Invertebrate Embryology and Reproduction

Invertebrate Embryology and Reproduction

Fatma El-Bawab
Professor in Department of Zoology,
Faculty of Science,
Alexandria University, Egypt

ELSEVIER

ACADEMIC PRESS
An imprint of Elsevier

Academic Press is an imprint of Elsevier
125 London Wall, London EC2Y 5AS, United Kingdom
525 B Street, Suite 1650, San Diego, CA 92101, United States
50 Hampshire Street, 5th Floor, Cambridge, MA 02139, United States
The Boulevard, Langford Lane, Kidlington, Oxford OX5 1GB, United Kingdom

Library of Congress Cataloging-in-Publication Data
A catalog record for this book is available from the Library of Congress

British Library Cataloguing-in-Publication Data
A catalogue record for this book is available from the British Library

ISBN: 978-0-12-814114-4

For information on all Academic Press publications
visit our website at https://www.elsevier.com/books-and-journals

Publisher: Charlotte Cockle
Acquisition Editor: Anna Valutkevich
Editorial Project Manager: Pat Gonzalez
Production Project Manager: Punithavathy Govindaradjane
Cover Designer: Matthew Limbert

Typeset by SPi Global, India

Dedication

To my dearly beloved father…

MAHMOUD EL-SAYED EL-BAWAB

and my highly valued teacher and professor.

PROFESSOR DR. MAHMOUD MOHAMED RAMADAN

Contents

Preface xi
Acknowledgements xiii

I Historical Review 3

Descriptive phase 4
Comparative phase 7
Experimental phase 9
References 15

II Introduction 17

Gametogenesis and gametes 17
Origin of germ cells 17
Formation of gametes 22
Fertilization 43
Development, Early development 60
References 64

KINGDOM PROTISTA 67

III Phylum Protozoa 69

Class 1- Mastigophora 69
Subclass Phytoflagellata 69
Order Volvocida 69
Chlamydomonas 69
Volvox 79
Class 2- Sarcodina 87
Subclass Rhizopoda 87
Order Foraminifera 87
Difflugia 88
Subclass Actinopoda 89
Order Heliozoa 89
Actinophrys 89
Class 3- Sporozoa 91
Subclass Telosporidea 91
Order Coccidiomorpha 91
Suborder Coccidia 91
Eimeria 91
Order Gregarinia 91
Monocystis 94
Class 4-Ciliophora 95
Subclass Ciliata 95
Order Holotricha 95
Suborder Hymenostomatida 95
Paramecium, Conjugation 96
References 101

KINGDOM ANIMALIA 103

Table of classification of Kingdom Animalia 103

Subkingdom Parazoa 105

IV Phylum Porifera (parazoa) 107

Definition 107
Historical review 107
General characteristics 107
Classification 109
Embryological characteristics of Porifera 110
Reproduction 110
Development 131
Class Calcarea 131
1- Subclass Calcaronea 131
2- Subclass Calcinea 137
Class Hexactinellida 138
Nutrition of the embryos 162
Larval release 162
Larval behaviour and settling 162
Metamorphosis 162
Oviparous sponges 162
Types of larvae 164
Biological associations 167
References 167

Diploblastica 171

V Phylum Cnidaria 173

Definition 173
General characteristics 173

Classification 175
Body plan of Cnidaria 177
Reproduction 177
Spawning 181
Embryogenesis 183
Class Hydrozoa 187
Class Scyphozoa 212
Class Cubozoa 239
Class Anthozoa 253
Representatives 253
References 303

VI Phylum Ctenophores 309

Definition 309
Distinctive features 309
Classification 310
Hormiphora plumosa 311
Reproduction 311
Regeneration 311
Brooding 313
Development 320
Mnemiopsis leidyi 326
Hatching 335
Brooding 335
References 336

Triploblastica 337
Acoelomata 337

VII Phylum Platyhelminthes, Turbellaria 339

Definition 339
Introduction 339
Distinctive features 339
Classification 340
Order Acoela 341
Reproduction 345
Order Polycladida (*Notoplana*) 354
Order Rhabdocoela 366
Order Tricladida 380
Development of composite eggs 380
References 385

VIII Phylum Nemertea 389

Definition 389
Introduction 389
Distinctive features 389
Classification 390
Reproduction 390
Reproductive system 390
Egg activation and the first block to polyspermy 397
Double eggs in Nemertinea 397
Micrura alaskensis 405
Procephalothrix stimulus 416
References 418

Pseudocoelomata 421

IX Phylum Nematoda 423

Definition 423
Historical review 423
General characteristics 424
Classification 424
Reproduction 426
Development 445
Chromatin diminution 447
Embryological peculiarities 452
Anabiosis 455
Caenorhabditis elegans 455
Historical review 455
Sex differentiation 457
Variation in nematode embryology 465
References 469

Coelomata 471
Supraphylum Arthropods
Mandibulata 473

X Phylum Crustacea 477

Definition 477
Historical review 477
A Branchiopoda 484
I Anostraca 484
II Diplostraca 493
B Cirripedia (Barnacles) 503
C Malacostraca 518
1 Hoplocarida (Stomatopoda) 518
2 Eucarida 530
Order Decapoda 530
Family Penaeidae 532
Penaeus kerathurus 532
Reptantia 590
Astacidea (Astacura) 590
Astacus fluviatilis 606
Brachyura 629
Portunus pelagicus 631

The male system 633

Formation of the spermatophore of *P. pelagicus* 641

Formation of the spermatophore of *C. danae* Smith 1869 653

Formation of the spermatophore of *Chionoectes opilio* 655

The female system 658

Copulation 660

Utilization of lipovitellin during Embryogenesis 700

References **704**

XI **Phylum Mollusca** 715

Definition **715**

General features of Mollusca **715**

Classification of Mollusca **716**

Class- Amphineura 716

Polyplacophora 716

Reproduction 716

Monoplacophora 733

Aplacophora 735

Class- Gastropoda 739

Reproduction 739

Eremina ehrenbergi 741

Differentiation of germ layer 793

References **810**

XII **Cephalopoda (Siphonopoda)** 815

General characters **815**

Classification **815**

Reproduction **816**

Spermatophores 822

Spermatophore of *Sepia officinalis* 823

Dehiscence of spermatophore 831

Formation of spermatophore 833

Hardening of spermatophore 839

Reproductive behaviour, mating, copulation, and spawning **850**

Development **851**

Organogenesis **855**

References **857**

Phylum Chordata **861**

XIII **Subphylum Urochordata, Tunicata** 863

Definition **863**

General characters **863**

Class Ascidiacea **864**

Order 1- Enterogona 865

Order 2- Pleurogona 865

Reproduction **865**

Spawning 874

Fertilization 874

Embryogenesis 877

Structure of fully formed larva 888

Experimental studies 890

References **898**

Further Reading 901

Index 905

Preface

Invertebrate Embryology and Reproduction is an essentially descriptive book. Although this branch has long been a core subject recognized as a necessary and valuable part of Zoology and Biology courses, it was not considered as a definite course in Egypt until the 1970s.

To give a brief account of my academic background, I ought to start by saying that I was an extremely curious student, fascinated by embryology, with an ambition to pursue an academic research career. I graduated from the Faculty of Science, Alexandria University in June 1955; at that time, unfortunately there were no available posts, and I was not able to join the faculty to fulfil my dream. However, I joined the Oceanography department to pursue my postgraduate studies. The Dean of the Faculty of Science at the time, the late Professor Doctor Abdel Fattah Mohammad, who happened to be teaching a course for this programme, became familiar with my work, and was quick to recognize my devotion to academe. Consequently, he offered me a post in the Zoology department; thus I became a university staff member in January 1956, and continue to be one. As a demonstrator in the practical laboratories, I was exposed to a variety of specialties in the science of zoology. However, I was always interested in embryology, and intended to pursue research in experimental embryology, but due to the lack of financial resources, my request to equip a suitable laboratory was rejected.

I was selected by the head of the Zoology department then, the late Prof. Dr Mahmoud Ramadan, who was the Professor of Invertebrates, as the demonstrator for the invertebrate laboratories. Consequently, my research was primarily concerned with invertebrates. My MSc and PhD theses were on Crustacea, and the PhD was on spermatophore, under the supervision of Prof. Ramadan. I feel privileged to have been his postgraduate student, and to have had the opportunity to work closely with him. His indelible influence on me, and on many others, is eternal. On the other hand, Prof. Dr Samy Riad, as a joint supervisor of my PhD, was the one who planned the course of the work, and basically taught me the principles of research, and the importance of integrity in carrying it out. I was incessantly guided and motivated by his discussions and decisive criticism. I was also well supported by the capable senior colleagues at this stage, Prof. Anga Ramadan and Prof. Milad Ishaa.

Eventually, I was considered the successor of Prof. Ramadan in Invertebrates. To follow on from his work, I created the 'Invertebrate school' in the Zoology department. Later, as a professor, I could confidently describe myself as an untraditional academic, working hand in hand with my students. I persistently examined the specimens myself, and aided in the actual writing of the theses, and in the interpretations and discussions.

By the time I had finished my PhD and started working as a lecturer on Invertebrates in the Zoology department at the Faculty of Science in Alexandria University, Oceanography had just started to be included as an undergraduate course. I was recommended by the head of the Zoology department at the time, the late Prof. Dr Shukralla Nakhla Sidra, to the head of the Oceanography department, Prof. Dr Saad El Wakil, to teach Invertebrate Embryology for the first time. I was the first, and the only one, to prepare and teach this course starting in 1970. I taught it to the general degree and special degree students, as well as for the pre-master's students. I also taught Experimental Embryology to the master's students. The great challenge then was that there were virtually no resources. I had to manage by using information out of the development and reproduction chapters from the available invertebrate books. For years, I had to ask any colleague or student travelling abroad to look for books specializing in invertebrate embryology.

It was only due to the genuine willingness of my departmental colleagues to help that I was supplied with the main available references in the field of invertebrate embryology over the years. The late Prof. Dr Mahmoud El-Zayat, Professor of Zoology, while in the United States ordered a book on invertebrate embryology edited by Kumé and Dan, translated from Japanese by J.C. Dan. After an unusually long wait, he had to place another order. Fortunately, I received both copies. Likewise, I can only be grateful to all other colleagues who travelled abroad over the years and went out of their way to provide me with the most-needed books: Dr Essam Abdulla, lecturer of Zoology, brought me *Experimental Embryology of Marine and Fresh Water Invertebrates,* by G. Reverberi, 1971, from the United States. Dr Desouky Mekawy, Professor of Zoology, brought me the valuable German book by O. Pflugfeder, 1970: *Lehrbuch dez Entwicklungsgeschichte und Entwicklungsphysiologie der Teire*. Dr Safaa Zaki, Lecturer of Zoology, brought me an unpublished MSc thesis, from the University of Maryland, United States, titled *A Histological Study of the Development of the Ovary and Accessory Reproductive Organs of the Blue Crab, Callinectes sapidus (Rabhdum)*, by L.E. Cronim, 1942. Dr Mohga El Abbady managed to send me different articles from the Congress Library in the United States, including an article about *Helix Pomatia* (E. Lind, 1973). Locally, people were equally supportive. Dr Mahmoud Ibrahim, Assistant Prof. of Molecular Biology, was always keen to provide me with all the related articles he found while searching on the Internet, in addition to his continual assistance with any article related to molecular biology.

The library in the Faculty of Science in Alexandria University has always offered me a fantastic service in terms of facilitating book borrowing. The late Prof. Dr Ibrahim El-Kholy, who was the Vice Dean of the Faculty of Science and Chair of the library, never rejected a purchase request that I submitted as a lecturer in Zoology. I therefore managed to get to the library a few purchases, including the series of *Reproduction of Marine Invertebrates* (volumes I–V). However, this series was not complete. I searched for it myself in UK bookshops during a visit there and realized that this series was not in circulation any more. I wrote to one of the editors, Giese, asking for the rest of the series, and received a reply from Pearse, the co-editor, who sent me volume VI, *Echinodermata and Lophophorates*, as well as all the related articles.

All those difficulties encouraged me into the idea of writing a comprehensive book about invertebrates. Having gone deeper in embryology through teaching, I found that books on the subject are quite rare in the market. For years, the update of the invertebrate embryology was confined to limited information from references dealing with general invertebrate reproduction. It did not cover all the phyla, and the books never included animals from the Mediterranean region. This does not fulfil the need for a comprehensive and exclusive update of the field. Even the recent publications, *Embryology, Constructing the Organism*, by Gilbert and Raunio, 1997, and *Developmental Biology* by Gilbert, 2000, are general and neither is exclusive to invertebrate embryology. In addition, the *Atlas of Comparative Invertebrate Embryology* by Dr Martin E.L. Scriba contains figures and not text. Its aim is to show clear figures for the invertebrate embryology, but not to describe and discuss the developmental stages for further research and teaching purposes. I therefore decided that it would be a pleasure and honour to write a comprehensive book as an Egyptian scholar, and to invest all the great effort spent preparing that course over the years.

This book therefore gathers all the information dealing with descriptive invertebrate embryology and updates this information in a single source, as a basic reference for teaching courses and research. The difficulty of obtaining teachable or research data in the field of descriptive embryology, not only in the Egyptian libraries but everywhere, demonstrated the need for a new work that collates all the available data concerning that field. The collection herein is sought from specialized references, in addition to general textbooks in reproduction, carried out by a professional expert in that field. It includes the description of embryology of newly studied species in most of the phyla, together with their systematic positions, and the results obtained from different researches using advanced techniques and different procedures, especially electron microscopical ones. It is specific to descriptive invertebrate embryology, including numerous invertebrate species. I hope that this effort will provide a coherent integrated account of the field that I attempted to cover, to the satisfaction of the reader.

Acknowledgements

It is only with the support of my family that this book was accomplished: both emotional support and unstinting encouragement as well as actual involvement with the work. I am immensely proud of my grandchildren, Ziyad and Zeinah Kalati, for their contribution in formatting the figures and revisions. The same goes to my daughter, Dr Siham Abdel Fattah, for her vital involvement with the work. My appreciation goes to my youngest daughter, Dr Azza Abdel Fattah, for the time that she generously invested in this work. She was involved in all stages of the work, and her thorough and meticulous revisions were invaluable. She handled all the required correspondence with the publisher, as well as the overwhelming copyright permissions requirements. It is in fact thanks to her zeal and keenness that this work was completed.

I would like to thank all who helped me during the planning and writing of this book, even the small efforts, which collectively made this project possible. I am truly grateful to Dr Amal Khafage, Lecturer of Invertebrates in the National Institute of Oceanography and Fisheries, who took time from her busy schedule to help in the searches for related articles to the points suggested. My sincere appreciation goes to Mr Mohammad Abdou, my nephew, who helped with the figure plates and Photoshop work, and to Dr Tarek Adnan, whose work in Bibliotheca Alexandria facilitated my obtaining copies of difficult-to-get articles. Finally, I would like to thank the staff at Elsevier for their patience and their help and expertise in editing and production.

Last, but not least, my sincere thanks to all members of the Department of Zoology, Faculty of Science, Alexandria University, for being a genuine family to me, sharing love, support, and care throughout my long journey of working in the department.

Chapter Outline

Descriptive phase **4**
Discovery of the microscope 5
Comparative phase **7**
Embryology and other subjects 8
Cell theory 8
Chromosome theory 8
Genetics and gene theory 8
Phylogeny and recapitulation theory 9
Retention theory 9
Experimental phase **9**
References **15**

Plato (left) and Aristotle (right), a detail of *The School of Athens*, a fresco by Raphael. Aristotle gestures to the earth, representing his belief in knowledge through empirical observation and experience, while Plato points up to the heavens showing his belief in the ultimate truth.

Chapter 1

Historical review

An invertebrate (nonvertebrate) refers specifically to an animal without a backbone, in contrast to the vertebrates represented by fish, amphibians, reptiles, birds, and mammals. An invertebrate animal has no central skeleton and has a ventral nervous system, while the heart is dorsal in position, with an open circulatory system, and its integument is formed of secretory substances, generally tanned proteins and sometimes chitinized.

At present, about 1.5 million species of invertebrates have been named and about 8000–10,000 new species are added to the list each year. With the exception of the protozoans (about 30,000 species) and the vertebrates (about 43,000 species), all other animals are invertebrates, comprising altogether about 95% of the animals in the animal kingdom. Strictly speaking, the protozoans are also invertebrates, but a distinction is sometimes made. If we also consider the extinct species, the total number of species that have lived would be much greater than that of the existing forms alone. The numbers of species are just estimations and vary enormously amongst zoologists. Evolution of new species is a continuous process; in addition, names of new species added annually to the list might result from the discovery of forms that have long been in existence but have only just come to light.

The invertebrates are grouped into about 30 phyla, the number of which varies among different zoologists. A phylum is somewhat difficult to define, but it may be considered as a basic structural pattern of evolutionary descent from a common ancestry, with its members more or less homogeneous and bearing a unique combination of distinct characteristics. The members of a phylum are all constructed along the same general plan. Some of the phyla are major ones (according to the constituting number of species), while others are minor phyla: a minor phylum can be a group separated (e.g. by class and/or order) from a major phylum such as Nemertinia from Platyhelminthes. Nemertinia was elevated to the rank of a phylum since its members have additional and/or different characters. A minor phylum can also be a newly discovered group of animals, which was established independently as a phylum, e.g. phylum Pogonophora.

According to embryology, the invertebrate phyla are grouped into two lines:

	Annelid line	Echinoderm line
Egg potency	Mosaic	Equipotent
Cleavage	Spiral	Radial
Blastomeres	Determinate	Indeterminate
Gastrulation	Delamination	Invagination
Coelom	Schizocoele	Enterocoele
Site of mouth	Protostome	Deuterostome
Larva	Trochophore (ciliated band around the equator)	Pluteus (ciliated band around the mouth)

Embryology (from Greek ἐμβρύον, *embryon*, 'unborn embryo'; and -λογία, *-logia*). It may be defined as that branch of biological science that deals with the development of the individual organisms. It is concerned with successive series of changes, in form and function, through which the initial germ of the new individual is transformed into a sexually mature adult. Normally, the germ with which development commences is an egg that has been fertilized by a sperm. The sexually mature adult is an individual that has developed to a point where it can produce mature eggs if a female, or sperm if a male. Sometimes the word ontogeny is used as a synonym for embryology, but more often it is defined to include the entire life history of an individual from its origin to its death.

Many principles of embryology apply to both invertebrate animals and vertebrates. Therefore, the study

Invertebrate Embryology and Reproduction. https://doi.org/10.1016/B978-0-12-814114-4.00001-1

of invertebrate embryology has advanced that of vertebrate embryology, despite the presence of many differences. The most obvious is the release of a larva in numerous invertebrate species before the completion of development.

The history of embryology has passed through three phases: fact-finding or descriptive, comparative, and experimental embryology. More recently, the molecular studies have been established.

DESCRIPTIVE PHASE

The first famous name associated with this period is Aristotle.

Aristotle (384–322 BC)

With the possible exception of Plato, Aristotle is by far the most influential philosopher in the history of Western thinking. Aristotle studied natural sciences, and these studies dominated until early modern times. Aristotle's influence was vast and extended far beyond philosophy. For example, he was the founder of biology, and Charles Darwin (1809–82) regarded him as the most important contributor to more than one subject.

The influence of Aristotle's family (his father was a physician) initiated his strong interest in anatomy, and the structure of living organisms in general. He described the reproduction and development of many kinds of animals. He was the first to study and write the accounts of development of the hen's egg. His Macedonian origin, being a student, and then a teacher in Plato's academy, and his connections with the group of outstanding scientists surrounding Plato (who worked in a wide variety of subjects, ranging from medicine and biology to mathematics and astronomy), contributed to him developing remarkable observation skills. These media had greatly influenced and characterized the scientific thinking of Aristotle. He was able to reach and formulate some scientifically significant findings as follows:

1. He noted the different methods by which animals originate from eggs, and described them as oviparous, viviparous, and ovoviviparous animals, and divided them into the following categories:
   - Mammals: because their young arose from a mixture of male and female fluids and are born alive. The eggs of mammals had not been discovered at that time.
   - Sharks: arose from eggs that are retained within the body of the mother and are also born alive.
   - Reptiles and birds: in which the eggs are complete ones.
   - Amphibia and bony fishes: in which the eggs are incomplete (complete eggs are those with an albumen coat and a shell).
2. He identified the two major cell division patterns by which embryos are formed: complete division of the egg (holoblastic), as in frogs and mammals; and incomplete division of the egg (meroblastic), as in chicks.
3. He was the first to figure out the functions of the placenta and the umbilical cord.
4. He also observed that the cephalic (anterior) part of the embryo grows faster than the posterior one (Fig. 1.1C and D). Therefore he discovered—as pointed out by Joseph Needham (one of the modern embryological pioneers) in 1934—the morphogenetic cephalocaudal gradient.
5. As a result of describing different embryos, he claimed the most valuable theory, which initiated massive arguments and discussions during later times, the **epigenesis**. This theory states that development proceeds from a simple formless beginning to the complex organization of the adult.

After Aristotle, there was hardly any progress in embryology for the next 2000 years. It was not until the 15th and 16th centuries that the science in general, and embryology in particular, was refreshed. The most famous man during that period was William Harvey.

The English Physician William Harvey (1578–1657)
(Before the invention of the microscope)

Harvey described the development of the chick in great detail. He also contributed a study of the development of the deer, and compared the two with each other. He was the first to observe the blastoderm of the chick embryo—that small region of the egg that contains the yolk-free cytoplasm which gives rise to the embryo. He was also the first to notice that islands of blood cells form before the heart does (Fig. 1.1A and B). Harvey suggested that the amniotic fluid might function as a shock absorber for the embryo. His most important finding was his conclusion that all animals arise from eggs, but this was based on theoretical considerations. He famously stated, 'Omne vivum ex ovo'. He finally published a book, following his numerous studies and results, *Exercitationes de generatione animalium* (*On Animal Generation*), published in 1651. His doctrine of *omne vivum ex ovo* (all life comes from the egg) was the first definite statement against the idea of spontaneous generation.

In addition to all this work, Harvey's main line of research was dealing with the blood circulation in the human body, and he became an embryologist, in addition to being the founder of modern physiology and medicine. He published his theories in a book titled *Exercitatio Anatomica de Motu Cordis et Sanguinis in Animalibus (An Anatomical Study of the Motion of the Heart and of the Blood in Animals)*, where he explained how the heart propelled the blood in a circular course through the body. His discovery was received with great interest in England, although it was greeted with some scepticism on the Continent (www. en.wikipedia.org/wiki/William_Harvey and www.bbc.co.uk/history/historic_figures/harvey_william.shtml).

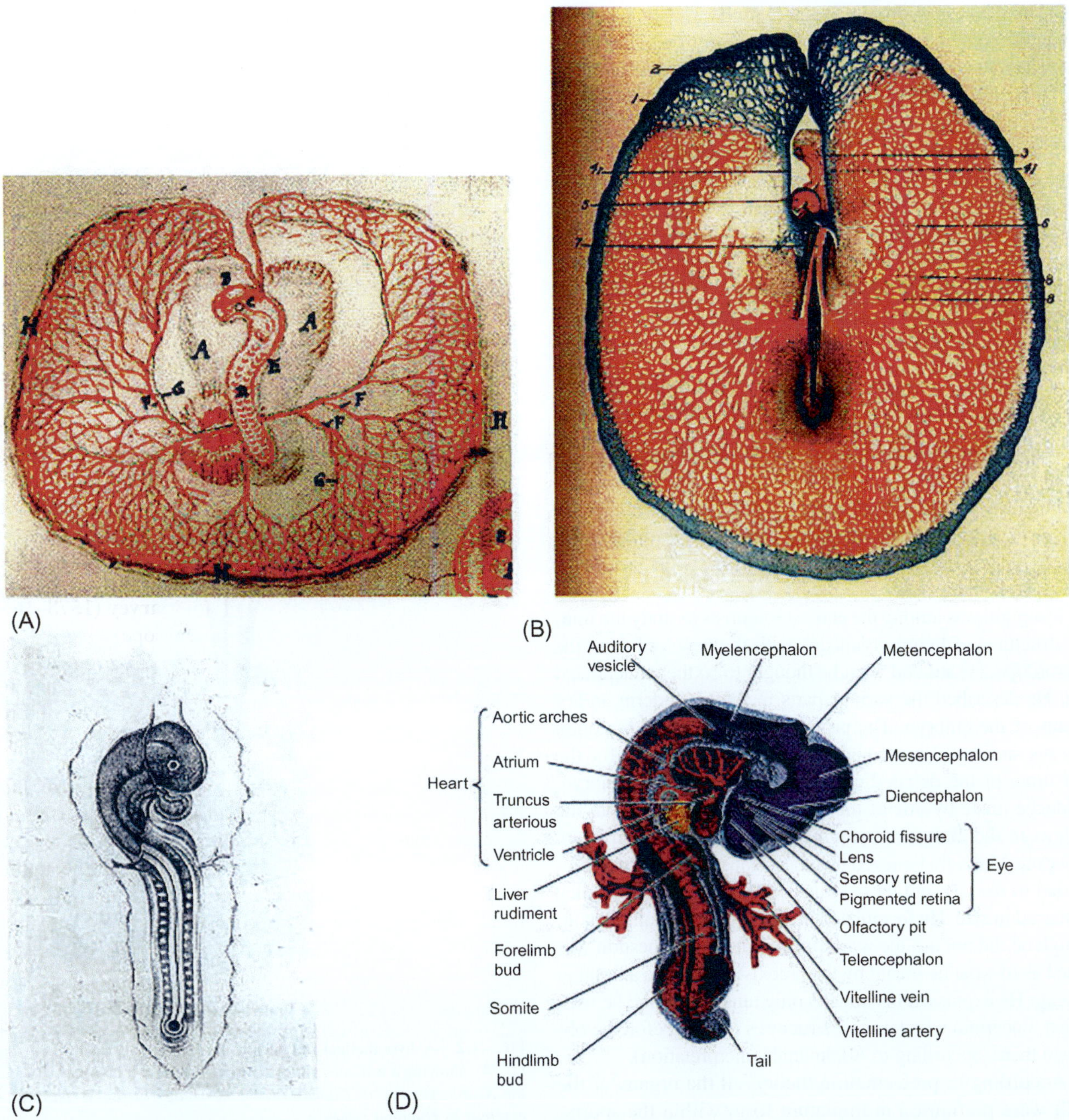

FIG. 1.1 Depictions of chick developmental anatomy. (A) Dorsal view (looking 'down' at what will become the back) of a 2-day chick embryo, as depicted by Marcello Malpighi in 1672. (B) Ventral view (looking 'up' at the prospective belly) of a chick embryo at a similar stage, seen through a dissecting microscope and rendered by F.R. Lillie in 1908. (C) Eduard D'Alton's depiction of a later stage 2-day chick embryo in Pander (1817). (D) Modern rendering of a 3-day chick embryo *after Carlson (1981)*. Details of the anatomy will be discussed in later chapters. *(From Gilbert (2000).)*

Discovery of the microscope

Leeuwenhoek, Anton van [lay'-vuhn-hook, ahn' -tohn vahn] **(1632–1723)**

Leeuwenhoek was a Dutch biologist and microscopist. He was a businessman, and became interested in science when he developed an interest in lenses and started building simple microscopes as a hobby. Each microscope consisted of a flat brass or copper plate in which a small, single glass lens was mounted. The lens was held up to the eye, and the object to be studied was placed on the head of a movable pin, just on the other side of the lens. Leeuwenhoek built over 400 microscopes, whose focuses were amazingly accurate and clear. Many of these microscopes still exist. The most powerful of these instruments can magnify objects up to 275 times. Although future microscopes were to contain more than one lens (compound microscopes), Leeuwenhoek's single lens had such perfection that he was able to make great advances and thus to draw attention to his field.

Leeuwenhoek was the first person to observe single-celled animals (protozoans) with a microscope. He described them in a letter to the Royal Society, which published his detailed pictures in 1683. Leeuwenhoek was also the first person to observe clearly, using a microscope, and to describe the red blood cells (corpuscles) in humans and other animals, as well as sperm cells. In addition, he studied the structure of plants, the compound eyes of insects and the life cycles of fleas, aphids, and ants.

(After the invention and general use of the microscope)

As might be expected, the study of embryology was based on speculation, until the invention of the microscope, which allowed detailed observations. One of the founders of the study of microscopic anatomy was Marcello Malpighi.

The Italian Physiologist Marcello Malpighi (1628–94)

Malpighi was among the earliest scientists to study the minute structures of living organisms with the new tool then, the microscope. He studied what he thought to be the unincubated egg. He described the various parts in the blastoderm and/or organs of the embryo. The primary contribution of Malpighi was his successful presentation of a visible evidence for the first time, of the detailed constitution of the young embryo. Evidence made by one of the new tools was as certain in the 17th as in the 20th century, which drew popular enthusiasm. However, Malpighi believed that the organs of the embryo preexisted in the egg and these structures became visible as they increased in size. He formulated his theory of development, although he did not use the word **preformation**. A question was raised as to what he meant by the preexistence of the animal in the egg. He expressed his notions only tentatively, and he was, in fact, uncertain whether new structures existed before he observed them (according to Adelmann's interpretation).

According to preformation theory, all the organs of the adult were prefigured in miniature form within the sperm, or, more usually the egg.

Consequently, many discussions and arguments took place during that period, and the scientists were split into ovists and animalculists. Leeuwenhoek, who co-discovered the sperm in 1678, and the other co-discoverer of sperm, Nicolas Hartsoeker, illustrated a drawing of what they hoped to find as a preformed human ('homunculus') within the human sperm (Fig. 1.2).

However, this theory of preformation opposed that of epigenesis by Aristotle, which resulted in a massive debate among the philosophically minded scientists of the 18th century. The most famous scientist who strongly disputed the preformation and revived Aristotle's epigenesis was Caspar Friedrich Wolff.

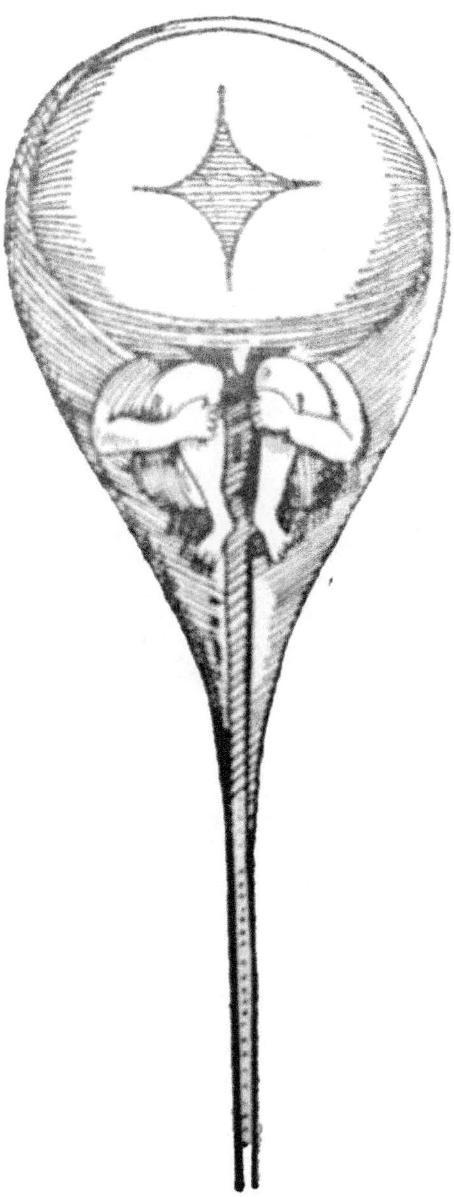

FIG. 1.2 A hypothetical drawing of the human sperm by Hartsoeker (1694), showing a miniature of the embryo preformed in the sperm. *(From the Arabic version of Comparative Embryology of the Vertebrates, the original by Huettner (1949).)*

The German Caspar Friedrich Wolff (1733–94)

Wolff was an anatomist and embryologist working in St Petersburg. He is often considered a founder of modern embryology at that time. He revived the view of one of the predecessors, Maître-Jan, who pointed out and explained, in 1722, the egg examined by Malpighi. His opinion was that Malpighi had studied an unincubated egg, after exposure to the heat of Bologna, not a fresh egg. After careful observation of the development of chick embryos, Wolff reached the first proof of epigenesis. He observed and identified that the blood vessels of the chick blastoderm are not present from the beginning. He also studied plant development; he was able to compare the development of some internal animal organs; and he interpreted the development of the intestine, and probably the central nervous system, which are formed by the folding of homogeneous tissue layers into tubes. This meant that the embryonic organs develop from tissues that have no counterpart in the adult organism, and he foreshadowed the germ layers. At the same time, Wolff supported epigenesis, and he is considered to be the father of epigenetic descriptive embryology. He marked the point when embryonic development began to be described in terms of the proliferation of cells, rather than preformed miniature. He postulated an unknown force, essential force, which acts like gravity or magnetism that organizes embryonic development. In his ***Theoria Generationis*** (1759), he tried to explain the emergence of the organism by the actions of a '*vis essentialis*', an organizing, formative force, and declared that 'all believers in epigenesis are vitalists'.

Several scientists and/or philosophers believed in this view, and added contributions to support it. Then, a mechanical, goal-directed force called developmental force (*Bildungstrieb*) was postulated—not theoretically, but it could be shown to exist by experimentation, e.g. regeneration of *Hydra* when cut. Thus, development could proceed through a predetermined force inherent in the matter of the embryo.

This hypothesis, in which epigenetic development is directed by preformed instructions, is not far from the view held by modern biologists that most of the instructions for forming the organism are already present in the egg.

COMPARATIVE PHASE

The end of preformationism did not come until the 1820s, when a revolution and evolution in descriptive embryology was created by the new microscopic investigators. Among the most talented of these was a group of three students (born in three consecutive years), of the philosophy of nature (*Naturphilosophie*) at Würzburg. They came from the Baltic region, but studied in northern Germany. They began to build constructively on Wolff's concepts, including the epigenesis and the results of his microscopic examinations. With the combination of new staining techniques, improved microscopes, and institutional reforms in the German universities, these students were able to document the epigenesis of anatomical structures and to carry on the work of their teachers. These students were the German Heinz Christian Pander (born 1794), German-Russian Karl Ernst von Baer (born 1792) and German-Prussian Martin Heinrich Rathke (born 1793), and they became known as the three Founders of Modern Embryology.

The German Heinrich Christian von Pander (1794–1865)

Von Pander was an anatomist. He was the first to discover and demonstrate the three germ layers, in the embryo of the chick. He had studied its development for 2 years (before becoming a palaeontologist). He thus realized that preformation could not be true. At the same time, his observations strengthened the belief in epigenesis since he discovered that organs are formed through interactions between simple structures. That is the view that we now call induction. Pander published a book, with glorious engravings (Fig. 1.1C), done by the artist Eduard D'Alton (a friend of Goethe who was one of the scientists of natural philosophy (*Naturphilosophen*), and whose influence on embryology was quite strong). However, Pander did not carry on, and von Baer took it upon himself to continue, and eventually broaden the base by additional examinations.

The German-Russian Karl Ernst von Baer (1792–1876)

Von Baer was a German-Russian biologist and the second of the three Founders of Modern Embryology of that time, but he was so famous that he was known as the 'father of the modern embryology'. In 1828, he reported: "I have two small embryos preserved in alcohol that I forgot to label. I am unable to determine the genus to which they belong. They may be lizards, small birds or even mammals." This quandary pushed von Baer to extend his study and examinations to many different vertebrates. Through his detailed study of chick development, and then comparing the results to the embryos of these different vertebrates, von Baer:

1. generalized Pander's germ layers to other animals, and thus generalized the science of embryology itself;
2. was the first famous embryologist to establish a new branch known as comparative embryology;
3. reached many valuable results through his detailed studies. He discovered the notochord, then the neural tube. The most important discovery was that of the mammalian egg in 1827. He observed and identified the egg cell, differentiating it from the fluid-filled sac or Graffian follicle that surrounds it. He therefore, fulfilled and revived the theoretical consideration of Harvey (1578–1657) that 'Life begins with an egg'; and
4. derived four generalizations (now often known as 'von Baer's laws'):
   - **(a)** The more general characteristics of any large group of animals appear in the embryo earlier than the more special characteristics, e.g. in all vertebrate embryos, the notochord appear first, shortly after gastrulation.
   - **(b)** After the more general characteristics, those that are less general arise, and so on until the most specific characters appear, e.g. all vertebrates initially have the same type of skin, and only later, the skin would develop: fish scales, reptilian scales, bird feathers, or the hair, claws, etc. of mammals.
   - **(c)** The embryo of any particular kind of animal grows unlike the forms of other species rather than passing

through them, e.g. human embryos never pass through a stage equivalent to an adult fish, frog, or bird.

(d) The embryo of a higher species may resemble the embryo of a lower species, but never resembles the adult of that species, e.g. human embryos initially share common characteristics with those of lower species, but none of them passes through the stages of the others.

Von Baer formulated these generalizations prior to Darwin's theory of evolution in 1856. 'Lower animals' would be those appearing earlier in life's history. His results initiated many discussions and arguments with others, especially philosophers. Afterwards, he understood the biogenetic law, which was clarified by Meyer (1935) as cited in Holmes (1944), and which was based on the recapitulation theory of Haeckel (1834–1919).

The German-Prussian Martin Heinrich Rathke (1793–1860)

Rathke was the last of the three Founders of Modern Embryology. His quality of mind, in particular, was in many ways similar to that of von Baer. He made distinguishable advances in demonstrating the existence of the germ layers in the invertebrates. He studied carefully the development of many different species especially the crayfish. On the other hand, Rathke looked at the development of some vertebrates (frogs, salamanders, fish, birds, and mammals). He emphasized the similarities in the development of all these vertebrate groups. These studies contributed significantly to a variety of topics. He was the first to discover the embryonic precursors of gills in embryos of higher animals that lack gills as adults, as well as, in mammals. In 1838, he published an important study of the pituitary gland. This embryologist is now known by his associated eponyms: Rathke's pouch—a depression in the roof of the embryonic mouth in front of the buccopharyngeal membrane; and Rathke's pouch cysts—groups of epithelial cells forming small colloid-filled cysts in the *pars intermedia* of the pituitary gland.

These were thus the three Founders of Modern Embryology: von Baer had been a student at Göttingen with von Pander. Rathke studied natural history and medicine at the University of Göttingen in 1814. Later, in 1818, he moved to Berlin to receive his MD degree. He became a physician, and was appointed in 1829 as a professor of physiology and pathology at the University of Dorpat. While in Dorpat, Rathke established contact with Karl Ernst von Baer, who was then a professor at Königsberg. When Baer left Königsberg for St Petersburg in 1834, Rathke succeeded him as a professor of zoology and anatomy. From there, these three scientists met, and started sharing their interest in embryology. They complemented each other's researches and added to their achievements.

Embryology and other subjects

Since the time of von Baer up to the present time, biological sciences, in general, have greatly improved. The comparative embryology has been influenced by different special studies and theories that are associated with it, such as:

- cell theory;
- chromosome theory;
- genetics and gene theory;
- phylogeny and recapitulation theory; and
- retention theory.

Cell theory

This theory was developed during the time of von Baer, by the German physiologist Theodor Schwann (1810–82) and the German botanist Matthias Jacob Schleiden (1804–81). They are both credited for formulating the cell theory, which believed that all living organisms are composed of cells. Schwann studied medicine at the University of Berlin. His most important work was the microscopic study of animal tissue, which led to the idea that all living organisms are made up of units-cells, and that each contains essential components such as a nucleus. He published this idea in *Microscopic Researches into accordance in the Structure and Growth of Animals and Plants* (1839).

Schleiden, after studying law, concentrated on botanical studies. In 1838, 1 year before Schwann, he developed the idea that the cell is the basic unit of plants and that growth consisted of production and development of new cells. Schleiden's two-volume textbook *Principles of Scientific Botany* (1842–43) was for a long time a model for modern botanical works. The cell theory was extended by other scientists, and it drew the attention of biologists to the study of the cell contents. Embryologists, on the other hand, started to base their field on the concept of the cell.

Chromosome theory

This theory, which deals with inheritance, was announced at the beginning of the 20th century by Walter Stanborough Sutton (1877–1916). He was an American surgeon and geneticist, and his work formed the basis of the chromosome theory of heredity. The latter greatly influenced the course of the field of genetics. In 1902, Sutton proposed that chromosomes during division make the basis of Mendelian inheritance.

Genetics and gene theory

The American zoologist Thomas Hunt Morgan (1866–1945) was a professor of experimental zoology at Columbia University (1904–28), where he conducted extensive breeding experiments on the fruit fly, *Drosophila melanogaster*. In 1927, he published a book entitled *Experimental Embryology*. From his research, he concluded that the individual units of heredity, the genes, are arranged in a line on the chromosomes. This theory, known as the chromosome theory of heredity, was presented in 1915, in *The Mechanism of Mendelian Heredity*, which he wrote with his co-workers, A.H. Sturtevant and C.B. Bridges. He claimed that embryology is fundamentally the study of heredity in action and he gave a support to Mendel's laws of inheritance; however, these were rejected for some time.

In 1926, Morgan wrote *The Theory of the Gene*, in which he stated that genes are held together in linkage groups, and that there is an interchange of alleles, genes, affecting the same trait. This interchange, known as crossing-over, allowed Morgan to construct chromosome maps showing the relative location of many genes.

For his work in the field of genetics, Morgan was awarded the Nobel Prize for physiology or medicine in 1933.

Phylogeny and recapitulation theory

This theory was published by Ernst Heinrich Haeckel (1834–1919), a German naturalist who made important discoveries in zoology. He is best known for his enthusiastic early support to Darwin's theory of evolution, and being its main expounder on the European continent. He presented strong evidence of evolution in 1866, in his book *General Morphology*, and is famous for popularizing the phrase 'ontogeny repeats phylogeny'. This phrase is a brief form of his recapitulation theory, later known as the biogenetic law. This theory states that ontogeny is a brief and incomplete recapitulation of phylogeny, or that an animal passes through stages in its development that are comparable to those which its ancestors passed through during their evolution. Haeckel also made significant contributions to knowledge of invertebrates and influenced the course of embryological research.

Retention theory

This theory has been proposed by the British biologist and Darwinian scholar, De Beer.

Sir Gavin De Beer (1899–1972)

De Beer is one of the modern embryology pioneers. In addition to being a zoologist and morphologist, he also contributed to experimental embryology, anatomy and evolution. He refuted the germ-layer theory, and developed the concept of paedomorphism: the retention of juvenile characteristics of ancestors in mature adults. This retention theory explains that such structures found in the bodies of some animals correspond to similar parts used for the same or different purposes in the body of some distant ancestor. Thus the vestigial tail of the human embryo, for example, arises in the same place and manner as the tails of other vertebrates, and therefore there is no doubt that some remote prehuman ancestor made use of a tail.

De Beer also developed the theory of piecemeal or mosaic evolution, on the basis of his work on the fossil *Archaeopteryx*, which had reptilian and avian traits, explaining the combination of both bird and reptile features. He elucidated the embryological origins of many tissues, wrote the authoritative *Charles Darwin: Scientific Biography* (1963) and served as a director of the British Natural History Museum (1950–60).

EXPERIMENTAL PHASE

A new generation of embryologists then felt that embryology should not merely be a guide to study anatomy and evolution, and that it should answer the question: 'How does an egg become an adult?' Embryologists were to study the mechanisms of organ formation (morphogenesis) and differentiation. This new branch of study was given a German name that means 'causal embryology', 'physiological embryology', or 'developmental mechanics'. These scientists began working with this new concept in the 1870s, and the first who led the group in this was Wilhelm His.

The Swiss-German Wilhelm His (1831–1904)

His was a researcher and teacher, who contributed to the fields of anatomy, histology and embryology. In addition to his investigations on some organs, he was the first to attempt a causal explanation of embryonic development. His was the most vocal against Haeckel theory and the most explicit. He claimed, with confidence, that during development of an embryo, any organ develops according to its future physiological function; and each stage—e.g. cleavage, blastulation, gastrulation, etc.—triggers the development of the following stage. According to this thinking, development is a mechanical and/or physiological process and never a recapitulation of the ancestral history.

One of His' remarkable achievements was his invention of the microtome (1870), a machine that can make very thin sections for the material that complements the micro-technique. It makes very accurate and better serial sections, even for a whole embryo that facilitates the detailed observation.

Around the same era, there was another scientist whose studies were related to and/or flourished the experimental embryology.

The German August Weismann (1834–1914)

Weismann was a German biologist who made fundamental contributions to the understanding of the processes of heredity and evolution. He proposed a hypothesis of development based on the fate map of the frog embryo. Weismann claimed that when the first cleavage division separated the future right half of the embryo from the future left half, there would be a separation of 'right' determinants from 'left' determinants in the resulting blastomeres. This led him to distinguish 'germplasm' (hereditary elements carried by sex cells) from 'somatoplasm' (the rest of the body). He thus proposed a model of inheritance that included a strict separation, in the early embryo, of a germ-line from the soma. He noted that the germplasm, on passing from generation to generation, remains largely unaffected by the environment (this work was based on the scant knowledge of fertilization available at that time).

Among Weismann's important works were *Studies in the Theory of Descent* (2 vols. 1882), *The Germ-plasm: A Theory of Heredity* (1893), and *The Evolution Theory* (2 vols. 1904).

One of the most eminent pioneers in the analysis of development at the same time as Weismann was Boveri.

The German Theodor Heinrich Boveri (1862–1915)

Boveri mostly preferred the sea urchin embryos, on which he experimented during the summer months at the Marine Biological Station in Naples-Italy. His work on sea urchins showed that it was necessary to have all chromosomes present, for proper embryonic development to take place. This discovery was an important part of the Boveri-Sutton chromosome theory (see above). His other significant discovery was the 'centrosome' (1888), which he described as the *especial organ of cell division.*

As some of Boveri's countrymen appreciated the importance of worms, he started in 1880, to study the embryogenesis of the nematode *Ascaris megalocepha* (=*Parascaris equorum*), a parasite found in the intestine of horses. The females produce millions of eggs, releasing more than 100,000 per day. In fixed and stained early cleavage stages, Boveri found a peculiar but reproducible division pattern accompanied by loss of chromatin in most cells. The first division generates two cells, one of which, at the following division, passes through a process called 'chromatin diminution' (see Chapter 9). Since Boveri was the first to discover this phenomenon, it should be his associated eponym.

Cell theory was formulated in 1838 and 1839, and was well accepted by scientists working on different lines of science. Wilhelm His disputed Haeckel's theory violently. Weismann proposed his germ-plasm theory with a new line of research. In the late 1800s, these three significant events explored different approaches of study in the field of embryology leading to the appearance of different schools of embryologists. Each school specialized in a definite major technique such as:

- defect, destroying a portion of the embryo;
- isolation, removing a portion of the embryo;
- recombination, replacing an original part with a part from a different region of the embryo;
- transplantation, whereby a portion of the embryo is replaced by a portion from a different embryo; and
- fate mapping experiments, tracing the fate of a portion and/or a cell of the embryo.

The work of these schools was parallel and/or succeeded each other, and their researches complemented each other, which greatly extended and flourished this new branch of experimental embryology. Each school had a leader or one famous scientist who planned and created a firm basis for the school's line of work. One of the most famous of these leaders was Wilhelm Roux.

The German Wilhelm Roux (1850–1924)

Roux was a young German embryologist and was considered the founder of the branch of experimental embryology. He strongly believed in the ideas and thoughts of His, in spite of the fact that Roux was a student of Haeckel, and considered himself his disciple. He unarguably acted as an intermediator between Darwinism and the new causal analytical embryology. The attempts at mechanical explanations of development were started by His and remained lacking in experimentation until Wilhelm Roux, who was able to succeed where His had apparently failed. This recalled a statement in the introduction of T.H. Morgan's book *Experimental Embryology*: 'The recognition of the advantage of applying the method of experiments to problems of development may be said to have begun with Wilhelm Roux in 1883'.

Roux was one of the first scientists to test Weismann's hypothesis. He followed the defect experiments. In 1888, Roux published the results of a series of defect experiments on frog embryos. He freed one blastomere in the two-celled amphibian egg, from the influence of the other, by killing the latter. He observed that the surviving cell developed into a half-larva. We now know that Roux's experiment was technically flawed. He thought that he had demonstrated the independent differentiation of the surviving cell. He interpreted the egg, for the first time, as a mechanism mechanically analyzable by outside interference. He concluded that the frog embryo was a mosaic of self-differentiating parts and that it was likely that each cell received a specific set of determinants and differentiated accordingly.

Roux not only performed the experiment, but was also able to generalize its significance. His ability to mobilize thoughts around it was due to his own qualities of mind and personality.

Some experimental embryologists, however, especially Driesch (see below), opposed Roux in his explanation that development simply proceeds mechanically. Eventually, the branch of experimental embryology was well established and differentiated into more specifications. The results and arguments tallied with the view of Roux. Many scientists, including cytologists, such as Boveri, Driesch, Delage, Brachet, and Flemming from Europe and Conklin, Harrison, Wilson, Loeb, Morgan, and Whiteman from the United States, agreed with the opinion and explanation of Roux.

The German Hans Adolf Eduard Driesch (1867–1941)

Driesch was a German experimental embryologist and philosopher. It is likely that nobody appreciated Roux's work and the experimental approach to embryology more than

Hans Driesch. However, contrary to Roux, Driesch's goal and attempts were to explain development, in terms of the laws of physics and mathematics. While Roux's studies were 'defect' experiments, those of Driesch were 'isolation' ones. In 1891, Driesch and Fiedler carried out, independently, the separation of the two-cell sea urchin blastomeres from each other by vigorous shaking (or later, by placing them in calcium-free seawater). To Driesch's surprise, each of the blastomeres of the embryo developed into a miniature, nearly complete larva. Similarly, when Driesch separated the blastomeres from four- and eight-cell embryos, some of the isolated cells produced entire pluteus larvae (Fig. 1.3). Driesch, then, reached a result that completely opposed the predictions of Weismann and Roux. He identified that each isolated blastomere regulated its development so as to produce a complete organism, rather than self-differentiating into its future embryonic part.

On the other hand, Driesch was a philosopher and one of the leading vitalists. He attempted to explain life processes through the theory of vitalism. At the same time, he was a teleologist. He believed that many of the basic problems of biology could not be solved simply by philosophy. In 1894, after publishing his papers on experiments on sea urchin eggs, Driesch wrote a theoretical essay, titled *Analytische Theorie der organischen Entwicklung*, in which he declared that his studies in developmental biology pointed to a 'blueprint' or teleology, an Aristotlean entelechy, a scientific demonstration of Immanuel Kant's notion that the organism develops as if it has a purposeful intelligence. *The History and Theory of Vitalism* (1905) was his main work.

In the same year of Driesch's experiments on sea urchin eggs, C. Chun reported the results of experiments he had started in 1877, in which he traced the development of separated blastomeres of two-cell ctenophore embryos. He chose the eggs of the ctenophore, e.g. *Beroë ovata*, because it is excellent for experimental work, as it is large, clear, and does not disintegrate when cut. He observed that each cell of the two developed, giving rise to a half-embryo that possessed exactly half of the normal set of adult structures. These results provided a view of embryos as being composed of a 'mosaic' of individual parts, which was completely different from Driesch's view. However, Driesch, being a powerful and arrogant man, refused and ridiculed Chun's results, and made personal attacks on Chun's integrity. T.H. Morgan collaborated with Driesch, repeated Chun's experiments and observed the same results as Chun in 1895.

There is an explanation that allows for each of the two experiments of Driesch and Chun and their results being accurate. The embryos of many animals (including those of amphibians) possess remarkable regulative abilities, while the embryos of many other animals do not. This is what is now known as indeterminate and determinate blastomeres.

The discovery of regulation led to the discovery of the 'organizers' by Spemann.

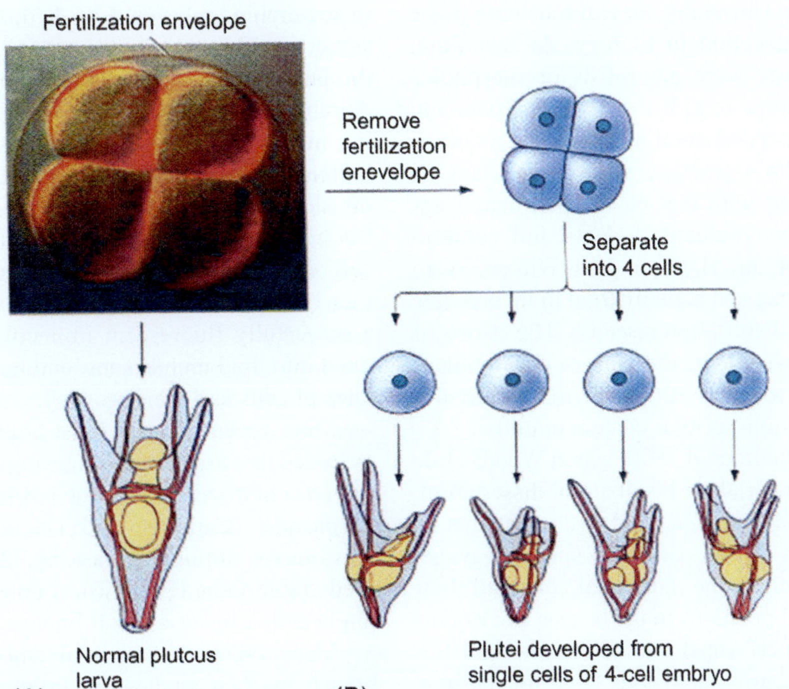

FIG. 1.3 Driesch's demonstration of regulative development. (A) An intact four-cell sea urchin embryo generates a normal pluteus larva. (B) When one removes the four-cell embryo from its fertilization envelope and isolates each of the four cells, each cell can form a smaller, but normal, pluteus larva. (All larvae are drawn to the same scale.) Note that the four larvae derived in this way are not identical, despite their ability to generate all the necessary cell types. Such variations are also seen in adult sea urchins formed in this way. *(From Gilbert (2000).)*

The German Hans Spemann (1869–1941)

Spemann was a German zoologist, the first graduate student of Boveri (see above) who completed his PhD thesis on nematodes. He won the 1935 Nobel Prize for physiology or medicine, for his work on the 'organizers' in the amphibian embryo. Using frog and newt embryos, Spemann found that development is determined by interactions between various parts of the embryo, with special areas serving as organizers for the development of nearby related tissues.

Another line of his experimental research was 'fate mapping the embryo', which was based on the cell theory. Following individual cells during development to see what they would become was one of the most important programs in embryological studies. Tracing the fate of the blastomeres of an embryo leads one to define the cell lineages of that embryo, and to construct a fate map to the gastrula or blastula of such embryo. What is most remarkable in this field is the zeal and commitment of those embryologists (Conklin, Wilson, Boveri, etc.) who patiently studied the 'cell lineage' of the invertebrate eggs without any other help than the microscope.

The American Edwin Grant Conklin (1863–1952)

Conklin spent most of his career as a distinguished academician and scientist in the Department of Biology at Princeton University. When he was a student, he was greatly interested in biology. Conklin decided to enter the graduate school at Johns Hopkins University, an extraordinary place to obtain a graduate education in biology. At that time, most American biologists were naturalists or morphologists, while those in Europe (e.g. Roux, Boveri, Spemann, Driesch) had applied experimental methods to embryological research. Conklin's graduate advisor, William K. Brooks, was not satisfied with that new experimental approach, while his fellow graduate students and contemporaries, including Morgan, Harrison and Wilson, were enthusiastic about this line, and actively tried to incorporate these methods into their dissertation research. The efforts of these graduate students were to be the roots of developmental biology in America. However, afterwards these individuals drifted in other directions as their careers matured.

Conklin, during the summer of 1890, was at Woods Hole in search of a suitable material for his doctoral dissertation. He chose the marine snail *Crepidula*, and discovered that its eggs divide according to a fixed pattern of spiral cleavage, which enabled him to follow the individual cells and their descendants to their final positions in the tissues and organs of the larva. In this he performed one of the first cell lineage analyses, which occurred at Woods Hole for the next decade. With this event began a lifelong love of the Marine Biological Laboratory at Woods Hole, where Conklin subsequently spent almost every summer to extend his cell lineage studies. In 1905, he studied the development of the egg of the tunicate sea squirt *Styela partita*. He patiently followed the fates of each cell in the embryo (the descendants of each particular cell) until they differentiated into particular structures. He was helped only by the peculiarity of the *Styela* egg, wherein the different regions contain different pigments. The most peculiar of these regions was the yellow crescent, or myoplasm, which was deposited in the tail muscle cells of the tadpole larva. Conklin had discovered a natural cell lineage marker.

He then constructed a linear version of the fate map of such species (Fig. 1.4). This was to be his most important work and his associated eponym. Conklin's fate map was confirmed by cell removal experiments, which is another line of studies. The removal of a certain cell in early cleavage produces an embryo without its descendants, i.e. the organ of that cell. For example, the removal of the B41 cell (which should produce all the tail musculature) resulted in the absence of tail muscles (Reverberi and Minganti, 1946).

Most embryos are not so accommodating as to have cells of different colours. In addition, not all embryos have as few cells as tunicates. The local staining method was introduced later by Vogt. In 1929, Vogt traced the fate of each blastomere in the developing amphibian embryo. He was thus able to establish a correct and precise 'map' of the cleaving embryo. A variety of methods have been used to study the fates of early blastomeres. One of the earliest studies was carried out by Hörstadius in 1939. He used the vital dye Nile Blue to mark the surface of individual cells in sea urchin embryos (Fig. 1.5). He found that all the mesomeres and some macromeres give rise to the ectoderm of the larva, and that micromeres become skeletogenic cells. Another fate mapping process was the radioactive labeling method. One of the problems with the vital dyes and radioactive labels is that they become diluted at each cell division. One way around this problem was the creation of fluorescent dyes, which could be injected into an individual cell of the early embryo. The descendants of that cell could then be seen under ultraviolet light. **'DiI'**, as an example, is a powerfully fluorescent molecule that becomes incorporated into lipid membranes, and has been used to follow the fates of cells and their progeny.

More recent studies have confirmed and considerably extended these observations, using intracellular injections of fluorescent tracer dyes in the sea urchin *Strongylocentrotus purpuratus* (Cameron and Davidson, 1991). Sea urchin blastomeres through at least the 32-cell stage have specific, predictable fates (Fig. 1.6), a condition common to many phyla called 'invariant cell lineage'.

Many other experimental embryologists extended this branch by their studies on different subjects, such as F.R. Lillie (1870–1947) on fertilization, and J. Loeb (1859–1924) on artificial parthenogenesis. Experimental embryology, which began in the late 19th century and reached its golden period during the early 20th century, changed the meaning

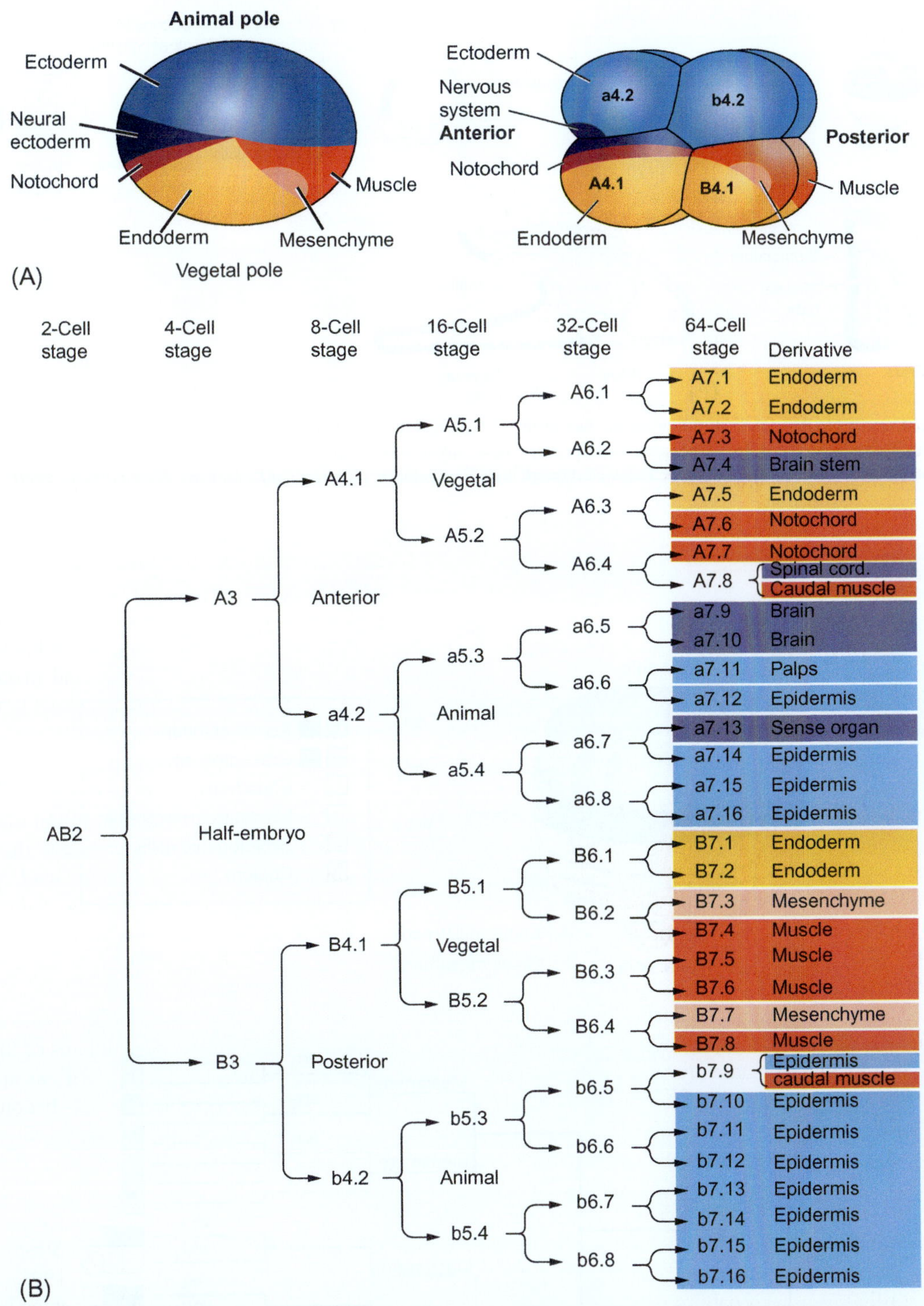

FIG. 1.4 Fate map of the tunicate embryo. (A) The one-cell embryo (left), shown shortly before the first cell division, with the fate of the cytoplasmic regions indicated. The eight-cell embryo on the right shows these regions after three cell divisions. (B) A linear version of the fate map, showing the fates of each cell of the embryo. *((A) After Nishida (1987) and Reverberi and Minganti (1946); (B) After Conklin (1905) and Nishida (1987). From Gilbert (2000).)*

of embryology. Not only did its aim become descriptive or comparative, but also it deals with much-detailed phenomena and processes. It deals with the protoplasmic organization of the egg and the sperm, their genetic constitution and the nature of the process by which the sperm initiates development in the egg. This modern embryology also traces the history of the different cleaved cells and their fates. It tries to explain the way in which the differentiation of each cell occurs. It is interested mainly in the mechanics of each process of each stage and how the different organs arise in the adult.

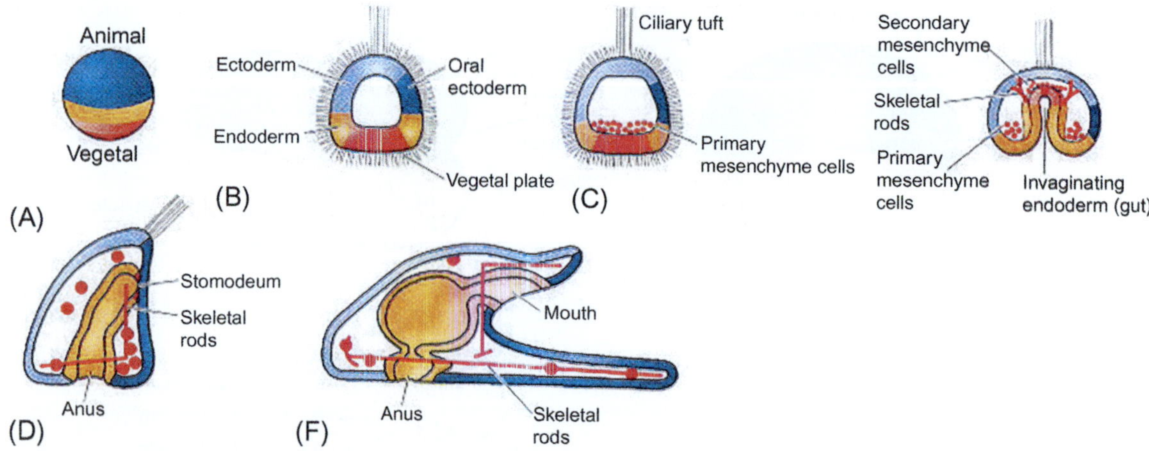

FIG. 1.5 Normal sea urchin development, following the fate of the cellular layers of the blastula. (A) Surface view of the blastula. (B) Late blastula with ciliary tuft and flattened vegetal plate. (C) Blastula with primary mesenchyme. (D) Gastrula with secondary mesenchyme. (E) Prism-stage larva. (F) Pluteus larva. Fates of the zygote cytoplasm can be followed through the *colour pattern*. *(Courtesy of D. McClay. From Gilbert (2000).)*

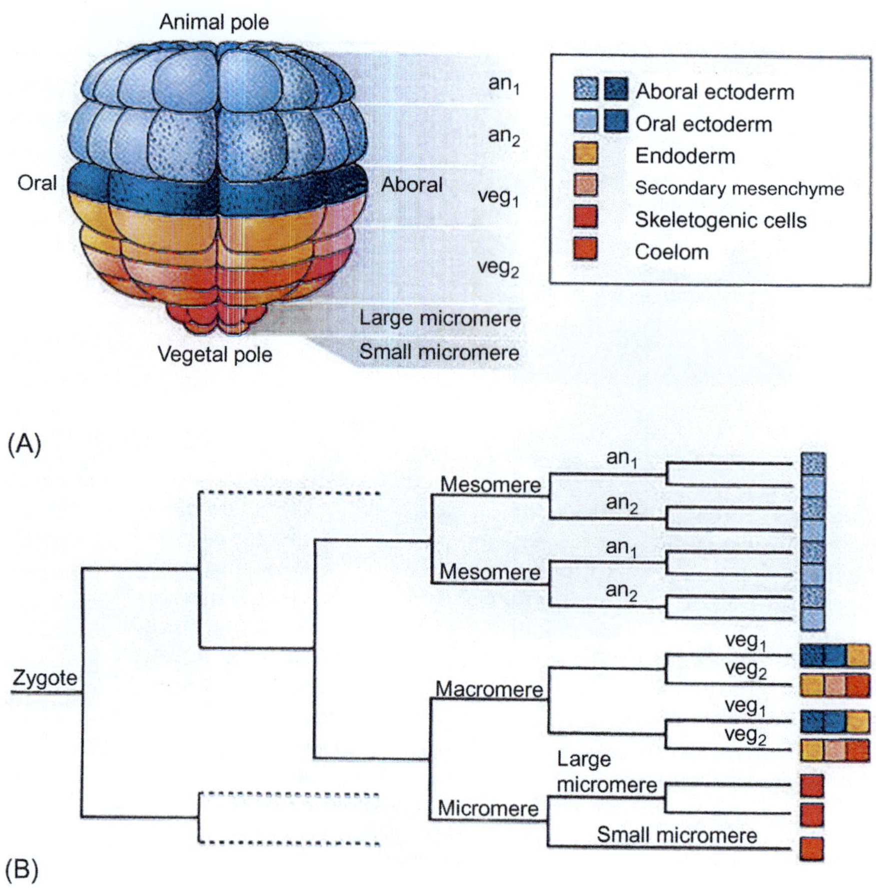

FIG. 1.6 Fate map and cell lineage of the sea urchin *Strongylocentrotus purpuratus*. (A) The 60-cell embryo is shown, with the left side facing the viewer. Blastomere fates are segregated along the animal-vegetal axis of the egg. (B) Cell lineage map of the embryo. For simplicity, only one-quarter of the embryo is shown beyond second cleavage. The veg_1 tier gives rise to both ectodermal and endodermal lineages, and the coelom comes from two sources: the second tier of micromeres, and some veg_2 cells. *(After Wray (1999). From Gilbert (2000).)*

Gradually, it became largely replaced by molecular embryology. Molecular biology, in general, began in 1967, when Francis Crick co-discovered the DNA and unravelled its helical structure. Since that time, most embryology researches have dealt with the ontogeny of molecular diversity and with how different biochemical substances direct and control development. The origin of cellular diversity (molecular differentiation) in an individual, which is composed of genetically identical cells, is one of the central themes of embryological research. Developmental biology has thus emerged as a field of study that attempts to correlate the genes with morphological change, and to determine which genes are responsible for each morphological change that takes place in an embryo, and how these genes are regulated.

REFERENCES

Cameron, R.A., Davidson, E.H., 1991. Cell type specification during sea urchin development. Trends Genet. 7 (7), 212–218.

Carlson, B.M., 1981. Patten's Foundations of Embryology. McGraw-Hill, New York. Cited in Gilbert (2000).

Conklin, E.G., 1905. The organisation and cell lineage of the ascidian egg. J. Acad. Nat. Sci. Phila. 13, 1–119.

Gilbert, S.F., 2000. Developmental Biology, sixth ed. Sinauer Association, Inc. Publishers, Sunderland, MA.

Hartsoeker, N., 1694. Essai de dioptrique, Paris.

Holmes, S.J., 1944. Recapitulation and its supposed causes. Quart. Rev. Biol. 19 (4), 319–331.

Huettner, A.F., 1949. Fundamentals of Comparative Embryology of the Vertebrates. Macmillan Company, New York. (Translated by Sidra, S. and Abdel Wahab, A. Dar Nahdat Misr for Print and Publishing.)

Needham, J., 1934. A History of Embryology. Cambridge University Press, Cambridge, England. Cited in Willier et al., 1956.

Nishida, H., 1987. Cell lineage analysis in ascidian embryos by intracellular injection of a tracer enzyme. III. Up to the tissue-restricted stage. Dev. Biol. 121, 526–541.

Pander, C., 1817. Beitrage zur Entwicklung-eschichte des Hungers im Eye. Bonner, Wurzburg. Cited in Gilbert (2000).

Reverberi, G., Minganti, A., 1946. Fenomeni di evocazione nello sviluppo dell'uovo di Ascidie. Risultati dell'indagine sperimentale sull'ouvo di Asidiella aspersa e di Ascidia malaca allo stadio di 8 blastomeri. Pubbl. Sta. Zoo. Napoli 20, 199–252.

Wray, G.A., 1999. Introduction to sea urchins. In: Moody, S.A. (Ed.), Cell Lineage and Fate Determination. Academic Press, New York, pp. 3–9.

Chapter Outline

Gametogenesis and gametes **17**
- Origin of the germ cells in some cnidarians 17
  - In *Obelia* sp. 17
  - In *Eudendrium* sp. 17
  - In *Hydra* sp. 17
- Natural phenomena indicating the identification of primordial gonocytes 18
- Features of the nucleus 19
  - Chromatin diminution in *Ascaris* 19
  - Chromosome elimination in *Sciara* 20
- Features of the cytoplasm 21
  - Localized cytoplasmic factors 21
- Formation of gametes (gametogenesis) 22
  - Oocyte growth (deutoplasmogenesis) or vitellogenesis 24
    - Deutoplasmogenesis in *Portunus pelagicus* 24
    - Deutoplasmogenesis in *Asterina pectinifera* 29
  - Meiotic arrest in animal oocytes 29
  - Life span of gametes 30
    - Senescence of spermatozoa 31
    - Senescence of egg 31
- Structure of gametes 31
  - Structure of the sperm 31
    - Evolution of the sperm 31
      - The typical marine spermatozoon 32
        - The acrosome 32
        - The nucleus 34
        - The centrioles and the anchoring structures 34
        - The axoneme 35
        - Mitochondria 35
  - Structure of the mature egg (the unfertilized egg) 36
    - Cortical granules 36
    - Exocytosis 37
    - Fate of cortical granule membrane after exocytosis 37
- Types of gametes 38
  - Types of sperm 38
  - Types of eggs 40
- Maturation of gametes 42
  - Oocyte maturation 42
  - Spermatozoon maturation 43

Fertilization **43**
- Description of fertilization reaction in sea urchin eggs 44
- Formation of the fertilization membrane 47
- Up-to-date concept of the process of fertilization 49
  - I-Cell adhesion (recognition and contact between sperm and egg) 49
  - II-Regulation of sperm entry into the egg and prevention of polyspermy 52
    - Ion-dependent changes in membrane potential 53
    - Other plasma membrane polyspermy blocks 53
    - Elevation of the fertilization envelope 55
    - Various secretory products released by cortical granules 55
    - Release of peroxide to inactivate sperm 56
- Other activities of the cortical granules 56
  - Hardening of the fertilization envelope 56
  - Role of Ca^{++} in the initiation of cortical granule reaction 57
  - III-Fusion of the haploid nuclei of the egg and sperm 58
  - IV-Activation of egg metabolism to start development 59

Development **60**
- Early development 60
  - Cleavage 60
  - Blastulation 61
  - Gastrulation 61

References **64**

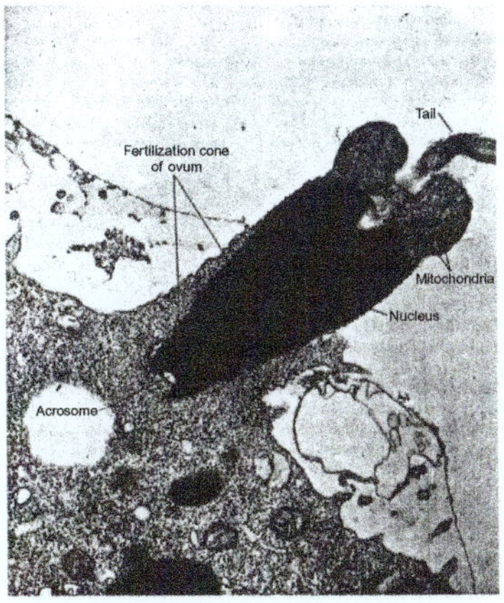

Chapter 2

Introduction

GAMETOGENESIS AND GAMETES

Gametogenesis (Fig. 2.8) is one of the vital activities through which the gametes are formed. This process is markedly similar among most animals. Generally, primordial germ cells (gametocytes) appear early in the development of an animal and either migrate to, or form the locus of, the gonads.

Origin of the germ cell (gametocyte) in some cnidarians

In Obelia sp.

The germ cells have been traced in the early development of the medusa on the blastostyle as originating in the ectoderm of the manubrium. The germ cells then pass through the endoderm and along the radial canals to the gonads where, before maturity, they leave the endoderm and lie between the ectoderm of the subumbrella surface, and the mesoglea. When they ripen, they burst through the ectoderm into the sea, where fertilization and development take place. Each medusa produces either male or female germ cells.

In fact, in *Obelia*, there is no true alternation between gametic and agametic reproduction 'generation'. The medusa is merely a modified zooid with free-swimming existence facilitating the possible dispersal of gametes, a factor of great importance to a sedentary animal. The gametes carried in the gonads of the medusa do not actually arise there, but may be observed in various stages of maturity in the ectoderm of the blastostyle. Later they migrate into the medusa and take up their position on the radial canals, making the differentiation between gametic and agametic generations impossible.

In Eudendrium sp. (loss of medusoid phase)

Among the hydroids with sessile medusoid or gonophores, there are many forms in which the medusoid structure is lost. A bud like structure is found in which a transverse section shows simply successive layers of ectoderm with generative cells; there is a structureless lamella and endoderm round an enteron, which does not open by a mouth. In such forms, the migration of germ cells, as in *Obelia*, is very noticeable. Thus, in *Eudendrium* (Fig. 2.1) the germ cells are often distinguished making their way along the coenosarc towards developing gonophores.

This is a striking example of a bushy, much branched colony. The male and female gonophores arise from the gastrozooid and differ from each other in appearance. No medusa generation occurs. If this degeneration of medusa is followed to its conclusion, a stage is reached at which there are no special reproductive buds at all, but the generative cells occur in the body of hydroid. This is a condition in *Hydra* where the multiplication of the interstitial cells at different positions produces either testes or ovaries.

In Hydra sp.

The germ cells arise from interstitial cells, which are migratory stem cells that divide to maintain the population and/or differentiate into several somatic cells (any type of cells) and the gametes. Thus interstitial cells are multipotent cells.

In *Hydra*, there is a population of interstitial cells composed of multipotent cells, unipotent stem cells, progenitor cells and cells about to enter a differentiation pathway (Fig. 2.2). Littlefield (1986) has provided strong evidence that there are subpopulations of interstitial cells specially committed to the germ line.

Invertebrate Embryology and Reproduction. https://doi.org/10.1016/B978-0-12-814114-4.00002-3

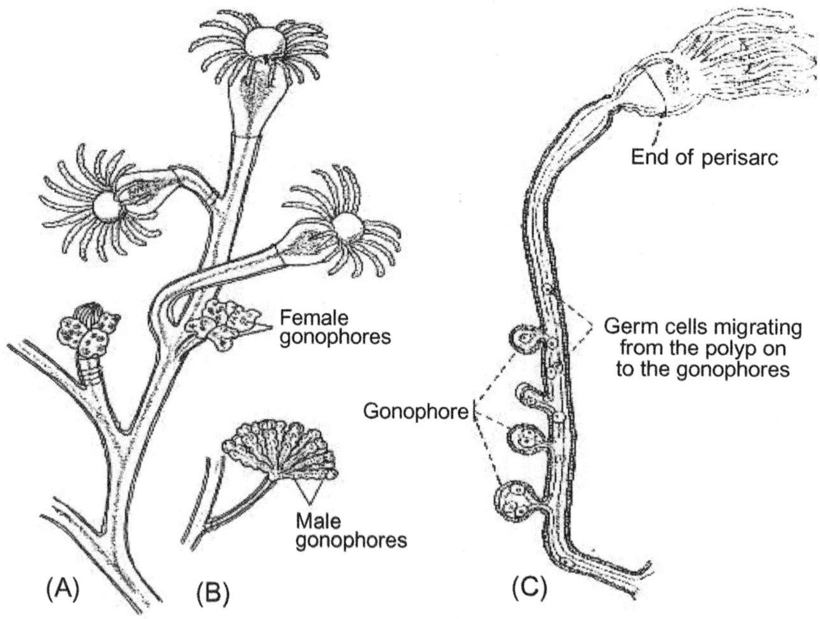

FIG. 2.1 Eudendrium (suborder Gymnoblastea). (A) Part of colony, with female gonophores (sketched from life). (B) Male gonophores *(A and B from Hickman, 1973)*. (C) Single polyp *(from Borradaile et al., 1977)*.

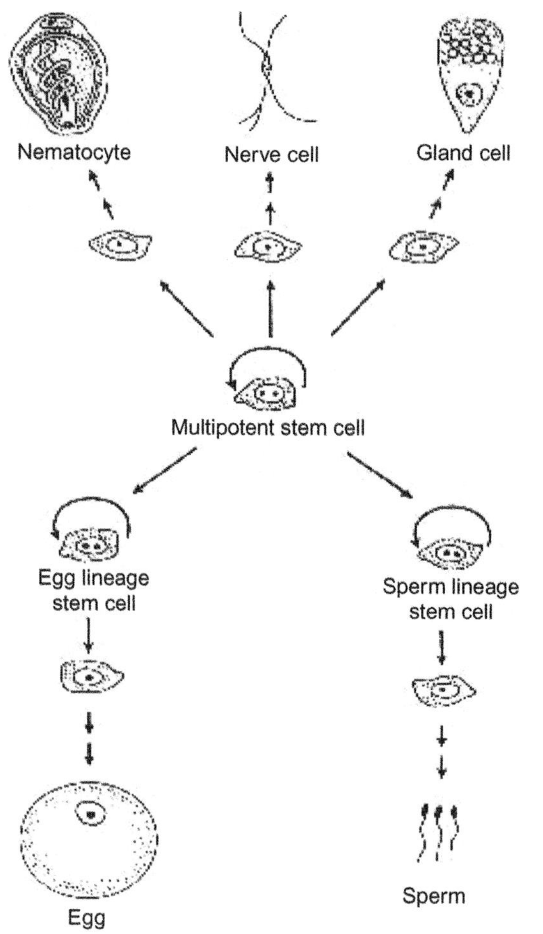

FIG. 2.2 Interstitial cell lineages is *Hydra*. Stem cells divide to replenish themselves and send committed cells down a differentiation pathway. *(After Littlefield (1994), from Gilbert and Raunio (1997).)*

With respect to the germ cells, some of the multipotent interstitial stem cells give rise to unipotent stem cells, which undergo both self-renewal and gamete differentiation. There is a unipotent stem cell for the egg lineage and another one for the sperm lineage. Using a fluorescent monoclonal antibody to label the cells, Littlefield (1986) clearly showed that there is a distinct subpopulation of interstitial cells developmentally restricted to sperm production. This antibody stains the entire sperm lineage, but does not stain egg or somatic cell lineages. In fact, stem cells committed to the sperm or egg lineages have been isolated from males and females of three *Hydra* species. Littlefield (1994) further demonstrated that the germ line cells, and not the somatic cells, are responsible for determining the sexual phenotype in *hydra*.

Interstitial cells preserve an embryonic character and may develop into germ cells, musculoepithelial cells and cnidoblasts. They are important in regeneration. There is much interest in the ultimate origin of the germ cells because they give rise to new individuals and serve to transmit hereditary characteristics between successive generations. The doctrine (sets of beliefs) of germinal continuity proposed by Weismann (1834–1914) held that the germ cells or germ plasm, constituted a substance apart from external influence and from the body, or somatoplasm.

Natural phenomena indicating the identification of primordial gonocytes (germ cells)

For each new generation, the parent germ plasm produces both the body (soma) and the germ plasm of new individuals. The continuity of the germ plasm is clear in some invertebrates (*Ascaris*, *Sagitta*, and *Miastor* (gallfly)), where one cell in the early cleavage of the egg can be traced as the ultimate origin of the future germ cells.

In these, the identification of the primordial germ cells is based on certain distinguishing features of the nucleus or the cytoplasm. There are some natural phenomena indicating the identification of primordial germ cells (gonocyte).

- **Features of the Nucleus:**
  1. **Chromatin Diminution** in *Ascaris*: the classic case is that of *Ascaris megalocephela* (=*Parascaris equorum*)
  2. **Chromosome Elimination** in *Sciara* (Fungus fly)
- **Features of the Cytoplasm:**
  1. **Localized Cytoplasmic Factors** termed germ cell determinants as in *Sagitta*.

Features of the nucleus

Chromatin diminution in Ascaris

Chromatin diminution is a fascinating exception to the general rule of the genome. Tadano (1968) reported that this phenomenon, in which selected parts of the chromatin are discarded into the cytoplasm, was first discovered in *Parascaris equarum* (Boveri, 1887c), and it was also he who made the later observations of the process (1890, 1892). His work was repeated and confirmed by Dostoiewsky (1888), Schneider (1891), and Herla (1893). Boveri (1899) adapted the term **diminution**, as used by Herla, to describe this phenomenon, calling it **chromatin diminution**. It was further shown that this process occurs in *Ascaris lumbricoids* (Mayer, 1895), *Ascaris rubicunda*, *A. labiata* (Mayer, 1895), and *Rhabditis Nigrovenosa* (Shcleip, 1911). The details of the process differ to some extent among these species; it has not yet been recognized as occurring in any other nematodes.

This phenomenon occurs during the cleavage in the cells that are to form somatic tissue but not in those that are to form the germ cells, i.e. separation and establishment of germ and somatic cell lineages. Recently, Esteban et al. (1995) reported that in *P. equarum univalens*, in addition to differences in cell size, cell cycle periods and spindle orientation, germline precursor cells and somatic line blastomeres greatly differ in both DNA content and organization of chromatin. They also concluded from their studies that chromatin diminution and somatic cell behaviour are not separable and constitute strictly correlated events in *Parascaris*.

In this process, the mid-portion of each of the chromosomes (two in *P. univalens*) breaks up into a number of small fragments (about 10 per chromosome), leaving two large terminal pieces. According to Boveri, this mid-portion (central part of the chromosome) forms a series of particles connected by a hyaline matrix. The particles then separate into granules, the hyaline matrix disappears, and the granules are distributed to the two new blastomeres to form the daughter nuclei. In the newly formed blastomeres, the daughter nucleus appears with the granules clustered in its centre, and the discarded chromatin of the terminal pieces remain in the cytoplasm and slowly disappear. The ability of the small fragments (or granules) to continue mitotic division in the daughter nucleus is evidently due to the fact that each possesses a centromere, the original chromosomes being polycentric. It is now well known that chromosomal terminals that are devoid of a centromere fail to attach to the spindles and to divide.

The diminution phenomena start at the second or the third division of the egg. In the former case, it occurs in the dorsal, somatic cells S_1 (AB) and not in the ventral stem cells P_1 (Fig. 2.3). In the latter case, it occurs in three cells at once but not in the stem cell P_2. In both cases, P_2 divides into a somatic cell, which undergoes diminution (of its next division) and another stem cell, P_3. The latter again divides (without diminution) into a somatic cell (that undergoes diminution at its next division) and the definitive primordial germ cell P_4. It was supposed that the differences in the cytoplasm of different regions of the egg were responsible for the occurrence or nonoccurrence of the diminution process and, at the same time, for the determination of the germ cells. This view was supported by results of studies on polyspermic and on centrifuged eggs. Centrifugation experiments provided the more direct evidence (Boveri).

This phenomena is well studied in *Ascaris megalocephala* (Cloquet) (=*Parascaris equorum* (Groza)) which is an intestinal parasite in horses with two different strains:

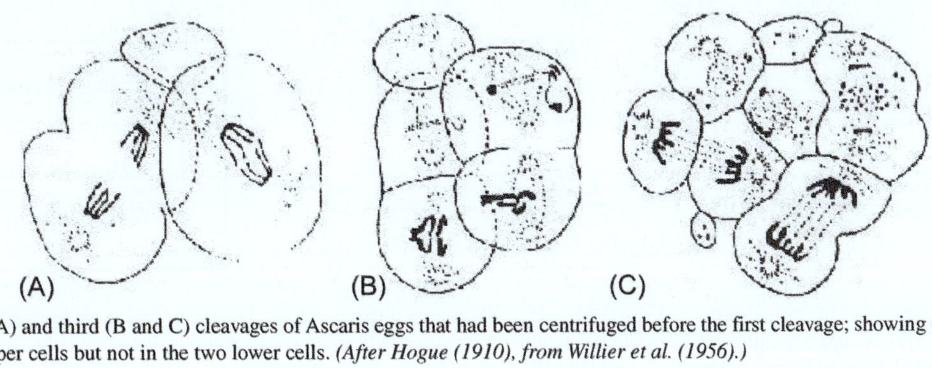

FIG. 2.3 Second (A) and third (B and C) cleavages of Ascaris eggs that had been centrifuged before the first cleavage; showing (in B and C) chromatin diminution in the upper cells but not in the two lower cells. *(After Hogue (1910), from Willier et al. (1956).)*

P. equorum bivalens in which the primordial germ cells have only four chromosomes and their gametes have two; and *P. equorum monovales*, in which the primordial germ cells have two chromosomes and the gametes one. The zygotes have four or two chromosomes, respectively.

While chromatin diminution was observed by Boveri over 100 years ago, it can now be studied on the molecular level. This has been accomplished in the roundworm *Ascaris suum* (=*Ascaris lumbricoids*), a parasite in the intestines of pigs (large quantities of this worm can be obtained cheaply from slaughterhouses all over the world). These studies reveal that chromatin diminution is not a haphazard ripping apart of the genome. Rather, this diminution is a carefully orchestrated genome rearrangement, where certain regions of the DNA are degraded and new telomeres are constructed.

Kubota et al. (1994) recorded that all the somatic cell precursors in *A. suum* undergo chromatin diminution. The process differs slightly from the one that Boveri observed in *Parascaris univalens*. First, in *A. suum*, diminution was not seen before the third cell division. Second, there is some chromosomal fragmentation in the germ line cells of *Ascaris*, and this results in the loss of some telomeric chromatin. Third, in *Parascaris*, about 85% of the total nuclear DNA is eliminated. In *Ascaris*, this figure is closer to 25%. In both species, all the heterochromatin is lost from the somatic cell nuclei. Chromatin diminution is not a general property of roundworms, neither it is in all species of parasitic roundworms. In the free-living nematode *Caenorhabditis elegans*, there is no known chromatin diminution.

Chromosome elimination in Sciara

Tyler (1956) recorded a remarkable example of differential behaviour of the chromosomes in germ cells and somatic cells, occurring in the fungus fly *Sciara*, studied extensively by C.W. Metz and his co-workers. The zygote starts development with three pairs of autosomes, three X-chromosomes, and (in 12 out of 14 sp. examined) one, two, or three large chromosomes called 'limited' chromosomes (Fig. 2.4). At the sixth division (sometimes the fifth) of the zygote, the 'limited' chromosomes are eliminated from all of the somatic nuclei. At this stage, the nuclei have migrated from the middle of the egg, where the zygote nucleus is originally located, nearly to the periphery. One or two of the nuclei at the posterior pole of the egg form the primordial germ cells, which retain for some time the full complement of chromosomes present in the zygote. Another elimination of chromosomes occurs in the somatic cells at the seventh or eighth division. At this time, the somatic nuclei of the female-producing eggs eliminate one of the three X-chromosomes and the male-producing eggs eliminate two X-chromosomes.

Schmuck (1934) studied *Sciara pauciseta* and reported that "in all species of *Sciara* that far adequately studied, the chromosome group of the germ-line in both sexes (i.e. spermatogonia and oogonia) consists of symmetrical pairs of chromosomes – except for variation in the number of 'limited' chromosomes which are found only in the germ line". He also added that the chromosome group of the male soma, e.g. in *Sciara coprophila*, is composed of three symmetrical pairs of chromosomes and an unpaired chromosome (Fig. 2.4). The eighth chromosome found in the

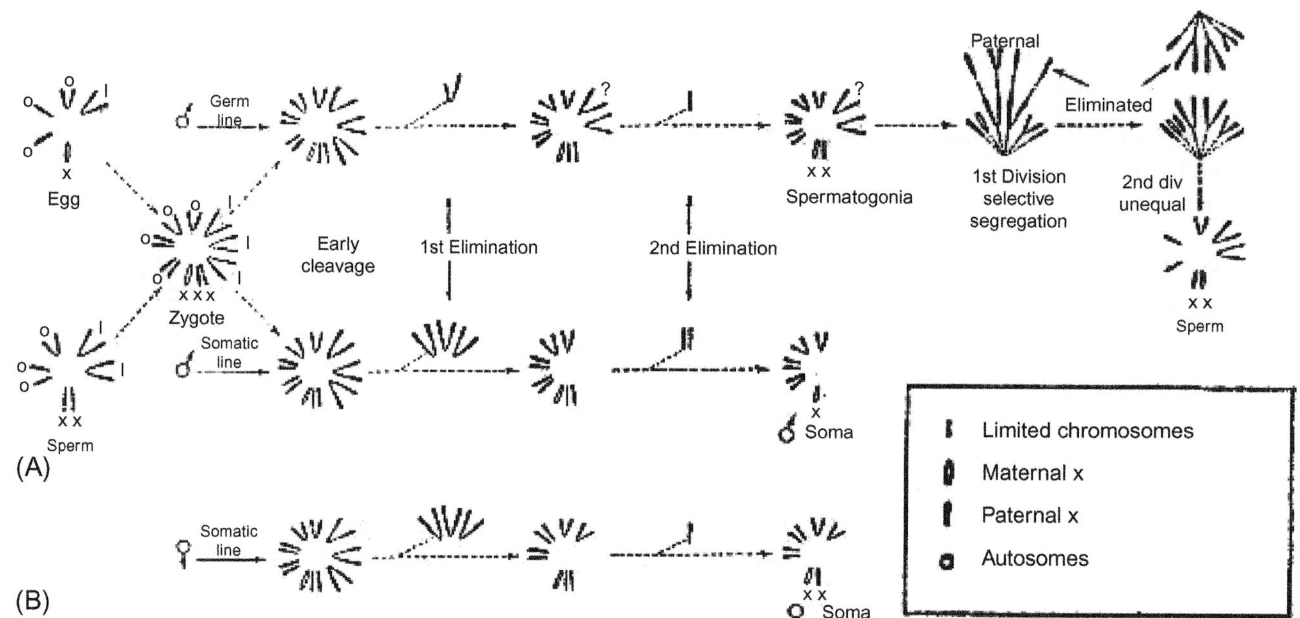

FIG. 2.4 Diagram of behaviour of chromosomes in *Sciara coprophila*, illustrating difference between somatic and germ line cells in regards to elimination of chromosomes. The female germ line (not illustrated) would be the same as in diagram (A) except that the meiotic divisions do not involve the cytological peculiarities exhibited in spermatogenesis. *(After Metz (1938), from Willier et al. (1956).)*

female soma is eliminated during cleavage in the somatic line of the male (Metz, 1931). Genetic evidence indicates that this unpaired chromosome is regularly the maternal X-chromosome, and therefore, it is the paternal sex chromosome that is regularly eliminated from the male soma. It seems probable that this parental sex chromosome is an X-chromosome not a Y, and that sex in the male is determined by the constitution of the soma.

How this elimination occurs? In these elimination divisions, the 'limited' and X-chromosomes that are to be discarded fail to divide, or divide incompletely. They are then, left on the middle of the spindle in the anaphase and do not become incorporated in the daughter nuclei, but slowly break up and disappear in the later embryo. In *Sciara*, an elimination also occurs later in the germ cells after they have migrated to the site of the gonads, one of the X-chromosomes (one of the two paternal X's) being extruded from each nucleus in both male and female larvae.

The observations on the chromosomes of the germ cells of the *Sciara ocellaris* Const. (Berry, 1939) have been made from the time of their differentiation to the formation of the gonads. These observations indicate that there is no chromosome elimination during cleavage stages of the germ cells comparable to the process that occurs in the somatic cells. However, there is an elimination of one chromosome from the germ cells of both the male and female embryos after these cells have migrated to their future site. This elimination occurs from a resting cell and is apparently effected by the moving of the chromosome from its position within the nucleus directly through the nuclear membrane into the cytoplasm (Fig. 2.5). The elimination occurs at a definite stage in embryonic development and approximately at the same time in all the germ cells of a single individual. The eliminated chromosomes remain in the cytoplasm for several days and then gradually degenerate. This elimination is described (Berry, 1941) as occurring in a resting nucleus, the chromosome being expelled through the nuclear membrane into the cytoplasm. The eliminated chromosomes cannot be distinguished morphologically from the other six rod-like chromosomes of the group, but the genetic evidence shows that it is one of the two sister sex chromosomes contributed by the sperm (Metz, 1938). Presumably, there is also an elimination of one or more 'limited' chromosomes from the germ cells in some cases, but this has not, as yet, been described.

Since the germ cells of *Sciara* have the same chromosome set in the male as in the female, the determination of whether they are to produce spermatozoa or eggs is evidently effected by the somatic cells which, as a result of the elimination of one or two X's at the seventh (or eighth) division, have the typical XO constitution in the male and XX in the female. This, in turn, is evidently determined in the female parent, which as a rule, produces unisexual families. The genetic evidence accumulated by Metz and his co-workers shows that the two X-chromosomes are alike (XX) in the male-producers but differ genetically (XX′) in the female-producers. The results of fertilization in these two types are diagrammed in Fig. 2.6.

Features of the cytoplasm

Localized cytoplasmic factors in germ-cell determination

Special cytoplasmic granules (termed germ-cell determinants) that are considered to be important for the formation of the germ cells, have been described among the scyphozoans (*Equorea*), cheatognathas (*Sagitta*), rotifers (*Asplanchna*), insects (*Chironomus*, *Calliphora*, *Calligrapha*, *Lepintotarsa*, *Copidosoma*, *Trichogroma*, *Apanteles*), crustaceans (*Cyclops*, *Diaptomus*, *Polyphemus*, *Moina*) and amphibians (*Rana*). According to the studies, the cells that receive these granules ultimately become germ cells. However, there is, as yet, no direct experimental evidence, such as might be derived from centrifugation experiments, to show that these particular granules are causative agents in germ cell determination.

Apart from the question of the particular granules and chromatin diminution, the importance of localized cytoplasmic factors in germ cell determination is very well

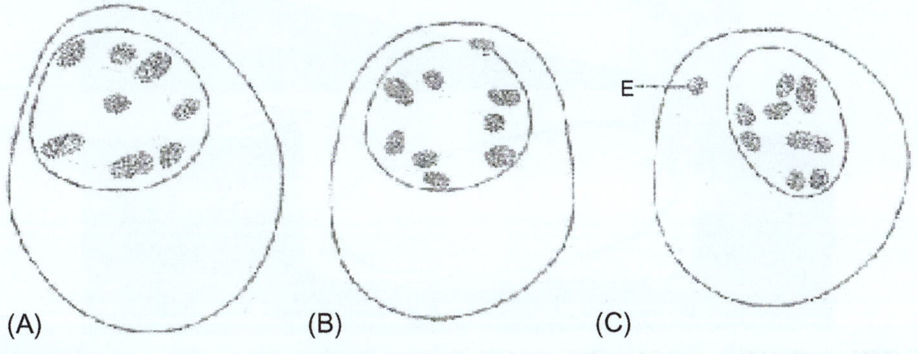

FIG. 2.5 Camera Lucida drawing from aceto-carmine preparations of cells in the germ line of *Sciara ocellaris* Const. 1800×. (A) A germ cell before migration, showing nine chromosomes. (B) A germ cell in which the one chromosome is being eliminated from the nucleus. (C) A germ cell after elimination has occurred. There are eight chromosomes in the nucleus and one in the cytoplasm (E).

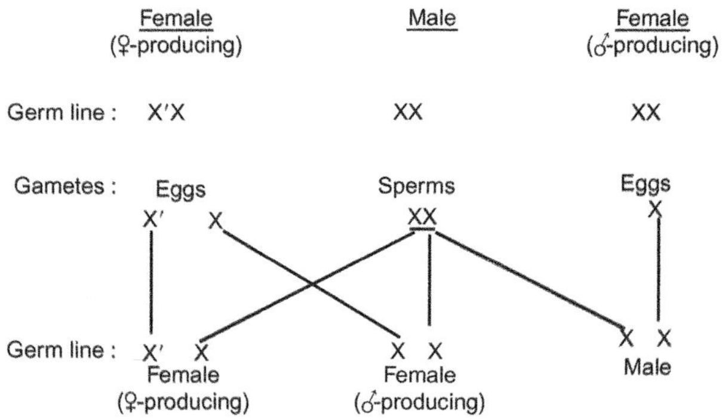

FIG. 2.6 Metz's interpretation of the production of 'unisexual' families in *Sciara coprophila* on the basis of genetically different (X and X') sex chromosomes. *(After Metz (1938), redrawing from Willier et al. (1956).)*

illustrated in the experiments on *Drosophila* (Geigy, 1931). These experiments were performed to inhibit the formation of germ cells by subjecting the posterior pole of the egg to ultraviolet irradiation (Fig. 2.7). After irradiation, the cleavage of the nuclei that wander into damaged pole plasm degenerate and the formation of pole cells are partially or completely suppressed. However, even in cases of complete suppression, adult flies are obtained that appear completely normal and have gonads of normal structure, although of reduced size. This work not only strikingly confirms the earlier indications that factors essential for germ-cell formation are localized in the posterior cytoplasm of the egg, but also furnishes further convincing evidence that the mesodermal parts of the gonad can develop normally in the absence of the germ cells.

Formation of gametes (gametogenesis)

Gametogenesis is a process, which is markedly similar among most animals (Fig. 2.8). In the gonads, the primordial germ cells multiply by mitosis and produce gonial cells. Gonial cells (spermatogonia in males, oogonia in females) can sometimes be separated into multiplying (primary) and terminal (secondary) stages. Primary gonial cells usually go through a limited number of mitotic divisions, usually two to six in invertebrates (Giese and Pearse, 1974), and the resulting secondary gonial cells transform into primary spermatocytes (in males) or primary oocytes (in females). In practice, primordial germ cells, primary and secondary gonial cells and young primary spermatocytes and primary oocytes are difficult to distinguish from each other; all are usually relatively large cells, each containing a conspicuous large nucleus. In some species, however, the primary germ cells have a conspicuous large nucleus relative to the size of the cell. Gradually, the cytoplasm increases, changing this proportion.

Both primary spermatocytes and primary oocytes undergo a number of characteristic nuclear changes during prophase of the first meiotic division. In spermatogenesis, these changes are followed by the first meiotic division of the primary spermatocytes to form the secondary spermatocytes. The second meiotic division occurs soon after the secondary spermatocytes are formed, and produces haploid spermatids (four spermatids for each primary spermatocyte). Differentiation of the spermatids into mature spermatozoa (sperm) is termed spermiogenesis or spermateleosis. Spermiogenesis is always completed, and usually large numbers of sperm are accumulated, before the sperm are shed.

On the other hand, nuclear changes are arrested when the diplotene stage is reached in the primary oocyte, and the paired, partially separated, chromosomes are dispersed in the nuclear sap of the enlarged nucleus (termed the germinal vesicle). Usually the nucleus contains one conspicuous nucleolus at this stage. Subsequent primary oocyte development goes through two distinct phases: First, there is a period of intense RNA synthesis and the large number of ribosomes are accumulated, typically causing the cytoplasm of the cell to be strongly basophilic. The second is the growth of the primary oocyte through the uptake and

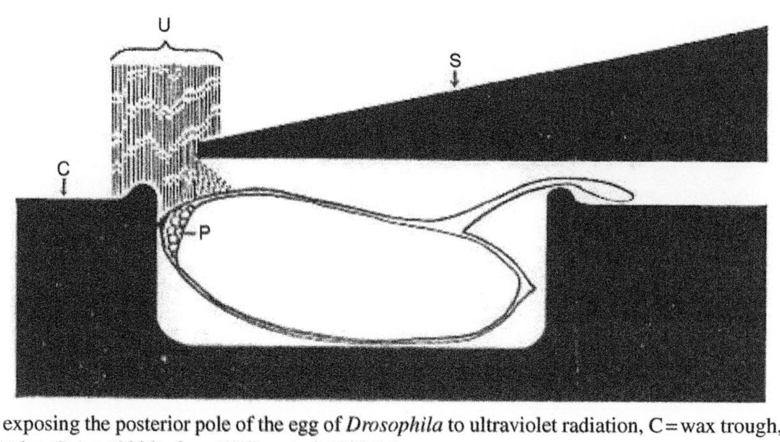

FIG. 2.7 Geigy's method for exposing the posterior pole of the egg of *Drosophila* to ultraviolet radiation, C = wax trough, U = ultraviolet beam, S = brass screen, E = egg, P = pole cells. *(After Geigy (1931), from Willier et al. (1956).)*

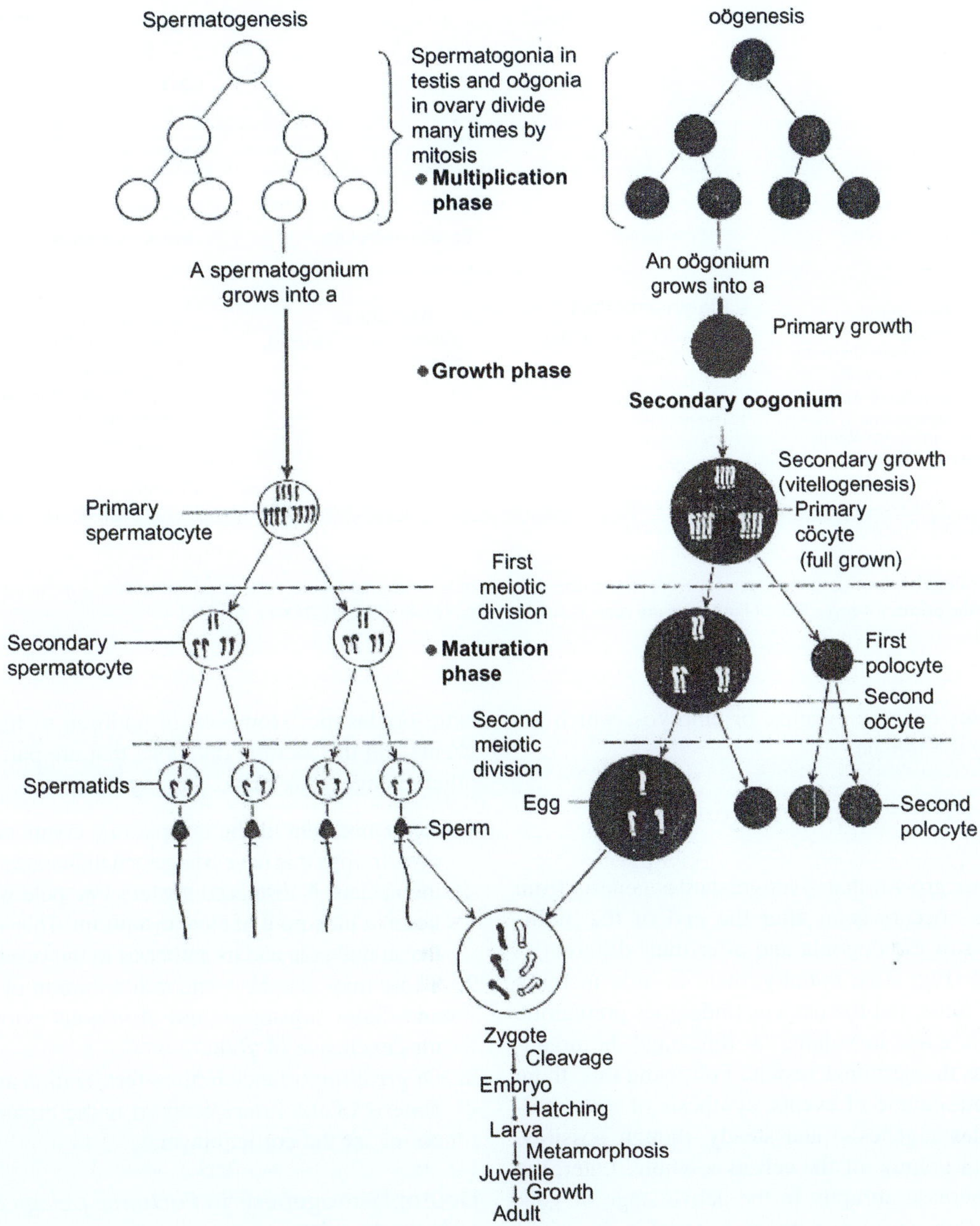

FIG. 2.8 Diagram of the general cellular changes that occur during gametogenesis. *(Modified from various sources.)*

accumulation of nutrients (due to plasmogenesis or vitellogenesis), and often the cytoplasm becomes weakly basophilic or even acidophilic. Primary oocyte growth is usually assisted by accessory cells in the ovary called nutrient, follicle, or nurse cells. Full-grown primary oocytes vary in size among different species, but usually they range from about 75 μm to several millimetres in diameter. After reaching full size, the primary oocytes may undergo the two meiotic divisions (maturation divisions) to form haploid ova.

Secondary oocytes, like secondary spermatocytes, are usually very transitory and are not often encountered. Some invertebrates (e.g. many echinoids) accumulate ova in the ovaries and these are shed during spawning. In many other invertebrates, however, 'eggs' are shed as full-grown, diploid primary oocytes, and the meiotic divisions are completed after spawning (e.g. asteroids) or fertilization (e.g. Platyhelminthes). Fig. 2.9 shows stages of egg maturation in some different invertebrate animals. These differences in the timing of the completion of meiosis in respect to fertilization have led to considerable confusion in terminology. In this treatise, the term *ova* refers to haploid female gametes only; the more general term *eggs*, is used for female

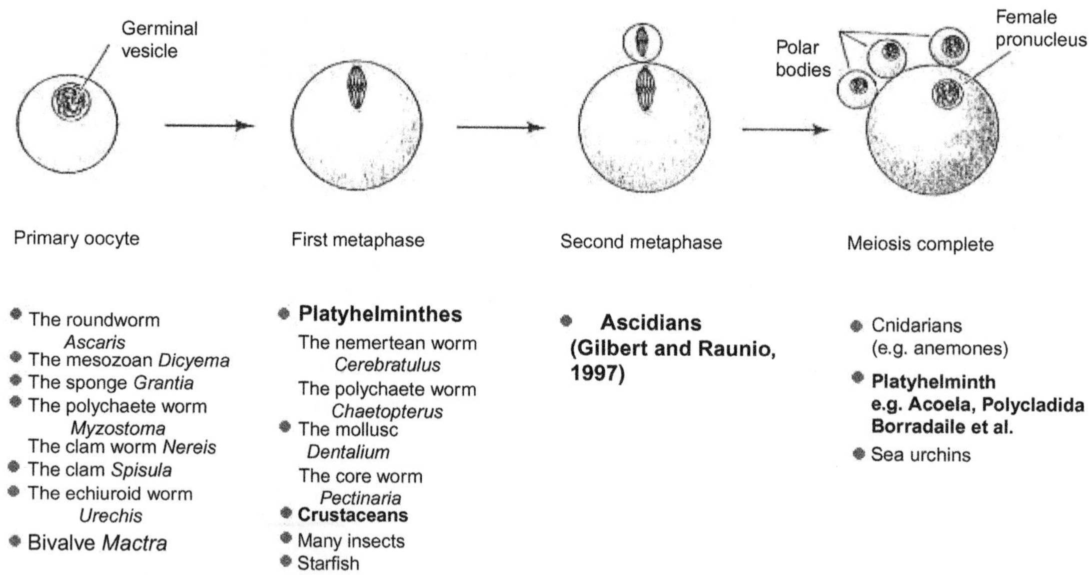

FIG. 2.9 Stages of egg maturation at the time of sperm entry in different invertebrate animal species. The germinal vesicle is the name given to the large diploid nucleus of the primary oocyte. The polar bodies are seen as smaller cells. *(From Gilbert (2000).)*

gametes (oocytes or ova), zygotes, or embryos, which are held or shed by the female.

Oocyte growth (deutoplasmogenesis) or vitellogenesis

The conspicuous growth that distinguishes oogenesis from spermatogenesis occurs soon after the end of the multiplication period of the oogonia and after their differentiation as oocytes (Fig. 2.8). Initially, there is little increase in cytoplasmic mass, but the nucleus undergoes premeiotic changes and increases in volume; at this stage the organelle is known as the germinal vesicle. Following this, there is, in the general course of events, synthesis of new cytoplasm (deutoplasmogenesis) and steady, though possibly, slow increase in volume of the cell as a whole. Later, the volume may increase abruptly in the active stage of yolk formation (vitellogenesis) as nutrient material is deposited in the cytoplasm and the oocyte reaches its definitive size (Fig. 2.11G).

Deutoplasmogenesis (vitellogenesis) is a time of intense synthetic activity on the part of growing oocyte, and one in which great drains are made upon the nutritive resources and reserves of the parent. This vital activity in the amphipod *Orchestia gammarella*, in natantian decapods and probably in other Malacostraca involves two phases, primary and secondary vitellogenesis (Charniaux-Cotton, 1985). Oogenesis from oogonia to the end of primary vitellogenesis is a continuous process. The primary proteic vitellus is endogenous. The secondary vitellogenesis takes place during the reproductive season. Therefore, it is now well established that the origin of yolk is intra-ooplasmic and extra-ooplasmic. However, in addition to the biochemical events and the meiotic divisions that are part of oogenesis, other changes take place. Among these are:

1. Rearrangement of the cytoplasmic components of the oocyte. In eggs that have a large or a moderate amount of yolk, the nucleus is displaced towards one pole where it usually lies in a little pool of clear cytoplasm. This is designated as the animal pole and its antithesis as the vegetal pole.
2. There may also be some redistribution of mitochondria and Golgi substance, and of various cytoplasmic granules exclusive of yolk.
3. Of great importance both to fertilization and to morphogenesis of the future embryo is the organization of the cortex, or the cortical layer.

Deutoplasmogenesis in *Portunus pelagicus* (in view of histochemistry)

Bawab and El-Sherief (1987a,b, 1988) published the study of the reproductive stages and the ovarian cycles of the female crab *Portunus pelagicus* (L., 1758). They interpreted, through histological examinations, that this crab has only one cycle in the year, with two ovarian cycles. The cells of each cycle undergo this important vital activity, deutoplasmogenesis (Bawab and El-Sherief, 1987a). This activity changes the oocytes into comparatively large ones, condensed with yolk (Fig. 2.11G and H).

Predeutoplasmic oocyte According to Charniaux-Cotton (1985) this is the primary vitellogenesis stage. The size of the ovary is comparatively very small and narrow (Fig. 2.10a). Its natural colour before deutoplasmogenesis

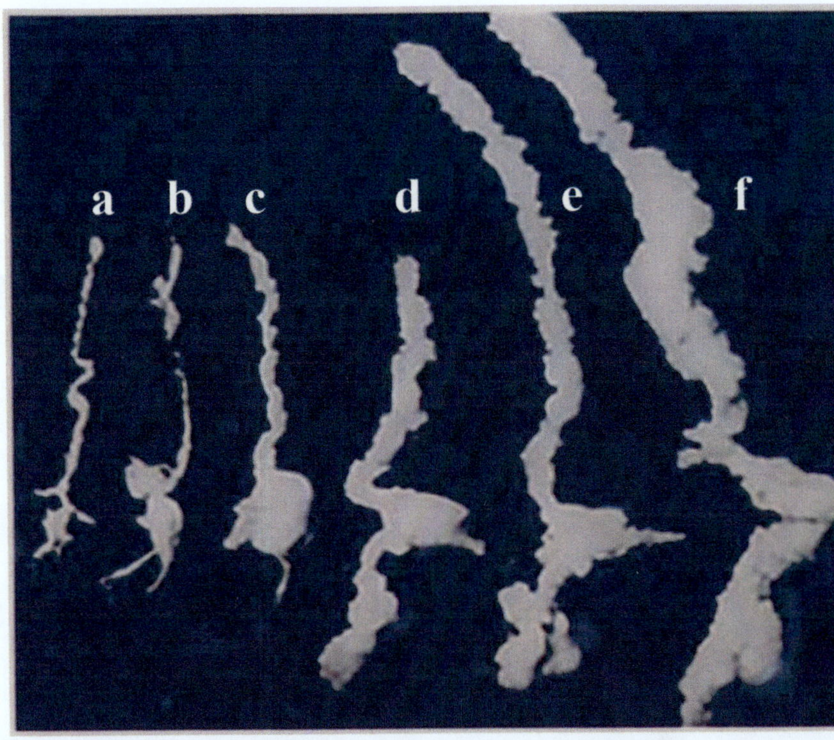

FIG. 2.10 Photomacrograph of the right side of the isolated female reproductive systems of adult *Portunus pelagicus*, showing the cyclic morphological changes during the breeding season, X = normal size, (a) immature ovary (virgin) early March; (b and c) ovary of copulated female; (b) beginning of maturation (late March–April); (c) late copulated stage; (d) early vitellogenic ovary (May–July); (e) ovary of prespent female (first spawning, May), still ovigerous female; (f) ovigerous ovary (September, second spawning), i.e. pre-September spawning. *(From Bawab and El-Sherief (1988).)*

is transparent-faint milky white (Bawab and El-Sherief, 1987b). Histologically each cell has a diameter less than 70 μm, with large nucleus and with one or more nucleoli (Fig. 2.11A and B). Histochemically, the cytoplasm proteins are, strikingly, with a complete absence of –s–s group, while the –S–H, reactive $-NH_2$ groups and the tryptophan are present in well detectable amounts (Table 2.1).

The carbohydrates are in the form of neutral and acidic mucopolysaccharides and glycogen, confirmed by Best's carmine, is also present. Goldstein (1962) suggested that strong basic groups such as arginine and lysine may stain with Best's carmine and this confirms the positive result of reaction of arginine. The lipid moiety is found in the form of the neutral and acidic nature while that of the steroid nature is absent completely.

Deutoplasmic oocytes (secondary vitellogenesis phase) The cell changes gradually both in size and chemical nature. All the changes occurring during this vital activity are grouped into four definite steps.

Step 1

This step begins nearly with the growth period of oogenesis. The ovary itself gradually enlarges (Fig. 2.10) and the diameter of each cell enlarges too. It ranges from 65 to 75 μm. Its cytoplasm becomes clearly loaded with homogeneously distributed fine droplets (Fig. 2.11C). They are not proteineous or of carbohydrate nature. They are strongly positive to Sudan black B and Nile blue sulphate. Their dominance obliterates the nature of this cytoplasm. Their nature ensures the basic or the first component of the yolk of this crab is of fatty nature composed of neutral and acidic lipids. This explains the natural pale yellow colouration accepted by the ovaries in the living state. Charniaux-Cotton (1985) reported that the 'fat body of malacostracans' synthesizes a precursor (vitellogenin) of the major egg yolk protein (vitellin). During this secondary phase, oocytes take up vitellogenin by endocytosis.

Step 2

The diameter of each cell at this step may reach up to 95 μm. The cytoplasm loaded with the fat droplets still appears with the same density and property. Additional to this picture, two somewhat large granules appear at one side of the cytoplasm (Fig. 2.11D). They are mainly of protein nature, but without tryptophan, –S–H or –S–S groups. Traces of neutral mucopolysaccharides can be detected, but both glycogen and fat have never been traced in these granules. They are given the name '**yolk nuclei**' (Bawab and El-Sherief, 1987b).

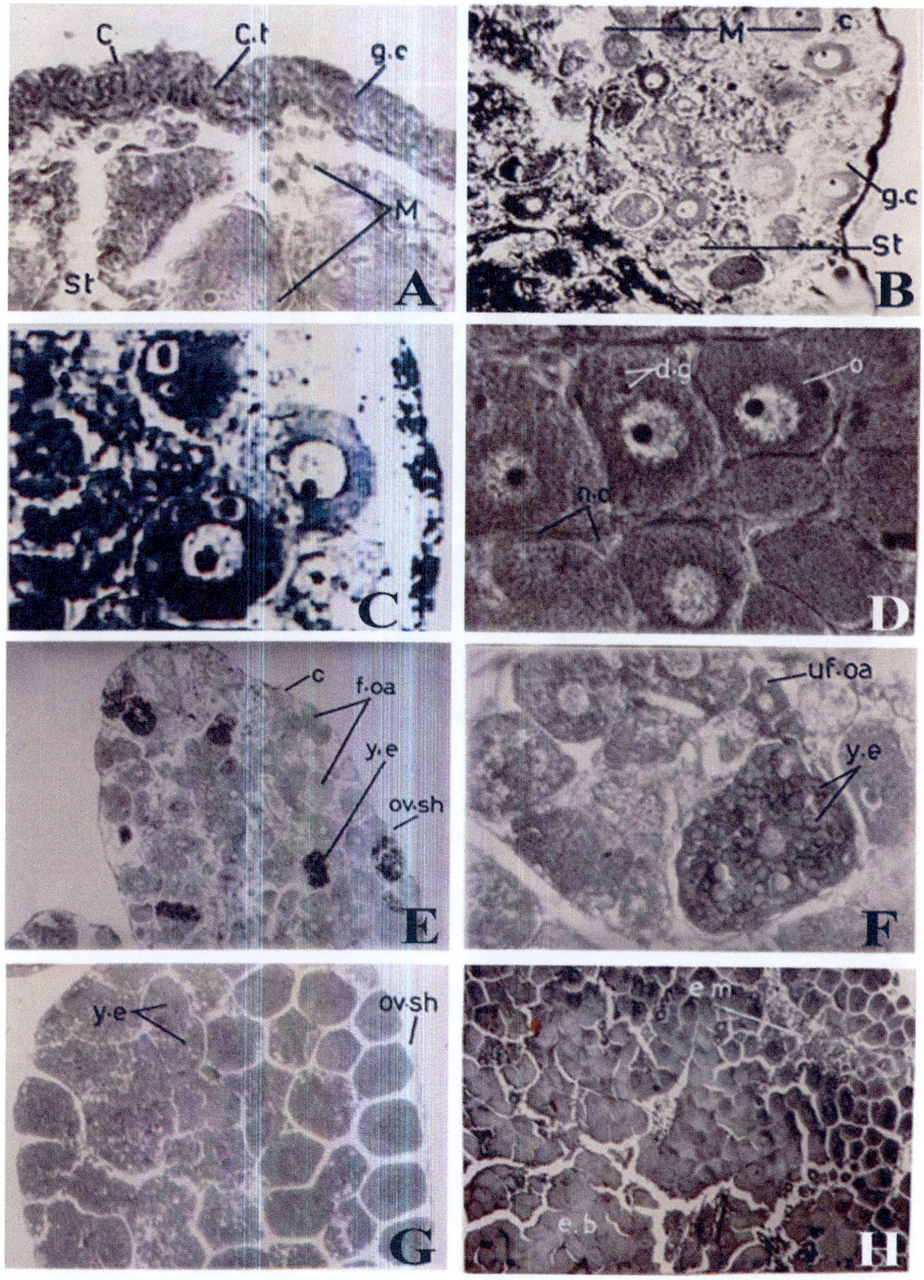

FIG. 2.11 Histological structure of the ovary of *Portunus pelagicus* during development: (A) magnified portion of a T.S. of the immature ovary, X=450; (B) magnified portion of a T.S. of ovary showing developing oocytes in the stroma, X=150; (C) highly magnified portion of stroma in figure (B) showing growing oocytes, X=300; (D) magnified portion of a T.S. of ovary showing oocytes at later stage of development (early vitellogenesis), X=400; (E) Portion of a T.S. of the ovary during the prespawn (May stage), showing a lobule with oocytes at different stages of vitellogenesis, X=150; (F) magnified vitellogenic oocyte from (E), X=600; (G) portion of a T.S. of ovary of prespent female showing oocytes at the end of vitellogenesis (yolk is a compact mass inside the oocyte, ovum), X=150; (H) T.S. of portion of ovary showing formation of egg brood, just before spawning, X=75; c=cortex, c.t=connective tissue, d.g=directive granules (yolk nuclei), e.b=egg brood, e.m=egg matrix, f.oa=fertilized egg, g.c=germinal cells, M=medulla, o=oocyte, ov. Sh=ovarian sheath, uf.oa=unfertilized egg, st=stroma; y.e=yolk mass inside oocyte (vitellogenic). *(From Bawab and El-Sherief (1987b).)*

Step 3

Histologically, it was proved that the previous two steps are oogonial while the following third and fourth steps are postoogonial and/or oocytes, during which each cell attains a conspicuous size (about 175 μm in diameter Fig. 2.11F). Thus, the ovary enlarges greatly and its natural colour gradually becomes orange (Fig. 2.10d). In addition, this step is characterized by the deposition of two additional globules (Fig. 2.11E and F), which are much larger than the yolk nuclei of step 2. One globule is larger and translucent. It consists of different substances, but the glycogen seems to be absent completely.

TABLE 2.1 The chemical nature of the different granules forming the yolk in the eggs of the crab *Portunus* during the process of deutoplasmogenesis

			Deutoplasmogenesis				
			Step 2		Step 3		
Reaction	Indication	Predeutoplasmogenesis	Cytoplasm	Directive granules	Granules	Translucent globules	Opaque globules
Mercuric bromophenol blue (Mazia et al., 1953)	Protein	bbb	bb	bb	bbbbb	bbbb	bbb
Biuret reaction (Gabe, 1976)	Peptide linkage	v	v	vv	vv	vv	vv
Aqueous bromophenol blue (Mazia et al., 1953)	Basic protein	bb	bb	bb	bb	bbb	bb
Toluidine blue (Kramer and Windrum) (Pearse, 1968)	Acidic groups (metachromatic sub.)	bb	bb	bb	bb	–	–
Xanthoproteic reaction (Pearse, 1968)	Pheny 1 group	oo	oo	ooo	oooo	o	oo
Millon reaction Benaley and Gersh modifn (Pearse, 1968)	Tyrosine	pp	pp	p	pp	p	–
Nitroso naphthol method (Udenfriend, Cooper, and Stoward) (Pearse, 1968)	Tyrosine	y	yy	y	yy	y	y
DAMB-nitrite method (Adams) (Pearse, 1968)	Tryptophan	b	–	–	–	–	–
Tryptophan reaction (Adams) (Pearse, 1968)	Tryptophan	B	–	–	–	–	–
Ninhydrin Schiff reaction (Yasuma and Itchikawa) (Pearse, 1968)	NH_2 group	p	p	p	p	pp	p
Oxidized Tannin Azo method (Dixon) (Pearse, 1968)	Amino group	p	p	p	p	pp	–
Sakaguchi reaction (Pearse,1968)	Arginine	oo	o	o	o	o	o
Ferric ferri-cyanide reaction (Pearse, 1968)	S–H	b	–	–	–	b	–
Performic acid-Alcian blue (Pearse, 1968)	S–S	–	–	–	–	–	–
PAS (Mc Manus) (Pearse, 1968)	Mucosubstance	R	R	R	R	RRRRR	RRR
Alcian blue method (Pearse, 1968)	Acid mucopolysaccharides	–	–	–	–	–	–
Alcian blue and PAS (Pearse, 1968)	Acidic and neutral MPS	bb RR	bbb RR	RR	bbb	RRRRR	RRR
Best's Carmine (Pearse, 1968)	Glycogen	pp	pp	–	ppp	p	pp

Continued

TABLE 2.1 The chemical nature of the different granules forming the yolk in the eggs of the crab *Portunus* during the process of deutoplasmogenesis—cont'd

			Deutoplasmogenesis				
			Step 2		Step 3		
Reaction	Indication	Predeutoplasmogenesis	Cytoplasm	Directive granules	Granules	Translucent globules	Opaque globules
Sudan black B (McManus) (Pearse, 1968)	Lipids	bb	bb	–	bbbb	bb	bbb
Sudan black B (Pearse, 1968)	Lipids	bb	–	–	bb	bbb	bbb
Nile blue sulphate (Cain, 1947)	Neutral and acidic lipids	b	bb	–	bbbb	bbbb	bbbb
Liebermann Bürchardt reaction (Pearse, 1968)	Cholesterols and its esters	–	–	–	–	–	–
Schultz reaction (Weber et al.) (Pearse, 1968)	Cholesterols and its esters	–	–	–	–	–	–

The symbol indicates the production of the colour, B=black; b=blue; g=green; o=orange; p=pink; R=red; v=violet; y=yellow. The repetition of the symbol indicates the intensity of the colour.

From Bawab and El-Sherief (1987a).

The translucent globules can be named lipovitelline, since they are similar in their composition to the vitelline of the yolk of the eggs of the hen. The other globule is smaller and opaque. It contains a specific amount of glycogen, in addition to the other substances. In comparison with the yolk of the eggs of hens (Thorpe, 1962), the present opaque globules are of the livetine type.

The deposition of these two types of globules is everywhere in the cell, but never on the expense or the displacement of the previous fatty droplets. The latter enlarge somewhat in their size and differ in their chemical nature. Their lipoid nature increases while conjugating probably with proteins especially of the basic nature and somewhat with carbohydrate substances of the directive granules. Therefore, the cytoplasmic appearance noticed in the predeutoplasmic stage disappears completely and the egg becomes heavily loaded with three types of spheres: enlarged fatty droplets, translucent and opaque globules. There is no definite sequence of the substances deposited in the oocytes of the different crustaceans. The fresh ovaries attain both the largest size and the deep orange colour at this step of deutoplasmogenesis. The directive granules disappear completely during this step. Their appearance during step 2 probably indicates the site at which begins the protein synthesis within the cell.

Step 4

During this step, the individuality of each of the three types of the deposited spheres disappears completely, i.e. all the deposited materials in the cell become a compact homogeneous substance (Fig. 2.11G and H) and accordingly, the size of the full-grown vitellogenic cell may decrease than that in the previous step. The ovary attains the reddish orange colour in the live state. This stage finishes both the deutoplasmogenesis and the ovarian cycle, after which the eggs start to discharge from the ovaries.

The natural colour of the ovaries that changes successively from yellow, orange to reddish orange may be due to the accumulation of the different substances mentioned above. More probably, it may be due to the deposition of a separate or conjugated carotenoid substance, which is essentially a type of lipid. Varadarajan and Subramoniam (1980) stated that in the anomuran crab *Clibanarius clibanarius,* the lipoprotein from extraovarian sources (which may be linked to carotenoid pigments) may serve as a carrier, facilitating entry of lipoprotein into oocytes. If this is also the case of Portunus, then it must occur during step 4 and this needs more study.

This picture of vitellogenesis shows that lipids constitute the major share in the yolk of the egg of *Portunus*. They are basic in the proteolipids (fatty granules) while a great part is in the lipovitelline and glycolipoprotein. According to the statement of Gardiner (1972), the yolk of the egg of *Portunus pelagicus* is of two types, fatty and proteid yolk.

Deutoplasmogenesis in *Asterina pectinifera* (in view of cytology)

In the publication of Metz and Monroy (1985a), the growth of the oocyte of the common Japanese starfish, *Asterina pectinifera*, was described. Firstly, the oogonia are clustered in and/or near the germinal epithelium. They are columnar in shape and about 10 µm in diameter, and are distinguished by the presence of small greenish granules in their cytoplasm after staining with toluidine blue (Fig. 2.12A). Takahashi and Kanatani (1981), divided the development (deutoplasmogenesis) of the oocytes of this species into the following five developmental stages on the basis of their cytological appearance.

- **Stage I:** Oocytes at the transforming period from oogonium to oocyte (about 10 µm in diameter). The chromatin in their nuclei shows synapsis (Fig. 2.12B).
- **Stage II:** Oocytes having a small, growing germinal vesicle with a single nucleolus in close contact with the nuclear membrane (10–30 µm in the nucleus). In the later part of this stage, yolk granules begin to accumulate among the small vesicles (Fig. 2.12C).
- **Stage III:** Oocytes having abundant yolk granules that stain deeply with toluidine blue. Oocytes at this stage are round or elliptical in shape and 30–70 µm in diameter. Lampbrush chromosomes are occasionally seen in some nuclei (Fig. 2.12D).
- **Stage IV:** Oocytes becoming large in size and having a pyriform shape. They are 70–150 µm in the long axis and 50–90 µm in the short axis; when isolated alive, they average ~90–100 µm in diameter (Fig. 2.12E).
- **Stage V:** Oocytes just before, or at the fully-grown state (up to~160 µm in diameter). The cortical granules in the oocytes at this stage stain very deeply with toluidine blue and are present just beneath the oocyte surface (Fig. 2.12F and G). At the beginning of stage V, the vitelline coat appears in the extracellular space between follicle cells and oocytes.

Meiotic arrest in animal oocytes

In animals that lack a haploid generation, completion of meiosis must be immediately followed by fusion of the gamete nuclei. In almost all species, the coupling between the two events has developed in such a way that oocyte meiosis is arrested at a certain stage of oogenesis and its resumption is induced by fertilization. This is, perhaps, the major reason that animals have developed devices that arrest meiosis during oogenesis in the course of evolution.

However, it is also noted that oocytes are arrested for a long period at prophase of the first meiosis (prophase 1 or diplotene) in all species, regardless of whether its resumption is induced by fertilization or not. Therefore, there may be another reason for the occurrence of meiotic arrest at prophase 1. The fact that synthetic activities of oocytes in

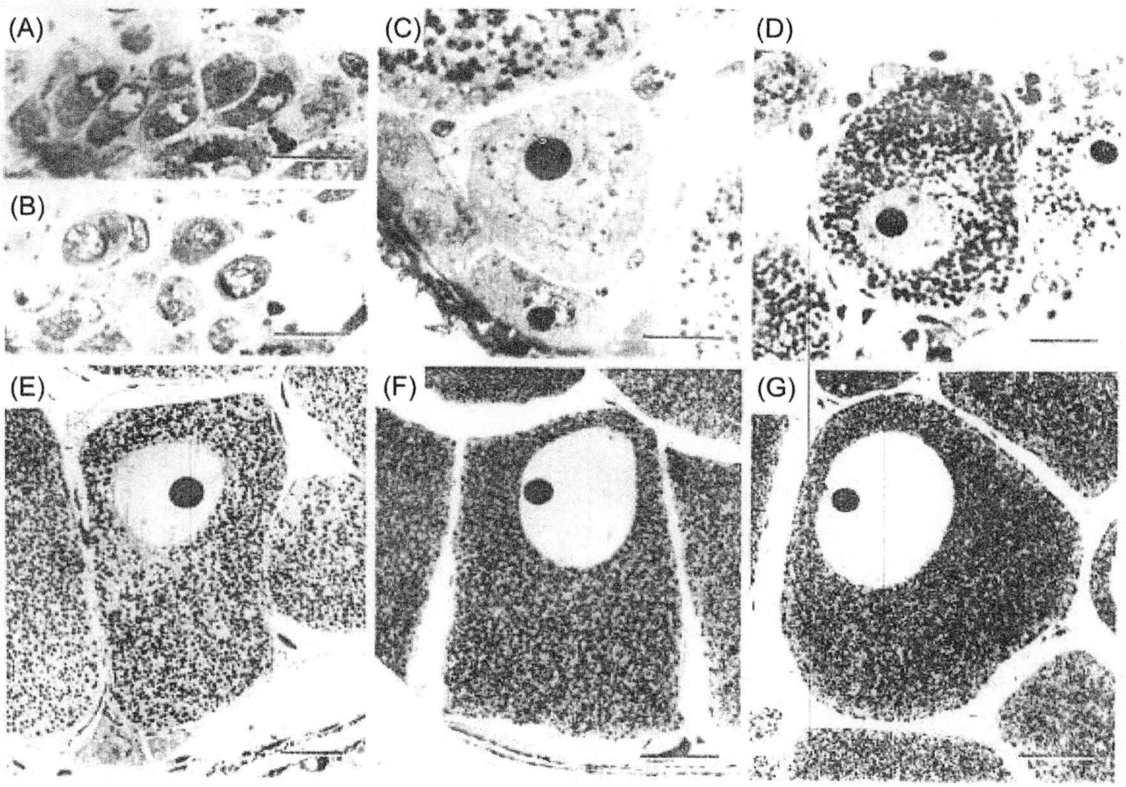

FIG. 2.12 Developmental stages of *Asterina pectinifera* oocyte. (A) Oogonia. (B) Stage I oocyte transforming from oogonia, (C) A late stage II oocyte. Yolk granules begin to accumulate. (D) A stage III oocyte characterized by the development of many yolk granules, (E) A pear-shaped stage IV oocyte. (F) An early stage V oocyte. (G) A late stage V oocyte characterized by the existence of small dense granules in the cortex of ooplasm. Toluidine blue stain; 1 μm section. Scale: 10 μm in (A) to (C); 20 μm in (D) and (E); 40 μm in (F) and (G). *(After Takahashi and Kanatani (1981), from Metz and Monroy, vol. 1., 1985a.)*

both the nucleus and the cytoplasm become very high at this stage of meiosis, causing the cell to grow enormously (for review, see Davidson, 1976), suggests a correlation between meiotic arrest and oocyte growth. In other words, the meiotic arrest at prophase 1 may be either required for, or caused by, oocyte metabolic activities for growth.

Although in many species resumption of prophase-arrested meiosis is caused by fertilization, in many other species it is resumed at the time when oocytes are released before fertilization. In this case, meiosis proceeds to the metaphase of the first or second meiosis, where it is again arrested until fertilization. Furthermore, in some animals, such as sea urchins, meiosis is completed even before oocytes are released from the ovary. These oocytes are then arrested after the pronucleus is formed, and await fertilization (Fig. 2.9).

In all cases, resumption of arrested meiosis in eggs is totally dependent on fertilization, regardless of the stage at which meiotic arrest occurred.

The above discussion may be summarized in the following hypothesis:

- First, female meiosis must be arrested at least once, and its resumption must be induced by fertilization to synchronize gamete chromosome cycles in the zygote.
- Second, this meiotic arrest must occur not only during oocyte growth but also at the stage at which oocytes attain maximum physiological tolerance to the optimum fertilized milieu.
- Third, the difference between the optimum condition for oocyte growth and that for fertilization acts as the major selection pressure that determines the stage at which meiotic arrest occurs prior to fertilization.

Then why is meiosis arrested only in oocytes, and not in spermatocytes? A conventional answer to this question may be that male gametes are the cells specialized for mobility to maximize the efficiency of fertilization. This specialization involves a great sacrifice of cytoplasm, and therefore spermatocytes are supplied with little cytoplasmic equipment necessary for any fine adjustment of the meiotic process, such as its arrest at specific stages.

Life span of gametes

Often the sperm or eggs are packaged, respectively, in spermatophores or egg cases for shedding. Sperm must be released from the spermatophores before they can fertilize the eggs and various degrees of embryonic development usually proceed in the egg cases before hatching the larvae or juvenile. These conditions are considered specialized cases. Generally, the sperm are discharged as semen and the eggs are embedded in a gelatinous substance. The conditions of insemination in most animals require that the gametes remain in a fertilizable state for a period of time following release from the gonads. It is of interest, therefore, to enquire about or examine the factors involved in the ageing of the gametes.

Senescence of spermatozoa

In some species of animals, such as bat, the honeybee and certain terrestrial isopods, the sperm may survive for periods of several months to several years in the female genital tract. Apart from such special cases, the usual life span of shed sperm is of the order of several hours to a few days. An important factor that is known to affect markedly the functional life span of sperm is dilution. Many researchers, such as Morgan (1927), have shown that the duration of life varies directly with the concentration of the suspension. The relative or complete lack of motility of sperm in undiluted semen of sea urchins and other animals has frequently been attributed to a relatively high tension of carbon dioxide. This means that increasing the oxygen tension activates the spermatozoa. It was proposed that immobility in semen, as well as subsequent senescence of sea urchin sperm, was due to a substance, termed **androgamone 1** that is liberated by the spermatozoa. A contradicting view showed that sea urchin sperm were as motile when diluted with seminal plasma as with seawater. This view was ascertained by the addition of the supernatant of a dense two-hour sperm suspension, which is supposed to contain androgamone 1, but had no inhibitory effect on respiration. Therefore, senescence is not attributable to substances such as the hypothetical androgamone 1, diffusing into the medium from the sperm.

In addition, the life span of sea urchin sperm can be considerably extended by addition of any of the amino acids to the suspension. The amino acid is not utilized metabolically, although it has effects on the respiratory metabolism of the sperm and even enables the sperm to remain motile anaerobically, whereas they ordinarily die at once in absence of oxygen. It was suggested that the amino acid acts by binding certain trace metals present in the seawater. The proteins of the seminal plasma can also bind heavy metals; so, the spermatozoa, in such denser suspensions, would be protected from the toxic action of the trace metals.

Another interesting effect of the amino acids is their ability to enable the sperm to induce a good fertilization-reaction under conditions in which the sperm, diluted and aged in ordinary seawater, call forth poor membrane elevation.

Senescence of egg

The unfertilized egg, likewise, has a relatively limited life-span under ordinary conditions, both in animals with external and with internal fertilization. Fertilizability is lost within a few seconds or minutes after spawning, in certain cases, as in amphibians. In such cases, marked visible changes occur, such as the elevation of the membrane at its contact with the new medium. A rapid loss of fertilizability may, however, occur without such visible changes, for example, as in *Platynereis*.

However, in most animals loss of fertilizability and cytolysis of eggs occur in a period of several hours to 1 or 2 days under ordinary conditions. This case cannot be attributed to depletion of endogenous nutrient, since, if fertilized, the eggs can survive considerably longer without any added nutrient. It was shown that various agents such as weak alcohol, slight acidity, and low calcium content of the medium can prolong the fertilizable life of eggs of marine animals to some extent. These agents, evidently, operate by delaying the onset of autolytic changes. On the other hand, the harmful effect of bacteria does not express itself clearly until after a certain immune period, since freshly shed eggs can survive in dense bacterial suspensions almost as long as in ordinary seawater.

Structure of gametes

The two gametes, the male gamete (spermatozoon) and the female gamete (ovum) differ completely from each other. The spermatozoon is produced from the testis in innumerable quantities, while eggs are produced from the ovary in comparatively small quantities. The size of the spermatozoon is always minute (microscopic), in contrast to the egg, which may be microscopic, e.g. that of *Grantia*, but its diameter may be measured in millimetres or even centimetres. Generally, the egg is provided with preserved nutrient material (yolk) for the development of the future embryo. In addition, the egg may be covered by one or more membranes for protection, while the sperm is protected with only the plasma membrane. A very important difference is that the sperm are movable while the eggs are stationary.

Structure of the sperm

Evolution of the sperm cell

The sperm cell is defined as the mature germinal cell of all multicellular organisms. An exception is the higher plants, in which this cell is extremely reduced and simplified. The sperm cell contains a haploid chromosomal complement that must be transferred into the egg cell. Therefore, the sperm cell must contain DNA in an easily transportable form, it must be capable of motility, and it must have receptors to recognize the female germinal cell of its own species. Lastly, the sperm cell must be able to penetrate through the egg investments and fuse its plasma membrane with that of the egg.

Anton van Leeuwenhoek (1632–1723) was a Dutch biologist and microscopist. He built over 400 microscopes, whose focuses are amazingly accurate and clear. By them, he observed and carefully described diverse things and minute organisms (protozoa). His most significant discovery, with his co-discoverer Nicolas Hartsocker, was the sperm cells, in 1678. Firstly, he believed them to be parasitic animals living within the semen (hence the term **spermatozoa**, meaning 'sperm animals'). He originally assumed that they had nothing at all to do with reproducing the organism in which they were found, but he later came to believe that each sperm contained a preformed embryo. Try as he might, Leeuwenhock was continually disappointed in his attempts to find the preformed embryo within the spermatozoa.

Life began in water, and all cells without exception are organized to live in an aqueous environment. The sperm cell is usually a highly evolved and differentiated cell that originated from a cell with an unspecialized structure by a complex morphogenetic process. The sperm structures specific for each type of spermatozoon derive from the spermatid organelles during spermiogenesis (Fig. 2.13). These organelles are rather uniform and relatively simple in aquatic species with external fertilization, and are substantially more complex in terrestrial species with internal fertilization (Fig. 2.14). For example, the human, a land animal with internal fertilization, and similar mammals, has a spermatozoon with characteristics selected for its highly viscous environment and egg investments. It is completely different from aquatic spermatozoa, e.g. jellyfish, sea urchin, fish and frogs, and more similar spermatozoa of highly evolved land animals, such as insects and snails, that have differentiated, independent structures corresponding to some of those found in mammalian spermatozoa.

The typical marine spermatozoon The typical marine spermatozoon (Metz and Monroy, 1985b), such as that of most bivalves, echinoderms and cephalochordates, is a roundish cell with a long, thin flagellum. It is surrounded by a plasma membrane about 10 nm thick. Enclosed in this membrane are the organelles typical to the classical spermatozoon: acrosome, nucleus, centrioles, axoneme, and mitochondria.

The acrosome Metz and Monroy (1985b, vol. 2) reported that Waldeyer is credited with the first description of the acrosome, a structure that he termed the **perforatorium**, in 1870. The term **acrosome**, or more accurately 'Akrosoma', was suggested by Von Lenhossek in 1898. It was thought that the acrosome was important in penetration and fusion with the egg, since it is often sharply pointed, screw-shaped or even barbed, in

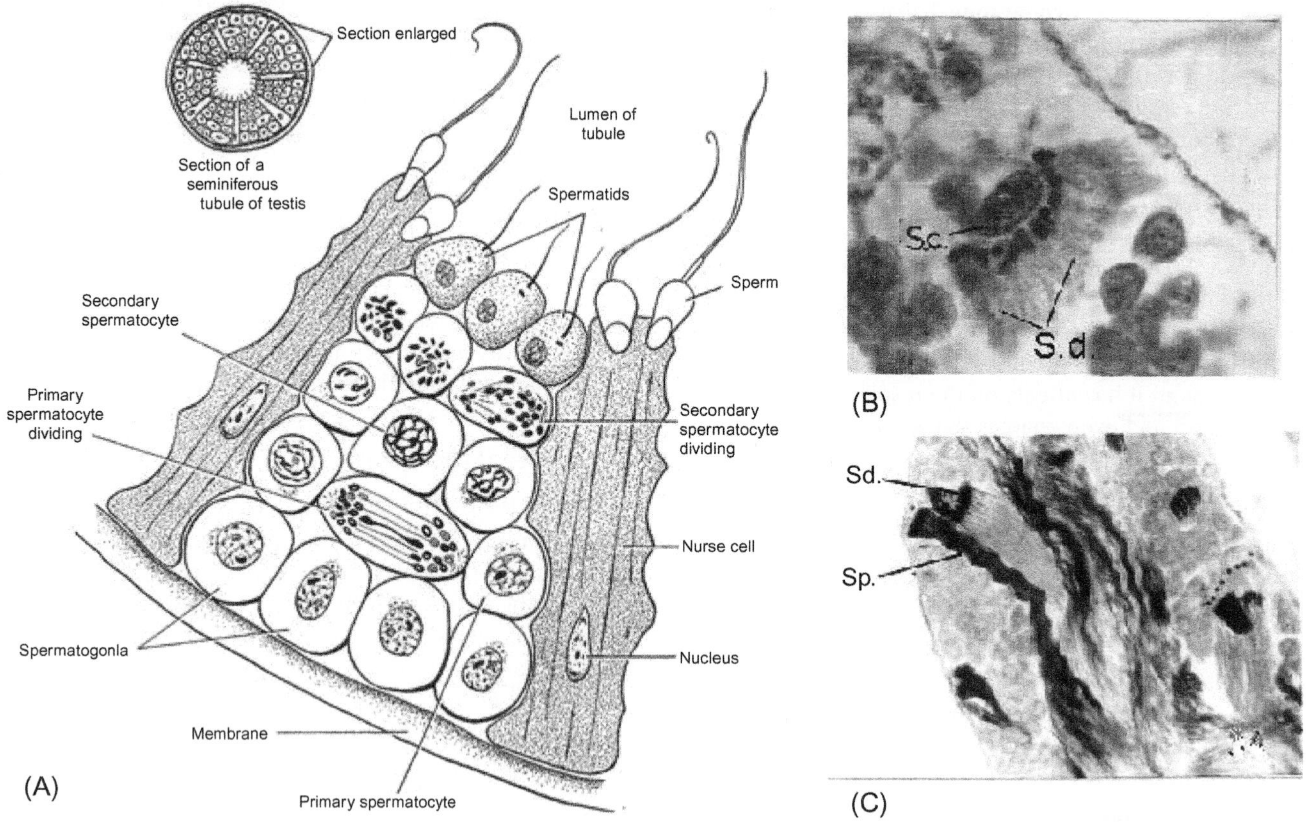

FIG. 2.13 (A) Diagram of a part of a section of a seminiferous tubule to show the stages in spermatogenesis and in the transformation of spermatid into a mature sperm. *(Villee: Biology, 4th ed.)*. (B and C) Microphotographs showing process of spermiogenesis in the snail *Eremina ehrenbergi*, *from Bawab et al. (1992a,b)*.

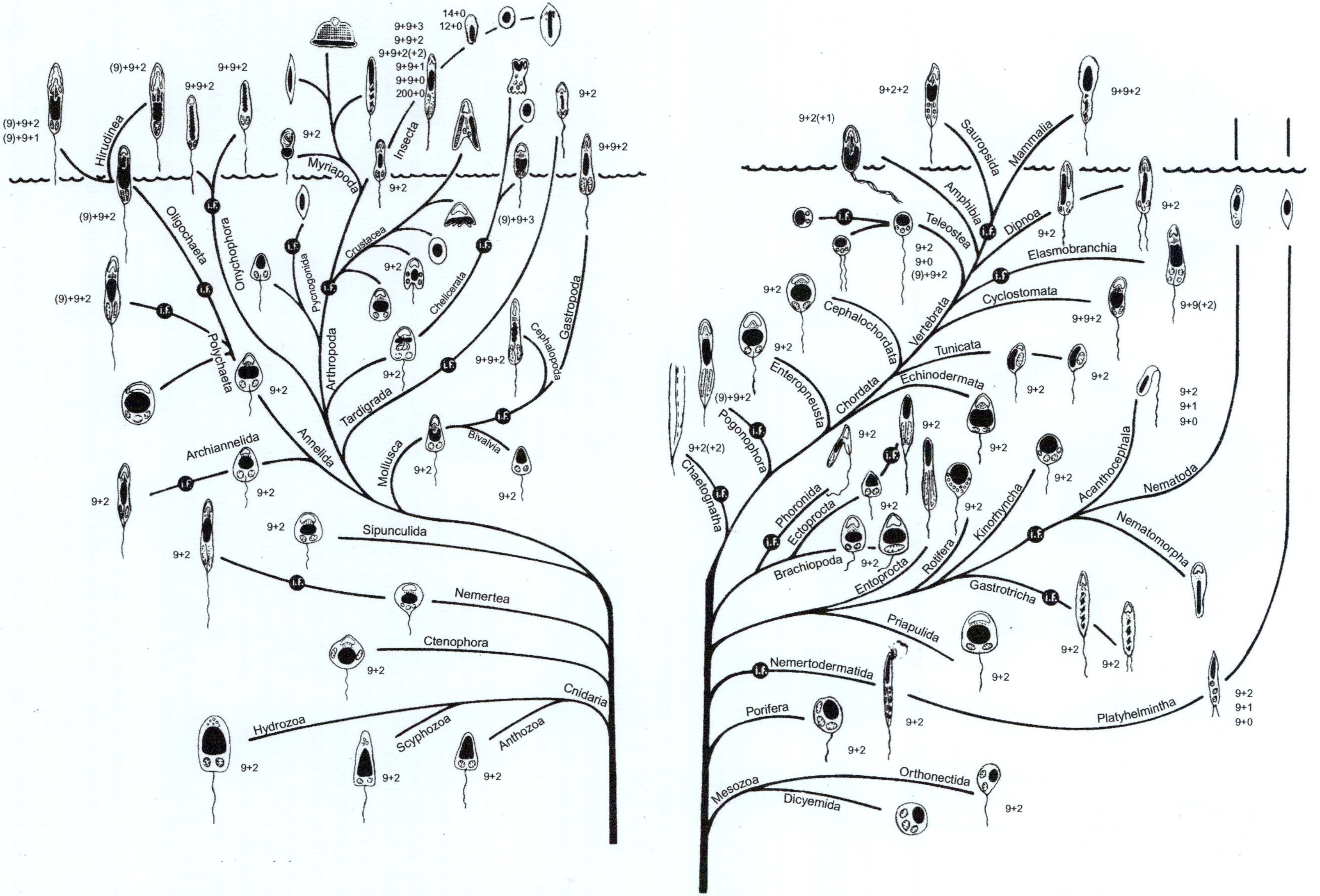

FIG. 2.14 Schematic drawing of the evolution of sperm structure during animal phylogeny. 'i.f.' = internal fertilization; 9 + 9 + 2 = polymorphic sperm. *(After Baccetti (1982), from Metz and Monroy, vol. 2., 1985b.)*

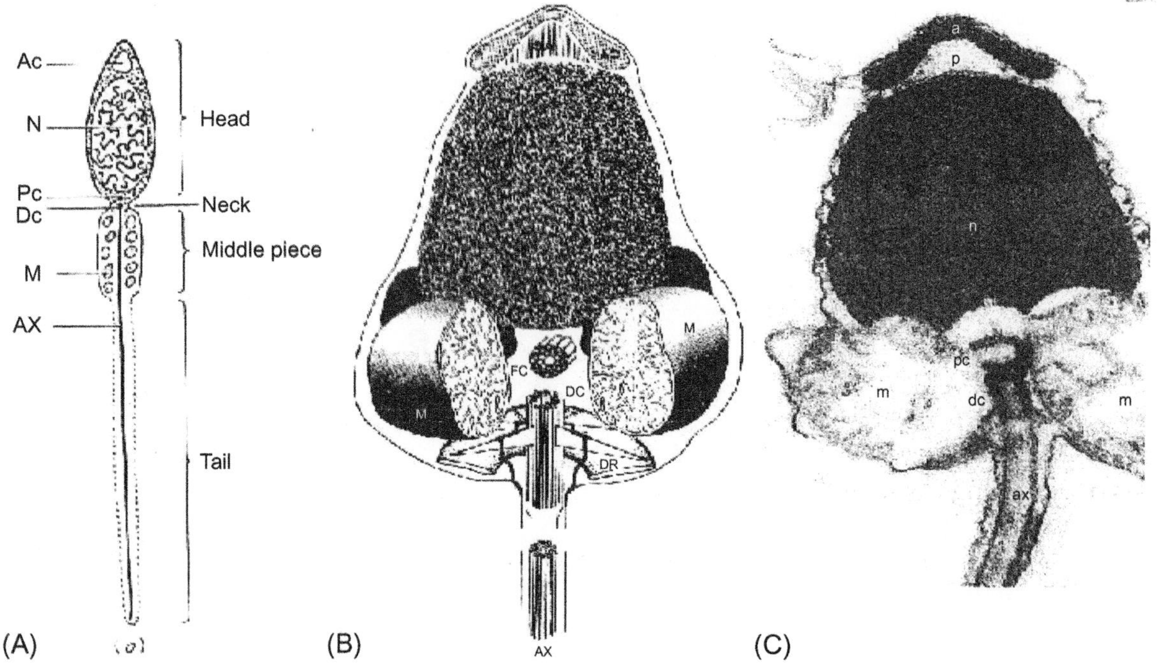

FIG. 2.15 Structure of spermatozoon. (A) Diagram of typical spermatozoon; (B) diagrammatic illustration showing the structure of marine spermatozoon of the sipunculid *Glodfingia gouldi* as seen by E.M.; (C) Electron macrograph of the above spermatozoon, AC = acrosome; AXE = axoneme; BR = basal plate; DC = distal centriole; M = mitochondria; N = nucleus; PC = proximal centriole; SA = perforatorium *(B and C, After Baccetti, 1977, from Metz and Monroy, vol. 2, 1985b).*

many sperm. Recently, it has become well known as the **acrosomal complex**. It is located in the anterior region of the cell, applied to the nucleus, and it is divided into two parts, the acrosome proper and the perforatorium (Fig. 2.15). The acrosome originates from the Golgi complex of the spermatid. It is surrounded by a membrane that is formed by the fusion of many Golgi vesicles and is cap-shaped, with its convex surface outward. It contains enzymes, e.g. 'lysine' in aquatic invertebrates. In sea urchins, there is a trypsin-like enzyme similar to the acrosin of mammals. Directly below the acrosome proper, there is a rod termed perforatorium or acrosomal process. It consists of actin and myosin. The actin, before its protrusion, is stored in a globular form or in helicoidal filaments associated with other proteins such as spectrin, profiling and α-actinin. An abrupt change in the crystalline organization of these actin and myosin causes the perforatorium to burst forwards and break through the acrosome, thereby releasing the acrosomal contents.

A comparative examination of all aquatic phyla (Fig. 2.16) shows that perforatoria containing G-actin are most primitive; perforatoria containing F-actin appear later in evolution and are found in terrestrial groups. In fact, in land animals, only filamentous perforatoria occur. On the other hand, the role of the myosin in perforatoria is not yet clear. From the acrosome of sea urchin and oyster sperm, a 35,000 molecular weight protein has been isolated that allows the sperm to adhere to the egg. This protein is termed 'bindin'. The egg surface has specific receptors for bindin. The acrosome does not function in fresh water and therefore becomes useless. Freshwater cnidarian sperm lack an acrosome (Fig. 2.16).

The nucleus The nucleus of the spermatozoon is more complex, more homogeneous and smaller than that of somatic cells; it has a haploid number of chromosomes, is highly dehydrated, and is not transcriptionally active. An important characteristic of the sperm nuclear membrane is its lack of pores. The chromatin is organized in helicoidal fibres. Instead of containing somatic-type histones, the helicoidal fibres are composed of other arginine-rich histones and protamines, which are synthesized in the spermatid cytoplasm and move into the nucleus when the acid proteins are released. This process has been described both in the higher land animals and in more primitive aquatic animals. In sea urchin sperm nuclei, the H_2 b histone also exhibits its unusual physico-chemical characteristics. Following sperm penetration into the egg, but prior to the substitution of sperm-type histones with egg-type histones, the chromatin becomes decondensed.

The centrioles and the anchoring structures The spermatid has two centrioles; one is called proximal, oriented parallel to the base of the nucleus, and the other is called distal, perpendicular to it. The nine doublets of the axoneme arise from the latter (Fig. 2.15). In the marine spermatozoon, both centrioles are retained until maturity. They are made up of nine triplet microtubules situated at the periphery of a cylinder ~0.2 μm in diameter and 1.0 μm long. The three

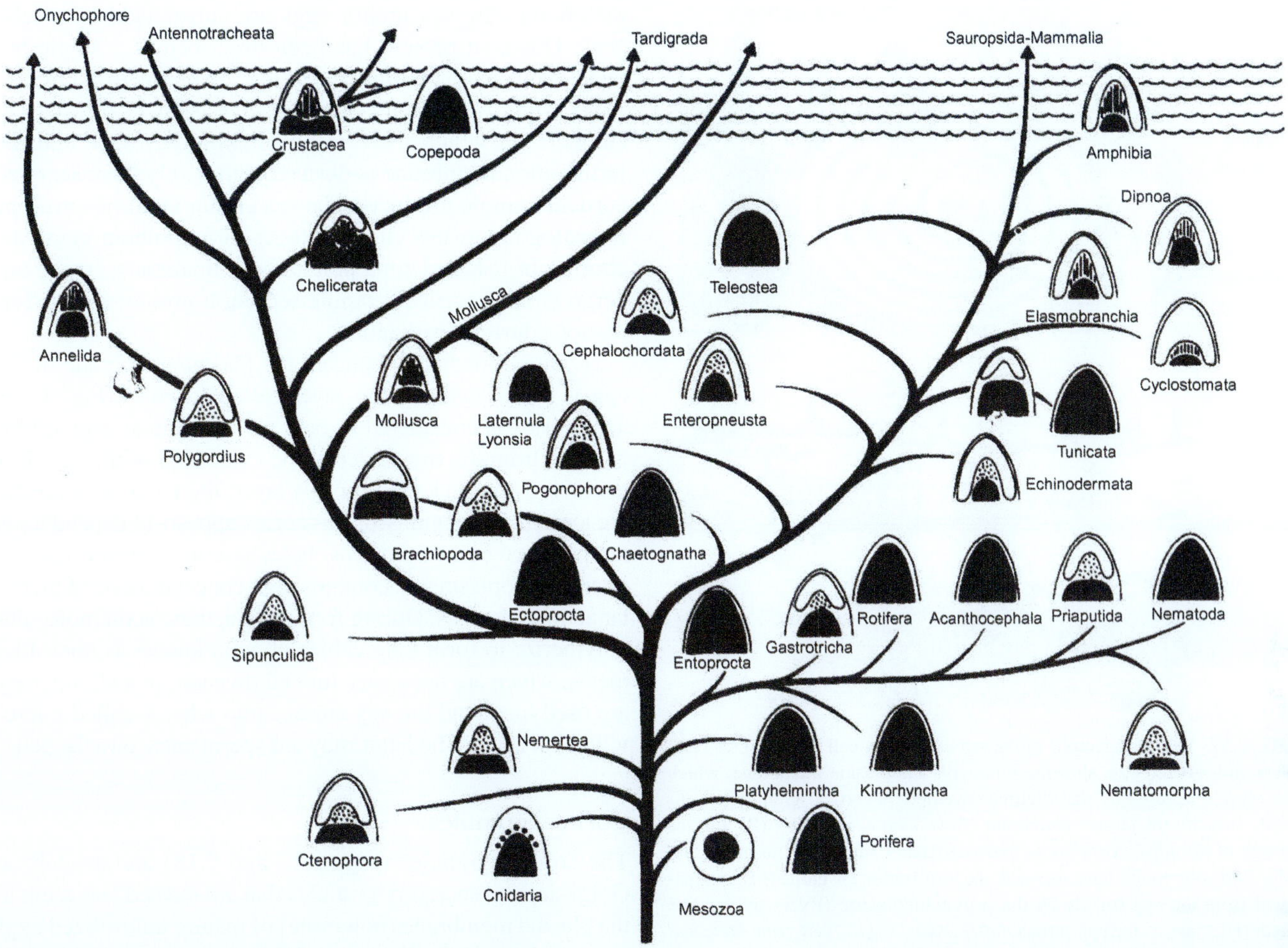

FIG. 2.16 Phylogenetic tree of aquatic acrosome evolution. The acrosome is depicted as a hollow vesicle, the subacrosomal substance or perforatorium as a group of spots or lines (according to the profilactin or filamentous actin models). The nucleus as a black body, and the plasma membrane as a peripheral line. *(After Baccetti (1979), from Metz and Monroy, vol. 2, 1985b.)*

microtubules are termed A, B, and C, beginning with the innermost one. The centrioles contain tubulin.

A complex radial anchoring structure surrounds the distal centriole (Fig. 2.15) and connects it to the plasma membrane at about the level at which the flagellum arises. In the simplest case, nine bridles extend radially from the nine triplets; in other cases they branch, and connective transversal bridles also occur. These bridles are proteineous and transversally striated. At the light microscopic level, actin, **alfa**-actinin, myosin, and tropomyosin have been found to be associated with the centriolar structure. With the use of the E. M., actin has been localized on the pericentriolar birdles.

The axoneme The axoneme, also known as the axial filament, is the motile flagellar component and is made up of microtubules and connecting elements. It contains the same elements typically found in cilia, arranged in the usual 9+2 pattern.

The doublets of peripheral microtubules are the most important elements of the axoneme. Almost all the axonemes contain nine doublets that are arranged in a regular pattern, 18 nm apart, forming a cylinder with a diameter of 0.18 μm. Two single microtubules, 9 nm apart, are found near the centre of the cylinder. For details, see Metz and Monroy (1985b).

Mitochondria Generally, in marine spermatozoa, the mitochondria are concentrated in the cell body, near the centrioles at the base of the flagellum (Fig. 2.15). This zone is called the 'mid-piece'. Mitochondria account for about 10% of the cell volume, with each cell containing one to four mitochondria, which are larger than those of somatic cells. Their structure is general, with deep cristae containing the cytochrome oxidase, but they lack the classical dense granules in their matrix. In the typical marine sperm of sea urchins, the mitochondria do not depend on exogenous substrates for respiration, and they utilize the phospholipids of their cristae, which in consequence progressively decrease in size. Lactic dehydrogenase is absent, probably because the pyruvate produced enters the Krebs cycle directly. Other marine spermatozoa also lack intramitochondrial glycogen.

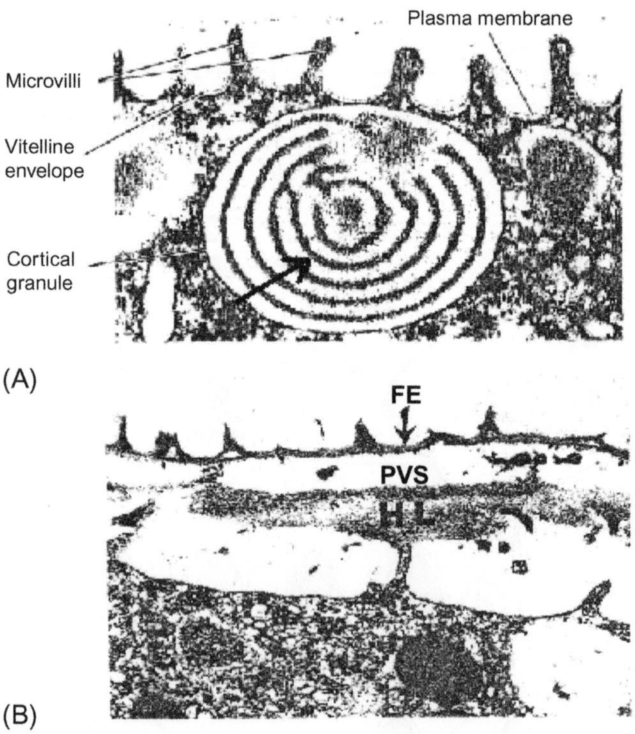

FIG. 2.17 TE micrographs of the sea urchin egg cell surface. (A) TEM of an unfertilized egg, showing microvilli and plasma membrane, which are closely covered by the vitelline envelope. A cortical granule lies directly beneath the plasma membrane *(from Schroeder, 1979)*. (B) Surface of egg of *Strongylocentrotus purpuratus* fixed 10 min after insemination. The cortical granules have secreted, the fertilization envelope (FE) is separated from the egg surface by the perivitelline space (PVS), the hyaline layer (HL) invests the egg surface *(after Schuel et al., 1982, from Metz and Monroy, vol. 3, 1985c)*.

The mature egg (the unfertilized egg)

The sea urchin egg (Fig. 2.17) was described ultrastructurally (Metz and Monroy, 1985c) as an example, but this is not necessarily applicable to all eggs. Generally, the egg is very large compared to the sperm. Its cytoplasm is enclosed by the **plasma membrane**, which is a gelation of the cytoplasm. This membrane has two functions:

a. it must regulate the flow of certain ions, especially during fertilization; and
b. must be capable of fusion with the sperm plasma membrane.

Outside the plasma membrane is the **vitelline membrane**, which forms a fibrous mat around the egg. In some electron microscopic preparations of sea urchin eggs, the vitelline envelope is found 20–30 nm above the plasma membrane. This envelope contains at least eight different glycoproteins and is often involved in sperm-egg recognition. It is supplemented by extensions of membrane glycoproteins from the plasma membrane and by proteineous vitelline posts that adhere the vitelline envelope to the plasma membrane. The vitelline envelope is essential for the species-specific binding of sperm. Outer to the vitelline membrane, the sea urchin eggs are surrounded by a jelly coat. This coat protein has been described as a single homogeneous band on ultracentrifugation or electrophoresis.

A slightly different situation may exist in the sea cucumbers, in which there is no vitelline envelope and the fertilization membrane is derived completely from secreted cortical granule products. The sea urchin vitelline envelope is analogous to the vitelline layer in amphibian eggs, the chorion in fish and zona pellucida in mammals, which undergo similar cortical granule secretion-promoted transformations during fertilization.

Lying immediately beneath the plasma membrane of the egg is a thin cytoplasmic layer called '**cortex**' (Fig. 2.17). This cortex is considered to be a gel found in an egg, differentiated from the remainder of the cytoplasm of the egg. It is about 1–5 μm thick. Within this layer, the **cortical granules** are located. The rigidity of this cortex appears to depend upon the presence of calcium ions. Its cytoplasm is stiffer than the internal cytoplasm and contains high concentrations of globular actin molecules. During fertilization, these actin molecules polymerize to form long cables of actin known as microfilaments, which are necessary for cell division. In addition, they are used to extend the egg surface into what is called microvilli (Fig. 2.17). The latter may aid sperm entry into the cell.

Cortical granules

The cortical granules (Figs 2.17 and 2.18) are specialized Golgi-derived secretory granules that are located subjacent to the plasma membrane (oolemma) of mature unfertilized eggs in many invertebrates and vertebrates (Anderson, 1968). The number of the cortical granules is about 15,000–18,000/cell. They are arranged within the cortex in interlaced rows and are connected to an underlying filamentous network. Each granule is about 0.9–1.3 μm in diameter (Nicosia et al., 1981). The structure and chemical composition of cortical granules of different eggs were studied ultrastructurally (Schuel, 1985).

1. Morphology: TEM examination revealed that the cortical granules in the eggs of most organisms are membrane-limited cytoplasmic particles with an internal matrix of high electron density. They closely resemble zymogen granules in the exocrine pancreas in this regard. However, in echinoderms, the matrix of the cortical granules contains electron-dense and amorphous electron-lucent components, with each genus showing a characteristic internal ultrastructure. This is clearly observed in Figs 2.17A and 2.18.
2. Molecular composition: considering this composition of these granules of sea urchin eggs, it was reported that each consists of:
   a. some different enzymes;
   b. acid mucopolysaccharides; and
   c. structural proteins, e.g. hyalin (major structural protein of the hyaline layer), which is secreted during the cortical reaction.

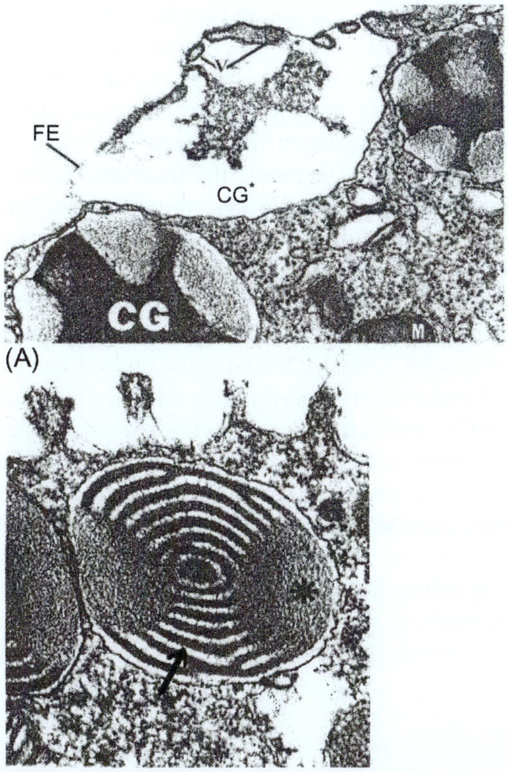

FIG. 2.18 TE micrographs of surface of eggs. (A) Section through a fertilized *Arbacia punctulata* egg showing discharging cortical granules (CG*). Vesicles (V) formed by fusion of membrane of the discharging granules with the plasma membrane are torn away from the egg surface in association with the elevating fertilization envelope (FE). Unreacted cortical granule (CG) shows the centrally located electron-dense stellate core and more peripheral electron-lucent components characteristic of this species. M = mitochondrion *(after Anderson, 1968, from Metz and Monroy, vol. 3)*. (B) Cortex of unfertilized egg of sea urchin *Strongylocentrotus purpuratus* showing undischarged cortical granule with electron-dense spiral lamellae *(arrow)* and amorphous electron-lucent (*) components characteristic of this species *(after Schuel et al., 1982)*.

The hyaline that is stored in small cytoplasmic vesicles of unfertilized eggs is not secreted at fertilization but it may be a reservoir released to regenerate the damaged layer, and

d. calcium ions (which has been localized to the limiting membrane of the cortical granules, but also found at other sites of the egg).

3. Function: these studies also led to the interpretation that the products derived from the electron-dense component of the matrix of the granules of the echinoderm eggs contribute to the fertilization envelope. Structural proteins destined for incorporation into the fertilization envelope are also secreted during the cortical reaction and contribute to the hardening of this fertilization envelope. On the other hand, the products derived from the electron-lucent component contribute to the hyaline layer. In addition, the calcium store may act to stabilize the structure of the cortical granule membrane and/or may be the source of intra-cellular calcium that is released at fertilization to initiate the cortical reaction and activate development.

It should be noted that certain molluscs, annelids and other invertebrate eggs contain cortical granules that do not secrete at fertilization. Little is known about these nonsecretory cortical granules.

Exocytosis

The cortical reaction (Fig. 2.18) in eggs is an attractive model system to study the mechanism of exocytosis as a general cellular phenomenon (Schuel, 1978). The contents of the cortical granules are normally discharged from the egg by means of exocytosis in response to stimulation by the fertilizing sperm. This process begins at the site of sperm fusion with the plasma membrane and then rapidly spreads around the entire surface of the egg. Exocytosis is a complicated process involving, in turn, fusion of the secretory granule membrane with the cell's plasma membrane, formation of an opening at this locus to the extracellular space, extrusion of the granule's contents into the external medium, recycling of the membrane of the secretory granule, and stimulation of somatic cells to synthesize and store more secretory product or, in the case of eggs, to initiate embryogenesis. In sea urchins, this secretory process (cortical reaction) is completed within 1–2 min postinsemination, depending upon species and temperature.

Cortical granules in unfertilized sea urchin eggs contain serine protease sensitive to soybean trypsin inhibitor (SBTI). This protease is inactive in unfertilized eggs and may be activated by calcium. Ionic calcium is released into the cytoplasm of the egg prior to the initiation of cortical granule discharge. The protease is transiently active in the eggs during fertilization and is secreted during the cortical reaction. Cortical granule discharge is retarded when sea urchin eggs are fertilized in the presence of SBTI. Retarded degranulation has been documented at the ultrastructural level by quantitative morphogenetic analysis of SBTI-treated *Arbacia* eggs and confirmed in *Strongylocentrotus* (Schuel, 1985). SBTI is specific enzymatic inhibitor of the cortical granule proteases and other serine proteases.

Fate of cortical granule membrane after exocytosis

Another important consequence of exocytosis is the insertion of the limiting membrane of the cortical granules into the oolemma to form the mosaic plasma membrane of the fertilized egg. This results in a large increase in the surface area of the egg's plasma membrane, as well as in changes in its electrical fluidity and permeability properties. The plasma membrane may also be modified by released cortical granule enzymes.

What is the fate of the cortical granule membrane after its insertion into the egg's plasma membrane? In somatic cells, secretory granule membranes are retrieved, presumably by endocytosis, and are reused in subsequent secretory cycles. It was proposed that the excess cortical granule membrane was accommodated only in part by elongation of microvilli and that the remainder must undergo a rapid and massive resorption.

Types of gametes

Types of sperm

The typical marine spermatozoon as described above (Fig. 2.15), but with a comparatively long flagellum, so that the whole spermatozoon cannot be seen in one field under the microscope, and of constant movement, is the general type of sperm in the animal kingdom, especially in vertebrates. That type is sometimes known as '**vertebrate sperm**', e.g. that of *Sepia* (Fig. 2.19H). However, the proportion, size and relations of the parts constituting the spermatozoon (Fig. 2.15A), (head, middle piece, and flagellum), are subjected to great variations. Also, the shape of the head differs with the species (Fig. 2.19), and so it is a taxonomic feature. In invertebrates, this vertebrate sperm is not a general type, it is found in, e.g. gastropods and cephalopods. On the other hand, in earthworms, the whole spermatozoon can be seen in one field under the microscope. To be differentiated from the long-flagellated vertebrate spermatozoon, that of the earthworm is termed '**short-flagellated**'.

In the majority of animals, the flagellum of the living spermatozoon exhibits cilia-like movements. This type is termed **flagellated spermatozoon**. In other two major groups, the nematodes and the crustacea, the spermatozoa are frequently present in very abnormal forms, and their movements are of amoeboid character. In Nematoda, they are pear-shaped, cylindrical and spine-shaped and they are mainly formed of protoplasm with a highly refracting nucleus. In Crustacea, they are sometimes simply spherical or they are composed of a nucleated body with stellate rays. These types are termed **aflagellated spermatozoons**.

There is another feature concerning the flagellated spermatozoon. Nearly all the Platyhelminthes are characterized by having **biflagellated sperm** (Fig. 2.20). Platyhelminthes

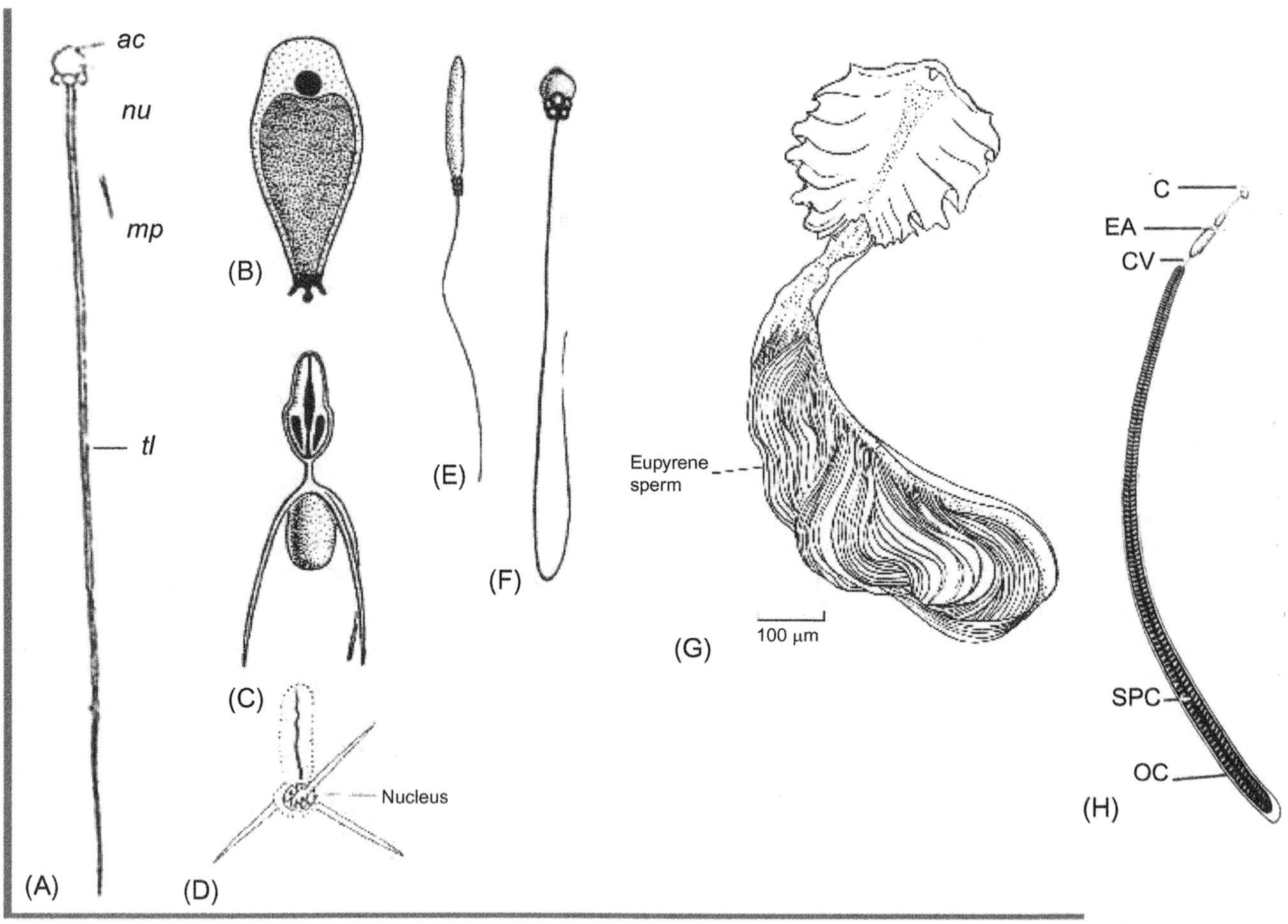

FIG. 2.19 Various types of spermatozoa. (A) Oyster, *Crassostrea*; (B) *Ascaris*; (C) Hemit crab; (D) *Homarus*; (E) Frog; (F) *Amphioxus*. (B), (C), and (D) are nonflagellate spermatozoa. (G) Spermatozeugma of *cerithiopsis tubercularis* carrying eupyrene spermatozoa *(after Fretter and Graham, 1962, from Giese and Pearse, vol. IV)*. (H) Camera lucida drawing of the fully formed spermatophore of *Sepia officinalis*, isolated from Needam's sac, X=40, C=cap, CON=connective, EA=ejaculatory apparatus, FC=free camber, OC=outer cover, SPC=spermatoporic cord *(from Bawab and Ismail, 1984)*.

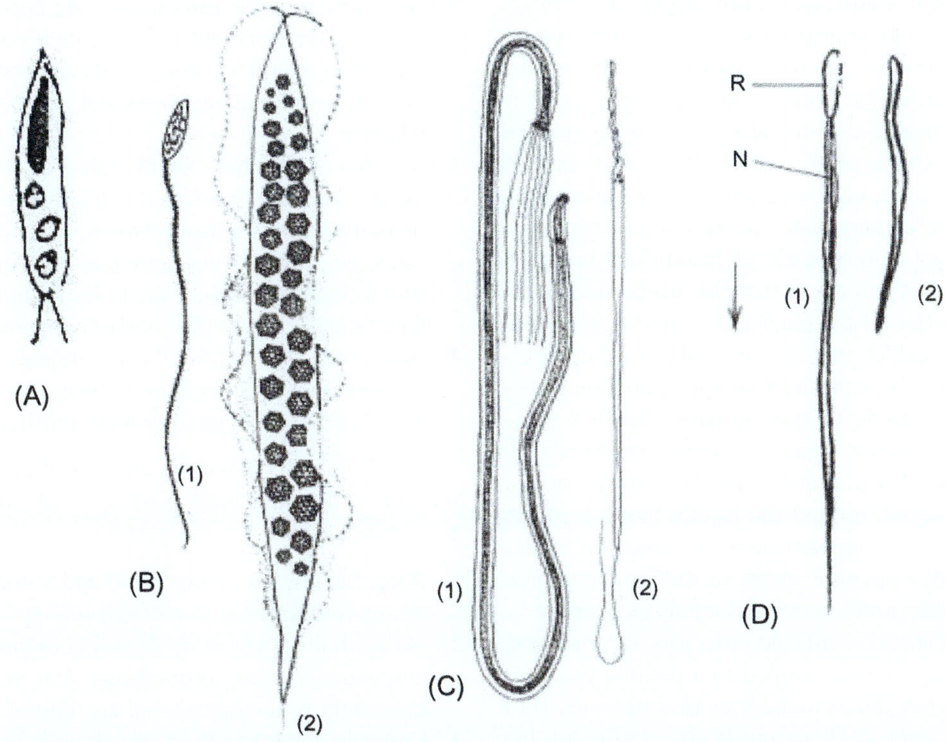

FIG. 2.20 Some types of spermatozoa in invertebrates: (A) Biflagellate sperm (Platyhelminth, refer to Fig. 2.14); (B, C, and D) polymorphic sperm, (1) Eupyrene sperm, (2) Oligopyrene (B) Apyrene (C and D), (B) of the snail *Strombus bituberculatus (after Galstoff, 1925, from Gardiner, 1972)*; (C) of the snail *Paludina vivipara (after Meves, 1903);* (D) of the acoelan flatworm *Otecelis rubropunctata.* R = Cytoplasmic residue; *arrow* = direction of swimming *(after Hendelberg, 1969), (C) and (D) from Roosen-Runge (1977).*

are hermaphrodites in which internal fertilization is firmly established, both in the free-living animals and in the parasitic forms. Of this phylum, only the primitive turbellarian Nemertodermatida and a few cestodes, such as *Echinococcus*, have spermatozoa with a single flagellum. Biflagellarity appears in all other turbellarians, in trematodes and in cestodes, probably having arisen by parallel evolution. Biflagellarity might be expected to result from the utilization of both centrioles as flagellar organizers and should, in principle, improve motility. However, an unusual 9 + 1 or 9 + 0 axonemal structure is found in all of these sperm (Fig. 2.6). Thus, in the case of biflagellarity as well, motility loses its importance with the refining of internal fertilization, and deviations from the classical complete 9 + 2 structure are found. In the trematode genus *Schistosoma*, the doublets in the two flagella lack dynein arms, and consequently the spermatozoa are immotile. This is clearly a step towards losing the flagellum. In Turbellaria higher than Acoela, there are many examples of aflagellate sperm, e.g. in Prolecithophora.

A completely different pattern is the formation of the **polymorphic spermatozoa** (Fig. 2.20). The most remarkable cases of special differentiation occur in molluscs and specifically in snails. In many prosobranchs such as the archaeogastropods [*Trochus*, *Patella*, *Bithynia*]; the mesogastropods (Taenioglossa) [*Viviparus* (= *Paludina*), *Pteroceros*]; and the neogastropods (Stenoglossa) [*Murex*, *Nassa*, *Purpura*], etc., the phenomenon of double sperm is encountered, i.e. the animal produces two kinds of spermatozoa: one is normal and filiform, and the other abnormal and vermiform. The classical case is that of the pond- or river-snail *Paludina* (Fig. 2.20), in which two types of spermatozoa were discovered by Von Siebold (1836), since that time, many cases of conspicuous polymorphism have been found in invertebrates and many have been thoroughly investigated as to development and morphology of the atypical spermatozoa. The latter are those sperm that during spermiogenesis cast off or diminish their chromatin material. Diminution or lack of chromatin was established as a frequent characteristic of atypical sperm by Meves (1903). Waldeyer to Meves used the term '**eupyrene**' (Gr. Pyrene, nucleus) for cells with a normal complement of chromatin, '**oligopyrene**' for cells with little chromatin, and '**apyrene**' for cells without chromatin. The atypical (abnormal) spermatozoa are generally morphologically distinct from eupyrene sperm. In many cases of polymorphism, the abnormal types are morphologically very characteristic and sometimes show a behaviour which suggests that they have become functional components of the reproductive process although they have never been seen to accomplish normal fertilization.

The abnormal spermatozoa differ in length and form according to species, and may possess an undulating membrane (Fig. 2.20), bristles, paired flagella, etc. By far, the most elaborate and, in fact, unique atypical sperm cell is the giant spermatozoon, also called a spermatozeugma, of the marine snail *Janthina* (Ankel, 1930). When fully mature it is ~750–900 µm long, equipped with a large undulating membrane, called a 'driving plate'; a mass of yolk granules; and a tail composed of thousands of fused flagella which have arisen through a process of extreme multiplication of centrioles. The surface of the giant tail is modified by longitudinal rows of small depressions into which typical spermatozoa may insert. Thus, the atypical spermatozoon serves as a carrier for hundreds of typical ones. Another function performed by smaller abnormal sperm has been suggested in Viviparus (Hanson et al., 1952). They described a polysaccharide sheath around the middle portion of the abnormal sperm. They suggested that this sheath provides nourishment for the eupyrene sperm in the female genital tract by disintegrating and liberating the polysaccharide.

Many different studies confirmed that this phenomenon, polymorphic spermy, is not confined to a definite group of animals and/or phylum, but is found even in vertebrates. (For more information, refer to Giese and Pearse, 1977, vol. IV; Roosen-Runge, 1977.)

Types of eggs

The very small egg (oogonium) has the character of a simple cell. It is formed of a mass of naked protoplasm, containing a nucleus within which there is a nucleolus. The nucleus and nucleolus are usually termed germinal vesicle and germinal spot. This oogonium is developed either from:

- one cell out of an aggregation or layer of cells all of which have the capacity of becoming ova; or
- one out of a number of cells segmented off from a nucleated mass of protoplasm, not divided into separate cells.

In both cases, the cells that have the capacity of becoming ova may be spoken of as germinal cells. In some cases, all the germinal cells eventually become ova, but as a rule only a small proportion of them have this fate, the remainder undergo various changes. The distinction of ova derived from each of the two cases is not an important matter. It is probable that ova of the second type are less common, it is stated to be found in most Platyhelminthes, Nemathelminthes, Insecta, and Crustacea.

During the growth phase of oogenesis, the oocyte may absorb other cells and so it is considered a compound structure. The growing oocyte is nourished by the nutritive fluids in which it is surrounded however, additional nutrition is provided as follows:

1. In the simplest case, as in many Hydrozoa, the germinal cells that do not become ova are assimilated by the ovum (Fig. 5.15C and D, Chapter 5).
2. In other cases, the growing oocyte becomes invested by a special layer of cells, which then constitutes what is known as a follicle. These cells are germinal ones. This case is found, e.g. in Styela and in Insecta (Fig.13.10D, Chapter 13).
3. In other cases, they seem rather to be adjoining connective tissue or epithelial cells, though it is sometimes difficult to draw a line between such cells and germinal ones, i.e. the growing oocyte is surrounded by connective tissue cells. This case is found in Chaetopoda and Crustacea (Fig. 2.11D) (and egg membranes Fig. 2.22).
4. In a fourth category, only a membrane surrounds the egg without a lining of cells, as in many arachnids in which eggs have no true analogy with a follicle.

The fully grown oocytes are classified according to the amount of the preserved nutrient (yolk substance) and its distribution within the cytoplasm:

1. Alecithal oocyte is very small and without yolk material or any nutrient, e.g. egg of *Grantia* sp.
2. Isolecithal oocyte is small and contains very little yolk embedded in their protoplasm. The yolk is distributed uniformly in the cytoplasm, e.g. egg of sea urchin.
3. Centrolecithal oocyte has much yolk concentrated in the centre surrounding the nucleus, e.g. eggs of crustaceans and insects (Fig. 2.21).
4. Telolecithal (moderately) oocyte has much yolk concentrated in one pole, which is the vegetative pole. The nucleus and cytoplasm occupy the other pole, called the animal pole, e.g. egg of Asterina pectinifera (Fig. 2.12D).
5. Telolecithal (heavily) oocyte in which the whole mass is yolk, the nucleus and the cytoplasm constitute a germinal small disc on the surface of the yolk, e.g. *Sepia*.

To differentiate between the two telolecithal oocytes, they are termed moderately telolecithal and heavily telolecithal oocytes, respectively. The differentiation of the oocytes according to the contained yolk leads to their differentiation according their cleavage. There are two types: **holoblastic** oocyte, which cleaves completely, e.g. the sea urchin egg; and **meroblastic** oocyte, which cleaves incompletely, e.g. the crustacean egg.

Egg membranes

The egg membranes constitute a third factor differentiating the oocytes. Most egg cells are surrounded by one or more membranes. However, different kinds of eggs when ready to be fertilized are 'naked eggs' devoid of any form of protecting coverings, as have been described in some sponges, cnidarians and lamellibranch molluscs. As a rule, the egg is invested by some form of membrane(s). These membranes are variable in their structure and origin and may be grouped into three groups:

1. Primary egg membranes: those are derived from the protoplasm of the egg itself. They are differentiated into:

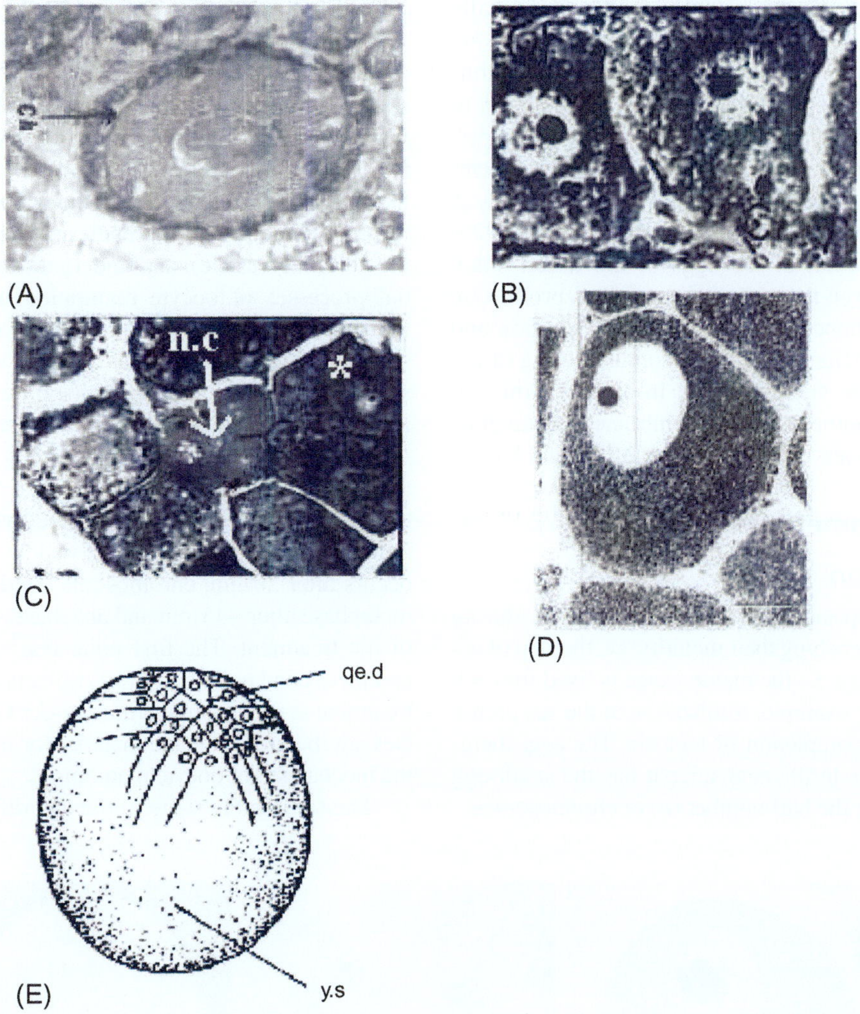

FIG. 2.21 Types of eggs according to Yolk: (A) Isolecithal egg of *Styela* (Bawab et al., 2012); (B and C) Centrolecithal egg of *Portunus pelagicus (Bawab and El-Sherief, 1987b)*; (B) early stage, and (C) fully grown egg (*); (D) Moderate telolecithal egg of *Asterina punctifera (after Takahashi and Kanatani, from Metz and Monroy, vol. 1, 1985a)*; (E) Heavy telolecithal egg of *Sepia (from Marshall and Williams, 1975)* (A, B, C, and D) microphotographs; (E) diagram, ch = chorion, ge.d = germinal disc, n.c = nurse cell, y.s = yolk sac.

(a) **Vitelline membrane** (Fig. 2.17A), which is formed essentially by gelation of the peripheral cytoplasm of the egg and some modification of its plasma membrane. The vitelline membrane may be thin and delicate or of relatively considerable thickness, and is sometimes composed of several layers. In some molluscs, worms and echinoderms, it shows radial striations, possibly representing microvilli extending from the egg surface, and is thus distinguished as a zona radiata (Gardiner, 1972);

(b) After fertilization, the vitelline membrane transforms into: **fertilization membrane** (Fig. 2.17B) by adding structural proteins from the cortical granules and so it is a primary membrane;

(c) External to the vitelline membrane there may be a jelly layer, which is also a primary membrane, as there is in the eggs of sea urchins. In many cases, there is a small passage or micropyle through the primary membranes; in echinoderms, a protoplasmic projection extends through this, reaching the surface of the jelly layer. The micropyle in other cases may represent the point of attachment of the egg stalk to the wall of the ovary and the scar left when this was broken and the egg set free.

2. Secondary egg membranes: these are formed by the follicle cells such as the chorion. This chorion is generally chitinous or horny in nature, and may be of complicated and intricate design. It is especially conspicuous in the eggs of cephalopods and insects. In cephalopods, Gardiner (1972) described that the follicle cells deposit tiny granules that enlarge and coalesce, at first into a rather coarse network, then into a finer one, and ultimately into a homogeneous sheet that makes a capsule around the egg. Bawab et al. (2012) evidenced that the follicular cells of the fully mature oocyte of *Styela plicata* precipitate the chorion during their degeneration.

Like the primary membranes, the chorion is usually pierced by a micropyle.

3. Tertiary egg membranes: those are derived as secretions from the wall, e.g. of the oviduct and/or other region of the female gonoduct. In some species, they are superimposed upon the primary and/or the secondary membranes and are acquired, generally after fertilization, as the egg passes along the gonoduct. Bawab et al. (1992b) studied the formation of the eggs of the desert snail, *Eremina ehrenbergi*, through studies of the reproductive system. They evidenced that this egg has a chorion and attains tertiary calcite membranes from the lining of the egg-channel of the spermoviduct. In addition, this region produces another tertiary membrane, the mucous in which the eggs are embedded (Fig. 2.22D and E).

Maturation of gametes

Oocyte maturation

Oocytes of different species arrest their nuclei before, during or following meiosis reaching their maturity, i.e. the step of acceptance of sperm entry. So the mature stage is fixed for each species (Fig. 2.9). For example, fertilization in the sea urchin takes place after the completion of meiosis. The egg, therefore, when it is laid is fertilizable since it has the small egg pronucleus containing the half number (n) of chromosomes.

Metz and Monroy (1985a) described how the fully grown starfish oocytes within the ovary still possess a large nucleus (germinal vesicle), which is arrested in the late prophase stage of the first maturation division. These oocytes are immature and fail to undergo normal fertilization when they are isolated and immediately inseminated. Just before spawning, the immature oocyte resume meiotic division, as indicated by the dissolution of the nuclear membrane (breakdown of the germinal vesicle), and undergoes the subsequent sequential processes of oocyte maturation until the discharge of the second polar body. Although the oocytes of species of starfishes including *Asterias* usually undergo the so-called spontaneous maturation when they are isolated and kept for a certain period in seawater without any hormone treatment, *Asterina pectinifera* oocytes generally fail to mature under the same conditions. In this species, when isolated immature oocytes are treated with 1-methyladenine (maturation-inducing substance), germinal vesicle breakdown generally occurs after 20 min, chromosome condensation after 35 min, metaphase after ~45 min and anaphase ~55 min, after the start of the treatment. The first polar body is formed 60–70 min and the second polar body 110–120 min after 1-methyladenine treatment at room temperature (20–35°C). Oocytes of this species are therefore suitable material for in vitro experiments on the mechanism of oocyte maturation.

The fertilizable stage of the egg differs with the species.

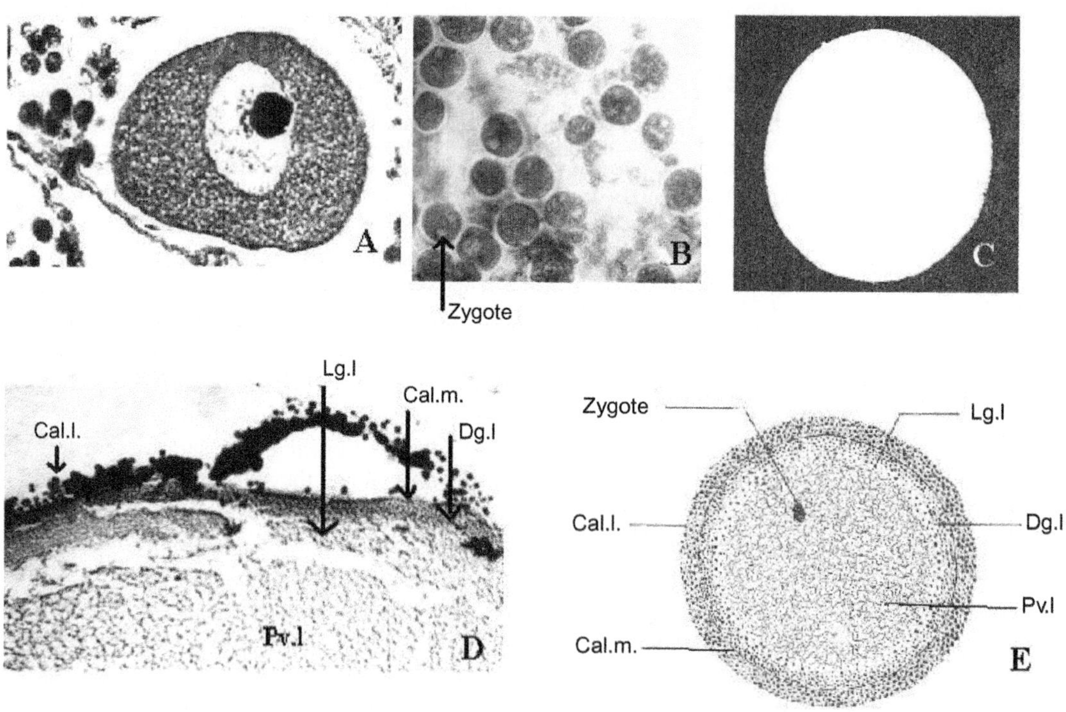

FIG. 2.22 Egg membranes, formation of tertiary membranes of *Eremina ehrenbergi*. (A) magnified portion of a transverse section of the gonad showing a spherical oocyte in its way to ovulate, 400×; (B) oil immersion magnification of eggs in the lumen of the end part of the egg-channel of the spermoviduct, 1000×; (C) a fixed freshly laid egg obtained from the field, showing its smooth surface, 10×; (D) a magnified portion of section of the laid egg, showing the surrounding layers, 100×; (E) a diagrammatic drawing of the upper section. Cal.l = calcified layer, Cal.m = calcified membrane, Dg.l = dense granulated layer, Lg.l = loose granulated layer, Pv.l = perivitelline layer *(Bawab et al., 1992b)*.

Spermatozoon maturation

Unlike eggs, the spermatozoon, in all the stages between the meiotic divisions until it reaches its final form, is incapable of fertilization. These stages are called 'stages of cytoplasmic metamorphosis'. Not every spermatozoon, which has achieved morphological perfection, has the fertilization capacity but the fertilization ability is parallel to the degree of motility, where vigorously moving spermatozoon has a high fertilizing capacity while immobile ones are generally useless.

The time limit of the fertilizing power of the sperm differs greatly in different animals. In a general way, it may be said that spermatozoa can function only so long as they are active. Unripe or inactive sperm will not fertilize unless excited to activity by external agents. As for how long the spermatozoa remain active, this depends partly on themselves and partly on external conditions. Morgan (1927) reported that as early as 1785, Spallanzani made some observations on the effect of temperature on the activity of spermatozoa. Prevost and Dumas in 1824 and Newport in 1854, were also interested in their activity. A. Koelliker, in 1856, carried out an elaborate series of experiments on the effect of external agents that excite sperm to activity or bring them to rest. More recently, when more accurate measurements began to be made and chemical relations to be discussed, sperm activity was enhanced by using external stimulation such as alkaline seawater. Alkaline seawater was also useful in cross-fertilization, where Loeb succeeded in causing fertilization of sea urchin eggs with starfish spermatozoa. Alkaline seawater is also useful for improving the percentage of cross-fertilization in difficult crosses. Loeb's (1903) success in causing fertilization of sea urchin, *Strongylocentrotus purpuratus* eggs with starfish, *Asterias ochracaea* spermatozoa by the use of alkaline seawater is a famous example of this. It is well known that spermatozoa, in general, will not adhere to, or penetrate eggs of other species or other tissues of the same species. Some exceptions to this are known. Thus sperm of the nemertean *Cerebratulus* penetrate eggs of the sand dollar *Echinarachnius* (Chambers, 1933). The sperm remain in the cortex without activating the egg, which can subsequently develop normally upon fertilization with the species sperm (Willier et al., 1956).

FERTILIZATION (L., *FERTILIS* = TO PRODUCE; TO BEAR; AND *FREVO* = TO BEAR)

Gilbert (2000) reported that Anton van Leeuwenhock, the co-discoverer of the sperm in 1678, originally assumed that they had nothing at all to do with reproducing the organism in which they were found. Leeuwenhock in 1685 wrote that the sperm were seeds (both sperma and semen mean 'seed') and that the female merely provided the nutrient soil in which the seeds were planted. In this, he was returning to a notion of procreation promulgated by Aristotle 2000 years earlier. Most investigators reached the same view and regarded the sperm as unimportant (see Correia, 1997 for details of this remarkable story).

The first evidence suggesting the importance of sperm in reproduction came from a series of experiments performed by Lazzaro Spallanzani in the late 1700s. Spallanzani demonstrated that filtered toad semen devoid of sperm would not fertilize eggs. He concluded, however, that the viscous fluid retained by the filter paper, and not the sperm, was the agent of fertilization. He, like many others, felt that the spermatic 'animals' were parasites. The combination of better microscopic lenses and the cell theory led to new appreciation of spermatic function. In 1824, J.L. Prevost and J.B. Dumas claimed that the sperm were not parasites, but rather the active agents of fertilization. They noted the universal existence of sperm in sexually mature males and their absence in immature and aged individuals. These observations, coupled with the known absence of spermatozoa in the sterile mule, convinced them that 'there exists an intimate relation between their presence in the organs and the fecundating capacity of the animal'. They proposed that the sperm entered the egg and contributed materially to the next generation. (For more clear and detailed information for this history, refer to Kume and Dan, 1968; Metz and Monroy, 1985a, vol. 1).

These claims were largely disregarded until the 1840s, when A. von Kolliker described the formation of sperm from cells within the adult testes. He ridiculed the idea that the semen could be normal and yet support such an enormous number OF parasites. Even so, von Kolliker denied that there was any physical contact between sperm and egg. He believed that the sperm excited the egg to develop, much as a magnet communicates its presence to iron.

It was only in 1877 that Oscar Hertwig and Herman Fol independently demonstrated sperm entry into the egg and the union of the two cells' nuclei. Fol carried on his experiments on the starfish *Asterias glacialis* (Balfour, 1880). Hertwig had sought an organism suitable for detailed microscopic observations, and he found that the Mediterranean sea urchin, *Toxopneustes lividus*, was perfect. Not only was it common throughout the region and sexually mature throughout most of the year, but also its eggs were available in large amounts and were transparent even at high magnifications. After mixing sperm and egg suspensions together, Hertwig repeatedly observed a sperm entering an egg and saw the two nuclei unite.

He also noted that only one sperm was seen to enter each egg, and that all the nuclei of the embryo were derived from the fused nucleus created at fertilization. Fol made similar observations and detailed the mechanism of sperm entry. He claimed that, in addition to the fact that the jelly of starfish eggs is relatively dense, the spermatozoa have rounded heads and are unable to penetrate the jelly layer. What Fol saw under these conditions at the time, as per Kume and Dan (1968) that a long, slender thread coming from the egg extended through the jelly layer to one spermatozoon, which was pulled through the jelly and to the egg surface. After sperm entrance the same sort of fertilization cone as that found in sea urchins and nereids, is formed (Fig. 2.23A and B). This thread is clearly of special interest, in that it is entirely different from the fertilization cone, and was therefore given the special name of '**reception cone**'. Almost 50 years later, Fol's conclusion was supported by Chambers (1923, 1930).

This process was, at last, recognized as the union of sperm and egg, and the union of sea urchin gametes remains one of the best studied examples of fertilization.

Description of fertilization reaction in sea urchin eggs

The unfertilized sea urchin egg is surrounded with transparent jelly. Since the refractive index of this jelly is the same as that of seawater, it cannot be seen by simple microscopic examination. Addition of some India ink colours the seawater and not the jelly, since the ink particles cannot enter the layer of jelly. The latter remains transparent, but it can be seen as a thick layer if it is stained by Janus Green. The eggs of such sand dollars as *Echinarachnius mirabitis* and *Astriclypeus manni* (Kume and Dan, 1968) have small pigment globules scattered throughout the jelly layer so that its approximate thickness is apparent without the use of any special measures. In the sea urchin egg, the point of sperm entrance is not predetermined, in this it differs from some other species. The sea urchin eggs at the time of sperm entry (after the discharge of the polar bodies) have a clear nucleus, which gradually travels towards the centre of the eggs and becomes known as the female pronucleus. If the spermatozoa are allowed to come into contact with such eggs, their heads soon become enveloped in the investing jelly layer surrounding the eggs. The first spermatozoon which comes up against the egg surface, applies its acrosome and rather simply penetrates its jelly, but those which arrive later and fail to enter, without exceptions, perform a clockwise rotational movement at the point of contact with the surface. The first spermatozoon can be carefully observed by using the oil immersion objective of the phase contrast microscope. Its head makes small oscillations by the aid

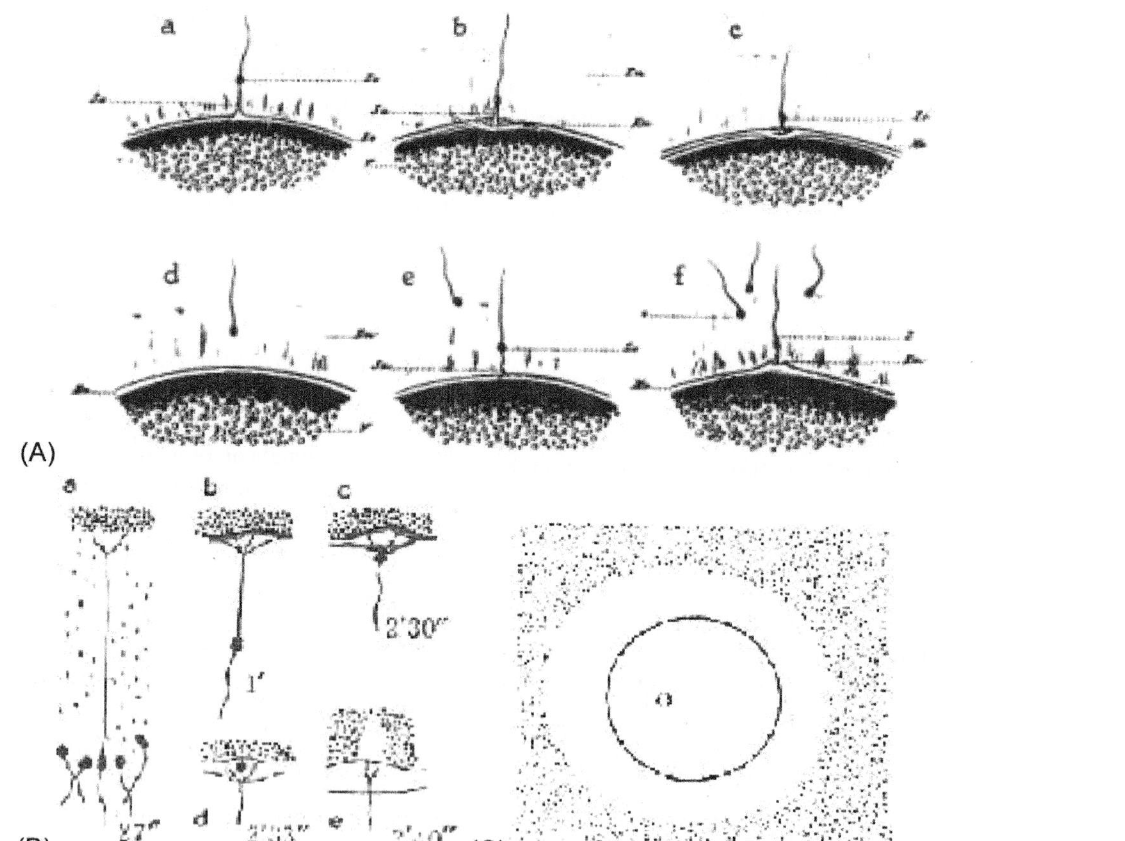

FIG. 2.23 (A) Reproduction of some of Fol's (1879) figures of fertilization in *Asterias glacialis*, showing filament connecting the spermatozoon with the surface of the egg. Three successive stages for one egg are illustrated in a, b, and c. and for another egg in d, e, and f *(from Willier et al., 1956)*; (B) Chambers' figure of fertilization in the star-fish *Asterisa forbesii (from Kume and Dan, 1968)*; (C) Jelly layer of sea-urchin egg demonstrated by adding India ink to egg suspension.

of the violent movements of its tail. The oscillations, then, suddenly cease and it looks as the acrosome of the head becomes fixed to the egg surface. This is the earliest indication of fertilization; this spermatozoon is the only one concerned in the process and is known as the **fertilizing spermatozoon**. It can be defined if one can quickly notice the spermatozoon whose acrosome stops oscillations. This means that this spermatozoon becomes motionless. The point of sperm contact causes the elevation of the superficial layers of the protoplasm of the egg forming a small prominence (a cone-shaped process) pointing towards and engulfing the acrosome of this nearest spermatozoon. This prominence is called the **fertilization cone** (exudation cone). The form and size of this cone differs with the species. This cone, in some cases, retracts within 2 or 3 min after the completion of sperm entrance and in others it remains a little longer. The sea urchin *Pseudocentrotus depressus* is exceptional in that its fertilization cone lasts for several hours; Endo (1954) recognized its existence in the 16-cell stage.

The head of the fertilizing spermatozoon is perpendicular to the fertilization cone while its acrosome is engulfed in it. Its tail remains visible for some time after the head has bored its way in. Its penetration to the inside of the egg is not effected by its mechanical activity, but the cone becomes responsible for pulling it inside. This fertilization cone is formed in many species of animals, but it takes different types and shapes. Examples of various types of cones may be found in studies by Chambers (1930, 1933). Kume and Dan (1968) described and illustrated the sperm entrance and their cones in different species. This is enough to emphasize that fertilization process differs markedly with the species. The most conspicuous is that which occurs in starfish (see above, Fig. 2.23).

At the moment and point of the slipping of the head of the perpendicular spermatozoon of sea urchin within the contour of the egg, the outermost layer of the protoplasm (vitelline membrane) is peeled off the surface. It changes into a **fertilization membrane** and expands gradually to enclose the whole egg, preventing the approach of other spermatozoa. During this event, the metamorphosed sperm tail can be observed projecting through a crater-like opening, which is left in the membrane. It must be able to slip to the inside just beneath the fertilization membrane. Actually, according to the still accepted opinion, the sperm tail was thought to remain outside of the membrane and be carried away from the egg surface as the membrane expands (Wilson, 1925; Chambers, 1933). However, phase contrast observation shows that it is usual for the tail to slip smoothly through the membrane. When the sperm head has proceeded a few microns from the surface into the egg, the tail, which is in a resting stage, begins to move. This phenomenon gives the impression that the egg, after once pulling the sperm through the surface, sets it free again inside this barrier—to such an extent does the spermatozoon move freely about. The head inside the egg ceases to move, and abruptly begins to swell and change into the spherical male pronucleus; from the mid piece, a centriole appears and is surrounded by a small aster. Once the sperm pronucleus is formed, the protoplasm in its neighbourhood assumes a radiate arrangement. Then, it gradually travels towards the female pronucleus, which remains motionless until the rays of the male pronucleus come into contact with it. The female pronucleus rapidly approaches the male pronucleus, apparently by means of its inherent amoeboid contractions, and eventually fuses with the male one (Fig. 2.24).

The product of the fusion of the two pronuclei forms the first segmentation nucleus, which soon, however, divides into the two nuclei of the two first segmentation blastomeres. In this act, the protoplasm of the two gametes, as well as their nuclei, coalesce, since the whole spermatozoon, which has been absorbed into the ovum, is a cell in which the head is the nucleus. It is clear that the ovum after fertilization is an entirely different body and hence the name **zygote** is provided (Latin *zygos*, union of two things, in biology it is union of two different gametes, zygos is from Arabic language, zegaa = zawag = marriage, for review, see Kume and Dan, 1968).

It does not, however, necessarily follow that the fertilization reaction begins simply by the attachment of a spermatozoon to the egg surface. It is entirely possible that sperm and eggs exert some decisive influence on each other before they come into spatial contact. Since the time of Fol and Hertwig, some other scientists were interested in demonstrating and performing experiments to observe the changes occurring to both sperm and egg of different species during their interaction; especially those that occur to the surface of the egg, as well as to its external investments. This interaction was defined as the **fertilization** process.

Loeb (1859–1924) was one famous (and the first) scientist who proposed that the sperm carried a substance located in its head that somehow activated the egg. He claimed this idea to explain his experiments on artificial parthenogenesis. This idea became of widespread interest following the introduction of the fertilizin theory of Lillie (1912), which was modified and expanded by the immunological studies of Tyler (1948), and others. The great discovery of Lillie, as well as the essential point of his fertilizin theory, was the phenomenon known as **sperm agglutination** (Lillie, 1923). As written in Kume and Dan (1968), this phenomenon is as follows: unfertilized eggs are suspended in seawater, allowed to settle, and the supernatant removed; if this supernatant is added to a sperm suspension, the cells that were swimming entirely separate and at random in pure seawater immediately come together to form clumps of various sizes. This is **agglutination**. On the other hand, the supernatant from immature or already fertilized eggs does not cause agglutination. Until the time of Lillie, the study of fertilization had been limited to a simple process of description,

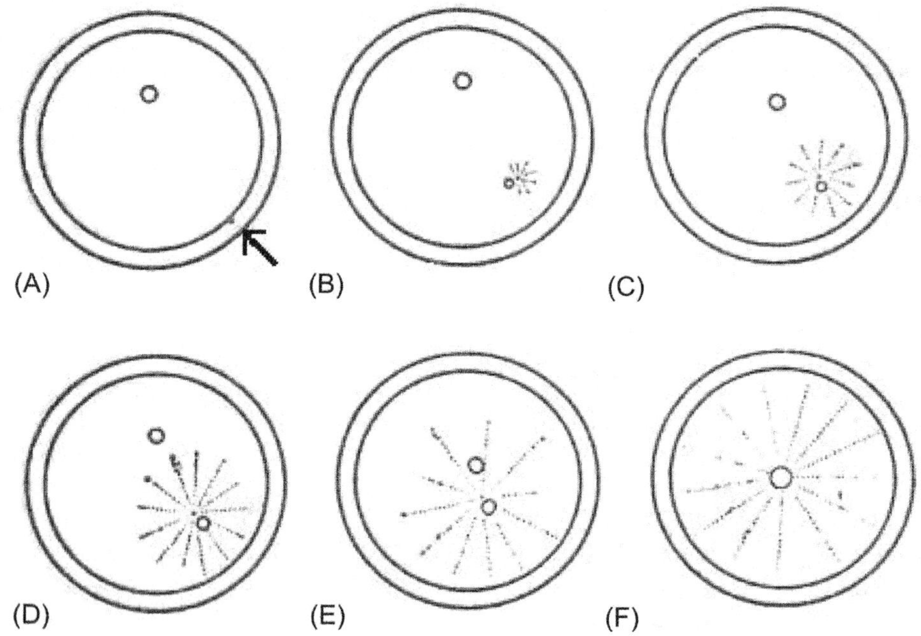

FIG. 2.24 Process of syngamy (A) entering sperm and fertilization cone *(arrow)*; (B) appearance of sperm monaster; (C) growth of sperm monaster; (D) migration of sperm monaster towards the egg centre; (E) migration of egg pronucleus; (F) fusion of nuclei and development of monaster. *(From Kume and Dan, 1968.)*

from an onlooker's point of view; this discovery for the first time started it on a materialistic course. Lillie expressly gave the name of **egg water** to the supernatant of unfertilized eggs, which had been endowed with the capacity to agglutinate spermatozoa, and the essential ingredient, which he believed to take part directly in agglutination. Lillie used the term **fertilizin** or agglutinin to designate the agent(s) in egg water that were responsible for agglutinating, activating and possibly chemotactic and other effects. However, it seems best to restrict the use of this term to the agglutinin. Lillie assigned it a central role in his theory of fertilization. In the early 1900s, fertilization research was framed by a dispute between Lillie and Jacques Loeb, who disagreed over whether the sperm recognized the egg through soluble factors or through cell–cell interactions. This early work in this field has been reviewed by Lillie (1919), Lillie and Just (1924), Morgan (1927), and Just (1930).

According to Lillie (1870–1947), there are three stages in the agglutination process. On encountering egg water, the spermatozoa first (i) become intensely active; (ii) come together, which he termed aggregation; and finally (iii) undergo agglutination—of the three, the agglutination process is the most important. In sea urchins, this process is defined as **head-to-head agglutination**, since the spermatozoa have their heads pointed towards the centre of the clump. A unique feature of the reaction in sea urchins is its spontaneous reversal after a period of time, ranging from a few seconds in dilute egg water to many minutes in concentrated egg water, whereas in other groups of animals the agglutination is essentially permanent or long-lasting. Following spontaneous reversal, in the sea urchins, the spermatozoa are incapable of being reagglutinated by the addition of fresh egg water and have lost their fertilizing capacity, although they are fully motile (Lillie, 1919; Tyler, 1941). Some support for the above interpretation of the spontaneous reversal of agglutination in sea urchins is obtained from experiments in which the ordinary agglutinin is converted into a nonagglutinating (univalent) form by the action of heat, proteolytic enzymes, ultraviolet or X-radiation (Tyler, 1941, 1942; Metz, 1942). This univalent antigen or antibody provides an interpretation for the apparent absence of fertilizin in many species, such as some starfish (Metz, 1945). Such failure to demonstrate fertilizin in all species of animals has been used as an argument against considering fertilizin to have any general significance for fertilization. However, that the egg water of such nonagglutinating species might contain a fertilizin-like substance was clearly established by Metz (1945), who was able to cause specific agglutination of sperm in species of starfish that normally do not give an agglutination reaction. Thus, the failure of agglutination to occur ordinarily in certain species of animals is not necessarily indicative of the absence of fertilizin in the egg water but can be interpreted on the basis of a univalent condition of either the fertilizin or the sperm.

Concerning the chemical nature of fertilizin, Lillie (1919) showed that the gelatinous coat of the sea urchin egg contained large amounts of fertilizin. He believed that it was being continuously secreted by the ripe unfertilized egg throughout its functional life. For chemical purposes, sea urchin fertilizin can be obtained in high titre by dissolving the coat with dilute acid, without damage to the rest of the egg. By this, the protein nature of fertilizin of the sea urchin and of the keyhole limpet was evidenced. However, later observations concluded that fertilizin is a complex of amino acids and sugars that may be termed a glycoprotein.

Not only fertilizin can be produced as a result of the interaction of eggs and sperm, but also three other groups of interacting substances. These four groups of substances are termed fertilizins, antifertilizins from sperm, antifertilizins from eggs, and lytic agents from sperm. The terms gynogamone and androgamone were introduced by Hartmann (1940) to designate the substances derived from eggs and sperm, but these terminologies are not sufficiently advantageous to warrant its adoption.

1. Fertilizins: Agglutination of sperm by homologous egg water has been reported in many species of animals among echinoderms, annelids, molluscs, tunicates and vertebrates.
2. Antifertilizins from sperm: The substance on the surface of the sperm with which fertilizin reacts. It is considered to be located on the lateral surface of the head between the acrosome and midpiece. It has been claimed that the jelly coat dissolves under the influence of such a substance. It is able to neutralize the agglutinating action of fertilizin on sperm. Another manifestation of its activity is its ability to agglutinate a suspension of eggs.
3. Antifertilizins from eggs: Lillie (1914) had postulated the presence of an antifertilizin within the egg. The effects of this agent are similar to those described for the antifertilizin from sperm. i.e. antifertilizin from within the egg can react with fertilizin of the same egg.
4. Lytic agents: Sperm can produce an agent which has the property of breaking down the membrane of viscous coat that normally surrounds the unfertilized egg. (For details of this subject at that time, see Willier et al. (1956). Follow later the evolution of this subject).

Formation of the fertilization membrane

Among the morphological changes that accompany fertilization in sea urchin eggs, and the most conspicuous, is the formation of the fertilization membrane. Extensive experimental work has been done to investigate the mechanism of the formation of the fertilization membrane. Kume and Dan (1968), described this vital activity in different species of Invertebrates. In 2000, Gilbert presented the more recent and advanced views around this process. In connection to sperm entrance, a single-layered membrane rises as though it was being peeled off, and gradually lifts away from the surface. It spreads around the egg to the side opposite to the point of sperm entrance so that it becomes an outer layer enclosing the whole egg. This event requires about 1 min after insemination to be completed. However, two or more facts are known about its physical properties. First, this fertilization membrane when firstly formed is a soft membrane, as the experiments evidenced; it can easily be torn by gentle shaking during the first minute after insemination. Within 2–3 min of insemination, it becomes hard and can no longer be torn easily (Lillie, 1921 and many others). Its hardening increases with time. It was reported that the fertilization membrane of sand dollar—*Dendraster excentricus*—eggs will be dissolved if they are transferred to 1 M urea within 5 min after insemination, but not later. Second, the hardening of the membrane does not occur when the Ca^{++} content of the medium is not at the proper level, even the fertilization process itself does not occur in such conditions. Loeb (1915) evidenced that fertilization is impossible in the absence of Ca^{++}. Eggs, after 30 s insemination have a fertilization membrane, if these eggs are shaken to remove the membrane and add Ca^{++}-free seawater, the surface precipitation reaction fails to occur because of the lack of Ca^{++}, and the eggs are injured or complete cytolysis takes place. Third, it is colloid osmotic pressure which lifts the fertilization membrane away from the egg surface. If egg white or some other colloid is added to the seawater in which eggs with fertilization membranes are suspended, the membranes deflate until they are again in contact with the egg surface; if they are washed with pure seawater, the membranes again inflate to their former dimensions.

From these facts, it is apparent that the fertilization membrane allows salts to pass freely but is impermeable to colloid molecules; that is, it is a so-called dialyzing membrane. Accordingly, it follows that under normal conditions a colloid released into the space between fertilization membrane and egg surface (perivitelline space), acts as the cause of membrane elevation; it is accepted that the resulting colloid osmotic pressure draws seawater from the medium into the perivitelline space, causing its increase in size (Loeb, 1908). Loeb's hypothesis has been demonstrated as a fact by later scientists.

About the mechanism of the formation of the fertilization membrane, Chambers (1921) performed experiments to determine the precursor of the fertilization membrane. He used starfish eggs since the unfertilized starfish egg has a surface layer which is recognizable from the beginning, and lifts off and hardens to form the fertilization membrane, so that the whole process is visible. He reported that when the outermost layer of this unfertilized starfish egg was torn off with a microneedle, the egg failed to form a fertilization membrane on subsequent insemination. Since it is usual to refer to the outermost layer in the case of the unfertilized starfish egg as the 'vitelline membrane', it is possible to state that the result of this work was that the vitelline membrane is the precursor of the fertilization membrane. Later on, Motomura (1941) traced the process of fertilization membrane formation in sections of sea urchin eggs, using a special Janus Green dye, which stained the cortical granules. He reported that as the membrane begins to rise, the granules, which at this time have been stained, become invisible, while the previously nonstaining vitelline membrane begins to take up the colour. This led him to think that when the cortical granules explode, this explosion product is extruded to the outside of the egg and adheres to the inner surface of the vitelline membrane, and he suggested that the addition of this substance is very likely the cause of the fertilization membrane hardening.

The work of Endo (1952, 1954) evidenced Motomura's idea by means of observations on living eggs, and carried it to completion. Endo found that when he observed the highly transparent eggs of *Clypeaster japonicus* and *Mespilia globulus* with dark contrast phase contrast objective, the cortical granules appeared dark; when these exploded and swelled as the result of fertilization, their contrast was reversed and they appeared bright. These bright-appearing masses of granule-explosion product are pushed outwards, becoming constricted in the process as though they were passing through small pores in the egg surface; since they are being extruded to the inner side of the vitelline membrane rather than to the true outside of the egg, they cause this membrane to rise away from the cytoplasmic surface (Fig. 2.25). As the vitelline membrane is thus stripped off and lifted away from the egg, the masses of the product stick to the inner surface of the vitelline membrane as flattened drops, together with the expansion of the membrane these drops spread out, and finally form a smooth inner lining of the completed fertilization membrane.

Metz and Monroy (1967, 1969) were very interested in the field of fertilization. In 1985, they edited a review and reported that the comparative study of gametes and their interactions in both unicellular and multicellular organisms is emerging as a fascinating field. It is an area that promises important contributions to the understanding of the evolution of reproductive processes. Interesting recent work has concentrated on the chemo-attraction of sperm and egg; the physiochemical mechanisms of the acrosome filament extrusion process, and so on.

This means that the recent tools and studies, especially molecular embryology, as they are developing the subject as a whole, have fulfilled the principal function of the process of fertilization; i.e. it transmits genes from parents to offspring and it initiates those reactions in the egg cytoplasm that permit development to proceed. So fertilization is described as a critical event in development. It is a very highly specialized example of the general phenomenon of cell–cell interaction with manifestations of its components. It can serve as a general model system for cell–cell interaction. These can include 'action-at-a-distance' (chemotaxis and other physiological effects on sperm). Much of the knowledge about the new view of the mechanism of fertilization has been learned from studies of marine invertebrates,

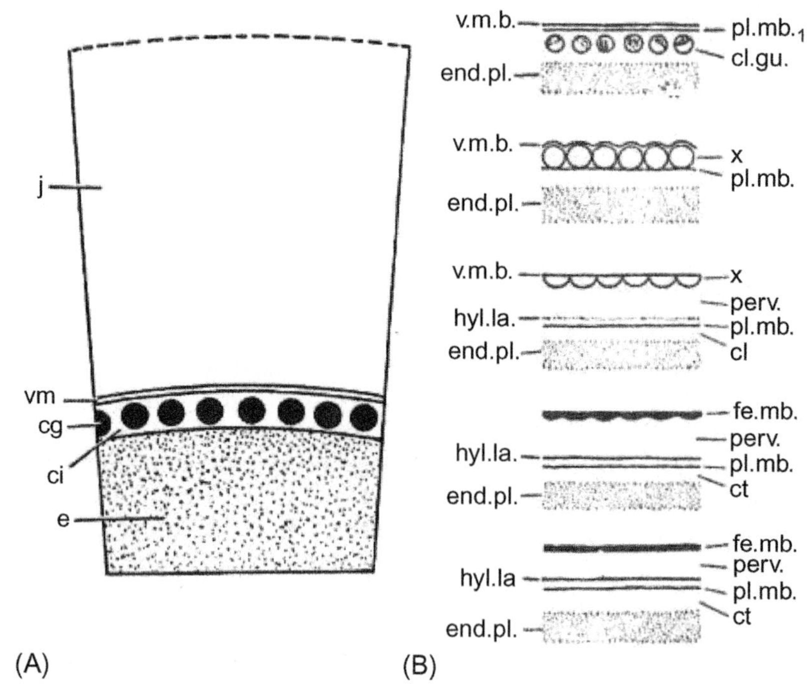

FIG. 2.25 (A) Diagram of part of surface of unfertilized sea-urchin egg *(after Runnström, 1949, from Willier et al., 1956)*; (B) Process of fertilization membrane formation *(after Endo, from Kume and Dan)*, ct, cortical layer—formerly 'granular layer', 'extra-granular zone'. In egg of sea-urchin *Hemicentrotus pulcherrimus*, yellow granular layer is present just beneath cortex, ct.gn = cortical granule, end.pl = endoplasm, fe.mb = fertilization membrane, hyl.la = hyaline layer, perv = perivitelline space, pl.mb = plasma membrane (surface of fertilized egg), x = products of breakdown of cortical granule.

especially sea urchins. In these organisms, fertilization occurs in seawater, into which huge numbers of both sperm and eggs are released. Such external fertilization is much more accessible to study than the internal fertilization of mammals. For this reason, most of the discussions of fertilization will deal with sea urchins. However, despite the great evolutionary distance between sea urchins and mammals, many of the cellular and molecular mechanisms involved in fertilization are similar in both.

Up-to-date concept of the process of fertilization

Although the details of fertilization vary from species to species, conception generally consists of four major mechanisms:

- cell adhesion (conjugation occurring in micro-organisms is also a model system for cell–cell interaction, see Chapter 3);
- regulation of sperm entry into the egg (block to polyspermy);
- fusion of the haploid nuclei of the egg and sperm (genomes) to create the diploid nucleus of the zygote; and
- initiation and activation of egg metabolism to start development.

I-Cell adhesion (recognition and contact between sperm and egg)

This event (interaction of sperm and egg) is a critical event in the development of an organism. It generally proceeds according to five basic steps:

1. the chemo-attraction of the sperm to the egg by soluble molecules secreted by the egg;
2. the exocytosis of the acrosomal vesicle to release its enzymes;
3. the binding of the sperm to the extracellular envelope (vitelline layer);
4. the passing of the sperm throughout this envelope; and
5. fusion of egg and sperm cell plasma membranes.

It is well known that many species of invertebrates and chordates are aquatic animals. The surrounding water may be as small as a tide pool, or as large as an ocean. In addition, fertilization of many of these animals occurs externally. Many of these animals release their gametes at the same time. This means that gametes of many different species are present in the same locality in water and at the same time. Thus, the meeting of sperm and egg of each species in such spacious medium becomes a difficult matter. These animals are then faced with two problems: (a) how can sperm meet their specific egg in such a dilute concentration?; and (b) how can sperm be prevented from trying to fertilize eggs of another species?

Step 1 (the chemo-attraction of the sperm to the egg) solves the first problem by **species-specific attraction**. This is a major mechanism of '**action-at-a-distance**'. This sperm attraction has been documented in numerous species, including cnidarians, molluscs, echinoderms, and urochordates. In many species, sperm are attracted towards eggs of their species by 'chemotaxis'—that is, by following a gradient of a chemical secreted by the egg. The mechanisms of chemotaxis differ with the different species. From the jelly coat of the sea urchin *Arbacia punctulata*, one chemotactic molecule, **a 14-amino acid peptide** called **resact**, has been isolated. Resact diffuses readily in seawater and has a profound effect at very low concentrations when added to a suspension of *Arbacia* sperm. Results of many different experiments confirmed that this resact is specific for *A. punctulata* and does not attract sperm of other species. This was explained through the fact that sperm of *A. punctulata* have receptors in their plasma membranes that bind resact, and can swim up a concentration gradient of this compound until they reach the egg.

The sperm receptor for resact is a transmembrane protein, and when it binds resact on the extracellular side, a conformational change on the cytoplasmic side activates the receptor's enzymatic activity. This activates the mitochondrial ATP-generating apparatus as well as the dynein ATPase that stimulates flagellar movement in the sperm. Thus, resact acts as a sperm-activating peptide, since it causes dramatic and immediate increases in mitochondrial respiration and sperm motility.

Step 2 (the exocytosis of the acrosomal vesicle) is a second interaction between sperm and egg known as the **acrosomal reaction** (the activation of the sperm) (Fig. 2.26 A and B). This is one of the major discoveries made during the latter half of the 20th century. Many details of this extraordinary process have been reviewed in earlier literature (e.g. Fertilization, vol. 1, 1967). This was first studied in echinoderms, but the principles discovered appear to apply to many other groups, including mammals. Sperm, after the action of resact, swim using long, whiplike flagella. A sperm cell approaching an egg (Fig. 2.26a) will burrow into the jelly coat (Fig. 2.26b), an action that triggers a dramatic response in the sperm called the acrosomal reaction with the egg jelly. The acrosomal reaction has two components: the fusion of the acrosomal vesicle with the sperm plasma membrane (Fig. 2.26b); and the extension of the acrosomal process (Fig. 2.26c and d). The acrosomal vesicle, a large membrane-bound vesicle near the tip of the sperm head, exocytoses, releasing a proteolytic enzyme called **sperm lysins** that digest a pathway for the sperm through the jelly coat to the egg surface. Meanwhile, actin monomers just under the surface of the acrosomal vesicle (Fig. 2.26a) polymerize to form a long actin-containing acrosomal process (filament) (Fig. 2.26C). This process involves a 1–75 μm long, thin, rigid, tube that projects or protrudes out from

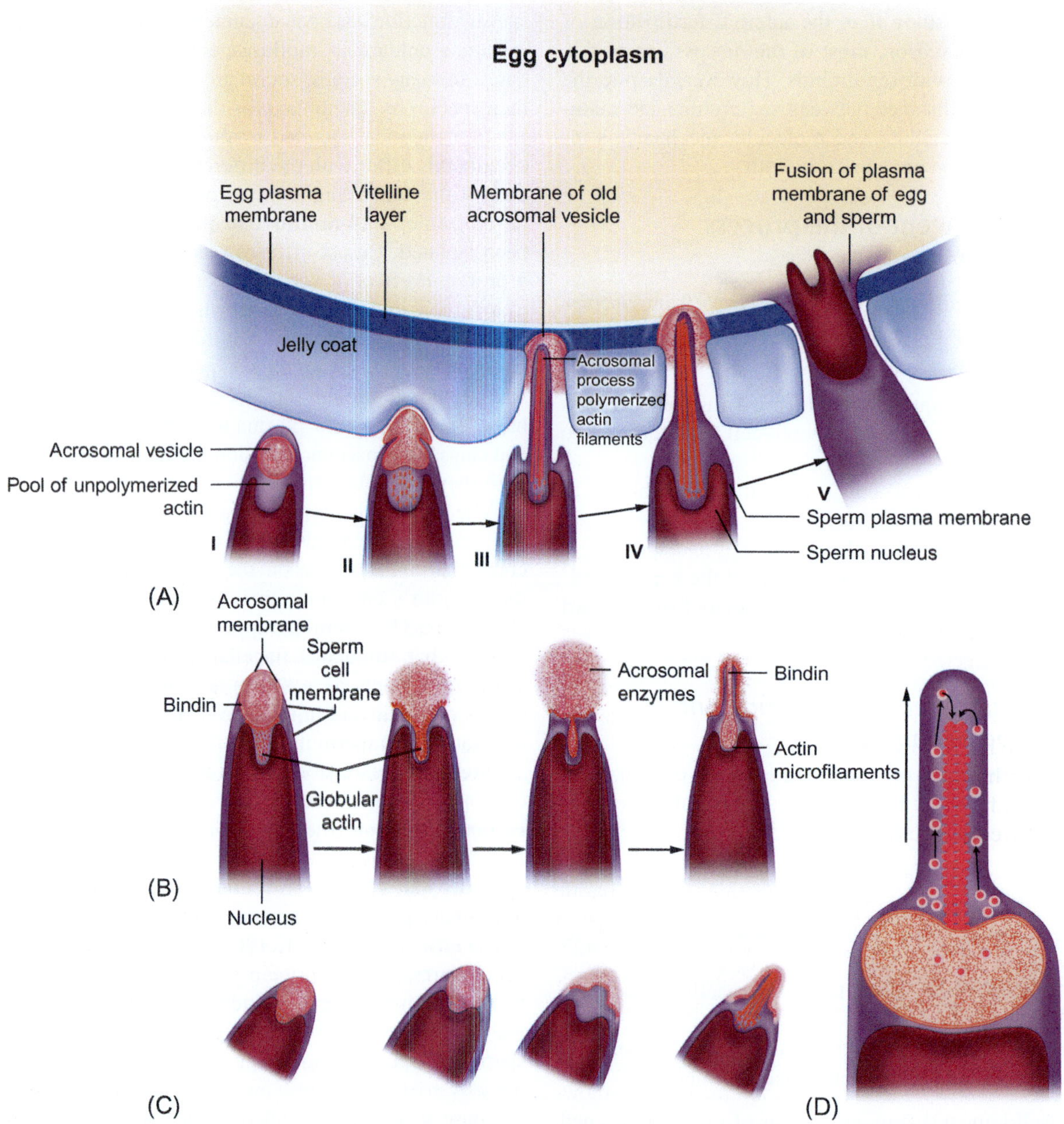

FIG. 2.26 Acrosomal reaction in sea-urchin sperm, (A and B) diagrammatic explanation; (C) actual photomacrographs of the acrosomal reaction. (a–c) Acrosomal membrane lying directly beneath the sperm plasma membrane; (d) the actin molecules extends the acrosomal process; (e) fusion of plasma membranes of sperm and egg *(B and C from Gilbert, 2000)*; (D) drawing illustrating the elongation of the actin filaments, actin monomers are liberated at the profilactin cup and diffuse to the tip of the acrosomal process *(from Metz and Monroy, vol. 2, 1985b)*.

the central anterior end of the sperm head. When the sperm exhibit such an acrosomal process, they are said to be activated for the ready penetration in the unfertilized egg. The tip of this process becomes covered by the components of the old acrosomal vesicle membrane (Fig. 2.26c). It also becomes coated with secreted contents of the acrosomal vesicle, including hydrolytic enzymes that help the sperm to penetrate the jelly coat and thus extends towards the egg surface (vitelline layer). As the surfaces of the gametes approach, a species-specific recognition (binding or interaction) takes place between the protein **bindin** (probably the fastest evolving protein known; closely related species may

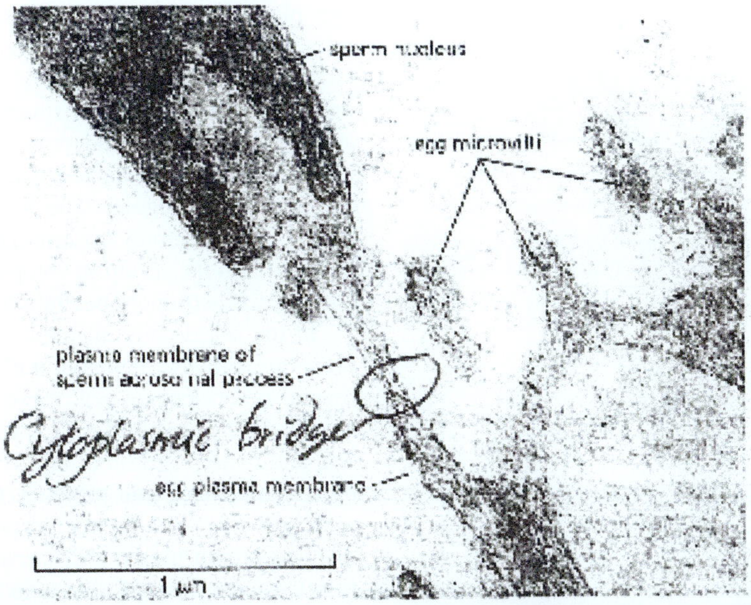

FIG. 2.27 Species-specific binding of acrosomal process to egg surface in sea-urchin. Electron micrograph showing actual contact of acrosomal process with an egg microvillus.

have near-identity of every other protein, but their bindins may have diverged significantly) on the sperm cell and a receptor on the egg surface (Figs 2.26d and 2.27). The membranes of the two gametes fuse, allowing the sperm pronucleus to begin moving inward (Fig. 2.26Ae).

In sea urchins, the acrosomal reaction is thought to be initiated by a fucose sulphate-containing polysaccharide in the egg jelly coat. The latter consists also of a highly sulphated glycoprotein. This jelly coat is insoluble in seawater but soluble in acidic water. When this polysaccharide is extracted from a sea urchin egg and added to a sperm suspension, it induces a normal acrosomal reaction within seconds. The jelly coat polysaccharide binds to a receptor glycoprotein in the sperm plasma membrane, causing the membrane to depolarize (changing the permeability of the acrosomal membrane). The depolarization is thought to open voltage-gated Ca^{++} channels in the membrane, allowing Ca^{++} ions in the seawater to enter to the sperm head. The jelly coat polysaccharide also activates a proton pump in the sperm plasma membrane that pumps H^+ ions out of the cell in exchange for Na^+ ions. This results in the rise of pH inside the sperm head, together with the increase in cytosolic Ca^{++}, which initiates the acrosomal reaction. The egg jelly factors that initiate the acrosomal reaction in sea urchins are often highly specific to each species.

Step 3 involves the binding of the sperm to the surface of the egg back to the **bindin**, which solves the second problem by **species-specific recognition**. Once the sea urchin sperm has penetrated the egg jelly, the acrosomal process of the sperm contacts the surface of the egg. A major species-specific recognition step occurs at this point. The acrosomal protein mediating this recognition is called bindin (Fig. 2.26Bd). In 1977, this bindin (**nonsoluble 30,500-Da protein**) was isolated from the acrosomal process of the purple sea urchin *Strongylocentrotus purpuratus* and found to be capable of binding to dejellied eggs of the same species. Further, its interaction with eggs is relatively species-specific. This means that bindin isolated from the acrosome of *S. purpuratus* binds to its own dejellied eggs, but not to those of *Arbacia punctulata*. For example, using immunological techniques, it was demonstrated that bindin is located specially on the acrosomal process—exactly where it should be for sperm-egg recognition (Fig. 2.27B).

Biochemical studies have shown that the bindins of closely related sea urchin species are indeed different. This finding implies the existence of species-specific bindin receptors on the egg, vitelline envelope or plasma membrane.

Step 4 and Step 5 consist of the sperm passing through the egg surface and fusion of egg and sperm cell plasma membranes. Once the species-specific binding proteins mediate the attachment (adhesion), the hydrolytic enzymes on the acrosomal process permit boring (digest) through the egg surface (vitelline membrane) to the egg's plasma membrane. Sperm egg binding appears to cause the extension of several microvilli to form the fertilization cone. An unfertilized sea urchin egg is covered with more than 100,000 microvilli (Fig. 2.17A). If the acrosomal reaction is successful, the vitelline envelope breaks down at the point of contact, and the egg extends a cytoplasmic fertilization cone (Fig. 2.28) towards the sperm head. This cone is formed

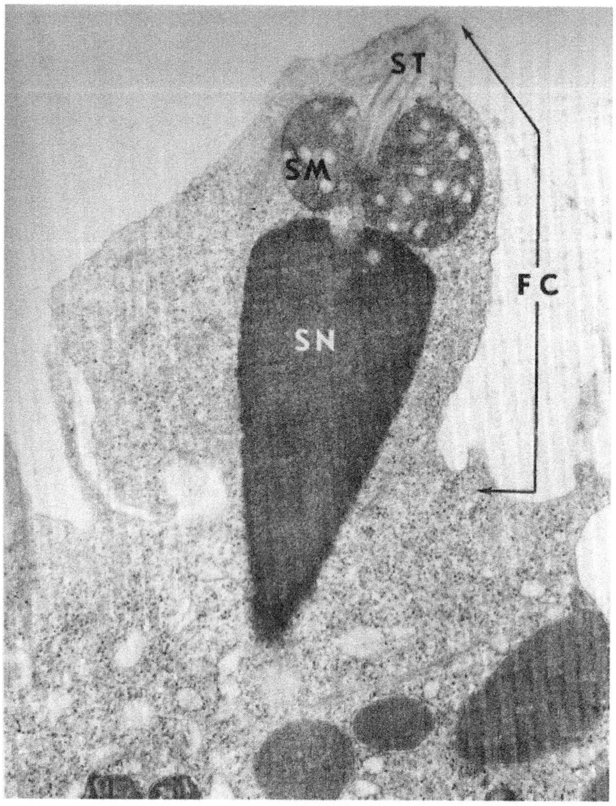

FIG. 2.28 Incorporated *Arbacia* spermatozoon, 1 min after insemination. The cytoplasm in the region of gamete fusion has engulfed the sperm, forming a fertilization cone (FC). ST=portion of the axonemal complex of the sperm tail, SM=sperm mitochondrion, SN=sperm nucleus. *(After Longo (1981), from Metz and Monroy, vol. 3, 1985c.)*

as soon as the tip of the acrosomal filament touches the egg plasma membrane; the latter extends several microvilli, which rapidly elongate and cluster around the sperm. Homology between the egg and the sperm is again demonstrated, because the transitory fertilization cone, like the acrosomal process, appears to be extended by the polymerization of actin. The fertilization cone may be conical, cylindrical or irregular. The fertilization cone engulfs the sperm, which moves inward. The sperm nucleus and tail pass through the resulting cytoplasmic ridge, which is widened by the actin polymerization.

In the sea urchin, all regions of the egg plasma membrane are capable of fusing with the sperm (Fig. 2.29). In several other species, e.g. those with a micropyle, certain regions of the membrane are specialized for sperm recognition and fusion. Fusion is an active process, often mediated by specific **fusogenic** proteins. It seems that bindin plays a second role as a fusogenic protein. It has been shown that sea urchin bindin will cause phospholipid vesicles to fuse together and that, like viral fusogenic proteins, bindin contains a long stretch of hydrophobic amino acids near its amino terminus. This region is able to fuse phospholipid vesicles.

The fusion of the sperm and egg plasma membranes ends the first major mechanism (cell adhesion). We now move on to the second mechanism.

II-Regulation of sperm entry into the egg and prevention of polyspermy

Normally, only one sperm enters the egg and fuses with the female pronucleus. Polyspermy, the fusion of more than

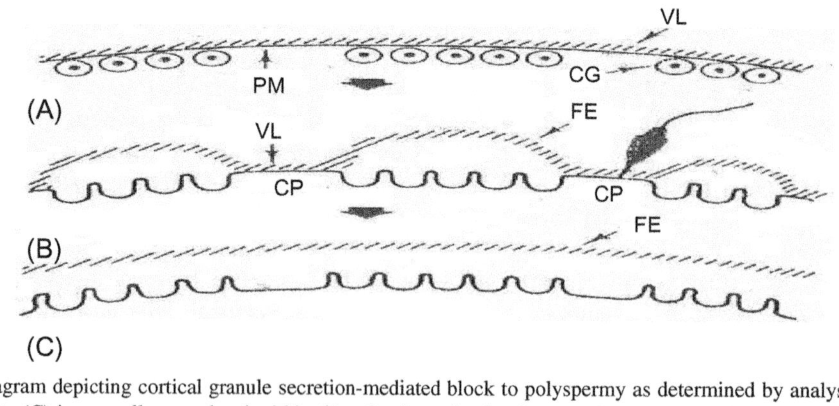

FIG. 2.29 Schematic diagram depicting cortical granule secretion-mediated block to polyspermy as determined by analysis of SBTI-treated *Arbacia* eggs. Transition from (A) to (C) is normally completed within 60 s after insemination in control eggs. (A) surface of unfertilized egg showing regions where cortical granules (CG) are tightly packed under the plasma membrane (PM) interspersed with regions devoid of subjacent cortical granules. The vitelline layer (VL) is attached to the outer surface of the egg's plasma membrane; (B) surface of SBTI-treated egg at 3 min after insemination. The fertilization envelope has elevated over regions previously occupied by the discharged cortical granules. The vitelline remains attached to the plasma membrane at the apex of the cortical projections. These sites correspond to regions where the plasma membrane was devoid of subjacent cortical granules prior to fertilization. Sperm can continue to fuse with and penetrate the egg at these sites as long as they are available; (C) surface of SBTI-treated egg at 20 min after insemination. The fertilization envelope has elevated from the entire egg surface and acts as a mechanical barrier to prevent entrance of additional sperm. *(After Schuel (1978), from Metz and Monroy, vol. 3, 1985c.)*

one sperm nucleus with the female pronucleus, invariably results in abnormal development and death of the embryo (Lillie, 1919 and many others). Thus, species have evolved ways to prevent the union of more than one male haploid nucleus with the female pronucleus, i.e. to ensure monospermic fertilization. However, several sperm normally may enter large yolky eggs (certain insects, molluscs, sharks, reptiles, birds, urodeles, etc.), but only one is allowed to fuse with the egg pronucleus. The most common way is to prevent the entry of more than one sperm into the egg. The prevention of polyspermy during fertilization is a critical event in the initiation of development. In sea urchins and most other animals, several processes act together to avoid polyspermy:

- ion-dependent changes in membrane potential;
- elevation of fertilization envelope as a mechanical barrier;
- various secretory products released by the egg's cortical granules; and
- release of peroxide to inactivate sperm.

Ion-dependent changes in membrane potential (polysperm block)

Elevation of the fertilization envelope is not the only process operating at the surface of the sea urchin eggs to prevent polyspermy. This was appreciated by early students over a century ago (Lillie, 1919 and others) regarding echinoid fertilization. The fertilization envelope does not elevate rapidly enough from the entire egg surface (1–2 min depending upon species and temperature) to be exclusively responsible for the prevention of polyspermy in excess sperm. The study of the fertilization rates indicated that a rapid block reduced the probability of polyspermy in the interval between the **first successful sperm-egg reaction** (fertilizing sperm-egg reaction) and completion of cortical reaction. In addition, study of the surface of the unfertilized sea urchin egg (presumably the vitelline layer) indicated that the sperm binding to receptors in the vitelline layer is a prerequisite for fertilization. There are 1500–6000 sperm binding sites on this surface. How many of these correspond to potential penetration sites is unknown. However, the potential penetration sites must be numerous since eggs can be fertilized upon insemination, with as few as 10 sperm per egg. Under these circumstances, polyspermy would be the rule rather than the exception even at low or moderate sperm densities in the absence of an effective rapid block.

Rapid prevention of polyspermy (fast block to polyspermy) is achieved by changing the electric potential of the egg plasma membrane. The possibility that an electrical mechanism may account for the fast block to polyspermy was recognized as early as 1922. However, technical difficulties delayed experiments to examine this idea. The first successful microelectrode recording during fertilization of any egg was made in 1956 by Tyler et al., in the eggs of starfish (Metz and Monroy, 1985c). Essentially, this membrane provides a selective barrier between the egg cytoplasm and the outside environment. The ionic concentration of the egg differs greatly from that of its surroundings (Fig. 2.30A). This concentration difference is especially significant for sodium and potassium ions. Seawater has particularly high sodium ion and low potassium ion concentrations, whereas the egg cytoplasm has relatively the reverse. This condition is maintained by the plasma membrane, which steadfastly inhibits the entry of sodium ions into the oocyte and prevents potassium ions from leaking out into the environment. If an electrode is inserted into an egg and another electrode is placed outside the egg, the constant difference in charge across the egg plasma membrane can be measured. This resting membrane potential is generally about 70 mV, usually expressed as −70 mV because the inside of the cell is negatively charged in respect to the exterior.

There are characteristic changes in membrane potential associated with fertilization. Within 1–3 s of the binding of the fertilizing sperm, the membrane potential shifts to a positive level, about +20 mV (this means that electrical depolarization occurs from −70 to +20 mV of the egg's plasma membrane). This change is caused by a small influx of sodium ions into the egg. Although sperm can fuse with membranes having a resting potential of −70 mV, they cannot fuse with membranes having positive resting potential, so no more sperm can fuse to the egg. The reason of the increase of sodium permeability is not quite known; it may be due to the binding of the first sperm or the fusion of the first sperm with the egg. Although the molecular mechanism is unknown, it seems likely that the membrane depolarization that normally accompanies fertilization alters the conformation of a crucial protein in the egg plasma membrane. Therefore, a sperm membrane can no longer fuse with the egg membrane. Experimentally, fertilization can be prevented entirely by artificially keeping the membrane potential of the egg positive. In addition, lowering the concentration of sodium ions in the water causes a fast block to polyspermy. On the other hand, when artificially supplying the egg with an electric current that keeps its membrane potential negative, the egg becomes polyspermic. This phenomenon, an electric block to polyspermy, also occurs in starfish, the marine worm *Urechis* (Phylum Echiura), clam and amphibian, but probably not in most mammals. However, fish eggs do not show a rapid electrical block to polyspermy owing to the presence of a chorion.

Other plasma membrane polyspermy blocks

Echiuroids, annelids, and molluscs

The electrical polyspermy block is not an absolute block; the change in potential simply reduces the probability of fertilization such that polyspermy does not occur when eggs are inseminated with a moderate concentration of sperm.

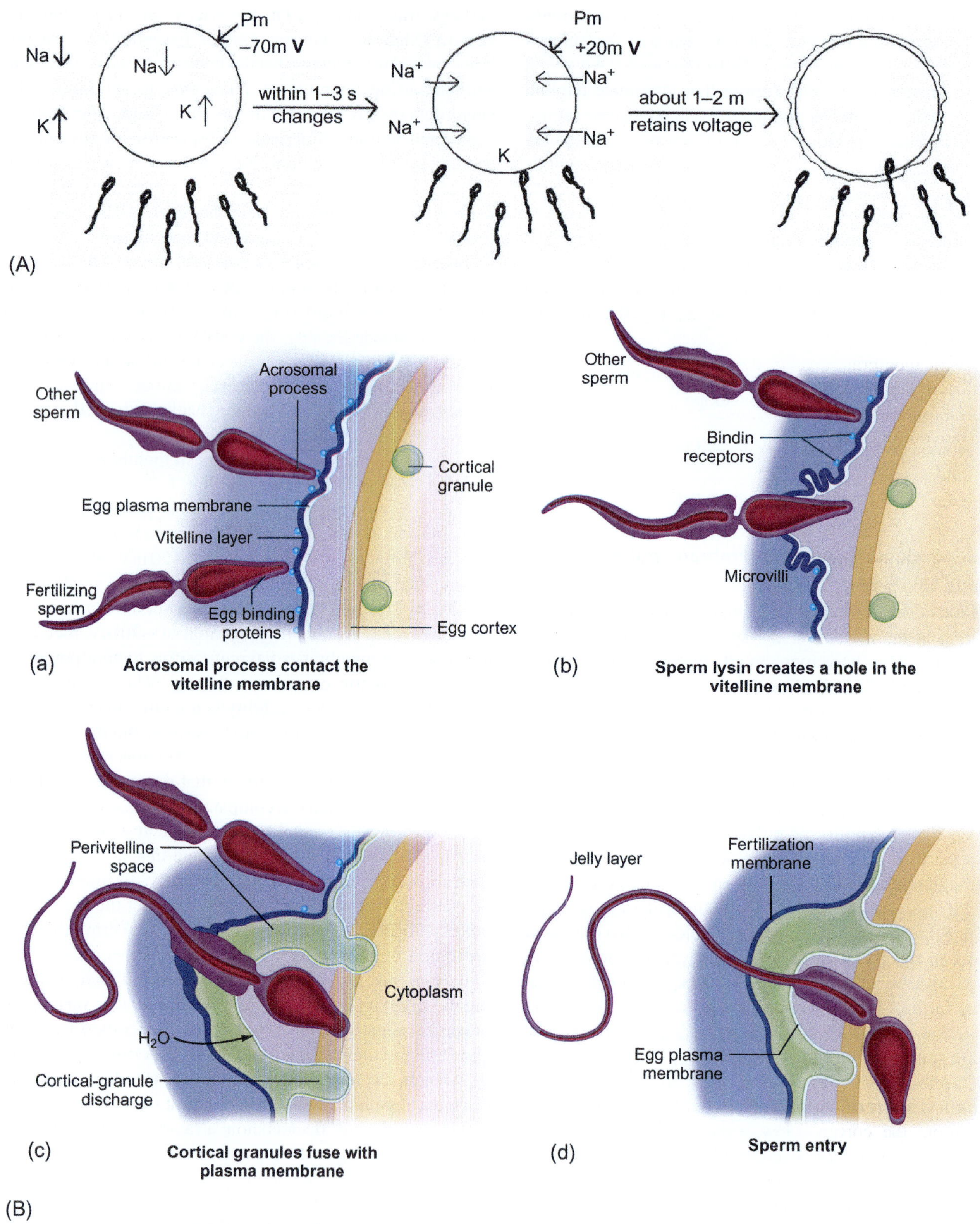

FIG. 2.30 Prevention of polyspermy. (A) Diagrammatic illustrations exemplifying the electric prevention. (B) Diagrammatic illustrations to show the mechanical prevention by formation of fertilization membrane.

It is understood that polyspermy block at the plasma membrane level is not always electrically mediated. One example is the echiuran worm *Urechis* (Borradaile et al., 1977), in which the first stage of polyspermy block is electrical, but it is transient and is followed by a permanent block, also at the level of the egg plasma membrane. Following fertilization, the surface coat elevates, but this does not stop penetration of sperm acrosomal tubules through the surface coat. Furthermore, by using trypsin at pH 7, the surface coat of the unfertilized egg can be elevated; subsequently, when the egg is inseminated, sperm can penetrate the elevated surface coat and fertilize the egg. The permanent polyspermy block is at the level of the plasma membrane. This is evidenced by the close contact between sperm and egg plasma membranes can be detected both by EM, and by electrical recording of small transient changes in membrane potential following reinsemination. The change in fusibility of the egg plasma membrane cannot be attributed to cortical granule secretion, since there is no significant decrease in the number of cortical granules in the egg up to the time of first cleavage.

This permanent polyspermy block at the level of the plasma membrane also occurs in *Spisula* (a marine bivalve mollusc), in other bivalve molluscs such as *Mytilus*, *Bankia*, and *Chama*, and in the polychaete annelids *Hydroides* and *Nereis*. However, it is not known whether these blocks are of electrical or nonelectrical nature (for details, see Metz and Monroy, 1985c, vol. 3).

Elevation of the fertilization envelope

It has been mentioned that the rapid prevention of polyspermy is transient, since the membrane potential of the sea urchin egg remains positive for only about a minute. This brief potential shift is therefore not sufficient to prevent polyspermy. The latter can still occur if the sperm bound to the vitelline membrane are not somehow removed. This removal is accomplished by the cortical granule reaction (exocytosis), a **slower mechanical block to polyspermy** that becomes active about a minute after the fertilizing sperm-egg attachment. This reaction results in the successive formation of both the fertilization envelope and the hyaline layer. Both of them change the morphology of the surface of the egg.

Formation of fertilization membrane At ~25 s postinsemination, the cortical granules begin to fuse with the plasma membrane, greatly increasing the egg's surface area (Fig. 2.30B), and they release their contents into the perivitelline space. The first released substances are the enzyme **proteases**, which dissolve the protein posts that connect the vitelline membrane proteins to the plasma membrane, and they clip off the bindin receptor of such membrane and any sperm attached to it. The second substances released are the **mucopolysaccharides**, which change the osmotic pressure inside the egg, causing the water to rush into the perivitelline space. This causes the expansion of the vitelline membrane, which becomes the fertilization envelope. The exocytosis of these granules initiates at the site of sperm fusion and continues around the egg in a wave that is completed within ~40 s (Just, 1919); fertilization envelope elevation follows this wave of exocytosis. Triggering of the exocytosis wave is tightly linked to fertilization-initiated changes in ion fluxes and membrane potentials, especially to elevated levels of free intracellular calcium.

It was demonstrated that the deposition of the cortical granule components onto or the reaction of these components with the vitelline layer result in the formation of a thin, less refractile intermediate layer termed the **transition membrane**. The incorporation of the vitelline membrane in the formation of this transition membrane (precursor of the fertilization envelope) was verified by two experiments:

- When the vitelline layer of a starfish's eggs was removed mechanically, it was found that no fertilization envelope was elevated.
- When the vitelline layer of a sea urchin egg was disturbed by reducing agents or proteases, this also prevented fertilization envelope formation.

During the next 10–30 min postinsemination, this transition membrane gradually hardens and becomes extremely difficult to remove or disperse. It thickens and changes its physical properties. It becomes inelastic and resistant to mechanical, chemical, enzymatic disruption, and it loses its affinity to bind sperm. However, it is not fully understood which of these changes in the vitelline membrane—elevation, thickening, hardening, or loss of sperm binding affinity—accounts for the change in sperm penetrability.

In mammals, the cortical granule reaction does not create a fertilization envelope, but its ultimate effect is the same.

Various secretory products released by cortical granules

Following the formation of the fertilization envelope, the second morphological change occurs to the surface of the sea urchin egg is the appearance of the **hyaline layer** (Fig. 2.30). The fourth substance released from the cortical granules is the hyalin protein, the major structural protein of the hyaline layer which forms a coat or layer around the egg. The egg itself extends elongated microvilli whose tips attach to this hyaline layer.

Recent studies have shown that the hyaline layer contains several other proteins in addition to hyalin. Some of these proteins are stored in granules other than the cortical granules in the unfertilized sea urchin eggs.

This layer, as its name may indicate, is translucent, a few micra in thickness and becomes clearly evident around the fertilized egg surface 5–7 min postinsemination. This, however, does not mean that it is suddenly formed at this time; actually, it gradually increases in thickness so that by 5–7 min it is easily visible to even casual observation.

This layer is not seen in eggs of other organisms and it may be unique to sea urchins and related echinoids. This hyaline layer acts as an extra-cellular cement to maintain blastomere adhesion during cleavage until they develop cell junctions in the blastula or in the early embryo. Medusan eggs, for example, are completely devoid of any sort of covering, and adhesion of blastomeres is achieved by a different method (see later).

The hyaline layer can be removed by placing embryos in calcium-free seawater. Cell division will continue under these circumstances, but an organized embryo will not form. In addition, weakening of the hyaline layer inhibits invagination of endodermal cells during gastrulation.

Release of peroxide to inactivate sperm

An interesting relationship has emerged from recent studies on polyspermy-preventing mechanisms in sea urchins. The sperm are active participants in both electrical and peroxide-mediated blocks. Sperm are preprogrammed to reduce their fertility in response to signals, electrical (see above) and chemical produced by the fertilized egg. The peroxide, which is produced within 15 s of insemination and reaches a maximum at about 1 min, is activated in sea urchin eggs at fertilization (production of H_2O_2 and secretion of ovoperoxidase), which is known to participate in cross-linking of the fertilization envelope. It was also suggested to contribute to the prevention of polyspermy. The postulated mechanism involved the combined effects of H_2O_2 and ovoperoxidase to constitute a spermicidal agent. This role was presumed to be analogous to the bactericidal peroxidatic system activated in phagocytic leucocytes as they ingest bacteria. If this is correct, then sea urchin eggs should become polyspermic:

- if H_2O_2 is removed from the system;
- if they are prevented from making H_2O_2; and
- if they are fertilized in the presence of ovoperoxidase inhibitors.

Extensive experimental studies (see Schuel, 1985) on sea urchin eggs (*Arbacia* and *Strongylocentrus*) proved that phenylhydrazine and 3-amino-1,2 and 4-triazole are inhibitors of ovoperoxidase. These two components are the most powerful inhibitors of ovoperoxidase in *Strongylocentrotus*. They are used to study the proposed role of this enzyme in the prevention of polyspermy in this species. Phenylhydrazine promoted polyspermy, while 3-amino-1,2 and 4-triazole did not. This result indicated that the egg's ovoperoxidase does not participate in the prevention of polyspermy, and suggested that the egg-derived peroxide may react directly with a peroxidase within the sperm to reduce their fertility (Boldt et al., 1981). On the other hand, it is well known that sea urchin sperm contain catalase (Rothschild, 1949). Recent biochemical studies have confirmed the presence of the proposed phenylhydrazine-sensitive peroxidase and 3-amino-1,2 and 4-triazole-sensitive catalase in sea urchin sperm (Boldt et al., 1981). Therefore, it was suggested and/or evidenced that during fertilization the fertilizing capacity of sperm would be regulated by the reactions of these enzyme systems with H_2O_2 that is released by the fertilized egg.

Other activities performed by cortical granules

To complete the picture and to understand perfectly the important role of the cortical granules in the process of fertilization, two items must be dealt with in some detail:

- hardening of the fertilization envelope; and
- the role of calcium ions in cortical reaction.

Hardening of the fertilization envelope

It is mentioned that the fertilization envelope first elevated as a thin, soft membrane. Some sperm remain attached to the elevated or fully elevated fertilization envelope for an extended period of time, yet they cannot penetrate it. How are they excluded? This is the critical transformation associated with the cortical reaction that results in establishing an absolute mechanical barrier to polyspermy (Fig. 2.31). It has been suggested that hardening of the fertilization envelope makes it resistant to sperm penetration. Gradually, this envelope hardens and becomes resistant by a complex process that occurs by the release of the **third** substance of the cortical granules which is the enzyme **peroxidase** (in such case it is called **ovoperoxidase**). This enzyme catalyses the cross-linking of the tyrosine residues on adjacent proteins. This complex process of hardening involves both **covalent cross-linking** and **structuralization**. The development of resistance to dissolution by organic dispersing agents probably reflects the covalent cross-linking of the fertilization envelope and is promoted by the cortical granule-derived ovoperoxidase. The latter is thought to use H_2O_2, which is also produced by the fertilized egg, to catalyse the formation of **dityrosine cross-links**. In addition, the fertilization envelope is impregnated by cortical granule-derived paracrystalline structural proteins to form a thickened trilaminar investment. This structuralization is responsible, at least in part, for the rigidity of the hardened envelope. Structuralization is modulated by cations in seawater, specially calcium and sodium. The initial to transient (I–T) transition in the configuration of the fertilization envelope is a characteristic of *Strongylocentrotus* eggs (Fig. 2.31). This transition is promoted by the insertion of structural proteins into the fertilization envelope, and is sodium dependent. The cortical granule protease appears to contribute to hardening by promoting the release of H_2O_2, as well as the dispersal and subsequent incorporation of secreted structural proteins into the fertilization envelope. These events are normally

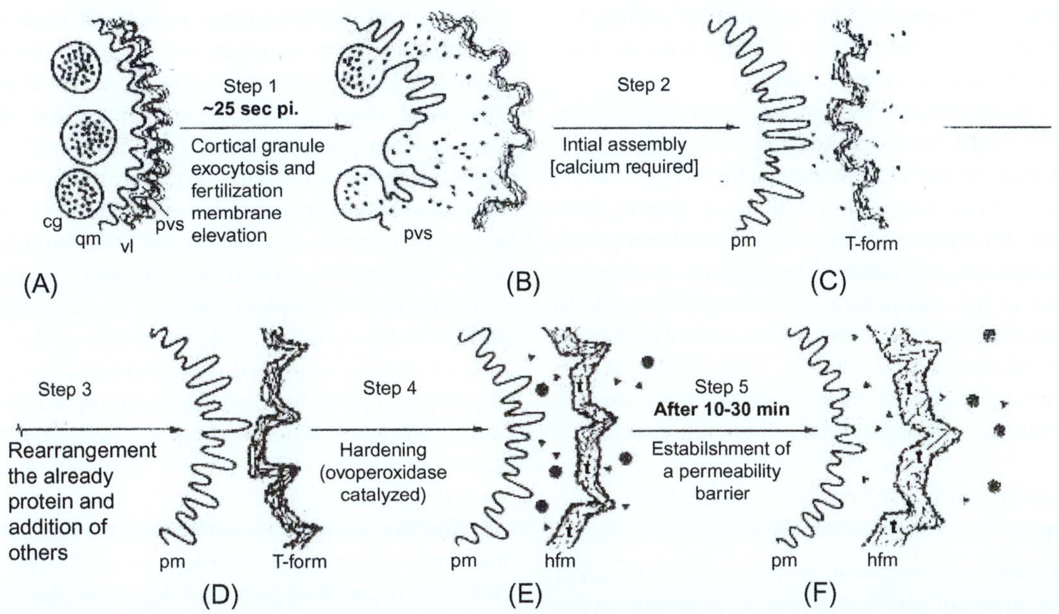

FIG. 2.31 Diagrammatic presentation illustrating the formation of the fertilization membrane and its hardening in the sea-urchin *Strongylocentrotus Purpuratus*. (A) Surface of unfertilized egg, cg = cortical granule, pm = plasma membrane, vl = vitelline layer; (B) One min after first contact of sperm. Discharge of mucopolysaccharide which increases the osmotic pressure causing the seawater to rush into the space and elevation of vitelline layer. (C) Incorporation of the discharged skeletal proteins especially tyrosine to the vitelline layer transforming it into fertilization membrane (initial 1, thin and soft); (D) Rearrangement of the already present proteins and addition of others; (E) The formation of the dityrosine cross-link using H_2O_2; (F) Incorporation of paracrystalline structural proteins leading to thick, hard and impermeable fertilization membrane, pi = postinsemination, t = dityrosine cross-link, ▲ = small molecule such as H_2O_2, ● = large molecule such as protein. *(From Metz and Monroy, vol. 3, with some modifications.)*

completed within 10 min of insemination. During the next 15–20 min, the permeability of the envelope to proteins is greatly reduced.

To summarize these events:

- Hardening: ovoperoxidase catalyses the formation of dityrosine cross-links.
- Structuralization: impregnation of paracrystalline structural proteins.
- Permeability: transition form of membrane is promoted by the insertion of structural proteins which is sodium dependent catalysed by protease.

The mature fertilization envelope appears by the light microscope as a negatively birefringent, highly refractile ring around the fertilized egg. At first, it is wispy, but later on, it becomes thicker and more refractile. The electron microscopy, on the other hand, shows that the completed fertilization envelope consists of trilamellar membrane ~50 nm thick. Two layers of highly organized, paracrystalline material ~15 nm thick sandwich the central 20 nm amorphous region.

Role of calcium in the initiation of the cortical granule reaction

The mechanism of the cortical granule reaction is similar to that of the acrosomal reaction. Upon fertilization, the intracellular calcium ion concentration of the egg increases greatly. An increase in the intracellular concentration of free calcium ions is now recognized to be a general feature of stimulus-secretion coupling by cells. The influx of calcium into sea urchin eggs at fertilization contributes to the rapid electrical depolarization of the plasma membrane. In addition, in a high-calcium environment, the cortical granule membranes fuse with the egg plasma membrane, releasing their contents (Figs 2.30B and 2.31). Once the fusion of the cortical granules begins near the point of sperm entry, a wave of cortical granule exocytosis propagates around the cortex to the opposite side of the egg.

In sea urchins and mammals, the rise in calcium concentration responsible for the cortical granule reaction is not due to an influx of calcium into the egg, but instead comes from within the egg itself. This hypothesis is based on more than one observation (for details, refer to Schuel, 1985). The release of calcium from intracellular storage can be monitored visually using calcium-activated luminescent dyes such as **aequorin** (isolated from luminescent jellyfish) or fluorescent dyes such as **fura-2**. These dyes emit light when they bind free calcium ions. When a sea urchin egg is injected with dye and then fertilized, a striking wave of calcium release propagates across the egg. Starting at the point of sperm entry, a band of light traverses the cell. The calcium ions do not merely diffuse across the egg from the point of sperm entry. Rather, the release of calcium ions starts at one end of the cell and proceeds actively to the other end.

The entire release of calcium ions is complete in roughly 30 s in sea urchin eggs, and the free calcium ions are resequestered shortly after they are released.

Several experiments have demonstrated that calcium ions are directly responsible for initiating and propagating the cortical granule reaction and that these calcium ions are stored within the egg itself. Electron-cytochemical studies have shown that calcium is present in the plasma membrane, cortical granule membrane and within mitochondrial; in addition, they are stored in the endoplasmic reticulum of the egg. In sea urchins and frogs, the reticulum is pronounced in the cortex and surrounds the cortical granules. Free calcium is able to release sequestered calcium from its storage sites, causing a wave of calcium ion release and cortical granule exocytosis.

III-Fusion of the haploid nuclei of the egg and sperm

This is the third major interaction in the process of fertilization (Fig. 2.24). For fusion of genetic material in sea urchins to occur, the sperm head enters the egg perpendicular to its surface. After the fusion of the sperm and egg plasma membranes and the incorporation of the sperm in the egg, the sperm nucleus and its centriole separate from the mitochondria and the flagellum. The mitochondria and flagellum disintegrate inside the egg, so very few (if any) sperm-derived mitochondria are found in developing or adult organisms. Thus, although each gamete contributes a haploid genome to the zygote, the mitochondrial genome is transmitted primarily by the maternal parent. In almost all animals studied (the mouse is an exception), the centrosome needed to produce the mitotic spindle of the subsequent divisions is derived from the centriole of the male gamete. Therefore, the sperm centrioles are crucial because the egg loses its own centrioles during the last mitotic division.

The egg nucleus, once it is haploid, is called the **female pronucleus**. Once inside the egg cytoplasm, the sperm head suddenly swells and begins to assume the characteristics of a **sperm pronucleus**. Therefore, 3 min after insemination, the sperm nucleus decondenses and undergoes a dramatic transformation. Although specific details involving the development of the male pronucleus vary from one species to another, there are three basic features common to all species studied so far:

1. breakdown of the sperm envelope;
2. dispersion of the condensed sperm chromatin; and
3. development of a nuclear envelope, the pronuclear envelope.

These processes are accompanied by a dramatic transformation in the shape, volume, chromatin conformation, nucleoprotein content and activity of the former sperm nucleus. It has been demonstrated that the elevation of intracellular pH that occurs in the egg of some species at fertilization is necessary for both egg activation and development of the male pronucleus. As to the first step, the nuclear envelope vesiculates into small packets, exposing the compact sperm chromatin to the egg cytoplasm. For the second step, the proteins holding the sperm chromatin in its condensed, inactive state are exchanged for other proteins derived from the egg cytoplasm. This exchange permits the decondensation of the sperm chromatin. In sea urchins, decondensation appears to be initiated by the phosphorylation of two sperm-specific histones that bind tightly to the DNA. The timing of the third step, which is the formation of the pronuclear envelope, appears to vary with the species studied so far. In *Arbacia*, development of the male pronuclear envelope is initiated during chromatin dispersion. In any case, morphological events involving this formation are similar. Vesicles coalesce along the periphery of the chromatin and fuse to form elongate cisternae, which develop pores (Fig. 2.32).

After the sea urchin sperm enters the egg cytoplasm, the male pronucleus rotates 180 degrees so that the sperm centriole is between the sperm and egg pronuclei. The centriole then acts as a microtubule organizing centre, extending its own microtubules and integrating them with egg microtubules to form an aster. These microtubules extend throughout the egg and contact the female pronucleus, and the two pronuclei migrate towards each other. Because the egg is large, the sperm and egg pronuclei have to migrate substantial distances to find each other. Not surprisingly, this migration depends on the cytoskeleton. In sea urchins and vertebrates, the two pronuclei come near to each other and close contact takes place between the two. The nuclear contents of both pronuclei are intermingled or fused. The endoplasmic reticulum forms a new common nuclear envelope around both pronuclei, and thus forms a zygote nucleus (from the Greek *zygotos*, yolked together), i.e. a diploid nucleus. The actual fusion of the sperm and egg pronuclei does not take place until some time after fertilization: about 1 h in sea urchins and about 12 h in mammals. The egg must complete its second meiotic division before the union of the two sperm and egg nuclei. When adjacent to one another, the pronuclei become associated in a characteristic fashion to establish the genome of the embryo and hence conclude the process of fertilization. Early studies of fertilization indicated that the association of the male and female pronuclei may take essentially two forms:

1. Fusion of the pronuclei to produce a single zygote nucleus (pronuclear fusion). Eggs exhibiting such a process have the **sea urchin type of fertilization**. This type is observed in eggs that are inseminated following meiotic maturation. This process has been studied in greatest detail in *Lytechinus* and *Arbacia*.
2. Both the male and female pronuclei give rise to a group of chromosomes for the ensuing cleavage division. In this form, there is an intermixing of maternally and paternally derived chromosomes without a fusion of the pronuclei. Eggs exhibiting these events have the ***Ascaris***

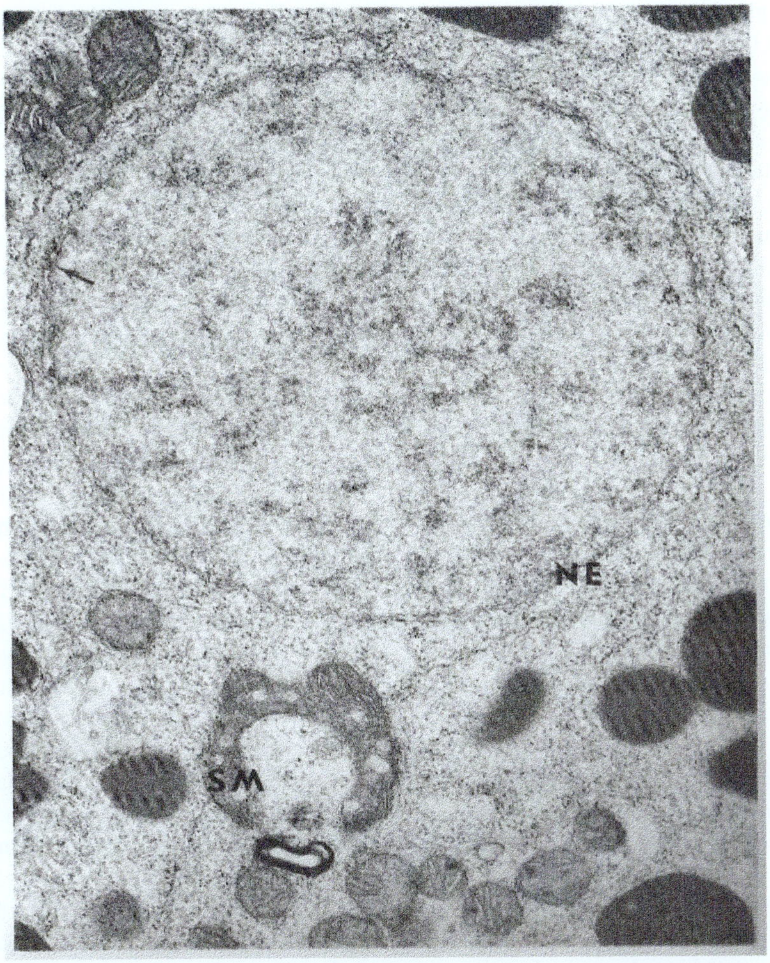

FIG. 2.32 Male pronucleus of an *Arbacia zygote*, 10 min postinsemination. The dispersed paternally derived chromatin is surrounded by a nuclear envelope (NE) possessing pores *(arrow)*, SM = sperm mitochondrion. *(From Metz and Monroy, vol. 3, 1985c.)*

type of fertilization. This type has been studied, in addition to Ascaris, in Spisula, Mtyelus and various mammals. Eggs of this type are inseminated prior to the completion to meiosis.

The initiation of DNA synthesis can occur either in the pronuclear stage (during migration) or after the formation of the zygote nucleus. In *Arbacia*, DNA synthesis normally follows the fusion of the pronuclei. Furthermore, there is evidence that the paternally derived genome may be active in RNA synthesis following formation and migration of the male pronucleus.

IV-Activation of egg metabolism to start development

Essentially, the function of fertilization is merely the means to merge the two haploid nuclei; in addition, it has an equally important role in initiating the processes that start development. These processes occur in the cytoplasm without the involvement of the nuclei. The mature sea urchin egg is metabolically sluggish cell that is activated by the sperm. This activation is a stimulus; however, it sets into action a preprogrammed set of metabolic events. Once the sperm contacts the egg surface, calcium ions are released within the egg that initiate the cortical reaction, and also activate a series of reactions that throws the egg's metabolic machinery into high gear. In addition, the intracellular pH of the cytoplasm increases, oxygen consumption rises and protein synthesis increases. However, the responses of the egg to the sperm can be divided into **early** responses, which occur within seconds of the cortical reaction, such as the two major blocks to polyspermy, the rapid block which is initiated by sodium influx into the egg cell and the slow block which is initiated by the intracellular release of calcium ions; and **late** responses, which take place several minutes after the beginning of fertilization. They begin with a rise in the intracellular pH, as a reflex to the increase of the intracellular calcium ions. This rise in intracellular pH begins with a second influx of sodium ions, which causes a 1:1 exchange between sodium ions from seawater and the hydrogen ions from the egg. It is thought that the pH increase and the calcium ion elevation act together to stimulate new protein and DNA synthesis. In sea urchins, this protein synthesis does not depend on the synthesis of new messenger RNA; rather, it utilizes mRNAs already present in the oocyte cytoplasm. These messages include mRNAs encoding proteins such as histones, tubulins, actins and morphogenetic factors that are utilized during early development.

It can be visualized that there are three different methodologies by which all the responses of the egg can be judged. These can be broadly described as the morphological, the physiological and the biochemical. Morphologically, the principal events of activation are the cortical reaction (secretion of cortical granules and elevation of the fertilization envelope), pronuclei formation and fusion, breakdown of the nuclear envelope and chromosome condensation, and the formation of the mitotic apparatus. All of these features are easily observed and studied with a phase contrast microscope, since the eggs of most sea urchin species are relatively transparent. Physiologically, early activation can be detected as the release of acid, changes in membrane potential, increases in oxygen uptake and activation of amino acid transport systems. Biochemically, the initiations of protein synthesis and DNA synthesis have been the usual parameters observed in judging the response to activation.

The activation of all eggs appears to depend on an increase in the concentration of free calcium ions within the egg. Such an increase can occur in two ways: calcium ions can enter the egg from outside, as in some snails and worms; or calcium ions can be released from the endoplasmic reticulum within the egg, as in sea urchins and mammals. Both mechanisms are used to different degrees in different species. In both cases, a wave of calcium ions sweeps across the egg, beginning at the site of sperm-egg fusion.

DEVELOPMENT

Early development

Following fertilization, the zygote undergoes developmental processes successively, beginning with cleavage followed by blastulation and then gastrulation. It is limited here to the types of these events, while the details will be described in appropriate positions, i.e. each with its species.

Cleavage

This event can be defined as the successive divisions of the zygote into a definite number of blastomeres specific for the species. All the blastomeres must divide, at each division, within the same time of that division. Cleavage ends if each blastomere behaves alone. There are two patterns of cleavage: radial and spiral cleavages.

Radial cleavage (Fig. 2.33A) occurs by cleavage planes (=furrows=sutures) parallel or perpendicular to the animal-vegetal axis of the egg in a regular manner, which alternate with each other. This means that the mitotic spindle of each cleavage takes a position at right angles to the position of the previous cleavage. The resulting blastomeres are arranged above each other, i.e. in tiers. The blastomeres may be equal or unequal according to the type of the egg. The furrows of the division may be complete or incomplete. The complete division is normal radial cleavage. If it is incomplete division, the cleavage is different type: discoid or superficial cleavage. Radial cleavage undergoes number of divisions more than the spiral cleavage, so the fate of each blastomere cannot be followed and hence it is described as **indeterminate**. Some irregular cleavages are introduced into the pattern, but the great majority of species conform to the type, at least as far as, the eight-cell stage.

Spiral cleavage (Fig. 2.33B) is holoblastic (=complete) and characteristic of several animal groups, including annelid worms, some flatworms and most molluscs. It differs from radial cleavage in numerous ways. First, the planes are neither parallel nor perpendicular to the animal-vegetal axis of the egg; rather, cleavage is at oblique angles. This is not particularly conspicuous in the two-cell stage, but at the four-cell stage, two of the diagonally opposite blastomeres lie up on the upper side, while the other two lie also on the lower side, i.e. four blastomeres appear on the animal pole, while only two appear on the vegetal pole. Second, at each successive division, the resulting four daughter blastomers (quartet=q) are unequal (Fig. 2.33Cc), lying on the animal pole, and shifted—according to the obliqueness—in a clockwise direction around the egg axis (*dextral*=*dexiotropic*=*dextrotropic*) or in an anticlockwise direction (*sinistral*=*laevotropic*=*laeiotropic*). This means that each daughter blastomere does not lie upper to its mother, thus forming a **spiral** arrangement of daughter blastomeres. Third, spirally cleaving embryos usually undergo fewer divisions during cleavage period, making it possible to follow the fate of each cell of the blastula. Hence, the blastomeres are described as **determinate**, so are labelled. The mother cells, the macromeres, are A, B, C, and D, and the daughter cells, the micromeres, are a, b, c, and d. Fourth, this pattern of cleavage divides the embryo longitudinally into genealogical lines corresponding to the descendants of the A, B, C, and D blastomeres. Since separating the embryo into its A, B, C, and D components is equivalent to dividing it into four lengthwise segments (along the egg axis), each of these is called a **quadrant**. In later developmental stages, the B quadrant forms the anterior part of the larva, the D quadrant forms the posterior part, and the A and C quadrants form the left and right sides, respectively. As a consequence of this fact, the anterior-posterior and lateral axes of the spiral-cleaved embryo are already determined in the four-cell stage.

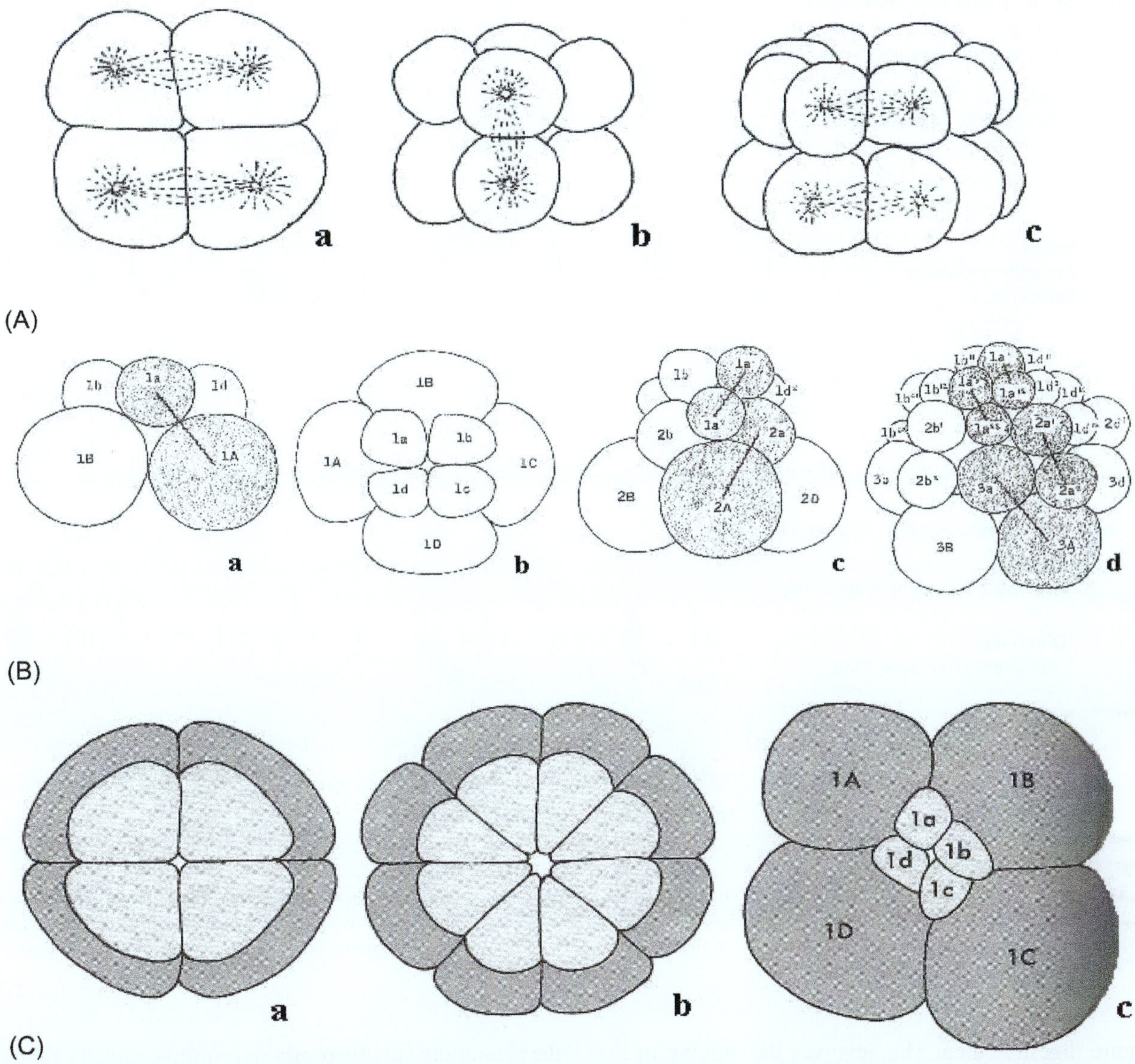

FIG. 2.33 Patterns of cleavage. (A) Radial cleavage in the sea cucumber, *Synapta*, (a) polar view, (b) and (c) lateral views; (B) Spiral cleavage, (a), (c), and (d) lateral views, (b) surface view. *Wavy black lines* indicate orientation of spindles at each cleavage *(A and B from Villee et al., 1964 and Barnes, 1974)*. (C) Comparison between radial and spiral cleavages, (a) radial cleavage at the eight-cell stage, as shown in the development of many echinoderms, (b) a similar egg after the fourth cleavage, 16-cell stage, (c) classic spiral cleavage at the eight-cell stage, as shown in a mollusc. In radial cleavage the daughter cell is located above its mother, while in spiral cleavage the daughter cell is shifted from its mother *(from Russell-Hunter, 1979)*.

Apart from these two major types, there are other types (Fig. 2.34), and each may be derived from one of the major ones. The amount and distribution of the yolk in the egg is the main factor that determines the type of cleavage.

Blastulation

The blastomeres of the cleaved embryo divide to increase their number and arrange themselves in one layer superficial of a spherical embryo, enclosing a cavity. This is the **blastula**. If this cavity (or blastocoele) is comparatively spacious, it is a **coeloblastula** (=holoblastula) (Fig. 2.35A–C), generally derived from radial cleavage. If the blastocoele is comparatively small or obliterated, it is a **stereoblastula** (Fig. 2.35D), generally derived from spiral cleavage.

Gastrulation

Gastrulation is the process of development of a double-layered embryo (**gastrula**) from a one-layered embryo (blastula). The term gastrula is derived from the Greek word *gaster*, meaning 'stomach'; gastrulation therefore implies segregation of gastrodermal from ectodermal cells and establishment of two functionally different cell layers and of

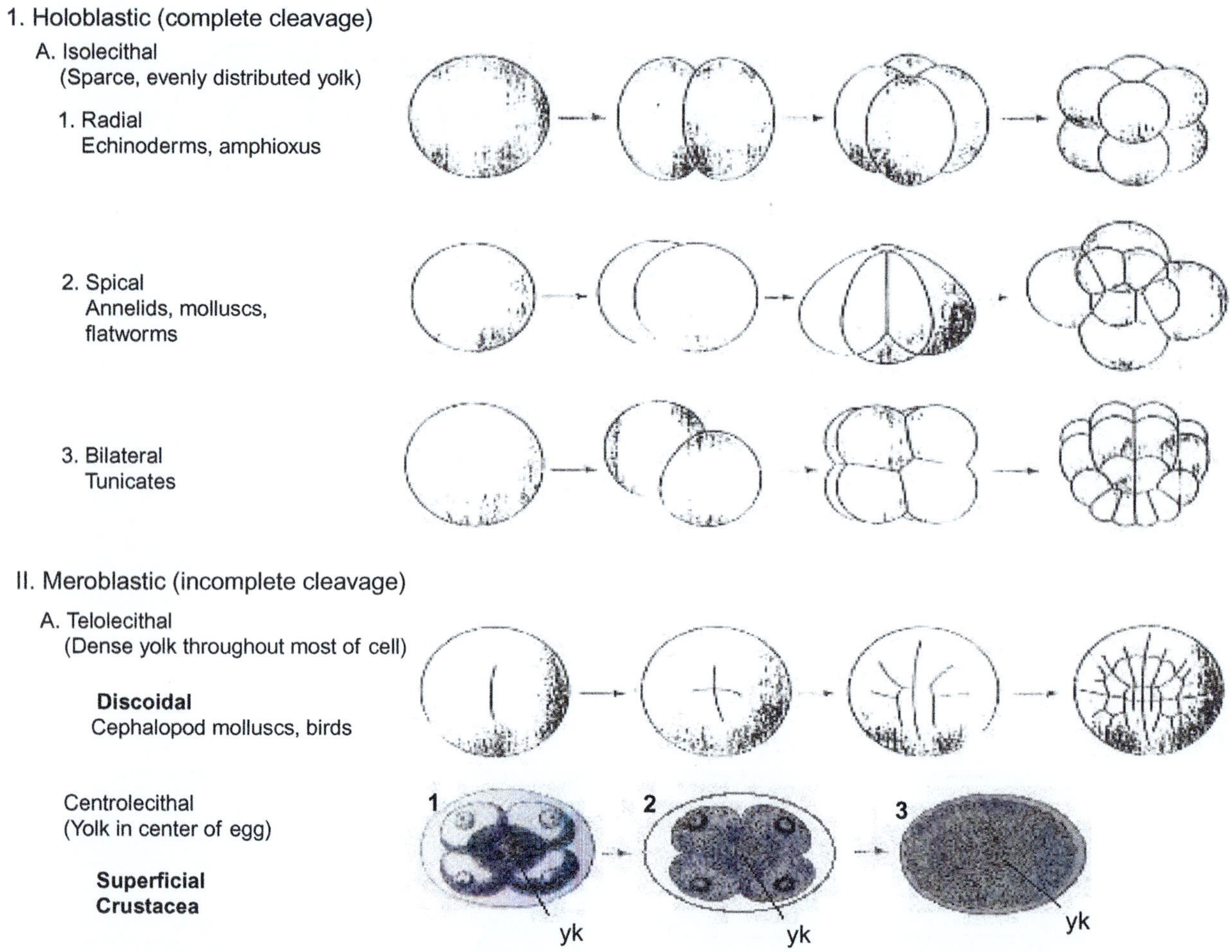

FIG. 2.34 Types of cleavage in invertebrate animals. Superficial cleavage of a crustacean ovum (*Penaeus*), the sections illustrate the yolk which is aggregated at the centre of the ovum. (1, 2) Surface view showing the stage with four segments, in (2) it is seen that the furrows visible on the surface do not penetrate to the centre of the ovum, (3) surface view of the ovum near the end of cleavage, the central yolk mass is very clear. *(After Haeckel, from Balfour (1880).)*

the future digestive area. This involves the moving of the blastomeres, which will develop into internal organs, to the inside of the embryo. This is another critical event in the development of the organism. There are different types of gastrulation processes. The type differs with the species, more precisely, with the type of the egg, cleavage and blastula (Fig. 2.35).

As the majority of animals show holoblastic cleavage giving rise to a hollow ball of cells or blastula, the most common form of gastrulation is by **invagination** (Fig. 2.35A), which may also be called **emboly** (Kume and Dan, 1968). Invagination is the inturning or inpushing of part of blastula, usually at or near the vegetal pole, to form a second layer or internal layer, the **endoderm** lining the outer layer, the **ectoderm**. The cavity formed by the endoderm is called the **archenteron** or primitive intestine. The site of initiation of the gastrulation process and the resulting hole into the inside of the gastrula, i.e. into the archentron, the blastopore (**protostomia** or primitive mouth), both have important places in the evolutionary picture. The fates of the blastopore and its relationship with the origin of the mouth or stomodeum are major taxonomical features.

The method of gastrulation is only possible when the blastula is a hollow ball of cells containing very little yolk. When the egg has a moderate amount of yolk concentrated at the vegetal pole and cleavage has produced large yolk-filled macromeres, the process of invagination is difficult or impossible. Under these circumstances, gastrulation is by epiboly (Fig. 2.35D), a modification of the invagination process. **Epiboly** (from the Greek *epiballein*, 'to throw upon') usually follows unequal holoblastic cleavage, with the smaller cells (micromeres) at the animal pole moving over the macromeres at the vegetal pole and ultimately enclosing them, so that they come to lie within the blastocoele and form either a coelo- or a stereogastrula. The part that is called the blastopore is a region at the vegetal pole where

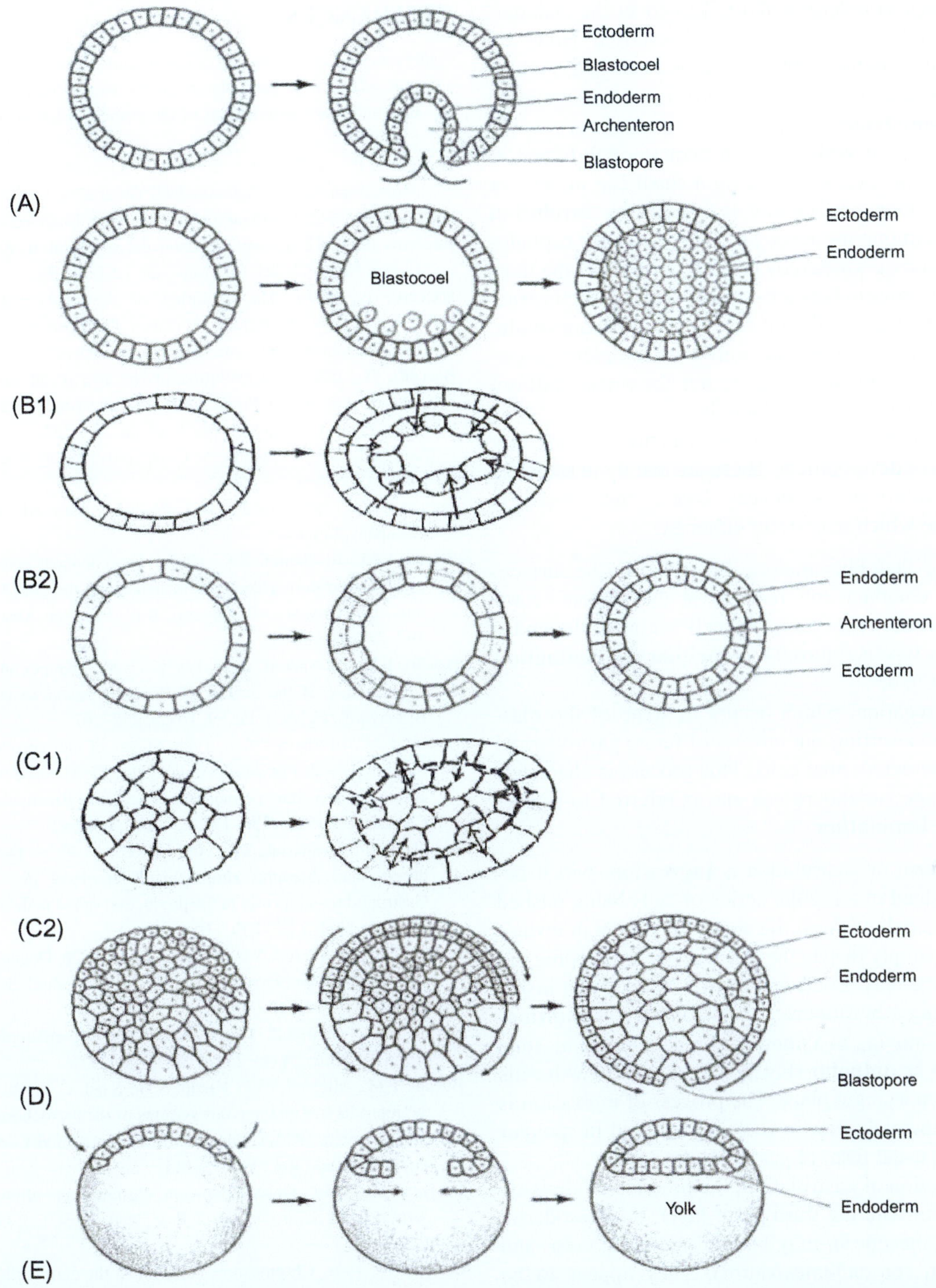

FIG. 2.35 Types of blastulae and types of gastrulation. (A) Invagination of a coeloblastula to form a coelogastrula, (B1) Unipolar ingression of a coeloblastula to form a stereogastrula; (B2) Polypolar ingression, migration of cells all around the blastula to form the inner layer; (C1) Primary delamination of a coeloblastula to form a double-layered coelogastrula, in this the outer cell divides parallel to the surface to give the inner cell; (C2) Secondary delamination of a stereoblastula to form a stereogastrula, selected cells migrate from the outer layer to form the endoderm; (D) Epiboly of a stereoblastula to form a stereogastrula, the outer layer expands on the surface to enclose the inside cells; (E) Involution of a discoblastula to form a discogastrula. *(From Marshall and Williams (1975) and Gilbert and Raunio (1997).)*

the micromeres (ectoderm cells) fail to cover the endodermal cells completely, leaving part of it exposed. However, as development proceeds, the ectodermal cell layer, which encircles this blastopore, sinks inward, forming a structure called the **stomodeum**.

If the amount of yolk has been great enough to necessitate meroblastic cleavage forming a small cap of cells or discoblastula, then gastrulation is usually by **involution** (Fig. 2.35E). Among invertebrates, it is seen in the cephalopods and is a process whereby cells in one area of the margin of the disc move in beneath the others and thus construct a two-layered disc. Involution results in a **discogastrula**, a special kind of coelogastrula with an incomplete gastrocoele. This is not a tubular structure but one whose wall is a sheet of cells spread out over the yolk.

There are many other forms of gastrulation seen in a few unusual types of development. These are mainly in response to special development sequences. One of these forms is **delamination**, which may occur either by:

- division of each blastomere cell parallel to the surface, the outer daughter cell remains at the surface (ectoderm), and the inner daughter cell becomes the endoderm. This type is referred to as **primary delamination** (Fig. 2.35C); or
- tissue segregation, which occurs to some of the blastula cells in a sorting-out process of future gastrodermal from future ectodermal cells. This process is characteristic of some stereoblastulae and is referred to as **secondary delamination**.

Another form of gastrulation is **ingression**, which occurs when instead of a regular series of cells being pushed into the blastocoele to form the second layer (as in invagination), cells simply invade the inner cavity by slipping out of the line to the blastocoele and form a second layer inside without forming a blastopore. Ingression may be **multipolar**, with cells moving in a number of places on the blastula wall, or it may be **unipolar** (Fig. 2.35B1 and B2), with cells moving in from a single place. The process of ingression is seen, among invertebrates, in coelenterates and in sponges where it is the usual form of gastrulation.

On completion of gastrulation, the products of division grow to differentiate the third germ layer, the mesoderm. Two kinds of mesoderm may be distinguished: ecto- and entomesoderm (or endomesoderm). They appear to be phylogenetically distinct. Ectomesoderm is derived from ectodermal cells that have wandered inwards; it is always mesenchymal. Entomesoderm includes all mesoderm originating within or from the endoderm, and is referred to as **true** or definitive mesoderm, since it characterizes all the higher animal phyla. After mesoderm formation, the embryos quickly take up the forms and modes of development characteristic of the various phyla to which they belong.

REFERENCES

Anderson, E., 1968. Oocyte differentiation in the sea urchin. *Arbacia punctulata*, with particular reference to the origin of cortical granules and their participation of the cortical reaction. J. Cell Biol. 39, 514–539.

Ankel, W.E., 1930. Die atypische Spermatogenese von Janthina (Prosobranchia, Ptenoglossa). Cell Tissue Res. 11 (3), 491–608.

Austin, C.R., 1965. Fertilization. Prentice-Hall, Englewood Cliffs.

Baccetti, B., 1977. Lo spermatozoo dei Sipunculidi. Atti. Acad. Naz̀. Lincei, Rend. Cl. Sci. Fis., Mat. Nat. LXII 89–92.

Baccetti, B., 1979. The evolution of the acrosomal complex. In: Fawcett, D.W., Bedford, J.M. (Eds.), The Spermatozoon. Urban and Schwarzenberg, Baltimore-Munich, pp. 305–329.

Baccetti, B., 1982. The evolution of the sperm tail. In: Amos, W.B., Puckett, J.G. (Eds.), Prokaryotic and Eukaryotic Flagella. Cambridge Univ. Press, London and New York, pp. 521–532.

Balfour, F.M., 1880. A Treatise on Comparative Embryology. I. Macmillan and Co, London.

Barnes, R.D., 1974. Invertebrate Zoology, third ed. W.H. Saunders Company, London.

Bawab, F.M., El-Sherief, S.S., 1987a. The chemical constituents of the yolk formed during deutoplasmogenesis in the crab *Portunus pelagicus* (L.) (Crustacea-Brachyura). Bull. Fac. Sci. Alex. Univ. 27 (2), 198–217 (chemistry).

Bawab, F.M., El-Sherief, S.S., 1987b. Cyclic changes in the histological structure of the ovaries of *Portunus pelagicus* (L.) (Crustacea-Brachyura). Egypt. J. Histol. 10 (2), 297–307.

Bawab, F.M., El-Sherief, S.S., 1988. Stages of the reproductive cycle of the female crab *Portunus pelagicus* (L., 1758) based on the anatomical changes in the spermatheca (Decapoda Brachyura, portunidae). Crustaceana 45 (2), 139–148. E.J. Brill, Leiden.

Bawab, F.M., El-Sherief, S.S., Abd-El-Kerim, 1992a. The gonad of the desert snail *Eremina ehrenbergi* Roth, 1839 (Stylommatophora-Gastropoda) and its role in the production of the male gametes. Funct. Dev. Morphol. 2 (2), 103–110.

Bawab, F.M., El-Sherief, S.S., Mehanney, S., 1992b. Oogenesis in the desert snail *Eremina ehrenbergi* Roth 1839. Proc. Zool. Soc. A. R. Egypt 23 (Part 1), 107–113.

Bawab, F.M., Ismail, S.Z., 1984. Structure of the spermatophore. Bull. Fac. Sci. Alex. Univ. 24 (4), 274–310.

Bawab, F.M., Michael, M.I., Khalil, S.H., Saad, G.A., 2012. Studies on the reproductive and nervous systems of the protochordate *Styela plicata* (Lesueur, 1823) to correlate the nervous activity on reproduction. Egypt. J. Aquat. Res. 38, 125–145.

Berry, R.O., 1939. Observations on chromosome elimination in the germ cells of *Sciara ocellaris*. Proc. Natl. Acad. Sci. U. S. A. 25 (3), 125–127.

Berry, R.O., 1941. Chromosome behavior in the germ cells and development of the gonads in *Sciara ocellari*. J. Morphol. 68, 547–583.

Boldt, J., Schuel, H., Schuel, R., Daradekar, P.V., Troll, W., 1981. Reaction of sperm with egg-derived hydrogen peroxide helps prevent polyspermy during fertilization in sea urchin. Gamete Res. 4, 365–377.

Borradaile, L.A., Potts, F.A., with chapters by Eastham, L.E., Saunders, J.T., 1977. The Invertebrata, fourth ed. Revised by Kerkut, G.A. Cambridge University Press.

Boveri, T., 1899. Die Entwicklung von *Ascaris megalocephala* mit besounderer Rücksieht auf die Kernverhältnisse. Festschr. f. C. von Kupfer, Jena.

Chambers, R., 1921. Microdissection studies, III. Some problems in the maturation and fertilization of the echinoderm egg. Biol. Bull. 41 (6), 318–350.

Chambers, R., 1923. The mechanism of the entrance of sperm into the starfish egg. J. Gen. Physiol. 5 (6), 821.

Chambers, R., 1930. The manner of sperm entry in the starfish egg. Biol. Bull. 58 (3), 344–369.

Chambers, R., 1933. The manner of sperm entry in various marine ova. J. Exp. Biol. 10 (2), 130–141.

Charniaux-Cotton, H., 1985. Vitellogenesis and its control in malacostracan Crustacea. Am. Zool. 25, 197–206.

Correia, C.P., 1997. The Ovary of Eve: Egg and Sperm Preformation. University of Chicago Press, Chicago.

Davidson, E.H., 1976. Gene Activity in Early Development, second ed. Academic Press, New York.

Endo, Y., 1952. The role of the cortical granules in the formation of the fertilization membrane in eggs from Japanese sea urchins. I. Exp. Cell Res. 3 (2), 406–418.

Endo, Y., 1954. Fertilization membrane. Cytochemical Symposium vol. 2, pp. 37–63. (in Japanese).

Esteban, M.R., Giovinazzo, G., Goday, C., 1995. Chromatin diminution is strictly correlated to somatic cell behavior in early development of the nematode *Parascaris univalens*. J. Cell Sci. 108, 2393–2404. Printed in Great Britain.

Fretter, V., Graham, A., 1962. British Prosobranch Molluscs, Their Functional Anatomy and Ecology. Ray Society, London.

Gabe, M., 1976. Histological Techniques. Springer-Verlag, Masson, New York, Paris.

Galstoff, P.E., 1925. Regeneration after dissociation (an experimental study on sponges). J. Exp. Zool. 42, 183–221.

Gardiner, M.S., 1972. The Biology of Invertebrates. McGraw-Hill Book Company, London, New York.

Geigy, R., 1931. Action de l'ultraviolet dur le pôle germinaldans l'oeuf de *Drosophila melanogaster* (Castration et Mutabilité). Rev. Suisse Zool. 38, 187–288.

Giese, A.C., Pearse, J.S., 1974. Reproduction of Marine Invertebrates. In: Acoelomate and Pseudocoelomate Metazoans. vol. I. Academic Press, New York, London.

Giese, A.C., Pearse, J.S., 1977. Reproduction of Marine Invertebrates. Molluscs: Gastropods and Cephalopods. vol. IV. Academic Press, New York, London.

Gilbert, S.F., 2000. Developmental Biology, sixth ed. Sinauer Association, Inc. Publishers, Sunderland, MA.

Gilbert, S.F., Raunio, A.M., 1997. Embryology. Constructing the Organism. Sinauer Associates, Inc. Publishers, Sunderland, MA.

Goldstein, D.J., 1962. Correlation of size of dye-particle and density of substrate, with special reference to mucin staining. Stain Technol 37, 71–79.

Hanson, J., Randall, J.T., Bayley, S.T., 1952. The microstructure of the spermatozoa of the snail *Viviparus*. Exp. Cell Res. 3 (1), 65–78.

Hartmann, M., 1940. Die stofflichen Grundlagen der Befruchtung und Sexualität im Pflanzen-und Tierreich. Naturwissenschaften 28 (51), 807–813.

Hendelberg, J., 1969. On the development of different types of spermatozoa from spermatids with two flagella the Turbellaria with remarks on the ultrastructure of the flagella. Zool. Bidr. Uppsala 38, 1–50.

Hickman, C.P., 1973. Biology of the Invertebrates, second ed. The C.V. Mosley Company, Saint Louis.

Just, E.E., 1919. The fertilization reaction in *Echinarachnius parma*: I. Cortical response of the egg to insemination. Biol. Bull. 36 (1), 1–10.

Just, E.E., 1930. The present status of the fertilizin theory of fertilization. Protoplasma 10 (1), 300–342.

Kubota, S., Nakai, Y., Sato, N., Kuro-o, M., Kohno, S., 1994. Chromosome elimination in Northeast Pacific hagfish, *Eptatretus stoutii* (Cyclostomata, Agnatha). J. Hered. 85 (5), 413–415.

Kume, H., Dan, K., 1968. Invertebrate Embryology. Translated by Dan, J.C. NOLIT Publishing House, Belgrade, Yugoslavia.

Lillie, F.R., 1912. The production of sperm asoagglutinins by ova. Science 36, 527–530.

Lillie, F.R., 1914. Studies of fertilization. VI. The mechanism of fertilization in *Arbacia*. J. Exp. Zool. 16, 523–590.

Lillie, F.R., 1919. Problems of Fertilization. University of Chicago Press, Chicago.

Lillie, F.R., 1921. Studies of fertilization: IX. On the question of superposition of fertilization on parthenogenesis in *Strongylocentrotus purpuratus*. Biol. Bull. 40 (1), 23–31.

Lillie, F.R., 1923. Problems of Fertilization. Chicago University Press, Chicago.

Lillie, F.R., Just, E.E., 1924. Fertilization, General Cytology. The University of Chicago Press, Chicago, IL.

Littlefield, L.C., 1986. Sex determination in hydra: control by a subpopulation of interstitial CELLS in Hydra oligactis males. Dev. Biol. 117 (2), 428–434.

Littlefield, L., 1994. Cell-cell interactions and the control of sex determination of hydra. Semin. Dev. Biol. 5, 13–20.

Loeb, J., 1903. Ueber die Befruchtung von Seeigeleiern durch Seesternsamen. Pflügers Archiv. Eur. J. Physiol. 99 (7), 323–356.

Loeb, J., 1915. Studies in artificial parthenogenesis. Biol. Bull. 29 (3), 149–203.

Longo, F.J., 1981. Regulation of pronuclear development. In: Jagiello, G., Vogel, C. (Eds.), Bioregulators of Reproduction. Academic Press, New York, pp. 529–557.

Marshall, A.J., Williams, W.D., 1975. A Text Book Zoology: The Invertebrates. ELBS and Macmillan.

Metz, C.W., 1931. Chromosomal differences between germ cells and Soma in *Sciara*. Biol. Zentr. 51, 119–124.

Metz, C.W., 1938. Chromosome behavior, inheritance and sex determination in Sciara. Am. Nat. 72, 485–520.

Metz, C.B., 1942. The inactivation of fertilizin and its conversion to the "univalent" form by x-rays and ultraviolet light. Biol. Bull. 82 (3), 446–454.

Metz, C.B., 1945. The agglutination of starfish sperm by fertilizin. Biol. Bull. 89 (1), 84–94.

Metz, C.B., Monroy, A. (Eds.), 1967. Fertilization, vol. 1. Academic Press, New York.

Metz, C.B., Monroy, A. (Eds.), 1969. Fertilization, vol. 2. Academic Press, New York.

Metz, C.B., Monroy, A., 1985a. Biology of Fertilization. In: Model Systems and Oogenesis, vol. I. Academic Press, Inc., New York, London.

Metz, C.B., Monroy, A., 1985b. Biology of Fertilization. In: Biology of the Sperm. vol. II. Academic Press, Inc., New York, London.

Metz, C.B., Monroy, A., 1985c. Biology of Fertilization. The Fertilization Response of the Egg. vol. III. Academic Press, Inc., New York, London.

Meves, F., 1903. Ueber oligopyrene und apyrene Spermien und uberihre Entsehung, nach Beobachtungen anPaludinaundPygaera. Archv für Mikroskopische Anatomie und Entwicklungsmechanik 61, 1–82.

Morgan, T.H., 1927. Experimental Embryology. Columbia University Press.

Motomura, I., 1941. Materials of the fertilization membrane in the eggs of echinoderms. J. Exp. Zoo. 16, 345–363.

Nicosia, R.F., Jeffery, W.R., Brandhorst, B., et al., 1981. The cytoskeleton of the *Lytechinus pictus* egg before and after fertilization. Biol. Bull. (Woods Hole, Mass.) 161, 309.

Pearse, A.G.E., 1968. Histochemistry. In: Theoretical and Applied, third ed., vol. I. Churchill, London.

Roosen-Runge, E.C., 1977. The Process of Spermatogenesis in Animals. Cambridge University Press, London.

Rothschild, L., 1949. The physiology of sea urchin spermatozoa: catalase. J. Exp. Biol. 26, 396–411.

Runnström, J., 1949. The mechanism of fertilisation in meats. Adv. Enzymol. 9, 241–327.

Russell-Hunter, W.D., 1979. A Life of Invertebrates. Macmillan Publishing Co., Inc, New York.

Schmuck, M.L., 1934. The male somatic chromosome group in *Sciara pauciseta*. Biol. Bull. 66 (2), 224–227.

Schroeder, T.E., 1979. Surface area change at fertilization: resorption of the mosaic membrane. Dev. Biol. 70, 306–326.

Schuel, H., 1978. Secretory Functions of Egg Cortical Granules. Cited in Biology of Fertilization, 1985, vol. 3, A. Academic Press.

Schuel, H., Schuel, R., Dandekar, P., et al., 1982. Sodium requirements in hardening of the fertilization envelope and embryonic development in sea urchins. Biol. Bull. (Woods Hole, Mass.) 162, 202–213.

Schuel, H., 1985. Functions of egg cortical granules. In: Metz, C.B., Monroy, A. (Eds.), Biology of Fertilization. vol. 3. Academic Press.

Tadano, M., 1968. Nemathelminthes. In: Kume, H., Dan, K. (Eds.), Invertebrate Embryology. NOLIT Publishing House, Yugoslavia.

Takahashi, N., Kanatani, H., 1981. Effect of 17β-estradiol on growth of oocytes in cultured ovarian fragments of the starfish, *Asterina Pectinifera*. Dev. Growth Differ. 23, 565–569.

Thorpe, S.R., 1962. A preliminary report on spawning and related phenomena in California Chitons. Veliger 4, 202–210. Cited in Metz and Monroy vol. 3.

Tyler, A., 1941. The role of fertilizin in the fertilization of eggs of the sea-urchin and other animals. Biol. Bull. 81 (2), 190–204.

Tyler, A., 1942. Specific interacting substances of eggs and sperm. West. J. Surg. Obstet. Gynecol. 50, 126–138.

Tyler, A., 1948. Fertilization and immunity. Physiol. Rev. 28 (2), 180–219.

Tyler, A., 1956. Physicochemical properties of the fertilizens of the sea urchin *Arbacia Punctulata* and the san dollar *Echinarachinus parma*. Exp. Cell. Res. 10, 377–386. Cited in Metz and Monroy vol. 3.

Tyler, A., Monroy, A., Kao, C.Y., Grundfest, H., 1956. Membrane potential and resistance of the starfish egg before and after fertilization. Biol. Bull. 111 (1), 153–177.

Varadarajan, S., Subramoniam, T., 1980. Histochemical investigations on vitellogenesis of an Anomuran Crab *Clibanarius clibanarius*. Int. J. Invertebr. Reprod. 2, 47–58.

Villee, C.A., Walker, J.R.W.F., Smith, F.E., 1964. General Zoology. W.B. Saunders Company, London.

von Lenhossek, M., 1898. Untersuchungen uber spermatozenese. Arch. Microsk. Anat. 51, 215–318. Cited in Metz and Monroy (Eds.), 1985, Biology of Fertilization, vol. 2.

Von Siebold, C.T., 1836. Fernere beobachtungen über dies spermatozoen der wirbellosen Tiiere. 2. Die spermatozoen der *Paludina vivipara*. Muller's Arch. Anat. Physiol. Wiss. Nat. 123, 232–256. Cited in Roosen-Runge (1977).

Weismann, A.F.L., 1834–1914. Embryo Project Encyclopedia. Arizona State University, School of Life Sciences, Center for Biology and Society.

Willier, B.H., Weiss, P.A., Hamburger, V., 1956. Analysis of Development. W.B. Saunders Company, Philadelphia and London.

Wilson, E.B., 1925. The Cell in Development and Heredity, third ed. Macmillan, New York.

Kingdom Protista

Chapter Outline

Class 1: Mastigophora [Flagellata] **69**
Subclass Phytomastigophorea 69
Order Volvocida 69
Chlamydomonas 69
Habitat 71
Morphology 71
Reproduction 72
Asexual reproduction 72
A- Zoospore formation 72
B- Aplanospore formation 74
C- Palmella stage 74
Sexual reproduction 75
A- Isogamous reproduction 76
B- Anisogamous reproduction 77
C- Oogamous reproduction 77
Zygote and zygospore 77
Volvox (family: Volvocidae) 79
Volvox habitat 81
Morphology 81
Reproduction 81
Asexual reproduction 81
Sexual reproduction 82
Wonderful features of *Volvox* 84
Class 2: Sarcodina (Amoebae) **87**
Subclass 1 Rhizopoda 87
Order Foraminifera 87
Suborder Monothalamia 87
Family: Difflugiidae 88
Difflugia sp. 88
Habitat 89
Reproduction 89
Asexual reproduction 89
Sexual reproduction 89
Amphimixis 89
Subclass 2 Actinopoda 89
Order Heliozoa 89
Habitat 89
Morphology 89
Family: Actinophyridae (Hartmann, 1913) 89
Actinophrys 89
Morphology 89
Reproduction 90
Asexual reproduction 90
Sexual reproduction 90
Autogamy (endomixis) 90
Amphimixis 91
Class 3: Sporozoa **91**
Subclass Telosporidea 91
Order Coccidiomorph 91
Suborder Coccidia 91
Example: *Eimeria* spp. 92
Order Gregarinidea 94
Suborder Eugregarinaria 94
Monocystis 94
Reproduction and/or life cycle 94
Class 4: Ciliophora **95**
Subclass Ciliata 95
Order Holotricha 95
Suborder Hymenostomatida 95
Habitat 96
Paramecium 96
Morphology 96
Reproduction 96
Asexual reproduction 96
Sexual reproduction 97
Conjugation in *Paramecium* 98
I- Cell adhesion 99
II- Fusion of meiotically reduced genome 99
III- Development of a new cell line 100
Endomixis; autogamy 100
References **101**

Chapter 3

Phylum Protozoa

The protozoans are unicellular organisms; strictly speaking, they are invertebrates, as mentioned above. They reproduce, generally, without forming gametes, i.e. agametically. The gametic reproduction that occurs in metazoa also occurs in protozoa in different ways. It occurs by the fragmentation of the body forming gametes, or the body itself acts as a gamete. Some protozoans with extremely striking life cycles and/or with conspicuous variety of modes of reproduction were selected to be represented here, while the following points were considered in the selection:

1. Protozoans that exemplify the types of agametic reproduction, on one hand, and describe the alternative(s) of the gametic reproduction, on the other.
2. They are mostly not familiar to the undergraduate students, and are not described in the usual invertebrate courses.
3. They are described and presented in the same order of their taxonomic positions. Taxonomy is according to different sources.

CLASS 1: MASTIGOPHORA (GREEK *MASTIX*, 'WHIP', AND *PHOREIN*, 'TO BEAR'), [FLAGELLATA] (LATIN *FLAGELLUM*, 'WHIP')

Subclass:	Phytomastigophorea (Greek *phyton*, 'plant')
	[Phytoflagellata] Usually with chromatophore; Mostly free-living organisms
Order:	Volvocida [Phytomonadida]

Small algae, solitary or swimming in colonies, some enclosed in cellulose membranes; with flagella, two or four (or eight); with no cytosome or gullet; chromatophores, if present, are bright green, usually large and cup-shaped; nutrition is holophytic or saprophytic; the reserve food is starch; mostly in fresh waters.

1. **Solitary forms** (Chlamydomonadidae) (Fig. 3.1A): oval with two flagella, green in colour, like *Chlamydomonas*, *Chlorogonium*, and *Haematococcus*, or colourless, like *Polytoma* and *Polytomella*. Some species may be coloured by haematochrome as *Chlamydomonas nivalis*, inhabiting melting snow (*nivalis* is from Latin and refers to snow), and *Haematococcus pluvialis,* which appears suddenly in rain pools, causing the myth of 'red rain'.
2. **Colonial forms** (Volvocidae) (Fig. 3.1B): in fresh waters; range in complexity from *Gonium*, as squares of 4 or 16 cells; *Pandorina,* globular, 16 cells; *Eudorina*, 32 cells near surface of gelatinous sphere; *Pleodorina*, 32, 64, or 128 cells in a solid gelatinous ball; *Volvox* with differentiated cells and complete division of labour, having about 15,000 cells on hollow gelatinous sphere, about 2 mm in diameter (Storer et al., 1983).

Chlamydomonas

C. parvula, *C. steini*, *C. euchlora*, *C. braunii*, *C. moewusii*, *C. media*, *C. angulosa*, *C. longistigma*, *C. caudata* Wille

The genus *Chlamydomonas* (Greek *chlamys*, 'a cloak or mantle'; *monas*, 'solitary'), was named by C.G. Ehrenberg, 1833–38. The particular species described by Ehrenberg is uncertain; Ettl (1976) suggested it might have been *C. pulvisculus* (Harris, 2009). It is a quite common microscopic, unicellular green plant (algae); more than 500 different species have been described, and have become popular research tools. However, most scientists work with only a few

Invertebrate Embryology and Reproduction. https://doi.org/10.1016/B978-0-12-814114-4.00003-5

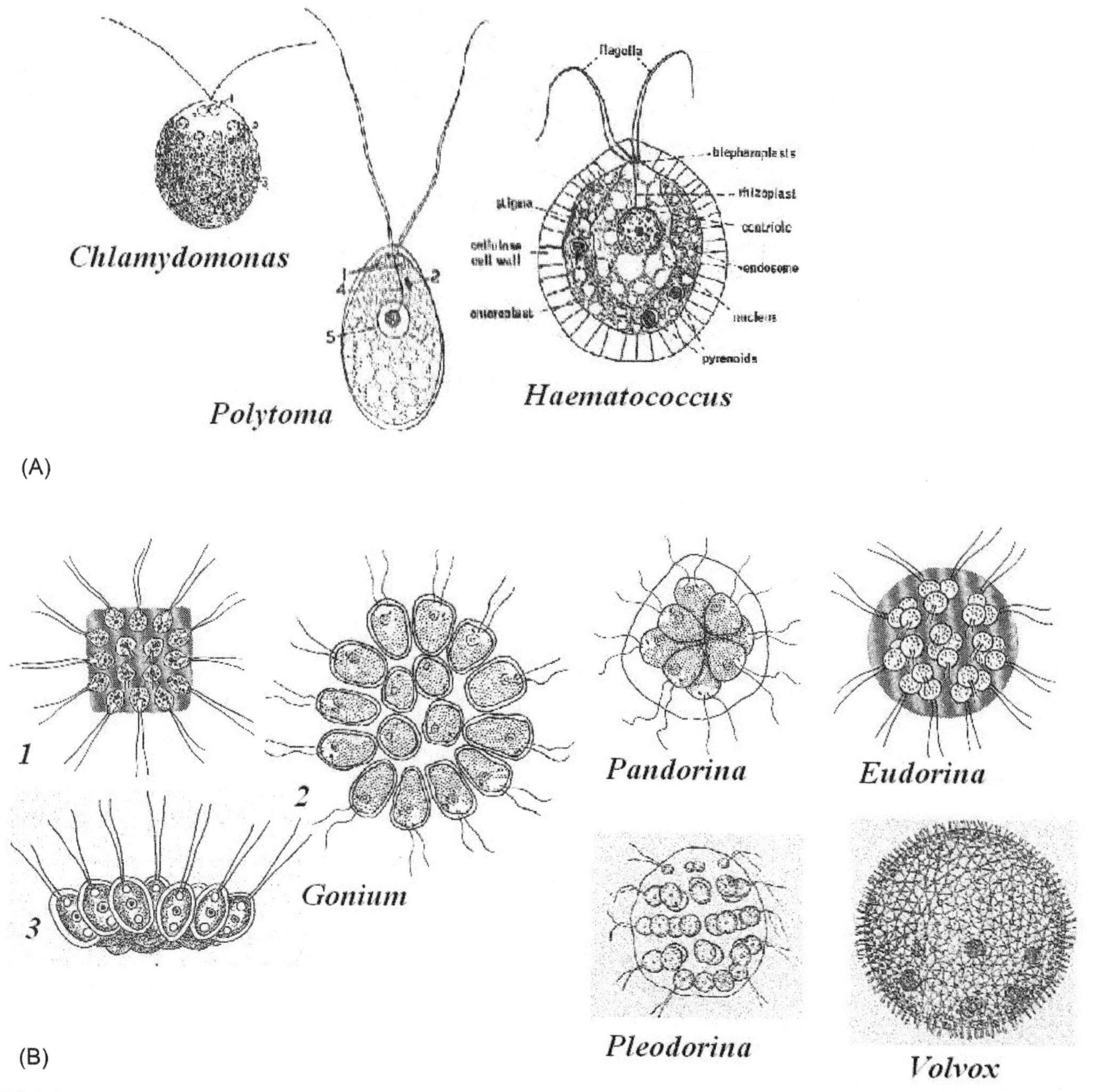

FIG. 3.1 Examples of the order Phytomonadina: (A) solitary forms, *Chlamydomonas*, from life; *Polytoma uvella (after Entz, 1918), the two from Hyman (1940)*. (1) contractile vacuole, (2) stigma, (3) cup-shaped chloroplast, (4) rhizoplast, (5) nucleus. *Haematococcus (after Elliot, from Barnes, 1974)*. (B) colonial forms, simple colony: *Gonium ((1) from Villee et al., 1964)*; (2),(3) *G. pectorale (from Hickman, 1973, (3) side view after Stein from Barnes, 1974)*; *Pandorina morum (from Hickman, 1973), Eudorina (from Villee et al., 1964)*, Pleodorina and the complex colony *Volvox (from Storer et al., 1983)*.

species. This genus is used extensively in research for the study of photosynthesis in eukaryotic cells, and as a model for plant genetics and physiology. *Chlamydomonas* has been used to study different aspects of evolutionary biology, molecular biology and ecology. Metz and Monroy (1985a,b,c) reported that this genus could serve as a general model system for **cell–cell interaction**, like other lower isogamous organisms, *Paramecium*, yeast, and the egg and sperm of metazoans. It is the organism of choice for many selection experiments because:

1. It has a short generation time
2. It is nutritionally simple and cheap to grow, since it is both a heterotroph and facultative autotroph, i.e. photosynthetically

3. It can reproduce both asexually and sexually; and
4. There is a wealth of genetic information already available

C. reinhardtii is widely distributed worldwide in soil and freshwater. It was not discovered until 1888 (at which point it was named after Ludwig Reinhardt), since it differs in sexual reproductive aspects from *C. pulvisculus*. According to the botanic taxonomic laws, Ettl (1976, in Harris, 2009) corrected its spelling to *C. reinhardtii*. By the 1960s, it became the most widely used species in laboratory work. It is primarily used as a model organism in biology, in a wide range of subfields. With illumination, *C. reinhardtii* can grow in a medium lacking carbon and energy sources, and can also grow in the dark when supplied with these. It is also of interest in the field of biofuel, as a source of hydrogen.

Habitat

Chlamydomonas species are found all over the world, in virtually every habitat on earth. They are found in freshwater such as pools, ponds, ditches, and lakes; in brackish water; in oceans (e.g. *C. ehrenbergi* is found in saline water); and even in snow on mountaintops, in alpine and arctic regions. Some terrestrial species live on damp soil, while others are simply in the air we breathe.

C. nivalis is a species of green algae that owes its red colour to a bright red carotenoid pigment, which gives the snow itself the red colouration. At the same time, the pigment protects the chloroplast from intense visible and ultraviolet radiation, as well as absorbing the heat. The alga is provided with liquid water as the snow melts around it.

During the winter months, when snow covers them, the algae become dormant. In spring, the nutrients and the increased levels of light and melted water stimulate germination. The resting cells, once germinated, release smaller green flagellate daughters, which travel towards the surface of the snow. There, they lose their flagella and form aplanospores, or thick-walled resting cells. Alternatively, they may function as gametes, fusing in pairs to form zygotes (see later).

Watermelon snow, also called **snow algae**, is snow that is reddish or pink in colour, with the slight scent of fresh watermelon. This type of snow is common during the summer in alpine and coastal polar regions. It is caused mainly by the presence of *C. nivalis*. Unlike most species of freshwater algae, it is cryophilic (cold-loving) and thrives in freezing water, hence its name.

Morphology

A typical solitary phytomonad, its body is a thallus, which consists of a single biflagellate **haploid cell**. It is spherical, pear-shaped or oval, about 20 μm in length and rarely exceeds 30 μm in major diameter (Fig. 3.2). The detailed internal organization of this alga is as follows:

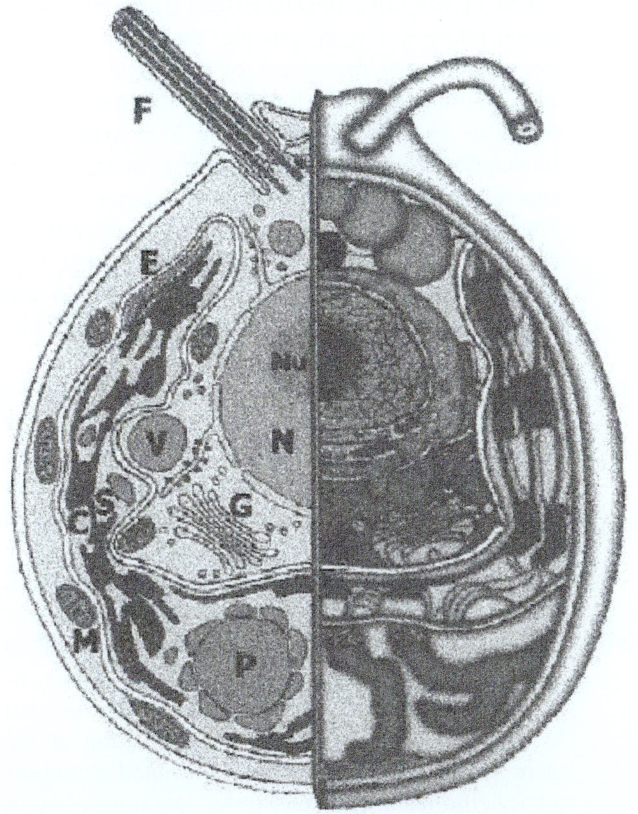

FIG. 3.2 Cell structure of *Chlamydomonas reinhardtii*, showing the central nucleus (N) with the nucleolus (Nu), the two isoform flagella (F), the cup shaped chloroplast (C) with the eye spot (E) and the starch-containing pyrenoid (P) and the mitochondria (M). In addition, one may distinguish the Golgi vesicle (G), starch grains (S), and vacuoles (V). *(After Nickelsen and Kück (2000), from Harris (2009).)*

(i) Early reports suggest a close-fitting cellulose cell wall, although studies of some species have shown that the chemical composition of the cell wall is not pure carbohydrate, but hydroxyproline-rich glycoprotein, e.g. *C. reinhardtii*. It is surrounded by a mucilaginous pectin layer, called the capsule layer.

(ii) The cytoplasm shows two distinct regions: the thin clear outer region with two anteriorly placed flagella, few strands of granular reticulum, free ribosomes, volutin granules and mitochondria; and the dense region possessing perinuclear dictosomes (Golgi complex), numerous mitochondria, free ribosomes, a few strands of granular endoplasmic reticulum, a single chloroplast and a nucleus.

(iii) The single nucleus is located centrally, sometimes near the anterior end, in the cup-shaped cavity of chloroplast, embedded in colourless cytoplasm. Its diameter is about 2–4 μm, with a distinct central nucleolus measuring about 0.5–0.9 μm in diameter (Pandey and Trivedi, 1995).

(iv) A pair of whip-like flagella extend from one end, to which attaches a basal granule, the blepharoplasty, that controls the movement of the flagella. Apical and equal in length, but longer than the length of the organism, each flagellum consists of an axial thread surrounded by a sheath. The axial is composed of nine doubled peripheral fibrils and two central fibrils. Their lashings pull the swimming organism through the water while rotating it, using a breaststroke type motion (Pandey and Trivedi, 1995). This aspect was studied in *C. reinhardtii* in a viscous medium. Flagella exhibit synchronous movement (see Fig. 3.3).

(v) The single large cup-shaped, double-membraned and sac-like chloroplast occupies the greater part of the cell, and is generally parietal in position. It may be star-shaped, H-shaped, or urn-shaped, or layered in appearance. In some species it has ridges, and appears laminate or stellate. It encloses a granular matrix (stroma), which contains ribosomes and is considered to be the site of chloroplast protein synthesis.

(vi) One or more protein pyrenoid bodies, spherical in shape, usually accompany the chloroplast and are involved in starch production, and may be surrounded by starch granules. The shape and position of the chloroplast and the number of associated pyrenoids are specific features. (Refer to the taxonomic key to the major groups of *Chlamydomonas* species (Ettl, 1976, in Harris, 2009).)

(vii) The most striking structure in the organism is the red stigma, or eyespot, which lies laterally in the anterior end of the cell inside the chloroplast; it is in fact a part of the chloroplast (Fig. 3.2). It is a carotenoid pigment called haematochrome. In light microscopy, it appears as a tiny orange or reddish dot. In electron microscopy, it can be seen as a linear arrangement of pigments associated with the chloroplast (Fig. 3.4). It consists of two parts: a curved pigmented plate (cup-shaped), located in the cup and overlain, in some forms, by a lens-like thickening, biconvex, hyaline photosensitive substance (it is 100 times faster than the human eye, as having **rhodopsin or visual purple**, a light sensitive substance also found in human eye (Keele and Neil, 1973)). In addition, it consists of mutually compressed 14–15 electron dense granules, each 75–100 μm in size. These have been referred to as fat droplets (Pandey and Trivedi, 1995). The hole constitutes a sensory organelle that receives light.

(viii) The organisms have no large central vacuole, but they do have one or more tiny, osmoregulatory contractile vacuoles. The various species exhibit isogamy, anisogamy, and intermediate conditions.

Reproduction

In **metazoa**, where true gametes are formed, it is preferable to refer to the types of reproduction as **agametic** and **gametic** reproduction. In **protozoa**, on the other hand, no true gametes are formed, and therefore the terms **asexual** and **sexual** types of reproduction are more suitable.

Asexual reproduction

Chlamydomonas reproduces asexually in more than one way. Vegetative cells are haploid, for example, *C. reinhardtii* has 17 small chromosomes.

Zoospore formation or binary fission (as in the majority of flagellates). Borradaile et al. (1977) reported that the term **spore** is applied to products of repeated or multiple fission of a zygote, and most often to products of any repeated or multiple fission. Most commonly this animal divides longitudinally (with respect to position of nucleus and chloroplast). Division is thus said to be **symmetrogenic**, producing mirror-image daughter cells, a minute miniature of the parent. In favourable conditions (sufficient moisture during the rainy season), *Chlamydomonas* individuals continue to grow and, at a certain size, become

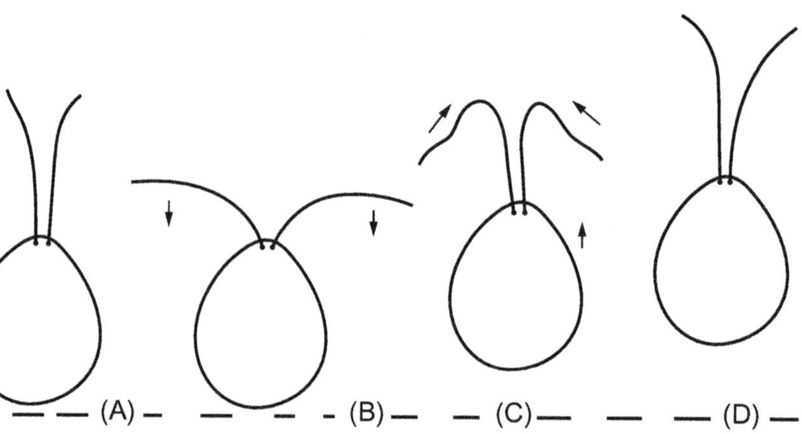

FIG. 3.3 A single complete action of flagellum of *C. moewusii* consists of: (A) start of power stroke; (B) power stroke; (C) recovery stroke; (D) cycle completed. *(Redrawn from Pandey and Trivedi (1995).)*

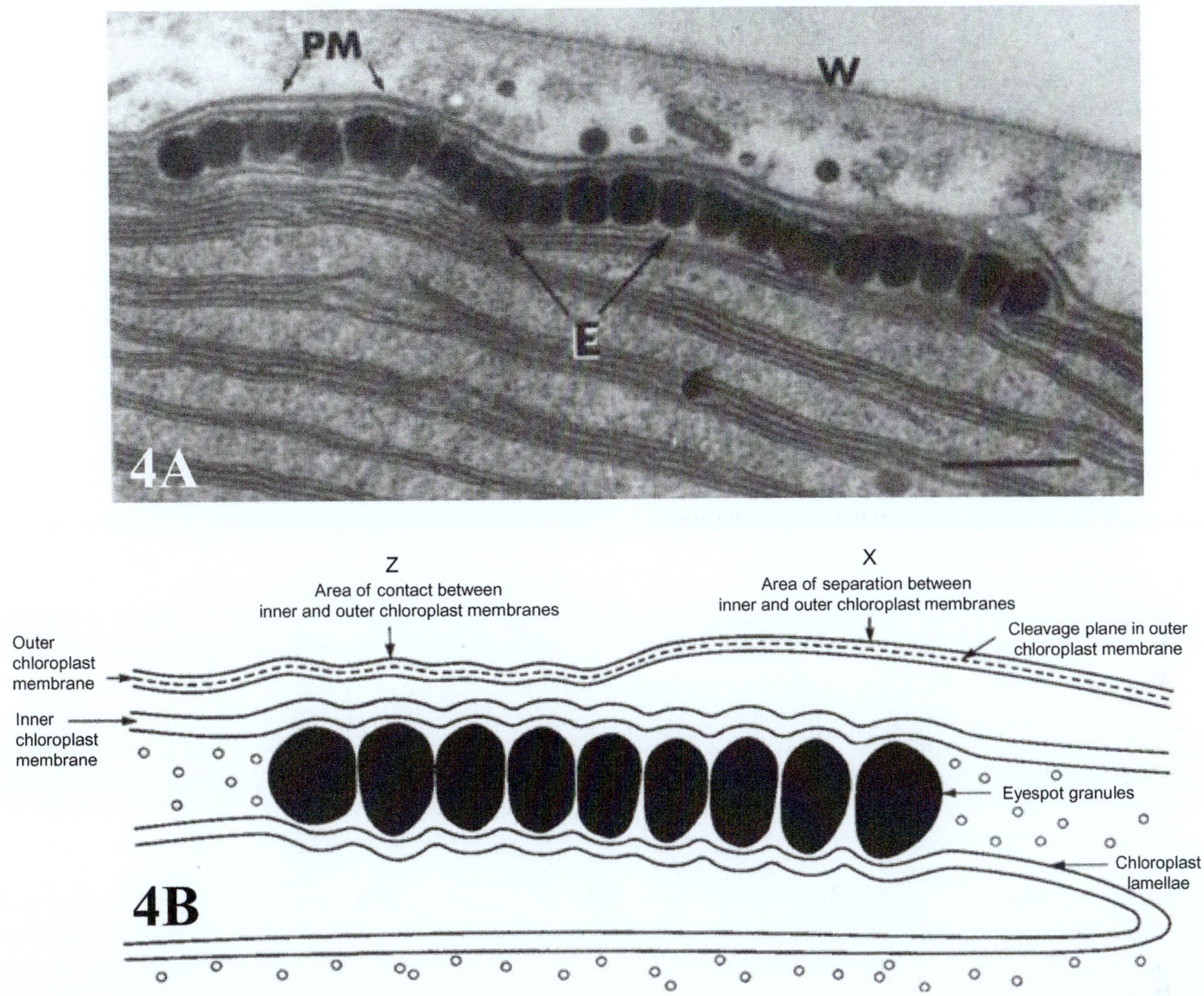

FIG. 3.4 Eyespot morphology in *C. eugametos*. (A) Micrograph showing the structure of the eyespot, (E) monolayer of approximately 23 granules lying below the plasma membrane (PM). Bar=0.2 μm. (B) Drawing showing the physical association of the eyespot granules with the chloroplast membrane system. *(From Harris, 2009.)*

inactive in respect to reproduction, frequently during the night. The flagella withdraw, the cytoplasm shrinks slightly within the cell wall and the contractile vacuoles disappear. The cytoplasm, chloroplast and nucleus divide once, twice or occasionally three times by longitudinal binary fission producing two, four or eight young daughters, which do not separate until the process is completed (Fig. 3.5). The division of the nucleus is mitotic and in some species, such as *C. longistigma* and even *C. reinhardtii*, the protoplast during or after the first division might rotate, until the plane of division becomes transverse. Fission of this kind is said to be **pseudotransverse** (Fig. 3.6). In most species under typical laboratory conditions, orientation of the second division in respect to the first might be a recognizable feature of the species (see Fig. 3.7). The second cleavage plane appears to be longitudinal in some cells of *C. reinhardtii* and equatorial in others. If equatorial division is to occur, the basal bodies need to migrate to the new anterior cleavage furrow. The basal bodies and flagellar root system are involved in the formation of the mitotic spindle (Harris, 2009). In *Polytoma* (Fig. 3.8), on the other hand, the only vestige of the longitudinal fission is a slight obliquity of the first division of the nucleus (Borradaile et al., 1977).

If the animal divides very rapidly, the dividing chromatophores may fail to be shared among the progeny. Those without these chromatophores are unable to develop new ones, and so remain colourless. In this way, Chlamydomonas can be transmuted into *Polytoma* (Borradaile et al., 1977).

Fission occurs repeatedly within the cell wall of the aflagellate; the daughters remain bound within the parent cell wall until conditions become favourable. Each daughter develops a pair of flagella and a cell wall. They produce specialized sporangial wall autolysins, which digests the parent wall but not their own. The parent wall bursts open and releases the product, commonly in the evening, to lead

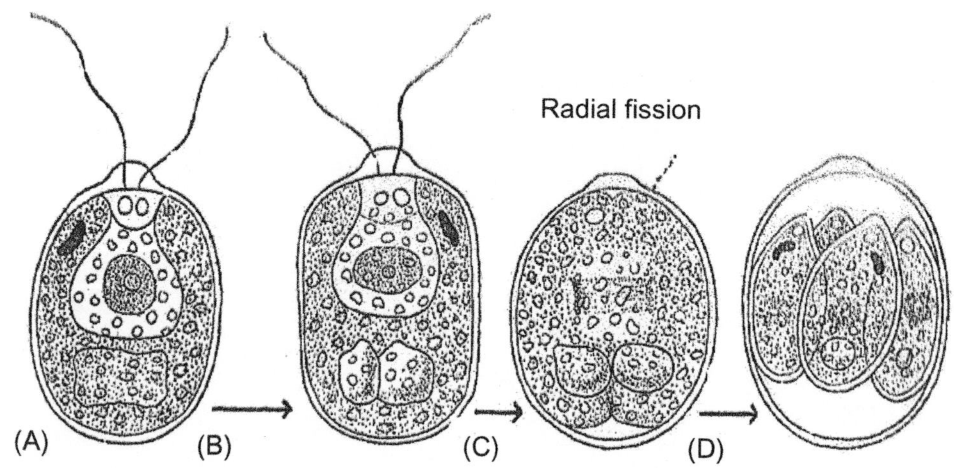

FIG. 3.5 Zoospore formation in *C. angulosa* 1000×, showing binary repeated fission (symmetrogenic division). *(From Borradaile et al. (1977).)*

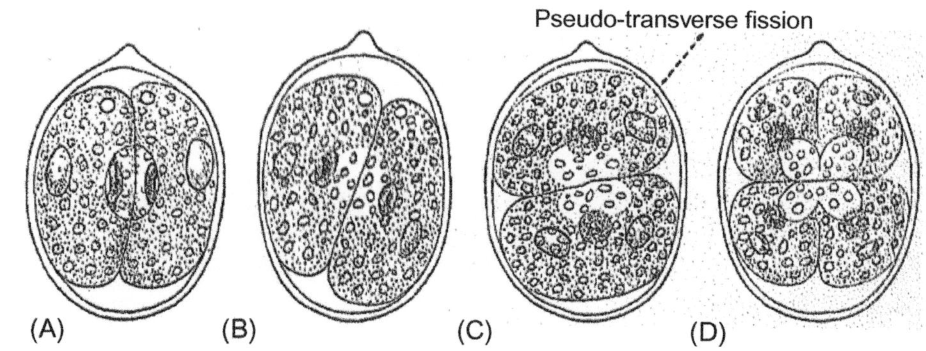

FIG. 3.6 Pseudo-transverse fission in *C. longistigma* 1000×. *(From Borradaile et al. (1977).)*

their own lives and grow to full size. This process is completed in few hours and takes place daily. Therefore great numbers of *Chlamydomonas* appear very rapidly, and usually make the water look green.

The flagella in *Chlamydomonas* and *Polytoma* are lost during fission, early or late in the process. Usually, a second flagellum grows out from the basal granule, which divides. Gardiner (1972) reported that this species produces isogametes by a series of mitotic divisions, as in their vegetative phase the individuals are haploid and meiosis occurs only at the zygote stage.

Aplanospore formation. When the conditions are not favourable to form zoospore, *Chlamydomonas* cell sometimes comes to rest and behaves as in forming a zoospore. However, in addition it rounds up as an envelope around itself to form a thin wall around, to become an aplanospore, e.g. *C. caudate* (Fig. 3.9A). It either germinates directly to give a new individual, or produce zoospore at favourable conditions. To survive severe conditions, the single spore secretes a thick wall around itself. Such resting spore is called **hypnospore**, e.g. *C. nivalis,* which develops strong red colouration due to the red pigment known as haematochrome. The hypnospore germinates to produce new daughters with the onset of favourable conditions.

Palmella stage. This stage is common in many species. It is a stage in the life history of some unicellular flagellate algae in which the cells lose their flagella and form a gelatinous aggregation. It becomes in a nonmotile reproductive phase, which helps the algae to survive during partial desiccation such as in damp soil conditions or agar culture medium. During this phase, the alga divides repeatedly into two, four, or eight daughter protoplasts, but they do not form flagella. Each continues division producing a group of two or four aplanospores within its own newly formed gelatinous wall. At the same time, the parent cell wall undergoes gelatinization, enclosing all in a mucilage large capsule resembling a large colony. This stage is called **palmella** (Fig. 3.9C), because earlier workers thought it was a separate species of algae. Physiologists generally refer to such palmellae as **little green balls**. On the return of favourable conditions, the mucilage wall dissolves, and

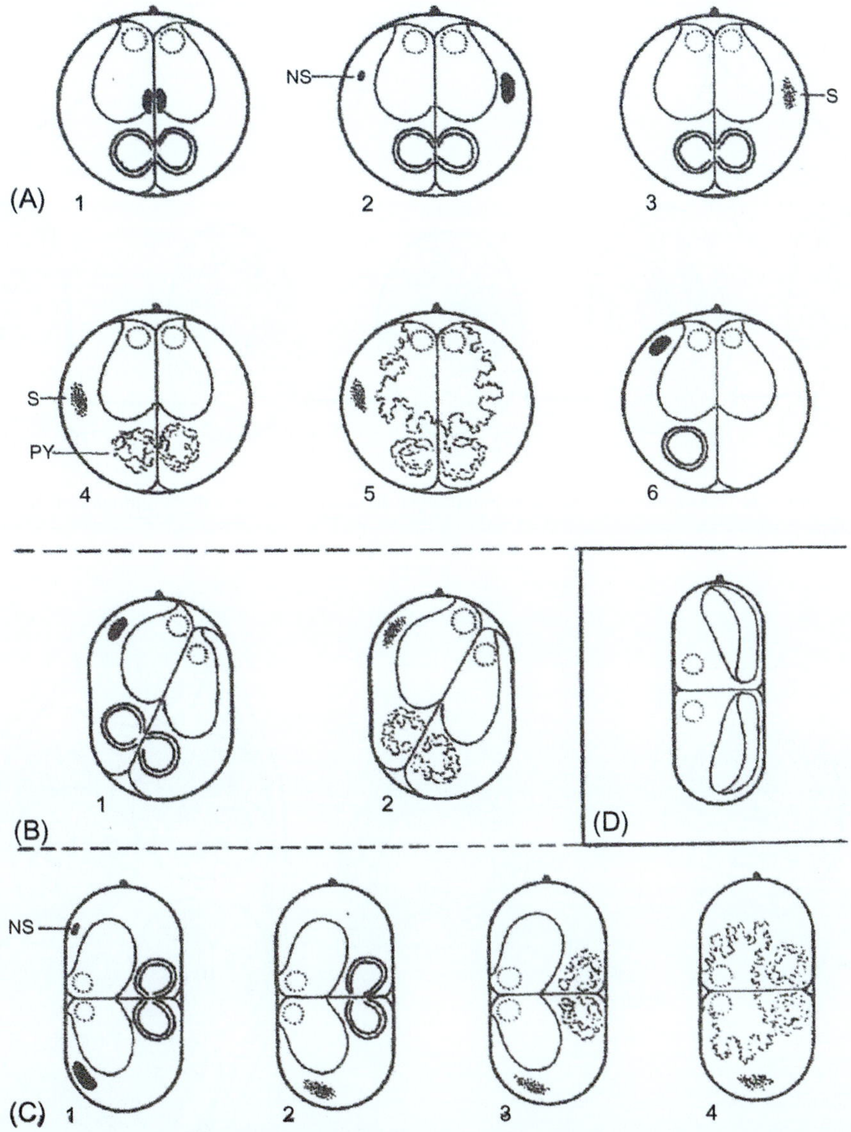

FIG. 3.7 Various types of protoplast division in Chlamydomonas species. (A) True longitudinal division: (1) eyespot and pyrenoid divided (e.g. *C. olifanii*); (2) chloroplast and pyrenoid divided, the original eyespot remains in one daughter, the other forms new eyespot (*C. angulosa*); (3) original eyespot degenerates and new ones form in the daughters (*C. proboscigera*); (4) both eyespot and pyrenoid degenerate and re-form later in the daughters (*C. concylindrus*); (5) eyespot, chloroplast, and pyrenoid lose discrete structure and re-form in daughters (*C. rotula*); (6) rare form, in which eyespot and pyrenoid persist in one daughter and the other re-form new ones (*C. praecox*). (B) Oblique divisions, with (1) (*C. ovalis*) or without (2) (species uncertain) persistence of original eyespot and pyrenoid. (C) Simulated transverse division, in which the protoplast rotates 90 degrees. Organelle division proceeds as in A, but perpendicular to the vertical axis of the mother: (1) *C. obverse*, (2) *C. lewinii*, (3) *C. svitaviensis*, (4) *C. geitleri*. (D) True transverse division, in which a lateral chloroplast in a narrow cell hinders protoplast rotation (*C. gloeophila*). S = eyespot; NS = new eyespot; PY = pyrenoid. *(From Harris (2009) after Ettl (1976).)*

the daughter protoplasts of the palmella (tetrasporales) become motile and develop into new individuals. Species of *Chlamydomonas* vary in terms of the extent to which they form palmelloid colonies of adherent, nonmotile cells in nature and in culture, with some species appearing to exist primarily in this state, e.g. *C. kleinii*. It was suggested that palmella formation has an adaptive advantage in protecting *Chlamydomonas* cells from predation by rotifers and other organisms (Harris, 2009).

Sexual reproduction

It is well known that individuals of genus *Chlamydomonas* are haploid organisms, which commonly reproduce sexually beginning with gametogenesis, i.e. transform or form gametes if they feel the approach of unfavourable conditions, or in the laboratory, if they are subjected to distilled water or low nitrogen content. This genus shows a considerable range of variation from isogamy, anisogamy to oogamy. These algae may be **homothallic**, e.g. *C. gynogama* and *C. media*,

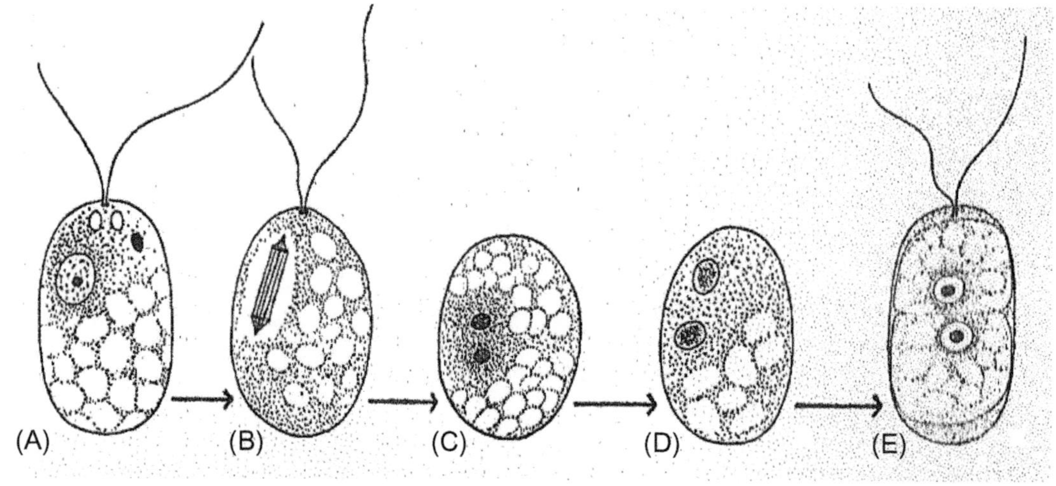

FIG. 3.8 Division of *Polytoma uvella* 1300×, the only vestige of longitudinal fission consists in a slight obliquity of the first division of the nucleus. *(From Borradaile et al. (1977).)*

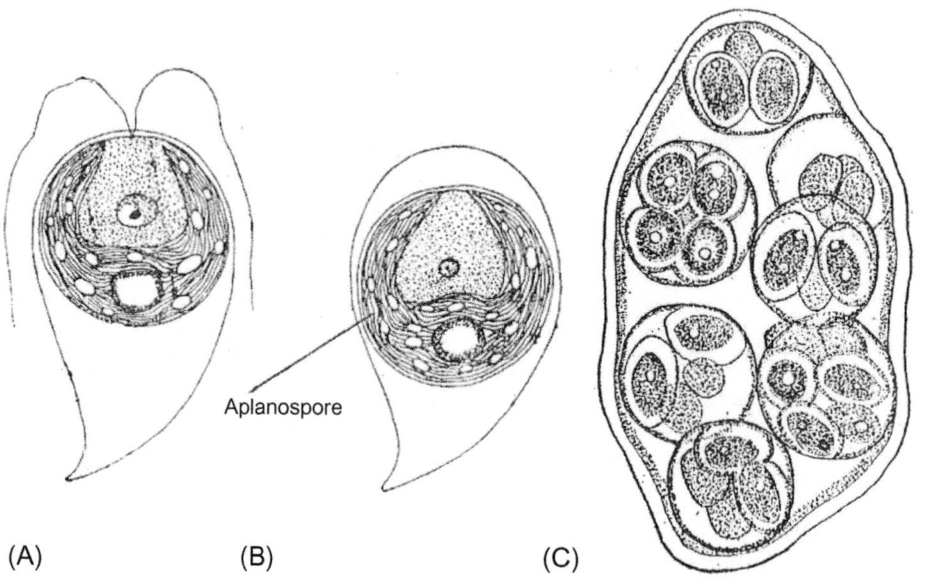

FIG. 3.9 *C. caudate* (A) normal plant; (B) withdrawal of flagella and formation of an Aplanospore, *from Pandey and Trivedi (1995)*; (C) *Chlamydomonas* in palmella state, *after Goroschankin from Hyman (1940)*.

in which both mating types can arise from a single progenitor cell, or may be **heterothallic**, e.g. *C. reinhardtii* and *C. moewusii*, in which mating type (plus or minus) is genetically determined in a cell line, behaving as a single Mendelian locus in crosses (Harris, 2009). The gametes formed are also of two types: gymnogametes, naked or without a cell wall, as in *C. longistigma*, or calyptogametes with a cell wall as in *C. eugametos* (Pandey and Trivedi, 1995).

Sexual reproduction is divided into three types according to the morphology and physiology of the gametes (Fig. 3.10):

Isogamous reproduction. Isogamy (Fig. 3.11A) is common in *Chlamydomonas*; this species is homothallic or heterothallic. The individual divides as described above, and undergoes two, three, or more rounds of mitosis producing up to 64 new individuals; the end number would differ with the species. This product does not develop cell walls, becoming gametes; all of them are morphologically similar but physiologically different. There are two mating types, referred to as **mt**(+) and **mt**(−). The plus gamete has a mating structure composed of three electron-dense layers that lie under the plasma membrane in an apical bud. The minus gamete has a mating structure composed of two electron-dense plates in a similar apical bud. Covering the plasma membrane of the apical buds of the two gametes is a **fringe** of fuzzy material that probably contains the recognition apparatus of the gametes. On release of the gametes from the parental cell, they swim about and may

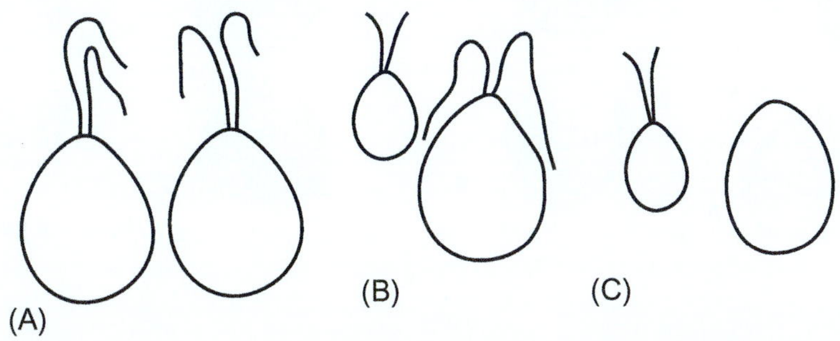

FIG. 3.10 Types of gametes in gametic reproduction of *Chlamydomonas* species; (A) isogamous; (B) anisogamous; (C) oogamous type. *(Redrawn from Pandey and Trivedi (1995).)*

meet other individuals of the same characters. Pairing of the mating types occurs: mt+ with mt− of the other parent. *C. snowiae*, an isogamous and heterothallic species, has been studied in detail. In this species, any vegetative cell may act as a gamete. However, the wall is shed prior to gametic fusion. In some species, such as *C. eugametos*, vegetative cells directly act as gametes and do not shed their wall.

As for how the two different mating types recognize each other, the two gametes come close by their anterior ends and recognition first occurs at the flagellar membranes. The flagella of two individuals twist around each other, enabling specific regions of the cell membranes to come together. These specialized regions contain **mating type-specific** components that enable the cytoplasm to fuse (for more molecular information about this aspect, refer to Harris, 2009). After flagellar agglutination, the mt+ gamete initiates fusion by extending a fertilization tube, which contacts and fuses with a dome-shaped specific site, on the mt− gamete (Fig. 3.11B). It was confirmed experimentally that the partner designated mt+ in *C. reinhardtii* and *C. moewusii* is the one that produces a fertilization tubule upon activation, and transmits its chloroplast genome to the other partner. Arguments based on anatomical grounds designated the +cell of *Chlamydomonas* as male. This is the swimming active gamete (Harris, 2009). Gametes fuse face-to-face in *C. moewusii* and **shoulder-to-shoulder** in *C. reinhardtii* (Fig. 3.12). The mechanism used to recognize and fuse the + and − gametes in a lower protist such as *Chlamydomonas* is amazingly similar to that occurring in the fertilization of sea urchins. This fulfils the view of Metz and Monroy (1985a,b,c), who stated that "fertilization is a very highly specialized example of the general phenomenon of cell-cell interaction which may be apparent among many lower isogamous organisms."

Soon after the completion of plasmogamy and karyogamy, each pair produces the **quadriflagellate zygote** with a diploid chromosomal set. It remains motile for several hours or even up to 15 days in some species, such as *C. pertusa*. It then comes to rest, rounds off and secretes a resistant wall, becoming a **zygospore**.

Anisogamous reproduction. In this sexual reproduction, the two types of gametes differ appreciably in their size (Fig. 3.13A) and in the number produced by the parent. One individual produces a larger number of smaller gametes, or **microgametes**, e.g. 8/cell; the other produces fewer and larger **macrogametes**, 4/cell. Both gametes are flagellated and only the microgamete is active. *Chlamydomonas brauni* is a heterothallic species and is a well-studied example of anisogamy.

The microgamete and macrogamete approach each other and fuse by their anterior ends, **vis-à-vis**. At the point of contact, they coalesce and the content of microgamete passes into the macrogamete, forming a **quadriflagellate** zygote. The latter becomes nonmotile and thick-walled, the **zygospore**.

Sexual reproduction involving fusion of similar or dissimilar gametic individuals is a common occurrence. In the author's opinion, this is amphimixis.

Oogamous reproduction. This is the true sexual reproduction since it occurs by male and female gametes like in *Volvox*. The female gamete is produced by maturation of a vegetative individual that enlarges and discards its flagella, becoming a nonmotile egg or ovum. It does not come out of the wall (Fig. 3.13-2 (D) and (E)). On the other hand, the male gametes are produced as 8, 16, or 32 in number by successive divisions of the cell (**antheridium**). They are biflagellate motile gametes; when the cell wall ruptures, these gametes are released into the medium. They approach the female gamete and fertilization occurs between a male gamete and an ovum, forming a **nonmotile zygote**. This is considered internal fertilization. Oogamy has been studied in detail in *C. coccifera* and *C. oogamum*.

Zygote and zygospore

The formation of the zygote in *Chlamydomonas* is summarized in the following steps:

1. gametic differentiation;
2. clumping;
3. pairing;

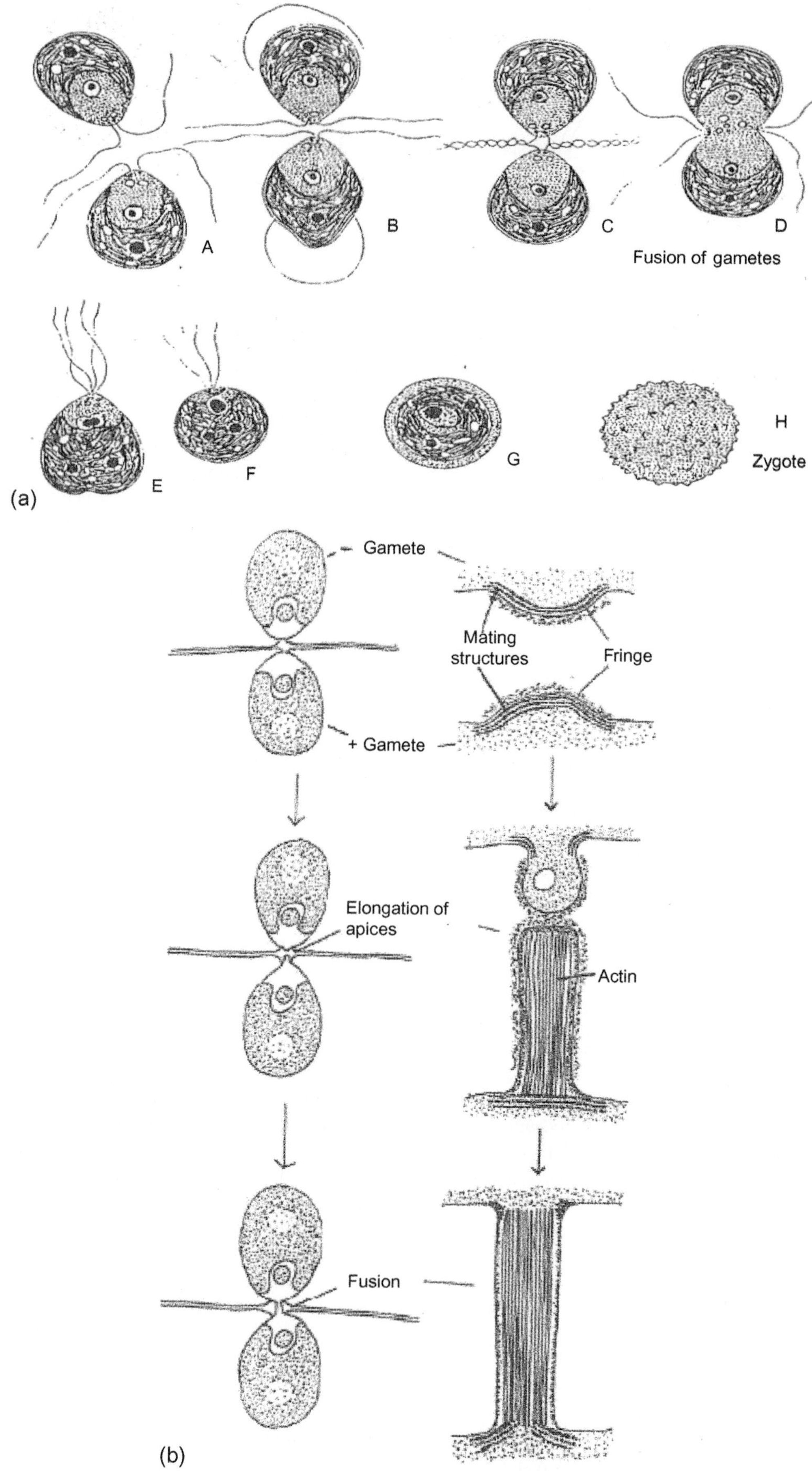

FIG. 3.11 (a) Isogamous reproduction in *C. snowiae*, (A and B) mutual approach of individuals and liberation of protoplasts; (C–E) early and late stages of fusion; (F) quadriflagellate zygote; (G and H) young and mature zygote. *From Pandey and Trivedi (1995).* (b) Drawings of fusing of individuals of *C. reinhardtii* on the left. On the right are enlargements of the apical areas of the fusing cells. *From Lee (2008).*

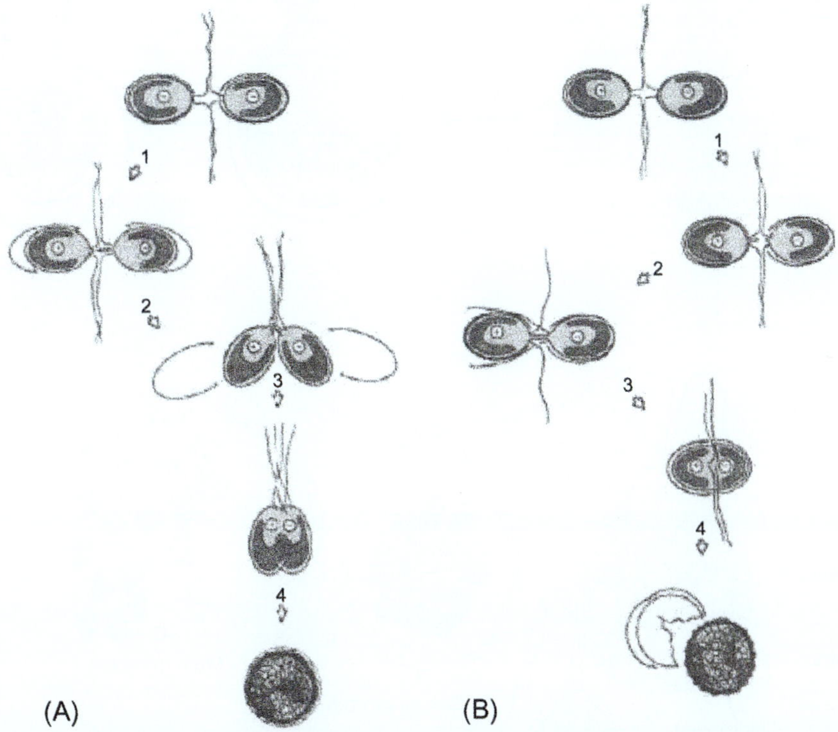

FIG. 3.12 Two styles of isogamous mating: (A) occurs in *C. reinhardtii* and *C. smithii*: (1) gametic flagellar agglutination between cells of opposite mating types, followed by flagellar tipping, activation of anterior mating structures, and activation of lytic enzyme to promote shedding of gametic cell walls; (2) activated mating structures fuse to form a cytoplasmic bridge, flagella de-adhere, and the cells immediately 'jackknife' to begin the final stages of cell fusion; (3) gametes fuse 'shoulder-to-shoulder' to produce the quadriflagellate zygote; (4) deposition of the multilayered zygospore wall completes the process. (B) Occurs in *C. eugametos, C. moewusii, C. monoica*: (1) gametic flagellar agglutination promotes activation of anterior mating structures and limited, anterior site-specific dissolution of the cell wall to allow extension of the fertilization tubule; (2) activated mating structures fuse to form a cytoplasmic bridge, gametic cell walls are retained, and mating-type specific flagellar paralysis allows directed movement of the tandem gamete pair ('vis-à-vis pair') which may persist for several hours; (3) the gametic cell walls loosen, and the cell bodies fuse following a shortening and widening of the cytoplasmic bridge; (4) the multilayered zygospore wall is assembled and a primary zygote wall is released carrying the loosely attached gametic cell. *(All from Harris (2009).)*

4. plasmogamy; and
5. karyogamy

The zygote becomes a zygospore. A mature zygospore is a spherical structure with a thick wall, which might be smooth or rough. The appearance of the surface of this wall is a specific feature. The zygospore has a diploid chromosomal set and often undergoes a period of rest, during which, it becomes red or orange in colour due to accumulation of fats. It can resist desiccation, cold and bad weather, and may be carried away to other positions by wind. It is therefore considered the dispersal stage of *Chlamydomonas*.

Upon the restoration of favourable conditions, the zygospore germinates. The zygote divides **meiotically** to produce **four haploid zoospores**, which are released in a vesicle by the rupture of the wall of the spore. The zoospores soon develop into *Chlamydomonas* algae to repeat the cycle (Fig. 3.14).

Volvox (family: Volvocidae)

V. aureus, V. carteri (V. nagariensis), V. globator, V. dissipatrix, V. tertius

The name '*Volvox*' comes from the Latin *volvere*, to roll and *-ox*, from *atrox*, fierce. Ever since Anton van Leeuwenhoek (1632–1723) first viewed these algal 'fierce rollers' with utter fascination in 1700, one biologist after the other has pointed to *Volvox* as a model organism that could be used to support or to disprove some important biological concepts of the day, such as spontaneous generation, preformation, epigenesis, the continuity of the germ plasm, etc. All attempts to exploit *Volvox* as a laboratory model system failed. However, in the 1960s, Richard Starr's group finally discovered a medium in which the organism would reproduce in captivity (Starr, 1968). By 1970, Starr concluded that a mating pair of isolates of *V. carteri* from Japan had the best combination of properties to serve as a genetic model system. Most studies of *Volvox* reported in the following 30 years have used those strains of *V. carteri* or their descendants as a model system for biochemical and physiological investigations. However, *Volvox* is often studied in introductory courses because its mode of colony formation is somewhat similar to embryonic development of some metazoa. It offers a unique opportunity to study the segregation of reproductive potential during the development of a multicellular organism.

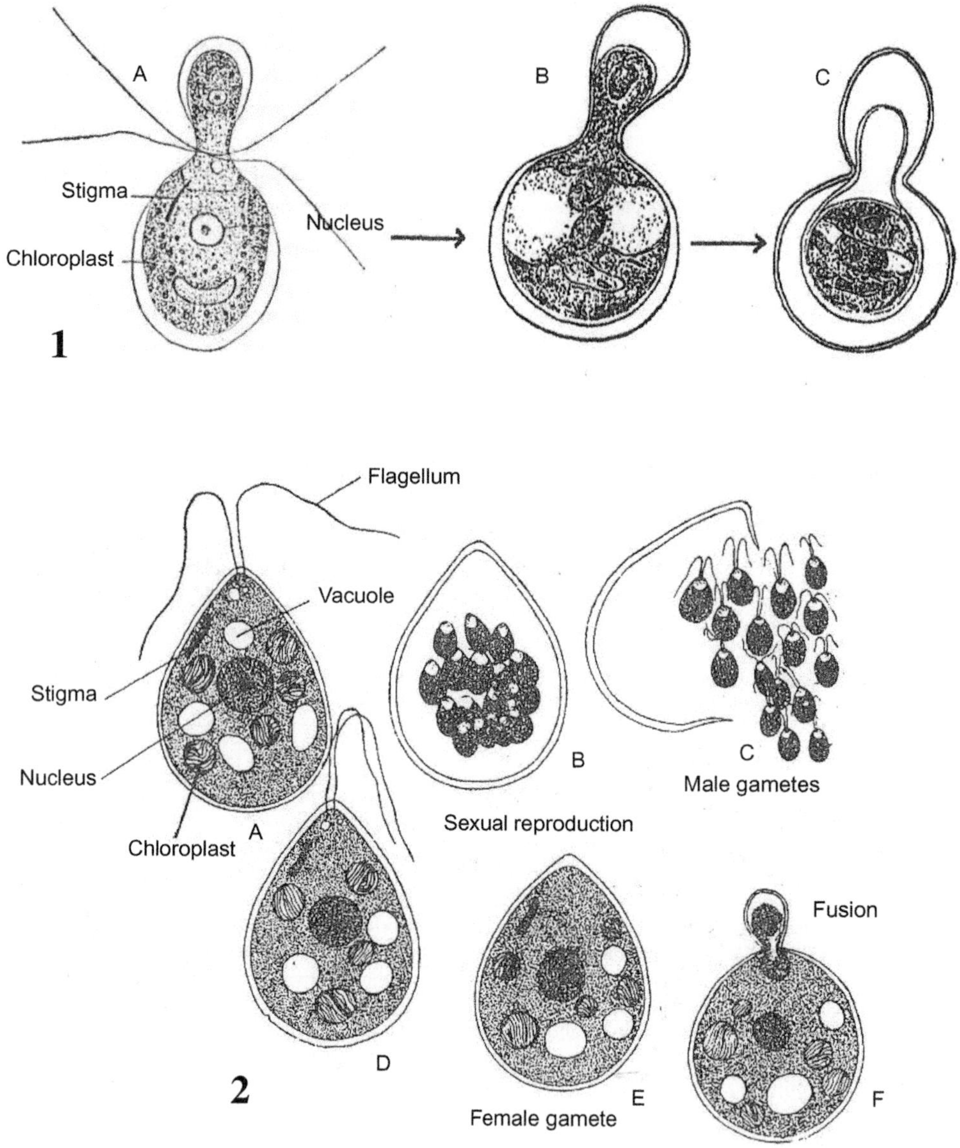

FIG. 3.13 (1) Anisogamous reproduction, anisogametes of *C. braunii*, *(A) from Hyman (1940)*, male gamete smaller and female gamete larger, the two in action of fertilization; *(B and C) from Borradaile et al. (1977)* showing the process of fusion. (2) Oogamous reproduction, *C. coccifera*: (A–C) stages in the development of male gametes; (D and E) development of female gamete; (F) act of fusion. *From Pandey and Trivedi (1995) and Vashishta (2004).*

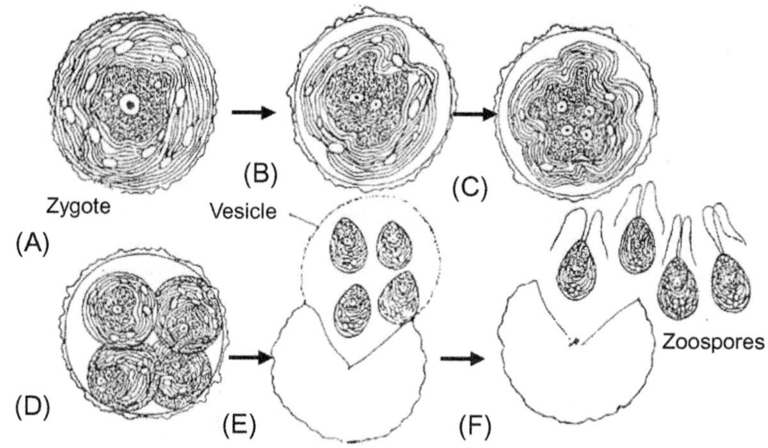

FIG. 3.14 Zygote and its germination in Chlamydomonas. (E) Liberation of zoospores in a vesicle. *(From Pandey and Trivedi (1995) and Vashishta (2004).)*

Volvox habitat

Volvox is found in ponds and ditches, and even in shallow puddles. The most favourable place to look for it is the deep ponds, lagoons and ditches, which receive an abundance of rainwater. *Volvox* is one of the typical genera of the group. It is a cosmopolitan, well-known type. It has even been suggested that *Volvox* is so cosmopolitan that new species are unlikely to be encountered (Powers, 1908).

Morphology

Volvox is a spherical multicellular (i.e. colonial) green alga, of worldwide distribution comprising about 20 species. It is one of the planktonic organisms, which help to colour the water of the pond green. The *Volvox* colony is of constant size and shape for a given species (Fig. 3.16A). Each colony consists of a definite number of cells arranged in a specific manner and forms an integrated whole, called '**coenobium**'. The *Volvox* coenobium is the largest and most highly differentiated of all the free-swimming colonial forms. It looks like a minute green ball, the size of just a small pinhead, varying from 0.5 to 2.0 mm in diameter, and therefore can easily be seen by the naked eye. It is a sphere (Fig. 3.16) of extracellular mucilage or matrix with a firm bounding layer. It contains many small biflagellate somatic, vegetative cells; each is much like *Chlamydomonas*, whose flagella protrude through the outer mucilage coat. In addition, there are a few large, nonmotile reproductive cells called **gonidia**, mostly in the posterior pole. The reproductive cells never develop functional flagella and never contribute to any somatic functions of the individual; they are entirely specialized for reproduction. In some species, the reproductive cells are derived from cells that originally look and function like somatic cells, before they enlarge and divide to form new progeny. However, in other members of the genus, such as *V. carteri*, there is a complete division of labour. The cells at the anterior pole are sterile and may also differ morphologically from the others. These numerous cells of the colony are embedded in the gelatinous matrix and are arranged in a single peripheral layer, with their flagella protruding through the outer mucilage coat that surrounds the sphere. The *Volvox* coenobium is thus a hollow sphere with the interior showing no cellular organization, and swims with a characteristic rolling motion.

Depending upon the species, the number of cells in the colony varies between 500 and 50,000. Each cell has its own gelatinous sheath or mucilaginous envelope, and is thus separated from its neighbours by a considerable expanse of the gelatinous material. This shows that *Volvox* is not an individual but an association of a number of independent cells. In the majority of species, each cell is connected to its neighbours by a series of broad protoplasmic strands (Fig. 3.16B) (a fine continuous, spherical network).

Reproduction

It is not only their **planetary movements** that make *Volvox* a definite candidate to be one of the seven wonders of the micro world; their interesting methods of reproduction are also really spectacular. *Volvox* (like other green algae) is haploid in all active phases of its life cycle. It has asexual as well as sexual reproduction. During the first half of the 20th century, all studies of both types of reproduction of this alga were carried out on material taken from natural populations (e.g. Powers, 1908). However, during the second half of the century, almost all investigations on this aspect were carried out in laboratory culture conditions (e.g. Starr, 1968, 1970). Only in the year 2000 did Desnitski publish his work on development and reproduction of two species of the genus *Volvox* in a shallow temporary pool in Russia. Further studies on natural populations of *Volvox* are still required.

All the cells of the young colony are green and vegetative in function. These are the somatic cells. In the adult colony, one of the first things that you notice in *Volvox* are the spheres inside. These are **daughter** colonies, called **gonidia**, which are the means of asexual reproduction. In each adult colony, a certain number of cells, especially in the posterior part of the colony, enlarge considerably and function either as asexual or sexual reproductive cells. These latter cells are recognizable by their large size, definite large nuclei and more dense granular protoplasmic contents. The adult *Volvox* colony exhibits a degree of polarity and differentiation of cells accompanied by division of labour. It is indicated by the position of the vegetative and reproductive cells. The anterior vegetative cells have larger stigma (eyespots) that guide the colony to light, while the reproductive cells, posteriorly, can reproduce. A coenobium has all its reproductive cells either entirely asexually when the season is favourable for growth, or sexually towards the end of the growing season.

Asexual reproduction

The daughter colony, or gonidium, grows from the reproductive cells around the equator of the colony. Fig. 3.17 shows how each special reproductive cell (A) gradually pushes back into the colony, withdraws its flagella, increases in size and becomes more or less round in form (B) (parthenogonidium or gonidium). Each undergoes about 11 successive longitudinal binary fissions, forming a daughter coenobium (autocolony) within the parent cell wall. The first five divisions are symmetrical, resulting in an embryo consisting of 32 cells of similar sizes. The first division of the cell is longitudinal and in the anterior–posterior plane in respect to the parent coenobium as a whole (C). The second is at right angles to the first (D). After that, each of the daughter cells divides again, forming a curved eight-celled plate (E). Then the 16-celled stage is arranged in a hemisphere (F), called **plakea**, in which the flagellated ends of the cells are directed towards the interior of the plakea. The fifth cleavage produces the 32 cells. The sixth division (G) is asymmetrical, since the 16 most anterior cells (one half of the 32 cells) typically cleave unequally to produce 16 cell pairs of unequal sizes (16 large gonidial, reproductive cells and 16 smaller somatic cells). The other half of the 32 cells cleave equally to produce additional somatic cells (Green and Kirk, 1982). The larger cells will become the new reproductive cells. Finally, at the end of the 11 cell divisions, the embryo is a hollow sphere with one layer of cells, whose flagella are directed inwards, and the sphere has a small aperture, or **phialopore**, which is directed to the exterior of the mother coenobium (F′). At the end of division stage, all the cells are in contact with each other. For the colony to be motile, the flagella must be directed to the exterior of the plakea. A fascinating process corrects this orientation in the life cycle of *Volvox*, called **inversion** (Kirk et al., 1982). In this process, the young daughter colony turns inside out through the phialopore to bring the flagella from its interior to its outer surface (H). The mechanism of this process (Fig. 3.17H and I), is a series of cell shape changes causing the multicellular sheet to bend outwards, and the bend propagates from the anterior to the posterior pole, eventually resulting in an inside-out spherical sheet of cells (Nishii and Ogihara, 1999).

Cleavage and inversion together take about 8 h, and the complete asexual cycle (Fig. 3.18) takes precisely 2 days, when it is synchronized by a suitable light–dark cycle (Kirk, 2004). About a day after this is done, the well-developed daughter colonies bulge into the hollow of the parent colony (Fig. 3.15), where they remain for some time before they are set free. Both the adult colony and the daughters within increase in size (without further cell division) by depositing large quantities of the matrix. These daughter colonies usually escape by rupturing of the parent wall, or they are enzymatically released from the parent and swim away. A number of asexual generations may follow most of the time, during the spring and summer and they would die in minutes once the pond dries up. *V. carteri*, for example, is able to survive by turning to sexual life cycle shortly before the pond dries up. Sexual reproduction occurs largely in autumn, producing dormant zygotes that survive the unfavourable weather.

Sexual reproduction

Sexual reproduction in the flagellates has rarely been observed, other than in the Volvocida, the symbiotic Hypermastigida and the Oxymonatida. The Volvocida display all gradations of sexual reproduction, from isogamy to highly developed heterogamy. In *Volvox*, it is of the advanced type, whereby true eggs and sperm develop from special reproductive cells. Therefore, the reproduction is **oogamous**. The initial inducing factor is not well understood; one hypothesis is that heat shock may be involved, i.e. gametic reproduction may occur when water temperature reaches a critical high level, particularly in shallow pools, temporary ponds or puddles. Kirk and Kirk (1986) experimented regarding this hypothesis, and discovered that the sexual cycle could be initiated by heating dishes of *V. carteri* to temperatures that might be expected in a shallow pond in late summer. The result was that the heat shock caused the somatic cells of the asexual *volvoxes* to produce the **sex-inducing pheromone**.

According to Graham and Wilcox (2006) in their excellent book *Algae*, the male colonies produce a chemical gametic attractant, or **pheromone**, a glycoprotein substance that diffuses through the water, inducing many other *volvox* colonies in the surrounding area to become gametic-competent. Hallmann et al. (1998) were interested in this subject and added that the level of pheromone produced by somatic cells is then further amplified by the ability of sperm cells to produce more sex-inducing pheromones. It has been established that the first biochemical responses to the sex-inducing pheromone are structural modifications within the extracellular matrix, which appears to play a key role in this unique signalling system (for details, see Hallmann et al., 1998; Hallmann, 2003).

Such pheromones can be effective at very low concentrations. Hallmann et al. (1998) reported that sexual reproduction in *V. carteri* is initiated by a glycoprotein pheromone, a **32-kDa glycoprotein** that acts at a concentration below 10^{-16} M. With such a view, a single sexual male releases enough pheromone to convert all the population of the entire pond to sexuality. This discovery explained a somewhat ancient observation made by Powers (1908) that 'in the full blaze of Nebraska sunshine, *Volvox* is able to appear, multiply, and riot in sexual reproduction in pools of rainwater of scarcely a fortnight's duration'.

Volvox, e.g. *V. carteri* (Fig. 3.19) reproduces asexually most of the time, and in nature, it reproduces sexually once each year. As in asexual reproduction, reproductive cells around the equator are specialized to develop into germ cells. Most species have male and female colonies. However, in some species the colonies are hermaphroditic, and are thus able to form sperm as well as ova, although not at the same time. The reproductive cells that will develop into male gamete (microgamete) are called **antheridia** (Fig. 3.16E). The antheridia undergo repeated fission to produce rounded or oval-shaped **sperm packets**, which are sort of loose aggregates (Mann, 1984). It is astonishing that such a lower organism has a device to preserve and protect its sperm for a longer time as in the case of spermatophores of metazoa. The sperm packet (Fig. 3.16) is a flat packet of 16, 64, or 128 considerably modified sperm. Each is slender, pale and bears a pair of flagella in the middle of its length (Fig. 3.16F). Upon the liberation of sperm packets from the parental colonies, they swim away as a unit, to a receptive colony, in search of eggs (Fig. 3.20).

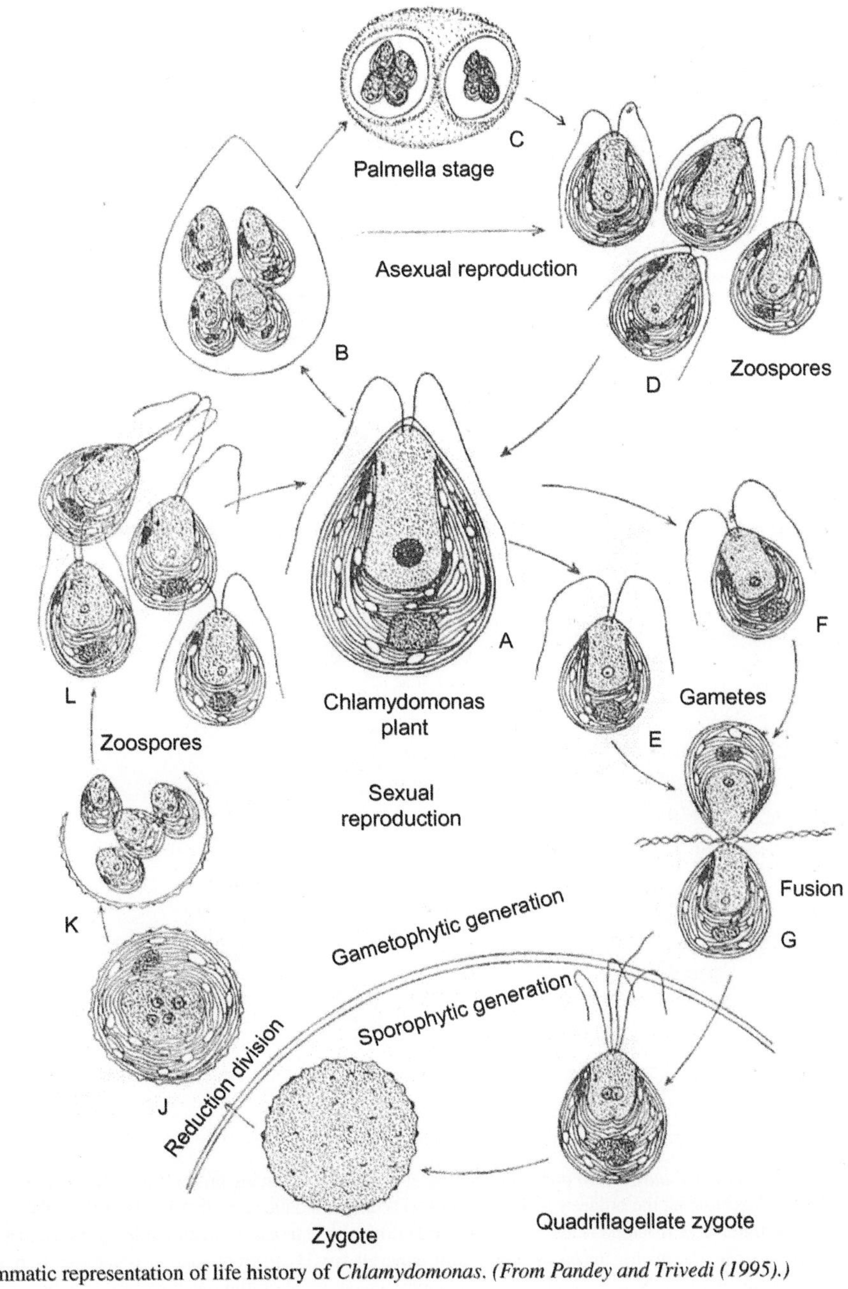

FIG. 3.15 Diagrammatic representation of life history of *Chlamydomonas*. *(From Pandey and Trivedi (1995).)*

On the other hand, eggs, or macrogametes, develop from the specialized reproductive cells that have not undergone division. The few macrogametes (Fig. 3.16) are large and loaded with food for nourishment of the young colony. When these macrogametes are mature, the sperm packet enzymatically lyses a hole in the colony mucilage, and dissociates into individual sperm, which swim slowly through the gelatinous oogonial wall. The sperm enters the egg probably through the beak-like receptive spot and fuses with the egg nucleus to form the zygote. The fertilization (Fig. 3.20) is thus internal, where the ovum is fertilized without becoming detached from the parent coenobium. The eggs that have not been fertilized might develop parthenogenetically, thus forming daughter colonies.

Soon after fertilization, the zygote secretes a thick, three-layered, smooth or spiny **protective** outer wall, thus becoming an **oospore** (Fig. 3.20). This functions as a '**resting**' stage. Meiosis occurs during this resting stage and the zygote produces four spores. The meiosis in volvocids is therefore **postzygotic** rather than prezygotic. The thick-walled oospores remain within the parent coenobium for some time and develop sufficient haematochrome or red pigment to colour their protoplast orange red. Eventually,

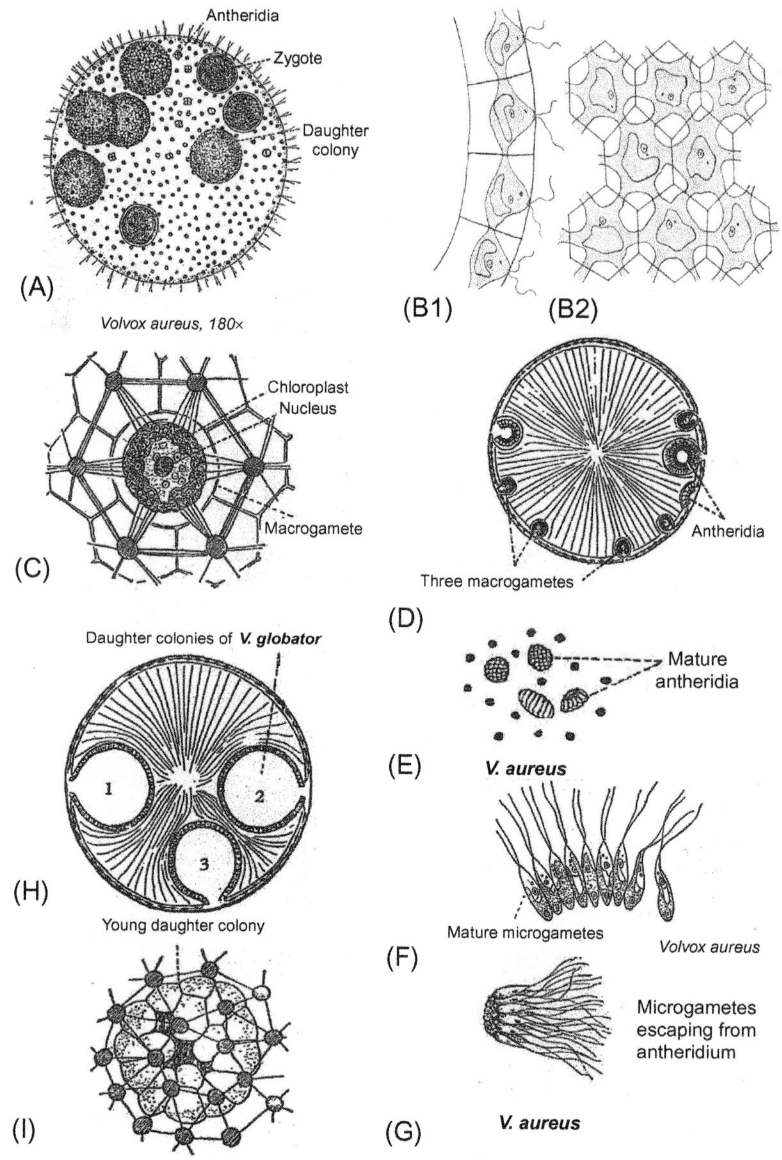

FIG. 3.16 Structure of *Volvox*: (A) whole mount of *V. aureus*; (B) portion of colony of *V. globator* showing protoplasmic connections: (1) side view, (2) surface view; (C) a single macrogamete of *V. aureus* among the ordinary somatic zooids; (D) diagrammatic section through colony of *V. globator* showing three antheridia in different stages of maturity and three large macrogametes; (E) mature antheridia of *V. aureus*; (F) mature microgametes of *V. aureus* just liberated from an antheridium and now beginning to separate; (G) microgametes escaping from antheridium; (H) diagrammatic section through an old colony of *V. globator* showing three large daughter-colonies projecting into the interior of the parent colony which is full of thin mucilage with a radiating structure; (I) a young daughter-colony of small size as seen through the layer of zooids of the parent colony. *(All from Borradaile et al. (1977) except (B) from Hickman (1973).)*

the parent colony decays and the oospores fall to the bottom of the pool, where they survive the winter (Fig. 3.21). Upon the restoration of favourable circumstances, the oospores germinate and each meiospore either develops to form a new small colony, or develops as a biflagellate cell that later divides into a small colony.

In fact, the morphological appearance of the oospores is a specific feature. For example, *Volvox aureus* has oospores with a smooth surface, while those of *Volvox globator* are star-like (Fig. 3.21).

Wonderful features of *Volvox*

Study of *Volvox* shows that, as simple as it is, it shares important scientific features that characterize the life cycles and developmental histories of more complex metazoans (Fig. 3.22). It is therefore considered **one of the seven wonders** of the world of the microorganisms. One of these features is cell differentiation and/or specification, which lead to the **division of labour**. Another feature is the interesting methods of reproduction, which are quite spectacular. *Volvox*, in

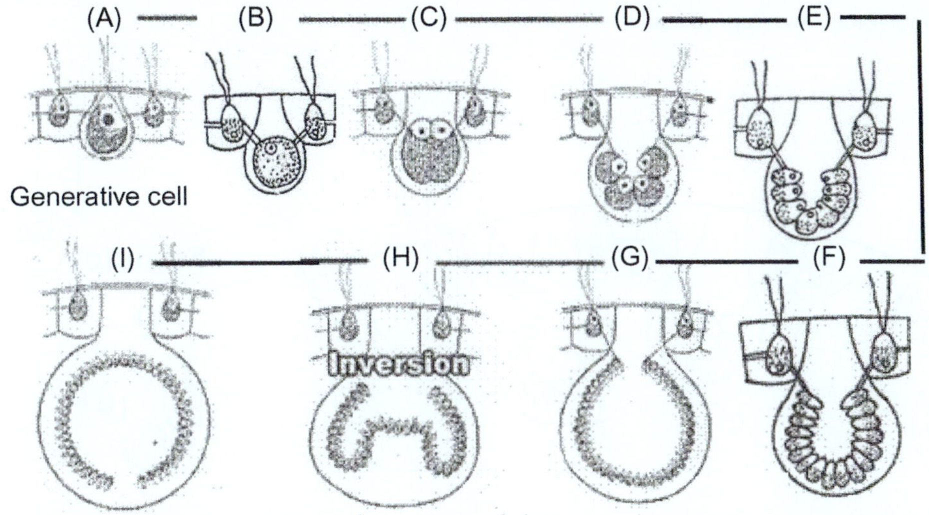

FIG. 3.17 Stages of asexual reproduction in *Volvox*: (A) generative cell; (B) gonidium; (C–G) cleavage steps to develop the gonidium into a daughter colony; (G) stage before inversion; (H) colony during inversion; (I) daughter colony after inversion and the phialopore still open, it will close later. *(Adapted and modified from Vashishta (2004).)*

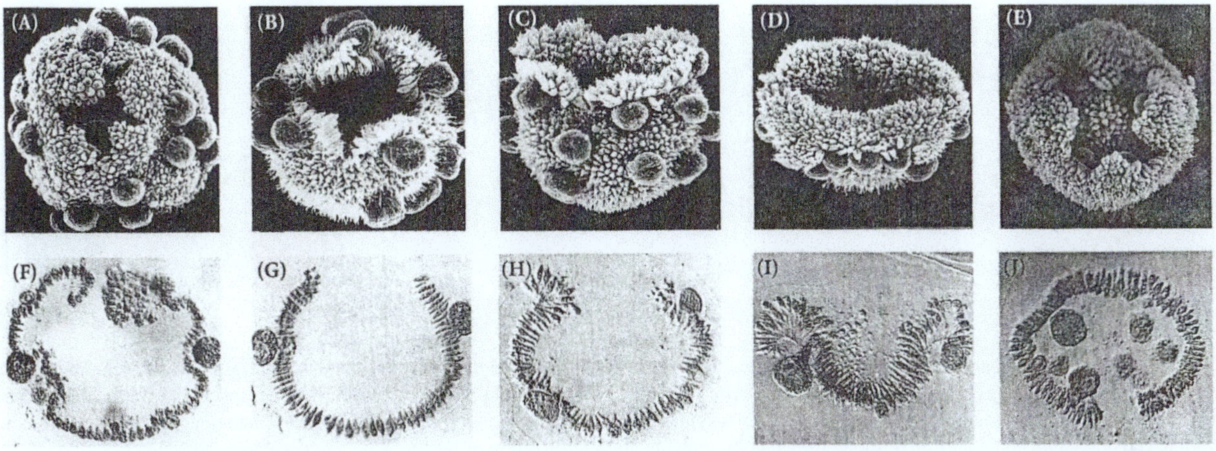

FIG. 3.18 Inversion of embryos of *V. carteri*, (A–E) scanning electron micrographs of whole mounts of embryos. (F–J) Sagittal sections through the centre of the embryos, visualized by differential interference microscope. *(After Kirk et al. (1982), from Gilbert (2000).)*

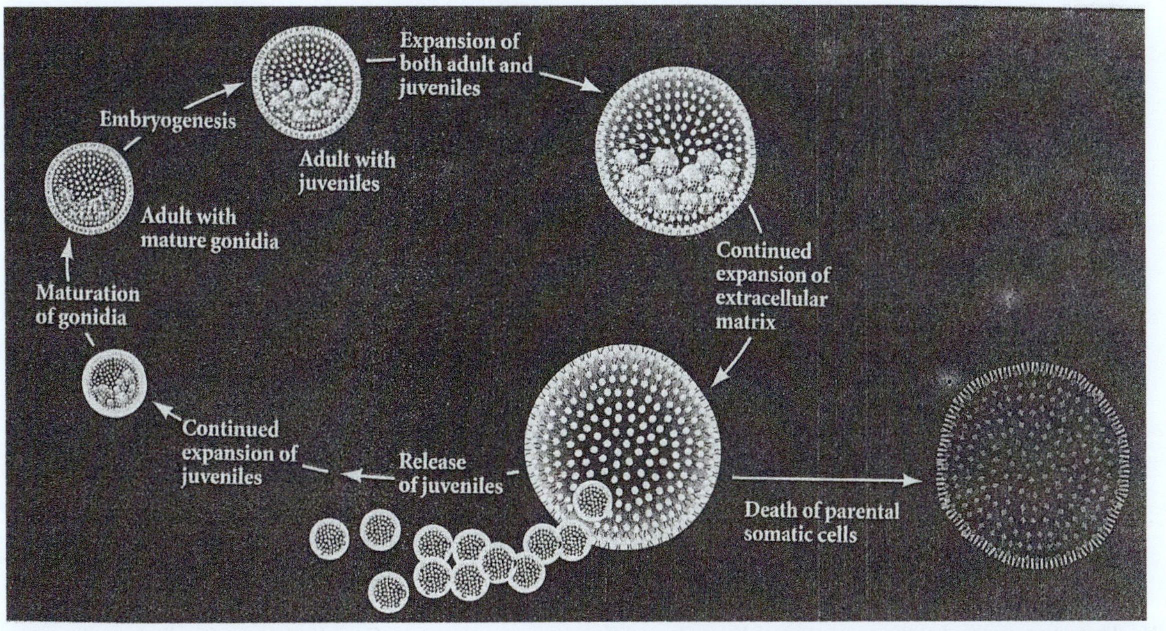

FIG. 3.19 Asexual reproductive cycle in *Volvox carteri*. The somatic cells of the parent lack gonidia, undergo senescence and die, while the juvenile colonies mature. The entire asexual cycle takes 2 days. *(After Kirk (1988), from Gilbert (2000).)*

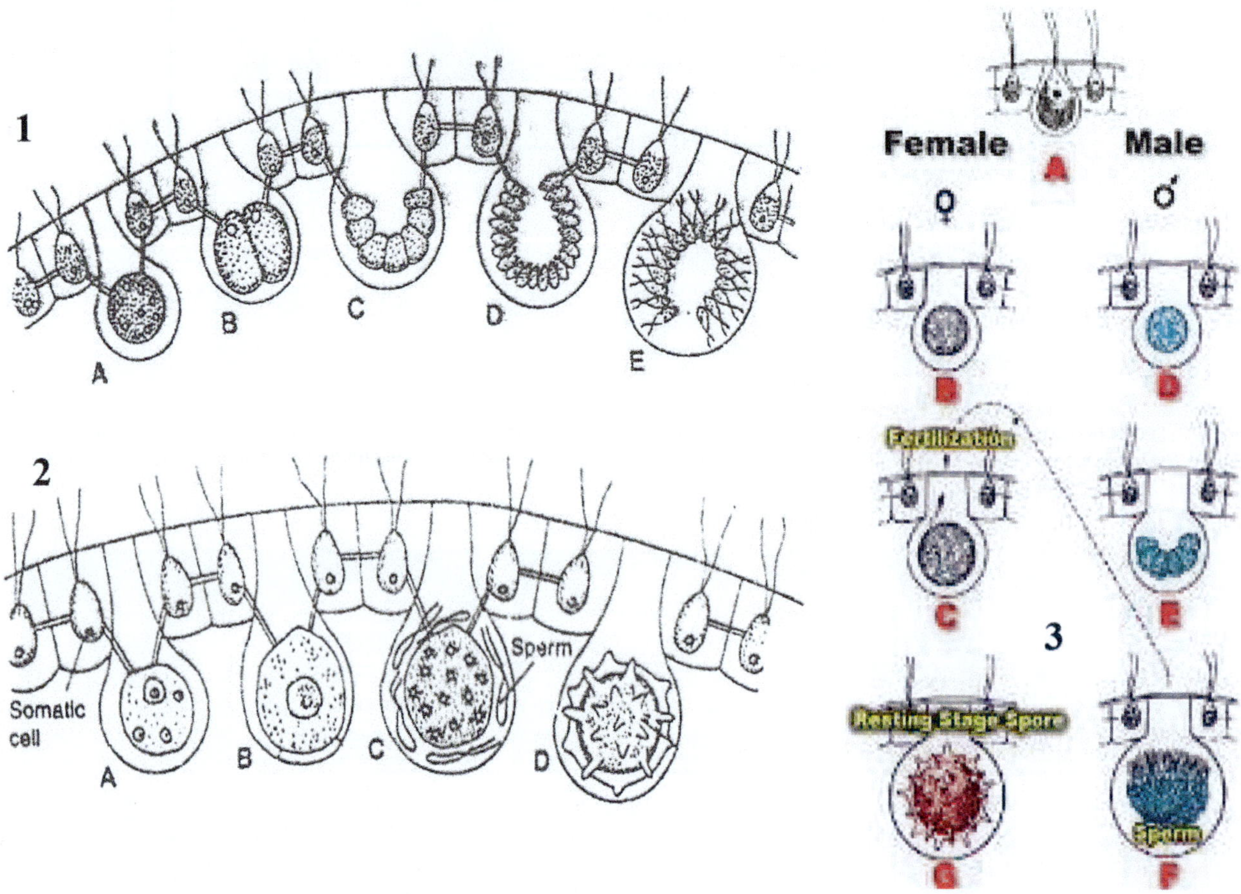

FIG. 3.20 Formation of gametes and fertilization in *Volvox*. (1) Sectional view of a portion of the colony showing stages of the development of sperm: (A) antheridium; (B and C) stages in the development of sperm; (D) antherozoidal bundle before inversion; (E) after inversion. (2) Sectional view of a portion of a colony showing development of oogonium: (A) oogonium with an immature egg; (B) oogonium with mature egg; (C) oogonium showing fertilization; (D) oogonium containing an oospore. (3) Diagram showing cross fertilization in *Volvox*. *Modified from different sources.*

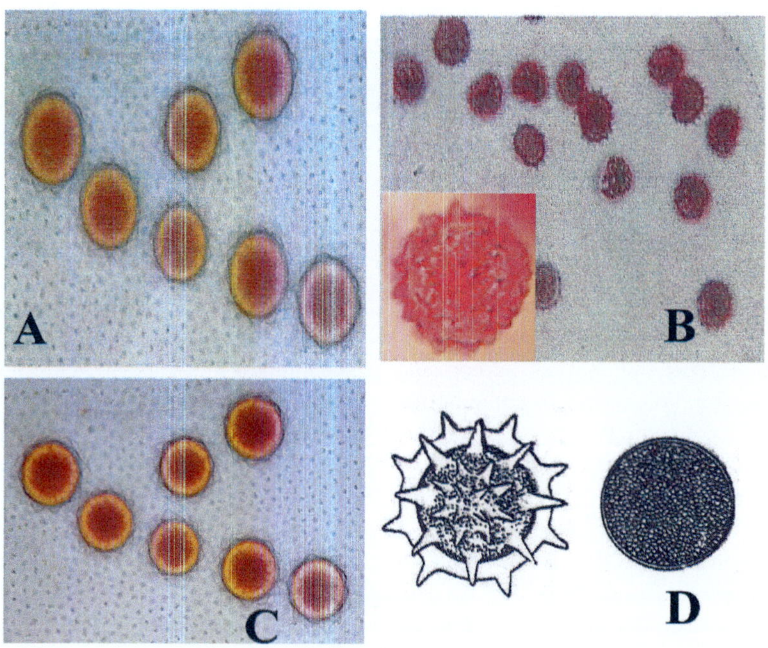

FIG. 3.21 Photomicrographs showing zygospores of *Volvox* from different sources. However, (C) of *V. globator*. (D) diagrams illustrating smooth-surfaced zygospore of *V. aureus* and spiny-surfaced zygospore.

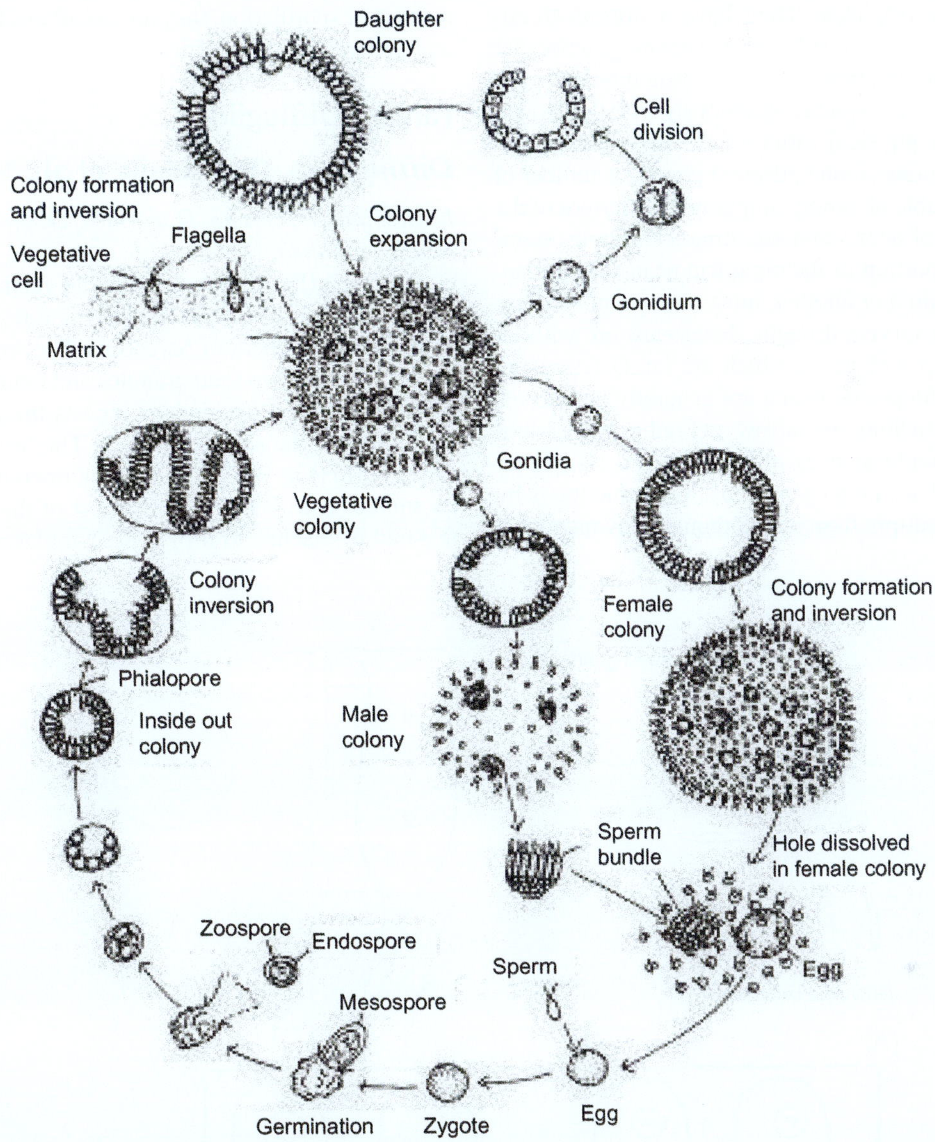

FIG. 3.22 The life history of *V. carteri*. *(Adapted from Smith (1944), from Lee (2008).)*

spite of its colonial nature, is still a simple organism, but produces true sperm and eggs. Moreover, it produces its sperm in protective aggregates to increase their lifespan. This is an alternative to e.g. spermatophores of advanced metazoans. Its development adds more features: it undergoes cleavage-like stage; moreover, during this process, segregation of germ cell line from somatic cell line occurs (Green and Kirk, 1982). The same phenomenon occurs in some nematodes. In addition, there is the fantastic process of inversion, which is a series of cell movements, resembling gastrulation, which occurs in some poriferans. The parent colony that has produced an offspring, and is incapable of further reproduction, as it lacks gonidia, undergoes senescence and its somatic cells die. Technically, these cells commit suicide, by synthesizing a set of proteins that cause the death (Kirk, 1988). This means that *Volvox*, in spite of its simplicity, includes death as a regular, genetically regulated part of its life history.

CLASS 2: SARCODINA (AMOEBAE) (GREEK *SARCOS*, 'FLESHY')

Subclass 1:	Rhizopoda (Greek *rhiza*, 'root' and *pous*, 'foot')
Order:	Foraminifera (Latin *foramen*, 'opening' and *ferre*, 'to bear')
Suborder:	Monothalamia (uniocular)

Sarcodin protozoans are amoeboid in the principal phase; they possess a test or shell which typically envelops all parts of the body except the pseudopodia and that part of the body from which these originate. They have a noncalcareous shell, a single-chambered shell with one opening; uniocular shells are of simple spherical, oval, pyriform, tubular, spiral or branched shape, and expand continuously with the growth of the animal. The physical nature of the shell and its embedded particles varies among different genera; a number of testaceans are capable of simply plugging the aperture of the shell with a mass of debris and secreting a chitinous membrane across that portion of the organism nearest to the aperture. By one method or another, most testate amoebae are effectively able to survive drought. Testaceans are without flagella, but with pseudopodia which are rarely reticulate. They range from filopodia, which are generally numerous, thread-like and branching, to narrow and rather blunt lobopodia, whose protoplasm does not extend as a layer over the shell and who have no meganucleus. Reproduction is by both binary and multiple fission, and though they may have a phase of sporulation, they are usually of freshwater habitat and generally not parasitic.

Family: Difflugiidae

Difflugia sp. (Borradaile et al., 1977, p. 78)

D. areolata, D. corona, D. garmen, D. globulus, D. lanceolata, D. oblonga, D. urceolata

Difflugia (Fig. 3.23A) is a familiar genus, which measures from 15 μm up to 500 μm long. It is a testate amoeba, one of the several genera of Sarcodina that produce egg-shaped **shells or tests** from sand granules and diatom frustules, over an organic base. The amoeba ingests the granules and embeds them in the secreted matrix. The test is usually ovoid or pyriform (Fig. 3.23B) and has an opening, the **pylome**, at its smaller end. The body fills most of the space within the test and is attached to it by long cytoplasmic strands. It has

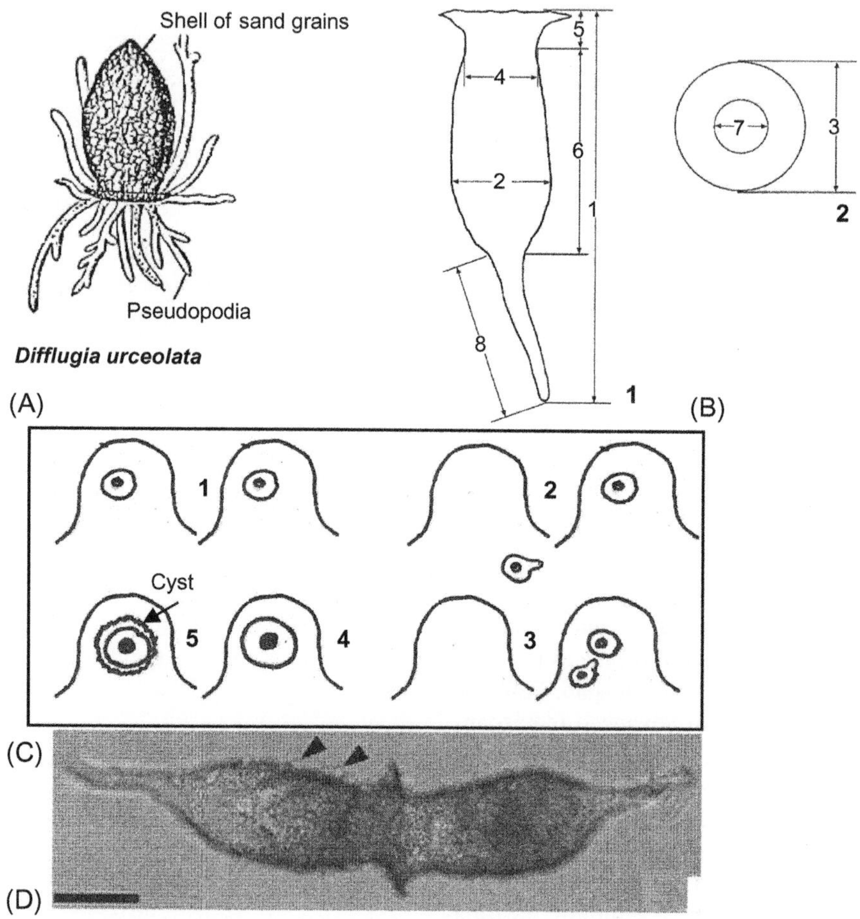

FIG. 3.23 *Difflugia*. (A) Whole mount of the organism *from Borradaile et al. (1977)*. (B) Shell outline and position of measured axes: (1) total length; (2) body width; (3) collar diameter; (4) neck width; (5) collar height; (6) body length; (7) aperture diameter; (8) aboral horn length. (C) Diagrammatic description of steps of amphimixis in *Difflugia* as written in Gardiner (1972): (1) two individuals may come together; (2) one leaves its shell moving towards the other, (3) the two individuals are inside the other's shell; (4) the two fuse; (5) the zygote encysts which withstands the conditions. (D) shows conjugation (shells of two organisms of *D. biwae* Kawamura, 1918 joined at their oral apertures) and some algae attached to the surface of the shell (fluorescence microscopy with DAPI stain) scale bar 50 μm. *(B and D) From Yang and Shen (2005).*

lobose pseudopodia, one or two nuclei, and a chromidium (this consists of numerous granules which, like chromatin, take basic stains). The nucleus is mostly ovular, and in some species vesicular. Contractile and food vacuoles are present in the cytoplasm. *Difflugia* can produce gas bubbles and utilize these to lift itself above the substratum. The taxonomy of this genus is based mainly on differences in the shape and size of their shells. As the shell is often opaque, cytoplasmic characters are rarely used. The small differences in shell size, shape or composition have been sufficient for many authors to describe more than 300 species and about 200 subspecies. In more recent works, other characters were considered, such as the appearance of the surface of the shell, and whether it is smooth. Some morphometric characters were also used, as that described in *Difflugia tuberspinifera* (Hu et al., 1997). Yang and Shen (2005) studied a Chinese species, *D. biwae* Kawamura 1918, and stated eight morphometric characters to portray and describe the shell of this species. These are illustrated in Fig. 3.23B.

Habitat

Difflugia, as other testate amoebae, appears to be restricted to freshwater and is not often found except where the water is acid. *Difflugia* is particularly common in marshes and generally common on the surface of sediments in lakes and ponds, and often occurs in aquarium sediments. It is heterotrophic and it engulfs its food.

Reproduction

Asexual reproduction

Difflugia reproduces by binary fission. The test of the parent individual is retained by one daughter, and a new test is built up around the other daughter; alternatively, the latter receives the necessary particles with which it forms a new test from the parental test.

Sexual reproduction

Sexual reproduction is common among all Sarcodina. In *Difflugia* (a shelled rhizopod), for instance, two individuals come together and attach by their openings and one leaves its shell, move bodily into that of the other, fusing with its partner to form a cyst-like zygote. Eventually, the zygote becomes active and a new individual emerges from the cyst. This is **amphimixis** (Fig. 3.23C), a type of reproduction that occurs among protozoans, in nearly all species, at some stage in the life cycle of an individual or of a population (Gardiner, 1972).

Reproduction in foraminifera is rather complicated and involves a definite alternation of generation (dimorphic). Thus an asexually produced uninucleate haploid generation (gamont), with a large **megalospheric** proloculus (chamber), alternates a sexually produced multinucleate diploid generation (agamont or schizont) with a much smaller **microspheric** proloculus, although often possessing a larger adult test.

The microspheric forms reproduce by multiple fission, and the products develop as megalospheric individuals (Fig. 3.24). This occurs by either a rapid series of divisions, after which, many small flagellate cells are set free, and as gametes, unite to form diploid zygotes and a new generation of microspheric forms, or the formation of fertilization cysts. In this latter event, characteristic of some species, megalospheric individuals come together in a common cyst and form gametes that unite within the cyst, from which, after varying periods of time depending largely on external conditions, a new microspheric individual emerges. There is, thus, in such dimorphic species, a regular alternation of generations of microspheric forms, reproducing asexually through multiple fission, and of megalospheric forms, reproducing sexually either by means of free, motile gametes, or by the formation of fertilization cysts, amphimixis.

Subclass 2:	Actinopoda (Greek *aktinos,* 'ray')
Order:	Heliozoa (Greek *helios,* 'sun') Sun animalcules

Habitat

Sarcodin actinopods occupy various aquatic freshwater habitats, such as pools, ponds, marshes, lakes, etc. They are generally of floating habit.

Morphology

These protozoans are without a shell or central capsule; sometimes with a siliceous skeleton; with spherical bodies, typical **axpodia** (pseudopodia, numerous and slender radiating from central spherical cytosome, so heliozoans are readily distinguishable from other amoeboid organisms); usually a highly vacuolated outer layer of protoplasm; contractile vacuoles are present; nutrition is heterotrophic.

Family: Actinophyridae (Hartmann, 1913)

Actinophrys

A. sol, A. vesiculata, A. tauryanini, A. eichhorni

Morphology

A well-known heliozoan is the **sun animalcule,** *Actinophrys sol* Ehrenberg, 1830 (Fig. 3.25A). This organism looks like a minute sun under the microscope, especially when using the phase contrast technique. Its body is nearly spherical, about 50 µm, and varies from 30 to 70 µm in diameter. It is unprotected, and has no internal skeleton, cilia or flagella. It has a large central nucleus, and the ectoplasm contains a number of conspicuous clear vacuoles, some of which are contractile. Small vacuoles are scattered throughout the endoplasm. Each of the pseudopodia, of axopodia type, has a rather stiff axial filament, which can be traced through the cytoplasm

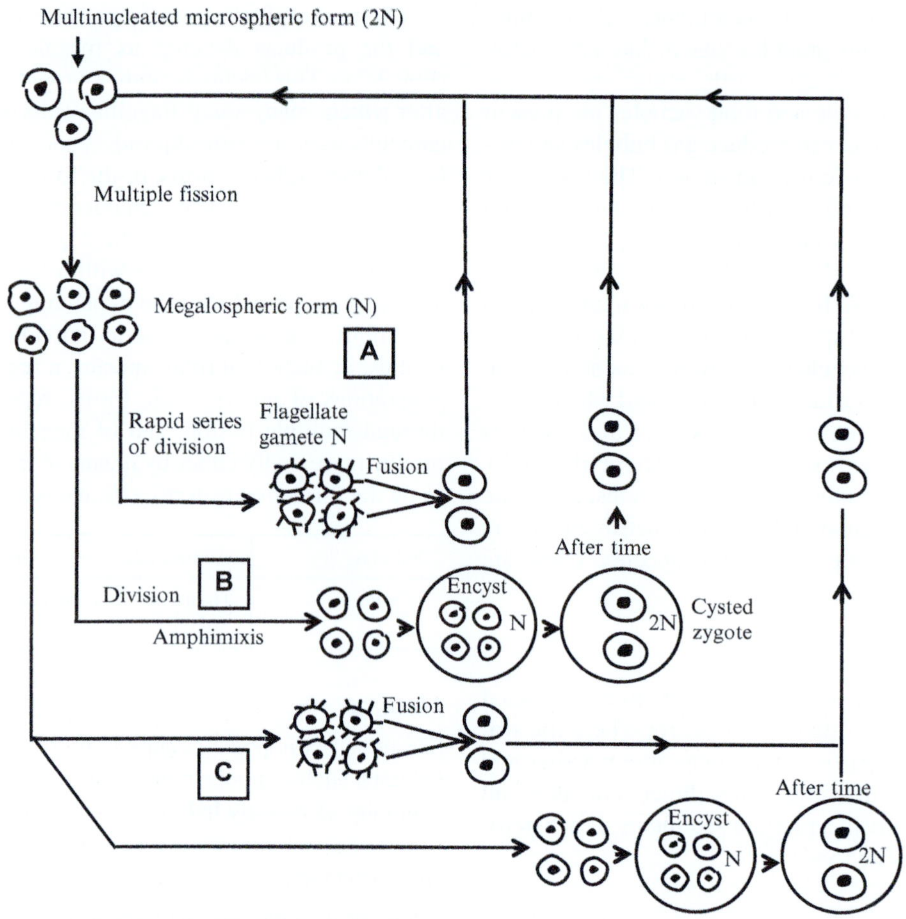

FIG. 3.24 Diagrammatic description of modes of reproduction of dimorphic foraminiferan, showing definite alternation of generations. The megalospheric form proceed in one of three modes: forming free gamete, every two fuse to give the other generation; **or** the megalospheric forms encyst and each two fuse inside the cyst withstanding the conditions, then give rise to the second generation; **the third**, the megalospheric forms, according to the weather, behaves as the first or as the second modes. According to what is written in Gardiner (1972).

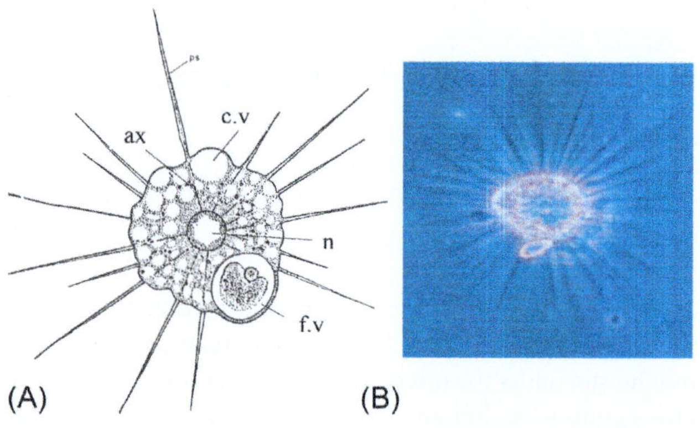

FIG. 3.25 The heliozoon *Actinophrys*; (A) whole mount of the organism about 50 µm, ax = axopodium; cv = contractile vacuole; fv = food vacuole; n = nucleus; ps = pseudopodium. *After Minchin from Marshall and Williams (1975)*; (B) a small ciliate was captured by *Actinophrys*. Size about 50 µm in diameter. (Phase contrast microscopy; objective 10×.) *From Chantangsi (2002).*

as far as the nucleus. Electron microscopy has shown that the axial filament consists of a bundle of tubular fibrils. Its axopodia have a number of indistinct extrusomes, which flow along its arms; these are special organelles for capturing or killing prey and for protection from its predators.

Reproduction

Like other sarcodines, or amoebae, the major reproduction in *Actinophrys* takes place **asexually** by binary fission, a division into two similarly equal daughters (Fig. 3.26). The heliozoan *Actinophrys* undergoes a different sexual reproduction called **autogamy** (nuclei divide into two and then fuse again). Gardiner (1972) reported that this heliozoan reproduces by both endomixis and amphimixis. **Endomixis**, or autogamy, is a process in which gametes or gamete nuclei, arising in a single individual, undergo fusion and, hence, through the antecedent meiosis, provide new combinations of

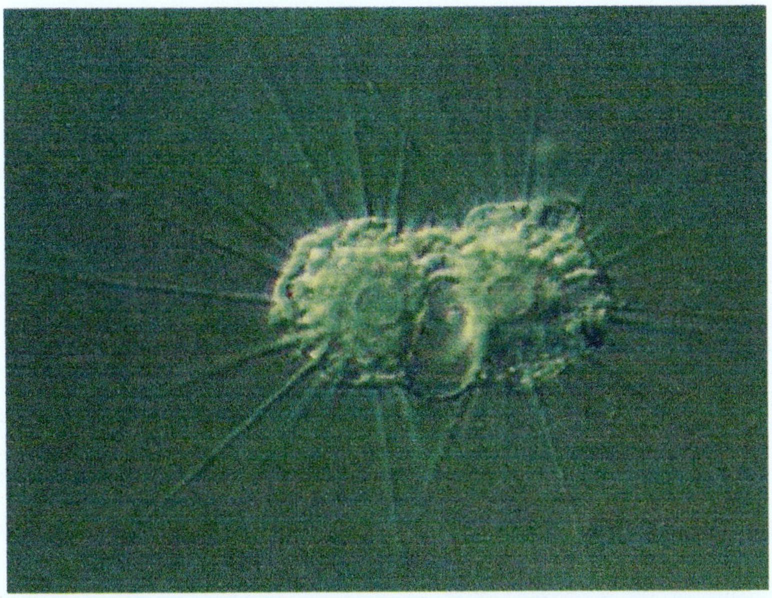

FIG. 3.26 *Actinophrys* undergoing binary fission and showing central nuclei, a large contractile vacuole, and some extrusomes, small convex objects within radiating pseudopods. Each cell is ~45 µm. (Taken by D.I.C.; objective 20×.) *(From Chantangsi, 2002.)*

genetic material. This is perhaps a parallel to self-fertilization in hermaphroditic metazoans. *Actinophrys* undergoes endomixis {autogamy or more correctly paedogamy (Borradaile et al., 1977)}. Paedogamy means production of premature individuals (Borradaile et al., 1977) and/or reproduction at a larval or early developmental stage (Gardiner, 1972). Storer et al. (1983) reported that parthenogenesis among larvae is termed paedogenesis such as sporocyst of the liver flukes, which produces stage that gives rise to radia. This process takes place as follows (Fig. 3.27): a single individual first retracts its axopodia and forms a cyst, within which it divides by binary fission. Thus, two diploid individuals called **gamonts** lie side by side in the cyst. The nucleus of each daughter cell, or gamont, passes through two typical meiotic divisions. At each, one of the resulting nuclei is expelled from the cell as a polar body; that is, it becomes pycnotic and degenerated, but the cell itself does not divide. This leaves one reduced nucleus in each cell, which becomes a gamete; the two morphologically identical isogametes unite, one behaving as a male by sending out a pseudopodium towards the other, and a diploid zygote results. A strong inner cyst is formed around the zygote within which it may remain dormant for some time. Ultimately, when the environment is suitable, the zygote divides by binary fission and the two products escape from the cyst as active individuals. This mode of reproduction, or autogamy, that occurs within the group is really more a process of genetic reorganization than reproduction.

In **amphimixis** (Fig. 3.28), occasionally two different individuals of *Actinophrys* encyst together and each divides by binary fission. There are four daughters within the cyst. The nuclei undergo the same series of divisions and expulsion of products as in endomixis. There are thus four haploid cells within the cyst. The cells derived from one individual may unite with a cell derived from the other, so that two zygotes are formed, or only two cells may be engaged in amphimixis and only one zygote is formed. In this case, the remaining cells develop **parthenogenetically** into new individuals (Gardiner, 1972). This specific mode of sexual reproduction in *Actinophrys* is referred to as conjugation in view of Borradaile et al. (1977).

CLASS 3: SPOROZOA (GREEK *SPORA*, 'SEED')

Subclass:	Telosporidea (Greek *telos*, 'end') Endoparasitic (inter- or intracellular) transmitted from host to host by encysted juvenile forms (sporocysts) or by naked ones (sporozooites)
Order:	Coccidiomorph (Greek *kokkos*, 'berry')
Suborder:	Coccidia gut parasites, zygote nonlocomotory
Example:	*Eimeria* spp.
Order:	Gregarinidea (Latin *grex*, 'hard')
Suborder:	Eugregarinaria No schizogony
Example:	*Monocystis* spp.

The adult of the vegetative stage of Sporozoa has only one nucleus, and it comes to one end with spore formation.

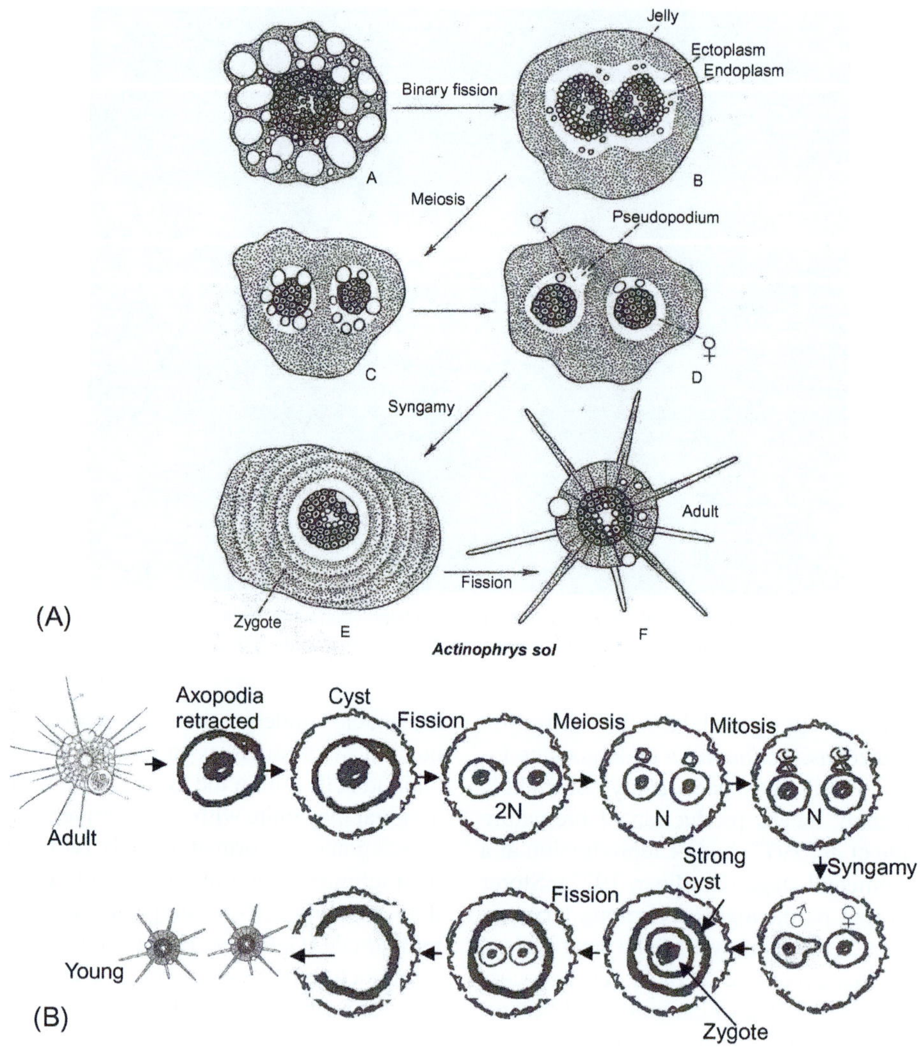

FIG. 3.27 (A) Successive stages in the autogamy (endomixis) of *Actinophrys sol, from Borradaile et al. (1977).* (B) Diagrammatic description of the process of endomixis as written in Gardiner (1972).

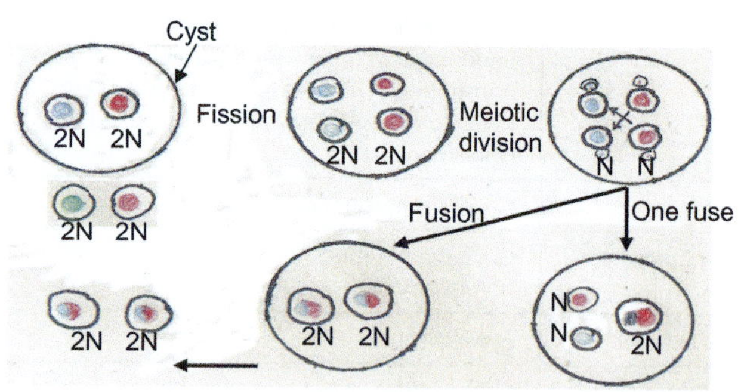

FIG. 3.28 Diagrammatic description of the process of **amphimixis** done by two different individuals of *Actinophrys* as written in Gardiner (1972).

The spores, if present, are simple in structures, and always contain several sporozoites. The vegetative stage, or **trophozoite**, usually has a definite shape. Its fission is multiple, and is usually known as schizogony.

Coccidiomorpha is usually a gut parasite. The zygote is nonlocomotory; the sporozoites are always encased; and the gamonts often form a syzygy. *Eimeria* is a commercially important sporozoite that lives as a parasite in the intestinal epithelium of various vertebrates and invertebrates.

Example: *Eimeria* spp.

Eimeria is parasitic in the intestinal epithelium of various invertebrates and vertebrates. *Eimeria schubbergi,* from the intestine of the centipede *Lithobius*, may be described as a type of the suborder. The spherical trophozoite (agamont) undergoes schizogony (agamogony) by multiple fission, within the epithelial cell that it inhabits. The spindle-shaped schizozoites (agametes) are set free into the cavity of the organ (intestine) (Fig. 3.29a (B)–(D)). Each infects another cell in which it grows similar to its parent (agametic

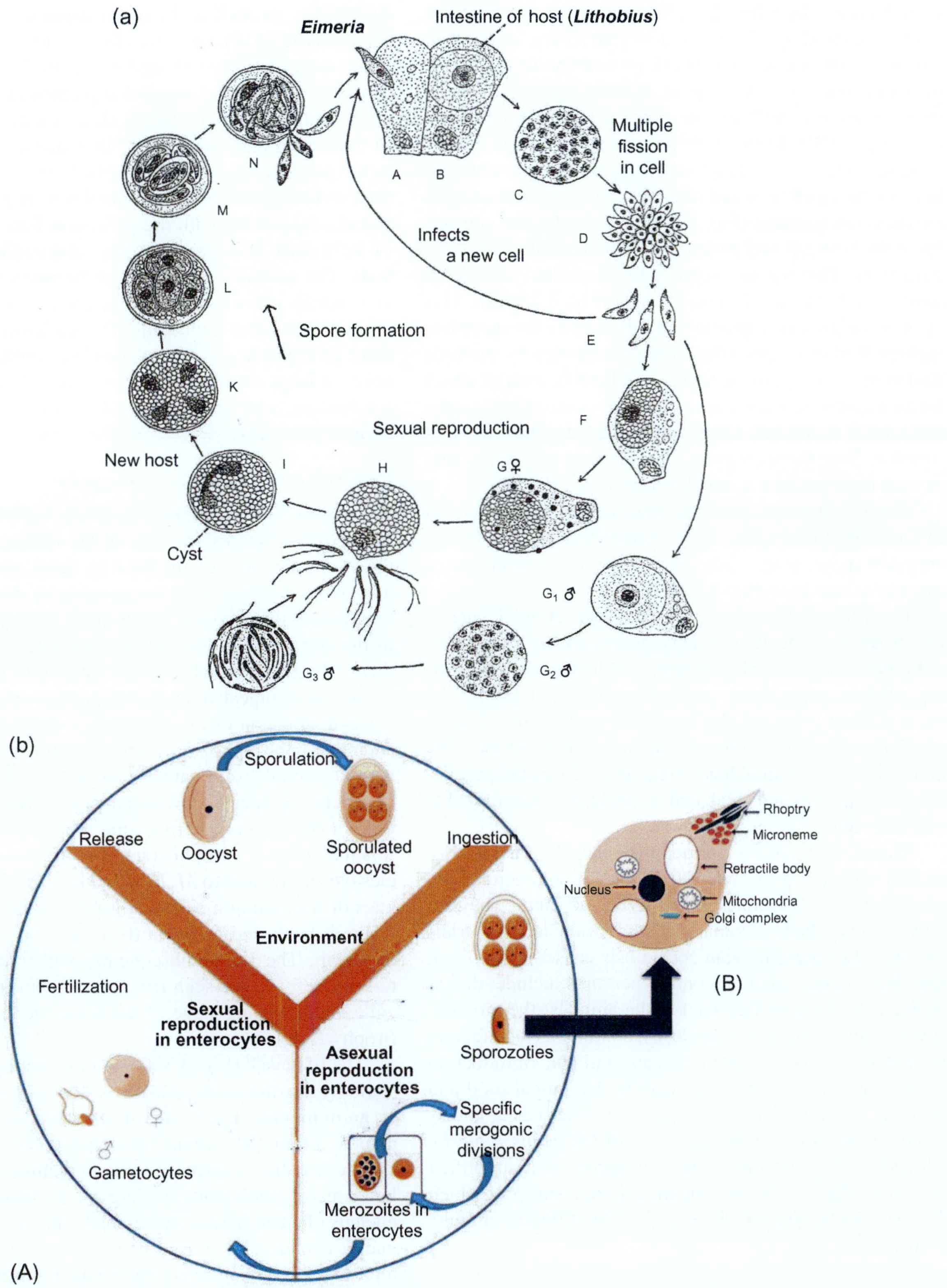

FIG. 3.29 (a) A diagram of the life cycle of *Eimeria schubergi (from Borradaile et al., 1977)*. (b) (A) Simplified life cycle of genus *Eimeria*. (B) A detailed structure of the sporozoite.

reproduction). In a few days, fissions occur in which the young, on invading a host cell grow into adults, unlike their parents, of two kinds, male (E, G1♂) and female (E, F, G♀) gamonts. Each female gamont extrudes stainable matter from its nucleus and thus becomes a female hologamete (Fig. 3.29a (H)). In the male gamont, the nucleus divides several times, and the daughter nuclei are set free with portions of the cytoplasm as biflagellate male gametes, which are thus merogametes (Fig. 3.29a G2♂, G3♂). The gametes leave the host cell and unite (fertilization) while free in the gut cavity. The zygote encysts and its nucleus undergoes what is probably a reduction division (Fig. 3.29a (I)). This cycle is the gametic reproduction followed by the spore formation. Within its cyst (the oocyst), it divides by multiple fission into four sporoblasts (Fig. 3.29a (K)), each of which forms a cyst of its own (Fig. 3.29a (M)) (a secondary sporocyst). Each sporoblast within the cyst divides into two sporozoites. Thus the sporogony takes place in two stages, and in each there is some residual protoplasm.

The oocyst passes out in the faeces of the host. Infection of a new host takes place by contamination of food by the encysted spores (Fig. 3.29a (M)), which hatch in the intestine of the new host (Fig. 3.29a (N)).

The different *Eimeria* spp. are the main etiologic agents for the disease Coccidiosis. This disease affects a variety of wild and domesticated vertebrates including chicken, turkey, rabbits, cattle, sheep, goats, pigs, fish and reptiles. The most virulent economically important types of that disease are those infecting poultry and mammalian livestock. This naturally affects animal husbandry, and reduces the production of milk, eggs and meat, and contributes to a remarkable mortality rate.

Ahmad et al. (2016) reported that since the early 1950s, several attempts were made to formulate commercial immunotherapies, but none proved to be sufficient. They reviewed, classified, summarized and evaluated the trials performed to prevent avian coccidiosis caused by *Eimeria* spp. The routine infection-control measures include: the introduction of oocyst-free feed to the animals; the biosecurity precautions to access the farms; continuous disinfection by potent anti-oocyst preparations; and the introduction of prophylactic anticoccidial drugs in the animal food and drinking water. Each of these measures needs a clear understanding and study of the life cycle of the genus *Eimeria*. Ahmad et al. (2016) presented a diagram of a simplified life cycle (Fig. 3.29b) showing the different stages at which *Eimeria* can be controlled, or targeted by different immune preparations.

Monocystis

M. magna, M. lumbrici (M. agilis)

The various species of *Monocystis* are parasites in the seminal vesicles of the common earthworms of the genus *Lumbricus,* as well as in other oligochaetes. The mature **trophozoite** of *Monocystis* (Fig. 3.30B) is characteristically elongated, cigar-shaped but worm-like, ranging from 10 μm to 16 mm, and somewhat pointed at both ends. It is extracellular, and it performs slow but strong movements of expansion and contraction, in a special manner known as the gregarine movement. The body is covered with a distinct striated pellicle. The outer layer of protoplasm (the ectoplasm) just beneath the pellicle is firm and rather free of inclusions. It is provided with contractile myoneme fibrils. The central protoplasm of the parasite is very fluid and rich in granules of glycogen-like carbohydrate, lipid droplets and other inclusions. The nucleus lies in the more fluid protoplasm, its boundaries are definite and it possesses a large endosome. There are no obvious organelles for feeding, and nutrition involves absorption of soluble organic materials.

Reproduction and/or life cycle

The **sporocysts** or **zygocysts** (often unfortunately incorrectly called 'spores') escape or are released, either down the vasa deferentia of the host by death and decay of the host, or by the host itself being eaten by a bird whose faeces contain them intact. When these zygocysts are present in the soil, the earthworms become infected by ingesting them. Presumably, when these zygocysts are swallowed, the action of digestive enzymes, perhaps coupled with mechanical abrasion of the zygocysts, causes them to rupture and release the sporozoites. The latter traverse the intestinal wall to reach the seminal vesicles and penetrate the clusters of developing sperm, in which they pass their earlier stages. Once a sporozoite penetrates the morula of an earthworm, it is then termed a trophozoite. The trophozoite of the species closely related to *M. lumbrici* lives in the seminal vesicles of the common earthworm *Allolobophora caliginosa*, where it feeds and grows at the expense of the protoplasm of sperm. The latter subsequently wither until only their tails are left and are seen fringing the parasite. The trophozoite is the feeding stage of the life cycle, hence the name (troph = food).

The life cycle (Fig. 3.30C), with no merogony involved, starts by pairing of trophozoites, now known as **gamonts** or **gametocytes**. Each pair of associated gamonts, which assume a rounded outline, secretes a common two-layered cyst wall called a **gamontocyst** or **gamocyst**. Within the latter, each trophozoite undergoes **multiple fission** (the nucleus of each divides successively into a large number of nuclei, and each becomes surrounded by a thin layer of cytoplasm), producing small cells called **gametes**. The gametes produced by one partner are similar to those produced by the other, and thus the term **isogametes** is used. These migrate to the periphery of the cyst, leaving the residual cytoplasm in the centre. The wall between the two gamonts disappears, then each gamete from one individual fuses

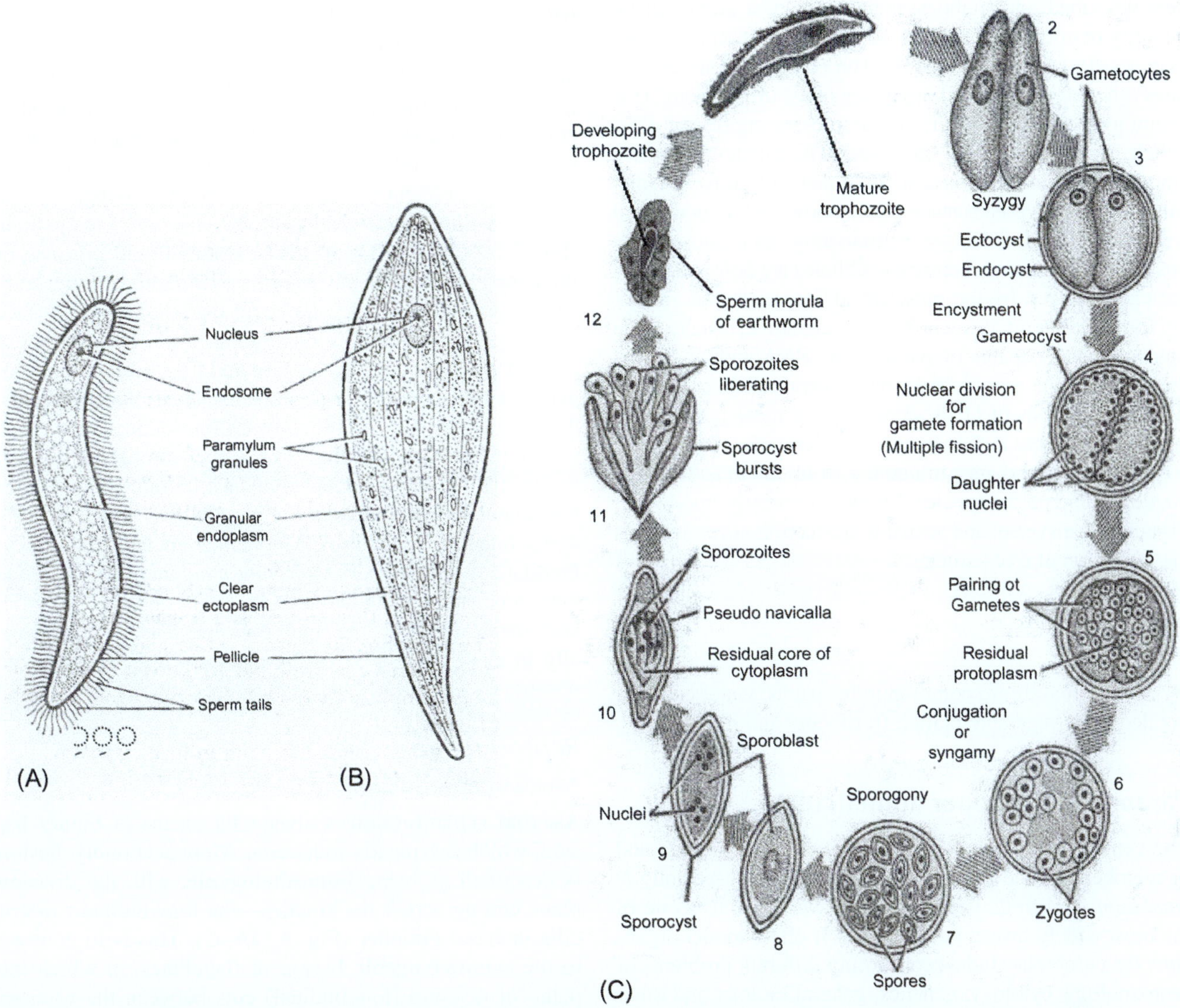

FIG. 3.30 (A) Trophozoite of *Monocystis*, free young. (B) Mature trophozoite. (C) Life history of *Monocystis*, *adapted from Verma (1977)*.

with a gamete from the other (**syngamy**: syn = together, gamy = gamete), forming a **zygote**. Thus, a large number of zygotes (**sporoblasts**) are produced. The zygote secretes a thin, but hard boat-shaped protective capsule (**zygocyst, oocyst**, or **sporocyst**). It is known as **pseudonavicella** (from its resemblance to the diatom *Navicella*). Within the sporocyst, the zygote undergoes two meiotic divisions followed by a mitotic division. Division of the cytoplasm results in the formation of eight young spindle-shaped gregarine zoites (generally still called **sporozoites**). These are haploid. The sporocyst, with the contained zoites, represents a spore. The mature gamonts, into which the zoites develop, and the gametes that they eventually produce, are also haploid. The zygote is the only diploid stage in the life cycle. Finally, the sporocysts find their way to the outside, as mentioned above, to repeat the cycle.

CLASS 4: CILIOPHORA (LATIN *CILIUM*, 'EYELID')

Subclass:	Ciliata (Infusoria Storer et al., 1983)
	Cilia and do not possess suctorial Tentacles
Order:	Holotricha (Greek *holos*, 'entire' and *thrix*, 'hair')
Suborder:	Hymenostomatida (Greek *hymen*, 'membrane')
	With mouth, complex gullet

The ciliates represent one of the most homogeneous groups among the protozoa. They are also distinguished by their wide distribution and the ease with which they can be found in water that will support any life. At least 10,000 species have been described, and doubtless many more exist. Although the majority of ciliates are solitary and free-swimming, there are both sessile and colonial forms. The ciliates are larger than most other protozoans, although they range in size from 10 or 12 μm to 3–4.5 mm long. Their locomotory organs are cilia, which persist throughout their life. The ciliates are never amoeboid. Cilia, in holotrichs, are equal and uniformly distributed on the whole surface of the body, with a mouth and a gullet (vestibule) that are permanently open and usually possess an undulating membrane. Ciliates are holozoic. They are characterized by the presence of heterokaryotic nuclei, i.e. formed of two types: one type is vegetative, usually large **macronucleus** capable of synthesizing RNA and DNA; and the other type is reproductive, the **micronucleus** concerned with the synthesis of DNA alone. One or more micronuclei are usually present in the ciliate. The micronuclei have small, rounded bodies and vary in number from one to as many as 80, depending on the species. Ciliates are mostly free-living. Fission is transverse, and sexual reproduction never involves the formation of free gametes.

Habitat

Hymenostomatids are common in infusions, water containing decaying vegetable matter, freshwater and marine.

Paramecium (slipper animalcule)

The genus *Paramecium* has been known for 250 years and its members are the commonest of protozoans, especially *P. caudatum*, *P. Aurelia* complex and *P. bursaria*. It is one of the most widely investigated ciliates. It offers model organisms for numerous studies concerning different problems of protozoology, cytology, genetics, general biology and other areas. The wide distribution of this genus (many species are cosmopolitan), ease of cultivation and the relative large sizes are important factors in allowing paramecia to become 'universal laboratory animals'. Fokin et al. (2001) used, in addition to the above three species, *P. putrinum*, *P. duboscqui*, *P. calkinsi*, *P. nephridiatum*, *P. woodruffi*, *P. multimicronucleatum*, *P. jenningsi*, *P. schewiakoffi*, *P. polycaryum*, and *P. triaurelia* to investigate the nuclear reorganization pathway during conjugation. They studied, for the first time, conjugation and nuclear reorganization for *P. woodruffi* and *P. schewiakoffi*. The main aim of this work was comparative analysis of the basic stages of nuclear reorganization in the *Paramecium* genus. Fokin et al. divided the nuclear reorganization into two phases: the **progamic** phase, which coincides with the behaviour and/or changes that occur during the period of fusion of meiotically reduced haploid genomes in **conjugation**; and the **metagamic** phase, which coincides with the period of development of new cell line in the same process.

Morphology

Paramecium caudatum (Fig. 3.31a) is an example of ciliates, which usually has a constant body shape, generally asymmetrical, although some have acquired radial or bilateral symmetry. *Paramecium* itself is cigar-shaped, with its posterior end slightly more pointed than the anterior end. The body is typically covered by a complex living, stiff but flexible **pellicle**, best studied in *Paramecium*. This pellicle (Fig. 3.31b (B)) is frequently sculptured, and consists of three membranes. The outermost is continuous over the body surface and cilia. These cilia cover the whole body surface and they are arranged in longitudinal rows. Some of these rows of cilia enter the oral groove and the vestibule, and, together with specialized membranelles, create water currents. This groove appears as a shallow oblique depression on one side of the body, which leads to a deeper conical depression: the vestibule. This vestibule, in turn, communicates with the funnel-like buccal cavity (cytopharynx), which extends down into the endoplasm. Myonemes encircle its wall to facilitate swallowing. *Paramecium caudatum* has only one macronucleus, a large ovoid body, easily observed in living specimens, and only one small micronucleus, which is usually in close apposition to the macronucleus, and may be obscure except in fixed and stained specimens.

Reproduction

Asexual reproduction

Asexual reproduction is always by means of binary fission, which is typically transverse. More accurately, fission is described as being **homothetogenic**, with the division plane cutting across the kineties—the longitudinal rows of cilia or basal granules (Fig. 3.32A–C). This is in contrast to the **symmetrogenic** fission of flagellates, in which the plane of division (longitudinal) cuts between the rows of basal granules. Fission involves both the micronucleus and macronucleus. The micronucleus divides by a typically **mitotic** process with spindle and chromosomes, while the behaviour of the macronucleus is more variable. It divides, usually, by simple elongation and constriction in the centre, i.e. a spindle does not form. (Alternatively, in other ciliates the macronucleus degenerates and arises anew from micronuclei. When a number of macronuclei are present, they may first combine as a single body before division.) However, the ciliate macronucleus is known to be highly polyploid, acquires diverse forms and contains many genomes produced by repeated endomitosis.

Modified fission in the form of budding occurs in some ciliate groups, notably the Suctoria. Small daughter cells arise from the outer surface of the parent body; alternatively, the budding may be internal in a brood pouch from which the young ciliated individuals are formed, unlike the adults, provided with several circlets of cilia. They then escape to lead a free existence. At the end, they develop to attain the features of the adult.

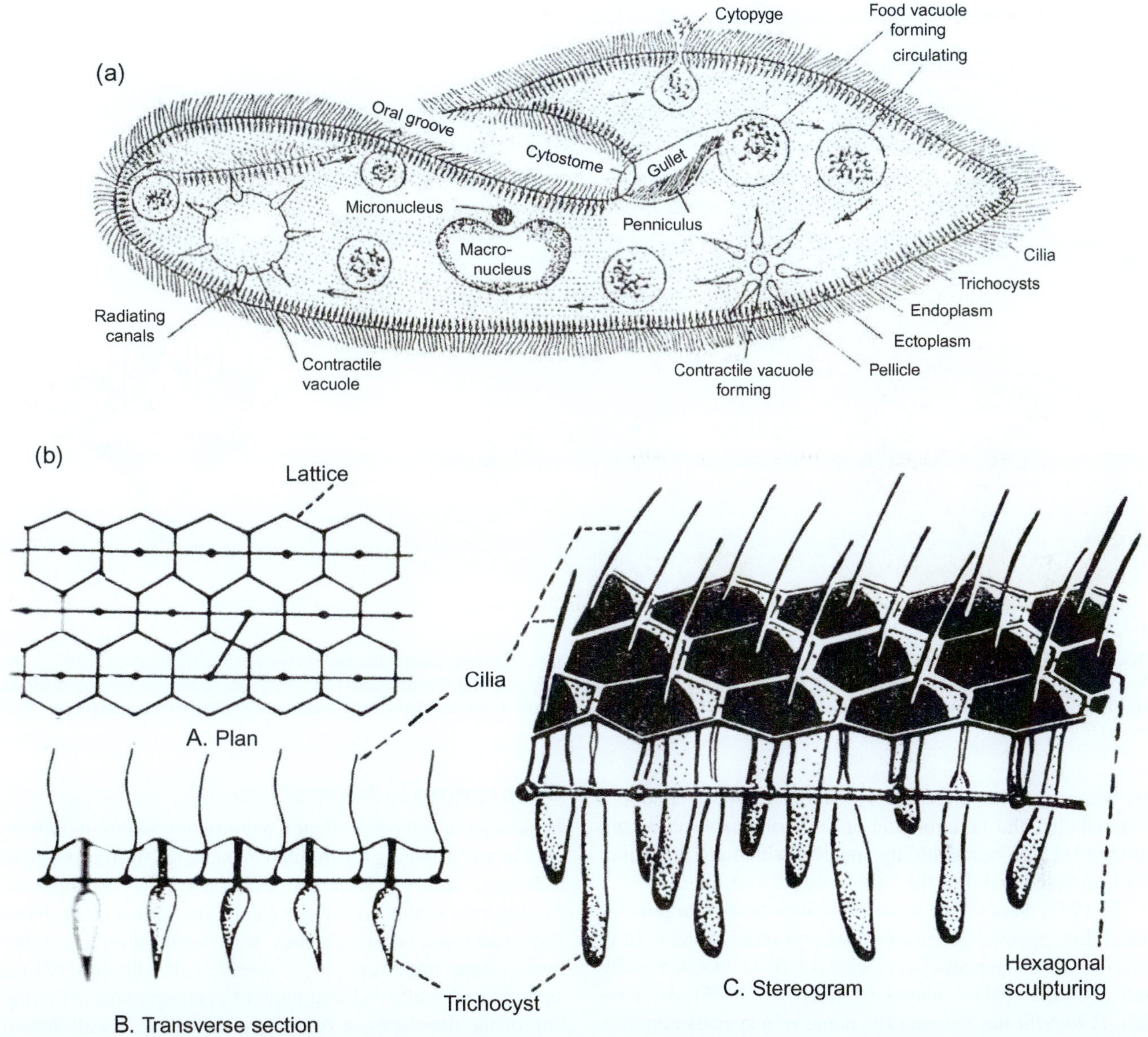

FIG. 3.31 (a) Diagram showing the structure of entire animal of *Paramecium caudatum, arrows* show path of food vacuoles in endoplasm. *From Storer et al. (1983)*; (b) structure of the pellicle of *Paramecium*; (A) plan view of the pellicle; (B) diagrammatic T.S. of the pellicular structure; (C) stereogram of the pellicle, *from Borradaile et al. (1977).*

Sexual reproduction

Sexual reproduction in *Paramecium* does not involve gamete formation, but only the exchange and fusion of micronuclear material by a process called **conjugation**. Conjugation in microorganisms, especially in *Paramecium*, is therefore the alternative process of fertilization in metazoa. It is well known that fertilization is a very highly specialized example of the general phenomenon of cell–cell interaction. This is especially apparent among many lower isogamous organisms. For example, the significant advances that have been made in the study of sexual phenomena (cell–cell interaction) in certain microorganisms (particularly *Paramecium*, *Chlamydomonas* and yeast). Conjugation can serve as a general model system for cell–cell interaction. The interpretation of conjugation in *Paramecium* as a phenomenon homologous to metazoan fertilization was first pointed out in the mid-19th century.

However, for years no differences were detected between the two paramecia joining in conjugation. Later on, the experimental work of Sonneborn (1937) discovered the **mating types** of *Paramecium aurelia*. He reported that there are stocks containing two kinds (**sexes**) of individuals; only members of opposite kinds would conjugate. This discovery was the start of the modern studies in *Paramecium* conjugation

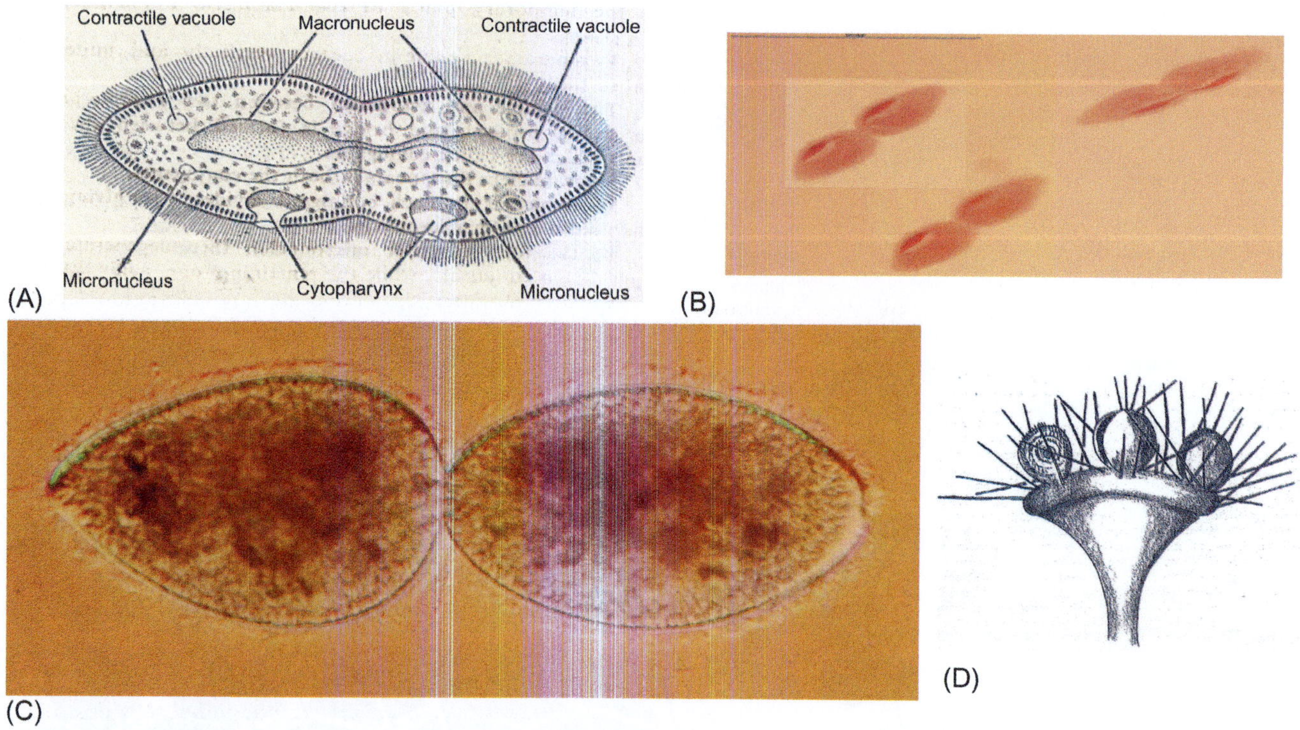

FIG. 3.32 Asexual reproduction in some protozoans. (A) Diagram showing transverse binary fission in *Paramecium (from Verma, 1977)*; (B) Photomicrograph of laboratory preparation of *P. caudatum* during the process of fission, scale bar 200 μm; (C) Live *Paramecium* in the process of fission which takes place in 2 h and is repeated four times a day (Beauté du Monde Animal Tome XII, 1974); (D) suctorian *Ephelota* with external buds, *from Hyman (1940)*.

by Metz and his colleagues (Metz, 1954). Sonneborn claimed that the taxonomical species of *Paramecium* are subdivided into many sibling species, which are reproductively isolated and thus they are biological species.

In 1957, Sonneborn provided a new term, **syngen**, for the sibling species of *Paramecium* and other ciliates. In *P. aurelia*, all 14 syngens were characterized biochemically and assigned species names (Sonneborn, 1975). As a result, *P. aurelia* has become the name of a species complex and is called the ***P. aurelia* complex**. In 2001, Fokin et al. studied the nuclear reorganization variety in *Paramecium* and reported that *P. Aurelia* complex is composed of 15 biological species (14 syngens of Sonneborn). In some species of *Paramecium* (e.g. *P. caudatum* and *P. Aurelia* complex), each syngen is composed of two complementary mating types, but in other species (e.g. *P. bursaria*), each syngen contains four or more types.

Conjugation is induced by the interaction of complementary mating substances, which are proteins or closely related to proteins. These substances exist in the ciliary membranes of the ventral surface of the organism. Evidence obtained from the chemical induction suggests that Ca^{++} plays an important role, as it does in the fertilization of higher organisms. Electrophysiological measurements before and during the process of activation revealed no membrane potential change comparable to the activation potential found in sea urchin eggs.

Conjugation in Paramecium

In paramecia and other ciliates, a temporary union of individuals in pairs (mating) with mutual exchange of micronuclear materials, occurs at intervals, and is known as conjugation. Conjugation in this unicellular eukaryote *Paramecium* shares important basic characteristics with fertilization in higher multicellular organisms: (1) a specific cell adhesion; (2) fusion of meiotically reduced haploid genomes; and (3) initiation of the development of a new cell line. It is well known that sperm and egg of metazoa are produced by the sexually mature individuals. Similarly, individuals of *Paramecium* attain their mating ability when they reach the **period of sexual maturity**. Before this, the individual must pass through an **immaturity period**, during which the individual has no ability to mate. This period is measured in terms of the number of cell divisions after the previous conjugation. The length of this immaturity period differs in different species. Even during the period of sexual maturity, individuals express the mating activity only in two conditions: (1) when they are in the stationary phase culture; or (2) when they are in the condition of so-called relative hunger. On the other hand, the individuals do not show mating ability in two different conditions: (1) when they are well fed; or (2) when they are extremely starved.

Using *Paramecium caudatum* (which has only one macronucleus and one micronucleus) as an example, the process of conjugation proceeds as follows.

Cell adhesion

Agglutination. When individuals of two different mating types of the same syngen (variety) are brought together under suitable conditions for conjugation, they undergo agglutination at once, forming clumps (Fig. 3.33B). This sexual cell agglutination in *Paramecium* is called the **mating reaction**, and is the first step in the process. Mating reaction is a specific cell contact by the cilia of the mating types, but **only** the cilia of the ventral surface have the ability to react. It was reported by Metz and Monroy (1985a,b,c) that this is clearly shown experimentally, as the cilia are detached from the mating reactive cells (individuals) of one mating type and are applied to reactive cells of the complementary mating type. The detached cilia adhere **only** to cilia of the ventral surface of the complementary cells.

Holdfast union. The continued cell interactions in the mating reaction (first step), for about an hour, are necessary for the appearance of the second step of cell adhesion. Afterwards, the individual paramecia begin to separate and conjugate together in pairs, by attachment of their ventral oral surfaces, by their anterior regions only (Fig. 3.33C). This is the **holdfast union**, which does not occur if the previous mating reaction is destroyed precociously by physical or chemical means. In the holdfast union pairs, cell adhesion occurs not at the ciliary surfaces, but at the surfaces of the cell bodies where cilia degenerate. The latter begins about 30 min after the mixing of mating types (in another description, there is a protoplasmic bridge between the conjugants). The holdfast union lasts for about an hour, during which the degeneration of cilia extends posteriorly, along the ventral surface reaching the paroral region (cytosome).

Paroral union. This is the last step of cell adhesion in conjugation (Fig. 3.33D). It is called paroral union as each pair of the conjugants adhere at this region, leaving the region in between the latter region and the holdfast region separate. Cell adhesion of this separate region probably comes last. In ordinary conjugation, however, it is difficult to distinguish that last stage of union from paroral union.

Degeneration of cilia does not end at the paroral region, but proceeds to the posterior end of the cell, followed by degeneration of trichocysts. During the completion of cell adhesion in conjugation, separation of conjugating pairs by physical or chemical means becomes difficult. In conjugation of paramecia, however, no complete fusion of cells occurs, in contrast to conjugation in unicellular algae or yeast, and also to fertilization of eggs with sperm. Electron microscopic observations show that, on the adhered membranes, many locally limited membrane fusions occur, making minute openings (0.2–0.5 μm in diameter) on the membranes.

Fusion of the meiotically reduced haploid genome

Immediately after the formation of the paroral union, a series of nuclear changes occurs in each conjugant.

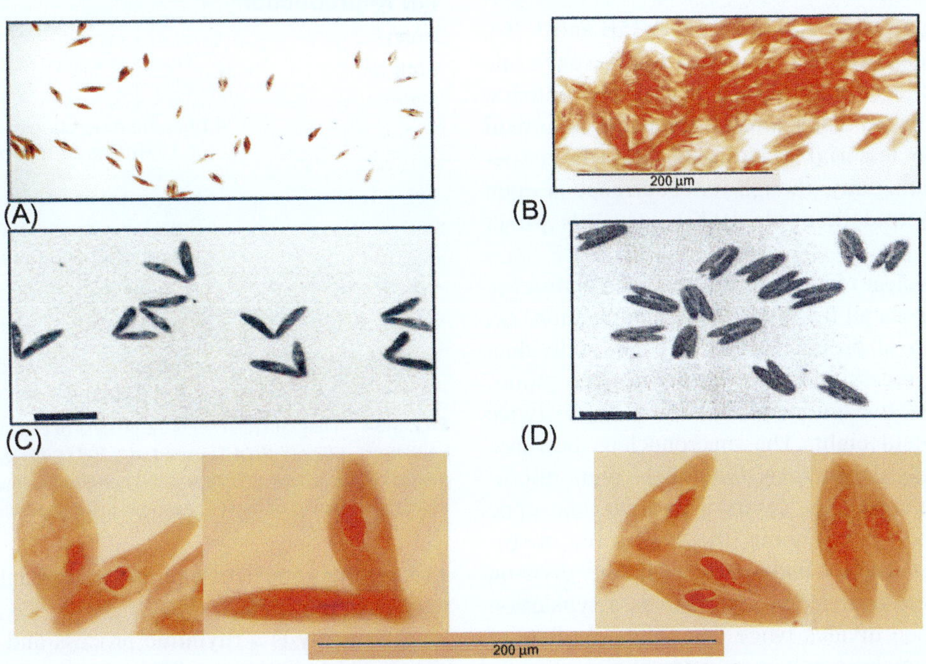

FIG. 3.33 Shows steps of stage cell adhesion (mating reaction) in the process of **conjugation** in *P. caudatum*: (A and B) steps of mating reaction; (C) holdfast union; (D) paroral union. (A, B, lower C, and lower D) Photomicrographs of laboratory preparations scale, bar 200 μm; *(upper C and D) from Metz and Monroy (1985a) vol. I,* scale 150 μm.

These changes involve the gradual fragmentation of the macronucleus and two meiotic divisions of the micronucleus to produce four haploid micronuclei (pronuclei), three of which disintegrate. The remaining haploid micronucleus undergoes a mitotic division to form one haploid stationary pronucleus (female) and one haploid wandering pronucleus (male). The male pronucleus of each conjugant migrates, probably through the opening of the membrane, into the partner and fuses with the stationary pronucleus to form a diploid zygote nucleus (synkaryon) (middle portion of Fig. 3.34).

Development of a new cell line

After the exchange of the male haploid pronuculei between the two conjugants, they separate as exconjugants. In each, the synkaryon divides successively into two, four and then eight micronuclei. Four of these enlarge to become macronuclei, and three others degenerate. The remaining micronucleus and the organisms themselves divide twice, so that from each exconjugant, there are four daughter paramecia produced, each with one macronucleus and one micronucleus (last portion of Fig. 3.34). These small cells then grow in size and undergo binary fission in the normal way.

Conjugation differs from the sexual union of gametes in other protozoa and in metazoa, because progeny are not a direct product of the fusion; after conjugation, each individual continues asexual fission. Conjugation is a process that provides hereditary transfer, as the two exconjugants are genetically altered by their exchange of micronuclear materials.

Endomixis occurs in *P. Aurelia* (Figs 3.35B and 3.36). It resembles conjugation but is carried out by only one individual, and it lacks meiosis and fusion of pronuclei. The resemblance lies in that a new micronucleus is formed from micronuclear material. Endomixis might be a specialized case of autogamy, though its occurrence is open to some doubt (Borradaile et al., 1977); see *Actinophrys* above.

Autogamy: Nuclear reorganization occurs in *P. Aurelia*. The animal undergoes all the processes of conjugation, i.e. meiosis and fusion of pronuclei, but only one individual takes part in the process and provides both of the pronuclei. *P. Aurelia* has two micronuclei, which divide twice (prezygotic), to yield eight. The macronucleus becomes fragmented and later degenerates. One of the eight micronuclei, as a gamete nucleus, enters a protoplasmic cone bulging through the cell cytosome (the other seven micronuclei disintegrate), where it undergoes yet another division yielding two gamete nuclei. The two fuse as a synkaryon (zygote), which then divides twice (postzygotic). Two of the four resulting nuclei continue as micronuclei and two become a macronucleus. The organism, with its micronuclei, divides to produce two daughter paramecia, each with a (new) macronucleus and two micronuclei. Thereafter, binary fission resumes.

Autogamy is a rhythmic process and one in which hereditary changes occur. The advantage of this, beside the formation of a new micronucleus, is that it establishes different combinations of allelomorphic genes.

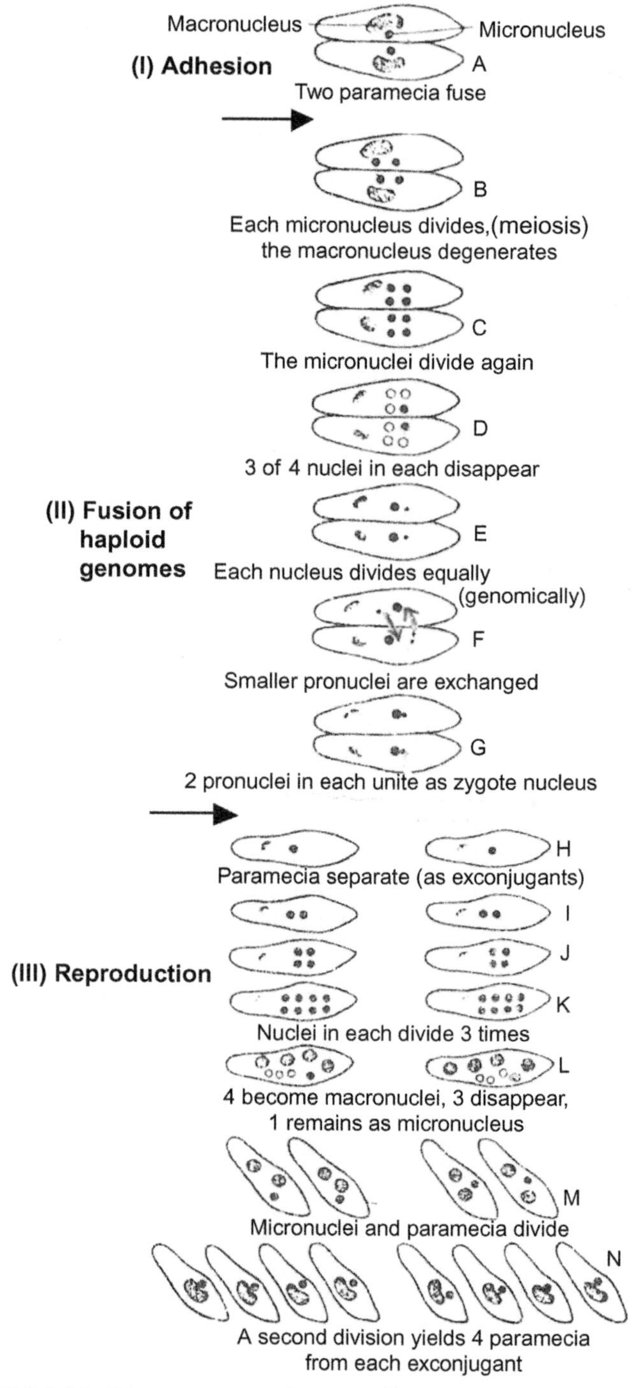

FIG. 3.34 Diagram represents the process of **conjugation** in *Paramecium caudatum, adapted from Storer et al. (1983). Long arrows* separate three phases of the process: (I) shows cell adhesion or mating reaction; (II) formation and fusion of haploid genomes; (III) production of new cell lines.

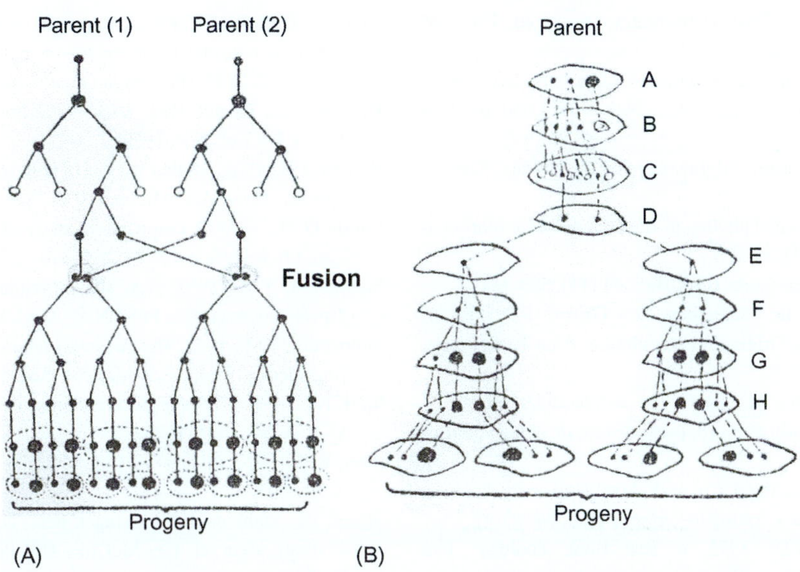

FIG. 3.35 Diagram showing comparison between the behaviour of the nuclei during conjugation and endomixis: (A) **conjugation** of *Paramecium caudatum*; (B) **endomixis** in *P. Aurelia. (From Borradaile et al. (1977).)*

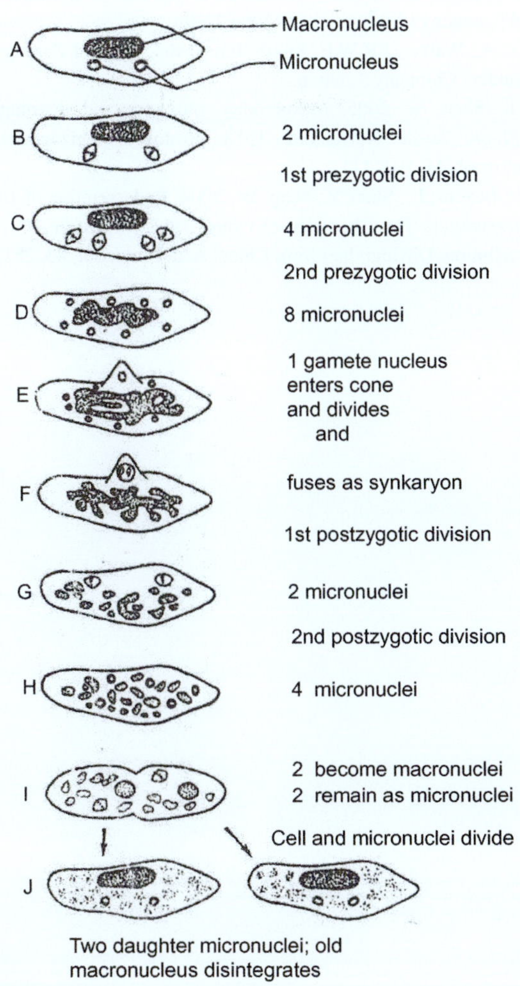

FIG. 3.36 Diagram showing behaviour of nuclei in the process of **autogamy (endomixis)** in *Paramecium Aurelia. (From Storer et al. (1983). See also Fig. 3.35B.)*

REFERENCES

Ahmad, T.A., El-Sayed, B.A., El-Sayed, L.H., 2016. Development of immunization trials against *Eimeria* spp. Trials Vaccinol. 5, 38–47.

Barnes, R.D., 1974. Invertebrate Zoology, third ed. W.H. Saunders Company, London.

Borradaile, L.A., Potts, F.A., with chapters by Eastham, L.E., Saunders, J.T., 1977. The Invertebrata, fourth ed. Revised by Kerkut, G.A. Cambridge University Press.

Chantangsi, C., 2002. Actinophrys. "A well-known sun animalcule". http://www.microscopy-uk.org.uk/mag/artfeb02/ccactino.html.

Desnitski, A.G., 2000. Development and reproduction of two species of the genus *Volvox* in a shallow temporary pool. Protistology 1 (4), 195–198.

Ettl, H., 1976. Die Gattung *Chlamydomonas Ehrenberg* (Chlamydomonas und die nachstverwandten Gattungen II). Nova Hedwig, Beih. 49, 1–1122.

Fokin, S.I., Przybos, E., Chivilev, S.M., 2001. Nuclear reorganization variety in *Paramecium* (Ciliophora, Peniculida) and its possible evolution. Acta Protozool. 40, 249–261.

Gardiner, M.S., 1972. The Biology of Invertebrates. McGraw-Hill Book Company, London, New York.

Gilbert, S.F., 2000. Developmental Biology, sixth ed. Sinauer Association, Inc., Publishers, Sunderland, MA.

Graham, L.E., Wilcox, L.W., 2006. Algae. Wisconsin University, Prentice Hall. ISBN: 0013-660333-5.

Green, K.J., Kirk, D.L., 1982. A revision of the cell lineages recently reported for *Volvox carteri* embryos. J. Cell Biol. 94, 741–742.

Hallmann, A., 2003. Extracellular matrix and sex-inducing pheromone in *Volvox*. Int. Rev. Cytol. 227, 131–182.

Hallmann, A., Godi, K., Wenzl, S., Sumper, M., 1998. The highly efficient sex-inducing pheromone system of *Volvox*. Trends Microbiol. 6 (5), 185–189.

Harris, E.H., 2009. In: Stern, D.B., Witman, G.B. (Eds.), The *Chlamydomonas* Sourcebook. vol. 1. Elsevier, San Diego, CA, pp. 293–302.

Hickman, C.P., 1973. Biology of the Invertebrates, second ed. The C.V. Mosley Company, Saint Louis.

Hyman, L.H., 1940. The Invertebrates, Protozoa Through Ctenophora, first ed. McGraw-Hill Book Company, Inc., New York, London. Third expression.

Keele, C.A., Neil, E., 1973. Sanson Wright's Applied Physiology, 12th ed. Oxford University Press.

Kirk, D.L., 1988. The ontogeny and phylogeny of cellular differentiation in *Volvox*. Trends Genet. 4 (2), 32–36.

Kirk, D.L., 2004. *Volvox*. Quick guide. Curr. Biol. 14 (15), 599–600.

Kirk, D.L., et al., 1982. In: Subtelney, S., Green, P.B. (Eds.), Developmental Order: Its Origin and Regulation. Alan R. Liss, pp. 247–274. inversion.

Kirk, D.L., Birchem, R., King, N., 1986. The extracellular matrix of *Volvox*: a comparative study and proposed system of nomenclature. J. Cell Sci. 80, 207–231.

Lee, R.E., 2008. Phycology, fourth ed. Colorado State University, USA.

Mann, T., 1984. Spermatophores. Springer-Verlag, Berlin, New York.

Marshall, A.J., Williams, W.D., 1975. A Text Book Zoology: The Invertebrates. ELBS and Macmillan.

Metz, C.B., 1954. In: Metz, C.B., Monroy, A. (Eds.), Mating Substances and the Physiology of Fertilization in Ciliates. vol. I. 1985a, Academic Press, Inc., New York, London.

Metz, C.B., Monroy, A., 1985a. Biology of Fertilization. Model Systems and Oogenesis. vol. I. Academic Press, Inc., New York, London.

Metz, C.B., Monroy, A., 1985b. Biology of Fertilization. Biology of the Sperm. vol. II. Academic Press, Inc., New York, London.

Metz, C.B., Monroy, A., 1985c. Biology of Fertilization. The Fertilization Response of the Egg. vol. III. Academic Press, Inc., New York, London.

Nickelsen, J., Kück, U., 2000. The unicellular green alga *Chlamydomonas reinhardtii* as an experimental system to study chloroplast RNA metabolism. Naturwissenschaften 87 (3), 97–107.

Nishii, I., Ogihara, S., 1999. Actomyosin contraction of the posterior hemisphere is required for inversion of the *Volvox* embryo. Development 126, 2117–2127. (Printed in Great Britain).

Pandey, S.N., Trivedi, P.S., 1995. A Textbook of Algae. Vikas Publishing House Pvt Ltd, New Delhi.

Powers, J.H., 1908. Further studies in *Volvox*, with description of three new species. Trans. Am. Microsc. Soc. 28, 141–175.

Smith, G.M., 1944. A comparative study of the species of *Volvox*. Trans. Am. Microsc. Soc. 63 (4), 265–310.

Sonneborn, T.M., 1937. Sex, sex inheritance and sex determination in *Paramecium aurelia*. Proc. Natl. Acad. Sci. U. S. A. 23 (7), 378.

Sonneborn, T.M., 1975. The *Paramecium aurelia* complex of fourteen sibling species. Trans. Am. Microsc. Soc. 94, 155–178.

Starr, R.C., 1968. Cellular differentiation in *Volvox*. Proc. Natl. Acad. Sci. U. S. A. 50, 1082–1088.

Starr, R.C., 1970. Control of differentiation in *Volvox*. Dev. Biol. Suppl. 4, 59–100.

Storer, T.I., Stebbins, R.C., Usinger, R.L., Nybakken, J.W., 1983. General Zoology, sixth ed. Tata McGraw-Hill Publishing Company Ltd, New Delhi.

Vashishta, B.R., 2004. Reprint: Botany for Degree Students, Part 1 Algae. S. Chand Group Company, Ltd, New Delhi.

Verma, P.S., 1977. A Manual of Practical Zoology: Invertebrates. S. Chand and Company Ltd, Ram Nagar, New Delhi.

Villee, C.A., Walker, J.R.W.F., Smith, F.E., 1964. General Zoology. W.B. Saunders Company, London.

Yang, J., Shen, Y., 2005. Morphology, biometry and distribution of *Difflugia biwae* Kawamura, 1918 (Protozoa: Rhizopoda). Acta Protozool. 44, 103–111.

Yang, J., Beyens, L., Shen, Y., Feng, W., 2004. Redescription of *Difflugia tuberspinifera* Hu, Shen, Gu et Gong, 1997 (Protozoa: Rhizopoda: Arcellinida: Difflugiidae) from China. Acta Protozool. 43, 281–289.

Kingdom Animalia

Table of classification of Kingdom Animalia

A brief synopsis of classification of invertebrate phyla according to various sources.

Kingdom	Animalia						
Subkingdom	Protozoa	Metazoa					
Division		Parazoa	Eumetazoa				
Section			Diploblastica	Triploblastica			
Subsection				Acoelomate	Pseudocoelomate	Coelomate	
Group						Schizocoelomate	Enterocoelomate
Phyla		1. Placozoa 2. Porifera	1. Cnidaria 2. Ctenophora	1.Platyhelminthes 2. Nemertinia 3. Endoprocta	1. Achanthocephala 2. Rotifera 3. Gastrotricha 4. Kinorhyncha 5. Nematoda 6. Nematomorpha	1. Annelida 2. Echiuroidea 3. Sipunculoidea 4. Onychophora 5. Arthropoda 6. Mollusca 7. Bryozoa 8. Phoronida 9. Brachiopoda 10. Pentastomida 11. Tardigrada 12. Ectoprocta?	1. Chaetognatha 2. Pognophora 3. Echinodermata 4. Chordata

Subkingdom Parazoa

Chapter Outline

Definition **107**
Historical review **107**
General characteristics **107**
Classification **109**
Embryological characteristics of Porifera **110**
Reproduction **110**
Agametic reproduction 111
Budding 111
External bud formation 111
Internal buds (gemmule formation) 114
Other agametic types of reproduction 116
Reduction bodies 116
Fragmentation 116
Factors influencing agametic reproduction 116
Regeneration 117
Gametic reproduction 117
Sexual differentiation 117
Gonads 117
Origin of the germ cells 117
Gametogenesis 118
Spermatogenesis 119
Structure of the fully formed spermatozoon 121
Oogenesis 123
Vitellogenesis 124
The mature oocyte 128
Reproductive behaviour and spawning 128
Fertilization 129
Development 131
Class Calcarea **131**
Subclass Calcaronea 131
Leucosolenia variabilis 131
Sycon raphanus 135
Subclass Calcinea 137
Hexactinellida **138**
Farrea sollasii 138
Oopsacas minuta 141
Demospongiae **148**
Tetractinomorpha 148
Donatia deformis 149
Viviparous sponges **149**
Plakina monolopha 150
Spongia officinalis 152
Stylocordyla borealis 155
Asbestopluma occidentalis 157
Nutrition of embryos **162**
Larval release **162**
Larval behaviour and settling **162**
Metamorphosis **162**
Oviparous sponges **162**
Types of larvae **164**
I Parenchymula larva 165
II Amphiblastula larva 166
III Simple blastula (undifferentiated blastula) (coeloblastula) 167
Biological associations **167**
References **167**

Chapter 4

Phylum Porifera

(Latin *porus*, "pore"; *ferro*, "to bear")

DEFINITION

Phylum Porifera comprises the sponges. These animals are asymmetrical or radially symmetrical multicellular organisms, without organs, mouth or nervous system; with a body perforated by pores, and with canals, and chambers through which water current flows; with one or many internal cavities lined with choanocytes.

HISTORICAL REVIEW

The animal nature of these sponges was defined in 1765 by Ellis, who saw the water currents and the movements of oscula. Linnaeus (1707–78), and Cuvier (1769–1882) with Lamarck, in 1809, proposed a theory of evolution, which placed the sponges under Zoophytes, or polyps, in their systems. They considered them allies to anthozoan coelenterates and many investigators sought to find in sponges the polyps they thought must be there. Later, de Blainville (1816) separated them into the group Spongiaria, allied to Protozoa. However, the discovery of the resemblance of sponges to choanoflagellata was made by James-Clark (1867), who, like most scientists at that time, considered sponges as protozoan colonies. Evidence in support of this view can be found in:

1. the presence of choanocytes;
2. the method of skeleton production (i.e. within single cell or small group of cells); and
3. the totypotency of cells (Marshall and Williams, 1975).

However, this idea gained little acceptance, and through most of the 19th century, more advanced studies on sponges were performed and sponges were placed with coelenterates. The latter sometimes referred to as Radiata.

The morphology and physiology of sponges were first adequately understood by R.E. Grant, beginning in 1825, who created the name Porifera for the group in 1836. This name is still generally used up to the present time. It was Leuckart (1854) who first drew attention to the structure of the sponge as a whole, and compared it to a coelenterate. Haeckel (1872) formalized this conception and termed the two layers constituting the body wall 'dermal' and 'gastral', respectively. Minchin et al. (1900) followed this terminology and it seems that these two terms are still used until now. Huxley (1875) and Sollas (1884) proposed the complete separation of sponges from other Metazoa, according to their many peculiarities. Since the 20th century, sponges have been recognized as constituting a separate branch of the metazoans, named Parazoa, after Sollas.

GENERAL CHARACTERISTICS

1. Metazoan organisms;
2. Asymmetrical or radially symmetrical animals;
3. Aquatic and sessile animals;
4. With numerous pores distributed on the whole body surface which are inhalant in function;
5. With one or more larger opening used as exhalant, osculum (Fig. 4.1);
6. Generally with skeleton (calcareous, siliceous, or horny); and
7. The sponge encloses a wide cavity (paragaster) lined wholly by choanocytes (the ascon grade), in others the

Invertebrate Embryology and Reproduction. https://doi.org/10.1016/B978-0-12-814114-4.00004-7

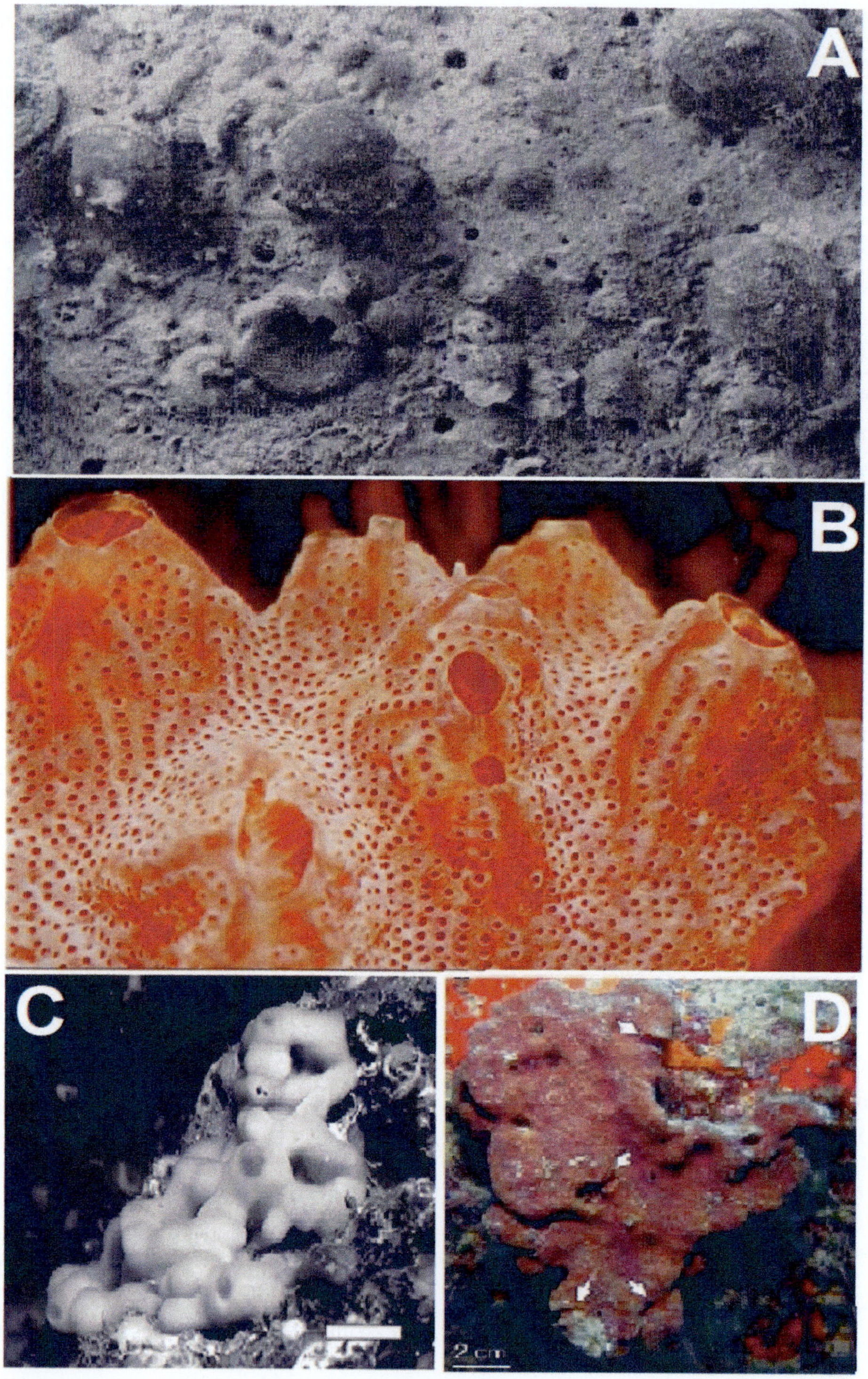

FIG. 4.1 (A) Surface of coral slab from Coralspecht, Curacao, showing bimodal size frequency of *Thecidellina*. The larger individuals are 4–5 mm wide. More characteristic of dark caves and crevices and at depths below 80 m, the thecidioid *Thecidellina barretti* (Davidson) is also an important brachiopod of the Jamaican fore-reef slope *(from Jackson et al., 1971)*; (B) Colony of ***Chondrosia reniformis*** showing details of inhalant pores and exhalant osculae *(from Invertébrés: Beauté du Monde Animal, Tome XII, La Rousse, 1974)*; (C) live *Oscarella tuberculata*, scale bar = 1 cm *(from Ereskovsky and Tokina, 2007)*; (D) a large live male of *Petrosia ficiformis* at 12 m depth, note the scars *(arrows)* produced by the periodical tissue sampling *(from Maldonado and Riesgo, 2009)*.

choanocytes are restricted to the flagellated chambers (the sycon grade), the choanocytes are further removed by the presence of exhalant canals (the leucon grade) (Fig. 4.2B–E). However, Minchin et al. (1900) differentiated between a simple type, with separate radial tubes and a more complex type, with radial tubes fused and thickened distally to form cortex and incurrent canals. Ereskovsky (2010) defined the simple type as the syconoid grade (Fig. 4.2C), and the complex syconoid with cortex to be named Sylleibid (Fig. 4.2D).

This phylum was a subject of dispute, as mentioned above, because their bodies consist of many cells, so they might be considered metazoans. However, they differ from all metazoans in the following:

- In no metazoa are choanocytes found;
- In none is the principal opening exhalant;
- Sponges have no nervous system coordinating their activities; and
- Lastly and perhaps most significant, in a sponge the 'cells' are less specialized and dependent upon each other than the cells of a metazoan.

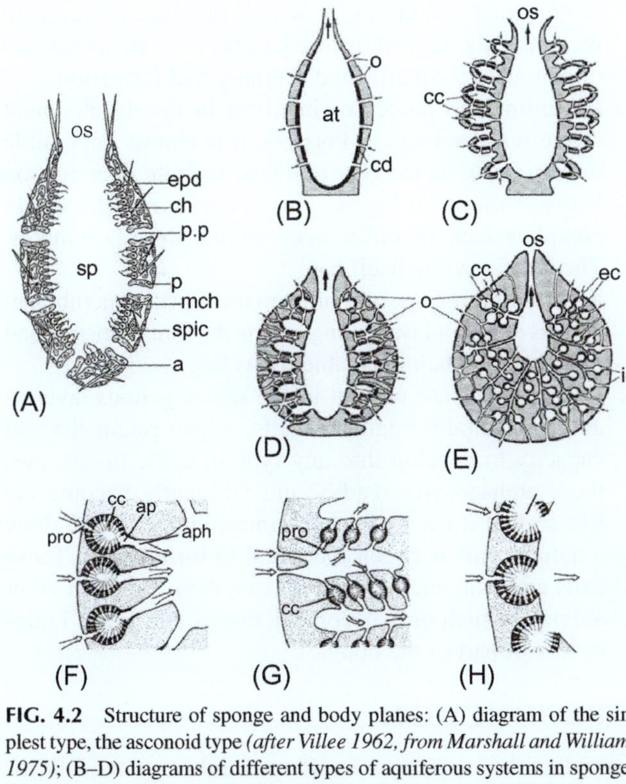

FIG. 4.2 Structure of sponge and body planes: (A) diagram of the simplest type, the asconoid type *(after Villee 1962, from Marshall and Williams, 1975)*; (B–D) diagrams of different types of aquiferous systems in sponges; (B) asconoid; (C) syconoid; (D) sylliebid; (E) leuconoid. (F–H) diagrams of types of choanocyte chambers: (F) aphodal, (G) diplodal, (H) eurypilous, *((B–H) from Ereskovsky, 2010)*, a=amoebocyte, ap=apopyle, aph=aphodus, at=atrium, cc=choanocyte chamber, cd=choanoderm, ec=exhalant canal, ic=inhalant canal, o=ostium, os=osculum, pro=prosopyle.

CLASSIFICATION (ACCORDING TO VARIOUS SOURCES)

Sponges are classified according to the nature and shape of the skeletal material, also, to the arrangement of the spicules and their position in the sponge. They are divided into three major classes:

1. Class Calcarea Bowerbank, 1864 (Calcispongiae) having only spicules of calcium carbonate ($CaCo_3$).
   a. Subclass Cacaronea: with amphiblastula larvae, leucoid structure is reached through syconoid stage. Ex, *Sycon, Leucosolenia, Grantia (Scypha).*
   b. Subclass Calcinea: with parenchymola larvae, leuconoid forms show no traces of syconoid ancestry, presumably evolved from reticulate asconoid forms. Ex., *Clathrina, Leucascus, Leucetta, Petrobiona, Minchinella, Ascandra, Leucandra.*
2. Class Hexactinellida Schmidt, 1870 (Hyalospongiae): having spicules or networks of siliceous material ($H_2Si_3O_7$).
   a. Subclass Hexasterophora, Ex, *Euplectella aspergillum* (Venus's flower basket); Farrea sollasi, Oopsacas minuta.
   b. Subclass Amphidiscophora, Ex, *Hyalonema* sp. The glass-rope sponge.
3. Class Demospongiae Sollas, 1885 (popular sponges): having skeleton of siliceous spicules, of spongin, of both, or none; spicules when present are not 6-rayed, mostly monaxon and tetraxon; leuconoid only.
   a. Subclass Tetractinomorpha: spicules are 4-rayed or none, no sponging. This group includes mainly oviparous species with stereogastrula larvae, and primitive families have incubated amphiblastula larvae. Ex. *Oscarella, Plakina, Haliclona.*
   b. Subclass Monaxonida: spicules monaxon, spongin abundant in some, most common of all sponges. Ex. *Suberites, Cliona* (boring sponge), *Halichondria, Microciona, Spongilla* (freshwater sponge), *Haliclona* (marine).
   c. Subclass Keratosa: horny sponges, skeleton is a network of spongin fibres, no silicate spicules, Ex. *Phyllospongia* (leaf-shaped), *Spongia* and *Euspongia* (bath sponges), *Hippospongia* (horse sponge), a sponge of the same kind with a coarser texture due to the inclusion of foreign matter in its skeleton, Carolline sponges. Recently Keratosa has been divided into: Keratosa, including two orders Di ctyoceratida + Dendroceratida; and Myxospongia, including the other orders, in other words, this is elevated to a subclass.
4. Class Sclerospongia: Hartman and Goreau (1970) have proposed this fourth Class. They differ in having:
   a. an internal skeleton of siliceous spicules and spongin fibres; and

b. an outer encasement of calcium carbonate crystals (aragonite).
Ex. *Ceratoporella nicholsoni* (Hickson), in Jamaica this sclerosponge was collected only from samples on the two traverses (Jackson et al., 1971).

Another view for class Demospongiae

The results of the extensive work of Lévi (1956) on the development of Demospongiae, clearly divided this class into two groups (subclasses) by their reproduction mode. The first group (subclass **Ceractinomorpha**) was characterized by the internal fertilization and the development of the embryo in the parent mesohyl (ovoviviparity and viviparity). The second group (subclass **Tetractinomorpha**) was characterized by external fertilization and the development of the larva in the environment (oviparity). Also, there are morphological differences between the two subclasses. The existence of the two traditional subclasses, the Tetractinomorpha and the Ceractinomorpha, has been challenged. They are now considered to be polyphyletic, and it is proposed that this division should be abandoned (Boury-Esnault, 2006).

Another recent view

The **Homoscleromorpha** is a very distinct group of Porifera. Homoscleromorphs are small sponges, encrusting or lumpy with a smooth surface. Lévi (1973) considered Homoscleromorpha to be a subclass of the class Demospongiae. Molecular studies opposed the phylogenetic relationships of Homoscleromorpha within Demospongiae, and alternative views have been proposed as follows:

1. they are closer to eumetazoan than to the other sponges in sponge paraphyletic hypothesis; or
2. they are considered as the sister group of the calcareous sponges in agreement with monophyletic view, as already postulated by some authors, based on an incorrect interpretation of the larval type.

Homoscleromorpha now forms a clade on its own, clearly separated from Demospongiae (Ereskovsky et al., 2014). An important morphological difference of the Homoscleromorpha from the Demospongiae is the lack of some cell types common in the Demospongiae, such as collencytes, spongocytes, and lophocytes. Instead, the mesohyl of homoscleromorphs contains mostly cells with specialized inclusions (spherulous, vacuolated and granular cells) and rare nucleolated amoebocytes (archaeocytes).

The Homoscleromorpha is not a subclass of the Demospongiae; this group of sponges shares several characteristics with the Eumetazoa (Ereskovsky, 2004).

Homoscleromorphs are characterized by the fact that the spicules may branch, and the aquiferous system is sylleibid or leuconoid (Fig. 4.2D and E), often with a vast basal exhalant chamber. Choanocyte chambers are large, usually eurypilous (Fig. 4.2H) (*Plakortis*, *Plakinastrella*, *Plakina*, and *Oscarella*), and aphodal (Fig. 4.2F) or diploid (Fig. 4.2G) (*Corticium*, *Pseudocorticium*); all the Homoscleromorpha are ovoviviparous sponges with the cinctoblastula larva. They can be both gonochoric and simultaneous hermaphrodites, i.e. male and female gametes mature in a sponge at the same time (Schulze, 1877; Tuzet and Paris, 1964). The representatives are the genera *Plakina*, *Corticium*, *Plakortis*, *Placinolopha*, and *Plakinastrella*. It was proposed that Homoscleromorpha is a definite class after class Demospongiae.

EMBRYOLOGICAL CHARACTERISTICS OF PORIFERA

It is a fact that sponges are animals of the simplest organization and hence they are found in a group outside the true Metazoa. In spite of this fact, their development is not so simple. There are many intricate problems in their development, which are not involved in the development of other metazoans. For example:

1. Gemmule formation that occurs as agametic reproduction of some sponges. It is believed that this process is a special adaptation for surviving during unfavourable environmental conditions. It is very difficult, however, to recognize the radical differences between the processes of gemmule formation and ordinary bud formation.
2. Regarding the processes involved in the development of the fertilized egg in Porifera, it is almost impossible to recognize, as in other animals, definite demarcation between the germ layers. This is because of the nearly complete lack of differences among the blastomeres. The poriferan egg itself has:
   a. no definite shape or characteristic capsular membrane;
   b. it is amoeboid wandering among the somatic tissue; and
   c. it is only distinguishable by its larger size.
3. Archaeocytes are present in the skeletogenous layer as amoeboid undifferentiated cells, which retain the full capacity to develop into any type of cells. In sponges, these archaeocytes readily and frequently become active so that it becomes, sometimes, impossible to draw a definite line between them and fertilized eggs. These cells are completely different from those found in other animals, which only become active as a result of injury or loss of part of the body.

REPRODUCTION

Reproduction in sponges is effected either gametically or agametically. In the latter, a bud is formed externally or internally; the former process is called budding and the second is gemmule formation. Gametic reproduction is accomplished by means of eggs and sperm as in other animals.

Agametic reproduction

In sponges, this mode of reproduction may serve a number of biological functions beyond reproduction in the narrowest sense; it may provide mechanisms for dispersal and for survival under extreme environmental conditions. This mode takes a variety of forms.

Budding

Budding is a common method of reproduction in sponges. These animals produce two types of buds, according to the process of formation of budding. The buds are external buds and internal buds (Gardiner, 1972). The latter is known as gemmule.

External bud formation

The bud always originates as a thin outgrowth of the sponge's cortex, supported by a spiculated stalk with few macroscleres. Separate cells or their groups migrate along the spicules to the terminal part of the outgrowth. There they form more or less spherical bodies. The main cell category is represented by cells with inclusions, which are scattered in the mesohyl and tend to concentrate along the distant surface of the bud. Masses of archaeocytes migrate to the tips of spicule clumps that project from the surface of the individual (Fig. 4.3A–E). During budding, cell proliferation or transdifferentiation was not observed either in the budding zone or in any special mitotically active region. Later, this outgrowth may be

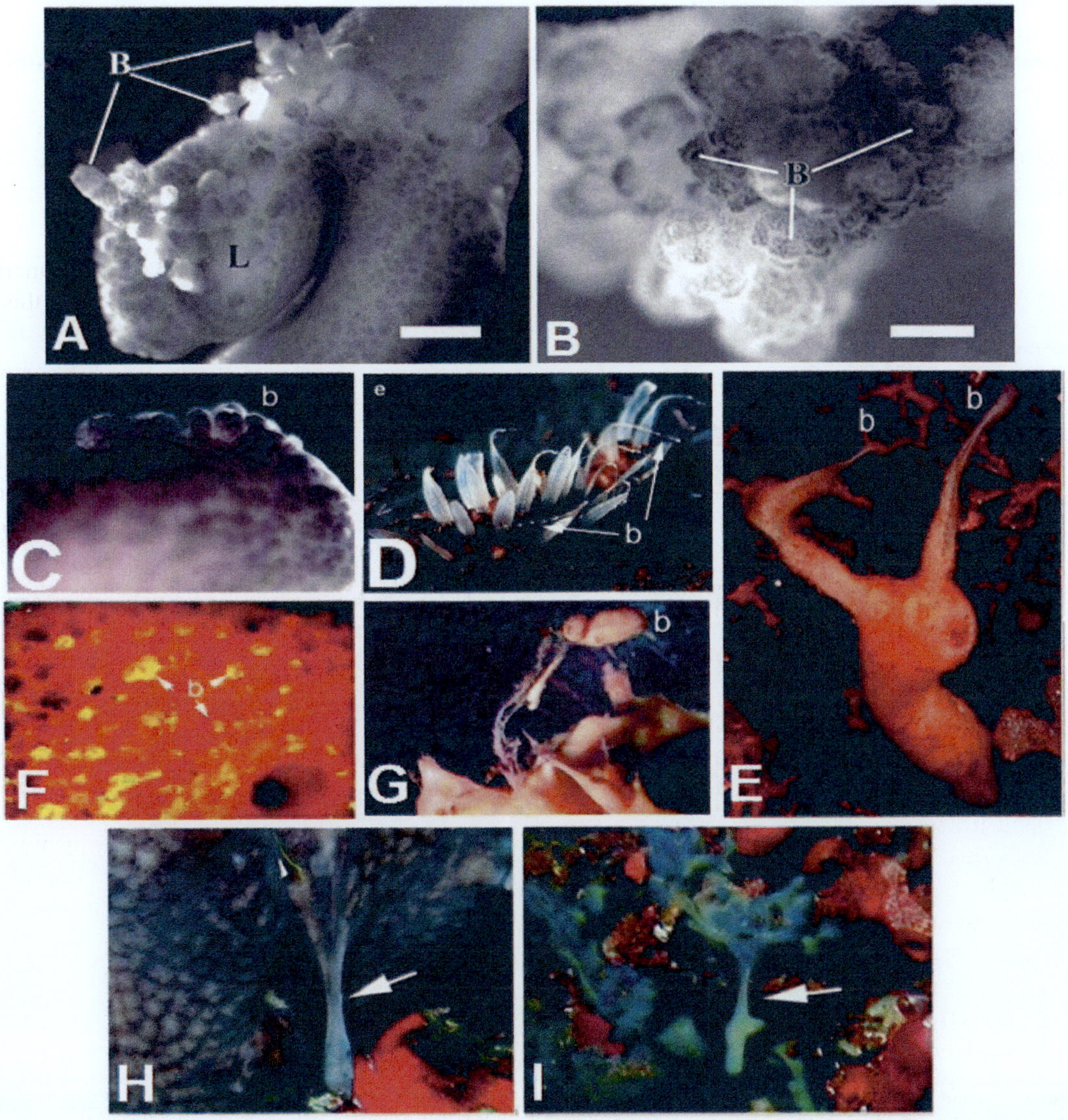

FIG. 4.3 In vivo agametic reproduction in sponges: (A and B) initial stage of process of budding in *Oscarella*: (A) *O. tuberculata*, bar = 1.5 mm; (B) *O. lobularis*, bar = 1 mm *(from Ereskovsky and Tokina, 2007)*; (C) Budding in *Oscarella lobularis*; (D) Papillar budding in *Polymastia arctica*; (E) Budding in *Haliclona* sp.; b = bud; (F) Budding of *Reniera fulva*; (G) Budding in *Aplysilla cavernicola*; (H) Fragmentation *(arrow)* in *Chondrosia reniformis*; (I) Fragmentation *(arrow)* in *Oscarella tuberculata*;

(Continued)

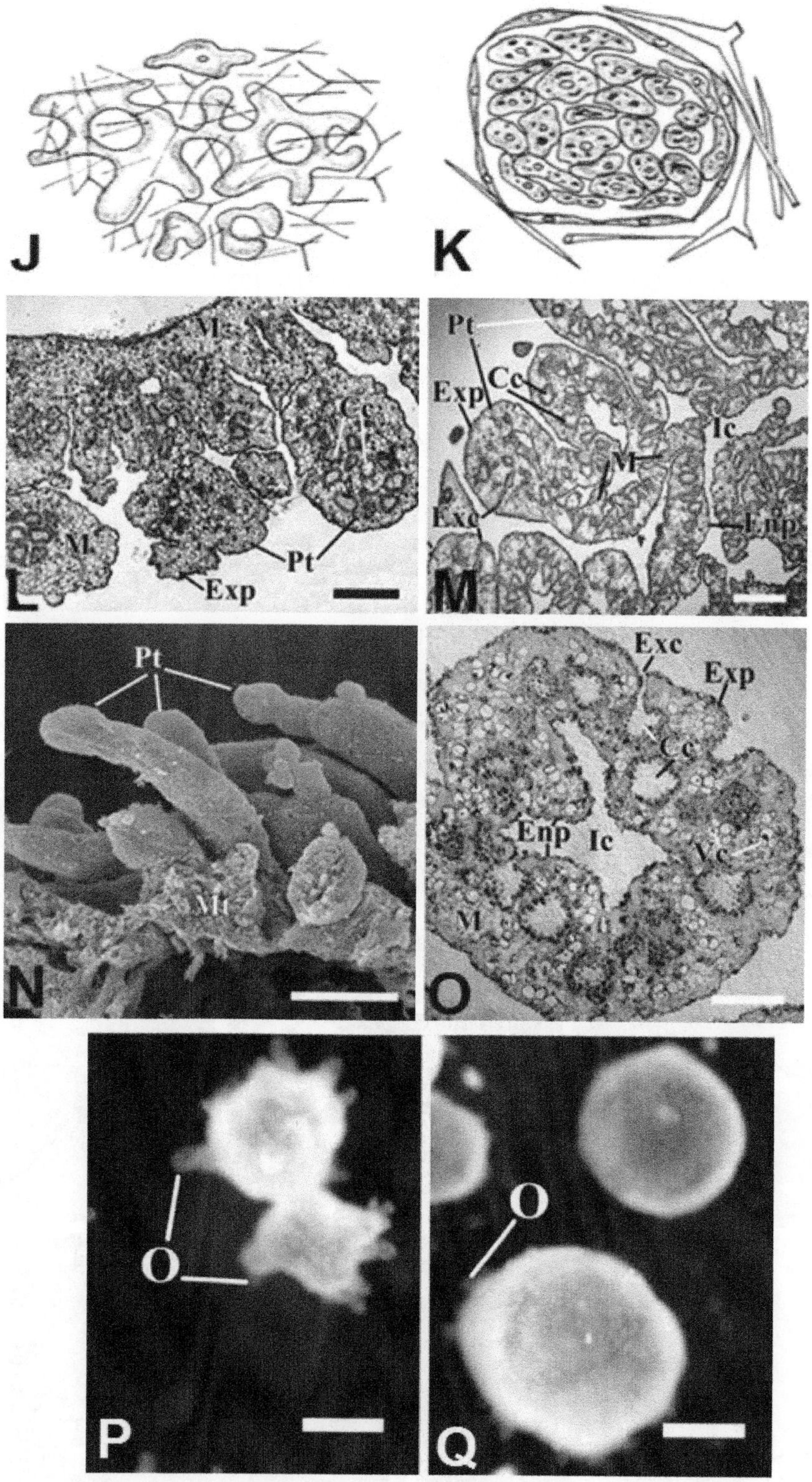

FIG. 4.3, CONT'D (J and K) reduction bodies in calcareous sponge, (K) section through reduction body showing mass of amoebocytes enclosed in 'epithelium'. This body differs from gemmule in the absence of the columnar layer with spicules *(from Hickman, 1973)* (L–O) budding in *Oscarella*, (L) *O. lobularis*, (M) *O. tuberculata* are longitudinal semi-thin sections showing the sponge ectosome (Exp) and protuberances (Pt), bar = 200 μm; (N) SEM micrograph of the nipple-like protuberances on second stage of bud development, bar = 25 μm; (O) perpendicular section of protuberance on second stage of budding, bar = 40 μm; Cc = choanocyte chamber, Enp = endopinacoderm, Exc = exhalant canal, Exp = exopinacoderm, Ic = inhalant canal, M = mesohyl, Mt. = maternal tissue, Pt = protuberances, Vc = vacuolar cells, scale bars L, M = 200 μm, N = 25 μm, O = 40 μm; (P and Q) floating buds, (P) of *O. tuberculata*, (Q) of *O. lobularis*, O = outgrowth, bar = 300 μm *((C–I) from Ereskovsky, 2010; (L–Q) from Ereskovsky and Tokina, 2007)*.

detached, differentiated and becomes a new stalked external bud. The bud attached to the substrate is similar to the rhagon developing. Among these are *Donatia (Tethya) deformis* and apparently *Cliona vastifica* (Fell, 1974). In *Donatia*, the buds which arise as club-shaped extensions of the sponge surface consist of bundles of spicules surrounded by masses of archaeocytes. The mature buds possess a long stalk, and their distal knob supports a varying number of long, slender rays. Under laboratory conditions, the buds are released from the parent sponge within 1 or 2 days following their initial appearance. The free bud develops into a small sponge, under suitable conditions.

Some other species of marine sponges produce buds without stalks, e.g. *Mycale contarenii*. Each bud in this species is situated in a depression in the external surface of the parent. The bud is formed by the pinching off of a small portion of the sponge containing all the types of cells (Fig. 4.4D and E). This portion develops into a bud by undergoing a

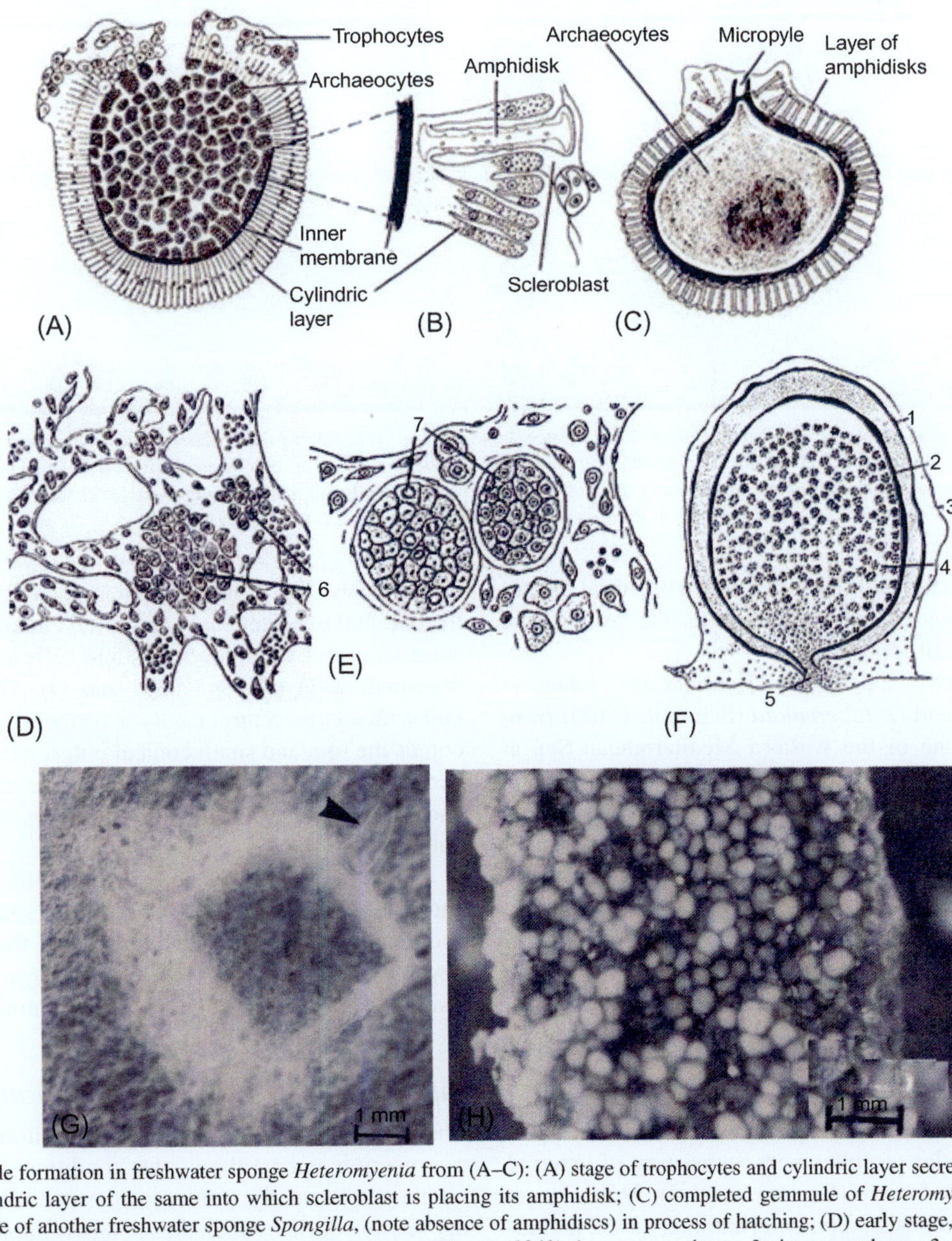

FIG. 4.4 Gemmule formation in freshwater sponge *Heteromyenia* from (A–C): (A) stage of trophocytes and cylindric layer secreting inner membrane; (B) portion of cylindric layer of the same into which scleroblast is placing its amphidisk; (C) completed gemmule of *Heteromyenia, (from Hickman, 1973)*. (F) Gemmule of another freshwater sponge *Spongilla*, (note absence of amphidiscs) in process of hatching; (D) early stage, (E) later stage of formation of gemmule of the marine sponge *Esperella* (=*Mycale*), *from Hyman (1940)*; 1=outer membrane, 2=inner membrane, 3='epidermis' of young sponge, 4=food-laden archaeocytes, 5=histoblasts streaming from micropyle, 6=assemblage of cells, 7=first layer encapsulated the mass of cells; (G) a living sponge (*Haliclona loosanoffi*) germinated from a mass of gemmules in the laboratory, note the oscular tube *(arrow)*; (H) living gemmules of *Haliclona loosanoffi* covering a blade of eel grass *((G and H) from Fell, 1974)*;

(Continued)

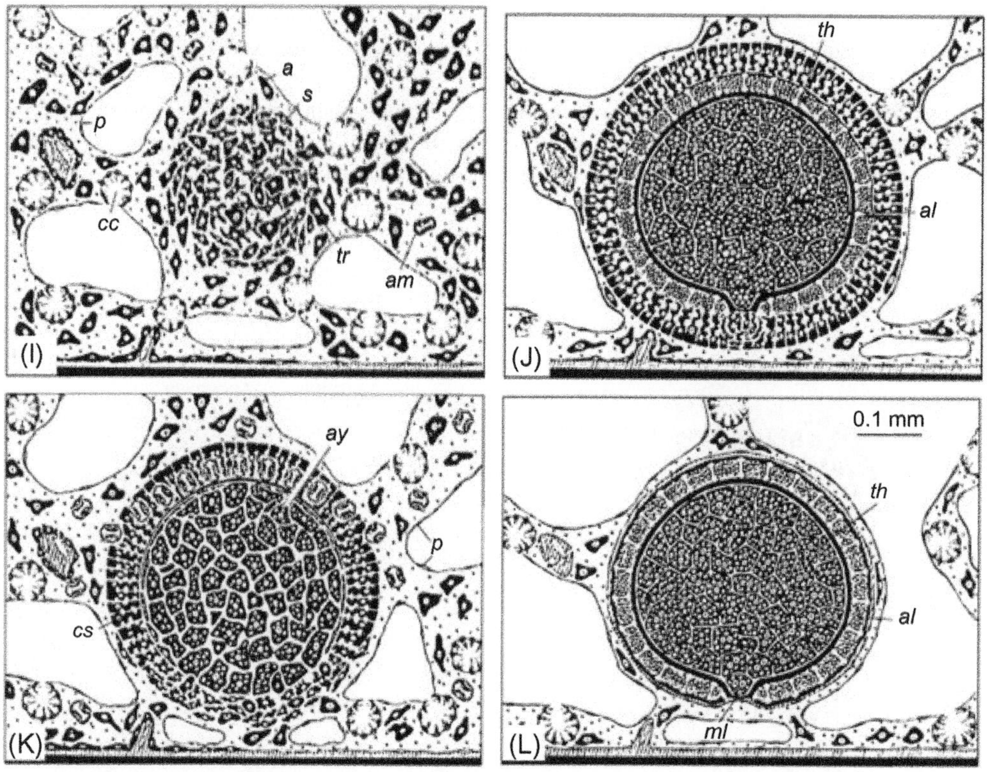

FIG. 4.4, CONT'D (I–L): scheme of gemmulogenesis of freshwater Haplosclerida: (I) beginning the process by the accumulation of cells (cell mass); (J) gemmule-envelope formation; (K) formation of alveolar layer, micopyle, and differentiation of thesocytes; (L) fully developed gemmule; a = archaeocyte, al = alveolar layer, am = amphidisoblast with a young amphidisc, ay = archaeocyte with yolk plates, cc = choanocyte chamber, cs = columnar layer of spongocyte, mi = micropyle = pinacocyte, s = spongocyte, th = thesocytes, tr = trophocyte *(from Ereskovsky, 1999)*.

complete reorganization of the cells, which begins at the periphery and progresses towards the centre. The result is the formation of a small sponge.

Budding in two species of *Oscarella* (*O. lobularis* (Schmidt, 1862) and *O. tuberculata* (Schmidt, 1868)) from the sublittoral zone of the western Mediterranean Sea at a depth of 5–25 m was studied (Ereskovsky and Tokina, 2007). These two species are similar in general shape and size (Fig. 4.1C), but the colour of *O. lobularis* is mostly purple or violet, whereas that of *O. tuberculata* is very variable. The authors reported that agametic reproduction is obligatory only in two Demospongiae orders. They reviewed the groups reproducing by gemmulation and those reproducing by external buds. Under laboratory conditions, the time required for bud development in these two species of *Oscarella* varies from 1 to 4 days, but is usually less than 2 days. Buds are formed in the marginal basal part of the sponge. The process proceeds in three well-defined stages. First, small irregular protuberances, consisting of external parental tissue, are formed (Fig. 4.3A and B). At this stage, the protuberances consist of parental external tissue, including the exopinacocytes, a thin layer of mesohyl, and some elements of aquiferous system (Fig. 4.3M and O). Second, they elongate and acquire a more regular, nipple-like shape; each becomes tube-like, with somewhat thicker walls in both species (Fig. 4.3O). The internal cavity is derived from the parental exhalant canal and small choanocyte chambers, short aquiferous system canals and cells are situated in the intermediate layer (Fig. 4.3M and O). Third, a spherical bud with a large central cavity is formed. The basal part becomes the foot and small conical outgrowths develop on the surface (Fig. 4.4A–D). The choanocyte chambers enlarge and open directly into the central cavity, without aphodus, but with an apopyle.

Cell composition of the buds is identical, quantitatively and qualitatively, to that in the upper parts of the adult sponge. Under laboratory conditions, the attached buds develop into small sponges in 4–10 days. It is similar to rhagon development, i.e. it has the syconoid organization.

Internal buds (gemmule formation)

The gemmules are formed in the late autumn in Europe, as a protection against the winter conditions, but in the tropics they are usually formed at the commencement of the dry season, during which the sponge is liable to desiccation. These internal buds occur in all fresh-water sponges and some marine forms of family Monaxonida, e.g. the *Tedania* and *Reniera* species. According to more recent studies, the gemmule of fresh-water sponges begins as an aggregation of archaeocytes, to which nurse cells (trophoblasts) are added (Fig. 4.4D and E). These cells contain granules of reserve food substances such as glycoproteins and lipoproteins. Other amoebocytes surround the sperm mass, forming layers of cylindrical cells. Meanwhile, scleroblasts secrete amphidisc spicules (Fig. 4.4A and B) that are arranged radially within the cylindrical layers; or between the inner and outer membranes surrounding the gemmule. A micropyle forms as an opening through all these enveloping layers, completing the formation of the gemmule (Fig. 4.4C).

Although the process of gemmule formation has long been studied, many points are yet to be clarified. Very early in the 20th century, a detailed study of the composition of gemmules concluded that the gemmule is most probably not formed from a single cell. There were no mitotic divisions observed during its formation. However, in its earliest stage it may be composed of a single cell or a mass of peripherally disposed cells.

In Autumn, fresh-water sponges belonging to the family Spongillidae produce a large number of gemmules within the body. The animals then disintegrate and some of the gemmules remain within the remnant of the parent sponge, while others fall to the bottom (Fig. 4.4H). These gemmules usually show a yellow tint, and are able to withstand freezing, cold, desiccation and drying during winter. When the temperature rises during spring, the gemmule germinates and the contained cells escape through the micropyle (Fig. 4.4F and G). The cells undergo histogenesis and morphogenesis to develop into young sponges (Okada, 1968).

There are two types of gemmules (Fell, 1974):

1. **Gemmules that develop directly into adult sponges:** It was believed that a number of marine sponges produce this type of gemmule. Two genera, *Haliclona* and *Suberites* have been studied in detail even in old references. This type of gemmule has an irregular hemispherical body. Following the degeneration of the parent sponge, the gemmule becomes flattened against the substratum to which it adheres. In many cases, nearly all the substratum that is occupied by the sponge, becomes covered by gemmules. These gemmules may attach to shells, rocks, eelgrass, or other similar objects. *Suberites domuncula* encrusts a hermit crab or dromiid crab. The gemmules of this type in different species vary in size from about 200 to ≈1000 μm in diameter. Each gemmule consists of a mass of cells, enclosed within a protective capsule composed of spongin substance. The capsule of some sponges is supported by spicules, e.g. *Haliclona*. In *Suberites domuncula* the capsule is aspiculous and it consists of two distinct layers. The inner layer is of flattened 'epithelial' cells, which are distinct from the cells of the gemmule. Others may be multilayered with spaces between the layers, e.g. *Laxosuberites lacustris*. In these gemmules there is no micropyle. The cells comprising a gemmule are of relatively uniform size and appearance. They are closely packed together and filled with numerous reserve granules. Small gemmules may be formed by unequal subdivision of large ones. In some species, gemmule formation may occur together with gametic reproduction.

   Dispersal

   Gemmules that are attached to eelgrass or other plant material may be dispersed by currents when the plants become fragmented, e.g. *Haliclona loosanoffi* (Fig. 4.4H). They may also be detached from firm substratum such as shells. Gemmules that are attached to rocks become exposed during the dry season, the gemmules become dry, and many groups of them peel away from the substratum and, apparently, are carried into new localities by wind and water currents. The air spaces in the capsule render the gemmules very light. In all cases, there is evidently a high mortality.

   Germination

   There is not enough information describing this aspect. Presumably, the capsule ruptures by an enzymatic process; and the cells emerge, divide, and progressively form a small adult. Apparently, in most cases, many gemmules give rise to a single sponge (Fig. 4.4F and G).

2. **Gemmules that develop into larvae**: this type of gemmule has been described in a number of marine sponges, including *Mycale (Esperella) fibrexilis*, *Euplectella marshalli*, *Callyspongia diffusa*, and *Spongia reticulata*. In most cases, the gemmule consists of a compact mass of arcaeocytes enclosed in a single layer of flattened follicle cells and/or a thin layer of spongin. This mass may increase in size by growth of its component cells (cell divisions and/or fusion with other small aggregates). In many of these species, the gemmules ultimately develop into free swimming parenchymula larvae. The formation of gemmules of this type is frequently associated with the disorganization of the endosome of the parent sponge. The gemmules of *Callyspongia diffusa* are formed by the aggregation of syncytial masses (disappearance of cell boundaries in the gemmule), which are produced by the fusion of a single amaebocyte with a number of choanocytes. These masses frequently fuse together, forming a single syncytium. Then, during the development of the gemmule into a larva, the syncytium breaks up into its component cells. If these syncitia are believed to occur, it is desirable to verify their existence by electro-microscopy. However, in some cases the agametic origin of these larvae has been disputed, especially in the descriptions recorded in the last years of the 19th century (1896, 97, 94).

A significant observation was recorded by Marshall (1884) (cited in Okada, 1968) that *Spongilla lacustris*, a fresh-water species, undergoes alternation of generations like that of Cnidaria; i.e. the animals which develop from gemmules produce reproductive cells, which unite to give rise to sponges. The latter, in turn, form gemmules.

Gemmule formation of haplosclerid sponges (Demospongiae, Ceractinomorpha) was studied in comparison of morphogenesis with gametic reproduction (Ereskovsky, 1999). Gemmulogenesis usually begins after the completion of gametic reproduction or during the period of embryogenesis. This vital activity was most successfully studied in freshwater Spongillidae. It proceeds in a series of stages (Figs 4.4I–L). It begins with movements of gemmular nurse cells, archaeocytes and spongocytes, particularly towards certain areas in the mesohyl (Fig. 4.4I). This study may give a definite view into the cells beginning the gemmule. The moved cells in a suitable situation make up dense aggregations 250–500 µm in diameter. These three types of cells might be considered to have common features: a large nucleus (about 6 µm) with a nucleolus; numerous Golgi complexes; and ribosomes, mitochondria, and cytoplasmic inclusions of different sizes. Within the aggregations, the archaeocytes actively phagocytose the nurse cells accumulating in their cytoplasm. This is the characteristic yolk plate of these cells. During the development of these yolk plates, two layers are formed around the aggregations. The inner layer consists of flattened archaeocytes, and the outer one of flattened spongocytes which makes up a columnar layer. During the formation of the latter, the microsclerocytes of the mesohyl bearing developed microsclerae (amphidiscs or rhabds), migrate to the developing gemmule to be incorporated into its columnar layer (Fig. 4.4J).

Gemmule-envelope formation begins at one pole and gradually extends towards the opposite pole. Later on, a micropyle arises in this area of the envelope, devoid of spicules (Fig. 4.4L). The fully developed gemmule consists of a homogeneous mass of cells, the thesocytes, enveloped in a dense three-layered wall. The thesocytes are characterized by the presence of a large number of yolk plates (in freshwater sponges) or yolk granules (in marine species). These thesocytes of spongillids are two nucleated, but in marine species are mononucleate. However, gemmulogenesis is similar in freshwater and marine Haplosclerids.

This study added evidence that gemmulogenesis and oogenesis are similar in certain stages (in species that have the two types of reproduction in their life cycle). For example, both gemmule and oogonium develop from nucleolate amoebocytes (archaeocytes); the pattern of development of yolk type inclusions is also similar in both processes, vitellogenesis of the oocyte and gemmulogenesis; the same nurse cells are involved in both processes; and there is phagocytosis of whole nurse cells or their parts in both processes. Ereskovsky (1999) stated that it is assumed that the movement of nurse cells towards the centre of gemmulogenesis is done by chemotaxis. During oogenesis, the nurse cells also show chemotaxis to chemical substances bound to the oocyte membrane. An old statement considered that 'the gemmule, in fact, is physiologically equivalent to the ovum at the close of segmentation', i.e. to a mass of blastomeres enclosed in a special capsule, and capable of developing into one or another form of cell (Minchin et al., 1900).

Other agametic types of reproduction

Reduction bodies

This type may occur during unfavourable conditions. It involves masses of many kinds of amoebocytes with 'epidermal' covering (Fig. 4.3J and K), that persist after the disintegration of the parent sponge. These reduction bodies can develop into complete sponges when conditions become favourable.

Fragmentation

The peripheral portions of sponges, especially those of branching structure, may be fragmented by water turbulence, predators or other causes. Probably, in most cases, the broken portions are carried away from the parent into other situations where they perish. However, such fragments may occasionally give rise to new sponges. It is well known that sponges have high regenerative power, so it leads finally to the dissociation of the cells of the organism. Wilson (1907) discovered that the cells of sponges can be readily dissociated by squeezing pieces of sponge through fine silk cloth and that the dissociated cells of some species can reaggregate and reform as a functional sponge. Among the sponges of which reconstitution has been studied are: *Microciona prolifra*; *Callyspongia diffusa*; *Spongilla lacustria*; and *Sycon raphanus* (Fell, 1974). Two species were photographed during the process of fragmentation of a portion from the body of the sponge, *Chondrosia reniformis* (Fig. 4.3H) and *Oscarella tuberculata* (Fig. 4.3I) (Ereskovsky, 1999). This phenomenon, which characterizes the sponge, is used to increase the production of commercial sponges naturally. It has also made them a good material for experimental studies of cellular adhesion. Many studies on the specificity and the mechanism of sponge cell adhesion have been performed by different scientists all over the world.

Factors influencing agametic reproduction

Little is known about the factors influencing agametic reproduction in marine sponges. However, consideration is only given to those which influence the production and germination of gemmules that develop directly into adult sponges. One important factor is water temperature. In temperate regions, gemmule production flourishes during spring; e.g. *Haliclona loosanoffi* survives the winter in the form of gemmules, which germinate when the water temperature rises in spring.

Regeneration

In sponges, the capacity for regeneration is highly developed, broken fragments of colonies being capable of re-attachment and soon becoming functional individuals. Regeneration is employed in the propagation of commercial sponges artificially in the Mediterranean Sea, e.g. pieces of sponges called 'cuttings' are attached to cement blocks and dumped into the water. Regeneration and several years' growth produce a sponge of marketable size. Cellular suspensions prepared by squeezing sponges through bolting silk into seawater are also capable of complete reorganization (Okada, 1968). Re-construction involves the random movement of individual cells, which adhere on contact and form cellular aggregates. These become surrounded by a peripheral layer of pinacocytes and the 'mesenchymal cells' move to take up their usual position. All cell types except scleroblasts have been identified in cellular suspensions, and it appears that these cells reform from archaeocytes.

Gametic reproduction

Sexual differentiation

Among the various marine sponges, there are examples of gonochorism (G), i.e. separate sex, e.g. the viviparous (V) sponge *Hippospongia communis*; successive hermaphroditism (SH), e.g. the (V) *Halisarca dujardini*; and contemporaneous hermaphrodism (simultaneous) (CH), e.g. (V) *Oscarella (= Halisarca) lobularis* and (V) *Spongilla lacustris*. There is no sexual dimorphism in the gonochoric sponges. Contemporaneous hermaphroditism can be readily detected in most cases. In these hermaphrodites, the male and female gametes are generally not restricted to different regions of the sponge, but both oocytes and spermatic cysts are intermingled. However, in (V) *Oscarella* the spermatic cysts tend to be situated superficial to the oocytes. Not all the specimens possess both kinds of gametes. Some contain oocytes and spermatic cysts and others contain only oocytes. Many specimens have been examined; none was found to contain only male gametes. A similar situation may possibly exist in *Grantia compressa*. On the other hand, the distinction between gonochorism and successive hermaphroditism is not always easy to make (Fell, 1974).

Gonads

No true gonads or reproductive ducts are found in sponges. In most cases, a major portion of the sponge body is involved in reproduction. Some types can be observed:

1. In a number of sponges including, e.g. *Polymastia mammilaris*; *Haliclona ecbasis*; and many calcareous sponges, the gametes are distributed more or less completely throughout the mesohyl and are surrounded by relatively normal somatic tissue containing flagellated chambers. This type of gonad is referred to as a 'diffuse gonad' (Gardiner, 1972). It appears that this type occurs in both oviparous and viviparous forms that produce relatively small eggs.
2. In other sponges, including *Microciona prolifera*, *Oscarella*, the fresh-water *Spongilla lacustris*, and *Ephydatia fluviatilis*, the mesohyl is differentiated into two, more or less, distinct zones. These zones are a superficial zone with canals and flagellated chambers, and a basal or central zone with no, or relatively few, flagellated chambers. Within the latter zone, the gametes and embryos develop. This type seems to occur largely in viviparous sponges that produce large, yolky eggs.
3. In *Halisarca*, in specific, the spermatic cysts occur in an intermediate zone displaced somewhat towards the external surface of the sponge (Lévi, 1956).

Origin of the germ cells

Histological studies of various types of cells of sponges indicate that the germ cells may arise from two different sources: from choanocytes and from archaeocytes. However, this view is still uncertain, and the origin of these germ cells has not been established (Fell, 1974). However, studies on calcareous sponges during the second half of the twentieth century suggest that both the oocytes and the spermatocytes derive either directly from choanocytes or flagellated larval cells; some authors described their origin in wandering amoebocyte cells; while others believe that they originate as choanocytes (Okada, 1968). Okada presented a diagram, after Dendy, describing the collar-cell of *Grantia compressa* transforming into oogonium. At first, it becomes rounded and slips from the others and attains the amoeboid form, then, the oogonium develops into oocyte which receives trophocyte caught by the nurse cell (Fig. 4.5D). Evidence was presented for the differentiation of oogonia from pinacocytes in *Ascandra minchini* and other calcinea (Borojević, 1969).

On the other hand, the oocytes of most acalcareous sponges appear to derive from archaeocytes, while the spermatocytes seem to develop in some cases from archaeocytes and in other cases from choanocytes. This latter case seems to be well established. In *Aplysilla rosea*, the transformation of flagellated chambers into spermatic cysts has been studied with the electron microscope, and the continuity between typical choanocytes and primary spermatocytes was demonstrated. If the transformation of choanocytes into archaeocytes was general among sponges, then the germ cells of acalcareous sponges would also appear to drive directly or indirectly from choanocytes. The recent ultra-structure studies may present convenient evidence that the choanocytes are the origin of the male gamete. Riesgo et al. (2007a,b) described in detail the spermatogenesis in *Corticium candelabrum* and documented this result by photos. Moreover, Maldonado and

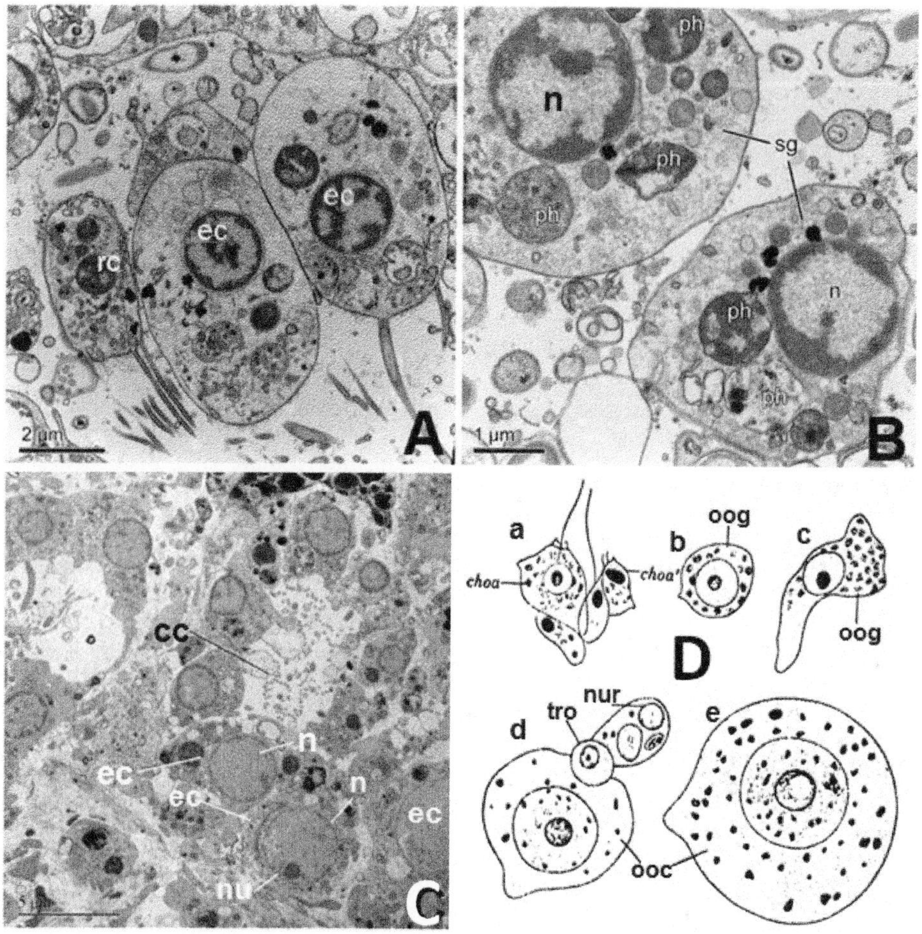

FIG. 4.5 Transformation of choanocytes to spermatogonia in spermatogenesis of the sponge *Petrosia ficiformis (from Maldonado and Riesgo, 2009)*: (A) regular choanocyte (rc) adjacent to choanocytes (ec) that are enlarging to become spermatogonia; (B) spermatogonia (sg) with a nucleus (n) having peripheral chromatin condensation and cytoplasm containing inclusions and phagosomes (pH) that are characteristic of the choanocyte cytoplasm; (C) showing choanocyte chamber in which the regular choanocyte becomes abnormally large with swollen nuclei and leaving the chamber to become spermatogonia; cc = choanocyte chamber, nu = nucleolus *(from Riesgo and Maldonado, 2009)*; (D) formation of egg from collar cell (choanocyte) in the sponge *Grantia compressa (after Dendy, from Okada, 1968)*; choa = choanocyte, choa' = collar cell beginning to change into oogonium, nur = nurse cell, ooc = oocyte, oog = oogonium, oog' = amoemboid oogonium, tro = trophocyte.

Riesgo (2009) confirmed this information by presenting photos showing the transformation of choanocytes to spermatogonia (Fig. 4.5A and B) in the demosponge *Petrosia ficiformis*. In the same year, 2009, the same authors presented more confirmation that spermatogonia are derived from choanocytes (Fig. 4.5C). Riesgo and Maldonado (2009) studied the formation of V-shaped spermatozoon in the demosponge *Crambe crambe* and observed the early stages of transformation of the choanocyte, which enlarges with swelling of its nucleus and then moves out of the choanocyte chamber (Fig. 4.5C). These results can be considered as more evidence to support the ultrastructural investigations of choanocytes as the origin of spermatogonia (Fig. 4.5D) in the spermatogenesis of the two species of family Spongillidae (Paulus, 1989).

Gametogenesis

There are no definite testes in sponges; the male gametes are formed inside spermatic cysts (Fig. 4.9). There are three patterns of cyst formation in acalcareous sponges:

1. In *Hippospongia communis* (V, G) all of the choanocytes of a flagellated chamber apparently lose their collar and flagellum and develop into spermatogonia. Entire chambers do not transform into cysts. Rather, some swollen choanocytes from different chambers migrate to the mesohyl for subsequent re-aggregation. The spermatogonia then constitute a more or less solid mass, which is surrounded by flattened follicle cells and/or spongin. The follicle cells may be derived from mesohyl amoebocytes.

2. In sponges including *Halisarca dujardini* and the fresh-water sponges, e.g. *Spongilla lacustris*, amoeboid cells aggregate at various points in the mesohyl, forming the primordial of the spermatic cysts. These masses subdivide into groups; each becomes enclosed within follicle cells comparable to that occurring in the first pattern.
3. In *Reniera simulans* (CH), a single sperm mother cell divides asymmetrically, producing one daughter cell with prominent nucleolus and another with no visible nucleolus. The former apparently becomes a follicle cell, which progressively envelops the mass of spermatogonia produced by the division of the latter. In this, each spermatic cyst contains 32 developing spermatocytes, which are surrounded by a single extended follicle cell. The size differs with the species.

Paulus (1989) illustrated a semi-diagram (Fig. 4.8) showing spermatogenesis and spermatocyst formation in *Spongilla lacustris* (left half of the illustration) and *Ephydatia fluviatilis* (right side) from the start of the transformation of the choanocyte in the flagellated chamber. More than one choanocyte from the chamber begins changing and moves to lie free in the mesohyl. They are bounded by pinacocyte-like cells (building cells) which are similar to the archaeocytes. These cells surround the spermatogonia, which undergo spermatogenesis. The cells inside the cyst may develop synchronously and the cyst itself develops. At first, the small spermatocyst has a diameter of 10–20 μm, but it develops by the uptake of free spermatogonia, extension of the cyst-building cells and by the fusion of spermatocysts.

There are three patterns for the development of the cysts:

1. All the cells in all the cysts of the sponge may be synchronized in their development, as it is formed in *Polymastia mammillaris*. This pattern is found in oviparous sponges.
2. In many sponges, including *Hippospongia communis*, the differentiation of the cells of a cyst is synchronous, but different cysts are at different stages of development. This case occurs in viviparous sponges.
3. In other sponges, such as *Oscarella lobularis*, the development of the cells in any particular cyst is asynchronous.

The orientation of the spermatozoa within the cyst differs. They may be oriented with their flagellae directed towards one pole of the cyst (Fig. 4.9E), e.g. *Corticium candelabrum*, and *Hippospongia comunis*. In other sponges, e.g. *Halisarca dujardini* and *Microciona microjoanna* (Fig. 4.9B), the spermatozoa are randomly oriented. In the fresh-water *Spongilla lacustris* and *Ephydatia fluviatilis*, the spermatozoa may be oriented with their heads central and their tails extended towards the periphery (Fig. 4.9C and D) or at random. This is according to whether the differentiation of the cells in the cyst is synchronous or asynchronous, respectively.

Spermatogenesis

Spermatogenesis in sponges has been studied in a number of forms, and appears to conform to one of two types (Minchin et al., 1900). Poléjaeff (1882) described the first type (Fig. 4.6A) in *Sycandra (= Sycon) raphanus* (see Roosen-Runge, 1977) and Fiedler described the same in *Spongilla* (see Minchin et al.). They described that each amoebocyte gives rise to and/or transforms into a primordial germ cell. Okada (1968) redescribed the process in *Sycon* species and reported that the nucleus of the germ cell divides into two dissimilar nuclei: one moves to the periphery, while the other remains in the centre (Fig. 4.6A (b)). The peripheral nucleus, with a portion of the cytoplasm, separates from the main bulk of the primordial cell, forming a smaller covering cell or spermatocyst. The main bulk is the spermatogonium proper, which undergoes spermatogenesis (Fig. 4.6A (c)–(f)). The covering cell remains undivided. The product of the sperm then aggregates in a mass covered with an envelope, formed of this covering cell, constituting a sperm ball (Fig. 4.9A). In ancient times, Poléjaeff described that most siliceous and horny sponges had a different mode of follicle formation. A covering of flat capsule cells developed secondarily from the surrounding mesohyl, and he speculated on the mechanism involved.

The second type of spermatogenesis is essentially similar, but differs in the absence of the covering cell. The whole spermatogonium gives rise to a sperm ball, which may be enclosed in an adventitious envelope or follicle derived from the cells of the mesohyl. It appears that the differentiation of the spermatozoa from primary spermatocytes may occur within approximately 2 weeks (temperature of the water was not recorded).

Very recent studies dealing with gametogenesis in sponges not only develop this aspect, but also clarify its fine details, especially the structure of the gametes themselves. Riesgo et al. (2007a,b) and Maldonado and Riesgo (2009) used Demospongiae sponges from the Catalan coast (Spain, north-western Mediterranean). The former work deals with *Corticium candelabrum*, which is a small sponge, orange-brown in colour, cushion-shaped to lobate in form, measuring up to 1.5 cm in thickness and up to 10 cm in its largest diameter. The latter is on *Petrosia ficiformis*, typically with sparsely scattered individuals, growing preferentially at semi-shaded rocky walls and overhangs. In both animals, gametogenesis proceeds in nearly the same stages. In *C. candelabrum*, both spermatogenesis and oogenesis occur next to each other in the same portion of the body of the mother (Fig. 4.6B). Histological examination of sections from about 150 individuals of *C. candelabrum* confirmed that this species is simultaneously hermaphroditic. Developing oocytes occur all over the year, while sperm production is restricted to late spring and early summer.

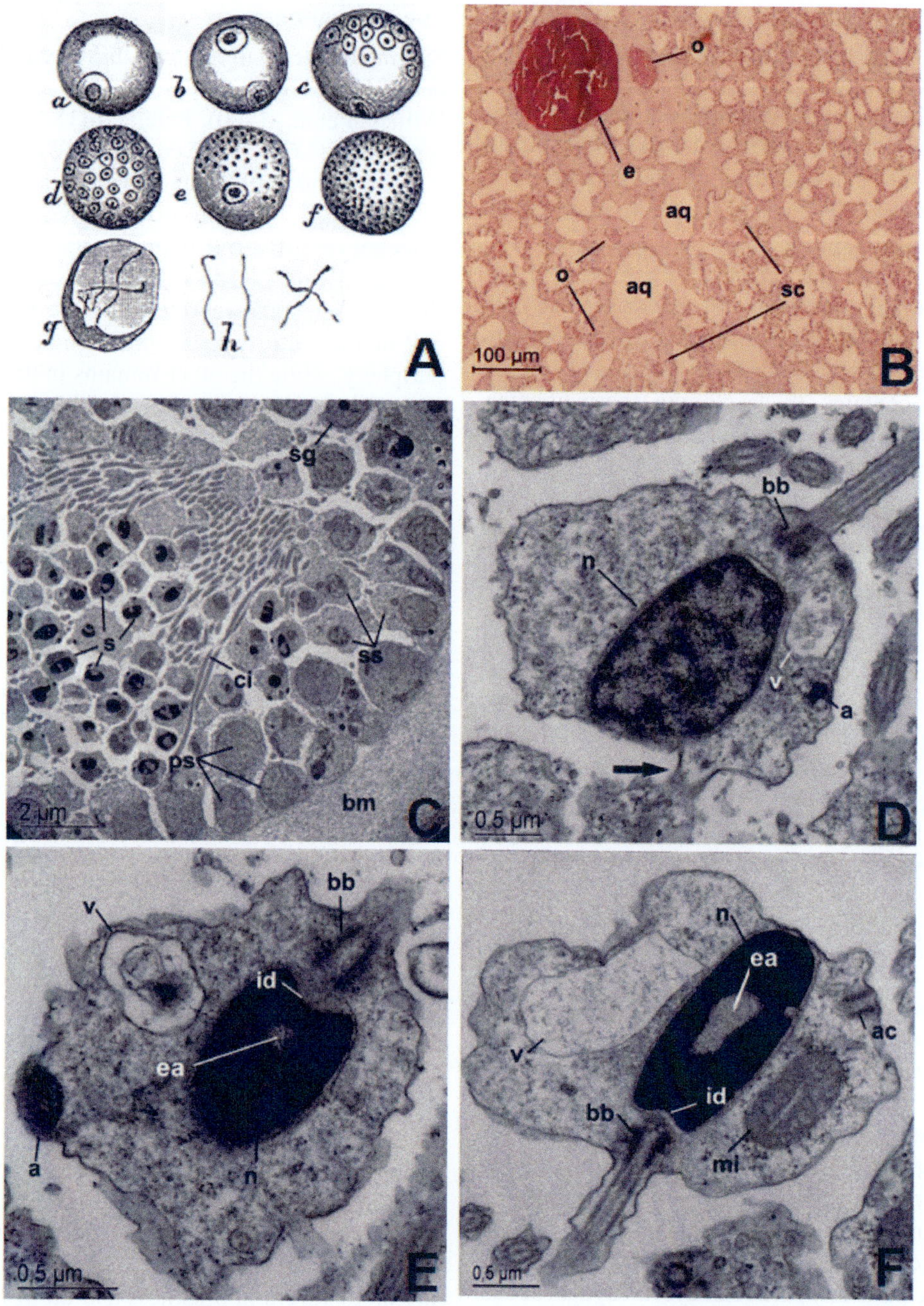

FIG. 4.6 (A) (a–h) Diagrams describing development of sperm of *Sycon raphanus*, h = matrure spermatozoa *(after Poléjaeff, 1882 from Minchin et al., 1900)*; (B) histological section showing co-occurrence of early stage oocytes (o), embryos (e), and spermatic cysts (sc) in the mesohyl of the sponge, *Corticium candelabrum*; (C) general view of a spermiocyst enveloped by a basement membrane-like structure (bm) and containing all the stages of spermatogenesis except the sperm, spermatogonia (sg), 1st spermatocyte (ps), 2nd spermatocyte (ss), spermatids (s) with the cilium (ci) directed towards the cyst lumen; (D) L.S. through 2nd spermatocyte showing a lengthened nucleus (n), a basal body of the cilium (bb), an incipient acrosome (a), an electron-clear vacuole (v), and a cytoplasmic bridge *(arrow)* with a sister spermatocyte; (E and F) L.S.s of spermatids showing advanced stages of spermiogenesis, more inclusions, nucleus with a peripheral indentation (id) and a central electron-clear area (ea), a single large mitochondrium (im), the accessory centrosome (ac) that has migrated to the cell side opposite the cilium insertion *((B–F) from Riesgo et al., 2007a,b)*.

Spermatocysts are found underneath the pinacoderm of the canals, usually around growing oocytes and embryos (Fig. 4.6B). On the other hand, the oocytes develop in the mesohyl immediately below the endopinacoderm of the exhalant canals (Figs 4.6B and 4.11B). Interestingly, production of oocytes in *C. candelabrum* is a continuous process, with a peak from October to January, i.e. during lower temperature season. When the oocytes are in maximum density, the spermatic cysts can be first detected in March during the same season. The spermatic cysts develop in those mesohyl areas where oocytes are already developing. In *Petrosia ficiformis*, oocytes and spermatocysts never co-occurred in the same individual (i.e. gonochorism), evidenced by histological studies. Observation and study for two successive years (2003–05) indicated that oogenesis extended for several months each year, with some interannual variability. There were no detectable spermatocysts (Fig. 4.11A and B) during this period; they were first noticed in 2006 as small, loose cell groups with poor affinity to stains. Spermatogenesis is a rapid process in this species, occurring in the mesohyl of males no longer than 2.5 weeks, a year, usually in late November.

Early spermiocysts of *C. candelabrum* are irregular in shape and externally limited by a basement membrane, lacking cellular follicle (Fig. 4.6C). The cells are stratified, arranged from the basement membrane to the lumen, with concentric, multilayered cell strata. The latter is comprised of pseudo-cubic to round spermatogonia derived from choanocytes, as proven by more than one piece of evidence (Fig. 4.5) from the same work; primary spermatocytes, as spermatogonia contain several mitochondria which engaged in fusion; secondary spermatocytes (Fig. 4.6C and D), are situated internal to the primary oocytes, the Golgi apparatus still clearly obvious, starting forming the acrosomal vesicle, a large electron-clear vacuole of unknown function also appeared in the cytoplasm; and spermatids respectively (Fig. 4.6C). The spermatids (Fig. 4.6E and F) have an oval nucleus and visible Golgi apparatus continue forming the acrosomal vesicle and they comprise multilayered stratum internal to the secondary spermatocytes (Fig. 4.6C). The nucleus has a small indentation in the region adjacent to the cilium insertion, and anchoring points linked the typical 9+2 basal body of the cilium to the surrounding cell membrane. All these cells are monociliated at all stages of spermatogenesis, with the cilium oriented towards the lumen. Again, all the cells are interconnected by cytoplasmic bridges (Fig. 4.7A) resulting from incomplete cytoplasm divisions during spermatogenesis, as occurs in other metazoa. The spermiocyst of *C. candelabrum* is membrane-bounded and consists of a 2×1 μm, very electron-dense body with an electron-clear area (Fig. 4.13B and C).

Structure of the fully formed spermatozoon. The sperm of sponges consists of the normal parts: head, middle piece, and tail. The head may be spherical, ovoid, conical, pear- or heart-shaped, depending on the species. The head contains a small nucleus with condensed chromatin. It measures about 0.5–1.5 μm, or maybe more, in length. The middle piece contains the mitochondria and may assume a variety of forms. The long tail (flagellum) extends from a pair of centrioles situated near the nucleus. The length of the entire spermatozoon is generally about 10–50 μm. It is a feature for the spermatozoon of Porifera that it lacks the acrosome, the acrosome is not clearly seen, or the acrosome is apparently absent (Fell, 1974, 1997). However, this acrosome may assume a variety of forms. Reiswig (1970) studied the sperm of three species of tropical Demospongiae (Fig. 4.10E). The spermatozoa of *Verongia archeri* (Higgin) (Dictyoceratida) (Fig. 4.10E (a)) possess a head 2.6 μm in diameter, which is almost completely filled by the uniformly dense-staining nucleus. This is followed by a tail 37–44 μm long, without a recognizable mid-piece. The sperm of *Geodia* (obtained from the diffuse turbid effluent emitted from the specimen) are slightly smaller but possess a more complicated structure (Fig. 4.10E (b)). The head, 0.88–1.4 μm in diameter, is again almost entirely filled by the nucleus and is capped anteriorly by a short acrosomal cone. A mid-piece sac is slightly smaller, and attached to the head at the junction of the 3–35 μm long flagellum. No structural details are visible within the mid-piece after normal staining with haematoxylin or toluidine blue. In another review, a definite acrosome is seen in the sperm of *Grantia compressa*, *Reniera simulans*, *Hippospongia communis*, and *Octavella galangaui* (Fell, 1974). Fell further confirmed its presence in 1997, in *Oscarella lobularis*, but *Suberites* possess a few Golgi-derived vesicles located anterior to the nucleus.

More recent studies proved that, generally, the acrosome is present in spermatozoa of sponges. The spermatozoon of *Petrosia ficiformis* (Maldonado and Riesgo, 2009) has a nucleus, three large mitochondria, flagellum provided by a basal body and accessory centriole. The section through this spermatozoon shows a large battery of proacrosomal vesicles (Fig. 4.9C and D). Another demosponge animal, *Corticium candelabrum* (Riesgo et al., 2007a,b) has a well-developed acrosome in its spermatozoon, which begins to be differentiated in the spermatid during spermiogenesis (Fig. 4.6E and F). The spermatids are still connected to each other with cytoplasmic bridges (Fig. 4.7A). The fully formed spermatozoa are no longer connected and the bridges disappear, so each can be seen swimming freely (Fig. 4.7B). On the other hand, Paulus (1989) described the details of spermiogenesis of *Spongilla lacustris*, especially the changes occurring in the nucleus and the flagellum (Fig. 4.7H), but no acrosome formation.

The unexpectedly sophisticated, V-shaped spermatozoon in the demosponge *Crambe crambe* was discovered and studied by Riesgo and Maldonado (2009). A large population of this animal is established on sublittoral rocky bottoms of the north-eastern Mediterranean coast of Spain. Early stages of spermatogenesis proceed as in other sponges; spermatogonia are derived from choanocytes that doubled their nuclear size (mean diameter 4.15 μm) and cell size (6.12 μm in mean height and 5.76 μm in mean width). These swollen choanocytes (Fig. 4.5C) were observed losing their collar and leaving the chamber. But spermiogenesis shows peculiar features. It appears that this activity begins from the secondary spermatocyte, as a process of chromatin condensation, starting at the inner side of the

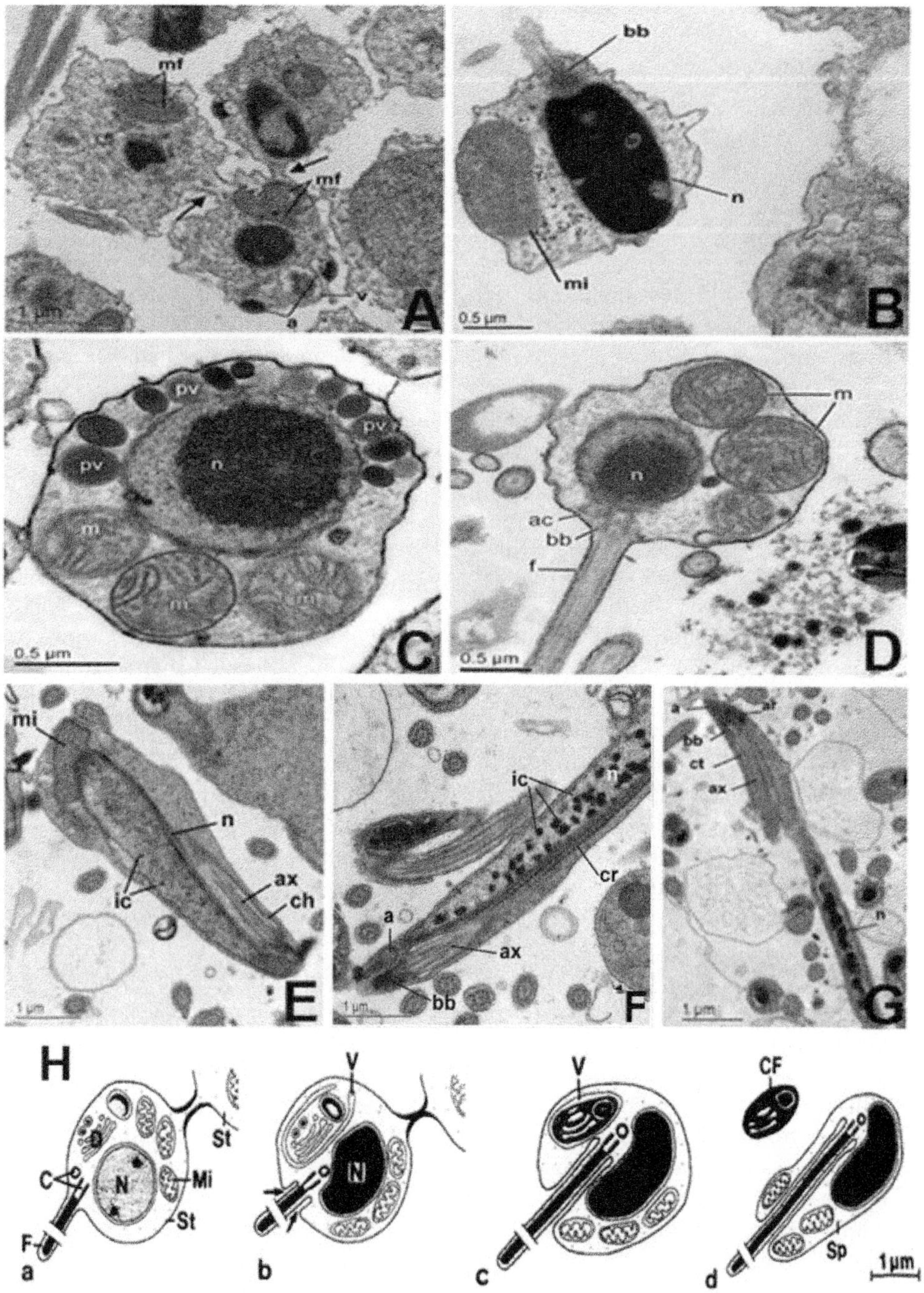

FIG. 4.7 Photomicrographs describing final stages of spermiogenesis in sponges. (A and B) in *Corticium candelabrum (from Riesgo et al., 2007a,b)*, (A) shows bridges *(arrows)* connecting sisters spermatids, notice the fusion of mitochondria (mf), the C-shaped acrosome represented by two acrosomal vesicles (a), electron-clear vacuole (v); (B) enlarged view of a free swimming spermatozoon of *C. candelabrum*; (C and D) in *Petrosia ficiformis (from Maldonado and Riesgo, 2009)*, (C) 'angling' section of the spermatozoon showing a large battery of proacrosomal vesicles (pv), nucleus (n), large mitochondria (m); (D) L.S. in the spermatozoon of *P.f.* ac = accessory centriole, bb = basal body, f = flagellum; (E–G) L.S.s of spermatids of *Crambe crambe* showing stages of spermiogenesis of V-shaped spermatozoon *(from Riesgo and Maldonado, 2009)*: (E) spermatid with lengthened nucleus; (F) shows axoneme (axe), the striated ciliary rootlets (cr), the nucleus (n), with multiple 'initiation centres' (ic), a developing acrosomal complex (a) occurs close to the basal body (bb); (G) shows the helical arrangement of the chromatin in the nucleus of fully formed spermatozoon, acrosomal rod (ar), and the cytoplasmic tunnel (ct); (H) semi-diagrammatic drawings of spermiocytogenesis of *Spongilla lacustris (from Paulus, 1989)* (a) the flagellum (F) of the spermatid (St) inserts in the vicinity of the circular nucleus (n) which is eccentric in position, (b) the flagellum is invaginated into the cell during the process *(arrows)*; (c) cytokinesis is completed; (d) mature spermatozoon (sp) gets its characteristic appearance, c = centriole, CF = cell fragment, Mi = mitochondria, v = vacuole.

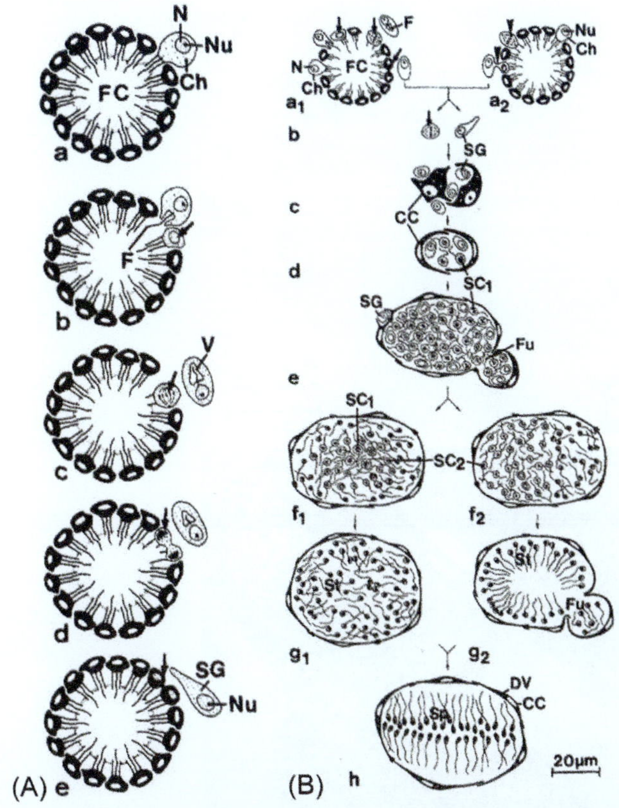

FIG. 4.8 (A) Semi-diagrammatic representation of spermatogonia development, (a) flagellated chamber with one enlarged choanocyte, (b) elimination of the collar and retraction of the flagellum into a vacuole, (c–e) transformation into spermatogonium; ch=chonocyte, F=flagellum, FC=flagellated chamber, N=nucleus, Nu=nucleolus, SG=spermatogonim, v=vacuole; (B) semi-diagramatic drawings showing stages of the process of spermatogenesis in *Spongilla lacastris (left side)* and *Ephydatia fluviatilis (right side)*, stages with common description are drawn only once, CC=cyst-building cell, Ch=choanocyte, DV digestive vacuole, F=flagellum, FC=flagellated chamber, Fu=fusion, N=nucleus, Nu=nucleolus, SC1=primary spermatocyte, SC2=secondary spermatocyte, SG=spermatogonia, Sp=spermatozoon, St=spermatid. *(From Paulus (1989).)*

nuclear membrane. Then, it becomes particularly evident at the area adjacent to the cilium insertion, where a transient nucleolus-like structure appears. At the same area, the basal body (principal centriole) and a perpendicular accessory centriole appear too. A well-developed Golgi apparatus is located in the cytoplasm.

Spermatids of the V-shaped spermatozoon change morphologically and become lengthened cells, and the nucleus stretches along the longitudinal axis (Fig. 4.7E). Chromatin compaction involves a peculiar process of 'multipolar' condensation, forming small, electron-dense clumps ('initiation centres'). The latter are scattered all over the nucleus (Fig. 4.7F). A longitudinal cytoplasmic tunnel, running from cilium insertion (i.e. the original posterior cell pole) to the anterior cell pole, is formed to harbour the proximal region of the cilium axoneme. The cilium is then bent at an 180-degree angle and, as a result, the cell becomes V-shaped. The first cyst appears in the deepest mesohyl, close to the basopinacocytes, while the late cysts occupy the remaining upper mesohyl regions, intermingled with functional choanochambers. However, no free-swimming spermatozoa were observed by Riesgo and Maldonado (2009).

The nearly mature sperm share the cysts with some late spermatids, still exocytosing part of their cytoplasm. Intercellular bridges, which appear connecting spermatids, disappear between the nearly mature sperm. The major difference between both stages is that the early chromatin clumps condense into a single helical structure in the spermatozoon (Fig. 4.10D). The spermatozoon is a long, narrow, and monociliated cell, with a cell body averaging 6.5 μm in length and 0.6 μm in width. There are neither Golgi apparatus nor accessory centriole. The proximal region of the cilium runs within a cytoplasmic tunnel. The most important inclusion in this spermatozoon is the electron-dense acrosomal rod located between the acrosome and the nucleus adjacent to the basal body. The single mitochondrion is located at the opposite pole (Fig. 4.10D).

Oogenesis

It appears that this vital activity was studied more in Calcispongiae. Many authors described the sequence of events occurring during the development of oogonia of *Sycon raphanus*, *S. elegans*, *Grantia compressa*, *Petrobiona massiliana*, *Clathrina coriacea*, *Leucosolenia botryoides* and several species of *Leucandra*. The egg-mother cells or oogonia move from beneath the choanoderm into the flagellated chamber, which is believed to arise from amoebocytes in the two genera *Sycon* and *Grantia*. Within the flagellated chamber they undergo two successive divisions and each oogonium results in the production of four oocytes. The latter then leave the chambers and assume positions under the choanocytes (some authors believe that the germ cells develop directly into oocytes in some calcareous sponges). The division of the oogonia within the flagellated chambers has also been described in two species of acalcareous sponges, *Reniera elegans* and *Octavella galangaui*.

Another unique story was told by Deny (1914) (cited in Roosen-Runge, 1977) who became convinced that in the calcareous sponge, *Grantia compressa*, the choanocytes directly transform into amoeboid germ cells. Later, these germ cells divide and form 'morulae' of 16 cells, which are apparently encapsulated.

During the growth phase of their development, the oocytes undergo two successive periods of growth. The first stage is the process of oogenesis, during which the oocyte undergoes a moderate increase in size. This is called initial growth. Following the period of initial growth, the oocytes begin a more extensive growth, which is accompanied by the accumulation of yolk. The oocyte may increase abruptly

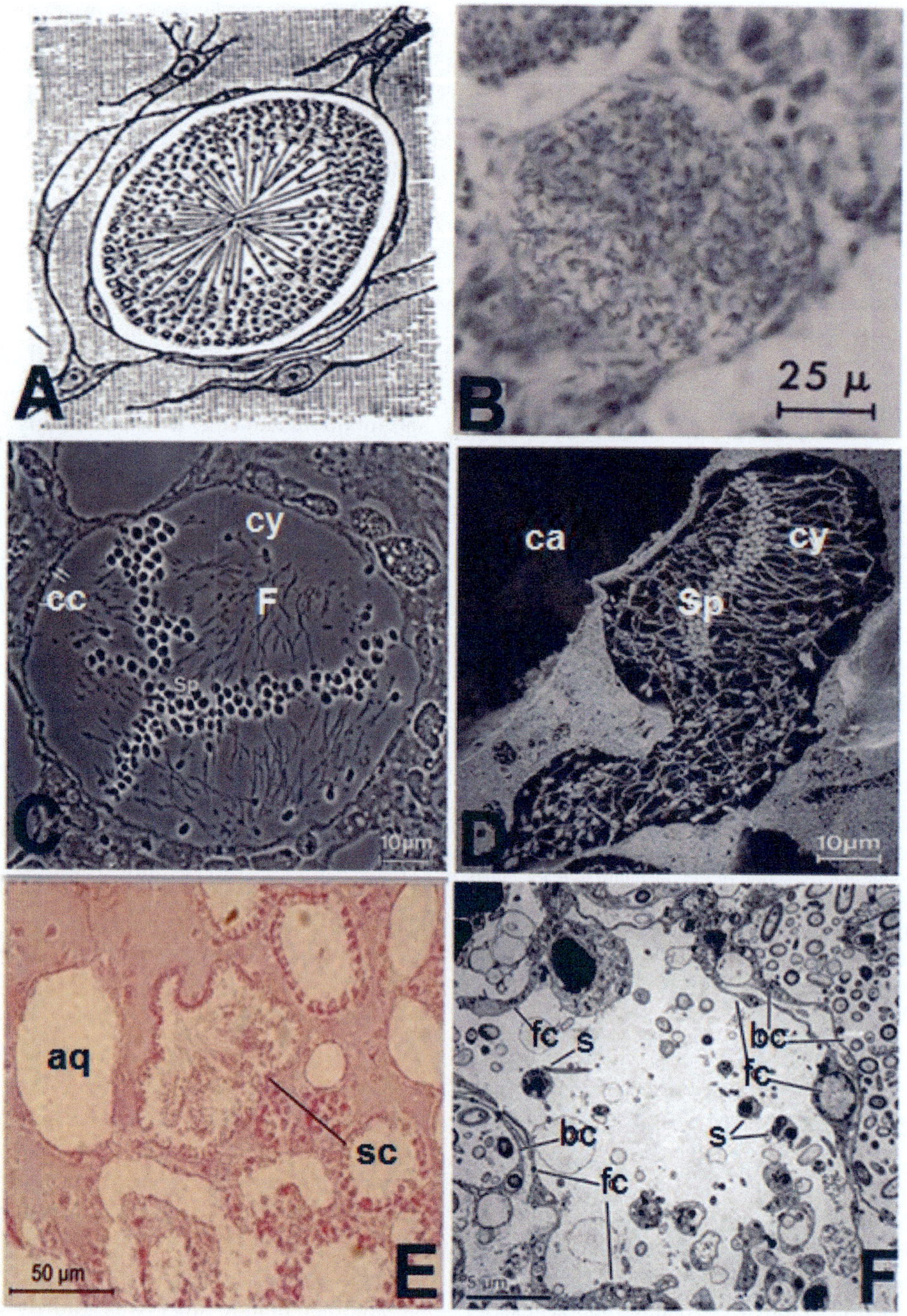

FIG. 4.9 Different types of spermiocyst in sponges: (A) spermatocyst of *Oscarella lobularis*, 500× *(from Minchin et al., 1900)* (B) spermatocyst of *Microciona microjoanna*, the spermatozoa in random *(from Fell, 1974)*; (C) of *Spongilla lacustris* and (D) of *Ephydatia fluviatilis*, heads arranged forming a pattern of bands in the centre *(from Paulus, 1989)*, (E) of *Corticium candelabrum*, flagella directed to one pole, the cyst within the tissue of mother *(from Riesgo et al., 2007a,b)*, (F) spawned spermatocyst of *Petrosia ficiformis* in which only three sperm remain *(from Maldonado and Riesgo, 2009)*; aq = aquiferous canal, bc = peripheral bacteriocytes, Ca = canal, CC = the cyst wall cells, Cy,sc = spermatocyst, F = flagellum, fc = continuous envelope of follicle cells, Sp,s = spermatozoon.

in size during this active stage of yolk formation (deutoplasmogenesis or **vitellogenesis** (Gardiner, 1972)). In the commercial bath sponges, *Hippospongia* and *Spongia*, after this activity they increase from ≈50 to about 300 μm in diameter (Fell, 1997). In most cases, this growth involves the activity of nurse cells (Fig. 4.12A). Among the calcareous sponges the nurse cells usually seem to be choanocytes or cells that are derived from choanocytes. In *Grantia compressa* and *Sycon raphanus* the oocytes apparently migrate from their position under the choanocytes to the vicinity of the paragaster, where the nurse cells are situated. Each oocyte becomes associated with two or three cells, which form a

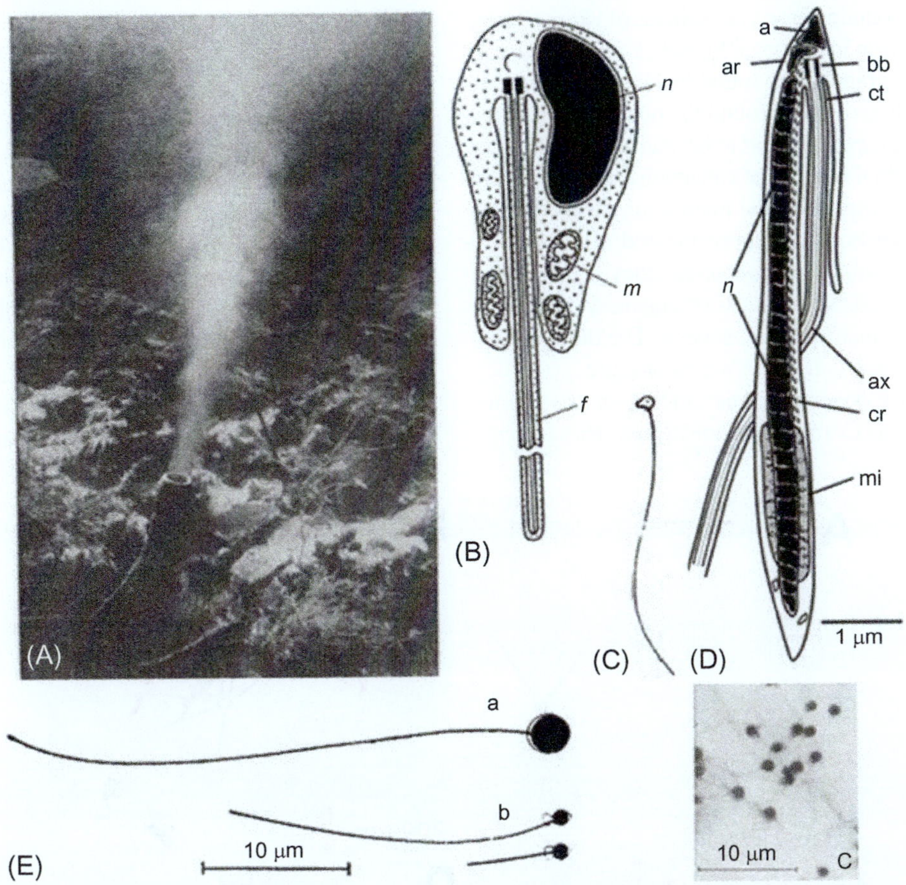

FIG. 4.10 (A) Sperm release from a single 1.5 m-long *Verongia archeri* at 49 m; (B) Scheme of a spermatozoid characteristic of the representatives of the order Haplosclerida: f = flagellum; m = mitochondrion; n = nucleus *(Ereskovsky, 1999)*; (C) mature isolated spermatozoon (may be from spermatic cyst of *Oscarella lobularis*), 800× *(after Schulze, from Minchin et al., 1900)*; (D) V-shaped spermatozoon of the poecilosclerid *Crambe crambe*, a = acrosome, ar = acrosomal rod, axe = axoneme, bb = basal body, ct = cytoplasmic tunnel, cr = ciliary rootlets; mi = mitochondrion, n = nucleus *(from Riesgo and Maldonado, 2009)*; (E) sponge sperm collected from effluent streams of (a) *Verongia archeri*; (b) *Geodia* sp.; and (c) photomicrograph of *Geodia* sp. *((A and E) from Reiswig, 1970)*.

nurse complex. The oocytes may then engulf at least one of the complexes, ingest all the nurse cells, or simply absorb material from them. After this period of growth, the oocytes move back under the choanocytes.

In clathrines, the nurse cells are special amoebocytes containing characteristic granules, which they transfer to the growing oocytes (Fell, 1974). This means that there is no engulfment of nurse cells and the granules are degraded within the oocyte. Granular porocytes may envelop the masses of cells that surround the oocytes, but they do not interact with the latter. In *Ascandra minchini*, the process is more complex. The oocytes undergo substantial growth by the phagocytosis of choanocytes. They then, move into nests of nurse cells to complete development. The nurse cells may be derived from pinacocytes, which grow by engulfing eosinophils and choanocytes.

In acalcareous sponges, the nurse cells are generally amoebocytes, which are described as granulated. The nurse cells aggregate around the oocytes of many sponges during their secondary period of growth. In most cases, the oocytes ingest these nurse cells, forming the yolk spheres (Fig. 4.12C and D). This process is well exemplified in *Spongilla lacustris*. The nurse cells involved in oogenesis and in gemmulation in the fresh-water sponges are identical, but the nature of the reserved material is very different. A number of the nurse cells accumulate within the oocyte, and the fully developed oocyte may attain 1300 cells. Whatever the process is that occurs to the nurse cells inside the oocyte, there exists the interesting possibility that there is more to these cells than simply representing nutritive stores. In *Haliclona ecbasis* and *H. permollis*, the nurse cells are engulfed by the oocytes but are not immediately degraded. Although the nucleus of the engulfed nurse cell disappears, the cytoplasm appears to remain essentially unchanged during oogenesis and the early stages of embryogenesis.

As the oocytes engulf, absorb and/or synthesis materials, yolk granules accumulate in their cytoplasm. The fully developed oocytes of a number of sponges, e.g. the fresh-water,

Spongilla lacustris, contain dense yolk spheres of various sizes, reaching up to 15–20 μm in diameter. In *Haliclona limbata*, e.g. only very small yolk granules are found in the oocyte. There is frequently a region free of yolk around the nucleus. Evidently, those of the oviparous sponges tend to be smaller than those of viviparous sponges. In a number of calcareous sponges and in few of the acalcareous sponges, the maturation divisions with the formation of polar bodies occur at the end of the growth of the oocyte. In *Hippospongia communis*, they occur before the major growth period. The process of oogenesis is a much more lengthy process than spermatogenesis. The duration differs with the species, and can be six weeks or more.

Ultra-structure of oogenesis was studied in the demosponges: the viviparous *Corticium candelabrum* (Riesgo et al., 2007a,b) and the oviparous *Petrosia ficiformis* (Maldonado and Riesgo, 2009). The process is almost the same in both, except that, in the former the animal is hermaphrodite, while the latter is gonochoristic. The observations and examination of the histological preparations during the 2 years (2003–05) indicated that oogenesis in *Petrosia* extended for several months each year. In this species, the authors were unable to identify the origin of oogonia. The smallest recognizable female reproductive cells are 10–20 μm, relatively amoeboid oocytes. At the beginning of oogenesis, small oocytes occurred sparsely scattered through the mesohyl (Fig. 4.11A and B). The mid-size oocyte (50–150 μm) is also amoeboid and in early stage of deutoplasmogenesis (Fig. 4.12E). It contains abundant large, electron-dense, membrane-bound yolk

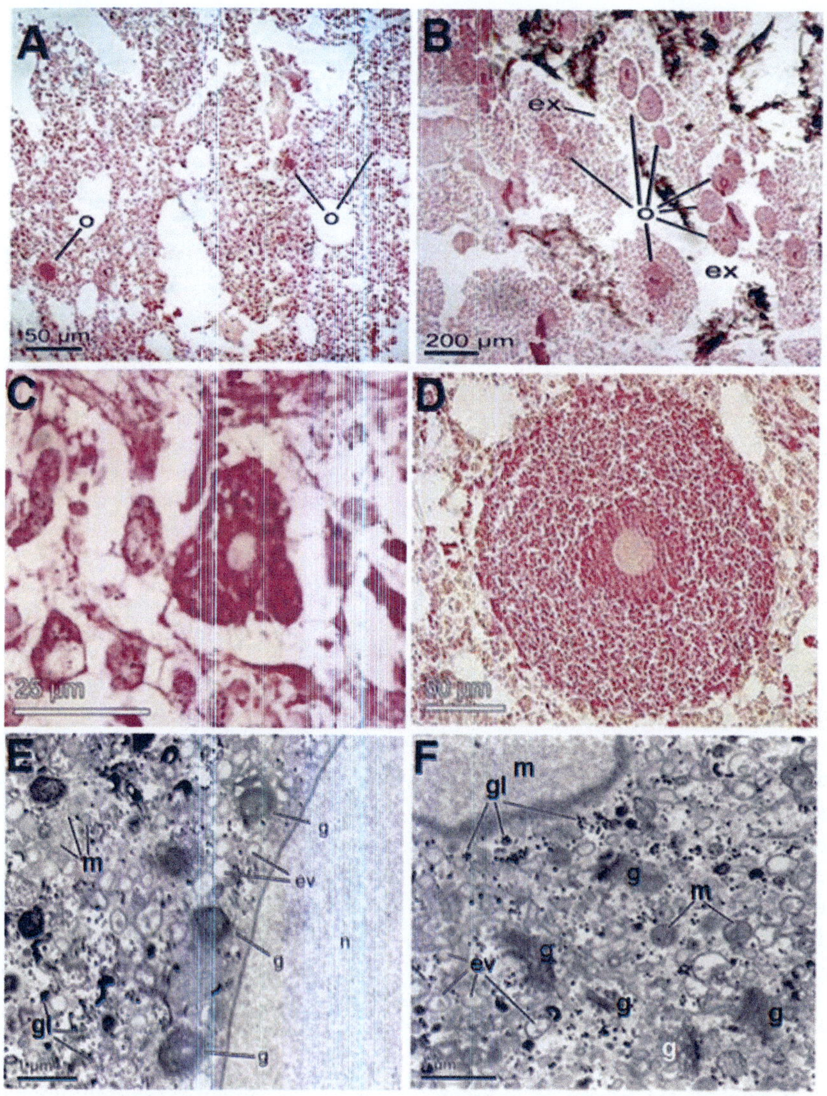

FIG. 4.11 Light microscopical preparations of *Petrosia ficiformis* showing steps of oogenesis: (A) early oocytes (o) scattered in the deep mesohyl; (B) late oocytes aggregated in the vicinity of excurrent canal (ex); (C) mid-stage oocyte after first growth; (D) full grown oocyte after second growth; (E and F) two stages of the cytoplasmic vicinity of perinuclear area, showing in (E) the nucleolate (nu) nucleus (n), which virtually lacks yolk bodies (y), but contains many electron-clear vesicles (ev), glycogen granules (gl), mitochondria (m), and numerous units of the Golgi apparatus (g); (F) abundant glycogen (gl), mitochondria (m), and electronclear vesicles (ev), as well as five units of the Golgi apparatus (g) *(from Maldonado and Riesgo, 2009).*

inclusions. The latter consist of a complex of heterogeneous materials, often abundant in glycogen granules (Fig. 4.3B). In this stage, clusters of small mitochondria are common among yolk bodies. In mature oocyte, a few yolk bodies are relatively homogeneous in substructure, mainly with glycogen (Fig. 4.12F). Rosettes of glycogen granules are commonly abundant in the late stage of mid-size oocyte (Fig. 4.12D). The nucleolate nucleus is surrounded by a 20 μm-wide perinuclear region in which yolk bodies are absent, but which contains multiple units of Golgi apparatus, clusters of mitochondria, and an abundance of small electron translucent vesicles and glycogen granules (Fig. 4.4B–D). The mid-size

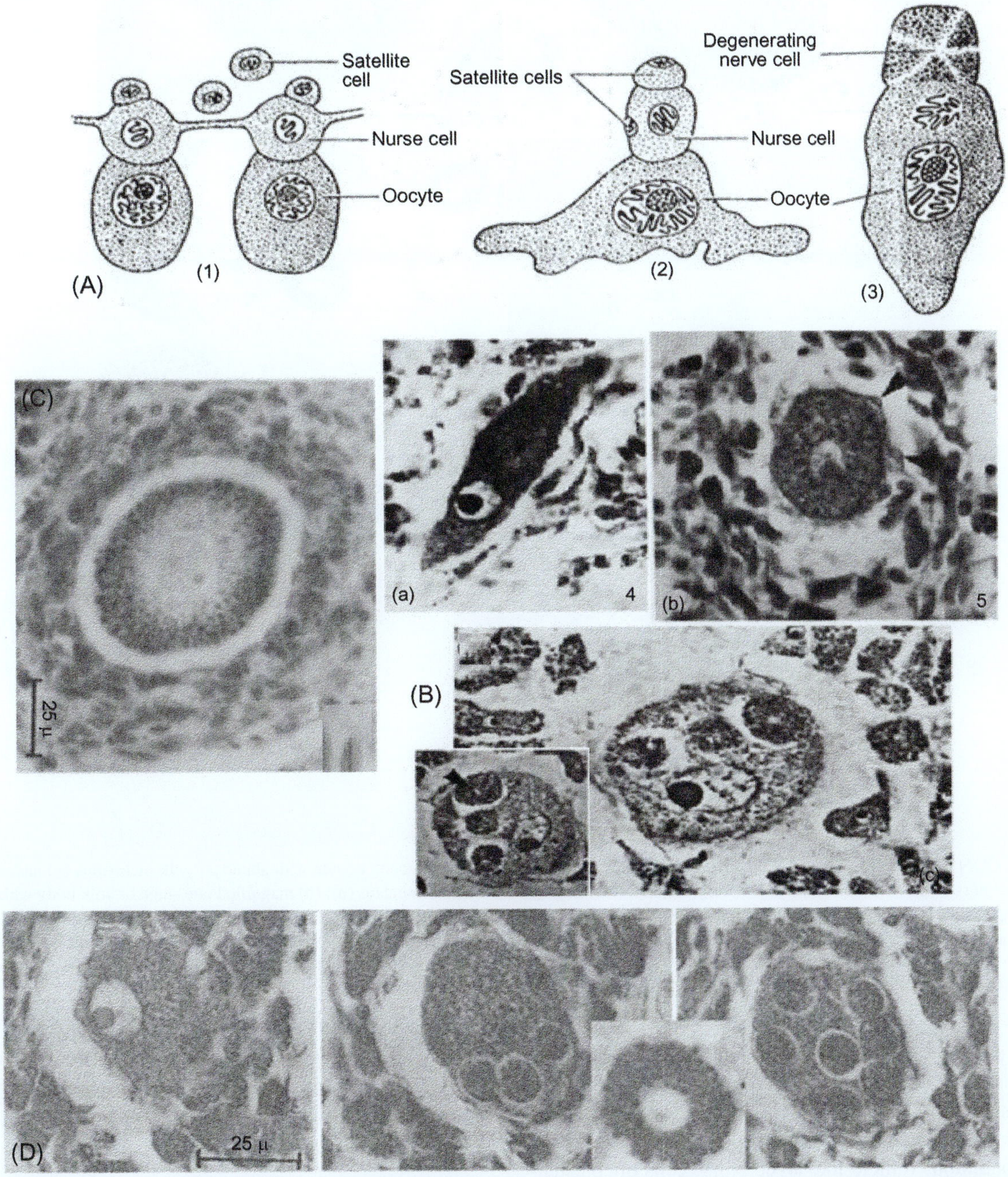

FIG. 4.12 Deutoplasmogenesis in sponges: (A) in the egg of the sponge *Sycon raphanus (after Duboscq and Tuzet, from Gardiner, 1972)*: (1) two oocytes with nurse cells and stellites; (2) larger oocyte with nurse cells and two stellite cells; (3) oocyte with degenerating nurse cells whose materials taken in by the oocyte; (B) phase contrast micrographs illustrating oogenesis in *Haliclona ecbasis*: (a) oocyte I without associated follicle cells, 1200×, (b) growing oocyte with follicle cells *(arrows)* and mass of surrounding nurse cells, 720×, (c) oocyte about full grown with associated nurse cells 1200×, inset shows an engulfed nurse cell 800× *(from Fell, 1969)*; (C) an oocyte of *Halichondria panacea* in early stage of yolk formation with surrounding nurse cells; (D) consecutive 8 μm sections through an oocyte of *Haliclona permollis* containing 10 nurse cells, inset shows a nurse cell filled with numerous granules *((C and D) from Fell, 1974)*; *(Continued)*

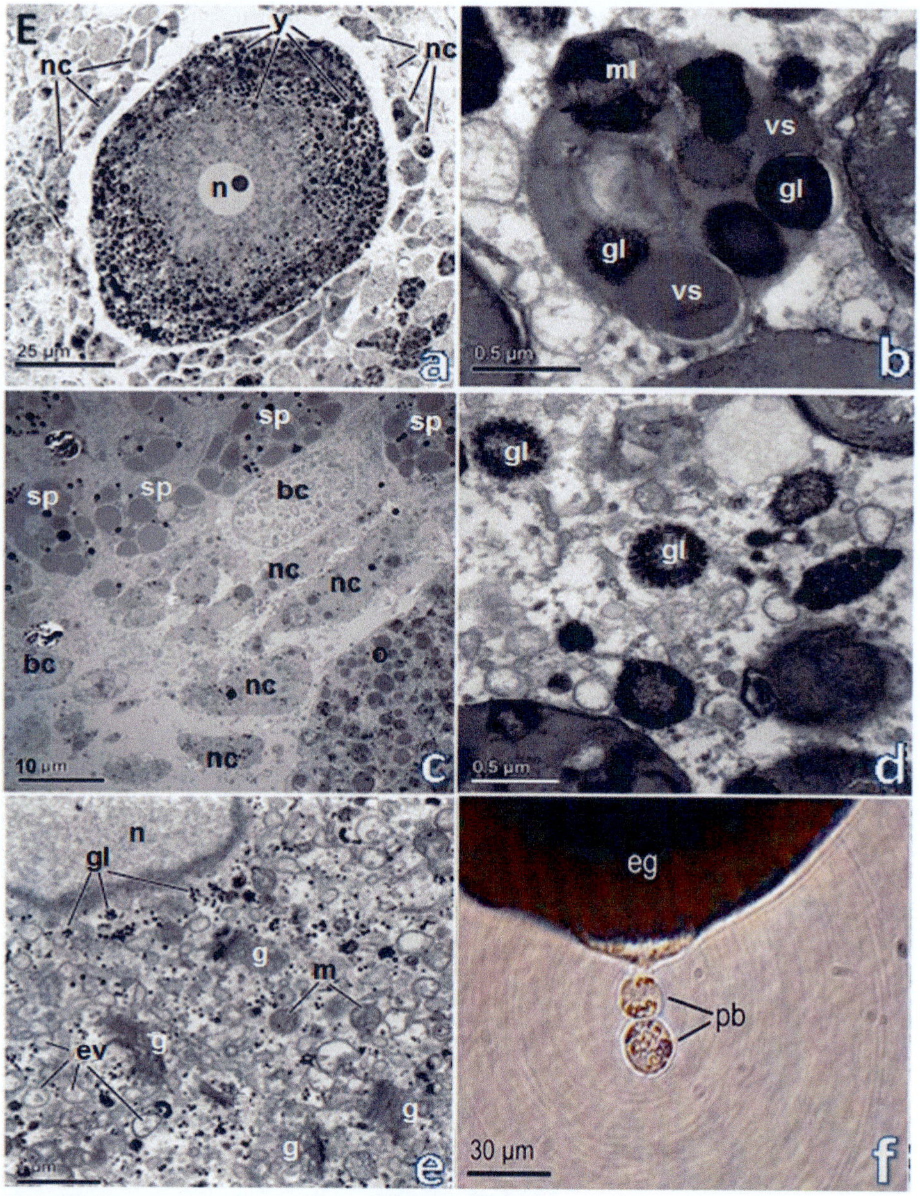

FIG. 4.12, CONT'D (E) vitellogenesis in *Petrosia ficiformis*: (a) mid-size amoeboid young oocyte with abundant yolk inclusions (y) and surrounded by numerous nurse cells (nc), notice the nonvitellogenic area surrounding the nucleolate nucleus (n); (b) magnified late stage of yolk body which appears more homogeneous, glycogen (gl), vesicular (vs), and multilamellar (ml) components can be observed; (c) oocyte (o) surrounded by nurse cells, bacteriocytes (bc), and more peripherally spherulous cells (sp); (d) mid-sized oocyte showing glycogen (gl) organized in rosettes; (e) detail of perinuclear region showing abundant glycogen (gl), mitochondria (m), electron-clear vesicles (ev), and five units of Golgi apparatus (g); (f) in vivo observations on spawned eggs, polar bodies (pb) attached to a released egg (eg) *(from Maldonado and Riesgo, 2009)*.

oocyte is usually surrounded by a 'lax' cell layer consisting of nurse cells, specialized pocket cells (i.e. bacteriocytes hosting enormous amounts of living bacteria) and most peripherally large spherulous cells (Fig. 4.5A and B). These cells never form a continuous follicle-like layer. Nurse cells often establish intimate contact with oocytes and they actively participate in oocyte deutoplasmogenesis.

The **mature oocyte** of *Petrosia ficiformis* (200–250 µm) has almost, except for the size and absence of nucleolus, the same cytological organization as the mid-size oocyte. The perinuclear region lacks yolk bodies but contains many units of Golgi apparatus, isolated round mitochondria, glycogen granules, free ribosomes and electron-clear vesicles (Fig. 4.7B).

Reproductive behaviour and spawning

The behaviour associated with reproduction in these sessile animals shows no peculiarity. Spermio-cysts are shed into the exhalant canals and are discharged through the oscula into the surrounding water. The rupture of the walls of the spermiocysts is probably an enzymatic process. Free sperm may fertilize similarly shed eggs. The result is that fertilization is external. In other oviparous species, it appears that fertilization is internal and that the zygotes are released into the water. The release of spermatozoa occurs either in a relatively dense column or in diffuse turbid clouds, which extend 2–3 m above the specimen. Reiswig (1970) observed and recorded that a single specimen of *Verongia archeri*, situated at a depth of 49 m on the deep fore-reef slope bordering the north coast of Jamaica at Discovery Bay, emitted a column of 'smoke', 3 m high for at least 10 min (Fig. 4.10A). A second single specimen, *Geodia* sp., situated at 29 m on the upper edge of the deep fore-reef slope, released a diffuse turbid effluent for at least 20 min. This specimen was active during this process. The third specimen, *Neofibularia nolitangere*, is a common species of the fore-reef and fore-reef slope at depths of 10–30 m. The colonies of this species were observed for a period of 30 min. Approximately, one-quarter of the surveyed colonies were emitting a diffuse cloud of material extending 2–3 m above the reef surface. Due to the toxic nature of the species, no attempt was made to collect any samples.

In some instances in which oocytes (or zygotes) were released in the laboratory but did not develop, it is not clear whether shedding was premature, precluding normal internal fertilization, or whether conditions for normal external fertilization and/or development were inadequate (Fell, 1974). On the other hand, many marine sponges are viviparous and consequently the female gametes are not spawned. Eggs are usually spawned singly or in small groups (Fig. 4.13A). Sometimes eggs are enclosed within a common gelatinous matrix that binds them together and to the adjacent substratum. The expulsion of the polar bodies occurs either during or immediately following the release of the eggs in *Cliona shetionis*. It was suggested from observations that light may coordinate spawning in some species.

Spawning in *Petrosia ficiformis* occurs in the morning. Mature eggs came out, in laboratory observations, bearing two (12–15 μm) attached translucent polar bodies, which were filled with turbid vesicles (Fig. 4.12E (f)).

Fertilization

In a majority of sponges, fertilization and embryonic development occur within the egg follicles, in the mesohyl. Generally, fertilization occurs either before or just after the period of the major growth of the oocyte, i.e. before the extrusion of the polar bodies. In *Clathrina (=Leucosolenia) coriacea* (Fell, 1974), fertilization occurs either before or soon after the oocytes enter the period of major growth. Once the incurrent water stream carrying the sperm comes inside the sponge-containing oocytes, the spermatozoon needs to penetrate one or more barriers in order to reach the oocyte. There are two basic problems related to the process of fertilization:

1. The site of the oocyte in the parent; and
2. Passing the barriers that exist between the incurrent water and the oocyte.

In all cases in which internal fertilization has been observed, it appears that the spermatozoon is transported from the water stream to the oocyte by a cellular intermediate known as a 'carrier cell' (Fig. 4.13D). This is a choanocyte that modifies, engulfs the sperm, loses its flagellum and collar and, leaving its layer, may increase in size and become amoeboid. The spermatozoon is enclosed within a vesicle, i.e. changes into 'spermiocyst', inside this carrier choanocyte, which applies itself against the surface of an oocyte and transfers the spermatozoon to it. In sponges where the oocytes are enclosed within a follicle, the carrier cell makes contact with the oocyte before the formation of the follicle. This sequence, or part of it, has been observed in a few calcareous sponges, e.g. *Grantia compressa* and *Sycon raphanus*; and in some acalcareous sponges, *Reniera elegans* and *Hippospongia communis*. In many of the calcareous and acalcareous sponges, the oocytes are situated immediately below the collar cells when the carrier cells establish contact with them. In other sponges, the oocytes are dispersed in the mesohyl, and so the mechanisms by which the spermatozoa may ultimately make contact with the oocytes are:

1. The sperm may enter the choanocytes at random, so there are two mechanisms:
   a. only the choanocytes which are initially associated with oocytes and/or which subsequently make chance contact with them, may function in fertilization; or
   b. these carrier choanocytes may locate the oocytes by chemotaxis.
2. The sperm may enter choanocytes neighbouring the oocytes by chemotaxis. Although there is no conclusive evidence for chemotaxis, there are a number of very suggestive observations. In *Sycon raphanus*, spermatozoa were seen only in choanocytes that are situated over oocytes.

In some sponges, including *Sycon raphanus*, the carrier choanocytes are associated with very small oocytes, but fertilization does not occur until the oocytes have nearly completed their growth. The carrier choanocyte apparently separates from the oocyte after the transfer of the sperm. In some sponges it is incorporated into the oocyte. In nearly all cases, the spermatozoon undergoes little change within the cytoplasm of the oocyte until the maturation divisions occur. Then the male and female pronuclei develop and fuse. The process of gamete fusion has not been studied in cases of external fertilization. However, an astonishing phenomenon was first observed and studied by Kumé (1952) in one oviparous sponge *Tetilla (=Tethya) serica*, and was

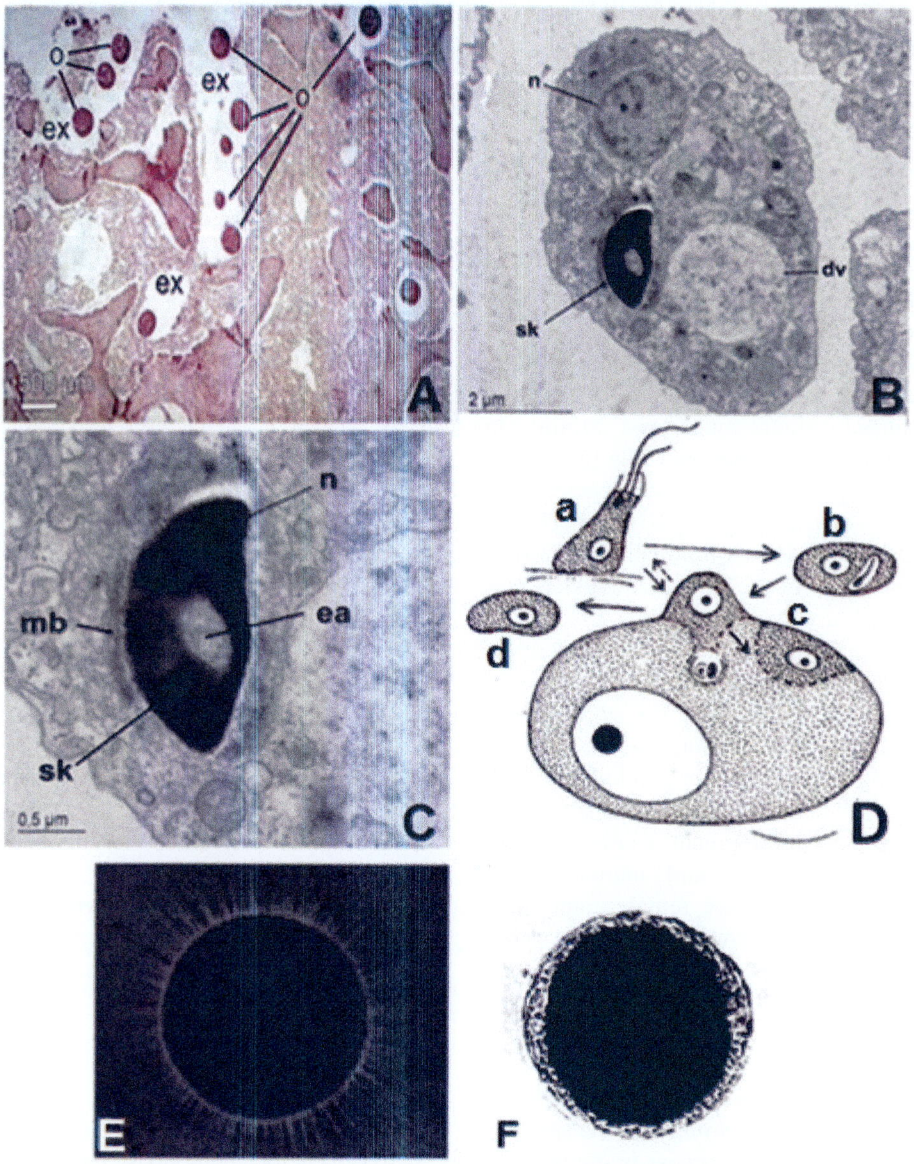

FIG. 4.13 Spawning and fertilization of sponges: (A) Oocytes of *Petrosia ficiformis* (o) after immigration to the excurrent canals (ex) for imminent spawning *(from Maldonado and Riesgo, 2009)*; (B–C) spemiocysts of *Corticium candelabrum*, (B) carrier cell with spermiocyst in its cytoplasm (sk), a large electron-clear digestive vacuole (dv) and a large nucleus (n), (C) detail of membrane-bounded (mb) spermiocyst (sk), note the highly condensed chromatin in the nucleus (n) and an electron-clear central area (ea) *(from Riesgo et al., 2007a,b)*; (D) transport of the spermatozoon to the oocyte by a carrier cell: (a) spermatozoon enters the choanocyte, (b) choanocyte loses collar and flagellum and become aoemoboid, (c) the latter may transfer it directly to the oocyte or move through the mesohyl to reach the oocyte, (d) the spermatozoon now in the oocyte, the carrier cell leaves the oocyte, or it is engulfed by the oocyte *(modified from Duboscq and Tuzet, 1937, from Fell, 1974)*; (E and F) spawned egg of *Tetilla japonica*, (E) unfertilized egg with radiating collagen fibre bundles as seen by a phase contrast microscope, (F) egg 2h after fertilization, with fibres retracted into the perivitelline space *(from Kumé, 1952)*.

recorded by Fell in both his chapters (1974 and 1997). The spawned unfertilized eggs are surrounded by a clear vitelline membrane and possess numerous radiating collagen fibre bundles. The latter may be longer than the diameter of the egg itself and extend through the clear vitelline membrane (Fig. 4.13E and F). Following fertilization, the vitelline membrane is elevated from the surface of the egg, presumably beginning at the point of entry of the sperm. This process is accomplished within about 1 min, at the same time the egg adheres to the substratum. After 10 min all the fibres are withdrawn into the perivitelline space and the egg then continues its embryonic development.

Experimentally, fertilization in *P. ficiformis* occurred externally in vitro in Petri-dishes. Male and female gametes were collected and mixed together in the Petri dish. After 6h, a membrane appeared external to the mucous layer surrounding each egg. This is a fertilization membrane. The polar bodies remain attached to this fertilization membrane

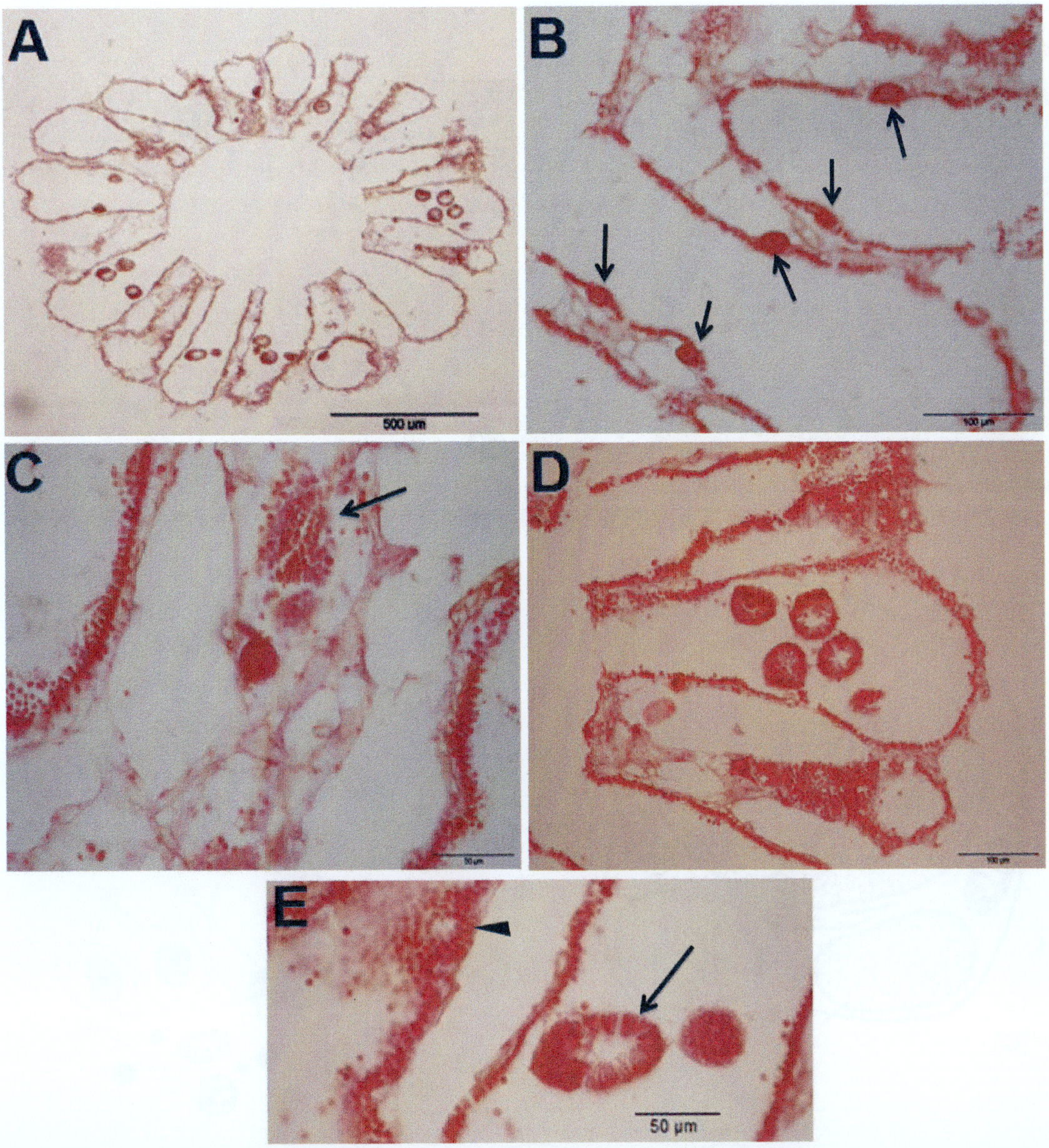

FIG. 4.14 T.S.s through the body of the Calcaronea sponge, *Grantia* sp., the contents are not correctly transverse *(laboratory preparations)*: (A) section shows embryos everywhere in the tissues; (B) enlarged portion of another section showing embryos in different early stages *(arrows)*; (C) another portion showing late stages, may be pre-larva *(arrows)*; (D) larvae in the aquiferous canal in their way to discharge; (E) completely formed amphiblastula larva in the canal *(arrow)* and another one still within the tissue *(headarrow)*.

(Fig. 4.12E (f)). Eggs started division in the laboratory 10h after spawning (Maldonado and Riesgo, 2009).

Development

Early development of sponges can be divided into three periods:

1. the embryonic period from the ova to the free swimming larvae. This period usually occurs within the maternal body (Fig. 4.14);
2. the larval or the free swimming period; and
3. the metamorphosing period from the settlement of the larvae to the formation of pores and osculum (Minchin et al., 1900).

CLASS CALCAREA

Subclass Calcaronea

Leucosolenia variabilis

Embryonic period. The development of *Leucosolenia variabilis* was described by Okada (1968). This description can also be applied to *L. complicata* (Fig. 4.15A), *Grantia compressa*, *Sycon raphanus*, and *Scypha ciliata* (Fell, 1997). Spermatogenic cells of *L. complicata* (Ereskovsky, 2010) were found to be diffusively distributed in the mesohyl. Single spermatocytes or small bunches were not surrounded by any specialized cells, did not form special

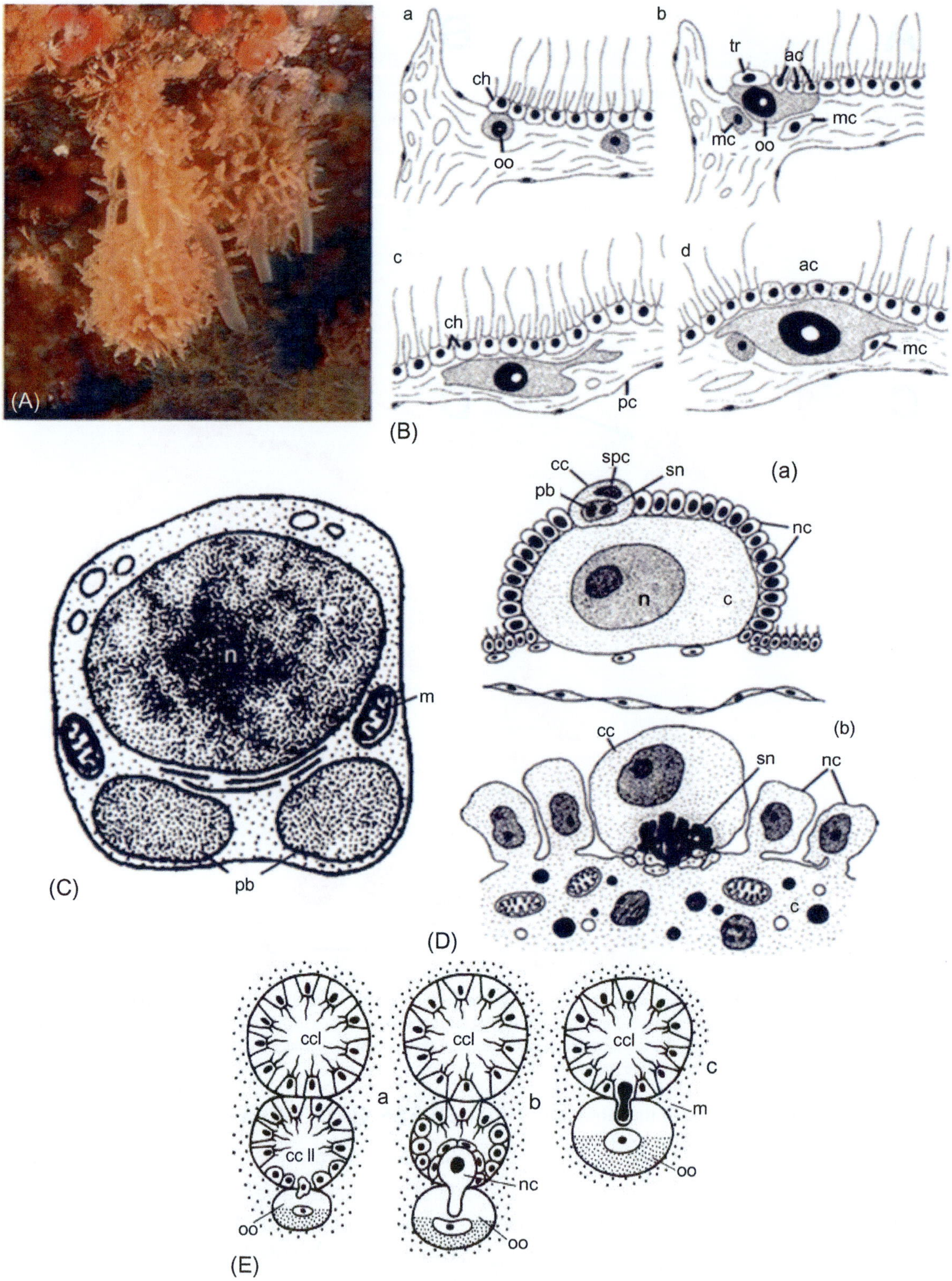

FIG. 4.15 (A) *Leucosolenia complicata* (Montagu 1818); (B–D) its embryonic stages (B) oogenesis in Calcaronea sponge, (a) oocyte at initial stage, (b and c) oocyte at primary growth (cytoplasmic growth), (d) oocyte at vitellogenesis, ac = accessory cells, ch = choanocyte, mc = mesohyl cells, oo = oocyte, pc = pinacocyte, tr = trophocyte; (C) Spermatozoon of *L. complicata*, m = mitochondria, n = nucleus; (D) illustrations showing process of fertilization in *L. complicata*, (a) formation of carrier cell, (b) penetration of the sperm nucleus from the carrier cell into egg cytoplasm, c = cytoplasm of the oocyte, cc = carrier cell, n = nucleus of oocyte, nc = nurse cell, pb = peptide body, sn = nucleus of spermium, spc = spermiocyst. (E) Oogenesis and vitellogenesis in *Petrobiona massiliana*, (a) oocyte (oo) located next to choanocyte chamber II (C II), (b) oocyte phagocytoses nurse cells (nc); (c) full grown oocyte with degenerated nurse cells, CCI = choanocyte chamber I, m = mesohyl. *(From Ereskovsky (2010).)*

spermatocyte, like in other sponges, and remained free under the choanoderm. Mature spermatozoa of *L. complicata* look like orbicular cells. Flagellum and acrosome are absent and the nucleus is formed by incomplete condensed chromatin. The latter is surrounded by a thin layer of cytoplasm (Fig. 4.15C). On the other hand, oogenesis in calcaronea, in general, is uniform and follows the usual pattern. However, the oocytes lack a vitelline envelope and possess amoeboid motility. At first, the oocyte takes its place beneath the choanoderm layer, and then it migrates by amoeboid motility through the mesohyl to the peripheral or basal parts of the sponge. There, it forms long pseudopodia and shows phagocytic activity to increase to its specific size (Fig. 4.15B). In *L. complicata* and during vitellogenesis, a complex of accessory cells is formed above each oocyte, derived from choanocytes that lose their flagellum and collar. The number of accessory cells does not exceed 6 for each of *Grantia* or *Sycon*, but may reach 30–40 for each of *L. complicata.* Ereskovsky (2010) reported that a unique condition is observed in *Petrobiona massiliana* (the only Cacaronea with a massive calcareous skeleton, in addition to the spicules). In this sponge, all the choanocytes in the choanocyte chamber, adjacent to a vitellogenic oocyte transform into cells of the feeding complex (Fig. 4.15E).

The cleavage of the fertilized egg (zygote) of sponges is peculiar, as the normal regular cleavage is somewhat modified. In *Leucosolenia* species (Fig. 4.16), during the first cleavage, the zygote divides along a meridional plane into two equal blastomeres. The second cleavage is along a meridional plane, perpendicular to the first, giving rise to four equal blastomeres. Each of these blastomeres modifies and develops a pointed end while the other end becomes broad (the pointed

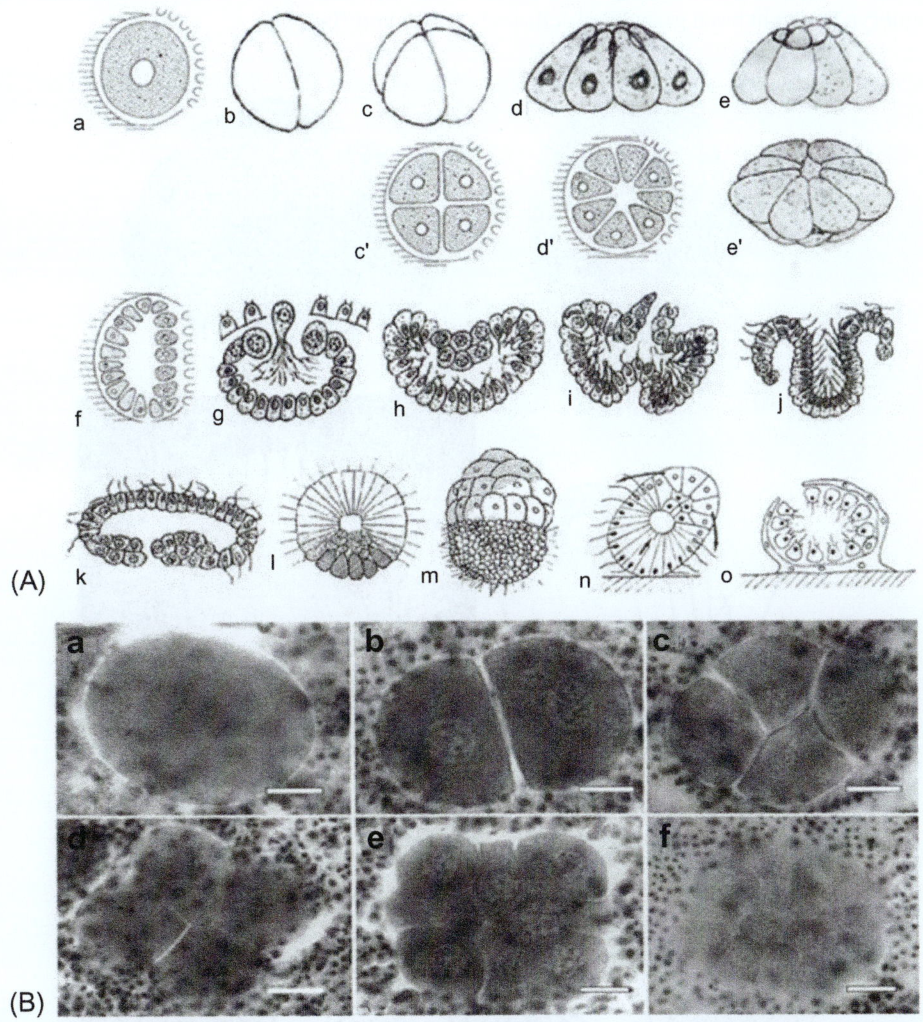

FIG. 4.16 (A) Schematic drawings showing embryological stages of Calcaronea (mostly of *Leucosolenia*), (b–e) and (e′) are original, others are collected from various sources, (a) zygote, (b–e) cleavage, (c′–e′) surface views for the upper lateral views, (f) blastula, (g) stomoblastula; (h–k) inversion of the blastula; (l) newly hatched larva of *Leucosolenia* (or pseudo-gastrula); (m) fully formed amphiblastula larva; (n and o) settlement of larva, *arrows* indicate direction of movement, (o) Olynthus stage; (B) Cleavage in the (asconoid) calcaronean sponge *Leucosolenia complicata*: (a) Anaphase of the first cleavage division; (b) two-cell stage; (c) second cleavage division; (d) four-cell stage; (e) third cleavage division; (f) sixteen-cell stage. Scale bars: 2 μm. *(After Anakina (1997), from Leys and Ereskovsky (2006).)*

ends will give rise to the ciliated cells of the future larva, and the broad ends to the granular cells). The third cleavage in sponges differs from the usual manner. It occurs in two vertical planes, intermediate in position between the first two planes of cleavage. The product is eight equal blastomeres, in one tier, each of which is like a pyramid. All of them are arranged as to give the whole growing embryo the form of a flat cone. The apex of the cone is formed by the pointed ends and the apices of the embryo do not meet, creating a central space. This space is an actual perforation through the axis of the growing embryo, open at both ends. The 4th cleavage is latitudinal (equatorial); it divides the embryo, giving rise to the 16-cell stage. This stage consists of two layers, eight upper blastomeres (apices) and eight lower, broad ones. Then the growing embryo assumes the form of a biconvex lens, in the axis of which is the central perforation (future blastocoel).

Probably, this 16-cell stage finishes cleavage of the zygote. The growing embryo has eight basal granulated cells and eight smaller upper cells, the central cavity becomes large, while its terminal apertures narrow and finally, at the end, they are closed. So, the axial perforation acquires the character of a closed blastocoele cavity. The growing embryo itself becomes a blastosphere. Its granulated eight cells abutting the wall of the radial canal, are the prospective granular cells, and the remainder are the prospective choanocytes (Fig. 4.17g).

The smaller cells divide rapidly and each acquires an internal flagella, while the granular cells do not divide for sometime, and form a distinctive group of eight large, rounded cells termed macromeres. In the centre of this group an opening appears that functions as a mouth and ingests the adjacent choanocytes of the mother (Fig. 4.16A (g)). This is the stomoblastula stage as designated by Duboscq and Tuzet (1937) (cited in Okada, 1968). The granulated cells then multiply, the mouth closes and the embryo is known as a blastula at this stage, with internally directed flagellae.

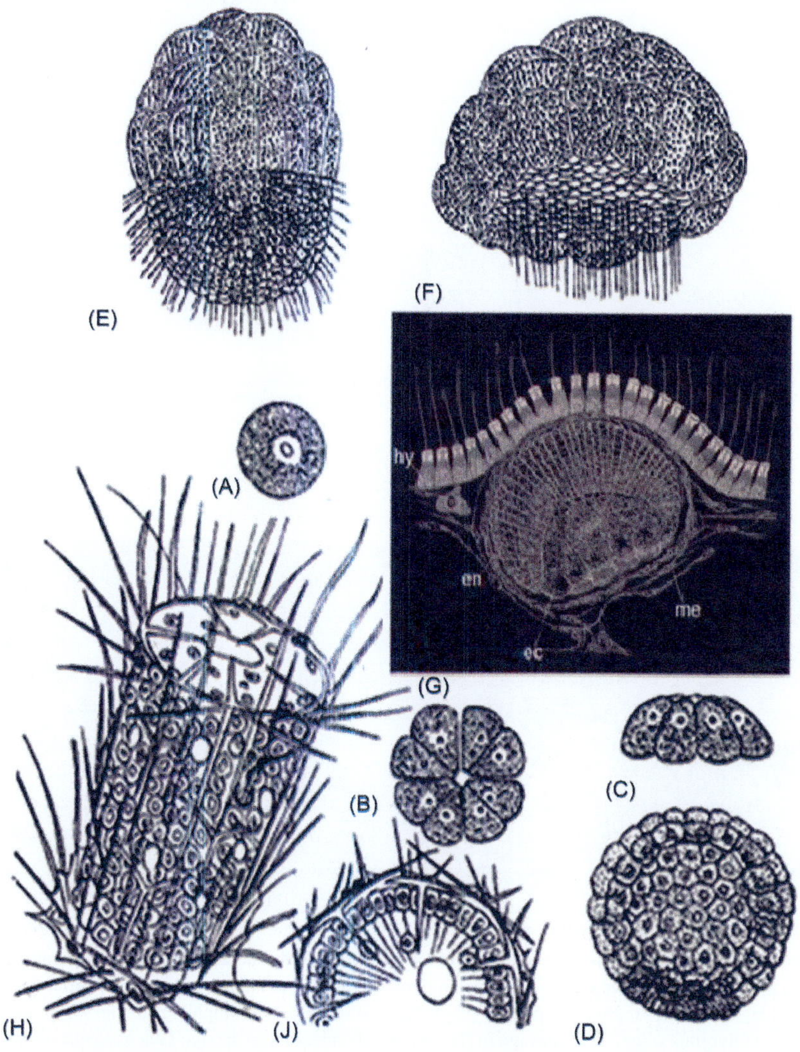

FIG. 4.17 Development of *Sycon raphanus* (*Sycandra raphanus*) *from Balfour (1880) and Minchin et al. (1900)*; (a) zygote; (b–d) cleavage; (g) growing amphiblastula larva brooded in mother, cc = ciliated cells of larva, gc = granulate cells of larva, hy = choanocytes, ms = mesohyl of mother; (e) swimming amphiblastula larva; (f) settlement of the larva; (h and j) young sponge, (h) Olynthus, (j) view of the upper surface of the former.

Inversion. The blastula undergoes a remarkable process of inversion, a process entirely comparable to that which takes place in the development of *Volvox*, but it has no counterpart in the Metazoa or indeed in siliceous sponges. This phenomenon was first discovered in *G. compressa* and *S. raphanus* by Duboscq and Tuzet (1935, 1937). They called this process 'inversion des surfaces'. It has been described only in the Calcaronea (Ereskovsky, 2010). This process (Figs 4.16A (h)–(k) and 4.18C) begins with the reopening of the blastula mouth, which enlarges to expose the flagellated cells. The embryo then turns inside out through this new opening by the detachment of the two lips of the opening away from each other. The flagellated region invaginates inwards and the lips extend upwards to meet in the opposite pole of the mouth. The result is a volvox blastula, with flagellae directed outwards. This is a growing amphiblastula larva. In *L. complicata*, before release, the embryo attains the pseudogastrula stage, in which granular cells enter and fill the larval cavity (Fig. 4.16A (l)). This stage is transitory, and this temporary invagination has nothing to do with gastrulation. Pseudogastrulation is merely a side effect of the separation of larval cells from the parent choanoderm prior to larval release (Ereskovsky, 2010).

Amphiblastula larva (free swimming period). This larva is characterized by the different structure of its two halves: the posterior half consists of large granulated cells, and the anterior half consists of smaller flagellated cells. Around the third and fourth tiers of the flagellated half, above the equator, usually situated crosswise at the border between the two types of cells, are found four cruciform cells, one on each quadrant, lacking flagellae (arranged as a cross). They are termed 'cellules en croix'. Each contains rows of chromatic corpuscles and has a prominent nucleus. They are differentiated from the columnar cells in the blastula. These cells apparently function as light receptors during larval life, and are expelled before metamorphosis. The embryo, from the blastula stage onwards, is enclosed in a trophic membrane maintained by the maternal choanocytes. This is the method of nourishment of the growing larva until it becomes fully formed. The larva forces its way through the choanocytes into the radial canal and swims to the exterior. Its orientation when swimming is with the flagellated hemisphere forward. The polar bodies are also in this area, which thus seems to correspond to the animal hemisphere of other metazoan embryos. When development of the larva is complete, the macromeres have multiplied to form ≈ half the embryo (Figs 4.16A (m), 4.17e, and 4.18D).

Metamorphosis and settlement. Postembryonic development of calcaronea may be divided into: a settling larva; larva undergoes metamorphosis; pre-olynthus stage; and Olynthus stage (juvenile with asconoid aquiferous system).

The amphiblastula larva develops only in calcareous sponges and a few primitive Demospongiae, such as *Oscarella*. After the free life span, the larva settles and a gastrulation process begins. Gastrulation in sponges is unusual. In these calcareous sponges, the macromeres overgrow the micromeres, whereas in other metazoans it is the macromeres that typically become internal. This means that the flagellated hemisphere invaginates into the blastocoele and the macromeres cover them. A typical gastrula with a blastopore results in *Leucosolenia* (Minchin et al., 1900; Fell, 1974). The larva, when settling, attaches by the blastoporal end to the substratum. A small tubular asconoid sponge, Olynthus, is formed (Figs 4.16A (n), (o), 4.17f, and 4.18E). The Olynthus stage was first established by Haeckel (Balfour, 1880). In some sponges, this asconoid construction is retained, e.g. *Leucosolenia.* In most cases, differentiation to the syconoid or leuconoid grades occurs. The nonflagellated cells of the larva give rise to the pinacocytes of the adult, also to the porocytes and sclerocytes. It is not possible to reconcile the type of development described above with conventional conceptions of germ layer formation in other metazoans.

Development of *Sycon raphanus*

The development processes in *Ascandra (=Sycon) raphanus* are somewhat different from those of *Leucosolenia*, in spite of the fact that the two genera belong to the same class: Calcarea. In 1880, Balfour described cleavage of the sponge *Sycandra (=Sycon) raphanus* (Figs 4.17 and 4.18). In *Sycon*, the first three cleavage planes are all meridional (as in *Leucosolenia*) and the 8-cells of the embryo lie in a single layer. The next two cleavages are both equatorial and the embryo becomes 32-cell in 4 tiers and with a small central cavity (Okada, 1968). Cleavage then proceeds irregularly, leading to the formation of a blastula composed of one layer of cells. The eight at the posterior pole are large, granulated and referred to as archaeocytes. They cleave to form a total of 32 cells, while the others of the blastula elongate greatly, becoming columnar, and each acquires a flagellum. The interior of the blastula is hollow, forming the blastocoel without an opening to the outside. A temporary aperture makes a fleeting appearance, between the granulated cells of the flattened posterior pole. The archaeocytes continue division and, as a solid mass, this invaginates inside the blastocoel. It becomes a 'pseudogastrula', and as a larva it moves, by means of its flagellae, through the maternal tissues, making its way to the outside through the osculum.

It is now hollow, with bright red pigmentation, swims for about 1–2 days and finally settles on a suitable substratum. In settling, it undergoes a striking transformation. The flagellated cells forming the anterior part of the larva shift to the inside of the posterior part. Accordingly, the anterior part flattens and the blastocoel becomes a narrow, slit-like space. So, a typical gastrula with a blastopore results in *Sycon.* This means that gastrulation in *Sycon* occurs by invagination. Leys and Eerkes-Medrano (2005) reviewed and examined the old works concerning this aspect (Haeckel, 1872; Hammer, 1908 and others) and restudied embryogenesis and metamorphosis in *Sycon raphanus* to

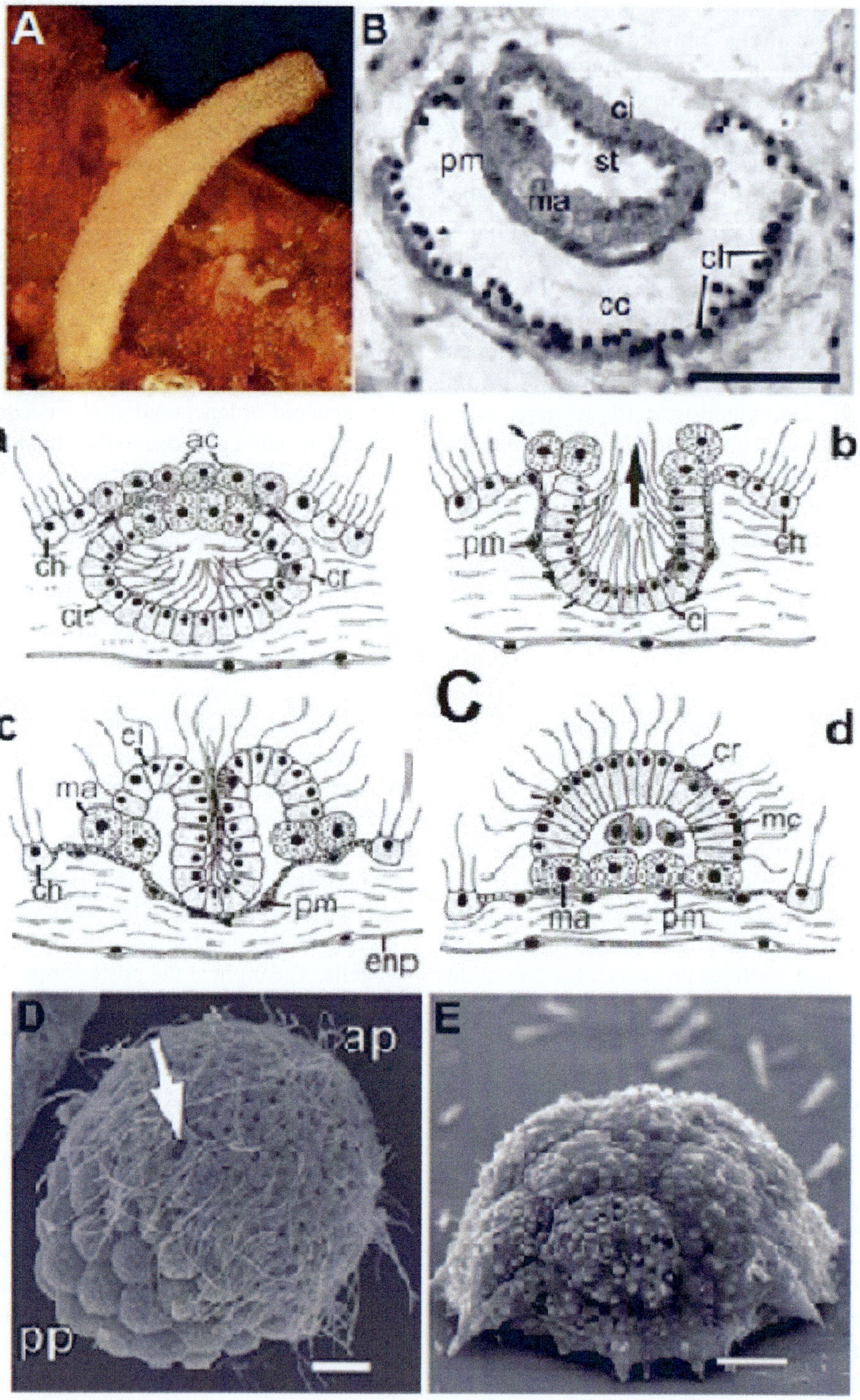

FIG. 4.18 Recent work of development of *Sycon* sp.: (A) photograph of whole mount *Sycon*; (B) semi-thin micrograph of *Sycon raphanus* stomoblastula (st) brooded in mother, cc=choanocyte chamber, ch=choanocyte, ci=ciliated cells, pm=placental membrane; (C) diagrams showing inversion of the stomoblastula of *Sycon*; (a) initial stage of formation of placental membrane *(arrows)* from the accessory–cell complex (ac); (b) early and (c) middle stages of inversion (*arrows* show growth of the membrane); (d) end of inversion; ci=ciliated cells, cr=cross cells, enp=endopinacocyte, ma=macromeres, mc=maternal cells; (D) SEM micrograph of 2-h old swimming amphiblastula larva of Calcaronea, *Sycon raphanus* showing the ciliated anterior hemisphere (ap) and the globular, granulated posterior hemisphere cells, *arrow* indicates one of the four 'cross cells'; (E) SEM of a newly adherent postlarva of *Sycon raphanus* and beginning metamorphosis, scale bar B=40 μm, D, E=10 μm. *((A–C) From Ereskovsky (2010) and (D and E) from Leys and Eerkes-Megrano (2005).)*

understand when and how gastrulation occurs. They concluded that the anterior cells 'invaginate' into the posterior half of the larva. The hole closes and it is not until some days later that the sponge forms an osculum at its apical pole. They also concluded that gastrulation occurs earlier, during formation of the two cellular regions of the larva, and metamorphosis involves the reorganization of these already differentiated regions.

Leys and Eerkes-Medrano (2005) described the larvae as seen by SEM, and thick and thin plastic sections by light microscopy. The larvae are 30–50 μm long, with ciliated columnar cells on the anterior pole and large globular granular cells on the posterior pole, as well as in the centre (Fig. 4.18D). There are four nonflagellated cells around the equator of the anterior hemisphere that correspond to the cross cells of Duboscq and Tuzet (1942). The larvae swim by rotating clockwise, as seen from the posterior pole. Leys and Eerkes-Medrano observed large numbers of live and fixed specimens, e.g. about 300 specimens as they underwent metamorphosis in Petri dishes. They stated that:

1. The larvae settle within 12 h of release from the parent, usually on their anterior pole (Fig. 4.18E) but occasionally on their side or posterior pole, and attach by their cilia which seem to be adhesive.
2. All settled postlarvae (12–48 h old) have a central mass of cells, enveloped by a single layer of the former granular cells (Fig. 4.18E). Thick sections show an outer layer of these granular cells, enveloping a mass of dense collagen and a mass of cells of different shapes.
3. In some 1–2 day old post larvae, choanocytes have already differentiated.
4. Spicules first appeared 2 days after release, and an osculum was first seen in 9-day old settled juveniles.

This is a young sponge with a simple canal system of the asconoid grade. It can then develop a highly complicated system of flagellated chambers.

Subclass Calcinea

Almost all information currently available on oogenesis and embryogenesis of Calcinea was obtained from light microscopic studies. Only their larval structure and metamorphosis were investigated at the electron microscopic level. Sexual differentiation in Calcinea has mostly remained unstudied and the investigated reproductive cycles of these sponges only describe the female gametes (Ereskovsky, 2010). As no sperm cells were described in Calcinea, only the origin of the female gamete can be described. Almost all the previous studies interpreted that oogonia originate either from choanocytes or from amoebocytes, e.g. Minchin et al. (1900). According to Borojević (1969), oogonium of *Clathrina* and *Ascandra* develop from pinacocytes (porocytes) which migrate into the mesohyl. However, Ereskovsky (2010) reported that oogonia divisions in the Calcinea have not been observed and the cytoplasmic growth of a young oocyte occurs in the thin layer of the mesohyl under the choanoderm (Fig. 4.20A (a) and (b)). At the beginning of the vitellogenic phase, the oocytes exhibit amoeboid motility and actively phagocytose various somatic cells (Fig. 4.20A (c)). At the end of vitellogenesis, the oocyte surface becomes smooth (Fig. 4.20A (d)) and its nutrition proceeds via cytoplasmic bridges connecting it to the phagocytes. In different Calcinea species, the oocytes may feed on different somatic cells. In some *Clathrina* species and *Ascandra* species, the oocytes first consume eosinophilic choanocytes and hyaline amoebocytes, then migrate to so-called 'nests', where they get nutrients from phagocytes via cytoplasmic bridges (Borojević, 1969). The mature egg of Calcinea is not polarized, its diameter varies from 85 to 120 μm and the nucleus is located centrally. It is isolecithal and heavily lecithal (rich in yolk granules).

During gametic reproduction, some Calcinea species (*A. minchini*, *A. falcat*, *C. contorta*) develop into peculiar temporary structures, referred to as 'nests' (Borojević, 1969). The nest is formed in the body wall, evenly throughout the sponge. A nest looks like an anarchized tissue with its own skeleton and a cylindrical cellular strand covered with 'pinacoderm'. Its long axis is perpendicular to the body wall. Some other Calcinea do not develop into nests, such as *C. coriacea* (Minchin et al., 1900). In *Clathrina*, each embryo is enveloped in a sheath of elongated parental cells.

Embryonic development was studied in several Calcinea species, e.g. *Ascandra minchini*. Cleavage of the egg is total and equal, and from the 8-cell stage, cleavage planes are radial and perpendicular to the embryo's surface (Fig. 4.20B (a) and (b)). At the end of the cleavage a hollow, one-layered, equal, nonpolarized blastula is formed, and retained from the 8-cell stage to the larval stage. Ereskovsky (2010) attributed this cleavage pattern to a new type, **polyaxial cleavage**, on the basis of characteristics mentioned in Ereskovsky and Boury-Esnault (2002). Polyaxial cleavage is intermediate between the chaotic cleavage and the radial one. It differs from the former in the regular position of cleavage planes, and from the latter, in the absence of an animal–vegetative axis determining the radial symmetry of the embryo. This cleavage is also associated with a polarized distribution of yolk granules in the egg.

Calcinea larva. Larval formation in Calcinea does not involve any morphogenetic movement of the cells or their layers. The cells of the embryo proliferate and differentiate. This

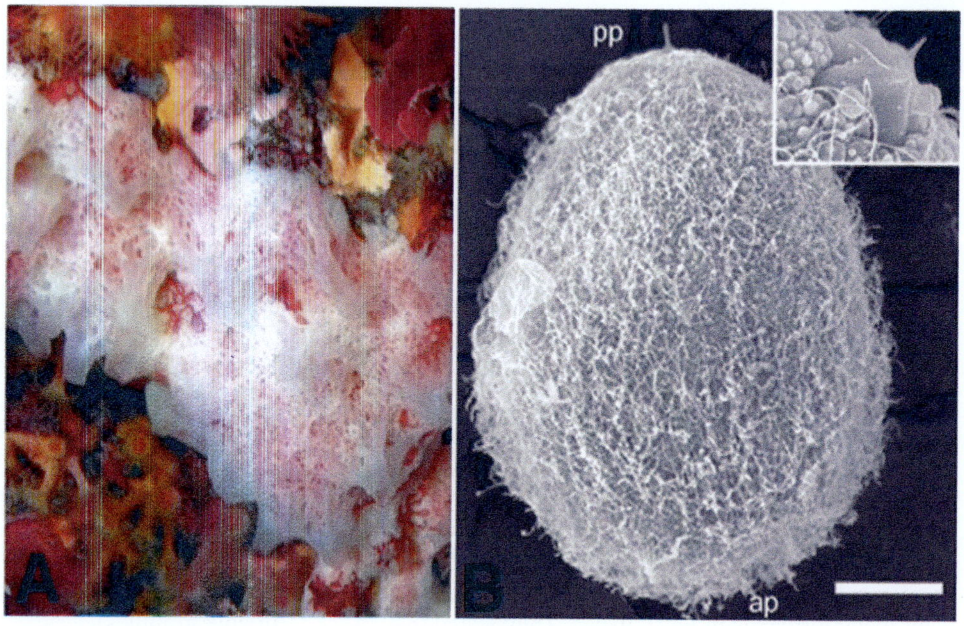

FIG. 4.19 (A) Calcinean sponge *Clathrina contorta* (Bowerbank 1866), Mediterranean Sea; (B) SEM graph of calciblastula larva of *Clathrina contorta*, inset shows portion of posterior pole (pp) with ciliated large cells, ap = anterior pole, scale bar 50 μm. *(From Ereskovsky (2010).)*

results in an increase in the surface area and change of the blastula to form plicate (Fig. 4.20C (d) and (e)). This larva of Calcinea is a calciblastula, previously called coeloblastula (Maldonado and Bergquist, 2002). Calciblastula is rounded or egg-shaped, 50–90 to 70–200 μm in length with a long antero-posterior axis. The body wall consists of a single layer of cells surrounding the larval cavity. The posterior pole has ciliated and granulated cells (Fig. 4.19B). Ciliated cells are the predominant cell type of the calciblastula. They do not have an underlying basement membrane and all cilia are of the same length. Bottle-shaped and vacuolar cells are present. Metamorphosis is mesenchymal and is accompanied by the migration of some of the flagellated cells into the blastocoel, to be transformed into choanocytes and amoebocytes. The remainder on the surface transform into pinachocytes (Borojević, 1969). The antero-posterior axis of the larva becomes the baso-apical axis of the Olynthus, juvenile sponge (Ereskovsky, 2004).

HEXACTINELLIDA (HYALOSPONGIAE)

Hexactinellid sponges, commonly called glass sponges, inhabit very deep sea waters, to depths reaching 6770 m. They are very variable in shape (Figs 4.21A and 4.22A). They may be tubular, cup-shaped, lumpy, branching, or lobulated, only encrusting forms are lacking. Owing to their habitat, studies of their reproductive behaviour, gametogenesis, and embryogenesis are lacking. In the literature, only three complete descriptions of development of two different species are found (*Vitrollula fertilis* Ijima, 1898 and *Farrea sollasii* Schulze, 1886). These descriptions were by Japanese workers in the late 19th and early 20th centuries (Ijima, 1901, cited in Fell, 1974; Okada, 1928). Later, only one species, *Oopsacas minuta* Topsent, 1927 was added. Fortunately, that of *Farrea sollasii* was available in the chapter of Okada (1968).

Reproduction. These glass sponges are thought to be hermaphrodites. They reproduce agametically and gametically. The **Agametic** method involves budding, as in other groups of sponges. It is suggested that some aggregation of archaeocytes present in the body of *Euplectella marshallii*, correspond to the gemmules in Calcarea. Budding in Hexactinellida seems to be epimorphic. In *Rhabdocalyptus dawsoni*, the bud's cavity is separated from the atrial cavity of the parent sponge, but the cytoplasm of the trabecular syncytium connecting the parent with the bud is continuous (Ereskovsky, 2010). In the **gametic** reproduction, the occurrence of eggs and sperm balls was observed in *Euplectella aspergillum* (Okada, 1968). Gametic reproduction takes place throughout the year, such as in the deep-sea *Farrea sollasii*. Okada (1928) studied reproduction of the glass sponge, *Farrea* (Fig. 4.21B) *sollasii*, the eggs are formed within a mass of archaeocytes. Each egg is large, measuring 0.7 mm, in diameter when mature, and has an alveolar nucleus. Surrounding the nucleus, there are numerous yolk granules and the peripheral cytoplasm contains lipid with vacuoles (Fig. 4.21B (a)). On the other hand, the spermatogonia are differentiated archaeocytes giving rise to spermatozoa with a spherical head and slender tail.

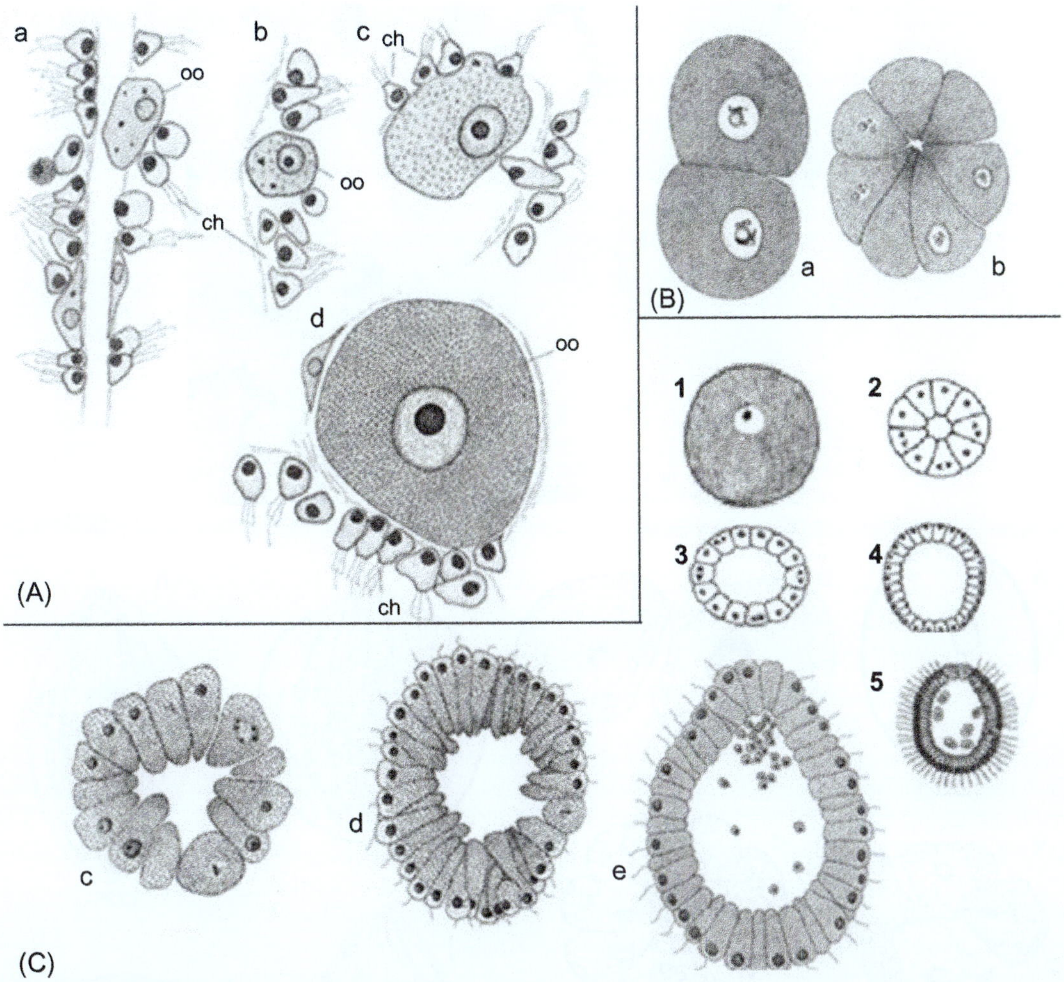

FIG. 4.20 Embryonic cycle of a calcinean sponge: 1–5 summary of the cycle *(from Ereskovsky, 2004)*, 1 = egg, 2 = cleavage, 3 = blastula, 4 = differentiation of blastula, 5 = larva; (A) drawings of oogenesis of *Ascandra minchini*: (a, b) early oocyte, (c) early vitellogenic oocyte, (d) fully grown oocyte; (B) cleavage of calcinea; (C) differentiation of the larva, (c) blastula, (d, e) larval morphogenesis (a–d *Ascandra minchini*). *((A–C) After Borojević (1969) from Ereskovsky (2010).)*

Fertilization is internal, taking place within the maternal body. Cleavage (Fig. 4.21B (a)–(d)) is total and, more or less, unequal and the resulting blastula (Fig. 4.21B (e)) is a parenchymula type. The embryo still develops within the maternal body. It is, at first, planula-like and spherical, later becoming ovoid. It is composed of an outer layer of cells, rich in yolk granules, surrounding a mass of mesohyl containing amoeboid cells. The latter arrange themselves in a layer surrounding a central cavity and change their form to become the future choanocytes. The flagellated chambers have not yet appeared, but six cross-shaped spicules (oxystauractines) are formed near the body surface (Fig. 4.21B (f)). Larvae of the later stages are oval, and become bluntly pointed at both ends, where the tips of the stauractines come together. The number of these spicules increases to 12 and their axes become thicker (Fig. 4.21B (g) and (h)). The cells constituting the posterior two-thirds of the larva are loaded with fatty material, responsible for the energy utilized during the later pelagic life. At this stage, the flagellated chambers are already beginning to be formed.

The most advanced larva is slightly pointed at the anterior pole and rounded at the posterior pole. While still within the maternal body, it develops two kinds of spicules (Fig. 4.21B (j)):

1. small discohexasters in the interior of the body; and
2. stauractines at the surface, which are thickened anteriorly and bear small processes.

It is probably at this stage that the larva acquires flagellae, escapes from the mother and settles onto some substrata. The smallest attached specimen is barrel-shaped,

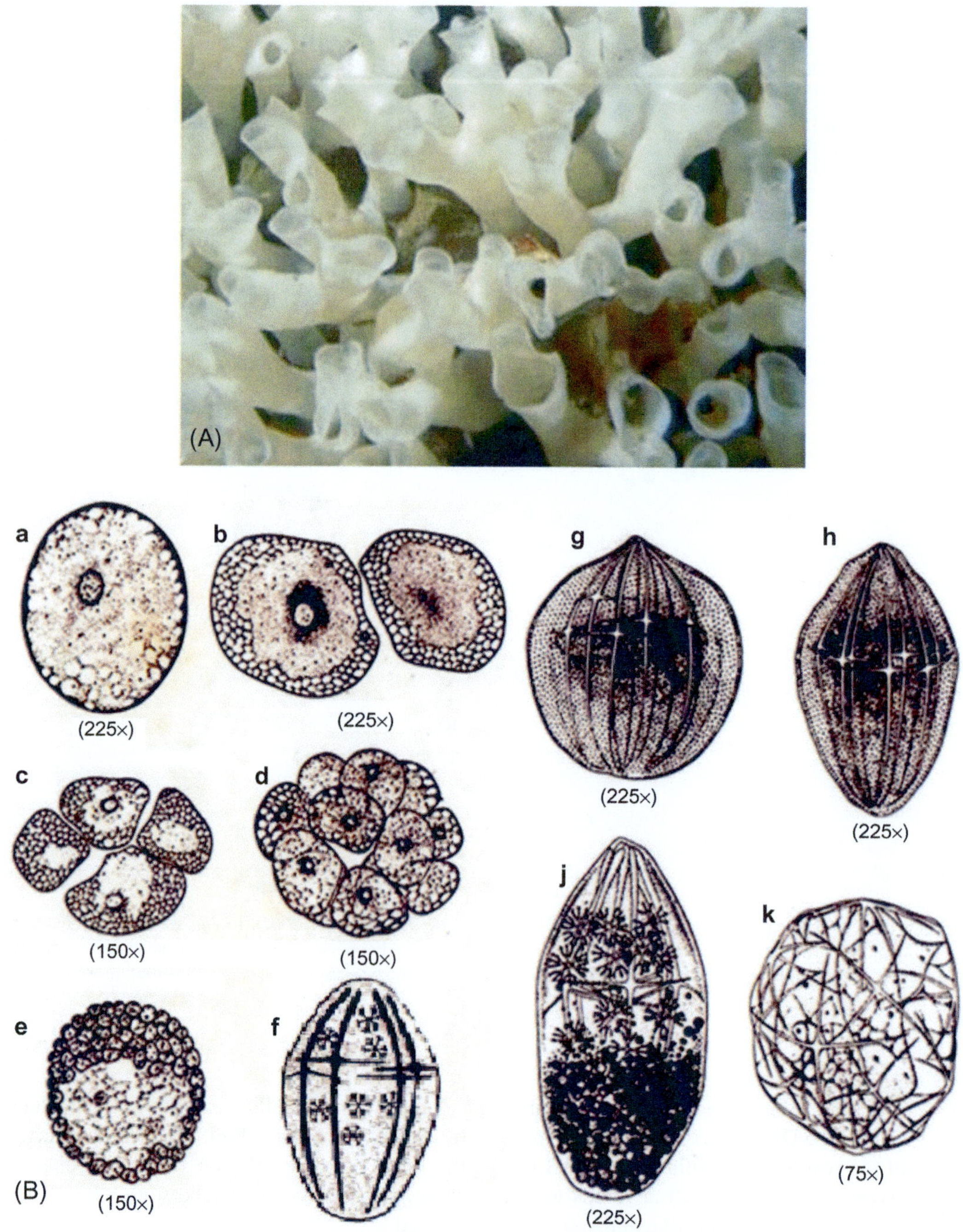

FIG. 4.21 (A) W.M. of a hexactinellid, *Farrea occa* Bowerbank 1862 *(from Ereskovsky, 2004)*; (B) (a–k) development of *Farrea sollasii (from Okada, 1968)*: (a) egg in body of adult sponge; (b) 2-cell stage; (c) 4-cell stage; (d) 16-cell stage; (e) section of larva at blastula stage; (f) early stage of larva of *Farrea*, stauractines are formed; (g) fairly well-developed larva; (h) elliptical larva with oxystauractines; (j) most advanced larva found within maternal body; (k) young sponge has escaped to outside, is about to attach. Spongocoel, gastric cavity, and mouth are formed.

0.54 mm long, 0.36 mm wide, somewhat flattened anteriorly and bluntly pointed posteriorly (Fig. 4.21B (k)). The flagellated chambers appear as several protuberances of the spongocoel towards the dermal layer. At the surface, small spicules of pentactines and stouractines develop, and the dictyonine net begins to be formed within the body near the point of attachment. Spicules are not yet formed in the gastral layer. A young sponge, measuring 0.68 mm long, 0.57 mm wide, is oval, and the openings of some of the flagellated chambers have been broken through, although the adult osculum is not yet formed. The paragastral cavity is formed with choanocytes lining its central portion, and the osculum breaks through to the outside.

Oopsacas minuta Topsent, 1927

Family:	Leucopsacasidae
Order:	Lyssacinosida
Subclass:	Hexasterophora
	Hexactinellida (Boury-Esnault et al., 1999)

In 1993, Boury-Esnault et al. and Vacelet et al. (1994) published that they discovered a population of the hexactinellid *Oopsacas minuta* in a cave in the Mediterranean Sea, near La Ciotat (France), a new habitat for this sponge, with a stable water temperature (13–14.5°C). It is easily accessible by SCUBA. This is a unique deep-sea habitat. These authors, in addition to Boury-Esnault et al. (1999) and Leys et al. (2006), collected adult specimens of *O. minuta* from this habitat. They studied gametogenesis, embryogenesis, especially gastrulation and formation of syncytia and structure of the larva of this species using up-to-date techniques and tools (TEM and SEM). These studies on the reproduction in a hexactinellid sponge, are the first since those made by Schulze (1880, 1887), Ijima (1901), and mostly by Okada (1928) on *Farrea sollasii*. This lack is probably because this group of sponges was considered to reproduce gametically, only rarely in the deep-sea. However, observations of Okada concluded that reproduction of this group is probably without seasonality, as was believed at that time for other deep-sea organisms. The discovery of a population of *Oopsacas minuta* in that cave mentioned above, which reproduce all the year round, proved that hexactinellids reproduce gametically all through the year. This discovery also provided, for the first time, a description of the hexactinellid larva using TEM and SEM.

Development of *Oopsacas*

Reproduction. Gametic reproduction in the cave population, as observed in all examined specimens, occurs all the year round. In hexactinellid sponges, the male and female gametes originate from archaeocytes, free nucleolar amoebocytes that are suspended within the trabecular reticulum between flagellated chambers (Boury-Esnault and Vacelet, 1994).

Spermatogenesis. The spermatic cysts are located between the choano-chambers close to the exhalant ducts. They are derived from large archaeocyte masses. The latter are groups of nucleolated cells present in the mesohyl just below the collar ducts. Archaeocytes are spherical or subspherical cells with densely granular cytoplasm and numerous mitochondria; the Golgi component and endoplasmic reticulum are not prominent and phagosomes are rare. Archaeocytes in conglomerates are frequently attached to one another and to the trabecular syncytium by plugged junctions (Leys et al., 2007). They are considered to be stem cells giving rise to other cell types. This view was suggested by Okada (1928). These cells are ovoid, about 28–30 μm long, 21–23 μm wide, and are surrounded by an envelope about 0.5 μm thick (Fig. 4.22B and C). In immature cysts, there is a gradient in the size of cells from the centre to the periphery: the largest with a 5.3 μm diameter, a 2.6 μm diameter of the nucleus and 1 μm diameter of the nucleolus, are present in the centre. The smallest, with a 2.7 μm diameter, 1.6 μm diameter of the nucleus and 0.5 μm diameter nucleolus are present at the periphery as spermatogonia. On the other hand, in the mature cysts this gradient disappears and all the spermatids have approximately the same size (Fig. 4.22C). Flagellae appear during the early stages of spermiogenesis. In the most advanced stages observed, the cytoplasm contains numerous mitochondria, a clear dictyosome and numerous small vesicles close to the cell membrane, and the nucleus is not condensed. Dense osmiophilic junctions are present in between the spermatogonia, and also between the spermatogonia and the thin envelope of the cyst (Fig. 4.22D). The mature sperm was not observed in such preparations. This may be because either the differentiation of sperm cells occurs over a short period, or the hexactinellid sperm cells never have a condensed nucleus, and never reach the stage considered as mature as in other invertebrates. Thus, the absence of the acrosome cannot be confirmed.

Oogenesis. Oogenesis also occurs within archaeocyte masses, as does the spermatogenesis. However, oogenesis is probably a relatively rapid process, since few oocytes have been observed even by light microscopy and paraffin sections. The smallest identified oocyte is 10 μm in diameter and is surrounded by archaeocytes that act as nurse cells. Only one oocyte is observed within each archaeocyte mass. The oocyte grows until it reaches a size of 100–120 μm (Fig. 4.23A). The only full-grown oocyte observed in this preparation using TEM, possesses numerous short pseudopodia, with a central nucleus of 45–50 μm diameter, and a

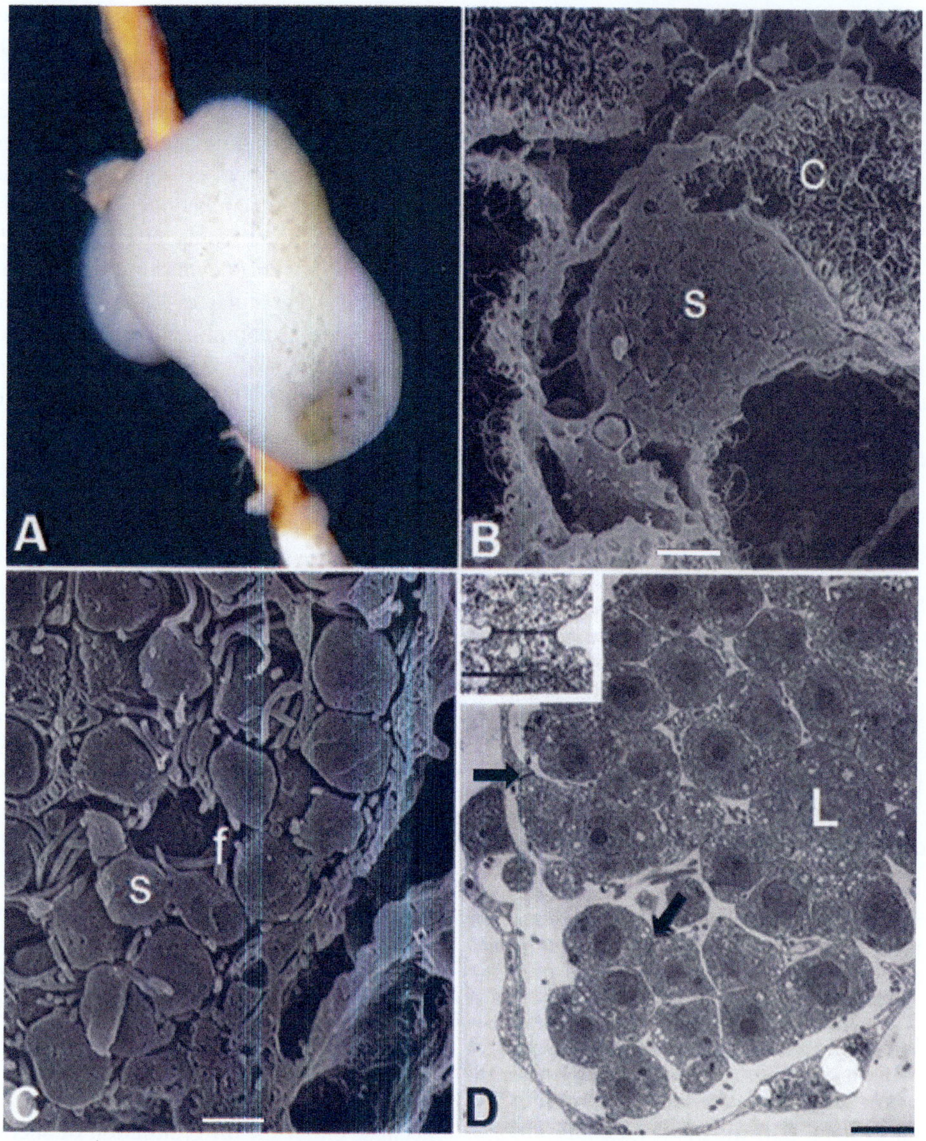

FIG. 4.22 (A) Whole mount of the sponge *Oopsacas minuta* Topsent 1927, from Mediterranean Sea *(from Ereskovsky, 2010)*; (B) SEM micrograph showing spermatic cyst (s), choanochamber (c); (C) SEM micrograph showing spermatids (s), flagella (f), bar=2.5 μm; (D) TEM micrograph of half part of immature spermatic cyst with large cell in the centre (L) on the right part of the graph, and smaller spermatogonia at the periphery, there are dark junctions *(arrows)* connecting the cells with each other and with the peripheral tissue, bar=2.4 μm *(from Boury-Esnault et al., 1999)*.

nucleolus 10 μm in diameter (Fig. 4.23A). The chromatin is homogeneous. The cytoplasm contains large, homogeneous lipid spheres localized mainly at the periphery of the cell. Membrane-bound vitelline inclusions with heterogeneous content are abundant in the centre, and small clear vesicles are distributed throughout the cytoplasm (Fig. 4.23B). The divisions of maturation of the egg have not been observed, but polar bodies have been found associated with the early embryos (Fig. 4.23C (a) and (b)).

Embryogenesis. Boury-Esnault et al. (1999) studied embryogenesis of *O. minuta* using light and electron microscopy. In 2006, Leys et al. restudied the same subject using TEM and SEM. The results of the two studies are similar, but the latter study of Leys et al. provided detailed ultrastructural information, especially regarding the formation of the syncytium that forms the inner and outer layers of this sponge.

Cleavage in *O. minuta* is radial, total, and somewhat equal until the 22-cell stage. The size of the embryo is the same as that of the largest observed oocyte (120 μm), and with the same cytoplasmic inclusions. The **first** cleavage is generally meridional, although it can differ in relation to the position of the polar body. The **second** cleavage is

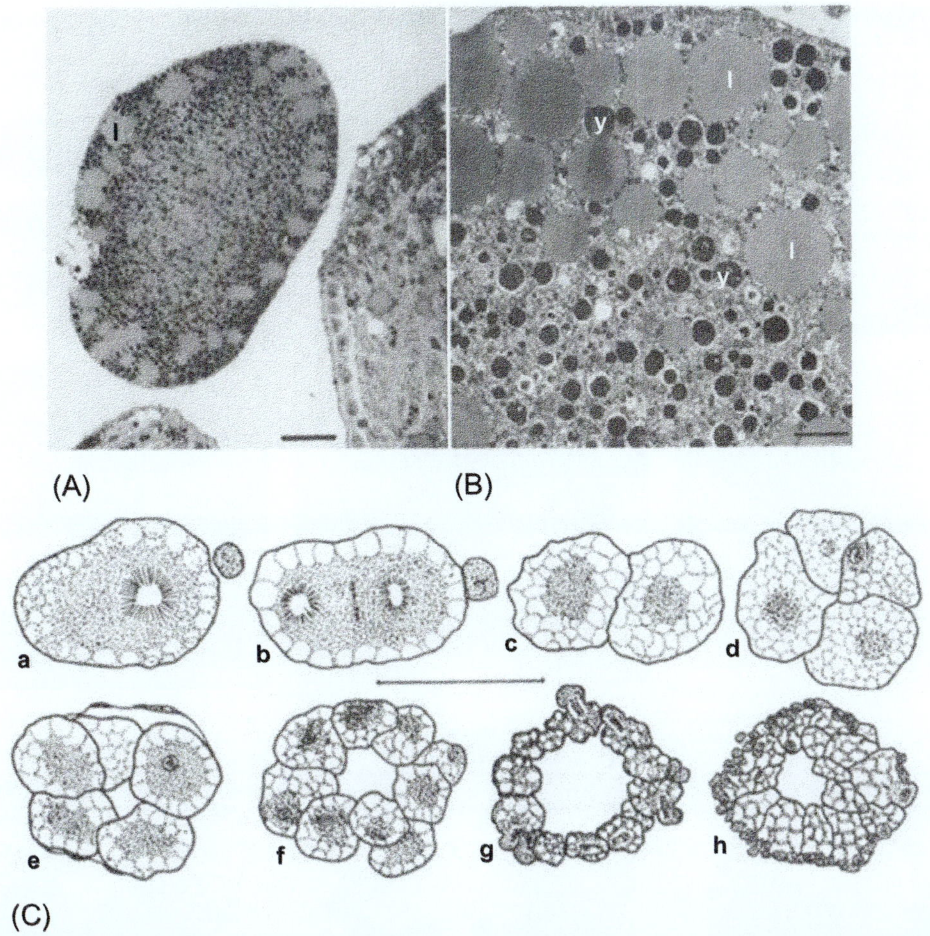

FIG. 4.23 (A) Semi-thin section of an oocyte isolated from the maternal tissue, lipid spherule (l), bar = 15 μm; (B) TEM micrograph of oocyte cytoplasm, vitelline inclusion (y), bar = 0.35 μm; (C) drawings show cleavage and embryogenesis in *O. minuta*, (a and b) embryos preparing for 1st division with the polar body at one pole; (c) 2-cell stage, the cells adhere tightly to each other; (d) 4-cell stage; (e) 8-cell stage, a blastocoel is clearly visible; (f) 16-cell stage, the cleavage is radial with the spindle parallel to the surface of the embryo and the blastocoel is growing; (g) stage after 6th division with unequal cells, every blastomere divides gives rise to a small outer cell and inner large one; (h) the gastrula is a double cell layered embryo, bar = 50 μm *(Boury-Esnault et al., 1999).*

also meridional, perpendicular to the first one producing the four equal blastomeres adhering tightly to each other. The **third** cleavage is also equal, but equatorial and results in 8-blastomeres, with displacement of the blastomeres which looks like a spiral cleavage. Embryos with 6-blastomeres were found, suggesting that early cleavage is asynchronous. At this stage, a small blastocoel appears, which increases in size during the subsequent divisions. The 16- and 32-blastomere embryos are hollow blastulae (Fig. 4.23C (g)) whose cells are held together by numerous phyllopodia that project from their surfaces. Yolk inclusions are distributed around the nucleus and lipid inclusions are at the periphery of each cell.

After the 32-blastomere stage, the division becomes unequal and results in a collection of irregularly shaped cells surrounding the blastcoel (Fig. 4.24G and H). Small cuboidal cells 4–6 μm diameter (micromeres) lie at the outer surface and large yolk- and lipid-rich cells (macromeres), also very irregular in shape, surround the blastocoel (Fig. 4.24G–L). Many of the embryos are polarized at this stage, one side of the embryo usually lacks micromeres altogether, whereas the opposite pole has fewer yolk- and lipid-rich macromeres (Fig. 4.24G–I). Some micromeres differentiate first a single, and subsequently multiple cilia. Importantly, all micromeres are connected to each other by cytoplasmic bridges with characteristic hexactinellid **'plugged junctions'** (Fig. 4.25A). It is assumed that plugs arise from the Golgi apparatus. This is a new finding, first observed and documented in the embryo of cave-dwelling glass sponge, *Oopsacas minuta* (Leys et al., 2006).

Initially, the macromeres still surround a hollow blastocoel, but sections of multiple embryos suggest that the

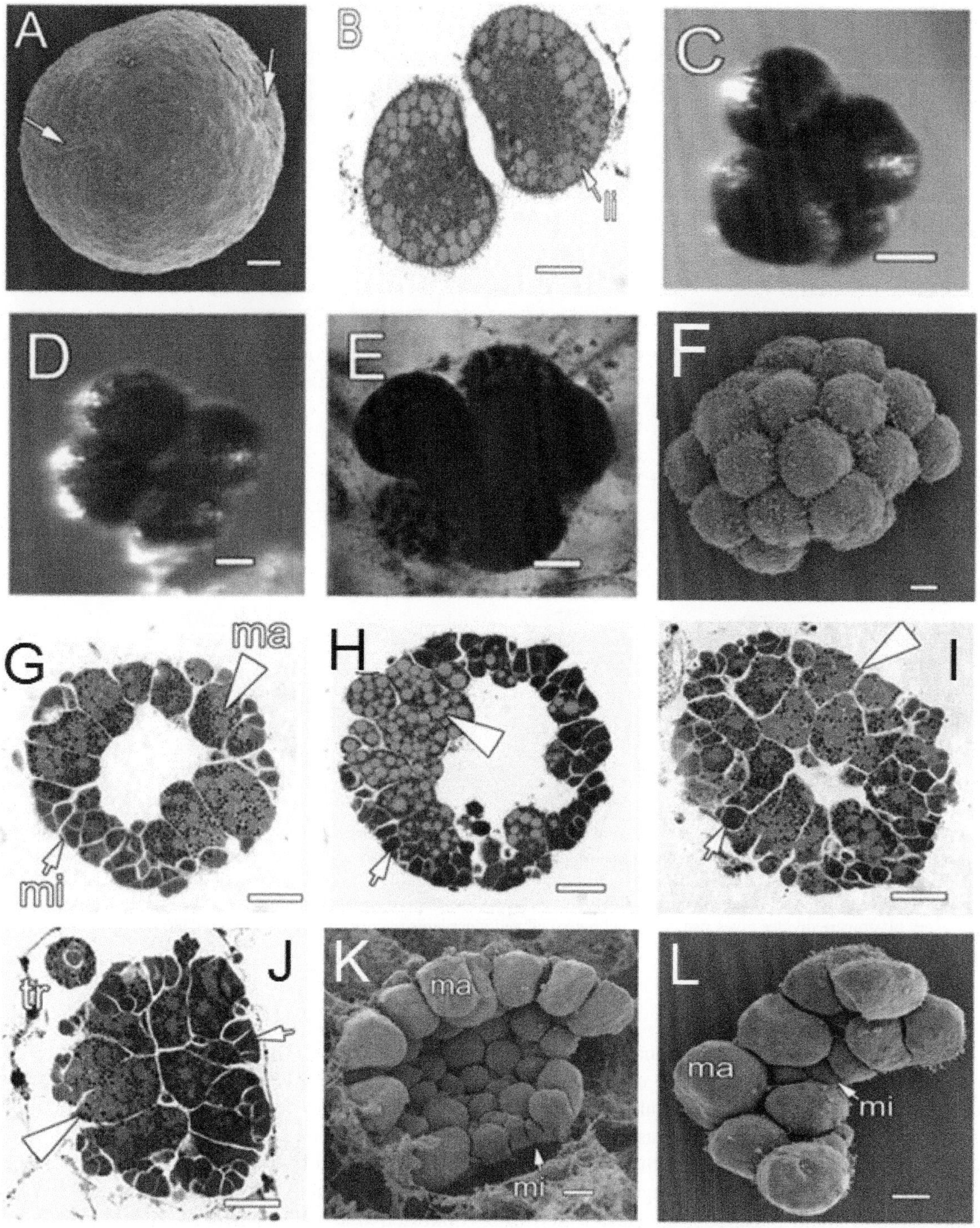

FIG. 4.24 Cleavage of *Oopsacus minuta*; (A) SEM micrograph showing depressions of the first cleavage of the oocyte *(arrows)* bar = 10 μm; (B) 2-cell stage with numerous filopodia, large lipid inclusions at the periphery (li); (C and D) stereomicrographs of 4-cell stage embryos dissected out of the mother sponge, (C) shows rotational and (D) equatorial cleavage patterns; (E) plastic section of 6-cell stage shows dark lipid inclusions; (F) SEM micrograph shows 16–32 embryo dissected out of the parent; (G–J) light micrographs of thick epoxy sections show embryos after gastrulation when micromeres (mi) are on the periphery and macromeres (ma) remain more central, blastocoel is gradually filled in; (K and L) showing views of the gastrula from the opened blastocoel; Bars, B–F = 20 μm *(Leys et al., 2006)*.

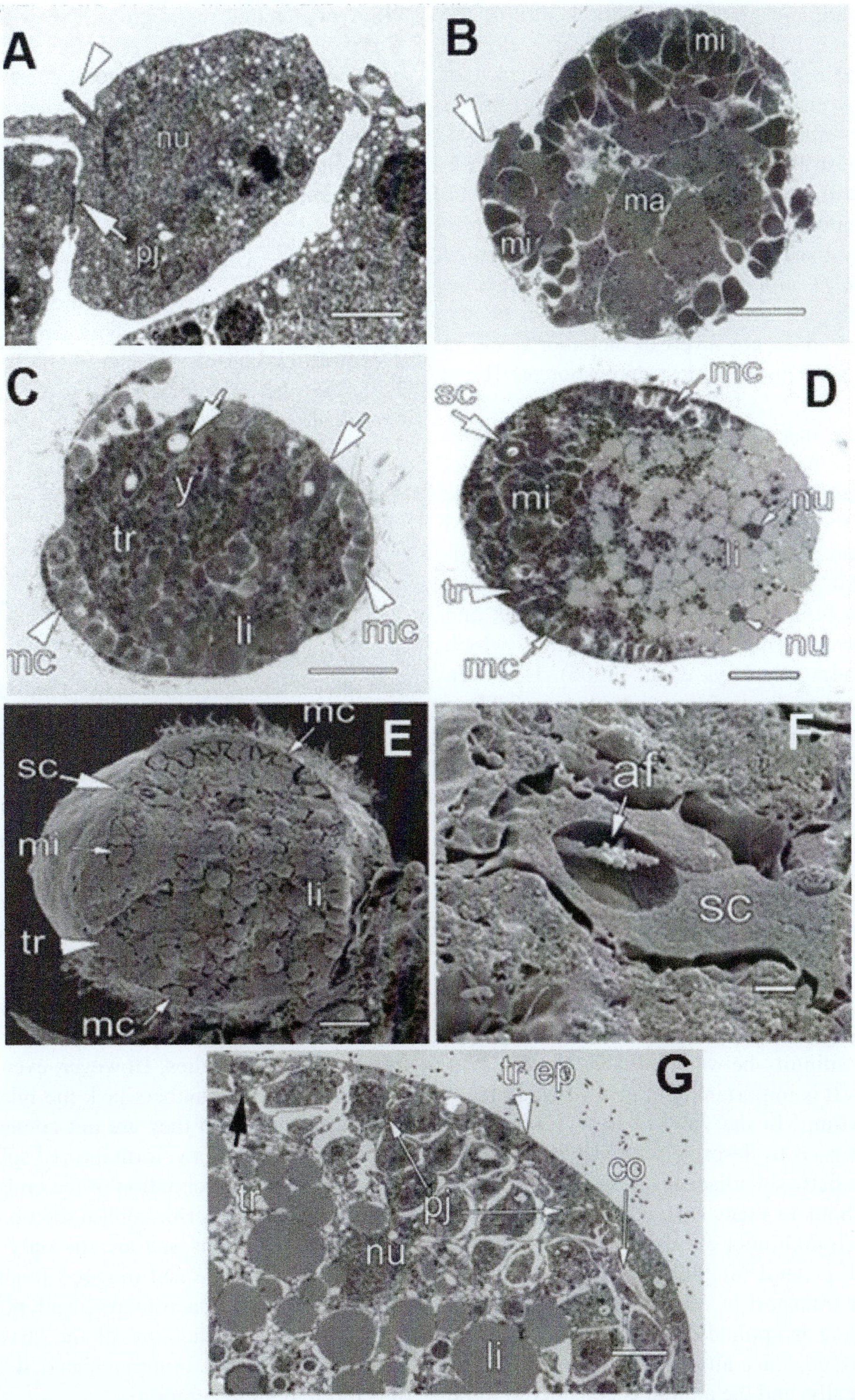

FIG. 4.25 Formation of the larva of *Oopsacus minuta*: (A) TEM micrograph shows micromeres (of the embryo in Fig. 4.24G) have already begun to differentiate, these cells are connected to others by a plugged cytoplasmic bridge *(Pj arrow)*, cilia *(arrowhead)*; (B and C) light microscopy of thick epoxy sections shows in (B) macromeres (ma) extend around micromeres *(mi arrow)* at the surface of the embryo and (C) an embryo in which the macromeres completed fusion and enveloping the micromeres within a continuous covering membrane, notice in (B and C) surface of the embryo was torn during isolation from the parent tissue, multiciliated cells (mc) and sclerocytes *(arrows)* are well differentiated, the remainder of the embryo consists of single syncytial trabecular (tr) tissue with yolk (y) and lipid (li); (D) light microscopy of a thick epoxy-embedded section shows the embryo becomes clearly polarized with the lipid (li) inclusions in the future anterior pole (right side), multiciliated cells (mc) round the equator, and sclerocytes (sc) towards the posterior pole (left side), the micromeres (mi) form the central-posterior portion of the embryo, the multinucleate (nu) trabecular tissue forms more than 70% of the embryo; (E) SEM micrograph shows the same of (D); (F) SEM micrograph shows enlarged view of the sclerocyte (sc) shown in (B) clarifying 4-rayed proteinaceous axial filament (af) of the stauractin spicule; (G) TEM micrograph shows enlargement view of the anterolateral region of the embryo shown in (D), clarifying the trabecular tissue (tr) with nuclei (nu) and lipid that reaches around and envelops *(arrow)* the multinucleated cells, bars A, C=2 μm, B, D, E=20 μm, F=1 μm, G=5 μm. *(From Leys et al. (2006).)*

macromeres elongate and gradually fill the centre of the blastula (Figs 4.24G, I, J and 4.25B). Once the embryo is solid, lamellipodia extend from the apical surface of the macromeres to envelop the micromeres (Fig. 4.25B and C), thereby forming the outer surface of the embryo. This means that gastrulation occurs, and this description confirms that gastrulation in hexactinellid embryos of *O. minuta* occurs by epiboly. This result differs from the preceding one (Boury-Esnault et al. (1999)) which claimed that gastrulation in *O. minuta* occurs by cellular delamination. Interestingly, the orientation of these two layers; 'dermal and gastral' does not invert at metamorphosis in this glass sponge as it does in most demosponges (Leys et al., 2006).

Furthermore, the macromeres also produce phyllopodia that interdigitate and appear to fuse. After this stage, each embryo consists of a single multinucleate tissue (the beginning of the formation of **trabecular reticulum** of the future larva and adult sponge) that envelops the micromeres (Fig. 4.25C–E). The syncytial tissue of the larva and of the adult is formed by the fusion of the yolk and lipid-rich micromeres at the time of gastrulation, which is another new finding by Leys et al. (2006). However, this macromere-derived tissue forms cytoplasmic bridges with the micromeres, and each of these bridges contains plugged junctions. The result is a fully cytoplasmic interconnected embryo and, afterwards, larva. The above description of the formation of the syncytial tissue in glass sponge, '*O. minuta*', during the formation of gastrulation and late embryo, is the principle difference from the earlier description of that of *Farrea sollsii* (Okada, 1928) and even from the very recent description of that of the same species *O. minuta* (Boury-Esnault et al., 1999).

Because no other metazoan possesses such extensive syncytial tissue, or such an unusual manner of maintaining cytoplasmic continuity between uninucleate and syncytial components, it is important to clarify what is meant by the term syncytium. In the view of Leys (2003), the term syncytium is used to describe any tissues that lack membrane barriers between adjacent nuclei. Syncytial tissues have come about in many different organisms, and the structure of each multinucleate tissue depends on how it was formed. The general function of syncytia is to enhance intracellular transport by way of cytoplasmic continuity. This meaning is applied to the terminology used for hexactinellid tissue, since all the components are connected cytoplasmically, and the animal is considered to be syncytial.

Formation of the larva. After the formation of the trabecular reticulum, the embryo undergoes the following steps to form a larva: **first**, the trabecular reticulum itself differentiates into two regions: one with more yolk, which will be at the future posterior pole, and another with more lipid, which will be at the future anterior pole of the future larva. **Second**, the micromeres also differentiate, but at all times they maintain cytoplasmic continuity with the trabecular reticulum and other cells via cytoplasmic bridges. They give rise to:

1. multiciliated cells that form a single (but comparatively broad) band around the equator of the embryo, with their cilia projecting through the syncytial outer layer (Fig. 4.25C–E). This can be identified as the **equatorial band**.
2. Some other micromeres also give rise to the sclerocytes, spicule-producing cells. Each cell produces a square axial filament within a small vacuole, while it is still at the periphery of the embryo. Each develops numerous long pseudopodia and migrates in at the posterior pole. At the same time, the vacuole enlarges with the secretion of silica around the proteinaceous axial filament. All the sclerocytes then elongate from the posterior region, and as they extend to the anterior pole, they become multinucleated (Figs 4.25G and 4.26B–D).
3. Other micromeres also ingress into the posterior central region and, as the embryo elongates, these cells form the choanocyte chambers (Fig. 4.26E). The Choanocyte begins as a single nucleated cell, with a single flagellum and a collar of microvilli, and all such cells are connected to similar cells and to the trabecular tissue by plugged cytoplasmic bridges. Subsequently, the portion of the cell containing the collar and flagellum (collar body) becomes separated by a long cytoplasmic bridge from the portion with the nucleus.

Third, in late embryos the trabecular reticulum envelops and penetrates the small spherical chambers that are formed by collar bodies. However, even in the fully developed larva, these chambers lack the bilayered reticulum of the adult sponge and they are not connected to the outside of the larva. **Fourth**, the formation of spherulous cells is the last step in the differentiation of the embryo. These become identified in the posterior-central region. They contain large spherulous inclusions and are the only cell type in which a cytoplasmic bridge and plugged junction have not been found. **Fifth**, the macromeres, and not the micromeres, from the outer 'epithelium' of the larva, and this layer is continuous with the multinucleated trabecular tissue that runs throughout the embryo.

The fully formed larva. The free-swimming mature larva is only 150–180 μm in length (Fig. 4.26A–C). It is syncytial, like the adult. It consists of a continuous multinucleated tissue that reaches from the anterior to the

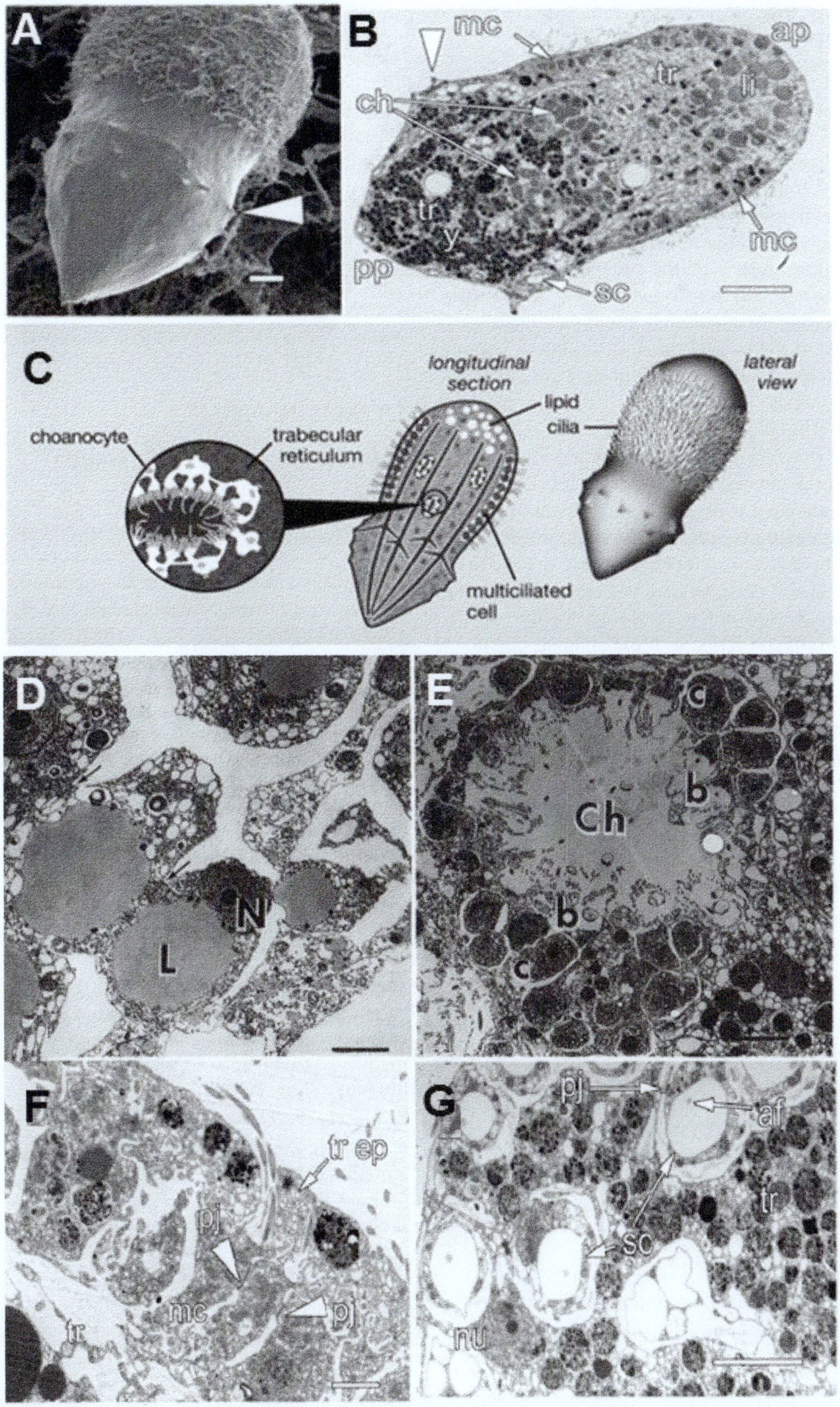

FIG. 4.26 Continuation of formation of the trichimella larva: (A and B) late larval differentiation; (A) SEM micrpgraph showing elongation of the larva at the anterior pole (ap), *arrowhead* indicates the protrusion of the transverse rays of spicules under the syncytial tissue, multiciliated cells form a band around the waist of the larva; (B) light microscopy of thick section of the larva at the stage shown in (A) indicates projections of spicules *(arrowhead)*, multiciliated cells (mc), lipid (li) at the anterior pole, yolk (y) at the posterior pole in the trabecular tissue (tr) choanocytes (ch) forming early flagellated chambers centrally, sc = sclerocytes, bar = 10 μm; (C) whole mount and L.S. section of the larva; (D, E, G) TEM micrographs, (D) shows undifferentiated macromeres with lipid granules (L) and often linked by dense junctions *(arrows)* bar = 1.7 μm; (E) shows choanochamber (Ch) collar bodies (b) choanoblasts (c) are differentiated bar = 6 μm; (F) SEM micrograph showing multiciliated cells are connected to each other and to the trabecular tissue (tr) by plugged cytoplasmic bridges (pi) bar = 2 μm; (G) shows differentiated sclerocytes (sc) at the posterior pole of the larva, axial filament of the spicule (af) and the trabecular tissue (tr) are separated by plugged cytoplasmic bridges (pj) bar = 5 μm. *((A–C and F–G) from Leys et al. (2006) and (D–E) from Boury-Esnault et al. (1999).)*

posterior pole, forming the outer layer and the bulk of the inner mass, and is connected to the cellular components via cytoplasmic bridges. It has a rounded anterior pole, which is made up of undifferentiated macromeres with lipid globules, and which is often linked by dense osmeophilic junctions (Fig. 4.26D). The posterior pole mainly consists of tissue rich in vitelline inclusions. Masses of choanoblasts differentiate at the posterior pole from the vitelline-rich tissue. Multiciliated mononucleated cells form a belt around the central third of the larva (equatorial band), each is about 5 µm height and 2.8 µm wide, with up to 50 cilia per cell. These cilia project through the syncytial layer (Fig. 4.26F and G). The choanochambers are situated in the centre-posterior half of the larva but are not connected to a canal system. Sclerocytes form a thin sheath of multinucleate cytoplasm that surrounds stauractines (4-rayed) spicules, whose long rays lie longitudinally from the lipid-dense anterior pole to the yolk-filled posterior pole, at the periphery of the larva (Fig. 4.26D and F). Within the trabecular reticulum in the centre-posterior portion of the larva, 4–6 choanocyte chambers lie, each of up to 30 µm diameter. They are smaller than those of the adult, which are 100–150 µm in diameter, and are composed exclusively of the collar bodies. There is no trace of the reticular tissue in which they are embedded in the adult (Fig. 4.26G).

Finally, the larva appears as wholly formed of the trabecular reticulum surrounding all the cellular components, and extends from the anterior to posterior poles. At the anterior pole, the trabecular tissue consists of thin strands with little yolk, and large spherical lipid inclusions (Fig. 4.26D). Towards the equator and posterior pole, portions of the trabecular reticulum are thicker, with many small yolk inclusions (Fig. 4.25F and G). Fine strands of this syncytial tissue interlaced with collagen are continuous throughout the larva. There are no membrane boundaries between lipid- and yolk-filled regions.

The new term **trichimella** (from the Greek 'bearing long threads') was proposed by Boury-Esnault and Vacelet (1994) for the *Oopsacas minuta* larva, and then extrapolated for the hexactinellid larvae. This terminology was based on the unusual features when compared with the demosponge parenchymula (Simpson, 1984).

1. During free living, the larva swims counter-clockwise, with the rounded, reserve rich pole forward.
2. Both in adult and embryo, individualized cellular units are linked by dense junctions, whose interpretation is still uncertain. They can be either intrasyncytial occlusions or intercellular junctions. These results suggest that the hexatinellid tissue needs further investigations (Boury-Esnault and Vacelet, 1994).
3. Trichimella larva is built up of three well-defined regions: a rounded anterior pole, a conical posterior pole, and a median flagellated zone.

As described by Okada (1928), a unique larval spicule skeleton, localized at the posterior pole, maintains the larval shape. In the larva of *Farrea sollasii* (Okada, 1928), both larval stauractines and adult microscleres are present, whereas only stauractines have been observed in *Oopsacus minuta* (Boury-Esnault and Rützler, 1997).

DEMOSPONGIAE

Tetractinomorpha (Tetraxonida)

The body form of demospongiae may be encrusting, massive, lobulated, tubular, filamentous and cup-shaped.

The sponges of this group reproduce agametically and gametically. In the former, the gemmule formation occurs as in all fresh water species, e.g. *Mycale*, *Chalina*, *Cladorhiza*, and in *Reniera*. Among these marine forms, the larva developing from the gemmule is similar to that developing from a fertilized egg. In gametic reproduction, the sperm are produced earlier than the eggs. Both kinds of gametes develop from archaeocytes. The egg has a finely granular cytoplasm and a large nucleus. It is surrounded by a cellular envelope originating from the neighbouring mesohyl. Probably, this envelope functions as a trophic membrane. The spermatogonium divides into two: one of them differentiates as a covering cell, while the other undergoes spermatogenesis. The details of oogenesis are not clear. The fertilized egg undergoes unequal holoblastic cleavage, forming a stereogastrula, or parenchymula larva. This larva corresponds to the gastrula of other animals. It swims by means of its flagellae for about 24 h, and it attaches by its anterior end to an object. It then undergoes the process of inversion of the cellular layers constituting the body wall. The outer flagellated cells become situated inside the body, and are differentiated into the paragasteral lining the flagellated chambers. At the same time, the amoebocytes from the inside emerge to the surface and arrange themselves to form the pinacoderm layer and the scleroblasts. The larva increases in size and becomes more complex to be an adult.

In freshwater sponges, it was found that the fertilized egg cleaves to form a stereoblastula, composed of cells of three sizes: micromeres, mesomeres, and macromeres. The micromeres, at first, are situated at the anterior pole, spread over the embryo and give rise to the flagellated surface cells. While the embryo is still enclosed in the maternal body, the cells of the inner mass transform into sclerocytes, choanocytes, and amoebocytes. The first two kinds of cells are derived from the macromeres. When the embryo escapes from the maternal body it is a flagellated larva, already provided with spicules and flagellated chambers. After swimming for several days, it becomes attached by its anterior pole. The developmental process of the freshwater sponges thus differs from that of the other monaxonid species in that there is no inversion of the embryonic cell layers.

Example: 1—*Donatia deformis* (Thiele)

This noncommercial sponge is one of the common forms found clinging to the under surface of stones close to the shores of the Hawaiian Islands. After discussion, the author Edmondson (1946), accepted the generic name as proposed by Nardo (1833) and adopted by Dendy (1905). The living animal is typically hemispherical in form, reaching up to about 1 in. in diameter, pale yellow in colour, sometimes white, and anchoring to a support by root-like extensions in the lower border.

The sponge *Donatia (Tethya) deformis* (Thiele, 1898), reproduces agametically by both external and internal budding. The chiasters (type of the microscleres) are distributed throughout both the medulla and cortex of the sponge, but more plentiful in the cortex. The external budding occurs by these spicules, which are carried out in large numbers into gemmules, filaments and other processes developed from the surface of the sponge (Edmondson, 1946). The buds move outward along the bundles of spicules and are released from the surface to form new sponge. It was investigated that this condition is the cause of the widespread occurrence of *Donatia deformis* observed in Hawaiian waters. The gemmules, the typical reproductive bodies of the species are frequently developed from the surface of the sponge under natural conditions, and perhaps to an even greater extent under subnormal conditions in the laboratory. The buds are pushed from the tips of the conical elevations of the surface. They are loose bundles of monaxon spicules surrounded by masses of archaeocytes, among which are scattered many chiasters. At early development, the thick stalk of the bud expands at the end into an irregular body from which pointed processes appear. Within 24 h the stalk elongates and becomes slender, and rays of variable length develop from the central nucleus of the gemmule. The rays are supporting spicules surrounded by archaeocytes, whose number ranges from a few to more than a dozen, and their length reaches up to 10 or 15 times the diameter of the body of the gemmule. In the laboratory, where the water is relatively quiet, the gemmules are released from the mother sponge within 1–2 days of their first appearance. Under favourable conditions, they may be transformed into young sponges. Much of the work of Edmondson (1946) was experimental in the aquarium, and he came to the conclusion that one gemmule develops into a young sponge, while large numbers of gemmules may come in contact with each other, and the result of this aggregation is a rapid conversion into a functional sponge.

In addition to gemmules, filaments of considerable length are often produced from the conical elevations of the surface of the sponge. They are flexible, retractile, extensible, and often fuse with each other. Their length may exceed twice the mother's diameter. Formation of filaments usually accompanies the formation of gemmules. Filaments are simple or branched, pointed or blunt at the end. To release from the mother, the filament may be completely retracted into the sponge body. However, under normal conditions, they do not release as do the gemmules. Experimentally, when they are cut off and kept under observation for a few days, they sharply contract and take on the form of a young sponge. There is no information about gametic reproduction in this work, however, the author, Edmondson (1946), recorded that 'it has long been known that the genus *Donatia* or *Tethya* propagate by a rather distinctive agametic method'. Also, he recorded that attention has been centred upon *D. deformis* in Hawaiian waters because of:

1. its widespread occurrence about the islands;
2. its unusual fertility in the Honolulu Aquarium where 'myriads of specimens' accumulate on the walls of the tanks; and
3. its remarkable behaviour as observed under controlled conditions.

VIVIPAROUS SPONGES

The majority of sponges are viviparous, that is the embryos develop within the maternal body. In many cases, these embryos are surrounded by the oocyte follicle. Cleavage is total in all the studied species, but there is considerable variation in the pattern of the cleavage among these studied species. In many of the calcareous sponges and also in many of the acalcarea, e.g. in *Oscarella = (Halisarca) lobularis* and *Hippospongia communis*, cleavage is equal or nearly so, during the early stages of development. In *Ephydatia fluviatilis*, cleavage is unequal from the beginning. In *spongia reticulata*, cleavage is irregular.

In most of the acalcareous sponges, the cleaved zygote develops giving rise to a parenchymula larva. In many sponges, these larvae are entirely covered by flagellated cells, and in some species the flagellae at the posterior pole are of different length and/or density than elsewhere. The larvae of many other species lack these posterior flagellae. In the larva of *Haliclona limbata*, the anterior pole lacks flagellae and there is only a transient existence of them at the posterior pole. In a number of species flask-shaped secretory cells appear among the flagellated cells, especially at the anterior pole.

Example: 2—*Plakina monolopha* (Maas, 1909)

The classic example of the group in which the gametic reproduction was studied in detail is *Plakina* (Fig. 4.27A) by Schulze (1880) and Maas (1909). It was also recorded in Okada (1968). The egg undergoes equal holoblastic cleavage. The blastula is ovoid to pear-shaped, its outer surface is composed of flagellated cells, while a very few granule-filled

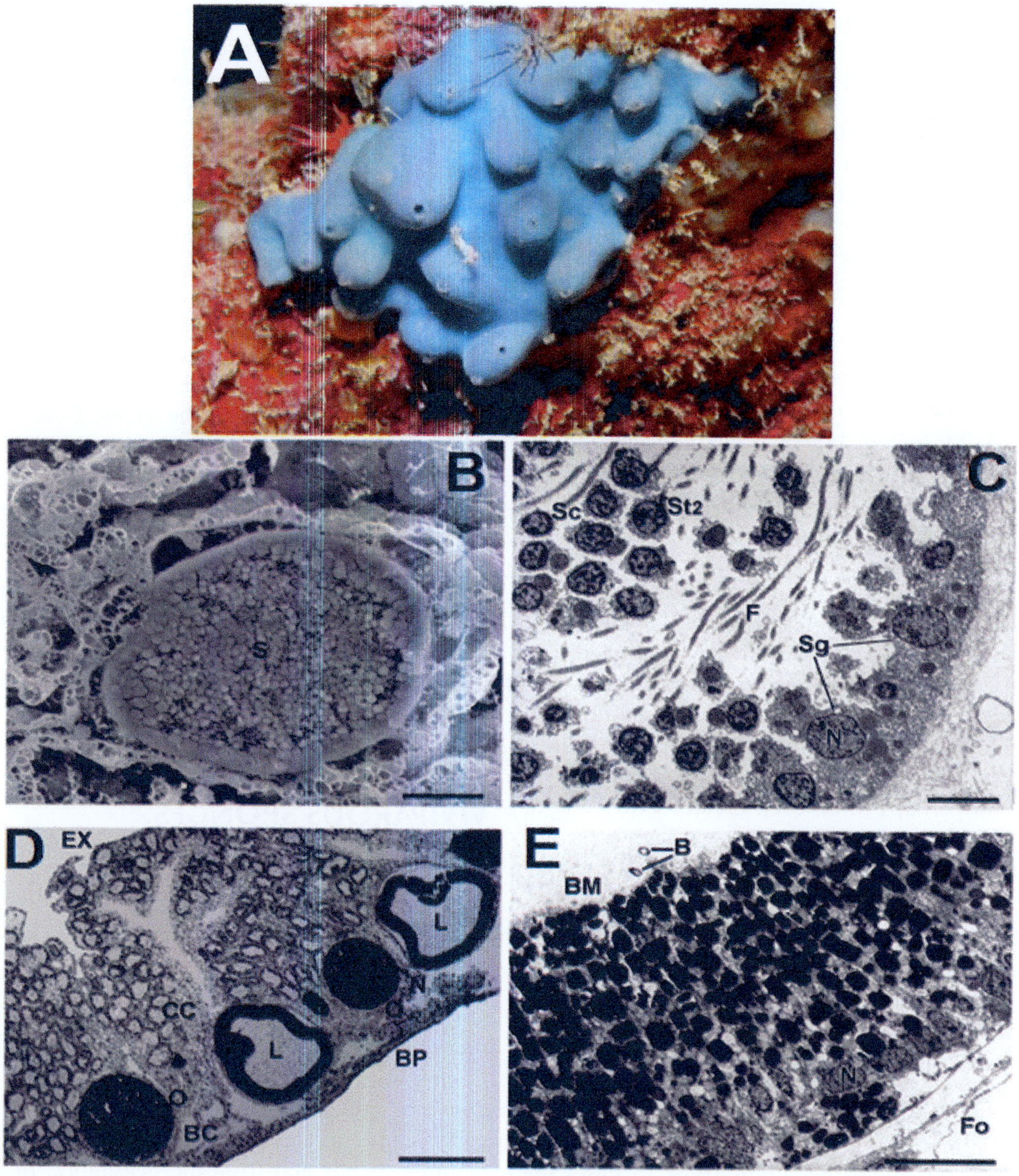

FIG. 4.27 (A) Small blue sponge, one of the four chromotypes of *Plakina kanaky* sp. nov. collected in New Caledonia, about 2–5×5–15 cm in size and 0.5–1 cm thick. Its surface is smooth with irregularly rounded lobes. Each lobe often harbours a tiny osculum, about 1 mm in diameter, bordered by a thin protruding membrane. The sponge is firmly attached to the substratum and its consistency is fleshy *(from Ruiz et al., 2015)*; (B–E) Reproduction in *Oscarella nathaliae* sp. nov.: (B) SEM micrograph showing spermatocyst in the choanosome; (C) TEM micrograph describing details of a spertmatocyst; (D) oocytes and cinctoblastulae larvae in the choanoderm (LM); (E) TEM micrograph describe details of cell wall of a larva; B = endobiotic bacteria inside a larva; BC = basal cavity; BP = basopinacoderm; BM = basement membrane; CC = choanocyte chamber; EX = exopinacoderm; F = flagella; Fo = follicle cell; L = larva; N = nucleus; O = oocyte; S = spermatic cyst; Sg = spermatogonia; St = spermatids; St2 = spermatocyte 2. Scale bars: B, 20 µm; C, 5 µm; D, 100 µm; E, 50 µm *(from Ereskovsky et al., 2014)*.

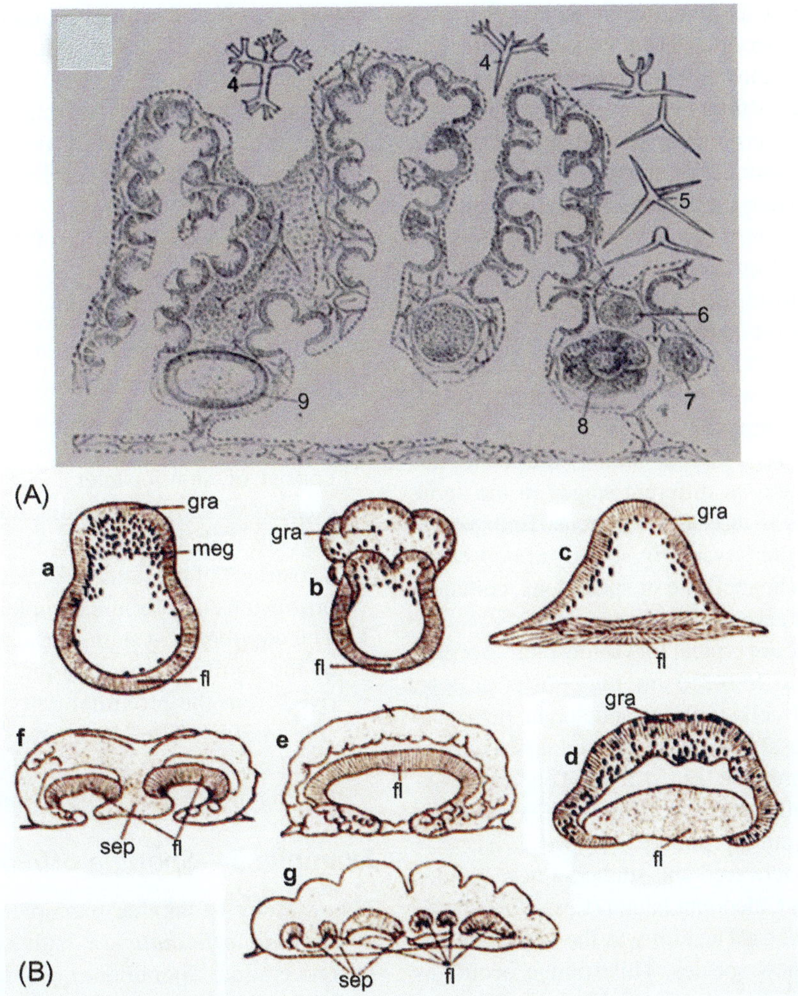

FIG. 4.28 (A) Part of Homoscleromorpha (Tetractinellid) sponge *Plakina*: 4,5 = spicules, candelabra and calthrops respectively, 6 = sperm ball, 7 = egg, 8 = cleaved embryo, 9 = larva *(from Hyman, 1940)*; (B) metamorphosis of *Plakina monolopha (after Maas, 1909 and from Okada, 1968)*; (a) amphiblastula larva, granulated cells (gra) present anteriorly and flagellated cells (fl) forming posterior part; (b) larva about to settle; (c) settled larva beginning metamorphosis; (d) settled larva with invaginating flagellated cells, (e) rhagon stage; (f and g) rhagon in two stages of formation of choanoflagellated chambers.

archaeocytes are formed in the blastocoel. At a later stage, the cells making up the posterior half of the body wall lose their flagellae and become filled with granules, and the mesohyl cells are pushed into the blastocoel (Fig. 4.28B (a) and (b)). The larva attaches by its anterior, flagellated half, which flattens and invaginates inwards. The flagellated cell layer becomes surrounded by the granule-containing columnar cells making up the posterior half of the body wall (Fig. 4.28B (c) and (d)). A blastopore is formed during this process but it is closed by the surrounding columnar cells, which meet at this position and eventually form the basal part of the sponge. The larva, at this stage, is called a **rhagon** (Fig. 4.28B (e)–(g)). The cavity of this rhagon, which was formed by the invagination of the flagellated layer, is divided into separate cavities, together with the corresponding parts of the flagellated layers, by the partitions originating in the dermal layer. These cavities become filled with flagellated cells, but later reappear, forming spherical flagellated chambers (Fig. 4.28B (f) and (g)).

The development of *Halisarca*, another primitive tetractinellid (recently Homoscleromorpha) sponge was studied, and the stages are more advanced than those studied in *Plakina*. The osculum breaks through the body wall only after completion of the rhagon stage. An excurrent canal system grows inwards from the osculum, making connection with the flagellated chambers through the paragaster cavity. The observation that the choanocytes and amoebocytes of the adult sponge are derived from the original posterior half of the larva was confirmed. Also, it was found that the inversion of the embryonic cell layers occurring in the Monaxonida does not take place in this form.

(The term 'dermal' is changed to 'outer' to avoid the terminology of the germ layers and tissues.)

Ereskovsky et al. (2014) discovered five new species of four genera of Homoscleromorpha, *Oscarella*, *Plakortis*, *Plakina*, and *Corticium*. These sponges were collected from Martinique and Jamaica, from vertical walls of reef caves at depths ranging from 23 to 28 m in the Caribbean Sea. The authors presented a re-description of ***Plakina jamaicensis*** based on newly collected material and the type of specimen. In sponges collected in Martinique, spermatogenesis was observed in early June. Reproduction in *Oscarella nathaliae* was described as an example of this group of animals. Spermatic cysts are surrounded by a thick layer of collagen located in the choanosome (Fig. 4.27B). Spermatogonia (Fig. 4.27C) are formed from choanocytes chambers which differentiate, become ovoid, about 65 μm in diameter, and undergo asynchronous spermatogenesis. Cysts, generally, with several generations of male germ cells, form clusters (Fig. 4.27C). Mature spermatozoa have long terminal flagellae and slightly elongated head containing acrosome and large mitochondria (Fig. 4.27C). On the other hand, oocytes occur simultaneously at different stages in the same individual, as oogenesis is also asynchronous. Embryos at different stages and mature larvae (Fig. 4.27D) are observed in the lower part of the choanosome of specimens, collected at the end of March and early April from Jamaica.

Cinctoblastula larvae are typical for Homoscleromorpha, with an oval shape, about 153–255 μm, pink posterior pole, surrounded by a belt of cells with intranucleolar paracrystalline inclusion. The basal side of the ciliated larval cells is lined by a basement membrane (Fig. 4.27E). The larval cavity is filled with extracellular matrix (Fig. 4.27D) including endobiotic bacteria of both morphotypes (Fig. 4.27E).

Ruiz et al. (2015) discovered and studied a new species of ***Plakina kanaky*** from Caledonia Island (East of Australia and North of New Zealand). The aim was the identification and classification of a new species. This sponge occurs as four similar sponges of different colours. They aimed to determine whether the four chromotypes represented different species, or phenotypic variations of a unique new species. No reproductive elements were recorded in specimens collected in November 2010. Gametogenesis was never observed, but coeloblastula (Fig. 4.29A (a)), pre-larvae, were found in blue (Fig. 4.27A), brick-red and green specimens collected in February 2013.

Prokaryotic transfer from the adult sponge to the larvae was observed inside the pre-larval cavity of the main prokaryote morphotypes, found in the adult mesohyl (Fig. 4.29A (a)).

In a third work (Ereskovsky et al., 2009), larval metamorphosis in ***Plakina trilopha*** was described. The cinctoblastula of this species (Fig. 4.29B) is egg-shaped with a broad anterior pole; they are about 180 μm in length and 250 μm in width. The cilia are distributed uniformly throughout the surface. The larva is built of columnar cells; its characteristic morphology divides the surface into three distinct areas: antero-lateral; postero-lateral; and posterior, enclosing a large central cavity. The latter contains an extracellular matrix and symbiotic bacteria, most abundant in the posterior region. Most remarkable is the postero-lateral area, the cells of which contain a spindle-shaped intranucleolar crystalloid, oriented parallel to the longer cell axis (Fig. 4.29B (a) and (d)). This structure can be used as a natural morphological marker when tracing the cell fate in the course of larval metamorphosis. Intermediate ciliated cells form a narrow belt above the posterior pole (Fig. 4.29B (a)).

The cinctoblastula larva of *Plakina trilopha* metamorphoses in three stages:

1. The settlement of the larva with its anterior pole, and its flattening on the substrate (Fig. 4.30C), so its antero-posterior axis is transformed into the baso-apical one;
2. The attached larva (the settler) acquires the shape of a hat or a bell (Fig. 4.30A and B). Its marginal and basal parts consist of anterior-lateral larval cells, the middle part consists of larval transition area cells; the apical parts of the cells include the former larval posterior pole. The apical surface of the latter cells bears numerous cytoplasmic protrusions that form a complex network (Fig. 4.30D);
3. The aquiferous system of rhagon is formed. The canals of the aquiferous system and endopinacoderm are derived from the proximal parts of the internal cell layer (Fig. 4.30E). The choanocyte chambers develop from distal parts of the internal folds (Fig. 4.30E). This signifies that the rhagon stage is completed.

Example: 3—*Spongia officinalis* Lenneaus

A new study of another example of **viviparous** sponge is of ***Spongia officinalis*** (a bath sponge) (Demospongiae, Dictyoceratida, Spongiidae) (Baldacconi et al., 2007). The specimens were obtained at a depth of 6–7 m, on the rocky bottom of La Pierta, a large embayment off the coast of Porto Cesareo (SE Italy, Ionian Sea), Mediterranean. Ten specimens were gonochoric (sex ratio 1:1), one specimen showed successive hermaphroditism, with alternate production of oocytes and spermatic cysts in the same reproductive season. All the examined specimens showed gametic reproduction, even though the process usually involved small portions of the sponge tissue. The examined specimens were obtained during June, which was concomitant with the occurrence of larvae in the sponge tissue. The released larvae in the field were trapped and were placed at 20°C in filtered seawater, and immediately preserved for histological and ultra-structural investigations. Statistical analysis proved that gametic elements (oocytes, spermatic cysts, embryos at different stages, and larvae) were found throughout the year with a peak occurring from September to November. The oocytes are found almost all the year round, with some fluctuations, but the maximum value recorded during June. **Gametogenesis.** Young oocytes of *Spongia officinalis* are fairly triangular in shape; each measures about 15 μm in diameter (Fig. 4.31A). Growing oocytes, they gradually change to become spherical, and show an eccentric

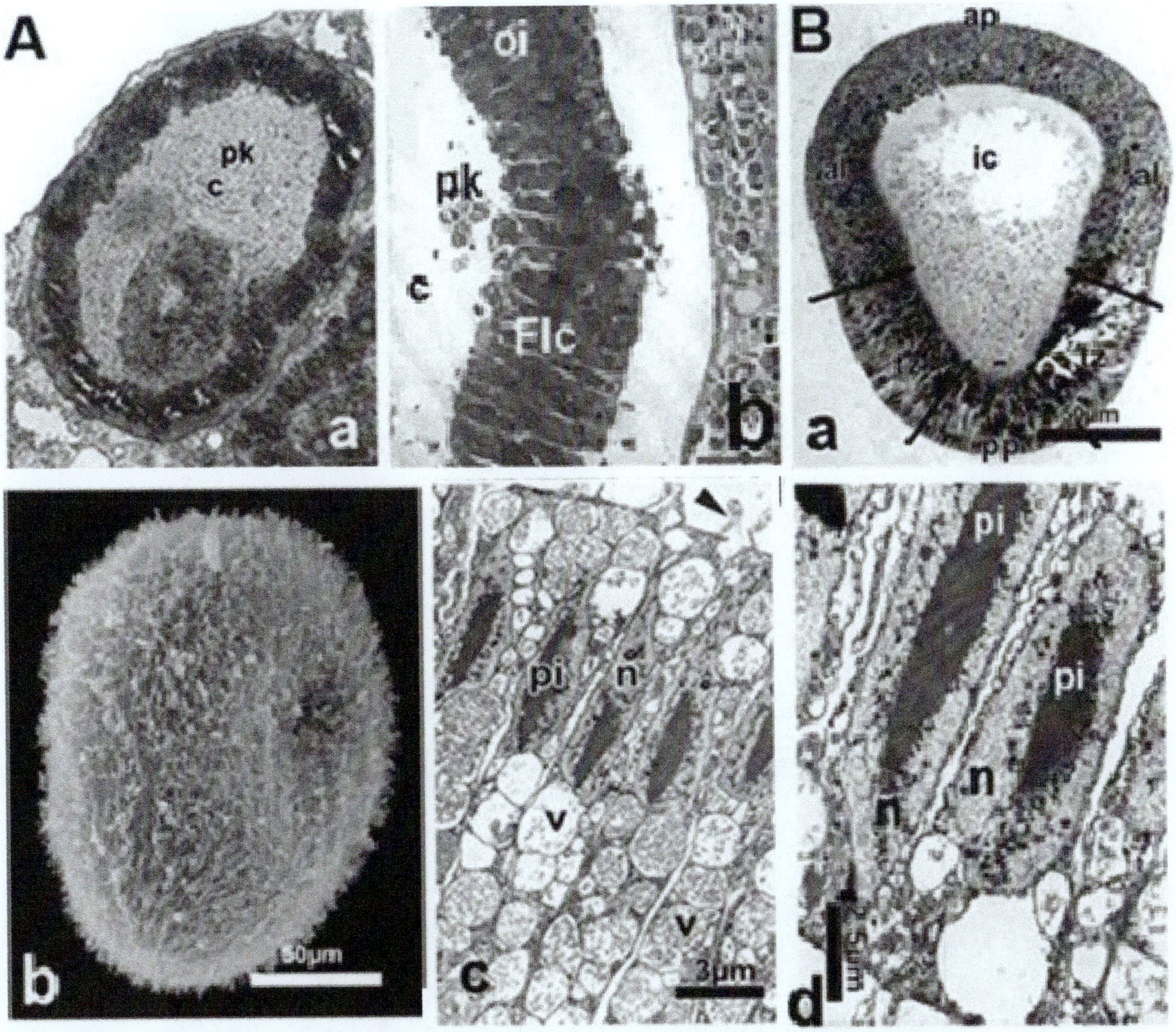

FIG. 4.29 (A) TEM micrographs of larva of *Plakina kanaky* sp. nov. (a) coeloblastula larva (scale bar = 50 µm); (b) detailed view of flagellated cells and prokaryote morphotype 1 inside sponge larvae and sponge mesohyl (scale bar = 10 µm); C = central cavity; Flc = flagellated cells; Oi = osmiophilic inclusion; Pk = prokaryot *(from Ruiz et al., 2015)*. (B) Free-swimming larva, cinctoblastula of *Plakina trilopha* (Homoscleromorpha); (a) Light microscopy, semi-thin longitudinal section of larva; (b) SEM of external view of a released, free-swimming, cinctoblastula; ap = anterior pole, al = antero-lateral zone, ic = internal cavity, tz = transitional zone of cells with intranuclear paracrystalline inclusions, pp. = posterior pole; (c) TEM of ciliated cells of the postero-lateral zone, each with an intranuclear paracrystalline inclusion (pi); (d) TEM of the nuclei (n) of cinctoblastular ciliated cell from transitional zone with intranuclear paracrystalline inclusions (pi). *(From Ereskovsky et al. (2009).)*

nucleolated nucleus and cytoplasm with little inclusions (Fig. 4.31B). They are found, in small numbers, throughout the choanosomal region, separated from one another by mesohyl elements and collagen. Each undergoes successive developmental stages and vitellogenesis (Fig. 4.31C); the cytoplasm gradually increases its contents of yolk elements. In the last stage of development, the oocyte becomes completely filled with yolk and reaches its maximum size (up to 200 µm) (Fig. 4.31D). On the other hand, the spermatic cysts are found in the choanosomal region (Fig. 4.31E). The germinal cells inside each cyst undergo development synchronously, 'whereas it is asynchronous in different cysts in the same specimen' (Baldacconi et al., 2007). The cysts flourish in autumn and early winter and they increase in number to about 100 times those estimated for female elements. In contrast with oocytes, the size of the spermatic cyst showed moderate oscillations (50–60 µm).

Embryogenesis appears to begin in November, and the cleavage is total and equal, giving rise to a solid 'morula' stage. The latter measures about 360 µm in diameter and consists of few cells (4–16 cells). Later, during December, the cells of the morula increase in number (Fig. 4.31F). The morula cell contains a nucleolate nucleus which tends to be eccentric (Fig. 4.31G). Although embryonic cells tend to maintain a certain distance from one another, on occasions a cell establishes contact by means of membrane bridges (Fig. 4.31H) that interact with the membrane of an adjacent cell. The development continues up to April, with resulting

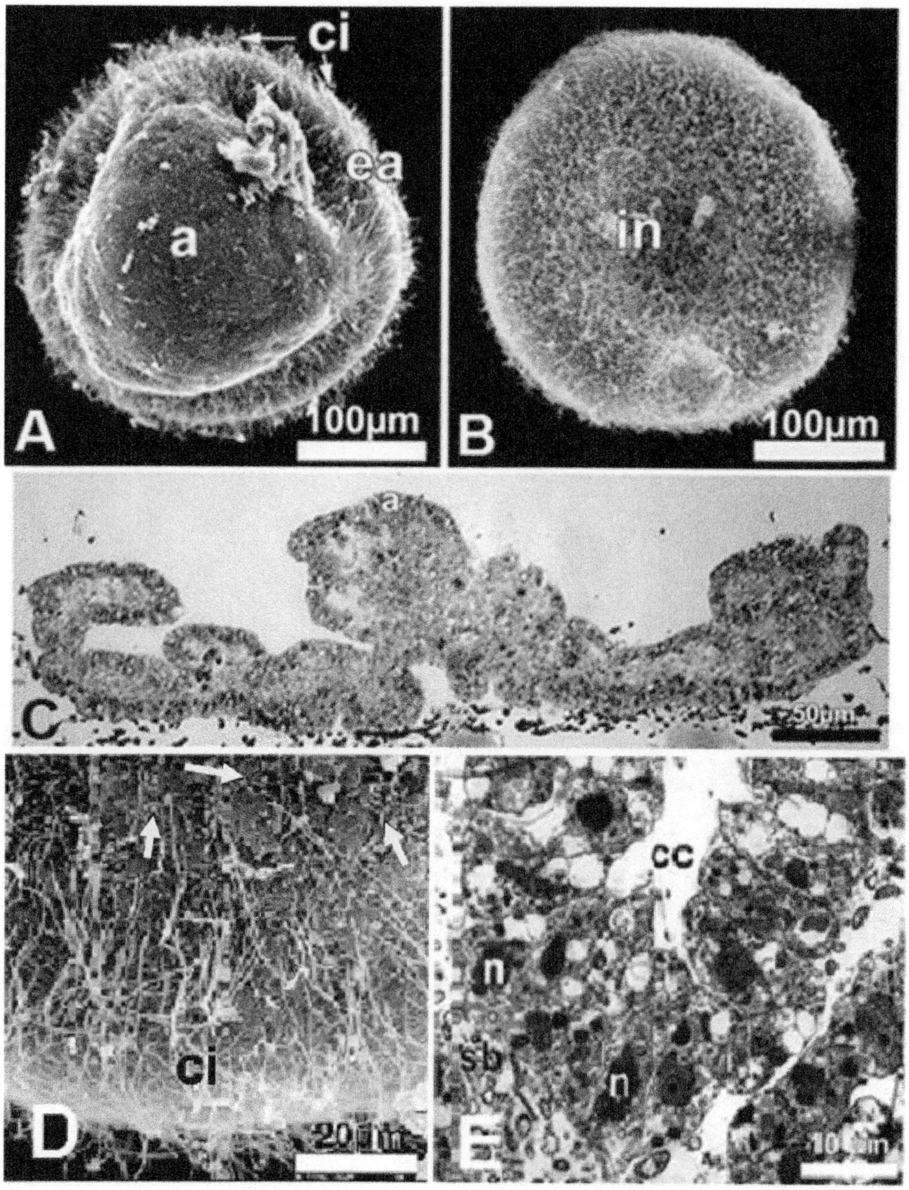

FIG. 4.30 Settler of *Plakina trilopha* (Homoscleromorpha). Fixed to substratum at the beginning of the second stage of metamorphosis by morphogenetic pathway, SEM. (A) Apical part of the settler. (B) Basal part of the same settler with dome-shaped invagination (in) of its basal layer, a = apical pole, ci = cilia, ea. = edge area; (C) Settler before the formation of an internal cavity during second stage of metamorphosis by alternative morphogenetic pathway; (D) Edge area of the settler. *Arrows* = cytoplasmic protrusions, a = apical pole, ci = cilia, ea. = edge area; (E) TEM. Aquiferous system formation in rhagon from the internal folds of larval ciliated cells at the third stage of metamorphosis, a = apical pole, ca = aquiferous system canal, cc = young choanocyte chamber, n = nucleus, sb = symbiotic bacteria. *(From Ereskovsky et al. (2009).)*

embryos measuring about 450 µm as a maximum. In May, the process ends, resulting in solid white-translucent stereoblastulae (Fig. 4.31I) and parenchymellae measuring about 400 µm. Larvae grow up to a final size of 480 µm in July. The cells of the solid stereoblastula differentiate into two types: peripheral nonflagellated cells, showing an elongated shape, and the more-rounded internal cells (Fig. 4.31J).

Within the brood chamber, mature parenchymellae and stereoblastulae can be found in clumps (Fig. 4.32A). The mature parenchymella, up to 600 µm in size along the main axis, are totally flagellated and their outermost surface protrude in numerous folds (Fig. 4.32D). The parenchymellae maintained the solid organization of the stereoblastula, but their superficial layer showed flagellated cells (Fig. 4.32E). The body of released larvae can be approximated to an oval (Fig. 4.32B), whose posterior pole had a slightly depressed region encircled by an uplifting ring (Fig. 4.32C) made of flagellated cells. These flagellae are very long, longer than any others on the surface of the larva. The parenchymellae of *Spongia officinalis* release, in the locality mentioned above, from June to July, asynchronously, either at the individual or at the population level, with a few days of de-phasing.

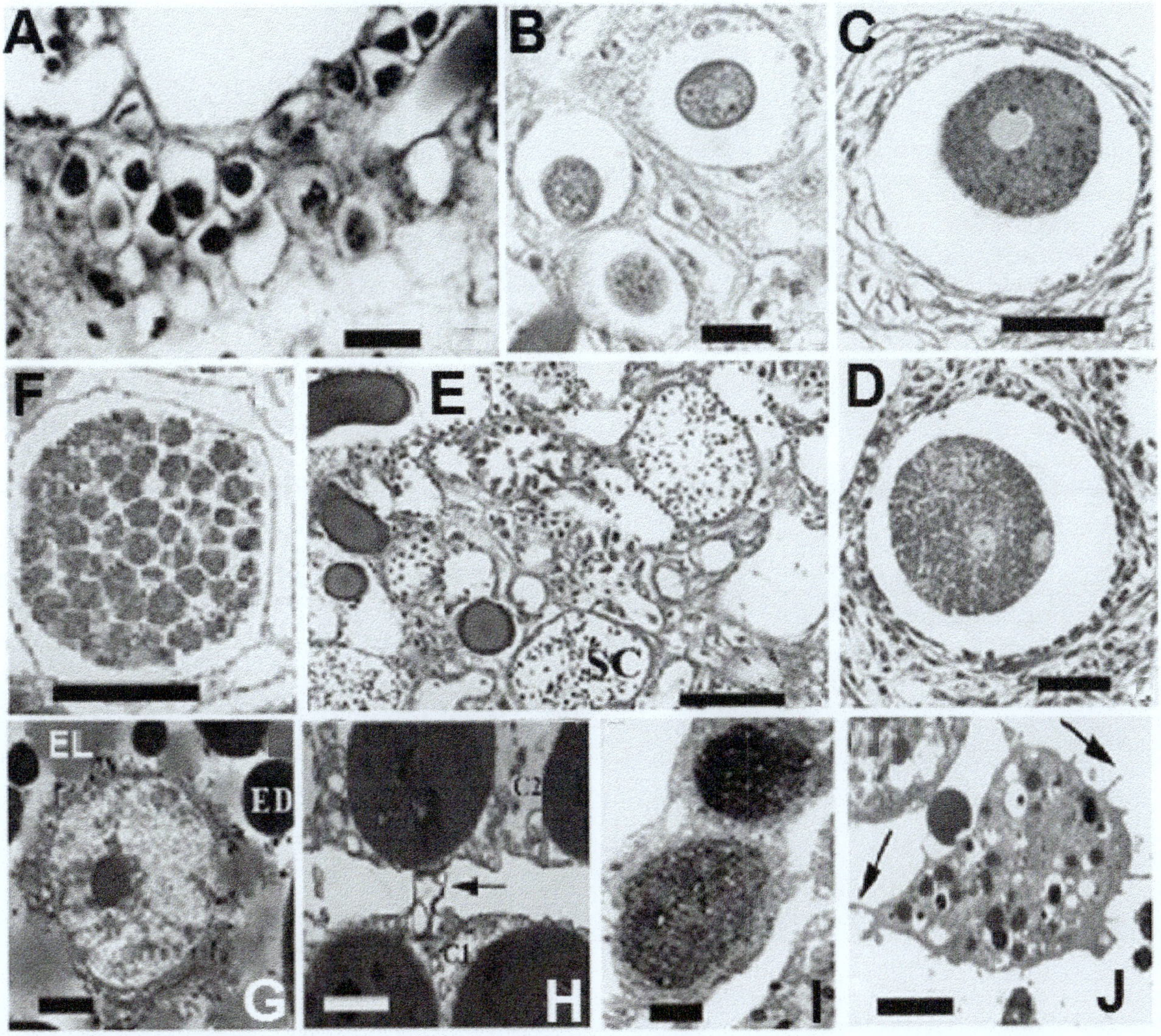

FIG. 4.31 Gametogenesis in *Spongia officinalis*: (A) undifferentiated oocytes (bar = 20 μm); (B) early differentiated oocytes, each with spherical shape and eccentric nucleolated nucleus (bar = 20 μm); (C) oocyte with nucleolated nucleus and in first stage of vitellogenesis (bar = 50 μm); (D) mature oocyte in last stage of vitellogenesis (full with stored yolk, bar = 50 μm); (E) spermatic cysts in choanosomal region containing sperm in different stages of differentiation, bar = 50 μm; (F–J) embryogenesis: (F) morula stage, bar = 150 μm; (G) detail of a nucleolated nucleus of a morula cell showing the surrounding electron-dense (ED) and electron-lucent (EL) inclusions, bar = 2.5 μm; (H) two adjacent cells (C1–C2) showing a contact by a membrane bridge protruding from C1 *(arrow)*, bar = 1 μm; (I) two solid stereoblastulae, bar = 100 μm; (J) a microgranular cell with evident microvilli-like marginal extensions *(arrows)*, inclusions of stereoblastula, bar = 1.5 μm. *(From Baldacconi et al. (2007).)*

Example: 4—*Stylocordyla borealis* Loven, 1868

This is another viviparous example, but the mother incubates the offspring from egg to young. The production of a young sponge, completely formed, inside the mother body is an unusual case of viviparity among sponges. Normally, viviparous sponges discharge larvae, which must undergo metamorphosis after fixation. Sarà et al. (2002) studied the development in the antarctic viviparous sponge ***Stylocordyla borealis*** which incubates its offspring till the young state. They collected two specimens in January from the Italian Terra Nova Bay, at depths of 100 and 150 m. (Mediterranean Sea). Both samples showed all the stages of development from the oocytes to the complete young sponge. The young oocytes are irregular in shape, may be rounded or elongated with some pseudopod-like outgrowths, ranging from 25 to 100 μm in diameter. Each has nucleolate nucleus, which is central in position (Fig. 4.33A). At the first growth phase, the oocyte is rounded or elliptical in shape and becomes larger, about 100–200 μm wide. All around the nucleus there is a granular crown of lipids (as indicated by Blue Nile sulphate) adherent to the nuclear membrane. The cytoplasm near the nucleus is hyaline and the oocyte is completely surrounded by nurse cells (Fig. 4.33B). After the second growth phase (Fig. 4.33C), the oocyte reaches a diameter of 300 μm and the granular crown surrounding the nucleus is

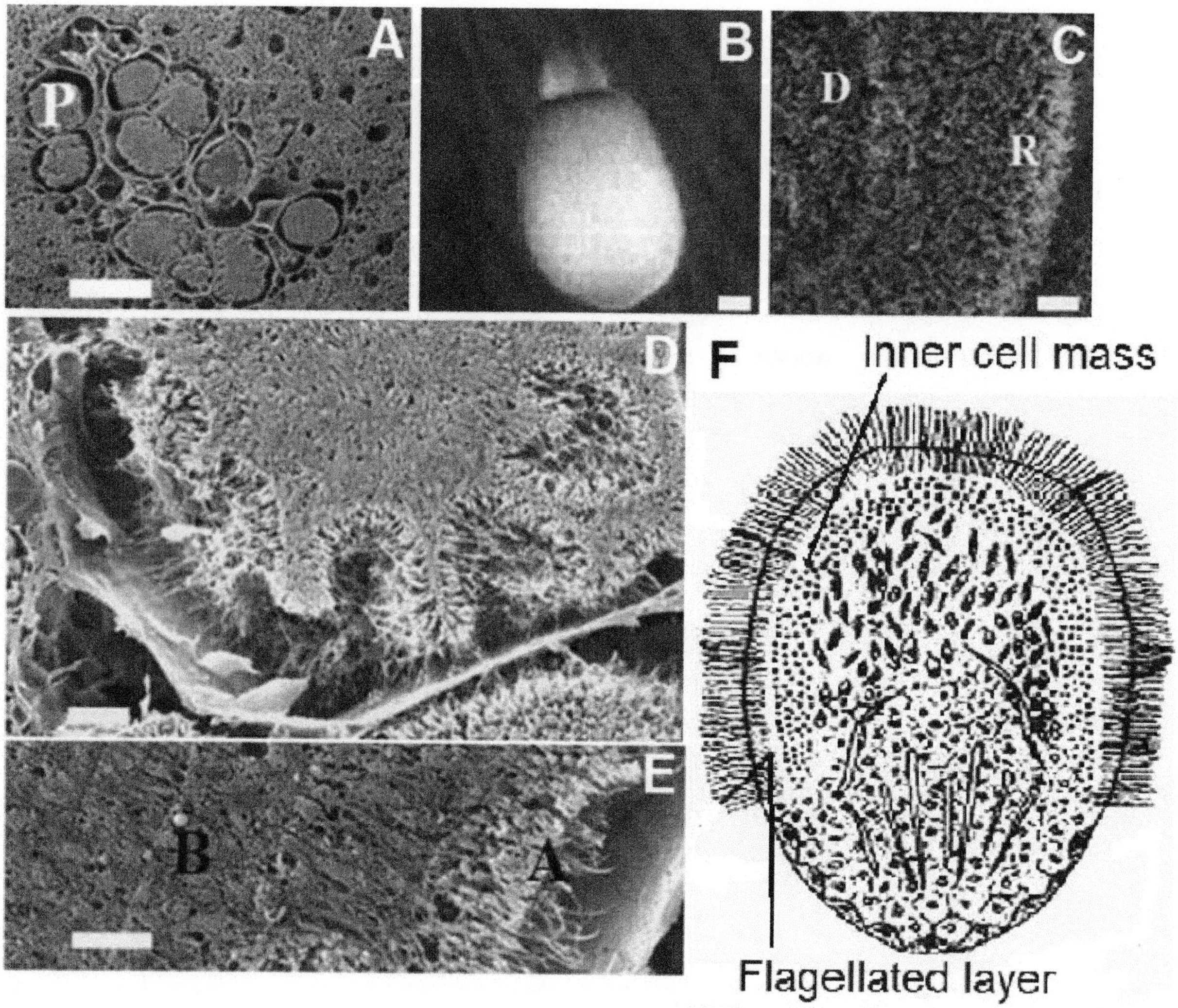

FIG. 4.32 SEM photographs of larva of *Spongia officinalis*: (A) showing stereoblastulae and parenchymella (*P*) grouped to form a clump within brood chambers scattered in the mesohyl of the sponge, bar = 500 μm; (B) of a newly released parenchymella, bar = 50 μm; (C) detail of the posterior larval pole showing the slight depression (*D*) delimited by an uplifted ring (*R*), bar = 25 μm; (D) of a section of a mature larva showing the folds of its outermost flagellated surface (bar = 25 μm; (E) of a portion of the flagellated surface (*A*) and internal region (*B*) of the larva, bar = 25 μm *((A–E) from Baldacconi et al., 2007)*; (F) general representation of parenchymella larva.

substituted by interconnecting bridges between the nucleus and cytoplasm. The cytoplasm contains several granular vacuoles. Oocyte II becomes surrounded by a thick layer of nurse cells. Both the crown and the interconnecting bridges are composed of neutral lipids. In the mature oocyte (oocyte II) numerous irregular vacuoles occur in the cytoplasm, which have interconnections with the nucleus (Fig. 4.33D).

The immature embryo is slightly ellipsoidal in shape and coated by a layer of nurse cells 100 μm thick. It has internal wide spaces, in which unequal cell masses are scattered. At this stage, it measures about 400–600 μm in diameter. During development, its internal structure changes and the spaces become filled and look denser. At the end of development, the large spaces disappear and only small single cells are clearly visible. At this stage, the embryo is mature, reaching a diameter of 700 μm and with a reduced nurse layer (about 30 μm thick), which might even be lacking (Fig. 4.34A).

After the disappearance of the nurse-cell coat, the embryo becomes a young sponge incubated inside the parental tissue and still with the cell masses. At this initial stage, it is characterized from the beginning by the development of an outer cortex and a spicular skeleton in which the spicules are disposed without a radial architecture (Fig. 4.34B). In a later stage, the choanocyte chambers are developed below the cortical layer. The young sponge shows polarity that at one pole there is a thick layer (about one-quarter of the whole sponge) of elongated cells, which build the cortical layer on the surface. This region lacks choanocyte chambers. The opposite pole is formed by the collagenous cortical layer (Fig. 4.34C). The young small sponge (Fig. 4.34D and

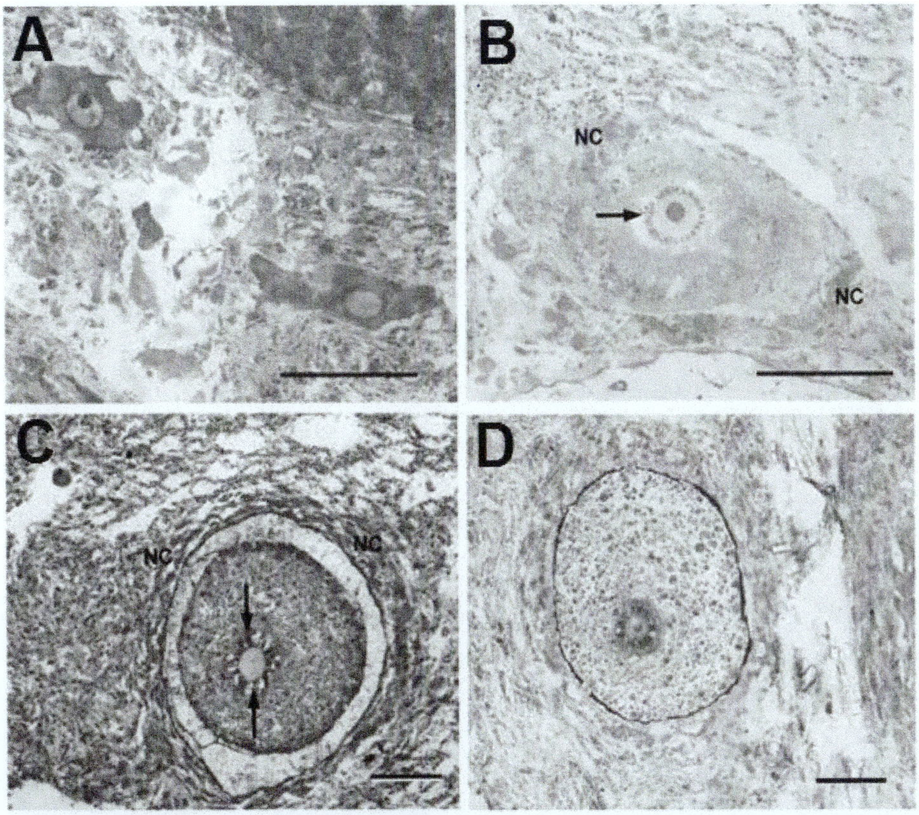

FIG. 4.33 Embryonic development of the antarctic sponge *Stylocordyla borealis*: (A) young oocytes with pseudopod-like outgrowths; (B) mature primary oocyte with granular ring of lipid adherent to the nuclear membrane *(arrow)* and nurse cells external to the cell (NC); (C) mature secondary oocyte with the bridges between the nucleus and cytoplasm *(arrows)* and nurse cells all around the oocyte; (D) Stage of transition between oocyte and embryo where there are still nuclear bridges present and there are dispersed vacuoles in the cytoplasm, scale bar = 100 μm. *(From Sarà et al. (2002).)*

E) reaches a diameter of 900–1000 μm and is similar in its tissue structure to an adult sponge.

Example: 5—Carnivorous *Asbestopluma occidentalis*

It is a major feature of sponges that they have an aquiferous system which connects the surface pores with the oscula. By this system, the sponge performs all its vital activities. A strange and astonishing sponge, which belongs to family Cladorhizidae (the order Poecilosclerida), has lost the aquiferous system and became carnivorous (Ereskovsky, 2010). They derive their nutrition from symbiotic bacteria. The **carnivorous *Asbestopluma occidentalis*** (Demospongia, Heterosleremorpha, Poeilosclerida), in which the aquiferous system and choanocytes are absent, was studied by Riesgo et al. (2007a,b). This sponge lives attached to rocks and glass sponge skeletons at 120 m depth. The specimens 'were collected by the remote-operated vehicle ROPOS from fjord walls at 120 m depth in Barkley Sound in July 2003, and by dredge at 100 m depth near Tahsis Inlet, Vancouver Island, in August 2004'. Riesgo et al. recorded that the basic structure of the adult *A. occidentalis* was described by Lambe (1893). The mature adult consists of a slender cylindrical trunk up to 6 cm long and 1.5 mm wide, from which long filaments arise (Fig. 4.35A). Cladorhizidae sponges from the deep sea are characterized by their skeletal composition and design (Fig. 4.35B) and unusual carnivorous habits. This unique feeding habit among sponges is likely due to the low nutrient levels present in abyssal depths (Vacelet and Boury-Esnault, 1995, 1996).

Asbestopluma occidentalis is a contemporaneous hermaphrodite, with oocytes, spermatic cysts and embryos simultaneously present in the tissue. Spermatic cysts are round to oval and enveloped by a thin layer of follicle cells. This layer gradually thickens as the cyst develops and becomes a complex interdigitated layer in the mature cyst (Fig. 4.35C). Early stages of spermatogenesis occur in the inner part of the core adjacent to the bundles of the spicules. Primary oocytes are found as loose groups of archaeocyte-like cells (Fig. 4.35C) surrounded by follicle cells (Fig. 4.35D). A few cells have a basal body indicating the formation of the flagellum. These are contained in the early stage cyst and may be spermatogonia or primary spermatocytes. Secondary spermatocytes are much smaller, flagellated and with numerous pseudopodia (Fig. 4.35F, G, and I). Spermatids, within later spermatocysts (Fig. 4.36A) are elongated cells

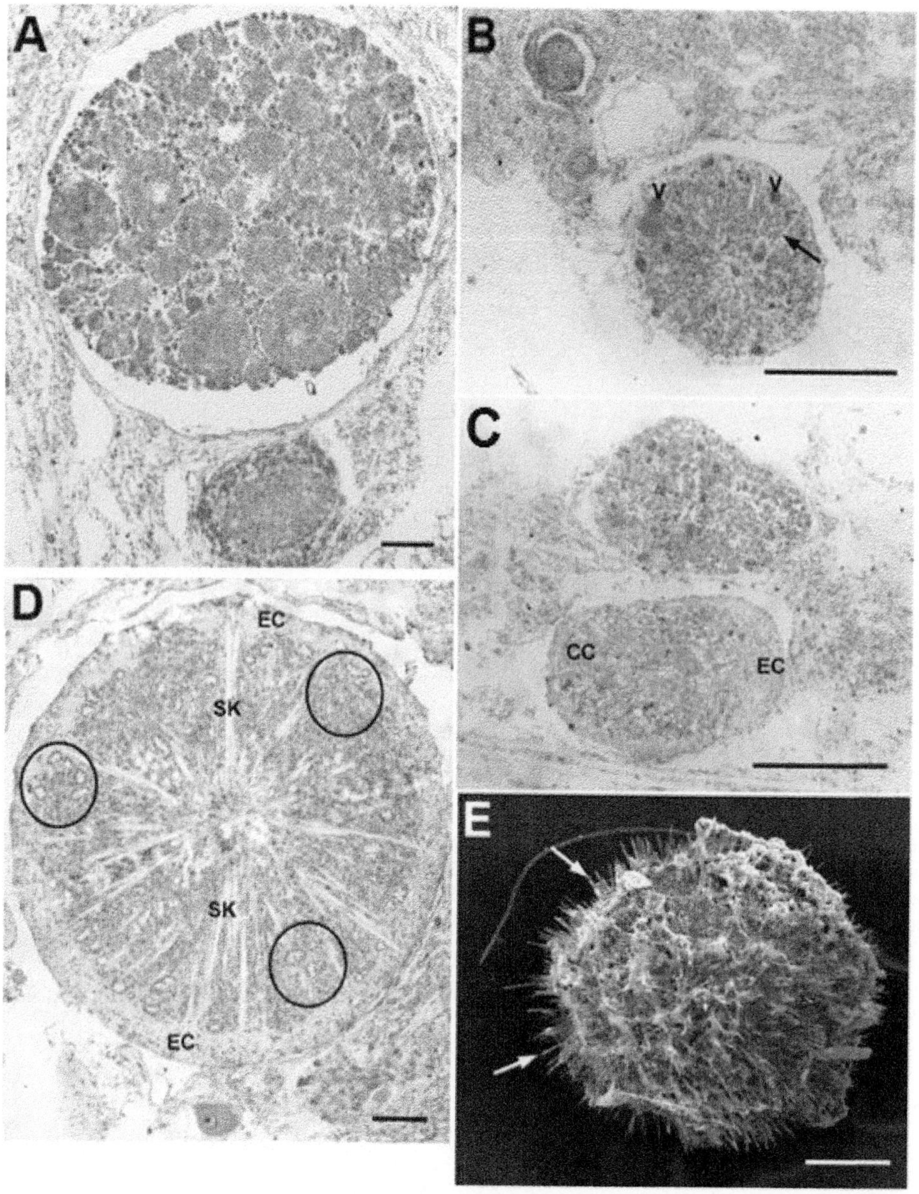

FIG. 4.34 Development of the larva of *Stylocordyla borealis*: (A) mature embryo without nurse cells, bar = 100 μm; (B) Young sponge with spicules *(arrow)* and vacuoles (V), bar = 500 μm; (C) Polarity in a young sponge. On one side the elongated cells are visible (EC) and on the opposite side the choanocyte chambers (CC), bar = 500; (D) Mature young sponge incubated inside the maternal tissue. Note the radial spicular traits of the skeleton (SK), the cortical layer of elongated cells (EC) and the choanocyte chambers *(circles)*, bar = 100 μm; (E) SEM photograph of young sponge with the spicules protruding from the sponge surface *(arrows)*, bar = 200 μm. *(From Sarà et al. (2002).)*

with a very long anterior extension containing the nucleus (Fig. 4.36B–D). Each cell has a flagellum inserted into the middle of the cell body. The proximal portion of the free flagellum is enclosed by a 1 μm-long cytoplasmic channel (a ciliary pit) (Fig. 4.36C, inset). The membrane, which is smooth in spermatocytes, becomes ridged in the spermatid, and retains the smooth feature in the mature sperm (Figs 4.35H and 4.36D, F). All the cells are connected by cytoplasmic bridges during the stages of spermatogenesis (Fig. 4.35I). The mature spermatozoon has a swollen anterior region containing the nucleus; the anterior-most end of the latter is like a hammer-head. At the very tip of the nucleus, 3–6 proacrosomal vesicles are found. Also in the mature sperm, parallel to the nucleus, bundles of microtubules are found (Fig. 4.36F and H). These later stages of spermatic cysts, especially the mature ones, are mostly located in the fine filaments projecting from the sponge stalk.

Oogenesis and embryogenesis, on the other hand, begin as small clusters of four or five oocytes in the outer part of the core with few inclusions and little yolk. Two or three nurse cells, each conical with long extensions, surround each oocyte (Fig. 4.37A and B). All developing embryos are found

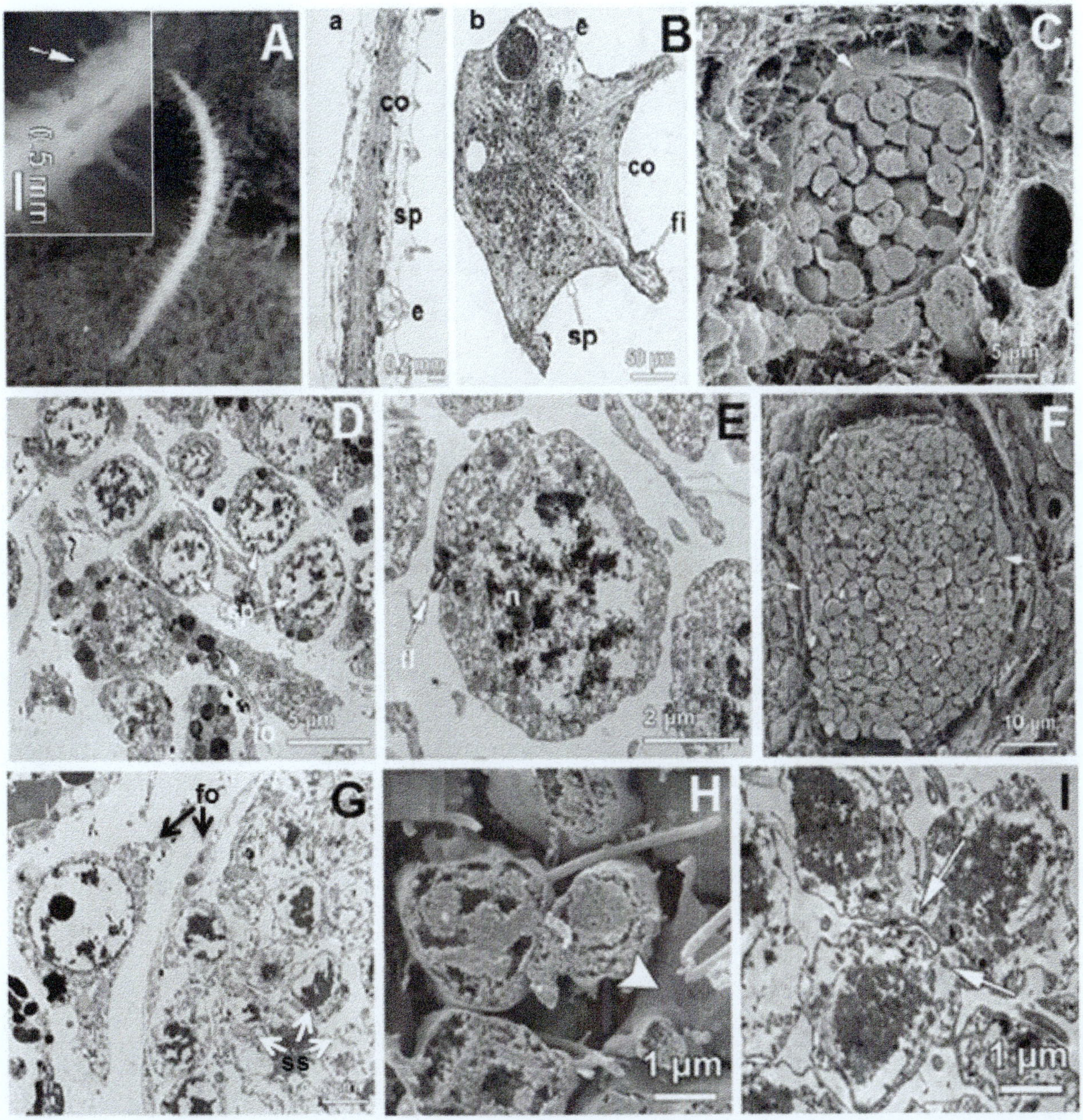

FIG. 4.35 (A) W.M. of live *Asbestopluma occidentalis* attached to the skeleton of a glass sponge in an aquarium after collection, **inset** shows detailed morphology of the sponge *(from Riesgo et al., 2007a,b)*, *A. occidentalis* is a contemporaneous hermaphrodite, with oocytes, spermatic cysts, and embryos simultaneously present in the tissue; (B) Longitudinal (a) and (b) cross sections of the sponge showing the core (co), subpinacoderm (sp), filaments (fi), and embryos (e); (C) Primary spermatocytes, surrounded by a thin layer of follicle cells *(arrows)*; (D) early primary spermatocytes (sp) adjacent to presumed future enveloping (follicle) cells (fo); (E) primary spermatocyte with a flagellum (fl) from the group shown in (D); (F) Spermatic cyst containing secondary spermatocytes, enveloped by a thin layer of cells *(arrows)*; (G) A spermatic cyst containing irregular secondary spermatocytes (ss) and surrounded by two layers of follicle cells *(black arrow)* (fo); (H and I) Secondary spermatocytes shown by SEM (H) and TEM (I) are connected by cytoplasmic bridges *(arrowhead and arrows)*. (D–I) Aspects of the outer surface of the late embryo.

in the outermost edge of the core. Clusters of the 2-cell stage prove that fertilization is synchronous (Fig. 4.37C). Early cleavage appears holoblastic, equal, and 4- to 8-cell stages are compact with cells tightly placed side by side. The 32-cell embryo is enveloped in two layers, of which the outer is flatter than the other (Fig. 4.37D). In the later stages of development, the cells of the embryo differentiate and those of the external layer develop multiple cilia (Fig. 4.37E). Cells in the inner region of early embryos are loosely arranged among filamentous bacteria and collagen (Fig. 4.37E). During all these stages, the embryo is surrounded by follicle cells separated from it by just a thin layer of collagen.

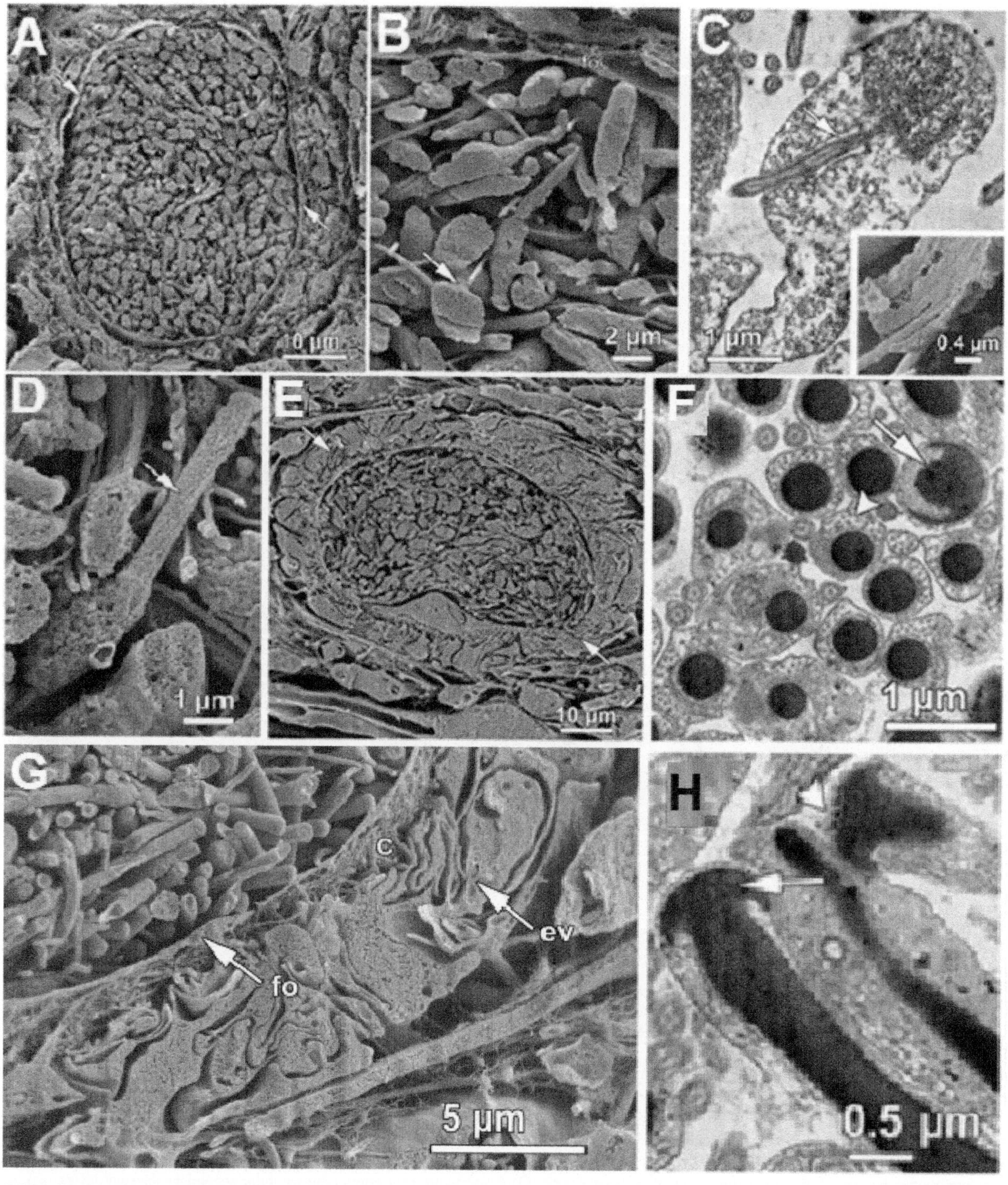

FIG. 4.36 Continued embryogenesis of carnivorous sponge: (A) Spermatids surrounded by a slightly thicker layer of cells *(arrows)*; (B) Sister spermatids connected by cytoplasmic bridges *(arrows)*. (fo), follicle cells enclosing the cyst; (C) and **inset** spermatids showing the cytoplasmic channel harbouring the flagellum; (D) An elongated spermatid showing the swelling of the head and the ridged surface typical of that stage *(arrow)*; (E) Mature spermatozoa in a late-stage spermatic cyst in one of the filaments. The envelope is now a complex layer of tightly interwoven cells *(arrows)*; (F) Cross section of a spermatic cyst containing mature spermatozoa. Note the hammer-shaped nucleus *(arrow)*, and the several bundles of longitudinal microtubules present in the cytoplasm *(arrow head)*; (G) portion of cross section of a spermatic cyst containing both spermatids and mature spermatozoa. Note the follicle cell layer (fo, *arrow*), the collagen envelope (c), and the thick enveloping layer of juxtaposed cells enclosing the cyst (ev); (H) Longitudinal section of two mature spermatozoa showing the hammer-shaped nucleus *(arrow)*, and the proacrosomal vesicles *(arrowhead)*. *(From Riesgo et al. (2007a,b).)*

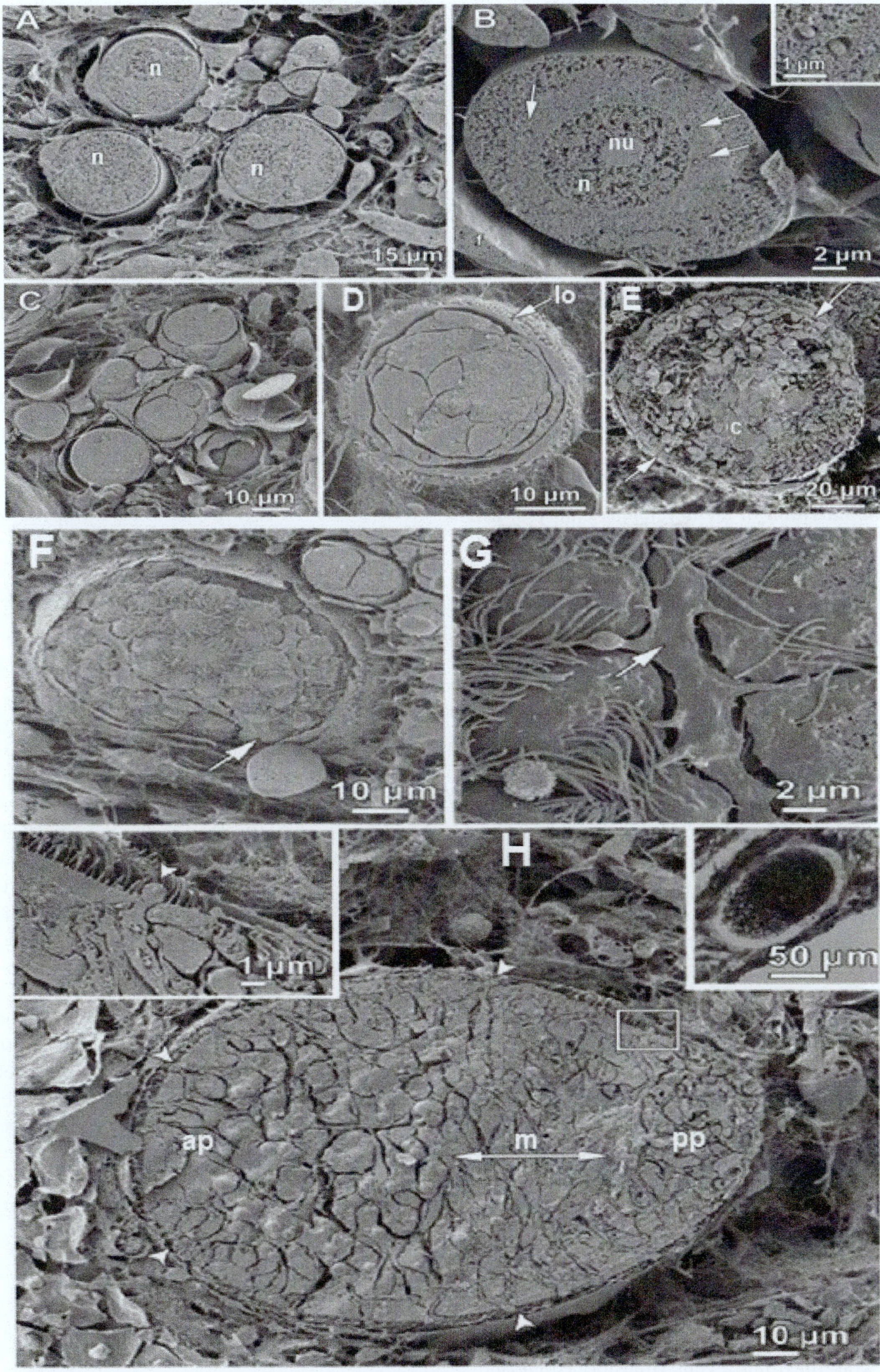

FIG. 4.37 Continuation of embryogenesis, formation of the larva of *Asbestopluma occidentalis*: (A) Oocyte cluster. Note the round nucleus (n) in all oocytes; (B) An oocyte with a nucleolated (nu) nucleus (n) and with bacterial symbionts cell (*arrows* and inset) in the cytoplasm; (C–F) cleavage, (C) A cluster of two-cell embryos; (D) 32-cell embryos. In the latter, two distinct layers have formed, and the follicle cells have numerous lobopodia (lo); (E) Late-stage embryo with multiciliated cells *(arrows)*, collagen (c) has been secreted in the centre of the embryo; (F and G) Multiciliated cells lie directly under the follicle cells: at one pole, there is an unusual nonciliated cell *(arrow)* that sends long extensions out between the ciliated cells; (H) A fracture of the sponge surface showing a pre-larva close to the pinacoderm. The larva shows three distinctive regions: the anterior pole (ap) with large cells, the mid-region (m and *double arrow*) with cells aligned perpendicular to the (A–P) axis and associated with dense bundles of collagen, and the posterior pole (pp) with small and vesiculated cells. Multiciliated cells cover all but the posterior pole and occasional nonciliated cells project through to the surface between the multiciliated cells *(arrowheads)*. The *boxed region* is shown in the **left inset** indicating the bulbous projection of the nonciliated cell *(arrowhead)* under the follicular envelope. (The **right inset** shows a longitudinal section of a larva viewed by light microscopy.) *(From Riesgo et al. (2007a,b).)*

The most fully differentiated embryos (pre-larvae) lie at the periphery of the sub-pinacoderm (Fig. 4.37H, right inset). This pre-larva is differentiated into three regions: anterior region with heterogeneous cells; mid region with cells arranged perpendicular to the A–P axis and associated with dense bundles of collagen; and cells at the posterior pole which are small and with numerous vesicles. All but the posterior pole cells are ciliated.

Absence of an aquiferous system in carnivorous sponges has allowed the evolution of features that are highly specialized for the habitat of this animal, but such modifications were not necessarily a prerequisite for the subsequent evolution of metazoans.

NUTRITION OF EMBRYOS

Concerning the **nutrition** of the developing embryos, it is recorded that the process of this vital activity distinguishes the so called viviparous sponges, whether the sponge is a true viviparous or it simply incubates the embryo. In many species, the oocytes contain large amounts of stored nutrients, and there appears to be little or no transfer of material from the parent to the developing embryos. The embryos, in many cases surrounded by the oocyte follicle, evidently gain protection from the parent constituents and/or from other factors, such as small predators and scavengers. In many other sponges, there are indications that cells and/or cellular products may be added to the developing embryos. For example, in *Microciona* some of the embryos are surrounded by masses of nurse cells which are identical to those surrounding the growing oocytes, and it has been suggested that some of these somatic cells may be added to the embryos. Another example, in *Grantia compressa* and *Sycon raphanus*, where the embryos are situated immediately under the maternal collar layer, and at all stages of development, even during the early cleavage stages, modified choanocytes appear to be incorporated into the embryo. Many of these are resorbed, but others are still present at the end of the larval development. Starting at the early blastula stage, a placental membrane is formed around the embryo by invaginated choanocytes, which have lost their collar and flagellum. The cells of this membrane, which lie adjacent to the granular cells of the amphiblastula larva, are filled with yolk; and it appears that there may be a transfer of material from the former to the latter. Placental cells may also be ingested by the cells of the embryo. The larva of *Clathrina* is enclosed within a placental membrane, which is similar to that of *Sycon* and *Grantia.* On the other hand, the larva of *Leucosolenia* rests on a layer of yolk-filled nurse cells (transformed choanocytes), some of which penetrate into the interior of the larva and others are engulfed by the posterior granular cells.

LARVAL RELEASE

There is little information concerning the process of releasing the larvae. However, it would not be surprising if it involves the production of an enzyme(s) by the larvae. In *Halisarca dujardini* the larvae rotate within their follicles, the walls of the follicle rupture, and the larvae leave the parent sponge by way of the excurrent canals and then oscula. Evidently, a similar sequence is followed in the release of most of the other sponge larvae.

LARVAL BEHAVIOUR AND SETTLING

The larvae of most sponges swim for a certain period of time and then creep on the bottom before leaving to the substratum. In some sponges, for example, *Callyspongia diffusa*, the larva swims for only 3 or 4 h, while in others, for example *Leucosolenia variabilis*, it may swim for 1 or 2 days. The benthic larvae, e.g. that of *Halichondria moorei*, creeps along the substratum for 20–60 h, and that of *Polymastia robusta* creeps for 18–20 days (Fig. 4.39B (d)). It appears that the behaviour of sponge larvae is relatively complex; and many factors, especially, light (phototaxis) and nature of the field (geotaxis) and currents, wave actions, turbidity, temperature, and the nature of substratum probably influence the movement of the larvae. Each factor may have influence on one species. For example, the larvae of *Callyspongia diffusa* are negatively phototactic; those of *Halichondria* moorei show no phototaxis, while those of *Sycon* = *(Sycandra) raphanus* are positively phototactic.

In *Petrosia ficiformis* (Maldonado and Riesgo, 2009), the larvae started attaching to the bottom of dishes about 2 weeks after spawning (Fig. 4.39C (c)). Cilia were still visible for several hours in the attached larvae (Fig. 4.39C (c) and inset). Many multilobate larvae remained unattached on the bottom of the Petri dishes for about two additional weeks before successful attachment, indicating a potential large variability in the duration of embryonic development and larval life, at least in the laboratory conditions. Settlers initially produced thin, pointing protrusions (Fig. 4.39C (d)). During attachment, settlers were able to fuse to each other producing large chimeric aggregates (Fig. 4.39C (e)). Juveniles developed obvious choanocyte chambers only after 1.5 months (not shown). After 3 months in the laboratory cultures, the young sponges had attained very little growth, although becoming more solid and nontranslucent.

METAMORPHOSIS

The time required for metamorphosis of the larva to a young sponge, differs with the different species, from 1 to 6 days. These larvae differ not only in their form, but also in the process(es) of development into the adult sponge.

OVIPAROUS SPONGES

The oviparous sponges shed their oocytes or zygotes into the sea where embryonic development occurs. In the studied

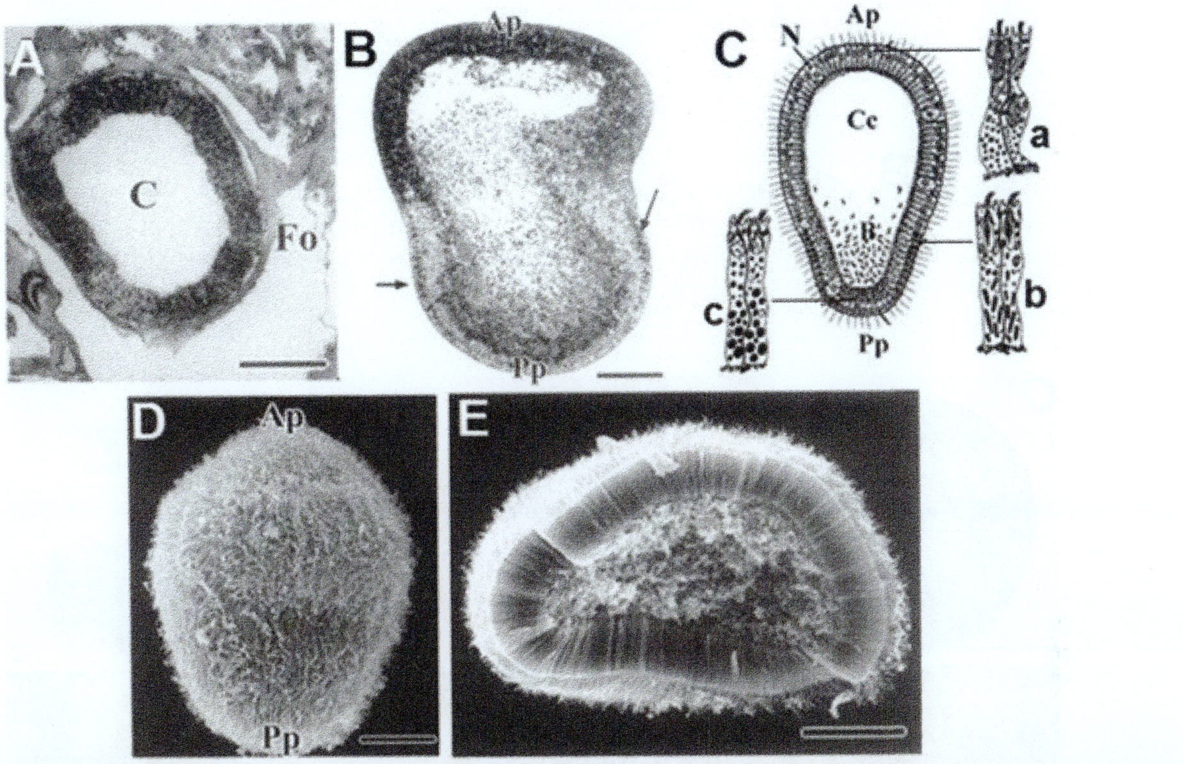

FIG. 4.38 (A) Coeloblastula (C) of *Corticium candelabrum* within the follicle (Fo). Semi-thin. Scale bar, 175 μM; (B) Longitudinal section through a cinctoblastula of *Corticium candelabrum*, free swimming stage. *Arrows* indicate the position of the postero-lateral belt of cells with intranuclear paracrystalline inclusions. Anterior pole (Ap), posterior pole **(Pp)**, semi-thin, Scale bar, 49 μm; (C) Diagram of homoscleromorph cinctoblastula (free-swimming larva). (a) = Flagellated cell of anterior pole, (b) = flagellated cell of postero-lateral zone, (c) = flagellated cell of posterior pole. Anterior pole (Ap), bacteria (B), central cavity (Cc), nonflagellated cells (N), posterior pole (Pp); (D) SEM showing external view of a released, free-swimming cinctoblastula of *Plakina trilopha*. Anterior pole (Ap), posterior pole (Pp), Scale bar, 48 μm; (E) SEM showing longitudinal section through a released, free-swimming cinctoblastula of *Plakina trilopha*, Scale bar, 46.3 μm. *(From Boury-Esnault et al. (2003).)*

species, it was observed that cleavage is total and nearly equal (Fig. 4.39A). Beyond cleavage, Fell (1974) reviewed the types that the embryo may follow in its development. Some are selected and summarized as follows:

1. In *Tethya aurantium*, the cleaved egg develops into a swimming parenchymula larva, which shows no differentiation of its two poles. Its central part consists of yolky cells, enclosed within columnar flagellated cells (Fig. 4.39B (a)). In a recent study of the gonochoric sponge *Petrosia ficiformis* (Maldonado and Riesgo, 2009), the life cycle produced, in vivo and in vitro observations, an entirely ciliated, solid larval stage (Fig. 4.39C (a)). It is hard to decide whether this larva, which is characterized by a simple cytology of undifferentiated cells, may correspond to the ancestral condition of the parenchymella larval type or to a secondary simplification of such a type.
2. In *Polymastia robusta*, the zygote adheres to the substratum in a mass of mucous produced by the maternal sponge. The cleaved zygote develops into a simple creeping larva, which has the form of a flattened blastula. It consists of a relatively small number of undifferentiated cells, each possessing a flagellum. Then, it moves out of the mucous mass and lives free for a long period with little change (Fig. 4.39B (d)).
3. In *Cliona*, the ova are extruded from the mother before cleavage, and undergo their development outside the maternal body (Minchin et al., 1900). In *Cliona stationis*, the young embryo can move over the substratum, and it soon develops into a swimming amphiblastula larva (Fig. 4.39B (b)).
4. In *Tetilla* = *(Tethya) serica*, the zygote adheres to the substratum, then after cleavage, it develops directly into an adult sponge (Fig. 4.39B (c)). The spawned eggs are surrounded by clear vitelline membrane, which becomes after fertilization, a fertilization membrane. Pseudopods and later single cells and groups of cells penetrate the fertilization membrane and move out radially on the substratum. The processes progressively come together under the growing embryo, and form the rudiment of the stalk. At the same time, the blastomeres of the cleaved zygote differentiate, giving rise to the different constituents of the adult sponge as spicules, canals, flagellated chambers and so on. This means that there is no larval stage.

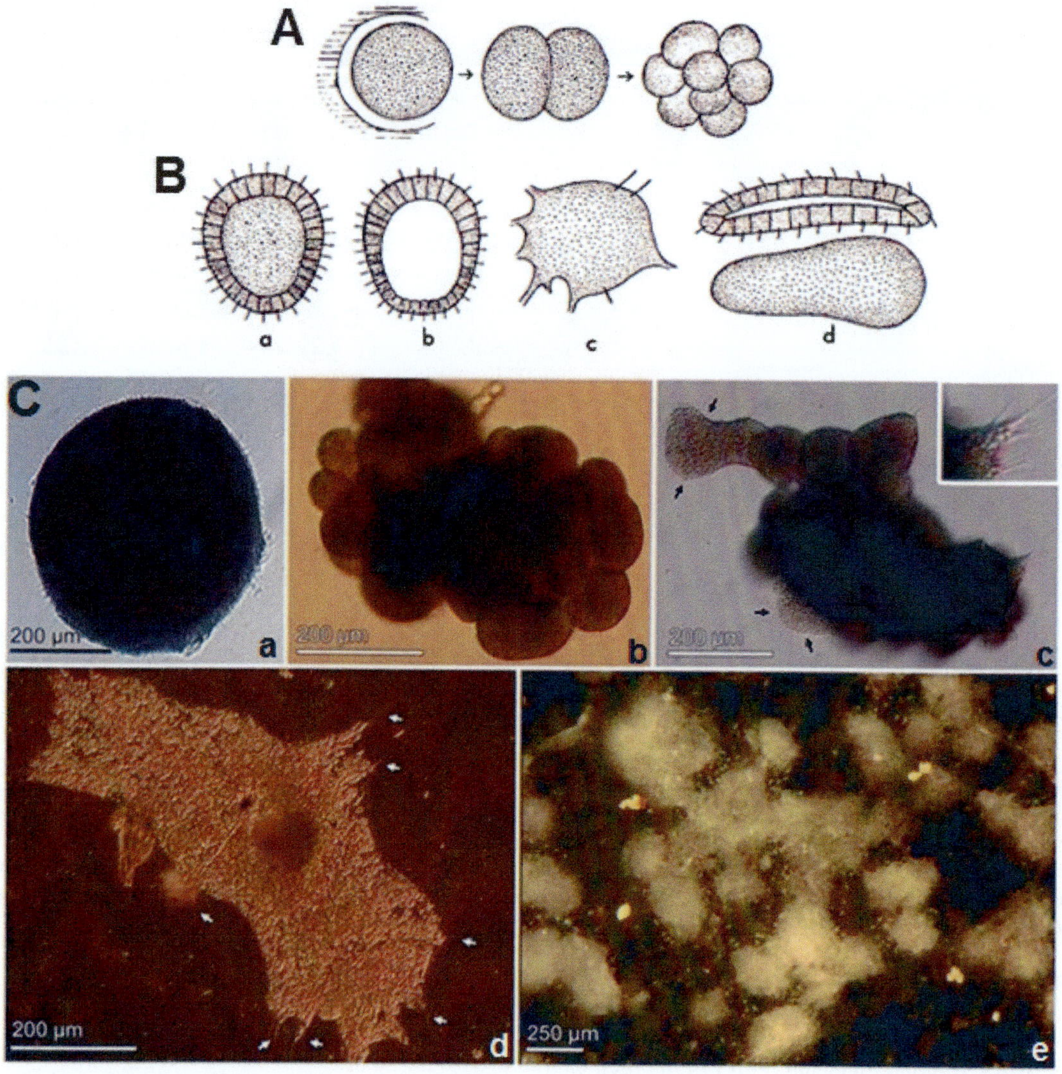

FIG. 4.39 (A) Cleavage of oviparous sponge; (B) settling stages of four oviparous Tetractinomorpha sponges: (a) in Tethya aerantiun, into a swimming parenchymula larva; (b) in *Cliona stationis*, into a swimming amphiblastula larva, (c) in *Tetilla (Tethya) serica*, directly without formation of larva (the dorsal aspect of an embryo flattened on the substratum is shown); (d) in *Polymastia robusta*, into a creeping benthic larva, (a sagittal section and dorsal aspect of the larva are shown), *(from Fell, 1974)*; (C) In vivo observations on spawned eggs, embryos and juveniles of *Petrosia ficiformis*: (a and b) view of the spherical multiciliated larva, the lobes do not correspond to a single cell, but are formed by dozens of them; (c) Multilobate settling larva, initiating attachment at several points *(arrows)*, **Inset** showing that cilia are still present at different regions of the larval body during the attachment process; (d) Early settler still lacking spicules and choanocyte chambers and characterized by thin, long protrusions *(arrows)*; (e) Aggregate of 1.5-month-old settlers, which remain fused and have developed choanocyte chambers. *(From Maldonado and Riesgo, 2009.)* Fig. 4.30 shows settling of *Plakina trilopha*.

TYPES OF LARVAE

Except for the recently discovered larva of *Oopsacus minuta*, a hexactinellid sponge, in 1993 and 1994, there are basically three different kinds of larvae. In all sponges that have been studied, these larvae are: parenchymula, amphiblastula, and simple blastula.

In 2003, Boury-Esnault et al. studied embryonic development from coeloblastula to fully formed larva of 8 Mediterranean Homoscleromorpha species: 4 *Oscarella*, 2 *Plakina*, 1 *Corticium candelabrum*, and 1 *Pseudocorticium jarrei*. The most distinctive feature of early development in homoscleromorphs is the formation of a coeloblastula from a stereoblastula by 'multipolar ingression'—the progressive migration of the internal cells to the periphery. This morphogenesis of the larva occurs within the maternal tissue, in the follicle where cleavage occurs. Morphogenesis of the larva is similar in all these species; however, cell proliferation is more active in species of *Oscarella* than in *Plakina* and *Corticium candelabrum*. The result of cell division is a wrinkled, flagellated larva, called a **cinctoblastula** (Fig. 4.38B and D). When the larvae have become fully developed, the follicle and the apopinacoderm of the excurrent canal of the adult merge together. The larvae unfold during release from the maternal sponge. The free swimming cinctoblastula is ovoid or pear-shaped, the anterior pole being larger than the posterior one, and the whole surface is flagellated (Figs 4.29B and 4.38B, D).

Ereskovsky (2004), on reviewing the works on embryonic development of Porifera, especially the recent ones, found seven types of gametic reproduction. Each of these types is referred to by the name of its resulting larva. He also defined the group of Porifera of each type. These are:

1. 'trichimella' (Hexactinellida) (Fig. 4.26A and B);
2. 'calciblastula' (Calcinea) (Fig. 4.19B);
3. 'amphiblastula' (Calcaronea) (Fig. 4.18D);
4. 'cinctoblastula' (Homoscleromorpha) (Fig. 4.29B);
5. 'disphaerula' (Halisarcida);
6. direct development (*Tetilla*, Spirophorida) (Fig. 4.39B (c)) without larva; and
7. 'parenchymella' (Fig. 4.32B and F) which can be subdivided into three subtypes.

All of these subtypes are for different groups of Demospongiae. Only the disphaerula larva is not described in the present work owing to the different schemes of classification. Ereskovsky tabulated the characters of each type of embryonic development from the cleavage to the larva and presented diagrammatic drawings describing each larva. However, each type of larva may have specific characteristics for the species

I Parenchymula larva

It is well known that the majority of sponges produce a parenchymula larva (Fig. 4.32F):

1. In many cases, e.g. *Callyspongia diffusa*, the larvae continue to develop during their free existence, and metamorphosis may actually begin during this period. The cells covering the anterior pole of the larva become disorganized during the later part of the motile phase. Not all the larvae of a species are released at the same degree of development, and there may be considerable variation in the extent of their differentiation.
2. In most species the larvae usually become attached to the substratum by the anterior pole or by a lateral region near the anterior pole. However, the larvae of some sponges appear to become fixed by their side, or even by the posterior pole. Following attachment, the larvae rapidly spread on the substratum. At the site of attachment, amaebocytes (collencytes) migrate away from the central mass of the larva to form a thin marginal membrane. This membrane eventually consists of a fold of the thin surface layer with a few amaebocytes inside. As the marginal membrane extends radially, the central mass becomes drawn out and flattened. At the same time, a network of the spongin (collagen) fibres is formed between the basal layer of cells of the sponge and the substratum. These events appear to be essential for the normal occurrence of the subsequent metamorphic processes.

The flattened cells lose their flagellae and migrate into the interior of the larva, while the amaebocytes migrate to the periphery where they spread on the surface and form a simple flattened layer. This migration of the two types of cells has been found to occur during the metamorphosis of all parenchymula larvae. It appears that the larval flagellated cells proceed to different fates in the different species. For example:

1. In *Ephydatia fluviatilis* and *Haliclona limbata*, in which choanocytes are found in the larvae, the flagellated cells have no direct role in the development of the adult sponge. They are engulfed by the archaeocytes and serve only as a nutritive reserve. The choanocytes, which develop precociously, derive from archaeocytes.
2. In most other species, larval flagellated cells develop into choanocytes in the interior of the sponge. In many cases, these flagellated cells are more numerous than the choanocytes of the young sponge, and a large number of them get phagocytized by the archaeocytes.
3. In *Halisarca* sp. and *Tethya aurantium* there is little, if any, cell loss; and apparently, all of the cells of the simple larvae participate directly in the formation of the adult sponges. In most cases, numerous flagellated chambers are formed but; in the young specimens of *Halisarca*, a single large chamber is produced.

Ultrastructural studies evidenced that continuity exists between larval flagellated cells and choanocytes during metamorphosis. It has been proved experimentally that the archaeocytes of the internal mass of the larva differentiate into choanocytes. While most, if not all, of the initial choanocytes of the young sponge normally develop from larval flagellated cells, the archaeocytes are clearly capable of differentiating into choanocytes under special conditions (Borojević, 1966).

Spaces, which become lined by pinacocytes, appear progressively. These give rise to the incurrent and excurrent canal system. The canals connect with the flagellated chambers, and pores and an osculum appears. Metamorphosis is then complete.

In some calcarea, e.g. *Leucandra blanca* and most demospongiae and Hyalospongiae, a parenchymula larva evolved. This larva is a stereogastrula. It is similar to the amphiblastula in being flagellated, but the exterior flagellated cells are more extensive and cover the entire outer surface except for the posterior pole. The interior of this larva is filled with archaeocytes. The larva breaks out of the mesohyl of the parent and has a brief free-swimming existence before settling. After gastrulation, the flagellated cells of the parenchymula form the flagellated chambers. In all calcareous sponges, development of the embryo from the egg to larval stage takes place within the body of the parent. In leuconoid calcareous sponges, the final stages of development after attachment of the larva recapitulate the more primitive asconoid and syconoid structures before reaching the leuconoid condition of the adult, e.g. *Leucosolenia*. In other leuconoid sponges (especially the demospongiae), the leuconoid condition is attained directly by a stage known as a rhagon (Fig. 4.28E–G). It resembles either the asconoid or syconoid structure, except that the walls are quite thick. The leuconoid plan of the adult develops directly from the rhagon by means of the formation of channels and flagellated chambers in the thick body wall (*Leucandra blanca*).

Typically the **parenchymula** larva is solid, however that of the freshwater sponge, e.g. *Ephydatia fluviatilis*, has a large anterior cavity lined by pinacocytes. The cells of the internal mass usually develop into collenocytes, archaeocytes and other kinds of cells of the adult mesohyl. The parenchymulae of many sponges contain spicules, which are frequently arranged in a bundle near or at the posterior pole. They appear during late cleavage, during the formation of the outer layer, or earlier. However, the larvae of some other sponges, e.g. that of *Haliclona limbata* possess no spicules at the time they are released from the parent sponge, and these spicules appear during the free-swimming stage of the larva.

The parenchymula of the majority of the sponges are highly differentiated, they contain most of the cell types of the adult except the choanocytes and reproductive elements, e.g. that of *Mycale contarenii.* The larvae of *Ephydatia fluviatilis* and *Spongilla moorei* are unique in that they contain flagellated chambers and the rudiments of the excurrent canal system. Few of the acalcareous sponges have very simple and differentiated parenchymula larvae.

II Amphiblastula larva

Oscarella (Halisarca) lobularis and some of the calcareous sponges (the Calcaronea) produce amphiblastula larva. These larvae attach to the substratum by the anterior pole:

1. *Oscarella* (leucon grade): the fixed larva flattens, the anterior half invaginates into the posterior half obliterating the central cavity to a large extent. The cells of the anterior half round up and lose their flagellae, and many of them enter the central cavity where they degenerate. Attachment of the larva is maintained by the marginal region, the inner part of which moves on the substratum towards a central point. In this way a new cavity, the primitive gastral cavity is enclosed. The cells of its walls progressively align, i.e. are arranged in a line, then folded and fragmented to form the flagellated chambers. Some cells of the superficial layer are also lost, but with less striking changes. A depression appears in the upper surface, and cords from the outer layer penetrate under the flagellated chambers. From the invaginated cells and cords, the osculum, paragaster and excurrent canals are formed. The incurrent canals also develop from cords of cells which penetrate between the flagellated chambers from the outer layer. Other cells from the outer layer and the canals move into the interior, forming mesohyl. Thus, the cells of the anterior half of the larva (flagellated cells) form the collared layer (choanocytes), and the cells of the posterior half (granulated cells) give rise to all the other elements of the adult sponge.
2. Calcareous sponges: the metamorphosis of the larvae is more or less similar to that of *Oscarella.*
   a. *Leucosolenia variabilis* (Ascon grade): following the attachment of the larva of this species to the substratum, the granular cells constituting the posterior half of the larva (in the free-swimming period) increase in number and then become much more prominent and extend around the flagellated cells (epipoly), which then form a solid internal mass. The granular cells then form a simple external layer (pinacocytes) that secretes monaxonid spicules and also gives rise to internal groups of cells that produce triradiate spicules. A central cavity appears in the mass of former flagellated cells, and the latter develop into choanocytes. Later an osculum appears.
   b. *Sycon raphanus* (Sycon grade): the metamorphosis of this larva is more advanced than that of *Leucosolenia.* During the free-swimming period, there is substantial outgrowth (true epiboly) of the flagellated cells by the granular cells (Fig. 4.17). Many of the flagellated cells degenerate during metamorphosis and are phagocytozed by the surviving flagellated cells. Experimental studies evidenced that the posterior cells alone can form almost a normal sponge (quasinormal) with a flagellated chamber, while isolated flagellated cells do not survive, and therefore the posterior granular cells of the amphiblastula larva is analogous to the internal mass of the parenchymula larva.

The calcareous sponges produce more or less differentiated blastulae larvae, e.g. various species of *Clathrina* and *Ascandra* (the calcinea). The larva is a simple flagellated blastula which, in some species, has a small membrane of large, non flagellated cells at the posterior pole.In *Grantia compressa*, *Sycon raphanus*, *Clathrina (=Leucosolenia) botryoides*, and *Petrobiona massiliana* (the calcaronea), the formation of the larva is more complex. During cleavage, two types of cells emerge: rounded granular cells with large nuclei, which are situated at the pole of the blastula that is directly beneath the collared layer, and elongated cells with small nuclei. The latter form flagellae which are directed into the blastular cavity. Then the embryo undergoes a process of inversion (Fig. 4.16A). An opening appears among the granular cells, and the blastula turns inside out as the opening enlarges and its edges fold down around the invaginating flagellated pole. The granular cells come together again at the opposite pole, and an amphiblastula larva with external flagellae is formed. In the final stages of differentiation, the cells of the larva increase in size and form tall columnar cells with the result that the blastular cavity is very much reduced. The granular non flagellated cells occupy the posterior half of the larva.

III Simple blastula (undifferentiated blastula) (coeloblastula)

Polymastia robusta and certain calcareous sponges (the Calcinea) release undifferentiated blastula; cytodifferentiation, apparently, begins only after settling of the larva. While in *Ascandra falcata* and *Clathrina reticulum*: during the free-swimming period of the larvae of these two species, individual cells from the wall migrate to the central cavity and then the cellular structure of the wall is completely lost. The blastula is converted into a solid, unorganized mass of cells, which lose their flagellae. This simple cluster of blastomeres settles on the substratum, and begins differentiation.

Experimentally, it is evidenced that the simple blastula lacks the anterior–posterior differentiation of the amphiblastula. When a recently released larva of *Ascandra falcata* was cut into anterior and posterior halves, each half developed into a small blastula, which later settled and proceeded to develop into a small but complete sponge. It seems that the cells differentiate according to their position in the fixed embryo.

According to the above, no mention has been made of gastrulation. Gastrulation is defined as rearrangement of the cells of the embryo, which establishes the basic organization of the young organism. It means that the movement of cells is the gastrulation and this occurs either during the free-swimming period of the larva or during its existence within the mother.

BIOLOGICAL ASSOCIATIONS

Sponges are not widely sought after as food by other marine animals, but some chitons, opisthobranchs and fishes are specialized sponge predators. Some of these predator–prey relationships are highly specific, e.g. the small red opisthobranch, *Rostangia rubicunda*, feeds on the encrusting sponge *Microciona coccinea*. The massive habitat and cavernous construction of many sponges renders them particularly suitable hosts for commensals. Sea anemones, zoanthids, polychaetes, amphipods, barnacles, ophiuroids, shrimps, crabs, and fish frequently inhabit sponges. The nature of the association has been investigated in very few cases.

Some animals use sponges as camouflage, for example, the small gastropod trochid *Herpetopoma bella*, which is commonly invested with a growth of *Microciona coccinea*. Zoochlorellae commonly live in the amoebocytes of freshwater and shallow-water marine sponges, and in this habitat do not accumulate starch reserves. It is assumed that the sugars they synthesize are utilized by the sponge. Zoochlorellae are not passed on in larvae or gemmules but re-infect young sponges. Filamentous blue-green algae commonly fill the mesoglea of marine sponges, particularly in tropical locations.

Of interest to human beings, *Spongia officinalis* Linnaeus is an example of a sponge known worldwide and much utilized for cosmetic purposes. Prolonged overfishing and some recent epidemics have caused its disappearance from wide areas of the West Mediterranean Basin, especially from the Ionian coasts of Apulia (SE Italy). Baldacconi et al. (2007) reported the inclusion of this bath sponge among the marine endangered species requiring specific management measures. Knowledge of the reproductive pattern and larval behaviour are essential for understanding the biological and ecological needs of this species, and may yield useful information for appropriate management.

In fact, in Egypt, there was marketable sponge fishing along the Mediterranean coast, west of Alexandria, up to the mid twentieth century. Later, sponge fishing became limited to the area from Marsa Matrouh to Saloum, using small fishing boats. The fished sponge was sold individually on a very small scale. It seemed that the reason was similar to what happened in Italy. However, in 1989, Ramadan et al. recorded, for the first time, the presence of four sponge species in the same region. One of which is the commercial *spongia agarcinia*, which indicates the commercial sponge abundance in that region. The actual reason for reduced fishing in Egypt is the industrial revolution in artificial sponge production.

REFERENCES

Baldacconi, R., Nonnis-Marzano, C., Gaino, E., Corriero, G., 2007. Sexual reproduction, larval development and release in *Spongia officinalis* L. (Porifera, Demospongiae) from the Apulian coast. Mar. Biol. 152, 969–979.

Balfour, F.M., 1880. A Treatise on Comparative Embryology. vol. I. Macmillan and Co, London.

Borojević, R., 1966. Etude expérimentale de la différenciation des cellules de l'éponge au cours de son développement. Dev. Biol. 14 (1), 130–153.

Borojević, R., 1969. Étude du dévelopment et de la différenciation cellulaire d' Eponges Calcaire Calcinéennes (Genre *Clathrina* et *Ascandra*). Ann. Emb. Morphogen. 2, 15–36.

Boury-Esnault, N., 2006. Systematics and evolution of *Demospongiae*. Can. J. Zool. 84 (2), 205–224.

Boury-Esnault, N., Rützler, K. (Eds.), 1997. Thesaurus of sponge morphology. Smithsonian Contrib. Zool. 596, 1–54.

Boury-Esnault, N., Vacelet, J., 1994. Preliminary studies on the organization and development of a hexactinellid sponge from a Mediterranean cave, *Oopsacas minuta*. In: van Soest, R.W.M., van Kempen, T.M.G., Braekman, J. (Eds.), Sponges in Time and Space: Proceedings of the Fourth International Porifera Congress. A.A. Balkema, Rotterdam, pp. 407–416.

Boury-Esnault, N., Harmelin, J.G., Vacelet, J., 1993. Les abysses méditerranéennes à 20 m de profondeur. La Recherche 256, 848–851.

Boury-Esnault, N., Efremova, S., Bézac, C., Vacelet, J., 1999. Reproduction of a hexactinellid sponge: first description of gastrulation by cellular delamination in the Porifera. Invertebr. Reprod. Dev. 35, 187–201.

Boury-Esnault, N., Ereskovsky, A., Bézac, C., Tokina, D., 2003. Larval development in the Homoscleromorpha (Porifera, Demospongiae). Invertebr. Biol. 122 (3), 187–202.

de Blainville, H.M.D., 1816. Quatrième mémoire sur les mollusques, de l'ordre des Cyclobranches. Bulletin des Sciences, par la Société Philomathique de Paris 1816, 93–97.

Dendy, A., 1905. Report on the sponges collected by Prof. Herdmanat Ceylon in 1902. In: Report to the Government of Ceylon on the Pearl Oyster Fisheries of the Gulf of Mannar. vol. 3. pp. 59–246, Plates 1–16.

Duboscq, O., Tuzet, O., 1935. Un nouveau stade du développement des éponges calcaires. C. R. Acad. Sci. Paris 200, 1788–1790.

Duboscq, O., Tuzet, O., 1942. Recherches complémentaires sur l'ovogenése, la fécundation, et les premiers stades du développement des éponges calcaires. Arch. Zool. Exp. Gen. 81, 395–466.

Edmondson, C.H., 1946. Reproduction in *Donatia deformis* (Thiele). Bernice P. Bishop Museum, Honolulu, Hawaii. Occas. Pap. XVIII (18), 271–282.

Ereskovsky, A.V., 1999. Development of sponges of the order Haplosclerida. Russ. J. Mar. Biol. 25 (5), 361–371.

Ereskovsky, A.V., 2004. Comparative embryology of sponges and its application for poriferan phylogeny. Boll. Mus. Ist. Biol. Univ. Genova 68, 301–318.

Ereskovsky, A.V., 2010. The Comparative Embryology of Sponges. Springer.

Ereskovsky, A.V., Boury-Esnault, N., 2002. Cleavage pattern in *Oscarella* species (Porifera, Demospongiae, Homoscleromorpha), transmission of material cells and symbiotic bacteria. J. Nat. Hist. 36, 1761–1775.

Ereskovsky, A.V., Tokina, D.B., 2007. Asexual reproduction in homoscleromorph sponges (Porifera; Homoscleromorpha). Mar. Biol. 151, 425–434.

Ereskovsky, A.V., Konyukov, P.Y., Tokina, D.B., 2009. Morphogenesis accompanying larval metamorphosis in *Plakina trilopha* (Porifera, Homoscleromorpha). Zoomorphology 129 (1), 21–31.

Ereskovsky, A.V., Lavrov, D.V., Willenz, P., 2014. Five new species of Homoscleromorpha (Porifera) from the Caribbean Sea and re-description of *Plakina jamaicensis*. J. Mar. Biol. Assoc. U. K. 94 (2), 285–307.

Fell, P.E., 1969. The involvement of nurse cells in oogenesis and embryonic development in the marine sponge, *Haliclona ecbasis*. J. Morphol. 127, 133–150.

Fell, P.E., 1974. Porifera. In: Giese, A.C., Pearse, J.S. (Eds.), Reproduction of Marine Invertebrates. Academic Press, New York, pp. 51–132.

Fell, P.E., 1997. Poriferans, the sponges. In: Gilbert, S.F., Raunio, A.M. (Eds.), Embryology, Constructing the Organism. Sinauer Associates, Inc. Publishers, U.S.A.

Gardiner, M.S., 1972. The Biology of Invertebrates. McGraw-Hill Book Company. London, New York.

Haeckel, E., 1872. Die Kalkschwarame. Eine Monographic in zwei Banden Text und einem Atlas mit 60 Tafeln Abbildungen. Berlin, 1872. Part translated Ann. Mag. Nat. Hist. 11 (4), 241–262, 421–430.

Hartman, W.D., Goreau, T.F., 1970. In: Fry, W.G. (Ed.), In the Biology of the Porifera. Symposia Series, Zoological Society of Londonvol. 25. Academic Press.

Hickman, C.P., 1973. Biology of the Invertebrates, second ed. The C.V. Mosley Company, Saint Louis.

Huxley, T.H., 1875. On the classification of the animal kingdom. Am. Nat. 9 (2), 65–70.

Hyman, L.H., 1940. The Invertebrates, Protozoa Through Ctenophora, first ed. McGraw-Hill Book Company, Inc., New York, London. Third expression.

Invertébrés: Beauté du Monde Animal, Tome XII, 1974. Librairie Larousse, Paris.

Jackson, J.B.C., Goreau, T.F., Hartman, W.D., 1971. Recent Brachiopod-Coralline communities and their paleoecological significance. Science 173 (3997), 623–625.

James-Clark, H., 1867. IV.—Conclusive proofs of the animality of the ciliate sponges, and of their affinities with the *Infusoria flagellata*. Ann. Mag. Nat. Hist. 19 (109), 13–18.

Kumé, 1952. Note on the early development of *Tethya serica* Lebwahl, a tetraxonian sponge. Nat. Sci. Rep. Ochanomizu Univ. 8, 97–104.

Lambe, L.M., 1893. Sponges from the Pacific coast of Canada and Behring sea. Trans. R. Soc. Can. 11, 25–43.

Leuckart, R., 1854. Zur nähern Kenntniss der Siphonophoren von Nizza: Mit drei Kupfertafeln. Nicolai.

Lévi, C., 1956. Etude des Halisarca de Roscoff. Embryologie et systématique des Démosponges. Arch. Zool. Exp. Gen. 93, 1–181.

Lévi, C., 1973. Systématique de la class des Demospongiaria (Demosponges). In: Trait de Zoologie. vol. 3, no. 1. Maison Cie, Paris, pp. 577–631.

Leys, S.P., 2003. The significance of syncytial tissues for the position of the Hexactinellidae in the Metazoa. Integr. Comp. Biol. 43 (1), 9–17.

Leys, S.P., Eerkes-Medrano, D., 2005. Gastrulation in calcareous sponges: in search of Haeckel's gastraea. Integr. Comp. Biol. 45 (2), 342–351.

Leys, S.P., Ereskovsky, A.V., 2006. Embryogenesis and larval differentiation in sponges. Can. J. Zool. 84 (2), 262–287.

Leys, S.P., Cheung, E., Boury-Esnault, N., 2006. Embryogenesis in the glass sponge *Oopsacas minuta*: formation of syncytia by fusion of blastomres. Integr. Comp. Biol. 1–14.

Leys, S.P., Mackie, G.O., Reiswig, H.M., 2007. The biology of glass sponges. Adv. Mar. Biol. 52, 1–145.

Maldonado, M., Bergquist, P.R., 2002. Phylum Porifera. In: Young, C.M., Sewell, M.A., Rice, M.E. (Eds.), Atlas of Marine Invertebrate Larvae. Academic Press, San Diego, pp. 21–50.

Maldonado, M., Riesgo, A., 2009. Gametogenesis, embryogenesis, and larval features of the oviparous sponge *Petrosia ficiformis* (Haplosclerida, Demospongiae). Mar. Biol. 156 (10), 2181–2197.

Marshall, A.J., Williams, W.D., 1975. A Text Book Zoology: The Invertebrates. ELBS and Macmillan.

Maas, O., 1909. Japanische Medusen. - Abhandlungen der bayerischen Akademie der Wissenschaften, mathematisch physikalische Classe, Suppl. Band 1, 1–53, plates 1–3.

Minchin, M.A., Fowler, B.A., Bourne, M.A., 1900. In: Lankester, E.R. (Ed.), A Treatise on Zoology. Part II: The Porifera and Coelen. Adam and Charles Black, London.

Nardo, G.D., 1833. Auszug aus einem neuen System der Spongiarien, wonach bereits die Aufstellung in der Universitäts-Sammlung zu Padua gemacht ist, Isis, oder Encyclopädische Zeitung Coll. (Oken: Jena) 519–523.

Okada, Y., 1928. On the development of a hexactinellid sponge, *Farrea sollasii*. J. Fac. Sci. Univ. Tokyo (Section IV Zoology) 2, 1–27.

Okada, Y.K., 1968. Porifera. In: Kumé, M., Dan, K. (Eds.), Invertebrate Embryology. NOLIT Publishing House, Belgrade, Yugoslavia.

Paulus, W., 1989. Ultrastructural investigation of spermatogenesis in *Spongilla lacustris* and *Ephydatia fluviatilis* (Porifera, Spongillidae). Zoomorphology 109, 123–130.

Poléjaeff, N., 1882. Uber das Sperma und die Spermatogenesis bei Sycandra raphanus Haeckel. Sitz Akad Wiss Wien, 86, 276–298. Cited in Roosen-Runge, E.C. (1977).

Ramadan, M.M., Kheirallah, A.M., Badawi, H.K., El-Beshbeeshy, M.M., 1989. A new record of sponges from the Egyptian Mediterranean waters. Bull. Nat. Inst. Oceanogr. Fish. ARE 15 (2), 63–71.

Reiswig, H.M., 1970. Porifera: sudden sperm release by tropical Demospongiae. Science 170 (3957), 538–539.

Riesgo, A., Maldonado, M., 2009. An unexpectedly sophisticated, V-shaped spermatozoon in Demospongiae (Porifera): reproductive and evolutionary implications. Biol. J. Linn. Soc. 97, 413–426.

Riesgo, A., Maldonado, M., Durfort, M., 2007a. Dynamics of gametogenesis, embryogenesis and larval release in a Mediterranean homosclerophorid demosponge. Mar. Freshw. Res. 58, 398–417.

Riesgo, A., Taylor, C., Leys, S.P., 2007b. Reproduction in a carnivorous sponge: the significance of the absence of an aquiferous system to the sponge body plan. Evol. Dev. 9 (6), 618–631.

Roosen-Runge, E.C., 1977. The Process of Spermatogenesis in Animals. Cambridge University Press, London.

Ruiz, C.E., Ivanisevic, J., Chevaldonne, P., Ereskovsky, A.V., Boury-Esnault, N., Vacelet, J., Thomas, O.P., Perez, T., 2015. Integrative taxonomic description of *Plakina kanaky*, a new polychromatic sponge species from New Caledoni (Porifera: Homoscleromorpha). Mar. Ecol. 36, 1129–1143.

Sarà, A., Cerrano, C., Sarà, M., 2002. Viviparous development in the Antarctic sponge *Stylocordyla borealis* Loven, 1868. Polar Biol. 25, 425–431.

Schulze, F.E., 1877. Untersuchungen über den Bau und die Entwicklung der spongien: die Familie der Chondrosidae. Engelmann.

Schulze, F.E., 1880. On the structure and arrangement of the soft parts in *Euplectella aspergillum*. R. Soc. Edin. Trans. 29, 661–673.

Schulze, F.E., 1887. Report on Hexactinellida. Challenger Rep. 21, 1–513.

Simpson, T.L., 1984. The Cell Biology of Sponges. Springer-Verlag, New York.

Sollas, W.J., 1884. Memoirs: on the development of *Halisarca lobularis* (O. Schmidt). J. Cell Sci. 2 (96), 603–621.

Tuzet, O., Paris, J., 1964. La spermatogenése, l'ovogenése, la fécondation et les premiers stades du développement chez *Octavella galangaui*. Vie Milieu 15, 309–327.

Vacelet, J., Boury-Esnault, N., 1995. Carnivorous sponges. Nature 373 (6512), 333.

Vacelet, J., Boury-Esnault, N., 1996. A new species of carnivorous sponge (Demospongiae: Cladorhizidae) from a Mediterranean cave. Bull. Inst. R. Sci. Nat. Belg. Biol. 66, 109–115.

Vacelet, J., Boury-Esnault, N., Harmelin, J., 1994. Hexactinellid Cave, a unique deep-sea habitat in the scuba zone. Deep-Sea Res. 41, 965–973.

Wilson, H.V., 1907. On some phenomena of coalescence and regeneration in sponges. J. Exp. Zool. 5, 245–258.

Diploblastica

Chapter Outline

Definition **173**
General characteristics **173**
Classification **175**
Body plan of Cnidaria **177**
Reproduction **177**
Agametic reproduction 177
Class Hydrozoa 177
Class Scyphozoa 179
Class Anthozoa 180
Regeneration 180
Gametic reproduction 180
Origin of germ cells 180
Gonadogenesis 180
Gametogenesis 181
Gametes 181
Spawning (release of gametes) 181
Brooding 183
Cnidarian embryogenesis 183
Fertilization 183
Cleavage 183
Blastulation 183
Gastrulation 183
Larval forms 186
Larval settling 186
Postembryonic development (metamorphosis) 186
Class Hydrozoa **187**
Obelia geniculata 187
Eucheilota paradoxica 189
Bougainvillea fructuosa 191
Tubularia mesembryanthemum 191
Eudendrium racemosum 195
Gonionemus 199
Hydra vulgaris 201
General development of Siphonophora, *Muggiaea, Hallistoma, Rhizophysa* 206
Bargmannia elongata 208
Nanomia bijuga 210
Class Scyphozoa **212**
Subclass Stauromedusa (*Lucernaria, Haliclystus*) 215
Stylocoronella riedli 218
Subclass Discomedusae 218
Order Coronatae 219
Ex. *Nausithoë aurea* 219
Order Semaeostomeae 220
Ex. *Aurelia aurita* 220
Other Semaeostomeae: *Pelagia noctiluca* 224
Chrysaora quinquecirrha 224
Order Rhizostomae 224
Mastigias 224
Cassiopea andromeda 228
Cotylorhiza tuberculata 233
Lychnorhiza lucerna 236
Class Cubozoa **239**
Tripedalia cystophora 239
Charybdaea marsupialis 241
Carybdea marsupialis 241
Carybdea sivickisi 243
Copula sivickisi 247
Alatina cf. *moseri* 248
Morbakka virulenta 251
Class Anthozoa **253**
Subclass Ceriantipatharia 253
Order Antipatharia 253
Ex. *Cirrhipathes anguina* 255
Ex. *Antipathes* sp. 258
Order Ceriantharia 260
Ex. *Pachycerianthus multiplicatus* 260
Ex. *Cerianthus lloydii* 262
Subclass Octocorallia 262
Order Stolonifera 262
Ex. *Clavularia hamra* 262
Order Alcyonacea 264
Ex. *Alcyonium digitatum* 264
Ex. *Alcyonium coralloides* 264
Ex. *Heteroxenia fuscescens* 266
Ex. *Xenia umbellata* 273
Ex. *Capnella gaboensis* 273
Order Gorgonacea (horny corals) 276
Ex. *Junceella fragilis* 276
Ex. *Anthoplexaura dimorpha* 279
Subclass Hexacorallia (Zoantharia) 280
(Sea anemones, Madrepore corals) 280
Order Actiniaria (sea anemones) 280
Ex. *Nematostella vectensis* 280
Ex. *Sagartia troglodytes* 285
Ex. *Halcampa duodecimcirrata* 286
Ex. *Epiactis japonica* 287
Ex. *Tealia crassicornis* 289
Cribrinopsis japonica 290
Order Scleractinia (Madreporaria) 291
True or stony corals 291
Ex. *Astrangia danae* 295
Ex. *Acropora* spp. 297
Galaxea fascicularis 300
Dipsastraea (Favia) speciosa 300
References **303**

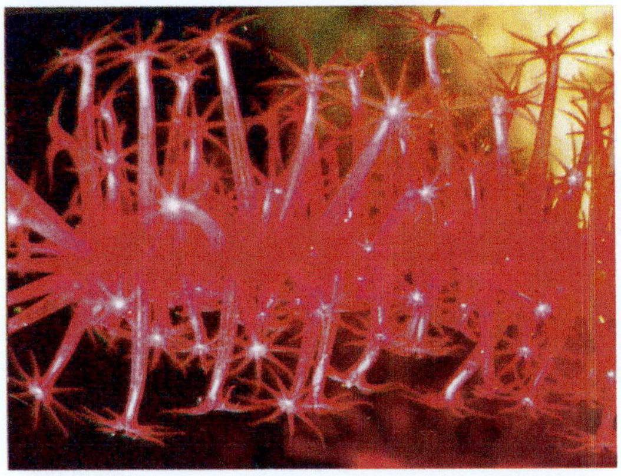

Chapter 5

Phylum Cnidaria

Verrill, 1865

(Greek *cnides*, 'nettle')

DEFINITION

Metazoans are either sedentary or free-swimming; they have primarily radial structure and are diploblastic animals, having a single opening. The phylum includes a variety of animals, such as hydras, jellyfish, sea anemones, and corals, as well as the less familiar hydroids, sea pens, sea rods, sea fans, siphonophores and zoanthids. There are about 9000 living species.

GENERAL CHARACTERISTICS

1. The majority of cnidarians are marine, except for hydras and some additional freshwater hydrozoans. Most cnidarians live in shallow water.
2. Cnidarians are sessile, sedentary or pelagic.
3. Radial symmetrical animals, the symmetry is often modified as biradial or quadriradial. There is one main axis: the oral-aboral axis, which extends from mouth to base.
4. They are solitary or colonial animals.
5. Many cnidarians are dimorphic; the animal exhibits two different adult forms during its life. This condition is referred to as alternation of generations, also called metagenesis. One form is the polyp, which is usually sessile and has the shape of an elongated cylinder. The other form, the medusa, is generally free-swimming and is a bell-, bowl- or saucer-shaped animal (Fig. 5.1B and C). Some cnidarians have only the polyp form, some have only the medusa form and others have both in their life cycle (Fig. 5.1).
6. The basic cnidarian body plan is simple, consisting of a solid body wall. The latter is composed of an outer ectoderm, and an inner endoderm enclosing the gastrovascular cavity, the enteron. The ectoderm consists mainly of epitheliomuscular cells, and the endoderm consists mainly of nutritive muscle cells. Between the two layers is a structureless layer, the mesoglea (Fig. 5.2).
7. From the above feature, it means that there is no defined muscular system. However, the muscular elements are included within the epithelial cells of both ectoderm and endoderm, as the movement of cnidarians is muscular.
8. Mesoglea is a secretion from the two layers. It varies in thickness depending on the organism. If it is a thick layer, it may contain wandering cells. Biochemical and immunohistochemical evidence show that the mesoglea is composed of type IV collagen, fibronectin, heparan sulphate proteoglycan and laminin (Sarras et al., 1991).
9. The body cavity is the enteron (see Fig. 5.5). It is often referred to as a coelenteron. It has a single opening at the oral end, which functions as inhalant and exhalant. It also functions for both digestion and circulation.
10. Cnidarians are characterized by the presence of specific cells, stinging cells called cnidocytes or nematoblasts. They are formed in the ectoderm layer (Fig. 5.4).
11. In between both the ectodermal and endodermal epithelial cells, there are the so-called interstitial cells (Fig. 5.2). These cells are multipotent cells, since they can differentiate into several somatic cells as well as into germ cells.
12. The nervous system is a primitive one, consisting of an independent nerve net for each layer (Fig. 5.2). The nerve net is composed of naked and largely nonpolar neurons (Fig. 5.3).
13. There are no distinct respiratory, excretory, circulatory or reproductive systems.
14. Gonads consist mainly of aggregations of germ cells, essentially differentiated interstitial cells.
15. Cnidarians reproduce both gametically and agametically by budding.
16. Cnidarians have a characteristic planula larval stage (Fig. 5.10D), without a cavity; this consists of ciliated free-swimming larva.

Invertebrate Embryology and Reproduction. https://doi.org/10.1016/B978-0-12-814114-4.00005-9

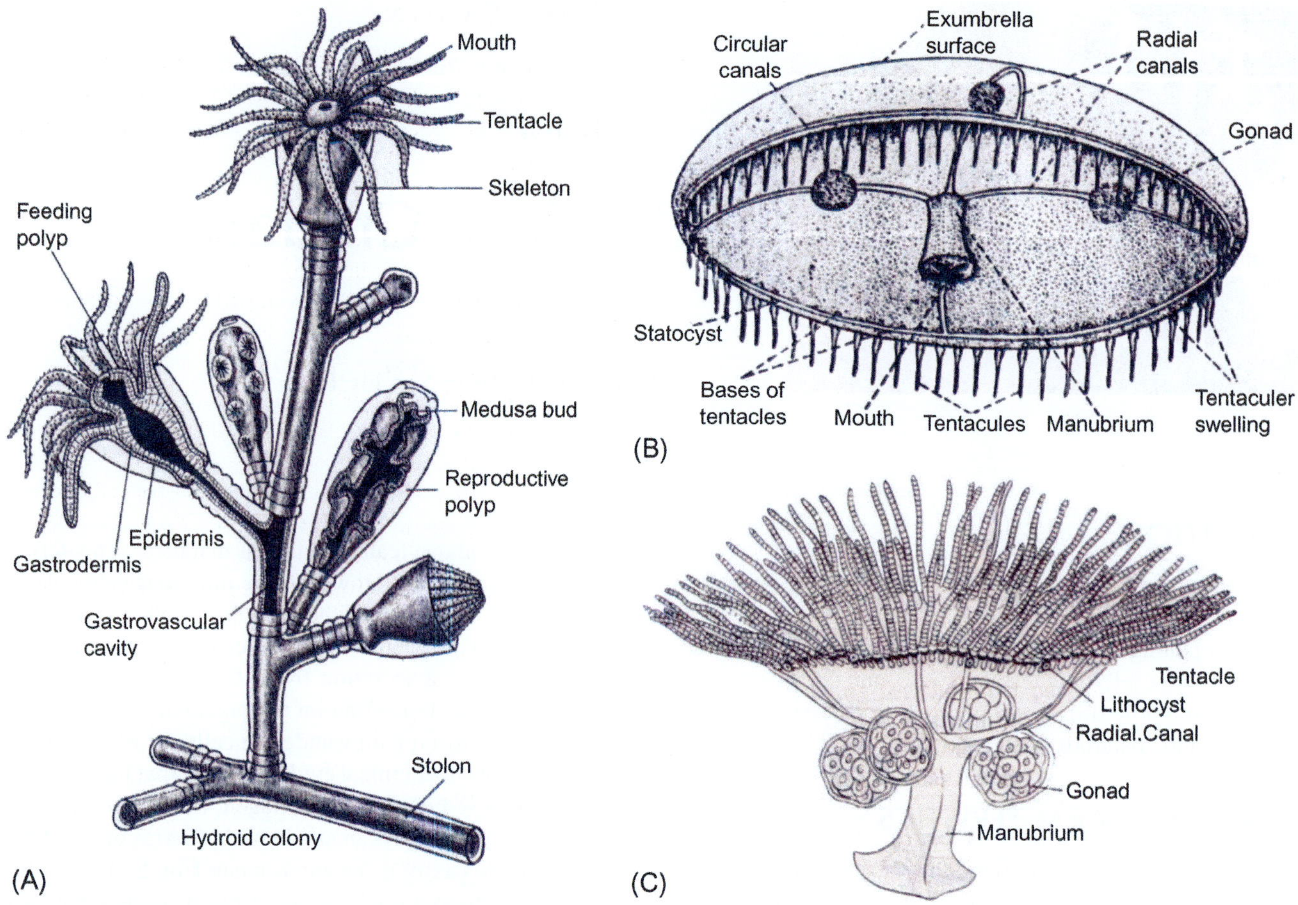

FIG. 5.1 Life cycle of *Obelia* showing the two generations: (A) polyp generation; (B and C) medusa generation: (B) medusa of *Obelia* seen partly from the subumbrella surface; (C) mature medusa swimming with everted umbrella surface *(Adapted from different sources.)*

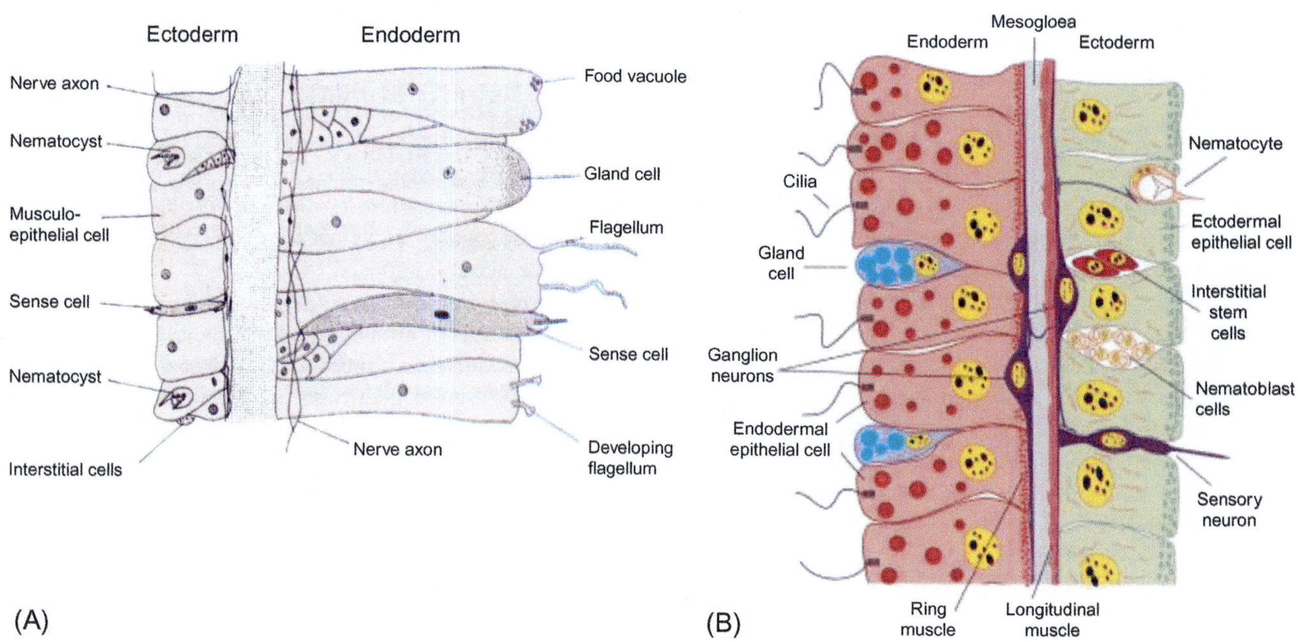

FIG. 5.2 Diagrammatic L.S. through the body of *Hydra* exemplified that in all cnidarians; (A) early presentation with its simplicity and clarity shows detailed structure; and (B) very recent coloured presentation, *Development 2011, 158 (8): 1447–1458 (Primer Series: Evolutionary crossroads in developmental biology: Cnidaria.). ((A) From Borradaile et al. (1977).)*

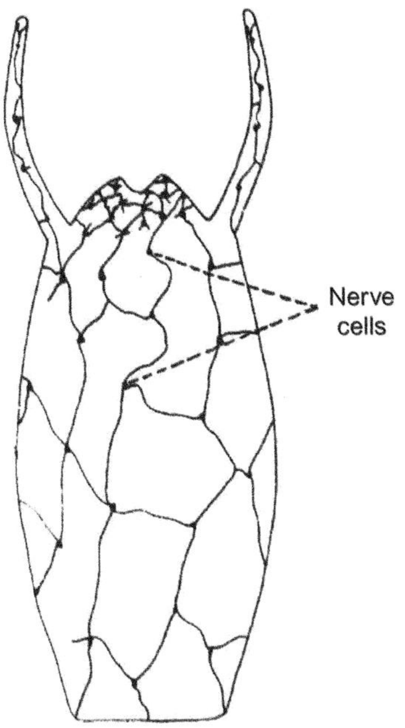

FIG. 5.3 Diagram of *Hydra* to show the nerve net *(From Borradaile et al. (1977).)*

CLASSIFICATION

Cnidarians are classified according to the phenomenon of alternation of generations and/or the form of the organisms. This phylum is classified into three large classes. They are as follows:

I. Class Hydrozoa

Typically the polyp alternates with the medusa. This class is divided into the following:

Hydrozoans: with a well-developed polyp stage and medusa (if present) is discoid or bell-shaped.

1. Order Hydroida
   a. Suborder Limnomedusae: solitary polyp and free medusa. Ex. *Gonionemus* sp.; the freshwater *Craspedacusta* sp.
   b. Suborder Anthomedusae (Gymnoblastea) (Athecata). Ex. *Bougainvillea, Ceratella, Tubularia mesembryanthemum, Pennaria, Sarsia, Syncoryne, Eudendrium, Hydractinia, Eleutheria, Clavatella, Branchiocerianthus*, the fresh water hydras and *Protohydra.*

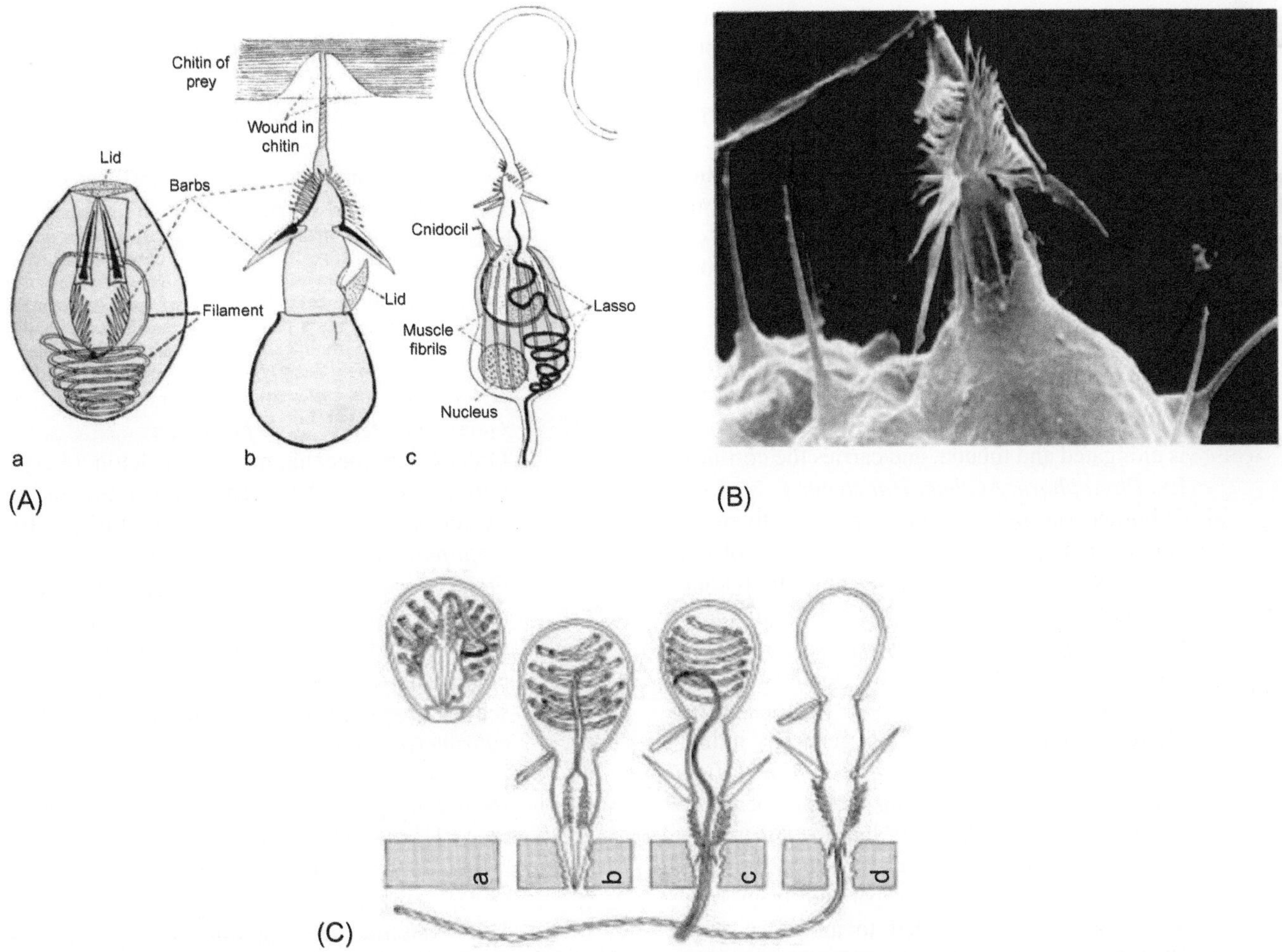

FIG. 5.4 Nematocysts of *Hydra*, large penetrating type: (A) (a) undischarged; (b) discharged; (c) discharged but retained within its thread cell; (B) SEM of a nematocyst after release. Note triggers in background of other nematocysts; (C) Steps of nematocyst discharge. *((A) From Borradaile et al. (1977), (B and C) after Tardent (1988), from Hale (1999).)*

c. Suborder Leptomedusae (Calyptoblastea) (Thecata), flat medusae. Ex. *Obelia, Sertularia, Plumularia, Campanularia, Polyorchis* (Barnes, 1974), *Diphasia* sp. *Clytia.*

2. Order Trachylina

The medusa is large and the polyp is reduced or lacking. The medusa develops directly from the egg (ovum) (hypogenesis).

a. Suborder Trachymedusae. Ex. *Glossocodon, Geryonia, Pelasus, Carmarima.*

b. Suborder/Narcomedusae. Ex. *Cunina, polycolpa, polyorchis, leptomedusa* (calyptomedusa).

3. Order Hydrocorallina (Hydroid corals)

There is a massive calcareous external skeleton. Ex. *Millepora* (Fire coral), *Stylaster.*

4. Order Chondrophora

The organism is a large *Tubularia*-like polyp floating at the surface. The adult is a free medusa. Ex. *Velella, Porpita* (Marshall and Williams, 1975).

5. Order Siphonophora

Organisms exist as free-swimming polymorphic colonies. Ex. *Physalia, Halistemma, Muggiaea kochi.*

a. Suborder Disconectae: Siphonophora with an apical pneumatophore, without nectophores (swimming polyp) or bracts (leaf-like protective polyp). A central single gastrozooid. Ex. *Discalia, Porpalia, pataria, Velella,* and *Porpita* (Minchin et al., 1900).

b. Suborder Calyconectae: Siphonophora without pneumatophore with one or more nectophores. The coenosarc is elongated and tubular, and carries the cormidia, which may become freed as Eudoxomes or Ersaeomes (= Monogastricae). Ex. *Diphyes, Muggiaea, Hippopodius.*

c. Suborder Physonectae: Siphonophora with an apical pneumatophore, followed by one or more coronae of nectophores or bracts. The coenosarc is elongated and tubular, and carries the cormidia. Ex. *Physophora, Agalma, Halistemma.*

d. Suborder Auronectae: Siphonophora with an apical pneumatophore, followed by coronae of nectophores, and carrying an aurophore. Individuals confined to lower surface of the colony, including numerous gastrozooids. Ex. *Stephalia.*

e. Suborder Cystonectae: Siphonophora with an apical hollow pneumatophore, without nectophores or bracts. Gastrozooids generally numerous, arranged either on the lower side of the pneumatophore or on a long tubular coenosarc. Ex. *Cystalia, Physalia* (The familiar Portuguese man-of-war), *Rhizophysa.* In some other systems Suborder Disconectae is separated from order Siphonophora and elevated to an order called Chondrophora.

II. Class Scyphozoa (Scyphomedusae) Jellyfish

The organism dominates as medusa while the polyp is reduced or absent.

1. Subclass Stauromedusae: sessile polyp-like scyphozoan. Ex. *Lucernaria, Haliclystus.*

2. Subclass: Discomedusae: mainly live as free-swimming medusae.

a. Order: Coronata (Peromedusae): the body is conical and divided transversely by a groove. Ex. *Nausithae, Periphylla, Pericolpa, Atolla.*

b. Order: Semaeostomeae: The body is discoid. Ex. *Aurelia, Cyanea, Pelagia, Stygiomedusa.*

c. Order: Rhizostomeae: The body is discoidal to nearly hemispherical, without tentacles, with sucking mouths. Ex. *Cassiopeia, Rhizostoma, Pilema.*

III. Class Cubozoa (Carybdeida)

Charybdeida—Cubomedusae—Sea wasps.

The medusa is cuboid shape with four flat sides (Tropical and subtropical oceans). Ex: *Carybdea, Chironex, Chiropsalmus, Tripedalia cystophora.*

IV. Class: Anthozoa (Actinozoa)

The organism dominates as a polyp stage, while the medusa is absent.

1. Subclass Ceriantipatharia; body enclosed in tube-like exoskeleton.

a. Order Anthipatharia: colonial black corals, thorny corals (100 sp.). Ex. *Antipathes.*

b. Order Ceriantharia: solitary (50 sp.). Ex. *Cerianthus, Pachycerianthus.*

2. Subclass Octocorallia (Alcyonaria) Soft corals (2500 sp.).

a. Order Stolonifera: *Tubipora* (organ pipe coral), *Clavularia, Hartea.*

b. Order Alcyonacea: Soft corals. Ex. *Alcyonium palmatum* (Dead Man's Fingers), *Xenia,* and *Gersemia, Heteroxenia fuscescens.*

c. Order Coenothecalia: massive skeleton, of crystalline calcareas fibres penetrated on the surface by tubes of two sizes. Contains only the Ex. *Heliopora* (Indo-pacific blue coral).

d. Order Gorgonacea (horny corals): colony usually of plantlike form; axial skeleton of calcareas spicules, of hornlike gorgonin, or both; polyps short; (1200 sp.) Ex. *Gorgonia* (beautiful sea fan) *Leptogorgia* (sea whip) and *Corallium rubrum* (precious red coral used for jewellery.

e. Order Pennatulacea (sea pens) (300 sp.). Ex. *Pennatula* sp. (sea pen), *Renilla* sp. (sea pansy).

3. Subclass Hexacorallia: (Zoantharia) sea anemones and stony corals. (Some thousands of species).

a. Order Zoanthidea: solitary or colonial, lacking skeleton. Ex. *Zoanthus, Epizoanthus, Parazoanthus* sp.

b. Order Corallimorpharia: organisms fail to form skeleton. Ex. *Corynactis* sp.
c. Order Actiniaria: sea anemones, solitary, no skeleton (1000 sp.). Ex. *Metridium, Adamsia, Calliactis, Edwardsia, Tealia, Minys.*
d. Order Scleractinia (Madreporaria) truce or stony corals, usually colonial. Ex. *Fungia* sp., *Galaxea.*

BODY PLAN OF CNIDARIA

Much of the interest of the cnidarians lies in the fact that the animal lives two modes of life: a sedentary existence (polyp) and a wandering or rather freely swimming life (medusa) (Fig. 5.1). The body plan of the two forms is almost the same (Fig. 5.5). The polyp reproduces agametically and the medusa can reproduce agametically, but this stage is responsible for the gametic reproduction. There are large divisions of the cnidarians in which only one type is present, while in others they may even be united in the same species or the same colony of that species. A survey of the phylum is very largely concerned with the variations of these types and their combinations in the life histories of the different cnidarians. Examples from the familiar animals: *Obelia* is typical in that it lives the two forms equally; *Tubularia* has a sessile medusa; *Hydra* lives as a polyp only and the medusa is absent completely; *Aurelia* exists as a medusa and the polyp is a short life in its cycle; for sea anemones, the polyp stage only represents their life cycle.

REPRODUCTION

Cnidarians reproduce agametically and gametically. Agametic reproduction occurs mostly by budding, while gametic reproduction by means of sperm and ova, in spite of the fact that the cnidarian animal has no reproductive system. Generally, the polyp stage is responsible for agametic reproduction. The medusa is dioecious, and at a specific time, the medusa produces sperm and ova. However, the medusa can also reproduce by budding.

Agametic reproduction

The work on this aspect is voluminous. Most histological and experimental work has concentrated on agametic reproduction of cnidarians, since they are excellent models for studying different aspects such as cyto-, morpho-, embryogenesis and others. Their simple forms, high degree of developmental reversibility and the feasibility to be controlled in the laboratory provide factors for successful investigation. It is an important vehicle (means) for propagation of the phylum.

The main process of agametic reproduction in cnidarians is the budding. Budding in cnidarians has been known ever since the pioneer observation of the Dutch microscopist Antony van Leeuwenhoek (1632–1723) and those of the Swiss biologist Abraham Trembley (1700–84), whose studies on *Hydra* are classics in the history of biology (Gardiner, 1972). *Hydra* (Fig. 5.6) has since provided much of the material for experimental investigations on budding, growth, and regeneration.

Class Hydrozoa

Within the hydrozoans, numerous agametic processes result in the creation of solitary individuals or the formation, growth and specialization of colonial forms. Most polyp stages reproduce themselves by budding, whereas medusae

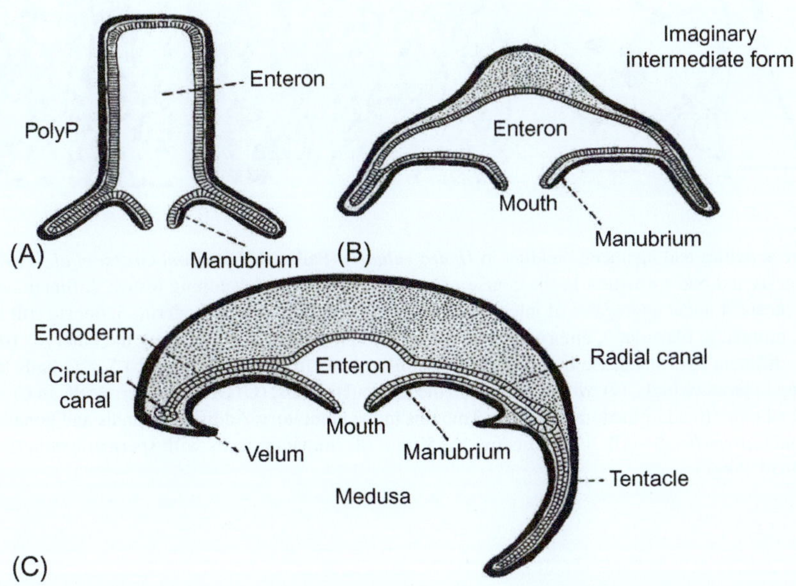

FIG. 5.5 Diagrams to show the relation between the polyp and medusa. The velum, present in many medusa, is absent in *Obelia (From Borradaile et al. (1977).)*

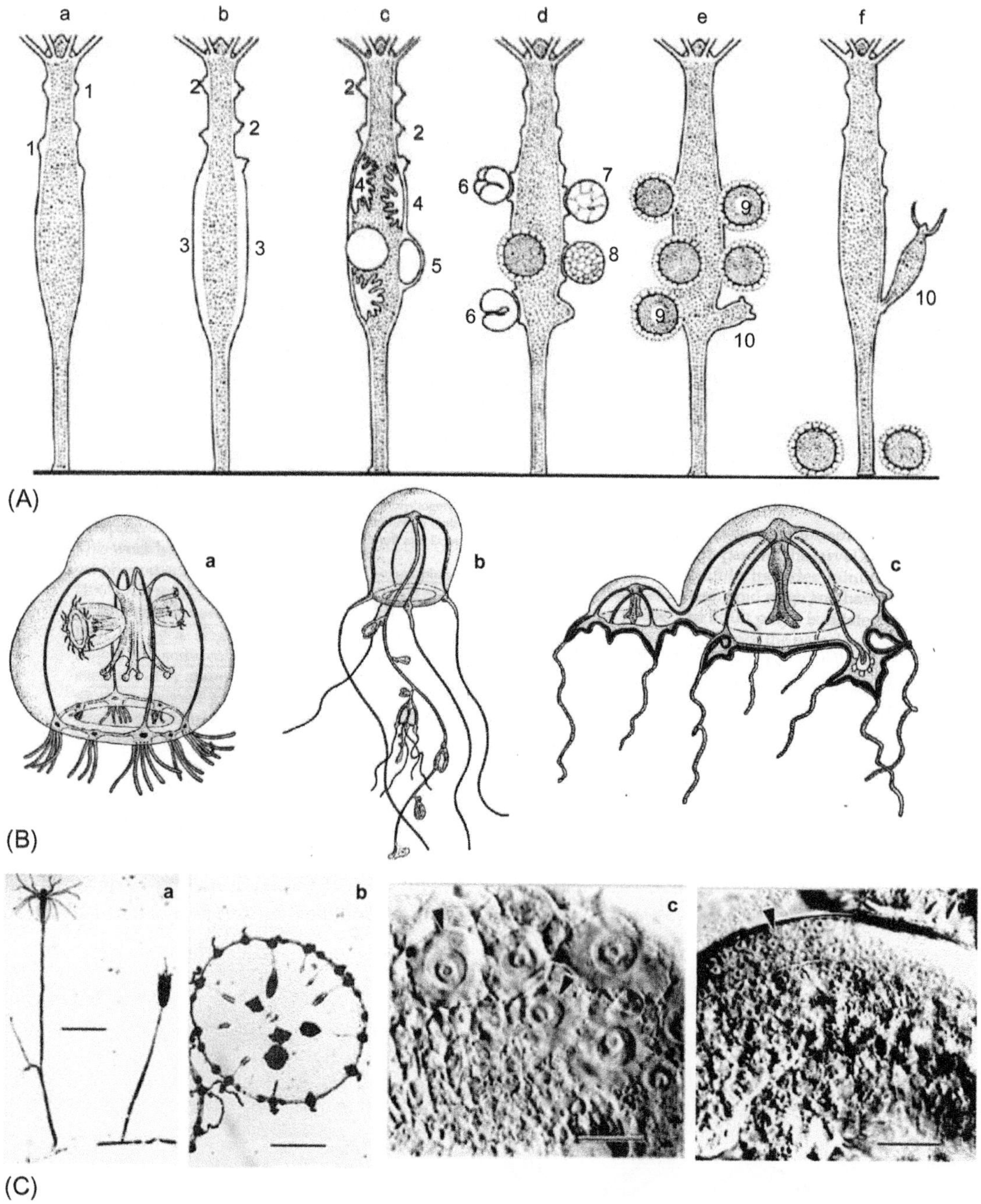

FIG. 5.6 (A) Sites of gametic activities and agametic budding in *Hydra vulgaris* (Pallas, 1766) *(from Grassi et al., 1995)*; External manifestations of sequential stages of gametogenesis and bud formation in the course of a sexual period (1, developing testes; 2, functional testes; 3, ectodermal swelling signalling the onset of oogenesis; 4, local aggregates of interstitial cells (oogonia) within the ectoderm; 5, oocyte still trapped within the ectodermal epithelium; 6, cleaving egg; 7, morula; 8, blastula; 9, embryos protected by the spiny ootheca secreted by the embryo; 10, asexual bud formation); (B) agametic reproduction in three different hydromedusae by budding: (a) *Rathkea* buds from manubrium; (b) *Sarsia* buds from its long thin manubrium. Daughter medusae are beginning to produce buds; (c) *Niobia* buds from the tentacular bulbs; (C) Origin of germ cells in *Clytia hemisphaerica* (Hydrozoa, Leptomedusa): (a) hydranth, X = 1 mm; (b) adult medusa raised for 3 months in the laboratory. Additional canals and gonads can be seen, X = 2 m; (c) immature ovary with young oogonia *(arrowheads)* (DIC) microscopy, X = 50 μm; (d) immature testis with spermatogonia *(arrowhead)* (DIC) microscopy, X = 50 μm. *(From Carré and Carré, 2000.)*

bud from polyps (Fig. 5.10C) and also from other medusae (Hyman, 1940) (Fig. 5.6B). All buds are genetically identical to their parent. In the colonial hydrozoans, budding usually occurs from stems and stolons (Figs 5.10–5.12). Thus the whole colony shares a single ectoderm, endoderm, mesoglea, and gastrovascular cavity. Such colonies may be sessile (hydroids) or floating (siphonophores) (Fig. 5.9).

In general, three kinds of buds may be formed:

- hydranth buds that remain attached to the parent colony as in *Obelia* (Fig. 5.10);
- planuloid buds, or frustules (Latin *frustulum*, 'little piece'), that separate from the parent and, like a planula larva, creep away, settle and develop into a new budding individual (Fig. 5.13J);
- and medusoid buds that develop into gametically reproducing medusa.

In species such as *Pennaria* and *Obelia,* they become detached from the parent as free-swimming individuals (Fig. 5.10) and in those like *Tubularia* and *sertularia*, they remain attached and reproduce in situ (Fig. 5.12B).

Some hydromedusae can bud new medusae from different parts of their bodies. Three different examples are given in Fig. 5.6B.

Other types of agametic reproduction

Budding is the primary mode of agametic propagation in hydrozoans. However, minor forms of agametic reproduction exist such as **transverse fission**, which occurs in *Protohydra*, the minute tentacleless hydroid, which cannot bud. Longitudinal fission occurs in *hydra*, and formation of frustules occurs in *Gonionemus* (Fig. 5.14A). Frustules are nonciliated planula like buds that creep or lie on the bottom and form polyps, they constrict from stolons and polyps (Hyman, 1940). Some siphonophores produce chains of individuals called cormidia, which may break free to begin a new colony, two types of siphonopore larvae, the calyconula and siphonula, bud (Carré, 1967, 1969). Budding usually takes place along stolons in colonial forms and from the polyp side in solitary forms. Some medusae bud directly from the manubrium, bell margins or tentacular bulbs. When a bud forms, it undergoes the following histological sequence (Gardiner, 1972):

1. One or both the epithelial layers thicken at appropriate position along the parent producing a bulge of tissue.
2. Sometimes ectodermal interstitial cells become more abundant in this bulge.
3. The bulge begins to evaginate at this position.
4. The endoderm acquires longitudinal ridges, frequently four in number, which establish the future radial symmetry of the young.
5. The outgrowth becomes papilli-form and then it undergoes morphogenesis, thus modelling the outgrowth to take the form of the mother.
6. The resulting young may separate from the mother or it may remain attached. The polarity of the bud is almost always with the future oral end, distal with respect to the mother. The proximal end of the bud is the attached part.

The entire process of formation of a bud takes 2 days. Variations of events of this histological sequence of the process of budding may occur in one species or the other, mainly in stages 4 and 5. However, the overall series within a species is so definite that a slight deviation even in a minute step in a stage of the sequence indicates abnormality. This overall sequence of events relates budding very closely to embryonic development and larval transformation, and sharply distinguishes budding from regeneration. By experimental studies, it was possible to help the process of budding, including epithelial cell behaviour (a principal factor) and gastrocoel fluid probably to some extent, and the form of external supporting structures such as the perisarc. Interstitial cells are thought to play only a little role in budding, but the accumulations of these cells at budding site is probably largely related to the production of the nondividing cells such as nerves and nematocysts.

Class Scyphozoa

Scyphozoans reproduce agametically by budding from the polyp or stolon by production of ciliated planuloid buds, or by formation of podocytes (encysted polyp rudiments). Scyphistomae may bud new polyps from the side of the stalk, like *Hydra*, or extend hollow stolons from the stalk or base that **bud** off new scyphistomae (e.g. *Aurelia* and *Chrysaora*) (Martin, 1997). In some scyphistomae, such as that of *Cassiopeia*, the head region pinches off free-swimming ciliated buds, the formation of the bud and its release may take 24 h. These buds are called **planuloid buds** since they resemble planula larva. Each bud swims with its distal/aboral end directed anteriorly, rotating from right to left upon its longitudinal axis. At metamorphosis, the planuloid bud attaches to the substratum by its distal/aboral end and then transforms into a young scyphistoma. The attached end of the bud forms the stalk of the young, while the proximal end forms the head of the young. This metamorphosis of the planuloid bud to be a young scyphistoma may occur in <24 h.

Scyphistomae of the members of order Semaeostomeae and Rhizostomeae produce **podocysts** from the foot region. These encapsulated structures, consist of epidermal and mesenchymal cells encased within a chitinous cyst. These structures remain dormant for a period of time, and then each cyst breaks to release either a small polyp or planula-like structure that settles to form a polyp.

Several other modes of agametic reproduction occur in cnidarians, but they are generally regarded as minor modes compared to budding. The most peculiar example is the transverse fission in cnidarians, which is known as the process of strobilation (Fig. 5.24). This process occurs in Semaeostomeae and Rhizostomeae and gives rise to the ephyrae (young medusa or larvae); see later for the development of Scyphozoa.

Class Anthozoa

Agametical reproduction in anthozoans occurs mainly by **fission**, but also by budding or by pedal laceration. Both transverse and longitudinal **fission** have been described for different species of sea anemones, e.g. *Nematostella* reproduces by transverse fission, while *Sagartia* reproduces by longitudinal fission. These two types of fission generally result in large groups or clones of genetically identical individuals (Hyman, 1940).

Budding is extremely important for the growth and formation of colonies of stony corals (Scleroctinia) and alcyonarians. In sea anemones, the individual buds from the column or pedal discs. In one rare example as reported by Martin (1997), little anemones actually bud from the tentacles. Colonies maybe formed by buds from polyps, which develop into polyps that remain connected to each other; from basal stolon which is found between polyps, or from a shared fleshy layer, the coenochyme.

Pedal laceration is an agametic reproduction which is common for many sea anemones, e.g. *Sagartia*, *Pherllia*, *Metridium*, *Aiptasia*, *Bunodes*, and *Haliactis*. Laceration occurs naturally or accidentally. In the former (Fig. 5.14E), a portion of the pedal disc extends as a lobe which fragments from the mother, developing into a new individual. In the latter process, when the individual moves about from the substratum to which it adheres, pieces of the pedal disc break and each portion develops into a tiny anemone.

Regeneration

The potential for regeneration is extensive in all classes of Cnidaria. Some species can be totally reconstituted from isolated groups of cells. In other individuals, some structures are incapable of regeneration. Isolated gonophores and tentacles of hydroids are incapable of regeneration and eventually die. In all cases, the regenerated animal is complete and correctly proportioned, but it may be smaller than the original one. It is not surprising that hydras, like many cnidarians, have considerable powers of regeneration. A classic experiment is that of Trembley, first performed in 1744; by means of a knotted thread inserted through the basal disc and drawn out through the mouth, Trembley turned a hydra the wrong side out. After a short period the gastrodermal cells reoriented themselves on the inner side of the mesoglea, and the ectodermal cells migrated to the outside (Barnes, 1974). As in sponges, dissociate cells will aggregate and from the larger aggregates an entire animal will develop. Entire animals have been produced from gastrodermis, and from ectoderm in other cnidarians.

Gametic reproduction

Origin of germ cells

In Cnidaria, beside the two epithelial cell lines that form and regenerate the ectoderm and endoderm, there are interstitial cells (stem cells) (Fig. 5.2) that, in *hydra*, make up 20%–30% of the cell population. At first, it was believed that these cells were totipotent and able to differentiate into either somatic cells (giving rise to cnidocytes and to neurons in particular) or into germ cells. These interstitial cells are migratory stem cells that divide to renew and/or to develop the different somatic cells, and to develop into gametes. Thus, the interstitial cells are multipotent cells. Some of the multipotent interstitial stem cells give rise to unipotent stem cells of each of two types of gametes. In hydras and apparently in all cnidarians, there is no double sex determination, first somatic and then germinal, but uniquely germinal determination. The male and female are identical and there are no somatic gonadal tissues. In addition, there are no primary or secondary sexual characters. In the absence of somatic sex determination, the sexuality of the gametes in hydras is the result of interactions between the precursors of the male and female germ cells. Carré and Carré (2000) reported that if interstitial cells from a male hydra are grafted onto a female hydra, sex change is induced, probably linked with the arrest of oogenesis, by a factor secreted or induced by the male gonocytes of the graft. A monoclonal antibody, AC2+, recognizes spermatozoids and their stages of differentiation specifically. However, the results of their experiments carried out on the hydrozoan *Clytia hemisphaerica* reared in their laboratory evidenced that temperature is the main factor influencing the germ cells. The medusae budded and cultured at 24°C were mainly female (80%). In contrast, if the medusae were released at 15°C, at whatever temperature they were raised to later, they were mainly male (85%). The female line will be dominant at 24°C but temperature-sensitive; inhibition of this line by a temperature drop to 15°C inverts the population sex ratio.

Gonadogenesis

Cnidarian gonads are not distinct organs, but are merely massive assemblages of gametes in the interstitial spaces of the body tissue (diffused gonads). Interstitial cells which will be sperm or egg stem cell, accumulate each in its specific area of the column and become a gonad. In Hydrozoa, the gonads are ectodermal, while in Scyphozoan, Cubozoa, and Anthozoa, the gonads are endodermal in origin. No accessory cells of somatic origin participate in the formation of these gonads.

Grassi et al. (1995) studied gametogenesis in the strain 'Scotland 1989' *Hydra vulgaris* (Pallas, 1766), which is well identified by Campbell (1989). The obtained specimens were successfully mass-cultured in 'finger bowls (260cm)' at 14°C in artificial culture medium, by careful keeping and good feeding, the culture covers an observation period of 4years (1989–1993). During this time, two gametogenically active specimens (No. 1 and No. 2) bearing side by side testes and eggs (Fig. 5.6A (c)) were isolated from the mass-culture and kept individually in 'wells (11cm)' to observe their behaviour regarding budding and gametogenic activities. This work provided data for the condition of the polyps as well as numbers of buds, testes, oocytes, eggs, and many others. In fact, only there are the sites of stages of gametogenesis and bud formation in one sexual period (Fig. 5.6A) recorded.

The onset of each sexual period is characterized by the appearance of an initially small number of testes, which emerge as conical outgrowths mostly of the subtentacular body ectoderm (Fig. 5.6A (a1)), well before signs of oogenesis are apparent. On the other hand, the beginning of oogenesis is signalled by a patchy milky thickening of the polyp's ectoderm within the middle region of the polyp's body (Fig. 5.6A (b3)), and shortly afterwards by the formation of one or more polymorphic aggregates of interstitial cells (Fig. 5.6A (c4)).

The growth and gametogenesis of each gonad causes this portion of the tissue to bulge, stretch and eventually rupture to release the gametes. Hydrozoans release their gametes directly to the outside, whereas the other three cnidarian groups release their gametes into the gastric cavity and later expel them to the outside through the mouth.

Gametogenesis

Spermatogenesis

During spermatogenesis, the interstitial cells that will be sperm lineage stem cells proliferate within the gonad forming large nests or clusters. The cells of these nests undergo division and morphological differentiation in synchrony. The spermatocytes undergo meiosis to form spermatids. Large number of sperm are produced and released during the breeding season. This process is similar to spermatogenesis of higher metazoans. Generally, the sperm are released in semen, but a conspicuous case is found in the cubozoan *Tripedalia cystophora* (Martin, 1997) which encapsulate the sperm, in what is called spermatozeugma (peculiar tubular sperm aggregates (Mann, 1984)) and transfer it directly to the female. It appears that this case of the cubozoan is a general character of the class (see later).

Oogenesis

Oogenesis is best described for the hydrozoans and some anthozoans. Interstitial cells concerned in the egg lineage divide to produce large groups of oogonial cells, which aggregate forming a large syncytial mass. One cell within this mass remains as the oocyte. All the other accompanied cells are often referred to as nurse cells. They gradually disintegrate and become incorporated into the cytoplasm of the growing oocyte (deutoplasmogenesis, i.e. yolk formation). Meiosis (the final step of maturation of the oocyte) may occur prior to or after spawning. However, there are representatives from all cnidarian classes in which the oogonia do not fuse, but simply grow, synthesize yolk, and enlarge.

The oocytes of Scyphozoans and anthozoans generally receive nutritive materials from the mother; specialized endodermal cells transfer nutrients to the oocyte.

Gametes

The mature cnidarian **sperm** is considered primitive, in spite of the fact that it resembles other spermatozoa. It consists of a small head that contains a nucleus, which is of conical or cylindrical form. The middle piece contains two to five small rounded mitochondria and a pair of centrioles. The latter are arranged linearly at right angles to each other. The lower centriole is the basal body of the flagellum of this spermatozoan. The flagellum shows the typical ultrastructure of other spermatozoans. Campbell (1974) reviewed the process of spermatogenesis and the structure of the sperm in Cnidaria (Fig. 5.7B). A conspicuous character of cnidarian sperm is that it lacks the acrosome. The only example of an acrosomal complex and an acrosomal reaction known to occur in the cnidarians is that of the siphonophore, *Muggiaea Kochi*. The sperm of this species undergoes an acrosomal reaction and Carré (1984) described a fully differentiated acrosomal complex in this species.

Cnidarian **eggs** differ in their morphological shape and in the amount of yolk distributed throughout their cytoplasm, i.e. they are isolecithal. Some eggs have a smooth surface, while others are covered with microvilli and a jelly coat. A special case is found in the eggs of the hydroid leptomedusan, *Bougainvillia multitentaculata* (Fig. 5.7A). These eggs are covered by thousands of nematocytes derived from the mother. These maternal nematocytes remain present in the egg during the embryological processes and then they are incorporated into the planula larva.

Spawning (release of gametes)

The breeding season of cnidarians occurs at a specific time of the year, but it differs from one species to another. Mostly, fertilization is external in this phylum and the majority of species release their gametes simply in the water where they live. Usually, the release of the two types of gametes is synchronized, providing a big chance for fertilization. This chance is assisted by the aggregation of the individuals of the same species, often just prior to reproduction. It was noticed that large swarms of medusae appear periodically in the water during breeding. I personally witnessed the swarm of *Cassiopiea* in the red sea during June. In addition, some

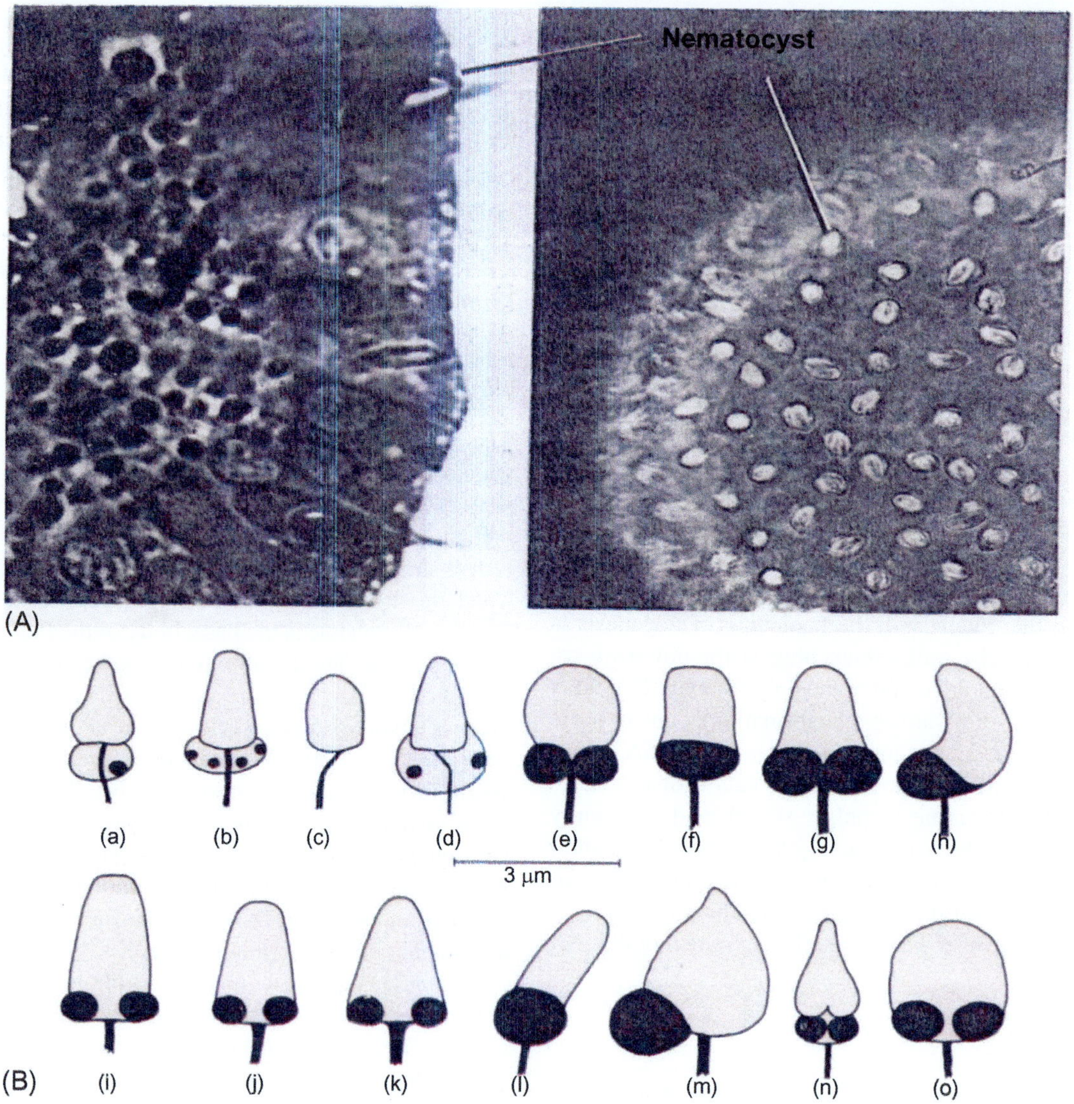

FIG. 5.7 (A) Armed egg of the hydromedusa, *Bougainvillia multitentaculata*: photo at right-hand has numerous nematocytes containing nematocysts on its surface, X=650; photo at left is section through a planula showing nematocytes which have been contributed maternally, X=2050 *(after Szollosi, 1969)*; (B) Cnidarian spermatozoa. Only heads are depicted: *black* indicates mitochondria. (a) *Caryophyllia smithii*, (b) *Bolocera tuediae*, (c) *Funiculina quadrangularis*, (d) *Sagartia troglodytes ((a–d) from Franzen, 1966)*, (e, f) *Polyorchis karafutoensis (Nagao, 1970)*, (g) *Phialidium* sp. *(Roosen Runge and Szollosi, 1965)*, (h) *Tubularia radiata (Nagao, 1965)*, (i) *Hydra fusca (Schincariol et al., 1967)*, (j) *Hydra attenuata (Stagni and Lucchi, 1970)*, (k) *Pennaria tiarella (Summers, 1970)*, (l) *Muricea fruticosa (Grigg, 1970)*, (m) *Tealia erassicornis (Rettzius, 1904)*, (n) *Limnoenida tanganyica (Bouillon, 1957)*, (o) *Corymorpha palma* (original). *(All from Campbell, 1974.)*

sea anemones aggregate just prior to spawning (Uchida and Yamada, 1968). Some sea anemones, however, have developed behavioural patterns associated with spawning. These sometimes involve the female moving for short distances to the male, stimulated by male spawning (Uchida and Yamada, 1968).

In many species, spawning is a response to changes in intensity of illumination, either lunar or solar, or darkness. Spawning occurs by rupture of the gonad wall in hydrozoans, and the gametes are released in the water, while in Scyphozoans, Cubozoans, and Anthozoans, gametes are released into the gastric cavity and then to the outside through the mouth. In some species, the gametes are shed to the outside through pores in tentacle tips or in the pedal disc.

Copulation is not a general feature of cnidarians. However, there is at least one known example, *Sagartia troglodytes* (European sp), in which copulation occurs. In this sea anemone, the pedal discs of two individuals are pressed together, forming a chamber in between, into which the gametes of the two individuals are released and so fertilization occurs. This copulatory position is maintained for several days, until the planula larvae are formed. The case of *Tripedalia cyclophora* or class cubozoa can be considered as copulation, since the male uses its tentacles to transfer its spermatozeugma to the female.

Brooding

Some species in all four classes of Cnidaria brood their young.

1. Hydroids lacking free-swimming medusae undergo internal fertilization and development in the female gonophore, then release planula or actinula larvae.
2. Many Scyphozoan, Cubozoan and anthozoan eggs are fertilized in the gonad or after release into the gastrovascular cavity. The zygotes may be retained in this cavity and undergo development until the stage of the planula larva, which is released to the outside.
3. Many Scyphozoans brood their young within their oral arms until the planula larva stage.
4. Some sea anemones and corals brood their young externally; the young adhere to the external surface of the mother or to the colony, e.g. *Epiactis*. The adults of *Epiactis japonica* (a species that lives in the northern part of Japan) very often have young attached to their outer surface (Uchida and Yamada, 1968).

Cnidarian embryogenesis

Fertilization

Generally, fertilization is an external event that occurs in the water where the cnidarians live. However, internal fertilization has evolved in some species of all four classes of the phylum. Internal fertilization occurs when the current of water containing sperm drains or flows into the polyp or medusa bearing eggs, e.g. sea anemones and *Obelia* respectively, or the current of water is attracted to the polyp, e.g. *Hydra*. In addition, internal fertilization occurs in the gonophore lumen of hydrozoans, e.g. *Tubularia* and in the gonad or gastrovascular cavity of the Scyphozoans, Cubozoans and anthozoans. In some hydroids, both Thecata and Athecata, chemotactic direction of sperm migration to the female gonad has been discovered (Miller, 1966a,b). This is particularly exciting since it represents the first well-documented case of spermatozoan chemotaxis in animals (Campbell, 1974). Chemical attractants, which can be isolated, are released by the female gonophores and have been shown to affect the spermatozoan swimming pattern. These attractants may be localized to the fertilization site or distributed over the entire surface of the egg. The substances were isolated and studied chemically, but show no perfect species specificity (Miller, 1966a,b).

The egg is ready to be fertilized after the extrusion of its polar bodies. Directly beneath the area of these polar bodies, the female pronucleus lies and this is the animal pole of the egg.

Cleavage

Four patterns of cleavage are seen in cnidarians:

1. Holoblastic, regular, radial and equal, many hydrozoans, e.g. Hydra, most scyphozoans and some sea anemone, several anthozoans, several stony corals, and various octocorals
2. Holoblastic, irregular, radial, many anthozoans, scyphozoans and some anthozoans tend to be much less regular in their divisions
3. Holoblastic, irregular, radial and unequal, some hydrozoans, such as *Pennaria tiarella* which produces blastomeres of unequal size and no one embryo cleaves in exactly the same pattern; and
4. Superficial, hydrozoans and anthozoans with very yolky eggs undergo this superficial cleavage (epiblastic). Variation, combination and modification of these patterns may exist throughout cnidarians. In many cnidarian embryos, the cleavage furrows are asymmetrical or unilateral.

Blastulation

Three types of blastulae are formed from the cleaved embryo.

1. A hollow coeloblastula that contains a simple layer of cells enclosing a large blastocoel;
2. A solid stereoblastula filled with cells and lacking a blastocoel; and
3. In superficially cleaving embryos a periblastula is produced, such as a blastula consisting of an outer layer of cells surrounding a central mass of yolk

Gastrulation

Gastrulation in cnidarians is diverse and different patterns are formed. These patterns or processes are unipolar and multipolar ingression, invagination and delamination may be primary or secondary, epiboly and mixed patterns (Hyman, 1940). In primary delamination, the endoderm arises by radial division of blastomeres such that one daughter cell remains on the surface (ectoderm) and the other daughter cell moves to the interior (endoderm). Apparently, this process occurs in the coeloblastula. In secondary delamination, which is characteristic of some stereoblastulae, tissue differentiation occurs by a simple histological rearrangement producing an outer ectoderm and an inner endoderm. A type of secondary delamination called syncytial delamination occurs in yolky egg that produces periblastulae. The gastrula develops into planula larva. The gastrula elongates, becoming a cylinder with two distinct ends. The anterior end is more or less flattened, while the posterior is narrow or pointed. It already consists of outer ectoderm and inner endoderm with a thin sheet of mesoglea separating the two layers. The larva creeps or swims in the water with the anterior end directed forward. It remains so for several hours to several days according to the species, during which time it increases in length and undergoes cell differentiation. The degree of differentiation varies with the species and depends on this species. A gastrovascular cavity may or may not be present. The mature planula consists of all the cells derived from both ecto- and endoderm specially the interstitial cells. Interstitial cells arise during gastrulation in the central core of endoderm along the entire length of the embryo; during embryogenesis and larval development, these cells emigrate to and populate the ectoderm. However, before this migration, many of these cells enter either the nematocyte or the neural differentiation pathway in the endoderm. During metamorphosis of the larva, its stem cells undergo changes to accommodate the adult nature. Larval interstitial cells and nematocytes are carried over into the adult; these cells constitute the adult stem cell system.

In *Pennaria tiarella*, gastrulation occurs 8h postfertilization, via invagination (Martin and Archer, 1986). Planula larva is formed by 10h postfertilization. This planula measures c. 325 μm in length and c. 200 μm in width. A morphological anterior–posterior axis has been established. The 10-h planula consists of an outer ectoderm and an inner endoderm separated by an acellular mesoglea. The ectoderm is composed of epitheliomuscular cells and mucous cells, while the endoderm contains interstitial cells, developing nematoblasts, developing digestive cells and developing glandular cells. The planula continues development and elongates, neurosensory cells differentiate in the ectoderm and cells migrate from the endoderm to the ectoderm. The mature planula of *Pinnaria* (24–96h postfertilization depending on temperature) is cylindrical in shape and has distinct anterior and posterior ends. It measures c. 400–1000 μm long and c. 70–90 μm wide in its anterior end. In *Pennaria*, the interstitial cells arise 8h postfertilization as a central core of cells located in the endoderm (Fig. 5.8A).

Another hydrozoan cnidarian, *Eudendrium racemosum* (Cavolini), was collected from the Mediterranean along the coast southeast of Genoa, Italy, at depths of 0.5–25 m. Sommer (1990) collected female colonies bearing numerous developing embryos to study larval development and metamorphosis. Sommer observed the process of hatching

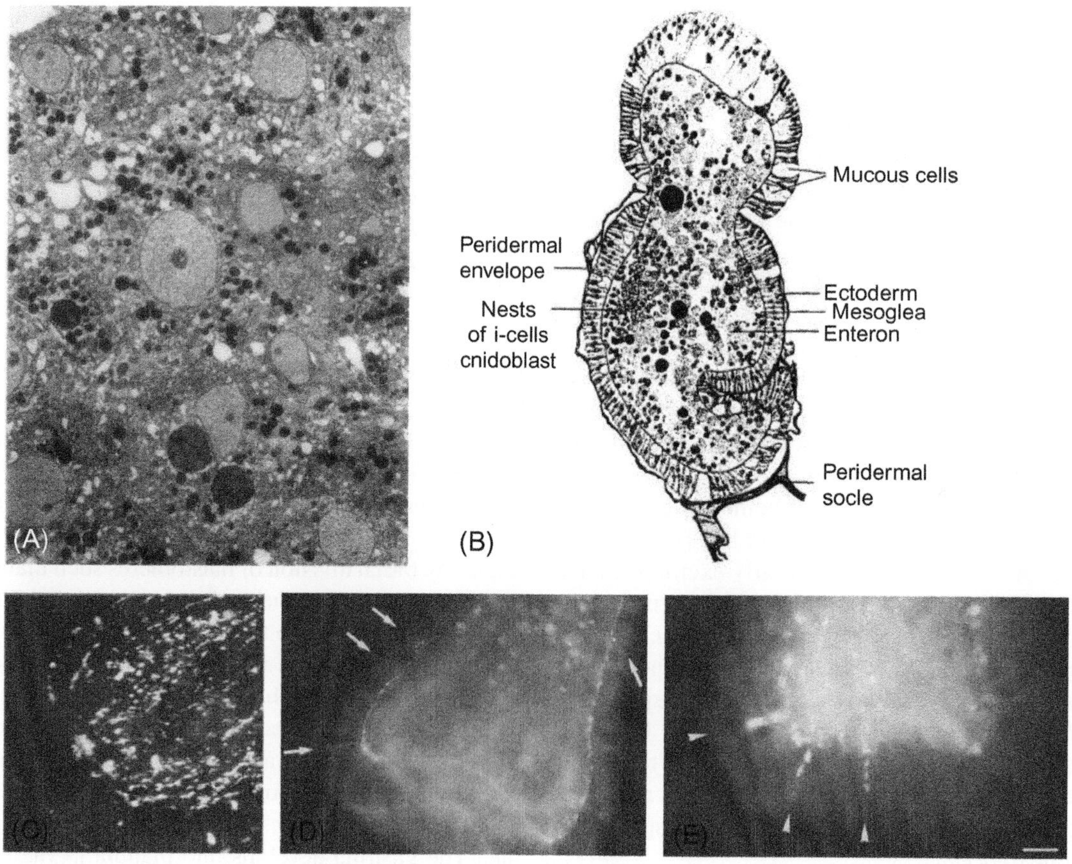

FIG. 5.8 (A) Central endodermal core of interstitial cells in an 8-h embryo of *Pennaria tiarella*, X = 4125 *(from Martin and Archer, 1986)*; (B) Hatching planula of *Eudendrium racemosum*, X = 50 μM *(from Sommer, 1990)*; (C) confocal laser scanning microscopical image of anterior part of *Hydractinia echinata* of a larva immunostained with the JPI (polyclonal rabbit antibody) antibody, X = 400 μM; (D) the anterior part of the same larva immunostained with JPI showing staining of granules in neurosensory cells *(arrows)*; (E) immunostaining with anti-RFamide visualineurocenscells *(arrowheads)* that probably are the same as those in (C), X of D and E = 20 μM *((C–E) from Leitz, 1993)*.

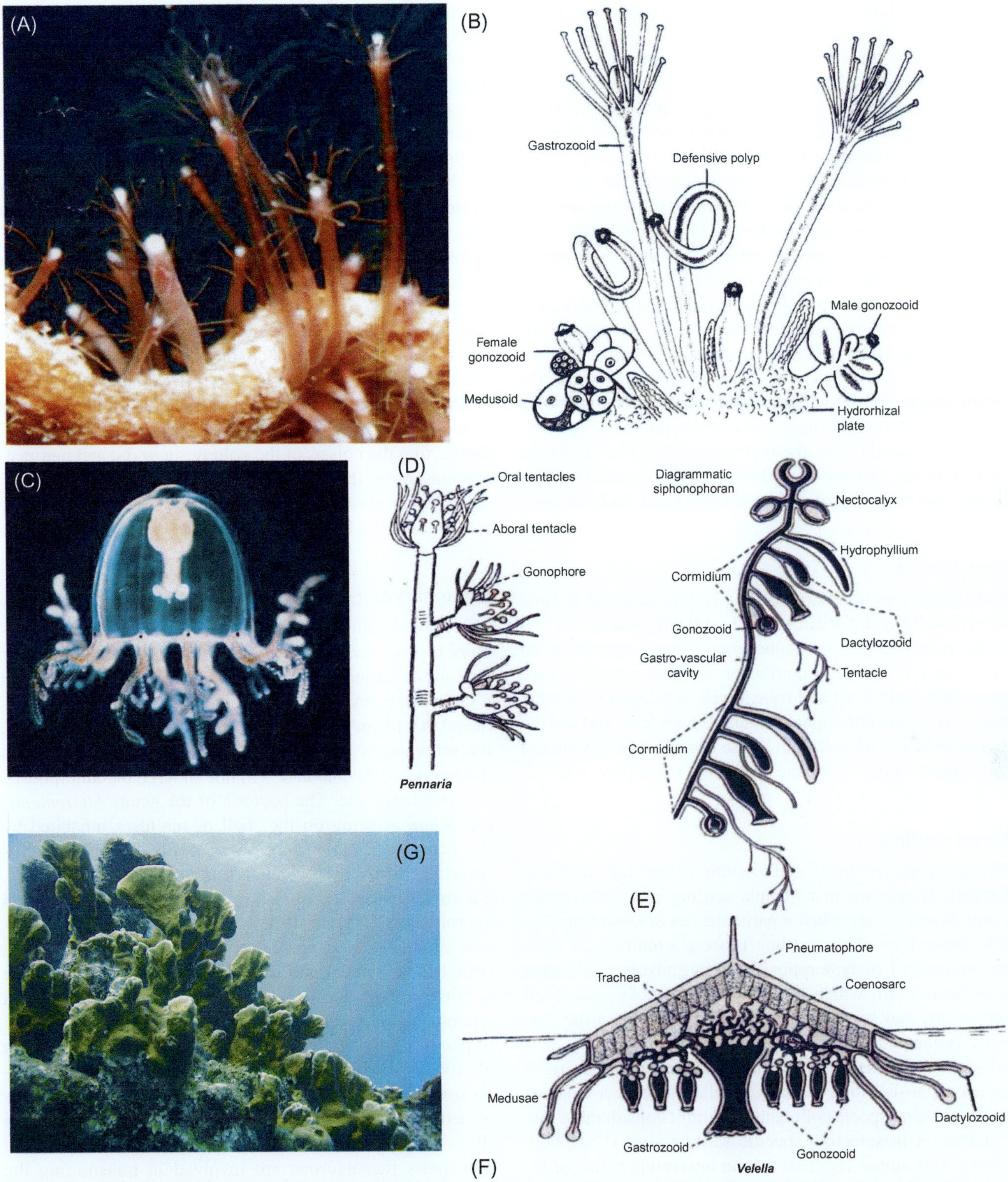

FIG. 5.9 Examples of cnidarian hydrozoans: (A) portion of *Clava multicornis* colony; (B) portion of *Hydractinia* colony, a tetramorphic hydroid. Polyps arise separately from mat of stolons, or hydrorhizae, skeleton limited to covering of stolons, and fusion of adjacent skeletons forming a plate *(after Hyman from Barnes, 1974)*; (C) medusa of *Cladonema radiatum*; (D) portion of *Pinnaria* colony; (E) portion of the siphonophore *Halistemma*; (F) a disc-shaped *Velella* colony has a superficial resemblance. *((A and C) From Beauté du Monde Animal, Tome XII, Larousse, 1974; (D–F) diagrammatic illustration from Borradaile et al., 1977)*; (G) Fire coral, *Millepora planata, © OCEANA/Houssine Kaddachi (The Corals of the Mediterranean).*

and reported that it is initiated by a striking decrease in the thickness of the embryonic envelope. The process culminates locally in the formation of a circular opening at the animal pole of the egg-capsule, which corresponds in position with the aboral pole of the hatching embryo (planula). At the opposite pole, the egg is fastened to the female 'blastostyle by a thick and multilayered periderm socle' (Fig. 5.8B). Planulae always hatch by leaving the egg aboral pole first. Then, they slowly wind the hind part of their body through the opening in the egg membrane. In particular, the ectoderm of the larva becomes highly constricted as it passes through the narrow opening (Fig. 5.8B).

The planulae are gastrula stage and the ectoderm is characterized by the presence of mucous cells in between the musculepithelial cells. The secretion of these cells is intense during the course of hatching. Thus, after leaving the egg, the larvae stay in contact with the empty capsule for some time. They are still attached by a slowly prolonging, sticky slime rope that extends from the oral pole of the planula to the surface of the egg-capsule. This behaviour, termed 'roping down', has been described in other species of *Eudendrium*.

Larval forms

The typical cnidarian larva is the planula, an ovoid to fusiform radially symmetrical, ciliated entity. Planulae are of two types: planktotrophic or lecithotrophic. The former feeds on plankton in the water or detritus on the bottom and they are potentially long-lived and have a well-developed gastrovascular cavity. Lecithotrophic planulae do not feed, and use the stored yolk for its nutrition. These planulae usually lack a gastrovascular cavity and are short-lived in the plankton.

Larval settling

Planula larvae probably settle within a few days of being formed. It appears that planula settling is a behavioural event which occurs when a particular set of conditions begin. Some part of the settling process stimulates at once the process of metamorphosis. Settling involves adhering to a substratum, which this is aided by sticky gland cell secretions, and sometimes by nematocytes, since the nervous system may be closely related to nematocytes. It is possible that a nervous impulse mediates a triggering of metamorphosis when nematocyst discharge accompanies settling. Many species of planulae exhibit considerable discrimination in selecting specific settling sites (Campbell, 1974). This author reported that an interesting series of investigations into planula settling concern *Hydractinia echinata*. This hydroid is a marine colonial hydroid frequently found in the North Sea. It grows almost exclusively on snail shells inhabited by hermit crabs. Fertile colonies, male and female, produce eggs and sperm, respectively, and within 3 days the fertilized egg develops into a mature planula larva (Walther et al., 1996). 'Mature' larva means a larva that is able to metamorphose into a polyp but not in the laboratory. To continue its development, it needs an external trigger that appears to be provided in the natural habitat by certain sedentary bacteria of the genus *Alteromonas*. A lipophilic substance produced by these bacteria is thought to act as the trigger (Leitz and Wagner, 1993).

It was concluded that *Hydractinia echinata* planula settling is determined by physical conditions, namely the presence of a hard substratum, across which water is briskly flowing. Some experimental studies were done to reexamine settling in this species. Four species of bacteria were isolated which, in pure culture, could stimulate *Hydractinia echinata* to settle and metamorphose. Naturally, the planula glides along the substrata in response to both light and oxygen conditions, and settles where bacterial density or activity is sufficient.

Metamorphosis into a primary polyp, the founder of a new colony, is completed within 1 day. Stolons, hollow tubes connecting the polyps of the colony, grow out and new polyps emerge on them at more or less regular distances. This new colony reaches sexual maturity, usually within a few months, and the life cycle is repeated (Fig. 5.10D and E).

Postembryonic development (metamorphosis)

Many free-swimming planulae change from active swimmers to slow swimmers or even crawlers as they are near the time of settlement and metamorphosis. Settlement is not a random process, nor is it determined internally. It is believed that planulae receive an external stimulus that signals the presence of an appropriate habitat. If this stimulus is not provided, the planulae will not undergo metamorphosis, and eventually die. The bacteria of the genus *Alteromonas espejiana*, growing on the shell of molluscs inhabited by the hermit crab *Eupagurus bernhardus*, are the stimulus received by the planulae of *Hydractinia echinata* Fleming, a colonial marine hydroid, triggering the settlement on an appropriate site (Leitz, 1993). Some part of the settling process stimulates metamorphosis at once. How such bacteria stimulate settlement and metamorphosis of these planulae is not completely known. Leitz and others proposed that the biochemical mechanism for metamorphic activation of planulae involved a signal transduction pathway using a phosphatidylinositol (PI)-cycle. They suggested that bacteria somehow activate this signal pathway in a small group of neurosensory cells in the anterior region of the planula (Fig. 5.8C). This view does not contradict the previous hypothesis that neurons are involved in transmitting the metamorphic stimulus throughout the larva. Again, Leitz stated that although there are few data at present about the biochemical mechanisms of induction of metamorphosis by exogenous inducers, there might be some similarities in species of very different taxa, including sensing the cues with specialized receptor cells, in which phosphorylation

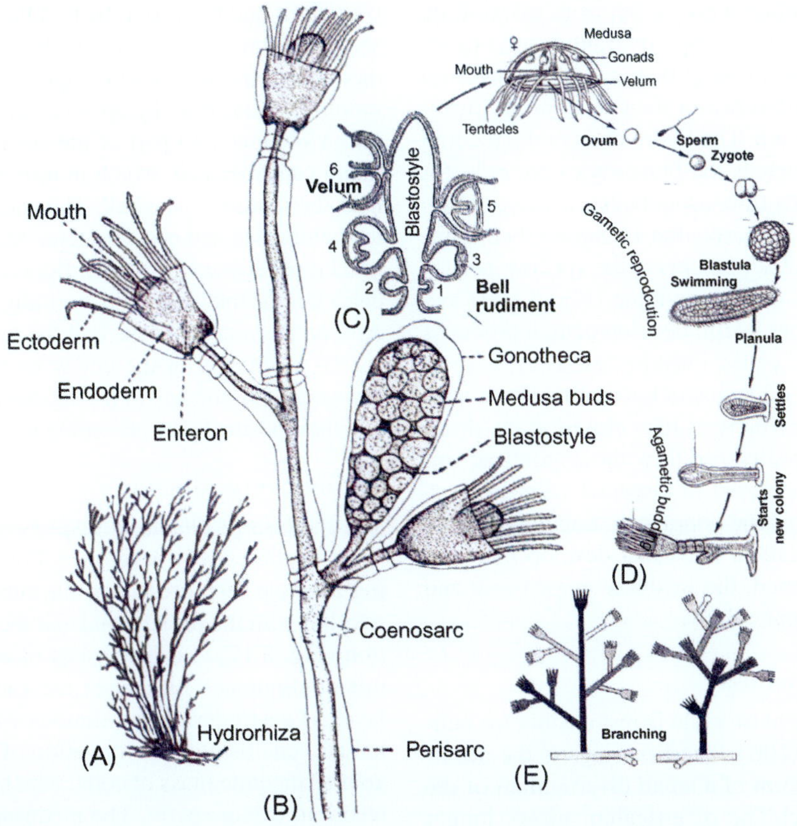

FIG. 5.10 Life cycle of *Obelia* showing the two generations and the two types of reproduction: (A) entire colony, life size; (B) branch of a mature colony showing the two types of zooids (hydranth and blastostyle); (C) stages of the formation of the medusa in the blastostyle; (D) gametic reproduction from the medusa until the planula larva, then agametic reproduction from the first polyp to that being mature and buds to form new colony; (E) hydroid growth patterns, black indicates the old polyp, youngest is clear, left-hand shows monopodial growth, the right-hand is the sympodial growth. *(Collected and adapted from various sources.)*

and depolarizations are part of the activated signal transduction mechanisms. The activated cells might then induce the intercellular communication and integration that is required for metamorphosis (Fig.5.8D and E).

When planulae undergo metamorphosis into polyps, some changes occur, such as:

1. the anterior (aboral end) becomes the base and stolon of the polyp;
2. the posterior (oral end) forms the head and tentacles of the polyp;
3. certain larval derivatives disappear;
4. new types of cells differentiate;
5. distribution patterns of cells change; and
6. Cellular migration patterns become more complex

Some representatives will be dealt with, in detail, to demonstrate how this phylum is very large and very variant, and includes very wonderful species. The development of a freshwater *Hydra vulgaris*, a marine hydroid *Tubularia mesembryanthemum* and an anthozoan *Nematostella vectensis* is presented in some detail to enable the reader to appreciate fully the beauty of their simplicity.

CLASS HYDROZOA

Origin of germ cells, sex determination, and sex inversion in medusae of the genus Clytia (Hydrozoa, Leptomedusae): The influence of temperature

Starting with *Obelia* (Calyptoblastea) and *Bougainvillea* (Gymnoblastea), both of these produce free-swimming medusae; *Tubularia* has sessile gonophores; Eudendrium, *Gonionema* and the freshwater *Hydra vulgaris* are without medusa. This series ends with Siphonophores.

Obelia geniculata

This species is a marine cnidarian which forms small, branching colonies attached to the surface of sea weeds, usually to *Laminaria*. Each colony (Fig. 5.10A) consists of a stem, or hydrocaulus, about 1 in. in length. It bears zooids and is attached to the weeds by a branched rooting portion, or hydrorhiza. The hydrocaulus, together with the hydrorhiza, forms the coenosarc. The individual zooids of the colony are attached alternately on opposite sides of the hydrocaulus. They are of two types: hydranths and blastostyles (Fig. 5.10B). Both are modifications of the hydroid type of zooid and the difference in their structure is correlated to the function of each. The hydranths are the feeding zooids of the colony, whereas the blastostyles are only for production of special buds (medusoid buds), which develop into medusa (Fig. 5.10C). These, after liberating their gametes in the sea, which helps to disperse the species, perish. The colony increases in size by budding (Fig. 5.10E) and when an erect stem reaches its full development, it produces special elongated zooids called blastostyles. They arise as buds in the axils of some of the lower hydranths. These buds grow to form a narrow cylindrical tube closed at the distal end. They are really simplified zooids without mouth or tentacles. A blastostyle is enclosed in a special vase-like portion of the perisarc termed the gonotheca. Each blastostyle gives rise to numerous lateral buds that develop into medusa. When fully developed, the medusae are set free and swim away from the parent colony.

Development of the medusa

As mentioned above, the blastostyle is responsible for budding the medusa (Fig. 5.10C). The first stage is the development of a bud in the form of a small diverticulum of the cavity of the blastostyle. The diverticulum grows longer and then enlarges at its extremity. It becomes a little vesicle enclosed by ectoderm, mesoglea and endoderm. It is connected to the blastostyle by a narrow stalk (Fig. 5.10C (3)). A cavity below the future subumbrella is developed in a peculiar way. The distal ectoderm separates into two layers, and the inner layer, in turn, acquires a cavity; this little sac is called the bell-rudiment (Fig. 5.10C (4)). The latter enlarges and assumes the shape of the subumbrella cavity with the manubrium in its centre. The cavity, therefore, although lined throughout by ectoderm, is at first closed by a layer of ectoderm. This ectoderm extends from one margin to the other. This is at last broken through and its remnant forms a circular shelf projecting inwards from the margin of the umbrella. This shelf is called the velum. The velum is permanent and conspicuous in most hydroid medusa, but in *Obelia* it diminishes and becomes quite vestigial.

When the medusa bud has acquired its final shape, the mouth breaks through at the extremity of the manubrium and finally the stalk is broken. The little fully formed medusa becomes free and escapes from the gonotheca through the aperture at its end.

Structure of the medusa

The medusa, a modified zooid (Fig. 5.5), has the form of a disc, which is concave on one side (inner surface known as the subumbrella) and convex on the other (outer surface, known as the exumbrella). The whole structure is referred to as a bell (Fig. 5.1B and C). From the middle of the concave surface, a short projection hangs; this is the manubrium. The structure then gives an umbrella-like appearance to the medusa. The mouth is situated at the apex of the manubrium, which leads into a small central part of the enteron. From the latter, four radial canals extend, which in turn open into a circular canal near the margin of the bell. The endoderm is not confined to the central sac and canal system, but is continuous as a fairly thin layer between the canals. Between the ectodem and endoderm lies the thick mesoglea forming the bulk of the organism. Around the margin of the bell the tentacles are found, which are 16 in number in the newly born medusa. Their number increases with the age of the medusa and becomes numerous and their positions are according to the radii of the medusa.

Gametic reproduction

The medusa of *Obelia* is responsible for this reproduction when becomes very developed. Four ovoid bodies are hanging at equal distances from the subumbrella, in immediate relation with the radial canal and therefore perradial in position (Fig. 5.1C). Each consists of an outer layer of epidermis continuous with that of the subumbrella, and an inner layer of gastrodermis continuous with that of the radial canal and enclosing a prolongation of the latter. There is also an intermediate mass of cells, which will become differentiated into ova or sperm. The medusae are dioecious.

Origin of the germ cells

Patches of the subumbrellar ectoderm are modified to form the gonads. The germ-mother cells originate in the ectoderm of the manubrium, pass through the endoderm and along the radial canals to the gonads and then migrate into the ectoderm again (Borradaile et al., 1977).

Development

When the gonads become ripe, the gametes are shed into the water by the rupture of the gonad and fertilization takes place. The zygote undergoes complete cleavage to form a hollow blastula. By the immigration of cells at one pole of the blastula (delamination), an elongated planula larva (Fig. 5.10D) with a solid core of endoderm is formed. It is ciliated and swims freely for some time, then settles down by its broader end. The other end develops a mouth and tentacles surrounding it. A cavity appears by splitting in the solid endoderm and the planula is easily converted into a single hydroid zooid. After swimming for a few days, the larva sinks to the bottom, drops its cilia and becomes dilated at one end, by which it attaches to stone or weeds. From the base of this first-formed polyp, there is an outgrowth along the surface of attachment, which is the beginning of the hydrorhiza. From this, the rest of the colony is developed. The opposite free end of the polyp enlarges to form the hypostome, on which the mouth will open. Then the circlet of tentacles grows out below the mouth. Over the whole exterior of the growing organism, a delicate pellicle, the future perisarc becomes secreted. At the hypostome edge, the perisarc ruptures and the tentacles become free. At the same time, the mouth opens and the creature becomes a single upright fully formed polyp.

From this first polyp, the branches develop through two different growth and budding patterns. Barnes (1974) described two patterns of growth. In **monopodial** growth, this first polyp continues to elongate as a result of a growth zone at the distal end of the stalk, or hydrocaulus (Fig. 5.10E, left-hand side). It may even lose its hydranth. As the primary hydrocaulus elongates, it also gives rise to new polyps by lateral budding. These secondary polyps may also elongate and produce buds, and the resulting colony may contain many individuals. In this colony, the main branch or axis of the colony is formed by hydrocaulus of the oldest polyp, and its hydranth, if still present, occupies the tip of the branch. The same is true of the secondary branches.

The colonies of other species of hydrozoans develop by **sympodial** growth. The primary polyp does not elongate, but produces one or more lateral polyps by budding (Fig. 5.10E, right-hand side). The secondary polyps do not continue to elongate either, but undergo lateral budding. In such a colony resulting from sympodial growth, the main branch or axis of the colony represents the combined hydrocauli of many polyps, and the youngest polyps are located at the tips of the branches. Thus, by budding, the colonial hydroid phase is produced and the cycle restarts.

Whether the colony develops by sympodial or monopodial growth, the form of the colony is quite variable, as seen in the figures. Most hydroid colonies are only a few inches or less in length. The colouration, which may be pink or orange, is usually not very striking because of the small size.

Eucheilota paradoxica

Eucheilota paradoxica Mayer, 1900 is a small leptomedusa (Calyptoblastea, Thecata) that occurs in the sea of Japan, and successively exhibits budding and reproduces gametically from late spring to early autumn. It was unknown in the Mediterranean Sea until 1977 (Carré and Carré, 1990). Since then, it has become endemic and occurs in great abundance in autumn in the bay of Villefranche-sur-mer (southeast France). Carré and Carré reported that no mature sexual stage has ever been collected neither from the Atlantic Ocean nor from the Mediterranean Sea. They presumed that the budding medusa of *E. paradoxica* had not yet attained the gamete differentiation stage at the time of collection from Villefranche-sur-mer. Therefore, they decided to investigate the still unknown life cycle of this species. They obtained the specimens from the bay of Villefranche-sur-mer from plankton hauls between 0 and 50 m depth during October and November 1986. The newly liberated medusa from the Mediterranean Sea displays the following characteristics: umbrella higher than one hemisphere (about 0.75 mm wide, 0.5 mm high) with cnidocysts on the exumbrellar surface; four radial canals; two diametrically opposed, hollow perradial tentacles, with large conical bulbs and with two short unequal lateral cirri each; two perradial hemispherical bulbs without tentacles but with two unequal lateral cirri; four long, interradial, full cirri with terminal knob (Fig. 5.11(I) A).

The authors established cultures and reared *Eucheilota paradoxica* to determine the mechanisms of such rapid adaptation to a new geographic area; to define optimal temperature conditions for agametic reproduction by medusa budding; and to determine the quantitative importance of this process in the settlement of *Eucheilota paradoxica* in the Mediterranean Sea. Their experiments were done on budding specimens collected from waters at 18–19°C, by varying the culture conditions and, especially, the temperature. One culture was incubated at six different temperatures (9°C 13°C, 15°C, 18°C, 21°C, and 24°C) in alternating 12 h light/12 h dark cycles. Another culture was incubated at 13°C in continuous darkness, i.e. under the conditions in deep Mediterranean waters. The medusae used for rearing measured an average of 1.8 mm diam. All were in a budding condition. Each had several buds, usually four of five, at different stages of development (Fig. 5.11(I) B). Over a period of 22 days, small medusae under different temperature conditions were liberated. The authors divided the results of their experiments into three phases:

1. **First** phase (days 0–10): at 18°C, 21°C, and 24°C, the liberation of medusae was substantial—426, 436, and 453, respectively—suggesting that at these temperatures, most of the induced medusa buds that were present at the beginning of the experiments continued to develop to the small free-medusa stage.
2. **Second** phase (after day 10): during this time, liberation of medusae ceased at 15°C, 18°C, 21°C, and 24°C in all cultures continued after day 10. However, under natural environmental conditions, budding appears to be a continued process, as all adult specimens of *Eucheilota paradoxica* found in plankton hauls were with more than one bud. The authors investigated that the resting-bud phase in their experiments, particularly at 18°C, was probably due to their culture conditions.
3. **Third** phase (after days 15–19): beginning at different times (days 17, 19, and 15 for the cultures maintained at 18°C, 21°C, and 24 °C, respectively). In each culture, a second liberation of medusae occurred. At 15°C, no new budding was observed.

During the time of these experiments, the mortality of the mother medusae was variable and the authors tabulated the daily budding rate of the surviving medusae.

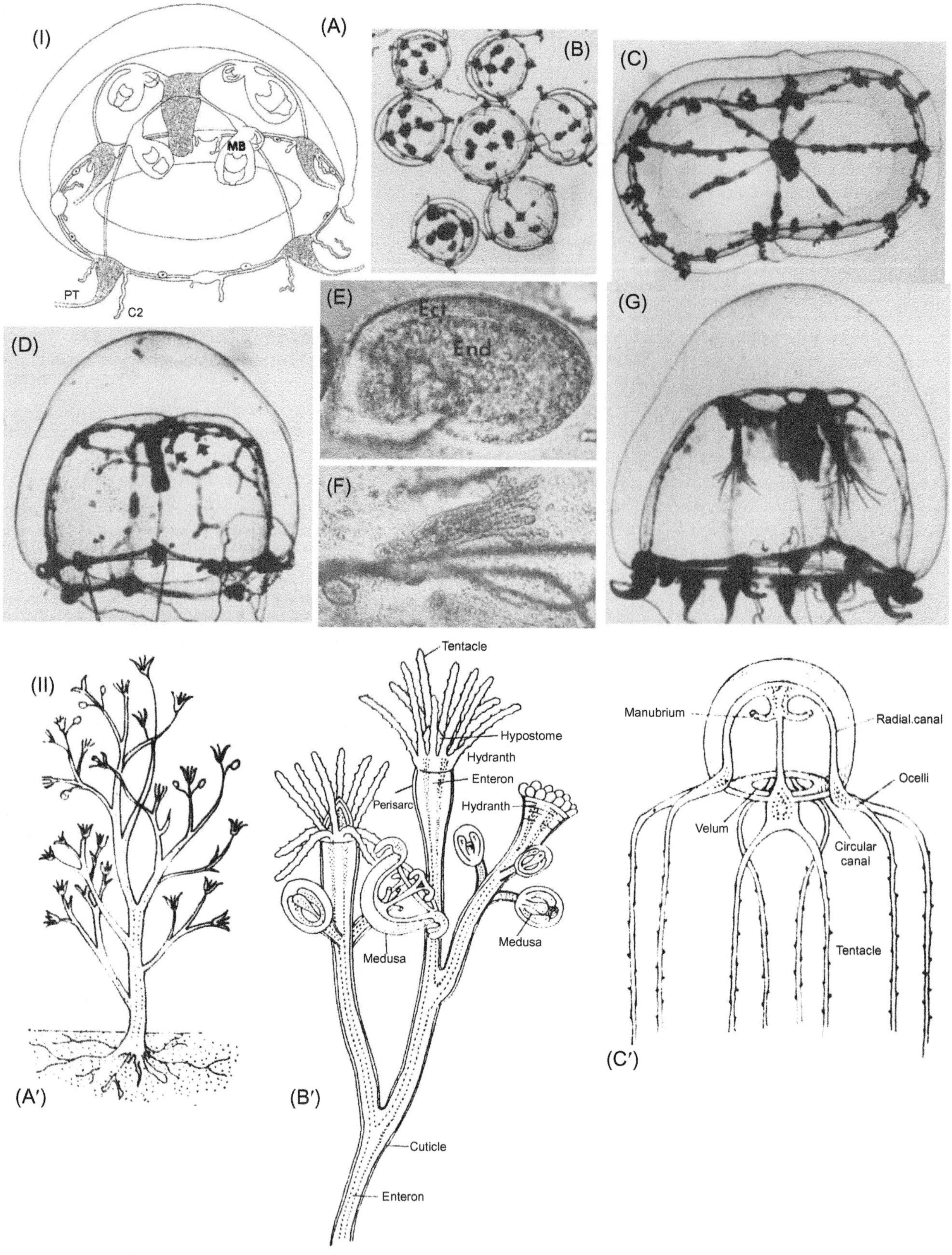

FIG. 5.11 (I) Leptomedusan *Eucheilota paradoxica* (Thecata): (A) diagrammatic drawing of adult medusa from Mediterranean sea, C2 = later cirri, MB = medusa bud, PT = perradial tentacle; (B) immediately after collection; all specimens including youngest bear several medusa buds (mean diameter = 1.74 mm) (stereomicroscope); (C) detail of specimen (4.2 mm diameter), showing emergence of four interradial canals. (Stereomicroscope); (D) medusa cultured at 17°C (3.5 mm diameter), small buds *(arrowed)* are forming along interradial canals and their branches (stereomicroscope); (E) frustule (0.25 mm long) maintained at 13°C; note didermal structure (Ect = ectoderm, End = endoderm) (differential interference-contrast microscopy); (F) polyp developed from a frustule transferred to 18°C (normal transmitted-light microscopy); (G) growth of polyp and progressive regression of medusa (stereomicroscope). *All from Carré and Carré (1990).* (II) Gymnoblastea *Bougainvillea ramosa* (Athecata), (A′) entire colony, natural size; (B′) portion of the colony magnified; (C′) immature medusa. *(From Marshall and Williams, 1975.)*

When cultures were maintained for over 1 month at temperatures >18°C, budding ceased. In all cultures, growth and morphological variations of medusae were observed with differences from the above-mentioned characteristics (Fig. 5.11(I) C):

1. the bell diameter increased from 1.5 to 4.5 mm, according to age;
2. relative flattening of the bell;
3. increased number of tentacles (from 4 to 20) and of marginal swellings (4–20);
4. increased number of statocysts (8–20) and of statoliths in each statocyst (1–4);
5. elaboration of new branches in the original canal system, i.e. formation of four centripetal interradial canals and irregular anastomoses between the four radial canals; and
6. irregular thickening along the greater portion of the canals

Formation of frustules, polyps, and persistent stages

At ~16–17°C, the subumbrella rapidly filled with abundant mucous secretions containing small detached portions (buds) of variable sizes. They were produced by short branchings on the radial canals and their ramifications, perpendicular to the subumbrella (Fig. 5.11(I) D). Histological examination of these portions revealed a cnidarial wall structure (Fig. 5.11(I) E). First, these fragments, in their mucus, were kept at 13°C (Mediterranean Sea temperature during winter) and rapidly became invested by microorganisms. Second, after 2 months no change occurred or degeneration, and so they were transferred gradually, still in their mucous envelope, to 18°C. Each fragment developed to a small polyp (Fig. 5.11(I) F). These fragments (buds) would seem to represent a persistent stage that is capable of resting at 13°C. It appears to be similar to frustules described in other cnidarians (see *Gonionmus* Fig. 5.14E). There is only a difference, that in *Gonionmus*, they are liberated from stolons of benthic colonies and not from medusal stages as in the present *Eucheilota paradoxica*.

The most surprising mode of agametic reproduction observed in *Eucheilota paradoxica* medusae was induced at temperatures close to 15°C in a culture maintained for 2 months. Only 12 medusae remained alive and all displayed. On each medusa, one or two polyps developed (Fig. 5.11(I) G) at the position usually occupied by the gonads and the medusa itself regressed. *Eucheilota paradoxica* is an example of a medusa budding polyps directly without sexual reproduction. The result is not proliferation of the species, but the occurrence of a benthic stage produced by a pelagic stage.

Bougainvillea fructuosa

This genus belongs to the Gymnoblastea (Anthomedusae, Athecata) as *Tubularia*, but resembles *Obelia* in having free-swimming medusa. The creeping hydrorhiza gives off branches (Fig. 5.11(II) A′), one of which becomes the first hydranth. Then numerous individuals are budded from it, most of which are polyps (hydranths) (Fig. 5.11(II) B′). These differ from the Calytoblastea in the fact that the perisarc stops short at the base of the polyp and does not form a hydrotheca. In addition, the medusoid individuals take their origin directly from the coenosarc. No blastostyle (modified polyp) develops in this genus. This condition, in which the polyp and medusa belong to the same grade of differentiation from the coenosarc, can be regarded as primitive, and that of *Obelia* as secondary. In the fully developed individual, the coenosarc gives off several buds, each as a simple bud within which a single medusa is developed (Fig. 5.11(II) B′). The latter emerges from the thin covering, breaks from its stalk and swims away. The medusa is conical or 'mitre-shaped' (Fig. 5.11(II) C′). Its walls are thick, owing to a great development of the gelatinous mesoglea of the exumbrella. The mesoglea of the subumbrella remains thin and the velum is considerably wider than in *Obelia.* The most important characteristics are the development of the gonads on the manubrium and the absence of lithocysts. However, sense organs are present as specks of red or black pigment at the basis of the tentacles. These ocelli are sensitive to light.

Campbell (1974) reported a very conspicuous phenomenon. The egg of *Bougainvillea multitentaculata* is invested with thousands of nematocytes, forming a solid layer over it (Fig. 5.7A). These presumably arise from the maternal tissue, and remain in place throughout cleavage. However, by the time of planula formation they have sunken down between adjacent ectodermal cells, in the histological position normal for nematocytes.

Tubularia mesembryanthemum

The development of *Tubularia* will be described as representative of a group, which lacks a free-swimming medusa phase and retains only the polyp form. *Tubularia* lives in shallow water on rocks and loose stones along the seashore, or attached to eel grass or other seaweeds (Fig. 5.12A). It is one of the species seen rather commonly on many coasts. The polyp reaches a height of 3–8 cm (Fig. 5.12B), and can be called one of the larger hydropolyps. Its embryology has been known for a very long time.

The reproductive cells are continuously produced agametically in gonophores formed on the polyp by budding. The gonophore does not separate from the polyp, but remains as a sporosac, which can be thought of as a sort of degenerate medusa (Fig. 5.12B). The gametes mature within these structures (sporosacs) and the eggs are fertilized within the female gonophores. The reproductive individuals of *Tubularia* originate from hollow branched structures springing from the polyp itself between the oral and aboral tentacles. Each polyp has several of these branches, and from each branch a number of reproductive individuals arise by budding (Fig. 5.12C). The branch is usually termed a blastostyle, although it is only a part of the polyp

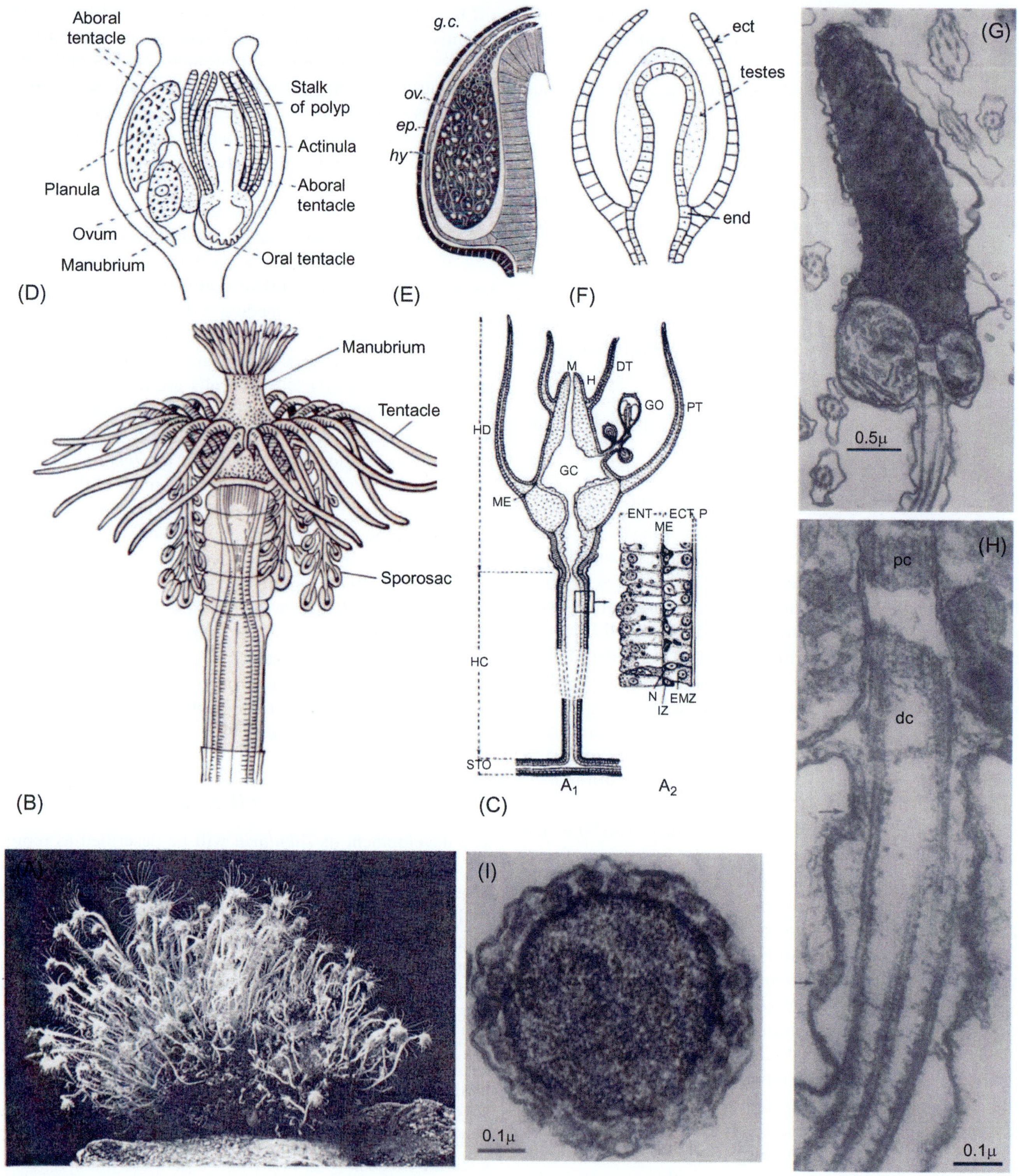

FIG. 5.12, CONT'D

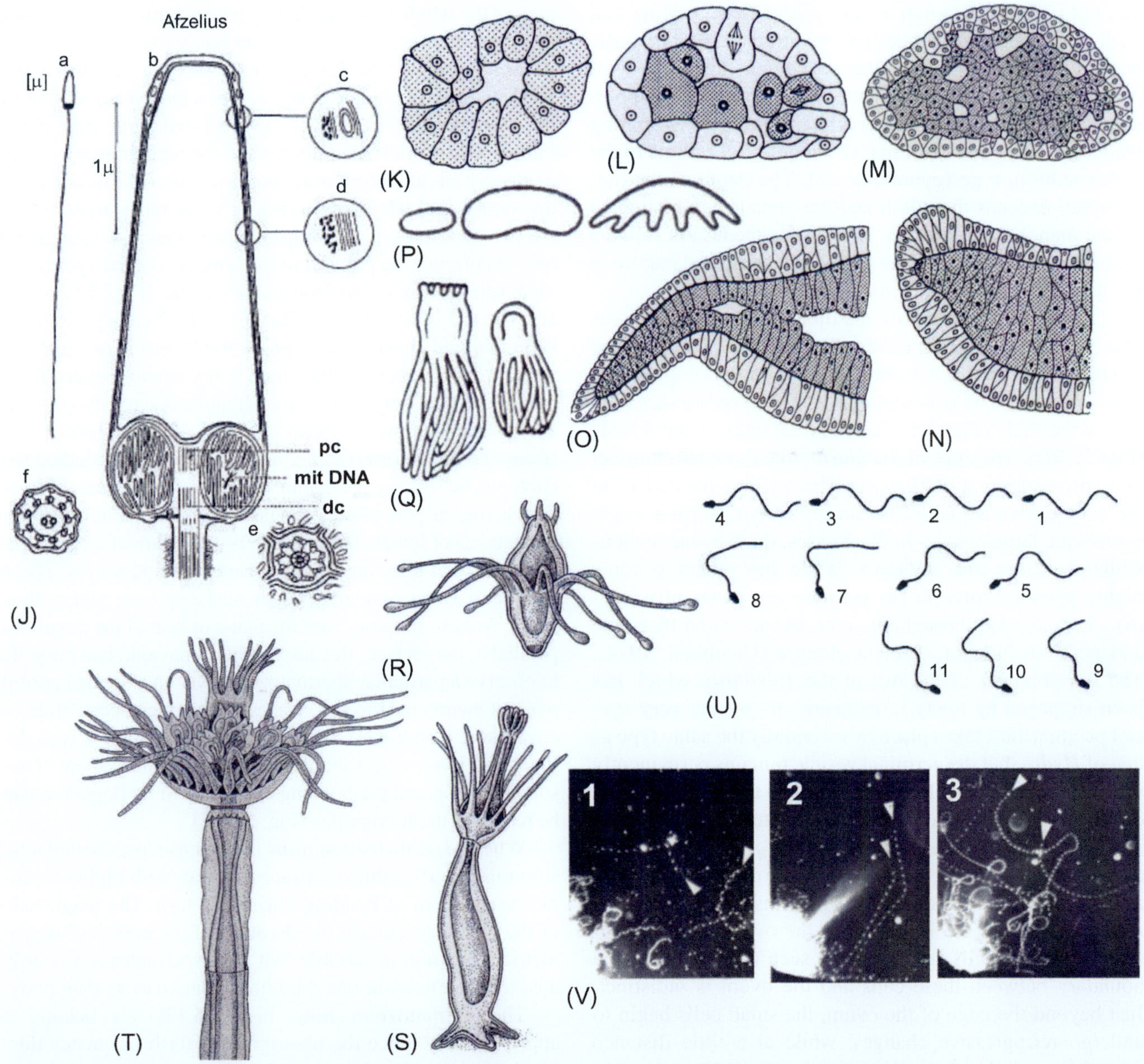

FIG. 5.12 Life cycle of *Tubularia mesembryanthemum*: (A) an expanded colony of the hyrozoan gymnoblastic *Tubularia*; (B) mature polyp; (C) anatomy of the polyp of *Tubularia larynx (from Tardent, 1963)*, DT = distal tentacle (oral), ECT = ectoderm, ENT = entoderm, ENZ = epitheliomuscular cells, GC = gastric cavity, GO = gonophore, H = hypostome, HC = hydrocaulus, HD = hydranth, IZ = interstitial cell, M = mouth, ME = mesoglea, N = nematocyte, P = perisarc, PT = proximal tentacle (aboral), STO = stolon; (D) L.S. through female gonophore with actinula larva; (E) female gonophore of *T.m.* containing one large ovum (ov) and a number of germinal cells (g.c.), ep = epiblast, hy = hypoblast; (F) L.S. through young male gonophore; (G) electron micrograph of L.S. of *Tubularia larynx* spermatozoon. Only a small portion of the tail is included (1 μm out of 40 μm). The nucleus appears dense and has indentations for the mitochondria of the midpiece, X = 37,000; (H) enlargement of a portion of (G) showing the distal centriole (dc), the proximal centriole (pc) and the proximal portion of the tail. Notice that a segment of the tail (between the *arrows*) is covered with a brush of fine fibrils, X = 95,000; (I) T.S. through the anterior portion of the head showing the granular appearance of the nuclear material in a narrow peripheral rim of cytoplasm, X = 100,000; (J) Diagrammatic drawing of the spermatozoon of *Tubularia larynx* ((A) The appearance of the living spermatozoon as seen with the phase contrast microscope (B) L.S. through the sperm head, midpiece and anterior portion of the tail. The proximal centriole (pc), the distal centriole (dc), and a bundle of mitochondrial DNA (mit. DNA) is shown. Approximate magnification X = 20,000 (C) A vesicle in the cytoplasmic rim covering the anterior portion of the nucleus (D) The membrane layers surrounding the main portion of the nucleus (E) T.S. of one of the centrioles (F) T.S. through the tail at its most anterior level) *((G–J) from Afzelius, 1971)*; (K–O) sections of blastula to gastrula with one pair of tentacles, show also the mode of gastrulation; (P) shows process of gastrulation in *T. indivisa*; (Q) two stages of the formation of actinula larva; (R) fully formed actinula larva (free living); (S) attached actinula (first polyp) *((P–S) from Uchida and Yamada, 1968)*; (T) mature polyp from which the colony (A) is formed; (U) sperm-tail patterns (this photographic sequence shows the development of asymmetrical beating and the return to normal beating throughout the turn at a particular phase of the beat cycle; (V) three photographs of sperm paths near micropipettes filled with hydranth extract. The micropipette is near the lower edge or lower left corner of each photograph. The flash frequency was 9 Hz. The *arrows* indicate the direction of sperm progression for paths. The direction is determined from the images of the sperm tails, which are visible in the original photographs and a few of these copies. *((U and V) from Miller and Brokaw, 1970), others are adapted from various sources.*

and not a modified polyp as the case of *Obelia*. However, each of the buds it produces has the structure of a medusa, but remains attached to the parent polyp as long as it lives (i.e. sessile medusa). Thus, the whole structure resembles a bunch of grapes. Like the structure of the free-swimming medusa of *Bougainvillea*, that of *Tubularia* conforms to the anthomedusan type (gymnoblastea). The depth of the medusa bell exceeds the width and the gonads being situated on the manubrium (spadix). This sessile medusa is called a 'gonophore' (sporosac). The gametes develop and mature in the gonads of these sessile medusae.

The species of *Tubularia* are dioecious, the gonophores of a given colony being either all male or all female. Very often, they exhibit a certain degree of morphological sex differentiation. The gonophore gives rise to sperm sacs (testes) in the male (Fig. 5.12F), or several eggs in the female (Fig. 5.12D). The eggs of *Tubularia*, like those in a number of hydrozoans (e.g. *Millepora*, *Hydra*), are formed from the oocytes that actually fuse successively to form a single symplastic large oocyte. In these cases, there is one nucleus which survives and enlarges, while the others become highly pycnotic (present but not live) and apparently inactive. They remain throughout embryogenesis and their fate and possible functions remain obscure (Campbell, 1974). The development of the ova of the *Tubularia*, which has been supposed by many investigators to present very special peculiarities, takes place by essentially the same type as that of *Hydra*, but the germinal vesicle remains permanently very small and difficult to observe. The mode of nutrition of the ovum may be very instructively studied in this type. The process is one of actual feeding; the germinal cells that do not become ova are assimilated by the ovum, much as an amoeba, which feeds on other organisms (Balfour, 1880). Adjoining one of the large ova of the ovary, a number of small germinal cells (g.c.) may be seen (Fig. 5.12E). The boundary between these cells and the ovum is indistinct. Just beyond the edge of the ovum, the small cells begin to undergo retrogressive changes, while at a little distance from the ovum they are quite normal.

The above description of the ova of the *Tubularia* is founded on the section of the gonophores of *Tubularia mesembryanthemum* (Fig. 5.12D). However, the absence of a distinct boundary between the germinal cells and the ovum is not usual.

The sperm of another species, *Tubularia larynx*, was studied by Afzelius (1971). The author collected his material in the neighbourhood of the Kristineberg zoological station at the Swedish west coast. The male gonophores were prepared for electron microscopical examination. The head of the sperm is conical, with a width of 1 μM at the base and a length about 3 μM and lacks an acrosome (Fig. 5.12G). The anterior 'surface' is flattened and the posterior end shows four shallow indentations for the four mitochondria of the midpiece, one of which is large. The nucleus occupies almost the whole head, with only a very narrow surrounding cytoplasmic layer. This cytoplasm has a finely granular appearance and the lateral sides have a layer of vesicles, about 10–15 vesicles, which can be seen in a transverse section of the sperm head (Fig. 5.12I). The most striking feature of the midpiece is the definite number and the regular arrangement of closely adjoined four mitochondria; three of this are about the same size, while the fourth is somewhat larger. There are two centrioles in the midpiece: proximal and distal. Both have an orientation parallel to the long axis of the sperm and their orientation seems unusually regular (Fig. 5.12G and H). The tail is 40 μM long. The set of 9 + 2 ciliary filaments is regular and constant throughout the length of the tail.

The eggs, which are large and yolky when liberated from the gonad, are fertilized in the subumbellar cavity (in situ) by spermatozoa, which swim out from the male gonophores (from another colony). Miller and Brokaw (1970) studied the chemotactic turning behaviour of *Tubularia* spermatozoan in the vicinity of gonophores or glass capillary pipettes containing extracts of female hydranths. The flagellum of a *Tubularia* spermatozoon generates nearly planar bending waves. These spermatozoa become trapped at surfaces over which they swim, without rotating, with the plane of beat of the flagellum parallel to the surface. Because of this behaviour, it is possible to observe a particular spermatozoon over an extended period while it swims next to the surface of the microscope slide or coverglass. Under these conditions, its orientation is well defined, and information about both the path of movement of the spermatozoon and the bending behaviour of its flagellum can be recorded in photographs (Fig. 5.12V).

When a spermatozoon turns in response to a chemotactic stimulus, the flagellum continues to beat, with highly asymmetrical pattern of bending during the turn. The magnitude of the turn, particularly the duration of the period of asymmetrical beating, is variable, but each spermatozoon is only able to make turns in one direction, relative to its own body.

The spermatozoon shown in Fig. 5.12U was beating at approximately twice the photographic flash frequency during the turn, so the photographic sequence shows the development of asymmetrical beating and the return to normal beating throughout the turn at a particular phase of the beat cycle. This photograph indicates that an asymmetrical pattern develops and is maintained for at least four complete beat cycles (positions 7, 8, and 9 in Fig. 5.12U).

The development of the fertilized egg occurs there, through the planula stage into an advanced larva called the actinula (Fig. 5.12R), which is really a polyp of *Tubularia* with a short stem.

Embryogenesis

Tubularia eggs are yolky, i.e. contain a large amount of yolk, and accordingly their cleavage is rather irregular. Furthermore, the presence of a number of eggs in one gonophore causes them to crowd each other (Fig. 5.12D) and so increases the irregularity to some extent. This type of egg of *Tubularia*, which cleaves irregularly, gives rise to multinucleate cells with the cytoplasm accumulated at the surface, while the yolk is left in the centre. This gives the impression or produces an effect like that of superficial cleavage. The ectodermal cells are arranged around the surface (Fig. 5.12K and L) forming the blastula (Reverberi, 1971), and few nuclei in the interior. The endoderm cells initially lie irregularly on the inner side of the ectoderm, but eventually they form a definite layer. This means that gastrulation occurs during cleavage stage, especially during the later steps of cleavage. In view of Reverberi (1971), young blastula at the 30-cell stage show incipient intoderm ingression by mixed delamination (Fig. 5.12L). Development of this leads to the solid 'pseudomorula' (Fig. 5.12M), which then flattens and forms the first pair of tentacles (Fig. 5.12N and O). These tentacles are of the actinula, grow and during which the gastric cavity is formed. However, another view described the process as follows: at about the 16-blastomere stage of the irregular cleavage, a small space is formed at the centre of the mass of the blastomeres. The embryo enters the blastula stage as the cleavage proceeds still further. This small blastocoel disappears and the embryo becomes a solid morula, as stereoblastula. This occurs by the repeated division of the blastomeres forming a part of the wall of the embryo, and the produced cells proliferate into the blastular cavity. The cells of this part of the wall later become the ectoderm, while proliferated cells that are now occupying the blastocoel become the endoderm. This means that a gastrula is formed by a process similar to delamination. In view of Campbell (1974), this is a secondary delamination present in the well-studied genus *Tubularia*. Afterwards, small empty spaces gradually appear within this interior mass of cells. The spaces become continuous with each other giving rise to a central cavity within the growing embryo and so this is a planula stage, and this cavity later is called the gastral cavity or coelenteron of the adult.

Formation of actinula larva

The embryo then flattens (Uchida and Yamada, 1968) or becomes somewhat compressed and appears roughly plate-shaped. A number of solid projections appear around the edge of this plate; these are the rudiments of the prospective aboral tentacles. As development proceeds, these begin to bend towards the concave side of the embryo and increase in length. At the same time, on the opposite convex side, the ectoderm and the endoderm of a definite portion of this side become thin and then a mouth opens. Gradually, the developing larva changes from plate-like to cylindrical form. Around the mouth opening at its upper end, a number of projections which appear to be the rudiments of the oral tentacles, and those of aboral ones become fairly long and slender. The aboral end of this stage is called an 'actinula' and so all the above described stages of development occur inside the gonophore until it reaches the actinula, which then escapes to the outside through the sporosac.

Actinula larva

Once the actinula larva becomes free, it immediately sinks towards the bottom, landing on a suitable substratum, where it spends a short time moving about by its tentacles, before finally becoming attached by its aboral end. The attached actinula increases in height and its part bearing the oral and aboral tentacles form the hydranth, while the attached part (aboral end) becomes the stalk (Fig. 5.12R and S). The latter secretes a substance forming a cover surrounding it, which is the chitinous perisac. At this stage, a new small hydropolyp of *Tubularia* is completely organized (Fig. 5.12S and T). From the attached base of this polyp, a number of stolons appear, and in turn they give rise to new polyps, forming a 'colony' (Fig. 5.12A). The hydranths of this new *Tubularia* would not yet have gonophores, being still sexually immature, and they would appear for the first time in the breeding season of the following spring.

The development of *Tubularia* is thus completed without inducing the free-swimming planula stage seen in the life histories of most other cnidarians, but with an actinula larva, which would be on its way to the adult state when it first escapes from the gonophores (sporosac).

As a rule, only one of three large eggs can be produced at one time and a ripe gonophore generally contains two larvae of different ages; one is a planula and the other an actinula, which may be seen protruding from the aperture of the bell (Fig. 5.12D).

Eudendrium racemosum (Cavolini, 1785)

This is a striking example of a bushy and quite branched colony (Fig. 5.13A). It is a species among the hydrozoons found in the red sea. Several Linnaean species were also found in the Mediterranean and in 1785, Cavolini reported on *Eudendrium racemosum* that it is probably the first species described as new to science from this sea. Schuchert (2008) reviewed all European hydroids belonging to the filiferan family Eudendriidae, occurring in the northeastern Atlantic and the Mediterranean Sea. Rarely, there is also a tentacle-like protrusion near the base of the hydranth, a nematophore (Fig. 5.13(a)). Male and female gonophores are normally produced in separate colonies, but a few species are hermaphrodites (e.g. *E. simplex*). The hydranth is provided with a single circlet of filiform tentacles (17–28) surrounding a funnel-shaped hypostome (Fig. 5.13B and C). Among the hydroids with sessile medusoids or gonophores, there are many forms in which the medusoid structure is lost. A bud-like structure is found, in which a transverse section shows simply successive layers of ectoderm with generative cells, structureless lamella and endoderm

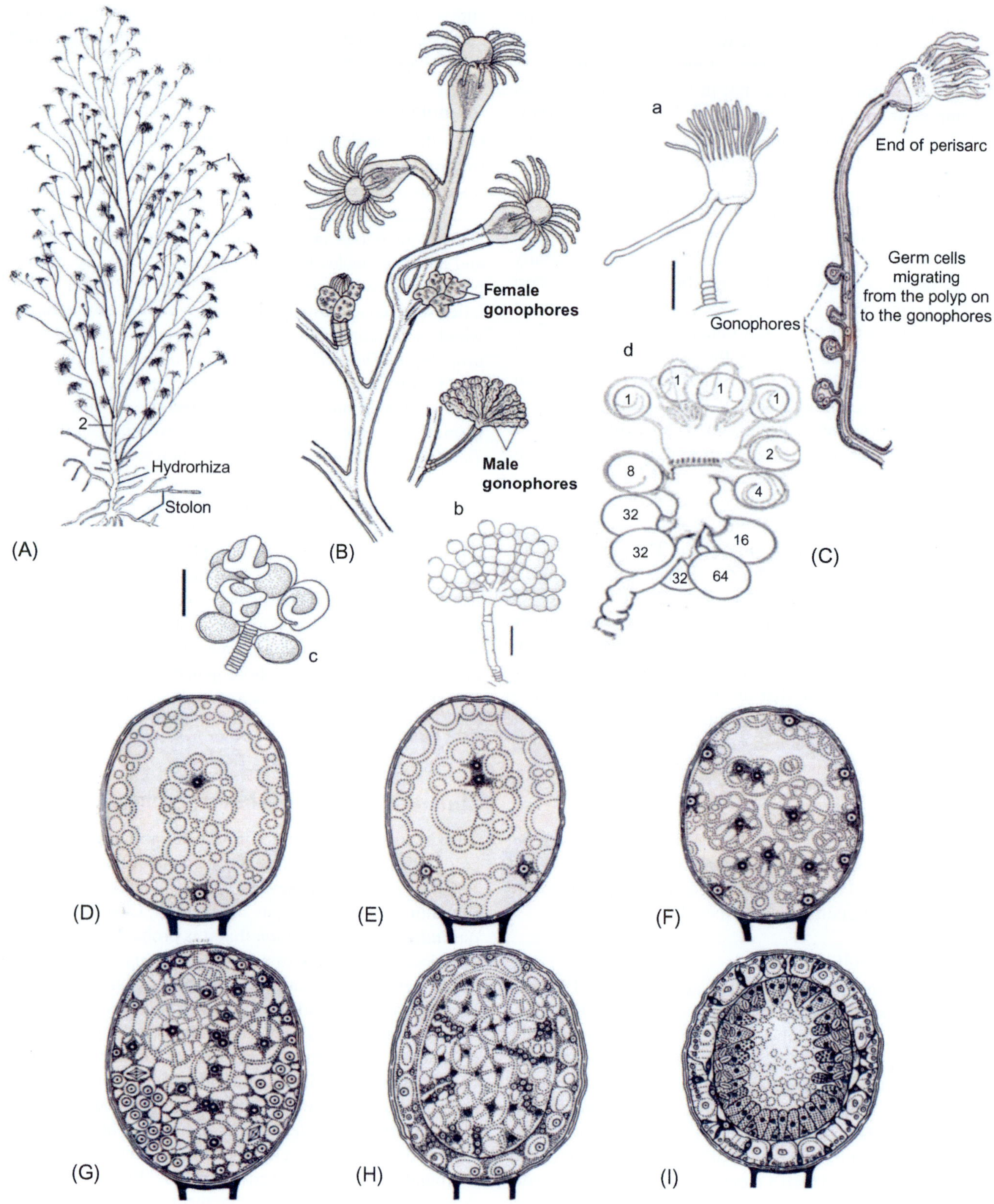

FIG. 5.13, CONT'D

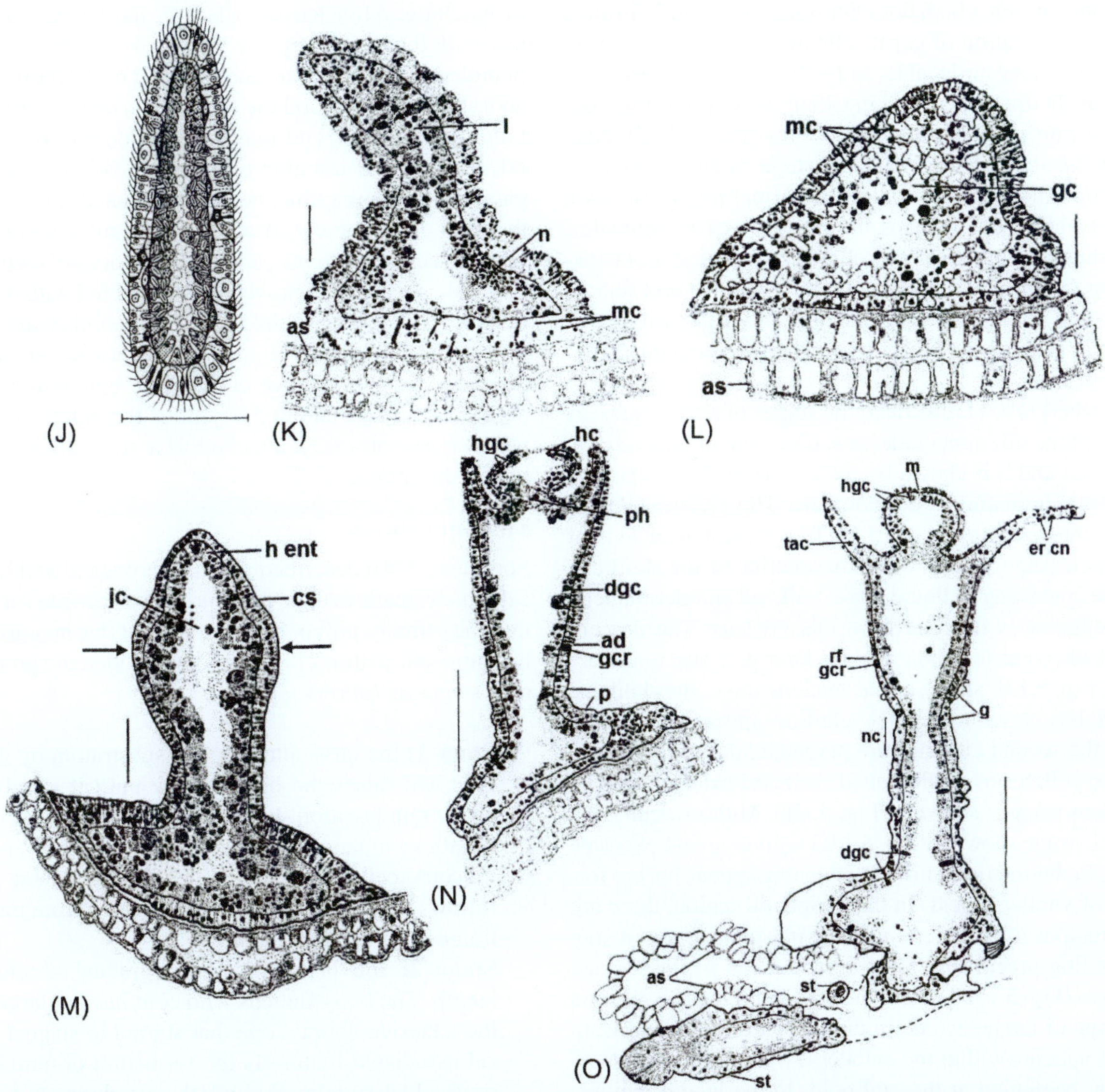

FIG. 5.13 Life cycle of a gymnoblastic hydroid *Eudendrium racemosum*, (A) portion of colony, from life; (B) part of the colony with female gonophore, beside which male gonophore is drawn (sketched from life) *(from Hickman, 1973)*; (C) one polyp showing formation of gonophores *(from Borradaile et al., 1977)*; (a)–(d) different polyps of *E. racemosum (from Schuchert, 2008)*: (a) hydranth with basal nematophore, (b) male blastostyle, (c) gonozooid with female sporosac, note bifid spadix (a–c, bar = 5 mm), (d) scheme of developmental stages on female blastostyle, the numbers indicate the number of nuclei during embryonic development; (D–I) sections illustrating cleavage to gastrulation, cleavage is syncytial: (D) 2-blastomere stage; (E) 4-blastomere stage; (F) 32-blastomere stage showing the yolk forms a ball around each cleaved nucleus; (G) multinuclear stage of late cleavage with incipient cleavage of yolk; (H) pre-gastrula stage in which the pre-ectoderm becomes delimited from the yolk spheres by formation of mesolamella; (I) represents old gastrula with differentiated cells; (J) primary polyp (planula larva), attachment of the fully differentiated larva at the aboral pole *((D–J) from Rverberi, 1971)*; (K–O) some selected stages of metamorphosis of *Eudendrium racemosum*: (K) the larva shortens gradually in length during this stage 1 of settling; (L) late phase of stage 2; (M) stage 6, the *arrows* indicate the position of tentacle budding (K–M, X = 50 μM); (N) stage 8 characterized by the differentiation of the hypostome; (O) stage 9 showing fully developed polyp of *Eudendrium racemosum*, scale of N and O = 100 μM, ad = annular depression, as = algal substratum, cs = constriction, dgs = digestive gland cell, gc = gastral cavity, gcr = gland cell ring, g = granula, hc = hypostomal cnidocyte, h ent = hypostomal entoderm, hgc = hypostomal granula cells, ic = i cells, jc = junction canal, l = lacunae in the gastral cavity, m = mouth, mc = mucous cells, n = nests of i-cells, cnidoblasts, and premature cnidocytes, nc = neck, p = periderm, pH = pharynx, rf = ring-furrow, st = stolon, tac = tentacular axial cell *((K–O) from Sommer, 1990)*.

around an enteron which does not open by a mouth. In such forms, the migration of germ cells mentioned as occurring in *Obelia*, is very noticeable. In *Eudendrium* the germ cells are often distinguished making their way along the coenosarc towards developing gonophores (Fig. 5.13C). Also in *Eudendrium*, an intermediate stage occurs. Medusae are budded off from the stalk of a normal polyp. As soon as this budding commences the polyp loses its tentacles, diminishes in length and considered to become a blastostyle (Fig. 5.13B and C, gonophores). The male and female gonophores arise from gastrozooides and differ from each other in appearance (Fig. 5.13(a) and (c)). No medusa generation occurs.

Reverberi (1971) illustrated the stages of embryogenesis of some of the different cnidarians. Cleavage in *Eudendrium* is syncytial and it is shown by sections (Fig. 5.13D–I). It is an extreme shortening of development. This cleavage begins as a two-nucleus stage (Fig. 5.13D); a daughter nucleus of the first cleavage wanders into the interior of the skull cap structure consisting of liquid phase yolk; all entoderm nuclei and vitellophages descend from this nucleus. The descendants of the outer daughter nucleus later give rise to the ectoderm. Fig. 5.13E shows a four-nucleus stage; the skull cap structure has closed to become a hollow sphere, the spindle axes of the second cleavage are perpendicular to that of the first. The yolk forms a ball around each cleaved nucleus. The 32-nucleus stage is shown in Fig. 5.13F. Multinuclear stage of late cleavage shown in Fig. 5.13G with incipient cleavage of yolk, in the interior, amitotic divisions appear for the formation of vitellophagess. In the peripheral region, there are accumulations of peripheral nucleus-plasma zones as a sign of beginning pregastrula formation. Section of the pregastrula stage (Fig. 5.13H) shows the peripheral nucleus-plasma zone stage of the preectoderm that becomes delimited from the yolk spheres within the entoderm by formation of mesolamella, beginning at the oral pole. In the lateral regions, the presumptive cnidoblast zone is prominent in the form of numerous nests of I-cell nuclei. The gastrula is histologically differentiated (Fig. 5.13I) into sensory nerve cell network, mucous cells and muscle cells in the ectoderm, and of the gastric epithelium, cnidoblasts in the entoderm and disintegrated yolk in the gastrovascular cavity.

At the time of hatching, the embryo has already attained the diploblastic organization characteristic of cnidarians. The embryo is surrounded by a peridermal envelope from an early stage of development, and hatching is initiated by a striking decrease in its thickness. This process leads to the formation of a circular opening at the animal pole of the egg-capsule, which corresponds with the aboral pole of the embryo. The opposite pole of the egg is fastened to the female blastostyle by a thick and multilayered periderm socle (Fig. 5.8B). Planulae hatch by leaving the egg aboral pole first.

Observations on living animals demonstrated that much secretion begins intensively even during the course of hatching. After leaving the egg, the larvae stay in contact with the empty capsule for some time. Newly hatched planulae are worm-like and about 0.6–0.7 mm long. The aboral pole is blunt and the oral pole is tapered (Fig. 5.13J). Later, it elongates and may reach a length of 1.3–1.6 mm after 1 day. It is not able to swim actively but rather float passively for some time, being 'distributed' by currents. At the time of settlement, it adheres to a substratum with the help of secreted mucous, and it glides forwards with the aboral pole ahead. This movement is coupled with a slow anticlockwise rotation of the body. The planula leaves a slime trail behind as that of a snail. This forward-movement is produced by the beating cilia of the ectoderm. Within a period of 2.5 h to about 1 day after hatching, most of the planulae are observed attached to a substratum and begin to metamorphose.

Metamorphosis

Sommer (1990) described the morphological and histological transformation of the planula of *Eudendrium racemosum* into the primary polyp. He reported that this metamorphosis is completed within 21–24 h and he divided the process into nine stages as follows:

Stage 1: the larva attaches to a substratum by its aboral pole and raises the body slowly until it stands free in an upright position. It adheres, at this step, probably by the sticky mucous secreted from the frontally condensed mucous cells (Fig. 5.13J). The permanent fixation follows by peridermal material secreted from the epitheliomuscular cells of the larva.

Stage 2: the larva undergoes gradual shortening in length. The body flattens with continued enlargement of the adhesive aboral zone that started in stage 1. A typical associated feature is the formation of annular folds or wrinkles running through the ectoderm (Fig. 5.13K).

Stage 3: the attached larva has flattened body to a maximum degree (Fig. 5.9). At about the same time, the enlargement of the adhesive disc stops and so the worm-like larva body changes into a lens- or button-like shape, with the adhesive zone representing the aboral end of the planula.

Stage 4: a blunt-ending column grows up from the centre of the flattened disc (Fig. 5.10).

Stage 5: the following slight constriction at the base of the column and a weak swelling of its distal part, thus dividing the body into three portions: the upper is the prospective head (=hydranth), the middle is the stem, and the lower is the foot of the developing primary polyp (Fig. 5.11). At the end of stage 5, the organism is completely surrounded by an acellular peridermal layer secreted by the cells of the upper side of the foot.

Stage 6: two morphogenetic events occur: the formation of an annular constriction, separating an oral cap (the future hypostome) from the lower hydranth body

(Fig. 5.13M), and the appearance of several vertical furrows that run through the hydranth wall and divide it into six or seven tentacle buds. The latter are well visible in transverse section (Fig. 5.13M). Most of the i-cells and cnidoblasts (Fig. 5.13M) are seen crowded in the upper basal disc.

Stage 7: the tentacle buds tips start to grow in an oral direction. New buds originate one by one from the same circle, that the tentacles become 9–13 in number.

Stage 8: characterized by the further outgrowth of the embryonic tentacles and the continuous formation of new ones that the number becomes 11–17. The base of the hypostome becomes fully constricted resulting in the separation of the oral (hypostomal) from the aboral gastric lumen by a tightly closing pharynx (Fig. 5.13N). This figure shows the developing polyp in which the mouth opening is about to be formed. Near the developing mouth, there is a new cell type in the endoderm, the hypostomal granular cell (Fig. 5.13N). Its cylindrical to cone-shaped form resembles that of the typical supporting cells of the hypostomal endoderm. It only differs in containing masses of minute acidophilic granules, which are mainly concentrated in the apical half of two-thirds of the cell body.

Stage 9: metamorphosis is completed. The tentacles grow to their final length of about three or four times that of Fig. 5.8. No new buds appear while the end number of tentacles in primary polyps varies from 13 to 19 (Fig. 5.13). During this process of tentacle, the mouth breaks open at the oral pole of the hypostome. This occurs apparently by rupture of the endodermal bridge described in stage 8. The most important change during this stage is the differentiation of the endoderm, which proceeds in a manner that enables the organism to change from lecithotrophic to planktotrophic nutrition. The hypostomal granular cells increase in number and become more widely distributed over the hypostomal endoderm than in stage 8.

Development of *Gonionemus depressum*

This species is common along the Pacific coast of Honshu, Japan. They can be collected through June by feeling for them among the eel grass or any marine plant, attached by adhesive pads on the tentacles. There are also *G. adriatic* (Fig. 5.14A) and *Cruspedacusta* (Fig. 5.14C); each is a polypoid stage, but this is tiny and solitary, and medusa bud off from the sides. Borradaile et al. (1977) noted that *Cruspedacusta sowerbyi* was first known from the Victoria Regia tank in the royal Botanic Gardens at Kew, but now has been discovered in various North American rivers and has even colonized ponds and canals in England. It has a polyp-like stage, *Microhydra*, which has a certain likeness to *Hydra*.

The gonads in *Gonionemus* are four in number; both male and female individuals have the gonads develop in the epithelium of the subumbrellar surface, and hang downwards as ruffles parallel to the four radial canals (Fig. 5.14B). Since the canals are close to the subumbrellar surface, the gonads are close to a nutrient source (Villee et al., 1964). They are somewhat different in colour, the testes being yellowish brown while the ovaries are dark brown. Since this species naturally spawns at night, one can obtain gametes of the medusa in the laboratory within 15 min after they are placed in a dark room, i.e. subjecting the animals to darkness.

The eggs are generally brownish, opaque, spherical and measure about 0.07 mm in diameter. They are released by the rupture of the ovaries, and fall to the bottom of the water where they are fertilized. Cleavage is somewhat radial, equal and total, but the furrows are unilateral, i.e. they begin at the animal pole and proceed to the vegetal pole. In the third cleavage, the furrow cuts equatorially, producing four upper and four lower blastomeres. The four upper ones are turned by 45 degrees and wedged by the four lower ones. This gives the pattern of spiral cleavage, but the embryo does not proceed as that mode of cleavage does. After the completion of the cleavage stage, the embryo enters blastulation, and then gastrulation. The differentiation of the germ layers is accomplished by the process of delamination.

During gastrulation, the ectodermal layer of the embryo acquires cilia and the embryo itself takes on the form of planula larva. By the cilia, the larva rotates within the fertilization membrane for about 1 h and then the membrane ruptures and the larva is set free. Then it elongates, swims with its broader end forwards. Twelve hours after fertilization are required for this creature to arrive at this stage, which is the free-swimming stage of development, at which time it attains a length of about 0.1 mm (Fig. 5.14D).

The larva leads this free stage for several days, and then sinks to the bottom, where it loses its cilia, but it does not develop directly into a medusa. It creeps about for a while and then attaches to some solid object before changing into a polyp. This first polyp (Fig. 5.14E) is tube-shaped with an outer epidermis and inner gastrodermis separated by a very thin meseglea. It develops a mouth opening, and then two tentacles are formed on opposite sides of the opening. Two more tentacles later appear beside the mouth opposite each other and at the same time at right angles to the first pair, making a total of four tentacles. The number of tentacles of the polyp may increase to five or six. During the four-tentacle stage, which is unusually small and squat, the polyp is only 1 mm in diameter. This tiny polyp reproduces agametically by budding. One side of the body thickens, becomes constricted as a separate tube and very slowly creeps away. These tubes, like planulae but lacking cilia (frustules), are budded off (Fig. 5.14E). This budding is repeated several times, and the frustules are believed to grow into polyps. The polyps then produce from one to several medusa buds, each of which develops radial canals, a manubrium, tentacles, statocysts, a velum and other structures. Finally,

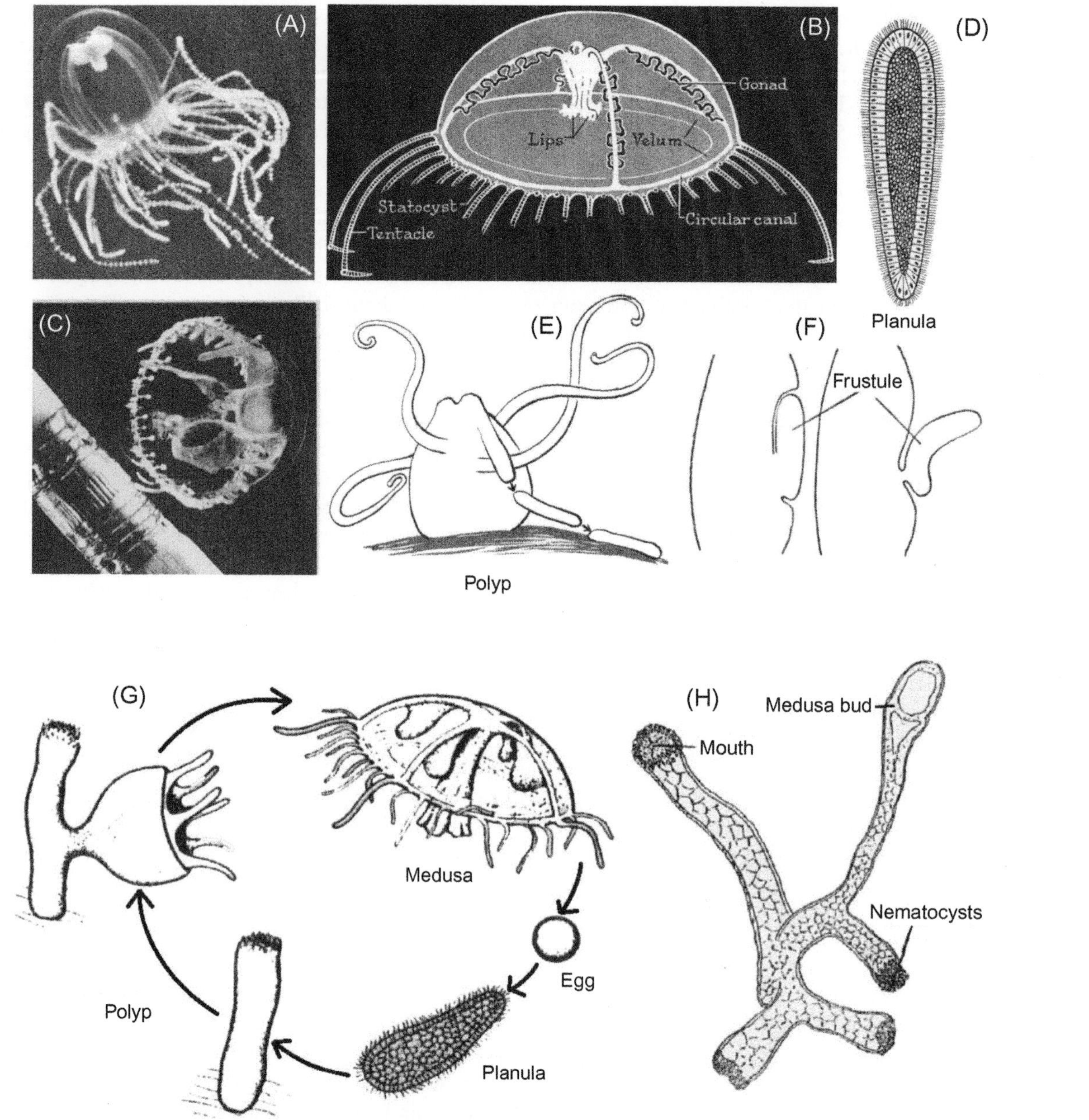

FIG. 5.14 (A) A hydrozoan medusa, *Gonionemus*, shows most of the features of the bell-shaped anthomedusae produced by many gymnoblastic hydroids; (B) side view of whole *Gonionemus* showing its structure, especially the gonads, tentacles incompletely drawn; (C) high-speed photograph of the freshwater medusa *Craspedacusta*; (D) planula larva of *Gonionemus*, (E) its polyp stage showing mouth and four tentacles. Notice frustule is formed on the side and the fully formed creeps away *((B, D, E) from Villee et al., 1964)*; (F) shows two stages of the formation of the frustule in *Microhydra* (hydroid stage of *Craspedacusta*); (G) life cycle of *Craspedacusta* (*Gonionemus* would be similar) *((C, G) from Barnes, 1974)*; (H) *Microhydra*, from life showing four polyps and one medusa.

the bud is set free as an independent medusa. The medusa, at this stage, has 12 tentacles and 8 statocysts, and these gradually increase in number (Fig. 5.14A). In Japanese species, the polyps of *G. depressum* live for at least 10 or 11 months.

The process of agametic reproduction has been followed in the other tiny hydrozoan, ***Cruspedacusta sowerbyi***. All the three types of buds are formed laterally near the mid region of the body of the polyp of this species (Gardiner, 1972). This polyp lives either as a solitary individual (Fig. 5.14G polyp) or as a member of a small colony of two to seven individuals (Fig. 5.14H), referred to as *Microhydra* (Hyman, 1940). The three types of buds, representing three types of morphogenesis, are formed successively. The hydranths are produced rapidly in the first 2 weeks. This young colony was observed again within the 12th–14th week. The frustules were produced most abundantly in the 3rd–6th week with a further increase in the 15th week. The medusae were produced in the 8th–16th week. The development of one kind of individual does not necessarily inhibit or prevent that of another.

Hydra vulgaris

The simplest members of the whole class (with the exception of one or two imperfectly known forms, which will be referred to below) are the freshwater polyps of the genus *Hydra* (found all over the world). The entire organism (Fig. 5.15A and B) consists of a simple cylindrical body with a conical hypostome and a circle of six or eight tentacles. It is usually attached to weeds, by a sticky secretion from the proximal end, but is able to detach itself and move about like a looping caterpillar. There is no perisarc and the tentacles are hollow communicating freely with the coelenteron. Both the body and tentacles are highly contractile.

Hydra has since provided much of the material for experimental investigations of growth and regeneration. Three regions of high metabolic activity, as measured by the relative capacity of the cells to reduce methylene blue,

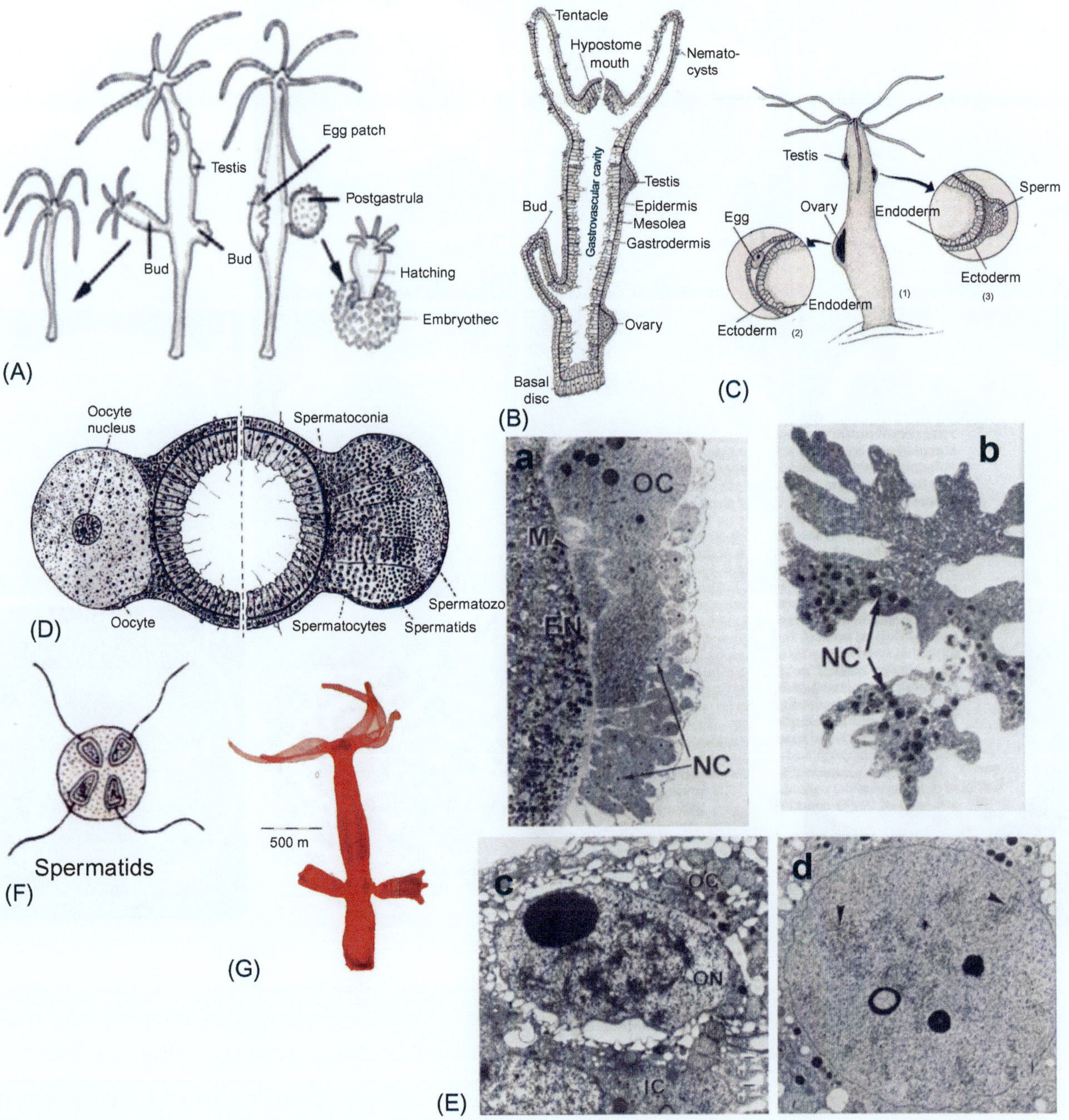

FIG. 5.15, CONT'D

(Continued)

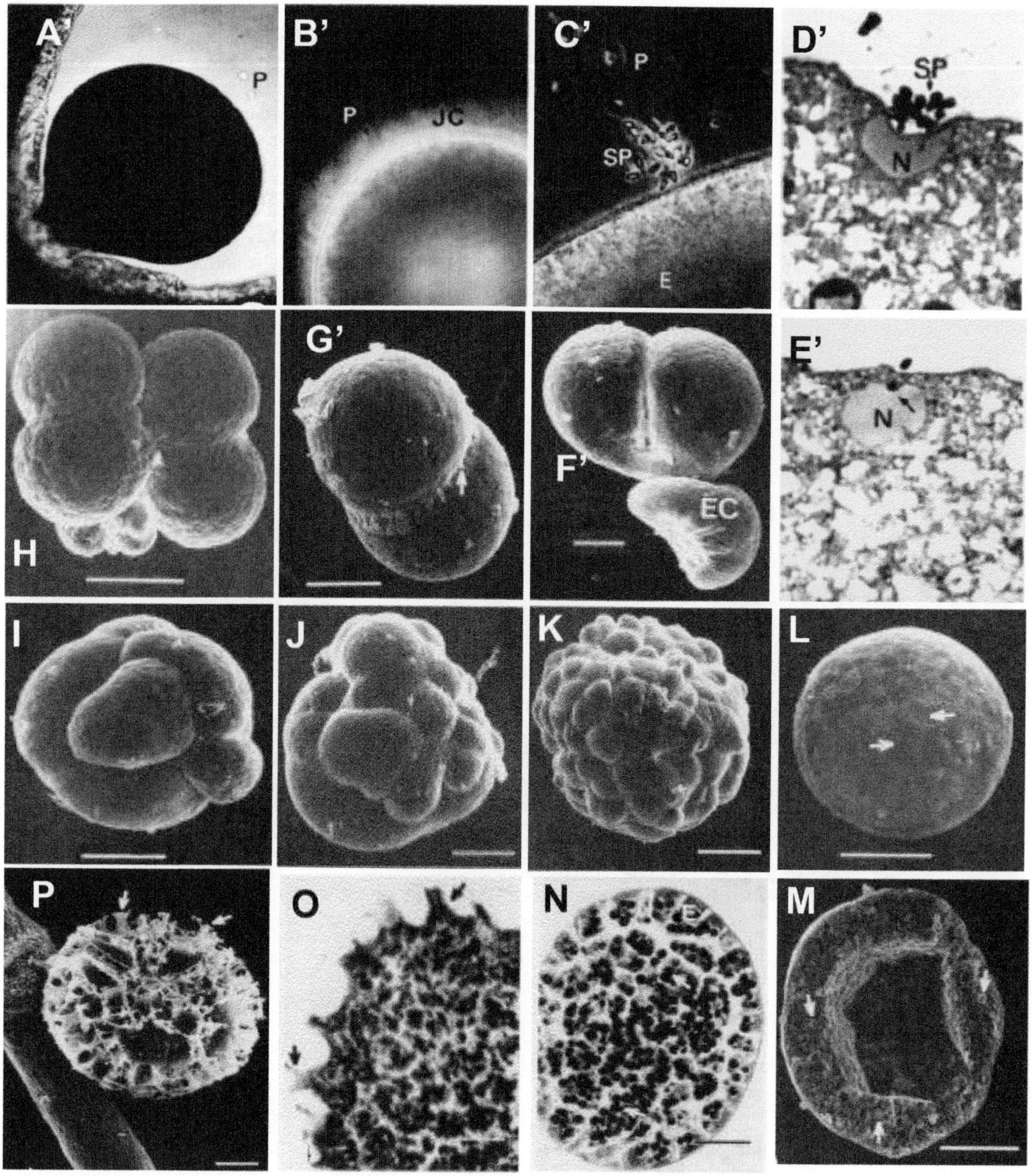

FIG. 5.15, CONT'D

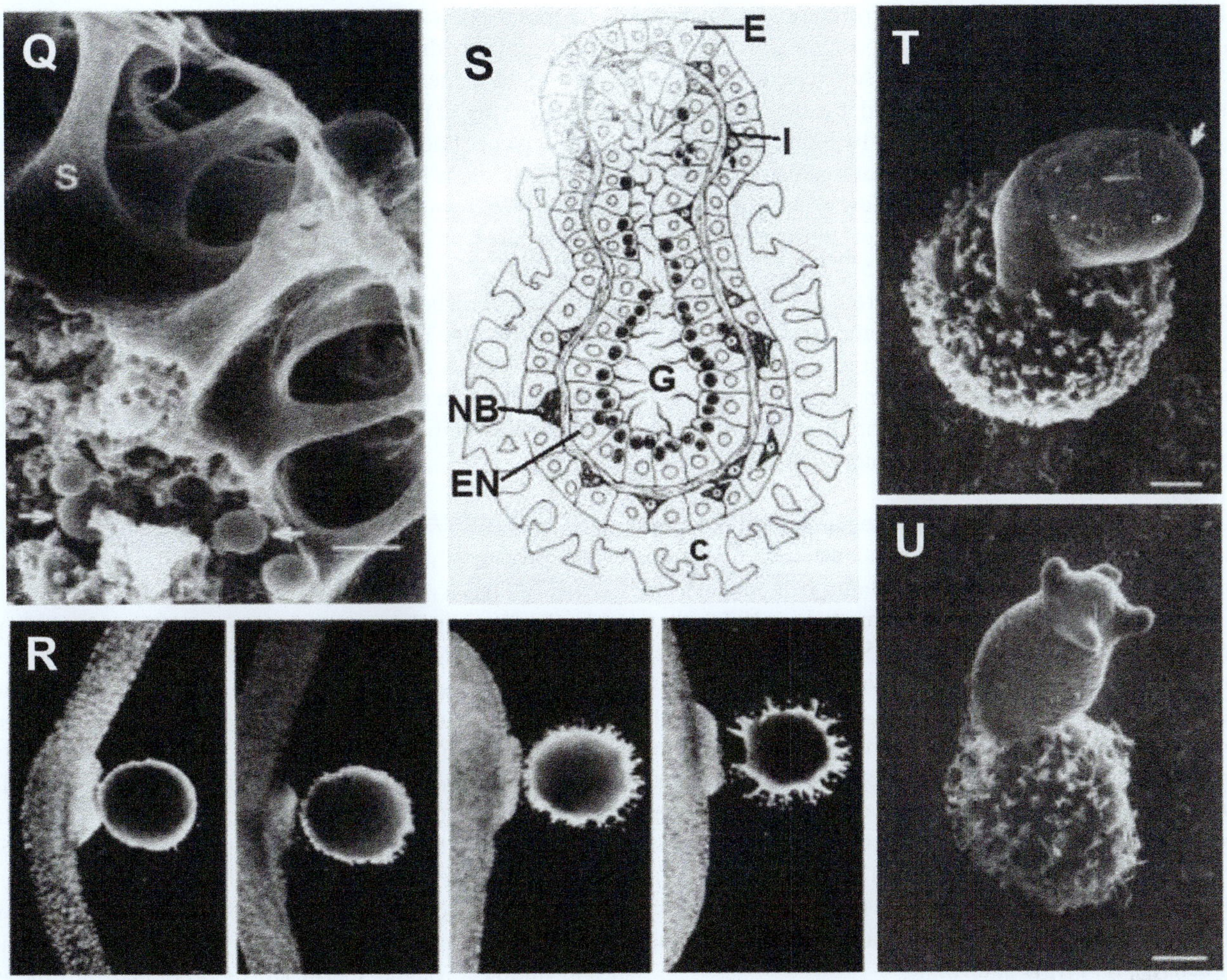

FIG. 5.15 Embryogenesis and development of *Hydra* sp. (A) whole mounts of dioecious *Hydra (from Technau and Steele, 2011)*; (B) L.S. illustrating whole mount of hermaphrodite *Hydra* sp. *(from Storer et al., 1983)*, (A and B) show the sites of gonads along the column of the animal; (C) drawing showing the initiation of gametogenesis in *Hydra*; (D) T.S. through the body of *Hydra*, showing the complete processes of gametogenesis, on the left through an ovary, while on the right through a testis; (E) development of *Hydra carnea* oocyte: (a) Egg forming region at early stage of differentiation showing the oocyte surrounded by nurse cells (NC). Endoderm (EN), mesoglea (M), X = 300, (b) semi thin section through isolated oocyte featuring vacuolated cytoplasm, incorporated nurse cells (NC). The centrally but more proximally located nucleus is not seen (cf. Fig. 5.8), X = 170; (c) oocyte (OC) at early stage of prophase I showing nucleus (ON) with dense nucleolus and chromosomes with synaptonemal complexes. Note the high amount of mitochondria and the highly vacuolated cytoplasm typical of the oocyte. IC, I-cell, X = 9700; (d) nucleus of oocyte in diplotene of prophase I with nucleoli *(arrowheads)*. Note the increase in size, X = 4700 *(from Honegger et al., 1989)*; (F) after the two maturation divisions in the process of spermatogenesis, four nuclei are produced but the cytoplasm does not separate and then each spermatid transforms into sperm *(from Storer et al., 1983)*; (G) *Hydra* with buds: on one side fully formed bud and on the other side it just begins, laboratory preparation; (A′–E′) photomicrographs illustrating the process of fertilization experimentally in *Hydra carnea (from Honegger, 1981)*: (A′) light photo showing 'released' egg, the process of release takes about 1 min (X = 100 μM); (B′) a 30-min old *Hydra carnea* egg in a suspension of indian ink to demonstrate the extent of the jelly layer (X = 150 μM); (C′) Phase-contrast photograph showing that sperm penetrate the jelly layer at a special site near the polar bodies (X = 300 μM); (D′) section of an egg fixed 2 min after insemination. Sperm lie densely packed in the jelly layer opposite the female pronucleus (X = 1000 μM); (E′) Section of an egg fixed about 3 min after insemination. A sperm *(arrow)* has fused with the egg and lies close to the female pronucleus (X = 1000 μM); (F′–M) cleavage in *Hydra vulgaris* egg: (F′) egg undergoing first cleavage, notice the distinctive heart-shaped created by unilateral furrowing (EC = egg cup); (G′) 2-cell stage, *arrow* indicates the microvilli in the region of the furrow which may help the attachment of cells together; (H) 4-cell stage; (I) 8-cell stage showing two unequal sized tiers of blastomeres; (J) mid-cleaved embryo, notice the unequal sized blastomeres; (K) late cleavage stage, the cells are more uniform in size and resemble a morula: (L) coeloblastula, *arrows* indicate the margins of cells; (M) T.S. through the coeloblastula showing nurse cells *(arrows)* within the one-layered wall; (N) gastrula of control embryo, a central mass of cells filled with darkly stained nurse cells *(arrows)* is surrounded by an outer epithelial layer (E); (O–Q) formation of the cyst; (O) 'cuticle' deposition in a control embryo, filipodia *(arrows)* extend from cells of the outer layer to the outside medium, cuticle material is deposited around the filipodia producing spines; (P) Hydra embryo with fully developed cyst, the cuticle material is ornamented by spines *(arrows)* (X = 100 μM for each photograph from F to P); (Q) enlargement of cuticular spine (S) and nurse cells *(arrows)*; (R) successive photographs of a *H. carnea* embryo 20, 22, 23, 24 h after insemination showing the formation of the species specific embryotheca, *from Honegger, 1981* (X = 50 μM); (S) emerging hatchiling, as the cuticle breaks open, the hatchiling exits head first. The cells of ectoderm (E) and endoderm (EN) expand and the gastrovascular cavity (G) increases in size, interstitial cells (I) are seen in great numbers in both layers. Clusters of nematoblasts (NB) are seen in the ectoderm; (T) Hydra hatchling emerging from the cuticle, *arrows* indicate the dome-shaped anterior head; (U) Hydra hatchiling showing four tentacle bumps (X for T and U = 100 μM) *((F–Q and S–U) from Martin et al., 1997)*.

are distinguishable in *Hydra*: the hypostome; the budding area, about midway between the hypostome and the basal disc (column); and the basal disc itself (Gardiner, 1972). These regions are also characterized by the presence of much interstitial and gland cells, and by high concentration of nutrients. Nutrition is important for agametic reproduction in *Hydra*. Well-fed *Hydra* buds every 3–4 days and in very well-fed hydras, two buds may develop simultaneously on opposite sides of the column (Fig. 5.15G). At times, multiple buds may arise simultaneously from the budding zone.

Of all the cnidarian species, different species of *Hydra* have been relatively studied in respect to reproduction. The majority of hydras are dioecious (Barnes, 1974; Villee et al., 1964), and some species are hermaphrodites. However, the general knowledge is that the freshwater *Hydra* is hermaphrodite (Fig. 5.15B).

Grassi et al. (1995) studied gametogenesis and development in *H. vulgaris*, a strain whose members are functional simultaneous hermaphrodites throughout their sexual periods, which however, was for particular clones, while others have remained sexually inactive for >5 years. In this species, the **testes** are conical protrusions provided with an opening, the nipple, through which the sperm exit (Fig. 5.6A (a1)); these testes are found along the entire upper body column from beneath the hypostome to the budding region (about the middle third of the column). **Ovaries** appear as large ectodermal bulges in the lower position of the body column near the budding region or just behind it (Fig. 5.6A (c 4, 5)). Each ovary contains one or more developing eggs covered by a thin layer of flattened ectodermal cells (epitheliomuscular cells) (Fig. 5.15D and E).

Gametogensis is induced in *H. oligactis* and *H. hymanae* by lowering the temperature, and in *H. magnipapillata* by decreasing food intake. Within the testes, distinct regions of maturation are found, with the young cells, the spermatocytes, lying closest to the mesoglea and the mature sperm located just beneath the ectodermal cells covering the testes. In *Hydra vulgaris*, as long as the testes persist, sperm are continuously released. The sperm of *Hydra* is the typical cnidarian type; it consists of a conical head with no distinct acrosome, a middle piece with five concentrically arranged mitochondria of equal size and a flagellar tail measuring about 30 μm in length. Once released from the testes, hydra sperm remain active for about 3 h (Fig. 5.15F).

During egg formation, one cell of the cluster that proliferated from the egg lineage stem cells becomes the oocyte (Fig. 5.15E), while its daughters begin to synthesize the yolk. These are called nurse cells. The oocyte increases in size by extending finger like processes in between the epitheliomuscular cells, which fuse with and/or engulf large numbers of nurse cells. As the oocyte grows, it assumes a flattened appearance and then gradually becomes spherical and so extending the overlying ectoderm. During the spherical shape of the egg, it undergoes meiosis and hence the extension of the polar bodies. The overlying ectoderm in the region of these polar bodies rupture exposing the egg to the medium and at the same time it slopes back around the egg forming a raised ring of remnants at the base of the egg referred to as the egg cup. Each mature egg then measures 100–500 μm in diameter and is loosely attached to the egg cup. This egg can be fertilized within 2 h; if it is not, it swells and degenerates within the day. The unfertilized egg of *Hydra* is surrounded by a jelly coat, and numerous short microvilli project from the surface. The egg pronucleus is found just beneath the egg membrane in the region of the polar bodies. Thousands of dead nuclei or remnants of the nuclei of the nurse cells are scattered in the cytoplasm. These remnants persist in the developing embryo until hatching.

Fertilization (*Hydra carnea*)

The portion of the egg surface at which the polar bodies are extruded has a minute cup-like depression referred to as the fertilization pit. Sperm penetration is restricted to this particular region of the egg surface (Honegger, 1983). It was observed that as many as 50 sperm can enter the fertilization pit, but only a single sperm fuses with the plasma membrane of the egg. The time required for a sperm to penetrate the jelly coat and fuse with the egg membrane is extremely short: about 30–40 s. Neither microvilli nor a fertilization cone is formed at this site of fertilization. In *Hydra carnea*, electron-dense material is released by the egg upon insemination, but no cortical granules are discharged and no fertilization envelope is formed (Honegger, 1983). Both the head and midpiece of the sperm are incorporated into the egg. After insemination for 8–10 min, the fusion of the male and female nuclei completes their fusion, and after 15 min, the fertilization pit disappears.

Honegger (1981) experimentally studied the sperm-egg interaction and fertilization of *Hydra carnea*. When a freshly 'released' (mature) egg (Fig. 5.15A′) is placed in a culture medium containing sperm, the sperm are observed moving over the jelly layer of the egg within a few seconds, and within a minute or two the egg is surrounded by a dense layer of these sperm. This picture is not found in the other regions of the egg. These observations strongly suggest that hydra sperm are attracted to the egg, as happens in many marine Cnidaria. Some of the motile sperm immediately penetrate the jelly layer in the region where the polar bodies are located (Fig. 5.15C′ and D′), and where directly under the egg membrane the female pronucleus is situated. No micropyle was observed in the jelly layer. Instead, a slight cup-like depression of the egg membrane and a circular indentation in the jelly layer are found in the same site. In addition, the sperm penetration is restricted to that particular region of the egg surface (Fig. 5.15D′).

The hydra eggs are fertilized by a single sperm, 1–2 min after being exposed to sperm culture medium. After about

3 min, the sperm has penetrated the jelly layer and fused with the egg membrane and is found lying close to the female pronucleus (Fig. 5.15E′). Ultrastructural and cytochemical investigations to clarify the interactions of the sperm with the jelly layer and egg membrane are in progress.

Cleavage and blastulation

Martin et al. (1997) described cleavage in *Hydra vulgaris* as holoblastic, radial, equal and synchronous, especially the first two cleavages, and it then becomes somewhat unequal and asynchronous. It is regular cleavage since it proceeds in the steps of typical radial cleavage 2, 4, 8, 32, 64 until the sixth cleavage (Fig. 5.15F′–M′). In hydras specifically, the cleavage is unipolar, meaning that the cleavage furrow proceeds inwards firstly from one side of the egg, usually at the animal pole, and then extends inwards across the cell. This is because of the specific site and attachment of the egg in respect to the column of the mother. This egg has a base attached to the mother with the egg cup and a naked opposite end, the apical end at which starts each cleavage. Up to the 32-blastomere stage, cleavage furrows occur roughly every 62 min, then the rate of cell division slows down until the formation of the blastula. During these steps, the blastomeres tend to become uniform in size. A coeloblastula is formed 6–8 h after fertilization (Fig. 5.15L′ and M′). The wall of the blastula is composed of approximately 76 rectangular cells arranged in one layer enclosing a fluid-filled blastocoel. The outlines of the cells are clearly distinct by rows of short microvilli.

Gastrulation

Shortly after the completion of the blastula, gastrulation begins. It occurs by multipolar ingression. Ingression begins at the apical end of the growing embryo, opposite to the egg cup, and progresses outwards from that point all around the embryo. Within about 4 h, the blastula develops into a gastrula, which then consists of an outer layer of columnar cells enclosing a central mass of unorganized cells (Fig. 5.15N). In the complete gastrula, the nuclei of the nurse cells are concentrated mainly in the central cells. At this stage, no cavity appears. Although the gastrula becomes somewhat smaller, yet the number of cells increases to about 3–5 cells. Gastrulation is completed 10–12 h after fertilization or 2–4 h after the blastula is formed. Afterwards, little activity is visible for the next 10–12 h.

Cyst formation

The gastrula remains as such for about 20–24 h when it starts to form a cyst, a thick protective layer surrounding it, which Martin (1997) referred to as 'embryotheca'. The cells of the outer layer of the gastrula extend fine filopodia into the surrounding medium (Fig. 5.15O) and then secrete a material deposited around the filopodia and in the apical substances of the cells themselves. This secretion starts from the region of the egg cup, and slowly progresses around the gastrula until it is wholly enclosed in this material, then a second very thin layer is deposited by the same cells beneath the former one, and thus the wall of the cyst is formed (Fig. 5.15O–Q). No information about the nature of this material is given. However, in previous works, the process occurs within about 20 h after complete gastrulation. By this time, the secretory cells flatten against the wall of the cyst. Three hours later, the material is built up to the point that the spines are visible. Over the following 18–24 h, material is continually deposited, producing a multilayered structure over the surface of the embryo, ornamented with spines. After the cyst wall is completely formed, the embryo, protected by the embryotheca, remains attached to the polyp (Fig. 5.15P) for a few days.

Honegger (1981) described the same process of cyst formation in the embryo of *Hydra carnea*, and he summarized the process in Fig. 5.15R.

In *Hydra fusca*, the development of the egg-capsule takes place differently. Korotneff (2009) investigated that here the mucous layer is formed by the parent animal itself. A perfectly smooth egg-capsule soon makes its appearance; at the same time, the 'epiblastic' nuclei shrivel up. In a short time we see that the 'hypoblastic' as well as the epiblastic cells form a common mass; the nuclei of the cells become pale and afterwards disappear, without inducing a degeneration of the cell-body itself. At this time, a vitelline membrane is secreted. After the setting-free of the primary hypoblastic cells, a division takes place in them; the newly formed cells then migrate to the periphery and constitute the definitive epiblast. Therefore, as regards *Hydra fusca*, it is established that the cell-body of the epiblastic elements is not entirely exhausted by the secretion of the egg-capsule, but, after a conversion, also takes part in the plastic development of the germ.

In *Hydra vulgaris*, the encysted embryo remains attached to the mother for few days, and then falls off and sinks to the bottom. The cyst detaches from the mother and the embryo enters a dormancy period during which the interstitial cells appear in between the outer cells (ectoderm) (Fig. 5.15S), but the rate of division of cells decreases and the wall of cells does not change. After the dormancy period, and 2 days before hatching, the outer layer of the cysted gastrula consists of low columnar epithelial cells, large numbers of interstitial cells, a few nematoblasts and some neurons. This composition closely resembles the ectoderm of the adult. The formation of this ectoderm precedes the formation of the endoderm. The cells of the central mass arrange themselves such that they form a layer lining the ectoderm and they become columnar rather than spherical. Cells that are not incorporated into the endodermal layer degenerate, producing an internal cavity that is the future gastrovascular cavity of the adult. Interstitial cells appear in the spaces in between the ectodermal cells. After the completion of the formation of the two layers, mesoglea is synthesized and the organism is ready to hatch. At this stage, the rate of division of cells increases and the interstitial cell lineage contributes about 40% of the total number of the organism.

Hatching of embryo

Martin (1997) and Martin et al. (1997) described this event after the embryo attains the bilayer form and begins to pulsate in a rhythmic contraction and relaxation. This pulsation continues until the hatching is completed. About 48 h after the formation of the bilayered embryo, the cuticle cracks open at its thinnest side, the original apical side of the embryo, where the head forms. The spherical embryo emerges from the cuticle through this crack (Fig. 5.15S). In the next 2.5 h, it changes from a compact sphere to an elongated cylinder. During this time, the hatching embryo is enveloped in an acellular accessory layer, which eventually ruptures as the hatching increases in length. Before the complete rupture of the accessory layer, the apical end of the hatching has the shape of a smooth dome (hypostome) (Fig. 5.15T). Within 15 min of rupture, the apical end narrows into a more conical shape, and one to five tentacle buds invaginate in a ring below the tip (Fig. 5.15U). At this point, the head of the hatching is morphologically fairly complete. The foot is also quite complete since, upon removing itself from the cuticle, the hatching is able to stick to the available surface (Petri dish), which is a sign of a functional foot.

The hatching has a full complement of the somatic cell types that an adult normally has. The number of cells doubles from about 5000 to 10,000 cells as the embryo progresses from the bilayer stage to the hatching stage. This increase is entirely due to increase in the cell populations of the interstitial cell lineage. The epithelial cell numbers do not change during the last two stages. The hatching has the capacity of regeneration as does the adult. In addition, the young hydra bud within 10 days of hatching and become mature (sexual) 4–8 weeks after hatching.

Within a 15-min period shortly after hatching, several small tentacles appear, evenly spaced in a ring in the tentacle zone of the head, directly below the hypostome or mouth region. The mouth has also formed by the time of hatching, as it is capable of opening and ingesting food shortly after hatching. Finally, as soon as the hatched organism leaves its cuticle, it is capable of secreting sticky material that allows it to adhere to the substrate, indicating that the foot is fully formed.

General development of siphonophora

Siphonophora (Fig. 5.16) is one of the specialized orders of Hydrozoa. The siphonophores are a group of about 170 described species of pelagic hydrozoans. Siphonophora are colonial animals, and have the most complex colony-level organization of all animals. They are arguably the most complex of all colonial animals. They exhibit the maximum development of polymorphism found in the Cnidaria or indeed in any group of the animal kingdom. Siphonophora include the longest animals in the world. Each animal or colony arises by a highly regulated budding process that arranges polyps and medusae in a precise, species-specific pattern (Dunn, 2005). These polyps and medusae, which are also called zooids, are physiologically integrated and fall into discrete functional categories. They are adapted for pelagic life (Figs 5.9E, F and 5.16). There are various constituents making up the colony, among which are those that develop the reproductive organs forming the gonophores. Most of these reach a degree of development similar to the sporosacs of the Hydrozoa, and the gonads forming from the part corresponding to the manubrium. The gonads of a single gonophore are either male or female, but the colony as a whole may have either gonophores all of one sex, or of both sexes, depending on the species.

In general, the eggs cleave totally and equally to form a morula, gastrulation usually takes place by secondary delamination (Campbell, 1974); the steriogastrula develops cilia on its outer surface and forms a planula, the shape of which may be spherical, oval or a modification characteristic of its species. Development after the planula stage varies to some extent from one species to another.

In the suborder *Calyconecta*, including the genera *Muggiaea*, *Diphyes*, *Hippopodius*, and others (Fig. 5.17 A(d)), the planula is solid and ovoid, with a ciliated surface. This ovoid planula buds from one side of its upper part a bell nucleus, which will become the under surface of the primary swimming bell tentacles that form on the lower part of this, and at the opposite side (lower) end of the planula, the primary gastrozooid is formed. When the primary swimming bell is completely developed, a stem grows out between the bell and gastrozooid, and buds extensively to produce the various other individuals, eventually forming a large colony.

In the suborder *Physonectae* (*Agalma, Physophora, Halistemma* (Fig. 5.9)) (Fig. 5.16B), the ectoderm cells on the upper side of ciliated planula increase in size and form a projecting structure like an umbrella; this becomes the pneumatophore, while tentacles arise from the side of the larva. A mouth opens at the lower side and the primary gastrozooid is formed (Fig. 5.17C). In *Halistemma* (Fig. 5.17B), the eggs give rise to a ciliated planula resembling that of the other Hydrozoa. At one pole the ectoderm becomes invaginated to form the float, the opposite extremity is gradually converted into the first polyp and a bud appears on one side and becomes the first tentacle. By gradual elongation, and the formation of new zooids as lateral buds, the adult form is produced; the various zooids are all formed between the first polyp and the float, so that the two become further and

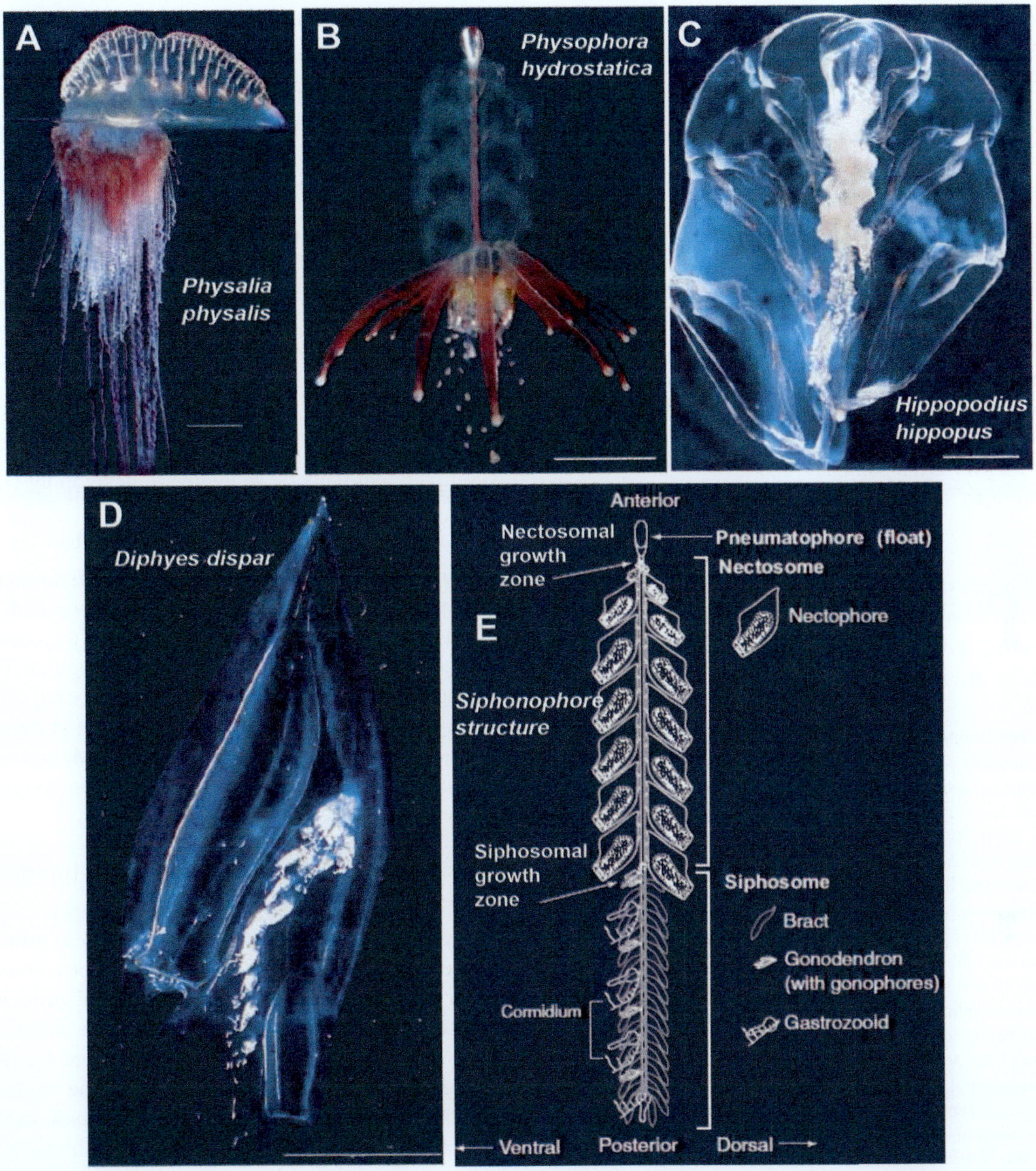

FIG. 5.16 Shows different whole mounts of siphonophores: (A) *Physalia physalis* (X = 5 cm), also known as the 'portuguese Man o' War', this familiar siphonophore is unique in that it lives at the air-water interface and has a hypertrophied pneumatophore that acts as a sail; (B) *Physophora hydrostatica* (X = 2 cm), a short-stemmed species with a conspicuous whorl of palpons above the gastrozooids, this species lacks bracts at maturity; (C) *Hippopodius hippopus* (X = 1 cm), a prayomorph calycophoran, the stem, which is white, is retracted between the six identical nectophores, the youngest of which are at the anterior end; (D) *Diphyes dispar* (X = 1 cm), this diphyomorph calycophoran has a well-differentiated anterior and posterior nectophore, the stemis retracted within the anterior nectophore. Pane (a) is a cystonect siphonophore *((A–D) from Dunn et al., 2005)*; (E) schematic overview of siphonophore structure. The depicted colony is a physonect and possesses a pneumatophore (gas-filled float), nectosome (a region with nectophores, the propulsive zooids), and siphosome (a region with zooids of mixed functional types). Cystonects lack a nectosome, whereas calycophorans lack a pneumatophore *((E) from Dunn and Wagner, 2006)*.

further apart, being always situated at the distal and proximal ends of the colony, respectively.

Suborder *Cystonectea* includes *Cystalia*, *Rhizophysa* and the famous siphonophore *Physalia*, the dreaded Portuguese man o' war. The latter species is of a brilliant peacock-blue colour, but orange-coloured specimens are sometimes formed. The development seems to be generally similar to that of the *Physonectae*; the pneumatophore develops to a still larger size and is separated from the gastrozooid by a septum, and no swimming bells are formed.

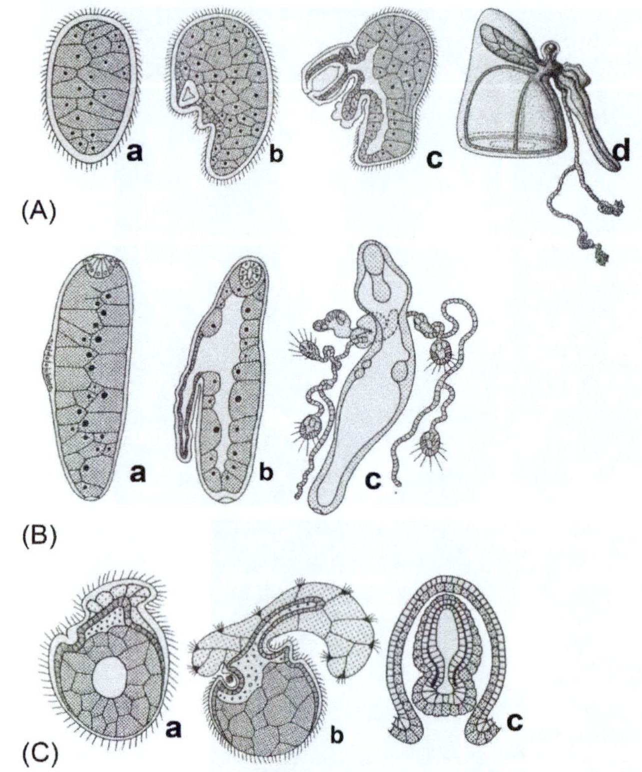

FIG. 5.17 Postembryonic development of three different genera of cnidarian Siphonophora. (A) of Galeolaria (a–c) and Muggiaea (d): (a) section of planula; (b) section of the larva showing the primordium of the first swimming bell and of the tentacle of the first nutritive polyp; (c) formation of the gastric cavity and further development of the bell and tentacle; (d) siphonula stage with completed first cormidium with swimming bell, nutritive polyp, tentacle, and bract, primordium of the second swimming bell on the budding zone, still embryonic; (B) of physophoride *Halistemma picta*, sections of: (a) planula with the anlagen of the pneumatophore at the upper end and that of the first tentacle at the side; (b) the tentacle grows and the gastric cavity develops; (c) older larva with two tentacles, nematocyst butteries, pneumatophore and first nutritive polyps; (C) of physophoride *Agalma sarsii* begins with the formation of the larval bract: (a) larva with the primordia of the primary bract and the pneumatophore, together with the aboral and central liquid cavity; (b) bract greatly enlarged and the pneumatophore sunken in; (c) after the regression of the primary bract, the definitive pneumatophore develops further. *(From Reverberi, 1971 but it was noticed that almost all were firstly studied in the 19th century.)*

The suborder *Disconectea* includes *Velella* and *Porpita*. The development shows rather wide divergence from the above. The youngest known larva of these species is called a '*Conaria*'. This is a hollow and nearly spherical larva with a slightly thickened area on its upper side, which later becomes the pneumatophore, while the nearly spherical portion forms the primary gastrozooid. The conaria is said to sink into deep water where it continues to develop, and only reappears at the surface after reaching the next stage, which is called a '*rataria*'. This larva *rataria* metamorphoses and gradually buds up the complex structure of the adult siphonophore.

Bargmannia elongata

The deep-sea siphonophore

The organization and development of *Bargmannia elongata* (Fig. 5.18A) is described by Dunn (2005) using 19 specimens collected by *ROV Tiburon* from depths of 350–2865 m. All collection sites were within 337 km of Moss Landing, California. All specimens of *Bargmannia elongata* were not visibly damaged during sampling, and survived the ascent to the surface intact. If not disturbed, they lasted for up to 2 days in a large volume of seawater (>6 L at 4°C in the dark) before sinking to the bottom of the vessel and disintegrating. They would not feed in these vessels, so dynamic observations of growth were not possible. Instead, the developmental process was inferred from the morphology of the growth zones (Fig. 5.18A–C), each of which had a complete ontogenetic sequence of developing cormidia.

The siphonosome cormidial organization

Each mature cormidium of *Bargmannia elongata* consisted of a gastrozooid, a tentaculozooid, a gonozooid, five mature bracts and several small buds. The organization of the zooids within the cormidia was entirely regular. They were arranged from posterior to anterior within each cormidium. All of the gonophores of each colony were of the same sex, indicating that this species is dioecious. All the zooids of the mature cormidium were attached to the stem independently (Fig. 5.18D).

The siphosome: The growth zone and budding sequence

The youngest cormidia were found on a protuberance, which in this work is referred to as the horn. It was curled into a sinister coil of about one and a half turns, with a radius in the order of 0.5 mm, and bore the siphosomal elements along the outer side (Fig. 5.18A and C). The spiral arrangement of the horn introduces some nomenclatural complexity for describing organizational axes. In this work, these axes are employed, in reference to the horn, as if the horn were straightened out into a linear anterior projection of the siphosomal stem. The outer zooid-bearing surface of the horn is continuous with the ventral zooid bearing side of the siphosomal stem, and is therefore, referred to as ventral. The tip of the horn is taken to be anterior, and left and right are defined in relation to these two axes (as they are for the colony as a whole).

The siphosomal elements were arranged in a linear ontogenetic sequence, with the posterior being more mature. There was no boundary or discrete transition zone between the zooids on the horn and those on the rest of the siphosomal stem. Pro-buds (P), all of the same type, are serially arranged at the anterior end of the horn (Fig. 5.18E (c)). The least developed pro-buds were at the very tip. Each pro-bud develops into a single cormidium. The complex organization

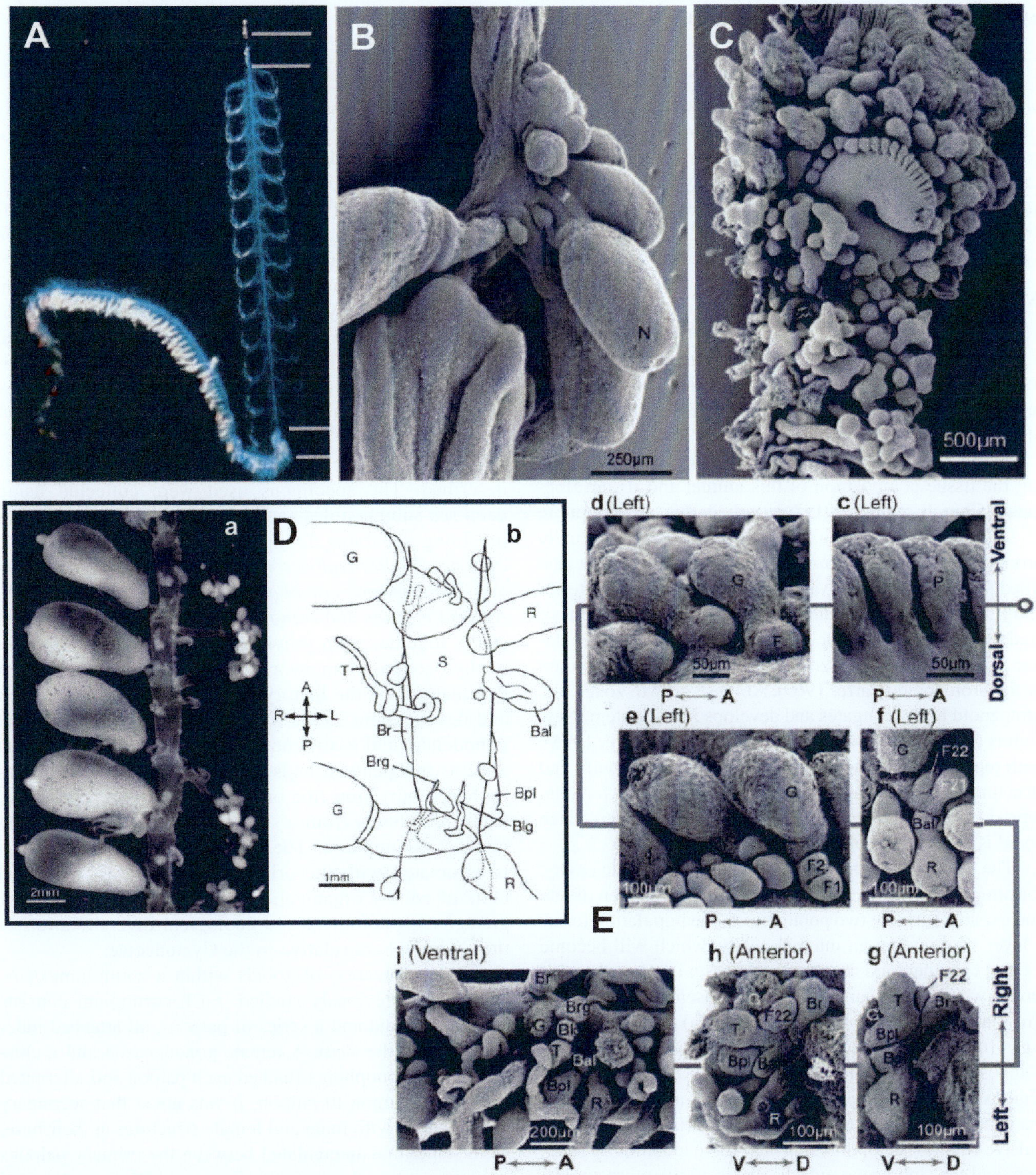

FIG. 5.18 Structure and development of the siphonophore *Bargmannia elongata*: (A) View of a living *Bargmannia elongata* colony, the entire colony is about 40 cm long; (B) scanning electron micrograph of the *Bargmannia elongata* nectosomal growth zone, the anterior portion shown between two lines in (A) (view from left, anterior up, dorsal to the right of the page). The *arrow* indicates one of the buds that form to the posterior of each nectophore (N) developing nectophore; (C) scanning electron micrograph of the siphosomal growth zone, portion shown between two lines in the posterior part of (A) (view from left, anterior up, ventral to the left of the page); (D) the organization of the mature cormidia of *Bargmannia elongata* (ventral view, anterior up, the left side of the stem faces the right of the page): (a) photograph of a length of mature stem. The white mature gonophores can be seen on the gonozooids. The largest structures are the gastrozooids; (b) illustration of the organization of a cormidium. The gastrozooid shown at the top of the illustration belongs to the next cormidium to the anterior, and the gonozooid shown at the bottom of the illustration belongs to the next cormidium to the posterior, S = stem, G = gastrozooid, T = tentaculozooid, R = gonozooid, Br = right lateral bract, Bpl = posterior left lateral bract, Bal = anterior left lateral bract, Brg = right gastrozooid-associated bract, Blg = left gastrozooid-associated bract (Blg, Brg, and Br have been dissected away, leaving only their lamellae) (unlabeled buds on the stem were inferred to be reserve bracts); (E) development of the cromidium, (c–i) cormidia in various stages of development. The *grey line* indicates the order of the ontogenetic sequence (with the youngest cormidia shown being the closest to the *circle*). The view is indicated in parentheses for each pane. The orientation of the axes is shown to the right of each row and below each figure. P = posterior; A = anterior; V = ventral; D = dorsal *(All from Dunn (2005).)*

of the cormidia of *Bargmannia elongata* results from a stereotypical series of divisions of the pro-bud into multiple zooid buds. The zooids of each cormidium were all attached to the stem by a common peduncle early in development, and only later in development (i.e. further to the posterior) do they come to be attached independently to the stem.

Nanomia bijuga

The siphonophore *Nanomia bijuga* (Delle Chiaje, 1841) is a pelagic hydrozoan Physonecte, Agalmidae studied histologically by Church et al. (2015). Specimens were collected in Monterey Bay via blue-water diving from depth of 10–20m.

Carré (1969) first provided a description for understanding the tissue composition of this animal and its development. Church et al. (2015) restudied this siphonophore *N. bijuga* and reported that it begins as a single sexually produced polyp, the protozooid. The pneumatophore, a gas-filled float, forms as an invagination at the aboral or anterior end (Carré, 1969). The authors followed the conventions of Haddock et al. (2005) for siphonophore axes and orientation. Two distinct growth zones arise along the body column of the protozooid (Carré, 1969). At these growth zones, the protozooid body elongates and develops into the stem. New zooids are added to form a colony (Carré and Carré, 1995), with repeated zooid morphologies that are functionally and structurally specialized for particular tasks such as feeding, reproduction, and defence (Dunn and Wagner, 2006). Each zooid is a modified polyp or medusa.

The bilayered epithelium pushes into the gastric cavity, forming an internal mass. The ectodermal cells then divide and expand, forming two populations: one population forms a layer surrounding an internal cavity, which will become the gas chamber; and the other population is internal to this cavity which are separated from the basement membrane. The mature pneumatophore (Fig. 5.19A (a) and B), which is a gas-filled float located at the anterior end of the animal, is composed of five distinct tissues (Fig. 5.19B (b)). The two regions that arise from these growth zones are referred to as the nectosome, which carries the pneumatophore, and propulsive zooids (nectophores), which serve locomotory function, and the siphosome, which carries all the other zooids. The siphosome contains cormidia, which are reiterated sequences of feeding zooids (gastrozooids), digestive zooids (palpons), male and female reproductive zooids (gonozooids), and protective bracts. The nectosomal growth zone is located posterior to the pneumatophore, and the siphosomal growth zone is located at the anterior end of the siphosome, adjacent to the posterior of the nectosome (Fig. 5.16E).

There are several terms that are specific to the group (Dunn, 2005). The term 'gonodendron' is used to describe any compound reproductive structure consisting of multiple zooids that are attached to the stem via a common peduncle. The structure and zooid composition of gonodendra vary widely across taxa; it is not clear if they are homologous across siphonophores or have independently arisen multiple times.

There are few studies of the colony-level development and organization of siphonophores, largely due to the difficulty of collecting them. Although they are often large and relatively abundant, they are also extremely fragile, and most species are found only in the open ocean. However, relatively recently, Dunn and Wagner (2006) studied colony-level development in the Siphonophora. They confirmed that siphonophores, a group of pelagic hydrozoans (Cnidaria), have the most complex colony-level organization of all animals. The colony-level development of five siphonophore species, strategically sampled across the siphonophore phylogeny, is described. The specimens used were collected using deep-sea submersibles and by self-contained underwater breathing apparatus diving. These species include three cystonects—*Bathyphysa sibogae*, *Rhizophysa filiformis* and *Rhizophysa eysenhardti*—and two 'physonects', *Agalma elegans* and *Nanomia bijuga*. The results of this study, together with previous findings, are analysed in a phylogenetic framework. The authors reconstructed key features of the history of colony-level organization and development in the Siphonophora. It is shown that gonodendra and gastrozooids of the examined cystonects arise as independent buds directly on the stem, whereas probud subdivision (the origin of feeding, reproductive, and other zooids from a single bud) is a synapomorphy of the Codonophora. The origin of probud subdivision is associated with the origin of cormidia as integrated units of colony organization, and may have allowed for greater morphological and ecological diversification in the Codonophora relative to the Cystonectae.

The organization of zooids within a cormidium of *N. bijuga* was previously studied. Each cormidium consists of a gastrozooid and a series of palpons, all attached independently to the stem. A female gonodendron and a cluster of male gonophores flanked each palpon and alternated sides from palpon to palpon. It was noted that secondary palpons, also with male and female structures at their base, were sometimes intercalated between the primary palpons in mature cormidia. In 2006, Dunn and Wagner observed and confirmed the earlier findings of *N. bijuga* (Fig. 5.19A). In addition, they were also able to describe the budding sequence by which the zooids of this species arise. There is a pronounced horn within the siphosomal growth zone, and each cormidium arises as a simple probud close to its tip (Fig. 5.19A (d)). The bulk of the probud gives rise to the gastrozooid, with most of the other zooids arising on the anterior side of its peduncle. A palpon and two flanking bracts are the first such zooids (Fig. 5.19A (e)). Additional bracts are added on both sides of the gastrozooid peduncle, with the youngest (i.e. the smallest) being the most distal

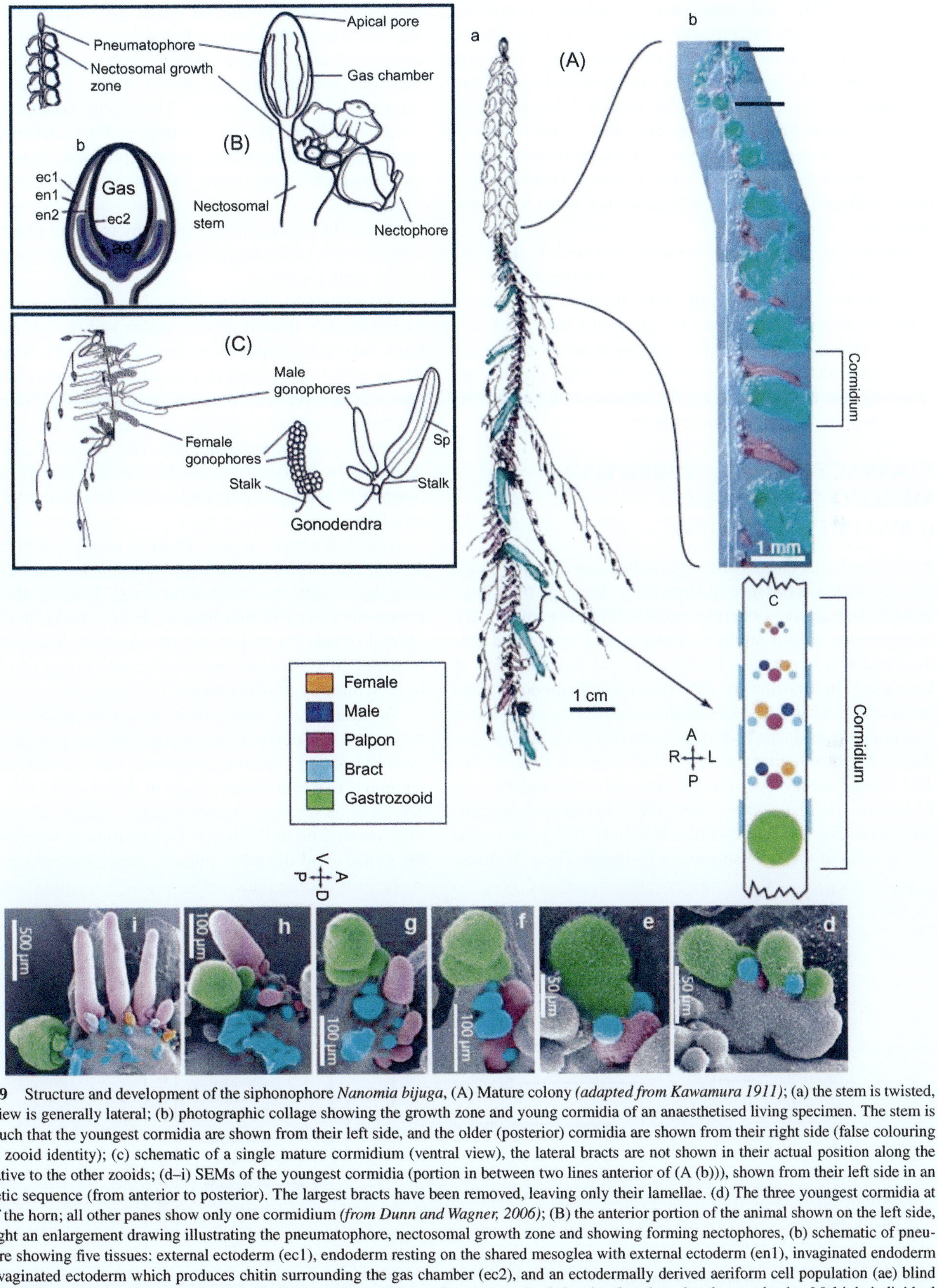

FIG. 5.19 Structure and development of the siphonophore *Nanomia bijuga*, (A) Mature colony *(adapted from Kawamura 1911)*; (a) the stem is twisted, but the view is generally lateral; (b) photographic collage showing the growth zone and young cormidia of an anaesthetised living specimen. The stem is twisted such that the youngest cormidia are shown from their left side, and the older (posterior) cormidia are shown from their right side (false colouring indicates zooid identity); (c) schematic of a single mature cormidium (ventral view), the lateral bracts are not shown in their actual position along the stem relative to the other zooids; (d–i) SEMs of the youngest cormidia (portion in between two lines anterior of (A (b))), shown from their left side in an ontogenetic sequence (from anterior to posterior). The largest bracts have been removed, leaving only their lamellae. (d) The three youngest cormidia at the tip of the horn; all other panes show only one cormidium *(from Dunn and Wagner, 2006)*; (B) the anterior portion of the animal shown on the left side, on the right an enlargement drawing illustrating the pneumatophore, nectosomal growth zone and showing forming nectophores, (b) schematic of pneumatophore showing five tissues: external ectoderm (ec1), endoderm resting on the shared mesoglea with external ectoderm (en1), invaginated endoderm (en2), invaginated ectoderm which produces chitin surrounding the gas chamber (ec2), and an ectodermally derived aeriform cell population (ae) blind tubes connected to the base of the gas chamber are shown; (C) cormidium of a mature animal showing female and male gonodendra. Multiple individual gonophores are borne on stalks which connect to the siphosomal stem. *((B and C) From Church et al., 2015.)*

(Fig. 5.19A (f)). Additional palpons bud on the anterior side of the primary palpon, and all palpons initially share a common base and form a fanlike structure (Fig. 5.19A (g)). The original palpon is the most distal in this structure, and the smallest (youngest) is the most proximal. During maturation, a lateral bract forms on each side of each palpon. The zooids of successively more mature cormidia come to be spread out along the stem such that each zooid is eventually attached to the stem independently (Fig. 5.19A (h) and (i)).

In spite of all these documented information, there are scanty published accounts on the early embryology of siphonophores, and almost all were in the 19th century. The only relatively recent study is that of Carré and Carré (1993). However, these studies stopped short of describing the origins and structure of the growth zones that are responsible for the colony-level development that continues throughout the life of a siphonophore.

CLASS SCYPHOZOA [SUBPHYLUM MEDUSOZOA PETERSO (LAWLEY ET AL., 2016)]

According to the classification presented above in this chapter, Scyphozoa is the second large class (see page 6) of phylum Cnidaria. Hale (1999) presented a different evolutionary relationships within phylum Cnidaria, and mentioned that there are two hypothesis for the origin of the stem line of Cnidaria. The first theory, the 'Basal Trachymedusae' hypothesis, held by many American scientists, suggests that a holoplanktic medusae-like stage gave rise to all known cnidarians by introducing a sessile polyp stage (Hyman, 1940). The second theory opposes the first and is held mostly by European scientists, whose view, the 'Medusozoa' hypothesis (Text Fig. 5.1A), assumes that only the polyp stage was present in the stem species of Cnidaria. Then, Technau and Steele (2011) believed in the growing interest in the use of cnidarian animals to investigate the evolution of key aspects of animal development, such as the formation of the third germ layer (mesoderm), the nervous system and the generation of bilaterality. The recent sequencing of the *Nematostella* and *Hydra* genomes, and the establishment of methods for manipulating gene expression, have inspired new research efforts using cnidarians. They evolved Text Fig. 5.1B.

Since the phylogenetic taxonomy is still not familiar, we arranged the following accounts according to the more familiar earlier scheme.

Members from this class, Scyphozoa, commonly referred to as jellyfish, inhabit all the oceans and seas of the world, from the tropics to the Arctic and Antarctic. The majority of the species live in shallow coastal waters, though there are species that have been found in deep seas. Regardless of the extensive distribution and varied diversity, there is relatively little known about these animals. The 'bloom' that occurred in the early 1980s in the Mediterranean, greatly affecting tourism in the area, generated more knowledge about the jellyfish (Hale, 1999).

Jellyfish range in size from a mere 12 mm to >2 m across, the largest is *Cyanea arctica* (Fig. 5.20D), which may have tentacles over 40 m long. Despite their often enormous size, jellyfish have no head, no skeleton and no special organs for respiration or excretion. Their life cycle involves an alternation between sessile polyp phase and a free-swimming medusa stage.

Alternation of generation occurs in scyphozoan animals, but the polyp is an inconspicuous part of the life cycle while the medusa, which may be as much as 1 m in diameter, is the dominant stage in Scyphozoa (Fig. 5.21F). The bell of *Cyanea capillata* (Fig. 5.20D), for example, may reach 2 m in diameter. Colouration is often striking, the gonads and the other internal structures, which may be

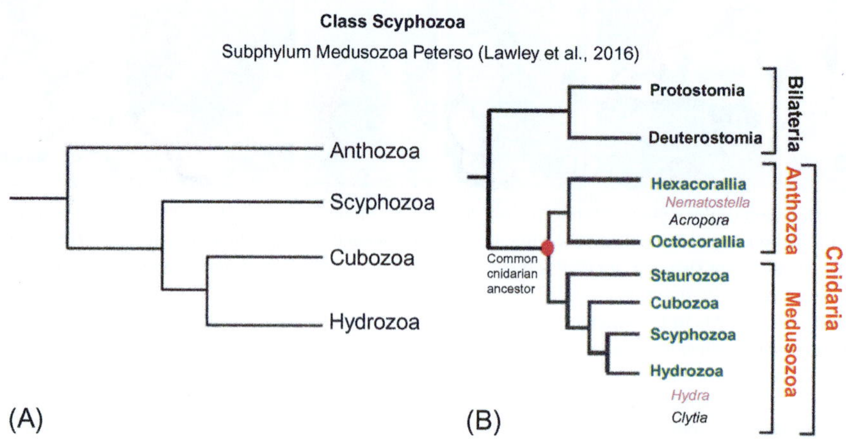

TEXT FIG. 5.1. (A) Evolutionary relationships of the Cnidaria, *(A) from Hale (1999)*; (B) a phylogenetic tree showing the relationships within the phylum Cnidaria.

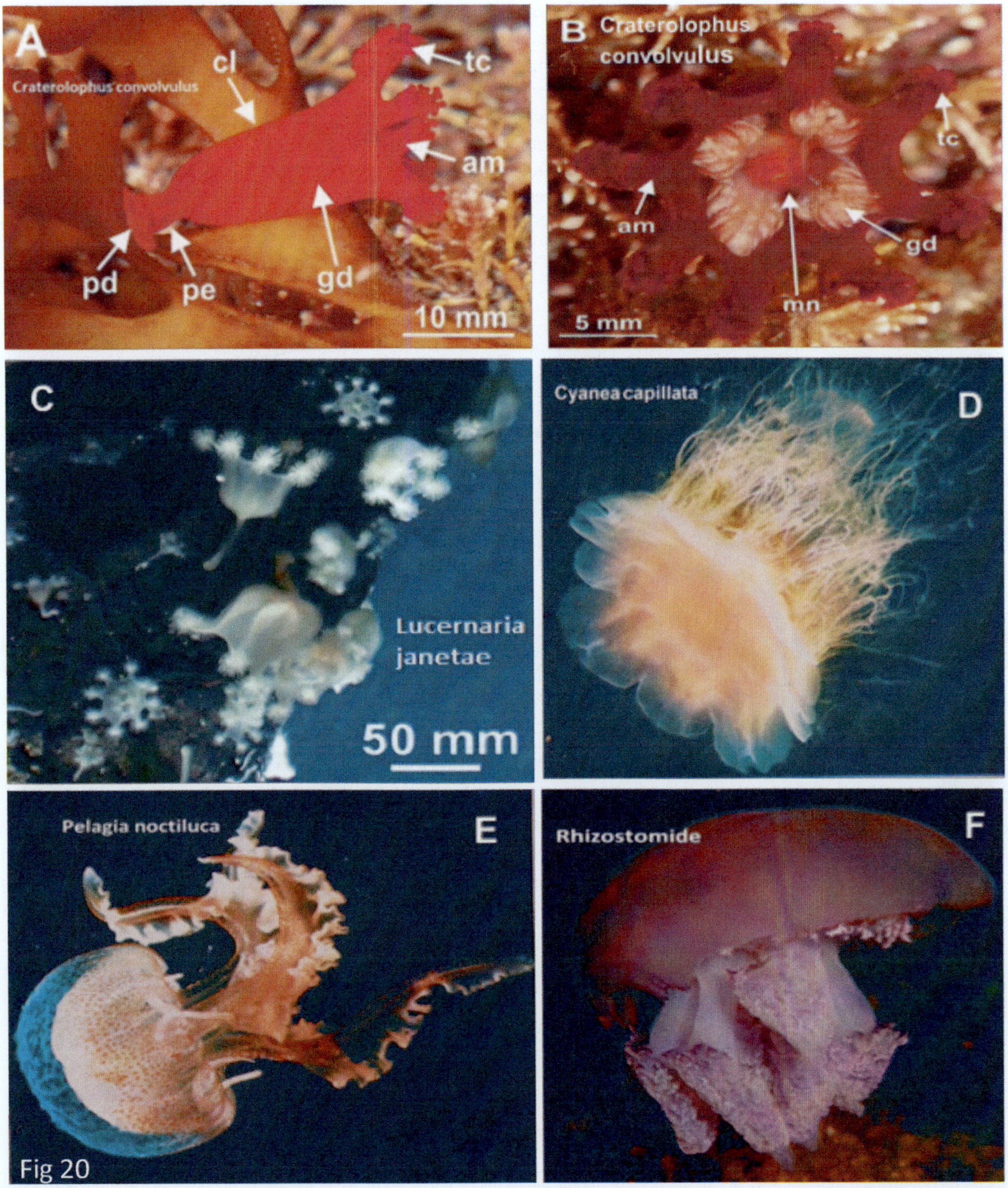

FIG. 5.20 Types of Scyphozoa: (A) lateral view; (B) oral view of the Stauromedusan *Craterolophus convolvulus*, am = arm, cl = calyx, gd = gonad, mn = manubrium, pd. = pedal disk, pe = peduncle, tc = tentacle cluster; (C) Living specimens of *Lucernaria janetae*, n. sp., in situ. Note variety of sizes represented in population *((A–C) from Miranda et al., 2016)*; (D) giant medusa; (E) mauve stinger *Pelagia noctiluca* (Forskål 1775) occurred in mass throughout the western and central basins including the Adriatic Sea since the early 1980s, becoming the most important jellyfish because of its widespread distribution, high abundance, and ecological role; (F) a Rhizostome with eight long oral arms of the manubrium; *(D–F) from Beauté du monde animal, Invertébrés Tome XII (1974).*

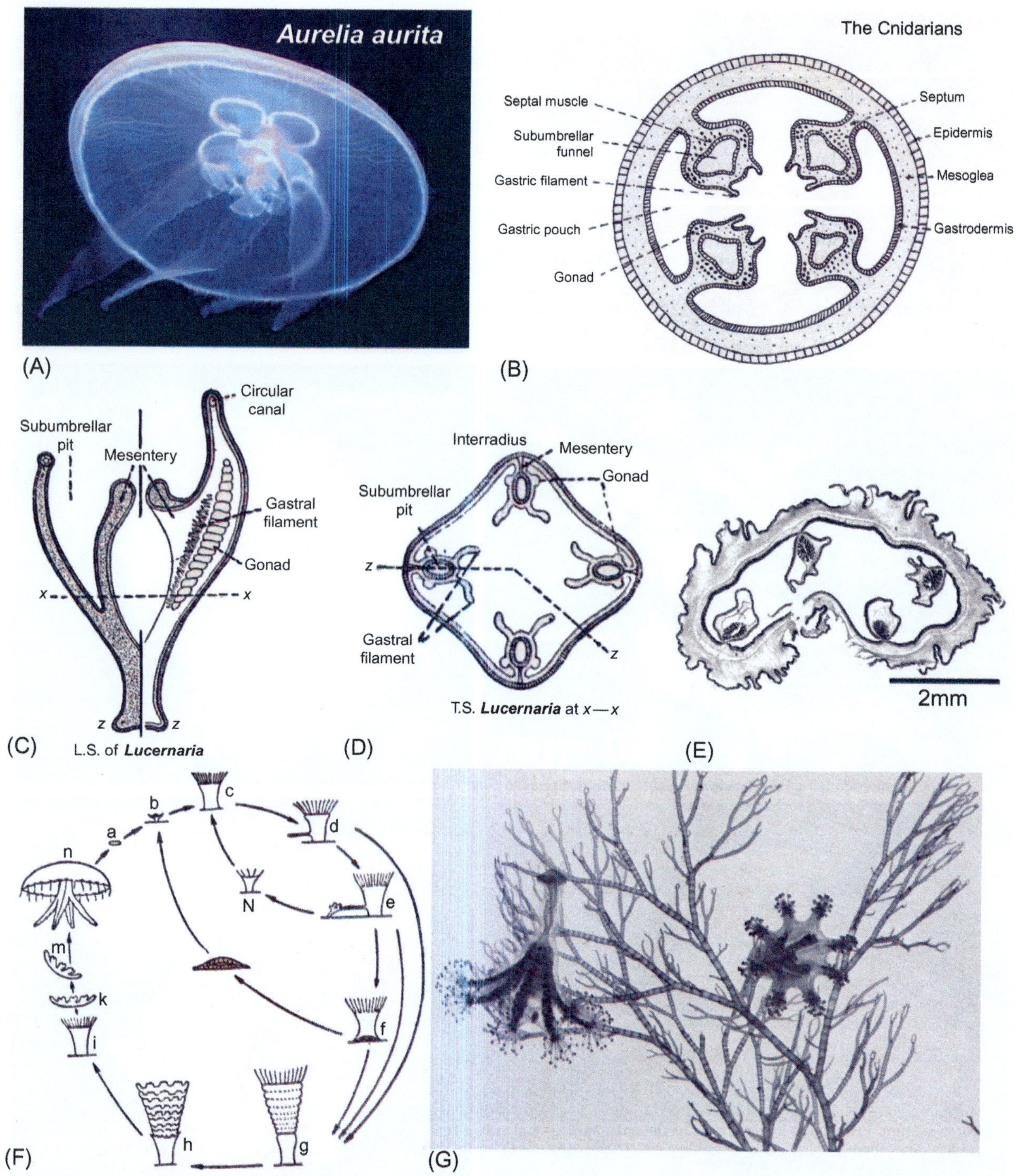

FIG. 5.21 (A) Photograph showing W.M. of live specimen of *Aurelia aurita* (moon jelly), has diameter measuring about 25–40 cm *(from Wikipedia)*; (B) diagrammatic drawing of T.S. of a scyphozoan through the gastric pouches *(after Hyman from Barnes, 1974)*; (C) L.S. and (D) T.S. through *Lucernaria* sp. showing internal anatomy *(from Borradaile et al., 1977)*; (E) internal anatomy of *Lucernaria janetae*, n. sp. showing T.S. through peduncle with four septal muscles *(from Collins and Daly, 2005)*; (F) life-history of *Aurelia aurita (from Uchida and Yamada, 1968)*, a = planula, b–f = scyphopoly, budding, g–m = strobilation, n = medusa; (G) living stauromedusans or 'attached jellyfishes', attached to branches of the red seaweed *Ceramium* are specimens of *Haliclystus auricula* (left) and *Craterolophus convolvulus* (right).

deep orange, pink or other colours, are visible through the colourless or more delicately tinted bell (Fig. 5.21A). With little exception, scyphozoans are dioecious, and the gonads are located in the gastrodermis, in contrast to the usual epidermal gonad in hydrozoans (Fig. 5.21B). In some groups, the scyphozoans have simple or elaborate septal structures, which function as gonads (Fig. 5.21C and D). In *Lucernaria* (Stauromedusae), the four gastric septa may be infolded by genital pits, with depressions arising from the subumbrellar surface and indenting the septa. In this case, cells arise and mature within the endoderm or mesoglea surrounding the pits. In some cases, the bottoms of the genital pits structure during the breeding season release gametes directly to the exterior. In semaeostome medusa (e.g. *Aurelia*) which lack septa, four horseshoe-shaped gonads lie on the floor of the stomach periphery (Fig. 5.21A). Generally, the gametes are released into the gastric pouches (gastric cavity) and may be either released through the mouth, before or after fertilization, or brooded somewhere within the complex gastric cavity.

Subclass Stauromedusa

The simplest type of Scyphozoa is formed in the division of Stauromedusa, ex. *Haliclystus*, *Lucernaria* (*Lucernaria quadricornis* Müller, 1776); its systematic position was uncertain, most assessments prior to the 1850s assumed that they were closely related to sea anemones until Sars (1846) noted that the presence of gastric cirri suggested that they were allied with jellyfish. Specimens of Stauromedusa are frequently found on algae, but they have also been reported attached to rocks, sea grasses, shells, mud, sand, coral/gorgonian, sea cucumbers and serpulid tubes. Most of the species are found in the intertidal and shallow subtidal regions, but species of Lucernaria have been reported at >3000 m deep (Miranda et al., 2017). They are not uncommon on the British coasts, adhering to the blades of *Zostera* or *Laminaria* (Borradaile et al., 1977). They are sessile animals (attached jellyfish, Fig. 5.21G). Generally, stauromedusae are very well camouflaged (colour and morphology) to match their surroundings (Miranda et al., 2017); they exist as individuals, being superficially more like a polyp than a medusa, but usually supposed to be a medusa. Their upper part is medusoid while their lower part is polypoid. Each has a narrow stem arising from its exumbrellar surface, by which it attaches itself temporarily to seaweeds. The edge of the bell is divided into eight lobes, on each of which there are several short tentacles and the adhesive organs which are called marginal anchors (absent in *Lucernaria* (Fig. 5.20C)). The mesenteries are vertical walls projecting from the body wall and composed of endoderm with an internal layer of mesogloea. They have a free edge centrally, while on each side a vertical series of gastric filaments projects into the enteron, and a parallel series of gonads stand nearer to the body wall (Fig. 5.21C).

Collins and Daly (2005) studied a new species, the deep-water stauromedusan *Lucernaria janetae* (Fig. 5.20C), collected from the East Pacific Rise. *Lucernaria janetae* is the first species in the genus recorded from the Pacific Ocean, and differs from its congeners in size and morphology. These spectacular fields of 'stalked jellyfish' (Stauromedusae) are up to 10 cm in height. They are typically small and solitary, and live in shallow near-shore habitats of temperate seas, highlighting the unusual nature of this deep-sea occurrence. It is a nonciliated creeping planula with 16 rectangular endodermal cells, a 4-chambered peduncle (Fig. 5.21E), an adhesive basal disc, 8 clusters of capitate tentacles and perhaps complex ovaries involving follicle cells. The generality of this last feature is somewhat tentative as it has been studied in only a single species, but it is dramatically different from what has been observed in coronates, rhizostomes and semaeostomes.

Collins and Daly's work used the molecular markers to favour the same topology of relationships among their samples. They clearly distinguished between two species, *Haliclystus sanjuanensis* and *H. octoradiatus*, which have been considered synonymous by many workers. They investigated a stable systematic framework for Stauromedusae that appears achievable through comprehensive study of both morphological and sequence data.

With respect to the embryological development of stauromedusos, they essentially differ from the other scyphozoans (i.e. Discomedusae) in the following: the hatched stage (planula) is without blastocoele or cilia and therefore it is nonswimming, instead it is a vermiform creeping organism (Fig. 5.22B (c)). This creature transforms into a scypho-polyp, which in turn becomes a single young medusa. This means that the fertilized egg develops into an individual somewhat resembling the parent, and so no alternation of generations occurs in this life cycle.

The developmental stages of ***Haliclystus octoradius*** were illustrated as an example of stauromedusan Scyphozoa without a free medusa stage (Reverberi, 1971). The early embryonic stages (cleavage) proceed as in any scyphozoon (Fig. 5.23(a)–(d)). Fig. 5.22B (a)–(g) describe the stages of postembryonic development (*H. octoradius*). Fig. 5.22B (a) is a T.S. that shows late blastula with some entoderm cells that ingress from one side, which means gastrulation occurs by polar immigration. The result is stereogastrula (Fig. 5.22B (b)) or preplanula with completed entoderm development and without cavity. This leads to solid planula with cuboid ectoderm cells, and a column of large vacuolated entoderm cells (Fig. 5.22B (c)). Metamorphosis begins with the attached larval stage (Fig. 5.22B (d)) with open mouth, a tentacle bud, gastric cavity and the appearance of the primordia of the pedal disc. The second stage of metamorphosis shows four tentacles, four tentacle primordia and the pedal disc deeply sunk in (Fig. 5.22B (e)). In the eight-tentacle stage, the mouth cone develops and appears,

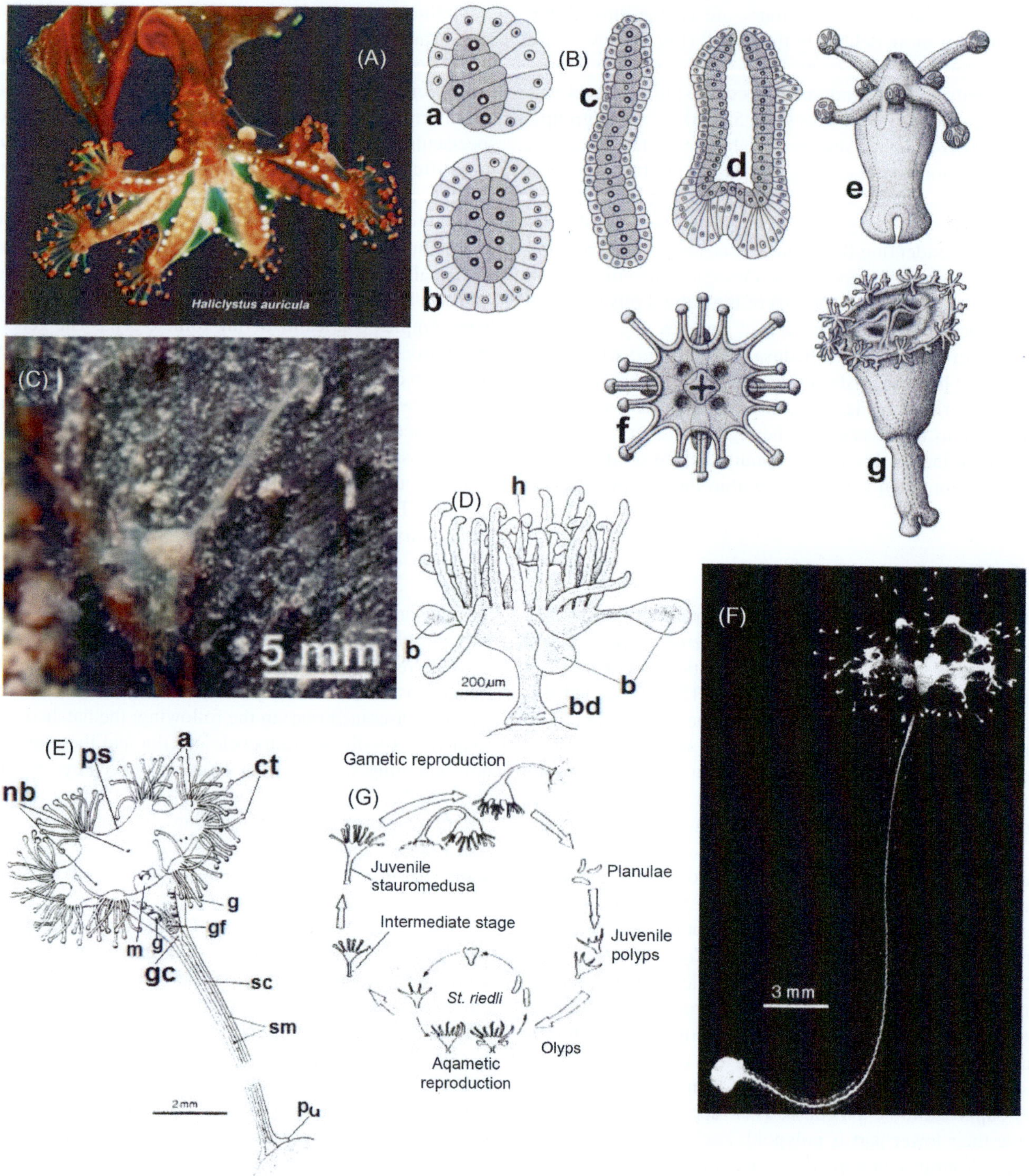

FIG. 5.22 Life cycle of the stauromedusan *Haliclystus*: (A) W.M. of live, attached specimen, *H. auricular (from Beauté du monde animal, Invertébrés Tome XII, 1974)*; (B) embryonic and postembryonic stages of *H. octoradius, from (a–h) (Reverberi, 1971)*: (a–d and f) (median sections), (e and h) (lateral view), (g) (surface view for upper end): (a) late blastula with immigration of few cells, (b) stereogastrula, (c) solid planula, (d) attached larva in metamorphosis with open mouth, (e) metamorphosed stage with 4 tentacles, (f) eight tentacle stage, (g) 16 tentacle stage, (h) young scyphopolyp; (C–G) structure and life cycle of the stauromedusan *Stylocoronella riedli (from Kikinger and von Salvini-Plawen, 1995)*: (C) photograph of lateral view of live specimen in nature *(from Miranda et al., 2016)*; (D) drawing of interstitial scyphopolyp stage with lateral buds, b = bud, h = hypostome, pd. = pedal disc; (E) schematic view of adult *Sfylocoronella riedli*. Total length: 20 mm. a = adradial arm, ct = capitate tentacles, g = gonad, gc = gastrocoel, gf = gastric filaments, m = manubrium, nb = nematocyst battery, pd. = pedal disc, ps = perradial pigment spot, sc = stalk canal, sm = septal muscle; (F) photograph of adult specimen with no anchors. The pedal disc is attached to a sand grain; (G) hypothetical reproductive cycle in *Stylocoronella riedli*. The sizes of the different developmental stages are drawn without scale.

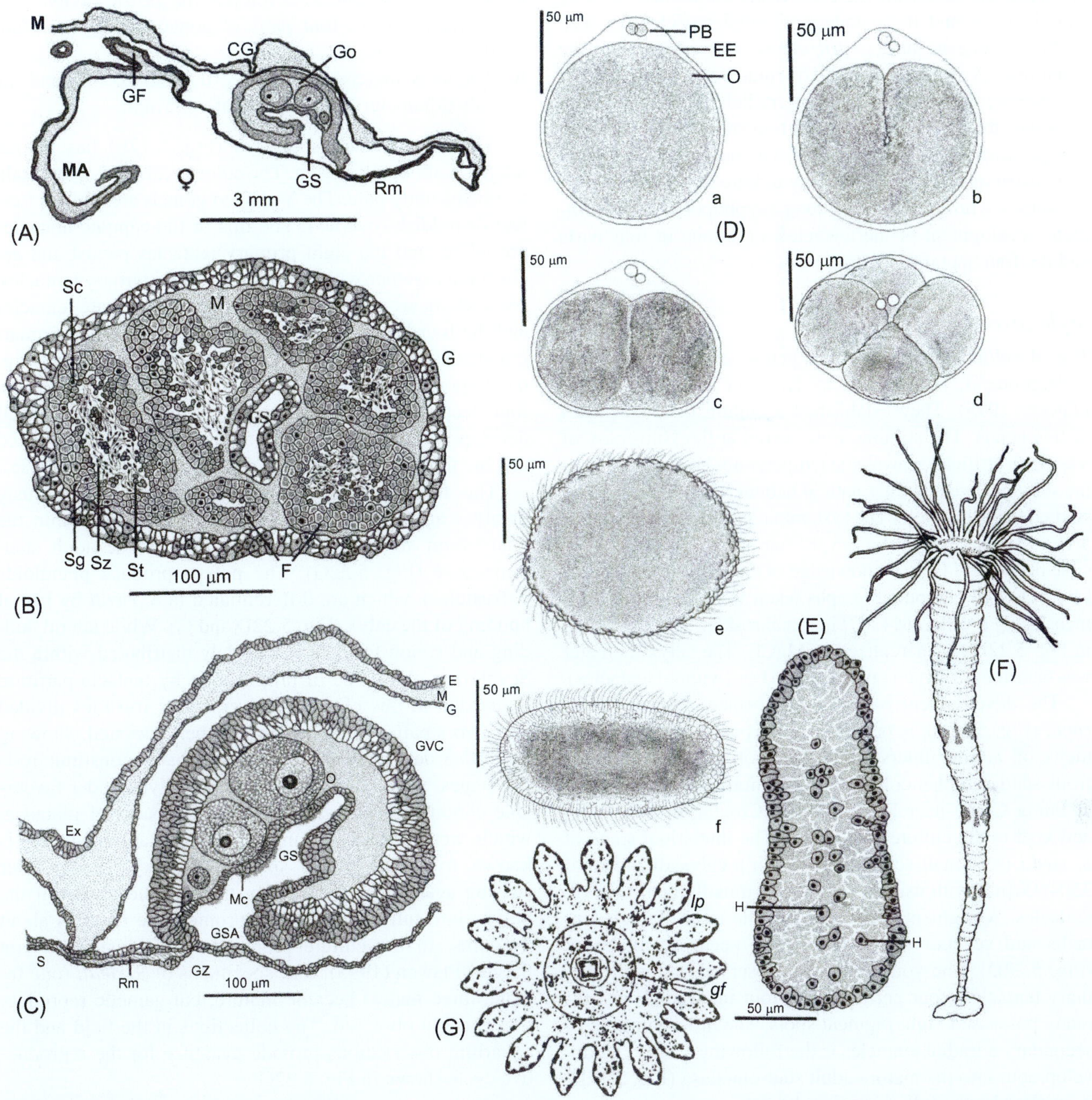

FIG. 5.23 Development of a scyphozoan discomedusan coronate *Nausithoë aurea*: (A) diagram of a histological section (oral-aboral) of the right half gonad of the female medusa, note that the gonads are evaginations of the gastrodermis; CG = coronal groove, GF = gastric filament, Go = gonad, GS = genital sinus, M = mouth, Ma = manubrium, Rm = ringmuscle, S = stomach. (B) Diagram of a T.S. of the histology of the testis, note the organization of the gonad with follicles and the development of the germ cells in the follicles, F = follicles, G = gastrodermis, GS = genital sinus, M = mesoglea, Sc = spermatocyte, Sg = spermatogonium, St = spermatid, Sz = spermatozoon; (C) diagram of a T.S. of the histology (of the oral-aboral plane and to the left of the stomach) of the ovary, note the oocytes in different stages of development, the differences between the gonad epithelia, the presence of mucous in the genital sinus, E = epidermis, Ex = exumbrella, G = gastrodermis, GS = genital sinus, GSA = genital sinus aperture, GVC = gastrovascular cavity, GZ = germinative zone, M = mesoglea, Mc = mucus, Rm = ringmuscle, O = oocyte, S = subumbrella; (D) early embryonic stages of *N. aurea*: (a) newly released oocyte, note the egg envelope and the polar bodies: O = oocyte, EE = egg envelope, PB = polar bodies, (b) first cleavage, (c) 2-cell stage, (d) 4-cell stage, (e) embryo 24 h after fertilization, (f) embryo 48 h after fertilization; (E) diagram of a L.S. of histological structure of the 48-h embryo of *N. aurea*, showing the gastrulation (multipolar ingression) process, H = migrating cells; (F) general habit of a solitary coronatae polyp. The basal disc attaches the tube to the substrate. Inside the tube there are whorls of internal cusps. At the apical pole the collar with its crown of tentacles above the rim of the skeleton; (G) ephyra of *N. punctate* resulting from strobilation of the giant polyp 'stephanoscyphus', gf = gastric filament, lb. = marginal lappet. *(From (A–E) after Morandini and Silveira (2001); (F) from Jarms et al. (2002); (G) after Komai (1935) from Uchida and Yamada (1968).)*

the septal funnels with the ectodermal cnidoblast forming sites beneath and the ectodermal septal muscles, as well as the pharyngeal tube, gastric cavity and sinking in of the pedal disc. A surface view of 16-tentacle stage (Fig. 5.22B (f)) shows four perradial, four interradial and eight adradial tentacles, four septal funnels, and four rhopaloid thickening and quadriradial mouth openings. The outlines of the septa and gastric pockets are shown by a dotted line. Fig. 5.22B (g) shows whole mount of young scyphopolyp with complete development of the tentacles, oral cone in four parts and the four septal funnels.

Stylocoronella riedli

The development of another genus of stauromedusae, *Stylocoronella*, was studied by Kikinger and von Salvini-Plawen (1995). They established a culture of two species of this genus. The cultures were reared at the University of Vienna in artificial seawater at temperatures corresponding to the seasonal cycle in their natural habitat. They observed and studied their life cycle. The systematic grouping of the genus *Stylocoronella* within the Scyphozoa has not yet been possible because of lack of knowledge of the gametic stage in its life cycle. All photomicrographs taken were of living specimens using a WILD M420 'Photomakroskop'; the individual in Fig. 5.22F was narcotized in $MgCl_2$. The calyx diameter was measured between opposing arm tips (without tentacles).

The development of only one species, *Stylocoronella riedli* (Fig. 5.22C), is presented in this account. The specimens of marine interstitial scyphopolyps were obtained from sand samples collected in the northern Adriatic Sea at Punta Croce near Rovinj, Istria (Croatia, east to Italy) and kept in the culture at 9–24°C). The interstitial stage of *S. riedli* is a small scyphopolyp with a calyx diameter of 350–600 µm, with usually up to 24 terminally bent filiform tentacles. A pigment spot occurs at the base of each tentacle, and vegetative buds sometimes occur on the calyx (Fig. 5.22D). The youngest polyp observed had eight primary tentacles (four perradial and four interradial) and already possessed eight pigment spots. The differentiation of secondary adradial tentacles is the following stage. The development into the mature adult stauromedusa (Fig. 5.22E) is reached by the following morphogenesis:

1. The filiform, distally bent tentacles, including the eight primary tentacles, transform into capitate tentacles.
2. All the tentacles become arranged into eight clusters, these become the eight adradial arms of the adult stauromedusa.
3. The four subumbrellar nematocyst batteries differentiate.
4. The formation of the oral disc and the four infundibulae.
5. The number of the tentacles are increased by adding tertiary tentacles up to 10–15 capitate ones. These additional tentacles develop at each of the eight equidistant arms.
6. The gastric filaments develop in the gastric cavity.
7. Development of four pairs of gonads along the infundibulae occurs.
8. The stalk increases in length to about 20 mm, and the calyx diameter (umbrella) to about 5 mm.

The fully mature specimen (Fig. 5.22F) has a total length of about 20–25 mm. The colour of the calyx and stalk is translucently pale. The V-shaped gonads are whitish (testes) or reddish (ovaries). The tips of the capitate tentacles are white, and the eight primary tentacles persist and are clustered together with the secondary and tertiary tentacles. No anchors are present at any stage. The capitate tentacles and the four larger perradial ones at the subumbrellar margin are characterized by the presence of numerous tiny pigment spots in their basal areas. The quadrate manubrium, four deep infundibulae and the gastric filaments are well developed. The long vermiform stalk is single-chambered and the four longitudinal muscles extend to the pedal disc.

The reproductive cycle of *Stylocoronella* obviously involves agametic reproduction by the polyps within the interstitium and gametic reproduction by the adult stauromedusae (Fig. 5.22G). The polyps produce planuloids (= frustules), which are differentiated in *S. riedli* by lateral budding of the calyx (Fig. 5.22D and G). While lateral budding and stolon formation is widely distributed within the Scyphozoa, the mode of propagation by tentacle-partition is unusual in this class. However, many frustules divided into two smaller ones, which are then encysted, showing no further development. Frustules probably constitute resting stages, which develop into small polyps under favourable conditions. The same encysting process of planulae, which were produced gametically by stauromedusae, is known from *Lucernaria* sp. (Kowalevsky, 1884). After ceasing agametic reproduction, the metamorphosis of the polyps into fully developed stauromedusae requires about 2 months. In the culture prepared by Kikinger and von Salvini-Plawen (1995), four specimens of *S. riedli* (one female, three males) became mature, but gametic reproduction was not observed. The collections in the field and the aquarium observations provide evidence for the reproductive cycle shown in Fig. 5.22G.

Subclass Discomedusae

The members of the three orders of this subclass (Coronatae, Semaeostomeae, and Rhizostomeae) are alike in having an ephyra stage, and generally exhibiting alternate gametic and agametic generations. With respect to the embryology of these three orders, there are doubtful points in the case of the deep-sea Coronatae, but the Semaeostomes and Rhizostomes have many aspects in common, although the metamorphosis of the rhizostome medusa is especially complex. In general, the course of development followed by the fertilized egg is the same for all orders.

Order Coronatae

Jarms et al. (2002) described and reviewed the behaviour and sites of living of coronatae species in general, and of genus *Nausithoë* specifically. These metagenetic cnidarians have traditionally been considered as deep-sea animals. Many medusa, such as some species of *Nausithoë*, have been found in mid- or deep waters. There are polyps from depths down to 7000m in the collection of Copenhagen and polyps with zooxanthellae have been described which can live only in the light-zone of shallow waters. These authors show more interest in three species from marine caves in shallow waters: *Nausithoë eumedusoides*, *N. planulophora*, and *Thecoscyphus zibrowii*. However, it may be concluded that coronates live in all zones of the ocean. Concerning the substrates to which the polyps need to be attached, these authors presented a very detailed description of the different hard substrates, as well as to the live animals or group of animals. Attachment to some substrates may be confusingly similar to the coronate attached portion. Therefore, specimens must be collected, together with a small piece of the substrate, and isolated in containers with fresh filtered seawater of the same temperature and salinity as at the sampling locality.

The work of Jarms et al. (2002) can be considered as a concise guide to the methods, techniques and equipment needed for the collection and transport of specimens, for arranging, maintaining and controlling cultures, for handling polyps, ephyrae, medusa and/or planuloids. In addition, for standardizing species description on the basis of life-cycle studies of Scyphozoa Coronatae.

The characteristic feature(s) of the members of this order is that the umbrella is divided by a horizontal coronary groove. They have 4–16 tentaculocysts (as confirmed by all available textbooks).

Ex. *Nausithoë aurea*

Morandini and Silveira (2001) studied the gametic reproduction of *Nausithoë aurea* and reported that this aspect and the microscopic anatomy of coronate medusa were studied early in the 19th century and beginning of the 20th. Later, Komai (1935, 1936) was interested in different aspects of gametic reproduction of *Nausithoë aurea*, and showed precocious gamete production during strobilation. The aim of the study of Morandini and Silveira was to update the knowledge of coronate gametogenes and to investigate early development of the species. They prepared for the study by the scyphistomae, which were sampled by scuba diving in August 1997 in the São Sebastião Channel at Ponta do Urubu (23°51.06′S 45°24.77′W) (São Sebastião Island), São Paulo State, Brazil. These scyphistoms were growing on the calcareous debris of the coral *Mussismilia hispida* at a depth of 3–7m. It remains to be answered which factors may account for the absence of ephyrae/medusae in plankton samples and intensify the mechanisms of asexual reproduction of coronates.

Twenty-four polyps were reared in the laboratory for 92days to obtain the medusa stage. Jarms et al. (2002) described the development of the polyp. The tentacles shorten, and then the whole head with collar, mouth-plate and resorbed tentacles is retracted into the tube (Fig. 5.23F). The uppermost tissue forms a plate, closing the aperture of the tube. Sometimes it produces a cover of a thin periderm layer. It is then not known what the result of the strobilation will be. Normally ephyrae are produced, ephyrae have to be handled in a special way. However, eumedusoids (*Nausithoë racemosa, N. eumedusoides*) or planuloids (*N. planulophora*) may also be the outcome of a strobilation. Fortunately, the culture of Morandini and Silveira (2001) produced a number of male and female medusa. A total of 63 female and 169 male medusae were produced by two and four strobilating scyphistomae, respectively.

Morandini and Silveira were also successful in performing and obtaining fertilized oocytes. Each morning (8–9a.m.), they separated oocytes from the females, and the medusae were fed for 5min or until their stomachs were completely filled. The medusae were kept in plastic cups at 22°C under a 2h light/22h dark regime. The oocytes were counted, measured and used for in vitro fertilization. The released oocytes were put together with five male medusae for a period no longer than 1h, to prevent medusae from eating the oocytes. This artificial fertilization also succeeded, since the authors continued the study of embryonic development in vitro. Their results were described mainly from histological preparations. The gonad is an evagination of the floor of the gastric cavity, peripheral to the gastric filaments within the central stomach. It consists of two gastrodermal tissue sheets, filled with mesoglea in which sperm in follicles or free oocytes develop (Fig. 5.23A).

Spermatogenesis The testis, as usual, appears in the section as follicles. Within each follicle, spermatogenesis occurs in a centripetal pattern, and the cell types form distinct layers (Fig. 5.23B). The margin of the cyst forms a thin layer of spermatogonia and the centre is occupied by spermatozoons with the tails pointing to the centre of the cyst before becoming mobile. The spermatozoon has a conical head (length 4.2–6.0μm), followed by an unstained region, which is the region of the mitochondria. The sperm are discharged without packages that characterize some other scyphizoans.

Oogenesis The ovary shows distinct areas of gastrodermal epithelium: an internal one, surrounding the genital sinus, with spherical cells and some vacuoles; and an external one, with columnar cells containing many vacuoles. The oocytes develop from the gastrodermis (Fig. 5.23C), and lie in the mesoglea showing different stages of maturation. From a limited region of the gastrodermis, small spherical cells migrate into the mesoglea, and show two distinct phases: basophilic and acidophilic oocytes. This is the germinative zone. In oral-aboral sections, a maturing oocyte gradient was observed towards the free side of the female gonad (Fig. 5.23C). One of the characteristic features of this coronatae scyphizoan, *Nausithoë aurea*, is the mode of the oocyte nutrition in the female gonad. This species follows the pattern defined for Coronatae; free oocytes in the mesoglea without close association to any cell. The oocytes are released through the mouth. Mature female medusa released oocytes during autumn. The total number of oocytes released varied throughout the period, ranging between 0 and 485 oocytes per day. However, the authors reported that the release of oocytes was continuous for the whole study period with variation in the number. Continuous gamete production may maximize the success of sexual reproduction. Thus, continuous egg release in this species is likely an adaptive trait that evolved in response to the natural selection of medusae.

Fertilization is external and the released oocytes remain grouped by a mucus strand. It was suggested that the large amount of mucus is peculiar to species with external fertilization and may provide a chemoattractant for the spermatozoa. Each released oocyte is enclosed by the egg envelope, and usually has two polar bodies (Fig. 5.23D (a)). It was stated that the polar bodies were formed prior to or at fertilization. The authors always observed polar bodies with released oocytes, which suggest that meiosis is completed prior to fertilization (following Campbell, 1974). In some cases, three or four polar bodies were observed and the spermatozoa were observed moving in the space between the egg envelope and the oocyte surface.

Embryogenesis Morandini and Silveira (2001) observed radial and holoblastic cleavage, and after 1.5–2 h of mixing the oocytes with male medusa, the first meridional cleavage occurs from the animal pole, producing a heart-shaped figure (Fig. 5.23D (b) and (c)). The second cleavage occurs almost 2.5–3 h after fertilization and up to the eight-cell stage the cleavage is regular. Afterwards, it becomes pseudospiral and after 24 h following fertilization, the embryo begins to rotate inside the egg envelope until it ruptures, and the embryo becomes free. At this stage, the embryo is almost spherical (Fig. 5.23D (e)) and actively moving in the water column of the rearing vessel (Fig. 5.23D (f)). These observed early stages of embryogenesis confirmed the absence of any zooxanthellae. At 24–48 h, gastrulation occurs through multipolar ingression (Fig. 5.23E). In the 48-h embryos, the basal lamina was not observed and the presumptive gastrodermis still had few cells. Most of the space inside the stereogastrula was filled with yolk granules. The ectodermal cells were very irregular in shape.

The mode of the oocyte nutrition in the female gonad differs in the different orders of Scyphozoa. *Nausithoë aurea* agrees with the pattern defined for Coronatae: free oocytes in the mesoglea without close association to any cell. As mentioned, the continuous egg release in this species is likely an adaptive trait evolved in response to the natural selection of medusae. Thus, the mean number of eggs released by *N. aurea* medusae fits the values for the deep-sea coronates. The authors suggested that *N. aurea* is indeed a neritic species and produces gametes throughout the year. This condition represents a new pattern for the biology of *N. aurea*.

Order Semaeostomeae

Generally, *Aurelia* is the representative of the subclass Discomedusae. *Aurelia* is a common genus abundant in Atlantic and Pacific waters as well as in the Mediterranean Sea.

Development of *Aurelia aurita*

This scyphomedusa is an object of morphological and embryological study almost all over the world. In Egypt, it is also a good example especially for the life cycle since it is among the forms with alternation of generation and a fixed larval stage (Fig. 5.21F).

Aurelia is dioecious; the gonads are purplish in the young medusa, but turn brown in older ones. They are located in the gastrodermis (endodermal in origin), and develop on the interradii in the floor of the stomach. Each is very folded, forming a horseshoe-shape pointing inwards (Fig. 5.21A). Under the gonads are the subgenital pits, the lower sides of which are thin and rupture during the breeding season, making a connection with the outside.

The spermatozoa are shed by the males through those apertures into the seawater. Water currents carrying these spermatozoa as well as the food enter the females through the mouth. The spermatozoa find their way into the gastral cavities of these females, where they fertilize the eggs as soon as they are released. In this, the fertilization is internal. The fertilized eggs pass within the female bodies to the grooves found in the oral arms, and embed themselves in pits along the sides of these grooves. These temporary brood chambers are the sites of early development through the planula. The fertilized eggs, as in other scyphozoans, undergo an undistinguished cleavage that resembles a radial type with the first two divisions always total and generally equal (Fig. 5.24(I) A (a)–(c)). Cleavage then becomes progressively less regular and synchronized (Campbell, 1974). However, the cleavage ends with the formation of a morula (Fig. 5.24(I) A (c)), and then a blastula (Fig. 5.24(I) B (a)). One end of this blastula becomes invaginated to form the gastrula (Fig. 5.24(I) B (b)). In other semaeostome, these pits may be the site of the fertilization area. The yolk granules contained in the cells of the blastula are absorbed gradually during the formation of the gastrula and its change from spherical to oval form. It has a somewhat wider anterior end, which is ready distinguished from the more pointed posterior end. The ectoderm of this creature acquires cilia and at this time the gastrula becomes a planula (Fig. 5.24(I) B (c)). It swims away from the body of the adult to the outside, where it leads free life. Little opaque patches can be seen by a lens along the sides of the oral grooves of the adult, and when dissected out, they prove to be masses of planulae.

The planula (Fig. 5.24(I) B (c)) has a small cavity and its ectoderm gives rise to nematocysts, sensory cells and gland cells. The endoderm contains large vacuoles. The planula swims for about 4–5 days, then loses its cilia, attached at the bottom by its broad end. It forms a pedal disc and its ectoderm cells secrete an adhesive substance that holds the planula to the substratum (Fig. 5.24(I) B (d)). This attached planula becomes somewhat flattened and the endoderm cells around the mouth proliferate rapidly and form a hypostome. The mouth opens in the centre of this hypostome. In the two opposite sides of this hypostome, small projections are formed; these are the first two tentacles. Two more tentacles are then formed perpendicular to the first pair. The tentacles are solid, composed of ectoderm cells including many nematocysts. During this time, four endodermal projections are also formed at positions alternating with the four tentacles becoming mesenteries (septa), which extend into the gasteral cavity (Fig. 5.24(I) B (e)). From the upper parts of these mesenteries, the ectoderm gives rise to muscular bands forming the longitudinal septal muscles. They extend to the base of the larva. This animal or this stage of development is called a scyphopolyp or scyphistoma (Fig. 5.24(I) B (f)). Alternating with each of the four tentacles of this scyphistoma, four new tentacles are formed, making a total of eight. The number of tentacles increases to 16, and in some cases to 20 or 24 (Uchida and Yamada, 1968). Nematocysts on the tentacles clump into many small, wart-like projections.

At this stage, scyphistoma feeds, increasing in size, and reproduces by budding. In addition, it forms new polyps by sending out hollow stolons from its base. This means that each scyphistoma produces, agametically, more

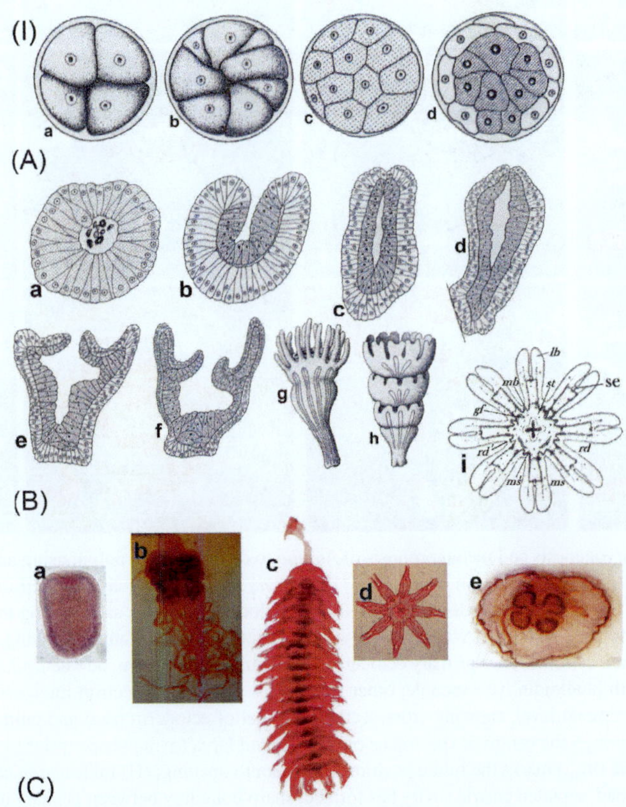

FIG. 5.24 (I) (A) Embryonic development of *Thaumaioscyphus distinctus* up to the morula, (a–d), as an example for scyphozoans: (a) 8-cell stage, (b) 16-cell stage, (c) morula, (d) few-celled polar ingression of the prospective endoderm; (B) development of scyphozoan *Aurelia aurita*: (a) section of blastula stage, (b) invaginated gastrula, (c) young planula, (d) attached larva, (e) young scyphopolyp, (f) older scyphopolyp with developing tentacles, (g) scyphostoma with the first constriction, (h) small strobili with three constrictions, (i) just released young ephyra, gf = gastral filament, lb = marginal lappet, mb = manubrium, ms = radial muscle, ms' = circular muscle, rd = radial sac, rd' = perradial sac, se = sensory organ, st = stomach *(from Uchida and Yamada, 1968)*; *(except (i) from Reverberi, 1971)*; (C) laboratory preparations of *Aurelia* sp.: (a) planula larva, highly magnified, (b) scyphistoma, (c) strobili, (d) ephyra larva, (e) adult medusa.

(Continued)

(II)

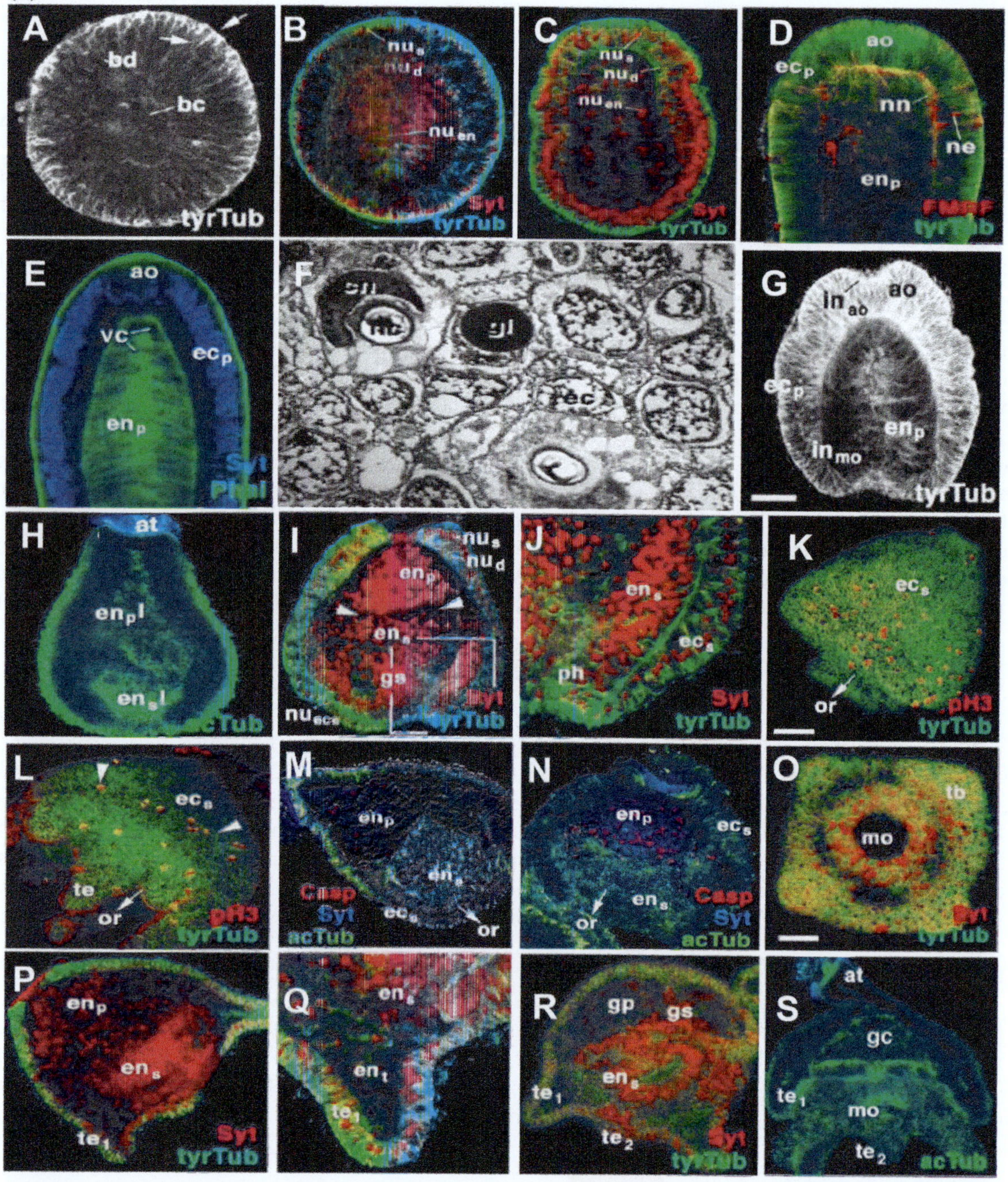

FIG. 5.24, CONT'D (II) Stages of embryogenesis and metamorphosis of *Aurelia* to develop into a polyp, using antibody markers combined with confocal against tyrosinated tubulin (tyrTub), and transmission electron microscopy *(Yuan et al., 2008)*: (A) equatorial section of late blastula, *arrows* indicate cortical cytoskeleton, bc = blastocoele; (B) T.S. of late gastrula, endodermal nuclei (nu_{en}) are located apically, surrounding the lumen of the primary endoderm (enl); (C) midstage (ellipsoid) planula ((B, C) nuclei are labelled with Sytox (Syt)); (D) section of late planula showing a subset of neurons is labelled with the anti-FMRFamide antibody, ao = apical organ, ci = cilia, ec_p = primary ectoderm, en_p = primary endoderm, ne = neuron, nn = nerve net; (E) section of late planula showing actin cytoskeleton is labelled with phalloidin, vc = vacuole, other lettering as in (D); (F) electron-microscopic section of the late planula showing ectodermal cells, tangential section at superficial level, showing cross-sectioned nuclei of ectoderm (rec) and cnidocytes (cn), as well as nematocysts (nc) and gland cells (gl) (X = 2 μm); (G) L.S. through the centre of specimens of early settled larva (ovoid shape, prior to formation of mouth opening), a shallow invagination of the ectoderm at the oral pole (in_{mo}) marks the future position of the mouth opening; (H) midstage of settled larva (ovoid shape, mouth opening has formed); (I) late settled larva (bell-shaped, septated gastric cavity has formed, sharp boundary between primary and secondary endoderm evident *(arrowheads)*, secondary endoderm forms ciliated cells that enclose an inner lumen (ensl) that is of the ectoderm has thinned out (ecs), presumably due to the fact that many cells have invaginated–delaminated from it to form the secondary endoderm. The cuboidal cells remaining at surface form the secondary ectoderm of the polyp; (K and L) in K and L, mitotic cells are labelled with antiphosphohistone 3 (pH3). During polyp stage, proliferation is still found in the oral belt *(arrowhead)* that now has come to lie at the base of the tentacles (te). Proliferation and apoptosis during early metamorphosis. (K) Confocal section of settled larvae, *arrow* point at oral pole (or), in settled larva, proliferating cells are concentrated in a belt-shaped region of secondary ectoderm (ecs) and secondary endoderm (ens) surrounding the oral pole; (L) during polyp stage, proliferation is still found in the oral belt *(arrowhead)* that now has come to lie at the base of the tentacles (te); (M and N) confocal sections of settled larvae, in these two figures, apoptotic cells are labelled by anti-caspase-3 (Casp). *Arrows* point at oral pole (or) cells of the primary endoderm (en_p) undergo apoptosis at the settled larva stage; (O) T.S. confocal section of *Aurelia* polyps, tentacle bud stage (tb). The latter consists of regularly spaced condensations of ectoderm and underlying secondary endoderm; (P and Q) L.S. of the polyp at four-tentacle stage; (R and S) sections cut at plane tilted between horizontal and longitudinal of eight-tentacle stage, at = attachment site of the polyp, en_s = secondary endoderm, gc = gastric cavity, gp = gastric pouch, gs = gastric septa, te1, te2 = first, second quartet of tentacle; (X = 25 μm). *(From Yuan et al., 2008.)*

scyphistoma (Fig. 5.21F). At the same time, some changes take place in the scyphistoma:

1. The hypostome (previously the mouth) further develops into a manubrium.
2. The margins of the septa thicken, giving rise to the rudiments of the gastral filaments in the gastral cavity.
3. Paired marginal lappets are formed on the perradii and interradii of the oral end, with the rudiments of the modified sense organs (rhopalia). The latter are lying in the cleft between the members of each pair.
4. At this time, a number of transverse constrictions appear in the long, slender, tubular polyp (Fig. 5.24(I) B (g) and (h)).

These constrictions gradually deepen, giving the polyp the appearance of a stack of inverted plates (Fig. 5.24(I) B (h)), i.e. these plates lie upon each other like a pile of saucers, connected by strands of tissue in which runs the muscle of the interradial mesenteries continuous through the pile of the developing individuals. This means that the polyp enters the strobila stage (Fig. 5.24(I) B (h)). The uppermost plate-like disc (saucer) undergoes the changes mentioned above and becomes provided with marginal lappets and rhopalia. It is thus ready to separate from the strobila by strobilation process. The muscles contract violently at intervals until the communicating strands snap and one by one ephyra swims away. This separated ephyra gives the next in the stack the time to prepare for its release. The basal part of the strobila remains and develops new tentacles, and after a resting period as a scyphistoma (essentially a hydratuba), it may strobilate again. This 'polydisc' strobilation is a secondary adaptation for the more effective spread of the species (Borradaile et al., 1977).

The newly released ephyra is a flower-shaped animal (Fig. 5.24(I) B (i)), pale reddish-brown in colour (Uchida and Yamada, 1968). It measures 0.3–0.4 mm in diameter, with eight pairs of marginal lappets, and with a rhopalium between each member of the lappet pairs. This rhopalium will become the statocyst in the adult. It has a short manubrium projecting from the centre of the subumbrella. The manubrium ends with a mouth opening with four lips. This ephyra swims free in the medium, and is regular found in spring plankton hauls. The ephyra then transforms into an adult by:

- filling up the adradial notches in the margins;
- the growth of the bell as a whole;
- the enormous increase of the mesoglea so as the two layers of the endoderm come together to form a solid lamella except where the canals occur; and
- the mesenteries losing their attachment and ceasing to exist as partitions with the collapse of the enteron; however, their position is marked by the gastric filaments.

In addition, during this process, new tentacles are formed on each adradius and their number continues to increase; additional marginal lappets are also formed in the adradial regions. The manubrium develops long oral arms and the subgenital pits, gastrovascular system, gonads and the other structures increase in size and so on. The young medusa that has completed these transformations is still only about 1 cm in diameter.

Embryonic development and metamorphosis of the scyphozoan *Aurelia* were studied by Yuan et al. (2008) using up-to-date tools (antibody markers combined with confocal and transmission electron microscopy). The authors collected their material (embryos and planula larvae) from oral arms of adult *Aurelia aurita* medusae obtained from the Cabrillo aquarium in San Pedro, California, United States. The authors noted that *Aurelia* sp.1 medusae have been housed with *Aurelia labiata* medusae, and it may be possible that some of the specimens are F1 hybrids between *A*. sp.1 and *A. labiata*. However, they found no evidence to support this. Planulae recovered from the oral arms of *A*. sp.1 were allowed to settle and develop in a bottle dish for a week in order to collect metamorphosing planulae and primary polyps.

In the view of Yuan et al. (2008), the fertilized eggs of *Aurelia* undergo equal and regular cleavage (Fig. 5.24(I) A (a)–(c)), resulting in a coeloblastula of columnar epithelial cells surrounding a small central blastocoel (Fig. 5.24(II) A). During the blastula stage, cells do not have cilia. Ciliation appears with the onset of gastrulation. Cytoplasmic microtubules radiate outwards and basally from around apically located nucleus of each blastomere, forming a conspicuous cortical cytoskeleton (Fig. 5.24(II) A). Gastrulation begins at a time when the blastoderm contains ~700–1000 cells. Cells on one side of the blastula invaginate and form an inner epithelial layer, the primary endoderm (Fig. 5.24(II) B) that surrounds a closed cleft-like central lumen. The blastopore is only transient and does not form a mouth opening in the planula. During the course of gastrulation, both ectodermal and endodermal cells start to produce cilia at their apical pole. Cell division is still frequent during and after gastrulation; there are estimated to be 2000–3000 cells.

Planula larva

The transition from embryogenesis into a freely motile larval planula stage is gradual. Early planulae are short and rounded (Fig. 5.24(II) C), and during later stages they gradually elongate (Fig. 5.24(II) D and E). The aboral pole is defined by the presence of the apical organ, consisting of elongated (sensory) ectodermal cells. This apical organ directs the movement of the organism. The ectoderm of the planula is pseudostratified, with superficial multilayer of densely packed nuclei and a deep layer of more widely spaced nuclei (Fig. 5.24(II) C). This ectoderm of the planula contains several types of specialized cells, including FMRFamido-immunoreactive neurons (Fig. 5.24(II) D). The endoderm of the planula consists of large regularly stacked epithelial cells and they are highly vacuolated. Their cytoplasm 'stands out' due to their high F-actin content, and vacuoles are visible as dark (F-actin-negative) spherical areas against the bright (F-actin-psitive) cytoplasm. The endodermal epithelium is ciliated at its apical surface that surrounds a closed cleft-like inner lumen (Fig. 5.24(II) E).

These detailed characteristics of the structure of the planula appeared by the transmission electron microscopy, which adds a number of other details such as cnidocytes and gland cells (Fig. 5.24(II) F) in between the ectodermal epithelium. Cnidocytes are easily observed as the egg-shaped nematocysts between the ectodermal cells.

Larval settlement and onset of metamorphosis

Planula settles by attaching their aboral end to the substrate. During this process, the apical organ 'indents' at its centre (Fig. 5.24(II) G). Afterwards, the solid planula body gradually changes into a polyp, which possesses tentacles and a gastric cavity. The latter opens through a mouth at the former posterior pole of the animal. The settled planula attains a bell-shape, shortening along the oral-aboral axis and widening in the transverse axis (Fig. 5.24(II) H and I). The densely packed layer of endodermal cells appears in the oral region and is called the secondary endoderm from the oral region to the aboral pole (Fig. 5.24(II) I). A lumen, lined by the ciliated apical surfaces of secondary endodermal cells, forms the gastric cavity and mouth (Fig. 5.24(II) G (h)). During settlement and early metamorphosis, the ectoderm changes. The pseudostratified epithelium contains in the order of 10,000–15,000 cells and the nuclei are densely packed, the ectoderm decreases in diameter, the nuclei become less densely packed and form one or two layers (Fig. 5.24(II) P and Q), i.e. the ectoderm of the early polyp contains a significantly lower number of cells than the planula per given area. Thus, during early metamorphosis when the secondary endoderm appears, the number of (primary) ectodermal cells seen in the planula decreases. No evidence of cell death was observed in the ectoderm (Fig. 5.24(II) M and N) and cell proliferation appears high during metamorphosis (Fig. 5.24(II) K and L). The authors' interpretation is that cells of ectodermal origin most likely contribute to the formation of the secondary endoderm.

Late metamorphosis: The formation of tentacles and gastric pouches

The settled larva prior to the appearance of tentacles has a square shape. At this stage, the secondary endoderm shows condensations that foreshadow the appearance of gastric septa and tentacles. Primordia of the gastric septa form four spikes, arranged in the shape of a cross, which radiate out from the endoderm surrounding the mouth opening (Fig. 5.24(II) O). These septa persist and subdivide the periphery of the gastric cavity into four gastric pouches. Tentacles, on the other hand, do not appear all at once, but the sequence is 4, 8, 16, and 32. The formation of the first four tentacles, which are arranged in a regular quadrant, is coordinated with the formation of gastric longitudinal septa, and is initiated in the secondary endoderm. At locations where these septa contact the perioral ectoderm, the gastric pouches produce small hemispherical evaginations for which the authors proposed the term 'tentacle papillae'. The ectoderm overlies the papillae evaginations and forms the tentacles. The papillae themselves are incorporated into the tentacle lumen and develop into the tentacle endoderm (Fig. 5.24(II) O).

At the stage when the first four tentacles have reached a length corresponding to approximately twice their diameter, the next four tentacles appear at positions between the existing ones (Fig. 5.24(II) R and S). The tentacle process repeats itself, resulting in a final number of 32 tentacles.

Other Semaeostomeae Scyphozoa

In addition to *Aurelia*, other genera of Semaeostomeae have been most thoroughly investigated. In general, their development is closely similar to that of *Aurelia*. ***Pelagia noctiluca*** (Fig. 5.20E) (a form famous for its phosphorescence (Marshall and Williams, 1975) is a small medusa with diameter that does not exceed 6 cm (Fig. 5.20E). *Pelagia* makes an exception to this general cycle of Aurelia. It differs in that its planula develops directly into an ephyra (Fig. 5.25A (g)), without passing through hydratuba and scyphistoma. The planula of *Pelagia* is ciliated, but a small mouth forms on its lower side and this region gives rise to endoderm tissue, which invaginates into the interior and forms a large gastral cavity. Eight lappets are first formed on the lower side; these soon become eight pairs, and a sense organ (short tentacle) develops in the spaces between them. After loss of the cilia, the larva (Fig. 5.25A (e)) moves by contraction of the bell.

Chrysaora quinquecirrha (Fig. 5.25B) is one of the common North Atlantic jellyfish. This species is variable in colour and attains a bell diameter of 10 in. It is a worldwide animal and lives in warm seas. It belongs to the same order of *Aurelia* (Semaeostomeae) and has gastric pouches and septa in the larval stage only, but the adults possess a system of radial canals, which result from a fusion of the upper and lower layers of the gastrodermis as the case of *Aurelia* (Fig. 5.21A). Reverberi (1971) reviewed and illustrated the embryonic and postembryonic development of *Chrysaora hyoscella* until the strobili stage (Fig. 5.25B).

Order Rhizostomae

Genus *Mastigias*

The rhizostome jellyfish *Mastigias* (Agassiz, 1862) is native to the western and central Indo-Pacific from Australia to Japan, and Micronesia to the Indian Ocean (Bayha and Graham, 2011). In the native range, these jellyfish occupy

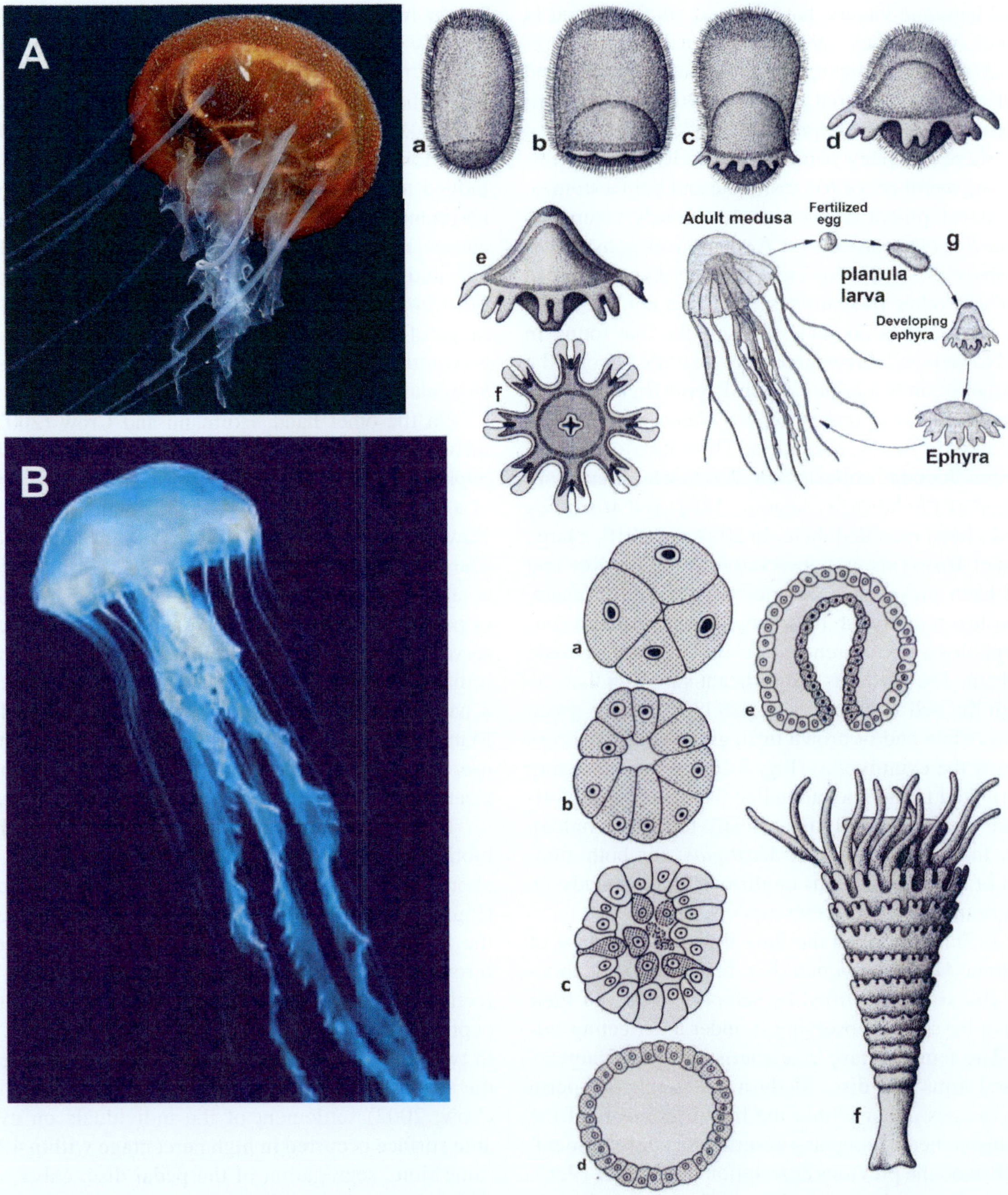

FIG. 5.25 (A) *Pelagia benovici* sp. nov., sublateral view, was recorded in the North Adriatic Sea from September 2013 to early 2014, collected and studied by Piraino et al. (2014); embryonic and post embryonic development of *Pelagia perla* from (a–f): (a) lateral view of planula with mouth cone at the oral end; (b) L.V. of planula with primordium of the eight marginal lappets; (c) the same with the growth of the marginal lappets; (d) shortening of the main axis, flattening of the subumbrella, and full development of the wing lappets, (e) full development of the mouth cone, lappets, sense clubs; (f) ephyra larva *(after Delap, 1905)*; (g) life cycle of *Pelagia*, a scyphomedusa lacking the polyploidy stage; (B) photograph of live *Chrysaora quinquecirrha* (Sea Nettle) a common species variable in colour, attains bell diameter of 10in. Warm seas worldwide (Keoki Stender); (a–f) embryonic and post embryonic development of *Chrysaora hyoscella* until the strobili stage: (a–e) are sections in cleaved eggs, (a) 4-cell stage, (b) later stage with beginning of the blastocoele, (c) blastula with numerous blastoderm cells in the cavity, (d) blastula starting gastrulation, (e) invaginated gastrula, (f) lateral view of polydisc strobili *(after Hadzi, 1907)*. *(All from Reverberi, 1971.)*

coastal and lagoonal waters, but are most notably found in marine lakes, where they can form famously large aggregations that are important tourist attractions. Uchida and Yamada (1968) recorded that Uchida (1926), as a native of Japan, studied ***Mastigias papua*** (the golden medusa) and stated that there are many points of similarity in the development of the members of Rhizostomae and Semaestomea. The early development of this species closely resembles that of *Aurelia*. (The process of formation of ephyra was not then observed in this species, but it is known that in ***Cassiopea*** and other rhizostomes, only one ephyra forms at a time, and there is no stalk of discs like that found in *Aurelia*.) Nowadays, rhizostomes are recorded from the western Atlantic, in a marine lake in Puerto Rico (Laguna Joyuda) and a working quarry on No Name Key, Florida (United States) in the Florida Keys. The single individual from Laguna Joyuda, collected in 2002, was previously misidentified as *Phyllorhiza* (Agassiz, 1862) and *Mastigias* has not since been recorded there. In 2009 and 2010, a large population of *Mastigias* was observed in No Name Key and may have been present for decades (Bayha and Graham, 2011). The lagoon jellyfish has many subspecies, and four main morphological differences are used to differentiate between them. The colour is a significant variation that exists, though the bell is usually greenish blue to olive green with yellow, white and/or brown oval, granular spots across the rim (over the exumbrella) (Fig. 5.26A). This colouring can be attributed to the zooxanthallae that reside symbiotically within the lagoon jellyfish (mostly in the mesoglea). Therefore, the identification of *Mastigias*, for both sampling sites of the Atlantic, was confirmed and the study on this genus flourished in different aspects.

Chang (2011) described the fully developed medusa of the rhizostome *Mastigias papua* (Fig. 5.26A) as being dioecious and able to be identified by sampling a part of their reproductive tissue and observing it under a dissecting microscope. The females have characteristic brood filaments on their oral arms and disc. Medusa males release sperm that swim to eggs either within the brood filaments of the female or inside her. Chang also described its development, which confirmed the previous description of Uchida (1926). The eggs hatch into tiny, oval-shaped, flat planula larvae, mouthless larvae that swim around until they find a suitable substrate to settle by their anterior ends. The process of settling takes between a few hours and a few days, during which the larva develops into a sessile polyp, or scyphistoma. The polyp feeds and gradually increases the number of tentacles from an original two to four and then eight (Uchida and Yamada, 1968). It grows to the degree that it can reproduce. At this point, it either can create motile buds agametically, or goes through monodisc strobilation yielding yellow-green ephyra. Thus strobilation in this group is mono-strobilation. Sugiura (1965) stated that water temperature has a great deal to do with the initiation of strobilation. Every year, the earliest appearance of ephyrae of *Mastigias papua* in the vicinity of the Misaki Laboratory (Japan) is in early summer. Considering seawater temperature to be related to strobilation must be that at the bottom of the bay, where the settled scyphistomae live, measurements were taken at 1.5–3.0 m in depth in the bays of Aburatsubo and Moroiso. The bottom temperature for strobilation was judged to be about 22°C. In addition, Sugiura studied experimentally the effect of temperature on strobilation. The author reared scyphistomae having zooxanthellae from late autumn until the next spring and concluded that for *Mastigias papua*, in the laboratory, the minimum effective range of temperature for induction of strobilation by giving zooxanthellae is found to be 20–22°C. It is about the same as in nature.

On the other hand, Hofmann and Crow (2002) experimentally induced metamorphosis of the larvae of *Mastigias papua*. They obtained the larvae from released egg masses of a medusa displayed at the Waikiki Aquarium in Honolulu Hawaii (originating from Jellyfish Lake in Paulo, Micronesia). The hatched planulae larvae maintained in filtered seawater containing antibiotics (ABS) made up with 100 μg/mL each of penicillin, streptomycin, and newmycin, were readily observed within about 48 h. The other planula larvae maintained in filtered seawater (through a 0.45 μm) at 26°C. To serve as a natural substratum, dark deteriorating leaves of the Red Mangrove (*Rhizophora mangle*) were collected and, before use, the surface biofilm was wiped off with filter paper. The cleaned leaves were rinsed thoroughly in filtered seawater.

The larvae maintained in filtered seawater without antibiotics were very low in development, since most embryos aborted and the live ones ran as controls in the experiments (Fig. 5.26C). They continued swimming around with the blunt end ahead and all remained in the larval state throughout the observation period. However, in the perfectly maintained cultures and with 15 μg/ml of the hexapeptide Z-Gly-Pro-Gly-Gly-Pro-Ala 'Z-GPGGPA' solution in two sets of larvae (for details of this substance forming the solution, refer to material and methods, Hofmann and Crow, 2002) settlement of the individuals on every suitable surface occurred in high percentage within 48 h. At the same time, segregation of the pedal disc, calyx, and stalk with the first tentacle anlagen appeared (Fig. 5.26D–F). Metamorphosis occurred in almost 100% in one set and about 70% in the other set within 24 h. The most obvious observation was that the deteriorating leaves of the Red Mangrove (*Rhizophora mangle*) proved to offer a very attractive biogenic substratum for *Mastigias papua* larvae to settle on. The planulae attached exclusively to the leaf surfaces (Fig. 5.26G), mostly on the shaded side. None was observed fixed to the glass walls, or hanging from the water surface pellicle.

This work of Hofmann and Crow (2002) shows that *Mastigias papua* polyps propagate agametically by strobilation and by lateral budding (Fig. 5.26H) of motile, larva-like buds, arising at one or more sites from the lower portion

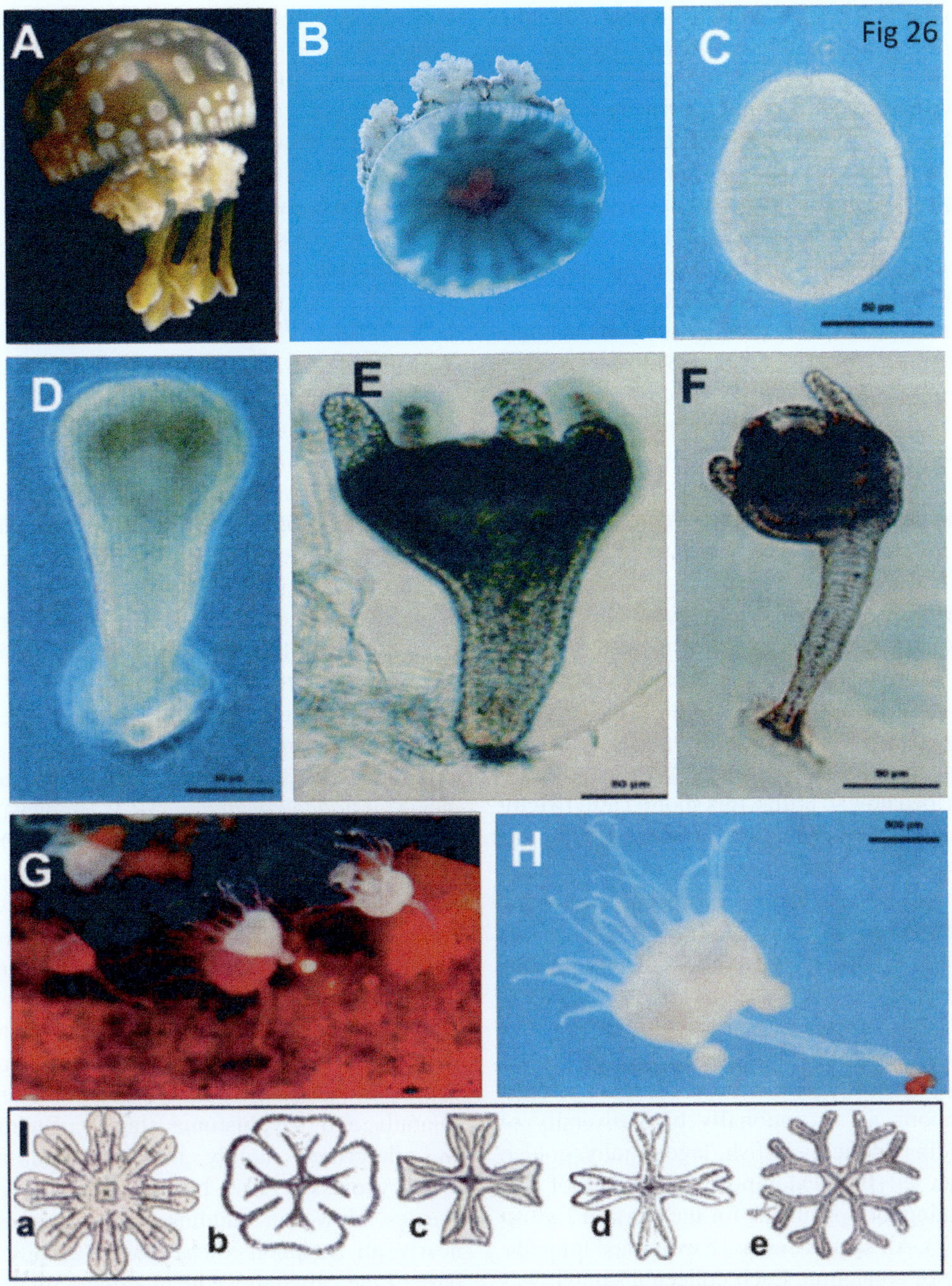

FIG. 5.26 (A) *Mastigias* spp. (golden medusa), collected from No Name Key, Florida (United States) in October 2009. Bell diameter is ~6 cm. *(Bayha and Graham, 2011)*; (B) *Cassiopea andromeda* (upside-down jellyfish), *Alamy Stock Photo*; (C–H) development of *Mastigias papua*; (C) photomicrograph of newly hatched planula larva of *Mastigias papua,* the larva will attach with the anterior blunt end; (D–F) early and successive advanced stages of planula larva undergoing settlement and polyp metamorphosis; (E) shows the fibrous material trapped accidentally in mucous secreted by the settling polyp; (G) *Mastigias papua* polyps developed from planula larva which had been induced to settle on a degrading *Rhizophora mangle* leaf. Each measures 3–4 mm in height with expanded tentacles; (H) polyp with spindle-shaped buds, each on either side of the calyx, just before release. *(C–H) From Hofmann and Crow (2002)*; (I) formation of manubrium of *Mastigias papua* after Uchida *from Kume and Dan (1968)*: (a) ephyra larva, immediately after liberation, (b and c) the edges of the four lips around the mouth opening spread horizontally, developing a fringe of small tentacles, (d) each adradial portion becomes deeply indented, while shallow constrictions appear in the perradii, (e) these arms later bifurcate, acquiring an extremely complex structure.

of the polyp's calyx. This work also states that some other scyphystomae, and *Cassiopea xamachana* share this type of bud formation.

Ephyra matures into free-swimming medusa with symbiotic zooxanthellae and sensory organs. Uchida and Yamada (1968) stated that, like the ephyrae of semaestomes, the rhizostome ephyra has a distinct mouth (Fig. 5.26I (a)). The transformation of ephyra into the adult medusa is quite different from that which occurs in semaestomes and is highly characteristic. The most obvious differences are the failure to form tentacles, and the complicated development and transformation of the gastrovascular septum and oral arms. In this process (manubrium formation), the edges of the four lips around the mouth opening spread horizontally, developing a fringe of small tentacles (Fig. 5.26I (b) and (c)). Each adradial portion becomes deeply indented, while shallow constructions appear in the perradii (Fig. 5.26I (d)). These arms later bifurcate again (Fig. 5.26I (e)). During the elaboration of the manubrium, the central mouth is closed off and many small mouths appear among the fringes of the oral arms, connected by brachial canals, which are subdivisions of the gastrovascular cavity formed by the repeated bifurcations of the oral arms. In *Mastigias* and similar species, the manubrium consists of four 'stomach pillars' supporting eight oral arms; each oral arm has two 'wings' and bears an elongated, cub-shaped appendage. Later in development, the gastrovascular canals branch to form a complicated network. *Mastigias papua* breeds once yearly. Breeding season is from May to June (Chang, 2011).

It appears that this species, *Mastigias papua*, has not only a scientific interest, but also an economic interest. They are also an indicator for water quality; as the oxygen rate in the water goes down, so does the respiratory rate of the jellyfish. It would also contain a variety of some important nutrients found in this kind of jellyfish. Egloso and Urboda (2017) studied and described how to utilize this spotted jellyfish for gelatin production, since this gelatin has less carbohydrate and fats and is perfect for food diet. Gelatin is also the basis for other different jellied products. A survey was conducted through ocular visitation. A sample was randomly taken and was brought to the lȧboratory for analysis. These spotted jellyfish are prevalent in both inshore seas and oceans in mostly warm tropical waters. Palompon (Philippines) is a community located along the coastal area and it is greatly known for its abundance of marine resources and other sea creatures. It supports an exceptionally high diversity of marine life including spotted jellyfish, lagoon jelly, golden medusa or papuan jellyfish. Palompon was the site of collecting and obtaining the specimens for the work of Egloso and Urboda (2017). All specimens were examined immediately at the place of collection. Each collected sample was examined quantitatively with the following morphological features in terms of diameter, size, visceral mass, physical structure, and colour dimorphism.

The process of making gelatin is usually derived from collagen, a natural protein substance present in mammals. So far, no studies have been conducted on the capabilities of producing gelatin compound from spotted jellyfish, however, Egloso and Urboda (2017) investigate the morphological characteristics and production rate of gelatin from utilization of spotted jellyfish (*Mastigias papua*). The method prepared 100 g samples of spotted jellyfish (*Mastigias papua*), 100 mL of vinegar (acetic acid), 20 g of seaweed and 10 mL of mineral water. The product yield of 0.35 g of gelatin per grams of raw jellyfish was produced containing 30% pure collagen and 70% seaweed extract. Furthermore, the researcher from Palompon Institute of Technology became interested in 'spotted jellyfish' to support the livelihood of their local fishermen. Thus this study proved that the production of gelatin using the collagen from spotted jellyfish was technically and economically feasible to invest in. Further studies are recommended for further development of this process.

Cassiopea andromeda Forskål, 1775

This genus, *Cassiopea*, has an almost circumtropical distribution, and its representatives are, at first sight, strikingly uniform (Hummelinck, 1968). They are usually to be found in shallow lagoons and marine pools lined with mangroves, where they lie with oral arms uppermost, scattered or sometimes closely together on the sandy or muddy bottom. The surface of the exumbrella (Fig. 5.26B) is adapted to this peculiar mode of life: the central part being, as a rule, slightly concave, while on the more peripheral part an inconspicuous raised ring has developed which enables a firmer contact with the bottom (Hummelinck, 1968). In the Caribbean region, only two species of *Cassiopea* are found: *C. frondosa* and *C. xamachana*. The Caribbean *andromedae* are generally considered as belonging to one and the same species, *Cassiopea xamachana* (R.P. Bigelow, 1892, described from Jamaica). Gohar and Eisawy (1960a) suggested *C. xamachana* not to represent a separate species but only a geographic variation of *C. andromeda*. However,*Cassiopea andromeda* can be recognized as having basically 16 rhopalial canals and 32 radial vessels (Hummelinck, 1968).

Agametic reproduction Gohar and Eisawy (1960b) reported that agametic reproduction in scyphomedusae occurs mainly by budding. It takes place in the larval generations (planula and scyphistoma stages). Budding of planulae was described mainly in Stauromedusae (*Haliclystus*, Wietrzykowski, 1912) but was not observed in any other species. However, budding of the scyphistoma occurs in nearly all Scyphomedusae. Buds usually grow from the base of the calyx above its junction with the stalk and grow as large protrusions without a stolon. This simplest type of budding is one of the six types that occur in scyphistomes and is associated with the rhizostome *Cotylorhiza tuberculata*. In the present species, budding is only a rarity and occurs in scyphistomes between the 22- and 28-tentacle stages. Gohar and Eisawy did not observe budding in scyphistomae reared in out-of-door aquaria. However, some buds appeared in the scyphistomes transferred into the laboratory. This procedure (moving scyphistomes from out-of-door aquaria into the laboratory) was repeated several times. Only about 12 buds were obtained out of the few hundreds of scyphistomes used. The authors investigated and reported that budding does not constitute an important method for the propagation of the species. It seems to occur only under unfavourable conditions.

The bud appears at first as a protrusion, gradually enlarges and becomes hemispherical. It then elongates into a spindle-shaped structure just before it separates (Fig. 5.27(I) E 2, 3). This process from the appearance until detachment of the bud takes about 10h. On becoming free, the bud looks like and swims as a planula, rotating on its long axis with the distal end forward. At the end of its free life span, it attaches to the substratum with the distal or anterior end. This end will form the stalk, while the posterior, originally proximal end becomes the oral side by the development of the mouth. Similar to the planula-like larva of *C. xamachana*, it contains greenish brown spots due to the presence of zooxanthellae. This planula-like larva differs from the normal planula in size, measuring about 360μm, which is greater than double the size of the normal one.

An astonishing and unusual case of agametic reproduction that was noticed only once, shown in (Fig. 5.27(I) E 4). This has not been described so far in scyphomedusae. This stage is one scyphistoma with one stalk and two calyces united at the bases. Each calyx has its own crown of tentacles and proboscis. The number of tentacles was 25 in one and 23 in the other. The authors investigated that this stage may be produced either by L. fission in the original scyphistoa, or by budding at an early scyphistoma stage (or possibly the two-tentacle stage). They proposed the more probability for the former view, as the buds never become equal in size to their parents. In addition, budding usually occurs in advanced stages (22- to 28-tentacle stages).

In some species of *Cassiopea* in the previous works, there was no definite view of the separation of sex (e.g. *C. xamachana*). In others, such as *C. frondosa*, they not only had separate sex, but also showed sexual dimorphism. Gohar and Eisawy (1960b) claimed that *C. andromeda* is unisexual, but there is no detectable sexual dimorphism. Only in the breeding season can one differentiate the males from the females: the males by the translucent white masses of spermatozoa at the base of the oral vesicles near the centre of the oral disc; the females, instead, have white opaque masses of eggs. In this species, the number of females is about double that of males, as among 108 mature specimens, 66 were females and 42 were males. In other species, females and males occur in approximately equal numbers, e.g. *C. frondosa*.

The fully formed eggs are usually spherical, with diameters ranging from 0.14 to 0.17mm, and they are usually cemented in masses (Fig. 5.27(I) A 1). They undergo cleavage inside the egg membranes. On the other hand, in fresh smears of testes, the spermatozoon consists of the normal parts, head, neck and tail (Fig. 5.27(I) A 3). The head is spindle-shaped, about 4μm and its anterior three-fourths are occupied by the nucleus. The tail is thin, whip-like and long, measuring about 35μm in length. The eggs are discharged into the coelenteron, and were not observed to pass out through the subgenital pits. The spermatozoa are discharged in the same way as the eggs. The whole germinal vesicle is discharged into the coelenteric cavity and then to the exterior through the axial canals. These vesicles rupture and the spermatozoa aggregate in masses around small pieces of the ruptured vesicles (Fig. 5.27(I) A 2). The aggregated spermatozoa move in the seawater until they reach the oral disc of the female, where fertilization occurs.

The authors obtained unsegmented eggs from the coelenteric cavity and reared them in filtered seawater; few underwent segmentation and the majority died and deteriorated. On the other hand, the eggs obtained from the oral disc segmented when reared. This proved that complete fertilization occurred mainly on the oral disc of females.

Development Gohar and Eisawy (1960b) reviewed the nature of development of the eggs of Scyphomedusae since it is correlated to the size of the egg and the type of gastrulation.

1. The smallest eggs produce vermiform planulae, which form scyphistomae after a series of aggregations and divisions;
2. The intermediate-sized eggs, form planulae, which metamorphose into scyphistomae in the usual way;
3. Larger eggs may form 'actinula' larvae, which give scyphistomes without passing an active planula existence;
4. The largest eggs develop directly into ephyra stages without forming scyphistomae.

It is appropriate to mention that this review was done before the separation of cubozoa from scyphozoan. The eggs of *C. andromeda* are of the intermediate size and thus form normal planula and scyphistoma stages.

The planula of *C. andromeda* is of ordinary type. It is oval, translucent, white and covered with very fine cilia. On the oral side, these cilia are somewhat longer (Fig. 5.27(I) B). It is compressed sideways so that it is oval in T.S. Its length ranges between 160μm and 190μm, rarely reaching 210μm. The largest width is about 130μm and the thickness is about 65μm. It has no mouth and during swimming, the narrower pole of the ovoid larva is directed forwards. The larva usually rotates upon its long axis from right to left (anticlockwise), rarely in the reverse direction. It is positively phototactic, swimming to the more lighted areas in the rearing basins. This life span is about 3–4days, after which the planula settles, and attaches to develop into the scyphistoma. In Scyphomedusae, in general, the individual undergoes two metamorphosis cycles in its life cycle: first from swimming planula to settled polyp; second from settled polyp to swimming medusa. Gohar and Eisawy described and illustrated these two cycles in detail for *C. Andromeda.* Fig. 5.27(I) B illustrated developmental stages (metamorphosis) of the planula to scyphistoma.

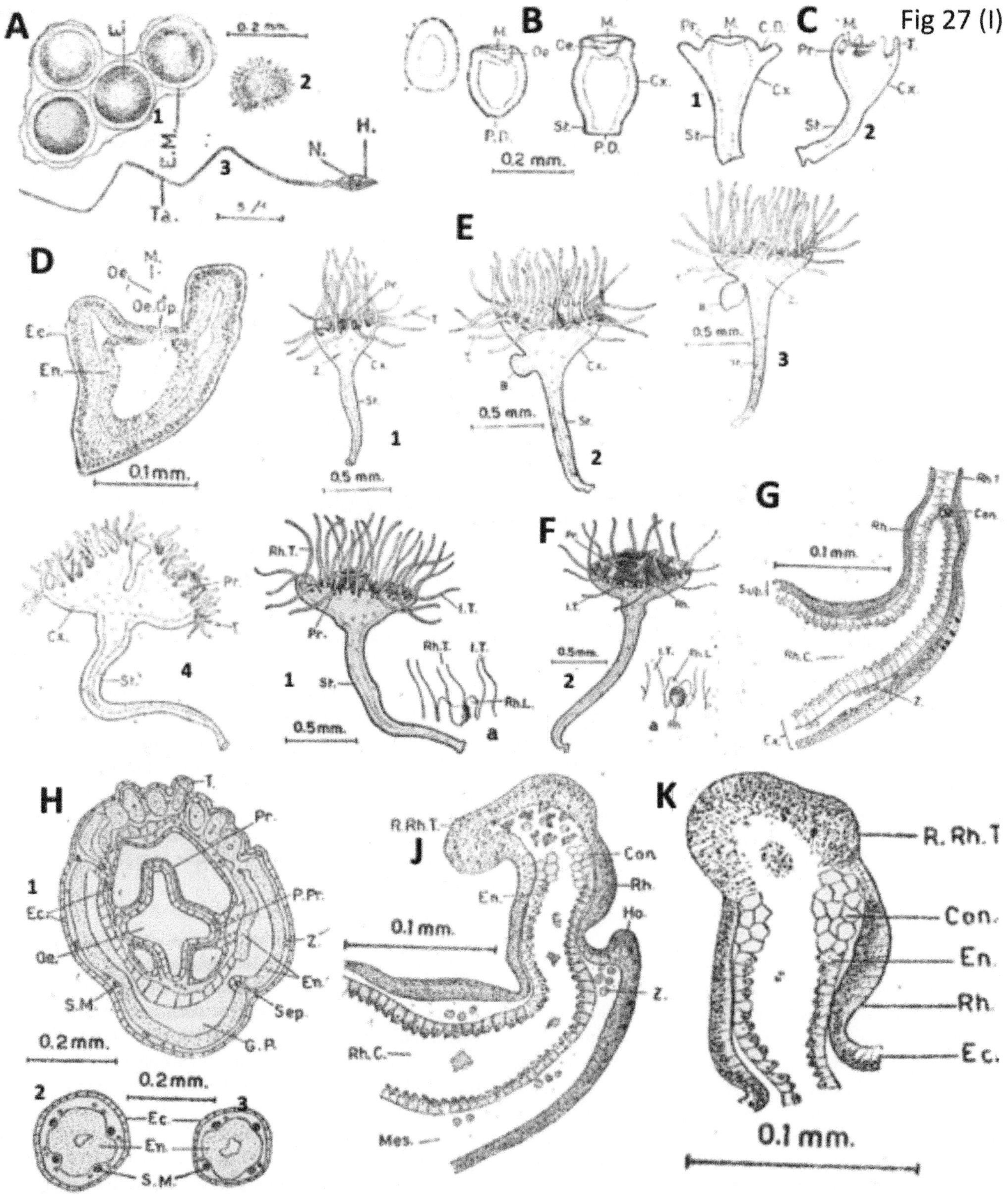

FIG. 5.27 (I) Development of the rhizostome *Cassiopea Andromeda (from Gohar and Eisawy, 1960b)*: (A) gametes: (1) egg mass, (2) sperm mass, (3) highly magnified spermatozoon (bar 0.2 mm between 172); (B) developmental stages of the planula: (1) free swimming planula, (2) just attached to substratum (P.D.) pedal disc, (3) slightly larger stage with cubical oesophagus (oe) and the endodermal evaginations disappear. Mouth (M) not open; (C) scyphistoma with 2 tentacles (1) and with 8 tentacles (2), C.D. = circum disc or peristome, Cx = calyx, Pr = 'prodocis', St = stalk, T = tentacle; (D) oblique LS of slightly older planula larva showing formation of oesophageal opening, Ec = ectoderm, En = endoderm, Op = oesophageal opening; (E) scyphistoma development and budding: (1) with 16 tentacles, (2) with 24 tentacles and developing a bud, (3) with 28 tentacles and with a fully developed bud just before separation, B = bud, (4) scyphistoma showing an abnormal case of binary fission; (F) strobili showing first traces of rhopalial structures: (1) early stage in strobilation in which rhopalial tentacles degenerating, a- part of margin of same specimen; (2) strobila with completely degenerated rhopalial tentacles, (a) portion of margin of same, enlarged; (G) rhopalial tentacle at first appearance of concretionary crystals, con = concretionary crystals or otolith, Ex = exumbrella, Rh = rhopalium, Rh.C. rhopalial canal, Rh.T. rhopalial tentacle, sub = subumbrella, Z = zooxanthellae; (H) oblique TSs in different regions of fully developed scyphistoma: (1) through upper part of calyx, (2) through junction between calyx and stalk, (3) through middle of stalk, G.P. = gastric pouch, P.Pr. = pillar of proboscis, Sep = septum, S.M. = septal muscle; (J and K) successive stages of degenerating rhopalial tentacle. Ho = hood, Mes = mesoglea. *(From Gohar and Eisawy, 1960.)*

Fig 27 (II)

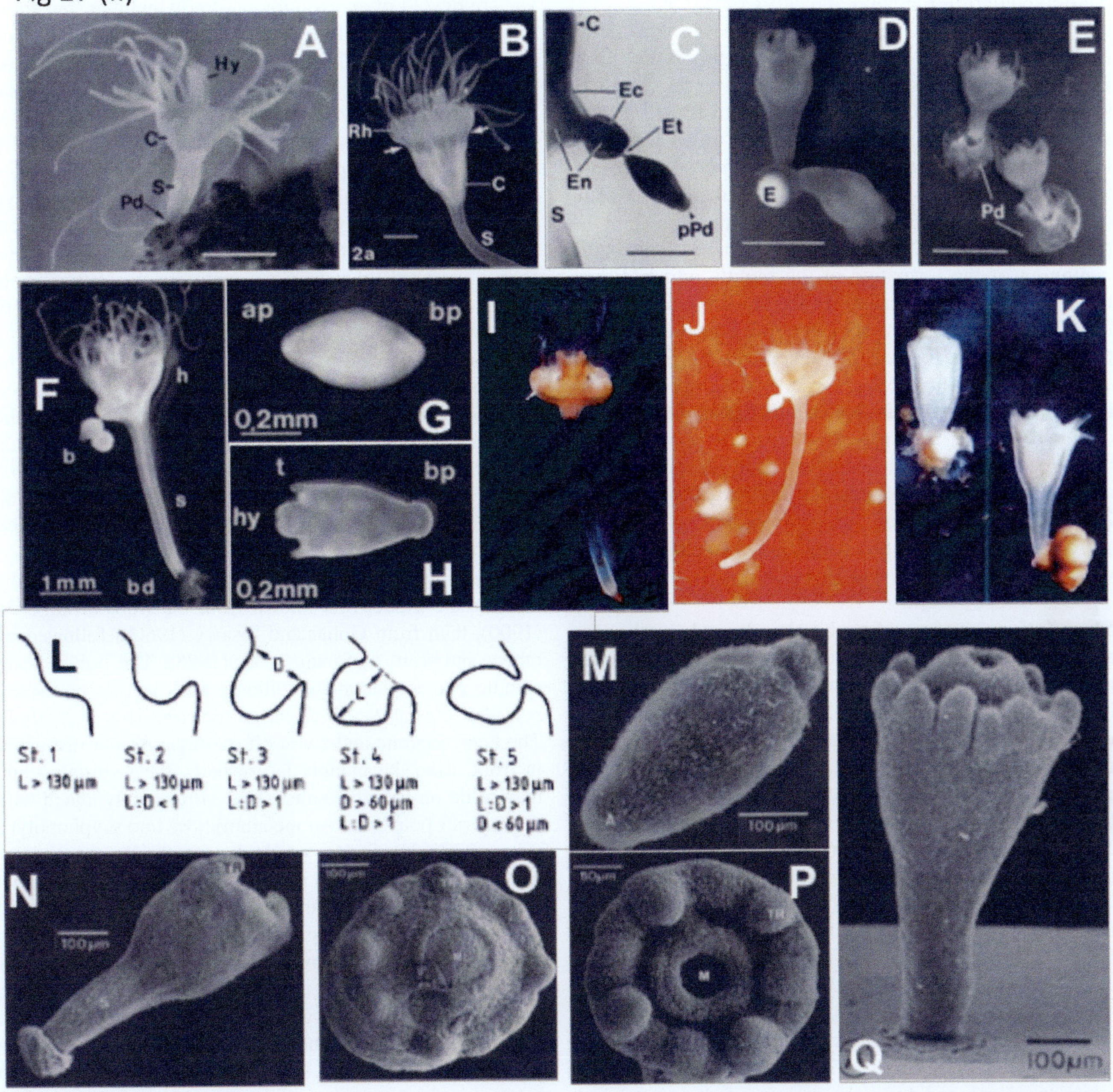

FIG. 5.27, CONT'D (II) Development and life cycle of the rhizostome *Cassiopea Andromeda*: (A–E) scyphistoma showing strobilation and budding *(from Hofmann et al., 1978)*; (A) mature scyphistoma; (B) early stage of strobilation, note rhopalial lobes developing at margin of the calyx and beginning of the constriction *(arrows)*; (C) budding polyp with one bud which will shortly be constricted off, (A–C) X = 1 mm; (D) budding region of scyphistoma with two buds at different stages of development, endoderm of proximal bud is continuous with endoderm of polyp, distal bud is connected only by an ectodermal tube; (E) two polyps developing from buds which settled on an *Artemia salina* egg shell (E), (D and E) X = 0.5 mm; (F–H) budding in *Cassiopiea Andromeda (from Neumann, 1977)*, (F) scyphistoma with three buds; (G) newly detached bud; (H) 4-day-old bud; (I–K) *other work of Hofmann et al. (1996)*, (I) full-grown polyp at an early stage of strobilation. Note the constriction of the calyx, regression of tentacles and formation of rhopalial and inter-rhopalial lappets *(arrowhead)*, upper diameter about 2.5 mm; (L–Q) other study of Hofmann and Honegger (1990), (L) definition of five bud stages (St.) according to profile, and ratio of bud length (L) to bud diameter (D); (M) Detached, ciliated-bud swims with its aboral end, presumptive foot region, directed forwards; (N) metamorphosing bud exhibiting tentacle primordia, stalk and prominent aboral knob; (O) oral view of metamorphosing bud featuring tentacle rudiments and prospective mouth; (F) oral view of cup-shaped head of young scyphistoma in which the mouth has been opened; A = aboral end, ap = apical pole, b = bud, bd = basal disc, bp = basal pole, C = calyx with tentacles ('head'), EC, Ec, EP = ectoderm, EN, En = endoderm, Et = ectodermal tube, h = head, Hy = hypostome with mouth opening, M = mouth, MG = mesoglea, Pd = pedal disc (covered by algal material), S = stalk (slightly contracted), Oa, Od = oral disc with oral arms developing from the perradial parts of the hypostome, Rh = rhopalial lobes with rhopalia, t = tentacle, TR = tentacle primordia.

During 1–2 days after attachment, the main parts of the polyp are differentiated: the calyx (head) and a small short, narrow stalk (Fig. 5.27(I) C). The growth of the larva continues, especially in the length of the stalk. During this period, the tentacles develop as small protrusions, in their characteristic sequence, to form a crown around the margin of the circum-disc, as will be seen later. In the fully formed scyphistoma, the stalk becomes four times the length of the calyx. After the development of the tentacles, shallow depressions are developed on the peristome in relation with the gastric pouches. This is the beginning of the formation of the mouth, which differs from the previous descriptions for other species. Gohar and Eisawy (1960b) described in detail how the mouth of *C. Andromeda* is formed, and discussed the process including the previous works. The fully developed scyphistomae are those 23–25 days after attachment. The number of tentacles is variable, typically 32, rarely less and often more reaching 42 tentacles. When fully expanded, they are one to two times the length of the body. At the 22- to 23-tentacle stage, the polyp begins reproduction agametically at the lower part of the calyx just above the junction with the stalk (Fig. 5.27(I) E 2–4).

Strobilation The type of strobilation depends on the size and shape of the calyx of the scyphistoma. When the latter has a shallow calyx and long stalk, usually a single ephyra is formed and thus strobilation is monodisc. The other type of strobilation, when the scyphistoma has a large columnar calyx and a short stalk, involves numerous constrictions appearing and several ephyrae being formed successively; the strobilation is polydisc. As described previously and as observed in the work of Gohar and Eisawy, strobilation is typically monodiscus in *C. Andromeda.* The first noticeable change in strobilation of this species starts 30–32 days after fixation of the planula and by the development of rhopalia. The developing polyp at this stage has calyx measuring about 2 mm in largest width. They appear as conical lobes at the bases of the erect series of tentacles and are referred to as rhopalial tentacles, in distinction from the inter-rhopalial ones (Fig. 5.27(I) F (1a)). These conical lobes are developed from the margin of the circumoral disc, and not as described before as widening of the basal portions of the tentacles.

Hofmann et al. (1996) were interested in the life cycle of *C. andromeda.* They worked on specimens obtained from population in the north of the red sea. They found the sex ratio to be 1:1. They did not observe gamete release and fertilization under natural conditions. However, they reported that female medusae are regularly found with masses of eggs or embryos enveloped in mucous and wrapped around the bases of sex specific vesicles in the centre of the oral disc. This is a simple mode of brood protection. This protection lasts until the ciliated planula larvae hatch from their egg envelopes. At this stage, the larvae are devoid of endosymbiotic dinoflagellates, but show morphological and physiological polarity. They swim with the blunt end ahead. These planulae may settle within hours or days on 'suitable substrate', attach irreversibly at their blunt anterior end. After settlement, they metamorphose into the benthic polyp stage (Gohar and Eisawy, 1960b). The developing polyp can acquire the algal symbionts after it forms its oral opening. The fully developed scyphopolyp (Fig. 5.27(II) I) may start forming a medusa by monodisc strobilation. This is the same as previously stated by Borradaile et al., that in *Cassiopeia* the 'hydratuba' only produces a single ephyra at a time, a condition that is obviously primitive compared with *Aurelia.* According to a progressive constriction of the calyx (Fig. 5.27(II) I), the upper portion, which bears tentacles, becomes separated and is transformed into a single medusa-anlage, and the ephyra, within about 1 week. The remaining basal portion of the polyp regenerates the lost tentacular and hypostomal region. It resumes its vital activities such as feeding and may strobilate again at intervals ranging from 3 to several weeks (Hofmann et al., 1978).

Cassiopea andromeda reproduces agametically by an alternate mode, first known from the work of Bigelow (1900), then from Gohar and Eisawy (1960b) followed by others and from Hofmann et al. (1978). The polyps form spindle-shaped, larva-like, ciliated buds at one or several perradial sites at the lower part of the calyx (Fig. 5.27(II) J). The buds separate and swim off, with the former distal end in front. Like the planula larvae buds, the swimming bud may settle on 'suitable substratum', irreversibly attach with the anterior pole and then metamorphose into scyphopolyps (Fig. 5.27(II) K). Contrary to buds of most other cnidarians, the spindle-shaped buds in *Cassiopea* spp. under normal conditions do not reach the juvenile polyp stage while still attached to the parent (Fig. 5.27(II) I). This type of larva-like swimming bud is exclusively found in species of the order Rhizostomea. It provides an important mode of agametic propagation of the polyp generation (Hofmann et al., 1996).

Neumann (1977) studied polyp morphogenesis in a scyphozoan, *Cassiopea andromeda,* and reported that it exhibits typical metagenetic development. Sessile scyphistomae representing the vegetative generation form medusa by monodiscous strobilation and these reproduce gametically to give rise to polyps via a planula stage (Gohar and Eisawy, 1960a,b). Neumann stated that ciliated vegetative buds produced in particular areas of the head region of the scyphistomae (Fig. 5.27(II) F) transform into polyps after settling down on a suitable substrate. When the buds detach from the polyps, their organization is already polarized, the site of attachment is the future apical region which gives rise to the head of the new polyp, while the opposite pole represents the presumptive foot region (Fig. 5.27(II) G). Four days after detachment from the polyp (mother), the free-swimming buds develop four or five short tentacles and a hyptome. Neumann concluded that this vital activity, production of a scyphistome from a free-swimming bud, is mediated by the influence of particular marine bacteria. Later, in 1978, Hofmann et al. evidenced experimentally that in sterile medium or in media containing antibiotics, the formation of polyps is inhibited. Both parts of longitudinal section of the spindle-shaped bud regenerate forming complete but smaller polyps, within 24h, when exposed to the above-mentioned bacterial influence.

As mentioned above, scyphopolyps of *Cassiopea andromeda* (Cnidaria: Scyphozoa) reproduce agametically either by strobilation of ephyrae, which grow into sexually mature medusae, or by releasing free-swimming buds which metamorphose into polyps. Recently, several aspects of these developmental processes have attained considerable attention. Hofmann and Honegger (1990) studied and reported that the formation of swimming buds, which metamorphose directly into polyps, plays the most important role in the propagation of *Cassiopea andromeda*. All the polyps of this species used in their study were originally supplied by the Löbbecke Museum and Aquarium, Düsseldorf, FRG. They were cultured according to the previously described methods (Hofmann et al., 1978). To study bud formation and the renewing of bud-forming cells, clusters of ectodermal polyp cells were marked by injecting Indian ink as described by Neumann et al. (1980). Budding and nonbudding polyps were injected at one or more sites, at various distances from the potential bud-site, then investigated with a Wild M5 dissecting microscope equipped for camera lucida drawing. Under laboratory conditions polyps raised from planula larvae start budding after 'c. 5–7 weeks'. The time required for bud development varies, but it usually takes less than 1 day until detachment, in well-fed individuals. Budding of the polyp occurs simultaneously at one, two, three, or even four sites, depending largely on the feeding regime and on environmental factors (Hofmann et al., 1978). Due to high propagative activity, new buds arise before the first one has separated. Five budding stages were defined (Fig. 5.27(II) A). The first stage was a protuberance formed by the ectoderm, mesoglea, and endoderm. Its development leads to a distinct roundish bud rudiment, then elongates and constricts at its base. Finally, it becomes spindle shaped, and the connection between the bud and polyp is reduced to a thin ectodermal strand ready to detach (Fig. 5.27(II) C). After detachment, the bud is propelled by the cilia of the ectodermal cells (Fig. 5.27(II) G). The spindle-shaped, uniformly ciliated bud (Fig. 5.27(II) G and H) is morphologically and physiologically polarized (Neumann et al., 1980).

Mainly two regions are affected by metamorphosis: (1) the basal knob, which is transformed into a basal or pedal disc for permanent adhesion. Formation of the pedal disc was investigated on buds which had settled on round coverslips (Fig. 5.27(II) E) or on Artemia spp. egg shells; and (2) the hypostomal region in which tentacles and mouth are differentiated. Hofmann and Honegger did not observe formation of the mouth before the appearance of tentacle rudiments. Metamorphosis of buds into scyphistomae is completed by the formation of mouth and tentacles.

Cotylorhiza tuberculata

Cotylorhiza tuberculata (Fig. 5.28A and B) is a common rhizostome in the Mediterranean like other scyphomedusae. Early reports on this medusa (e.g. Claus, 1877) were on its polyps and on its ephyrae in 1890, and also described the morphological aspects on its development in 1884. In recent years, the morphogenesis from ephyra to young medusa and the biochemical composition have been studied. However, this species is much less studied than *Aurelia*, which is the most known species in Scyphozoa. Kikinger (1992) tried to increase information about this species to be known scientifically like *Aurelia*. He obtained the specimens from the Mediterranean waters of Lefkada, a Greek Ionian island.

Beginning with the time of strobilation, as indicated by the first appearance of ephyrae, 8–10 weeks were required until the young medusae attained a mean diameter of 3 cm. At this stage, growth rates became much more rapid, resulting in weekly diameter increases of 3–4 cm. Rapid growth was maintained over a wide size range; only upon attaining a diameter of 25 cm or more did growth rates decrease. There were no sex-specific differences in growth rates, maximum size or abundance. Mean medusa diameter to be adult was 35 cm (maximum 40 cm) (Fig. 5.28A and B). However, in all of these medusae, gonads were always fully developed and gametes were present.

Reproductive system The gonads of the gonochoristic *Cotylorhiza* medusae are band-like endodermal epithelia originating from the gastrogenital membrane. This thin membrane separates the gastric cavity from the subgenital space (Fig. 5.28C). Tissues of both gastrogenital membrane and gonads contain zooxanthellae, mostly being situated in the mesoglea. They are especially numerous in the region of the gastric filaments.

Spermatogenesis *Cotylorhiza* medusae are dioecious; the sperm in the male individual develop in follicles produced by immigration from the genital epithelium into the gonadial mesoglea. This cell immigration takes place near the branching of the gonad from the gastrogenital membrane (Fig. 5.28D). With increasing distance from this area, the size and developmental stage of the follicles increase. Ripe follicles average 200–400 μm in diameter. The progressive spermatogenesis inside each follicle is similar to *Mastigias papua* (Uchida, 1926), *Cassiopea frondosa* (Smith, 1936) and *Cassiopea andromeda* (Gohar and Eisawy, 1960a),

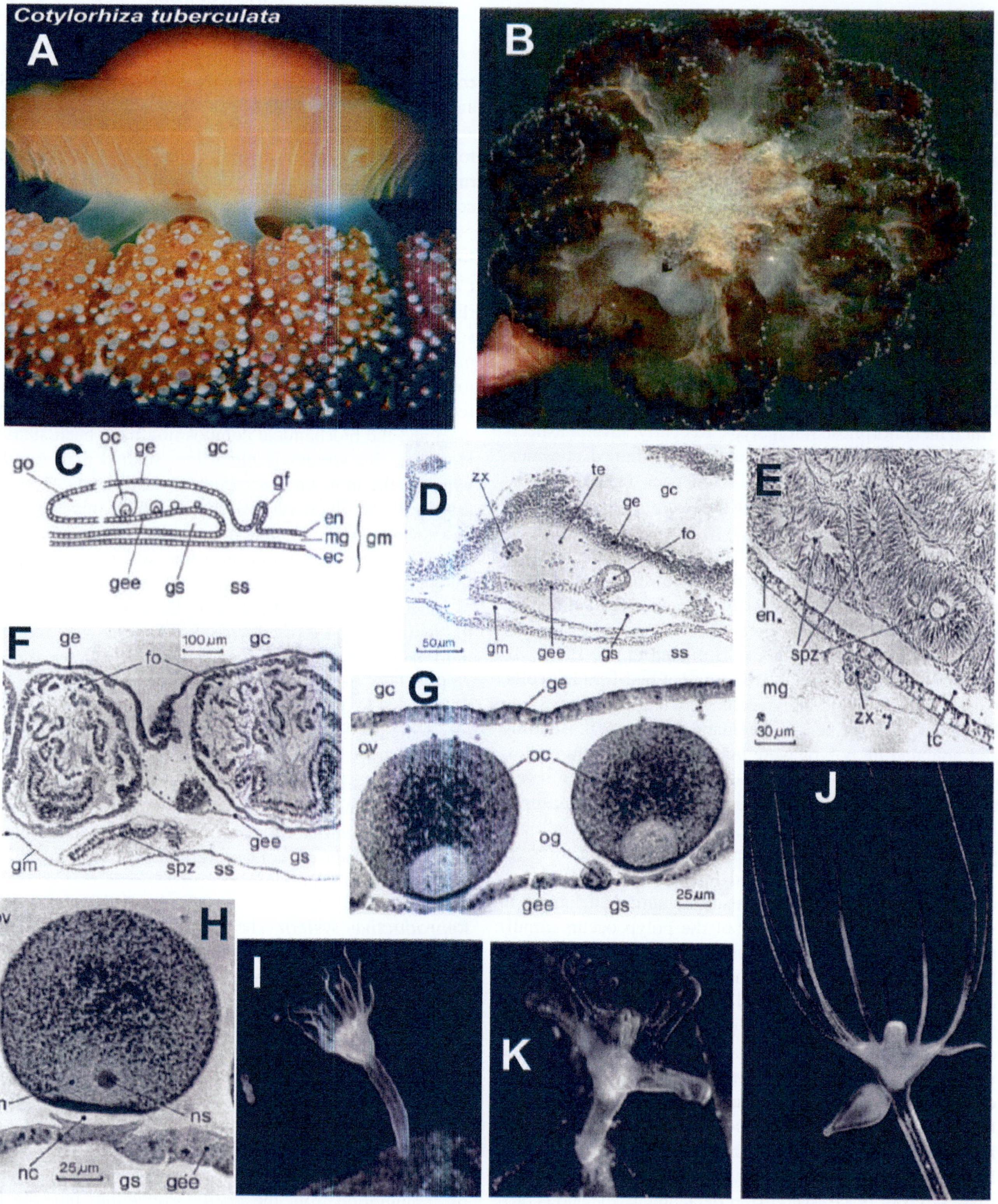

FIG. 5.28 Photographs of live W.M. of the Rhizostome *Cotylorhiza tuberculata* showing their natural colours: (A) of lateral view *(from Beauté du monde animal, Invertébrés Tome XII, 1974)*; (B) of oral view of mature female with embryo-filled moutharm canals (with diameter 34 cm) *(from Kikinger, 1992)*; (C) L.S. through gonad, ec = ectoderm, en = endoderm, gc = gastric cavity, ge = gastric epithelium, gee = genital epithelium, gf = gastric filament, gm = gastrogenital membrane, gs = genital sinus, go = gonad, ir = interradius, ma = moutharm appendage, mc = moutharm canal, rnd = moutharm disc, mg = mesoglea, oc = oocyte, ss = subgenital space; (D–F) spermatogenesis: (D) section of the testis with developing follicle, other lettering: fo = follicle, te = testis, zx = xanthellae; (E) T.S. through sperm-filled male moutharm appendage, other lettering: spz = spermatozeugma, tc = tentacle canal; (F) section through testis with free spermatozeugma in genital sinus, lettering as before; (G and H) oogenesis: (G) section through ovary with oogonia and oocytes, oc = oocyte, og = oogonium, ov = ovary, (H) mature oocyte with nucleus, nucleolus, and nurse cells, n = nucleus, nc = nurse cell, ns = nucleolus; (I–K) scyphistomae: (I) Scyphistoma of *C. tuberculata*, slightly contracted; (J) scyphistoma budding; (K) monodisc strobilation in *C. tuberculata*. *(After Kikinger, 1992.)*

resulting in spermatozeugmata (Werner, 1973). These sperm packages with somatic cells in the centre consist of centripetally oriented sperm with the tails radiating outwards. These packages make their way out through the follicle wall and through the genital epithelium into the genital sinus, which functions as a seminal vesicle (Fig. 5.28F). They reach the open water through the moutharm canals (Fig. 5.28E). Sperm-filled canals of the moutharm appendages of male medusae are white due to the densely packed spermatozeugmata. The presence of sperm-filled appendages indicated a spawning period, which is from August to October in these waters.

Oogenesis The size and shape of the ovary is the same as that of the testis in this species, *Cotylorhiza tuberculata*. The oogonia originate from the genital epithelium and protrude during their growth into the ovarian mesoglea, similar to *Mastigias papua* (Uchida, 1926), *Cassiopea frondosa* (Smith, 1936) and *Cassiopea andromeda* (Gohar and Eisawy, 1960a). In contrast, in *Cassiopea* species, the oocytes always maintain contact with the genital epithelium via nurse cells (Fig. 5.28G and H). As opposed to *Mastigias papua*, the oocytes are arranged in the order of their age. Fully developed eggs measure 140–160 μm and are enveloped by a very thin membrane. Zooxanthellae are abundant in the ovarian mesoglea, most being aggregated in small groups near the gastric epithelium. They do not immigrate into the cytoplasm of the oocytes.

Fertilization Fertilization is internal in this species, since the spermatozeugmata make their way from the open water into the female genital sinus through the moutharm canals. On the way, the spermatozeugmata disintegrate and only separate sperm are found near the ovary. Histological sections of the ovary show sperm quite visible inside the genital sinus and at the nurse cells of the oocytes. This observation indicates that fertilization is not only internal but occurs inside the ovary.

Embryology Just after fertilization, the expulsion of the eggs from the ovary takes place and the transport through the genital sinus into the moutharm canals (where the embryology occurs). Cleavage is equal and regular. The duration of embryogenesis—from fertilization to the freshly hatched planula—is 30–35 h in vitro. Most of this development takes place inside the moutharm canals. The path of the embryos inside these canals is usually looped and extends into the distal parts of the canal (Fig. 5.26B).

Release of the embryos The female medusa possesses a great number of slender, brood-carrying filaments. At the time of the exit of the embryos from the canal system, most of them have attained the ciliated planula stage, but are still inside a thin egg membrane. They frequently rotate around their longitudinal axes within this transparent envelope. "Of 137 females with planulae between their central filaments, 22% had embryo-filled moutharm canals."

The whitish brood-carrying filaments are developed only in females, so become organ of sexual dimorphism, macroscopically, at umbrella diameters of 12–14 cm. These appendages differ in length, but seldom exceed 20 mm. Their morphology is the same as that of the larger moutharm appendages. Their ends are disc-like covered with nematocyst warts. These warts detach in great numbers from the disc-like ends under strong mechanical stimuli.

Planula Newly hatched planulae from the egg membrane have a length of 270–330 μm and are brownish to reddish in colour. They lack zooxanthellae. They aggregate in groups between the brood-carrying filaments; their great numbers give the central moutharm disc of the mature female its reddish colouration. The number of the planulae in one specimen is great and the ovaries of this specimen contain all stages of oogenesis. These females occurred in the above-mentioned waters from early August.

The planulae themselves are elongated with a rounded anterior end, which is broader than the somewhat laterally compressed posterior end. They are mouthless and swim by ciliary action with the broader end forward, rotating around their longitudinal axis. A reversal of the ciliary beat is not possible; upon contact with a barrier, the planula must swim around it. The duration of the free living could not be determined in situ. Most planulae, freshly isolated from medusae and incubated in a 1-L jar with natural seawater, settled within 5 days.

Settlement The planula settles with the broader anterior end attached to the substrate. The opposite pole develops the apical organs of the scyphistoma. In the laboratory, the greatest number settled on the undersides of hard substrates. The muddy sediment was avoided for settlement, which was often in great numbers.

The polyp Due to their small size and apparently cryptic life habitat, no *Cotylorhiza tuberculata* polyps have been found in the field during the season. However, the author found a small number of *Cotylorhiza* scyphistomae in the field during August. Five specimens settled on submerged artificial substrata (glass slides in an open Perspex jar inoculated with planulae and positioned on the sea bottom in 2 m depth). A sixth specimen was found in its natural habitat, on a leaf of *Zostera* sp. All of these scyphistomae possessed zooxanthellae; one polyp was in the process of agametic reproduction by lateral budding (Fig. 5.28J).

The scyphistoma In the aquaria, metamorphosis from settled planula to fully formed scyphistoma with 16 tentacles was completed within 8–10 days at temperature 23–24°C of the water. The scyphistoma is whitish and is about 5 mm in height from the pedal disc to the mouth. The slender long stalk, 3–4mm in length, bears the cone-shaped calyx. The latter ends with one whorl of 16 filiform, contractile tentacles (Fig. 5.24(I) B (g)). The ectoderm of the whole creature is ciliated and propagates slight water current from the pedal disc in the direction of the hypostome (feeding process). Fully developed scyphistomae reproduce agametically by budding. At the beginning, a roundish protrusion appears in the inferior region of the calyx. The protrusion elongates and develops into a pyriform bud (Fig. 5.28J). In the early stage of bud formation, the endoderm of the polyp and the bud is continuous. Then, the endoderm bridge becomes constricted, and older bud stages remain connected to the mother polyp only by a thin ectodermal tube. The bud itself is clearly polarized. Its stalk-like distal end is the presumptive pedal disc and stalk of a future polyp. The broad proximal end connected to the mother polyp develops into peristome and proboscis after separation. *Cotylorhiza tuberculata* scyphistomae always produce a single, laterally flattened bud, never bud chains of two or more as *Cassiopea* polyps occasionally do. Bud separation occurs by rotation of the ciliated bud around its longitudinal axis. At the moment of separation, it is 0.5–1.0mm long. The entire differentiation process requires 40–48h. After separation, a new bud may be generated. If the food supply is sufficient, scyphistomae are highly reproductive (up to six buds per month). The ciliated bud swims with its stalk-like end directed forward. After a short time it settles on its prospective pedal disc, in many cases near the mother. Two days after separation the 8-tentacle stage is attained, and all the 16 tentacles are developed after 5 days. While reproduction by budding occurs in the absence of zooxanthellae, none of these polyps strobilate in their absence. After 19days of infection of 10 scyphistomae with zooxanthellae by feeding algae-containing tissue of *Anemonia sulcata*, one polyp started strobilation (Fig. 5.28K). Three days later, one zooxanthellate ephyra (diameter 1.8mm) was liberated by monodisc strobilation. During strobilation, the calyx of the scyphistoma already developed new marginal tentacles, and 6 days after the liberation of the ephyra, the 16-tentacle stage was attained.

Lychnorhiza lucerna

The life cycle of the Rhizostome *Lychnorhiza lucerna* Haeckel 1880, has not been previously described. However, Schiariti et al. (2008) presented a somewhat recent work dealing with the life cycle of this Rhizostome animal, using laboratory-based rearing experiments. The emphasis was focused on the benthic stages, namely the settlement preferences of planulae, the agametic reproduction of scyphistomae, the process of strobilation and the development of the ephyra. First of all, this species displayed the typical metagenetic (i.e. medusoid and polypoid) life cycle (Fig. 5.29K) reported for other rhizostomes. *Lychnorhiza lucerna* is the most abundant rhizostome in the southern Atlantic coasts of South America. It is distributed from Guyana (5°N) to Buenos Aires Province in Argentina (38°S). While there has been no monitoring of populations of *L. lucerna*, there are several accounts of large numbers of this species in the Río de la Plata estuary (35–37°S) and the adjacent marine area. Twenty-eight 'stranded' *L. lucerna* were obtained in March 2006 from beaches along the Argentinean coast of the Rio de la Plata estuary. All the specimens were mature, having coloured gonads (testis: greenish, ovary: light to dark brown); 16 were males and 12 females. The testes in the males were ripe containing follicles with mobile spermatozoa. On the other hand, the ovaries in the females contained oocytes ranging from 30 to 120µm in diameter. The large oocytes (60–120µm diameter) had a germinal vesicle, indicating that oogenesis stopped at the Prophase I stage of meiosis. Fertilized eggs and planulae were not found in the ovaries, or in the oral arms of the females.

In the laboratory, Schiariti et al. (2008) performed artificial fertilization to examine the settlement preferences of the planulae (one of their aims) by the following experiment: gonadal tissue (c. 10g) from 10 to 15 specimens, including both males and females, was isolated and incubated with weak aeration in 500mL plastic buckets containing 25µm mesh filtered seawater. The incubation was at a salinity of 17psu, which was the approximate salinity at locations where the medusae were obtained. During 1–3days after incubation, the authors collected fertilized eggs; each was approximately the same size as the large oocytes. In addition, each had two polar bodies in the transparent perivitelline space, which was about 4µm thick (Fig. 5.29B).

After 24h, at temperature (19–22°C), free actively ciliary-swimming larvae were found. The larva was elongated with a rounded, broader anterior end and a somewhat laterally compressed posterior end. Planulae varied in size from 95 to 207µm in length and 39–54µm in width. The authors suggested that these planulae are lecithotrophic, since no mouth appeared until this stage. To observe and study settlement, the authors isolated, under a dissecting microscope, about 100 planulae and transferred them to each of 13 buckets containing 250mL filtered seawater (17psu). Different types of substrata were randomly placed in each of the 13 buckets. The majority of planulae settled on the air-water interface ($P < .001$). Of those that settled on the settlement plates provided (which were selected from a range of possible substrata common in the study area), no significant differences were observed between styrene slides, glass slides and shells of the bivalve *Mactra isabelleana* ($P > .05$). No planulae settled on stones. The larvae settled on a variety of natural and artificial substrata, indicating that they do not have a preferred type of substratum for settlement. On the other hand, the attachment of large number of scyphistomae to the water's surface could be due to the experimental

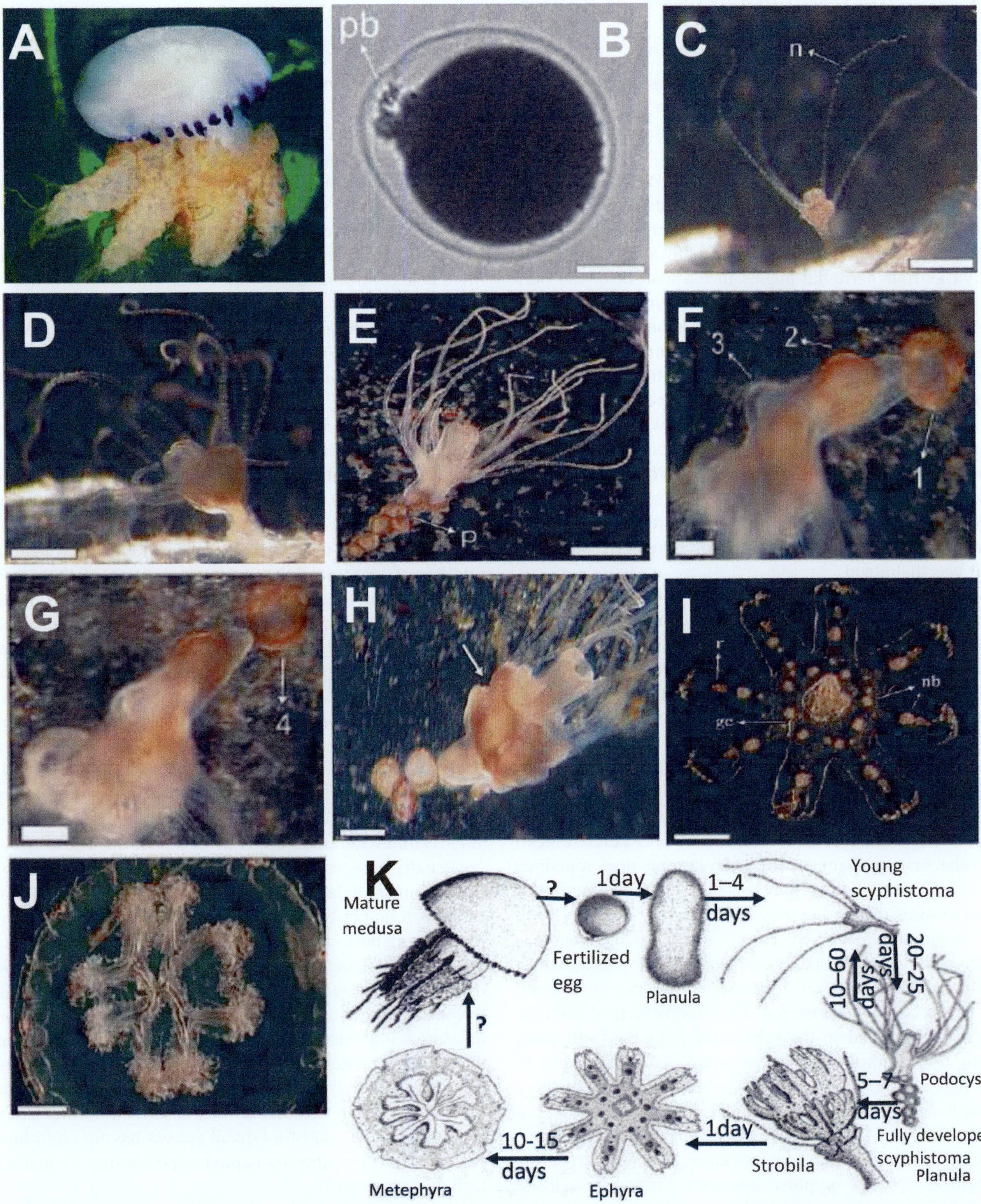

FIG. 5.29 Schiariti et al. (2008) reared and studied the life cycle of the Rhizostome *Lychnorhiza lucerna*: (A) photograph of W.M. of the animal with *whitish or purplish-blue marginal lobes*; (B) fertilized egg with polar bodies (pb) and X=60 μm; (C) a Newly metamorphosed, 4-tentacled scyphistoma, n=nematocyst batteries, X=200 μm; (D) intermediate 8–10-tentacled scyphistoma, ~6 days postsettlement, X=400; (E) fully developed, 18–22-tentacled scyphistoma, ~25 days postsettlement, producing podocyst (P) by stolon formation, X=650 μm; (F) simultaneous production of three podocysts, indicating the order of formation by 1, 2, and 3, X=200 μm; (G) the first formed podocyst (4) which has been liberated from the stolon and left behind; (H) strobilation, *arrows* show calyx incisions that indicate the incipient ephyra formation X=300 μm; (I) oral view of newly released ephyra with its typical arrangement of nematocyst batteries (nb), gastric cirri (gc), rhopalium=(r), X=300 μm; (J) oral arms in the young metephyra bifurcated into 8. Each arm branched into two wings at the tip bearing papillae, X=1 mm; (K) schematic representation of *Lychnorhiza lucerna* life cycle. Interstage durations are also shown (stippling indicates shadows, not pigmentation).

conditions (i.e. the small volume of still water in the buckets) and this probably does not occur in nature.

After 24h of settlement of the planulae, scyphistomae appeared. The base of each was attached to the substratum by a stalk enclosed within a transparent cuticle. Each was cone-shaped, translucent-white and typically had four filiform tentacles bearing nematocyst batteries (Fig. 5.29C). These initial tentacles arose from the distal end of the calyx as four perradial angles in the margin of the peristome. Newly developed scyphistomae measured 220μm in height (from the pedal disc to the mouth) with a calyx width measuring 90μm. Other four tentacles developed one by one, to be eight perradial and interradial tentacles (Fig. 5.29D), and then new tentacles appeared in between to form a fully developed scyphistoma with up to 22 tentacles. The oral disc was largely occupied by a prominent dome-shaped proboscis. The mouth was capable of extensive expansion and could ingest prey of similar size to the scyphistomae themselves. The scyphistomae can renew their base or form a new pedal stolon to reattach perfectly, if it was dislodged.

The scyphistomae attained full growth 20–25days after settlement. They reached 1500μm in length with calyx diameter between 550 and 800μm and bearing 18–22 tentacles (Fig. 5.29E). The flexible knob-shaped mouth (proboscis) occupied most of the oral disc. The stalk was somewhat variable in length and width depending on stolon formation and/or the shape and size of ingested prey. Schiariti et al. (2008) reported that a sort of tentacle 'cleaning' behaviour, i.e. drawing the tentacles to the mouth, was repeatedly observed in most scyphistomae while not feeding. When the animal captured a second prey before digesting the first, the tentacle with the second prey remained contracted until the first prey was ingested. The mouth then bent towards the second prey to digest it.

In the culture of Schiariti et al. (2008), scyphistomae were observed reproducing agametically by production of yellowish, crater-shaped podocysts at the base of the stalk by means of stolon formation. The stolon emerged from the stalk and attached to the substratum 0.23–0.85mm away from the scyphistoma. The scyphistoma then moved to the new site leaving its podocyst behind (Fig. 5.29E–G). Simultaneous formation of 2–3 podocysts was observed (Fig. 5.29F and G). A single scyphistoma was observed producing up to 17 podocysts, which extended c. 3.4mm in row length. Three new scyphistomae were formed from those 17 podocysts, and so on. Forty-six days after settlement at constant temperature (22°C), polydisc strobilation was observed in some buckets, but no ephyrae were observed. Early strobila (Fig. 5.29H) with segmentation of the calyx occurred 5days after the first noticeable sign of strobilation (Fig. 5.29H) appeared. The circular incision became progressively deeper and more pronounced within the following 30h and the second incision appeared proximal to the first one. Each forms a segment, which will be an ephyra. One day after the appearance of the second incision, rhopalia, marginal lappets and conspicuous, shiny statocysts appeared. At this stage, the ephyra pulsate by contracting their marginal lappets, ingest food and produce podocysts. The whole process of strobilation, from the appearance of the first incision to the liberation of the third ephyra, occurred within 8days. Under suitable conditions, scyphistomae were able to strobilate five times over 4months.

Forty-six days after settlement at constant temperature (22°C), strobilation was observed in some of the buckets. Early strobilae (Fig. 5.29H) with segmented calyx were observed. A first circular incision was formed at the base of the tentacles c. 5days after the first noticeable sign of strobilation (Fig. 5.29H). The incision became more pronounced within the following 30h. The second incision occurred proximal to the first one. Each segment became an ephyra with rhopalia, marginal lappets and conspicuous, shiny statocysts appeared 1day after the second incision. The authors did not find any correlation between the number of marginal lobes and the position of the ephyrae in the strobila. They observed the release of a typical ephyra having eight lappets followed by other two ephyrae possessing 11 and 12 lappets, respectively. At this stage, the ephyrae begin pulsation by contracting their marginal lappets and begin feeding and production of podocysts. The whole process of strobilation, from the appearance of the first incision to the liberation of the third ephyra occurred within 8days. Under suitable conditions, scyphistomae can strobilate five times over 4months. Newly liberated ephyra, and when extended, measured about 1.4–1.9mm wide from lappet tip to lappet tip. These ephyrae usually possessed eight marginal lobes; each lobe contained a pair of lappets and a rhopalium (Fig. 5.29I). In turn, each lappet was provided with a variable number of tips ranging between two and nine. The manubrium of this ephyra at this stage had circular shape or irregular shape (Fig. 5.29J), elongated as the ephyra developed, and began to constrict longitudinally in four positions. Older ephyrae measured 3–4mm diameter, secondary lappets emerged at the adradial planes and the nematocysts, any correlation between the number of marginal lobes and the position of the ephyrae in the strobila as was described before. They observed the release of a typical ephyra having eight lappets followed by other two ephyrae possessing 11 and 12 lappets, respectively.

Advanced ephyrae became metephyra stage, 7–12days postliberation. Metephyra were 6–8mm and its secondary lappets reached the length of the primary lappets, making the bell margin polygonal (Fig. 5.29J). Twenty days after liberation, the metephyra reached up to 14mm width.

Scyphomedusan development is known to the point at which a polyp is produced, but the manner of formation of medusa remains undiscovered.

CLASS CUBOZOA WERNER, 1973 (LAWLEY ET AL., 2016)

Werner et al. (1976) established the Cubozoa as a new class within the Cnidaria. This is a small class that stands alone. Cubomedusae are commonly known as box jellyfish. They possess image-forming eyes (Coates et al., 2001), which certainly evolved independently from other metazoans. Cubomedusae therefore offer a unique perspective on the evolution of image-forming eyes.

Until now, class Cubozoa has included two orders having about 50 species described worldwide and grouped in seven families. Its members, the cubomedusans, are commonly called 'box jellies' or 'sea wasps' because of their shape and sometimes their severe sting. They are inhabitants of neritic tropical regions of the oceans. They have been known for >200 years (Werner et al., 1971). They are the most highly adapted for a pelagic life. The morphology of several species has been investigated. Previously the members of this class were included as an order of the subclass Discomedusa of the class Scyphozoa. The nature of their nematocytes, the possession of a velum, and their life cycle present evidence that they are not closely related to the scyphozoans. In addition, each medusa has four rhopalia located around the margin of the bell (Fig. 5.30(I)) (arrows) which alternate with the four groups of tentacles. The complex eyes point towards the centre of the bell (Coates, 2003).

Piatigorsky and Kozmik (2004) showed that **cubomedusa**, such as ***Tripedalia cystophora***, contain a well-developed visual system located on specialized sensory structures (rhopalia) with eyes and statocyts. In addition, these authors have been exploring that each rhopalium in this species has a large and a small complex, camera-type eye with a cellular lens containing three distinct families of 'crystallins' which apparently serve nonlenticular functions (Fig. 5.30(II) (b)). Another distinctive feature characterizing this cubozoan jellyfish is that it discharges its sperm encapsulated in spermatozeugmata and/or spermatophores, like all the known animals of this group. Thus this group of animals should be separated and elevated to a different class, Cubozoa (Werner, 1975; Calder and Peters, 1975). These cubomedusans are medusoid cnidarians with bells having four flattened sides (Fig. 5.30(I)). The bell margin is simple and bears four tentacles or tentacle clusters. Gonads are four pairs of narrow plate-like organs, attached one along each side of each interradial septum. An attached polypoid larva follows the planula larva.

Practically speaking, not much is known about the development of this group. This means that the developmental history of the cubomedusae has still not been thoroughly investigated, but it is known that no ephyra is formed, and therefore alternation of generations cannot be taking place. However, it is a well known information that cubozoans (box jellyfish) undergo remarkable body reorganization throughout their life cycle when, first, they metamorphose from swimming larvae to sessile polyps, and second, through the metamorphosis from sessile polyps to free swimming medusae. In the latter, they develop complex structures like the central nervous system (CNS) and visual organs. In some species, the scyphistoma (Fig. 5.30(III) (b)) transforms directly into a small medusa. In the one species of Cubomedusae (*Tripedalia cystophora*) for which the life cycle is known, the scyphistoma transforms directly into small medusa.

Tripedalia cystophora

Tripedalia cystophora is unusual among cubozoans in having a rounded, not box-shaped bell (Coates, 2003). In addition, Werner et al. (1971) reported that the fertilized eggs of *Tripedalia*, unlike other Cubomedusae, develop into planulae in the gastral pockets of the female medusa. Then the mature planulae are released within 2 days. Each is slightly pyriform, has a diameter measuring 0.13 mm and contains a yellow-brown pigment. Within another 2 days, the planula settles preferably on bivalve shells and in darkness, but would settle on glass or plastic. The planula develops into a polyp with a mouth and four tentacles within 2 or 3 days after settlement. Although naked at first, the polyp's base is enveloped by a thin transparent periderm at the four-tentacle stage. By feeding, the polyp grows rapidly.

The polyp becomes mature and can reproduce agametically when it reaches 0.5 mm long and with five or six tentacles. It buds from the basal body wall (Fig. 5.30(III) (a)) and the bud resembles that of *Hydra* since it acquires a functional mouth and tentacles during brooding. The released two- or four-tentacle bud or the secondary polyp elongates and one of its tentacles becomes much longer than the others. By this long tentacle, the secondary (daughter) polyp glides over the substrate and then attaches becoming pyriform, and develops peridermal sheath around the base. It buds off more polyps in the same manner as the primary (mother) polyp. When 8–10 weeks old, the mother and daughter polyps reach a body length of 0.8–1.0 mm, having a single whorl of 7–9 tentacles and thus become fully grown.

The solitary polyp has long, contractile, cone-shaped peristome and solid, completely retractile, capitate tentacles. Each of the latter bears nematocysts only at its tip. Other scyphopolyp features are lacking, such as: a cruciform-shaped mouth; peristomial pits; gastric septa and pockets; septal funnels and muscles; and thick mesoglea. However, at the time of transformation to a medusa, the characteristic feature of cubozoa, the tetraradiate symmetry, appears. This polyp of *Tripedalia* is unique in that it metamorphoses directly into a single medusa. Gurska and Garm (2014) studied this second metamorphosis from sessile polyps to free-swimming medusa in the life cycle of the box jellyfish

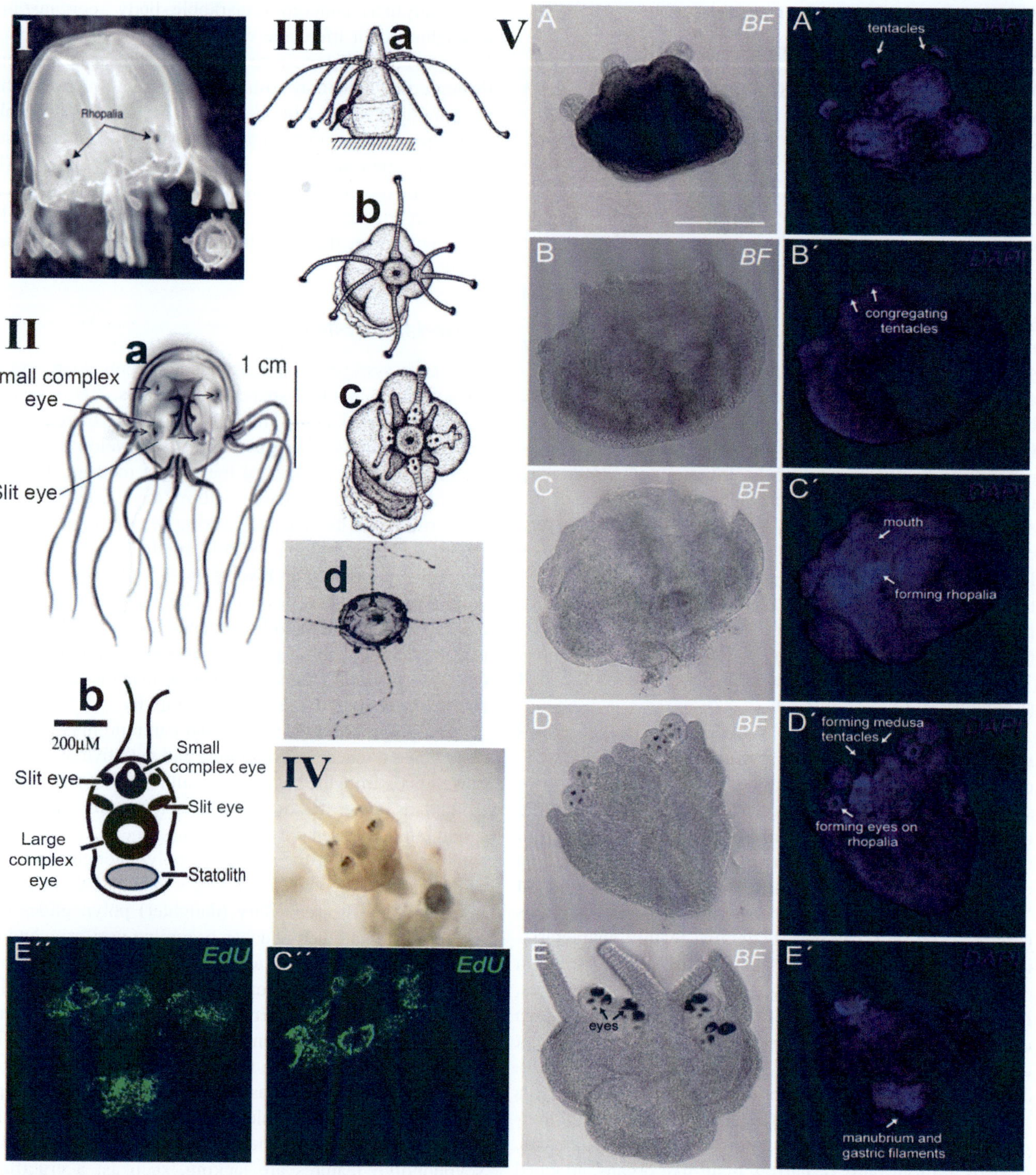

FIG. 5.30 Cubozoa, *Tripedalia cystophora*, (I) adult medusa *(from Piatigorsky and Kozmik, 2004)*; (II) drawings of *T. cystophora (courtesy of F. Sommer)*: (a) note that *T. cystophora* is unusual among cubozoans in having a rounded, not box-shaped bell. Each medusa has four rhopalia located around the margin of the bell *(arrows)* which alternate with the four groups of tentacles. The complex eyes point towards the centre of the bell. (b) Schematic diagram of the rhopalium. The large and small complex eyes lie along the medial line while the pit and slit ocelli are paired laterally *((II a and b) from Coates, 2003)*; (III) life cycle of *Tripedalia cystophora (from Werner et al., 1971)*, (a) adult polyp with a secondary young polyp, basally on the left, just before detachment, length of the mother is 1 mm; (b) adult polyp in the tetraradiate stage of the beginning of the metamorphosis into the medusa; (c) same polyp 2 days later *((D–F) drawn by Heckmann)*; (d) young medusa just after metamorphosis from polyp reared in the laboratory, bell margin above, diameter of the bell is 1.3 mm *(photograph by B. Werner)*; (IV) newly metamorphosed *Tripedalia* medusa with the rhopalia situated behind the 1^{ry} tentacles *(from Piatigorsky and Kozmik, 2004)*; (V) proliferation zones during the metamorphosis of *Tripedalia cystophora* polyps: polyps during different stages of metamorphosis unstained from (A)–(E); stained with DAPI (A′, B′, C′, D′, E′) and S phase cells visualized with EdU (C″, E″), X = 300 µm. *(All photographs of (V) from Gurska and Garm, 2014.)*

of both species *Tripedalia cystophora* and *Alatina moseri*. During this metamorphosis, the organism develops complex structures like the central nervous system (CNS) and visual organs. Gurska and Garm examined the cell proliferation at different stages of metamorphosis, through in vivo labelling of cells in the 'synthetic' phase (S phase) of the cell cycle. The authors described the cultures and the stains in detail (the nuclei stain (DAPI, 0.1 mg/mL and the EdU that is being incorporated into DNA), and the procedures. Fig. 5.30(V) A–E show the nonstained successive stages of metamorphosis of the sessile polyp to free-swimming medusa of *Tripedalia cystophora.* Fig. 5.30(V) B shows the beginning of metamorphosis and the early appearance of the proliferation zones (high density of S phase cells) around the mouth and at the bases of the tentacles (Fig. 5.30(V) B′). This stage is known as stage of 'congregating tentacles'. In the second step in metamorphosis (Fig. 5.30(V) C) and the early stage of rhopalia formation, a proliferation zone is found in the oral end of the polyp and especially in the forming rhopalia. During the late stage of rhopalia formation (Fig. 5.30(V) D), the eyes are visible on the rhopalia. In the polyp, in the late stage of metamorphosis (Fig. 5.30(V) E), an additional proliferation zone is observed in the forming manubrium and gastric filaments.

The medusa differs from the polyp since the crown of the latter changes to become a quadrangular shape; the tentacles congregate at four corners (Fig. 5.30(III) (b) and (c)) while their bases thicken and develop small pigment spots (Fig. 5.30(III) (c)). At the same time, the distal parts of the tentacles resorb. The tentacles, either singly or fused, become the four rhopaloids. Before the resorption of the polyp tentacle, the four primary medusa tentacles appear, de novo, between the four rhopaloids (Fig. 5.30(III) (d)).

Charybdaea marsupialis

Lawley et al. (2016) revised systematically some of the known cubozoans. They described species *Alatina alata* (Reynaud, 1830, with a synonymy list that included; *Charybdea alata* Haekel, 1880, p. 441); and *Charybdea moseri*: Mayer, 1906, pp. 1135–1136, pl. 1, fig. 2–2c; *n. sp.*, description and illustrations; Bigelow, 1909, pp. 19–20, young stage of *C. grandis*; Bigelow, 1938, p. 144, junior synonym of *C. alata*; Kramp, 1961, p. 304, in synonymy of *C. alata.* It is understood from Lawley et al. that Charybdia and Carybdia are different genera.

Gueroun et al. (2015) recorded, for the first time, the cubozoan jellyfish *Carybdea marsupialis* Linnaeus, 1758, from the eastern Tunisian coast (Hammamet beach), south central of the Mediterranean Sea, in spite of the fact that there are records of *C. marsupialis* from the tropical and subtropical regions of the Atlantic Ocean, the Caribbean Sea and Californian waters. In the light of recent morphologic and genetic analyses results, *C. marsupialis* seems to be an endemic (familiar) species for the Mediterranean Sea, with its type locality being the Northern Adriatic (coastal waters in close proximity to Rimini along the Italian coastline). The present work deals with these two species separately.

Previous developmental studies of *Charybdaea marsupialis* (Fig. 5.31B2) were in the accounts of Okada (1927). (The best known species as a scyphomedusan at that time.) Uchida and Yamada (1968) recorded that fertilization occurs in the gastrovascular cavity (internal fertilization). Cleavage is total and equal and follows about the same sequence as that in the other scyphomedusae, as far as the fourth cleavage (Fig. 5.31A). Then the morula stage is formed, which is succeeded by the formation of the hollow blastula. A gastrula is formed by multipolar ingression or delamination. The gastrula then develops cilia and changes into a small, pear-shaped planula with pigment granules in its central part. This planula swims for 2–3 days, then becomes sessile. It loses its cilia and becomes a plate-like mass of cells. A pair of tentacles appear from this mass. The tentacles elongate and the body increases in length. A small hypostome is formed and two other tentacles appear. This is all that is known in these accounts.

The larval nature of the polyp and its direct transformation into medusa in cubomedusae seems to support the belief that the medusoid form is primitive in cnidarians and that the polypoid form evolved as a larval adaptation. There are zoologists such as Werner et al. (1971) and Werner (1973), who took the opposite view. It should be noted that stauromedusae and cubomedusae exhibit many morphological as well as embryological differences from the other scyphozoans; at the same time, they have many important characteristics in common, such as the absence of ephyra stage and the absence of alternation of generation.

Carybdea marsupialis

Carybdea marsupialis (Linnaeus, 1758) is a small transparent jellyfish, known as sea wasp. It is distinguished from other species of Order Carybdeida Gegenbaur, 1857 (Lawley et al., 2016) by its four elongated tentacles at the four lower corners of the box-shaped bell (Fig. 5.31B1). The tentacle reaches up to 30 cm (the bell about 3 cm), characterized by the red banding. The sexes are separate and sexual reproduction takes place with the emission of gametes into the open water. This is the first difference—that the fertilization is external. After fertilization, a planula larva forms, which later develops into a cubopolyp with a few tentacles, and this can reproduce agametically by budding.

After the detachment of the daughters, at first, they creep across the substrate with the tentacular region leading. During this mobile stage, pieces may become detached or the cubopolyp may split transversely. These fragments regenerate their missing parts within 72 h. The resulting polyps then attach themselves to the substrate and grow a full

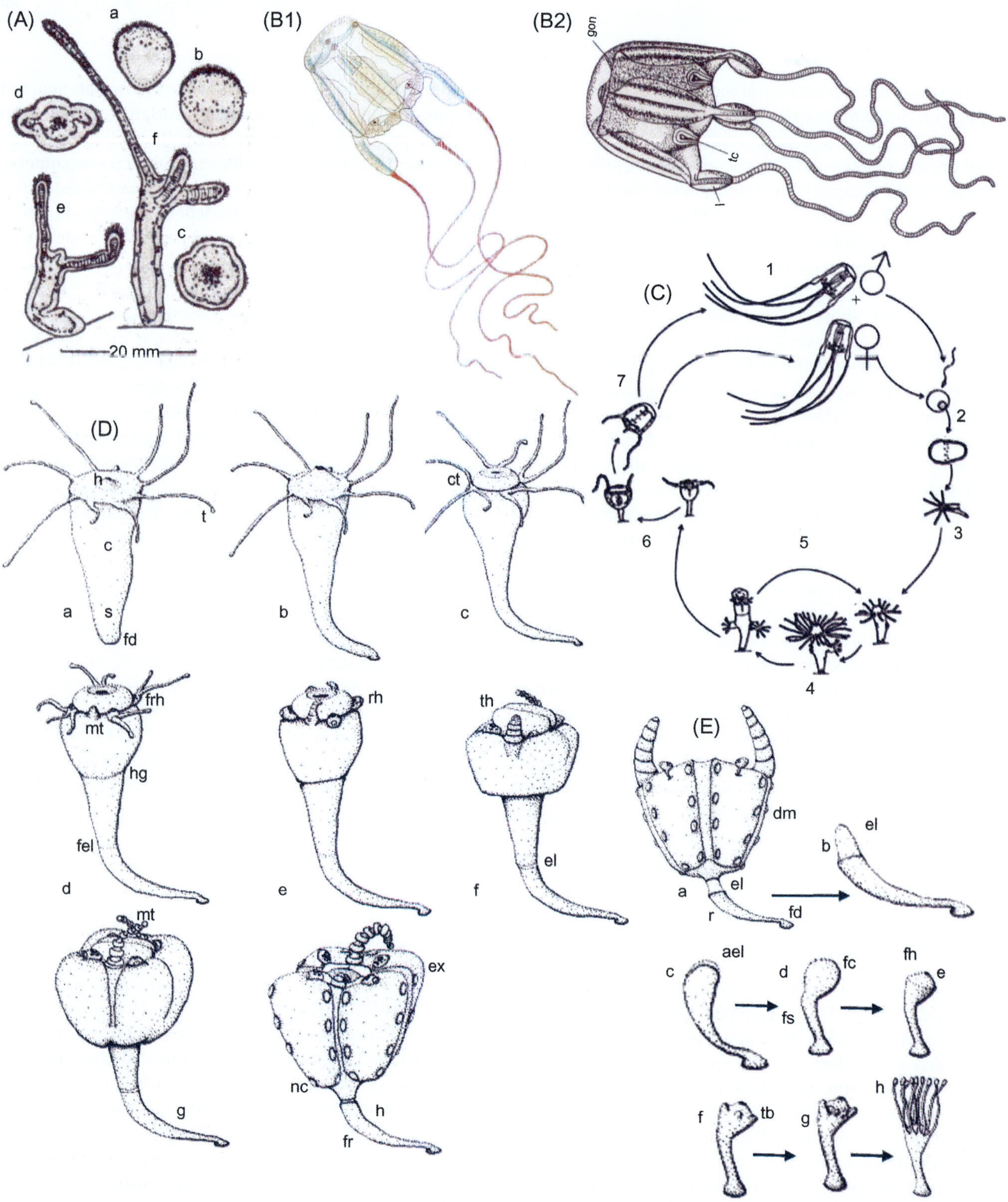

FIG. 5.31 (A) Early development of *Charybdea rastonii* (now it is Cubozoa): (a and b) successive stages of planula, (c) attached planula, (d) formation of two primordial tentacles, (e and f) formation of scyphopolyp *(after Okada, 1927 from Kumé and Dan, 1968)*; (B1) illustration of *Carybdea marsupialis*, from Wiki Loves Africa, (B2) illustration of *Charybdea*; (C) life cycle of *Carybdea*: (1) male and female mating (2) release of the fertilized egg into sea water (3) settlement of planula larva on substrate after ~2 days (4) benthic polyp phase (5) new polyps budding from existing polyps (6) polyp metamorphosing into juvenile medusa (7) release of juvenile cubomedusa *(from Dudley, 2016)*; (D) metamorphosis of *Carybdea marsupialis*, (a–h) forming a regenerating remnant: c = calyx, ct = congreate tentacles, el = ectodermal lamella, ex = exumbrella, fd = footdisc, fel = forming ectodermal lamella, frh = forming rhopalia, fr = forming remnant, h = hypostome, hg = horizontal groove, mt = medusoid tentacle, nc = nematocyst cluster, rh = rhopalium, s = stalk, t = tentacle, th = transforming hypostome; (E) (a–h) regenerating of the remnant into a fully developed, small polyp, ael = ectodermal lamella being absorbed, el = ectodermal lamella, dm = detatching medusa, fc = forming calyx, fd = footdisc, fh = forming hypostome, fs = forming stalk, r = remnant, tb = tentacle buds. *((D and E) from Straehler-Pohl and Jarms, 2005.)*

complement of 24 tentacles. About a month later they become detached, the multiple tentacles are reabsorbed and the four tentacles typical of the medusa phase develop.

The partially published PhD thesis (2016) by Melissa J.A. Dudley, dealing with the biology of the same above species, *Carybdea marsupialis*, reported that the Caribbean Carybdea sp. (= *C. xaymacana*) is reported to be ovoviviparous. The spermatozoa are released into the water, enter the gastrovascular cavity of female via the manubrium, and fertilize the ova. Zygotes remain there until blastulae are released and develop into swimming planulae.

Straehler-Pohl and Jarms (2005) reported that unlike scyphozoans, the polyps of *C. marsupialis* did not strobilate but went through a metamorphosis, in that the whole polyp transformed into a single medusa. They obtained the polyps used in their study, from the cultures established by Werner in 1973 and continued by Jarms. The authors carefully kept the culture to observe and record data about the development during the two types of metamorphosis in the cubopolyp *C. marsupialis*, and discuss the implications these results have on the classification of this group of Cnidaria. The authors induced metamorphosis by raising temperature up to 28°C and applying definite feeding regime. Two types of metamorphosis were observed: in type one, a single polyp transformed into a single medusa without leaving any regenerative remnants. This metamorphosis was described several times in the literature and in the present work (Fig. 5.30(V)). A second type of metamorphosis was investigated, in which the polyp transformed into a single medusa, but left a remnant (Fig. 5.31D), named residuum by Werner (1973). This remnant regenerated a small polyp (Fig. 5.31E). This second type resembles the first one except that the fate of the stalk differs.

In the second type, the polyp grows and at a definite length of the stalk (Fig. 5.31D (b)), its terminal portion begins to separate from the calyx. The detail of how this occurs was described histologically by Straehler-Pohl and Jarms (2005). The stalk was finally divided by an ectodermal lamella into two still connected parts. Both parts stayed as opaque as the calyx, and the ectodermal lamella became apparent as a transparent horizontal circle around the middle part of the stalk (Fig. 5.31D (d)–(h)). After the detachment of the young medusa, the thick ectodermal lamella was absorbed (Fig. 5.31E (b) and (c)). For 1–2 days, there was no visible change and the remnant did not react to mechanical stimuli. After about 2 days, a little cone formed on the top of the remnant, followed by the appearance of a little mouth-like opening at the upper end (Fig. 5.31E (d) and (e)). At this stage, the remnant reacted to stimulation by contraction and enlargement of the mouth opening. Twenty-four hours later, four tentacle-buds appeared around the mouth-cone, each containing a single 'stenotele' (Fig. 5.31E (f)). Six hours after that, four more tentacle-buds appeared (Fig. 5.31E (g)). The tentacles grew and became functional (Fig. 5.31E (h)). During the next 2 days, the number of tentacles increased until it reached the original number of tentacles of the mother. The regenerated polyp resembled the mother in appearance, but measured only one-fifth of its original size. The whole regeneration process took about 7–10 days. The regenerated polyp grew to the former size in 2–3 weeks and propagated agametically afterwards, creeping polyps and medusae as well.

Carybdea sivickisi

Toshino et al. (2015) tabulated the current names of some Japanese Cubozoa. *Carybdea sivickisi* is the former name and is currently known as *Copula sivickisi* (Stiasny, 1926). This information is according to Uchida (1929, 1970) and Bentlage et al. (2010). This Japanese species, *Carybdea sivickisi*, was recently studied in terms of a different, but related aspect. Lewis and Long (2005) were interested in the courtship and fertilization in Cubozoa. They collected their animals at four sites around Okinawa Prefectural, Japan, one of which is Motubu Shinko port, between May and June or August. This species was actually first described taxonomically by Stiansy (1922). It is a small (≤14 mm diameter) usually dimorphic. (Fig. 5.32C) shows the full bell (diameter of 8.5 mm), mature adult male and female, of *Carybdea sivickisi*. The female possesses 'velar spots' pigmented terminal regions of structures that branch off from the palmate velar canals (four per quadrant). These spots are microscopical, dark-orange in colour and are believed to have a specific dimorphic character. In both male and female individuals, there is a pair of hemispherical sacs below each of the gastric 'phacellae', continuous to the gastric ostium. These gastric sacs have a dark-orange pigment, and contain sperm. This characteristic, in addition to their location, suggests that these sacs function as sperm-storage organs, as spermathecae in females and seminal vesicles in males.

Sexual maturation

Juvenile medusae of *Carybdea sivickisi* (from the field) were observed as for how to be mature. When the medusa reached ~4 mm in bell diameter, gonadal development begins in the region flanking the interradial canals. Males developed two hemispherical patches of pale-orange pigment at each interradius (a total of eight). Male medusa becomes sexual mature at a bell diameter of 5 mm, and the orange patches at the interadii develop into a pair of flattened hemispherical gonads (hemigonads) which are pinched at the perradius. Paired gonads contained a triangular sperm-filled structure that was red or dark-orange in colour. In female medusae, pairs of gonads (gastric pockets) developed at each interradius as narrow, pale sheets of tissue (Fig. 5.32C). Velar spots were faint in females 4 mm in bell diameter, but were clearly defined in mature females 5 mm in diameter. Of the 99 females collected, 77 had conspicuous spots. Males did

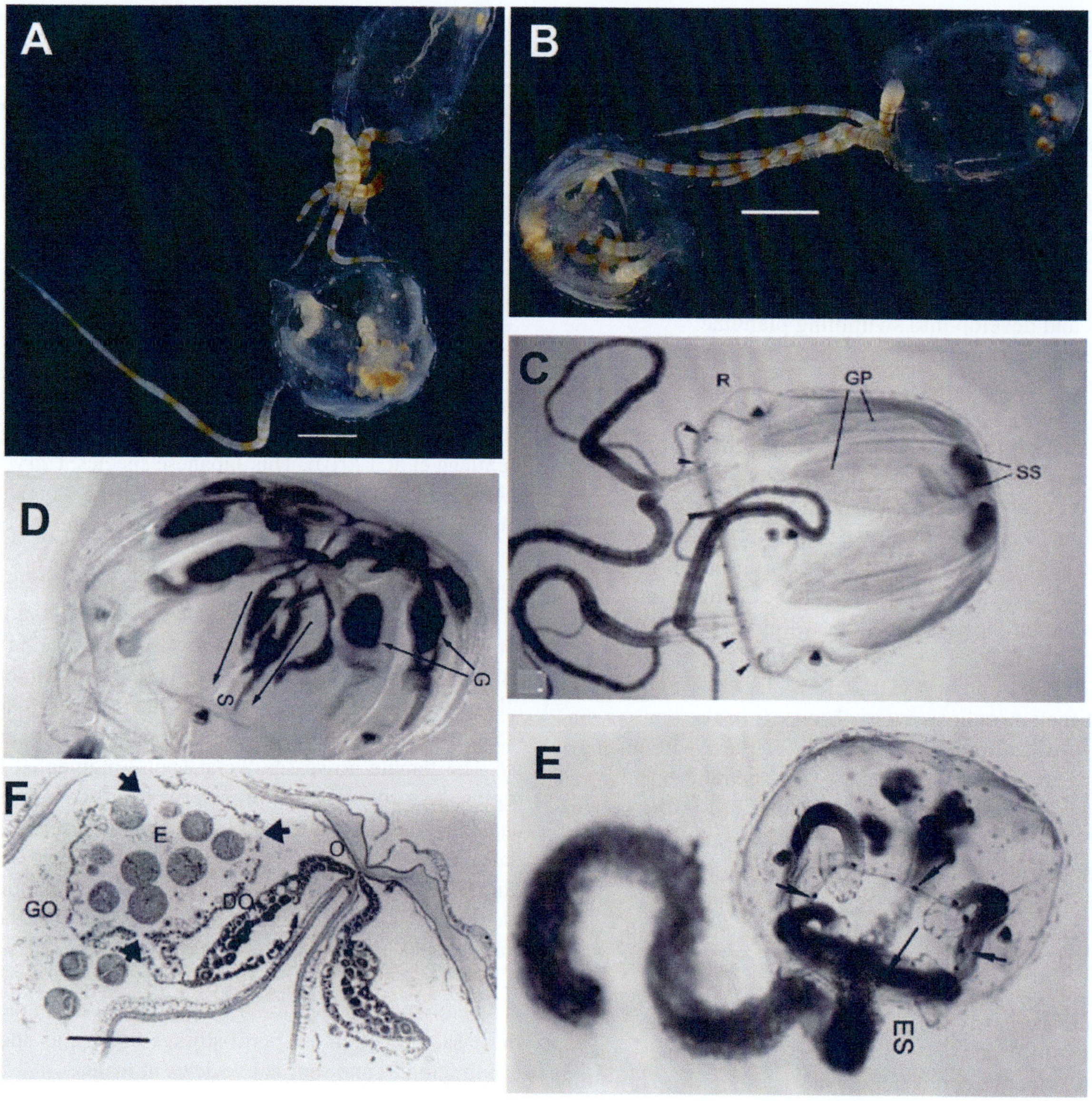

FIG. 5.32, CONT'D

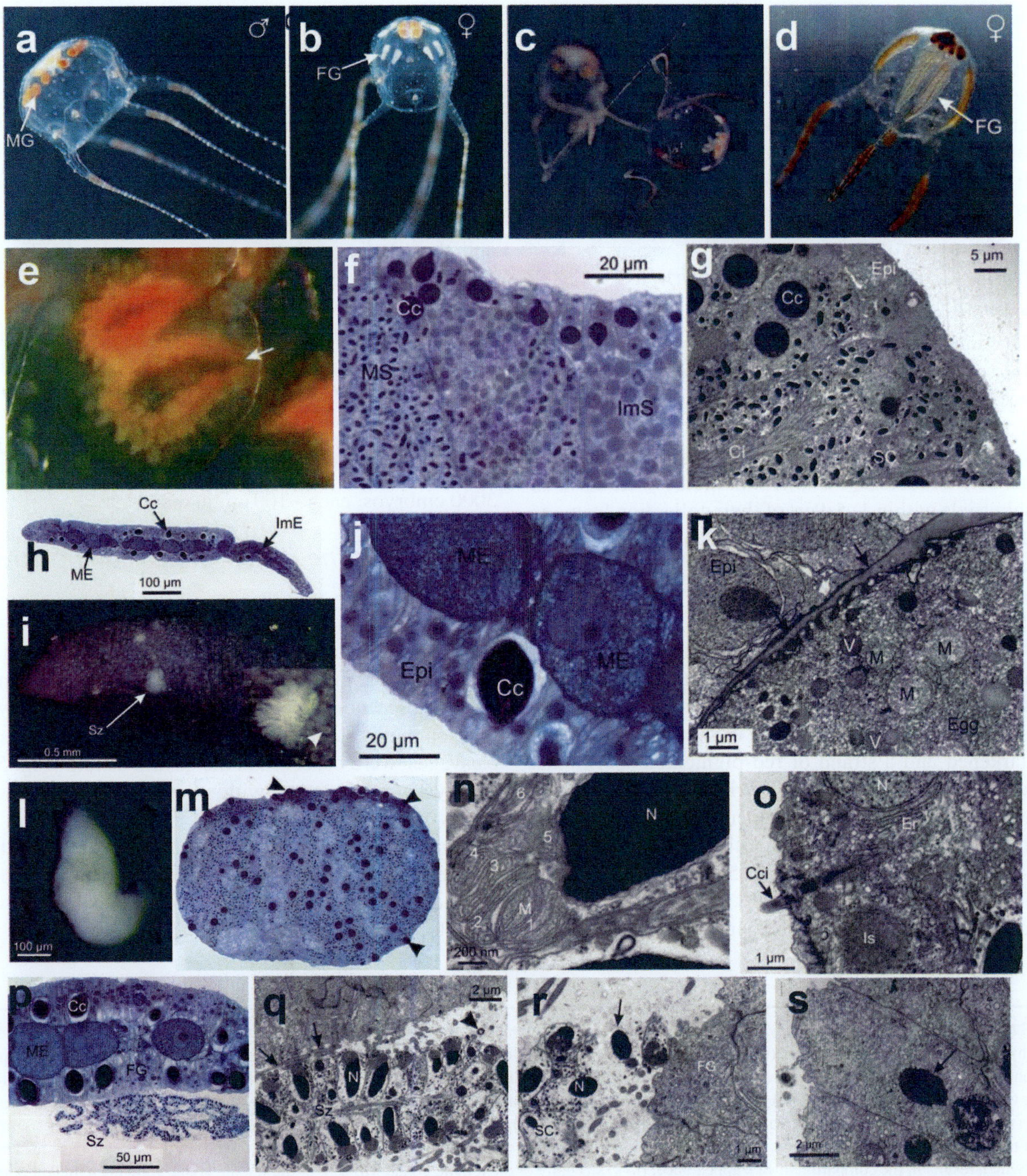

FIG. 5.32 Reproductive behaviour and fertilization in Cubozoa: (A and B) copulation in *Copula sivickisi (from Marques et al., 2015)*; (A) pair formation, female (bottom) and the male (top) during the ejaculation of the 'spermatophores', the latter are visible as opaque central strands inside the male, while a 'spermatophore' from a previous mate is visible as an orange bundle in the centre of the female's stomach, X = 2 mm; (B) female (left) inserting spermatophore into her manubrium using her tentacles as the male (right) releases her from his tentacle grip, X = 2 mm; (C–F) fertilization in *Carybdea sivickisi*: (C) lateral view of a mature adult female medusa of *Carybdea sivickisi*, bell diameter of 8.5 mm, rhopalia (R), subgastric sacs (SS), male hemigonads (G), gastric pockets (GP) which encase the female gonads, and ovaries (O). *Arrows* point to the velar spots on the female's bell margin; (D) lateral view of mature adult male *Carybdea sivickisi* discharging a spermatophore during courtship. Each of the eight hemigonads releases sperm strands that merge into four strands in the subgastric sacs, gonads (G), and spermatophore strands (S), photo reproduced from video; (E) oral view of a mature adult female medusa extruding a larviparous embryo strand (ES) from her manubrium, *small arrows* indicate to conspicuous velar spots on the bell margin, photo reproduced from video; (F) T.S. through the gastric ostium region at the interradius of a recently spentmature adult female medusa showing the ovaries (O), developing oocytes (DO) and residual embryos (E). *Arrows* indicate where the gastric pockets have ruptured liberating embryos throughout the gastric ostium (GO), X = 100 µm. ((C–F) from Lewis and Long, 2005) (a–d) Photographs showing mating in *Copula sivickisi*, in view of Garm et al. (2015):

(Continued)

not initiate courtship with females lacking velar spots, and were mostly unresponsive to females with pale spots. In one instance, courtship was observed between a male and a female with pale spots, and four spermatophores were transferred, but this did not produce any viable embryos. This female might have been perceived by the male as being sexually mature, as her velar spots became conspicuous the following day.

Courtship, spermatophore formation and sperm-transfer

In order simultaneously to examine courtship behaviour and male choice in *Carybdea sivickisi* and to test for the presence of sperm competition between large and small males, the authors conducted 25 in vitro courtship trials. They used mature adults (two males and one female, all with bell diameter ≥5 mm). Individuals were used in courtship trials within 1–5 days of their collection from the field. In the in vitro trials, courtship began with the female and one of the males swimming around in the jar, with their tentacles fully extended. The male would then attach one of his tentacles to one of the females, gaining control of the female and pulling her around. The male then contracted his tentacle in order to bring the oral opening (manubriums) of the two individuals into direct contact. Upon contact, the area below each of the male's hemigonads darkened and a strand of red-pigmented sperm was released from the perradial puckering into the subgastric sacs. These eight strands of sperm were merged into four strands in the stomach and finally coalesced into one thick strand in the manubrium (Fig. 5.32D), where a spermatophore was formed at the bell's margin. It was then transferred to one of the female's tentacles. Following this process, the male released the female, who inserted the spermatophore into her manubrium.

Dehiscence of the spermatophore and fertilization

Following the insertion of the spermatophore(s) into the female, they were transferred from the manubrium into the subgastric sacs (putative spermathecae) after ~1–2 h and the gastric pockets soon became cloudy. The gestation period lasted ~55 h, after which time thousands of embryos began to circulate through the gastric ostia giving the bell an opaque brown colour. Just after this point, the female extruded embryos by rapid pulsation that the embryos crowded at the bell's four interradii. After an hour, the female extruded a gelatinous cream-coloured embryo strand (Fig. 5.32E). Five of the 18 females that mated in the trials produced embryo strands; each strand contained over 3000 embryos, each ~100 μm in diameter.

Histology of the female gonad and putative spermathecae

A histological picture of a female *Carybdea sivickisi* with conspicuous velar spots (~250 μm in diameter) and a bell diameter of 7.5 mm revealed a ruptured gastric ostium, suggesting a recent discharge of embryos. Blastulae at various developmental stages, some having undergone three or four cleavages, were observed in the gastric ostium (Fig. 5.32F). Embryos were observed within the gastric ostium and velar canals of the velarium, as well as near the opening of the sinus that connects the subgastric sacs. The ovaries exhibited a development gradient, with the most-developed oocytes located near the gastric pockets and the least-developed oocytes located near the interradius.

FIG. 5.32, CONT'D (a) an adult male with clearly visible *orange male gonads* (MG); (b) an adult female with the gonads (FG) not yet fertilized as indicated by the *pale colour* and small size; (c) mating medusa. The male transfers a spermatozeugmata using its tentacles and the female picks it up using a single tentacle; (d) a fertilized female approx. 24 h after mating. The gonads are now enlarged and have a *yellow brownish colour*; (e–g) TEM photographs showing the structure of male gonad: (e) a pair of hemigonads from a live male. The *arrow* indicates the pore by which the spermatozeugmata leaves the gonad; (f) magnified portion of T.S. showing the transition from mature to immature sperm arranged in compartments; (g) the cnidocytes are imbedded between the sperm cells (SC) and are never placed at the very surface of the epithelium (Epi). The cilia (flagellae) (Ci) of the sperm cells are arranged in bundles which results in the compartment-like appearance; (h–k) structure of the female gonad: (h) T.S. of the female gonad showing a thicker part having mature eggs (ME) and a thinner 'wing' having immature eggs (ImE). Cnidocytes (Cc) are distributed throughout the gonad; (i) T.S. of female gonad with an attached spermatozeugmata (Sz). Insert shows attached spermatozeugmata magnified. The *arrowhead* indicates a cnidocyte; (j) magnification of the epithelium (Epi) with an embedded cnidocyte; (k) magnified portion of the egg cytoplasm showing dense aggregations of mitochondria (M) and different vacuoles (V). Between the egg cell and the epithelium, there is a thin layer of mesoglea *(arrows)*; (l–o) structure of the spermatozeugmata: (l) photograph showing one retrieved from a female stomach; (m) T.S. of the spermatozeugmata which still has the cnidocytes (Cc) but now several are placed at the very surface of the epithelium *(arrowheads)*. All sperm cells (SC) appear mature; (n) magnification of a sperm cell showing at least six mitochondria in the proximal end; (o) the only cilia found on the surface of the spermatozeugmata are the cnidocils (Cci) from the cnidocytes, a cnidocyte of the isorhiza type (Is) placed at the very surface of the spermatozeugmata, Er = endoplasmatic reticulum; (p–s) interaction of a female gonad with an attached spermatozeugmata: (p) T.S. of the female gonad (FG) midway through the spermatozeugmata (Sz). The cnidocytes (Cc) of the gonad is still pulled back from the surface of the epithelium. The spermatozeugmata is attached in the region of the mature egg cells and has no cnidocytes; (q) close up of the area with contact between the spermatozeugmata and the female gonad. This area has free cilia from both the gonad and from the sperm cells. In this area, another structure about twice the diameter of the cilia is also seen *(arrowhead)*; (r) partly digested sperm cells (SC) with apparently intact nucle *(arrow)* lying free close to the female gonad (FG); (s) in the area where the spermatozeugmata lies close to the gonad, there are large electron dense vacuoles present in the epithelial cells of the gonad *(arrows)*. These vacuoles are similar in shape and size to the sperm nucleus.

Settlement and development of embryos

The female deposited the embryo strand onto the substrate (day 0), and the spherical embryos began to develop. The gelatinous membrane that surrounds the embryo strand dissolved after 4.5 days. After 6 days, the embryo developed into pear-shaped motile planula, 140 µm in diameter. After 9.5 days, the motile planula settled on the substrate and developed into polyps (200–700 µm in diameter) with 1–10 tentacles by day 14.5. The next step was the elongation of the polyps to become creepers by day 17.5. These creepers often aggregated into long chains (>1000 µm long) that appeared to function as a single unit. These aggregations are in accordance with those made by other carybdeids. One notable exception is that no mention is made in the cubozoan literature of the aggregations seen in this study. Whether the development of these aggregations is a unique and normal characteristic of *Carybdea sivickisi* life history, or just an artefact of holding conditions, requires further study.

Copula sivickisi

It is appropriate to record the available information dealing with mating and internal fertilization that occur within cubozoons. One species, *Copula sivickisi*, is found in a large area of the Indo-Pacific. They have separate sexes and when mature males and females meet, entangle their tentacles and the males transfer a sperm package, or spermatozeugmata (Fig. 5.32(a)–(d)). The latter is ingested by the female, to fertilize her eggs internally (Fig. 5.32(p)–(s)). Garm et al. (2015) examined the gonads and spermatozeugmata ultrastructurally to understand the structural adaptations to this specialized way of reproduction, and to understand how the fertilization takes place. The authors discovered that the male gonads were heavily packed with cnidocytes of the 'isorhiza' type and surprisingly, they are transferred to the spermatozeugmata. The latter does not dissolve in the female gastrovascular cavity but is attached to the female gonad, probably using the isorhizas.

Structure of the gonads of *C. sivicksi*

The **male gonad** is disc shaped with a diameter of ~1 mm and the sperm producing compartments arranged in grape-like clusters (Fig. 5.32(e) and (f)). These compartments are also seen in the sections and they contain sperm cells at different developmental stages (Fig. 5.32(f)). The youngest sperm cells are found in the periphery of the gonad and proceeding towards the centre, the sperm cells become more mature (Fig. 5.32(f) and (g)). Generally, the spermatogonia are in the periphery and maturation proceeds to the centre. The mature sperm are found in the centre of the gonad, with their cilia (flagellae) gathered in bundles in cavities of the gonad (Fig. 5.32(g)). The male gonads were heavily packed with cnidocytes of the isorhiza type. Cnidocytes are found in the epithelium, especially of the central part of the male gonad and they appear nonfunctional, but they are transferred to the spermatozeugmata.

The **female gonad** of *C. sivickisi* is made of two flattened and elongated hemigonads (Fig. 5.32(b) and (d)). In the mature female, each hemigonad is about 2 mm long and 700 µm wide (Fig. 5.32(h)). The edge facing the neighbouring hemigonad has a band of about 200 µm, which is thinner than the rest of the gonad (Fig. 5.32(h)). From the sections, it is seen that this thinner part is occupied by immature eggs, whereas mature eggs are found in a monolayer in the rest of the gonad (Fig. 5.32(h)). The eggs are surrounded by a monolayered epithelium with columnar cells 20–30 µm long (Fig. 5.32(j)), but between the epithelium and the egg cells, there is a thin layer of mesoglea (Fig. 5.32(k)). The epithelial cells are monociliated with short cilia and cnidocytes, which seem to be nonfunctional, and found embedded in the epithelium.

Spermatozeugmata

The spermatozeugmata removed from the stomach of a female was ~500 µm long (Fig. 5.32(l)) and is assumed to contain only mature sperm cells, which is supported by them being very similar in morphology (Fig. 5.32(m) and (n)). They are also similar to the central sperm cells in the male gonad, with the only difference being a high number of small spherical acrosomal vesicles (~100 nm). The sperm cells here contain up to six mitochondria (Fig. 5.32(n)). A monolayer of cells lacking cilia covers the entire spermatozeugmata, which means that all the cilia of the sperm cells (flagellae) are 'internalized'. The cnidocytes are still present in the spermatozeugmata and are fully functional.

Spermatozoon

During mating (Fig. 5.32(a)–(d)), a sperm package is transferred to the female and the fertilization occurs at the female gonad, the sperm cells seem to be motile. Each has a highly condensed nucleus, 4–6 mitochondria (Fig. 5.32(n)) placed in a cluster at the proximal end of the nucleus, and a number of other organelles like a golgi apparatus. Each has a single fully developed flagellum, with a well-developed basal body complex. The flagellum including the basal bodies appear to have all the elements of a motile cilium. In general, they appear very similar to sperm cells of other cnidarians. Like in many other cnidarian species, the sperm cells of *C. sivickisi* are missing a well-defined acrosome. Acrosomal vesicles are not concentrated in the apical end and seem rather randomly arranged in the cytoplasm. After the sperm has been delivered to the female, it is transported (possibly by their own flagella) to the female gonads in the gastrovascular cavity.

Female gonad with attached spermatozeugmata

This discussed part of female gonad is identical to the female gonad described above (compare Fig. 5.32(h) and (p)). The attached spermatozeugmata were considerably smaller than the one isolated from the stomach, just 100–200 µm across (Fig. 5.32(i) and (p)). From the sections, it is seen that these spermatozeugmatas are no longer compact when attached to the female gonad but rather a mesh of strings of sperm cells (Fig. 5.32(p) and (q)). The epithelial cells are almost completely digested and no cnidocytes could be found (Fig. 5.32(p)). The sperm cells are thus the only remaining cells and some of them appear partly digested, as indicated by a ruptured cell membrane and few organelles (Fig. 5.32(r)). In the area close to the spermatozeugmata, some of the epithelial cells of the female gonad have large electron-dense vacuoles that are similar in size (and to some degree in shape) to the sperm nucleus (Fig. 5.32(s)). In one case, such a vacuole was found inside an egg cell. These vacuoles were not seen in the female gonads without attached spermatozeugmatas.

Fertilization

When inside the female gastrovascular system, the sperm cells are initially covered by unciliated cells, but this is probably digested by the female gut enzymes. When the sperm package reaches the female gonad, the sperm cells are uncovered, and the sperm flagella are no longer covered. They are transported (possibly by their own flagellae) to the female gonads. As mentioned before, the spermatozeugmata does not dissolve in the female gastrovascular cavity but is attached to the female gonad, probably using the isorhizas. On the female gonad, the sperm cells are partly digested and the nuclei are released. The actual fertilization seems to happen through phagocytosis of the released nuclei by the epithelial cells. In addition, the female gonads are heavily loaded with cnidocytes, of the 'eurytele' type. They putatively serve to protect the developing larvae once the embryo strand is laid. This specialized mode of fertilization was quite new information to Garm et al. (2015) and so becomes the first account of cnidocytes being directly involved in cnidarian reproduction.

The authors reported that there are no indications that the sperm cells swim and single sperm cells were never seen. This needs smear preparation of the spermatozeugmata. In the case of *C. sivickisi*, the acrosomal vesicles are randomly arranged in the cytoplasm, which fits nicely with the apparent specialized way the fertilization happens. The sperm attaches to the gonad and starts to break down, which finally releases the nucleus of the sperm. The authors do not have direct evidence for the following steps, but their observations of electron-dense vacuoles similar in size, in terms of both the epithelial cells and to some degree the shape, to the sperm nucleus inside the egg cells suggests that the nucleus is phagocytozed by the epithelial cells and transported to the egg cells. In addition, these vacuoles were only found in female gonads carrying a spermatozeugmata. This way of fertilization differs from that known in other cnidarians where the sperm and egg fuse in the normal way. However, it is similar to fertilization in some poriferans (described in all the available textbooks, see Chapter 4). In the poriferans, a specialized carrier cell phagocytozes the entire sperm cell not just the nucleus. In *C. sivickisi*, no specializations were found in the cells in the areas of sperm attachment, indicating that any part of the gonad can perform the nucleus transport (Garm et al., 2015).

Cubozoan species such as the carybdeid *Carybdea marsupialis* and the alatinid *Alatina alata* are ovoviviparous, a reproductive mode in which ova are internally fertilized by 'sperm' taken up by the female medusae during spawning aggregations with males (Marques et al., 2015). The complex cubozoan sexual reproduction is the elaborate behaviour seen only in the species of family Tripedaliidae—*Copula sivickisi* and *Tripedalia cystophora*. These tiny box jellyfish (i.e. venomous box jellyfish) display sexual dimorphism and courtship behaviour in which 'spermatophores' (spermatozeugmata) are formed by coalescing sperm packets ejaculated from 'seminal receptacles' of the male testes (called hemigonads). Testes are exhausted with successive ejaculations during breeding experiments, becoming visibly paler. However, the next day, the testes regained activity and became rich in sperm for subsequent copulations. During mating, the male uses his tentacles to control the female (Fig. 5.32A), bringing her in close, to transfer the spermatozeugmata that she invariably ingests (Fig. 5.32B). These spermatozeugmata disperse from the stomach into the gastrovascular cavity, facilitating internal fertilization within hours to days, after which time the female deposits an embryo strand (unique among cnidarians) onto the substrate. This feature was described in the above accounts of *Carybdea sivickisi*. One documented change induced by copulation in *Copula sivickisi* is the presence of conspicuously dark pigmented velarial spots in sexually active females, which are absent in juvenile females and all males (Marques et al., 2015).

Alatina cf. *moseri* (Mayer, 1906)

The life cycle for the majority of the cubozoan species is missing. However, only one complete life cycle, from spawning medusa through sessile stages to spawning medusa, has been published previously for the Caribbean cubozoan *Tripedalia cystophora* by Conant, 1897 (Carrette et al., 2014). Carrette et al. studied the embryological life stages of the cubomedusan, *Alatina* cf. *moseri* (Fig. 5.33A). This cubomedusan was known previously as *Charybdea moseri* Mayer, 1906 or *Carybdea alata* Reynaud, 1830 (Lawley et al., 2016). It has been documented in Hawaii

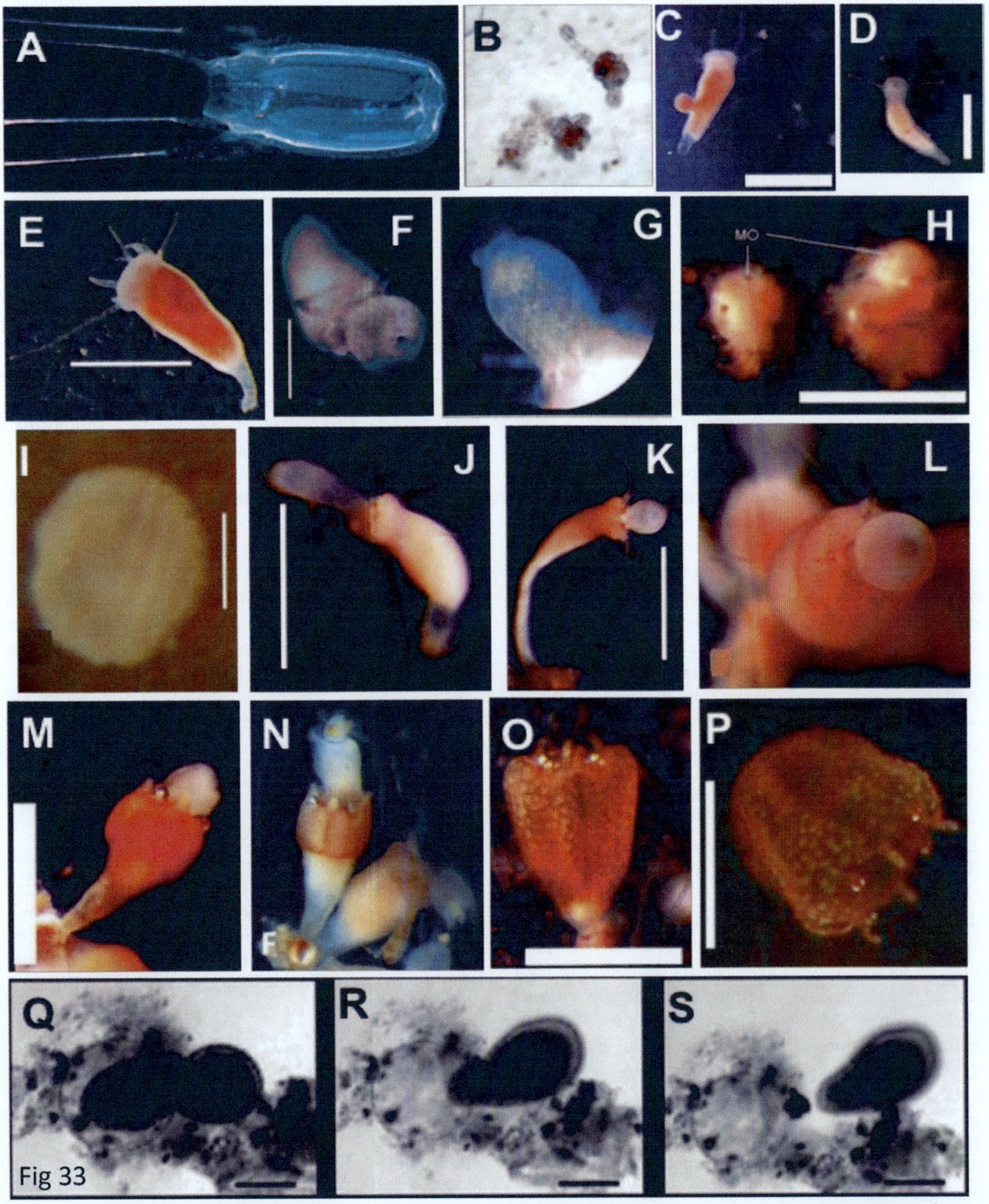

FIG. 5.33 (A) Photograph of the live winged box jellyfish *Alatina moseri* from Hawaii **(Marinelife photography. Com, Keoki Stender)**; (B) primary polyps (only few μm in height, no scale, picture was taken by camera through 400× microscope objective); (C) budding adult polyp (Hawaiian population), X = 1 mm; (D) creeping polyp (bud) X = 0.5 mm; (E) adult polyp (>21 days postfertilization), lateral view (Australian population) X = 1 mm; (F) adult polyp, oral view—note lips (Australian population) X = 0.5 mm; (G) hypostome, note lips (hypostome c. 0.5 mm in length) (Hawaiian population); (H) cyst, stage right after encystment, note the still visible mouth opening (MO) X = 0.5 mm; (I) cyst, stage after a week, note the unstructured tissue inside the mucus shell, X = 0.25 mm; (J–P) metamorphosis: (J) Stage 1, elongation of hypostome, calyx and stalk, tentacles cluster at four spots; (K) (lateral view); (L) (oral view close up of same animal as K), Stage 2 clustered tentacles fuse at base, note the *red-violet pigmentation* around hypostome and eye spots at the fused tentacles bases; (M) stage 3, medusa tentacles appear in space between rhopalia; (N) (same animal as M), note that this stage is still feeding on *Artemia* nauplii; (O) stage 4, nematocyst clusters appear on the developing exumbrella; (P) stage 5, after reabsorption of the remaining stalk tissue, the newly detached medusa free swimming. *(All after Carrette et al., 2014)*; (Q–S) series of digital frame grabs taken from video footage of *A. alata* planula emerging from a cyst, X = 50 μm. *(From Lawley et al., 2016.)*

since the late 1980s with their predictable arrival on Waikiki beach on the 8th to 10th days after every full moon. Carrette et al. reported that many specialized scientists observed aggregations of sexually mature *Alatina* from this site on consecutive monthly cycles from Waikiki Beach in Hawaii on various occasions over the last 10 years, leading to the hypothesis of a spawning population.

Carrette et al. (2014) described how they obtained their specimens. On November 15, 2006, 10 days after the full moon, specimens of *Alatina* cf. *moseri* were collected between 0000 h and 0230 h at Osprey Reef (Northern Queensland, Australia). To attract medusae for collection, high power underwater lights were set up off the diving platform of a research dive vessel anchored at the reef's edge. Collection of over 250 individual animals occurred; they were taken out of the water with dip nets and transferred into large collection bins filled with fresh seawater. Visual inspections of animals revealed ~60% of individuals were in spawning condition with milky white gonads evident. However, the sexes of individual medusae (from Hawaii) were determined by light microscopy observation. For fertilization, ~0.5 mL of egg solution was extracted from each spawning female with a glass pipette and transferred into individual Petri dishes filled with clean seawater. A single drop of the 'sperm mixture' taken from the water, in which spawning males were kept, was added to each of the egg solution dishes. Regular observations were made and, when back on land, planulae were observed in the plastic containers. They were extracted and placed in new 650 mL containers with filtered seawater.

Lawley et al. (2016), while examining developing embryos of another species of box Jellyfish *Alatina alata* in the lab ~24 h after their release into the aquarium water, discovered multiple cysts (~150 µm in diameter), each containing a single planula bearing characteristic equatorial eye-spots, rotating on its longitudinal axis. During the course of their microscopic examination, one planula successfully bored through the outer, 'shell-like' perisarc (within 5–50 min), emerging as a swimming planula (Fig. 5.33Q–S). Lawley et al. reported that this is the first documented occurrence of a planula hatching in a cnidarian. Whether this encysting stage is a form of diapause (i.e. a dormancy stage or delay in development in response to adverse environmental conditions) remains to be confirmed. Another open question is whether all or just some of the maturing embryos develop a perisarc.

Turning back to the planulae in the work of Carrette et al. (2014), both populations were observed from 24 to 48 h post gamete mixing in Petri dishes, and at this stage showed the presence of 'eye spots'. The planulae were then extracted and placed in clean plastic containers with seawater, and were held in a constant temperature aquarium room (24–26°C). Observations were made regularly over the following 18 months. Seventy-two hours postfertilization, the planulae were observed on the bottom of the Petri dishes and had attained irregular shapes. The planulae attained two to four round tentacle buds, which protruded from their circumference containing two to four small 'eurytele' nematocysts in their tips. The tentacle buds continued to grow over the next 24 h and on the fifth day, primary polyps were visible at the bottom of the container. Each polyp is cone shaped, with four long tentacles. They creep with one of the long tentacles, which were longer than the body length and extended out on the substrate. About 50% of the primary polyps settled down on the bottom of the plastic containers on the sixth day. Each polyp had well-developed hypostome (Fig. 5.33B). From days 10–13, the morphology of the primary polyps changed and appeared to flatten against the substrate, while their four tentacles were extended out to the sides. Their tentacles elongated, each having just a single 'stenoteles'.

Polyps increased in size from day 18 onwards and by day 21, each of 40% of the polyps developed six or eight tentacles. The polyps of this species were successfully reared from spawning adults. Polyps of *Alatina* cf. *moseri* were cultured at temperatures of 23–28°C, developed up to 19 tentacles and reached up to 1.7 mm in height.

Agametic reproduction

These polyps were capable of reproducing agametically. Many of them were observed with buds at their calyx base (Fig. 5.33C), and each could continue this reproduction until they have as many as five buds at any time. The bud was in two forms: it either remained next to the base of the original 'parent' polyp after detachment, or crept for some time and then relocated to another position. The latter form did not resemble the parent polyp or the primary polyps during the creeping phase (Fig. 5.33D). The creeping polyps showed four to eight very short, still filiform tentacles around a cone-shaped hypostome and a high number of large stenotele nematocysts at the rear end of the body. Single stenoteles were located at the bases of the tentacles that migrated to the tentacle tips as soon as the polyps settled. The creeping polyps markedly differed from the primary polyps by having a distinctly larger body size (up to 2 mm), and by the 'daisy flower'-like 'head region'. This form resulted from the very short, nematocyst-lacking tentacles surrounding the mouth cone. These creeping polyps attained free life for 2–3 days over the substrate before they settled. They then grew to a 9- to 12-tentacle stage before starting budding themselves.

Cyst formation

If the conditions changed and became unfavourable, such as drastic changes in temperature, salinity or absence of food for prolonged periods, the polyps of both populations encysted. The polyps retracted their tentacles completely into their calyx and contracted the whole body into a ball-shape (Fig. 5.33H). The different parts of the body like tentacle tips, mouth opening and calyx fused into an indistinguishable white tissue ball, enclosed by a transparent and flexible mucous coat (Fig. 5.33I). These cysts were attached to the substrate and survived for up to 3 months. Three days after starting feeding, polyps of both populations (from Australia and Hawaii) regenerated. Carrette et al. (2014) reported that the longest observed cyst state was by the Australian culture and this was a 12-month period, before recovery was commenced.

Gueroun et al. (2015) added that within the life cycle of other cubomedusae, such as *C. morandinii*, the polyp can encyst into a resting stage to overcome unfavourable environmental situations; the cyst can then detach and float, giving rise to new polyps after its reattachment in a different habitat (Straehler-Pohl and Jarms, 2011).

Metamorphosis

Metamorphosis in *Alatina* cf. *moseri* polyps cultured in Germany occurred only between the months of November and March, with or without temperature change. The metamorphosis in Australian cultures occurred spontaneously after their original collection. However, not in the same numbers, as was initially observed 6 weeks after initial fertilization took place. Metamorphosis in both populations completed after about 14 days from the start. The first signs of metamorphosis were apparent in all cultures on days 29–31 with an observed change in polyp shape. The polyp pigment darkened and the tentacles became four distinct clusters. In addition, the hypostome of the polyp and its stalk lengthened. The polyps of the two observed *Alatina* cf. *moseri* populations showed no distinct differences. Polyps of both populations were solitary (Fig. 5.33E). The body of the polyp is divided into three parts: the hypostome, the calyx and a stalk region. The latter includes a basal disc and a tiny periderm beaker enveloping the pedal region. The hypostome of polyps of both populations was four-lipped (Fig. 5.33F and G), balloon-shaped and ~14%–15% of the total body length (TBL). It was not completely contractible into the body.

Carrette et al. (2014) divided the process of metamorphosis in both populations of *Alatina* cf. *moseri* into five stages, documented photographically and with detailed description:

- Stage 1 was characterized by the elongation of calyx, stalk and specially the hypostome (Fig. 5.33J), in some cases, the hypostome being as long as the calyx. The tentacles formed clusters at four points around the peristome. At the base of the hypostome, reddish violet pigmentation appeared at four spots between the tentacles.
- Stage 2 was characterized by the clustered tentacles fused at the bases, and the development of sensory structures (Fig. 5.33K and L). Rhopalia developed at the fused tentacle bases. Statolith formation occurred initially while the tentacles were still reabsorbed.
- During stage 3 (Fig. 5.33M and N), four medusa tentacle buds appeared. A deep cleft around the hypostome marked the transformation of the polypoid hypostome into a medusoid manubrium. The red-violet pigmented region transformed into a velarium. At this stage, the metamorphosing polyps were still catching and consuming prey (Fig. 5.33N).
- Stage 4 was characterized by the change of the colour of the calyx from reddish dark brown to yellowish light brown, and the calyx itself grew further in size, forming a slightly transparent umbrella. Small nematocyst batteries appeared, densely covering the developing exumbrella. The medusoid tentacle buds, which ahd appeared during the previous stage, grew out to very short, hollow and moniliform tentacles. Each developed thick, yellowish-brown rolls of nematocyst battery (Fig. 5.33O).
- The remaining stalk was reabsorbed in stage 5 (Fig. 5.33P), during which the developing medusa completed its formation and initiated its pulsating, at least 48 h prior to detachment.

During all the stages of metamorphosis, no feeding behaviour was observed.

Young medusa

The young medusae of these two populations (Hawaiian and Australian populations) are similar; no differences were found, as was described by Carrette et al. (2014). This young medusa had tetraradial symmetry and it was yellowish-brown to yellowish-olive in colour. Its umbrella was spheroidally pyramidal with diameter measuring (=UD) 1.1–1.5 mm (mean: 1.24 mm). The manubrium of the young medusa was short, measuring 0.24–0.3 mm (mean: 0.27; <20% of UH) and four-lipped. One gastric filament per quadrant was visible through the apex of the umbrella. No additional medusa tentacles developed after the four that already formed before detachment. Carrette et al. reported that the extended tentacles resembled short pearl strings because of the up to four thick, round, yellowish-brown nematocyst batteries. In their well-kept culture, the longest life span of medusae ($n = 6$) kept alive was for 29 days; at their time of death, medusae ranged in size from 15 to 25 mm.

Morbakka virulenta

Another Japanese giant cubozoan, *Morbakka virulenta* (Fig. 5.34A), collected from the central part of the Seto Inland Sea, western Japan in October 2009, was studied by Toshino et al. (2013). The bell height of medusa ranged

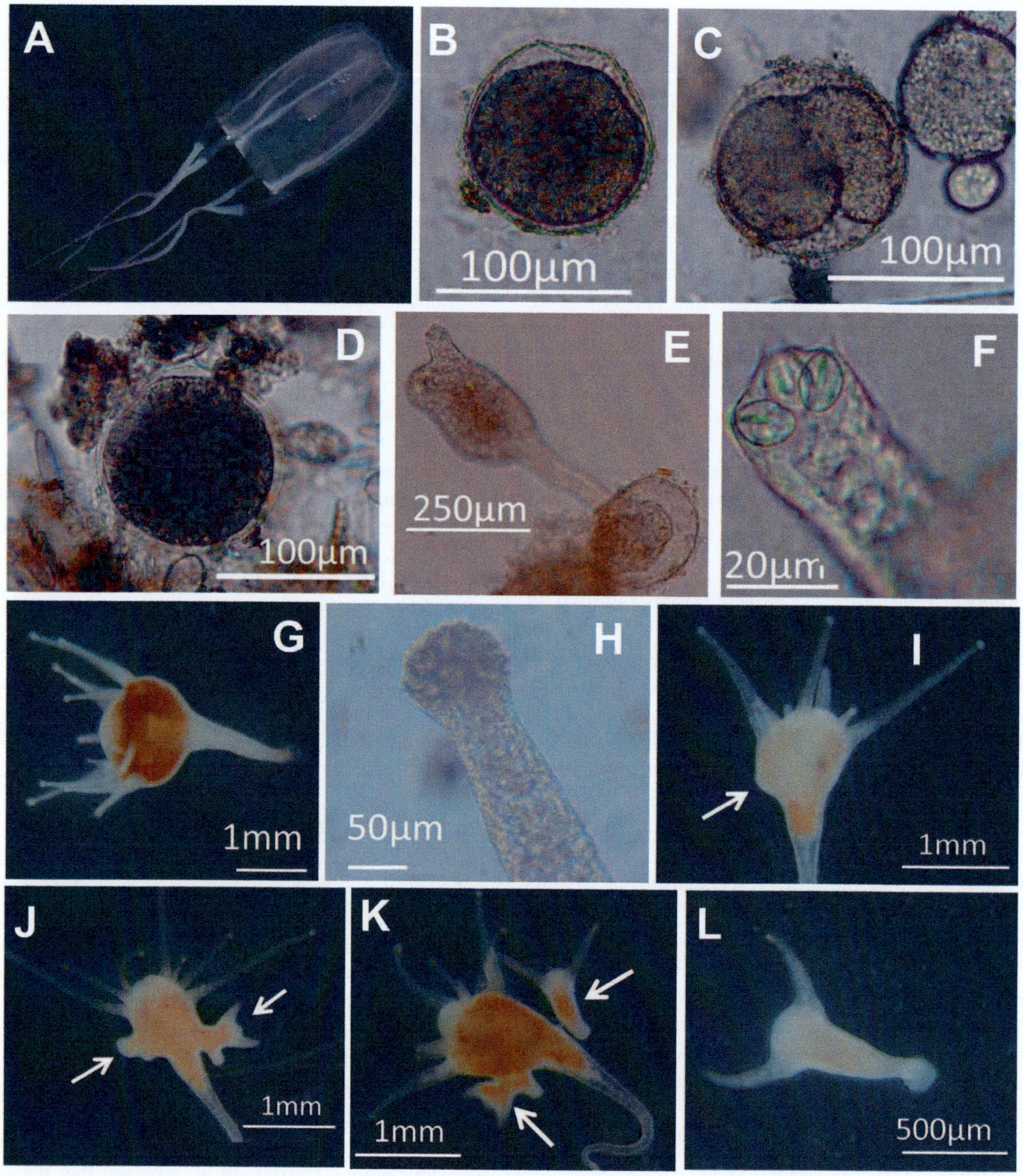

FIG. 5.34 (A) Lateral view of the giant box jellyfish *Morbakka virulenta* (Kishinouye, 1910), live, in situ and bell height of 200mm; (B–K) development and polyp formation; (B) fertilized egg; (C) planula rotating in egg membrane; (D) planula; (E) one-tentacled polyp; (F) nematocyst of one-tentacled polyp; (G) 16-tentacled polyp; (H) nematocysts of 16-tentacled polyp; (I) just at the beginning of the process of polyp budding, an *arrow* indicates a bud; (J) after 6days; (K) after 9days, the right bud just released; (L) swimming polyp. *(From Toshino et al. (2013).)*

from 100 to 250mm. The sex of the medusa was determined under a microscope by observing the condition of the gonads. The gonads of mature medusae were whitish to whitish-yellow in males and females. Fertilized eggs were not found in the gonads of females, indicating that fertilization occurred externally. Few eggs were observed in the aquarium on the night of collection. Two days after collection, thousands of eggs were obtained. These eggs were about 0.10mm in diameter (Fig. 5.34B) and most of them were at the two-cell stage (Fig. 5.34C). Within 2h, they developed into the 8- or 16-cell stage. These eggs developed into blastulae (Fig. 5.34D) within the following 2h, and developed into blastocysts within 30h from the blastulae stage. Blastocysts could play an important role in adaptation to low temperatures and unfavourable settlement conditions without solid substrate. Under the conditions of this work (Toshino et al., 2013), embryos of *M. virulenta* formed blastocysts at low temperature (17°C) and blastocysts excyst at a higher temperature (21°C). Development stopped during these days and then the blastulae finally metamorphosed into actively rotating planulae within the egg membrane. Planulae were about 0.15 mm in diameter and had no band of larval ocelli. Lacking this feature is a characteristic of cubozoan planulae (Werner et al., 1971).

Planulae stayed within the egg membrane until they became 5 days old, after which they developed into polyps, with their basal discs attached to the egg membrane. At this stage, they hatched out of the eggs. They looked like tulips with long stalks, and from the ovoid calyx, only a single tentacle protruded (Fig. 5.34E). The calyx of the polyps was about 0.06 mm in mouth disc diameter and the polyps were about 0.90 mm in total body length. At the tip of the tentacle of one-tentacle polyp, three nematocysts developed (Fig. 5.34F). The second tentacle appeared in the 13-day-old polyp. The polyps developed four tentacles in 3–4 weeks, eight in about 5 weeks, and 16 in about 3 months (Fig. 5.34G). Polyps reached 1.08 mm in maximum mouth disc diameter and 4.66 mm in total body length when fully developed. The number of their tentacles reached 17, and the tip of each had >30 nematocysts (Fig. 5.34H).

Polyps of *M. virulenta* were distinguished from all described polyps of other cubozoans by shape and size. These polyps of *M. virulenta* were characterized by being tulip-shaped. They are the largest of all known cubozoan polyps, and are as large as known scyphozoan scyphistomae (Morandini et al., 2004). Toshino et al. (2013) tabulated these two features for comparison between all studied cubozoons.

The eight-tentacle polyps became mature and could reproduce agametically by budding. Buds were formed in the lower part of the calyx (Fig. 5.34I). The new bud developed two tentacles and a basal disc after 4 days (Fig. 5.34J). The new polyp was released from the original polyp (Fig. 5.34K) after 9 days from the start of the budding process. Swimming polyps were 0.7 mm in length and had two tentacles (Fig. 5.34L). They swam near the bottom or at the water surface. These polyps settled on the wall or near the edge of a Petri dish, and transformed into adult polyps within 2–3 days. In contrast, other cubozoan polyps formed from agametic buds are worm-like creeping polyps with a tiny hypostome cone surrounded by two to eight tentacles. This type of polyp is found in Scyphozoa (Gohar and Eisawy, 1960a,b; Sugiura, 1965). Polyps of *M. virulenta* can be distinguished from polyps of other cubozoan species by having a much longer stalk and a tulip-shaped body. Their long stalks bear, at the base, some traces of egg membrane that the polyp can use as a substrate in areas without hard substrate. The long stalk of the polyp may diminish the possibility of being buried by soft sediment. However, this character of long stalks suggests that *M. vilurenta* is a link species between Cubozoa and Scyphozoa (Toshino et al., 2013).

CLASS ANTHOZOA

Anthozoans (Fig. 5.35) are either solitary or colonial polypoid cnidarians, in which the medusoid stage is completely absent. Many familiar forms, such as the sea anemones, corals, sea fans and sea pansies, are members of this class, and so it is a somewhat heterogeneous group. Although the anthozoans are polypoid, they differ considerably from hydrozoan polyps (Fig. 5.9).

Since the anthozoans have no medusan phase, their life history is therefore relatively simpler than other cnidarians. The fertilized egg develops into a planula, which attaches and becomes a polyp. Tentacles are formed, septa become prominent and agametically reproducing species in most cases form branches, plate-like or clumped colonies. The reproductive cells of the anthozoa are of endodermal origin and there are many species in which the eggs contain large amounts of yolk and undergo superficial cleavage, so that it is difficult to observe their mode of germ layer formation. In species with only small amounts of yolk, cleavage is total and radial, although there are some species that show temporary appearance of spiral cleavage in the eight-cell stage.

In most cases, germ layer formation takes place by invagination. Some yolk usually remains in the blastocoel, which is mostly absorbed by the time the embryo becomes a planula and begins its swimming life. This lasts for several days. Some species develop into colonies by budding or by fission.

In solitary forms, the septa become highly developed. Motile species, whether solitary or colonial, form chiefly muscular bodies, while the sessile species produce a calcareous or keratinoid skeleton. These skeletons are secreted by mesogleal cells, which arise as modified ectodermal cells and gradually migrate into the interior.

Both dioecious and monocious forms are found. The gonads consist merely of these cells accumulated either at the base of the endodermal 'epithelium' or within the mesoglea of the gastric mesenteries (septa). These gonads develop on all septa except the two 'asuleal' ones (directives). Gametogenesis only occurs in the band between retractor muscles and septal filament. Accumulation and growth of the gametogenic cells causes the local epithelia to rupture, liberating the gametes directly through the mouth, although there are reports of gamete release through pores in the tentacle tips or base. Antipatharians are best known as black corals.

Subclass Ceriantipatharia

Order Antipatharia

Antipatharians are commonly known as black corals. Although its tissue is brilliantly coloured, the black coral takes its name from the distinctive black or dark brown skeleton. Black corals are treasured by many cultures for medicinal purposes and to produce jewellery. They are used to manufacture precious coral jewellery, and as a result are

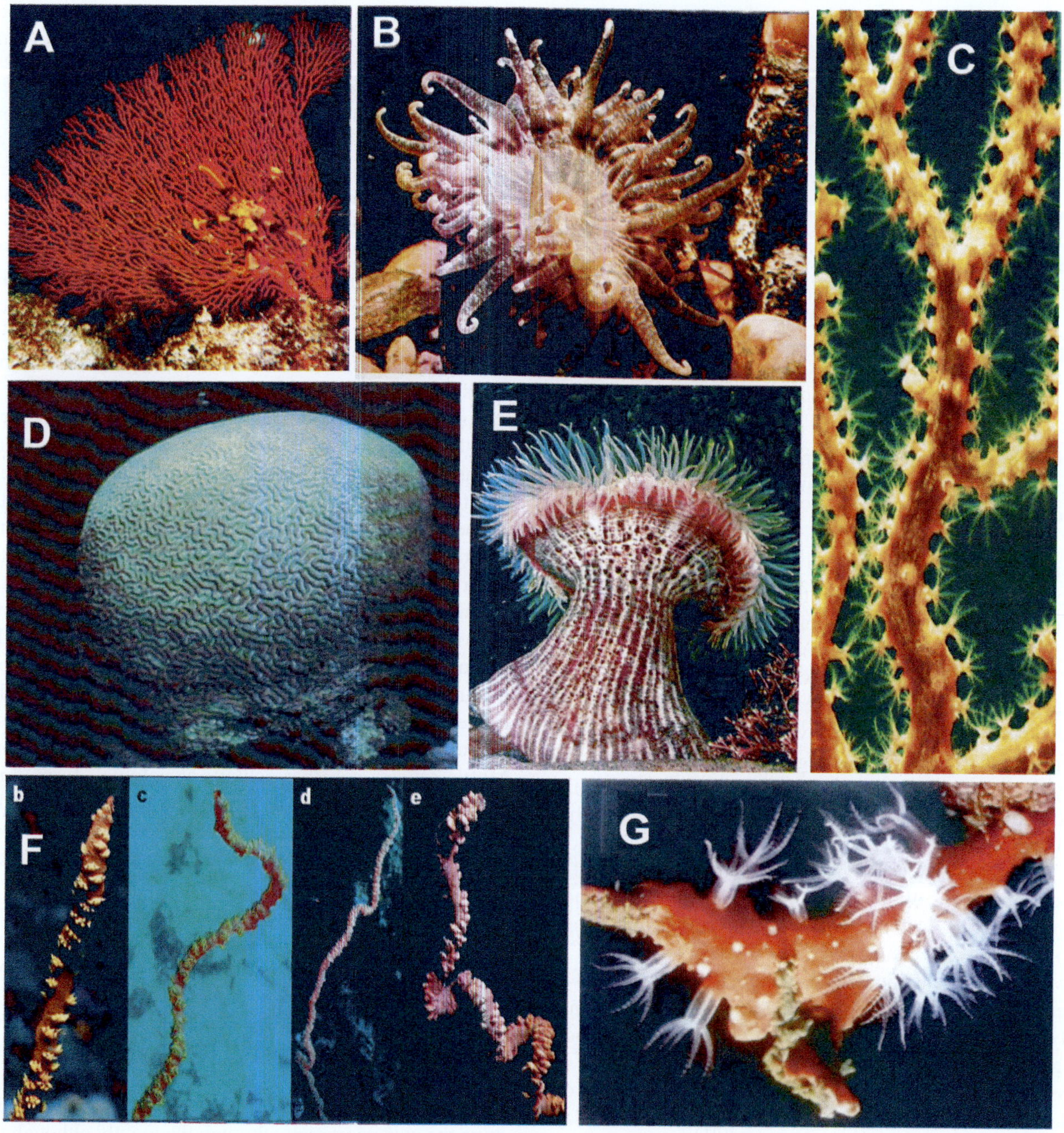

FIG. 5.35 Different types of Anthozoa: (A) a gorgonian of the great Barrier; (B) sea anemone exotic, *Anthopleura balii*; (C) the gorgonian, *Eunicella cavolini*; (D) photograph of whole mount of the living brain coral, *Platypyra lamellina*; (E) sea anemone, *Actinia equina* with cylindrical body, 7cm in diameter; (F) in situ photographs of *Cirrhipathes* cf. *anguina* colonies from Hawaiian waters, *photos courtesy of Jim Maragos*, bar = 1 mm. The corallum diameter is generally 1.39–6.12 mm at midheight. Polyps are of variable size, ranging from 0.67 to 4.45 mm in transverse diameter, and are arranged irregularly on all sides of the corallum; (G) portion of the *Octocorallia Pennatulacea* colony, *Pennatula phosphorea*. *((A–C, E, and G) from Beauté du monde animal, Invertébrés Tome XII (1974); (D) from Saudi Arabian, Red Sea by H. Schmid and P. Vine; (F) from Wagner (2011).)*

targeted by commercial fisheries in several regions around the world. Most species inhabit deeper-water environments (>50 m). Although black corals occur throughout all oceans, and from subtidal to abyssal depths, they are particularly common in tropical and subtropical regions at depths below 50 m (Wagner et al., 2012a,b). Despite this wide bathymetric range, over 75% of described antipatharian species are restricted to depths below 50 m. Additionally, there are no records of black corals from intertidal zones, perhaps reflecting that antipatharians are sensitive to air exposure (Wagner, 2011). Antipatharians reproduce through both gametic and agametic processes. In general, polyps and colonies are gonochoric, with fertilization and larval development likely occurring externally. However, to date antipatharian larvae have only been observed for a single species. Antipatharians are generally slow-growing and long-lived organisms with maximum longevities ranging from decades to millennia.

Agametic reproduction

Antipatharians, like other cnidarians, reproduce agametically, and Wagner (2011) enumerated the types of this mode of reproduction that have been reported for black corals which include: budding of new polyps; breakage of fragments which can subsequently reattach to the substrate; and production of agametic larvae in aquaria under stressful conditions. The latter two modes of agametic reproduction result in the formation of new colonies, whereas the former leads to new polyps on existing colonies.

Black corals are exclusively colonial organisms, and budding of new polyps is part of their growth. Budding regularly occurs in between polyps, and leads to small polyps interspersed within branches containing larger polyps. Also, within several antipatharian species, the existing polyps may form buds directly. The Hawaiian black corals *Antipathes griggi* and *A. grandis* often break away their terminal branches from their colonies. However, this phenomenon is rarely successful in producing new colonies because broken portions hardly ever reattach to the substrate. Regeneration of anchorages has not been observed in broken portions, suggesting that this mode of agametic reproduction is rarely successful. Wagner (2011) reported that the artificial transplantation of black coral fragments has been attempted for *Antipathes* sp. in Cuba, *Stichopathes* cf. *maldivensis* and *Cirrhipathes* cf. *anguina* in Indonesia, and *Antipathes griggi* in Hawaii, with mean survival rates of 93%, 80%, 68%, and 45%, respectively. These high survival rates indicate that agametic reproduction via fragmentation can be successful under favourable conditions.

Gametic reproduction

Gaino and Scoccia (2010) stated that gametic reproduction of the black coral is far from being fully explored, mainly because of the bathymetric range where the colonies live, which limits sampling procedures and subsequent investigations. Wagner (2011) reported that most information on the gametic reproduction of antipatharians is derived from scarce notes on the anatomy of reproductive tissues accompanying taxonomic descriptions. Even though the majority of this information is derived from the examination of a limited number of specimens, previously published data on gametic reproduction exists for at least 56 nominal species, representing >20% of the described antipatharian fauna. Wagner (2011) tabulated these previous data to be available to date. It appears that a number of generalities within the order Antipatharia were clarified:

1. Out of the six primary mesenteries, only the two in the T.S. bear the filaments and gametes.
2. Individual polyps are strictly gonochoric because polyps with both oocytes and spermatocysts have not been found within any antipatharian species examined to date.
3. Entire colonies are either female or male, with the exception of *Stichopathes saccula*, which has mixed colonies with both male and female polyps.
4. There is no evidence for internal fertilization within the Antipatharia, indicating that fertilization and larval development likely occur externally in the water column and not internally within the polyps.

Very little is known about the larval biology and reproductive seasonality of antipatharians. Only larvae of the species *Antipathella fiordensis* cultured in the laboratory that have been observed (Miller, 1996). The ciliated planulae were 200 µm in length and developed within 36 h of fertilization. These larvae were negatively buoyant, weak swimmers, likely nonfeeding and stayed alive for a maximum of 10 days, but none settled.

Gaino and Scoccia (2010) examined gametic reproduction in different species of family Antipathidae and others of family Myriopathidae. They obtained their material by means of scuba diving at various depths along the Indonesian coral reefs. They considered the results of the study of the polyps of the colonies of *Cirrhipathes anguina* (Fig. 5.35F) as an example to highlight the general features of reproduction in the black coral. They sampled this species in January and August 2007, and in February 2008. Not only was the process of gametogenesis described, but also particular attention was paid to the detailed organization of the polyp itself and the mature gametes.

Ex. *Cirrhipathes anguina*

Mature male polyp

T.S. through mature polyp showed that the two primary mesenteric transversal septa, each separating the lateral tentacles from one another, were involved in spermatogenesis. The male gametes were contained within spermatocytes, as other anthozoons, which occupied the entire width of each septum (Fig. 5.36A). During spermatogenesis, the spermatic cysts increased in size enormously, with obvious enlargement of these septa. Owing to this activity, the mesenteries occupied the whole of the cavity of the lateral tentacles (Fig. 5.36B). The gastrodermal cells penetrated deeply inside the mesentery, forming cytoplasmic extensions that separated spermatocyts from one another (Fig. 5.36E). Ultrastructural examination showed that the spermatocyst is covered by a thin layer consisting of outer flattened cells rich in electron-lucent inclusions, and inner fibrillar thin mesoglea. Developing spermatozoa were arranged in rows (Fig. 5.36C). Within each

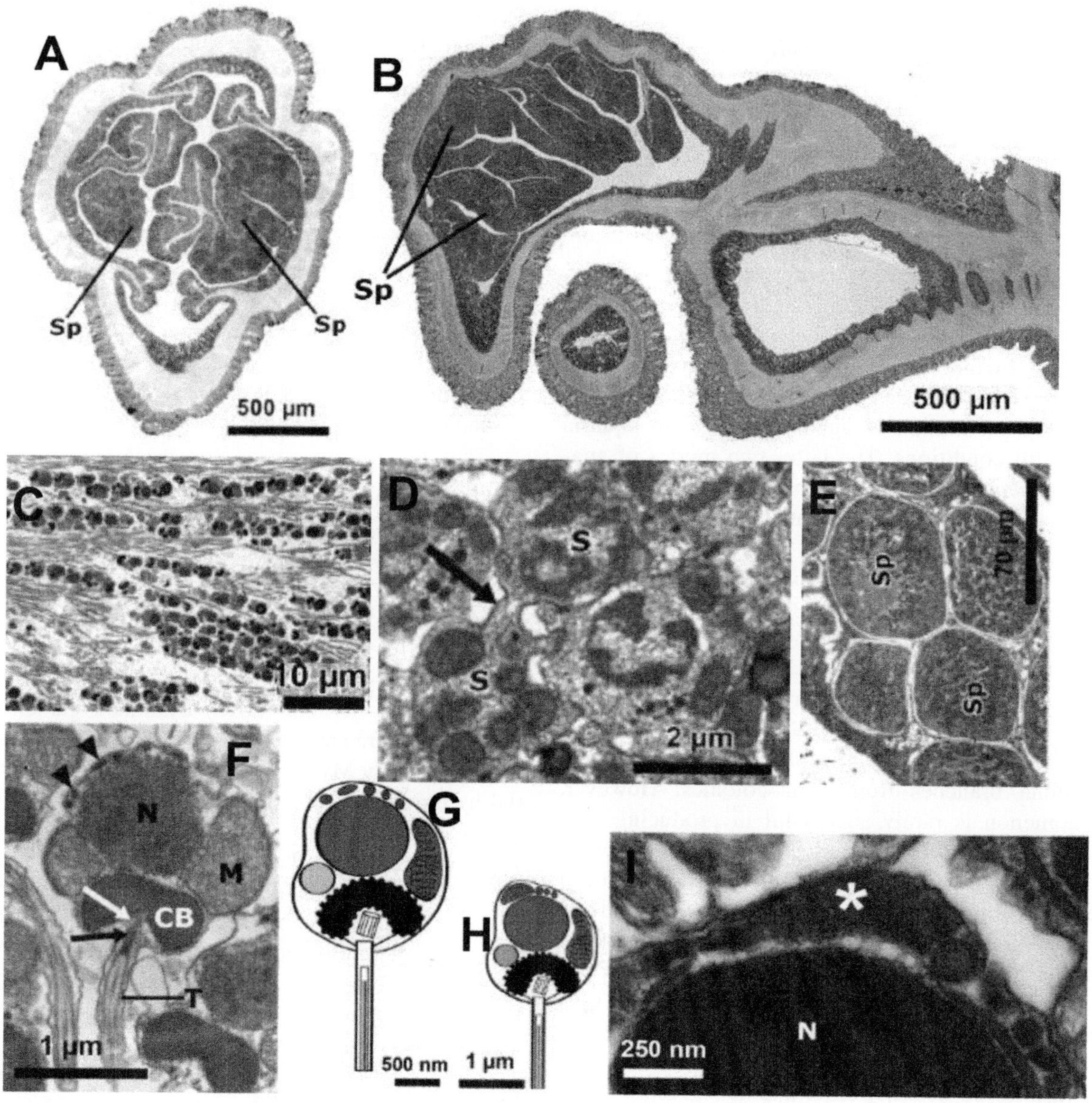

FIG. 5.36, CONT'D

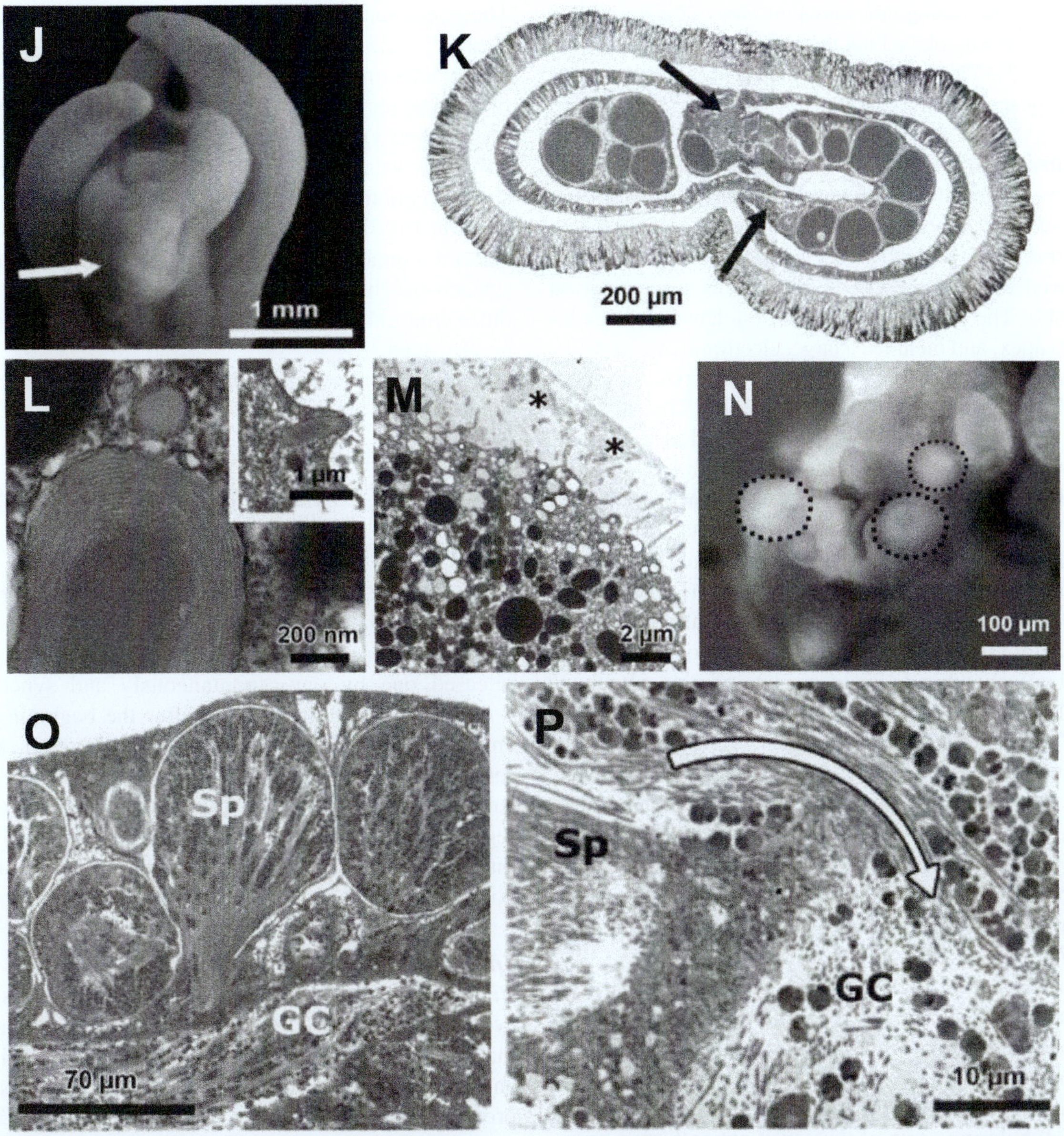

FIG. 5.36 Spermatogenesis and sperm structure of *Cirrhipathes anguina*: (A) section of a mature male polyp showing 1^{ry} transversal mesenteries loaded by spermatocysts (Sp); (B) mesentery occupying the cavity of a lateral tentacle owing to the numerous spermatocysts (Sp); (C) sperm arranged in rows within a spermatocyst; (D) cytoplasmic bridge *(arrow)* connecting two spermatocytes (S); (E) spermatocysts (Sp) separated by a network of the cytoplasmic extensions produced by the gastrodermal cells bordering the mesentery; (F) mature sperm structure. Note: apical pro-acrosomal vesicles *(arrowheads)*, nucleus (N), mitochondrion (M), cup-like body (CB), centrioles *(arrows)*, tail (T); (G) sperm model derived from the ultrastructural images of Antipathidae and of the Myriopathidae *Cupressopathes pumila*; (H) sperm model derived from the ultrastructural images of *Myriopathes japonica*; (I) acrosome-like structure *(asterisk)* located above the nucleus (N). *After Gaino and Scoccia (2010)*. Oogenesis in *Cirrhipathes anguina* and spawning, *after Gaino and Scoccia (2010)*: (J) mature female polyp showing the base of the lateral tentacles, loaded *(arrow)* by the presence of oocytes inside the primary transversal mesentery; (K) semi-thin section showing the gradual development of the oocytes inside the mesentery. Note the young oocytes *(arrows)* located close to the point of emergence of the mesentery from the gastroderm; (L) inclusions scattered in the cytoplasm or inside the pseudopodia (inset) and characterized by an electron-dense core surrounded by a sequence of tightly packed fibrils; (M) portion of a released egg surrounded by a fibrillar coat *(asterisks)*; (N) spheres *(dotted circles)* protruding from the mouth of a polyp; (O) a spermatocyst (Sp) emptying into the gastric cavity (GC); (P) sperm flowing *(arrow)* from the spermatocyst (Sp) into the gastric cavity (GC). In situ photographs of *Cirrhipathes* cf. *anguina* colonies from Hawaiian waters *(photos courtesy of Jim Maragos)*.

spermatocyst, spermatogonia were confined to the peripheral region, and some displayed interconnections via cytoplasmic bridges (Fig. 5.36D). It was established that this is the stage of secondary spermatocytes.

The mature sperm Fig. 5.36F shows SEM of the structure of a mature sperm of *Cirrhipathes anguina*. The sperm has a large and fairly round nucleus (diameter ~1–1.5 μm) with one mitochondrion on each side. However, in serial sections, these proved to be two portions of a single mitochondrion (Fig. 5.36G). The sperm showed electron-dense vesicles located at its apex, and included a large, electron-dense cup-like body, measuring about 800 nm along its long axis and about 500 nm along the shorter one (Fig. 5.36F–H). This structure occupies just underneath the nucleus and it has a concave shape in the region adjacent to the nucleus, while on the opposite side it is bounded by an electron-lucent cytoplasmic region. The two centrioles, proximal and distal, are almost coaxial (Fig. 5.36F). The axoneme of the tail, which was characterized by a 9+2 microtubular array, originated from the distal centriole (Fig. 5.36F). Ultrastructural observations enabled the different phases of the formation of the cup-like body, and mitochondria do not merge into a single mitochondrion to illustrate two schematic drawings of the sperm organization. The sketch illustrated in Fig. 5.36G is derived from the ultrastructural observations carried out on the representatives of Anthipathidae, namely *Cirrhipathes anguina*. The sketch illustrated in Fig. 5.36H is derived from the ultrastructural analysis carried out on *Myriopathes japonica* belonging to Myriopathidae. The two sketches differ in terms of the presence of an acrosome-like structure, which is only observed in *M. japonica* (Fig. 5.36H).

Gaino and Scoccia (2010) concluded that the structure of the black coral sperm is therefore fairly simple, like that of the other members of Anthozoa, in which an actual acrosome is lacking. In addition, the cup-like body is restricted to Antipatharia. Examiniation of its presence in the members, which belong to two of the seven antipatharian families, confirms the hypothesis that the cup-like body is an apomorphic characteristic.

Mature female polyp During oogenesis, the female polyp shows, like the male polyp, obvious enlargement of the base of its lateral tentacles (Fig. 5.36J), since the two primary transversal mesenteries are occupied by the developing oocytes (Fig. 5.36K). The oocyte measures from 30 to 125 μm according to the stage of maturation (Fig. 5.36K). Maturation of oocytes (oogenesis) begins at the point where the mesentery emerges from the gastroderm, which includes smaller eggs, and proceeds towards the central region of the mesentery, where the oocytes reach their maximum size (Fig. 5.36K). The stage of oocyte maturation is determined by the amount of material accumulated in the cytoplasm (steps of the process of vitellogenesis). These contents of the cytoplasm may push the nucleus slightly aside. These oocytes are separated by cytoplasmic extensions, extending from the cells of the wall of the mesentery. These extensions inside the mesentery form a dense network, as observed in the case of the male spermatocysts (Fig. 5.36E). The developing oocyte has short irregular microvilli, which extend into the surrounding thin fibrillar matrix (Fig. 5.36G). Occasionally, some cellular elements of the mesoglea approach the oocyte surface and their cytoplasmic extensions interdigitate with those emerging from the oocytes. Lobose pseudopodia can be also seen.

The elements contained in the cytoplasm resulting from vitellogenesis are:

1. lipids, which appear as electron-lucent vesicles;
2. protein yolk, appearing as electron-dense granules of different sizes and shapes; and
3. rod- or ovoid-shaped bodies (Fig. 5.36L).

A striking observation reported by Gaino and Scoccia (2010), which has attracted attention, is that from the fragments of some reared colonies, pigmented eggs were released one by one, spontaneously and synchronously 1 day after sampling. Shortly after the beginning of their emission, the eggs underwent depigmentation, a process followed by their rapid degeneration and desegregation. A subsequent inspection of the polyps cut from the different regions of the colonies (basal, central and apical) showed that reproduction takes place synchronously in the whole colony.

Spawning Ultrastructural sections showed that in *Cirrhipathes anguina*, sperm were released from spermatocysts after the breakage of the mesentery wall, a process that enables sperm to enter the gastric cavity (Fig. 5.36O and P) before spawning. In some modified polyps, spheres protruded directly from the mouth (Fig. 5.36N). The largest spheres had a maximum diameter of 300 μm. The reared polyps released the spheres, which settled on the bottom of the tanks where fragments of the black coral were reared. Semi-thin-sections showed that each released sphere consisted of multiple spermatocysts, spherical or oval in shape, containing sperm arranged in rows, a feature confirmed at the ultrastructural level, as observed in *C. anguina* (Fig. 5.36C).

Ex. *Antipathes* Pallas, 1766

The Hawaiian species ***Antipathes griggi*** Opresko, 2009 (Fig. 5.37A–C) was previously identified as *A. dichotoma* (Bayer, 1961), a species originally described from waters off the coast of Marseilles in the Mediterranean Sea (Pallas, 1766). Colonies of *Antipathes griggi* were sampled as part of the study of Wagner (2011). A total of 220 specimens

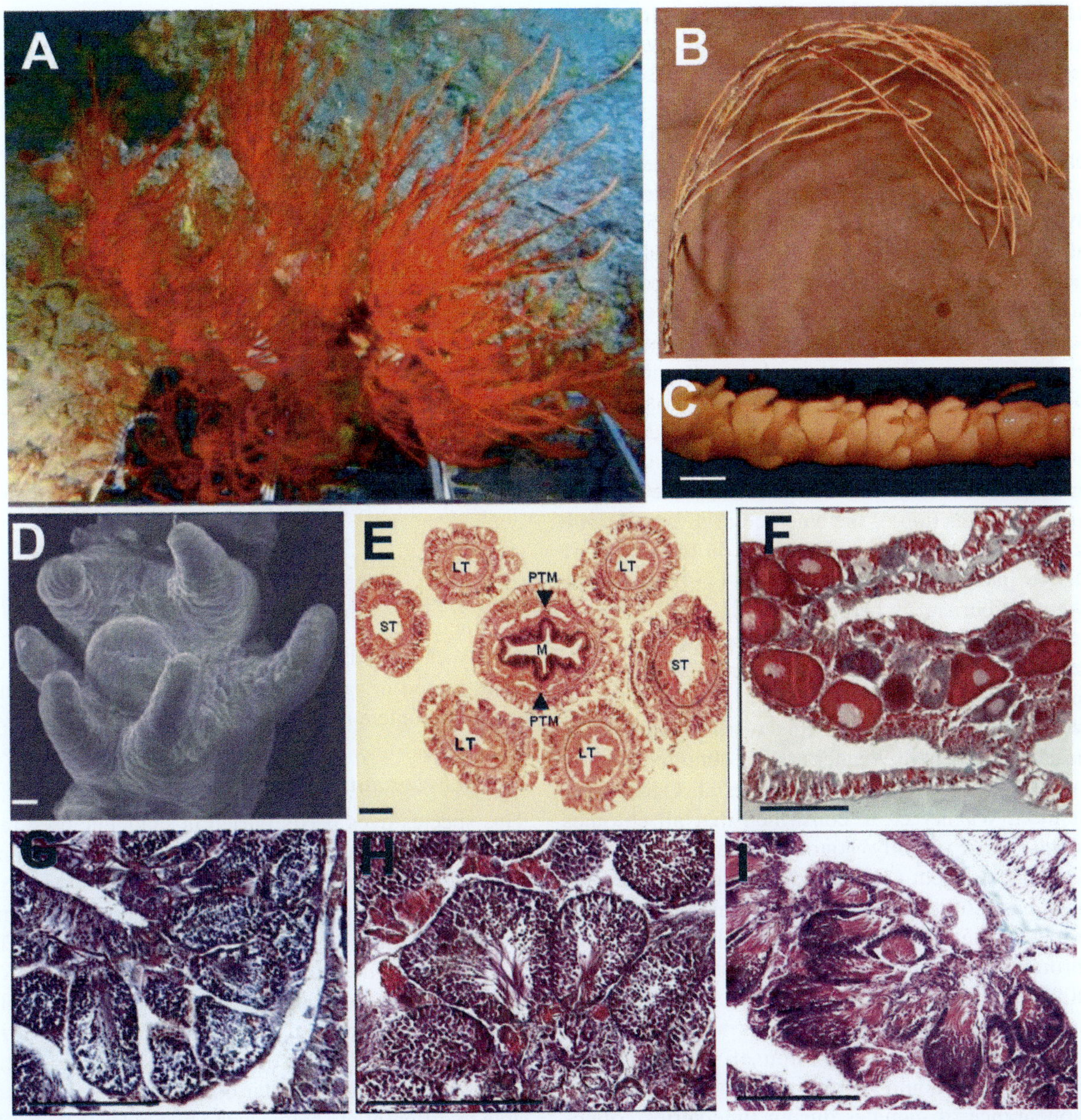

FIG. 5.37 Photographs describing development of *Antipathes griggi from Wagner (2011)*: (A) whole mount of a colony in situ; (B) portion of the colony, as a specimen; (C) polyps on terminal branch under light microscopy, bar = 1 mm; (D) SEM of *Antipathes griggi* polyp; (E–I) histological T.S.s through polyps of *A. griggi* showing: (E) locations of primary transverse mesenteries which bear the filaments and gametes, M = mouth, PTM = primary transverse mesentery, LT = lateral tentacles, ST = sagittal tentacles; (F) vitellogenic and previtellogenic oocytes in close proximity to each other; (G) stage 1 spermatocysts; (H) stage 2 spermatocysts; and (I) stage 3 Spermatocysts, bars = 100 μm.

were collected throughout the known range of *A. griggi*, from the islands of Hawaii to Laysan at depths of 10–100 m. Although *A. griggi* can be found at depths of 100 m, it is rare below the 75 m depth limit, at which commercial harvest occurs in Hawaii. Thus, the supposed depth refuge from harvest does not really exist, thereby calling into question population models used.

Wagner examined gametic reproductive processes within *Antipathes griggi*, the dominant species targeted by the Hawaiian black coral fishery, using histological techniques. The results indicated that all colonies were strictly gonochoric with a 1:1 sex ratio, and this species has an annual reproductive cycle. Externally, there were no apparent morphological differences between males and females in any of the samples, and the presence of oocytes or spermatocysts was the only character that allowed for the distinction between sexes. The gametes would always develop in association with the primary transverse mesenteries, which extended into the cavity of lateral tentacles in many cases (Fig. 5.37E).

There were no brooding signs observed in any of the examined polyps (Fig. 5.37D), and none of the oocytes exhibited any signs of fertilization (Fig. 5.37F). Externally, none of the examined polyps contained brooded larvae or eggs on their surfaces. Spawning was not observed in situ or in histological sections, and could therefore only be inferred by the disappearance of gametes during time-series collections. Wagner (2011) statistically studied the reproductive cycle of *A. griggi* and concluded that this species spawns continuously between June and December, when temperatures are highest. Furthermore, Wagner et al. (2012a,b) reported that their histological results showed that the percentage of polyps containing gametes dropped continuously throughout the reproductive season, indicating that spawning occurs in successive events with greatest intensity between November and December.

Wagner et al. (2012a,b) were interested in the study of the gametic reproduction of the black coral *A. griggi*, with specific objectives to obtain information on the following reproductive parameters of this species: (1) reproductive strategy (gonochorism vs hermaphroditism); (2) mode of reproduction (spawner vs brooder); (3) reproductive cycle; (4) minimum size of sexual maturity; and (5) maximum depth of reproduction.

First of all, the material of this work was obtained from the same site where Wagner (2011) obtained his material or it might be even the same material. Concerning the first item, Wagner et al. confirmed that all colonies of *A. griggi* were strictly gonochoric and added that out of the 10 tagged colonies that were sampled on a monthly basis off South Kaua'i, 5 were males and 5 were females. The sex of each tagged colony stayed the same throughout the duration of the study (July 2008–July 2009). No brooding signs were observed internally or externally, and histological sections showed rupture of the mesenteries and accumulation of gametes in the gastric cavity. This could therefore only be inferred by the disappearance of gametes during time-series collections (see below). The authors examined the reproductive cycle statistically and stated that both male and female colonies had a similar seasonal pattern of reproductive output, although there were slight temporal differences between sexes. All male polyps were completely void of spermatocysts by January, whereas few female polyps (<3%) still contained oocytes through February. Gamete maturity stages generally increased from June through to December. Spermatocysts in male colonies were mostly in early stages of development (stages 1 and 2, Fig. 5.37G and H) in June–August, and then progressively became dominated by spermatocysts in late developmental stages through December (stage 3, Fig. 5.37I). On the other hand, female colonies had polyps containing a higher proportion of previtellogenic oocytes in June–August, which gradually progressed to a high proportion of vitellogenic oocytes of larger size in January–February (Fig. 5.37F). Generally, female colonies contained oocytes of a large range of size and maturity stage throughout the reproductive season, with immature oocytes often occurring within close proximity to mature oocytes even within the same mesentery (Fig. 5.37F).

Wagner et al. (2012a,b) studied the fourth reproductive parameter, the minimum size of sexual maturity, also statistically. They stated that the proportion of sexually mature colonies increased with increasing colony size until a height of 130 cm, at which 100% of colonies were mature. However, the smallest *A. griggi* colony containing gametes was 40 cm in height, and the sexually mature colonies measured 64–80 cm. Current state regulations prohibit commercial harvesting of *A. griggi* colonies that are smaller than 90 cm.

Order Ceriantharia

The members of this order (Ceriantharia) are similar to sea anemones in size and general appearance. They have greatly elongated bodies adapted for living within mucous tubes buried in sand or mud. They have one siphonoglyph and all complete mesenteries (Fig. 5.40C).

Ex. *Pachycerianthus multiplicatus*

This species lives partly buried in sand, enclosed in a tube formed of discharged stinging capsule, the oral end with its crown of tentacles alone being exposed (Marshall and Williams, 1975) (Fig. 5.38A). This species is hermaphrodite, the sperm being shed first, and then the eggs. The eggs are yolky and the polar bodies are extruded before fertilization, and the egg does not form a special fertilization membrane (Uchida and Yamada, 1968). The fertilized egg undergoes normal total, somewhat equal cleavage. These authors recorded that the polar bodies are lost and it is impossible to differentiate between the two poles (Fig. 5.38A (a)), since the fertilization furrows advance from both the animal and vegetal poles. The first cleavage occurs 15–20 min after fertilization, and about 10 min later, the second cleavage follows at right angles to it, forming four blastomeres (Fig. 5.38A (c)). After this stage, a shifting in the relation of the blastomeres takes place (Fig. 5.38A (e) and (f)) that one blastomere moves to a different level. This returns later to its original position and the next cleavage takes place, giving rise to the eight-cell stage (Fig. 5.38(g)). Cleavage continues regularly forming a blastula, the centre of which is solidly filled with yolk, but it does not gastrulate by invagination. Reverberi (1971) redrew the illustrations of the stages of embryonic and postembryonic development from the cleavage to the young (Fig. 5.38(a)–(l)). Fig. 5.38(a)–(c) are T.S.s showing first and second cleavages: (a) metaphase

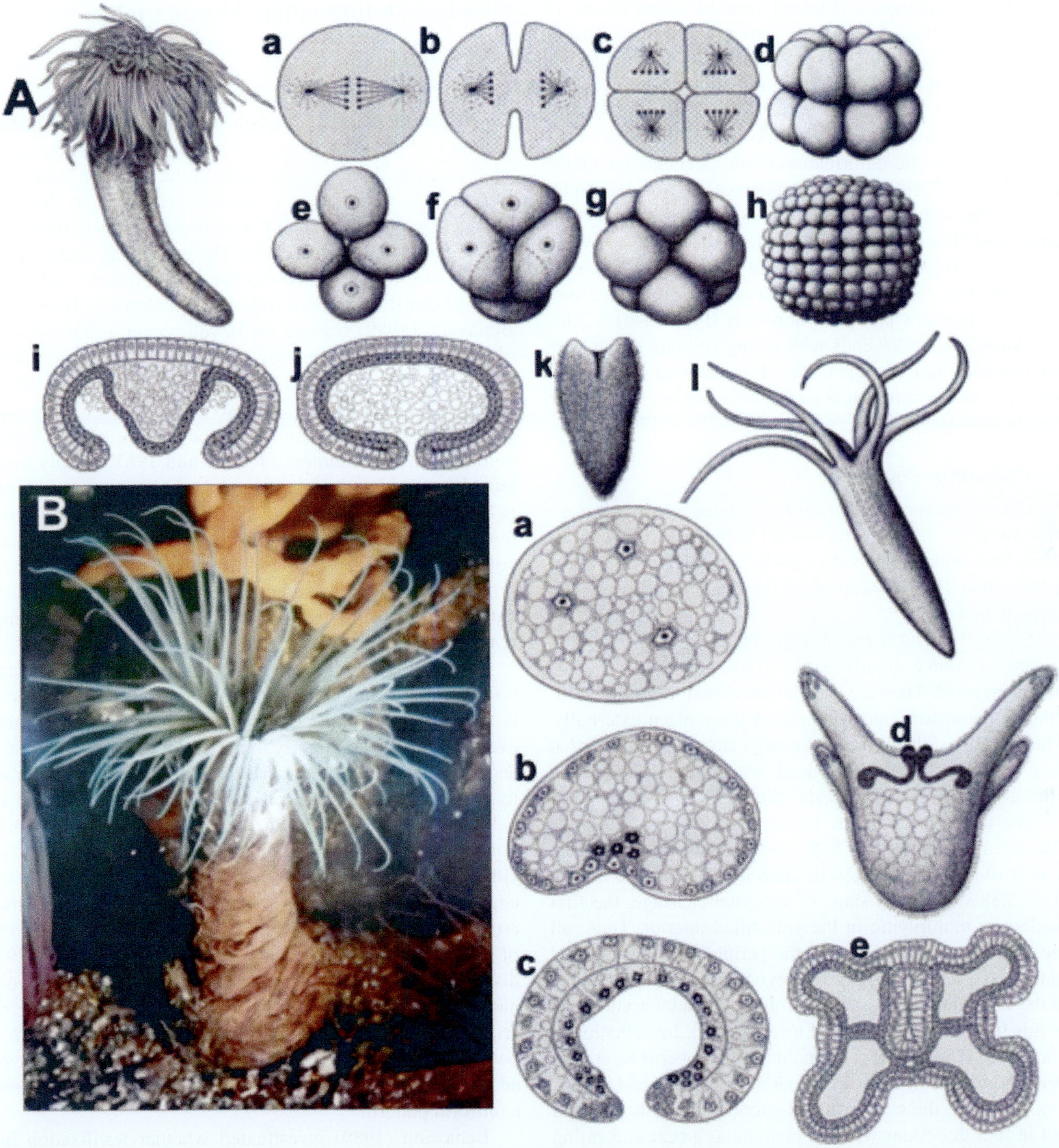

FIG. 5.38 (A) Illustration showing the ceriantharian *Pachycerianthus multiplicatus after Kükenthal from Marshall and Williams (1975)*; (a–i) embryological and postembryonic development of *P. multiplicatus,* example of direct development without passing through the Arachnactis stage: a–c and I, j are median sections; d, h, k, l are side views; e–g polar views; (a) first cleavage (step metaphase), (d) 16-cell stage; (e–g) pseudospiral cleavage as a variant of the normal radial cleavage occurs from a–d, (e) 4-cell stage with two cells displaced by ~45 degrees, (f) displacement of a cell in the 4-cell stage; (g) 8-cell stage, (h) young blastula, (i) invagination (modified due to yolky blastocoel), (j) completed invagination, all the yolk passed into the lumen of the archenteron, (k) young larva with broadened oral and initiation of pharynx formation, (l) young actinian polyp with 6 tentacles and 6 septa. *(All after Nyholm, 1943 from Reverberi, 1971)*; (B) photograph of a large burrowing anthozoan *Cerianthus membranaceus, from Beauté du monde animal, Invertébrés Tome XII (1974)*; (a–e) embryonic and postembryonic development of *C. iloydii*, an example of early development by way of superficial cleavage and invagination: (a) median section of early stage of syncytial cleavage (the first cleavage nuclei are multiplying in the yolk-filled interior), (b) invagination, all cleavage nuclei have reached the periphery of the embryo, (c) invagination is completed, development of cell boundaries in both germ layers, (d) young 'Arachnactis' larva, with 4 tentacles filled with yolk, (e) section across the region of the pharyngeal tube of the larva showing 2 formed and 2 still incompletely developed septa *(after Nyholm, 1943 and Beneden, 1891 from Reverberi, 1971) ((A a–j) and (B d, e) are found in Uchida and Yamada, 1968).*

step of first cleavage; (d) lateral view of whole mount of 16-cell stage. Fig. 5.38(e)–(g) shows pseudospiral cleavage, as a variant of the normal radial cleavage of (a)–(d). This somewhat modified normal, radial cleavage gives rise to the young blastula (Fig. 5.38(h)), filled with yolk. This type of blastula is the cause of the modified course of invagination and the deformation of the archenteron of the gastrula (Fig. 5.38(i)). After the passage of yolk into the lumen of the archenteron (Fig. 5.38(k)), the invagination is completed. The body surface becomes ciliated; pharynx and septa are formed internally and the larva begins to swim. Subsequently, there is a young larva with broadened oral end and incipient pharynx formation (Fig. 5.38(l)). This developmental series ends with young actinian polyp with six tentacles and six septa. This species is an example of the type that develops directly, without passing through an Arachnactis stage.

Ex. *Cerianthus lloydii*

This European species, *Cerianthus*, is a large burrowing anthozoan (Fig. 5.38B). The animal secretes a soft mucous tube, into which the body can be retracted. *Cerianthus* differs from *Pachycerianthus* in the mode of cleavage and invagination.

Cerianthus is a hermaphrodite that breeds in February, and the eggs have already extruded their polar bodies before being spawned. They are shed about an hour after the shedding of the sperm, and fertilization takes place externally. There is no visible change in the eggs after fertilization, that they might be thought to be dead, until invagination begins after 24 h (Uchida and Yamada, 1968).

The egg of *Cerianthus* undergoes superficial cleavage (Reverberi, 1971). Fig. 5.38B (a)–(e) shows the T.S.s from a to c of the early stages of cleavage. The first cleavage (Fig. 5.38B (a)) early stage of syncytial cleavage, the first nuclei are multiplying in the yolk-filled interior. Then, all cleavage nuclei have reached the periphery of the cleaved embryo (Fig. 5.38B (b)); the embryo has somewhat flattened at the vegetal pole; prospective entoderm nuclei accumulate at this location (primordial of invagination). Invagination is completed (Fig. 5.38B (c)), and the cells of germ layers develop boundaries. After the endoderm is formed, and a pharynx develops, the ectoderm and endoderm together extend out in four directions with the pharynx as a centre, forming a rectangular figure (Fig. 5.38(d) and (e)). These extensions are the rudiments of the 'marginal' tentacles. After this, two more 'oral' tentacles appear at the sides of the mouth, making a total of six tentacles. Internally, septa are formed, and the so-called **Arachnactis larva** is reached and begins its pelagic life (Uchida and Yamada, 1968) with four tentacles filled with yolk (Fig. 5.38B (d)), and (Fig. 5.38B (e)) is a T.S. across this larva at the level of the pharyngeal tube. It shows two formed and two still incompletely developed septa.

Subclass Octocorallia (Alcyonaria)

Soft corals

Order Stolonifera

Ex. Tubipora Another example of this order is ***Clavularia hamra*** Gohar, 1948 (Fig. 5.39A), which is an abundant species at the shallow coral reefs of the northern Red Sea. In this area, colonies of *C. hamra* are found in shallow crevices on the reef flat and on hard substrates in the back reef lagoon (0.5–1.0 m depth). The encrusting and patchy growth form of *C. hamra* (Gohar, 1948) made it difficult to distinguish between individual colonies. However, Benayahu (1989) studied the development and life cycle of this species, by collecting samples of 10 large colonies at random, monthly from August 1985 and August 1987. During observations in August 1985, orange spawned eggs of *C. hamra* remained on the outer surface of the colonies. In the summers of 1986 and 1987, freshly collected fragments were examined for large orange oocytes in the polyp cavities. A few weeks prior to the spawning date, predicted on the basis of the 1985 observations, colonies were carefully removed from the sea and transferred to running seawater aquaria for spawning observations. Fertilized eggs, obtained from the spawns, were reared in aerated aquaria.

Clavularia hamra reproduces gametically by spawning (Gohar, 1948); and maturation of male and female gonads is highly synchronized (Benayahu, 1989). The spawning event was observed directly in midsummers of 1985, 1986, and 1987, both in the field and in aquaria, a few days prior to the new moon. After spawning, the orange eggs were held together by mucus and remained attached among the polyps on the outside of the female colonies (Fig. 5.39B). Gohar (1948) observed a case of *C. hamra* spawning white eggs in an aquarium late in May at Hurghada, the northern Red Sea. However, Gohar did not mention retention of spawned eggs by the female colonies. Benayahu (1989) discussed and reviewed the brooding habit within alcyonaceans and mentioned that released eggs of *Clavularia crassa* are attached under the tentacles on the outside of the polyps. Benayahu reported that these alcyonaceans have adopted a variety of modes of external brooding rather than a uniform pattern.

Benayahu (1989) investigated whether fertilization in this studied species is external, since neither embryos nor sperm were ever found in female polyps. Embryogenesis is also external, and occurs on the outside of female colonies (Fig. 5.39C); the eggs start to cleave within 5–8 h of spawning. Many of the eggs cleave synchronously, with holoblastic and equal cleavage, resulting in embryo with a regular blastomere arrangement (Fig. 5.39D). However, 35%–40% of the embryos exhibit abnormal shapes (Fig. 5.30E), due to a pronounced unequal and asynchronous cleavage.

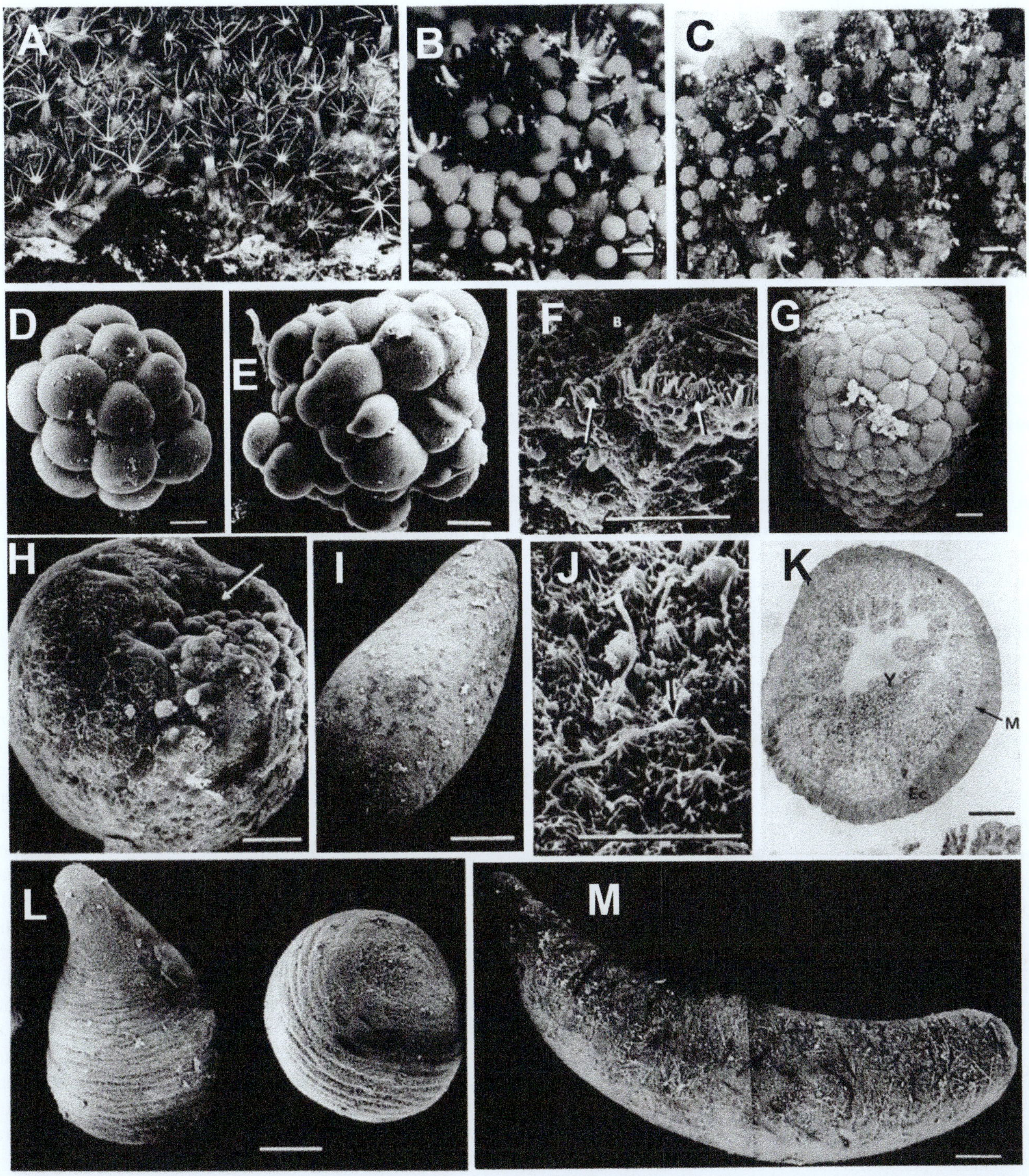

FIG. 5.39 (A) Photograph of colony of *Clavularia hamra*; (B) spawned eggs among polyps of a living colony of *Clavularia hamra*. Bar = 750 μm; (C) early cleaved embryos on a living colony. Bar = 750 μm; (D) SEM of a cleaved embryo (6 h after spawning). Bar = 100 μm; (E) SEM of an irregular and unequal cleavage pattern of embryo (6 h after spawning). Bar = 100 μm; (F) SEM of a fractured blastula (6 h after spawning). The surface of the blastomeres (B) is covered with microvilli *(arrows)*. Bar = 10 μm; (G) SEM of a morula (30 h after spawning) with equal sized blastomeres. Bar = 100 μm; (H) SEM of an early gastrula (36 h old). *Arrow* indicates the blastopore. Bar = I00 μm; (I) SEM of a 3-day-old young planula. Bar = 100 μm; (J) SEM of a surface of 3-day-old planula. Microvilli are common, some of them arranged in a collar structure, associated with a cilium *(arrow)*. Bar = 10 μm; (K) photomicrograph of a T.S. of 4-day-old planula with ectoderm (Ec), thin mesogloea (M) and yolk laden material (Y). Bar = 100 μm; (L) SEM of 6–7-day-old planulae exhibiting body contractions. Bar = 100 μm; (M) SEM of 8-day-old mature planula. Bar = 100 μm. *(All from Benayahu (1989).)*

Microvilli are uniformly scattered on the surface of the blastomeres (Fig. 5.39F). This cleavage process is rather rapid and occurs within 5–16h of spawning. Within 24–36h of fertilization, morula and/or solid blastula are formed (Fig. 5.39G). Gastrulation occurs by delamination and at this stage, all embryos, both regular and irregular-shaped, have blastomeres of an equal size. One day later, the embryos are smooth and a blastopore is apparent, indicating a fully formed gastrula, between 48 and 72h of spawning (Fig. 5.39H). The gastrula elongates to form an early planula stage (Fig. 5.39I) measuring about 600μm in length and 400μm in width at its anterior pole. It is densely covered by microvilli and with sporadically scattered short cilia (Fig. 5.39J). The mouth opening is not visible at this stage.

Four days after spawning, the planula differentiates histologically and develops an outer ectodermal layer, a thin mesogloea and yolk-laden material fills its interior (Fig. 5.39K). Two to three days later, the planula grows in size and numerous planulae are found on the surface of the parent colonies. These larvae gradually change their shape by body contractions from pear-like to round, and vice versa (Fig. 5.39L). During this, the larvae detach and sink down to the bottom, and by day 8, mature planulae are obtained. Most of them are scattered on the surrounding substrates and exhibit slow movements by crawling. Mature planulae are orange, elongated and when fully extended their maximal length reaches 1.8–3.0mm (Fig. 5.39M).

Further studies on gametic reproduction and embryogenesis of alcyonaceans are required to draw generalities on their morphogenesis and their relation to other cnidarians.

Order Alcyonacea

Actinozoa have eight mesenteries and eight pinnate tentacles; stomodaeum have a single siphonoglyph (ciliated groove) (Fig. 5.40C). The skeleton is internal, consisting of spicules in the mesoglea (Fig. 5.40B), and occasionally supplemented by an external skeleton; longitudinal muscles are present on the ventral faces of the mesenteries (Fig. 5.40C).

Alcyonium digitatum (dead man's fingers)

This is a colonial form, which occurs below low-tide mark, attached to stones, in various sizes and shapes, but usually in the broad-lobed masses (Fig. 5.40A) in the Red Sea. A transverse section through a portion of a colony (Fig. 5.40B) shows extended polyps with pinnate tentacles and coenosarc. It can be seen that the polyps project, in life, from the general surface of the colony. The polyps are delicate and withdraw on the slightest stimulus; the oral disc with its crown of tentacles is pulled inside the enteron by the contraction of longitudinal muscles running in the mesenteries and attached to a tube lined with ectoderm, the stomodaeum or the gullet, which communicates with that cavity. The whole of the stomodaeum is ciliated, and at one end of it there is a groove, which is lined with especially strong cilia, which draw water into the mouth. This is the siphonoglyph, and it occupies a ventral position (Fig. 5.40C). Below the level of the stomodaeum, the mesenteries end in an enlarged free edge: the mesenteric filament (Fig. 5.40D).

The gonads are developed at the breeding season, from groups of endodermal cells near the filaments, but they only occur on the six ventral mesenteries (Fig. 5.40D). The eggs are comparatively large and pass very slowly up the enteron and out of the stomodaeum, get fertilized outside the polyp and develop into a planula larva. After a free-swimming period, this fixes and becomes a single individual, which by budding gives rise to a colony.

Alcyonium coralloides

Alcyonium coralloides is an epibiotic soft coral overgrowing living colonies of Mediterranean gorgonian, octocoral *Paramuricea clavata*. Quintanilla et al. (2013) studied its gametic reproduction, and they obtained the samples from a population found at 18–22m depth off the Medes Islands (north-west Mediterranean), in Spanish waters. The sampling site is a large tunnel connecting two sides of one of the islets, and harbours a population of *P. clavata*. This tunnel is among the tourist attractants where the corals might be subjected to injury or damage. All the samples used for the study were obtained by scuba diving between March 2008 and March 2009, usually monthly when weather conditions allowed the fieldwork. The morphological pattern of *A. coralloides* is a small colony, about 10mm wide (coenenchyme covering the dead gorgonian axis) and c. 50mm long (Fig. 5.40E). The smallest fragments possible of *A. coralloides* are from the apical part of the gorgonian branches. The coenenchyme is red, the tentacles and the proximal part of the polyp are yellow (Fig. 5.40E and F), while the distal part of the polyp is red. The pattern may differ with the different sites of the populations. *A. coralloides* is gonochoric, its reproductive cycle is annual and gametogenesis occurs over 5–6months. This means that male and female colonies have similar duration, and oogenesis is shorter than that of other octocorals. Quintanilla et al. (2013) studied gametogenesis and the fecundity of this species statistically. They reported that the development of the reproductive products probably begins in December (2008) and ends with the release of sperm and larvae in May (2009).

In **male** colonies, the first spermatic cysts were detected in January 2009, and they achieved a final diameter of 311 ± 97μm in March 2009. No spermatic cysts were found in samples between June and November 2008. The histological study confirms that spermatogenesis starts in December, since colonies sampled in November lacked reproductive products, both male and female. On the other hand, in the following sampled month (January), in males, spermatic cysts (Fig. 5.40K) showed numerous heads of

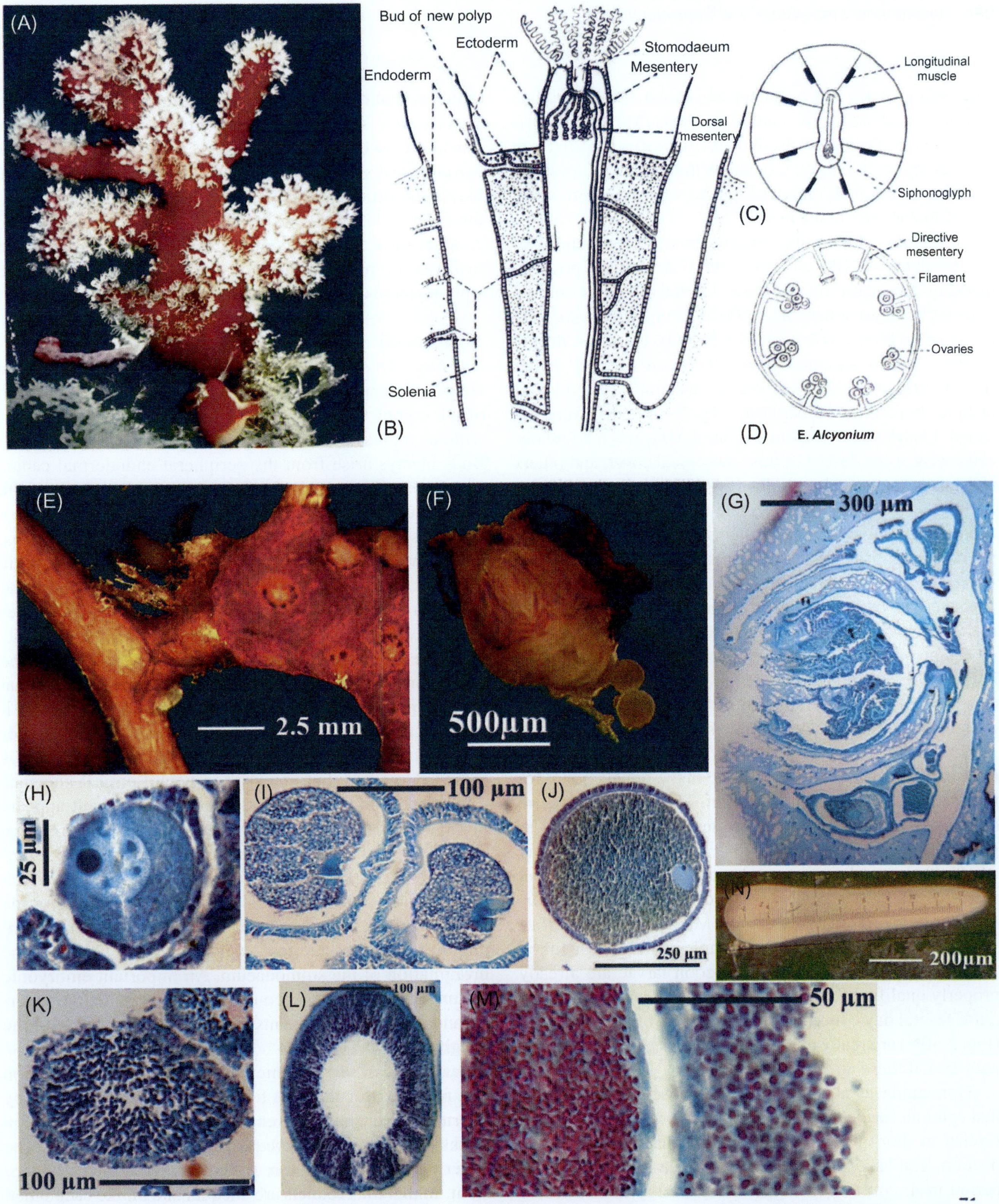

FIG. 5.40 (A) Photograph of whole mount of a small colony of *Alcyonium palmatum*, *from Beauté du monde animal, Invertébrés Tome XII (1974)*; (B) diagram of section through colony of *Alcyonium*, showing extended polyp with pinnate tentacles and coenosarc. The direction of the water current is shown by *arrows*; (C and D) diagrammatic sections through *Alcyonium* at the level of the gullet (C), below the gullet (D), *(B–D) from Borradaile et al. (1977)*. (E–N) *Alcyonium coralloides*, histological sections of colonies: (E) colony of the epibiont *A. coralloides* on a branch of a gorgonian, *Paramuricea clavata*; (F) polyp from a male colony with spermatic cysts; (G–J) histological sections of female colonies showing oogenesis: (G) retracted polyp with oocytes in the gastrovascular cavity; (H) previtellogenic oocyte c. 50 μm in diameter; (I) vitellogenic oocytes with the ooplasm increasing in drops of yolk; (J) oocyte (mature or close to maturity), still inside mesogleal capsule; (K–M) histological sections of male colonies showing spermatogenesis: (K) early spermatic cysts, without clear distinct lumen and disorganized spermaries; (L) developing spermatic cyst with typical lumen and future sperm flagellae and head of immature sperm still rounded and radially arranged; (M) two adjacent spermatic cysts from a colony collected in April, showing the asynchrony in the development of cysts (the one on the right still contains rounded immature sperm heads, while the one on the left is fully developed, with disorganized sperm heads reaching the parental mesogloeal capsular layer); (N) photograph of W.M. of larva of a female colony collected in May 2008. *(All are adapted from Quintanilla et al., 2013.)*

developing sperm and a thinner mesogleal envelope. Male colonies examined at the end of the cycle (May), were heterogeneous: the gastrovascular cavities of some of them were completely empty, indicating that spermatic cysts or sperm were emitted; others had gastrovascular cavities still full of mature spermatic cysts.

In **female** colonies, the first evidence of small reproductive products (0–100 μm) was found in January, possibly marking the onset of oogenesis in the previous month. Oocytes of intermediate size (300–400 μm) were also present in some female colonies (Fig. 5.40G). However, during January, previtellogenic oocytes between 50 and 70 μm (Fig. 5.40H) and vitellogenic oocytes with increasing amount of lipid droplets in the ooplasm (Fig. 5.40I and J) were detected. Oocytes increased in size until May, when a few oocytes were some devoid of their mesogleal cover, and others still inside mesogleal capsules (Fig. 5.40J). Gastrovascular cavities of other polyps of female colonies were observed occupied mainly by larvae, showing epidermis with columnar cells and disaggregated internal yolk granular reserves. *A. coralloides* is considered an internal brooder, and thus is characterized by internal fertilization and retention of larvae in the gastrovascular cavity of the polyps.

Quintanilla et al. (2013) reported that internal brooding is the most common phenomenon among *Alcyonium* species, whereas broadcast spawning is most common in tropical soft corals, and surface brooders are not common.

The release of sperm in *A. coralloides* occurs from mid-May onwards, while the oocytes are in the female polyp for maturation and internal fertilization. This release of sperm coincides with increase in water temperature by two degrees (from 14.8°C to 16.7°C) between May 14 and May 21, 2009. Internal fertilization and planulae development might occur for a short period, and the release of *A. coralloides* larvae probably occurs in May to early June (larvae were completely absent in samples collected in June). Some of the newly released anthozoon larvae are already properly qualified for life 2 h after release, while others require several days. It is plausible that those of *A. coralloides* (Fig. 5.40N) are ready to settle on naked segments of *P. clavata* axes if these are available.

Quintanilla et al. (2013), in their discussion, reported that epibiont species show fast growth, gametic reproduction limited to short and seasonal periods, and agametic reproduction. The lack of reproductive products during 6 months period in *A. coralloides* suggests that energy is invested in processes other than gametic reproduction (such as growth and maintenance), or environmental conditions do not permit investments in reproductive tasks for this epibiotic species.

Heteroxenia fuscescens (Ehrenberg, 1834)

Another example of order Alcyonacea that is also found in the Mediterranean Sea is *Heteroxenia fuscescens*, a hermaphroditic and dimorphic anthozoan that commonly inhabits shallow-reef xeniid in the northern Red Sea. All *Heteroxenia* species are characterized by polyp dimorphism, i.e. two kinds of polyps are present in a colony, autozooids and siphonozooids. Some *Heteroxenia* species are hermaphroditic and have four methods of reproduction—three agametic and one gametic. During the latter, the Xeniids have a rather complex brooding mechanism that involves the retention of the expelled planulae among the siphonozooids (Gohar, 1940).

Agametic reproduction

Budding. This is the commonest method of reproduction among Cnidaria in general, which only adds to the size of the colony by increasing the number of its individuals, without increasing the number of colonies (Gohar, 1940). Buds always arise from the peripheral endodermal canals (see Fig. 5.40B), particularly those at the periphery of the summit of the syndete of *Xenia* and *Heteroxenia*, e.g. a few buds also arise in between the fully developed polyps. Buds never arise in direct connection with the coelenteric cavities of adult polyps. The first or the primary polyp that develops from the larva acquires a thickening at its lower portion, which presents the primordial syndete. The bud itself arises at first as a widening of the peripheral end of one of the endodermal tubes, and grows towards the external surface, pushing the mesoglea and endoderm in front of it to the exterior. This part of the mesoglea becomes thinner as the process continues. An outgrowth results, which is at first a low pad, but increases in height until it becomes a finger-shaped bud. The distal end flattens, a pit appears in the centre and eight radiating grooves emerge from the central pit to the periphery separating the primordial tentacles. The pit deepens, forming the stomodaeum, which finally opens into the internal cavity. The primordia of the tentacles grow, first into simple tentacles, and later acquire pinnules. The mesenteries develop in connection with the invagination of the stomodaeum. An important embryonic feature is that there is no connection between the mesenteries and mesenteric filaments of the mother and those of the daughter polyp.

Transverse fission. This method of agametic reproduction was recorded for the first time in 1961 in Alcyonaria, by Gohar and Roushdy. In Xeniidae, a transverse constriction marks the upper part of the polyp from the lower part. The upper part separates before the lower has started development, so that the products are unequal and unlike each other. In fact, the lower part does not possess tentacles, mouth or stomodaeum at the time of separation. In all the Xeniidae, particularly during the summer months, this method of propagation by T.F. is common and has been observed both in the sea and in the aquaria. A short time (a few hours to about 1 day) before division takes place, the polyps expand and become thinner and more transparent. A faint constriction appears in the wall of the 'anthocodia' or polyp nearer to the base than the tentacles and grows deeper and deeper. The **upper part** of the polyp eventually falls and is carried far off by water movements. It finally settles down, attaches itself by its lower end and starts a new colony (Fig. 5.41A). The **lower part** of the polyp that is left on the colony heals and rounds its tip and undergoes the same course of development of the buds mentioned above.

It is of interest to note that further development of colonies from such fallen-off polyps completely resembles the development of colonies from a single polyp that arises from larva (Fig. 5.41B).

In one case, a fallen part (upper part) of *Heteroxenia fuscescens* was rolled about in the aquarium for some days by water movements, so that it failed to attach itself to the bottom. It could not form a syndete at the lower end, but developed a set of tentacles, a mouth and a stomodaeum, etc. The product was an abnormal specimen with two continuous polyps in one line and opposite directions (Fig. 5.41C). However, it settled down and attached itself to the bottom on the side of the common to form the syndete and the normal state of affairs was restored.

Gohar (1940) reported that he observed the T.F. in a species of *Alcyonium,* but he did not study this observation further.

Pinnitomy. This third type of agametic propagation, which Gohar called 'Reproduction by pinnitomy', is very interesting and has been observed only in *Anthelia*. He mentioned that this type has no described parallel in the Actinozoa as a whole. He isolated a number of colonies of *Anthelia glanca* Savigny in a separate aquarium. After a few weeks, numerous very small solitary polyps, in different stages of development, were found attached to the bottom of the aquarium. Some were sac-like, short, of practically uniform breadth and width, with knob-like tentacles; others were longer with longer tentacles with or without pinnules, etc. At first, Gohar thought they had developed from larvae, but further and careful investigations proved that some of the pinnules were detached, attached themselves to the bottom and started growing into new polyps. This occurs in the same way that the polyps fall off and grow in other species. In *Anthelia glanca*, and probably in other *Anthelia* sp., this method of reproduction is more common than T.F. There are accounts for a great abundance of small solitary polyps of these species in nature, especially during summer months.

Gametic reproduction. Gohar and Roushdy (1961) selected *Heteroxenia fuscescens* to study as an example for the Xeniidae for many reasons:

1. The colonies of this species were abundant and widely distributed in the vicinity of the Marine Biological Station, Hurghada (Red Sea) at that time.
2. These colonies usually attain the largest size of all species of *Heteroxenia* and *Zenia* known in the Red Sea.
3. These colonies are hermaphrodites with relatively large gonads, genital cells and embryos.
4. The breeding season is long lasting throughout the greater part of the year.

Gohar first studied the development of the germ cells; in this dimorphic, hermaphrodite species, the male and female gonads lie side by side, sometimes touching one another in the coelentric cavities of the autozooids (Fig. 5.41F). Monomorphic colonies are usually unisexual and are males. Siphonozooids develop in connection with the female gonad (Gohar, 1940). In *H. fuscescens*, as in other alcyonarian animals, gonads develop in the cavity of the autozooids on the six ventral and lateral mesenteries, being absent on the dorsal pair. Siphonozooids are always sterile.

The primordial germ cells (generative cells) originate from the endoderm covering the mesenteries and migrate to the mesoglea, so as to lie near the free edges of the mesenteries (Fig. 5.41D (a)). From these cells, the genital structures, either ova or spermaries, are developed. After several successive divisions, they give rise to the larger auxocytes: either oogonia or spermatogonia. At the early stages of development, both are closely similar. The auxocyte appears as small, almost rounded with a relatively large central nucleus, embedded in a scanty but homogeneous cytoplasm. Later on, the oocytes and spermatocytes become distinguishable through the enormous growth of the former. The auxocytes migrate gradually to the periphery of the inner edges of the mesenteries and then lie underneath their endodermal coverings: the site of gametogenesis. The free inner edges of the ventral and lateral mesenteries in the upper two-thirds of the polyp, especially during the breeding season, swell gradually due to the progressive development of the genital cells.

Spermaries. Due to the increase in size of the primary spermatocyte, this portion (primordium of a spermary) bulges slightly into the cavity of the polyp, pushing around itself two coatings: an inner one formed of a layer of denser mesogloeal substance, derived from the mesoglea of the mesentery and completely surrounding the developing spermary; and an outer coat formed of endodermal cells, derived from the endodermal covering of the mesentery and still continuous with it. Due to the rapid successive divisions of the spermatocytes (spermatogenesis), the spermary gradually increases, bulging into the coelenteron. At first, the young spermary is usually solid, and later on as it ripens and increases in size, a central cavity gradually develops (due to the process of gametogenesis) (Fig. 5.41D (b)–(f)). At the end of the process of spermatogenesis and spermeogenesis (Fig. 5.41(g)–(i)), the central cavity of the fully mature spermary becomes occupied with masses of thousands of ripe spermatozoa. After maturation, the spermary breaks loose from the mesentery and comes to lie free in the coelenteric cavity.

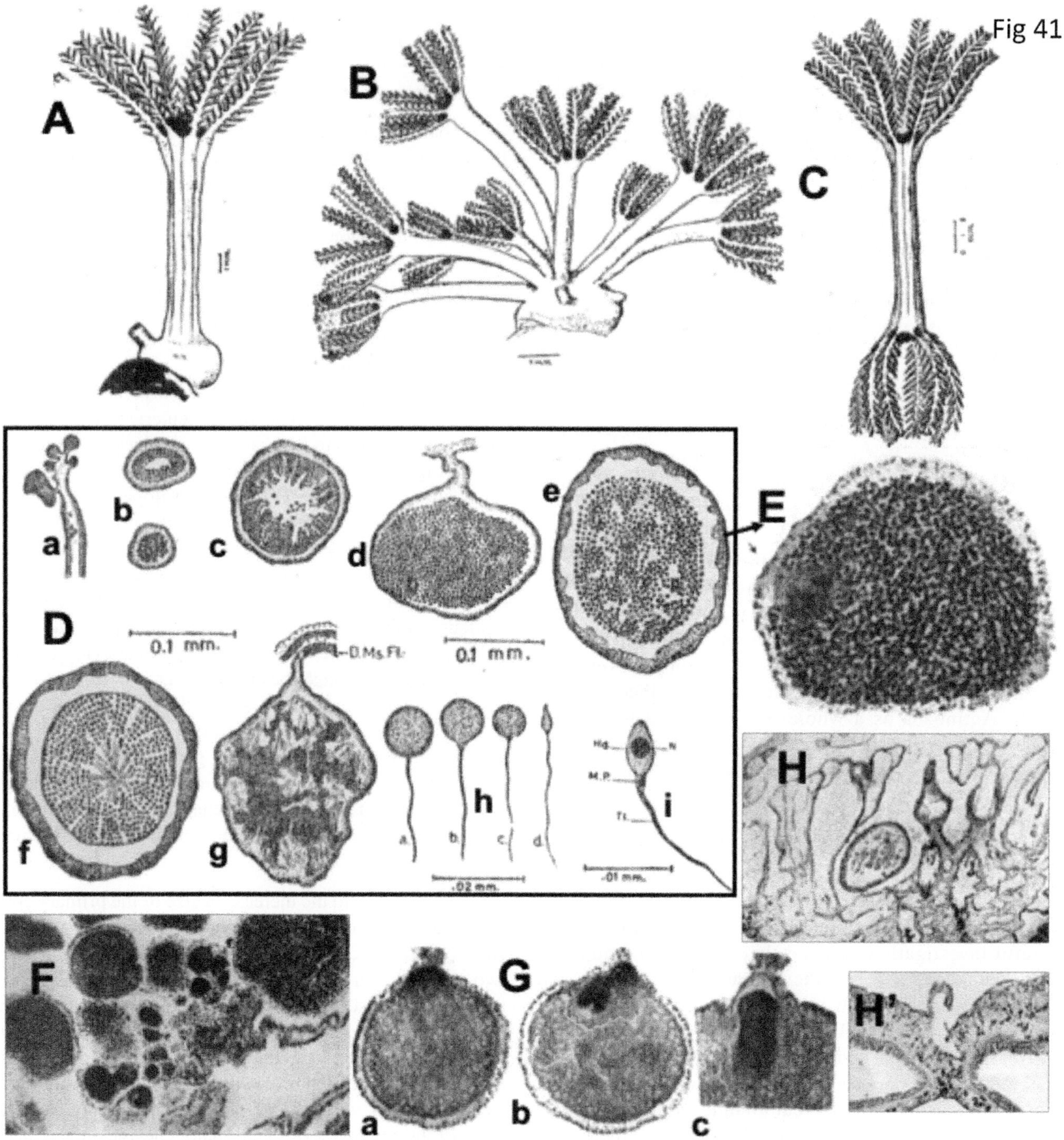

FIG. 5.41, CONT'D

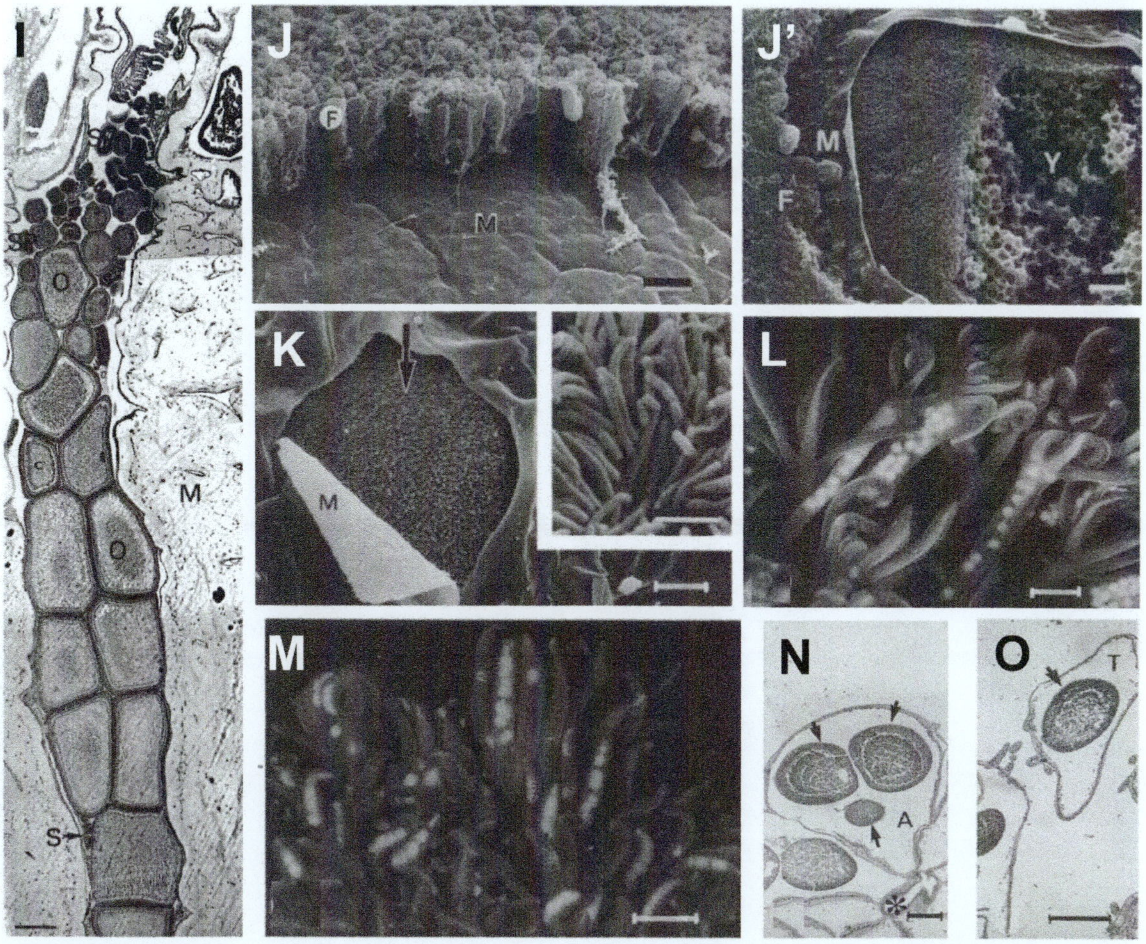

FIG. 5.41, CONT'D

(Continued)

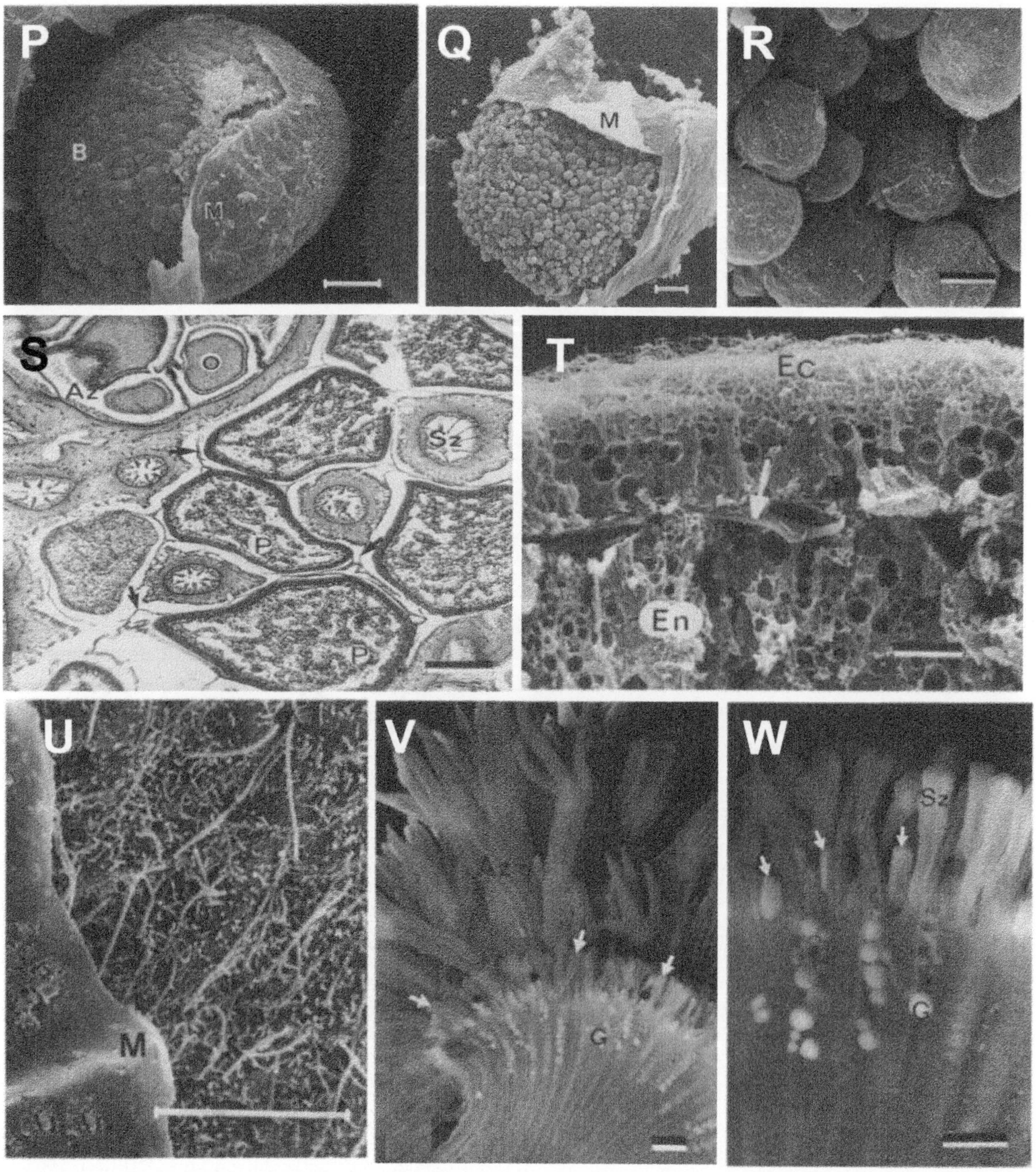

FIG. 5.41 Reproduction in *Heteroxenia fuscescens*: (A–C) agametic reproduction *(Gohar, 1940)*; (A) a polyp of *H. fuscescens*, separated by T.F., regenerating normally and showing the early formation of a syndetial portion from which buds are developed; (B) a small colony formed by T.F. to show the relatively small and low syndete and the absence of siphonozooids; (C) a polyp separated by T.F., was rolled about in the aquarium by water movements and developed another set of tentacles, mouth, stomodaeum, etc. at the end opposite to the original head; (D–G) gametogenesis: (D) sections of spermaries within coelenteric cavities of autozooids showing: (a) early development of spermaries in free end of mesentery; (b–e) stages of spermatogenesis, (f–h) stages of spermiogenesis, (i) mature sperm; (E) spermary abnormally enclosing three immature 'ova' oocytes among the developing sperm pointed by an *arrow*; (F) section through gonad showing development of male and female genital cells at free edge of mesentery (hermaphrodite); (G) stages of fertilization: (a) formation of fertilization cone, (b) female nucleus migrated to the cone region, (c) fusion of the two nuclei; (H) section in summit of syndete of *H. fuscescens,* showing larva in between siphonoids; (H′) section in summit of syndete of *X. coerulea* showing temporary gonopore. *(D–H′) from Gohar and Roushdy, 1961*. Gametogenesis in *Heteroxenia fuscescens*: (I) L.S. through autozooid cavity and gonads of a mature dimorphic colony showing arrangement of male and female gonads, bar = 200 μm *(from Achituv and Benayahu, 1990)*; (J, J′–K) SEM describing the developing oocyte: (J) portion of an oocyte surrounded by ciliated endodermal follicular layer (F) and mesoglea (M) (bar = 10 μm); (J′) a fractured oocyte with follicular layer (F), mesoglea (M) and yolk material (Y) (bar = 10 μm); (K) of an oocyte with partly broken mesoglea (M) and exposed oocyte surface *(arrow)*, bar = 1 μm. Inset: SEM showing microvilli of the oocyte surface, bar = 10 μm; (L and M) SEM showing brooded embryos, bar = 5 mm; (L) inside anthoeodiae of living colony; (M) inside the tentacles of living colony; (N and O) photomicrographs of fertilized egg *(arrows)*, bar = 200 μm: (N) inside anthocodium (A), at the level of the pharynx (asterisk); (O) inside a tentacle (T). *(I–O) From Benayahu et al. (1989)*. Postembryonic development of *Heteroxenia fuscescens*: (P) SEM of a stereoblastula with partly removed parental mesogleal coat (M) exposing the blastomeres (B), bar = 100 μm; (Q) SEM of an advanced cleaved stereoblastula with partly removed mesogleal coat (M), bar = 100 μm; (R) SEM of exposed blastomeres indicating no junctional filipodia between cells. Scale bar = 10 μm; (S) photomicrograph of cross section through autozooid (Az) with oocytes (O) and through the siphonozooids (Sz). Immature planulae (P) are found in the intersiphonozooid spaces. Each planula is surrounded by the parental mesogleal coat, indicated by *arrows*, bar = 100 μm; (T) SEM of a fractured mature planula showing ciliated ectoderm (Ec), mesoglea *(arrow)* and endoderm (En). Scale bar = 10; (U) SEM of a partly removed mesogleal coat (M) showing microvilli and cilia on the embryo surface, bar = 10 μm; (V) L.S. through living colony showing autozooids (Az), siphonozooids *(arrows)*, gonads (G) in autozooids, planulae between siphonozooids *(asterisks)*, bar = 5 mm; (W) L.S. through living colony showing clusters of gonads (G) in autozooids and brood planulae *(arrows)* between siphonozooids (Sz), bar = 2 mm. *(All from Benayahu, 1989.)*

The mature spermatozoon (Fig. 5.41D (i)) is a vertebrate sperm, i.e. it is differentiated into a head, middle piece and a tail. The spherical head measures about 3.4 μm in diameter and the middle piece is about 1 μm in length. The spherical head gradually assumes a conical shape with a somewhat pointed end. It finally measures in average 3.4 μm in length and 1.7 μm in breadth at its base.

After maturation, spermaries of *H. fuscescens* float freely in the cavity of the autozooids, and enter into the endodermic canal system of the 'syndete'. Under the surface of the summit of the syndete, they bulge out and finally rupture and the spermaries become free in seawater, intact, via the so-called temporary gonopores, which heal afterwards. These ripe spermaries float for some time in seawater before rupturing and liberating their spermatozoa.

The ovum. The auxocytes, which become oocytes, absorb a good deal of nutriment from the neighbouring mesogloeal and endodermal cells. They gradually increase in size and bulge slightly into the coelentric cavity, pushing around two coats, as in the case of the spermaries (Fig. 5.41F), but here the mesogloeal coat is relatively thick and prominent, while the inner endodermal coat forms a follicle of more or less cubical cells. The bulge appears as a feeble elevation from the surface of the mesentery, then slowly increases in size with the gradual development of the oocyte, which undergoes gametogenesis and vitellogenesis. The oocyte becomes loaded with yolk granules and the bulge attains a more or less alveolar appearance in T.S. With further increase in size, the ovum (developed oocyte) protrudes more into the autozooid, until it becomes delimited from the mesentery. Its attachment becomes constricted to a thin stalk consisting of central mesoglea surrounded by an endodermal tube. As the young ovum increases in size, the nucleus moves gradually towards the periphery at the part where the follicle is attached to the mesentery (Fig. 5.41G (a)–(c)). As the ovum becomes ripe and increases in size, the mesogloeal coat gradually becomes thinner until in the fully mature ovum, it is too thin to be recognized. At the same time, the follicular cells multiply actively, and gradually increase in length, forming long columns.

Fertilization. Mature ova remain for sometime moving up and down the cavity of the autozooids. Fertilization and early stages of development usually occur in this early mobile stage of the ovum. The spermatozoa, liberated into the seawater, are sucked into the polyps with the inhalant current. It is interesting to note that the ovum of *H. fuscescens* is fertilized while still enclosed within its follicular cells of the mesentery. The sperm moves towards the outer surface of the follicular covering of the fertilization-cone (Fig. 5.41G (a) and (b)), underneath which the female nucleus is found. Gohar and Roushdy (1961) described histologically the event of fertilization at that early time, and reported that it is monospermic. However, in rare cases more than one sperm were seen to penetrate the follicular coat and travel to come close to the site of the nucleus. However, all except one sperm failed completely to enter, probably prevented by the vitelline membrane. The tail of the sperm falls off. Fig. 5.41G (c) shows the fusion of the two nuclei.

Embryology. After fertilization, the mesogloeal coat thickens forming a firm capsule with the outer follicular coat, to protect the embryo during its early stages of development. After the zygotic nucleus finishes its cycle and disappears, the capsule protecting the early embryo, separates from the embryo at definite points, and then completely detaches along the whole surface, leaving a conspicuous space surrounding the embryo. Later, the coats break off leaving the embryo free in the cavity of the autozooids. In the breeding season, especially in the climax, the embryos increase in size and number, and may block the cavity of the autozooid in the upper two-thirds of the syndete, thus pressing each other and becoming not spherical. However, the embryos may be driven up to other cavities even those of the tentacles, where they appear in a row, one behind the other. Tentacles carrying embryos, or sometimes spermaries, pulsate slightly slower than normal.

The developing embryos leave the autozooid cavity to the superficial canals of the summit of the syndete, where they complete their development. The cytoplasm differentiates into macromeres, larger in size and loaded with yolk, and small micromeres, which rapidly arrange themselves on the periphery of the embryo, giving rise to the ectoderm. The nuclei become detectable in these micromeres. These formed immature larvae are solid, being full of yolk cells fused into an ill-determined mass (Fig. 5.41H). These larvae increase gradually in size at the expense of the yolk substance. The nuclei of the endodermal cells become detectable and the larvae become mature. Before liberation, the larvae attain thin structureless membranes, foreshadowing the mesoglea, which makes its appearance in between the ectodermal and endodermal cells. At the same time, the bases of the ectodermal cells are produced into fine muscular processes while their free surfaces develop a coat of cilia to render the larvae mature.

Gohar and Roushdy (1961) observed the liberation of larvae of *H. fuscescens* both in the aquaria and in the sea. This occurs by day and night. Direct sunlight seems to stimulate this process. In the aquaria, the liberation of the larvae may also be due to the shock of collection and removal of the mother colonies from the reef. *H. fuscescens* larvae are not freed immediately after getting out from the gonopores, but are retained for some time in the spaces between the siphonozooids (Fig. 5.41H). From the time a larva is just visible among the siphonophores until it is completely liberated takes 2–5 months. No mucous secretion or zooxanthellae accompany the larvae on extrusion (Gohar, 1940).

It is appropriate to record how the temporary gonopores are formed in the Xeniidae, in general, and in *H. fuscescens* in particular. Gohar (1940) described for the first time the peculiar way of the extrusion of the larvae and spermaries from temporary gonopores in the Xeniidae. Xeniidae, as all Cnidaria, are lower animals that do not have organs, and therefore no definite passages to liberate reproductive products. The superficial endodermic canals of the summit of the syndete become greatly distended through the progressive increase in size of the developing larvae within them. The endoderm-lining of the so-formed cavities does not keep pace with the rapid swelling and becomes discontinuous due to rupturing at numerous points. Through further widening of the cavities, the ruptured endoderm-lining shrinks and becomes confined to small areas. Some parts detach and fall into the cavity, and later the greater part becomes without an endodermal lining. Around the cavity, the mesogloeal cells increase rapidly both in size and number, particularly at 'Anlagen' of the temporary gonopores. The ectodermal cells become progressively loose and the underlying mesogloeal substance gradually dissociates. Finally, through the increased pressure within the cavities and the progressive dissociation of both the ectoderm and the underlying mesogloeal substance, the gonopore is formed (Fig. 5.41H′) and the larvae make their way to the exterior. In *H. fuscescens*, this description applies; however, in this species, the gonopores are completely hidden by the siphonozooids.

After the ejection of the larvae, the opening narrows and healing consequently takes place. The superficial endodermic cavities also shrink and restore their natural size. The endodermic lining of the canals is regenerated (Fig. 5.41H).

Benayahu et al. (1989) collected colonies of *Heteroxenia fuscescens* from the coral reef east of the Mediterranean Sea at depths of 4–12 m. All colonies smaller than 10 mm in diameter are monomorphic. The appearance of siphonozooids is size-dependent and the percentage of dimorphic colonies gradually increases with size. All colonies larger than 36 mm are dimorphic (Achituv and Benayahu, 1990). These colonies are easily recognized among other xeniids by their large size and long pulsating polyps. The colonies of *Heteroxenia fuscescens* of this work were collected on the reefs of the Red Sea, between August 1985 and August 1987. Each month, six to eight dimorphic colonies were sampled randomly, between 2 and 6 m depth. In addition, every 4–5 months, monomorphic and dimorphic colonies of various sizes were collected, including juveniles with a diameter of <5 mm. This location is the same from which Gohar studied his animals. Benayahu et al. (1989) studied its embryology, focusing on the development of planulae within the mesogleal coat. Since this species is hermaphrodite, the two types of gonads appear in the autozooid (Fig. 5.41I) on the four lateral and two ventral mesenteries of the polyp, excluding their anthocodial extensions. Clusters of gonadal primordia appear at the uppermost part of the autozooid. Both male and female gonads are opaque and creamy-white when alive and present on the same mesentery. Sex determination is only possible by histological preparations during the early stages of development.

The mature oocyte reaches a maximum diameter of 750–900 µm and is surrounded by a follicular layer of ciliated columnar cells with an underlying mesogleal layer (Fig. 5.41J and J′). Both layers are continuous with the parental tissues, which form the gonadal pedicle that attaches them to the mesentery. The outer surface of the oocyte is covered by microvilli (Fig. 5.41K); each extends 1–2 µm from the surface, towards the mesogleal coat (Fig. 5.41K, inset). Mature oocytes of *H. fuscescens* become detached from the mesenteries and are fertilized internally, within the gastrovascular cavity of the autozooids. They undergo first stages of embryology inside this cavity; they resemble white-coloured strings and are easily observed underwater through the transparent wall of the polyp (Fig. 5.41L and M). The embryos are very often driven up into the anthocodiae (Fig. 5.41N) and they are conspicuous in the tentacle cavities (Fig. 5.41O) where they undergo the second stages of embryology (Benayahu et al., 1989).

All these stages of embryogenesis are carried out while the embryo is enveloped within a mesogleal coat. This coat is apparently parental mesoglea that surrounds the oocyte (Fig. 5.41P). The young embryo is about 1.0 mm long, and still covered with the mesogleal coat. Embryos at the stereoblastula stage (Fig. 5.41P) were isolated from the autozooids to study. After further cell divisions, the blastomeres are loosely arranged inside the mesogleal coat (Fig. 5.41Q) without any recognizable filipodia between them (Fig. 5.41R). The blastomeres are nearly smooth and have no surface specializations. Gradually, the microvilli appear on the surface of the cells, initially short and sparsely scattered and later they become longer and more numerous. Cilia also develop among the microvilli beneath the mesogleal coat (Fig. 5.41U). Immature planulae of *H. fuscescens* are extruded from the autozooids through temporary gonopores that opened on the surface of the capitullum. These planulae remain attached externally on the capitulum, in spaces between the bases of the siphonozooids (Fig. 5.41S, V, and W). The immature planula is elongated or pear-shaped (Fig. 5.41W), about 1.5 mm and surrounded by the parental mesoglea coat (Fig. 5.41S). The inter-siphonozooid spaces are used as external brooding sites where the final stage of larval maturation takes place. In the laboratory, numerous aborted young white embryos and immature planulae fail to complete their metamorphosis into mature larvae. Release of mature planulae occurs at dusk and this process is rapid within the colony. Often, it takes a few minutes for hundreds of larvae to emerge from the inter-siphonozooid spaces in this process. Planulation takes place when the planulae hatch from the mesogleal coat and escapes from these spaces.

Mature planulae of *Heteroxenia fuscescens* are elongated, 3.5–4 mm long, ciliated (Fig. 5.41T) and opaque white because their tissues lack symbiotic algae. These planulae can change their shape because they are capable of contractions. In addition, they move slowly by these contractions, coupled by their ciliary movement to glide on the substratum.

Xenia umbellata

Embryogenesis and acquisition of algal symbionts by planulae of ***Xenia umbellata*** (Octocorallia: Alcyonacea) were studied by Benayahu et al. (1988). This internally brooding soft coral, like *Heteroxenia*, inhabits the Red Sea. The authors collected colonies of *Xenia umbellata* during the summer of 1986 from the reef in front of the Marine Biological Laboratory found north of the Red Sea. Colonies were sampled rondamly by scuba diving at depths of 3–6 m. In the laboratory, live colonies were dissected longitudinally and by using binocular stereoscope, the gonads and developing larvae in the brooding chambers were examined. Other colonies were carefully detached from the natural substratum and perfectly transferred to the laboratory to be prepared for SEM and TEM and/or light microscope studies.

Results of studies of Benayahu et al. (1988) revealed that the ripe eggs of *X. umbellata* detach from the mesenteries and enter the gastrovascular cavities of the polyps. After fertilization, the zygotes pass into endodermal brooding chambers, which are located at the distal part of the polyp cavities (Fig. 5.42A). Within these chambers, embryogenesis occurs. Like in *Heteroxenia*, the embryo is enveloped within a mesogleal coat (Fig. 5.42B), but in *X. umbellate*, this coat is loaded with symbiotic algae (Fig. 5.42C). Later, this mesogleal integument disappears and a naked round stereoblastula is observed (Fig. 5.42F). At this stage, algal cells appear to be on the surface of the cells of the newly formed blastula (Fig. 5.42D). Thus, symbionts are incorporated by the embryo very early in development. The following stages lead to the formation of young planulae within the brood chambers (Fig. 5.42E). The peripheral layer of the planula becomes the ectoderm, which is attached to a thin layer of mesoglea. These immature planulae contain an inner mass of undifferentiated yolk-laden with intracellular algal symbionts embedded in the yolk (Fig. 5.42G). The larvae feed on this yolk, which is gradually consumed, leaving a cavity enclosed by a wall of two tissue layers, separated by a thin mesoglea (Fig. 5.42H).

The mature planula of *X. umbellata* is slender, reaching 2.0 mm or more in length, and has a broad oral end (Fig. 5.42G, inset). Gradually, the body tapers towards the aboral end. The live planula is brownish due to the presence of symbionts, and its outer ectoderm is covered with cilia, and collars surround many of them. Endodermal cells of this mature planula contain algae, which adhere to the thin mesoglea and are ciliated together with microvilli on the surface lining the polyp cavity (Fig. 5.42C).

Capnella gaboensis Verseveldt, 1977

Capnella gaboensis is a soft coral (Octocorallia, order Alcyonacea), abundant and widespread, and grows in temperate Australian waters. It is dioecious species and is a surface-brooder. Farrant (1986) studied and reported a detailed account of gonad development, and the development and settlement of the planulae of *Capnella gaboensis*. He investigated the annual changes in gonad number and size, variation in gonad development between branches of colonies, between colonies and between populations of *C. gaboensis*. He described the developmental stages of the female and male gonads, the time of year and timespan of such stages, and described the development of the planulae. Farrant (1986) collected the material for his study from Fairlight, Cobblers, Dobroyd and Manly Point (four study sites in Sydney Harbour) in the years 1981–84. Seawater temperature was measured at Fairlight (at 6 m depth, using a thermometer) at fortnightly intervals during the first year of the study. Seawater temperatures for Manly ocean beach, 1 km away, were obtained from the local newspaper for the entire study period.

The gonads of *Capnella gaboensis* develop on the ventral and lateral mesenterial filaments, in the part of the polyps below the level of the 'coenenchyme' (see Fig. 5.40D). The study of Farrant described that these gonads develop on two or three (rarely more) of the filaments in any one polyp. The development of both female ovaries (each consisting of a single oocyte) and male spermaries follows an annual cycle. Ovaries in *C. gaboensis* never contain two oocytes in their later stages in the manner described for other corals. The eggs are attached by a stalk to the mesentery, as for other octocoral eggs, and they contain nuclei up until the time they are spawned. Statistically, the number of spermaries per filament is much greater than the number of oocytes per filament, especially in the period from November to May. Gonad abundance is greatest in May, and least in July to November. Spermaries are smaller than ovaries throughout most of the year. The variation in gonad size, at any one time, is due to the difference of several gametogenic stages in the two types (e.g. vitellogenesis in oogenesis). The number of ripe ova or spermaries in the polyps at spawning is not great, and female colonies generally spawn about 5–10 ova per polyp, while males spawn about 10–20 spermaries per polyp. Much of the results of the studies of Farrant (1986) were obtained by applying the statistical methods of Sokal and Rohlf (1969) and the statistical tables of Rohlf and Sokal (1969).

The gonads develop during the warm months, and colonies spawn their gametes in late autumn and early winter.

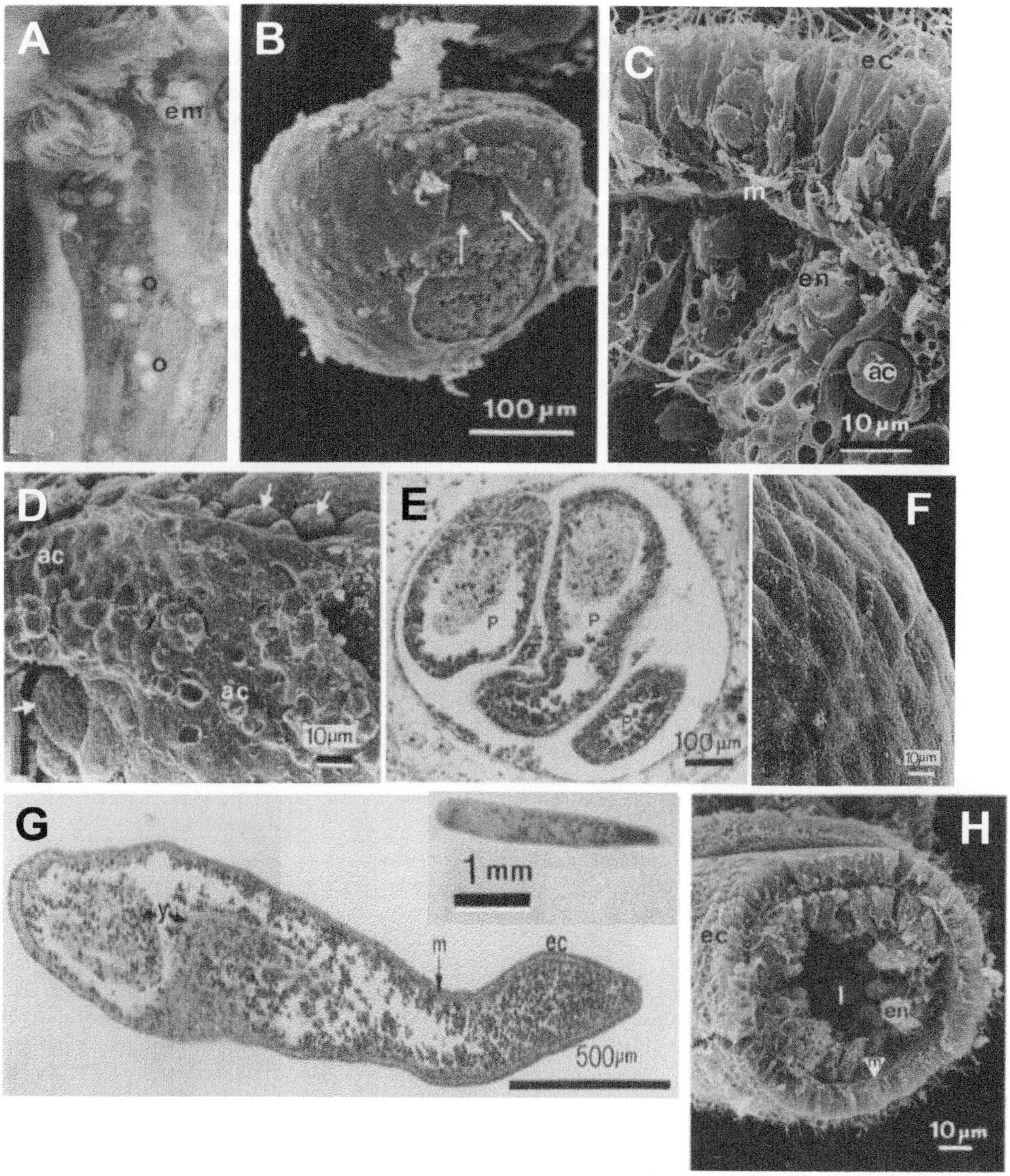

FIG. 5.42 Development of *Xenia umbellata* by *Benayahu et al. (1988)*: (A) Dissected living female colony with oocyte (o) and brooding chamber with developing embryos (em); (B) SEM of a near cleavage embryo coated by mesogleal cover isolated from brood chamber. *Double arrows* mark exposed blastomeres, where mesoglea was removed; (C) SEM of fractured mature planula showing ciliated columnar ectoderm (ec), mesoglea (m) and endoderm (en) inhabited by algal cells (ac); (D) SEM of mesogleal sheath around embryo, with embedded algal cells (ac), *arrow* points to exposed blastomeres; (E) photomicrograph of sectioned brood chamber with developing young planulae (p); (F) SEM of portion of blastula after detachment of mesoglea; (G) photomicrograph of immature planula showing ectoderm (ec), mesoglea (m) and inner yolk laden mass (y) with numerous darkly stained algal cells (ac). Inset: Living planula with zooxanthellae visible as *dark patches*; (H) SEM of transversely fractured mature planula prior to liberation, differentiated into ectoderm (ec), thin mesoglea (m), and endoderm (en) with interior lumen (I).

Gonad development is not synchronous within colonies or within populations, possibly reflecting the protracted nature of spawning.

In addition, the histological studies resulted in detailed figured description of spermatogenesis and oogenesis, and described the same of the related species already mentioned above.

Description of larval development was one of the aims of Farrant (1986). *Capnella gaboensis* is a surface-brooder, and the polyps of the parent colony bend over to conceal the embryos during their development. The eggs of this species are fertilized on the outside of the female colonies, and not inside the parent colony. Sperm were never seen along the mesenterial filaments of recently spawned female colonies, although they were plentiful around embryos on the outside of the colonies (Fig. 5.43A). Embryos were never observed inside polyps in sections. The larva of *C. gaboensis* measures 1.0–2.0 mm in length and up to 0.4 mm wide, and it contains numerous zooxanthellae (Fig. 5.43B). Abnormal-shaped larvae were always observed, including several that

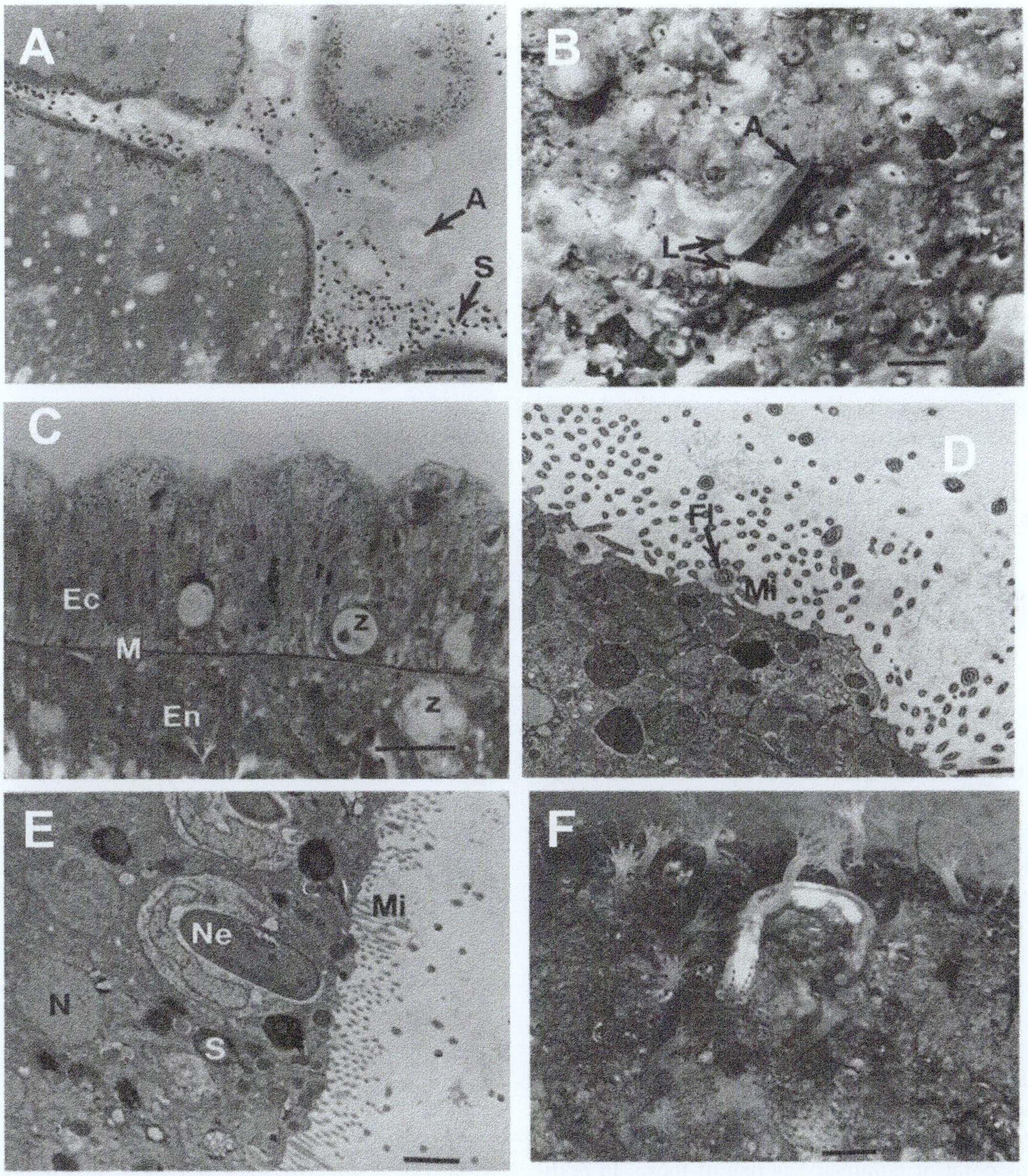

FIG. 5.43 Development of larvae of *Capnella dimorpha after Farrant (1986)*: (A) Section through edge of an early embryo showing sperm (S) and algae (zooxanthellae (A) on outside, bar = 10 μm; (B) larvae (L) with algae (A) visible as dark patches inside, bar = 1 mm; (C) section through edge of a larva showing endoderm (En), mesogloea (M), and ectoderm (Ec) (z) zooxanthellae, bar = 10 μm; (D) TEM showing flagellate collar region of a cell in the larval ectoderm, flagellum (F1) surrounded by collar of microvilli (Mi), bar = 1 μm; (E) TEM showing nematocysts (Ne), secretory granules (S), nuclei (N), and microvilli (Mi) in larval ectoderm, bar = 2 μm; (F) polyps 13 days after settling, with well developed tentacles and pinnules, bar = 2 mm.

appeared to have resulted from either two fused eggs or two fused larvae. Each larva consists of an endoderm, which is continuous with the central yolk material, a thin mesogloea and an ectoderm (Fig. 5.43C). TEM studies show that the ectodermal cells are columnar with microvilli, flagellate collar cells (Fig. 5.43D), secretory cells and nematocysts (Fig. 5.42E). Zooxanthellae occur only in the endoderm of mature larvae, but are frequently seen in the ectoderm of immature larvae (Fig. 5.43C). Zooxanthellae adhering to the sticky spawned eggs may be trapped in the bulges of dividing embryos (Fig. 5.43A).

Capnella gaboensis planulae are fully benthic. However, in the laboratory they begin to settle within 1 or 2 days of leaving the parent colony, and they remain capable of settling for at least 1 month after leaving the parent colony. The larvae crawl with the blunt (aboral) end forward, and they are slow moving and settle quickly, they would not normally disperse far from parent colonies. They attach to the substratum by the aboral end. They are also able to contract and change shape. They then become wrinkled, flatten and metamorphose into polyps with mouths after several days. After ~1 week or more, the polyps develop spicules and tentacles. In the laboratory, the polyps did not grow as rapidly as those in the field and did not reproduce agametically. However, one small colony in the field was observed to consist of four polyps ~10 weeks after spawning.

Order Gorgonacea (horny corals)

Sea fan and *Corallium* The compound tree-like Octocorallia (Fig. 5.35A and C) has a skeleton that is internal and is an integral part of the animal. This skeleton is ectodermal in origin, and may be calcareous separate or fused spicules, or of horny proteinaceous material, forming a branched axis throughout the colony. The skeleton is present in the mesoglea (coenenchyme). There is no siphonoglyph. The beautiful 'sea fans' and the precious 'red coral' (*Corallium*) belong to this group. In the red coral *Corallium rubrum* (Fig. 5.44G), there is an upright branched colony with a rigid axis (Fig. 5.44H) composed of originally separate spicules of different shapes and colours. Spicules are compacted together and are embedded in a cement-like deposit of carbonate of lime. It is the colour of the calcareous skeletal components that accounts for the yellow, orange or lavender colour of some species. The yellow-brown colour of many reef species results from the presence of symbiotic zooxanthellae. The result is the production of an extremely hard and dense branched rod, which extends as an axis through the coenosarc, from which the short polyps arise. The precious coral of commerce is one of these. In addition, it contains a network of endodermal tubes, some of which run along the parallel grooves, which are sometimes seen on the surface of a piece of the precious coral. The mesogloea contains spicule-forming cells derived from the ectoderm, and these travel inwards and add their secretion to the central skeleton.

Secondary polyps are budded off from endodermal tubes, which ramify in the much-thickened mesoglea of the body wall of the primary polyp (see Fig. 5.40B). The latter are of two types: the normal 'autozooids', which feed the colony; and the 'siphonozooids', with reduced mesenteries and enlarged siphonoglyph (Fig. 5.44H). This siphonozooid has no tentacles and its function is only to maintain the circulation of water in the canals of the colony (Borradaile et al., 1977).

Anthozoons, in general, reproduce gametically and the gametes are lodged in the gastrodermis. When ripe, they are discharged into the coelenteron. The fertilized egg develops into a planula which, after a short free existence, settles down and undergoes metamorphosis into the adult form (Marshall and Williams, 1975). In most members of these anthozoons, extensive processes of agametic reproduction take place. The result is the formation of colonies of various forms and often of great size. Budding is the normal mode, and a kind of transverse fission also occurs. It is difficult, in many cases, to distinguish between incomplete longitudinal fission and budding. However, some of these animals reproduce agametically by other types, e.g. by fragmentation. Walker and Bull (1983) reported that in the Antipatharia, the precious coral, *Antipathes dichotoma*, undergoes natural fragmentation of branch ends. Other octocorals, particularly Alcyonacea, reproduce by vegetative fragmentation but few details are known. A new method of agametic reproduction was discovered in the gorgonian coral, the unbranched seawhip *Junceella fragilis* Ridley (Walker and Bull, 1983).

Junceella fragilis Ridley

The gorgonian *Junceella fragilis* (Fig. 5.44K) is a common and conspicuous inhabitant of reefs along the Great Barrier Reef. It is a widespread species throughout the Indo-West Pacific. In Great Barrier Reef waters, two species occur side by side in the same habitats: *J. fragilis* and *J. juncea*. They are readily identified in situ by colouration and anatomical features. *J. fragilis* has a white, yellow or orange coenenchyme with a brown zooxanthellae-containing polyp. Walker and Bull (1983) studied the reproduction of this species and all their methods and observations were done in the field. Consequently they discovered a new method of reproduction. The *Junceella fragilis* reproduces vegetatively by a process of autotomy of terminal portions of a colony (Fig. 5.44L). Prior to autotomy, the polyps are uniformly distributed over the surface of the colony (Fig. 5.44L (a)). Autotomy begins by the resorption of the polyps over a subterminal region of the colony 0.5–1 cm wide (Fig. 5.44L (b)). The polyps of this region disappear completely (Fig. 5.44L (c)). The coenenchyme is resorbed in a transverse band until the daughter colony is connected to the mother colony (Fig. 5.44L (d)). This connection is only by a hair-like axial core, which means that no living

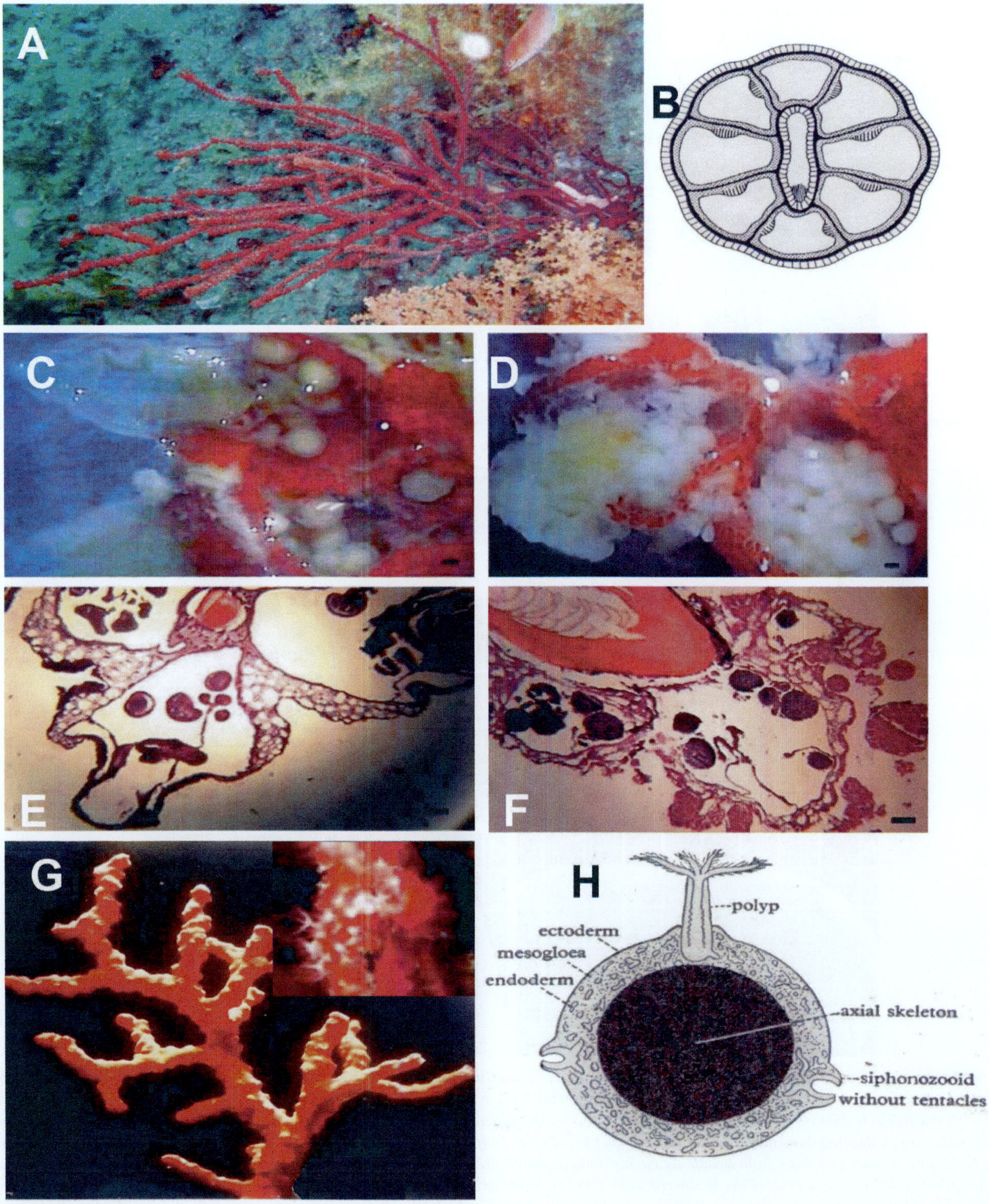

FIG. 5.44, CONT'D

(Continued)

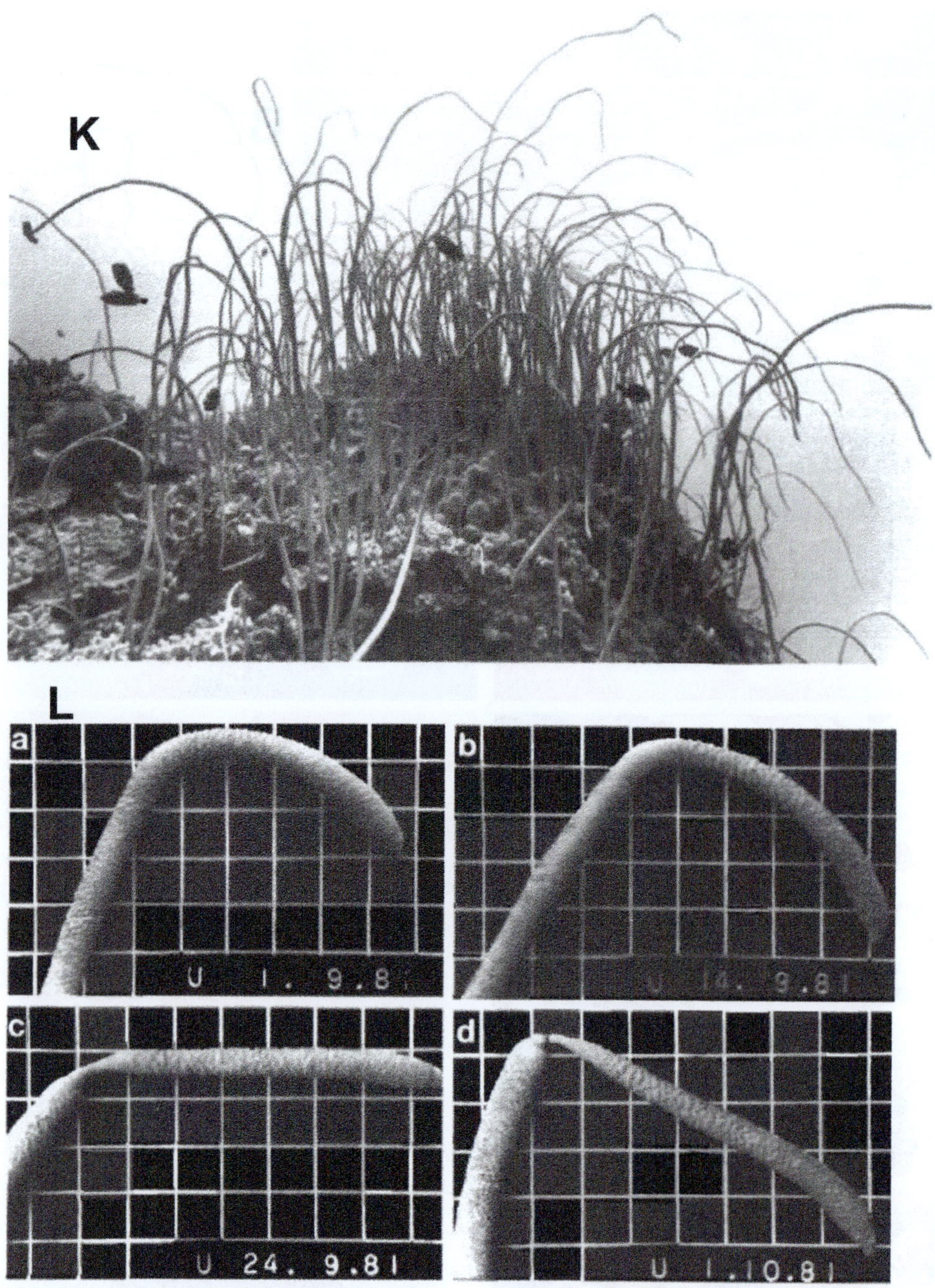

FIG. 5.44 Development of gorgonaceans: (A) a colony of *Anthoplexaura dimorpha* from Munseom, Jejudo island, Korea; (B) T.S. at the level of the pharyngeal tube of primary polyp of *Anthoplexaura dimorpha*, showing 8 septa (mesenteries) with muscle bundles, *after Kinoshita (1910) from Reverberi (1971)*; (C) oocytes in female polyps; (D) spermaries in male polyps; (E) general view of female polyps showing the arrangement of oocytes; (F) general view of male polyps showing the arrangement of spermaries. *(A and C–F) from Seo et al. (2008)*, scale bars = 100 μm; (G) portion of a colony of the red coral, *Corallium rubrum* (not live), *from Beauté du monde animal, Invertébrés Tome XII, 1974*. Inset: portion of live colony of *Corallium rubrum*, © *OCEANA/Carlos Suárez (The Corals of the Mediterranean)*; (H) T.S. to the axis of *Corallium*, *from Borradaile et al. (1977)*. (K and L) Reproduction in the gorgonian unbranched sea-whip *Junceella fragilis*: (K) a dense clump at Lizard Island; (L) daughter colony formation of *Junceella fragilis* on a colony at Magnetic Island: (a) Day 1; (b) Day 14; (c) Day 24; (d) Day 31. Scale lines are 1 cm apart. *(From Walker and Bull, 1983.)*

tissue connects the daughter with the mother. Complete retraction of the polyps of the mother colony is caused by mechanical stimulation. However, this event has no effect on the daughter colony and vice versa. Final separation of the axial core and release of the daughter needs only gentle water currents. The released daughter colony drops rapidly to the bottom near the base of the mother, but heavy currents may take them somewhat beyond the mother. Following separation, both parent and daughter colonies regrow coenenchyme tissue over the small ends of the exposed axial core. When epigrowth occurs on the exposed axial core, the colonies are capable of overgrowing or displacing it.

However, the daughter length is largely independent of the mother colony size.

The widespread occurrence of dense, isolated clumps of *Junceella fragilis* (Fig. 5.44K) demonstrates that autotomy is a relatively successful method of reproduction. This may be described that clumps are clones derived from one initial colony rather than amphimicts resulting from aggregated larval settlement. Walker and Bull (1983) proposed that *Junceella fragilis* demonstrates "the effectiveness of non-dispersive, vegetative reproduction in a sessile, uniaxial (or sparsely branched) animal. Reproductive autotomy is ideally suited to fast growing species."

Anthoplexaura dimorpha Kûkenthal, 1908

Gorgonian corals belong to the subclass Octocorallia, and are characterized by three modes of their reproduction: broadcasting of gametes; internal brooding; and external surface brooding of planulae. Seo et al. (2008) studied gametic reproduction of *Anthoplexaura dimorpha* (Plexauridae: Holaxonia). This octocorallian species is a fan-shaped gorgonian (Fig. 5.44A) abundant in the soft coral community on the southern coast of Jejudo Islands, Korea. The authors collected samples, 5–6 cm long fragments from 3 to 7 randomly chosen colonies (height >30 cm), at a depth of 12–24 m via scuba diving, between August 2003 and August 2005. These colonies are normally distributed on the horizontal and vertical rocky substrata from 10 to 30 m in depth (authors' observation). In the laboratory, the samples were prepared for dissection, for histological studies, to describe the morphological aspects of gametogenesis, to determine the annual reproductive cycle and to record quantitative data regarding sex ratio, including the comparison of gonadal development with environmental factors.

The microscopic and histological studies determined that *Anthoplexaura dimorpha* is a gonochoric species. Gonochorism is known to be the dominant reproductive pattern in the soft corals belonging to octocorals together with gorgonians, although a few species have a hermaphroditic sexuality. However, in *A. dimorpha*, oocytes developing in female colonies were continuously observed in the all samples during the study periods, while spermaries developing in male colonies were only surveyed during summer and autumn seasons, from June to November. Statistical analysis shows that the sex ratio is 1.86:1 (female to male), and this species has one annual reproductive cycle affected by temperature of seawater. Soft corals, including *A. dimorpha*, have reproductive cycles correlated with fluctuation of seawater temperature and release the planulae or gametes at the same periods.

Gonads and gametogenesis. The gonads of both sexes develop from the gastrodermal mesenteries within the calyx of autozooid, and move into the gastrovascular cavity (Fig. 5.44C and D). This means that the gonads are attached to the mesenteries by pedicles. During maturation, they gradually detach from the mesenteries (Fig. 5.44E and F). Gametogenesis, in both sexes, is divided into four stages. In the case of **oogenesis**, the oocyte is ensheathed in a follicular sheath, and it has a large nucleus with a definite nucleolus. During stage 1, the oocytes are embedded in the gastrodermal mesenteries and clustered together. During the following stages, they gradually undergo the following changes: the cytoplasm increases in amount; vitellogenesis occurs; and due to this, it undergoes cyclic colour change: transparent, opaque, faint yellow, orange, then red or reddish orange according to the species. The oocyte at stage 4 attains the optimum size characteristic of the species. The oocyte at its full size has sharp edges and numerous granules containing yolk, with a diameter over 276 μm. It thus appears that the oocyte size may be related to the storage of energy for dispersal and early development of planulae rather than reproductive mode, because nonfeeding and azooxanthellate planulae are common among octocorals. The fully mature red oocytes of *A. dimorpha* are visible through the external surface of polyps having transparent calyxes, which may be useful to infer the level of maturation in oocytes and consequently the timing of spawning.

On the other hand, **spermatogenesis** is also divided into four stages, and stage 1 appears as small transparent spermaries containing small clusters of spermatogonia embedded in the mesenteries. In stage 2, the opaque spermaries contain spermatocytes that move into the gastrovascular cavity, and are attached to mesenteries by pedicles. At this stage, spermaries become with distinct boundaries and increase in size. The spermatocytes in stage 3 arrange themselves in the periphery of spermaries, and maturation proceeds from the periphery to the centre of the spermary. The centre of the maturing spermary is occupied by small spermatids. The stage 4 spermaries were fully matured since the spermatids differentiated into spermatozoa with sperm flagellae arranged in arrays. At this stage, spermaries become dark opaque cream in colour and attains somewhat larger size, with a maximum of 425 μm in diameter.

This gorgonian coral, *Anthoplexaura dimorpha*, has one annual reproductive cycle affected by elevation of seawater temperature. Oogenesis is longer than spermatogenesis: 12 months in females and 6 months in males. The possible spawning events may occur after the seawater temperature reaches the annual peak in September. In general, gametogenesis of cnidarians may be affected by environmental factors such as seawater temperature, lunar cycle and energy resources. Therefore, the gametogenic period may differ in the same species, in the different geological regions. The planulae, in the work of Seo et al. (2008), were not observed from the field survey and histological sections, which suggested broadcast spawning as the reproductive mode of *A. dimorpha*, although the actual spawning events were not observed. Brooding of embryos or larvae, internal or external of polyps was not found from the histological sections and dissected mother colonies.

Subclass Hexacorallia (Zoantharia)

Sea anemones and stony corals

Actinozoa have mesenteries varying greatly in number, typically arranged in pairs, the longitudinal muscles of which face each other (see Fig. 5.48(III)), except in the two opposite pairs, directives, in which the muscles are on opposite sides. Tentacles are usually simple: six or multiples of six. Mesenteric filaments are trefoil-shaped in section (see Fig. 5.48(II) B); the stomodaeum has two ciliated grooves. They typically have a calcareous exoskeleton (in corals) but this may be entirely absent (in sea anemones).

This group of cnidarians includes two apparently different categories: the sea anemones and the madreporarian corals. However, the polyps may all be referred to the same type of structure, the alcyonarian type. The stomodaeum is elongated in the same plane, but possesses two siphonoglyphs. There are hollow unbranched tentacles, and often very numerous. Numbers and grouping of the mesenteries vary greatly within the limits of the Zoantharia itself. The simplest form, and the most like *Alcyonium* (Fig. 5.40C), is found in the small burrowing sea anemone, *Edwardsia* (Fig. 5.48(III) B). In the typical sea anemone, such as *Actinia*, and in coral polyps, the mesenteries are arranged in cycles (or generations) (see Fig. 5.48(III) F).

Order: *Actiniaria* (sea anemones)

Sea anemones Sea anemones constitute a large group of the Hexacorallia (Barnes, 1974; Borradaile et al., 1977). They are solitary polyps and are considerably larger and heavier than the polyps of hydrozoans (Figs 5.1 and 5.15). Most sea anemones range from 1.5 to 5.0cm in length, with a diameter varying from the size of a dime to that of a half dollar. However, there are species of *Goniactinia*, *Edwardsia* and *Nematostella* that are as small as 4mm in diameter (Fig. 5.45A), and others that attain great sizes. Specimens of *Tealia columbiana* (Marshall and Williams, 1975) on the North Pacific coast of the United States and Stoichactis on the Great Barrier Reef of Australia may have a diameter of more than a metre at the oral end. The colourful Tealia are found in temperate regions. The colour of sea anemones is bright white, green, blue, orange or red, or combinations of colours. Sea anemones inhabit coastal waters throughout the world, but are particularly abundant in tropical oceans. They commonly live attached to rocks, shells, and submerged timbers; a few forms burrow in mud or sand.

Agametic reproduction is not uncommon in sea anemones. One method is by pedal laceration, in which parts of the pedal disc are left behind as the animal moves. In some instances, the disc puts out lobes that pinch off. These detached portions then regenerate into small sea anemones. Many sea anemones reproduce agametically by longitudinal fission, and a few species do so by transverse fission.

Sea anemones may be hermaphroditic or dioecious. The gonads are located in the gastrodermis of certain mesenteries, as mentioned above, in the form of longitudinal band-like cushions behind the mesenterial filament. In hermaphroditic species, eggs and sperm are produced as in invertebrates.

Nematostella vectensis

The starlet sea anemone, *Nematostella vectensis* (Fig. 5.45A), has recently emerged as an important cnidarian developmental model system for use in embryological and especially experimental studies. It has now generated interest for the fact that, unlike most other cnidarians, it can be readily spawned in the laboratory, allowing for the experimental comparisons of patterning during the developmental events of embryogenesis, agametic reproduction and regeneration. Layden et al. (2016) reported that *Nematostella vectensis* lives in estuarine environments like salt marshes or brackish water pools, and is thus normally exposed to significant fluctuations in temperature and salinity. The physiological tolerance required for life in an unsteady environment may contribute to the relative ease with which *Nematostella* can be cultured. Martindale et al. (2009) stated that, in the laboratory, animals can be cultured in 1/3 X artificial seawater, fed newly hatched *Artemia* and can be induced to spawn all year round using light and temperature cues. Embryonic development takes 5days from the time of the egg spawning to the development of the planula and finally the settled polyp stage. The life cycle of *Nematostella* takes ~12weeks in culture. They are dioecious, and fertilization occurs in the external environment after spawning. Hand and Uhlinger (1992) cultured *Nematostella vectensis* collected subtidally from the Rhode River, a subestuary of the Chesapeake Bay in Maryland, United States. *Nematostella vectensis*, a widely distributed, burrowing sea anemone, was successful as a culture and produced successive gametic generations at room temperature in noncirculating seawater. These anemones become sexually mature at 3–4months of age and at column lengths of 1.5–3.5cm. The sex is separate. Individuals that have been isolated for >2years continued on as males or females. *Spawning.* By these cultures, Hand and Uhlinger observed spawning of this species. In cultures of mixed sexes, or on comparable feeding regimes, they tended to spawn at the same time. The females extrude egg masses from the mouth (Fig. 5.45B). The eggs are opaque and creamy white, and they have diameter that varies from 170 to 240μm. The mass consists of a gelatinous-like material, which adheres to near objects when first extruded. The production of gelatinous egg masses by *Nematostella* is a unique feature of this sea anemone. The mass may contain many more; the authors had the largest egg masses containing >2000 ova. Additional to the ova, egg masses contain hundreds, even thousands, of nematosomes. These can be seen rotating

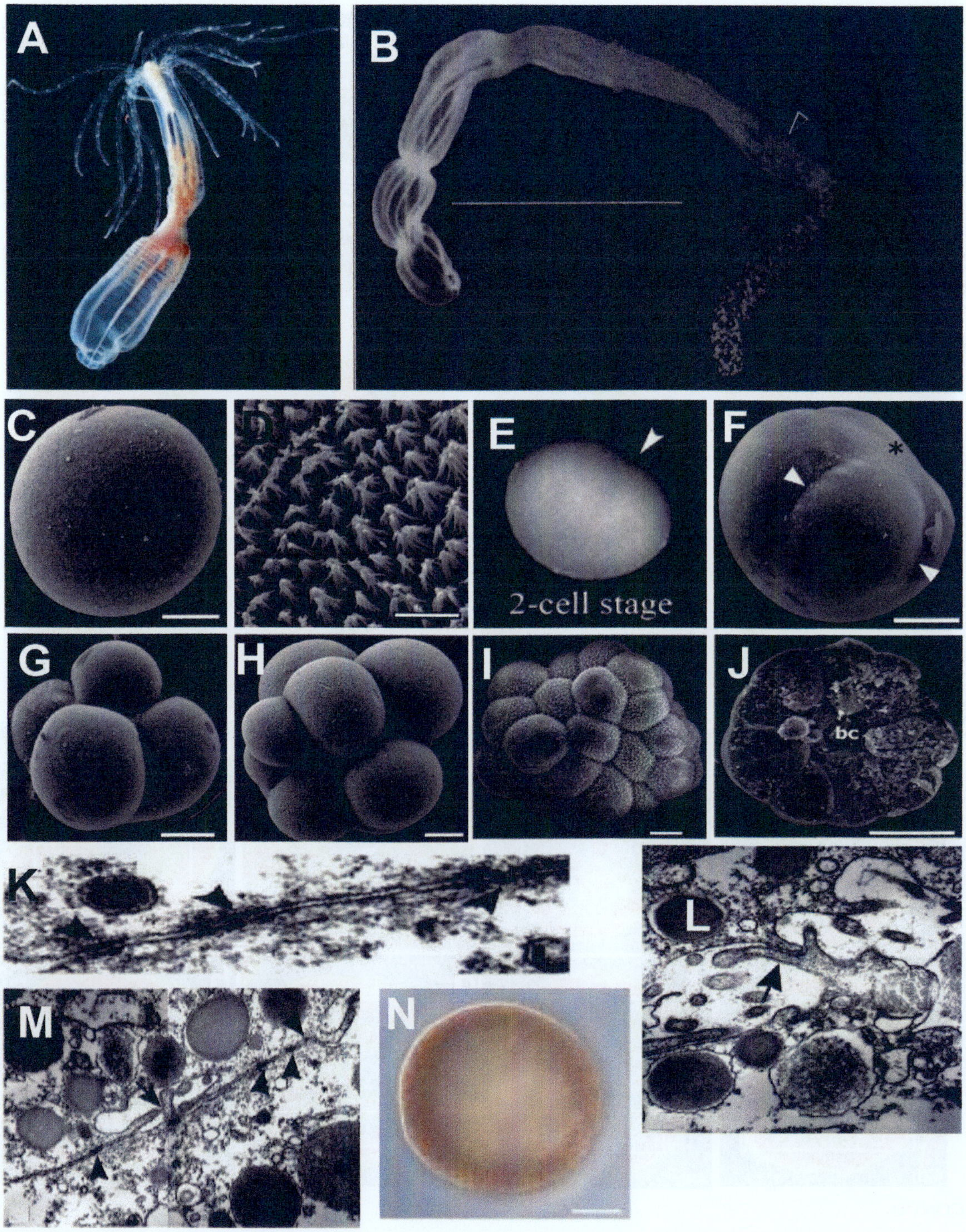

FIG. 5.45, CONT'D

(Continued)

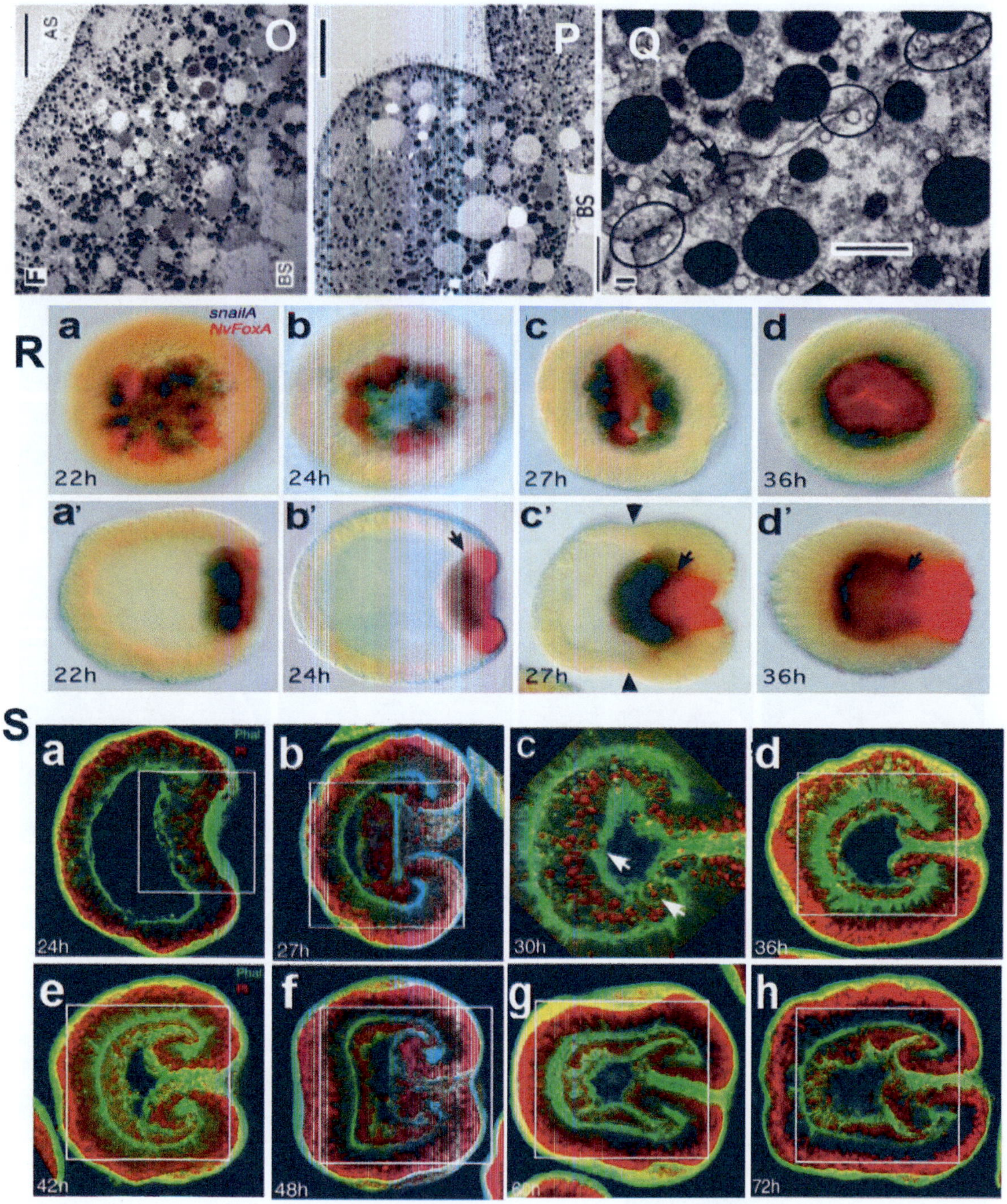

FIG. 5.45, CONT'D

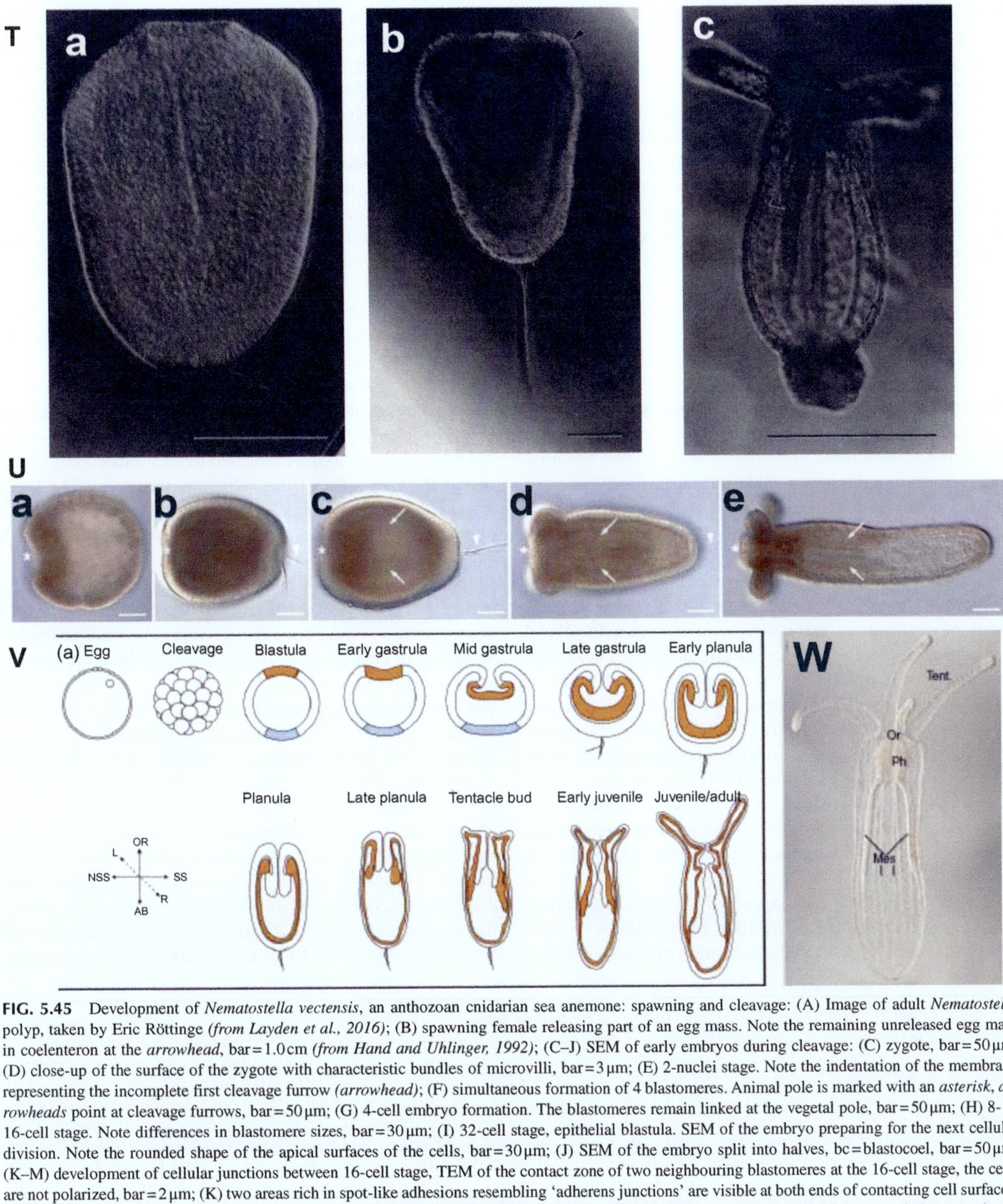

FIG. 5.45 Development of *Nematostella vectensis*, an anthozoan cnidarian sea anemone: spawning and cleavage: (A) Image of adult *Nematostella* polyp, taken by Eric Röttinge *(from Layden et al., 2016)*; (B) spawning female releasing part of an egg mass. Note the remaining unreleased egg mass in coelenteron at the *arrowhead*, bar = 1.0 cm *(from Hand and Uhlinger, 1992)*; (C–J) SEM of early embryos during cleavage: (C) zygote, bar = 50 μm; (D) close-up of the surface of the zygote with characteristic bundles of microvilli, bar = 3 μm; (E) 2-nuclei stage. Note the indentation of the membrane representing the incomplete first cleavage furrow *(arrowhead)*; (F) simultaneous formation of 4 blastomeres. Animal pole is marked with an *asterisk*, *arrowheads* point at cleavage furrows, bar = 50 μm; (G) 4-cell embryo formation. The blastomeres remain linked at the vegetal pole, bar = 50 μm; (H) 8- to 16-cell stage. Note differences in blastomere sizes, bar = 30 μm; (I) 32-cell stage, epithelial blastula. SEM of the embryo preparing for the next cellular division. Note the rounded shape of the apical surfaces of the cells, bar = 30 μm; (J) SEM of the embryo split into halves, bc = blastocoel, bar = 50 μm; (K–M) development of cellular junctions between 16-cell stage, TEM of the contact zone of two neighbouring blastomeres at the 16-cell stage, the cells are not polarized, bar = 2 μm; (K) two areas rich in spot-like adhesions resembling 'adherens junctions' are visible at both ends of contacting cell surfaces. Spot 'adherens junctions' are indicated with *arrowheads* (K and M); filopodial contacts are marked with *arrows*. Rectangular frames mark the areas shown on close-up (L, M), K–M bars = 10 μm; (N) ciliated coeloblastula (4–6 h fertilization) *(from Martindale et al., 2009)*; (O) TEM of the columnar cells in the spherical blastula, bars = 10 μm; (P) rounded cells; (Q) subapical contact zone of the two rounded cells. Areas of 'adherens junctions' are marked with ovals, *arrows* point at filopodial contacts; (R) gastrulation: a–d are oral views, a′–d′ lateral views of comparably staged embryos of *Nematostella vectensis*. SnailA is expressed in all endoderm. The endoderm–ectoderm boundary coincides with the boundary of the snailA and NvFoxA expression domains.

(Continued)

in place within the egg mass. In this culture of Hand and Uhlinger, gametic reproduction occurred in every month of the year, with no sign of seasonality or correlation with moon phases.

In the first experiment, the authors recorded that both sexes spawned within a few hours of one another, between mid-afternoon and early evening. The females spawned from one to three egg-masses each spawning, and males released varying amounts of sperm. In the second experiment, the authors recorded larger numbers of animals.

Fertilization. Sperm produced in both of the two experiments were added to the extruded egg masses, and development was observed.

Another work on *Nematostella vectensis* (Fritzenwanker et al., 2007) followed Hand and Uhlinger (1992), to establish a culture to study early development and axis specification. Fritzenwanker et al. induced gametogenesis and observed and described cleavage in detail. They reported that gametes in anthozoans cnidarians are formed between the two layers of the mesenteries, which are endodermal folds, reaching into the gastric cavity of the adult polyp. Upon maturation, the eggs are squeezed through the endodermal epithelium of the mesenteries into the gastric cavity and are, in *Nematostella*, collectively embedded in gelatinous substances secreted by endodermal cells (Hand and Uhlinger, 1992). The eggs of this species are rounded, having a diameter of $230 \pm 40\,\mu m$ (Fig. 5.45C). The surface of the egg is covered with microvilli, which are arranged on the surface in bundles (Fig. 5.45D). Such arrangement of microvilli on the surface is a characteristic of the early development of many anthozoans (Fritzenwanker et al., 2007). These authors examined the early embryonic development of *N. vectensis* by means of time-lapse microscopy, experimental and histological techniques, and electron microscopy.

Cleavage. Early cleavage stages in *N. vectensis* are characterized by substantial variability from embryo to embryo, yet invariably lead to the formation of a coeloblastula (Fritzenwanker et al., 2007). In *N. vectensis* the first two cleavage furrows originate at the animal pole (Fig. 5.45E and F), and a high degree of variability was observed in the cleavage pattern of *Nematostella* embryos. The two-nuclei stage is nearly indistinguishable from the egg itself, since no cytokinesis occurs. Only a slight indentation of the cell membrane on one side of the embryo is visible (Fig. 5.45E). In a few cases, this also holds true for the four-cell stage embryos. Between the two-cell and the four-cell stage, as well as between the four-cell and the eight-cell stage, cells can fuse again and become visibly indistinguishable from the zygote (Fig. 5.45E). In most cases, the blastomeres of the four-stage embryo are not completely isolated from one another. However, from the eight-cell stage on, the blastomeres become completely separated. From the third cell cycle, the cleavage is holoblastic; the blastomeres are usually not equal in size and vary from embryo to embryo (Fig. 5.45H). The cells of the 8- to 16-cell stage embryo are not polarized and loosely attached to each other, only meeting at the sites of spot-like adherens junctions. The latter are randomly distributed along the neighbouring cells and contacting by filopodia forming on the adjacent lateral cell surfaces of the cells (Fig. 5.45K–M).

Fritzenwanker et al. (2007) reported that despite the highly variable cleavage pattern, cell division remains largely synchronous until the pre-gastrula stage. However, extensive variations of cleavage patterns, even within the same species, are not uncommon among cnidarians. This study indicates that *N. vectensis* embryos cleave in a variety of ways, but always reach a coeloblastula stage by 32–64 cells.

Blastula. The blastocoel becomes first visible after the 16- to 32-cell stage (Fig. 5.45J). The process of cell division continues with further development of the blastula from the 32- to 256-cell stage (seventh cleavage), forming the coeloblastula stage (Fig. 5.45N). The lateral surfaces of the cells are brought into close proximity to one another in their apical thirds, and the adhesion contacts, which can be morphologically identified as adherens junctions, become concentrated apically and form belts (Fig. 5.45O–Q). Early embryonic development

FIG. 5.45, CONT'D (a–d′) 2-colour in situ hybridizations of gastrulating *Nematostella* embryos labelling the snailA *(blue)* and NvFoxA *(red)* expression domains; (S) description of developmental stages of *Nematostella* embryos from 24 h of development (early gastrula, a to 72 planula formation, h). All images are projections of confocal z-stacks taken of *Nematostella* embryos stained with phalloidin *(green)* to label actin, and propidium iodide (PI; *red*) to label nuclei. Hours of development at 16°C are indicated in the lower left corner of each figure *((C–M and O–Q) from Fritzenwanker et al., 2007); ((R and S) from Magie et al., 2007).* (T) (a) Three day postfertilization planula with early apical tuft, bar = 100 μm; (b) 5-day postfertilization planula with fully developed apical tuft and developing tentacle buds. Tentacle bud at *arrow*, bar = 100 μm; (c) 2-week-old four tentacle juvenile anemone. *Arrow* points to one member of the first couple of mesenteries. Note its size compared to the adjacent smaller primary mesenteries, bar = 1.0 mm, *from Hand and Uhlinger*; (U) Late embryonic development in *Nematostella vectensis* from gastrula to the polyp: (a) early gastrula (12–15 h.p.f.), gastrulation initiates at one pole of the coeloblastula and proceeds via invagination; (b) early planula (24 h.p.f.). ectodermal and endodermal layers are well-organized and the planula exhibits directional swimming with the aboral end, distinguished by the apical tuft forward *(arrowhead)*; (c) late planula which elongates along oral-aboral axis and the primary mesenteries *(arrows)* begin to develop, the apical tuft persists *(arrowhead)*; (d) early tentacle-bud stage of polyp. The first four tentacle buds develop around the mouth at the oral pole; (e) juvenile polyp (p.f. at 25°C); *low asterisk* indicates site of gastrulation/oral pole; bar for a = 60 μm, b–e bar = 90 μm *(from Martindale et al., 2009)*; (V) diagrammatic *coloured illustrations* exemplified the development of *Nematostella vectensis* and can be used for other anemones, *orange* represents presumptive endoderm in blastula and the endoderm in all subsequent stages. The outer ectoderm is white, and the aboral ectodermal domain is in *light blue*; (W) DIC image taken by Aldine Amiel of juvenile polyp, Or = the mouth (oral opening), tent = the tentacles, pH = the pharynx, mes = the mesenteries, oral is up in all images. *((V and W) from Layden et al., 2016.)*

shows that starting from the 64-cell stage, *Nematostella* embryos undergo a series of four or five invagination–evagination cycles, which are temporally linked to cell division. These cycles take place between 7 and 12h of development, during which the embryo remains spherical until the onset of gastrulation, about 20h postfertilization. Invagination–evagination movements are strongly correlated to the cell cycle. In fact, inhibition of the replication by Aphidicolin showed that no invaginations occur in the absence of cell division. Mitotic divisions occur in flattened blastulae, while the spherical blastulae contain mostly interphase nuclei.

Magie et al. (2007) studied the process of gastrulation in *Nematostella vectensis* using a diversity of techniques, including expression analysis, cell labelling, confocal microscopy and TEM. They stained the embryos with propidium iodide to label nuclei (red), and phalloidin to label filamentous or f-actin (green) and so to outline cell boundaries. This work and that of Fritzenwanker et al. (2007) evidenced each other, that the cleavage pattern is highly variable from embryo to embryo, forming a spherical blastula, and gastrulation begins ~24h postfertilization at 16°C. Magie et al. (2007) stated that gastrulation in the anthozoan cnidarian *Nematostella vectensis* has been described as a combination of invagination and unipolar ingression through epithelial to mesenchymal transitions (EMT), possibly controlled by snail genes, which are important regulators of EMT in other organisms (Fig. 5.45R). Magie et al. described the process of gastrulation, which begins with the apical constriction of the presumptive endodermal cells at one pole of the embryo (Fig. 5.45S (a)). This apical constriction results in the 'buckling' of the presumptive endoderm and the subsequent initiation of invagination at the future oral pole of the embryo. This event leads to the internalization of the endodermal cells (Fig. 5.45S (b)). During this process, the archenteron forms a mushroom-like shape, in which the lateral edges are pulled towards the side of the internal lateral surface of the embryo (Fig. 5.45S (b) and (c)). Thirty hours postfertilization, the internalization of the endoderm is complete and it lies as a layer beneath the overlying ectoderm (Fig. 5.45S (c)). The archenteron becomes filled with a mass of mesenchymal and endodermal cells, and the authors identified that EMTs (epithelial to mesenchymal transitions) were occurring. At 36h postfertilization, the endodermal cells flatten against the overlying ectoderm (Fig. 5.45S (d)). This vital event results in the opening up of the 'coelenteron', allowing space for the internalization of the pharyngeal ectoderm (Fig. 5.45S (d) and (e)). This movement is involution of the pharyngeal ectoderm through the open blastopore. This is quite distinct from the initial invagination event. However, the formation of the pharynx is similar in behaviour to the pharyngeal endoderm. The prospective pharyngeal endodermal cells must flatten against the body wall endodermal cells, much the same way the endoderm flattens against the overlying ectoderm. The involution of the pharyngeal ectoderm continues for 48h postfertilization (Fig. 5.45S (f)), and by 60h the embryo has elongated into a planula larva (Fig. 5.45S (g)). The elongation of the planula continues for 72h to attain the form of the polyp (Fig. 5.45S (h)).

Hand and Uhlinger (1992) observed that the gastrulae emerge from the egg mass as 200–250μm spherical, ciliated planulae 36–48h postfertilization. The planulae alternate between periods of swimming and resting, then develop an apical tuft of large cilia that becomes obvious by day 3. They change their shape from spherical, to pear-shaped, to elongate. By 5days, some develop four tentacle buds around the mouth (Fig. 5.45T (a) and (b)). At 4–5days, there are two thickened areas of tissue internally that represent the first mesenteric couple. On day 7, many planulae cease swimming, settle to the bottom and metamorphose giving rise to long juveniles measuring 250–500μm with four tentacles (Fig. 5.45T (c)).

Martindale et al. (2009) presented detailed micrographs of the embryonic development in *Nematostella vectensis*, especially from gastrula to polyp (Fig. 5.45U (a)–(e)). Gastrulation initiates at one pole of the sterioblastula and proceeds via invagination (Fig. 5.45U (a)). In early planula (Fig. 5.45U (b)), 24h postfertilization, ectodermal and endodermal layers are well organized and the planula exhibits directional swimming with the aboral end, distinguished by the apical tuft (arrowhead) forward (Fig. 5.45U (c)). Late planula elongates along the oral–aboral axis and the primary mesenteries (arrows) begin to form and the apical tuft persists (arrowhead) (Fig. 5.45U (d)). The early tentacle-bud stage of polyp (Fig. 5.45U (e)) follows, and then the first four tentacle primordia develop around the mouth at the oral pole, the juvenile polyp.

More than one work has summarized the description of the developmental stages of *Nematostella vectensis* by simple illustrations. The description presented in the work of Layden et al. (2016) (Fig. 5.45V) is a model not only for this species, but also for any sea anemone, in these accounts.

Before *Nematostella vectensis*, accounts describing the gametic development of sea anemones were based mainly on that of the European species, since no Japanese species was described at that time. *Sagartia troglodytes*, the English sp. *Tealia crassicornis* and some others were studied.

Sagartia troglodytes

Uchida and Yamada (1968) reported that this sea anemone breeds every year in the beginning of August, when the temperature of the seawater reaches 15°C. The sperm are shed into the seawater through pores in the tips of the tentacles or in the pedal disc of the males. The females respond to this stimulus by moving towards the source of the sperm until they are in contact with the male animals. They described and illustrated its early embryological stages (Fig. 5.46). The mature eggs, which are 0.08–0.09mm in diameter, are fertilized as soon as they are spawned, and form a

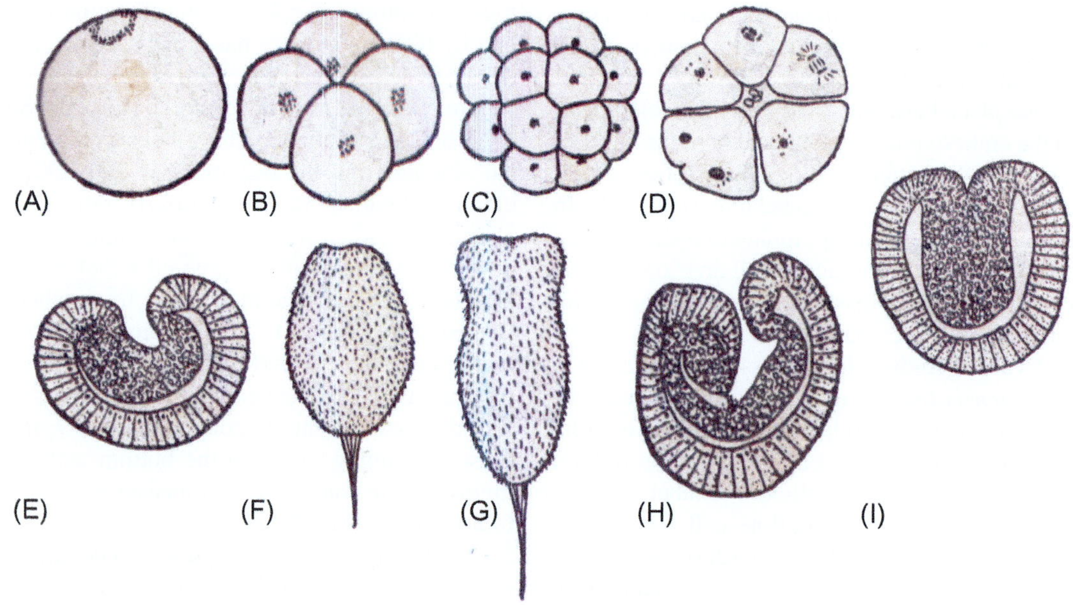

FIG. 5.46 Early development of the sea anemone *Sagartia troglodytes*: (A) fertilized egg, fusion of nuclei at animal pole; (B) 4-cell stage; (C) 16-cell stage; (D) T.S. through embryo at 16-cell stage; (E) L.S. of embryo 8-h after fertilization; (F and G) successive stages in formation of swimming planula; (H) L.S. through free-swimming larva with completely formed oral canal; (I) L.S. through embryo derived from two fused eggs. *(After Nyholm from Uchida and Yamada (1968).)*

fertilization membrane. About 1 h postfertilization, cleavage begins. The first cleavage is normal radial, and all the cleavages begin at the animal pole and cuts through towards the vegetal pole. The second cleavage furrow is perpendicular to the first, forming four blastomeres. The next furrow arises in the equatorial plane dividing the embryo into four upper and four lower blastomeres. At this eight-cell stage a blastocoel has already formed, and becomes larger as cleavage proceeds forwards. The embryo continues cleaving regularly with radial symmetry, until it finally reaches the 128-cell stage (seventh cleavage), about 4 h postfertilization. The cleavage produces a blastula with a blastocoel, which is not extensive, and rather large numbers of yolk cells fill one end of its narrow lumen. These yolk cells serve as a store of nutriment for the growing embryo.

At this stage, the embryo still has no cilia and, like the cleaved eggs, lies motionless on the bottom. Gastrulation is formed by invagination; when this process is practically complete, the ectoderm invaginates and forms the future stomodaeum. While the endoderm is invaginating, cilia develop on the body surface of the developing planula, which begins to move within the fertilization membrane, about 15 h postfertilization. After the beginning of the swimming stage, the body becomes longer and thinner, and a bunch of especially long cilia develops at the end opposite to that of the stomodaeum. This is the planula larva with an aboral tuft of cilia. The possession of this bundle of long cilia is a characteristic feature of sea anemones. The planula swims with this end forward and later attaches with this same end. In this stage, the so-called *Edwardsia* stage, the septa also become conspicuous. After 6–7 days of life as a planula, the larva sinks to the bottom and becomes attached. The size of the body increases and tentacles appear in a regular order; internally the number of septa increases and the gonads approach sexual maturity.

In this species, two eggs sometimes fuse after being spawned and develop as a single large embryo, with certain slight differences in the developmental process.

Halcampa duodecimcirrata

Another European species, elongated and living in the coasts of Sweden, buries its body in the sand. *Halcampa duodecimcirrata* breeds in October and November; the males first come out of the sand and spawn, the females then come out in response to this stimulus and spawn. The eggs (Fig. 5.47A) are shed from the mouth, travelling along the tentacles, and then disperse into the seawater. The eggs are large, about 0.3 mm in diameter, and opaque due to the large amount of yolk. In many cases, they are fertilized within the coelenteron. Such eggs have extruded the polar bodies and are surrounded by a rather thick jelly layer to which sand grains and other particles adhere. Under the fertilization membrane, there is a thin cortical layer of cytoplasm surrounding the whole bulk of the endoplasm. The nucleus lies in this cortical layer where it undergoes cleavage (Fig. 5.47B) and the daughter nuclei are in the peripheral layer, presenting an appearance like that of the superficial cleavage of arthropod eggs. The interior cytoplasm is completely filled with yolk granules.

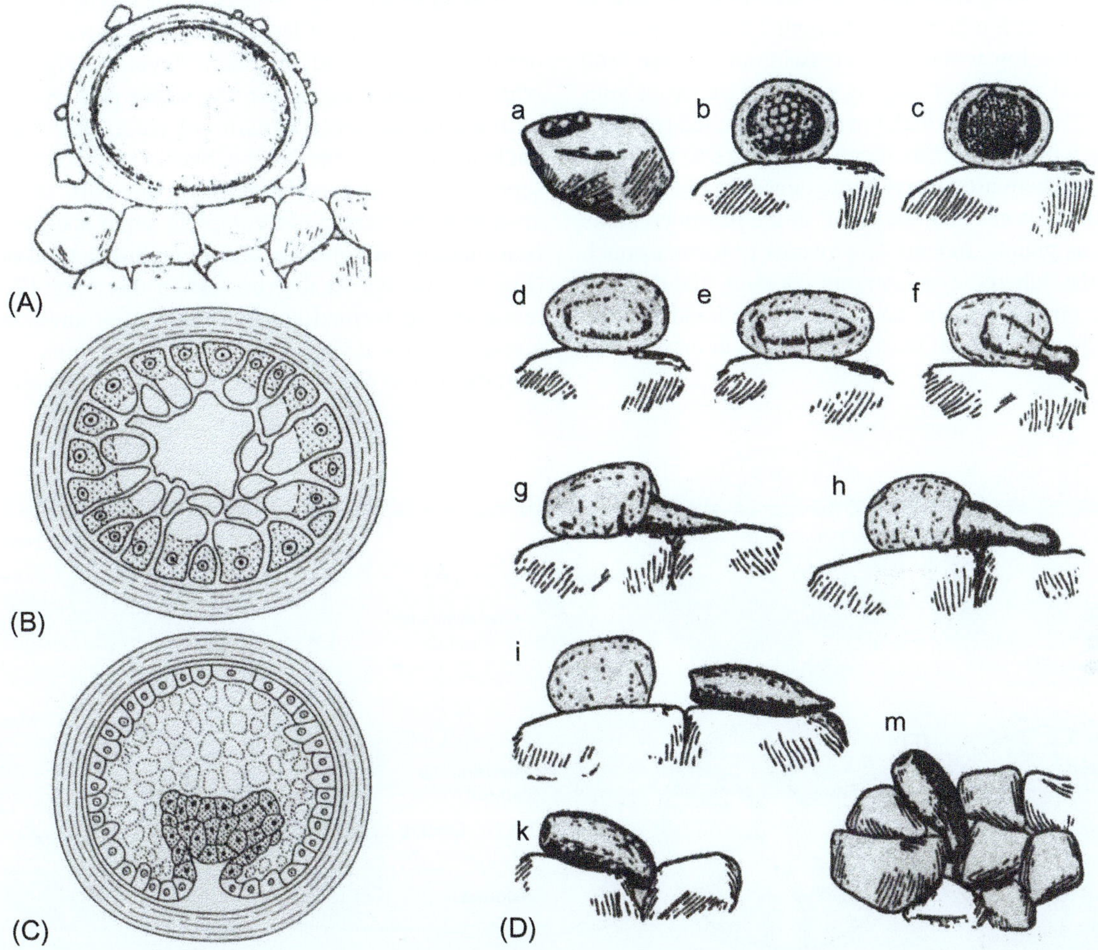

FIG. 5.47 Development of the sea anemone *Halcampa duodecimcirrata*: (A) Section through fertilized egg with thick jelly layer; (B) section through blastula (embryo at 25 h after fertilization) with central yolk mass and peripheral blastoderm cells at the end of superficial cleavage; (C) section through embryo at 35 h showing endoderm development by greatly modified invagination, while the blastomeres eliminate the yolk before their ingression. *After Nyholm from Uchida and Yamada (1968) and Reverberi (1971).* (D) Several diagrams showing the course of changes from the freshly laid eggs to the adult about to burrow into the sand (in (C) the embryo is still solid even after the completion of endoderm formation, (D, a–c)), (d and e) the part which will become the oral plate flattens and a mouth opens in its centre, while the opposite end becomes pointed, (f–h) the case is torn and the larva emerges, the pointed end foremost, (k and m) the larva digs its way into the sand. *(After Nyholm from Uchida and Yamada, 1968.)*

Gastrulation (Fig. 5.47C) occurs by invagination at about 35 h postfertilization at 7.5°C. The embryo is still solid by the yolk inside, even after the endoderm has been completely formed. Gradually this yolk is absorbed and the stomodaeum is formed; the septa are developed gradually, and on their completion, the yolk disappears completely, i.e. completely absorbed. The larva at this stage is elongated and oval; the oral plate flattens and the mouth opens in the centre of this plate. The opposite end of the larva (aboral pole) narrows and becomes pointed. All the described development has taken place within the jelly case (Fig. 5.47D (a)–(m)) and then the larva begins to move inside this case. The case is torn and the larva emerges with its pointed end forward. Immediately it digs its way into the sand and continues its development. In this species, the planula has no swimming stage since it has no cilia and transforms immediately and directly into the polyp.

Epiactis japonica

A Japanese sea anemone, *Epiactis japonica*, is found in the northern part of Japan. The adults of this species very often have small sea anemones attached to the outer surface of their bodies. This means that the animal broods its youngsters. Its breeding season is from October to December. It is hermaphrodite, having the ovaries on the younger septa and the testes on the older ones. The sperm are shed first, and several hours later, the eggs are shed. The two gametes are expelled from the gullet together with whitish mucous secretion, as a result of rhythmical muscular contractions of the body wall.

The spermatozoon has a head, a midpiece (measuring together 2.5 μm) and a long tail (measuring 45 μm). The sperm live for 5 days in seawater at a temperature of 7.5°C. The eggs are reddish-brown in colour and are apparently fertilized before being spawned. They are among the largest cnidarian eggs, with a diameter of about 1 mm. In general, species that develop without a free-swimming phase tend to have a small number of large eggs with a generous supply of yolk. These eggs, which have been expelled from the gullet along with the viscous secretion, adhere to the body of the parent. There are always some eggs that fail to stick, and fall to the bottom. The body wall of the parent contains many mucous glands, indents downwards to form a pouch to envelop the adhered eggs. Among the eggs that develop under these conditions, the youngsters are already in the gastrula stage with a small blastocoel visible at one pole of the embryo.

The gastrula (Fig. 5.48(I) A) of this species has blastocoel completely filled with yolk; however, the endoderm is formed and the septa develop (Fig. 5.48(I) B), while the gullet appears in the upper part. As long as the outlines of the septa remain not clear, there is still yolk packed into one end of the blastocoel. By the time the septa become distinctly defined, the yolk is completely absorbed. At first, eight complete septa are formed; later four incomplete septa are added, making a total of 12 septa (Fig. 5.48(I) B). At the time when there are 12 septa, 12 tentacles are formed and the young sea anemone has 24 tentacles when it leaves the parent and begins an independent existence. Fig. 5.48(I) C shows the halcampula stage,

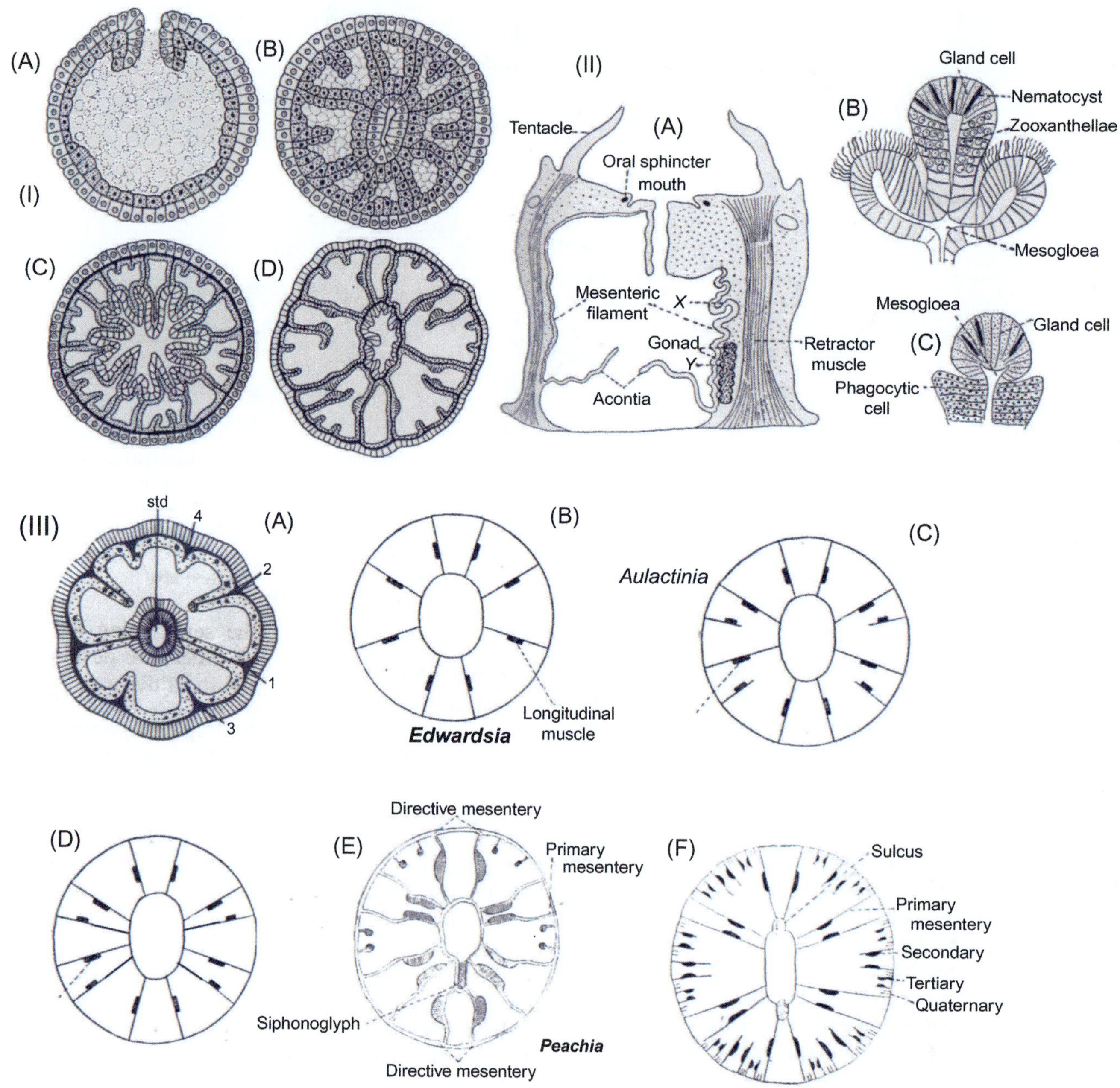

FIG. 5.48, CONT'D

FIG. 5.48 (I) Development of *Epiactis prolifera*: (A) median section through late gastrula with yolk-filled gastrovascular cavity; (B) T.S. through late Edwardsia stage, upper part with pharyngeal tube, and with 8 complete and 4 incomplete septa. The septal pouches filled with yolk; (C) T.S. through Halcampula stage, upper part with densely folded pharyngeal tube, with 12 complete and 12 incomplete septa. The yolk has already been absorbed; (D) T.S. through 24-tentacle stage, lower part with 8 complete and 16 incomplete septa. Formation of muscle bundles on the septa (young actinian). *(After Uchida and Iwata, 1954 from Reverberi and Kume and Dan)*. (II) (A) Diagrammatic drawing of dissected L.S. anemone showing internal organs, especially mesenteries; (B) section through filament at level X; (C) section through filament at level Y. *(From Borradaile et al., 1977.)* (III) Diagrammatic T.S.s of corals and sea anemone to describe the sequence of development of the cycles (or regenerations) of the mesenteries: (A) early embryonic stage of an anemone (*Actinia*) (*after Korschelt and Heider, from Marshall and Williams, 1975*); (B) an Edwardsia stage with 8 complete mesenteries; (C) from B the hexactinian type is developed quite simply by the subsequent development of 4 additional incomplete mesenteries (*Aulactinia*); (D) the completion of the 4 pairs, i.e. the incomplete mesenteries join up with the stomodaeum and complete the first cycle. This is *Halcampa* with these four pairs never develop mesenteric filaments but with muscles facing each other; (E) the simplest form of *Peachia*, no second siphonoglyph and the second cycle of mesenteries is incomplete and without mesenteric filaments; (F) this complete adult arrangement in *Actinia equina* with successive three cycles. *((B–F) From Borradaile et al., 1977.)* (IV) Fluorescence in *Cribrinopsis japonica*: (A) Under a fluorescent lamp; (B) through the cut-off filter under blue LED light; (C) tentacles under a fluorescent lamp; (D) tentacles observed through the filter under blue light; bar = 1 cm; increase in the fluorescence-expressing area of *C. japonica* after blue-light exposure: (E–E′) 0 min after exposure; (F–F′) 10 min after, bar = 1 cm; (E and F) photographs were taken without the cut-off filter under the fluorescent lamp, which was turned on only during photographing; (E′ and F′) photographs were taken through the cut-off filter under blue LED light; (G) *C. japonica* and the deep-sea shrimp in the laboratory. The photograph was taken through the cut-off filter under fluorescent lamp and blue LED light. *Arrows* indicate the shrimp, bar = 1 cm, *adapted from Tsutsui et al., 2016*.

its upper part with densely folded pharyngeal tube, with 12 complete and 12 incomplete septa, the yolk has already been absorbed. Then, the 24-tentacle stage (Fig. 5.48(I) D) has the lower part with eight complete and 16 incomplete septa with formation of muscle bundles on the septa (young actinians).

Tealia crassicornis

This sea anemone, a conspicuous English species, usually has a flower-like form and brilliant colour, and undergoes gametic reproduction more or less similar to that described above for *Sagartia*. The only difference is that fertilization occurs inside the gastric cavity of the mother. The gonads, when mature, often form very noticeable structures. The reproductive products are lodged in the gastrodermis, as in the scyphozoan. The sperm, when ripe, are discharged into the stomach and escape by the mouth. They are then carried, partly by their own movement, and partly by ciliary action, down the gullet of a female to the ovaries where they fertilize the eggs. The fertilized eggs are brooded inside and the planulae escape to the outside through the mouth of the mother.

The fertilized ovum undergoes more or less regular radial division, the details differing considerably in individual cases. It becomes converted into a planula, an elongated ovoidal body with an outer layer of ciliated ectoderm and inner layer of large endoderm cells surrounding a closed enteric cavity. Usually, this cavity is filled with a mass of yolk serving as a store of nutrients. In this condition, the planula escapes from the mother through the mouth, swims about for some time, and then settles down, becoming attached by its broader or anterior end. At the opposite or narrow end a pit appears; this is the rudiment of the stomodaeum. This pit deepens, and its lower or blind end opens, so it can communicate with the enteric cavity.

The mesenteries are developed in regular order, but in a way that certainly would not be suspected from their arrangement in the adult. Firstly, a single couple of mesenteries (Fig. 5.48(III) A1) grow from the body wall to the gullet, being situated one on each side of the vertical plane, at right angles to the long diameter of the stomodaeum, and near to one end of that tube. These mesenteries, which connect the body wall on one side and the wall of the gullet on the other, are called 'complete' mesenteric. This first couple divides the enteric cavity, which becomes divided into two chambers: a larger or dorsal and a smaller or ventral one. In this condition, the embryo acquires a distinct bilateral symmetry. Secondly, a couple of mesenteries appear in the dorsal chamber (Fig. 5.48(III) A2), dividing it into a media and two lateral compartments, firstly this couple appear incomplete, i.e. do not reach the wall of the gullet, and later become complete. Thirdly, another couple appear in the ventral chamber (Fig. 5.48(III) A3), producing a similar division and also begins as incomplete and later become complete. Fourthly, in the middle compartment of the dorsal chamber a fourth couple appear, in the same way as the second and third couples. In this condition, the larva has eight complete mesenteries (Fig. 5.48(III) B). This stage of the complete eight mesenteries is called the Edwardsia stage. Then a fifth couple in the lateral compartments of the dorsal chamber, and a sixth couple appear in the lateral compartments of the ventral chamber. The fifth and sixth couples are incomplete mesenteries (Fig. 5.48(III) C). This means that each couple of mesentery develop beside each other. This means that each mesentery on one side has a mate of the same age on the opposite side, the two together forming a couple, and this condition with 12 mesenteries, four of which are incomplete, is called late Edwardsia (Marshall and Williams, 1975), while another view called it the Aulactinia stage (Borradaile et al., 1977).

The order in which the mesenteries appear differs in the different Anthozoa. However, the most common temporal sequence from dorsal to ventral is 2, 4, 1, 3, instead of 4, 2, 1, 3, as in the example of Fig. 5.48(III) A. Soon the longitudinal muscles are developed and then in four pairs of these mesenteries, the muscles face each other. Primitive couples of mesenteries can be seen, the third and fourth couples become the two directive pairs of the adult, since they attach the tapered ends of the gullet, their muscles face away from each other (Fig. 5.48(III) B). Later, the fifth and sixth pairs of mesenteries become complete, i.e. the mesenteries attach the gullet wall to give a condition of six pairs of complete mesenteries called primary mesenteries, two pairs of which are the directives (Fig. 5.48(III) D). This is the *Halcampa* stage.

The mesenteries usually occur in multiples of 12. When only 12 mesenteries are present, such as primitive Halcampoides, they are complete and are called the primary cycle. The addition of a cycle of 12 are called secondary mesenteries, and each pair develop in the spaces between two adjacent pairs (exocoeles) and never between two members of a pair (entocoeles) (Fig. 5.48(III) E). The secondary mesenteries bring the total to 24. Finally, there may be tertiary and even quaternary mesenteries, always in exocoelec spaces of the preceding generation, making third and fourth cycles. This condition is found in a typical sea anemone such as *Actinia equina*, the commonest of the British anemones (Fig. 5.48(III) F). The subsequent cycles to the secondary or even to the primary cycles are usually incomplete and successively smaller. Many exceptions to the numerical symmetry just described exist. Moreover, agametic reproduction produces considerable regularity particularly in *Metridium*.

The tentacles are developed in a somewhat similar order to that of the development of the mesenteries. The first to appear is connected to the layer or dorsal enteric chamber mentioned above; for some time, it remains much longer than any of its successors, and this adds a marked degree of the bilateral symmetry to the embryo.

It is appropriate to mention that the two intermediate stages in the metamorphosis of the planula of Anthozoa; namely the Edwardisia and that later (halcampula) (Fig. 5.48(I) B and C) compare two prevalent larval types of embryo, with certain slight differences of the developmental process (Fig. 5.46).

Cribrinopsis japonica

It seems that some sea anemones, if not all, like the scyphozoon *Mastigia papua* (as mentioned before) attract the attention of scientists and/or economists. Tsutsui et al. (2016) reported that light is present in deep waters, via organisms producing light known as bioluminescence, and this is used for predation, defence and communication.

With technological advances, it is now possible to observe bioluminescence in some deep-sea organisms. Tsutsui et al. (2016) identified and cloned a novel fluorescent protein from the deep-sea anemone *Cribrinopsis japonica* Tsutsui et al., 2014. Their specimens were collected at a depth of 600m in Toyama Bay, Sea of Japan, and were maintained in artificial seawater at 2°C in the dark. About one-third of the aquarium water was replaced every week to keep the anemones healthy. The anemones were fed with dried krill once a week.

This anemone is photosensitive, perhaps as a reflection of the abundance of light in the deep-sea. It is a pink anemone under fluorescent light (Fig. 5.48(IV) A). The authors never observed bioluminescence in this anemone. Even the physical stimuli, such as poking with a glass rod and massaging the anemone, induced only nonluminous secretion of mucous and internal fluids. However, when the anemone was exposed to blue light, green fluorescence was observed (Fig. 5.48(IV) B) in the tentacles and was more intense at the tips (Fig. 5.48(IV) C and D). The intensity of the fluorescence varied greatly among the individuals, even though the specimens were maintained in the same aquarium under the same rearing conditions. Behavioural observations revealed that a blue light excited the fluorescence in the tentacles of the anemone. Initially, the oral disc expands, and then fluorescent tentacles become fully exposed to the blue light (Fig. 5.48(IV) E, E′ and F, F′). These studies of Tsutsui et al. (2016) revealed a novel green fluorescent protein, cjFP510, in the tentacles of this anemone. They mentioned that this is the second report following that which was isolated from a deep-sea ceriantharian. However, there was little similarity between the primary structures of these two deep-sea fluorescent proteins. The excitation and emission peaks of *C. japonica*'s fluorescent protein were at 500 and 510nm, respectively, which were greener than those reported in homologues. Furthermore, this protein was highly tolerant of increased temperatures and repeated freeze–thaw treatments.

The biological role of this fluorescent protein, in the view of these authors, is to react with free radicals, and thus have the potential to act as antioxidants in living animals. However, localization of concentrated cjFP510 in tentacles makes their role as antioxidants unlikely in *C. japonica*. Another important biological role of fluorescent proteins is prey attraction. The anemone, *C. japonica*, is usually surrounded by a co-living shrimp under laboratory conditions (Fig. 5.48(IV) G) and often captured along with this shrimp at the depth of 800m. The anemone probably attracts the shrimp via visual or chemical cues, but the glowing tentacles of *C. japonica* could function as a lure for prey attraction.

Order Scleractinia

True or stony corals

The other large group of the Hexacorallia is order Scleractinia (Madreporaria), or true corals. They closely resemble sea anemones in structural anatomy of the mesenteries and gastrovascular cavity. A major difference between these corals and the sea anemones is the colonial pattern in the majority, with very small polyps averaging 1–3mm in diameter. However, the entire colony can become very large (Fig. 5.35D). Some other corals are solitary, with polyps reaching 25cm in diameter (Fig. 5.50E). Another major difference is the secretion of a massive calcium carbonate skeleton. Corals have a relatively simple life cycle involving a dominant benthic polyp phase and a shorter planula larval phase. Essentially, the polyp develops from a planula that is usually planktonic and dispersive to some degree. If the planula survives and successfully attaches and settles permanently on hard substratum (Fig. 5.50A), it metamorphoses from the larval form into a juvenile polyp. From this first individual initiates the formation of a calcium carbonate exoskeleton (Harrison, 2011). Subsequent growth is by hundreds of thousands of polyps in a large colony that are formed by division or gemmation. The polyp is identical to an actinarian except for the absence of a differentiated pedal disc, and it does not possess siphonoglyphs (Barnes, 1974). Initially a prototheca is secreted by the ectoderm as a basal plate. Instantly, the under surface of the juvenile polyp develops radial folds, which secrete septa; at the same time, a rim is formed and built up as a thecal wall around the polyp (Fig. 5.49B). Actually, the septa of the skeleton are built up from the bottom, and the animal is lying on top of these as they are secreted by the ectoderm. Then further skeletal material is secreted into the gaps between the septa. Thus, while the skeleton grows, the septa of the skeleton usually alternate with the mesenteries of the living cnidarian.

There are two other patterns of formation of coral colonies (Borradaile et al., 1977). For the first pattern, an example of division is when the polyp has reached a certain size, the oral disc becomes elongated in the direction of the long axis of the mouth, tentacles and mesenteries increase in number, and finally a transverse constriction divides the mouth first, then the disc and lastly the whole polyp (Fig. 5.49D). The second pattern is shown by Fig. 5.49C, the part of the coral polyp that overlaps the theca, the extrathecal zone, which gives rise to young polyps when a colony is formed by gemmation (Fig. 5.49E) and the parent remain connected by the extrathecal portions, and this constitutes the coenosarc of the colony. The gaps between the thecae of the colony are filled up by calcareous material secreted by the coenosarc and called coenenchyme.

In a few forms, there are solitary individuals—the so-called cup-corals—but the majority of species are colonial. By budding, these colonies increase in size and grow into diverse shapes and forms. There are two major patterns of branching (Russell-Hunter, 1979): (1) there is always a primary polyp at the top of the colony with lateral branches on either side (Fig. 5.49F (a)) (e.g. stags-horn-coral, *Ctenopora*); and (2) every budded individual is potentially

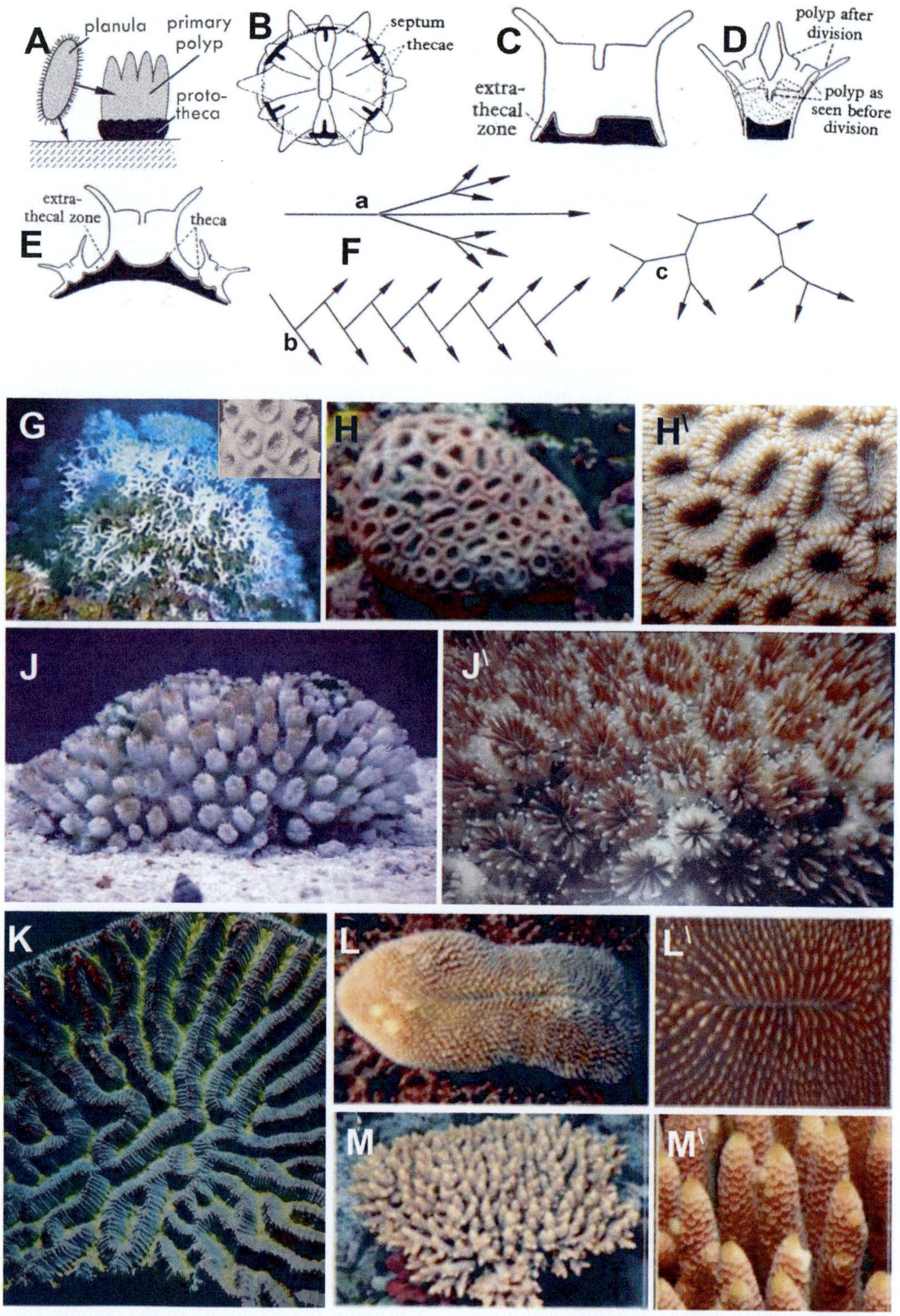

FIG. 5.49 Skeleton formation in corals: (A) the initial settlement of a coral, formation of a prototheca; (B) oral view of a young coral polyp with the beginning of the skeleton seen through the transparent tissues. Shows the position of the formation of the septa; (C) vertical section through a later stage of A; (D) division of polyp by fission; (E) development of a colony showing budding from the extra-thecal zone; (F) shows patterns of branching in the growth of corals: (a) there is always a primary polyp at the top of the colony with lateral branches on either side (e.g. stags-horn-coral); (b) every budded individual is potentially equal to others. This gives rise to forms with solid colonies which may be encrusting or foliaceous (e.g. *Oculina, Favia*); (c) a subtype, again involving equal division but with alternate growth giving rise to cymose branching; (B–E) *after Von Koch, from Borradaile et al. (1977)*. (G–M) different corals mentioned in the text: (G) *Oculina* Lamarck, 1816, colonial mostly straggly branching structure and are mostly pale yellow *(from Wikipedia)*, inset shows magnification of some of cups; (H) small colony of *Favia lizardensis* has closely packed, but clearly separated polyps; (H′) S.V. of highly magnified portion of *Favia* sp.; (J) *Galaxea (from Wet Web Media, courtesy of Robert Fenner)*; (J′) S.V. of the coral *Galaxea*, is a common species in the Red Sea, the polyps are clearly separate and the delicate transparent feeding tentacles are clearly observed; (K) S.V. of magnified portion of 'Cerveau de Netune' genus *Meandrina, from Beauté du monde animal, Invertébrés Tome XII, 1974*; (L) *Ctenactis echinata*, a solitary coral, consisting of a single large polyp; (L′) S.V. of the central portion, magnified showing the slit from which the polyp emerges to feed; (M) *Acropora millepora* colony, composed of many branches emerging from a single base; (M′) a few branches magnified, each covered in many polyps. Each scale-like structure on the branch marks the location of a single polyp. *(H, L, L′, M, and M′) from Okubo et al. (2016); (H′ and J′) from Saudi Arabian, Red Sea by H. Schmid and P. Vine.*

equal to the others with persistent growth. This gives rise to forms with solid colonies, which may be encrusting or foliaceous (Fig. 5.49F (b) and (c)). A subtype, again involving equal division but in this case alternate growth, gives rise to cymose branching. The degree to which the daughter polyps remain together after agametic division also determines the form of the coral. They may retain a common skeletal base but have widely spaced cups as in *Oculina* (Fig. 5.49G and inset), or the thecae may be placed closer and share common walls as in *Favia* (Fig. 5.49H and H′), or the polyps may be confluent into rows not separated by skeletal materials as in the meandrine group of corals (Fig. 5.49K). In these, the winding grooves or valleys on the surface of the calcareous mass are occupied in life by fused lines of polyp, with each row of mouths opening into a common gastrovascular cavity. These meandrines are the characteristic 'brain-corals' (Fig. 5.35D) which can grow into massive coral boulders or amalgamate into reefs (Russell-Hunter, 1979).

Harrison (2011) discussed this aspect of budding in corals with a different view. Agametic budding of polyps leads to the formation of coral colonies, while various forms of agametic reproduction result in the production of new modules that form physically separate but genetically identical clones. Most 'hermatypic' coral species and about 26% of the known 'ahermatypic' azooxanthellate coral species form colonies by agametic budding of polyps and are therefore modular, 'iterative' colonial organisms.

Budded polyps usually form by growth and internal division of existing polyps (intratentacular budding) or development of new polyps from tissues adjacent to, or between, existing polyps (extra-tentacular budding). In most colonies, these budded polyps remain interconnected (Fig. 5.50B), and the colony is partly integrated via nerve and muscular networks within the thin veneer of tissues that overlie the skeleton they secrete. Budding produces genetically identical polyps within each colony; however, DNA damage and somatic mutations can genetically alter cell lineages and may induce development of neoplasms (tumours) within colonies. Colonial growth enables corals to grow much larger than most single polyps; thus colonies can occupy more space and more effectively compete for resources by growing above the reef substratum or over other benthos. Thus, the colonies can survive the death of individual polyps and partial colony mortality. As the number of gravid polyps and fecundity increase with size and age (Fig. 5.50C); hence larger colonies can dominate gamete production within coral populations.

Corals exhibit a wide range of agametic reproductive processes that produce new colonal solitary corals or colonies. These processes include colony fragmentation resulting from storm and wave impacts or other damage, colony fission, L. and T. division, polyp expulsion or polyp 'bail-out', growth and detachment of polyp balls in some Goniopora colonies, and budding of polyps from an anthocaulus or regenerating tissues in fungiids and some other corals.

In these corals, agametic reproduction produces genetically identical modules that may prolong the survival of the genotype. On the other hand, gametic reproduction enables genetic recombination and production of new coral genotypes that may enhance fitness and survival of the species (Harrison, 2011): hermaphroditic broadcast spawners, hermaphroditic brooders, in which corals have both sexes, develop in their polyps and colonies. The other two patterns are gonochoric broadcast spawners, or gonochoric brooders in which corals are separate sexes. However, not all coral species belong to one of these basic patterns, as mixed patterns or both modes of development are known to occur in some species. However, hermaphroditic broadcast spawners remain the dominant group of corals. Hermaphroditic coral species occur in 13 different families, e.g. Acroporidae (155 species) (Fig. 5.50C), and faviids and associated corals (74 species) (Fig. 5.50I) in clade group XVII. The vast majority of known hermaphroditic coral species are simultaneous hermaphrodites that develop both ova and sperm within each polyp (Fig. 5.50D). On the other hand, in other corals that are generally recorded as gonochoric, their different polyps are typically only female or male (monoecious pattern) such as fungiids (Fig. 5.50E and F). Gonochoric coral species are less commonly reported than hermaphroditic coral species. They are characterized by colonies or solitary corals that have separate sexes, and are therefore functionally only male or female during their lifetime. Fungiidae (22 gonochoric species) is one of the families included in this group of the gonochoric corals.

As for the mixed patterns, more unusual mixed gametic patterns have been recorded in a few other coral species. One of these unusual patterns is that explained in *Galaxea* species. For the first example, *G. fascicularis* and *G. astreata* were originally described as simultaneous hermaphrodites. However, subsequent research demonstrated that these species have populations composed of female colonies that spawn pinkish-red eggs, and hermaphroditic colonies that produce sperm and lipid-filled white eggs (Harrison, 1988). In the second example, hermaphroditic *G. fascicularis* colonies produce functional sperm that can fertilize the spawned pigmented eggs from female colonies (Fig. 5.50G and H). The third example is the white eggs that contain unusually large lipid spheres and are not able to be fertilized. These function to lift the sperm bundles up to the water surface where the buoyant pigmented eggs accumulate, thus potentially increasing fertilization success (Harrison, 1988). This pseudogynodioecious sexual pattern, in at least some *Galaxea* species, is therefore functionally gonochoric.

Coral have two different primary modes of development: broadcast spawning of gametes from their polyps into the

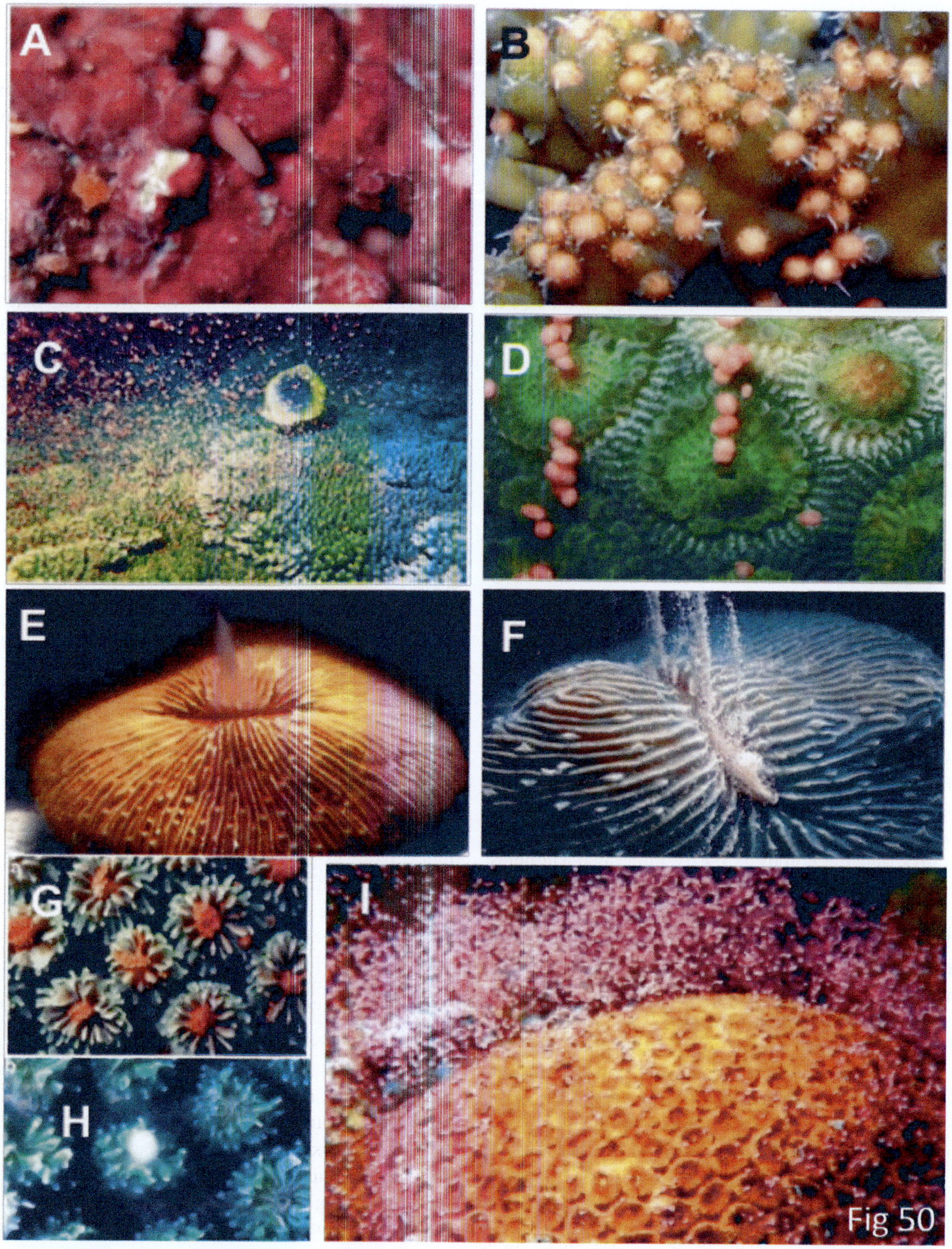

FIG. 5.50 (A) A brooded *Isopora cuneata* planula searching reef substratum covered with crustose coralline algae for a suitable site for initial attachment; (B–I) photographs showing different modes of spawning activity in corals; (B) colony of *Acropora arabensis*, where bundles of eggs and sperm bulge under the oral disc of each polyp in the 'settling' phase just prior to synchronized spawning. This photo shows integration of polyp tissues leading to coordinated behaviour of gametic reproduction of colonies that develop from agametic budding; (C) *Acropora tenuis* colonies seen in this photo, spawning many thousands of buoyant egg–sperm bundles on the Great Barrier Reef (GBR) during the 'crepuscular' period just after sunset; (D) the simultaneous hermaphrodite, *Favites* brain corals that develop both eggs and sperm within each polyp, which are spawned in buoyant egg–sperm bundles; (E and F) the gonochoric Fungiid mushroom corals have some individuals that spawn sperm (E), and some that spawn eggs (F); (G and H) *Galaxea fascicularis* populations have: female colonies that spawn pigmented eggs (G), and other colonies that spawn functional sperm and white eggs that cannot be fertilized (H), so this species is a pseudo-gynodioecious sexual pattern; (I) the broadcast spawning simultaneous hermaphrodite, *Favia* brain corals that forcibly eject egg–sperm bundles from the polyps during brief annual spawning events. *(All from Harrison (2011).)*

seawater for external fertilization; development (spawners) or brooding of embryos and planulae within their polyps that are subsequently released, usually at an advanced stage of development (brooders). The vast majority of corals studied are broadcast spawning species (Fig. 5.50I), so external fertilization leading to planktonic larval development is the dominant mode of development.

Astrangia danae

Szmant-Froelich et al. (1980) reported that there is great lack in the study of the reproduction, especially the embryogenesis, of corals in general. They studied the reproductive activities of this temperate scleractinian coral *Astrangia danae*. They collected their material from the southern end of Narragansett Bay, Rhode Island, where this species lives. It occurs as small encrusting colonies on rocky surfaces and shells. It is found in nature both with and without the endosymbiotic dinoflagellates, Zooxanthellae, and is remarkable among corals in tolerating an annual temperature range of −2°C to 22°C. Szmant-Froelich et al. collected the colonies of *A. danae* via scuba divers using knives and others to remove the coral from their rocky substrate and taking them to the laboratory in plastic bags containing seawater.

As in other anthozoans, the gonads of *Astrangia danae* are found on the mesenteries, between the mesenterial filaments and the muscle bands (Fig. 5.51A). Each polyp has up to 24 mesenteries, and all are capable of bearing gonads. The smallest colonies, of six to nine polyps (estimated age of 1–3 years, based on the growth rates in the laboratory), were already capable of gametic reproduction. All polyps of a colony were of the same sex; only one hermaphrodite polyp was found out of 5000 polyps examined. From the colonies examined, 77 males were found and 73 females, which indicate that the sex ratio of this species is 1:1. Gametogenesis started in the early spring (March–April) and proceeded through June and July. The spawning season is from early August through September, when water temperatures are at their peak.

Spermatogenesis. Spermary formation began in the early spring with the appearance of interstitial cells in the mesoglea. In this site, the interstitial cells formed small membrane-bound aggregates (stage I) (Fig. 5.51A). The number and size of the cells in these early spermaries gradually increased, but the spermatocytes still retained an appearance similar to that of the initial interstitial cells (stage II). The spermatocytes advanced in maturation and their nuclei accepted more hamatoxylin (stage III). The spermatocytes became much more numerous and smaller, with condensed nuclei (stage IV) (Fig. 5.51B). The spermatozoan tails became evident and the sperm appeared ready to be spawned (stage V) (Fig. 5.51B).

Oogenesis. Oogenesis does not begin as aggregates of cells observed in male gonads, but it begins as only one interstitial cell, which it is difficult to recognize. However, stage I oogonia were observed along the mesogleal lamellae and known by their larger nuclei and nucleoli (Fig. 5.51C). The smallest oogonium observed measured about 5 μm in diameter, with very little cytoplasm surrounding the nucleus. Vitellogenesis begins as stage II and the oocytes continue to grow that each of the cytoplasm and nucleus increase in size (stage III). During this stage, the oocyte attains its full size and so it is the growth stage. In this work, the increase in size of the oocytes was not identified, but the authors investigated that small oocytes may have been resorbed as the larger oocytes grew (Fig. 5.51D). The oocytes attain stage IV by the formation of the vitelline membrane, which consists of cortical vesicles just beneath the cell membrane (Fig. 5.51D). The oocytes remained at this stage IV as mature eggs until spawned some time in August and early September.

Spawning. Spawning was experimentally induced by inserting fine-tipped forceps through the mouths of male polyps and isolated small pieces of male gonadal tissue, and placing each inside the female polyp and vice versa. Fertilization of eggs was accomplished by pipetting the freshly spawned eggs into small bowls, and adding water from the bowls of spawned males. Samples of the sperm, unfertilized and fertilized eggs, and various stages of embryonic development were observed. However, colonies in the laboratory were observed to be spawned when they were advertently subjected to heat and possibly low-oxygen stress. In all cases, a certain behaviour pattern was noticed. The polyps first expanded to an extreme degree, but the tentacles remained short and stubby, unlike the expansion in a feeding posture. The oral disc was very prominent, with the mouth tightly closed and 'puckered' in the centre. At the moment of spawning, the polyp suddenly shortened the column and a narrow stream of eggs or sperm ejected out of the mouth. Within a few minutes, the polyp would be fully expanded again, and the contraction with gamete ejection is quickly repeated. Individual polyps were observed to go through this cycle up to six or seven times.

Gametes. The mature spermatozoan consists of the usual parts: head, midpiece and a tail or flagellum. The large head with the very small middle piece measures about 2–3 μm in length, not including the tail. This sperm is characterized by having a single large mitochondrion and a large, electron-dense lipid body. The latter is located at the base of the conically elongated nucleus (Fig. 5.51E). No acrosome was observed, but in Fig. 5.51E, one can observe an accumulation of moderately dense material anterior to the tip of the nucleus, similar to that of *Aurelia*. On the other hand, the mature ovum measures about 100–130 μm in diameter and has an asymmetrically located nucleus. The surface of

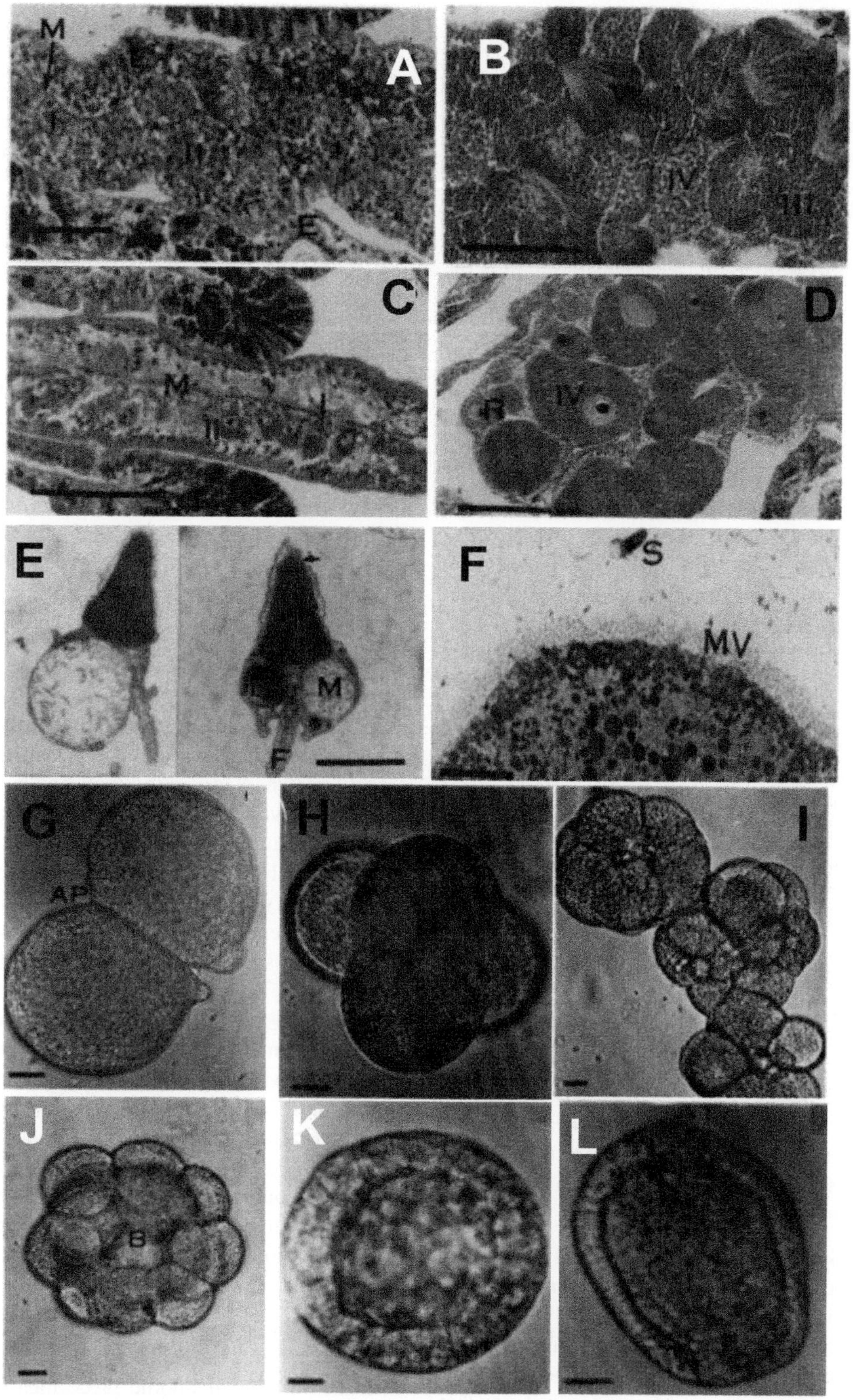

FIG. 5.51 Photomicrographs of histological preparations of tissues of *Astrangia danae* (Helly's fixative and stained with Heidenhain's aniline-blue) illustrating the gametogenic sequences; (A and B) TSs of gonad-bearing portion of a mesentery with spermaries in developmental stages I–V, bar = 50 μm; (C) early oogenesis seen in L.S. of a mesentery. Stage I and II oocytes are strung along the mesogleal lamella, bar = 100 μm; (D) section through ovary showing mature oocytes with well-developed vitelline membrane (stage IV). Notice that some of the smaller oocytes seem to be partially resorbed, bar = 100 μm. Abb: E = endoderm, F = mesenterial filament, M = mesoglea; (E) TEM of LS through spermatozoa show single mitochondrion and large lipid body. Notice less dense material anterior to the tip of the nucleus *(arrow)*, bar = 1 μm; (F) TEM of TS of outer portion of an unfertilized spawned egg. A layer of subcortical vesicles lies beneath the microvillous oolema. Yolk granules and mitochondria are abundant within the cell, bar = 5 μm; (G) first division complete, notice small protuberances being pinched off at vegetal pole, bar = 10 μm; (H) pseudospiral arrangement of blastomeres after second division, bar = 10 μm; (I) irregularly shaped embryos resulting from nonsynchronous third and fourth divisions, bar = 10 μm; (J) embryo of around 20 blastomeres, with large blastocoele, bar = 10 μm; (K) blastula stage, blastomeres compressed, bar = 10 μm; (L) planula stage, showing apical tuft of cilia and oral pore, bar = 10 μm. Abb: AP = animal pole, AT = apical ciliary tuft, B = blastocoele, CG = cortical granules, F = flagella, L = lipid body, M = mitochondrion, MV = microvilli, N = nucleus, OP = oral pore, S = spermatozoan, Y = yolk granule. *(From Szmant-Froelich et al. (1980).)*

the ovum is microvillous, with a subsurface layer of cortical granules (Fig. 5.51F). The ooplasm is densely populated with yolk bodies and mitochondria.

Fertilization and development. Most of the spawned eggs quickly settled to the bottom of the culture where they could be pipetted and concentrated before fertilization to observe and photograph the sequence of development from the beginning. No fertilization membrane could be distinguished by light microscope for the egg of *Astrangia danae*. Cleavage begins within 1 h of fertilization. The first cleavage furrow started at the animal pole and proceeds inward, towards the vegetal pole giving rise to two complete and equal blastomeres. A phenomenon that frequently happened at this time was the pinching off of two small portions of the cytoplasmic material at the vegetal end of the two cells (Fig. 5.51G). No acceptable reason was given. The timing of subsequent divisions varied between batches of eggs obtained on different occasions, but they were generally 30–60 min apart. The second division was always synchronous between the two blastomeres. The cleavage plane was radial but the blastomeres rearranged themselves after the division, into a pseudospiral formation (Fig. 5.51H). Subsequent divisions were not always synchronous, resulting in a variety of embryos of different cell numbers and forms (Fig. 5.51I). A blastocoele was formed early during development (Fig. 5.51J). It was observed in embryos with as few as 12 blastomeres. At the 64-cell stage, the blastomeres were still fairly rounded, but with further development and at 128-cell stage they became closely spaced and columnar in shape (Fig. 5.51K). They attained cilia and at this time, they remained on the bottom of the culture. Ciliation increased and the embryos elongated and gastrulated and began to swim and congregate on the surface of the culture. At this stage, 6–8 h postfertilization, they became strongly attracted to light.

There was no information about the process of gastrulation, as it was not observed and no evidence of invagination in any of the embryos was observed.

Within 12–15 h of fertilization, the elongated gastrulae developed oral pores and apical ciliary tufts, planula stage (Fig. 5.51L). The young planulae swam and spun on their longitudinal axes quite rapidly, but remained near the surface of the aquarium. In the next 24 h, the larvae became more elongated and their mouth and gullet were formed. From 48 to 72 h postfertilization, the larvae no longer congregated at the surface, or were attracted to light, but they swam in the mid-water of the aquarium. In a few of the larvae, the first signs of mesenterial development were noticed. By time, the development proceeded and the larvae reached the planulae stage, they were about 75 μm long, only half to three-quarters of the diameter of the original egg. Some of the cultures were contaminated by small flagellate, diatom and bacteria, but the planulae were never observed feeding on this material. Planulae in all the cultures shrank at about the same rate. Different attempts to settle the planulae on different substrates failed and eventually all of the larvae died.

The reef-building coral *Acropora* spp.

The coral genus *Acropora* contains >300 defined species. They are widespread throughout the Indo-Pacific, and typically dominate the shallow reef assemblage. Kenyon (1992) reported that the existence of *Acropora* in Hawaii had long been doubted until the late 1970s, when three species were reported from the Northwest Hawaiian Islands. Since the discovery of *Acropora* in these Islands, the study of reproduction in corals has undergone a renaissance. Kenyon (1992) presented new evidence showing that *Acropora valida*, *A. cytherea* and *A. humilis* are sexually reproductive at French Frigate Shoals. The author collected samples of these three corals, at this site, from 7 to 9 m depth around La Perouse Pinnacle. The collection occurred roughly at biweekly intervals from late May through to July in 1989, with consideration given to the time of the full moon.

Field observations and comparison of sizes of fixed and fresh oocytes obtained from the same colony were factors to evaluate the sexual reproductive maturity of the species. The results of the study of Kenyon (1992) proved that during two consecutive summers, *Acropora valida* gonads matured at different rates in different habitats in both years. Spawning was induced by the sudden disappearance of gonads of mature size. Histological sections of fertile polyps confirmed the maturity of gonads prior to spawning. In the second coral, *A. humilis*, the isolated colony spawned in early summer. There were strong indications of sexually mature colonies of *A. cytherea*, but clear temporal patterns were not apparent. The three species did not spawn at the same time. The lunar period of spawning in reef flat *A. valida* and *A. humilis* differed from that reported for these species from other regions.

This new evidence of Kenyon (1992) was useful for the re-evaluation of the allochthonous origin of larval recruits, as well as the absence of *Acropora* from high islands of the Hawaiian chain.

Okubo and Motokawa (2007) studied embryogenesis (the main aim of the present work) in the reef-building coral of five species of *Acropora*. A few days before the predicted mass-spawning event, the authors collected fragments from five colonies from Japan waters: *A. tenuis* and *A. digitifera* from the reefs around Akajima Island, Okinawa at depths of 3–6 m in May and June; *A. hyacinthus* and *A. solitaryensis* from Wakayama in July and August; and *A. intermedia* fragments from Okinawa Churaumi Aquarium in June. The fragments were placed in the harbour or aquariums and examined every night, until bundles appeared at the mouth of the polyps. At that time, each fragment was placed in a plastic container and brought to the laboratory. Fragments of the same species were transferred to a plastic container filled with artificial seawater (Fig. 5.20L) and kept in the dark. When spawning occurred, the contents of the container were gently stirred to mix the bundles and to ensure that insemination occurred. The water temperature was maintained at 26.0–26.5°C throughout the observations.

Spawning The egg and sperm produced by a single polyp were compressed into a buoyant bundle (Fig. 5.52A), which had no membrane. Instead, the eggs cohered via a type of sulfated mucosubstance, which stained by Alcian blue (pH 1.0). The cohered eggs had a central cavity in which the sperm were located (Fig. 5.52B). The egg had no germinal vesicle, indicating that its breakdown had occurred before the egg was released from the mouth. A polar body was observed 30 min postspawning (Fig. 5.52C). It was unknown if it was the first or the second polar body. With magnification and with special treatment, it was well observed (Fig. 5.52D). After the bundles were broken, the eggs attained various shapes, but became round after 1 h. They were isolecithal and filled with large amount of yolk granules. Its nucleus or possibly the pronucleus was visible.

Cleavage. In spite of the yolk granules contained in the egg, the cleavage is holoblastic, and the first cleavage furrow initiated at the animal pole, which contained the nucleus. The result was a heart-shaped 'zygote' at 2 h (Fig. 5.52E and F). The cleavage divided the egg symmetrically, although some zygotes were asymmetrical (Fig. 5.52E, arrow). The second cleavage furrow also initiated at the animal pole, but the two blastomeres twisted (Fig. 5.52G and H) and so the cleavage plane for each blastomere was not at a right angle to the first cleavage furrow. The result was four equal-sized blastomeres at 3 h (Fig. 5.52I). This is similar to what occurred in the previous example, *Astrangia danae*. The third cleavage was equatorial and perpendicular to the first two cleavages (Fig. 5.52J) and resulted in eight blastomeres. The fourth cleavage was meridional and resulted in 16 blastomeres (Fig. 5.52K). The blastomeres after this division were displaced obliquely (Fig. 5.52L) as shown from the side view. At 4 h, about half of the developing eggs were at the eight-cell stage and the other half at the 16-cell stage, indicating that the synchronous cleavage was gradually lost. The blastomeres continued to divide irregularly after the 16-cell stage. In the morula stage, the nucleus of each blastomere shifted from the inner side to the centre. No blastocoel appeared during these stages of cleavage (Fig. 5.52).

Gastrulation. At 7 h to 10–12 h, the embryos changed in form, in terms of arrangement of blastomeres and in position of the nuclei in the blastomeres, until each embryo was one-layered and the cells were columnar. At 12 h, the blastocoel was formed and numerous yolk granules were simultaneously 'excreted' from the inner side of the cell membrane into the blastocoel (Fig. 5.52M). After 10 h, the blastula gradually thickened until it was roughly spherical at 18–20 h. This stage was the gastrula.

At 36 h, the round-shaped larvae developed cilia on the outer surface and began rotary motion. The blastopore was closed in all but a few embryos at this time with open blastopore. The outer cells became a single columnar layer characteristic of the ectodermal epidermis (Fig. 5.52N). The cells inside of the mesoglea and the yolk granules accumulated in the blastocoel. The cells dispersed randomly and assumed irregular shapes among the yolk granules. Before the closure of the blastopore, the boundary between the outer and inner cells of the larva became clear, indicating the formation of the mesoglea (Fig. 5.52N, arrowhead).

Formation of the stomodaeum and gastrodermis. At 42–54 h, the larva became elongated. The region near the animal pole, near the blastopore, invaginated for a short distance, forming the stomodaeum (Fig. 5.52O, arrow). At the same time, the formation of another pore was often observed in one side of the developing planulae (Fig. 5.52O, circle). Most of the embryos of *A. tenuis* and *A. digitifera* formed this pore, around which some ectodermal cells accumulated. A concavity at one side was observed externally, and in the other species of *Acropora*. There was no explanation or significance for these observations known. At 86 h, the larva elongated further, the formation of the stomodaeum completed and the inner cells became arranged along the inside of the mesoglea. This was the first sign of an endodermal cell layer and at 98 h, the aboral region became thicker and slightly concave. Spirocysts appeared only in the aboral region. At 170 h, the concave region in the aboral region became more conspicuous, and numerous spirocysts were observed; in contrast, other regions of the planula had few spirocysts. At this stage, numerous mucous cells were observed. At 218 h, mesenterial filaments appeared to form via continuous invagination of ectodermal cells from the oral region. Around this stage, the larvae frequently touched the substrate with their concave aboral end and then stretched out onto the substrate.

Before the above study, in 2002 Ball et al. studied the embryological development of *Acropora millepora* to investigate the molecular control of this vital activity. They reviewed the descriptive embryology of this species, and provided a diagrammatic summary of the major morphological stages of embryonic development of *A. millepora* (Fig. 5.52P). They then reviewed their recent work on cnidarian Pax genes, nuclear receptor genes and Hox-like genes. They restudied and selected transcription factor gene families in *A. millepora*. They selected *A. millepora* because their knowledge about *Acropora* is particularly advanced relative to other cnidarians. They discovered and sequenced four Pax genes, nuclear receptor genes, and Hox-like and Other Homeobox Genes in the coral *Acropora*.

Studies of cnidarian development and its molecular control will provide information about the origins of the major bilaterian body axes, the origin of the third tissue layer, the mesoderm and the evolution of nervous system

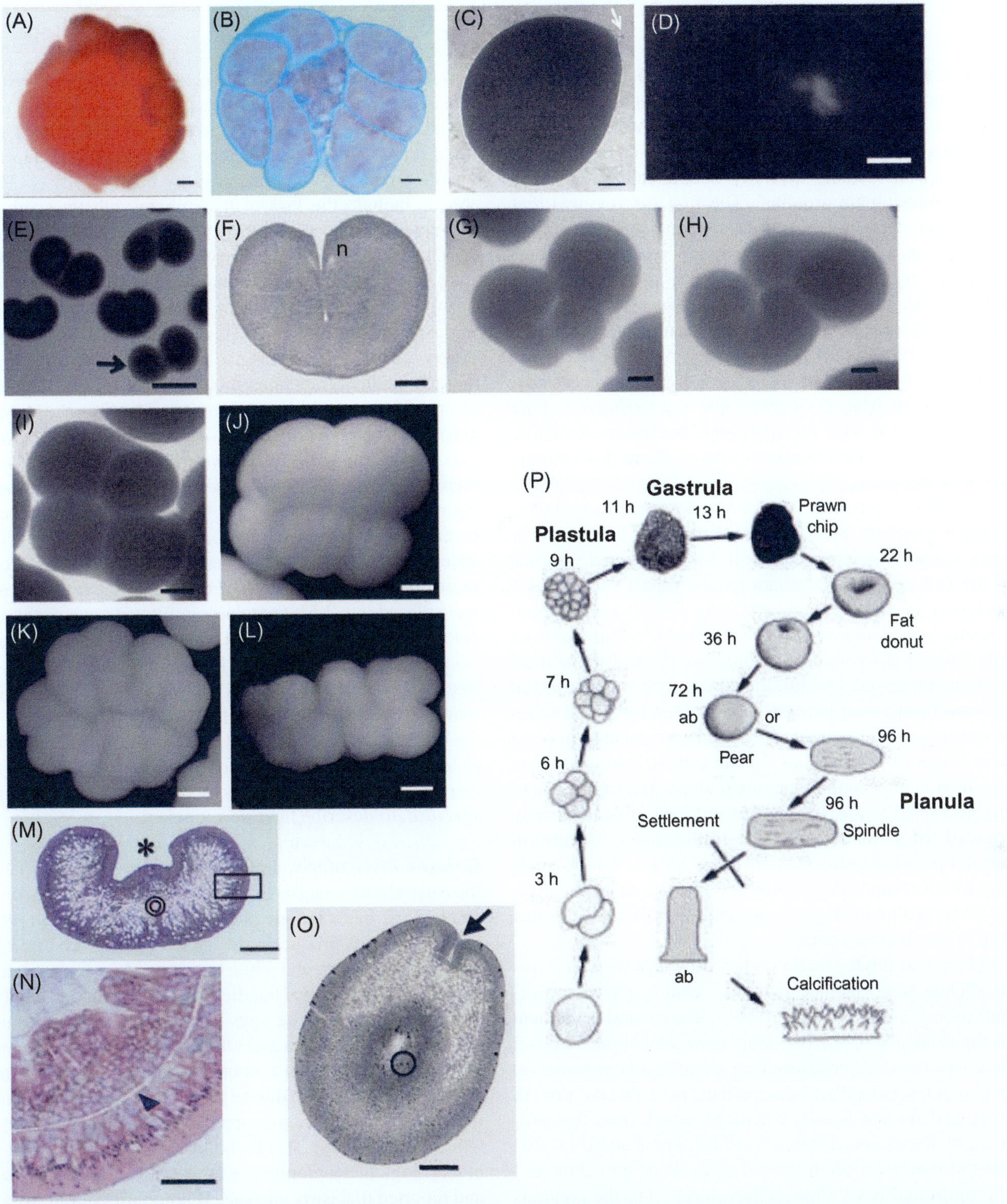

FIG. 5.52 Embryological development in *Acropora tenuis.* (A and B) Egg-sperm bundle released, 0h: (A) external view just after spawning; (B) cross-sectional view, eggs cohered via sulfated muco-substances (stained by Alcian blue (pH 1.0)) and sperm were located in the central cavity. Cells are stained with Alcian blue (pH 1.0) and Kernechrot stain. Bar = 100 μm. (See wide view of spawning of this species Fig. 5.50C); (C) external view of the egg and polar body *(white arrow)* (30 min). Fertilization may or may not have occurred, bar 100 μm; (D) DAPI staining of the polar body after its magnification, bar = 5 μm; (E and F) early cleavage, (E) two-blastomere stage (2 h). An asymmetrically cleaved zygote is indicated *(black arrow)*, bar = 50 μm; (F) T.S. of the 2-blastomere stage zygote, bar = 100 μm; (G) twisting blastomeres during the second cleavage, bar = 100 μm; (H) cleavage initiated at the animal pole (AP), bar = 100 μm; (I) four-blastomere stage (3 h), bar = 100 μm; (J) eight-blastomere stage (4 h), bar = 100 μm; (K) 16-blastomere stage formed by meridional division (5 h), bar = 100 μm; (L) side view of the 16-blastomere stage, bar = 100 μm; (M) gastrula showing excretion of the yolk granules into the blastocoel and gradual thickening of the blastula (18–20 h), bar = 100 μm; (N) the inner cells and yolk granules are excreted into the blastocoel, *headarrow* indicates the mesoglea, bar = 100 μm; (O) planula larva shows formation of the stomodaeum via invagination (*thick arrow*, 54 h), bar = 100 μm. *From Okubo and Motokawa (2007)*; (P) diagrammatic summary of the major morphological stages in the embryonic development of *A. millepora*. *(From Ball et al., 2002.)*

patterning. These studies gave rise to one clear message that a surprising diversity of genes has been identified in *Acropora*, at least four genes encoding Pax proteins, 10 encoding nuclear receptors and a range of other genes encoding a considerable diversity of 'Antennapedia superclass homeodomain proteins'. The first surprise was that, despite the fact that the Cnidaria are equally diverged from the chordate, nematode and arthropod lineages, coral genes frequently match significantly better to their chordate orthologs than to the *Drosophila* or *Caenorhabditis* gene. The second surprise was that a number of genes previously thought to be vertebrate-specific are present in *Acropora*. Genes that have been characterized in vertebrates but are clearly absent from *Drosophila* and *Caenorhabditis* have been assumed to have recent origins, but the coral data indicate that there are a number of cases where this assumption was incorrect.

Another work (Okubo et al., 2016) described early embryology of *Acropora millepora* (Fig. 5.49M and M′) the previously characterized complex coral, for comparison with that of the two previously uncharacterized robust corals, *Favia lizardensis* (a massive brain coral) (Fig. 5.49H) and *Ctenactis echinata* (a solitary coral) (Fig. 5.49L and L′). This study not only adds to the embryology, but also allowed the authors to carry out, in situ, hybridizations revealing both conserved and novel patterns of gene expression. Developmental series of images of whole mounts and sections revealed that the early stages of both species contain a blastocoel, consistent with their membership of the robust clade. The expression of *brachyury* and *forkhead* was consistent with that previously reported for *Acropora*, and allowed the authors to confirm that the pseudo-blastopore sometimes seen in robust corals such as *Favia* spp. is not directly associated with gastrulation. *C. echinata* chordin expression, however, differed from that seen in the other two corals.

Molecular phylogenetic studies show that the stony corals (Scleractinia) include two large clades: the 'complex' and 'robust' corals. There seems to be a general agreement on the existence of the two large clades mentioned above; the robust (so-called because they are plate-like or massive) and complex (so-called because they have diverse growth forms and are less heavily calcified), which were first proposed by Romano and Palumbi (1996). These authors stated that the molecular phylogenetic analysis of mitochondrial 16S ribosomal RNA showed departures from the previous hypotheses of coral evolution. These fossil and molecular data suggest multiple origins of scleractinians skeleton.

Nevertheless, another school of Okubo in 2013 described the embryonic development of four 'complex' species: *Pseudosiderastrea tayamai*, *Galaxea fascicularis*, *Montipora hispida* and *Pavona decussata*, and seven 'robust' species: *Oulastrea crispata*, *Platygyra contorta*, *Favites abdita*, *Echinophyllia aspera*, *Goniastrea favulus*, *Dipsastraea speciose* (previously *Favia speciosa*) and *Phymastrea valenciennesi* (previously *Montastrea valenciennesi*). In these authors' view, a comprehensive understanding of coral reproduction and development is needed because corals are threatened in many ways by human activity. Major threats include the loss of their photosynthetic symbionts (Symbiodinium) caused by rising temperatures (bleaching), reduced ability to calcify caused by ocean acidification, increased storm severity associated with global climate change and an increase in predators caused by runoff from human agricultural activity. However, detailed description of embryonic development is not available for many coral species. The authors obtained their material from different regions of Japanese and/or Australia waters. In some species, colonies were brought to the laboratory, where spawning occurred in 'tubs'. Once spawning had occurred, gametes were stirred gently to mix the bundles and ensure insemination. In other species, gametes were collected by scuba divers from the surface of colonies as they spawned. These species are marked with an asterisk, and in all cases, three or more colonies were sampled. For all species, fertilized eggs were placed into 2.5 L containers in filtered seawater and development allowed to proceed. Gametes and the early stages of development were sampled and examined hourly for the first 24 h after spawning, and every 4 or 6 h thereafter. The materials were then prepared for histological studies.

The present work selected two animals that are familiar, at least for the author: one complex species, *Galaxea fascicularis*, and another robust species, *Dipsastraea (Favia) speciosa*, to describe their development.

Galaxea fascicularis. In Okinawa (Heyward et al., 1987), the Australian Great Barrier Reef (Harrison, 1988), *Galaxea fascicularis* has a unique pattern of reproduction, in which female colonies produce red eggs and hermaphroditic colonies produce sperm and white eggs (pseudo-eggs) which are spawned as egg-sperm bundles (Fig. 5.50G and H) and are more buoyant than true eggs because of their greater content of wax-ester (Mita and Okubo, unpublished). The eggs have a natural reddish colour and the red colour is more intense in Fig. 5.50G, due to the technical preparation. On the other hand, there are opposing views on whether the white eggs are natural. Okubo et al. (2013) described the embryological development of this species, *G. fascicularis*, and reported that early cleavage is holoblastic, in spite of the abundance of yolk granules. In the proceeding stages, after ~10 h, the embryo assumed a concave-convex dish shape (Fig. 5.53F and G). The embryo then gradually thickened and attained the spherical form enclosing a core packed with yolk-containing cells (Fig. 5.53J and K).

Dipsastraea (Favia) speciosa. The second animal to describe its development is ***Dipsastraea (Favia) speciosa***.

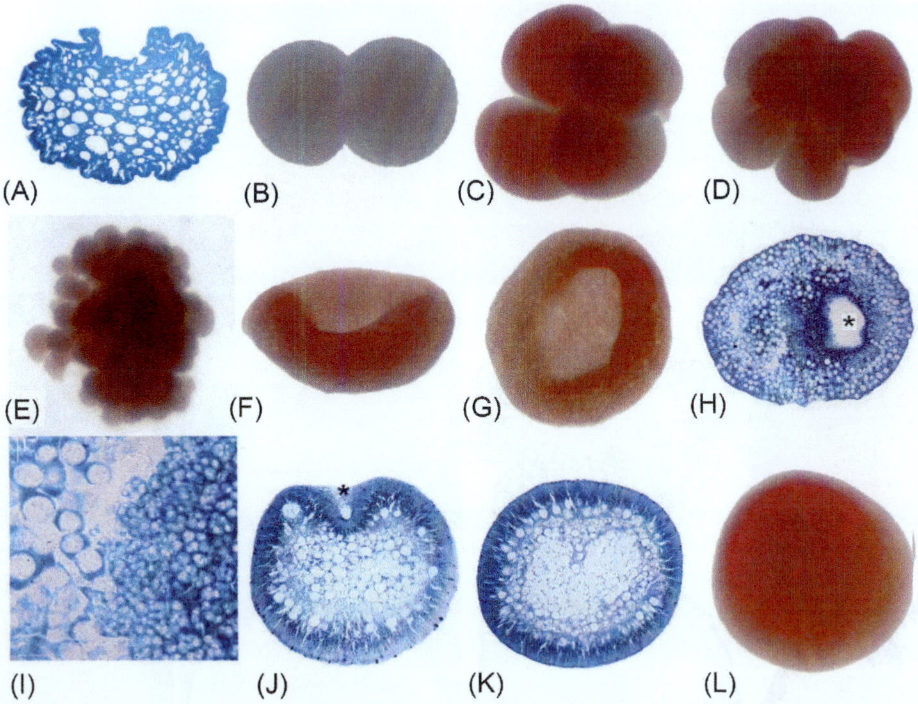

FIG. 5.53 Embryological development of *Galaxea fascicularis*: (A) pseudo-eggs appear vacuolated and the lipid is localized to much larger bodies; (B) 2-cell stage; (C) 4-cell stage; (D) 8–16 cell stage; (E) morula stage; (F) bowl stage (concave side up); (G) bowl stage (embryo shown in F-viewed from above); (H) glancing section of an embryo comparable to (F) and (G), the *asterisk* marks the blastopore; (I) magnification of a portion of the embryo shown in (H), with two sizes of lipid-containing bodies. It appears that the large arise by fusion of the small; (J) spheroidal embryo with closing blastopore showing the centre filled with lipid, the asterisk marks the blastopore; (K) mesoglea is gradually forming at this stage; (L) W.M. of stage similar to (K). Red pigment localized at the animal pole as the result of a marking experiment persists in C. *(Adapted from Okubo et al. (2013).)*

In addition, Okubo et al. (2013) described its embryology and the present work followed this description. A polar body (pb, Fig. 5.54A) started to appear in several eggs ~1 h after spawning while the eggs were still spheroidal. The egg then became spherical and the polar body disappeared. The first cleavage resulted in two equal blastomeres (Fig. 5.54B), and the second cleavage was observed ~1 h after the first (Fig. 5.54C and D). The cleavage proceeded and the embryo became hollow and flattened (Fig. 5.54E–K). It then became more spherical and attained a concavity, the pseudo-blastopore, on one side (Fig. 5.54L). This remained for 4 h (Fig. 5.54M and N) and then disappeared before the beginning of the invagination after 13 h (asterisk in Fig. 5.54J and subsequent figures). This cavity was established as irregularly shaped, and elongated (Fig. 5.54K) and then became round (Fig. 5.54L), while the embryo began to swim in the upper or intermediate water layer of the container. It was held at about 14–17 h, while the cavity was still opening. The pore then gradually reduced in size (Fig. 5.54M) and it appears that endoderm formation occurred by unipolar ingression (Fig. 5.54N). The pore eventually became the mouth of the planula larva. After ~24 h, some of the planulae began swimming towards the bottom of the bowl, assuming a pear (Fig. 5.54O) or barrel shape (Fig. 5.54P) after 39 h.

It is well-known that most corals spawn at night, and in some cases only once per year, so descriptions of the development of many species are still fragmentary. In addition, Okubo et al. (2013) reported that the descriptive developmental data available for corals are quite limited, considering the number of species and their morphological diversity. Their study was to provide information on the development of a number of species, additional to those that have previously been described. Secondly, the study aimed to see whether the prawn chip stage typical of the *Acropora* species that they had previously studied was a characteristic feature of the development of complex corals. They found that this was not the case, since the genus *Pavona*, which is listed among the complex corals in all recent phylogenies, lacks a prawn chip and has a well-developed blastocoel comparable to that of the robust corals.

Fig. 5.55 summarizes the divergent patterns of development seen in complex corals (as exemplified by *Acropora*) and robust corals (as exemplified by *Dipsastraea (Favia) speciosa*), as well as illustrating some of the descriptive terms used in their text.

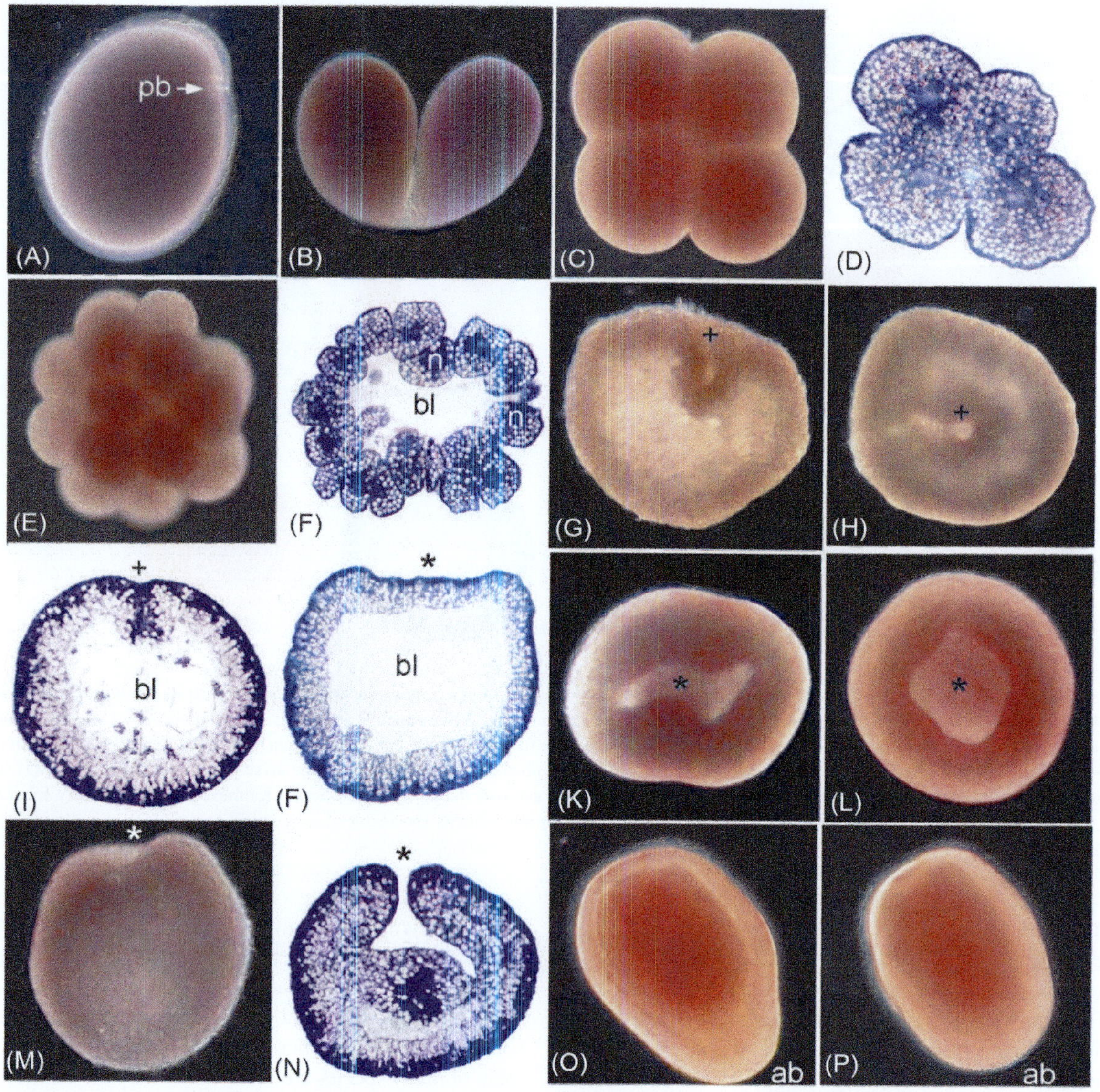

FIG. 5.54 Development of *Dipsastraea (Favia) speciosa*: (A) spawned egg, showing polar body (pb); (B) 2-cell stage, the blastomeres are almost fully separated; (C–D) 4-cell stage. The cells are tightly adherent; (E) 16-cell stage; (F) section of 16-cell stage with blastocoel (bl in this and subsequent photos). Nuclei (n) occupy a substantial portion of the volume of each cell; (G) the pseudo-blastopore has appeared *(plus sign)*; (H) the embryo has now become spherical with a smooth surface except for the invagination associated with the pseudo-blastopore; (I) section of spherical embryo with the pseudo-blastopore, within the periphery of the embryo is a layer of lipid globules. The blastocoel is starting to fill; (J) this section reveals a flattening where invagination of the blastopore is about to begin *(asterisk)*; (K) invagination of the blastopore has now started; (L) the blastopore has first become circular and then started to shrink in diameter; (M) the embryo has now begun to elongate in the oral-aboral axis; (N) this section shows the blastopore closing and two germ layers being formed; (O) a pear shaped planula with aboral end (ab in this and the next figure) becoming thicker; (P) barrel-shaped planula. *(Adapted from Okubo et al. (2013).)*

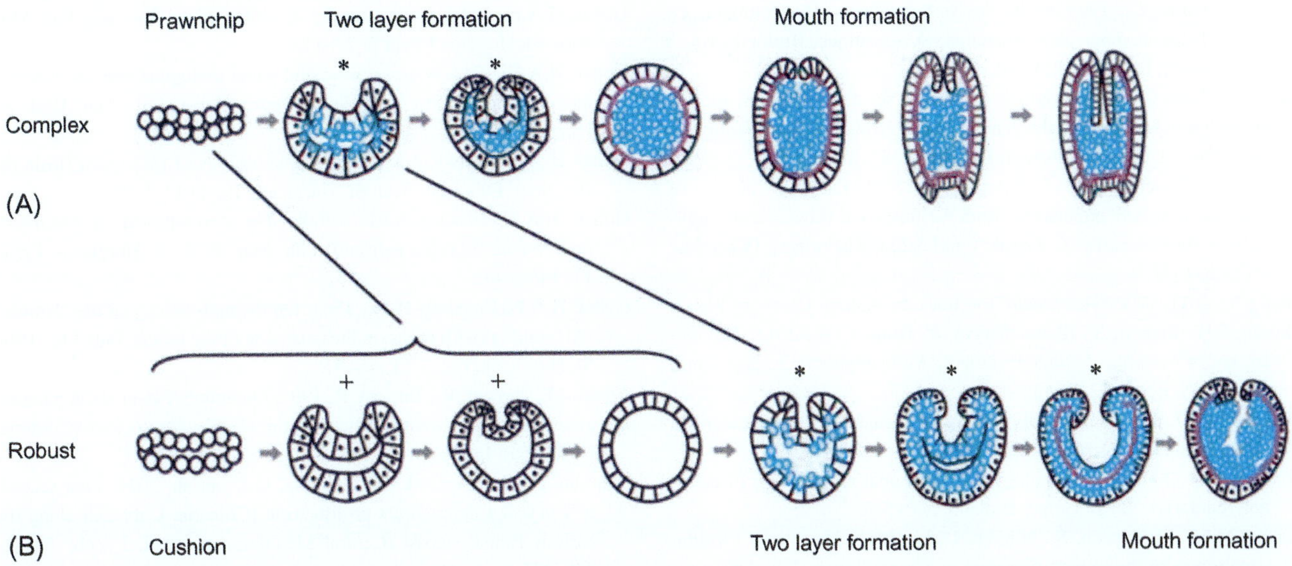

FIG. 5.55 Diagrammatic representation of the two extreme forms of coral development. (A) Early in their development *Acropora* spp. embryos pass through a prawn chip stage consisting of an extended cellular bilayer lacking a blastocoel. (B) which is that seen in robust corals such as *Dipsastraea (Favia) speciosa*. Rather than a spatially extended prawn chip lacking a blastocoel, these corals pass through a cushion stage, which is flattened but always retains a blastocoel. This cushion then rounds up and develops a depression in its side which we have termed a 'pseudo-blastopore', but no material enters the blastocoel.

REFERENCES

Achituv, Y., Benayahu, Y., 1990. Polyp dimorphism and functional, sequential hermaphroditism in the soft coral *Heteroxenia fuscescens* (Octocorallia). Mar. Ecol. Prog. Ser. 54, 263–269.

Afzelius, B.A., 1971. Fine structure of the spermatozoon of *Tubularia larynx* (Hydrozoa, Coelentrata). J. Ultrastruct. Res. 37, 679–689.

Balfour, F.M., 1880. A Treatise on Comparative Embryology. vol. I. Macmillan and Co, London.

Ball, E.E., Hayward, D.C., Reece-Hoyes, J.S., et al., 2002. Coral development: from classical embryology to molecular control. Int. J. Dev. Biol. 46, 671–678.

Barnes, R.D., 1974. Invertebrate Zoology, third ed. W.H. Saunders Company, London.

Bayer, F.M., 1961. Summary of black coral investigations to date. Hawaiian Shell News 9 (12), 8.

Bayha, K.M., Graham, W.M., 2011. First confirmed reports of the rhizostome jellyfish *Mastigias* (Cnidaria: Rhizostomeae) in the Atlantic basin. Aquat. Invasions 6 (3), 361–366.

Benayahu, Y., 1989. Reproductive cycle and developmental processes during embryogenesis of *Clavularia hamra* (Cnidaria, Octocorallia). Acta Zool. (Stockholm) 70 (I), 29–36. Printed in Great Britain.

Benayahu, Y., Achituv, Y., Berner, T., 1988. Embryogenesis and acquisition of algal symbionts by planulae of *Xenia umbellata* (Octocorallia; Alcyonacea). Mar. Biol. 100 (1), 93–101.

Benayahu, Y., Berner, T., Achituv, Y., 1989. Development of planulae within a mesogleal coat in the soft coral *Heteroxenia fuscescens*. Mar. Biol. 100, 203–210.

Bentlage, B., Cartwright, P., Yanagihara, A.A., et al., 2010. Evolution of box jellyfish (Cnidaria: Cubozoa), a group of highly toxic invertebrates. Proc. R. Soc. Lond. B Biol. Sci. https://doi.org/10.1098/rspb.2009.1707.

Bigelow, A.P., 1900. The anatomy and development of *Cassiopea xamachana*. Mem. Boston Soc. Nat. Hist. 5, 193–236 (cited in Hofmann, 1996).

Borradaile, L.A., Potts, F.A., with chapters by Eastham, L.E., Saunders, J.T., 1977. The Invertebrata, fourth ed. Revised by Kerkut, G.A.Cambridge University Press.

Calder, D.R., Peters, E.C., 1975. Nematocysts of Chiropsalmus quadrumanus with comments on the systematic status of the Cubomedusae. Helgoländer Meeresun. 27 (3), 364.

Campbell, R.D., 1974. Cnidaria in Giese and Pearse: Reproduction of Marine Invertebrates. vol. 1. Academic Press, Inc.

Campbell, R.D., 1989. Taxonomy of the European *Hydra* (Cnidaria. Hydrozoa). A reexamination of its history with emphasis on the species *Hydra vulgaris* (Pall.), *Hydra attenuate* Pallas and *Hydra circumcincta* Schulze. J. Linn. Soc. Lond. 95, 219–244.

Carré, D., 1967. Etude du développement larvaire de deux siphonophores: *Lensia conoide* (Calycophore) et *Forskalia edwardsi* (Physonecte). Cah. Biol. Mar. 8, 233–251.

Carré, D., 1969. Etude histologique du dévelopment de *Nanomia bijuga*. Cah. Biol. Mar. 10, 325–341.

Carré, D., 1984. Existence d'un complexe acrosomal chez les spermatozoïdes du cnidaire Muggiaea kochi (Siphonophore, Calycophore): différenciation et réaction acrosomale/Existence of an acrosomal complex in the spermatozoids of the cnidarian *Muggiaea kochi* (Siphonophora, Calycophora): differentiation and acrosomal reaction. Int. J. Invertebr. Reprod. Dev. 7 (2), 95–103.

Carré, D., Carré, C., 1990. Complex reproductive cycle in *Eucheilota paradoxica* (Hydrozoa: Leptomedusae): medusae, polyps and frustules produced from medusa stage. Mar. Biol. 104, 303–310.

Carré, C., Carré, D., 1993. Ordre des Siphonophores. In: Doumenc, D. (Ed.), Traité de Zoologies: Anatomie, Systematique, Biologie. vol. 3 (2). Masson, Paris, pp. 523–596.

Carré, D., Carré, C., 1995. Ordre des siphonophores. In: Doumenc, D. (Ed.), Traité de Zoologies: Anatomia, Systematique, Biologie. vol. 3 (2). Masson, Paris, pp. 523–596.

Carré, D., Carré, C., 2000. Origin of germ cells, sex determination, and sex inversion in Medusae of the genus *Clytia* (Hydrozoa, Leptomedusae): the influence of temperature. J. Exp. Biol. 287, 233–242.

Carrette, T., Straehler-Pohl, I., Seymour, J., 2014. Early life history of *Alatina cf. moseri* populations from Australia and Hawaii with implications for taxonomy (Cubozoa: Carybdeida, Alatinidae). PLoS One 9 (1), e84377.

Chang, Y., 2011. '*Mastigias papua*' (on-line). In: Animal Diversity Web.

Church, S.H., Siebert, S., Bhattacharyya, P., Dunn, C.W., 2015. The histology of *Nanomia bijuga* (Hydrozoa: Siphonophora). J. Exp. Zool. (Mol. Dev. Evol.) 324B, 435–449.

Claus, C., 1877. Studien uber Polypen und Quallen der Adria. I. Acalephen (Discomedusen). Denkschr. Akad. Wiss. Wien. 38, 1–64.

Coates, M.M., 2003. Visual ecology and functional morphology of cubozoa (cnidaria). Integr. Comp. Biol. 43, 542–548.

Coates, M.M., Thompson, S., Nilsson, D.-E., 2001. The image forming capabilities of the camera-type eyes of a box jellyfish, *Tripedalia cystophora*. Am. Zool. 41, 1640.

Collins, A.G., Daly, M., 2005. A new deepwater species of Stauromedusae, *Lucernaria janetae* (Cnidaria, Staurozoa, Lucernariidae), and a preliminary investigation of Stauromedusan phylogeny based on nuclear and mitochondrial rDNA data. Biol. Bull. 208, 221–230.

Delap, M.J., 1905. Notes on the rearing, in an aquarium, of Aurelia aurita, L. and Pelagia perla (Slabber). Fisheries, Ireland, Sci. Invest. 7, 22–26.

Dudley, M.J., 2016. Biology, Ecology and Ecophysiology of the Box Jellyfish *Carybdea marsupialis* (Cnidaria: Cubozoa). Universitat Politècnica de Catalunya.

Dunn, C.W., 2005. Complex colony-level organization of the deep-sea siphonophore *Bargmannia elongata* (Cnidaria, Hydrozoa) is directionally asymmetric and arises by the subdivision of pro-buds. Dev. Dyn. 234, 835–845.

Dunn, C.W., Wagner, G.P., 2006. The evolution of colony-level development in the Siphonophora (Cnidaria: Hydrozoa). Dev. Genes Evol. 216, 743–754.

Dunn, C.W., Pugh, P.R., Haddock, S.H., 2005. Molecular phylogenetics of the siphonophora (Cnidaria), with implications for the evolution of functional specialization. Syst. Biol. 54 (6), 916–935.

Egloso, N.L., Urboda, E.M.V., 2017. Utilization of spotted jellyfish (*Mastigias Papua*) for gelatin production in Palompon, Leyte Phils: basis for product commercialization. In: International Conference on Chemical, Agricultural, Biological and Medical Sciences (CABMS-17) Jan. 23–24, Manila (Philippines).

Farrant, P.A., 1986. Gonad development and the planulae of the temperate Australian soft coral *Capnella gaboensis*. Mar. Biol. 92, 381–392.

Fritzenwanker, J.H., Genikhovich, G., Kraus, Y., Technau, U., 2007. Early development and axis specification in the sea anemone *Nematostella vectensis*. Dev. Biol. 310, 264–279.

Gaino, E., Scoccia, F., 2010. Ultrastructural investigation of the sexual reproduction of some members of the black coral: a review. In: Méndez-Vilas, A., Díaz, J. (Eds.), Microscopy: Science, Technology, Applications and Education. Formatex, pp. 936–945.

Gardiner, M.S., 1972. The Biology of Invertebrates. In: McGraw-Hill Book Company. London, New York.

Garm, A., Lebouvier, M., Tolunay, D., 2015. Mating in the box jellyfish *Copula sivickisi* – novel function of cnidocytes. J. Morphol. 276 (9), 1055–1064.

Gohar, H.A.F., 1940. Studies on the Xeniidae of the Red Sea. Pub. Mar. Biol. St. Hurgada Egypt 2, 27–118.

Gohar, H.A.F., 1948. A description and some biological studies of a new alcyonacean species *Clavularia hamra* Gohar. Pub. Mar. Biol. St. Hurgada, Egypt 6, 3–33.

Gohar, H.A.F., Eisawy, A.M., 1960a. The biology of *Cassiopea* (from the Red Sea). Pub. Mar. Biol. St. Hurghada, Egypt 11, 5–42.

Gohar, H.A.F., Eisawy, A.M., 1960b. The development of *Cassiopea andromeda* (Scyphomedusae). Pub. Mar. Biol. St. Hurghada, Egypt 11, 148–190.

Gohar, H.A.F., Roushdy, H.M., 1961. On the embryology of the Xeniidae (Alcyonaria) with notes on the extrusion of the larvae. Pub. Mar. Biol. St. Hurghada, Egypt 11, 45–70.

Grassi, M., Tardent, R., Tardent, P., 1995. Quantitative data about gametogenesis and embryonic development in *Hydra vulgaris* Pall. (Cnidaria, Hydrozoa). Invertebr. Reprod. Dev. 27 (3), 219–232.

Gueroun, S.K.M., Acevedo, M.J., Yahia, O.K., et al., 2015. First records of *Carybdea marsupialis* proliferation (Cnidaria: Cubozoa) along the eastern Tunisian coast (Central Mediterranean). Ital. J. Zool. 82 (3), 430–435.

Gurska, D., Garm, A., 2014. Cell proliferation in Cubozoan jellyfish *Tripedalia cystophora* and *Alatina moseri*. PLoS One 9 (7), 1–11. Open access, freely available online.

Haddock, S.H., Dunn, C.W., Pugh, P.R., 2005. A re-examination of siphonophore terminology and morphology, applied to the description of two new prayine species with remarkable biooptical properties. J. Mar. Biol. Assoc. UK 85, 695–707.

Hale, G., 1999. The classification and distribution of the class Scyphozoa. In: BI 375-Biological Diversity. University of Oregon. Available at http://gladstone.uoregon.edu/~ghale/pdf/scyphozoa.pdf (Accessed May 2, 2011).

Hand, C., Uhlinger, K.R., 1992. The culture, sexual and asexual reproduction, and growth of the sea anemone *Nematostella vectensis*. Biol. Bull. 182, 169–176.

Harrison, P.L., 1988. Pseudo-gynodioecy: an unusual breeding system in the scleractinian coral *Galaxea fascicularis*. In: Proceedings of the 6th International Coral Reef Symposium, vol. 2, Townsville, pp. 699–705.

Harrison, P.L., 2011. Sexual Reproduction of Scleractinian Corals. Springer Science+Business Media B.V.

Heyward, A., Yamazato, K., Yeeman, T., Minei, M., 1987. Sexual reproduction of corals in Okinawa. Galaxea 6, 331–343.

Hickman, C.P., 1973. Biology of the Invertebrates, second ed. The C.V. Mosley Company, Saint Louis.

Hofmann, D.K., Crow, G., 2002. Induction of larval metamorphosis in the tropical scyphozoan *Mastigias papua*. Striking similarities with upside down-jellyfish *Cassiopea* pps. (with notes on related species). Vie Milieu. 52 (4), 141–147.

Hofmann, D.K., Hellmann, M., 1995. Studies in the reef-dwelling cnidarian *Cassiopea* spp.: RF amide positive elements of the nervous system at different stages of development. Beitr. Paläontol. 20, 21–29.

Hofmann, D.K., Honegger, T.G., 1990. Bud formation and metamorphosis in *Cassiopea Andromeda* (Cnidaria: Scyphozoa): a developmental and ultrastructural study. Mar. Biol. 105, 509–518.

Hofmann, D.K., Neumann, R., Henne, K., 1978. Strobilation, budding and initiation of scyphistoma morphogenesis in the rhizostome *Cassiopea andromeda* (Cnidaria: Scyphozoa). Mar. Biol. 47, 161–176.

Hofmann, D.K., Fitt, W.K., Fleck, J., 1996. Checkpoints in the life-cycle of *Cassiopea* spp., control of metagenesis and metamorphosis in a tropical jellyfish. Int. J. Dev. Biol. 40, 331–338.

Honegger, T.G., 1981. Light and scanning electron microscopic investigations of sexual reproduction in *Hydra carnea*. Int. J. Invertebr. 3 (5), 245–255.

Honegger, T.G., 1983. Ultrastructural and experimental investigations of sperm-egg interactions in fertilization of *Hydra carnea*. Roux's Arch. Dev. Biol. 192, 13–20.

Honegger, T.G., Zurrer, D., Tardent, P., 1989. Oogenesis in Hydra carnea: a new model based on light and electron microscopic analyses of oocyte and nurse cell differentiation. Tissue Cell 21 (3), 381–393. Longman Group UK Ltd.

Hummelinck, P.W., 1968. Caribbean scyphomedusae of the genus Cassiopea. In: Studies on the Fauna of Curacao and Other Caribbean Islands. vol. 25 (1). M. Nijhoff, Hague, pp. 1–57.

Hyman, L.H., 1940. The Invertebrates, Protozoa Through Ctenophora, first ed. McGraw-Hill Book Company, Inc., New York, London. Third expression.

Invertébrés: Beauté du Monde Animal, Tome XII, 1974. Librairie Larousse. Paris.

Jarms, G., Morandini, A.C., da Silveira, F.L., 2002. Cultivation of polyps and medusae of Coronatae (Cnidaria, Scyphozoa) with a brief review of important characters. Helgol. Mar. Res. 56 (3), 203–210.

Kenyon, J.C., 1992. Sexual reproduction in Hawaiian *Acropora*. Coral Reefs 11, 37–43.

Kikinger, R., 1992. *Cotylorhiza tuberculate* (Cnidaria: Scyphozoa) – life history of a stationary population. Mar. Ecol. 13 (4), 333–362.

Kikinger, R., von Salvini-Plawen, L., 1995. Development from polyp to stauromedusa in *Stylocoronella* (Cnidaria: Scyphozoa). J. Mar. Biol. Äss. U.K 899–912.

Komai, T., 1935. On *Stephanoscyphus* and *Nausithoë*. Mem. Sci. Kyoto Imp. Univ. Ser. B 10, 289–339.

Komai, T., 1936. On *Stephanoscyphus*. Zool. Mag. 48, 535–544 (in Japanese).

Korotneff, A., 2009. On the embryology of Hydra. Ann. Mag. Nat. Hist. Ser. 5 (published online).

Kume, H., Dan, K., 1968. Invertebrate Embryology. NOLIT Publishing House, Belgrade, Yugoslavia. Translated by Dan. J.C.

Lawley, J.W., Ames, C.L., Bentlage, B., et al., 2016. Box jellyfish *Alatina alata* has a circumtropical distribution. Biol. Bull. 231 (2), 152–169.

Layden, M.J., Rentzsch, F., Röttinger, E., 2016. The rise of the starlet sea anemone *Nematostella vectensis* as a model system to investigate development and regeneration. Wiley Interdiscip. Rev. Dev. Biol. 5 (4), 408–428.

Leitz, T., 1993. Biochemical and cytological bases of metamorphosis in *Hydractinia ehinata*. Mar. Biol. 116, 559–564.

Leitz, T., Wagner, T., 1993. The marine bacterium Alteromonas espejiana induces metamorphosis of the hydroid Hydractinia echinata. Mar. Biol. 115 (2), 173–178.

Lewis, C., Long, T.A.F., 2005. Courtship and reproduction in *Carybdea sivickisi* (Cnidaria: Cubozoa). Mar. Biol. 147, 477–483.

Magie, C.R., Daly, M., Martindale, M.Q., 2007. Gastrulation in the cnidarian *Nematostella vectensis* occurs via invagination not ingression. Dev. Biol. 305, 483–497.

Mann, T., 1984. Spermatophores. Springer-Verlag, Berlin, New York.

Marques, A.C., García, J., Ames, C.L., 2015. Internal fertilization and sperm storage in cnidarians: a response to Orr and Brennan. Trends Ecol. Evol. 30, 435–436.

Martin, V.J., 1997. Cnidarians, the jellyfish and hydrans. In: Gilbert, S.F., Raunio, A.M. (Eds.), Embryology, Constructing the Organism. Sinauer Associates, Inc. Publishers.

Martin, V.J., Archer, W.E., 1986. Migration of interstitial cells and their derivatives in a Hydrozoan Planula. Dev. Biol. 116, 486–496.

Martin, V.J., Littlefield, C.L., Archer, W.E., Bode, H.R., 1997. Embryogenesis in hydra. Biol. Bult. 192, 345–363.

Martindale, M.Q., Heather, T.A., Marlow, Q., 2009. Anthozoan Cnidarian development: the Starlet sea anemone, *Nematostella vectensis*. In: Woods Hole Embryology.

Marshall, A.J., Williams, W.D., 1975. A Text Book Zoology: The Invertebrates. ELBS and Macmillan.

Miller, R.L., 1966a. Chemotaxis during fertilization in the hydroid *Campanularia*. J. Exp. Zool. 162, 23–44.

Miller, R.L., 1966b. Chemotaxis during fertilization in the hydroid *Tubularia* and *Gonothyrea*. Am. Zool. 6, 509 (Abstract).

Miller, K., 1996. Piecing together the reproductive habits of New Zealand's endemic black corals. Water Atmos 4, 18–19.

Miller, R.L., Brokaw, C.J., 1970. Chemotactic turning behavior of Tubularia spermatozoa. J. Exp. Biol. 52, 699–706.

Minchin, M.A., Fowler, B.A., Bourne, M.A., 1900. In: Lankester, E.R. (Ed.), A Treatise on Zoology. Part II: The Porifera and Coelen. Adamand Charles Black, London.

Miranda, L.S., Hirano, Y.M., Mills, C.E., Falconer, A., Fenwick, D., Marques, A.C., Collins, A.G., 2016. Systematics of stalked jellyfishes (Cnidaria: Staurozoa). Peer J. 4, e1951.

Miranda, L.S., Mills, C.E., Hirano, Y.M., Collins, A.G., Marques, A.C., 2017. A review of the global diversity and natural history of stalked jellyfishes (Cnidaria, Staurozoa). Mar. Biodivers. https://doi.org/10.1007/s12526-017-0721-4.

Morandini, A.C., Silveira, F.L., 2001. Sexual reproduction of *Nausithoe aurea* (Scyphozoa, Coronatae). Gametogenesis, egg release, embryonic development, and gastrulation. Sci. Mar. 65 (2), 139–149.

Morandini, A.C., Silveira, F.L., Jarms, G., 2004. The life cycle of Chrysaora lactea Eschscholtz, 1829 (Cnidaria, Scyphozoa) with notes on the scyphistoma stage of three other species. Hydrobiologia 530/531, 347–354.

Neumann, R., 1977. Polyp morphogenesis in a scyphozoan: evidence for a head inhibitor from the presumptive foot end in vegetative buds of *Cassiopeia Andromeda*. Wilhelm Roux's Arch. Dev. Biol. 183, 79–83.

Neumann, R., 1979. Bacterial induction of settlement and metamorphosis in the planula larvae of *Cassiopea andromeda* (Cnidaria: Scyphozoa, Rhizostomeae). Mar. Ecol. Prog. Ser. 1, 21–28.

Neumann, R., Schmahl, G., Hofmann, D.K., 1980. Bud formation and control of polyp morphogenesis in *Cassiopea andromeda*. In: Tardent, P., Tardent, R. (Eds.), Developmental and Cellular Biology of Coelenterates. Elsevier/North-Holland Biomedical Press, Amsterdam, pp. 217–223. Oceana the Corals of the Mediterranean.

Okada, Y.K., 1927. Note sur l'pontogenic de *Charybdea rastonii* Haacke. Bull. Biol. Fr. Belg. 61, 148–241.

Okubo, N., Motokawa, T., 2007. Embryogenesis in the reef-building coral *Acropora* spp. Zool. Soc. Jpn. 24, 1169–1177.

Okubo, N., Mezaki, T., Nozawa, Y., Nakano, Y., Lien, Y.-T., et al., 2013. Comparative embryology of eleven species of stony corals (Scleractinia). PLoS One 8 (12), e84115.

Okubo, N., Hayward, D.C., Forêt, S., Ball, E.E., 2016. A comparative view of early development in the corals *Favia lizardensis*, *Ctenactis echinata*, and *Acropora millepora* – morphology, transcriptome, and developmental gene expression. BMC Evol. Biol. 16, 48.

Pallas, P.S., 1766. Elenchus zoophytorum sistens generum adumbrationes generaliores et specierum cognitarum succinctas descriptiones cum selectis auctorum synonymis. Frankfurt am Main, Hagae-Comitum, p. 451.

Piatigorsky, J., Kozmik, Z., 2004. Cubozoan jellyfish: an Evo/Devo model for eyes and other sensory systems. Int. J. Dev. Biol. 48, 719–729.

Piraino, S., Aglieri, G., Martell, L., Mazzoldi, C., Melli, V., Milisenda, G., … Boero, F., 2014. *Pelagia benovici* sp. nov. (Cnidaria, Scyphozoa): a new jellyfish in the Mediterranean Sea. Zootaxa 3794 (3), 455–468.

Postelnicu, T., Rohlf, F.J., Sokal, R.R., 1970. Statistical tables. WH Freeman & Comp., San Francisco 1969. XI+ 253 S., 5 Abb., Preis 641—. Biom. J. 12 (3), 192–193.

Quintanilla, E., Gili, J.M., López-González, P.J., 2013. Sexual reproductive cycle of the epibiotic soft coral *Alcyonium coralloides* (Octocorallia, Alcyonacea). Aquat. Biol. 18, 113–124.

Reverberi, G., 1971. Experimental Embryology of Marine and Fresh-Water Invertebrates. North-Holland Publishing Company. Amsterdam, London.

Romano, S.L., Palumbi, S.R., 1996. Evolution of scleractinian corals inferred from molecular systematics. Science 271, 640–642.

Russell-Hunter, W.D., 1979. A Life of Invertebrates. Macmillan Publishing Co., Inc, New York.

Sars, M., 1846. Ueber Arachnactis albida, einen schwimmenden Polypen. In: Sars, M. (Ed.), Fauna littoralis Norvegiae oder Beschreibung und Abbildungen neuer oder wenig bekannten Seethiere, nebst Beobachtungen über die Organisation. Leb-ensweise und Entwickelung derselben. Druck und Veblag von Johann Dahl, Christiana, pp. 28–30.

Sarras, M.P., Madden, M.E., Zhang, X., Gunwar, S., Huff, J.K., Hudson, B.G., 1991. Extra cellular matrix (mesoglea) of *Hydra vulgaris*. Dev. Biol. 148 (2), 481–494.

Schiariti, A., Kawahara, M., Uye, S., Mianzan, H.W., 2008. Life cycle of the jellyfish *Lychnorhiza lucerna* (Scyphozoa: Rhizostomeae). Mar. Biol. 156, 1–12.

Schuchert, P., 2008. The European athecate hydroids and their medusae (Hydrozoa, Cnidaria): Filifera Part 4. Rev. Suisse Zool. 115 (4), 677–757.

Seo, S.Y., Hwang, S.J., Song, J.I., 2008. Sexual reproduction of *Anthoplexaura dimorpha* (Gorgonacea: Octocorallia) from Munseom, Jejudo Islands, Korea. Anim. Cells Syst. 12, 231–240.

Smith, G., 1936. Contribution to the Anatomy and Physiology of *Cussiopea frondosa*. Pap. Tortugas Lab. vol. XXXI, pp. 18–52.

Sokal, R.R., Rohlf, F.J., 1969. Introduction to Biostatistics. Dover Publications, Inc., Mincola, NY.

Sommer, C., 1990. Post-embryonic larval development and metamorphosis of the hydroid *Eudendrium racemosum* (Cavolini) (Hydrozoa, Cnidaria). Helgoländer Meeresun. 44, 425–444.

Stiasny, G., 1922. Die Scyphomedusen-Sammlung von Dr. Th. Mortensen nebst anderen Medusen aus dem zoologischen Museum der Universitat in Kopenhagen. Vidensk Medd Dan Naturhist Foren Khobenhavn 77, 33–558.

Storer, T.I., Stebbins, R.C., Usinger, R.L., Nybakken, J.W., 1983. General Zoology, sixth ed. Tata McGraw-Hill Publishing Company Ltd, New Delhi.

Straehler-Pohl, I., Jarms, G., 2005. Life cycle of *Carybdea marsupialis* Linnaeus, 1758 (Cubozoa, Carybdeidae) reveals metamorphosis to be a modified strobilation. Mar. Biol. 147, 1271–1277.

Straehler-Pohl, I., Jarms, G., 2011. Morphology and life cycle of *Carybdea morandinii,* sp. nov. (Cnidaria), a cubozoan with zooxanthellae and peculiar polyp anatomy. Zootaxa 2755 (2), 36–56.

Sugiura, Y., 1963. On the life-history of Rhizostome medusae I. *Mastigias papua* L. Agassiz. Annot. Zool. Jpn. 36, 194–202.

Sugiura, Y., 1964. On the life-history of rhizostome medusae. Embryologia, 8 (3) 223–233. https://doi.org/10.1111/j.1440-169X.1964.tb00200.x (Accessed 02 May 2011).

Sugiura, Y., 1965. On the life-history of Rhizostome Medusae. III. On the effects of temperature on the strobilation of *Mastigias papua*. Biol. Bull. 128 (3), 493–496.

Szmant-Froelich, A., Yevich, P., Pilson, M.E.Q., 1980. Gametogenesis and early development of the temperate coral *Astrangia danae* (Anthozoa: Scleractinia). Biol. Bull. 158, 257–269.

Tardent, P., 1963. Regeneration in the Hydrozoa. Biol. Rev. 38, 293–333.

Technau, U., Steele, R.E., 2011. Evolutionary crossroads in developmental biology: Cnidaria. Development 138 (8), 1447–1458.

Toshino, S., Miyake, H., Ohtsuka, S., et al., 2013. Development and polyp formation of the giant box jellyfish *Morbakka virulenta* (Kishinouye, 1910) (Cnidaria: Cubozoa) collected from the Seto Inland Sea, western Japan. Plankton Benthos Res. **8** (1), 1–8.

Toshino, S., Miyake, H., Shibata, H., 2015. *Meteorona kishinouyei*, a new family, genus and species (Cnidaria, Cubozoa, Chirodropida) from Japanese waters. ZooKeys 503, 1–21.

Tsutsui, K., Hatada, Y., Tsuruwaka, Y., 2014. A new species of sea anemone (Anthozoa: Actiniaria) from the sea of Japan: *Cribrinopsis japonica* sp. nov. Plankton Benthos Res. 9 (4), 197–202.

Tsutsui, K., Shimada, E., Ogawa, T., Tsuruwaka, Y., 2016. A novel fluorescent protein from the deep-sea anemone *Cribrinopsis japonica* (Anthozoa: Actiniaria). Sci. Rep. 6, 23493. https://doi.org/10.1038/srep23493.

Uchida, T., 1926. The anatomy and development of rhizostome medusa, *Mastigias papua L.* Agassiz, with observations on the phylogeny of Rhizostomae. J. Pac. Imp. Univ. Tokyo, Sec. Zool. 1, 45–95.

Uchida, T., 1929. Studies on the Stauromedusae and Cubomedusae, with special reference to their metamorphosis. Jpn. J. Zool. 2, 103–193.

Uchida, T., 1970. Revision of Japanese Cubomedusae. vol. 17. Publications of the Seto Marine Biological Laboratory, pp. 289–297.

Uchida, T., Iwata, F., 1954. On the development of a broodcaring actinian. J. Fac. Sci. Hokkaido Univ. Scr. 6, Zool. 12, 220–224 (cited in Kumé and Dan, 1968).

Uchida, T., Yamada, M., 1968. Coelenterata, I- Cnidaria. In: Kumé, Dan (Eds.), Invertebrate Embryology. Reproduced by National Technical Information Service.

Villee, C.A., Walker, J.R.W.F., Smith, F.E., 1964. General Zoology. W.B. Saunders Company, London.

Wagner, D., 2011. The Biology and Ecology of Hawaiian Black Corals (CNIDARIA: ANTHOZOA: HEXACORALLIA: ANTIPATHARIA). (a dissertation submitted for the Degree of Doctor of Philosophy. Partially published).

Wagner, D., Luck, D.G., Toonen, R.J., 2012a. The biology and ecology of black corals (Cnidaria: Anthozoa: Hexacorallia: Antipatharia). Adv. Mar. Biol. 63, 67–132.

Wagner, D., Waller, R.G., Montgomery, A.D., Kelley, C.D., Toonen, R.J., 2012b. Sexual reproduction of the Hawaiian black coral *Antipathes griggi* (Cnidaria: Antipatharia). Communicated by Biology Editor Dr. Mark Vermeij, In: Coral Reefs. Springer-Verlag.

Walker, T.A., Bull, G.D., 1983. A newly discovered method of reproduction in gorgonian coral. Mar. Ecol. Prog. Ser. 12, 137–143.

Walther, M., Ulrich, R., Kroiher, M., Berking, S., 1996. Metamorphosis and pattern formation in *Hydractinia echinata*, a colonial hydroid. Int. J. Dev. Biol. 40, 313–322.

Werner, B., 1973. Spermatozeugmen und Paarungsverhalten bei *Tripedalia cystopohra* (Cubomedusae). Mar. Biol. 18, 212–217 (cited in Carrette et al., 2014).

Werner, B., 1975. Bau und Lebensgeschichte des pollypen von *Tripedalia cystophora* (Cubozoa, class nov. Carybdeidae) und seine Bedeutung für die Evolution der Cnidaria. Helgol Wiss Meeresunters 27, 461–504 (cited in Jarms et al., 2002).

Werner, B., Cutress, C., Studebaker, J., 1971. Life cycle of *Tripedalia cystophora* Conant (Cubomedusae). Nature 232, 582–583.

Werner, B., Chapman, D.M., Cutress, C.E., 1976. Muscular and nervous systems of the cubopolyp (Cnidaria). Experientia 32 (8), 1047–1049.

Yuan, D., Nakanishi, N., Jacobs, D.K., Hartenstein, V., 2008. Embryonic development and metamorphosis of the scyphozoan *Aurelia*. Dev. Genes Evol. 218, 525–539.

Chapter Outline

Definition **309**
Distinctive features **309**
Classification **310**
- ***Hormiphora plumosa*** 311
- **Reproduction** 311
  - Regeneration 311
  - Brooding 313
  - Gametic reproduction 313
    - Origin of the germ cells and gonads 313
    - Gonads 314
    - Gametogenesis 314
      - Spermatogenesis 314
      - Mature sperm 315
      - Oogenesis 315
      - Oogenesis in *Gastrodes* 320
    - Development 320
      - Embryonic development 320
      - Fertilization 320
      - Cleavage 323
      - Gastrulation 326
      - Larval types 326
  - ***Mnemiopsis leidyi*** 326
    - Cell lineage experiments 326
  - Organogenesis 327
    - Comb plates 327
    - Stomodeal pharynx 331
    - Sense organ 334
    - Nervous system 335
    - The gastrovascular system 335
  - Hatching 335
  - Brooding 335

References **336**

Chapter 6

Phylum Ctenophora

The word Ctenophora is derived from the Greek for 'Comb-Bearers'.

DEFINITION

Ctenophores are free and solitary animals. They are biradially symmetrical organisms, lacking nematocytes, but possessing colloblasts (lasso cells), ciliary plates in eight rows used for movement and a gelatinous ectomesoderm containing mesenchymal muscle fibres.

Martindale and Henry (1997a,b) reported that the statocyst-ctene row complex, mesenchymal muscle cells, colloblasts, life history strategies (plan) and the stereotyped plan of embryogenesis all set the ctenophores apart from cnidarians. Molecular analysis may provide some clues as to the evolutionary position of the ctenophores. Some initial suggestions indicate that the ctenophores diverged from other major metazoan phyla prior to its divergence from the cnidarians. The clear absence of ctenophores in the fossil record has made it difficult to resolve the relationships within the phylum and with other groups.

The Ctenophora is a small phylum of marine animals. Ctenophores (pronounced 'teen-o-for' or 'ten-o-for') are more commonly known as sea walnuts, comb jellies, moon jellies, or sea gooseberries (Fig. 6.1). They comprise a group of gelatinous zooplankton found in all the world's seas. There are about 150–200 described species; most are holopelagic (living in the open ocean), but new species continue to be discovered in bluewater and deep-sea regions. Most ctenophores are transparent or translucent, and range in size from millimetres up to 2 m in length, although most are in the few centimetres range (Pang and Martindale, 2008). Some of the more common animals are the sea gooseberry (genus Pleurobrachia), the sea walnut (genus Mnemiopsis) and the Venus' girdle (genus Cestum). Most of these beautiful and exquisite animals are not very well studied because they are either difficult to obtain or extremely delicate (or both). These genera are abundant in coastal waters and are common genera along the north Atlantic coast.

DISTINCTIVE FEATURES

- Free-living very well defined and successful phylum of marine predators, primarily pelagic and planktonic.
- Solitary; the formation of colonies is entirely unknown.
- No indication of a polyp stage (characteristic of the Cnidaria), neither in the adult nor in the embryonic condition (i.e. there is no alternation of generations).
- Cnidarian medusae are said to be radially symmetrical around their oral–aboral axis, while ctenophores display a special form of biradial symmetry, with no planes of mirror symmetry but rather an infinite number of planes of rotational symmetry (Pang and Martindale, 2008).
- The ectodermal and endodermal tissues are separated by a thick mesoglea of uncharacterized glucopolysaccharides, and the mesoglea is sparsely populated with several types of individual muscle and mesenchymal cells (Martindale, 2016).
- Ciliary movement, cilia fused to form comb-like structures arranged in eight meridional rows or swimming plates (costae).
- The eight part system of comb plates (with evidence for eight part of distribution colloblasts of nervous elements) is, to some biologists, a foreshadowing of the eight nerve strands in certain flat worms (Storer et al., 1983).
- Tentacles, when present, are usually two in number, on the aboral pole, situated in opposite (right and left) perradii, and retractile in pouches.
- Instead of the cnidoblasts in cnidarians, the tentacles of Ctenaria are provided with specific adhesive cells known as lasso-cells (colloblasts).
- On the aboral pole, there is a single sense organ which is like a peculiarly modified lithocyst, situated in opposite (right and left) perradii, and retractile in its pouches.

Invertebrate Embryology and Reproduction. https://doi.org/10.1016/B978-0-12-814114-4.00006-0

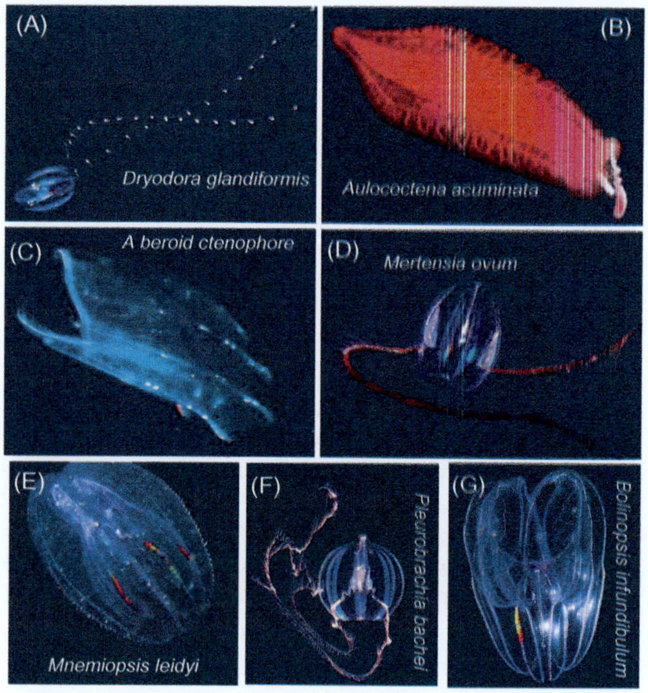

FIG. 6.1 Photos of whole mounts of almost live examples of ctenophores: *(A–E from Ocean Portal) ((A and D) couretsy of Alexander Semenov, (E) courtesy of Marco Faasse, (C) courtesy of NOAA office of Ocean Exploration and Research)* and (F and G) *P. bachei*, the sea gooseberry, and *B. infundibulum*, the lobed comb jelly. *(From Quanta magazine.)*

- The enteron communicates with the exterior by a large stomodaeum which functions as the chief digestive cavity.
- From the enteron extends a system of canals, the ultimate branches of which are adradial and have meridional position, lying beneath the swimming plates.
- A single axial canal is continued to the aboral pole, where it commonly opens by two excretory pores.
- There are no gastric filaments.
- Nervous system is a subepidermal nerve network, well developed beneath the comb plates (rows). As a further support for the existence of a definite nervous system, evidence from electron microscopical investigations revealed numerous synapses in the floor of the apical organ, underlying the ctene rows. In addition, it is associated with the tentacles and various sensory structures. These indicate that this nervous system regulates both; ciliary and muscular effectors.
- No agametic reproduction.
- The gonads of both sexes are found in the same individual.
- All ctenophores are hermaphrodites.
- Testes and ovaries are found on opposite sides of the meridional canals.
- The fertilized eggs undergo unequal cleavage.
- Gastrulation occurs by epiboly or overgrowth.
- There is no planula larva; however, the parasitic platyctenean (Tenteculata), *Gastrodes,* forms a planula larva. Only one member of this group, Gastrodes, is a parasite in the body of Salpa (a tunicate, Thaliacea; Borradaile et al., 1977). Its chief interest, however, is in the fact that it has a planula larva, found nowhere else among the ctenophores. Thus this provides the strongest piece of evidence for the close relationship of the Ctenophora with the Cnidaria (Borradaile et al., 1977).
- In some cases, development is accompanied by a well-marked metamorphosis.

CLASSIFICATION

Bases of taxonomy

- No agametic reproduction.
- Aboral sensory organ.
- The only sense organ is an apical simple depression.

Class: 1. Tentaculata (Micropharyngea), Tentacles present.

Order: 1. Cydippidea, sea gooseberries (primitive order) (Kydippe = 'a Nereid')

- Body rounded or ovoid in form.
- Two tentacles, branched, retractile into sheaths.
- Unbranched stomodaeal vessel (meridional).
- Ex. *Hormiphora, Pleurobrachia, Callianira, Dryodora*.

Order: 2. Cestidea, Venus' girdle (cestus = girdle)

- Body band-like in form or ribbon owing to the extreme compression of the body in the sagittal plane (tentacular plane).
- Tentacles reduced, with sheaths, bases of the two principal tentacles are present, enclosed in sheaths, and these are also numerous lateral tentacles contained in a groove.
- Union of the meridional and stomodaeal vessels.
- Four of the rows of ciliary plates are rudimentary.
- Ex. *Cestus veneris* (smaller *Velamen*) ('Vexillum').

Order: 3. Lobata (lobes = lobe)

- Body laterally compressed, with oral end expanded into two contractile lobes or lappets and into four pointed processes or auricles.
- Numerous nonretractile lateral tentacles contained in a groove.
- Two principal tentacles also present, but have no sheath.
- Ex. *Polinopsis vitrea 'Bolina hydatina'*
- Stomodaeal and meridional vessels unite.
- Ex. *Deiopea; Mnemiopsis* (Gardiner, 1972).
- *Leucothea* (= *Eucharis*); *Deiopea; Bolinopsis*.

Order: 4. Platyctenea (Platys = broad, flat)

- Dorsoventrally flattened.
- Creeping habit or sessile.
- Pair of retractile lateral tentacles with sheaths.
- The (costae) ciliary plates may be present in larvae only, short and deeply sunk.
- No meridional canals.
- Has a system of branching peripheral vessels.
- Ex. *Ctenoplana, Tjalfiella, Coeloplana, Lyrocteis, Planoctena, Savangia, Vallicula*.

Class: 2. Nuda (Atenteculata, Macropharyngea), Tentacles absent.

Order: 1. Beroidea, Beroë = 'a nymph'

- Mouth is very wide, and the gullet occupies the greater part of the interior of the body.
- Body conical or thimble-shaped.
- Meridional vessels are produced into a complex system of anastomosing branches.
- Ex. *Beroë* sp.

Hormiphora plumosa

Hormiphora (Fig. 6.2D) is a pear-shaped organism about 5–20mm in diameter, and of glassy transparency. This species is found in the Mediterranean. Allied forms belonging either to the same genus or to the closely allied genus Pleurobrachia are common pelagic forms all over the world.

Structurally, from opposite sides of the broad end, two long tentacles hang provided with numerous little tag-like processes. Each tentacle springs from a deep cavity or sheath into which it can be completely retracted (Fig. 6.2F). At the narrow end of the stalk of the pear-shaped animal, there is a slit-like aperture: the mouth. This end is therefore the oral pole, and at the opposite aboral pole there is a slight depression, in which lies a prominent sense organ.

The most striking and characteristic feature in the external morphology of *Hormiphora*, and in all ctenophores, is the presence of eight equidistant meridional bands. Each band starts from near the aboral pole, and extends about two-thirds of the distance towards the oral pole. Each band is constituted by a row of transversely arranged comb-like structures, consisting of a narrow plate frayed at their outer ends. During life, the frayed ends are in constant movement, lashing to and fro (costae), or propelling the animal through the water. The combs are, in fact, rows of immense cilia. Now, the electron microscopy shows these to be compound cilia, fused at their proximal ends. Their presence and mode of occurrence, arranged in meridional comb ribs, costae, or swimming plates, are strictly characteristic of the phylum, and indeed give it its name.

Apart from all considerations of internal structure, *Hormiphora* presents a similar combination of radial with bilateral symmetry as in some Hydrozoa, such as *Ctenaria* (Fig. 6.2E) and the majority of the Anthozoa. The swimming plates are radially arranged, and mark the eight radii. The slit-like mouth and the two tentacles indicate a very marked and characteristic bilateral symmetry. An imaginary line passing from the middle of the mouth to the sense organ is the primary axis. A plane passing through the longitudinal axis of the body, parallel to the long axis of the mouth, is called (as in the Anthozoa) the vertical or sagittal plane. It includes two perradii, the dorsal and ventral ones. A plane at right angles to this, passing through both tentacles, and including right and left perradii, is called the transverse or lateral plane. It is along these two planes alone that the body is capable of division into approximately equal halves, and the term biradial best describes this (Marshall and Williams, 1975).

Reproduction

Regeneration

These fragile animals are virtually impossible to collect intact in plankton nets, and thus it is not surprising that adults of most species examined have a high capacity to regenerate (Martindale, 2016). It is well-known that all adult ctenophores are able to regenerate lost parts, and many can regenerate completely from a small fragment. This ability is known to be used in agametic reproduction only in the platyctenid genera *Coeloplana*, *Planoctena*, and *Vallicula*. The process is similar in all species, and has been compared with pedal laceration in anemones (Hyman, 1940). Small pieces break off from the edge of the adult body, usually as the animal is creeping about, and often at night. Each piece regenerates completely to form a new individual. This laceration may occur all year-round. In other species, it may occur only during a definite time, e.g. from June to August, when the water temperature ranges from 22°C to 29°C. Spontaneous laceration is inhibited in the laboratory at temperature above 30°C. Not only the adult ctenophores possess this phenomenon, but the larval stages can also regenerate. Martindale and Henry (1997a,b) reported that one interesting feature of some ctenophore species is their ability to generate progeny at very early stages of 'larval' life. This process was termed 'dissogeny' by Chun (1880–1892) and has been reported for members of the Lobata and Cydippida. Dunlap-Pianka (1974) confirmed this phenomenon and reported that 'paedogenesis', or reproduction by larvae or juveniles, has been documented in several ctenophores, and may be of wide occurrence throughout the phylum. Chun (1880–1892) first observed this phenomenon in two lobate species from the Mediterranean. Larvae of these species develop ovaries and testes, similar to adult gonads, although occurring only in subsagittal canals (Fig. 6.5D). Chun, at that time, employed the term 'Dissogeny' to describe the phenomenon of two periods of sexuality, one in the larva and one in the adult, separated by a period surrounding metamorphosis during which the gonads regress.

Martindale (2016) studied the onset of regenerative properties in ctenophores. It is not surprising that adults of most species examined have a high capacity to regenerate, but the regenerative ability varies throughout the Ctenophora, with some interesting variations to the pattern. The platyctenes, one group of ctenophores, lose their comb plates after embryogenesis and crawl along the substrate. These animals can, not only regenerate but have also evolved modes of agametic reproduction, either by undergoing symmetrical fission through their esophageal axis, leaving each piece with four comb rows and one tentacle, or by shedding small pieces of peripheral tissue that contain neither tentacle,

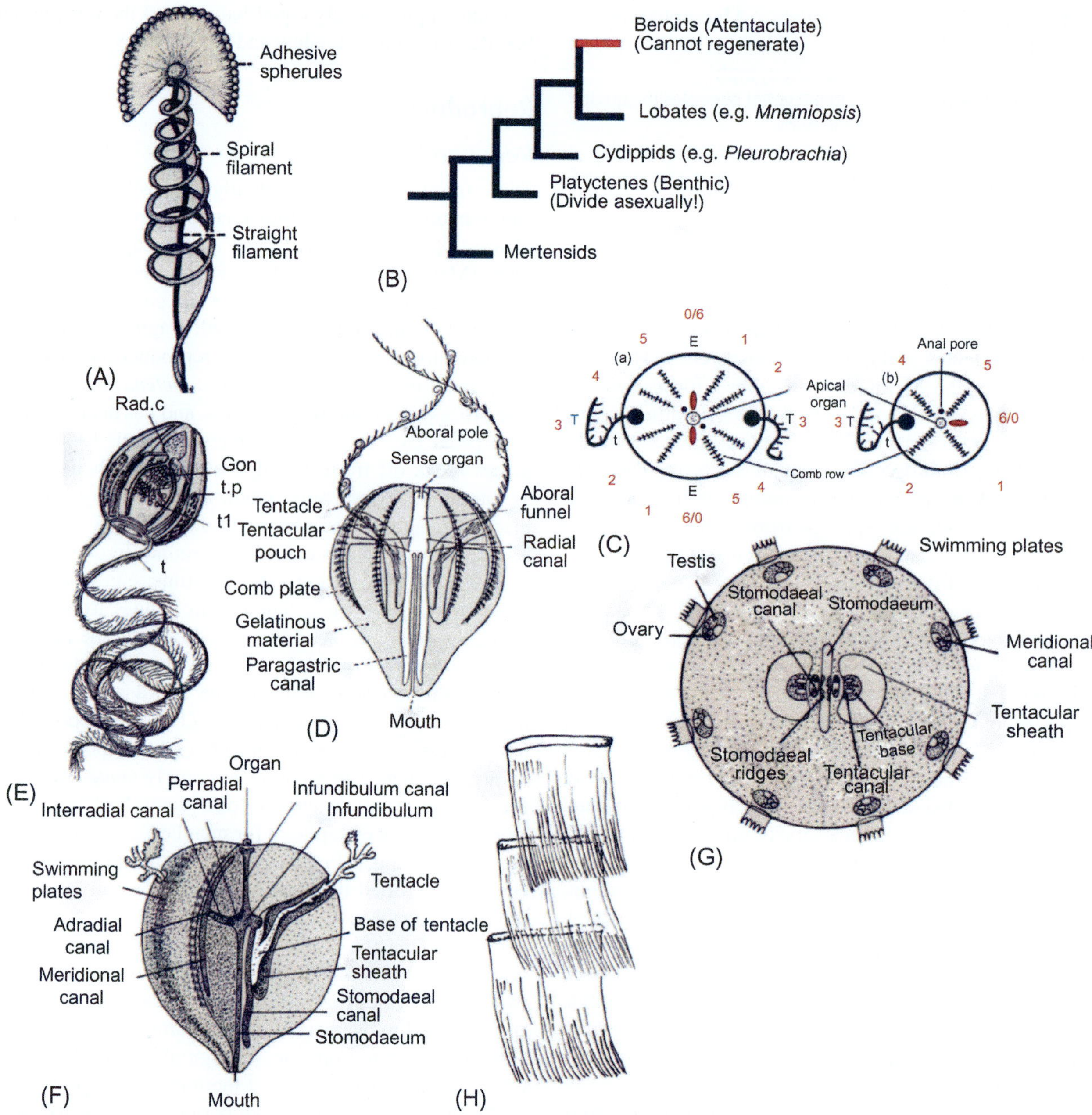

FIG. 6.2 (A) Colloblast, *after Hyman from Gardiner (1972)*; (B) Simplified internal phylogeny of ctenophores to indicate that regenerative ability varies within the group. Parsimoniously, ancestral ctenophores were able to regenerate, while agametic reproduction evolved in the benthic platyctenes. Regenerative ability was lost in the atentaculate Beroids *(red line)*; (C) Body plan and model of positional information of ctenophores viewed from the aboral pole: (a) an intact ctenophore is composed of four quadrants defined by two planes, the tentacular (T) and the esophageal (E). Each quadrant possesses two rows of comb plates, and half of a tentacle (t) and a quarter of the apical organ (located on the aboral pole). The apical organ is composed of four groups of balancing cilia (not shown) and two polar fields *(red elipses)*. However, oppositely opposed quadrants are more similar to one another than any pair of adjacent quadrants. Note the position of the two anal pores in the opposite quadrants *(blue circles)*; (b) Ctenophores can remain as stable adult 'half animals', possessing exactly half of the normal complement of adult structures. The stability of these morphs can be explained by a modified positional information system *(red numbers)*. The positional informational system predicts that gaps in the continuity of normal tissues initiate a regenerative response that will replace the intervening structures *((B and C) from Martindale, 2016)*; (D) diagrammatic drawing for the side view of the ctenophore *Hormiphora plumosa (after Chun from Borradaile et al., 1977)*; (E) *Ctenaria* (a hydrozoon showing a combination of radial with bilateral symmetry), gon = gonad, rad.c = radial canal, t = marginal tentacle, t^1 = oral tentacle, tp = tentacle pouch; (F) dissected specimen of *Hormiphora plumosa* having rather more than one quarter of the body cut away; (G) diagram of section at right angles to long axis of the same *((E–G) from Marshall and Williams, 1975)*; (H) Portion of comb row showing three plates, each composed of transverse rows of long fused cilia. *(From Hickman, 1973.)*

comb plates, nor apical organ tissue. Phylogenetic reconstruction of the Ctenophora (Fig. 6.2B) demonstrates that regenerative ability was likely shared by all ctenophores, lost in the Beroids and amplified to facilitate agametic fission in the platyctenes.

The variability in regenerative potential in different ctenophore groups provides unique opportunity to investigate the stability of cell fate and the origins of regenerative capacities in these animals. Unfortunately, only a handful of ctenophore species have been studied in detail. This review summarizes the key features of the regenerative properties of adults, and the characteristics of the embryological onset of regenerative abilities. One of the experiments is that juvenile or adult ctenophores are bisected along the oral–aboral axis (in any plane), and an incomplete regenerative event will occur in a substantial percentage of the cases (20%–40%). Wound healing proceeds normally, but instead of replacing the missing parts, a stable 'half-animal' forms. These animals possess one tentacle, four ctene rows and 'half' of an apical organ (i.e. two groups of balancing cilia rather than four, and a single polar field) (Fig. 6.2C (b)). These half animals are capable of feeding and will live as stable 'half-animals' well through their reproductive adult stages. These 'half-animals' can themselves be challenged to regenerate. If the apical organ of a 'half-animal' is amputated, a new 'half' apical organ will be formed, indicating that there is feedback from the periphery that regulates regenerative output. Interestingly, if 'half-animals' are created that possess two opposite, rather than adjacent, quadrants, they regenerate directly to 'whole-animals' (eight ctene rows, two tentacles and a complete apical organ) in nearly all of the surviving cases.

Henry and Martindale (2000) defined the process of regeneration as follows: "Regeneration is typically associated with the replacement of structures removed during the larval or adult phases of life through processes of epimorphosis and/or morpholaxis." They studied regulation and regeneration in the ctenophore *Mnemiopsis leidyi* and they collected adult specimens from Quissett Harbour, Eel Pond, and off the rock jetty at the National Oceanographic and Atmospheric Administration research facility in Woods Hole, Massachusetts, using small plastic dippers. They reported that during regeneration, populations of existing adult cells (including differentiated, or satellite 'stem' cells) generate the missing structures, which may involve the initial process of de-differentiation.

Not only do the adults lobate ctenophores (tentaculates) exhibit a remarkable ability to regenerate missing structures, but also their embryos exhibit a highly mosaic behaviour when cut into halves or when specific cells are ablated. These deficient embryos do not exhibit embryonic regulation, and generate incomplete adult body plans. Under certain conditions, however, these deficient animals are subsequently able to replace the missing structures during the adult phase in a process referred to as 'postregeneration'. To study the phenomenon of postregeneration, Henry and Martindale (2000) chose to examine the regeneration of one distinctive set of structures, the comb rows. Unlike the regeneration of other structures such as the tentacles, comb plate regeneration occurs quickly once the adult body plan is established (24–72 h following fertilization), and their regeneration occurs without the need for prolonged feeding. The authors have found that the progeny of three specific cell lineages are required to support postregeneration of the comb rows (the e1, e2 and m1 micromeres). Furthermore, postregeneration of the comb rows involves a suite of cell–cell inductive interactions, which are similar to those that take place during their embryonic formation.

Brooding

Internal brood chambers have been observed in the platyctenids *Tjalfiella tristoma* (Fig. 6.3A) and *Lyrocteis imperatoris*. Those in *Tjalfiella* (Dunlap-Pianka, 1974) are arranged regularly beneath the lateral epidermis, sometimes in a double layer, and each is a thin-walled sac touching (abutting against) a blind end of the ramifying gastrovascular system (Fig. 6.3B). Thus the brood chambers are separated from the gonads by much of the canal system. In *Lyrocteis imperatoris*, on the other hand, the chambers occur in rows paralleling the series of ovaries and each is evidently an expansion of the distal part of an ovarian diverticulum (Fig. 6.3C and D). At least in this *Lyrocteis* the brood chambers may well be homologues of oviducts in other ctenophores. Ovoviviparity could be accomplished by having sperm enter through the ovarian gonopores, and fertilization occurs within the oviducts, which would then function as uteri. These oviducts might be the site of brooding, as in *Coeloplana mitsukurii*. Brooded embryos, when fully developed, hatch by rupture of the ectoderm.

Gametic reproduction

Origin of the germ cells and gonads

Dunlap-Pianka (1974) reported that germ cells in ctenophores are always seen closely associated with the endoderm. They could have migrated there during early embryonic development, from either the mesoderm or the ectoderm, but there is no convincing evidence of either. There is, however, considerable support for an endodermal origin of the gametes and of all gonadal tissues from the early studies. One of these studies (Chun, 1880–1898) investigated that germ cells are first recognized in the cydippid larva as clusters of cells associated with the outer endodermal lining of the meridional canal primordia. The development of larval gonads from these cells has been described briefly by Chun (1898) and in greater detail by Garbe (1901). Garbe observed three initial germ cells clusters in the oral extremes of each of the two meridional canal primordia. These clusters proliferate aborally into three germ bands (Fig. 6.5A).

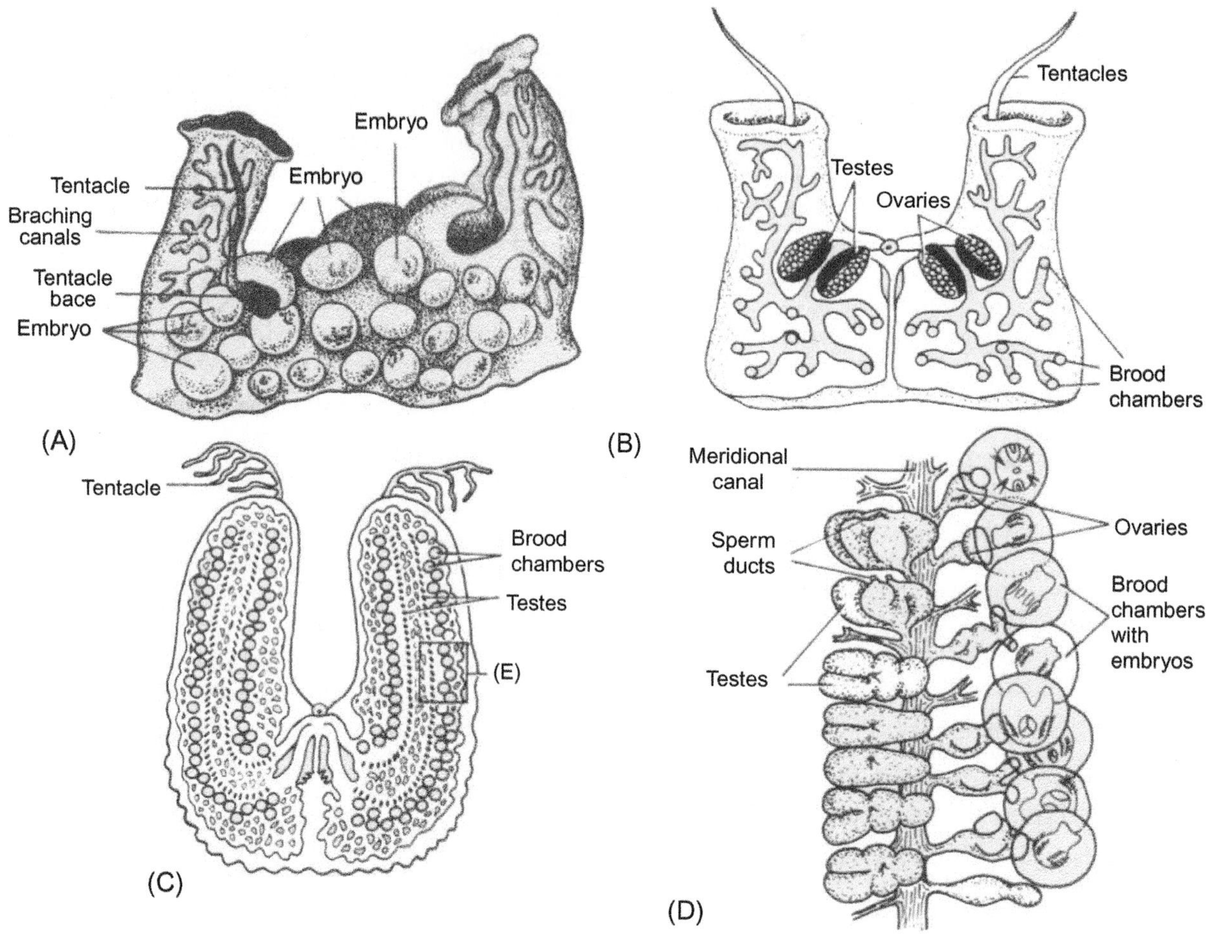

FIG. 6.3 Brooding in Ctenophora: (A) side view of adult specimen of *Tjalfiella tristoma* with many embryos, oral (attached) surface upwards *(After Mortensen from Marshall and Williams, 1975)*; (B) lateral view of *T. tristoma* showing four of the eight genital organs. Brood chambers shown as circles at the blind ends of the peripheral gastrovascular branches; (C) lateral view of *Lyrocteis imperatoris* showing numerous testicular diverticulae and brood chambers along meridional canals; (D) enlargement of part of one meridional canal, with testicular and ovarian diverticulae, and brood chambers containing developing embryos, shown as dissected away from surrounding tissues. *((B–D) From Dunlap-Pianka, 1974.)*

Gonads

As mentioned above, most but not all ctenophores are hermaphrodites, possessing both male and female gonads, which occur close to each other, underneath each of the ctene rows. Within a pair of ctene rows of the same quadrant, the testes face each other, while the ovaries are located on the 'outside' of the mirror plane of symmetry (Figs 6.5C, D and 6.2G). Chun (1880–1892) and Martindale (1987) noted that it is also of interest that only four of the eight possible gonads—the four associated with the ctene rows adjacent to the oesophagus—become precociously reproductive. The idea is that these ad-oesophageal gonads differ from the ad-tentacular gonads in respect to larval reproduction, which is supported by additional findings. 'Half-animals' generated by bisecting early ctenophore embryos become reproductive at a higher rate than normal animals, and in these partial embryos, only the ad-oesophageal gonads generate functional gametes.

Gametogenesis

Spermatogenesis. As was described in more than one work, the spermatogonia in the ctenophore testis occur in a localized part and give rise to batches of developing spermatocytes (Fig. 6.6). These occur in large packets in which all component cells are in the same stage of development. Near the onset of spermeogenesis, each packet becomes separated from the less mature packets by a cellular septum formed by interstitial processes of the digestive epithelium in some species, e.g. *Pleurobrachia bachei*, or by separate small cells in others (*Bolinopsis microptera*) (Fig. 6.6B). These septa are continuous with the testicular membrane surrounding the outer part of the testis.

Electron micrographs reveal narrow open intercellular bridges connecting developing spermatocytes of *Bolinopsis* (Fig. 6.6D), similar to those observed in other animal testes (Fawcett, 1961).

Mature sperm often form large wedge-shaped bundles, in which the gametes are arranged parallel to each other.

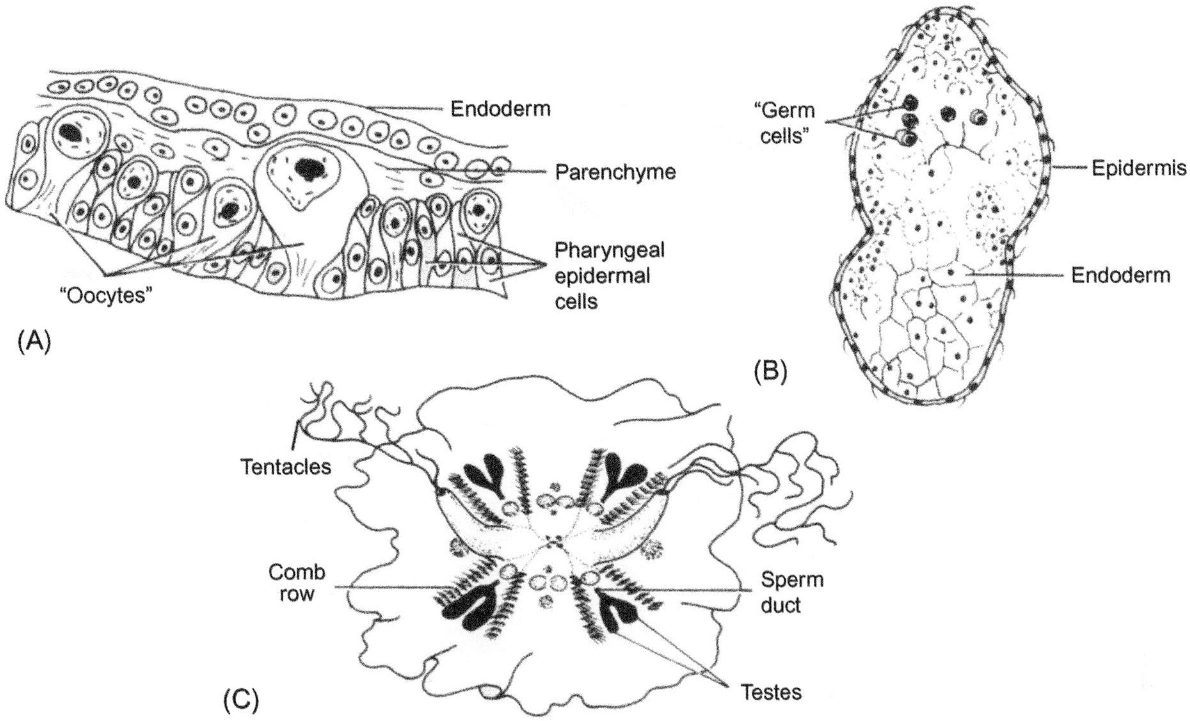

FIG. 6.4 (A and B) *Gastrodes parasiticum*: (A) part of a section of the ventral pharyngeal epidermis, showing possible oocytes in several stages of development; (B) L.S. of a planula larva of *Gastrodes* showing ciliated epidermis, and 'primordial germ cells' among the cells of the solid endoderm. *(After Komai, 1922 from Dunlap-Pianka, 1974.)* (C) Aboral view of *Planoctena yuri* showing the four pairs of testes, each joined by a common sperm duct, lying along the interradii between comb rows. Ovaries have not been observed in this species. *(From Dunlap-Pianka, 1974, redrawn from Dawydoff, 1936.)*

In Ctenophora *Ctenoplana korotneffi*, mature sperm are arranged as to form large pyriform bundles, with tails pointing both towards the centre of the bundle and radiating outwards.

Mature sperm. Sperm are considered to be primitive in form (Fig. 6.6E). The head is usually spherical to pyriform and contain the haploid nucleus, a Golgi-derived acrosomal vesicle, two centrioles and a small number of mitochondria (Fig. 6.6E). Sperm heads are connected to the distal centriole, giving rise to a normal flagellum composed of the standard 9 + 2 microtubular array. There is no definite midpiece, however, with the distal centriole being attached to a star-shaped structure at the base of the sperm head (Martindale and Henry, 1997a,b). Motile sperm are shed into a testicular sinus before discharging to the sea through the gonopores.

Oogenesis. Scanty studies of oogenesis found in the previous literature revealed that in all major respects, the process is similar in the *Pleurobrachia bachei*, *P. pigmentata*, *Dryodora* and *Beroë*. In addition, these studies indicate that oogenesis is similar throughout the phylum. In respect to the nurse cells, they have been identified in many ctenophores, and are interpreted as young oocytes in some species, which are sacrificed in development of the definitive oocytes. Another view proposed that these oocytes are 'phagocytized' by their more vigorous neighbours.

Then, in 1966, Dunlap perfectly studied oogenesis in *Bolinopsis microptera* and documented the results by electron micrographs. An adequate and detailed picture of the process was provided, especially the structure and role of the nurse cells (Fig. 6.7). During oogenesis, the *Bolinopsis* oocyte is associated with approximately 100 nurse cells, divided into three unequal clusters. Cells of each cluster form a syncytium, i.e. they are connected by cytoplasmic bridges, basically similar to those between spermatocytes. The author reported that this picture looks like the ovarian nurse cells and oocytes of some insects and annelids.

This pattern and the approximate number of nurse cells in each of the three clusters suggest a sequence of eight synchronous divisions resulting in an oocyte-nurse cell complex, starting with an oogonium. One of the two oldest cells, being central in position, becomes the definitive oocyte. Consistent changes in the intercellular bridges correlate with major stages in development of the oocyte-nurse cell complex. These changes may be divided into four stages: previtellogenesis, early vitellogenesis, late vitellogenesis and corticogenesis (Fig. 6.7):

1. Previtellogenesis involves growth of the most proximal nurse cells, but not of the oocyte (Fig. 6.7A and B).
2. Early vitellogenesis begins with the onset of vitellogenesis. This is a simultaneous process with initial inflow

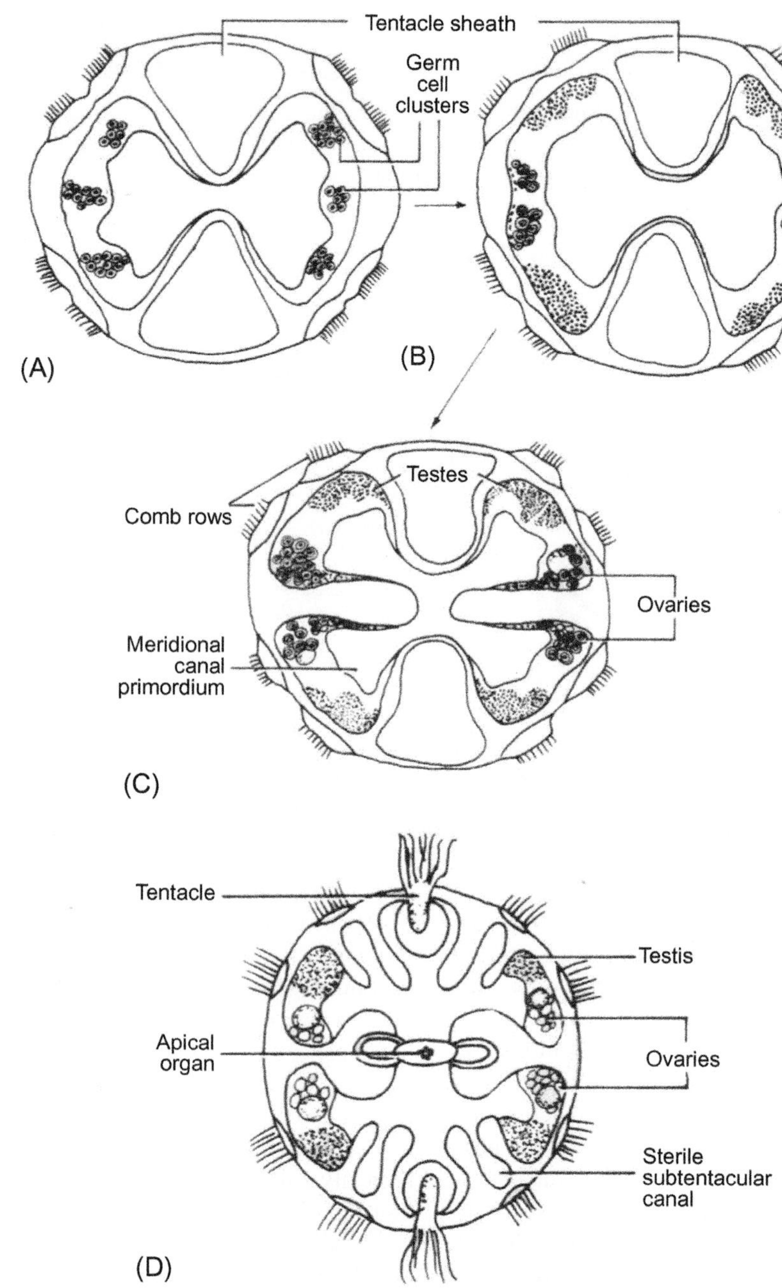

FIG. 6.5 Diagrammatic representations of development and structure of larval gonads in two *Bolinopsis* species, shown as in equatorial cross sections of cydippid larvae: (A–C) stages in development of larval gonads of a *Bolinopsis* larva ('*Pleurobrachia rhodopsis*'). *(Redrawn from Garbe, 1901.)* Subtentacular canals were not described in this form; (D) sexual larva of *Bolinopsis vitrea* ('*Bolina hydatina*') showing gonads in four subsagittal canals *(redrawn from Chun, 1892). (From Dunlap-Pianka, 1974.)*

of nurse cell cytoplasm (Fig. 6.7C) (in this figure, yolk is not preserved). Early vitellogenesis is essentially the growing of the oocyte to a sphere of about 100 μm in diameter. During this growth phase, the oocyte becomes filled with small yolk bodies (Fig. 6.7D) and only the nine most proximal nurse cells (three per cluster) contribute cytoplasm. This is the endoplasm of the fully grown oocytes.

3. Late vitellogenesis involves further growth of the oocyte to about 15 μm in diameter (Fig. 6.7E). During this phase of growth, the central (old) yolk spheres become differentiated from the peripheral (recently formed) yolk ones. There is further addition of nurse cell cytoplasm from 12 more nurse cells (four in each cluster), and the yolk formation evidently continues in the oocyte periphery.

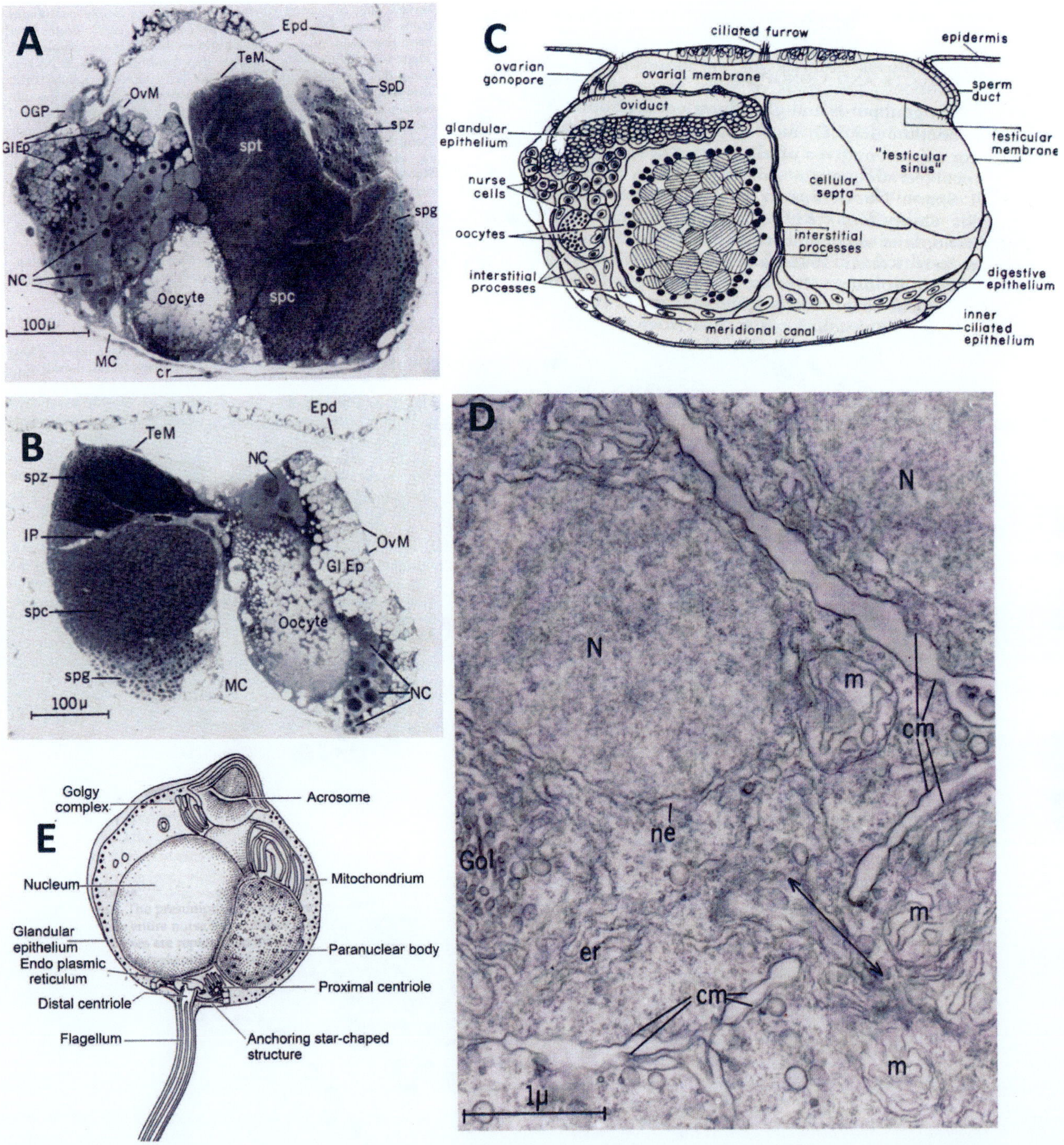

FIG. 6.6 Photomicrographs of cross sections of: (A) a meridional canal and gonadal tissues of the lobate *Bolinopsis microptera* and (B) the cydippid *Dryodora glandiformis*. Material fixed with Osium tetroxide, hence yolk in the oocytes is not preserved; OGP = ovarian gonopore cells, OvM = ovarian membrane, GlEp = glandular epithelium of the oviduct, NC = nurse cells, SpD = sperm duct, TeM = testicular membrane, spg = spermatogonia, spc = spermatocytes, spt = spermatids, spz = spermatozoa, IP = interstitial process of digestive cells, MC = lumen of meridional canal, cr = cell rosette, Epd = epidermis; (C) diagram of a representative cross section of a meridional canal and gonadal tissues, at the level of the gonopores, in *Bolinopsis microptera*. (Compare with (A)). Ovary on the left separated from testis on the right by the long central interstitial processes; (D) electron micrograph of an intercellular bridge connecting two developing spermatocytes of *Bolinopsis microptera*. *Arrow* passes through the cytoplasmic connection; N = spermatocyte nucleus, ne = nuclear envelope, Gol = Golgi body, er = endoplasmic reticulum, m = mitochondria, cm = cell membranes; (E) ultrastructural reconstruction of a Beroë sperm. *(From Gilbert and Raunio, 1997.)*

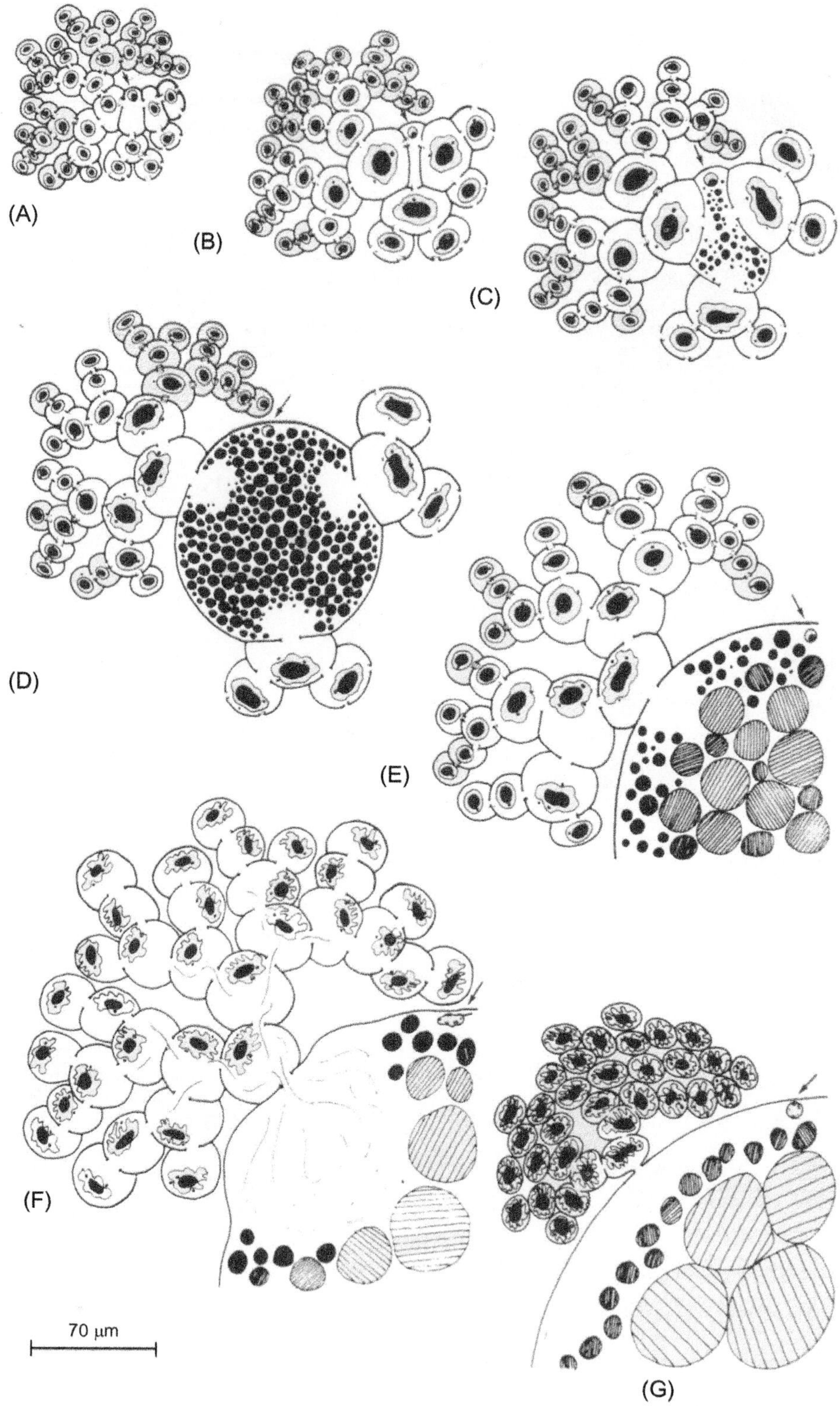

FIG. 6.7 Diagrams of major stages of oogenesis in *Bolinopsis*: (A–B) previtellogenesis; (C–D) early vitellogenesis; (E) late vitellogenesis; (F–G) corticogenesis. *Small arrows* point towards the oocyte nucleus. In (A) through, in (D) one entire nurse cell cluster is shown with the oocyte, while each of the other two clusters is represented by its three most proximal cells. Only one complete cluster is depicted in (F) and (G), *from Dunlap-Pinaka (1974)*.

4. Corticogenesis involves the time of the final oocyte growth, during which the oocyte attains a diameter of 200–210 μm; the approximately 80 remaining distal nurse cells which have only begun to differentiate until now, mature rapidly and apparently simultaneously. The cytoplasm of these inflows to the developing oocyte and spreads out evenly over the peripheral yolk, forming the definitive cortical plasm (Figs 6.7G and 6.8A). The nurse cells then shrink sequentially and gradually disintegrate among the ovarian tissue.

Thus, the fully grown *Bolinopsis* oocyte, and generally the ctenophore oocyte, has a centrolecithal organization.

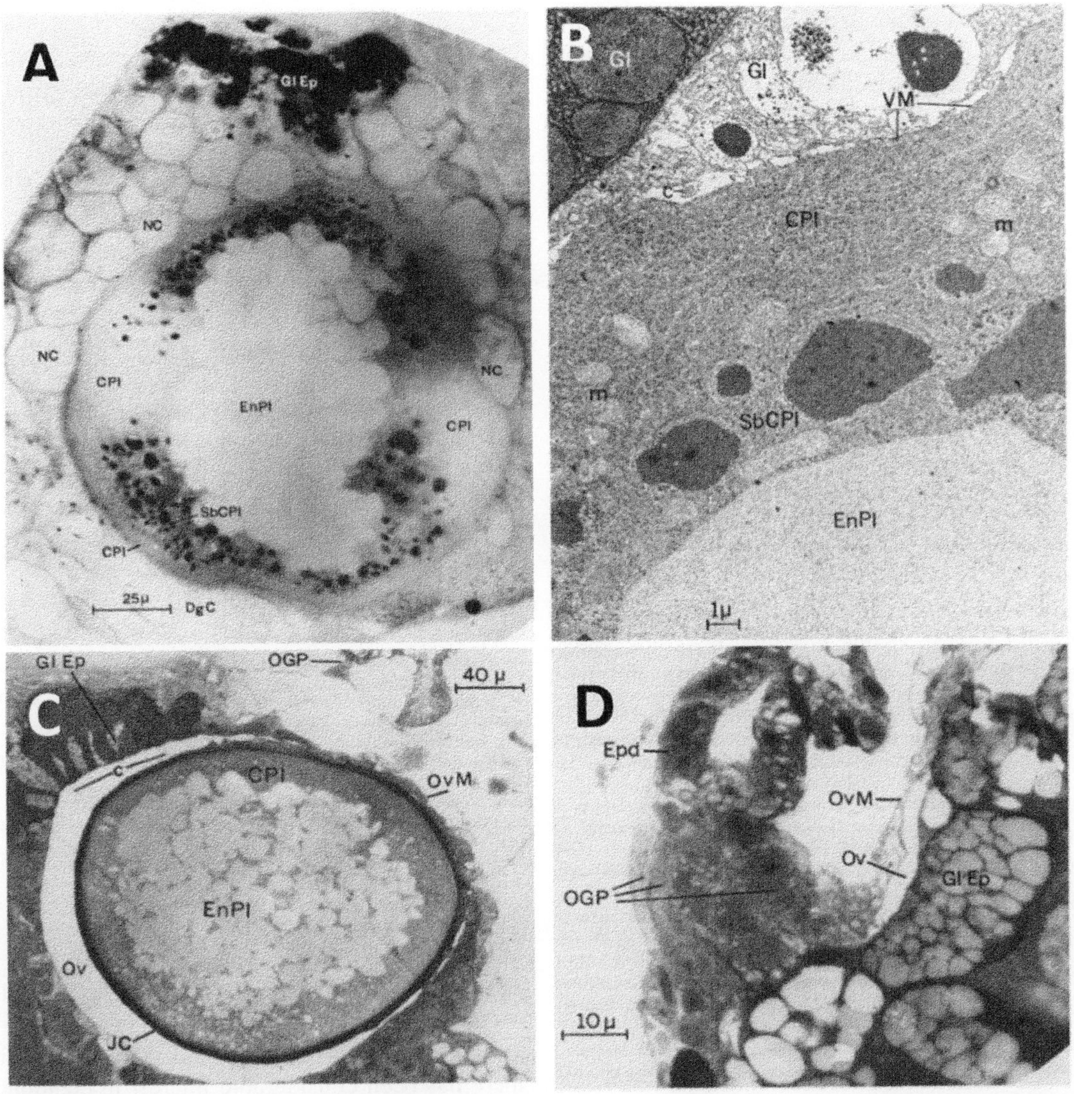

FIG. 6.8 Photomicrographs of oocyte of *Bolinopsis microptera*: (A) during corticogenesis, comparable to the stages (F) and (G), Fig. 6.7, the material was fixed with glutaraldehyde and stained with PAS procedure to demonstrate carbohydrates. The endoplasm (EnPl) is unstained, while the peripheral yolk bodies in the subcortical plasm (SbCPl) are strongly PAS+ve and thus have high carbohydrate content. The mucous secretion of the oviduct's glandular epithelium (GlEp) also stains strongly. Note the absence of a reaction in the nurse cells (NC) and in the cortical plasm (CPl). DgC, digestive cells lining the meridional canal; (B) electron micrograph of the edge of a mature oocyte in the oviduct, comparable to that in (D) except that this oocyte has been preserved with glutaraldehyde, and hence the difference between the yolk in the endoplasm and in the subcortical plasm is apparent. The cortical plasm contains mainly lamellae of endoplasmic reticulum oriented more or less parallel to the oocyte cell membrane (oolemma). Mitochondria (m) occur among these lamellae and also among the endoplasm. The vitelline membrane (VM) is visible as a layer of fibrillary material on the outer surface of the oolemma. Two gland cells (Gl), one of each type composing the inner oviducal epithelium are in the upper part of the micrograph. Several cilia (c) are also included; (C) mature oocyte within the oviduct, covered with a jelly coat (JC). The two main zones of the ooplasm: endoplasm and cortical plasm are evident. Yolk in the former is not preserved. Ov = lumen of oviduct, c = cilia in oviduct, GlEp = glandular epithelium, OGP = part of ovarian gonopore cells near their connection with the epidermis; (D) enlarged section through the ovarian gonopore cells (see Fig. 6.6A and C). The ciliated epidermal invagination (Epd) connects with one end of the gonopore cells (OGP), while the inner ends of these cells border the oviduct lumen (Ov). The oviduct is lined with an inner glandular epithelium (GlEp) and an outer ovarian membrane (OvM). *(From Dunlap-Pianka (1974).)*

It is composed of three distinct zones: (1) a central mass of large, pale yolk spheres with few intervening cytoplasm, this is the endoplasm of the oocyte; (2) a narrow intermediate zone surrounding the first zone and composed of dense, small periodic acid-Schiff stain (PAS)-positive peripheral yolk bodies are dispersed in relatively abundant cytoplasm; (3) and an outermost layer of yolk-free cytoplasm (Figs 6.7G and 6.8B); this is the 'cortex' or 'ectoplasm' of the oocyte. The intermediate zone is described for the first time, since it is not readily visible in living eggs, and is clearly distinguished only after fixation with glutaraldehyde. This zone is also present in mature oocytes of *Dryodora* and *pleurobrachia*, and hence probably occurs in all ctenophore eggs. Because of its association with the outer cortical plasm during embryological development, it is referred to a subcortical plasm.

Meiosis (maturation of the fully grown oocyte) occurs some time after spawning in *Beroë ovata*, while in *Bolinopsis* the first polar body is formed during ovulation in the oviduct. This polar body appears at the apical end of the oocyte, where the nucleus has been located throughout oogenesis. The second polar body is given off either in the oviduct or during extrusion through the gonopore.

During this process of oogenesis in *Bolinopsis*, two important characteristics must be recorded:

1. Nurse cell development begins prior to any oocyte development. All nurse cells undergo the same general structural changes, although the more distal cells do not become quite as large as the proximal cells, and their cytoplasm is never as densely packed with organelles.
2. Nuclei of oocytes associated with nurse cells are smaller and less active than nuclei of solitary oocytes. *Bolinopsis*, then, affords the first reported example of an oocyte nucleus that does not grow at all.
3. In fact, all the major synthetic functions of oogenesis in *Bolinopsis* are performed by the nurse cells, and not by the oocyte.

Oogenesis in Gastrodes. In view of the parasitic nature of *Gastrodes*, and the occurrence of the planula larva (Fig. 6.4B), both egg structure and oogenesis might be expected to differ from that in other ctenophores. However, gonads have not been observed in *Gastrodes*, but certain large cells occurring among the endodermal cells of the planula larva and later among the ectodermal epidermal cells of the ventral everted pharynx (Fig. 6.4A). The presence of the egg cells in the ectoderm seems to be a secondary adaptation to its parasitic life. The primordial germ cells initially found in the endoderm of the planula larva migrate into the pharyngeal ectoderm where they can get nourishment directly from the host. Since sperm have not been observed, it has been proposed that *Gastrodes* is parthenogenetic.

Spawning. In many ctenophores, spawning is controlled by light, e.g. adult specimens of the lobate ctenophore *Mnemiopsis* will spawn approximately 8 h after the lights are turned off. However, the timing of spawning varies considerably for different species. Some species spawn in response to the onset of daylight. Generally, sperm are shed first and can be seen as white lines in the testicular sinuses underneath the ctene row. On the other hand, the oocytes can be seen in the gonads soon after, and then they are released into the ovarian sinus before being shed through the gonopores. However, in view of Dunlap-Pianka (1974) who studied the lobate ctenophore *Bolinopsis microptera*, the oocyte during spawning becomes amoeboid and ovulate through the glandular epithelium into the oviduct. Within the latter, secretions of the gland cells adhere to the vitelline membrane and form the distinct circular aperture, 50–100 μm in diameter. In *Dryodora*, one gonopore forms for each oviduct and is about 200 μm in diameter. Oocytes become constricted and evidently amoeboid as they leave the pores.

In most and perhaps all ctenophores, except certain ovoviviparous species, gametes are released through the epidermis above the gonads, probably always through ducts and pores, and the first observation of egg release through the epidermis, was observed in *Beroë ovata*, a *Pleurobrachia* species, and the cydippid *Lampetia pancerina*. At Friday Harbour, Dunlap-Pianka repeatedly observed the release of both oocytes and sperm through consistently located pores above the gonads in *Bolinopsis microptera*, *Dryodora glandiformis*, and *Pleurobrachia bachei*, and oocyte release alone in *P. pigmentata* (Dunlap, 1966).

The comb plates overlying the gonads stop beating during the process of spawning for both sperm and eggs. However, they quickly resume, to help disperse gametes. The oocytes complete their maturation within the ovarian sinus. This means that the first and the second maturation divisions occur regardless of the presence of the sperm. In some species, the polar bodies remain on the surface and mark the site of the underlying oocyte nucleus.

Development

Embryonic development. Because of the unique segregated ooplasm, the equally unique biradial pattern of cleavage and the highly determinate development of the embryo, there has been considerable interest in ctenophore embryology. Cleavage, gastrulation and early larval development are remarkably similar throughout the phylum, with differences becoming increasingly apparent during postgastrular differentiation, and being most striking in postlarval transformations.

Fertilization. Dunlap recorded that fertilization apparently occurs when an oocyte is extruded through the ovarian gonopore into the seawater, and in the view of Yatsu (1912), more than one sperm may enter the egg. A small projection, probably representing a fertilization cone, has been observed on the egg surface near the site of the polar body

(Fig. 6.9F and G) emission in both *Coeloplana bocki* and *Bolinopsis microptera*. Pelagic ctenophores are self-fertile. Nearly 100% normal development has been repeatedly obtained from self-fertilized eggs of a single *Bolinopsis microptera*, *Pleurobrachia bachei* and *Dryodora glandiformis* isolated at the time of collection (Dunlap, 1966).

Certainly, the hermaphrodite nature of the ctenophores and the sequence of events of spawning constitute the main factor causing self-fertilization: an individual first sheds sperm, briefly disperses them and then releases oocytes into its own sperm 'suspension'. Self-fertilization may be of widespread occurrence in nature among pelagic ctenophores. While in swarms, cross-fertilization might also occur, perhaps synchronized by reaction to light.

Carré et al. (1991) reared and studied the event of fertilization in the ctenophore, *Beroë ovata*. They collected mature *B. ovata* (5–15 cm) between February and June in surface waters of the bay of Villefranche-sur-Mer. In the laboratory, the animals were kept under equipped conditions. Ripe animals were observed with gonads beneath

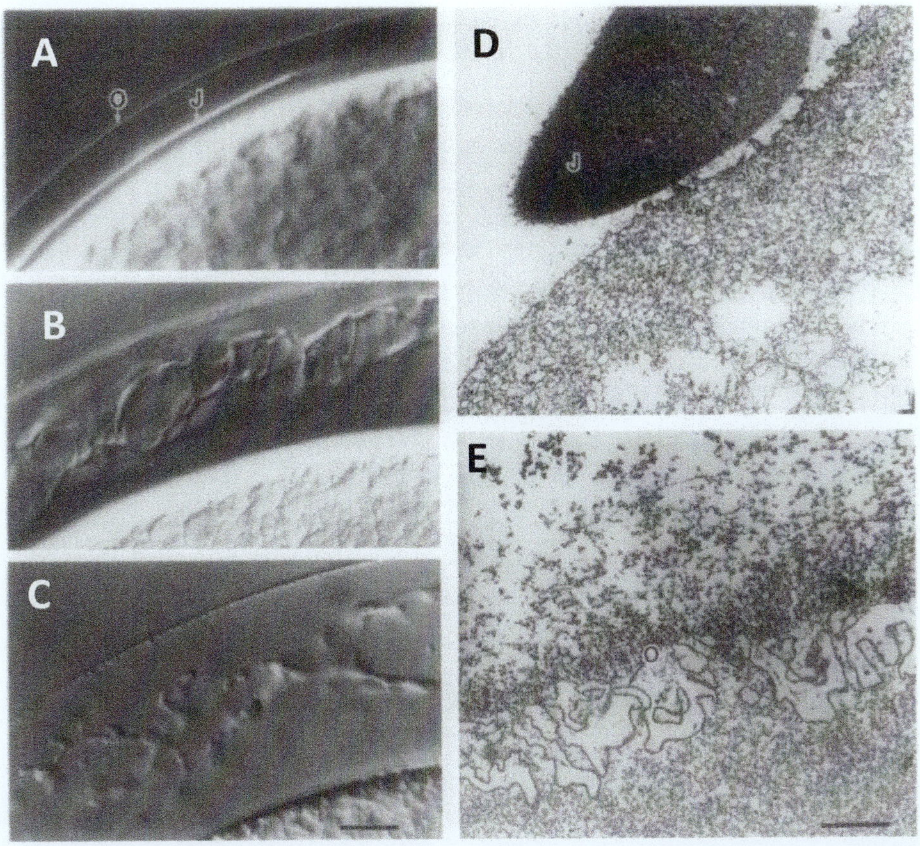

FIG. 6.9 Fertilization of the egg of *Beroë ovata, adapted from Carré et al. (1991)*. DIC images photographed from an optical memory disc recorder (OMDR) video sequence: (A–E) **the envelope** and its transformations after fertilization, (A)–(C) are DIC images and scale bar = 30 μm; (A) In freshly shed eggs the peripheral layer envelope composed of an outer layer (0) and a refractile jelly layer (J) are still closely apposed to the egg surface; (B and C) after fertilization, the envelope swells and the refractile jelly meshwork is modified; (D) electron photograph of the jelly (J) on top of the freshly laid egg surface (corresponds to the situation shown in (a)); (E) electron photograph of the fully swollen outer layer (0) 30 min after fertilization, scale bar of D and E = 2 μm; (F) **Sperm entry** in the egg of *Beroë ovata* (some selected photographs): **Left column**: showing sperm entry seen from above (times are indicated in minutes after fertilization). Right column: Growth and resorption of the fertilization cone (c) seen from the side (the times indicated are approximate times after fertilization in minutes), scale bars = 20 μm; (G) The **migration of the female pronucleus** and its fusion with the male pronucleus, DIC images photographed from an (OMDR) video sequence: only one photo was selected from the left column showing the beginning of the migration of the female pronucleus (FN and *white arrow*) towards the immobile sperm pronucleus (SN and white arrowhead). The female pronucleus originates from the site of meiosis indicated by the position of the polar body (Pb) and bulges the egg surface as it moves; (f–i) Last part of the trajectory of the female pronucleus (FN and *white arrow*) towards the male pronucleus (SN *black triangle*) surrounded with its locally differentiated zone (smooth area), (i) Corresponding mitotic apparatus with the long axis of the mitotic spindle (MS, *arrows*) perpendicular to the female pronucleus trajectory. (f)–(I) are DIC optics, images taken respectively at times 0 (f), 2 min 20 s (g), 5 min (h), and 85 min (i) from an OMDR video sequence. Scale bar = 20 μm; sequence showing the fusion of the male (SN) and the female (FN) pronuclei: in (m) fusion occurs; (n and o) the male chromosomal material (SC) decondenses, images taken respectively at times (8 min for m), (9 min n) and (10 min o) from an (OMDR) video sequence, scale bar = 5 μm; (H–K) **First mitosis** in the egg of *Beroë ovata*. Selected photos from DIC images which are taken minutes apart as indicated from (OMDR) video sequence showing mitosis taking place beneath the egg surface. The chromosomes condense (C) within the zygote nucleus (ZN) surrounded by mitochondria (m). From anaphase through telophase (H and I), the spindle pole (P). *(From Carré et al. (1991).)*

(Continued)

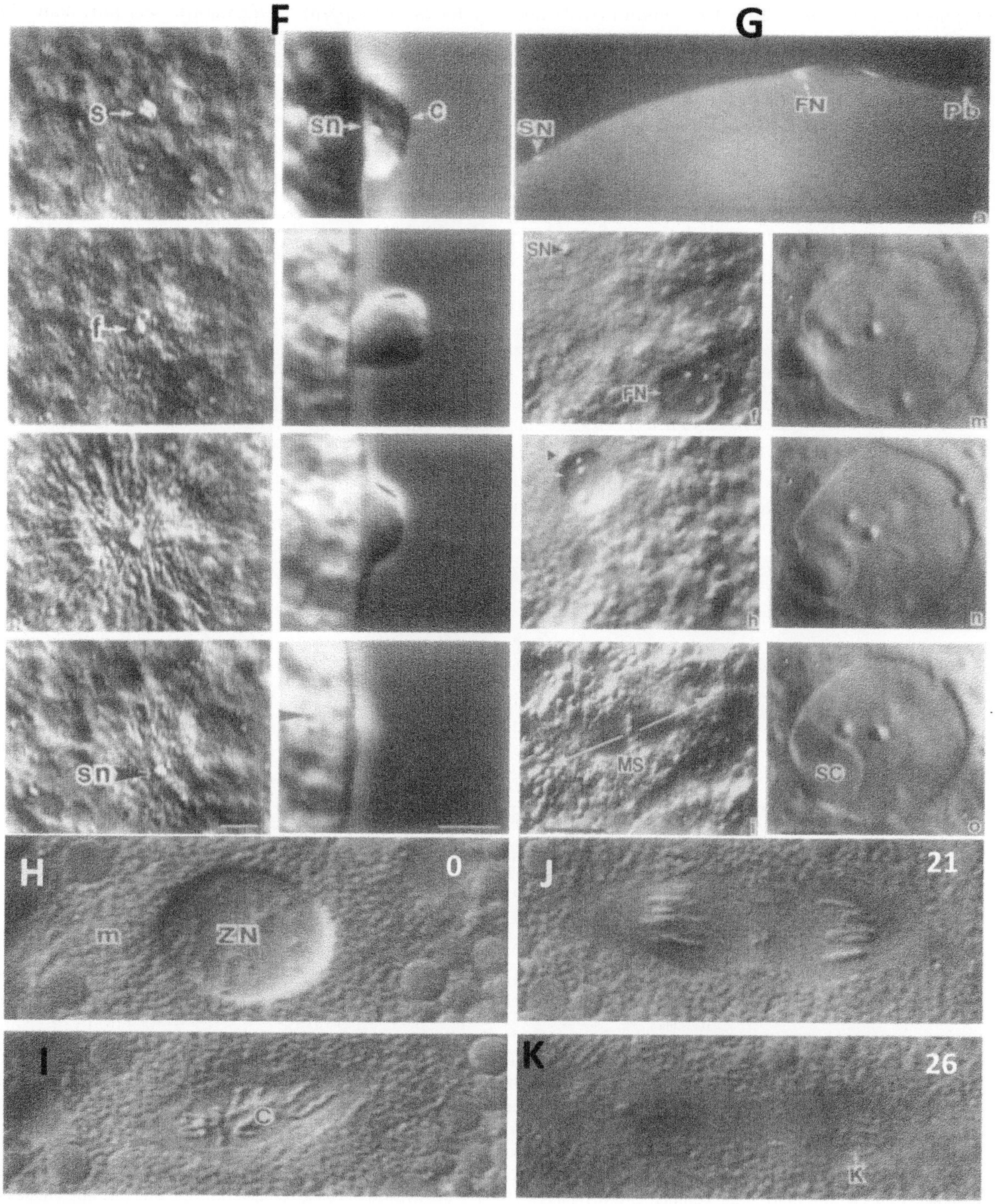

FIG. 6.9, CONT'D

the rows of comb plates. They were placed for spawning in isolation, kept in a large volume of seawater (1–5 L/animal). The authors had the opportunity to observe details of sperm entry with exceptional clarity. In addition, they have obtained new information on polyspermy (Yatsu, 1911), reorganization of the egg cortex, pronuclear fusion and mitosis. *B. ovata* can produce eggs every day or every 2 days for up to 2 weeks.

Most observations were with differential interference contrast (DIC) and epifluorescence optics using an imaging station consisting of a Zeiss Axiophot microscope. Sperm is emitted through spermiducts situated at the edge of the eight rows of comb plates. The shedding lasts for several minutes. During that period, the beating of comb rows ceases momentarily four or five times and then resumes, thereby dispersing the spermatozoids. The laying of eggs follows through separate gonopores (Fig. 6.8D). Naturally, in daylight, animals seem to shed gametes without any predictable periodicity about once a day or every 2 days. By isolating animals immediately after collection, the authors could obtain unfertilized eggs and motile sperm from the same animals. They also achieved in vitro fertilization. When kept in isolation, most animals become self-fertile after several days. Naturally, where freshly collected animals are kept together, all egg batches are fertilized within 30 min of being released. If they are kept in the dark for long periods (48–72 h), spawning can be induced in most mature animals within 2–3 h of exposure to bright sunlight.

Carré et al. (1991) reported that experimental fertilization in ctenophores has never been achieved (at that time). It has been reported that self-fertilization occurs in some species favoured by the simultaneous shedding of sperm and egg (Dunlap-Pianka, 1974). However, Carré et al. found that when freshly collected *B. ovata* were isolated from each other, only 21% of the animals were self-fertile (9 out of 41).

When the egg is just shed, it is already undergoing the first meiotic division and completes maturation (30 min) whether it is fertilized or not. The oocyte is surrounded by an envelope, consisting of a refractile meshwork (jelly layer) limited by a tougher outer layer. This envelope swells upon contact with seawater. It is gradually modified after fertilization, as illustrated in Fig. 6.9A–E. The ability to control fertilization in *B. ovata* allows these authors to observe sperm entry in real time and in time lapse with extraordinary clarity (Fig. 6.9F). The spherical sperm head first becomes modified when in contact with the egg (Fig. 6.9G and H). A flange forms in the anterior part, probably due to an acrosomal reaction. The flagellum then becomes rigid. It contracts around the sperm entry site; radiating membrane folds develop in a 70-μm zone around the spermatozoid during the first minute after sperm fusion and then subside during the following minute (Fig. 6.9F). About 40–60 min after the initial sperm entry, the zones have reached their final appearance and male pronuclei have decondensed slightly and moved closer to the egg surface within the dense accumulation of mitochondria and autofluorescent organelles.

By the end of this period, the sperm nucleus is situated 15 μm beneath the egg surface, but remains within a few micrometres of its initial entry site until pronuclear fusion. In the natural situation, where freshly collected animals are kept together, all egg batches are fertilized within 30 min of being released.

The female nuclear migration proceeds at approximately 0.4–0.5 μm/s and can last from a few minutes, if the fusion occurs with a sperm pronucleus close to the meiotic site, to hours if the sperm pronucleus (or several successively visited sperm pronuclei) is located far away from the meiotic site (up to 1500 μm in this very large egg). As soon as the male and female pronuclei meet, fusion occurs and is followed by a visible decondensation of the male chromatin within the zygote nucleus (Fig. 6.9G (m)–(o)). In this large egg, mitosis occurred just beneath the plasma membrane, enabling the authors to record its progression in detail (Fig. 6.9G (i) and H). The unipolar cleavage furrow is initiated at this site perpendicular to the long axis of the spindle. In eggs kept with minimum compression between slide and coverslip, the trajectory of the female pronucleus, the fusion of pronuclei, and the position and orientation of the mitotic spindle could be followed and recorded. The authors have noted that the long axis of the spindle formed perpendicular to the trajectory of the female pronucleus.

Cleavage. Ctenophore development is unique and does not bear obvious similarities to embryogenesis in any other extant organisms (Martindale and Henry, 1997a,b). However, Cleavage is unilateral, holoblastic, unequal and (after the second division) biradial and highly determinate. Early divisions are usually quite rapid, each requiring 5–10 min. The first two cleavages begin at the animal pole, and are meridional and equal. The third cleavage begins near the animal pole and divides each of the four blastomeres along a plane of 45 degrees and hence oblique to the first division plane. It is also unequal, resulting in an eight-cell stage composed of two rows, of four blastomeres each, paralleling the future tentacular plane (Dunlap-Pianka, 1974) (Fig. 6.10P and A).

By the eight-cell stage, the finding characteristics of biradial symmetry of the embryo have been established: the oral–aboral axis, the tentacular plane and the sagittal plane. This symmetry is reflected not only in the configuration of blastomeres, but also in their reproductive developmental potentials (Fig. 6.10A). During oogenesis in *Bolinopsis*, the three nurse cell clusters join the oocyte at three points: two lateral and one more or less basal to the apical nucleus (Fig. 6.7D) in biradial symmetry of the oocyte-nurse cell complex, which could well correspond with that of the embryo and adult (Fig. 6.10P (d)), although a causative correlation with developmental potentials of the blastomeres is not being suggested. The fourth cleavage is perpendicular to the first three, hence is horizontal, and is also distinctly unequal, with each of the eight blastomeres giving off a micromere towards the aboral pole. Until then, the endoplasm and cortical plasm have been generally equally distributed among blastomeres, each being a small edition of the zygote. However, with the fourth cleavage there is a partial separation of the plasms, and the eight micromeres contain predominantly cortical plasm, while the eight macromeres contain mainly endoplasm. The fate of the 'subcortical plasm' has not been adequately studied, but preliminary observations of *Bolinopsis* development suggest that much of it comes to lie at the 15-cell stage in the micromeres.

Through the 16-cell stage, divisions are synchronous in all quadrants, but during the rest of cleavage, divisions of

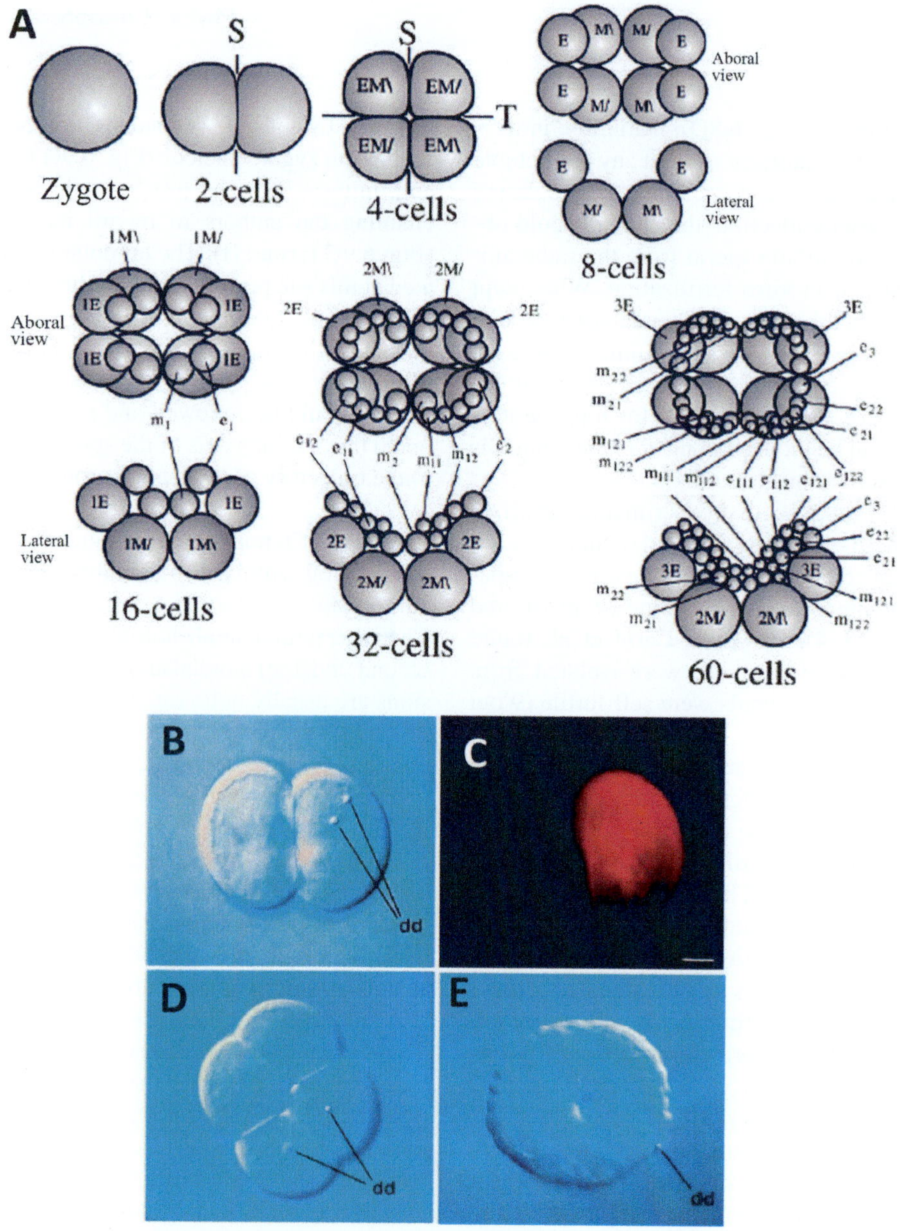

FIG. 6.10 (A) Diagrams of the ctenophore cleavage process: The first cleavage plane corresponds to the future adult sagittal plane (S), while the second cleavage plane corresponds to the tentacular plane (T). Cell lineage experiments done by Martindale and Henry, revealed the existence of two distinct sets of diagonally opposed quadrants referred to as the EM/ and EM\ blastomeres in the 4-cell-stage embryo. The third cleavage shows two distinct cell types which are: 4 middle cells (M cells) and 4 end (E cells). At the 16-cell stage (the 4th cleavage) each E cell gives rise to a small e1 micromere at the aboral pole and each M cell forms a similar m1 micromere. At the 32-cell stage (5th cleavage), 1 M and 1E macromeres give rise to m2 and e2 micromeres at the aboral pole, respectively. e1 micromeres divide to form e11 and e12 cells, and m1 cells divide to form m11 and m12. The sixth cleavage, the 2E macromeres give rise to another round of micromeres (called e3) to generate the 60-cell-stage embryo. By definition, e11 and m11 lie closest to the sagittal plane. Additional micromeres are given off at the oral pole during gastrulation (not shown). (B–E) These photos show that the intracellular injection of DiI into ctenophore blastomeres is a reliable cell lineage marker. (B) DIC image of a two-cell-stage *Mnemiopsis* embryo approximately 1 min after one cell was injected with two DiI-containing oil droplets (dd); (C) Fluorescent micrograph of the embryo seen in (B). Note that the DiI quickly labels internal and external membranes; (D) The same embryo seen in (B) after one subsequent cleavage division; (E) gastrula stage embryo seen from the oral pole in which one cell at the four-cell stage was injected with DiI. (F–K) Identified blastomeres of the e1 lineage labelled with DiI. The photographs on the left side (F, H, J) are DIC images corresponding to the fluorescent images shown on the right side of each row (G, I, K). (F and G) Aboral view of a cydippid showing that e1 micromeres contribute to both the adesophageal (ae) and adtentacular (at) ctene row within each quadrant. The ciliated grooves, portions of the apical organ (ao) and tentacle bulb (tb), are also labelled. There is a small amount of epidermal staining below the tentacle bulb not visible in this photo; (H and I) Lateral view of a cydippid showing the DiI labelling pattern of an e11 micromere. The apical organ, tentacle bulb, ciliated grooves (cg) and epidermis (ep) are all labelled. Note that, as expected, the adesophageal ctene row is labelled but, in addition, the oral-most ctene plate located in the adtentacular row is also labelled; (J and K) Aboral view of an e12 labelling pattern. The adtentacular ctene row and ciliated groove are labelled. The adesophageal row is also brightly labelled. Thus, the two daughter cells of an e1 micromere do not consistently give rise to complimentary clones. (Scale bar = 25 μm); (L–O) Identified blastomeres of the m1 lineage labelled with DiI. The photographs on the left side are DIC images, of the same cydippid stage adults seen on the right side of each row. The middle and bottom rows indicate that both descendants of the m1 micromere, m11 (L and M) and m12 (N and O) give rise to epidermis, subcomb row tissue and parts of the apical organ but that ctene row and the tentacle bulb staining is segregated to m12. Abbreviations

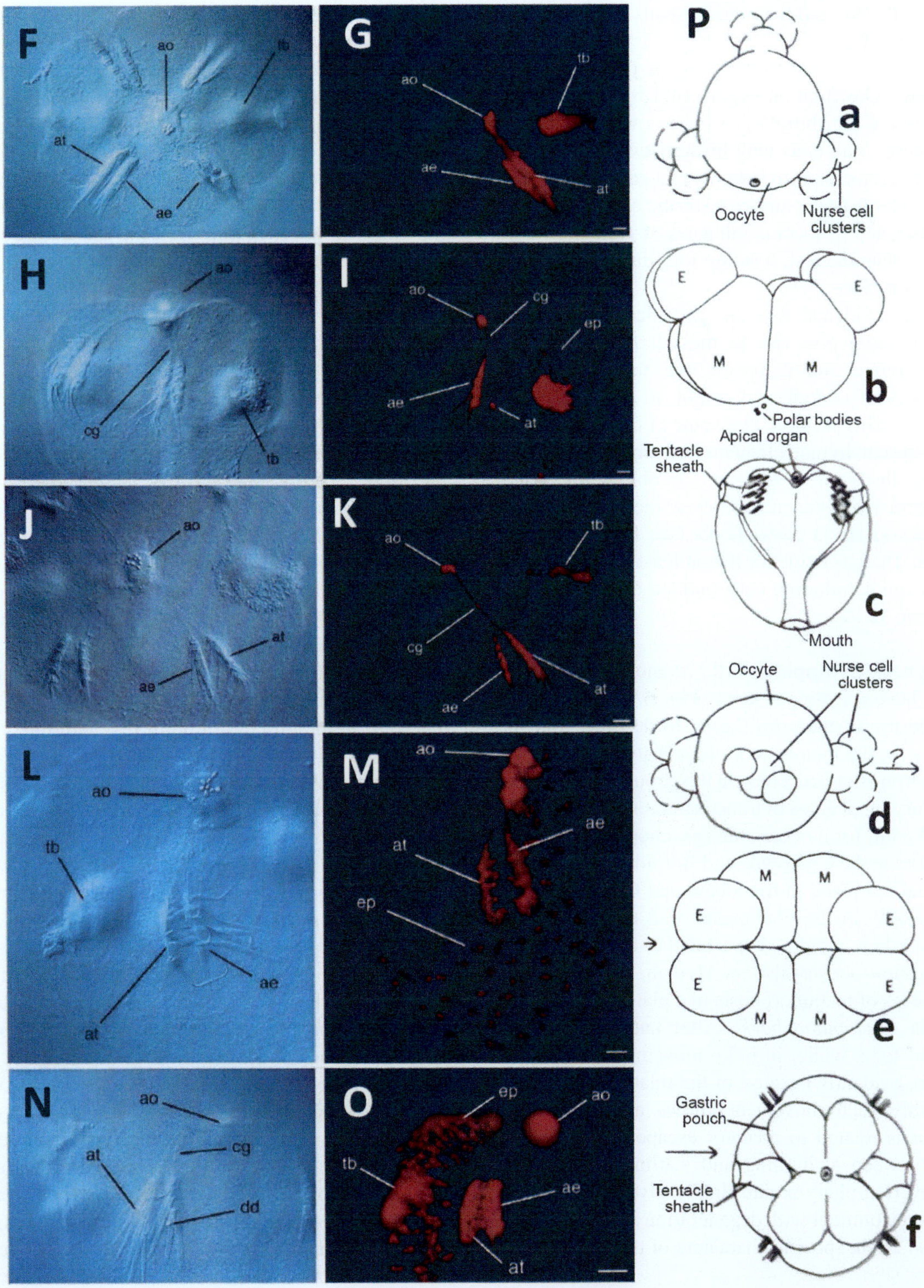

FIG. 6.10, CONT'D are as above; bar=25 μm. *(A, cleavage from Martindale and Henry, 1999. All the photographs (B)–(O) from Martindale and Henry, 1997a,b.)* (P) Diagrams of the possible correlation between the biradial symmetry of the oocyte-nurse cell complex (a) and (d) and that of the 8-cell stage of the embryo (b) and (e) and of the cydippid larva (about 1-day-old) (c) and (f) of *Bolinopsis microptera (from Dunlap-Pianka, 1974)*. (A), (B), (C) viewed from the sagittal axis; (D) and (E) viewed from the aboral pole. The polarity relationships between the embryo and larva (B) to (C) and (E) to (F) have been established (Reverberi and Ortolani, 1963); Whether the loci of the two lateral nurse cell clusters bear any relationships to the ends of either the tentacular (as shown) or sagittal axis has not been determined; and thus (A) to (B) and (D) to (E) are partly hypothetical; the 'ends' and 'middle' blastomeres of the embryo are labelled, respectively 'E' and 'M'.

the middle cells (M) and their descendants lag behind those of the end cells (E).

Gastrulation. Gastrulation occurs first by epiboly, or an oral migration of the aboral micromeres over the surface of the macromeres, and ends with invagination, or emboly, in which the macromeres turn inwards and carry some micromeres to the inside of the embryo. During epiboly, these micromeres destined to form comb rows, divide and migrate more rapidly than the rest, forming four thickened ribs over the embryo's surface.

Hence the evidence that the gastrular micromeres in *Mnemiopsis leidyi* give rise to mesodermal musculature and collenchyma, and therefore that the mesoderm in ctenophores is of endodermal origin, is placed in doubt, although it remains possible that some of these micromeres may form mesenchymal elements. In any case, although ectoderm is clearly derived from the aboral micromeres and endoderm from the macromeres, the problem of the origin of mesoderm in ctenophores (see Hyman, 1940) is still with us. There is evidence that at least some mesoderm may arise from ectodermal cells budded off at the oral region (Hyman, 1940).

Larval types. **Cydippid:** All ctenophores, except *Gastrodes*, have a cydippid larva (Fig. 6.10P (c) and (f)), although the use of the term 'larva' to describe this stage is valid only for the members of order Platyctenea, because true metamorphosis occurs only in this group of Ctenophora. However, varying degrees of transformation of cydippid larvae into the adult form occur among ctenophores. The least change occurs in the Cydippida and Beroida, with considerably more taking place in the lobata and Cestida (Hyman, 1940). It is only in the Platyctenea that the change from larva to juvenile is rapid and spectacular, and deserving of the term metamorphosis (also see Hyman, 1940).

The process of metamorphosis in a platyctenid has been described in *Coeloplana bocki*. After hatching, the larva swims about for a while, in jerky movements reminiscent of copepods, evidently because of the relatively long comb cilia. Possibly such movement allows a more effective dispersal, or is related to predator escape. The larva then alternates between swimming and settling briefly on its expanded everted pharynx and finally gives up swimming altogether, concomitant with degeneration and eventual loss of its combs. Then, specific characters of the adult begin to develop.

Planula: The youngest stage of the platyctenid *Gastrodes* (Fig. 6.4B) so far observed bears a remarkable resemblance to the planula larva of the Cnidaria Dunlap-Pianka (1974) reported that Komai (1922) described that this larva has a solid endodermal core, containing possible primordial germ cells, and is covered with uniformly ciliated epidermis. Komai observed one such larva in the process of boring through the test of the tunicate *Salpa*. The planuloid larva of *Gastrodes parasiticum* having entered the mantle of the host *Salpa*, becomes cup-shaped and develops eight comb rows, an apical organ, rudimentary tentacles and a gastrovascular system. This is the first of the four pouches, and later of eight subequal pouches reminiscent of those of a freshly settled juvenile *Coeloplana bocki* (Komai, 1922; Hyman, 1940). Komai (1922) considered this last stage, the oldest to occur in *Salpa*, as the sexual stage, and interpreted as oocytes certain large cells in the ventral epidermis (Fig. 6.4A).

Mnemiopsis leidyi

Cell lineage experiments

The works of Komai (1922), Hyman (1940), and Dunlap-Pinaka (1974) on ctenophores were mainly on *Bolinopsis*, but more recently, *Mnemiopsis* became the favourite animal. Martindale and Henry (1995–2000) worked experimentally on embryogenesis of this lobate ctenophore, *Mnemiopsis leidyi*. They collected their animals at Quissett Harbour or off the rock jetty at NOAA in Woods Hole, Massachusetts, United States during the months of July and August. They published a series of papers, where they illustrated and described the early developmental stages in detail (Fig. 6.10A). In the 1996 study, they discovered that, during the process of adult regeneration, ctene plates are generated by m1 micromere derivatives (Martindale and Henry, 1996). They have therefore reinvestigated the embryological origins of ctene rows in the ctenophore *Mnemiopsis leidyi* using intracellular lineage tracer DiI. They showed that the use of DiI dissolved in soybean oil is a reliable cell lineage tracer in ctenophore embryos that is more accurate than previous labelling methods. Previous cell lineage experiments, using chalk particles, revealed that each pair of rows is derived from one of the four e1 micromeres (Fig. 6.10A, 16-cell stage). Those classical cell lineage experiments indicated that comb rows are derived exclusively from the four e1 micromeres at the 16-cell stage (Fig. 6.12A). This conclusion was also supported by the fact that no ctene rows (or their underlying endodermal canals) form when all four e1 micromeres were deleted. They have used intracellular DiI cell lineage tracing to determine that, in addition to e1 micromeres, the four m1 micromeres also make significant contributions to the ctene rows. In another work, in 1997, they injected identified blastomeres, in order to determine the fidelity of the lipophilic dye DiI as a lineage tracer. They then watched the domains of fluorescent labelling through subsequent development. Injected droplets of DiI quickly labelled the internal and plasma membranes of the injected, but never the adjacent, blastomeres (Fig. 6.10B–E). Labelled clones stay together in a contiguous domain through gastrulation (Fig. 6.10E). Development proceeded normally in both injected and uninjected cells through cydippid formation. These results indicate that the intracellular injection of DiI dissolved in soybean oil is an effective and reliable cell lineage tracer in the *Mnemiopsis* embryo.

These injection experiments confirm previous cell lineage experiments of ctenophore embryogenesis that each e1 micromere generates a pair of ctene rows. Labelled e1 clones ($n=32$) always include both ad-oesophageal and ad-tentacular ctene rows and their respective ciliated grooves (Fig. 6.10F–K), as well as contributions to the floor of the apical organ, tentacle apparatus and a small amount of epidermis. The presence of unlabelled portions of the ctene rows following the injection of e1 micromeres suggested that other cells contribute to these structures. Previous reports indicated that e11 micromeres generate the ad-oesophageal ctene row while e12 micromeres generate the ad-tentacular row. Martindale and Henry's injections indicate that this is not entirely correct. They injected each of the two daughter cells of the e1 micromeres, the e11 ($n=14$) and e12 ($n=11$) cells in order to see how their fates contributed to each ctene row. Fig. 6.10H and I show an example in which an e11 micromere clone has generated the ad-oesophageal ctene row and ciliated groove, as well as the oralmost ctene plate of the ad-tentacular row. Fig. 6.10J and K illustrates another embryo in which the sister cell of the e11 blastomere called e12, has generated the ad-tentacular row and ciliated groove plus a large portion of the ad-tentacular region of the ad-oesophageal row and groove. Both e11 and e12 cells generated cells in the apical organ, tentacle sheath and epidermis as well, but did not generate consistently unique cell-specific domains of labelled tissue. There is normally variation in the partitioning of adult descendants in labelled e1-derived clones. No contribution to the endodermal canal system was observed.

In contrast to what is seen in the e1 lineage, the daughters of m1 micromeres generally give rise to distinctly different subclones (Fig. 6.10L–O). Both m1 descendants contributed to epidermis, the polar fields, the floor of the apical organ and sub-ctene row material (possibly including the somatic gonad). m11 descendants ($n=13$) gave rise predominantly to epidermal and sub-ctene row material associated with the somatic gonad, while m12 cells made ctene plates ($n=18/22$), sub-ctene tissue, dome cilia ($n=11/15$) and components of the tentacular system ($n=12/22$). These results reveal a picture of ctenophore embryogenesis, which is in marked contrast to the former rigid mosaic reputation of ctenophore development, and invite speculation as to the role of the cleavage programme in embryonic patterning in the lower Metazoa.

Martindale and Henry continued their experimental research work on embryogenesis of the lobate ctenophore, *Mnemiopsis leidyi*, and in 1999 reported that the third cleavage is oblique and allows the overt identification of the two planes and the oral–aboral axis. The four E ('end') and M ('middle') blastomeres present at this stage (eight-blastmere stage) represent the major stem cell lineages in ctenophore embryos. Each M macromere gives rise to two rounds and each E macromere to three rounds of micromeres at the aboral pole (Fig. 6.10A, 8-cell stage). These micromere derivatives proliferate and move over the macromeres by epiboly during gastrulation. The macromeres remaining at the oral pole generate a second set of oral micromeres later during development (Reverberi, 1971; Martindale and Henry, 1997a).

The authors also injected individual E and M blastomeres at the eight-cell stage (data not shown), but their contributions are best understood by examining their sub-lineages, which are described below. In addition to generating major components of the ectodermal ctene rows, all four e1 micromeres also contribute to discrete regions of the apical organ and tentacle apparatus. In particular, the e1 micromeres contribute to cells in the floor of the apical organ (including the balancing cilia; see Fig. 6.11B) and cells of the ciliated groove (Fig. 6.11A, C–G), but not to dome cilia or lithocytes.

Martindale and Henry (1999) reported that Reverberi and Ortolani (1963) used chalk particles to trace the fates of cells from two different Mediterranean ctenophores, the atentaculate *Beroë forskalii* and the lobate *Bolina hydatina*. Due to the limitations of the techniques of that era, they were unable to make detailed statements regarding the origins of adult structures. In their view, the two comb rows located in each quadrant are derived separately from each daughter of the e1 micromeres (i.e. e11 and e12). Their results clearly revealed that both e1 daughters normally make contributions to both comb rows within a quadrant and that individual m12 micromeres also contribute to both comb rows (Martindale and Henry, 1997b).

In this work (Martindale and Henry, 1999), they also reported in their discussion that endoderm, and thus endomesodermal derivatives, originate from the animal and not the vegetal pole of the ctenophore embryo, which is apparently unique in the Metazoa and clouds the relationship of the oral–aboral axis of diploblasts to the anterior–posterior axis of bilaterians (Martindale and Henry, 1998).

Organogenesis

The main structures of the cydippid larva are the combs (motility), tentacles (prey-catching), an apical sense organ (orientation), a stomodeal pharynx (feeding, settling) and a simple gastrovascular system (digestion).

Comb plates

The combs constitute one characteristic set of structures displayed by most adult ctenophores that are the eight longitudinal rows of comb plates (comb or ctene rows) that are used for movement. Comb plates (Fig. 6.2D and F) arise

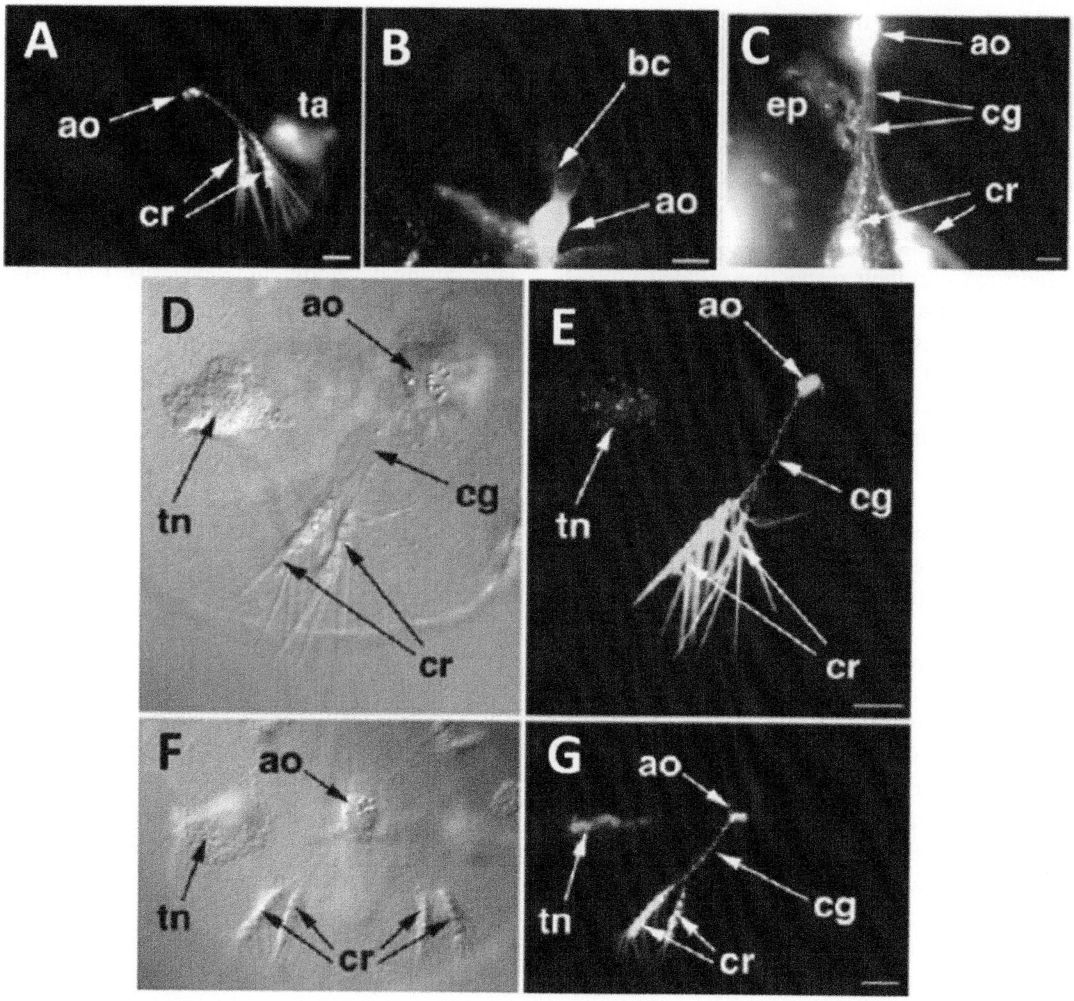

FIG. 6.11 (A–C) Fluorescent labelling patterns observed following the injection of a single e1 micromere with DiI in a 16-cell stage *Mnemiopsis leidyi* embryo: (A) Fluorescence micrograph of the embryo, scale bar, 40 μm. The e1 micromere has given rise to both ctene rows (cr), the ciliated grooves, a portion of the floor of the apical organ (ao), and the tentacle apparatus in the corresponding \ quadrant; (B) The floor of the apical organ and one bundle of balancing cilia, but not dome cilia or lithocytes, are fluorescently labelled in the \ quadrant that contains the injected e1 micromere, bar = 10 μm; (C) High magnification view showing the labelled ciliated grooves running between the apical organ and the pair of labelled ctene rows. Ctene row beating is thought to be controlled by signals sent from the apical organ to the ctene rows via the ciliated grooves (cg). The floor of the apical organ and a variable group of epidermal cells are also labelled following e1 injections; (D–G) The two daughters of e1 micromeres (e11 and e22) most often generate the same complement of cell types. (D and E) Aboral views of a cydippid in which an /e11 micromere was labelled with DiI. e11 micromeres normally contribute to both ctene rows within each quadrant; ciliated grooves, the floor of the apical organ, and tentacle are also labelled; (F and G) Aboral views of a cydippid following the labelling of an e12/ micromere (D). The apical organ, tentacle, and ciliated grooves are labelled. Labels are the same as those used above. Scale bar = 10. *(From Martindale and Henry (1999).)*

from polster cells of the outer epidermis and are composed of hundreds of thousands of individual cilia bound together by compartmenting lamellae (Martindale and Henry, 1996). Ctenophores have a highly stereotyped cleavage pattern, so individual blastomeres can be identified and their fates can be followed throughout embryogenesis. It was known, from previous cell lineage experiments, using chalk particles, that each pair of comb rows is derived from one of the four e1 micromeres (Fig. 6.10A). Martindale and Henry (1996) killed two adjacent ad-oesophaeal e1 micromeres of the ctenophore *Mnemiopsis leidyi* with glass needles at the 16-cell stage (Fig. 6.10A, 16-cell stage). They then injected two other identified blastomeres in the surviving embryo with the fluorescent lineage tracer DiI dissolved in soybean oil. The injected blastomeres included the two 1 E macromeres, two 1 M macromeres and two m1 micromeres, which were all located on the same side as the two deleted e1 micromeres. The authors also injected the two surviving e1 micromeres on the side of the embryo opposite the two deleted e_1 micromeres. Another part of the experiment was to see whether the same lineage that makes ctene plates during embryogenesis also makes them during 'postgeneration'. This operation generated postembryonic juveniles (cydippid larvae) that were completely deficient in four of their eight ctene rows.

Experimental embryos were examined 24–72h after the operation for the presence of DiI in their newly formed (postgenerated) comb plates. Comb plates that appear during the larval period do not form in the same way that they do during embryogenesis, they do not become functionally connected to the apical organ until multiple plates in each row subsequently appear. The results of this experiment show that 1 E macromeres (12 cases) and 1 M macromeres (11 cases) do not generate cells in new comb plates formed during the larval period. Furthermore, the surviving two e1 micromeres which make the comb rows on the opposite side of the embryo (Fig. 6.12A and B) also do not participate in the formation of 'postgenerated' comb plates (15 cases). However, in 18 cases, when the two m1 micromeres were labelled in the same hemisphere of the embryo as the missing e1 micromeres, the newly formed comb plates were labelled (Fig. 6.12C and D). The authors investigated an alternative explanation; that previous ctenophore embryonic fate maps may be incorrect, and that m1 descendants normally contribute to comb plate formation during embryogenesis.

Genome sequencing of *Mnemiopsis* has revealed two genomic clusters containing 10 distinct copies of photoproteins, closely resembling the luciferase-type photoproteins found at the base of the Metazoa. In situ hybridization studies have shown that at least subsets of these mRNAs are expressed in photocytes prior to when these embryos are bioluminescent. Thus, the presence or absence of bioluminescence is a strong indicator of the developmental fate of a differentiated photocyte. In *Mnemiopsis*, bioluminescence is generated by the expression of luciferase-like photoproteins in a specific cell type, the photocytes. Fate-mapping experiments have shown that during development, the 2 M macromeres give rise to photocytes (Fig. 6.13(I)). Photocytes are normally located in the endodermal canals underneath the comb row (Fig. 6.13(II)). To visualize the photoprotein, embryos were acclimatized to the dark and imaged under a GFP (green fluorescence protein) filter, set on a Zeiss compound microscope. The blue fluorescent light excites the photoprotein in these cells, causing them to emit light at around 485–496nm (Fischer et al., 2014).

Conclusions of the work of Fischer et al. (2014) are as follows:

1. Embryos of the comb jelly *Mnemiopsis leidyi* undergoes a mosaic development and their embryos show differentiation of selected cell fates even when cytokinesis is arrested using cytochalasin B during the course of early embryogenesis (Fig. 6.14A).
2. Fischer et al. demonstrated that embryos arrested prior to the eight-cell stage do not show visible signs of cell fate differentiation from either the E or M lineages. However, if cytokinesis is blocked at the eight-cell stage or later, comb-plate-like cilia, which are derived

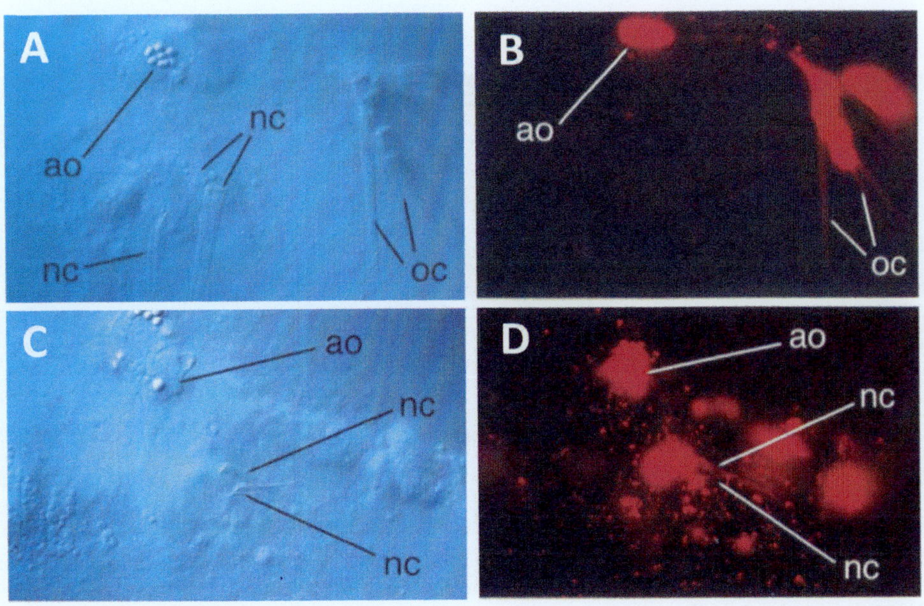

FIG. 6.12 (A) DIC micrograph a 48-h-old *Mnemiopsis* cydippid larva in the process of generating new comb plates (nc) following the destruction of two adjacent e1 micromeres at the 16-cell stage. The apical organ (ao) marks the aboral pole. The original well-organized comb plates (nc) can be seen on the right side of the photo (B) fluorescent micrograph of the cydippid larva seen in photo (A). The two surviving e1 micromeres were labelled with DiI at the 16-cell stage following the removal of the opposite two e1 micromeres. Note the absence of staining in the vicinity of the new comb plates; (C) DIC micrograph of 36-h-old cydippid larva in which two adjacent adesophageal e1 micromeres were destroyed at the 16-cell stage, and the two m1 micromeres on the same side were labelled with DiI; (D) fluorescent micrograph of the cydippid larva seen in photo (C) showing newly formed comb plates were labelled as adjacent and subjacent tissue including the apical organ. None of the existing comb plates on the unoperated side of the embryo contain labelled m-derived cells (not seen in this photo). *(From Martindale and Henry (1996).)*

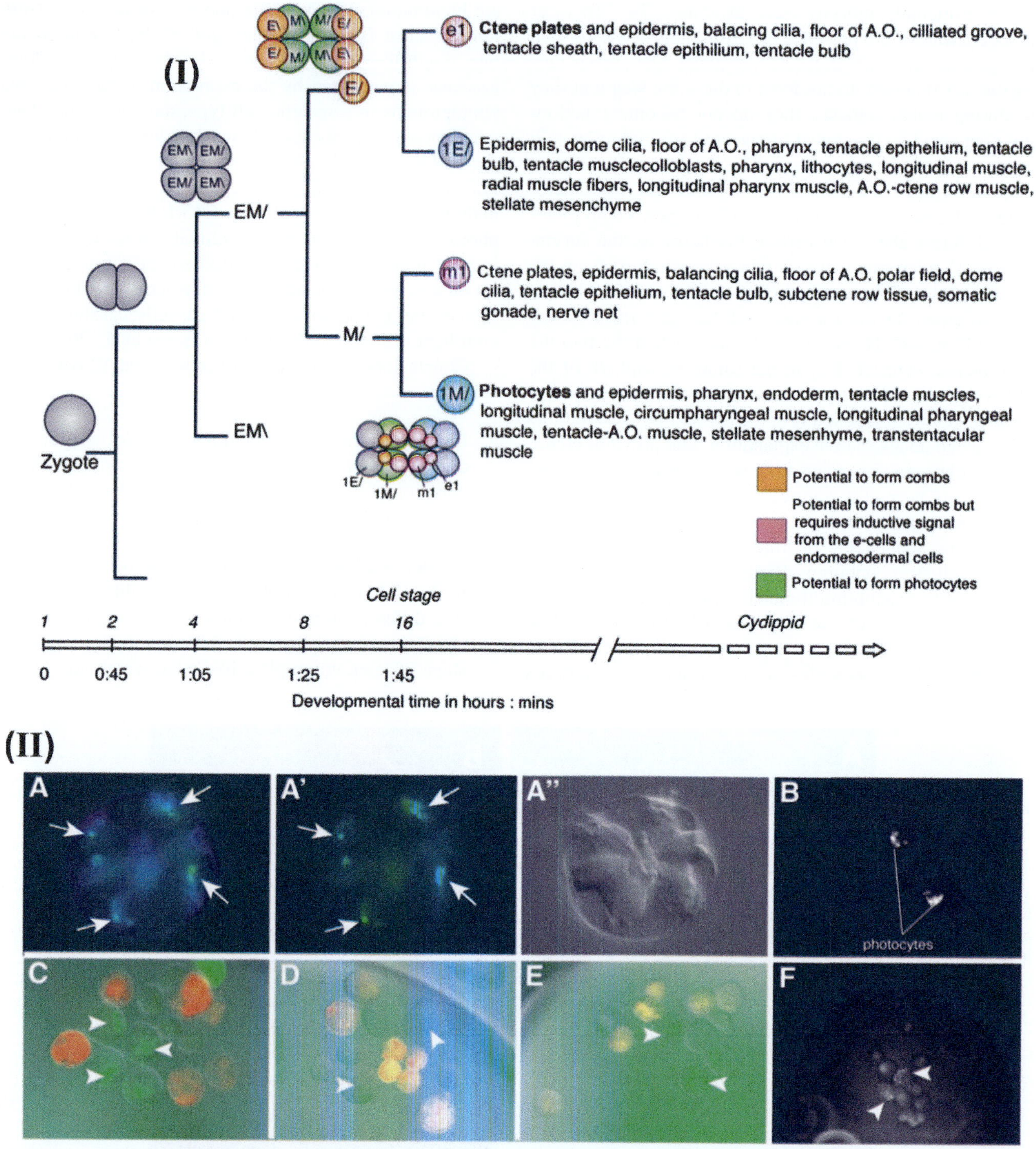

FIG. 6.13 (I) Simplified cell lineage fate map showing the e1, m1, 1E/ and 1 M/ micromere sublineages in the ctenophore *Mnemiopsis leidyi*. For each of the different early cleavage stages there are diagrams showing the embryo from the aboral pole; (II) Photocytes in control embryos and cleavage-arrested embryos. (A, A′, A″) In normal embryos, photocytes *(white arrows)* are visible at 9–10 hpf (hours post fertilization) in the endoderm beneath the comb rows. In this live animal, viewed from the aboral pole; (A) shows nuclei in *blue* (**Hoechst 33342 staining**) overlaid with photoprotein fluorescence *(green)*; (A′) Photoprotein fluorescence as viewed under the GFP channel; (A″) DIC image of the same embryo shown in (A) and (A′); (B) Photoprotein fluorescence of a control animal; (C–E) E macromeres were labelled with DiI at the 8-cell stage and embryos were cleavage arrested at the 16-cell stage. Photoprotein is visible as *green dots (arrowheads)*, in cells that are not labelled; (F) Photoprotein in an embryo with cleavage arrested at the 16-cell stage. *(From Fischer et al. (2014).)*

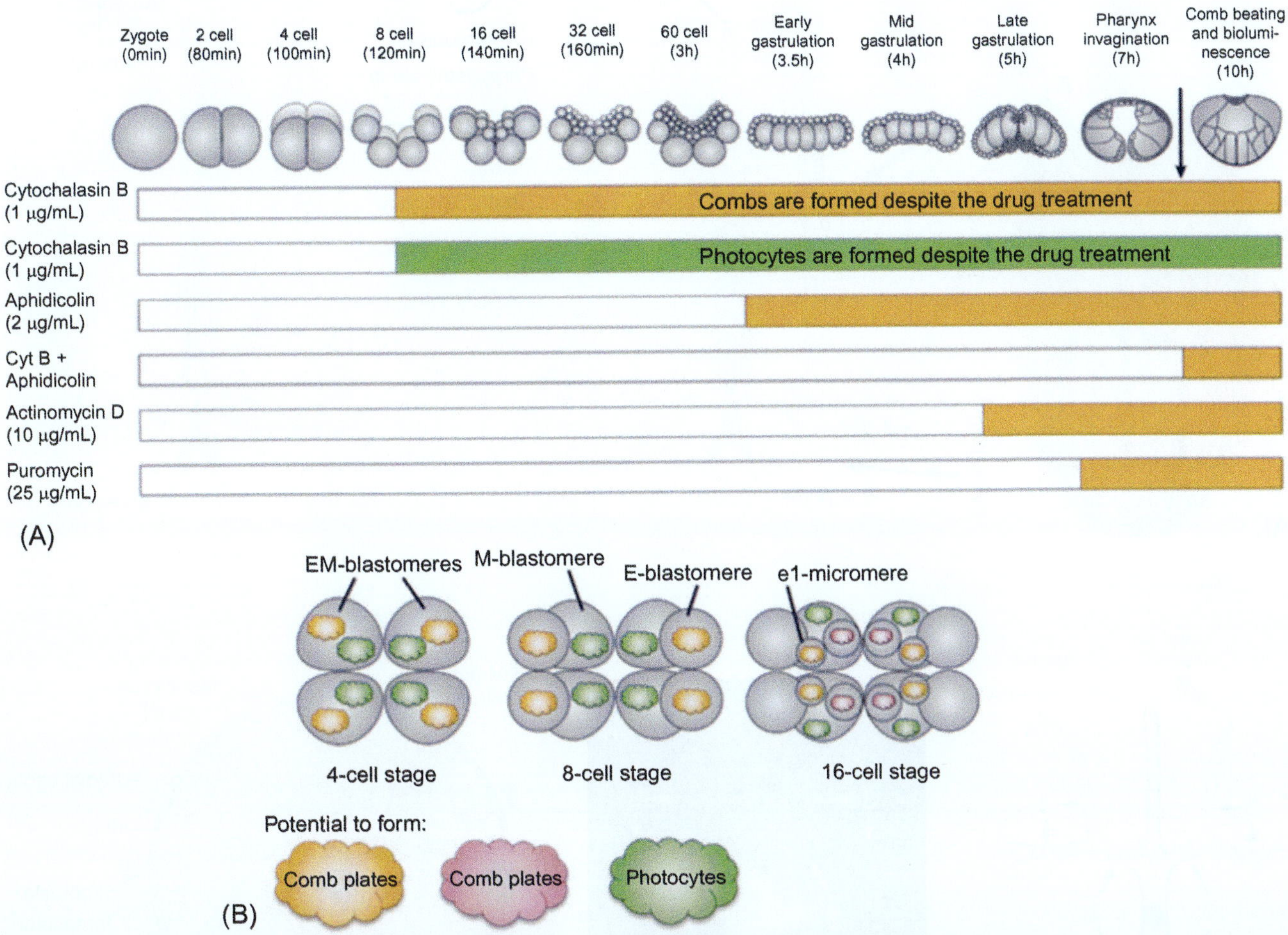

FIG. 6.14 Summary of mechanisms controlling the development of comb plates and photocytes in *Mnemiopsis leidyi*: (A) Overview of the normal embryonic development and the different treatments with cytochalasin B, aphidicolin, cytochalasin B plus aphidicolin, actinomycin D, and puromycin. The *black arrow* indicates when combs form during normal development. The *yellow part* of each bar indicates when a treatment starts that permits comb formation, and the *green part* indicates when a treatment starts that permits photocyte formation, in those cases where it has been tested. The *white part* of each bar indicates the developmental period, when treatment with the inhibitor blocks comb or photocyte formation; (B) Overview of the segregation of developmental determinants throughout early cleavages. All embryos are shown from the aboral pole. At the 4-cell stage, the developmental determinants are not yet segregated. At the 8-cell stage, the M blastomeres inherit factors that specify the photocyte lineage (*green clouds* labelled with 'photo') and the E blastomeres inherit factors that specify the comb-cell lineage (*orange clouds* labelled with 'comb'). At the 16-cell stage, factors that specify the comb-cell lineage are segregated to the e1 micromeres. The m1 micromeres can form combs but they require an inductive signal from the e cells *(pink clouds)* in addition to signals provided by the endomesoderm. *(From Fischer et al. (2014).)*

autonomously from the E lineage, and the bioluminescence characteristic of photocytes form at the correct time compared with control embryos (Fig. 6.14A).

3. Treatment with the DNA synthesis inhibitor aphidicolin reveals that the number of nuclear divisions is essential for the proper timing of the differentiation of the comb-plate-like cilia (Fig. 6.14A).
4. Lineage tracing experiments show that cells producing comb-plate-like cilia are formed autonomously from the E lineage, particularly the e1 micromeres, and not any of the M or m1 lineages.
5. Bioluminescence is only observed in the M lineage and specifically in M macromeres; additionally, Fischer et al. showed that developmental determinants that are required to generate motile comb-plate-like cilia and photoprotein expression are already present and localized at the eight-cell stage and further localize to the appropriate cells in the subsequent cell divisions (Fig. 6.14B).
6. Fischer et al. also presented evidence that suggests that zygotic transcription in *Mnemiopsis* begins around the 60-cell stage, just before the onset of gastrulation.

Stomodeal pharynx

It was commonly known that the nonbilaterian metazoan lineages either lack a gut (Porifera and Placozoa) or have a sac-like gut (Ctenophora and Cnidaria). Presnell et al. (2016) reported that the idea of Ctenophora (comb jellies), which diverged very early from the animal stem lineage, possessing a sac-like blind gut with only one major opening remains a commonly held misconception. Dating to Agassiz

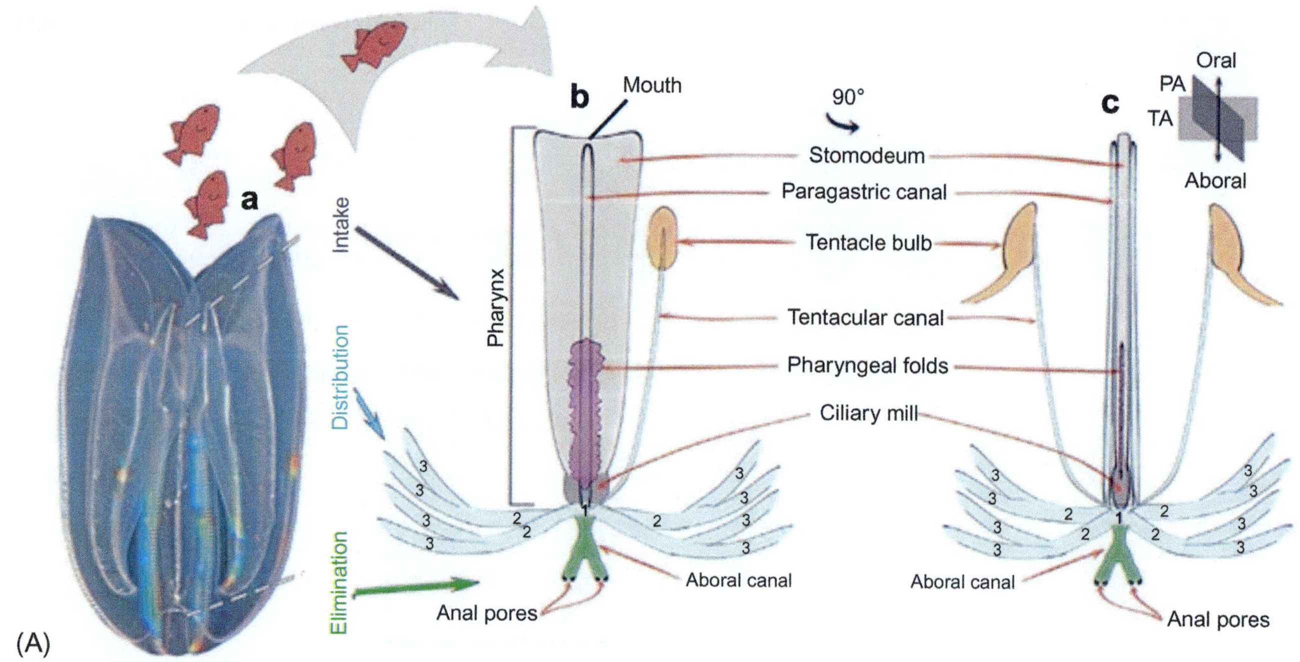

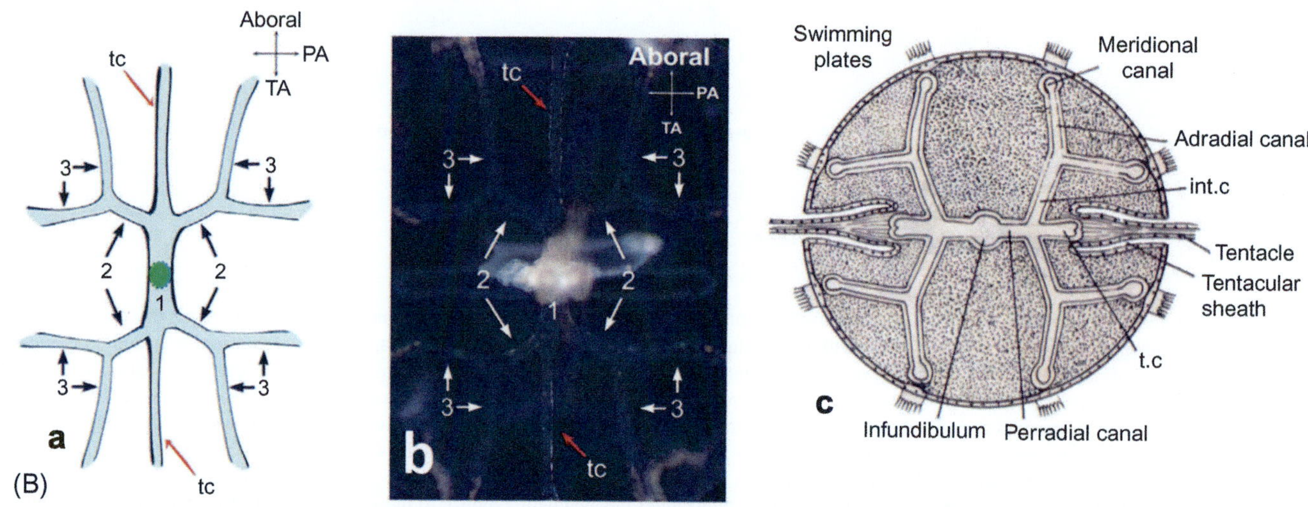

FIG. 6.15 (A) Schematic of the major features of the ctenophore digestive system. The pharyngeal axis (PA) is to the left, and the tentacular axis (TA) is to the right. Food enters the stomodeum and moves aborally through the pharynx *(light grey)*, where digestive enzymes are secreted by the pharyngeal folds *(purple)*. Large particles are then mechanically processed by a region of stiffened cilia, the ciliary mill *(dark grey)*. Small food articles move to the infundibulum (1 in the endodermal canal system), where isodynamic ciliary beating distributes particles for absorption in the radial canals (2), the paired adradial canals (3), and the paragastric canals and tentacular canals. Waste (solid particles and fluid) is eliminated from the branched endodermal canal system by transport through the aboral canal and anal canals *(green)* and evacuation through the anal pores. (B) (a) Aboral schematic highlighting the major endodermal canal system branches. The infundibulum, or transverse canal (1), serves as the central hub from which two tentacular canals (tc) and four radial canals (2) branch. Each radial canal bifurcates into adradial canals (3). The *green circle* marks the aboral canal branch; (b) Aboral view of a juvenile cydippid stage *Mnemiopsis leidyi* showing the principal branches of the endodermal canal system as schematized in (B-a); (c) diagrammatic section at right angles to long axis (horizontal) of *Hormiphora plumosa* showing the endodermal system: the epidermis is dotted, endoderm not dotted, the mesoglea stippled, and the muscular axis of the tentacles striated *(all from Presnell et al., 2016 except (Bc) from Marshall and Williams, 1975)*; (C–F) photos showing waste evacuation: (C–F) Waste Evacuation (C) and (D) in *Mnemiopsis leidyi* and (E) and (F) in Pleurobrachia bachei: (C) showing waste material being ejected from open anal pore; (D) showing an aboral view of a dilated anal pore during waste ejection; (E) showing a lateral view of a fluorescein plume emerging from an open anal pore; (F) showing an aboral view of a fluorescein plume during waste material ejection from a dilated anal pore; (C–F) photographs from Movies S1, S2, and S3; *asterisks* mark the position of the aboral apical organ, *arrowheads* mark the position of the open anal pore, and *arrows* denote waste plumes. (C) and (D) are lateral views, with the oral end to the left; (E) and (F) are aboral views; (G–J) (G) DIC image of aboral region of *Mnemiopsis*. *Arrows* highlight organized giant smooth muscle bundles that contract and regulate the diameter of the aboral and anal canals; (H) aboral region of *Mnemiopsis* stained with phalloidin in *red* and DAPI in *blue*. The *arrowhead* and *dashed circle* mark the ring of F-actin-rich cells that contribute to the circumference of the contractile sphincter associated with the anal pore; (I) aboral region of *Pleurobrachia* stained with phalloidin in *red*. The *arrowhead* and *dashed circle* mark the position of the ring of F-actin-rich cells contributing to the contractile sphincter associated with the anal pore; (J) aboral view of *Mnemiopsis* from merged brightfield and fluorescence microscopy. The *arrowhead* and *dashed circle* mark the position of a fully dilated anal pore. Digested fluorescent fish particles appear as bright spots in the anal canals. *(All from Presnell et al., 2016.)*

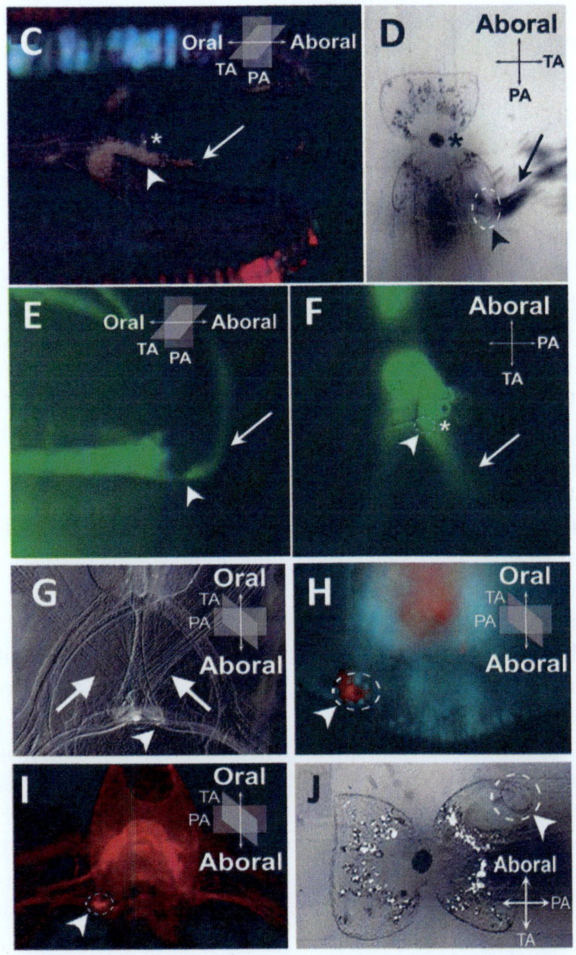

FIG. 6.15, CONT'D

(1850) who identified two openings of the digestive system opposite of the mouth—called 'excretory pores' by Chun (1880), referred to as an 'anus' by Main (1928), and coined 'anal pores' by Hyman (1940)—contradictory reports, particularly prominent in recent literature, posit that waste products are primarily expelled via the mouth. Although recent evidence supports Ctenophora as the earliest-branching extant metazoan phylum, ctenophores possess many 'complex' traits found in other animals, such as definitive muscles and a nervous system. Thus, the revised phylogenetic position of Ctenophora has also led to novel hypotheses reconsidering the origin of both muscles and nerves.

The incongruence among these previous observations motivated Presnell et al.'s experiments, which were explicitly designed to explore ctenophore gut functional anatomy. The authors used the ctenophores *Mnemiopsis leidyi* and *Pleurobrachia bachei* to provide an update to previous literature on the ctenophore digestive system using high-resolution in vivo time-lapse videography combined with fluorescent labelling techniques and primary cell culture. The main longitudinal body axis in ctenophores is the oral–aboral axis, with the mouth at the oral pole and two anal pores at the aboral pole (Fig. 6.15). The unidirectional ctenophore gut is highly differentiated and organized into three functional regions (intake, distribution and elimination) (Fig. 6.15A). The results resolve conflicting reports on the function of the ctenophore anal pores, confirming their role in relation to the overall digestive system as secondary gut openings functionally equivalent to an anus. The authors unequivocally show that ctenophores possess a functional through-gut from which digestion waste products and material distributed via the endodermal canals are expelled to the exterior environment through terminal anal pores that are anatomically and physiologically specialized to control outflow from the branched endodermal canal system (Fig. 6.15C–F).

Presnell et al. (2016) followed the pathway of the food inside the body of the ctenophores *Mnemiopsis leidyi* and *Pleurobrachia bachei*. The transparency of the animals allowed for easy in vivo observation of internal processes. The authors provide an update to previous literature on the ctenophore digestive system using high-resolution in vivo time-lapse videography combined with fluorescent labelling techniques and primary cell culture. The authors fed

M. leidyi Brainbow fluorescent zebrafish larvae and filmed on a Zeiss SteREO Discovery.V8 with Zeiss AxioCam MRm rev3. For *P. bachei*, marine rotifers and *Artemia* were added to the sterile sea water with 0.1 mg/mL fluorescein (Sigma Aldrich). The animals were fed in the dark for 90 min. Imaging was performed on a Leica M205FA stereoscope with DFC360FX CCD and DFC425C CCD. In contrast to prior studies, the authors did not observe food waste expulsion or regurgitation through the mouth during their feeding experiments with either *Mnemiopsis* or *Pleurobrachia*. Within the pharynx, combined mechanical and enzymatic action results in the predigestion of food into large particles that then enter the morphologically distinct ciliary mill (Fig. 6.15A (b) and (c)). The ciliary mill (Fig. 6.15A (b) and (c)) is uniquely characterized by dense arrays of stiffened cilia that act as a physical sieve and disrupt large particles, allowing only small particles to pass through into the infundibulum (transverse canal), the first chamber of the endodermal canal system (Fig. 6.15B). Both nutrient distribution and waste elimination are aided by peristaltic activity within the pharynx and the endodermal canals (Fig. 6.15B (a)–(c)) that direct flow of material (Fig. 6.15G). During waste removal, ciliary beating reverses direction, collecting waste material at the centrally located infundibulum, prior to moving into the aboral canals.

The results presented from these experiments show that ingested food processed by the pharynx and distributed through the endodermal canal system is subsequently expelled 'en masse' at regular intervals through the anal pores (Fig. 6.15C–F). The anal pores are composed of actin-rich cells in a ring configuration, consistent with a role in contraction (Fig. 6.15H–J). This actin distribution, coupled with the previous report of myosin heavy chain expression in *Pleurobrachia pilius* anal pores, provides molecular evidence that the anal pores function as contractile, ring-like sphincters.

Comparatively, the functional anatomy of the *Mnemiopsis* and *Pleurobrachia* digestive systems is typical of Ctenophora, although some ctenophore body plans have modifications to endodermal canal peripheral branching patterns. The authors have observed the same digestive system function across a range of ctenophore body plans, including representatives from the genera *Beroë*, *Bolinopsis*, *Ocyropsis*, *Cestum*, *Coeloplana*, and *Vallicula*.

Thus, Presnell et al. (2016) revised the phylogenetic position of Ctenophora, which also leads to novel hypothesis reconsidering the origin of both muscles and nerves. They provide new data for and discuss previous evidence of the presence of a through-gut in ctenophores and the broader implications for evolution of the metazoan through-gut and anus.

Sense organ

All the available textbooks and Hickman (1973) described the sense organ, or statocyst of Ctenophora (Fig. 6.16B). It is located at the aboral pole and is somewhat complicated in structure. Generally, it is a cyst containing centrally located statolith, which is composed of fused calcareous bodies. The floor of the statocyst is composed of ciliated epidermal and secretory cells. The statolith is supported on four pillars of fused cilia (the balancers) located interradially on the sensory floor. Two ciliated furrows lead from the base of each balancer to the pair of comb rows in the same quadrant. These tracts may carry impulses to the comb rows

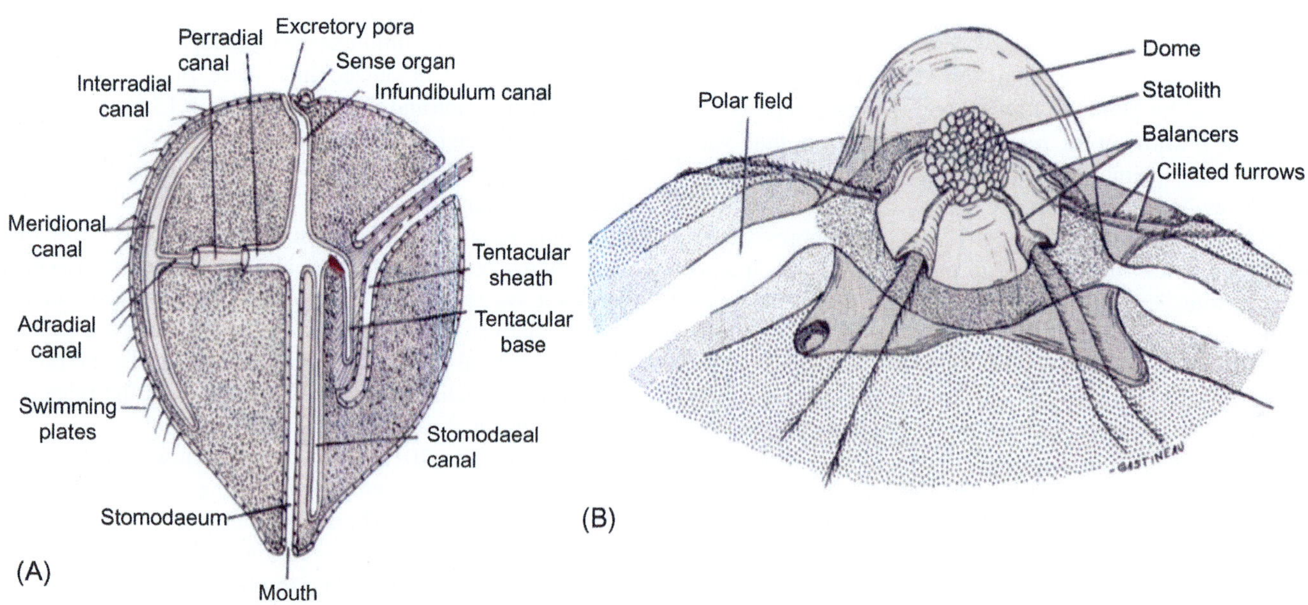

FIG. 6.16 (A) Drawing of dissected *Hormiphora plumosa* with more than one quarter of the body cut away. The epidermis is *dotted*, the gastrodermis *not dotted*, the mesoglea stippled, and muscular axis of the tentacles striated. The sensory organ and the excretory pores are shown *(from Marshall and Williams, 1975)*; (B) diagrammatic drawing showing details of enlarged statocyst *(modified from Chun, from Hickman, 1973.)*

from the sense organ. The dome-shaped cover of the statocyst is formed of fused cilia. On each side of the statocyst in the sagittal plane, there is a ciliated area of unknown function called the polar field.

Nervous system

Previous accounts describing Ctenophora did not consider a definite nervous organ. However, Marshall and Williams (1975) with their description of the sense organ reported that a subepithelial plexus of nerve fibres with nerve cells extends all over the surface of the body. Thus, nerve elements can also be traced in the mesoglea. Recently, Ryan (2014), discussing the evolution of ctenophore nervous system, reported that phylogenetic analysis of transcriptomic data from many animals and genomic data from the ctenophores *Mnemiopsis leidyi* (Ryan et al., 2013) and *Pleurobrachia bachei* (Moroz et al., 2014) have suggested that ctenophores—rather than sponges—are the most distant relatives to all other animals. Accordingly, the new view described that the ctenophore nervous system is organized as an epithelial nerve net consisting of short 'nerve chords' arranged in a polygonal mesh, as well as a less organized mesogleal nerve net made of single neurites and the apical organ is underlined by its own nerve net. Many ctenophores possess two retractile tentacles, which are innervated by a pair of parallel longitudinal nerve chords. Ctenophore neurons can be both isopolar and multipolar; they can also have multiple effectors, e.g. giant axons in the lips of *Beröe* make synaptic connections with both muscle fibres and adhesive cells.

The gastrovascular system

The gastrovascular system consists of an extensive canal system that is closely involved with the biradial symmetry of the group. This system begins with the mouth which leads to a flattened tube (Figs 6.2F and 6.16A) often called the stomach, but more correctly the gullet or stomodaeum. This tube elongates inside the animal and reaches about two-thirds of the way towards the aboral pole. The walls of this tube internally produce ridges that increase the area for the absorption of digested food. Living prey is seized by the tentacles (Fig. 6.15A (b) and (c)), ingested by the aid of the mobile edges of the mouth and digested in the stomodaeum, which is thus physiologically (though not morphologically) a stomach (Marshall and Williams, 1975). The products of digestion move into the various parts of the canal system at that time, it was described that the digestible food particles are passed out at the mouth.

Towards its upper or aboral pole, the stomodaeum gradually narrows and opens into a cavity called the infundibulum, which is probably equivalent to the stomach of an anthozoan or a medusa. It is flattened in a direction at right angles to the stomodaeum, i.e. in the transverse plane. Three tubes are given off from the infundibulum: one, the infundibular canal, which passes directly upwards, and immediately beneath the aboral pole divides into four short branches, two of which open on the exterior by minute apertures. These are the excretory pores (Fig. 6.16A). The two other canals given off from the infundibulum are the perradial canals, which pass directly outwards in the transverse plane, and each divides into two interradial canals. Then each of the latter, in turn, divides into the adradial canals. These successive bifurcations of the canal system all take place in a horizontal plane (Fig. 6.15C). Each of the adradial canals opens into a meridional canal, which extends upwards and downwards beneath the corresponding swimming plate. Furthermore, each perradial canal gives off a stomodaeal canal (Fig. 6.16A), which passes downwards parallel to and in close contact with the stomodaeum, and a tentacular canal (Fig. 6.15B (c)), which extends outwards and downwards into the base of the corresponding tentacle. Each tentacle presents a thickened base, closely attached to the wall of the sheath, and gives off a long flexible filament beset with processes of two kinds: one is simple and colourless; the other leaf-life with branchlets and of a yellow colour (Marshall and Williams, 1975). This canal system in other sites and in view of others (e.g. Presnell et al., 2016) who consider these canals collectively to be the endodermal canals, function in both nutrient absorption and distribution (Fig. 6.15B (a)).

Hatching

Larvae hatch by rupturing of the vitelline membrane. At least in *Coeloplana*, larvae leave the membrane with the oral end first. The time of hatching in pelagic ctenophores is usually about 1–2 days after fertilization.

Brooding

Only platyctenids have been reported to brood their embryos. Ovoviviparity, or retention of the embryos within the adult organism, occurs in *Tjalfiella*, *Coeloplana mitsukurii*, *Lyrocteis* and *Vallicula* (Fig. 6.3). Brood chambers of *Tjalfiella* and *Lyrocteis* are described in *Coeloplana bocki* and *Coeloplana gonoctena*; embryos are brooded under the creeping sole of the adult, within a mass of gelatinous material which is sticky in the former animal and relatively solid in the latter. Embryos of *C. mitsukurii* are retained within the 'gastrovascular spaces' of the adult. This latter process of brooding is also observed in *Vallicula*. In addition, brooding is observed outside this animal attached to the ventral epidermis.

In respect to the developmental stage of the brooded embryos, there are two cases: all the brooded embryos in one individual are in approximately the same stage of development. For example, in *Coeloplana bocki* and hence in this case the eggs were probably fertilized at the same time; the second case is that the brooded embryo in a single individual are in different stages of development, e.g. *Lyrocteis* and *Tjalfiella*.

The number of the brooded embryos within the individual varies with the species. A single *Tjalfiella* broods up to 35 embryos at a time, while a *Coeloplana bocki* broods from 10 to 200 but usually not more than 50. One *Lyrocteis* may contain up to 400 developing embryos. However, these numbers of the brooded embryos by a platyctenid individual is generally fewer than the number of eggs of any one spawning by a single pelagic ctenophore.

REFERENCES

Agassiz, L., 1850. Contributions to the natural history of the Acalephæ of North America. Part II: on the Beroid medusæ of the shores of Massachusetts, in their perfect state of development. Arts Sci. 4, 313–374 (cited in Presnell et al., 2016).

Borradaile, L.A., Potts, F.A. with chapters by Eastham, L.E., Saunders, J.T., 1977. The Invertebrata, fourth ed. Revised by Kerkut, G.A., Cambridge University Press.

Carré, D., Rouviiére, C., Sardet, C., 1991. *In vitro* fertilization in ctenophores: sperm entry, mitosis, and the establishment of bilateral symmetry in *Beroe ovata*. Dev. Biol. 147, 381–391.

Chun, C., 1880. Die ctenophoren des Golfes von Neapel und der angrenzenden Meeres-abschnitte (Vol. 1). W. Engelmann.

Chun, C., 1892. Die Dissogonie, eine neue Form der geschlechtli-chen Xeugung. In: Festsch. Zum siebenzigten Geburtstage RudorfLeuckarts. Engelmann, Leipzig, pp. 77–108.

Chun, C., 1898. Die Ctenophoren der Plankton Expedition. Ergeb-nisse Plankton Expedit. Humboldt–Stiftung 2, 1–32.

Dunlap, H.L., 1966. Oogenesis in the Ctenophora (Ph.D. thesis). Univ. of Washington, Seattle, WA. 230 pp. (cited in Dunlap-Pianka).

Dunlap-Pianka, H., 1974. Ctenophora. In: Giese, A.C., Pearse, J.S. (Eds.), Reproduction of Marine Invertebrates. Acoelomate and Pseudocoelomate Metazoans, vol 1. Academic, New York, pp. 201–265.

Fawcett, D.W., 1961. Intercellular bridges. Exp. Cell Res. 8, 174–187.

Fischer, A.H.L., Pang, K., Henry, J.Q., Martindale, M.Q., 2014. A cleavage clock regulates features of lineage-specific differentiation in the development of a basal branching metazoan, the ctenophore *Mnemiopsis leidyi*. EvoDevo 5, 4. Licensee BioMed Central Ltd.

Garbe, A., 1901. Untersuchungen über die Entstehung der Geschlechtsorgane bei der Ctenophoren. Z Wiss. Zool. 69, 472–491. pl. 36, 37.

Gardiner, M.S., 1972. The Biology of Invertebrates. McGraw-Hill Book Company, London, New York.

Gilbert, S.F., Raunio, A.M., 1997. Embryology, Constructing the Organism. Sinauer Associates, Inc. Publishers, Sunderland, MA.

Henry, J.Q., Martindale, M.Q., 2000. Regulation and regeneration in the ctenophore *Mnemiopsis leidyi*. Dev. Biol. 227 (2), 720–733.

Hickman, C.P., 1973. Biology of the Invertebrates, second ed. The C.V. Mosley Company, Saint Louis.

Hyman, L.H., 1940. The Invertebrates, Protozoa Through Ctenophora, first ed. McGraw-Hill Book Company, Inc., New York, London. Third expression.

Komai, T., 1922. Studies on Two Aberrant Ctenophores, *Coeloplana* and *Gastrodes*. Author, Kyoto. 102 pp., 9 pL (cited in Dunlap-Pianka, 1974).

Main, R.J., 1928. Observations of the feeding mechanism of a ctenophore. *Mnemiopsis leidyi*. Biol. Bull. 55 (2), 69–78.

Marshall, A.J., Williams, W.D., 1975. A Text Book Zoology: The Invertebrates. ELBS and Macmillan.

Martindale, M.Q., 1987. Larval reproduction in the ctenophore *Mnemiopsis mccradyi* (order Lobata). Mar. Biol. 94 (3), 409–414.

Martindale, M.Q., 2016. The onset of regenerative properties in ctenophores. Curr. Opin. Genet. Dev. 40, 113–119.

Martindale, M.Q., Henry, J.Q., 1996. Development and regeneration of comb plates in the ctenophore *Mnemiopsis leidyi*. Biol. Bull. 191, 290–292.

Martindale, M.Q., Henry, J., 1997a. Ctenophores, the comb jellies. In: Gilbert, S.F., Raunio, A.M. (Eds.), Embryology, Constructing the Organism. Sinauer Associates, Inc. Publishers.

Martindale, M.Q., Henry, J., 1997b. Reassessing embryogenesis in the Ctenophora: the inductive role of e1 micromeres in organizing ctene row formation in the 'mosaic' embryo *Mnemiopsis leidyi*. Development 124, 1999–2006 (1997) 1999 Printed in Great Britain © The Company of Biologists Limited 1997 DEV3636.

Martindale, M.Q., Henry, J., 1998. The development of radial and biradial symmetry: the evolution of bilaterality. Am. Zool. 38, 672–684 (cited in Martindale and Henry, 1999).

Martindale, M.Q., Henry, J., 1999. Intracellular fate mapping in a basal metazoan the Ctenophore *Mnemiopsis leidyi*, reveals the origins of mesoderm and the existence of indeterminate cell lineages. Dev. Biol. 214, 243–257.

Moroz, L.L., Kocot, K.M., Citarella, M.R., Dosung, S., Norekian, T.P., Povolotskaya, I.S., Grigorenko, A.P., Dailey, C., Berezikov, E., Buckley, K.M., Ptitsyn, A., Reshetov, D., Mukherjee, K., Moroz, T.P., Bobkova, Y., Yu, F., Kapitonov, V.V., Jurka, J., Bobkov, Y.V., Swore, J.J., Girardo, D.O., Fodor, A., Gusev, F., Sanford, R., Bruders, R., Kittler, E., Mills, C.E., Rast, J.P., Derelle, R., Solovyev, V.V., Kondrashov, F.A., Swalla, B.J., Sweedler, J.V., Rogaev, E.I., Halanych, K.M., Kohn, A.B., 2014. The ctenophore genome and the evolutionary origins of neural systems. Nature 510, 109–114. https://doi.org/10.1038/nature13400.

Pang, K., Martindale, M.Q., 2008. Ctenophores. Curr. Biol. 18 (24), R1120.

Presnell, S.J., Vandepas, E.L., Warren, J.K., et al., 2016. The presence of a functionally tripartite through-gut in Ctenophora has implications for metazoan character trait evolution. Curr. Biol. 26, 2814–2820.

Reverberi, G., 1971. Experimental Embryology of Marine and Fresh-Water Invertebrates. North-Holland Publishing Company, Amsterdam, London.

Reverberi, G., Ortolani, G., 1963. On the origin of the ciliated plates and mesoderm in the Ctenophore. Acta Embryol. Morphol. Exp. 6, 175–199 (cited in Martindale and Henry, 1999).

Ryan, J.F., 2014. Did the ctenophore nervous system evolve independently? Zoology 117, 225–226.

Ryan, J.F., Pang, K., Schnitzler, C.E., Nguyen, A.D., Moreland, R.T., Simmons, D.K., Koch, B.J., Francis, W.R., Havlak, P.NISC Comparative Sequencing ProgramSmith, S.A., Putnam, N.H., Haddock, S.H.D., Dunn, C.W., Wolfsberg, T.G., Mullikin, J.C., 2013. The genome of the ctenophore *Mnemiopsis leidyi* and its implications for cell type evolution. Science 342 (6164), 1242592.

Storer, T.I., Stebbins, R.C., Usinger, R.L., Nybakken, J.W., 1983. General Zoology, sixth ed. Tata McGraw-Hill Publishing Company Ltd, New Delhi.

Yatsu, N., 1911. Observations and experiments on the ctenophore egg. Annot. Zool. Jpn. 7, 333–346 (cited in Carré et al., 1991).

Triploblastica
Acoelomata

Chapter Outline

Definition **339**
Introduction **339**
Distinctive features **339**
Classification **340**
Origin of Platyhelminthes 341
Order Acoela **341**
Acoel *Polychoerus carmelensis* 341
Morphology of gametes 341
Mating 342
Development of simple egg (Acoela) 342
Polychoerus carmelensis 342
Neochildia fusca 343
Convolutriloba longifissura 343
Reproduction **345**
Agametic reproduction 345
Gametic reproduction 348
Reproductive system 348
Method for obtaining eggs 350
How to prepare embryos 350
Gametogenesis 350
Spermatogenesis 350
Spermatozoa 350
Polymorphism of sperm 352
Mechanisms of sperm transfer 352
Oogenesis 352
Oocytes 353
Copulation and/or fertilization 353
Order Polycladida **354**
Development of another simple egg 354
Cleavage 354
Gastrulation 356
Indirect development 356
Formation of Müller's larva 356
Metamorphosis of Müller's larva 358
Hoploplana inquilina (Polycladida) 358
Götte's larva (*Stylochus* sp.) 359
Larvae of Polycladida 359
Direct development in Polycladida 363
Notoplana humilis 363
Order Rhabdocoela **366**
Gieysztoria superba 366
Juvenile of *Gieysztoria* 368
Mesostoma lingua 369
Comparison between *Gieysztoria* and *Mesostoma* 372
Macrostomum sp. 372
Juvenile of *Macrostomum* sp. 376
Order Tricladida **377**
Epimorphosis 377
Morphallaxis 377
Development of composite egg (Tricladida) 380
Dendrocoelum lacteum 380
Girardia tigrina (Girard, 1850) 380
References **385**

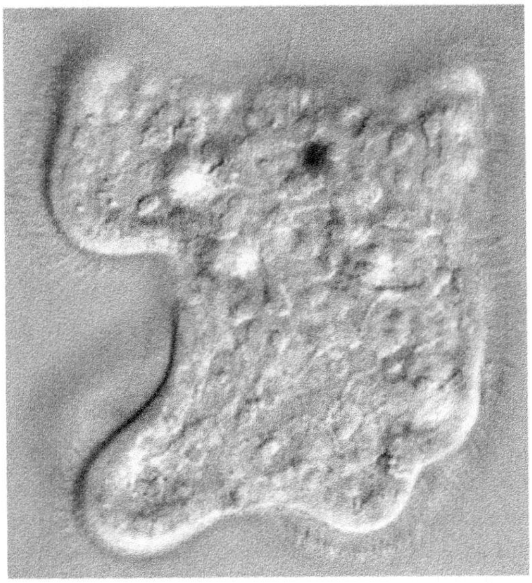

Chapter 7

Phylum Platyhelminthes, Turbellaria

(Greek *platys*, 'flat' and *helmins*, 'worm')

DEFINITION

The animals in this phylum are flat worms, as their name should suggest, dorsoventrally compressed, unsegmented, enteron (with only one opening functioning as inlet and exit, i.e. without anus) and they include the planarians, flukes, and tapeworms.

INTRODUCTION

Henley in her chapter (1974) informed that collection of living turbellarians is somewhat difficult, at least for the beginner, because some of the worms are quite small and because turbellarians live in a wide variety of habitats. These range from sand or mud bottoms, wharf-pilings (frequently in association with colonial cnidarians), under rocks, or on algae, to the gills of horseshoe crabs (*Bdelloura*), the interior of molluscan shells inhabited also by hermit crabs (*Stylochus*) and the branchial chambers of the whelk *Busycon* (*Hoploplana*). Hyman (1944a) described a polyclad, *Toenioplana teredini*, which lives in the burrows of the shipworm *Teredo*; she suggested that as the worm is always found in burrows lacking shipworms, it might be a useful enemy of these pests. Hyman (1944b) also reported that *Ectocotylea* was an ectoparasite of the crab *Pagurus*. Another turbellarian *Stylochoplana parasitica* lives in the pallial groove of a chiton (Kato, 1940), and there are many such examples known of rather bizarre association of turbellarians with other animals. Some acoels and polyclads are pelagic as adults, and many polylcads have pelagic larval and juvenile forms (Hyman, 1951). Turbellarian worms are found at a wide variety of depths, from very shallow to very deep water and in psammon.

DISTINCTIVE FEATURES

- Bilaterally symmetrical animals.
- Usually dorsoventrally compressed metazoan.
- Triploblastic animals, not metameric.
- Acoelomate animals (without body cavity).
- The well-developed organ systems being separated by a special parenchymatous connective tissue.
- Devoid of a hard general body skeleton, but some parasites have hard supports associated with their attaching organs.
- A system of flame cells and excretory vessels is nearly always present.
- No blood vascular system.
- A gut may be entirely absent, or it may consist of a loose syncytium, or there may be a well-developed alimentary canal with only one opening.
- Anus is absent, or present only very rarely.
- Hermaphrodites.
- Ovary is divided into germarium and vitellarium.
- Egg is surrounded with a shell (from the zygote or yolk cells).
- Development is direct or accompanied by a metamorphosis.
- Nervous system consists of nerve ring from which radiate a number of N. Cords.

Invertebrate Embryology and Reproduction. https://doi.org/10.1016/B978-0-12-814114-4.00007-2

CLASSIFICATION

Class 1. Turbellaria (Latin *turba*, 'disturbance') includes Planarians, about 3000 species

- Mostly nonparasitic.
- Nearly always with ciliated epidermis and rhabdites.
- Gut cavity present in all except the Acoela.
- Excretory pores variable in number and position or excretory system absent.
- Simple life cycle usually direct with juvenile worm which resembles the adult emerging from egg or cocoon, but occasionally a free-living larva is present.

Subclass 1. Archeophora (Greek *arche*, 'beginning', *oion*, 'egg' and *phorein*, 'to bear')

- Female reproductive system; primitive with gonads undivided.
- Eggs; endolecithal.
- Some form of spiral cleavage.
- Exclusively marine.

Order 1. Acoela (Greek *a*, 'not' and *koilos*, 'hollow').

- Gut without pharynx and devoid of cavity.
- Instead has a loose digestive syncytium which can partly engulf prey.
- No protonephridia or oviducts.
- Three to six pairs of nerve cords.
- Statocysts.
- Exclusively marine.
- Ex. *Convoluta* sp. *Polychoerus*, *Anaperus*, *Childia*, *Oligochoerus*, '*Palmenia*' (=*Anaperus*)

Order 2. Polycladida (Greek *poly*, 'many' and *klados*, 'branch').

- Body broad and flattened.
- Gut with pharynx and numerous radiating branches.
- Nervous system with numerous radiating nerve cords.
- Numerous eyes (there are a few eyeless species).
- Ovaries and testes; numerous and scattered, one or two gonopores, yolk glands lacking.
- Marine (few exceptions).

Suborder 1. Acotylea, no sucker; Ex. *Planocera*, *Hoploplana*, *Notoplana*.

Suborder 2. Cotylea, have a sucker; Ex. *Stylochus*, *Thysanozoon*, *Natocoylus*, *Yungia*.

Subclass 2. Neoophora (Greek *neos*, 'new')

- Ovary divided to germarium and vitellarium.
- Eggs; ectolecithal.
- Irregular cleavage.
- Marine, freshwater and occasionally terrestrial.

Order 1. Rhabdocoela

- Many free-living.
- Gut with pharynx and usually a simple undivided intestine.
- Protonephridia paired.
- One pair of nerve cords.
- Ex. *Rhynchoscolex* sp. *Macrostomum*, *Mesostoma*, *Microdalyellia*.

Order 2. Temnocephala (Greek *temnein*, 'to cut' and *kephale*, 'head').

- Cilia few or lacking; epidermis is syncytial.
- Anterior end extended into 2–12 tentacles.
- Posterior end with one or two adhesive discs.
- Single gonopore.
- Ectocommensal.
- Freshwater.
- Ex. *Temnocephala*, *Actinodadylellia*.
- Live on outer surfaces of various freshwater animals, e.g. crustaceans, molluscs and turtles, with which they have a specific relationship

Order 3. Tricladida (Greek *tria*, 'three' and *klados*, 'branch').

- Elongate.
- Pharynx; usually directed posteriorly.
- Gut with three primary branches: one directed anteriorly and two posteriorly.
- With many diverticula.
- Testes numerous.
- Yolk glands.
- Single gonopore.
- Three or four pairs of nerve cords.
- Marine, freshwater and terrestrial.
- Ex. *Bipalium*, *Polycelis*, *Procerodes*, *Rhynchodemus*, *Dugesia*, sp. *Dendrocoelum*, *Bdelloura*, *Planaria*.

Order 4. Seriata.

- Pharynx plicate.
- Enteron with lateral diverticula.
- Paired yolk glands, arranged in row along yolk duct.
- Enteron not forked, four pairs of nerve cords usually.
- Ex *Monocelis*, *Otoplana*, *Nematoplana*.

Class 2. Trematoda (Greek *Trema*, 'hole' and *eidos*, 'form').

- Fluks 6250 species.
- No epidermis, rhabdites or cilia.
- One or more suckers.
- Gut usually bifurcate, but sometimes simple with many branches.
- Single ovary.
- Life cycle simple or complex, ecto or endoparasites.

Class 3. Cestoda (tapeworms): 3400 species

- No epidermis, rhabdites or external cilia.
- Body with cuticular covering, differentiated into or made up of scolex (head) and a strobilae, a series of segments (proglettides), of which the youngest are the most anterior, each containing a complete reproductive system of both sexes.
- No mouth or enteron.

- Endoparasitic, principally in the intestines of vertebrates, with complex life cycle, usually involving two or most hosts.

Two theories concerning the origin of Platyhelminthes

1. Ctenophore theory (A. Lang): Based on the fact that there are certain structural similarities between the polycladida and the ctenophore *coeloplana*.
2. Planula theory (Ludwig von Graff): Holds that the turbellarian order Acoela resembles the ctenarian planula stage and suggests that the present Acoela had their origin in some acoele ancestor, which evolved from a planuloid form. According to this idea, there are two suggestions: (1) the acoele represent the link between the radiata and the bilateral metazoans; all the Platyhelminth worms originate from Acoela, which gave rise to the Alloeocoela and Rhabdocoela on one hand, and the Tricladida and Nemertini on the other; and (2) the Alloeocoela are believed to have produced the polycladida, while the Rhabdocoela gave rise to the parasitic Trematoda and Cestoidea (Kato, 1968).

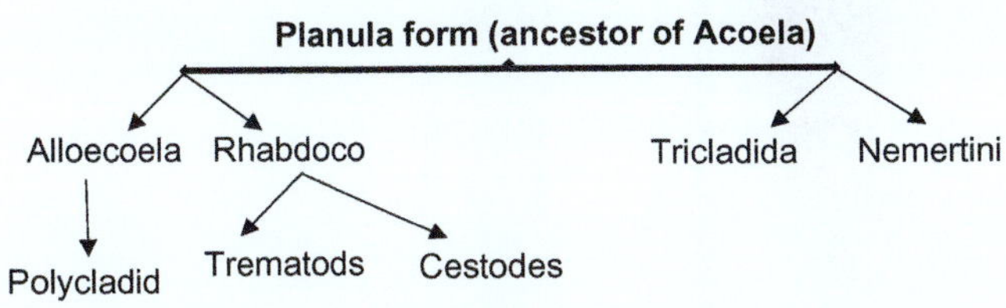

The members of the trematods and the cestods are parasitic, and are better considered within the context of parasitology. This chapter focuses on the development of members of class Turbellaria among the phylum of the Platyhelminthes, in order to make clear the relation between the Turbellaria and the other members of all phyla.

The Turbellaria (Fig. 7.1), in turn, are further divided into five orders: Acoela, Rhabdocoela, Alloeocoela, Tricladida, and Polycladida. Previously, a sixth order, the Temnocephala, was recognized, but this has recently been included among the Alloeocoela. Of these groups, the Polycladida and Nemertini are like the Acoela in forming simple eggs, while the Rhabdocoela and Alloeocoela, like Trematoda and Cestoda, have composite eggs. The reproductive system of the latter is complicated. This system of some turbellarians is shown in Fig. 7.7.

ORDER ACOELA

It is appropriate to pay much attention to order **Acoela**. Acoels are small, less than 15mm (Fig. 7.3A (a)–(d)), only marine and are found in all the world's oceans (Bourlat and Hejnol, 2009). They live between sand grains or gravel on beaches and on the sea bottom, forming part of the benthic 'meiofauna'. Some species, such as *Hofstenia miamia*, live on submerged mangrove leaves and roots. Most acoels are free-living, but can also live as commensals—a type of nonparasitic interaction—on or in other marine invertebrates. Henry et al. (2000) reported that the acoel flatworms, which are considered to be members of the Turbellaria, have the simplest morphology of the group (Figs 7.1A and 7.3A). In 1882, Ludwig von Graff already considered acoels as possibly representing the bilaterian ancestor. The Acoela derive their name from their lack of a defined digestive cavity where the midventral mouth opens into the central parenchyma (endoderm), as represented by a mass of nonepithelial tissue with both syncytial and cellular elements. In this tissue, they digest unicellular algae or other prey. The body wall, consisting of the ciliated epidermis and complex layers of circular, longitudinal and diagonal muscles (Figs 7.3A (b) and 7.6H), surrounds the peripheral cellular parenchyma, which includes mesodermal tissues other than muscles and reproductive structures.

Henry et al. (2000) collected sand bearing pink bacterial deposits from Sippewisset Marsh, Massachusetts, and kept them in the laboratory until specimens of *Neochildia fusca* migrated to the surface. Adults were harvested and maintained in culture dishes containing seawater and a few sand grains, where they laid eggs in gelatinous masses for about a week.

Acoel *Polychoerus carmelensis*

Costello (unpublished observations) has found that in the acoel *Polychoerus* the earliest oogonia are characterized by large nuclei with prominent nucleoli. As the primary oocytes grow, these nucleoli become very enlarged, with two distinctly different regions.

Morphology of gametes

In living oocytes of *Polychoerus carmelensis* observed by phase constant microscopy (Fig. 7.8C (a)), a large spherical germinal vesicle is present, with a conspicuous nucleolus. The yolk is rather uniformly distributed in the cytoplasm of oocytes of this form, as in those of most acoels and many alloeocoels, where as in the more advanced groups the nutritive materials are provided from an exogenous source: the yolk cells (Henley, 1974).

Pigment granules of various types are found in the eggs of many turbellarians. Those of *Polychoerus* contain orange granules, superficially resembling the surface structures on ova of *childia,* but lacking their content of DNA, as determined by the Feulgen procedure (Costello, personal communication). Kato (1940) stated that the cytoplasm of eggs of many of the Japanese polyclads that he studied had dense masses of granules in their cytoplasm, which were colourless or slightly pink.

Costello and Henley (unpublished data) have found that the cytoplasm of the ripe ovum of the acoel *Polychoerus* is relatively free of cell inclusions, at both the light and electron microscope levels. However, the oocyte of the ocoel *Childia* (Fig. 7.8C (b)) has an abundance of rather dense

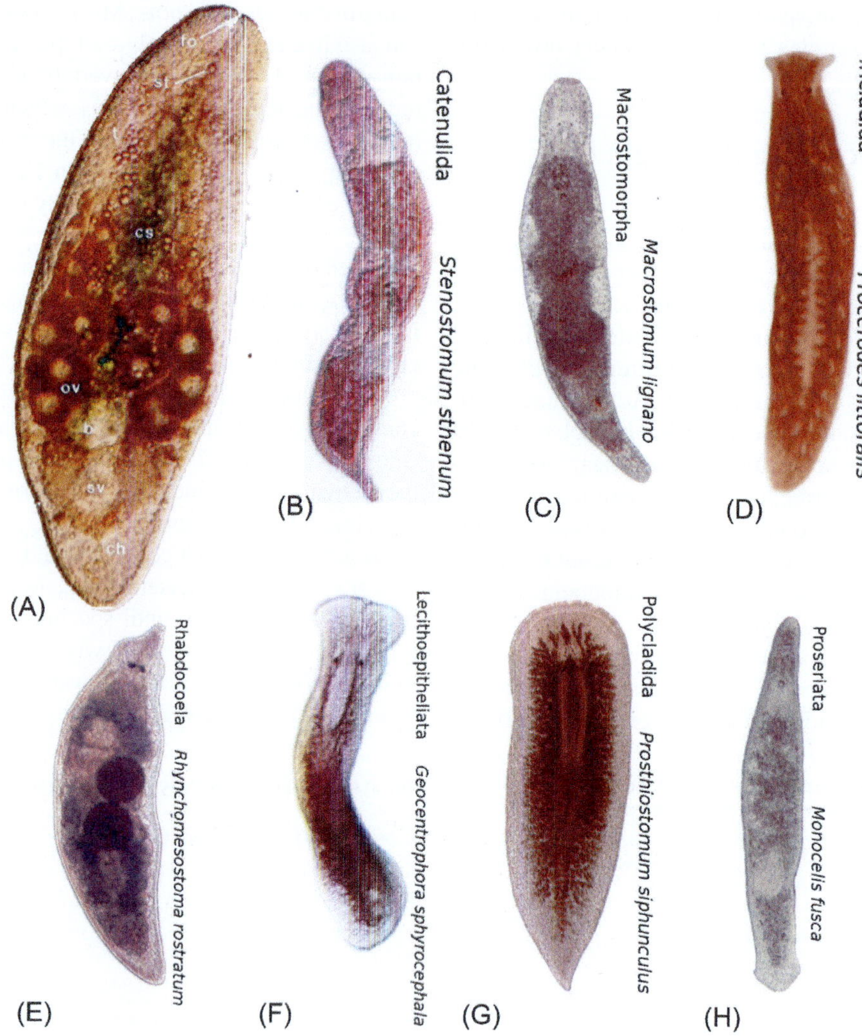

FIG. 7.1 Photographs of some live turbellarian animals: (A) live *Convoluta pulchra*, with paired rows of developing eggs (ov) with large nucleoli and green digestive central syncytium (cs). b=bursa, ch=choroid vacuoles, fo=frontal organ, st=statocyst, sv=seminal vesicle with penis, t=testes; bar=100mm *(from Ladurner and Rieger, 2000)*, at present this animal belongs to its own clade; (B–H) live turbellarians *(from Martín-Durán and Egger, 2012)*.

granules (some of which are presumably yolk), vacuoles and a fair number of mitochondria.

Henley (1974) reviewed and described the spermatozoa of Turbellaria and reported that the most striking differences in the morphology and/or length of the spermatozoa are found within those of acoels. Those of *Polychoerus carmelensis*, for example, are 300–400 µm long (Henley et al., 1968), slender and with paired undulating membranes. The tail is provided with one or more rows of retractile granules.

Mating (Costello and Costello, 1938)

During copulation of *Polychoerus carmelensis*, spermatozoa were deposited directly by the penis of each partner into the dorsal portion of the bursa of its mate. They likewise observed a form of courtship behaviour, in which prospective copulants exchanged head contacts, then faced each other with their head ends elevated and their ventral surfaces in contact. The pair then moved slowly backwards and forwards, 'swayed' from side to side in opposite directions; one partner would then bend its anterior half down under its partner, with the ventral surfaces of the two animals being in contact. Eventually, the two formed a ball and remained in contact with each other for 40–50 s, after which one partner would move away; the entire process including courtship required about 1–1.5 min. This character is in agreement with the view reported by Hyman (1951) that self-fertilization usually does not occur in these hermaphrodites, because spermatozoa are motionless in the male system and do not become motile until they are ejaculated during copulation.

Development of simple egg (Acoela)

Polychoerus carmelensis

Previously, Gardiner (1895) studied the development of the acoel *Polychoerus caudatus*, while Breslau (1909) studied that of *Convoluta roscoffensis*. After these studies, Kato (1968) and Henley (1974) described the development of the egg of the acoel, *Polychoerus carmelensis*. The general mode of development of this group, acoela, is similar to that of the polyclads, as it will be seen later, with the exception that: the spiral cleavage is duet, not quartet; with stereogastrula; there is no intestinal cavity; there are no free larvae; and the nervous system is formed by the first set of micromeres.

In *Polychoerus carmelensis*, the first cleavage is meridional, and separates the zygote into two blastomeres of approximately equal size (Fig. 7.2A (a) and B (a)). The cleavage proceeds in a spiral fashion, producing the first to the fourth micromere sets, but with only two micromeres in each set. The second cleavage cuts off two smaller cells (micromeres Ia and Ib, the first duet micromeres), from the two macromeres (IA and IB). The two micromeres are nearer to the animal pole. This cleavage is therefore designated as leiotropic (left-handed or counterclockwise). The Ia and Ib next divide by a dexiotropic (right-handed or clockwise) division, producing Ia^1 and Ia^2 and Ib^1 and Ib^2. At about the same time, the first generation macromeres (IA and IB) divide again, this time almost bilaterally but dexiotropically, to produce the second pair of micromeres (2a and 2b) and a pair of second generation macromeres (2A and 2B).

In *Convoluta*, according to Bresslau (1909), Kato (1968) reported that the macromeres 3A, 3B become in the interior of the cleaved embryo, surrounded by the overlapping micromeres. By time, they form the fourth set of micromeres (Fig. 7.2D (c) and C). This process occurs somewhat later in *Polychoerus*, but in either case a stereogastrula is formed. The cells filling the interior of this gastrula are chiefly 4A, 4B and the descendants of 4a and 4b. It is believed likely (at that time) that there are also ectomesodermal cells derived from the third set of micromeres. When epiboly is complete and the blastopore is closed, the cells on the surface remain there as the ectodermal layer. Internally, a clear distinction has been established between the cells of the endoderm and mesoderm.

The eventual fates of the cells of acoel embryos, according to Bresslau (1933) and Hyman (1951), are as follows: the first duet of micromeres will give rise to ectodermal structures including the nervous system; and the second and possibly also the third duet will form ectomesoderm (epidermis and peripheral parenchyma). Part of the internal mass of the embryo will be formed by progeny of the fourth duet of micromeres. The fourth-generation macromeres and third descendants also contribute to this interior mass, which, in the gastrula stage, is not clearly separable into mesoderm proper and mesendoderm (Henley, 1974). Kato (1968) added that the mouth is formed by an invagination of the ectoderm in the region of the former blastopore. No intestinal cavity appears. There is no free-swimming larval stage, and the definitive adult carries over the structure of the stereogastrula.

The acoel *Neochildia fusca*

Henry et al. (2000) studied the embryological development of this flatworm and evidenced that this species exhibits the unique form referred to as 'duet spiral cleavage' (Fig. 7.2A (a)–(d) and B–D). This unusual cleavage pattern, which is also unique among the bilaterians, is a stereotyped cleavage programme. It allows tracing back the expression of developmental genes to individual blastomeres with known cell fate (Bourlat and Hejnol, 2009). In *Neochildia fusca*, Henry et al. (2000) described this different cleavage as follows: the **first** cleavage passes through the animal–vegetal axis and gives rise to two indistinguishable, equal-sized, blastomeres called A and B. In the **second** cleavage, unlike in the typical pattern of quartet spiral cleavage, the cleavage spindle forms at an oblique 45 degrees angle to the animal–vegetal axis with a smaller 'micromere' being generated towards the animal pole. This is the first 'duet of micromeres' generated in counterclockwise, or leiotropic, orientation when viewed from the animal pole (Fig. 7.3B and C). To the six-cell stage, the spindles do not reorient, so that a second set of micromeres is born from the macromeres, again in a leiotropic direction. These cells of the second duet become shifted somewhat more laterally due to the presence of the first duet micromeres following the cleavage division. The fourth cleavage (eight-cell stage) arises as the first duet micromeres divide, and unlike the macromeres, the mitotic spindles of these blastomeres reorient by 90 degrees at the next division to give rise to a dexiotropic division. The 10-cell stage is reached when the second duet micromeres divide in a bilaterally symmetrical fashion through the future frontal plane. Close observation reveals that this division occurs in a slightly dexiotropic fashion. The third duet of micromeres is generated at the 12-cell stage when the macromeres undergo yet another slightly leiotropic division (Fig. 7.3B and C). Following that, the third duet macromeres are closely apposed at the vegetal pole and ultimately become covered by the dividing overlying micromeres.

The fourth duet of micromeres has been reported for two other species (Bresslau, 1909). The authors have not observed this stage in living embryos of *Neochildia fusca*. However, division of the third duet micromeres is also equal and bilateral in nature. To determine the fates of the micromeres, the authors injected individual cells, or sometimes pairs of cells, with fluorescent lineage tracers. Both the first A and B cells are equal in size, and the authors were not able to distinguish derivatives of the A cell from the B cell at the time of the injection. However, since labelling patterns were relatively consistent, it is clear that these two cells and their progeny are fated to give rise to different regions of the adult worm.

The acoel *Convolutriloba longifissura*

Hejnol and Martindale (2008) aimed to highlight the development and/or the evolution of the opening(s) of the digestive

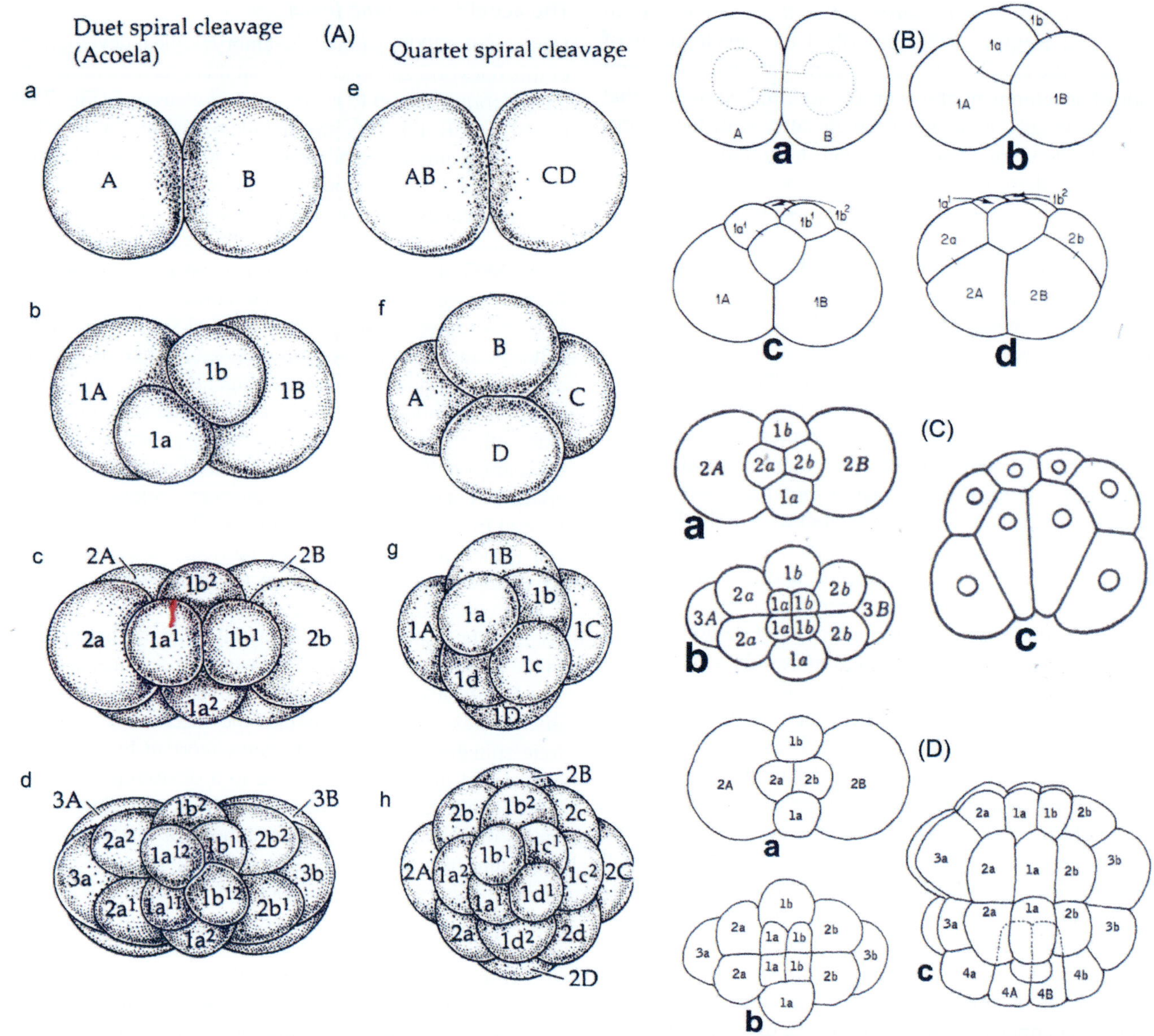

FIG. 7.2 (A) Duet versus quartet patterns of archoophoran cleavage: (a–d) duet spiral cleavage (acoela); (e–h) quartet spiral cleavage; (B) duet spiral cleavage (in acoel *polychoerus*): (a) two-cell stage, (b) four-cell stage, (c) six-cell stage, (d) eight-cell stage; (C) (a) six-cell stage, (b) 16-cell stage (in *Polychoerus caudatus*) (Gardiner), (c) 12-cell stage (in *Convoluta roskoffensis*) (Bresslau), note that macromeres have already been surrounded by micromeres; (D) duet spiral cleavage (in *polychoerus*): (a and b) views from animal pole, (c) lateral view *(after Gardiner from Hyman). (A) After Gardiner, 1895, from Ellis and Fausto-Sterling (1997); (B) from Henley (1974); (C) from Kato (1968); (D) from Barnes (1974).*

system. They postulated that most animals begin with either a radially symmetric larva or a cnidarian polyp-like organism that elongated its body along the future anterior–posterior axis, followed by a lateral closure of this slit-like blastoporal opening, with the ends giving rise to a mouth and anus (Fig. 7.5A). They studied the molecular phylogenesis of the flatworm acoel *Convolutriloba longifissura* (Fig. 7.3A (d)), which has only one opening to its digestive system. They studied the expression of genes involved in bilaterian foregut and hindgut patterning during the development of the acoel *Convolutriloba longifissura*. The results show that the genes brachyury and goosecoid are expressed in association with the acoel mouth. Therefore, the authors suggested that this single opening is homologous to the mouth of other bilaterians. To determine whether there is an embryonic relationship between the formation of the mouth and the expression of these two genes, they compared their results (Fig. 7.5B (a)–(c)) to the embryonic fate map of acoel flatworms. That of *Neochildia fusca* indicates that the mouth is formed at the border of the micromeres 1b, 2a, 2b, 3a and 3b (Fig. 7.4) at the anterior–ventral edge of the site of gastrulation. On the other hand, the authors find that the genes caudal, orthopedia

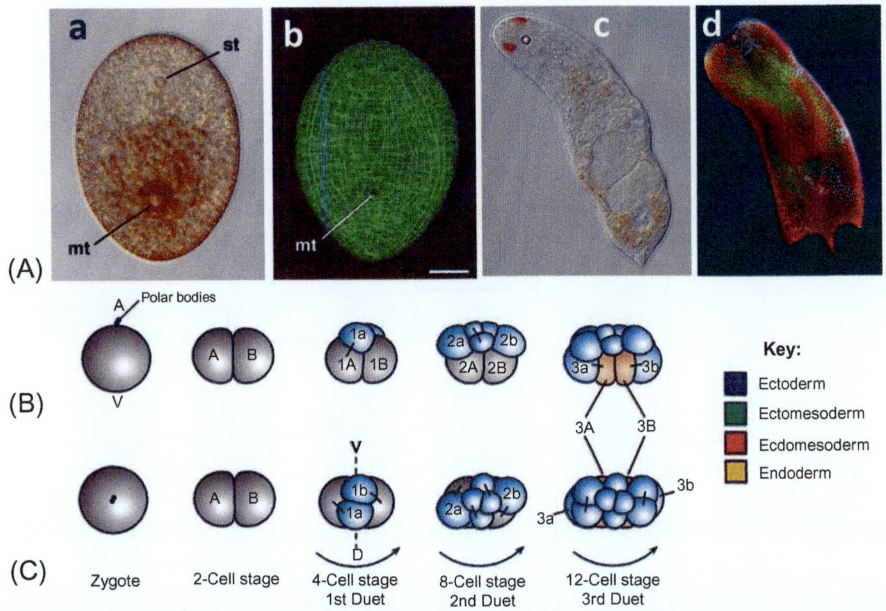

FIG. 7.3 (A) Photomicrographs of WMs and cleavage of acoels: (a and b) DIC photomicrographs of the ventral views of juvenile worm *Neochildia fusca*, the mouth (mt) and statocyst (st) are clearly seen; (b) bodipy green phallicidin (Molecular Probes, Eugene, Oregon), F-actin-stained juvenile worm revealing the extensive array of circular, spiral, and longitudinal body wall muscle fibres, bar = 50 μm; (c) the acoel *Notocelis gullmarensis*, is 800 μm in length. The statocyst is visible between the *red eye fields* and two large egg cells can be seen in the lower half; (d) adult acoel *Convolutriloba longifissura*, the *green colour* of the body comes from algal symbionts integrated in its skin and the *red stripes* in the middle of the body of this 3 mm long worm indicate mature oocytes; (B and C) lateral and animal pole views, respectively, of duet 'spiral' cleavage in the acoel *N. fusca* through the formation of the 12-cell stage when the third duet of micromeres has formed (not to scale). Note that *N. fusca* does not exhibit the characteristic alternating clockwise and counterclockwise generation of the various micromere duets. Note also that the acoel does not form ectomesoderm *(green cells)* and that there are two rather than one sources of endomesoderm *(orange cells)*, the dorsoventral axis is shown as a *dashed line* labelled D–V at the four-cell stage. *((A a, b, B, and C) From Henry et al. (2000); (A c and d) from Bourlat and Hejnol (2009).)*

and brachyury (bra), which are expressed in various bilaterian hindguts, are expressed in a small region at the posterior end of the animal (Fig. 7.5C (j)–(l)). They suggested that the anal opening of protostomes and deuterostomes could have arisen independently in distinct lineages, and the posterior expression domain of (bra) in acoels corresponds to the adult gonopore (Fig. 7.5D (o)–(q)).

The findings of the work of Hejnol and Martindale (2008) contradict the hypothesis that the bilaterian mouth and anus evolved simultaneously from a common blastoporal opening, and suggest that a through-gut might have evolved independently in different animal lineages. In addition, they stated that the morphology and recent molecular phylogenies place the Acoela and the 'nemertodermatid' flatworms as basal branches of the Bilateria. For the details of this aspect, see Hejnol and Martín-Durán (2015).

REPRODUCTION

Agametic reproduction

Occurrence and types

Agametic reproduction, with a very few exceptions, is restricted to the freshwater rhabdocoels of the catenulid group, the microstomids (which are both marine and freshwater), some land triclads and freshwater planarians (Hyman, 1951). However, a few instances of agametic reproduction have been described for acoels (Marshall and Williams, 1975). A few turbellarians multiply by budding, e.g. *Microstomum* (Fig. 7.6A), and the buds form long chains that give a superficial resemblance to the chain of segments or proglottids present in many tapeworms. Indeed, this resemblance has been held to reflect an evolutionary relationship between the two groups, but if it is accepted that the 'unsegmented' condition of cestodes like *Caryophyllaeus* (Fig. 7.6E) is the survival of a primitive character, then there is obviously no direct relationship between budding turbellarians and segmented tapeworms. Henley (1974) proposed that the process has been studied quite extensively in the freshwater groups, less so in the marine. In all, however, the process appears to involve either transverse fission or the formation of chains of zooids or fragmentation. In the case of fission, the region where division will occur (Fig. 7.6D) is often marked first by an indentation (small hollow) and then by the appearance of ciliated pits (Hyman, 1951). During the process, the near end of the worm is firmly attached to the substratum, while the anterior end moves away and breaks off. Both portions produce new worms. In the freshwater worm *Phagocata*, there is fragmentation into a number of pieces, which subsequently encyst; fission does not occur.

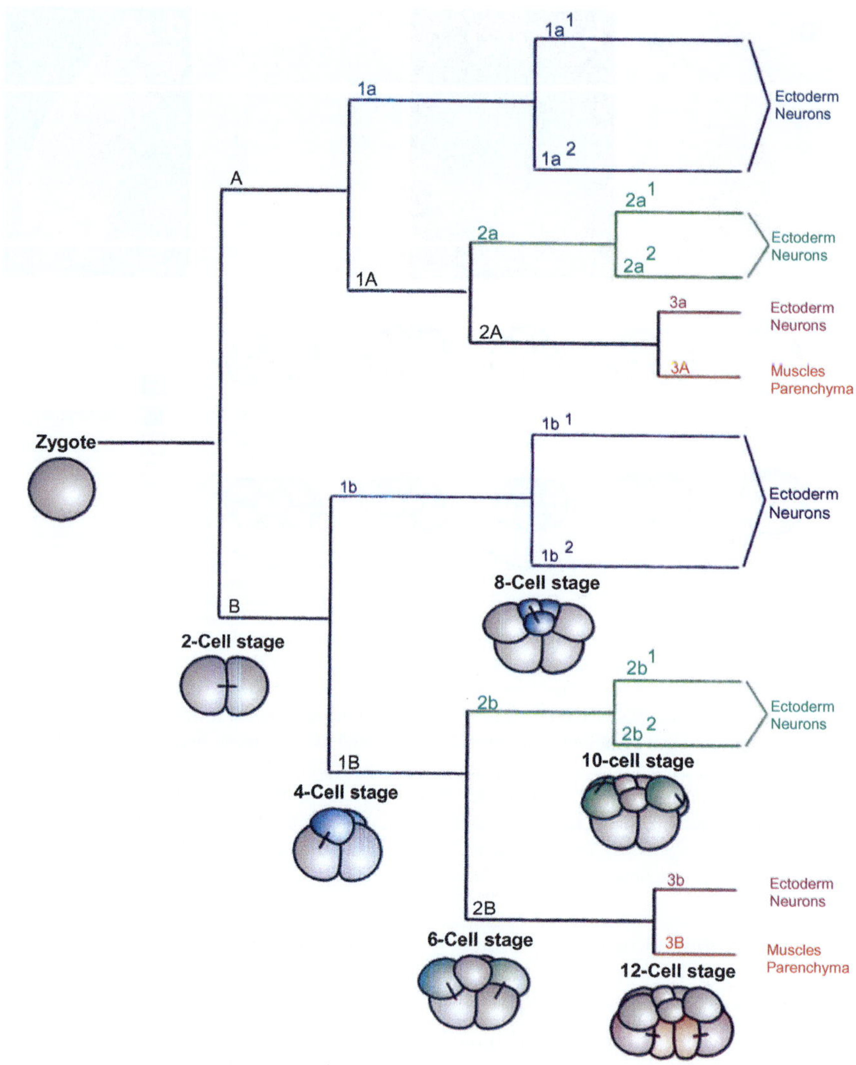

FIG. 7.4 Summary lineage diagram illustrating the derivatives of the acoel duet cleavage programme (of *N. fusca*) including the fates of identified blastomeres; colours distinguish the ectodermal derivatives of the first *(blue)*; second *(green)*; and third *(purple)* duet micromeres from the mesodermal and endodermal derivative (parenchyma) of the third duet macromeres *(orange)*, as indicated in the diagram. *(From Henry et al. (2000).)*

Architomy (Fig. 7.6C), in which fission occurs before regeneration of the parts to be lost has occurred, is opposed to **paratomy**, which occurs in the Rhabdocoela. In the latter, the organs that are due to be lost in fission are first formed, and division takes place afterwards (Fig. 7.6B).

In general, the flatworms have a strong capacity for regeneration, and agametic reproduction by transverse fission is widely practised. This form of agametic reproduction takes place repeatedly over a rather long period, before the body attains full growth and the gonads and copulatory apparatus are developed so that gametic reproduction can occur.

In ordinary planarians, after the body has been divided posterior to the pharynx, a tail region develops on the isolated anterior part, and a head region regenerates on the posterior part, forming two complete new animals (Fig. 7.6F). However, Ellis and Fausto-Sterling (1997) reported that planarians have a strong sense of head and tail (anterior–posterior polarity). If one makes extremely thin transverse cuts (Fig. 7.6G (a)), the normal antero-posterior polarity is sometimes disrupted, often leading to the production of a double-headed Janus-like organism. If, however, one makes a single transverse cut, and allows a short time to elapse before making a second cut (Fig. 7.6G (b)), this thin transverse fragment regenerates with normal polarity.

Borradaile et al. (1977) reported that agametic reproduction occurs commonly in the Turbellaria. In *Microstomum lineare*, the hinder end buds off new individuals, which remain attached for some time, so that chains of three or four individuals in different stages of development are often seen. Planarians undergo autotomy, cutting themselves into two by a ragged line, which traverses the middle of the body. Lost parts are easily regenerated in the Tricladida and the group is a favourite one for experimental work on regeneration.

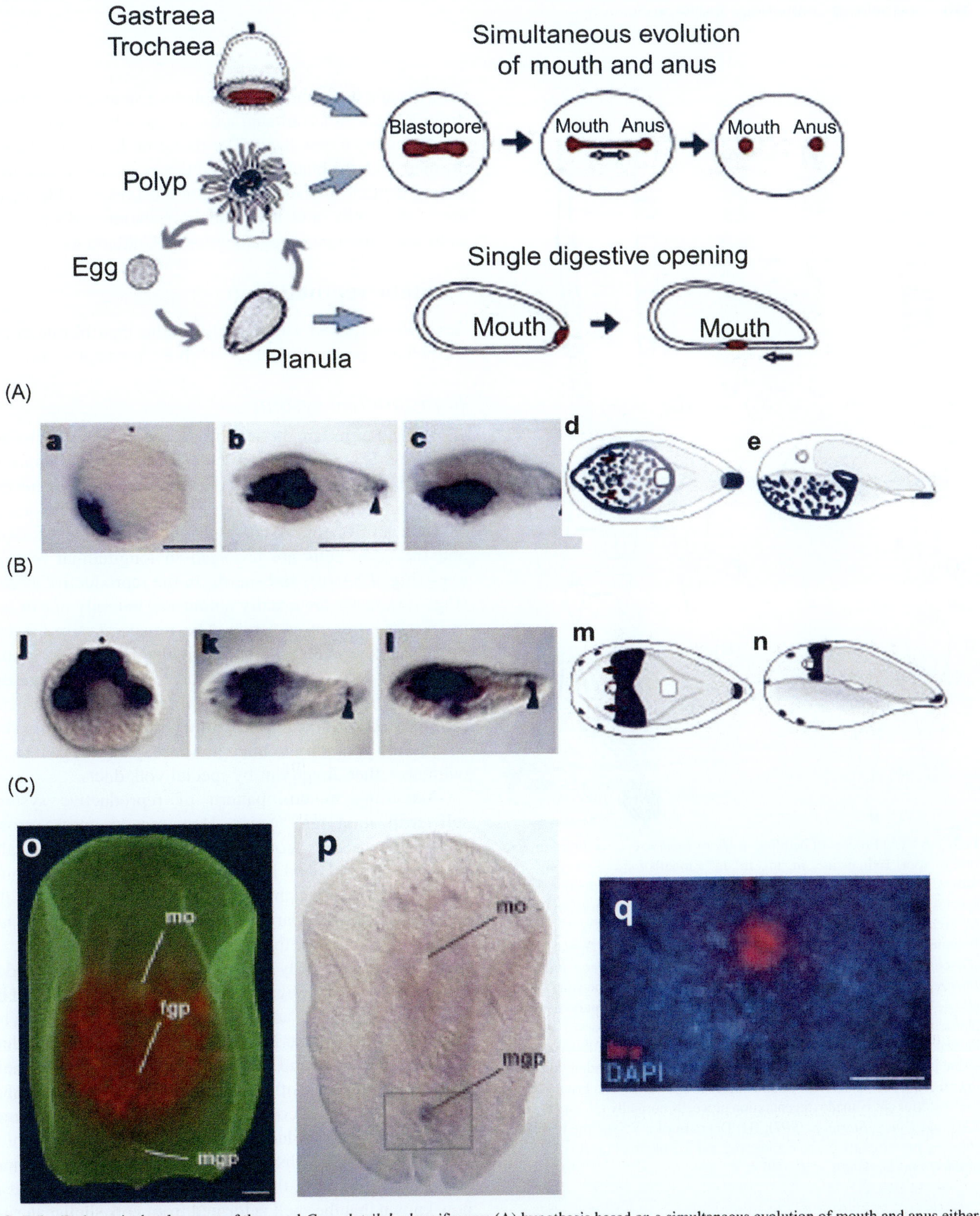

FIG. 7.5 Embryonic development of the acoel *Convolutriloba longifissura*: (A) hypothesis based on a simultaneous evolution of mouth and anus either start with a pelagic hypothetical ancestor (gastraea and trochaea) or start with a sessile cnidarian polyp-like ancestor. The common assumption is that the single opening, arising from the blastopore, gave rise to mouth and anus simultaneously by a lateral closure. Alternative hypothesis suggests a cnidarian planula-like ancestor with a single opening, which shifts ventrally during the transition to bilateral symmetry with an anus forming secondarily; (B and C) Gene expression during development of the acoel *C. longifissura*, asterisks indicate the animal pole, which gives rise to the anterior end, anterior is to the left, dorsal is to the top: (a–c) ***bra*** (the brachyury orthologue) expression in the embryos; (a) lateral view of the postgastrula showing the expression in the future mouth; (b) expression in ectoderm around and anterior to the mouth, and in the posterior tip *(arrowhead)*; (c) lateral view of (b); (d and e) illustrations of (b) and (c), respectively; (j) ***otp*** expression in the postgastrula in the presumptive neural precursor cells; (k) in the juvenile in the brain-commissure and in a posterior spot below the outer ectoderm *(arrowhead)*; (l) Lateral view of (k); (m and n) illustrations clarifying (k) and (l), respectively; (o) Ventral view of an adult stained with Alexa488-labelled phalloidin, red fluorescence shows the oocytes. Posterior to the mouth (mo) and female gonopore (fgp) is the *bra*-expressing male gonopore (mgp); (p) in the young adult, *bra* is expressed in the male gonopore; (q) *bra* gonopore expression of (p), counterstained with nuclear stain (DAPI); expression shown in *red false-colour*; bars of o–q = 100 μm. *(All from Hejnol and Martindale (2008).)*

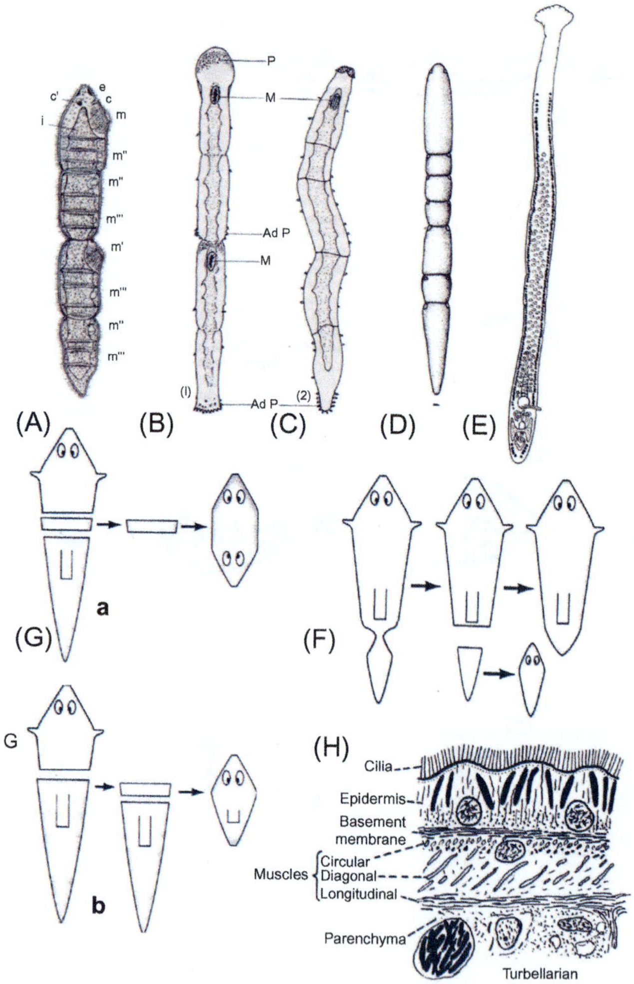

FIG. 7.6 (A) Process of budding in *Microstomum*: c, c′ = ciliated groove, e = eye spot, I = intestine, m, m′, m″ m‴ = mouth *(after von Graff from Marshall and Williams, 1975)*; (B and C) agametic reproduction in two species of the *Microstomum*, by paratomy in *M. rubromaculatum* (B) and by architomy in *M. papillosum* (C): AdP = adhesive papilla, M = mouth, P = anterior pigment *(redrawn after Westblad, 1952, from Henley, 1974)*; (D) chain of zooids in *Stenostomum (Hyman redrew after Child, from Barnes, 1974)*; (E) adult *Caryophyllaeus laticepus* from intestine of carp *(from Marshall and Williams, 1975)*; (F) spontaneous binary fission in planaria. At 24°C, the small tail piece differentiates a pharynx and will become a feeding worm within 2 weeks; (G) (a) a narrow, transverse anterior piece often leads to the production of a double-headed janus-like organism; (b) if the anterior wound border is allowed to heal briefly before the posterior cut is made, regeneration proceeds normally *((F and G) from Ellis and Fausto-Sterling, 1997)*; (H) Diagram of a V.S. of the body wall showing the structure of the ectodermal and mesodermal layers of a tubellarian *(from Borradaile et al., 1977)*.

The interstitial cells play an important role in regeneration of planarians. If one takes white planaria and cuts off its head, a new head will soon be regenerated. If one repeats the experiment but this time irradiates the body with X-rays, then a new head will not grow. If one takes such an irradiated decapitated animal and implants a small piece of tissue from a black nonirradiated animal, the white animal starts to regenerate a new head. The tissue in the head is largely formed from the black implanted tissue, as can be seen from its pigmentation. The cells that are most responsible are the interstitial cells, and it should be remembered that these cells are important in regeneration in Cnidaria too.

Gametic reproduction

Sexual dimorphism is the exception rather than the rule among Turbellaria, practically all of which are hermaphroditic.

Reproductive system

Most flatworms are hermaphroditic, and their reproductive systems are the most complex in the animal kingdom. Sex cells are developed from certain free cells of mesenchyme and migrate to regions where gonads are formed. In the primitive form Acoela no gonads or oviducts exist, and the germ cells are arranged in longitudinal rows of eggs (Fig. 7.3A (c)) and sperm. In the reproductive system (Fig. 7.7), testes are usually numerous, but only one or two ovaries are present as a rule. The female system has one unique feature is that the yolk is added by vitelline glands to the eggs after the latter leave the ovary. This feature differs from the other animals in which the yolk substance becomes a part of the egg structure within the ovary. In flatworms, the yolk, or vitelline, glands may discharge into the oviducts either directly or by special yolk ducts.

According to this pattern of reproductive system, self-fertilization may occur. However, to ensure cross-fertilization, a copulatory apparatus or complex is usually present (Fig. 7.7H and I). This consists mainly of a genital antrum for receiving the male penis and sacs to store the sperm from the sex partner. The oviducts may or may not unite before entering the complex. The female system, on the other hand, usually consists of a branched or rounded ovary from which elongates a short oviduct that is joined by a common yolk duct, and a short duct from the seminal receptacle. The muscular Laurer's canal also joins the oviduct between the ovary and the entrance of the common yolk duct. Laurer's canal (often absent) may open to the outside and may function in copulation in some unknown manner (at that time). The oviduct widens to form the ootype, which is surrounded by Mehlis' glands (unknown function). These glands were formerly thought to secrete the shell, but it is now known that the shell is formed from yolk shell material.

In 1974, Henley described additional recent observations, the present account selected only few of them: Dahm (1950) described another case, in the alloeocoel *Bothrioplana semperi* M. Braun (Turbellaria, Alloeocoela), in which the male copulatory organ, if present, is very much reduced and serves merely as a storage point for

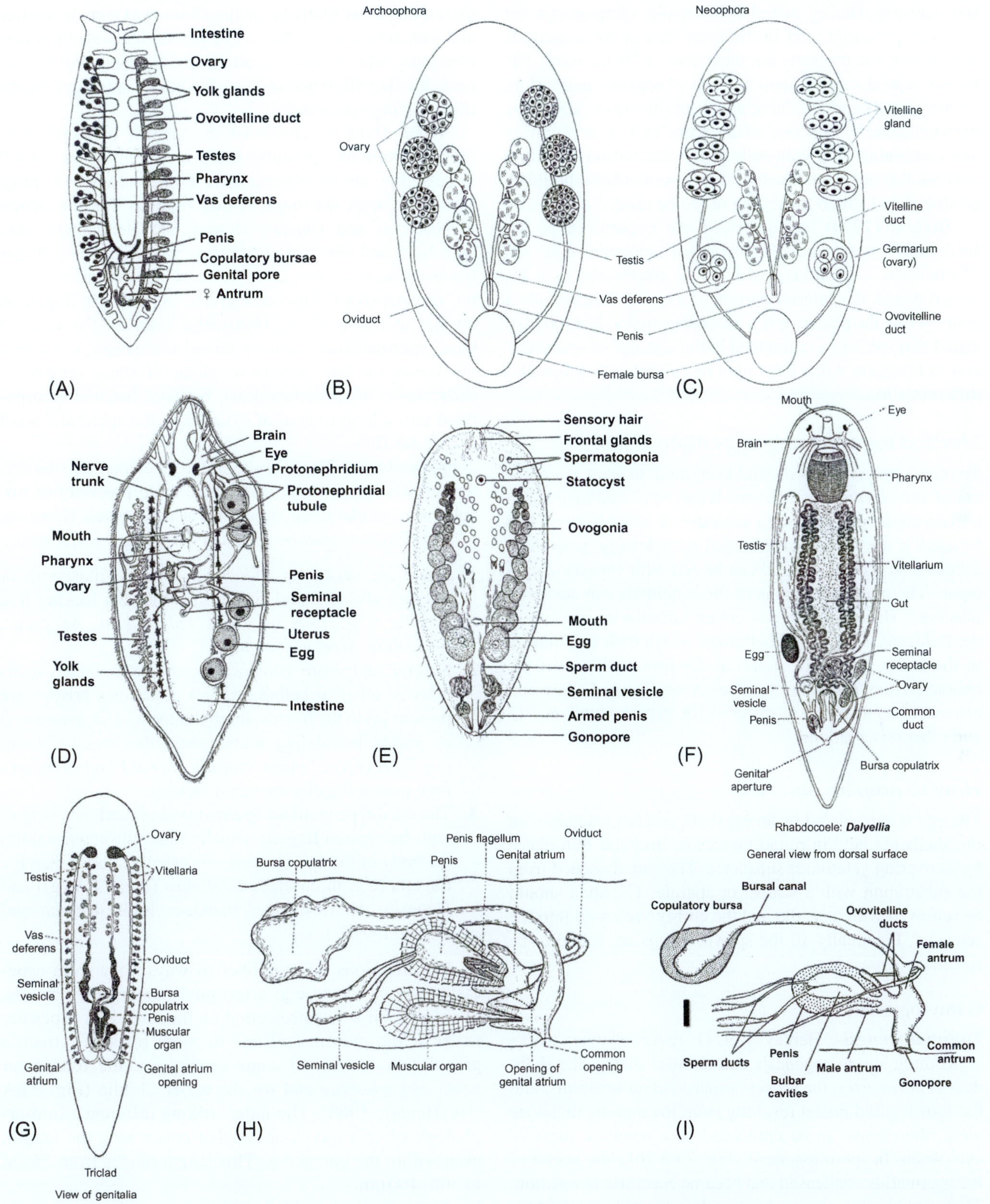

FIG. 7.7 Diagrams of dissected different turbellarian flatworms showing reproductive system: (A) model shows male structures on left and female on right, but in life both systems are paired as shown below; (B and C) diagrams of dissected animals: (B) archoophoran polyclad; (C) Neoophoran proseriate, triclad; (D) a rhabdocoel *Mesostoma*, a common freshwater form whose summer eggs hatch within the body, testes shown only on left, uterus only on right; (E) acoel *Childia spinosa*, primitive form has epidermal musculature similar to that of coelenterates *(modified from von Graff, 1911)*; (F) dorsal view of a rhabdocoel *Dalyellia viridis (from Bresslau)*; (G) view of genitalia of a planarian triclad *(after Steinmann)*; (H) L.V. section through region of genital atrium in *Dendrocoelum lacteum (after Ullyott and Beauchamp)*; (I) L.S. of a planarian showing copulatory organs. *(A, D, E, and I) From Hickman (1973); (B and C) from Ellis and Fausto-Sterling (1997); (F–H) from Borradaile et al. (1977).*

spermatozoa. Henley stated that usually spermatozoa are not even produced, and in the event that spermatogenesis does occur, the products are incapable of fertilization. The female reproductive system consists of separate germarium (ovary gonad) and vitellaria, and reproduction is said to be parthenogenetic with one ovum from each ovary uniting and combining with yolk cells of the vitellarium to form a cocoon. It is not clear, however, why spermatozoa should be produced and stored if they are not to be used.

Borkott (1970) has described the occurrence in the freshwater rhabdocoel *Stenostomum sthenum* nov. sp. (Turbellaria, Catenulida), of what he interpreted as a bisexual gonad, the anterior portion of which is functionally a testis, while the posterior is a secondary ovary. Borkott also stated that some ova segmented in the absence of spermatozoa and suggested that this might represent parthenogenetic differentiation.

Method for obtaining eggs (Kato, 1968)

To obtain polyclad worms and keep them to spawn, a number of animals with mature gonads can be placed together in a glass container of running seawater. A suitable cover can be used. If the seawater is changed occasionally, animals in a Petri dish will spawn and can be fed with various simple algae. The mating reactions of these animals can easily be observed. However, one can obtain turbellarian eggs from the field, since various turbellarians attach their egg masses to the under-surface of stones at the low tidal line; if the characteristics of these egg masses are learned, they can also serve as a source of material for experimental studies and other observations.

How to prepare embryos

The egg is surrounded by an eggshell, and large numbers of the shells are held in round masses or irregular thin plates by a covering gelatinous substance. They are loosened from the substratum with a scalpel or spatula. The shell should be removed with a needle and the embryo released into the seawater. Practically all the spawned eggs are found to be fertilized.

Gametogenesis

Spermatogenesis. Henley (1974) reviewed the works concerning spermatogenesis and showed that in the rhabdocoel *Paravortex*, the process proceeded quite rapidly, and the four haploid nuclei resulting from the meiotic divisions were often found to be embedded in a common mass of cytoplasm. In spermiogenesis (Fig. 7.8A (b)), the nuclei of the spermatids condensed and became eccentric in position. Their cytoplasm was more hyaline than at earlier stages. One axial filament was borne by each centrosome (presumably a single one for each spermatid, although this is not made clear); this axial filament, or flagellum, was sometimes free and sometimes embedded in cytoplasm. The flagellum then elongates, and the nucleus attains the S-shape. During this time, the flagellum was observed to have a spiral configuration in living spermatids.

When Henley considered the studies of Hendelberg (1965, 1969) on spermiogenesis in a number of turbellarians, the above observations took on new meaning, especially with the unpublished findings for the acoels *Polychoerus* and *Childia*. Henley and Hendelberg used both light and electron microscopy in their studies. There has been no evidence, at the level of electron microscopy, for spermatids of flatworms giving off a single flagellum during spermiogenesis. Invariably, even in the case of those spermatozoa having a thread-like shape in the mature condition, two flagella are given off; these are free, in early stages of spermiogenesis, but later become incorporated into a long tongue of cytoplasm, the spermatid shaft (Fig. 7.8A (b)).

Spermatozoa. The morphology of the spermatozoa differs greatly in the different groups, more precisely of turbellarians, of platyhelminthes and/or the species. However, these differences may be confined to three general forms:

1. Some are very long, slender and thread like, with or without visible undulating membranes but lacking free flagella (Fig. 7.8A (d)) (e.g. *Polychoerus*, *Anaperus*, *Convoluta*, *Mecynostomum*).
2. Shorter and more compact spermatozoa, which characterize other turbellarians, and sometimes bristles are present (as in the freshwater rhabodocoel *Macrostomum*) or paired undulating membranes (the acoel *Childia*, Fig. 7.8A (e)) or 'wings' (the *alloeacoel Plagiostomum*). Free paired flagella are again lacking.
3. The third type includes spermatozoa of moderate length, with free paired flagella, usually inserted subterminally; the body of the spermatozoon is usually shorter than the flagella (e.g. the triclads *Bdelloura* (Fig. 7.8A (g)) and *Dugesia*, the freshwater rhabdocoels *Mesostoma* and *Microdalyellia*).

Modifications, in a number of ways, may form especially within acoels e.g. a terminal bundle of filaments, or delicate cilia-like projection on the thread-like spermatozoa or the presence of one or more rows of retractile granules in the tail of some acoels, e.g. the freshwater acoel *oligochoerus* and for the acoel Childia (Fig. 7.8A (f)) (Henley, 1968). The most striking difference in morphology of spermatozoons is that concerning the length, even within the one genus. This length ranges from 35 μm to 300–400 μm.

At the beginning of the 20th century, it was thought that the acoel *Amphiscolops* had dimorphic spermatozoon, one type being 8–10 times larger than the other and with a

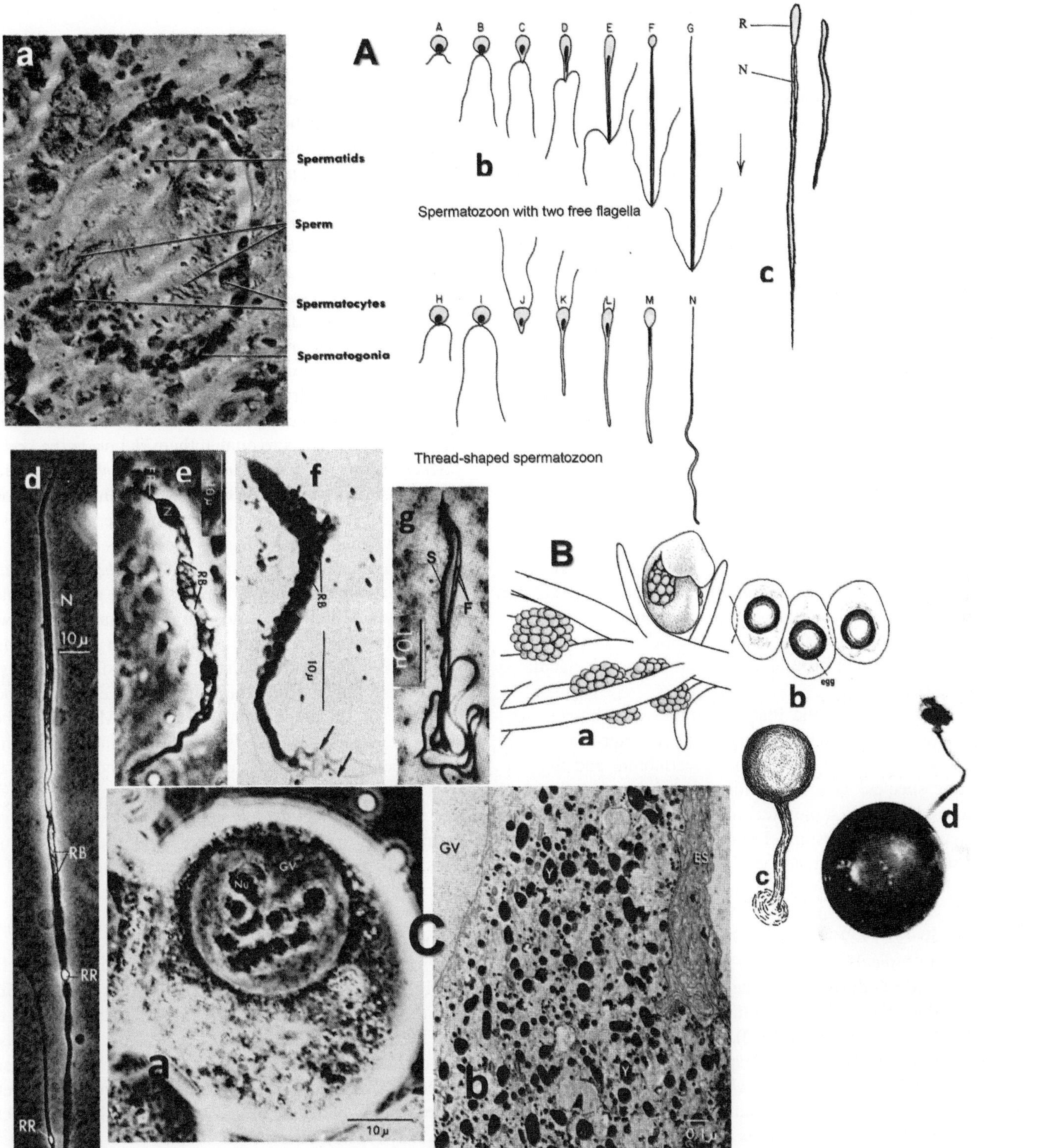

FIG. 7.8 (A) Spermatogenesis in Turbellaria: (a) photomicrograph of section through testis of the triclad *Dugesia* showing different stages of the process; (b) illustrations of the possible steps in spermiogenesis of spermatozoan with free flagella and with incorporated axonemes *(redrawn and modified from Hendelberg, 1969)*; (c) typical eupyrene (left, 95 μm long) and atypical apyrene (right, 45 μm long) spermatozoan of the acoelan flatworm *Otocelis rubropunctata*. N = nucleus (Feulgen +ve), R = cytoplasmic residue, *arrow* = direction of swimming; (d) phase contrast micrograph of living spermatozoon of the acoel *Anaperus*, the nucleus (N) is anterior and behind it, is a double row of small refractile bodies (RB). There are also two very conspicuous refractile regions (RR) in the tail; (e) phase contrast micrograph of living late spermatid of the acoel *Childia*, the nucleus (N) has still not taken on the mature oval shape. A double row of refractile bodies (RB) is found along the length of the tail behind the nucleus, and a short filamentous process (FP) extends forwards from the tip of the cell; (f) light micrograph of fixed and stained mature spermatozoon of *Childia* with heavily stained refractile bodies (RB), the *arrows* indicate a region where the double undulating membranes are flattened out *(from Henley, 1968)*; (g) phase contrast micrograph of living spermatozoon of the triclad *Bdelloura* with two long free flagella (F) are inserted subterminally at the anterior end of the spermatozoon (s); (B) Gametes of Turbellaria: (a) egg masses of *Convoluta (from Apelt, 1969)*; (b) section of egg ribbon of polyclad *Stylochus (after kato, from Hyman)*; (c) stalked cocoon of a fresh-water planarian *(modified from Pennak)*; (d) egg capsule (cocoon) of planarian, contains 1 to several eggs and yolk cells; (C) oocytes of Turbellaria: (a) phase contrast micrograph of living oocyte of acoel *Polychorerus carmelensis*. GV = conspicuous germinal vesicle, N = a skeinlike nucleolus; (b) Electron micrograph of section of oocyte of *Childia* showing cytoplasm with the margin of GV at left and the egg surface = ES at right, Y = large yolk granules, and there are other types of cytoplasmic inclusions. *(A a and B d) From Hickman (1973); (A b and d–g and C) from Henley (1974); (A c) from Roosen-Runge (1977); (B a–c) from Barnes (1974); (d) from Hickman (1973).*

'contractile spiral flange'. The anterior end of the shorter spermatozoon terminates in an abrupt, nearly transverse face, and there was a short filamentous portion. Both types had the granules referred to above, arranged along the length of the tail. On the basis of observations (Henley, 1974) for a number of forms on spermiogenesis in the living condition, it seems likely that this might be the explanation for the apparent dimorphism.

As long ago as 1916, it was reported that the spermatozoa of the rhabdocoel *Paravortex* are motionless in the male system, and then the fluid inside the female system stimulates them to become motile. The release of spermatozoa into seawater also stimulates them to become active. It appears that this is true for the turbellarian spermatozoa, since the later foregoing studies, e.g. those 'extensive studies of living intact specimens of *Polychaerus carmelensis*', proved that the spermatozoa of this form are almost invariably motionless when still in the various parts of the male system. However, when contained in the seminal bursa of the female tract, they become very active. This character is in agreement with the view reported by Hyman (1951) that self-fertilization usually does not occur in these hermaphrodites, because spermatozoa are motionless in the male system and do not become motile until they are ejaculated during copulation.

Spermatozoa of the Platyhelminthes are characterized by the absence of the acrosome at the anterior-most tip, in contrast to these of many other invertebrate and the vertebrate species. This was evidenced by the ultrastructural studies. However, some exceptions are present and/or to the structure of the acrosome. Detailed information concerning the structure of spermatozoa of Platyhelminthes is found in the chapter written by Henley (1974).

Polymorphism of sperm. In an acoelan flat worm, Hendelberg (1969) found two kinds of spermatozoa: a very peculiar 'typical' one with two flagella, and a smaller atypical one which during spermatogenesis casts off its nucleus with residual cytoplasm and remains in the semen as short cytoplasmic rod (Fig. 7.8A (c)). Diminution or lack of chromatin was established as a frequent characteristic of atypical sperm by Meves (1903), who used the term 'eupyrene' for cells with normal chromatin content, 'oligopyrene' for those with subnormal amounts and 'apyrene' for cells without chromatin. The eupyrene is 95 μm and the apyrene is 45 μm long in acoelan flatworm *Otocelis rubropunctata* (Fig. 7.8A (c)).

The spermatozoa may possess undulating membranes, bristles, paired flagella, etc., and vary in length from less than 5 to 400 μm. According to Lentati (1970), the fertilizing spermatozoa are usually haploid, although polyploid biotypes exist in which, as for instance in *Dugesia benazzii*, excess sets of chromosomes may gradually be eliminated during spermatogonial or meiotic divisions.

Mann (1984) reviewed the structure and development of the spermatophore and/or the sperm in animals. He then reported that Lang (1884), an early authority of Platyhelminthes, especially of the turbellarians, was the first to apprehend clearly that the peculiar small white threads attached to the bodies of *Leptoplana* and certain other Polyclads inhabiting the Gulf of Naples are not some of thread-like packages of spermatozoa, wrapped in tough membranes. He also deserved the credit for pointing out that in Polycladida, the so-called male copulatory organ is not necessarily inserted into the female genital pore, but may serve the purpose of attaching the spermatophores to the surface of the partner's body. The fixed spermatophore, having penetrated the 'skin' and musculature, then releases the spermatozoa so as to enable them to invade the body.

Mechanisms of sperm transfer in Turbellaria. Many years after Lang's observations, it was investigated that while in *Macrostomum Romanicum* (Macrostomida), the cork-screw-shaped spermatozoa are pumped during copulation straight into the vagina, in *Archaphanostoma agile* (Acoela), the copulating partners exchange their spermatozoa by hypodermal injection into the 'bursal tissue', with the whole copulation lasting no longer than 3 s.

Mann (1984) reported that sperm transfer mechanisms in Platyhelminthes (Turbellaria and Monogenea) have been receiving special attention in recent years. Members of both groups are hermaphrodites (see Fig. 7.7), and most of them have a 'penis' (Fig. 7.7H and I). Among the relatively few species in which copulation has been actually observed, only some reproduce by exchange of spermatophores; others transfer their spermatozoa either by intromission of a penis into a vagina, or by hypodermic impregnation. All the described examples were for Monogenea.

Oogenesis. One very conspicuous characteristic of the Platyhelminthes is their possession of a special gland, called the yolk gland, which forms yolk cells. These are combined with the egg cells to make up composite eggs. Within one shell there may be included as many as several thousands of yolk cells, and from 1 to 10 egg cells. Such 'eggs' are large in general, usually being several millimetres in diameter, while the especially large eggs of some of the terrestrial triclads may reach a diameter of 1 mm.

This characteristic originated with the Turbellaria, but no yolk gland develops in Acoela, Polycladida and some species among Rhabdocoela and Alloeocoela. Consequently they lay eggs called simple eggs, to differentiate them from the composite eggs. These turbellarian groups are known as more primitive archoophoran (in which there are no yolk glands and the eggs contain yolk distributed in the cytoplasm). Tricladida and some of Rhabdocoela and Alloeocoela develop yolk glands and form composite eggs (Neoophora) (Barnes, 1974). Since the yolk is accumulated outside the egg cell in composite eggs, these are sometimes called ectolecithal neoophora eggs, while the simple eggs are called endolecithal archoophora since their yolk is located inside the egg.

Oocytes (ova). Henley (1974) reviewed the morphology of the flatworm ova, which has received very little attention in the literature, and apart from some investigations on their development after fertilization, she reported that in living oocytes of *Polychoerus carmelensis*, studied by phase contrast microscopy (Fig. 7.8C (a)), a large spherical germinal vesicle is present, with a conspicuous nucleolus. The yolk is rather uniformly distributed in the cytoplasm of the oocyte of this form, e.g. those of most acoels and many alloeocoels. In the more advanced groups, the yolk is provided from an exogenous source: yolk cells produced from yolk glands.

Surface (1907) studied the development of the polyclad '*Planocera*' (=*Hoploplana*), and Papi (1953) (cited in Henley, 1974), described the embryology of the rhabdocoel *Macrostomum*. Both described the striking amoeboid surface changes in the fertilized ova preceding polar body formation and early cleavages. Another bit of evidence concerning the nonrigidity of the eggs of some turbellarians was also reported.

As mentioned before, Kato (1940) stated that the cytoplasm of eggs of many of the Japanese polyclads he had studied had dense masses of granules in their cytoplasm, which were colourless or slightly pink. He also reported a dark brown pigment in ova of the polyclad *Thysanozoon*; this pigment accumulates at the animal pole after the eggs have been laid.

The size of full-grown turbellarian ova varies greatly. Among the polyclads, for example, Kato (1940) studied the development of ova varying from 85 μm in diameter (*Stylochus*) to 330 μm (Planocera). Some terrestrial triclads are said by Kato (1968) to have ova 20 μm in diameter; fixed and sectioned ova of *Polychoerus carmelensis* were reported by Costello (1961) to vary between 200 and 280 μm in diameter, while those of *Hoploplana* are 100 μm in diameter (Costello et al., 1957).

Copulation and/or fertilization

Kato (1968) reported that gametic reproduction begins when the reproductive organs are fully developed; although these species are almost without exception hermaphroditic, self-fertilization does not take place as a rule. There is a well-developed copulatory pouch (seminal receptacle) and a true copulation takes place between any two mature individuals by the union of the respective male and female reproductive organs. In the rhabdocoels, some of the freshwater triclads and the polyclads, the spermatozoa are bundled in **spermatophores**, which are placed in the copulatory pouch (bursa copulatrix) of the partner. The most simplified mode of copulation, **hypodermic injection**, occurs widely in many species among the Acoela, Polycladida, several of the Rhabdocoela, Alloeocoela and marine Tricladida. In this method of impregnation, one animal punctures the epidermis of the other by means of the copulatory apparatus, apparently at any point on the body, and injects its spermatozoa into the parenchyma. The sperm travel through the tissue spaces and enter the uterus, where fertilization takes place. This means that fertilization occurs just before spawning. In exceptional cases in *Thysanozoon brocchi*, fertilization may take place after spawning (Selenka, 1881), but Kato (1940) believed this to be an error.

Kato (1940) studied copulation or mating in a number of polyclads including *Notoplana*, *Stylochus*, and *Pseuodostylochus*. In all cases, he found that pairing occurred during the day as well as at night. Other reports described mating behaviour, including those of Surface (1907) for the polyclad *Hoploplana*, and for the freshwater rhabdocoel *Mesostoma* and the alloeocoela *Monocelis*. The general pattern was comparable for all and the duration of the process was usually short.

The view of Henley (1974) was that in many cases, self-fertilization of these hermaphroditic animals is anatomically possible, but apparently it rarely occurs. There are said to be two general methods of cross-insemination: hypodermic impregnation and true copulation (Hyman, 1951). The first method was formerly considered to prevail among the acoels, but more recent observations have shown that this is not the case. It does seem to occur in a few acoels, rhabdocoels and polyclads: the spermatozoa are injected via the male copulatory apparatus as described above; they then wander through the parenchyma to the oocytes. Reciprocal copulation, often with deposition of spermatozoa into the partner's copulatory or seminal bursa or into the parenchyma, is more commonly the rule, particularly among the triclads. Rather complex patterns of courtship behaviour often precede the event (see Hyman, 1937).

Oviposition. Fertilization of most turbellarian ova occurs internally, and the eggs are then encased in delicate capsules, which are embedded in a jelly mass as for most acoels and polyclads, or in a shelled cocoon. In the latter, there may be one to several ova within a single shell (Fig. 7.8B (d)). A few cases are known where ova develop within the body of the parent (e.g. the rhabdocoels *Paravortex* and *Bresslauilla*). Cement glands, usually found in proximity to the gonopore, produce an adhesive material, which in many cases attaches the deposited eggs to the substratum (Fig. 7.8B (a) and (b)). Some forms have stalked cocoons (Fig. 7.8B (c)), while others do not. There is attachment of the cocoon by an elongated elastic stalk to sand grains in beach surf, and in even more violent environments, additional protective coats may be present in the capsule. In the work of Henley (1974), a very strange and astonishing cocoon was described.

This cocoon, produced by the rhabdocoel *Kronborgia caridicola*, is a very long tubi-form, dextrally coiled, 46–71 cm long, with closely packed egg capsules within. One end (which presumably had been anchored in the mud substratum) was irregularly thickened and twisted, while the other had a funnel-shaped opening with a diameter of about 1 mm, through which the hatched young emerged.

The process of egg-deposition varies with the group. In those forms lacking an oviduct, notably some of the acoels, von Graff (1911) and others concluded that fertilized ova would be deposited either via the mouth or through rupture of the body wall. However, the deposition through the mouth was not actually observed, and in a careful study of the process in the acoel *Polychoerus*, Costello and Costello (1939), using both living and technically prepared material, found no evidence that the eggs were extruded through the mouth. However, they were able to demonstrate clearly in their sectioned material that these eggs were extruded through an extensive rupture of the body wall.

ORDER POLYCLADIDA

Development of another simple egg

The Polycladida are characterized by highly typical spiral cleavage. Their development is the object of the historically significant discovery that homology of blastomere determination, which had been found among the annelids and molluscs, exists even in such a phylogenetically low group of animals as the Turbellaria (Lang, 1884; Wilson, 1898). Since then, extremely detailed investigation into the embryology of this group has been continued by Surface (1907) and Kato (1940). The size of the ovum varies with the species; in *Planocera reticulata* it measures 310–330 μm in diameter, but that of *Notoplana humilis* measures 130–150 μm; there are some eggs as small as that of *Stylochus uniporus* (88–95 μm). The larger and more yolky the egg, the more slowly it develops. The time required until hatching is 3 weeks in *Planocera*, 2 weeks in *Notoplana* and 1 week in *Stylochus*. The meiotic divisions begin about 1 h after the egg is laid, when a depression appears at the animal pole, where the first polar body is formed. After this, the egg seems to shrivel and its outline becomes extremely irregular for some time; it then regains its spherical shape and gives off the second polar body.

Cleavage of this simple egg (Kato, 1968)

This egg undergoes spiral cleavage. Spiral cleavage is a derivative of radial cleavage. The word spiral was first used in relation to cleavage, by Selenka (1881) in his description of the rotation of blastomeres in the cleavage of polyclad flatworms. Wilson (1892) proposed the recognition of a spiral type of cleavage as a derivative of radial cleavage. Spiral cleavage is holoblastic, total, determinate and unequal. The planes are not parallel or perpendicular to the animal-vegetal axis of the egg; rather, cleavage is at oblique angles. This results in the blastomeres having a spiral arrangement, with any one cell being located between the two blastomeres above or below it. This one tier or set of cells alternates position with the next tier. The cells touch one another at more places than those of radially cleaving embryos do. Spirally cleaving embryos usually undergo fewer divisions before they begin gastrulation, making it possible to follow the fate of each cell of the blastula. When the fates of the individual blastomeres from annelid, flatworm and mollusc embryos were compared, many of the same cells were seen in the same places, and their general fates were identical (Wilson, 1898).

Fig. 7.9 shows illustrations of stages of the cleavage of *Notoplana humilis* (Kato) where the egg divides totally and unequally, and as stated above, in a typical pattern of spiral cleavage. The **first** cleavage plane cuts vertically through the animal and vegetal poles and divides the egg into two nearly equal-sized blastomeres (AB, CD) (Fig. 7.9A). CD is always larger than AB. The **second** cleavage is also vertical, but unequal, dividing the blastomeres into two fairly large ones (B, D) and two small ones (A, C) (Fig. 7.9B). At this time, the D blastomere is the largest, and its position indicates the future posterior side of the embryo. As a consequence, the B blastomere marks the anterior side. Viewed from the animal pole, the A blastomere lies on the left and the C blastomere on the right side. Seen from the side at the time of the second cleavage, the positions of the two spindles intersect each other practically at right angles. As a result, two of the four blastomeres lie above the other two; a pair of smaller blastomeres (A and C) are formed on the animal pole, and a pair of larger ones (B and D) make contact at the vegetal side, forming the so-called 'cross furrow'. This cleavage is clearly laevotropic.

The **third** cleavage is horizontal and unequal, dividing the blastomeres in a strongly *dexiotropic* direction. In general, the larger a blastomere is, the more rapidly it cleaves; in this case D thus cleaves first, followed by B, C, and A. The blastomeres cut off towards the animal pole side (micromeres) are smaller than those left on the vegetal side (macromeres). The difference in size between micromeres and macromeres differs with the species. This third division gives (1A–1D) + (1a–1d). 1a–1d are the first quartet (Fig. 7.9C). The **fourth** cleavage, which follows next to produce the (8–16-cell stage), is laevotropic; first the macromeres at the vegetal side divide, forming the second quartet of micromeres 2a–2d, but before these are completely separated, the first micromere quartet divides into $1a^2$–$1d^2$ on the vegetal side and $1a^1$–$1d^1$ on the animal pole side. The $1a^2$–$1d^2$ blastomeres correspond to the trochoblast of annelids and molluscs. The embryo at this stage constitutes the blastomeres: (2A–2D) + (2a–2d) + ($1a^1$–$1d^1$ + $1a^2$–$1d^2$) 1st q (Fig. 7.9D–F).

The **fifth** cleavage (16–32-cell stage) is dexiotropic, leading to the formation of a typical spiral cleavage 32-cell

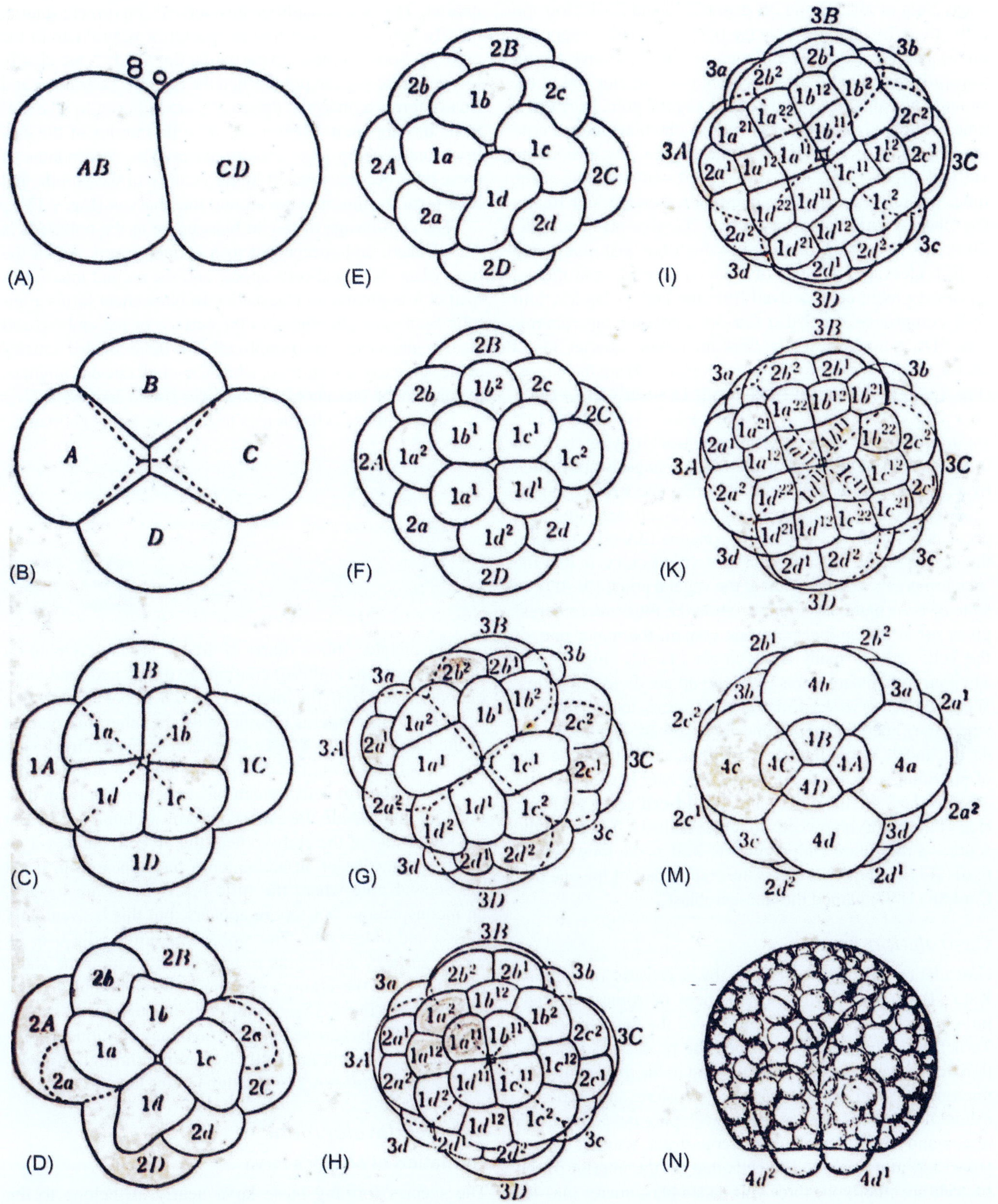

FIG. 7.9 Diagrams seen from animal pole of cleavage in the polyclad *Notoplana humilis* (Kato): (A) two-cell stage (lateral view); (B) four-cell stage; (C) eight-cell; (D) 8–10 to 12 cell; (E) 12-cell; (F) 16-cell; (G) 24-cell; (H) 28-cell; (I) 28–32-cell; (K) 32-cell; (M) 36-cell (vegetal view); (N) cleavage of mesentoblasts. *(From Kato (1968).)*

stage. First of all, 2D and 2B give off 3d and 3b dexiotropically; there is a considerable lag before 2A and 2C separate off 3a and 3c. In the meantime, 2d and 2b cleave dexiotropically, producing $2d^1$–$2b^1$, which are smaller and at the animal pole, and $2d^2$–$2b^2$ are on the vegetal pole. During the time of the division of the 2d and 2b, the blastomeres constituting the first quartet divide dexiotropically, producing $1a^{11}$–$1d^{11}$ + $1a^{12}$–$1d^{12}$ + $1a^{21}$–$1d^{21}$ + $1a^{22}$–$1d^{22}$. At the same time, the remaining 2a, 2c micromeres divide and finally the other two macromeres, 2A and 2C, give off 3a and 3c. Thus the 32-cell stage embryo, which has just completed its fifth cleavage, consists of 16 cells derived from the 1st q (1a–1d); eight cells derived from the 2nd q (2a–2d), four cells comprising the 3rd q (3a–3d) and four macromeres (3A–3D). Among these, the third micromere quartet 3a–3d are the smallest and the macromeres (3A–3D) are the largest (Fig. 7.9G–K). (3A–3D) + (3a–3d) 3rd q + ($2a^1$–$2d^1$) + ($2a^2$–$2d^2$) 2nd q + $1a^{11}$–$1d^{11}$ + $1a^{12}$–$1d^{12}$ + $1a^{21}$–$1d^{21}$ + $1a^{22}$–$1d^{22}$ 1st q. In the subsequent division, the isochrony of cleavage is increasingly lost, so that the 32-cell stage is succeeded by a 36-cell, and then 40-, 44-, 45-, 53-, and 61-cell stages.

After the embryo has reached the 32-cell stage, the next event is the cleavage of the macromeres (3A–3D) to form the 36-cell stage. This division is highly peculiar in that the 'macromeres', which are left at the vegetal pole (4A–4D), in spite of their name are minute, while the four 'micromeres' given off at the animal pole side contain the major part of the yolk and are many times larger (4a–4d) (Fig. 7.9M). The four little macromeres (4A–4D) do not divide after this. When the embryo gastrulates, they move to the interior, where they later degenerate and are absorbed. The three cells 4a, 4b, and 4c also fail to cleave further, and in the same role of providing yolk for the embryo, they later degenerate and are absorbed. On the other hand, the **4d-cell** at the posterior side of the embryo proceeds to divide actively, becoming the source of the future endodermal as well as the mesodermal tissue (Fig. 7.9N). This cell is the counterpart of the one that Conklin (1897) named the 'mesentoblast'.

Gastrulation

Gastrulation takes place chiefly by a process of epiboly. Kato (1940) studied the development of some polyclads turbellarians and, in 1969, he reviewed the development of Turbellaria. This author described the process of gastrulation that the cells derived from the first to the third quartet of micromeres proliferate by repeated divisions and gradually extend towards the vegetal pole, enveloping the cells of this region and finally forming a stereogastrula, without a blastocoel left in its centre. The cells that fill the interior of the gastrula are chiefly the three yolk-laden blastomeres (4a–4c) of the fourth quartet and the descendants resulting from the division of the 4d cell; in addition there are $1a^{111}$–$1d^{111}$ the brain cells derived from the 1st micromere quartet and the ectomesoderm cells which arise from the second micromere quartet. The descendants of this second micromere quartet chiefly form the ectoderm of the posterior ventral part of the embryo. Some of these, in particular $2a^{222}$–$2d^{222}$, are clearly located at the vegetal pole. When the embryo gastrulates and finally forms a pharynx, these cells are believed to take part in its development. They move into the interior of the embryo and develop into ectomesoderm. The descendants of these cells become located in the vicinity of the mouth, and later form the musculature around the pharynx (Fig. 7.13).

The cell-lineage of the 4d blastomere in the polyclads is complicated, and specialized to a certain extent. About the time when the apical cells appear and the second micromere quartet is beginning to cleave, the **4d** blastomere forms a mitotic figure at right angles to the surface of the embryo and cleaves into a large and a small cell. The large cell **$4d^1$** remains at the surface of the embryo, while the small one **$4d^2$** migrates inwards to the mesentoblast, and gives rise to endoderm 2.

Each of these cells divides to form the left–right pairs:

$4d^2$
Divide
↙ ↘
(Smaller) (Larger)
4 d^{22} (Smaller) and 4 d^{21}
↙ ↘
Endoderm Mesoderm

$4d^1$
↙ ↘
$4d^{11}$ $4d^{12}$
Remain on the surface

Planocera	Müller's
Stylochus	Göttes
Notoplana	direct

The daughter blastomeres of $4d^{11}$ 4^{12} (which remain on the surface of the embryo) continue to divide and send more endoderm cells into the interior (Fig. 7.13), but at about this time, these blastomeres together with the other micromeres of the fourth quartet (4a, 4b, 4c) become covered by the ectoderm cells as a result of the epiboly which is taking place.

The endoderm cells, which have been multiplying during this time inside the embryo, form a clump located at the lower side of the embryo near the 4b cell, but as yet no cavity suggestive of an archenteron has appeared in it. At the vegetal pole where the little macromeres are located, an inconspicuous blastopore appears, but this is eventually closed and obliterated. The embryo acquires cilia over its whole surface, and begins to turn slowly within the shell. At this stage, the embryo continues development directly (*Notoplana*) (Fig. 7.15), or indirectly through larval stages. The larva that has arrived at this stage may continue its process of development indirectly by passing through Müller's larva stage (*Planocera*), or through Göttes larva (*Stylochus*).

Indirect development

Formation of Müller's larva

The species forming these larva nearly all belong to the *Cotylea* although among the Acotylea, Müller's larvae have been found by Surface in *Hoploplana inquilina* and by Kato in *Planocera multitentaculata*. In the latter *Planocera multitentaculata* (Fig. 7.10), the general course of development

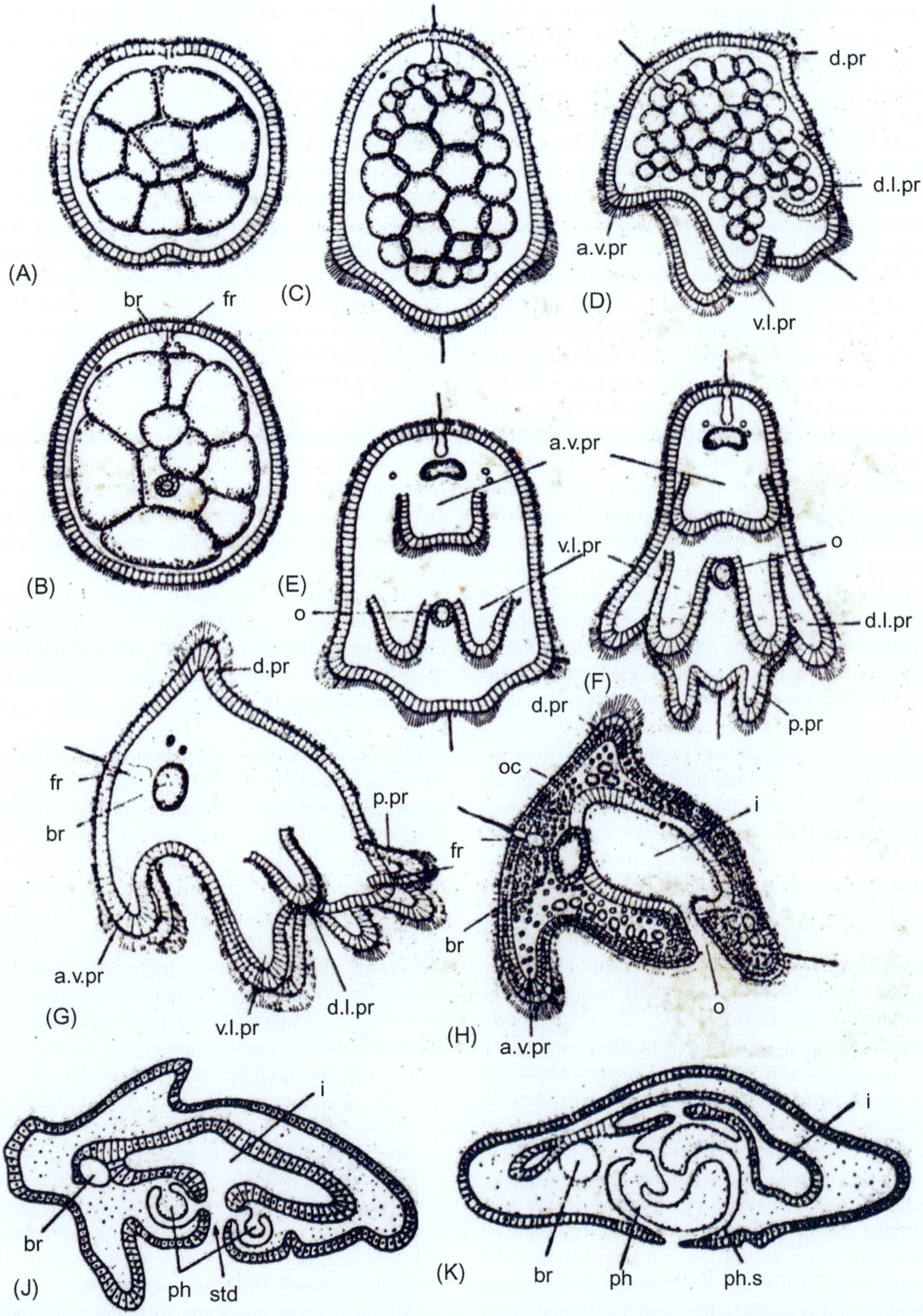

FIG. 7.10 Diagrams showing indirect development in the polyclad *Planocera multitentaculata* (Kato): (A) gastrula; (B) formation of procerebral organ rudiment; (C) two-eye-spot stage; (D) side view of three-eye-spot stage embryo; (E) V. view of d; (F) V. view of Müller's larva; (G) side view of f; (H) L.S. of Müller's larva. a.v.pr.=antero-ventral process, br=brain, d.l.pr.=dorso-lateral process, d.pr.=dorsal process, fr.=procerebral organ, i.=intestine, o.=mouth, oc.=eye-spot, p.pr.=posterior process, v.l.pr.=ventro-lateral process; (J and K) metamorphosis in *Yungia aurantica* (Lang): (J) L.S. of Müller's larva; (K) L.S. after metamorphosis. ph=pharynx of adult, ph.s=pharyngeal sheath, other letters as above. *(From Kato (1968).)*

is similar to that of the directly developing forms (see later). Five days after spawning, epiboly is complete and the blastopore is closed over. On the sixth day, as in *Hoploplana*, an ectodermal invagination takes place in the vicinity of the original blastopore and a broad, short pharynx, which connects with the intestinal lumen, is formed. Cilia develop on the interior surface of the pharynx and intestine, and at about this time, an invagination at the aboral pole gives rise to the rudiments of the brain and frontal gland. Only a single eye-spot appears, however, rather far to the left of the body midline. As in direct development, rhabdite rudiments are formed in the ectodermal cells, the embryo bends anteriorly and a bundle of sensory filaments, the apical sensory tuft, appears at the anterior end. Internally, as in the former case, the yolk mass is subdivided into small cells (Fig. 7.10C).

Next, the ventral side of the embryo becomes rather flattened, and longitudinal depressions appear at the sides of the mouth. The epidermis at the anterior end of these depressions protrudes, giving rise to the rudiment of an antero-ventral process. Following this, a ventro-lateral process is formed on each side of the mouth and posteriorly, a pair of dorso-lateral processes and finally a single small dorsal process is formed. The cells making up these processes are tall columnar, with long cilia. It is considered most probable that these cells arise from the $2a^2$–$2d^2$ quartet (corresponding to the annelid trochoblasts). At about the tenth day, another eye-spot is added on the right of the mid-line and a second sensory tuft appears at the posterior end. Meantime the rotational speed increases, and the left eye-spot divides, producing a larva with three eye-spots (Fig. 7.10E). Each of the processes grows in size. The antero-ventral one does especially, increasing markedly in width. A new pair, the posterior processes, are added at the posterior end, making a total of eight such processes or lobes (lappets), and completing the full development to the Müller's larva (Fig. 7.10F–H).

On day 14 or 15, the larva escapes from its shell and begins a free-swimming life. It is about 0.2–0.3 mm in length, and forms (seen from the side) a rounded triangle with the antero-ventral process as the front angle. The dorsal process forms the top angle and the posterior processes complete the triangle at the rear. Besides the unpaired antero-ventral and dorsal processes, there are one pair each of ventro-lateral, dorso-lateral and posterior processes; each process bears long cilia. There is a circular mouth behind the antero-ventral process. From this mouth, a narrow stomodaeum leads into a spacious intestine (Fig. 7.10H). A mass of mesoderm cells is found around the stomodaeum. This is the rudiment of the pharynx. There are sensory tufts on both anterior and posterior ends (Fig. 7.10F), and the brain and frontal gland are well developed (Fig. 7.10G). However, the frontal gland will degenerate a short time after the larva hatches.

Metamorphosis of Müller's larva. Kato (1968) reported that according to Lang (1884), who found Müller's larvae of *Thysanozoon* and *Yungia* in his plankton, the larva flattens as it increases in length, the various processes diminish in size and finally disappear, and for the first time the larval shape begins to resemble that of an adult polyclad. The most striking internal change that accompanies this metamorphosis is the formation of the pharynx. The innermost part of the stomodaeum swells to a spherical shape, and a mass of mesoderm cells collects to form a protuberance that will become the pharynx. The pharynx sheath forms in the empty space around the protuberance. The connection between the pharynx sheath and the stomodaeum consists of an extremely narrow slit. At the time of metamorphosis, the stomodaeum is pulled outwards and its wall forms part of the body wall (Fig. 7.10J and K).

Kato (1968) reported that *Planocera reticulata* (Japanese sp.) reaches Müller's larva inside the eggshell, and the young animal emerges only after metamorphosis. This is called intracapsular Müller's larva (Kato) and this mode of development is **intermediate development**. This larva is flattened, its dorsal process is lacking and instead, the tip of the antero-ventral process is bilobed. The intestine is restricted in size.

Another interesting point recorded by Kato is that *Graffizoon lobata* (Cotylea) a turbellarian found in California, has larvae less than 1 mm in length that closely resemble Müller larvae and become sexually mature. This is regarded as a case of neoteny (Heath, 1928).

Hoploplana inquilina* (Polycladida, Acotylea).** A somewhat recent study of the development of this platyhelminth species, ***Hoploplana inquilina, not only highlighted the description of the quartet spiral cleavage pattern of the polyclad embryo, but also mainly discussed the relationship of the Platyhelminthes to other metazoan groups. Boyer et al. (1998) studied the cell lineage of this polyclad, *H. inquilina*, based on recent techniques using fluorescent lineage tracers. These tracers have made it possible to refine and extend the previous studies. By fluorescently labelling identified cells with DiI, the authors have followed the fates of individual blastomeres of the polyclad *H. inquilina*, providing essential background information on development of a probable basal spiralian. However, the present work copied the cell linage from Kato (1968), since it is very detailed, in addition to the names of the organs written in Boyer et al.'s lineage (Fig. 7.13).

Boyer et al. (1998) obtained the specimens of this polyclad, *H. inquilina*, from the mantle cavities of the whelk *Busycoptypus cannaliculatus*. The latter were obtained from the Aquatic Resources Division of the Marine Biological Laboratory (Woods Hole, Massachusetts).

Although there are some differences in the interpretations between this study of Boyer et al. (1998) and the previous studies of Surface and Kato, the former evidenced the previous ones (Fig. 7.9). At the four-cell stage, the *H. inquilina* embryo consists of two large vegetally positioned blastomeres which meet at the vegetal cross-furrow, and two somewhat smaller animal positioned noncross-furrow cells. Thus, while it is not possible to identify specifically each quadrant in an unlabelled embryo, the vegetal cross-furrow cells and their descendants are easily distinguished from the noncross-furrow cells. Fig. 7.12 illustrates, by colours, the fates of the blastomeres of each quartet.

The results of this study of Boyer et al. demonstrate a striking similarity between the cell lineages of polyclad and the higher spiralian embryos, in which the four quadrants also bear the same relationships to the larval axes. They also give rise to comparable larval structures, including derivation of mesoderm from both ectodermal (2b) and endodermal precursors (4d) (see Figs 7.9 and 7.13).

At 22°C, *H. inquilina* embryos reach a well-differentiated Müller's larva (Fig. 7.11(I) G) in ~5 days. The larva consists of a series of ciliated lobes, including the oral hood and middorsal lobe. They are separated by the apical tuft and two eyes. There are paired ventrolateral and lateral lobes, and the posterior-dorsal lobe, which bears the posterior tuft. The midventral mouth leading into a blind gut is surrounded by a ring of 'peculiar ciliated cells' at the junction of the stomodaeum and gut. The appearance of the larval form, especially the size of the individual lobes, and the labelling patterns observed varied slightly from embryo to embryo.

The results of the study of Boyer et al. (1998) on the development of *Hoploplana inquilina* confirm largely those of Surface (1907). One of the most significant results of this cell lineage study indicates that the axial positions of the alternating cleavage pattern are in the polyclad embryos. In addition, the blastomere fates as well are very similar to those of the previously described polychaete, nemertean and molluscan embryos (Spiralians). These embryological features support the concept, and provide further evidence, that the turbellarian platyhelminthes such as the polyclads, which exhibit the quartet spiral cleavage, represent the extant basal spiralians (Text Fig. 7.1B). Another possible scenario for the relationship of the Platyhelminthes to other metazoan groups has been argued to be the sister group of the coelomate Bilateria, or a sister group of the coelomate protostomes (Text Fig. 7.1A) (Boyer et al., 1998).

Götte's larva (*Stylochus* sp.)

The Götte's larva was so named by Lang (1884) in honour of its first discoverer. This larva has only four processes instead of the eight of Müller's one (Kato, 1968). There are antero-ventral, a dorsal, and a pair of latero-ventral processes (Fig. 7.11(II) E). This larva is the free-living stage of genus *Stylochus uniporus* (Kato), a Japanese species. Its developmental process (Fig. 7.11(II)) is essentially like that of *Planocera* (Fig. 7.10). The embryo that reaches this stage is 0.1 mm in length. As stated above, it has four processes. Its brain is differentiated near the anterior end of the body, and the frontal gland is not very highly developed. It metamorphoses several days after hatching. Lang believed that this larval form was an intermediate stage in the development of a Müller's larva, but the results of Kato's (1940) rearing experiments show that metamorphosis occurs without any increase in the number of processes, and it is thus clear that this is an independent larval form.

Larvae of flatworms Polycladida

Rawlinson (2014) paid attention to a related subject: the polyclad larvae. First of all, Rawlinson begins the article by schemes (Fig. 7.14) representing the present phylogenetic relationships of the Platyhelminthes. The flatworms of this phylum are an extremely diverse spiralian clade (over 100,000 known species (Park et al., 2007), that includes the catenulids and rhabditophorans, but excludes the acoels and nemertodermatids (Larsson and Jondelius, 2008) (Fig. 7.14A). Then, the members of the Platyhelminthes were differentiated into definite groups (Fig. 7.14B). It is appropriate to copy these schemes, as they may help and/or extend knowledge. The article itself deals with polyclad larval structures, diversity, development and evolution. It summarizes classical literature, synthetizes recent work and presents new data; it also establishes a foundation for future studies of metazoan larval character evolution. One of the attractive and significant inclusions of this article is the table in which the author tabulates the distribution of the larvae in the polyclads with their taxonomy, the type of larva, the length of the larva and the reference. Rawlinson stated, the well-known information that polyclad larval stages (known as Götte's and Müller's larvae) swim using ciliary bands, or ciliary tufts, found on protrusions of the body wall. The other flatworms have direct development. The exception is one freshwater species of catenulid, which swims at hatching using several anterior rings of elongated cilia, but it resembles the adult in other respects; this is known as a **Luther's larva**, a swimming stage.

Classically, polyclad larval stages have been divided into two groups based on the number of lobes: the Götte's larva having four lobes (Fig. 7.11(II) E), and the Müller's larva with eight lobes (Fig. 7.11(I) G and H). However, as larvae with different numbers of lobes were subsequently discovered, they were assigned to one group or the other; consequently there is a Götte's larva with five lobes (*Stylochus tauricus*), and Müller's larvae with 6, 7, and 10 lobes, e.g. at hatching, the cotylean (*Pseudoceros Canadensis* has 6 lobes but a pair of dorso-lateral lobes develops later (Fig. 7.14(I) C); *Planocera reticulata* has seven lobes, intracapsular and free-swimming (length at hatching 310–330 μm). This is known as **Kato's larva** (Fig. 7.14(I) D–F). In respect to the structure of the polyclad flatworms, in the early stages, the most obvious larva-specific features are the lobes and ciliary bands or tufts.

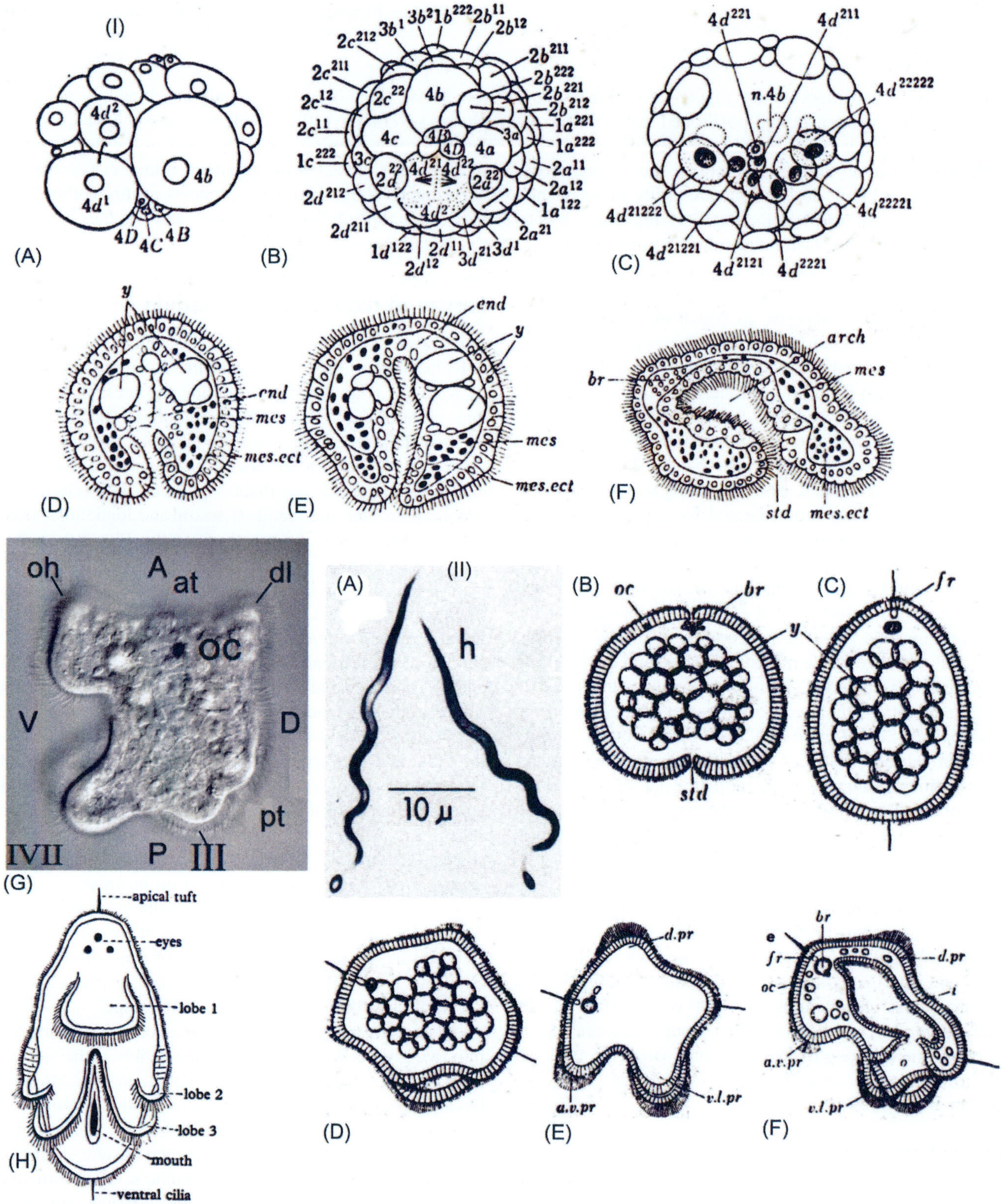

FIG. 7.11 (I) Gastrulation and formation of Müller's larva of *Hoploplana inquilina* (Surface, 1907): (A) T.S. of embryo at 45-cell stage, $4d^2$- blastomere moves into segmentation cavity; (B) vegetal polar view of embryo at 66-cell stage, $4d^2$ divides into right and left cells; (C) subsequent division of $4d^2$ (cross-section) near vegetal pole; (D) invagination of stomodaeum; (E) opening of archenteron after stomodaeum invagination; (F) early stage of Müller's larva. Arch = archenteron, br = cerebral ganglion, end = endoderm, mes = mesoderm, mes.ect = ectomesoderm, std. = stomodaeum, y = yolk; (G) left lateral view of a 5-day-old Müller's larva of *H. inquilina*, indicating the larval axes, A = anterior is up, P = posterior, D = dorsal, V = ventral. The larva is ~ 120 μm in length, but most cases have been flattened for photomicrographs. Tracing of the larva shown in (G): at = apical tuft, dl = dorsal lobe, III = left lateral lobe, IVII = left ventrolateral lobe, oc = ocellus, oh = oral hood, pt. = posterior tuft; (H) diagram of Müller's larva of a polyclad *Cycloporus papillosus* Lang. *(A–F) From Kato (1968); (G) from Boyer et al. (1998); (H) from Borradaile et al. (1977).* (II) Formation of Götte's larva of *Stylochus uniporus* (Kato): (A) phase contrast micrograph of living spermatozoon of the polyclad *Stylochus*, the anterior end is indicated by h *(from Henley, 1974)*; (B) young embryo; (C) one-eye-spot stage; (D) two-eye-spot stage, beginning of process of formation; (E) Götte's larva; (F) L.S. of the larva. d.v.pr. = antero-ventral process, br = brain, d.pr. = dorsal process, fr. = frontal organ, I = intestine, o = mouth, oc. = eye-spot, std. = stomodaeum, v.l.pr. = ventro-lateral process, y = yolk. *(B–F) From Kato (1968).*

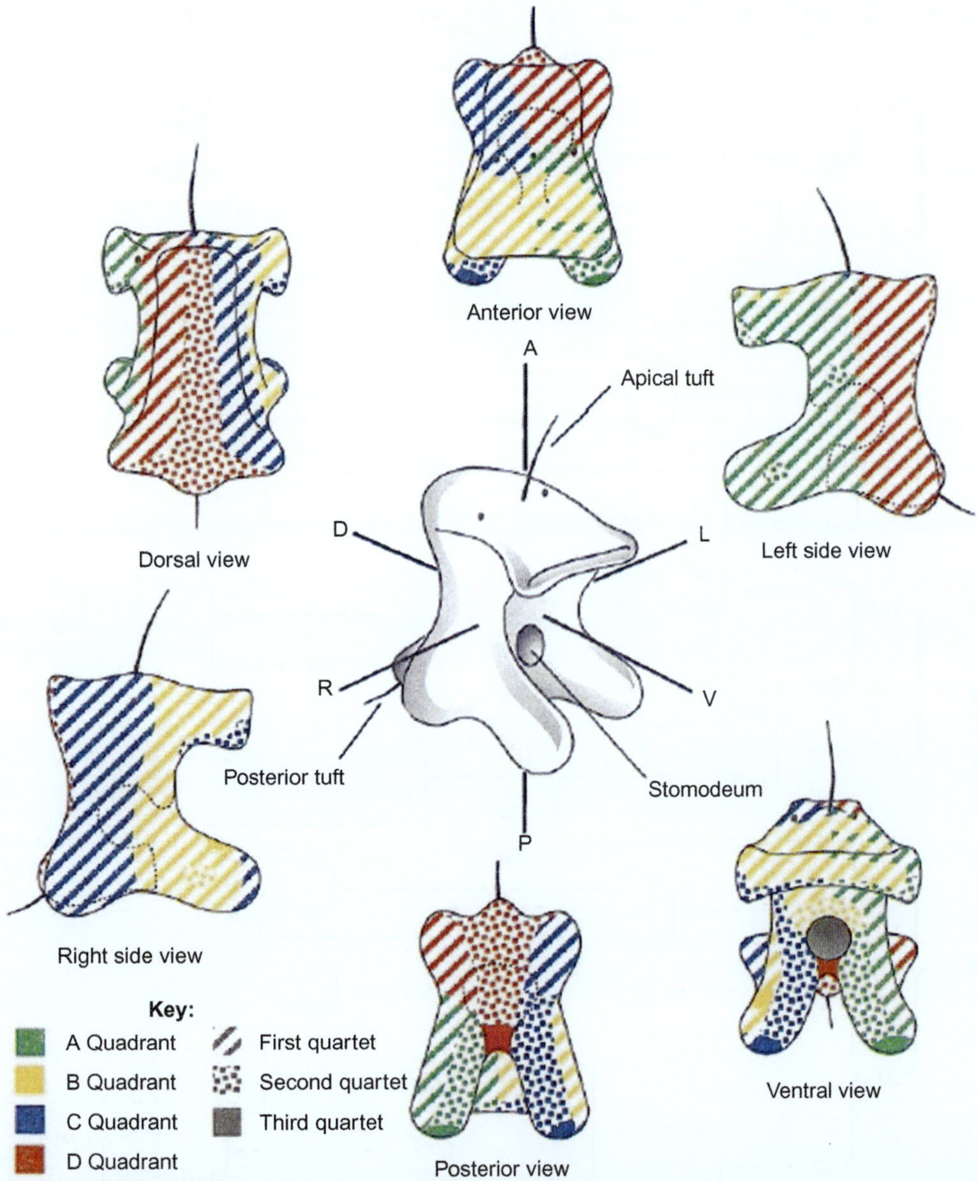

FIG. 7.12 Diagrams summarizing the typical contributions of each cell, through the formation of the third quartet of micromeres to the ectoderm of the müller's larva in the polyclad flatworm *H. inquilina*. Specific contributions are as indicated by the *colour-coded key*. Axes shown on the central three-dimensional view are as follows: A–P, anterior–posterior; D–V, dorsoventral; L–R, left–right. Other views are as labelled. *(Modified from Henry et al. (1995), presented here from Boyer et al. (1998).)*

Other characters, such as the frontal (or apical) organ, are also reported to be lost during metamorphosis. More detailed studies on the metamorphosis of indirect-developing species, and comparative analyses of indirect and direct-developing taxa, are needed to distinguish between characters that are unique to the larval stage, and those that persist through the larval stage and into the juvenile/adult. Rawlinson (2014) described the larval lobes as temporary protrusions made up of epidermal cells, such as prototrochal cells, ciliated epidermal cells, monociliated sensory cells and neurons, body wall muscle and connective tissue constituting an extracellular matrix (Fig. 7.6H). The lobes lack epidermal rhabdites and yolk (Fig. 7.14(I) A and B), unlike the rest of the animal body. The long lobes found in larvae collected from the plankton would suggest that lobe growth continues in the water column (Fig. 7.14(I) D–F). At the peripheral margins of the lobes, cells with longer cilia form continuous, preoral ciliary bands (often called prototroch), discontinuous bands or numerous distinct bands or tufts (Fig. 7.14(I) C).

In *Maritigrella crozieri*, the ciliary band is formed by hundreds of prototrochal cells, positioned in rows and

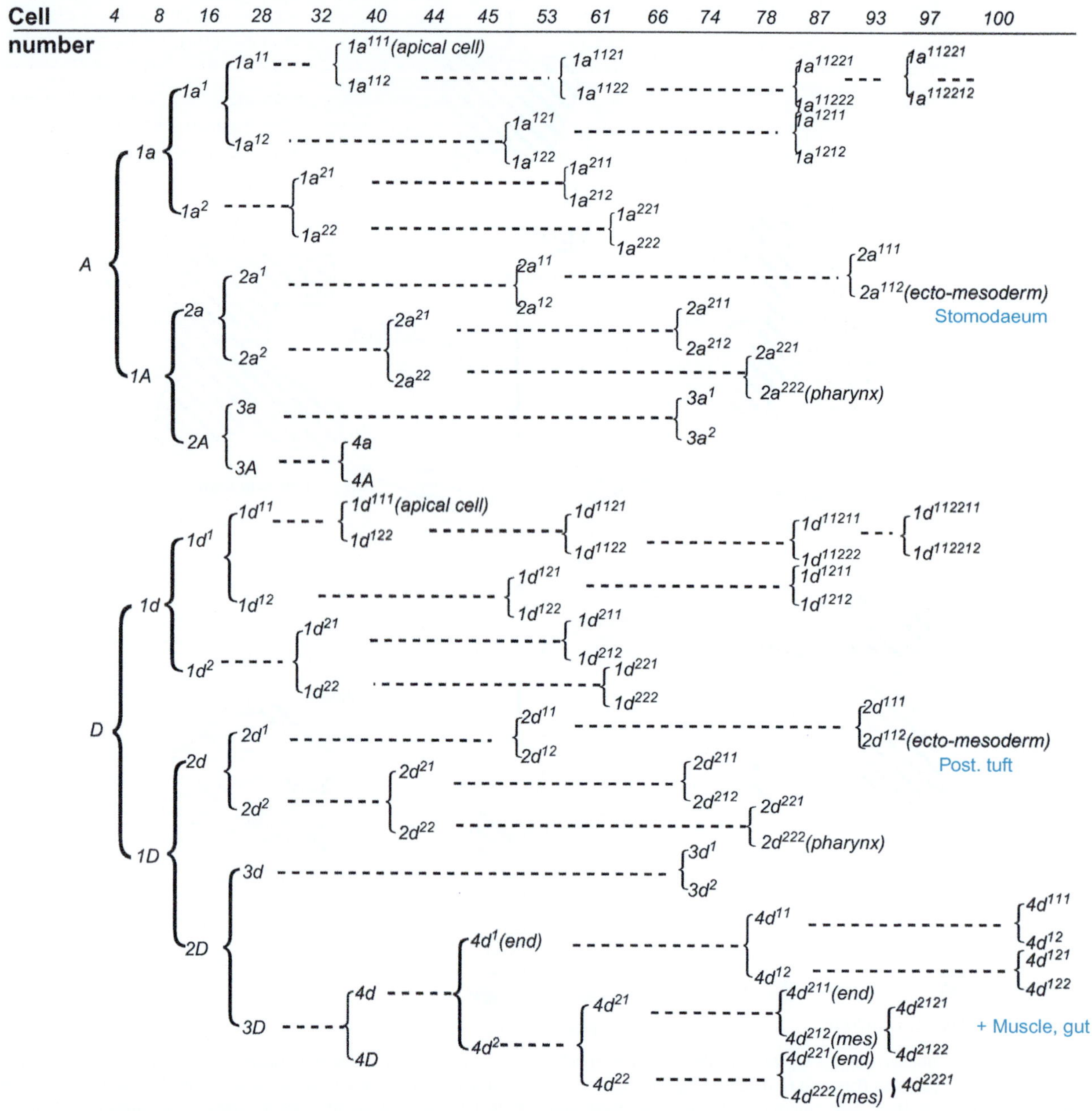

FIG. 7.13 Cell lineage in *Hoploplana inquilina* (Surface, 1907). Division of B and C blastomeres, same as that of A. end = endoderm, mes = mesoderm *(from Kato, 1968)*. Names of organs written by *different colour* are *after Boyer et al. (1998).*

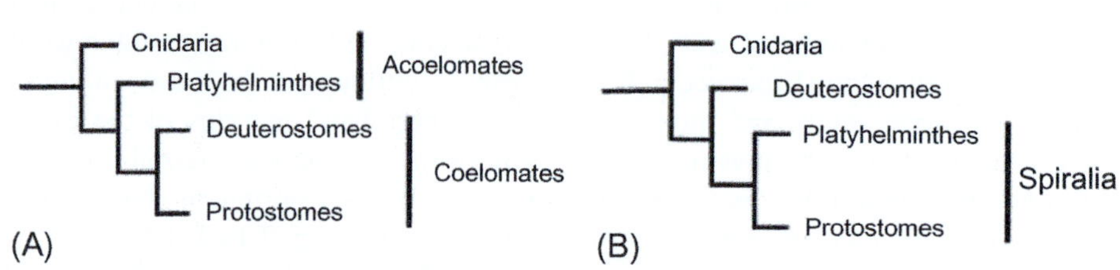

TEXT FIG. 7.1 Two possible scenarios for the relationship of the Platyhelminthes to other metazoan groups (assuming Platyhelminth monophyly). (A) Phylogeny indicating a sister group relationship with the coelomate Bilateria. (B) Phylogeny supported by molecular and developmental data, with the Platyhelminthes as the basal taxon of the "Spiralia," a large group of protostome animals displaying spiral cleavage. *(From Boyer et al., 1998.)*

innervated by a dense subcilial plexus of serotonin and FMRFamide immunoreactive cells. The ciliary band forms shortly after gastrulation as a ring around the developing stomodaeum at the posterior end of the embryo (in *M. crozieri*). As the stomodaeum migrates ventrally, the ciliary band becomes convoluted and migrates to the position of the future larval lobes. The ciliary bands are swimming organs and are also used for feeding.

At the anterior pole of polyclad larvae is a tuft of cilia and associated gland cells. These have been named a frontal organ, apical organ or apical plate, and a 'glandulo-sensory' function has been suggested. Rawlinson (2014) calls this organ a frontal organ, for specification. Histologically, the frontal organ looks like a columnar mass of connective tissue between the brain and the apical cilia (Fig. 7.14(I) B(ii)). Its ultrastructure shows that the frontal organ consists of a cluster of monociliated or multicilated sensory cells surrounded by a circle of gland-cell necks supported peripherally by microtubules. The bases of the monociliated sensory cells in *Pseudoceros canadensis* are located on top of the brain, but no direct connection to the brain has been observed. It was determined that these cells were not nerve cells. Kato (1940) suggested that the frontal organ is used to break the eggshell at hatching and noted that it degenerates 1–2 days posthatching in *Planocera reticulata* (which has already undergone metamorphosis inside the egg capsule). In 1968, Kato suggested that the frontal gland breaks the eggshell of *Notoplana humilis*.

Direct development in Polycladida

Acotylea *Notoplana humilis*

Kato (1968) presented this species as an example for direct development in Polycladida. The author reviewed this process and reported that as the ciliated gastrula begins to rotate within its shell, with the animal–vegetal (aboral–oral) axis as the main axis. Little by little, it tends to incline towards the anterior side. Between this ectoderm, the endoderm

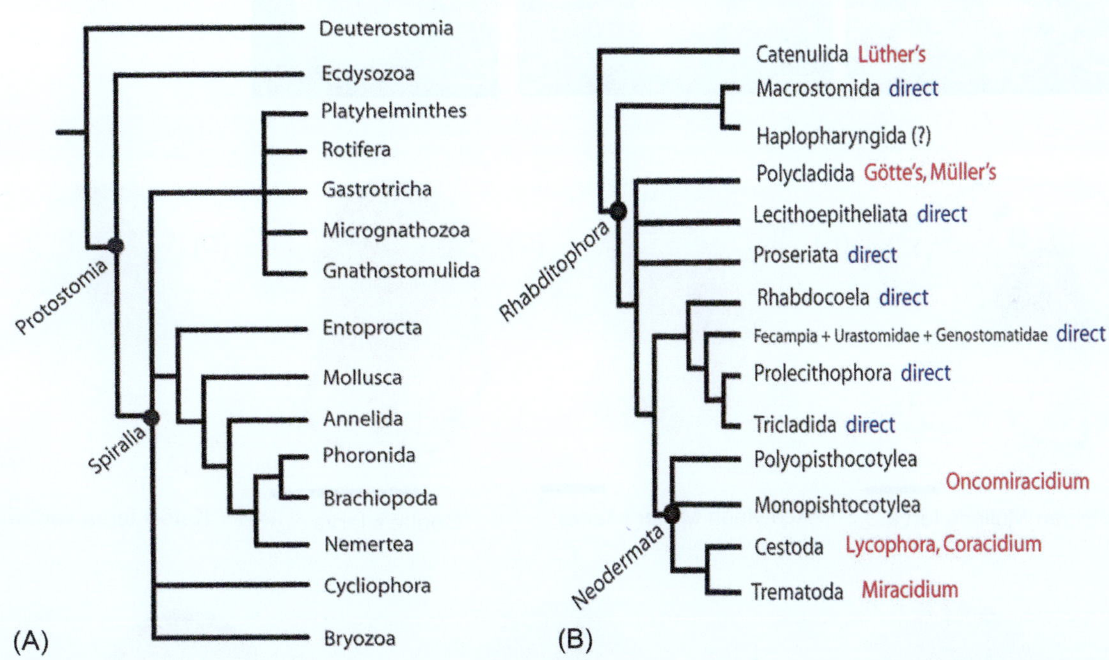

FIG. 7.14 Phylogenetic relationships of the Spiralia and the Platyhelminthes (A) and the different groups of the Platyhelminthes (B). *From Rawlinson (2014).* (I) Schematic diagrams and histological sections showing the gross anatomy of an eight-lobed larva and variation in ciliary band and ciliary tuft configurations *(red)* among polyclad larvae: (A and B) *Maritigrella crozieri* (2 days post hatching). (A) (ii) Frontal section; (B) (ii) sagittal section. Masson's trichrome stain, scale = 100 μm; (C) *Pseudoceros canadensis* Müller's larva with a single but discontinuous band, breaks in ciliary band indicated with arrowheads, it has six lobes, but a pair of dorso-lateral lobes develop later; (D–F) Photomicrographs of live polyclad larvae, 8-lobed Müller's larvae collected from plankton offshore of Fort Pierce, Florida; (E and F) 10-lobed larvae, the late larval stage of this worm is recognizable by elongation in the anterior–posterior axis, dorso-ventral flattening and the development of a sucker on ventral surface. The lateral lobes are relatively short and are slowly being incorporated back into the body. br = brain, cci = caudal cilia, epr = putative extraocular photoreceptor cells, dl = dorsal lobe, dll = dorsolateral lobe, fo = frontal organ, in = intestine, ll = lateral lobe, m = mouth, oh = oral hood, ph = pharynx, rc = putative rejectory cells, rh = rhabdites, rpr = rhabdomeric photoreceptor cells, vll = ventrolateral lobes, y = yolk. *From Rawlinson (2014).* (II) Larval types and juveniles of Polycladida: (A) of *Prosthiostomum siphunculus*; (B) of *Planocera multitentaculata*, both (A) and (B) hatching with eight lobes and three eyes (two cerebral eyes and one *epi*dermal eye; (C) of *Imogine mediterranea*, hatching with four lobes and a cerebral and an epidermal eye; (D) ventral view of Kato's larva of the acotylean *Planocera reticulata*, hatching with eight lobes and 12 eyes and being dorsoventrally flattened; (E) dorsal view of Kato's larva; (F) directly developing juvenile of the acotylean *Pseudostylochus obscurus*, hatching with no lobes and four eyes; (G) directly developing juvenile of an undetermined acotylean, hatching with no lobes and 12 eyes; all scale bars = 50 μm. *(A–G) From Martín-Durán and Egger (2012).*

(Continued)

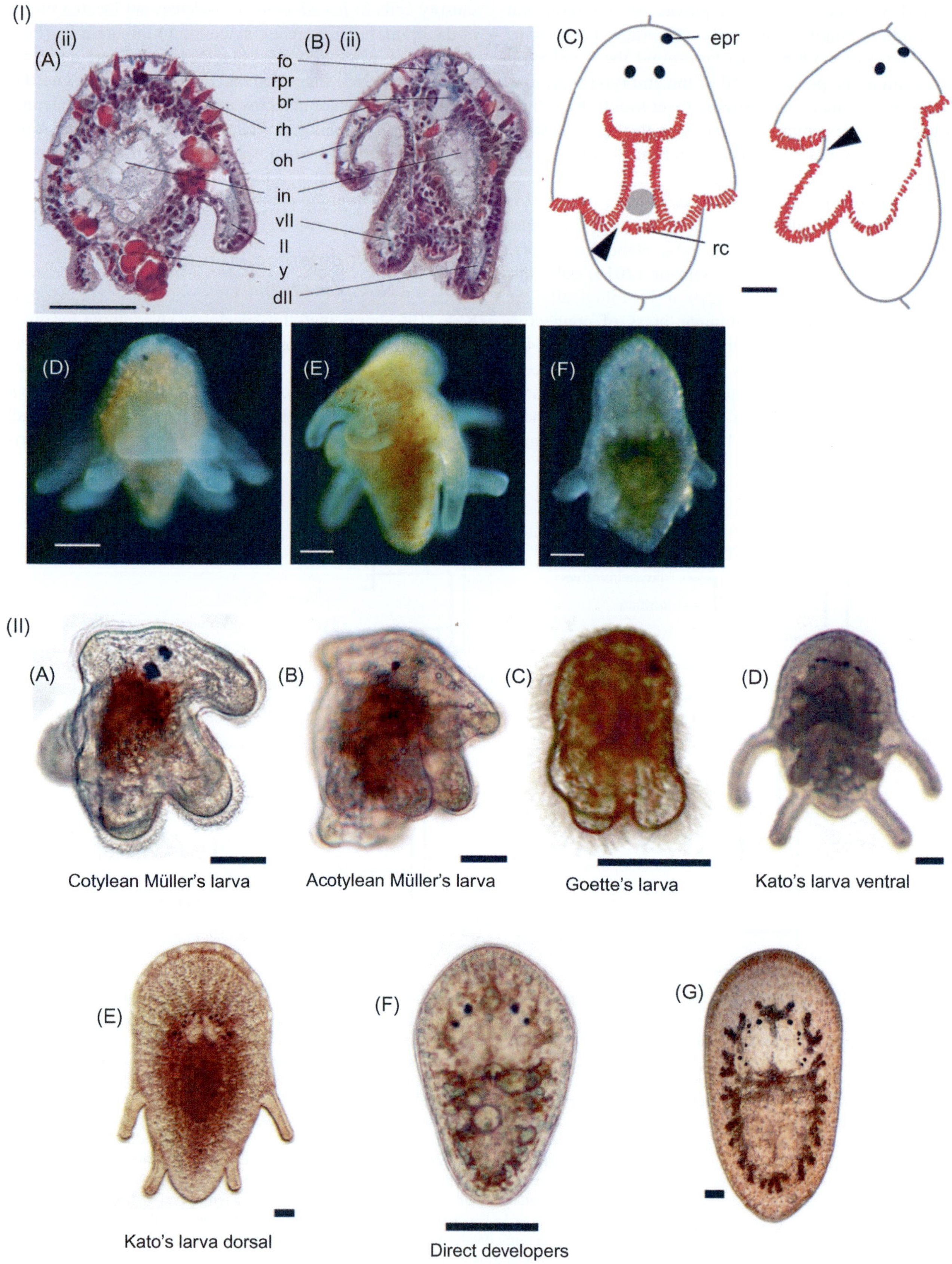

FIG. 7.14, CONT'D

cells can be seen; a pair of mesodermal strands formed by proliferation of cells arising from the mesentoblasts expand laterally and hence it becomes possible to define the bilateral symmetry of the embryo. The epithilial cilia increase considerably in length, the rotation becomes faster and the rudiments of rhabdites begin to appear among the ectoderm cells. Internally, the yolk-laden cells of the fourth micromere quarter fuse with each other, their cell boundaries disappear and they form a single mass to divide into a large number of cells (Fig. 7.15A and B). The endoderm cells proliferate, spreading in a network around this mass of yolk and repeatedly extend their processes into the yolk to be absorbed, and thus to nourish.

The embryo bends anteriorly, so that the aboral pole shifts to its front end, and the oral pole finishes in a position near its posterior end. The embryo itself becomes an oval body, having a pair of eye-spots that appear in the ectoderm near the aboral pole (2 weeks after spawning). These advanced embryos sink from the surface into a deeper layer, while the rudiments of the brain and frontal gland make their appearance at the aboral pole. The endoderm continues development, becoming sac-shaped and begins to undergo extensive ramification. The mesoderm is well developed in the anterior part of the embryo, and in the form of a large mass posteriorly. This will become the pharynx rudiment. During this stage, the ectoderm in the former blastopore region invaginates in the form of a slender cord which passes through the pharynx rudiment and reaches the lumen of the intestine. A slit next appears in this cord, extending through it and connecting with the intestinal cavity. The whole structure becomes tubular, and develops into the epidermis of the pharyngeal sheath.

As development continues, annular folds arise in the pharynx rudiment, and the pharyngeal sheath is formed. In *Hoploplana villosa*, however, this mode of formation of pharynx is rather different, in spite of the same direct

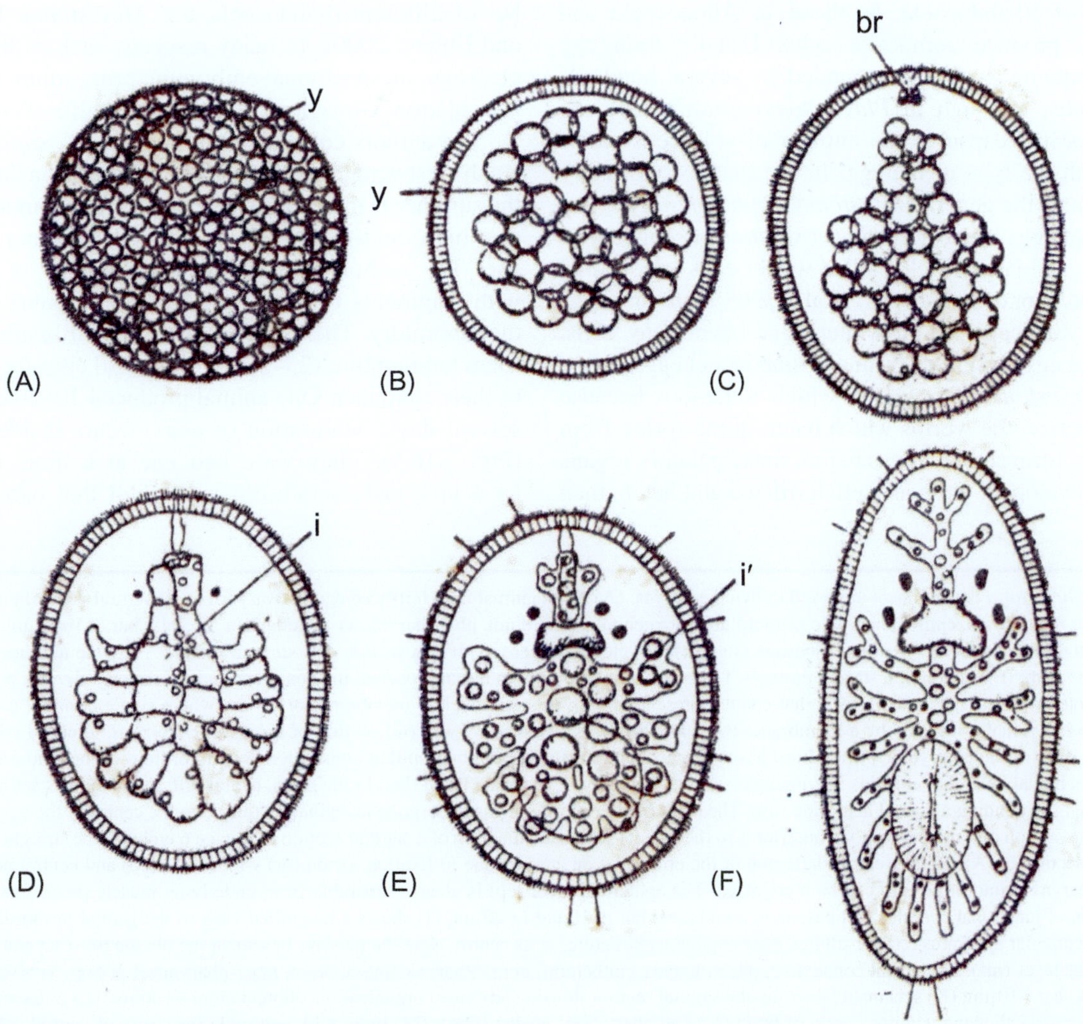

FIG. 7.15 Direct development in *Notoplana humilis* (Kato): (A) gastrula; (B) embryo with the interior yolk divided into small clumps; (C) embryo with the beginning of ectodermal invagination at the anterior end; (D) embryo at the two-eye-spot stage; (E) embryo at four-eye-spot stage; (F) embryo immediately after hatching. br = cerebral ganglion, I = intestine, y = yolk. *(From Kato (1968).)*

development being the type of development. Within another week, the eye-spots divide into two pairs, the frontal gland is clearly defined and the musculature of the pharyngeal wall takes its radial arrangement. At this stage, the mouth is not yet open to the outside. A small number of immobile sensory filaments appear around the edge of the body. Hatching takes place at about 21 days; it is believed that the breaking through of the eggshell is the function of the frontal gland. The larva has already nearly attained the adult shape when the mouth opens 5 days after hatching.

Martín-Durán and Egger (2012) showed that the turbellarian polyclads develop through different larvae. They presented a number of photographed larvae and juveniles of this group of Platyhelminthes (Fig. 7.14(II) A–G).

ORDER RHABDOCOELA

Dalyellid flatworm, *Gieysztoria superba*

Members of Rhabdocoela, as those of Alloeocoela and Tricladida, produce composite eggs. Usually each egg capsule contains one egg surrounded by several hundreds of yolk cells, although in *Plagiostoma girardi* there are 10–12 eggs in a capsule. The number of yolk cells varies widely with the type of the egg. In the rhabdocoel family Typhloplanae, the genera *Mesostoma, Typhloplana*, etc. lay two types of eggs: small summer (or *subitaneous*) eggs with thin shells, and large, thick-shelled winter eggs (also called *dauerei*, latent or dormant eggs) (Kato, 1968). Summer eggs have few yolk cells, while winter eggs have many cells. These latter eggs resemble ordinary rhabdocoel eggs. In the species *Mesostoma Ehrenbergi*, which is famous because of its large size, the worms which hatch in the spring from winter eggs form summer eggs before the copulatory organs are fully developed; these are self-fertilized and hatch, then grow to maturity within the parental body, finally rupturing it to merge. This occurs repeatedly during the summer, until the reproductive system reach maturity in the autumn. The animals then copulate and form winter eggs (Kato, 1968).

Younossi-Hartenstein and Hartenstein (2000) studied histologically and electron microscopically, and analysed the embryonic development of a new dalyellid species, *Gieysztoria superba* (Fig. 7.16A). This species is a small rhabdocoel flatworm that occurs in freshwater habitats all over the world. These authors initiated a series of studies on embryonic development of various representatives of the phylum platyhelminthes (flatworms). This rhabdocoel group, to which *G. superba* belongs, possess a highly evolved pharynx ('pharynx doliiformis') and have developed a peculiar mechanism of adding yolk to their eggs. Thus, instead of 'dumping' yolk into the oocyte itself, they invest the oocyte with a thick layer of specialized yolk cells. This 'ectolecithal' pattern of yolk distribution has a strong effect on early embryonic development, as shown for a number of different rhabdocoels, e.g. *Mesostoma* (Hartenstein and Ehlers, 2000). In many respects, such as the irregular cleavage, mesenchymal embryonic primordium and lack of gastrulation, *Gieysztoria* embryos resemble *Mesostoma*.

The authors collected individuals of *G. superba* from a small freshwater stream near Brisbane, Queensland. During the summer (January and February), these animals occurred at a high density in the fine grained mud and around algae. The authors transferred small groups of specimens with fragments of algae and minute amounts of mud to the laboratory. These specimens survived several weeks at room temperature (23–25°C). They laid eggs on the bottom of their container. One animal produced 10–20 eggs within several days. Maturation of eggs occurs inside the ovary (Fig. 7.16A). They were laid one at a time, surrounded by a brownish, semitransparent shell that barely allowed

FIG. 7.16 *Gieysztoria* development observed in living embryos: (A) Adult animal with fertilized egg in ovary (ov); (B) ventral view of embryo at stage 2; (C) at stage 8. br=brain, epm=embryonic primordium, gp=genital primordium, ph=pharynx, vt=vitellarium, yk yolk; bar (A)=50 µm, (B,C) 20 µm; (D) parasagittal section of embryo at late cleavage (stage 2); (E) electron micrograph of a section of a stage 2 embryo. Note the absence of junctional contacts in the epm; (F) section of a stage 3 embryo. The embryonic primordium has approached the ventral surface of the egg. *Arrows* point at cells in telophase. Note arrays of spherical vesicles that could represent an early stage in nuclear membrane formation, where single or small groups of chromosomes are individually invested by a membrane; Bars: D=20 µm, E=2 mm, F=5 µm; (G) section of a stage 5 *Gieysztoria* embryo with embryonic invagination, cells of the embryonic primordium have segregated into a surface epithelium that constitutes the primordium of the epidermis (ep) and a deeper mesenchymal layer that contains the precursors of the musculature (mp) and brain (br). In the brain, neural differentiation has set in, resulting in an outer cortex surrounding a central neuropile (np). The epidermal primordium has undergone invagination (ein) into the centre of the egg; bar=20 µm; (H) an oblique section of a stage 6 embryo; anterior is to the left; (I) schematic diagram of a sagittal section of a stage 6 embryo, the straight line indicates the plane of section in (A). Dorso-ventral flattening of the embryo is an artefact due to fixation. Brain (br) with cortex (co) and central neuropile (np), as well as pharynx primordium (ph) can be recognized. The epidermal layer (ep) is clearly separable from underlying muscle precursors (mp), due to differentiation of junctional complex and basement membrane; bar of H and I=20 µm; (J) shows a magnified view of the genital primordium (gp) with rudimentary cuticular apparatus (cta constitutes male copulatory structure) at its centre. Note the massive basement membrane (bm) separating epidermis from muscular layer (ms). dc=dorsal connective, phe=pharynx epithelium, phm=pharynx musculature, php=pharyngeal pocket, vc=ventral connective, yk=yolk; bar=10 µm; (K) schematic diagram of a sagittal section showing advanced organization, abbreviations as above; (L) cross section through juvenile *Gieysztoria* showing anterior cortex of brain (br) and pharyngeal pocket (php); (M) brain with neuropile (np), root of ventral connective (vc), and pharynx musculature (phm); bar=20 µm; (N) ultrastructure of adult body wall with epidermis (ep), basement membrane (bm), musculature (ms) and protonephridium (pnp); bar=2 µm; (O) ultrastructure of internal organs of juvenile *Gieysztoria* showing gut epithelium (int) surrounding irregular lumen (lu); bar=1 µm. *From Younossi-Hartenstein and Hartenstein (2000).*

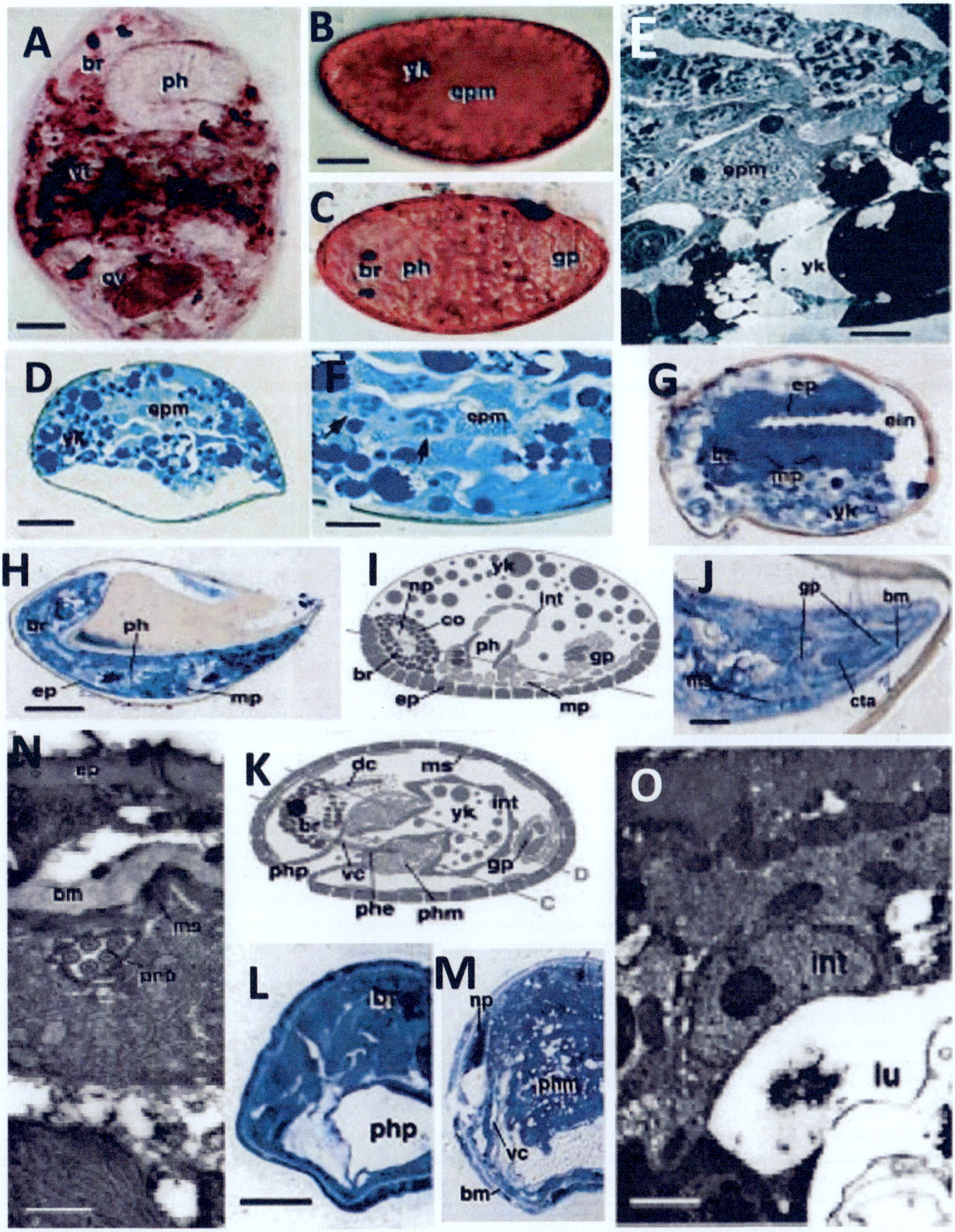

FIG. 7.16 See Figure Legend in previous page.

one to observe some landmark events, such as the ventral migration of the embryonic primordium, the formation of the brain and pharynx and eye pigmentation in the living embryo (Fig. 7.16C). Eggs are oval shaped, measuring app. 100 μm in length and 75 μm in width. The oocyte cannot be observed in the living embryo during the first cleavage division. However, it is small and found in the centre of the yolk mass (**stage 1**, 10 h of development). Cleavage leads to the formation of an apparently irregular, mesenchymal cluster of cells: the embryonic primordium. This structure is similar to *Mesostoma* embryo by Bresslau (1904) and thus the term. This embryonic primordium in living material forms barely visible hyaline disc surrounded by granular yolk, in the centre of the egg (Fig. 7.16B) (**stage 2**, second half of first day of development). In prepared material, the histological sections (Fig. 7.16D) show the embryonic primordium containing flattened, irregularly shaped cells with abundant mitotic figures. The embryo is surrounded by more irregularly shaped nuclei of yolk cells. Some cells contain several small, spherical vesicles (Fig. 7.16F). This probably represents an early stage in nuclear membrane formation. Despite close contact of the cleaving cells, cellular contacts such as adherens junctions are sparse, if present at all (Fig. 7.16E). The embryonic primordium then migrates to one surface, which will be the future ventral side of the embryo (Fig. 7.16F) (**stage 3**, the end of the first day of development). This movement of the embryonic primordium from the centre to the ventral side appears to be typical for rhabdocoel development; it has also been described in *Mesostoma* (Bresslau, 1904; Hartenstein and Ehlers, 2000) and other groups (Younossi-Hartenstein and Hartenstein, 2000). The disc-shaped embryonic primordium increases in size and, at this stage, it is ~5–7 cell diameters in thickness (dorso-ventrally) and ~15 cells in diameter. Histologically, cells appear homogenous, and mitotic activity is scattered throughout the primordium.

During the middle stages, as the embryonic primordium expands, it invaginates into the interior of the embryo, forming a deep furrow along the antero-posterior axis (Fig. 7.16G). This movement characterizes *Gieysztoria*, but is absent in *Mesostoma*. The primordia of the main organs, such as the brain and pharynx, appear as local condensation of cells in the anterior wall of this invagination (**stage 4/5**, second day of development). There is no gastrulation—that is, no tissue differentiation (no layer formation). Instead, the surface layer of the embryonic primordium, facing the lumen of the invagination, differentiates into a monolayered epithelium that will give rise to the epidermis. The remaining 'deep' cells of the primordium give rise to different lineages such as the nervous system, pharynx, somatic musculature, gut, and gonads. The primordium of the brain is the first distinct structure that becomes visible under the light microscope as a condensation of 'deep' cells at the anterior of the embryo. Differentiation of nerve cells proceeds very fast and results in the formation of a two-layered brain, consisting of an outer cortex of small, round neurons that surround an inner neuropile (Fig. 7.16M). While the organs grow and differentiate, the embryo itself evaginates to cover the ventral surface of the egg (Fig. 7.16H and I). The epidermis becomes very flat, yet maintains its cellular structure. In *Gieysztoria*, at the basal surface of the epidermis, spindle-shaped muscle precursors appear. The basement membrane that separates these two layers is visible as a light zone light microscopically from stage 6 onwards. At the end of the process of embryogenesis, the epidermis attains its enormous thickness (Fig. 7.16J).

Starting from stage 5, and during stage 6 onwards, oscillating movement in the centre of the embryo can be observed. This movement probably corresponds to myogenic twitching of somatic and/or pharyngeal musculature. The primordia of the brain and pharynx occupy the anterior half of the embryo. The pharynx develops right next to the brain. In both these rhabdocoels, *Gieysztoria* and *Mesostoma*, the pharynx epithelium does not arise by invagination from the outer ectoderm. It develops from a group of 'deep' cells differentiating into an epithelium surrounding an inner lumen (Fig. 7.16H and I). Initially, this emergent pharynx lumen has no connection to the outside. The definitive mouth is formed during later stages. Pigmentation of the eyes that develop in the anterior cortex of the brain indicates the onset of **stage 7/8** (middle of third day of development). The epidermis has now grown around the yolk and the embryo shows rapid muscular movement in the egg. The growth of the brain, especially the neuropile core, increases to the degree that its size becomes greater, compared to the size of the embryo itself. On the other hand, the pharynx elongates greatly so that it fills almost the whole space in the embryo, ventrally and posterior to the brain (Fig. 7.16I and K). The rudimentary gut develops from a population of 'deep' cells, which are organized into a loose epithelial layer that surrounds the remaining yolk mass (Fig. 7.16K). The primordium of the reproductive system develops from another group of 'deep' cells, posterior to the gut. It consists of an outer muscular layer surrounding a group of large cells differentiating into male reproductive, cuticular organs (Fig. 7.16J and K). This early development of more or less complete gut and reproductive organs distinguishes the embryo of *Gieysztoria*.

Juvenile of *Gieysztoria*

Younossi-Hartenstein and Hartenstein (2000) described this stage ultrastructurally in much detail, especially the body wall (Fig. 7.16N). That description recalls the typical structure of a turbellarian (Borradaile et al., 1977). It consists of the major constituents: epidermis of ciliated flat cells with rhabdites; basement membrane; and complete muscular layer (Fig. 7.6H). The body of the juvenile is filled with the brain and pharynx anteriorly, and with the gut posteriorly (Fig. 7.16L and O). Fig. 7.16O shows a loosely organized layer of flattened cells that are devoid of apical cilia and a prominent junctional complex constitutes the intestinal wall. At the posterior tail of the juvenile, a characteristic structure, the male reproductive system, develops (Fig. 7.16J).

In this analysis the authors have employed a morphological staging system recently introduced for the rhabdocoel species *Mesostoma lingua* (see the foregoing data).

The rhabdocoel *Mesostoma lingua*

Hartenstein and Ehlers published in the same year (2000) the study on the embryonic development of the rhabdocoel *Mesostoma lingua*. It was understood that this study is prior to that of the other rhabdocoel *Gieysztoria*. *Mesostoma* is a small flatworm that lives in freshwater ponds and was collected from pond near Göttingen, Germany, and raised in pond water at 23°C. During the summer months, this species produces eggs (called subitan eggs) that mature inside the mother. Their thin transparent shell and through the body wall of the live mother enabled the observation of their development. Their serial developmental stages were easily defined by morphological criteria and were applied on that of *Gieysztoria*. These stages (Fig. 7.18A) are also applicable to other platyhelminth taxa that are currently under investigation in the same laboratory. By life observation, whole mount, and light and electron microscopical preparations, these developmental stages were analysed in detail.

Mesostoma lingua produces 20–30 subitan eggs (summer eggs) that are arranged linearly on either side of the body (Fig. 7.17G). Eggs develop synchronously inside the female; juveniles hatch through the body wall of the mother after 7–8 days (at 23–25°C). At production, the eggs are disc-shaped; their diameter initially is 60 μm. They increase to about twice that size towards the end of embryogenesis. Development proceeds through the eight stages, each lasting for approximately 24 h (Fig. 7.17A–G). Those eight stages that were described above in the development of *Gieysztoria* are according to those defined in the development of *Mesostoma*, and also become applicable to other platyhelminth taxa that are currently under investigation in the same laboratory. Hence, it will be summarized for *M. lingua*: **stage 1**, the embryo itself is too small to be visible in the live specimen. By the end of this stage, the embryo forms a roughly circular disc, 4–5 cells in diameter and 1–2 cells thick. By the end of **stage 2**, a bilateral symmetry and anterior–posterior axis becomes established. The embryo itself, attains the shape of a 'trapezoid' disc, 2–3 cells thick, 7–9 cells wide and 5–7 cells long (Fig. 7.17H and I). In **stage 3**, in the live animal, the clear area corresponding to the embryo has increased in size and migrated to the periphery of the egg. It still forms a trapezoid disc, 2–3 cell diameters thick, 15–20 cell diameters wide and 12–15 cell diameters long (Fig. 7.17J and K). During **stage 4**, the size of the embryonic disc has increased to cover 60%–80% of the ventral hemisphere of the egg (Fig. 7.18A). Anteriorly, a lateral bulge as the brain primordia, and posterior bulge corresponding to the pharynx primordium are visible. Surrounding the brain and pharynx primordia, the thickness of the embryonic disc decreases to 2–3 cell diameters; here, mitotic cells are abundant. At about the end of stage 4, the embryo, viewed ventrally, attains a circular shape with three concentrically arranged regions, and the embryonic disc forms a thin layer measuring 1–2 cell diameters in thickness. Moving on to **stage 5**, the clear area forming the embryonic disc has expanded and reaches the cording equator of the embryo (Fig. 7.18A). The brain and pharynx primordia become prominent bulges. The most characteristic properties of stage 5 in prepared embryos are the formation of an epithelial outer layer that will form the ventral epidermis, and the appearance of a small neuropile region in the centre of the brain hemispheres (Fig. 7.17L and M). This epidermis shows a midventral gap along the ventral midline, which will not close until late in development. The subepidermal layer of mesenchymal cells of the epidermis contains progenitors of muscle cells, protonephridia and gland cells.

In **stage 6**, during which the embryo elongates in the anterior–posterior axis, a wider gap between brain and pharynx can be observed. These two organs develop; in addition, the epidermis becomes more advanced and differentiated. Protonephridial cells begin to be observed, by the electron microscope, to form elongated clusters of ciliated cells with an inner lumen. These clusters indicate the beginning of **stage 7** during which the three organs—epidermis, brain, and pharynx—grow further in development and differentiation. The brain hemispheres have approached each other further and condensed into a single brain mass. The condensed brain and increased neuropile become surrounded by a cortex of neuronal cell bodies 3–4 cell diameters thick (Fig. 7.17O). Located anteriorly in the brain, the eyes develop at the interface of the anteriorly directed process of the neuropile and cortex. They begin to distinguish as two weakly pigmented spots. Behind the pharynx, a transverse ridge of densely packed mesenchymal cells forms the primordium of the genitalia (Fig. 7.17O). Later, other scattered subepidermal cells that remain large and undifferentiated represent the population of stem cells (neoblasts) that keep proliferating postembryonically; these cells give rise to gonads and replace epidermal cells. During this stage, the body wall and pharyngeal muscles begin to contract. When embryos are freed from their egg case, they appear lemon shaped and glide unidirectionally by ciliary movement. They do not stretch or bend, indicating that their muscular and neural apparatus has not yet differentiated completely. In **stage 8**, the end stage in embryological development of *M. lingua*, the embryo attains its definitive length. It folds back inside the eggshell, and moves rapidly (both ciliary and muscular movement) because it is ready to hatch.

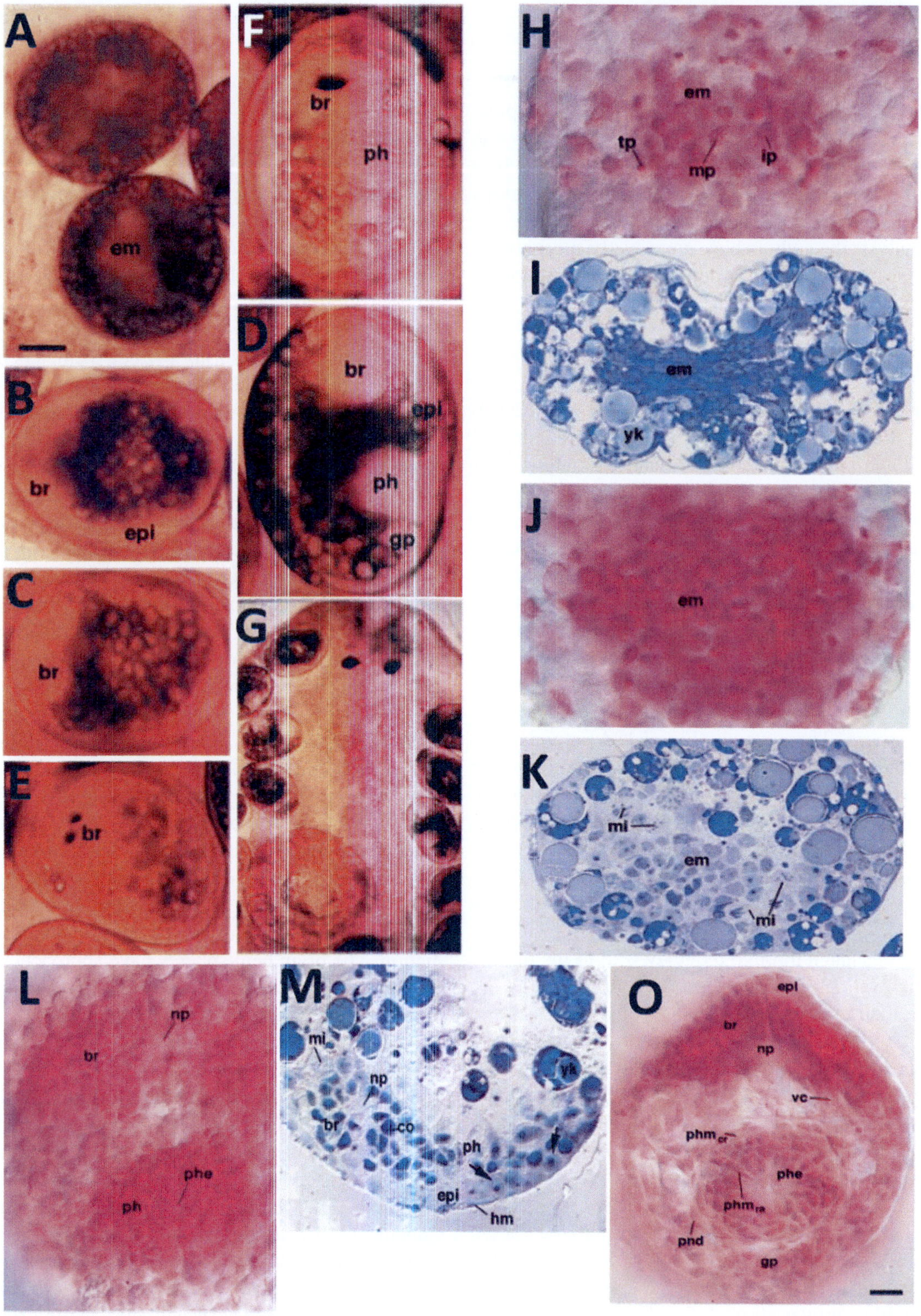

FIG. 7.17 (A–G) Photographs of living *Mesostoma* embryos at different stages of development: (A) lateral view of late stage 2 showing light embryonic disc (em) surrounded by dark yolk; (B) dorsal view of late stage 5, with symmetrical brain hemispheres (br) and epidermal layer (epi); (C) dorsal view and (D) lateral view of late stage 6 showing brain, epidermis, pharynx (ph) and genital primordium (gp); (E) dorsal view and (F) lateral view showing stage 8 embryo; (G) low magnification view of part of mother animal (ph = pharynx, br = brain, with eye spots) containing one row of stage 6 embryos on each side; bar = 25 µm; (H, J) W.M. of early stage of ventral view of *Mesostoma* development stained with basic fuchsine; (I, K) Parasagittal section of embryo at corresponding stage; (H and I) stage 2, late cleavage. Embryo is formed by irregularly shaped disc several layers thick. Dividing cells at different stages of mitosis (mp, metaphase; tp, telophase; ip, interphase) can be seen in all layers; (J–K) stage 3. Embryonic disc has grown and contains ~400–600 cells. It resembles a bilaterally symmetric trapezoid with the long side facing anteriorly (J, top). As in previous stages, mitotic figures (mi) are scattered throughout the embryo; (L) W. M. and (M) Parasagittal section showing early organogenesis in stage 5. A clear neuropile (np) has formed in the centre of the brain primordium (br). The epidermis forms a lightly staining layer of cuboidal cells, some of which (arrows) are shown in telophase. Mitotic activity (mi) becomes sparse and is restricted to the inner, peripheral regions of the embryo. A group of large cells in the centre of the pharynx primordium (ph) represent the precursors of the pharynx epithelium (phe); (O) shows W.M. of Late organogenesis of stage 7 embryos stained with basic fuchsine in ventral view. br = brain, epi = epidermal layer, np = neuropile, phm_{cr} = thinner circular layer, phm_{re} = massive radial layer (these two layers are pharyngeal muscles); bar 10 µm. *(From Hartenstein and Ehlers (2000).)*

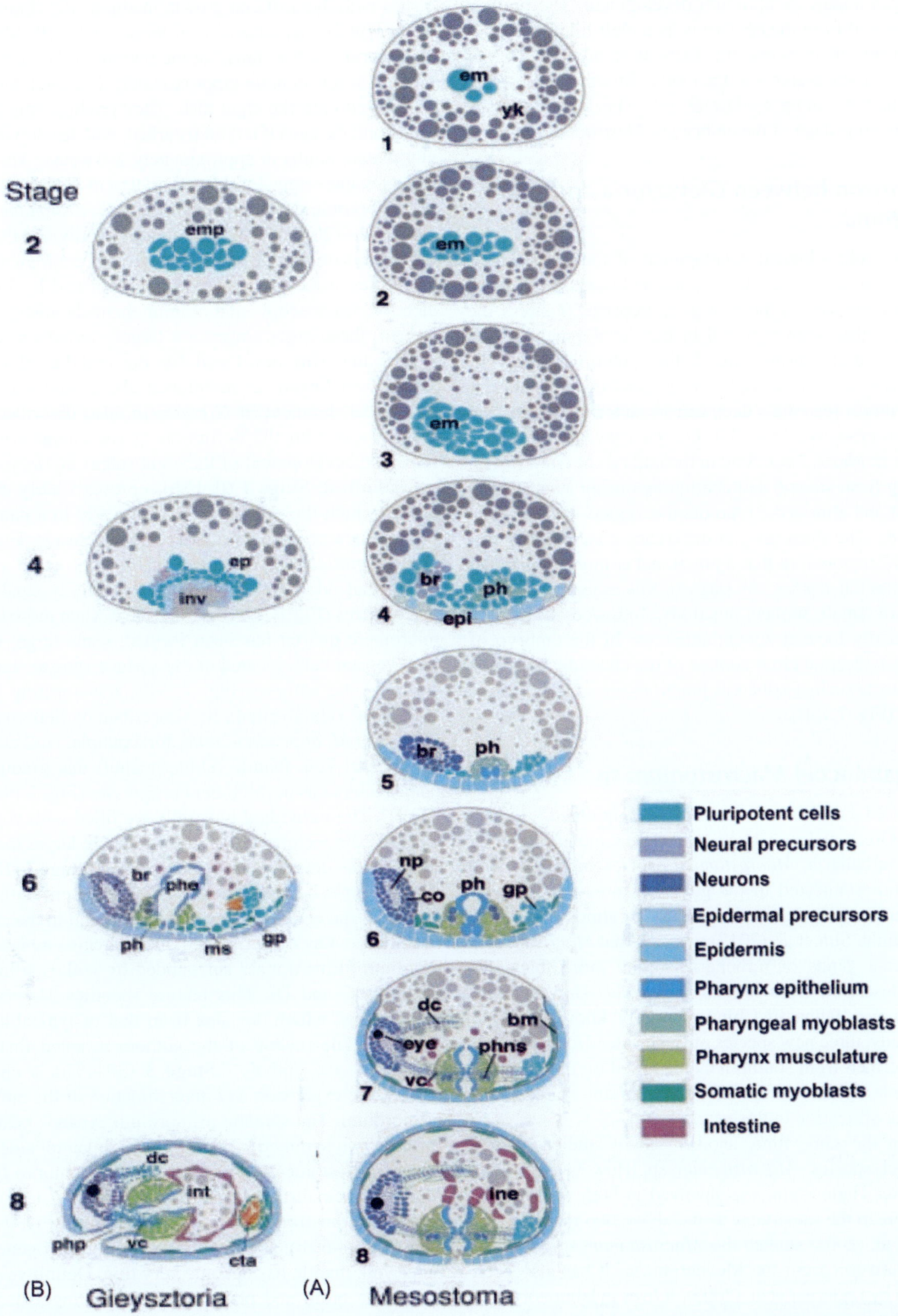

FIG. 7.18 Comparison of morphogenesis in *Mesostoma*, a typhloplanoid flatworm, and *Gieysztoria*: (A) Schematic drawings of lateral views of embryos of *Mesostoma* development, illustrating major events of embryogenesis. Anterior is to the left and dorsal at the top. The number at the bottom left of each drawing indicates the embryonic stage. co=cortex, cta=cuticular apparatus, dc=dorsal connective, eye=eye pigmentation, vc=ventral connective *(from Hartenstein and Ehlers, 2000)*; (B) the same for *Gieysztoria (from Younossi-Hartenstein and Hartenstein, 2000)*.

Each organ attains its optimum development. The brain is now separated from the epidermis by a cleft filled with processes of muscle cells and the ducts of gland cells. One of the characteristic features of the Platyhelminthes is the presence of rhabdites in the epidermal cells. This feature appears in the finishing stage of the embryo of *Mesostoma lingua*.

Comparison between *Gieysztoria* and *Mesostoma*

Since the embryological development of the two rhabdocoel turbellarian species, *Mesostoma* and *Gieysztoria*, were studied by the same school, it is appropriate to stress the differences that were reported in the development of the latter. First of all, during stage 5, the epidermis of the embryo of *Gieysztoria* ventrally invaginates into the interior of the embryo, forming a deep furrow, anteriorly, along the antero-posterior axis (Fig. 7.16G). This invagination does not occur in *Mesostoma*. Also in the embryo of *Gieysztoria*, beginning from stage 6 and continuing during stages 7 and 8, the gut and almost the reproductive organs are complete (Fig. 7.18). The third distinct difference concerns the embryo of *Mesostoma*, in that a junctional complex is formed in all epithelial tissues. At stage 6, this complex consists initially of small septate junctions, followed later by a more apically located zonula adherens. In the embryo of *Gieysztoria*, despite close contact of the cleaving cells, cellular contacts such as adherens junctions are sparse, if present at all (Fig. 7.16E).

The Rhabdocoel *Macrostomum* sp.

The term 'Macrostomum', meaning 'big-mouthed', derives from the Greek makros, 'large', and stoma, mouth. Relative to other turbellaria, *Macrostomum* species have indeed a long mouth connected to the gut by a muscular pharynx which can expand, in certain species, to almost the width of the animal. Sun et al. (2017) reported that *Macrostomum* is a diverse genus of turbellarians with more than 180 species described from around the world. However, the *Macrostomum* fauna in China is poorly known. In Sun et al.'s study, three new species of freshwater *Macrostomum* were described from southern China based on morphology of the penis stylet (see Fig. 7.7), an important character for species identification in this genus.

Within the same time, another school studied a third rhabdocoel species, *Macrostomum* sp. However, one author, Volker Hartenstein, was involved in both schools and the work is in the same view as the above two rhabdocoels. Morris et al. (2004) studied this *Macrostomum* sp. as a new marine flatworm from the Mediterranean. It has also been described by Ladurner et al. (2005). It lives in laboratory cultures of the diatom *Nitzschia curvilineata*. Morris et al. cultured this *Microstomum* in their laboratory, preparing Petri dishes with the artificial growth medium 'F2' (Guillard and Rythe, 1962), 'inoculated' with *Nitzschia*. By 10–14 days, the algae form a dense 'lawn' at the bottom of the dish. Groups of adult *Macrostomum* (approximately 1.5 mm in length) are transferred onto the algal dish. They produce eggs continuously, and the eggs if left undisturbed, will develop into sexually mature adults in approximately 2–3 weeks. The authors collected these eggs for the preparation of their work.

Embryonic development of this *Macrostomum* sp. proceeds for ~120 h at 20°C. The authors subdivided this development into eight stages according to morphological criteria since they used live observation, confirmed by histology, electron microscopy and immunohistochemistry studies. Indeed, these eight stages are largely based on a similar system that was developed for the rhabdocoel flatworm *Mesostoma lingua*, as mentioned above, and was applied on the development of *Gieysztoria*, also described above. Briefly and with the definition of the characteristic features, the development of the eight stages of *Macrostomum* are described: **Stage 1** (0–15 h) represents early cleavage, during which the zygote divides basically in a pattern that shows characteristics of quartet spiral cleavage (Fig. 7.19). Some significant deviations from typical spiral cleavage occur, that micromeres are not substantially smaller than macromeres (Fig. 7.19A). At the stage when more than one micromere quartet has been formed, some large, presumably vegetal cells located at the surface start to flatten and surround the embryo (Fig. 7.19B), transforming into the 'hull cells' (Huellzellen), first described by Seilern-Aspang (1957) for *M. appendiculatum*, for example. Hull cells form an external yolk mantle (Dottermantel) that gradually becomes thinner as development progresses (Fig. 7.19D, E, G, and H). The individual hull cells are filled with yolk granules and show elongated nuclei which are larger than those of most blastomeres (Fig. 7.19B). During **stage 2** (15–30 h), a yolk mantle begins to form. It forms a proliferating mass known as the embryonic primordium found in the centre of the embryo. Once internalized, the cleaving embryonic primordium forms a mass surrounded by yolk-rich hull cells (Fig. 7.19C and D). This feature specifies *Macrostomum* embryology, which deviates from that of typical spiralian embryos. The review of the authors revealed that this is called 'inverse epiboly'. **Stage 3** (30–45 h) is characterized by the expansion and diversification of the embryonic primordium. The somatic primordium expands posteriorly, forming two symmetric plates along the lateral sides of the embryo. Blastomeres in the centre of the embryo (gut primordium) remain large and yolk-rich (Fig. 7.19D). These cells later transform into the gut epithelium (Fig. 7.13 4d). **Stage 4** (45–60 h), about the middle of embryogenesis, the outer yolk mantle has become very thin. Definitive primordia of the brain and pharynx can be distinguished at the anterior pole. Later, these two organs develop further. The gut primordium fills the posterior-central part of the embryo

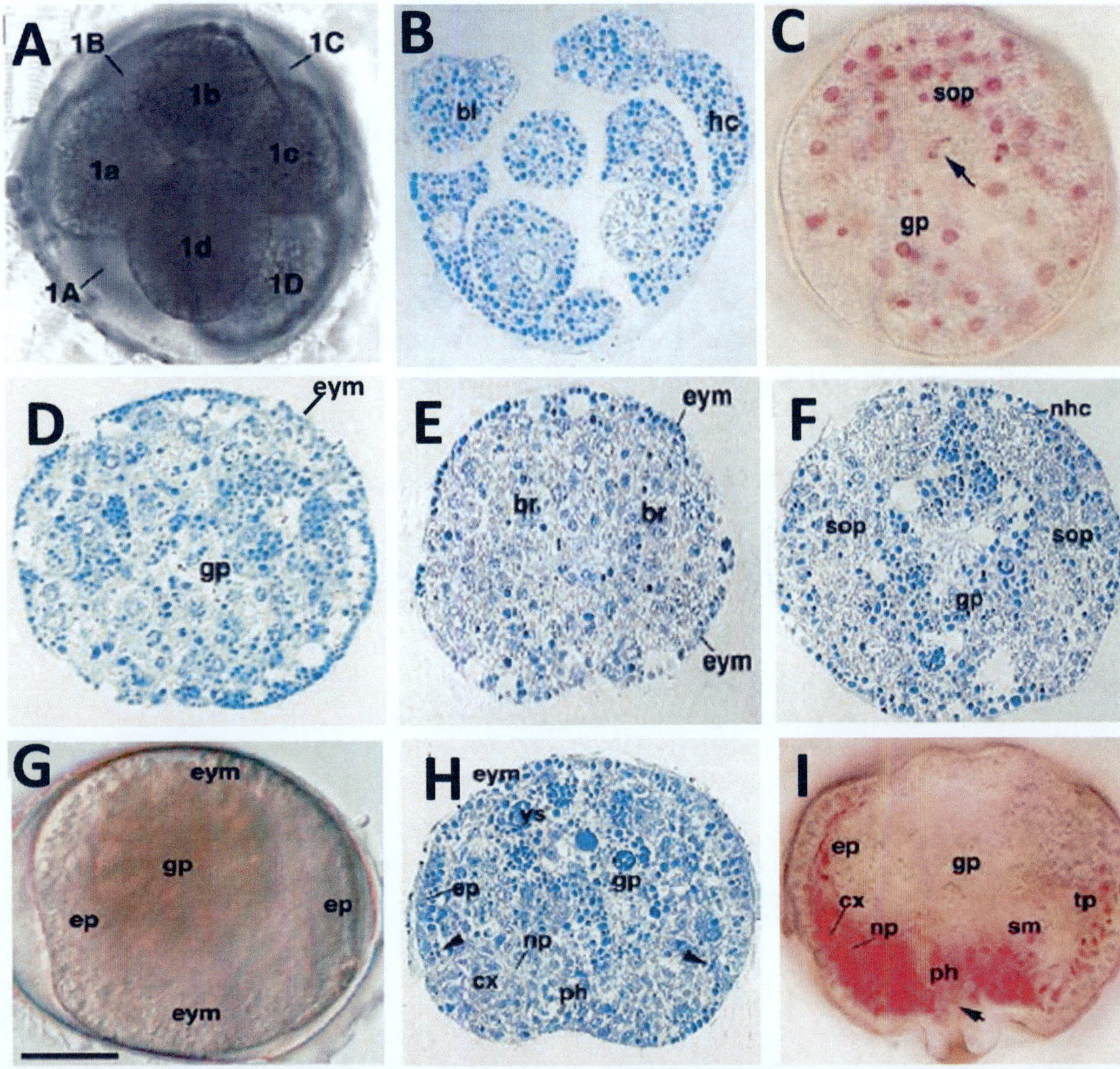

FIG. 7.19 Photographs of living embryos showing early stages of embryogenesis of *Macrostomum* sp.: (A) view from animal pole of eight-cell stage, focal plane through micromere quartet 1a–d (b) and macromere quartet (c). Note larger size of 1D macromere and relatively large size of micromeres, especially 1d; (B) Parasagittal section showing yolk-rich animal blastomeres (bl) and incipient hull cells (hc); (C) ventral view of W.M. embryo stained with basic fuchsine, *arrow* indicates one of numerous mitotic figures that occur throughout early development; (D) cross section near posterior pole, hull cells have extended over the surface of the embryo and form the external yolk mantle (eym), large-sized cells will form the gut (gut primordium, gp); (E–I) early organogenesis: (E) Cross sections of somatic primordium near anterior pole; (F) Cross section at mid-levels of embryo. At the anterior of the embryo, the primordium of the brain (br) appears as a bilateral condensation within the somatic primordium (sop). The somatic primordium forms a belt around the equator of the embryo. It measures 20–25 cell diameters in length (AP axis), 10–15 cell diameters in width (DV axis) and three to four cell diameters in thickness, bringing the overall cell number up to 2–3000. The gut primordium (gp) lies in the centre of the embryo and contains large, yolk-rich cells (nhc hull cell nucleus); (G) onset of organ differentiation (stage 5). A photograph of living embryo, anterior view; (H) cross section of embryo near anterior pole, *arrowheads* indicate precursors of muscle cells; (I) lateral view of W.M. of fuchsine-labelled embryo, arrow indicates cells that close the pharynx lumen ventrally. Note the folding of the eggshell ventral of the *arrow*, which probably constitutes an artefact related to shrinkage of the preparation. ep = primordium of epidermis, ex = cortex, eym = external yolk spheres, gp = gut primordium, np = central neuropile, ph = pharynx primordium, sm = muscle cells, tp = tail plate, ys = yolk spheres; bar 20 μm. (J–P) Electron micrographs of details of organ primordia of organogenesis in *Macrostomum* sp. (stage 5): (J and K) magnified views of part of somatic primordium, containing superficial layer of epidermal precursors (ep) and deep layer of muscle precursors and neural precursors (ne), showing details of bundles of cell processes (cp), probably axons, located at the border between neural precursors (ne) and epidermal cells (ep); (L) Apical part of epidermal cell, showing cilia (ci), zonula adherens (aj) within septate junction (sj); (M) Electron micrograph of details of L.S. of a stage 6 embryo showing overview of body wall, with ciliated epidermal layer (ep), subepidermal muscle fibres (mf), main longitudinal cord (mln) and gut primordium (gp). Epidermal cells still contain yolk granules (yg) near their basal pole; bar = 1 μm; (N) electron micrograph of cross section of body wall of hatching juvenile (magnified portion) showing epidermis (ep), subepidermal gland cells (glc) with rhabdites (rhb), muscle fibres (mf), sm = muscle cells. Epidermal cells are covered apically by microvilli (mv) and cilia (ci). Pronounced junctional complex (jc) interconnects neighbouring cells. Microfilaments form terminal web (tw); bar = 1 μm;

(Continued)

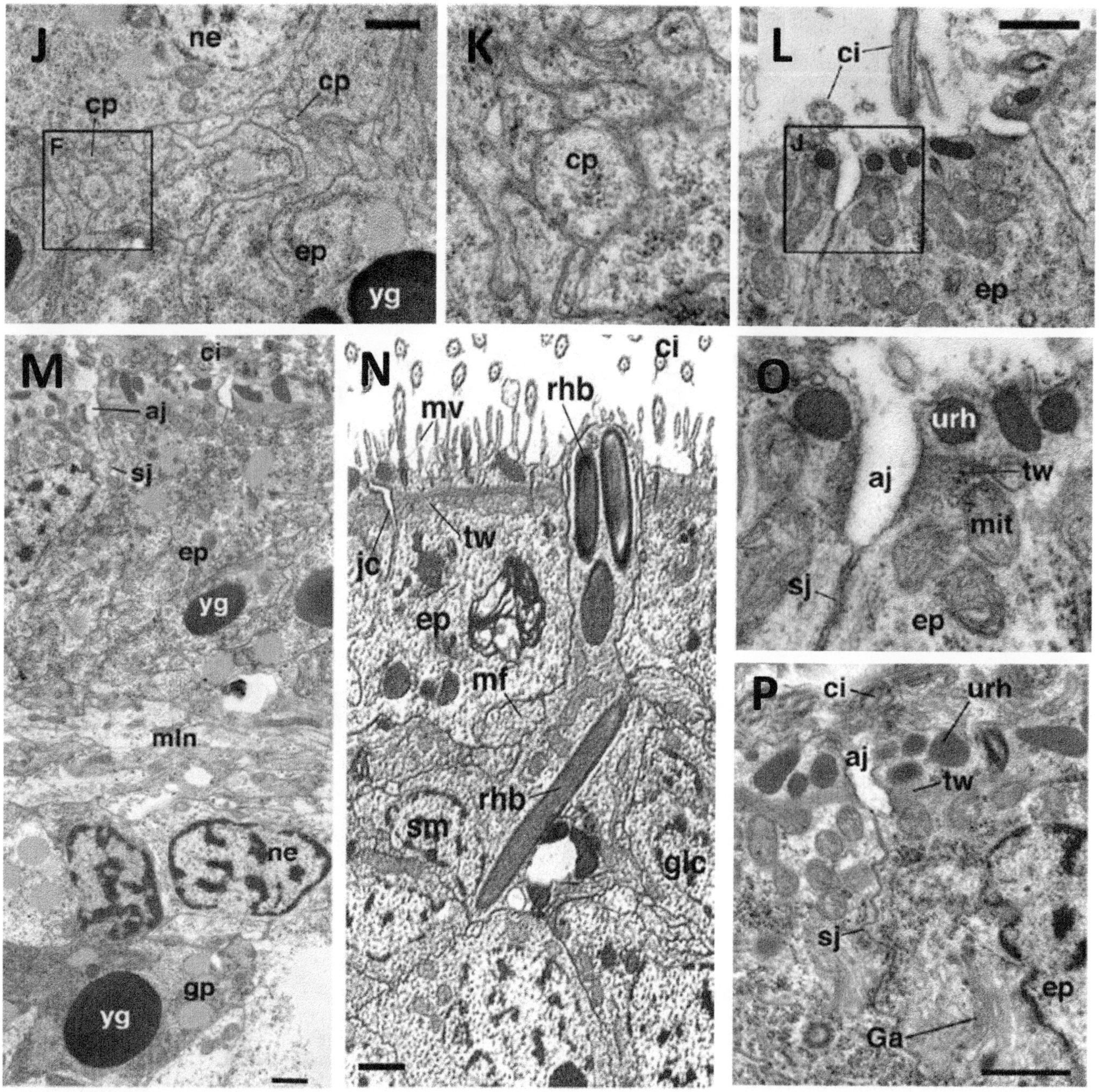

FIG. 7.19, CONT'D (O) Apical part of epidermal cell (squared portion in L), showing zonula adherens (aj), septate junction (sj), and terminal web (tw; microfilaments inserting at the zonula adherens). Apical of the terminal web is a layer of ultrarhabdites (urh), which are small secretory granules of epidermal cells. Mitochondria (mit) are clustered underneath the terminal web; (P) magnified view of apical part of epidermal cell with cilia (ci), ultrarhabdites (urh), terminal web (tw), zonula adherens (aj) and septate junction (sj). Epidermal cells contain prominent Golgi complexes (Ga); bar = 1 μm; (Q) a photograph of ventral view of living embryo, during organogenesis and differentiation of the body wall (stage 6); (R) parasagittal section of embryo showing the gut primordium has reorganized into a population of large, yolk-rich epithelial cells (ge) loosely arranged around a central lumen. Some of the yolk still forms dense yolk spheres (ys). At the posterior end of the embryo, a subepidermal condensation of cells forms the primordium of the male copulatory apparatus and the tail plate (tp); bar = 20 μm; (S) photograph of ventral view of living embryo (late phase of organogenesis, stage 7), the embryo has flattened and elongated and moves actively in the eggshell. A characteristic feature of stage 7 are the pigmented eyespots (eye) embedded in the brain (br); bar = 20 μm; (T) photograph of lateral view of living embryo at late phase of organogenesis (stage 8), the embryo has elongated further and adopted shape of juvenile; bar = 20 μm; (U) W.M. of stage 5 embryo labelled with acTub antibody labelling cilia of pharnx (ph), gut lumen (gl) and protonephridia (pn); bar = 20 μm; (V) confocal section (1 μm) of W.M. of juvenile showing muscle pattern, deep systems of muscle fibres, consisting of brain-related muscles (brm), pharyngeal suspensors (psm), and internal pharyngeal muscles (phm); (W) Schematic diagram of lateral view of juvenile, showing location of muscle systems, somatic muscle fibres (sm) of ventral body wall, gt, gut; ph, pharynx; vm, visceral muscle, pharyngeal suspensors (psm); bar 20 μm; (X) schematic drawings of lateral view of embryos of increasing age, illustrating the major events of *Macrostomum* sp. embryogenesis. Anterior is to the left and dorsal to the top. The number at the top left of each drawing indicates the embryonic stage, lettering as above. *(All are selected and adapted from Morris et al. (2004).)*

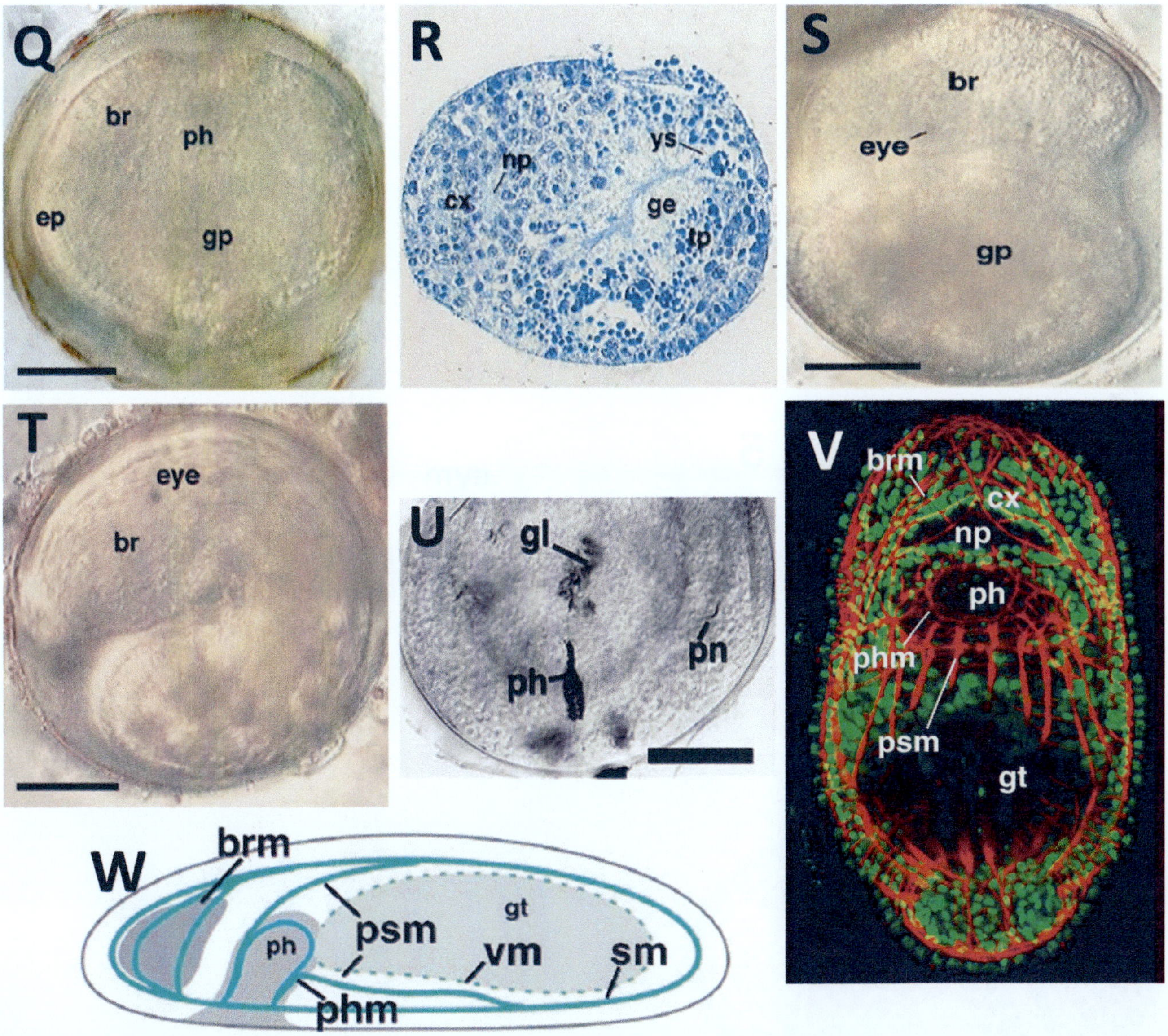

FIG. 7.19, CONT'D

(Continued)

(Fig. 7.19D and F). Starting towards the end of stage 4, peculiar spherical structures can be distinguished within the gut primordium (Fig. 7.19F and I). These 'yolk spheres' stand out in histological sections, in electron micrographs (not shown) and can even be seen in living material. **Stage 5** (60–75 h) is defined by the onset of tissue and organ differentiation. Most prominent during this stage are the epidermis, brain, pharynx and gut. Outer cells along the lateral margins of the embryonic primordium transform into a ciliated epithelial epidermal layer that displaces the external yolk mantle (Fig. 7.19M). During stage 5, the epidermis covers only the embryo laterally, whereas yolk still reaches the periphery of the embryo at dorsal and ventral levels (Fig. 7.19G and H). Later, the developing epidermis fully encloses the embryo by the end of stage 6. However, the protonephridial system is formed by a pair of longitudinal tubes, which give off several evenly space side branches. The tube and side branches are lined by flat epithelium. Whole-mounts of embryo preparations reveal the appearance of cilia in protonephridia for the first time during late stage 5 (Fig. 7.19U). **Stages 6–8** (during the last 2 days of the 5-day embryonic period) are the organ differentiation period (Morris et al., 2004). Fig. 7.19Q, R, and V show, in much detail, the histological and ultrastructure development of each organ, epidermis, brain, and gut, in addition to the reproductive organs.

The formation of an external yolk mantle by hull cells characterizes the embryological development of *Macrostomum*

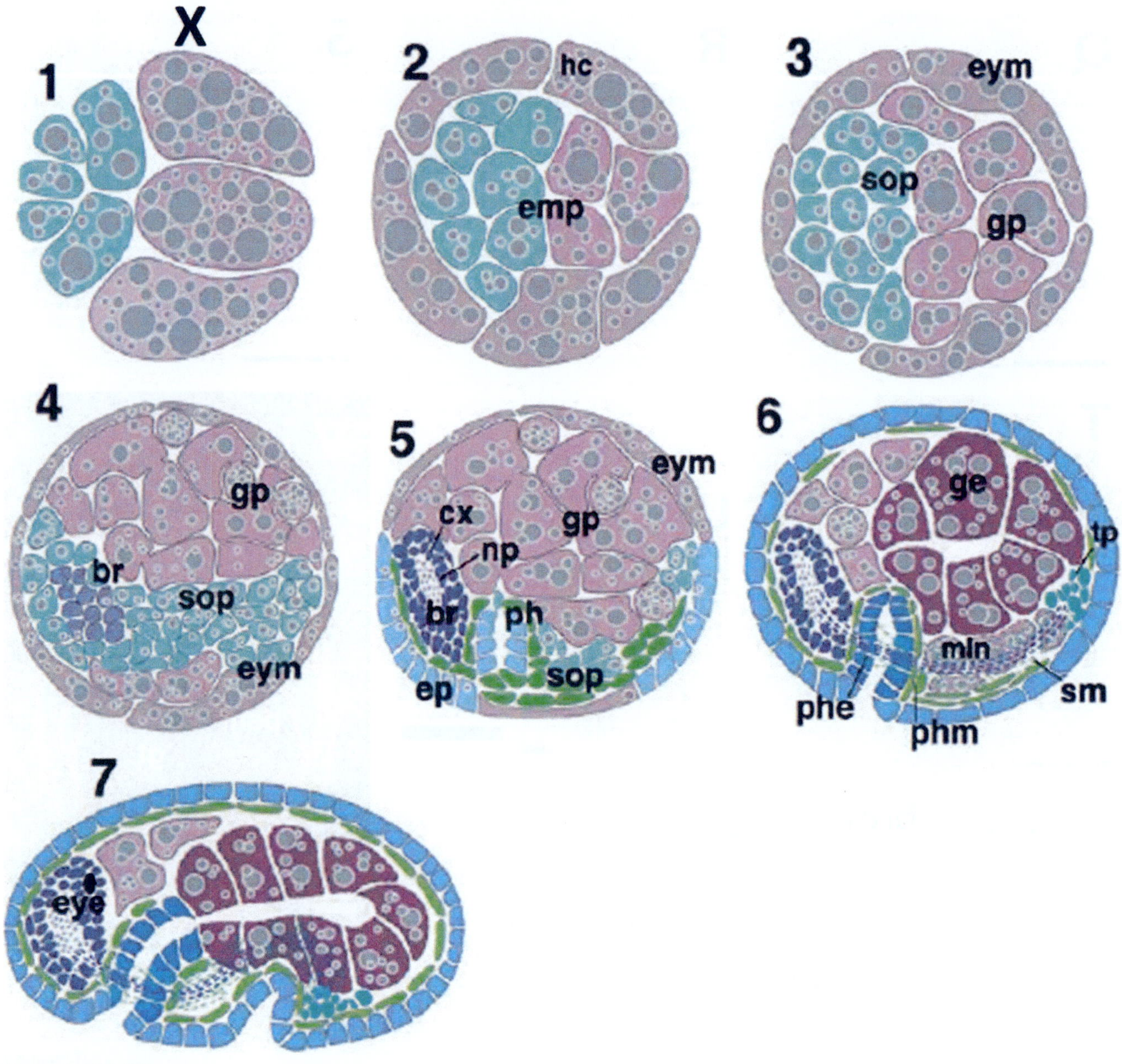

FIG. 7.19, CONT'D

sp., and the absence of a characteristic gastrulation is another criterion that would place macrostomids between the more 'basal' archoophorans (e.g. polyclads) and neoophorans. Gastrulation in this species deviates significantly from the typical spiralian mode. In living embryos, this process cannot be observed due to the presence of the yolk mantle. The data presented by Morris et al. (2004) suggest a morphogenetic process that leaves out typical gastrulation movements and, like the formation of the external yolk mantle, bears resemblance to early developmental stages in neoophorans.

Juvenile of *Macrostomum* sp.

The embryonic stages that lead to the early embryonic primordium of *Macrostomum* sp. to the body of the hatching juvenile correspond closely, in most important aspects, to what has been described for other flatworm taxa. The first organ primordia are invariably those of the **brain** and the pharynx; in *Macrostomum* sp., as in the dalyellid (rhabdocoel) *Gieysztoria superba*, the **pharynx** primordium appears directly posterior of the brain. The position of the pharynx primordium at early stages thus closely corresponds to the position where the pharynx ends up later. In *Macrostomum* sp., pharynx muscle precursors in the pharynx simplex form only a thin, inconspicuous layer around the pharynx epithelium. The **epidermis** in freshly hatched juveniles measures ~30–35 cells in perimeter (at midbody level), and possesses the same number of longitudinal muscle fibres. In addition, they possess a thin layer of muscles surrounding the **gut** which are undetectable in embryo preparations. The authors assume that formation of this visceral muscle layer must take place shortly before hatching. The musculature surrounding the **genitalia** develops postembryonically. Recently the homologue of the muscle specification gene Mef 2 (Morris et al., 2004) was cloned which, in both vertebrates and Drosophila, is expressed in mesoderm, followed by muscle cells.

Specific molecular markers will also be needed to make progress in understanding the origin of the central nervous system in *Macrostomum* sp. or any other flatworm.

ORDER TRICLADIDA

Paludicola Tricladida, the freshwater planarians, sometimes brackish water, is divided into three families: Dugesiidae, Planariidae, and Dendrocoelidae (Gardiner, 1972). They are easily cultured in the laboratory, and their phyletic position and capability for gametic and agametic reproduction, among other characteristics, have attracted the attention of researchers of developmental biology (da Vara et al., 2008), and also for other aspects. For example, with respect to planariidae, there is an old saying that believes that the planaria eats itself. This means that planaria never dies because of absence of food and/or unsuitable conditions. Very early, Stoppenbrink in 1903 reported that the Turbellaria are able to go without food for long periods, but during starvation they grow smaller and smaller. He studied the effect of starvation on *Planaria alpina*. He kept a group of animals entirely without food, while as a control he kept a similar group supplied with food. His results are given in Table 7.1. The measurements are in millimetres.

The reduction in size of the starved animals is accompanied by the absorption and digestion of the internal organs. They disappear in a regular order and the animal uses these organs as food. The first things to go are the eggs that are ready for laying, then follow the yolk glands and the remainder of the generative apparatus. Finally, the ovaries and the testes disappear, so that the animal is reduced to sexual immaturity. Next, the parenchyma, the gut and the muscles of the body wall are reduced and consumed. The nervous system alone holds out and is not reduced, so that starved planarians differ in shape from the normal forms in having a disproportionately large head end, the bulk of which is the unreduced cerebral ganglion. On feeding, these starved animals will regenerate all the lost organs and return to their normal size. This is another mode of regeneration.

The mutilated body, on the other hand, cannot move, cannot find food and cannot eat it. Its organs of locomotion and organs of cognition no longer exist. However, the damaged animal does not die. The process of regeneration starts just at the change of the unsuitable conditions. It begins with the assembly of regeneration cells into a regeneration blastema in the region of the wound. The first act of regeneration is wound closure; the wound is drawn together by muscle contraction (Kühn, 1971/421).

Epimorphosis

> *The regeneration blastema grows out from the edges of the wound as a cone or ridge, according to the form and size of the wound, and it gradually replaces in outer form and inner organization the lost body parts. These events have been called epimorphosis by Morgan.*
>
> (Kühn, 1971/422).

Morphallaxis

In addition to the new construction around the edge of the wound, however, other important changes take place in the intact remainder of the body, and these changes are called, collectively, morphallaxis. If the form of a little piece is so changed that the normal proportions must be recreated on a smaller scale, some organs that are normally far apart must be crowded together. Organ parts or whole organ complexes, such as the reproductive apparatus, are dismantled if they are now disproportionately large. They are then formed anew.

This phenomenon (consume the organs and regenerate them) becomes an attractive field to study the effect by different biological techniques. The present work selected that of Bowen et al. (1976), which employed a combination of

TABLE 7.1 Stoppenbrink's experiment

	Fed				Starve			
	Largest		Smallest		Largest		Smallest	
Date	Length	Breadth	Length	Breadth	Length	Breadth	Length	Breadth
16.iii.03	13	2	10	1	13	2	10	1
15.vi.03	17	2.5	12	1⅓	10	1⅓	6	⅔
15.ix.03	17	2.5	13	2	7	1	4	½
15.xii.03	17	2.5	14	2	3½	½	2⅓	⅓

From Saunders, J.T., 1963. Acoelomata. In: Kerkut, G.A. (Ed.), The Invertebrata. Cambridge University Press, pp. 196–197.

microscopical, biochemical and autoradiographic techniques to detect the primary effects of starvation on adult *Polycelis tenuis*. In this paper, particular attention was given to changes in the activity and distribution of the so-called 'lysosomal enzymes'. The authors cultured adult *Polycelis tenuis* in shallow perspex trays containing aerated tap water at room temperature (18°C). The planaria were fed a standard diet of boiled chopped earthworm at a given time and subsequently starved for periods up to 36 days. This short-term starvation (36 days) results in a substantial decrease in size (Fig. 7.20A). This decrease is accompanied by a steep drop in mitosis (Fig. 7.20B) and an initial rise in RNA synthesis (Fig. 7.20C). Mitotic activity appears to be confined to neoblasts, the highest population of which occurs in the pharyngeal region. Increased RNA synthesis in response to starvation is confined to the parenchyma and most of the labelling occurs in the gland cells. Fine structural, biochemical and cytochemical changes have been followed over these initial 36 days of starvation.

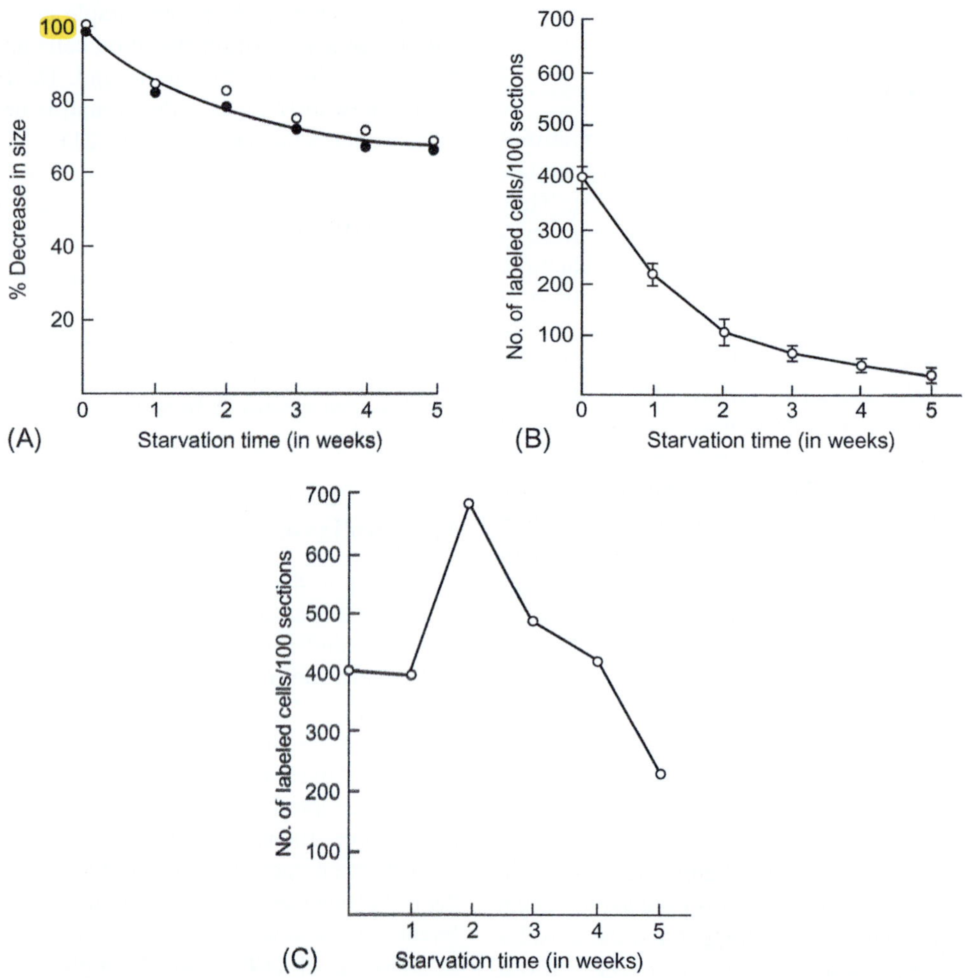

FIG. 7.20 (A) The graph illustrates the decrease in size of *Polycelis tenuis* over a 5-week period of starvation. The results are mean values obtained from a population of fifty animals and are expressed as a percentage of the dimensions of the unstarved group. The mean standard deviation was ±10%. Key, ○ = length, ● = width; (B) The pattern of incorporation of 6^3H thymidine over 5 weeks starvation. The number of labelled cells, per hundred sections per animal were counted. The results are the mean values from five replicates; (C) The pattern of incorporation of 5.6^3H Uridine during 5 weeks starvation. The results are expressed as the mean number of labelled nuclei per hundred sections per animal; (D–G) TEM photographs showing effect of starvation on different tissues of *Polycelis tenuis*: (D) Part of the parenchyma after 36 days starvation. Note the vacuities *(headarrows)* containing evidence of lysis in the form of broken subcellular fragments. Pigment cells (P) in an advanced state of lysis can be seen, the pigment granules (PG) are still recognizable, as are the large empty vacuoles (V) that normally contain lipid-like material. Part of dedifferentiating gland cell (GC) can also be seen; (E) Part of a muscle cell (M) after 21 days starvation. Note that in this instance, acid phosphatase activity has invaded the fibrillar structure of the muscle. Individual fibrils (f) surrounded by enzymatically released reaction product can be seen. The surrounding sarcoplasm (SC) is full of vacuities; (F) Remains of part of the gastrodermis after 36 days' starvation. Hardly any subcellular structures survive in the lysed area *(headarrows)* which is infused with reaction product. One thin phagocytic cell remains above the basement membrane (BM); (G) A lysing Sertoli-type cell from the reproductive system. The whole cell is lysing and contains free acid phosphatase activity (sodium fl-glycerophosphatase). A reabsorbed sperm head can be seen *(arrow)* in the disintegrating cytoplasm. The cell is surrounded by transverse sections through sperm cells, the microtubules of which bear acid phosphatase activity. This, however, is a normal feature; bar = 1.0 μm, unless otherwise stated. *(From Bowen et al. (1976).)*

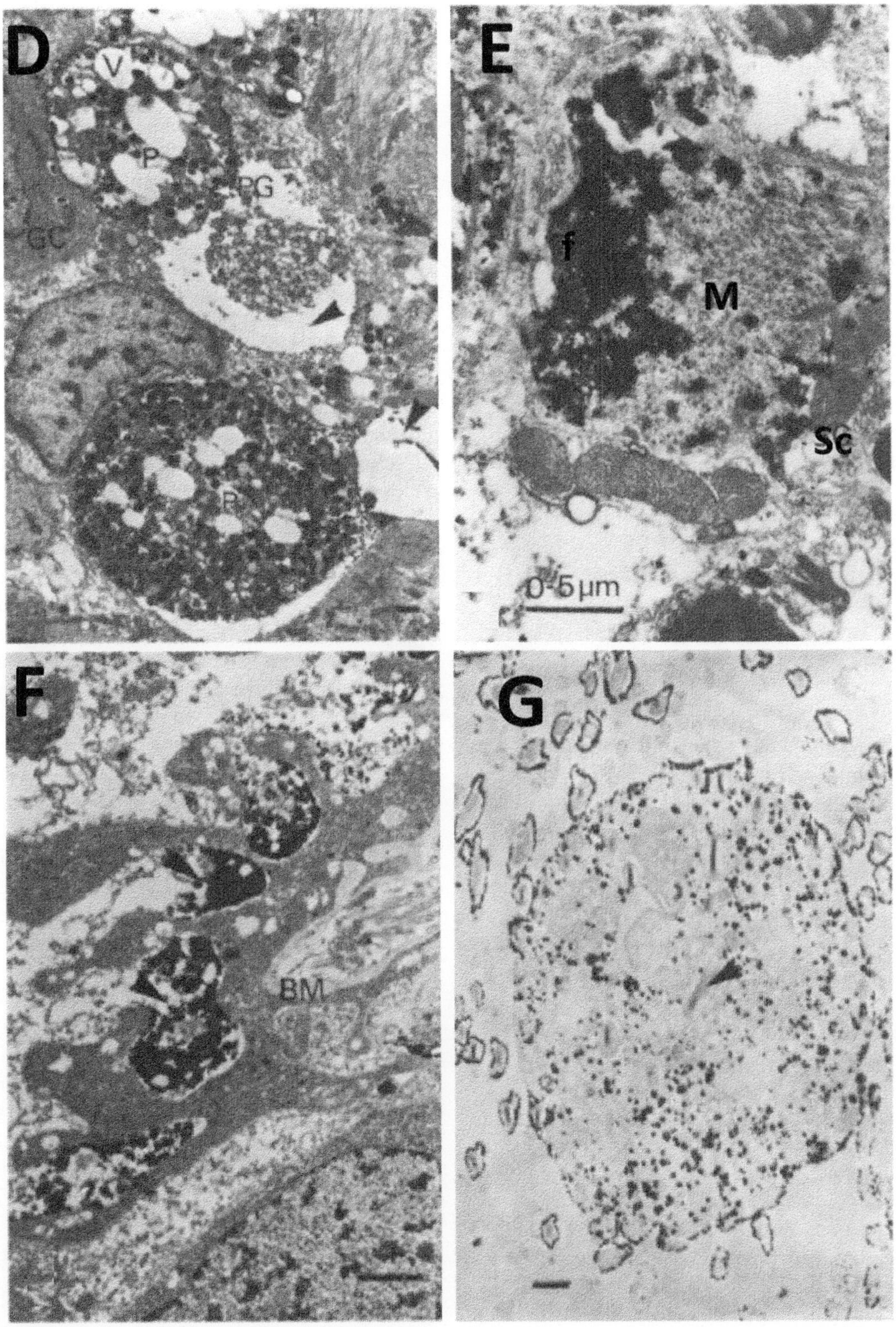

FIG. 7.20, CONT'D

The initial fine structural effect of starvation appears to be an increased vacuolation within the gland cells and particularly in the pigmented 'fixed parenchymal cells'. Pigment granules appear to be broken down within lysosome-like vacuoles. The pigment granules themselves appear to be distended due to fusion with lysosomes. There is a marked loss of lipid-like material characteristic of this cell type, leaving many empty vacuoles (Fig. 7.20D). Quite large areas of autolysis can be observed in the epithelial and subepithelial rhabdite containing cells. Between the seventh and the fourteenth day of starvation, marked morphological changes have taken place. Some muscle blocks are completely lysed, the muscle fibres becoming digested and consolidated into a dense mass (Fig. 7.20E). Changes in the gland cells are followed by gross histolysis of large parts of the gut (14–16 days starvation) (Fig. 7.20F) and the reproductive system (24–28 days starvation; see Fig. 7.20G). Pharyngeal musculature remains largely unaffected and nerve cells also remain structurally intact.

The general decrease in size brought about by starvation appears to be due to a combination of factors, which include a reduction in mitosis and an increased cell death and cell shrinkage, the latter, leading to dedifferentiation through autophagy and crinophagy. By the 36th day of starvation, although there has been an extensive loss of tissue, all cell types are still represented by survivors. The continued existence of survivors may in turn depend on a controlled histolysis as a source of recycled nutrient.

Descriptive embryology of members of Planariidae (*Planaria* and *Polycelis*) and Dendrocoelidae (*Dendrocoelum*) were studied and, more recently, that of *Schmidtea*. The following is one of these studies.

Development of composite egg (Tricladida)

Dendrocoelum lacteum

The embryos of triclads have rigid capsules: the cocoons. It is usual to find within one eggshell (capsule) of triclad species a large number of yolk cells (sometimes as many as several thousands) surrounding a few egg cells. In the ordinary planaria, there are usually 4–6 eggs, while in *Dendrocoelum lacteum* there are said to be 20–40 (Kato, 1968). The cleavage of these eggs is extremely peculiar in that the blastomeres become isolated from each other and distribute themselves among the yolk cells. Under these conditions, it is difficult to find evidence of spiral cleavage; in addition, regular germ layer formation does not take place. The yolk cells surrounding the blastomeres proceed to fuse together, forming a syncytium, while some of the blastomeres transform into wandering amoeboid cells and, separating from the other blastomere masses, migrate to the vicinity of the yolk syncytium. They form a thin membrane, which acts as a boundary between the ordinary yolk cell masses outside and the yolk syncytium inside. This membrane is called the 'outer membrane' or 'provisional ectoderm' (Fig. 7.21).

At the same time, another group of wandering cells assembles at one point on this membrane; eventually a cavity appears in this cell mass and the whole thing takes the shape of a small sac. Next, a mouth opens to the outside of the membrane and an **embryonic pharynx** is formed, at the inner end of which is a sac that functions as a temporary intestine. The wall of this structure is also called the internal membrane. As the embryonic pharynx sucks in the yolk from the outside, the temporary intestine swells enormously and its wall, pushing away the blastomeres and yolk which lie outside of it, forms the embryonic wall (Fig. 7.21H and I). Within the embryonic wall, the blastomeres, suspended in yolk, proliferate actively, and the resulting cells eventually gather at the ventral side, which is determined by the embryonic pharynx, and form the germinal cord along the mid-ventral line. This embryonic end consists of three cell masses: the middle one near the embryonic pharynx, and the others lying anterior and posterior to it.

The embryo eventually becomes flattened, and the embryonic pharynx as well as the inner and outer membranes begin to degenerate. However, the cells of the part forming the inner surfaces of the germinal cord are destined to become the endoderm of the adult; they spread out in the region where the inner membrane is degenerating, and there form the wall of the intestine. The other parts of the germinal cord give rise to the ectoderm and endoderm; in particular the anterior cell mass forms chiefly the ectoderm of the anterior part of the body, taking the place of the degenerating outer membrane to become the adult epithelium. Furthermore, two groups of cells, in close contact with this epithelium, differentiate into cerebral ganglia. These two ganglia eventually fuse into a single unit from which the other nerves of the body grow out. In the central part of the embryo, the embryonic pharynx degenerates; shortly afterwards, spaces appear in the cellular mass and the definitive pharynx takes shape around them. In this species, there is no clear-cut stomodaeal invagination.

Recent studies of embryonic development of Tricladida paid attention to prepare histological sections of the cocoons. da Vara et al. (2001) used Dugesiidae cocoons of *Girardia tigrina* (Girard, 1850). The specimens were maintained in the laboratory at a constant temperature of 22 °C. The cocoon (Fig. 7.22A) consists of sclerotize and keratinized proteins. This constitution is the cause of the difficulties encountered during the histological preparation. The fixatives and infiltrating solutions cannot penetrate the tissues. To study the embryo development, the histological processes of the cocoons should be adapted for light microscopy. Thus, the studies of these authors were to adapt techniques for the histological processing of these cocoons. Two embedding techniques were tested: Paraplast (Fig. 7.22B) and glycol methacrylate (Historesin) (Fig. 7.22C). Their results were tabulated to show and state the most suitable processing techniques to obtain good histological sections.

Girardia tigrina (Girard, 1850)

The embryonic development in the genus *Girardia* is almost unknown; only a note is given regarding the formation of the pharynx of *Girardia tigrina* (Curtis, 1902). The same authors, da Vara et al. in 2008, adapted suitable techniques and prepared histological sections, and studied embryological development of *Girardia tigrina* by means of

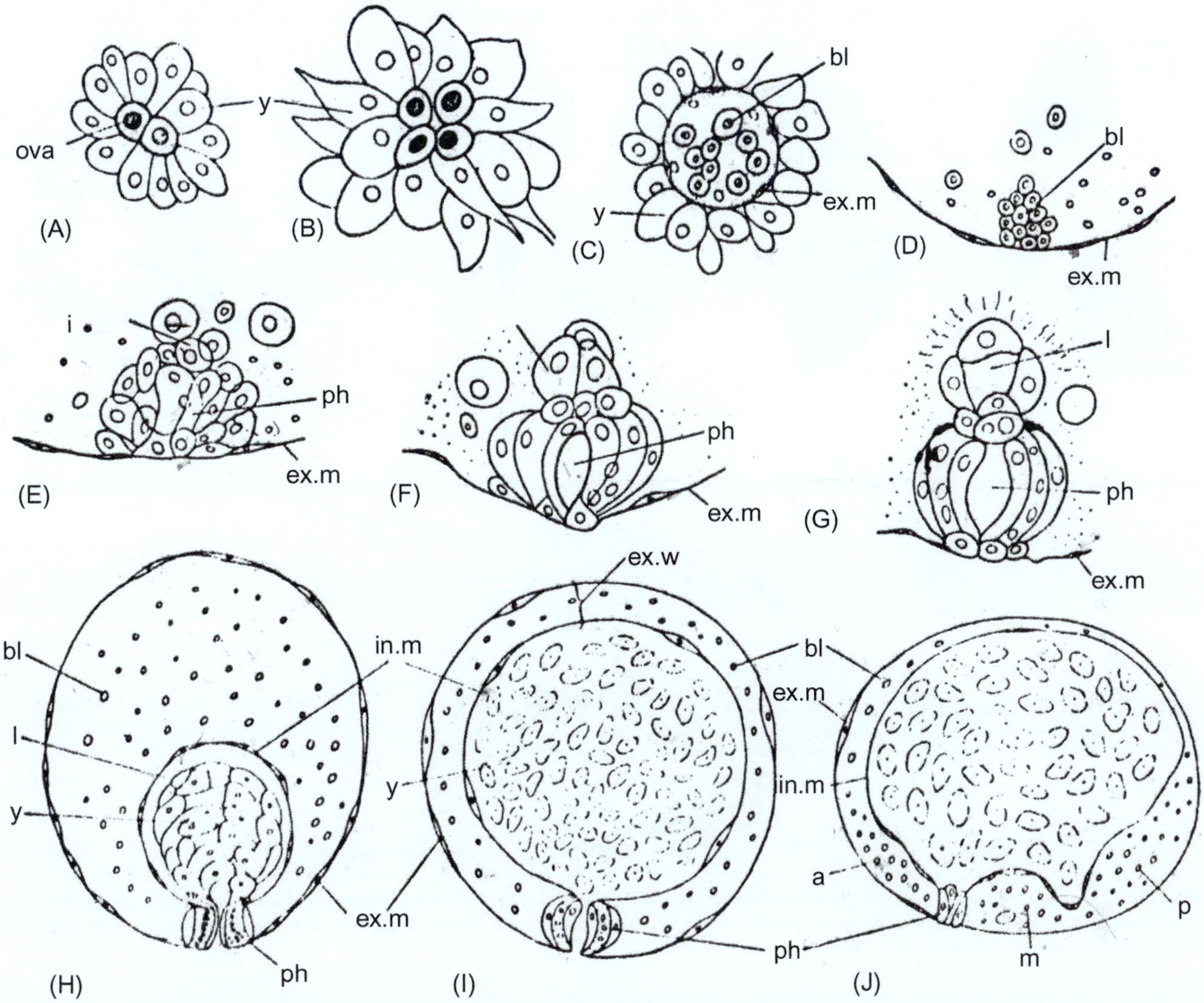

FIG. 7.21 Development of the Tricladida (Ijima, 1884; Mattiesen, 1904; Fulinski, 1914–1916): (A and B) early cleavage stages; (C) formation of external membrane (temporary ectoderm); (D) aggregation of blastomeres towards ventral side; (E–G) differentiation of embryonic pharynx and internal membrane (temporary intestine); (H) absorption of yolk into temporary intestine; (I) formation of embryonic wall by extension of temporary intestine; (J) germinal cord formed at ventral side and anterior, mid- and posterior cell masses appear. a = anterior cell mass, bl = blastomere, em.w = embryonic wall, ex.m = external membrane, I = temporary intestine, in.m = internal membrane, m = mid-cell mass, ova = ovocyte, p = posterior cell mass, ph = temporary pharynx, y = yolk cell. *(From Kato (1968).)*

optical microscopy. The authors kept 80 adult specimens at 20 ± 1°C in Petri dishes with 'well-water' and cocoons were obtained from this laboratory culture. The cocoons were fixed immediately after laying and at 2, 4, 6, 12, 18, and 24 h after laying during the first day. There were about 10 cocoons from each stage; however, 94 cocoons were collected in the period from laying to hatching.

A freshly deposited cocoon has a light brown colour, and is ovoid, with the largest axis measuring 0.7–1.3 mm (Fig. 7.22A). Within few hours, the colour darkens. The cocoon contains 4–16 oocytes, which all go synchronously through the embryological stages. This vital activity is differentiated into nine stages: in **stage 1** (0–12 h after laying), egg cells can be observed within the huge number of the yolk cells (Fig. 7.23A). The egg cell contains round or ovoid nucleus and one to two prominent nucleoli. The yolk cells are of two types, type I and type II, and both are larger in size compared to the egg cell. Type I cells are more numerous, while type II are larger in size (Fig. 7.23A–C). At the end of this stage, the egg cell starts its development. In **stage 2** (12–24 h after laying), the egg cell divides mitotically giving rise to varied number of blastomeres, some of which form aggregates. The triclads, and some species of other neophoran orders, reveal a highly modified cleavage compared to the spiral quartet or duet form occurring in most arcoophorans as seen above. However, all the

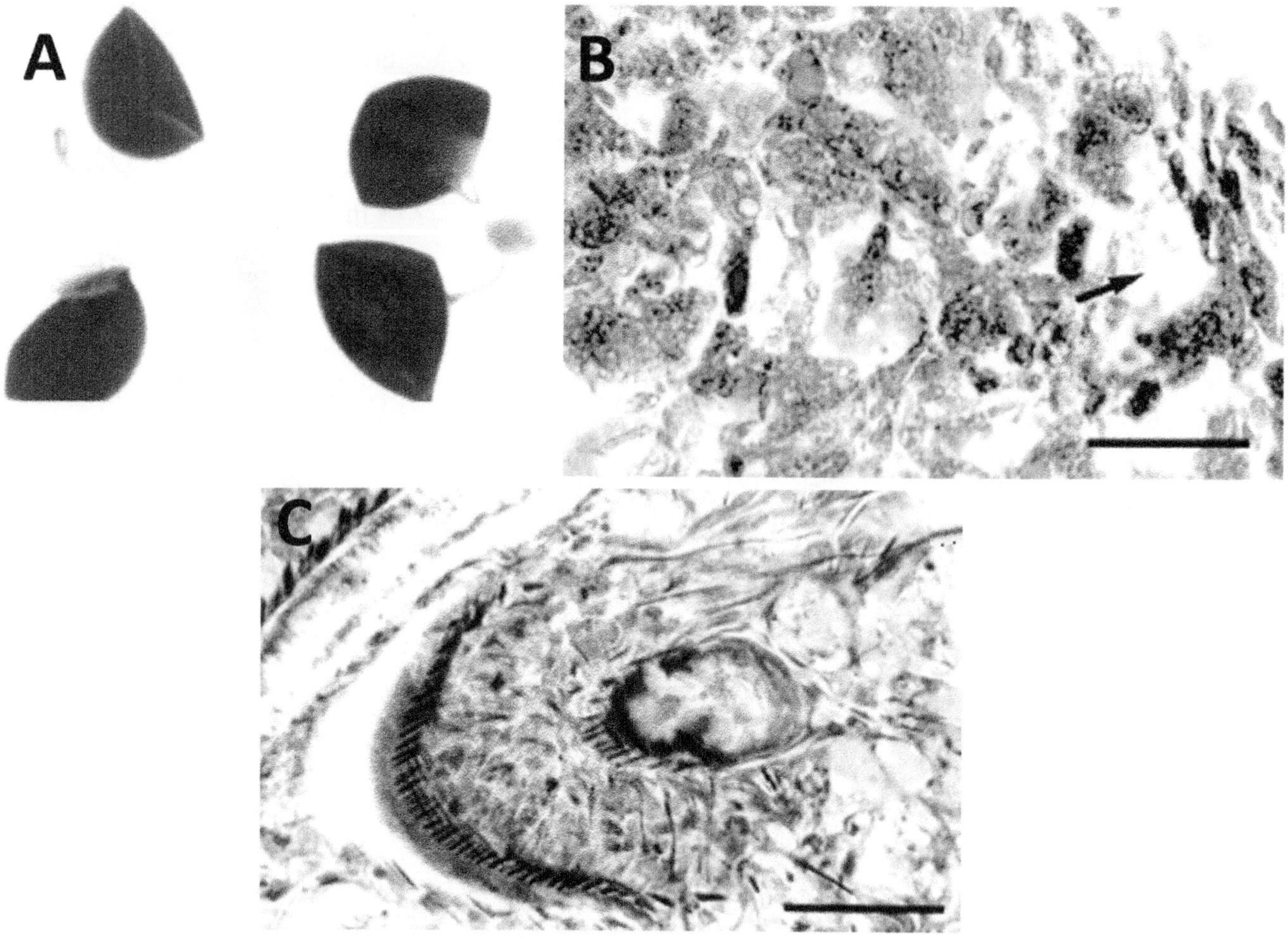

FIG. 7.22 (A) Photograph of WM of cocoons of *Girardia tigrina*; (B) Histological section (7 μm) of material fixed in formaldehyde, treated with formic acid, and embedded in Paraplast. The arrow shows artefacts. Bar = 50 μm; (S) Histological section (2 μm) of material fixed in paraformaldehyde/glutaraldehyde, and embedded in Historesin; bar: 50 μm. *(From da Vara et al. (2001).)*

blastomeres and the aggregates in *Girardia* are dispersed among yolk cells. At the end of this stage, the aggregates form a spherical blastema with eccentric small cavity, enteroblastula (Fig. 7.23C). This means that cleavage is so irregular that the fate of blastomeres, the precise origin of cell lines and organs, and a gastrula stage cannot be identified. **Stage 3** (~2 days after laying) begins with the formation of the embryonic epidermis which consists of flattened cells with an underlying fibrous tissue (Fig. 7.23D). The yolk cells form a syncytium surrounding each developing embryo, although these cells are observed dispersed within the cocoon. Later, the formation of the pharynx begins at one pole of the developing embryo (arrow) and intestine, embryo, embryonic cells (Fig. 7.23E). It begins as a tubular organ lined with cuboidal epithelium, and coated with a fibrous tissue. Small round cells delimit the communication of the pharyngeal lumen with the interior of the embryo. Distal to the embryonic pharynx, the fibrous tissue forms a sacculi-form structure, which accumulates yolk cells, similar to an embryonic intestine. In **stage 4** (~3 days after laying), the embryo absorbs the yolk cells surrounding it and the embryonic epidermis is comprised of a single layer of cubic to flattened cells (Fig. 7.23F). During this stage, the embryonic pharynx degenerates and is not observed from this stage onwards (Fig. 7.23G). In **stage 5** (until the fourth day after laying), the embryonic epidermis becomes thicker and multilayered due to the accumulation of undifferentiated cells. This is the origin of the definitive epidermis. Some of these cells delaminate determining the site of the definitive pharynx, while internally some other cells forms a split (Fig. 7.23H) corresponding to the site of the definitive intestine. **Stage 6** (4–6 days after laying) involves the beginning of organogenesis of most of the definitive organs. First is the pharynx, which appears in the ventral region of the embryo. In an early stage of formation, it has an internal cavity lined by an epithelium with an underlying connective tissue, besides a layer of radially arranged muscle fibres. In addition, the definitive intestine appears as a blind saccular organ containing small amount of yolk. The embryo itself becomes spherical in form, with

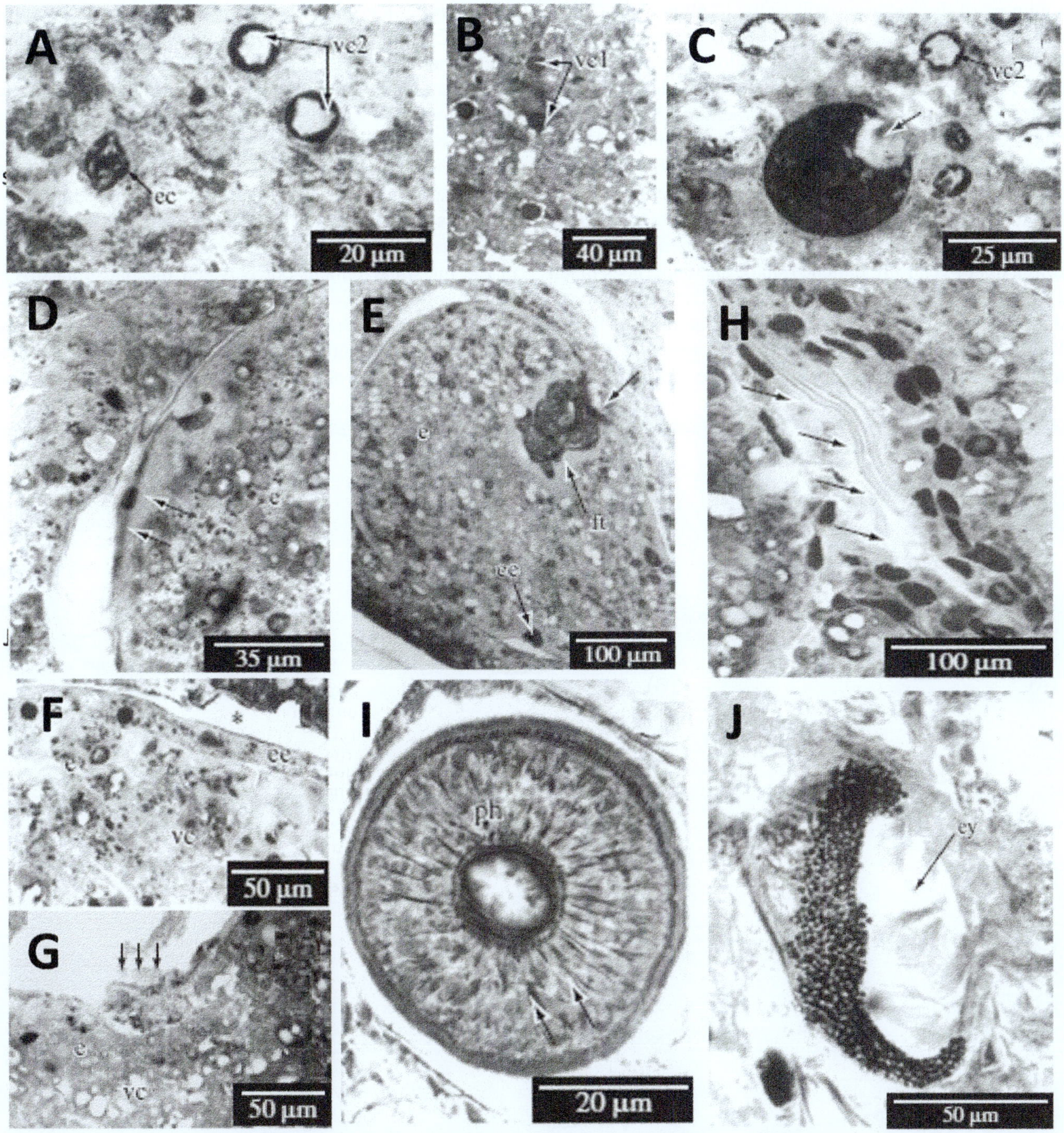

FIG. 7.23 Semi-thin sections of cocoons of *Girardia tigrina* showing developmental stages: (A) stained with methylene blue/basic fuchsine showing early phase of stage I with embryonic cells (ec) and one type of yolk cells (vc2) *arrows*; (B) another section stained with toluidine blue showing the other type of yolk cells (vc1) *arrows*; (C) stained with methylene blue/basic fuchsine showing advanced phase of stage II with the blastomeres aggregating in a spherical blastema (enteroblastula) presenting an eccentric small cavity *(arrow)*, the vc2 appears; (D) early phase of stage III, showing the formation of embryonic epidermis *(arrows)*; (E) advanced phase shows the embryonic pharynx *(arrow)* and intestine; (F) stained with toluidine blue showing the embryonic epidermis during stage IV. The *asterisk* indicates the space between embryos without yolk; (G) stained with hematoxyline/eosine showing the degeneration of the embryonic pharynx *(arrows)* during stage IV, (H) stage V, the delamination of some epidermal cells *(arrows)* giving rise to a split; (I) section of stage VIII stained with toluidine blue shows the definitive pharynx containing necks of the pharyngeal glands *(arrows)*; (J) shows a completely formed eye during stage VIII. e=embryo, ec=embryonic cells, ee=embryonic epidermis, ei=embryonic intestine, ft.=fibrous tissue, vc vitelline cells, vc1=vitelline cells of type I, vc2=vitelline cells of type II, ph=pharynx, ey=eye. *(From da Vara et al. (2008).)*

a number of degenerating yolk cells. In **stage 7** (6–8 days after laying), the embryos no longer have their definitive form, but become more elongated along the prospective antero-posterior axis. The pharynx develops in ~ the third quarter of the body. It is lined by epithelium with an underlying connective tissue, internal and external muscle layers and radial muscles (Fig. 7.23I). Within the embryonic mesenchyme, also, muscle fibres directed differently are observed. The yolk cells become an amorphous mass. During **stage 8** (at an average of 8 days after laying), the embryo attains a dorso-ventrally flattened shape and undergoes much differentiation preparing for hatching. The mesenchyme shows incompletely formed dorso-ventral muscle fibres. The epidermis differentiates: the dorsal becomes thicker and attains rhabdites seen clearly in the juvenile, while the ventral epidermis presents cilia. The pharynx contains necks of the pharyngeal glands between external and internal epithelia (Fig. 7.23I). Only one main intestinal branch with small lateral diverticulae becomes well developed and has lumen filled with degenerating yolk cells. The nervous system represented by the cerebral ganglia appears. The eye pigment cup begins to be formed (Fig. 7.23J). In **stage 9** (at an average of 14 days after laying), the intestine completes its formation. The anterior branch becomes longer and the two shorter posterior ones appear. No yolk appears in the intestinal lumen, but between the intestinal diverticulae appear numerous mesenchymal cells and free undifferentiated cells. The nervous system attains its definitive formation; the longitudinal nerve cords and the cerebral ganglion and the eyes are also completed (Fig. 7.24H). The

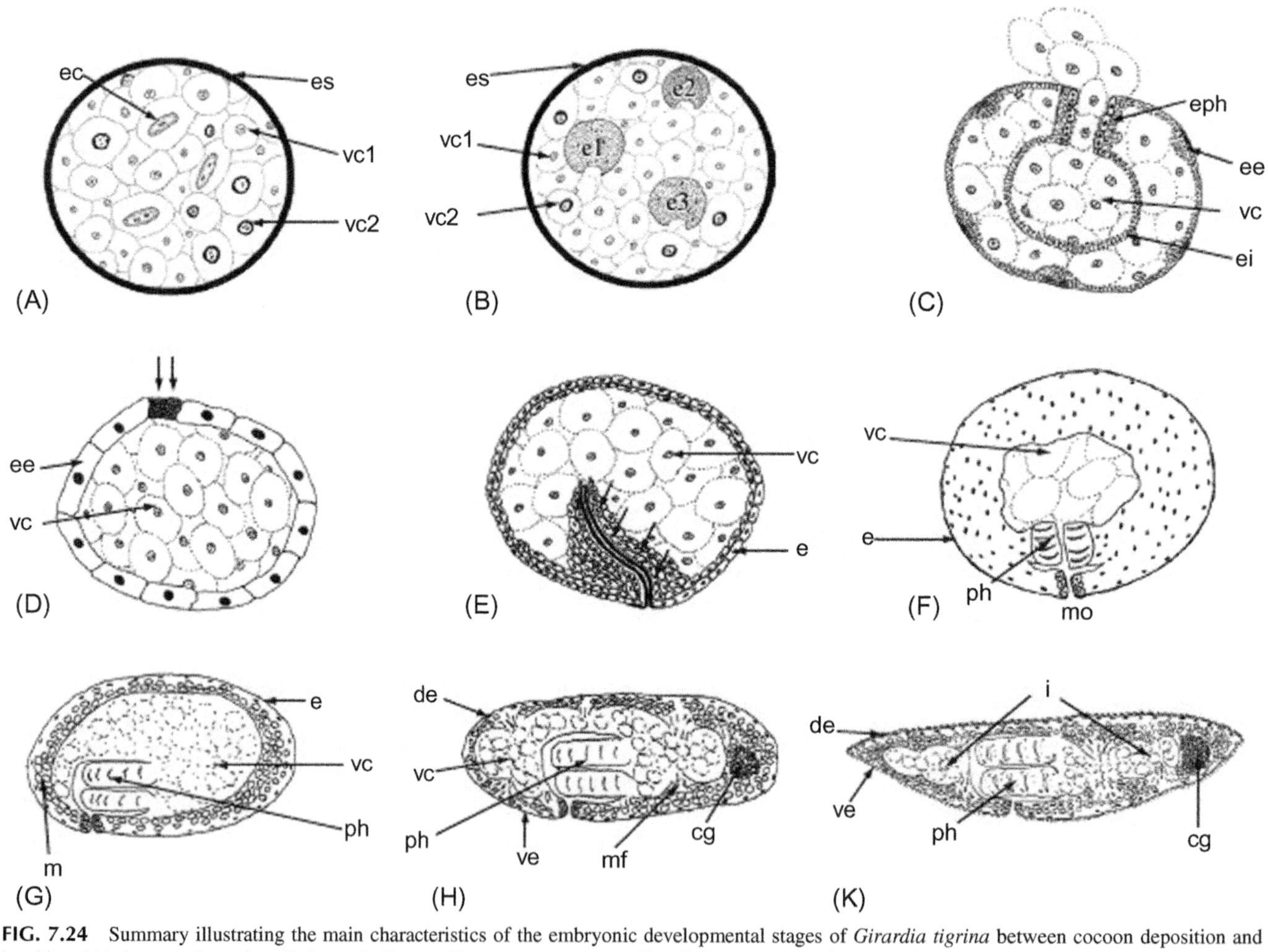

FIG. 7.24 Summary illustrating the main characteristics of the embryonic developmental stages of *Girardia tigrina* between cocoon deposition and hatching: (A) shows embryonic cells (stage I) dispersed between numerous yolk cells in an egg capsule; (B) embryonic cells (stage II) in a capsule, aggregate forming a round blastema with an eccentric small cavity; (C) an embryo (stage III) forming embryonic epidermis and pharynx; (D) stage IV embryo showing degeneration of the embryonic pharynx *(double arrow)*; (E) stage V embryo forming the definitive epidermis and delamination of a group of epidermic cells *(arrows)* determining the site of the definitive pharynx; (F) stage VI embryo in early organogenesis with formation of definitive pharynx and intestine (g); (G) stage VII embryo with definitive pharynx completed and the embryo begins to elongate; (H) stage VIII embryo showing the formation of the nervous system; (K) stage IX embryo, late differentiation with the embryo showing similar characteristics to those of the juveniles. cg = cerebral ganglium, de = dorsal epidermis, e = definitive epidermis, e1–e3 = embryos, ec = embryonic cells, ee = embryonic epidermis, ei = embryonic intestine, eph = embryonic pharynx, es = eggshell, I = definitive intestine, m = mesenchyme, mf = muscle fibres, mo = mouth, ph = definitive pharynx, vc = vitelline cells, vc1 = vitelline cells of type I, vc2 = vitelline cells of type II, and ve = ventral epidermis. *(From da Vara et al. (2008).)*

embryos (Fig. 7.24K), already 0.8–4.5 mm long, show all the juvenile characteristics.

Juveniles hatch between 12 and 16 days after cocoon laying. However, a period of up to 22 days before hatching has been known, in some cases. The maximal hatching frequency occurring between 12 and 16 days after cocoon deposition (at $20 \pm 1°C$). The newly hatched juvenile measures 2–3 mm in length. Fig. 7.24 summarizes the main morphological changes occurring during embryogenesis of *Girardia tigrina* (da Vara et al., 2008).

REFERENCES

Apelt, G., 1969. Fortpflanzungsbiologie, Entwicklungszyklen und vergleichende Frühentwicklung acoeler Turbellarien. Mar. Biol. 4 (4), 267–325.

Barnes, R.D., 1974. Invertebrate Zoology, third ed. W.H. Saunders Company, London.

Borkott, H., 1970. Geschlecheehtliche Organisation. Fortpflanzungsverhalten und Ursachen Der sexuellen Vermehrung von *Stenostomum sthenum* nov. sp. (Turbellaria, Catenulida), mit Beschreibung von 3 neuen *Stenostomum*-Arten. Z. Morphol. Tiere 67, 183–262. (cited in Henley, 1974).

Borradaile, L.A., Potts, F.A., with chapters by Eastham, L.E., Saunders, J.T., 1977. The Invertebrata, fourth ed. Revised by Kerkut, G.A. Cambridge University Press.

Bourlat, S.J., Hejnol, A., 2009. Acoels. Curr. Biol. 19 (7), R279–R280.

Bowen, I.D., Rider, T.A., Dark, C., 1976. The effects of starvation on the planarian worm *Polycelis tenuis* Iijima. Cell Tissue Res. 169 (2), 193–209.

Boyer, B.C., Henry, J.J., Martindale, M.Q., 1998. The cell lineage of a polyclad turbellarian embryo reveals close similarity to coelomate spiralians. Dev. Biol. 204, 111–123.

Bresslau, E., 1904. Beitraege zur Entwicklungsgeschichte der Turbellarien. I. Die Entwicklung der Rhabdocoelen und Alloiocoelen. Z. Wiss. Zool. 76, 213–332.

Bresslau, E., 1909. Die Entwicklung der Acoelen. Verh. Dtsch. Zool. Gesamte 19, 314–324. (cited in Henry et al., 2008).

Bresslau, E., 1933. Erste Klasse des Cladus Plathelminthes. In: Kükenthal, W., Krumbach, T. (Eds.), Handbuch der Zoologie. vol. 2, no. 1. Gruyter, Berlin, pp. 52–320.

Conklin, E.G., 1897. The embryology of Crepidula, a contribution to the cell lineage and early development of some marine gasteropods. J. Morphol. 13 (1), 1–226.

Costello, D.P., 1961. On the orientation of centrioles in dividing cells and its significance. A new contribution to spindle mechanics. Biol. Bull. 120, 286–312.

Costello, H.M., Costello, D.P., 1938. Copulation in the acoelous turbellarian *Polychoerus carmelensis*. Biol. Bull. 75, 85–98. (cited in Henley, 1974).

Costello, H.M., Costello, D.P., 1939. Egg laying in the acoelous turbellarian *Polychoerus carmelensis*. Biol. Bull. 76, 80–89. (cited in Henley, 1974).

Costello, D.P., Davidson, M.E., Eggers, A., Fox, M.H., Henley, C., 1957. Methods for Obtaining and Handling Marine Eggs and Embryos. Marine Biological Laboratory, Woods Hole, MA.

Curtis, W.C., 1902. The life history, the normal fission, and the reproductive organs of *Planaria maculata*. Proc. Bost. Soc. Nat. Hist. 30, 515–583. (cited in Vara et al., 2008).

da Vara, D.C., Leal-Zanchet, A.M., Lizardo-Daudt, H.M., 2001. Histological processing techniques for the study of Dugesiidae development (Platyhelminthes, Tricladida, Paludicola). Rev. Bras. Biol. 61 (2), 341–345.

da Vara, D.C., Leal-Zanchet, A.M., Lizardo-Daudt, H.M., 2008. Embryonic development of *Girardia tigrina* (Girard, 1850) (Platyhelminthes, Tricladida, Paludicola). Braz. J. Biol. 68 (4), 889–895.

Dahm, G., 1950. On *Bothrioplana semperi* M. Braun (Turbellaria Alloeocoela Cyclocoela). The taxonomy of the 'genera' and of the known 'species' of the family Bothrioplanidae. Ark. Zool. I, 503–510. (cited in Henley, 1974).

Ellis, C.H., Fausto-Sterling, A., 1997. Platyhelminthes, the flatworms. In: Gilbert, S.F., Raunio, A.M. (Eds.), Embryology, Constructing the Organism. Sinauer Associates, Inc.

Gardiner, E.G., 1895. Early development of *Polycoeeus caudatus*. J. Morphol. 11 (cited in Kato, 1968).

Gardiner, M.S., 1972. The Biology of Invertebrates. McGraw-Hill Book Company, London, New York.

Guillard, R., Rythe, L.H., 1962. Studies on marine planktonic diatoms. I. Cyclotella nana Hustedt and *Detonula confervacea* (Cleve). Gran. Can. J. Microbiol. 8, 229–239. (cited in Morris et al., 2004).

Hartenstein, V., Ehlers, U., 2000. The embryonic development of the rhabdocoel flatworm *Mesostoma lingua* (Abildgaard, 1789). Dev. Genes Evol. 210, 399–415.

Heath, H., 1928. A sexually mature turbellarian resembling Müller's larva. J. Morphol. Physiol. 45, 187–207.

Hejnol, A., Martindale, M.Q., 2008. Acoel development indicates the independent evolution of the bilaterian mouth and anus. Nature 456, 382–386. (letter).

Hejnol, A., Martín-Durán, J.M., 2015. Getting to the bottom of anal evolution. Zool. Anz. 256, 61–74.

Hendelberg, J., 1965. On different types of spermatozoa in Polycladida, Turbellaria. Ark. Zool. 18, 267–304.

Hendelberg, J., 1969. On the development of different types of spermatozoa from spermatids with two flagella in the Turbellaria with remarks on the ultrastructure of the flagella. Zool. Bidrg. Uppsala 38, 1–50.

Henley, C., 1968. Refractile bodies in the developing and mature spermatozoan of *Childia groenlandica* (Turbellaria, Acoela) and their possible significance. Biol. Bull. 134, 382–397. (cited in Henley, 1974).

Henley, C., 1974. Platyhelminthes (turbellaria). In: Giese, Pearse (Eds.), Reproduction of Marine Invertebrates. vol. I. Academic Press, London.

Henley, C., Costello, D., Ault, C.R., 1968. Microtubules in the axial filament complexes of acoel Turbellarian spermatozoa, as revealed by negative staining. Biol. Bull. 135, 422–423. (cited In Henley, 1974).

Henry, J.Q., Martindale, M.Q., Boyer, B.C., 2000. The unique developmental program of the acoel flatworm, *Neochildia fusca*. Dev. Biol. 220, 285–295 (Acoels).

Hickman, C.P., 1973. Biology of the Invertebrates, second ed The C.V. Mosley Company, Saint Louis.

Hyman, L.H., 1937. Reproductive system and copulation in *Amphiscolops langerhansi* (Turbellaria Acoela). Biol. Bull. 72, 319–326.

Hyman, L.H., 1944a. A new Hawaiian polyclad flatworm associated with Teredo. Occas. Pap. Bernice P. Bishop Mus. Honolulu 18, 73–75.

Hyman, L.H., 1944b. Marine Turbellaria from the Atlantic coast of North America. Am. Mus. Novit. 1266, 1–15.

Hyman, L.H., 1951. The Invertebrates: Acanthocephala, Ashelminthes and Entoprocta. The Pseudocoelomate Bilateria. vol. III. McGraw-Hill Book Company, Inc., New York, Toronto, London.

Kato, K., 1940. On the development of some Japanese polyclads. Jap. J. Zool. 8, 537–573. (cited in Henley, 1974).

Kato, K., 1968. Platyhelminthes. In: Kumé, Dan (Eds.), Invertebrate Embryology. Molit Publ. House, Belgrade, Yugoslavia. Translated by J.C. Dan.

Kuhn, A., 1971. Lectures on Developmental Physiology. Springer Verlag, Berlin.

Ladurner, P., Rieger, R., 2000. Embryonic muscle development of *Convolute pulchra* (Turbellaria-Acoelomorpha, Platyhelminthes). Dev. Biol. 222 (2), 359–375.

Ladurner, P., Schärer, L., Salvenmoser, W., Rieger, R.M., 2005. A new model organism among the lower Bilateria and the use of digital microscopy in taxonomy of meiobenthic Platyhelminthes: *Macrostomum lignano*, n. sp. (Rhabditophora, Macrostomorpha). JZS 43 (2), 114–126. https://doi.org/10.1111/j.1439-0469.2005.00299.

Lang, A., 1884. Polycladen. Fauna u. Flora v. Neapel 11. (cited in Kojiro Kato, 1968).

Larsson, K., Jondelius, U., 2008. Phylogeny of Catenulida and support for Platyhelminthes. Org. Divers. Evol. 8, 378–387. (cited in Rawlinson, 2014).

Lentati, G.B., 1970. Gametogenesis and egg fertilization in Planarians. Int. Rev. Cytol. 27, 101–179.

Mann, T., 1984. Spermatophores. Springer-Verlag, Berlin, New York.

Marshall, A.J., Williams, W.D., 1975. A Text Book Zoology: The Invertebrates. ELBS and Macmillan.

Martín-Durán, J.M., Egger, B., 2012. Developmental diversity in free-living flatworms. EvoDevo 3, 7.

Meves, F., 1903. Über oligopyrene und apyrene Spermien und ihre Entstehung nach Beobachtungen Paludina and Pygaera. Arch, mikrosk. Anat. Entw. Mech 61, 1. (cited in Roosen-Runge, 1977).

Morris, J., Nallur, R., Ladurner, P., Egger, B., Rieger, R., Hartenstein, V., 2004. The embryonic development of the flatworm *Macrostomum* sp. Dev. Genes Evol. 214, 220–239.

Park, J.K., Kim, K.H., Kang, S., Kim, W., Eom, K.S., Littlewood, D.T.J., 2007. A common origin of complex life cycles in parasitic flatworms: evidence from the complete mitochondrial genome of *Microcotyle sebastis* (Monogenea: Platyhelminthes). BMC Evol. Biol. 7, 11. (cited in Rawlinson, 2014).

Rawlinson, K.A., 2014. The diversity, development and evolution of polyclad flatworm larvae. EvoDevo 5, 9.

Seilern-Aspang, F., 1957. Die Entwicklung von *Macrostomum appendiculatum* (Fabricius). Zool. Jahrb. Anat. 76, 311–330.

Selenka, E., 1881. Zoologische Studien. II. Zur Enwicklungsgeschichte der Seeplanarien. In: Ein Beitrag zur Keimbla tterlehre und Descendenztheorie. Engelmann, Leipzig, Germany. (cited in Rawlinson, 2014, p. 2192).

Sun, T., Zhang, L.V., Wang, A.T., Zhang, Y., 2017. Three new species of freshwater *Macrostomum* (Platyhelminthes, Macrostomida) from southern China. Zootaxa 4012 (1).

Surface, F.M., 1907. Early development of a polyclads, *Planocera (Hoploplana) inquilina*. Proc. Acad. Nat. Sci. Phila. 59. (cited in Kato, 1968).

von Graff, L., 1882. Monographie der Turbellarien: Rhabdocoelida. I. vol. 1. W. Engelmann.

von Graff, L., 1911. Acoela, Rhabdocoela und Alloeocoela des Ostens der Vereinigten Staaten von Amerika mit Nachträgen zu den 'marinen Turbellarien un der Küsten Europas'. Arb. Zool. Inst. Graz. 9, 321–428. (cited in Henley, 1974).

Wilson, E.B., 1892. The cell-lineage of *Nereis*. A contribution to the cytogeny of the annelid body. J. Morphol. 6, 361–480.

Wilson, E.B., 1898. Considerations on cell lineage and ancestral reminiscence. Ann. N. Y. Acad. Sci. 11. (cited in Kato, 1968).

Younossi-Hartenstein, A., Hartenstein, V., 2000. Comparative approach to developmental analysis: the case of the dalyellid flatworm, *Gieysztoria superba*. Int. J. Dev. Biol. 44, 499–506.

Chapter Outline

Definition **389**
Introduction **389**
Distinctive features **389**
Classification **390**
Reproduction **390**
Reproductive system 390
Gametic reproduction 394
Gametogenesis 394
Spermatogenesis 394
Oogenesis 395
Spawning 396
Fertilization 396
Egg activation and the first block to polyspermy 397
Obtaining material 397
Double eggs in Nemertinea 397
Early development 398
Cleavage 399
Blastula 401
Gastrulation 401
Indirect development 401
Pilidium larva 401
The structure of pilidium larva 403
Metamorphosis of pilidium larva 403
Micrura alaskensis 405
Embryonic development 406
Cleavage 406
Pilidium larva 406
Juvenile development in *M. alaskensis* 406
Juvenile rudiment development 406
Desor larva 410
Schmidt's larva 410
Iwata's larva 410
Lineus ruber (Schmidt's larva) 411
Structure of the Schmidt's larva 411
Metamorphosis of the Schmidt's larva 411
Development of the proboscis 414
Early juvenile 415
Direct development 415
Procephalothrix simulus 416
References **418**

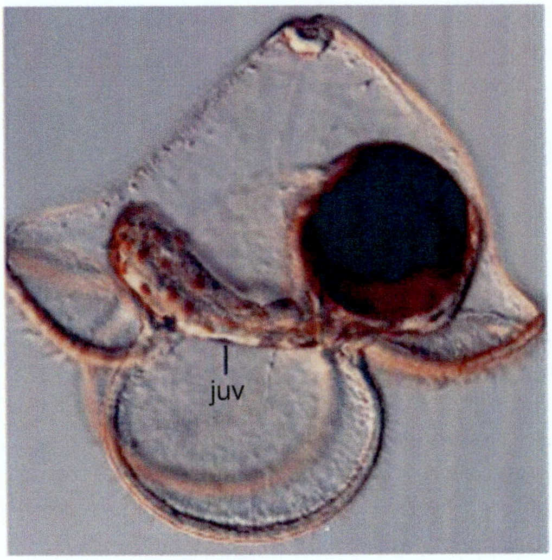

Chapter 8

Phylum Nemertinea

(Greek: *Nemertes;* one of the Nereids)
Rhynchocoela (Greek: *Rhynchos* = snout and *koilos* = cavity)

DEFINITION

Nonparasitic, elongated, flattened, unsegmented worms with a ciliated ectoderm and an eversible proboscis. This proboscis lies in a sheath on the dorsal side of the alimentary canal, but is not connected. The alimentary canal has two openings: mouth and anus. There is an excretory system with flame cells and there is a blood vascular system. Gonads are simple, repeated and sexes separate.

INTRODUCTION

Phylum Nemertinea was one of the minor 'Acoelomata', also known as Nemertini, Nemertinae or Rhynchocoela. Members of this phylum are commonly referred to as 'ribbon worms' or 'proboscis worms' due to their characteristic shape and to the presence of a eversible proboscis, located within a fluid-filled cavity called the 'rhynchocoel'. This cavity opens anteriorly via a thin canal referred to as 'rhynchodaeum', through which the proboscis is thrust to be used in feeding.

This phylum contains over 900 species, most of which are marine; only a few forms live in fresh water (e.g. *Prostoma*); and a single subtropical genus live on land (*Geonemertes*). Gravid females range in length from ~3 mm (*Carcinonemertes epialti*) to over 2 m (*Tubulanus polymorpkus*) (Stricker et al., 2001). Nemerteans, the ribbon worms, are commonly looked at, as they are closely related to the Turbellaria and were formerly included in that class. However, they differ in having a complete digestive tract, in being dioecious and having a simplified reproductive system without accessory organs and/or copulatory apparatus. They possess a closed blood vascular system and a unique proboscis, and are higher in organization than turbellarians. In addition, they exhibit certain special features distinguishing them from the rest of the lower worms. Previously, it was thought that the nemerteans were acoelomate, lacking a true body cavity. It is now clear that the nemerteans are, in fact, true coelomate. Recent careful observations revealed that the lateral blood vessels present in these organisms form within the mesoderm through a process of 'schizocoely'. The rhynococoel also forms in a similar manner. Henry and Martindale (1997) dealt with this phylum in Section V 'Bilateral Animal Phyla, Protostome Coelomates' in Gilbert and Raunio (1997). They reported that members of the Heteronemertea display indirect development involving the formation of a feeding, planktonic larval stage. This mode of development (which is strikingly different) is considered by many to represent the typical ancestral life history of many metazoan phyla.

Previously, it was thought that the nemerteans were 'acoelomate', lacking a true body cavity. Thus, some investigators argued that the nemerteans are closely related to the flatworms (turbellarians). Nemerteans also possess 'protonephridia' and in some species, red-shaped secretory cells called 'pseudocnids' or 'rhabdoids', which were thought to be homologous to rhabdites found in turbellarian flatworms. Henry and Martindale recorded (in Gilbert and Raunio) that in 1992, in addition, comparisons of 185 rRNA sequences indicated that the nemerteans are more closely related to the annelids and molluscs than to the turbellarians.

DISTINCTIVE FEATURES

- Bilaterally symmetrical.
- Unsegmented, flattened, elongated, with ciliated body surface.
- Without distinct coelom.

Invertebrate Embryology and Reproduction. https://doi.org/10.1016/B978-0-12-814114-4.00008-4

- Eversible proboscis enclosed in a sheath and capable of being protruded to a great length.
- Proboscis is found in a cavity called **rhynocoel** dorsal to the alimentary canal. In many species the rhynocoel opens separately from the mouth. In others (Hoplonemerteans), the oesophagus opens into the rhynocoel.
- A complete digestive tract with a separate mouth and anus.
- **Intestine** usually has distinct lateral **diverticulae.**
- Replicated simple saccular **gonads**, sexes **separate**.
- Development may be direct or through a larva (pilidium).
- Blood vascular system.
- Excretory vessel, with **ciliary** flames.
- Nemerteans move via ciliary gliding and rhythmic muscular contraction of the body.

CLASSIFICATION (ACCORDING TO VARIOUS SOURCES)

1. **Class Anopla** (Greek, *anoplos*, unarmed)
   - Proboscis without stylets.
   - Mouth posterior to ganglionic mass.
   - Central nervous system beneath epidermis or embedded in musculature of body wall.

   Order Palaeonemertini (Greek, *palaios*, ancient)
   - Musculature of body wall of two or three layers, third (innermost), if present, always circular.
   - Dermis gelatinous.
   - Direct development.
     Ex: *Cephalothrix, Tubulanus.*

   Order Heteronemertini (Greek, *heteros*, other or different)
   - Musculature of body wall in three layers, innermost longitudinal.
   - Dermis fibrous.
   - Indirect development, feeding larva.
     Ex: *Lineus, Cerebratulus Micrura.*
2. **Class Enopla** (Greek, *enoplos*, armed)
   - Proboscis usually with stylets.
   - Mouth **anterior** to ganglionic mass.
   - Central nervous system internal to body musculature.

   Order Hoplonemertini (Greek, *hoplos*, weapon)
   - Includes the freshwater and terrestrial species.
   - Proboscis with one or more stylets.
   - Enteron straight with one or more diverticula.
   - Direct development.
     Ex. *Prosorhochmus, Geonemertes, Emplectonema, Ptostoma.*

   Order Bdellomorph (Greek, *bdellu*, leach and morphe; form)
   [Bdellonemertini] (Bdellonemertea)
   - Includes a relatively small number of **commensal** species that live in the **mantle** cavity of certain molluscs or inside tunicates.
   - Proboscis without stylets.
   - Enteron sinuous.
   - Ectocommensals of bivalve molluscs.
   - Direct development.
     Ex: *Malacobdella.*

REPRODUCTION

Reproductive system

Stricker et al. (2001) studied oogenesis in 16 species of nemerteans that represent virtually all of the major subdivisions of the phylum. Nemertean gonads, in general, and so the ovaries, in all species, comprise serially repeated sacs (Figs 8.1A and 8.8(I) A, B) with an ovarian wall that characteristically includes myofilament-containing cells 'interspersed' among the germinal epithelium. The ovaries tend to be restricted to the ventrolateral parts of the worms (Fig. 8.2E) and are interspersed among the lateral diverticula of the intestine (Fig. 8.2F). The female nemertean system is simple in organization, as it comprises ovaries with or without short oviducts and lacks accessory organs for storing or processing gametes. Fully gravid specimens of *Micrura alaskensis* and *Cerebratulus lacteus*, have ovaries that fill most of the body (Fig. 8.2H and J). These two species may have several hundreds to a few thousands of pairs of ovaries down the length of their body. In few specimens, the ovary connects a short tubular gonoduct (Fig. 8.2G) that traverses the body wall musculature and integument to connect with a gonopore. The latter is typically found on the ventral or lateral side of the body.

The immature ovaries, prior to the breeding season, consist of several cells of undetermined embryological origin (Stricker et al., 2001) that form accumulations measuring only a few hundreds of micromeres wide. No lumen or ovarian wall appears at this stage. On developing, the outermost edge of each ovary characteristically constitutes a relatively thin wall (Fig. 8.2K). The developing wall comprises a simple squamous to cuboidal germinal epithelium that possesses a basal lamina and junctional complexes. In between these epithelial layers, nurse cells are found. In addition, the germinal epithelium remains connecting the oocytes via a stalk-like projection. This stalk indicates the future vegetal end of the oocyte and it ranges in size from only a few micrometres to more than 50 μm wide and, at least in benthic species, tends to be situated opposite to the germinal vesicle marking the animal pole. Stricker et al. (2001) reported that previous investigations showed that the nurse cells are connected with each developing oocyte via multiple 'cytoplasmic bridges' (Fig. 8.2L) that are numbered from three to 150 in the different species studied.

In the male specimens, as in the female specimens, the testes are serially repeated sacs, arranged along the side of the whole body (Stricker and Folsom, 1998). Except in *Nectonemertes*, the saccular testes are restricted to a more anterior region. The number of testes per worm differs with the different species, but characteristically total more than

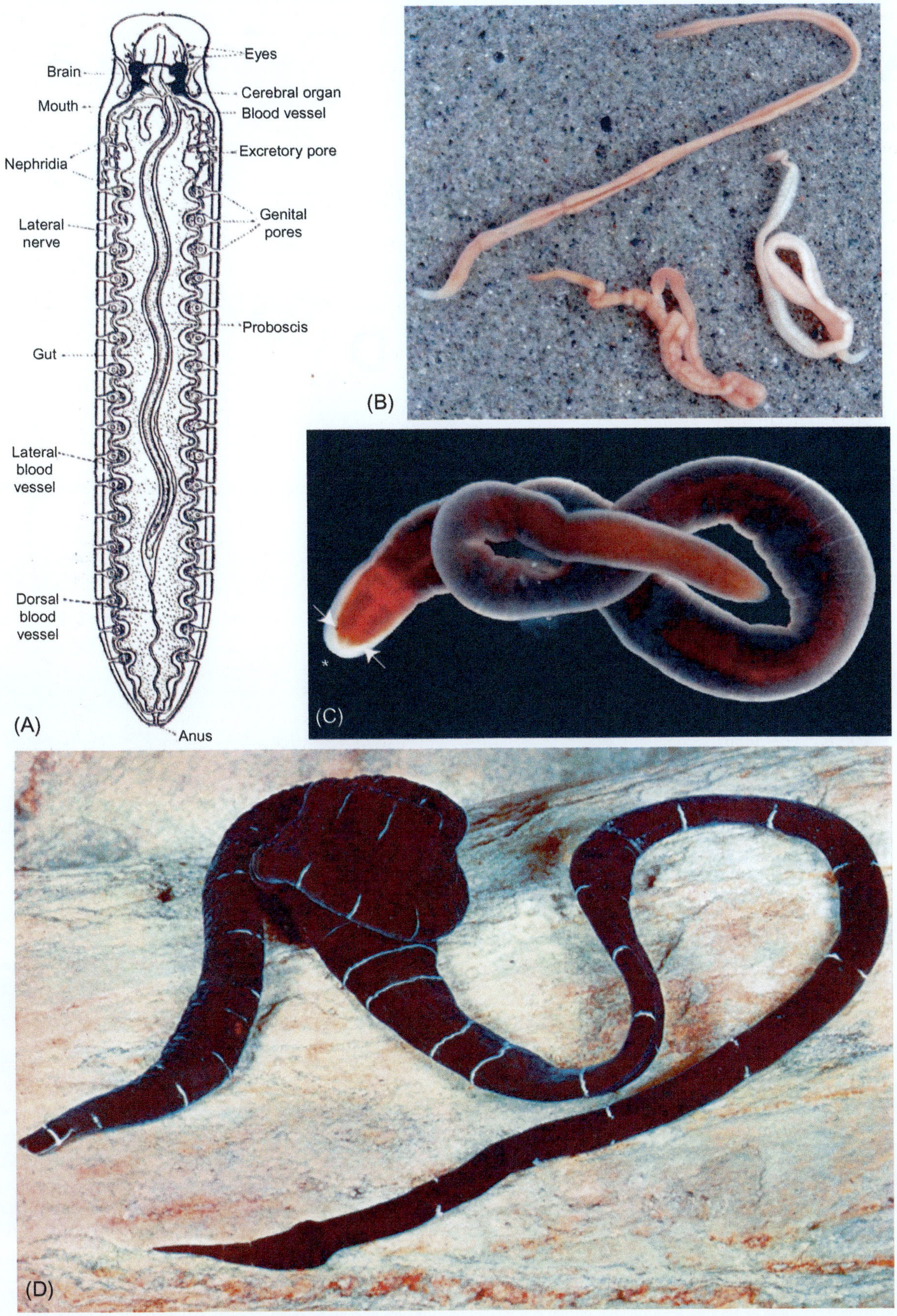

FIG. 8.1 (A) Diagram showing dorsal view of a nemertine *(after Kükenthal, from Borradaile et al., 1977)*; (B) External appearance of live adults of *Micrura alaskensis*. The worms are about 5–10 cm long and 2–3 mm wide, *pinkish* with *lighter coloured head (from Maslakova, 2010)*; (C) photograph of live adult specimen of *Lineus ruber* showing the typical elongated ribbon-like appearance of nemertean worms, with a *dark red pigmentation* throughout the body. The head region is of *lighter red colour*, and contains several pairs of eyespots *(white arrows)* and a terminal anterior proboscis *(asterisk) (from Martín-Durán et al., 2015)*; (D) photograph of live *Lineus geniculatus*, common in the Mediterranean Sea, measures 40 cm *(from Invertébrés: Beauté du Monde Animal, Tome XII, 1974)*.

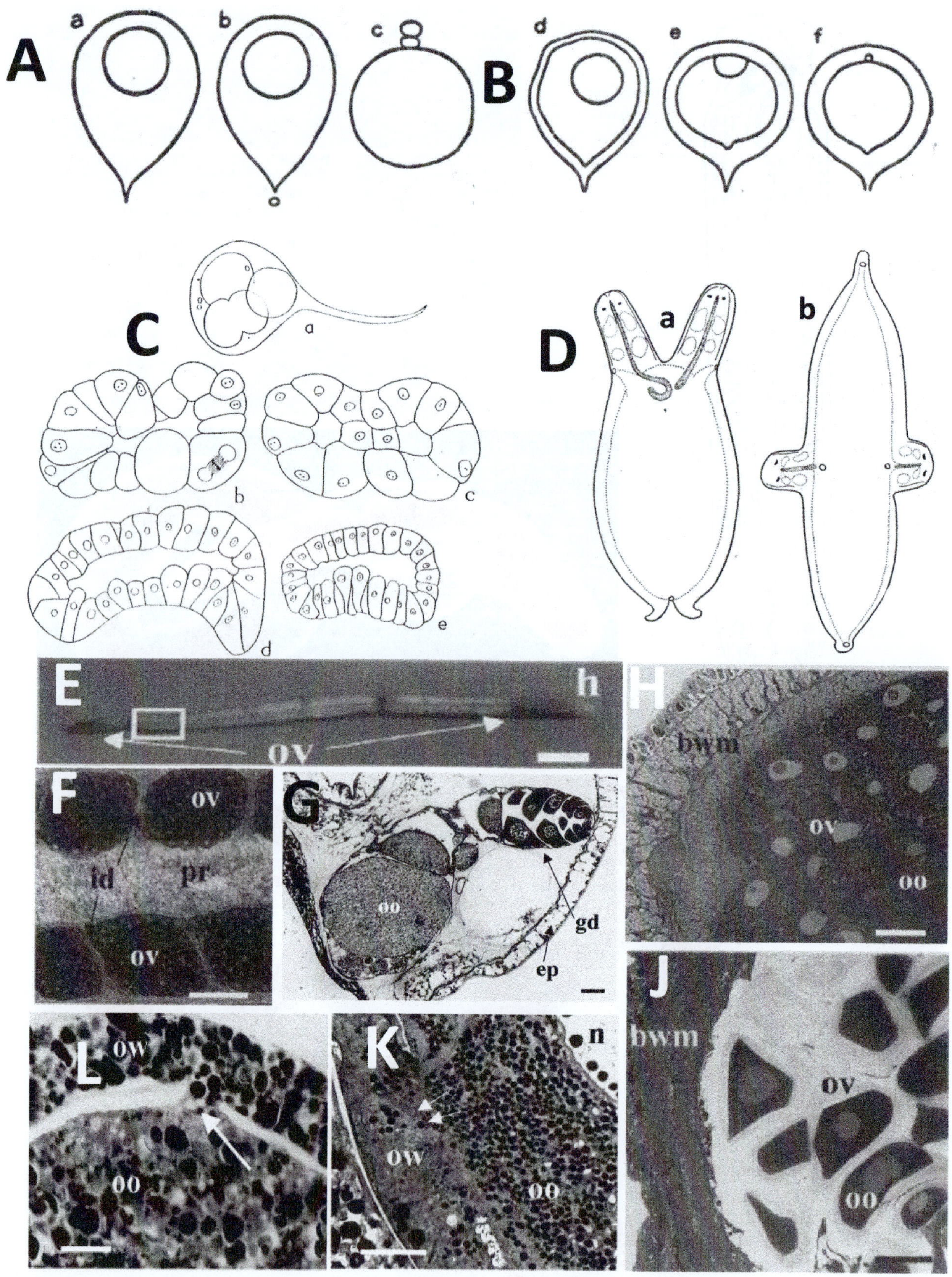

FIG. 8.2 Diagrams illustrating the eggs in nemertins: (A) (a–c) show egg of *Lineus torquatus* (egg membranes removed), X = 140, a = just after oviposition, (b) protrusion separates as vacuole, c = formation of first and second polar bodies after fertilization (Iwata); (B) (d–f) egg of *Cerebratulus lacteus*, d = just after oviposition, e = at 1 h formation of first polar spindle, f = formation of first polar body after fertilization *((A and B) after Wilson, 1903, from Iwata, 1968)*; (C) egg and/or capsule of *Lineus*: (a) capsule containing two normal and one fused egg, (b and c) fused segmentation stages, (d and e) later stages of fused eggs; (D) two double-headed embryos of *Lineus* produced from fused eggs *((C and D) after Nusbaum and Oxner, from Morgan, 1927)*; (E) Photograph of gravid *Micrura alaskensis*, showing reproductive system; (F) magnification of the portion shown in the rectangular in (E); (G) section of the ovary of *Cerebratulus* sp. showing the gonoduct; (H) T.S. of a gravid *Micrura alaskensis*; (J) T.S. of a gravid *Cerebratulus lacteus*; (K) light microscopic T.S. showing the wall of the ovary of a nemertean; (L) high magnification view of cytoplasmic bridge *(arrow)*; (M–Q) Vitellogenesis in nemertean worms: (M) T.S. of ovarian wall in *Cerebratulus lacteus*, showing oogonia and previtellogenic oocytes *(arrow)*; (M′) TEM of a *Paranemertes peregrina*

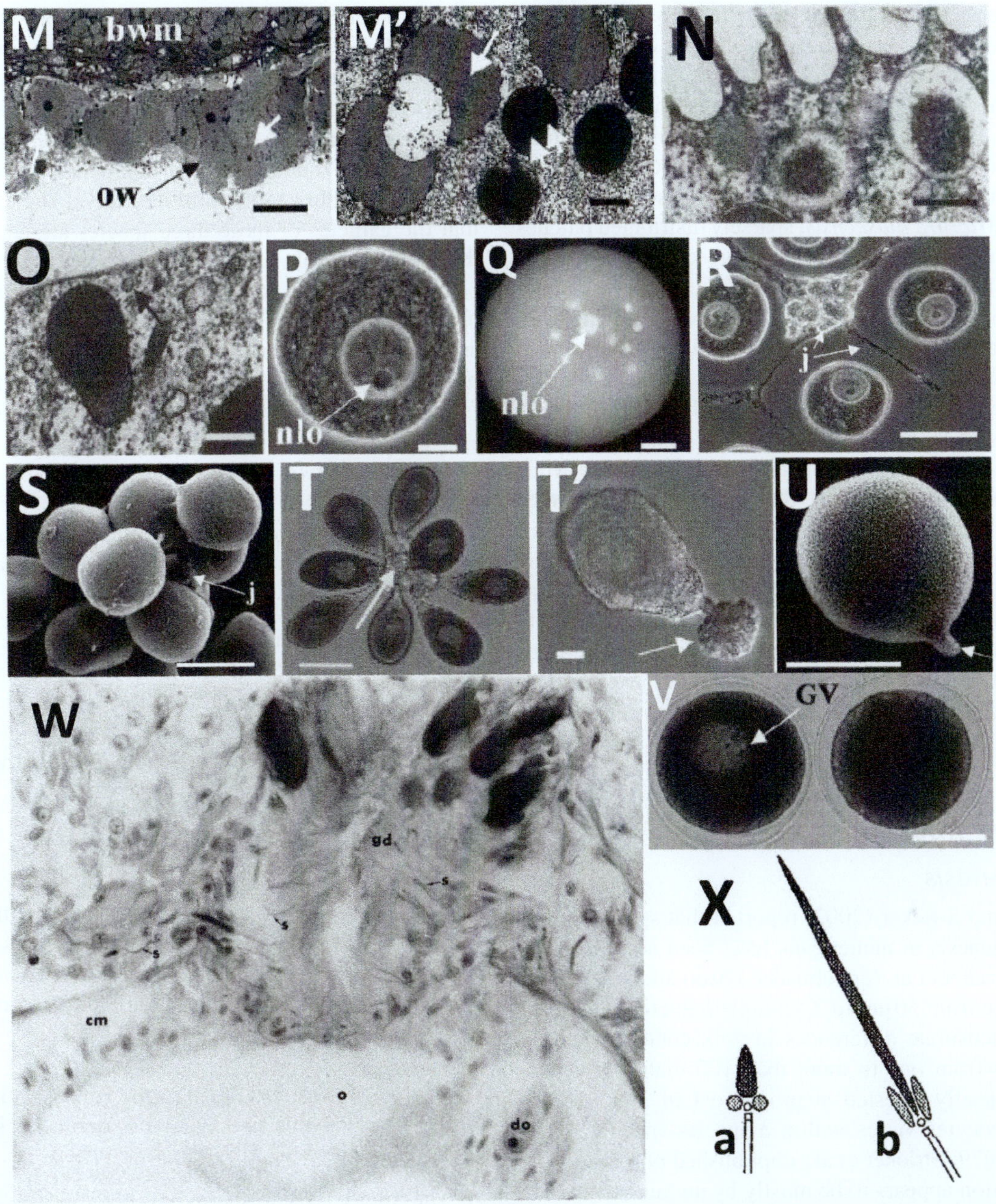

FIG. 8.2, CONT'D oocyte showing lipid droplets *(arrow)* and yolk granules *(double arrows)*; (N) High magnification EM of the oolemma in vitellogenic oocytes from *Cerebratulus* sp.; (O) from *M. alaskensis*, showing relatively few putative endocytotic vesicles (*arrow* in O); (P) Phase contrast micrograph of a *Micrura alaskensis* oocyte with a single nucleolus; (Q) Fluorescence photomicrograph of a *M. alaskensis* oocyte stained with Hoechst 33342 to show a single nucleolus surrounded by numerous patches of heterochromatin; (R) Phase contrast photomicrograh of several discharged oocytes of *M. alaskensis* that are held loosely together by jelly; (S) SEM of *M. alaskensis* oocytes with surrounding jelly; (T) Photomicrograph of interconnected oocytes stripped from an ovary of *Micrura alaskensis*, *arrow* points to attached tissue that probably represents part of ovarian wall; (T′) Photomicrograph of *Micrura alaskensis* oocyte showing vegetal attachment stalk *(arrow)*; (U) SEM of vegetal attachment stalk in oocytes of *Micrura alaskensis (arrow)*; (V) Oocyte maturation and reorganizations of the endoplasmic reticulum, Photomicrograph of immature *Cerebratulus lacteus* oocyte (left) immediately after removal from ovary, versus mature oocyte (right) that had undergone spontaneous maturation within ~1 h after contacting seawater; (W) section through gonoduct of a female *Lineus viridis* immediately after spawning showing sperm (s) migrating through epidermis and in the cutis (cm) circular muscle; (X) diagrams of compact-headed sperm (a) of *Cerebratulus* sp. and elongate-headed sperm (b) of *Cerebratulus lacteus*; bwm = body wall muscles, do = degenerating oogonium, ep = epidermis, gd = gonoduct, GV = germinal vesicle, h = head, id = intestinal diverticulum, in = intestine, n = nucleus, nlo = nucleolus, oo or o = oocyte, ov = ovary, ow = ovarian wall, pr = proboscis; scale bars: E = 5 mm; F = 500 μm; H,J = 500 μm; M, P, Q = 10 μm; N, O = 1 μm; R, S = 50 μm. *All are selected from Stricker et al. (2001), except Fig. W from Riser (1974), and Fig. X from Stricker and Folsom (1998).*

several dozens, and in the even smaller *Carcenonemertes epialti*, a single transverse 1 μm section often contains more than 10 testes. Each ripe testis connects to the exterior via a simple, short gonoduct. This system is without copulatory organs. Histologically, the testicular wall is delimited by a basal lamina, which lies directly beneath a thin layer of muscles in many species. In the anoplans *Cerebratulus lacteus* and *Micrura alaskensis*, sparsely distributed patches of myofilament-containing cells are present in the wall of the testes. These cells might represent the muscles that actually lie above the basal lamina constituting part of true intratesticular set of scattered muscle cells. In addition to those myofilament cells, the basal lamina underlies numerous germ cells. The latter are aggregated into several distinct clusters or clones that apparently originate around the edges of the testis. In all the species examined in the work of Stricker and Folsom (1998), the testes possessed neither a well-developed, stratified germinal epithelium around the entire periphery of the testis, nor a continuous gradient of sperm maturation from the periphery to the centre of the testis. Scattered among the developing germ cells are Sertoli-like cells, referred to in nemerteans by the more generalized term of somatic cells. The role of these intratesticular somatic cells, until the time of Stricker and Folsom (1998), was unknown.

Gametic reproduction

Gametogenesis

Norenburg and Stricker (2002) reported that spermatogenesis and oogenesis in nemerteans have been addressed recently by Stricker et al. (unpublished observations). Mature oocytes range from 50 μm to 2.5 mm in diameter, with generally commensurate differences in yolk content. Ovaries may possess from one to more than 100 mature oocytes, which are usually arrested at prophase I of meiosis; the number of oocytes varies within a species-specific range (Friedrich, 1979; Stricker et al., unpublished observations). Yolk production appears to be mostly by an 'autosynthetic' mode, but several exceptions are suspected—for example, *Carcinonemertes epialti* (Stricker and Reed, 1981), *Tubutanus polymorphus* (Stricker et al., unpublished observations) and most pelagic nemerteans (Norenburg and Roc, 1998) exhibit some exogenous contribution to yolk synthesis.

Spermatogenesis

Stricker and Folsom (1998) studied this vital activity in 11 species of nemertean worms. The authors obtained specimens of *Cerebratulus lacteus* from Wood Hole laboratories, and a single ripe male of the pelagic *Nectonemertes mirabilis* from Drs. J. Norenburg and P. Roc. All the other studied nemerteans were collected during spring and summer months near San Juan Island, Washington, United States. The authors used electron microscopy to determine if sperm morphology is related to the type of the extracellular coats surrounding the oocytes. They examined oocytes of *Cerebratulus* sp., *C. lacteus* and *Micrura alaskensis* in in vitro fertilization and these animals were continued in laboratory culture to the pilidium larval stage. This means that the gametes of these three species are capable of producing viable embryos.

Spermatogonia are well known from their large size and peripheral position. Each contains numerous mitochondria, and a single dense nucleolus. Even in the ultrastructural preparations, the 1^{ry} spermatocytes are not distinguished from 2^{ry} spermatocytes. However, the nucleus of the 2^{ry} spermatocytes is smaller, due to the lack of an intervening period of DNA synthesis during meiosis (Stricker and Folsom, 1998). In addition, these cells are flagellated. During the maturation of the spermatids to be spermatozoa (spermiogenesis), some cytological reorganizations occur:

1. condensation of the chromatin;
2. the Golgi apparatus forms the acrosome;
3. aggregation of the mitochondria forming several relatively large mitochondria in the middle region of the developing sperm; and
4. 'sloughing' and/or redistribution of the cytoplasm to become more fluid (stream-shape).

In the spermatid, which will produce long-headed sperm, two other reorganizations occur:

1. elongation of the nucleus, in association with 'a longitudinally arranged manchette of microtubules'; and
2. further fusion of mitochondria to form either a single mitochondrion or up to several elongated ones extending anteriorly along the nucleus.

Structure of the spermatozoon. Henry and Martindale (1997) reported that two types of sperm are identified in nemerteans.

1. The more typical sperm morphologically, with small heads and mid bodies, which may reach 8 μm in length, attached to an elongated flagellum. This type is produced by species that shed their gametes freely into the surrounding water, e.g. *Cerebratulus lacteus* (this species possesses elongate-head sperm; see the following paragraph).
2. Modified sperm that possess an elongated head and midpiece which may reach 25 μm in length, e.g. *Lineus ruber*. This modified sperm is produced by species that do not release the sperm freely in the medium, but fertilize their eggs internally or in close proximity to the adults, while encased in mucous sheaths.

Stricker and Folsom (1998) studied nemertean sperm in a number of species and ascertained that it is the same as other sperm; it consists of a head containing a condensed nucleus, a mitochondrial component (middle piece) and a single tail. Their ultrastructural examinations evidenced that the tip of the head of all types contain a typical acrosomal vesicle and the flagellum has the typical 9+2 arrangement of the microtubules. In the sperm of *Cerebratulus* sp. and *Micrura alaskensis*, the mitochondria are in a distinct midpiece posterior to the nucleus. In *Cerebratulus lacteus*, the mitochondria are fused and found partly along the side of the nucleus (Fig. 8.2X (b)). The acrosome comprises a more or less bullet-shaped structure that lacks a deep central indentation. In this elongate-headed sperm *C. lacteus*, the sperm head is straight or slightly curved (Fig. 8.2X (b)), whereas in *Carcinonemeretes epialti*, the head is twisted to form a slightly helical structure.

In all species examined ultrastructurally by Stricker and Folsom (1998), the tail consisted of a single flagellum that typically measures about 30–50 μm long. Anteriorly, the tail connects to a centriolar apparatus. The latter contains a perpendicularly arranged, pair of basal bodies known as the proximal and distal centrioles. The former lies within a slightly invaginated nuclear fossa and is situated between the nuclear envelope and distal centriole. Extending from the distal centriole is the flagellar axoneme. The axoneme of all mature spermatozoa examined consists of the '9+2' arrangement of microtubules that is characteristic of other motile flagella and cilia.

As mentioned above, this work examined experimentally the correlation of sperm morphology with the extracellular coat surrounding the female gamete. In *C. lacteus*, which produces elongate-headed sperm, the oocyte averages 11.5+3.8 μm in diameter (N=10) and is directly ensheathed by an electron-dense layer, referred to by the general term 'vitelline envelope'. In *Cerebratulus* sp. and *Micrura alaskensis*, on the other hand, the sperm has a compact head and the oocytes lack a vitelline envelope or any other prominent coat covering the microvilli of the oocyte. Oocytes from each of the three above species were inseminated in the laboratory (in vitro fertilization) with its proper identified species. In all three cases, normal cleavage occurred and the embryos developed to pilidium larvae. The authors interpreted that the results of this experiment are as follows:

1. The gametes were either correctly matched with its species or with different species that readily hybridize.
2. All three species could produce viable embryos without either internal fertilization or paeudocopulation.
3. The oocytes of *Cerebratulus* sp. and *Micrura alaskensis* were capable of developing normally without a detectable vitelline envelope.

Oogenesis

Most nemerteans are dioecious; however, there are a few hermaphroditic species, particularly among the terrestrial and fresh water worms, such as *Prostoma*, and in some commensal species that occupy unusual ecological habitats (Henry and Martindale, 1997). Most nemerteans are oviparous, though some are ovoviparous or viviparous (Stricker et al., 2001). Many nemertean species possess a small peduncle or stalk located at the vegetal pole, which apparently represents the point where the oocytes were connected in the ovaries (Fig. 8.2T and U). The opposite, animal pole, represents the site where the polar bodies will form during the meiotic maturation divisions. In many species, the eggs are surrounded by an extracellular **chorion** and a surrounding **jelly coat**. In some species, the latter may be sticky and may help to attach the eggs and embryos to the substrate. Animal-vegetal polarity may also be evident in the chorion.

Just before the above accounts, Reverberi in 1971 mentioned that Hörstadius evidenced experimentally that the eggs of *Cerebratulus lacteus* (the only one used for embryological experiments) do not form a fertilization membrane. They are surrounded by jelly and soft membranes.

In the studies of oogenesis of nemerteans, Stricker et al. (2001) used light and electron microscopy to compare the biology of oogenesis in the 16 different species they worked on. They examined fixed material and observed living specimens. Oogenesis of only one or two species will be followed to simplify this aspect, that of *Cerebratulus lacteus* and *Micrura alaskensis*. Adult specimens of the former species were purchased from the Marine Biology Laboratory (Woods Hole, Massachusetts), while those of the latter, like the other studied species, were obtained from intertidal and shallow subtidal habitats near San Juan Island, Washington, United States. To collect oocytes, gravid specimens of *M. alaskensis* were teased apart with fine forceps, and the released oocytes were subjected to the technical processes. For analysis of living oocytes, time-lapse video studies of maturation events were conducted using an inverted microscope equipped with a thermoelectric cooling stage.

In living oocytes of *M. alaskensis* that were removed from the ovary at various times during the breeding season, the size and persistence of the attachment stalk to the germinal epithelium varied depending upon the apparent age of the oocyte. Oocytes that are relatively young, based on their smaller size and less spherical shape, were difficult to remove from the ovary and often remained permanently attached to an irregular mass of tissue that presumably represented part of the ruptured ovarian wall (Fig. 8.2T). The larger and more spherical the oocyte, the less pronounced the stalk becomes, and it eventually becomes either fully absent in oocytes suspended in the ovarian lumen, or it is resorbed after removal from the ovary. This isolated oocyte may undergo cytoskeletal reorganization and becomes round in seawater. In most examined nemertean species, each ovary contained only a single fully grown oocyte that ranged in size from ~60 μm to 1 mm. However, within the ovarian lumen of *Cerebratulus lacteus* and *Micrura alaskensis*, at least several dozens of fully-grown oocytes exist per ovary (Fig. 8.2H and J). Within the ovary, the oocytes lack an enveloping layer of follicle cells.

At the beginning of oogenesis, each oocyte is a small cell in the ovarian wall, measuring ~5–20 μm in diameter (Fig. 8.2M). The previtellogenic oocyte is larger in size, basophilic stained and, as vitellogenesis begins, the oocyte gradually attains the acidophilic staining pattern and larger size (Fig. 8.2N and O). These oocytes are within the lumen of the ovary, while the more advanced stages of vitellogeneic oocytes are shifted away from the so-called gonoduct region. The younger oocytes are situated towards the more medial portions of the ovary. Vitellogenesis itself (deutoplasmogenesis) starts by the deposition of membrane-bound yolk bodies and lipid droplets lacking a delimiting membrane in the ooplasm of the developing oocyte (Fig. 8.2M′). This process of vitellogenesis occurs via an 'autosynthetic' mode whereby the precursors are formed within the oocyte itself. However, possible exceptions to this autosynthesis type of vitellogenesis were observed in *Carcinonemertes epialti*. Light microscopic examinations of several pelagic species have also suggested that yolk precursors are transferred from nurse cells to the developing oocyte via interconnecting cytoplasmic bridges (Fig. 8.2L).

The nucleus of the vitellogenic oocytes, in all the species examined in the work of Stricker et al. (2001), underwent a gradual increase in size to form a hypertrophied 'germinal vesicle' as the oocyte remained arrested at prophase I of meiosis. In most nemerteans, the germinal vesicle of the yolky oocytes measures ~30–60 μm in diameter and contains multiple nucleolus-like bodies that tended to aggregate at the nuclear periphery. On the other hand, some other yolky oocytes and the microlecithal oocytes produced by *Cerebratulus lacteus* and *Micrura alaskensis* contain only a single large nucleolus (Fig. 8.2P and Q).

Spawning

The sexes are separate in almost all the nemertinean worms, except for a few hermaphroditic genera among the Hoplonemertea (Ejnopla). In the Anopla species, both the Paleonemertea and Heteronemertea, the small eggs are shed almost simultaneously from the ovaries. The mode of spawning is known for relatively few nemerteans, mostly those from littoral habitats. Among littoral forms, spawning ranges from broadcast release of gametes into the surrounding sea, to pseudocopulation with eggs attached in a gelatinous matrix to a benthic substratum (Riser, 1974). The eggs may be shed directly into the surrounding water or deposited on various substrates in gelatinous masses. In most of the Hoplonemertea, spawning takes place continuously; at intervals of a day or a week, one large egg matures and is released from each ovary.

However, Stricker et al. (2001) cultured 16 nemertean species in the laboratory, and reported that egg release was not observed in the field for any of the species examined in their study. However, ripe female of *Cerebratulus lacteus* and *Micrura alaskensis* occasionally discharged hundreds to thousands of oocytes in their laboratory aquaria. These oocytes were either spawned freely or deposited within some loosely adherent jelly that was easily dispersed by currents in the aquarium (Fig. 8.2R and S). The oocytes of these two species and those of *Tubulanus polymorphic* were spawned from the female body apparently lacking a discrete extracellular coat other than a dispersed layer of jelly (Stricker and Folsom, 1998). Directly following removal from the ovary, fully developed oocytes possessed a prominent germinal vesicle that indicated a prophase-I arrest (Fig. 8.2V). Within just ~20–40 min after contacting natural seawater, isolated oocytes typically resumed meiotic maturation. This is evidenced by the germinal vesicle breakdown (Fig. 8.2V). After the completion of this vital activity, most unfertilized oocytes reached a secondary arrest at metaphase I and remained at this stage of meiosis until fertilization.

The breeding season differs according to species and locality. The two species of one genus breed at different times in the same locality. The one species breeds at different times in the different localities. For example, *Lineus alborostratus* breeds during August at a locality, and in April in another locality. Generally, this species breeds during summer.

Fertilization

Fertilization is usually external, although there are few viviparous or ovoviviparous species, in which the young larva is formed inside the body of the mother, e.g. *Prosorhochmus viviparous*, *Lineus viviparous*, *L. biliniatus*, *Geonemertes australiensis*, *G. Agricola*, *Prostoma lacustra*, *Stichostemma graecense* (Iwata, 1968). In *Lineus ruber* and *L. viridis*, a form of external pseudocopulation occurs by close contact of two or more individuals, typically inside a secreted mucous mass or cocoon. In such cases, sperm may take their way into the ovaries where internal fertilization can take place. Riser (1974) reported that sperm in the gonoducts of *L. ruber* were described and figured, and this author found large numbers of sperm throughout the epidermis and dermis in the region of the gonoduct of postspawning female *L. viridis* (Fig. 8.2W).

Accessory organs are developed in some species to be involved in the process of copulation. For example, the bathy pelagic (deep ocean dwelling) *Nectonemertes* has a pair of cephalic clasping tentacles and eversible penis facilitating copulation and therefore internal fertilization. There are a number of species with viviparous or ovoviviparous development that display internal fertilization, e.g. *Cephalothrix galathea*, *C. rufifrons* and *Carcinonemertes carcinophila*.

Fertilization occurs before the maturation division (extrusion of the polar bodies). The eggs become arrested in metaphase of the first meiotic division.

Egg activation and the first block to polyspermy

Henry and Martindale (1997), interested in this vital phenomenon, reported that in the nemertean egg, as many other eggs, the cell membrane becomes electrically polarized as a result of fertilization. The egg cell membrane of *Cerebratulus lacteus* shifts from a negative resting potential to a positive one that persists for approximately 1 h following fertilization. Experiments have demonstrated that this electrical 'first block to polyspermy' prevents additional sperm from entering the zygote. Studies indicate that an artificially imposed positive membrane potential will inhibit oocytes from becoming fertilized. On the other hand, if the positive membrane potential is reversed by placing zygotes in low Na^+ seawater, they can become 'polyspermic'.

Obtaining material

For the study of the development and for experimentation in the laboratory, the heteronemerteans *Lineus* and *Cerebratulus* are the best material. Reverberi (1971) added that a further advantage of the *Cerebratulus* egg as an object for experiments is the fact that the egg-axis can be accurately determined from the moment of its discharge from the ovary. This is because of the presence of the conical protuberance (Fig. 8.2A and B). Iwata collected females during the breeding season and kept them in seawater tanks, that spawned, and if males and females are placed together, it is very easy to secure material for observation of developmental stages. In 1968, Iwata also reported that to obtain unfertilized eggs, a piece of the body is cut off, and by longitudinal incisions along the sides and by the muscular movements of this piece, the eggs will then be expelled. With a pipette, these eggs should be transferred to fresh seawater. In such eggs, the germinal vesicle is still intact and clearly visible, but normal fertilization does not take place. These eggs become fertilized in one of the two following conditions: (a) immediate insemination usually results in **polyspermy**; or (b) if the eggs are left for some time, the germinal vesicle soon breaks down in response to the seawater stimulus, and the first polar spindle is formed. This is completed in 10–30 min, depending on the temperature and the degree of maturation of the eggs. These eggs are then ready to be fertilized normally if inseminated. This means that normal fertilization occurs after the extrusion of the first polar body of the egg.

The insemination can be done if a small incision is made in the body wall of a male worm to get sperm (Iwata, 1968), which exudes as a white, viscous thread-like mass. A very small amount of this stirred into the egg suspension with a pipette will bring normal fertilization of these eggs. In the case of the development of these eggs and reaching the pilidium larva, it is necessary to provide food in the form of small diatoms, ciliates, etc., for this swimming stage that precedes metamorphosis.

By this method described, the above authors studied the early cleavage stages and they followed the developmental 'capacity' of each blastomere. According to their results, each of the two blastomeres resulting from the first cleavage is able to give rise to a complete larva of half size, while defects appear in larvae reared from separated four-cell stage blastomeres.

Double eggs in Nemertinea

In 1927, Morgan in his book *Experimental Embryology* identified that giant embryos of the nemertean, *Lineus ruber*, arise from the fusion of two eggs (which had already been described by Nusbaum and Oxner, 1913). This condition was also mentioned much later by Henry and Martindale (1997). The eggs are laid in cocoons, with two or more eggs in each (Fig. 8.2C (a)). They may fuse before cleavage (Fig. 8.2C (a)), during cleavage or in the blastula stage (Fig. 8.2C (b)). The eggs that fuse before cleavage may contain two or more nuclei, but sometimes only one. When more than two eggs fuse, the cleavage is so irregular that embryos do not develop. The two fused eggs develop reaching the gastrulation stage (Fig. 8.2C (d)), and a giant embryo may be formed. This may have a common blastocoel and a single archenteric invagination. These appear to give rise to a single giant embryo at times (Fig. 8.2D). When partial fusion takes place between the two blastulae, each of them may invaginate separately. The double-headed embryos that have been found (Fig. 8.2D (b)) may be produced in this way.

The development of whole embryos from parts of eggs, or from isolated blastomeres, or even from pieces of blastulae of another nemertean (*Cerebratulus*, e.g.), indicates that the blastomeres of these worms are little differentiated, or else have extensive powers of 'regulation'. Hence, the results of fusion of two eggs or embryos are entirely consistent with the known possibilities of these eggs.

Turbeville (2002), in his reinvestigation of the cell lineage of *Cerebratulus lacteus* using fluorescent tracers, revealed that endomesoderm forms from the 4d cell as in other spiralians, and the ectomesoderm is derived from the 3a and 3b cells as in annelids, echiurans and molluscs. Studies examining blastomere specification show that cell fates are established precociously in direct developers, and later in indirect developers (cell lineage of *Cerebratulus lacteus*, Table 8.1). Previously, Iwata (1968) recorded that nemerteans develop either directly or indirectly. Species of Paleonemertea of the group Anopla and those of the Enopla develop directly, i.e. without metamorphosis of a larval stage. Studies concerning this direct development, besides that of Ikeda (1915) on *Stichostemma grandis*, have been published by more than one author since 1873. However, the number of species that has been adequately investigated is very small. On the other hand, the indirect development of the **Heteonemertea** has

TABLE 8.1 Cell lineage of *Cerabratulus lacteus* as determined using a fluorescent tracer

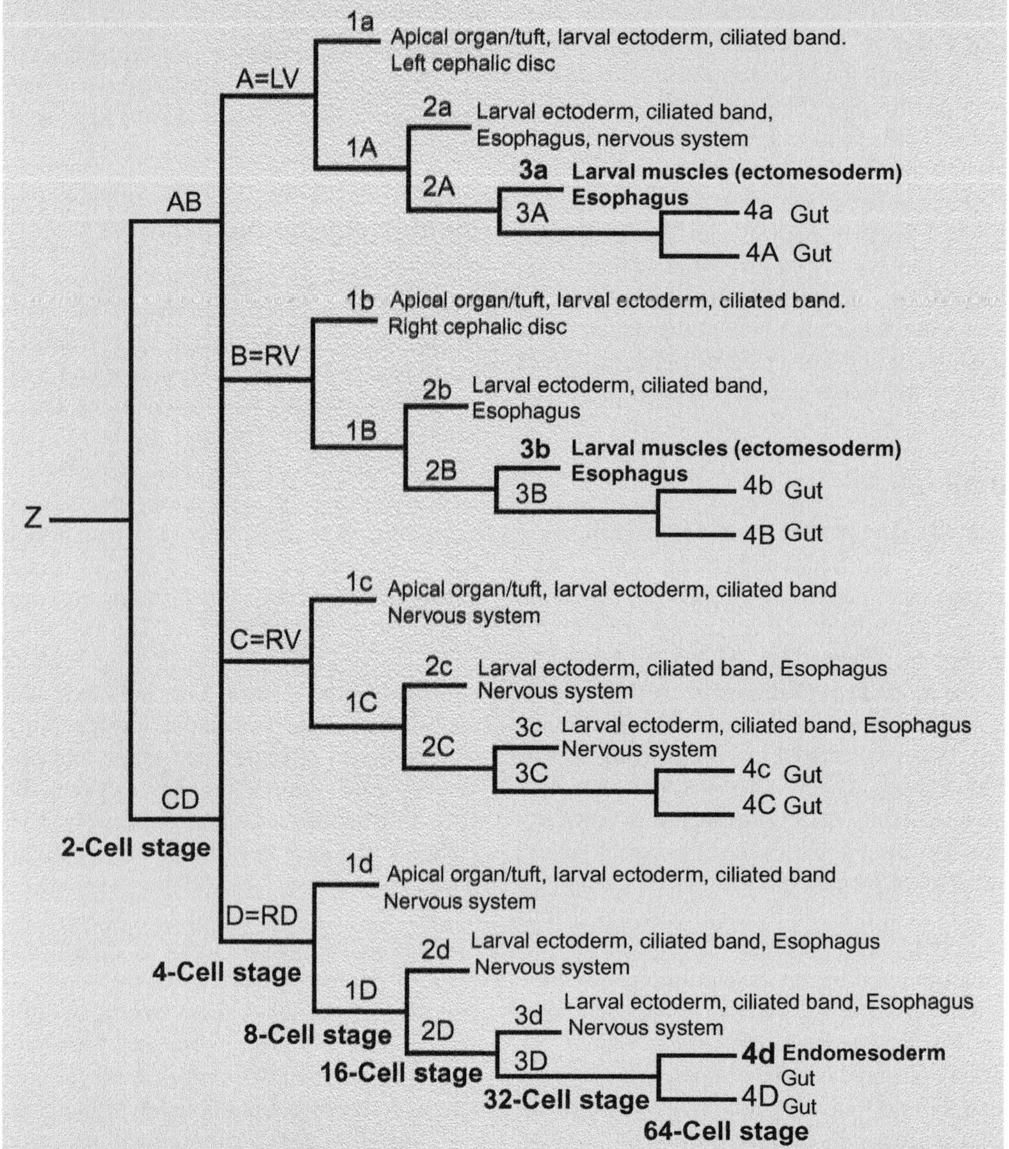

Redrawn with permission from Henry and Martindale (1998). Z=zygote, from Turbeville (2002).

been studied for the longest time. Müller (1847) applied the name *Pilidium gyrans* to a little swimming animal found in the plankton of the North Sea nearest to Heligoland Island. He himself afterwards, in 1854, observed that this Pilidium passed through metamorphosis to become a young nemertinean worm. Since that time, reports concerning the development of this helmet-shaped or Pilidium larva and its metamorphosis were published.

Another conspicuous larva that is formed by the development of the *Lineus ruber* egg was observed by Desor (1848). It is not a swimming larva, but it undergoes a shortened development inside a membrane. Investigations concerning the development of the larva of this genus, which is called Desor's larva, have been published by Barrois (1877) and others.

In both cases, for Pilidium and Desor larvae, the body wall of the worm is formed by the inner wall, the imaginal discs, of the amniotic invaginations of the larval ectoderm.

Early development

Iwata (1968) described the nemertean eggs as being of two types:

1. eggs containing large amount of yolk, spherical when spawned; and

2. eggs containing small amount of yolk, rather irregular in shape like those of *Cerebratulus* and *Lineus* (Fig. 8.2A and B).

This egg has a conical protuberance at one side and a germinal vesicle with a clearly visible outline. Within about 10–20 min of shedding, whether fertilization has taken place or not, the germinal vesicle breaks down and its material, distinguishable as a clear mass, begins to move towards the side of the egg, opposite to the conical protuberance. By that time the first polar body is formed, while the clear mass is reaching the opposite pole. In *Celebratulus lactues*, the cone-shaped protuberance disappears at this time in most cases, but in *Lineus toquatus* it pinches off from the egg as a small vesicle before the egg becomes spherical. In these eggs, as mentioned above, the egg axis is determined before fertilization. The side with the conical protuberance is the vegetal pole, at which the egg is attached to the ovarian wall, while the side to which the germinal vesicle substance moves, and where the first and the second polar bodies are given off after fertilization, is the animal pole.

Cleavage

Cleavage in nemertean eggs is of the typical spiral type. Riser (1974) confirmed the view of Iwata that cleavage is spiral and added that it becomes 'determinant' after the four-cell stage. The initial four blastomeres are capable of producing miniature normal larvae, but larvae obtained from blastomeres of the eight-cell stage are abnormal. Iwata (1968) described the process of cleavage in the eggs of both *Celebratulus* and *Lineus* (Fig. 8.2A and B), which have little yolk and are rather irregular in shape. The **first** cleavage occurs after the expulsion of the polar bodies and divides the egg into two equal-sized blastomeres (Figs 8.3 and 8.4A) by the vertical plane which passes close to the polar bodies and parallel to the egg axis. The **second** cleavage also occurs by a longitudinal plane forming a right angle with the first plane and produces four blastomeres of equal size (Figs 8.3 and 8.4B). The **third** cleavage divides the egg along the equatorial plane forming four equal-sized macromeres (1A–1D) at the vegetal pole, and dextrally rotated, in respect to these macromeres, four equal-sized 'micromeres')

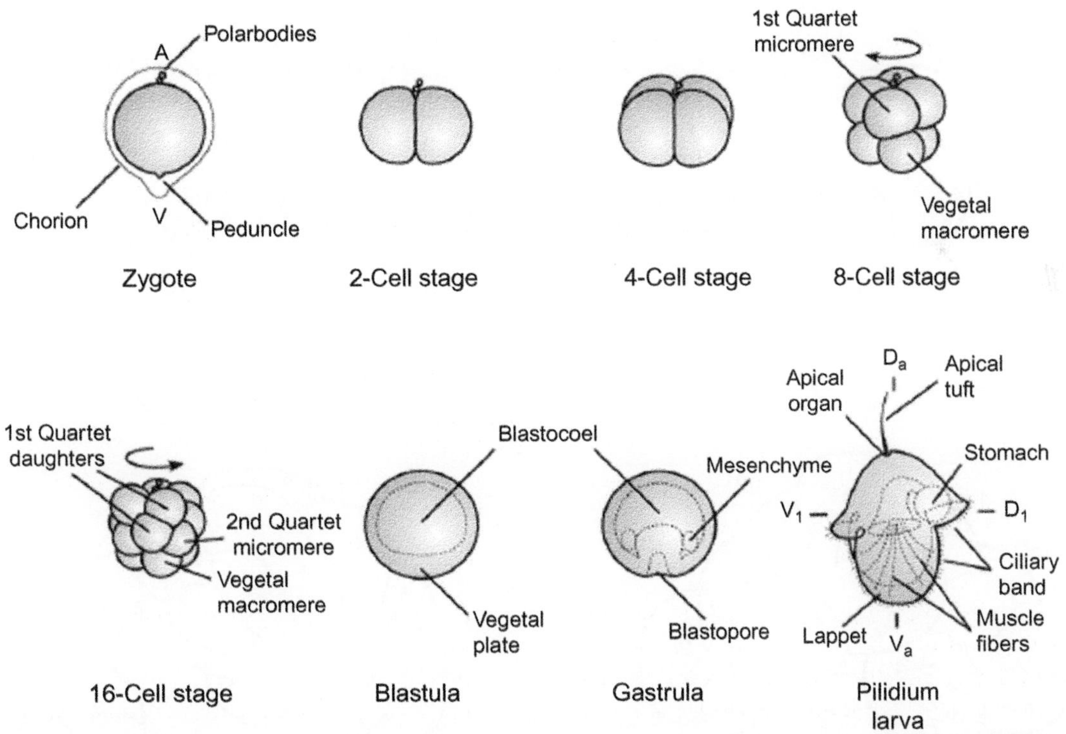

FIG. 8.3 Diagram summarizing development from the fertilized egg to the formation of the feeding pilidium larva of *Cerebratulus lacteus*. All embryonic stages are illustrated with the animal pole (A) directed upwards, which is also indicated by the polar bodies. The first two cell divisions are equal. Note that the first quartet micromeres of the 8-celled embryo are larger than the vegetal macromeres due to the subequatorial plane of the third cleavage division. Also note the oblique arrangement of the blastomeres in the 8- and 16-celled embryos that results from the shifting relationship of the cleavage spindles. The first quartet micromeres are given off in a clockwise direction, while the second quartet is formed in a counterclockwise direction *(arrows)*, as viewed from the animal pole. Bilaterally situated populations of mesenchyme lie within the blastocoel during the early gastrula stage along the sides of the invaginating archenteron. Subsequent development leads to the formation of the swimming pilidium larva with a blind digestive tract, numerous muscle fibres, lappets, ciliary band and an anterior apical organ and ciliary tuft (located at the animal pole). The pilidium larva is shown as a left lateral view with the apical ends directed upwards (animal pole). Ultimately, the pilidium larva will undergo a radical metamorphosis to give rise to the juvenile worm. The larval dorsoventral axis is labelled Vl–Dl, while the future adult dorsoventral axis is labelled Va–Da. *(From Henry and Martindale (1998).)*

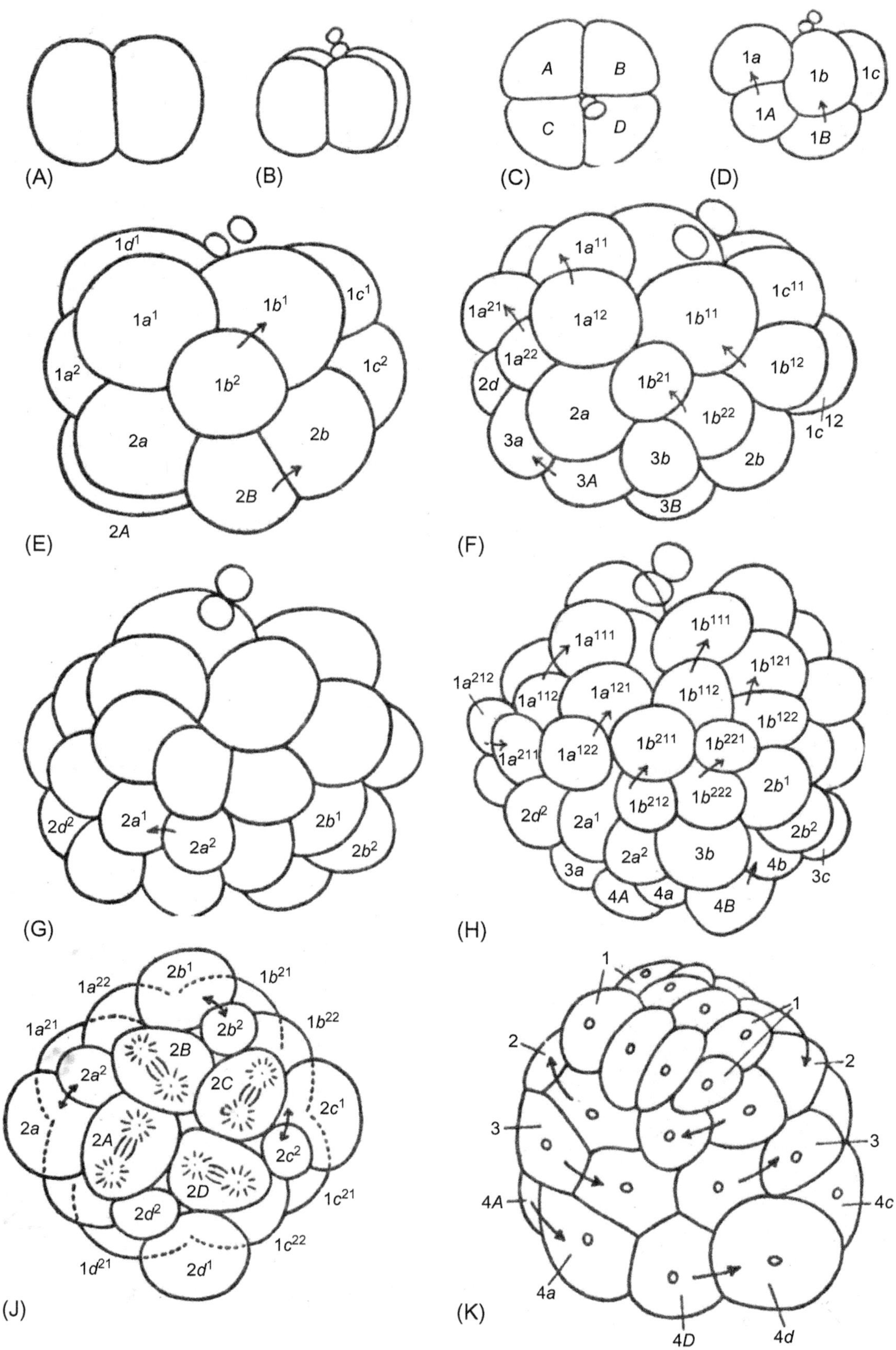

FIG. 8.4 Diagrams showing cleavage stages in nemertins: (A–H) in *Lineus torquatus*; (A) 2-cell stage; (B) side view of 4-cell stage; (C) 4-cell stage from animal pole; (D) 8-cell stage; (E) 16-cell stage; (F) 28-cell stage; (G) 32-cell stage; (H) 52-cell stage; (J) in *Malacobdella grossa*, 28-cell stage from vegetal pole; (K) in *Lineus ruber* at 64-cell stage, lateral view. *(After Nusbau and Oxner.)*

(1a–1d) (first quartet). Riser (1974) reported that the term 'micromere', in nemertinean embryology, is utilized topographically and does not refer to cell size. The marked size distinction between the first eight blastomeres does not occur in most species, although the D cell of *Lineus ruber* was described as larger than the three cells of the first quartet. At this point, it is important to note that the equal size of the blastomeres occurring here in nemertineans differs from that which occurs in the Platyhelminthes described before. Also, the so-called 'micromeres' here are larger than the 'macromeres' (Fig. 8.4D and E), a characteristic that is not found in any of the other spirally cleaving groups. In the **fourth** cleavage, all the blastomeres are divided by furrows parallel to the equatorial plane, but in this cleavage with a sinistral rotation as opposed to the dextral rotation of the third cleavage. This produces eight blastomeres in the animal hemisphere: an upper ring $1a^1$–$1d^1$ and a lower ring $1a^2$–$1d^2$, and eight blastomeres in the vegetal half: four secondary micromeres (2a–2d) second quartet above and four macromeres (2A–2D) below. This cleavage makes up the **16-cell stage**. In the **fifth** cleavage, a second dextral twist occurs in this cleavage. The blastomeres of the first quartet in the animal hemisphere divide almost simultaneously, forming $1a^{11}$–$1d^{11}$, $1a^{12}$–$1d^{12}$, $1a^{21}$–$1d^{21}$ and $1a^{22}$–$1d^{22}$.

The blastomeres in the vegetal pole, namely 2A–2D and 2a–2d, vary in the cleavage at this stage according to the species. In *Lineus* and *Cerebratulus*, the lowest group (2A–2D) divides first to form the third quartet of micromeres (3a–3d) and the third quartet of macromeres (3A–3D). This cleavage results in the **28-cell stage** (Fig. 8.4F). In *Emplectonema* and *Malacobdella*, the cells of the second quartet 2a–2d (Fig. 8.4J) divide first, followed by the division of the blastomeres 2A–2D, resulting in **the 32-cell stage** (Fig. 8.4G) In this stage in *Malacobdella*, the blastomeres of the $2a^1$–$2d^1$ quartet project outwards, giving the embryo a square outline when seen from the vegetal side.

Reports of later cleavages are available to the **sixth** cleavage in *Malacobdella* and *Lineus*, and to the **eighth** cleavage in *Emplectonema* and *Cerebratulus*; in all cases, right and left torsion continue to alternate.

Blastula

Blastulation takes place very early in the cleavage period by the appearance of the blastocoel, which may be large, or small, depending on the species (coeloblastula: Riser, 1974). Those with little yolk, which will be eventually pilidium larvae, have large blastocoels, while the yolky eggs have small ones. In the **blastula** stage of pilidium larvae, a dish-shaped indentation appears at the animal pole. This part is formed of closely packed columnar cells, which are smaller than those of the surrounding ectoderm. A little later, this will become the **apical plate** and gives rise to the long, slender **cilia** forming a bundle. At this time, a pair of mesoderm mother cells appear at two sides of the vegetal pole and hence the embryonic symmetry changes from radial to bilateral. When the embryo reaches the blastula, its surface becomes thickly covered with delicate cilia, and it begins to rotate slowly within the egg membrane. The speed of rotation increases with time and the apical **tuft**, which looks like one long flagellum can be detected at the animal pole. At this stage, the embryo makes its way out of the egg membrane and swims free in the seawater (hatched), while the animal pole region which bears the **apical tuft** becomes pointed and the vegetal side flattens like a flask shortened in its vertical axis (Fig. 8.5(I) A–C).

Gastrulation

This process takes place by invagination (In view of Riser, 1974; it occurs by emboly, unipolar ingression, or a combination of the two processes) and the endoderm cells invaginate to form a pouch-shaped archenteron. By that time, a circular blastopore can be detected in the centre of the flattened vegetal side. According to the invagination of the endoderm, the blastocoel becomes smaller. The embryo then becomes a **gastrula**, and still swims but rapidly in a spiral fashion with the animal pole directed forward.

Indirect development

Pilidium larva

The following stage to the gastrula is the pilidium larva. The ectoderm of the vegetal side extends down around the blastopore forming the lateral lappets, and the larva acquires complete bilateral symmetry. Accordingly, it assumes the shape characteristic of this type (Fig. 8.5(I) D). The animal pole side will be the dorsal side of the adult form, while the vegetal region will become the ventral side. The most notable change that occurs to the gastrula to form this larva is that the blastopore moves inwards together with the neighbouring ectoderm cells, meaning that the larval mouth does not coincide with the blastopore. In this pilidium larva, the blastopore remains open during the larval period. Another change is that the blind end of the **archenlerm** turns towards one side of the embryo as it elongates. This side will be the future posterior end, and the opposite side will form the anterior or head end.

About the mesoderm, more than one view was encountered concerning the formation of the mesoderm. The only full accounts concerning the formation and development of mesoderm are those that have been published as early as the 20th century on *Lineus ruber* and *Malacobdella grossa*. No agreement in respect to the development of the **coelom** has been put forward. However, Iwata (1968) reported that the formation of mesoderm in *Lineus ruber* was chosen to be recorded. At the 64-cell stage, one blastomere of the fourth micromere quartet, the 4d blastomere, which is especially large and further distinguishable by its colour, is recognized to be the mesoderm mother cell (Fig. 8.4K). This blastomere later

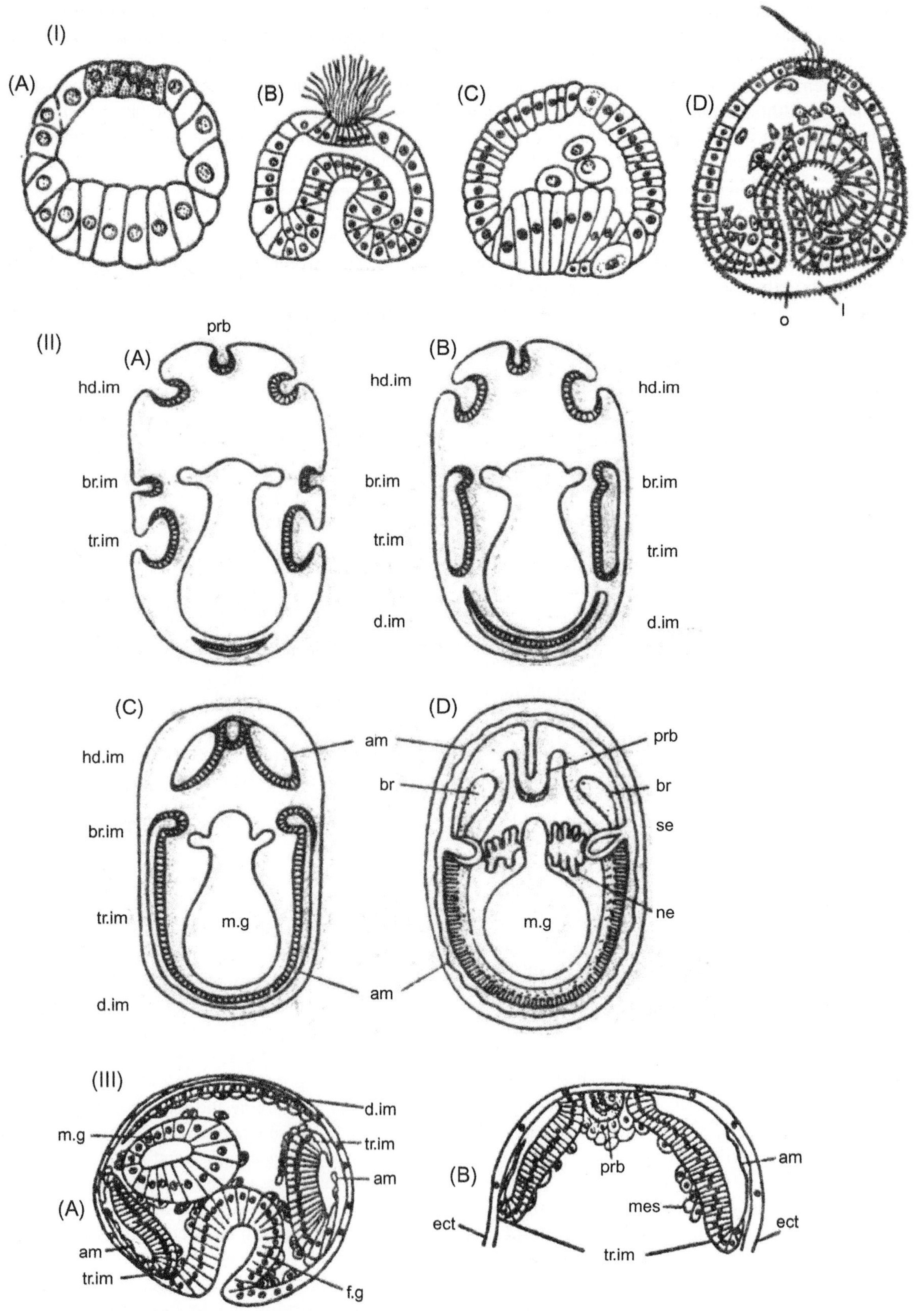

FIG. 8.5 (I) (A) blastula; (B) gastrula of *Cerebratulus lacteus (after C.B. Wilson)*; (C) beginning of gastrulation; (D) early pilidium larva of *C. marginatus* (Coe), ap=apical plate, l=lateral lobe, o=mouth. (II) diagrammatic drawings of pilidium larva during metamorphosis (Korschelt u. Heider), am=amnion, br=brain, br.im=imaginal disc of brain, d.im=dorsal imaginal disc, hd.im.=imaginal disc of head, m.g=mid gut, ne=nephridium, prb=proboscis rudiment, se=head sensory organ, tr.im=imaginal disc of trunk. *(From Iwata, 1968.)* (III) formation of blastodisc in Desor's larva (Arnold) am=amnion, d.im=dorsal imaginal disc, ect=ectoderm, f.g=foregut, m.g=mid-gut, mes=mesoderm, prb=proboscis rudiment, tr.im=imaginal disc of trunk. *(All are from Iwata, 1968.)*

becomes two cells situated on opposite sides of the blastopore. Like their counterparts in annelids and moluscs, they proliferate inwards to form strands, extending into the blastocoel, in which the body cavity later develops. Iwata added that according to Wilson (1900), in *Cerebratulus arginatus* a pair of large cells can be distinguished in the late blastula stage before gastrulation begins, lying on opposite sides of the flattened vegetal surface. These later proliferate inwards and become mesenchyme, instead of forming strands. Moreover, they do not arise by division of one mesoderm mother cell, but are mesoderm cells which originate from widely separated ectoderm cells. From this, it appears that Iwata reviewed the formation of the mesoderm to recognize the development of any cavity, but no definite result was obtained.

However, Iwata described the structure of the larva and its metamorphosis, but because of the great differences between the tools and technical procedures at the time of Iwata and the present days, only the pilidium larva is described.

The structure of pilidium larva

The pilidium (Helmet Larva), first described by Müller in 1847, looks nothing like the juvenile nemertean worm. The species that have been found to pass through the pilidium stage are: two species of *Cerebratulus*; two species of ***Micrura***; and one species of *Lineus*. In addition, four species of this type of larva discovered in plankton hauls have been named, although the adult species remains unidentified (Fig. 8.6A–F). However, Norenburg and Stricker (2002) added descriptive illustrations of other different Pilidia (Fig. 8.6G–K). Pilidia belonging to different species vary in shape, size and colour, but few have been matched to their respective adults.

The Pilidium, 1 day after the larva had escaped from the egg membrane and begun to swim, is characterized by the following: the blastopore sinks inwards and the mouth is formed external to it. The region between the mouth and the blastopore forms the **foregut**, which will become the oesophagus of the adult. The part extending crookedly posterior is the gut, which will be the intestine of the adult. The cells of the foregut are flat, like those of the ectoderm and carry short cilia. A valve develops near the blastopore and regulates the entrance and exit of food particles. This region is formed of a small number of nonciliated **amoeboid** cells, which send out processes to make the valve aperture narrower, or widen it by retracting them.

The pilidium shape is achieved within 1–2 weeks of escaping from the egg membrane and it is characterized by the following:

1. The ectoderm on the two sides of the oval mouth opening extends downwards to form the lateral lappets. The cells along the lower edges of these are large and crowded with long cilia. In some species such as *Cerebratulus lacteus* and *Pilidium gyrans*, the lateral lappets are extremely conspicuous. In other species such as *Micrura caeca*, the lateral lappets are not so noticeable.
2. In *Micrura caeca*, the dorsal ectoderm is umbrella-shaped, causing the back of the larva to be broader than the ventral side (Fig. 8.6C).
3. On the posterior wall of the mouth, a protruding lip bearing especially long cilia extends forwards and downwards. The ingested food enters the mouth through the anterior part, while the undigested material is expelled to the outside, along the back wall in the region of the lip.
4. Mesenchyme cells can be recognized, at an early stage, suspended in the transparent body fluid filling the body cavity between the wall of the body and that of the gut. By developing of the pilidium stage, they have differentiated into amoeboid mesenchyme and muscle cells bearing long muscles fibres. These long slender fibres of the muscle cells join together the various parts of the larval body and direct its movements.

Metamorphosis of pilidium larva

The description of this vital activity in the work of Iwata (1968) was written in the present accounts very briefly (in points) to simplify the understanding. Metamorphosis begins by the invagination of the ectoderm at **seven or eight** positions around the circumference of the lower part of the umbrella. Six of these are paired and two are unpaired (Fig. 8.5(II) A). The paired ones are: the **cephalic discs** lying at the two sides of the head end; the **trunk** discs beside the **gut**; and the **cerebral discs** located between the other two pairs beside the foregut. The unpaired ones are the **proboscis disc** at the anterior-most part of the head end; and the **dorsal disc** at the posterior end. The proboscis disc had been recognized only in the Desor larva, not having been confirmed in pilidium larvae at that time.

Formation of the prospective body wall. During metamorphosis, these invaginations (discs) become connected to form another secondary new body wall surrounding the gut. This process is completed when cilia appear on this new body wall and the larval body wall is burst and cast off. How does this occur?

Firstly, the invaginations appear as hollows in the larval body wall, which gradually become deeper and at the same time elongate parallel to the body wall, while their openings eventually close. Then new body wall appears as pouches inside the larval wall (Fig. 8.5(II) B).

- The inner walls of the pouches are made up of deeply columnar cells. Later, they become many-layered and form the adult **epithelium**.
- The pouch is then called the **imaginal** disc or blastodisc, the outer wall of which is composed of a very thin simple layer of flattened cells called **amnion** (Fig. 8.5(II) C and D).
- Each blastodisc encloses a cavity called amniotic cavity.
- The dorsal disc is not formed by invagination, but it is formed by **delamination**. A few cells of the posterior body wall divide and proliferate, and arrange themselves forming this dorsal blastodisc.

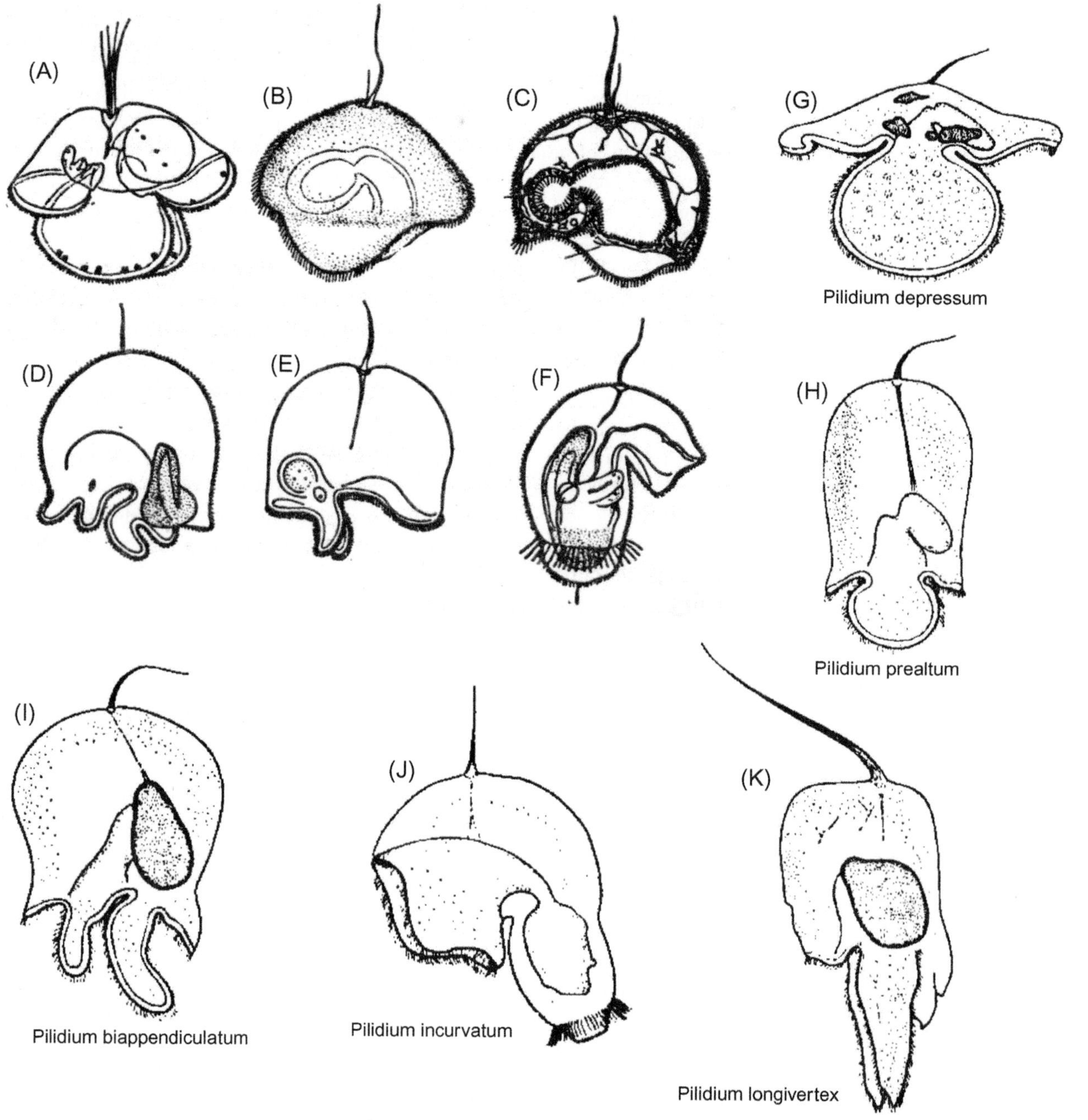

FIG. 8.6 Diagrams showing pilidium larvae of different species: (A) *Cerebratulus lacteus*; (B) *C. marginatus*; (C) *Micrura caeca*; (D) *Pilidium brachiatum*; (E) *P. auriculatum*; (F) *P. recvatum*; g=gut, l=lateral lobe, o=mouth *((A–F) from Iwata, 1968)*; (G–K) the name not according to species but to the form of the larva *(from Norenburg and Stricker, 2002)*.

- Early during the formation of these blastodiscs, mesenchyme cells arrange themselves in a single layer over the inner face of the blastodisc facing the body cavity. The latter occupies the distance between the blastodiscs and the gut. As development proceeds, this mesenchyme layer becomes multilayered, giving rise to the musculature of the body wall, muscles of the proboscis and the mesenchyme of the body fluid.

Fusion of the discs. The development of the several blastodiscs and the order of their fusion are as follows:

- The left and right cephalic discs appear first, grow anteriorly and fuse together. Simultaneously, they unite with the proboscis disc.
- The **trunk** discs gradually extend posteriorly and **ventrally**. They then join each other at the mid-ventral line.

During this time, they also fuse anteriorly with the cerebral discs.

- The dorsal blastodisc appears **late**, then unites with the trunk disc since it advances upwards on either side.
- At the same time the cephalic and cerebral unite at the sides of the body in the anterior part.
- As this process continues, the dorsal disc spreads forwards and sidewise. Finally, it joins the cephalic discs on either side.

Development and/or structure of the organs

Proboscis. Independent of the body wall formation process, the proboscis disc forms a pouch in the upper part of the body cavity and gradually grows **anteriorly**. In some species, this proboscis disc is not recognized, and Salensky proposed that the proboscis takes its origin in an invagination of the upper layer of the **blastodisc**, formed by the union of the left and right cephalic discs. At any rate, the proboscis disc becomes the simple-layered proboscis epithelium. The mesodermal mesenchyme attached to its inner side forms a many-layered **proboscis sheath**. This undergoes delamination and differentiates inwards as proboscis muscle. The space between the layers develops into the rhynchocoel.

Nervous system. The brain, formed on both left and right sides by the proliferation of ectodermal cells, which appear in the cephalic discs, sinks into the body cavity. It gives off the ventral ganglion below, and at the same time the left and right **ganglia** are connected by the dorsal and ventral **commiseres**. Since the lateral nerves are formed well before the union of the dorsal with the cephalic blastodiscs, they grow out of the ventral ganglia and gradually extend posteriorly, running between the blastodisc and the mesenchyme. The cerebral discs do not participate in the formation of the body wall. Later, they become tubular and extend into the body cavity. The cells at their tips proliferate, forming the cerebral sense organs and making connection with the dorsal ganglia.

Body wall. The regions of the adult epithelium formed by each of the respective blastodiscs are:

- anterior tip of head to posterior end of lateral cephalic grooves-cephalic discs;
- ventral side-trunk discs; and
- dorsal and lateral body walls-dorsal discs.

Nephridia. A pair of ectodermal invaginations appear between the larval mouth and the cerebral discs (Fig. 8.5(II) D). The openings persist until a late developmental stage. Later, they are closed off and **efferent** ducts are secondarily formed. A recent view (Norenburg and Stricker, 2002) mentioned that Pilidia and related larvae lack nephridia until these are formed in the juvenile, as reported for *Micrura alaskensis* (Iwata, 1958). It is not known when the nephridia of palaeo- and hoplonemertean larvae develop.

Muscles. The mesenchyme which is massed against the blastodiscs becomes multilayered and forms the circular and longitudinal muscle layers of the body wall as well as the mesenchyme of the body fluid. However, there are different views concerning the origin of the outer longitudinal muscle layer and the dermis (cutis), which lies between this and the epithelium. Some believe that they have an ectodermal origin in the pilidium larva, while others state that they arise from the mesoderm in their larvae.

Body cavity. From the above description, it can be thought that the body cavity of the adult stage takes its origin from the cavity of the blastula, i.e. **blastocoele**. There is no general conclusion concerning the relation between the body cavity and the rhynococoel or the circulatory system.

Alimentary canal. The central part of the foregut, between the larval mouth and the blastopore which makes the border between the foregut and the intestine, is constricted by the body muscles, and after **moulting** the adult mouth is formed there. A considerable time after the completion of adult epithelium formation, the intestine unites with the body wall and an opening is formed, which is the anus. Norenburg and Stricker (2002) reviewed and discussed the formation of the anus. They reported that although all adult nemerteans have an anus, and its presence has been of central importance in discussions of nemertean evolution, little is known about timing of anus formation in planuliform larvae or pilidia Juveniles. An anus is not evident in planuliform larvae until the other adult structures such as cerebral ganglia and proboscis are clearly differentiated.

Micrura alaskensis

Very recent studies, in respect to that of Iwata, on the development of the pilidium larva of the heteronemertean *Micrura alaskensis* from fertilization to metamorphosis were carried out by Maslakova (2010). This species, *M. alaskensis*, is a common Northwest Pacific coast intertidal nemertean (Fig. 8.1B), and its larva, like those of other nemerteans, is ciliated planktotrophic organism. It is one of the most characteristic and unmistakable representatives of marine plankton. Maslakova collected reproductive adults of *M. alaskensis* during negative low tides in False Bay (San Juan Island, Washington, United States) in July–August 2005, June–August 2006 and July 2007. Occasionally, ripe individuals were encountered outside the usual summer reproductive period, e.g. in February 2007 in False Bay in Charleston. Through the body wall, males and females can be distinguished under the dissecting microscope. Several hundreds to thousands of eggs can be obtained from a single female. Upon contact with seawater, naturally spawned eggs soon undergo germinal vesicle breakdown. These eggs can be fertilized by a dilute suspension of sperm obtained by dissecting male individual. Developing embryos were cultured and Maslakova successfully reared dozens of cohorts of pilidium larvae of *M. alaskensis* to metamorphosis.

Embryonic development

Maslakova (2010) described the oocyte of *M. alaskensis* as 'frisbee-shaped', relatively opaque and possess conspicuous nucleus without a distinct egg envelope, measuring about 76 µm in diameter (Fig. 8.7(I) A). It rounds up and undergoes maturation division within 30–40 min of contact with seawater. At 11°C the first polar body is formed (Fig. 8.7(I) B), and the second polar body is formed about 1 h after fertilization (Fig. 8.7(I) C). It is surrounded, as mentioned above, by a transparent jelly layer. Maslakova mentioned that the latter was easily removed by pipetting or repeatedly passing the eggs through a 'nitex mesh' with a mesh size slightly larger than the egg diameter.

Cleavage

The fertilized egg undergoes, as in other nemerteans, spiral cleavage, equal and holoblastic (Fig. 8.7(I)). The **first** cleavage occurs about 2 h and 15 min after fertilization (Fig. 8.7(I) D–F). The initial cleavage furrow is broad (Fig. 8.7(I) D) and the blastomeres separate almost completely (Fig. 8.7(I) E), then re-compact (Fig. 8.7(I) F). The second cleavage occurs about once an hour. The third division is dextral (Fig. 8.7(I) H) and, as in other nemerteans, the first quartet micromeres are slightly larger than the macromeres in the eight-celled embryos (Fig. 8.7(I) I).

Blastula. The blastula of *M. alaskensis* is flattened along the animal-vegetal axis, and square-shaped, that the author referred to as a 'blastosquare' (Fig. 8.7(II) A). The blastosquare has a small blastocoel; this becomes ciliated and starts swimming as early as 16 h after fertilization at ambient sea temperature (11–14 °C). By 24 h of development, **swimming gastrulae** are radially symmetrical, and possess an apical tuft at the animal pole and a prominent invagination at the vegetal pole (Fig. 8.7(II) B). By the end of the second day of development, the **young pilidium** larva exhibits bilateral symmetry, as the developing gut is inclined towards the dorsal side (Fig. 8.7(II) C).

Pilidium larva

The young helmet-like pilidium larva has a long apical tuft, two lateral lappets and a band of longer cilia spanning the lappets. Its gut is a blind 'sack' differentiated into the funnel-like thin-walled oesophagus (or vestibule). The posterior wall of the oesophagus (with respect to the axis of the future juvenile worm) has a pair of ciliated ridges (Fig. 8.7(II) E). The oesophagus leads to the thick-walled stomach, with a muscle sphincter in between (Fig. 8.7(II) D). This pilidium is capable of feeding on unicellular algae as early as 66 h after fertilization. During the growing of the pilidium, the lateral lappets also grow more and more, and two additional lobes develop, which are usually referred to as the anterior and the posterior (in respect to the prospective juvenile) (Fig. 8.7(II) F).

Juvenile development in *M. alaskensis*

The typical pilidium larva possesses three pairs of imaginal discs, which develop as ectodermal invaginations: the cephalic discs, the trunk discs and the cerebral organ discs. The juvenile rudiments always appear in the same sequence: first a pair of cephalic discs, then a second pair of trunk discs. The third step is the unpaired proboscis rudiment. Finally, a pair of cerebral organ discs and the unpaired dorsal rudiment appear (Fig. 8.7(II) G). Maslakova observed differences in the timing of the appearance of these discs, and not in the sequence of their appearance. The cephalic and the trunk discs appear as early as 9 days after fertilization, although, more typically, they appear at 14–15 days. The proboscis rudiment appears at about the same time as the cerebral organ discs, as early as 14 days of development, but more typically around 20 days. However, fusion of discs occurs: first the proboscis rudiment fuses with the cephalic discs, forming the head rudiment; second, the cerebral organ discs fuse with the trunk discs (Fig. 8.7(II) H). At the same time, a small dorsal rudiment appears just below the stomach and dorsal to the trunk discs (Fig. 8.7(II) G). Initially, this rudiment is clearly separate from the trunk discs, but later fuses with the trunk discs contributing to the trunk rudiment. As another fusion, this trunk rudiment fuses with the head rudiment, forming a 'toroid' juvenile rudiment (Fig. 8.7(II) I). Maslakova observed this 'torus' stage as early as 28 days after fertilization, although development to this point may take 5 weeks or longer. In addition, the proboscis rudiment elongates and reaches beyond the margin of the juvenile head, stretching out along the dorsal side of larval oesophagus towards the stomach (Fig. 8.7(II) J). The dorso-posterior wall of the juvenile rudiment gradually grows over the larval stomach and overlaps the proboscis. The juvenile rudiment then most resembles a 'slipper shoe' with a hole in the middle of its 'sole' (Fig. 8.7(II) K), and its dorsal wall edge continues growing anteriorly, covering up the proboscis. It then fuses with the margin of the head rudiment, thus closing the dorsal gap and completing the juvenile (Fig. 8.7(II) L).

Typically, metamorphosis of pilidia occurs at 6–7 weeks, but in cultured pilidia of *M. alaskensis* occurred as early as 35 days after fertilization.

Juvenile rudiment development

As mentioned above, larvae of *M. alaskensis* have three pairs of imaginal discs. The cephalic discs develop first as invaginations of pilidial epidermis just apical of the ciliated band at the junction of pilidial lateral lappets and the anterior lobe—one on each side. The cerebral organ discs and the trunk discs invaginate below the ciliated band, and so separate from the larval epidermis. Once the invaginated rudiments of cephalic and trunk discs separate from the larval epidermis, they consist of two layers separated by a narrow

I

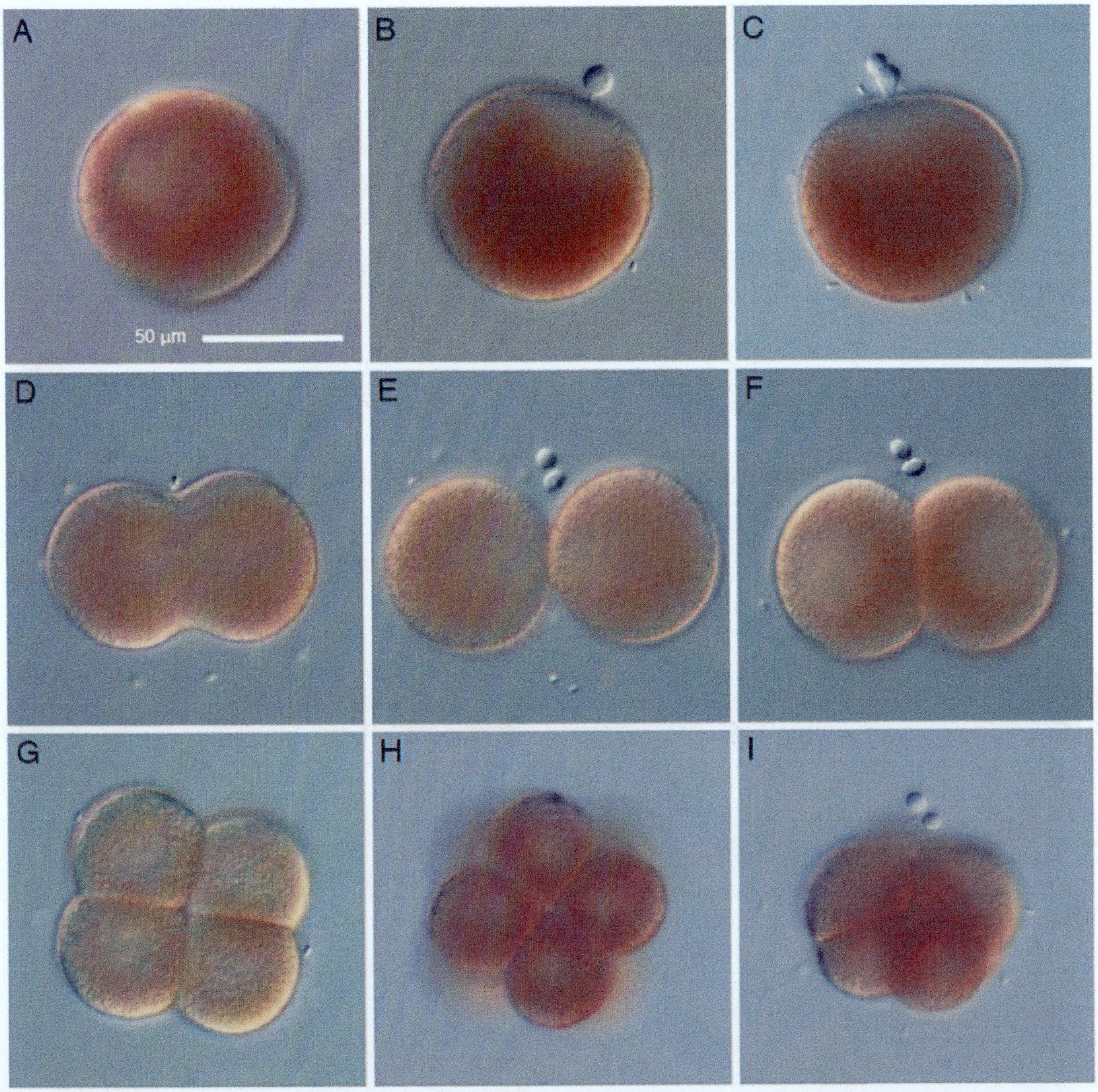

FIG. 8.7 (I) DIC images showing early cleavage in *Micrura alaskensis*: (A) unfertilized egg undergoing germinal vesicle breakdown; (B) fertilized egg with the first polar body; (C) second polar body appears; (D–F) first cleavage producing two-cell stage, there is no egg chorion and so the blastomeres are separate; (F) the blastomeres recompact; (G) polar view of the 4-cell stage; (H) vegetal view of an 8-cell stage with dextral spiral cleavage; (I) lateral V. of 8-cell stage. Apical pole, marked by polar bodies, is up. Animal micromeres are larger than the vegetal macromeres. *(From Maslakova, 2010.)* (II) (A–D) DIC images of development of *Micrura alaskensis* from blastosquare to young pilidium: (A) ciliated blastosquare (18–20 h after fertilization); (B) gastrula with an apical tuft (ap) (27-h-old); (C) 2-day-old pilidi um (49–53 h old) showing bilaterally symmetrical gut (gt) and a long apical tuft; (D) 3-day-old pilidium has a blind gut differentiated into oesophagus(es) and stomach (st), a prominent apical tuft, ciliated band (cb), and nascent lateral lappets (lp). The young pilidium increases in size even before onset of feeding due to the expansion of the blastocoel; (E–I) DIC images of lateral views showing onset of feeding, and early juvenile development in *M. alaskensis*, apical plate up: (E) 4-day-old feeding pilidium features an apical plate (ap), extensive blastocoel (bl), oesophagus(es) with a pair of ciliated ridges (cr), stomach (st), and circumoral ciliated band (dotted line) spanning the lappets (lp); (F) the second pair of imaginal discs, called trunk discs (td), appear in vicinity of the ciliated band at the posterior margin of the lateral lappets as early as 9 days of development. At this stage, the anterior larval lobe (al) and posterior lobe (pl) are well developed; (G) the unpaired dorsal rudiment (dr) appears between the stomach and the pilidial epidermis in diametrically opposite position to the proboscis rudiment as early as 24 days of development. Note the pilidial posterior cirrus (pc); (H) head and trunk stage. The cephalic discs fuse with each other and the proboscis rudiment to form the head rudiment (hr). The cerebral organ discs together with the trunk discs and the dorsal rudiment form the trunk rudiment (tr); (I) torus stage is characterized by a doughnut of juvenile tissue (juv) around the larval gut formed by the fused head and trunk rudiments; (J–L) DIC images of lateral view of Advanced stages of juvenile development inside pilidium of *M. alaskensis*: (apical plate up, juvenile anterior left) (J) complete proboscis stage is characterized by proboscis (pb) reaching the stomach (st); (K) hood stage is characterized by the anterior margin of the juvenile dorsal epidermis overlapping the proboscis *(arrowhead)*; (L) a complete juvenile inside the pilidium larva. At this stage the larva is competent to undergo metamorphosis. *(From Maslakova, 2010.)* (III) Cerebral organ

(Continued)

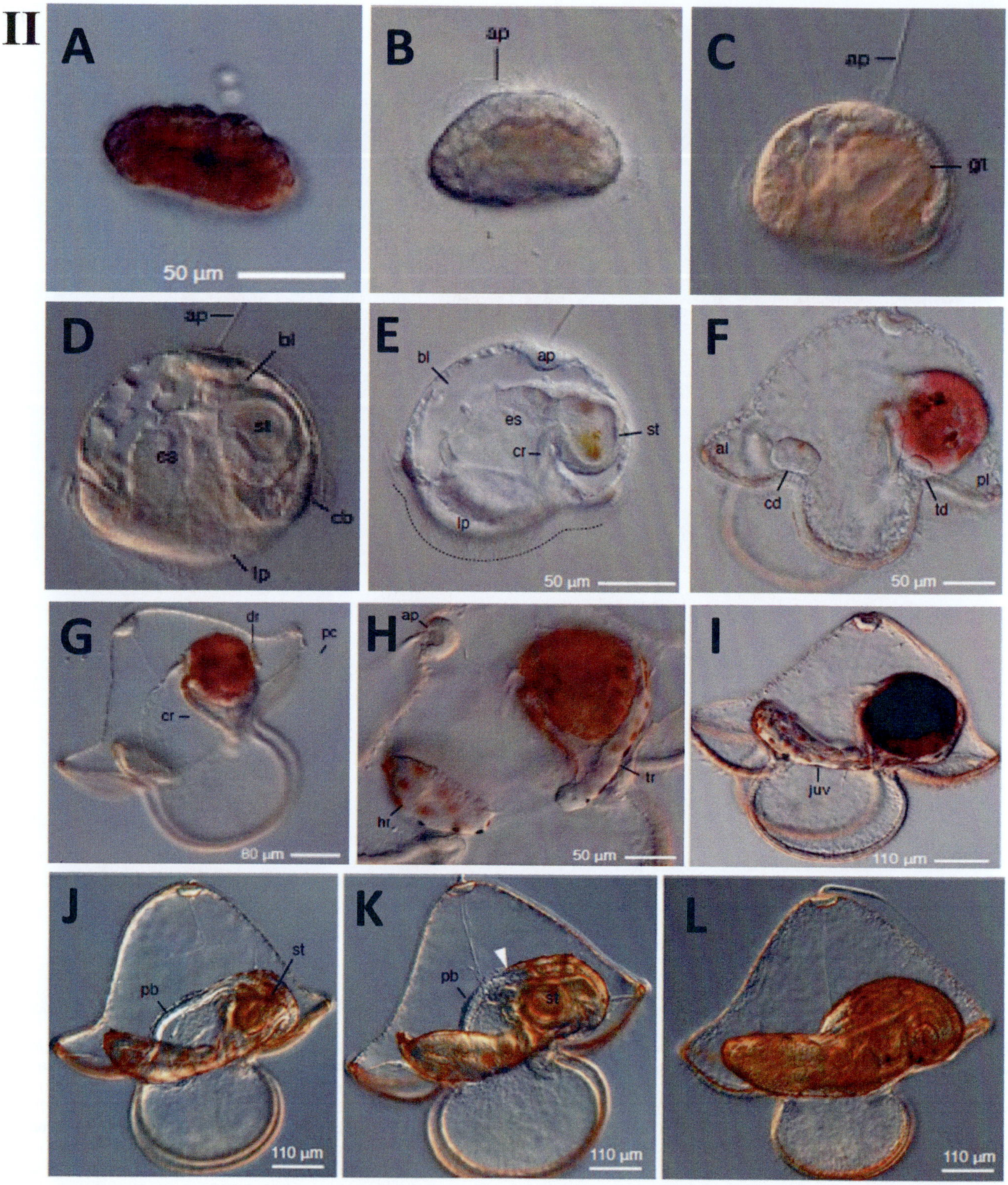

FIG. 8.7, CONT'D disc invaginations in a 17-day-old pilidium of *M. alaskensis*: (A, A′) Confocal images of two different pilidium larvae stained with phalloidin *(white)* and hoechst *(orange)*. (A) A z-projection of the entire larva; side view. Apical organ (ap) is at upper right, anterior lobe to the left, stomach (st) to the right. The left cephalic disc (cd), the left cerebral organ disc *(arrowhead)*, and the left trunk disc (td) are discernible as dense clusters of nuclei. (A′) A substack (6 μm thick) of the same larva, that highlights the left cerebral organ disc *(arrowhead)*, which originates as an invagination of subumbrellar pilidial epidermis, on the inner side of the lateral lappet (lp) in close proximity to the ciliated ridge (cr). The section also passes through the left trunk disc, the left cephalic disc, the proboscis rudiment (pb), which at this stage already consists of multiple cells, and the stomach; (B) diagram of the pilidium larva in lateral aspect; (C, D) DIC images showing development of the proboscis in *M. alaskensis*: (C) proboscis rudiment *(arrowhead)* first appears as a small and discrete cluster of cells in the plane of bilateral symmetry of the larva between the two cephalic discs (cd); (D) similar developmental stage as on (A): the proboscis rudiment *(arrowhead)* is separate from the cephalic discs; (E) the proboscis (pb) continues to elongate along the larval oesophagus(es) and reaches beyond the margin of the juvenile head (jh) at extended proboscis stage; (F) the distal tip of proboscis *(arrowhead)* reaches the stomach *(dark magenta shape)* at complete proboscis stage. *(From Maslakova, 2010.)*

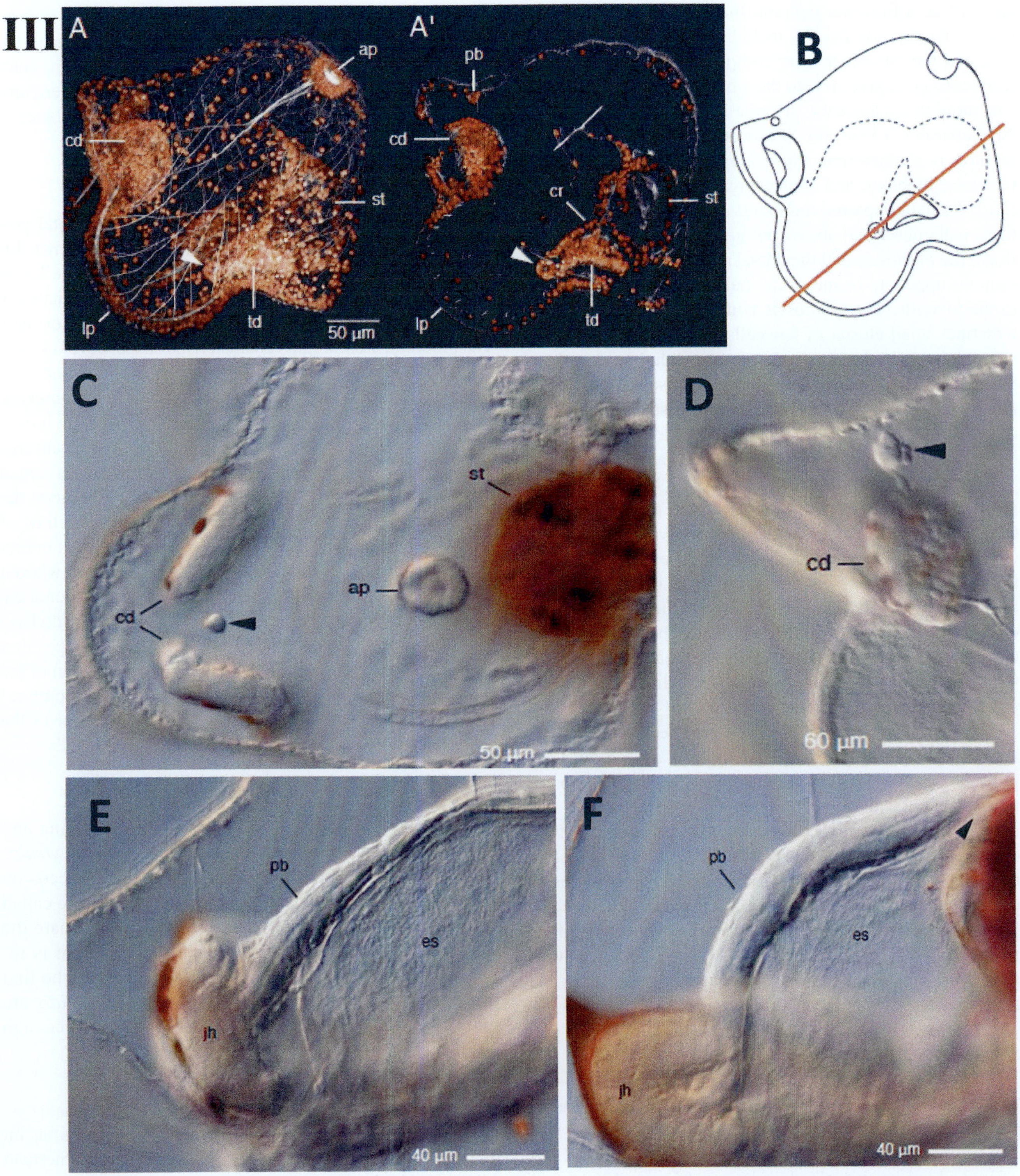
III
A
ap
cd
st
lp
td
50 µm
A'
pb
cd
cr
st
lp
td
B
C
st
ap
cd
50 µm
D
cd
60 µm
E
pb
es
jh
40 µm
F
pb
es
jh
40 µm

FIG. 8.7, CONT'D

cavity, like a flattened balloon: the thick inner layer gives rise to the juvenile tissues, including the definitive epidermis; and the thin outer layer composed of a single layer of squamous cells gives rise to the amniotic 'sack', or amnion. The amnion encloses the juvenile inside the pilidium larva and separates it from the blastocoel. In *M. alaskensis*, the amnion is characteristically decorated and defined by the presence of black and brown pigment granules (Fig. 8.7 (II) J–L). Two unpaired rudiments appear, in addition to the three pairs described above, but not as epidermal invaginations: the proboscis and the dorsal rudiments. These two rudiments appear to be mesenchymal in origin and contribute to the juvenile. The **proboscis rudiment** is first evident as a distinct small cluster of few cells in the plane of bilateral symmetry of the larva—between the pilidial epidermis of the anterior lobe and the two cephalic discs (Fig. 8.7(III) A′, B, C, D). The proboscis rudiment fuses with the two cephalic discs, and forms a distinctly bilayered proboscis bud inside the head rudiment at the 'head and trunk' stage. The proboscis bud elongates and protrudes beyond the margin of the head rudiment at the 'extended proboscis' stage (Fig. 8.7(III) E). The proboscis follows along the larval oesophagus in the larval mid-plane and reaches the larval stomach at the 'complete proboscis' stage (Fig. 8.7(III) F). At the diametrically opposite end from the proboscis, the **dorsal rudiment** appears. It is enclosed between the epidermis of the posterior lobe and the stomach (Fig. 8.7(II) G). It subsequently fuses with the trunk discs to form the trunk rudiment (Fig. 8.7(II) H).

The nervous system. 'Serotonergic' neurons were observed in the young pilidium larva as early as 40h after fertilization, even before the onset of feeding. Two groups of neurons can be distinguished: several apical neurons, and several associated with the developing ciliated band. In a fully developed pilidium larva, this system includes numerous serotonergic neurons connected by the marginal ciliary nerve associated with the ciliated band, as well as numerous neurons that form an extensive subepidermal nerve net. On each side of the apical plate, two monociliated serotonergic neurons are always associated. In the sphincter between the larval oesophagus and the stomach, the circular oral nerve can be distinguished.

This work of Maslakova (2010) gives scientifically significant results:

1. New information concerning metamorphosis of pilidium of *M. alaskensis*, with images and photos of live specimens, taken by advanced procedures, that clarified and added much to the description of Iwata (1968).
2. The data presented by Maslakova stated new identification of the sequence of the developmental stages of a nemertean as follows: blastosquare, gastrula, young pilidium, feeding pilidium, cephalic discs, trunk discs, cerebral organ discs, head and trunk, torus, extended proboscis, complete proboscis, hood, metamorphosis.
3. Because the development of pilidia in culture is typically asynchronous, it makes little sense to refer to the larval stages by absolute age. To facilitate future studies of pilidial development in this and other nemerteans, Maslakova proposed and tabulated the characteristics of each stage to facilitate their identification.

Desor larva

Iwata (1968) reported that although the developmental processes of the pilidium and Desor larvae are somewhat different, their mode of metamorphosis is the same. The Desor larva grows by digesting and absorbing the yolk of its own, differing from the pilidium in that it does not swim or feed. The pilidium, assuming its characteristic shape, develops as it swims free in the sea. It feeds and **moults** after metamorphosis, taking on the adult form. The author (Iwata) has observed that the larva of *Micrura akkeshiensis* swims although it does not have the pilidium morphology, and digests and absorbs the egg yolk (like the Desor larvae). It forms the various organ systems and results. However, unlike the pilidium larva, the Desor larva does not form a complete **amnion** at the time of blastodisc formation. After the several amniotic membranes separate from the larval body wall, they disappear without uniting into a common membrane (Fig. 8.5(III)). Second, the archenteron, which is formed by invagination of the endoderm at gastrulation, closes temporarily. Following this, the position of the blastopore is moved inwards by the invagination of the ectoderm. Then, the foregut develops between the mouth region and the blastopore. The oesophagus and intestine of the adult are formed secondarily, at the expense of the yolk.

Schmidt's larva

As mentioned above, the Desor's larva is nonfeeding and is a larval stage of the heteronemerteans *Lineus viridis*. However, *L. ruber* forms a Desor's larva, which feeds on the undeveloped eggs inside the capsule, so they are called 'nurse eggs', or it feeds on aborted embryos that share the common cocoon. This unusual feeding larval stage is referred to as Schmidt's larva, after the investigator who first discovered this behaviour. The eggs of *Lineus viridis* are much larger than those of *L. ruber* and so they have enough nutrients to supply the developing larva.

Iwata's larva

Iwata's larva was first discovered in 1958 by Iwata, as an intermediate stage in the life cycle of the heternemertean *Micrura akkeshiensis*. It is nonfeeding similar to the Desor's larva, and also forms imaginal discs. However, it differs in forming only five ectodermal invaginations. These include two cephalic, two trunk and a single dorsal imaginal discs. Iwata's larva differs from Desor's in having an apical organ and an elongated apical tuft. In this, it is similar to the pilidium larva, and also in being a free-swimming larva. While the juvenile worm maintains the same plane of bilateral symmetry and dorso-ventral axis as that of Iwata's larva, the worm's anterior–posterior axis is reversed by 180 degrees. During metamorphosis, the juvenile worm sheds the larval integument as it crawls out of the amniotic cavity.

Other members of the genus *Micrura* form pilidium larva. Thus Iwata himself argued that the larva of *M. akkeshiensis* represents an intermediate between that of the pilidium and Desor's larva.

Lineus ruber (Schmidt's larva)

The Nemertea is a group of spiralians that exhibits a broad diversity of larval forms, including the iconic pilidium. A pelagic planktotrophic pilidium is the ancestral form in the Pilidiophora, but several lineages exhibit deviations of this condition, mostly as a transition to pelagic lecithotrophy. The most extreme case occurs, however, in the Pilidiophoran ***Lineus ruber***, which exhibits an adelphophagic intracapsular pilidium, the so-called Schmidt's larva. Martín-Durán et al. (2015) studied development and metamorphosis of the Schmidt's larva of *L. ruber* by confocal laser scanning microscopy and gene expression. The authors found gravid specimens under stones in the coast near Bergen (Fanafjorden; GPS coordinates) Norway. They collected the specimens during day low tide in March and April 2014 and kept them in aquaria with constantly aerated filtered seawater (FSW) at 14°C, under a day-night cycle. By suitable feeding and under careful maintenance of the culture, the animals laid egg masses spontaneously, and the authors could obtain them daily for their studies.

L. ruber is a dioecious species and the mature female animal has multiple individual ovaries distributed along two lateral rows (see Fig. 8.8(I) A) (typically extended from shortly behind the head region to near the posterior end of the body). Each ovary connects to the epidermis through a small gonoduct and a gonopore (Fig. 8.8(I) A). Martín-Durán et al. (2015) did not observe the process of fertilization, but they reported that it occurs internally after pseudocopulation, and oviposition occurs spontaneously. Eggs are released through the lateral gonopores packed into multiple pyriform egg capsules. During the release of the eggs, epithelial glands secrete an enclosing jelly. The female individual progressively glides out of the jelly and lays the gelatinous cocoon. Within these cocoons, the eggs undergo development (Fig. 8.8(I) B). The authors reported that this culture of *L. ruber* was kept under their lab conditions over a whole year and the animals were still able to reproduce. This means that *L. ruber* can display iteroparous reproduction, at least in captivity. The eggs are rich in yolk content and the authors observed that the eggs undergo a stereotypical spiral cleavage that results in the formation of a blastula 4 days after oviposition (Fig. 8.8(II) A). However, they reported that during cleavage, the eggs underwent abnormal cell division and they arrested at this early stage of embryogenesis. Gastrulation begins at 6 days of development, and by invagination, the embryo attains an archenteron and a rounded blastopore (Fig. 8.8(II) B′, B″). The embryo firstly becomes bilateral symmetrical after 10 days of development (Fig. 8.8(II) C′, C″) with somewhat elongated anterior end. The intracapsular Schmidt's larva forms 12 days after oviposition (Fig. 8.8(II) D′, D″).

Structure of Schmidt's larva

The larva is a bilateral symmetrical organism with a more pointed anterior end, the mouth replaces the blastopore and the ectodermal rim cells of the latter appear more developed. On either side (left and right) of both the anterior and posterior ends of the larva, an ectodermal disc appears, and a thin ciliated epidermis connects these four discs, so encloses the larval body. Internally, the archenteron becomes a blind gut. Inside the capsule, the larvae feed on the unviable arrested embryos with which they often cohabitate. As a result of the ingestion of other siblings, the larva increases in size and the blind gut becomes filled with nutrients (Fig. 8.8(II) E′, E″). About 18 days after deposition, Schmidt's larva is elongated along the anteroposterior axis and is more developed with the ectodermal discs extending over the whole body, and it shows a more worm-like shape (Fig. 8.8(II) G′). This stage of development shows the mouth still opens in an antero-ventral position; the original larval epidermis covers the body, but starts to detach from the developing definitive epidermis in some regions of the larva.

Metamorphosis of Schmidt's larva

About 20 days after oviposition, most of the larvae metamorphose into early juveniles (Fig. 8.8(II) G′), which lack the larval epidermis. The authors could not follow the fate of this epidermis and they put three possible cases of its fate:

1. The larval epidermis is ingested by the developing juvenile, as in the sister species *L. viridis* (von Döhren, 2011).
2. It is resorbed by the definitive epidermis.
3. It is simply discarded.

The imaginal discs are no longer obvious in this worm-shaped early juvenile. The head has an anterior terminal proboscis, but the eyes are not yet formed. The gut is still blind but full of yolk and the mouth occupies the same position. These juveniles actively move inside the capsules. Further development and maturation of the tissues and organs are at the expense of the yolk content, and the juvenile progressively adopts the morphology of a small adult (Fig. 8.8(II) H). Juveniles can escape out of the capsule, and glide within the jelly enclosing the egg mass.

About 30 days after oviposition, the free-living juvenile attains the first signs of eyespots and after 40 days, the body becomes pigmented. However, at the moment of hatching, the size and morphology of the juveniles can vary, but most

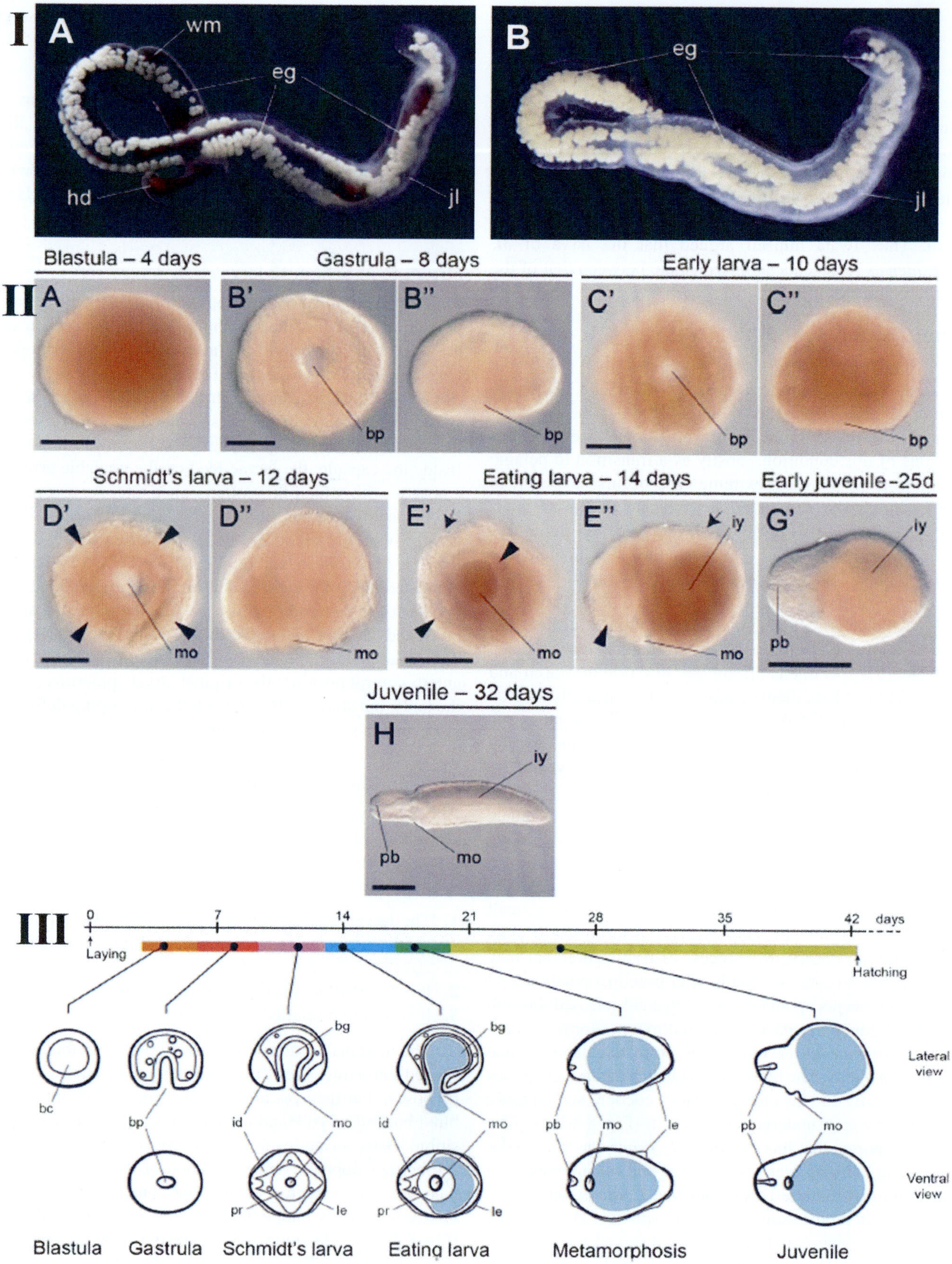

FIG. 8.8 (I) (A, B) Photographs of live specimens taken under the stereomicroscope showing oviposition in *Lineus ruber*: (A) oviposition occurs through the multiple gonopores located on both lateral sides of the female. Several fertilized oocytes are packed within individual pyriform capsules, which are bounded together inside a jelly mass. As the female releases the eggs and secretes the jelly, it glides out of the forming egg string; (B) Laid egg mass containing multiple egg capsules. The entire development occurs inside this mass; e.g. = egg capsules, hd = head, jl = jelly cover, wm = worm. *(From Martín-Durán et al., 2015.)* (II) (A–H) Photographs of fixed dissected embryos of *L. ruber* at representative stages of development (some selected stages):

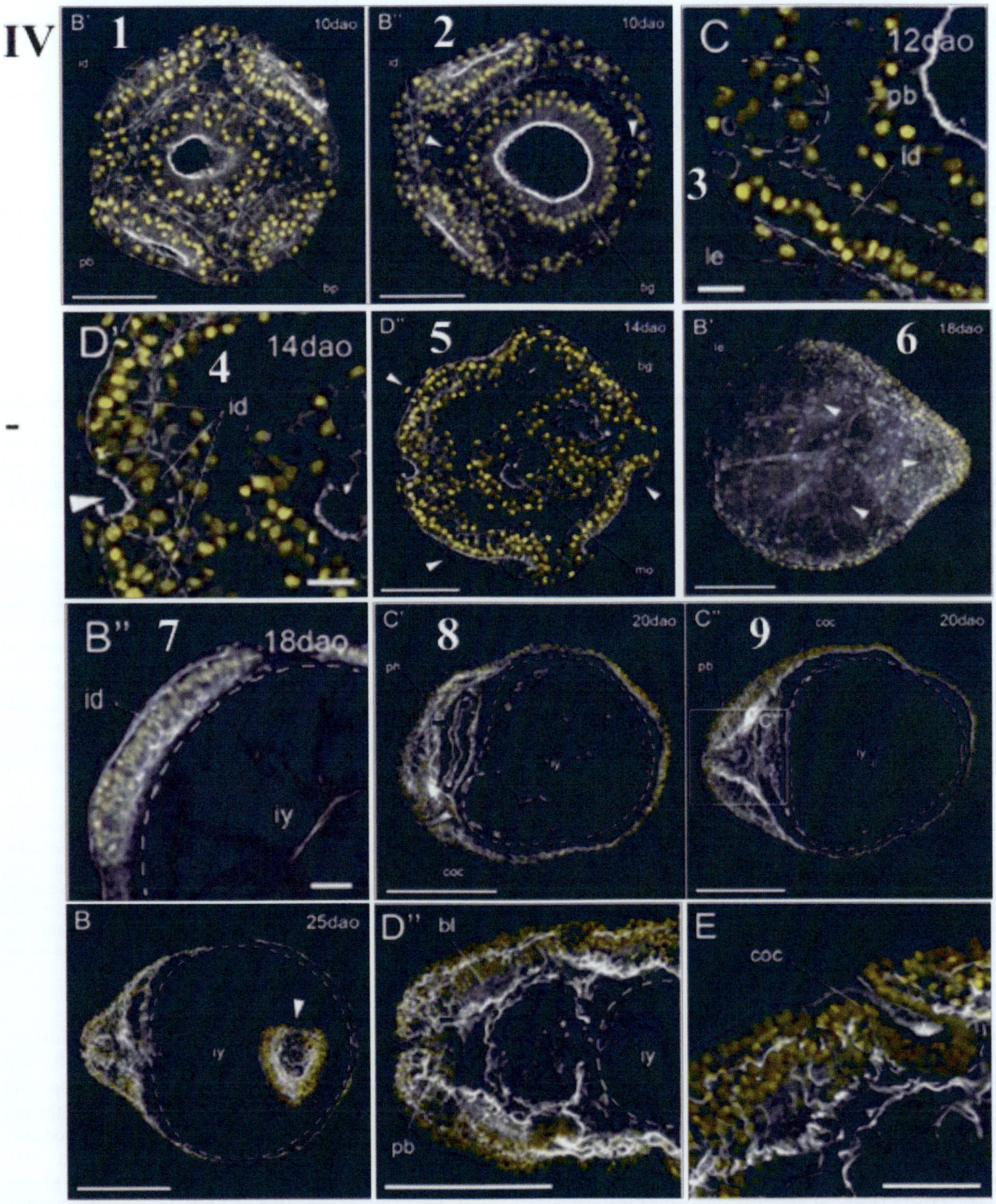

FIG. 8.8, CONT'D (A) a stereotypical quartet spiral cleavage results in the formation of a yolky coeloblastula; (B′, B″) Gastrulation occurs at one pole of the embryo and results in the formation of a central round blastopore and an internal archenteron; (C′, C″) The late gastrula shows the first signs of bilateral symmetry, with the future anterior end pointier than the posterior side, which is more round (more evident in C″); (D′, D″) The Schmidt's larva forms after 12 days of development. It exhibits an anterior pair and a posterior pair of ectodermal imaginal discs *(arrowheads)*, and the blastopore has become the mouth of the larva. A thick ectodermal tissue lines the mouth opening; (E′–E‴) the Schmidt's larva of *L. ruber* can feed on other siblings that occupy the same egg capsule. The ingestion of this yolk material increases the size of the larva and fills up the blind gut. At this stage, the imaginal discs are still visible *(arrowheads)* and a conspicuous epidermis covers the larva *(arrows)*; (G′) After metamorphosis, the juvenile shows a worm-like shape, elongated along the anteroposterior axis but not along the dorsoventral axis, mostly because of the high amounts of yolk present in the blind gut.

(Continued)

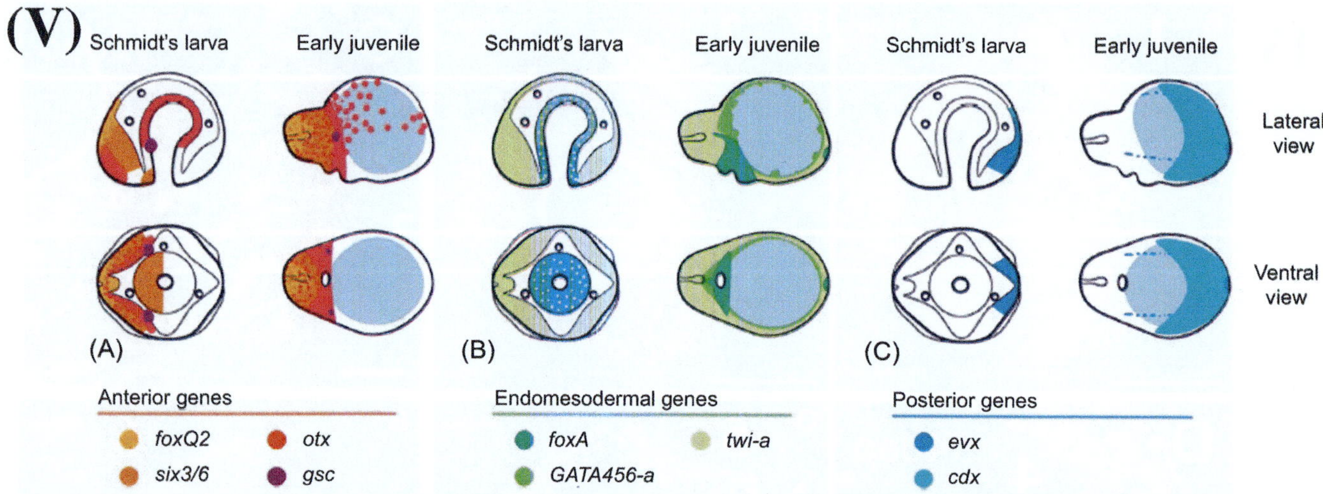

FIG. 8.8, CONT'D The head region is more developed and the rudiment of the proboscis is now evident; (H) 8 weeks after oviposition, the juveniles are morphologically similar to the adults, and have absorbed part of the ingested yolk; bc = blastocoel, bg = blind gut, bp = blastopore, id = imaginal discs, iy = ingested yolk, le = larval epidermis, mo = mouth, pb = proboscis, pr = pharyngeal rudiment. Scale bars (A–D″) 50 μm; (E′, E‴) 100 μm; (G′, H) 500 μm. (III) Schematic summary of the embryonic development of *L. ruber* at 14°C. The light blue depicts the ingested yolk. Drawings are not to scale, lettering as above. *(From Martín-Durán et al., 2015.)* (IV) Formation of the Schmidt's larva in *Lineus ruber*. (1–5) z projections of confocal scans of embryos and larvae labelled against F-actin (*grey*: (B′–B″) After 10 days of development, the embryo shows 4 paired ectodermal discs (marked by *dashed lines* (B′)), an incipient proboscis rudiment, and the ventral ectodermal ring around the blastopore (*dashed circle* (B′)). Internally, the archenteron widens and bends backwards, forming a blind gut. There are still numerous cells in the blastocoel *(arrowheads)*; (C) After 12 days, an anterior indentation forms between the two anterior imaginal discs, and the whole embryo becomes covered by a thin epidermis; (D′–D‴) The mature Schmidt's larva shows 7 separate imaginal discs covered by a larval epidermis (*arrowheads* (D″)); (6–9) metamorphosis of the Schmidt's larva of *Lineus ruber*. z projections of confocal scans of larvae and early juvenile labelled against F-actin *(grey)*: (B′–B″) After 18 days of development, the larvae are more elongated along the anteroposterior axis, and circular and longitudinal fibres extend below the epidermis (*arrowheads* (B′)). The imaginal discs are more developed (B″); (C′–C″) After 20 days of development, the imaginal discs have formed the basic anatomical features of the juvenile. The rhynchocoel occupies the central region of the head, and the mouth opening occupies the antero-ventral side of the animal; (10–12) the early juvenile of *Lineus ruber*. Projections of confocal scans of juveniles labelled against F-actin *(grey)* counterstained with the nuclear marker Sytox Green *(yellow)* 25 days after oviposition (dao): (B) Juveniles with a large volume of ingested yolk keep a spherical shape, although they exhibit mature tissues. The adelphophagy of the Schmidt's larva not only affects unviable embryos, but also developing siblings *(arrowhead)*: (D″) a central rhynchocoel with the two brain lobes at each side (D″–D‴). Posterior to the brain, the ectoderm on the left and right sides of the head invaginates and forms the cerebral organ canal; (E) the connection of the cephalic organ with the exterior. Note in (E) the particular ciliation of the inner canal compared with the outer epidermis; all photos are ventral views, anterior is to the left; bl = brain lobes, coc = cerebral organ canal, iy = ingested yolk, le = larval epidermis, mo = mouth, pb = proboscis; bars, B, D″ = 100 μm; E = 25 μm. *All are selected and adapted from Martín-Durán et al. (2015).* (V) Summary of gene expression during embryonic development of *Lineus ruber*; (A–C) Schematic representation of the expression patterns of anterior, endomesodermal, and posterior markers, respectively, in the Schmidt's larva and early juvenile of *L. ruber*; (A) The anterior/head markers *foxQ2*, *six3/6*, and *otx* are expressed in the cephalic imaginal discs and head of the juvenile. Additionally, otx is expressed in the gut of the Schmidt's larva and scattered dorsal cells of the juvenile trunk. The gene *gsc* is expressed in two antero-lateral clusters of the Schmidt's larva and in the cerebral organs of the juvenile; (B) The endodermal genes *foxA* and *GATA456* are complementary, expressed in the blind gut of the Schmidt's larva and early juvenile. The mesodermal gene *twist-a* is expressed broadly in all imaginal discs and juvenile tissues; (C) The posterior markers *evx* and *cdx* are expressed in the posterior end of the Schmidt's larva and juvenile. The depicted expression patterns are for guidance and do not necessarily represent exact expression domains. *Dots of colour* indicate overlapping expression. The ingested yolk is depicted with a *light-blue circle*. Drawings are not to scale. *(From Martin-Duran et al., 2015.)*

of them exhibit a normal behaviour. Fig. 8.8(III) presents a summary of the main developmental events during *L. ruber* embryogenesis.

Development of the proboscis

By confocal laser scanning microscopy and Phalloidin labelling (actin labelling) from gastrula to the early juvenile stage, Martín-Durán et al. (2015) described the metamorphosis and organogenesis of Schmidt's larva. Here, only that of an eversible proboscis characterizing this phylum (Nemertini) is presented. The gastrula has archenteron cavity and rounded blastopore. After 10 days, the embryo has anterior cephalic and posterior trunk pairs of discoidal ectodermal concentrations, clearly segregated from the ring of ectodermal cells that surround the blastoporal opening (Fig. 8.8(IV) B′ [1]). In this figure, the first signs of the unpaired proboscis rudiment are visible. The archenteron cavity is expanded and bent backwards (Fig. 8.8(IV) B″ [2]), and isolated cells are still present inside the former blastocoel. After 12 days and at Schmidt's larva stage, the ectodermal discs are monostratified epithelium (Fig. 8.8(IV) C [3]) and the proboscis rudiment is more visible; the whole body is covered by a thin ciliated epithelium. Two days later, the indentation of the proboscis rudiment is more pronounced

(Fig. 8.8(IV) D′ [4]). At this stage, the larva has at least seven discs as rudiments of the organs of the juvenile: an unpaired anterior proboscis rudiment, two cephalic discs, the ventral mouth/pharynx rudiment, the internal endodermal blind gut and two trunk discs (Fig. 8.8(IV) D″ [5]). In addition, this metamorphosing larva feeds on other 'siblings', filling the blind gut with yolk.

Early juvenile

After 16 days of development, the larva increases in size, while the blastocoel becomes obliterated and is still spherical in form. Two days later, the larva attains a worm-like morphology, with the posterior tip of the trunk more elongated (Fig. 8.8(IV) B′ [6]) and the cephalic discs are much more developed (Fig. 8.8(IV) B″ [7]). After 20 days of development, the larva has metamorphosed into a juvenile (Fig. 8.8(IV) C′, C″ [8, 9]). At this stage, the mouth has a clear antero-ventral position, with a well-developed musculature (Fig. 8.8(IV) C′ [8]). The head shows two bilateral lobes, with a conspicuous longitudinal and transverse musculature, and the cerebral organ canals become visible (Fig. 8.8(IV) C″ [9]). According to the amount of yolk consumed during the larval stage, the form of the juvenile can vary considerably after 25 days of development. Those that retain more yolk attain a more rounded form. This developing Schmidt's larva not only feeds on arrested embryos, but can also ingest apparently viable embryos (Fig. 8.8(IV) B). The anterior median region is occupied by the proboscis (Fig. 8.8(IV) D″), and the dorsal ectoderm posterior to the brain exhibits well-formed cerebral organ canals on each side of the head (Fig. 8.8(IV) E). The results of this work (Martín-Durán et al., 2015) indicate that the juvenile of ***Lineus ruber*** acquires the basic anatomy of the adult after 25 days of development at 14°C.

This work also identified and studied in detail the expression pattern of genes involved in the specification of anterior and cephalic tissues (Fig. 8.8(V)). Anterior (*foxQ2*, *six3/6*, *otx*, *gsc*) and posterior (*evx*, *cdx*) gene markers are expressed in the cephalic and trunk discs, respectively, and endomesodermal genes are detected in the pharynx and gut rudiments (*foxA*, *GATA456-a*) and broadly in all discs (*twi-a*). In addition, posterior (*evx*, *cdx*) markers demonstrate that the juvenile retains the molecular patterning of Schmidt's larva. Importantly, the gene *gsc*, which is related to the formation of the oral nervous system of many spiralian larvae, is expressed in two lateral clusters of cells of Schmidt's larva, and later on the developing cerebral organs of the juvenile (Fig. 8.8(V) A). Therefore, the molecular mechanism patterning the juvenile is only active in the imaginal discs and not during the early development of the larva, suggesting two separate molecular programmes during nemertean embryogenesis. After metamorphosis, the juveniles stay within the egg masses for more than 20 days, until they are fully mature and hatch.

Direct development

Iwata (1960) reported that direct development has been observed in *Cephalothrix rufifrons* (Smith, 1935) in the order Palaeonemertea and in five species, namely *Prosorochmus viviparus* (Salensky, 1914), *Tetrastemma vermiculus* and *Drepanophorus spectabilis* (Lebedin-sky, 1897), *Prostoma graecense* (Reisinger, 1926; Reinhard, 1941) and *Malacobdella grossa* (Hammarsten, 1918) in the orders Plonemertea and Bdellonemertea. However, this direct development in nemerteans has been defined as 'a path of development that leads to the establishment of the adult body plan (bauplan) without the formation of an intermediate feeding larval stage that possesses a distinct body plan' (Norenburg and Stricker, 2002). These authors also reported that direct developing nemerteans may hatch from the egg capsule beginning at the gastrula stage, or at a stage up to a fully functional, nonsexual juvenile (Iwata, 1960). Among those hatching as precrawling juvenile forms, some remain within a cocoon, as in some supralittoral hoplonemerteans (e.g. *Pantinonemertes*, cf. Norenburg, 1986).

In directly developing species (Fig. 8.9), the characteristics of the later pilidium larva do not exhibit and the larva attains the following characters (Iwata, 1968): the larva become oval in shape and thickly covered with cilia; a **tuft** of long cilia develops at the anterior end; and the larva enters a free-swimming stage. Beside the **tufts** at the head and tail ends, in *Procephalothrix simulus* two additional pairs of **tufts** develop at the anterior region.

The larval epithelium generally changes into that of the adult, although it is known that in *Cephalothrix* and *Emplectonema*, a ciliated young worm bursts out of the larval epithelium. The **proboscis** develops from an ectodermal invagination at the anterior end, exactly as in the pilidium larva. As in the indirect development, the **blastopore** moves inwards. However, in *Cephalothrix*, this blastopore closes for a time, and later reopens and unites with the intestine.

In Enopla, after the larval mouth, the foregut shifts towards the ciliated anterior end joining proboscis sheath and opening into the rhynochocoel. Together with the proboscis, it forms a common passage leading into the anterior most part of the head. The mouth, oesophagus, stomach and **pylorus** are formed from the foregut.

The **brain** develops from a pair of ectodermal proliferations in the head region. They grow and sink inwards. The cerebral sense organs are formed from a pair of invaginations on the ventral side of the head. At the same time, the cephalic gland develops from an invagination in the anterior most part.

The **anus** in the **Hoplonemertea** appears early as an ectodermal invagination of the posterior body wall. In the Paleonemertean *Cephalothrix*, it opens after the intestine has been completely formed, even later than in the ***Heteronemertea***, the opening appearing after the intestine and the ectoderm had united.

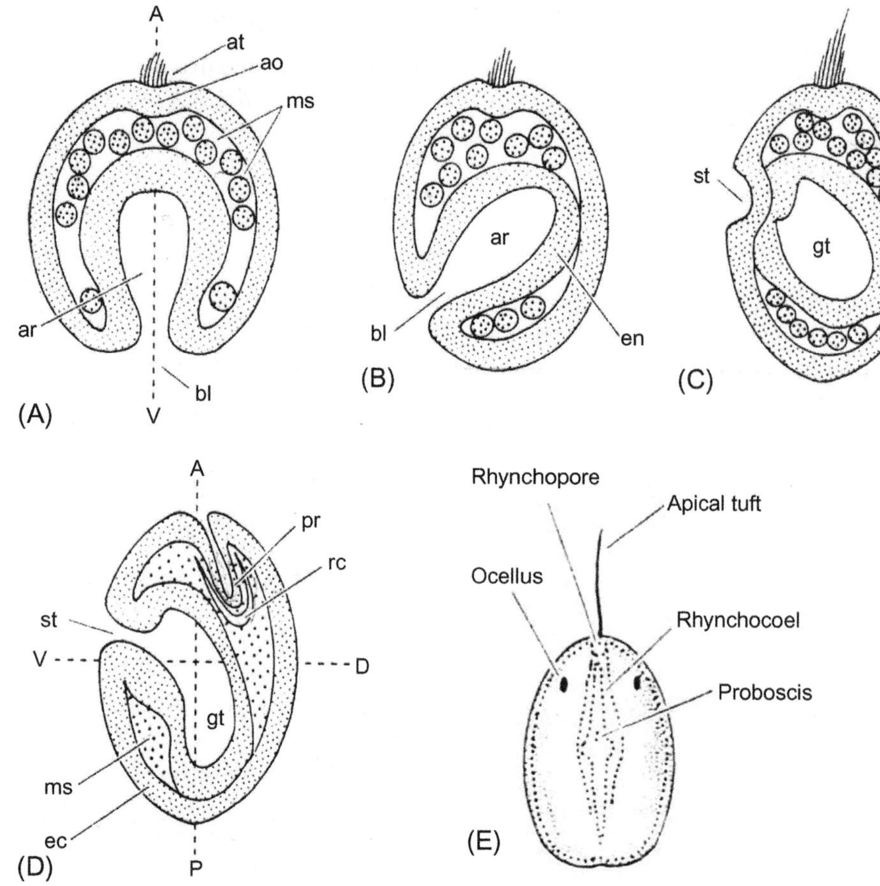

FIG. 8.9 Sagittal sections representing four stages of direct development: (A) gastrula stage, the embryo has the invaginated archenteron (ar) which will be giving rise to the gut. The blastopore (bl), located at the vegetal pole, numerous mesenchyme cells (ms) are located within the blastocoel. The thickened apical organ (ao) is located within the ectoderm at the animal pole. An elongated apical tuft (at) of cilia emerges from the cells of the apical organ; (B) an older embryo after expansion of the dorsal ectoderm has begun reposition of the blastopore to the ventral side of the embryo. The endoderm (en) encloses the archenteron; (C) embryo after closure of the blastopore. The stomodaeum (st) or mouth appears as an invagination of the ectoderm on the ventral surface of the embryo; (D) late swimming stage after the stomodaeum forms connection with the developing gut (gt). The proboscis rudiment (pr) appears as an invagination of anterior ectoderm. The proboscis lies within the rhyncocoel (rc), which has formed within the mesoderm (ms). The apical tuft is no longer present at this stage; ec=ectoderm. The adult axes shown in D apply for all four stages, the axes A=anterior, P=posterior, V=ventral, D=dorsal; (E) direct developing larva is basically a miniature version of a juvenile worm. *(From Henry and Martindale (1997).)*

The mesenchyme of mesodermal origin, within the body cavity, adheres to the inner side of the body wall and to the intestine. According to Salonsky, this occurs in ***Prosorhochmus***, where it forms a layer. Later develops into the body wall musculature and the mesenchyme, which lies between this layer and the intestine.

The circulatory system first appears as tubules surrounded by mesenchyme cells, which lie in the inner of the body wall and become the lateral blood vessels. The dorsal blood vessel, at the same time, is formed from proboscis sheath mesenchyme. In *Malacobdella*, Hammarsten found that the ectoderm at the posterior body wall grows inwards and forms a sucker. At the same time, it proliferates inside the body and forms mesenchyme of ectodermal origin.

Procephalothrix simulus

Procephalothrix and *Tubulanus* exhibit the direct type of development (Iwata, 1960). This work is greatly voluminous and scientifically significant. It worked on many nemertean species. The author obtained the mature adult worms from between the tide marks from the shores of Akkeshi and Daikokujima, a small island 2 miles off the Akkeshi coast, Sapporo, Japan. The worms were brought to the laboratory, and the males and females of each species were kept separately in shallow finger bowls containing seawater. Those of *P. simulus* were maintained from the middle of 1953 to about 1960. The author obtained the eggs artificially by cutting the gonads transversely and they were inseminated, also artificially. Naturally, the eggs are discharged simultaneously through the gonoducts; they are embedded in a loose jelly substance secreted from the epidermis of the females. The jelly dissolves after spawning. The egg is yellow-orange in colour, opaque with rich yolk and measures about 0.19 mm in diameter. The germinal vesicle is represented as a small clear zone. The egg membrane is thin.

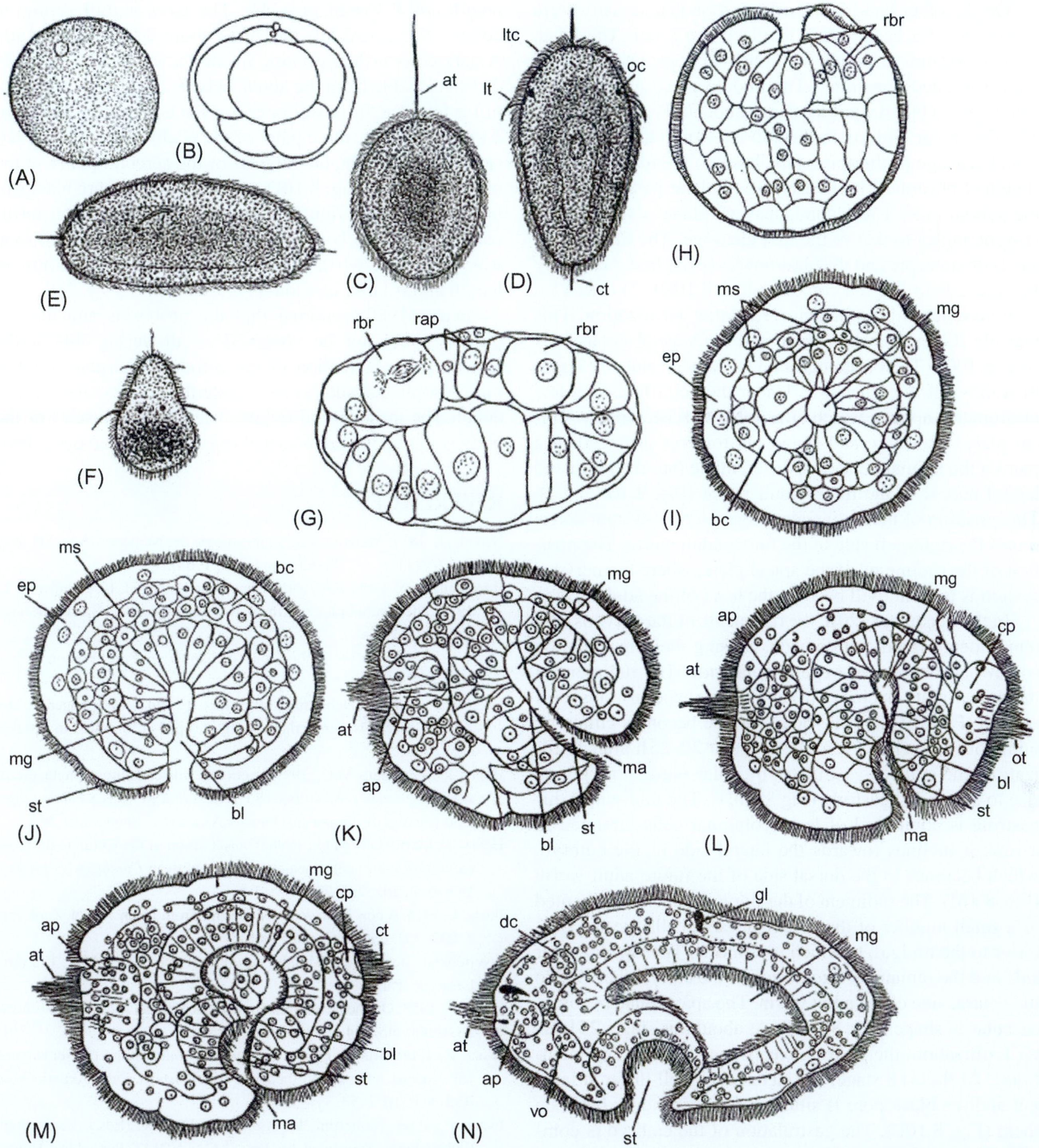

FIG. 8.10 Embryological development of *Procephalothrix simulus* Iwata: (A) Egg immediately after liberation. 105×; (B) 8-cell stage. 105×; (C) 46-h larva. 105×; (D) 165-h larva. 85×; (E) 165-h larva from the lateral view. 85×; at = apical tuft, lt = lateral tuft, ltc = lateral tuft derived from the cerebral plate, ct = caudal tuft, oc = ocellus; (F) 41-day larva, 105×; (G) section of the 13-h blastula through the polar body, rap = rudiments of the apical plate, 340×; (H) sagittal section of the 25-h gastrula, showing the rudiments of the brain and the lateral nerves (rbr), 340×; (I) transverse section of the 31-h-larva, showing the rudiment of the brain and lateral nerves extending inside the epidermis, bc = blastocoel, ep = epidermis, mg = mid gut, ms = mesoderm cells, 340×; (J) horizontal section of the 43-h larva, showing the blastopore (bl), the rudiment of the brain and lateral nerves (rbr) and the stomodaeum (st), 340×; (K) sagittal sections of the 55-h larva through the apical plate and the rudiments of the brain and lateral nerves and of the lateral plate, 340×, ap = apical plate, at = apical tuft, bi = blastopore, ma = mouth of the future adult worm; (L) sagittal section of the 67-h larva through the apical plate, the nervous system, and the caudal plate (cp), caudal tuft (ct) 340×; (M) sagittal section of the 79-h larva through the apical plate, 340×; (N) sagittal section of the 85-h larva through the apical plate, dg = dorsal ganglion, gl = gland cell, vc = ventral commissure of the brain, 340×. *(Selected from Iwata, 1960.)*

On the other hand, the spermatozoon is a nematosperm composed of a head, a middle piece and a tail. The head, about 0.003 mm in length, is elliptical in shape and provided with a pointed acrosome. The middle piece appears as a small spot, while the tail measures about 0.04 mm in length. The first polar body is formed about 40 min after fertilization. Cleavage of the spiral type leads to the formation of a flattened blastula. The first cleavage plane passes through the animal pole. The second cleavage plane is vertical and at right angles to that of the first cleavage. The third cleavage is dexiotropic and the blastomeres of the first quartet are far larger than the macromeres (Fig. 8.10B). The cleaved embryo reaches blastula stage 13 h after fertilization. This blastula flattens on the animal and the vegetal polar sides (Fig. 8.10G). The blastomeres on the vegetal side are larger than those of the other side. Under the polar bodies, a few blastomeres appear, which afterwards will become the apical plate. The large and round blastomeres arranged in a pair on the animal polar side differentiate into the brain and lateral nerves of the future adult worm (Fig. 8.10I and J). This position of the rudiments of the nervous system determines the right–left side of the future adult worm. The position of the rudiments of the apical plate, where the nervous system is present, will become the head of the adult worm.

In the blastula, at 20 h, the rudiments of the nervous system undergo rapid cell division, causing their invagination somewhat towards the vegetal polar side. The rudiment of the apical plate becomes triangular in shape in transverse section (Fig. 8.10K). The body surface becomes ciliate by numerous fine cilia. From this stage at 20–25 h after fertilization, the embryo gastrulates by polar ingression, giving rise to a rounded gastrula (Fig. 8.10H). The mid-gut in this gastrula is composed of large columnar cells arranged in a row; it inclines towards the lateral side of the embryo, which becomes to the dorsal side of the future adult worm (Fig. 8.10J). The rudiment of the stomodaeum is represented as a small number of the columnar large cells located posterior to the midgut, to which it connects. The vegetal polar side and the remainder lateral side of the embryo become to the ventral side of the adult worm. The apical plate becomes saccular in shape (Fig. 8.10M). At about this stage, 25 h after fertilization, the gastrula creeps out from the egg membrane. At the 31 h stage, the larva has a well-differentiated gut and the blastopore is situated at the boundary between them (Fig. 8.10L). The gastrulation of the embryo is completed in the 43-h larva. However, at 31 h, a small number of the mesoderm cells are differentiated in the blastocoel, then it widens and the mesoderm cells increase in number. The apical plate produces a bundle of long flagella forming an apical tuft in the 31-h larva.

The young larva just escaped from the egg membrane is spherical in shape and is provided with a long apical tuft. The larva swims in a straight line along the long axis of the body. At 46 h, it elongates measuring about 0.22 mm in length and 0.16 mm in width. The larva is dull yellow in colour. The apical tuft measures about 0.15 mm in length. A dark contour line running inside the epidermis is clearly distinguishable from the above (Fig. 8.10C). A short caudal tuft is found at the posterior end of the body at the 50th hour. At the 85th hour, three pairs of the ocelli appear, of the lateral tufts and of the lateral tufts of the cerebral plates in the cephalic region (Fig. 8.10D and E). The ocelli are blackish-brown. These two figures show the larvae 165 h after fertilization. The larva becomes smaller in size because no food was given; the 41-day larva measures about 0.16 mm in length and 0.1 mm in width (Fig. 8.10F).

Iwata (1960) reported that the proboscis and the nephridia could not be observed at all during this study. However, the formation of the ectodermal organs, epidermis, oesophagus and nervous system were described in full detail. The mesodermal organs, such as the muscles of the body wall and the endodermal organs, were also described.

REFERENCES

Barrois, J., 1877. Mémoire sur l'embryologie der Nemertes. Ann. Sci. Nat. Ser. 6 Zool. 6, 1–232. (cited in Iwata, 1968).

Borradaile, L.A., Potts, F.A., with chapters by Eastham, L.E., Saunders, J.T., 1977. The Invertebrata, fourth ed. Revised by Kerkut, G.A.Cambridge University Press.

Desor, E., 1848. On the embryology of Nemertes. Proc. Boston J. Nat. Hist. Soc. 6, 1–18.

Friedrich, H., 1979. Nemertini. In: Seidel, F. (Ed.), Morphogenese der Tiere. Gustav Fischer-Verlag, Stuttgart, pp. 1–136. (cited in Stricker et al., 2001).

Henry, J., Martindale, M.Q., 1997. Section V Bilateral animal phyla, protostome coelomates. In: Gilbert, S.F., Raunio, A.M. (Eds.), Embryology, Constructing the Organism. Sinauer Associates, Sunderland, MA.

Henry, J., Martindale, M.Q., 1998. Conservation of the spiralian developmental program: cell lineage of the nemertean *Cerebratulus lacteus*. Dev. Biol. 201, 253–269.

Ikeda, I., 1915. A new fresh water Nemertine from Japan. Annot. Zool. Jap. 8, 239–255. (cited in Iwata, 1968).

Invertébrés: Beauté du Monde Animal, Tome XII, 1974. Librairie Larousse, Paris.

Iwata, F., 1958. On the development of the nemertean *Micrura akkeshiensis*. Embryologia 4 (2), 103–131.

Iwata, F., 1960. Studies on the comparative embryology of Nemerteans with special reference to their interrelationships. Publ. Akkeshi Mar. Biol. Stat. 10, 1–53.

Iwata, F., 1968. Nemertini. In: Kume, H., Dan, K. (Eds.), Invertebrate Embryology (Translated by Dan, J.C.). NOLIT Publishing House, Belgrade, Yugoslavia.

Martín-Durán, J.M., Vellutini, B.C., Hejnol, A., 2015. Evolution and development of the adelphophagic, intracapsular Schmidt's larva of the nemertean *Lineus ruber*. EvoDevo 6, 28. BioMed Central.

Maslakova, S.A., 2010. Development to metamorphosis of the nemertean pilidium larva. Front. Zool. 7, 30.

Morgan, T.H., 1927. Experimental Embryology. Columbia University Press.

Müller, J., 1847. Fortsetzung des Berichts über einige neue Thierformen der Nordsee. Arch. Anat. Physiol. 45 (2), 157–179.

Norenburg, J.L., 1986. Redescription of a brooding nemertine, *Cyanophthalma obscura* (Schultze) gen. Et comb. n., with observations on its biology and discussion of the species of *Prostomatella* and related taxa. Zool. Scr. 15 (4), 275–293.

Norenburg, J.L., Roc, P., 1998. Reproductive biology of several species of recently collected nemertean worms. Hydrobiologia 365, 73–91. (cited in Stricker et al., 2001).

Norenburg, J.L., Stricker, S.A., 2002. Phylum Nemertea. In: Atlas of Marine Invertebrate Larvae. Academic Press (Chapter 7).

Nusbaum, J.U., Oxner, M., 1913. Die embryonalentwicklung des *Lineus ruber*. Zeitschr. wiss. Zool. 107, 78–191. (cited in Iwata, 1968).

Reverberi, G., 1971. Experimental Embryology of Marine and Fresh-Water Invertebrates. North-Holland Publishing Company, Amsterdam, London.

Riser, N.W., 1974. Nemertinea. In: Giese, Pearse (Eds.), Reproduction of Marine Invertebrates. vol. 1. Academic Press, London.

Smith, J.E., 1935. Early development of *Cephalothrix rufifrons*. Q. J. Microsc. Sci. 77, 335–378. (cited In Iwata, 1068).

Stricker, S.A., Folsom, M.W., 1998. A comparative ultrastructural analysis of spermatogenesis in Nemertean worms. Hydrobiologia 365, 55–72.

Stricker, S.A., Reed, C.G., 1981. Larval morphology of the nemertean *Carcinonemertes epialti* (Nemertea: Hoplonemertea). J. Morphol. 169, 61–70.

Stricker, S.A., Smythe, T.L., Miller, L., Norenburg, J.L., 2001. Comparative biology of oogenesis in Nemertean worms. Acta Zool. (Stockholm) 82, 213–230.

Turbeville, J.M., 2002. Progress in nemertean biology: development and phylogeny. Integr. Comp. Biol. 42, 692–703.

von Döhren, J., 2011. The fate of the larval epidermis in the Desor-larva of *Lineus viridis* (Pilidiophora, Nemertea) displays a historically constrained functional shift from planktotrophy to lecithotrophy. Zoomorphology 130, 189–196. (cited in Martín-Durán et al., EvoDevo, 2015).

Wilson, C.B., 1900. The habits and early development of *Cerebratulus lacteus*. Q. J. Microsc. Sci. 43 (2), 99–191.

Wilson, C.B., 1903. Experiments on cleavage and localization in the Nemertine-egg. Arch. Entw. Mech. 16, 411–458.

Pseudocoelomata

Chapter Outline

Definition **423**
Historical review **423**
General characters **424**
Classification **424**
Reproduction **426**
- Agametic reproduction 426
- Parthenogenesis 427
- Gametic reproduction 430
  - Sexual dimorphism 430
  - Spicules 430
  - Intersexes 431
  - Hermaphroditism 431
  - Reproductive system 432
  - Origin of germ cells and gonads 433
  - Gametogenesis 434
    - Spermatogenesis 434
    - Oogenesis 437
  - Reproductive behaviour 441
    - Sex attractant 441
  - Copulation 441
  - Fertilization 441

Development **445**
- Early embryology 447
- Cleavage 447
  - Chromatin diminution 447
- Gastrulation 450

Hatching **451**
Embryological peculiarities **452**
Postembryonic development **452**
- Life cycles 453
- Anabiosis 455

Caenorhabditis elegans **455**
- Historical review 455
- Sex differentiation 457
- Reproductive system 457
- Gonadogenesis 458
- Gametogenesis 458
  - Gametes 462
- Development 462
  - Embryogenesis 462
  - Gastrulation 464
  - Polarity 464
    - P-granules 465
  - Postembryonic development 465
  - Dauerlarva 465

Variation in nematode embryology **465**
References **469**

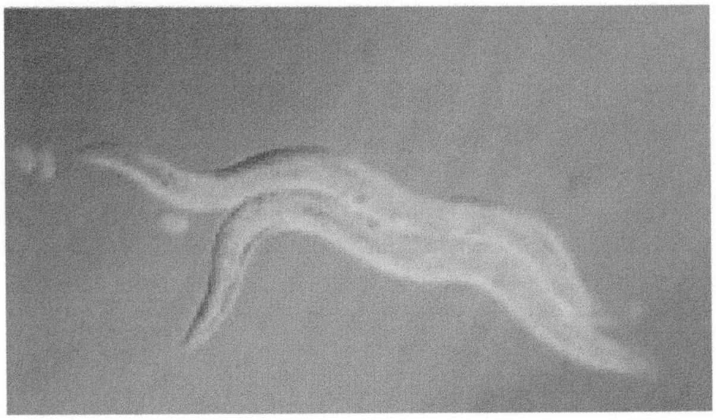

Chapter 9

Phylum Nematoda

(Greek *nematos*, 'thread')

DEFINITION

Roundworms; elongated, unsegmented, and cylindrical animals. The body is fusiform or filiform and mostly with tapered two ends; often with bristles but without cilia. Subepidermal musculature of longitudinal fibres only. Due to this musculature, movement is a characteristic whipping flexure in a snake-like movement. They are aquatic, terrestrial and parasitic; differ in length from minute less than 1 mm, and even microscopic (e.g. many free living nematodes), to several metres in length (e.g. a parasite in a whale).

HISTORICAL REVIEW

The nematodes known nowadays were not originally well differentiated from other worms in the *Systema Naturae* of Linnaeus (1758). This author defined all the parasitic worms infecting the intestine of hosts collectively as **order Intestina**, under the class Vermes. This class is one of the classes in the classification of Linnaeus: Mammalia, Aves, Amphibia, Pisces, Insecta, and Vermes (Chitwood et al., 1950; Hyman, 1951). The free-living nematodes, which are mostly microscopic, were not discovered until the invention of the microscope, when the vinegar eel and the other so-called eelworms were visualized by early microscopists. Goeze (1782) was the first to study the vinegar eel and other nematodes. He observed the differences among these free-living worms and distinguished them into different groups. This 18th-century step was so perfect and convenient that Zederin in 1800 (cited in Chitwood et al., 1950) was able to give each group a definite name, e.g. the nematodes were named 'round worms' (Rundwürmer). Later, Rudolphi (1808) created a scientific name for each group and applied the name 'Nematoidea' to these round worms. This name is scientifically significant since it gives the proper meaning of the shape of the worm. In his *Entozoorum synoosis* published in 1819, Rudolphi listed 11 genera and about 350 species of nematodes, but he grouped the two recently known groups, Nematoda and Nematomorpha, under this name Nematoidea.

No progress was made towards a more natural grouping of the parasitic worms, until Vogt (1851) divided the Vermes into Annelid, Rotatoria and differentiated between the flat and round worms. He created the two names 'Platyelmia' to refer to the flat worms and 'Nematelmia' to refer to the round worms. Gegenbaur in 1859, proposed a scheme for classification of the parasitic worms and he defined each group as an order. Therefore the Nematoidea of Rudolphi became an order in this scheme of Gegenbaur. The latter altered Vogt's names to Platyhelminthes and Nemathelminthes. He also placed the two orders Acanthocephala and Nematoidea in one group and elevated it to the class rank, named 'Nemathelminthes'. Later on, he himself separated them into three classes: Acanthocephala, Nematoda, and Gordiacea (=Nematomorpha), since he elevated the class Vermes, as a whole, into phylum status.

From the above, one can understand that the name Nemathelminthes proposed by Gegenbaur (1859) replaced the name Nematoidea of Rudolphi (1808). Hyman (1951) mentioned that the name Nematoidea was retained by some zoologists and it is usually altered to Nematoda. During the 19th century, different parasitic worms were studied in more detail and more features were added to their description and differentiation, most importantly the body cavity. In nematodes, it is evidenced that it is a pseudocoel. Grobben (1910) grouped all the pseudocoelomate in one phylum under the name Aschelminthes. This phylum included the six classes: Rotifera, Gastrotricha, Kinorhynca, Priapulida, Nematoda, and Nematomorpha. Hyman (1951) followed this view.

On the other hand, in the work of Chitwood et al. (1950), each of these classes was raised to the status of phylum. Some other zoologists followed this view, e.g. Borradaile et al. (1977) and Gardiner (1972). This view is probably more convenient because the evidence shows that the

Invertebrate Embryology and Reproduction. https://doi.org/10.1016/B978-0-12-814114-4.00009-6

relationship between these six groups is weaker than in case of the classes of most other phyla. Accordingly, the nematodes constitute a definite phylum named Nematoda, which belongs to the pseudocoelomata under grade Bilateria, branch Eumetazoa, subkingdom Metazoa, kingdom Animalia.

GENERAL CHARACTERS

1. Nematodes are triploblastic, bilateral symmetrical, and pseudocoelomate animals.
2. Nematodes in general lack colouration, being transparent, with a whitish or yellowish tint.
3. Elongated and cylindrical; the body tapers in both ends, there is no distinct head, the terminal mouth is usually guarded by lips (number, structure and position of the lips are specific characters) and the posterior end may be curved (Fig. 9.2A and B).
4. The majority of free-living nematodes are microscopic or very small, i.e. a fraction of 1 mm, although some marine species may attain a length of 5 cm. While some parasitic species are also of small size, others are excessively long. Smyth (1962) recorded that the kidney worm *Dioctophyma renale* measures more than 1 m length. Another example is *Philometra amazonica*, which measures about 95 cm (Rasheed, 1963).
5. The body wall consists of syncytial hypodermis covered with an elastic, somewhat thick, nonchitinous cuticle (Fig. 9.2D). The hypodermis is thickened to form two main lateral lines and less evident mid-dorsal and mid-ventral lines (hypodermal ridges), due to which the body wall is divided into four equal quadrants (Fig. 9.2C).
6. Musculature of the wall consists of longitudinal muscle fibres only. In Nematoda, these muscles are primitive in structure, they are arranged in the quadrants, and their number in the quadrant is a specific feature. Gibbons (2002) reported that one to two rows are referred to as holomyarian, two to five rows as meromyarian (Oxyuridae, Ancylostomatidae) and larger number of rows as polymyarian, as in the Ascarididae (Fig. 9.2C). Each muscle cell is spindle-shaped and divided into contractile portion and noncontractile portion. There are three types of cells according to the shape of their contractile portion: platymyarian with a wide shallow contractile portion close to the epidermis; coelomyarian in which the contractile portion extends up the lateral sides of the cell; and circomyarian where the striated muscle fibres envelop the noncontractile portion. In addition, the nematodes are peculiar in that each muscle cell sends a long axon-like 'innervation process', a slender arm or tail (Soliman, 1996) which extends from the sarcoplasm and communicates with the corresponding portion of the nervous system to receive synaptic input from the moto-neurons (Fig. 9.2D and E). In the view of Chitwood et al. (1950), this characteristic feature is unique in the animal kingdom. Specialized muscle cells are mainly associated with the digestive system, especially the pharynx, and the male and female reproductive tracts.
7. The gut is a simple tube, with long oesophagus having triradiate lumen (Fig. 9.2G). The length of the oesophagus is a specific character. The gut opens posteriorly through an anus which is subterminal (Fig. 9.2A).
8. The space between the body wall and the gut is sometimes filled with a small number of highly vacuolated cells, the vacuoles joining together and simulating a perivisceral cavity (pseudocoele) (Fig. 9.2C).
9. The excretory system typically comprises two laterally placed intra-cellular types. Flame cells are absent.
10. The nerve cells are aggregated at the anterior end of the body forming a circum-oesophageal ring. One or more nonganglionated nerve cords extend from this ring. Sense organs are of the simplest type.
11. Sexes are usually separate, dimorphism usually occurs, also parthenogenic or hermaphroditic.
12. The posterior end of the two sexes frequently differs in shape and external features, being generally curved in the male, and provided with papillae, caudal alae, spicules, and bursa, which are lacking in the female (Fig. 9.4).
13. The gonads are tubular and continue as gonoducts. The female system is paired and opens anteriorly. The distance length between the anterior end and the female genital opening is a specific character. The male system is unpaired and opens posteriorly into the hind-gut through a diverticulum on which, in most cases, one or two protrusible copulatory spicules lie, i.e. penial setae (Soliman, 1996).
14. Spermatozoa are rounded and amoeboid; fertilization is internal.
15. The egg hatches as a larva, very similar to the adult. There are four larval stages.
16. Almost all the known nematodes are parasitic, often with complex life history.

CLASSIFICATION

Nematodes are able to adapt to almost every habitat available to animal life. They occur anywhere from the polar regions to the tropics, in all types of environment including deserts, hot springs, high mountain elevations and great ocean depths. They are free-living in freshwater, sea, soil or decomposed organic matter, or parasitic on or in plants and animals. It was recorded in different sources that the vinegar eel *Turbatrix aceti* (Müller, 1783) (=*Anguillula aceti*) (Rhabditoidea, Rhabditina, Rhabditida, Secernentea; Hope, 1974) lives in the sediment of nonpasteurized vinegar. They often occur in enormous numbers, and millions or even billions could exist in limited areas of the muddy sea bottom, sandy beaches or cultivated land. Storer et al. (1983) recorded that some marine bottom muds have been estimated to contain more than 4 million nematodes per square metre, while good soil may have billions per acre. The human body harbours more than 50 species causing variable degrees of harm. Many nematodes show an extraordinary capacity to survive conditions suboptimal to viability, expressing variable degrees of anabiosis. Many free-living and plant parasitic species show an extreme state of dormancy (cryptobiosis). When unfavourable conditions become absent, dormant nematodes shortly regain their activity. It appears that these nematodes are greatly successful in any habitat, and this recalls the view of Stunkard (1953) that the arthropods are only first to the nematodes in this respect. Another study believes that among metazoan animals, nematodes are probably second only to insects in number of individuals (Storer et al., 1983). About 20,000 nematode species have been described so far, but the estimate of how many species exist in total, varies dramatically between 80,000 and 1 million (Schierenberg, 1997). Independent of which number comes closer to reality, nematodes are the animals to look for if you want to find a so-far undescribed species (and to become immortal by giving it your name).

It seems that these diverse habitats lead to different interesting biological branches: some investigators were experts only on free-living, phytoparasitic or zooparasitic nematodes. Accordingly, convenient forms of classification have been proposed for each of these groups. On the other hand, there have been few zoologists who have studied all nematodes, with the result that a definite schematic classification is not well established. The similarity of the individuals of nematodes is great enough that the general description of one nematode can be applied to all others, and this is the main factor that the classification of lower grades in phylum Nematoda was very difficult. However, Chitwood (1933) divided the nematodes into two subclasses, Phasmidia and Aphasmidia, according to the presence or absence of the phasmidial organs at the tail region of the worm. (Phasmids are a pair of prominent—not only microscopical but ultrascopical—structures located laterally, one on each side of and at a distance greatly close to the spiny distal process (Fig. 9.3).) Chitwood et al. (1950) raised the subclasses to the rank of classes since the nematodes themselves were raised into a Phylum, as mentioned above. Dougherty (1958) noted that Phasmidia had been used much earlier as a name for walking sticks (Orthoptera), and he preferred to use the name Secernentea to replace Phasmidia. At the same time, Chitwood (1958) preferred the name Adenophorea to replace Aphasmidia (the two new names were previously proposed by von Linstow, 1905). Some authors followed this preferable nomenclature for the two classes of Phylum Nematoda, e.g. Gardiner (1972) and Hope (1974). The latter presented a very detailed scheme of classification of Nematoda up to the genera, including the ecological habitat of each group and/or genus. A more recent work (Schierenberg, 2006) presented a phylogenetic tree of nematodes, based on 18s RNA gene sequences (Fig. 9.1), but it is still not familiar.

Many other authors used minor characteristics to arrange the nematodes in lower grades, in spite of its difficulty owing to the uniformity of the nematodes. The lips, if present or absent, their number, position, and structure constitute a somewhat reasonable character. Therefore, classification of nematodes is highly variable and there is no convenient and/or simple scheme to be recorded. This work will present a simple scheme for major taxa and the taxonomy of each encountered species will be mentioned next to it.

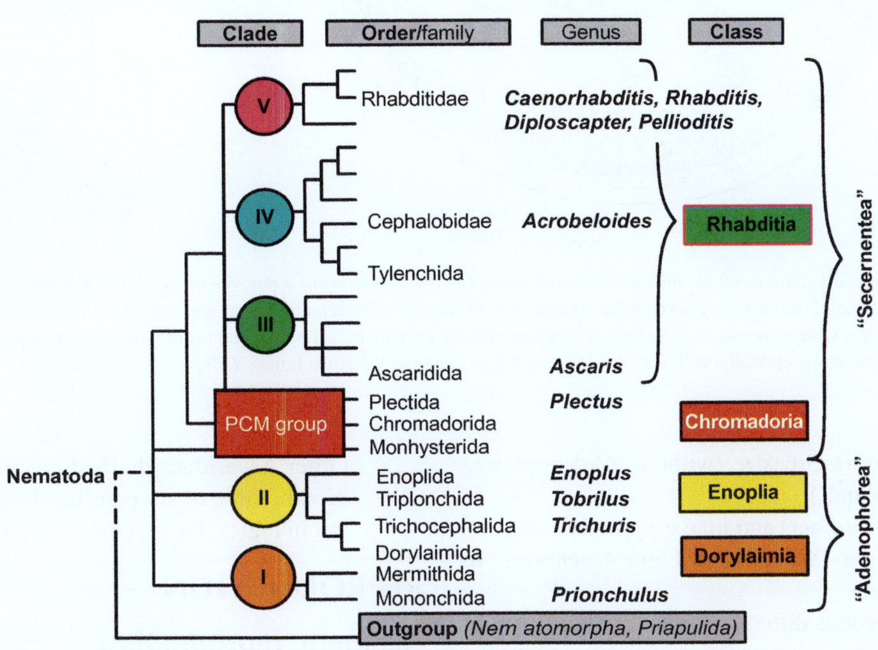

FIG. 9.1 Phylogenetic tree of nematodes. Based on 18s RNA gene sequences, five different clades (I–V) have been defined. Branches that could not be assigned to one of these clades are grouped collectively and named here as 'PCM group' (for Plectida, Chromadorida, Monhysterida; Schierenberg and Lahl, 2004). With Nematoda as a phylum, four classes are defined by combining suggestions of Malakhov (1994) and De Ley and Blaxter (2002) because such a subdivision fits best with our embryological data. *(Adapted from Blaxter et al. (1998), from Schierenberg (2006).)*

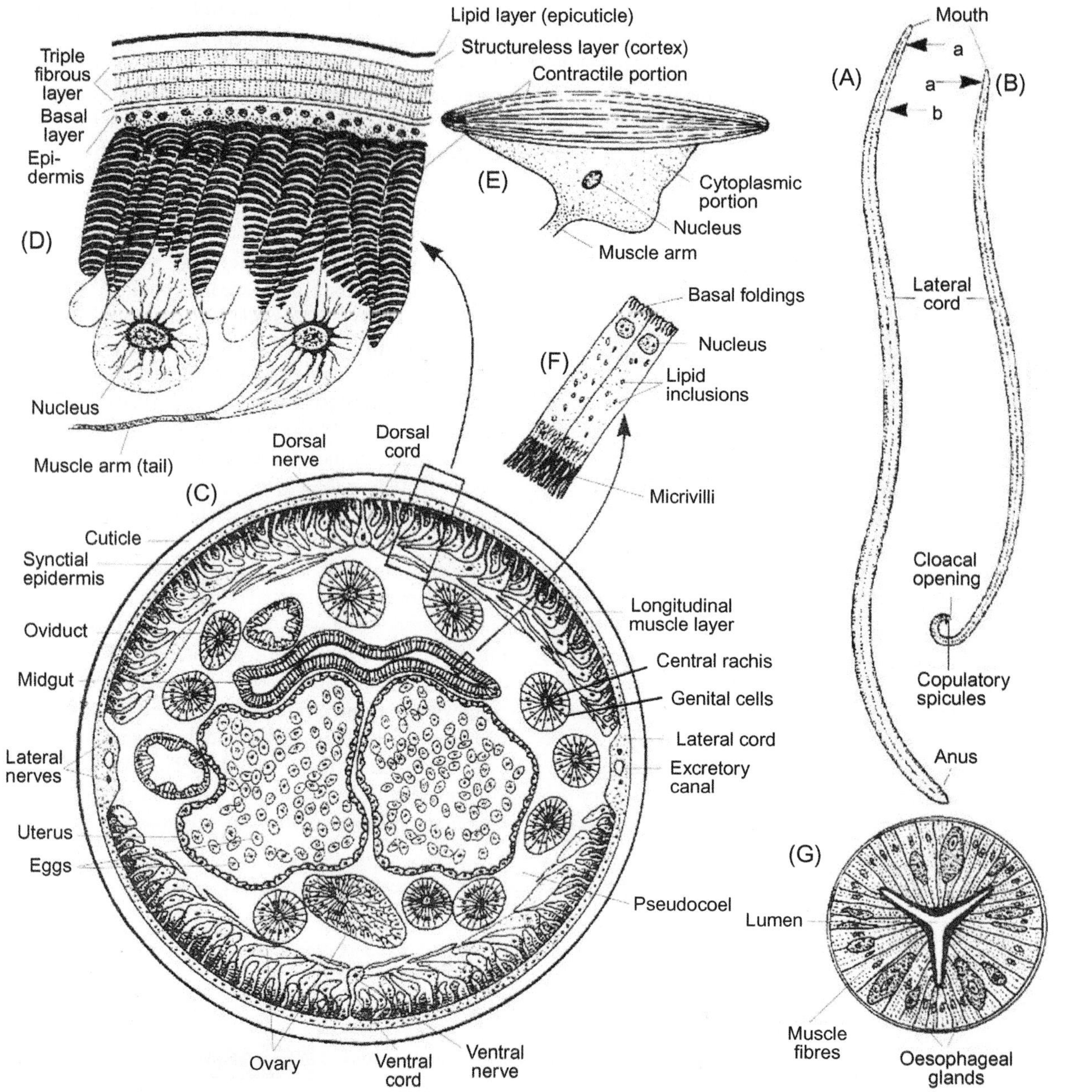

FIG. 9.2 (A) and (B) show diagrams of whole mount of *Neoascaris* (=*Toxocara*) *vitulorum*, a parasite of cattle. (A) Female side view and (B) Male side view. (a) Points to the position of the excretory pore on the ventral side; (b) points to the female genital aperture; (C) T.S. in female *Neoascaris*, note poly-myarian type of muscle layer with numerous muscle cells projecting into the pseudocoel; (D) enlarged portion of the body wall; (E) a single longitudinal muscle cell; (F) enlarged intestinal epithelial cell; (G) T.S. in oesophagus showing Y-shaped lumen. *(All from Soliman (1996).)*

Class 1. Aphasmidia Greek *a*, 'without' (Adenophorea Greek *adenos*, 'gland').

- Five orders (Gardiner) and nine groups (Hope).

Class 2. Phasmidia (Secernentea Latin *secenrere*, 'to separate' 'secrete').

- Number of orders differs with the author (five to eight).

  Order Rhabditata (small, free-living, simple mouth or six papillae. ex. *Rhabditis*).

  Order Ascaridata (bulbed oesophagus, long vagina, mouth, three to six papillae, two spicules and curled tail in male). Ex. *Ascaris*, *Enterobius*).

REPRODUCTION

Agametic reproduction

Hope (1974) reported that there is no evidence that nematodes reproduce by fission or by polyembryony. However,

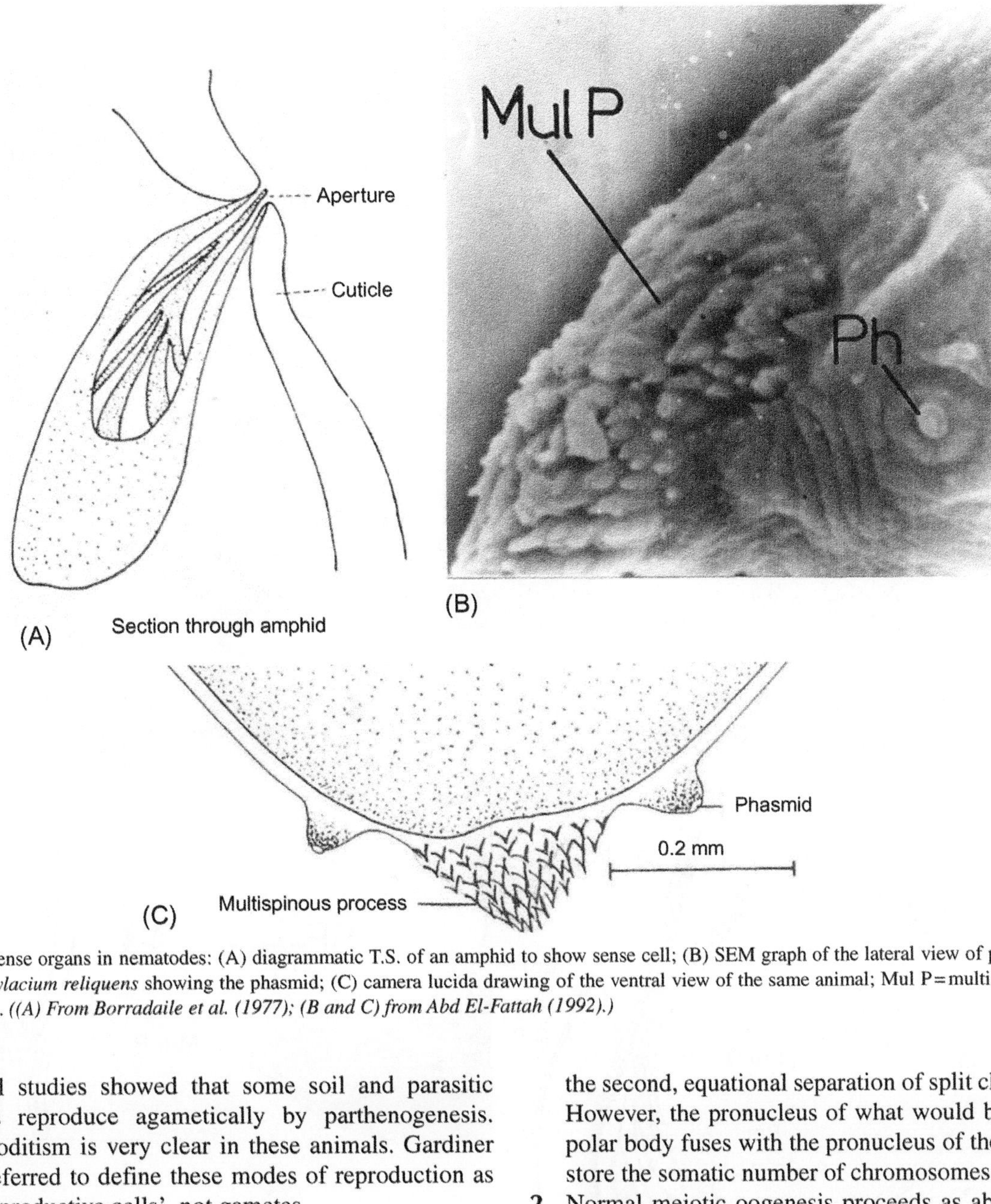

FIG. 9.3 Sense organs in nematodes: (A) diagrammatic T.S. of an amphid to show sense cell; (B) SEM graph of the lateral view of posterior portion of *Hysterothylacium reliquens* showing the phasmid; (C) camera lucida drawing of the ventral view of the same animal; Mul P = multispinous process, Ph = phasmid. *((A) From Borradaile et al. (1977); (B and C) from Abd El-Fattah (1992).)*

cytological studies showed that some soil and parasitic nematodes reproduce agametically by parthenogenesis. Hermaphroditism is very clear in these animals. Gardiner (1972) preferred to define these modes of reproduction as 'special reproductive cells', not gametes.

Parthenogenesis

The occurrence of parthenogenesis has been proved for several terrestrial nematodes (Hyman, 1951), e.g. *Mermis subnigrescens* (Mermithoidea, Dorylaimida, Adenophorea; Hope, 1974). Parthenogenesis in these nematodes is accomplished through one of three variations in oogenesis; two represent meiotic and the third mitotic parthenogenesis. These are as follows:

1. Oogenesis proceeds as in normal amphimictic species, with synapsis and two maturation divisions: the first, a disjunction of homologous (paired) chromosomes, and the second, equational separation of split chromosomes. However, the pronucleus of what would be the second polar body fuses with the pronucleus of the ovum to restore the somatic number of chromosomes.
2. Normal meiotic oogenesis proceeds as above until the telophase II, which stops short of the formation of the second polar body pronucleus, and the two groups of telophase chromosomes become enclosed in a single pronucleus, thus re-establishing the 2N condition.
3. No synapsis of homologous chromosomes takes place and the somatic (2N) condition is maintained throughout the first division. A second maturation division does not occur.

Despite the absence of sperm, parthenogenetic nematodes express an anterior–posterior (a–p) polarity similar to *C. elegans*, as both generate, with their first cleavage, a larger somatic cell AB and a smaller germline cell P (Schierenberg, 2006).

The majority of the species that are known to reproduce by parthenogenesis appear to reproduce solely by obligatory mitotic parthenogenesis. In other instances, different populations within the same nominal species may reproduce by different methods. For example, one population of *Paratylenchus scribneri* (Tylenchida, Secernentea; Hope, 1974) reproduces by meiotic parthenogenesis, while another population reproduces by mitotic parthenogenesis. At least one instance is known in which a hermaphroditic species reproduces by mitotic parthenogenesis and pseudogamy. Pseudogamy is the necessity for sperm to penetrate the egg to initiate cleavage divisions, but without karyogamy, it occurs in several species of the family Rhabditidae. There is no direct evidence that any marine nematode may reproduce agametically. The ratio of males to females may be an indirect indication of the mode of reproduction for a given species, as amphimictic nematodes usually have a fairly equal proportion of males and females. On the other hand, parthenogenetically and hermaphroditically reproducing species would be expected to have fewer males, if any. In

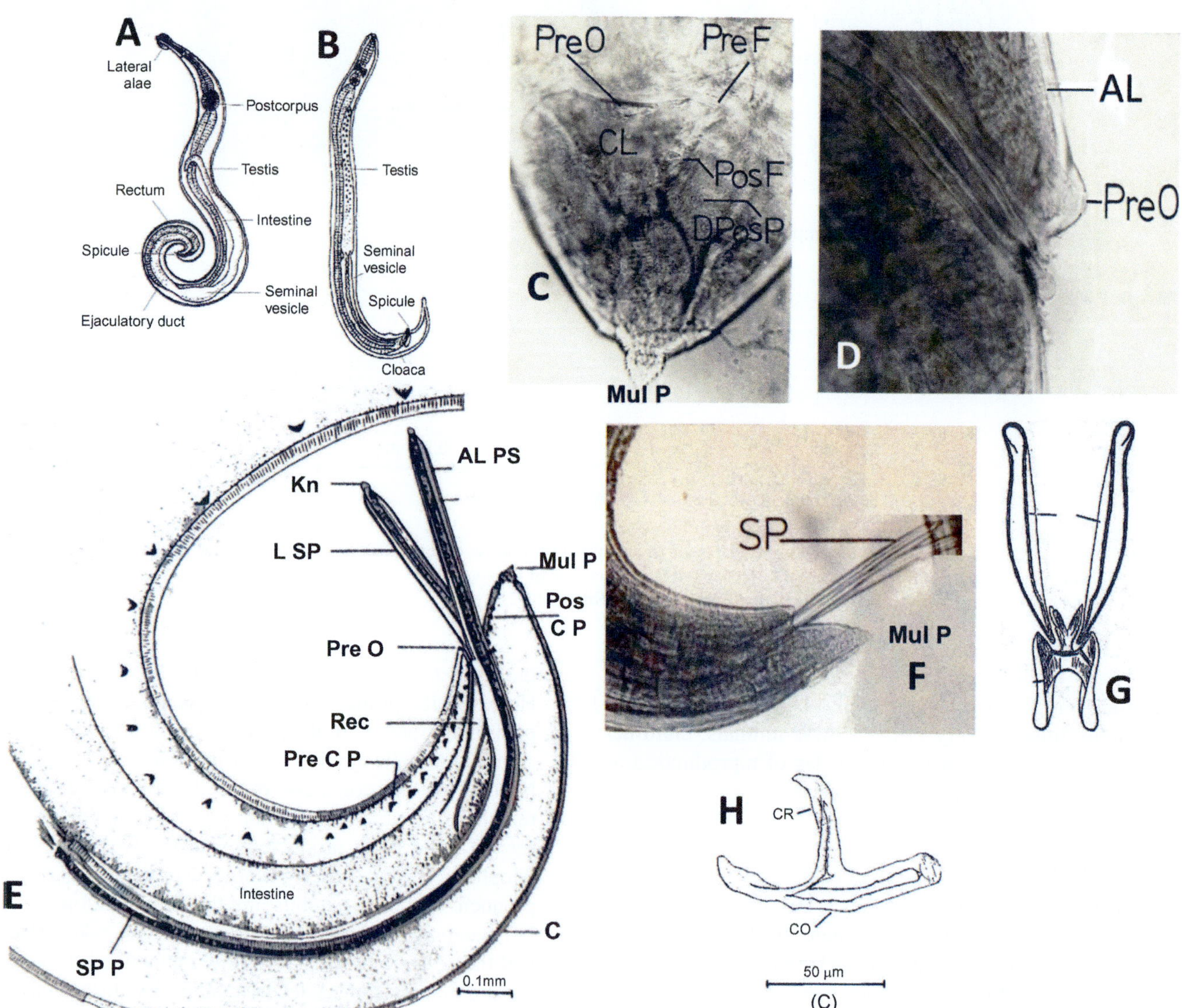

FIG. 9.4 Dimorphism in nematodes: (A) and (B) show the male coiled tail, (A) *Stercoraliss tercoralis* and (B) *Enterobius vermicularis (from Soliman, 1996)*; (C–F) tail of adult *Hysterothylacium reliquens* showing its structure *(from Abd El-Fattah, 1992)*; (C) photomacrograph of the ventral view; (D) photomacrograph of the lateral view to show the preanal organ; (E) camera lucida drawing of the lateral view, (F) photomacrograph of the lateral view; AL=ala, CL=cloaca, D Pos P=double postcloacal papilla, Pos F=postanal fold, Prec p=Precloacal papillae, Pre F=preanal fold, Pre O=Medioventrapreanal organ, Pos C P=postcloacal papillae, SP=spicule, SP P=spicule pouch; (G) spicules and gubernaculum of a chromadoroid, G=gubernaculum *(from Hyman, 1951)*; (H) right lateral view of gubernaculum of *Deontostoma californicum*, CO=corpus, CR=crus *(from Hope, 1974)*;

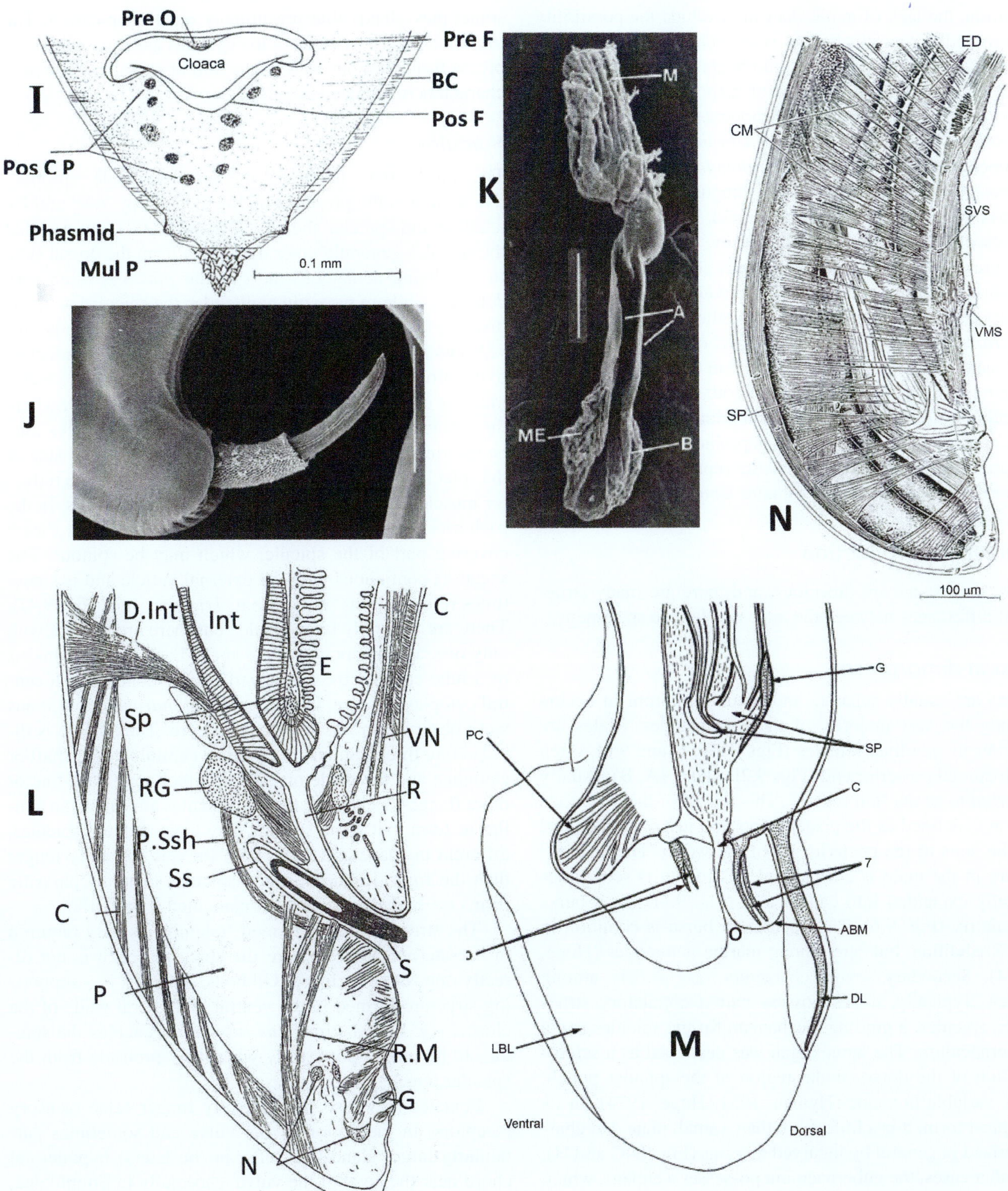

FIG. 9.4, CONT'D (I) camera lucida drawing of the tail of the male of *H. reliquens* showing the distribution of the postcloacal papillae; (J) posterior end of male *Trichuris suis* showing the single spicule protruding through the cloaca and part of the spinous sheath or cirrus; (K) one of the two spicules dissected out of the body of *Gazellostrongylus lerouxi* showing the muscles (M), branch (B), alae (A) and additional membranes (ME), scale bar of J and K = 100 μm, *after Gibbons and Khalil (1983), from Gibbons (2002)* (L–N) Tail region of male nematodes: (L) sagittal section through the tail of *Ascaris*, showing the relation of spicule to rectum *(from Hyman, 1951)*; (M) diagram of the lateral view of the male tail of an ostertagid nematode (Trichostrongyloidea) showing the components of the genital 'cone' *(after Gibbons and Khalil (1983), from Gibbons (2002))*; (N) male tail of *Deontostoma californicum (from Hope, 1974)*.

addition, the lack of males does not exclude the possibility of protandric reproduction; it is in such populations that a search for parthenogenesis would begin.

More convincing evidence of disproportionate sex ratios and the possibility of parthenogenesis was obtained by the study of the marine nematode *Chromadorina epidemos*, Hopper and Myers, 1967 (*Chromadorida*, Aphasmidia; Hope, 1974). It was cultured on fungus-infested cellulose matrix and was also found in the field. This nematode was represented in the field by both sexes, but the males were not present in the cultures. This was considered as an indication that this species might reproduce both amphimictically and parthenogenetically. Another species, *Viscosia macramphida* Chitwood, 1951 (Oncholaimina, Enoplida, Aphasmidia; Hope, 1974), are found in their natural habitat with relatively equal occurrence of both sexes. After 10 days in culture, all sexually mature individuals were females. It was concluded that this species is probably capable of both amphimictic and parthenogenetic reproduction, because several generations of only females were produced.

Gametic reproduction

Most nematodes are dioecious, and there are many structural differences between the male and female specimens.

Sexual dimorphism

Sexes are usually separate and sexual dimorphism occurs among the vast majority of these nematodes. Males are smaller in size than females (Figs 9.2A, B and 9.6), often with curved posterior end (Figs 9.2B and 9.4A, B), usually referred to as the 'tail region'. This region of the nematode worm is defined as the posterior portion just from the level of the anus to the posterior tip of the worm. The anal aperture in the male is cloacal and this region is sometimes distally expanded into caudal alae (Fig. 9.4D) or a bursa copulatrix (Fig. 9.6). The copulatory bursa is common in the Rhabditida, but rare among marine nematodes (Hope, 1974). Secondary sexual characters vary widely among males. Typically, males possess paired copulatory structures, spicules, a guiding mechanism for the spicules and a gubernaculum. The latter organ was described as a sclerotization of the dorso-caudal region of the spicular pouch, very variable in shape (Hyman, 1951; Hope, 1974). In its simplest form, it is a little more than a small plate, but characterized in general by incurved margins (Fig. 9.4G and H). In other cases, the gubernaculum possesses a cuneus, which is a medial process that extends into the spicular pouch between the spicula. According to Chitwood et al. (1950), the posterior part of the gubernaculum is then termed the corpus (Fig. 9.4H). It directs the spicules towards the anus and prevents them from piercing the cloacal wall. In addition, the males frequently possess various sensory receptors and sometimes glands that presumably aid in copulation. The morphological variations of the spicules and their accessory pieces furnish important generic and/or specific taxonomic characters in nematodes.

Spicules

The characteristic feature of the nematode male reproductive system is the presence of spicule(s) (Fig. 9.4E and F), telamon and spicular sheath. Gibbons (2002) reported that the spicules generally enter the cloaca from the dorsal side. In free-living nematodes the spicular pouch is often very delicate, while in parasitic nematodes the spicular pouch is distinct. In the majority of the species, each spicule is provided with two retractor and two protractor muscles attached to the proximal end of the spicule (Fig. 9.4L). The retractor muscles extend anteriorly towards the lateral chord, where they attach to the body wall. The protractor muscles may be inserted postanally to the body wall or the dorsal side of the spicular pouch. If only one spicule is present, the retractor muscle extends to both left and right body walls. In the trichurids, the cloacal lining is eversible, forming a sheath covering part of the spicule, which may be spinous. The sheath is continuous with the external cuticle and is sometimes referred to as 'cirrus', e.g. *Trichuris suis* (Fig. 9.4J). There are generally two spicules but there are species with only one, and others with none at all. Each spicule consists of a tube covered by sclerotized cuticle containing a central cytoplasmic core. In *Heterakis gallinarum*, nerve axons were observed in this cytoplasmic core. The spicule is divided into three regions: the head or capitulum; the shaft or calomus; and the blade or lamina, which may have one or more flanges. Generally, the two spicules are equal and similar in form, but some species may have unequal spicules, different in shape and form, with the left generally longer than the right. In addition, the spicules of many parasitic forms are branched, alate or ornamented (Fig. 9.4K).

The structure that is termed 'telamon' is only required in bursate nematodes where the spicular pouch is not directly opposite the cloaca (Gibbons, 2002). It is a supporting structure formed in the ventral and lateral walls of the cloaca, e.g. of *Hyostrongylus rubidus*. It enables the spicules to be turned posteriorly when they protrude from the spicular pouch.

Females may also have slightly longer setae (sensory receptors) in the region of the vulva and sometimes particularly large, gland-like cells, in the lateral hypodermal chord near the level of the vulva, especially in Enoploidea, Aphasmidia.

Hysterothylacium reliquens (Deardorff and Overstreet, 1980) (Phasmidia, Ascarida), a parasite in fishes, is a nematode that has been studied relatively recently (Abd El-Fattah, 1992) under our supervision (unpublished work) and has evidenced this generalization of the tail region. This worm has a short, conical, multispinous tail (Figs 9.4C, E, F, I and 9.5A, B). In the two sexes, the anal aperture has preanal and postanal folds. In the female these folds are simple, while in the male the middle portion of the preanal fold is thickened, papillated and elevated, forming a distinct, large projection referred to as a medioventral preanal organ (Fig. 9.4C, D, I). In the male, this region is characterized by the possession of a number of caudal papillae. These are sensory prominences found on the surface and arranged in two rows, each on either side of the mid-ventral line, anterior and posterior to the cloacal aperture (Fig. 9.4E and I). The total number of these caudal papillae differs in the different specimens of the present species and ranges from 31 to 37 pairs (Fig. 9.4E and I).

It appears that the tail region is very important to identify not only the sex, but also the genus and/or the species. In addition, some investigators study its interior structure by dissection or by histology as longitudinal sections (Fig. 9.4L–N).

A notable exception to the above generalization is found in the deep-sea nematode, *Rhaptothyreus typicus* (Hope and Murphy, 1969) (Enoploidea, Enoplida, Adenophorea; Hope, 1974). Males are characterized by such unusual features as the absence of a well-developed digestive system and the amphids (Fig. 9.3A) are exceptionally large and shield-shaped. Females have not been identified in the same samples. The females *are* present with the males; however, they may be parasitic, or are yet to be found in some other specialized but unsampled habitat. They have not been recognized due to extensive sexual dimorphism.

Intersexes

Intersexes are known to occur in species of all major groups of nematodes. Intersexes are found in some genera and should not be confused with hermaphrodites because only one set of reproductive organs is functional, the other appearing as a vestige. Intersexes are most commonly functional females with male secondary characters and/or occasionally with rudimentary testes. Being a functional female is frequently evidenced by the presence of eggs in the uterus. The male secondary characters that may appear on intersexes are the spicules, gubernaculum, ventromedian supplements, subventral supplements or copulatory muscles. A single male character may appear on female intersexes or they may occur in several combinations. Various degrees of maleness with different combinations of male features may occur between individual intersexes of the same species. Hope (1974) tabulated intersexes of marine and freshwater species. One of these is *Deontostoma californicum* (Enoplina, Adenophorea), which lives as female with normal female reproductive system and with all the male features mentioned above, but without testes (Hope, unpublished).

Anderson and Kimpinski (1977) studied a female intersex of a soil inhabitant species, *Aphelenchoides*, from a cornfield in Prince Edward Islands, Canada. It is a sterile intersex of a mycophagus species, identified as *A. composticola* (Aphelenchoidea, Aphelenchina, Tylenchida, Secernentea; Hope, 1974). Its description shows that the body, tail and reproductive system are female in character. The only male sexual structures evident are well-developed paired spicula. This intersex differs from all other female intersexes in having an underdeveloped gonad, gonoduct and postuterine sac. The underdeveloped reproductive system in the intersex from Canada is possibly due to a genetic factor expressed late in the development of the germinal primordium. If associated with factors responsible for proliferation of male tissue, this intersex would represent another grade of intersexuality in nematodes. The arrested maturation of germ cells in the *Aphelenchoides* intersex from Canada may be characteristic of other intersexes in the Tylenchida.

Male intersexes with female secondary characters have also been reported, but are less common. Hope reported that maleness could be produced in females by certain environmental forces, and when the force was insufficient to transform the females completely into a functional male, they became females. Hope also reported that some genetic factors are believed to be operative in the production of intersexes in other animals, but in nematodes these need further study.

Hermaphroditism

Protandric hermaphroditism has been well documented among soil nematodes. The gonad first produces sperm that are stored, and later fertilize the eggs that are subsequently developed by the same gonad. Hermaphroditism is usually called 'syngony'. Syngonic nematodes have the external appearance of females (Hyman, 1951). The sperm produced by syngonic females are usually insufficient to fertilize all the eggs matured by the ovary, meaning that following the production of fertilized eggs, there is a period of emission of sterile eggs. Sometimes the ovary reverts to sperm production again. In some hermaphroditic nematodes, the sperm production continues throughout the egg-laying period, simultaneously with the production of eggs.

It is generally regarded that among nematodes, hermaphroditism has evolved from gonochorism. This belief is based largely on the numerous degrees of hermaphroditism extant among the rhabditids (Hope, 1974). In many species of *Rhabditis*, there is a remarkable disparity in numbers of the sexes. For 1000 females there may be only 10 or 20 males, and they are lethargic in their sexual activities. The females, on the other hand, have developed an interesting kind of hermaphroditism. When the gonad first becomes ripe, a number of spermatozoa are produced. Afterwards the gonad produces only eggs, which are fertilized by the individual's spermatozoa, and after these are exhausted, only sterile eggs are laid. Experiment has proved that in these animals, self-fertilization may occur for an immense number of generations without any deterioration of the species (Borradaile et al., 1977). Nigon (1949) studied the hermaphrodites of *Caenorhabditis elegans* (Rhabditoidea, Rhabditida, Secernentea) and postulated that they are derived from normal bisexual species through mutation or hybridization.

Reproductive system

Chitwood et al. (1950) reviewed the anatomy of nematodes, including the reproductive system, adding original observations of their own. Hyman (1951) presented a review, with exceptions, clarifying the description with good illustrations. Something similar was done by Hope (1974); in addition, he contributed in detail that of *Deontostoma californicum*. Basically, there are two gonads in all nematodes. They are of tubular shape, varying greatly in length, and may be straight, sinuous, reflexed or coiled back and forth. There are two types of gonads according to the site of the production of the gonocytes: **hologonic** gonad, in which the gonocytes are produced throughout the length of the gonad; and the **telogonic** gonad, in which the production is limited to a short region at the distal end of the gonad. Each gonad is continuous with a separate gonoduct.

In males, a single testis is usually present (monorchic) as in *Rhabditis*, and extends anteriorly (Fig. 9.6). More frequently two testes are present (diorchic), each of which is continuous with a separate seminal vesicle. The number of testes is not necessarily correlated with the number of ovaries in females of the same species. Generally, the testes are oppositely oriented (Fig. 9.7 except A, C, H) except in *Heterodera* (Heteroderoidea, Tylenchida, Secernentea) *marioni*, where they are parallel (Fig. 9.7H), and in *Anticoma typica* (Enoploidea, Enoplina, Adenophorea) in which they have a tandem arrangement (Fig. 9.7A) (Hyman, 1951). At the point where the two seminal vesicles converge begins a common vas deferens and/or ejaculatory duct. The latter opens into the hind-gut (cloaca) and then to the exterior through a slit-like transverse vent (Hope, 1974).

In monorchic types (e.g. genus *Labiduris*), the system consists of a testis, seminal vesicle and vas deferens (Fig. 9.6). In some forms, a vas efferent is present and separates the testis from the seminal vesicle (Gibbons, 2002). The posterior end of the vas deferens, if provided with a muscle layer, may act as an ejaculatory duct. In addition, ejaculatory glands, which open into the posterior end of the vas deferens, may be present. The vas deferens enters the rectum from the ventral side in all species except in the trichinelloids, where it enters on the dorso-lateral side. In the phasmidian nematode species, the junction of the vas deferens with the hind-gut is very close to the intestine-rectal valve. In these species, there is almost no rectum and the whole of the hind-gut forms part of the cloaca. In the adenophorean species, particularly the Enoplida, the vas deferens may join the rectum more posteriorly so that there is a rectum and a cloaca (Gibbons, 2002).

In **females**, there are usually two ovaries (didelphic), oriented in opposite directions (amphidelphic) (Fig. 9.8A). They may be straight, sinuous, reflexed or, if very long, wound back and forth. The ovaries and ducts may be arranged in a parallel position (Fig. 9.8B) extending either anteriorly or posteriorly (Hyman, 1951). In the female *Hysterothylacium reliquens* (Abd El-Fattah, 1992), the whole system occupies the posterior portion of the body (Fig. 9.5A). It was described as didelphic and opisthodelphic by Yamaguti (1961). It shows a peculiarity that one of its ovaries extends anteriorly for a short distance, just above the vulva (Fig. 9.5C). The reduction of the female system to a single ovary and duct (monodelphic) is not uncommon (Fig. 9.8C). The 'polydelphic' condition, in which there are more than two (up to 10 or 11), also occurs, especially in the Spiruroidea. Gibbons (2002) reported that in some animal-parastitic nematodes there may be up to four (*Polydelphis*), six (*Hexametra*) or 32 (*Placentonema gigantissima*). Each ovary, which is a blind ending sac, continues distally as the oviduct, which soon widens into a broad tube, the uterus. The beginning of the uterus usually functions as a seminal receptacle in which sperm are stored and fertilization occurs (Fig. 9.8B and C). In addition, the uterus functions in the shell formation, and some degree of embryonic development, and in some cases as far as the formation of juveniles (Fig. 9.5D). The two uteri unite to form a common short vagina (Figs 9.5B and 9.8C) that opens to the exterior by the female gonopore or vulva on the ventral surface of the body. The uterus and vagina have a highly developed musculature designed for the efficient ejection of the eggs and referred to as 'ovejector' (Fig. 9.8A). The morphology of this organ in the Strongyloidea is divided into two types: 'ovejector I' is usually Y-shaped; and 'ovejector II' is generally J-shaped (Gibbons, 2002).

The female gonopore is a mid-ventral transverse slit with bulging lips that can be anywhere along the length of the body. Usually, however, it is situated along the middle third of the body but may be shifted anteriorly, or to the vicinity of the anus, or it may protrude from the body surface near the tail (*Pseudalius inflexus*, Fig. 9.5D) (Gibbons, 2002). Extremes in the posterior displacement of the vulva are found in the marine nematode family Lauratonematidae (Enoploidea, Enoplina, Adenophorea). In the genus *Lauratonemoides*, the vulva is located just anterior to the anus, while in the genus *Lauratonema* the vulva actually opens into the rectum, being a true cloaca (Hope, 1974). However, Hyman (1951) reported that *Rondonia rondoni*, a parasite of fishes, is apparently the only nematode in which the vagina opens into the rectum. Both vaginal and vulvar

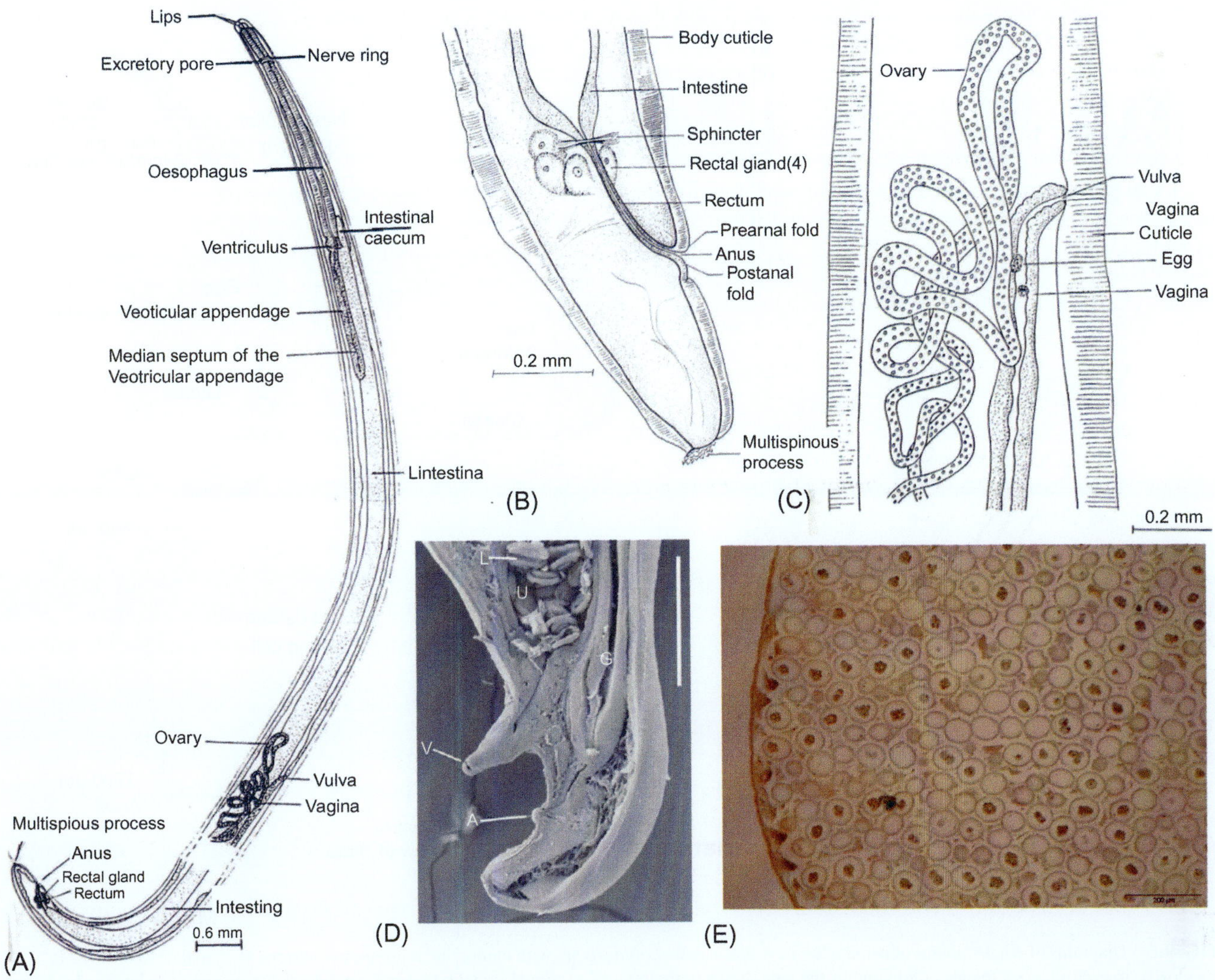

FIG. 9.5 Camera lucida drawings of the female adult *Hysterothylacium reliquens*: (A) lateral view of the whole Animal showing general features; (B) posterior portion of the animal; (C) lateral view of the region of the vulva, notice the eggs inside the vagina; (D) body of *Pseudalius inflexus* cut open to show the uterus (U) with larvae (L) and the vulva (V) opens on a projection just anterior to the anus (A) near the tail, scale bar = 231 μm *(after Gibbons (1986), from Gibbons (2002))*; (E) portion of L.S. of the uterus of *Ascaris megalocephela* (=*Parascaris equorum*) showing different stages of development of the eggs, scale bar = 200 μm (laboratory preparation).

glands (Fig. 9.8D) may occur, chiefly in free-living marine nematodes. The vulvar glands consist of one to several pairs of gland cells situated anterior and posterior to the vulva. Observations that eggs of *Enoplus communis* (Enoploidea, Enoplida, Adenophorea; Hope, 1974) are deposited in a string-like mass evidenced that gland cells are associated with the female reproductive system (Hope, 1974). Dilator muscles in the form of radiating bands encircle the vulva in most nematodes.

Origin of germ cells and gonads

Hyman (1951), followed by Hope (1974), reviewed the embryogeny of the reproductive system of nematodes from the early classic studies in the cytology and germ cell lineage of certain parasites such as the studies of Boveri (1892–1910). The early success in following the germ cell lineage is substantially attributable to the facts that the germ cell line, is in some ascarids segregated from the somatic cells early in development, and these lines can be distinguished readily from one another. The distinction lies in the occurrence of chromatin diminution in the somatic line, a phenomenon that is absent in the germ cell line. By this distinction, it has been determined that the stem line cell classically designated P3 divides to produce stem line cell P4 and somatic cell S4. P4 then divides to produce the paired primordial germ cells GI and GII which, after gastrulation, come to lie in the blastocoel at or slightly posterior to the mid-body length. A pair of somatic cells, one on each opposite end of the paired primordial cells, is invariably present, and these give rise to the epithelium of the gonad and gonoduct (see

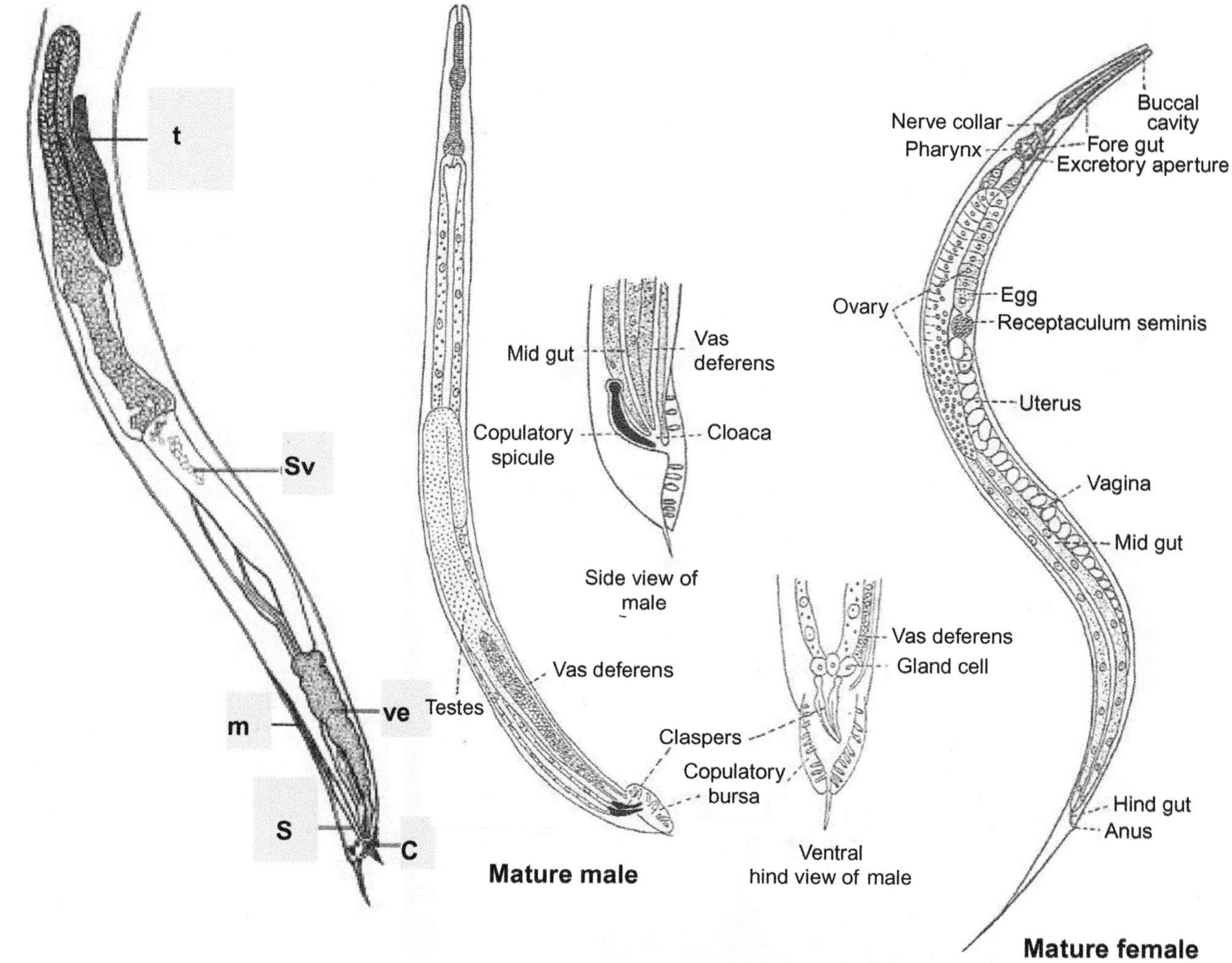

FIG. 9.6 Diagrams of whole mounts of nematodes: Left side is male *Labiduris* sp. with monorchic reproductive system. Its female is shown in Fig. 9.8C. c=cloaca, m=muscles, s=spicules (only one of the pair shown complete), sv=seminal vesicle, t=testis, ve=vas deferens and ejaculatory gland *(from Gibbons, 2002)*; others are male and female *Rhabditis (after Maupas, from Borradaile et al., 1977).*

later, Fig. 9.15). This pair of somatic cells may be derived from somatic cell S4. Hope (1974) reported that in some other nematodes the somatic cell S4 is ectodermal, giving rise to the hind-gut, while P4 divides to produce P5 and S5. The former is the definitive primordial germ cell and the latter gives rise to the epithelium of the gonad and gonoduct. As in the case of the ascarids, P5 and S5, each divide one more time during early embryogeny and become paired. Further development of the gonads and gonoducts is arrested until the first stage juvenile is fully formed. With the completed development of this juvenile and, usually, its hatching from the egg, the genital primordial resumes development.

Gametogenesis

Hope (1974) mentioned that gametogenesis of nematode parasites of plants and animals has been studied extensively and is usually by normal meiosis, and amphimixis is the common mode of reproduction. Spermatogonia and oogonia undergo mitotic divisions in the germinal zone, and the beginning of the growth zone occurs at the level where the mitotic divisions cease. The chromatin material of the gametocytes becomes compacted into a densely staining network or mass, which indicates synapsis.

Spermatogenesis

Spermatogenesis in *Ascaris megalocephala* (=*Parascais equorum*) is one of the most famous subjects of cytology and has been extensively investigated ever since Van Beneden and Julin (1884) (Roosen-Runge, 1977). This author reviewed, in detail, all the works concerning this vital activity although without illustrations or photos. However, he reported that spermatogenesis in nematodes has some unusual and variable features. He summarized the outstanding phenomena that occur in *Ascaris*, as the only genus that has been thoroughly and frequently examined, are the

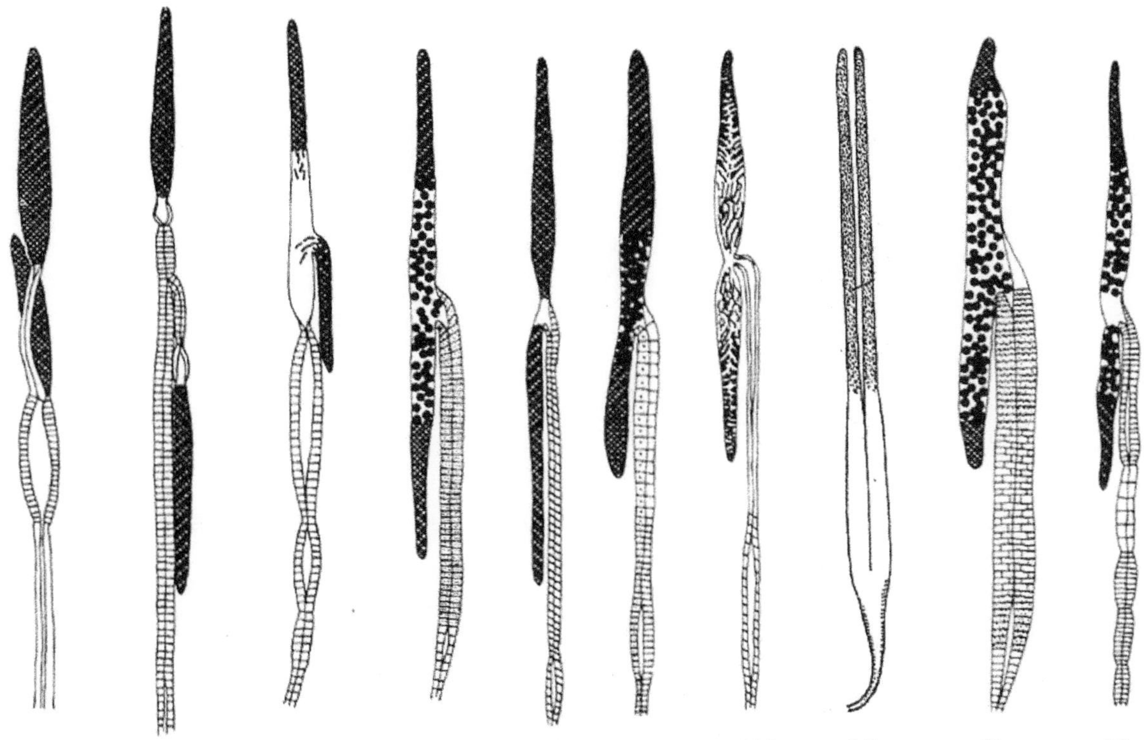

FIG. 9.7 Male reproductive system in nematodes showing different modes of diorchic species: (A) *Anticoma typica*, both testes are directed anteriorly and their positions are tandem to one another; (B) *Agamermis decaudata* with two opposite testes and seminal vesicle; (C) *Enoplus communis* with opposite testes but common sperm duct; (D–G) with only the testes are opposite to each other; (D) *Desmolaimus zeelandicus*; (E) *Metoncholaimu spritiuris*; (F) *Anaplectus granulosus*; (G) *Tobrilus gracilis*; (H) *Heterodera marioni* with parallel system; (I) *Halichoanolainus robustus* with system like that of D–G; (J) *Sabatieria hilarula* has the same of J. *(All from Hope (1974) except (H) after Atkinson from Hyman (1951).)*

formation of a rachis in the spermatogonial stage. The electron microscopic study concluded that spermatogonia, in contrast to oogonia, detach from the rachis after their last divisions. All stages of the development of the male gamete of *Ascaris* are characterized by the formation of granules. Histo- and cyto-chemical studies show that these granules are proteinaceous in nature and in the spermatid stage, they fuse to form the 'refringent cone', which later constitutes the tip of the amoeboid nonflagellar spermatozoon (Fig. 9.10(I) (a)–(d)). The development of a cytophore, not yet entirely clarified, connects a quartet of spermatids, and there is an unusual accumulation of proteinaceous 'yolk granules' in the spermatocytes. Many authors have pointed to the great similarity between the formation of these unique granules in the male gametes of *Ascaris* and the formation of proteinaceous yolk in its oocytes and in other animals.

In many other nematodes, these features have not been seen. A rachis is not a general feature of nematodes. It appears that nurse cells of some kind play a prominent role in the species without rachis.

Deontostoma californicum (Hope, 1974) has two testes; the posterior is 2.46 mm long, while the anterior testis is 1.90 mm. Histologically, the testis is enveloped in an extremely thin glycocalyx (tunica propria). Underneath this envelope there is a single layer of squamous epithelium, and each cell has a single nucleus. The electron microscope shows that this epithelium is not syncytium. The epithelium encloses a germinal chord which is telogonic, but without a rachis (a feature of marine nematodes). The distal end of the germinal chord includes spermatogonia (Fig. 9.9A); these undergo mitotic divisions near the distal end of the testis. The division gives rise to spermatocytes, each has little cytoplasm and contains a nucleus that stains darkly. Nearly 37% of the posterior testis is occupied by these cells. Meiotic divisions occur in the following region, which begins about 1.2 mm from the distal end of the testis and extends proximally for a distance of about 150 μm. Within a single section taken near the middle of this region, several stages of meiosis can be observed (Fig. 9.9B). The second division follows producing relatively small spermatids, each with a nucleus, densely packed with chromatin material.

Spermiogenesis occurs in the remainder of the testis. This process produces mature sperm; each is spindle-shaped, measuring ~8 × 18 μm with a nucleus 3 × 15 μm (Fig. 9.9C).

The sperm of nematodes is usually quite unlike that of other animals. It may have a rounded, conical, lobulated, or elongated body (Fig. 9.10(I)) that moves slowly in amoeboid

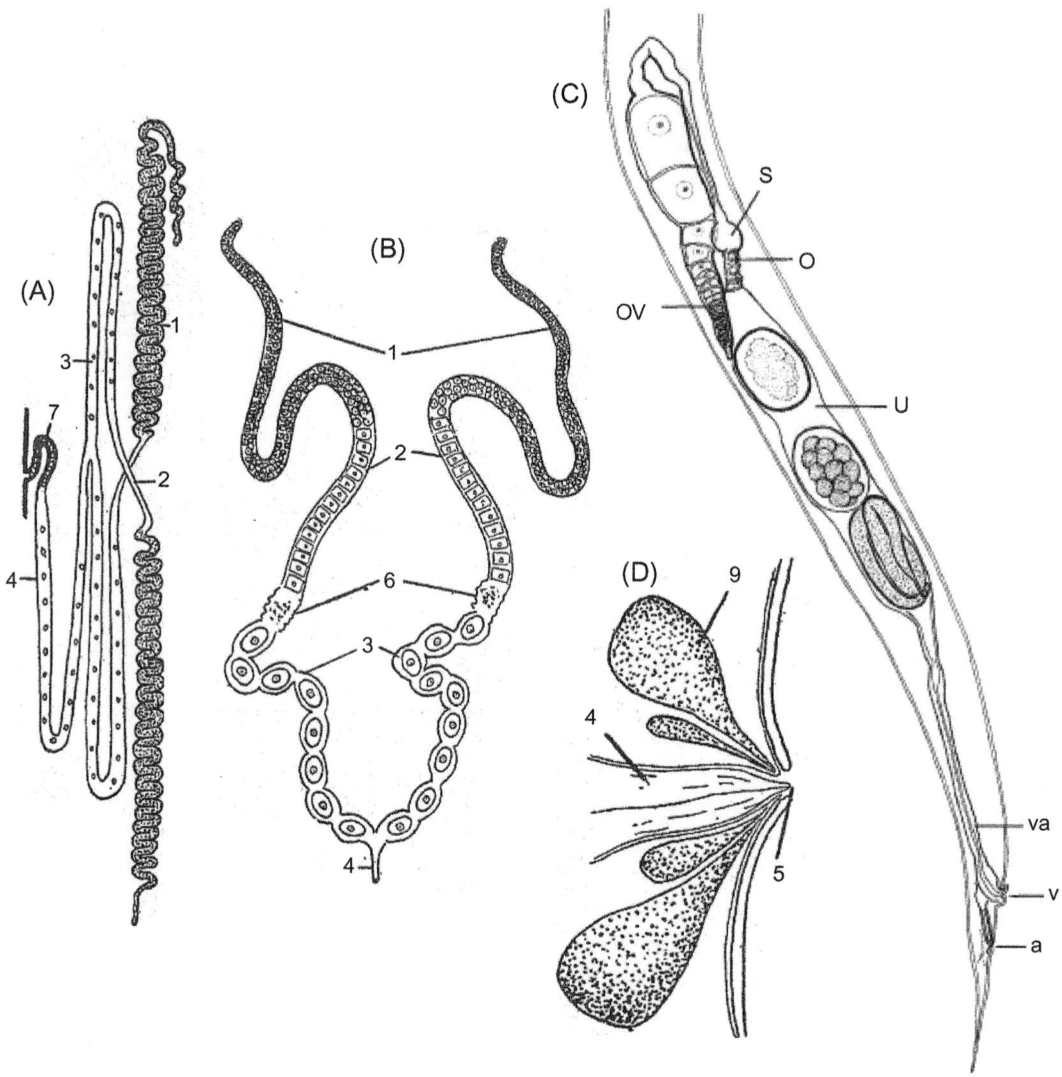

FIG. 9.8 Different types of female reproductive system in nematodes: (A) opposite sinuous ovaries and simple ovejector in *Heterakis gallinae*; (B) anterior parallel ovaries in *Heterodera*; (C) monodelphic female system in the nematode genus *Labiduris*; (D) vulvar glands of *Halichoanolaimus*. 1, ov=ovary, 2, o=oviduct, 3, u=uterus, 4, va=vagina, 5, v=vulva, 6, sr=seminal receptacle, 7=ovejector, 9=vulvar glands, a=anus. *((A, B, and D) From Hyman (1951); (C) after Gibbons (1986) from Gibbons (2002).)*

fashion. Even when the external shape shows the flagellate type, the cytological structure characteristic of flagellate sperm is wanting (Fig. 9.10(I)). In *Deontostoma californicum*, the lumen at the proximal end of each testis contains, in addition to the mature sperm, a substance of uncertain identity (Fig. 9.9C). Hope (1974) suggested that this substance might be a secretion of the epithelium. Up to that time, Nematoda was considered as one of the phyla whose members do not form spermatophores, and the male discharges the sperm to the female as semen. On reviewing the literature, we discovered that some nematode species form spermatophores. The newly discovered genera of nematodes, *Rhynchonema* Cobb, 1920 and *Prorhynchonema* (Xyalidae) obtained from the intertidal zone, and adjacent seagrass, *Zostera marina*, Linne bed in Amakusa, south Japan, are peculiar in the production of one or two spermatophores. The latter are transferred to the female by the spicular system of the male (Gourbault, 1982; Gourbault and Renaud-Mornant, 1982a,b).

Justine (2002) reviewed the ultrastructure studies of nematode sperm. By that time, a relatively large number of species sperm had been studied; about 73 species belonging to around 56 genera. Accordingly, the nematode sperm can be broadly divided into round and elongate types (Fig. 9.10(II)) and morphologically it is generally characterized by:

1. Constant absence of cilia or flagella
2. Development of pseudopodia, at least in utero, and amoeboid motility
3. Presence of a special protein, 'major-special-protein' (MSP), associated with amoeboid motility in most sperm

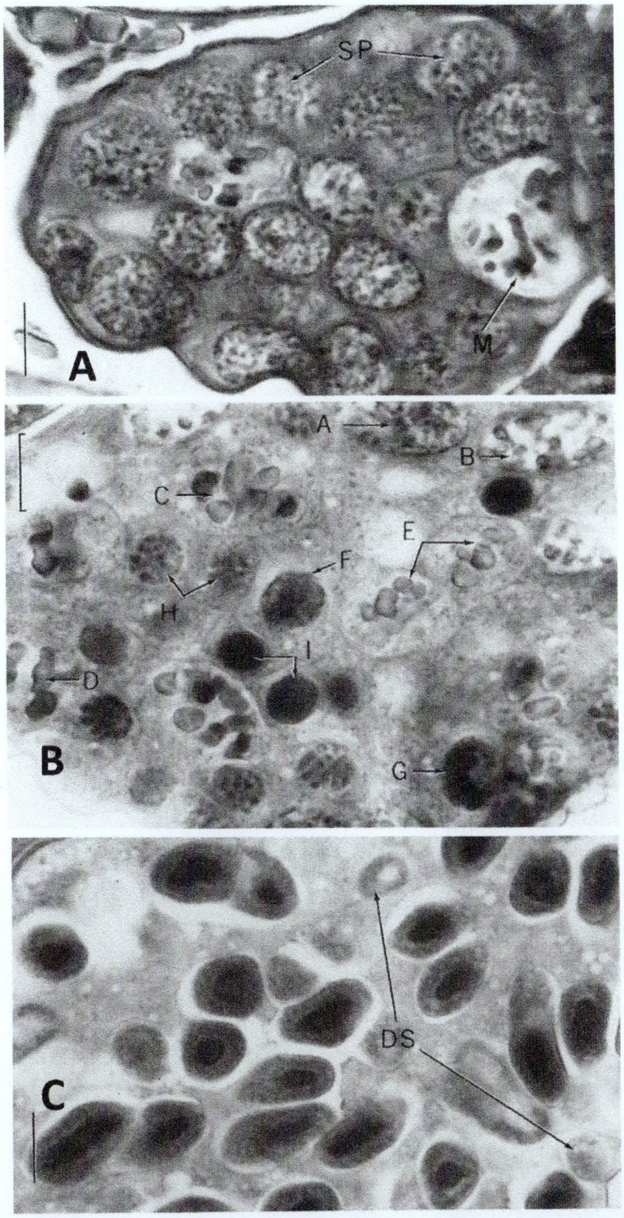

FIG. 9.9 T.S.s through different regions of the testis of *Deontostoma californicum*: (A) distal end showing the germinal chord is comprised of spermatogonia (SP) which are proliferated by mitotic divisions (M); (B) posterior region (the region of meiosis), leptonema or zygonema of prophase I (A), pachynema or diplonema of prophase I (B), prometaphase or metaphase I (D), telophase I (E), metaphase II (F), early anaphase II (G), telophase II (H), and young spermatids (I). As spermiogenesis proceeds, the nucleus becomes densely packed with chromatin material; (C) proximal region in which the spermatozoa are sectioned transversely and obliquely. The lumen is filled with amorphic substance, which envelops degenerating sperm (DS). Scale line indicates 10 μm. *(From Hope (1974).)*

studied. This is correlated with the presence of 'fibrous bodied' in early stages (Fig. 9.10(III)), and a fibrillar cytoskeleton in mature spermatozoa (Fig. 9.10(II)). Exceptions may be due to insufficient data

4. Presence of unusual membranous organelles in mature spermatozoa of most nematodes studied (Fig. 9.10(II) A).
5. Presence of modified centriole(s) with triplets replaced by singlets
6. Absence of a nuclear envelope at maturity (except in the Enoplea); and
7. A variable form, from rounded to elongated

It appears that most of these ultrastructure descriptions evidenced the observations of the light microscope studies. The only addition is the description of the fine details of the inclusions of the sperm of some species. The most characteristic were part of the motile cellular system, which was described for only two species: *Coenorhabditis elegans* (see later) and *Ascaris suum* (Fig. 9.10(II), (IV) A). The MSP is responsible for this amoeboid nematode sperm. In most amoeboid cells, such as leucocytes, motility is based on actin and its associated proteins. In nematode amoeboid sperm, this actin is replaced by an extensive filament system known as MSP. *A. suum* has been used for studies on the protein itself due to the large size of the worm and its large quantities of sperm, and therefore of the protein.

The ultrastructure description of the sperm in the uterus of a female *A. suum* (Justine, 2002) is shown in Fig. 9.10 (IV) A. It has a clear anterior region, pseudopod, and a denser posterior region, which contains the nucleus. The latter is rounded, dense and lacks a nuclear membrane. It also contains a characteristic refringent body, mitochondria and membraneous body organelles. The cytoplasm of the pseudopod possesses fibrillar elements consisting of the active form of MSP and forms the pseudopodial processes. The posterior region contains a large 'refringent body', which has a glistening appearance under the light microscope. It appears that this body is present only in ascarid nematodes.

Oogenesis

Most of nematodes have telogonic ovaries, in which the production of oogonia is limited to a short region at the distal end of the ovary nearest to the vulva. A narrow cylindrical structure termed the rachis commonly exists along the median axis of the ovaries in ascarids, oxyurids, strongylids, and spirurids. The developing oogonia are radially arranged around this rachis (Figs 9.2C and 9.23C). It is an important structure despite its presence in some species, and its absence in other closely related forms. It consists of nonnucleated tissue, containing such organelles as lamellar bodies, lipid droplets, dense granules and what might be ribosomes. The oocytes are attached to the rachis up to a short distance before the beginning of the oviduct. The rachis maintains the spatial arrangement of the oocytes. Within the lumen of the oviduct, the oocytes lie free. Ovaries of marine nematodes, although telogonic, are not known to possess a rachis (Hope, 1974).

The ovary of *Ascaris lumbricoides* and *Parascaris equorum* is telogonic, has an epithelial cell at its blind proximal end, which functions as an undifferentiated germinal cell giving rise to the oogonia. According to Chitwood et al.

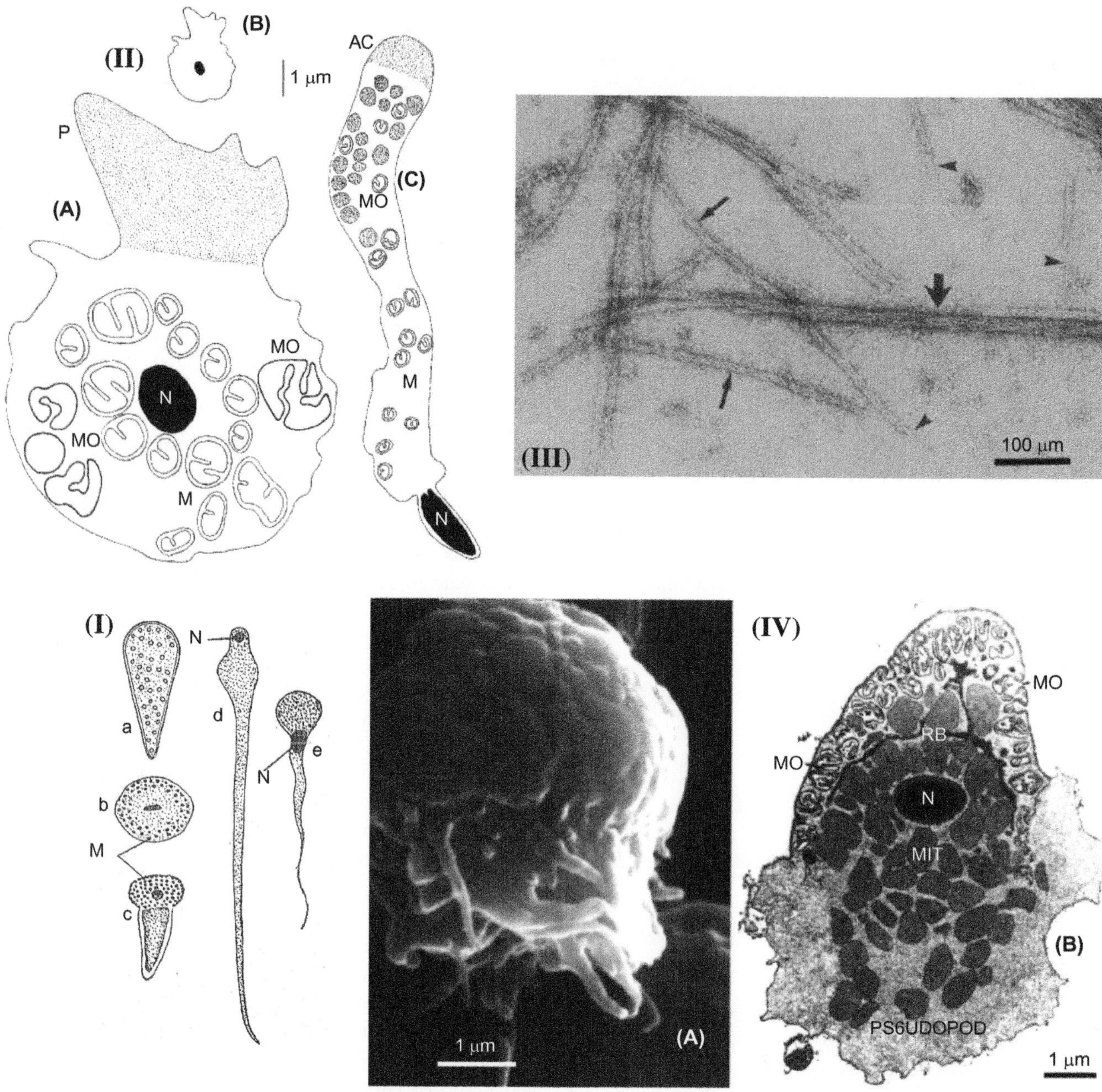

FIG. 9.10 (I) Light microscopical drawings showing types of nematode sperm, c of *Ascaris megalocephala*, d of *Tilobus*, e of *Passalurus (from Hyman)*; (II) SEM of variety of form and size of the same, (A) of *Caenorhabditis elegans* small rounded, (B) diagram at the same magnification as the other species, (C) of *Heligmosomoides polygyrus* large elongate sperm, scale bar valid for (B), (C); (III) structure of MSP filaments of *Ascaris suum* negatively stained with uranyl acetate, *small arrows* show filaments have a characteristic axial pattern of dots repeated every 11 nm, *arrowheads* indicate subfilaments projecting from filaments or in the background, *thick arrows* indicate filament bundle forming macrofibres commonly containing three filaments; (IV) (A) SEM of spermatozoon of *Ascaris suum* after addition of sperm activating substances extracted from the male accessory glands. Note filipodia at the anterior end; (B) TEM of mature spermatozoon of *Ascaris lumbricoides* dissected from the uterine-oviduct junction of a female. AC=anterior cap, MIT &M=mitochondria, MO=membraneous organelle, N=nucleus, P=pseudopodia, RB=residual body. *(II, III, and IV) From Justine (2002), after Abbas and Cain (1979), Stewart et al. (1994) and Nelson and Ward (1981).*

(1950), there was no evidence of nuclear division observed in this cell. Such evidence is absent in the case of *D. californicum* as well. Hyman (1951) termed this as the germinal zone (Fig. 9.11B). It is succeeded by a growth zone. The germinal zone occupies a small distal region (Fig. 9.11C) where small spherical oogonia are proliferated. The beginning of the growth zone is identified by the abrupt transition of the oocytes from the spherical to the disc shape, each with its flat surface facing the adjacent oocytes.

In *D. californicum* and the other leptosomatids, the ovary is in reality an elongated epithelial sac, closed at both ends, but with an opening into the oviduct. This ovary is

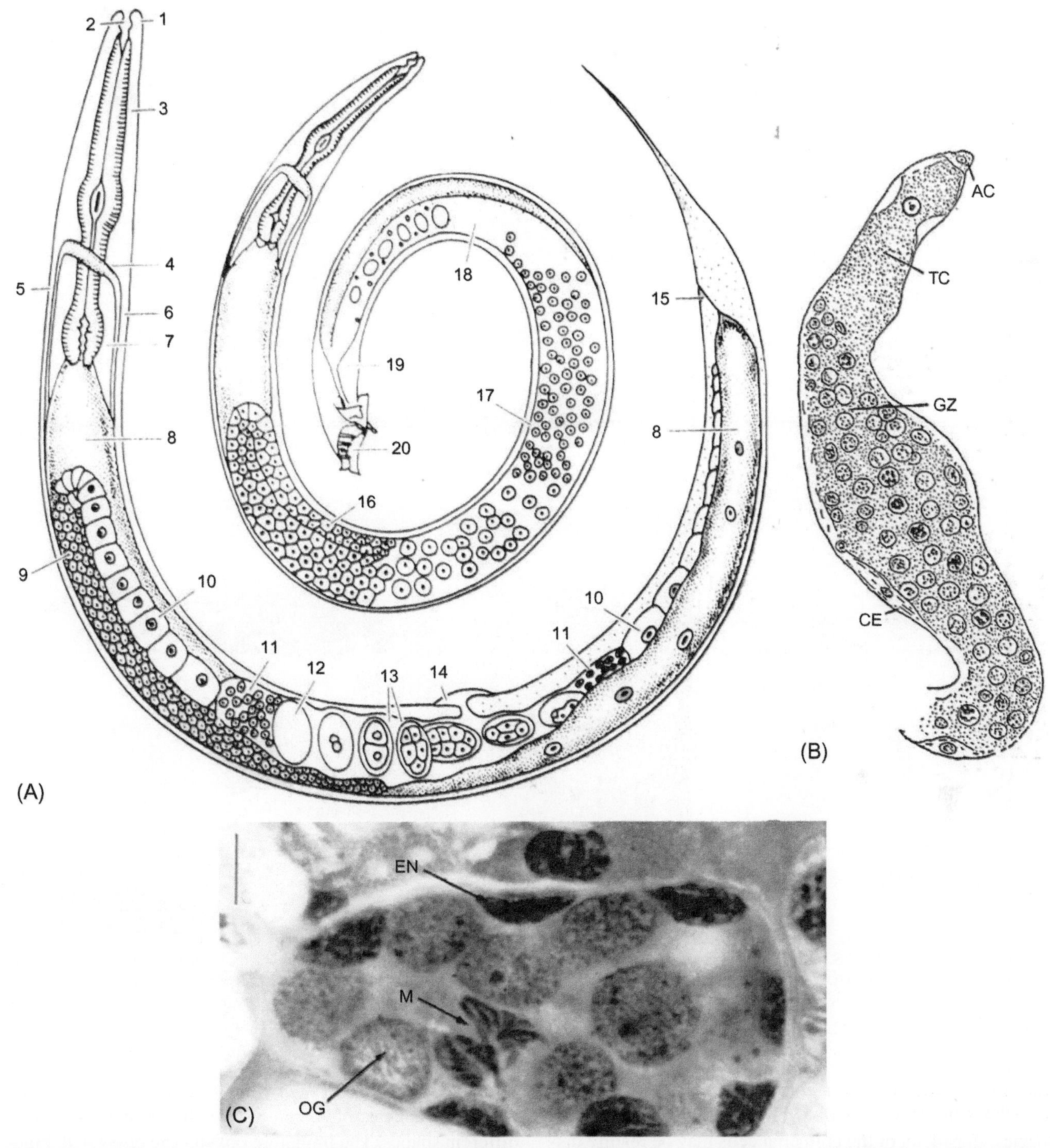

FIG. 9.11 (A) General anatomy of a nematode. Larger animal outside female; smaller animal inside, male. 1=one of six lips, 2=mouth, 3=pharynx, 4=nerve ring, 5=dorsal nerve cord, 6=ventral nerve cord, 7=posterior pharynx bulb, 8=intestine, 9=oogonia, 10=oocytes, 11=sperm, 12=freshly fertilized eggs, 13=early cleaved stages, 14=vulva, 15=anus, 16=spermatogonia, 17=sperm, 18=vas deferens, 19=cloaca, 20=copulatory bursa *(after Meglitsch and Schram, 1991, from Schierenberg, 1997)*; (B) diagram showing the tip of gonad of pig ascaris *(after Musso from Hyman (1951))*; AC=apical cell, CE=covering epithelium, GZ=germinal zone, TC=terminal cell; (C) T.S. through the germinal zone of the ovary of Deontostoma californicum showing the proliferation of oogonia (OG) in the germinal chord by mitotic divisions (M). The entire ovary is enclosed by epithelial cells (EN), scale bar=10 μm;

(Continued)

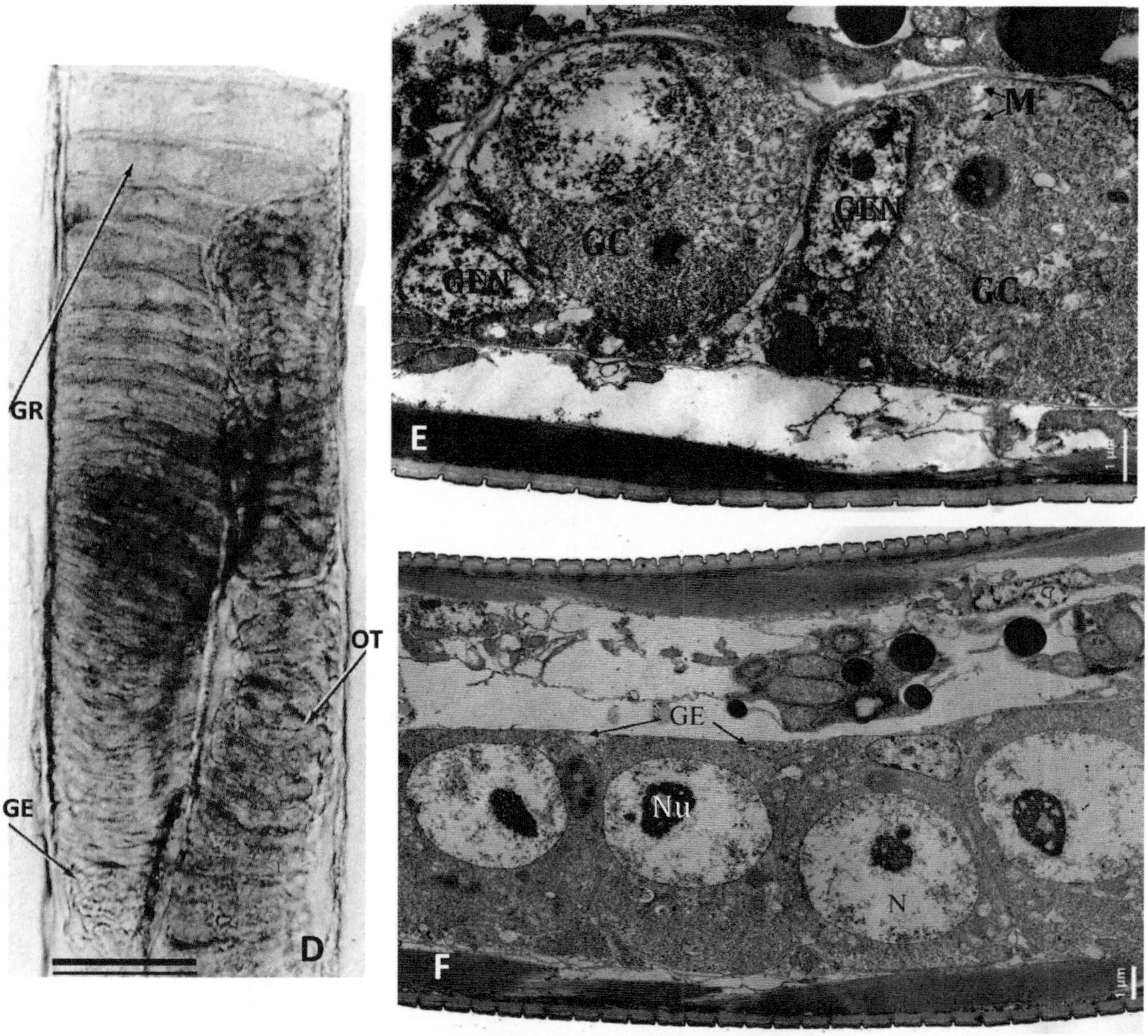

FIG. 9.11, CONT'D (D) T.S. of the same ovary showing the oviduct (OT) is connected to the dorsal side of the ovary not far from the germinal zone, scale bar=50 μm *((C and D) from Hope, 1974)*; (E) T.S. through the distal region of the ovary of *Pratylenchus penetrans* showing three distal cells; (F) T.S. through the proximal region of the same ovary showing growth stage of the oocytes; GC=germinal cell, GE=gonad epithelium, GEN=nucleus of gonad epithelium, M=mitochondria, N=nucleus, Nu=nucleolus, *(E and F) from Justine, 2002.*

telogonic, and consists of a chord of germinal tissue enveloped in a single layer of epithelial cells (Fig. 9.11F). In this species, the nuclei of the epithelial cells may be readily distinguished from those of the oogonia by their smaller size, flattened form and being more basophilic with coarser nucleoplasm. The gonoduct is located on one side of the ovary fairly close to the juncture between germinal and growth zones (Fig. 9.11D).

Oocyte development has been studied in detail in *Ascaris lumbricoides* (Justine, 2002). The oogonia and the rachis in the region of early oocytes contain numerous smaller bodies. The oocytes are attached to the rachis by cytoplasmic bridges, which contain many microtubules. The SEM studies on the oocytes of *A. suum* have shown that they are of different shapes. The immature oocytes are elongated, bipolar cells, with an atypical region, which contains numerous lobate cytoplasmic processes, and a slender opposite pole corresponding to the basal region previously attached to the rachis. Later, they gradually round up. Justine (2002) presented an extremely richly illustrated study of oogenes in *Pratylenchus penetrans* (Tylenchoidea, Tylenchida, Secernentea; Hope, 1974). The germinal cells (oogonia) in the 'distal' region of the ovary have spheroid nuclei, numerous polyribosomes, high concentrations of rough endoplasmic reticulum, mitochondria, dictyosomes, and electron-dense granules (Fig. 9.11E). The oocytes, in the middle of the ovary, enlarge and accumulate in a single row (Fig. 9.11F). They have similar organelles as in the oogonia. Oocytes, prior to their entry into the oviduct, show relatively few cytoplasmic inclusions. In the spermatheca, they have numerous, large lipid droplets and protein granules.

Borradaile et al. (1977) reported (e.g. in *Ascaris*) that a single ovum is discharged at a time into the oviduct; as soon as this happens, another ovum ripens in its place. To reach the uterus, the egg first has to pass through the very short region, the seminal receptacle, where fertilization occurs.

Reproductive behaviour

Sex attractants

Although parthenogenetic and syngonic reproduction occur in nematodes, most species are dioecious and fertilization is internal. Both sexes are commonly mobile, and all appear to have sensory receptors. It is not unreasonable to assume that for a given species, one or both individuals would be attracted to the other for purposes of reproduction. Investigations on *Ancylostoma caninum* and different other nematodes have provided evidence that this attraction is chemically mediated, and that members of one or both sexes will migrate to the general locality of the other. Females are attracted to males in *A. caninum.* In addition, attraction of females to males was greater than males to females in the case of *Trichinella spiralis*. The females of some nematodes are known to produce a pheromone that attracts males. This chemical attractant seems to be species-specific, and it is likely that chemical attraction between opposite sexes of the same species will prove to be the rule rather than the exception.

Copulation

In nematodes, copulation generally takes place with the tail and posterior region of the body of the male. The male is usually oriented at an angle to the female and often coils about its vulval region (Fig. 9.12A). Borradaile et al. (1977) described the process as follows: special muscles (as many as five muscles may be involved in moving each speculum; Hope, 1974) operate to protrude spicules from the anus (cloacal aperture). To each side of this aperture, there is a lateral cuticular flange, supported by ribs, which meets its fellow at the root of the drawn-out tail. This acts as a sucker (copulatory bursa), by which the male retains its position on the body of the female until the spicules are thrust through the female aperture, and keep both the female and male apertures apposed and open. The spicules are inserted in the vulva. Then, by the contractions of the muscles of the cloaca, the spermatozoa are expelled and passed into the vagina of the female. Presumably, the various accessory copulatory structures, as well as adhesive secretions, assist the process. Exceptions might be the stouter-bodied species such as *Desmoscolex* Claparède, 1863 (=*Prodesmoscolex* Stauter, 1924, Desmoscolecida, Adenophorea). The insertion of the spicula into the vagina ensures the alignment and close adherence of the female gonopore to the male cloacal vent. In bursate males, the bursa folds over the female body, and may also aid in holding and sealing the open gonopore from the external environment. In parasitic nematodes, the bursa may be large with muscular rays, which enable the male to grasp the female firmly; or it may be too small to function as a hold-fast organ. Union between the mating pairs is assured by copious secretions from the male, which cement the bursa over the vulva.

Hyman (1951) reported that the trichuroid worm, *Trichosomoides crassicauda* (Trichostrongylina, Strongylida, Secernentea), which is parasitic in the urinary tract of rats, is notable in the animal kingdom in that the minute males, 1.5–3.4 mm long, dwell in the vagina of the female, 10–13 mm long, where they were originally taken to be young worms.

Fertilization

Nematodes are fundamentally bisexual, parthenogenetic, hermaphroditic or are paeudogamenous, in which development of the ovum is stimulated by a sperm. The nucleus of the sperm does not fuse with that of the ovum and contributes nothing to the hereditary composition of the embryo (Justine, 2002). In bisexual species, internal fertilization through female genital ducts is the general rule. After sperm enter the female gonoduct, they migrate presumably by amoeboid movement, to the seminal receptacle (spermatheca, short portion in between the oviduct and uterus; Fig. 9.8B) where fertilization occurs. Sperm entrance occurs preceding the meiotic divisions of the oocyte (immature oocyte) (Fig. 9.12B). Syngamy (fusion of gametes, not nuclei) occurs simultaneously with the formation of a perivitelline space, but the prospective point of sperm entrance is not predetermined. Tadano studied this aspect thoroughly, and reviewed and reported its details in his work (Tadano, 1968). He stated that the eggs of *Parascaris equorum*, *Ascaris suum*, *A. lumbricoides*, etc., might begin to develop as a result of sperm entrance.

In *Parascaris*, granules collect in the centre of the egg, the vacuoles begin to enlarge, and a thick egg membrane is formed. The spot where the spermatozoon is attached becomes hyaline. At this time, the cytoplasmic granules in the vicinity of the sperm entrance point begin to move about energetically and swell. The sperm pronucleus is immediately surrounded by these granules. The plasma membrane in the region where the sperm entered becomes indistinct, but few vacuoles collect there and dissolve to repair the membrane.

The eggs of the oviparous species have a thick and resistant egg membrane. Egg membranes of *Ascaris*, depending on the mode of formation and properties of the membrane,

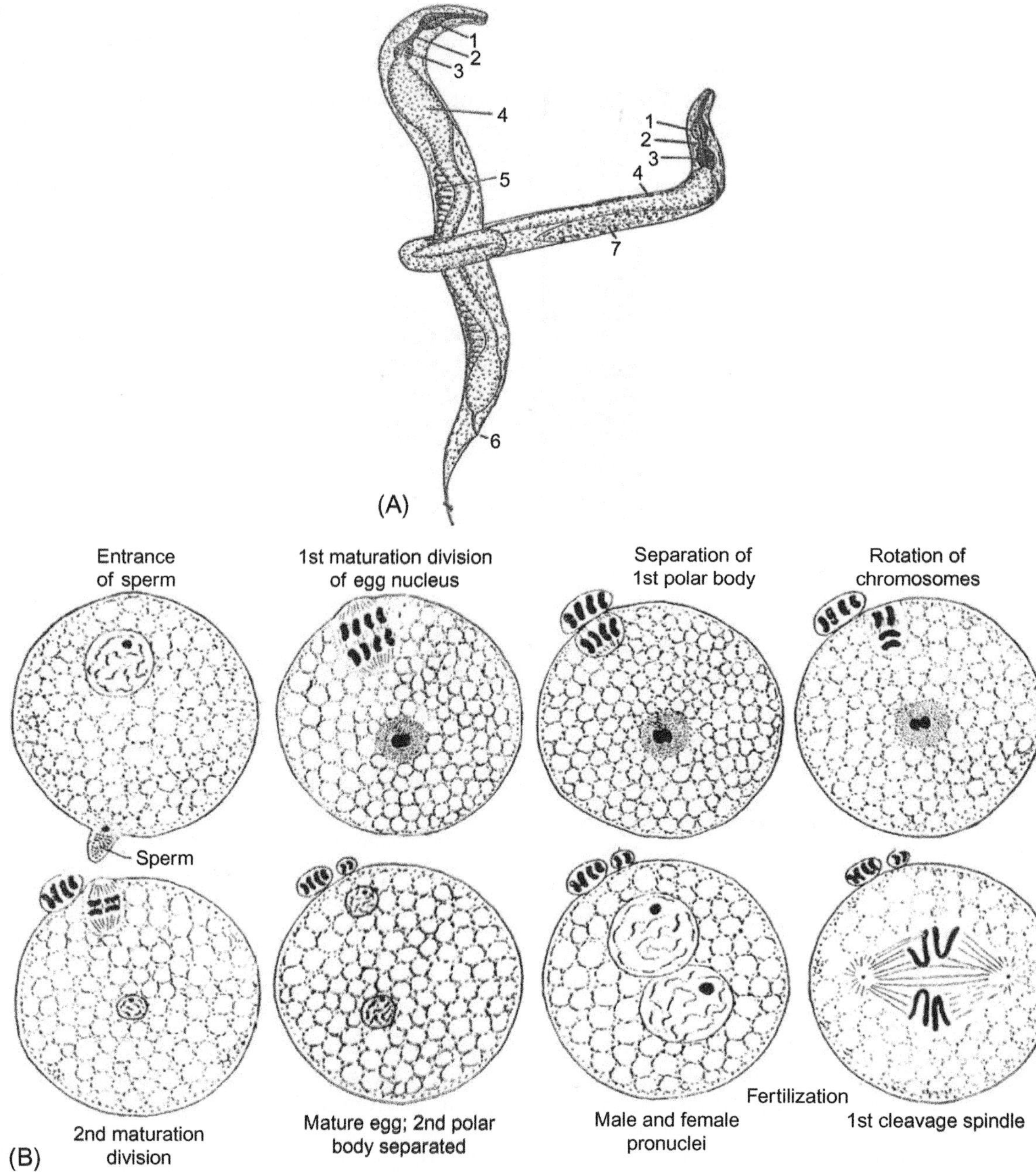

FIG. 9.12 (A) *Pristionchus aerivora* in copulation, 1 = median bulb, 2 = isthmus, 3 = end bulb, 4 = intestine, 5 = uterus, 6 = anus, 7 = testis; (B) Maturation of the ovum, sperm entrance, junction of pronuclei, and preparation for the first cleavage in the nematode *Ascaris megalocephala*. (just at the expulsion of the first polar body, the fertilization membrane elevates) *(from Storer et al., 1983).*

show five layers; the outermost is referred to as the first layer. The first layer is formed by a secretion product of the uterine wall (tertiary membrane). The second to the fifth layers are produced by the egg cytoplasm (primary membranes); each layer is composed of a basal membrane and an additional membrane. The fifth layer, in particular, has properties different from those of the other layers. In viviparous species such as *Rhabditis* or in ovoviviparous species, the egg membrane is generally thin. Tadano (1968) described in detail the three major events by means of which fertilization takes place; the first and second are in succession, while the third occurs parallel to the second.

First is the formation of the egg membrane and the expulsion of the first polar body. During this process, five definite layers are formed. The **first layer** (protein) is formed of granules secreted by the wall of the uterus. They swell and adhere to the egg surface, then fuse together. This layer is sticky, contains a large amount of albumen and is easily peeled off the egg. It is usually lacking in ovoviviparous and viviparous species like *Rhabditis*. This layer may be formed before or after sperm entrance. The **second layer** (lipoid) follows the formation of the first layer, but begins after sperm entrance. Large granules from the centre of the egg migrate to the periphery and swell into vacuoles; at this time, the medium-sized granules adhere around the vacuoles and join together to form a lipoid layer (inner to the first layer). In the beginning, it is irregular, but during the formation of the third layer it becomes smooth (Fig. 9.13A (a)).

This description of the formation of nematode egg membrane recalls the view of Hyman (1951), in which this egg is 'typically surrounded by three definite membranes, the lipoid membrane, the shell, and the protein membrane'. In Tadano's view, the egg is surrounded by a membrane consisting of five layers. The second lipoid layer of this membrane corresponds to the first lipoid membrane of Hyman.

Just after the formation of the second layer, small granules move from the central region to the periphery of the egg. They swell, and together with a hyaline substance, become attached to the inner surface of the second layer forming the **third layer**. This layer thickens by the addition of vacuolated large granules. It is compressed and smoothened by the following fourth layer. Compared with the second and fourth layers, the third layer is definitely thick (Fig. 9.13A (b2)). In the same way as the third layer, the **fourth layer** is formed, but the addition of vacuolated large granules gives the effect of 'soap bubbles'. In this stage, the number of the cytoplasmic granules diminishes and the interior of the cell appears brighter. The colour of the cytoplasmic granules changes from yellowish-black to brownish-yellow. Just as the fifth layer of the egg membrane begins, the oocyte nucleus, accompanied by granules, moves to the periphery. There, the first polar body extrudes and acquires a flattened shape as it is pressed against the inner wall of the fourth layer (Fig. 9.13A (b) and (c)). After this extrusion, the nucleus moves back to the centre of the oocyte. Around the periphery of the cytoplasm, a homogeneous region appears. Swollen vacuoles and small granules move into this region, fusing and dissolving together to give rise to very delicate lamellae. The gradual increase in the amount of a fibrous material on the inner side of the fourth layer completes the formation of the rough and thick **fifth layer**. The formation of the perivitelline space makes it smooth and regular in thickness.

Especially in the egg of *Parascaris equorum*, the inner side of the fifth layer is denser than the other regions (Fig. 9.13A (d)). In *Ascaris suum*, the fifth layer is closely attached to the fourth layer, and its inner surface is undulatory.

It is appropriate to mention that Hyman (1951) reported that the chemical composition of the egg membranes identified in her work has been investigated. It was shown that the shell membrane consists of chitin. The ripe 'ovocyte' contains glycogen and lipoid globules. Immediately after fertilization, the glycogen globules move to the surface where they break down into glucose from which glucosamine, the chief constituent of chitin, is synthesized. Tadano (1968) also reported that the second to the fourth layers of the *Parascaris equorum* egg are composed of chitin. Tadano presented his own table showing histochemical properties of the eggshell of *Ascaris*. Chemically, chitin is mainly a sugar derivative known as a glucosamine polysaccharide. Structurally, chitin consists of *N*-acetyl-D-glucosamine units joined by ß (1–4)-glycosidic linkages (Fig. 9.13F) (Murray et al., 1996). The table of Tadano did not contain specific reactions to verify the presence of chitin.

Second is the formation of hyaline membrane, plasma membrane and perivitelline space. Following the formation of the fifth layer, the plasma membrane is formed in contact with the inner surface of the egg cortex. It is separated only very slightly from the fifth layer. This membrane is irregular at first but becomes smooth when the perivitelline space and external cytoplasmic zone are formed. In a parallel process, a thin hyaline membrane appears outside the plasma membrane. The perivitelline space is formed between the hyaline layer and the inner surface of the fifth layer. It is at this time that the hyaline and plasma membranes become thinner and denser.

Third is the formation of the second polar body, and the fusion of the nuclei. It is a parallel process to the formation of the perivitelline space. After the extrusion of the second polar body, the egg pronucleus returns to the centre of the egg where it unites with the sperm pronucleus. Granules gather around the fused nucleus (zygote). The volume of the mature *Parascaris* egg is reduced to about 70% of that of the unfertilized egg. The same process takes place in *Rhabditis* and *Dipolagaster* (Diplogasteroidea, Rhabditina, Rhabditida Secernentea; Hope, 1974).

The upper detailed description of the fertilization process was done by the aid of the light microscope, and it appears that it is evidenced by the following more recent information.

Perry (2002) reviewed the structure of the 'eggshell' of nematodes and presented a diagrammatic illustration to describe a generalized form (Fig. 9.13C). It consists of one to five layers, depending on the species. However, the most common form has three layers. The outer vitelline layer is derived from the oolemma and it retains the unit membrane-like structure. The middle chitinous layer (corresponding to the third layer of Tadano) is often the thickest and the strongest since it has a chitin microfibril core. In addition, it has a collagen-like protein coat, which provides its rigidity. It appears that chitin has been demonstrated exclusively in the eggshell of nematodes. The innermost lipid layer represents the main permeability barrier of the eggshell. However, it was suggested that in *Ancylostoma* spp. (Ancylostomatoidea, Strongtlina, Strongylida, Secernentea), the chitinous layer

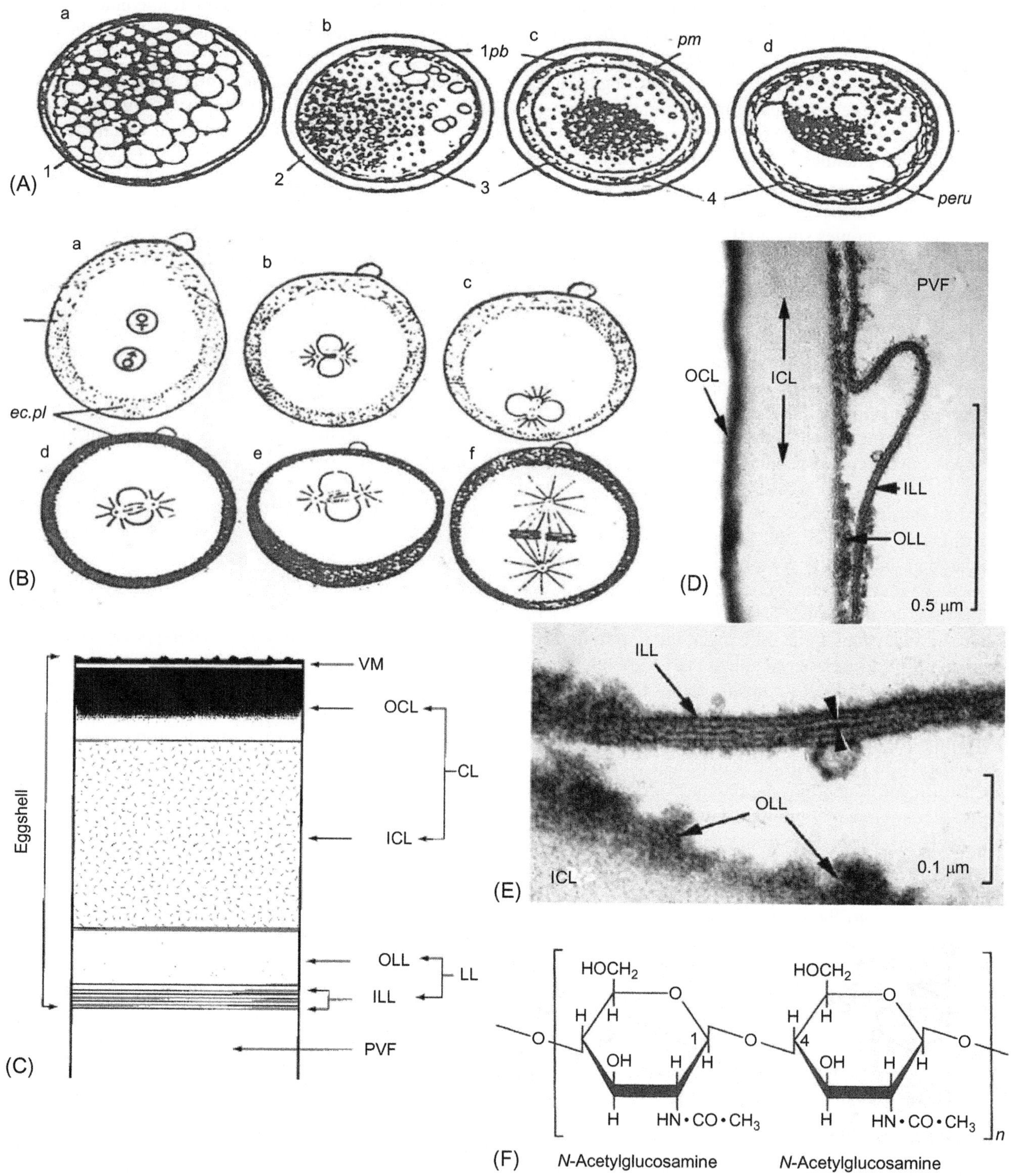

FIG. 9.13 (A) Egg membrane formation in *Ascaris megalocephala* (Tadano); (a) formation of second layer in centrifuged egg (1), (b) formation of first polar body (1pb) and fifth layer in centrifuged egg (4), (c) formation of plasma membrane (pm), (d) formation of perivitelline space in centrifuged egg (perv), 2 = third layer, 3 = fourth layer; (B) localization of mitotic apparatus and behaviour of ectoplasm, 2pb = second polar body, ec. pl = ectoplasm; (C) diagrammatic reconstruction of a generalized embryonated tylenchid egg; (D) section through the eggshell of Heterodera schachtii showing the inner lipid layer detached from the amorphous outer lipid layer; (E) detail of the same inner lipid layer showing the tetralaminate structure with a mean lamina separation of 30 nm *(arrow heads)*; CL = chitin layer, ICL = inner chitin layer, OCL = outer chitin layer, LL = lipid layer, ILL = inner lipid layer, OLL = Outer lipid layer, PVF = pervitelline fluid, VM = vitelline membrane, *(D–F) from Perry (2002)*; (F) chemical formula of chitin.

might act as a sieve, preventing larger molecules reaching the lipid layer.

In plant-parasitic nematodes, the inner lipid layer usually consists of two or three lipoprotein membranes, although in *Heterodera schachtii* this layer is tetralaminate (Fig. 9.13D and E). In *Ascaris* spp., the lipid layer has been termed the 'ascaroside layer'. This layer contains 25% protein and 75% lipid and it belongs to a unique class called ascarosides that is not found in any other eukaryote group. However, the more general term 'lipid layer' is used for the innermost layer of the eggshell.

In addition to these three layers (primary membranes, since they are synthesized by the egg itself), the eggshell of some species of nematodes, possesses one or two layers secreted by the cells of the uterus, e.g. the eggshell of *Ascaris lumbricoides*. In aquatic species, the uterine layer may form filaments by which the egg attaches to the substrate.

Justine (2002) reported that the ultrastructural aspects of fertilization have been studied in a small number of nematode species, and that *Ascaris lumbricoides* was one of them. As already stated, the nematode sperm has no acrosome (Metz and Monroy, 1985a,b) and no flagellum. The process of fertilization therefore differs from that of other more familiar animals. In *Ascaris lumbricoides*, pseudopods projecting from the anterior cytoplasm make first contact with the primary oocyte. The gamete membranes then interdigitate and finally fuse. After this fusion, the sperm nucleus becomes disorganized and the refringent body, a special structure typical of *Ascaris* sperm, disappears, and the sperm reaches the centre of the egg.

Polyspermy frequently occurs, and up to six spermatozoa have been observed either penetrating, or within the oocyte of *Acanthocheilonema (Dipetalonema) viteae*. The anterior region of the sperm is the first to contact the oocyte and causes a deep invagination of the oocyte membrane. No pseudopods were observed. The surface coat around the oocyte terminates where the membranes of the two gametes are in contact.

The mature eggs of most nematodes are usually covered with a chitinous cuticle (Borradaile et al., 1977) (Fig. 9.13C). The form of the eggs (Fig. 9.14(I) (a)) may be ovoid (*Ascaris lumbricoids*), but occasionally they are spindle-shaped (*Trichocephalus dispar*). There are eggs with polar filaments (*Tetrameres*). In addition, there are eggs with lid-shaped structures or plugs at one pole (Oxyuroid egg of *Dermatoxys veligera*) (Fig. 9.14(I) (c)), or at the two poles (*Trichuris suis*) (Fig. 9.14(I) (e)). Sometimes the egg encloses the larva (*Enterobius vermicularis*) (Fig. 9.14(I) (b)). Therefore, the precise stage at which the egg leaves the parent varies among the nematodes. In *Ancylostoma* the eggs are already at a state of early cleavage. In *Enterobius* the eggs are at a late embryonic stage, while those of *Wuchereria* are fully developed little embryos.

Development does not begin until the eggs have been oviposited and environmental conditions are favourable (oviparous). When conditions are unfavourable, *Ascaris* eggs, e.g. may remain dormant for long periods of time. Some marine nematodes, such as *Viscosia carnleyensis*, are known to be in temperate latitudes as adults and juveniles from February to October, while these stages are absent through the remaining winter months. This species may survive this period as resistant eggs. Adult (including gravid females) and juvenile *Deontostoma californicum* are present during all months of the year on the northern coast of California, where temperatures vary little from 15°C throughout the year.

DEVELOPMENT

Tadano (1968) reported that representing this group of animals, the nematode *Parascaris equorum* (Goeze, 1782) (=*Ascaris megalocephala*, Cloquet, 1824) and *Rhabditis ikedai* and *Rhabditis* sp. (Fig. 9.6) were widely studied. *Rhabditis* sp., which is a species parasitizing *Armadillidium vulgare* (Latreille), the common wood louse, is completely terrestrial in its habitat (Isopoda, Crustacea; Borradaile et al., 1977). *Ascaris megalocephala*, is an intestinal parasite of horses with two different strains: *A. m.* bivalens in which the primordial 'germ cells' have only four chromosomes and therefore the gametes have two; and *A. m.* monovalens, in which the primordial 'germ cells' have two chromosomes and the gametes have only one chromosome. The zygotes have four or two chromosomes (Gardiner, 1972).

Nematodes proved to be a particularly suitable material for detailed embryology and cell lineage studies, since their development is highly determinate and the number of the somatic cells is relatively small (Hope, 1974). The textbook *The Biology of Nematoda* (Hope, 2002) added that the transparency and availability of nematode embryos made them attractive subjects to study, especially their development. Classically, embryonic development was described for many nematode species, leading to several key concepts in developmental biology. A number of zoologists have investigated the embryology of these animals, and in fact the studies in cytology and germ-cell lineage have flourished. However, the most eminent pioneer in the analysis of nematode development was Theodor Boveri (1862–1915), whose first graduate student Hans Spemann (doing his PhD thesis on nematodes) won the Nobel Prize in 1935 for his work on the 'organizer' in the amphibian embryo. In 1880, Boveri started to study embryogenesis of the nematode *Ascaris megalocephala* (*Parascaris equorum*), a parasite found in the intestine of horses. The females produce millions of eggs, releasing more than 100,000 per day. After fertilization the eggs start to cleave, only after being released from the anaerobic environment.

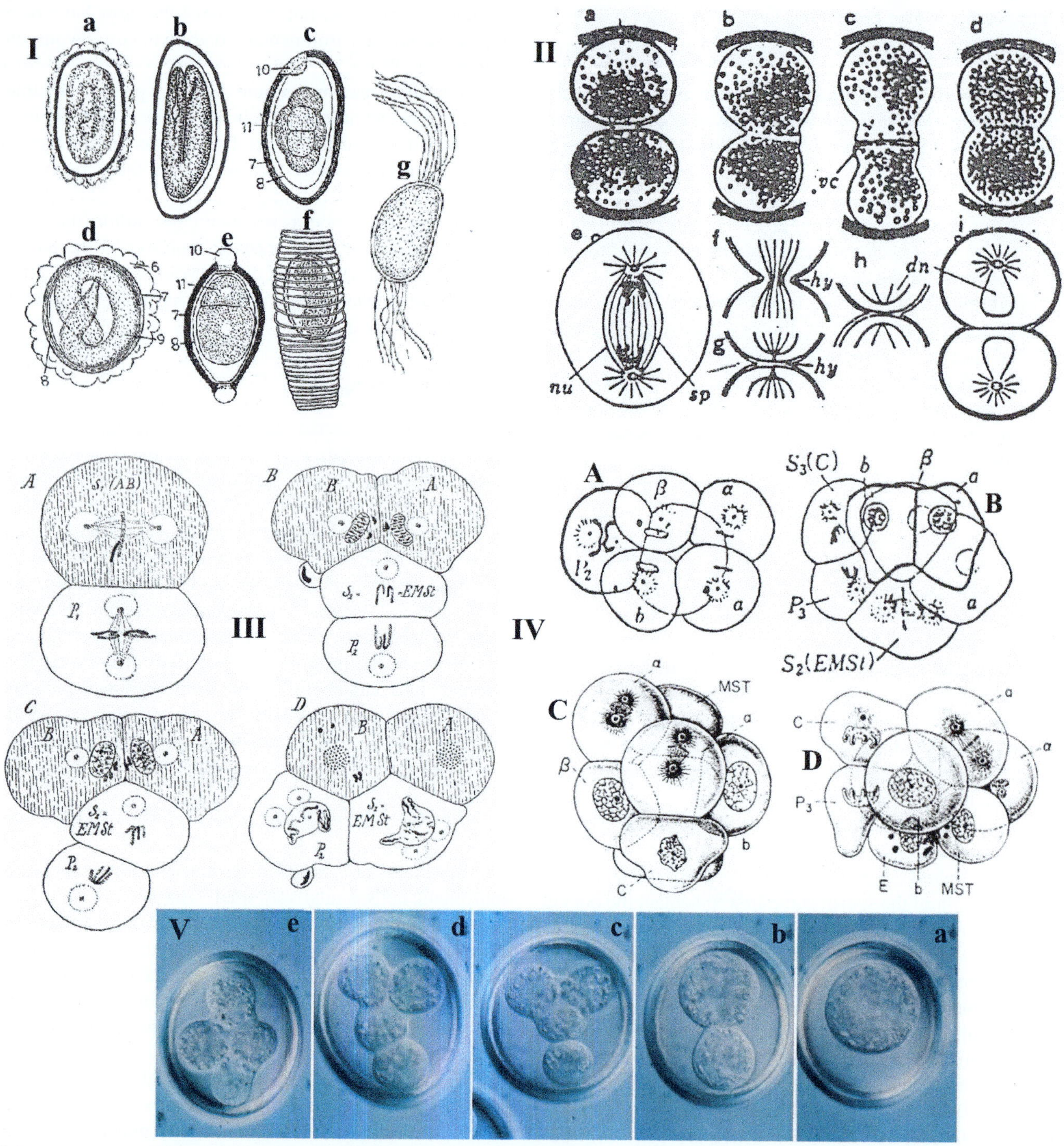

FIG. 9.14 (I) Types of eggs of nematodes: (a) fertilized egg of *Ascaris lumbricoides*; (b) egg of *Enterobius vermicularis*, enclosing larva *((a and b) from Cheng, 1964)*; (c) egg of *dermatoxys veligera* (Oxyuroid) with one opercular plug; (d) Ascaris egg; (e) egg of *Trichuris suis* (Trichuroid) with two opercular plugs; (f) Pseudonymus egg enclosed in spiral filaments; (g) egg of Tetrameres with polar filaments *((c–g) from Hyman, 1951)*; (II) (a, e–g) 'early process' of first cleavage, (b–d) and (h–l) 'late process', dn = daughter nucleus, hy = hyaline membrane, nu = nuclear region, sp = spindle fibre, vc = vacuole; (III) two- to four-cell stage, the second cleavage plane is parallel to the egg axis in S1 (AB) and the division is with chromatin diminution, while in P1 the plane is perpendicular and without chromatin diminution: (A) two-cell stage, (B) four-cell stage (T), (C) four-cell with P2 shifting, (D) four-cell (rhombic) stage *(after Boveri from Pflugfeder, 1970)*; (IV) four- to eight-cell stage: (A) six-cell stage (dorsal) showing right–left pairs, a–α and b–ß; (B) seven-cell stage (right side) ((A, B) from Tadano); (C) (dorsal surface) and (D) (lateral view of eight-cell stage *(from Hope, 1974)*); (V) photomicrographs showing embryological stages from the egg (a) to the four-cell (rhomboid stage, e) of *Parascaris equorum* (=*A. megalocephala*), using Nomarski optics, at 1000×, *(from Hope, 2002)*.

Early embryology

Immediately after the maturation divisions, the *Parascaris* egg is fairly oval, with a rather thick peripheral zone of pale yellowish-green hyaline substance (Fig. 9.13B (a)). The latter is clearly distinct from the interior cytoplasm, which is crowded with a large number of variously sized granules. Large vacuoles can be observed scattered throughout the cytoplasm. Granules of all sizes change to vacuoles and then dissolve into this hyaline zone, increasing its refringence and increasing the contrast between the external and internal cytoplasm. In addition, the egg becomes spherical (Fig. 9.13B (b)). The egg and sperm pronuclei make contact in the centre. The contact surfaces of the pronuclei disappear and they fuse, forming the transparent fused nucleus, in which the chromosomes are clearly visible (Fig. 9.13B (c)). As soon as the nuclei join, they begin to migrate towards the centre of the zygote where the mitotic figure develops. Then, it moves towards the periphery, at the animal pole, and the egg takes an asymmetrical oval shape (Fig. 9.13B (d) and (e)). Commonly, the mitotic figure rotates from 90 degrees to as much as 360 degrees. Finally, the spindle elongates and the cleavage furrow forms (Fig. 9.13B (d) and (f)).

Cleavage

One- to two-cell stage (first cleavage)

Where the zygote is oval, which seems to be the usual case, the first division is typically transverse to the long axis, i.e. the spindle axis is parallel to the polar axis (Fig. 9.14(II)). This results in two blastomeres somewhat unequal in *Parascaris equorum* and in *Toxocara canis* (Werner, 1782) (Ascaridiodea, Ascaridina, Strongylida, Secernentea). This is generally because the mitotic spindle is more in the vegetal region of the zygote (Tadano, 1968). However, they are equal in *Deontostoma californicum* (Hope, 1974). The animal pole becomes the anterior S1 blastomere and the vegetal pole the posterior P1. The stem cell P1 is usually the smaller one. The letters used in designating these blastomeres are after Boveri's nomenclature system (1899), which is still widely used today. The letter S or sometimes AB designates the blastomere that produces the somatic tissues (somatic-cell line). The letter P designates the blastomere that gives rise to the primordial germ cells (germ-cell line). These reflect the highly determinate nature of nematode development.

Two- to four-cell stage (second cleavage)

The second cleavage in various species of *Ascaris* and in *Rhabdias bufonis* (=*Rhabditis nigrovenosus*) (Rhabdiasoidea, Rhabditina, Rhabditida, Secernentea) is brought about by a mitotic division. It is different in S1 cell as compared to P1. The second cleavage plane is parallel to the egg axis in S1 (AB) cell producing A and B blastomeres (the primary somatoblasts as shown in Fig. 9.15), and perpendicular in the P1 cell, producing P2 and S2 (EMSt) blastomeres. The net-result of this cleavage is a T-shaped four-cell stage. Later, it becomes rhombic-shaped by a shift in the position of the P2 blastomere (Fig. 9.14(III)). Fig. 9.14(V) shows photographs taken by Nomarski optics to show this stage in the cleavage (Hope, 2002). In this rhomboid, A is anterior, B dorsal, S2 ventral, and P2 posterior. At the second cleavage, the ends of the chromosomes of the S1 blastomere are cut off and thrown out into the cytoplasm. This means that the S1 blastomere undergoes the phenomena called chromosome fragmentation and chromatin diminution. The P1 blastomere does not undergo these phenomena. The characteristic shift from the T-shape to the rhombic, which appears in early nematode development, parallels a movement of the nuclei and a change (migration, new formation) in the condition of the peripheral cytoplasm in the 'later phase' of the second cleavage.

This second cleavage is characterized by two phenomena: chromatin diminution and the T-shaped arrangement of the four cells. This transitory T-shaped arrangement is not known among nematodes other than ascarids. Hope (1974) reported that in several other species, cleavage patterns lead to the rhomboid arrangement without the T-shaped configuration (Fig. 9.16C). It was described for *Rhabdias bufonis* and *Turbatrix aceti* that S1 and P1 change shape to have a diagonal common surface, and then cleave in parallel to form a rhomboid configuration. Another example is in *Pratylenchus* sp. and some other species, which show that S1 and its descendants divide in advance of P1 to form six-celled embryos, and the resulting somatoblasts shift positions to form a rhomboid (Fig. 9.16C).

Variation in the arrangement of the blastomeres of the four-cell stage was also observed in *Deontostoma californicum*. Basically, two cleavage patterns are evident. In the first type, which occurred in 20 of 26 developing embryos, the first cleavage was transverse and the cleavage planes of the daughter blastomeres, S1 and P1, were longitudinal and at right angles to each other. In the six remaining cases, the first cleavage was as before and the second cleavages are also at a right angle to the first, but parallel to each other. With two exceptions, the positions of the blastomeres remained unchanged through the subsequent cleavages (Hope, 1974).

Thus, it is clear that early ascarid development is not typical of all nematodes and that the patterns of early cleavage are numerous, and more than is generally recognized.

Chromatin diminution

This phenomenon, in which part of the chromatin is discharged into the cytoplasm, was first discovered in *Parascaris equorum* by Boveri (1887e). In addition, Boveri himself made the later detailed observations of the process,

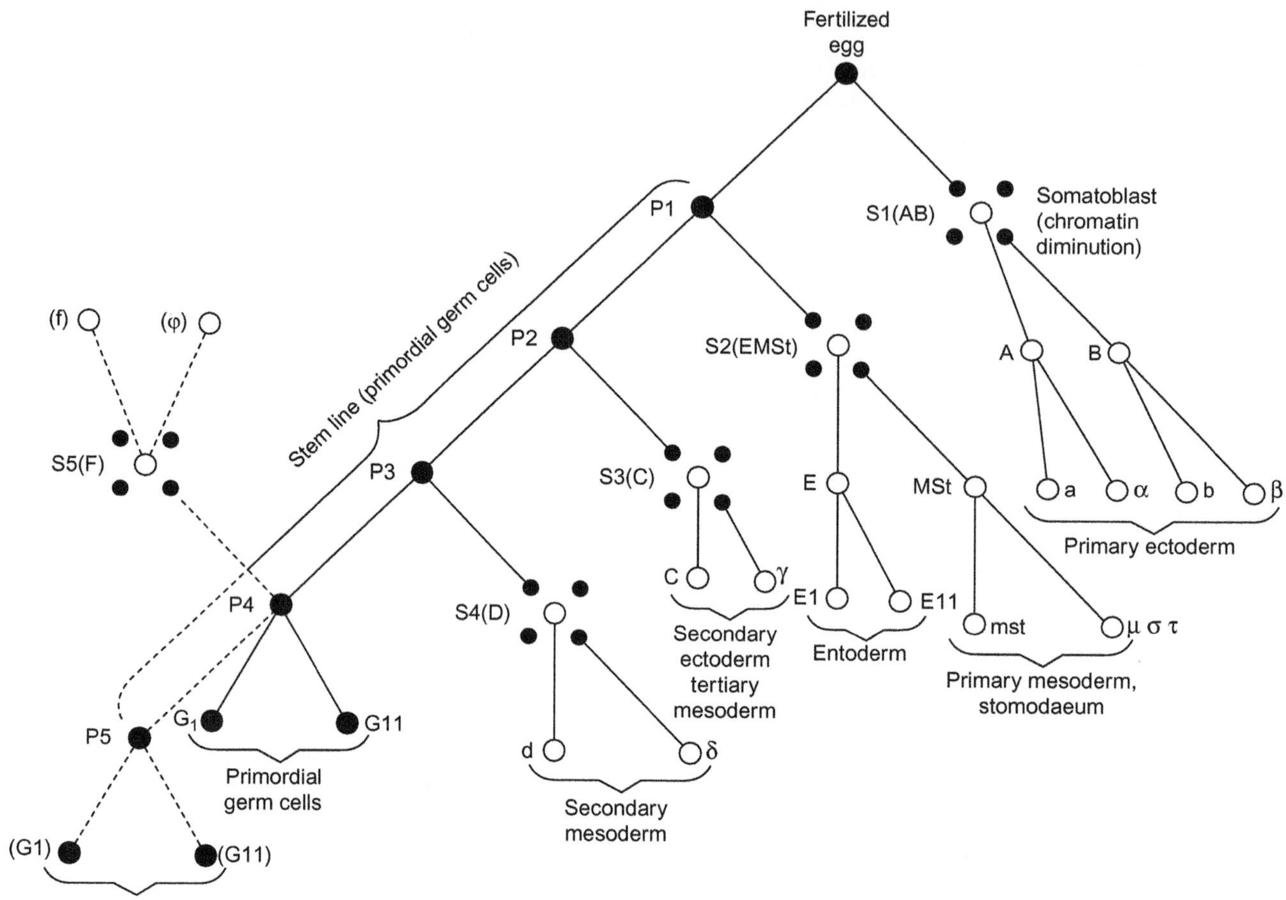

FIG. 9.15 Cell-lineage in the nematode *Parascaris equorum*. *(From Tadano (1968) based on Boveri (1899, 1910).)*

followed by others who studied the process in *Ascaris lumbricoides*, *As. rubicunda*, *As. Labiate* and *Rh. nigrovenosa*. The details of the process differ to some extent among these species. It has not yet been shown to occur in any other nematodes.

As has been stated above, nematode cleavage proceeds in a determinate fashion, and the differentiation of the somatic cell line from the germ cell line begins as early as the two-cell stage (S1 and P1). P1 possesses the capacity to form nearly the whole embryo. It can therefore be described as 'universal'. By contrast, the S1 blastomere can be called 'partial'. The nuclear division leading to this situation is of profound significance from the embryological point of view. In *Parascaris equorum*, the ends of the S1 (AB) chromosomes thicken into a flask shape by the metaphase (Fig. 9.14(III) A and B), and during the period between metaphase and telophase, they are cast out of the nuclear region into the cytoplasm (Fig. 9.14(III) C–D) where they lie scattered, in a row or chain, or all clumped at one side. In *Toxocara canis* this process occurs during the anaphase and the pieces are distributed proportionately to the daughter blastomeres. While most investigators have found that chromatin diminution takes place during the division of the S1 blastomere into its daughter cells A and B, it may be delayed until the next cleavage when A and B give rise to a, α, b, and β (Hope, 1974). It has been thought that the end pieces do not contain genes, although Tadano (1968) found that they give a very strong Feulgen reaction.

As mentioned above, *Parascaris equorum* includes two varieties. The gamete of *P. equorum* var. univalens has a single large compound chromosome composed of an X-*complex* is made up of eight components, and an *autosomic complex* containing 22 components. Since *P. equorum* var. bivalens discards a larger amount of chromatin than *univalens*, the somatic resting nuclei of the two varieties after diminution are about the same size. On the other hand, the resting nuclei of the germline cells (P1–P4) and of the primordial germ cells (GI and GII) are clearly larger in var. *bivalens* (Tadano, 1968).

This chromatin diminution phenomenon takes place during each cleavage of the somatic line blastomeres from S1 to S4 (or S5), so that only the germ cells descending from the P4 (or P5) blastomeres retain a set of complete chromosomes.

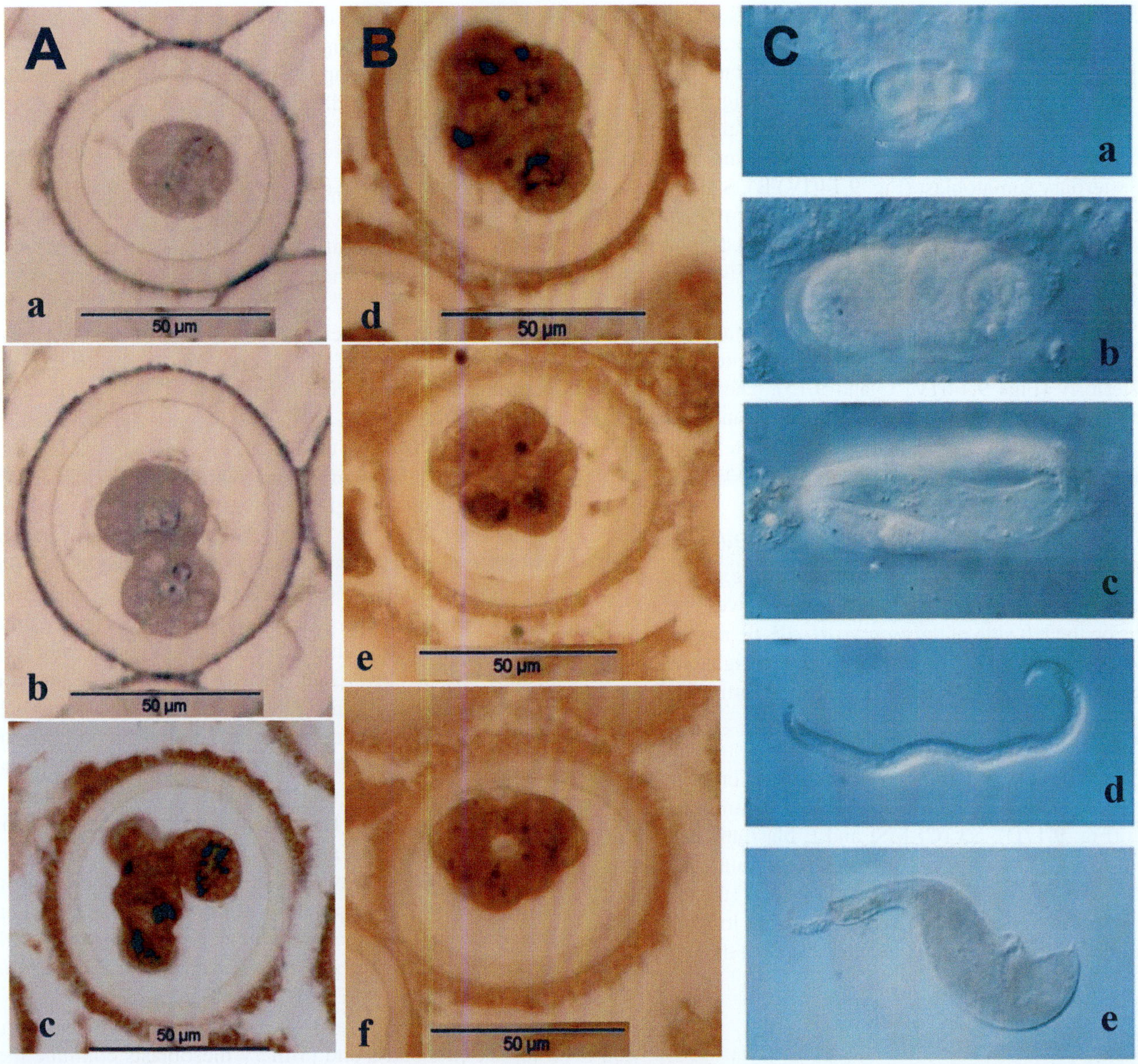

FIG. 9.16 (A and B) Light microscopic photographs of early embryological stages of *Ascaris megalocephala* from the egg to the blastula (laboratory preparations), (e) shows four- to eight-cell stage; each photo is selected (from Fig. 9.5E) and magnified. Note the internal egg membranes; (C) photographs of embryological stages of *Rotylenchulus reniformis* taken by Nomarski optics, anterior end is to the left throughout, (a and b) four-cell stage viewed at 400× and 1000×. The blastomeres are linearly arranged and the egg is elongated and 75 µm long; (c) an egg containing a fully elongated worm with a head in focus, viewed at 1000×; (d) a juvenile viewed at 400×; (e) an adult viewed at 200×. The adult swells from the typical nematode, but the vulva is visible along the upper surface in the figure and the tail is apparent to the right, the anterior is specialized for interfacing with the plant host *((C) from Hope, 2002)*. Notice the difference in the arrangement of the four-cells in the two animals.

Four- to eight-cell stage (third cleavage)

Embryonic development beyond the eight-cell stage is best known for some species, e.g. *Turbatrix aceti*, and especially *Parascaris equorum*. Accordingly, the following account of further nematode development is based on studies of these species. In the third cleavage, the mitotic spindles of the ectoblasts A and B lie perpendicular to the embryonic axis. The daughter cells derived from these A and B blastomeres in this third division are arranged in right–left pairs, a–α and b–β, respectively (Fig. 9.14(IV) A). In accordance with the same system of nomenclature, Greek letters are assigned to daughter blastomeres on the left and Roman letters to those on the right. These four cells are asymmetrical when first formed, but become more or less symmetrical during the 'second process' (this means that these divisions establish the bilateral symmetry of the embryo). The a, β blastomeres lie somewhat forward and below a and b, and more or less divergent. The a and α blastomeres lie slightly to the left of the midline, and since a and ß come to lie in contact with each other, an asymmetry makes its appearance between the left and right halves of the embryo. These four blastomeres (a, α, b, β), all give rise to ectoderm only (Fig. 9.15).

The P2 blastomere, which lies above the midline in the posterior part of the embryo, divides into P3 and S3 (C)=tertiary somatoblast (Fig. 9.14(IV) B). S3 gives rise to the epidermis covering the posterior half of the embryonic body (i.e. =secondary ectoblasts). However, Müller (1903) and Boveri (1910) established that it also gives off part of the mesoderm.

S2 (EMSt) divides, with chromatin diminution, and in the same direction as P2, forming MSt anteriorly and E posteriorly (Figs 9.14(IV) C and 9.15). The embryo now has eight cells, with a 'ventral familic' arranged in an arc along the ventral midline, consisting of MSt, E, P3, and S3 (Fig. 9.14(IV) D).

Eight-cell to 16-cell stage (fourth cleavage)

Like the third cleavage, the fourth cleavage in *Parascaris equorum* begins with the ectoblasts. This gives rise to four cells on the left side of the embryo αI, αII, βI, and βII, which form a rhomboid configuration. On the right side, their counterparts, aI, aII, bI, and bII, are in the form of a T-shape (Fig. 9.17A and B).

P3 cleaves to produce S4 (D, the fourth propagative cell and the secondary mesoderm) at the posterior extremity of the embryo, and P4 on the ventral side continues the stem line (Fig. 9.17D). S3 (C) MSt and E blastomeres begin to cleave. S3 (C) undergoes chromatin diminution, forms c and γ and MSt divides into mst and μστ. E (the endoblast) divides transversely in the ventral mid-line to form EI and EII, the former anterior to the latter and both lie in the sagittal plane (Fig. 9.17C).

Fifth cleavage—Blastula

The ectoblasts also begin the fifth cleavage and their daughter cells assume a symmetrical arrangement. The aI and αI give aI_1 and aI_2, αI_1 and αI_2, and so on, and the primary ectoderm-producing blastomeres become 16 cells.

S4 (D) (mesoblast), which lies anterior to P4 (Fig. 9.17D), divides transversely while undergoing chromatin diminution to form the bilateral pair of blastomeres d and δ (Fig. 9.17F). Endoblasts EI and EII then divide transversely; their daughter cells form a somewhat rhomboid arrangement. The fifth cleavage is not completed until the division of P4, which is delayed. In the meantime, the ectoblasts have undergone their **sixth cleavage** and the embryo has, by this time, become a blastula with a small blastocoel (**coeloblastula**; Fig. 9.17G). This coeloblastula is generally dorsoventrally flattened and elongated in the anteroposterior direction. The dorsal and lateral walls of the blastula consist chiefly of the descendants of S1 with some additions from S3 and S4. The flat or even concave ventral surface is composed of EMSt and P4 or their offspring. Fig. 9.16A and B show micrographs of stages of development of *A. megalocephala* from the egg to blastula stage prepared for laboratory studies.

Gastrulation

Gastrulation is generally believed to be by the process of epiboly, since an archenteron is absent. However, an old opinion stated that it occurs by invagination. When the primary mesoblast cells, particularly the stomatoblasts, form the two sides of the embryo and approach the mid-ventral line, gastrulation begins. Next, the stomatoblasts divide, slightly ahead of the mesoblasts. All their descendants move towards the midline, gradually causing the endodermal blastomeres (descendants of EI and EII) to slip into the blastocoele. Eventually the mesodermal cells follow them (Fig. 9.17H–J), and the increase in number of blastomeres causes a reduction in the size of the blastopore. Gastrulation continues as the primary mesoblasts enter the blastocoele. The secondary mesoblasts (d and δ descendants of S4) then divide and their daughter cells occupy the posterior and postero-lateral lip of the blastopore, thus embracing P4 (Fig. 9.17F). Blastomere P4 then divides to produce the primordial germ cells GI and GII, which along with the secondary mesoderm cells, are subsequently carried into the blastopore as invagination processes (Fig. 9.17H). In *Turbatrix*, the fate of P4 has been followed more precisely than in other nematodes, this P4 divides into two cells, S5 and P5. The offspring of S5 produces the epithelium that covers the gonad and forms the wall of the gonoduct and P5 proliferates the germ cells. Similar observations were made for *Rhabdias* (Hyman, 1951).

Following the division of the P4 blastomere, the stomatoblasts, endoblasts and primary mesoblasts continue their division. Some descendants of the stomatoblasts come into contact with the most anterior of the secondary ectoblasts, and these two groups of cells form the lip of the blastopore. The descendants of the mesoblasts contribute to the lateral wall of the embryo. Next, the stomatoblasts begin to form a pocket like invagination, stomodaeal pocket, in the anterior region of the blastopore (Fig. 9.17H–J). Concurrently, the proliferating ectoblasts begin closing the posterior and lateral regions of the blastopore, covering the GI and GII. As the blastopore closes and the ventral surface of the embryo becomes smooth, the stomodaeal invagination is reduced to a small opening. Later, this opening shifts to the anterior end of the embryo and becomes the stoma. Posteriorly, the stomodaeam makes contact with the endodermal tissue (primordial of the midgut) (Fig. 9.17J). As development continues and gastrulation is completed, the previously flattened embryo becomes a cylinder and then elongates. The embryo

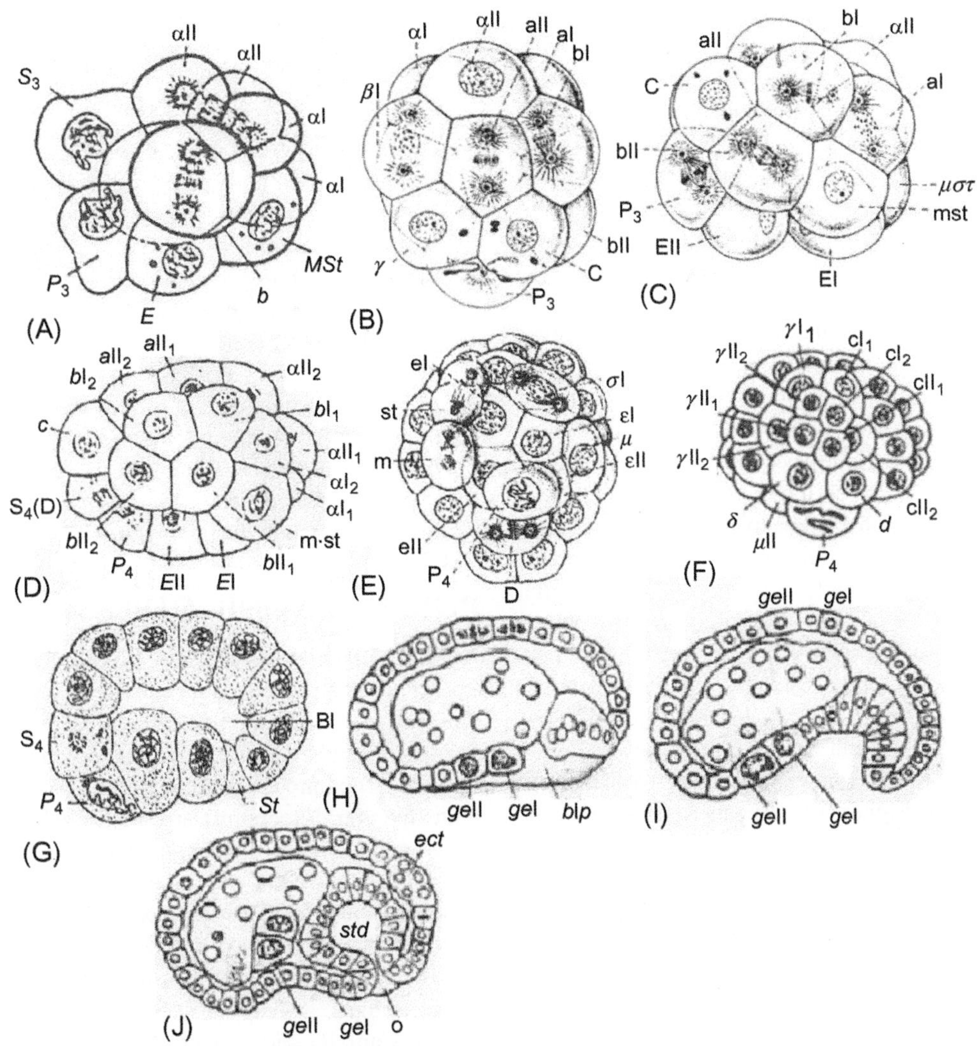

FIG. 9.17 Later embryonic development in *A. megalocephala*: (A) 10-cell stage (right side); (B) 15-cell stage (dorsal surface); (C) 15-cell stage (right side); (D) 32-cell stage (right side); (E) ventral view of later stage; (F) gastrula (posterior); (G–J) sections through the embryo showing gastrulation: (G) blastula; (H) formation of stomodaeum (mid-plane); (I) appearance of stomodaeal invagination (mid-plane); (J) more advanced, invagination of germ cells and closure of blastpore. *(A, D, F, H–J) From Tadano (1968); (B, C, E) from Hope (1974); (G) from Pflugfeder (1970).* (For letters, see Fig. 9.15.)

begins to assume the form of the juvenile worm curved inside the egg membranes.

In summary, the principal organs of the juvenile, and the blastomeres that contribute to their formation, are as follows: in *Parascaris*, S1 (AB) forms most of the epidermis (ectoderm), and the remainder comes from S3 (C). The primary mesoblasts M, which separates from S2, form most of the mesoderm; the E blastomere gives rise to the endoderm, forming the stomodaeal rudiment. S3 (C) supplies the primary endoblast cells and forms part of the mesoderm; S4 (D) also forms the mesoderm. Concerning the germline cells, Boveri (1899) held that P4 divides into P5 and S5 (F), P5 then divides into GI and GII, while S4 forms the f and φ at the posterior end of the embryo. On the other hand, Tadano (1968) reported that other investigators considered that GI and GII are derived from P4. Later, Boveri (1910) maintained that the production of P4 is normal in *Parascaris*, but that P5 and S5 (F) may be formed as an exception.

HATCHING

The majority of the nematode adult females lay eggs, and each egg contains a single juvenile. Despite the vast difference in size of adults of various species of nematodes, the majority of eggs are of similar size and morphology. In many species, it is the first-stage juvenile that hatches. Therefore, 'hatching' means the emergence of a fully developed, but sexually immature juvenile, from the confines of the egg membranes in which it develops. In most plant-parasitic nematodes, the juvenile moults inside the egg, and the resulting second-stage juvenile hatches. In some animal-parasitic species, there is a further moult in the egg and it is the third-stage juvenile that hatches. In many species, the egg is the main survival stage of the life cycle, and in species such as *Ascaris suum*, the unhatched juvenile can remain viable for many years. The eggshell affords protection to the enclosed juvenile and hatching only occurs when environmental conditions are favourable. (In nematodes, the egg hatches a juvenile not larva, since the larva differs markedly from the parent and attains the adult stage by a process of metamorphosis.)

In Hope's opinion (1974), the hatching process involves mechanical force and/or chemical breakdown of at least a portion of the chorion. More or less the same meaning was encountered in the work of Perry (2002), but the process proceeds in three definite successive phases: changes in the eggshell; activation of the juvenile; and eclosion. In many species, activation of the juvenile appears to precede, and may even cause changes in the eggshell structure. Mechanical force corresponds to activation of the juvenile, i.e. body movements of the first- or second-stage juveniles. Changes in the eggshell or its chemical breakdown have been demonstrated early in the hatching of *Ascaris lumbricoides* (Hope, 1974). At the time of hatching, juveniles of this species secrete lipase, chitinase and protease. If an embryo has achieved the development necessary for hatching, it may not do so until triggered by the appropriate environmental stimuli. This delay may be prolonged and the stimuli may be complex. In suitable environmental conditions such as appropriate temperature and oxygen availability, and for soil-dwelling forms, soil-moisture levels and absence of physiological barriers, such as diapause, hatching of most species occurs without requiring specific cues emanating from the food source. In the case of *Ascaris lumbricoides*, hatching does not occur until stimulated by gaseous carbon dioxide and/or undissociated carbonic acid. This case was verified experimentally (Perry, 2002). *A. lumbricoides* was stimulated to hatch in vitro by treatment with carbon dioxide in the presence of a reducing agent at neutral pH and 30°C temperature. These conditions are probably similar to those present within the intestine of the host. In some animal-parasitic and insect-parasitic nematodes, eggs are ingested by the host and the host gut provides the conditions for hatching.

Temperature extremes are also relevant to hatching biology. High temperatures can suppress hatching or are lethal to nematodes. Dormancy can occur at most stages of the nematode's life cycle, and where the unhatched juvenile is the affected stage, it is of importance in mediating the hatching response. The transition from a fluctuating external environment to a stable increased temperature of 37°C can often be the stimulus for cessation of dormancy. Substantial hatching of *Ascaris lumbricoides* only occurs between 30°C and 40°C, and they do not respond to temperatures below 30°C. Temperature is an important environmental cue for the termination of diapause.

EMBRYOLOGICAL PECULIARITIES

The above embryological description of nematodes show that this group of animals is peculiar in the following:

1. development is highly determinate and the fate of each cell can be definitely followed;
2. the phenomena of chromosome fragmentation and chromatin diminution;
3. the segregation of gametic-cell line from somatic-cell line in the early embryo, the two-cell stage;
4. the four-blastomere stage in the form of T-shape (short period)
5. the T-shape changes into rhomboidal form;
6. the embryo from the beginning appears bilaterally symmetrical; and
7. except for the gonads, the number of cells in other organs remains constant during postembryonic morphogenesis

POSTEMBRYONIC DEVELOPMENT

After the juvenile has hatched, it is usually fully developed except in size, and the reproductive system together with the accessory reproductive structures. Before the juvenile reaches sexual maturity, it undergoes a phase of growth. Growth in nematodes is not generally accomplished by cellular division. Exceptions are the growth and development of the gonads in all species. At hatching, the gonad usually consists of four cells, two inner ones (product of P4 or P5) which are the primordial germ cells, and two outer encircling ones (product of S4 or S5), which give rise to the gonoducts and the epithelial covering of the gonad. Another exception is the midgut, at least in some cases. In *Paracamallanus sweeti* (Camallanoidea, Spirurida, Secernentea), the intestinal cells are 35 in young juvenile, and increase to about 200 in the adult (Hope, 1974). The number of cells remains constant in other organs during postembryonic morphogenesis and throughout the life of the nematode. For example, in *Rhaditis* there are 200 nerve cells and 120 epidermal cells, and the digestive tract is composed of 172 cells (Barnes, 1974). However, the number of the intestinal cells is known to increase with growth in some species. Growth results from an increase in cell size; consequently, certain cells in the larger nematodes may reach considerable sizes, up to 10mm as in case of muscle cells.

A remarkable feature of the postembryonic growth of nematodes is the occurrence of moults, in which the entire cuticle including the lining of the buccal capsule, pharynx, rectum, and vagina is shed. There are always four moults that divide the postembryonic period of the life cycle into four juvenile stages. The nematode emerges from the fourth moult as a fully developed adult, although growth may continue. Sexual maturity is attained with the last moult, when the copulatory organs are formed. The gonads also reach maturity at this stage. Some difficulty was encountered to differentiate between the adult and the fourth-juvenile stage of the fish-parasitic nematode, *Hysterothylacium reliquens*, in spite of the fact that the site of infection is the lumen of the intestine of the fish host, *Sparus auratus*. An extensive study of the morphology, morphometric and histology of all the organs was carried out to establish the adult stage (Abd El-Fattah, 1992).

Life cycles

Life cycles and the reproductive habits of many species have been studied, e.g. that of the free-living nematode *Ditylenchus triformis* (Hirschmann, 1962). Stock cultures of the dioecious and intersexual lines of this species were maintained on fungus cultures in the laboratory at 24–26°C, which is the optimal temperature range for development and reproduction. Twenty females or intersexes and 20 males were added to 8-day-old Petri dish cultures of the fungus growing on potato dextrose agar. Within 1.5–2 months, each Petri dish culture yielded large numbers of nematodes in all stages of development. This indicates that the fungus has high degree of antibiotic activity. However, in these studies, the nematodes were transferred to new fungus cultures every 5 days.

The average number of eggs laid by a single female of *Ditylenchus triformis* or intersex was 79 (maximum 168). The uterus usually contained only one well-formed egg at a time. In general, the adult female remains sexually functional for three-quarters of its lifetime. The eggs are laid singly before undergoing cleavage. Four moults occur during the life cycle. The first moult occurs within the egg (Fig. 9.18B) and the second stage 'larva' hatches (it was preferred to refer to the stages of nematodes as juveniles). The genital primordium of the young second stage consists of one large germinal cell and two small somatic cells (Fig. 9.18C). The germinal cell could be traced to the late embryonic stages. In *Ditylenchus triformis*, it does not divide before the fourth larval stage in males and the fourth moult in females. The first indication of sex differentiation occurs in the second larval stage, when two to four specialized ventral-chord cells appear in the female and intersex larvae, but are lacking in the male larvae. Early in the third stage, the morphological differences between larvae of different sexes become more pronounced. In the male larva, the anterior part of the gonad consists of oval-shaped group of 9–12 small cells (Fig. 9.19A). Four specialized cells of the ventral-chord will form the vagina by invagination, which appear during the second moult of the female larva (early third stage, Fig. 9.19B). Two large spicule primordia, derived from the rectal epithelium, give rise to spicules and gubernaculum. Each of the three moults outside the egg was clearly defined based on body and gonad length.

With the beginning of the third moult, the gonad extends in length. In the male larva (Fig. 9.19C), the anterior

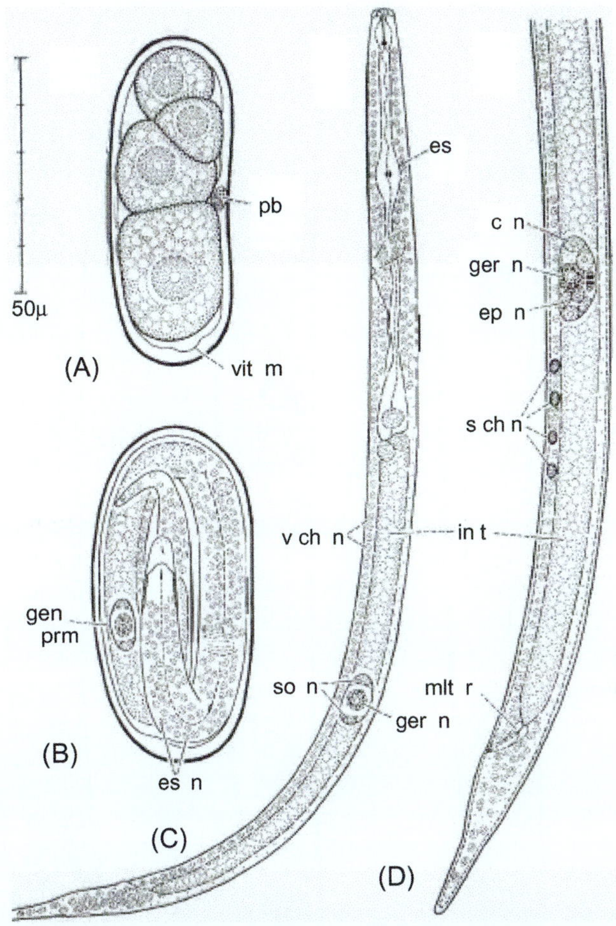

FIG. 9.18 Development of *Ditylenchus triformis*: (A) egg with cleaved embryo (four blastomeres); (B) egg with moulted juvenile (first stage); (C) second stage juvenile (freshly hatched); (D) second moult (posterior part of female juvenile); c n = cap cell nucleus, ep n = epithelial nuclei, es = oesophagus, es n = nuclei of oesophagus, gen prm = genital primordium, ger n = germinal nucleus, int = intestine, mlt r = moulting rectum, pb = polar body, s ch n = specialized ventral cord, so n = somatic nuclei of genital primordium, v ch n = ventral cord nuclei, vit m = vitelline membrane. *(From Hirschmann (1962).)*

oval-shaped group of cells of the gonad (9–12 small cells) differentiates into two double rows of small spherical cells, which turn backwards and grow posterior along the ventral side. The remaining gonad with the germinal cell, cap cell and four large epithelial cells is pushed dorsally and gradually shifts anteriorly. The rectal cells also divide, resulting in the enlargement of the spicule primordium. On the other hand, the posterior cell group of the gonad of the female larva (Fig. 9.19D) increases in size. The gonad grows out anteriorly by forming two single rows of cells between this posterior cell group and the two anterior large epithelial cells. The four specialized ventral-chord cells undergo further divisions, and by completion of the third moult, they are arranged in two groups of eight cells, each in the vicinity of the posterior region of the gonad. Later, they take part in the formation of the vagina. The distance between the

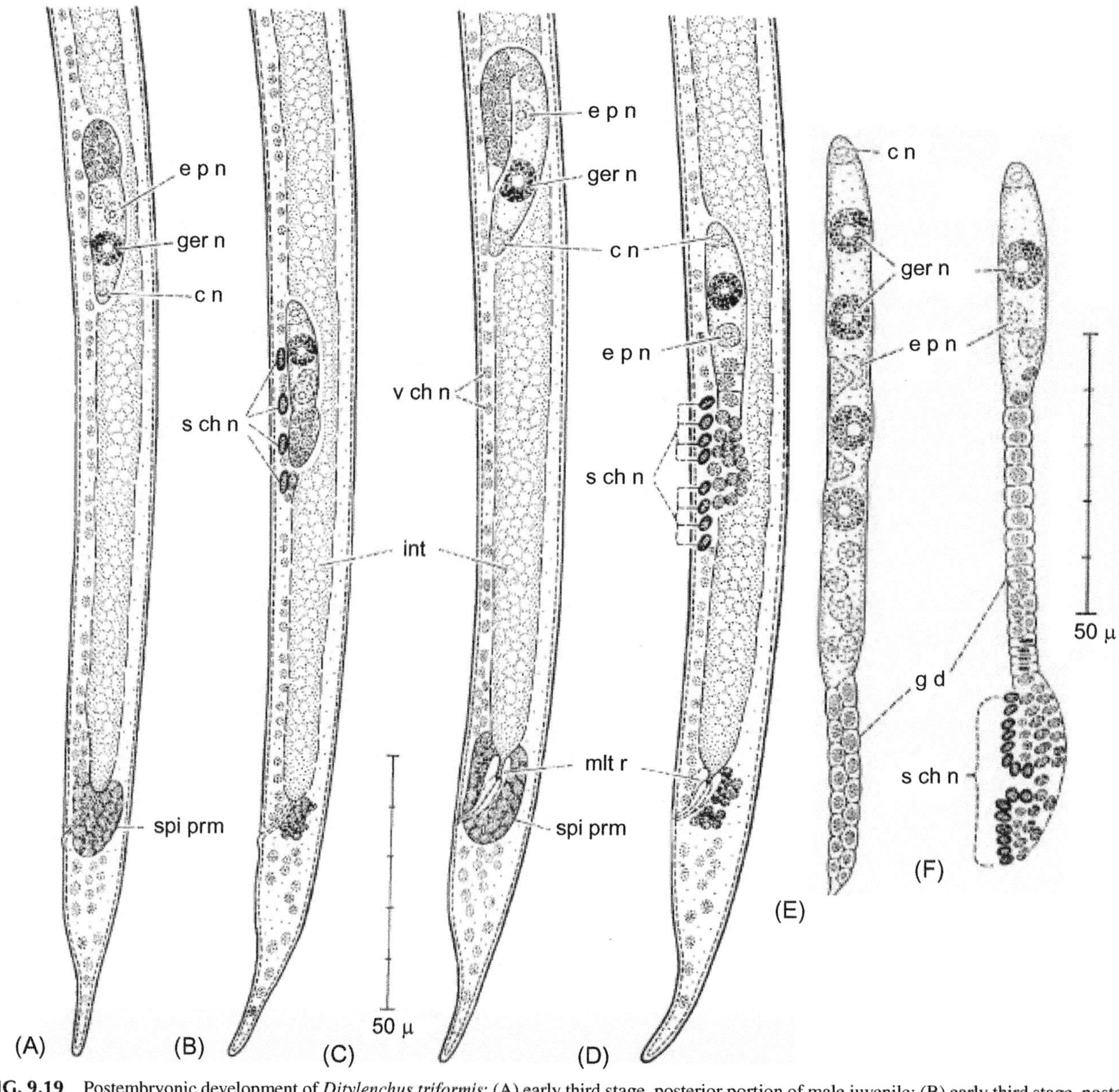

FIG. 9.19 Postembryonic development of *Ditylenchus triformis*: (A) early third stage, posterior portion of male juvenile; (B) early third stage, posterior region of female juvenile; (C) third moult, posterior portion of male juvenile, (D) third moult, posterior portion of female juvenile; (E) fourth stage, gonad of male juvenile; (F) fourth stage, gonad of female juvenile; gd = gonad, spi prm = spicule primordial, other lettering as in Fig. 9.18. *(From Hirschmann (1962).)*

posterior end of the gonad in female larva is about half of that in the male larva. During the fourth stage, the gonads of the two sexes continue to increase in length (Fig. 9.19E and F). The anterior testicular and ovarian parts are clearly distinguishable from the posterior somatic parts, which will form the gonoducts. In both sexes, the gonad gradually differentiates into its different regions.

During the fourth moult (from fourth larvae to adults), the gonad of the male larva clearly differentiates into testis and vas deferens. The freshly moulted males may have as many as 13 spermatogonial cells. The most mature ones are preparing for the first maturation division. In the gonad of the early fourth stage female larva (Fig. 9.19F), the germinal cell does not divide, while the two single rows of small epithelial cells extend in length by repeated divisions. They will eventually differentiate into the various parts of the female gonoduct. The germinal cell in the freshly moulted female retains its activity and begins to divide for the first time, producing four or fewer germinal cells. A slight constriction marks the end of the ovarian part and the beginning the gonoduct. The anterior narrow part of this gonoduct is the oviduct, the end part of which is the spermatheca. At the same time, this spermatheca constitutes the anterior modified end of the uterus. The vagina and vulva complete their development during this fourth moult.

The life cycle of *Ditylenchus triformis* from egg to egg was completed in these experiments (Hirschmann, 1962) within 16–21 days at 24–26 °C, when the nematodes fed on cultures of an unidentified fungus.

The intersex larvae proceed in development as the female larvae, but combine characters of the two sexes. They resemble the female larvae in respect to gonad morphology and location, and exhibit a large group of spicule primordial cells in the rectal region as in male larvae.

Anabiosis

Experiments carried out by Hirschmann (1962) resulted in the observation of the ability of this species, *Ditylenchus triformis*, to survive unfavourable conditions under dormancy. In heavily populated fungus cultures, this nematode in all stages tended to swarm together and form clumps. In older cultures, these clumps or aggregates of motionless, tightly packed nematodes were easily visible to the naked eye as white patches in the agar medium. Ten heavily populated fungus cultures were allowed to dry out slowly at room temperature. The thin sheets of dried agar medium, containing the nematodes, were then stored for 2.5 years at 24–26°C. At the end of this period, most of the specimens revived when the desiccated medium was soaked in water for 48 h. Larvae of all stages and adults were capable of anabiosis. When transferred to new fungus cultures, they resumed feeding, and reproduction was normal. This anabiosis is a common phenomenon in the genus *Ditylenchus*.

It is well known that embryogenesis in nematodes takes place completely or partially inside the mother, or cleavage may start only after eggs have been laid. The more or less transparent, rigid eggshell (mechanical protection) and an underlying thin lipid layer ('vitelline membrane', a chemical protection) make the early embryo independent of a supporting maternal environment. The duration of embryogenesis is species-specific and temperature-dependent. Under optimal conditions, it may be as short as 12 h in one species and as long as several weeks in another. Schierenberg (1997) summarized four different nematode life cycles by the illustrations shown in Fig. 9.20.

CAENORHABDITIS ELEGANS

Nomenclature

Caenorhabditis elegans is a free-living (not parasite) transparent nematode (round worm), about 1 mm in length, that lives in temperate soil environment (Fig. 9.16C). The name is a blend of Greek (*Caeno*—recent, *rhabditis*—rod-like) and Latin (*elegans*—elegant). In 1900, Maupas initially named it *Rhabditides elegans*; Osch placed it in the subgenus *Caenorhabditis* in 1952, and in 1955, Dougherty raised it to the status of genus.

Historical review

Caenorhabditis elegans was first discovered in 1897 by Émile Maupas, a French zoologist and botanist. He described this species as a nematode that lives in rich humus, in the surroundings of Algeria, which he visited twice (Maupas, 1900). He isolated specimens from the soil to study. His sketches show remarkable details for both forms of *C. elegans*: the hermaphrodite producing both sperm and oocytes, and the male producing sperm only. He described the hermaphrodite soma as having a female form with a vulva that is behind the middle part of the body (Fig. 9.21B and E). He also described the male form as having a specialized tail structure with a flattened, somewhat 'heart-shaped bursa' which contains '9 pairs of papillae'. In 1948, Ellsworth Dougherty proposed that free-living nematodes of the suborder Rhabditina might be useful for genetic study, noting their relative structural simplicity and invariant cell lineage. Dougherty and Victor Nigon in the 1940s obtained the first mutant, from a laboratory culture of the nematode *Caenorhabditis briggsae*. However, much of the early laboratory work on *Caenorhabditis* nematodes was directed towards the establishment of a defined axenic culture medium. By the early 1960s, Sydney Brenner had made several important contributions to molecular biology, notably a demonstration (with Francis Crick and other colleagues (1961)) that 'the genetic code is triplet in nature'. In June 1963, he wrote to Max Perutz, who was the head of the MRC laboratory of Molecular Biology, Cambridge (Medical Research Council, United Kingdom) at that time, proposing future research to study development.

The success of the researches on bacteria inspired Brenner to believe that the same approach could be used to study specification and control of more complex processes in cells of higher organisms. Brenner looked for a simple metazoan organism, with a short life cycle, that can be controlled genetically, in which molecular genetics is analogous to that of bacteria. He succeeded in finding a small free-living nematode *Caenorhabditis elegans*, which is 1 mm long and has a rapid period of embryogenesis (about 16 h). It has relatively few cell types. Moreover, it can be cultured in a Petri dish on agar plates with a thin layer of bacteria as food. Brenner isolated and characterized mutants, localized about a hundred genes on the six linkage groups and then published his pioneering paper 'The genetics of *Caenorhabditis elegans*' (Brenner, 1974). Since then, this nematode has become the most favourite organism for different aspects of research, and has proved to be superior to the sea urchins and *Drosophila*.

For a long time, sea urchins were the favourite organisms of embryogenesis because the gametes are readily

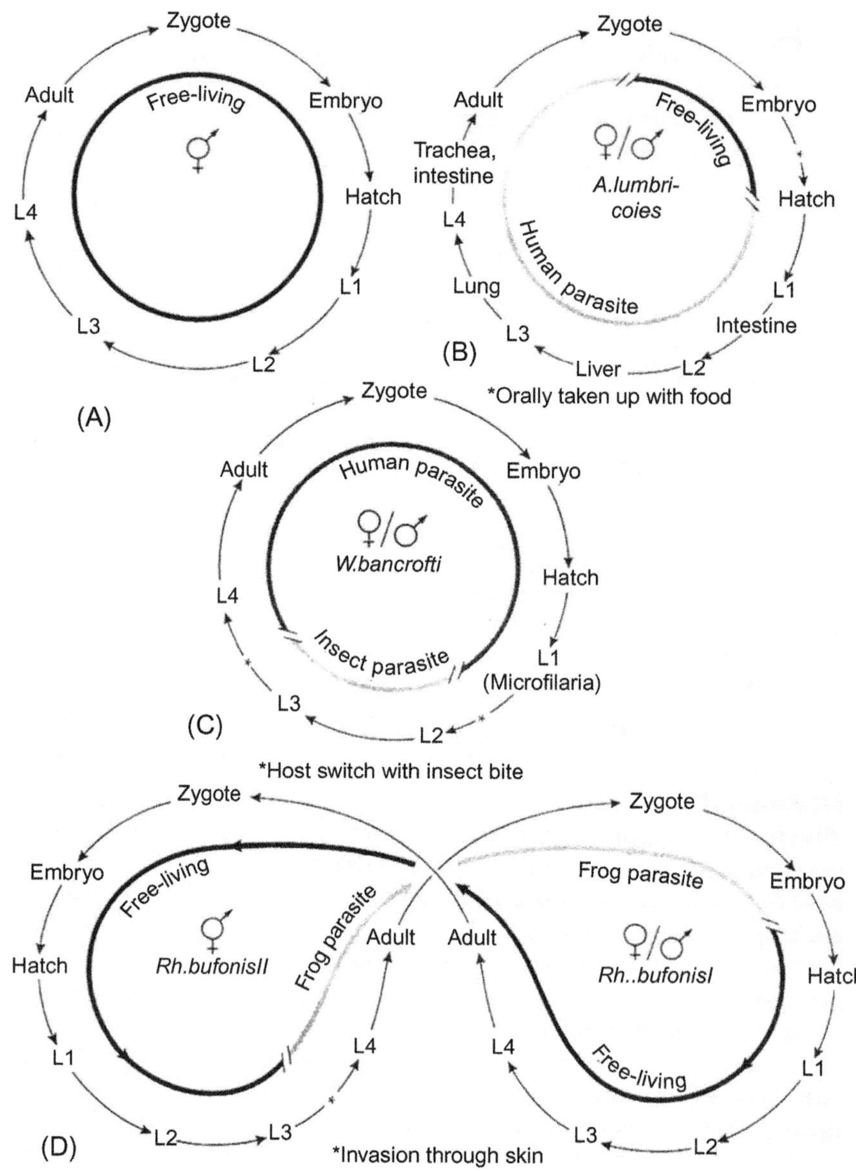

FIG. 9.20 Nematode life cycles: (A) *Caenorhabditis elegans*, free-living; (B) *Ascaris lumbricoides*, parasitic except during embryogenesis, the eggs pass out of the host with faeces; (C) *Wuchereria bancrofti*, microfilaria blocking lymph glands may cause elephantiasis, parasite in two successive hosts; (D) *Rhabdias bufonis*, alternating between amphimictic, free-living and hermaphroditic, parasitic generations. L1–L4, four juvenile stages. *(From Schierenberg (1997).)*

obtainable in large numbers, the eggs and embryos are transparent, and fertilization and development can occur under laboratory conditions.

However, sea urchins are difficult to rear in the laboratory for more than one generation, making their genetics difficult to study. On the other hand, geneticists (at least those working with multicellular eukaryotes), favour *Drosophila* because of its rapid life cycle, its readiness to breed and the polytene chromosomes of the larva (which allow gene localization), which make this fruit fly superbly suited for hereditary analysis. However, the development of *Drosophila* is complex and difficult to study.

Today, *Caenorhabditis elegans* is probably the most completely described metazoan organism in terms of anatomy, germ cell development and differentiation, behaviour, genetic control and molecular events. In the 1980s, John Sulston and co-workers identified the lineage of all 959 cells in the adult hermaphrodite. The first genes were cloned, and the physical map began to be constructed in 1998. The worm became the first multicellular organism to have its genome sequenced. Notable research using *C. elegans* includes the discoveries of caspases, RNA interference and microRNAs. Six scientists have won the Nobel Prize for their work on *C. elegans*.

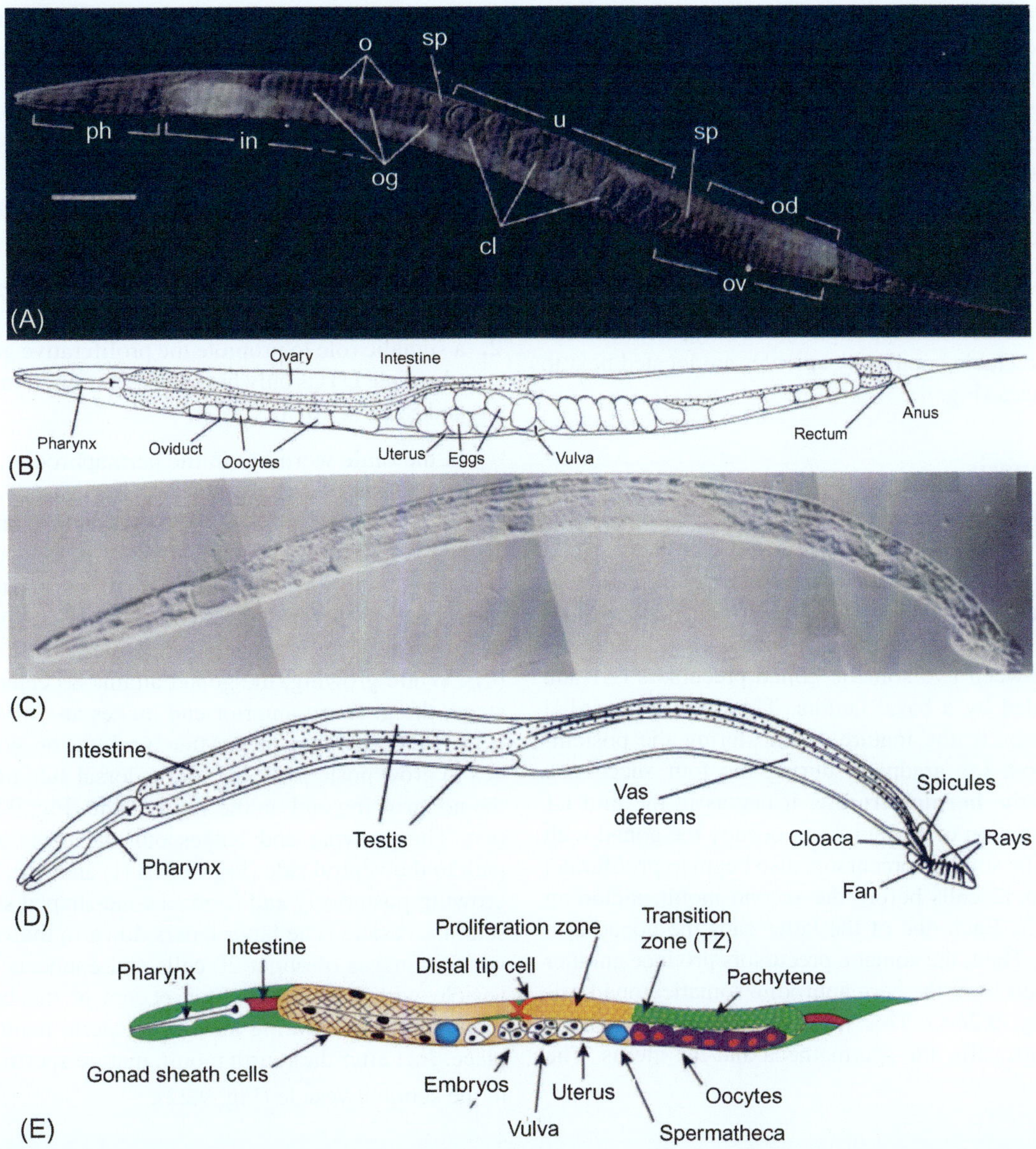

FIG. 9.21 (A) Young, gravid adult hermaphrodite *C. elegans*, bar = 100 μm *(from Schierenberg, 1986)*, cl = cleaving eggs, in = intestine, o = mature oocytes, od = oviduct, og = oogonia, ov = ovary, ph = pharynx, sp. = spermatheca, u = uterus; (C) photomicrograph of a wild-type male taken with Nomarski optics *(from Klass et al., 1976)*; (B and D) Bright field illumination of hermaphrodite (B) and male (D) of *C. elegans*, 137× *(from Sulston and Horvitz, 1977)*; (E) schematic drawing of whole mount of hermaphrodite *C. elegans*, highlighting the reproductive system *(from Pazdernik and Schedl, 2013)*.

Sex differentiation

The free-living nematode *Caenorhabditis elegans* is usually found as a hermaphrodite (Fig. 9.21A and B), while males occur spontaneously in populations with a frequency of about one per 700 worms. Males (Fig. 9.21C and D) are XO in karyotype and are most easily distinguished from hermaphrodites by the copulatory bursae on their tails. The adult male is ~1000 μm long and 50 μm in diameter (Fig. 9.21C and D). Hermaphrodites are XX with a diploid set of autosomes (2X.2A).

Reproductive system

In the **hermaphrodite** *C. elegans*, this system consists of two U-shaped tubes, the ovaries, facing each other and joined at a common uterus ventrally (Fig. 9.21E). The vulva lies ventrally, midway along the length of the worm. It is a flat slit against the external cuticle (Figs 9.21B, E and 9.23A, B). The ovary (U-tube) contains germ cells in different stages of development, arranged linearly with the proliferative germ cells in the distal region, just beyond the somatic distal tip cell (DTC) (Fig. 9.23A). The latter caps the distal end of the tube. The proliferative zone is limited by a very narrow region, the transition zone. The tube continues to the loop, in which the cells enter and progress through meiotic prophase (Fig. 9.23A). The fully formed oocytes are found in the proximal gonad. The hermaphrodite specifies male germ cells in the L3 stage, which develop into sperm in the L3/L4 stage. It then specifies female germ cells from L4 throughout adulthood, which develop into oocytes.

The adult **male** gonad is an asymmetrical structure located in the coelomic cavity of the worm (Fig. 9.21D). The testis is found on the ventral side of the worm and it loops giving rise to a longer and wider branch, and continues as a small region, the seminal vesicle (Fig. 9.23D). The latter extends while narrowing to be the vas deferens, consisting of about 20 cells. It is the narrow tube connecting the seminal vesicle to the cloaca and eventually to the exterior. The copulatory bursa is the posterior end of the male system and opens to the ventral side (Fig. 9.21C and D). It is composed of a membrane-like fan supported by nine rays on each side. A pair of retractable copulatory spicules are located next to the small cloaca (Fig. 9.22).

Gonadogenesis

The hermaphrodite and male gonad of *C. elegans* develop from morphologically identical primordia. This small primordial gonad lies ventrally in the coelomic cavity of the newly hatched L1 in both worms (Fig. 9.24). It consists of four cells: two primordial germ cells (Z2 and Z3) embraced between two somatic gonad precursors (Z1 and Z2) surrounded by a basal lamina. This small primordial gonad develops to the mature gonad during the postembryonic period, i.e. gradually during the four successive juveniles. In the **hermaphrodite**, it begins in the mid L1 by the proliferation of Z2 and Z3 to occupy the gonad with germ cells. The somatic precursors also begin to proliferate, giving rise to 12 cells before the second moult, including the two DTCs. Each one of the latter caps the gonad arm (Fig. 9.24A). Then, the somatic precursors produce another 10 cells, which form the hermaphrodite somatic gonad primordium (Fig. 9.24A). This will give rise to the somatic gonadal sheath cells, the spermatheca and the uterus. The sheath cells have more than one role; they are necessary for oocyte maturation and ovulation, promoting the meiotic prophase progression and activating the germ cell proliferation. From the L4 stage, the hermaphrodite gonad begins to look like its adult counterpart, with the gonad arm on the dorsal side of each U-shaped tube capped by the DTC (Figs 9.23A and 9.24A). The hermaphrodite DTC has two functions:

1. a migratory/morphogenetic leader function that gives rise to the U-shape of each gonad arm; and
2. a signalic role to promote the proliferative germ cell fate (the male DTCs only have the latter function) (Pazdernik and Schedl, 2013)

In the **male** worm, as in the hermaphrodite, the primordium consists of four cells. They begin to divide posteriorly and anteriorly along the longitudinal axis of the worm. After 14–18 h from hatching at 25°C, the developing gonad contains about 30 cells. At this stage, the posterior end of this gonad stops growing, while the anterior end continues to grow anteriorly along the ventral side (Fig. 9.24B (c)). While growing, the gonad attains 80 cells, and at this stage, the growing anterior end makes an 180 degrees turn (Fig. 9.24B (c) and (d)). After turning, the gonad continues to grow posteriorly, along the dorsal side until it passes the nongrowing end on the ventral side (Fig. 9.24B (d) and (e)). The growing end bulges out and takes a 45 degrees path to the ventral side (Fig. 9.24B (f) and (g)). It continues growing posteriorly and forms a cone-shaped structure, the seminal vesicle. The latter tapers down to the vas deferens, which consists of about 20 cells and connects the seminal vesicle with the cloaca. This occurs at the beginning of the L4 stage and during this stage, sperm maturation takes place. Just after the fourth moult, mature sperm can be seen in the seminal vesicle (Fig. 9.22).

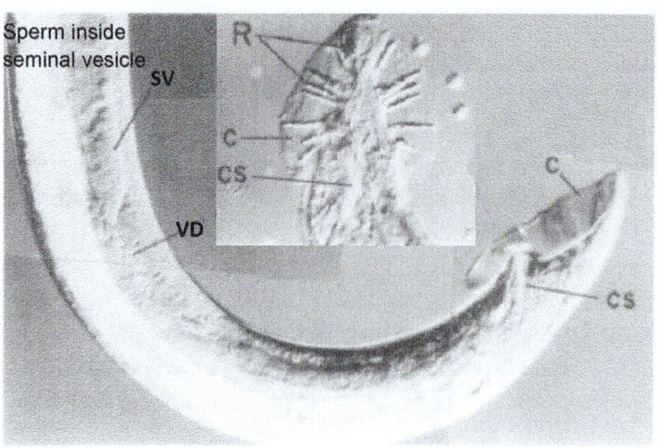

FIG. 9.22 Photomicrograph of wild-type male tail of *C. elegans* taken by Nomarski optics at 440×, C = copulatory bursa, cs = copulatory spicules, sv = seminal vesicle (notice fully formed sperm inside), vd = vas deferens, inset shows the supporting rays (R) of the bursa. *(From Klass et al. (1976).)*

Gametogenesis

The male individuals of *C. elegans* produce sperm, while the self-fertile hermaphrodites produce sperm and eggs. All the germ cells in the male differentiate, giving rise to sperm. In hermaphrodites, the first germ cells to enter meiosis give rise to sperm, and later ones become oocytes. This transition occurs near the beginning of the adulthood of the worm. In both sexes, all germ cells in mitosis or early stages of meiosis are part of a large syncytium. These immature cells are similar in both individuals, and there is no visible difference at the end of the pachytene stage of meiosis I.

In male individuals, the isolated testis stained with Feulgen (Klass et al., 1976) shows many mitotic figures (Fig. 9.25A and B), and pachytene stage of meiosis are present through and beyond the loop (Fig. 9.25A and C). Then follows a continuous progression of stages of pachytene, diplotene and diakinesis (Fig. 9.25D). After the latter

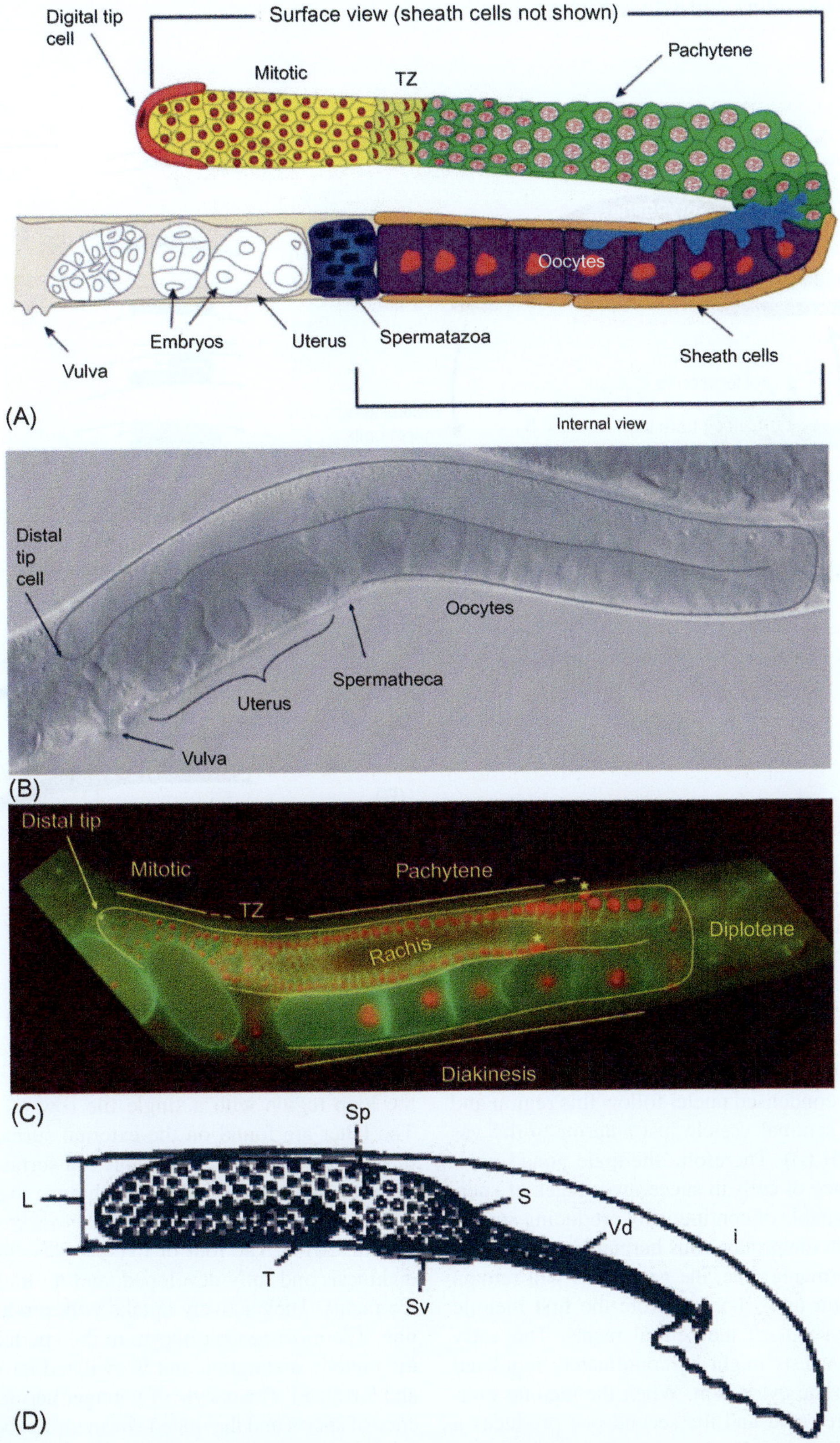

FIG. 9.23 (A) Detailed view of one adult hermaphrodite gonad arm of *C. elegans*, the upper part of the arm shows a surface view without covering sheath cells (this is the proliferative zone, *yellow*), the transition zone (a very small zone is visible as a *light green colour*), followed by a long portion in *green colour* is the growth zone, the lower part of the gonad shows the interior of the proximal region covered by the sheath cells, the oocytes closest to the loop are connected to the central rachis (appeared in *light blue*), while the following oocytes nearest to the spermatheca are fully developed, (B) An interior view of live *C. elegans* adult hermaphrodite from Nomarski Optics showing the U-shaped gonad arm in the posterior half of the worm *C. elegans*, the distal tip cell caps the distal end of the gonad and the developing oocytes inhabit the proximal end. In the distal half of the gonad, on the dorsal surface, germ cell nuclei are situated along the outer surface of the gonadal tube. The rachis is devoid of germ cells nuclei. By a bulk flow, germ cells move from the very distal end to the proximal end. Developing oocytes are found from the loop to the proximal half of the gonad ventrally. (C) Fluorescence image of live *C. elegans* adult hermaphrodite highlighting the U-shaped gonad arm in the posterior half of the worm, the gonad is outlined in *yellow*, the distal tip cell caps the distal end, the developing oocytes populate the proximal end, the *green colour* visualized the plasma membranes, the red colour indicates the nuclei, the rachis is devoid of germ cells/nuclei, (D) Diagrammatic drawing showing the structure of the testis of adult male *C. elegans*, magnified from Fig. 9.24B i, L=loop, S=mature sperm, Sp=spermatocytes, Sv=seminal vesicle, T=testis, Vd=vas deferens *(from Klass et al., 1976). (A–C) From Pazdernik and Schedl (2013).*

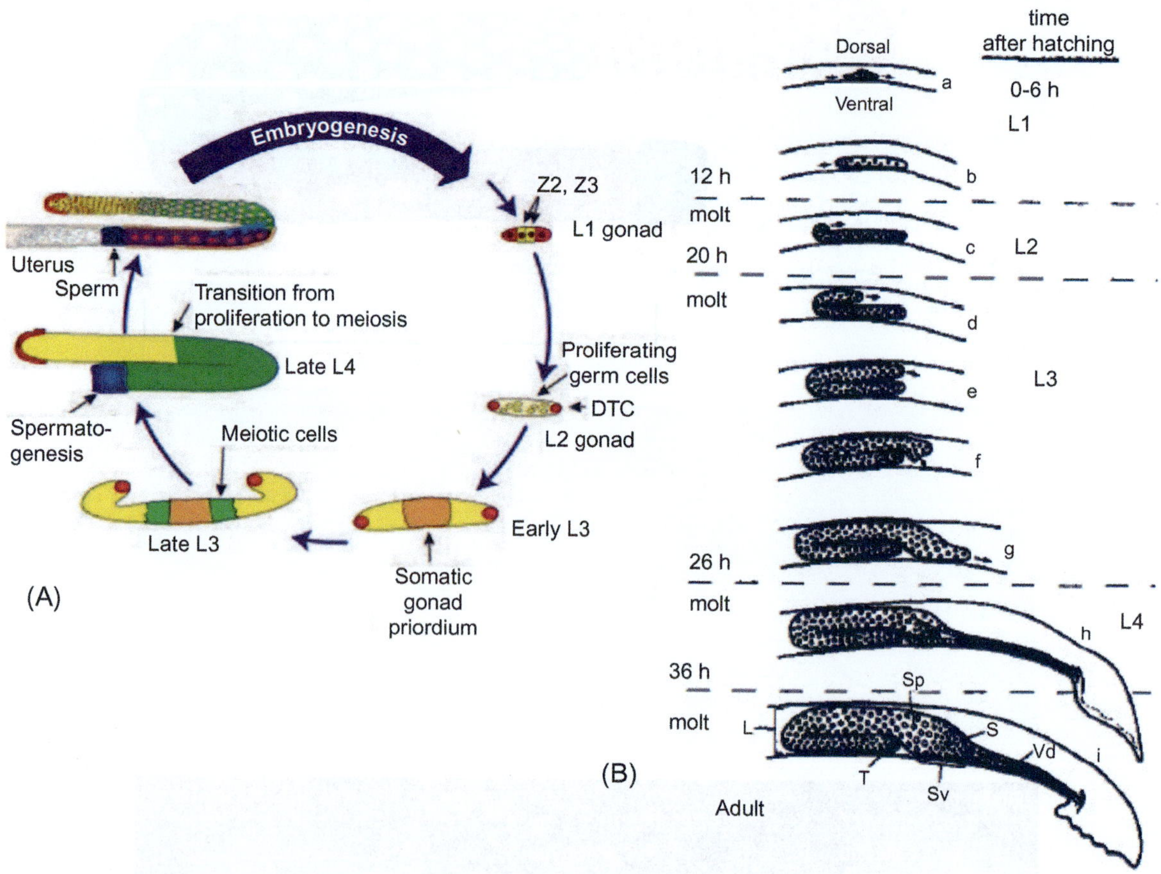

FIG. 9.24 Illustrates the process of gonadogenesis in *C. elegans*: (A) hermaphrodite worm *(from Pazdernik and Schedl, 2013)*; (B) male worm, S = mature sperm, Sp = spermatocytes, Sv = seminal vesicle, T = testis, Vd = vas deferens *(from Klass et al., 1976)*.

region, there is a small region (the only region) in which one can see meiotic metaphase figures. Klass et al. interpreted that both meiotic divisions occur in this small region. Mature sperm with condensed nuclei follow this region and can be seen in the seminal vesicle just anterior to the vas deferens (Fig. 9.24B (i)). Therefore, the male gonad presents a linear sequence of cells in successive stages of spermatogenesis. It is capable of continuously producing sperm, and thus differs from the protandrous hermaphrodite worm.

In self-fertile hermaphrodite, the germ cells will remain part of the syncytium until shortly before the first meiotic division, when they bud off the central rachis. The early stages of spermatogenesis might be coordinately regulated by factors shared in the cytoplasm. When the meiotic divisions begin, they are rapid and the second one produces a residual body and four round spermatids (Fig. 9.26A). The spermatids remain inactive until they are pushed into the spermatheca during ovulation (Fig. 9.26B). Oogenesis in the self-fertile hermaphrodite has an extended pachytene stage, during which the germ cells synthesize RNAs and proteins needed for the growth of the oocyte. Much of the germline is a syncytium, and during the pachytene stage, the germ cells are on the surface of the gonadal tube with an interior region called the rachis or core (Fig. 9.23C). As germ cells progress from pachytene to diplotene, oocyte differentiation begins in the loop region with a single file row of growing oocytes. The latter are found on the external surface of the gonadal tube, and the rachis is on the internal surface. The RDNs and proteins synthesized during pachytene stage are deposited into the rachis and delivered to the oocytes (Pazdernik and Schedl, 2013). The four or five oocytes after the loop are in diakinesis and fully developed, and no longer connecting to the rachis. They actively uptake yolk produced by the intestine. The most nearest oocyte to the spermatheca undergoes the meiotic maturation and is ovulated into the spermatheca and fertilized. The oocyte of younger hermaphrodite, in presence of sperm and the mated worm undergoes maturation and ovulation every ~23 min, and oogenesis is continuous.

The unmated hermaphrodite is able to produce about 300 embryos, while the mated worm can produce up to 1000 embryos. However, the limiting factor for self-fertility is not oocyte production, but the amount of self-sperm production by the hermaphrodite. Thus, if mating occurs, 50% of cross progeny will be male (Pazdernik and Schedl, 2013).

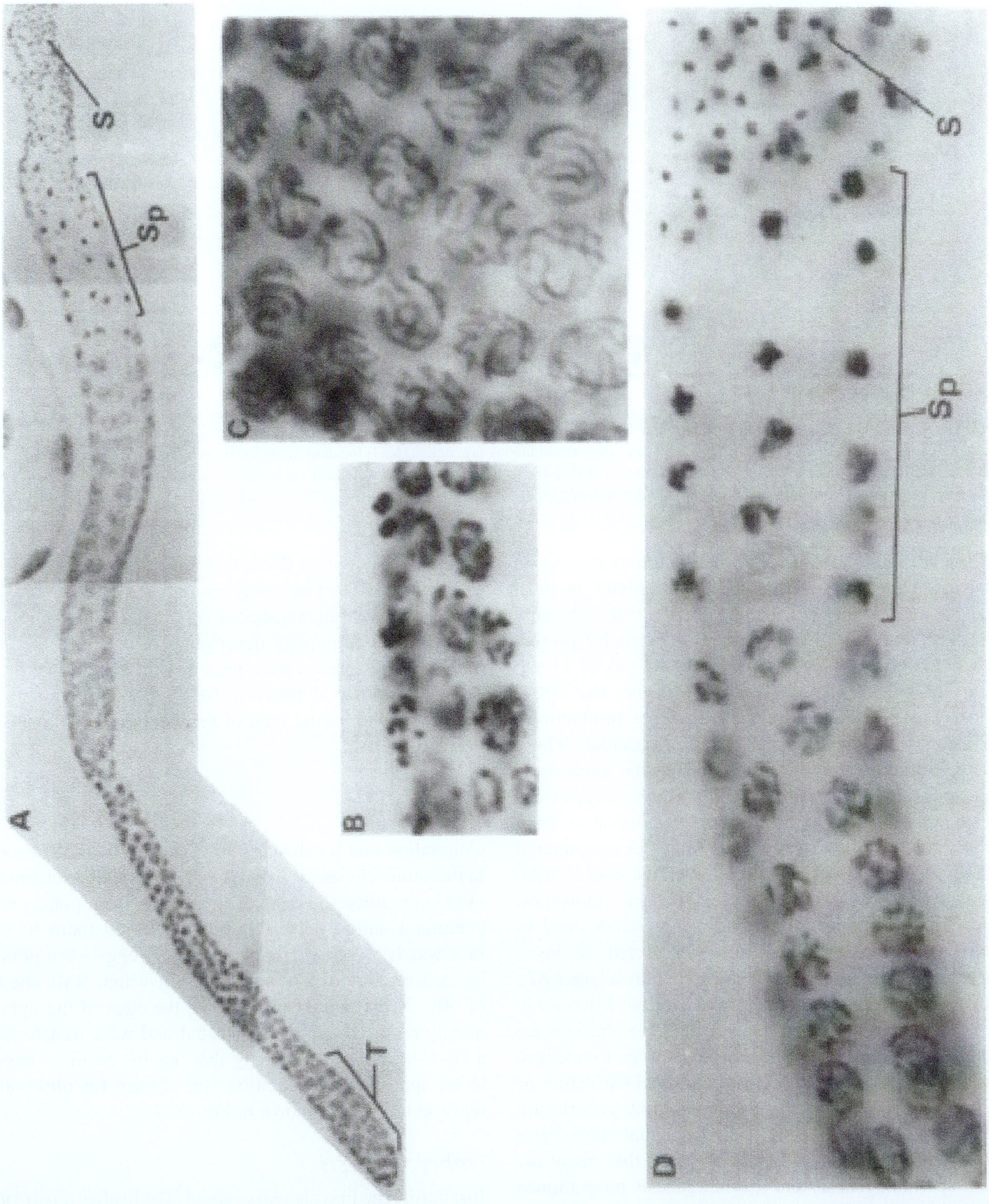

FIG. 9.25 Histology of the process of spermatogenesis of male *C. elegans*, the testis was stained by Feulgen reaction: (A) whole mount of isolated testis showing the linear arrangement of the stages of the process (magnification 387×); (B) enlargement of the mitotic region of the testis (magnification 3060×), (C) enlargement of the loop region showing the pachytene stage (magnification 20,430×); (D) enlargement of the region preceding the seminal vesicle showing the linear arrangement from pachytene, diplotene, and diakinasis (magnification 1530×); S = mature sperm, Sp = spermatocytes, T = testis. *(From Klass et al. (1976).)*

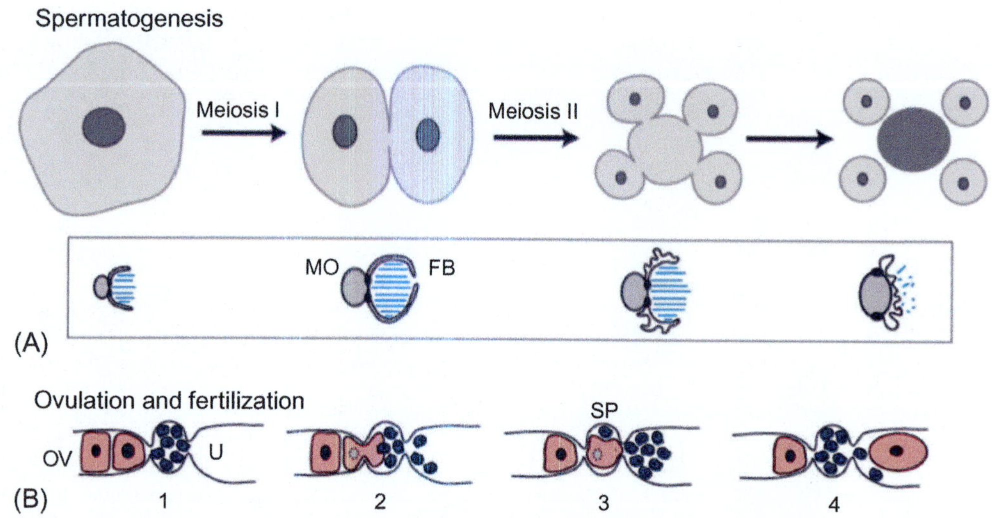

FIG. 9.26 (A) Spermatogenesis in the hermaphrodite *C. elegans*, the spermatocyte undergoes two meiotic divisions, producing four spermatids *(light grey)* and a residual body *(dark grey)*. During this process, the membraneous organelles (MO) localize many sperm proteins, and their associate fibrous bodies (FB) are made of polymerized major sperm protein (MSP, *blue*), the differentiation of each organelle is shown in the inset below each. (B) ov=ovary, sp.=spermatheca, u=uterus. *(From Ellis and Stanfield (2014).)*

Ellis and Stanfield (2014) studied spermatogenesis in *C. elegans*. They reported that this vital activity, in most animals, functions under a major constraint: the chromatin must be reorganized into a compact form for protection. This unique chromatin structure uses special Sperm Nuclear Basic Proteins. Sperm of *C. elegans* use SPCH proteins, which resemble some invertebrate ptrotamines, and a histone 2A variant like that found in mammals. These begin to restructure the chromatin during the karyosome stage of meiosis I, just after diplotene. This process makes DNA inaccessible, so transcriptional regulation is only possible early in spermatogenesis. Many developmental decisions in spermatogenesis occur after the end of transcription regulation. Mature spermatids lack ribosomes, which segregate to the residual body (Fig. 9.26A), and so protein–protein interactions would be expected to play a major role in spermatogenesis, and also to play a larger role in the subsequent process of sperm activation. Microarray analysis indicates that protein kinases and phosphatases are strikingly overrepresented in sperm, in contrast to oocytes. Transcripts for proteins that control chromatin structure or RNA translation are under-represented. Similar patterns are found in transcripts from the parasitic nematode *Ascaris suum*. Therefore, Ellis and Stanfield stated that many aspects of spermatogenesis and spermiogenesis programmes are probably controlled by the phosphorylation condition of proteins.

Gametes

Mature sperm of *C. elegans* are stable, polarized cells (Fig. 9.10(II) A).They crawl rather than swim. Their pseudopodia contain fibres of Major Sperm Protein (MSP) (Fig. 9.10(III)). The body contains a stereotypical arrangement of membraneous organelles and chromatin (Fig. 9.26B). Complete description in page 19. In *C. elegans* genetic screens have identified at least 50 genes that define a biological pathway for sperm development. The mature oocyte, at the time of fertilization, is elongate oval and 50 μm long.

Development

Copulation and fertilization in *C. elegans* can be described in the same way as for *Ascaris megalocephala*. Sulston et al. (1983) obtained the eggs of *C. elegans* from plates, which contained ample bacteria, and transferred them to water in a watch glass. Alternatively, young eggs were obtained by cutting gravid hermaphrodites in water. With the help of 50% X dissecting microscope, the eggs of the approximately required age were selected and were transferred to a layer of 5% agar. The egg that has the required age and in an appropriate orientation was chosen for observation; representatives are shown in Fig. 9.27.

Embryogenesis

Just before the first cleavage, about 30 min after fertilization, the *C. elegans* egg is somewhat oval in shape. It contains the two pronuclei apposed in the centre (the zygote) and the first polar body is visible beneath the eggshell at the anterior end (Fig. 9.27B). Embryogenesis begins with the **first** cleavage, which is unequal, producing an anterior larger somatic founder cell (AB) and a smaller germline (P1),

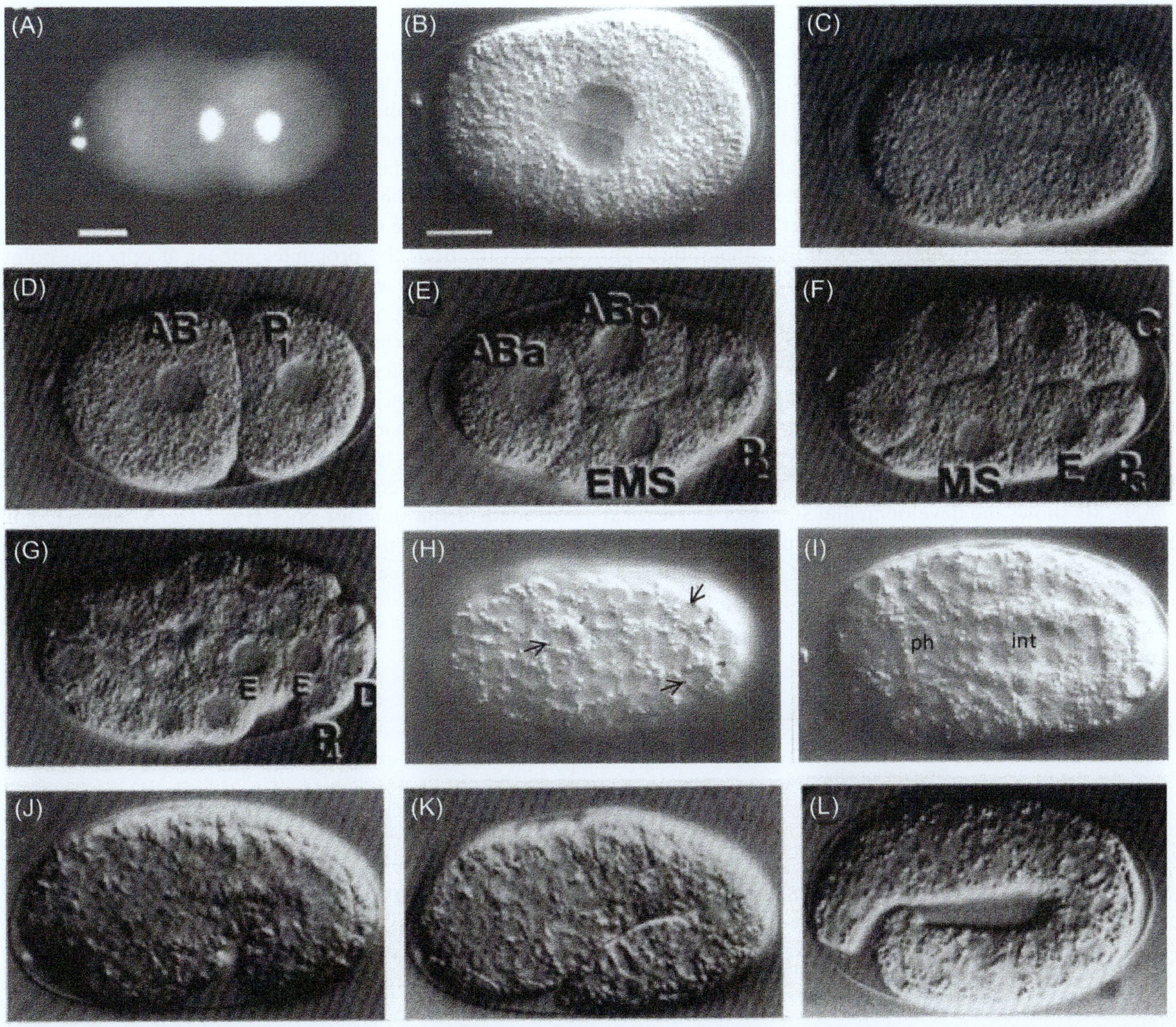

FIG. 9.27 Photomicrographs showing embryonic development of *C. elegans*: (A) one-cell embryo before fusion of the pronuclei, two PBs, are seen at the anterior pole (epifluorescence visualizing DAPI-stained nuclei); (B) just before first cleavage, PB is visible beneath the eggshell at anterior end; (C) preparation of first cleavage; (D) two-cell stage with larger anterior somatic founder cell AB and a germline cell P1; (E) four-cell stage with the second somatic founder cell EMS and a new germline cell P2; (F) eight-cell stage with a third somatic founder cell C and a new germline cell P3; (G) all the founder cells divide, AB-cell, at this stage (fourth cleavage) produce eight cells, the last unequal cleavage in the embryonic germline produces the somatic founder D-cell and the primordial germ cell P4, soon afterwards gastrulation starts with the immigration of the two E-cells; (H and I) later stages of gastrulation; (H) ventral view, superficial plane, small neuroblasts over the entire surface *(head arrows)*, dying cells are engulfed by their sisters *(arrows)*; (I) dorsal view, midplane, intestinal cells (nine nuclei in this focal plane) and the cytoplasm heavily loaded with granules, Ph = pharyngeal cells; (J–L) completion of morphogenesis which starts with the indentation at the ventral side, then elongation of the body leads to the formation of the worm. L1 hatches about 12 h after fertilization at 25°C. Bar = 10 μm. *((A) From Schierenberg (2006); (B, H, I) from Sulston et al. (1983); (C–G and J–L) from Schierenberg (1986).)*

which are beside each other (two-cell stage, Fig. 9.27D). The P1 is the primordial germline. The **second** cleavage divides the AB into Ab a and AB p, while P1 divides unequally into a larger somatic cell EMS and a new germline P2 (four-cell stage, Fig. 9.27E). **third** cleavage; the division of all cells is followed by unequal cleavage of P2 into a somatic founder cell C and a new germline P3, and the EMS gives rise to E and MS (eight-cell stage, Fig. 9.27F). During the **fourth** cleavage, all the somatic cells divide in the order AB, MS, E, and C, with the synchronous duplication of the eight AB cells (16 cells). The last unequal cleavage in the embryonic germline generates the somatic founder cell D and the primordial germ cell P4. Soon afterwards, gastrulation begins.

Gastrulation

Schierenberg (2006) divided the process of gastrulation in *C. elegans* into two distinct phases: phase 1 starts at 26-cell stage by the immigration of the precursors of the gut, two E-cells (Fig. 9.27G); and phase 2, during which a larger number of mesodermal progenitors migrate, cells of which derived from different lineages at a much higher stage. However, just after the immigration of the E-cells, the daughters of P4 follow them, through the same ventral opening (blastopore), just at the beginning of phase 2. Morphogenesis takes place and begins with the indentation at the ventral side (Fig. 9.27J). After completion of morphogenesis, the embryo elongates and takes the shape of the worm. The latter hatches as L1, 12 h after fertilization at 25°C (Fig. 9.27L). Newly hatched juvenile consists of about 550 cells, AB 389, MS 80, E 20, D (body muscles) 20, P4 2, and C 47 (Fig. 9.28A). Fig. 9.28B shows the developmental stages of *C. elegans* from the two-cell stage to the adult by photographs taken by Nomarski optics.

Polarity

Before the fertilization of the *C. elegans* oocytes, no signs of polarity are seen under the light microscope, with the exception of the nucleus of the most mature oocyte, which can be more or less displaced from the centre of the cell. The egg of this worm is fertilized at the pole that enters the spermatheca first, which will become the posterior pole of the embryo. About half an hour after fertilization, the pronuclei appear at opposite poles. At that time, the egg clearly shows a visible anterior–posterior (a–p) polarity. However, it remains unclear whether sperm penetration induces the appearance of the a–p polarity (Schierenberg, 1986). In 2006, Schierenberg went on to evidence that the sperm entrance into the egg induces the primary polarity, i.e. defined the posterior pole. In addition to its essential function, it contributes by its DNA and centrosome. Both polar bodies usually extruded at the anterior pole (Fig. 9.27A and B). This means that the a–p polarity in *C. elegans* is defined very early in embryogenesis.

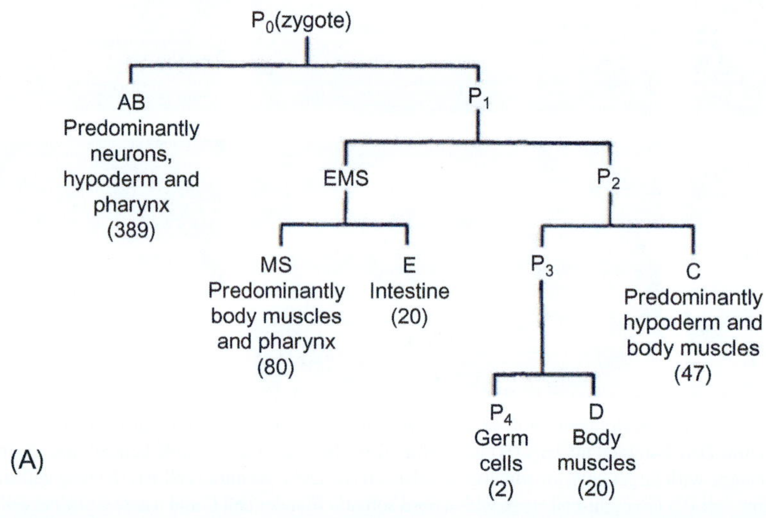

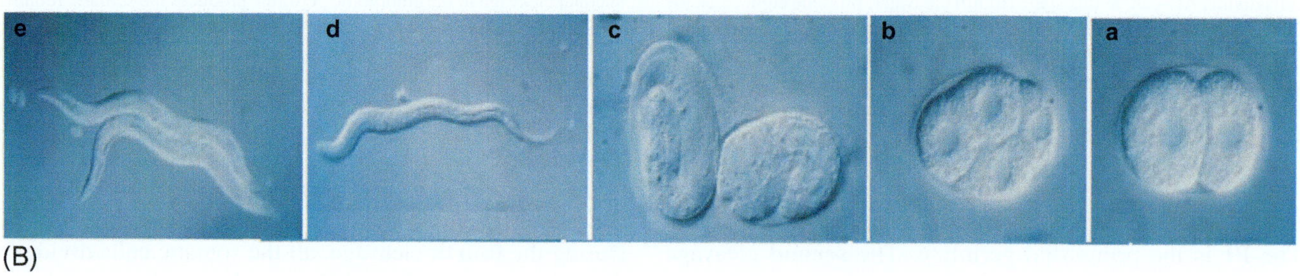

(B)

FIG. 9.28 (A) Schematic early cell lineage tree of *C. elegans*, the numbers of surviving cells produced in each Cell lineage during embryogenesis are given in parenthesis *(from Schierenberg, 1986)*; (B) Photomicrographs taken using Nomarski Optics showing embryonic development of *Caenorhabditis elegans*: (a) two-cell stage at 1000×; (b) four-cell stage at 1000×, dorsal is to the top, EMS is ventral and P2 is posterior; (c) the embryo on the left is fully elongated, the other is elongated to the 1½-fold stage (at 1000×); (d) L1 larva at 400×; (e) adult hermaphrodites and three eggs at 100× *(from Hope, 2002)*.

Prior to fertilization, the germline specific structures (**'P-granules'**), as identified by fluorescent antibodies, are randomly distributed in the cytoplasm of the most mature oocyte of *C. elegans* (Schierenberg, 1986). These granules play an important role for the preservation of germline identity. They gather at one pole only, prior to mitosis, but are distributed all over the cell during interphase. Around the time when the two pronuclei meet (Fig. 9.27B), the P-granules accumulate in the posterior region of the egg and are thus incorporated only into the germline cell P (Fig. 9.30H). The alteration from apparently random cytoplasmic distribution to prelocalization at one pole is repeated during each cell cycle of the germline cell, until the primordial germ cell P4 is formed. When this cell divides, both daughters receive P-granules in approximately equal amounts. Thus, the P-granules mark the pathway of the germline in *C. elegans* with the same precision as chromatin nondiminution does in *Ascaris*.

Postembryonic development

The life cycle of *C. elegans* is rapid; during 3.5 days at 20°C, it develops from a fertilized egg and through four larval stages to a mature adult. Superficially, the newly hatched juvenile appears to be similar to the adult; and the hermaphrodite and male juvenile are almost identical morphologically (Fig. 9.29A). This juvenile, at this stage, consists of about 550 cells (Fig. 9.29A). Its ectoderm, which comprises the majority of cells (389), includes the nervous system with its supporting cells, and the hypodermis. The postembryonic development, does not only increase the size (Fig. 9.29B L1-adult) and number of cells in some organs, but also differentiate its sex. The cells that make this change are initially identical in the male and in the hermaphrodite, but later diverge. These cells are referred to as 'blast cells' (Sulston and Horvitz, 1977) which produce, in the hermaphrodite, structures used in egg-laying and mating. In male they produce morphologically different structures used in copulation. In addition, development of the male involves a number of lineages derived from cells that do not divide in the hermaphrodite. The primordium gonad of the newly hatched juvenile contains four cells. Its early development is the same in the two sexes. Later, the result differs (Fig. 9.24) and its cells increase in number to about 2500 in the mature adult.

Dauerlarva

The dauerlarva (Fig. 9.29C) is a postembryonic developmental variant of the nematode *C. elegans*. It is a dormancy state that may occur at the end of the developmental stage L1 or L2. A similar stage has been observed for the related nematode *C. briggsae* (Cassada and Russell, 1975). It is well known that many other animals, during their lives, enter an inactive state, known as hibernation, during the winter when the conditions are unfavourable such as lack of food, high temperature, cold, or dryness. Likewise, *C. elegans* is able to enter a quiescent state, designated as dauer (a German word meaning 'enduring or persisting') stage. The fertilized eggs of soil nematode *C. elegans* would normally develop into the adult reproductively mature stage within 3–5 days, after quickly going through the L1 to L4 larval stages, moulting at the end of each stage (Fig. 9.29B). However, in postembryological development of this free-living nematode, a nongrowing stage may arise, which is a dauer stage. It is formed by a facultative, reversible arrest at a specific point in the life cycle in response to external conditions (Cassada and Russell, 1975). Usually at the end of the L2 (or less commonly of the L1) larval stage, under unfavourable conditions such as high-population density, or other conditions as mentioned above (Fig. 9.29D), the dauer state commonly allows *C. elegans* to survive for up to 120 days, instead of the normal 20 days. The dauer state also allows it to disperse effectively via **'phoresy'** (relationship in which one organism transports another organism of a different species) to a more favourable condition place (Wang et al., 2009).

Morphology and ultrastructural studies proved that the dauerlarva has several distinctive properties. It is very thin, and has unusual internal appearance, e.g. no open space in the gut. Its specific gravity is increased and it has an altered cuticle. Most, if not all, have a loose outer sheath when first formed. They are less desiccated and greatly resistant to any harsh condition.

VARIATION IN NEMATODE EMBRYOLOGY

Since the early cell lineages and arrangement of blastomeres in *C. elegans* (Clade V Fig. 9.1) are similar to the pattern found in *Ascaris* (Clade III) and other studied nematodes, embryonic development shows little variation within phylum Nematoda.

However, the development of *Ascaris* is considered the classical example or mode, so *C. elegans* differs in developing without chromatin diminution, which is most important and most significant difference between *Ascaris* and other nematodes. Schierenberg (2006) studied three parthenogenetic species, and Schulze and Schierenberg (2011) studied 12 species, belonging to different clades of the phylogenetic tree (Fig. 9.1), to compare to *C. elegans*. These somewhat recent studies of a variety of species, on a descriptive level, demonstrate more prominent variations in critical steps of early embryogenesis. Schierenberg (2006) focuses his studies on the initial phase between egg activation and gastrulation. The present work chooses some examples of these variations.

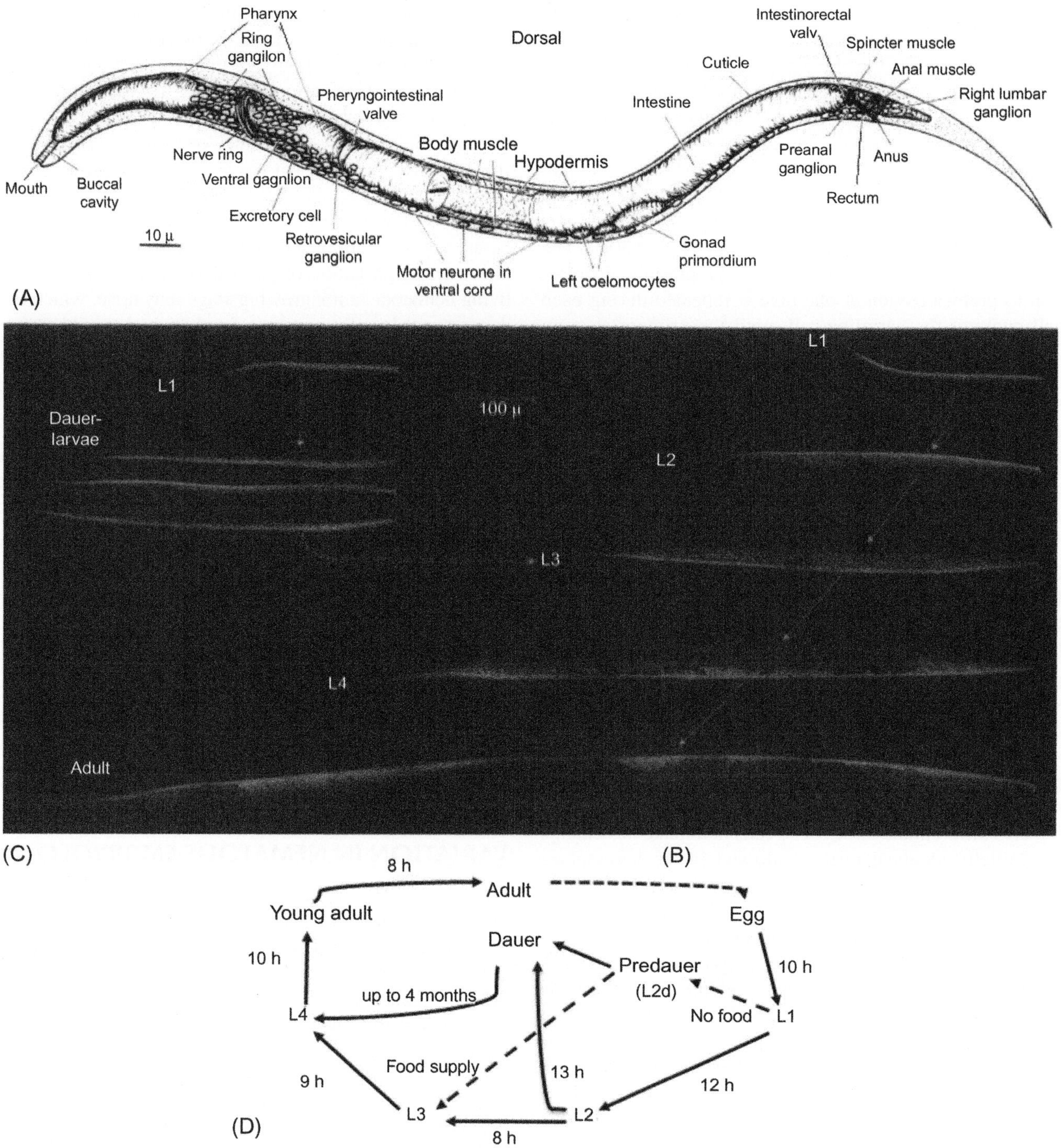

FIG. 9.29 (A) General anatomy of newly hatched juvenile L1 of *C. elegans*, left lateral aspect, body wall was cut at mid-line and also part of the intestine cut away *(from Sulston et al., 1983)*. (B) photomicrographs showing normal postembryonic development of *C. elegans* from L1–L4 and adult *(solid arrows)* taken by Nomarski optics, with (C) dauerlarvae at the same magnification *(dashed arrows) (from Cassada and Russell, 1975)*; (D) The life cycle of *C. elegans*. Dauer is an alternative larval stage, taken when the environmental conditions are unfavourable for further growth. Dauerlarvae appear to be nonaging and can live nearly 10 times their normal lifespan (about 20 days). The length of time the animal spends at a certain stage is indicated.

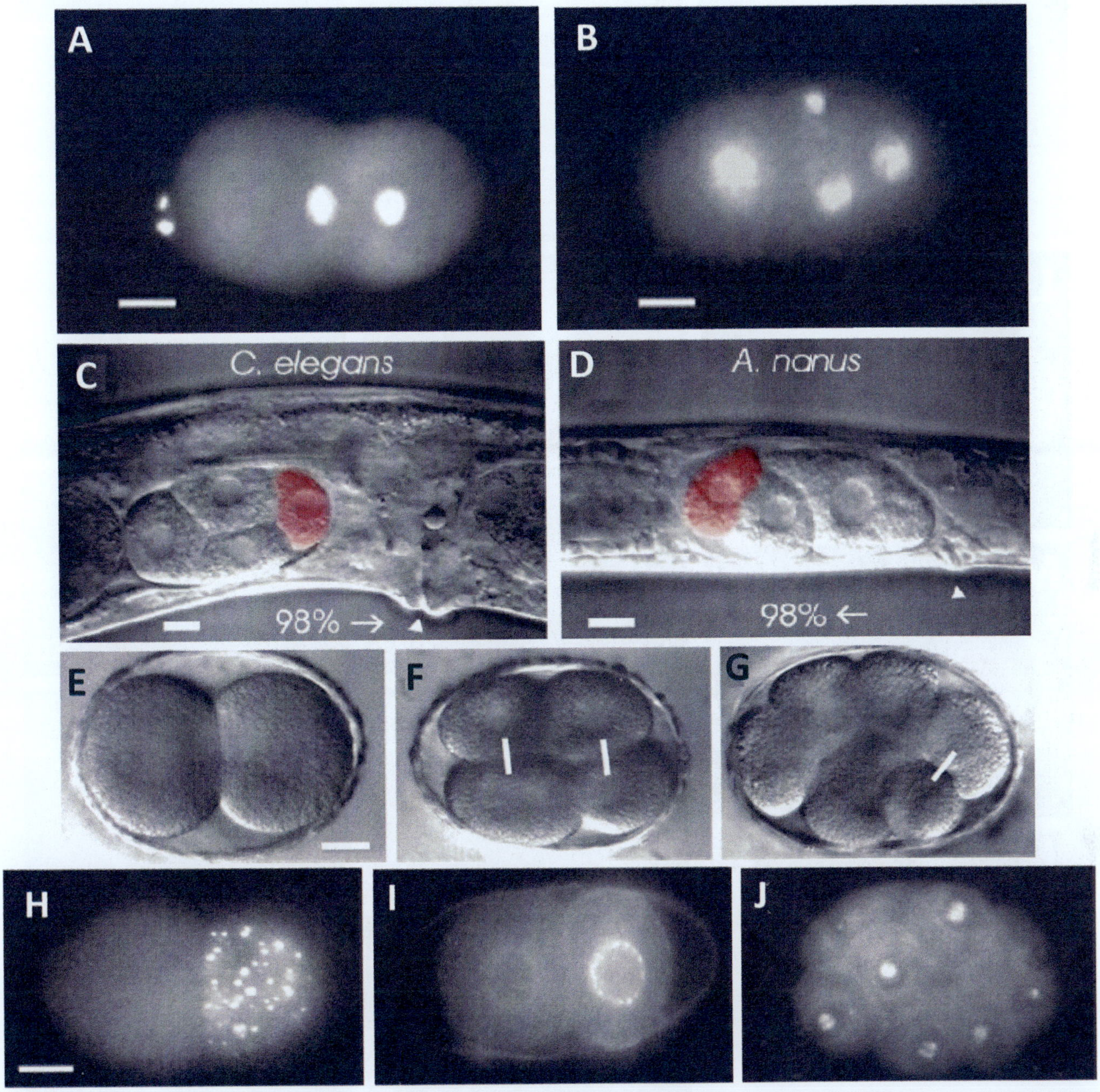

FIG. 9.30 Examples of embryological variation in nematodes: (A and B) variation in meiosis and polar body (PB), Epifluorescence visualizing DAPI-stained nuclei, (A) *C. elegans* (see Fig. 9.27A), (B) *Acrobeloides nanus*, three-cell stage and one PB; (C and D) variation in polarity, in (C) the posterior pole of *C. elegans* directed to the vulva, while in (D) the anterior pole of *Acrobeloides nanus* directed to the vulva; (E–G) variation in cleavage, *Prionchulus* sp., (E) two-cell stage derived from a symmetric first cleavage, (F) four-cell stage (both daughter cells have divided symmetrically) with transverse spindle; (G) eight-cell stage, one cell has performed a germline-like asymmetric division; (H–J). Variant staining with antibodies against *C. elegans* P granules. (H) *C. elegans*, two-cell stage, P granules are only present in the germline cell P; (I) *A. nanus*, two-cell stage, strong perinuclear staining in the germline cell P1, weak staining in the somatic AB cell; (J) *Rhabditis terricola*, eight-cell stage, intranuclear staining in all blastomeres. Orientation: anterior, left. Scale bar, 10 mm. *(From Schierenberg, 2006.)*

(Continued)

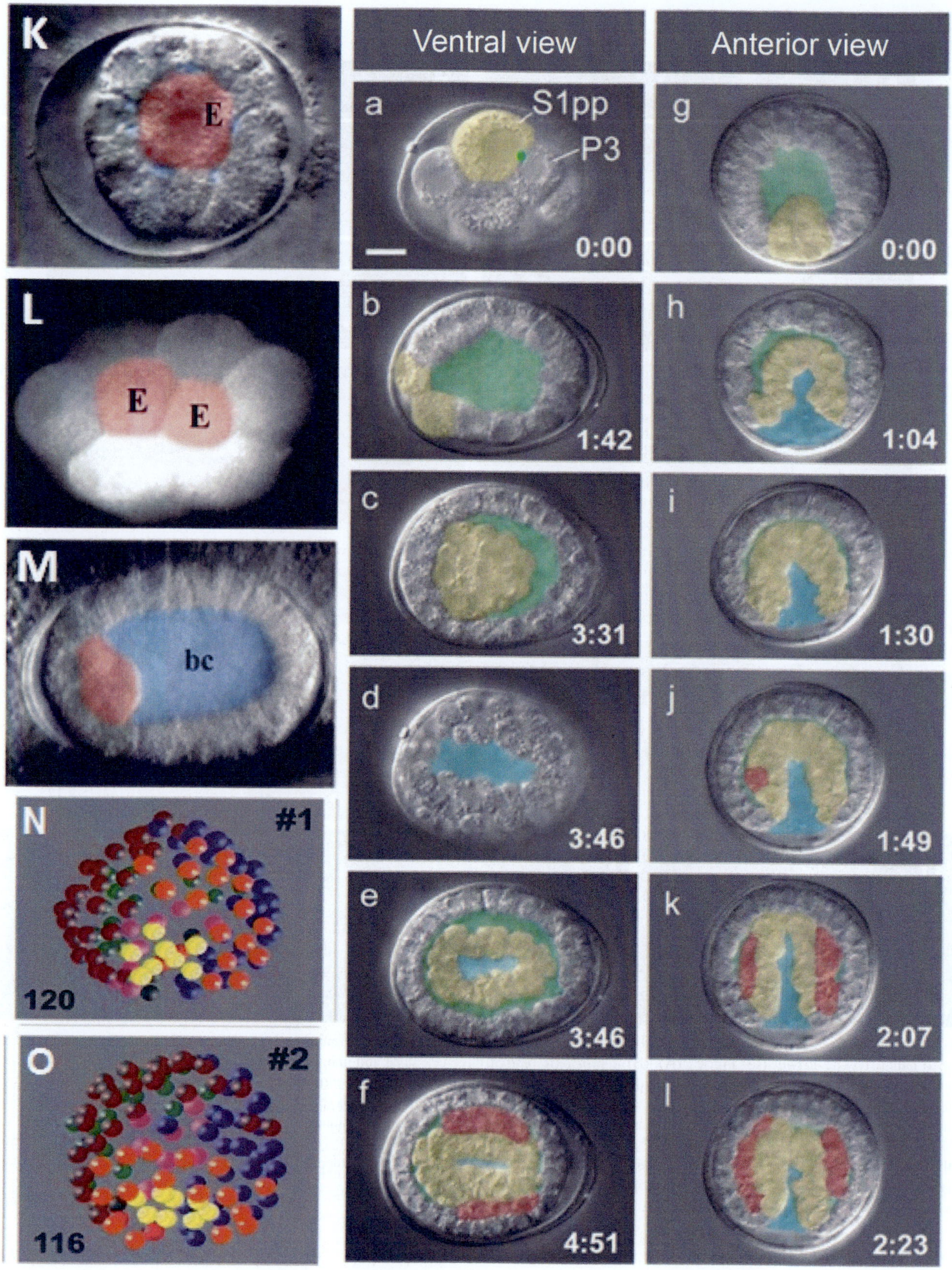

FIG. 9.30, CONT'D (K–M) Variation in gastrulation; (K) *Plectus* sp. (ES 601). The E cell has moved inside the embryo after the 12-cell stage; (L) *Enoplus brevis*, 16-cell stage, two gut precursor cells have started to immigrate; (M) *Tobrilus diversipapillatus*. After the 64-cell stage gut precursors have started to move into the large blastocoel (bc) from the anterior region; colour code, *blue*, blastocoel; *red*, gut precursor cells (endoderm); *green*, intermediate layer ('mesoderm'); orientation: anterior, left. Bars, 10 μm *(from Schierenberg, 2006)*; (a–f and g–l) Embryogenesis of two embryos of *Tobrilus stefanskii* showing cleavage, blastocoel formation, gastrulation, and tissue formation; colour code, *Yellow*, alimentary tract precursor(s); *green*, blastocoel; *blue*, slit-like blastopore; *red*, body muscle cells, *green dot*, primary organizing centre (POC); time in hours and minutes; scale bar, 10 μm *(from Schulze and Schierenberg, 2011).*

One of the presented species, *Acrobeloides nanus* (clade IV, Fig. 9.1), shows more than one difference. First is in preservation of its diploidy, compared to *C. elegans*. Parthenogenesis needs preservation or restoration of diploidy, in the absence of sperm. In *A. nunus*, usually only large PB is formed, and diploidy is reached by a truncated meiosis (Fig. 9.30A and B). The *sec*ond example concerns the polarity. It was described above that the oocytes of *C. elegans* are usually fertilized at the pole that enters the spermatheca first, developing embryos in the uterus, directing the same pole (posterior pole) towards the vulva (with few exceptions). *A. nanus* directed its anterior pole towards the vulva (Fig. 9.30C and D). The third example shows the difference in cleavage patterns. In *Ascaris* and *C. elegans*, the cleavage causes the separation of soma/germlines and produces a small number of somatic founder cells and the primordial germ cell P4. This type of cleavage is referred to as 'strong expression of the germline' (Schierenberg, 2006). In other nematodes, Schulze and Schierenberg (2011) found six different early cleavage patterns: two among Enoplea and four among Chromadorea. In *Prionchulus* (Clade I), clear asymmetric cleavages can only be observed at later stages (Fig. 9.30E–G), and early stages completely differ from that of *C. elegans*. This difference was also encountered in Fig. 9.16C.

The localization of P granules is the fourth example. They are found distributed in the cytoplasm of the mature oocyte of *C. elegans*, and accumulated in one pole only prior to mitosis (see above). This localization differs in different nematodes (Fig. 9.30H–J). Variations in gastrulation occur only in the first step. In *C. elegans*, this process begins at the 26-cell stage and this appears to be similar in members of clades I, III, IV, and V (Fig. 9.1). In *Plectida* (PCM group), gastrulation starts as early as the eight-cell stage, and it is always the gut founder cell itself that immigrates (Fig. 9.30K). In *Enoplus brevis*, this process begins at the 16-cell stage (Fig. 9.30L). The most striking and most different from *C. elegans* and from *Ascaris* is the development of the coeloblastula (Fig. 9.30M). In three different species of one genus, *Tobrilus*, the blastocoel is large during early embryogenesis. In *T. diversipapillatus*, gastrulation occurs after the 64-cell stage by the movement of the precursors of the gut into the large blastocoel from the anterior region (Schierenberg, 2006) (Fig. 9.30M). In the three species, the blastocoel starts to form as early as the four-cell stage and grows to its full size before the 100-cell stage. Gastrulation starts with the invagination of the eight gut precursors (Fig. 9.30N and O) (Schulze and Schierenberg, 2011). As divisions continue, the immigrated cells fill the blastocoels (Fig. 9.30(c), (g)–(i)), leaving an oval-shaped blastopore furrow on the ventral side (Fig. 9.30(d)). After immigration of the future alimentary canal cells, leaving only small remnants of the blastocoels, body muscle precursors of various lineage origin move between these and the outer layer (Fig. 9.30(f) and (j)–(l)).

REFERENCES

Abd El-Fattah, A.M., 1992. Studies on Some Stages in the Life Cycle of Nematodes Infecting Three Marine Fish Species (M.Sc. thesis) Faculty of Science, Alexandria University.

Anderson, R.V., Kimpinski, J., 1977. Sexual anomalies in a female intersex of a species of *Aphelenchoides* (Nematoda: Aphelenchoididae). Can. J. Zool. 55, 1209–1211.

Borradaile, L.A., Potts, F.A., with chapters by Eastham, L.E., Saunders, J.T., 1977. The Invertebrata, fourth ed. Revised by Kerkut, G.A. Cambridge University Press.

Boveri, T., 1892. Befruchtung. In: Merkel, F.S., Bonnet, R. (Eds.), Ergebnisse der Anatomie und Entwicklungsgeschichte, (Wiesbaden: Joseph Friedrich Bergmann), pp. 386–485.

Boveri, T., 1899. Die Entwicklung von Ascaris megalocephala mit besonderer Rücksicht auf die Kernverhältnisse. Festschrift für C v Kupffer, Fischer, Jena, pp. 383–430.

Boveri, T., 1910. Die Potenzen der Ascaris Blastomeren bei abgeänderter Furchung. Festschrift für R Hertwig, Fischer, Jena, pp. 133–214.

Brenner, S., 1974. The genetics of *Caenorhabditis elegans*. Genetics 77, 71–94.

Cassada, R.C., Russell, R.L., 1975. The dauerlarva, a post-embryonic developmental variant nematode *C. elegans*. Dev. Biol. 46, 326–342.

Cheng, T.C., 1964. The Biology of Animal Parasites. W.B. Saunders Company, Philadelphia, London.

Chitwood, B.G., 1933. A revised classification of the Nematoda. J. Parasitol. Urbana 20, 131.

Chitwood, B.G., 1958. The designation of official names for higher taxa of invertebrates. Bull. Zool. Nomencl. 15, 860–895.

Chitwood, B.G., Chitwood, M.B., Christenson, R.O., Jacobs, L., Wallace, F.G., 1950. An Introduction to Nematology, Section 1: Anatomy. Monumental Printing Company, Baltimore.

Crick, F.H., Barnett, L., Brenner, S., Watts-Tobin, R.J., 1961. General nature of the genetic code for proteins. Nature 192 (4809), 1227–1232.

Deardorff, T.L., Overstreet, R.M., 1980. Review of *Hysterothylacium* and *Iheringascaris* (both previously= *Thynnascaris*)(Nematoda: Anisakidae) from the northern Gulf of Mexico. Proc. Biol. Soc. Wash. 93 (4), 1035–1079.

Dougherty, E.C., 1958. Notes on the naming of higher taxa with special reference to the phylum (or class) Nematoda. Bull. Zool. Nomencl. 15, 896–906.

Ellis, R.E., Stanfield, G.M., 2014. The regulation of spermatogenesis and sperm function in nematodes. Semin. Cell Dev. Biol. 17–30.

Gardiner, M.S., 1972. The Biology of Invertebrates. In: McGraw-Hill Book Company. London, New York.

Gegenbaur, C., 1859. Grundzüge der vergleichenden Anatomie. 1870, second ed. (cited in Hyman, L.H., 1940, 1951).

Gibbons, L.M., 2002. General organization. In: Lee, D.L. (Ed.), The Biology of Nematoda. CRC Press (Taylor & Francis Group).

Goeze, J.A.E., 1782. Versuch einer Naturgeschichte der Eingeweidewürmer tierischer Körper. Mit 44 Kupfertafeln. – pp. I–XI [= 1–11], [1–5], 1–471. Blankenburg. (Pape).

Gourbault, N., 1982. Nematodes marins de Guadeloupe. I. Xyalidae nouveaux des genres *Rhynchonema* Cobb et *Prorhynchonema* nov. gen. Bull. Mus. Natn. Hist. Nat. Paris, 4e Sér. 4, sec. A (1–2), 75–87.

Gourbault, N., 1983. Systeme reproducteur d'un Nématode marin à fécundation par spermatophore. Rev. Nematol. 6, 51–56.

Gourbault, N., Renaud-Mornant, J., 1982a. Un nouveau mode de fécundation, par spermatophore, chez les Nématodes. C. R. Acad. Sci. Paris 294 (ser III), 285–287.

Gourbault, N., Renaud-Mornant, J., 1982b. Un nouveau mode de fécondation, par spermatophore, chez les Nématodes. Comptes rendus de l'Académie des Sciences 285–287.

Grobben, K., 1910. Claus-Grobben: Lehrbuch der Zoologie, second ed. Elvert'sche Verlagsbuchhandlung, Marburg.

Hirschmann, H., 1962. The life cycle of Ditylenchus triformis (Nematoda, Tylenchida) with emphasis on post-embryonic development. Proc. Helminthol. Soc. Wash. 29 (1), 30–43.

Hope, W.D., 1974. Nematoda. In: Giese, Pearse (Eds.), Reproduction of Marine Invertebrates. vol. I. Academic Press, New York/London.

Hope, I.A., 2002. Embryology, developmental biology and the genome. In: Lee, D.L. (Ed.), The Biology of Nematoda. CRC Press (Taylor & Francis Group).

Hopper, B.E., Meyers, S.P., 1967. Foliicolous marine nematodes on turtle grass, Thalassia testudinum König, in Biscayne Bay, Florida. Bull. Mar. Sci. 17 (2), 471–517.

Hyman, L.H., 1951. The Invertebrates: Acanthocephala, Aschelminthes and Entoprocta. The Pseudocoelomate Bilateria. vol. III McGraw-Hill Book Company, Inc., New York, Toronto, London.

Justine, J.-L., 2002. Male and female gametes and fertilization. In: Lee, D.L. (Ed.), The Biology of Nematoda. CRC Press (Taylor & Francis Group).

Klass, M., Wolf, N., Hirsh, D., 1976. Development of the male reproductive system and sexual transformation in the nematode *Caenorhabditis elegans*. Dev. Biol. 51, 1–18.

Linnaeus, C., 1758. Systema Naturae, 10th ed., vol. 1. Salvii, Holmiae. 824 pp.

Maupas, E., 1900. Modes et formes de reproduction des nematodes. Archives de Zoologie Experimentale et Generale, 8. Translation provided by Marie-Anne Felix, pp. 463–624.

Metz, C.B., Monroy, A., 1985a. Biology of Fertilization. Model Systems and Oogenesis, vol. I. Academic Press, Inc, New York, London.

Metz, C.B., Monroy, A., 1985b. Biology of Fertilization. Biology of the Sperm, vol. II. Academic Press, Inc., New York, London.

Müller, H., 1903. Beitrag zur embryonalentwicklung von Ascaris megalocephala. Zoologica 41, 1–30.

Murray, R.K., Granner, D.K., Mayes, P.A., Rodwell, V.W., 1996. Harper's Biochemistry, 24th ed., Middle East edition Librairie de Liban, Appleton and Lange, Norwalk, CT/Los Altos, CA.

Nigon, V., 1949. Les modallités de la reproduction et le déterminisme du sexe chez quelques nématodes libres. Ann. Sci. Nat. 11, 1–132. (cited in Hope, 1974).

Pazdernik, N., Schedl, T., 2013. Introduction to germ cell development in *C. elegans*. Adv. Exp. Med. Biol. 757, 1–16.

Perry, R.N., 2002. Hatching. In: Lee, D.L. (Ed.), The Biology of Nematoda. CRC Press (Taylor & Francis Group).

Pflugfeder, O., 1970. Lehrbuch dez Entwicklungsgeschichte und Entwicklungsphysiologie der Teire. Veb Gustav Fischer Verlog Jena.

Rasheed, S., 1963. A revision of the genus *Philometra costa*, 1845. J. Helminthol. 37 (1–2), 89–130.

Roosen-Runge, E.C., 1977. The Process of Spermatogenesis in Animals. Cambridge University Press, London.

Rudolphi, K.A., 1808. Entozoorum, sive vermium intestinalium historia naturalis. Volumen I. - pp. I–XXVI [1–26], 1–527, Tab. I–VI [= 1–6]. Amstelaedami.

Rudolphi, C.A., 1819. Entozoorum Synopsis Cui Accedunt Mantissa Duplex et Indices Locupletissimi. Sumtibus A. Rücker, Berolini.

Schierenberg, E., 1986. Developmental strategies during early embryogenesis of *Caenorhabditis elegans*. J. Embryol. Exp. Morphol. 97 (Suppl), 31–44.

Schierenberg, E., 1997. Nematoda, the round worms. In: Gilbert, S.F., Raunio, A.M. (Eds.), Embryology: Constructing the Organism. Singauer Associates, Inc, USA. (Chapter 8).

Schierenberg, E., 2006. Embryological variation during nematode development. In: Worm Book. The *C. elegans* Research Community.

Schulze, J., Schierenberg, E., 2011. Evolution of embryonic development in nematodes. EvoDevo 2 (1), 18.

Smyth, L.D., 1962. Introduction to Animal Parasitology. The English Universities Press, London.

Soliman, G.N., 1996. Invertebrate Zoology, The Mid-eastern Invertebrate Fauna. Part I. The Palm Press, Cairo.

Storer, T.I., Stebbins, R.C., Usinger, R.L., Nybakken, J.W., 1983. General Zoology, sixth ed. Tata McGraw-Hill Publishing Company Ltd, New Delhi.

Stunkard, H.W., 1953. Life histories and systematics of parasitic worms. Syst. Zool. 2, 7–18.

Sulston, J.E., Horvitz, H.R., 1977. Post-embryonic cell lineages of the nematode, *Caenorhabditis elegans*. Dev. Biol. 65, 110–156.

Sulston, J.E., Schierenberg, E., White, J.G., Thomson, J.N., 1983. The embryonic cell lineage of the nematode *Caenorhabditis elegans*. Dev. Biol. 100, 64–119.

Tadano, M., 1968. Nemathelminthes. In: Kumé, M., Dan, K. (Eds.), Invertebrate Embryology. Nolit, Belgrade, pp. 159–191. (J.C. Dan, translator from Japanese).

Vogt, C. 1851. Zoologische Briefe, vol. 1. WIEGMANN, A. F. A. 1835. Bericht iiber dieFortschritte der Zoologie im Jahre 1834. Arch. Naturg., Berlin, I.J., I(3):301–361.

von Linstow, O., 1905. Helminthen der Russischen Polar-Expedition 1900–1903. Imperatorskaja Akademija Nauk.

Wang, Y., Ezemaduka, A.N., Tang, Y., Chang, Z., 2009. Understanding the mechanism of the dormant dauer formation of *C. elegans*: from genetics to biochemistry. IUBMB Life 61 (6), 607–612.

Yamaguti, S., 1961. Systema Helminthum'. Vol. III: The Nematodes of Vertebrates. Part I: Nematodes of Fishes. Interscience Publishers, London.

Coelomata

Supraphylum Arthropods Mandibulata

These are the only major invertebrate group with many members adapted for life in all areas: on land, and the insects are the only invertebrates capable of flight (Storer et al., 1983). Artropods occur at attitudes of over 6000 m (20,000 ft) on mountains and crustaceans to depths of more than 9750 m (32,000 ft) in the sea. Different species are adapted for life in the air, on land, in soil, and in fresh water, brackish water, or salt water.

Most arthropods are important economically. The larger crabs, lobsters, and shrimps are eaten by humans, small crustaceans are the major herbivores in the sea. Most crustaceans are marine, occurring in all oceanic habitats, but many live in fresh water or brackish water, and a few, like the sea bugs are in moist places on land. The great majority are free living, some planktonic species occur in vast schools. The barnacles are either sessile or parasitic, certain other crustaceans are commensals or parasitic on various aquatic animals, from hydroids to whales.

It is necessary to give a brief exoposition of the principal groups into which the arthropods fall: One small section stands apart from the rest, The **Onychophora**, of annelid type: with thin cuticle, without joints and with a continuous muscular body wall (Borradaile et al., 1977). The remaining groups of the arthropods fall into two sharply differentiated sections; the **mandibulate** or **crustacean-insect-myriapod** section; and the **chlicerate** or **arachnid** section. These are sometimes referred to as the **Mandibulata** and the **Chelicerata**. This division of arthropods maybe the most stable, but the lower always changes according to different aspects, especially the molecular biology.

Chapter Outline

Definition **477**
Historical review **477**
General characters **482**
Classification **483**
Branchiopoda 484
1- Anostraca 484
Morphology 484
Reproductive systems 484
Ex. *Artemia* sp. 485
Male reproductive system 485
Female reproductive system 486
Early development (embryology) 489
Cleavage 489
Presumptive areas of the blastula 489
Gastrulation 490
Differentiation of the embryo 491
The hatched nauplius 491
Parthenogenesis 493
2- Diplostraca 493
Morphology 494
Cyclomorphism 495
Reproductive system 495
Ex. *Daphnia, Polyphemus* 495
Early development (embryology) 499
Cleavage 499
Presumptive areas of the blastula 501
Gastrulation 501
Differentiation of the embryo 502
Cirripedia (barnacles) 503
Ex. *Lepas* sp., *Balanus* sp. 503
Reproductive systems 503
Dwarf males 505
Copulation 508
Fertilization and brooding 509
Development 511
Gametes 511
Embryology of *Mitella mitella, Tetraclita rosea* 511
Cleavage 511
Presumptive areas of the embryo 514
Gastrulation 516
Differentiation of the embryo 516
Metamorphosis of the larvae 516
Malacostraca 518
1- Hoplocarida (Stomatopoda) 518
Ex. *Squilla* 518
Definition 518
History 518
Habitat and behaviour 520
Morphology 520
Sexual dimorphism 521
Reproductive system 521
Formation of the sperm cord 522
Courtship and copulation 523
Development 524
Gametes 524
Early embryology 528
Postembryology (larval development) 528
2- Eucarida 530
Order Decapoda 530
Penaeoidea 531
Definition 531
Habitat 531
Family Penaeidae 532
Morphology and sexual dimorphism 532
Reproductive systems 532
Male reproductive system 532
Formation of the spermatophore of *P. kerathurus* 541
The external appearance of the spermatophore of *P. kerathurus* 553
Chemical composition of the spermatophore of *P. kerathurus* 554
Female reproductive system 557
Copulation 560
Development 563
Gametes, spermatogenesis and oogenesis 563
Spawning 574
Fertilization 575
Early embryology 575
Larval development 588
Reptantia 590
Astacidea (Astacura) 590
Definition 590
Habits and habitat 590
Morphology and sexual dimorphism 590
Reproductive systems 590
Formation of the spermatophore of *Astacus leptodactylus* 593
The fully formed spermatophore of Astacidea 595
The ultrastructure of astacid sperm 597
The female reproductive system 598
Copulation 598
Spawning 602
Brooding 603
Development 604

Formation of gametes 604
Spermatogenesis 604
Oogenesis 605
Annual cycle of ovary development in sexually mature crayfish 605
Early embryonic studies 606
Astacus fluviatilis 606
Up-to-date embryological studies on Astacidea 609
The Japanese *Cambaroides japonicus* 611
The parthenogenetic marmorkrebs *Procambarus fallax* 613
Hatching 620
Postembryonic development 620
Maternal care in Astacidea 624
Parthenogenesis 628
Brachyura 629
Definition 631
Habits and habitat 631
Morphology and sexual dimorphism 632
Reproductive systems 633
1- The male system 633
The testes 633
The vasa deferentia 633
The proximal vas deferens 634
The middle vas deferens 637
The distal vas deferens 641
The intromittent organ 641
Formation of the spermatophore of *P. pelagicus* 641
A- Formation of the sperm masses 641
The spermatozoa 643
B- Deposition of the main layers of the spermatophore 647
C- Role of the MVD 650
D- Role of the DVD 650
Moulding of the spermatophore of *P. pelagicus* 650
Structure of the spermatophore of *P. pelagicus* 651
Chemical composition of the spermatophore of *P. pelagicus* 653
Absence of chitin 653
Distribution of lipids 653
Distribution of proteins 653
Formation of the spermatophore of *C. danae* Smith 1869 653
Formation of the spermatophore of *Chionoecetes opilio* 655
Spermatid differentiation 658
2- The female system 658
Copulation 660
Duration of copulation in Brachyura 664
Formation of the spermatheca 664
1- Precopulation stage 667
2- Fresh-copulation stage 667
Plug of the female *P. pelagicus* 667
Formation of the plug 668
3- Late-copulation stage 668
Chemical structure of the plug 668
Hardening of the plug 670
Maturation of the ovaries 671
1- Precopulation stage 672
2- Fresh-copulation stage 673
3- Late-copulation stage 674
4- Prespawning stage 675
Deutoplasmogenesis 675
5- May-spawn stage 676
6- Inter-spawn stage 676
7- September-spawn stage 679
Egg-membranes 679
Embryology 683
Hatching 695
Postembryonic development 696
Artificial insemination 697
Embryology of artificially inseminated egg 697
Utilization of lipovitellin during embryogenesis 700
References **704**

Chapter 10

Phylum Crustacea, Pennant (1777)

DEFINITION

Arthropods for the most part of aquatic habit and mode of respiration; whose second and third somites bear antennae; and their fourth somite a pair of mandibles.

HISTORICAL REVIEW

The adjective 'Crustata' or Crustacea (L. crusta, hard surface) had been applied to animals possessing hard skeleton, and Brisson (1756) used it 'as the designation of the group'. **Calman** (1909) in his treatise *Crustacea* recorded that **Linnaeus** (1767) placed most of the crustaceans then known, in his group 'Insecta'. Linnaeus divided the animals, at his time into groups and called them '**classes**'. These are: mammals, birds, reptiles, amphibians, fishes, insects, and worms. Such division is somewhat similar to that established by **Aristotle** (384–322 BC). It is well known that Linnaeus (Karl von Linné, 1707–1778) is usually called the 'father' of taxonomy, since he made notable contributions to the taxonomic system, *Systema Naturae*. He began, by following Ray (John Ray, 1627–1705), in the use of the genus and he established the second name of the species. Ray and Linnaeus brought '**order**' into the classification of animals and plants, to be a category between the class and genus. Therefore, Linnaeus divided the class Insecta into two orders: one of them was **Aptera** that includes the crustaceans and arachnids so far studied; another was **Ptera**, which includes insects proper. The class Insecta of Linnaeus is somewhat equivalent to subphylum Arthropoda of Calman (1909).

In those days of Linnaeus, the class Crustacea did not include the Isopoda or the Cirripedia (Pelseneer, 1906; Calman, 1909; Smith et al., 1923), but it included the Pycnogonida, Xiphosura, and Trilobita (Calman, 1909; Smith et al., 1923). The Isopodes were classified with the order Ptera (insects), while cirripedes were classified with molluscs under the class worms.

Both the old work (Calman, 1909) and the more recent work (Schram,1986) recorded that **Pennant** (1777) seemed to be the first post-Linnaeus author who suggested the separation of order Aptera to be a distinct class under the name Crustacea. Just before 1800, other workers introduced the '**Family**' as a category between the order and the genus, and soon thereafter, classes were grouped into higher categories, the '**Phyla**'. However, in 1829, Cuvier divided the animals into four 'branches' (Vertebrata, Mollusca, Articulata, and Radiata). These may be the phyla concerned above in the work of Calman (1909). Crustacea belongs to subphylum Arthropoda, class Articulata, which included, at that time, the annelids, crustaceans, insects and spiders. The term Arthropoda was not known before Cuvier's time. This view was definitely established in Cuvier's *Lecons d'Anatomie comparée* (1800). Crustaceans were still without the Isopoda and Cirripedia. Just 1 year later, in 1801, **Lamarck** included the isopods in that well-established class Crustacea and then **Latreille** in 1806 was the author who applied the name 'Isopoda' to refer to these crustaceans.

Concerning the latter group 'Cirripedia', the earlier taxonomists, e.g. Cuvier classified the barnacles and the acorn shells as Mollusca (Pelseneer, 1906; Calman, 1909; Smith et al., 1923; Schram, 1986). It seems strange that Cuvier's anatomical investigations (the book mentioned above) survived that view. Again, Lamarck (1801) was the first author who gave those early studied barnacles the name 'cirripedes' (later changed to 'Cirripedia'). He also referred, more or

Invertebrate Embryology and Reproduction. https://doi.org/10.1016/B978-0-12-814114-4.00010-2

less, vaguely to their affinities to Crustacea (Calman, 1909; Schram, 1986). That indistinct view was adopted by some subsequent authors, but without definitely removing them from Mollusca. It was not until 1830 that the developmental stages of one of the barnacles were carefully studied by **Thompson** (Calman, 1909). The discovered *Cypris* larva of *Balanus* (Thompson, 1930) strengthened the vague view of Lamarck and placed the crustacean nature of these barnacles beyond dispute. Later, Thompson in 1835 completed the outline of the life history by discovering the earlier *Nauplius* stage. That discovery confirmed the long-forgotten observations of **Slabber**, who had figured the *Nauplius* of *Lepas* as early as 1767 (Calman, 1909). In spite of those documented contributions, the taxonomy of this group remained in the hands of conchologists, perhaps because of the new genera and species that were established on the superficial characters of the shell alone. It is notable that even at that date, **Milne-Edwards** (1834–1840), who was considered the most prominent author working in that field, did not include the Cirripedia in his survey of the Crustacea. **Darwin**'s series of monographs on fossil and recent barnacles (1851–1854) placed the study of these animals on a sound foundation and within the province of carcinologists (Schram, 1986). Indeed, Darwin's work remains the standard reference on such subject.

However, the above picture of the group Crustacea is not the one accepted nowadays. The limits of this group as was adopted by Milne-Edwards (1834–1840) were without Cirripedia, but included Pycnogonida, Xiphosura and Trilobita. During that time, there was little doubt that the affinities of these groups are with the Arachnida. Therefore, each of these groups received as much doubt in its taxonomic position as the comparative studies occurred on their appendages and those of the Crustacea. Lamarck (1802), who formulated the group Arachnida, allowed it to embrace the pycnogons. Latreille (1804–1810), who immediately followed Lamarck, defined more clearly the **Pycnogonida** as a subdivision of the greater group Arachnida. The second group **Xiphosura** remained crustacean from the time of Latreille until the late 1800s. In 1881, **Lankster** in his classification of Arachnida effectively demonstrated the affinities of *Limulus* sp. with scorpions, and thus all are arachnids (Smith et al., 1923). The third group **Trilobita** attained this name very early in 1771 from **Walch** (Smith et al., 1923), due to the marked longitudinal division of the dorsoventral flattened body into an axial part and a pleural part on each side. It was thought that the affinities of this group of fossils were with Xiphosura. However, after the careful study of their ventral surface together with the comparative study of their appendages and those of Crustacea, they were included within the phyllopod crustaceans. This means that both Trilobita and Crustacea might have some ancestral connection. However, approximately at the end of the 19th century, **Kingsley** in 1892 did not admit this relationship, and he regarded the Trilobita as a group quite distinct from all other crustaceans. During the 20th century, the investigators more clearly defined each of these three groups as an independent group, side by side, with each of the group Crustacea and the group Arachnida. However, the somewhat recent publication of Borradaile et al. (1977) kept the independency of the Trilobita and placed it as a class of phylum Arthropoda; at the same time, both the Xiphosura and Pycnogonida were placed as orders of class Arachnida of this phylum Arthropoda.

With the separation of the above three groups from Crustacea, the limits of the latter were specifically defined whatever its taxon was. Back again to the time of Latreille (1806) who made many changes in the classification of Crustacea and created more than one significant view in this respect (Table 10.1). He divided the Crustacea into Entomostraca and Malacostraca, and he outlined the features which define the members of each group. At that time, the Cirripedia was still out of the Crustacea. The classification of Latreille, which was accepted during that time, is still known at the present time. All the subsequent investigators followed this scheme of classification. The most stable was the division of the Malacostraca, but the name of its grade differs with the different systems. To define a malacostracan refer to Calman (1909, pp. 143–144) and Schram (1986, Chapter 43, p. 525). The publication of Calman, which is still a standard reference of Crustacea, rejected the division Entomostraca, and elevated each of its subdivisions into a subclass side by side with subclass Malacostraca.

A further step of classification was made by **Leach** (1815), who divided the Malacostraca into the Edriophthalma (sessile-eyed) and Podophthalma (stalked-eyed) crustaceans. However, the most scientifically significant change that occurred to the Malacostraca was what concerned *Nebalia* and its group. The systematic position of this latter group faced much dispute by different investigators. **Claus** in 1889 restudied *Nebalia* and evidenced that it has affinities to Malacostraca rather than Branchiopoda. He ended the dispute, since he proposed the name **Leptostraca** as a definite group for this *Nebalia.* In Borradaile et al. (1977), this Leptostraca is an order included within the subclass Malacostraca. However, according to Claus, **Grobben** in 1892 separated the Leptostraca and grouped the other orders together in a new category and proposed its name **Eumalacostraca**. Therefore, Calman (1909) divided the Malacostraca into two series: Leptostraca and Eumalacostraca. To differentiate between these two groups, refer to the characters of each in Calman (1909, p. 148).

During the above period of time, the research on Crustacea was in continual progress. One of the pioneers was **Boas**, who in 1883 rearranged the Malacostraca by the following: (1) rejection of its division into Edriophthalma and Podophthalma; (2) separation of Schizopoda into two orders: Mysidacea and the Euphausiacea; and (3) division

TABLE 10.1 Scheme of classification proposed by **Latreille** (1810) as it was recorded after **Milne-Edwards** (210, 211)

■ Sous-classe: Entomostracés
♦ Section: Operculés.
◘ Ordre: Xyhosures. (Genre: *Limule.*)
◘ Ordre: Pneumonures. (Genre: *Calige Binocle, Ozole.*)
◘ Ordre: Phyllopodes. (Genre: *Apus.*)
◘ Ordre: Ostrachodes. (Genre: *Lyncè, Cypris, Daphnie, Cythérée.*)
♦ Section: Nues.
◘ Ordre: pseudopodes. (Genre: *Cyclope, Argule.*)
◘ Ordre: Céphalotes. (Genre: *Polyphéme, Zoé, Branchiopode.*)
■ Sous-classe: Malocostracés
◘ Ordre: Décapodés.
♦ Section: Brachyures.
• Famille: Cancérides
* Platymatiens.
□ Littoraux. (Genres: *Calappe, Hépate, Dromie, Crabe.*)
□□ Pelagiens. (Geners: *Matute, Fortune, Podophthalme.*)
** Vigilans. (Genres: *Porcellane, Ocypode, Grapse, Pinnothére.*)
• Famille: Oxyrynques.
(Genres: *Orithie, Ranine, Dorippe, Coryste, Leucosie, Macropode, Maïa.*)
♦ Section: Macroures.
• Famille: Paguriens.
(Genres: *Pagure, Albunée, Hippe*)
• Famille: Langoustines.
(Genres: *Scyllare, Langouste, Galathée.*)
• Famille: Homardiens.
(Genres: *Écrevisse, Alphée, Penée, Palémon, Crangon.*)

of the Malacostraca into seven orders: the above two and the Cumacea, Isopoda, Amphipoda, Decapoda, and Squilliacea (Stomatopoda). This means that he separated the Leptostraca from Malacostraca, i.e. the Crustacea was then divided into Entomostraca, Leptostraca, and Malacostraca. The classification of the last group became attractive and many different views were proposed. Another pioneer was **Hansen,** in spite of the fact that his work was greatly affected by that of Boas. Hansen in 1895 carried this classification a step further. He placed the stomatopoda and Leptostraca apart and put the remaining Malacostraca in two well-defined groups. He coined the Euphausiacea with the Decapoda as one of these groups. The present work stresses the research of the above two authors, since their studies and views were the basis of the work of Calman afterwards.

Calman (1904) greatly followed Hansen but with considerable modifications. He greatly extended the classification of Malacostraca. First of all, Calman followed Grobben (1892) in that Leptostraca is a division of Malacostraca, while all the malacostracans then known are referred to as '**Eumalacostraca**'. He divided the latter into four divisions and proposed a name for each: Syncarida, Peracarida, Eucarida, and Hoplocarida. This plan of division was so significant that it solved the problem of the systematic positions of both the Stomatopoda (Squilliacea) and the Euphausiacea. He placed the former in one of his divisions named Hoplocarida, while he included the Euphausiacea and Decapoda as two separate orders in another division—the Eucarida. The features characterizing the crustaceans of these divisions were described in his textbook (Calman, 1909).

Concerning order Decapoda, many animals were described very early as far back as the time of Aristotle (384–322 BC) Some of these animals were described with accuracy and also with surprising details since the time of Linnaeus (1767) forward. These descriptions were the material that aided **Fabricius** in 1793 to divide the early described very famous genus *Cancer* into a large number of genera (the majority of which are still recognized). Accordingly, this author was considered the first to lay the foundations of the classification of this group of animals. Few years later, Latreille, in 1806, proposed the name **Decapoda** to refer to this group and included it in his Malacostraca. He determined the features characterizing the decapod crustaceans and which were recorded in the work of Calman (1909), Smith et al. (1923), and Borradaile et al. (1977). Latreille divided the group Decapoda into **Macrura** (long-tailed) and **Brachyura** (short-tailed) and subdivided each into families: Macrura three families and Brachyura two families. He placed all the shrimps and shrimp-like crustaceans (palaemonids and penaeids) so far studied, together in one macruran family: **Homardiens**. At that time, only the word shrimp was known, which means caris in Greek, subsequent authors used the word carides to define the group, while the word prawn has no origin (for details, see **Holthuis**, 1980, pp. xv, xvi). The scheme of classification proposed by Latreille (1810), as was recorded after **Milne-Edwards** was re-recorded in the work of **Khafage** (2000). It seems that there was some difficulty to define a crustacean, if it is long-tailed or short-tailed. Milne-Edwards (1834–1840) was greatly interested in this problem and so he established a third group and proposed its name—**Anomura**—which becomes an intermediate one in-between Macrura and Brachyura (Milne-Edwards, 1834–1840, p. 247).

To define a decapod crustacean and/or an animal of its subdivisions, one should refer to Smith et al. (1923). It is appropriate to mention that this group Decapoda is the most stable with respect to its taxonomic position from the moment of its establishment up to the present time.

Milne-Edwards (1834–1840, 1837) himself again changed much in the system of Latreille without any change in the basic frame of the classification of this group Crustacea as a whole.

Since the system of Latreille did not differentiate between penaeids and palaemonids, Milne-Edwards divided the Macrura into four and not three families. In addition, he differentiated between the penaeids (**Pénées**) and the palaemonids (**Palemoniens**) and elevated each in an equal grade named **tribe**, but still both were included in one of the four families named **'Salicoques'** (Table 10.2). At that time, the tribe was a subdivision of the family, opposite to what is known nowadays, and also in Calman's treatise, the tribe was a grade of the order. This means that 'Salicoques' of Milne-Edwards is equivalent to 'Homardiens' of Latreille. He planned keys in the form of tables to define each of the categories of the Crustacea down to the genus. In this more than in any other group of Crustacea, the work of Milne-Edwards, and especially his *Histoire Naturelle de*

TABLE 10.2 Scheme of classification proposed by **Milne-Edwards** (1834) (pp. 236, 247, 269, 270) (1837) (pp. 305, 328, 338, 339, 405, 413, and 418)

Class: Crustacés
▫ Subclass: Crustacea Maxillés
♦ Légion: Podophthalmiens
• Order: I. Décapodes
○ Section: Brachyures
• Family: 1. Oxyrhinques
2. Cyclometopes
3. Catomètopes
4. Oxystomes
○ Section: Anomures
• Family: 1. Apterures
2. Ptérygures
▫ Tribe: a. Porcellaniens
b. Hippiens
c. Paguriens
○ Section: Macroures
• Family: 1. Cuirassés
▫ Tribe: a. Galatheides
b. Eryons
c. Scyllarides
d. Langoustiens
• Family: 2. Thalassiniens
▫ Tribe: a. Cryptobranchides
b. Gastrobranchides
• Family: 3. Astaciens
Ecrevisses
Homards
Nephrops
• Family: 4. Salicoques
▫ Tribe: a. Crangoniens
b. Alphéens
c. Palémoniens
d. Penéens
Penée = Penaeus
P. caramote
P. canaliculata
P. monoceros
P. indicus
P. mondon
P. affinis
P. membranaceus
P. brevicornis
Sicyonie
Euphème
Oplophore
Ephyre
Pasiphée
Sergeste
Acète
• Order: II. Stomapodes
♦ Légion: Édriophthalmes
• Order: I. Amphipodes
II. Isopodes
III. Laemipodes

Crustacae (1834–1840, 1837), may be taken as a mark indicating the beginning of the modern period. In addition, his classification of the Decapoda has been considered the most generally accepted until very recently. Moreover, he was considered to be the first to establish the foundations of the taxonomy of penaeids in specific.

It appears that this field of work was very attractive during that period. Therefore, there were other carcinologists whose investigations were so valuable and significant that became useful in the progress of the taxonomy of this group, Crustacea.

One of these was **de Haan**, whose investigations were published during the period 1833–1850. He agreed with Latreille on the one hand, and considered the proposed ideas of Milne-Edwards on the other. The most important addition of de Haan was that he considered the numerical value of the 'branchiae' as a main factor in his taxonomy, in spite of the fact that he described the structure of the branchia itself and differentiated between two forms. His system of classification evidenced that of Milne-Edwards. The only difference might be the new factor or might be the names which he used. For example, he applied the name '**Carides**' to refer to the Salicoques of Milne-Edwards. **Dana** was another carcinologist whose investigations in 1852 were so scientifically significant. It was of great probability that Dana considered the difference in the structure of the branchiae described by de Haan. He not only differentiated between the penaeids and the palaemonids, but also separated them and elevated each in an equal grade. At the same time, each was equal to the 'Homardiens' of Latreille and the 'Salicoques' of Milne-Edwards. This means that he rejected these names and applied the names '**Penaeidea**' and '**Caridea**' instead of both the penaeids and palaemonids, respectively, and placed each as a direct division from the Decapoda. He then divided the Penaeidea into **Penaeidae**, **Sergestidae**, and **Eucopidae**. The information of classification of Dana was recorded in the work of Bate (1888). The features characterizing any member of Penaeidea were recorded in Calman (1909) and Smith et al. (1923).

The system of Dana was not stable, especially the separation of the penaeids and palaemonids, since some other investigators, e.g. **Huxley** (1883), stated different scheme of taxonomy for this group, Decapoda. Huxley considered and restudied the structure of the branchiae previously described by de Haan and applied a name for each type. Accordingly, he divided the Decapoda into: (1) '**Trichobranchiata**' to refer to those having branchia which bears series of cylindrical filaments; (2) '**Phyllobranchiata**' to refer to those having branchia which bears series of leaf-like plates, i.e. foliaceous; and (3) '**Abranchiata**'. In addition, he changed the names of the subdivisions to introduce the word '**caris**' which means shrimp. Therefore, he introduced the name '**Caridomorpha**' instead of 'Penaeidae'. However, unfortunately, he placed these penaeids under the Trichobranchiata. **Bate** (1888) restudied the branchiae of Penaeidae and evidenced that its structure and form differ from both types of Huxley. It has a distinct plume-like structure. Therefore, he added a third type to that of Huxley and applied the term '**Dendrobranchiata**', which corresponds closely with the Penaeidea of Dana. This means that Bate followed the systems where the penaeids were separated from palaemonids and he added the feature of the branchiae to the keys of Milne-Edwards.

During about the same time, **Boas** in 1880 added some changes to the classification of Malacostraca, but he made a complete departure from Latreille's system. He did not accept Latreille's division of Decapoda into Macrura and Brachyura, since he evidenced that the Brachyura and Anomura (the system of Milne-Edwards), which constitute branches of decapod stock, were by no means equal in systematic value to Macrura. The latter included several other branches not more closely connected with each other. Accordingly, Boas proposed an alternative division to the order Decapoda. He divided it into two primary groups of **Natantia** (swimmers) and **Reptantia** (crawlers/walkers). This division is undoubtedly a more natural one than that formerly employed. Calman (1909, p. 310) reported several features describing a natantian decapod crustacean. This system of Boas has been further elaborated by the following carcinologists. The most famous in this field were **Ortmann** and **Borradaile**. The former was greatly interested in the study of crabs (Ortmann, 1894, 1896), while the investigations of the latter were extensively on the classification of the Malacostraca, in general, and the decapods, in specific (Borradaile, 1901–1904, 1907).

The above review shows that Latreille was a pioneer in the beginning of the 19th century and Milne-Edwards was another one in the middle of the same century. **Calman** was a third one whose investigations were published at the beginning of the 20th century. His fossil record was one of his main studies (Calman, 1896, 1899). This record facilitates the interpretation of the phylogenetic relationship and the correct classification of the group Crustacea. In addition to his extensive studies on the classification of Malacostraca, he reviewed all the well-accepted and scientifically significant views concerning the classification of Crustacea, in general. Then, he in 1909 formulated the first monograph on Crustacea as a part of **Lankster's Treatise on Zoology**. This classic volume is still a standard reference even though quite out of date, as it is reported in Schram (1986). Calman in his book established the framework upon which the discipline has developed even to the present. He summarized and enumerated the definitive features for each group. He did not accept the division of Decapoda into Macrura and Brachyura. Instead, he followed Boas in his division into Natantia and Reptantia.

It is worth noting that the classification of Crustacea as a group is somewhat stable from the moment of its establishment up to the present. The additions of the carcinologists of the 20th century are only elucidations and/or detailed descriptions of the features especially of the newly discovered genera and species. Despite all that happened in the field of Arthropoda during this century, the only observed change concerning Crustacea was the work of Schram, especially his book published in 1986. The latter is the first monograph of all fossils and living crustaceans grouped and published since 1909, i.e. that of Calman. It is considered the first systematic phylogenetic review. In spite of the fact that some of the works, especially the book is available to the current author, the book still needs further studying and further comparison with that of Calman. However, the most obvious difference is that Schram and before him Burkenroad (1981, 1984) who followed Dana with some changes reject the group Natantia but not Reptantia. In addition, they changed the name of the group Penaeidea into Dendrobranchiata according to Bate (1888). At the same time, they elevated the group into suborder as a direct division from Decapoda with changes in its lower grades. The most significant scientific addition is the key to differentiate any genus of adult dendrobranchiate crustacean living anywhere (Schram, 1983a,b). Moreover and the most scientific change Schram made in the classification of Crustacea, is the elevation of this class into the grade of **phylum** (Schram, 1986).

GENERAL CHARACTERS

It is usually possible to distinguish a crustacean with more characters than the mere possession of two pairs of antennae. A more complete description, both as a crustacean and as an arthropod, might be as follows:

1. Shortening of the metameric body.
2. Few numbers of metameres, ranging between 16 and 60.
3. Since it is an arthropod, a crustacean **head** has five fused segments with the presegmental region.
4. The other body segments are typically divided into two regions, a thorax and an abdomen terminated by a postsegmental region (embryonic segment) or **telson**.
5. Often, some or all the thoracic segments fuse with the head to form a **cephalothorax**.
6. Usually the dorsal shield of the head extends backwards to cover the fused segments forming the **carapace**. The latter undergoes varying degrees of development, which becomes diagnostic of many orders.
7. Each metamere of the body bears a pair of primitively and typically biramous appendages.
8. Anteriorly, the appendages of the head: two pairs of antennae.
9. The first antennae (antennules) belong to the second metamere.
10. The second antennae (simply antennae) belong to the third metamere; this is the most characteristic character of the crustaceans.
11. The mandibles belong to the fourth metamere.
12. The first maxillae (maxillules) belong to the fifth metamere.
13. The second maxillae (simply maxillae) belong to the sixth metamere.
14. The thoracic and abdominal appendages modify, forming the following:
15. Often some of the anterior thoracic appendages modify as feeding, and become closely associated with the head appendages.
16. Posterior ones modify as walking, swimming, or respiratory organs.
17. Or as accessory reproductive organs.
18. As in all arthropods, the body is covered by an **exoskeleton**, which differs by calcification hardening.
19. Typically, external respiration takes place on the surface of **gills**, which are usually associated with the appendages.
20. Sometimes it occurs over the whole body surface.
21. There is a great variety of feeding mechanisms, but the digestive tract is almost always straight. It is characterized by the following:
22. A very short mesenteron in between two long stomodaeum anteriorly and proctodaeum posteriorly.
23. The mesentron is not lined by cuticle and often gives rise to digestive glands; while the other two components are lined by a cuticle.
24. In Malacostraca, these elements become more complex. The stomodaeum is provided with a '**gastric mill**' and a '**filtering apparatus**'.
25. The coelom is much reduced and represented only by small cavities in the gonad, excretory glands, and some ducts.
26. The open vascular system nearly always includes a contractile heart, which is formed as a muscular dilation of the dorsal vessel, and communicates by valvular ostia with a pericardial sinus.
27. One pair of excretory organs, rarely two pairs metamerically situated in the coxopodites of the second antennae and those of the second maxillae. In rgw case of the former, they are '**green glands**'; in the latter, they are referred to as '**shell glands**'.
28. The nervous system is built on the advanced plan of invertebrates. It exhibits a very complete series of stages from the ideal arthropod condition to the most extreme concentration. Usually, the first four pairs of ganglia unite to form the brain.
29. (Supraesophageal ganglia) to connect the subesophageal ganglia via the circumesophageal connectives. Then, connects:

30. The sense organs that comprise **eyes**, usually compound and frequently stalked; **statocysts** which occur in certain malacostracans; proprioceptors; chemoreseptors and general tactile receptors.
31. The sexes are usually separate, and often display **sexual dimorphism**, but few forms such as the Cirripedia, which are sessile, certain parasitic Isopods, and a few exceptional species in other groups, are hermaphroditic. Reproduction is typically gametic, but parthenogenesis also occurs in many of the Branchiopoda and Ostracoda. **Sperm** are very varied in form and often of complex structure; usually, but not always, they are immobile. They are transferred to the female, often in sperm-packets (**spermatophores**); or mobile and provided with bristles or spines.
32. These animals possess some noteworthy characteristics in common, in respect to embryological development: (a) Eggs are typically centrolecithal; some, however, are almost alecithal, others telolecithal. (b) Cleavage is ordinarily **superficial**, but it may be complete or discoid. It retains vestiges of the spiral type found in Platyhelminthes, Annelida, and Mollusca. Although the Arthropoda, in general, undergo superficial cleavage, only the Malacostraca among the Crustacea conform to this rule, the majority of the others cleave totally. Not only do the totally cleaving eggs follow the spiral pattern, but the superficially cleaving ones also show traces of this type to certain extent, in their modes of cleavage furrow formation. (c) The teloblasts also play an important role in forming the body region posterior to the first maxillary segment, at least in the Malacostraca. (d) Crustaceans, without exception are characterized by the production of a **nauplius stage**, which may or may not be released as a free-swimming form. (e) The various complicated structures of all the adults are built up conforming to the general ontological principle that their bodies are metameric. (f) In addition, the appearance of such a habit as brooding the youngs exerts a definite influence in the amount of yolk in the ova, and in turn the mode of embryological development and the stage of hatching.

CLASSIFICATION

The classification of crustaceans is built on the shortening of the body; number of segments; specification of the appendages and the development of the carapace. Different books seldom contain identical schemes of classification.

Therefore, the classification herein is not built according to one definite book, but from different sources, especially Calman (1909), Gardiner (1972), and Borradaile et al. (1977), in which the group is considered (like many others) a class of phylum Arthropoda. The textbook of Schram (1986) elevated the class Crustacea into phylum Crustacea, and so the subclasses automatically become classes. Apart from the grades and because of the lengthy classification of this group, only its basic frame is recorded. The detailed taxonomy of each species will be dealt with in appropriate site.

Class — **Subclass** — **Order**

- I. Cephalocarida
- II. Branchiopoda
  - 1. Lipostraca (fossils)
  - 2. **Anostraca**
  - 3. Notostraca
  - 4. Diplostraca
    - Conchostraca
    - **Cladocera**
- III. Ostracoda
- IV. Mystacocarida
- V. Copepoda
- VI. Branchiura
- VII. Cirripedia
  - 1. **Thoracica**
  - 2. Acrothoracica
  - 3. Ascothoracica
  - 4. Rhizocephala
- VIII. Malacostraca
  - 1. Leptostraca
  - 2. **Hoplocarida** (Stomatopoda [Schram, 1986])
  - 3. Syncarida
  - 4. Peracarida
  - 5. Eucarida
    - Euphausiacea
    - **Decapod**

It is well known that crustaceans are among the most widespread species on earth, but we know very little about their embryology. There are very few studies following the development of any one species in detail. Since there is no general plan of reproduction and/or embryology, each species will be dealt with separately.

Branchiopoda (Greek *branchia*, 'gill')

Includes: fairy shrimp, brine shrimp, Daphnia, clam shrimp, tadepole shrimp.

Branchiopods live almost exclusively in fresh water, only some cladocerans are marine. The majority of branchiopods are only a few millimeters in length, and some are as small as 0.25 mm.

The largest are the fairy shrimps (Anostraca), which are usually larger than 1 cm. Most branchiopods are pale and transparent, but rose or red colours and some cladocerans and clam shrimp (Conchostraca) are brown or black.

Anostraca (Greek a, *'not' and* ostracon, *'shell')*

Fairy shrimp and brine shrimp.

Ex. *Chirocephalus* and *Artemia*.

The anostracans are found worldwide. *Chirocephalus diaphanous* and *Artemia salina* are two British representatives. *Artemia* differs from *Chirocephalus* in having only six abdominal segments and in the form of the antennae of the male. *Chirocephalus* turns up from time to time in temporary pools of water in various districts. *Artemia* occurs in various parts of Europe in salt lakes and marshes and in pans in which brine is concentrated, but never the ocean. In Egypt, like in Europe it occurs in such places, especially along the shores west Portsaid and west of Alexandria. Our laboratories always obtain *Artemia salina* from the pans of El Mex Salines Co. *Artemia* can endure a very high concentration of salt, and some of its minor features change with the degree of the concentration, so that it has been described under different specific names.

Morphology

Anostracans are small to medium in size, up to approximately 10 cm in maximum length (though more typically only about 1 cm). The body is clearly circular in cross section, and does not possess any pleural lobes. The body is clearly tagmatized into three regions: shieldless head (cephalon) with prominent stalked compound eyes and the second antennae are comparatively large, modified into claspers in males; the thorax consists of 11 segments, with metameric foliaceous appendages (phyllopodia); and the thin abdomen consists of seven simple, limbless segments and ends with a telson. The latter bears a pair of caudal rami as narrow, pointed plates, fringed with bristles.

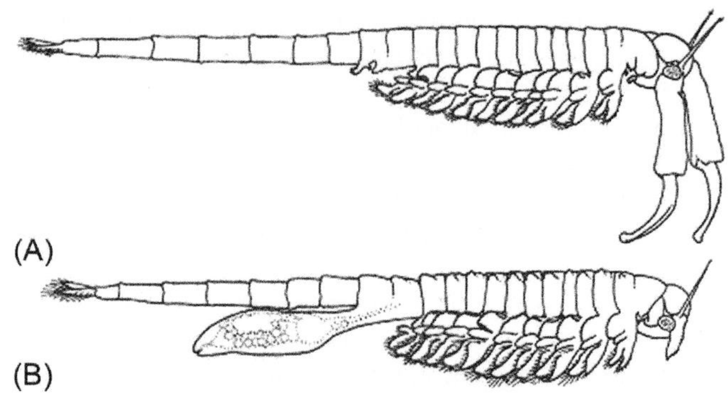

FIG. 10.1 Body plan of anostracan, with (A) male and (B) female of *Branchinecta campestris*. *(From Schram (1973), after Lynch (1960).)*

The 12th thoracic segment upon which are the genital openings is fused ventrally with the first abdominal segment (Calman, 1909; Borradaile et al., 1977) forming a 'genital segment'. In the male, in addition to the antennal claspers, this fused segment bears a pair of ventrolateral processes, in each of which is the terminal portion of a vas deferens, with a protrusible penis. The latter probably represents an appendage. In the female, on the other hand, this fused segment has a median, ventral, projecting egg pouch, which like the penis, is held to represent a pair of appendages (Fig. 10.1).

Reproductive systems

Each of a pair of tubular gonads lies on either side of the alimentary canal in the abdomen and is continuous in front, with a short duct. The female reproductive system of *Streptocephalus dichotomus* Baird, 1860 is morphologically as well as functionally similar to that of *Artemia* and *Chirocephalopsis bundyi* (Munuswamy and Subramoniam, 1985). The oviducts are pouch-like structures appearing temporarily at the time of ovulation. Each originates from the lateral part of the ovary and the two oviducts combine into a single median uterus (ovisac) in the egg pouch (Fig. 10.3A). After ovulation, the eggs are temporarily stored in the dilated oviducts (oviduct pouches). During the early stage of growth, the oviducts may be seen as two finger-like outgrowths from the ovary. Interestingly, the oviduct shows a shutter-like structure between the oviduct and ovisac. This structure may aid in the regulation of the passage of eggs. It has also been reported in the oviduct of *Artemia*. The ovisac receives two pairs of shell glands: one anterior and the other posterior.

In the male, the vasa deferentia lead to the penis. An enlargement near the entry of each vas deferens into the penis acts as a seminal vesicle. Small accessory glands open independently on the tip of the penis adjacent to the seminal vesicle and seem to produce a lubricant or sperm activator (Schram, 1986).

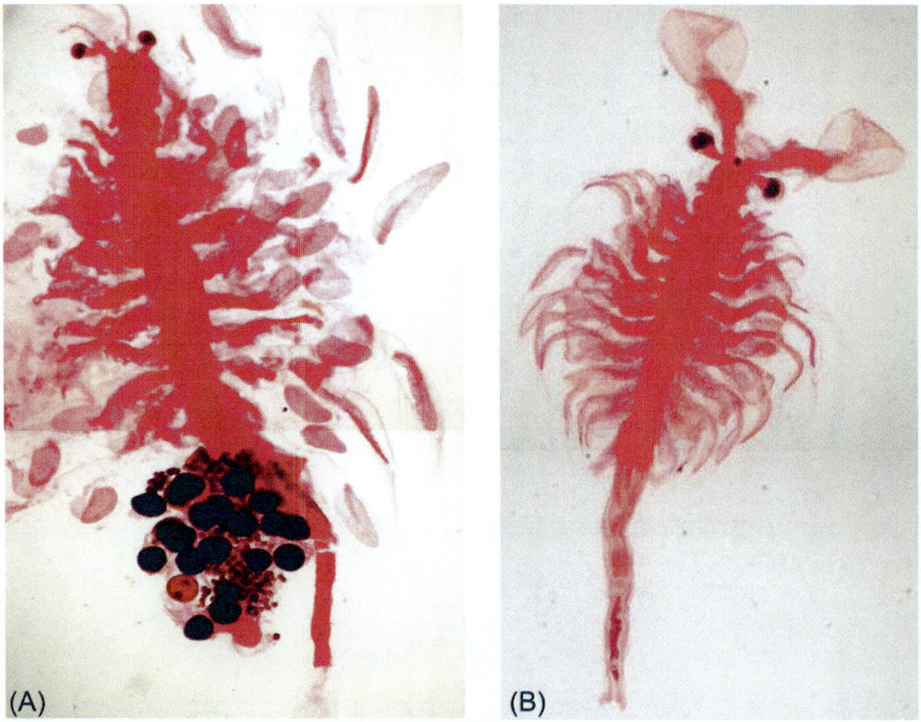

FIG. 10.2 Montage of micrographs of whole mounts of adult *Artemia salina*: (A) female showing egg pouch; (B) male showing geniculate antennae. *Laboratory preparations.*

Example Artemia *sp.*

The first descriptions of *Artemia* were with reference to bisexual strains, which were classified as *Artemia salina*, *A. tunisiana*, *A. franciscana*, *A. persimilis*, *A. urmiana*, and *A. monica* (Sorgeloos et al., 1986). *Artemia salina* (Fig. 10.2) is also known as brine shrimp. It is a primitive crustacean, which thrives in such salty conditions mainly because there is an abundance of food (phytoplankton) and there are no predators. Their use in the aquarium hobby is very widespread. Newly hatched brine shrimps are viewed as a nutritious food for both fish and corals. They are particularly useful in the fish culture business where they are used as larval food of a wide variety of fishes ranging from freshwater to saltwater. Adults are usually sold in frozen form as fish food.

The male reproductive system

In the **male *Artemia*** (Wolfe, 1971), this system consists of paired testes, vasa deferentia, accessory glands, and penis. The testes are paired tubes, slightly flattened laterally, begin from about the second or third abdominal segment (Fig. 10.3B) and elongate forwards to the first of the genital segment, each on either side of the intestine. However, this length is variable. Each testis in the first genital segment continues as a vas deferens, which enlarges and passes posteriorly to the end of the second genital segment. Then, it turns and extends anteriorly within the second genital segment to join the accessory gland. Afterwords, it takes its way posteriorly to the tip of the penis (Fig. 10.3B). The histology of the testis is similar throughout its entire length. In a cross section (Fig. 10.4(I) A), it is surrounded by a membrane (perhaps cortex and/or membrana propria, in the opinion of the author), beneath which there are two types of cells: supporting cells (alternative to sertoli cells) and germ cells. The latter undergo spermatogenesis and spermiogenesis, and the spermiation occurs centrally in the wide cavity. The histological structure of the vas deferens is well differentiated from that of the testis by lacking germinal tissue and it is lined by large squamous epithelial cells. Its wall consists of longitudinal and circular muscle layers. Sperm are stored along the entire length of this duct and the rhythmic contraction of the muscles keeps the appropriate movement of these sperm inside.

The vas deferens leads into the eversible penis. In the retracted state, the lumen of the penis is lined with epithelium and a cuticle (Fig. 10.4(I) D). The epithelium is surrounded by several bands of muscles. The penis is expanded all around the papilla-like ending of the vas that protrudes into the lumen. In the everted penis, three to five spines appear located near the junction of the eversible and noneversible penis. During eversion of the penis, the papilla-like ending of the vas deferens becomes the distal most part of the reproductive tract. Located near the junction of the vas deferens and eversible penis is the accessory gland. In the mature animal, this gland becomes situated antero-laterally to the tip of the eversible penis. The gland is composed

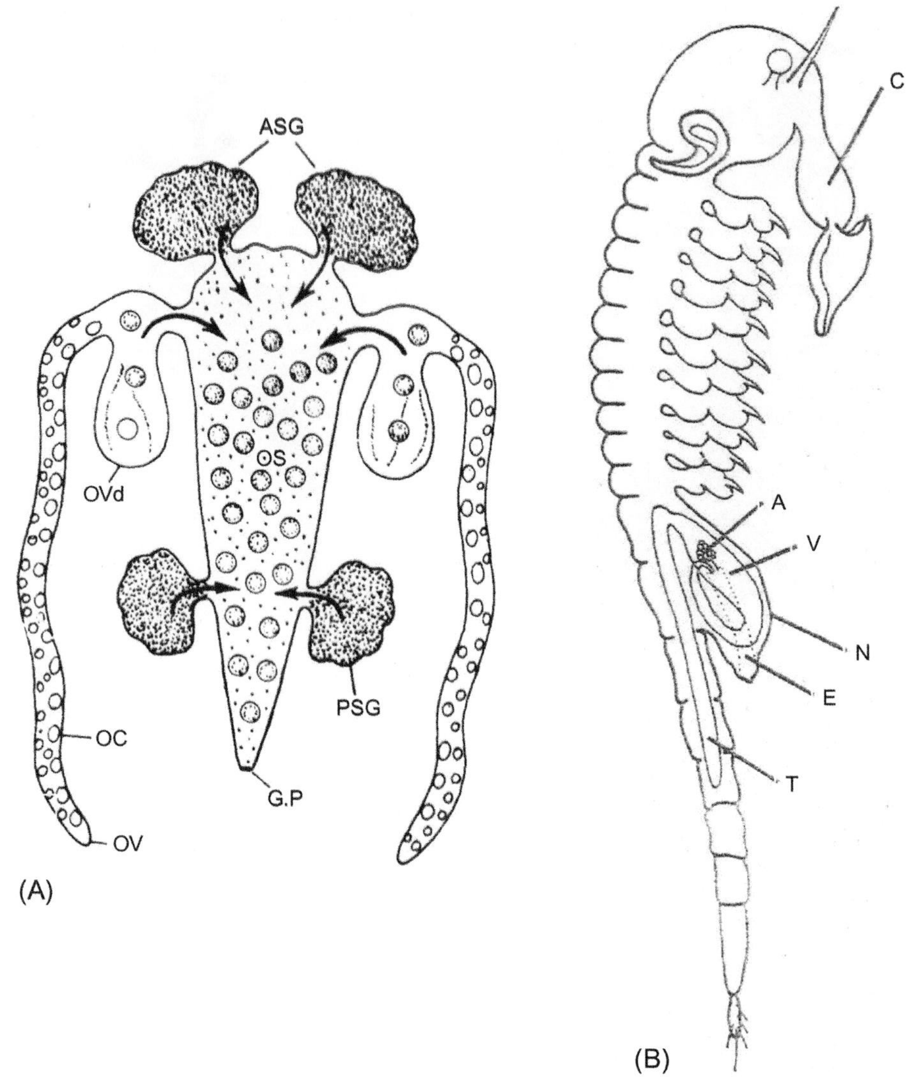

FIG. 10.3 (A) Diagrammatic drawing of the female reproductive system in *Streptocephalus dicotomus* Baird, 1860. ASG = anterior shell gland, GP = genital opening, OC = oocyte, OS = ovisac, OV = ovary, Ovd = oviduct in pouched phase, PSG = posterior shell gland, *arrows* indicate the passage of oocytes from the oviduct and secretion from the shell gland *(from Munuswamy and Subramoniam, 1985)*; (B) Diagram of lateral view of male *Artemia* showing the parts of the reproductive system. A = accessory gland, C = clasper, E = eversible penis, N = the ventral cuticle of the noneversible penis, T = testis, V = vas deferens. *From Wolfe (1971)*.

of approximately 20 pairs of gland cells (Fig. 10.4(I) B). Each pair of gland cells is surrounded by connective tissue capsule. Attached to each pair of gland cells is a pyramidal neck cell, whose apex is directed away from the gland cells. A duct cell connects the neck cell to the collecting duct (Fig. 10.4(I) F). The latter extends throughout the middle of the gland and varies in diameter along its extension, becoming narrowest at its opening in the penis.

The spermatozoa are immobile and usually spherical (Calman, 1909) and those of *Artemia* are aflagellated (Wolfe and Dillman, 1990). These authors studied the sperm of *Artemia* from the testis, vas deferens, and in the medium immediately after release from the penis using phase, Namarski, fluorescence, and electron microscopy. They described those extruded from the everted penis as spherical cells bearing many projections; some are spike-like resembling the cytoplasmic arms of malacostracan sperm. Some others are blunt and look like pseudopodia, but no actual movement was observed. However, the presence of actin and other filaments within the cytoplasm suggests the possibility of motility.

Female reproductive system

The **female reproductive system** is similar in form and location to the male system. However, the two oviducts unite to form median uterine chamber, with groups of gland cells on its wall comprising the shell gland (Calman, 1909; Schram, 1986). This chamber ends as an ovisac (Fig. 10.2A). Therefore, the genital opening is unpaired. As the eggs mature in the ovary they pass to the lateral pouches of the oviduct and then to the median uterus where they attain their hard shells. Before attaining this last step, two vital activities must occur (see later).

Generally, temperature has little effect on breeding, though breeding behaviour seems to be triggered by rising temprature. When the female becomes ready to mate, she undergoes some unusual behaviour to attract males. The male then approaches his partner and grasps the middle of the female dorsum with his strong antennal claspers (Fig. 10.5). Then periodically, sometimes over the duration

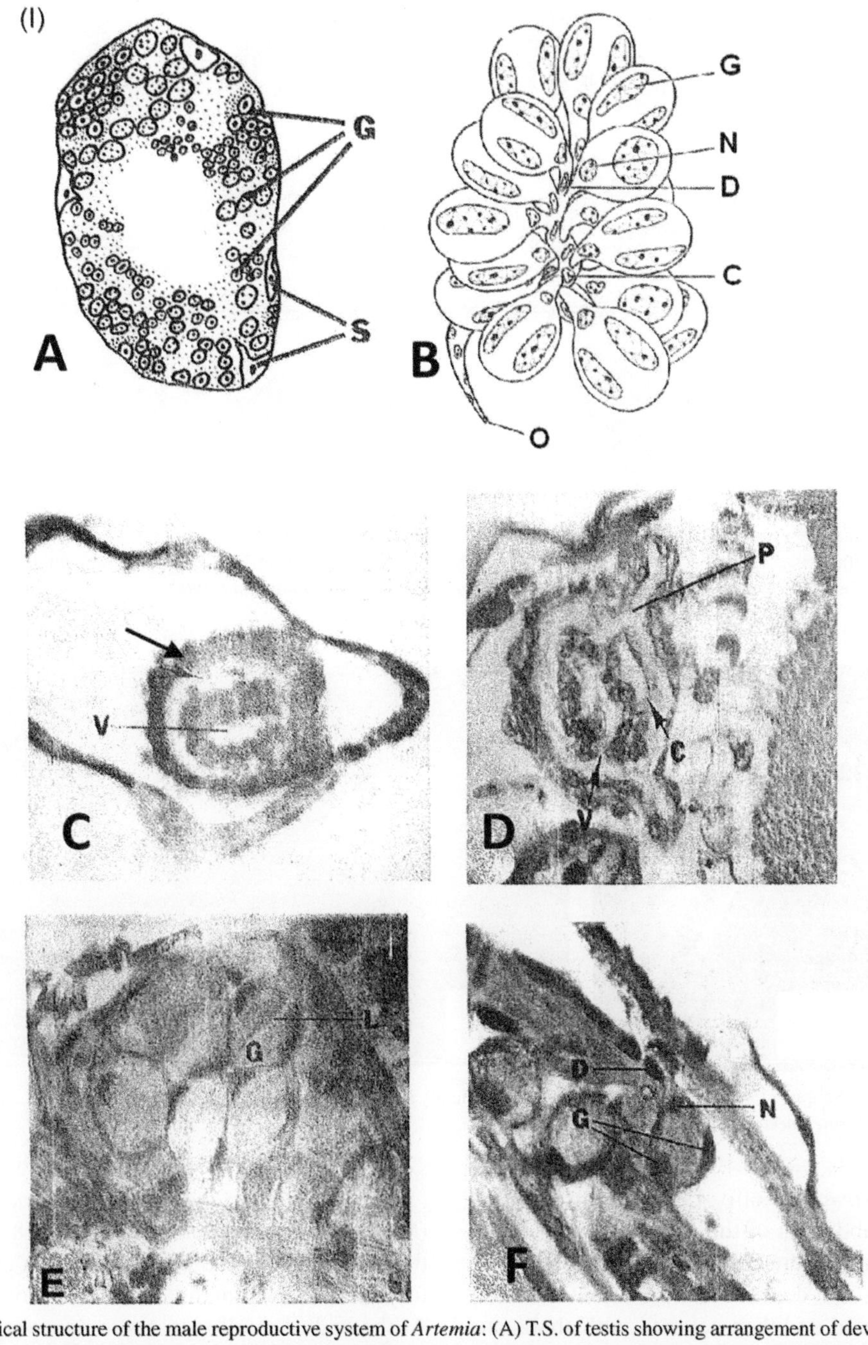

FIG. 10.4 (I) Histological structure of the male reproductive system of *Artemia*: (A) T.S. of testis showing arrangement of developing germ cells (G) and nuclei of supporting cells (S). 220×; (B) Camera Lucida Drawing of reconstruction of an entire accessory gland. Note that each pair of gland cells drain by a neck cell (N) and a duct cell (D) into a collecting duct (C) which opens to the outside (O). 500×; (C) T.S. through the penis, the *arrow* indicates the opening of the collecting duct to the outside. 630×, lettering as before; (D) T.S. through the retracted penis of an adult animal showing the irregular lumen of the eversible penis (P), *arrows* indicate the separate openings of the Deferens (V). 630×; (E) Sagittal S. of an accessory gland showing their common intercellular lumen (L). 450×; (F) Sagittal S. of an accessory gland cell unit. 450×, lettering as before. *From Wolfe (1971).*

(Continued)

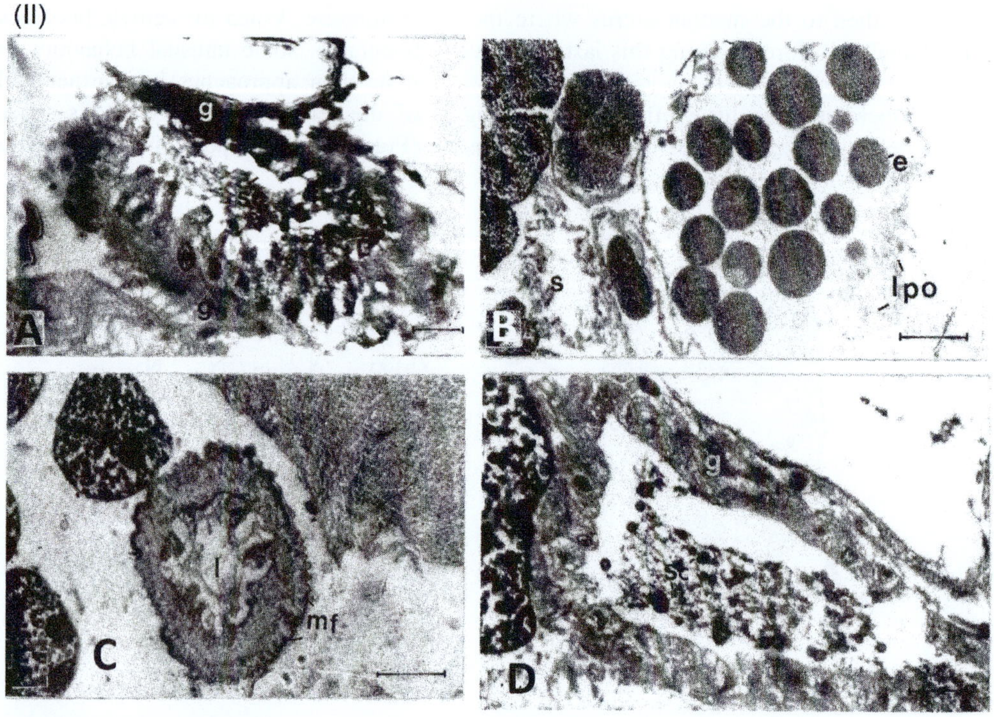

FIG. 10.4, CONT'D (II) Photomicrographs showing T.S.s of the oviduct of *Streptocephalus dichotomus* at different levels during the reproductive cycle: (A) one hour after descent of eggs (uterus). Scale bar=50 μm; (B) frontal section through the oviduct. Scale bar=100 μm; (C) oviduct 30 min after the descent of the eggs into the ovisac. Scale bar=200 μm; (D) 10 h after the descent. Scale bar=50 μm. e=eggs, g=glandular cells, l=lumen of the oviduct, mf=muscle fibre, S=shutter, Sc=secretion in the lumen. *From Munuswamy and Subramoniam (1985).*

FIG. 10.5 *Artemia* mating, Schram (1986) reported that anostracans typically swim with dorsal side down. *WoRMS (2012). 'Artemia salina* (Linnaeus, 1758)'. *World Register of Marine Species.*

of several hours in *Artemia* (Wolfe, 1973), the male bends his body around the abdomen of the female so their gonopores are opposed and inserts his penis into her ovisac. Therefore, the sperm are deposited in the female ovisac where eggs are fertilized and then covered with shells and stored. This means that copulation in *Artemia* (first activity) occurs, during which internal fertilization (second activity) occurs. Once fertilized, eggs quickly undergo cleavage and development through the gastrula stage. After one to a few days, oviposition takes place.

A study of the oviduct of *Streptocephalus dichotomus* (Munuswamy and Subramoniam, 1985) during the ovulation cycle revealed the existence of a corresponding secretory cycle in the glandular cells of the oviduct (Fig. 10.4(II)). At the end of vitellogenesis in the ovary, the oocytes begin to pass to the lateral pouch of the oviduct. This leads to an enormous dilation of the lumen and flattening of the wall (Fig. 10.4(II) A). The oocytes in the oviduct measure about 0.20–0.22 mm and remain without any distinct envelope (Fig. 10.4(II) B). The lining epithelium of the lateral pouch is highly secretory, and the secretion appears soon after it becomes filled with oocytes (Fig. 10.4(II) D). The contraction of this region is observable in the living animal through the transparent body wall. Such a peristaltic movement facilitates the downward movement of the eggs embedded in the mucus into the ovisac, where the eggs are fertilized and shelled.

Artemia, as other anostracans, produce different-sized eggs depending on habitat of the populations involved. Two types of eggs may be laid: thin-shelled (summer eggs) that continue development and hatch quickly; or thick-shelled, brown (winter eggs) in which development is arrested at about early gastrula stage. Such **winter eggs** in their dried and encysted form, survive dry conditions for years, up to 10 years or more, while retaining the ability to survive severe environmental conditions. The egg of *Artemia franciscana* (= the North American version of *Artemia salina*) is one example, that may remain viable after heating to 80°C for 1 h; cooling to – 190°C for 24 h; or reducing air pressure to 0.000001 mm (Drewes, 2002).

Early development (embryology)

Cleavage. Basic crustacean development, unlike that of other arthropods, originates from a small egg. The latter passes through a total cleavage, which in some species has a spiral basis. Anderson (1973) recorded that intimations of spirality in the cleavage of cladocerans, copepods, and cirripeds were clear in the results of a number of cell lineage studies, e.g. Kühn (1913). The egg then proceeds to hatching as a small planktonic larva, the nauplius. Little is known of cleavage in orders other than the above groups. However, Schram (1986) recorded that the most complete description of anostracan early ontogeny is given by Benesch (1969) for *Artemia salina*, which clarifies and supplements the work of Heath (1924). There is, however, enough fragmentary information to indicate that *Artemia* exemplifies cleavage in the noncladoceran branchiopods. The egg of *Artemia* (about 150 μm in diameter) has equally distributed yolk (isolecithal) and so cleavage is total (complete), equal and radial (Anderson, 1973). However, it has a uniformity which prevents the identification of the axes of development. Consequently, spindle orientations and cell lineages cannot be followed. However, Fig. 10.6 shows that the first cleavage (A) is latitudinal (transverse) division, while the second cleavage is meridional (longitudinal) as shown in the same figure. The cleavage divisions continue until at least the ninth cleavage, 512 tall, narrow cells make up the wall of the hollow, spherical blastula (E).

Presumptive areas of the blastula. The crustacean comparative embryological studies (Anderson, 1973) show that there is a wide range of cleavage types and morphological forms of blastula or blastoderm, but Crustacea exhibits the same stability in their presumptive areas (Fig. 10.7A). The presumptive rudiments of crustacean embryos are segregated from one another in a spatial pattern. The major variations are found in the presumptive midgut and extra-embryonic ectoderm, since they are most affected by changes in the volume of yolk in the egg. In *Artemia*, which has hollow blastula of numerous, narrow, columnar cells, the **presumptive midgut** occupies an extensive area of the

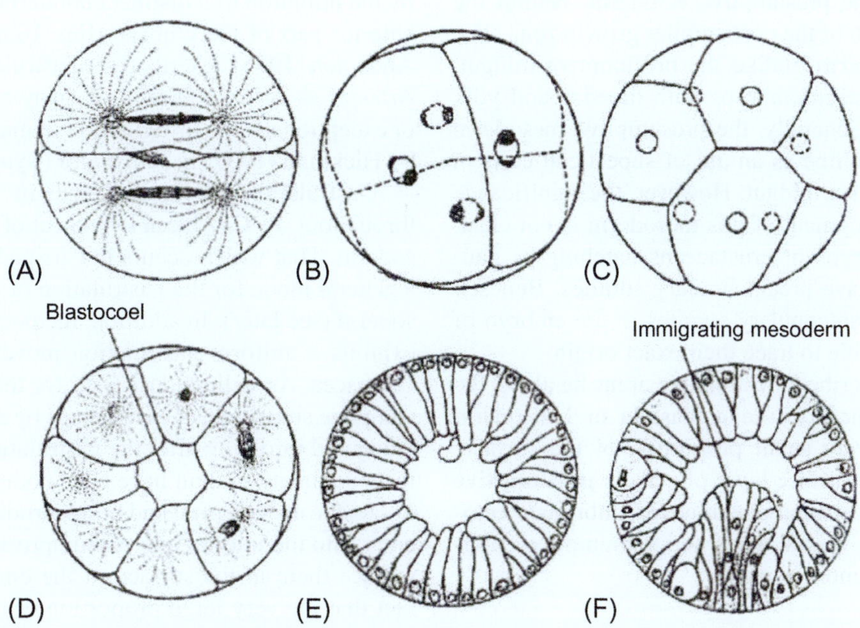

FIG. 10.6 Diagrammatic drawings presenting the cleavage of *Artemia salina*: (A) second cleavage division; (B) 4-cell stage; (C) 16-cell stage; (D) onset of fifth cleavage division; (E) 512-cell stage; (F) first phase of gastrulation. *(After Benesch (1969), from Anderson (1973).)*

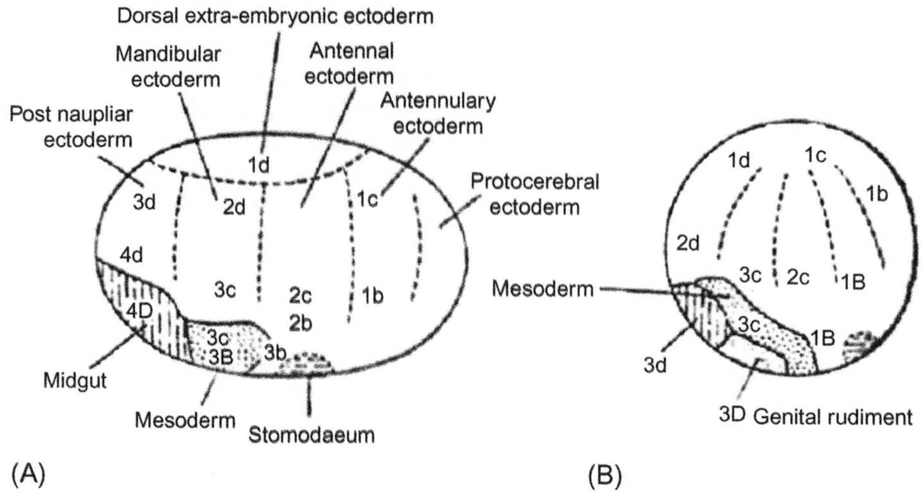

FIG. 10.7 Fate maps of the crustacean blastula or blastoderm, right lateral view: (A) Cirripedia; (B) Cladocera and Copepoda. *(Based on data of Kühn and others and also Anderson (1969). From Anderson (1973).)*

blastula wall. Their entry into the interior is by immigration or invagination of mass of cells as a whole. Unfortunately, it is not possible on present evidence to establish a precise fate map of *Artemia* or of any relative species. Thus, Fig. 10.7B may be an alternative.

The **presumptive stomodaeum**, as in all crustacean embryos, is a circular area of superficial cells in the ventral midline, anterior to and widely separated from the presumptive midgut. In later development, the presumptive stomodaeum makes an independent invagination into the interior and gives rise to the lining epithelium of the foregut. On the other hand, the **presumptive proctodaeum**, in all crustaceans, is a small group of superficial cells in the postero-ventral midline of the presumptive ectoderm, behind the presumptive ectoderm of the postnaupliar growth zone. The **presumptive mesoderm**, unlike the presumptive midgut, varies little in crustacean embryos with the size and yolk content of the egg. Generally, the presumptive mesoderm lies in the ventral midline as an arc of superficial cells in front of the presumptive midgut. However, the significance of the separate development of this mesoderm is not clear, since the small embryos of crustaceans hatching as nauplii do not usually have preantennulary somites. Benesch (1969) identified preantennulary somites in the embryo of *Artemia*, but was unable to trace their exact origin.

All the above-described presumptive areas lie along the ventral midline of the crustacean blastula or blastoderm and occupy a relatively small proportion of the surface. The remainder of the surface layer of cells is **presumptive ectoderm**, divisible between presumptive embryonic ectoderm adjacent to the ventral areas and presumptive extra-embryonic ectoderm more dorsally.

Gastrulation. As in all yolky arthropod embryos, gastrulation in the Crustacea is a relatively minor aspect of development (Arderson, 1973). The patterns of gastrulation movements in crustacean embryos are highly variable, proportionate with the variations in the composition and mode of segregation of presumptive areas in different species. It is advantageous to deal with these movements of each presumptive area separately. However, gastrulation in *Artemia* occurs in two distinct phases (Schram, 1986). The first begins with the formation of a posteriorly located blastopore and results in the invagination of the mesoderm and primary germ cells (Fig. 10.8A), a process that begins about 32 h after fertilization and is completed after about 50 h. The second phase begins with the differentiation of naupliar buds (anlagen), shortly followed later (about 60 h) by the initiation of a distinct endodermal migration near the anterior part of the embryo (Fig. 10.8B and C). However, Anderson (1973) described the gastrulation of the midgut in *Artemia* as an immigration of many cells (Benesch, 1969) or exceptionally, an invagination of many cells, as described by Hickman (1937) in *Anaspides* (Syncarida).

Gastrulation movements remain much more distinct throughout the Crustacea in respect of the presumptive mesoderm. That which occurs in Cirripedia can be considered a general mode for the gastrulation of this presumptive mesoderm (see later). In addition, the presumptive stomoaeum exhibits a uniform gastrulation movement throughout the Crustacea. As will be described for the Cirripedia, the presumptive stomodaeum invaginates by an independent thickening and infolding after the gastrulation movements of the mesoderm and midgut have been completed.

As the mesodermal and midgut rudiments complete their entry into the interior, the presumptive ectoderm spreads to replace them at the surface of the embryo. In spite of the fact that the very large proportion of the surface is already occupied by the presumptive ectoderm in the crustacean blastula or blastoderm (Fig. 10.7), little ectodermal spread

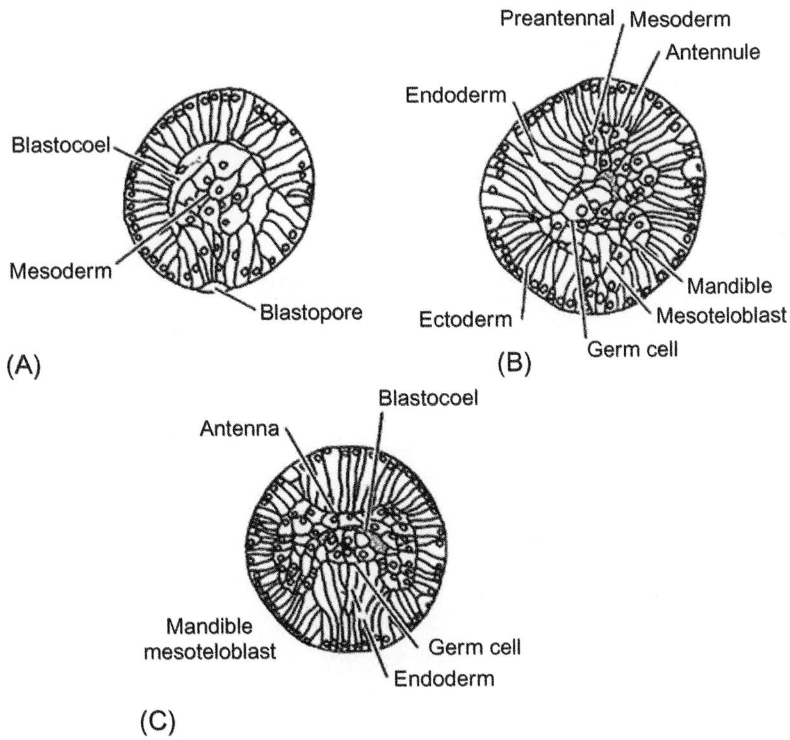

FIG. 10.8 Diagrammatic drawings of late ontogeny of *Artemia salina*: (A) second phase of gastrulation (advanced stage of mesoderm formation); (B) L.S. of A (inward migration of endoderm cells); (C) cross section of (B). *(Modified from Benesch (1969), From Schram (1986).)*

is involved in any species in covering the superficial, posteroventral areas previously occupied by mesoderm and midgut cells. Gastrulation in respect to ectoderm is thus minimal in Crustacea.

Differentiation of the embryo. After the above-described step, gastrulation is completed and the differentiation of the naupliar organs begins. The external development of crustaceans after gastrulation is a highly complex and varied process (Anderson, 1973). However, in those crustaceans, which hatch as a nauplius larva, the external development of the embryo before hatching is, in general terms, a relatively simple process. So in *Artemia* (Schram, 1986), the first notable change is the development of the three pairs of naupliar buds which will give rise to the antennules, antennae, and mandibles, but anterior to these is an unpaired mass of preantennal mesoderm (Fig. 10.9A). This mass behaves like any of the more posterior metameres and produces a segmentally organized mass of muscles. This latter will eventually give rise to the labral, esophageal, and eye muscles, as well as lipid storage cells and connective tissue in the labrum (Fig. 10.9B). The teloblasts of the posterior end of the body (Fig. 10.9) appear to be unpaired and indistinct and, as a result, the earliest stages of segment formation cannot be determined in *Artemia.* During the course of naupliar development, Benesch noted the appearance of cell masses derived from the somites in the mandibular and thoracic segments (Schram, 1986). These are the so-called 'blood glands' that gave rise to hemocytes (Fig. 10.9C). However, Anderson (1973) doubted whether these should be interpreted as segmental organs.

Eggs of *Artemia* hatch as nauplii (Fig. 10.10) in which the naupliar region (as in all Branchiopoda) is well developed with a conspicuous postnaupliar region (Anderson, 1973; Schram, 1986). The naupliar region is structurally and functionally specialized in relation to active swimming and planktotropfic feeding, while the postnaupliar region is not externally evident. The structural and functional elaboration of the naupliar region is completed during the first three larval stages during which the larva lives on yolk reserves, and feeding begins during stage 3 or 4. The newly hatched nauplius of *Artemia* (Fig. 10.10C) is 450 μm; when feeding begins, this length increases to 725 μm. At the third moult, a series of postnaupliar appendage buds suddenly appear. The trunk limbs continue to develop either essentially in blocks or in small gradual steps. Naupliar development in *Artemia* continues through 10 stages (Fig. 10.11), through which the postnaupliar region increases in length and becomes further elaborated. Metamorphosis to the juvenile adult form then takes place at the ninth moult, and this juvenile is about 2 mm long.

The hatched nauplius. Stage 1 nauplius of *Artemia* (Fig. 10.10C) is brownish-orange in colour due to the presence of yolk material. It is somewhat elongated larva with three pairs of appendages belonging to the naupliar region. A nauplius eye is present, but it is not easily distinguished

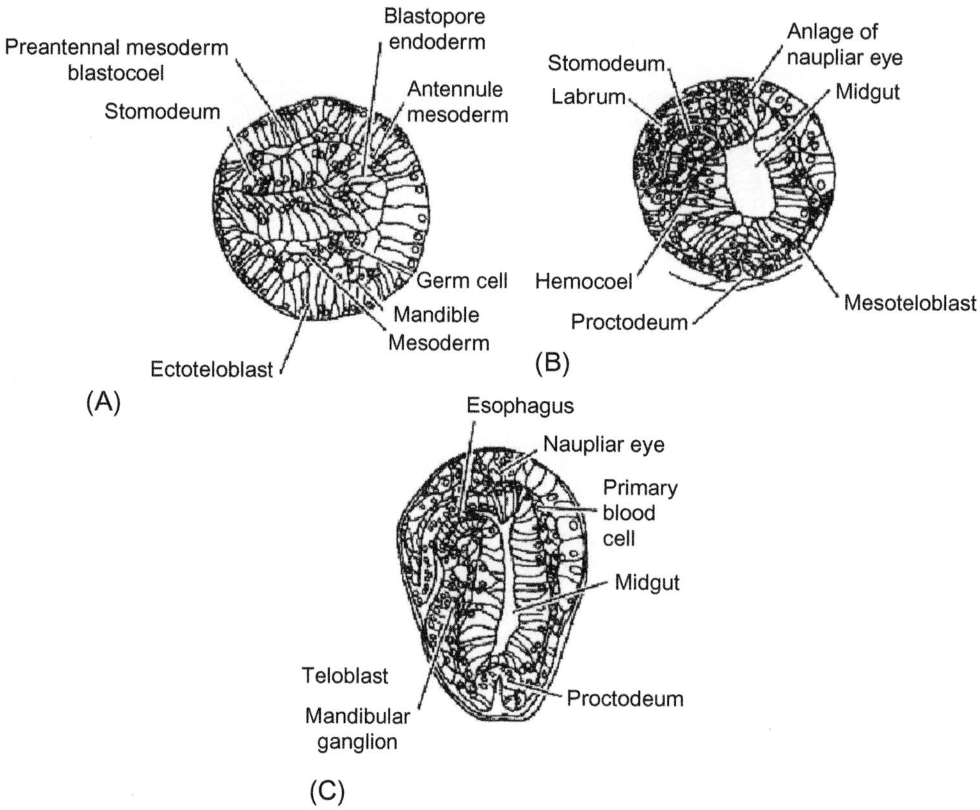

FIG. 10.9 Differentiation of the larva: (A) initiation of stomodeum, completion of gastrulation; (B) anterior gut and labrum complete, initiation of proctodeum; (C) advanced naupliar differentiation. *(Modification of Benesch (1969), from Schram (1986).)*

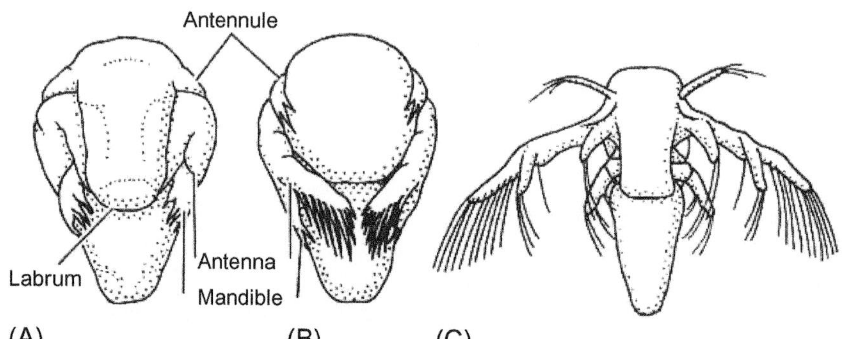

FIG. 10.10 Illustrations showing nauplius of *A. salina* at hatching: (A) ventral; (B) dorsal views of the larva *(modified from Benesch, 1969)*; (C) hatched and free swimming nauplius (Heath, 1924). *From Schram (1986)*.

at this stage. The first pair of appendages are uniramous antennules; they are small and project forwards as sensory processes. The second pair are the well-developed antennae which, in contrast, are long, biramous with long hinged setae on the exopode and endopode. The mandibles are the third pair, short and uniramous, but also carry terminal hinged setae. There is a large lip-like structure (the labrum) covering a ventral mouth. At this stage, the larva does not have a complete digestive tract and so it does not immediately feed. After the third moult, the larva subsists by naupliar feeding, that is, the antennae with their terminal fans of setae and with the mandibles sweep food into the sublabral space. Therefore, the antennae are locomotory and feeding organs during larval stages.

The specialization of development of the branchiopod postnaupliar region is evident even at hatching. Although not marked externally by intersegmental annuli or limb buds at this stage, the region already includes the rudiments

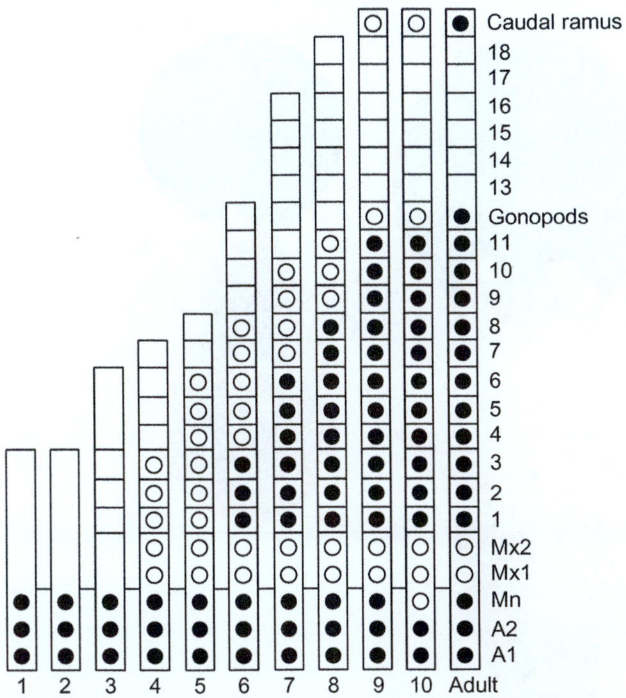

FIG. 10.11 Diagram describing postembryonic development in *A. salina*. ● Mature and functioning limbs; ○ immature and/or reduced limb. *(From Schram (1986), modified from Anderson (1967).)*

of the first eight postnaupliar segments (the maxillulary, maxillary, and first six trunk segments) together with a pretelsonic growth zone and a terminal telson which carries the anus. The maxillulary and maxillary segments remain vestigial, but the first six trunk segments develop rapidly and more or less synchronously, exhibiting limb buds at stage 4. The six pairs of limb buds become functional, coincident with metamorphosis at stage 10 in Anostraca.

Parthenogenesis

Typically, *Artemia* reproduces gametically and so sexes are separate, as shown above. However, in some species and populations of *Artemia* (e.g. Europe), males may be rare and females reproduce parthenogenetically. Parthenogenetic reproduction has been reported in *Artemia salina* and in some other anostracans (Wolfe, 1971) and also in some Ostracoda and Copepoda, and in some malacostracan Isopoda. In these, parthenogenesis is referred to as geographic type (Charniaux-Cotton, 1960a,b). In this geographic parthenogenesis, in the single species or in two closely occurring species, exists in two forms: bisexual and parthenogenetic. The two forms generally have different distribution. This geographic parthenogenesis differs from cyclic parthenogenesis, which occurs in *Daphnia* (Charniaux-Cotton, 1960a,b) (see later). It was recommended that all parthenogenetic *Artemia* be referred to as *Artemia parthenogenetica* followed by the vernacular or Latin name of the locality of origin (Kuruppu, and Ekaratne, 1995). These authors proved experimentally, for the first time, that *Artemia parthenogenetica* Sri Lanka was a parthenogenetic strain. Kuruppu and Ekaratne obtained cysts (Fig. 10.12A) from Mahalewaya and incubated them under controlled conditions and studied development and life cycle of *Artemia parthenogenetica* Sri Lanka (Fig. 10.12). They found that this species was able to develop and reproduce ovoviviparously within a period of 15 days after hatching. Preadult females, complete with eight abdominal segments, appeared on day 12. They attained the brood pouch and a pair of ovaries on day 13, and on day 14 the ova were visible within the brood pouch. Free-swimming nauplii were released on the 15th day. No males were observed during the experiment.

Diplostraca

Diplostraca (Branchiopods with bivalved and laterally compressed shell or carapace). It is divided into two suborders (Borradaile et al., 1977):

a. Conchostraca (Greek *konche*, 'shell'): hinged carapace with adductor muscle. Ex. Clam shrimps. (*Limnadia*).

b. Cladocera (Greek *clados*, 'sprout' and *keras*, 'horn'): carapace without hinge or muscle. Ex. Water fleas (*Daphnia*, *Leptodora*, *Sida*, *Polyphemus*).

Branchiopods live almost exclusively in fresh water; only cladocerans inhabit streams, large ponds, vernal pools, and lakes. Moreover, only some cladocerans are marine with thought to be restricted to neritic habitats. In fact, they can in some areas and some seasons be very abundant and occur in a high proportion of samples. Therefore, cladocerans were thought to be common representatives of the group. *Daphnia* is obtainable from the fountains found in the plant gardens of the college and also from the duck and goose ponds of the Zoo in Alexandria, Egypt.

Because of its ease of handling in the laboratory, and ability to reproduce by both parthenogenesis and genetic reproduction, genus *Daphnia* was used to study some different aspects especially ecological ones. For example, Chandini in 1989 used the changes in food (Chlorella) levels as a factor to study the chronic toxicity of cadmium on the various life-history parameters in *D. carinata.* The study indicated that at low food levels, these parameters, such as survival, growth, and fecundity, were reduced by 50%; at high food levels, the toxic effect of cadmium was greatly reduced. Therefore, the study stresses the need to incorporate in laboratory tests other relevant factors that might modify pollutant toxicity. Another species, *D. magna* Straus, 1820, is a native in the state of Rio Grande do Sul, southern Brazil. It lives in regions where water is hard and the rivers have soft water. Terra and Feiden (2003) exposed this animal to water hardness variations for 21 days to evaluate the existence or not of stress in this species, due to changes in hardness. The results obtained indicate that there

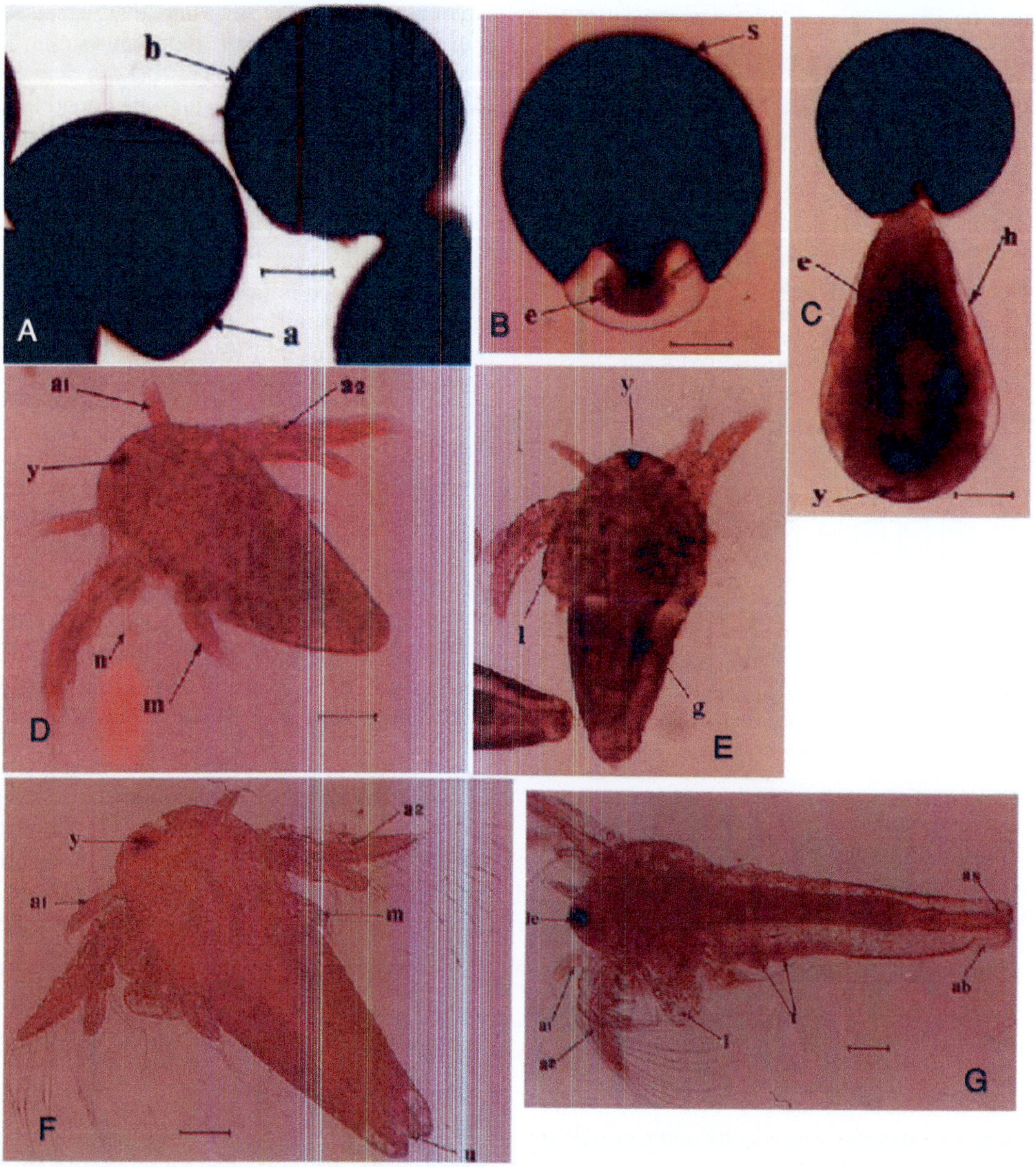

FIG. 10.12 Life cycle of *A. parthenogenetica* Sri Lanka (*Kuruppu and Ekaratne, 1995*) showing lateral views of: (A) biconcave nature of cysts before hydration (a) and spherical shape (b) one hour of hydration; (B) embryo (e) beginning to protrude through fissure of cyst shell (s); (C) oval-shaped pre-nauplius stage embryo (e). h= thin hatching membrane, y=red eye; (D) instar I nauplius stage, after release from hatching membrane in dorso-lateral view. a1=antenna I, a2=antenna II, m=mandible, n=gnathobase. Scale bar for (A), (B), (C), and (D) is 76 μm; (E) instar II nauplius stage (the first feeding stage). g=gut, I=ventral labrum, y=naupliar eye. x 96 μm; (F) instar III nauplius stage in dorso-lateral view. u=anus; (G) instar IV nauplius stage. ab=short abdomen, as=initial abdominal segmentation, le=developing lateral eye, t=developing thracopods. Scale bar of (F) and (G) is 100 μm.

is no significant difference in the reproduction of *D. magna* between the least and most water hardness. Survival at different hardness was always within the percentage required for control, and thus considered normal. The same species, *D. magna*, was used in another ecotoxicological study by Hayashi et al. (2008). They exposed the animal (5 days old) to different concentrations of ibuprofen (IB) for 10 days followed by a 10-day recovery period in uncontaminated water. They recorded that IB-stressed individuals produced offspring faster during recovery, and so the effect of IB is reversible in *D. magna* as it is in mammals.

Morphology

The 600 or so species of water fleas are among the most distinctive of crustacean body plans (Fig. 10.13). The small, familiar organisms of Cladocera are 0.15–3 mm in length (Hickman, 1973). The most familiar genus is *Daphnia* (Fig. 10.14), in which the body is reduced to, at most, four

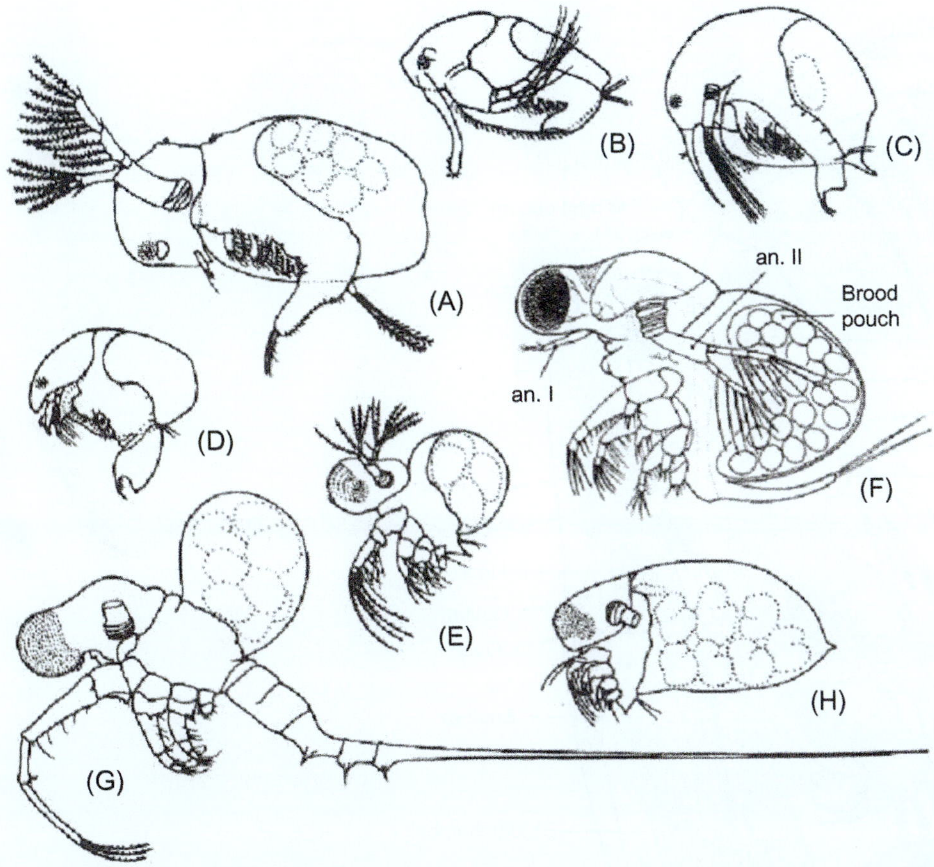

FIG. 10.13 Diagrams of different cladocerans: (A) *Sida crystalline*; (B) *Moina brachycephala*; (C) *Bosmina longispina*; (D) *Eurycercus glacialis*; (E) *Podon intermedius*; (F) *Polyphemus pediculus (Barnes, 1974)*; (G) *Bythotrephes longimanus*; (H) *Evadne nordmanni*. *(From Schram (1986).)*

to six limb-bearing trunk segments and a recurved postabdomen. The segmentation of the trunk is obscure. The first two somites are fused with the head, as is shown by the position of their appendages. Behind these are three fairly distinct limb-bearing somites (in all five pairs of trunk limbs), and then three limbless and hardly distinguishable somites, and a telson, which is compressed and produced on each side of the anus into a toothed plate, bearing a spine terminally that may represent a furcal ramus. The carapace is bivalved shell, but no muscle or hinge, and not covering the head. The head is bent downwards, so the median eye and the small antennules are ventral to the antennae. A large, sessile compound eye, formed by the fusion of a pair, stands in front. Rarely, there is a nauplius larva, only in *Leptodora*.

One of the more confusing aspects to cladoceran studies is the phenomenon of **cyclomorphosis**. In some cladoceran species, such as the lake-inhabiting *Daphnia dubia* and *D. retrocurva*, the head progressively changes from a round to helmet-like shape between spring and midsummer (Barnes, 1974) (Fig. 10.15). This phenomenon is still poorly understood. It may result, as in rotifers, from internal factors; or may result from interaction between external conditions (temperature) and internal factors (perhaps genetic) (Barnes, 1974). However, Schram (1986) reported that it was demonstrated that this phenomenon was related to differential rates of growth between the head and trunk regulated by temperature and food availability. For example, in *D. recurvata* and *D. galeata*, a temperature of less than 7°C resulted in animals with round heads, while a temperature in excess of 7°C produced spiked-head animals (Fig. 10.15C).

Reproductive system

In *Daphnia* and *Simocephalus*, as examples of Cladocera, the one pair of gonads are simple elongated sacs lying in the trunk, on either side of the midgut (Fig. 10.14). Each gonad is continuous with its duct, which opens on the telson in the male, near the anus or the postabdomen. In some cladocerans, the male gonopores, if united or separate, may be modified in the form of copulatory organ (Barnes, 1974). Among the Cladocera, the spermatozoa (Fig. 10.16A–D) are not very different in their constitution from ordinary cells. However, they present a remarkable variety of form, differing greatly sometimes even in the species of one genus (Calman, 1909). They have abundant cytoplasm, in which the large nucleus is embedded. They exhibit rod-like,

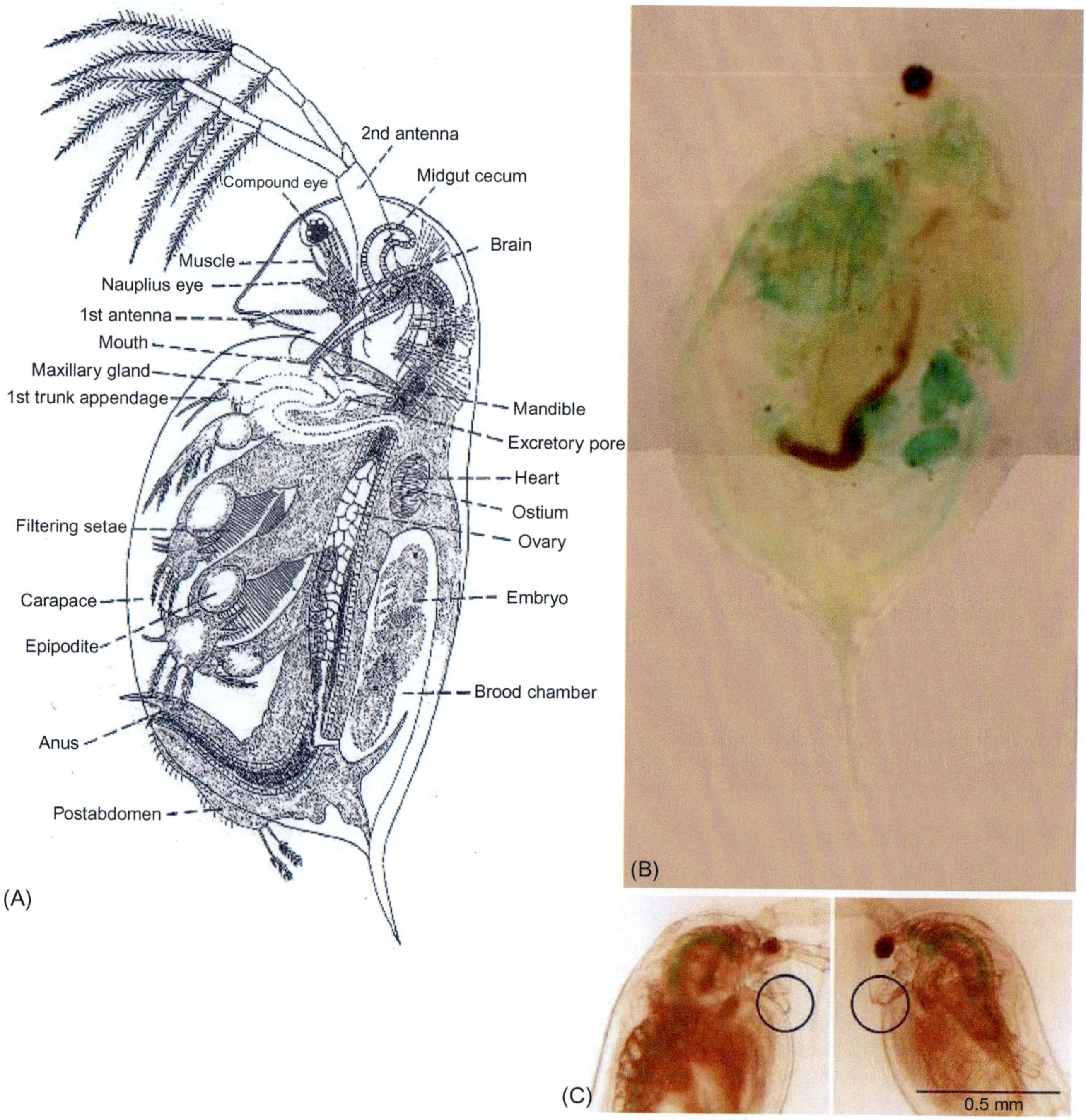

FIG. 10.14 Lateral view of female cladoceran *Daphnia pulix*: (A) diagrammatic drawing *(Barnes, 1974)*; (B) montage of photomicrographs of whole mount of *Daphnia* sp. *(laboratory preparation)*; (C) heads of reared *D. magna* aged c. 24 h showing male (left) with longer antennule and female (right) short antennule as appear within the *circles*. *(After Tatarazako et al. (2004).)*

sickle-like, or other similar shapes, to which are sometimes added a number of protoplasmic processes. In *Polyphemus* (Fig. 10.16D), they are even able to perform amoeboid movement (Kumé and Dan, 1968). In the female, on the other hand, the oviducts open dorsally into a brood chamber (Fig. 10.14) located beneath the carapace. The female produces centrolecithal eggs in which it is difficult to distinguish the egg axis except by the location of the polar bodies (Kumé and Dan, 1968). In *Moina*, two kinds of differentially staining yolk granules surround the nucleus concentrically. The *Holopedium* egg has a large, eccentric oil drop as well as a concentric arrangement of oil drops of much smaller but varying sizes and a similar arrangement of yolk granules. The summer egg of *Polyphemus* is accompanied by three nutritive cells during its formation. When they have been almost completely absorbed by the

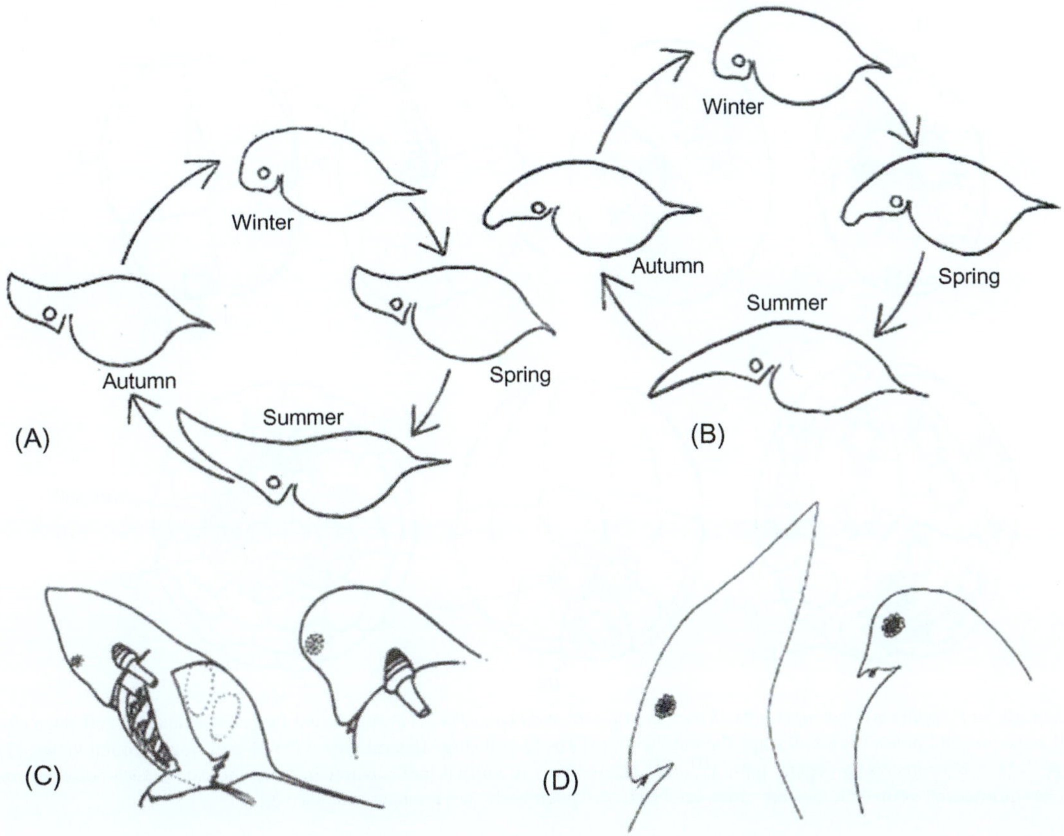

FIG. 10.15 Cyclomorphosis in *Daphnia*. The animals often show seasonal variations in form. Four different examples are shown. *((A and B) from Borradaile et al., 1977)*; (C) cyclomorph of *D. galeata (from Schram, 1986)*; (D) summer long-head form and spring and fall round-head form *(from Barnes, 1974)*.

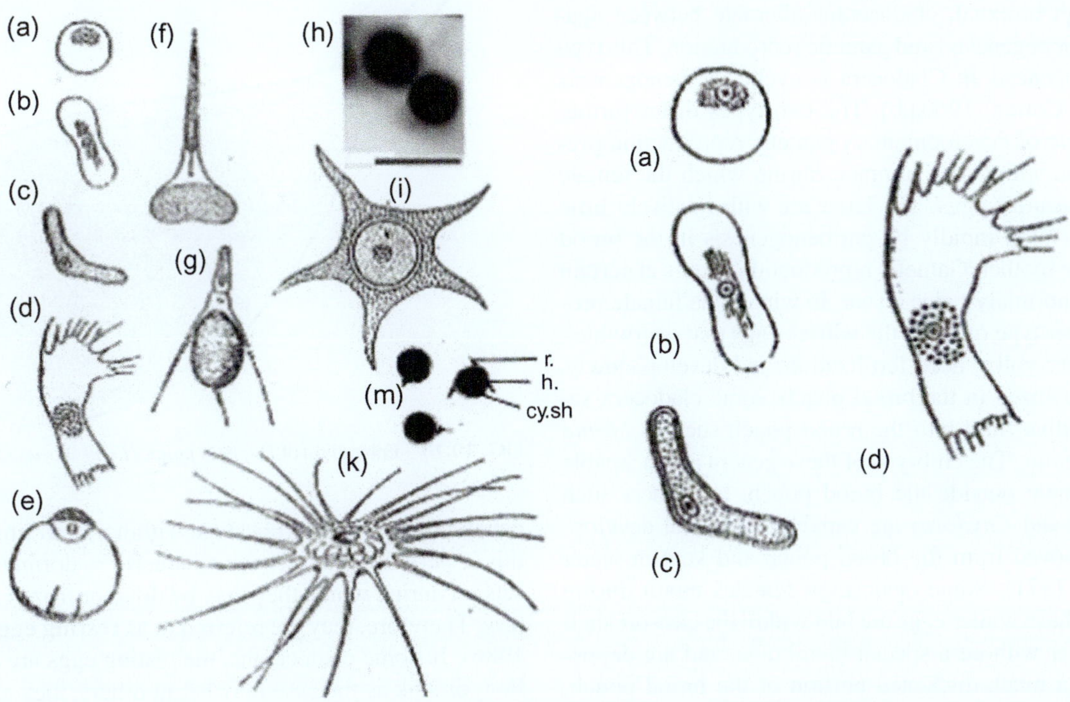

FIG. 10.16 Spermatozoa in several crustaceans. (a)–(d) Cladocerans: (a) *Latona setifera*; (b) *Daphnella brachyuran*; (c) *Moina paradoxa*; (d) *Polyphemus pediculus*; (e) *Squilla oratoria*; (f) *Leander aspersus*; (g) *Galathea squamifera*; (h) *Oratosquilla oratoria* (section) *(from Kodama et al., 2009)*; (i) *Ethusa mascarone* (upper side); (k) *Astacus fluviatilis*; (m) *Portunus pelagicus*, camera Lucida drawing of a smear of the most posterior region of the testis. cy.sh. = cytoplasmic sheath, h = head and r = ray *((m) from Bawab and El-Sherief, 1981, others from Kumé and Dan, 1968)*.

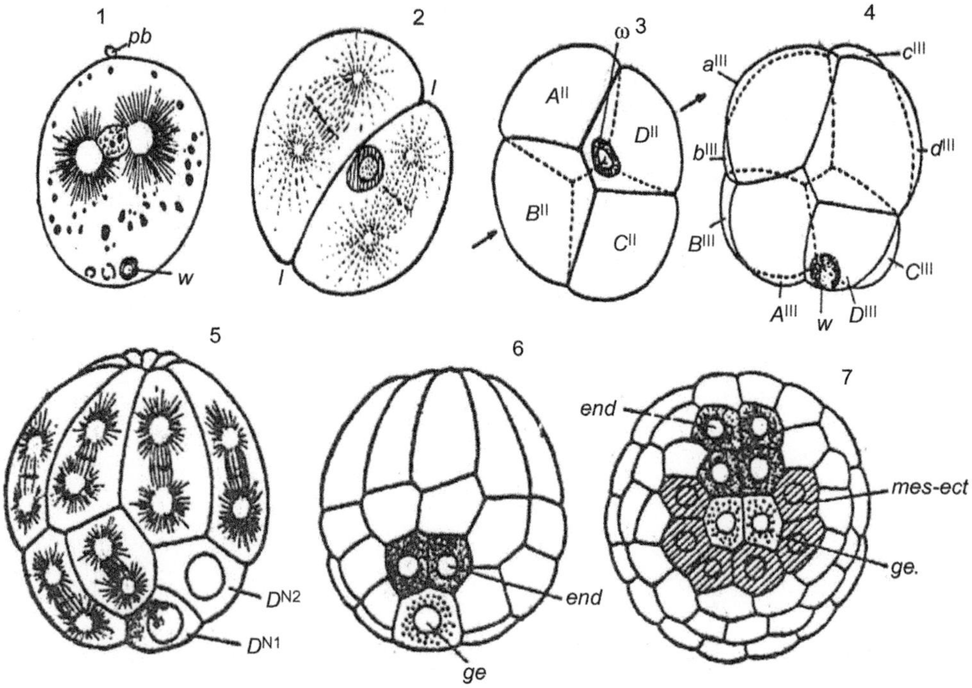

FIG. 10.17 Cleavage in *Polyphemus pediculus (after Kühn, from Kumé and Dan, 1968)*: (1) egg during first cleavage; (2) 2-cell stage *(from Pflugfeder, 1970)*; (3) 4-cell stage (vegetal pole); (4) 8-cell stage (lateral view); (5) 16–32-cell stage (lateral view); (6) 31-cell stage (frontal view); (7) 62-cell stage (vegetal pole). A^{III}–D^{III} blastomeres in vegetal half, a^{III}–d^{III} blastomeres in animal half, *arrows* in three indicate antero-posterior axis of embryo, end = endoblast, ge = primordial germ cell, mes-ect = mes-ectoblast, pb = polar body, w = remnant of nurse cell.

egg, their remnants remain as a small mass within the egg at one side (Fig. 10.17(1)), and the polar bodies emerge at the opposite side (Kumé and Dan, 1968).

Although bisexual, cladocerans alternate between agametic (parthenogenesis) and gametic reproduction. This type of parthenogenesis in Cladocera is cyclic parthenogenesis (Charniaux-Cotton, 1960a,b). The two types differ further in their mode of development. A gametic reproduction predominates in spring and summer, during which the female produces **summer eggs**. The latter are with relatively little yolk and develop rapidly by parthenogenesis in the brood pouch of the mother. Gametic reproduction occurs at certain times only, normally twice a year. In winter, the female produces another type of eggs, the **winter eggs** that overwinter. They are very yolky, need fertilization, and develop slowly. They are fertilized in the brood pouch, some cladocera secrete a nutritive fluid into the brood pouch such as *Moina* and *Polyphemus*. The embryos of these genera are incapable of development outside the brood pouch, but others such as *Daphnia* and *Chydorus* are capable of normal development if removed from the brood pouch and kept in water (Reverberi, 1971). Some cladoceran females moult during brooding. These winter eggs are laid within the cast-off shell of the mother without a special membrane, and are deposited within a much-thickened portion of the brood pouch. The membrane is called the **ephippium** (Fig. 10.18), within which they are able to withstand freezing and desiccation (Kumé and Dan, 1968). They go through the early stages of

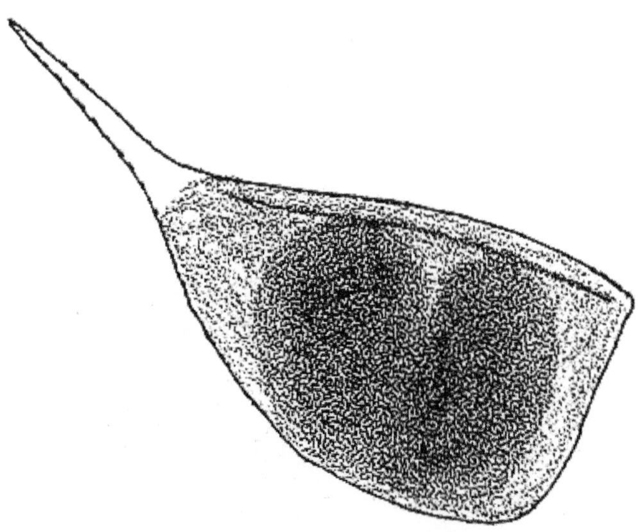

FIG. 10.18 Ephippium of *Daphnia pulex*. *(From Barnes (1974).)*

development to a gastrula stage within a short time, but after this a period of quiescence (diapause = dormancy period) sets in, during which they may be dried or frozen without injury. Therefore, they are referred to as **resting eggs** (Schram, 1986). In some cladocerans, the resting eggs are released to float or sink as the case may be; in others, they are attached to the bottom or to the surrounding macrophytes. Even the oceanic *Evadne* produces resting eggs, but their fate is currently not known (Schram, 1986).

Daphnids switch their mode of reproduction from parthenogenesis to gametic reproduction in response to environmental changes such as short day length, food depletion, and high population density. Several studies revealed that one of the major hormone groups in insects and crustaceans, juvenile hormones, are involved in this shift of reproductive mode from parthenogenesis to gametic reproduction. Accordingly, Tatarazako and Oda (2007) used offspring sex ratio of *Daphnia magna* as a new endpoint to identify chemicals with juvenile hormone-like effects on crustaceans. After males are produced, mating behaviour, in *Daphnia* in general appears to depend on random contact between these males and sexual females rather than diffusible pheromones. In addition, males may be able to discriminate sexually, the receptive females from females in other developmental stages and increase their mating efficiency (Winsor and Innes, 2002).

The number of eggs produced during the life of a female varies. Generally, in nature, the number of eggs per brood increases with the age of the female. In *Daphnia*, e.g. with increasing age of the mother, the proportion of young matures at the fifth instar, rather than the sixth, increases. These can be seen as modes of reproduction meant to increase the probability of overwintering success in the population (Schram, 1986). Previously, Green (1954) studied egg production and size (length) in natural populations. He examined four populations of *Daphnia magna* Straus, 1820 from different localities. He reported that in favourable conditions, reproduction is purely parthenogenetic. Eggs are laid into the brood pouch and are carried until they are liberated as miniature immature adults. Youngs are liberated a few hours before the mother moults. Soon after moulting, another clutch of eggs is laid into the brood pouch. The number of parthenogenetic eggs produced during the life of a female increases in the early broods and decreases in later ones. However, during the gamogenetic reproduction, only two eggs (resting eggs) are produced in each brood, e.g. *Daphnia*. Some genera produce only one resting egg at a time, and *Eurycercus* produces a variable number, up to about 10 (Reverberi, 1971). From his studies, Green (1954) came to the following conclusions:

1. In natural populations, parthenogenetic egg production in this species is directly related to length. He measured each individual from the crown of the head to the base of the tail spine, with microscope fitted with a calibrated eyepiece micrometer.
2. Under constant conditions and with varying food, egg production and size are positively correlated in the first adult instar.
3. The size of the young liberated varies with maternal age, the third brood consisting of young of a larger average size than earlier or later broods.
4. The maternal age may affect growth, maturation and longevity of the young.

Development (embryology)

Gametes: Oogenesis In *Daphnia*, groups of four cells are formed in the ovary, three of which break down and are ingested by the fourth, which becomes the definitive oocyte. Sometimes all four cells break down and are ingested by a neighbouring oocyte (Reverberi, 1971). This process enables the rapid build-up of relatively large cells in a short time. This oogenesis cycle in *Daphnia* is closely related to the moulting cycle. In good conditions, an adult female *Daphnia* moults once every 2 or 3 days. After each moult, a batch of eggs is laid into the brood pouch, and the ovary then has to regenerate rapidly, ready for the next moult. The next batch of oocytes enters vitellogenesis about a day after the last batch of eggs has been laid.

Cleavage. Schram (1986) reported that detailed knowledge of early development in cladocerans is largely restricted to the earliest stages described by Grobben (1879) and Kühn (1908, 1911, 1913). However, Anderson (1973) recorded that his work (Anderson, 1969) and that of von Baldass (1937) show that the cladoceran *Holopedium*, with spherical eggs 120 μm in diameter, has total and spiral pattern of cleavage more distinct than that of cirripeds. The cleavage in *Daphnia*, on the other hand, follows the same sequence as those of *Holopedium*, but evolved one step further due to its larger and more yolky egg. Unfortunately, the diagrams describing the cleavage *of Daphnia* are not available. However, Pflugfeder (1970) reported that the egg of *Daphnia* was used for experimental studies. After centrifugation, the distribution of yolk in the egg changes. However, at low speed, no change occurs and the egg develops normally since the cleavage proceeds as normal. But higher speeds affect the process of cleavage and cause deformation of the embryo. For example, if the yolk accumulates in the vegetal pole after centrifugation, the head of the produced embryo becomes very large. The opposite case produces an embryo without a head.

The total radial cleavage of the summer egg of *Polyphemus pediculus*, with approximately the same diameter as that of *Holopedium*, is used as a general pattern for cladocerans (Fig. 10.17). The **first** cleavage is meridional; it divides the egg into two somewhat unequal blastomeres. The **second** cleavage is also meridional but perpendicular to the first, forming a cross at the poles. It yields four blastomeres. At the end of this stage, the embryo consists of three equal blastomers A, B, and C, and somewhat smaller D, which contains the remnants of the nutritive cells as granular material persisting in this cell during early cleavages (first observed by Kühn, 1913). This is one well-known example for the recognition of primordial germ cells in metazoans recorded by Biffis et al. (2009). B, lying opposite D, is in contact with it at the vegetal side, but separated at the animal pole. Since D is smaller than the others, the bilateral symmetry of the embryo is already determined at this 4-blastomere stage.

The **third** cleavage is a horizontal furrow lying nearer to the vegetal pole and yields four larger blastomeres (1a–1d) in the animal hemisphere and four smaller vegetal ones (1A–1D). The descendants of 1a–1d give rise to ectoderm (Fig. 10.17 and Table 10.3) and are called primary ectoblast, to be differentiated from the ectoderm cells, which will be derived from 1A–1C. These latter blastomeres are called mesectoblasts, and later form ectoderm and mesoderm. The 1D blastomere will give rise to the endoderm and primordial germ cell. At the end of this cleavage, all the blastomeres are completely separated from each other and a small segmentation cavity is formed at their centre. At this stage, the vestiges of spiral cleavage are hardly perceptible, either in the direction of the spindles or the arrangement of the blastomeres. The **fourth** cleavage (the 16-cell stage) is meridional in all the blastomeres except 1D. The latter divides later latitudinally into $2D^1$ and $2D^2$; the former includes all the remnants of the nutritive cells and lies nearer to the vegetal pole.

TABLE 10.3 Cell-lineage of *Polyphemus pediculus*

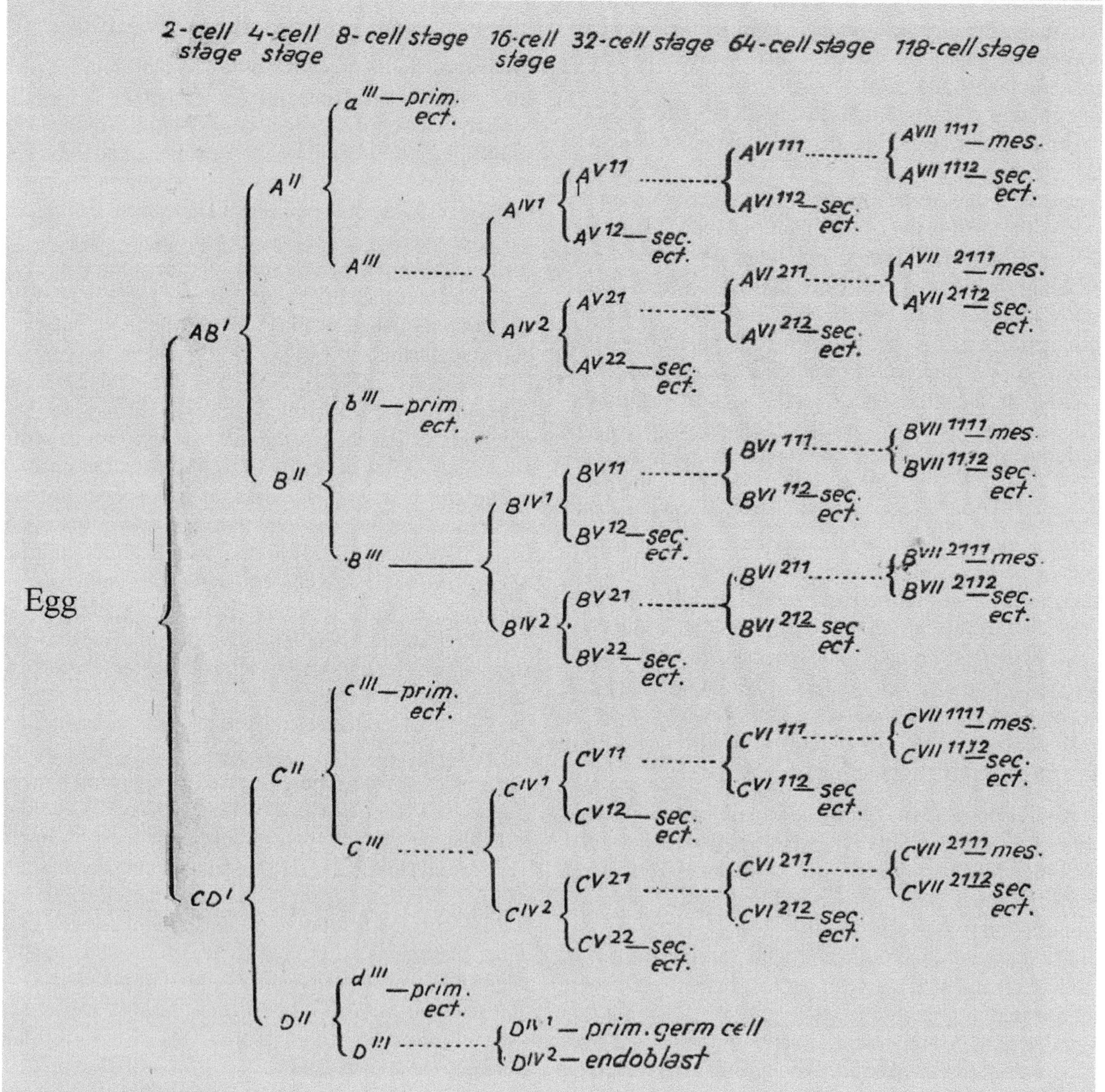

From Kumé and Dan (1968).

It is differentiated into the primordial germ cell, which will give rise to the generative cells. $2D^2$ represents the primordial endoderm cell. The **fifth** cleavage is latitudinal results in the 32-cell stage. The tendency shown by the descendants of 1D to lag behind those of 1A–1C, and the still more retarded cleavage of the primordial germ cell, lead to the appearance of intermediate stages (e.g. 30-, 31-, 32-, 62-cell). After the **sixth** cleavage, the blastomeres of the animal hemisphere divide before the vegetal hemisphere, and in later stages it becomes irregular. In the 118-cell stage, six mesoblasts are differentiated for the first time from the descendants of 1A–1C. The period of cleavage of the egg comes to its end by the eighth step and is described as holoblastic cleavage. The embryo then may be called a blastula.

The blastula consists of 128 primary ectoblasts, 84 secondary ectoblasts, 12 mesoblasts, 8 endoblasts, and 4 germ cells, making a total of 236 cells enclosing a relatively spacious blastocoel.

The radial cleavage in *Polyphemus*, which yields a hollow spherical blastula of relatively few cells, is clearly secondary to an ancestral condition of spiral, unequal cleavage of early segregation of presumptive areas. Details concerning the fates of the blastomeres in *Polyphemus pediculus* are given in Table 10.3.

Presumptive areas of the blastula. Schram (1986) recorded that the presumptive cells are already recognized at the 16-blastomere stage (Fig. 10.7B). Anderson (1973) presented Fig. 10.7B to show the presumptive areas of cladocerans and copepods. He described those of *Polyphemus* and Cyclops as crustaceans that retain total cleavage and a blastula of 32–64 cells. The present work mentions *Polyphemus* only. It has a single presumptive midgut cell. It divides into a pair of presumptive midgut cells located posterventrally in the blastula wall (Fig. 10.17). This presumptive midgut cell has a different division history and a different enumeration from that of cirripeds. It is 2d, which is homologous with the 4D presumptive midgut cell of cirripeds. In *Polyphemus*, the D stem cell cuts off one presumptive ectoderm cell, 1d, and then divides into a posteroventral presumptive midgut cell, 2d, and a midventral primordial germ cell, 2D. After its initial subdivision into two cells, the presumptive midgut of *Polyphemus* immigrates into the blastocoel and divides repeatedly to form the midgut epithelium. In the Cladocera, in general, the presumptive midgut cells invade the yolk mass, proliferate within the yolk as vitellophages, and then assume the definitive epithelium midgut form.

The presumptive stomodaeum and proctodaeum

These areas are similar to what is written for *Artemia.* Since the proctodaeum always develops late, after the naupliar segments have undergone preliminary development, it has a definite location at this time, midventrally behind the ectoderm of the pretelsonic, segmental zone. Thus its presumptive area in the blastula can be limited in relation to the presumptive ectoderm.

The presumptive mesoderm. In *Polyphemus*, as in other embryos of *Holopedium* and *Cyclops*, when their lineage can be identified, the arc of presumptive mesoderm cells is found to have the same ventral A-, B-, and C-quadrant lineage (Fig. 10.7A and Table 10.3), as in Cirripedia. Though the number of cells in the arc is greater than three and further development of this arc is the same in all of these cases. The cells migrate into the interior, undergo further divisions, and move in a posterior direction to lie behind the midgut rudiment at the posterior end of the embryo. They proliferate a pair of lateral bands of mesoderm cells forwards on either side of the midgut. These cells then aggregate as three pairs of naupliar somites, leaving a residual posterior mass, which gives rise to all of the postnaupliar mesoderm.

The presumptive ectoderm. The presumptive ectoderm of the Crustacea takes a number of forms, depending of the form of cleavage. In small crustacean eggs, the presumptive ectoderm has the same relative distribution and pattern of subareas, though it may be laid down as a wall of yolky cells around a blastocoel as in *Polyphemus*, or as the major part of a blastoderm around a yolk mass as in *Holopedium*, *Daphnia* and other cladocerans (Anderson, 1973). The presumptive naupliar embryonic ectoderm extends forwards as a pair of divergent short bands, each comprising presumptive mandibular, antennal and antennulary segmental ectoderm, and a terminal area of presumptive preantennulary ectoderm. The presumptive ectoderm occupies less than half of the surface of the blastoderm. The remainder, covering the yolk mass dorsally and laterally, is an extensive presumptive extra-embryonic ectoderm.

Gastrulation. Gastrulation in Crustacea, as in all yolky arthropod embryos, is a relatively minor event. Gastrulation in *Polyphemus* egg, which begins about the time of the ninth cleavage, is effected by sinking of the mesoblasts, endoblasts, and germ cells into the blastocoel, rather than by a clear-cut invagination (Kumé and Dan, 1968). However, it is convenient to describe the gastrulation movements of each presumptive area separately. *Polyphemus*, which develops a hollow blastula and two cells as a presumptive midgut, exhibits a clear gastrulation movement. This occurs by the immigration of the two cells from the blastula wall into the blastocoel. Then, they enter into a proliferative organogeny at the yolk surface and their products invade the yolk mass. In respect to the presumptive mesoderm, its gastrulation movement remains more distinct throughout the Crustacea

(Fig. 10.19). As in Cirripedia (see later), the cells of the presumptive mesodermal area migrate beneath the surface of the blastoderm and are overgrown by ectoderm from the front. Gastrulation in respect to ectoderm is thus minimal in Crustacea, reflecting the fact that the presumptive ectoderm is laid down in extended form during all types of crustacean cleavage (Anderson, 1973). Epiboly of the ectoderm eventually serves to close the site of the blastopore, which is marked by the primordial germ cells (Schram, 1986).

Differentiation of the embryo. The external development of crustaceans after gastrulation is a highly complex and varied process, since it includes the details of functional larval elaborations and metamorphosis in species with indirect development and the retention of embryonic anamorphosis, even in those species with direct development (Anderson, 1973). The first notable change is the development of the three pairs of naupliar limb buds ventrolaterally along the sides of the body (Fig. 10.20A and C). Most of the embryo is made up of the rudiments of the three naupliar segments: antennulary, antennal, and mandibular. A relatively short, preantennulary region completes the anterior end, and an even smaller postnaupliar region completes the posterior end. Other than the limb buds, however, the embryo does not manifest any signs of external subdivision. Associated with the ventral concentration of the limb buds and labrum, the dorsal surface of the embryo becomes more convex. In the majority of nauplii, there is no sign of the development of a carapace fold until hatching has occurred and naupliar development is proceeding.

In *Limnadia*, as in *Artemia* (see above), the newly hatched nauplius is structurally and functionally specialized in relation to active swimming and planctotrophic feeding. The newly hatched larva of *Limnadia* is 400 μm long (Fig. 10.21A). After feeding, which occurs during stage III, and metamorphosis which takes place at the fifth moult and proceeding through few moults, it yields a juvenile adult 1 mm long (Fig. 10.21D).

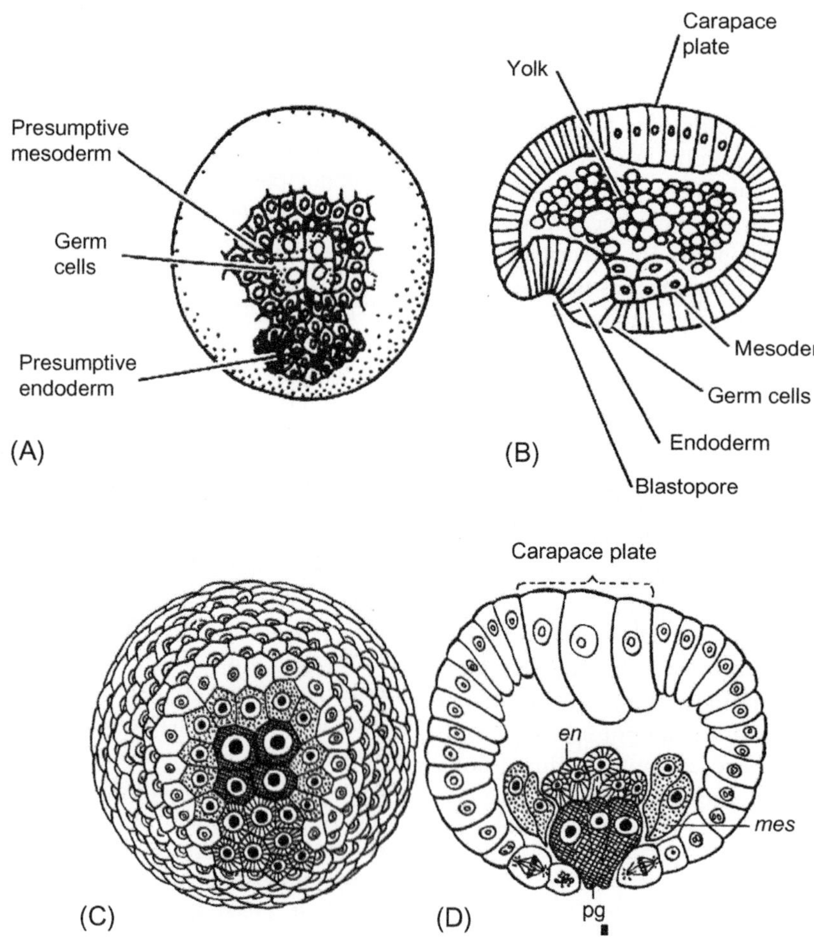

FIG. 10.19 Gastrulation in a cladoceran: (A) blastopore region as gastrulation is about to commence; (B) L.S. of embryo during gastrulation *((A and B) after Grobben, 1879, from Schram, 1986)*; (C and D) gastrulation in *Polyphemus*: (C) at the beginning; (D) median section of the gastrula *((C and D) after Köhn, from Pflugfeder, 1970)*. en=endoderm, mes=mesoderm, pg=primordial germ cells.

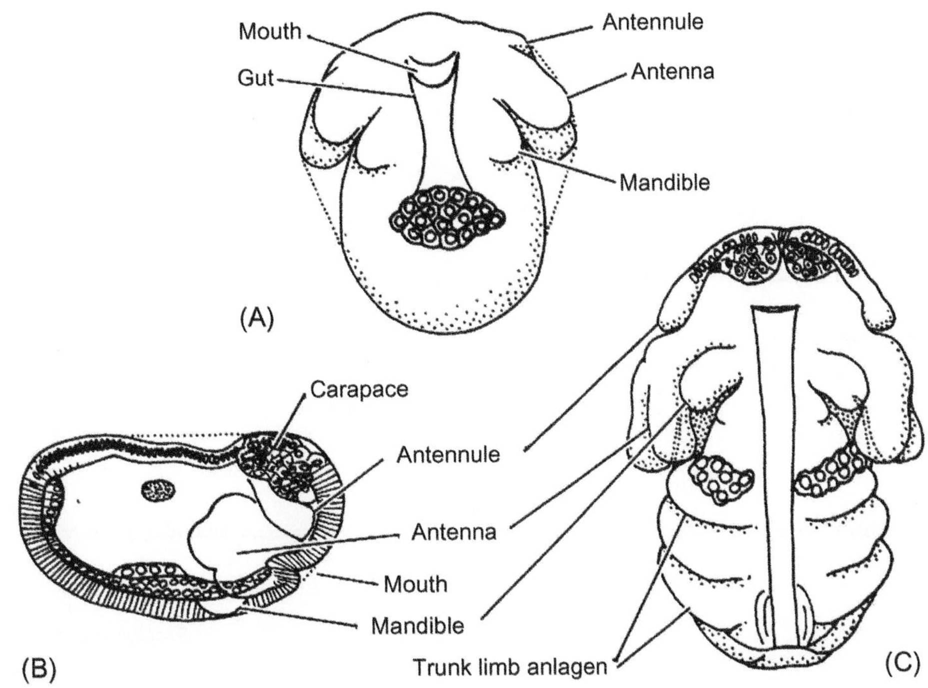

FIG. 10.20 (A) External ventral view of naupliar stage of a cladoceran; (B) L.S. of (A); (C) advanced embryo in a metanaupliar stage. *(After Grobben (1879), from Schram (1986).)*

Except for the great majority of the Cladocera and a few conchostraca (*Cyclestheria*) in which the development is embryonic, the Branchiopoda have a free-swimming nauplius or metanauplius stage (Calman, 1909). This means that the Cladocera also have a direct development in most species. Anderson (1973) recorded that among the conchostraca, there is one species (*Cyclestheria hislopi*) in which the embryo does not hatch as a nauplius larva, but undergoes a more direct development which yields, as a first hatched stage, a juvenile conchostracan with six pairs of functional trunk limbs. This unusual mode of development was described by Sars (1896a). After hatching, juvenile development continues in the usual way, with the addition of further trunk segments and limbs. This unusual mode of embryonization is the dominant feature of development in the Cladocera, *Polyphemus* and *Leptodora* (Anderson, 1973).

Cirripedia (Latin *cirrus*, 'curl', and *pous*, 'foot', includes barnacles)

Fixed animals and for the most part hermaphrodite crustaceans; without compound eyes in the adult; with a carapace (except in rare instances) as a mantle; with usually a mandibular palp, which is never biramous; and typically with six pairs of biramous thoracic limbs.

Thoracica (Greek thorax, *'chest')*

Cirripedia with an alimentary canal; six pairs of biramous thoracic limbs; no abdominal somites; free-living with permanent attachment by the preoral region; and are normally filter-feeders. They are divided into the following:

According to Calman (1909)

Pedunculata, e.g. *Lepas* (the common goose barnacle)
Operculata, e.g. *Balanus* (rock or acorn barnacle)

According to Schram (1986)

Lepadomorpha, *Lepas*
Balanomorpha, *Balanus*

Thoracican barnacles, despite their deviation of form, or perhaps because of it, inhabit almost every conceivable marine habitat from the deep sea to the intertidal, from pole to pole, and from rock solid substrates, to mud bottoms, and even to living tissues (Schram, 1986). These barnacles are usually found in groups attached to such objects as algae, rocks, ships, whales, turtles, decapod crustaceans, etc. Therefore, *Lepas* is found all over the world floating on objects in the sea (Fig. 10.22A and B). It hangs by a stalk or peduncle, which represents the fore-most part of the head, greatly elongated but still bearing at its far end the vestiges of the antennules embedded in a cement by which it is held fast. The glands that produce the cement are contained in the peduncle, and open on the antennules.

Reproductive system

In Cirripedia, the ovaries and testes of an individual animal are separate; hermaphroditism is simultaneous, but self-fertilization does not occur except in the Rhizocephala;

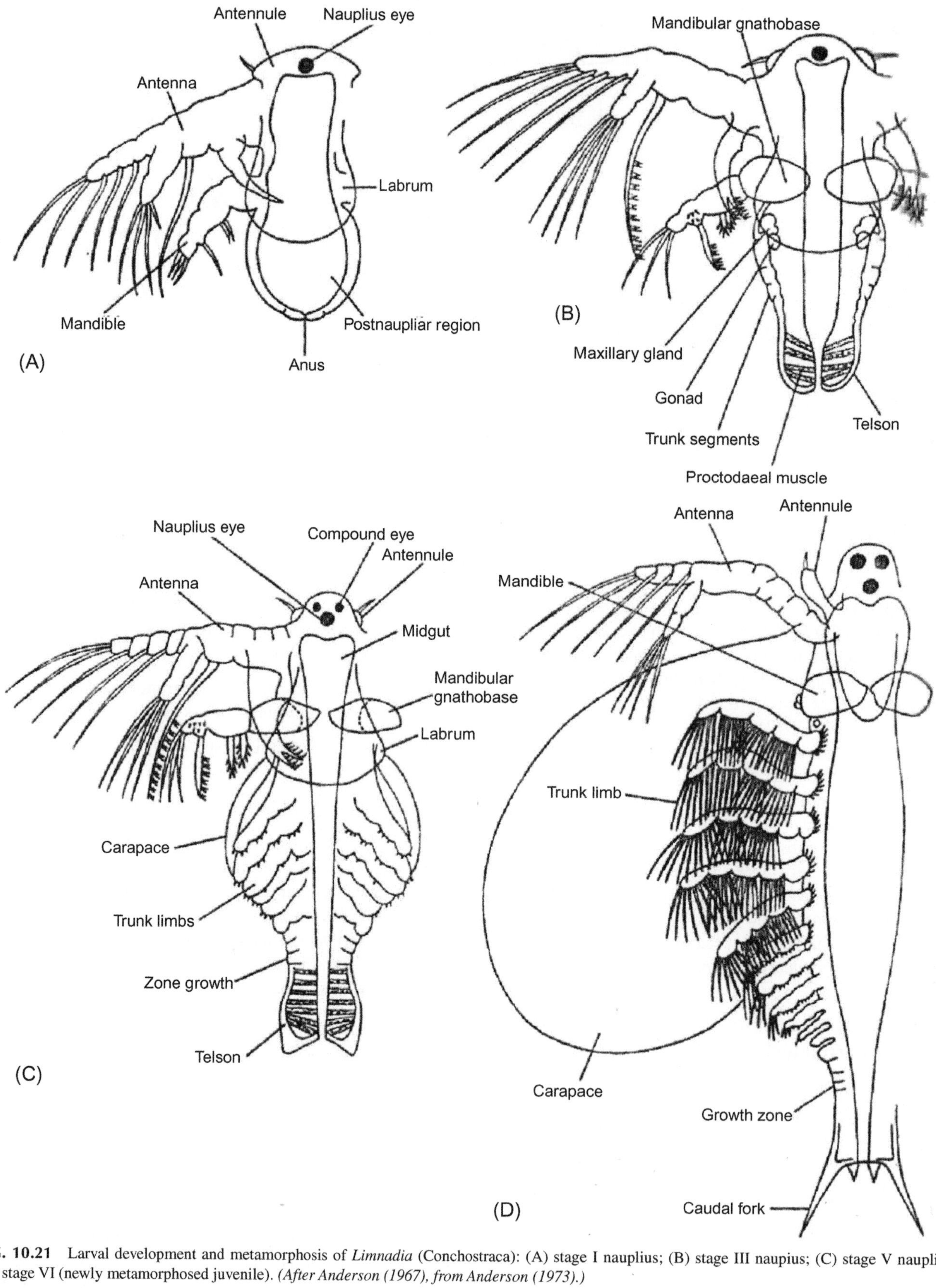

FIG. 10.21 Larval development and metamorphosis of *Limnadia* (Conchostraca): (A) stage I nauplius; (B) stage III naupius; (C) stage V nauplius; (D) stage VI (newly metamorphosed juvenile). *(After Anderson (1967), from Anderson (1973).)*

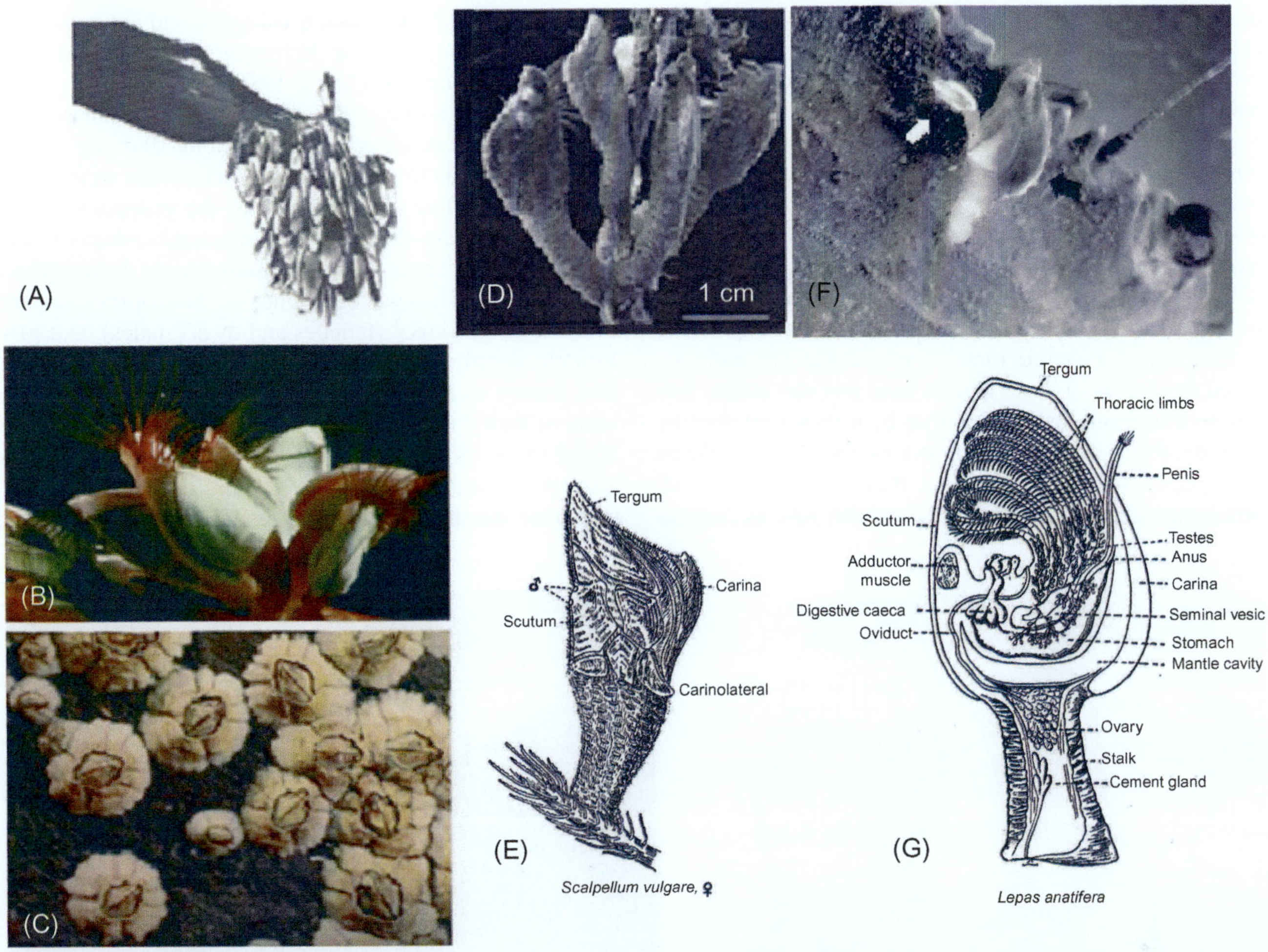

FIG. 10.22 Different species of barnacles: (A) a bunch of stalked or ship's barnacles, *Lepas anatifera*, live specimens attached to a floating bottle *(photo by Wilson, D.P. from Russel-Hunter, 1979)*; (B) individuals of the same species have extended feeding cirri; (C) group of *Balanus tintinnbulum* fixed on submerged, e.g. bottom of boats *((B and C) from Beauté du Monde Animal, vol. 12, Librairie Larousse)*; (D) Photo of group of *Scalpellum scalpellum* hermaphrodites attached to hydroida mainly *Tubularia indivisa (from Buhl-Mortensen and Høeg, 2006)*; (E) female *Scalpellum vulgare* showing site of the males as parasites; (F) Two individuals of *Octolasmis warwickii*, one attached directly on the carapace of the crab *Charybdis japonica (black arrow)*, and the other on the scutum of the former individual *(white arrow) (photo by Ishimura M., from Yusa et al., 2010)*; (G) A view of *Lepas. anatifera* cut longitudinally to show the different organs *(partly after Claus)*. *(E and G) From Borradaile et al. (1977).*

certain genera exhibit a more or less complete reversion towards unisexuality (*Scalpellum*, *Ibla*) (Charniaux-Cotton, 1960a,b).

In the pedunculate cirripedes, the ovaries lie in the peduncle (Fig. 10.22G). The ovaries of *Balanus amphitrite* and *B. eburneus* consist of branched tubules (ovarioles) located in the connective tissue between the mantle cavity and the basal membrane (Fyhn and Costlow, 1977). The oviducts open on the bases of the first pair of thoracic limbs, much farther forwards than is usual in Crustacea. The testes are branched tubes, which lie at the sides of the alimentary canal and in the basal parts of the cirri. Each vas deferens enlarges into a vesicular seminalis, whose duct joins that of its fellow in the penis (Borradaile et al., 1977). The latter is long and thread-like (Calman, 1909), located posterior to the last pair of cirri. It is whiplike organ capable of extending beyond the mantle cavity. Darwin (1851) originally thought this structure is an atrophied abdomen, but Newman et al. (1969) indicated probable homology with the penis cone on the first abdominal segment of ascothoracicans (Schram, 1986).

Dwarf males

In the Operculata and in the great majority of the Pedunculata, all the individuals of a species are similar and are hermaphrodite. In two genera of Pedunculata, *Scalpellum* (Fig. 10.22D and E) and *Ibla*, in a few species of Balanus (Fig. 10.22C) and in the boring Acrothoracica, tiny, mostly nonfeeding male individuals occur (Calman, 1909; Barnes, 1974; Foster, 1983). They are permanently attached within the mantle-cavity of the larger feeding female or hermaphrodite individuals. Darwin (1851) was the first to recognize these tiny cirripede males (Fig. 10.23) and differentiated between those that live within pure females and within hermaphrodites. In the former case, they are **dwarf males** and the latter, which are attached to the hermaphrodite individual, present almost a unique sexual relation, and were termed **complemental males**. Darwin and his successors believed that the primary function of thoracican dwarf males was to complement the reproduction of hermaphrodites at low densities, and hence their name. Both males exhibit successive degrees of degeneration that range from complete but miniature version of the host hermaphrodite to those with degenerate mouth parts. It was generally assumed that larvae of these species have two characters, hermaphroditism and bisexuality; that is, if they settle on the substrate, they become hermaphrodites, and if they settle on a conspecific (of a same species), they become males. All complemental males are small and exhibit various degrees of degeneration. Some have mouthparts and are essentially miniature versions of the hermaphrodite, e.g. males of *Bathylasma alearum* (Schram, 1986).

Since Darwin's investigations on dwarfism in barnacles, from about 150 years ago, there is a long-standing interest among biologists in this aspect. Therefore, the tremendous studies proved that these barnacles—the sedentary marine crustaceans—have the most diverse sexual systems in animal kingdom. These include **simultaneous hermaphroditism**, **androdioecy** (coexistence of hermaphrodites and dwarf males), **dioecy** (females and dwarf males), and **protandric simultaneous hermaphroditism** (the individuals first mature as males and then acquire the female function without losing the male function). This fourth system was added by Kelly and Sanford (2010) and then recorded by Yamaguchi et al. (2012, 2013). Again, these dwarf males, which are attached to hermaphrodites and have the potential

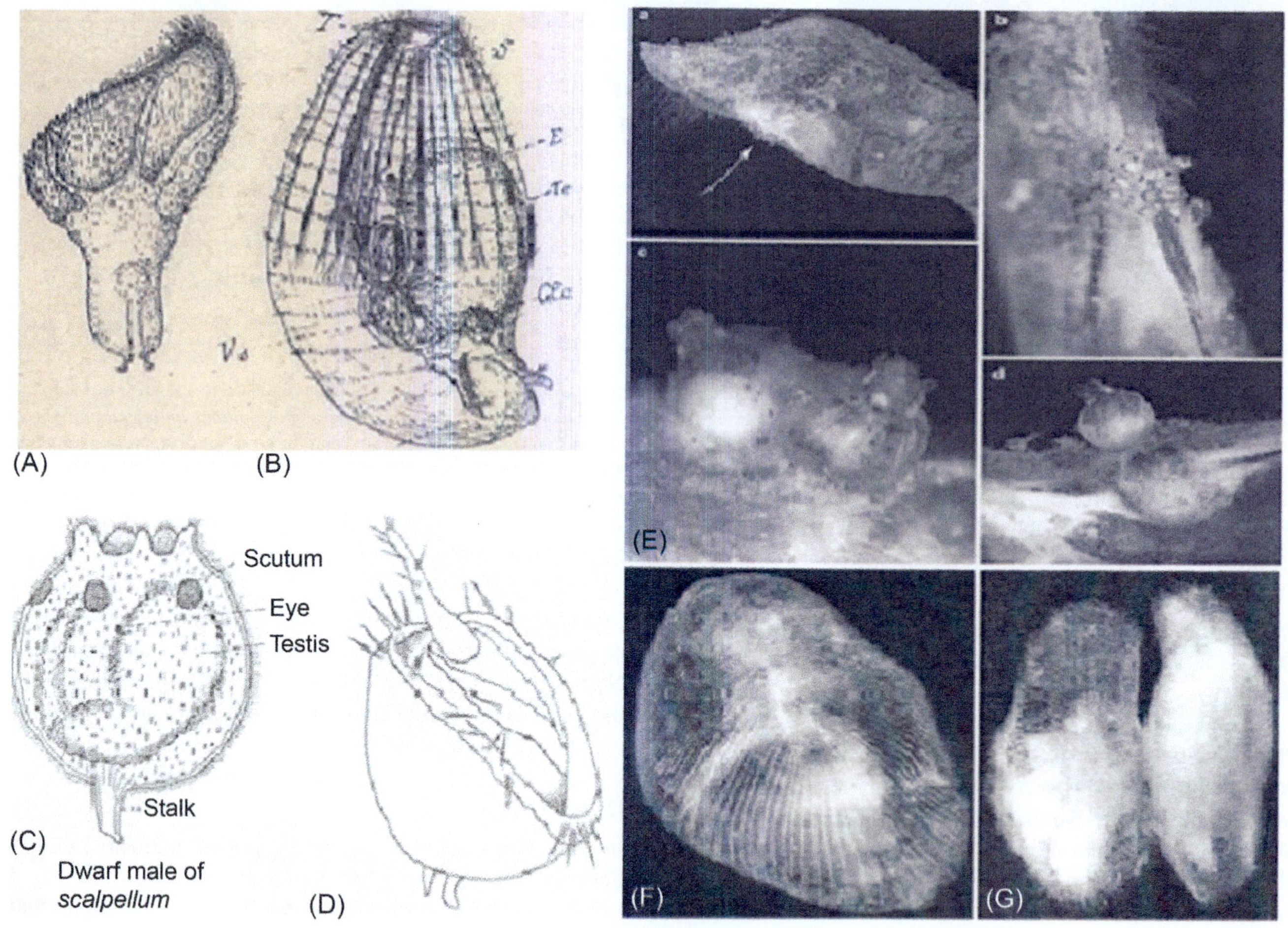

FIG. 10.23 Dwarf males of barnacles: (A) *Scalpellum peronii* (=*Similium peronii* (*Buhl-Mortensen and Høeg, 2006*)); (B) *S. gigas* (=*Trianguloscalpellum gigas* (*Buhl-Mortensen and Høeg, 2006*)). E=stomach, Gl.c=cement gland, S=scutum, T=tergum, Te=testis *((A and B) from Calman, 1909)*; (C) *S. vulgare*, the attachment of males on female is indicated in Fig. 10.22E *(from Borradaile et al., 1977)*; (D) complemental male of a species of *Balanus (from Barnes, 1974)*; (E) Photos of dwarf males of *S. sclpellum*: (a) hermaphrodite with cypris larvae settling close to the receptacle indicated by an *arrow*, (b) magnification of the above area, (c) two dwarf males with four primordial shell plates, (d) two metamorphosed dwarf males in the receptacle of the hermaphrodite; (F) photo showing large-scale line structures on surface of dwarf male of *Ornatoscalpellum stroemii*; (G) two dwarf males, front and side views of *Arcoscalpellum michelottianum*. *(E–G) From Buhl-Mortensen and Høeg (2006).*

to become hermaphroditic, are called **apertural** males (Yamaguchi et al., 2012). Androdioecy is rare among animals, but has been described in more than 30 species of barnacles. Kelly and Sanford (2010) tabulated these androdioecious barnacles with information on habitat/substrate and distribution of each species with the references. They added that the table lists all 13 known androdioecious species outside the Scalpellidae, plus an additional 19 scalpellids where androdioecy has been well described.

In the same year, 2010, Yusa et al. published their studies on *Octolasmis warwickii* (Fig. 10.22F). The larger individuals of this gooseneck barnacle are frequently attached externally on large arthropod hosts (mainly crabs of the family Portunidae) as sublittoral symbiotic species. The authors observed smaller individuals attached to the scutum of larger ones that attached directly on the carapace of the crab *Charybdis japonica*. They proved that these smaller individuals are dwarf males (Fig. 10.22F) having greatly developed testis and penis than that of the larger ones. Therefore, they considered this species, *Octolasmis warwickii*, the first example of androdioecy in the suborder Lepadomorpha, since they elevated its families into suborders. Schram (1986) in his textbook *Crustacea* divided this suborder Lepadomorpa into superfamilies and families, including Iblidae and Scalpellidae. Accordingly, *O. warwickii* becomes additional genus to Lepadomorpha. In addition, the title of their work indicates that the sexual system of this species is androdioecy, while they reported that these 'dwarf males are with a tendency to become hermaphroditic', i.e. the sexual system is protandric simultaneous hermaphroditisn. Therefore, as per their opinion, this needs more elucidation.

As regards the structure of the dwarf males, Calman (1909) reported that these males show great differences even in closely allied species of the same genus. Therefore, the male of *Scalpellum peronii* (= *Smilium peronii* (Buhl-Mortensen and Høeg, 2006)) is an almost perfect miniature of the large hermaphrodite to which it is attached (Fig. 10.23A). The peduncle is distinct, though short, and the capitular plates are well developed. The mouth parts are normal, and the cirri are all present, but are composed of a restricted number of segments. The alimentary canal is complete, and in some cases, even vestiges of the ovaries have been found. In *S. gigas* (= *Trianguloscalpellum gigas* (Buhl-Mortensen and Høeg, 2006)), the peduncle and capitulum are no longer distinct, and the capitular plates are vestigial. The mantle cavity is greatly reduced and the cirri are six pairs of unsegmented processes, each carrying two or three long setae (Fig. 10.23B). A third example is *S. stromii* (= *Ornatoscalpellum stroemii* (Buhl-Mortensen and Høeg, 2006)); the capitular plates have entirely disappeared and the structure further degenerates (Fig. 10.23F). The degree of degeneration of the male appears to be correlated, to some extent, with the place of attachment to the female or hermaphrodite individuals. The less modified males are lodged in 'fossettes' in the margin of the mantle, while those of more pronounced modification are attached within the mantle-cavity below or behind the adductor muscle. More recent studies (Buhl-Mortensen and Høeg, 2006) on reproduction and larval development in three species of pedunculated barnacles (Lepadomorpha) with different depth and distribution added much information. In addition to the ecological data, the sexual system and the dwarf male with its position in the mother animal were strictly defined. In the three species, the males are very small and lack any kind of thoracic appendages. They are essentially sac-shaped with an apical opening that represents the mantle aperture from which the penis probably extends at spawning. The first species, *Scalpellum scalpellum*, is androdioecious with maximum size of its adult hermaphrodite of 22 mm, while its male measures ~300 µm (Fig. 10.23E). Around the aperture of the latter, it carries four tiny shell plate primordia, representing the paired scuta and tergal plates, which never develop beyond this rudimentary state. The second species, *Ornatoscalpellum stroemii* is dioecious, with maximum size of adult female of 11 mm, and that of its male is 800–900 µm (Fig. 10.23F). The latter has a thin cuticle with structures that resemble combs sensu. The third species, *Arcoscalpellum michelottianum* is also dioecious and its adult female of maximum size ~43 mm, and male with size almost twice that of *O. stroemii* ~1.600 mm (Fig. 10.23G). The cuticle of this male gives a 'furry' impression, because it is covered with scales and spines.

All three species have a pair of receptacles (fossettes of Calman) for hosting males, situated one on each side of the orifice at the margins of the scutal plates. The receptacle of *S. scalpellum* is a shallow and furrow-like pocket and rather open to the exterior. The receptacles of this species are preformed and present before settlement of any cyprid males. The maximum recorded number of males in a single receptacle of a sexually mature hermaphrodite was five, and the maximum number for a single hermaphrodite was five plus three. The position of the dwarf male of *O. stroemii* is often marked by a scar or light area on the scutal plate, clearly visible from the outside. However, the pouch-like receptacle is marked by the protruded apertural end of the male, which is completely embedded in the female tissue. The average number recorded of males on mature female was one, since the smallest females, measuring 0.49 and 0.51 cm, lacked males altogether. *A. michelottianum* has very large, deep, and spacious receptacles; a single sexually mature female hosted as many as 12 dwarf males. In both *S. scalpellum* and *A. michelottianum*, a single receptacle can host several males of different ages. In the former species, the multiple males can sometimes sit directly above each other with the topmost one being almost fully exposed and hardly protected at all by the pocket. Therefore, the older and deeper situated males are sometimes dead or at least nonfunctional. However, in *O. stroemii*, the authors never found receptacles hosting more than a single male, but they observed that the receptacle fits tightly around the single male leaving no space for any additional cyprid that might try to settle.

This dwarfism and the various patterns of sexuality in barnacles provide an opportunity for developing and testing theories about evolution of sexuality (Yamaguchi et al., 2008; Kelly and Sanford, 2010; Yusa et al., 2012). The presence or absence of dwarf female/hermaphrodite—the large feeding individual hosting the dwarf males (Buhl-Mortensen and Høeg, 2006)—is the key to the diversity in sexuality among barnacles. Yamaguchi et al. (2008) developed a model that predicts the pattern of sexuality in barnacles by two key environmental factors: food availability and the fraction of larvae that settle on the sea floor. They defined the pattern of sexuality by the combination of the optimal resource allocations. They added that sexuality pattern changes with the food availability through the mating group size, e.g. simultaneous hermaphroditism appears in food-rich environments, where the mating group size is large. On the other hand, androdioecy is observed in food-poor environments and small mating group size. Mating group size (in which barnacles mate with each other) is probably one of the most important factors in animals, that is when males compete for a small number of mates, increasing sperm production (Kelly and Sanford, 2010). The recent theoretical studies of Yamaguchi et al. (2007, 2008) have stressed the role of reduced sperm competition in the evolution of dwarf males in barnacles. In addition, these theoretical studies developed two other theories:

1. **Sex allocation theory**, first developed by Charnov (1980, 1982, 1987) and recently built upon by Yamaguchi et al. (2008), studies allocation of reproductive resources between male and female functions (Yamaguchi et al., 2012). Because sex allocation often depends on body size, this theory was proposed (Yamaguchi et al., 2013).
2. **Life history theory**, another branch of evolutionary ecology, studies allocation of resources between growth and reproduction. Dwarf males are animals that adopt a life history strategy that does not invest growth so much, as well as a sex allocation strategy that does not invest female function.

Therefore, both sex allocation and life history theories are needed to understand male dwarfism, and hence sexual systems in barnacles (Yamaguchi et al., 2012). In addition, life history strategy may affect sexuality even without dwarf males, since it determines the structure of mating groups.

Copulation

Like most crustaceans, Barnacles reproduce by copulation, but they are unique among crustaceans in that the adults are entirely sessile. Contrary to many other marine filtrators, cirripedes do not spawn freely but have copulation between sessile mating partners. The eggs are fertilized within limited brood chambers in the mantle cavity where they are kept until they hatch (Buhl-Mortensen and Høeg, 2006). Also, in Another old view (Calman, 1909), the penis (Fig. 10.24A) can be protruded from the opening of the shell to deposit spermatozoa within the mantle-cavity of an adjacent individual. From Darwin's monograph (1854), the cirripede penis can extend up to eight times an individual's basal diameter (Kelly and Sanford, 2010). Gilbert and Raunio (1997) reported that other species of barnacles have evolved other elaborate ways of getting the sperm to the eggs. In species where males and females may live apart from each other, the male can send out a 'groping penis', which moves about the colony, eventually entering a female barnacle and depositing its sperm (Fig. 10.24B). However, self-fertilization may probably occur in isolated individuals.

Barnes et al. (1977) described experimentally the act of copulation—or, strictly speaking, pseudo-copulation—since the penis only enters the mantle cavity to transfer the semen and not the terminal part of the female reproductive tract. Previously, copulation was considered in relation to cirral beat. However, these authors extended the observations, largely on *B. balanoides* (L.), and reported that the first indication of copulation is penis activity. At the time of copulation, the penis extended, even in its relaxed state, is extremely long—at least as long as the larger cirri, the operculum of the functional male was raised to a high level. Consequently, two events must occur: (1) increasing the volume of the mantle cavity; and (2) cirral activity, initiated by the first action initiates, through relaxation of the depressor muscles. The cirri protruded and the penis was extruded, giving the appearance of being 'unrolled'. During 'searching' and copulation, the cirri were thrust even further back against the carina than in normal extension. Subsequent searching movements over virtually 360 degrees in the horizontal plane and considerable vertical movement are brought about by the activity of the muscles of the penis. In its searching behaviour the penis, fully or partially extended, is moved over the surrounding substrata in its 'search' for a functional female. During this behaviour, the terminal portion of the penis is repeatedly bent downwards to touch the substratum in an apparent testing action. For this, Gilbert and Raunio (1997) described this penis as a 'groping penis'. This searching behaviour would, however, appear to be random rather than immediately directed by any characteristic chemosensory material. A potentially functional female would release this material, which would be perceived by the functional male on first extrusion of its penis. The setae of this penis play an essential role in this recognition. In accepting a penis for copulation, the functional female raises its valves high as during the extrusion of the penis.

In contrast, her cirri are not extruded and may be seen through the opercular opening to remain lightly curled over the body as if the opercular valves are closed. Remembering that these animals are hermaphrodite, these authors stated that mutual copulation in pairs was never observed—an animal acting as a functional male is unable to accept the penis of another functional male. Contrarily, a single functional female will accept several penes at the same time and receive semen from all of them. The observations of these authors proved that single copulation is the most frequent, but the other case is also common and as many as six were seen in one functional female. If the animals are disturbed during copulation, the penis is withdrawn and both male and female will close completely.

Fertilization and brooding

During copulation or just at its end, the 'receptive' *B. balanoides* females have greatly swollen oviducal glands and the oocyte nuclei are in the metaphase of the first meiotic division (Barnes et al., 1977). Each oviduct, just before reaching the exterior, expands to form a genital atrium (described by Darwin as an 'acoustic organ'), with glandular walls within which a sac or sheath is secreted for the reception of the eggs (Calman, 1909). After copulation, the oocytes pass up the paired oviducts and enter the sacs where they are fertilized. This activity causes the expansion of the sac that is pushed through the exit of the oviducal atrium and breaks away from the oviduct. The extruded eggs contained in these sacs

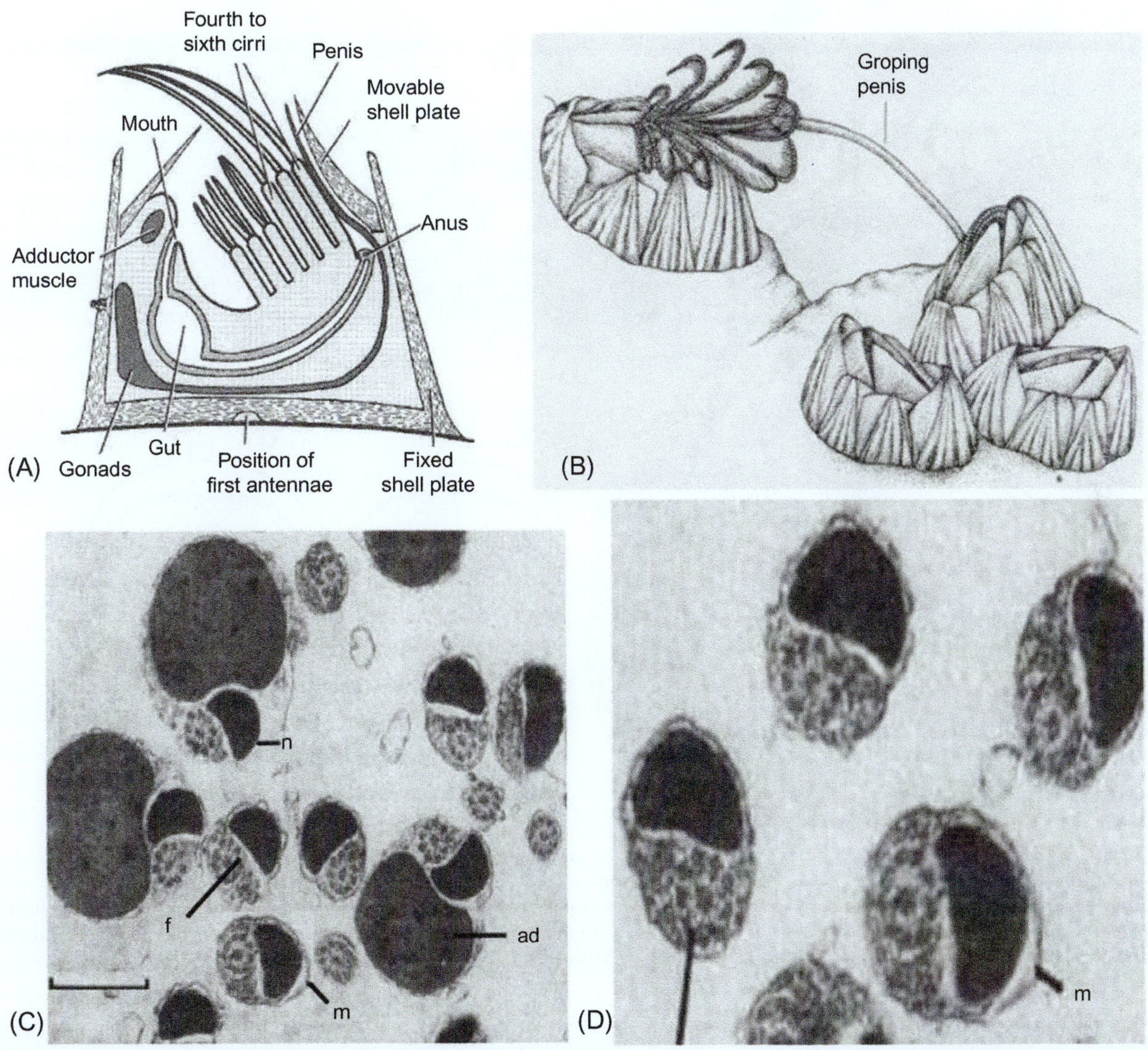

FIG. 10.24 Vital activities of *Balanus*: (A) general anatomy of acorn barnacle, *B. crenata* with penis in a nonfunctional condition *(after Douglas Wilson from Russel-Hunter, 1979)*; (B) groping penis of barnacle *(after Brusca and Brusca, 1990, from Gilbert and Raunio, 1997)*; (C) portion of a TEM section of spermatozoa in normal vesicular semen of *B. balanus*. Scale bar 0.5 μm (Barnes et al., 1977); (D) magnified sperm from the previous figure to show that it is biflagellate. ad = accessory droplet, f = axoneme, m = membrane, n = nucleus.

(Continued)

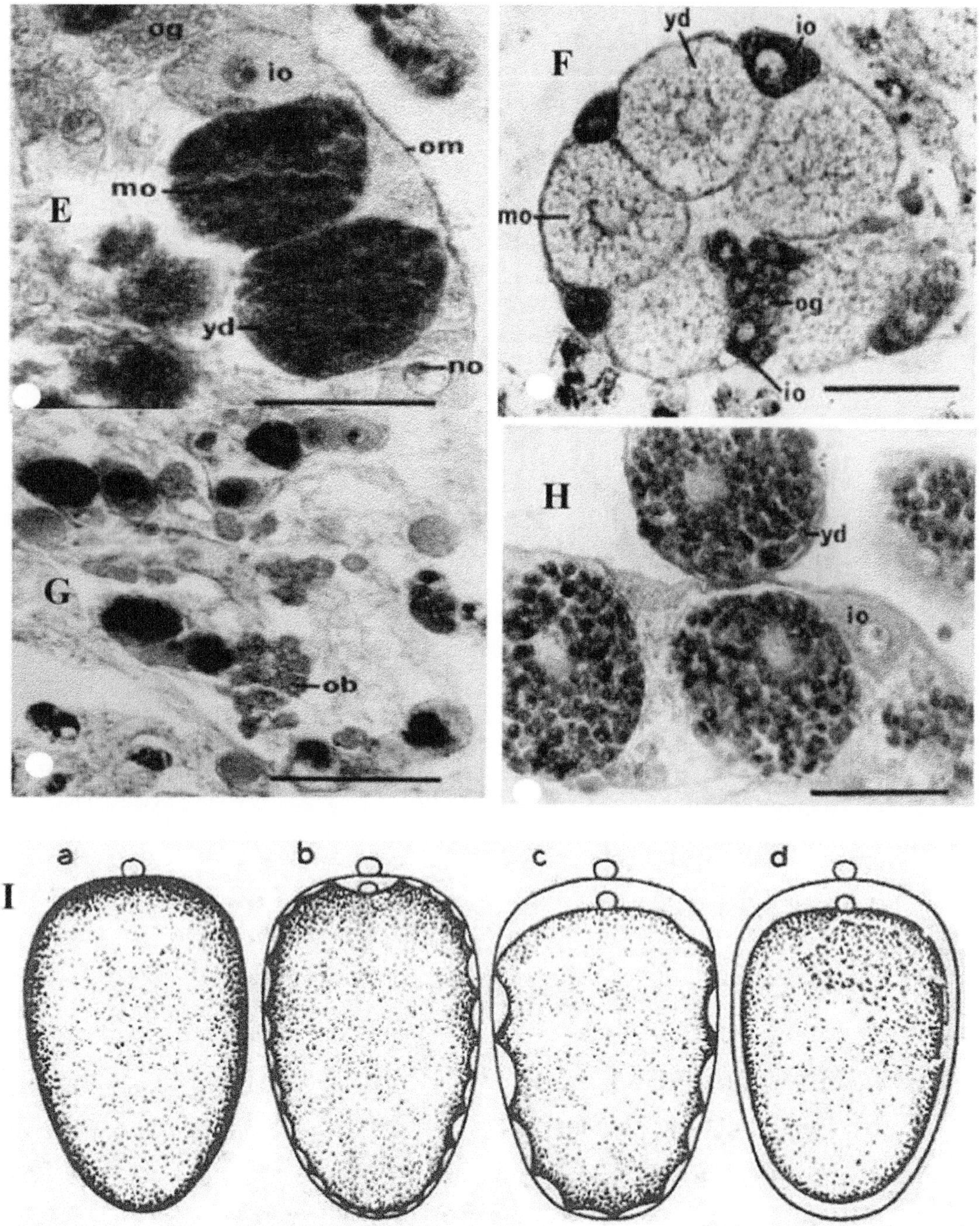

FIG. 10.24, CONT'D (E–G) Histology of ovariole of *B. amphitrite* L.; and (H) of *B. eburneus* Gould *(from Fyhn and Costlow, 1977)*. Scale 50 μm except (G) 150 μm. io = immature oocyte, ob = ovariole buds, og = oogonia, om = ovariole membrane, mo = mature oocyte, no = nucleus, yd = yolk droplets. ((G) shows oocytes in various stages of development); (I) Formation of polar bodies and fertilization membrane in *B. amphitrite albicostatus (after Hudinaga from Kumé and Dan, 1968)*: (a) formation of first polar body, (b) formation of second polar body and beginning of separation of fertilization membrane, (c) appearance of perivitelline space, (d) completion of elevation of fertilization membrane.

form the 'ovigerous lamellae' frequently found within the mantle cavity, where they are retained until hatching takes place (Calman, 1909). Alternatively, these sacs form paired egg lamellae, which are manipulated down the sides of the animal and come to lie free in the mantle cavity (Barnes et al., 1977). These authors observed the events within the mantle cavity after copulation, and the animal after four penis insertions, 3 min total time of insemination: (1) mantle cavity full of semen, oviducal sac formed inside its atrium, no oocytes; (2) after 20 min, lamellae only just visible, eggs pass into the sac, and still semen found in the mantle cavity; (3) after 60 min, excess spermatozoa still expelling, egg lamellae hour-glass shaped being extruded, eggs rounded with first polar body, and egg membrane not raised; after 2 h egg lamellae with different cases, eggs with wavy outline not divided, and first polar body often present; and (4) after 3 h, egg lamellae free and soft, eggs still not divided, membrane quite distinct, and first polar body still present. However, the egg of *Balanus amphitrite albicostatus* (Kumé and Dan, 1968) has no surrounding membrane in the stage immediately following its deposition. The fertilization membrane appears all around the egg at once, after the formation of the first polar body.

Development

Gametes

Roosen-Runge in his book *The Process of Spermatogenesis in Animals* (1977) did not mention the thoracican example. However, he recorded that in Rhizopoda, a parasitic order, a predominant part of spermatogenesis occurs in the female tract. It was revealed that the so-called 'testes' occurred in animals of female type. The larval males (cyprids) deposit masses of germ cells in the female where they develop into functional spermatozoa within sacs, which should be called 'seminal receptacles' or 'spermatothecae'. These transferred cells appear to be in spermatogonial stages. They complete the process in the receptacles, but not all the cyprid germ cells transform into spermatozoa. Some remain on the wall of the receptacle in the manner of supporting cells. The sperm cells are spherical or ovoid, in the Cirripedia they are filiform and motile (Calman, 1909), with a spherical head and a long flagellum. In *Balanus balanus* (Barnes et al., 1977), the nucleus has the shape of a half-sphere and is provided by a very large accessory droplet (Fig. 10.24C), but differs considerably in morphology from ordinary flagellated spermatozoa in that the nucleus has two flagella (Kumé and Dan, 1968). Careful examination of Fig. 10.24C shows two flagella of the sperm of *B. balanus* (Fig. 10.24D).

On the other hand, oogenesis was studied in two balanids, *Balanus amphitrite* and *B. eburneus* (Fyhn and Costlow, 1977). Four types of cells can be distinguished in the developing ovariole. (1) Cells of the ovariole membrane make up a continuous lining of the ovariole and are of squamous epithelium (Fig. 10.24E). Mitotic configurations were not observed in these cells. (2) The oogonia are closely packed cells, making up a continuous string of cells frequently seen in mitosis. Each is large rounded cell with a diameter of about 2 μm (Fig. 10.24F). (3) Immature oocytes (primary oocytes) have a large nucleus with one nucleolus (Fig. 10.24F). Chromosome structures or premeiotic configurations were not clearly seen. (4) The largest immature oocytes have a diameter larger than 40 μm and cytoplasm that is more coarsely granulated. Mature oocytes are approximately spherical with a diameter of 45–70 μm. The nucleus has one nucleolus, but meiotic chromosomes were not observed. The cytoplasm is filled with spherical yolk droplets (Fig. 10.24F and H). In most oocytes, the droplets were evenly distributed throughout the cytoplasm. In some other oocytes, the periphery of the cytoplasm was free of droplets and they were significantly smaller and less clear in appearance. No accessory cells, a nurse-cell function, or a degeneration of immature oocytes were described. In this work, some histochemical tests were applied to identify the yolk droplets and their structure. It was investigated that the barnacle oocyte takes up nutrients through the cell membrane directly from its surroundings complying with the definition of an autonomous yolk formation. Therefore, vitellogenesis seems to be initiated predominantly in postecdysis and is completed during one intermoult cycle. Barnes et al. (1977) confirmed experimentally that in *Balanus amphitrite* and *B. balanoides*, and perhaps in Cirripedia in general, the oocytes do not leave the ovary before copulation.

Kumé and Dan (1968) illustrated and described how the egg of *Balanus amphitrite albicostatus* has no surrounding membrane in the stage immediately following its deposition. The fertilization membrane makes its appearance all around the egg at once, after the formation of the first polar body (Fig. 10.24I). In contrast to the smoothness of the membrane, the peripheral layer of cytoplasm has an irregular surface, retaining points of contact with the membrane. Following the emission of the second polar body, this layer is completely detached from the membrane and its surface again becomes smooth.

Among the Cirripedia, the newly laid egg is oval, and has one blunt and one more pointed end (Figs 10.25 A and 10.26(I)), like a hen's egg; the blunt end includes the animal pole and the pointed end will be posterior in the developing embryo. The length of the egg varies (e.g. *Tetraclita rosea* 215 μm; *Balanus balanoides* 300 μm (Anderson, 1969)), the egg size of *S. scalpellum* is 300–500 μm, and in *O. stroemii* the eggs or embryos contained in the mantle cavities of berried females had a mean length of 520 μm (Buhl-Mortensen and Høeg, 2006). However, the long axis is always the anteroposterior axis of development (Anderson, 1973). Before the maturation divisions end, the oocyte of *Balanus amphitrite rosa* is centrolecithal, and the granular yolk fill the space between the peripheral protoplasm and the central distributed oil spherules. About the time when the polar bodies are given off, the two protoplasmic regions unite and the confluent mass flows towards the animal pole. Afterwards, the nucleus returns to the centre of the egg and the oil spherules disperse and disappear, so the yolk structure becomes homogeneous. In *Mitella mitella* (Fig. 10.25), when cleavage begins, the oil spherules concentrate towards the vegetal pole so that the egg appears to be telolecithal.

Cleavage. The most recent and detailed study of thoracican development is that of Anderson (1969). Anderson reexamined and carefully analysed the cell lineages of three

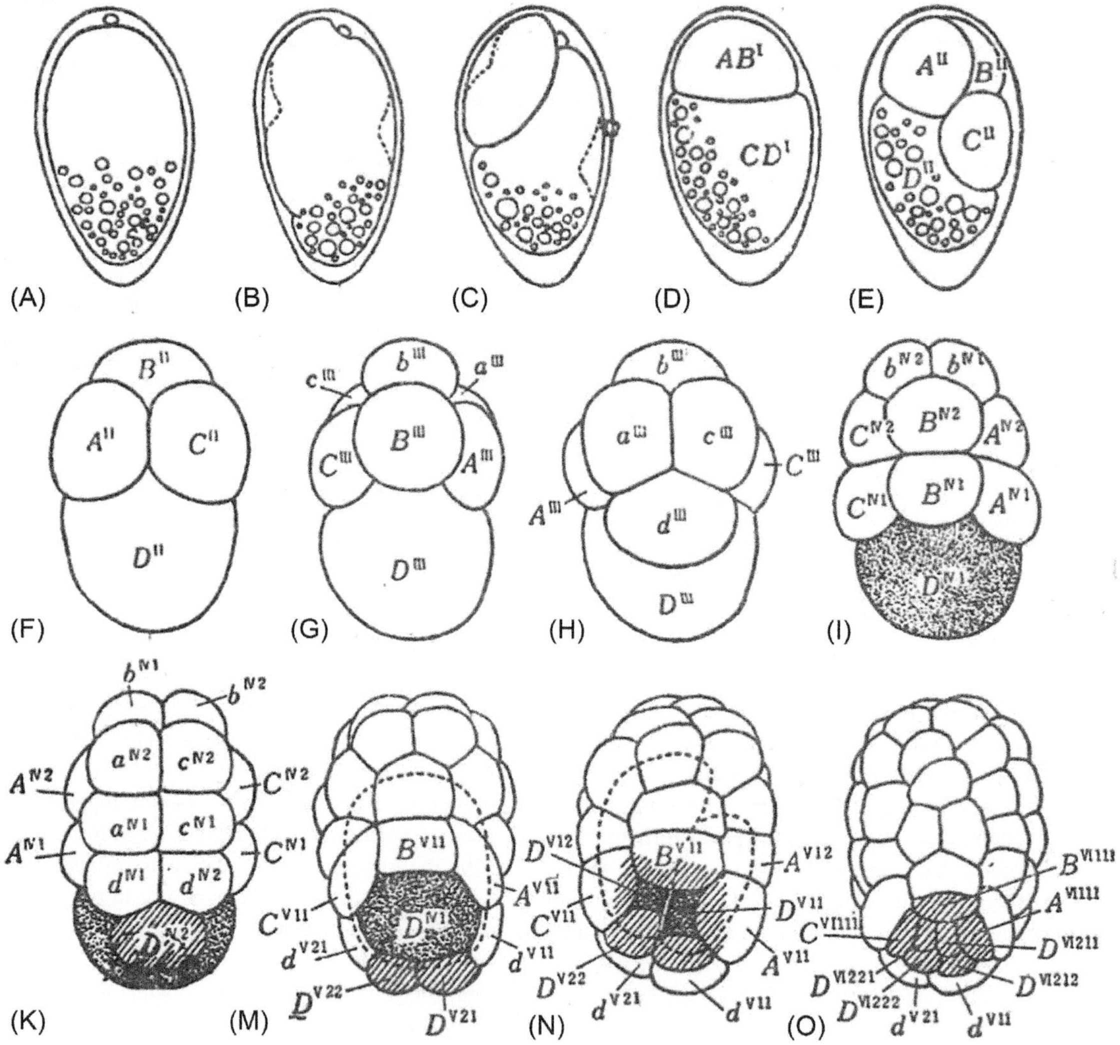

FIG. 10.25 Cleavage in *Mitella mitella (after Shiino, from Kumé and Dan, 1968)*: (A) uncleaved egg; (B) beginning of first cleavage; (C) early 2-cell stage; (D) late 2-cell stage; (A–D lateral views); (E) early 4-cell stage (animal pole); (F) late 4-cell stage; (G) 8-cell stage (vegetal pole); (H) 8-cell stage (animal pole); (I) 16-cell stage (vegetal pole); (K)16-cell stage (animal pole); (M) 31-cell stage (beginning of gastrulation, vegetal pole); (N) 32-cell stage (beginning of closure of blastopore); (O) 60-cell stage (sinking of mesoderm). Cross-hatched area = endoderm, Oblique-lines = mesoderm.

balanomorph barnacles; *Tetraclita rosea* is one of them. The study of this early ontogeny has revealed that cleveage is not only total (as in *Scalpellum*), but it is also uniquely primitive in that it is of a modified spiral type. Two examples will be presented in this work, that of *Mitella mitella* (Fig. 10.25) and *Tetraclita rosea* (Fig. 10.26). All the cirripedes brood their embryos. The fertilized eggs are kept within the receptacles until they hatch (Buhl-Mortensen and Høeg, 2006). The **first** cleavage begins with the rotation of the egg within the egg membrane, as evidenced by the displacement of the polar body along the egg surface from the blunt to the pointed end. This division is also accompanied by a segregation of the yolk towards the posterior half of the egg. A third change is the rotation of the mitotic spindle to appear oblique to the long axis, becomes vertical to it by the end of this stage. The cleavage thus appears to be equatorial, cuts the egg to separate an anterior at the blunter end yolk-free cell (AB) from a posterior pointed end, yolk-filled cell (CD) as Wilsonian spiral cleavage terminology (Figs 10.25A–D and 10.26(I), (II)). The **second** cleavage is also meridional, divides AB equally into A, which pushes backwards to a left lateral position, and B, which retains an anterior position. CD, in contrast is divided into a second yolk-free cell C anteriorly in the right and D, which receives a much greater part of the amount of yolk, so it becomes the large yolk-filled cell in the posterior position (Figs 10.25E, F and 10.26(II)–(V)). The furrows cleaving AB and CD are not in a single line, but meet the first furrow at separate points. Following the second cleavage, A and C move around the polar axis in a counter-clockwise direction as viewed from the animal pole, so they become bilaterally arranged and then come in contact with each other dorsally

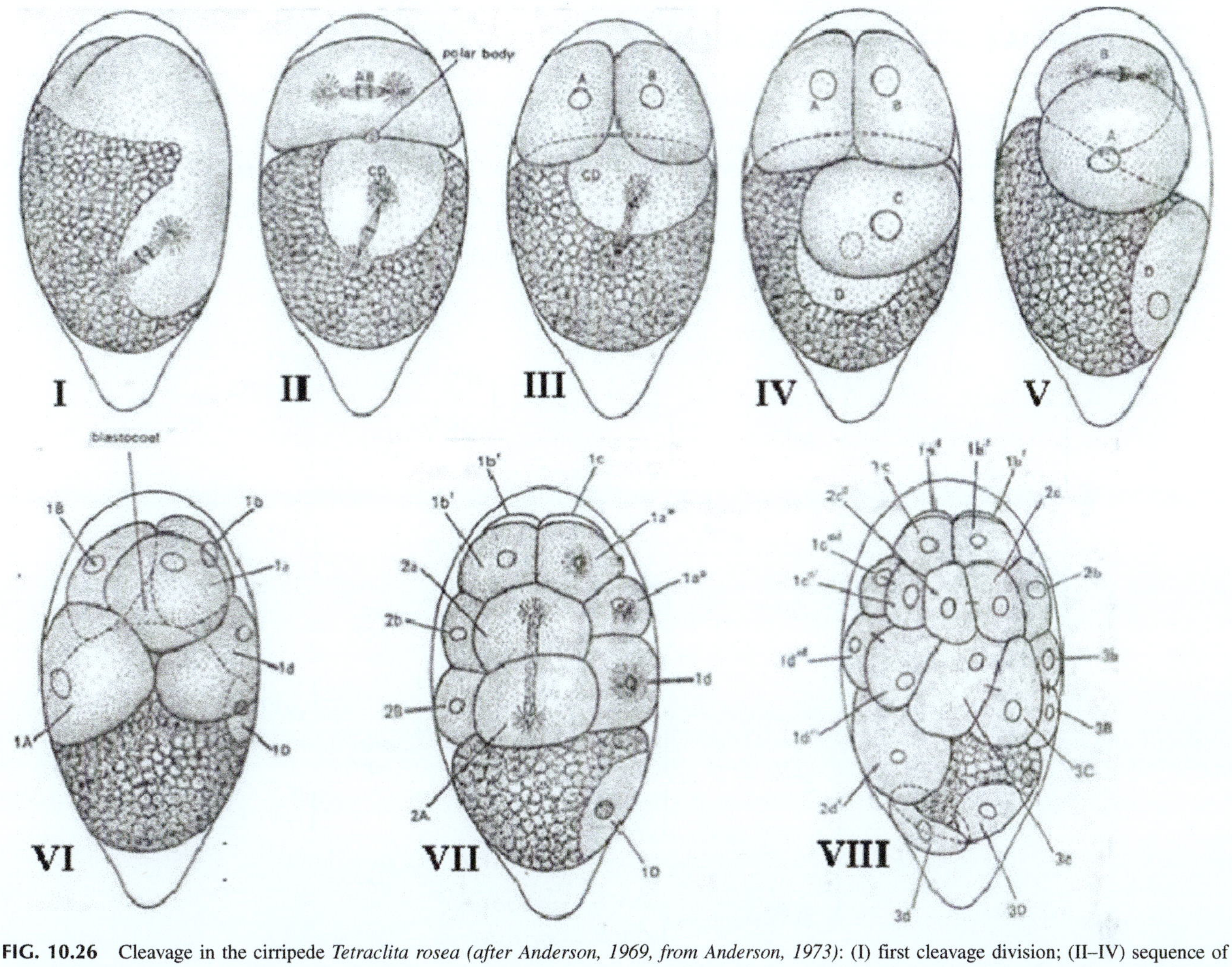

FIG. 10.26 Cleavage in the cirripede *Tetraclita rosea (after Anderson, 1969, from Anderson, 1973)*: (I) first cleavage division; (II–IV) sequence of second cleavage, dorsal view; (V) 4-cell stage, left lateral view; (VI) 8-cell stage, left lateral view; (VII) fourth cleavage, left lateral view; (VIII) 28-cell stage, right lateral view.

in animal pole. The B and D are in contact at the vegetal pole ventrally. All four quadrants proceed through three division cycles: the third, fourth, and fifth cleavage divisions. The furrow that divides each cell is perpendicular to the previous one in the same sequence of typical spiral cleavage. The **third** cleavage is primarily latitudinal. A divides almost synchronously, giving rise to 1A–1C + 1a–1c, while D divides somewhat later and unequally, cutting off dorsally 1d which is similar in size to the other micromeres, all forming the 1a–1d (first quartette) (Figs 10.25G, H and 10.26(VI)). In this 8-blastomere stage, 1a–1c are located at the anterior end of the embryo and they are ectoblasts. The 1A–1C as well as 1d are mesectoblasts. In the **fourth** cleavage, the second quartette is produced (2a–2d), but 1D again lags behind the others, producing the dorsal micromere 2d and the ventral macromere 2D (Figs 10.25I, K and 10.26(VII)). The descendants of 2d later become the mesoderm and are called the primary mesoblast to be distinguished from the mesoderm, which will be derived from the mesectoblasts. The 2D is the primordial endoblast (Kumé and Dan, 1968). At the end of this division, the micromeres begin to move over the macromere 2D towards the posterior. This change is more obvious on the dorsal than on the ventral side.

During the divisions, third to fifth, other features of spiral cleavage are maintained. The most characteristic feature, which appears at the end of the **fifth** cleavage, is the segregation of the ectodermal, mesodermal and midgut rudiments from each other (Table 10.4). Therefore, in cirripede cleavage, in fact, is a functional modification of spiral cleavage appropriate to a small but densely yolky egg. Cleavage proceeds in such a manner that, during the first five division cycles, no cleavage furrows pass through the yolk. During the third to fifth division cycles, the division products of the yolk-free (B-cell) spread ventrally backwards over the D-quadrant yolk cell to cover the anteroventral and ventral surfaces of the yolk cell (Fig. 10.26(VIII)). Accompanying the division of B cell, the daughter cells of the yolk-free

TABLE 10.4 Cell-lineage of *Mitella mitella*

	2-cell stage	cell stage	8-cell stage	16-cell stage	32-cell stage	62-cell stage
egg	AB'	A''	a''' — prim. ect.			
			A'''	AIV^1	AV^{11}	AVI^{111} — sec. mes.
						AVI^{112} — sec. ect.
					AV^{12} — sec. ect.	
				AIV^2 — sec. ect.		
		B''	b''' — prim. ect.			
			B'''	BIV^1	BV^{11}	BVI^{111} — sec. mes.
						BVI^{112} — sec. ect.
					BV^{12} — sec. ect.	
				BIV^2 — sec. ect.		
	CD'	C''	c''' — prim. ect.			
			C'''	CIV^1	CV^{11}	CVI^{111} — sec. mes.
						CVI^{112} — sec. ect.
					CV^{12} — sec. ect.	
				CIV^2 — sec. ect.		
		D''	d'''	dIV^1	dV^{11}	dVI^{111} — sec. mes.
						dVI^{112} — sec. ect.
					dV^{12} — sec. ect.	
				dIV^2	dV^{21}	dVI^{211} — sec. mes.
						dVI^{212} — sec. ect.
					dV^{22} — sec. ect.	
			D'''	DIV^1 — endoblast		
				DIV^2 — mesoblast		

After Shiino from Kumé and Dan (1986).

A and C cells also spread backwards along the sides of the yolk cell. At the end of fifth cleavage division, these cells cover the yolk cell anterodorsally and laterally. During these events, this yolk cell undergoes the same divisions to cleave in such a way as to cover its dorsal and posterior surfaces with its own yolk-free daughter cells. Then it becomes 4D in the **sixth** cleavage. No cleavage furrows have passed through this cell and it has lost little in size as compared with the initial zygote. At the same time, 4D has become covered, except posteroventrally, by yolk-free blastomeres. The result is a **blastula** in which all cells lie at the surface but one cell, the posteroventral 4D, extends inwards to fill the blastocoel.

Lepas anatifera is very similar to *Mitella* in its mode of cleavage, but differs only in that all the descendants of 1d develop into ectoderm. The differences between the mode of cleavage of the Cirripedia and of *Polyphemus* appear in the following: (1) 1D gives rise to a primordial endoblast and primary mesoblast in Cirripedia but in *Polyphemus* it produces a primordial endoblast and a primordial germ cell; (2) the mesoderm cells differentiate in the sixth cleavage in the former, and with the seventh in the latter (Kumé and Dan, 1968).

The major variations in the presumptive areas of Crustacea are found in that of midgut and extra-embryonic ectoderm, are the two areas most affected by changes in the volume of yolk in the egg. The mesoderm, stomodaeum, and proctodaeum are conservative in their formation and fates throughout the Crustacea. An early segregation of primordial germ cells as a definite presumptive area does not occur in cirripedes or other Malacostraca. Cirripedes provide the most basic example of the composition, origins, and relative juxtaposition of presumptive areas in the crustacean blastula (Fig. 10.7A).

Presumptive midgut. The presumptive midgut comprises the single large, yolky cell 4D. This cell fills the interior of the blastula, but is exposed at the surface posteroventrally (Fig. 10.27(I)). Its descendants constitute the rudiments of the anterior and posterior midgut.

Presumptive stomodaeum and proctodaeum. In the Cirripedia, the origin of the presumptive stomodaeal cells can be traced to the ventral, superficial cells of the B quadrant, 2b and 3b. The presumptive stomadaeum in all crustaceans is formed by the equivalent of the blastula or blastoderm. The proctodaeum always develops late, after the naupliar segments have undergone preliminary development. In all crustaceans, the presumptive proctodaeum is a small group of superficial cells in the posteroventral midline of the presumptive ectoderm, behind the presumptive ectoderm of the postnaupliar growth zone.

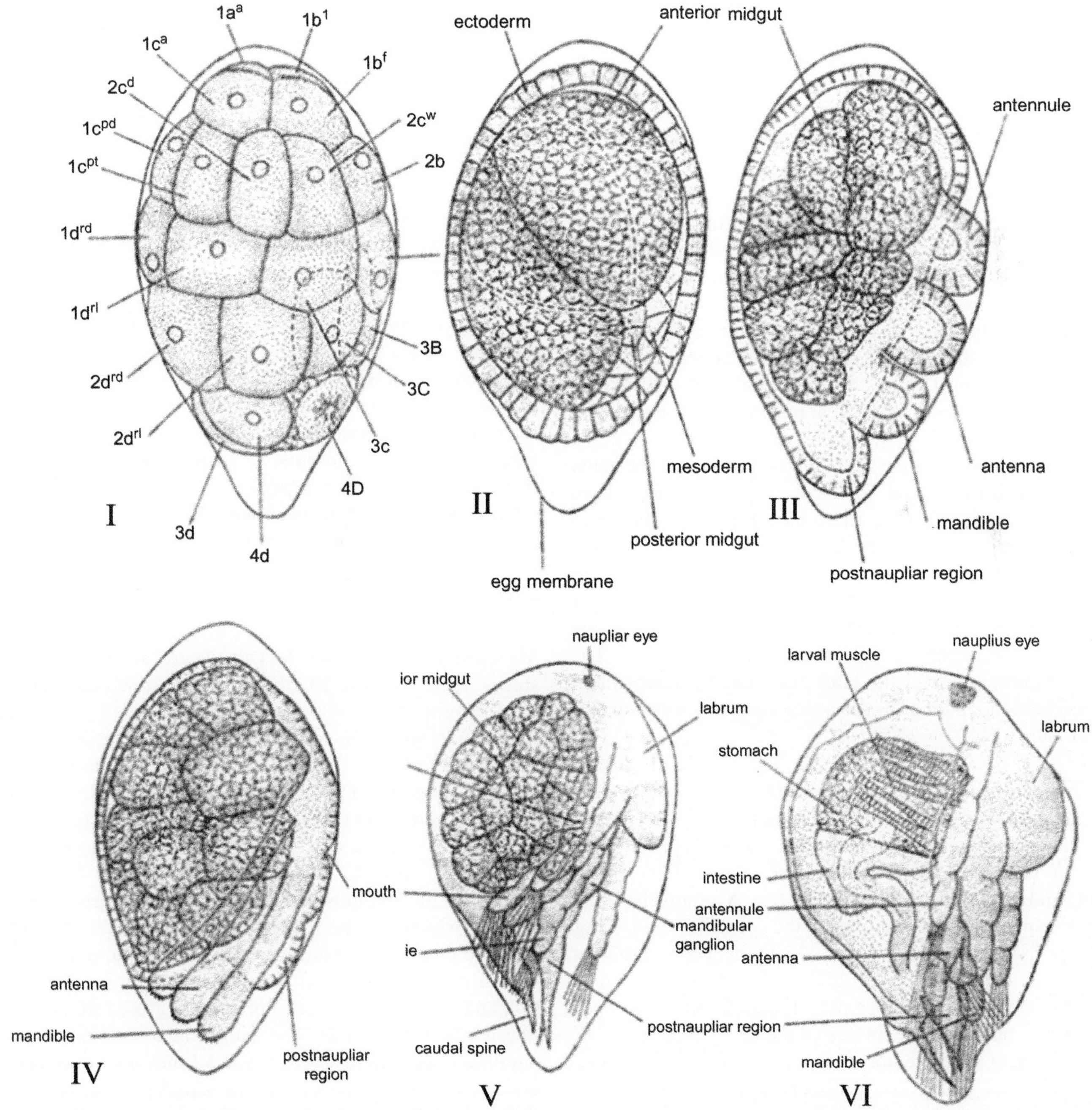

FIG. 10.27 Gastrulation of the embryo of *Tetraclita rosea (after Anderson, 1969 from Anderson, 1973)*: (I) 31–33-cell stage, right lateral view showing onset of mesodermal immigration; (II–VI) external development of the embryo after the completion of gastrulation.

Presumptive mesoderm. The presumptive mesoderm lies in the ventral midline as an arc of superficial cells in front of the presumptive midgut (Anderson, 1969). The three cells which comprise this arc are the cells 3A, 3B, and 3C (Fig. 10.25I). In Cirripedia, the 4d cell has no role in presumptive mesoderm formation and is merely one of the many cells that constitute the presumptive ectoderm.

Presumptive ectoderm. In the Cirripedia, it is a layer of superficial, yolk-free cells laid down around the yolk cell 4D (presumptive midgut) during cleavage. The basic cells that constitute the presumptive ectoderm are the cells of the first three quartettes 1a–1d, 2a–2d and 3a–3d, together with 4d. The presumptive stomodaeum is also of 2b and 3b origin; and probably the presumptive proctodaeum is of 4d origin (Anderson, 1969).

Gastrulation. As was mentioned above, gastrulation in the Crustacea is a relatively minor event. Again, gastrulation in Cirripedia can be taken as a starting point. In *Tetraclita* and other similar thoracican embryos, it begins when the fifth cleavage division has been completed. The first step of gastrulation (Fig. 10.27(I)) is an inward and forward migration of the three presumptive mesodermal cells 3A, 3B, and 3C beneath the ventral ectoderm cells in front of them. During this event, the yolk-cell 4D divides equally into two, by an approximately longitudinal division. The two presumptive posterior midgut cells, which lie behind the presumptive mesoderm, at the same time, slip inwards and forwards to join the mesoderm cells in the interior. The three mesoderm cells begin to divide, and the presumptive embryonic ectoderm bordering both mesoderm and posterior midgut spreads to cover their small area of the ventral surface. The major events of gastrulation are then complete and the external surface of the embryo retains no surface openings. The presumptive stomodaeum later gastrulate by independent invagination, and that of the proctodaeum is independent and later still.

Differentiation of the embryo. In Cirripedia, the naupliar limb buds occupy the major part of the body length of the embryo, which means that the embryo is made up of the three naupliar segments: antennulary, antennal, and mandibular (Fig. 10.27(III)). A relatively small postnaupliar region completes the posterior end. This region is the rudiment from which the postmandibular segments develope anamorphically during postembryonic development. No intersegmental annuli are developed over the convex dorsal surface of the developing larva. As development of the embryo continues, the naupliar limb buds elongate and protrude, usually in a posteroventral direction. At an early stage of their elongation, their bases rotate and/or move so as to become crowded around the mouth. At the same time, the labrum arises as a protrusion in front of the mouth. Afterwards and during the elongation of the limb buds, they begin to show the external annulations indicating their subdivision into podomeres, and the antennae and mandibles become biramous (Fig. 10.27(V) and (VI)). Secretion of the stage I nauplius cuticle begins at about this stage, accompanied by the onset of pigmentation of the median eye (nauplius eye). The cirripedes are exceptional in developing a pair of projecting frontal horns at the sides of the preantennulary region. The postnaupliar region then subdivides into a narrow dorsal protrusion forming the median caudal spine and a large caudal papilla. Barnes (1974) and Schram (1986) recorded that the developing eggs are brooded within an ovisac in the mantle cavity in all barnacles, while the view of Buhl-Mortensen and Høeg (2006) is that 'all cirripedes brood their embryos'. Hatching is achieved when the adults produce a diffusible factor (probably a lactone or lactan) that stimulates the embryo to gyrate and break typically **I nauplii**. A single gravid individual may release over 13,000 larvae.

Following hatching, the larvae are almost immediately expelled into the plankton. In the ground pattern, larval development comprises a series of planktotrophic naupliar stages followed by the nonfeeding settlement stage, called the sypris (Calman, 1909). However, many species have nonfeeding (lecithotrophic) nauplii, e.g. the three scalpellid barnacles studied by Buhl-Mortensen and Høeg (2006), *S. scalpellum*, *O. stroemii* and *A. michelottianum*. However, *S. scalpellum* and *A. michelottianum* release pelagic nauplii that spend more than 10–15 days at 15–16 °C in the plankton in the former, and several weeks, while *O. stroemii* release cyprids that settle soon after release.

Metamorphosis of the larvae. There are six naupliar stages (Fig. 10.28A–F), the first stage, which usually has a length of 200–400 μm subsisting on yolk. After the first moult, which occurs within 2 h of hatching, and the development of the gut are completed, the nauplii take up active feeding. The naupliar form is quite distinctive, characterized by: (1) prolonged caudal process; (2) paired anterolateral horns; (3) frontal filaments anterior to the antennules; and (4) triangular shield-shaped carapace. The horns have glands that produce a water-insoluble secretion, which may be for protection, or acts as a wetting agent to prevent the nauplius from getting trapped in surface films. The frontal filaments have complex innervation and a possible ion receptor was suggested that in turn might affect phototaxis. The nauplii bear well-developed naupliar eyes, and in stages V and VI they also have compound eyes. The stage VI nauplius usually varies between 500 and 1200 μm in length in different species. By stages

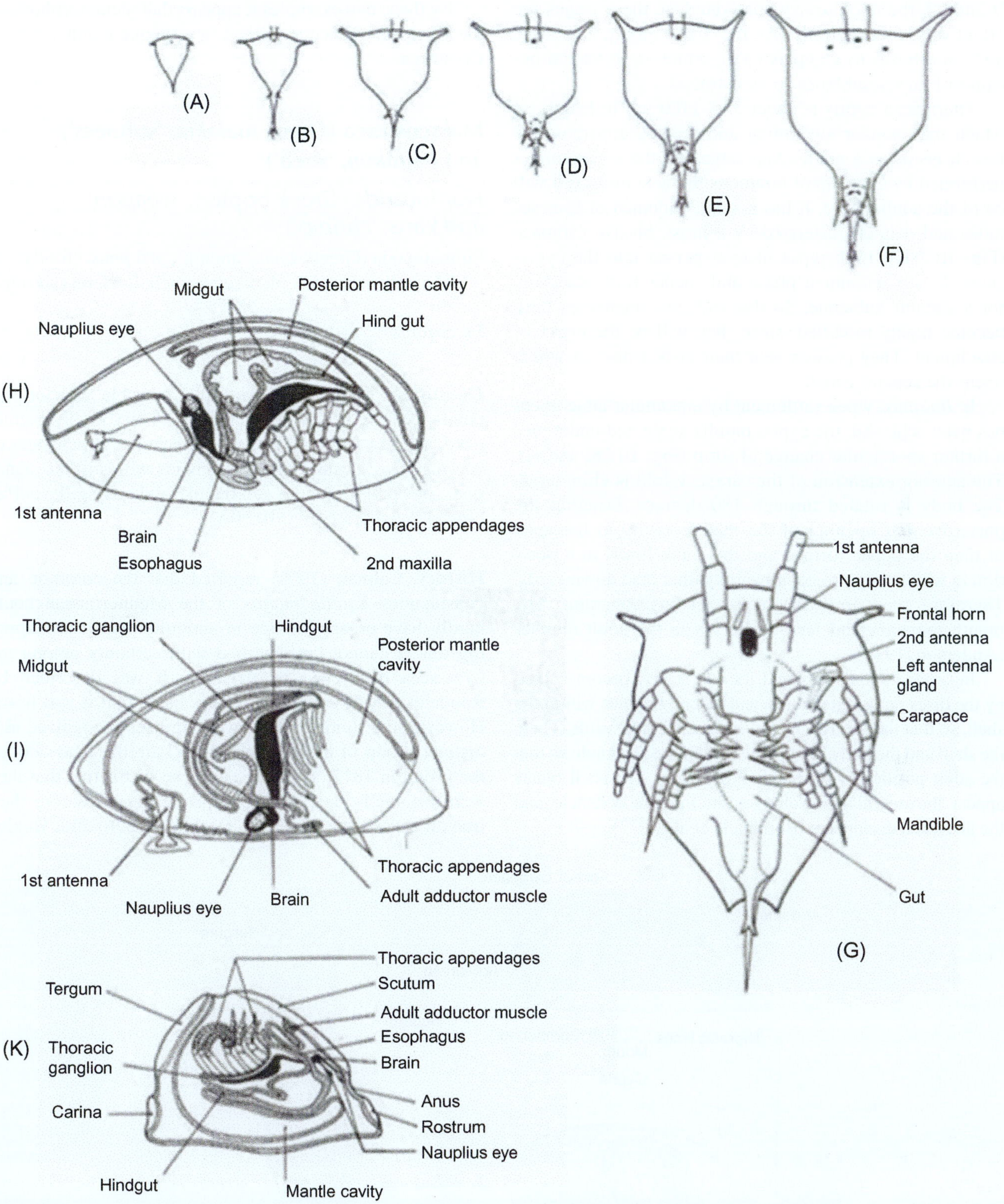

FIG. 10.28 Larval development: (A–F) body outline of six naupliar stages of *Balanus nubilis (from Schram, 1986)*; (G) nauplius larva of *Balanus*, setae omitted; (H) free-swimming cypris larva; (I) cypris shortly after attachment; (K) young barnacle *((G–K) from Barnes, 1974)*.

V and VI, the teloblasts are activated, so these stages are effectively **metanauplii**. At the sixth moult, the larva metamorphoses to a **cypris** larva, which is so named because of its resemblance to an ostracod.

Thoracican cypris is about 500–1100 μm in length, in which the naupliar swimming and feeding apparatus are lost. It becomes a nonfeeding stage and the swimming is performed by six pairs of biramous thoracic limbs (all will be of the adult stage). It has a small abdomen of four somites and it is characterized by a large, bivalve carapace (Fig. 10.28H). Both types of eyes persist into the cypris stage. It is a settlement phase and spends time searching for a suitable substrate. To this end, the antennules have become highly modified, since they will be the organ of attachment. They possess near their ends a disc on which opens the cement gland.

In *Balanus*, when settlement by antennular attachment has been achieved, the cypris moults again and undergoes a further spectacular change of form (Fig. 10.28I and K). The anterior extension of the carapace fold is eliminated. The body is rotated through 180 degrees, bringing the posteroventral opening of the mantle cavity to the apex of then the upper surface and the trunk limbs to a position in which they project upwards rather than downwards. Therefore, these limbs become a feeding apparatus. The mouth reopenes and feeding begins in the adult manner (Anderson, 1973).

In *Lepas*, on the other hand, the cypris larva becomes fixed by the discs on its antennules, and its body rotates within the shell, so that the ventral surface is directed backwards. Then, the shell and the body rotate upwards on the antennule so that the adult position is assumed (Fig. 10.29). The shell plates appear, the preoral region elongates to form the peduncle, and the abdomen disappears (Borradaile et al., 1977).

By these two examples it appears that metamorphosis of the Cirripedia is accomplished in a unique manner among Crustacea.

Malacostraca (Greek *malakia*, 'softness', and *ostrakon*, 'shell')

Hoplocarida (Greek hoplon, *'weapon', and* karis, *'shrimp')*

Stomatopoda (Greek *stoma*, 'mouth', and *pous*, 'foot').

The only order of hoplocaridans include Mantis shrimps.

Ex. *Squilla*

Definition. Marine; head with two movable anterior segments, bearing eyes and antennules; thorax highly tagmatized, anterior five segments with subchelate appendages of varying sizes; posterior three segments with stiltlike stenopods; pleopods biramous and flaplike and carry gills; tailfan typically elaborate (Fig. 10.30A).

History. Calman (1909) reported that the common and conspicuous *Squilla mantis* of the Mediterranean could hardly have escaped notice in antiquity, and it is surprising that it cannot be identified with certainty among the Crustacea mentioned by Aristotle. It was described by Rondelet (1555) under the generic name that it still bears. However, the stomatopods have long been recognized as a distinct group of crustaceans, since Latreille who defined the group in 1817. Their larvae are so distinctive that they were originally described as separate genera before their true affinities were discerned. Grobben (1876) after Huxley

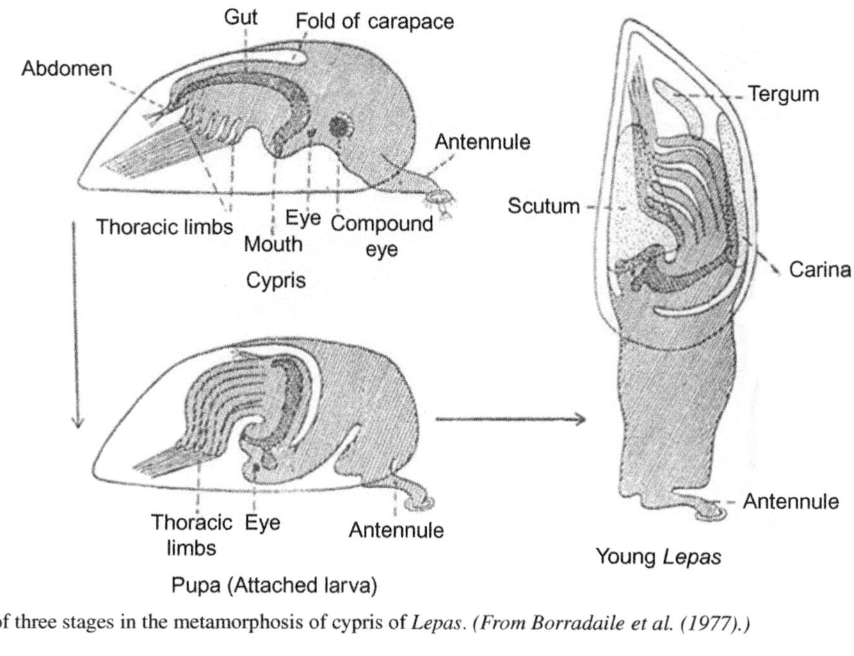

FIG. 10.29 Diagrams of three stages in the metamorphosis of cypris of *Lepas*. *(From Borradaile et al. (1977).)*

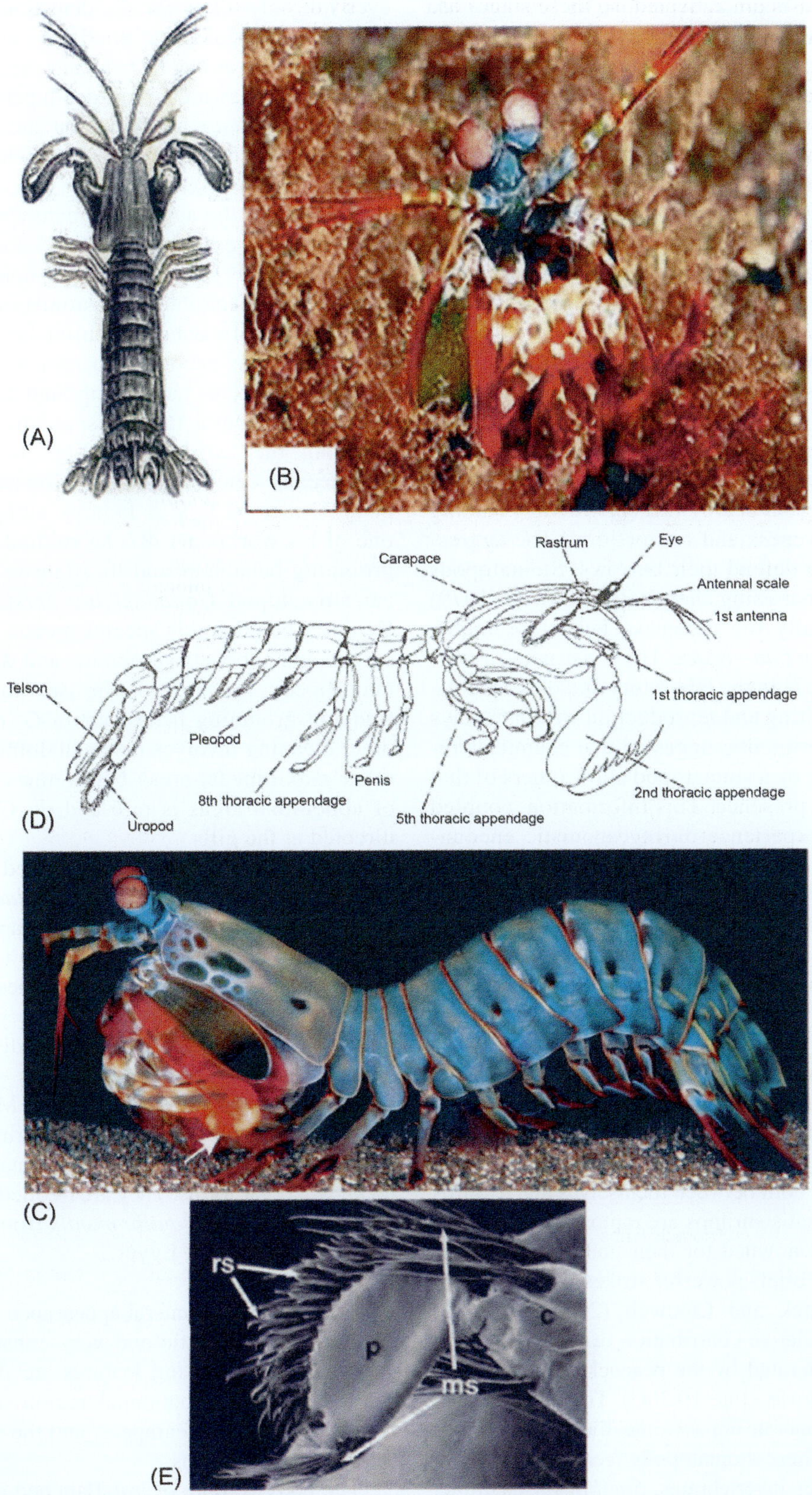

FIG. 10.30 (A) Example of a stomatopod; (B) *Squilla empusa* in a burrow potential—impacts of trawling in a marine ecosystem of the southeast coast of India; (C) Peacock mantis shrimp *Odontodactylus scyllarus* uses a pair of large raptorial appendages *(white arrow)* to strike hard objects with such high speeds that cavitation bubbles from between the appendages and striking surface *(Patek and Caldwell, 2005)*; (D) Male *Squilla mantis*, lateral view *(from Barnes, 1974)*; (E) Lateral view of first maxilliped of *Gonodactylus oerstedii* showing grooming setae of terminal segments. c = carpus, ms = multiscaled setae, p = propodus, rs = rasp setae. Scale bar = 520 μm. *From Bauer (1987).*

pointed out the great differences separating the Stomatopoda from the other stalk-eyed groups. Calman (1904) finally formalized the separate position of stomatopods when he established the grade Hoplocarida. At that time, they were generally concerned to be an early and distinct branch of the Eumalacostraca. However, Stomatopoda are the basalmost Malacostraca, a position supported by most phylogenetic analyses, e.g. Schram (1986). An extensive and ongoing taxonomic revision of the living stomatopods has been under way for several years, focused around the extensive work of Manning (Schram, 1986).

Habitat and behaviour. There are ~350 species of mantis shrimp. Most stomatopods live in burrows, they construct simple burrows (Fig. 10.30B) comparable to marine astacidean ones. The burrows are excavated in the bottom or in rock or coral crevices. Burrows of different individuals are often nearby (20 burrows/m^2) leading to a lot of aggressiveness and competition. The aggressiveness is solely to defend their burrows. Stomatopods can identify each other using chemicals. Caldwell (1979) studied experimentally the defensive behaviour in the stomatopod *Gonodactylus festai*. He constructed artificial burrows in aquaria in the laboratory and observed the animals during moulting and reproduction within 60 days of capture. His experiments suggested that chemicals released into the water by a stomatopod allow others of that species to detect its presence. This information, coupled with his previous experience during agonistic encounters with the resident, allows the intruder to identify the individual and determine the probability of successfully evicting it. In this case, they (stomatopods) are only one of a handful of crustaceans known to be able to do so joining the hermit crabs and shrimp. Stomatopods take it one step further, by being able to remember each other, and especially the results of previous fights (Atema, 1995), and thus a sort of dominance hierarchy can be thought to exist.

Various species of stomatopod are known to exhibit agonistic and aggressive behaviour in natural populations. They are very antagonistic both between themselves and towards other species. All mantis shrimps are raptorial, eating only live prey. They are renowned for their unusual method of breaking shells with brief, powerful strikes of their raptorial appendages. Patek and Caldwell (2005) examined the magnitude and relative contribution of the impact and cavitation forces generated by the peacock mantis shrimp *Odontodactylus scyllarus* (Fig. 10.30C). They investigated that this species can generate impact forces thousands of times their body weight. These stomatopods feed on small fish, crustaceans, and other invertebrates. *Squilla*, for example, is active and predatory, and hunts prey; others lie in wait at the mouth of the burrow. The highly developed stalked compound eyes play an important role in prey capture. The eyes not only allow for the detection of moving objects, but also for judging the proximity, and with the position in which the eyes can be held and the accuracy with which the animal can swim to prey, depth perception must be possible. The antennae are important sites of chemoreception, and are also used in prey detection when the range is short enough.

In spite of the aggressive behaviour of stomatopods, these animals exhibit specialized social, courtship, and more extensive parental care of their young. Moreover, they exihibit excitable and astonishing grooming behaviour. The importance of grooming behaviour in the life of crustaceans has become apparent in recent years. Many crustacean species have compound setae, organized into brushes and combs that are specialized for scraping and brushing the exoskeleton (Fig. 10.30E). Stomatopod crustaceans can frequently be observed grooming the body. Bauer, R.T. is a famous author in this field. In one of his works in 1987, he studied experimentally the grooming behaviour and its adaptive value in the tropical stomatopod *Gonodactylus oerstedii* Hansen, 1895. He described that this species preens body parts with the carpus and subchela (propodus and dactylus) of the first maxillipeds. Preening of the two antennae is the most frequent grooming behaviour in *G. oerstedii*, and most other cleaning involves the cephalothoracic region (eyes, other maxillipeds, areas below the carapace). Cleaning of abdominal areas is rare and, in *G. oerstedii*, mainly directed at the gills.

Some species are easily recognized in the field by their colour and appearance, e.g. *Odontodactylus scyllarus*, but more scientifically by their flat carapace that leaves the sides of the thorax exposed and don't even cover the last four thoracic segments. Many species are brilliantly coloured. Green, blue, and red with deep mottling are common, and some species are striped or display other patterns. Their size ranges from small species ~5 cm long to giant forms greater than 36 cm in length. Most mantis shrimps are essentially tropical and live in warm temperature water, but some species live in temperate waters. To date there are no known polar forms. They are frequently caught by prawn trawlers, especially *Squilla mantis* from the Mediterranean shores of Alexandria, Egypt.

Morphology. The general appearance of the Stomatopoda is highly characteristic and very constant throughout the group. Its most striking features are due to the great development of the abdominal region and its appendages, the small size of the carapace, and the large and peculiarly formed raptorial limbs.

The body is more or less flattened dorso-ventrally. The carapace is fused dorsally with at least two of the thoracic segments; two others are represented by indistinct vestiges, overlapped by its hinder edge, while the last four are free and completely developed. Anteriorly, the carapace does not extend to the front of the head leaving uncovered two movably articulated segments, which carry the eyes and the antennules, respectively. A small rostral plate, movably articulated with the front edge of the carapace, overlies the antennular segment. The exopodite of the antenna is scale-like. The first five pairs of the thoracic limbs are subchelate, the second pair of which are very massive forming powerful weapons; the last three are biramous. There are no oostegites. The abdomen is large with the first five pairs of limbs bearing gills on the exopodites; the sixth pair form a tail-fan with the telson.

Sexual dimorphism. Individuals of this group are dioecious as almost all the crustaceans, but without well-developed dimorphic organs. However, Deecaraman and Subramoniam (1983), working on *Squilla holoschista*, reported that the male and female are similar except for variation in colour and size, and the dimorphic organ is the intromittent organ of the male shown in Figs 10.30D and 10.31B (penis of Calman, described above). These authors added that the appendages of the first abdominal segment have two curved processes. In *Oratosquilla oratoria* (Kodama et al., 2009), sex was determined by the presence or absence of a penis.

Reproductive systems. In the male, the paired testes are delicate convoluted tubules that lie in the abdomen; extend only from the fourth abdominal segment to the telson. They unite posteriorly in an unpaired piece, which lies in the telson and passes anteriorly into the vasa deferentia. Each vas deferens extends forwards to the eighth thoracic segment (Fig. 10.31A) to open to the exterior at the end of a long penis (Calman, 1909). Deecaraman and Subramoniam (1980) referred to this penis as the intromittant organ, and described its tip in *Squilla holoschista* as blunt, with the genital orifice located at its centre (Fig. 10.31B). It differs from the corresponding organs of other Malacostraca not only in its great length, but also in the fact that it is more or less strongly chitinized and it is divided by a movable articulation about the middle of its length (Calman, 1909). A more recent view is that each vas deferens opens on the precoxa of the eighth thoracic appendage by means of a long process called the genital papilla (Schram, 1986). In the posterior thoracic somites, a pair of convoluted tubular glands lie, the accessory glands, very similar in form to the testes. In *Squilla holoschista* (Deecaraman and Subramoniam, 1980), these glands are situated in the cephalothorax and extend to the eighth thoracic segment. Each is a white, transparent, convoluted gland whose anterior portion is enlarged and club-shaped, the left is longer than the right and lies slightly over it (Fig. 10.32A). The ducts from the accessory glands run on each side of the midline to the eighth thoracic segment. Each is extensively coiled, floats in the haemocele, and opens separately at the tip of the intromittant organ, parallel to the vas deferens. These glands and their ducts never contain spermatozoa (Calman, 1909). In another species, *Oratosquilla oratoria*, Kodama et al. (2009) described the penis as

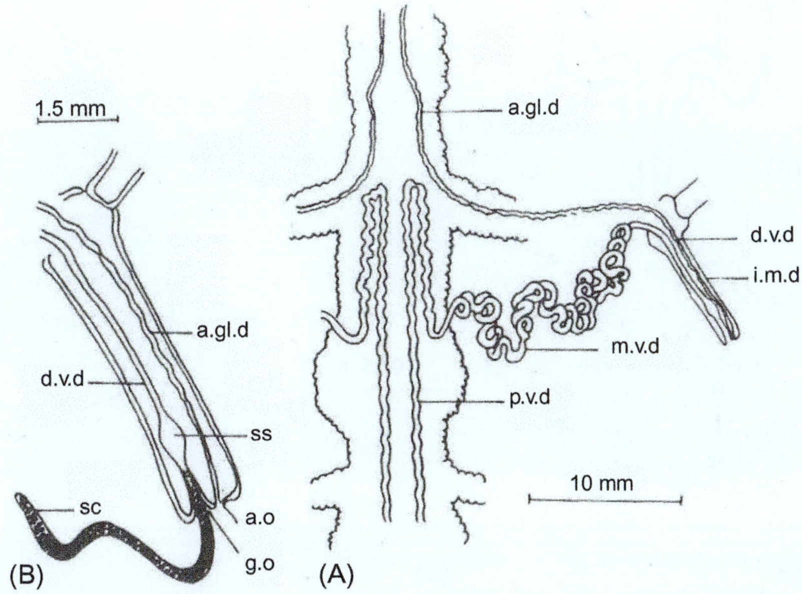

FIG. 10.31 Reproductive system of *Squilla holoschista*: (A) diagram showing the three regions of the vas deferens with the accessory gland ducts; (B) intromittant organ. a.gl.d.=accessory gland duct, a.o.=accessory gland opening, d.v.d.=distal vas deferens, g.o.=genital orifice, i.m.o.=intromittant organ, m.v.d.=middle vas deferens, p.v.d.=proximal vas deferens, s.c.=sperm cord, s.s.=sperm sac. *(From Deecaraman and Subramoniam (1980).)*

includes two ducts, the accessory gland duct, which never contains sperm, and the seminal duct. They presented figures of histological sections to prove this description (Fig. 10.32B). In addition, Charniaux-Cotton (1960a,b) reported that a set of androgenic glands are found on coxal muscles of the eighth thoracic appendage of the stomatopos *Squilla mantis*. In Stomatopoda, Calman described the mature spermatozoa as spherical in form, without processes of any kind, and appear to be simple nucleated cells (Fig. 10.16(e) and (h)). They resemble superficially those of the decapods. In *S. holoschista* (Deecaraman and Subramoniam, 1980), perhaps as in all the members of the group, they are transferred to the female, not as spermatophores, but as sperm cords (Fig. 10.31B).

Formation of the sperm cord. No direct information describing this process was available in the literature. However, from the work of Deecaraman and Subramoniam (1980), one can understand that the sperm cells coming from the testis intermingle with the viscous fluid secreted in the lumen of the proximal vas deferens (Fig. 10.32C). The sperm cell measures about 1.2–2.4 μm in diameter at this step. It is nonflagellate, spherical or elliptical in shape, and contains many deeply staining granular structures. The secretory product of PVD may be utilized by the developing spermatozoa. The latter become of a larger size ranging from 4.8 to 5.4 μm and completely occupy the lumen of the middle vas deferens forming a compact mass. The granular structure seen in the sperm cell in the PVD is no longer recognizable. It was suggested that the sperm finish their maturation, becoming a compact cord, and are stored in this region, MVD (Fig. 10.32D). Their final mould and separation into the length specific for the species occur in the distal vas deferens (Fig. 10.32E). It is equipped by its thick muscular wall, quite narrow lumen and with its four typhlosoles for such function. Deecaraman and Subramoniam (1980) identified histochemically and discussed the role of the acid musopolysaccharide secreted in the DVD. They recorded that the 'cementing' and rigidity of the sperm cord has been attributed to this substance. Further, they pulled 'mechanically'

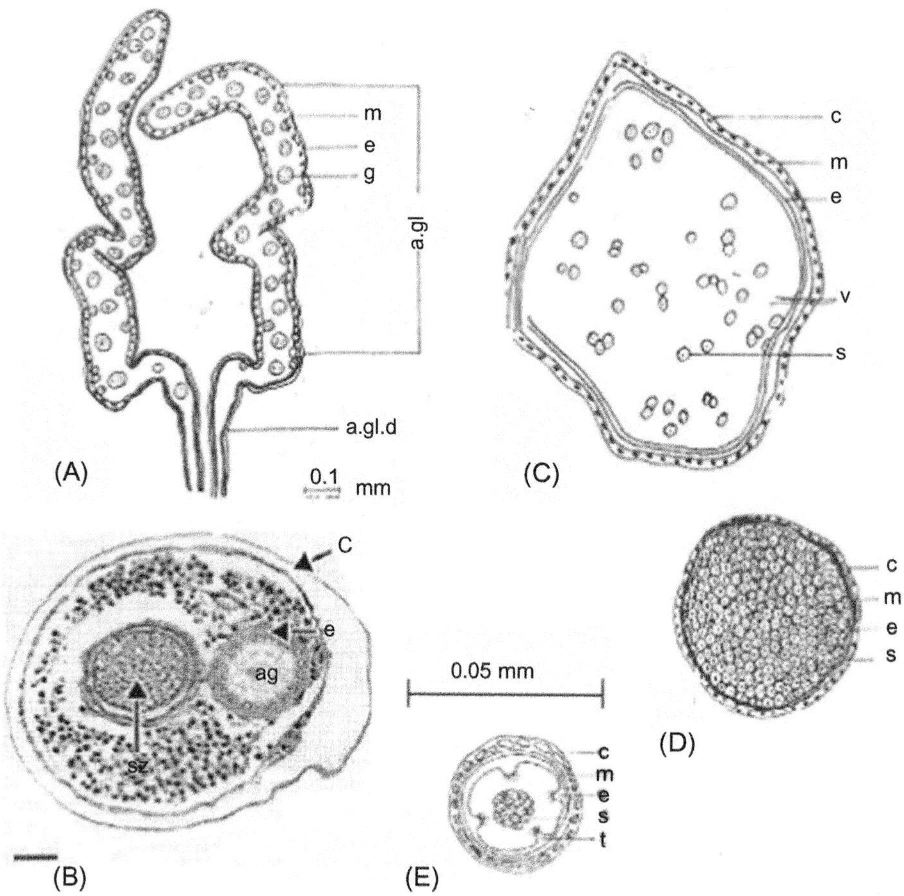

FIG. 10.32 Histology of the male reproductive tract of *S. holoschista* showing camera Lucida drawings of: (A) accessory glands and their ducts; (C) PVD; (D) MVD; (E) DVD *(From Deecaraman and Subramoniam, 1980)*; (B) penis in which sperm are present of *Oratosquilla oratoria (from Kodama et al., 2009)*. Scale bar indicates 50 μm. a.gl. = accessory gland, ag & a.gl.d. = accessory gland duct, c = connective tissue and in B cuticle, e = epithelium, g = granule, m m = muscle layer, s & sz = sperm, t = typhlosole, v = viscous fluid.

the rigid sperm cord out of the intromittant organ, supposing that a kind of hardening occurs in this DVD.

Since the accessory glands have separate openings to the exterior, it appears that their secretion does not share in the formation of the sperm cords. However, their secretion may have some postcopulatory role inside the oviuct. The preliminary observations of Deecaraman and Subramoniam (1980) support this role. They performed a test by mixing crude extract of accessory gland with a sperm cord, the result was the dissolution of the latter, but the spermatozoa remained alive.

In the mature female, the paired ovaries are very voluminous and closely approximated, so that they appear to form a single-lobed mass. It is not easy to separate the lobes from each other in the mantis shrimp *Oratosquilla massavensis* (El-Shereif et al., 2012). They extend far forward enough to the hinder limit of the carapace. Posteriorly, they extend through the abdomen and connect in the telson. It seems that this long extension of the ovaries is a feature for this group of the mantis shrimps. It is also observed in the female *Oratosquilla oratoria* (Kodama et al., 2009) and in that of *Harpiosquilla raphidi*a (Yusli and Ali, 2010). The oviducts open near the middle line on the sternal surface of the sixth thoracic somite, and in the same area there is a pouch-like seminal receptacle. On the ventral surface (sternite) of each of the three last thoracic somites of this female, a glandular mass lies (cement-gland), sending numerous ducts to the exterior (Calman, 1909). In the mantis shrimp *Erugosquilla massavensis* obtained from the Mediterranean Sea at Port Said, Egypt, Sallam (2005) observed and described synchronous development and activity of the cement glands with gonadal maturation of the females. By means of the adhesive secretion of these glands, the eggs are agglutinated, forming a globular mass (Barnes, 1974). The eggs in *Squilla* are picked up by the maxillipeds as they are extruded and then kneaded for hours with the anterior thoracic appendages (Fig. 10.33A).

Courtship and copulation. Before the above vital activity (spawning), stomatopods undergo an excitable and astonishing courtship. In sexual displays, the maxillipeds will be whirled around; whether this is a dance-like display or merely a way to waft attractant chemicals around is not known. An interesting twist to the usual sex stories can be found in at least some stomatopods. In *Pseudosquilla ciliata*, the female does the courtship rituals, not the male, and the male is the one that picks and chooses which female to mate with, choosing the larger females. It is thought that this is due to the ejaculate being nutrient-rich (Hatziolos and Caldwell, 1983).

During the fighting ritual itself, one individual assumes the telson curl position (Fig. 10.33B), in which the telson is curled up (Taylor and Patek, 2010). The aggressor will then strike blows on the curled-up telson, while the stomatopod will hit its aggressor from its rather uncomfortable position. This ritual is fascinating from a functional and mechanical point of view. The raptorial appendage is immensely powerful, yet the stomatopods can bash each others' telsons without breaking the telsons. Schram (1986) has analysed and summarized courtship and copulation described by others especially that described by Dingle and Caldwell (1972). He recorded that mating in such naturally aggressive creatures can have its little anxieties. Males generally are always ready to mate, except just after a previous liaison. The female is very particular in her receptivity, and it is not always evident to a male whether he will be received with indifference or will receive a chop to the head. After complete mounting and more male stroking of the carapace, the male curls his abdomen around the female (Fig. 10.33C), causing her to roll over. With ventral surfaces opposed, the male inserts his genital papillae into her seminal receptacle. For 10–50 s the male thrusts his abdomen up and down injecting the sperm cords into the seminal receptacle, while the relaxed female raises her head and tail up and down. The male then either moves slowly off while the female rights herself, or she finally manages to throw him off.

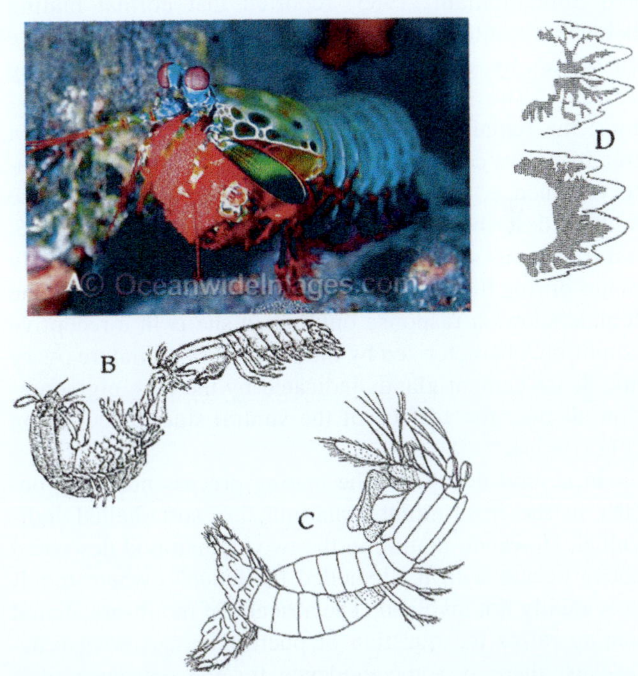

FIG. 10.33 Shows stomatopod behaviour: (A) female *Odontodactylus scyllarus* carrying egg mass, found on sand and rubble throughout the Indopacific; (B) Ritualized fighting in mantis shrimp. The animal on the left is in a coiled position, presenting its telson to the opponent. The animal on the right is striking the telson of the opponent with both raptorial appendages *(diagrammed by Taylor and Patek, 2010, adapted from Caldwell and Dingle, 1976)*; (C) Mating in *Gonodactylus bredini* with male above clasping carapace of female below (stippled); (D) ventral view of telson: above nonreceptive female of *G. bredini*, below receptive female *((C and D) from Dingle and Caldwell, 1972)*.

The work of Dingle and Caldwell (1972) described in detail these vital activities in the mantis shrimp *Gonodactylus bredini* Manning, in both cases; the male approaches first or the female approaches first. If the male approaches, with or without antennular palpation, he grasps the female with his maxillipeds most often at the telson. However, if the female makes the initial approach, she does so from the front or side of the male, not from the rear. When she reaches him, she pushes her head and carapace beneath his cephalothorax, which usually induces him to grasp her carapace and copulation takes place (Fig. 10.33C). At any case, the courtship-copulation sequence takes place only if the female is receptive, which can be distinguished morphologically by darkened internal organs (ovary-hepatopancreas) visible through the telson (Fig. 10.33D), and it is distinguished behaviourally by its acceptance of a male approach. On the other hand, it seems that there are no morphological criteria to distinguish receptive males. At the completion of copulation, either the male releases and swims slowly away while the female turns back over on to her ventral surface, or the female pushes the male off with her telson and struggles free.

It appears that courtship sequences of the mating are not alike in all species. In *Squilla holoschista*, Deecaraman and Subramoniam (1983) reported that normal mating behaviour could easily be differentiated from aggressive encounter by closely observing the various premating gestures. The male usually exhibits aggressive behaviour; the female also behaves similarly when she is not in a receptive condition and also at the end of the copulatory sequence. Under laboratory rearing conditions, these authors described the mating process for this species, *S. holoschista* as shown in Fig. 10.34. This species generally mates during the evening, preferably in diffused light. The female shows a response only when she is in a receptive condition, characterized by the possession of mature ovary and dense cement glands indicated by the dark pigmentation all over the surface of the ventral side of the telson (Fig. 10.34).

In several decapods, the mating process normally occurs in the fresh-moult condition, i.e. soft-shelled individual. However, mating in the two stomatopod described above occurs with hard-shelled individuals, where moult is evidently not involved. The absence of moult-associated mating raises the question of pheromones. Among crustaceans, there is some evidence for the existence of a sex pheromonal attraction as in the decapod brachyuran *Portunuus sanguinolentus* (Ryan, 1966). Again, in both species described above, and perhaps in all members of the group stomatopoda, sex pheromonal attraction has not been demonstrated. The results of the work of Deecaraman and Subramoniam (1983) have not provided any evidence in support of a sex pheromonal attraction in mating in *Squilla holoschista.*

The male and female may share the same den, repeatedly copulating for as long as a week, in behaviour probably designed to ensure that no other male mates with the female. The female becomes gradually more aggressive until the male is evicted (Schram, 1986). Some species of mantis shrimp can actually mate with a partner, and stay together for their entire life. However, most mantis shrimps mate with many different partners. After the mating process is finished, the female spawns the fertilized eggs. She can either retain them, or lay them immediately in her burrow or crevice, or can keep them on the forelimbs (Fig. 10.33A). The entire process of spawning may take 4 h, during which time she stands on the legs and the telson with her body arched upward. The interval between first copulation and egg-laying was about 10 days at a water temperature of 24–26°C in *Gonodactylus bredini* (Dingle and Caldwell, 1972). The agglutinated egg mass is about the size of a walnut and may contain as many as 50,000 eggs (e.g. *Squilla*), and is carried by the smaller chelate appendages. The female guards, cleans, and constantly turns to aerate the developing eggs and embryos. The female does not feed while she is brooding (Barnes, 1974). During this time, she would only rarely make short excursions outside the den, returning immediately to guard the eggs if danger threatened. In summary, the reproductive behaviour of *G. bredini* seems to reflect the co-evolution of four elements: cavity dwelling; defence of the cavity; courtship and copulation; and maternal care.

Development (embryology)

Gametes. Roosen-Runge (1977) recorded that an extensive account of spermatogenesis in a stomatopod species, *Squilla oratoria*, demonstrates many similarities between this order and the Amphipoda and Decapoda (Komai, 1920) (see later). The germinal zone in the tubular testes is a narrow longitudinal band in the mid dorsal line, where a few large spermatogonia and some 'nutritive cells' with polymorphic nuclei are situated. Spermatogonia and spermatocytes are intimately associated with nurse cells, but Komai did not discuss the relationship between spermatids and nurse cells. More recently, Kodama et al. (2009) studied the reproductive biology of the Japanese mantis shrimp *Oratosquilla oratoria* with stress on spermatogenesis (Fig. 10.35). They only added the detailed description of the histological structure of the four types of cells of the development of the testis. Spermatogonia, the largest of all has a nucleus with several strongly basophilic chromatin particles (Fig. 10.35A). In the following stage, the spermatocyte is about the same size as the spermatogonia, with a nucleus that consists of fine thread-like strongly basophilic chromatin material (Fig. 10.35B). The third stage is the spermatids, which are about half-size of the previous stage, with a nucleus that usually contain crescent-shaped basophilic material concentrated on one side (Fig. 10.35C).

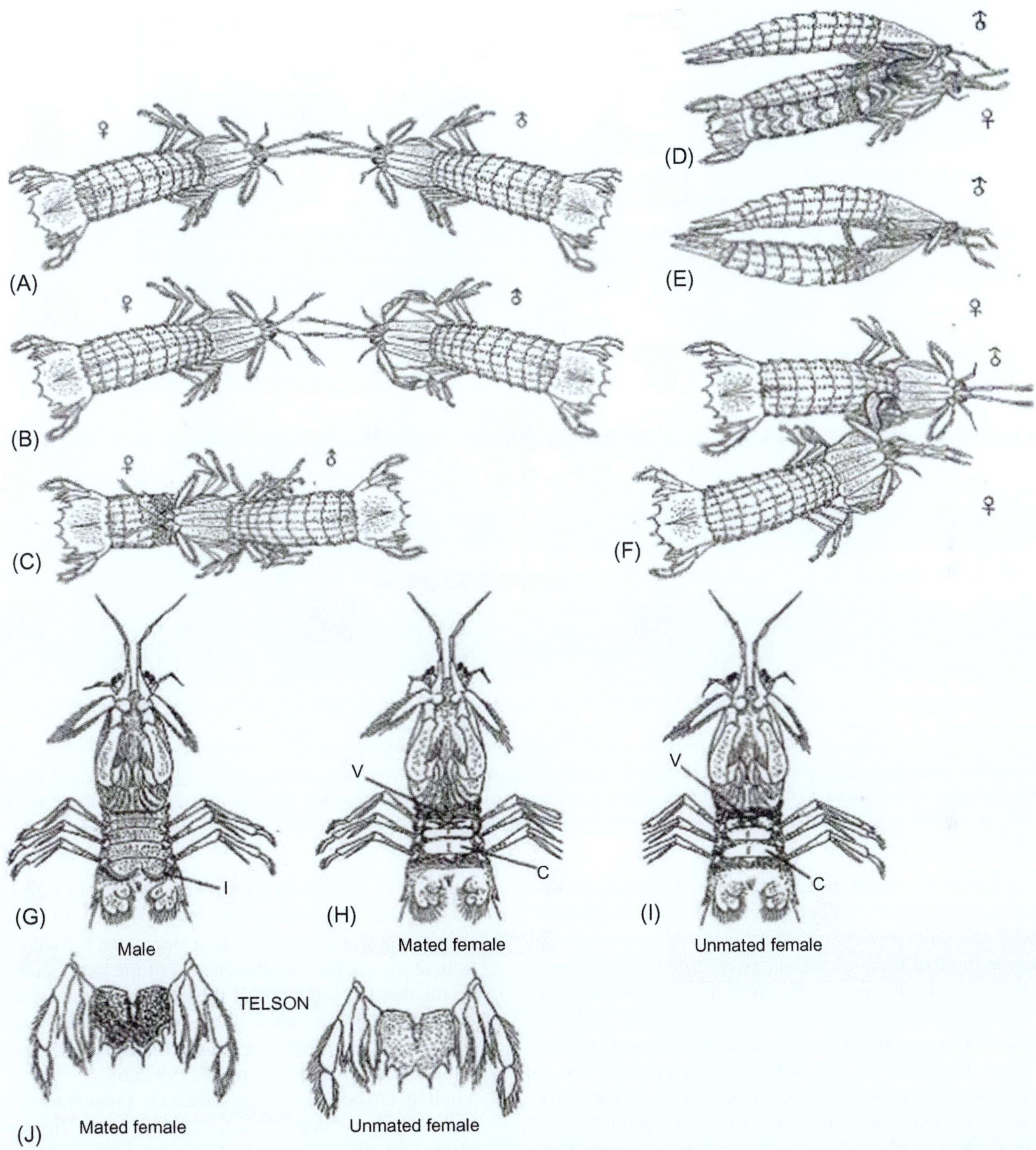

FIG. 10.34 Mating behaviour in *S. holoschista*: (A) antennule contact between the male and female; (B) spreading of the raptorial meri by the male; (C) movement of the male over the female; (D) tilting posture of the female by the male; (E) male-erection of paired intromittent organs; (F) final mating process, during which the female strikes the male; (G) male ventral view; (H and I) unmated and mated females—note the difference at the vulval region; (J) telson of the mated and unmated females. C=cement glands, I=intromittent organ, V=vulva. *(From Deecaraman and Subramoniam (1983).)*

These spermatids develop to give rise to the spermatozoa, which occupy the centre of the follicle (Fig. 10.35D). Each spermatozoon is spherical, nearly ball-like, with a smaller size than that of the spermatid (Fig. 10.16h). Its nucleus is very large, strongly basophilic, and fills almost the whole cell. No details for the spermatozoon were given in the study mentioned.

On the other hand, Kodama et al. (2004) studied oogenesis in *Oratosquilla oratoria* and El-Shereif et al. (2012) studied the same process in *Oratosquilla massavensis*.

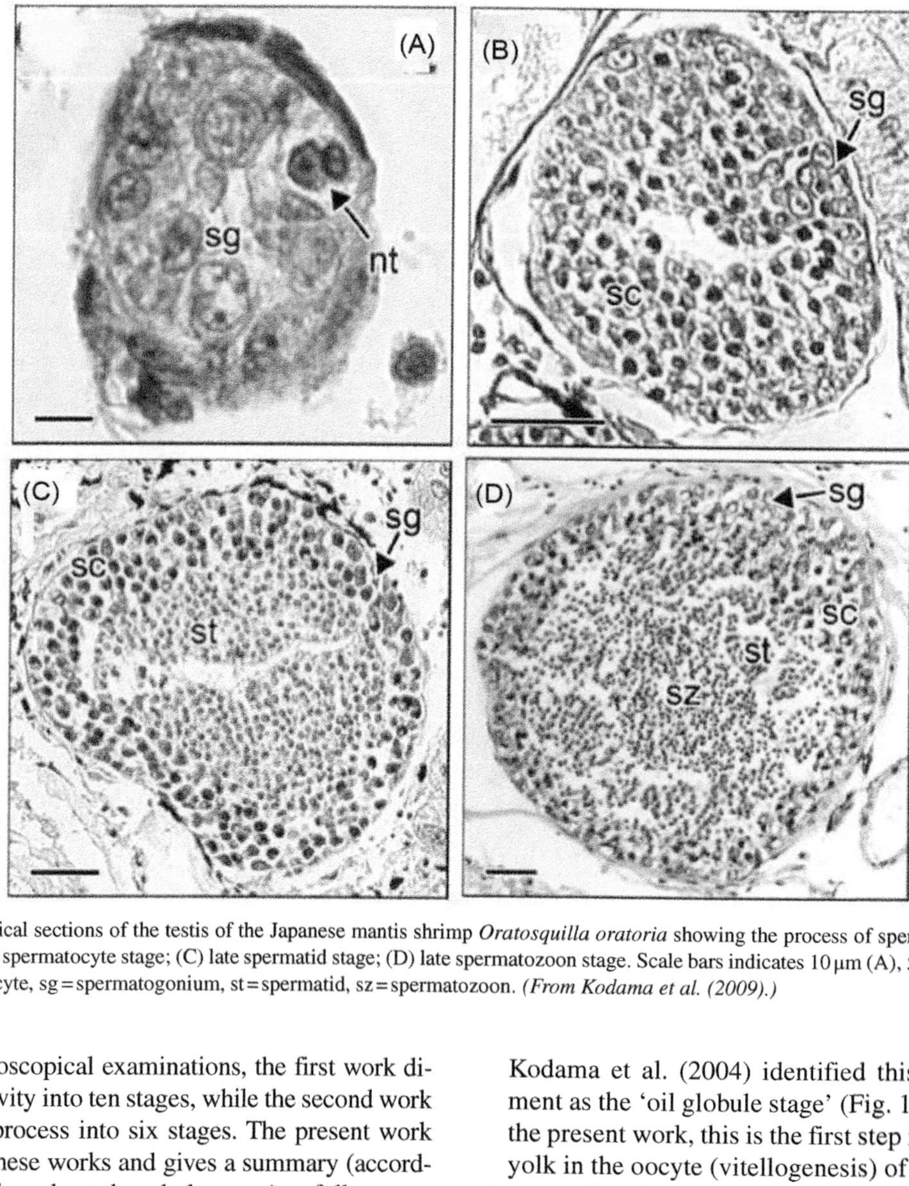

FIG. 10.35 Histological sections of the testis of the Japanese mantis shrimp *Oratosquilla oratoria* showing the process of spermatogenesis: (A) spermatogonium stage; (B) spermatocyte stage; (C) late spermatid stage; (D) late spermatozoon stage. Scale bars indicates 10 μm (A), 50 μm (B–D). nt = nutritive cell, sc = spermatocyte, sg = spermatogonium, st = spermatid, sz = spermatozoon. *(From Kodama et al. (2009).)*

According to microscopical examinations, the first work divided this vital activity into ten stages, while the second work divided the same process into six stages. The present work discusses both of these works and gives a summary (according to the germ cells and not the whole ovary) as follows:

1. In *O. oratoria*, the oogonium at its maximum development, measures 55.5 ± 18.8 μm in diameter. It has a large, round, strongly basophilic nucleolus that appears in the nucleus which measures 30.6 ± 5.8 μm in diameter. The cytoplasm becomes visible, but in proportion to the size of the nucleus, is very little. Smaller follicle cells surround the oogonium and gradually increase in number; each is round or oval according to the species (Fig. 10.36A and D).
2. Oocyte stage (previtellogenic stage), in both species the cytoplasm increases gradually in amount and so the size of the cell itself. The cytoplasm is weakly basophilic. The outline of the cell may change and the nucleus may be eccentric. At its maximum size specific to the species, droplets appear in the periphery of the cell. Kodama et al. (2004) identified this step of development as the 'oil globule stage' (Fig. 10.36B). In view of the present work, this is the first step in the formation of yolk in the oocyte (vitellogenesis) of the two species. It seems that the sequence of this process is the same, at least in crustaceans, since the beginning in these species is similar to that which happens in the crab *Portunus pelagicus* (Bawab and El-Sherief, 1987a,b).
3. Vitellogenic oocyte, during which the precursors of the yolk substance deposit successively inside the cytoplasm and hence the gradual increase of the cell itself. Acidophilic 'yolk' granules appear from the circumference of the cytoplasm and increase gradually in number and occupy the cytoplasm completely. The nucleus has lesser basophilic affinity than the nucleolus. The shape of the oocyte becomes more irregular and polygonal. Follicle cells still form a flat sheath around the vitellogenic oocyte (Fig. 10.36C).
4. The remature oocyte reaches its maximum size specific for the species at this stage by the increase in the acidophilic yolk granules. The nucleus shrinks and

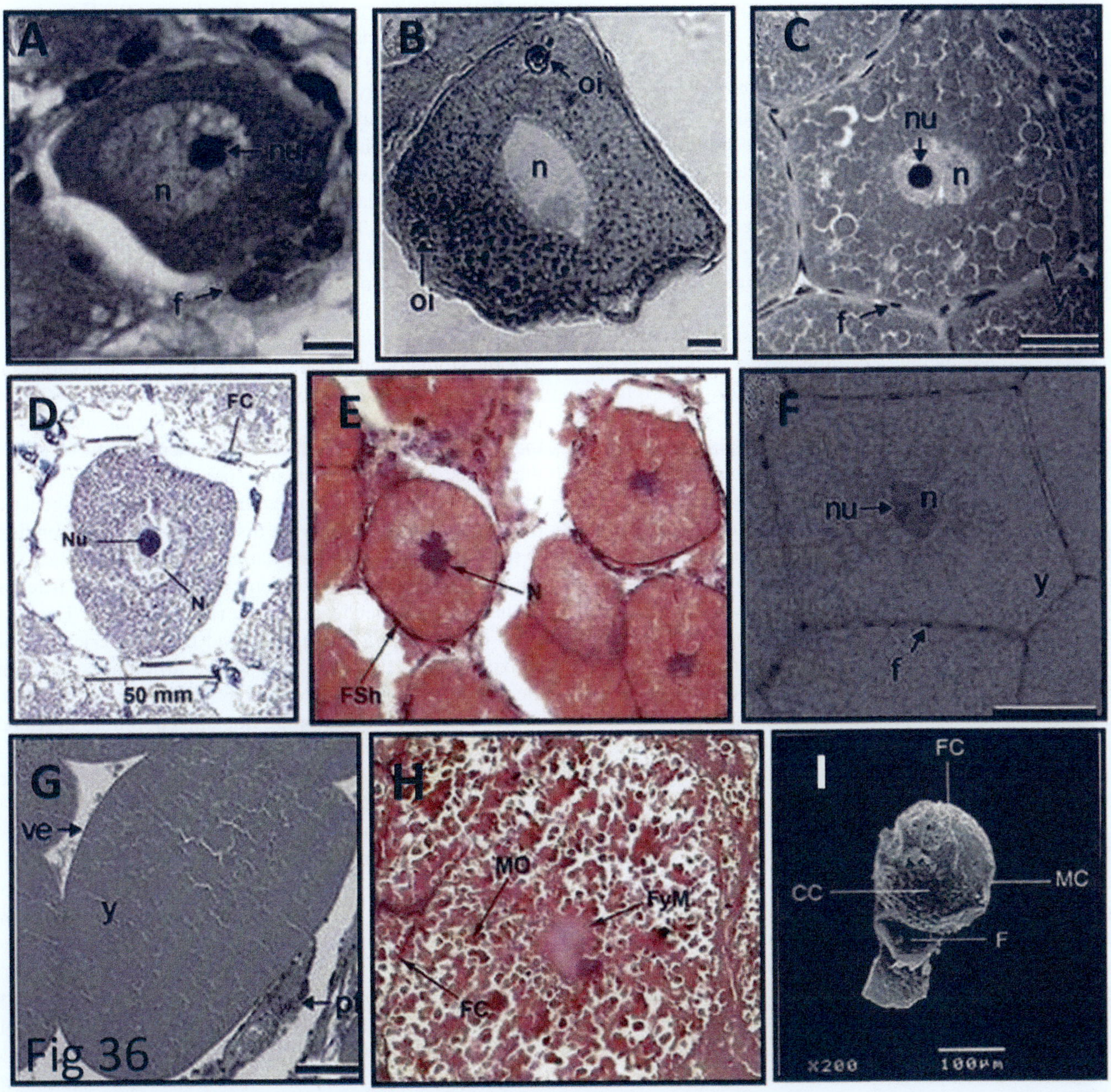

FIG. 10.36 Histological sections of the ovary of the female Japanese mantis shrimp showing the process of oogenesis in *Oratosquilla oratoria ((A–C and F, G) from Kodama et al., 2004)* and in *Oratosquilla massavensis ((D, E, H, and I) from El-Sherief et al., 2012)*. (A) and (D) showing pre-vitellogenic stage. f or FC = follicle cell. Bar in A = 10 μm; (B) shows oil globule stage. Bar = 10 μm; (C) primary yolk granule stage. y = yolk granule. Bar = 50 μm; (E and F) secondary yolk granule stage. FSH or f = sheath of follicle cells. Bar in F = 100 μm; (G and H) maturation stage. FYM = fluid yolk matrix, pf = postovulatory follicles, ve = vitelline envelope. Bar in G = 100 μm; (I) SE micrograph of the fertilized spawned egg of *O. massavensis* hung by a well noticed stalk or funiculus (F). The chorion is divided into three regions: central chorion (CC), marginal chorion (MC), and frontal chorion (FC).

moves to the periphery of the oocyte, the individuality of the granules can be identified with difficulty. The follicle cells persist (Fig. 10.36E and F).

5. Mature stage, at which the vitellogenic oocyte reaches its specific size (in *O. oratoria* the diameter is 589.5 ± 65.9 μm and in *O. massavensis* it is 425 μm). The cytoplasm becomes condensely loaded by a homogeneous substance. The loaded oocytes become densely packed to each other, and the follicle cells are rarely observed around the fully mature oocytes (Fig. 10.36G and H). In *O. oratoria*, a vitelline envelope with a considerable thickness coats the periphery of the oocyte, but in *O. massavensis*, a chorion was identified and studied in a spawned egg (Fig. 10.36I).

Early embryology. The stomatopod egg is very rich in yolk. The nuclei make their way to the surface of the egg and undergo only partial cleavage (superficial cleavage) (Kumé and Dan, 1968; Schram, 1986). With each division the cleaved nucleus, together with the surrounding cytoplasm, moves towards the egg surface until it reaches the peripheral cytoplasm. Then all the cells unite with this peripheral cytoplasm to form the blastoderm. In the egg of *Squilla oratoria*, all the cleaved nuclei reach the peripheral cytoplasm by the 128-cell stage and the cleavage furrows appear towards the end of this stage (Fig. 10.37A and B). The yolk pyramids, however, are very short in this species; the furrows extending for only a short distance into the yolk and advancing very little farther even in the succeeding stages. Therefore, the central yolk mass remains homogeneous. At the stage with about 500 cells, regional differences in the distribution of nuclei almost obliterate the yolk pyramids, although the cell boundaries remain visible on the surface. Eventually an area on the surface emerges where the nuclei are slightly more crowded, the germinal disc (Fig. 10.37C and D). The earliest differentiation on the disc results in the formation of the rudiment of the optic lobes and a posterior area known as the ventral plate (Fig. 10.37E and F). The latter develops the blastopore, from whose anterior end the primary mesoderm delaminates (Fig. 10.37D). This mesoderm extends forwards to join a secondary preantennular mesoderm and thus forms a mesodermal ring around the edge of the germinal disc (Fig. 10.37I). The preantennular mesoderm later disintegrates and the posterior region of the blastopore gives rise to the endoderm. The latter, in turn, form a compact mass of yolk-laden cells (Fig. 10.37D). Thus, an egg nauplius stage is formed (Fig. 10.37G and H). The three primary appendages arise while the ventral plate develops into a protruding thoracoabdominal rudiment. The naupliar appendage bud thickens and rises up above the germinal disc (Fig. 10.37H and K). A teloblast ring develops around the blastopore prior to the initiation of the egg–metanauplius stage (Fig. 10.37L), which then begins to bud off the maxillulary and maxillary segments in sequence.

Stomatopod development, as will be seen later, shares several features with that of decapods on the one hand; several other features of mantis shrimp development are shared with Leptostraca on the other hand.

In *O. oratoria* (Kodama et al., 2004), the female spawns in the field or under rearing conditions, mostly at night. The female probably lays almost all of its oocytes at one time and broods the egg mass for 3–4 weeks until hatching, during which time this brooding female ceases feeding. Although multiple spawning of *O. oratoria* under rearing conditions has been reported, these female shrimps failed to brood the egg mass until hatching and resumed feeding for the next spawning.

Postembryology (larval development). The larval stages of Crustacea usually display a remarkable series of form changes. The stage in which the larva becomes free-living is intimately related to the brooding habit of the mother. When a larva hatches in a fairly advanced stage, it always passes through a stage corresponding to the nauplius at some time in its embryonic life. There is, however, no form in which the larva becomes free-living before it attains the nauplius stage. This larval stage may be regarded as representing one of the primitive features of the Crustacea. The larvae of the Stomatopoda pass through morphologically peculiar stages, which are not comparable with those of the other malacostracans. Very early, Calman (1909) recorded that the stomatopod larval life is extremely remarkable, on account of its prolongation and its complicated metamorphosis. The most obvious feature is that the larval forms of the various species differ from each other more widely than the adults do. Larvae of many species were described by the older observers as adult animals, under several generic names. Two main types were distinguished, corresponding to the old genera *Erichthus*, Latreille, and *Alima*, Leach. In *Squilla*, *Gonodactylus*, and some others, the larva hatches in the alima or pseudozoea stage (Fig. 10.38(I)). The first larva of *Lysiosquilla*, *Coronida* and others is erichthus or antizoea (Fig. 10.38(II)). One of the differences between the two larvae is that the former has stalked compound eyes, while the eyes of the latter are not stalked. Shiino in Kumé and Dan (1968) presented diagrammatic illustrations explaining the metamorphosis in several malacostracan larvae and so Fig. 10.38(III) shows that of Stomatopoda.

It appears that studies on embryology and/or larval development of Stomatopoda are scarce, since Schram (1986) recorded that only those of few species have been completely worked out e.g. of *Gonodactylus oerstedii*, *Squilla armata* and *S. empusa*. However, Dingle (1969) studied the responses of the newly hatched larvae of the mantis shrimp *G. Bredini* Manning. He observed that the larvae of this species pass through propelagic stages, lasting a total of 7–8 days, before becoming pelagic in stage IV. These stages are spent in the den occupied and defended by the spawning female. These three stages are characterized by a lack of developed mandibles, initial development of maxillae and maxillipedes, and a great deal of yolk. At the fourth stage the yolk is consumed, appendage development completed, and the pigment in the eyes is then green instead of black as previously. The anatomical development of *G. Bredini* larvae seems to be identical to that of the *G. oerstedii* larvae (Schram, 1986).

Two elements of behaviour seem to be characteristic of the three propelagic stages: a strong thigmokinetic and a negative phototaxis. With respect to the first, the larvae will grasp with the large second maxillipeds any suitable surface with which they come in contact. In the field, larvae with

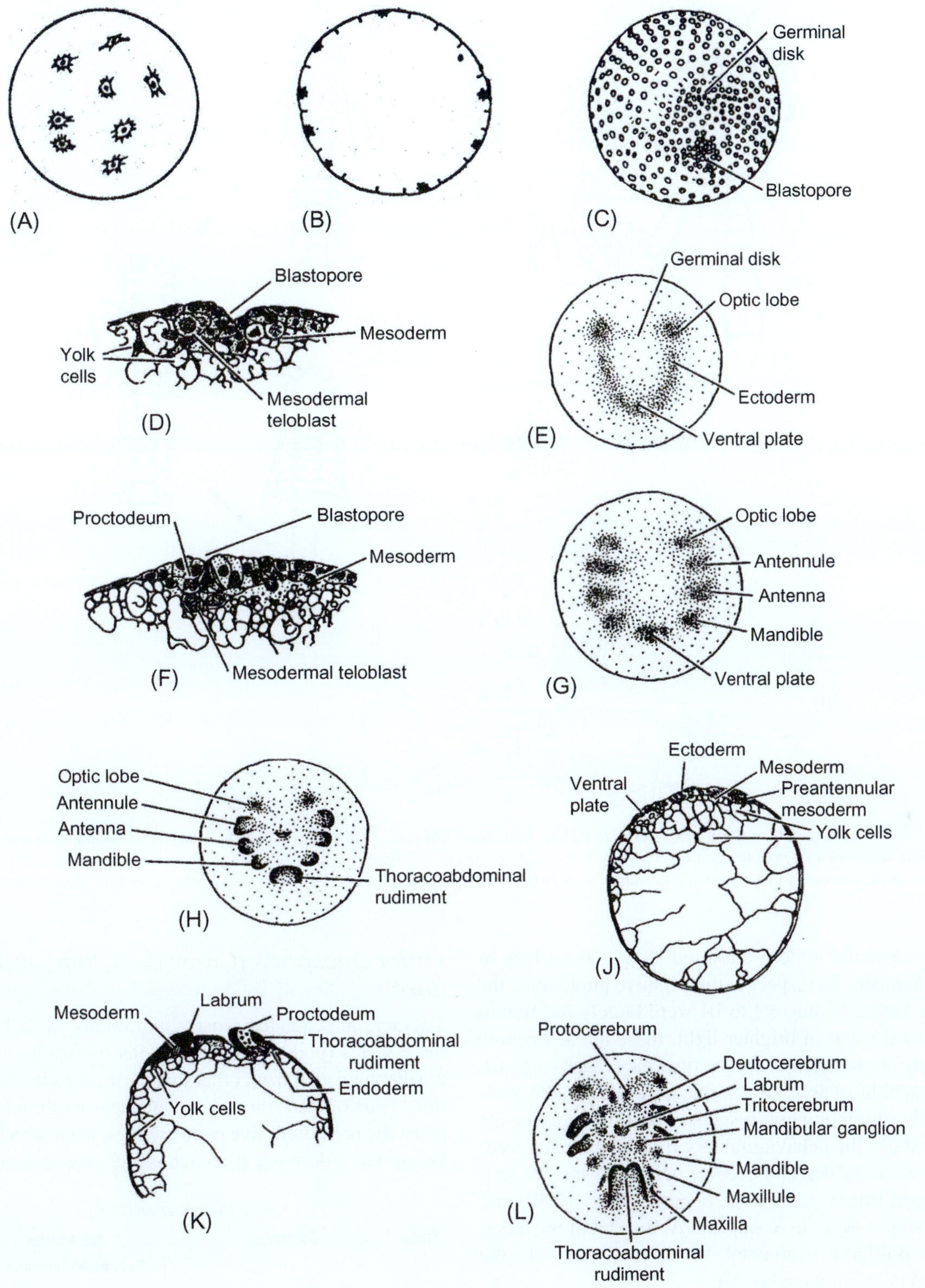

FIG. 10.37 Early embryology of stomatopods: (A and B) cross sections of cleaved specimens of *Squilla oratoria (after Shiino from Kumé and Dan, 1968)*; (A) 8-cell stage, cleared specimen showing all nuclei and protoplasmic islands; (B) 200-cell stage, arrival of protoplasmic islands in periphery; *((C–L) after Schram, 1986, modified from Shiino, 1942)* (C) late blastula showing the initiation of the germinal disc and blastopore; (D) L.S. through stage C; (E) early differentiation of germinal disc; (F) L.S. through stage E; (G) beginning of egg-naupius stage; (H) development of raised limb buds and the thoracoabdominal rudiment; (I) diagrammatic L.S. through stage E; (K) diagrammatic medial L.S. through stage H; (L) early egg-metanauplius.

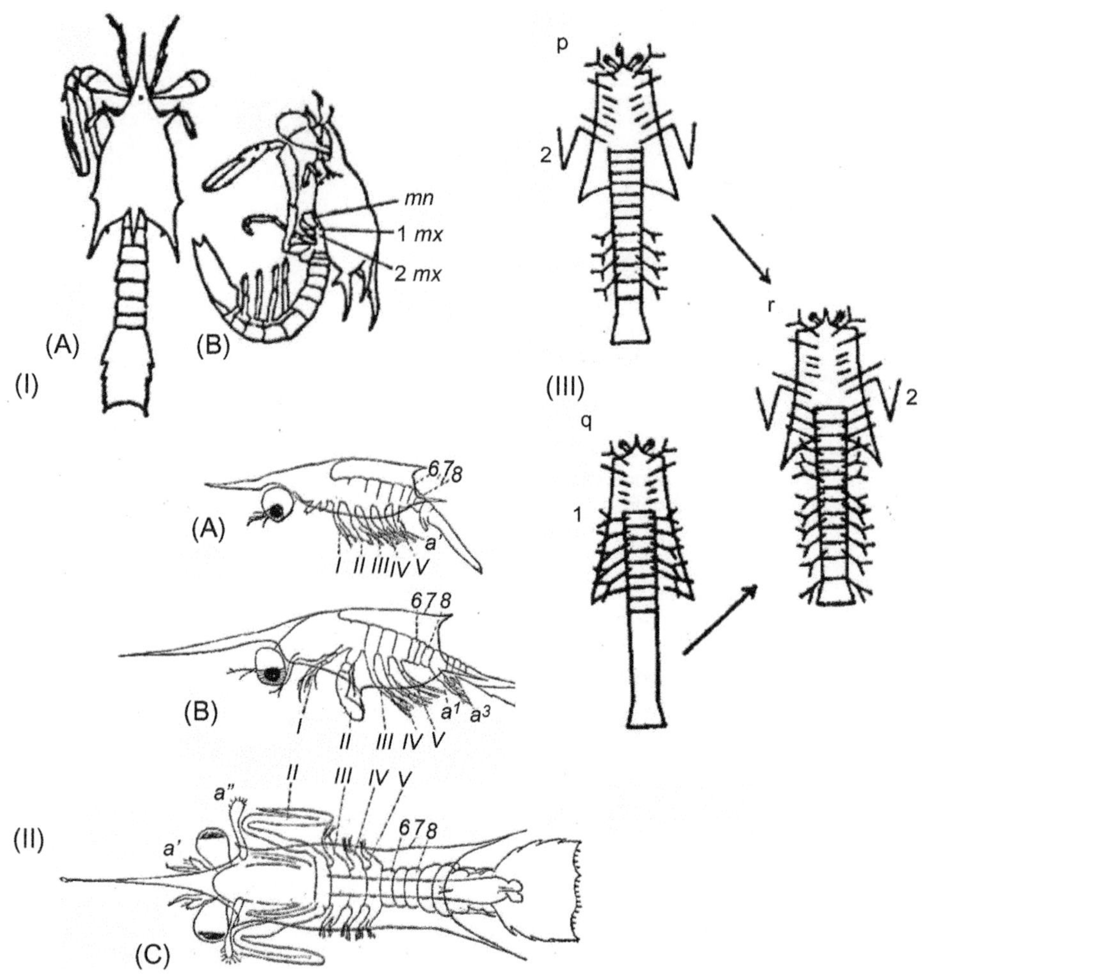

FIG. 10.38 (I) Alima larva of *S. oratoria (after Komai and Tung, from Kumé and Dan, 1968)*. mn = mandible, 1 mx = first maxilla, 2 mx = second maxilla; (II) consecutive stages of a larva of the first *Erichthus*-type, C-stage belongs to a different species *(after Claus, from Calman, 1909)*; (III) diagrammatic representation of the process of metamorphosis in stomatopods *(after Shiino from Kumé and Dan, 1968)*. a = nauplius, p = alima, q = erichthus, r = synzoëa.

females were found in dens clustered along one surface of the rock chambers. In respect to the negative phototaxis, the propelagic larvae of stages I to III were largely indifferent to the dim light, but in brighter light, these larvae become increasingly photonegative. By the time they reach stage III, they are capable of free swimming for considerable periods, and the thigmokinetic responses are still present. At the fourth stage, the behavioural modifications are induced; the larvae leave the den and become pelagic. A total of four pelagic larval stages occur in *G. oerstedii* (Fig. 10.39) and nine such stages occur in *S. empusa*. At the end of the larval series, the postlarva is attained, which then settles to the bottom and takes up the adult life.

Eucarida (Greek eu, *'true', and* karis, *'shrimp')*

Carapace large, covering all the thorax, pedunculate eyes, thoracic appendages with branchia.

Order Decapoda (Greek deca, 'ten', and pous, 'foot')

Thoracic appendages mostly uniramous, with branchia in three series (podo-, arthro- and pleurobranchia). Decapods are distinguished from other malacostracans in that their first three pairs of these thoracic appendages modified as maxillipeds; the remaining five pairs are legs, from which the name Decapoda is derived; abdomen is elongate or short.

According to Calman (1909).

Order — **Suborder** — **Infra-order**

Decapoda
- Natantia (shrimps) (swimming)
  - **Penaeidea** (primitive prawns)
  - Caridea (prawns & shrimps)
- Reptantia (crabs, etc.) (crawling)
  - Palinura (crawfishes & bear-crabs)
  - **Astacura** (crayfishes & lobsters)
  - Anomura (hermit crabs)
  - **Brachyura (crabs)**

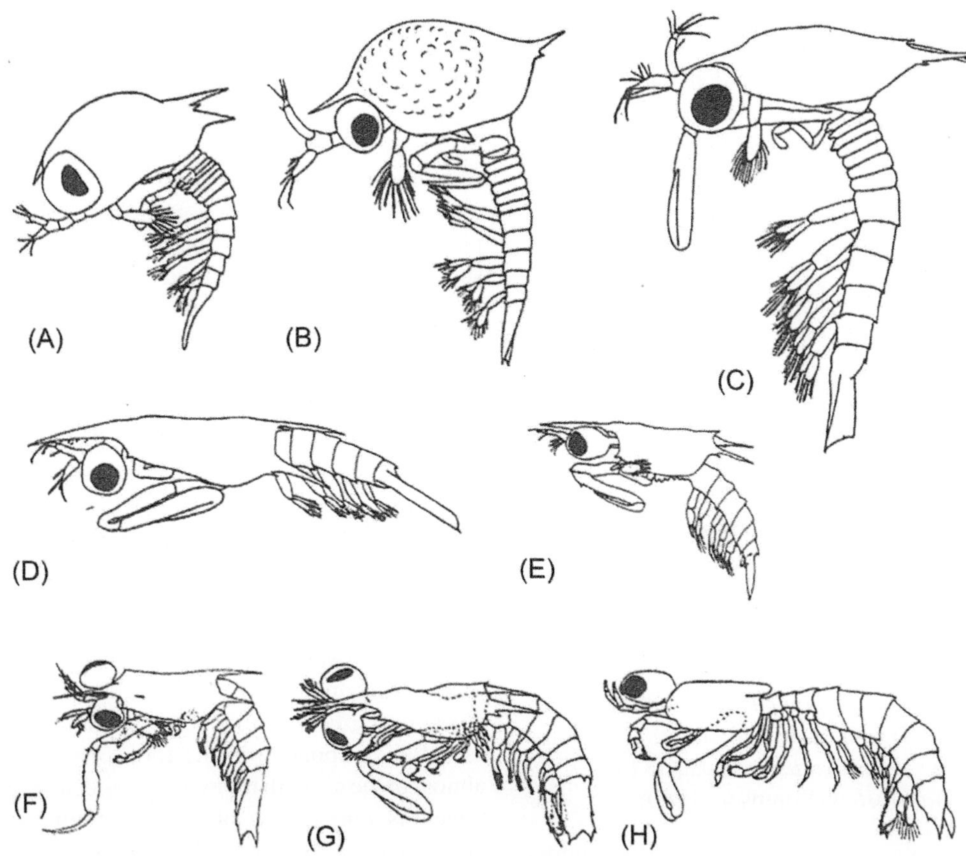

FIG. 10.39 Larval development in *Gonodactylus oerstedii*: (A–G) the first through seven stages; (H) postlarval. *(From Schram (1986), after Manning and Provenzano (1963); Provenzano and Manning (1978).)*

ACCORDING TO BORRADAILE ET AL. (1977).

Decapoda	Macrura, abdomen hard, long, and extended.
	Anomura, abdomen soft.
	Brachyura, abdomen hard, short, folded beneath the body.

More recently, Burkenroad, a modern carcinologist, worked on taxonomy of Crustacea, especially Decapoda, since 1934. In 1981, he tabulated a new scheme of classification of Decapoda depending on: the adult morphology of characters of gill arrangement, modes of pleopoda incubation, patterns of overlap of abdominal pleura, pleural lock and hinge arrangements, chelation of legs, presence of appendix interna, and form of spermatozoa; the fossil record; and ontogeny. Schram (1986), also a modern carcinologist, followed Burkenroad and so the scheme is as follows:

Order Decapoda (Latreille, 1803)
- Suborder Dendrobranchiata Bate, 1888
  - Superfamily **Penaeoidea** Rafinesque, 1815
  - Superfamily Sergestoidea Dana, 1852
- Suborder Euzygida nov.
  - Family Stenopodidae Huxley, 1879
- Suborder Eukyphida Boas, 1880
  - Infraorder Caridea Dana, 1852 (from Schram, 1986)
  - And several superfamilies with about 20 living families
- Suborder Reptantia Boas, 1880
  - Supersection Thalassinida Boas, 1880
  - Supersection **Astacina** H. Milne-Edwards, 1839
  - Supersection Palinura Borradaile, 1907
  - Supersection Anomura de Haan, 1841
  - Supersection **Brachyura** Boas, 1880

Penaeoidea

Definition. Pleura of the second abdominal somite not overlapping those in front; third maxillipeds with seven segments; third legs chelate (except when legs are much reduced); first pleopods of male with petasma; the abdomen is without a sharp bend; gills dendrobranchiate (different sources).

Habitat. Benthic shrimps are found in all sorts of bottom habitats. Species of *Penaeus*, which are the most important shrimp for commercial fisheries throughout the world, live on sand bottoms in 5–30 m of water. Some shrimp are shallow burrowers in soft bottoms, and the sand shrimp *Crangon* (Caridea) uses the beating pleopods for excavating. Many small species live within algae, beneath stones and shells, and within holes and crevices in corals and rocks (Barnes, 1974).

Family Penaeidae Bate, 1881

The last two pairs of legs are well developed; there is a nearly complete series of gills.

In the Mediterranean Sea, along the Egyptian shores extending from about Alexandria to Port Said, seven species of prawns constitute the commercial catches nowadays. Holthuis in FAO Species Cataloge (1980) renamed some of these species, e.g. *Penaeus (Melicertus) kerathurus* (Forskål, 1775), then Türkay (2001) confirmed this name in World Register of Marine Species (WoRMS).

1. *Metapenaeus monoceros* (Fabricius, 1798).
2. *Metapenaeus stebbingi* Nobili, 1904.
3. *Parapenaeus longirostris* (Lucas, 1846).
4. *Penaeus (Marsupenaeus) japonicus* Bate, 1888.
5. *Penaeus (Melicertus) kerathurux* (Forskål, 1775).
6. *Penaeus (Penaeus) semisulcatus* de Haan, 1844.
7. *Trachypenaeus curvirostris* (Stimpson, 1860).

During the sixties of the 20th century, the current author prepared her MSc and then the PhD theses on *Penaeus kerathurus*, only the three species of genus *Penaeus* and the *Metapenaeus monoceros* were found in these shores. This means that the fauna is not stable—at least in the Mediteranean and Red seas because of the Suez Canal—and the habitat of the species can or may change according to many and different factors. When these seven species of prawns were used for studying and preparing by Khafage (2000) for an MSc degree, the current author planned and supervised this thesis and suggested the construction of a specific key (Table 10.5) to facilitate the identification of the species for future and other workers. Therefore, this is the appropriate place to record this key.

Penaeus kerathurus

Morphology and sexual dimorphism. Many different available sources presented drawings and photographs of their local penaeid species. Fig. 10.40A is a photograph of a live adult *Penaeus kerathurus*, the most delicious, favourite and famous prawn among customers along the seashore of the Mediterranean in Egypt. Calman (1909) reported that phosphorescent organs are known in a number of deep-sea Decapoda, but the nature of the organs differs widely in the different groups. Alcock (1900) observed in *Aristeus coruscans* Duvernoy (Fig. 10.41) a luminous fluid to be omitted from the base of the antennae, apparently from the orifices of the antennal glands. This case may be compared with that of some species among the Mysidacea. More recently, Barnes (1974) reported that 17 genera of shrimp contain species known to be luminescent. The luminescent organs are most commonly photophores with lenses and reflectors, and may arise anywhere on the body, even the roof of the branchial chamber (as in *Sergestes*). On the other hand, species of the deep meso-pelagic zone are red and possessed dermal photophores, when photophores are present. Some deep-sea shrimp emit a luminous cloud-like secretion into the water.

Dimorphism is a common feature in penaeids. There are very important external organs characterizing each sex. The male specimen is defined by the petasma (modified first abdominal appendages) (Fig. 10.40D); while the female specimen is characterized by its thelycum (a structure modified of seventh and eighth thoracic sternites). Khafage and Bawab (2013) presented an intensive study on the structure of the closed thelycum of seven penaeids from the Mediterranean region of Alexandria, Egypt (oral presentation), where *P. (Melicertus) kerathurus* was one of them. The structure of this type of thelycum is differentiated into three components (Fig. 10.40B): (1) posterior basal ridge, a transverse, hard and strong plate that extends inbetween the coxae of the fifth pereiopods; (2) middle component is the essential component, in the form of wide pocket(s) found almost in the seventh thoracic sternite; and (3) anterior component is a proximal plate, similar to the posterior ridge. It is very hard, stiff and extends longitudinally in front of the pocket(s). It reaches the coxae of the fourth pereiopods, i.e. the left and right coxae embrace this anterior plate which faces the female genital openings. The latter are found on the coxae of the third pereiopods. The form and dimensions of the components change with the species. The thelycum has no connection with the internal genitalia.

The most important and scientifically significant result of this study is that the SEM shows that the coxae of the fourth pereiopods are strikingly modified. Each coxa has a wide circular opening, directed underside knowledgeable with the movement of the coxa. It is guarded by numerous dense, stiff, and long bristles. It is as a complementary component to the thelycum. It was referred to as a 'coxal vesicle' (Fig. 10.40C). In *P. kerathurus*, the vesicle is spherical in shape but it is triangular in the species of genus *Metapenaeus*. It is appropriate to mention that this vesicle was overlooked by previous authors, which might be because of its very tiny size, or it might have been considered as a normal extension of the pereiopod. However, Heldt (1938) observed and illustrated a very small white protrusion at the base of the fourth pereiopods of *P. kerathurus* without any additional information, but she referred to the tiny openings.

Reproductive systems

Male reproductive system. The male reproductive system of *Penaeus kerathurus* as described by Bawab (1964) can be regarded as conforming to a basic pattern common

TABLE 10.5 Taxonomic features used in diagnosis of the Mediterranean (Alex. Egypt) species

	Penaeus			*Metapenaeus*		Parapenaeus	Trachypenaeus
	P. kerathurus	*P. japonicus*	*P. semisulcatus*	*M. monoceros*	*M. stebbingi*	*P. longirostris*	*T. curvirostris*
	Fig.						
-Common Name	Gambari Azzazi	Gambari Azzazi yabani	Gambari Nylon Swissy	Gambari Ahmar	Gambari Abiad	Gambari Englizy Cuba	Gambari Okr
-Colour	Bluish purple bands	Reddish brown bands	Pale brown	Pink	Bluish grey	Red	Pale grey
Carapace L./Total L. in mm	65/165	70/190	60/160	55/135	45/120	40/100	35/90
Carapace grooves:							
• Rostral grooves	Well developed End at carapace	Well developed End before carapace edge	Well developed End at middle of carapace	–	–	–	Very shallow
	Posterior edge			–	–	–	End at epigastric tooth
• Middorsal g.	Well developed	Well developed	Well developed	Present	Present	–	–
• Cervical g.	Deep wide	Deep wide	Deep wide	Shallow narrow	Shallow narrow	–	Very faint
• Hepatic spine	Present	Present	Present	Present	Present	Present	Present
• Gastrofrontal g.	Present	Present	–	–	–	Present	Present
• Gastrofrontal s.	Small	Small	–	Very small	Very small	–	Very small
• Gastroorbital g.	Deep	Deep	Deep	–	–	–	–
• Antennary s.	Present	Present	Present	Present	Present	Present	Present
• Accessory g.	–	Present	–	–	–	–	–
• L. suture	–	–	–	–	–	Long (from carapace anterior edge to posterior one) above first pereipod	–
• Transverse suture	–	–	–	–	–		–
• Branchiocardiac g.	–	–	–	Present	Present	–	–

Continued

TABLE 10.5 Taxonomic features used in diagnosis of the Mediterranean (Alex. Egypt) species—cont'd

	Penaeus			*Metapenaeus*		**Parapenaeus**	**Trachypenaeus**
	P. kerathurus	*P. japonicus*	*P. semisulcatus*	*M. monoceros*	*M. stebbingi*	*P. longirostris*	*T. curvirostris*
	Fig.						
• Branchiostegal s.	–	–	–	–	–	Present	–
• Rostal pathway	Somewhat convex	Straight	Straight	Straight directed upward	Straight directed upward	Long wavy passage	Short sharply curved upward
-Rostal teeth							
• Dorsal teeth	8–13	9–11	5–8	9–12	7–10	8	8
• Ventral teeth	1	1	2–4	–	–	–	–
-Telson spines	3 mobile pairs	3 mobile pairs	–	Microscopic spinules	Microscopic spinules	Tridentate	spinules

From Khafage (2000).

FIG. 10.40 (A) Shows photograph of whole mount of a live specimen *P. kerathurus*, the specific characters are very clear; (B) left hand-side is a photograph and the right hand-side a camera Lucida drawing showing the structure of the thelycum of the female. A.C, M.C, and P.C = anterior, middle and posterior components, respectively, cp = coxal process, cv = coxal vesicle, cx3, cx4, cx5 = coxae of third, fourth and fifth pereiopods, respectively, fl = flap, go = genital opening, ms = median suture, p = proximal plate, pbr = posterior basal ridge, sp = sternal process. X = 7; (C) SEM photograph showing the coxal vesicle of fourth pereiopod with circular wide opening beside part of the proximal plate. cv = coxal vesicle, hs = hairy spines, op = opening, p = proximal plate. X = 45; (D) photograph showing the petasma of the male (modification of the endopodites of the first abdominal appendages). *(Photographs by Khafage and the drawing from Bawab (1964).)*

to all Crustacea Decapoda (Calman, 1909). It consists of paired testes and paired vasa deferentia that occupy dorsally the posterior region of the cephalothorax (Fig. 10.42A and B). The **testes** rest on the posterior half of the hepatopancreas, being partly concealed by the heart, which rides over them like a saddle. They appear as two unpigmented, lobulated, delicate, and translucent sheets. The inner edge of each component is completely separated from the other, except the first most anterior lobes which extend medially. They are completely attached to each other, where the other lobes from the second to the eighth are completely separated from each other at their outer edges, but completely fused in their inner edges. This case of the testes in *P. kerathurus* may be considered as a general feature for all

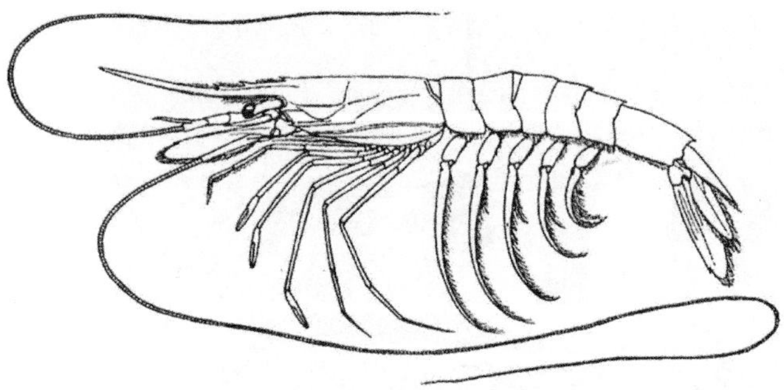

FIG. 10.41 *Aristeus coruscans* (Penaeidae). *(After Alcock, Naturalist in Indian Seas. From Calman (1909).)*

penaeids, since those of *P. setiferus* (Linnaeus) are more or less united along their entire length (King, 1948). However, the lobulation and number of lobes in both species are the same. In more recent study, Bauer and Cash (1991) reported that the reproductive tract dissected from a male of *P. setiferus* (Fig. 10.43B) is illustrated (in their work) to show the general features of that system for *P. duorarum* and *P. aztecus*. Champion (1987) presented a schematic diagram describing the male reproductive system of *Penaeus indicus* Milne Edwards (Fig. 10.43A). The appearance and lobulation of the testis in this diagram greatly differs from that of *P. kerathurus*. A very recent study by Muhammad et al. (2014) described the testis of *Parapenaeopsis stylifera* (H. Milne Edwards)—a penaeid but Tribe Parapeneae (Schram, 1986)—as comprised of two to mostly three short, broad, and flattened lobes (Fig. 10.43D and E).

In *P. kerathurus*, the inner edge of each testis was considered as the axis of the testis (Bawab, 1964). The first lobe is the smallest, the third is the longest, and the others show gradual reduction in length in a posterior succession. It is noteworthy, that the external limits of the lobes from the third to the seventh conform to the margin of the hepatopancreas (Fig. 10.42B). The last, which is the eighth lobe projects posteriorly as a short, narrow extension. The testes in live and fresh specimens of different sizes, collected at different periods of the year, are transparent when immature. Maturity is indicated by the degree of their opacity. The highly mature testis attains the milky white colour, a character that was reported very early by Huxley (1880) to describe the mature testis of *Astacus fluviatilis*.

Histologically (Fig. 10.44) each testis is surrounded by the cortex which is a very thin membraneous wall; consisting of an outer layer of squamous epithelium and an inner layer of connective tissue (CT). This cortex is devoid of any muscle fibres (Fig. 10.44C and D). The two testes, while they are separated morphologically, are connected with each other by a very thin sheet of CT (Fig 10.44A–C). The testis itself is composed of convoluted seminiferous tubules, which appear, in a transverse section, as follicles of different sizes, outlines, and degrees of maturity. They are closely packed together in a comparatively loose CT in which the blood supply is found. The follicle is covered by a very thin wall of CT, referred to as membrana propria. The immature follicle is somewhat blocked with the germ cells, while the mature follicle has a large acentric cavity formed as a result of the process of spermiogenesis. It appears that the maturation of the seminiferous tubules of the testis of *P. kerathurus* proceeds from the outward, towards the inner edge of the testis. Bawab (1964) examined serial preparations of the two testes and described a collecting tubule that proceeds all along the inner edge of each testis forming its main axis, or more precisely piercing this axis (Fig. 10.44A and B). Each collecting tubule receives seminiferous tubules during its passage along the testis (Fig. 10.45A and B). At certain levels, other extra smaller tubules were observed receiving seminiferous tubules (Fig. 10.45C and D). In the opinion of the author, the seminiferous tubules receive each other until they ultimately reach the main collecting tubule.

At the most posterior region of the testis, the collecting tubule is outside the cortex, and proceeds independently forming the initial part of the vas deferens (Fig. 10.42A and B).

Opposite to the closely approximated testes, the **vasa deferentia** are widely separated (Fig. 10.42B). A great proportion of each of the two vasa deferentia runs beneath its corresponding testis (Fig. 10.42A). Each vas deferens begins as a continuation of the inner axis of the testis (collecting tubule). It separates from the latter at the level of the eighth lobe, and runs as a very narrow short duct that increases gradually and slowly in diameter. This is the **proximal vas deferens** (PVD), which is naturally concealed by the eighth lobe. Beyond the level of this lobe, the PVD increases abruptly in diameter and at the same time flexes to run forwards underneath the corresponding testis. This flexed portion with large diameter projects as a pouch beyond the level of the eighth lobe and so referred to as **blind pouch** (Malek and Bawab, 1974a). Just at about the level of the second lobe of the testis, it flexes again to run backwards to the level of the eighth lobe forming an inverted U-tube. This is the **main or medial vas deferens** (MVD), which is completely concealed by the corresponding testis. The continuation of the blind pouch is the ascending limb of the MVD, after the second flexion it is the descending limb of MVD. The three terms—blind pouch, ascending limbs, and descending limbs—were used throughout the work of Bawab (1964) and that of Malek and Bawab (1974a). The larger diameter attained at the first flexion, is uniformly shown by the whole length of the ascending limb, and is further maintained along the greater part of the descending limb. Normally, the main region attains a milky white colour in adult mature specimens. However, two hyaline streaks run a particular course along the two limbs of the vas deferens (Bawab, 1964, 1970).

The descending limb, beyond the level of the testis, tapers gradually so as to merge with the succeeding region. Just before this tapering, the descending limb rotates

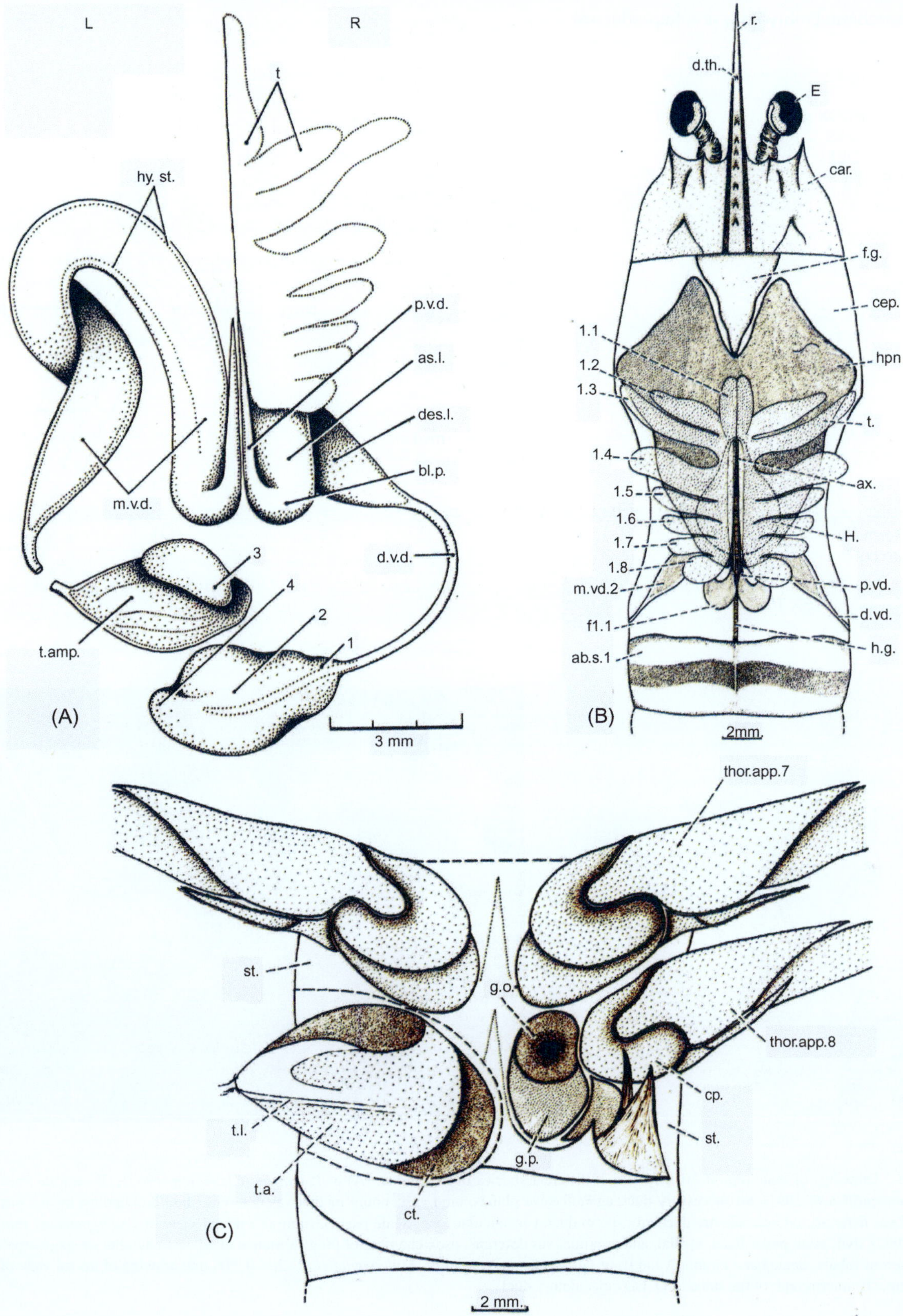

FIG. 10.42 Male reproductive system of *Penaeus kerathurus*: (A) the right (R) as seen in situ with the position of the corresponding testis (t) in outline, the left (L) displaced to expose its medial region *(from Malek and Bawab, 1974a)*. As. L. = ascending limb, bl. P. = blind pouch, des. L. = descending limb, d.v.d. = distal vas deferens, hy. St. = hyaline streaks, m.v.d. = medial vas deferens, p.v.d. = proximal vas deferens, t. amp. = terminal ampoule with its regions: (1) apex; (2) trunk; (3) pocket; (4) base; (B) Camera Lucida drawing showing dorsal view of a dissected male adult specimen *P. k.* showing the reproductive system in situ. ax. = main axis of the testis, ab. s.1 = first abdominal segment, car. = carapace, cep. = cephalothorax, d. th. = dorsal tooth, E. = eye, f. g. = fore gut, H. = heart, h. g. = hind gut, hpn. = hepatopancreas, l. 1–8 = lobes of the testis, r. = rostrum, t. = testis; (C) Camera Lucida drawing of the ventral posterior region of the cephalothorax of the male specimen showing the genital opening on the right side and the exposed terminal ampoule, in situ, on the left side. c. p. = coxopodite, ct. = crest of the terminal ampoule, g. o. = male genital opening, g. p. = genital papilla, st. = sternum, t. a. = terminal ampoule, t. l. = transparent line of the spermatophore, thor. App 7, 8 = seventh and eighth thoracic appendages. *(B and C) From Bawab (1964)*.

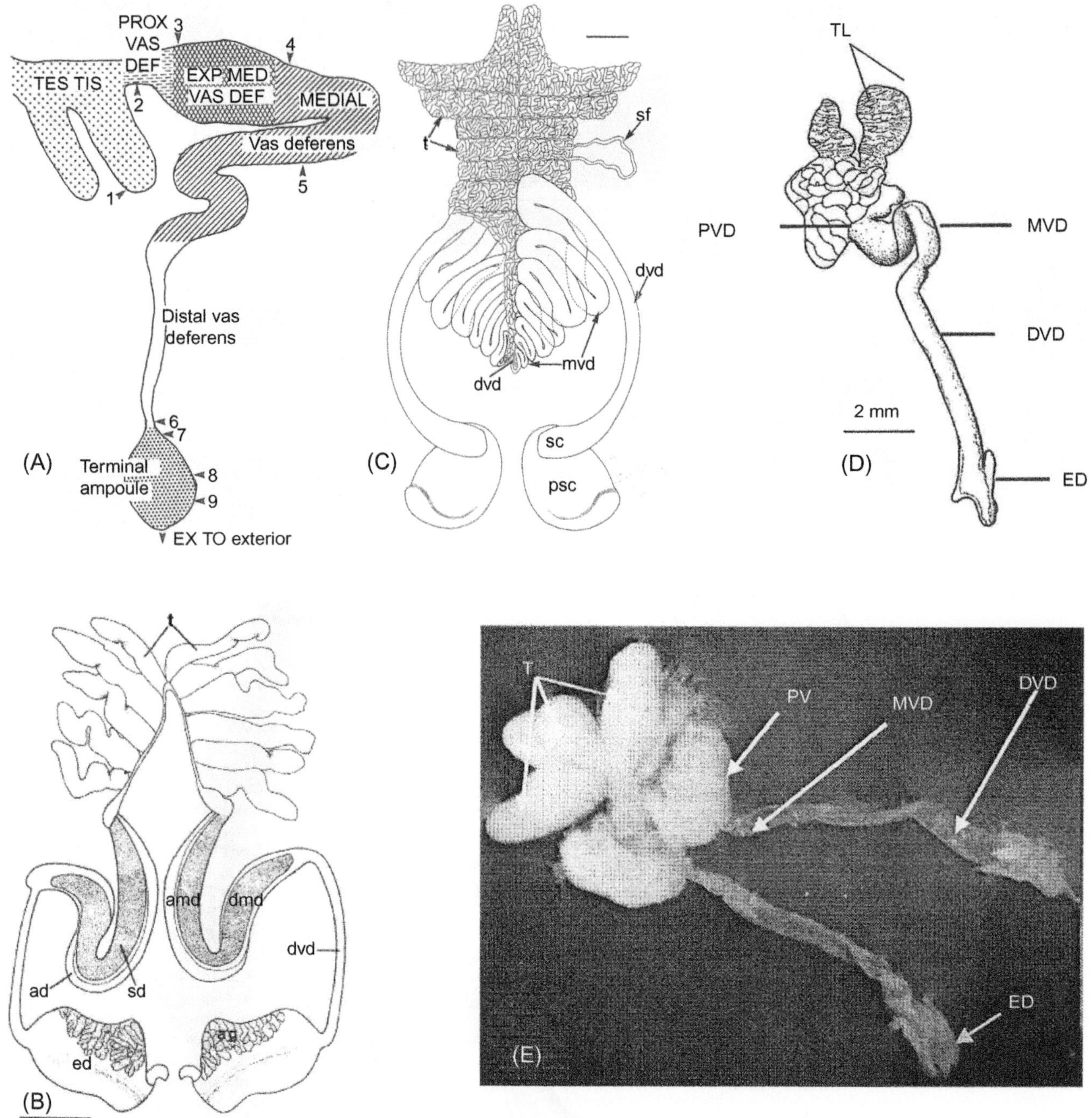

FIG. 10.43 Drawings of male reproductive system of different species of Penaeidae: (A) of *P. indicus (Champion, 1987)*; (B) ventral view of *P. setiferus (Bauer and Cash, 1991)*. ad = accessory duct, ag = adhesive glands, amd = ascending medial vas deferens, dmd descending medial vas deferens, dvd = distal vas deferens, ed = ejaculatory duct, sd = sperm duct, t = testicular lobe. Scale bar = 5.0 mm; (C) dorsal view of *Trachypenaeus similis (Bauer and Min, 1993)*. dvd; mvd; pvd = distal, medial, and proximal vas deferens, psc = chamber of plug substance of ed, sc = chamber of spermatophore of ed, sf = seminiferous tubule. Scale bar = 1 mm; (D and E) of *Parapenaeopsis stylifera (Muhammad et al., 2014)*: (D) line drawing of dorsal view of right half of the system; (E) micrograph of the same of D. ED = ejaculatory duct.

around its longitudinal axis through 90 degrees. As may be expected, the rotation of the vasa deferentia occurs in opposite directions. The tapered descending limb becomes a narrower duct and proceeds uniformly and takes its passage around the body of the specimen to be ventrally reaching the coxa of the corresponding fifth pereiopod. This narrow, straight duct, of uniform diameter, is the **distal vas deferens** (DVD). A noticeable observation is that the gradual decreased portion of the descending limb seen in between the seventh and eighth lobes of the testes, and the blind

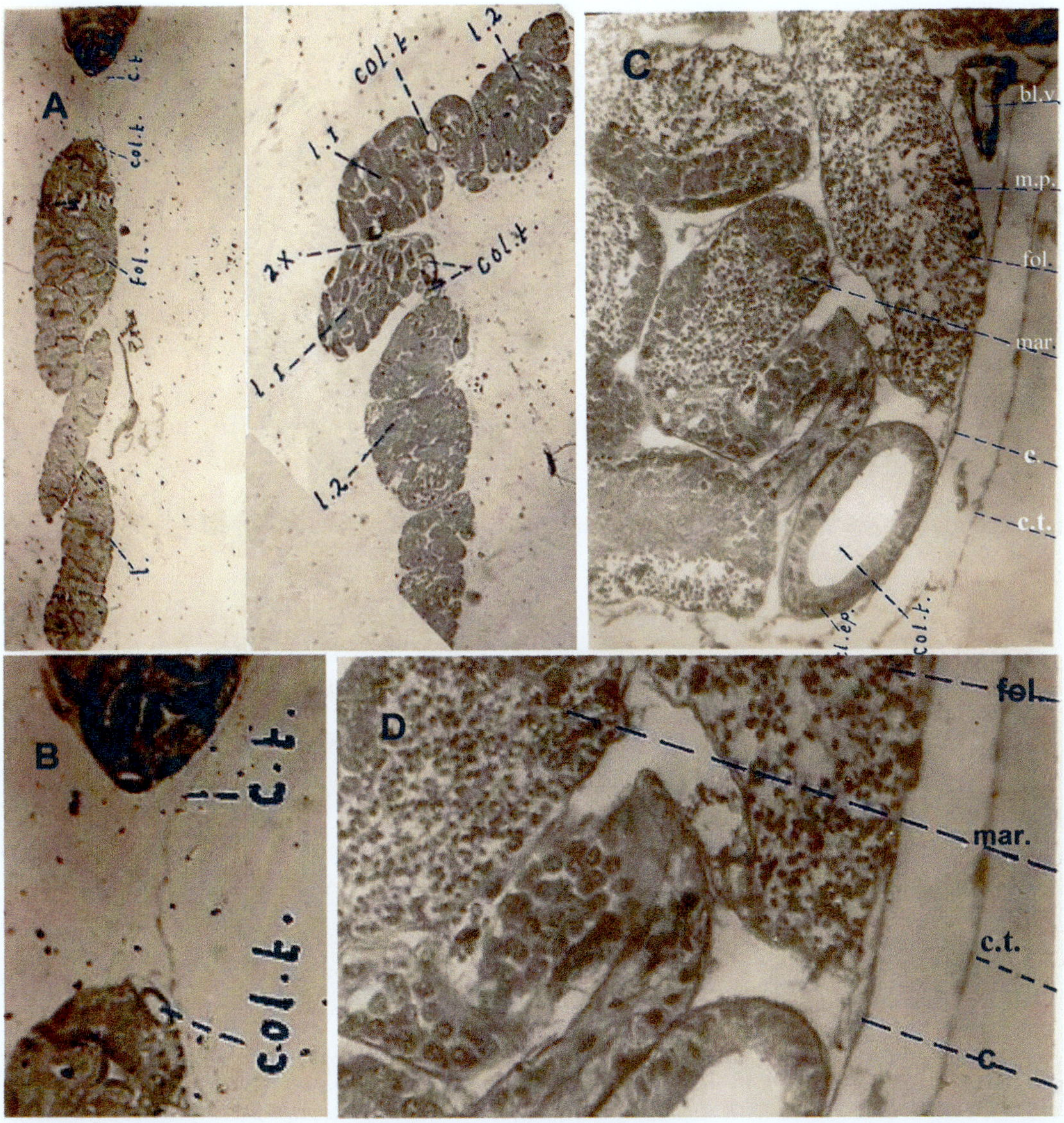

FIG. 10.44 (A) Photomicrographs of T.S. through the testes, right photo through the anterior most region showing undivided collecting tube in the right testis and divided one in the left testis. l.1 and l.2 = first and second lobes of the testis, 2x = axes of the testes; left photo through the middle region of the testis showing three lobes (l.) and the connective tissue (c.t.) sheet connecting the two testes; (B) enlarged portion of (A) to clarify the connective tissue sheet; (C) photomicrograph of a T.S. through the middle region of the testis. bl. V. = blood vessel, c. = cortex of the testis, c. t. = connective tissue, cl. Ep. = columnar epithelium, col. t. = collecting tubule, fol. = follicle, m. p. = membrane propria of the follicle, mar. = marginal spermatocytes; (D) enlarged portion of (C) to clarify c, c.t. and mar. *(From Bawab (1964).)*

pouches, are the only portions of the vasa deferentia that appear dorsally in a dissected specimen (Fig. 10.42B). Each vas deferens ends by a conspicuous enlarged pear-shaped sac, the terminal ampoule (TA). The DVD, when reaching the corresponding coxa, suddenly enlarges greatly in diameter. Owing to its extension ventrally, the volumenous TA takes transverse position in relation to the main axis of the animal (Fig. 10.42A and C). Each TA opens to the exterior by a comparatively large rounded genital opening, which lies on a soft genital papilla found inside its coxa (Fig. 10.42C).

This above description shows that the vas deferens of *P. kerathurus* is divided morphologically and functionally (see later) into four distinct regions: (1) the PVD originates as a continuation of the main axis of the testis, proceeds shortly as a comparatively very narrow duct; (2) the MVD begins by the flexion of the PVD, proceeds as an enlarged inverted U-shaped duct concealed underneath the testis; (3) the DVD begins after the gradual tapering of the descending limb of the MVD, proceeds as uniform narrow, somewhat long duct and ends ventrally in the fourth region; and (4) the TA, the enlarged pear-shaped structure and the end region

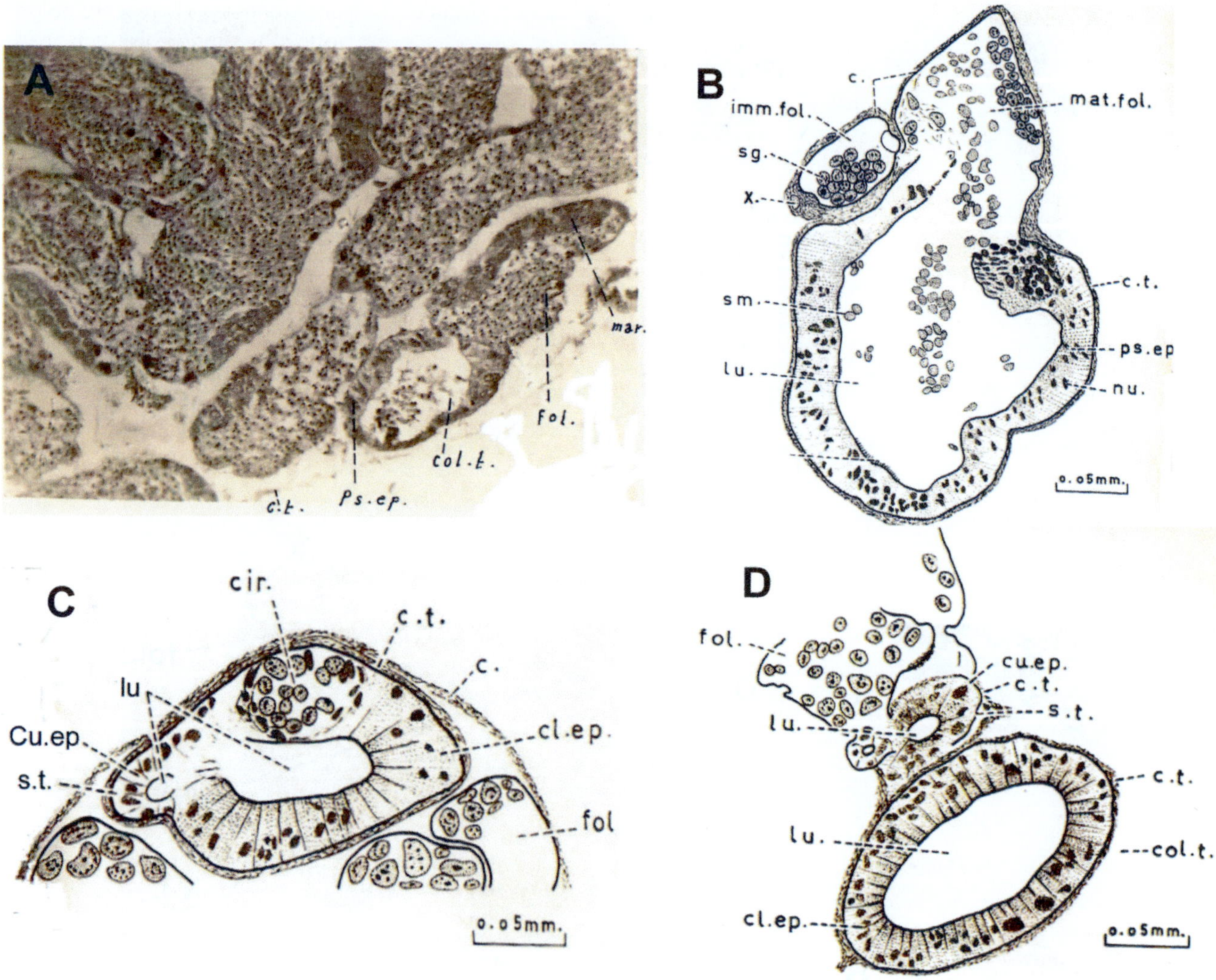

FIG. 10.45 (A) Photomicrograph of a T.S. through the middle region of the testis showing the collecting tubule receiving follicle filled with spermatids; (B–D) Camera Lucida drawings: (B) of a T.S. through the anterior region of the testis showing a collecting tubule receiving a follicle; (C) showing the main collecting tubule receiving a smaller one; (D) showing the collecting tubule with a neighbouring smaller one receiving a follicle. cir. = circle of specialized cells, cu ep. = cubical epithelium, imm. Fol. = immature follicle, lu. = lumen, mat. Fol. = mature follicle, nu. = nucleus, ps. ep. = pseudo-stratified epithelium, sm. = sperm, other lettering as before. *(After Bawab (1964).)*

of the vas deferens, concealed completely within the coxa of the fifth pereiopod. This division is greatly conformed to the old typical division of the vas deferens of Decapoda in general as viewed by Calman (1909). It was divided into three divisions: (1) a narrow efferent duct leading from the testis; (2) a glandular part, with a wider lumen often convoluted; and (3) a terminal ductus ejaculatorius with muscular walls. Only the DVD was missing. However, in Penaeidae in earliar works, e.g. the present species (Heldt, 1938) and *P. setiferus* (King, 1948), the vas deferens was divided almost the same as the present division and described with even the two flexions of the MVD. Even the subsequent works such as the spermatophore formation in the American white shrimps *P. setiferus* and *P. vannamei* (Chow et al., 1991) prove that the gross morphology of the male reproductive tract of different species of the genus *Penaeus* is virtually the same. In addition, this fact can be applied on the three species of *Penaeus* studied by Bauer and Cash (1991), except for the different terminology of some regions. On the other hand, in *P. indicus* (Champion, 1987) the vas deferens has more or less the same morphology with the same terminology (Fig. 10.43A). Muhammad et al. (2014) divided the vas deferens of *Parapenaeopsis stylifera* into four distinct parts as that of *P. kerathurus* with the same terminology, but not with the same morphology (Fig. 10.43D and E). Likewise is that of *Trachypenaeus similis* (Bauer and Min, 1993). In another somewhat recent work, Subramoniam (1995) divided the vas deferens of *Sicyonia ingentis* (Family Penaeidae, Subfamily Sicyoninae (Schram, 1986)) into proximal secretory, mid storage, and distal ejaculatory regions. It appears that the structure and division of the vas deferens into definite parts differ with its product. Each of that of *Penaeus kerathurus* produces only one highly complex spermatophore, that of *Parapenaeopsis stylifera* and of *Trachypenaeus similis* produce a large number of simple and small spermatophores (Fig. 10.46), while that of *Sicyonia ingentis* does not produce spermatophores, but transfers the sperm suspended in seminal plasm. Therefore, the structure and regions of the vas deferens of Penaeidae is correlated with its function.

Formation of the spermatophore of P. kerathurus. Bawab (1970) and Malek and Bawab (1974a,b) studied in detail the functions of the different regions of the male reproductive system of *P. kerathurus* and the mechanism of the formation of its spermatophore. There were no preceding works or evern following studies as such views were added. In any mature specimen of this species at any time of the year, the male reproductive system contains three spermatophores in three successive stages of formation. The first one is the somewhat compact sperm mass in the blind pouch; the second is in the inverted U-shaped MVD as an incomplete but unified spermatophore, i.e. a ribbon of enveloped sperm mass in one duct and another solid ribbon (the precursor of the wing) in the other complementary duct; the third one is the completed spermatophore inside the TA, moulded to its final form, but not completely finished, i.e. perhaps not hardened (Bawab, 1970).

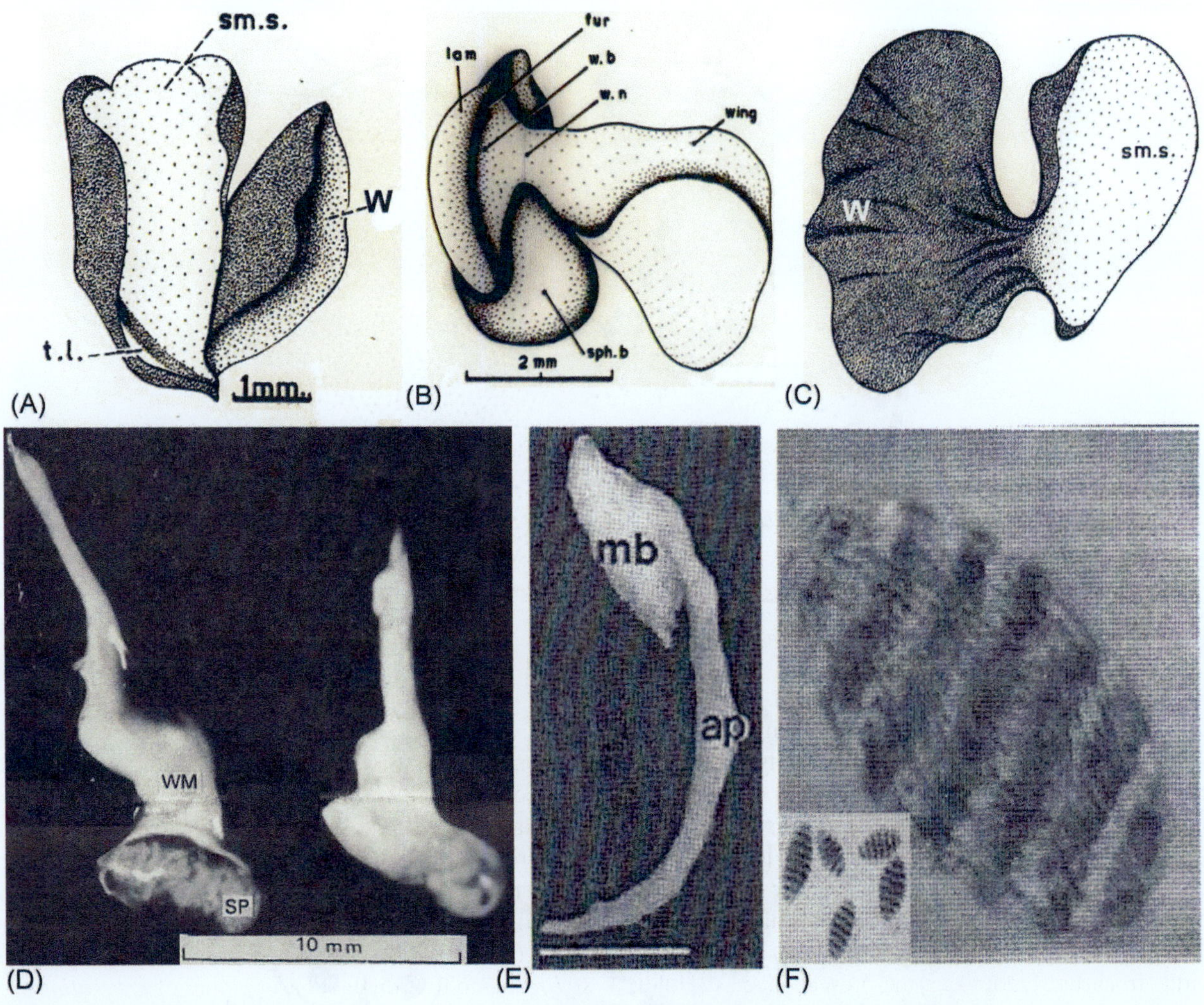

FIG. 10.46 Drawings and photographs of whole mounts of spermatophores of different penaeids: (A–C) of *P. kerathurus*; (A) spermatophore just after discharging from the terminal ampoule; (C) the same with completely unfolded wing *((A and C) from Bawab, 1964)*; (B) the same with partly folded wing *(from Bawab, 1970)*. fur = furrow, lam = lamina, sph. b = spermatophoric body, sm. s. = sperm sac, t. l = transparent line, w = wing, w. b = wing base, w. n = wing neck; (D) photograph of the extruded spermatophoric masses from the left and right terminal ampoules of *P. indicus (from Champion, 1987)*. WM = the viscous white wing matrix, SP = oval, semitranslucent spermatophore; (E) spermatophore of *P. duorarum* placed in 100% formalin immediately following ejaculation. Scale bar represents 4.2 mm *(from Bauer and Cash, 1991)*. ap = appendage, mb = main body; (F) photographs of spermatophores of *Parapenaeopsis stylifera* (base of the figure) and a magnified one showing striations *(from Muhammad et al., 2014)*.

Initial formation of a sperm mass. The morphology of the PVD denotes that it may be regarded merely as a conducting tube, but it is also secretory. Its wall is of considerable thickness due to the incorporation of irregularly distributed multicellular glands. The cells of these glands appear loaded with highly refractile granules (Fig. 10.47A). These loaded cells, when ruptured, extrude their product (Fig. 10.47B); the secretion is of holocrine type (Garven, 1965). A copious secretion is seen spreading, rather unevenly, among the individually separated sperm cells. The latter reach the PVD from the testis via collecting tubules whose lumens are always occupied by sperm somewhat difficult to discern, not accompanied by any secretion (Fig. 10.45). Therefore, the role of the PVD in the formation of the spermatophore is not only to conduct the sperm cells from the testis to the succeeding MVD, but also to assist in coalescing these individually separated cells.

Fig. 10.42 shows that the PVD joining the succeeding region at one side, rather than terminally just at the first flexion, so the MVD begins as a blind pouch. This latter pouch acts as a reservoir for storing the sperm cells, which are presumably delivered over a long period of time by the PVD. The much wider lumen of the pouch (compared with that of the PVD) is a necessity for accommodating greater amounts of sperm cells (Fig. 10.47C). Moreover, it participates in converting these cells into a coherent mass. Indeed, critical examination of stained paraffin sections reveals the presence of innumerable extremely fine thread-like filaments distributed everywhere among the sperm cells (Fig. 10.47D). It was evidenced that the properties of these filaments are the same as the free secretion of the PVD and they are produced by the much regular glandular wall of this pouch. However, no such filaments can be seen in fresh

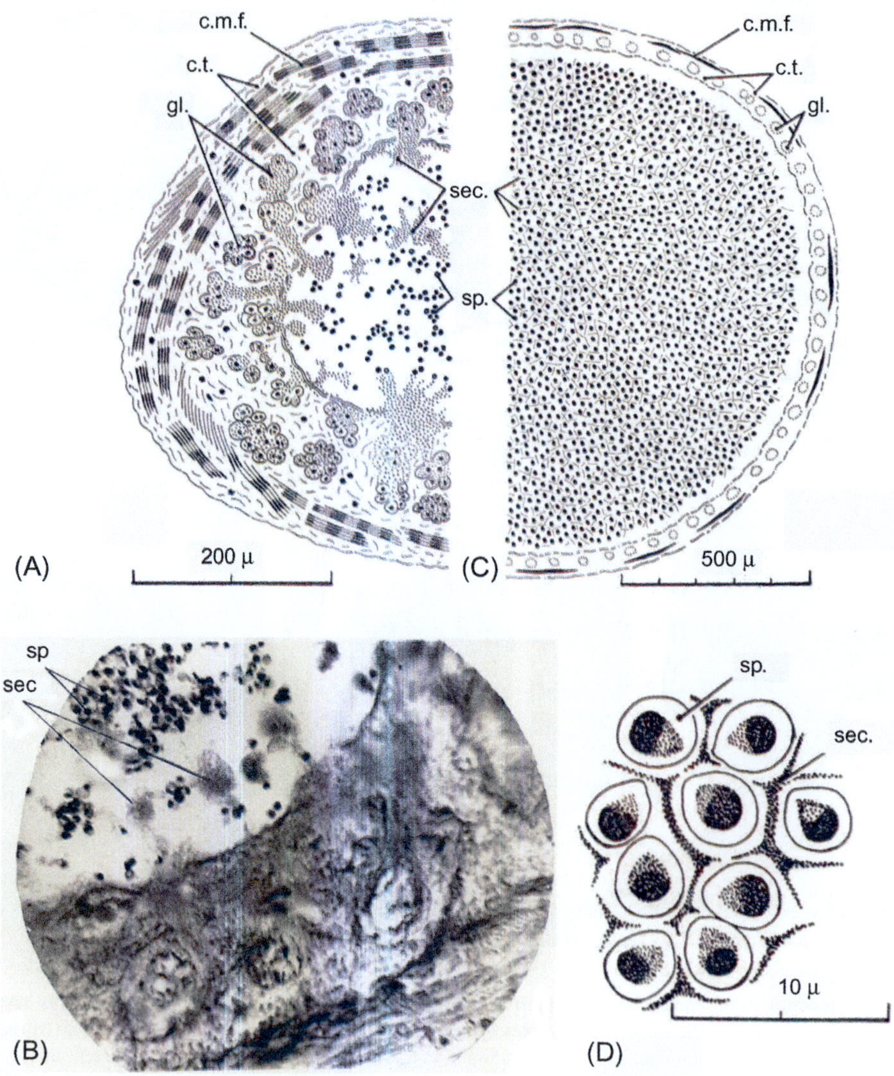

FIG. 10.47 Camera Lucida drawings of (A), (C), and (D) of T.S. through parts of the vas deferens of *P. k.* showing in: (A) the proximal vas deferens; (B) photomicrograph of an enlarged portion of the secretory wall of (A). 500×; (C) the blind pouch; (D) enlarged part of the contents of the pouch. c.m.f. = circular muscle fibres, c.t. = connective tissue, gl. = glands, sec. = secretion, sp. sperm cells. *(From Malek and Bawab (1974a), except (B) from Bawab (1970).)*

preparations, since when smeared some of the contents of the pouch on clean slides, a viscous substance appeared bathing and/or embedding the sperm cells. It was concluded that it is in this pouch that a true sperm mass is ultimately formed.

The importance of this viscous substance remains to be discussed; whether in *Penaeus* it would serve other functions is not quite clear. However, it may be compared with the fluid part of the semen of vertebrate reproductive organs. In man, for example, the semen is known to contain various inorganic and organic components, a high concentration of fructose, which is apparently metabolized by the sperm, and, on the whole, to provide the sperm with the necessary medium for their viability and motility (see Keele and Neil, 1973). The presence of a nourishing material may seem a necessity in the analogous substance in *Penaeus.* However, it cannot be accepted without reserve whether or not this substance in a crustacean should truly be necessary for the motility of sperm cells. Therefore, the nature and function of this substance in the sperm mass, at least in *Penaeus*, is an open question.

The two ducts of the ascending limb. With very few exceptions, Calman (1909) considered the possession of a wing to be characteristic of the spermatophores of macruran decapods. In *Penaeus*, this wing is a foliaceous accessory structure attached to a main globular body, which contains the sperm mass.

Unlike the preceding regions of the vas deferens (the PVD and the blind pouch), the interior of the ascending limb, into which the sperm mass moves next, is divided into two independent ducts (Fig. 10.48). These are of unequal capacity, and only the wider of them is continuous with the cavity of the blind pouch. Inside this wider duct, the sperm mass attains its outer layers to become the body of the spermatophore, and hence its name is the **spermatophoric duct** (Malek and Bawab, 1974b). On the other hand, the narrow duct begins blindly from the wall of the ascending limb at its most proximal end. The examination of serial preparations of the system proves that it begins as a thickening in the wall of the ascending limb at the position nearest to the main axis of the animal. An enormous increase in the connective tissue separating the muscle layer from that of the glands results in this thickening, and it is there that lies the blind lowermost end of a new duct (Fig. 10.49). At this level, this duct is narrow, and its lumen appears lined with irregular clusters of glandular cells. This new blind duct is merely a side channel inside which the fundamental layers of the wing are laid down and hence its name, **wing duct** (Malek and Bawab, 1974b). In the work of Bauer and Cash (1991), these two ducts are referred to as the 'sperm duct' and 'accessory duct'.

However, all along the ascending limb of *P. kerathurus* extends a streak of connective tissue that subdivides its cavity into two independent ducts. This is the septum, which joins the inner walls of the ascending limb at varying planes and can be observed as hyaline streaks in fresh specimens (Bawab, 1964, 1970). This septum remains basically made of connective tissue as it was when it first began, with glands on either side. A few muscle fibres invade its matrix here and there and, of more importance, blood islands for conveying the material which the glands apparently metabolize (Fig. 10.48).

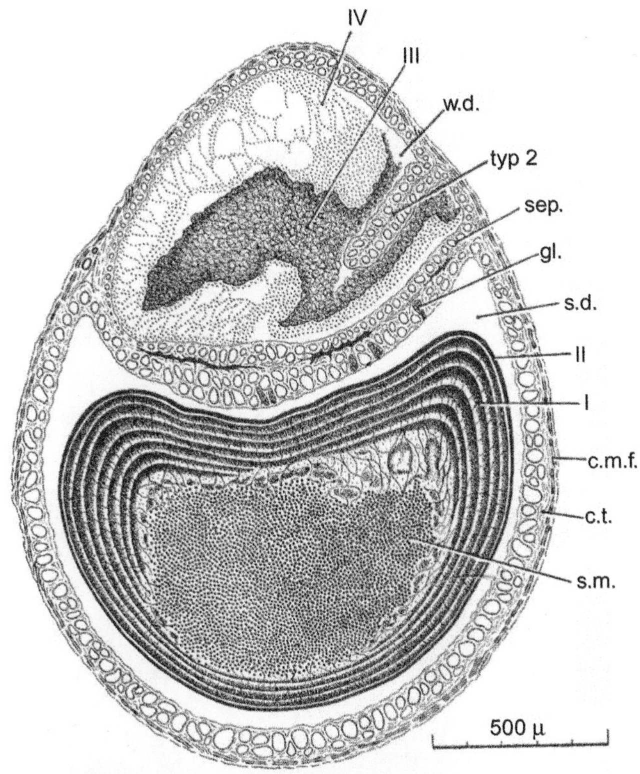

FIG. 10.48 Camera Lucida drawing of a T.S. through the ascending limb slightly beyond its middle, with the septum (sep) completely separating the wing duct (w.d.) and the spermatophoric duct (s.d.), layers I, II surround the sperm mass (sm), layer III appears applied to typhlosole 2 (typ2) and layer IV fitting around III. c.m.f. = circular muscle fibres, c.t. = connective tissue, gl. = glands. *(From Malek and Bawab (1974b).)*

The deposition of the main layers of the body and the wing. The most prominent feature in the histology of both ducts, spermatophoric and wing ducts, is the extraordinary activity of the glands present in their walls. No other region throughout the entire length of the vas deferens of *Penaeus kerathurus* contains as highly active glands. This is not surprising, since in these two ducts as mentioned above, the basical structure of the spermatophore takes place.

All along the whole length of the ascending limb, three phases of activity can readily be recognized in the lumen of the spermatophoric duct. At the most proximal portion of this limb, just above the blind pouch, this lumen is practically free from secretion. It is only occupied by the coherent sperm mass passed from the blind pouch (Fig. 10.49B and C). A little beyond this level, the glands of the

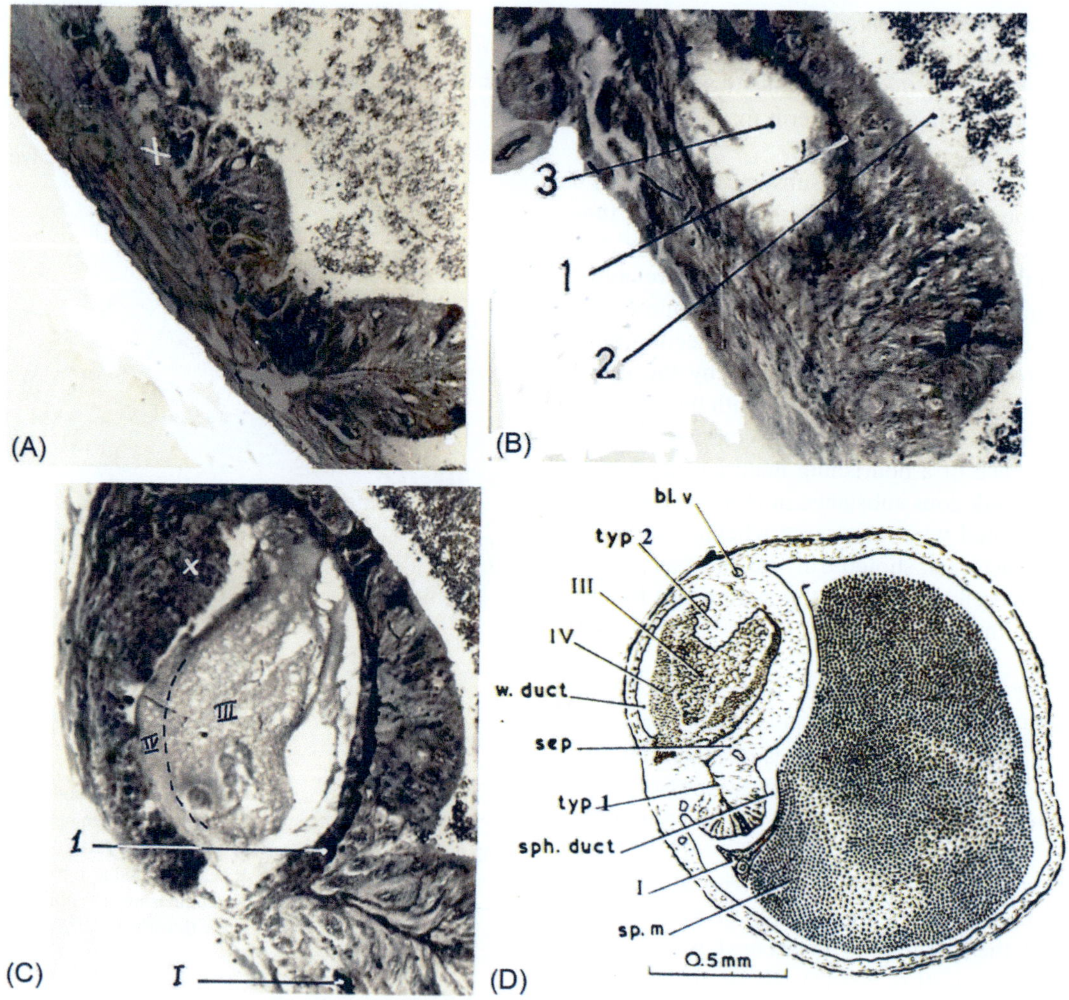

FIG. 10.49 Photomicrographs of successive T.S.s of the portion of the ascending limb showing the formation of the wing duct: (A) the thickened portion indicated by X where the duct originates. 110×; (B) cavity appears within the thickened portion, i.e. lumen of the new wing duct (3). 150×; (C) beginning of the activity of the new duct. 150×; (D) the complete formation of the wing duct separated from the spermatophoric duct (sph.duct). bl.v.=blood vessel, 1 and sep=septum, sp.m=sperm mass, typ=typhlosole, I, III, IV=secretions I, III, IV, respectively. *(From Bawab (1970).)*

spermatophoric duct show extraordinary activity (Fig. 10.49D). Owing to a marked increase in their number, they project into the lumen as a prominent typhlosole. This is, however, only a temporary structure disappearing halfway up the length of the spermatophoric duct (Fig. 10.48). It was termed typhlosole 1 (Malek and Bawab, 1974b). Its glands, active as they remain, may be regarded as forming the principal secretory centre throughout. At the level of the figure (Fig. 10.49D), the **first phase** of activity occurs by the glands of typhlosole 1, which appear highly loaded with discrete granules. The latter is discharged as a secretion, which stains, as the granules, clear green with Masson's trichorome stain, and a pale blue with the combination of Mallory. Hitherto, it shows no affinity towards Hiedenhain's haematoxylin. This secretion is first seen covering the surface of the portion of the sperm mass, just opposite to the typhlosole. At a slightly higher level, both the glands and secretion change reactions and become pronouncedly haematoxylinophil. This is because of the impregnation of the initial material with another secretion of a different kind. This is the **second phase** of activity. Still further up the spermatophoric duct, the formed conjugate settles down beginning the formation of layer I of the body of the spermatophore. This layer is as a network of haematoxylinophil material embedded in a continuous green-staining matrix. The only addition to this description is the electron microscopic study of the secretions of this layer in *P. setiferus* and *P. vannamei* (Fig. 10.50A and B) (Chow et al., 1991).

These activities in the spermatophoric duct of *P. kerathurus*, hitherto confined to the region of typhlosole 1, are repeated in the same sequence at advancing levels, where adjacent areas in the septum and then areas of the duct wall become similarly active. Continuation of the process leads to a gradual spreading of the deposited material of layer I around the sperm mass, and to its advancing edges eventually fusing together in the manner represented in

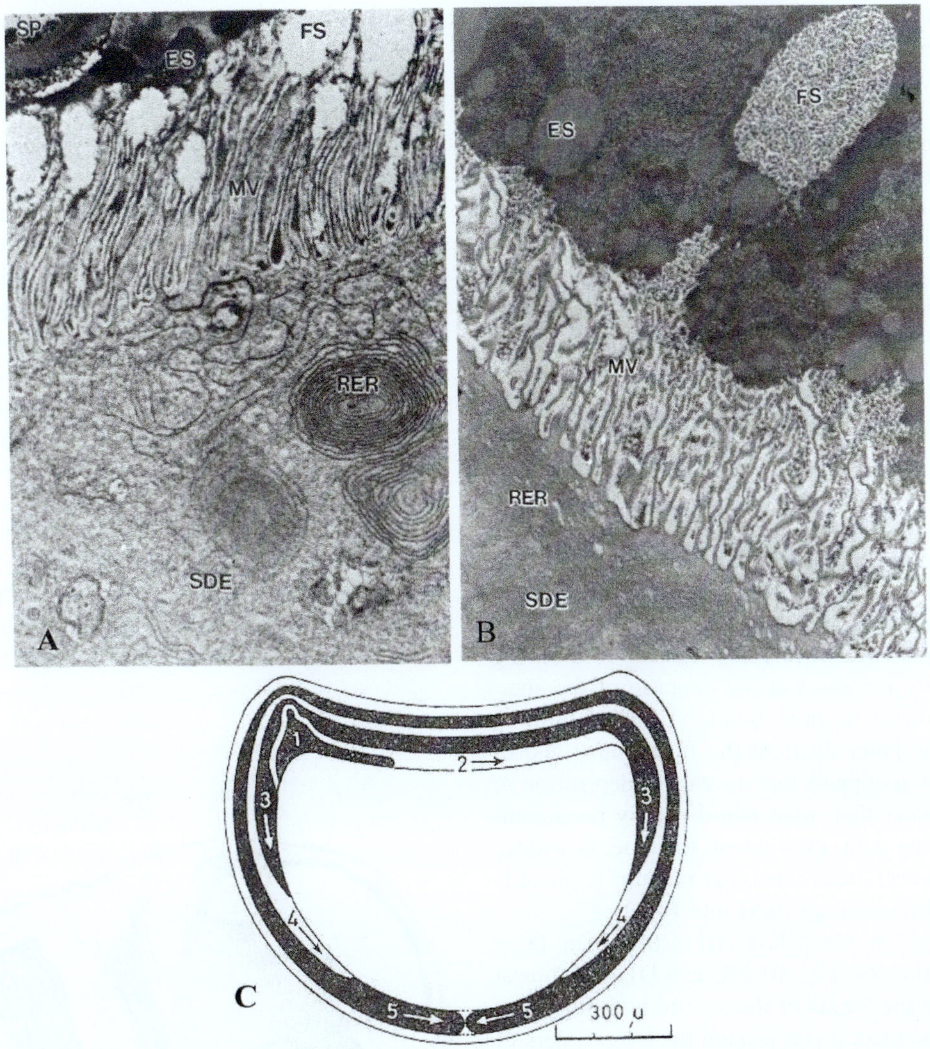

FIG. 10.50 (A and B) *from Chow et al. (1991)* present SEM sections showing cells of spermatophoric duct epithelium (SDE) that exhibit abundant rough endoplasmic reticulum (RER) in *Penaeus setiferus* ((A) 6.450×) and *P. vannamri* ((B) 12.900×). Secretory product is observed at cell apices, which possess numerous microvilli (MV). Primary spermatophore layer consists of electron-dense substance (ES) and flocculent substance (FS); (C) Diagrammatic representation of five successive stages (1–5) in the deposition of Layer I in the spermatophoric duct. *From Malek and Bawab (1974b).*

Fig. 10.50C. This process of formation of layer I of the body of the spermatophore is completed along the lower half of the ascending limb. It is seen from Fig. 10.48 that the fully formed layer reaches a maximum thickness of about 200 μm next to the septum where secretion first began, but thins out in a gradual manner to as little as 70 μm at the opposite side. Anywhere along its length, layer I has a laminar appearance. The laminae are widely spaced in gelatin sections, but are more closely packed together in paraffin ones. These differing appearances of the laminae can only be attributed to a difference in the nature of the fixatives employed. With Gilson's mercuro-nitric mixture, used in preparing paraffin sections, a differential fixing effect results in the separation of the components of which layer I is built. No such effect seems to be produced by the formol-saline with which is treated the gelatin-embedded material.

The glands responsible for the secretion of the materials forming layer I, change their secretory activity along the upper half of the ascending limb. They metabolize a new product, which in due course sets down around layer I a continuous layer: layer II (Fig. 10.48). The formation of the latter represents the **third** and final phase in the series of activities shown by the glands of the spermatophoric duct. Although this layer is much thinner than layer I, it is homogeneous and more easily identified in gelatin than in paraffin sections. In common with layer I, it shows strong affinities for the acid fuchsin and haematoxylin of Masson, but it further colours to some extent with Sudan black B, to which layer I is completely refractory.

Like the spermatophoric duct, the wing duct which is regarded as complementary to the former (Fig. 10.49), contains a typhlosole with apparently as highly active glands.

For distinction, it is referred to as typhlosole 2. It is not merely secretory as is typhlosole 1, but further participates in adjusting the position of the wing relative to the accompanying body until the entire spermatophore attains its final form in the terminal ampoule. At a very low level of the ascending limb, shortly after the wing duct has appeared, a frothy secretion can already be seen occupying a large area of the lumen of this duct. This secretion constitutes layer III (Fig. 10.49C and D) which, being a product of typhlosole 2 alone, it remains closely applied to its secretory border. During the formation of this layer, a finely granular matrix is first laid down which stains merely green with the combination of Masson. However, owing to secondary impregnation with a different material, this matrix soon changes reactions. It becomes purple with haematoxylin and acid fuchsin at nearly the same level, and in just the same manner, as does layer I in the accompanying spermatophoric duct. However, layer III is not laminated and both layers are closely similar in their chemical composition. More of layer III continues to be added at increasingly higher levels of the ascending limb. Meanwhile, the capacity of the wing duct itself gradually increases at the expense of the spermatophoric duct, owing to an inclination of the delimiting septum towards the latter duct. At the furthermost end of the ascending limb, it appears that the rate of deposition of secretion III is greater than what could readily be accommodated by the wing duct, even where this duct is widest. The secretion thus becomes folded, but remains resolvable into an elongated lamina (Figs 10.51 and 10.52).

At the same level at which layer III is first seen, layer IV begins to be deposited (Fig. 10.49C and D), but stops at about two-thirds of the length of the ascending limb. From the moment it is secreted, it sets around the free borders of layer III. Apart from its homogeneous appearance, which strongly contrasts with the frothy condition of layer III, layer IV exhibits two peculiarities. It always stains green at all levels of the ascending limb, with no affinity for haematoxylin. Moreover, it takes up Sudan black much more pronouncedly than does layer III. In this latter respect, it more closely resembles layer II of the spermatophoric duct than layer III of its own duct.

Preliminary connection of the two parts of the spermatophore. It is clear from the above description that the two parts of the spermatophore—the body and the wing—are independently formed in separate ducts of the ascending limb of the main vas deferens (MVD). Just at the level where layer IV stops deposition, a new material is then metabolized and discharged along the remaining third of the wing duct. Copious globules of differing sizes are first yielded (Fig. 10.51); a few stain either with acid fuchsin or further with haematoxylin, and many more with light green. All of these eventually run together into a continuous layer, referred to as layer V, which sets around layer IV. Like the latter, the newly deposited layer is markedly sudanophil. Because of this similarity, the two layers are hardly distinguishable from each other in paraffin sections. In gelatin sections, however, layer V is considerably thicker, and is seen clearly to be laminated.

Fig. 10.51 is a T.S. through the distal end of the ascending limb. It shows the septum intervening between the wing duct and the spermatophric duct to have detached from the opposing wall nearer to the typhlosole. Excess material of layer V runs across the resulting opening into the spermatophoric duct where it branches into two unequal and widely separated limbs. Each extends on either of the two nearest sides of the semilunarly shaped spermatophoric body (Fig. 10.52). Both limbs apparently adhere firmly to the underlying layer II; it fuses with layer II of the body, which also contains lipid. This is the only change that appears to have taken place after the separation of the two ducts.

Therefore, the role of the main vas deferens of *Penaeus kerathurus* is to secrete the materials and form the essential components of the spermatophore. At this stage, the properties of its constituents are given in Table 10.6.

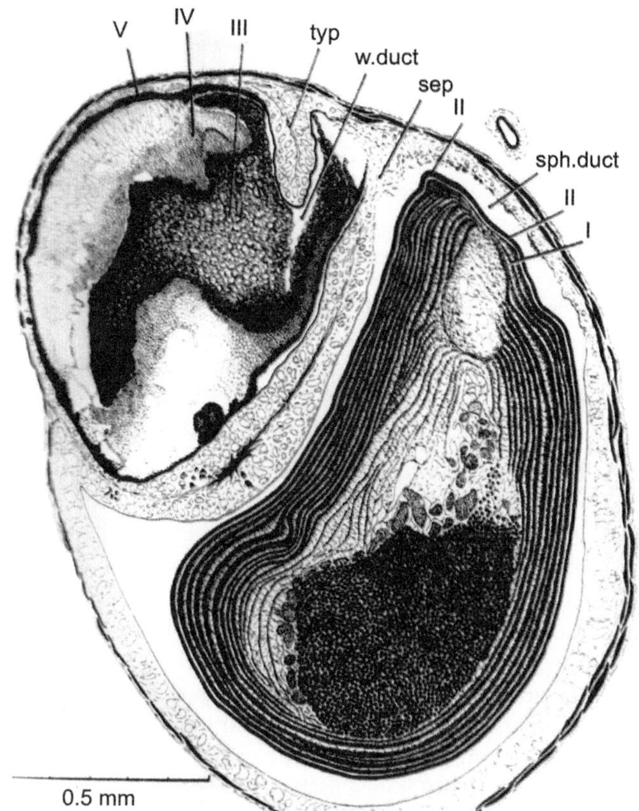

FIG. 10.51 Camera Lucida drawing of T.S. through the distal end of the ascending limb, the septum is about to detach from the opposing wall, and the deposition of secretion V (V) in the wing duct in progression, lettering as before. *(From Bawab (1970).)*

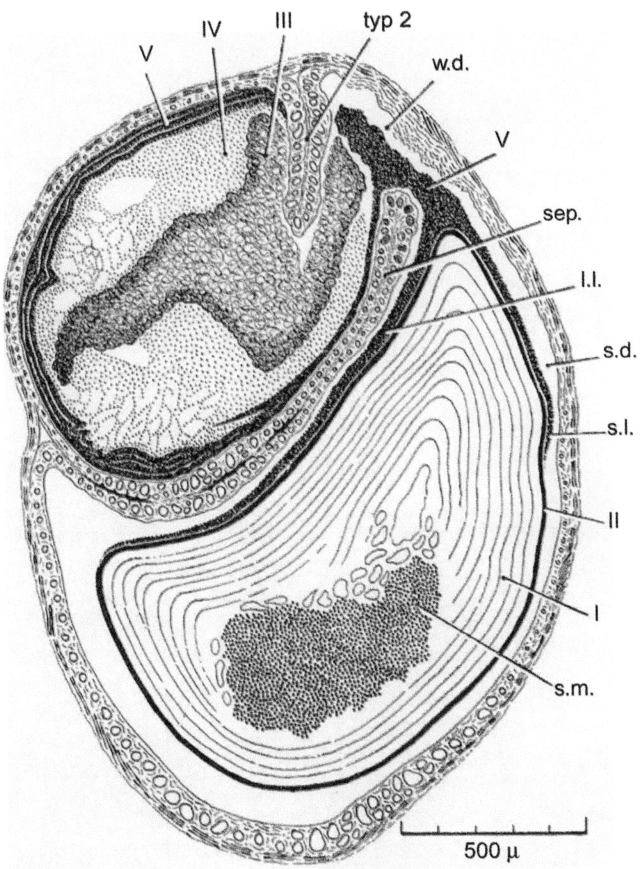

FIG. 10.52 T.S. through the ascending limb at its farthermost end, with the septum open at one side. Secretion V of the wing duct is seen flowing into the spermatophoric duct across the open part of the septum. l.l.=long limb and s.l.=small limb of secretion V, other lettering as before. *(From Malek and Bawab (1970b).)*

The descending limb. The descending limb bears the same general properties as the adjoining ascending limb, since they form one region of the vas deferens. Therefore, it contains a septum and a typhlosole continuous with those already recognized in the preceding region. However, in contrast with the ascending limb, the secretory activity of the descending limb is very limited. However, only the glands of the typhlosole and of the septum remain active in the accompanying wing duct. Even these yield no new secretions; they only continue to add more of the same materials that they produced in the ascending limb. In spite of this, the descending limb performs an important function. It keeps the two components of the spermatophore—the body and the wing—attached together into one structure without any distortion in spite of the abrupt and acute bending at its beginning course (Fig. 10.53A). During this process, if both components are to undergo distortion, as is theoretically to be expected, then it is the body rather than the wing that would be liable to undergo more. This bending results in the following.

1. The spermatophoric channel becomes wider at the expense of that of the wing. And the latter becomes more round owing to the septum undergoing a marked curvature (Fig. 10.53B). This may be a safeguard against the wing itself undergoing severe distortion.
2. The typhlosole gradually shifts position in the direction of the base of the free septum to be perpendicular to it.
3. Still more important than either of the above two changes is the rotation which takes place almost halfway down the course of the descending limb (see Fig. 10.42A). This results in the exchange in the relative positions of the channels of the wing and of the body of the spermatophore (Fig. 10.53A); compare this with Fig. 10.52.

TABLE 10.6 The appearance and staining reaction of the layers constituting the spermatophore of *Penaeus kerathurus* (Forskal, 1775)

Layer no.	Site of formation	Part of spermatophore	Appearance	Heidenihain's haematoxylin	Masson's trichrome stain	Mallory's triple stain	Sudan black B	
							Gelatin	paraffin
I	Sperm. d.	Body	Laminated	+++	Purple	Red	−ve	−ve
II	Sperm. d.	Body	Not	++++	Purple	Red	++++	+
III	Wing d.	Wing	Feebly laminated	++	Purple	Red	++	+
IV	Wing d.	Wing	Not	−ve	Green	Pale blue	++	++++
V	Wing d.	Wing and body	Laminated	Variable	Variable	Variable	++	++++
VI	Terminal ampoule	Body	Not	++++	Purple	Red	-ve	-ve
VII	Terminal ampoule	Wing	Not	−ve	Green	Pale red	−ve	−ve

From Malek and Bawab (1974a,b).

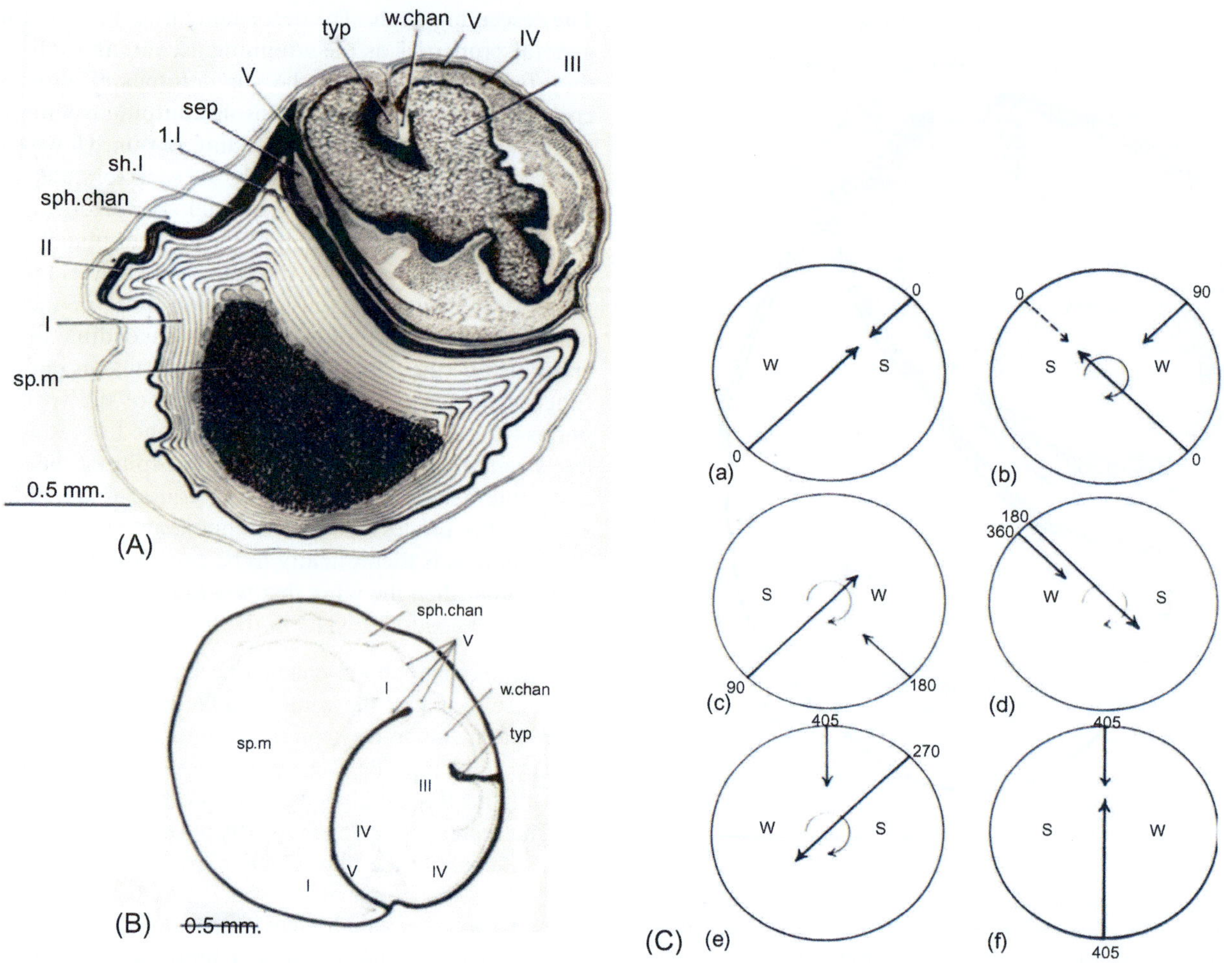

FIG. 10.53 (A) T.S. through the upper (proximal) region of the descending limb of the right vas deferens. Secretion passing from the wing channel forms two limbs, short and long limb around the sperm mass in the spermatophoric channel. l.l. and sh.l.=long and short limbs, sph. Chan and w. chan=spermatophoric and wing channels; (B) T.S. after the external coiling of the ascending limb through 90° has taken place, lettering as before. *(From Bawab, 1970)*; (C) Diagrams of selected sections from a series across the right vas deferens of *Penaeus* to illustrate the shifts in the positions of the typhlosole *(short arrow)* of the wing duct (W), and of the septum *(long arrow)* separating this duct from the spermatophoric duct (S). (a) At the distal end of the ascending limb (see Fig. 10.42A); (b) a mirror-image of (a) as appears at the proximal end of the descending limb, the typhlosole has moved in a clock-wise direction through 90°; (c) along the descending limb just after its external coiling through 90°, the septum has moved through 90°, and the typhlosole through 90° additional to the 90° it has already gained at (b) (i.e. 90° + 90° = 180°); (d) at the tapering end of the ascending limb, the typhlosole has shifted 180° (i.e. 180° + 180° = 360°), while the septum has shifted only 90° (i.e. 90° + 90° = 180°); (e) at the apex of the terminal ampoule, the septum gaining 90° (90° + 180° = 270°), while the typhlosole gains only 45° (45° + 360° = 405°); (f) at the base of the trunk of the terminal ampoule, no change in the positions of the typhlosole, but the septum has further moved through 135° (i.e. 135° + 270° = 405°).

Notice that the relative positions of the septum and typhlosole do not alter. Such a spiral coiling, starting immediately after the bending of the main vas deferens and extending halfway down the length of its descending limb, is undoubtedly an important device, which appears to have been designed to minimize distortion of the soft secretions passing within (Fig. 10.53B).

The main region of the vas deferens has maintained a uniform thickness all along its two limbs. Thereafter, beyond the spiral coiling of the descending limb, the latter begins to taper gradually in the directon of the DVD and to merge smoothly with it (see Fig. 10.42). No change in the histological appearance of this tapering portion except: (1) the septum becomes longer compared to the narrow diameter of this portion; (2) the capacity of the wing channel is wider; and (3) the wall gradually increases in thickness with increase in its muscles (Fig. 10.54).

Role of the distal vas deferens. The remarkable length of this region is in justification of the function that it serves. It connects the dorsally situated main region with the TA, which lies ventrally inside the coxa of the fifth periopod. It has a uniform appearance with narrow diameter. However, histologically, it has the thickest wall with well developed muscles, compared with that of all the regions of the vas deferens. Unlike these regions, its muscle layer is characterized by its great thickness and differentiation into outer circular

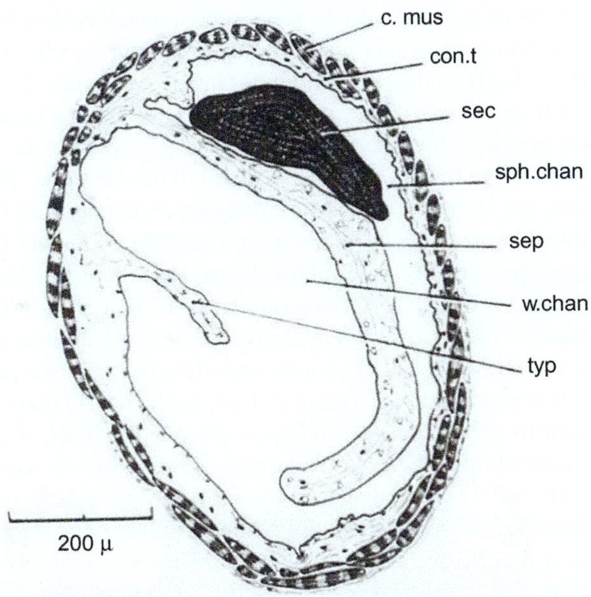

FIG. 10.54 T.S. through the tapering end of the descending limb. Notice the reappearance of the circular muscles and the secretion occupying the spermatophoric channel is that closing the advancing end of the spermatophoric body, lettering as before. *(From Bawab (1970).)*

and inner longitudinal sublayers (Fig. 10.55). While the circular fibres run singly and show discrete cross striations as that in the PVD, the longitudinal ones appear arranged in bundles. The lumen of this region is lined by the same multicellular glands, but typically they are rosette-shaped. No secretory granules are observed within their cells or freely in the lumen. Whether in this state these glands would serve any function is therefore doubted. This is true at least at the time the preparations examined in this work were made, when a spermatophore was being secreted in the MVD and another elaborated in the TA. Practically, the typhlosole and the septum continuous with those in the immediately preceding descending limb occupy the narrower lumen. They are much shorter and thicker. Nevertheless, the lumen is hardly distinguishable into distinct spermatophoric and wing channels. Both the typhlosole and the septum have a connective tissue core and rosette-shaped glands continuous with that present in the surrounding walls, but invading this connective tissue only longitudinal muscles, arranged in somewhat smaller bundles.

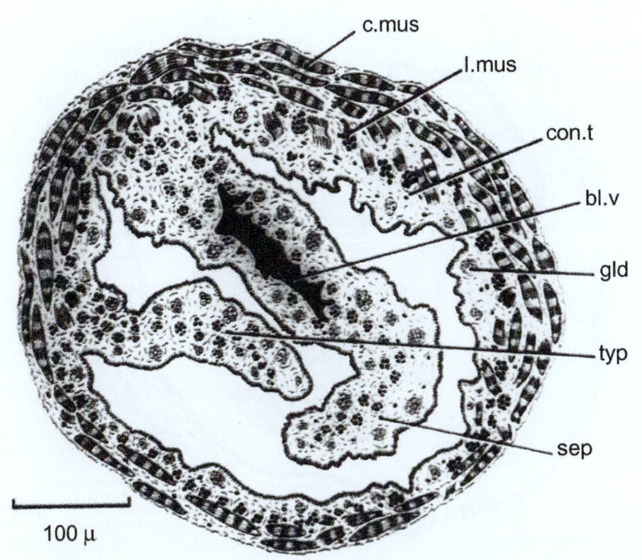

FIG. 10.55 T.S. through the distal vas deferens at about the level nearer to the main vas deferens of the right side. bl.v. = blood vessel, c. mus. = circular muscles, con.t. = connective tissue, gld = glands, l. mus. = longitudinal muscles, sep = septum, typ = typhlosole. *(From Bawab (1970).)*

From the above description, one may conclude that the prime, and perhaps the only, function of the DVD of the vas deferens is that it acts as a duct in the strict sense. It has to conduct the weakly connected unified spermatophore—body and wing—from the main region into the TA, where it will undergo further changes. The typhlosole and the septum inside the DVD facilitate this passage: the former to maintain the laminar form of the wing; and the latter to avoid unnecessary mixing of the soft materials of the two components. However, it cannot be accepted that such soft material arranged in layers as such and given particular shape would pass undistorted from a wide chamber as the main region, into such a narrow duct as the DVD. An earlier observation by Heldt (1938) seems to provide the right clue. Heldt studied the same species of *Penaeus* of that of the present work. She noticed that evacuation of the TA takes place whenever the adult male moulted its old cuticle. By this time, a dissected specimen revealed the DVD to be highly contracted—shortened and widened—so as to allow a new spermatophore to pass to the TA. The present detailed histological observations give full support to Heldt's observations. The well-developed longitudinal muscles of the walls, typhlosole, and septum would effect the required shortening. The lumen of this DVD, presumably then widened as much as the main and the ampoule are, would allow the spermatophore to pass easily, and moreover with minimal distortion, from one chamber to another. The highly folded connective tissue immediately lining the lumen of DVD (Fig. 10.55) is significant. Equally significant is the complete absence of circular muscles from the typhlosole and from the septum. All participate in widening of the duct. Once the spermatophore has passed into the TA, the circular muscles, confined to the walls of the duct, would ensure a return to their normal positions of all such parts involved.

The terminal ampoule. The TA is the most complex region of the vas deferens of *P. kerathurus*. It seems that it is because of this complexity that such competent workers, preceding Bawab and Malek and Bawab, as Heldt (1938) and King (1948), who examined several species of the Penaeidae, did not attempt any detailed study of this structure. The subsequent workers, such as Champion

(1987), Chow et al. (1991), and Bauer and Cash (1991), that studied *P. setiferus* and different species other than *P. kerathurus*, did not give an adequate description of that TA, in spite of the advanced tools used. That might be because its structure differs from that of *P. kerathurus*, or because their spermatophores differ in structure (see later).

The complexity of the TA of *P. kerathurus* appears in its morphological appearance. It is pear-shaped but retains the tubular pattern shown by its preceding regions. However, it is the most dilated region. It is the shortest region of all of this vas deferens. It measures about 4 mm in length in specimens measuring 15–18 cm and 3 mm where it is widest (Bawab, 1964). It is thus open at both ends: at the apical side continuous with the DVD and at the basal side leading to the exterior: the genital pore (Fig. 10.42A and C). It is opaque except for a fine transparent line extending along its whole length on the ventral surface. Again, along the latero-ventral side of the body of this ampoule extends obliquely a bulging part, which marks the position of an internal pocket (Fig. 10.42A and C) communicating with the cavity of the ampoule. The pocket thus adds further width to the ampoule where it is already widest. Owing to the presence of this pocket, another septum appears which runs almost parallel to the original one, but is complete at its upper and lower parts. As this pocket protrudes outside the normal contour of the ampoule, its cavity may be regarded as being added to, rather than cut out, from the original cavity of the ampoule. The cavity of this latter region increases gradually in diameter in the direction of the base. It is widest at the level of the side pocket (Fig. 10.58). The ampoule retains the incomplete septum of the DVD. However, the length of this septum is not constant since it increases with the increase of the cavity of spermatophoric channel. Oppositely, the wall of the latter maintains a uniform thickness except in the region of the pocket. It consists of the same three layers recognized elsewhere in the vas deferens. An innermost layer rich in the rosette-shaped glands is found in the DVD.

As was mentioned above, the ampoule is lodged within the coxopodite, so its body cannot be seen neither externally, nor after careful dissection. However, an intact ampoule can readily be pulled loose by a pair of forceps applied gently to the DVD. If the ampoule of a living animal, freed in this way, is placed in tap water, an intact spermatophore gradually comes out. However, if the whole coxopodite containing the ampoule is similarly treated, nothing happens. The precise nature of this mechanism remains unknown.

The histological studies of the successive levels of the TA provided detailed description for a clear understanding of its various activities in the formation of the spermatophore. The apex of the ampoule (Fig. 10.42B) merely represents a transitional region between the DVD and the trunk of the ampoule, which embodies almost the whole mass of the spermatophore. Fig. 10.56 shows the **upper region** of the trunk of the ampoule. In spite of the considerable width of the ampoule at this level, the spermatophoric channel, with the body of the spermatophore inside, occupies almost the whole cavity of the ampoule. The septum has undergone considerable shortening with the greater part of its length having fused with the inner wall of the ampoule. This causes the shift of its position to be perpendicular with the typhlosole (previously of the wing channel). Compare the relative positions of the typhlosole and septum at this level with that in Fig. 10.55. A consequence of this shift is the formation of a narrow wing channel extending around the spermatophoric channel. Another result is that the free part of this septum, as short as it is at this level, is nearly perpendicular to the typhlosole, and faces directly the body of the spermatophore. The typhlosole then yields more of secretion III; this is pressed against the globular spermatophore, and a 'furrow' is formed. Both the addition of secretion III and the furrow, extend the length of the spermatophore forming an important feature of this structure: the **wing neck**. Many of the secretions are involved in the formation of the spermatophore, but none is more chemically resistant than secretion III, even long before this secretion is eventually hardened in the TA. Its addition in the latter region is to provide a '**base**' to the wing, firmly fixed on one side to the body of the spermatophore (see Fig. 10.46B). It is also continuous with what is previously secreted of the wing material in the MVD, and then becomes properly elaborated in the ampoule. It should be recalled that only secretion V has so far provided connection, as weak as it must be, between

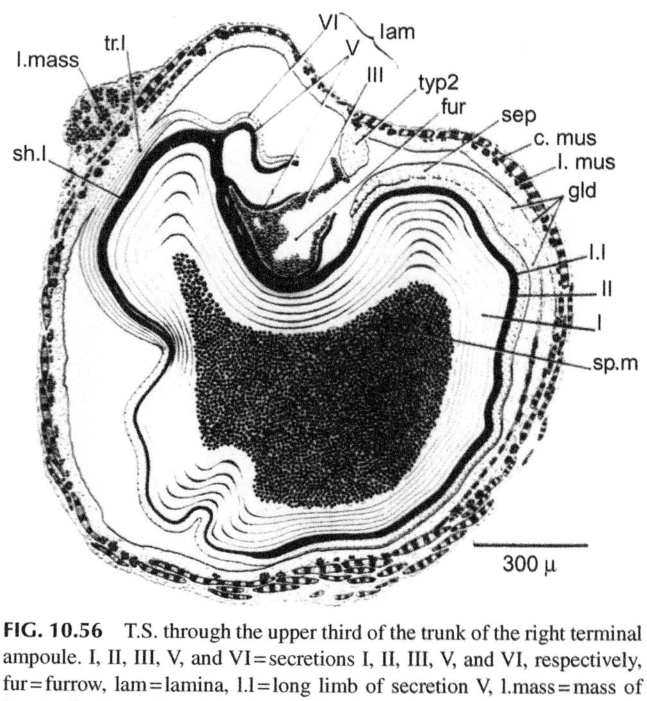

FIG. 10.56 T.S. through the upper third of the trunk of the right terminal ampoule. I, II, III, V, and VI = secretions I, II, III, V, and VI, respectively, fur = furrow, lam = lamina, l.l = long limb of secretion V, l.mass = mass of longitudinal muscles, sh.l = short limb of secretion V, tr.l = transparent line of layer, other letters as before. *(From Bawab (1970).)*

the body and the wing of the spermatophore. This secretion, as mentioned above, is sudanophil and it will be indicated later that it consists of lipoprotein; it provides no sufficient attachment between the two parts of the spermatophore. This is satisfied by secretion III, so the formation of a wing neck is considered as a second connection between the body and the wing of the spermatophore.

At this level (Fig. 10.56), a new layer is also added in the ampoule forming (layer VI), like layer V and outer to it, an incomplete ring around the body of the spermatophore. Unlike layer V, the exception is the region of the furrow. It differs also in that its thickness is not the same throughout. A definite small region applied to the internal wall of the ampoule, where it is externally overlain by the longitudinal mass of muscles, appears relatively broad and clearly laminated. It takes up light green of Masson's and aniline blue of Mallory's triple stains. In a freshly discharged spermatophore from a living specimen, this region appears transparent, extends along the length of the spermatophore as a transparent line (Fig. 10.42C). The remainder of layer VI attains the same colours shown by layers I and II with the above two stains (see Table 10.9). For this reason, layer VI is largely indistinguishable from the others.

One more point is worth mentioning before shifting to the next level in this description of the TA. Fig. 10.56 shows, in addition to the furrow with secretion III, that another part continuous with it remains applied to the side of the typhlosole far away from the wing channel. Unlike the furrow, it is surrounded by layer VI, in the same manner as happens around the body of the spermatophore. It will be seen from Figs 10.57 and 10.58 that this is the 'lamina' consisting of layers V and VI in addition to III. This lamina forms, besides the furrow, an important landmark, which remains visible at the external surface of the fully formed spermatophore (Fig. 10.46B).

At the **middle region** of the trunk of the ampoule (Fig. 10.57), it becomes wider, and the free part of the septum increases in length owing to a gradual separation of its basal part from the opposing wall of the ampoule. Then two important changes appear:

1. The septum shifts 90 degrees in a clockwise direction. No further shifting of this direction occurs at lower levels, and the relative position of the septum to the typhlosole remains unchanged.
2. The wing channel extends deeper around the spermatophoric mass to accommodate, at this level, a larger part of the wing.

In spite of this, still none of the wing material originally manufactured in the MVD is visible at this level. Oppositely, the new secretion III, which appeared in the previous level extends into the wing channel. To this newly extending part, the term '**neck**' will be applied. Participating in the formation of the 'neck' are apparently the typhlosole (for layer III) and the walls of the wing channel (for layers IV and V). In this way, the wing is considered to consist of three parts: a base, a neck, and a main foliaceous part (Fig. 10.46B). Only the latter part is manufactured in the MVD; the other two parts are made in the ampoule. They provide a firmer connection between the foliaceous part of the wing and the body of the spermatophore.

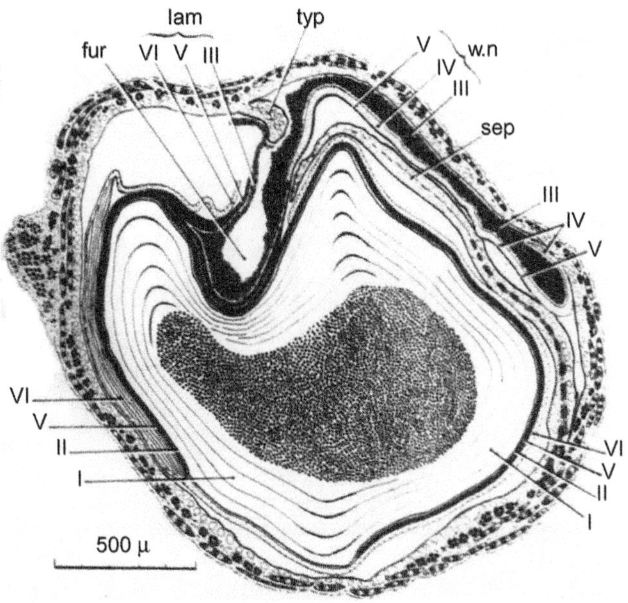

FIG. 10.57 T.S. through the middle region of the trunk of the terminal ampoule showing the wing neck complete (w. n). w. chan = wing channel, other lettering as before. *(From Bawab (1970).)*

Where the ampoule appears widest and the spermatophore inside largest, this is the **lower region** of the trunk of the ampoule (Fig. 10.58). The wing channel is at maximum length; it extends as a full crescent embracing one side of the body of the spermatophore, but remains separated from it by the septum. The latter is completely free except at the basal point where it remains fixed to the ventral wall of the ampoule. As mentioned above, the new septum (wall of the pocket) is fixed at both its ends and so the two cavities (ampoule and pocket) are completely separated at this level. Then, the gradual withdrawal of this new septum from the dorsal side of the ampoule leads to their continuity (Fig. 10.59A). This does not take place, however, until a slightly lower level. Still further down the ampoule the new septum disappears, the wing channel is once more reduced, and the old septum shuts off. The remainder of the cavity of the pocket is occupied by the body of the spermatophore.

Looking at Fig. 10.58 again, the relatively thick gelatin sections, cut with freezing microtome, reveal a U-shaped wing loop occupying the wing channel. A similar picture is visible in paraffin sections except that the fixatives employed appear to cause distortion of the wing and to produce unnecessary complications. The proximal limb of this loop is continuous through the neck part with the base

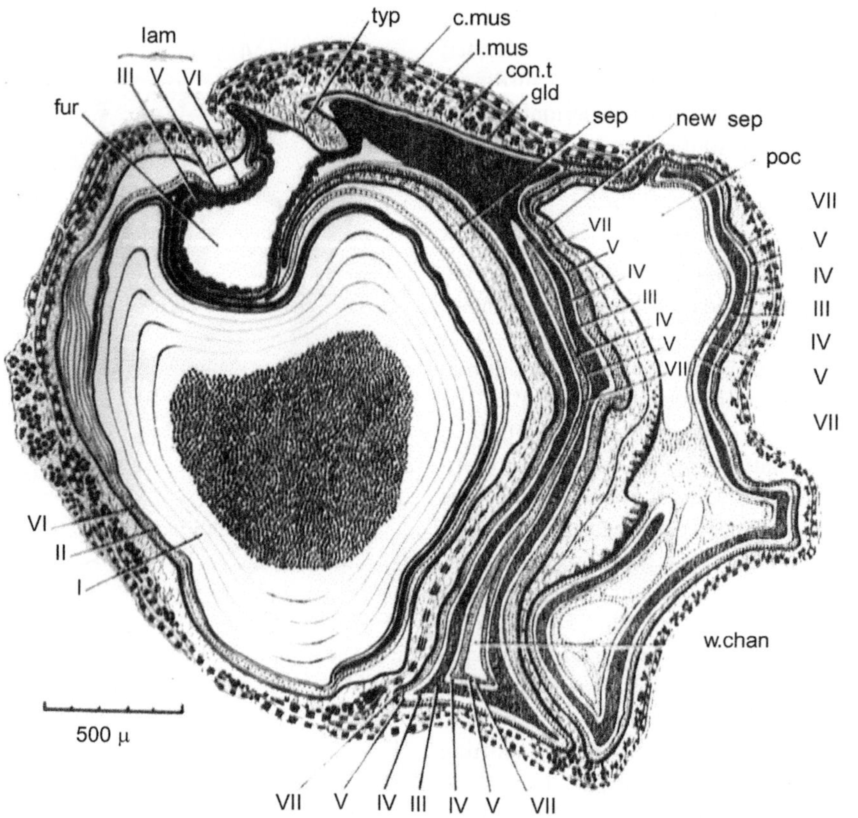

FIG. 10.58 T.S. through the lower region of the trunk of the terminal ampoule. The new septum completely separates the wing channel from the pocket. The wing is slightly idealized. pok=pocket, VII=secretion VII. *(From Bawab (1970).)*

of the wing, which occupies the furrow; the distal limb is free. Another loop of the wing is enclosed in the pocket, which also has a proximal limb much shorter than the distal one. At a slightly lower level (Fig. 10.59A), the two loops are continuous with each other through their distal limbs. Fig. 10.59B represents the complex structure of the wing material which has been manufactured in the MVD, and which in the TA spreads out into an enormous, thin, and highly folded flap.

At this lower level of the trunk of the TA, an additional wing layer appears external to layer V secreted by the highly active rosette-shaped glands, lining the wing channel and the pocket. This new layer is referred to as layer VII (Figs 10.58 and 10.59B). It stains a clear green with Masson and a pale blue with Mallory, and shows a very characteristic granular appearance. The existence of this layer VII at the outermost surfaces of the wing raises an important problem. As described above, the wing always attains the form of the letter U, whether in its own channel or in the pocket where only the outer surfaces of each loop are facing the glandular lining. The inner surfaces are quite distant from any secretory glands, but they are similarly covered by this layer. Bawab (1970) discussed this problem and found possibilities. For the wing channel, the glandular lining extending next to the typhlosole as a roof overhanging this channel

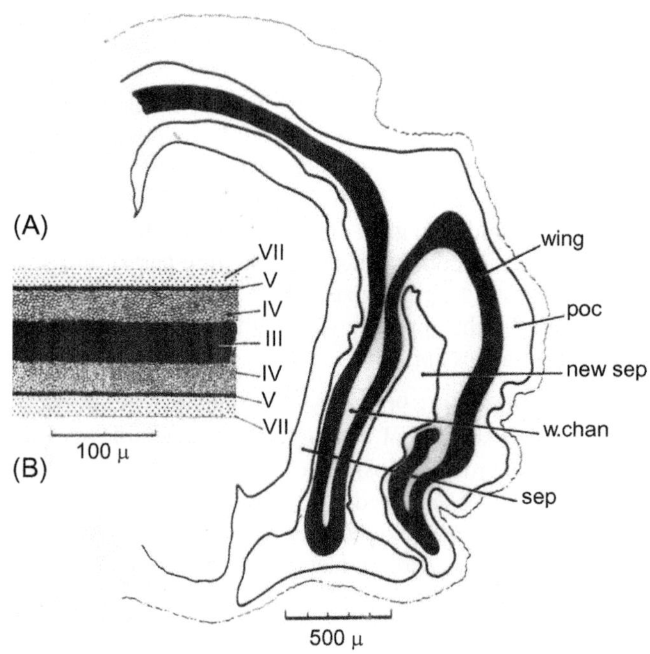

FIG. 10.59 (A) Part of the trunk of the terminal ampoule in outline as appears in T.S. at a slightly lower level than in Fig. 10.58. The new septum here is not complete allowing continuity of the wing material in both the wing channel and in the pocket. To be very clear, the wing is drawn from gelatin section, sep=old septum; (B) a part of the wing shown in (A), highly enlarged to illustrate the seven constituting layers.

may be responsible, or the closed bottom of the channel instead of its roof. Either of these two regions, or both, may yield an adequate amount of material VII to pass in the little space left in between the facing surfaces of the U-shaped wing loop. For the pocket, on the other hand, in Fig. 10.58 and more clearly in Fig. 10.59A, the glandular lining of the new septum immediately overhangs the shorter limb of the wing loop, thickens greatly and its secretion flows easily between the inner surfaces of the two limbs of this loop.

The end part of the TA or more precisely its **base** (Fig. 10.42B) is occupied by none of the components of the spermatophore. Its cavity is more elongated than round (Fig. 10.60). It is lined with a continuous epithelium that completely differs from that lining the whole length of the ampoule. This epithelium consists of unicellular glands arranged in crypts. Immediately surrounding the latter, there is a broad layer of connective tissue with very well developed longitudinal muscles at its outer surface. A continuous secretion appears to be yielded by these unicellular glands (Fig. 10.60A). Neither the nature nor the precise function of this secretion is known with certainty in the work of Bawab (1970), but she investigated whether it may facilitate the passage of the spermatophore whenever this takes place. Therefore, it cannot be compared with the 'glutinous' material. King (1948) described it as 'an important factor in its adherence to the female' *P. setiferus*, which has open thelycum.

As complicated as the ampoule is in structure, it performs several functions. In the light of the above description, one can summarize these functions. It is conducting such as P and DVD, storing as the blind pouch, or otherwise secretory as the MVD. In addition, it provides the firm connection by forming a base and layer VI and its wing attains its outermost and waterproofing layer VII. The most significant and important function of the TA in the formation of the spermatophore of *P. kerathurus* is the moulding and hardening of this structure to attain its final form. To refer to it as an 'ejaculatory duct' (Bauer and Cash, 1991) is neither appropriate nor accurate terminology, at least in *P. kerathurus*, since only the base of this TA is responsible for the discharging of the spermatophore.

The external appearance of the fully formed spermatophore of P. kerathurus. The spermatophore of *P. kerathurus* is nonpedunculate, and comparatively large. It consists of a pear-shaped spermatophoric body, which takes the form and size of the TA (Fig. 10.46). It measures about 5 mm in length in specimens measuring 15–19 cm. This length is within the range previously recorded by Heldt (1931). Its wall is comparatively very thin except a definite thickened region with different colour. It is limited from the rest of the body by a fine transparent line. This line can be seen along the spermatophore inside the TA (Bawab, 1964). In the work of Bawab (1970), it was evidenced that the thickened region is the

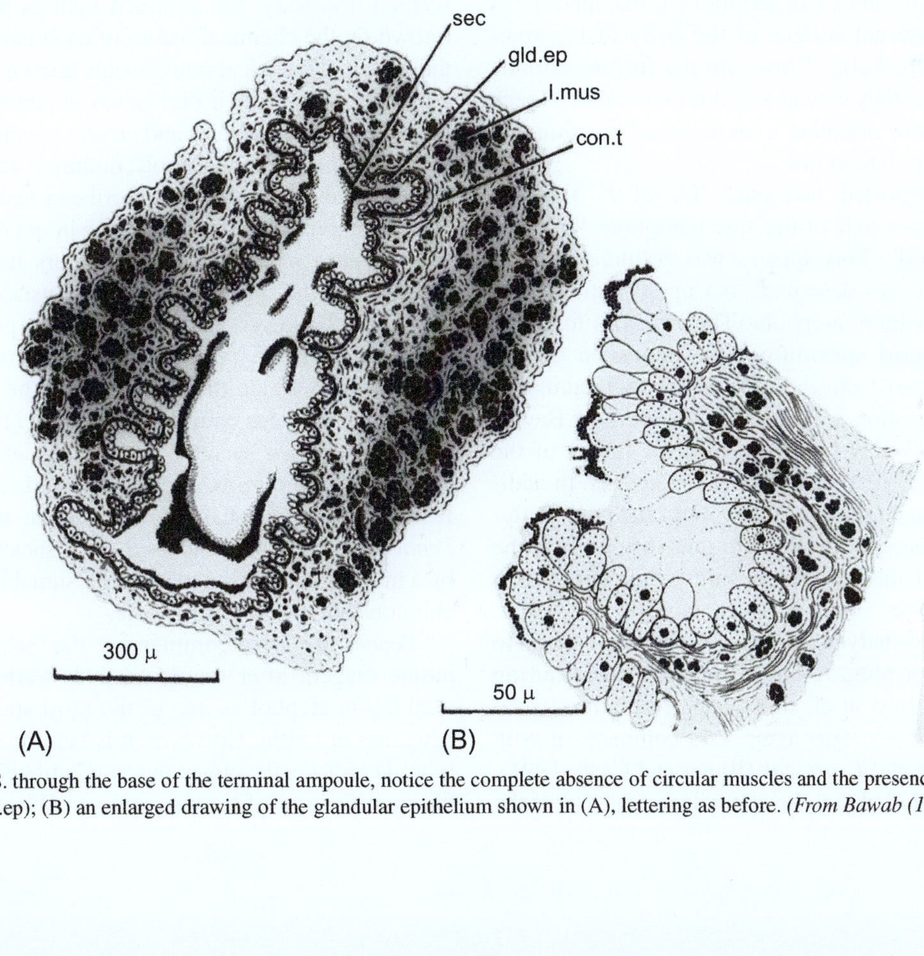

FIG. 10.60 (A) T.S. through the base of the terminal ampoule, notice the complete absence of circular muscles and the presence of a continuous glandular epithelium (gld.ep); (B) an enlarged drawing of the glandular epithelium shown in (A), lettering as before. *(From Bawab (1970).)*

furrow (base) of the wing, which connects the latter with the body of the spermatophore (Fig. 10.46B).

Fresh spermatophores can be obtained when isolated ampoules of living males are immersed in fresh or tap water. Malek as a supervisor and Bawab as a PhD student (1970) observed under the binocular microscope the discharge of the spermatophores from such treated ampoules. The process was described as follows: within 15–60 min, depending on the vividity of the specimen used, an intact spermatophore gradually passes out from the opening at the base of the ampoule. Its body is the first to protrude, but the discharge of the attached wing is withheld until it has first undergone gradual unfolding in the side pocket of the ampoule. When this has occurred, the wing is trailed behind its body. The spermatophore remains in the water for some time unchanged. Thereafter, the wing completely spreads out, as illustrated in Fig. 10.46C. This wing assumes a foliaceous form, rather like the petal of a plant flower. Calman (1909) reported that in the thelycum of Penaeidae lies 'the large foliaceous spermatophores deposited by the male'. The length of the wing of the spermatophore of *P. kerathurus* is somewhat greater than that of the accompanying body (5.5 mm compared with 4 mm). It has the same consistency as the thickened region of the body. It begins as a short narrow neck, consisting of the same wing material. Freely hanging down from the other side of the furrow is a lamina, which runs parallel to, and normally overhangs, the transparent line (Fig. 10.46B).

According to the above description, three landmarks characterize the external surface of the body of the spermatophore of *P. kerathurus*. These are the furrow, lamina, and transparent line; they extend lengthwise parallel to each other, with the furrow attaining a more dorsal position and the other two a more lateral one.

King (1948) reported that each TA of *P. setiferus* (Linnaeus) discharges half of the spermatophore 'with the winged end forward'. This species was restudied (Chow et al., 1991) and it was described that upon mating, each of the TA expels a spermatophore. The two join together, forming a **compound spermatophore** deposited on the open thelycum. Apart from the type of the thelycum, this evidenced the suggestion mentioned in the work of Bawab (1970) that layer VI may be involved in the fusion of the two bodies of the spermatophores of *P. kerathurus*. In addition, Chow et al. concluded that the wing of the open thelycum species is formed in the TA, and thus differs from the so-called 'wing' in the closed-thelycum species, e.g. that of *P. kerathurus*. They also suggested that the 'wing' described in the closed-thelycum species should be referred to as part of stopper or plug. Another work published during the same time of Chow et al. studied the formation of the spermatophore of *P. setiferus* again, and compared it with that of *P. duorarum* and *P. aztecus* (Bauer and Cash, 1991). These authors concluded that the appendage material of the two latter species appears to be homologous to the glutinous material of the spermatophore of *P. setiferus*. This appendage is formed in the channel with typhlosole 2 (wing duct of Malek and Bawab, 1974b; accessory duct of Chow et al., 1991). Therefore, the appendage of the spermatophore of *P. duorarum* (Fig. 10.46E) and *P. aztecus* is homologous to the wing of the spermatophore of *P. kerathurus*. The wing of the spermatophore of *P. setiferus* is completely formed in a chamber in the ejaculatory duct (TA), it is neither homologous to the appendage of *P. duorarum* and *P. aztecus* nor to the wing of the spermsatophore of *P. kerathurus* nor *P. indicus*. The latter species was studied previously by Champion (1987), who concluded that the entire extruded mass of *P. indicus*, including the wing and globular matrices, is termed the spermatophoric mass (Fig. 10.46D). This is a picture that differs completely from those of the above *Penaeids*.

Chemical composition of the spermatophore of P. kerathurus. In total, seven different secretions have been shown to be involved in the formation of this spermatophore. Four of these secretions constitute the body (I, II, V, and VI), and four constitute the wing (III, IV, V, and VII). Only layer V is shared between these two parts and joins them together at the earlier stage in the MVD. At a later stage in the TA, they join together firmly by the added parts of the wing (base and neck).

The appearance and staining reactions of these secretions after they set down as independent constituent layers of the spermatophore are summarized in Table 10.6. However, it seemed necessary that as much light as possible should be thrown on the chemical nature of each constituting layer. Of the principal chemical components known to occur in the arthropod tissues or their derivatives in general are: chitin, certain amino acids; proteins and protein conjugates; true fats and sterols; free dihydroxyphenols; quinines; and polyphenol oxidase enzyme. Davey (1959) described a neutral mucopolysaccharide in the spermatophore of the insect *Rhodnius prolixus*.

The absence of chitin. Many authors have contributed to the idea that the spermatophores of crustaceans are chitinous structures. However, none of them has presented adequate chemical evidence in support of this contention. Jackson (1913) simply stated that the whole of the spermatophore of *Eupagurus* contains chitin, and Spalding (1942) believed that this component is secreted by the vas deferens of *Carcinus maenas* in a fluid form, in addition to protein. More recently, King (1948) made the blunt statement that the sperm of *Penaeus setiferus* are enclosed 'in a sheath and surrounded by a mass of chitinous material'. Presumably the sheath is not chitinous but the outer layers are.

Pearse (1968) recommended the Schultze chloro-zinc iodine reagent after treatment with warm KOH 50% and washing in alcohol as one of the most specific tests for the detection of chitin. However, it is better to be done in a test tube. Therefore, the chitosan test (Campbell, 1929), which is the modification of the original van Wisselingh procedure, based on the conversion of this substance to chitosan was applied. Intact whole spermatophores artificially released from the TAs of *P. kerathurus* were heated at 155°C with a solution of KOH saturated at room temperature. The results obtained are unequivocal; the entire material disappeared completely within a few minutes, except for a few oily droplets that remained floating on the surface of the solution. The experiment was repeated using whole ampoules or even whole vasa deferentia. The results were the same. A control experiment was done, portions of chitinous cuticle, such as of the cockroach *Periplaneta* survived this treatment for at least 2 h, and after washing with water developed with iodine and dilute sulphuric acid, the characteristic intense violet colour of a positive chitosan reaction was observed. It becomes evident that chitin is entirely absent from the entire spermatophore of *P. kerathurus*. Whether the spermatophore of this species, as such found to be free from chitin, would be exceptional to other crustacean spermatophores is highly improbable. For as early as 1931, Mouchet, on the basis of the less specific diaphanol-naphthol test of Schulze and Kunike, failed to detect the presence of chitin in the spermatophores of several species of Anomura and Brachyura.

Further, the persistence of oily droplets during the alkali treatment described above suggests that they are unsaponifiable, i.e. they are sterols. This means that sterols incorporate in the formation of the spermatophore of *P. kerathurus* (see later).

Distribution of lipids. Pure lipid may not be present in any of the components of the spermatophore. However, if it is present, it cannot stand the lengthy treatment of organic solvents through which paraffin sections pass. For this, gelatin sections in which these organic solvents are completely avoided were used for the detection of lipids. Both types of sections were subjected to the Sudan black B stain, the most recommended for fat detection (Pearse, 1968). The results obtained were shown in Table 10.7. Considering these results, the following remarks are of importance:

1. Some of the layers constituting the spermatophore are lipid free, layers I, VI, and VII; while the other layers, II–V, are lipid-containing but they show different affinities for Sudan in gelatin and in paraffin sections.
2. The sudanophil affinity of layers II and III is more pronounced in gelatin sections. This is because some of the lipid naturally incorporated in these layers must have been lost during the course of preparing paraffin sections. However, the Baker's fixative employed in preparing gelatin sections preserve this lipid. In this respect, layer II appears to have lost more lipid than layer III has.
3. Layers IV and V have an opposite situation, i.e. their sudanophil affinity is intensive in paraffin sections than in gelatin ones. It appears that their lipid is naturally more stabilized by polymerization and when subjected to the nitric acid of Gilson's fixative, which is used in preparing the paraffin sections, some oxidation occur. Therefore, they show more affinity to Sudan stain in these sections.

Application of the Liebermann-Burchardt test indicates that the lipids in the spermatophore of *P. kerathurus* are of steroid type or at least contain sterols (Table 10.7). It appears that it is this sterol, which by nature is unsaponifiable, that persisted in alkali treatment during the chitosan transformation described before.

Distribution of proteins. If it is accepted that lipids impregnate certain layers of this spermatophore, then proteins must also be present. No layer consisting purely of lipid,

TABLE 10.7 Responses to ammoniacal silver hydroxide, protein, carbohydrate, and lipid tests given by the layers constituting the spermatophore of *Penaeus kerathurus* (Forskal, 1775)

						Sudan black B	
Layer no.	Argentaffin	Biuret	Xanthoproteic	Millon	PAS	Gelatin	paraffin
I	−ve	+	+	+	+	−ve	−ve
II	+++	+++	+++	+++	−ve	++++	+
III	+	+++	++	++	+	++	+
IV	+++	++++	+++	+++	−ve	++	++++
V	+++++	++++	++++	+++	−ve	++	++++
VI[a]	−ve	+++++	+++++	+++++	−ve	−ve	−ve
VII	±	−ve	−ve	−ve	+	−ve	−ve

Biuret gives purple colour; Xanthoproteic gives orange colour; Millon gives red colour. The repletion of + indicates intensity of colouration.
[a]*Except the transparent line, which is PAS-Positive and Millon-negative.*
From Bawab (1970).

polymerized or unpolymerized, can stand the lengthy treatment involving repeated use of organic solvents through which paraffin sections pass. Application of protein colour tests indicates that this is true. All the layers except layer VII contain protein, but with different intensities (Table 10.7). It is evident from this table that proteins occur in abundance in layer VI, and to lesser degrees in the other layers and comparatively little amount in layer I. It appears that these proteins are rich in the tyrosine as indicated by Millon's test. This tyrosine grouping is the only hydroxyphenol (Oser, 1965; Murray et al., 1996) present in the protein molecule. However, it is not clear whether the same protein in different concentrations is present in the responding layers of the spermatophore or whether different proteins are involved.

Bawab (1970) studied the isoelectric point (I.E.P.) of the various layers of the spermatophore of *P. kerathurus*. The method adopted was that employed by Yonge (1932). The results of this study are recorded in Table 10.8. They evidenced that at least three proteins with different I.E.P. (pH 6.8, 5.4, and 7.2) contribute to the formation of this spermatophore. The value of pH 7.2 for the I.E.P. of layer I, and that value of 5.4 of layer III, are not shown by any other layer of the spermatophore. It is highly interesting that the secretions responsible for the formation of these two layers I and III are yielded at the same time and at nearly the same level of the main region, though in independent ducts of its ascending limb. The value of the I.E.P. of these two layers is maintained in the TA. Likewise, layers II, IV, and V show identical I.E.P., pH 6.8 in the main region, but differ in that the value rises to pH 7.8 in the TA.

TABLE 10.8 The isoelectric points of the layers of the spermatophore of *Penaeus kerathurus* (Forskal, 1775)

Layer no.	I. E. P. in main region	I. E. P. in terminal ampoule
I	pH 7.2	pH 7.2
II	pH 6.8	pH 7.8
III	pH 5.4	pH 5.4
IV	pH 6.8	pH 7.8
V	pH 6.8	pH 7.8
VI	Absent	pH 6.8[a]
VII	Absent	Not measurable owing to absence of protein

[a]*No value could be measured for the transparent line forming part of this layer owing to the absence of protein from it.*

From Bawab (1970).

Layer VI has an I.E.P. of pH 6.8, again the same as that of layer II when freshly deposited, but remains unchanged since it is secreted only in the TA. The proteins that have an I.E.P. of pH 6.8 and 5.4 mix with lipid (see Table 10.7) and the produced lipoprotein conjugate are consequently of two types: in one type the protein moiety is relatively rich in tyrosine (layers II, IV, and V); the other type is restricted to layer III but has a relatively lower content of tyrosine. Layer VI has an I.E.P. of pH 6.8, but this protein is without lipid, and its higher tyrosine content indicates that it is simply more concentrated. It is worth mentioning that the transparent line, which forms a special part of this layer VI, is free from protein.

Distribution of carbohydrates. Davey (1959) first recognized what he believed to be a mucoprotein or a neutral mucopolysaccharide in the spermatophore of the insect *Rhodnius prolixus*. The periodic acid-Schiff (PAS) is widely employed in histochemical examination to detect such substances. The results obtained by the application of this test are interesting in that the layers of the spermatophore that developed the deepest colours are those poorest in protein as already shown in Table 10.7. These are layers I, III, and VII and the transparent line of VI. Of these, the last two have been shown to be protein-free, while the protein content of layers I and III is considerably less than the remaining layers of the spermatophore.

The fact that layers II, IV, and V remain PAS-negative, in spite of a high content of lipid, immediately exclude them from containing unsaturated lipids, glycolipids, or phospholipids. This is because Pearse (1968) categorically listed these substances among the PAS-positive.

Only layer III is remarkable for its moderate content of lipid and for a positive reaction with the PAS reagent. It may consist of glycolipids and/or mucoproteins.

Argentaffin material in the spermatophore. An important outcome of the forgoing observations on the distribution of the essential substances in the spermatophore of *P. kerathurus* is the very close similarity in chemical composition between layers II, IV, and V. All of them are strongly sudanophil and Millom-positive, and all are free of carbohydrate. Although layer III also consists of lipid and of protein, it differs in being further PAS-positive. These four layers are also similar in that their impregnating phenolic substance is capable of reducing ammoniacal silver hydroxide (the argentaffin reaction; Lison, 1953). The results obtained from this reagent, shown in Table 10.7, reveal a close correspondence between the intensity of the argentaffin reaction given by these layers and the concentration of the phenolic substance as indicated by the Millon reaction. It becomes evident that the reducing substance is either the phenolic substance itself, or is bound to it.

Two objections, however, are immediately raised concerning this conclusion. **Firstly**, the correspondence between Millon-positivity and the argentaffin reaction indicated in layers II–V does not apply to either layer I or layer VI. In fact, of all the layers of the spermatophore, layer VI gives the strongest reaction with Millon's reagent. At the same time, it completely fails to reduce ammoniacal silver hydroxide. Lison (1953) stated that structurally, tyrosine does not fall into any of the groups that are capable of reducing this substance, and added that it is not reducing when cold. In spite of Lison's remark, Dennell and Malek (1955) found that authentic samples of tyrosine are argentaffin when cold. Therefore, they explained this contradiction with Lison on the basis that tyrosine may undergo oxidation to silver-reducing quinones even in the absence of enzymes. In the light of this explanation, it seems possible that tyrosine in the layers of the spermatophore may undergo oxidation. The combination of the resultant quinone nuclei with the amino portions of other molecules may result in an argentaffin product. To confirm that tyrosine is the reducing substance in the layers of the spermatophore, Bawab (1970) evidenced experimentally, by prolonged extraction of about 300 TAs of fresh specimens with suitable solvents, the complete absence of dihydroxyphenols.

Secondly, a closer correspondence between the argentaffin substance and the lipids incorporated, not only in some of the layers of the spermatophore, but towards them all. Lison (1953) does not include lipids among argentaffin substances. However, unsaturated lipids would theoretically be reducing agents, but it is completely lacking from this spermatophore, since it is PAS-positive. Likewise, although some carbohydrates are reducing, no correspondence is shown between their distribution in the spermatophore and the ability to reduce the ammoniacal silver solution (see Table 10.7).

Hardening of the spermatophore of *P. kerathurus*. Concerning the above observations together with the natural colour of some layers of the spermatophore of *P. kerathurus* (Table 10.6), Bawab (1970) and Malek and Bawab (1971) decided to study the tanning process in this spermatophore. Layers II and III have a natural colour tinge of orange or brown, respectively, when freshly laid in the ascending limb, and these colours are retained after they become hardened. Layers IV and V become decidedly yellow in the terminal ampoule when appear colourless in the ascending limb. It seems that any yellowish colour due to tanning secondarily developing in Layers II and III would be masked by their natural colours. The above results proved that these layers consist of lipoprotein; its protein moiety is characteristically rich in phenolic groups owing to the presence of tyrosine.

No free dihydroxyphenols could be detected, either histochemically or after prolonged extraction with suitable solvents. However, the enzyme phenolase was readily detected by 'Nadi' reagent (dimethyl-*p*-phenylene-diamine and α-naphthol). The reagent prepared with the addition of alkali according to Lison (1953) has been applied to gelatin sections of formol fixed ascending limb (spermatophores fixed during the course of formation) and TA. Layers II, IV, and V were coloured intense blue within few seconds, and layer III paler blue after some time, but layer I remained entirely uncoloured. The development of blue colouration indicates the presence of the enzyme. Confirmation of this result has been given by the use of catechol (ortho-dihydroxyphenol) as a substitute of the naturally occurring substrate. The results of this treatment were in full agreement with those utilizing the 'Nadi' reagent. Since this effect of the above two reagents was not produced in previously boiled sections or after treatment with cyanide, it can be concluded that this enzyme complex, in the layers mentioned above, oxidizes the tyrosyl residues of the naturally occurring substrate in situ.

It was evident, then, and for the first time, that a process of phenolic tanning is operative in the spermatophore of a crustacean. Therefore, this spermatophore owes its hardness to a definite enzymically catalysed chemical transformation and not to the mere exposure to seawater as has hitherto been thought (e.g. Dahlgreen and Kepner, 1908; Allen, 1916; Heldt, 1932; Matthews, 1951). It must be added that the process of tanning taking place in the spermatophore of *P. kerathurus* may not be completed in the TA. As long as the phenolase appears active in this region, the process is continued for some time after the spermatophore has been transferred to the female. This may even be seen as a necessity, for in the female this component first has to acquire the shape of the thelycum into which it is lodged, and over the surface of this spreads the wing in its final form before it is finally hardened.

Female reproductive system. The ovaries in *P. kerathurus*, as in all penaeids, are similar in structure and location to the testes. They are composed of two bilaterally symmetrical components, which are distinctly separated medially throughout their length. However, they differ from the testes in that they extend posteriorly along the whole length of the abdomen. They are colourless, unpigmented, translucent, and very delicate when immature. Therefore, it is very difficult to identify them in fresh and/or live specimens. However, during their development, their appearance and size change that their definition into three regions becomes very clear (Fig. 10.61A and B). The anterior region is a pair of very long lobes, each one on either side of the fore gut, i.e. embracing this organ; the median region, which is short and relobed (about eight lobes in *P. kerathurus*), rests on the hepatopancreas, as in the testes being partly concealed by the heart; the posterior region is the elongation of the last lobules of the median region forming a pair of highly elongated lobes embracing the hind gut, until approximately the telson where they fuse for a very short distance (Fig. 10.61A). It is worth mentioning that the structure and site of the ovaries in this dissected live specimen is the same as the previous illustration (Heldt, 1938).

From each ovary, at about the level of the sixth lobule of the middle region of *P. kerathurus*, a simple tube, thin-walled oviduct proceeds ventrally. The latter passes around the body of the specimen, downwards, without convolutions, to the ventral surface to the coxa of the sixth thoracic appendage (third leg), which is current author own observation. However, the same information was recorded in Calman (1909). The histological structure and function of this oviduct in penaeids have not been well studied. However, in 2012, Lamsaard et al. studied the structure of the oviduct (Fig. 10.61C–H) in the black tiger shrimp, *Penaeus monodon*, and its changes during ovarian maturation. They concluded that the changes of the oviduct 'were related to ovarian maturation' and prior to spawning, only at ovary stage IV, the oviduct epithelium secreted much of spherical substances into the lumen (Fig. 10.61E–H). They proposed that these substances might act as lubricants to facilitate the spawning process and/or play a significant role in the process of fertilization. The author confirms thoroughly both and especially the latter.

In the Mediterranean Alexandria waters, the ovaries of *P. kerathurus*, as those of other penaeids, begin maturation in May just at or after the female receives the spermatophores, i.e. being mated (personal communication, Khafage, A.R.). In open thelycum shrimp, such as *P. vannamei*, Yano et al. (1988) concluded that mating and spermatophore transfer closely follow ovarian maturation and take place immediately before spawning, while in penaeid shrimp with a closed thelycum, mating and spermatophore transfer take place before ovarian maturation, and mating does not accelerate ovarian maturation directly. Kao et al. (1999) evidenced that development of the ovary of *Aristaeomorpha foliacea* Risso, 1826 (Fam. Aristaeidae, Superfam. Penaeoidea (Holthuis, 1980)) is induced by copulation. During the maturation of the ovaries of this species and those in all the penaeids available, they undergo a chain of colour change correlated with increase in size (Fig. 10.62). This chain of colour change is not specific for the species, since it may change with the food taken. Heldt (1931) described the ripe females of *P. caramote* (= *P. kerathurus*) obtained from Tunisia waters. She reported that they have the ovaries full of eggs, the colouration of which vary with the individual, sometimes yellowish white, but more frequently clear olive green. Very recently, it was reported that the gravids (stage IV) of the Pacific green tiger prawn, *P. semiculcatus* de Haan, obtained from Japanese waters, have well-developed olive green ovaries (Ronquillo et al., 2006).

The first studies examining yolk protein of eggs of Crustacea focused on shrimps and prawns of the genus *Penaeus*, but especially the giant freshwater prawn of the Southeast Asia, *Macrobrachium rosenbergii* (Wilder et al., 2010). This yolk protein consists of vitellogenin and vitellin. Several groups of authors studied those of the egg of *M. rosenbergii* and all concluded that vitellogenin is composed of three subunits of molecular weight of around 200, 100, and 90 kDa (kilo Dalton, unit for molecular weight). Further, vitellin consists of two subunits only, both of which have identical molecular weights and are immunologically equivalent to smaller subunits of vitellogenin. Wilder et al. in a previous work postulated that 'vitellogenin is initially secreted into the haemolymph from its synthetic site (likely the hepatopancreas) as a large precursor, which is cleaved into the 102 and 90 kDa subunits before being taken into the ovary'. This information was later confirmed by more authors. Moreover, amino acid composition of vitellin was seen to be comparable among shrimp and prawns (Wilder et al., 2010).

In addition, the eggs of many crustaceans are pigmented by carotenoids (Reverberi, 1971). Opposite to the proteins, it is generally accepted that crustaceans do not synthesize carotenoids from noncarotenoid precursors, but need a source of carotenoid in the food. These pigments may be free and the colour ranges from yellow to red, or they may be linked to a protein and the colour produced may be green, blue, brown, and purple. The colour of the carotenoprotein in the eggs of, e.g. *Cyclops vernalis* varies according to the food eaten by the animal. The carotenoid does not vary, but is linked in at least two ways, so that a range of colours can be produced by varying the relative proportions of each linkage. However, the carotenoids taken in with the food can be modified and changed into other carotenoids. There is some evidence that β-carotene is the main precursor of other carotenoids in the Crustacea.

Returning to the maturation of the ovaries, it is divided into stages according to these two factors mentioned above, but more precisely according to the development of the oocyte itself. Kao et al. (1999) divided the maturation of the ovaries of the deep-water giant red shrimp *Aristaeomorpha foliacea* (Risso, 1826) (obtained from a deep-sea fishing port, Tashi, in northeastern Taiwan, Pacific region) into four distinct types (Fig. 10.62(I)). These types appear to be correlated with the chain colour change mentioned above. Ovaries of the same species, *A. foliacea*, obtained from another region (Tyrrhenian Sea, between the Gulf of Patti and Aeolian Islands, Italy region of the Mediterranean Sea) were studied (Perdichizzi et al., 2012). According to the histology of the ovaries, these authors divided the maturation process into five stages with different terminology (Fig. 10.62(II)). On the other hand, the maturation of the ovaries of *P. Melicertus kerathurus* was divided into five distinct stages according to the process of vitellogenesis (Medina et al., 1996; Kücükdermenci and Lok, 2012). It seems that this latter division is more appropriate, since it includes both the colour and the histology of the oocytes. That of *P. kerathurus* (Fig. 10.70) is typical of that described previously for other penaeids. These stages are as follows:

- stage 1: previtellogenic can be seen in late December and January;
- stage 2: early vitellogenic in June;

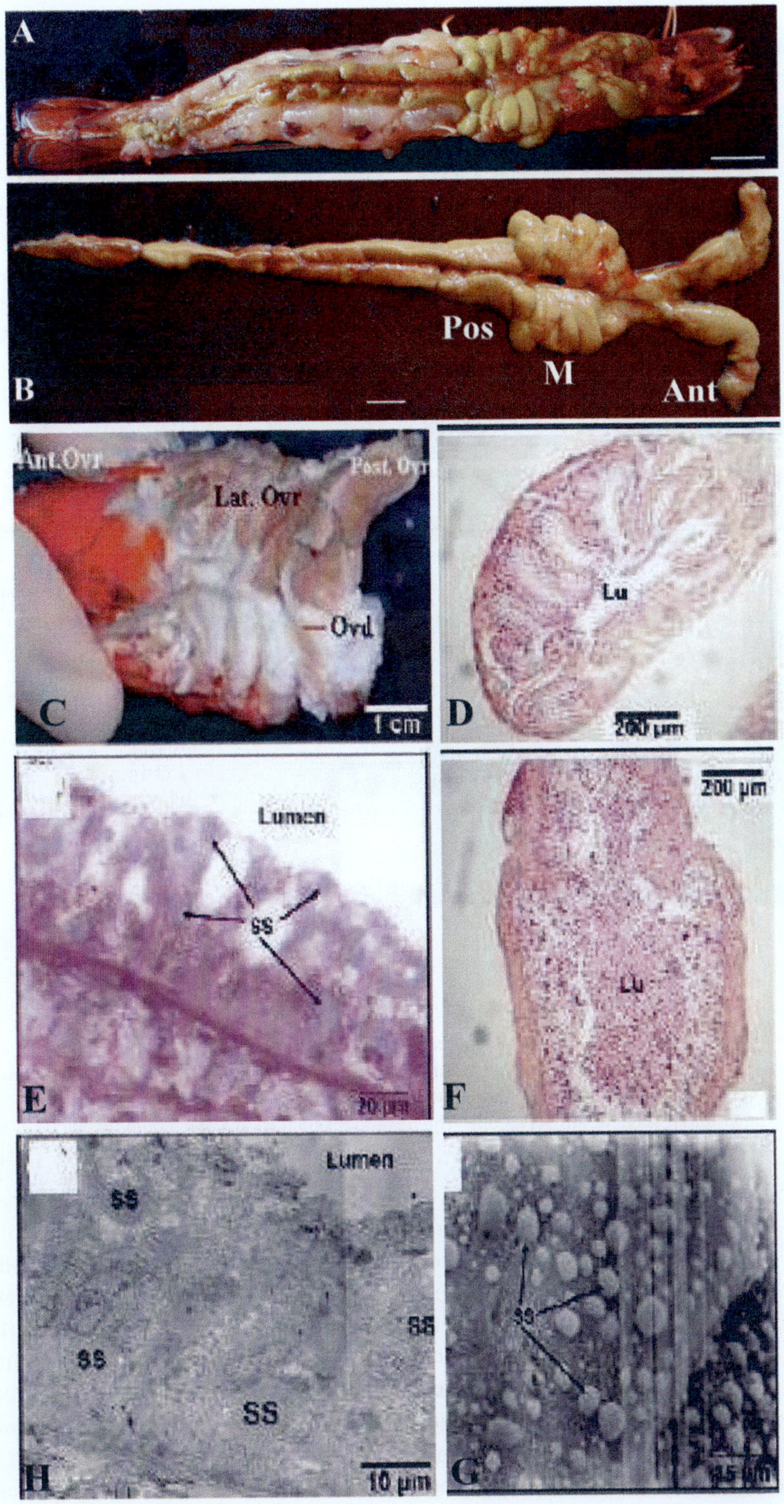

FIG. 10.61 (A) Dissected specimen of the prawn *Penaeus kerathurus* showing the ovaries in situ; (B) the isolated system of the same species *((A and B) prepared by Khafage, A. R.)*; (C) photograph showing the gross anatomy of the oviduct connecting to the ovary of the black tiger shrimp *P. monodon*; (D) T.S. showing the histological structure of the oviduct of *P. monodon* during maturation of the ovaries; (E) portion of T.S. of the oviducal secretory epithelium (basic fuchsin-stained semithin); (F) T.S. through the oviduct during stage IV mature ovary (H&E stain); (G) TEM graph; and (H) SEM of the secretion of the oviduct *((D–H) from Lamsaard et al., 2012)*. Ant., Lat., and Post. Ovr = parts of ovary, Lu = lumen, SS = secretory substance.

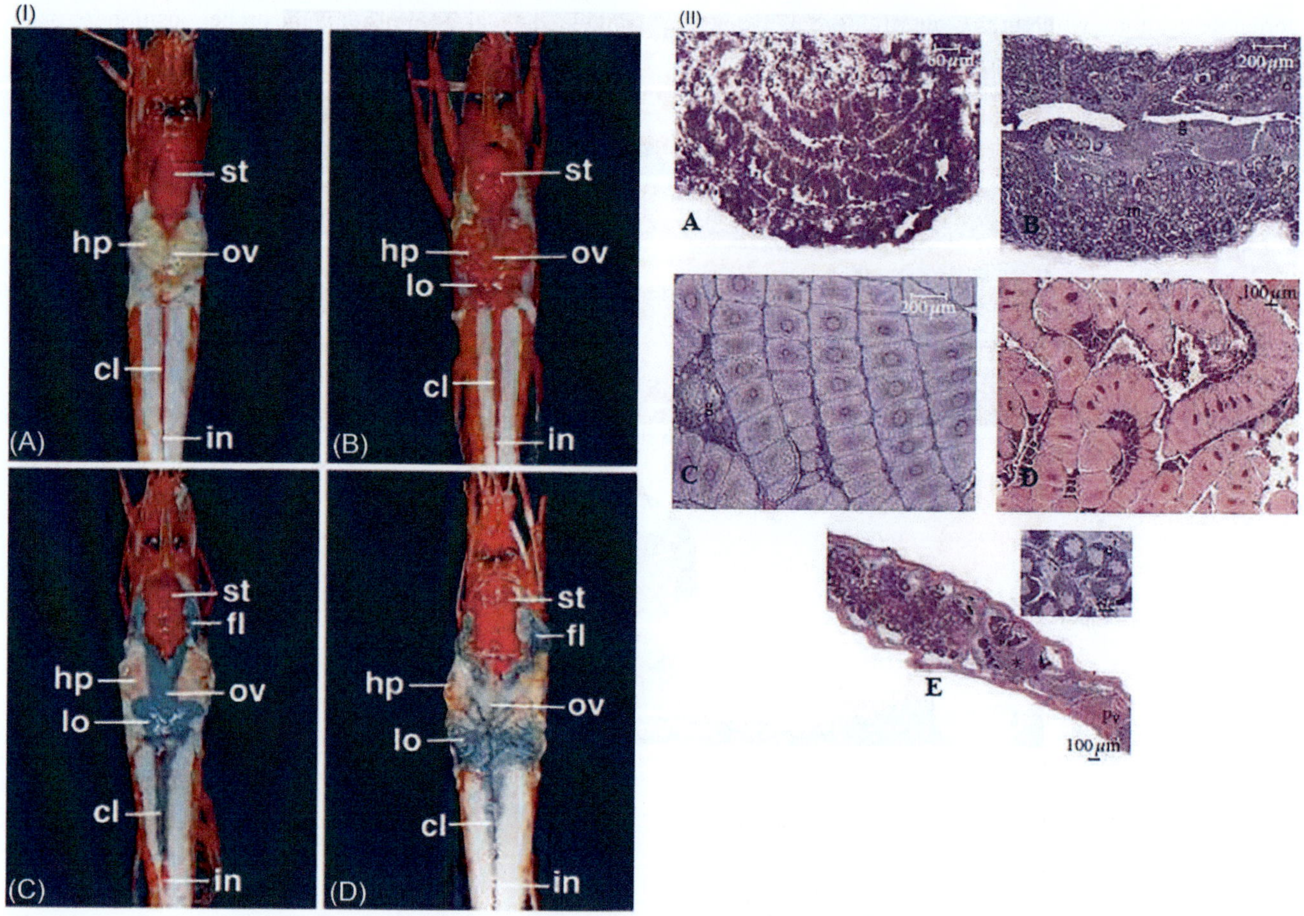

FIG. 10.62 (I) Photographs of dissected specimens during maturation of the ovaries of *Aristaeomorpha foliacea* showing changes in size and colour of the ovaries: (A) type I ovary, translucent and colourless; (B) type II ovary, semitranslucent and orange-red or light grey; (C) type III ovary, large and bluish grey; (D) type IV or ripe ovary, reaches maximum size and *pale black colour*. cl = caudal lobe of ovary, fl = frontal lobe of ovary, hp = hepatopancreas, in = intestine, ov = ovary, st = stomach. *From Kao et al. (1999).* (II) Ovarian maturity stages of *Aristaeomorpha foliacea* according to the histological structure (haematoxylin-eosin): (A) stage 1, immature ovary with oogonia, early primary oocytes scattered in the connective stroma; (B) stage 2, recovering ovary showing distinct germinative (g) with og and Ep, and maturative (m) zones, the latter occupied by late primary oocytes; (C) stage 3, maturing ovary showing g-zone with og & Ep and m-zone with a single batch of Ev oocytes; (D) mature ovary showing m-zone with Lv oocytes piled up in tubular units; (E) resting ovary (spent) with Lp at an initial stage of atresia among which some residual Pv oocytes could be detected. Ao = atresial oocyte, Ep = early primary oocyte, Ev = early vitellogenic oocyte, Lp = late primary oocyte, Lv = late vitellogenic oocyte, og = oogonia, Pv = postvitellogenic oocyte. *From Perdichizzi et al. (2012).*

- stage 3: late vitellogenic from February to May;
- stage 4: mature from late June to October; and
- stage 5: spent or degenerating in the middle of September (Kücükdermenci and Lok, 2012).

Copulation. Spermatophore transfer to the female is the essential final step of a series of events comprising mating behaviour (copulation) in decapod crustaceans. There was no observation of this vital activity of *P. kerathurus* during the present work, neither were any specimens coincidentally obtained in action. However, Yano et al. (1988) studied mating behaviour in the open thelycum species, *P. vannamei*. They reported that these events were described in a variety of caridean prawns, but little information is available on mating behaviour in the penaeid shrimps. However, mating behaviour has been described for the closed thelycum species in *P. japonicus* Bate, 1888 (Hudinaga, 1942), *P. monodon* Fabricius (Primavera, 1979), and *P. paulensis* Pérez-Farfante, 1967 (De Saint-Brisson, 1985). Primavera and De Saint-Brisson mentioned, in addition, *P. stylirostris* Stimpson (Aquacop, 1977). The textbook '*The Biology of the Penaeidae. Advances in Marine Biology*' (Dall et al., 1990) reviewed these works with others, where *P. kerathurus* was one of them and presented the illustrations of mating of *P. monodon*. Discussing these available descriptions, in the opinion of the author, no significant difference can be observed between the courtship and mating process itself in open and closed thelyca. Only in the former case, the male *P. vannamei* mates with parturial females (very ripe just before spawning) that are in intermoult stage, and the male attaches the spermatophores to the thelycum (external deposition), while in the latter (closed thelycum), the male mates with newly moulted females with immature ovaries and inserts the spermatophores into the thelycum (internal deposition). This evidenced the very early observation of Heldt (1931), who reported that the spermatophores are introduced in the 'seminal receptacle' of the female of *P. caramote* Risso (= *P. kerathurus*) at the period of the moult (prespawn moult). This means that copulation occurs in this species when the female is soft-shelled.

Copulation (Fig. 10.63), as a vital activity, generally takes place during the night-time in decapod crustaceans. It was divided into four stages in the open thelycum species: approach, crawling, chasing, and mating (Yano et al., 1988). On the other hand, in the closed thelycum species, *P. monodon*, involving both unablated and unilateral ablated females (Primavera, 1979), it was divided into three stages. The courtship was described in much more detail in the first three stages in the former case (Fig. 10.64), while in the second case, only the first stage described this courtship. The two descriptions are almost the same, even down to the expression that a single swimming or resting female is chased by one to as many as three males. Apparently females are not always passive and some resist the males' approach. This courtship is the longest step; it can last up to 2 h if, for example, the male cannot attain the proper position for mating. It may be as quick as 20 min if the male immediately attains the mating position described below. The male turns ventral side up and attaches to the female (Fig. 10.63C). The ventral-to-ventral position is maintained for 1–2 s in open thelycum and the spermatophore transfer is achieved. Usually a face-to-face position is assumed but occasionally an inverted position is observed (Fig. 10.63D). In both types of thelyca, once the ventral-to-ventral position is achieved, it is difficult for other males to displace the first male, and copulation is certain. If unsuccessful, the male immediately returns to the former upright position, still trying to swim parallel to the female. In the closed thelycum, the ventral-to-ventral position is further followed by another position. The male turns perpendicular to the female, rotating at the point of the posterior end of the thorax (Fig. 10.63E). At this position, the pair of individuals may either maintain their position in the water or slowly settle to the bottom. In *P. paulensis*, at the beginning or during the sexual act the animals did not swim as described for *P. monodon* or *P. japonicus*, but stayed on the bottom, the male practically touching it with his back. Immediately after assuming a position perpendicular to the female, the male curves his body in a U-shaped around the thorax of the female (Fig. 10.63F)

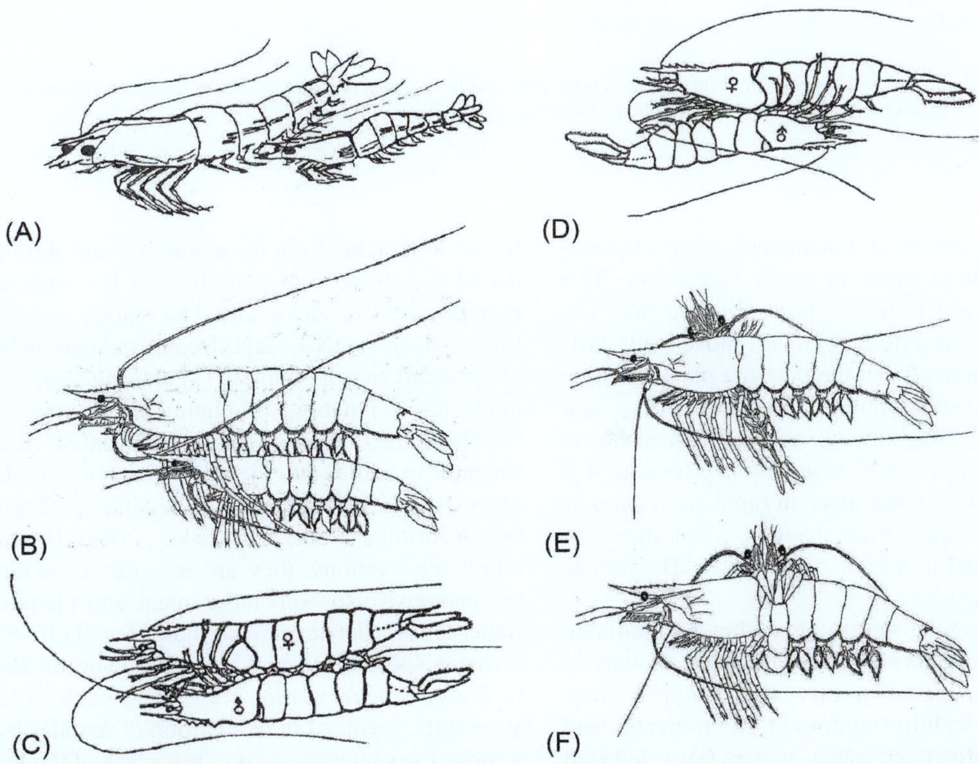

FIG. 10.63 Since the courtship and mating process are the same in open and closed thelyca species, these illustrations represent the processes in different species: (A) male *Penaeus paulensis* at the side of the female under an angle of 45° *(from De Saint-Brisson, 1985)*; (B) male *P. monodon* below the female in parallel swimming *(from Primavera, 1979)*; (C) male *P. vannamei* turns ventral side up and attaches to female *(from Yano et al., 1988)*; and (D) an inverted position in mating of *P. vannamei* (open thelycum); (E) male *P. monodon* turns perpendicular to female; and (F) male curves body around the female and flicks head and tail simultaneously.

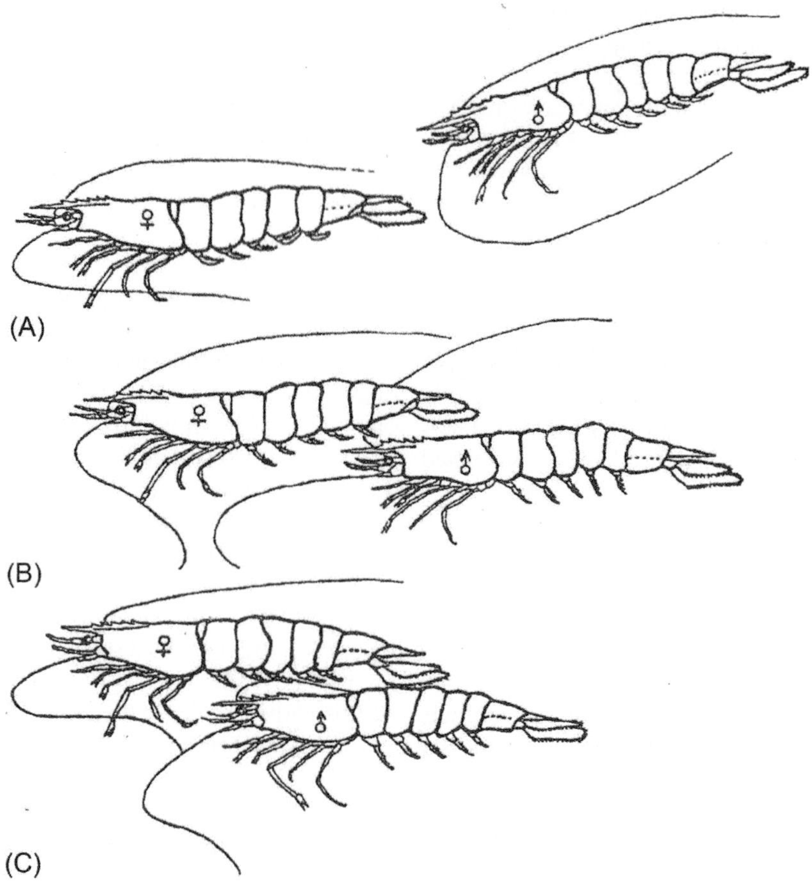

FIG. 10.64 These illustrations represent courtship behaviour in *P. vannamei*: (A) male walks close to female from behind; (B) male crawls his head under the tail of female; (C) male chases female in parallel position. *(From Yano et al. (1988).)*

and flicks both head and tail simultaneously, as in a squeezing action, up to three times in quick succession. This last position is very quick, lasting just a few seconds. The whole process from the initial courtship movements lasts 0.5–3 h. It is hypothesized that the petasma plays a role in the transfer of the spermatophores from the terminal ampoules (Primavera, 1979). Although no specific mechanism for spermatophore transfer in *P. monodon* is proposed, it is probable that this action takes place in rapid succession as the male achieves an appropriate position. Soon after this, the male separates and moves or swims away. The female may also move away.

It is worth mentioning that these studies on courtship and mating of penaeid shrimps were done on animals in captivity. They were reared in very well-equipped tanks under highly and carefully controlled enviromental and biological factors. Moreover, many authors have deduced the presence of sex pheromones, which attract a mate and initiate the appropriate copulation in decapod crustaceans. Yano et al. (unpublished data) evidenced experimentally that male *P. vannamei* apparently recognize mature females by sex attractants from the mature female shrimp. They deduced the presence of two different sex attractants, which may be produced and released by mature female for signalling sexual receptivity, and chasing and mating behaviour in the penaeid shrimp, namely (1) chasing-stimulating pheromone, and (2) mating-stimulating pheromone.

The long flagella of the second antennae of decapod shrimps, in addition to their important role in identification of food items, have been implicated in sex recognition in caridean shrimps (Bauer and Caskey, 2006). Based on behavioural observations, they are recognized as chemotactile, i.e. responsive to both mechanical and chemical stimuli. Bauer and Caskey examined these flagella by SEM in four caridean species and in the penaeoid shrimp *Rimapenaeus* (= *Trachpenaeus*) *similis*. They proposed that their setae are as sensilla, involved in recognition of females by males via a contact sex pheromone on the surface of the female. They also tested the hypothesis of sexual dimorphism in antennal flagellar setae by observations on the type and number of basic setal types. They examined the distal part of the antennal flagellum, the most likely to first touch another

FIG. 10.65 Antennal flagellar setae of *Rimapenaeus similis*: (A) arrangement of arched plumose setae on the antennal articles. Scale bar 110 μm; (B) long simple seta, indicated by *arrow*, along side of plumose setae. Scale bar 25 μm. *(From Bauer and Caskey (2006).)*

object in the environment. Fig. 10.65 illustrates that of the penaeoid *Rimapenaeus*. Dimorphism in antennal flagellum setae across species studied indicates that male capabilities to detect female contact pheromones in decapod shrimps will be found in the ultrastructure and physiology of setal neurons or at neurological levels.

Just after copulation and separation of the pair, the mated female has its thelycum occupied with spermatophores (Fig. 10.66). The mated females of *P. kerathurus* are always obtainable during the season from May to September in Alexandria waters. Only the wings of the spermatophores appear external to the thelycum (Fig. 10.66A), opposite to the open thelycum (e.g. that of *P. setiferus* (King, 1948)) which wholly appears external (Fig. 10.66B).

In *P. vannamei* with open thelycum, mating and spermatophore transfer closely follow ovarian maturation and take place immediately before spawning (Yano et al., 1988).

Development

Gametes. Roosen-Runge in his textbook *The Process of Spermatogenesis in Animals* (1977) reported that the first comprehensive report of the male reproductive organs of decapods was given by Grobben (1878). This work involved no cytological details, but it is still of interest. It noted that only two-thirds of the wall of the tubular testes of decapods is covered by germinal epithelium, while the remainder is occupied by normal epithelium. This means that up to the time of this book, there was no information about spermatogenesis in penaeids. However, the process of **spermatogenesis** was studied by examination of serial transverse sections through successive levels of the testes of *P. kerathurus* (Bawab, 1964). It appears that, at any level, the follicles exhibit various outlines and various degrees of maturity (Fig. 10.44). Maturity here is clearly shown by the regions where the spermatogonia undergo developmental stages to be transformed into sperm (Fig. 10.67). However, the follicles in the middle, as well as the distal parts of the testes, show a more advanced stage of maturity. The immature follicle (Fig. 10.67A) always has a relatively large cavity. It is surrounded by a wall as described above, but the portion that bounds the cavity is a comparatively thick layer of connective tissue. Scattered in the latter, there are a few sharply defined cells, oval in outline with nuclei of different shapes and with the same affinity to stains as the spermatogonia. In the opinion of the author, these might be the original germ cells, but from where they first appear is not clear. Each spermatogonium in this immature follicle is a large rounded cell with granulated cytoplasm and a nucleus occupying nearly the whole cell. In this studied specimen, their nuclei are in different stages of mitosis, and they gradually increase in number to obliterate its cavity (Fig. 10.67A). In the follicles where the cavity has disappeared, other stages of maturity follow. As it is described and illustrated in Carleton (1944), the primary spermatocytes appear as the result of the activity of spermatogonia. They are smaller in size, and are at first interspersed between them. Gradually, they increase in number and nearly completely replace the spermatogonia, except for several neighbouring cells forming marginal region of two layers of three to five cells in each (Fig. 10.67B). This marginal region gradually increases in thickness during successive stages of maturity and may correspond to the marginal epithelium described by Grobben. The primary spermatocytes have nuclei also in different phases of division (Fig. 10.67B). In this preparation, the whole bulk of the follicle consists mostly of one type of cells, but few number of nurse cells are confined to the periphery of the follicle. They are elongated and well demarcated, with the nuclei not sharply distinguished. In maturing follicles just as those constituting spermatids, the nurse cells increase in number (Fig. 10.67C) and are crowded

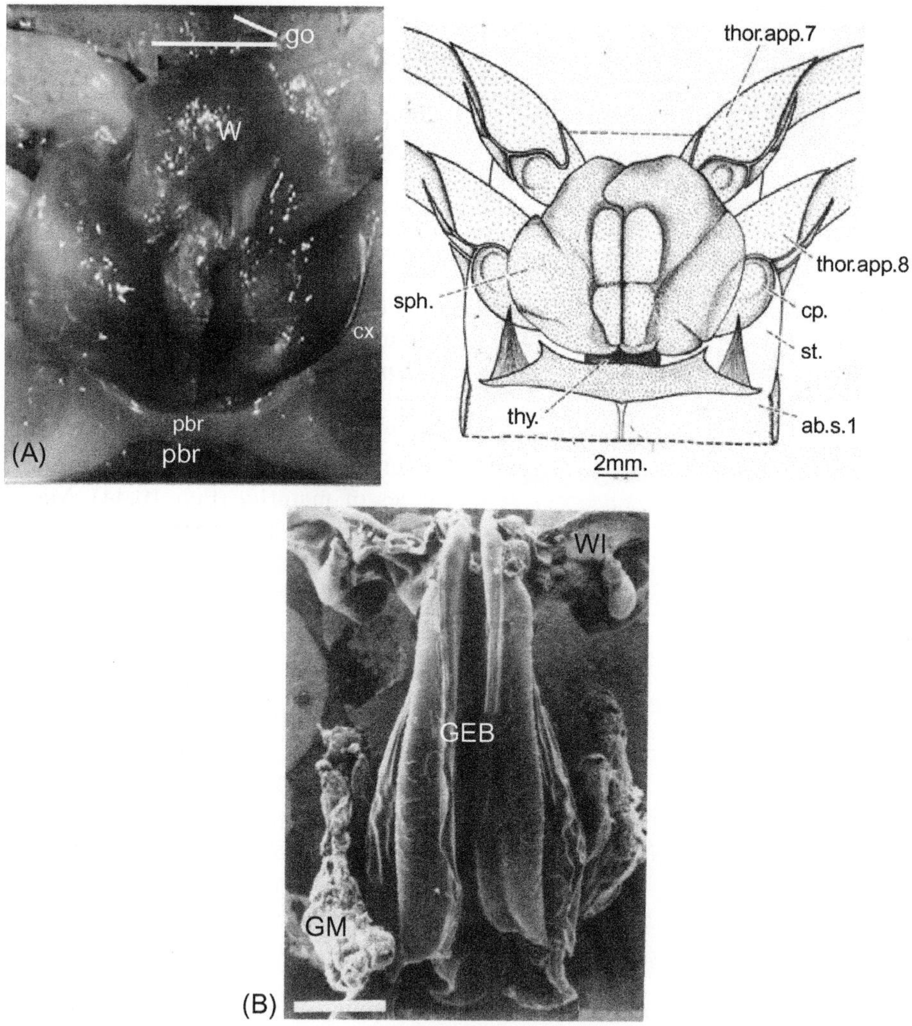

FIG. 10.66 (A) Occupied (impregnated) thelycum of female *P. kerathurus*, the left-hand-side photograph and the right-hand-side camera Lucida drawing. ab.s.1 = sternum of first abdominal segment, cp = coxopodite, pbr = posterior basal ridge, sph = spermatophore, st = sternum, thor.app. 7 and 8 = seventh and eighth thoracic appendages, thy = thelycum, W = wing; (B) SE micrograph illustrating ventral view of artificially assembled compound spermatophore of *P. setiferus*. GEB = germinate body (main trunk), GM = glutinous material, WI = pair of wings at anterior portion. *(From Chow et al. (1991).)*

along the faint line separating the marginal band from the bulk of contents. However, in the fully mature follicle that contains developing sperm (Fig. 10.67D), the nurse cells almost completely disappear. It is worth mentioning that the structure and appearance of this mature follicle is similar to the mature follicle of the testis of the crab *Menippe mercenaria* present in Roosen-Runge's textbook (1977) (see later).

It is well known that reduction division takes place during spermatogenesis in the last stage of primary spermatocytes (see Figs 2.8 and 2.13, Chapter 2), producing secondary spermatocytes. Histologically, no difference can be observed in this preparation between 1^{ry} and 2^{ry} spermatocytes except that the latter are smaller and they produce spermatids. The spermatids are the smallest cells and have small, rounded nuclei (Fig. 10.67C). They segregate from each other and become no longer a tissue, then undergo the process of spermiogenesis. This process occurs in an acentric fashion, not central as described in textbooks—that is, the cavity produced is not central (Fig. 10.67). Since this process needs cytological studies, it was not in consideration in the work of Bawab (1964). Yet later, Medina (1994) studied this process ultrastructurally in the penaeid shrimp *Parapenaeus longirostris* (Lucas, 1846). Early spermatids of this species are polarized cells from the cytoplasm by the nuclear envelope. The perinuclear cytoplasm includes numerous small vesicles, some of which seemingly coalesce with the large proacrosomal vesicle. Thin cisternae of endoplasmic reticulum frequently form parallel concentric arrays around the nucleus (Fig. 10.68G), or constitute large areas of membrane accumulations in close contact with the nucleus and proacrosomal vesicle (Fig. 10.68H). Other organelles that are present in the perinuclear cytoplasm are granular ribosome-like particles and a few

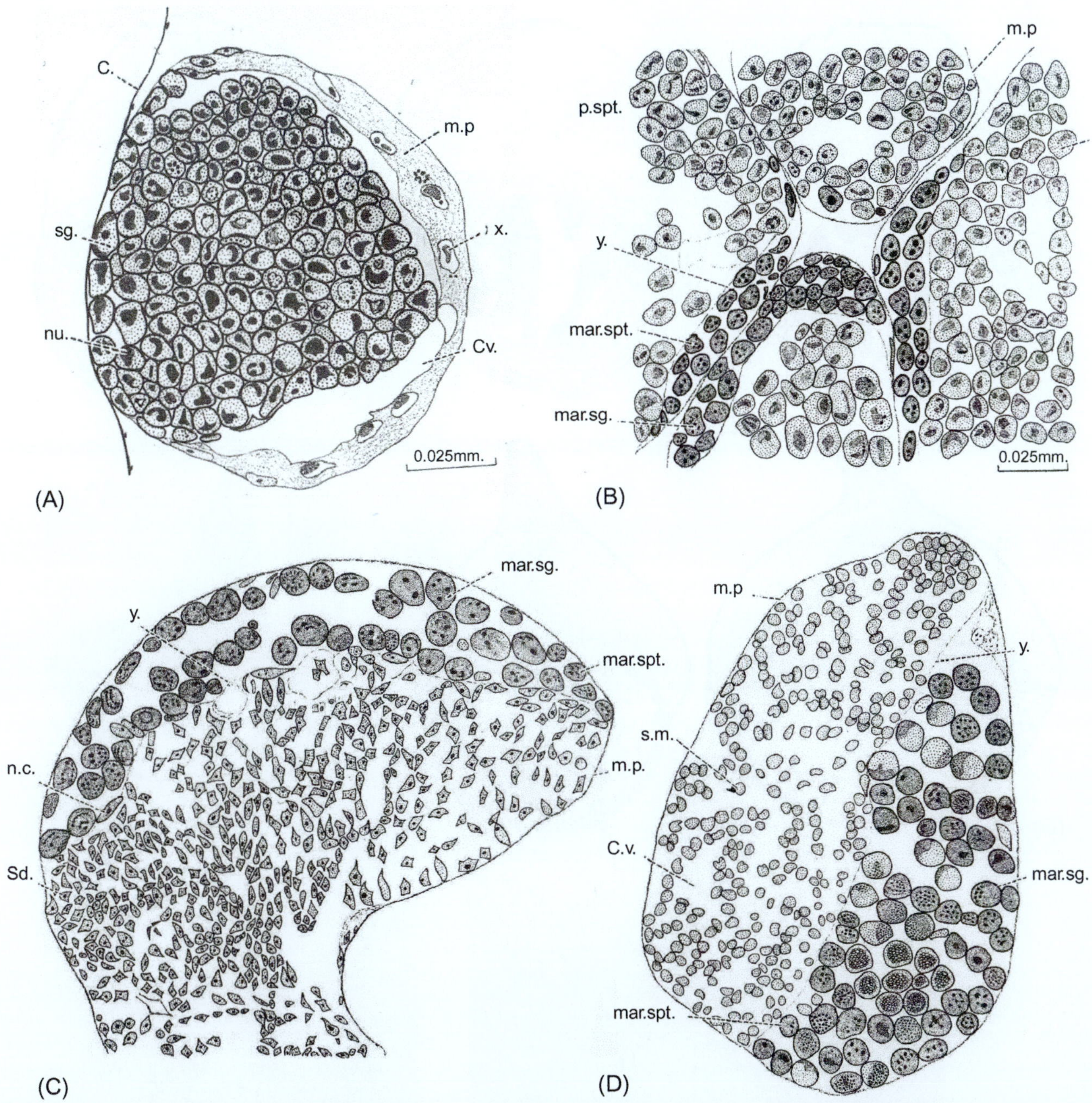

FIG. 10.67 Camera Lucida drawing of portions of T.S.s of the testis of adult *P. kerathurus* showing: (A) immature follicle with spermatogonia in stages of mitosis; (B) neighbouring follicles with specialized margins; (C) mature follicle full of spermatids; (D) highly mature follicle with acentric cavity full of developing and/or mature sperm. (C and D) X=0.025. C=cortex, Cv=cavity, m.P.=membrane propria, mar.sg.=marginal spermatogonia, mar.spt.=marginal spermatocytes, n.c.=nurse cell, Nu=nucleus, p.spt.=primary spermatocyte, sd=spermatid, sg=spermatogonia, s.m.=mature sperm, x=cells in m.p., y=faint line separating the marginal from the main bulk of the follicle. *(From Bawab (1964).)*

spherical mitochondria with irregular cristae (Fig. 10.68H). Centrioles are seldom found, and no trace of Golgi bodies or any other similar membranous organelles were observed.

Fig. 10.67D shows a mature follicle of the testis of *P. kerathurus* full of metamorphosing and/or mature sperm. At the time of this study, the electron microscope was not in common use, and the mature sperm was studied with the oil immersion magnification of the light microscope (Fig. 10.69(I) A). The general structure and size were in accordance with previous studies (Heldt, 1938; King, 1948). The electron microscopic studies (Medina et al., 1994) evidenced the size and the general appearance but not the structure of the sperm of *P. kerathurus* (Fig. 10.69(I) B). The EM studies evidenced that the structure and form of the spermatozoon of different

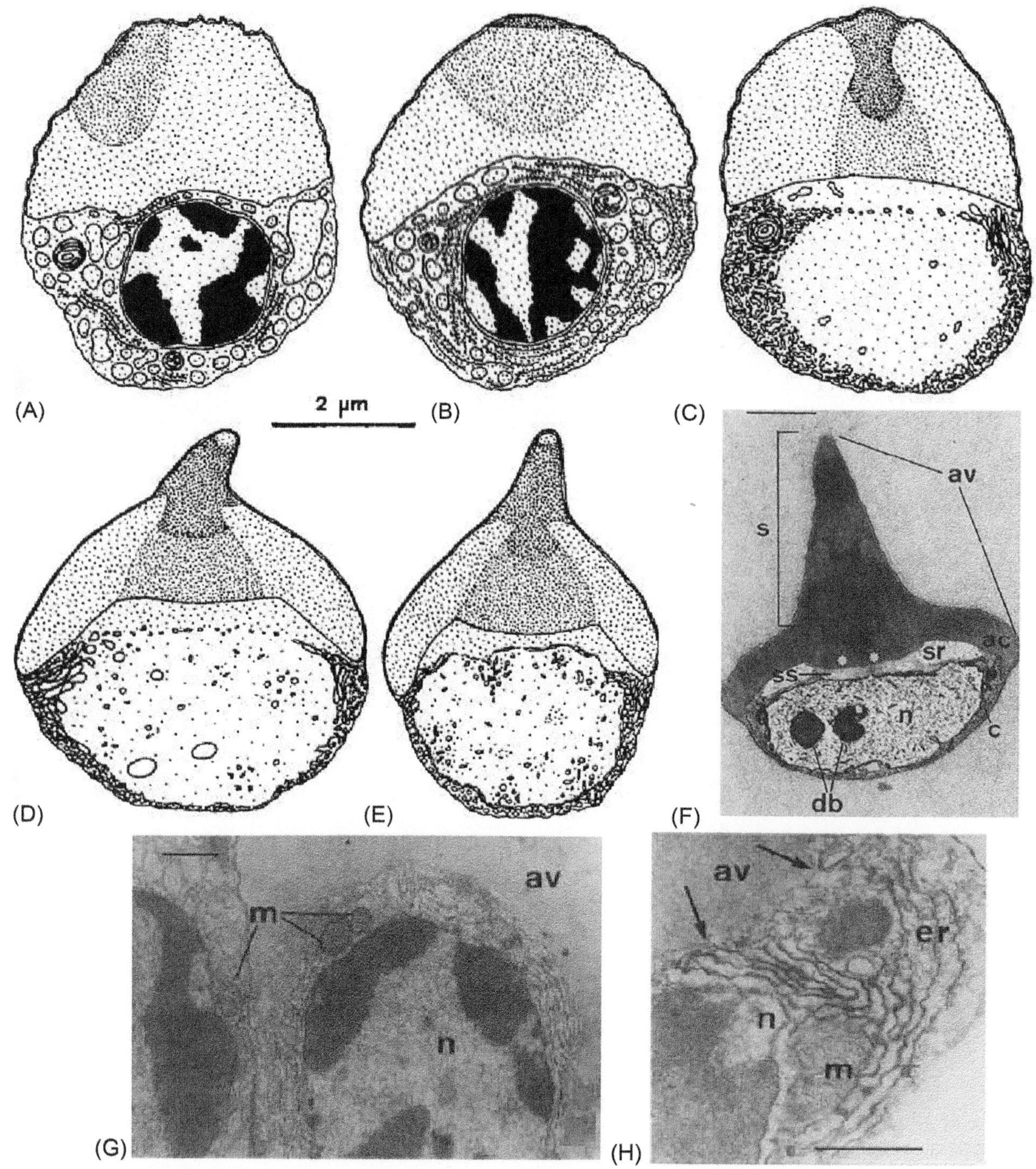

FIG. 10.68 (A–E) Schematic drawings of SEMicrographs summarizing spermiogenic process of *Parapenaeus longirostris*: (A and B) early spermatids; (C) mid spermatid; (D and F) late spermatid; (F–H) TEM: (F) longitudinal section; (G) nucleus and perinuclear region containing concentric arrays of endoplasmic reticulum; (H) parallel membrane accumulations of endoplasmic reticulum closely associated with proacrosomal vesicle *(arrows)* and nucleus. av = acrosomal vesicle, c = perinuclear cytoplasm, db = dense bodies derived from membranes, er = endoplasmic reticulum, m = mitochondria, n = nucleus, s = spike, sr = subacrosomal region, ss = subacrosomal substance. Scale bars F = 1 µm, G, H = 500 nm. *(From Medina (1994).)*

(I)

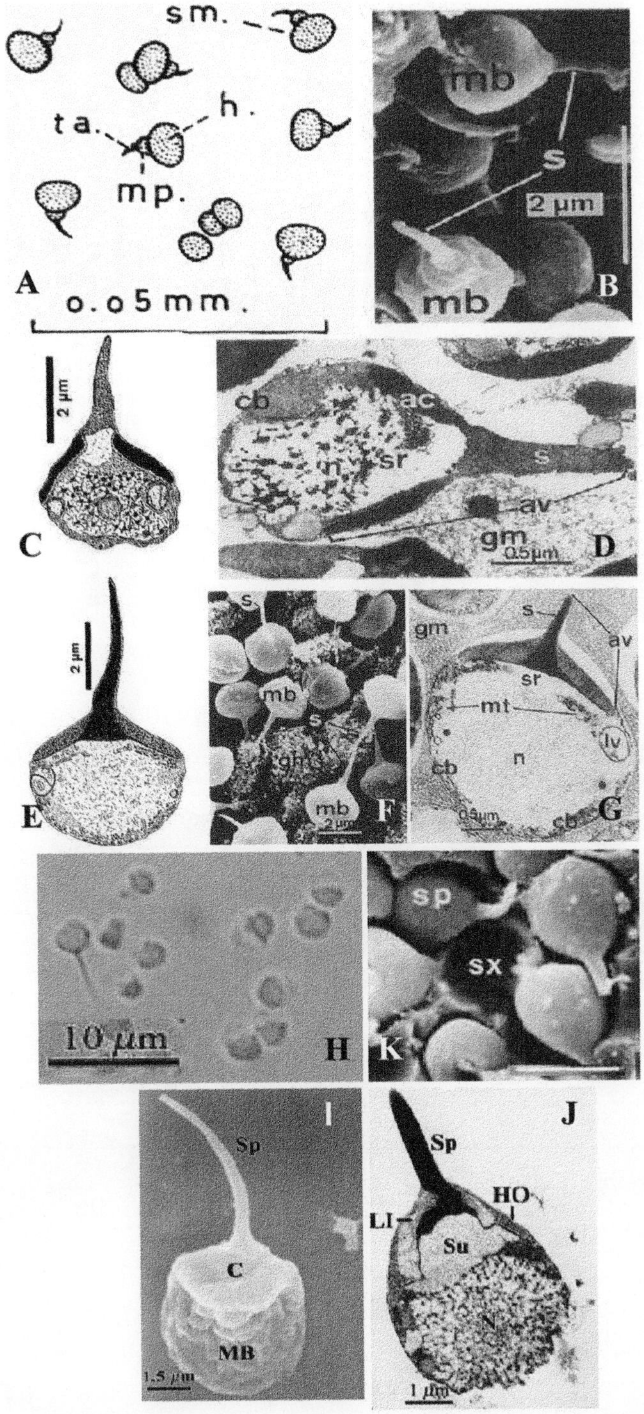

FIG. 10.69 (I) Illustration and photographs showing structure of the sperm of different species of genus *Penaeus*, (A–D) of *P. kerathurus*: (A) smear of the content of the PVD *(from Bawab, 1964)*; (B) SEM; (C) line drawing of D; (D) TE micrograph of sagittal section; (E–G) of *P. japonicus*: (E) line drawing of (G); (F) SEM; (G) TE micrograph of sagittal section; (H–J) of *P. monodon*: (H) light microscopic photograph of smear; (I) SE micrograph of the sperm; (J) TE micrograph of the sperm; (K) SE micrograph of spermatozoa of *P. setiferus* (sp) surrounded by seminal matrix (sx) contained in descending vas deferens. Scale bar = 150 μm. *(B–G) From Medina et al. (1994), (H–J) from Pongtippatee et al. (2007) and (K) from Bauer and Cash (1991).* ac = acrosomal cap, av = acrosomal vesicle, C = cap, cb = cytoplasmic band, gm = gelatinous matrix enveloping the spermatozoa, HO = high-density outer layer, LI = low-density inner layer, lv = large vesicle, mb & MB = main body sperm, mt = microtubules, n = nuclear region, s & sp = spike, sr & Su = subacrosomal region.

(Continued)

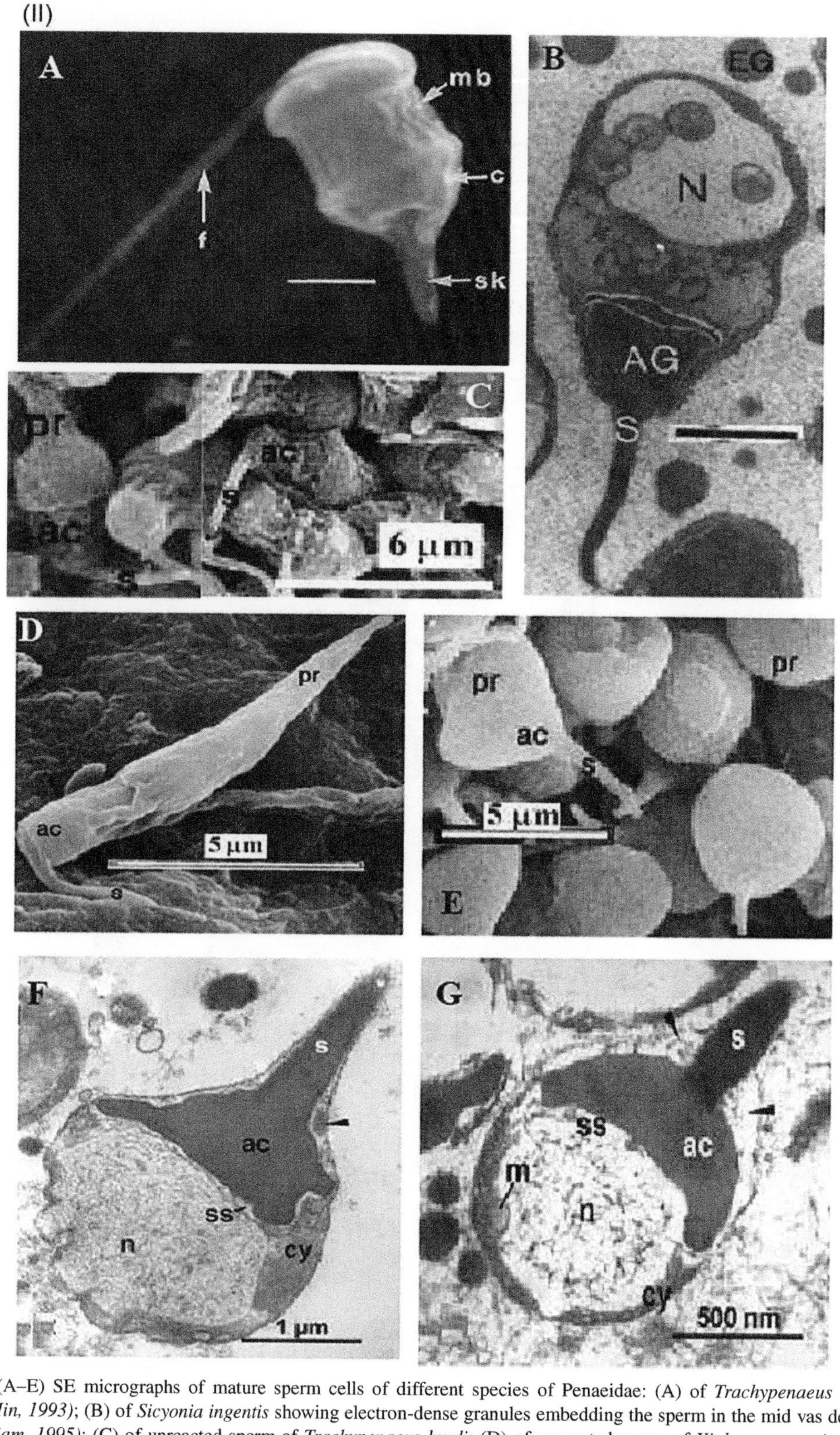

FIG. 10.69, CONT'D (II) (A–E) SE micrographs of mature sperm cells of different species of Penaeidae: (A) of *Trachypenaeus similis*. Scale bar=35 µm *(from Bauer and Min, 1993)*; (B) of *Sicyonia ingentis* showing electron-dense granules embedding the sperm in the mid vas deferens. Scale bar=10.0 µm *(from Subramoniam, 1995)*; (C) of unreacted sperm of *Trachypenaeus byrdi*; (D) of unreacted sperm of *Xiphopenaeus riveti*; (E) of unreacted sperm of *Litopenaeus occidentalis ((C–E) from Alfaro et al., 2003)*; (F & G) TE micrographs of solenocerid shrimps (Family Aristeidea *after Schram, 1986*): (F) of *Solenocera membranacea;* and (G) of *S. Africana (from Medina et al., 2006)*. ac&c=cap, ac=acrosomal region which includes the spike, ag=acrosome granule, av=acrosome vesicle, cy=cytoplasm, m=mitochondria, mb=main sperm body, n=nuclear region, pr=posterior region contains the nucleus, s&sk=spike, ss=subacrosomal space.

species of genus *Penaeus* are almost the same, differing only in the dimensions of the components (Fig. 10.69(I)). However, structure and form of different penaeids differs (Fig. 10.69(II)). The EM corrects the origin and structure of the organelles of the sperm of penaeids in particular and of Crustacea in general. There is no axoneme in this sperm, and this evidences the general form of decapod sperm described in textbooks (e.g. Barnes, 1974).

During the last few decades, there have been many electeron microscopic studies dealing with the decapod sperm (for a review, see López-Camps et al., 1981). Accordingly, the structure of decapod sperm was divided into at least two morphological groups:

1. Dendrobrachiatan, formerly known as natantian and including shrimps and prawns. The sperm of this group have an incomplete nuclear envelope and a single spike, which in some species (e.g. *Sicyonia*) is believed to be the acrosome. The above description of penaeid sperm evidenced this information.
2. Pleocyematan, formerly known as reptantian, including crayfishes, lobsters, and crabs (López-Camps et al., 1981 and Beach and Talbot, 1987).

Oogenesis: to the author's knowledge, there are no studies available specifically describing the process of oogenesis or vitellogenesis in *P. kerathurus*. However, through the histological studies of the ovarian development, one can describe this process. Both the works of Medina et al. (1996) and Kücükdermenci and Lok (2012) divided ovarian development into five distinct stages:

1. Previtellogenesis: according to Medina et al., the ovaries during March contain only oogonia, but previtellogenic oocytes with two distinct sizes can be recognized; the smaller with 17–26 μm diameter, lie at the centre of the ovary; whereas the larger ones with 26–65 μm diameter, and appear in follicles located more peripherally (Fig. 10.70A).
2. Early vitellogenesis: the ovaries during April contain mainly previtellogenic oocytes and some early vitellogenic oocytes, with 80–100 μm diameter, display in this stage. Their cytoplasm becomes more acidophilic with increase in size (Fig. 10.70B).
3. Late vitellogenesis: the ovaries during June contain large oocytes with 100–200 μm diameter and the cytoplasm is filled with yolk granules. These yolky oocytes are located in the outer regions of the ovary (Fig. 10.70C).
4. Mature stage: the ovaries during July are characterized by the presence of fully mature oocytes of diameters that exceed 200 μm. This mature ocyte shows conspicuous **cortical rods** at the cell periphery (Fig. 10.70D).
5. Spent stage: the ovaries during July are distinguished by the presence of atretic (degenerating) oocytes (Fig. 10.70E).

There is a difference between these two works describing the development of the ovaries of the same species, *P. kerathurus*, such as the duration of the cycle as well as the timing of each stage. The localities from which the animals used in both works were obtained are in the same region, between 35° and 40° north of the equator, but those used in the work of Medina et al. (1996) were obtained from the waters of the ocean south-west Spain, while those of the work of Kücükdermenci and Lok (2012) were obtained from the waters of the east of the Mediterranean in Turkey, very far from Spain. The waters differ, the environmental conditions must differ, and as such the results of these two works differ.

However, before the time of both the above authors, oocyte development in the kuruma prawn *P. japonicus* Bate was studied by Yano (1988). This author obtained his specimens from Ise Bay, Japan and divided the oocyte development into 10 stages according to morphological characters (Fig. 10.71(I)).

1. Synapsis stage (multiplication period), oogonia and oocytes are spherical with the same size (10 μm in diameter). The largest nucleolus is typically half-moon shaped and the cell membrane is indistinct (Fig. 10.71(I) A).
2. Chromatin nucleolus stage in which the oocyte measures 16–34 μm and the nucleus has basophilic chromatin materials (Fig. 10.71(I) B).
3. Early perinucleolus stage in which the oocyte increases in size to 30–80 μm and its nucleus has central basophilic nucleolus, with six or seven circular nucleoli situated peripherally. Very few follicle cells appear in this stage (Fig. 10.71(I) C).
4. Late perinucleolus stage with oocyte measuring 68–88 μm and has a structure as in the previous stage. The cytoplasm, which increases in amount in the previous stage, is strongly basophilic. Follicle cells increase in number around the oocyte (Fig. 10.71(I) D). Stages 2, 3, and 4 constitute the period of previtellogenesis.
5. Oil globule stage I, in which the oocyte may reach 160 μm and its cytoplasm becomes loaded with numerous acidophilic globules strongly positive with Sudan black B indicating that they are oil globules. The nucleus still central in position but shrinks rapidly and becomes less visible. The follicle layer consists of 28–34 cells; each expands rapidly on the surface of the oocyte (Fig. 10.71(I) E and F). The follicle cells are implicated as the possible cell type responsible for ovarian vitellogenin synthesis in *P. japonicus*. However, yolk protein is not accumulated in the oocyte during this stage.
6. Oil globule stage II differs from the previous stage in that a large oil globule appears in the peripheral cytoplasm (Fig. 10.71(I) F).

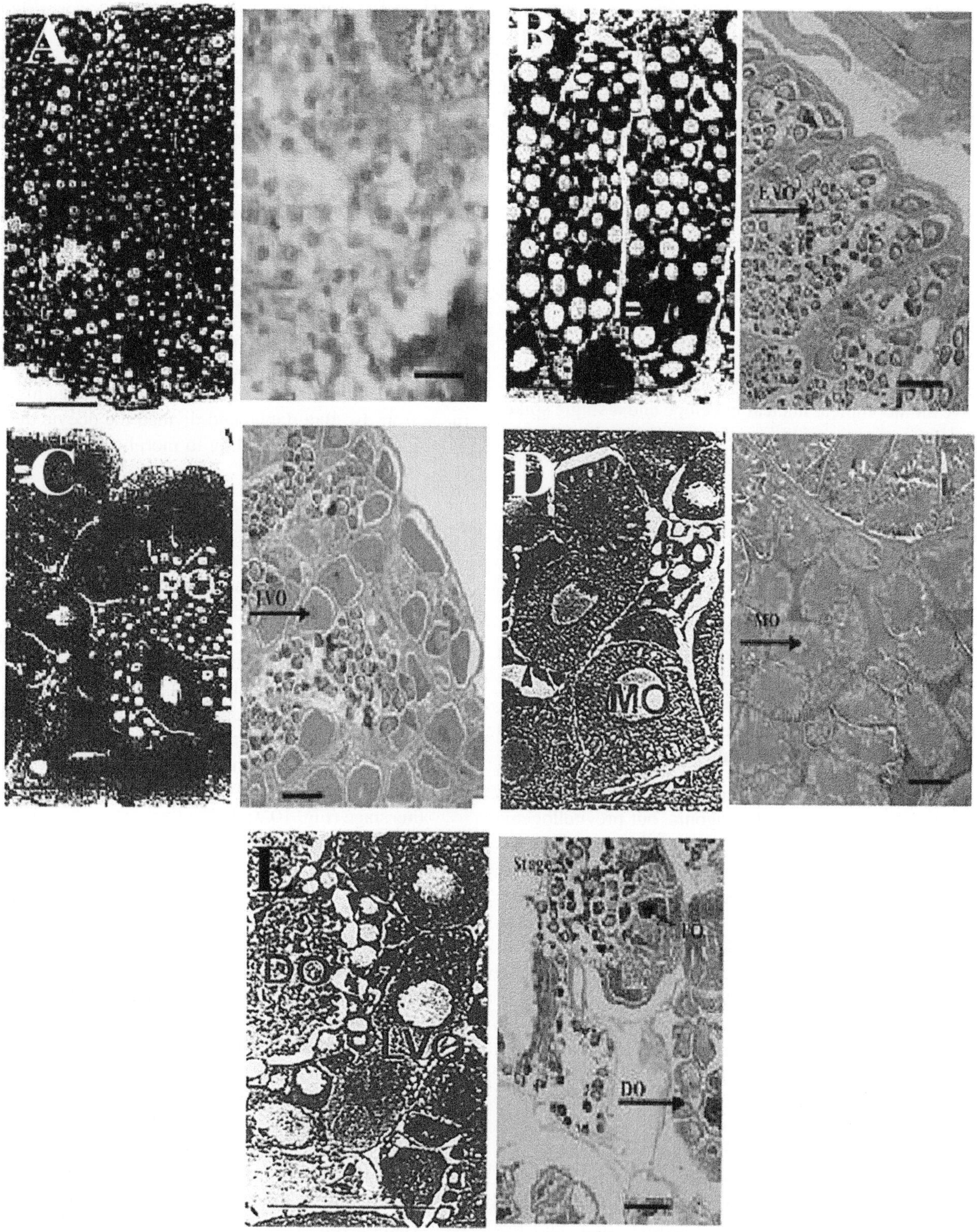

FIG. 10.70 Oogenesis in wild-caught *Penaeus (Melicertus) kerathurus* (Forskål, 1775). Each figure is presented by two halves: **left** are sections of the ovaries of specimens obtained from the Gulf of Cádiz (off Sanlúcar de Barrameda, southern Spain), the sections stained with toluidine blue. Scale bars = 200 μm *(from Medina et al., 1996)*; **right** are sections of ovaries of specimens obtained from Izmir Bay, southern Aegean Sea, Turkey. Scale bars = 100 μm *(from Kücükdermenci and Lok, 2012)*. (A) Stage I (left during March, right during December–January) contain only oogonia, small and large previtellogenic oocytes, the former in the centre and the latter peripherally in the follicle; (B) stage II with early vitellogenic oocytes (left during April, right during June); (C) stage III with late vitellogenic oocytes (left during June, right during February–May); (D) stage IV shows mature oocytes with peripheral cortical rods (left during July, right during June–October); and (E) stage V appears in late July (left during July, right during September). DO = degenerating oocytes, EVO = early vitellogenetic oocytes, LVO = late vitellogenic oocytes, MO = mature oocyte, *arrowhead* indicates cortical rods, PO = previtellogenic oocyte.

7. Yolkless stage during which the oocyte measures 110–146 μm and the basophilic cytoplasm contains, in addition, strongly basophilic vesicles (Fig. 10.71(I) H). The possible function of this basophilic vesicle is that it is the frame and matrix for accumulation of yolk granules in the cytoplasm of oocytes.
8. Yolk granule stage, oocyte of which may reach 237 μm and in its cytoplasm appear strongly acidophilic yolk granules within the basophilic vesicles. At first, they are concentrated centrally, then gradually spread peripherally (Fig. 10.71(I) I).
9. Prematuration stage has oocytes measuring 200–264 μm and with cytoplasm still with yolk granules. The strongly basophilic nucleus shrinks gradually and its outline becomes irregular. At the late phase of this stage, it begins to migrate to the periphery. Follicle cells are still observed surrounding the oocyte. In the early phase of this stage, acidophilic, round cortical crypts are found in the peripheral part of the oocyte (Fig. 10.71(I) J). These crypts are separated from the cytoplasm by the cytoplasmic membrane, but there is a space between them and the membrane. In the late phase of this stage, the cortical crypts are situated in a club-shaped structure (Fig. 10.71(I) K). The vitelline envelope is visible just at the top of the cortical crypts. This stage thus lasts for relatively a long period of time during oogenesis in the penaeid shrimp. Germinal vesicle breakdown (GVBD) is initiated in the late phase of this stage and continues until the late phase of the maturation stage, immediately prior to spawning.
10. Maturation stage in which the cocyte reaches 288 μm in diameter and the surrounding follicle cells disappear. This indicates that the oocytes are ovulated. A large number of these ovulated maturation stage

(I)

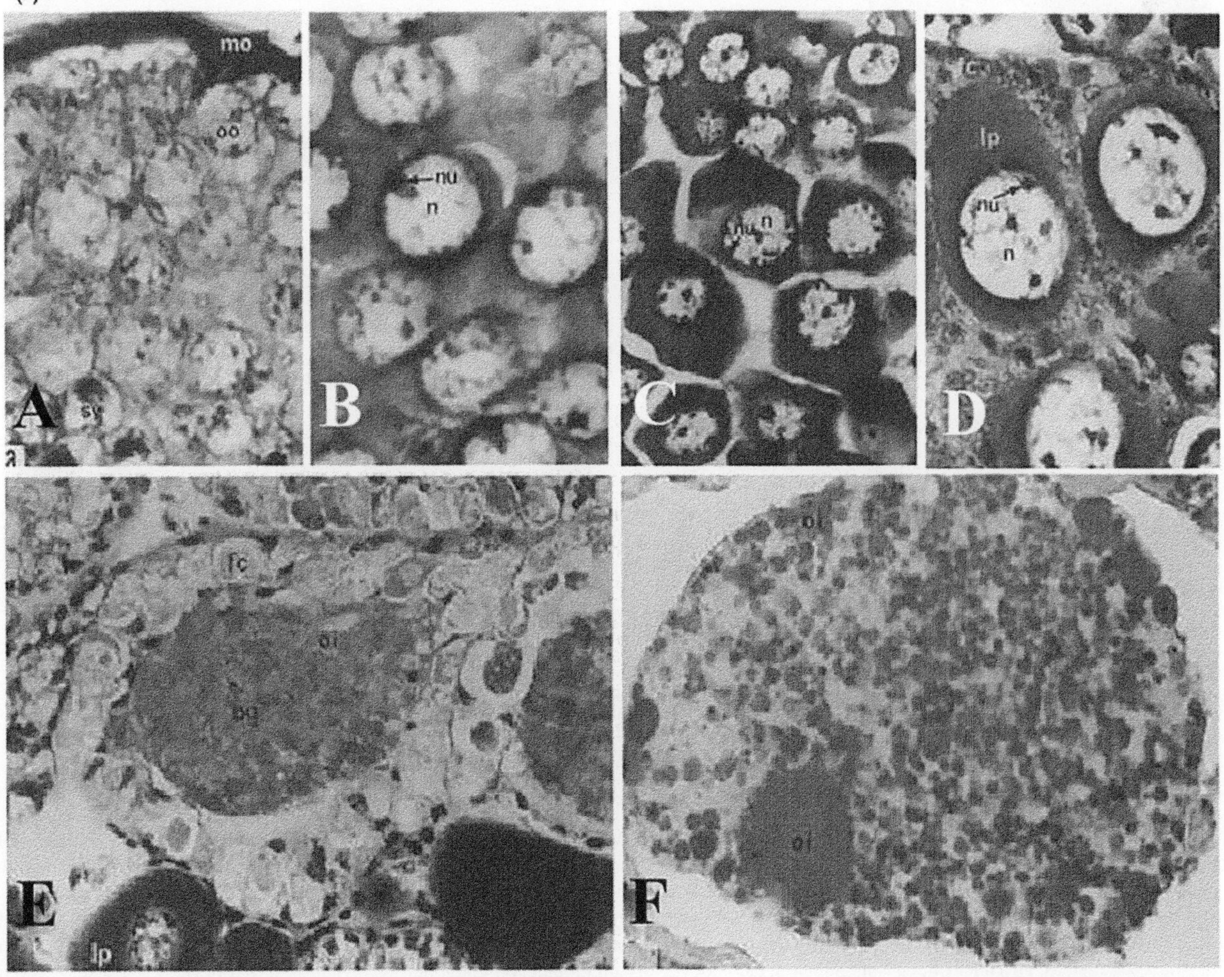

FIG. 10.71 (I) *Penaeus japonicus*, oocyte development: (A) Oogonia and synapsis stage oocytes near membrane of ovary; (B) Chromatin nucleolus stage oocyte; (C) Early perinucleolus stage oocytes; (D) Late perinucleolus stage oocytes, follicle cells are visible on outer surface of oocytes; (E) Oil globule stage I oocytes and expanded follicle cells on outer surface of oocytes and numerous oil globules are found in cytoplasm; (F) An oil globule stage I oocyte showing positive-stained oil globules with Sudan black B. X=810, (A and B) X=1180.

(Continued)

(I)

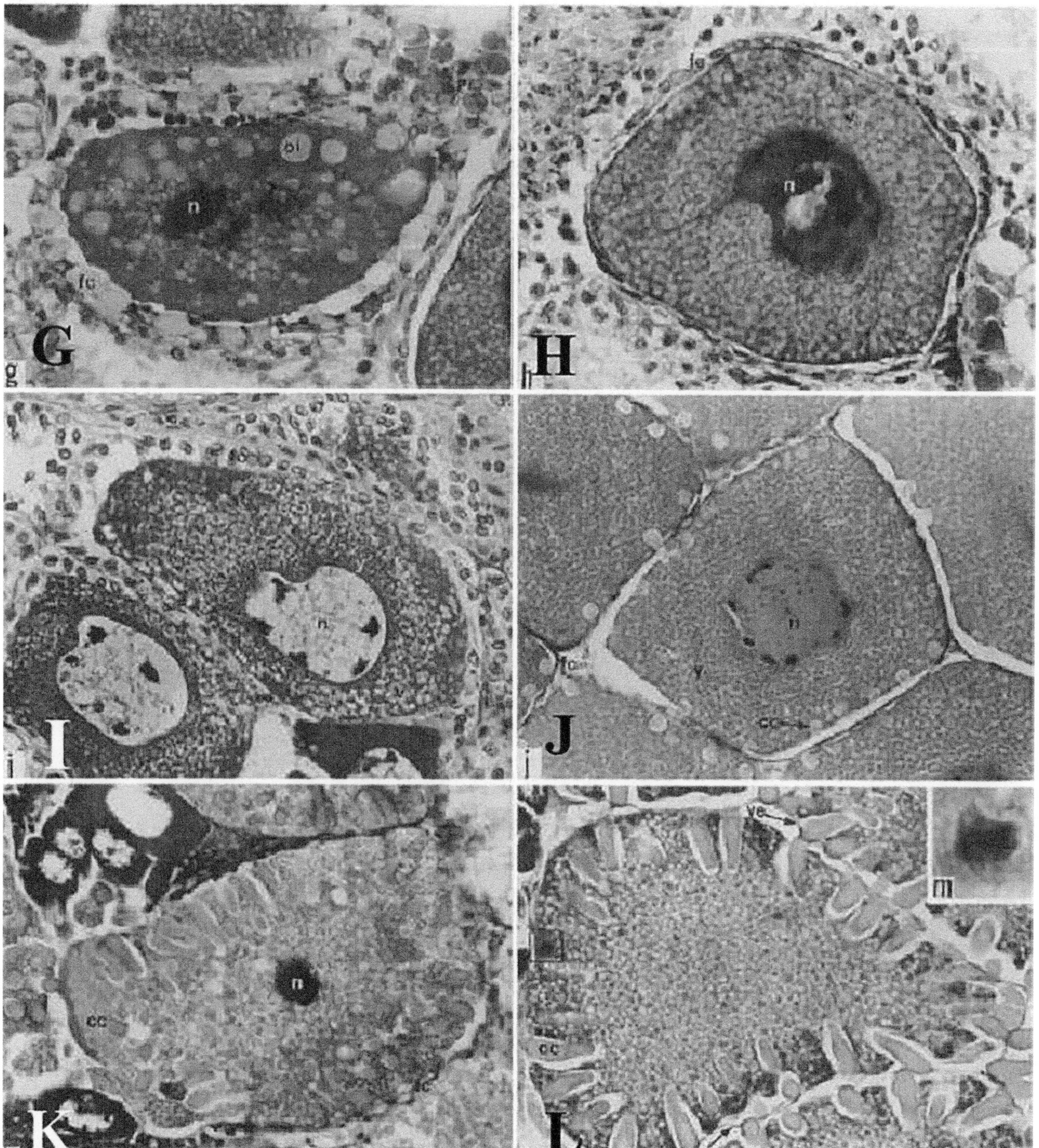

FIG. 10.71, CONT'D (G) Oil globule stage II oocyte with expanded follicle cells; (H) Yolkless stage oocyte and shrinked follicle cells. The basophilic cytoplasm contains strongly basophilic vesicles; (I) Yolk granule stage, yolk granules are first concentrated centrally then spread to periphery; (J) Prematuration stage oocytes, acidophilic round cortical crypts appear in early stage of this phase; (K) Prematuration stage oocytes with shrinked nucleus and cortical crypts in club-shaped structure; (L) Maturation stage oocyte during which follicle cells disappear, indicating that oocytes are ovulated, vitelline envelope appears; (M) Higher magnification of enclosed area (*small square* at the left-hand side of the figure) showing meiotic metaphase. X=1180. All stained with haematoxylin & eosin except F. all X=405 except A, B, F & m. cc=cortical crypts, lp=late perinucleolus oocyte, mo=membrane of ovary, n=nucleus, nu=nucleoli, og=oil globule stage I oocyte, oi=oil globule, oo=oogonia, sy=synapsis stage oocyte, v=vesicle, ve=vetilline envelop, y=yolk granule. *After Yano (1988).*

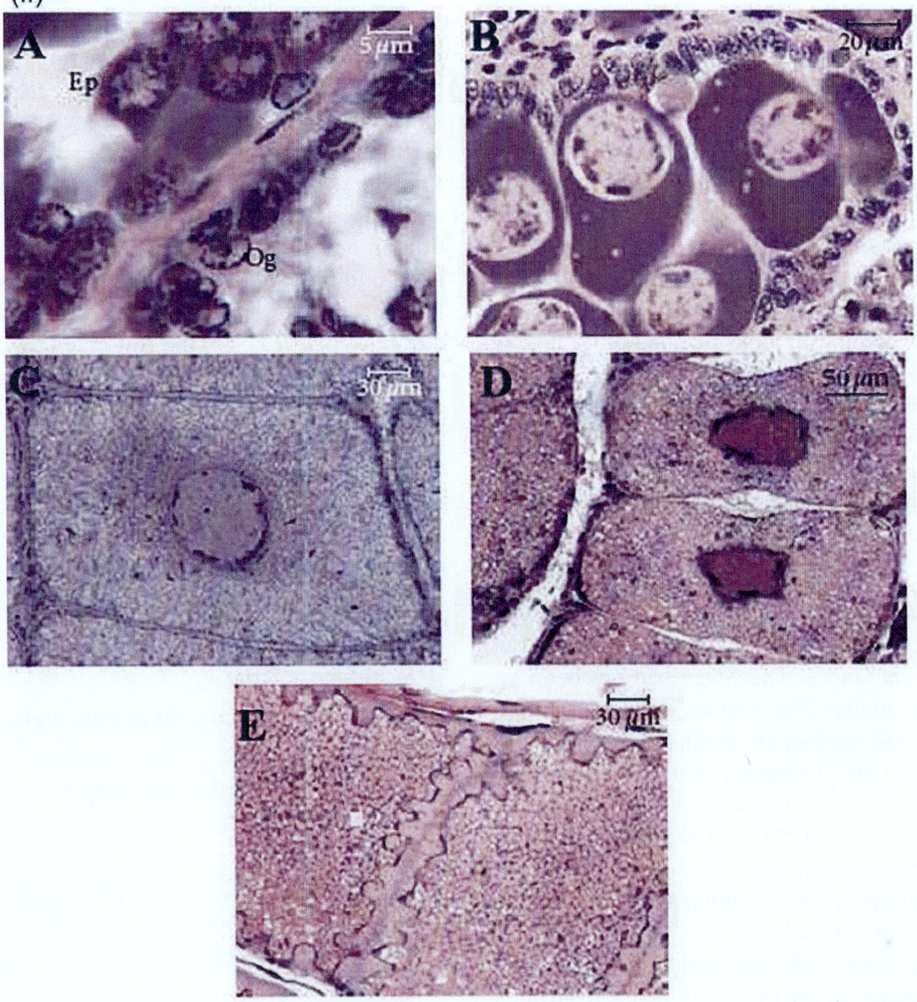

FIG. 10.71, CONT'D (II) Oocyte developmental stages of *Aristaeomorpha foliacea* (haematoxylin-eosin): (A) Oogonia (Og) with large clear nucleus and early primary stage (Ep) cell with large clear nucleus and chromatin clusters; (B) Late primary stage, cell has nucleus with many peripheral nucleoli; (C) Early vitellogenic stage, cytoplasm filled with lipid vesicles and yolk granules, the cell is rectangular in section; (D) Late vitellogenic stage, cell with small central nucleus; (E) Postvitellogenic stage, nucleus no longer visible and cytoplasm has columnar cristae protruding to the cortex. *From Perdichizzi et al. (2012).*

oocytes were observed in the same ripe ovary, implying that ovulation is synchronous within the same ovary. After ovulation and in the late phase of this stage, the nuclei shrink even further and migrate towards the cytoplasmic membrane and their nuclear membrane become indistinct. Meiotic divisions of the ovarian oocytes were found to progress to the metaphase of primary maturation division. The review of Dall et al. (1990) certained this observation and reported that penaeid eggs, when released are in thelate prophaseor early metaphase of the first maturation division. Thus, after ovulation, GVBD in *P. japonicus* is not completely synchronized within the same ovary. Polar bodies were not observed in the ovarian oocytes. These oocytes have many club-shaped cortical crypts and contain extensive accumulations of yolk granule dispersed throughout the cytoplasm and the whole oocyte is coated by the vitelline membrane (Fig. 10.71(I) L).

Through this detailed description of the process of oogenesis in this species, *P. japonicus*, one can conclude the evidence of the same four principal stages described above.

A satisfactory histological description of oogenesis is that of the giant red shrimp *Aristaeomorpha foliacea* (Risso, 1827) (another family Aristeidea (Schram, 1986)) (Fig. 10.71(II)) obtained from the southern Tyrrhenian Sea, Italy (Perdichizzi et al., 2012). This process in this species is divided into the following:

1. oogonia and early primary stage (Ep) in which the oogonium measures about 10 μm and the Ep measures 15–25 μm. The oogonium has large, clear nucleus containing many strongly basophilic nucleoli, one of which

is more evident. The cytoplasm is barely visible. The Ep has large, clear nucleus with chromatin clusters and filaments;

2. late primary stage (Lp) in which each oocyte measures 25–85 μm and has many peripheral nucleoli, strongly basophilic;
3. early vitellogenic stage (Ev) in which each oocyte measures 90–300 μm and with visible cytoplasm filled with lipid vesicles and yolk granules;
4. late vitellogenic stage (Lv), with oocytes rectangular in section and each measures 200–350 μm, with small central nucleus. The cytoplasm looks like that of Ev oocyte;
5. postvitellogenic stage (Pv) in which the oocyte enlarges greatly in diameter measuring 350–470 μm. Its nucleus becomes no longer visible and the cytoplasm contains columnar cristae protruding into the outer cortex; and
6. atresic stage with the oocytes appear like late primary oocytes, but show massive vacualization of cytoplasm.

Therefore, this ovarian development shows six stages (Fig. 10.62(II)) (Perdichizzi et al., 2012). These authors add a new stage, postvitellogenic stage, in between the late vitellogenic and spent stages. This stage had not been described in previous studies. The oocytes at this stage have the same size and yolk inclusions as those of the late vitellogenic stage. They are characterized by the presence of **cortical cristae**, resembling the cortical rods of penaeid shrimps. In this study of Perdichizzi et al. (2012), light microscopy showed that cortical cristae are housed in crypts formed by invaginations of the oolemma, and therefore they are extracellular, as cortical rods.

After the completion of yolk accumulation in the fourth stage of the development of the ovaries of penaeid shrimps, the large oocytes become characterized by the appearance of cortical rods. The latter are formed radially around the periphery of the oocyte plasma membrane, but under the vitelline membrane. Perdichizzi et al. (2012) reported that they are involved in the cortical reaction by contact with seawater by forming a jelly layer around the egg, preventing polyspermy. Wilder et al. (2010) reported that two major proteins comprising the cortical rods have been purified and characterized: shrimp ovarian peritrophin (SOP) in *P. semisulcatus* or cortical rod protein (CRP) in *Mar. japonicus*; and *Mar. japonicus* thrombospondin (MjTSP). Except for their molecular weights, these three proteins show similar molecular structure in terms of repeated cysteine-rich domains and proteolytic sites for posttranslational modulation and glycosylation, as well as similar deduced amino acid sequences. Further studies suggested that they are the products of a single gene or are encoded by orthologous genes.

Spawning. Heldt (1931) reported that the period of laying eggs of *Penaeus caramote* (= *P. (Melicertus) kerathurus*) in aquariums extends from June to September in the region of Tunisia (Mediterranean), with the peak during hot periods in July and August. On July 24, 2014, Khafage obtained two live specimens of *P. kerathurus* selected as gravid females from the fisheries of the Gulf of Abu-kir, 15 miles east of Alexandria, Egypt. She dissected one straight away and kept the other live in a suitable aquarium in the National Institute of Oceanography and Fisheries, Alexandria. The first one was used to expose the female reproductive system and then isolated (Fig. 10.61A and B). In addition, the portion of the ventral surface containing the thelycum was isolated and prepared for histological study of the role of the female in the dehiscence of the spermatophores (Khafage and Bawab, 2014). On the following day, a change was observed in the tank. The examination of a drop of water of the tank revealed that it embedded many eggs and few very tiny larvae. This means that the animal spawned but unfortunately photography was not available then. However, Fig. 10.72A shows female *P. monodon* during spawning. This figure greatly evidences the description. The diagram (Fig. 10.72B) illustrates the spawning 'posture', as it swims, of *P. japonicus*.

The study of these two gravids evidenced the following:

1. they are parturients;
2. the spawning occurs during the night;

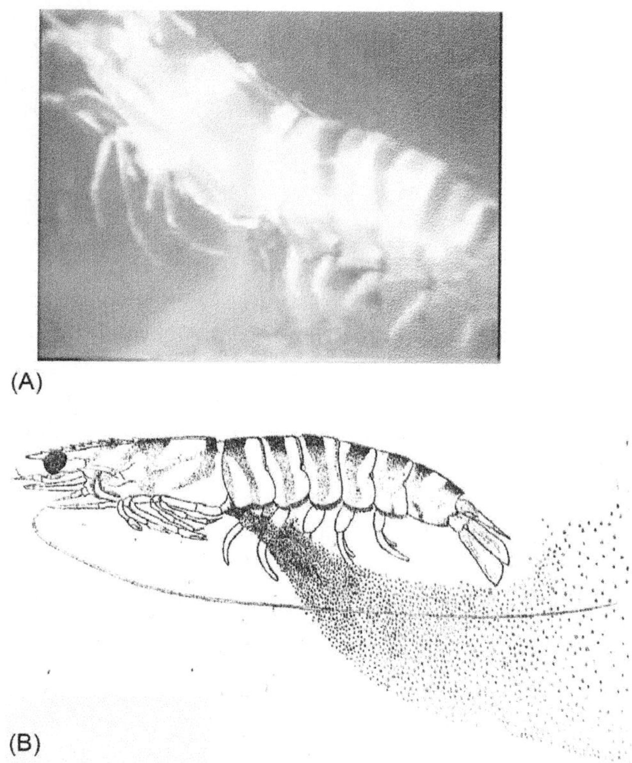
(A)
(B)

FIG. 10.72 (A) Shows *Penaeus monodon* during spawning taken from video prepared by the Southeast Asian Fisheries, Development Center, Department Aquaculture (Biology & ecology), 1988; (B) A copy from the work of Hudinaga, 1942 showing *P. japonicus* during spawning. One wonders that the recent photo greatly evidenced the earlier illustration.

3. the peak of spawning occurs during the hot period which extends from June to September, and the maturation of the ovaries occurs during the spring and summer in the Mediterranean (first reported by Heldt, 1931);
4. the colour of the mature ovaries is olive green;
5. the female must have a role in the dehiscence of the spermatophores; and
6. fertilization is synchronized with spawning.

At the time of spawning, the spermatophores inside the thelycum in the female must dehisce to free their spermatozoa. Heldt in 1931 recorded and, in 1938 illustrated by figure, two little openings, each at the base of the fourth pereiopod of *P. kerathurus*. From these openings, the 'spermatic flow' discharges. In addition, Hudinaga (1942) reported that at the time of spawning of *P. japonicus*, the spermatozoa within the spermatophores inside the thelycum are discharged into the seawater through a pair of very tiny openings located at the base of the fourth pereiopods shown in one of his figure. Khafage (2000) studied the thelycum of the seven penaeid species of the fisheries of Alexandria waters. She presented SEM photo showing a very wide opening, which belongs not to the base of the coxa of *P. kerathurus*, but to an extension of this coxa referred to as the coxal vesicle (Fig. 10.40C). This vesicle is so tiny that it was overlooked previously. Its form and size change with the species in the seven penaeids studied. The histological study of the thelycum just prior to spawning (see above) evidenced that fertilization occurs during spawning.

Fertilization. Studies on the process of fertilization in Crustacea are rather limited because of the opacity of their eggs. This opacity makes the observation very difficult owing to the abundance of yolk. In addition, it is more difficult owing to the rare occurrence of free egg-laying as a result of the highly developed brooding habit. However, in penaeids where there is no brooding and the fertilization is external, the latter occurs just at spawning. Hudinaga (1942) studied reproduction, development, and rearing of *Penaeus japonicus* Bate. He observed, described, discussed, and illustrated the process of fertilization that occurs in this species. According to this information of Hudinaga, the present work is illustrated Fig. 10.73. It is as follows:

1. Sperm entrance occurs almost simultaneously with the formation of the first polar body, which appears immediately after egg laying.
2. When a number of spermatozoa reach the egg after penetrating the jelly layer (Fig. 10.73A), the contact of each results in the elevation of a fertilization (or entrance) cone (Fig. 10.73B and C).
3. The sperm heads are in contact with the apices of the cones, which are at first transparent cylindrical elevations.
4. Soon they become filled with inflowing ooplasm (Fig. 10.7D).
5. The first cytoplasm reaches the sperm head, this cone begins to contract (Fig. 10.73E), drawing the entire sperm including the spike into the interior (Fig. 10.73F and G);

6. As the original smooth contour of the egg surface is restored, all the other cones then disappear without engulfing their spermatozoa (Fig. 10.73H). The second polar body discharges after the formation of the fertilization membrane, which separates it from the first polar body.

Early embryology. It appears that Heldt (1938) was the first to describe the early development of *P. caramote* (*P. (Melicertus) kerathurus*) and *Parapenaeus longirostris*. In addition, Hudinaga (1942) was the first to describe the early development in *P. japonicus*. It appears that this aspect concerning penaeids was not studied enough at that time. However, very recently during the first years of the 21st century, the early development (cleavage and gastrulation) and larval stages of some penaeids were studied using very recent tools. Ronquillo et al. (2006) studied the embryonic development and larval stages of *P. semisulcatus*. They reported that they are generally similar to those of other closely related species in the family Penaeidae, just as that of *Marsupenaeus japonicus*. Later, Pawlak et al. (2010) verified this information. Ronquillo et al. (2006) obtained specimens of *P. semisulcatus* from the southern Yatsuhiro Sea, southwest of Kyushu Island, Japan (Pacific region). Gravid shrimps were spawned by natural means in well-aerated fibreglass tanks with filtered seawater and suitable capacity. The early embryology was studied and photographed (Fig. 10.74), and a table was presented summarizing the chronology of the embryonic development of this species, *P. semisulcatus* (Fig. 10.75A). Its larval development was thoroughly studied and illustrated in detail (see later).

The early embryology of the black tiger shrimp *P. monodon* (Biffis et al., 2009) and of the kuruma shrimp *P.* (*Marsupenaeus*) *japonicus* (Bate) (Pawlak et al., 2010) were thoroughly studied using very advanced tools and stains (nucleic acid stain Sytox Green and confocal microscopy). The present work followed that of *P. japonicus*, since this species is one of the common prawns in the commercial catches in Egypt. At the same time used the work of Biffis et al. for comparison and/or discussion. Pawlak et al. (2010) selected gravid females (ovary stage IV) from tank-reared stocks by shining a flashlight through their dorsal exoskeleton during dark hours. Gravid females that were 'impregnated' with spermatophores were unilaterally eyestalk ablated using hot forceps. Spawning occurred in suitable tank fitted with an automated spawning detection and alarm system. Coman et al. (2003) described and constructed this system. It has a false mesh base and three glass trays permitting timely

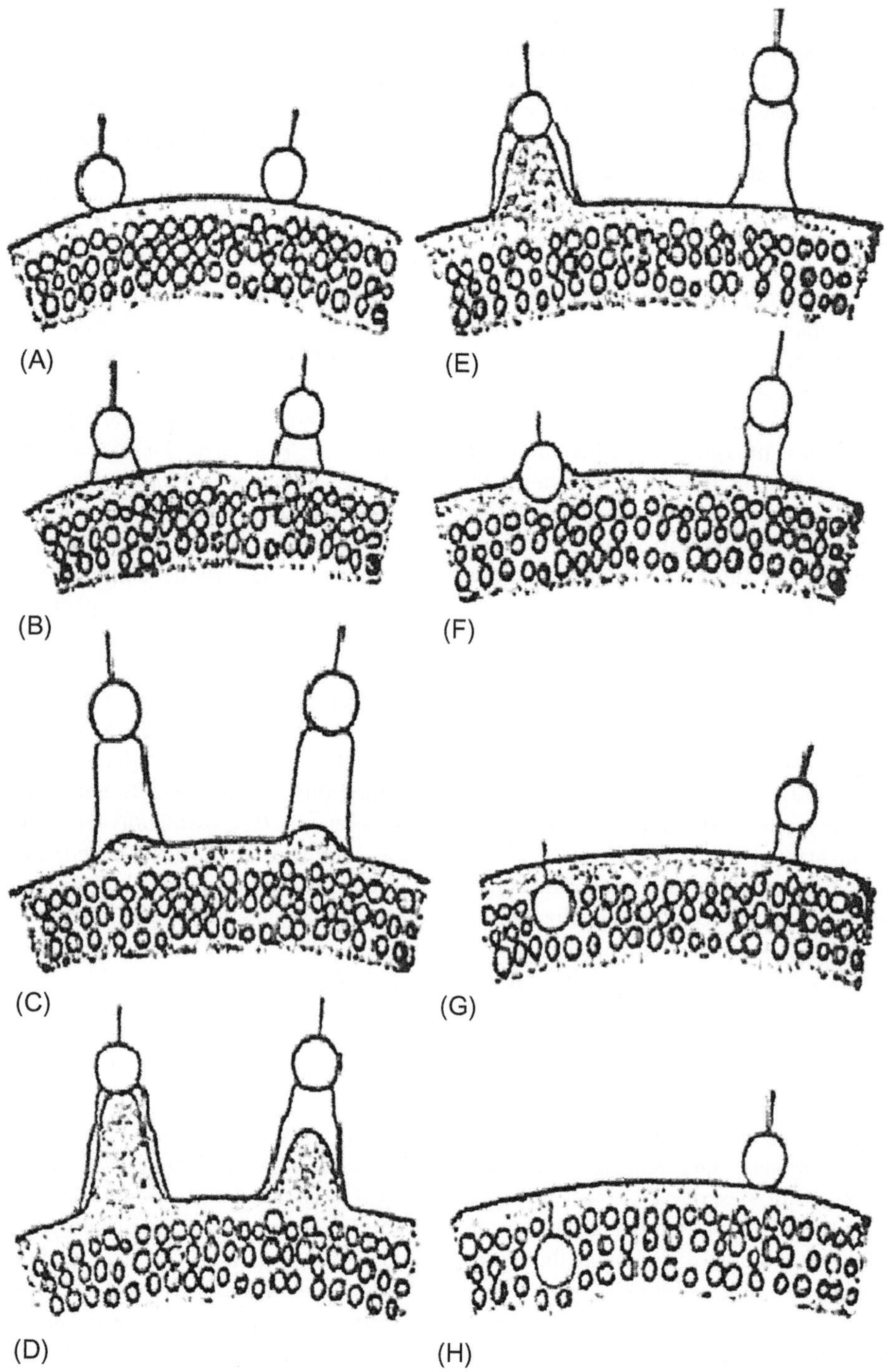

FIG. 10.73 Diagrammatic drawings illustrating the process of fertilization in *Penaeus japonicus* according to the description of Hudinaga (1942): (A) Sperm comes into contact with egg surface; (B) Beginning of formation of fertilization cone; (C) Maximum size of fertilization cone; (D) Influx of cytoplasm into fertilization cone and becomes in contact with sperm head; (E) Contraction of the fertilization cone in the left-hand side; (F) Complete contraction of fertilization cone with engulfing the sperm head; (G and H) Entry of only one spermatozoon into the egg, while the rest remain on its surface.

collection and manipulation of penaeid shrimp embryos. The material was prepared in Cleveland, Queensland, Australia, and subsequently shipped from Queensland to Mount Pleasant, Michigan, United States to be analysed.

The fertilized egg of *P. japonicus* is approximately 200 μm in diameter and round in shape with uniformly distributed yolk granules (Fig. 10.76A). The migrating pronuclei are visible in this figure. Normal development occurs within the 'hatching envelope' and proceeds for 12–13 h postspawning detection (psd). Cleavage occurs in the same stereotyped pattern described for other dendrobranchiates, including mirror-image variations. The cleavage pattern in *P.*

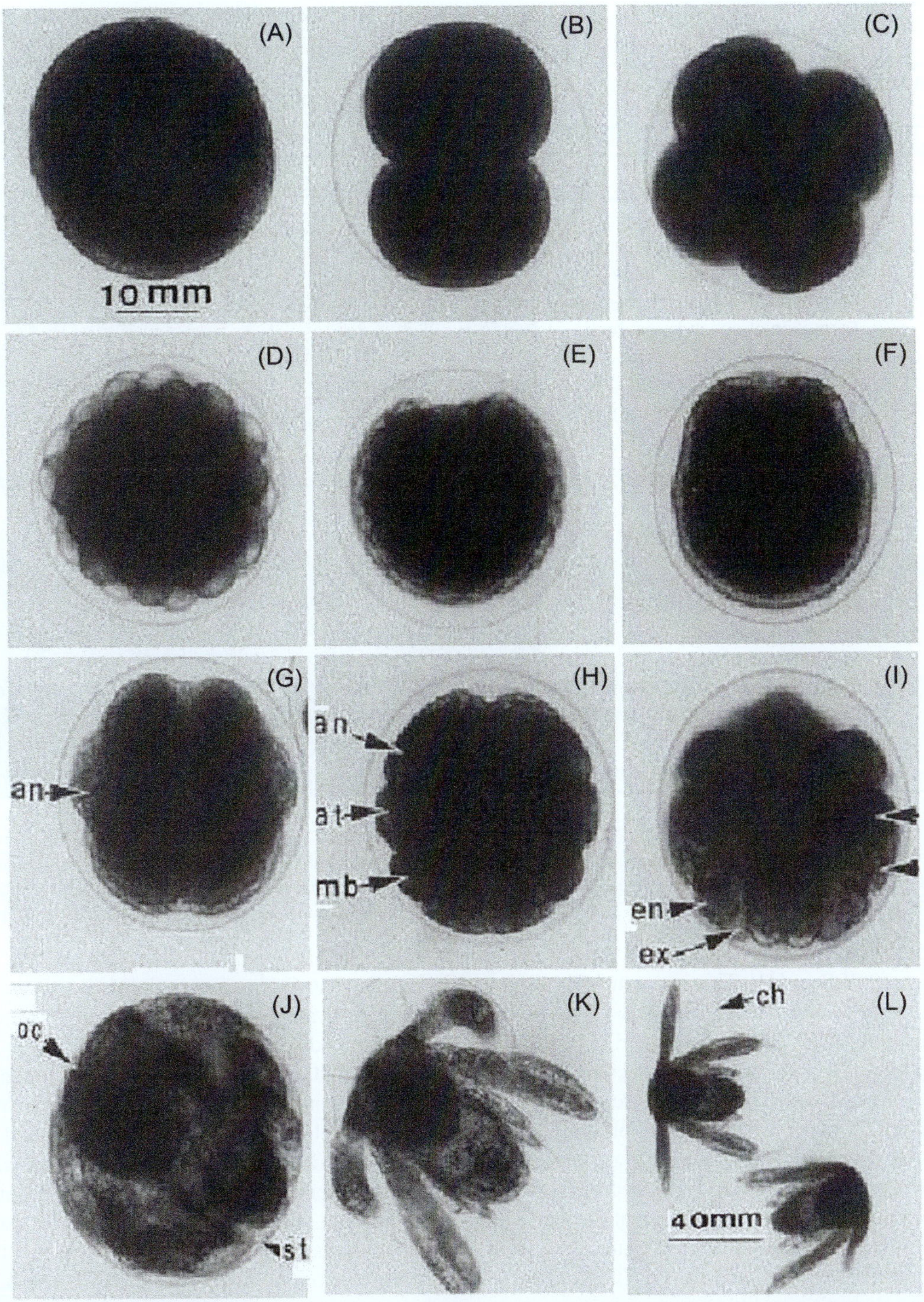

FIG. 10.74 Photomicrographs of stages of embryonic development of *P. semisulcatus* at 28.2±0.2°C, pH 8.20–8.32, and 34.5 g kg^{-1} salinity: (A) newly spawned egg; (B) 2-cell stage; (C) 4-cell stage; (D) morula stage; (E) blastula stage; (F) gastrula stage; (G) antennal limb-bud stage; (H) mandibular limb-bud stage; (I) biramous antenna and mandible (*head arrows* in the right hand side are the same of those in the left-hand side); (J) elongation of setae and spine; (K) hatching nauplius; (L) newly hatched nauplius beside empty chorion. an=antenna, at=antennule, ch=chorion, en=endopod, ex=exopod, la=labrum, mb=mandible, oc=ocellus, st=seta. *(After Ronquillo et al. (2006).)*

japonicus described in the work of Pawlak et al. (2010) does not support the view that it is of a spiral pattern. The cell lineage of this species to the 62-cell stage and two mesendoderm (ME) divisions is shown in Fig. 10.75B. The **first** cleavage, about 35 m psd in fixed samples, resulted in two similar-sized blastomeres: AB and CD (Fig. 10.76D and F). Therefore, it was presumed that the CD blastomere contains the vegetal pole as the previous studied penaeids. Polar bodies were not tracked throughout development in *P. japonicus*, but one polar body was well observed in

Chronology of embryonic development of *Penaeus semisulcatus* at 28.2±0.2 °C, pH 8.20–8.32, and 34.5 g kg^{-1} salinity

Cumulative time (h:min)	Stage
00:00	Spawn
00:08	Sperm penetration stage
00:08	Formation of first polar body
00:19	Formation of fertilization membrane
00:23	Formation of second polar body
00:32	First cleavage (2-cell stage)
00:48	Second cleavage (4-cell stage)
00:58	Third cleavage (8-cell stage)
01:15	Fourth cleavage (16-cell stage)
01:30	Fifth cleavage
01:48	Sixth cleavage
02:07	Seventh cleavage
02:36	Blastula
04:32	Gastrula
05:12	Antennal limb bud
06:20	Mandibular limb bud
06:35	Antennular limb bud
08:05	Biramous antenna and mandible
11:10	Elongation of spines and setae
14:05	Movement of limbs
14:10	Emergence from egg membrane

(A)

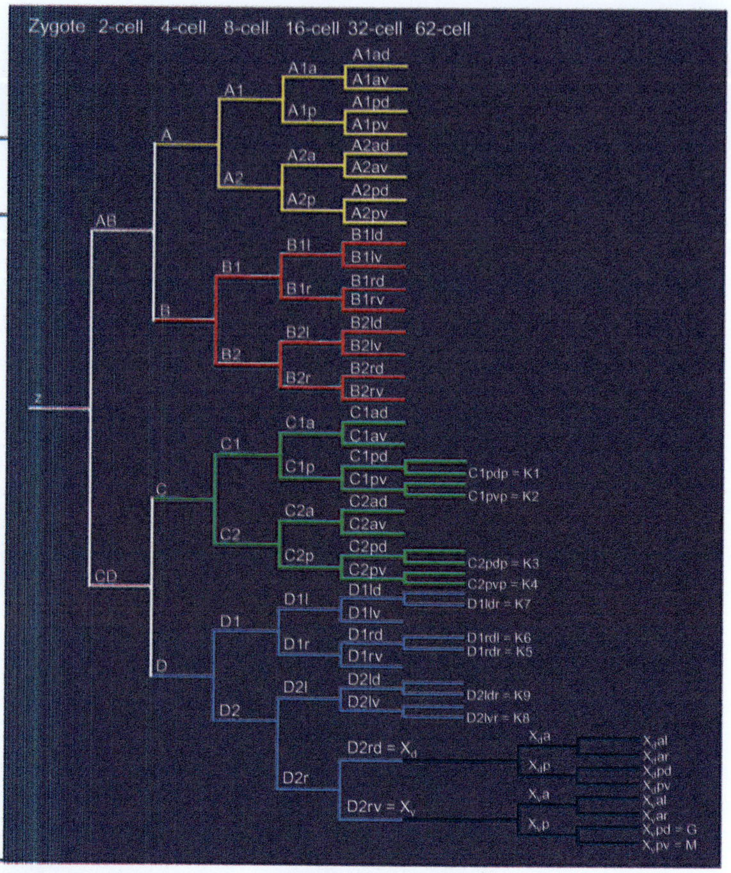

(B)

FIG. 10.75 (A) Copy of Table 1 of the work of Ronquillo et al. (2006) showing the timing of stages of development of *P. semisulcatus*. That of *P. japonicus* proceeded for 12–13 h *(Pawlak et al., 2010)* about the same time; (B) A copy from Pawlak et al. (2010) showing the cell lineage of *P. japonicus* to the 62-cell stage and two mesendoderm (ME) divisions. The A blastomere derivatives are *yellow*, B are *red*, C are *green* and D are *blue*.

P. monodon (Fig. 10.76G and H). This is not the only difference during the early development of these two species, but the intracellular body (ICB) was well detected in this 2-cell stage of *P. monodon* (Fig. 10.76G–I). The **second** cleavage resulted in the four-cell embryo, ABCD, in close packing arrangement. The CD blastomere divided obliquely in direction to the AB division. The B and D blastomeres lay ventrally to the anterior and posterior, respectively, while A and C lay dorsally (Fig. 10.77). All four blastomeres of this stage were of similar size and not distinguishable from each other, except for the presence of the (ICB) in the D blastomere. The orientation and blastomere identities were determined by the location of the ICB at this early stage, 4-cell stage to 62-cell stage. The ICB were observed consistently by Sytox green and SYTO RNAaSelect staining in the cleaved embryo of *P. japonicus* like that of *P. monodon* (Biffis et al., 2009). In *P. japonicus*, this ICB typically had an oval shape and an average diameter of 6.3 μm ($n = 16$). It was deduced that the presence of ICB in the D blastomere of both species is a natural marker of the germ line (Biffis et al., 2009), since this blastomere gives rise to the ME cells. The C blastomere derivatives were distinguished from A derivatives by their proximity to the D derivative that contained the ICB.

The **third** cleavage is to the 8-cell stage, where the B and D blastomeres divided towards each other, resulting in a band of blastomeres four cells long and one cell wide (Fig. 10.78C and C′). The B2 and D2 blastomeres were located at the ends of the band while the B1 and the D1 met at the middle of the band. The A and C blastomeres also divided towards each other resulting in another band, the same as the above one (Fig. 10.78D and D′). The A2 and C2 blastomeres were located at the ends of the AC band, and the A1 and C1 met at the middle of the band. This third orthogonal division produces the interlocking bands observed in cleavage of other penaeoidean species. This pattern starting from the 8-cell stage also exists in *P. kerathurus* and *Parapenaeus longirostris* (Heldt, 1938; Figs 10.36 and 10.38). The ICB was located in the D2 blastomere at the 8-cell stage in many examples. Biffis et al. (2009) described the form of these blastomeres at this stage in *P. monodon* as two bands of four cells each, which are interlocked as two horseshoe-like

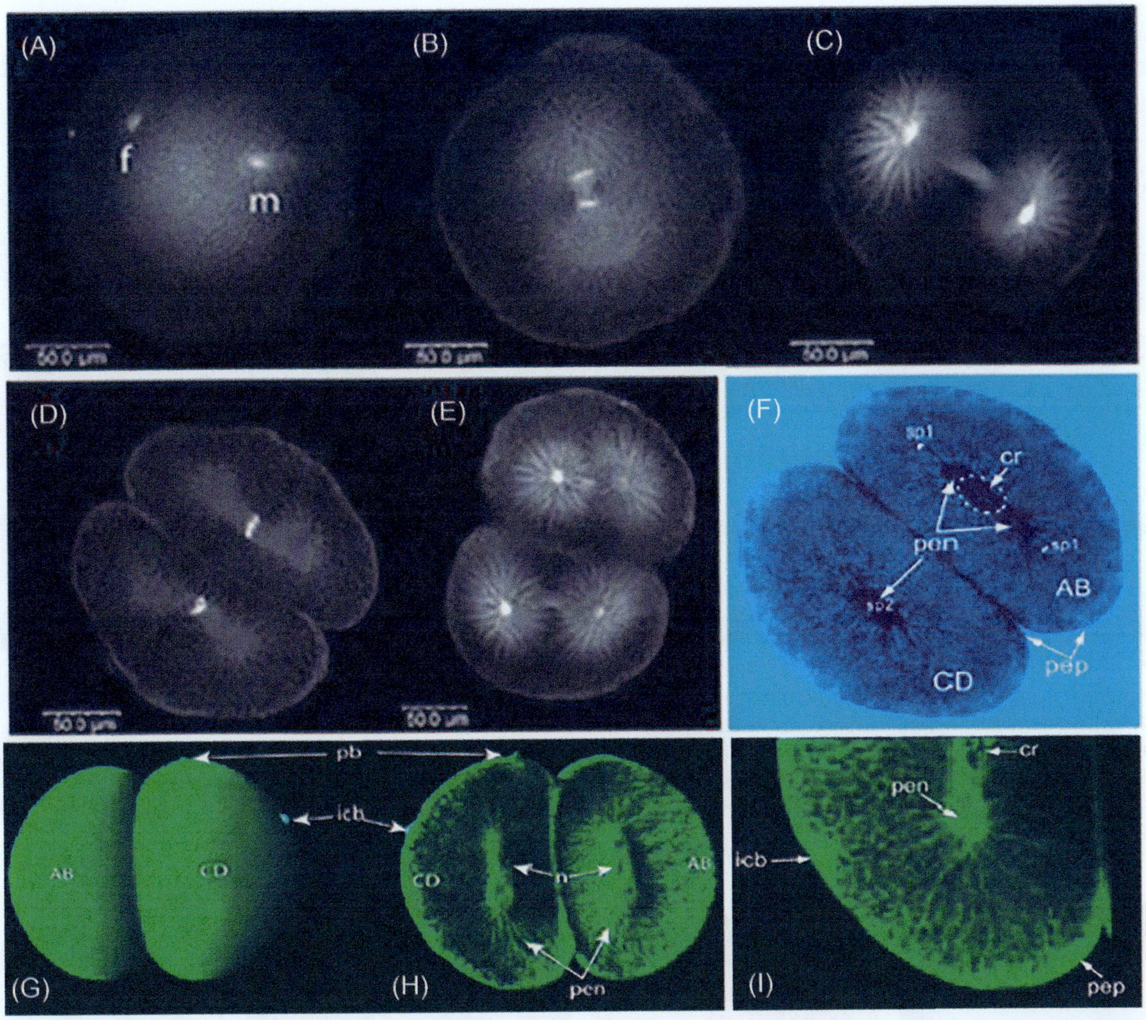

FIG. 10.76 (A–E) One- and two-cell stage embryo of *P. japonicus*: (A) composite optical section showing pronuclear migration after fertilization, the male m and the female f pronuclei are shown; (B) anaphase of first mitotic division; (C) metaphase spindle of the same; (D) metaphase of the second mitotic division in 2-cell stage; (E) metaphase spindles of the latter stage. Bars = 50 μm *(from Pawlak et al., 2010)*; (F–I) two-cell stage of *P. monodon (from Biffis et al., 2009)*: (F) semi-thin section showing uniformly distributed yolk droplets which are surrounded by the periplasm, the perinuclear cytoplasm links the periplasm through a mesh-work of threads. The chromosomes are surrounded by the perinuclear cytoplasm. The section is parallel to the spindles of AB cell and orthogonal to CD cell; (G) surface view of the cells; (H) inside view of the cells, in this the egg is rotated 180° around the vertical axis; (I) enlarged portion of the CD cell showing the ICB stained brightly within the periplasm. (C and E) stained by antitubulin antibody, (F) stained by Toluidine-blue, all the others stained by Sytox Green, all are scanning confocal microscopical images. cr = chromosomes, icb = intracellular body, n = nucleus, pb = polar body (on the surface of the egg membrane), pen = perinuclear cytoplasm, pep = periplasm, sp1 and sp2 = spindle of AB cell and of CD cell.

sheets (Fig. 10.79). The beginning with the 8-cell stage in this species forms a small blastocoel (Fig. 10.80) and at the same time, the division of D, is slightly unequal. The somewhat larger cell extends deeper into the blastocoel and includes the ICB. In *P. japonicus*, the ICB was located in the left-posterior quadrant associated with the mitotic spindles near the spindle pole at the start of the fourth division (Fig. 10.78D and D′).

The **fourth** cleavage is to the 16-cell stage, where the blastomeres divided perpendicular to the direction of the previous division. The two bands, AC and BD, expand to a new band size of four cells long by two cells wide (Fig. 10.78E–F′). A2a and A2p are located at one end of the band while C2a and C2p are located at the other end. A1a met C1a at the middle of the band on the anterior side while A1p and C1p met at the middle on the posterior side. Similarly, the blastomeres of the BD band divided in the direction of the left-right axis. In this stage, the ICB is located in blastomere **D2r** in the ventral-left quadrant (Fig. 10.78E and E′) and associated with the spindle poles at the start of the fifth division. In addition, it was slightly delayed in its division (at metaphase) compared with the rest of the blastomeres at anaphase/telophase. The **fifth** cleavage occurs to the 32-cell stage, during which all the blastomeres divided perpendicular to the previous division. Each of the two bands extends to eight cells in length in two rows of cells (Fig. 10.81J–N). The blastomere

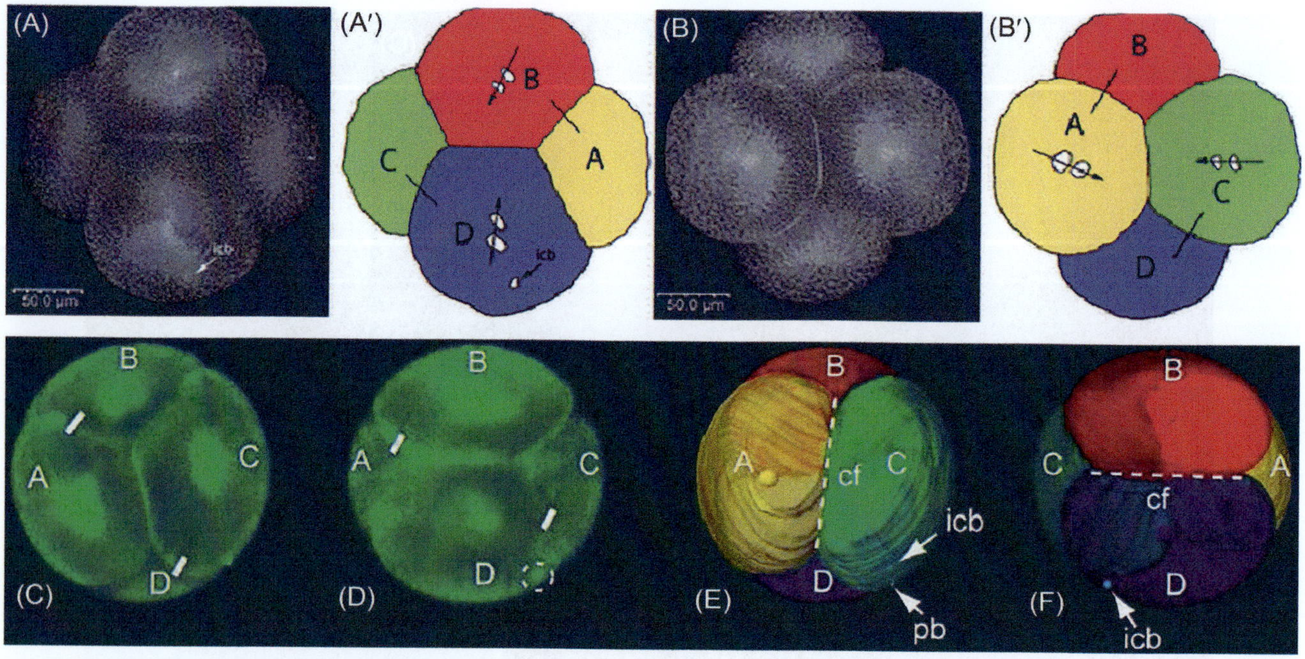

FIG. 10.77 Sytox Green-stained embryos at 4-cell stage (second cleavage) of *P. japonicus* (A–B′) and of *P. monodon* (C–F); (A and B) optical sections; (A′) (ventral) and (B′) (dorsal) corresponding drawings, a *one-sided arrow* represents an oblique cleavage direction; notice the ICB in the blastomere (D) seen from the ventral view *(From Pawlak et al., 2010)*. Confocal image in (C) upper surface and (D) bottom surface, the lines indicate the sister cell relationship, the ICB *(white broken line)* is located in (D) shifted towards the sister cell (C); 3D reconstruction of the blastomeres in (E) and (F) (3D reconstruction) the volume of each blastomere is represented by a transparent surface which allows the observation of the position of each nucleus inside represented by coloured spheres, the ICB and the polar body are pointed out by *white arrows* and are represented by *light blue spheres* smaller than the nucleus spheres, the ICB is always shifted towards the cells derived from (C), the cross furrows *(white broken line)* between (A and C) and (B and D) blastomeres are shown in (E) and (F), respectively. *(C–F) From Biffis et al. (2009).*

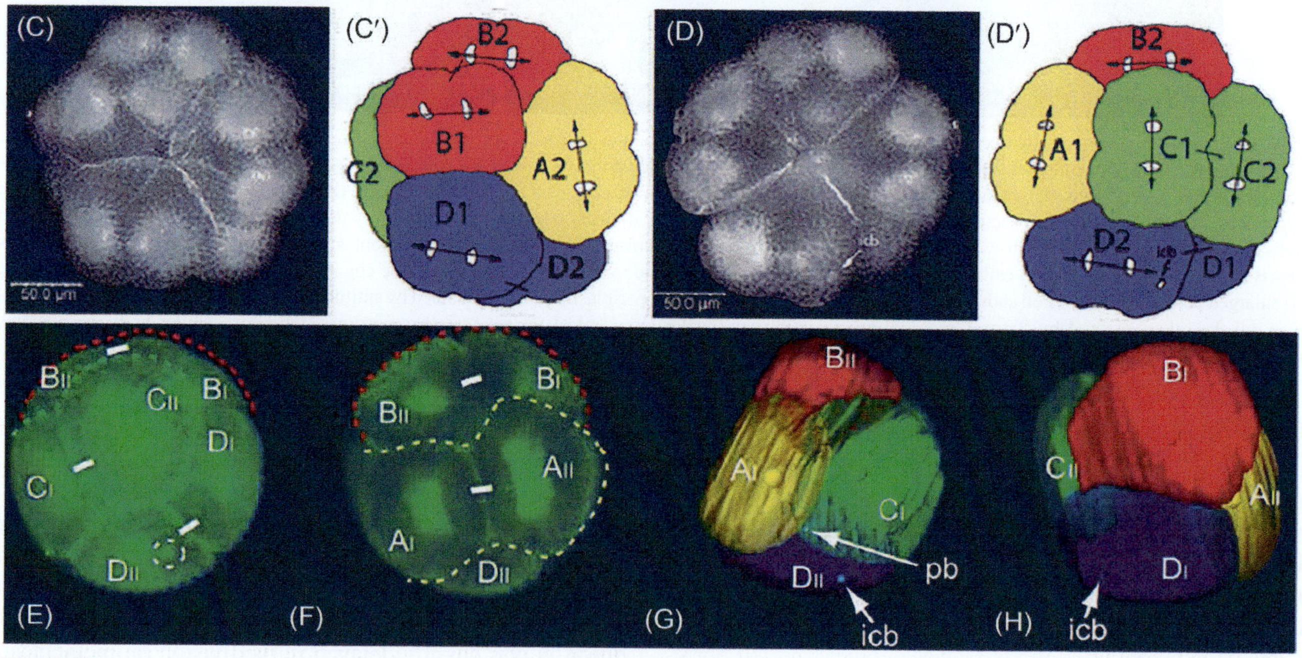

FIG. 10.78 Sytox Green-stained embryos at 8-cell stage (third cleavage) of *P. japonicus* (C–D′): (C, C′) ventral views, (D, D′) dorsal views; and of *P. monodon* (E–H), the *two-head arrows* represent the direction of division; (C, D) optical sections; while (C′, D′) interpretive drawings; notice the ICB located in (D2). *((C–D′) From Pawlak et al., 2010)*. Confocal image in (E) upper surface and (F) bottom surface, the lines indicate the sister cell relationship, the ICB in (E) *(white broken line)* is located in $\mathbf{D_{II}}$ shifted towards the sister cell $\mathbf{D_I}$, view of one end of the **AC**-band *(green broken line)* with $\mathbf{C_{II}}$ and $\mathbf{C_I}$ on focus at the upper surface of the embryo, while the **BD**-band is viewed by the red and blue broken line interlocking the **AC**-band, $\mathbf{D_I}$ contacts $\mathbf{B_I}$ within the band, $\mathbf{D_{II}}$ and $\mathbf{B_{II}}$ define the two ends of **BD**-band; in (F) view of one end of the **AC**-band *(yellow broken line)* with $\mathbf{A_{II}}$ and $\mathbf{A_I}$ on focus at the bottom surface, $\mathbf{B_{II}}$ is on focus at the end of the **BD**-band *(red and blue broken lines)*; (G and H) (3D reconstruction): in (G), $\mathbf{A_I}$ contacts $\mathbf{C_I}$ within the **AC**-band while $\mathbf{B_{II}}$ and $\mathbf{D_{II}}$ are at the two ends of the **BD**-band; in (H), $\mathbf{D_I}$ contacts $\mathbf{B_I}$ within **BD**-band while $\mathbf{C_{II}}$ and $\mathbf{A_{II}}$ are at the two ends of **AC**-band, the two bands are interlocked as two horseshoe-like sheets, the ICB and the polar body indicated by *white arrows* in (G) and (H). *(E–H) From Biffis et al. (2009).*

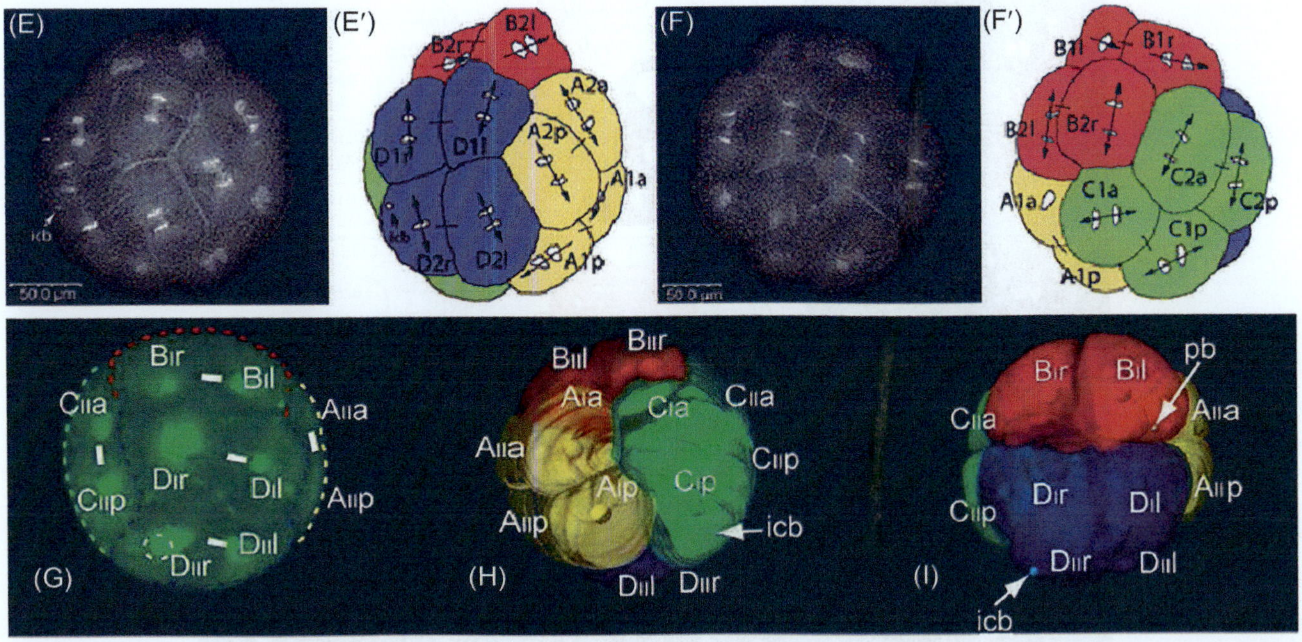

FIG. 10.79 Sytox Green-stained embryos at 16-cell stage (fourth cleavage) of *P. japonicus* (E–F′). (E, E′) Ventral views, (F, F′) dorsal views; and of *P. monodon* (G–I); the *two-head arrows* in (E′), (F′) represent the direction of division; (E and F) optical sections; (E′, F′) interpretive drawings, notice the ICB located in **D2r** blastomere in (E, E′); (G) view of the upper surface of the image stack of the confocal scanned egg showing a part of the **BD**-band *(red and blue broken line)*, the sister cells $\mathbf{B_Ir}$ and $\mathbf{B_Il}$ contact $\mathbf{D_Ir}$ and $\mathbf{D_Il}$ within the **BD**-band (on focus at the upper surface of the egg), the sister cells $\mathbf{D_{II}r}$ and $\mathbf{D_{II}l}$ are located at one end of the **BD**-band. $\mathbf{C_{II}a}$ and $\mathbf{C_{II}p}$ (in focus on the left) define one end of the **AC**-band *(green and yellow broken line)*. The ICB *(white broken line)* is located in $\mathbf{D_{II}r}$ shifted towards the derivative **C** cells; (H and I) 3D reconstruction of the 16-blastomere stage, in (H) the sister cells $\mathbf{C_Ia}$ and $\mathbf{C_Ip}$ contact $\mathbf{A_Ia}$ and $\mathbf{A_Ip}$ within the **AC**-band while the sister cells $\mathbf{B_{II}l}$ $\mathbf{B_{II}r}$ and $\mathbf{D_{II}r}$ $\mathbf{D_{II}l}$ are at the two ends of the **BD**-band; in (I) the sister cells $\mathbf{B_Ir}$ and $\mathbf{B_Il}$ contact $\mathbf{D_Ir}$ and $\mathbf{D_Il}$ within the **BD**-band while the sister cells $\mathbf{C_{II}a}$ and $\mathbf{C_{II}p}$ and $\mathbf{A_{II}a}$ and $\mathbf{A_{II}p}$ are at the two ends of the **AC**-band. *((E–F′) From Pawlak et al., 2010, and (G–I) from Biffis et al. (2009).)*

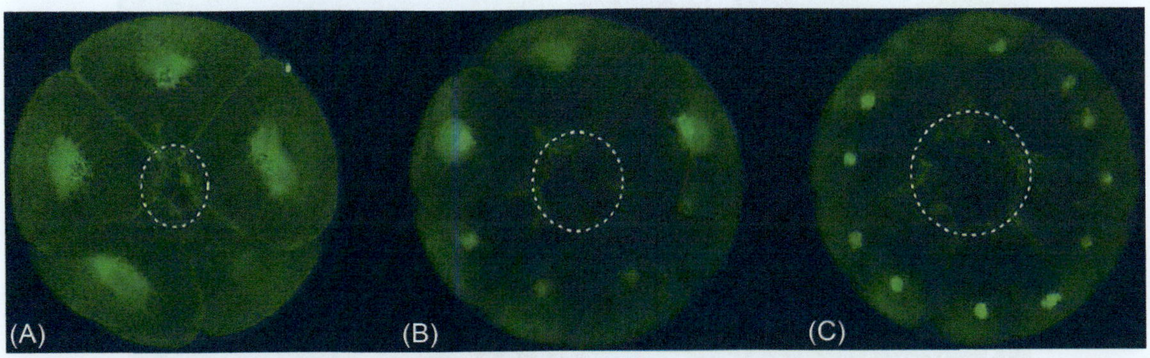

FIG. 10.80 Confocal image stacks processed by software Imaris, section mode showing formation of the blastocoel in the embryo of *P. monodon*: (A) view of different embryos at the 8-cell stage; (B) the 16-cell stage; (C) the 32-cell stage, the blastocoel is pointed out within the *white circled broken line*. *(From Biffis et al. (2009).)*

D2r divided into D2rv and D2rd (Fig. 10.81C), the ME cells, which will be known as Xv (ventral) and Xd (dorsal) ME cells, respectively. The division of the D blastomere descendants, which contained the ICB, was delayed and temporarily resulted in a **31-cell embryo**, with the ICB associated with one of the spindle poles (Fig. 10.81G). The delay in this fifth division is more pronounced than that which occurs in the previous division. The IBC then became allocated to one of the two ME cells (Fig. 10.81H).

Cleavage is complete and synchronous in *P. japonicus* to the 32-cell stage at which the ME cells arrest division while the ectoderm continue to divide for three cycles. This phenomenon has also been observed in other penaeoidean shrimp species. It was reported that in *P. kerathurus*, the ME cells arrest at the 16-cell stage rather than the 32-cell stage, resulting in their arrested division until the 114-cell stage (seventh division). Older literature described ME cell arrest in *P. japonicus* at a stage other than 32-cells stage.

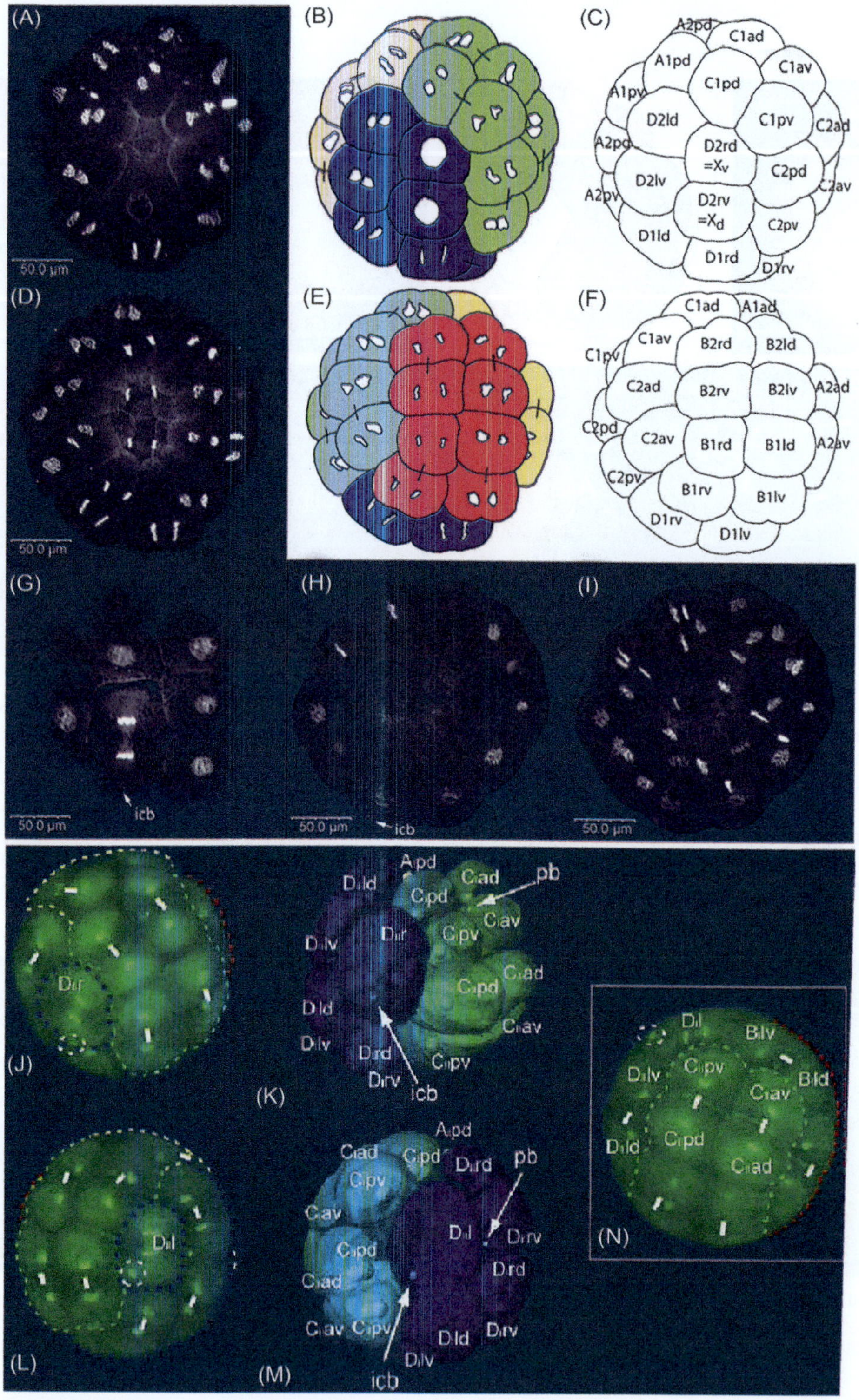

FIG. 10.81 Sytox Green-stained embryos at 31- or 32-cell stage (fifth cleavage) of *P. japonicus* (A–I) (A–C) posterior views, (D–F) ventral views; and of *P. monodon* (J–N). (A and D) optical sections; (B and E) corresponding drawings showing cleavage; and (C and F) labelled cells; (G) shows optical section of a 31-cell stage with a division delayed ME progenitor cell containing the ICB; (H) is an optical section showing the ICB in X_v at the 32-cell stage; (I) is a projection of all optical sections of the same embryo in (B). In (J)–(N) the lines indicate the sister cell relationship: (J) view of vegetal pole of the image stack of the 32-cell stage showing view of a part of the **AC**-band *(green and yellow broken line)* and of the **BD**-band *(red and blue broken line)*, $\mathbf{D_{II}r}$ at one end of the **BD**-band shows a delay in its division with respect to the other blastomeres, it contains the ICB and its nuclei are still completing their mitosis; (K) 3D reconstruction of the same stage, the $\mathbf{C_{I}pd}$ and $\mathbf{C_{I}ad}$ contact $\mathbf{A_{I}pd}$ and $\mathbf{A_{I}ad}$ (not visible in the image) within the **AC**-band while $\mathbf{D_{II}ld}$ and $\mathbf{D_{II}r}$ define the end of the **BD**-band, $\mathbf{D_{II}r}$ is delayed in its division and contains the ICB; (L) view of the vegetal pole of the image stack of the confocal scanned egg showing a part of the **AC**-band *(green and yellow broken line)* and of the **BD-band** *(red and blue broken line)* $\mathbf{D_{II}l}$ at one end of the **BD**-band shows a delay in its division; (K and M) 3D reconstruction that the volume of each blastomere is represented by a transparent surface that allows the view of the nucleus inside, represented by *coloured spheres*, the ICB and the polar body (pb) are indicated by *white arrows* and represented by *light blue spheres* smaller than those of the nuclei, the ICB is always shifted towards the C derivative cells; (N) view of another egg showing the pattern T2a M2 (see Fig. 10.82), the ICB *(white broken line)* is located in $\mathbf{D_{I}l}$. *((A–I) From Pawlak et al. (2010), (J–N) from Biffis et al. (2009).)*

The improvements in techniques and technology permit the re-examination of the embryology of the dendrobranchiates such as that of *P. monodon* and *P. japonicus*. These recent studies proved the 32-cell stage ME arrest is the case of penaeoideans and this remains as a basal condition to this group of animals. In addition, Pawlak et al. (2010) noticed that from the 4- to 32-cell stage, the most common path by which the ICB segregated was D2 to D2r to D2rv. Other paths were observed (see Fig. 10.82).

The **sixth** cleavage is to the 62-cell stage during which the ME cells remain in interphase while the rest of the embryo continues division. These ME cells ingress anteriorly into the blastopore (first step of gastrulation) and become surrounded by the rest of the embryo leaving a blastopore opening at the posterior end (Fig. 10.83A). During this activity, the ME cells elongate while their nuclei remain at the posterior end of the cells (Fig. 10.83C). The ICB in Xv remains at the posterior end of the cell near the cell membrane. In the embryo where the ME cells derived from D2r, nine cells (five D derivatives and four C derivatives; see Fig. 10.75B) contact the ME cells and divide radially into the blastopore (Fig. 10.83A, D, and F). Their names are shown in Fig. 10.83F and they form a ring around the ingressed ME cells. Often, they are not shown in a single plane of focus. These cells are considered the **crown cells** and their divisions are delayed compared with the ectoderm and are partially asynchronous to each other. This is a characteristic feature common in Dendrobranchiata. These crown cells in *P. japonicus* are homologous to the nine crown cells found in *P. monodon* (Biffis et al., 2009). They give rise to the naupliar mesoderm as described in other penaeoidean shrimps. If *P. kerathurus* is confirmed to have eight crown cells, it is likely that the number of crown cells is related to the stage of ME arrest, as all penaeoidean species known to have nine crown cells also have a 32-cell stage ME arrest. Nine crown cells may be the basal character state for Dendrobranchiata along with 32-cell stage ME arrest based on the out-group to the penaeoideans, *S. ingentis*, which also share those character states.

During the **seventh** cleavage, the ME cells remain in interphase while the rest of the embryo divides to 122 cells (Fig. 10.84). They remain arrested while the ectoderm cells divide once more for their eighth division. The divisions of the crown cells are retarded compared with the ectoderm in both the seventh and eighth divisions. The ICB in Xv is no longer visible during the seventh division and is not detected at any later stages. Oppositely, in *P. monodon* (Biffis et al., 2009), the ICB is still visible within the Xv during the seventh division and its position is towards the opening of the blastopore (Fig. 10.84H and J). It is no longer visible during the following division (Fig. 10.85(II)). During the **eighth** cleavage in *P. japonicus*, the ME cells lay dorsal and ventral, but are not distinguished from each other. Following the eighth ectodermal cleavage, the ME cells (mesendoderm) resume their division during this stage after three stages of arrest at 32-, 62-, and 122-cell stages (Fig. 10.85(I)). ME cells continue division beyond the second one, which result in eight ME derivatives, two of which are the presumptive primordial germ cell and the primordial mesoteloblast (Fig. 10.86A–G). The latter gives rise to the posterior (postnaupliar) mesoderm by undergoing teloblastic cell division. The presumptive primordial germ cell can be identified by its smaller size and large nucleus with distinct chromosomes (Fig. 10.86C and G). On the other hand, Biffis et al. (2009) described the behaviour of the crown cells beyond the seventh cleavage in *P. monodon*. During the subsequent divisions, these crown cells show a radial orientation towards the blastopore, and their derivatives move into the blastopore, forming a tier of the 'archenteron' wall, posterior to the mesendoderm cell (Fig. 10.85(II)). This is the advanced gastrula in which the two mesendoderm cells show their enlarged nuclei and they do not participate in the next division. They appear flattened against the basal wall of the ectoderm.

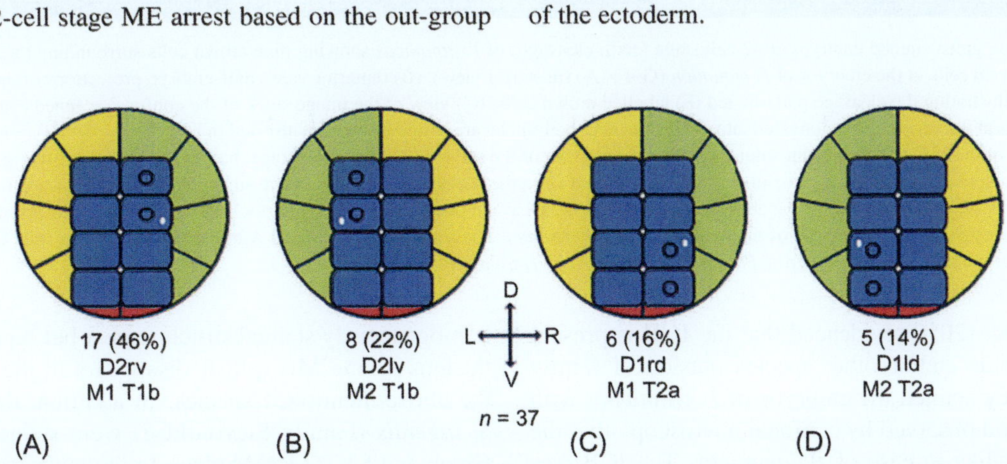

FIG. 10.82 Illustration and rate of occurrence of the four most common cleavage types at the 32-cell stage of the embryo of *P. japonicus*. (A) D2r divides into mesendoderm cells ME and D2rv receives the ICB and D2rd, this is cleavage type M1 T1b; (B) D2l divides into the ME cells and D2lv receives the ICB (type M2 T1b); (C) D1r divides into the ME cells and D21d receives the ICB (type M1 T2a); (D) D1l divides into the ME cells and D1ld receives the ICB (type M2 T2a). *(From Pawlak et al. (2010).)*

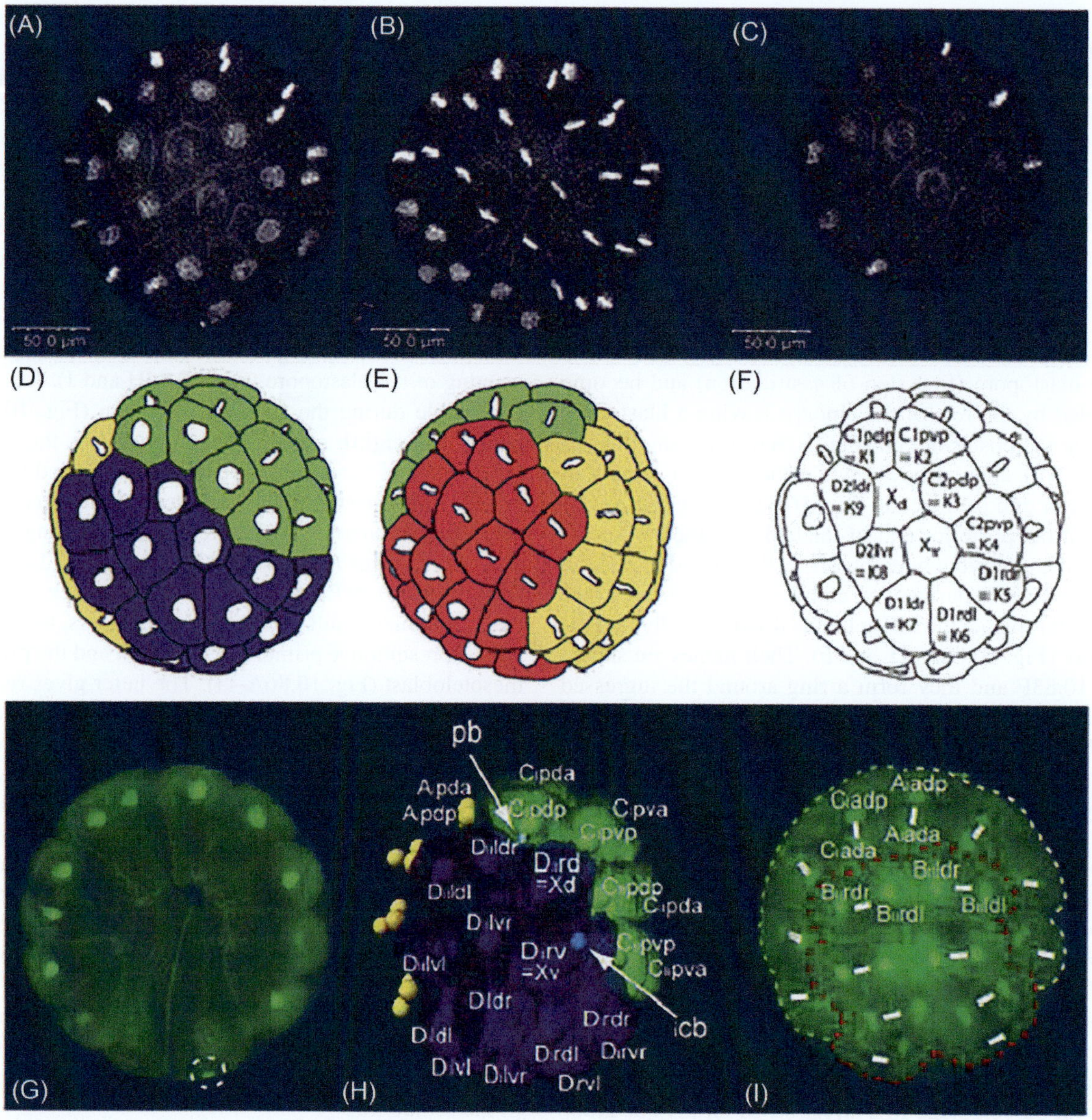

FIG. 10.83 Sytox green stained embryos at 62-cell stage (sixth cleavage) of *P. japonicus* showing nine crown cells surrounding the blastopore (A–F) and the mesendoderm cells in the embryos of *P. monodon* (G–I): (A) (posterior view), (B) (anterior view) half-embryo projections of optical sections; (D and E) drawings illustrating the cleavage pattern; and (F) labelled crown cells; (G) view of the image stack of the confocal scanned egg showing the two mesendoderm cells at the vegetal pole elongated into the blastocoel, their nuclei are in apical position arrested in interphase, the ICB *(white broken line)* is located on the top of the ME cell at the right side; (H) 3D reconstruction of the same egg of the same stage, but rotated pointing the vegetal pole showing the X_v and X_d correspond to the arrested blastomeres **D_{II}rv** and **D_{II}rd** from the previous division, they are surrounded by the crown cells, X_v contains the ICB; (I) the same as (G) showing view of the animal pole in which the cleavage pattern of the two interlocking cell bands is preserved at this pole. The sister cells derivatives of **B_{II}** define one end of the **BD**-band *(red broken line)*. The sister cells **$A_I$adp** and **$A_I$ada** contact the sister cells **$C_I$adp** and **$C_I$ada**, respectively, within the **AC**-band. *((A–F) From Pawlak et al. (2010), (G–I) from Biffis et al. (2009).)*

Pawlak et al. (2010) evidenced that the **ICB** is present in other penaeids and/or other species outside of family Penaeidae. They stained the embryos of *P. vannamei* with Sytox Green and observed by confocal microscopy that the ICB is present like in case of *P. japonicus*. The ICB was found in the 32- and 62-cell stage embryos (Fig. 10.87A–C). It has the same appearance as in both the above species, *P. japonicus* and *P. monodon*. It consists of an ovoid, homogeneously stained structure near but separate from the nucleus in one ME cell. It disappears in the same way as the above-mentioned species. In addition, the embryos of *S. ingentis* (Family Sicyoniidae) were stained with Sytox Green and SYTO RNASelect. In the embryos with the former stain, a slightly enhanced fluorescence was observed with the nuclear envelope of one ME cell. With the latter stain, a string-like body was observed in close association

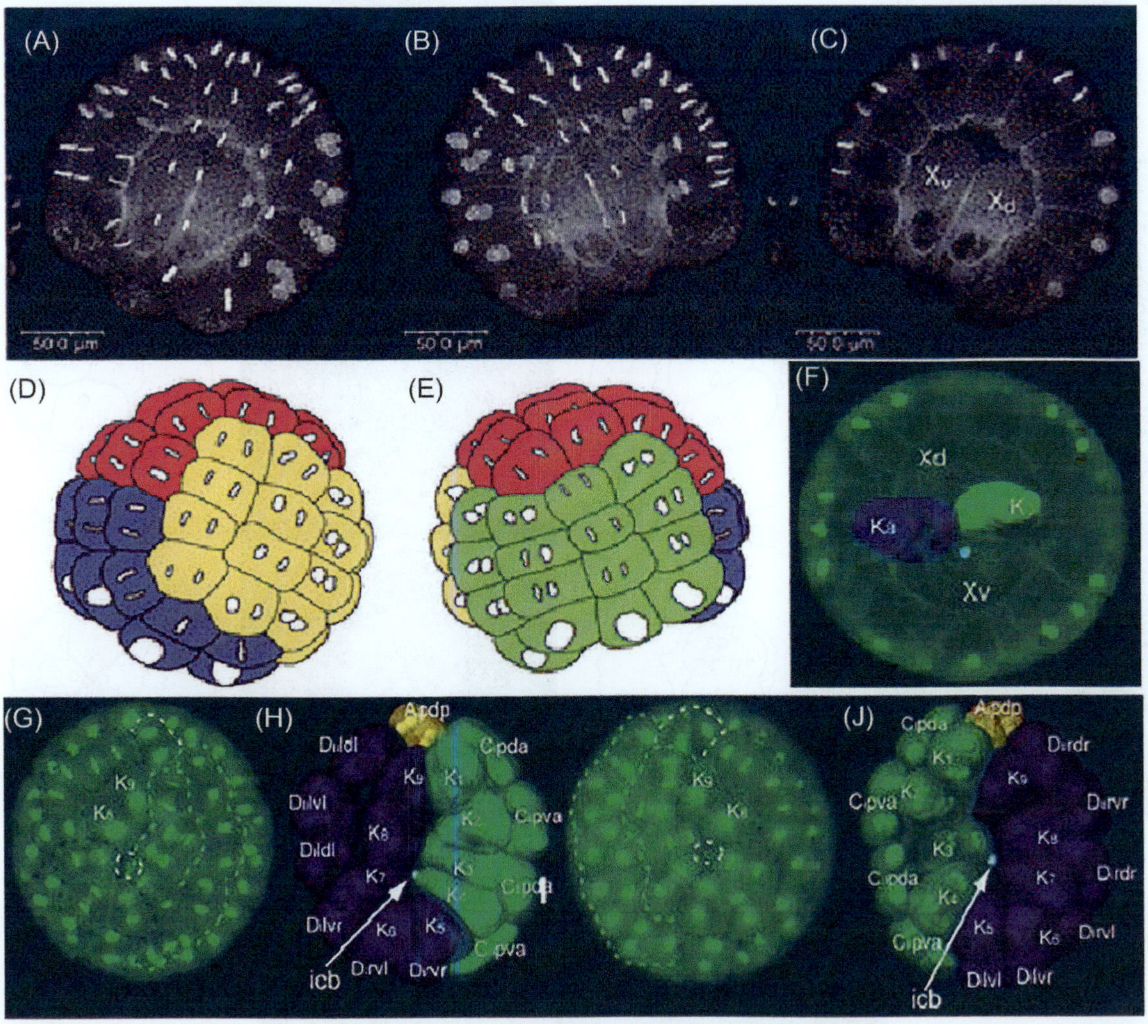

FIG. 10.84 Sytox green stained embryos at 122-cell stage (seventh cleavage) of *P. japonicus* showing anterior embryo in left (A and D) and right side view (B and E), (A–C) optical sections while (D) and (E) drawings illustrating the cleavage pattern, there are nine crown cells surrounding the blastopore in interphase while the remaining cells are in metaphase of the next cell division, the ME cells (Xd, Xv) have ingressed into the blastocoel with the blastopore (b) seen at their apical side in (C); (F–J) the same cleavage stage of *P. monodon*: (F) view of the image stack of the confocal scanned egg associated with the 3D reconstruction showing the vegetal pole with focus plane on the two enlarged ME Xd and Xv (arrested in the interphase) extending more deeply into the blastocoel, the ICB is inside Xv, represented by a *light blue sphere*, the two crown cells K8 and K3 are visible in contact to each other in their 3D reconstruction; (G) confocal image showing view of the vegetal pole, the inner ring of the crown cells and the corresponding outer ring are shown *(blue and green broken line)*, K_9 and K_8 (on the left side) are delayed in their division, the ICB *(white broken line)* is visible in the middle through the blastopore opening; (H) 3D reconstruction of the same stage of the same egg of the same pole showing the nine crown cells are composed of four **C** derivative cells *(green)* and five **D** derivative cells *(blue)*, the M1 pattern is shown which correspond to the crown cells (K_1–K_9) arranged clockwise around the blastopore; (I) confocal image of another egg showing the mirror image pattern M2 view of the vegetal pole, the ring of the crown cells and the corresponding outer ring are shown *(blue and green broken line)*, K_9 and K_8 (on the right side) are delayed in their division with respect to the others, the ICB *(white broken line)* is visible in the middle through the blastopore opening; (J) 3D reconstruction of the same stage of the same egg in previous image showing the nine crown cells from the vegetal pole composed of four **C** derivative cells and five **D** derivative cells; the M2 pattern is shown which correspond to the crown cells (K_1–K_9) arranged clockwise around the blastopore. *((A–E) From Pawlak et al., 2010; (F–J) from Biffis et al. (2009).)*

with the nuclear envelope in 62-cell stage embryo (Fig. 10.87D–F), but not later. Therefore, it becomes a fact that the ICB is a germ granule in penaeoidean shrimp, since it has been observed in embryos of three species of family Penaeidae and one species of family Sicyoniidae within the Dendrobranchiata. Therefore, it is likely to be found generally among the Dendrobranchiata.

By nucleic acid staining and sensitivity to nucleases, the ICB contains RNA. In the work of Pawlak et al., the ICB was also detected with SYTO RNASelect, which is an RNA-preferential stain. By their staining experiments, these authors provided stronger evidence that the ICB contains RNA but not DNA (Fig. 10.88). The ICB formed in a vegetal blastomere at the 2–4-cell stage, is inherited by

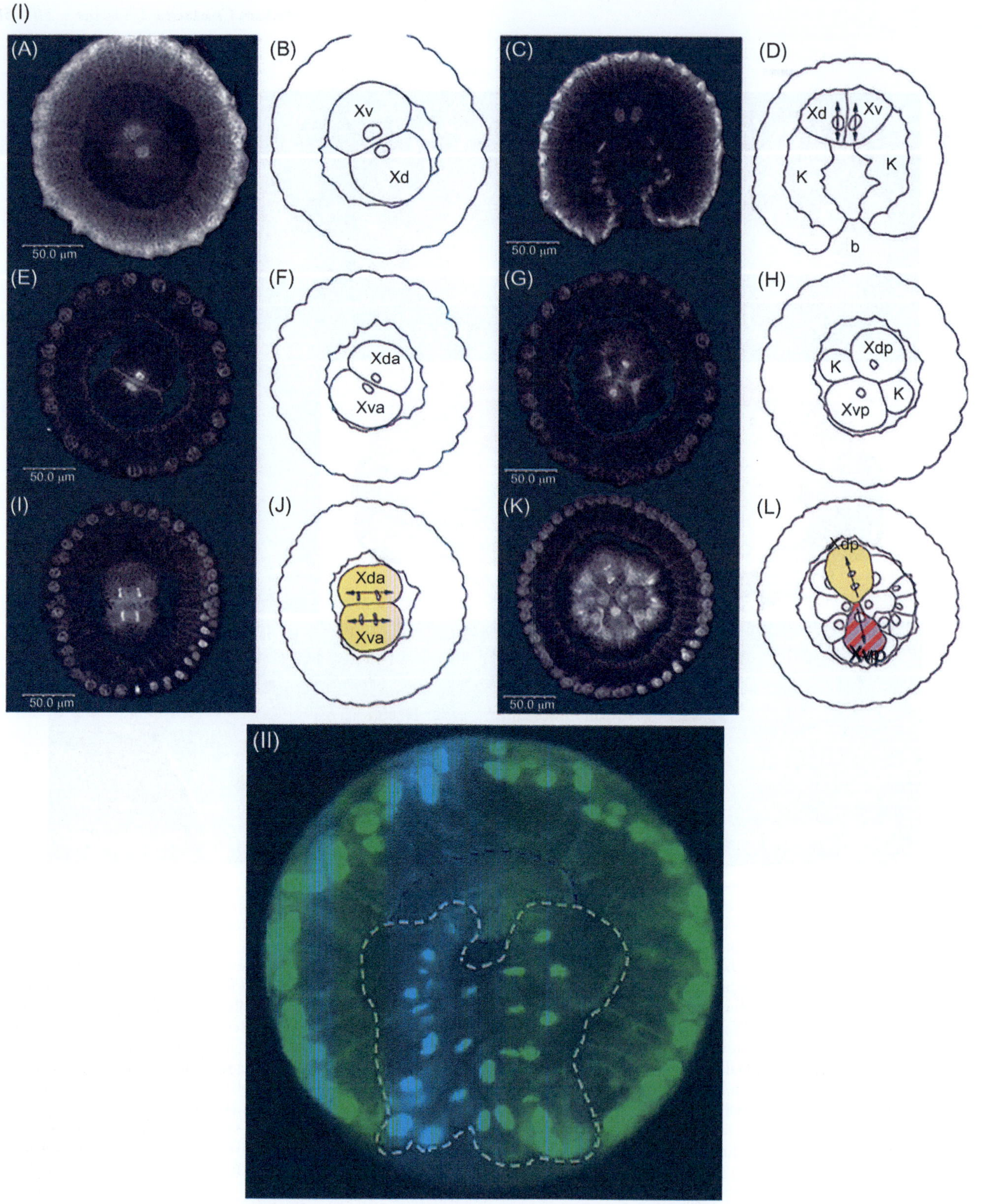

FIG. 10.85 (I) Sytox green stained embryo of *P. japonicus* showing late gastrula: (A and B) mid-transverse optical section of the ME cells (X) with condensed chromosomes following arrested division; (C and D) mid longitudinal optical section of the ME cells as chromosomes condense for anterior-posterior division (indicated by *arrows*), crown cell derivatives (K) form an invagination from the blastopore (b) at the posterior; (E–H) single transverse optical sections (E and G) and interpretive drawings (F and H) of an embryo with four ME derivatives (Xa, Xp) following the first ME division, (E) and (F) are more anterior to (G) and (H) and the crown cell derivatives (K) are also in the plane of section; (I–L) single optical sections (I and K) and interpretive drawings (J and L) of an embryo during second ME division. (I and J) Anterior region showing the two most anterior ME derivatives dividing along the left-right axis, (K and L) more posterior showing the posterior ME derivatives dividing in an oblique plane dorsal-ventral/anterior-posterior. Xvp *(striped in red and purple)* is hypothesized to give rise to the primordial mesoteloblast and germ cell on the ventral side. Bars 50 μm. *From Pawlak et al. (2010).* (II) Sytox Green-stained egg of *P. monodon* during gastrulation, confocal image stack (Imaris Section mode), the division of the crown cells *(grey broken line)* occur along the antero-posterior axis maintaining the same radial orientation into the blastopore, they form the archenteron wall posterior to the mesendoderm cell. One of the two mesendoderm cells *(blue resp. black broken line)* shows its enlarged nucleus which appears flattened against the basal wall of the ectoderm. The ICB is no longer visible at this stage. *From Biffis et al. (2009).*

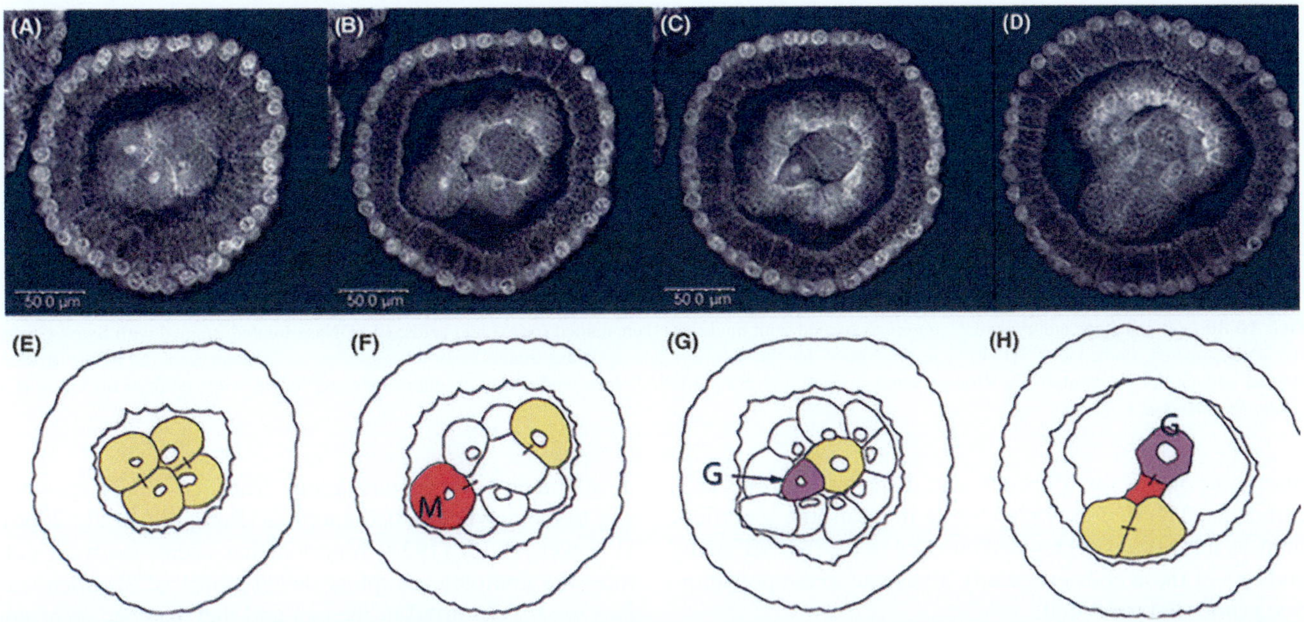

FIG. 10.86 Development of mesendoderm (ME) derivatives in the embryo of *P. japonicus* stained with Sytox Green to observe chromosomes and nuclei. (A–D) Single optical sections, (E–H) interpretive drawings: (A and E) an anterior transverse section showing the four most anterior ME derivatives, which are endodermal yolk cells; (B and F) a more posterior section showing two more ME derivatives, the primordial mesoteloblast (M) is located on the ventral side of the embryo; (C and G) an even more posterior section showing two more ME derivatives, the presumptive primordial germ cell (G) is a sister cell to the primordial mesoteloblast in the previous section; (D and H) a single optical section from a different embryo at a slightly later time point, displaying the presumptive primordial germ cell (G) with a characteristically large nucleus. Bars 50 µm. *(From Pawlak et al. (2010).)*

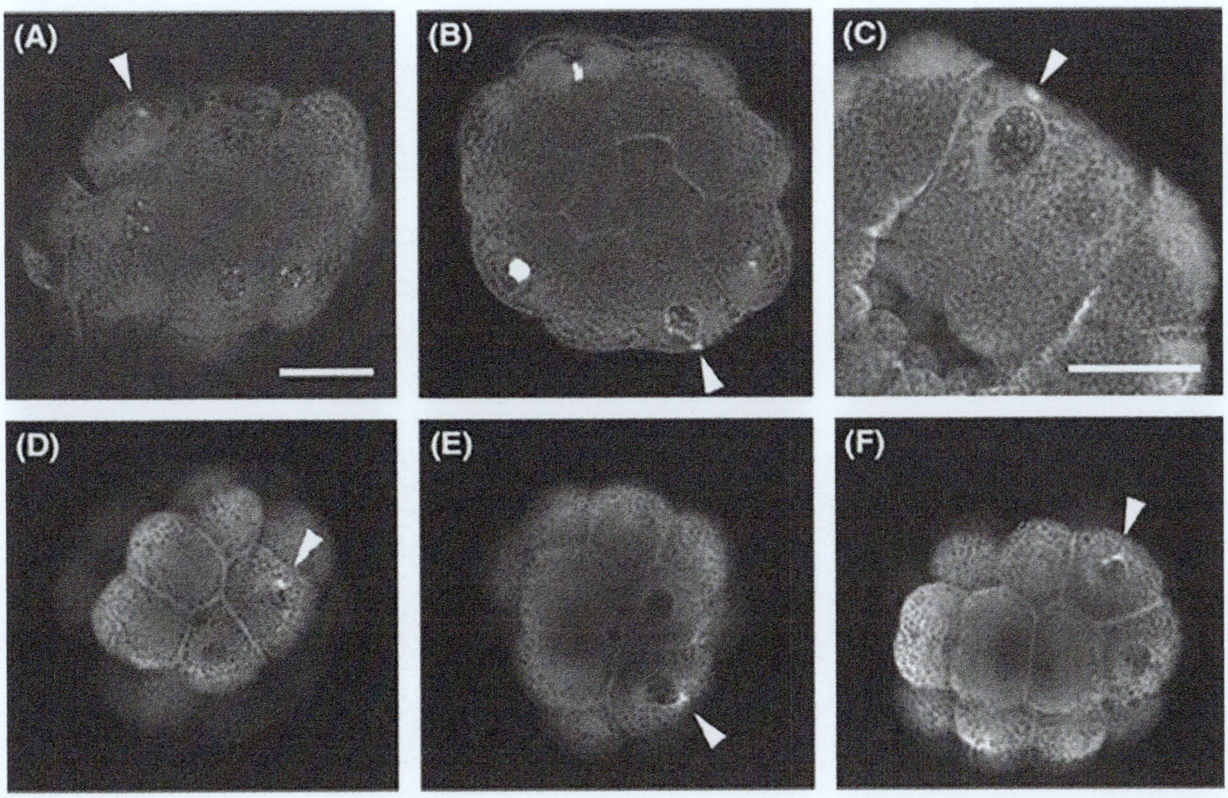

FIG. 10.87 (A) The intracellular body (ICB) in *P. japonicus* 32-cell stage interphase at higher magnification; (B and C) Sytox Green stained embryos of *P. vannamei* at 32-cell stage showing mitosis in (B) and early 62-cell stage at higher magnification in (C); (D–F) three different Syto RNASelect-stained *S. ingentis* embryos at the 32-cell stage. The ICB indicated by *head arrows*. Scale bars in A, B = 20 µm and in C, D–F = 50 µm. *(From Pawlak et al. (2010).)*

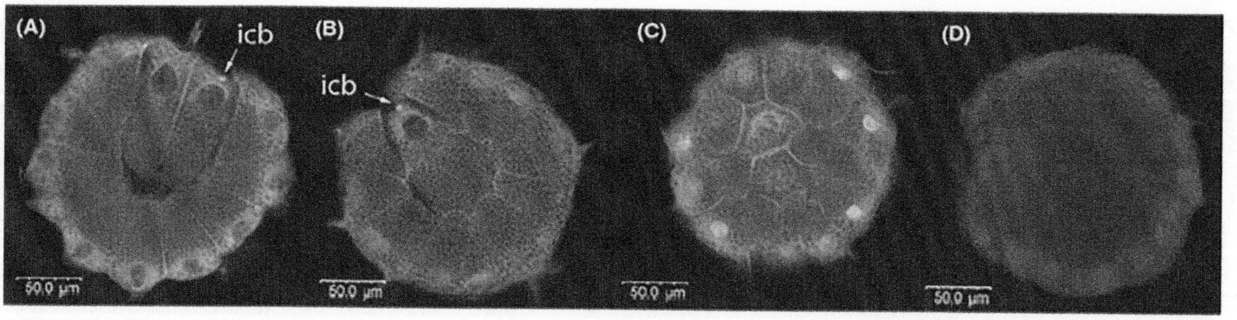

FIG. 10.88 62-cell stage embryos of *P. japonicus* treated with nucleases then stained for nucleic acids: (A) DNase-treated, stained with Sytox Green; (B) DNase-treated, stained with Syto RNASelect. Notice that the ICB is visible in DNase-treated embryos regardless of the stain used; (C) RNase-treated, stained with Sytox Green and; (D) RNase-treated, stained with SYTO RNASelect. Notice that chromosomes are visible in the embryo treated in (C). *(From Pawlak et al. (2010).)*

lineage to one of the ME cells, and disappears by the 122-cell stage. The ICB may play a role in germ line specifications as it is always segregated into one of the two ME cells, and one of these cells eventually gives rise to the presumptive primordial germ cell.

Larval development. The review of Dall et al. (1990) presented illustrations showing late gastrula of a penaeid (mostly of *Parapenaeus longirostris*) with the differentiated buds of the naupliar organs just before hatching (Fig. 10.89). They described the process of hatching in much detail in some species other than those mentioned in the present work. However, Fig. 10.74 shows the stages of development of *P. semisulcatus* from the newly spawned egg to the newly hatched nauplius (Ronquillo et al., 2006). However, Heldt (1938) reported that occasionally in culture, the complete nauplius development of *Parapenaeus longirostris* occurred in the egg and they hatched as protozoa. The embryonic and larval stages of *P. semisulcatus* are generally similar to those of other closely related species in the family Penaeidae, such as *P. (Marsupenaeus) japonicus.* The larvae of *P. semisulcatus* passed through six naupliar, three protozoeal, and three mysis stages before reaching the first postlarval stage (Fig. 10.90) in a minimum of about 9 days (224 h) from hatching at 28.2 ± 0.2°C, pH 8.20–8.32,

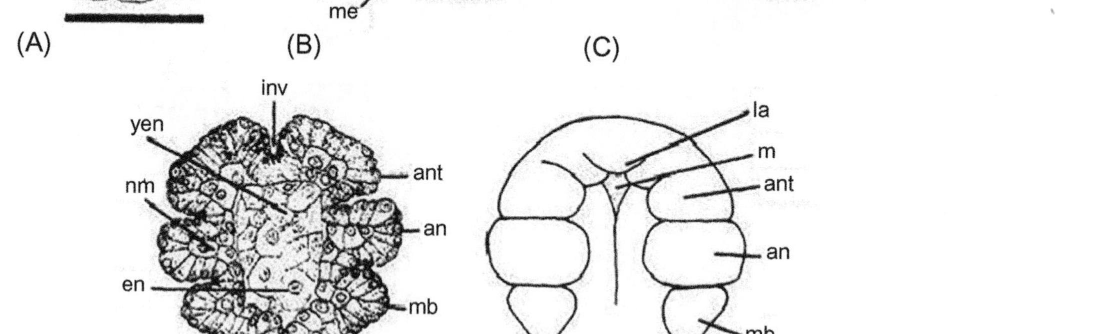

FIG. 10.89 Embryological development of a penaeid showing the differentiation of the gastrula before hatching *(from Dall et al., 1990)*: (A) section through 30-cell stage of cleavage; (B) sagittal section through the gastrula; (C) T.S. through the anterior region of the embryo; (D) frontal section through the embryo; (E) frontal view of a whole mount embryo. an = second antenna, ant = first antenna, b = blastopore, bl = blastocael, ec & et = ectoderm, en = endoderm, gc = germ cells, inv = invagination of neuro-ectoderm, la = labrum, m = mouth, mb = mandible, me = mesoderm, mes = mesendoderm, mg = midgut, nm = naupliar mesoderm, pnr = postnaupliar region, yen = yolk endoderm.

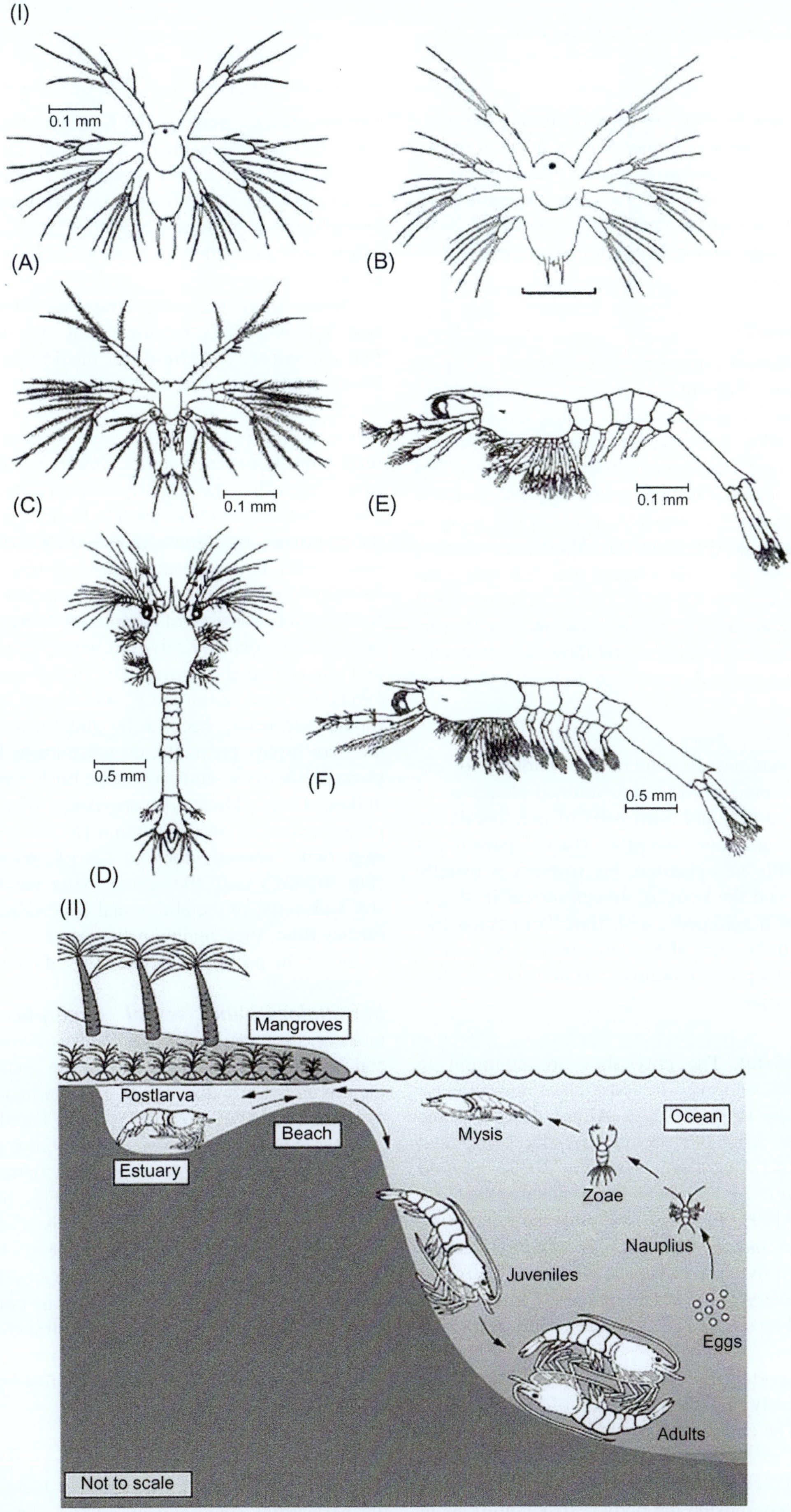

FIG. 10.90 (I) Larval stages in the life cycle of *P. semisulcatus* de Haan: (A) ventral view of first nauplius; (B) ventral view of first nauplius of the kishi velvet shrimp *Metapenaeopsis dalei* (Rathbun) for differentiation; (C) ventral view of six naupllius; (D) ventral view of third protozoea; (E) lateral view of third mysis; (F) lateral view of first postlarva. *From Ronquillo et al. (2006), except (B) from Choi and Hong (2001).* (II) Ecological life cycle of Indian white prawn, *Penaeus indicus* H. Milne Edwards, 1837. *From Aquaculture Feed and Fertilizer Resources Information System, FAO official common names.*

and 34.5 g kg^{-1} salinity. The drawings illustrating the structure of these larvae are almost the same of those of the kishi velvet shrimp, *Metapenaeopsis dalei* (Rathbun) (Choi and Hong, 2001). However, there are some specific characters for the differentiation, such as the size of the larva, setation of appendages and duration of metamorphosis (see Fig. 10.90(I) A and B).

Astacidea (Astacura)

Almost all the authors consider this crustacean group as a subdivision from Reptantia (suborder). The group includes crayfish and clawed lobsters, which are variously known as Astacidea Latreille, or Macroures Cuirassés Milne Edwards, or Homaridea Boas, or Nephropidea Ortman (Schram, 1986). The latter author applied the name Astacina to this group. A common example of this group is the famous ***Astacus*** sp. (Fig. 10.91B). There are a number of kinds of this crayfish, quite similar, that bear that common surname of *Astacus*. By way of distinction, one kind is called *fluviatile*, another *slender-handed*, another *Dauric*, from the region in which it lives, and these double names are rendered *A. fluviatilis, A. leptodactylus*, and *A. dauricus* (Huxley, 1880).

Definition. Reptantians in which the abdomen is extended, well-armoured, with well-developed pleura and a broad tail fan. The first and third pairs of pereiopods are similar, and there are numerous gills. The carapace is not fused at sides with the epistome; the rostrum is usually well developed, and the body is subcylindrical in shape. The Expodites of maxillipeds; with flagella directed forwards. The first three pairs of legs are chelate. No appendix interna on pleopods. Examples: *Homarus, Astacus, Nephrops, Cambarus.*

Habits and habitat. The crayfishes are common in streams and rivers, and even in some lakes and marshes throughout Europe (Fig. 10.91A). Allied genera of the same family, the Astacidae, occur in Asia, North and South America and Australasia. Holdich (2002) reported that *Procambarus clarkii* has been in Egypt since the early 1980s. Hamadi (1995) has been studying the population in the River Nile and reports that it is a highly tolerant invader capable of living anywhere in water and cultivated ground. Nearly every possible site in Giza, Cairo and the Nile Delta is colonized. However, there is little exploitation of the species.

The most important favourable factor for their presence is a sufficiently high concentration of dissolved calcareous matter. The cuticle of a crustacean animal is often strengthened by calcification. The major portion of the endocuticle is a calcified zone (Fig. 10.91E). The cuticle does not lie entirely outside the body (exoskeleton), as in certain regions, such as the head and thorax, the epidermis sinks inwards beneath the muscles to form a system of rods or plates called apodemes. These apodemes can be so extensive to form a definite arrangement termed the endophragmal skeleton. In *A. fluviatilis*, this system was extensively described and carefully illustrated by Huxley (1880).

Some crayfishes may attain very large size. Beach and Talbot (1987) reported that the Tasmanian crayfish *Astacopsis franklinii* and the Murray River crayfish *Euastacus serratus* are the largest species of crayfish in the world, and attain weights of ~2 kg. *Cherax tenuimanus* (a native to the large streams of southwestern Australia) is the third largest species. *Cherax albidus* is smaller, yet is still much larger than the native American genera (*Pacifastacus, Procambarus*) which weigh less than 20 g. During the day, the crayfishes avoid the light by hiding in crevices or in burrows which they excavate themselves (Fig. 10.91C and D). Activity begins in the evening, when they search for their food, which consists of almost anything organic, animal or vegetable, dead or alive. In winter they burrow deeper, and remain inactive during the coldest weather (Bullough, 1981). *C. tenuimanus* and *C. albidus* are excellent animals for aquaculturing, because in addition to their large size, they are highly palatable, do not burrow, have an uncomplicated life cycle, and grow large fairly rapidly (Beach and Talbot, 1987). Mating of crayfishes (Fig. 10.92D) takes place commonly in the autumn (Bullough, 1981) and the eggs being carried glued to the pleopods of the female (Fig. 10.92C) until the spring. After hatching, the young still lack some of the abdominal appendages, and for some further time, they maintain their grip on the pleopods by means of the peculiarly hooked tips of their chelae.

Morphology and sexual dimorphism. *Astacus*, as most crustaceans is dioecious and is characterizd by special organ(s) defining each sex. The abdomen of the female is much broader than that of the male. The pleopods (Marshall and Williams, 1975) in the female are all similar (and serve to carry the eggs, Fig. 10.92B (c)), the elongated basipodite, to which are hinged the flattened endopodite and exopodite fringed with setae (Fig. 10.92B (c)). The first pair is more or less vestigial (it is absent altogether in crayfish from the southern hemisphere). In the male, the first and second pairs of pleopods are modified into incomplete tubes or spout-like which act as copulatory organs (Fig. 10.92B (a) and (b)), serving to transfer the spermatophores to the female. There are never any pleopods on the first male abdominal segment in crayfish from the southern hemisphere.

Reproductive systems. The reproductive system (Fig. 10.93) lies in the thorax, just beneath the floor of the pericardial sinus and above the alimentary canal. In

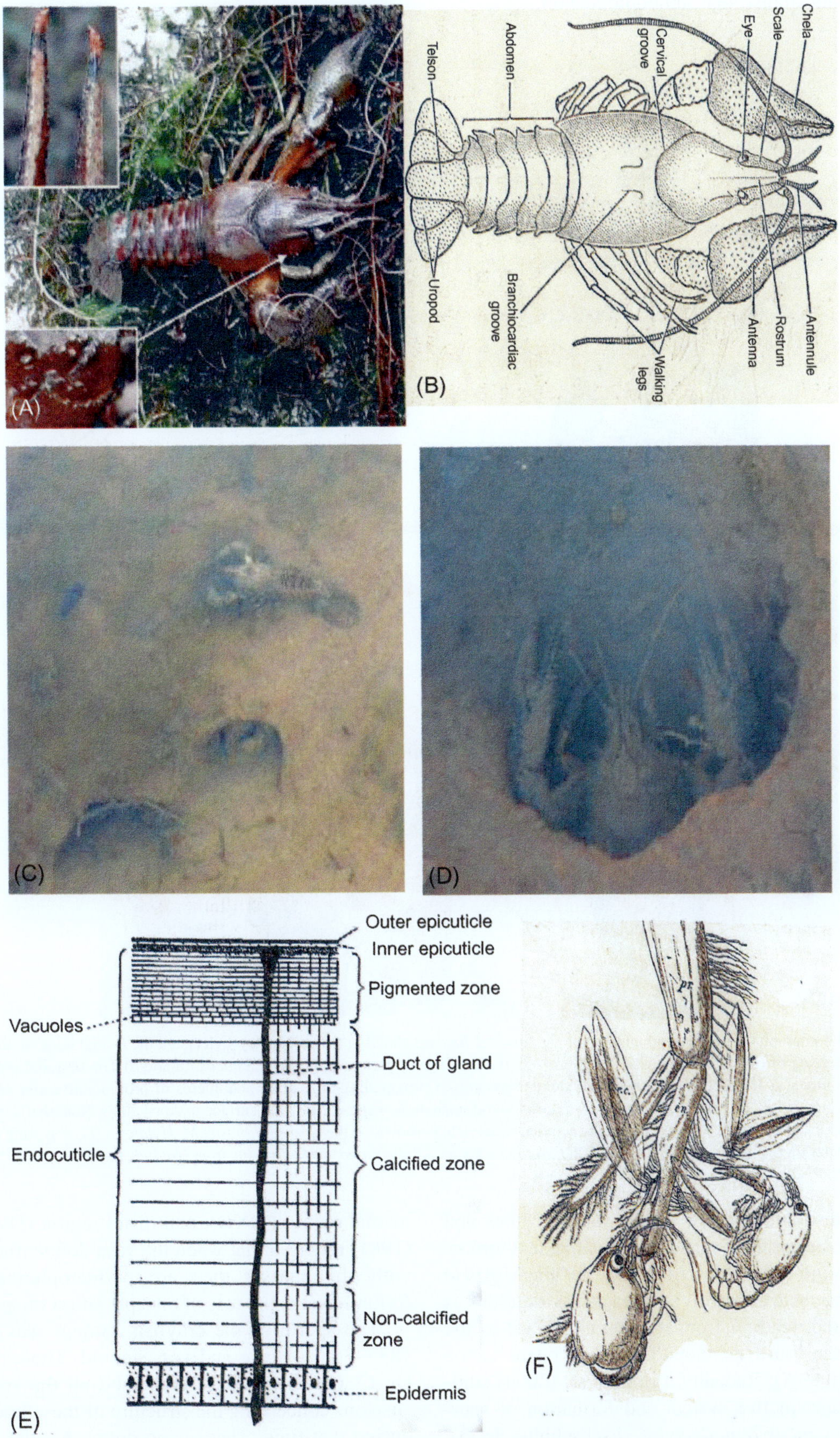

FIG. 10.91 (A) An adult male *Orconectes limosus* (Rafinesque, 1817) with its characteristic spiny cheeks *(arrow and lower inset)*, orange tipped legs *(upper inset)*, and abdominal stripes *(photos D.M. Holdich)*; (B) Diagrammatic drawing illustrating the features of genus *Astacus (from Verma, 1977)*; (C) An *O. limosus* on the pond floor close to a number of burrows; (D) A crayfish about to emerge from its burrow. Note that it is covered in silt (photos (C and D): J. Black; *(A, C, and D) from Holdich and Black, 2007)*; (E) Schematic drawing illustrating the structure of crustacean cuticle *(after Dennel, from Borradaile et al., 1977)*; (F) Illustration shows two recently hatched crayfish (*A. fluviatilis*) attached to one of the swimmerets of the mother. 4×. pr = protopodite, en = endopodite. ex = exopodite, ec = ruptured egg-cases. *From Huxley (1880)*.

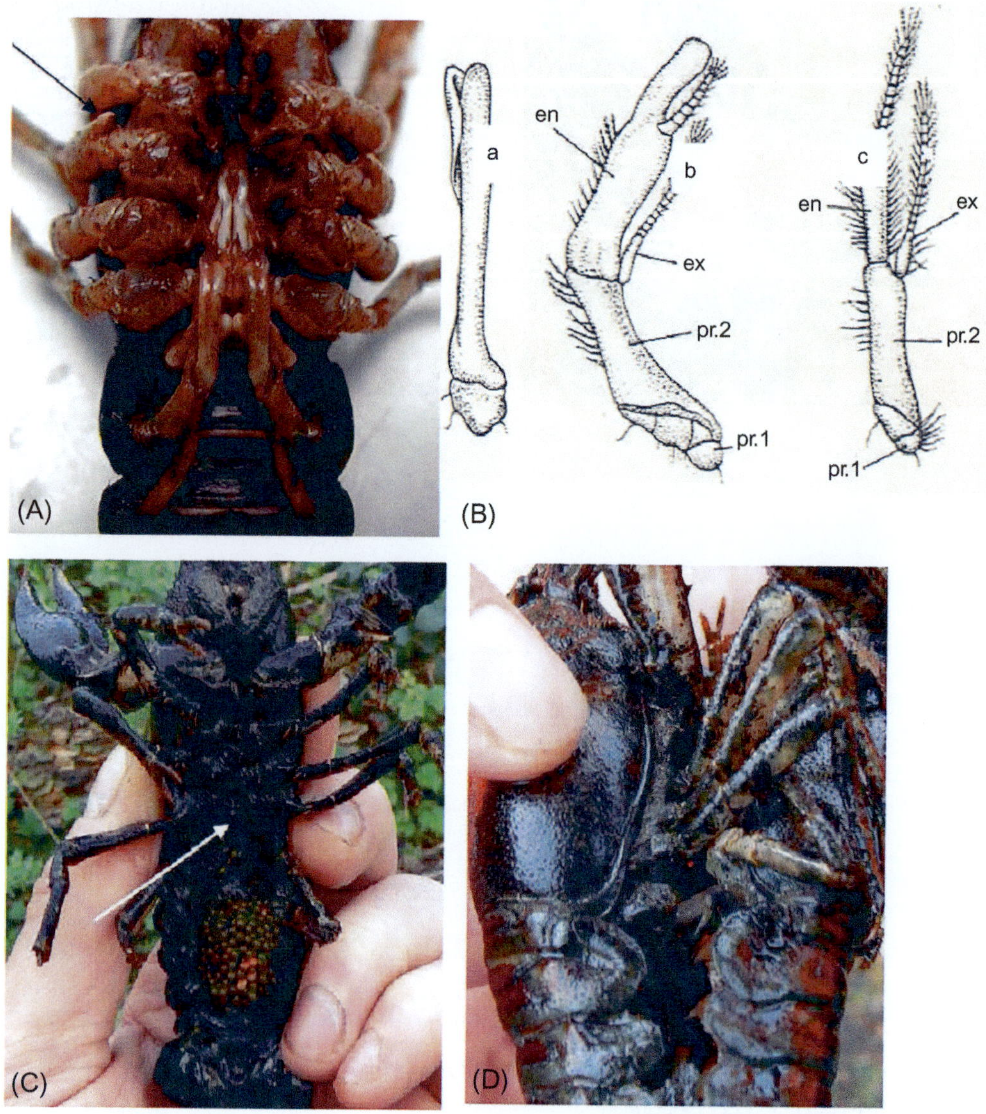

FIG. 10.92 (A) Photograph of male individual crayfish, *Orconectes limosus* showing the copulatory stylets and the ischial hook on the second walking leg *(arrow)* characteristic of F1 males *(photos: D.M. Holdich)*; (B) drawings of pleopods of genus *Astacus*. (a and b) The first and second are modified as male stylets, (c) pleopod of the female unmodified. en=endopodite, ex=exopodite, pr.1, pr.2=podomeres of protopodite *(after Huxley, 1880, from Marshall and Williams, 1975)*; (C) A berried female *O. limosus* found in shallow water on the mud surface in April 2006. Note that some of the eggs are dead having been attacked by fungi. The position of the seminal receptacle is shown by the *arrow (photos D.M. Holdich)*; (D) A mating pair of *O. limosus* found in shallow water on the mud surface in April 2006 *(photos J. Black). All the photos of O. limosus from Holdich and Black (2007).*

the **male**, the testis consists of paired anterior lobes and an unpaired posterior lobe (all the available textbooks). However, Bullough (1981) described that the single gonad of *Astacus* is bilobed anteriorly and each vas deferens is highly convoluted, leading from its ventral surface to the coxopodite of the fourth pereiopod where it opens to the exterior (Fig. 10.93A). Recently, Erkan et al. (2009) studied the male reproductive system and formation of spermatophores in *Astacus leptodactylus* (Eschscholtz, 1823). They presented photos of the fresh isolated male system (Fig. 10.93C). They only added the differentiation of the vas deferens into definite regions: proximal vas deferens (PVD), medial region (MVD), and distal region (DVD). Bullough (1981) reported that when the vasa deferentia are distended with white masses, these are spermatophores in process of formation (i.e. process of encapsulation of spermatozoa).

The sperm of the crayfish *Astacus* was discovered in 1835–36 by Henle and von Siebold. Then, the memoir of Dr. C. Grobben (1878) contained all the studies and discussions concerning the structure of these bodies published during that time. There is no doubt that the spermatozoon consists of 'a flattened or hemispherical body, produced at its circumference into a greater or less number of long tapering curved processes' (Huxley, 1880). It is a curious

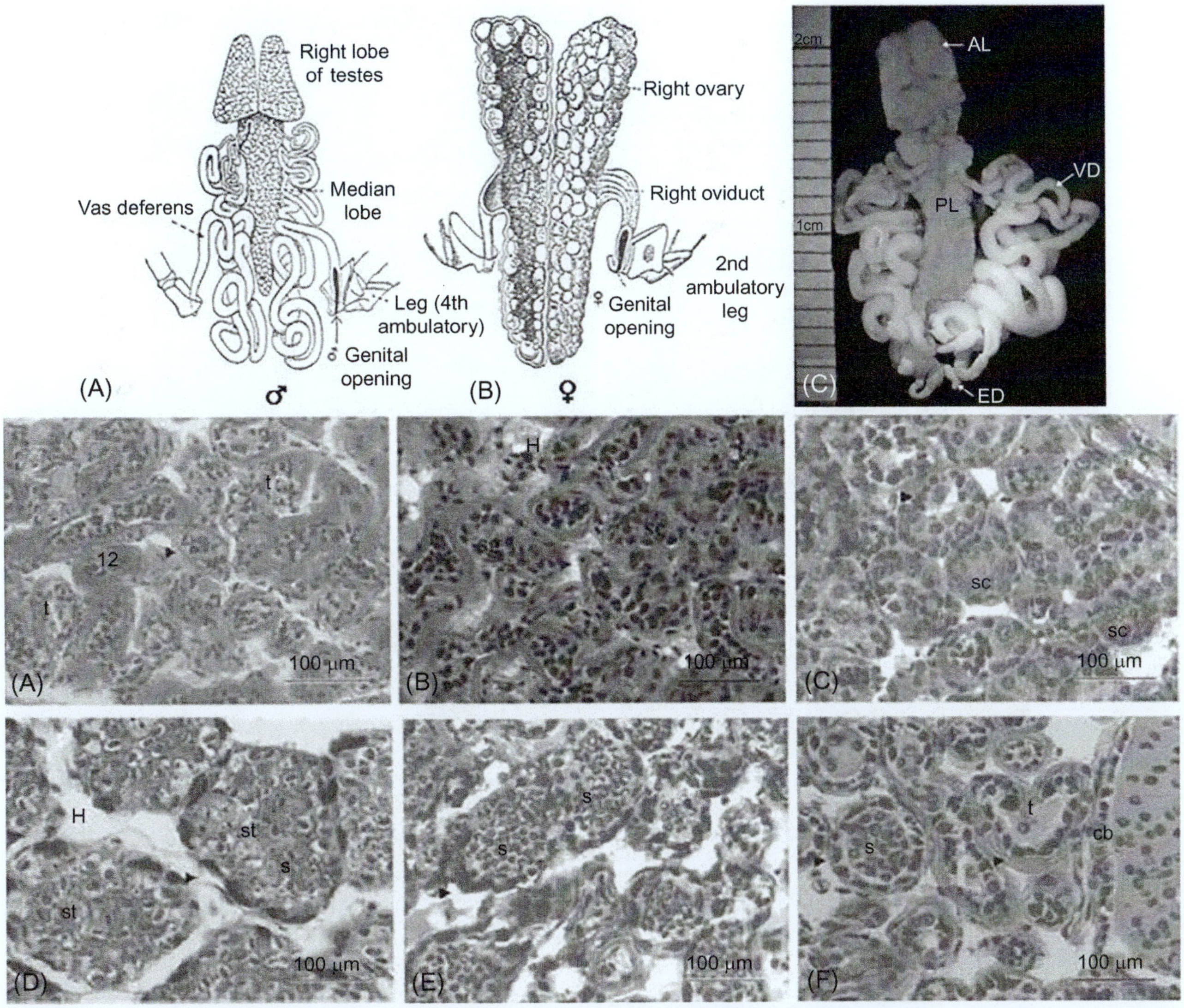

FIG. 10.93 (A) and (B) show diagrammatic drawings of the male and female reproductive systems of *Astacus fluviatilis*; (C) shows freshly isolated male reproductive system of *A. leptodactylus*. AL & PL = anterior and posterior lobe of testis, ED = ejaculatory duct, VD = vas deferens *(from Erkan et al., 2009)*. Lower histological photos show **spermatogenesis** in *A. astacus*: (A) testis of animal caught during May; (B) during July; (C) during August; (D) during September; (E) during October; (F) during November. Cb = cuboidal epithelium, ct = collecting tubules, H= heamal sinuses, s = spermatozoa, sc = spermatocytes, sg = spermatogonia, st = spermatids, t = seminiferous tubule, *head arrow* = Sertoli cell, H&E method. *From Lucić et al. (2006).*

amoeboid disc-like centre produced into a number of stiff processes or pseudopodia (see Fig. 10.16k).

The histological structure of the **testis** in *Astacus leptodactylus* (Fig. 10.93A–F) is basically composed of acini (follicles) and collecting ducts. The T.S. of the testis at different regions shows clusters of acini of similar stages of maturity (Fig. 10.93A). Each acinus is surrounded by a layer of connective tissue (as all the regions of the system), internal to which, there is a layer of generative cells and/or their progeny (spermatogonia, spermatocytes, spermatids, and spermatozoa) along with accessory cells. The latter are located in the wall of the acinus and can be easily distinguished by their flattened nuclei, which are more deeply stained than those of other cells in the acinus (Fig. 10.93B). As spermatogenesis progresses within an acinus and reaches spermatozoa stage (Fig. 10.93D and E), they accumulate in the centre of collecting tubules in syncytial clusters (Erkan et al., 2009).

Formation of the spermatophore of A. leptodactylus. The epithelial lining of the collecting tubule is of cuboidal cells (Fig. 10.94A and C), the cytoplasm of which stains with eosin considerably well. This epithelium secretes PAS-positive material, in which the sperm are embedded. This means that the sperm produced in the testis reach the vas deferens via the collecting tubules embedded in a material. The PVD is lined by a layer of

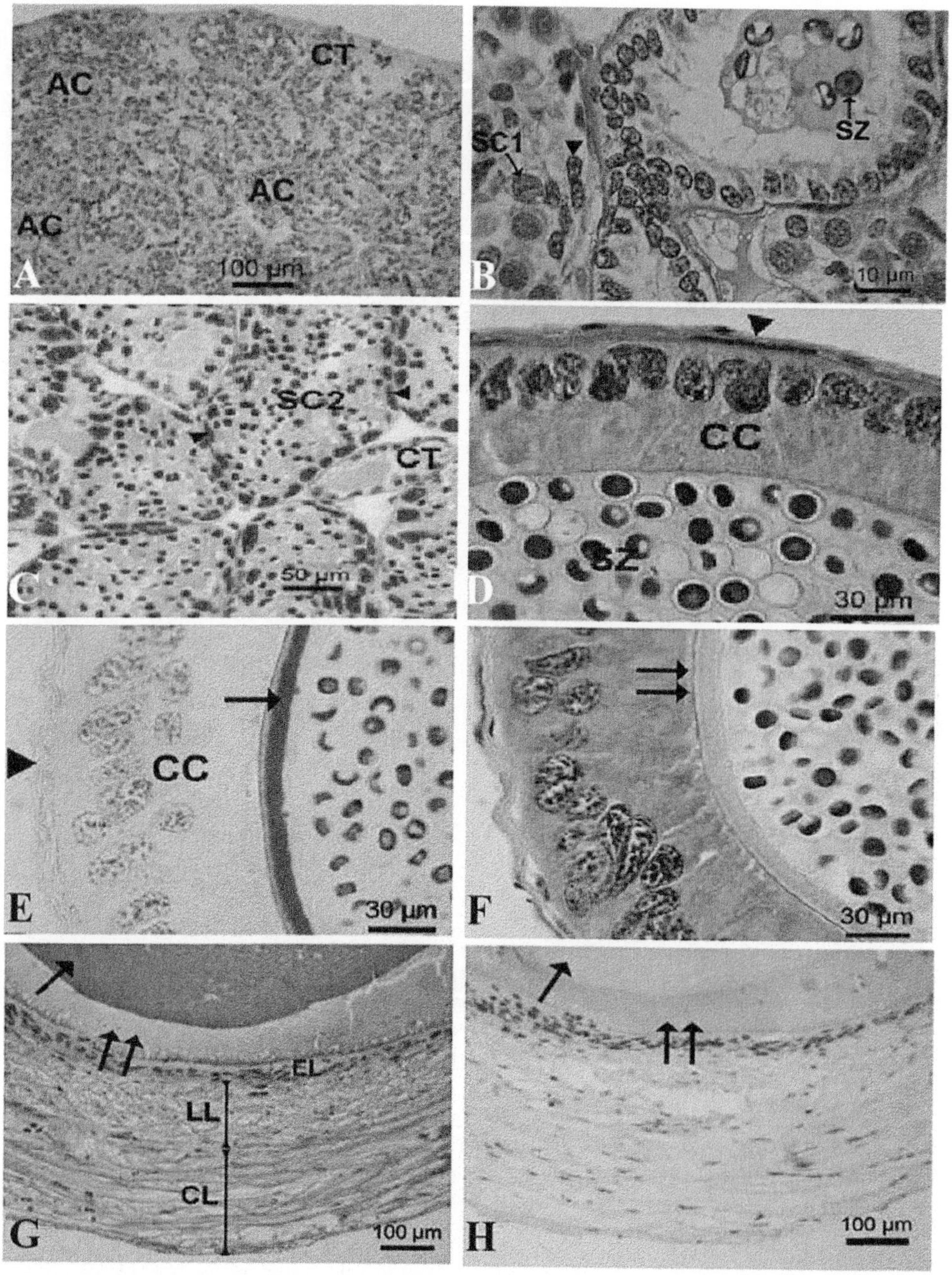

FIG. 10.94 Formation of the spermatophore through T.S.s of the reproductive system of *A. leptodactylus (from Erkan et al., 2009)*; (A) general view of the testis showing the acini (AC) and collecting ducts (CT); (B) the acini showing primary oocytes (SC1), accessory cells with flattened nuclei *(arrowhead)*, and spermatozoa (SZ); (C) the acini showing secondary spermatocytes (SC2); (D–H) T.S.s through regions of the vas deferens (VD): (D) the lumen of PVD occupied with spermatozoa embedded in a material; (E) middle region of MVD with high secretory activity, showing primary layer of spermatophore *(arrow)* stained deeply with bromephenol blue; (F) near the end of MVD showing secondary layer of spermatophore *(double arrow)* as a thin layer, stained with Alcian blue; (G) ejaculatory duct showing its thick muscular wall, primary layer of spermatophore *(arrow)* surrounds the sperm mass and secondary layer of spermatophore *(double arrow)* around the primary one reaching its thickest form in this region; (H) secondary layer stained with Alcian blue while primary one is not stained with the same stain. *Arrowhead* indicates fibrous connective tissue layer. CC = columnar epithelium, CL = circular muscle layer, EL = narrow epithelial layer, LL = longitudinal muscle layer. A–D,G stained with H&E.

columnar epithelium; their apical cytoplasm is stained with Alcian blue. The lumen of the PVD is occupied by the sperm, still embedded in the PAS-positive material with no definite layer surrounding them (Fig. 10.94D). The MVD is highly convoluted and is lined by a single layer of secretory columnar epithelium, taller than those of PVD, and also more basophilic. Its lumen is occupied by the sperm mass with its embedding material. At the beginning of the MVD, this mass becomes surrounded by somewhat thick layer (the 1^{ry} layer of the spermatophore).

Obviously, the material of this layer is secreted in the MVD and is strongly stained with bromephenol blue, i.e. protein in nature (Fig. 10.94E). At about the end of the MVD, a comparatively very thin layer appears surrounding the 1^{ry} layer (Fig. 10.94F). Opposite to the latter, the 2^{ry} layer is stained with Alcian blue, i.e. acidic in nature. Erkan et al. (2009) interpreted that this material contains hyaluronic acid, which is also a product of the MVD. The last region is the DVD, which is the widest compared with the previous two regions. It is lined by columnar epithelium similar to the MVD, but it differs in being folded, that it becomes more active. Within this last region, the layers constituting the spermatophore are prominently thicker and completely formed. Thus the spermatophore completes its maturation and it will be transferred to the ejaculatory duct.

The end part of the DVD changes gradually in its histological structure to lead to the ejaculatory duct. The latter differs greatly from the regions of the vas deferens. Its epithelial lining becomes very narrow and flattened, while its outer layer is very thick and rich in muscles arranged circularly and longitudinally (Fig. 10.94G and H). The spermatophore inside the lumen of this ejaculatory duct almost completes its formation. Its two covering layers reach the thickest form within this ejaculatory duct.

The fully formed spermatophore of Astacidea. The completely formed spermatophore of *A. leptodactylus* is nonpedunculate, tubular, 2–10 mm long, 1 mm in diameter (Fig. 10.95) and the sperm mass is enclosed in three-layered wall (Fig. 10.95) (Niksirat et al., 2014). It is in the form of an interrupted column as in *Pasifastacus leniusculus*, *Homarus americanus*, and *Enoplometopus occidentalis* (J.W. Randall, 1840) and is covered by several layers made of materials secreted by the vas deferens (Erkan et al., 2009). In that of *A. leptodactylus*, the sperm are embedded in a PAS-positive material, which is composed of variable carbohydrate molecules (in the opinion of the author, it contains nutritive and/or hormones and enzymes to reactivate the sperm at fertilization). The primary layer is formed by proteins and may have a protective role in maintaining the sperm alive (Erkan et al., 2009), while the second layer (which is the peripheral layer) is a highly Alcian blue-positive component that was reported to be hyaluronic acid. In other related species, it is reported to be chondroitin sulphate. Both substances are antimicrobial. The existence of hyaluronic acid in the secondary layer of the spermatophore of *A. leptodactylus* may be associated with hardening of the spermatophore and/or adhesion to ventral surface of the female during copulation (Erkan et al., 2009). In *A. astacus*, the spermatophore is covered by a wall that consists of three acellular layers. Just the outer layer stained positively with the PAS reaction, which indicates its polysaccharide content.

The freshly extruded spermatophore of the mature male lobster, *Homarus americanus*, weighs 500–900 g, and measures 1–2 cm in length with a circumference of 0.5–1.0 cm. The animals were obtained from a local market of California, Riverside, United States (Kooda-Cisco and Talbot, 1982). Spermatophore extrusion was induced by electrically stimulating the base of the fifth walking leg (pereiopod) with 12 V delivered through an alternating current variable transformer (Kooda-Cisco and Talbot, 1983). A single spermatophore is extruded through the gonopore; the sperm inside are morphlogically normal and undergo normal acrosome reactions. The spermatophore is tubular, 'trifoil' in shape with two lobes tapered laterally from a third lobe situated ventrally (Fig. 10.95A). The dorsal and ventral surfaces are labelled with respect to the spermatophore's orientation during extrusion. The sperm are densely packed into a continuous tube, contained largely within the ventral lobe. The sperm mass, which appears white, is easily seen through the translucent layers of the spermatophore wall. Most of the sperm mass is located within the first half of the extruded spermatophore. In cross section through the spermatophore, four regions can be identified by light microscopy (Fig. 10.95B). The TEM and SEM (Fig. 10.95C) show the following:

1. The sperm are embedded in a matrix of moderate electron density.
2. The primary spermatophore layer (1 mm in width) is uniformly thick around the sperm mass and contains at its peripheral margins both ring structures and crystals in an amorphous matrix.
3. The intermediate layer is thicker dorsally than ventrally with dense granules that dominate ventrally, while inclusions dominate the dorsal half. Both react positively to the PAS technique.
4. The outer bounding layer (~ 13.3 μm in width) has the innermost portion composed of parallel fibrils; the peripheral portion composed of flocculent material which is presumed to impart stickiness to freshly extruded spermatophores.

Dudenhausen and Talbot (1983) compared ultrastructurally between the soft and hardened spermatophores of the crayfish *Pacifastacus leniusculus* Dana (Fig. 10.95E). No significant difference was observed in the structure of the spermatophore of both species—*H. americanus* and *P. leniusculus.* The authors discussed the hardening of the spermatophore and supported the conclusion that the middle layer of the spermatophore wall in these species is the site of hardening. This layer forms the bulk of the spermatophore and is structurally modified after attachment on the female, and is stable during prolonged storage. However, the present author, Bawab, F.M. noticed that they recorded that "hardening of the spermatophore of *Penaeus trisulcatus* (*P. kerathurus*) is the result of phenolic tanning---associated

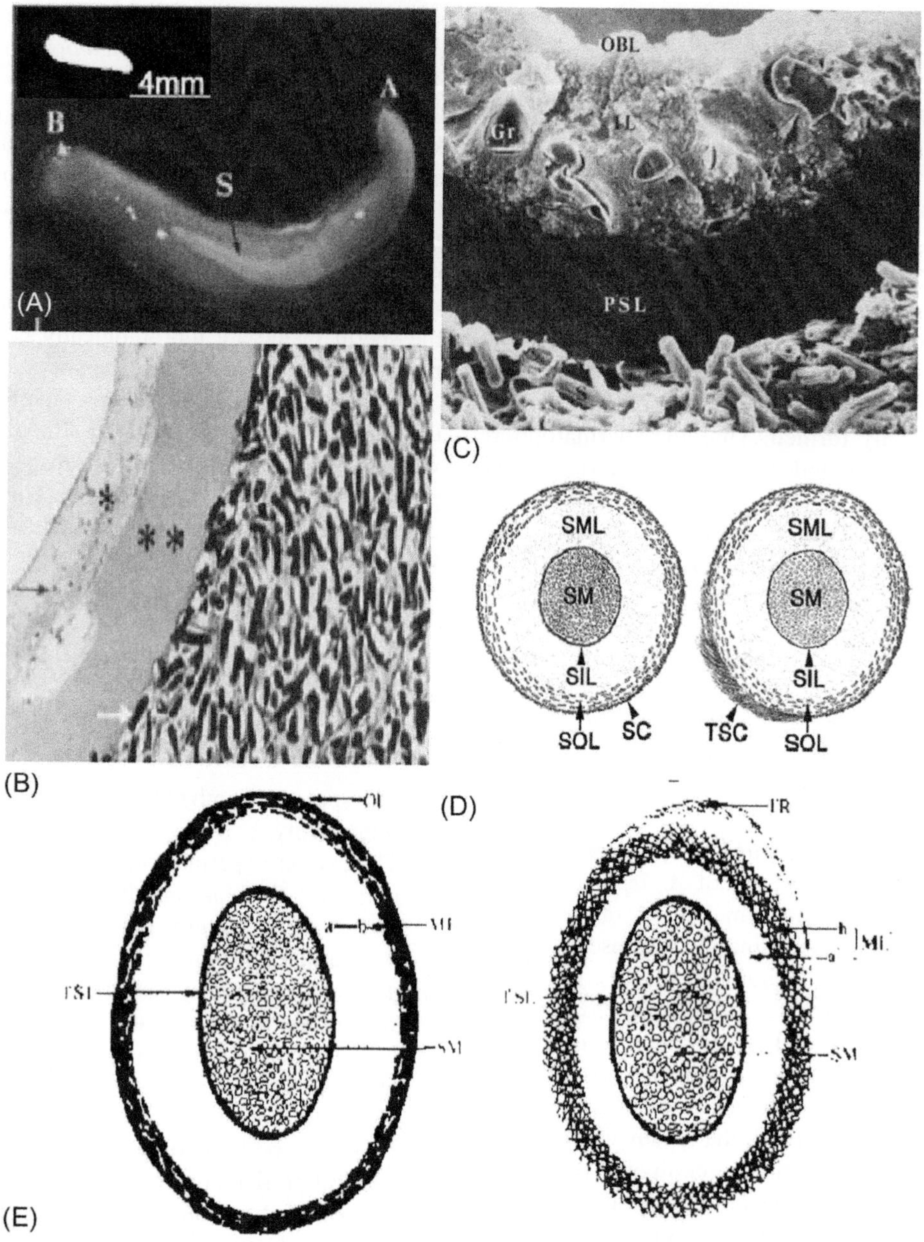

FIG. 10.95 (A–C) Spermatophore of the lobster *Homarus americanus (from Kooda-Cisco and Talbot, 1982)*: (A) Freshly extruded spermatophore; the sperm are clearly seen through the wall of the spermatophore. (A) indicates the end first appearing upon extrusion and (B) indicates the end appearing last. 4×. Upper inset shows schematic drawing of whole mount spermatophore of *A. leptodactylus (from Nikstrat et al., 2014)*; (B) Section of the spermatophore following the PAS method showing strong +ve reaction of the sperm acrosome *(white arrow)*, a moderate reaction of the 1ry spermatophore layer (**) and granules *(black arrow)* of the intermediate layer, and a weak reaction (*) of the intermediate layer's matrix. 400×; (C) SE micrograph of a fractured spermatophore showing the sperm mass (S), 1ry spermatophore layer (PSL), intermediate layer (IL) granules (Gr), and the outer bounding layer (OBL). 46,000×; (D) Drawings of cross sections of the spermatophore of *Astacus leptodactylus* (upper inset in (A)) showing before (left-hand sec) and after (right-hand sec) mating. The sticky coat of the spermatophore is thickened at the site of attachment. SC = sticky coat, SM = sperm mass (SIL = inner layer, SML = middle layer, SOL = outer layer of spermatophore), TSC = thickened sticky coat; (E) Schematic drawings of sections of the spermatophore of *Pacifastacus leniusculus* before (left-hand) and after (right-hand) hardening *(from Dudenhausen and Talbot, 1983)*. The former shows the structure of the unextruded spermatophore (before), the latter shows that of the hardened one. SM = sperm mass, 1° = primary spermatophore layer, ML = middle layer, OL = outer layer, a′ = compact region and b′ = outer reticulated region of the middle layer of the hardened spermatophore correspond roughly in size to regions a and b of the middle layer of the unextruded spermatophore, TR = thickened ridge. Note that the outer globular layer is absent in the hardened spermatophore.

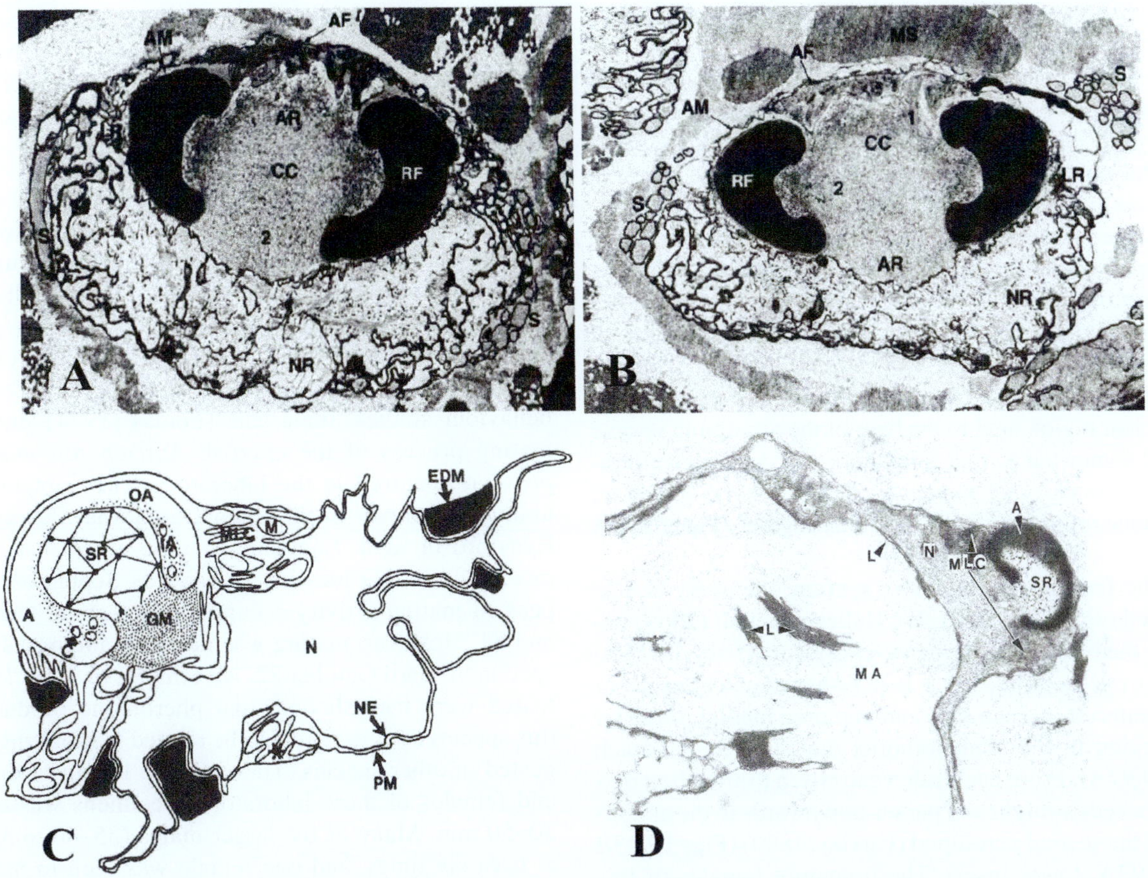

FIG. 10.96 TEM longitudinal sections of mature sperm of crayfishes, (A, B, and D) photos while (C) schematic diagram: (A and B) of *A. astacus* at two different levels or of two different sperm. The acrosomic region of the spermatozoon (AR) consists of a central canal (CC), which is delimited by the acrosome vesicle, formed in turn by the apical formation (AF), the ring-like formation (RF) and an amorphous matrix (AM). Peripheral material (1) and internal material (2) of the central canal. The gametes are covered by a mucoid substance (MS) which determines that the spines (s) arising from the sperm are maintained encircling the cell body. X of A = 9000 and of B = 6000; (C) of *Cherax tenuimanus* and (D) of *C. albidus*. The acrosome (A), the subacrosomal region (SR), the collar containing granular matrix (GM) and mitochondria (M) in a membrane lamellar complex (MLC), the nucleus (N), NE = nuclear envelope, PM = plasma membrane. In *C. tenuimanus*, extracellular electron-dense material (EDM) only and its acrosome includes a homogeneous outer region (OA) and a granular inner region (IA) that has electron-translucent canals (C) containing an electron-dense component at their periphery. Some of the membrane lamellar complex (MLC) extends around the periphery of the sperm (*). In *C. albidus* there is extracellular lamellae (L), the acrosome contains electron-translucent canals arranged in whorling pattern (C), a portion of the membrane lamellar complex (MLC) extends around the periphery of the sperm *(arrow)*. X = 13,700. *Sperm of A. astacus from López-Camps et al. (1981), the sperm of the two Cherax from Beach and Talbot (1987).*

with chitin complexes." Malek and Bawab (1971) did not mention the presence of chitin in this spermatophore, since its absence was evidenced chemically (see page 86). However, Dudenhausen and Talbot (1983) added a table of the decapod animals whose spematophores were studied by light microscope, up until the time of their work.

The ultrastructure of astacid sperm. Ultrastructural studies on the sperm of crayfishes were performed by sectioning spermatophores. These sperm belong to the type pleocyematan (reptantian) as mentioned above. The sperm of this type have a complex acrosome, a membrane lamellar complex, uncondensed chromatin, nuclear envelope, and number of spikes containing microtubules. Beach and Talbot (1987) studied the sperm of two species of *Cherax*: *C. tenuimanus* (Fig. 10.96C) and *C. albidus* (Fig. 10.96D). The extracellular materials associated with the sperm surfaces of these two species are morphologically different. That of *C. tenuimanus* has a very electron-dense substance associated with infoldings of the plasma membrane (Fig. 10.96C). This material may keep the sperm compact and may be analogous to a capsule, though incomplete. Sperm of *C. albidus*, in contrast, do not have this substance, but rather appear to bind long extracellular lamellae to their surfaces (Fig. 10.96D).

In addition, there is a major difference between the sperm of these two species of *Cherax* and those of other crayfishes in regards: (1) packing in the spermatophore and the extracellular materials associated with them. Sperm of *Cherax* are not tightly packed and do not have long rigid spikes, thus they may not require elaboration of a discrete capsule for packing. Astacinid and cambarid sperm are tightly packed in the spermatophore and are surrounded by complete capsules; presumably, the capsules keep the spikes wrapped around the sperm body and facilitate packing in the male tract. (2) The unusual feature of the sperm of *C. albidus* and especially of *C. tenuimanus* is the extensiveness of the membrane lamellar regions. It occupies much of the sperm volume, and in some cases extends around the periphery of the sperm. In other crayfish sperm, these are small and lateral to the acrosome. This latter feature is well observed in the sperm of *Astacus astacus* (Fig. 10.96A and B) (López-Camps et al., 1981). These authors reported that in a nuclear region next to the base of the acrosomic vesicle situated somewhat to one side, there is a certain proliferation of 'laminae' and membranes, which have amyelinic appearance referred to as the 'laminar region' (Fig. 10.96A and B).

In the **female reproductive system**, the ovary is also a trilobed organ (Fig. 10.93B). Hubenova et al. (2009) described the ovary of the narrow-clawed crayfish, *A. leptodactylus* Eschscholtz, 1823. It is Y-shaped and consists of paired anterior ovarian sacs and unpaired medial posterior sac, located in the cephalothorax dorsal to the stomach (Fig. 10.97A). From each side ventrally, a thin-walled oviduct proceeds straight and passes downwards to the coxopodite of the second pereiopod (Huxley, 1880) (Figs 10.93B and 10.97A, lower inset). The immature females of this narrow-clawed crayfish and/or juveniles reared in carp ponds, south Bulgaria, can be observed all the year round. From June, the ovary starts maturity and the eggs attain yellow colouration. In August–September, they change their colour to orange or orange-brownish and enlarge in size, reaching definitive sizes (Fig. 10.97B–I). The fully ripe yellowish orange oocyte of this species, *A. leptodactylus* (collected from Aras Dam Lake, western Azerbaijan province, Iran) has a maximum diameter length of 1250 μm (Mirheydari et al., 2014). From a third location, Lake Terkos, Turkey, the oviducts of the ovaries of this species (Fig. 10.97J) show activity synchronized with the development of the ovaries (Ünis and Erkan, 2012). Their epithelial lining produce two different types of secretion: one is in a 'band' form, which lies through the epithelium that surrounds the lumen, stains with bromphenol blue, and also gives a PAS-positive reaction (Fig. 10.97K); the other type spreads throughout the inside of the lumen, and stains only with the bromphenol blue (Fig. 10.97L). Neither the first nor the second types stain with Alcian blue.

Hubenova et al. identified, as a result of their observations, that the female narrow-clawed crayfish, at conditions of carp pond rearing, can reach sexual maturity during the first year of their life (6–7 months old), at body weight above 9.47 g and body length above 7.0 cm. The copulation in this crayfish can be observed at the end of November and the beginning of December. The ovulation of the mature eggs occurs immediately after it or within 1 month after spermatophore deposition.

Copulation. The earliest description of mating behaviour of a crayfish belonging to the astacids is that by Roesel von Rosenhof (1755). This and subsequent accounts dealing with the same subject until early in the 20th century, probably all refer to the one species *Astacus astacus*. Later descriptions of mating, even of other species, lack sufficient detail that would be of value as comparative behaviour studies. Ingle and Thomas (1974) studied the mating process of the crayfish *Austropotamobius pallipes* that occurred in the laboratory. They obtained their animals from part of the River Darent that flows through Eynsford in Kent, England. They observed that *A. pallipes* mates in this location as early as September 24, the peak of mating activity occurs between about October 24 and 31. However, mating was not seen among laboratory specimens until October 22. All females of *A. pallipes* that mated were hard-shelled, that pheromone production in this species does not seem to be related to moulting as suggested in other species. The carapace length of the males and females of these laboratory specimens was of about 30–50 mm. Many of the larger males (35–40 mm) mated at least six times, and one female was seen to mate with five different males.

According to Ingle and Thomas (1974), the following phases were recognised.

1. Contact in which the male meets the female from a frontal position and grasps her chelipeds, pereiopods, and sometimes her antennae. The female remains stationary or retreats by walking backwards (Fig. 10.98A). In eight observed matings, this contact phase was omitted and the male began by the second phase.
2. 'Seizure' of the female where the male moves rapidly around the female from left to the right through 180 degrees, retaining his grasp of her or pressing his chelipeds onto her carapace. Grasping of the female's antennae or rostrum is not uncommon during this phase. The male then turns onto his side and prepares to encircle the female's body (Fig. 10.98B).
3. Turning of female, the male encircles the female's body with his pereiopods, interlocking his third pair (Fig. 10.98C) and rotates her within the confines of this embrace using his fifth, fourth and sometimes the first pair of pereiopods. Thus, their ventral surfaces become face to face (see before Fig. 10.92D) and the male is beneath the female. The male usually establishes himself as the uppermost of the pair, but in four matings the male remains beneath the female.

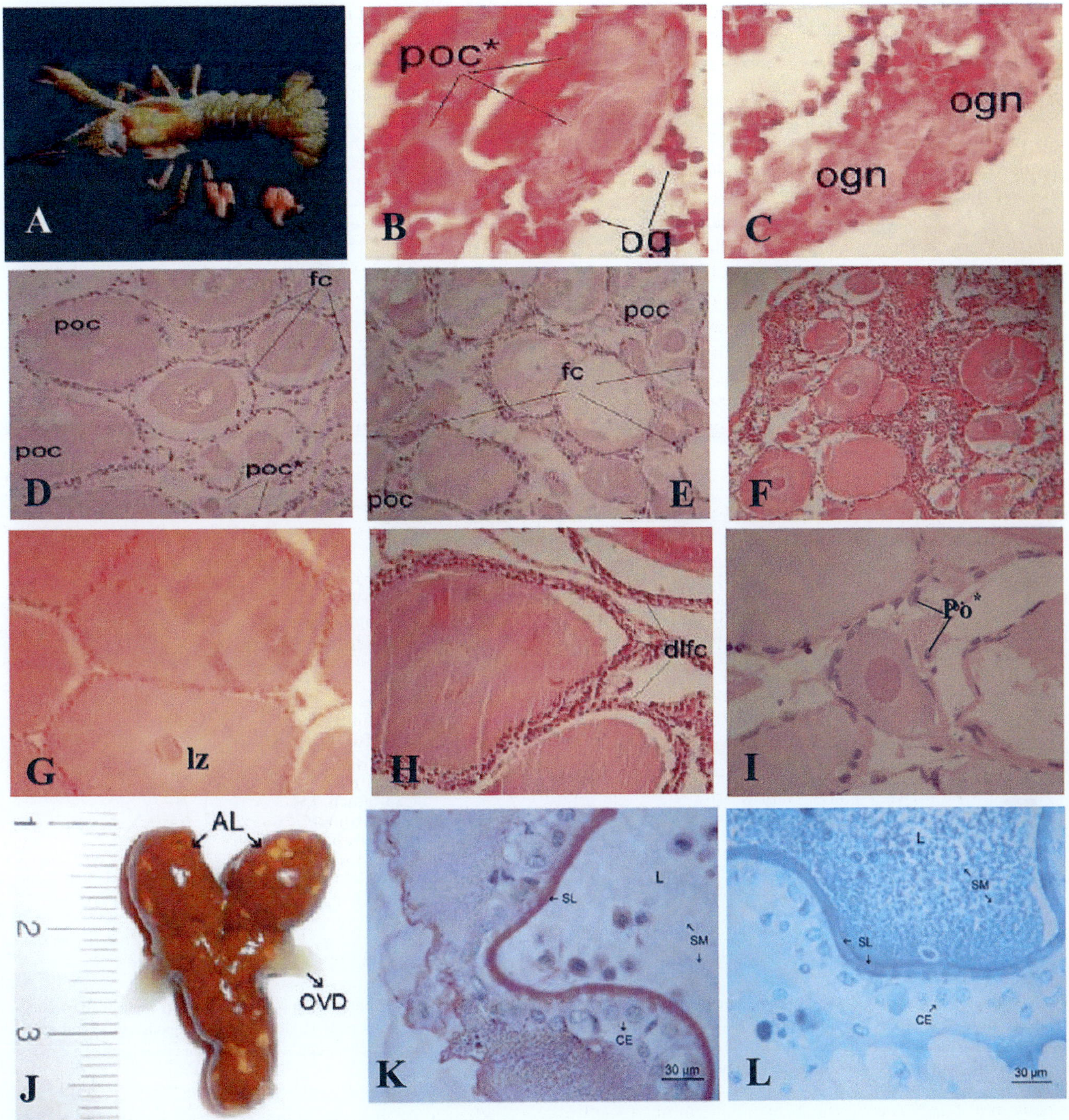

FIG. 10.97 Female reproductive system of the narrow-clawed crayfish *A. leptodactylus* Ech. 1823 *(from Hubenova et al., 2009)*: (A) Photograph of a dissected animal showing the ovary in situ and the isolated system in the lower inset, the ovary after ovulation and the oviduct empty of oocytes (February); (B–I) histological sections showing oogenesis in the same animal (H&E); (B–G) each shows part of ovary of one-summer old crayfish; (B) August, early previtellogenic oocytes (poc*) and oogonia (og). 400×; (C) September, nests of oogonia (ogn). 400×; (D) September, previtellogenic oocytes (poc) and (poc*). Fc=foliclle cell. 200×; (E) January, different location of poc, primary follicle oocyte (fc). 200×; (F) January, ovary wall. 200×; (G) Part of ovary of one-summer old crayfish, May, endogenic vitellogenesis, light zone (lz). 200×; (H) part of ovary of two-summer old crayfish, September, vitellogenic oocytes with secondary follicle cell layer (dlfc). 200×; (I) part of ovary of mature crayfish, February, early previtellogenic oocytes (po*). 200×; (J–L) ovary of the same species *(from Ünis and Erkan, 2012)*, (J) isolated ovary showing oviduct (ovd) begins laterally from the base of the anterior lobe (AL); (K and L) show the secretions of the oviduct histochemically. CE=columnar epithelium, L=secretion inside the lumen, SL=secretion layer, SM=secretion material, SM=secretion stained with bromphenol blue.

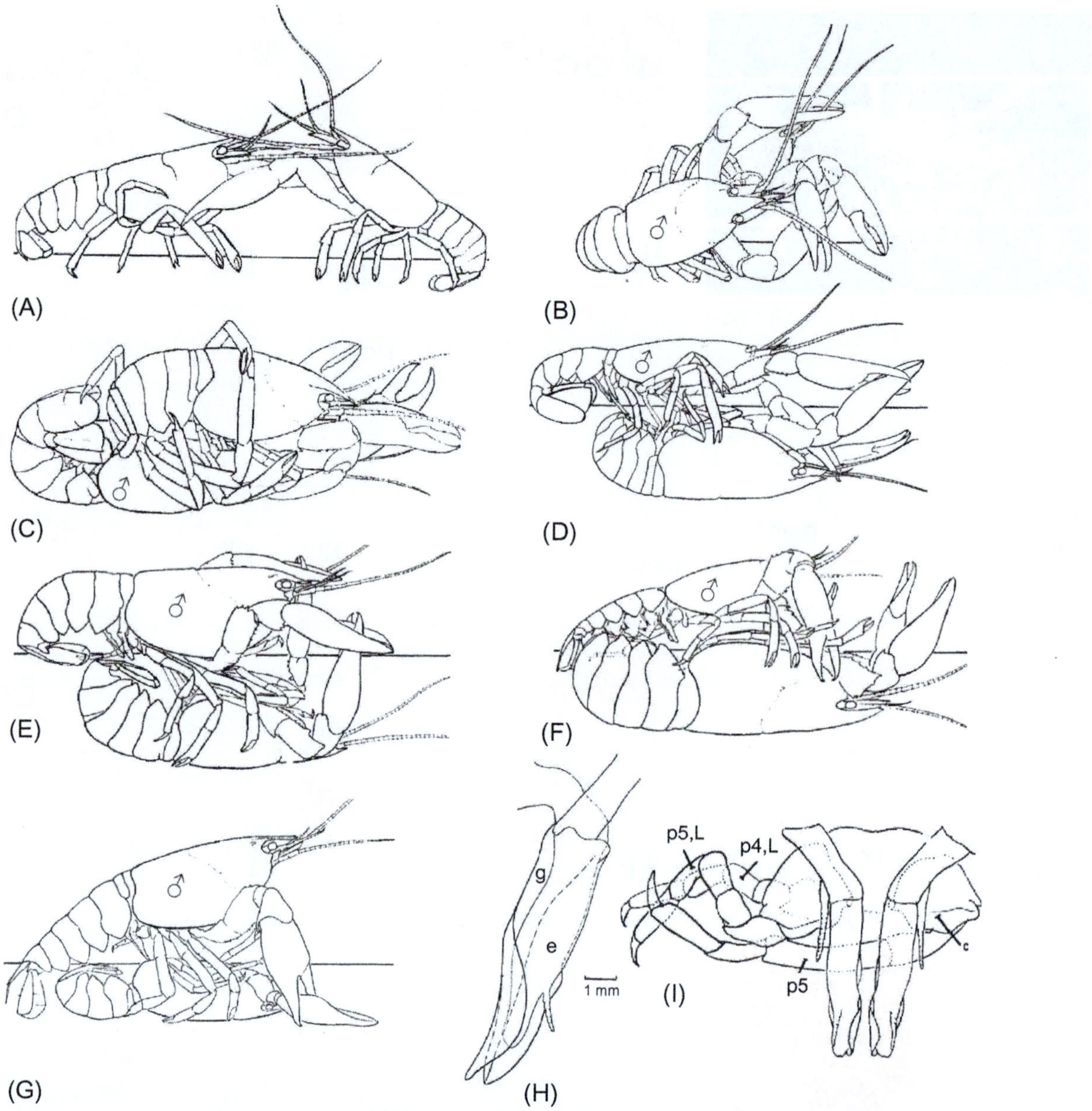

FIG. 10.98 Mating process shown by *Austropotamobius pallipes (from Ingle and Thomas, 1974)* proceeds in seven phases: (A) contact; (B) seizure of female; (C) turning of female; (D) mounting by male; (E) fifth leg crossing prior to interlocking of gonopods; (F) locking of Gonopods; (G) spermatophore deposition; (H) structure of right pair of interlocked gonopods; (I) position of limbs during spermatophore deposition. c = right coxal papilla, e = endopodite, g = groove, p4,l and p5,l = fourth & fifth left peraeopod, p5 = fifth right peraeopod.

4. Mounting by male who then relaxes his firm grip on the female's chelipds or pereiopods and repositions himself (Fig. 10.98D) to be uppermost.
5. Fifth leg crossing and locking of gonopods, in this phase the right or left fifth pereiopod is positioned transversely across the sternum (Fig. 10.98E). The distal part of each second gonopod (see Fig 10.92B 2) endopod is inserted into the grooved portion of its respective first gonopod. Both pairs of gonopods, now interlocked, are swung forward. The 'coxal papilla' on each fifth pereiopod (where the male gonopore is found) is pressed into the groove formed by the interlocked gonopods. The male then straightens his abdomen attaining the final position of this phase (Fig. 10.98F).
6. The spermatophore deposition occurs by the pressing of the tips of the interlocked gonopods of the male onto the female's sternum (portion between the bases of her fourth and fifth pereiopods, i.e. annulus ventralis), thus attains the position shown in Fig. 10.98G.

A spermatophore mass flows from each gonopore as a hyaline fluid passing through the 'tubes' formed by the gonopods and emerging onto the female's sternum (annulus ventralis) as opaque, white, short tubular deposits (Fig. 10.99(I) H and I). Fig. 10.98H shows the structure of the right pair of interlocked gonopods. The endopod of the second not only completes the inner 'wall' of the tubular first gonopod, but also contributes to the formation of a proximal groove into which the coxal papilla of the right fifth pereiopod is inserted. The endopod of the second gonopod is moved lengthwise rapidly as the spermatophore mass passes down the 'tube'. This action cuts the mass into small lengths as it is extruded onto the annulus ventralis. The accurate position of the appendages responsible at the moment of spermatophore deposition is shown in

(I)

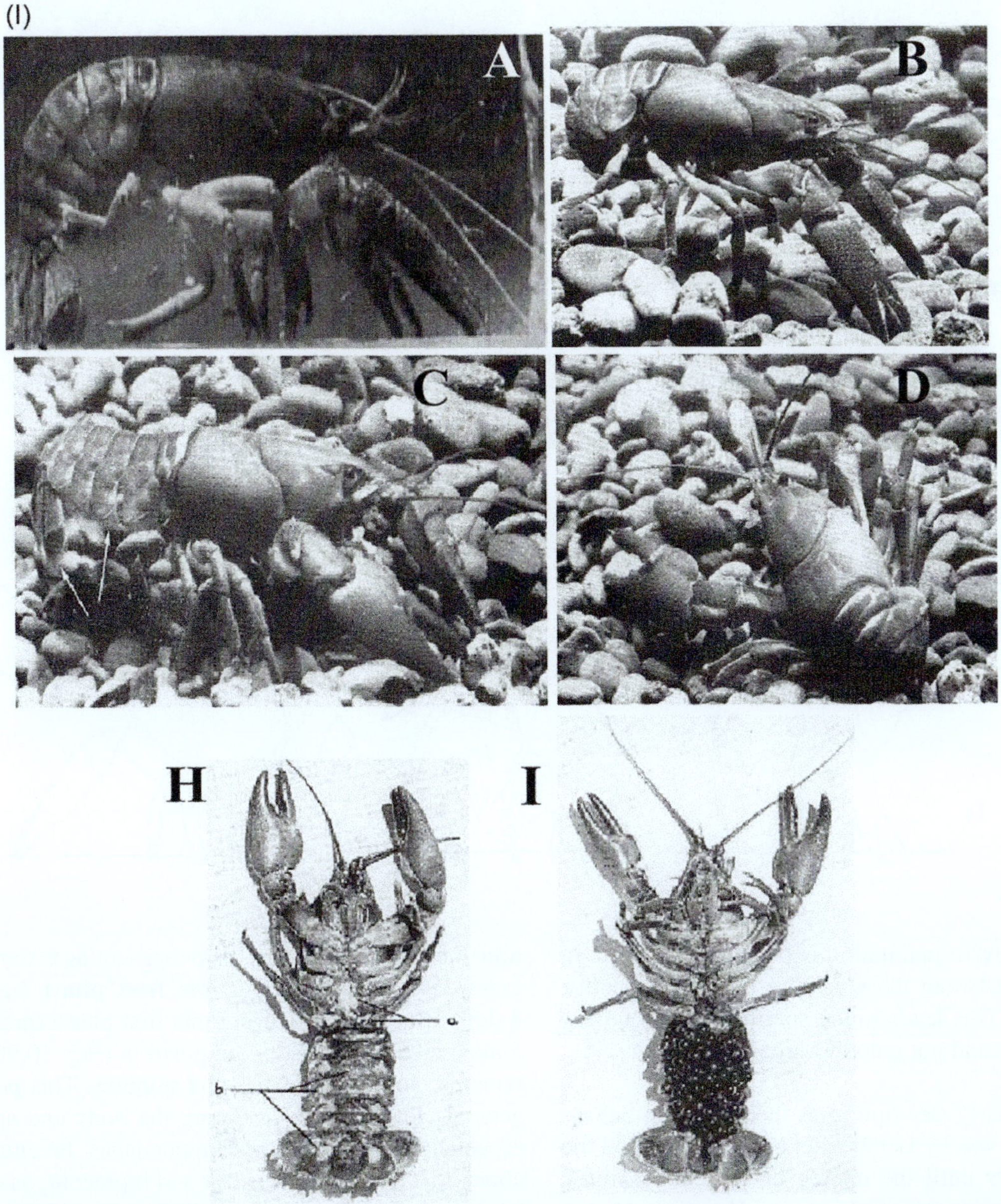

FIG. 10.99 (I) Photographs showing stages of spawning process by two animals *((B–D and H, I) from Mason, 1970; (A, E–G and a–d) from Ingle and Thomas, 1974)*: (A) preening; (B) vigorous abdominal contraction; (C) slow unfolding of the tail fan becoming slightly convex, *arrows* point to the clean fringes of setae on the fan; (D) initial stage of the assumption of thawning posture, the walking legs of the right side are elevated, with the aid of the right main cheliped, the female can turn onto her back; (E) initial regaining of the upright stance; (F and G) rolling; (a–d) illustrations showing regaining of upright stance, i.e. explain the (E) and (G) photos; (H) female crayfish about to spawn, showing the pattern of spermatophore deposition (a) and the distribution of the mucous secretory glands (b) on the ventral abdomen; (I) recently spawned female carrying a full complement of eggs. The spermatophores have disappeared completely. (II) Egg membranes in *Astacus pallipes (from Cheung, 1966)*: (A–C) show photographs of three stages in course of formation of the trichrome membrane. 1, 2, and 3 are membrane layers; (D) and (E) are T.S. through the uropod of a female *Astacus* showing cement glands; (D) the specimen with mature ovary; (E) the same with spent ovary and degenerating cement glands; (F) diagrammatic drawing of an *Astacus* egg with funiculus on a row of hairs (A) and (B) section at *S-S*. 1,2,3 are membrane layers, h = hair.

(Continued)

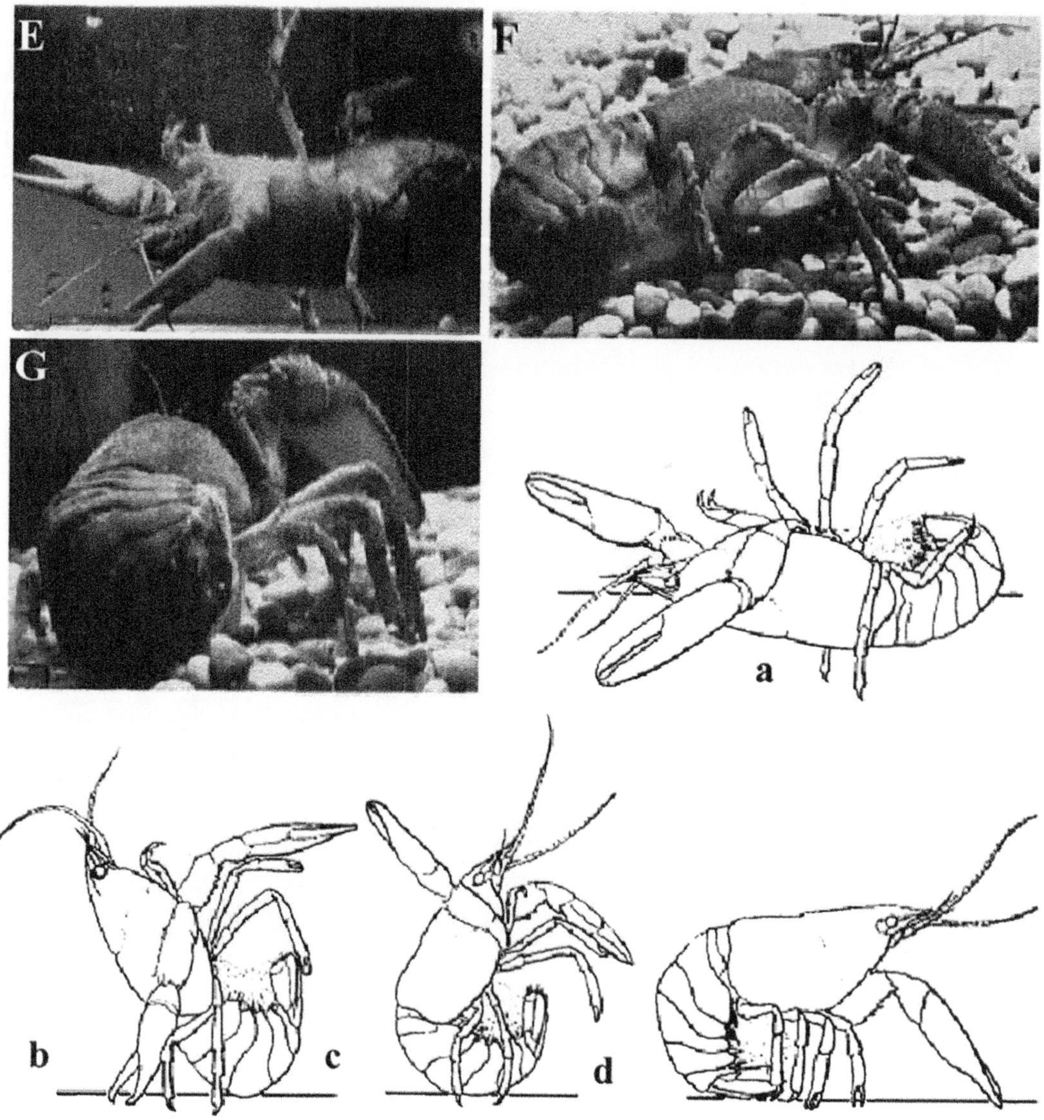

FIG. 10.99, CONT'D

Fig. 10.98I. Failure to maintain this position results in an incomplete seal between the gonopore and its respective gonopod groove. This leads to the spermatic mass falling directly elsewhere and not onto the suitable position.

Spawning. The early description of the spawning behaviour of *A. astacus* was by Lereboullet (1860). This, and the following accounts until the middle of the 20th century, described only part of this vital activity. However, Mason (1970) and Ingle and Thomas (1974) provided a detailed description of spawning in the eastern American astacid crayfish *Pacifastacus trowbridgii* (Stimpson) by the former, and in the astacid crayfish *Austropotamobius pallipes* by the latter. Observations of this vital activity in both species were made in the laboratory and they were almost the same. It was divided into five phases. Except in *P. trowbridgii* the first phase was described as a prespawning behaviour (the time between copulation and egg-laying), during which grooming occur. However, the **first phase** begins about 3 days after mating. During this first phase (preening), the female takes a characteristic position (Fig. 10.99(I) A) and remains motionless for several minutes. This period is for general cleaning and grooming the body and appendages, especially the setae of these appendages. Preening lasts for about 36 h and during the last 3–4 h preening is continuous while the female itself shows, throughout this phase and the next, extreme sensitivity to all external stimuli. The **second phase** is the vigorous abdominal contraction, one of a series just prior to spawning, whereby the posterior tip of the telson surpass the oviduct openings on the coxopodites of the second pair of walking legs (Fig. 10.99(I) B). This posture is maintained for about 10 min and is terminated by a slow straightening out of the tail (Fig. 10.99(I) C). The female makes several movements to assume the spawning

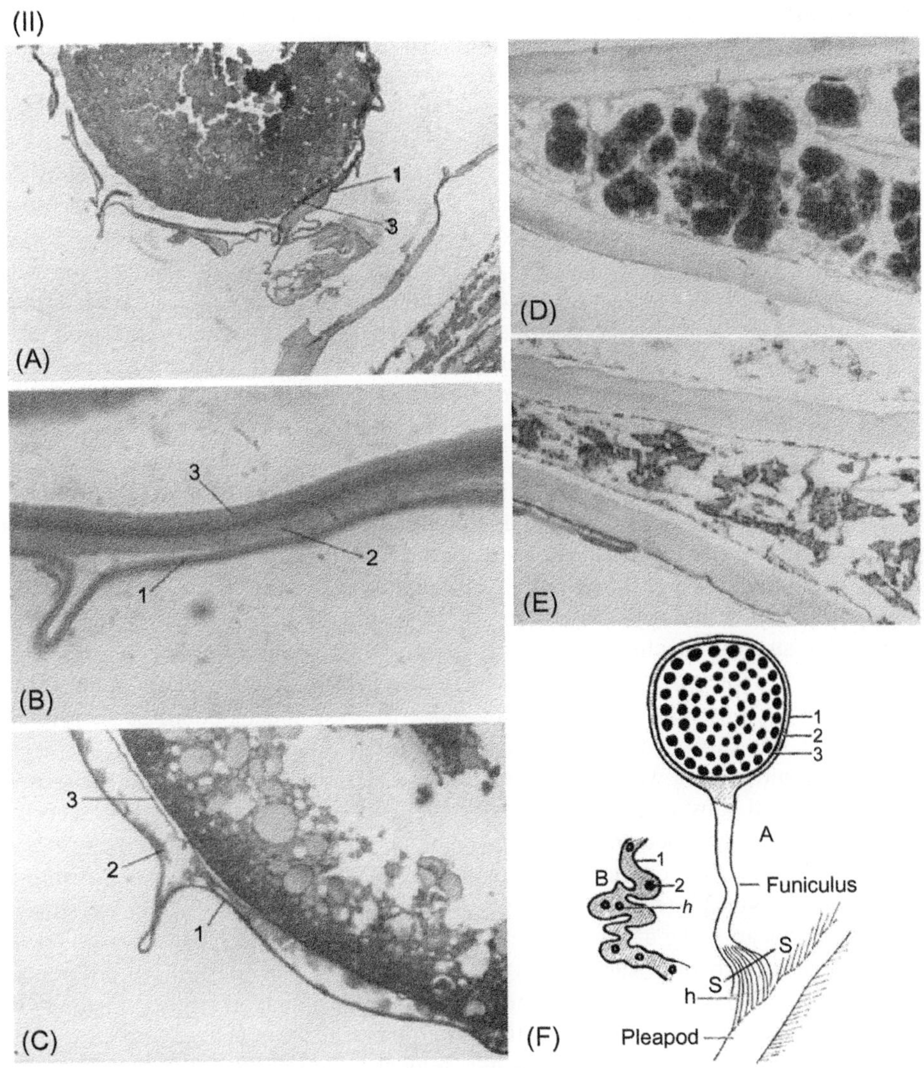

FIG. 10.99, CONT'D

position. During the **third phase**, spawning occurs. The desired body angle of the female is maintained by the first pair of chelipeds, aided by the walking legs (Fig. 10.99(I) D). The abdomen remains tightly folded until the full spawning posture is attained, then relaxes slightly. Therefore, the tail fan constitutes a basket-like receptacle for the extruded eggs. Stream of eggs can be observed leaving the oviducts. Recently laid eggs appear soft light brown, immersed in the almost transparent substance (mucous substance). Folding of the abdomen (rolling) alternates with periods of rest for about 30 min. In the **fourth phase**, the female regains the upright position rapidly, in about 12 s (Fig. 10.99(I) E) with the abdomen held at almost a right angle to the thorax, and with the telson cupped. The **fifth phase** is termed 'turning' in *P. trowbridgii* and 'resting and rolling' in *A. pallipes* (Fig. 10.99(I) F, G and (a)–(d)). After resting for about 30 min, the spawned female slowly constitutes a regular series of alternating positions taken by lying first on one side and then on the other. This phase continues for about 12 h. The spawned female then assumes her normal position and the eggs appear to be firmly attached to the pleopod setae (Fig. 10.99(I) I). Then, the egg mass is preened frequently and the residual mucous removed.

Brooding. The female *A. fluviatilis* Fabricius, 1775, after accepting the spermatic matter of the male, retires to a burrow and the process of laying eggs commences. These eggs, as they leave the apertures of the oviducts, are coated with a viscid substance. The latter is readily drawn out into a short thread, the end of which attaches itself to one of the long 'hairs' of the swimmerets (Huxley, 1880). At subjecting to water, the substance rapidly hardens and the egg thus becomes attached to the limb by a stalk. The movements of the swimmerets keep these eggs clean and well aerated. Subsequently, the newly hatched youngsters are carried about, protected by their mother's abdomen, as a kind of nursery, for as long as they need maternal care (Fig. 10.91F). Both vital activities—brooding and maternal care—were described in such amazing details by Huxley (1880).

Shiino (1968) mentioned that the members of the Decapoda carry their offspring attached to their pleopods; a secretion from the dermal glands on these pleopods, adheres to the egg-membranes and coagulates, forming threads which fasten the eggs to the setae of the pleopods. Such glands are found in *Astacus* on the sternal surface of the abdomen with ducts extending to the interior of the pleopods and opening to the exterior at the bases of the setae of these pleopods.

Cheung (1966) was greatly interested in the development of egg membranes and their attachment in decapods. He studied this aspect mainly in *Carcinus maenas*, and also in other decapods, *Nephrops norvegicus*, *Homarus vulgaris*, and *Astacus pallipes* (freshwater crayfish) (see later in *Portunus*). In all the eggs already attached to hairs of pleopods of any of these species, a thick trichromatic membrane was found. The stages of its formation were observed (Fig. 10.99(II) A–C). Layer 1 was supposed to be derived from the vitelline membrane, and layer 2 was found to be formed by solidification of a fluid substance produced by the egg. In *Astacus pallipes*, this layer becomes uniform and comparatively very thick after the completion of its solidification (the formation of layer 2 by solidification of a viscous fluid from the egg appears to be a rather special feature in the decapod Crustacea (Cheung, 1966)). Layer 3 arose under the fluid that forms layer 2 and is formed by highly PAS-positive substance from the egg. When the *Astacus* egg was first laid, it passed immediately into a jelly-like secretion, which had appeared a short time previously in the incubation chamber, formed by the curved abdomen and pleopods. By histological examination, Cheung stated that the oviducal epithelium undergoes cyclic activities corresponding to the spawning cycle. No function could be attributed to the oviducal secretion from his work. However, his discussion shows that it may contribute in part to the material for sticking the eggs to the pleopodal setae. In any case, it seems apparent that such secretion would be advantageous to the process of egg-laying because of its lubricating effect. It is natural to believe that this secretion must somehow be related to the physiology of spawning. It was thought that the oviducal secretion was responsible for the formation of a waterproof layer around the egg and for the funiculus attaching the egg to the pleopod hairs.

In *Astacus*, it is suggested that additional membranes must be acquired from external sources, i.e. from the oviduct (secretion that surrounds the spawned eggs) and from the cement glands, which also show activities corresponding to spawning. Concerning the former, it was observed that these eggs on being laid were surrounded by the secretion actually not in contact with the water. These eggs were neither adhering to each other, nor attached to any setae. They were only held to the animal by the secretion. After a few days, the secretion was greatly diminished in quantity and the eggs became attached to the nonplumose hairs in their proper way. Nearly every egg was successfully attached. Later, each attached egg formed a stalk (funiculus) with the outermost layer of the trichromatic egg membrane. The movements of the pleopods may play an important role in this process. The funiculus of each egg of *Astacus* is attached to a number of adjacent hairs running parallel from the limb (Fig. 10.99(II) F). This view of Cheung (1966) greatly evidenced the very old description of Huxley (1880). On the other hand and concerning the cement glands, the opinion of Yonge (1937) was that the cement glands alone contributed the cement for egg attachment, and for the funiculus. It is worth noting that Yonge worked on *Homarus vulgaris* whose pleopods contain these cement glands. Cement glands are present in *Astacus*, mostly in uropods and telson (Fig. 10.99(II) D and E). Macroscopically, the size of these glands could be seen to correspond with the developing stage of the ovary. These glands, with a cycle of secretion corresponding to the spawning cycle, did not occur in sections of pleopod on whose nonplumose hairs, the eggs were attached. However, in the aquarium, practically all the eggs laid by *Astacus* into the substance(s) in the incubating chamber became attached successfully to the pleopodal hairs. This view is in accordance with that of Shiino (1968), in spite of the difference in the position of the cement glands. However, Cheung (1966) indicated that neither of these views is correct, as the funiculus and the sement secretion are not the same substance.

Development

Formation of gametes

Spermatogenesis. in *Astacus astacus* (lower histological photos, Fig. 10.93) obtained from Vukovina gravel pit, located at northwest Croatia (Mediterranean region), was studied through morphological changes in the gonads during the reproductive cycle (Lucić et al., 2006). The seminiferous tubules of the testis of this species during May and June were observed filled with numerous spermatogonia (Fig. 10.93A). During August, most of the seminiferous tubules were filled with the spermatocytes (Fig. 10.93C). Primary spermatocytes could not be distinguished from secondary ones with light microscopy, but both cell types were filled with chromosomes at different stages of meiosis. For the first time, mature spermatozoa were noticed together with spermatids in seminiferous tubules of specimens obtained in September (Fig. 10.93D). Spermatids have very dense, basophilic, and elongated nuclei. They are mostly located centrally in the follicle (T.S. of the seminiferous tubule).

A month later, during November, most of the follicles were observed filled with spermatozoa (Fig. 10.93E and F). During this process of maturation of the testis, its size gradually enlarges greatly. The radius of seminiferous tubules changes from approximately 50 μm measured in May to approximately 200 μm measured in October–November (Fig. 10.93A and E). Because of the enlargement of the radius of the tubules during spermatogenesis, the connective tissue surrounding them becomes thinner and this affects the intensity of the PAS colouration. In addition, the acrosomal vesicle of the sperm inside these tubules is basophilic and stained positive with PAS (Fig. 10.93D–F).

Oogenesis. was described through the histological analysis of ovary development in narrow-clawed *Astacus leptodactylus* Eschscholtz, 1823 reared in carp ponds in South Bulgaria (Hubenova et al., 2009). The body length of female individuals (of the one-summer-old, 6–7 months) at first maturation is from 7 to 8.9 cm. The annual cycle of the ovary development in these immature one-summer-old narrow-clawed crayfish (August–September), is as follows: oogonia have been established in the ovary on a mass scale, as well as single previtellogenic oocytes. The primary oocytes have ~one and the same size, their nucleus is oval, centrally located, and fills up almost the whole ooplasm (Fig. 10.97B). The oogonia are observed in nests among the primary oocytes (Fig. 10.97C). During September, oocytes with smaller and bigger sizes have been observed. Their ooplasm is vacuolizated by one or more layers of vesicles (vacuoles) located at the periphery of the oocyte. In January, differentiated distribution of previtellogenic oocytes in the ovary has been established; the bigger ones are located in its periphery, while the smaller ones in the interior part of the ovary (Fig. 10.97E and F). The oocytes are surrounded by well-shaped primary follicle cells. The ovarian wall in the one-summer-old crayfish is narrower, built of one layer of cells, free of muscles (Fig. 10.97F). In 1-year-old crayfish (BW = 12–16 g), from March until August, the changes in the ovary concern mainly oocyte preparation for vitellogenin accumulation in the ooplasm (process of vitellogenesis). Accordingly, the oocytes start enlarging in size (Fig. 10.97G). A characteristic feature of the oocytes during this stage is the differentiation of a lighter zone around the nucleus, which is located centrally in the ooplasm (Fig. 10.97G).

In two-summer-old crayfish (BW = 24.86 g), during September, vitellogenic oocytes with an already formed secondary follicle cell layer are in abundance in the ovary (Fig. 10.97H).

A characteristic feature of all one-summer-old females is that the individuals, after reaching a body weight of 10 g or more in September, have reached maturity during the first year of their life. By forming vitellogenic oocytes, which after copulation with the male during the autumn-winter period, have ovulated.

The earlier maturity of the narrow clawed crayfish, reared in ponds, according to the opinion of the authors, Hubenova et al., has been the result of the rearing conditions effect—mainly the abundance of food in the ponds. Moreover, the higher water temperature in the ponds has a considerable influence on the development of their reproductive system.

Annual cycle of ovary development in sexually mature crayfish. Concerning aouth Bulgaria, the copulation in *A. leptodactylus*, in these rearing conditions, is carried out from the middle of November until the beginning of December. Immediately after copulation, within some days to 2–3 weeks, the ovulation and the fertilization of the ovulated eggs has begun. After the ovulation and until February, the previtellogenic oocytes have approximately the same size; and represent the main fraction in the ovary. Each oocyte is surrounded by one layer of follicle cells with elongated shape and with big, oval nucleus provided with many peripheral nucleoli (Fig. 10.97I). Oogonia nests located centrally in the ovary, close to the early previtellogenic oocytes are present at that time in the ovary as well (Fig. 10.97I). Some nonovulated oocytes and empty follicles, both in process of atresia and desorption, is a characteristic feature of the ovary during February. In March, the most characteristic feature of the ovary is the great number of the previtellogenic oocytes with various sizes; the bigger ones occupy the periphery of the ovary, and the smaller ones its interior parts. Proliferation of oogonia continues. No atretic follicles have been found and no great change in the picture of the ovary during April. Then, the oocytes continue development. The previtellogenic oocytes and oogonia, which represent the new wave of cells and will be differentiated during the following period, are closer to each other and gradually change their shape from elongated (in February–April) to a spherical one (Fig. 10.97G). They attain this spherical shape when they begin to fulfil their typical function, namely, to contact with the blood vessels to transfer the vitellogenin into the oocyte. During June and July, the endogenic synthesis of vitellogenin continues. The follicle cells layer becomes thicker around the oocytes and in August, all of the cells become completely spherical ones, and the light zone around the nucleus decreases. An active accumulation of vitellogenin has been carried out in the maturing oocytes.

The presence of two-layer follicle cells determines the differentiation of the primary vitellogenic oocytes into secondary ones. Gradually, the oocytes enlarge in size, change colour towards yellow-orange, and are located very close to each other.

Ovary development in the narrow-clawed crayfish, *A. leptodactylus*, can be divided into two clearly defined stages: the first one occupies the greater part of the year (from January to August), when vitellogenic or early vitellogenic oocytes are observed in the ovary; and the second is characteristic of the period August–November, when vitellogenin accumulation in the oocyte is observed.

Early embryological studies. During the 19th century, certain aspects of the egg development of some crayfish species: *Astacus astacus* (Linnaeus, 1758), *Austropotamobius pallipes* (Lereboullet, 1858), *A. torrentium* (Schrank, 1803), and *Astacus leptodactylus* (Eschscholtz, 1823) were studied by several authors (Rathke, 1829; Lereboullet, 1862; Reichenbach, 1877, 1886; Huxley, 1881; Schimkewitsch, 1885; Morin, 1886; Fulinsky, 1908). All these works are cited in Celada et al. (1991) and are in German, except that of Lereboullet, which is in French. However, almost all the textbooks described the embryology of genus *Astacus*, especially *A. fluviatilis*. Early studies on the embryonic development of this species were performed by Rathke (1829), who described the structure of the 'exochorion' and the primary yolk pyramids. Lereboullet (1862) studied the same aspect on the same species as well as other species. He described the formation of the blastomeres, whose number increases while their size decreases, and he also pointed out the strong influence of temperature on the rhythm of the embryonic development. Reichenback (1877) established 11 different phases in the egg development of the freshwater crayfish. Until that moment, the different authors considered that the egg segmentation (cleavage) of these crayfishes was of discoidal type, though Morin (1886) discovered the superficial character of this cleavage. Zehnder (1934) restudied and described the embryology of *A. fluviatilis* and *A. torrentium* from prior spawning to the first postembryonic moult. He increased their developmental phases to 16 and established their chronology.

Astacus fluviatilis Fabricius, 1775. *Astacus* is a scavenger and is found in fresh-water ponds, lakes, streams, and swamps all over the world. Therefore, it was available for these previous studies. The egg is generally yolky, of considerable size and is centrolecithal (Fig. 10.100(I)). The amount of yolk influences the segmentation and further development of the embryo. The yolk is enclosed in a layer of cytoplasm lying beneath the egg membrane. The nucleus lies in the centre of the egg, and is enclosed in a thin capsule of cytoplasm (centrolecithal type). It measures about 2.5 mm in diameter. Shortly after fertilization, the nucleus begins to undergo repeated mitotic divisions without any corresponding protoplasmic divisions, to form a syncytial mass (Fig. 10.100(I) B). The daughter nuclei then migrate into the superficial cytoplasm, on the upper part of the egg, converting it into a sort of **blastoderm** above the still undivided yolk (Fig. 10.100(I) C). The cleavage then reaches the 128-cell stage (seventh cleavage). Within this blastoderm, it soon becomes possible to distinguish groups of cells, which are the presumptive areas of, e.g. ectoderm, endoderm and mesoderm (Fig. 10.100(I) L). The central part of the egg consists only of yolk material. Cleavage of this nature in *Astacus* is said to be **superficial** (typically partial or meroblastic). The meroblastic ovum does not segment completely; only part of the ovum, called the germinal disc undergoes segmentation while the remainder of the ovum usually forms the yolk sac. Shiino (1968) described that subsequent to the seventh cleavage, blastomere boundaries, surrounding each nucleus make their appearance abruptly and synchronously over the whole surface in the 512-cell stage (ninth cleavage). Under the influence of these cells, the yolk itself is simultaneously divided into a corresponding number of radiating yolk pyramids called **primary yolk pyramids** (Fig. 10.100(I) C and G). Each has its base in contact with one of the blastoderm cells and its apex pointing towards the centre of the egg. The yolk, somewhat altered by a digestive process beginning about the 16-cell stage (fourth cleavage), becomes granular; the granules are concentrated near the apices of the yolk pyramids (Fig. 10.100(I) H).

The first indication of the future crayfish takes the form of five thickenings on what will become its ventral surface: an anterior pair, the cephalic lobes (Fig. 10.100(III) A) on which the eyes subsequently appear; a more posterior pair, the thoracico-abdominal rudiments; and one, posterior and unpaired, the endodermal disc. On the latter an invagination of the blastoderm takes place, giving rise to a small sac, the archenteron which communicates with the exterior by an aperture, the blastopore.

The cells of the blastoderm are at first uniform, but in most decapods as in *Astacus*, they soon become more columnar for a small area forming a circular patch. The edge of the patch invaginates, as a roughly speaking circular groove, deeper anteriorly than posteriorly. This is the first step of the process of **gastrulation**, which begins at the 1024-cell stage (10th cleavage) (Reichenbach, 1886; Zehnder, 1934). Within this groove, the remainder of the patch forms a kind of central plug, which does not become invaginated until a somewhat later period (Fig. 10.100(I) I). After the invagination of this patch, the remainder of the blastoderm cells forms the ectoderm. In the gastrula of *A. fluviatilis*, the blastopore makes its first appearance as a large circular area, contracts and this causes the epithelial layer to invaginate without losing its cellular continuity. This results in the formation of a distinctly invaginated archenteron. Very soon, the embryo becomes triploblastic, or three-layered. The primordial mesodermal cells 'are not derived directly from either the ectoderm or the endoderm, but have a *quasi*-independent origin, from a mass of cells appear in the neighbourhood of the blastopore (Fig. 10.100(I) J), between the endoderm and the ectoderm. Though they are probably derived from the former' (Huxley, 1880).

By the gradual incorporation of yolk granules, the cells constituting the archenteron grow taller and cylindrical, forming a yolk sac enclosing a central cavity. After they have taken in the whole yolk mass, they finally

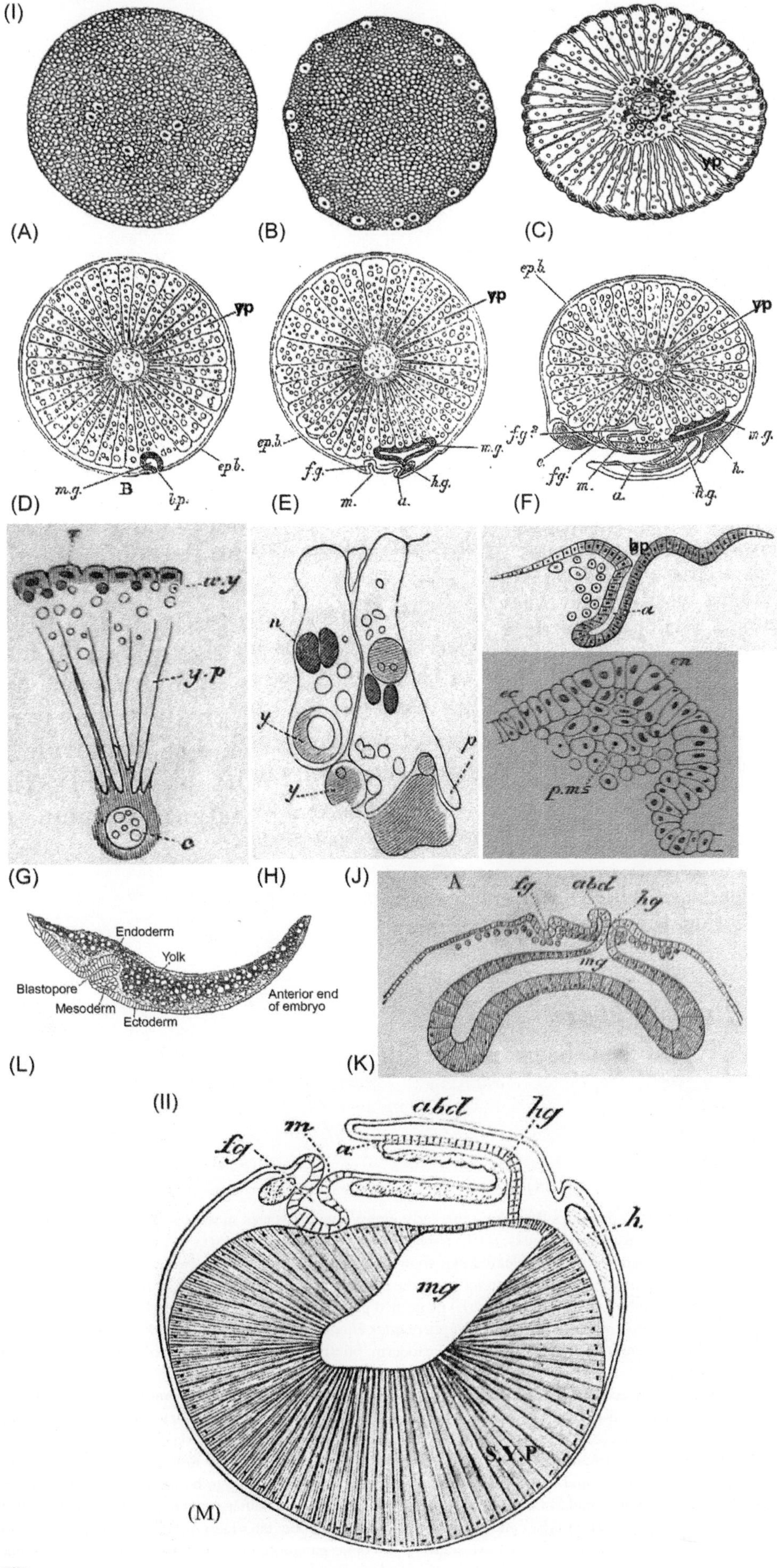

FIG. 10.100, CONT'D.

(Continued)

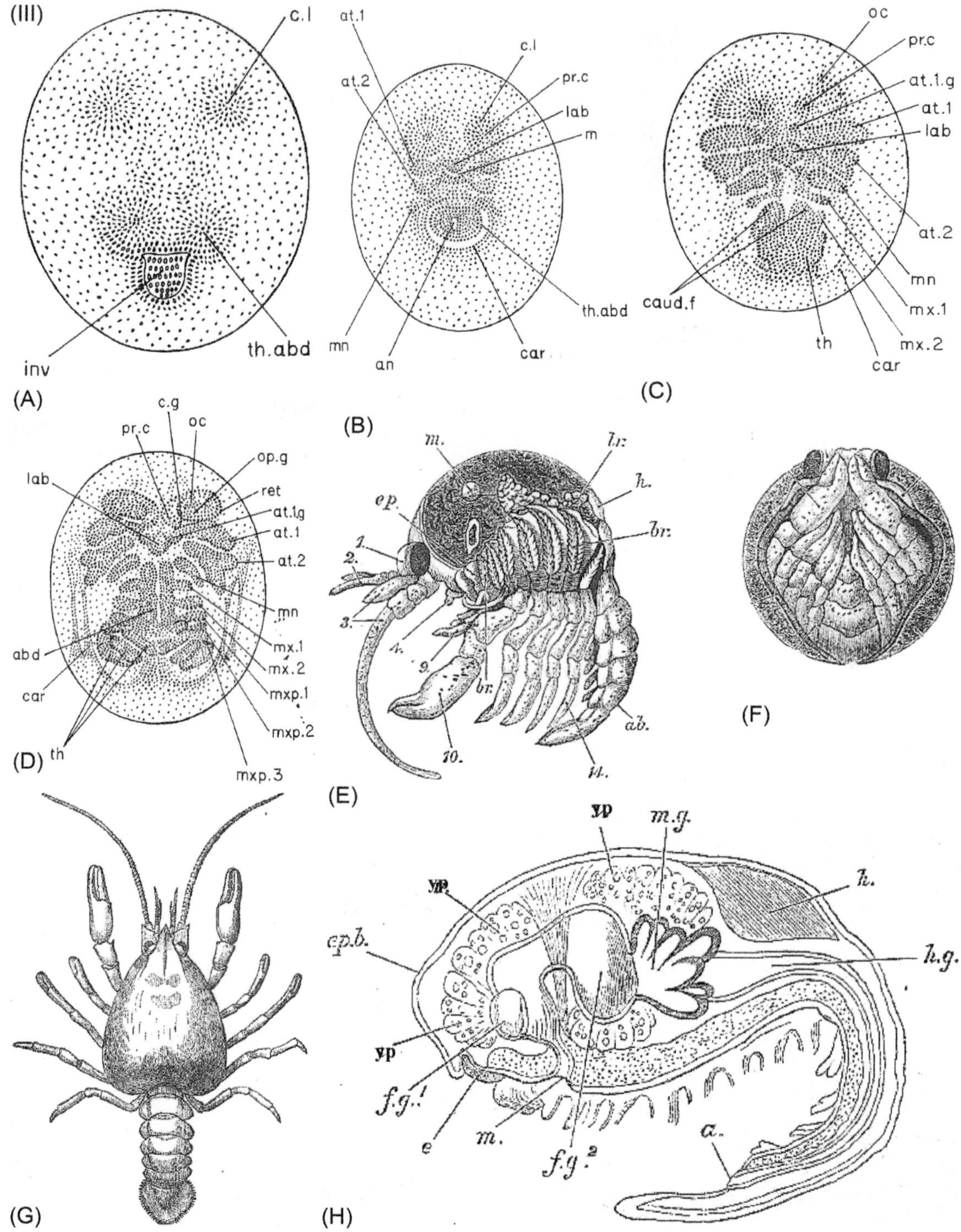

FIG. 10.100, CONT'D (I) Diagrammatic drawings describing the early embryological stages in *Astacus fluviatilis*. (A–C) showing three stages of superficial cleavage and formation of primary yolk pyramids; (D–F) *(from Huxley, 1880)* showing stages of gastrulation and formation of the gut; (G–K and M) *(from Balfour, 1880)*: (G) magnified portion of a cleaved egg showing the yolk pyramids; (H) two endodermal cells showing the amaeba-like absorption of yolk spheres; (I) highly magnified view of the anterior lip of the blastopore showing the origin of the primary mesoblast from the wall of the archenteron; (K) portion of a L.S. of the embryo at nauplius stage; (L) magnified portion of the embryo at an early stage of gastrulation *(from Borradaile)*. *(All after Reichenbach except (D)–(F) partly after Reichenbach and party after Huxley and (K) after Bobretzky.)* (II) (M) L.S. of the embryo at a later stage *(after Bobretzky)*. a=anus and arcenteron, abd=abdomen, bl=blastoderm, bp=blastopore, c=central yolk mass, e=eye, en & ep.b=epiblast (ectoderm), fg=fore gut, fg^1=oesophagus, fg^2=gastric portion, h=heart, hg=hind gut, m=mouth, mg=mid gut (arcenteron), n=nuclei, p=pseudopodial process, p.ms=primary mesoblast, S.Y.P.=secondary yolk pyramids, w.y=white yolk, y=yolk, y.p=yolk pyramids. (III) (A–D) Diagrammatic drawings showing surface views of embryos in successive stages of *Astacus fluviatilis (from Marshall and Williams, 1975 after Reichenbach)*: (A) ventral view of gastrula stage; (B) ventral view of nauplius stage; (C) stage showing the rudiments of maxillae and the forked abdomen; (D) stage showing rudiments of thoracic appendages and the segmented abdomen; (E–G) whole mounts of embryos *(from Huxley, after Rathke)*; (E) lateral view with removed carapace and the appendages were spread out; and (F) ventral view of the young embryo when nearly ready to be hatched. 15×; (G) newly hatched young. 6×. abd=abdomen, an=anus, at.1=rudiment of antennule, at.2=rudiment of antenna, at.1.g.=antennulary ganglion, car=carapace, caud.f.=forked extremity of abdomen, c.g.=cerebral groove which gives rise to the optic ganglion, c.l.=cephalic lobe, lab=labrum, m=mouth, mn=rudiment of mandible, mx.1 and 2=first and second maxilla, mxp.1, 2 3=maxillipeds, oc=eye stalk, op.g.=optic ganglion, pr.c.=protocerebrum, ret=retinulae, th.abd.=thoracico-abdominal rudiments, th=rudiments of thoracic appendages; (H) L.S. of an embryo just hatched *(from Huxley, 1880)*. *Dotted portions* in (F) and (H) (III) represent the nervous system.

become completely typically **secondary yolk pyramids** (Fig. 10.100(II) M) They have very thin protoplasmic membranes as cell boundaries, and their nuclei are situated on the periphery of the yolk mass as in the primary yolk pyramids. The yolk sac represents a transient mesentron, which is later replaced by epithelial endoderm. The archenteron gradually grows forwards; its opening is wide at first, but becomes continuously narrowed and is finally closed. Very shortly after the latter step, slightly in front of the position of the closed blastopore, the portion of the ectoderm invaginates, giving rise to the proctodaeum. The opening of this proctodaeum remains as the definite anus and very soon it communicates with the mesenteron (midgut) (Fig. 10.100(I) K). The stomodaeum is formed during the same stage as the proctodaeum. The ectoderm invaginates in the region between the antennules and antennae, giving rise to the oesophagus and stomach. The interior wall of this stomodaeum grows backward over the mouth opening, covering it as a labrum.

Before long the blastopore closes, converting the archenteron into a blind sac, the enteron, and the thoracico-abdominal rudiments unite with one another, forming an obvious oval elevation (Fig. 10.100(III) B). Three pairs of elevations appear between it and the cephalic lobes. Since the stage with these three pairs of appendages is equivalent to the free-living nauplius larva, it is called the egg-nauplius. The germinal disc of this embryo corresponds to the ventral surface of the nauplius. The crayfish continues its embryonic development within the egg case. The thoracico-abdominal rudiment soon begins to increase rapidly in length, not backwards but forwards, being in fact folded upon the anterior part of the body in much the same way as the adult abdomen during extreme flexion. In the meantime, postmandibular appendages are formed in regular order from before backwards (Fig. 10.100(III) C and D); in addition, the eye stalks appear, as well as the labrum.

The rudiment of the carapace is formed as a fold on each side of the thorax; each gradually extends dorsally, until it meets its fellow of the opposite side and covers the cephalothorax. The embryo then consists of a nearly globular cephalothorax with a small abdomen and a nearly complete set of appendages. All these appendages are tucked in under the cephalothorax and closely packed together within the egg membranes. The embryo is hatched in this condition (Fig. 10.100(III) E and F). It differs from the adult in the form of the cephalothoracic region and in the size of the latter relative to the abdomen. For some time, the hatched crayfish clings to the pleopods of the mother or to the empty egg shells, usually by means of the peculiarly hooked chelae of the first pair of legs.

The newly hatched young (Fig. 10.100(III) G) differs from the adult not only in the two above features, but also the rostrum is short and bent down between the eyes. The sterna of the thorax are relatively wider, and hence there is a greater interval between the bases of the legs than in the adult. The proportion of the limbs to one another and to the body is nearly the same as in the adult.

Up-to-date embryological studies on Astacidae. The recent studies on the embryology of different species of crayfishes and lobsters used improved techniques as stereoscopic microscope and techniques of high resolving power (electron microscopes). These studies presented photographs of the whole mount of the developed eggs at successive stages, such as of *Pacifastacus leniusculus* (Celada et al., 1987) and of *Austropotamobius pallipes* (Celada et al., 1991). In natural habitats of Europe, these two species spawn in autumn and hatching takes place in spring. Fig. 10.101 shows more focused and coloured photographs of some stages of the embryonic development of *A. leptodactylus* (Koca et al., 2013), and another indigenous crayfish species of Europe. As a coldwater species, it spawns in autumn-winter and the eggs hatch in spring/summer. Incubation processes for these species are long. That of *A. leptodactylus* was 4.5 months (Koca et al., 2013). The females carrying sperm tubes (being ovigerous) were obtained from their natural habitat in Egirdir Lake (Isparta, Turkey) in February 2011 and maintained in the laboratory in equipped aquaria. The conditions were controlled similar to the natural habitat (9–15°C). The live eggs were taken from the 'pleon' of the mother with tweezers every day, prepared for observation, and photographed with stereomicroscope. The embryonic stages were determined according to Celada et al. (1987, 1991). The embryological stages of the three species are thus almost identical (compare photos in Fig. 10.101).

Koca et al. (2013) continued their studies to the postembryonic development of *A. leptodactylus* from the juvenile stage I (hatched stage) to juvenile stage III (Fig. 10.102). **Juvenile I** hatched from the egg at the end of the incubation period (134 days). It has sessile eyes, a small rostrum curved ventrally, and the carapace is globose. The telson is large and roundish. The chelipeds are large and equipped with strongly recurved hooks on the propodus and dactylus (Fig. 10.102A). The juvenile crayfish is connected to her mother via the **telson thread**; one end is attached to the posterior hooked spines of its telson, while the other end is connected to the egg shell which is attached to the maternal pleopods (Fig. 10.102B). There is no feeding activity during stage I juvenile, and it moults after 7–8 days, reaching stage II.

Juvenile stage II (Fig. 10.102C and D) has less yolk in its body, and its carapace is not globose as in the preceding stage. The eyes attain short stalks and the rostrum emerges between them. The body surface is covered with many setae, and the telson, in particular, is equipped with long plumose setae. This stage has feeding activity. Its total length was determined as 9–11 mm and its weight as

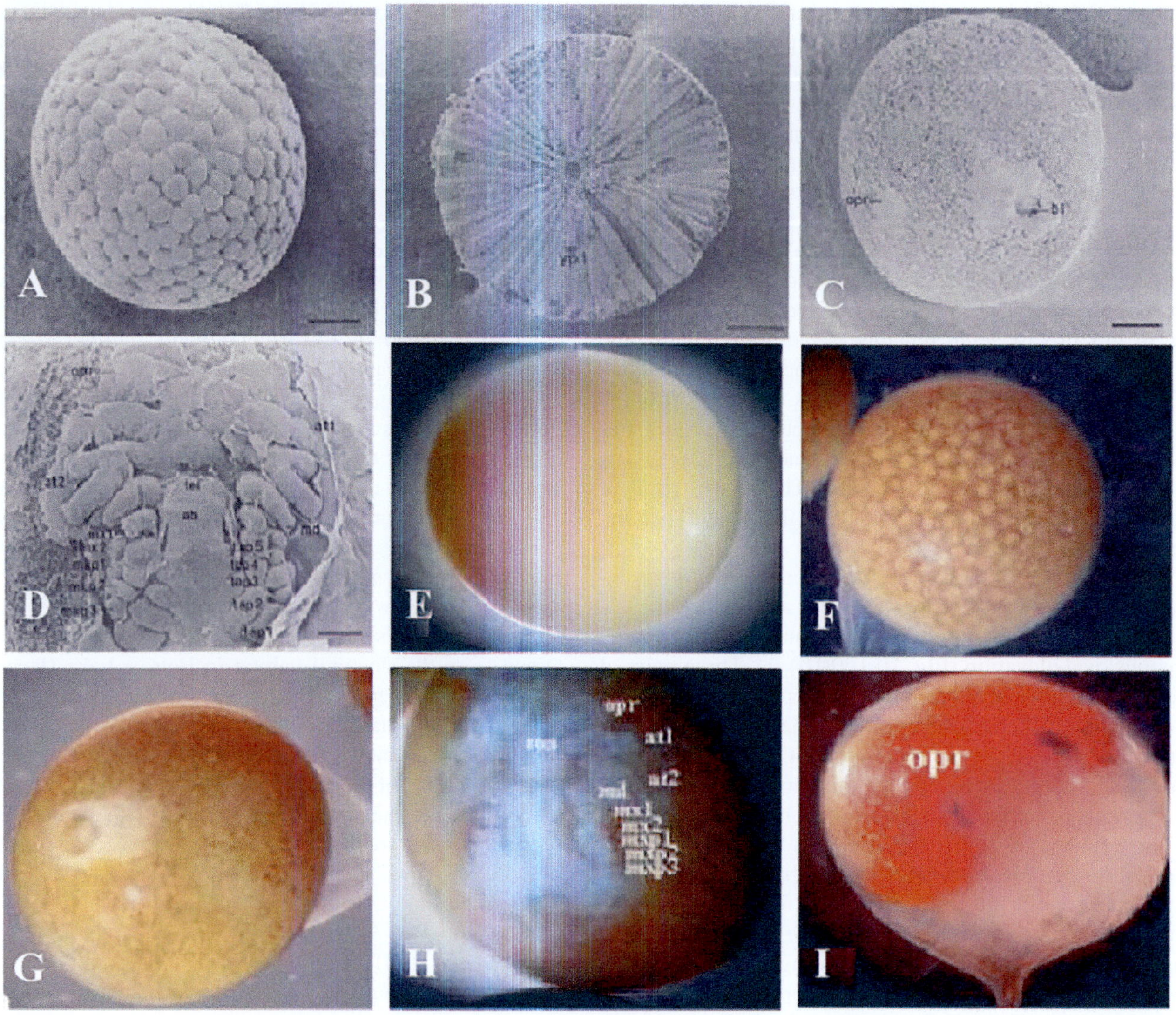

FIG. 10.101 (A–D) SEM photographs showing embryonic stages of *Austropotamobius pallipes (from Celada et al., 1991)*: (A) 8-day-old egg (phase II), formation of the blastoderm; (B) section of 9-day-old egg (phase III) showing primary yolk pyramids; (C) 31-day-old egg (phase VII) showing embryo with existing blastopore (bl). Bar = 100 μm for the three stages; (D) 45-day-old egg (phase XI) showing anlagen of all appendages. Bar = 45 μm; (E–I) stereomicroscopical photos showing embryological stages of *Astacus leptodactylus (from Koca et al., 2013, stages according to Celada et al.)*: (E) new spawn egg; (F) surface view of cleaved egg (phase III); (G) embryo with existing blastopore; (H) embryo with anlagen of pereiopods (phase XII); (I) shortly before hatching.

30–60 mg. The second postembryonic moult occurs after 9–10 days reaching stage III. After this moult, the **juvenile stage III** looks more or less like miniature version of the adult (Fig. 10.102E and F). The uropods become unfolded, and the tail fan is complete. The gonopores become invisible under stereomicroscope. Its total length was determined as 12–14 mm and its weight as 60–90 mg.

It is appropriate to present the postembryonic stages of *Homarus americanus* Milne-Edwads, 1837 (Helluy and Beltz, 1991). These authors are mainly interested in the neural organization and the control of behaviour. Studies on lobsters and other crustaceans have made a significant contribution to their understanding of this field. A deeper knowledge of lobster embryology could provide more insight and understanding of many particular aspects of development such as the differentiation of the nervous system. They obtained egg-bearing female lobsters from Massachusetts State Lobster Hatchery on Martha's Vineyard, Massachusetts. In addition, eggs detached from the mother's abdomen were provided by the New England Aquarium in Boston, where lobsters were reared under equipped and controlled conditions. In New England waters, egg development spans about 10 months, from spawning in July or August to hatching the following May or June. Studies on the embryonic development of this lobster, *Homarus americanus* (Helluy and Beltz)

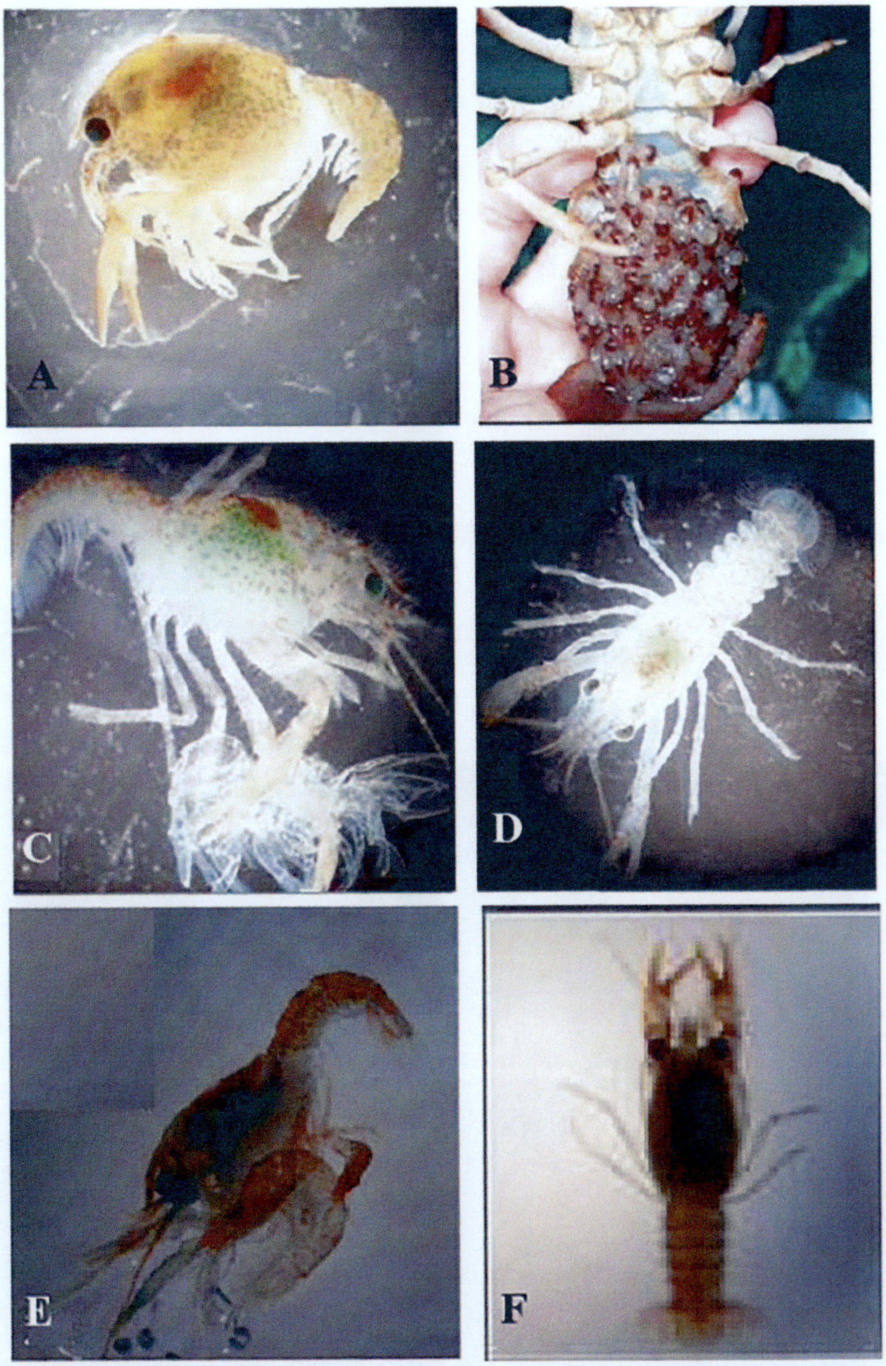

FIG. 10.102 Stereomicroscopical photos showing postembryonic development of *A. leptodactylus (from Koca et al. 2013)*: (A and B) juvenile stage I; (C and D) juvenile stage II; (E) juvenile stage III just after moulting; (F) dorsal view of juvenile stage III.

show no observative difference than that described above (Fig. 10.103). However, the perihatching development differs completely. *Homarus* hatches as a mature metanauplius (prelarva, prezoea) (Fig. 10.103F) (the identification of the larvae according to Shiino, 1988) that moults rapidly into the first larval stage. It is necessary for the present work to describe different examples of embryonic and postembryonic development in Astacidae.

The Japanese Cambaroides japonicus. Embryonic and postembryonic developmental studies provide characteristics that constitute important, rather essential factors to identify a foreign and/or unknown species, systematic position, phylogeny, and evolution of adaptations of different aspects such as maternal care of freshwater astacids. This was the aim of Scholtz and Kawai (2002) to describe aspects of embryonic and postembryonic

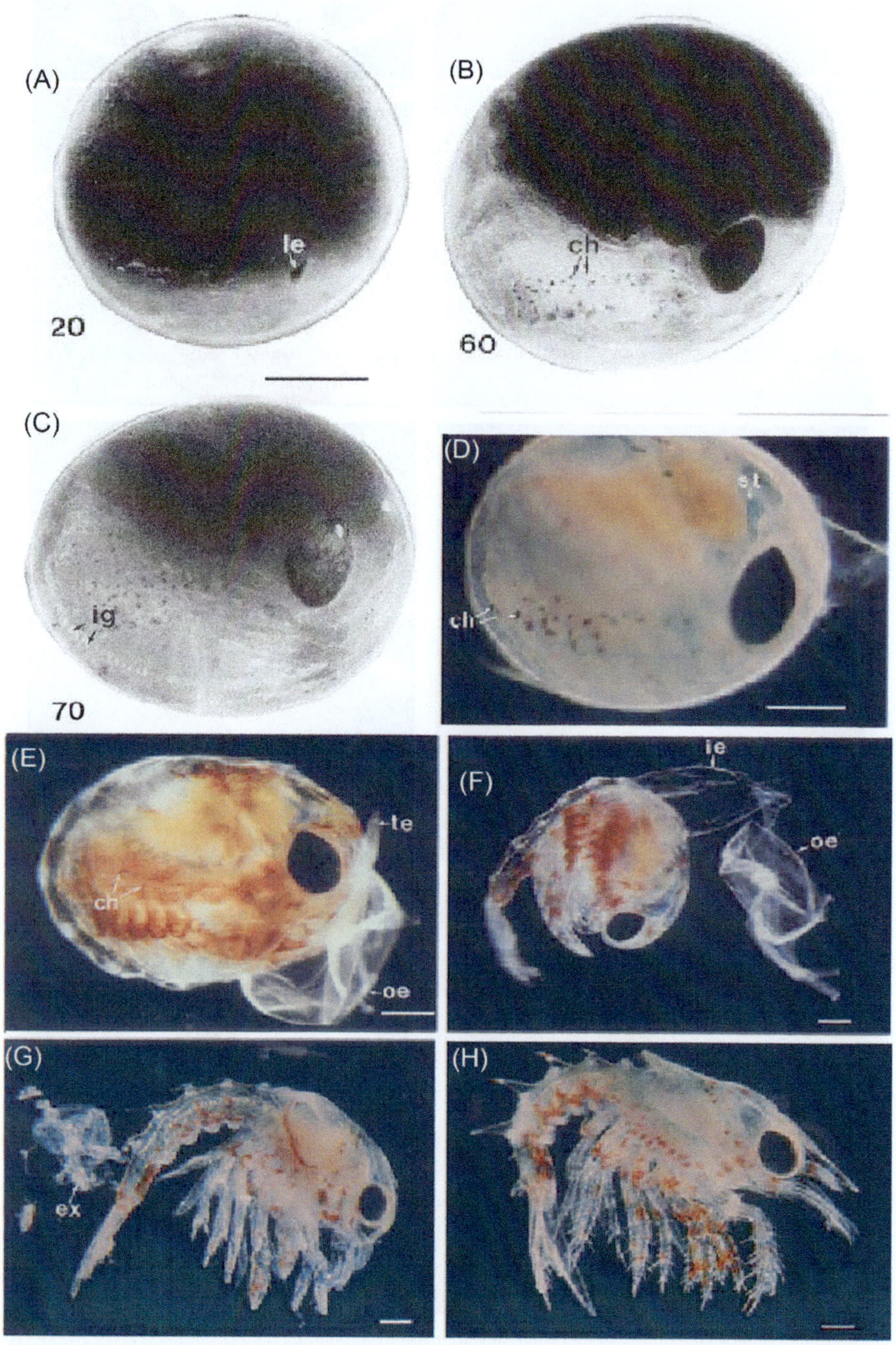

FIG. 10.103 Unfixed, intact eggs of *Homarus americanus* at successive embryonic development *(from Helluy and Beltz, 1991)*. The figures in the lower left corners refer to the percentage of development. All the stereomicroscopical photographs show the dorsal side at the top and the head on the right; (A) the eye pigment is visible in the lateral eyes (le); (B) the red chromatophores (ch) already present in this stage; (C) the intestinal granules (ig) are particularly clear; (D) embryo just prior to hatching (100%, blue embryo), note the blue tinge of the haemolymph, the blue stomach (st) and the red chromatophores (ch) in which the pigment is still concentrated; (E) hatching stage: the outer egg envelope (oe) has burst and the telson (te) is piercing the inner egg envelope, the red pigment has spread in the star-shaped ch; (F) the metanauplius (prelarva, prezoea) is now free of both outer and inner (ie) egg envelopes; (G) early first larval stage (first zoea), the exuvia (ex) of the metanauplius has been sloughed; (H) mature first larval stage, rostrum, abdominal spines, and other acuminate structures are now erect. Scale bars = 500 μm.

development of the Japanese freshwater crayfish *Cambaroides japonicus* (de Haan, 1841). Genus *Cambaroides*, East Asian freshwater crayfish, shows some controversy concerning its systematics and phylogenetic relationships between its species and between other crayfish. Little is known about its biology; even its morphological features have not been described adequately. The authors collected females of *C. japonicus* with eggs and young specimens (23.4–28.6 mm postorbital carapace length) from Otaru City Hokkaido, Japan during May 1997. All juvenile specimens were deposited at the Humboldt University of Berlin. As with almost all other crayfish species, *C. japonicus* undergoes direct development and the eggs are yolky. Only a few aspects of its embryonic development, which are relevant in the context of phylogeny and evolution of freshwater crayfish, were described. For example, the characteristic ventrally folded caudal papilla appears in the general appearance of the embryo (the surface view of the 'germ band'; Fig. 10.104A). The growth zone (Fig. 10.104B) consists of about 40 ectoteloblasts, arranged in a ring surrounding the caudal papilla anterior to the telson primordium. The eight mesoteloblasts form an inner ring showing a specific arrangement of a pair of median mesoblasts and three pairs of lateral mesoteloblasts. These findings, the shape of the germ band, and the characteristics of the growth zone agree with those described for other freshwater crayfish representatives—in particular, the number and arrangement of the ectoteloblasts, which fits well as described so far for representatives of the Astacidae. In contrast, the eight mesoteloblasts show the conserved pattern found in all malacostracans.

The postembryonic development of *C. japonicus* also shows some other characteristics in common with that described for representatives of the Astacidae. This development comprises three juvenile stages. Except for the size, which is specific for each species, each juvenile is almost the same as that described above. However, Scholtz and Kawai (2002) described, with much interest, some characteristics as follows.

1. the shape of the telson of the juvenile stage I. That of *C. japonicus* is large, roundish, and equipped wth curved hooks around its free edge (Fig. 10.104D). These hooks seem to be setal precursors. There is a small notch in the centre of the posterior telson margin. After hatching, this juvenile is connected to its mother via the telson thread (Fig. 10.104C). One end of this thread is attached to the posterior hooked spines of the telson of this juvenile. The other end is connected to the egg shell, which is attached to the maternal pleopods. The authors investigated that the telson thread is probably formed by the embryonic cuticle, which is shed during hatching. They could not determine for how long the stage I juveniles are connected to the mother with the telson thread. They observed that this thread is lost during this stage and the juveniles exclusively use the hooked chelipeds to cling to the maternal pleopods.
2. The occurrence of plumose telson setae in juvenile stage II. As in stage I, this stage II (Fig. 10.104E) appears the same as described above, setae occur on the body surface. The telson, in particular, is equipped with long plumose setae (Fig. 10.104F). First pleopods and uropods are still absent and the recurved cheliped hooks are no longer present (Fig. 10.104G). The loss of these hooks indicates an early independence of the juvenile from the mother.
3. The early appearance of a spermatheca (*annulus ventralis*) is a feature of the juvenile stage III. In this juvenile, the first pleopods are present as small anlagen in both sexes. In addition, this juvenile is characterized by being dimorphic. The sternites of the seventh and eighth of the females are wider, and bear the putative anlagen of the annulus ventralis (Fig. 10.104J). The gonopores are visible under the fluorescent microscope but not in living animals.

The spermatheca as an apomorphic morphological feature is shared between *Cambaroides* and the American representatives of the Cambaridae. However, all the above characters indicate that, like those of Astacidae, *C. japonicus* juveniles become mobile and independent from their mother after the first posembryonic moult. These characters differ from the corresponding characters of American Cambaridae.

It remains to mention that the findings gained from the work of Scholtz and Kawai (2002) gave them the opportunity to review the evolution of the extended maternal care in Cambaridae in specific and in Astacidae in general. That concerning the *C. japonicus*, the so-called **anal thread** which is formed by the shed cuticle of the hind gut connecting the juvenile to the mother via the cuticle of the previous juvenile stage. The latter is attached to the maternal pleopods with the hooks of the chelipeds. This case is found in cambarids but not in *C. japonicus* or in any representative of Astacidae.

Parthenogenetic Marmorkrebs Procambarus fallax. The subject of the heading, which refers to the first parthenogenetic Decapod Crustacean Marmorkrebs (Marbeled Crayfish) (Vogt et al., 2004), was quite intriguing. It was reported that this crayfish (Fig. 10.105A) derives its name from its beautiful green, brown, and white typical colouration that resembles marble (Fig. 10.105B). The background colour of the cuticle and the colouration of the marbled spots can vary and seem to depend on rearing conditions and food (Vogt et al., 2004). It was observed and/or discovered in the mid-1990s in the German aquarium trade (Scholtz et al., 2003; Vogt et al., 2004). It was unidentified and of uncertain geographical origin. At the beginning of

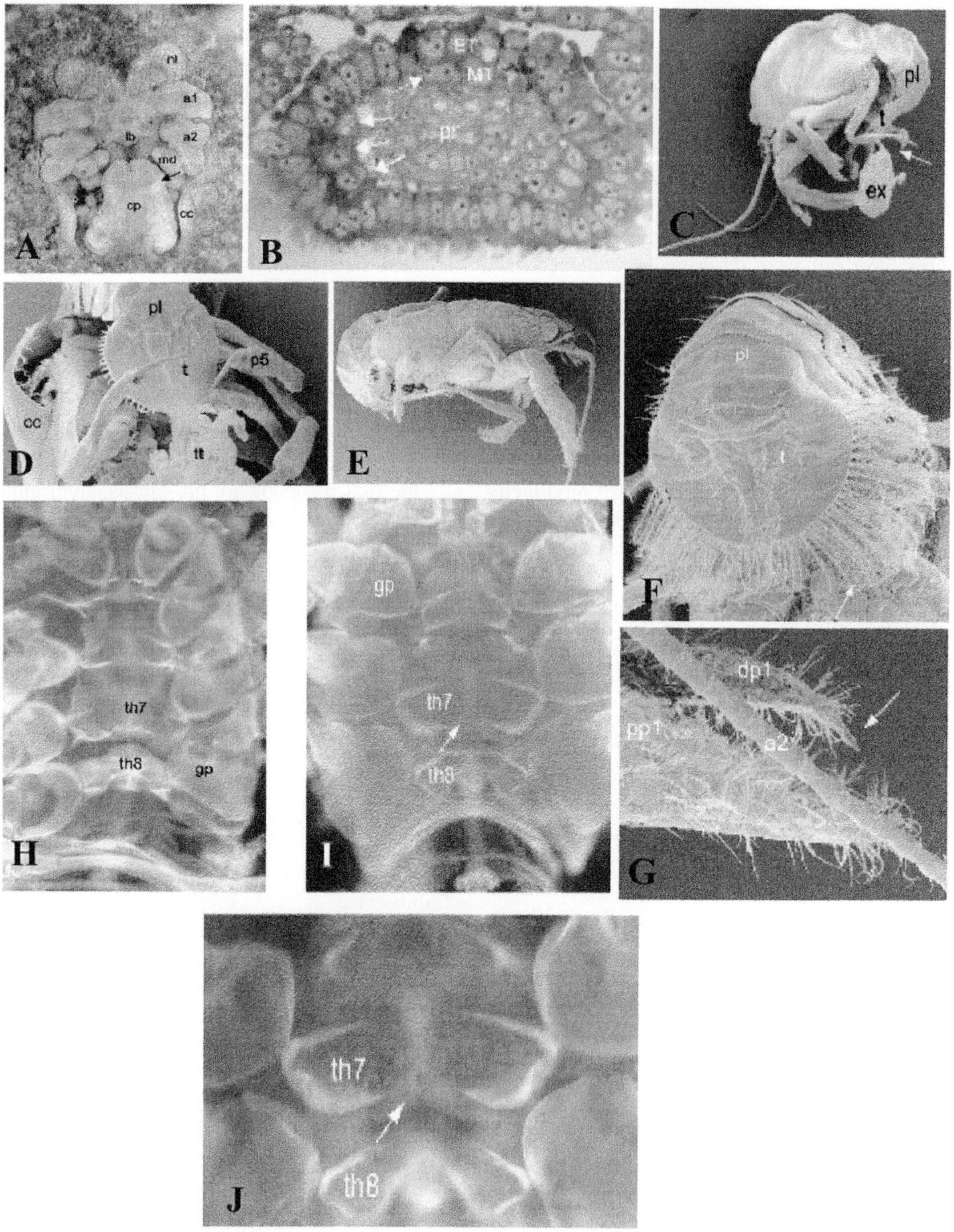

FIG. 10.104 Photographs of the embryo of *Cambaroides japonicus (Scholtz and Kawai, 2002)*: (A) whole mount of the embryo stained by bisbenzimide fluorescent dye showing surface view of the embryonic germ band in the nauplius stage. The anterior-most part is formed by the headlobes (hl) which are the anlagen of the eyes and the lateral brain parts. a1 = first antenna, a2 = second antenna, md = mandibles, cp = the ventrally folded caudal papilla contains the superficial ring of ectoblasts *(arrow)* in its posterior region; (B) T.S. through the caudal papilla of (A), stained with toluidine blue. ET = outer ring of ~40 ectoblasts, MT = the inner ring of four pairs of mesoteloblasts, arranged in a specific pattern *(arrows)* surrounding the pr = proctodaeum, cc = carapace, lb = labrum; (C–G) SEM views of the juvenile stages: (C) the newly hatched juvenile I which is connected to the mother via a telson thread *(arrow)* which is connected to the telson (t) of the hatchling and which is mainly formed by the shed embryonic cuticle (ex), no setae occur; (D) The telson of the stage juvenile I with its characteristic round shape and the hooks around its margin. The posterior end of the telson is connected to the telson thread (tt). The carapace (cc) margin is equiped with setae; (E) SEM overall view of juvenile stage II showing the advanced general shape of this stage and the occurrence of setae all over the body; (F) the telson (t) at the end of the pleon (pl) which is still round and equipped with long plumose setae around its margin *(arrow)*, no uropods are yet present; (G) the tip of the cheliped (first pereiopod) has the spines *(arrow)* of the dactyl (dp 1) and the propodus (pp 1) are no longer recurved at this stage. a2 = second antenna; (H) the sternal region of a male stage 3 juvenile (fluorescent staining), the gonopore (gp) is found on the coxa of the last thoracopod, compare the shape of the sternites of the seventh and eighth thoracic segments (th7, th8) with those of the female of the same stage (I); (I) the sternal region of a female stage 3 juvenile (fluorescent staining), the gonopore (gp) is found on the coxa of the sixth thoracopod, notice that the sternites of the seventh and eighth thoracic segments (th7, th8) are wider and shorter than in the male; (J) close up of the last two thoracic sternites (th7, th8) of the female, the *arrow* points to the posterior part of the seventh thoracic sternite (th7) in which the annulus ventralis is formed, the slight invagination and the density of cells in this region resembles what has been described in the American cambarids (Andrews, 1906).

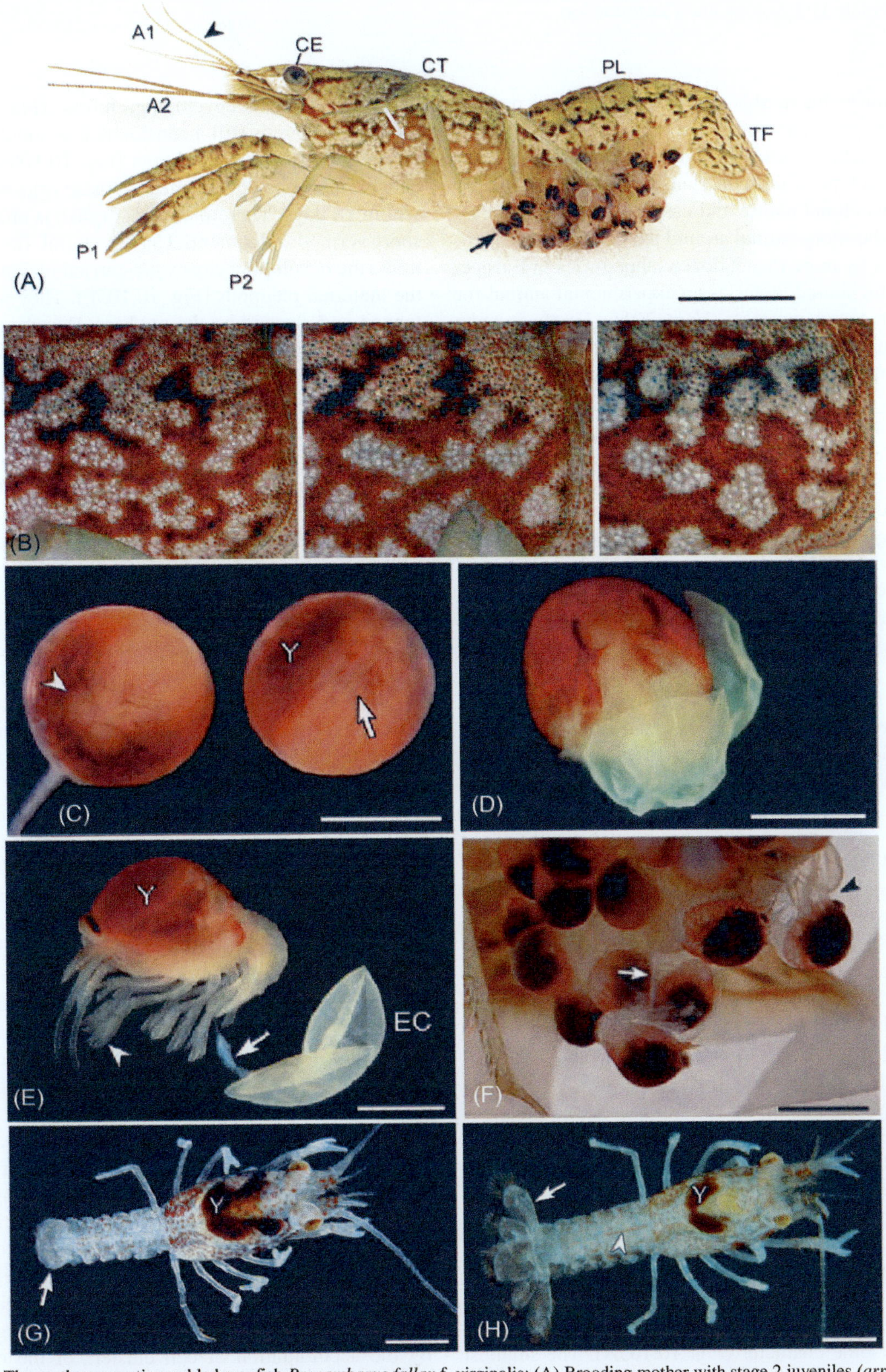

FIG. 10.105 The parthenogenetic marbled crayfish *Procambarus fallax* f. virginalis: (A) Brooding mother with stage 2 juveniles *(arrow)* under pleon (pl) showing eponymous marmoration. A1=first antenna with aesthetasces bearing outer flagellum in upright position *(arrowhead)*, A2=second antenna, CE=compound eye, CT=cephalothorax, P1=first pereiopod, P2=second pereiopod, TF=tail fan *(from Vogt and Tolley, 2004)*. Scale bar=1 cm; (B) variation of marmoration pattern between three genetically identical and communally reared batchmates. Shown in the postero-lateral part of the cephalothorax indicated by an *arrowhead (from Vogt, 2011)*; (C–H) except (F) microphotographs *(from Vogt et al., 2004)*. Scale bar=1 mm; (C) embryonic stages at 85% development, displaying anlagen of eyes *(arrowhead)* and appendages *(arrow)*. y=yolk; (D) hatching in vitro, showing emerging hatchiling in frontal view and split egg case; (E) first postembryonic life stage (stage I juvenile) linked to the egg case (EC) by a telson thread *(arrow)*, first pereiopod with hooks *(arrowhead)*; (F) macrophotograph of cluster of hatchlings underneath the pleon showing one bending upwards to attach with its chelipeds while securing by the telson thread *(arrow)*, *arrowhead* denotes hatchling successfully hooked in an egg case *(from Vogt, 2008b)*; (G) first juvenile life stage with well-developed external sense organs (stage 2 juvenile), recognized by an incomplete tail fan *(arrow)*; (H) first free-lancing and feeding stage (stage 3 juvenile) with yolk remnants, characterized by a complete tail fan with uropods *(arrow)*, *arrowhead* denotes food in intestine.

the 21st century, this marble crayfish became very popular among hobby aquarists and researchers, since it is a suitable material adding much knowledge concerning parthenogenesis as a type of reproduction and phylogeny. Mainly because of its clonal nature and easy culture, it was introduced as a laboratory animal around the year 2000. At present, it is used by more than a dozen laboratories in Europe, Japan, and the United States as an experimental animal for a variety of research purposes (Vogt, 2011). Firstly, it had a German name, then an English name (Scholtz et al., 2003), but no scientific name. Their morphological and molecular analyses, under laboratory conditions at the Heidelberg University of Berlin, indicate that it belongs to the North American Cambaridae, and more specifically to the large and diverse genus *Procambarus* (Scholtz et al., 2003). Of the cambarid species included in their molecular relationship, it was closest to the similar-looking *Procambarus fallax*. It is only recently that a precise determination based on detailed molecular analysis has shown a high degree of DNA sequence correspondence between *Procambarus fallax* and Marmorkrebs (Martin et al., 2010a). Therefore, it was identified as the parthenogenetic form of the 'slough' crayfish *Procambarus fallax* (Hagen, 1870) and the tentative scientific name *Procambarus fallax* f. virginalis was proposed. It is native to Georgia and Florida, United States (Martin et al., 2010a). This identification was again confirmed by conducting a molecular analysis based on two mitochondrial genes, CO I (cytochrome oxidase subunit I gene) and 12 S (ribosomal RNA gene) (Martin et al., 2010b). These two genes are well-established markers for identifying species. They were used to check the identity of the wild cambarid collected in Saxony is the above Marmorkrebs in the laboratory culture at the Humboldt University of Berlin. Chucholl and Pfeiffer (2010) studied ecologically the populations of this marbled crayfish in different regions of Germany, in specific, and in central Europe in general. They reported that, In addition to the recently suggested hypothesis that Marmorkrebs might be temperature limited on most parts of Europe, it is presumably able to colonize summer-warm, lentic habitats in most parts of central Europe.

The molecular analysis is required to identify the phylogeny of a species, and embryogenesis is equally required, especially postembryonic development. Vogt et al. (2004) investigated the embryological developmet of this marbled crayfish. The latter, when berried, the underside of the abdomen of the female is full of eggs, which gives the individual a cumbersome character (Fig. 10.105A). The smallest berried female investigated had a total length (tip of rostrum to end of telson) of 3.9 cm and a weight of 1.34 g. The eggs, brooded on the pleopods, had an average diameter of ~1.5 mm. Like other freshwater crayfish, marbled crayfish develops directly within the eggs. At the beginning of embryonic development, the eggs appear completely filled with yolk, but later, more and more space of the egg is occupied by the developing embryo (Fig. 10.105C). In the larvae, there is still a considerable amount of yolk present in the dorsal cephalothorax (Fig. 10.105D and F). The entire embryonic development of these eggs was completed within 18 days. Therefore, this crayfish is characterized by short-germ development. Under natural (in vivo) conditions, the developing eggs were carried in dense clusters on the maternal pleopods (Fig. 10.105F). They were regularly fanned and cleaned by the mothers. Decaying eggs, which were orange to yellow instead of dark grey to brown, were mostly removed by the mother, but occasionally they remained for several days in the clutch (Vogt, 2008b).

Alwes and Scholtz (2006) studied, in detail, the embryonic development of this parthenogenetic marbled crayfish until hatching. The specimens were obtained and worked on in the laboratory of the Heidelberg University of Berlin. Freshly laid eggs were taken from the pleon of the mother with tweezers and prepared for observation under a dissecting microscope. The egg membrane itself is enclosed by a strong envelope, composed of an embryonic endochorion and a maternal exochorion. The egg has a strictly spherical shape of about 1300–1500 μm in diameter. The colour of the yolk varies between dark brown and yellowish brown (Fig. 10.106). The similarity of the embryological stages of the marbled crayfish with those of *A. leptodactylus* (Fig. 10.101E–I) is well noticed.

Alwes and Scholtz (2006) moreover studied and described the morphogenesis in more detail with emphasis on limb development, segment formation, and cell division patterns of the ectoderm using improved stains and microscopses. It is difficult to observe the 'germ disc' (blastoderm) in live eggs; however, the first mitotic divisions take place in the centre of the egg without external appearance (Fig. 10.106A). The first appearance of the nuclei as whitish spots occurs at the 128-cell stage (Fig. 10.107B). Due to the synchronous mitotic divisions in all nuclei, the blastoderm cells uniformly cover the egg surface (Fig. 10.107A). At the beginning of gastrulation, as soon as the blastopore has formed, the early caudal papilla anlage appears as a white cell cluster adjacent to this blastopore. The latter looks like a horseshoe furrow with the opening to the posterior. The furrow deepens more and more until it becomes an orbital furrow of about 350 μm in diameter, surrounding the transparent mesendodermal tissue (Fig. 10.106B). Anterior to the caudal papilla anlage, the beginning of the V-shaped germ disc (beginning of nauplius stage) is formed by the accumulation of cell clusters of the optical lobes (Fig. 10.107C). The preparations of the fixed embryos reveal no regular ectodermal cell division pattern during the formation of the V-shaped germ disc. Fig. 10.107D shows the nauplius stage with the blastopore sill open and appears as a dark ovoid patch (Fig.10.106C). During this stage, the formation of the buds of the naupliar appendages results in a well-defined protuberances, more or less arranged in a

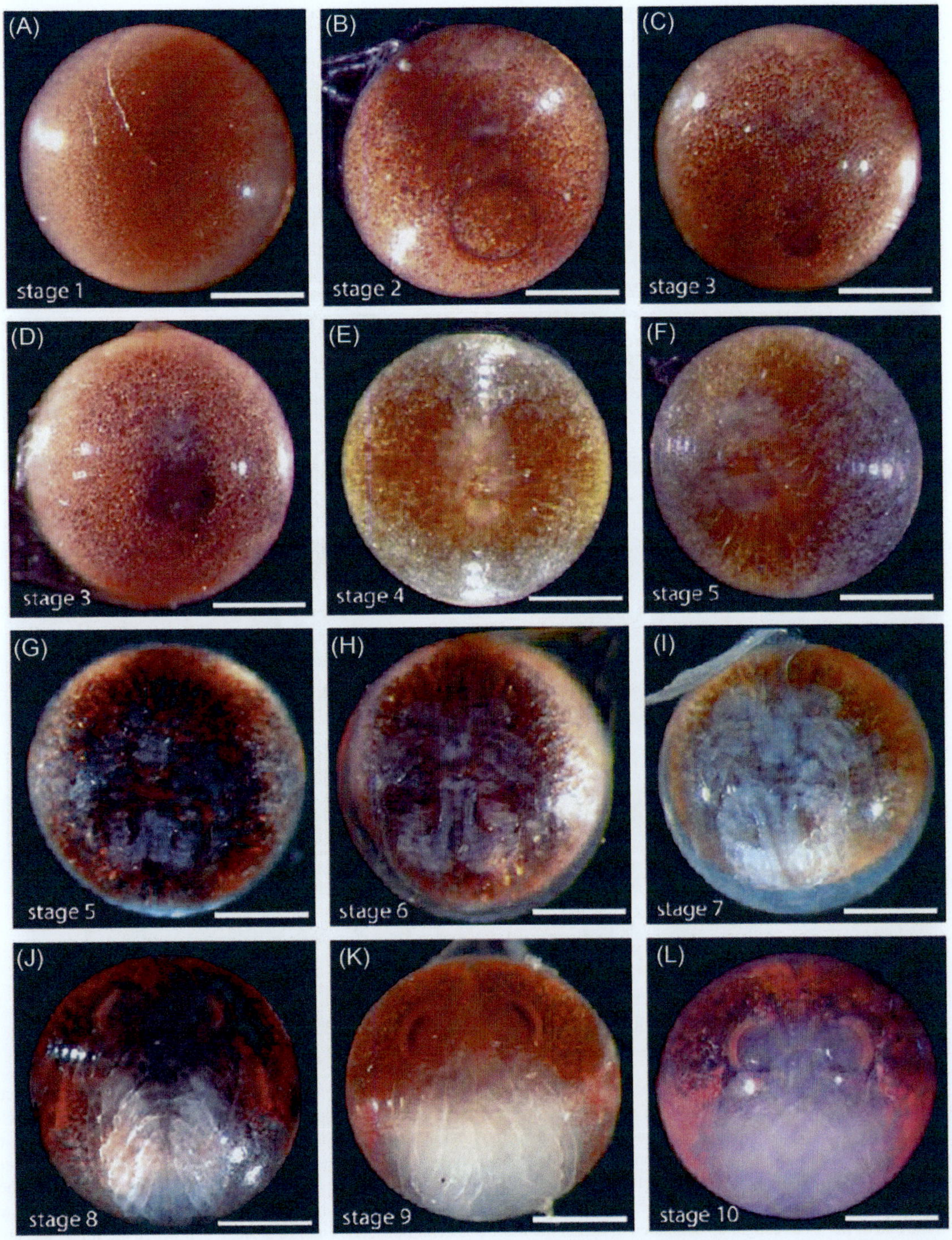

FIG. 10.106 Developmental stages of unfixed embryos exemplified the characteristic stages. Scale bar = 500 μm *(from Alwes and Scholtz, 2006)*: (A) Stage 1 showing the egg without external signs of cleavage; (B) stage 2 the beginning of gastrulation results in the formation of a circular blastopore of about 300 μm in diameter; (C) stage 3 in which the caudal papilla adjacent to the closing blastopore and the two optical lobes are visible as **whitish regions**. The closing blastopore appears as a contracting **dark patch** posterior to the germ disc; (D) embryo at the end of stage 3, the recruitment of cell material of the naupliar anlagen is advanced and the blastopore is now closed, beneath the posterior half of the germ, there structuring yolk appears as a **dark diffuse cloud**; (E) embryo at the end of stage 4, the dark patch has disappeared and the yolk is recognized, the caudal papilla is bent ventrally from now on; (F) stage 5, the caudal papilla is flexed and its tip reaches the margin of the second maxillary buds; (G) embryo at the beginning of stage 6. The tip of the caudal papilla reaches the segment of the second maxillae; (H) stage 6 in which the tip of the caudal papilla now reaches the posterior margin of the mandibles; (I) stage 7 the thoracopods are arranged like a basket; (J) stage 8 the pereiopods cover the caudal papilla, but do not meet at its midline. The retinal pigmentation begins as **yellowish stripes** below the outer eye margin. The **orange** chromatophores of the carapace appear; (K) stage 9 in which the tips of the chelae of the first pereiopods nearly overlap the rostrum between the eyes. The advanced retinal pigmentation becomes distinct **brown** retinal layers; (L) stage 10 is the prehatching.

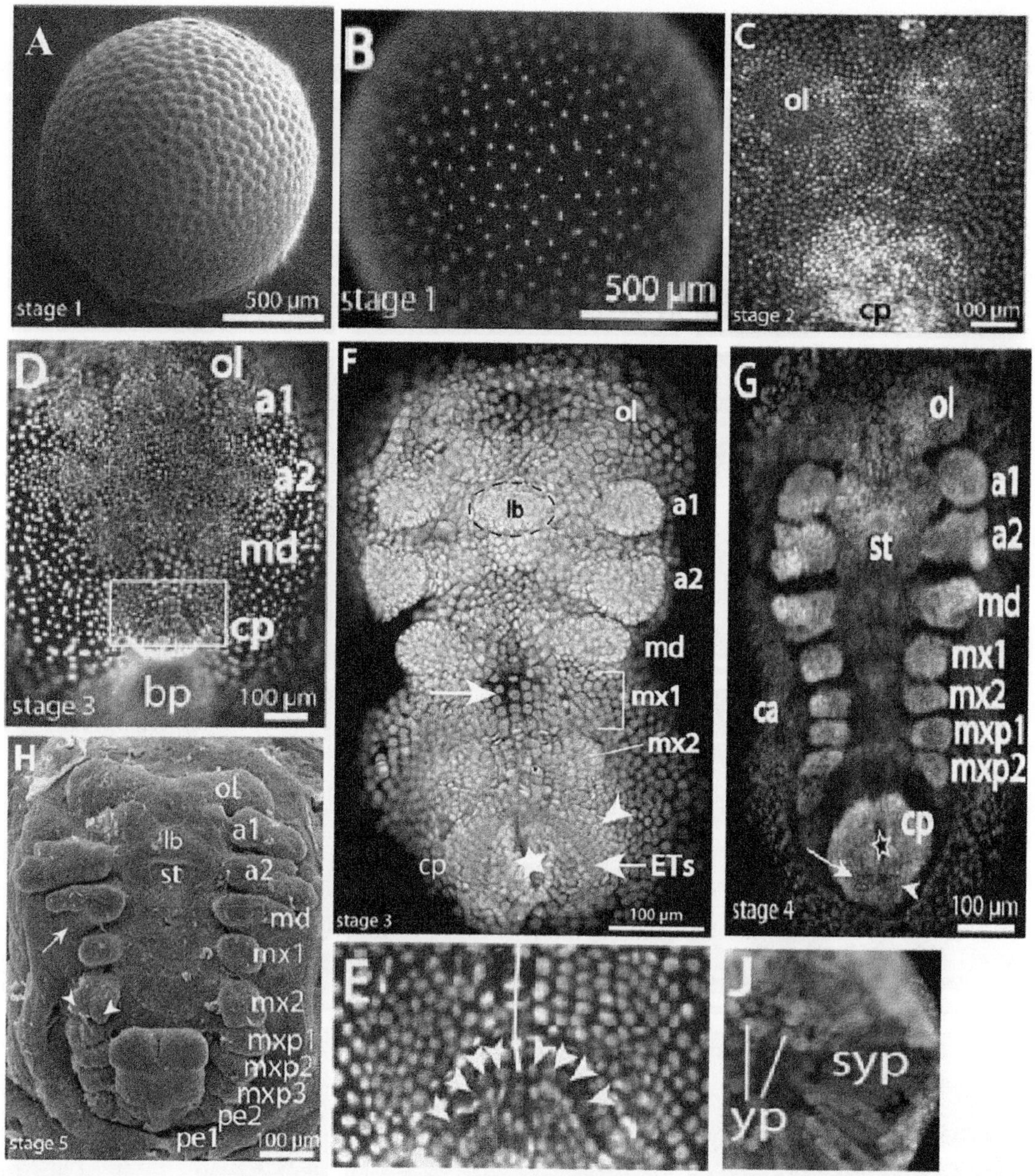

FIG. 10.107 SEM photographs of the blastoderm stages *(from Alwes and Scholtz, 2006)*: (A) blastoderm cells uniformly cover the surface of the egg at the end of stage 1; (B) blastoderm at the same stage of A stained with fluorescent nuclei dye Hoechst. The nuclei appear as small **spots** within the comparatively large blastoderm cells; (C) early **V-shaped** naupliar anlagen without regular cell division pattern at stage 2, stained with the same dye; (D) early formation of naupliar segments at stage 3, the blastopore is still open; (E) details of the growth zone anterior to the caudal papilla (*white rectangle* in D), *white arrowheads* point to the first differentiated ectoteloblasts around the ventral midline *(white line)*; (F) beginning of the formation of the postnaupliar segments at the end of stage 3; (G) advanced teloblastic ectoderm formation of an embryo at a stage 4, the buds of mxp 1 and 2 are visible as round buds, the formation of the bud of mx 1 has caught up compared to that of the more posterior buds; (H) embryo at stage 5 in which the limb buds up to the second pereiopod appear (pe1, pe2 = first and second pereiopods, respectively), the first maxillae still uniramous, the limb buds of the second maxillae and first two maxillipeds start to split into endopodite and exopodite (*arrowheads* for right second maxillae), *arrow* points to the basal socket of the mandible; (J) view of the inside of an egg at stage 5, showing formed secondary yolk pyramids (syp), except in the centre part where pellet-shaped yolk (py) is still left.

circle (Fig. 10.107F). A characteristic feature of this stage is that the buds of the second maxillae are visible anterior to the caudal papilla. The first maxillary segment shows no buds at this stage. Near the midline of the first maxillary segment, the ectodermal cells slightly drift aside. This means that the cephalic body part of the embryo from the optic lobes to the mandible anlagen is simultaneously generated. The following postnaupliar segments, leading to the eight thoracomeres and the six pleomeres, are sequentially formed by the activity of ectoteloblasts and mesoteloblasts (Fig. 10.107E) that are located within superimposed rings in the caudal papilla (see later Fig.10.108E). The caudal

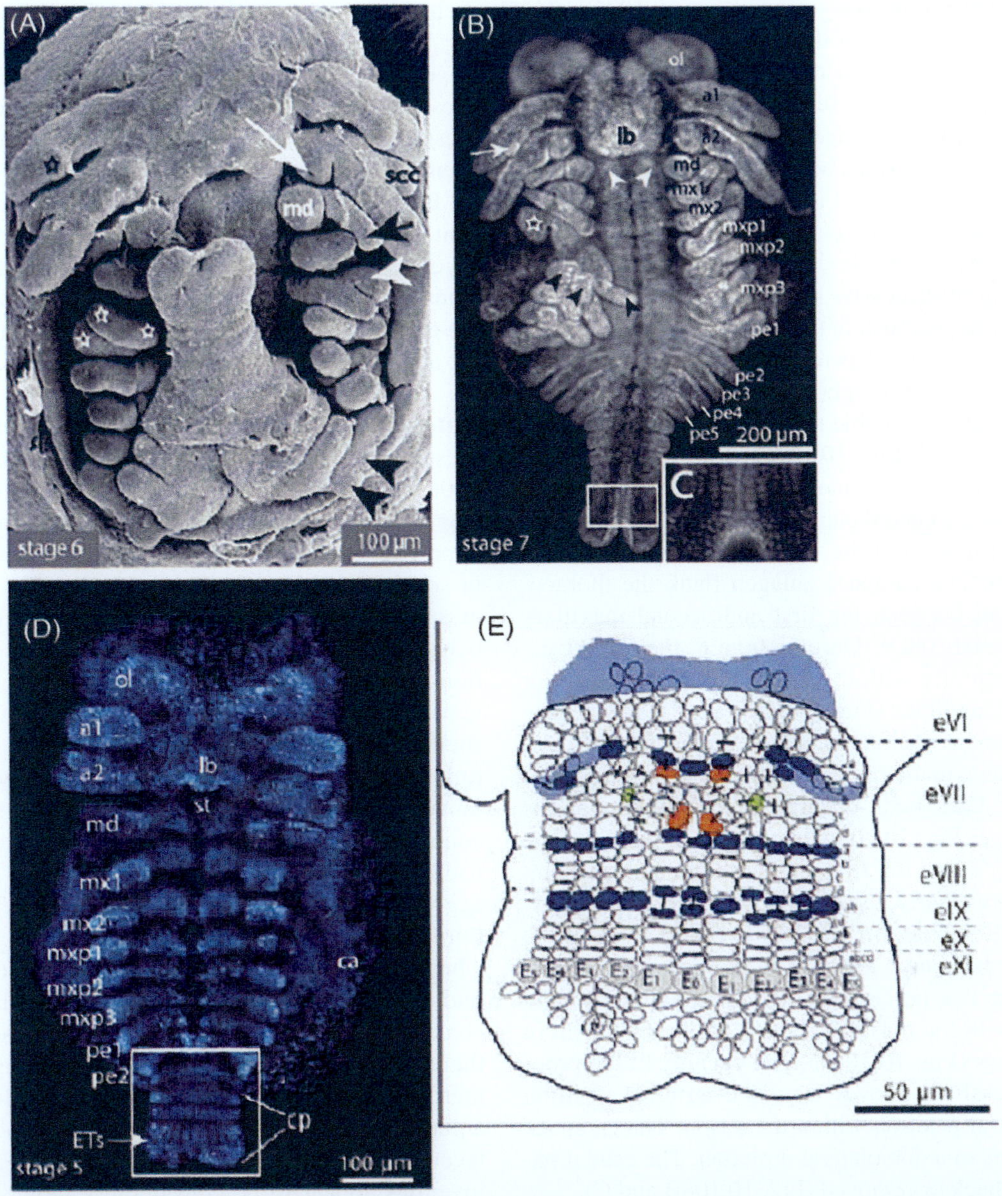

FIG. 10.108 SEM photographs *(from Alwes and Scholtz, 2006)*: (A) shows embryo at stage 6, **black star** points to the bud of the inner branch of the first antenna, **white arrow** points to the opening of the antennal gland in the most proximal joint of the antenna. **scc** = small bud of anntenal scaphocerite, **white arrowhead** points to scaphognathite anlage, **black arrowheads** point to the protrusions of the first pereiopod giving rise to the podobranchial gills, **stars** mark the three buds of mxps; (B) embryo at the beginning of stage 7 with the caudal papilla folded back in which the ectodermal segment formation is nearly finished (Hoechst stained), the first to third maxillipeds of the right side are folded to show a view of the three-lobed first maxilliped with external gill anlage, the exopodite in the midline and the endopodite, here pointing to the midline *(black arrowheads)*, the *white arrowheads* point to buds of paragnaths, *white arrow points* to the inner branch of first antennae, white star = scaphognathite; (C) detail of ventral view of caudal papilla (*white rectangle* in (B)) showing the end of teloblastic segment formation; (D) ventral view of an embryo at late stage 5, in which the engrailed expression pattern is counterstained with the nuclei fluorescent dye Hoechst (fluorescence quenched), engrailed is expressed in the posterior region of each segment; (E) schematic detail of the ventral view of the caudal papilla during the formation of rows eVI to eXI (projected from the embryo) shown in (D).

papilla is not yet flexed forward (Fig. 10.107F). Its shape is almost triangular with its apex directed anteriorly. A median groove ends posteriorly into the opening of the proctodaeum. A dark diffuse region occurs in the living eggs beneath the rear of the embryo (Fig. 10.106D), caused by the restructuring of the yolk to the so-called secondary yolk pyramids of a later embryo. At the end of this stage 3, the blastopore is closed.

During the **fourth stage**, the first postnaupliar segments are formed. The dark region mentioned above enlarges, until it spreads under the entire embryo. The naupliar region increases in size, the stomodaeum invaginates, and the appendages undergo morphogenesis (Fig. 10.107G). The most characteristic feature of this stage is the starting flexion of the caudal papilla to the anterior, concealing the medio-ventral parts of the second and third maxilliped segments. The carapace anlagen (ca) occurs as small longitudinal elevations laterally, beginning from the first maxillary segment posteriorly (Fig. 10.107G). Development occurs with two processes occurring side by side: morphogenesis of the earlier developed appendages; and the gradual addition of successive segments with their buds. At the stage of the appearance of second maxillae, three maxillipeds and early buds of pereiopods (**stage 5**), the yolk becomes restructured throughout the egg and the secondary yolk pyramids are visible through the nearly transparent embryonic tissue (Fig. 10.107J). The opening of the proctodaeum successively moves from the more dorsal side to the tip of the caudal papilla. The latter is now flexed to the posterior margin of the second maxillary limb buds (Fig. 10.106D). The carapace anlagen flank the thoracopods, narrowing between the first and second maxillary segments (Fig. 10.108D). During **stage 6**, the caudal papilla is bent ventrally with the tip median to the posterior margin of the mandibles (Figs 10.106E and 10.108A). The proctodaeum opens terminally. The heart, which was visible in living eggs in the previous stage, beats strongly and regularly. The limb buds undergo much morphogenesis during this stage (Fig. 10.108A) and continue during **stage 7**. In the living eggs, the first and second antennae point posteriorly more clearly, the biramous identity of the maxillipeds are still visible, the tips of the endopodites of the third maxillipeds enlarge and start to grow over the caudal papilla, and the five pereiopods increasingly elongate and are arranged with the maxillipeds in a basket-like form in these living eggs (Fig. 10.106F). The opening of the proctodaeum successively moves from the terminal position to a more ventral position within the telson. The carapace anlage broadens to a dorsolateral direction. The caudal papilla is folded back or removed (Fig. 10.108B and C).

With the obvious morphogenesis of the appendages, the caudal papilla elongates during **stage 8** and its tip reaches the basis of the first antennae. The pereiopods cover the caudal papilla, but do not meet yet in the midline (Fig. 10.106J). The first pereiopods are differentiated as large chelipeds, and the second and third pereiopods also show a first differentiation of chelae. Fig. 10.106J also shows the first pigmentation of the retinal cells in the anterior margin of the eyes as an incipient yellow stripe. This pigmentation is more pronounced during **stage 9** as the pigments of the retinal cells occur as a dense brown stripe (Fig. 10.106K). The chelipeds more or less reach the tip of the pronounced and elongated rostrum between the eyes. The pereiopods cover the caudal papilla completely. The amount of yolk decreases to nearly half of the egg, owing to the increase in size of the developing embryo. During **stage 10**, the prehatching stage, the amount of yolk that remains at hatching becomes the stock for the nonfeeding postembryonic stages. The first pereiopods cover the rostrum between the eyes (Fig. 10.106L). The 'prehatchling' appears to be compressed within the egg, indicating the moment of hatching event.

In the opinion of the present author, there is no significant difference in this very recent description of the embryonic development of Marmorkrebs *Procambarus fallax*, from that very ancient one of *Astacus fluviatilis*. However, the former is better explained by magnificent photos due to the advanced technology.

Hatching. Eggs ready for hatching are recognized by the presence of a fully developed embryo (Alwes and Scholtz, 2006) and a weakened egg shell. Softening of the rigid egg shell starts approximately 1 day prior to eclosion and is probably caused by a hatching enzyme as it was found in some other crayfish (Vogt, 2008b). The hatching process itself is rather uniform in respect to time course and behavioural sequence, both in vivo (under natural conditions) and in vitro (experimentally in the net culture system) (Vogt, 2008b). After the softening of the rigid egg shell, it becomes easily dented by forceps. During this time, there are no rotations of the embryo or flapping movements of its abdomen, but there are slow rhythmic contractions and expansions of the yolk sac. Hatching itself lasts for 10–12 min. It begins with rupture of the egg shell at the back of the embryo. Suddenly, the dorsal third of the embro appears (Fig. 10.109B). After ~5 min a beak appears and the scaphognathites (exopodite of second maxilla) beat rapidly to propel water through the gill chamber. During the following 3 min, the second third of the embryo slowly pushes out by stretching of the abdomen and the appendages. In the last step, the long second antennae move intensely and the pereiopods are used as levers to push the hatchling out of the egg. Freshly hatched juveniles appear to be considerably larger than the eggs (Fig. 10.109C). Usually, the egg case ruptures along the midline of the dorsal cephalothorax. If this portion of the embryo is positioned directly underneath the egg stalk, the egg shell ruptures in the opposite side and the hatchling pushes out with the pereiopods first. Such hatchlings mostly cannot complete hatching and usually die within a few hours.

Postembryonic development. Brood care in this freshwater marbled crayfish was further studied by Vogt and Tolley (2004) (Figs 10.109D–F and 10.110). At hatching, the egg capsule generally splits on the opposite side of the egg stalk over the back of the embryo. Then the hatchling (stage 1 juvenile) glides out backwards into the water (Fig. 10.110A and B). Freshly hatched juveniles have a spherical body shape resembling the embryos (Fig. 10.110A). Each soon becomes much larger than the

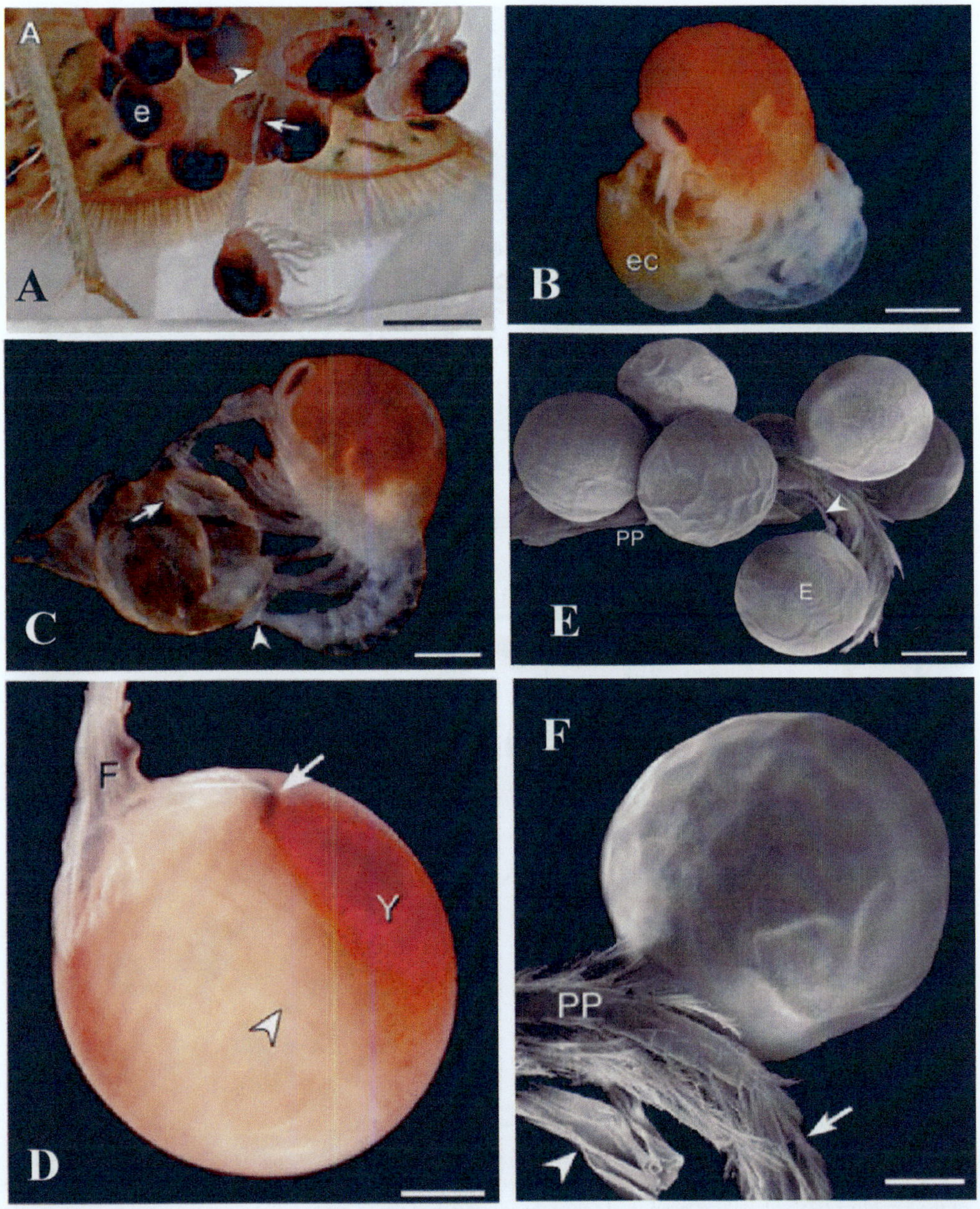

FIG. 10.109 (A–C) Hatching and early postembryonic development of the marbled crayfish: (A) on the maternal pleopods; (B and C) in vitro *(from Vogt, 2008b)*. (A) Macrophotograph showing cluster of hatchlings and late embryo (e) underneath the pleon, note freshly hatched juvenile dangling on a telson thread *(arrow)*, which keeps the new born firmly roped to its case *(arrowhead)*. Bar = 2 mm; (B) first phase of hatching (sudden appearance of the cephalothorax). ec = egg case; (C) hatched juvenile in a side position on bottom of culture vessel, gripping its egg case with the cheliped *(arrow)*, the telson *(arrowhead)* is attached to the inner face of the egg shell by an unexpanded telson thread. Bar B and C = 500 μm; (D–F) showing brooding in the marbled crayfish *(from Vogt and Tolley, 2004)*: (D) microphotograph showing embryonized egg at 90% development, eyes *(arrow)*, appendages *(arrowhead)*, yolk deposits (Y) of the embryo are well discernable. F = funiculus or egg stalk. Bar = 300 μm; (E) group of eggs (E) adhered to distal part of pleopod (PP), *arrowhead* denotes egg stalk. Bar = 600 μm; (F) close-up of egg attached to pleopod, *arrow* indicates terminal oosetae, *arrowhead* indicates broken egg stalk. Bar = 300 μm.

egg and reaches more stretched form, probably by uptake of water (Fig. 10.110B). The studies of Vogt et al. (2004) also concerned the postembryonic development of the Marmorkrebs *Procambarus fallax.* The first two postembryonic juvenile stages (Fig. 10.105E and G) are nonfeeding and permanently carried under the mother's abdomen. They live exclusively on the remainder yolk stored under their carapace. The **stage 1** juvenile (Figs 10.105E and 10.110A) has a crescent-shaped body form with an average carapace length (CL) of 1.8 mm and a total length (TL) of 3.5 mm. During this period, the hatchling is largely immobile and helpless. It is safeguarded by a relatively long rope-like structure, the telson thread that keeps the hatchling firmly attached to the egg case (Fig. 10.110A–C). After

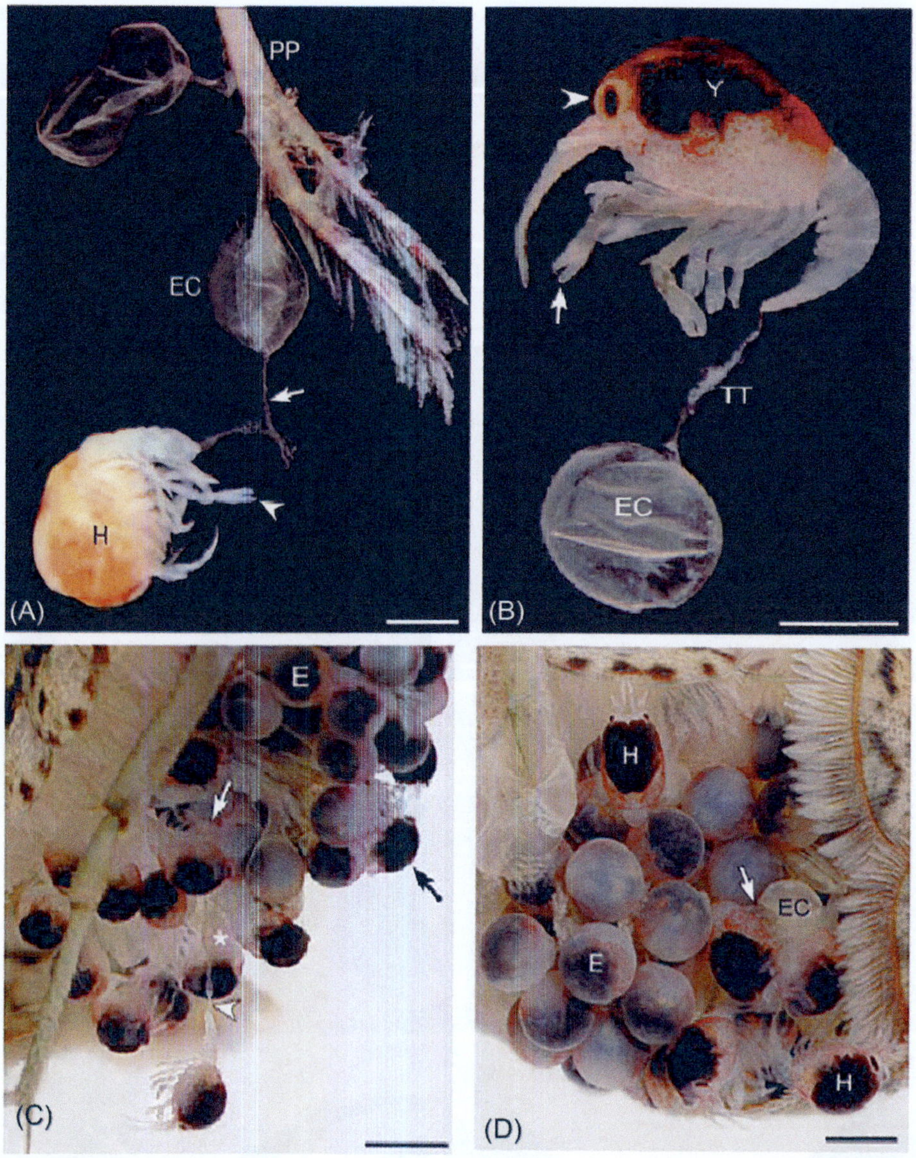

FIG. 10.110 Safeguarding of hatching in the marbled crayfish *(from Vogt and Tolley, 2004)*, (A and B) microphotography and (C and D) macrophotography: (A) newly hatched juvenile passively attached to a maternal pleopod (PP) via telson thread *(arrow)* and stalked egg case (EC), note rolled-up embryonic habitus of the hatchling (H) with its first pereiopod *(arrowhead)* stretched out to grasp for structures suited to cling to. Bar = 1 mm; (B) older hatchling with more straightened body form, still connected to the egg case by a telson thread (TT), this stage 1 juvenile was removed from a pleopod where it was passively adhered to via telson thread and egg case and additionally hooked upon with its pereiopods *(arrow)*. *Arrowhead* indicates pigmented eye, Y = yolk. Bar = 1 mm; (C) free-hanging hatchling secured by the telson thread *(arrowhead)*, other hatchlings *(arrows)* have already attached on pleopodal structures, asterisk points to egg case. E = embryonized egg. Bar = 2 mm; (D) newly hatched juveniles among eggs in final stage of development under the mother's pleon, *arrow* denotes hatchling emerging from the egg case. Bar = 1.5 mm.

a while, the hatchling tries to get a hold on the maternal pleopods by clinging itself by its hook-bearing chelae of its first pereiopods (Fig. 10.110C). Sometimes a newly hatched juvenile is found in a free-hanging position followed by upward movements and attempts to get a hold on the pleopods. This indicates that in vivo, the telson thread functions as a safety line that prevents the hatchling from being lost. The eyes appear well developed; they are large and pigmented (Fig. 10.110B). The first and second antennae of the hatchling are glabrous, except for a few setal precursor buds at their apices (Fig. 10.111(I) A), indicating that the hatchling is devoid of aesthetascs and hydrodynamic receptor hairs. The stage 1 juvenile is kept inactive on the maternal pleopods for 4–5 days until moulting. The **stage 2** juvenile is relatively transparent and its abdomen is bent like in the stage 1 juvenile. Sometimes it changes position being away from mother's pleopods. It takes a typical crayfish-like form (Fig. 10.105G) and starts to walk or swim, but swimming is less effective than in later stages. This is because its tail fan still lacks the uropods. In this life stage, the juvenile has much longer antennae; the vital external sense organs like the olfactory aesthetascs on the first antennae and the gustatory fringed setae on the chelae of pereiopods 1–3 emerge for the first time (Vogt et al., 2004). The compound eyes are stalked and start functioning, since they are partitioned into ommatidia (Fig. 10.111(I) A and B). However, in stage 2 juvenile the ommatidia of the eyes, the aesthetascs, hydrodynamic receptor hairs, and fringed setae appear fully developed but are still relatively few in number (Fig. 10.111(II) A–E). This stage 2 juvenile lasts 6–8 days.

In adults, the aesthetasces are located on the ventral side of the outer branch of the first antenna (Fig. 10.111(I) D and E) and the hydrodynamic receptor hairs are on the inner branch of the first antenna (Fig. 10.111(I) C) and on the second antenna. The gustatory fringed setae are found in adults along the inner margins of the propodus and dactylus of the chelae of pereiopods 1–3 (Fig. 10.111(I) F and G).

All the following life stages are free-feeding and 'lancing'. However, some times stage 3 juveniles, and even stages 4 and 5, may regularly return to adhere to the pleopods of

(I)

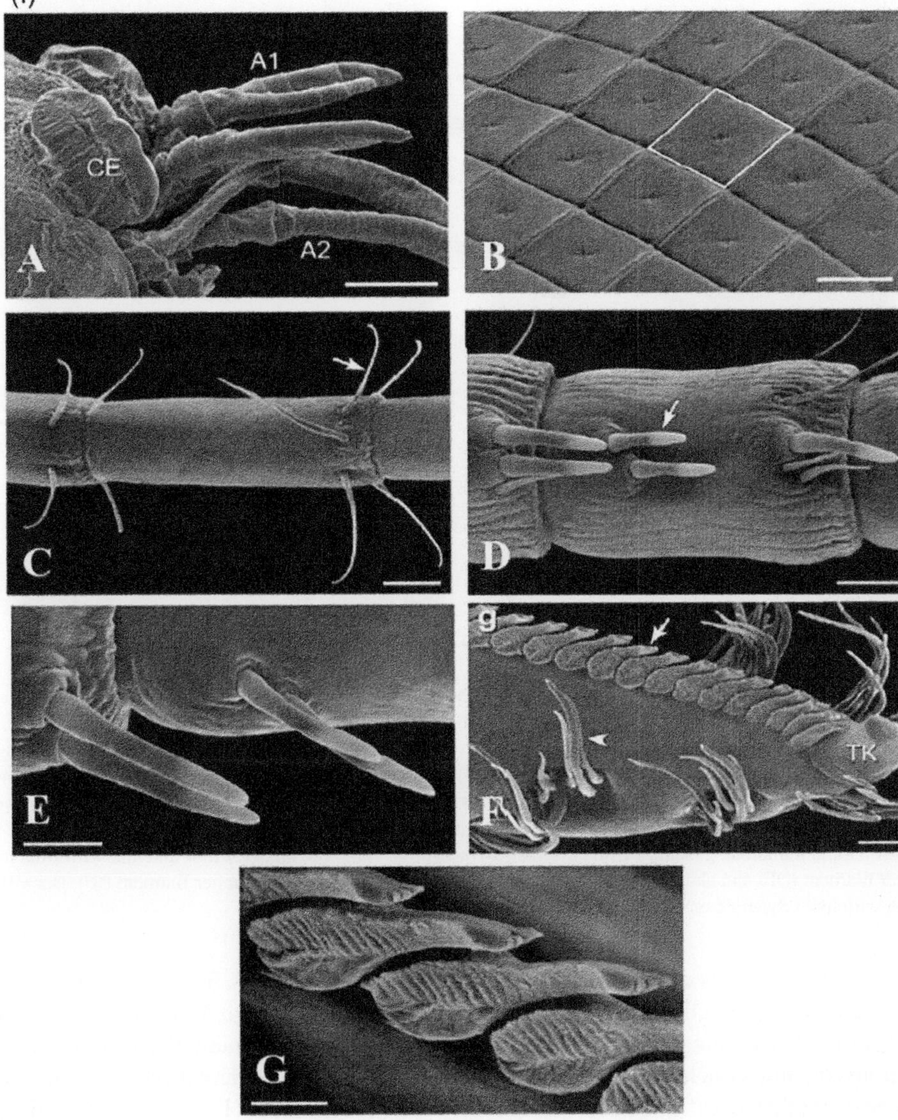

FIG. 10.111 (I) SEM photos showing comparison of major sense organs of hatchlings (A) and adults (B–G) of the marbled crayfish *(from Vogt and Tolley, 2004)*: (A) anterior cephalothorax of hatchling with incompletely developed compound eyes (CE) and glabrous first and second antennae (A1, A2). Scale bar = 200 μm; (B) surface view of eye of adult characterized by rectangular ommatidia (marked by *white lines*). Bar = 50 μm; (C) inner branch of first antenna of adult with whorls of hydrodynamic receptor hairs *(arrow)*. Bar = 50 μm; (D) ventral side of outer branch of first antenna of adult showing olfactory aesthetascs *(arrow)*. Bar = 50 μm; (E) lateral view of aesthetascs. Bar = 30 μm; (F) propodus of second pereiopod of adult with numerous gustatory fringed setae *(arrow)* at inner margin, instead of terminal hook adults have a blunt terminal knob (TK), *arrowhead* denotes groups of nongustatory sensory hairs. Bar = 50 μm; (G) close up of fringed setae. Bar = 20 μm.

(Continued)

(II)

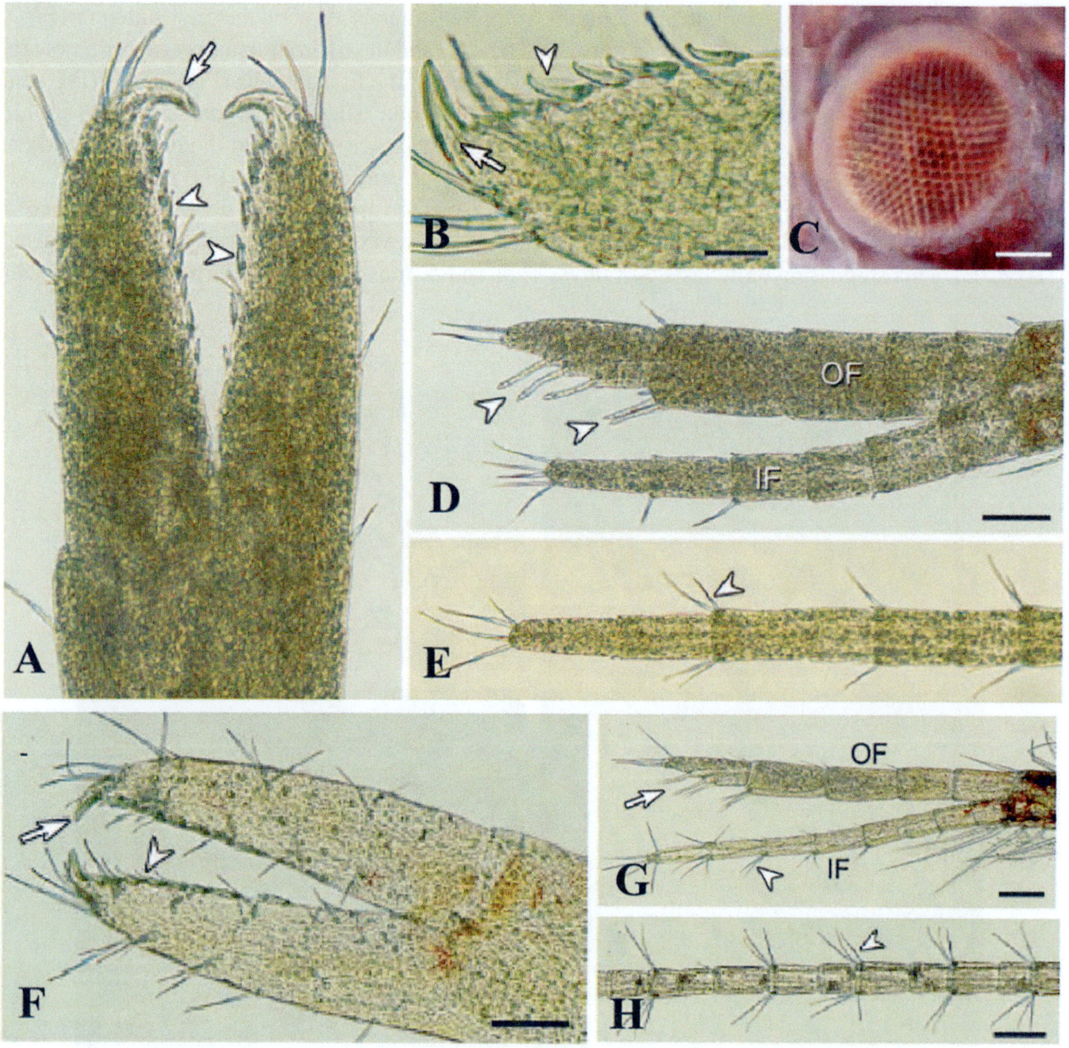

FIG. 10.111, CONT'D (II) Brood care, attachment mechanisms, and sensory development of stage 2 juvenile of the marbled crayfish *(from Vogt and Tolley, 2004)*: (A–H) light microscopy except (C) microphotography; (A) first pereiopod with recurved terminal hooks *(arrow)* and gustatory fringed setae *(arrowheads)*. Scale bar = 100 μm; (B) propodus of second pereiopod with slightly curved terminal spike *(arrow)* and fringed setae *(arrowhead)*. Bar = 40 μm; (C) compound eye clearly subdivided into ommatidia. Bar = 100 μm; (D) first antenna with olfactory aesthetascs *(arrowheads)* on terminal articles of outer filament (OF), the inner filament is studded with a few hydrodynamic receptor hairs. Bar = 100 μm; (E) terminal part of second antenna with hydrodynamic receptor hairs *(arrowhead)*. Bar = 100 μm. (F–H) in stage 3 and 4 juveniles: (F) first pereiopod of stage 3 juvenile with short and slightly curved terminal spikes *(arrow)* and numerous gustatory fringed setae *(arrowheads)*. Bar = 100 μm; (G) first antenna with olfactory aesthetascs *(arrow)* on terminal articles of outer filament (OF) and abundant hydrodynamic receptor hairs *(arrowhead)* on inner filament (IF). Bar = 100 μm; (H) subterminal segment of second antenna with hydrodynamic receptor hairs *(arrowhead)*. Bar = 100 μm.

their mother for shelter. Vogt and Tolley evidenced that their sense organs are enlarged and their functional units are multiplied (Fig. 10.111(II) F–H). Concerning the statocysts, the mentioned investigations do not conclude any structural data, but their effect may be deduced from the behaviour of the various juvenile stages. Therefore, the sense of orientation seems fully developed only from stage 3 on. The **stage 3** juvenile can be distinguished from stage 2 by its complete tail fan (Fig. 10.114G). This juvenile has a CL of 2.7 mm and a TL of 5.7 mm, and it feeds on the yolk remainder at least in the first half of its life period. The latter lasts 10–22 days. The following stages are morphologically similar to stage 3 juveniles and look more or less like miniature adults. Gradually, each increases in size with each moult and the pigmentation increases too. The **stage 4** juvenile (Fig. 10.114H), e.g. has a CL of 3.1 mm and a TL of 7.2 mm, and so on (Fig. 10.112).

Maternal care in Astacidea. In the decapod Crustacea, parental care is the rule but is usually confined to carrying, cleaning, and aeration of the eggs that are attached to ovigerous setae on the pleopods of the abdomen of the female. Only the phylogenetically ancestral dendrobranchiate

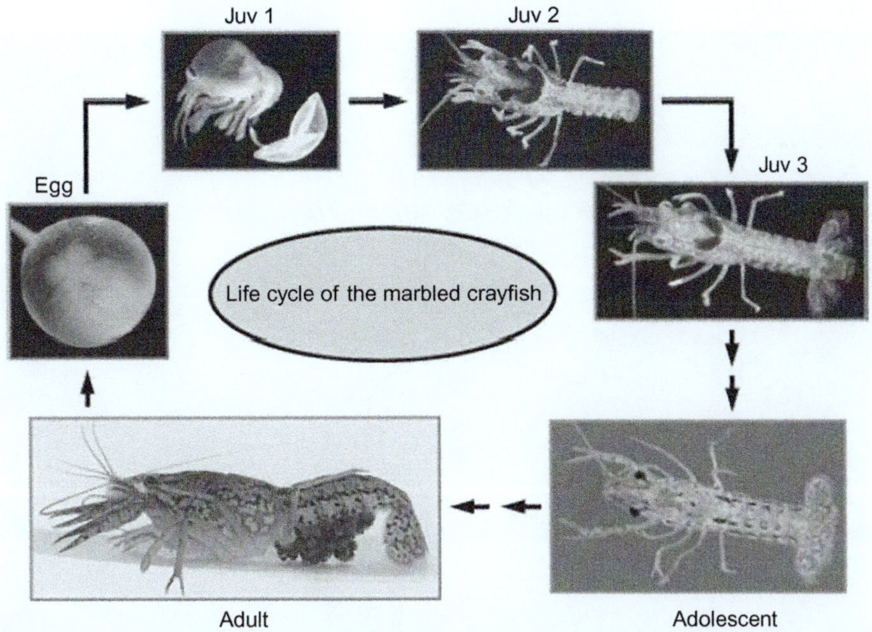

FIG. 10.112 Life-cycle of the marbled crayfish, includes embryonic stages that are passed within the egg, three dependent (juvenile 1–3), and 4–5 independent juvenile stages, 7–10 adolescent stages, and several reproducing adult stages. *(From Vogt (2008a).)*

shrimps (Vogt and Tolley, 2004) and family Penaeidae lack brood care, releasing their eggs directly into water (Shiino, 1968). Brooding of eggs and postembryonic brood care are to some extent found in freshwater Decapoda like freshwater crabs and aeglid anomurans. Prolonged parental care that includes care of the hatched juveniles is rare in Decapoda. It is evolved mainly in animals that live in open freshwater habitat (limnetic animals) and terrestrial ones. The most advanced example of this brood care is found in freshwater crayfish. As mentioned above, the first postembryonic stages (particularly stage 1 juveniles) remain attached to their mother's abdomen by the telson thread and then by its pereiopodal hooks. This unusual mode of parental care is found in all crayfish families, the Astacidae, Cambaridae, and Parastacidae, but with various details. Fig. 10.110A–C shows that the telson thread can persist until moulting to the next juvenile stage but often detaches from the egg case earlier. When moulting into stage 2, the juvenile of the Cambaridae and Parastacidae remain linked to the hooked exuvia of the first stage for a short time by an anal thread that is composed of the inner lining of the hind gut. Vogt and Tolley have not observed such anal thread in the marbled crayfish, perhaps because it persisted for only a very short period. The role of holding the hatchling to the mother by the telson thread was an acceptable view by earlier authors. Vogt and Tolley (2004) were able to document, by SEM studies, this function of the telson thread of the marbled crayfish for the first time in vivo (Fig. 10.113(I) A–D). As in the other species, the telson thread of the marbled crayfish often appears as a twisted rope that includes larger pieces of tissue. It was found 'to span from the inner surface of the egg case (Fig. 10.113(I) C) to the end of the abdomen' of the hatchling. In the latter it is tightly fixed to spines (14 spines) at its posterior margin (Fig. 10.113(I) B). The telson spines seem to release a secretion, as may be concluded from tiny filaments that are occasionally visible on the tips of the spines and interwoven in the first segment of the telson thread (Fig. 10.113(I) B). In the exuvia of Stage 1 juvenile, the telson thread was found still to be attached to the telson, but detached from the egg case (Fig. 10.113(I) D). Vogt and Tolley concluded that the telson thread of the marbled crayfish resembles more that of astacid than cambarid crayfish. Previous studies believed this structure to be derived from glandular secretion of the telson, the inner lining of the egg capsule or the exuvia of a moulting embryo. The results of Vogt and Tolley (2004) evidenced the second origin and the thread is attached to the telson by a narrow string originating from a hardened secretion of telson glands.

Formation of the telson thread was investigated in vitro in embryos of 95% development (Vogt, 2008b). In embryos ready to hatch, there was a drop of transparent material visible between the pleon and the inner layer of the egg shell. When such specimens were carefully pulled out from the egg and lifted by forceps, a telson thread appeared while the transparent drop disappeared (Fig. 10.109C). This indicates that this material is used to form the thread or at least part of it. When the pleon of such specimens is bent forwards on the ventral body side, like in earlier stages, it is almost completely covered by the pereiopods, but if this specimen

(I)

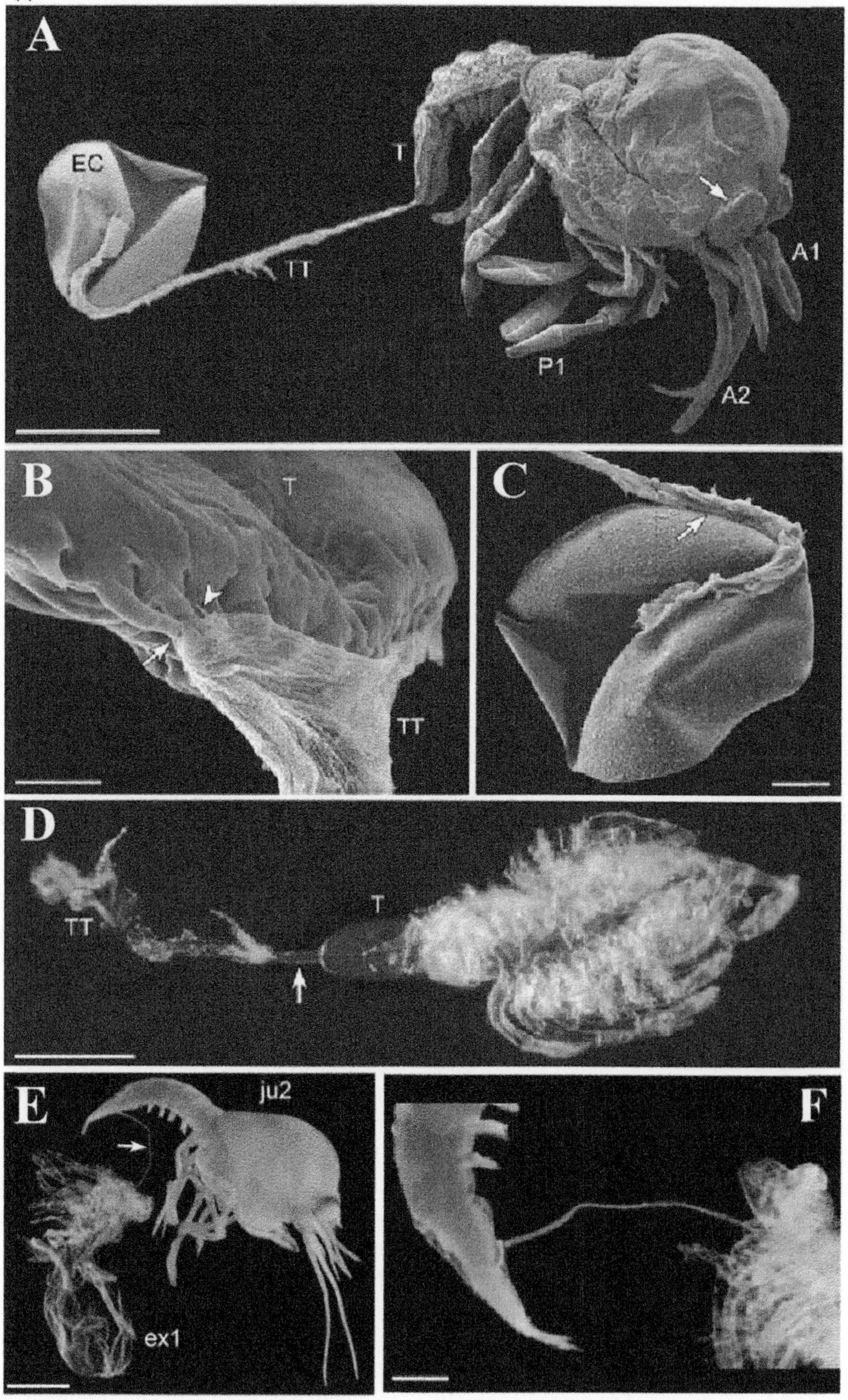

FIG. 10.113 (I) SEM photographs showing passive attachment of hatchlings to the mother in the marbled crayfish *((A–D) from Vogt and Tolley, 2004)*: (A) hatchling connected to empty egg case (EC) by telson thread (tt), *arrow* points to the eye. A1 = first antenna, A2 = second antenna, P1 = first pereiopod, T = telson. Scale bar = 1 mm; (B) ventral view of posterior part of hatchling's abdomen showing attachment of telson thread to spines *(arrow)* at margin of the telson, *arrowhead* denotes filamentous material to telson thread. Bar = 20 μm; (C) attachment of telson thread to inner surface of egg shell *(arrow)*. Bar = 200 μm; (D) exuvia of stage 1 juvenile with attached telson thread, notice division of telson thread into two parts, a short and narrow unstructured string close to the telson *(arrow)* and a broader heterogeneous part farther posterior. Bar = 1 mm. (E and F) Photomacrographs of anal thread in vivo-raised specimens *(from Vogt, 2008b)*; (E) stage 2 juvenile (ju 2) linked to its stage-1 exuvia (ex 1) by an anal thread *(arrow)*; (F) details of anal thread connection.

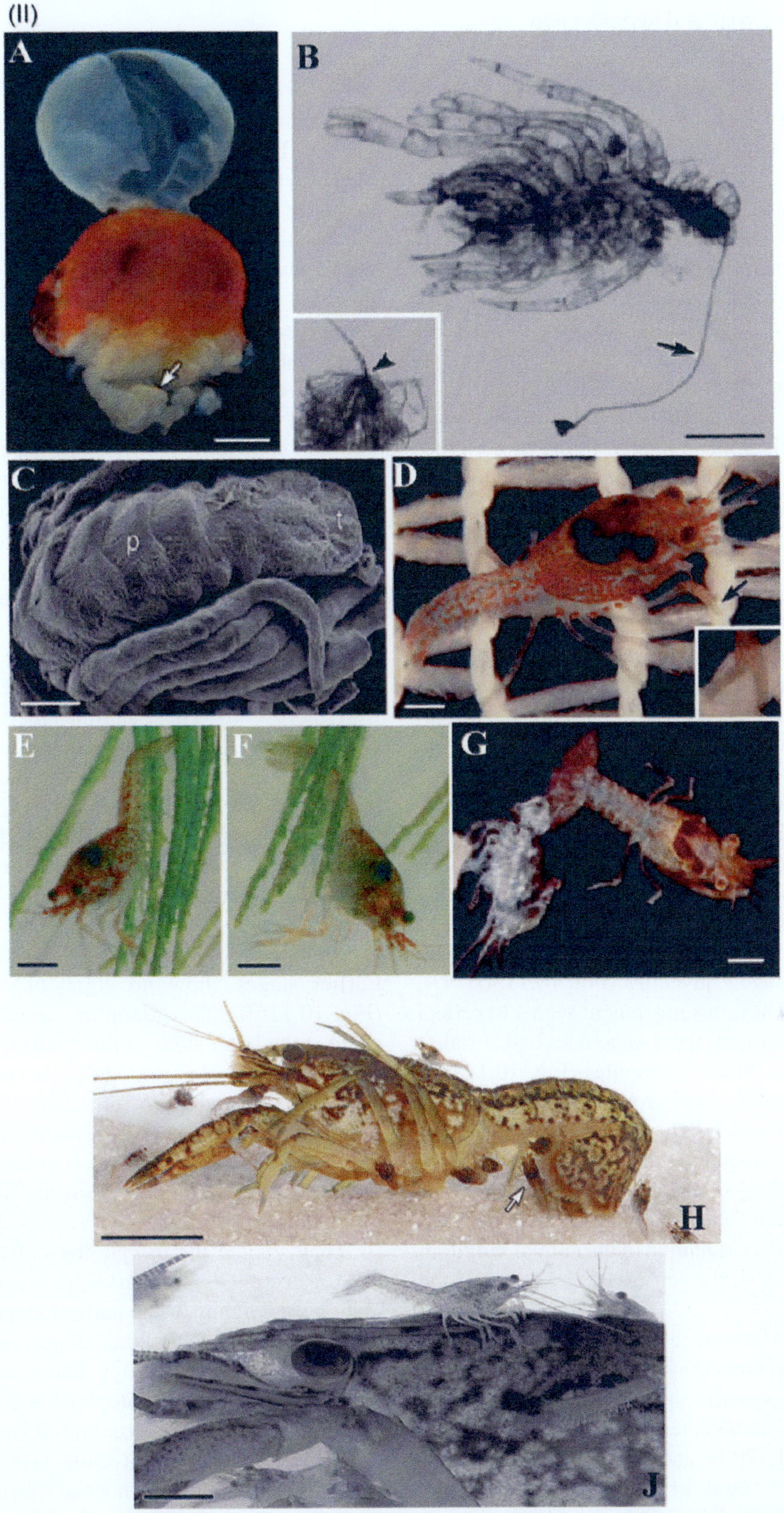

FIG. 10.113, CONT'D (II) Formation of the telson thread in marbled crayfish *(from Vogt, 2008b)*: (A) microphotograph showing late embryo with egg case, *arrow* denotes posterior margin of telson lacking a telson thread; (B) macrophotograph showing stage 1 exuvia showing an anal thread in total length *(arrow)*, inset shows attachment site *(arrowhead)* of anal thread; (C) pleon (p) and telson (t) of late embryo lacking terminal spines and a telson thread; (D–F) attachment behaviour and holding structures in early juvenile stages; (D) microphotograph showing stage 2 juvenile gripping the net with all pereiopods, *arrow* denotes cheliped tightly hooked in the net, inset shows hooked-in cheliped in detail. Bar = 40 μm; (E and F) macrophotographs. Bar = 1 mm; (E) stage 3 juvenile crawling on *Ceratophyllum demersum*, the branches of the plant are grasped with all pereiopods; (F) resting stage 3 juvenile keeping hold with pereiopods 2 and 3 only; (G) freshly moulted stage 3 juvenile crawling on bottom of petri-dish, the exuvia is sill firmly attached to the net. Bar = 1 mm; (H–J) mother-juvenile association, enabling investigation of learned behaviours and potential transmission of maternal effects to the next generation *((H) from Vogt and Tolley, 2004)* shows stage 3 juveniles mounting their mother from all sides, *arrow* denotes juvenile grasping a pleopod. Bar = 1 cm; (J) bar = 3 mm *(from Vogt, 2008a)*.

is removed from the egg, it is without this thread. The telson thread-linkage between such embryo and the egg shell was not yet established (Fig. 10.113(II) A and C).

On the other hand, the second safety line, the anal thread (Fig. 10.113(I) E and F) which secures the stage 2 juvenile during moulting and emerging from the stage 1 juvenile, differs completely in origin and structure from the telson thread. In addition, it is difficult to be sampled in vivo because the exuviae are usually torn into pieces by crawling movements of the stage 2 juvenile. In contrast, in vitro it can easily be observed, sampled, or manipulated with forceps (Vogt, 2008b). It consists of the inner cuticular lining of the hindgut of the stage 1 juvenile. The anal thread of the marbled crayfish is about 2 mm in length when completely pulled out from the intestine of a stage 2 juvenile (Fig. 10.113(II) B). On the exuvial side, the thread ends at the anus and is continuous with the cuticle of the telson (Fig. 10.113(II) B, inset). This anal thread can easily hold the weight of a stage 2 juvenile in the water and resist moderate tearing forces, as testing by lifting experiments with forceps. In vitro (in net culture system), this structure kept the emerging stage 2 juvenile passively connected to the net via the stage 1 exuvia, which remained hooked in the net with the first pereiopods.

Again, in vivo (Pleo of mother) it is very difficult to study how the juveniles attach to the maternal pleopods and how motile they are on the pleopods. However, in vitro (Figs 10.113(II) D and 10.114A) it was possible to take photos during short observation intervals. This enables to suggest that stage 1 juveniles are usually stationary and do not change position. Stage 2 juveniles (Fig. 10.113(II) D), generally move in clutches, this movement seems to reflect turnings of the bodies around fixed attachment sites rather than positional changes. Stage 3 juveniles (Fig. 10.113(II) E–G) and later stages often crawl around and change position (Fig. 10.113(II) H and J). They mostly fled by rapid, backward or even more frequently upward swimming.

However, safeguarding of hatching is confined to the Astacida only (Vogt and Tolley, 2004). This sophisticated mode of passive brood care is unique in the animal kingdom and is apparently related to the sensory deficiencies of the first juvenile stage

Parthenogenesis. Marmorkrebs is one of the 12 currently known nonindigenous crayfish species to be found in German waters, in specific, and in central Europe in general. It is unique in the manner that only females exist. Therefore, Scholtz et al. (2003) and Vogt et al. (2004) tested whether Marmorkrebs could be parthenogenetic. They obtained this species from Frank Steuerwald (Ostringen-Tiefenback, Germany) and defined it by a German and English name, but there as no scientific name at this time. Due to its beautiful colouration (Fig. 10.105A), rapid reproduction, and easy handling, the authors were interested in this striking species, especially its parthenogenetic nature. Five days after spawning, they removed one female to be reared separately. First, they studied its general life periods. Under favourable conditions, Marmorkrebs (marbled crayfish) can grow to a total length (TL: tip of rostrum to end of telson) of ~ 12 cm, but mostly the adults remain much smaller. The maximum age of the marbled crayfish is not known, but may be considerably more than the 2 years recorded by those authors so far. The first spawning occurs on average at an age of 25 weeks (about 6 months) when reared at 25°C. Under very good conditions, marbled crayfish can reproduce all year round, with intervals of only 8–9 weeks between spawnings. The embryonic period between spawning and hatching lasts 2–3 weeks, depending on temperature. The maternal care remains for another 2–5 weeks until the juveniles become fully independent. This results in a total breeding period of 4–8 weeks (Vogt and Tolley, 2004).

They studied the external sexual characters and the internal reproductive organs. They examined more than 100 specimens of their population in Heidelberg under the dissection microscope and by means of light and electron microscopy. These samples included individuals from hatchlings (stage 1 juvenile) to repeatedly spawned adults having the lengths of 3.5 mm to 6.8 cm. All the specimens investigated were purely female as shown by the presence of the unpaired annulus ventralis (Fig. 10.115A) at the ventral side of the cephalothorax between the pereiopod pairs 4 and 5 (as usual in females of all crayfish families). This annulus ventralis is first detectable in stage 4 juvenile, but only with the SEM technique (Fig. 10.115C). Earlier stages show no signs of sexual differentiation (Fig. 10.115B). The gonopores in adult specimens are located at the inner sides of the coxae of the third pereiopods and are easily observed as depressions (Fig. 10.115A). First signs of gonopore formation are detected in stage 5 juveniles, where a small crescent-shaped slit appears. In the following stages, this slit is gradually enlarged and the area of the later gonopore becomes more and more visible as a cuticular depression.

They studied the morphology of the internal reproductive organs of 93 specimens from seven mothers of two generations, beginning at the earliest stage at which sex can be determined. They found that all specimens (BL 0.8–8.0 cm) showed female morphology that excludes protandric hermaphroditism. No spermatophore was detected in this study. The histology and ultrastructure of the reproductive system of 39 specimens from juvenile (1.9 cm) to postbrooding (6.8 cm) stages, all show normal ovaries (Figs 10.115D and 10.116) with oviducts. Histological studies show that the nuclei were found to be relatively uniform through oocyte maturation (Fig. 10.115E) and showed no meiotic features. They are characterized by a high degree of chromatin decondensation and include numerous nucleoli (Fig. 10.115F), particularly when grown to a diameter exceeding 100 μm. Fig. 10.116 represents the maturation of this ovary.

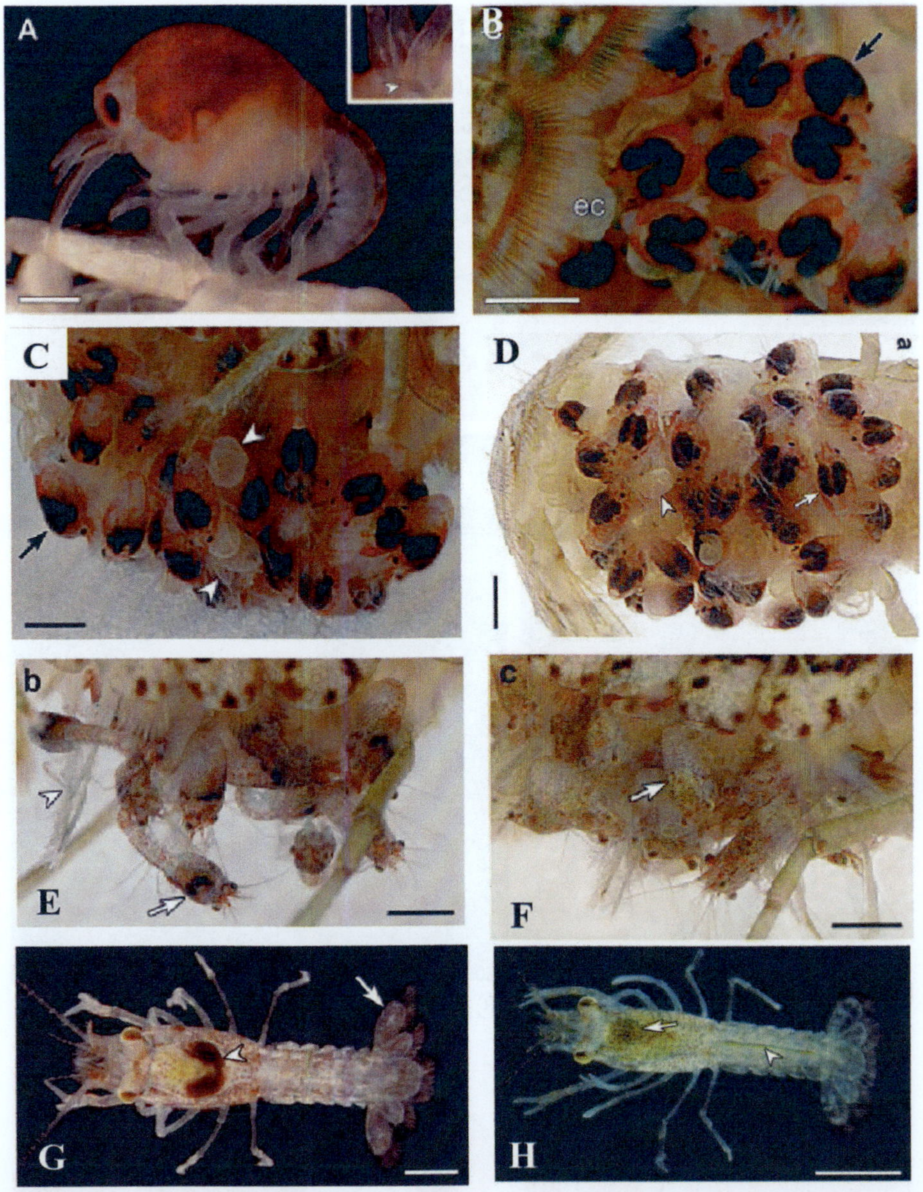

FIG. 10.114 Macrophotography showing post embryonic development of the marbled crayfish on the maternal pleopods *((A–C) from Vogt, 2008b, (D–H) from Vogt and Tolley, 2004)*: (A) microphotography shows stage 1 juvenile in typical attachment posture on nylon net, inset shows only the cheliped *(arrowhead)* hooked in the net. Bar=500 µm; (B) cluster of stage-1 juveniles under the pleon, *arrow* denotes extensive yolk deposit. ec=egg case; (C–D) dense aggregation of stage-2 juveniles *(arrow)* with interspersed egg cases *(arrowheads)*. Bar of C=2 mm, of D=2 µm; (E) stage-3 juveniles under their mother's abdomen, *arrowhead* denotes adherence of juvenile to pleopod and *arrow* points at yolk reserves in cephalothorax of juvenile; (F) stages-3 and 4 juveniles *(arrow)* under mother's pleon. Bar of both E and F=3 mm; (G) free-lancing early stage-3 juvenile with typical tail fan *(arrow)*, despite extensive yolk reserves *(arrowhead)*, such juveniles feed. Bar=1 mm; (H) free-lancing stage-4 juvenile with food in stomach *(arrow)* and intestine *(arrowhead)*. Bar=2 mm.

This marbled crayfish fulfils the basic requirements for laboratory culture such as robustness against handling stress, high fertility, relatively short generation time, and adaptability to a broad range of culture conditions. It is large enough to allow individual physiological and biochemical analyses. In this it becomes a model organism, more favourable than the fly *Drosophila melanogaster* and the nematode *Caenorhabditis elegans*. For details and review, refer to Vogt (2011).

Brachyura Latreille, 1803

The infraorder Brachyura of suborder Reptantia Boas, 1880 of order Decapoda Latreille, 1803 is divided into four sections; one of them is section Eubrachyura de St. Laurent, 1980. Many families belong to this section, and one of them is family Portunidae Rafinesque, 1815 (Schram, 1986). The members of the Brachyura, the true crabs, have the most highly specialized short body form and, judging from the number of species (about 4500), are probably most successful decapods.

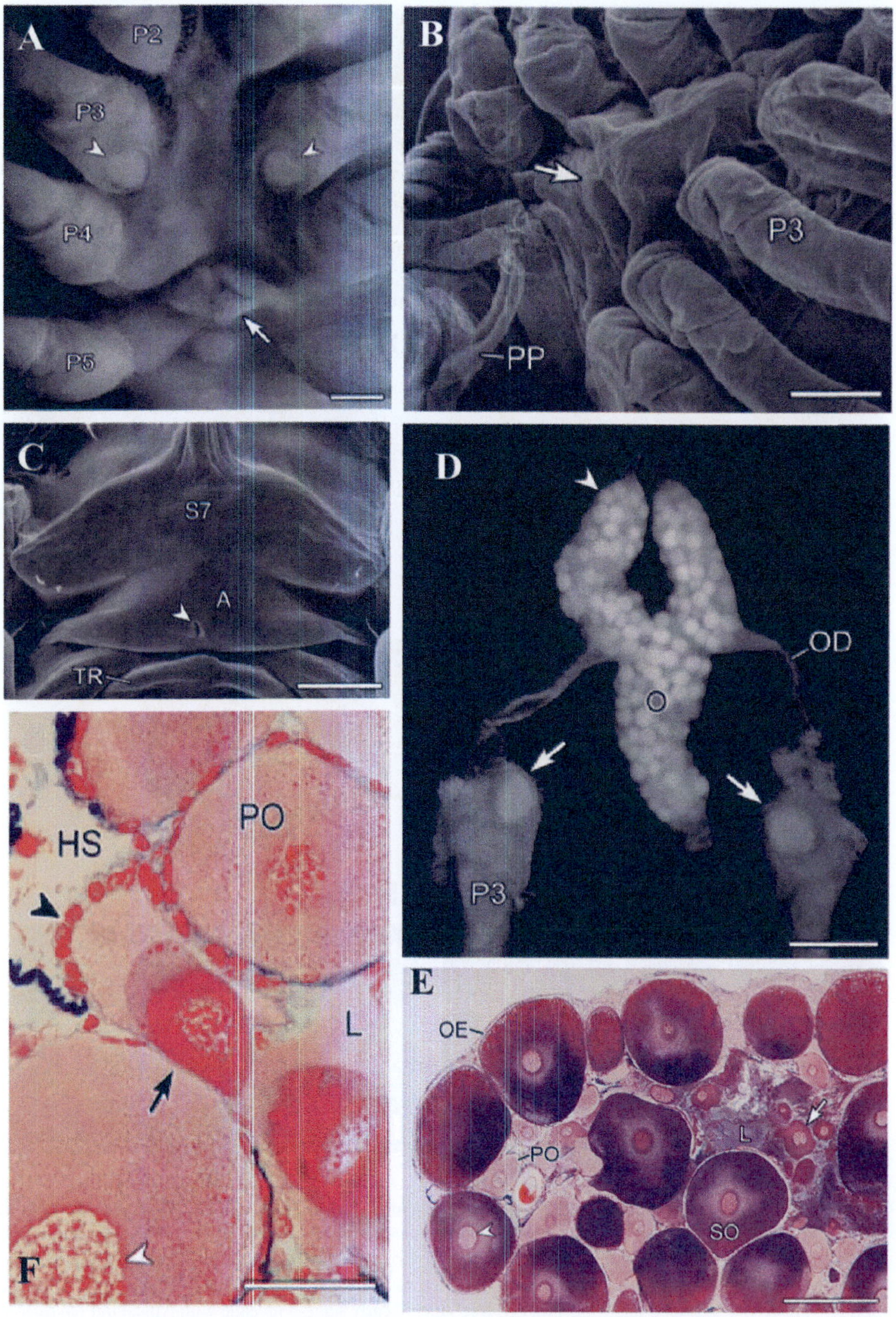

FIG. 10.115 External sexual characters on ventral side of cephalothorax of immature and mature marbled crayfish: (A) annulus ventralis *(arrow)* and gonopores *(arrowheads)* in adult of 4.9 cm TL. P2–P5 = pereiopods 2–5. Scale bar = 1 mm; (B) stage 2 juvenile lacking sexual characters, *arrow* points to the site where annulus appears later. PP = pleopods. Bar = 200 μm; (C) first appearance of the annulus (A), a posterior structural derivative of the seventh sternite (S7), in stage juvenile 4, *arrowhead* denotes spot-like primordial central groove. TR = transversal ridge of eighth sternite. Bar = 100 μm; (D) isolated reproductive system from specimen 5.5 cm TL showing y-shaped ovary (O). Bar = 2 mm; (E) section of the ovary of first-time maturing female containing previtellogenic *(arrow)*, primary vitellogenic (PO), and secondary vitellogenic oocytes (SO), note thin ovarian envelope (OE). *Arrowhead* = nucleus of oocyte, L = ovarian lumen. Bar = 400 μm; (F) previtellogenic oocyte *(arrow)* invading an oogenetic pouch, black *arrowhead* = follicular epithelium, *white arrowhead* = nucleolus, HS = interstitial hemal sinus. Bar = 50 μm. *(All from Vogt et al. (2004).)*

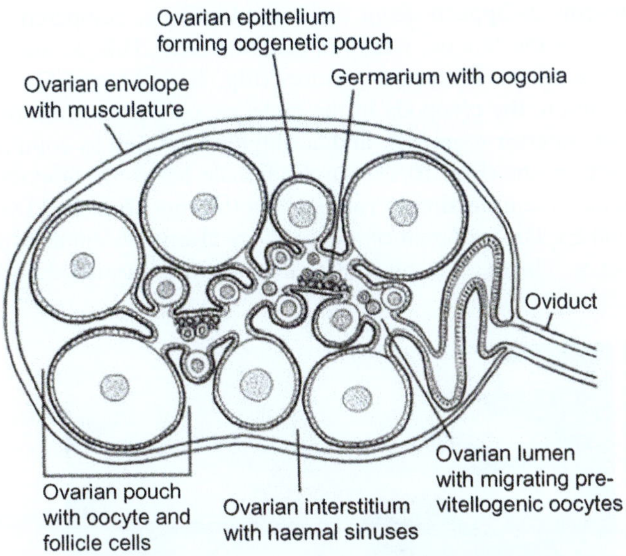

FIG. 10.116 Schematic representation of maturing ovary exemplifying architecture and function of the crayfish ovary, describing an ovarian lobe. *(From Vogt et al. (2004).)*

Definition. The abdomen is greatly reduced and fits tightly beneath the cephalothorax, so as to be invisible from the dorsal surface. Uropods have disappeared. In the female, pleopods are retained for brooding eggs; in the male, only the anterior copulatory pleopods are present. The carapace is very broad, at least as wide as it is long and commonly much wider, increasing the flattened appearance of the body (Barnes, 1974; Marshall and Williams, 1975).

Habits and habitat. Brachyuran crabs as reptantians are benthic animals found in all types of habitats and to great depths. Crabs can crawl forward slowly, but they commonly move sideways, especially when crawling rapidly. Most crabs cannot swim, but members of family Portunidae, which includes the common edible red crab *Portunus pelagicus* (Linnaeus, 1758) (Fig. 10.117) and the blue crab *Callinectes sapidus* (Rathbun) are the most powerful and agile swimmers of all crustaceans. The last pair of legs in members of this family terminates in broad flattened paddles. Members of this family and family Cancridae have a premoult attendance of the females by the males, in which

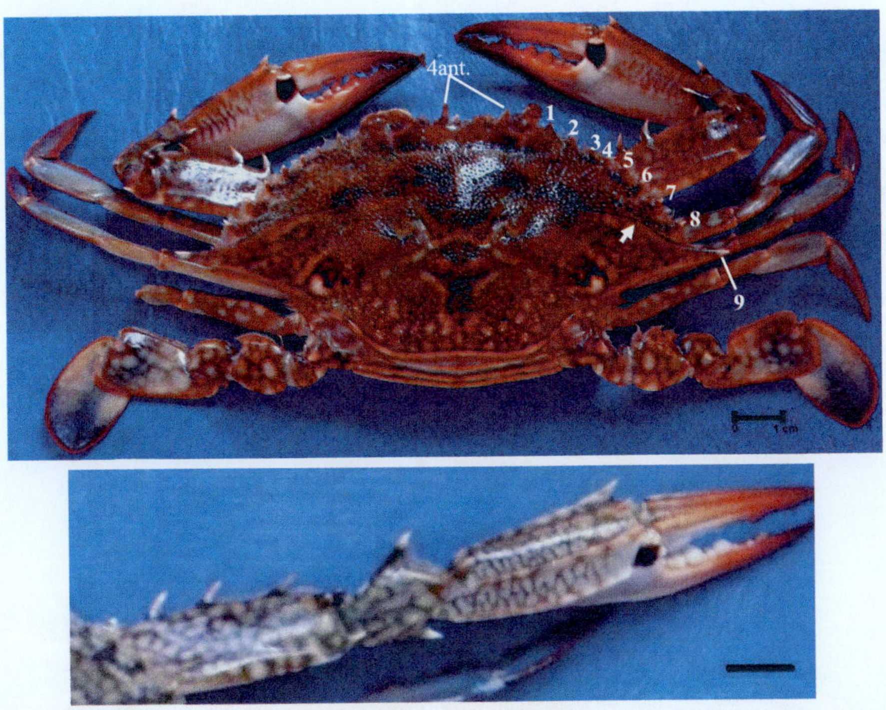

FIG. 10.117 Photograph of the crab *Portunus pelagicus* showing the specific characters. Its identification was based on the characteristics specified by Stephenson and Campbell (1959) and before them Holthuis (1956) and Balss (1936). More recently, the same characteristics were illustrated in the FAO species identification guide for fishery purposes (1987) and (2012). Formerly, it was recognized as *Neptunus pelagicus* Linnaeus (1758). Although this change of name has been suggested by Weber as early as 1795, the name (*Neptunus*) remained in use by many authors, particularly by those who studied Egyptian collections (Balss, 1936; Ramadan, 1936). It was not until 1956 that the present generic name, *Portunus*, has been finally established by the International Commission of Zoological Nomenclature at the suggestion of Holthuis (1956), and confusion with other species belonging to the same genus has since been totally avoided. It is known among the population in Alexandria and Port Said, EGYPT (east of the Mediteranean sea) as the 'red crab' to be differentiated from the 'blue crab' *Callinectes sapidus* (other one in the catch). It was necessary to elaborate on the history of nomenclature and identification of this species, having personally had few contributions to research in its reproductive biology. The lower photo shows an expanded chela of another specimen exposing the three spines of the merus *(arrows)*. 4 ant.s = 4 prominent frontal spines, 1–9 = 9 triangular spines on anterolateral margin, 9 = greatly expanded ninth spine, *arrow* indicates transverse ridge on the carapace reaching the ninth spine. *(Preparation and photo A.R. Khafage.)*

the male carries the female about beneath him, her carapace beneath his sternum. He releases her so that she can moult. Copulation occurs shortly afterwards (Barnes, 1974) (see later).

Morphology and sexual dimorphism. In brachyurans, the first antennae and the stalked eyes can be retracted into special pits excavated in the carapace. The gills are usually few in number. The reduced abdomen in the male is very narrow as appears from the ventral surface, compared to that of the female, which appears broader. This is one of the external dimorphic features (Fig. 10.118A and C). In addition, the pleopods in the male are only present on the two anterior segments, and are highly modified as copulatory organs (Fig. 10.118B); the female has four pleopods, which are used simply for carrying the eggs (Fig. 10.118D and E). The sixth pair of pleopods are always absent in both sexes. The egg hatches out as a zoaea; it is furnished with

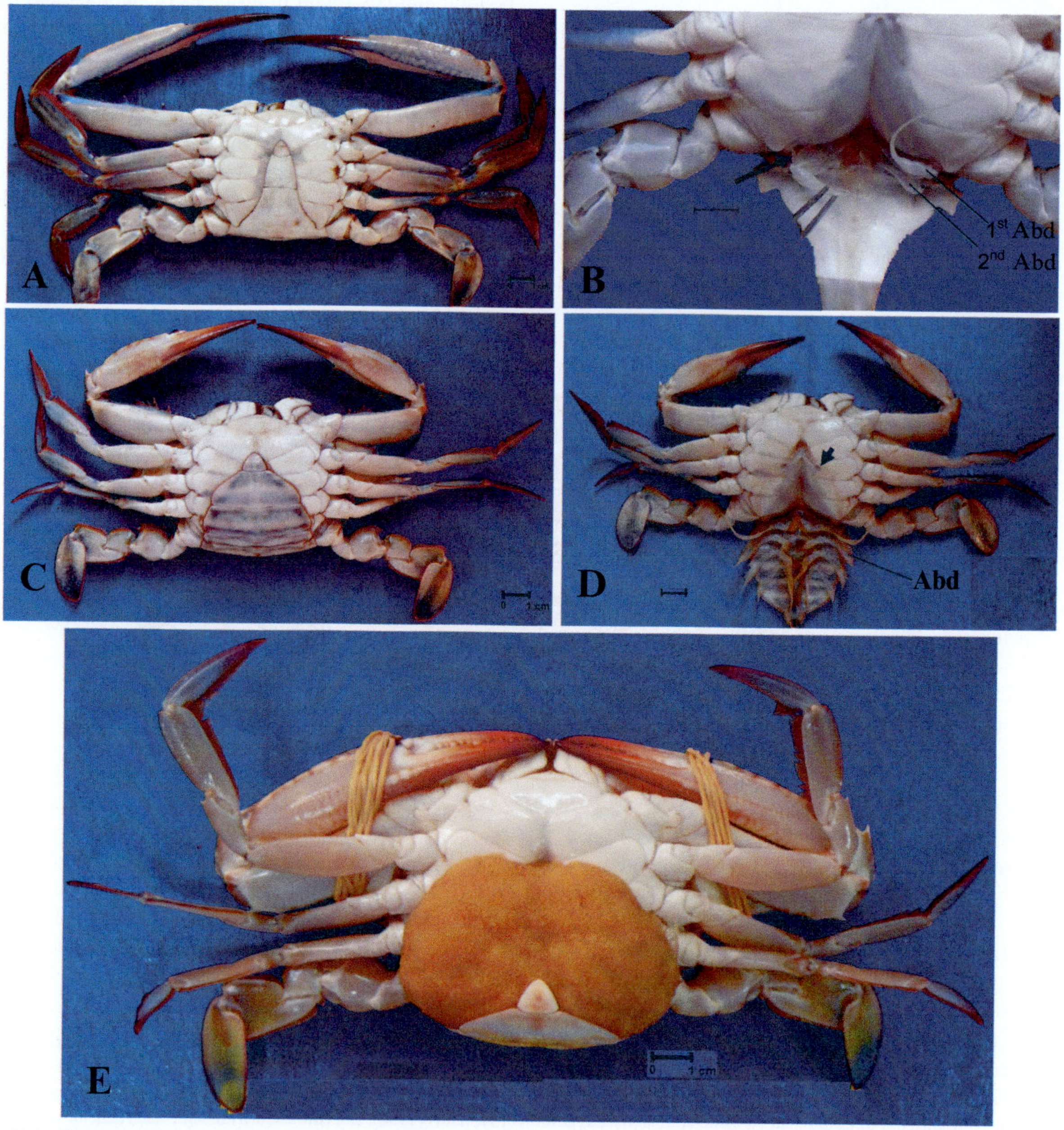

FIG. 10.118 Photographs of fresh whole mounts of ventral view of crabs, *Portunus pelagicus*, showing the dimorphic organs: (A) male crab with narrow overlapped abdomen; (B) live male crab with unfolded abdomen showing first and second Abd = first and second abdominal appendages, the second abdominal appendage displaced to expose the intromittent organ protruding from the genital opening *(arrow)*; (C) female crab with broad abdomen; (D) female crab with unfolded abdomen exposing the Abd. = 4 pairs of abdominal appendages, *black arrow* indicates the female genital opening; (E) ventral view of ovigerous (berried) live female. *(Preparation and photos by A.R. Khafage.)*

an anterior and posterior spine on the carapace. This larva is characterized by the definitive expansion of its third maxillipede, and it is never a biramous swimming organ (Harmer and Shipley, 1923).

Reproductive systems. In Brachyura, in general, these systems always occupy the posterior region of the cephalothorax dorsally (Fig. 10.119). Each system is H-shaped and the anterior horns of the H are widely separated from each other. The two sides of this system, as symmetrical as they might be, are always joined together medially by a short transverse bridge of gonadal tissue. The constituents of the horns of the H differ from the male to the female animal. As an example of Portunidae, the reproductive systems of the red crab *Portunus pelagicus* (Linnaeus, 1758) (its identification according to Stephenson and Campbell, 1959) will be described since this species is one of the two common edible crabs in Egyptian waters. The other one is the blue crab *Callinectes sapidus* (Rathbn, 1896).

The male system. The anterior horns of the H-shaped are the testes, while the adjoining vasa deferentia constitute its posterior horns (Fig. 10.119B and C). In a dissected specimen of *P. pelagicus* examined in situ (Fig. 10.119A), the connecting transverse bridge may be exposed either after the heart has been completely excised, or after the pyloric tip of the stomach is displaced from its position. In that it thus assumes an H-shaped form, the reproductive system of specifically male crabs recalls that of the anomures described by Matthews (*Panulirus penicillatus*, 1951; *Parribacus antracticus*, 1954a; *Enoplometopus occidentalis*, 1954c). It appears that in spite of the remarkable length of the reproductive system of *Portunus*, the whole system occupies a limited area of the haemocoele. This results in the great coiling of the vasa deferentia, a feature that may have significant implications in the manner of production of spermatophores. The latter of *Portunus* are remarkably simple and nonpedunculate, as they are in other brachyures (Mouchet, 1931; Spalding, 1942; George, 1963). In other crustaceans, which produce pedunculate spermatophores, the complexity in structure of the peduncle seems directly dependent upon the presence of coilings and folds in the concerned vasa deferentia (Mouchet, 1931; Matthews, 1953, 1954b).

The **testes** in *P. pelagicus* are widely separated from each other, contrary to the situation in prawns where they remain closely applied together at their median borders. Each testis is located on the hepatopancreas, which furnishes a suitable background for the immature translucent stage. It begins from a point far away in the lateral region of the cephalothorax just beneath the lateral spine (Fig. 10.119A). It proceeds forwards, as a narrow folded ribbon, along that narrow lateral area which is limited between the anterolateral margin of the specimen and the outer border of the branchial chamber. It curves around this chamber and then proceeds to the middle region of the cephalothorax, passing between the anterior border of the branchial chamber on one hand and the pyloric stomach on the other. Just beneath the tip of the latter, the two inner ends of the two testes connect by the transverse bridge.

The present author supervised the work of the MSc presented by El-Sherief (1975), and noticed during the period from March 1971 to about the end of December 1973 that the male crabs (*P. pelagicus*) have no definite breeding season, and they are always sexually mature. However, the peak of maturity is reached during January, February, and March. Maturity of the testis is indicated morphologicaly by its milky white appearance, and histologically by the contents of its seminiferous tubules. The monthly serial histological preparations of the testes evidenced the above information. The different levels along the whole length of the testis exhibit somewhat different degrees of maturity. This means that one can find the two extremes of maturity in one testis (Fig. 10.120B and I). Maturation commences in the base of the testes where each joins its corresponding vas deferens. This picture is in agreement with the findings of other works (Bawab, 1964; Huxley, 1880; Spalding, 1942; Matthews, 1951, 1953, 1954a,b,c).

Each testis is surrounded by a very thin membraneous wall, the cortex (Fig. 10.120A and B). Masson's triple stain enables the differentiation of this wall into an outer squamous epithelial layer and an inner connective tissue layer devoid of muscle fibres. This connective tissue extends inwards in the form of two very thin septa subdividing the testis into three lobes (Fig. 10.120A) running longitudinal, parallel to each other throughout the whole length of the testis. Each is composed of convoluted seminiferous tubules. A transverse section, at any level, gives the appearance of follicles with various outlines, the wall of which is a very thin layer of connective tissue, the membrana propria (Fig. 10.120B, C, G, and I).

The **vasa deferentia** begin just behind the transverse bridge. Each testis leads to its corresponding vas deferens (Fig. 10.119A and B) which is highly elongated and convoluted. The vasa deferentia differ from the testes in that they are situated close to each other and are confined to a limited area. Morphologically and functionally, the vas deferens of *P. pelagicus* is differentiated into three specific regions: the proximal, middle, and distal regions (El-Shereif, 1975). This division is in agreement, only morphologically, with the basic pattern in decapodes stated by Calman (1909). This pioneer carcinologist divided the decapod vas deferens into three definite regions, the second glandular, and the third ejaculatory. This division cannot apply on that of *P. pelagicus*. However, that of *Carcinus maenas* (Spalding, 1942) was divided into four secretory regions and so, apart from the nomenclature, it can be in agreement with that of

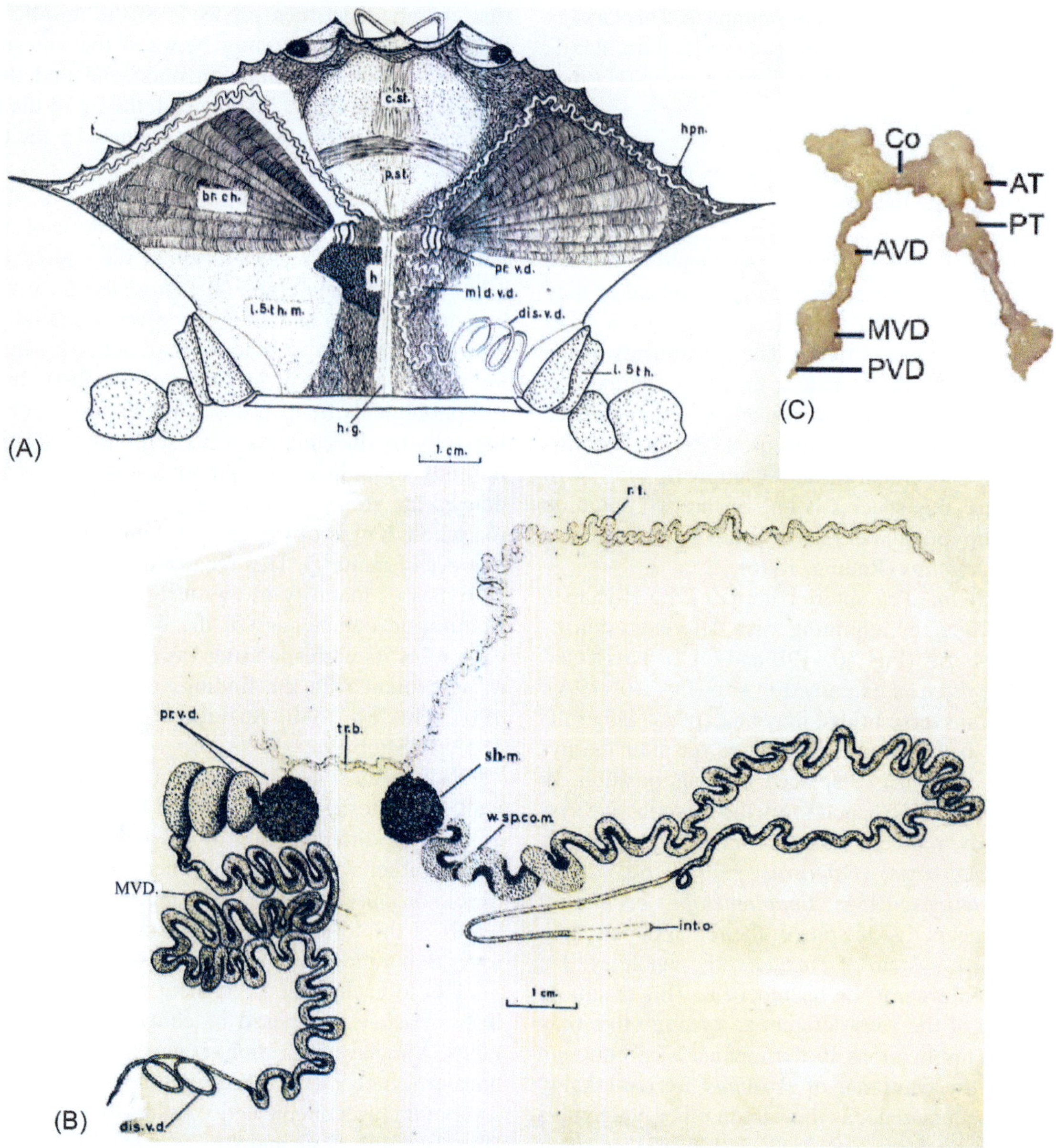

FIG. 10.119 Diagrammatic drawings of the male reproductive system of adult crab *P. pelagicus (from El-Sherief, 1975)*: (A) A dissected male measuring 6–8 cm in length and 12–16 cm in width showing the system in situ. Half of the heart and the muscles of the fifth pereiopod were removed from the right side to expose the course of the system; (B) The isolated system showing the different regions. The left testis was removed, while its corresponding vas deferens remains as seen in nature. The right vas deferens was somewhat expanded to show its different regions. br.ch. = branchial chamber, c.st. = cardiac stomach, co.m. = coiled mass of the vas deferens, dis.v.d.= distal vas deferens, h = heart, h.g. = hind gut, hpn. = hepatopancreas, int.o. = intromittent organ, l.5th. = fifth thoracic leg, L.5th.m. = muscles of the fifth thoracic leg, M.v.d. = middle region of the vas deferens, p.st. = pyloric stomach, pr.v.d. = proximal region of the vas deferens, r.t. = right testis, t = testis, t.b. = transverse bridge, w.sp.c. = wide spirally coiled part; (C) Photo of isolated male reproductive system of *P. pelagicus (from Stewart et al., 2010)*. AT = two anterior lobes of the testis, PT = two posterior lobes, Co = central commissure, AVD, MVD, PVD = anterior, middle, and posterior regions of the vas deferens, respectively.

P. pelagicus. Fig. 10.119B shows the distinct regions of the latter, both raveled and unravelled.

The **proximal vas deferens** (PVD), in turn, is divided into the ball-like finely coiled mass (referred to as the **spherical mass**), and the wide spirally coiled region (referred to as the **spiral region**) (Fig. 10.119B). The portion of the continuation of the testis to its corresponding vas deferens cannot be easily distinguished morphologically in *P. pelagicus*. However, critical observation reveals that this very short part at which the testis connects the transverse bridge (transitional part) abruptly changes from the ribbon-like testicular tissue into fine tubular structure of the spherical mass of the vas deferens (Fig. 10.121C). This part is about 0.08 mm, along which the ductules of

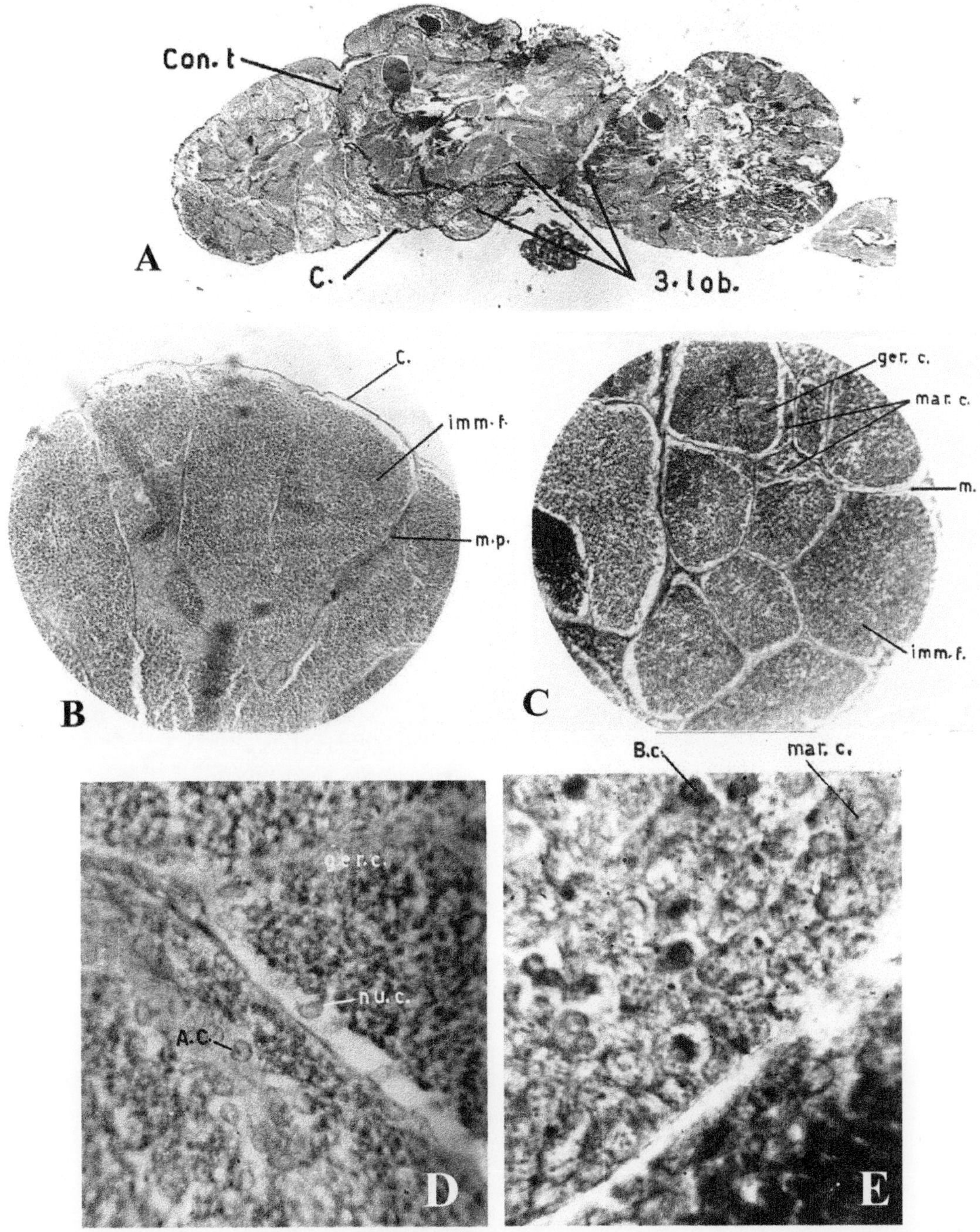

FIG. 10.120 Photomicrographs of T.S.s through different levels of the testis of *P. pelagicus* showing its histological structure *(from Bawab and El-Sherief, 1981)*: (A) through the anterior region showing the three lobes. X=30; (B) as A showing immature follicles; notice the cortex (c) surrounding the testis and the membrane propria (m.p.) separating the follicles. X=400; (C) through the middle region showing the beginning of maturation of the follicles, notice the few marginal cells (mar.c.). X=300; (D and E) magnified portions of section (C). X=1260; (D) shows germinal cells (ger.c.), notice the nurse cells (nu.c.) and A-cell (A-c.) scattered at the periphery of the germinal mass; (E) shows the B-cells scattered between the marginal cells and A-cell (A-c.);

(Continued)

the seminiferous tubules are transformed into fine tubules of the proximal vas deferens. This is ascertained by examining serial histological sections of this transitional part. The steps through which each ductule of the follicle is transformed into a tube were described in Fig. 10.122. The histological beginning of this transitional part cannot be sharply defined owing to two facts: (1) each tubule is a continuation to the temporary ductule of the testis; and (2) the tubules are highly coiled and the coils are firmly applied to each other. The histological structure of the tubules of the spherical mass (Figs 10.121C and 10.123A–C) changes from region to region. During the course of this

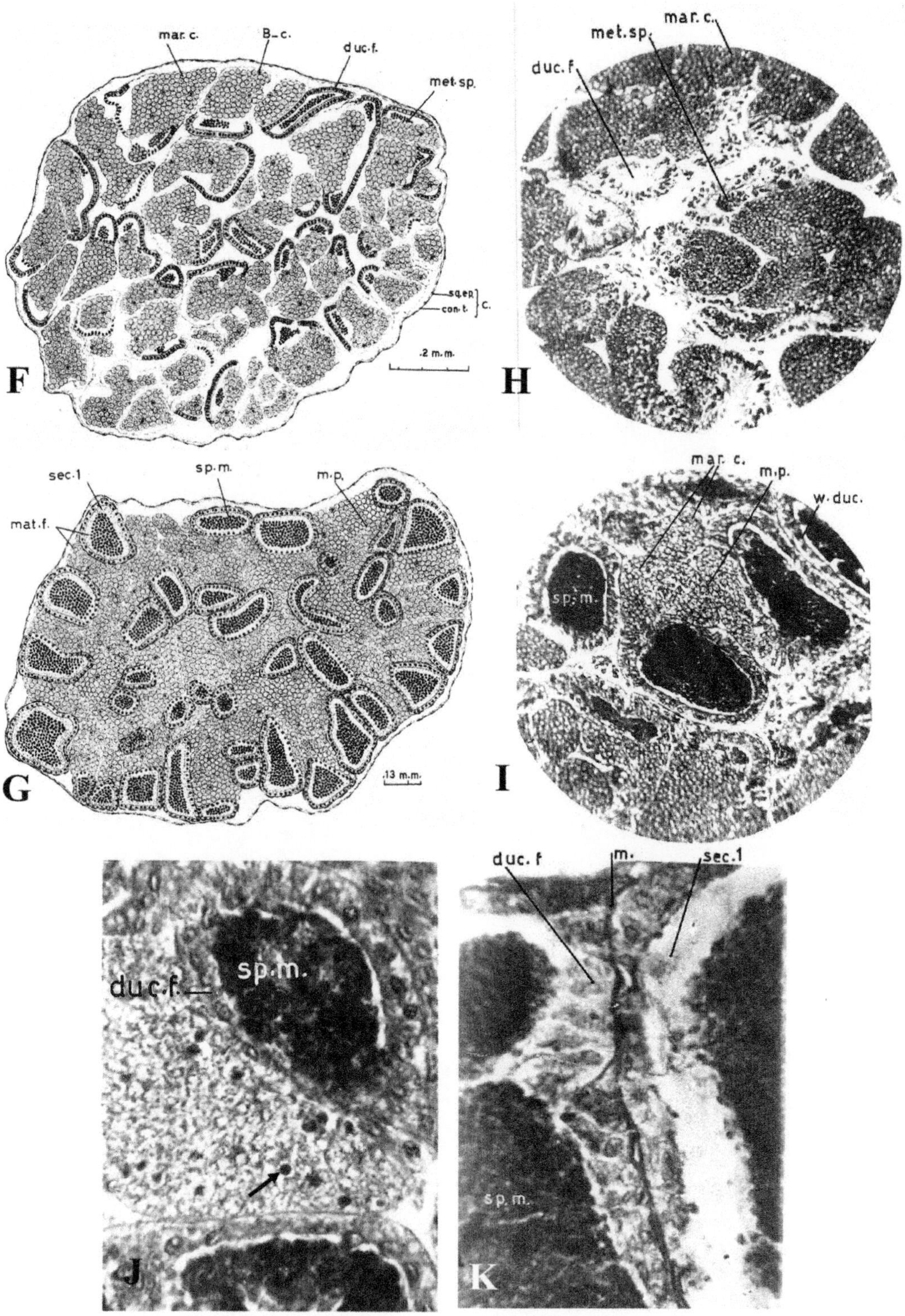

FIG. 10.120, CONT'D (F & G) Camera Lucida drawings of T.S.s through the posterior region of the testis for two successive levels; (F) B-cells still existing between the marginal ones; (G) shows the mature follicles with complete ductules (duc.f.) enclosing metamorphosing sperm (met.sp.), notice the compressed membrane propria between the marginal areas; (H–K) photomicrographs showing details of (F) and (G). X=250; (H) portion of (F) showing the formation of the ductule; (I) portion of (G) showing the completely formed ductules and the continued marginal areas of the neighbouring follicles; (J) magnified follicle with incomplete ductule enclosing metamorphosing sperm mass, notice the scattered B-cells within the marginal cells *(arrow)*. X=400; (K) highly magnified portion showing the secretory walls of two adjacent ductules, notice the muscle fibers within the connective tissue surrounding the ductule. X=1200. m=muscles, sec. 1=secretion 1, sp.m.=sperm mass.

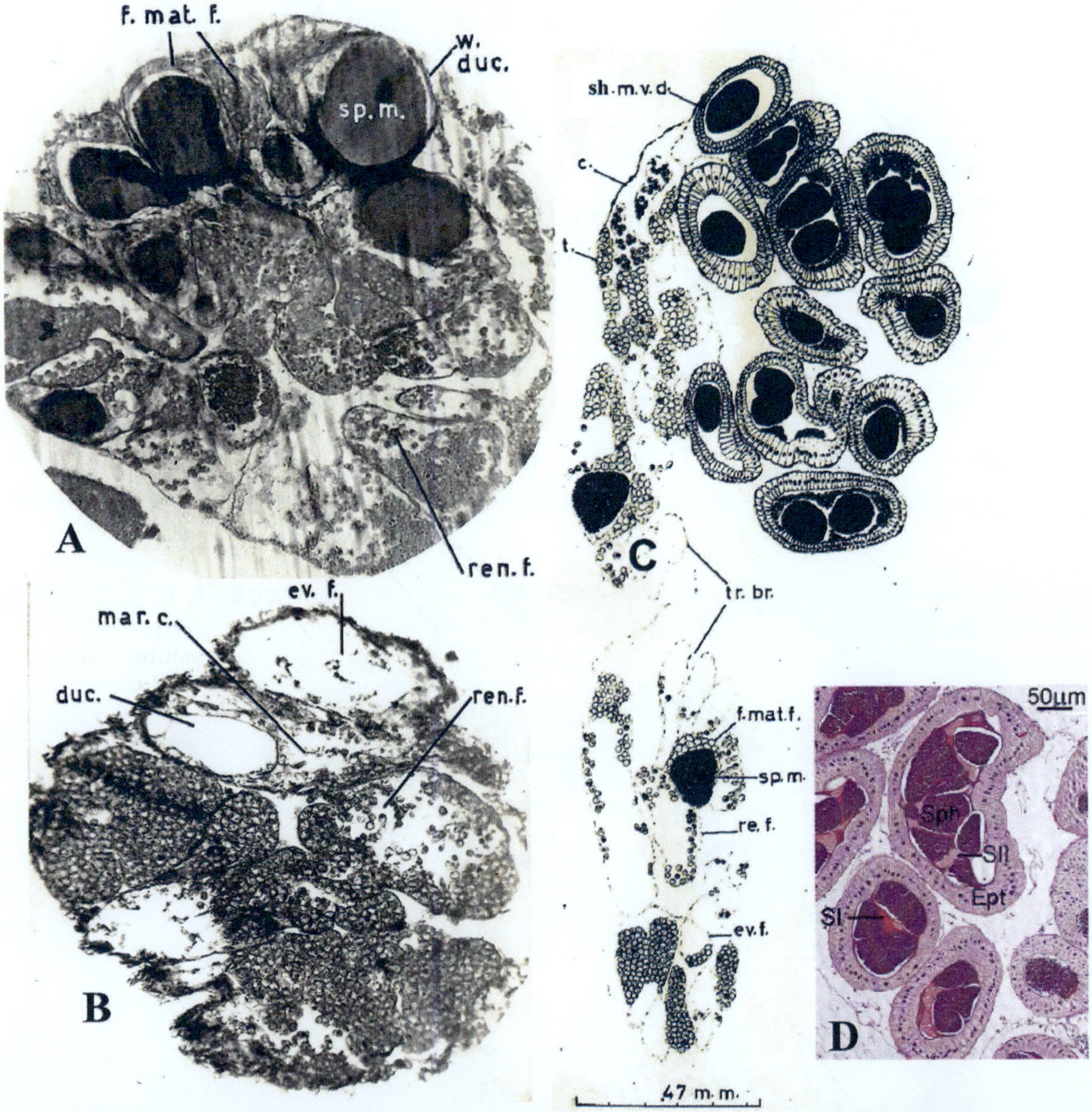

FIG. 10.121 (A and B) Photomicrographs of T.S.s of two successive levels of the testis posterior to Fig. 10.120 *(from Bawab and El-Sherief, 1981)*; (A) shows different stages of fully mature follicles. X=300; (B) shows the evacuated and renewed follicles. X=250; (C) Camera Lucida drawing of a T.S. of the testis at the level of transverse bridge. Notice the tubes of the spherical mass of PVD in the right side, that of left side omitted. ant.co.v.d.=spherical mass, c=cortex, duc=ductule, ev.f.=evacuated follicle, f.mat.f.=fully mature follicle, mar.c.=marginal cells, re.f.=renewing follicle, sp.m.=sperm mass, t=testis, tr.br.=transverse bridge, w.duc.=wall of the ductule; (D) T.S. of paraffin-embedded anterior vas deferens of *P. pelagicus* stained with H&E *(from Stewart et al., 2010)*, to show the similarity between the two preparations. Ept=columnar epithelium, SI=substance 1 stained red, S II=substance 2 stained pale blue, Sph=spermatophores.

mass, the tubules gradually and continually fuse with each other, ending as almost one large tube in the posterior region to continue to the spiral region of the proximal vas deferens. This end region of the spherical mass increases gradually but rapidly in diameter, to lead to the spiral region (Fig. 10.119B). The course of the latter appears as three large, well-defined coils that presented great difficulty in sectioning as it hardens to a great degree after fixation. Owing to its coils, its outline in cross section never attains a circular appearance (Fig. 10.124).

Naturally, the spherical mass appears more or less translucent, while the spiral region has a whitish colour owing to its internal contents. Because of its great distension with secreted material, it is delicate and is easily ruptured by any mild touch, even after fixation. The examination of its gelatinous material reveals that it is heavily loaded with spermatophores, mostly at the fully formed stage (Fig. 10.124B).

The **middle vas deferens** (MVD) is the continuation of the third coil of the preceding region, following the gradual decrease of its diameter. Thus, it is narrower and proceeds

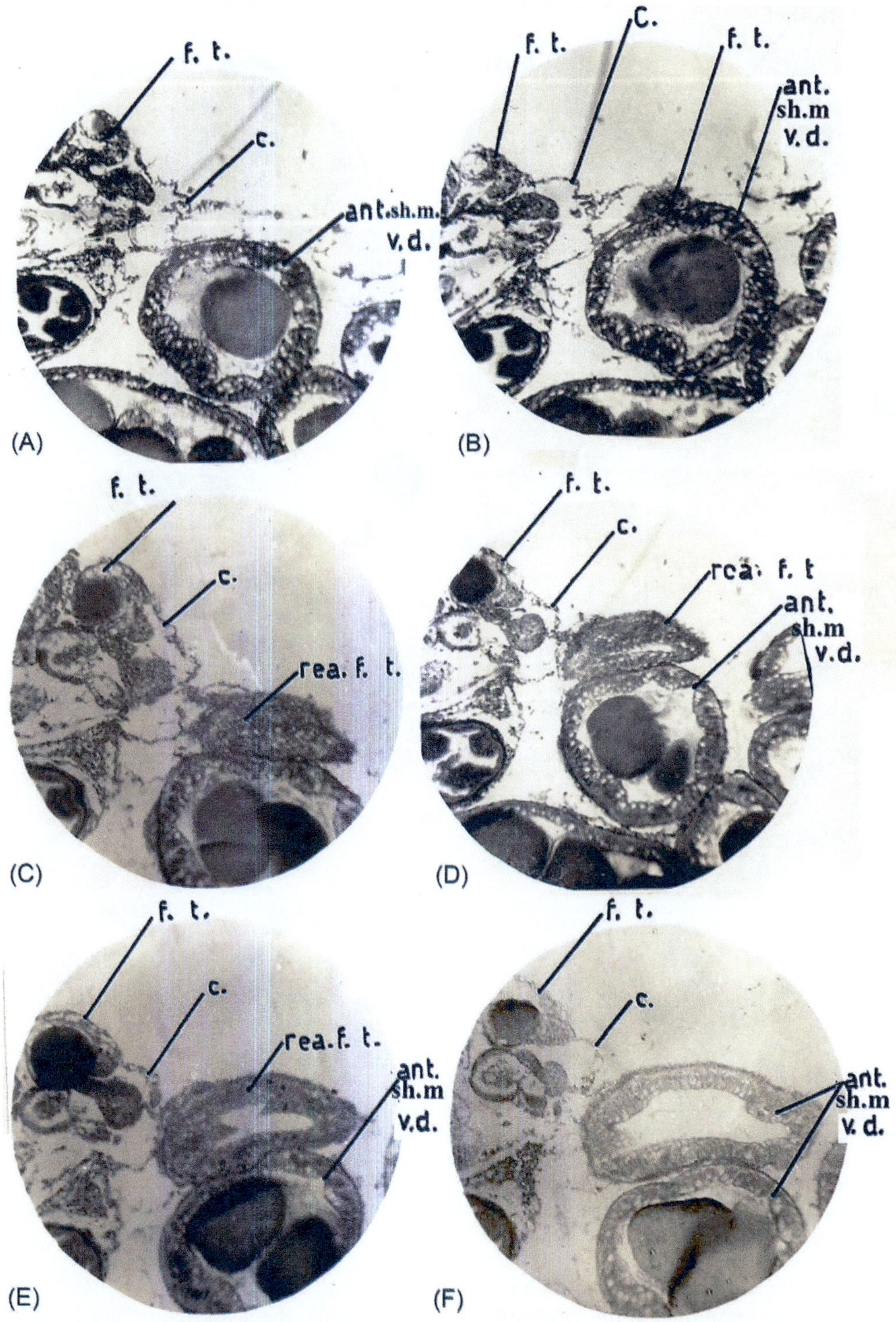

FIG. 10.122 Photomicrographs of T.S.s of the AVD just below the transverse bridge showing how the testicular ductules continued as tubes of the spherical mass of AVD. X=150 *(from El-Sherief, 1975)*: (A) the testis with ruptured cortex; (B) follicle approaching the tubules of the spherical mass; (C) the approaching follicle with its contents in activity; (D) the appearance of a lumen inside the approached follicle; (E) the central lumen of the approached follicle enlarges and the cells rearrange themselves forming a definite wall; (F) the approached follicle fully becomes tubule of the spherical mass, notice the approach of another follicle. Ant.sh.m.v.d.=anterior tubule of the spherical mass of PVD, c=cortex, f.t.=follicle of the testis, rea.f.t.=re-arranged follicle of the testis.

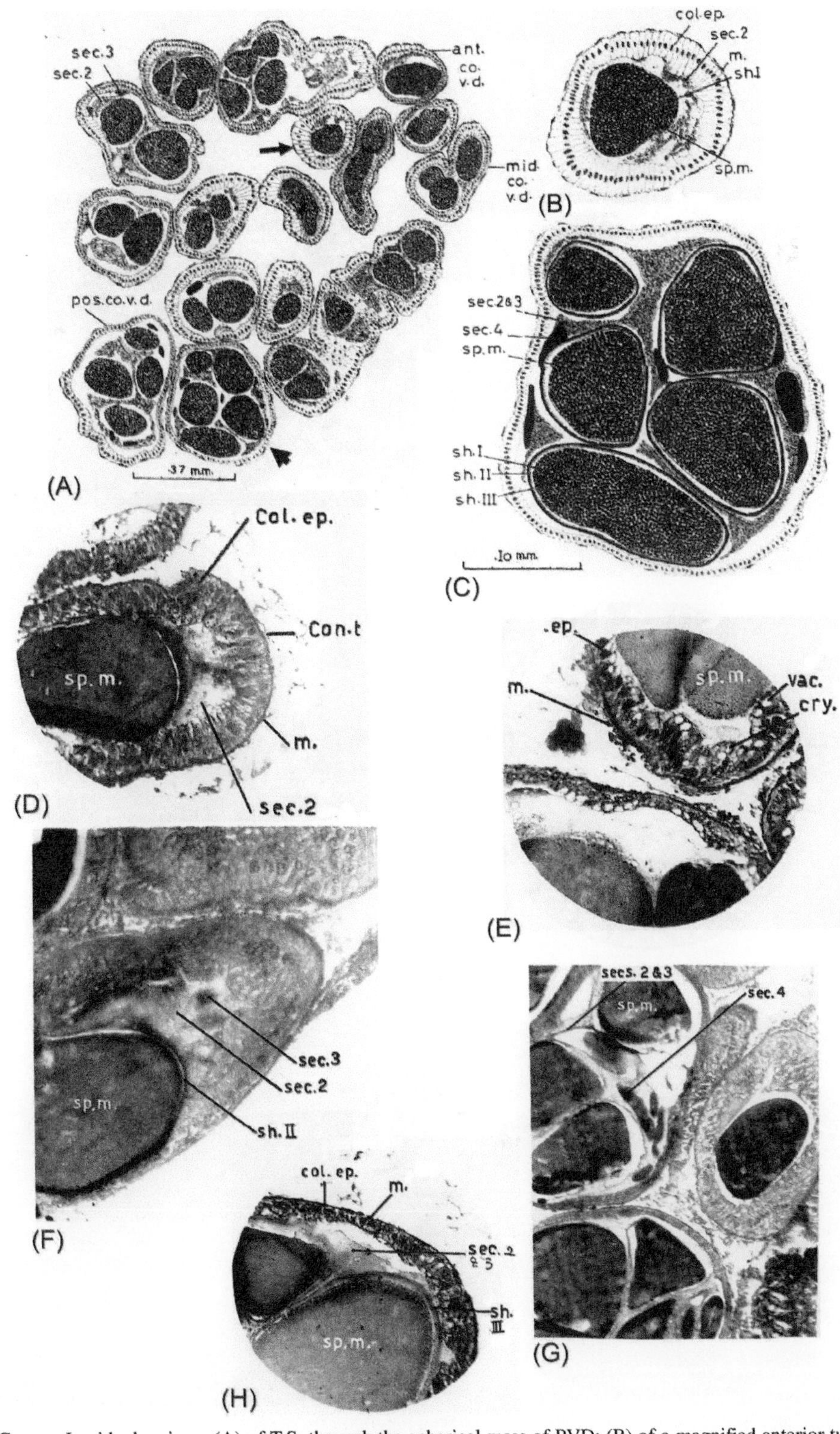

FIG. 10.123 (A–C) Camera Lucida drawings: (A) of T.S. through the spherical mass of PVD; (B) of a magnified anterior tubule (indicated by an *arrow*) of the spherical mass showing the activity of its wall and the formation of sheath I of the spermatophore; (C) of a magnified posterior tubule of the spherical mass (indicated by *arrowhead*); (D–G) photomicrographs of the secretory walls of different tubules of spherical mass of the PVD; (D) tubule of anterior region showing secretion 2. X=200; (E) tubule of the middle region showing the temporary crypts of the epithelial wall and the holocrine secretion 3. X=100; (F) tubule of the middle region showing secretion 3 mixing with secretion 2 and formation of sheath II. X=500; (G) tube of the posterior region showing secretions 2, 3 and 4. X=250. col.ep.=columnar epithelium, con.t.=connective tissue, cr.y.=crypts of the epithelial wall, m=muscles, sec. 2=secretion 2, sec. 3=secretion 3, sp.m.=sperm mass, sh. II=sheath II, vac.=vacuole. *(From El-Sherief (1975).)*

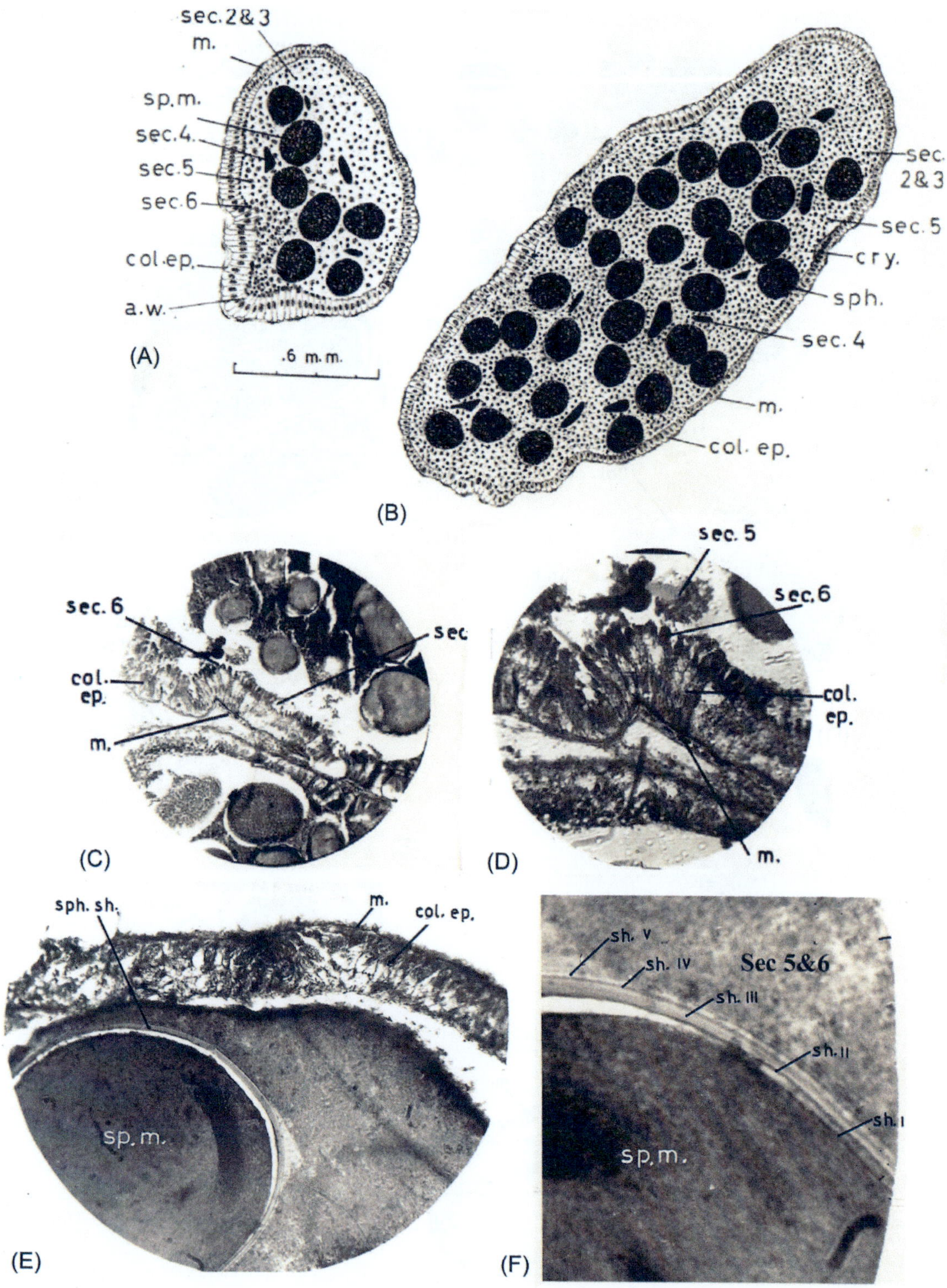

FIG. 10.124 (A and B) Camera Lucida drawings of T.S.s of the spiral region of the PVD *(from El-Sherief, 1975)*: (A) through the anterior part showing the activity of its wall and secretions 5 and 6; (B) through the posterior part showing the great number of fully formed spermatophores embedded in a great amount of secretion 5; (C) photomicrograph of a portion of the wall of the spiral region showing its activity. X=150; (D) photomicrograph of a magnified part of the previous active wall showing holocrine and apocrine secretions 5 and 6. X=300; (E) photomicrograph of a magnified portion of a T.S. through the posterior spiral region showing the fully formed spermatophore. X=350; (F) photomicrograph of a greatly magnified portion of a fully formed spermatophore shown in the previous photo to distinguish the five sheaths surrounding the sperm mass. X=950. a.w.=active wall, col.ep.=columnar epithelium, cry.=crypts, m=muscles, sec. 2&3=secretions 2 and 3, sec. 4=secretion 4, sec. 5=secretion 5, sec. 6=secretion 6, sh. I, II, III, IV, V,=first, second, third, fourth and fifth sheathes, respectively, sp.m.=sperm mass, sph.=spermatophore, sph.sh.=spermatophoric sheath.

differently: as a regular folding tube. In situ, this folded tube takes a definite folding passage (Fig. 10.119A and B), since it flexes three times underneath the heart making an S-shaped structure to fit in this limited area. The S-tube proceeds, with less folding, along the side of the hind gut beneath the hepatopancreas to reach nearly the posterior margin of the cephalothorax, to lead to the distal vas deferens. Unlike the preceding region, the spiral region, this middle vas deferens withstands touch, perhaps because of its thick wall or owing to its internal contents. It appears opaque and its gelatinous contents rarely embed fully formed spermatophores.

The distal vas deferens (DVD) is found posteriorly, in the cephalothoracic region and underneath the hepatopancreas. Unlike the previous region, it is a straight smooth tube with a smaller diameter and is completely concealed among the muscle bundles of the last thoracic leg (Fig. 10.119A). Along its passage, it coils twice: the first coil is around the dorsal muscle of the fifth leg; the second is around the ventral muscle of the same leg. Its length is roughly half that of the folded middle region. It is difficult to be differentiated from the muscles around which it coils, but it becomes clearly visible if any fixative is added. It opens ventrally in the male genital opening (Fig. 10.118B) found on the coxopodite of the fifth leg without elevation or genital papilla. It is appropriate to mention that the corresponding region of the vas deferens of a related crab *Neptunus sanguinolentus* Herbst (George, 1963) is called the ejaculatory duct.

The **intromittent organ** is the end portion of the DVD. It protrudes from the genital opening and always rests in a groove found on the adjacent surface of the coxopodite of the first pleopod (Fig. 10.118B). It is delicate, transparent, and does not exceed 1 cm in length. The most characteristic feature of this organ in *P. pelagicus* is that its outer cuticular covering is continuous with the exoskeleton of the body wall of the animal. George (1963) referred to the corresponding portion of the crab *N. sanguinolentus* as a penis. Since it is not introduced directly into the female but into the first abdominal appendage of the same specimen, the present work preferred the term 'intromittent organ', following Spalding (1942).

Formation of the spermatophore of P. pelagicus

Formation of the sperm masses (role of the testis). Bawab and El-Sherief (1981) studied and published the role of the testis in the formation of the spermatophore of *P. pelagicus*. They noticed that the T.S. through the anterior region of the testis shows the **immature follicles** closely applied to each other to the degree that one cannot define the limits of any follicle. However, each follicle is blocked with a homogeneous mass of small rounded cells with comparatively large dense nuclei (Fig. 10.120B). These cells were counterstained by light green and aniline blue of Masson's and Mallory's triple stains, respectively. In the course of maturation, which proceeds gradually posteriorly, each follicle undergoes some definite changes as follows: (1) the membrana propria gradually thickens and hence the gradual distinction of the follicles (Fig. 10.120C). (2) Very few larger cells with diffused nuclei appear along one side of each follicle, i.e. different from the whole bulk, called **marginal cells** to differentiate them from the original bulk. Owing to their gradual increase in number (high activity) and the coiling of the seminiferous tubules, these marginal cells appear in T.S. as patches occupying the different corners of the adjacent follicles (Fig. 10.120C and H). These cells were densely stained with acid fuchsin and refuse the counterstains of Masson's and Mallory's triple stains. In spite of their well-marked activity, they never undergo spermatogenesis and the spermatozoa are produced from the original cells of the bulk. Thus the latter cells are truely **germinal cells** since they undergo spermatogenesis (no cytological study). It is worth mentioning that in 2010, Stewart et al. studied spermatogenesis in the same species from another locality, in Thailand (Indian Ocean waters). In spite of the fact that they did not present a detailed study of the testis, but the photos indicate that the maturation of the testes proceeds as that described by Bawab and El-Sherief (1981). However, Stewart et al. presented TEM micrographs describing in detail 13 stages of this process, based on cytological characteristics, two stages of spermatogonia, six of primary spermatocytes, a secondary spermatocyte stage, three of spermatids and mature sperm (Fig. 10.125). Histologically in *Portunus*, this is indicated by the decrease of the size of the individual cells and the increase of their affinity to stains. Accordingly, the area occupied by them decreases too and is replaced by the product of the activity of the marginal cells. (3) few well-marked cells appear scattered in between the maturing germinal cells. They are round with very small nuclei and contrary to the germinal cells, they are faintly stained with acid fuchsin. They are termed **A-cells** (Fig. 10.120D). At the same time, few other cells deeply stained with acid fuchsin are distinguished between the newly formed marginal ones (Fig. 10.120E, F, and J) but smaller in size with large dense rounded nuclei. For differentiation, they are called **B-cells**.

At this stage of development, each follicle becomes differentiated into two distinct areas: germinal and marginal areas. The wider the marginal area is, the higher the degree of maturity of the follicle (Fig. 10.120G–I). It is worth mentioning that this mode of development of the testis of *P. pelagicus* is more or less the same as that of *Penaeus kerathurus* (see Fig. 10.67). The difference is found in the later stages. In addition, it recalls the description and illustration of the testis of the crab *Menippe mercenaria* (Binford, 1913).

The mature follicle. As the spermatids of *P. pelagicus* become released, the germinal area becomes sharply separated within the follicle, then gradually changes into a definite ductule. Some marginal cells become cubical and arrange themselves one by one to surround this germinal area (Fig. 10.120F–J), forming its wall. At the same time, the membrana propria invades the interior of the follicle to

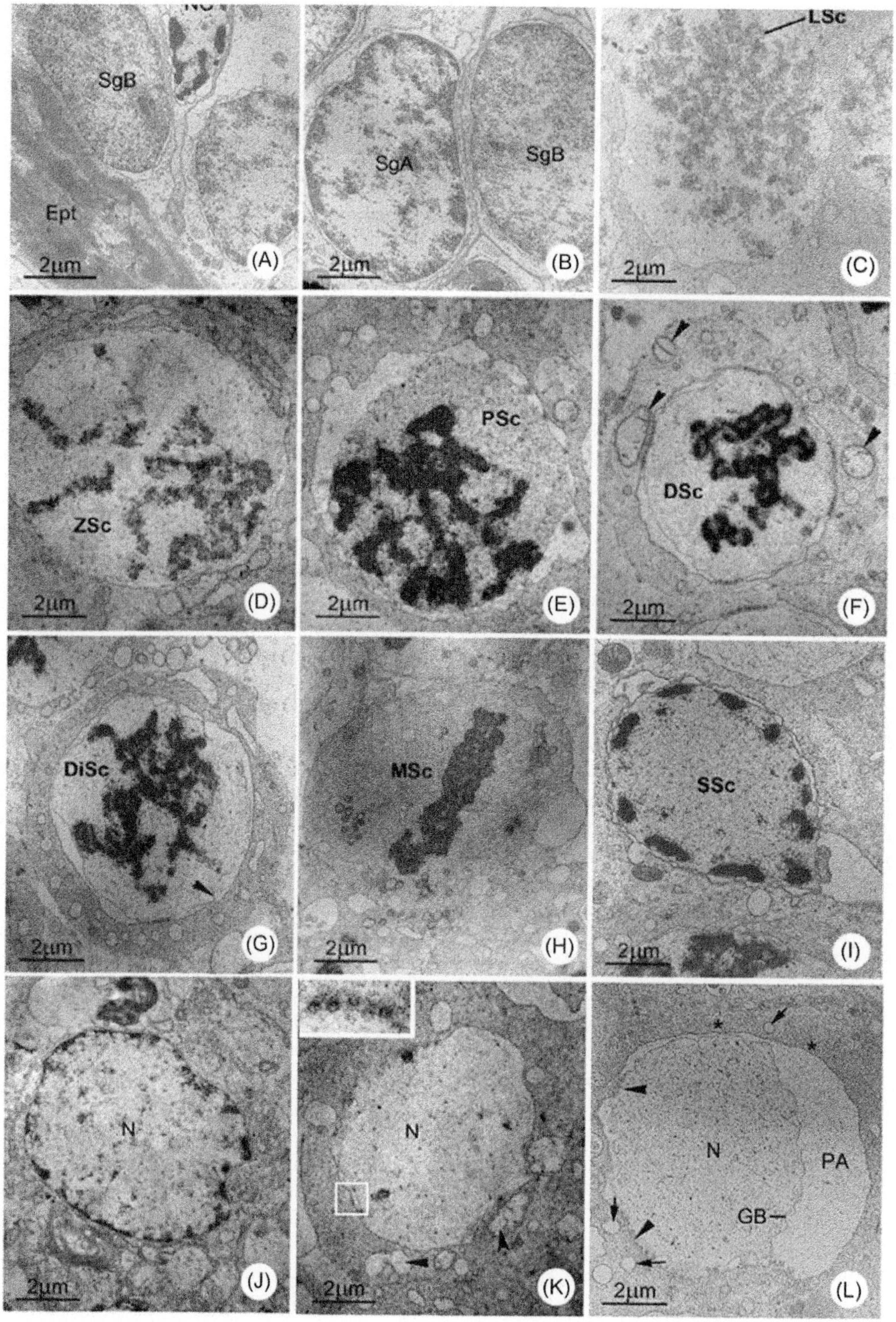

FIG. 10.125 TEM micrographs of spermatogenesis in *P. pelagicus (from Stewart et al., 2010)*: (A) and (B) show spermatogonia (SgA) and (SgB) are located directly on the basement membrane, accessory nurse cells (NC) are seen in intercellular space between the spermatogonia; (C–H) prophase stages of primary spermatocytes: (C) leptotene spermatocyte (LSc) shows the forming of chromatin cords which contain loosely packed fibers, while the rest of the chromatin is still in euchromatic form; (D) zygotene spermatocyte (ZSc) shows thick cords of heterochromatin that are interwined through the nucleus, and some cords are paired; (E) pachytene spermatocyte (PSc) shows thicker and denser cords of heterochromatin that occupy the whole nucleus; (F) diplotene spermatocyte (Dsc) shows thick strands of chromatin starting to clump, *arrowheads* indicate few Golgi complexes and some large mitochondria; (G) diakinesis spermatocyte (DiSc) shows clusters of thick chromatids in the centre of the cell and the dispersion of the nuclear envelope *(arrowhead)*; (H) metaphase (MSc) shows fully formed chromosomes aligned at the equatorial region; (I) secondary spermatocyte (SSc) is smaller than primary spermatocyte and the heterochromatin blocks bordering the nuclear envelope are arranged in a clock-face pattern; (J) an early spermatid (St1), nucleus (N); (K) the middle stage spermatid (St2) the heterochromatin disappears almost entirely from the nucleus, apart from two opposite regions where there are still blocks of heterochromatin in contact with the nuclear envelope, most cytoplasmic organelles are aggregated on one side, there is formation of nucleopores (inset); (L) late stage spermatid (St3) there is partitioning of nucleus by granular belt (GB) and formation of a large proacrosomal vesicle (PA), nucleopores are now more numerous *(arrowheads)* along the nuclear envelope and few vesiculated mitochondria *(arrows)* in the cytoplasm matrix *(asterisk)*.

surround this wall and thickens to become more obvious. Fine muscle fibres that are circularly arranged can be seen embedded in this connective tissue coat. The histological picture of the testis, at this stage, has a number of scattered ductules with different sizes and outlines, referred to as **testicular ductules** (Fig. 10.120G, I, and J). This histological description of the testis of *P. pelagicus* can also be applied to that of the adult blue crab, *Callinestes Danae* Smith, 1869 (Zara et al., 2012). Thus, the maturation of the testes of these two species is identical (Fig. 10.126). At a later stage of maturity, it becomes clear that these testicular ductules in *P. pelagicus* are not only for protecting spermatids and/or spermatozoa during spermiogenesis and conducting the testicular product, but also for producing a peculiar secretion. Its epithelial wall secretes, by eccrine method and later on by holocrine (Garven, 1965; Leonhardt, 1977), a fluid substance (Fig. 10.120K), which diffuses in between the developing spermatozoa. Since this secretion is the first to be deposited as a constituent of the spermatophore, it is named **secretion 1** (sec. 1). Histochemical reactions show that it is mostly of carbohydrates with some lipids, while the proteins take no share in this substance (Table 10.10). The most noticeable and interesting observation is the complete disappearance of A-cells at the beginning of the secretory function of the ductule. However, the assumption that these A-cells are nutritive ones is improbable, although their persistence is limited to the period before spermiation. Therefore, the testicular ductule is formed to protect the developing sperm, nourish them, and convey them to the vas deferens. This means that spermiogenesis in *P. pelagicus* occurs within the testis, specifically within the testicular ductules. This observation agrees with previous information (Huxley, 1880; Pearson, 1908; Heldt, 1938; Bawab, 1964; Malek and Bawab, 1974a). Moreover, Stewart et al. (2010) recorded that the 'spermatozoa from the testis enter the AVD as a densely homogeneous mass'. Zara et al. (2012) observed and stated that the 'seminiferous ducts' of *C. danae* receive the spermatozoa from the 'testicular lobules'. This means that spermatogenesis and spermiogenesis occur within the testicular tissue, and not in definite ducts of the testis. On the contrary, Sainte-Marie and Sainte-Marie (1999a,b) observed and reported that the testicular products of the adult snow crab, *Chionoecetes opilio* (O. Fabricius), which enter the vas deferens, are not spermatozoa. They enter the vas deferens as precursors of immature spermatids. This means that the site of spermiogenesis of *Chionoecetes opilio* is the initial portion of the vas deferens (see later the formation of its spermatophore). This means that there are three sites of spermiogenesis in Brachyura: (1) within the tissue of the testes, e.g. *C. danae* (Zara et al., 2012); (2) within definite testicular ductules of the testis, e.g. *P. pelagicus* (Bawab and El-Sherief, 1981); and (3) within the anterior portions of the vasa deferentia, e.g. *Chionoecetes opilio* (Sainte-Marie and Sainte-Marie, 1999a,b).

Bawab and El-Sherief (1981) identified that two other functions are performed by these testicular ductules in *P. pelagicus*. **First**, its secretion (sec. 1) is also to accompany the testicular product to the vas deferens. Hamon (1941a,b) stated that the sperm coming from the testes of *Eupagurus prideauxi* Leach are surrounded by a fluid material, and Matthews added that the secretion of the collecting tubules of the testes of *Panulirus penicillatus* Oliver (1951) and *Dardanus asper* (De Haan) (1953) may serve as a lubricant and a nutrient. To be a lubricant in *Portunus* is a reasonable assumption since it contains lipids, but as a nutrient, might be mainly for retaining activity at the proper time. **Second**, the testicular ductules of *Portunus* lead directly to the tubules of the spherical mass of the PVD (Fig. 10.122). This is ascertained from the fact that the number of the tubules of the VD ± equals those of the testis.

A more advanced stage of maturation shows that the membrana propria surrounding the testicular ductule persists, but the secretory epithelial wall gradually fades (Fig. 10.121A). Its content becomes more condensed to the degree that the individuality of the metamorphosing sperm is completely lost, i.e. the content of each ductule is converted into coherent mass called sperm mass. This means that the sperm are inactive but live. The marginal area of each seminiferous tubule still has the same histological pattern, except that the B-cells increase greatly in number.

This work of Zara et al. (2012) added cytological studies of spermatogenesis and spermiogenesis in *C. danae*. What concerns the development of the sperm of *Portunus*, might be the same in *C. danae*. The early spermatids are difficult to identify before the beginning of production of the acrosomal vesicle. This α-metachromatic vesicle arises in the cytoplasm and facilitates the identification of early spermatids, even when nuclei are still round (Fig. 10.126D). Nuclei are less dense than those of secondary spermatocytes. Middle spermatids are characterized mainly by changes in the shape of the nucleus, from round to C-shaped over the acrosomal vesicle (Fig. 10.126E). In late spermatids, the nuclear cap (3.2 ± 0.2 μm) becomes more evident and surrounds most of the α-metachromatic vesicle. Nuclei are also narrower than those observed in previous stages (Fig. 10.126F). No differences were found between the diameters of these nuclei compared to those of secondary spermatocytes. Sperm are characterized by nuclei surrounding the acrosome almost completely (Fig. 10.126G). However, despite morphological changes, the average diameter (3.0 ± 0.2 μm) was not significantly different from those of late spermatids.

The spermatozoa. In *P. pelagicus*, at this level of study (light microscopy), the mature spermatozoa are indistinguishable from the spermatids in the ductule of the mature follicle; owing to two facts. Firstly, the metamorphosing spermatids are very small rounded cells, and the same as the fully grown sperm (Fig. 10.120G–K). Secondly, the secretory substance (sec. I) coaleses the sperm forming a compact mass, which renders the identification of their individuality

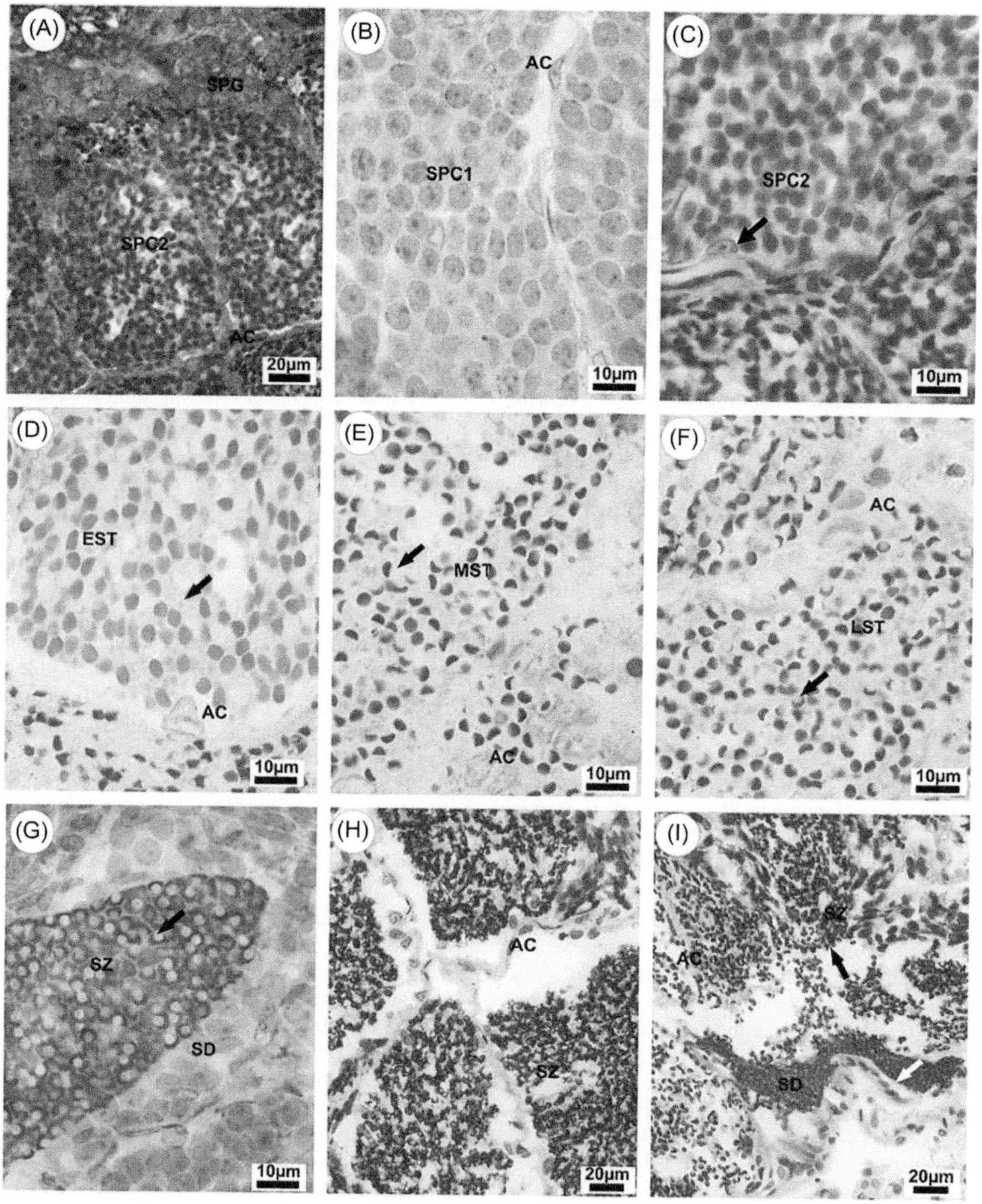

FIG. 10.126 Spermatogenesis in developing males of *C. danae*, Toluidine blue staining (pH 4.0) *(from Zara et al., 2012)*: (A) germinal centre of spermatogonia (SPG) in the periphery of 2 seminiferous lobules filled by spermatocyte II (SPC2), accessory cells (AC) delimited both lobules; (B) seminiferous lobule of spermatocytes I (SPC1) with characteristics of meiotic prophase delimited by AC with flat nuclei; (C) spermatocytes II with small and homogeneous nuclei and elliptical AC *(arrow)*; (D) early spermatids (EST) with round nuclei and proacrosomal vesicle with α metachromasia *(arrow)*, note increase in size of AC during spermiogenesis; (E) intermediary spermatids (MST) with nuclei forming cup on proacrosomal vesicle *(arrow)*; (F) late spermatids (LST) with compact nuclear cup on acrosome *(arrow)*; (G) seminiferous duct (SD) filled with spermatozoa (SZ) with α metachromatic acrosome *(arrow)*; (H) general view of several testicular lobules in mature males filled with mature spermatozoa and accessory cells with condensed nuclei; (I) testicular lobule *(arrow)* with flat accessory cells releasing spermatozoa into seminiferous duct formed with cubical/columnar epithelium *(white arrow)*.

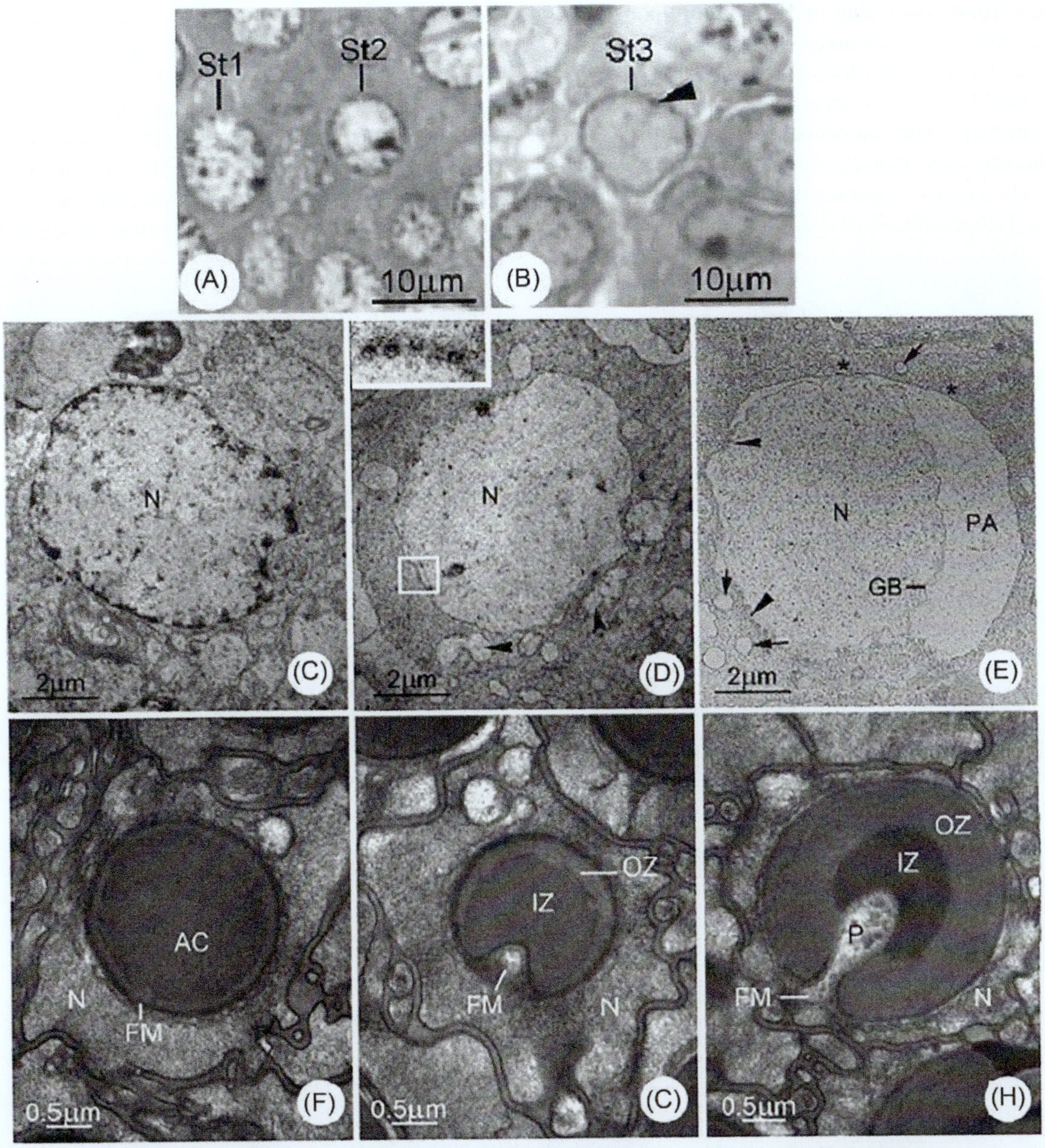

FIG. 10.127 Spermatogenesis in *P. pelagicus*, semithin sections stained with methylene blue: (A) and (B) show spermatids (St 1–3). Early stage spermatids St 1 possess nuclei with clusters of small blocks of heterochromatin (A), whereas the heterochromatin of middle St 2 and late St 3 (B) is dispersed within the nuclei, the cytoplasm contains dense granule is a part of the acrosome vesicle (B, *arrow*); (C–H) TEM micrographs of three stages of spermatids: (C) an early spermatid St 1, nucleus (N) contains small blocks of heterochromatin disposed along the inner surface of the nuclear membrane; (D) middle spermatid St 2, in which most cytoplasmic organelles are aggregated in one side. There is formation of nucleopores (inset); (E) late stage spermatid St 3, there is partition of nucleus by nuclear granular belt (GB) and formation of a large proacrosomal vesicle (PA), nucleopores are more numerous *(arrowheads)* and few vesiculated mitochondria *(arrows)* in the cytoplasm matrix *(asterisk)*; (F–H) maturation of the spermatozoa showing the invagination of the acrosome vesicle (AC) from the nuclear side, the loose fibrous material (FM) which originates from the cytoplasm occupies the space within the invaginated pouch which continues to grow inwards to form a tube-like structure forming the innermost core of the acrosome, equivalent to a subacrosomal space or perforatorium (P), the nucleus (N) surrounds much of the acrosome which contains a homogeneous dense material, which divides into a denser inner zone (IZ) and a less outer zone (OZ), the inner zone later reduces in size but its matrix becomes denser and attains a horse-shoe shaped configuration in (H). *(From Stewart et al. (2010).)*

difficult (Fig. 10.121A). However, Stewart et al. (2010) studied spermiogenesis (Fig. 10.127) in the blue swimming crab *Portunus pelagicus* (the same species worked on by Bawab and El-Sherief (1981)). Afterwards, Zara et al. (2012) studied cytologically and histochemically the same aspect in *C. danae* and reported that spermatid maturation in this species (described above), undergoes the same three stages described by Stewart et al. (2010). To study the structure of the full-grown sperm, Bawab and El-Sherief (1981) prepared successive smears from: the posterior region of the testis; the portion connecting the testis with the transverse bridge; and the succeeding anterior portion of the spherical mass of the PVD. It was found that the second preparation contained fully grown sperm, while the third one was loaded with them. The full-grown spermatozoon is a very small spherical mass densely stained with haematoxyline, the head. The latter is ensheathed with a thin membrane of cytoplasm. It measures about 2.7 μm in diameter. From the cytoplasmic sheath, two or three processes, called rays, radiate at different levels and can only be identified by oil immersion magnification (Fig. 10.128A).

TEM studies evidenced the morphology of this sperm, but not the details. El-Shereif (1991) studied this sperm (Fig. 10.128B–H) from the contents of the anterior spherical mass of the PVD and from the preserved spermatophores in the spiral PVD. The sperm of *P. pelagicus* appears globular

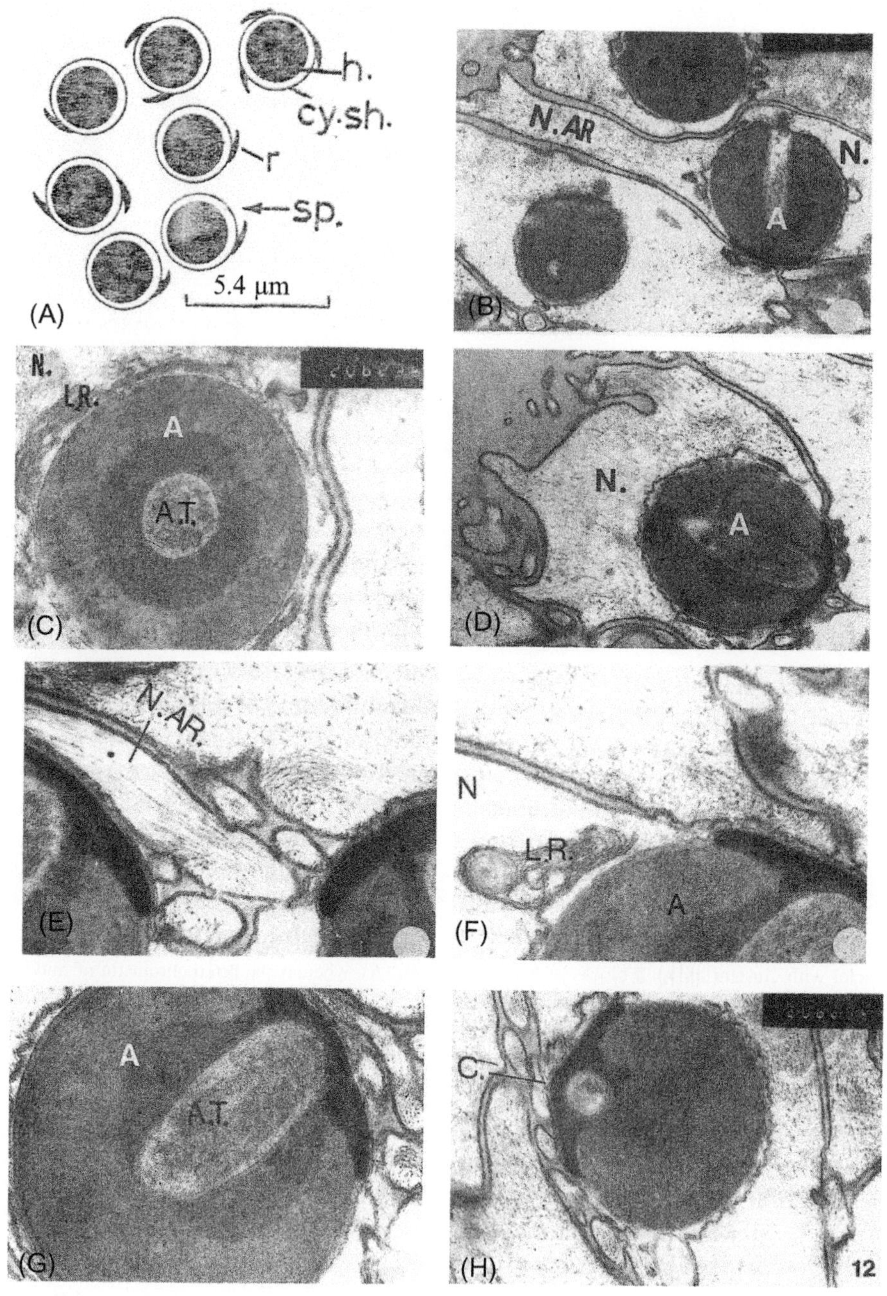

FIG. 10.128 TEM micrographs showing the fine structure of the sperm of *P. pelagicus* from spermatophores in spherical mass of the PVD *(from El-Sherief, 1991)*: (A) portion of the sperm shows the nucleus and nuclear arm. X = 10,000; (B) shows the acrosome and lamellar region, notice central acrosomal tube and two different areas forming acrosome. X = 20,000; (C) shows the acrosome and nucleus. X = 10,000; (D) shows radial nuclear arm with its fibrillar nature. X = 4000; (E) shows middle lamellar region in between nucleus and acrosome. X = 40,000; (F) shows different areas of acrosome and acrosomal tubule. X = 30,000; (G) shows anterior part of acrosome with raised centre of circular apical cap. X = 20,000. A = acrosome, AT = acrosomal tubule, C = apical cap, LR = lamellar region, MT = microtubules, N = nucleus, N.AR = nuclear arm.

in outline. It is composed of three different parts: the outer Feulgen positive **nucleus** with long **nuclear arms** (rays) (Fig. 10.128B); the middle **lamellar or cytoplasmic region** (Fig. 10.128C); and the antero-inner large PAS-positive, rounded **acrosome** (Fig. 10.128D). The nuclear arms contain a fibrillar, uncondensed chromatin (Fig. 10.128E) and no microtubules were found in either the nucleus or the nuclear arms. The lamellar region is found in between the nucleus and the acrosome (Fig. 10.128C and F). It consists of cytoplasmic remnants of a thin layer of membranes that expanded in some areas by the aggregation of mitochondria. This lamellar region is found to be continuous with the central acrosomal tubules. The acrosome consists of two main portions of different electron density. The inner portion is circular in outline, two-thirds of which are occupied by concentrated acrosomal tubules (Fig. 10.128G), while the remaining third is electron dense material that appears as two lateral lobes around the central tubules. The outer portion is the widest and is made of homogeneous, light electron dense material. In addition, a very narrow outer lamellar region is observed to cover the entire acrosome. Anteriorly and beneath the plasma membrane is the **apical cap**, that is circular in outline but slightly raised in the centre (Fig. 10.128H). It is composed of electron dense material, distinctly differentiated into an outer highly condensed area and an innermost, less condensed area.

The TEM studies description by Stewart et al. (2010) of the structure of the sperm is almost the same as described above i.e. evidencing that of El-Shereif (1991). They just added the stages of the differentiation of the spermatid into mature sperm (Fig. 10.127F–H).

The rayed sperm is a characteristic feature for many brachyurs (Calman, 1909), and they are only differentiated by the number of these rays. Binford (1911) found five or more pseudopodial rays in the sperm of *Callinectus sapidus* (Rathbun). The sperm of the crab *Carcinus maenas* are similar to those of other crabs and have three or four rays (Spalding, 1942). Those of the sperm of another crab *Neptunus sanguinolentus* studied by George (1963) are three or four in number. The sperm of *P. pelagicus* is similar to those of these crabs, but with only two or three rays (Bawab and El-Sherief, 1981).

The flow of the product of the seminiferous tubules of the testes of *P. pelagicus* to the vas deferens occurs directly from the testicular ductules to the fine tubules of the spherical mass of the PVD, i.e. no collecting tubules as in *Carcinus maenus* (Spalding, 1942) and in *Penaeus kerathurus* (Bawab, 1964). The direct flow to the vas deferens was early information, since Pearson (1908) described the testes of *Cancer* without collecting tubules and the follicles of the testes that lead directly to the vas deferens. As mentioned above, the portion connecting the testis with the transverse bridge in *P. pelagicus* (Fig. 10.121C) shows the close relation and similarity between the testicular ductules and the spherical mass of the PVD. In addition, Fig. 10.122 illustrates the changes take place in this short transitional portion. These critical examinations of this transitional portion during February reveals that the true testis (as a tissue) ends just with the posterior margin of the transverse bridge, beyond which each testicular ductule is converted quickly into another type of tubule.

As mentioned above, the testicular ductules of the fully mature follicles contain coherent sperm masses. The latter differ in size and are conveyed to the proceeding region by the rhythmic contractions of the muscles found in the wall of these ductules. At this stage, the histological picture of the testis appears different from both the immature and mature follicles and is referred to as an **evacuated follicle** (Fig. 10.121B). The membrana propria surrounding the follicles becomes distinct again, and the portion forming the ductule becomes empty but with some debris. Its epithelial wall fades and the cells of the marginal area disintegrates except the B-cells which persist, increase in number, and try to fill or renew the follicle.

Deposition of the main layers of the spermatophore (role of the PVD). Study of serial T.S.s and L.S. of histological preparations through the spherical mass of VD reveals that this region, in addition to being conducting, is secretory. However, the latter function differs from one level to the other along this region. **Anteriorly**, just beneath the transverse bridge, each fine tubule looks like the occupied testicular ductule with respect to the contents (Fig. 10.121C). However, the wall is comparatively very thick, and consists of an outer thin connective tissue layer with very fine muscle fibres (continues along this region), and an inner layer of highly elongated columnar epithelial cells with central granulated nuclei (Fig. 10.123B) (this histological picture of the anterior region of the spherical mass of the PVD was evidenced by Stewart et al., 2010 (see Fig. 10.121D)). From the first glance, one can recognize the high activity of this glandular epithelium, which is increased by forming temporary crypts (Fig. 10.123D). The product is a copious frothy secretion discharged merocrine and has the same affinity to stains as sec. 1, but it differs in the amount and the appearance in fixed condition. Since its site of secretion differs, and it follows sec. 1, it is referred to as **secretion 2** (sec. 2). It is adhesive that it further coalesces the individuality of the sperm masses and, in addition, it ensheaths each mass by a definite very thin membrane that can be identified as **sheath I** (Fig. 10.123B and C). Therefore, the coherence of the masses becomes more pronounced and this was examined by a simple experiment. The spherical mass was isolated from a live crab; its wall was ruptured by a pin and put in a small petri dish with seawater. The compactness of the extruded sperm masses persist for a long time and their consistency as coherent masses was preserved. However, one of the smear preparations to study the structure of the fully formed sperm, as mentioned above, was done from these masses. This ascertained that these masses are firmly coalesced but not hardened.

Gradually and beyond the transverse bridge, at about its **middle** level, the fine tubules fuse with each other forming larger tubes (Fig. 10.123A). Their contents may fuse forming the final size of sperm masses of the spermatophores. In addition, the histological structure of the wall changes. The epithelial lining becomes comparatively smaller but still secretory. The active parts are thicker than the nonactive and their cells are arranged in groups as rosettes. A large round vacuole is clearly observed at the base of each active cell (Fig. 10.123E). Their product is **secretion 3** (sec. 3); it is little in the amount and discharged apocrine. It differs in its affinity to stains from sec. 2, since it is positively stained with Heidenhains-haematoxylin (H-h) and gives faint red colouration with the complex stains of Mallory's and Masson's combinations (Table 10.9). Sec. 3 diffuses smoothly throughout the abundant sec. 2, intermingled with each other, and the resulting substance surrounds the sperm masses, forming **sheath II** (Fig. 10.123F). However, sec. 2 remains in excess among the developing sperm masses.

Fusion of the tubes of the spherical mass of PVD precedes that at its **posterior** level, only few large tubes are found; each contains more than one sperm mass. Its epithelial lining is very thin compared to that of the anterior and middle levels, but the active portions become thicker, measuring about one-quarter that of the anterior level (Fig. 10.123G). Its holocrine secretion, **secretion 4** (sec. 4), differs in its appearance from the former ones. It is in the form of dense patches that stain deep green and bright blue with Masson's and Mallory's triple stains, respectively. It is consumed in the formation of the third sheath of the spermatophores, **sheath III** (Fig. 10.123H).

As the **spiral region** of the PVD differs in morphology from the spherical mass, it also differs in its T.S. (Fig. 10.124A and B), and so in its function. Along its large three coils, the outer connective tissue layer with the fine muscle fibres constitute its outer coat, which is continuous to that of the spherical mass. The epithelial lining is still secretory but differs in its thickness and the nature of its product. Not all the parts of this epithelium are seen in the same physiological state at the same time. It appears that they act in succession (Fig. 10.124C and D). That part in activity has columnar cells heavily loaded with secretion. The latter discharged holocrine and referred to as **secretion 5** (sec. 5), which increases greatly in amount along the whole length of this spiral region that fills all the spaces between the developing spermatophores. It is in the form of large translucent granules, heavily stained with H-h, deep bright red with acid fuchsin in Masson's, while it is orange-red with Mallory's triple stain. Along just a short distance of the first coil of this spiral region, and after the discharge of sec. 5, the epithelium secretes what is called **secretion 6** (sec. 6). The latter resembles sec. 5 in being in the form of translucent granules, but differs in many respects. Firstly, its secretion is confined to small portion of the first coil; it is discharged apocrine; and it accepts bright deep green and blue colours with Masson's and Mallory's stains, respectively, while it is negatively stained with H-h (Table 10.10). Sec. 6 was well observed impregnating the granules of sec. 5 (Fig. 10.124D). The two fuse together and the resulting substance ensheaths each spermatophore with a complete sheath, referred to as **sheath IV**. This sheath IV appears as a homogeneous substance (Fig. 10.124F). After sec. 6 stops, the field along the whole length of this spiral region is seen to be loaded with only sec. 5, which changes to become a homogeneous matrix embedding the fully formed spermatophores. Sec. 5 not only furnishes the matrix, but also forms **sheath V** around each of the spermatophores.

Summarizing the role of the PVD in the formation of the spermatophore of *P. pelagicus*, one can say that the spherical

TABLE 10.9 The appearance and staining reactions of the secretions of the spherical mass of the proximal vas deferens of *Portunus pelagicus*

No. of secretion	Site of secretion	Appearance	Affinity to stains: Heid.	Affinity to stains: Masson	Affinity to stains: Mallory	Sudan Black B	Sheath
(2)	Anterior spherical mass	Frothy	–	Bright green	Bright blue	–	I
(3)	Middle spherical mass	Clear	Faint +	Faint red	Faint red	+	II
(4)	Posterior spherical mass	Dense patches	–	Deep green	Deep blue	–	III

From El-Sherief (1975).

TABLE 10.10 The appearance and staining reactions of the secreted material taking a part in the formation of the spermatophore of ***Portunus pelagicus***

							Gelatin				
				Paraffin			Sudan black B		Proteins		
No. of secretion	Part of the systems	Method of secretion	Nature of secretion	Heid haen.	Masson's	Mallory's	Paraffin	Gelatine	Millon	Xanthop.	Function of secretion
(1)	Ductule of testis	Holocrine	Fluid	–	Green	Blue	–	++	–	–	Nourishing
(2)	Anterior coiled mass	Merocrine	Frothy	–	Bright green	Blue	±	+++	+++	++	Adhesive, sheaths I and II
(3)	Middle coiled mass	Holocrine	Clear	+	Faint red	Faint red	–	–	+++	+++	Sheath II
(4)	Posterior coiled mass	Holocrine	Dense patches	–	Deep bright green	Bright blue	+	+++	++	+	Sheath III
(5)	Wide spiral region	Holocrine	Large translucent granules	+	Deep red	Orange red	–	–	++++	++++	Sheaths IV and V
(6)	First coil of the spiral region	Apocrine	Small translucent granules	–	Dense green	Dense blue	–	–	–	+	Sheath IV
(7)	MVD crypts	Holocrine	Fluid	+	Red	Orange red					
(8)	MVD	Apocrine	Fluid		Green	Blue					

From EL-Sherief (1975).

mass is responsible for further coalescence of the coherent sperm masses coming from the testicular ductules. In addition, it is responsible for the formation of the three successive sheaths around each spermatophore. It seems that the corresponding region in other crabs is the site of formation of the spermatophores. For example: (1) Beninger et al. (1988) reported that the spermatozoa of *Chionoecetes opilio* (O. Fabricius) and the matrix of the AVD are packed into spermatophores and surrounded by a pellicle; (2) the simple ellopsid spermatophores of the deep water crab *Charybdis smithii* McLeay (Indian specimen) are formed and lodged in the AVD (Balasubramanian and Suseelan, 2000); and (3) Zara et al. (2012) described in detail the formation of the spermatophores of *C. danae* in the anterior part of AVD and stated that it is the site of secretions forming their walls. In *Portunus*, the large three coils of the spiral region of the AVD add the IV and V sheaths, and in addition form the copious matrix in between the spermatophores. It is the site of preservation of the fully formed spermatophores and the hardening of the latter. Therefore, this is the cause of the difficulty encountered during sectioning this region. However, the PVD in general, and its three coils in specific, is the most important region of the male reproductive system of *Portunus* in respect to spermatophore formation and storage. Storage function of the corresponding region was mentioned before in *Carcinus meanas* by Mouchet (1930). In addition, in *Eupagurus prideauxi*, Hamon (1941a) stated that this region is an enlarged one, and can be considered as a reservoir for the storage of the spermatophores before being ejected to the outside. The corresponding region of *Neptunus sanguinolentus* was described as a massive part filled with white spermatophores (George, 1963). More recently, Zara et al. (2012) evidenced this view by their findings that the 'mature round spermatophores are found in the posterior part of AVD'.

Role of the MVD. Shortly beyond the spiral region *P. pelagicus*, the section through the MVD (folded tube) has different outline with an oval lumen, but it is a narrower tube. Its outer coat is thicker and its circularly arranged muscle fibres are larger in size and greater in number. It is interesting that its epithelial lining is still secretory in spite of the fact that the spermatophores are completely formed within the PVD. It is composed of highly elongated columnar cells (Fig. 10.129), which seem to be the most active cells in this reproductive system. This is because of the numerous invaginations and evaginations of the epithelial lining forming what is called crypts. Owing to the latter, the outline of the section is never continuous and its lumen is greatly reduced (Fig. 10.129A). The crypts secrete a holocrine fluid substance, **secretion 7** (sec. 7), which stains red and orange-red with Masson's and Mallory's triple stains and positively stained with H-h. At the same time, the noncrypt epithelium secretes another substance, **secretion 8** (sec. 8), which differs in being apocrine and stains green and blue with the two above triple stains, respectively. It is interesting to note that the epithelial lining of the corresponding region in the crab *Neptunus sanguinolentus* has almost the same description. George (1963) stated that there are numerous glandular pockets found along this epithelial wall. In *Portunus* posteriorly, when the folds of this region fade morphologically, the internal crypts fade too, but its secretory function persists until its end (Fig. 10.129D). The most noticeable observation is that the spermatophores are rarely found in any portion of this MVD, so it is only secretory and never being for storage of these spermatophores. However, its secretions 7 and 8 may accompany the fully formed spermatophores as a matrix, to be transferred to the female during copulation.

Role of the DVD. In *Portunus*, this region of the VD is characterized by its simple structure, both morphologically and histologically. All the above-mentioned authors agree that this last region of the VD of their investigated animals is made mainly of thick muscular layers, differently arranged and embedded in a thick connective tissue layer. In *Portunus*, Fig. 10.129E and F shows that the thickness of the muscle layer in this straight tube is approximately equal to its epithelial lining. Most of these muscles run singly and all are of the stripped type characteristic of the arthropods. In spite of this, the wall of this narrowest region of the VD is comparatively thin and the lumen is wide. Its epithelial lining appears inactive compared to the preceding epithelia. Yet the free edges of the cells show continuous secretion, which in stained preparations and inside the lumen shows affinity to the light green. At the level of this study, neither the nature of this secretion nor its precise function was known with certainty. However, it is intermingled with the previous fuchsinophil substance. The wall's musculature equips it to ejaculate the huge numbers of the spermatophores during copulation, a function previously mentioned by George (1963).

Moulding of the spermatophore of P. pelagicus. Since almost all the regions of the male reproductive system of *P. pelagicus* share in the formation of the spermatophores, their moulding proceeds in more than one step and can be outlined as follows: (1) Inside the testicular ductules, the metamorphosing sperm are mixed and coalesced with sec. 1. The contraction of the muscles of the wall of these ductules is not continuous, that the product is produced in masses and not in a continuous column. (2) This contraction also facilitates the passage of these unequal masses to the spherical PVD. Within the latter anteriorly, the masses are mixed together and by the muscle contractions and sec. 2, masses of nearly equal size are produced. Owing to the high coiling of the narrow tubes of this region, these masses undergo rolling throughout its length that they become spheres. However, the typical spheroidal shape of

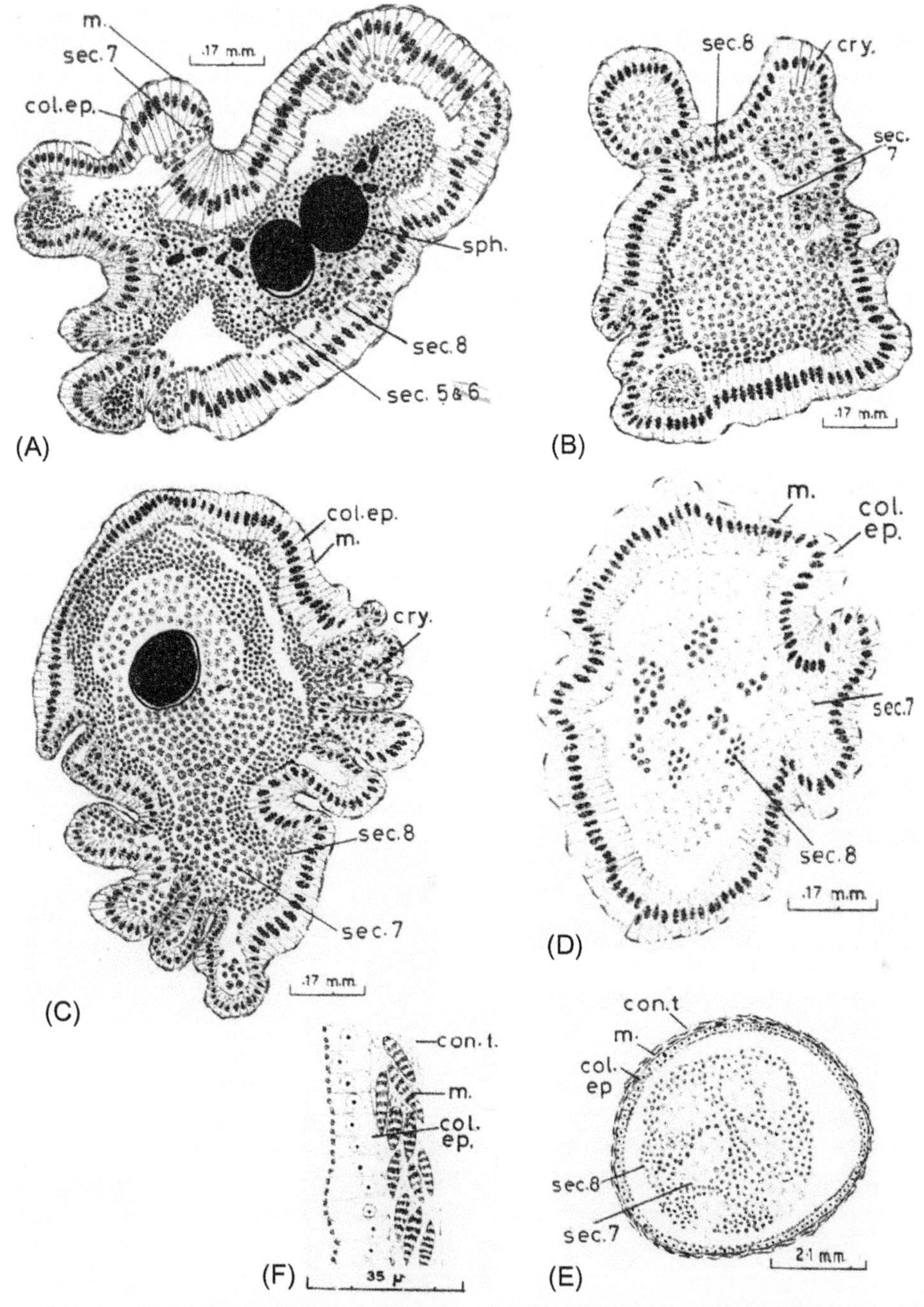

FIG. 10.129 Camera Lucida drawings of T.S.s of different regions of vas deferens of *P. pelagicus*: (A) through the most anterior part of the MVD; (B) through the anterior part of the MVD; (C) through the middle part of the MVD; (D) through the posterior part of the MVD; (E) through the DVD; (F) magnified portion of the wall of the DVD. Col.ep. = columnar epithelium, con.t. = connective tissue, cry. = crypts, m. = muscles, sec. 5&6 = secretions 5 and 6, sec. 7 = secretion 7, sec. 8 = secretion 8, sph = spermatophore. *(From El-Sherief (1975).)*

the prospective spermatophore is not attained in this region, because they are not yet hardened. In addition, the sperm masses acquire sheath I. Then successive two secretions are added forming sheaths II and III. (3) The three sheathed spermatophores pass to the spiral PVD where they accept sheaths IV and V in excess of sec. 5, which forms the matrix embedding these fully formed spermatophores. This spiral PVD is also the reservoir keeping these elements, until the proper time of ejaculation. During their storage and after attaining their definite spheroidal shape, they become hardened elements.

Structure of the spermatophore of P. pelagicus. The posterior portion of the spiral PVD preserves great number of very minute sphere-shaped fully formed spermatophores, which are readily distinguished by the naked eye. They are embedded in a huge somewhat homogeneous gelatinous substance. Macroscopic examination reveals that they are simple small nonpedunculate spheres with an average diameter of about 0.36 mm. Each fresh spermatophore appears as a homogeneously compact spherical mass ensheathed with a distinct sheath (Fig. 10.130A). Fixed spermatophores inside the spiral PVD show that this sheath is composed of a number of very thin sheaths (Fig. 10.124F). SEM and TEM studies of these contents of the spiral PVD (El-Shereif, 1991) show that the adjacent spermatophores are surrounded by a spermatophoric matrix (Fig. 10.130B). The latter appears as an electron dense, flocculent material with large globules of different sizes (Fig. 10.130C). The globules consist of a homogeneous, fairly intense electron dense material and small, scattered, intense electron dense material. The outer surface of the preserved spermatophores (wall) is distinguished by the presence of numerous deep folds distributed along the whole surface (Fig. 10.130B and D). The wall of the spermatophore is about 4 μm thick. It is apparently formed from the thick spermatophoric sheath externally and the sperm matrix internally. The spermatophoric sheath measures

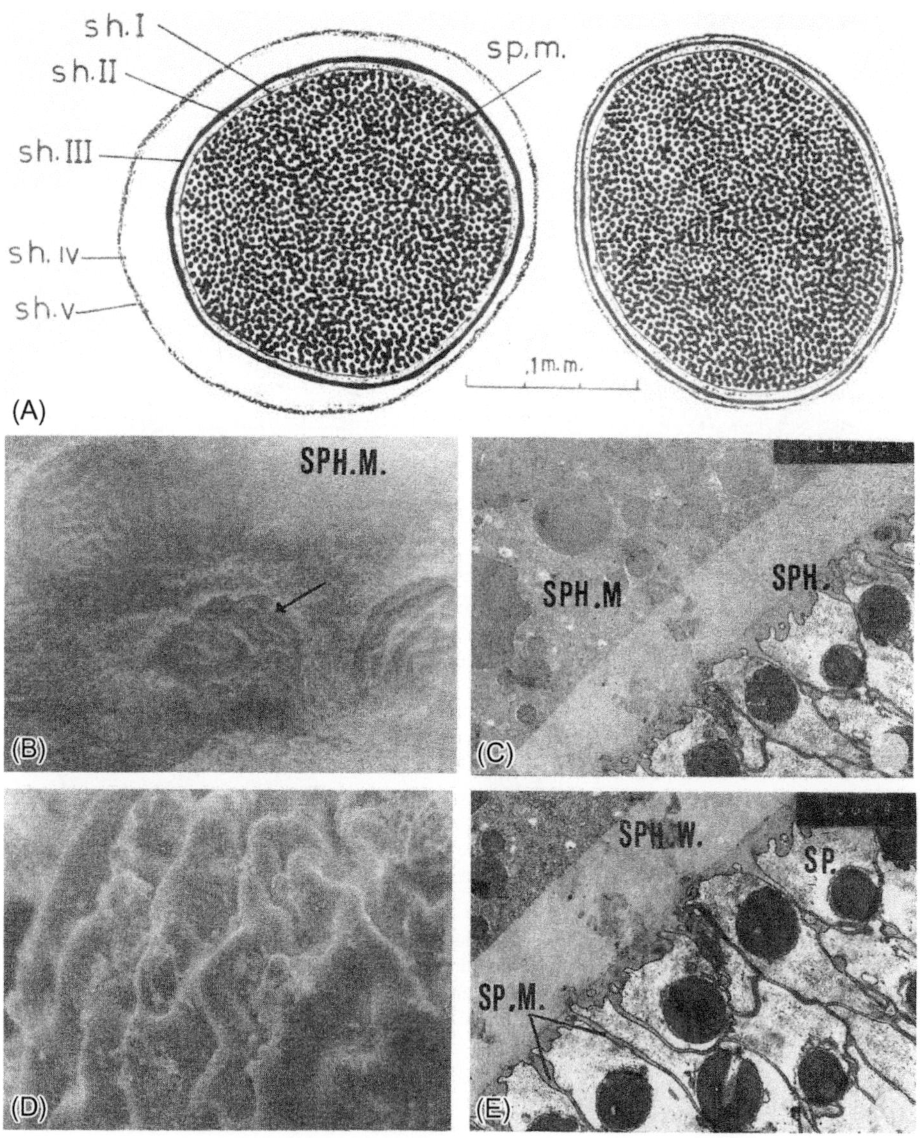

FIG. 10.130 Structure of the fully formed spermatophore of *P. pelagicus:* (A) from a smear of the contents of the spiral region of PVD; (B–E) TEM micrographs of the spermatophore of the same species: (B) of contents of spiral region of PVD showing spermatophores *(arrow)* embedded in spermatophoric matrix. X = 150; (C) section through spermatophore showing spermatophoric matrix, notice associated globules of secretion. X = 4000; (D) of spermatophore and associated globules of spermatophoric matrix, notice sculpturing surface of spermatophoric wall. X = 800; (E) section through spermatophore showing sperm, sperm matrix and spermatophoric wall. X = 5000. sh. I–V = sheaths of the spermatophore, SP = sperm, sp.m. = sperm mass, SP.M. = sperm matrix, SPH = spermatophore, SPH.M. = spermatophoric matrix, SPH.W. = spermatophoric wall. *(From (A) El-Sherief (1975), (B–E) El-Sherief (1991).)*

about 3 μm and shows homogeneous, slightly electron dense material (Fig. 10.130E). Interiorly, the sperm are not closely packed, but there is much homogeneous, electron dense matrix in between the individuals.

Chemical composition of the spermatophore of P. pelagicus

Absence of chitin. As for the chemical nature of the decapod spermatophore, discrete information was encountered. Jackson (1913) stated that the whole spermatophore of *Eupagurus* contains chitin. Mouchet (1931) studied the spermatophores of a number of decapod anomures and brachyures, and reviewed the works concerning the nature of the substance forming these spermatophores. She recorded that in the lobster it is an amorphous envelope, in *Carcinus maenus* it is made of albuminous liquid, and in the pagures it is of chitinous nature. Later, Hamon (1937) mentioned that the spermatophore of *Eupagurus* is made up of sperm immersed in a hyaline fluid and forms an elastic elongated mass enclosed in an ampoule. Spalding in 1942 believed that a chitinous component is secreted by the VD of *Carcinus maenas* in a fluid form, in addition to protein molecules. On the other hand, King (1948) stated that the sperm of *Penaeus setiferus* are enclosed in a sheath surrounded by a mass of chitinous material found inside the spermatophore. However, Bawab (1970) gave a good description of the secretory activity of the VD of *Penaeus (Melicertus) kerathurus* correlated with the formation of the spermatophore. She gave, for the first time, a full description of the chemical nature of the seven sheaths around the main body of the spermatophore. She also evidenced experimentally that chitin is entirely absent from this element. The procedure adopted by Bawab (1970) was described here in this chapter for the spermatopores of *P. kerathurus*. El-Shereif (1975) applied this procedure on released fully formed spermatophores obtained from the spiral PVD of a live specimen of *Portunus*. The result was the complete disappearance of the material within few minutes. This result was obtained when the experiment was repeated on the whole spiral region and even on the whole VD. Therefore, it is evident that chitin is entirely absent from the spermatophors of *P. pelagicus*.

Distribution of lipids. As Table 10.10 shows, the application of the reaction of Sudan black B on paraffin and gelatin sections of the posterior region of the testis and PVD differs in its affinity in both paraffin and gelatin sections, for the same region. When it appears positive in gelatin section, it is negative in paraffin one, this is because Baker's fixative employed in preparing the gelatin sections is the best for the preservation of lipids. Accordingly, sec. 1, 2, and 4 contain lipids while sec. 3, 5, and 6 are free of lipids in either paraffin or gelatin sections.

Distribution of proteins. It is evident from the chemical reactions applied in Table 10.10 that the proteins occur in abundance in almost all the secretions, either free or conjugated. Rather than the preliminary indication of the presense of protein by the stains used, xanthoproteic test indicates the phenyl group in the protein. Millon's reaction was used to specify the presence of tyrosine (β-hydroxy phenyl-α-aminopropionic acid). Only sec. 1 was found to be free from proteins, while sec. 6 may contain protein. Although the chemical constituents of the spermatophore of *P. pelagicus* were studied in full detail, yet its hardening was not. However, previous experience in studying hardening of the spermatophore of *Penaeus kerathurus* (Bawab, 1970; Malek and Bawab, 1971) and studying hardening of the plug in the female crab *Portunus pelagicus* (Bawab and El-Sherief, 1989), in addition to what was encountered in the literature, suggest that the process of hardening of this spermatophore occurs in a similar manner. The presence of tyrosine (phenolic group) and the presence of lipids with probability of conjugated substances (lipoprotein), make the phenolic tanning the most probable mechanism.

Formation of the spermatophore of C. danae Smith, 1869. Zara et al. (2012) studied histologically and histochemically the formation of the spermatophore of *C. danae*. This formation begins as follows: the 'seminiferous duct receives the spermatozoa from the testicular lobule (Fig. 10.126I). The spermatozoa move within the seminiferous ducts and converge to a single **collecting duct or vas efferens** (Fig. 10.131B) which opens at the most anterior part of convoluted vas deferens'. This is a difference from that of *P. pelagicus*. Despite the nomenclature of the vas deferens in the two species, the first region (AVD, Fig. 10.131A) is divided into two portions: AVD proximal and AVD distal. In *C. danae*, immediately after the vas efferens (Fig. 10.131B) receive the sperm from the seminiferous ducts, open into the AVDp. Both vas efferens and AVDp show the same type of columnar epithelial lining (Fig. 10.131B and C). A thin layer of musculature covers this inner epithelium, and an outermost layer of connective tissue with fibroblast-like cells (Fig. 10.131C). In the lumen of AVDp, many free spermatozoa are surrounded by a type I secretion which is highly basophilic and fluid (Fig. 10.131B). This secretion separates the spermatozoa into large distinct masses, by forming small ridges in between. This is another difference from that of *P. pelagicus*. A second type II secretion, which is strongly acidophilic, penetrates into the ridges formed by secretion I (Fig. 10.131B and C) which will be part of the wall of the spermatophore. The second portion, the AVDd, is the same as the preceding portion but it is lined by cubical epithelium. Its lumen exhibits the same two types of secretion but in greater amount. In this lumen, completely formed ovoid spermatophores are observed. Secretion II is stained for neutral polysaccharides and seems to be added gradually to the spermatophores wall (Fig. 10.131D–F). It is also positive for proteins (Fig. 10.131G) and forms the glycoproteinaceous wall surrounding the sperm masses. This wall appears as only one layer at light microscopy

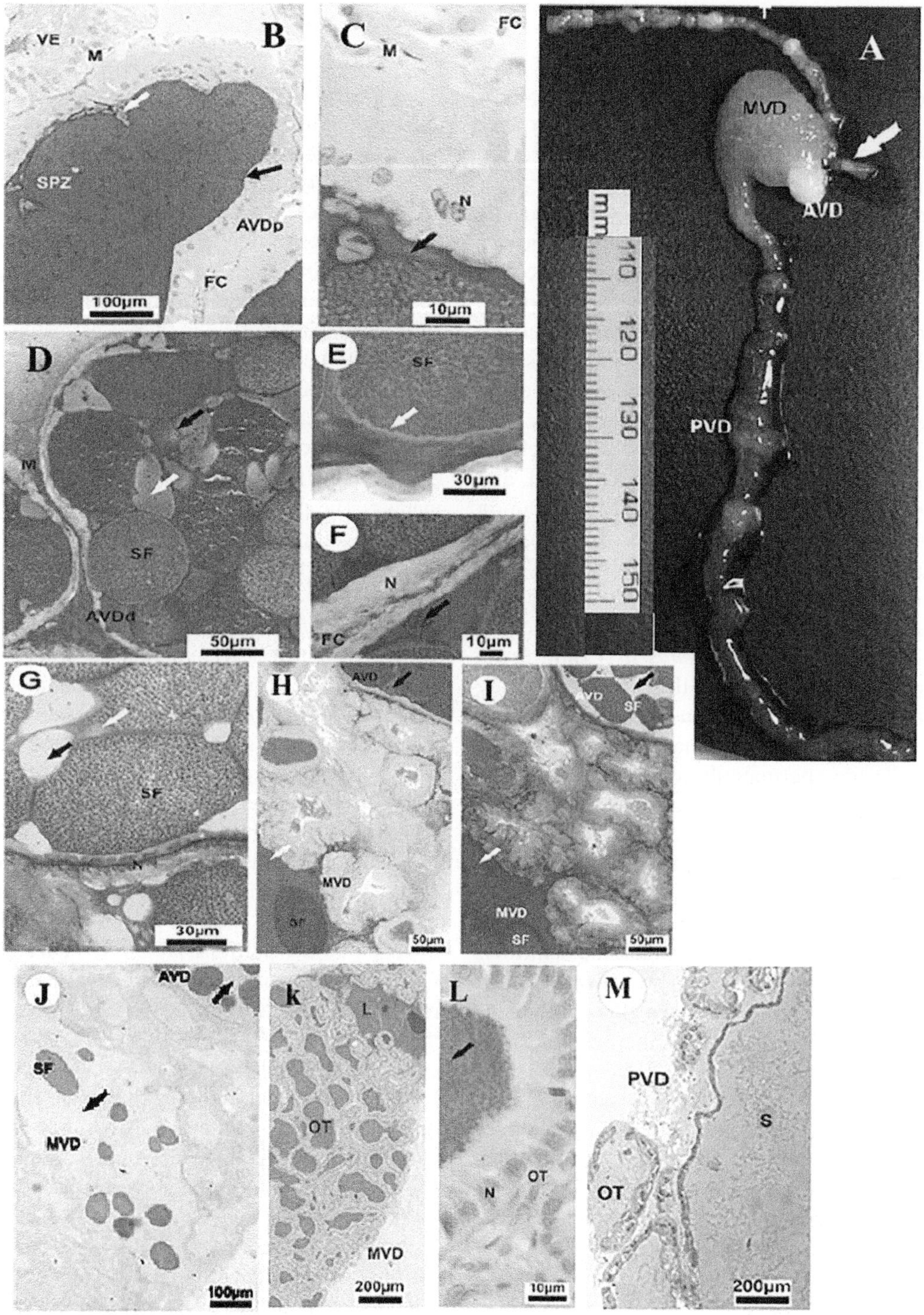

FIG. 10.131 Formation of the spermatophore of developing *C. danae (from Zara et al., 2012)*: (A) general view of left reproductive system isolated from mature male (*arrow* depicts commissure joining both testes); (B–G) T.S.s through AVD in developing male *C. danae:* (B) showing general view of vas efferens filled with sperm (VE) and proximal portion of AVD (AVDp) showing lumen filled with basophilic secretion *(black arrow)*, which apparently compacts and separates the mass of free spermatozoa (SPZ), thick acidophilic secretion penetrates between masses of spermatozoa *(white arrows)*, H (C) showing thick acidophilic secretion penetrating between spaces formed by basophilic secretion *(arrow)*; (D) distal portion of AVD showing in lumen, acidophilic secretion strongly positive for acidic polysaccharides *(white arrow)*, and basophilic secretion positive for neutral polysaccharides *(black arrow)*; (E) details of the PAS stained wall of the spermatophore *(arrow)*; (F) details of luminal secretion strongly stained for neutral polysaccharides *(arrow)*, within an alcianophilic matrix, (C–F) PAS/Alcian blue pH 2.5 staining; (G) lumen of distal region of AVD (AVDd), basophilic secretion negative for proteins *(black arrow)*, while acidophilic secretion positive *(white arrow)*, note that spermatophore (SF) walls positive for protein, mercuric bromophenol blue staining; (H–M) T.S.s through the transition portion between AVD and MVD in mature male: (H) differences in luminal secretion of distal portion of AVD rich in acidic polysaccharides *(black arrow)*, while in MVD secretion positive for neutral polysaccharides *(white arrow)* PAS/Alcian blue pH 2.5 staining; (I) luminal secretion of the same portion of H negative for proteins *(black arrow)*, except for that associated with spermatophores, MVD with secretion strongly stained for proteins *(white arrow)*, mercuric bromophenol blue staining; (J) both AVD and MVD negative for total lipids *(arrows)*, Sudan black B staining; (K and L) T.S. through MVD in mature male: (K) MVD with several lateral out-pocketing (OT) and central duct (L) both showing same monostratified columnar epithelium; (L) detail of lateral out-pocketings columnar epithelium intermingled with connective tissue (luminal secretion composed of acidophilic granules exhibiting different staining patterns *(arrows)*), H (M) general aspect of the fluid secretion (S) without granules in the lumen of PVD, H&E (PVD with lateral out-pocketing show same cubical epithelium of central duct). FC = fibroblast-like cells, M = muscles, N = nucleus.

(Fig. 10.131C–K). Secretion I is negative to proteins, in contrast with secretion II. Neither secretions present in the AVD contain lipids (Fig. 10.131J). Concerning the number of secretions in AVD, this is a third difference from that of *P. pelagicus*, it may be that the latter was studied in more detail.

The MVD in *C. danae* is characterized by several well-developed lateral 'out-pocketings' filled with secretions that are discharged into a larger central duct (Fig. 10.131K). There are no histological differences between the lining epithelium of the lateral out-pocketings and the central duct. It is columnar cells with basal nuclei, which lie upon a connective tissue layer. The luminal secretion is very different from that found in AVD. This difference is clearly observed in the transition area between the two regions (Fig. 10.131H–J). The acidophilic secretion in MVD contains granules, although basophilic granules are also present (Fig. 10.131L). This difference is also observed in that the MVD secretion is strongly stained for neutral polysaccharides and proteins. Ovoid spermatophores are observed in this duct between the acidophilic and granular secretion. The average diameter of spermatophores of developing specimens is 270 ± 41 μm; that of mature specimens is 277 ± 32 μm.

The PVD also exhibits lateral out-pocketings as in MVD, but in a smaller amount. Its lining epithelium is flatter and the secretion is less basophilic. This secretion becomes more fluid and lacks granules (Fig. 10.131M), it is negative for acidic polysaccharides, positive for neutral polysaccharides and proteins, and negative for lipids. The most characteristic feature of these secretions forming the sprmatophore of *C. danae* is the absence of lipids. This is a fourth difference from that of *P. pelagicus*. The histochemical results of the vas deferens and spermatophore are summarized in Table 10.11.

Formation of the spermatophore of Chionoecetes opilio. It was mentioned above that spermiogenesis in *Chionoecetes opilio* occurs not in the testis, but in the vas deferens (Sainte-Marie and Sainte-Marie, 1999a). These authors described, by light microscopy, the process of formation of spermatophores, which begins with the formation of sperm, the spermiogenesis (Fig. 10.132). The sections of the vas deferens nearest to the testis were loaded with a dense and homogeneous population of 'young sperm cells' (Fig. 10.132A). These cells are rounded, pinkish, and very pale except for a blue 'dark ring' present at the surface of one cell pole. The ring delimited a minute intrusive colourless area that apparently corresponded to a shallow cell-surface invagination (Fig. 10.132B). The other following sections showed comparable but slightly more evolved cells in which the dark ring is more prominent (Fig. 10.132C). Therefore, half of the ring appears to be homogeneously dark, whereas the other half appears as a row of small and slightly lighter dots (Fig. 10.132D). Sections taken further along the vas deferens showed additional changes that occurred to the sperm cells before their enclosure in spermatophores (Fig. 10.132F). First, under the dark ring of these cells appears a small, light-brownish 'hemispherical component', below which a thin, colourless crescentic area was detected (Fig. 10.132F). The cell invagination extending within the hemispherical component became more obvious, while the ring, hemispherical component, and invagination widened and the related cell pole became flatter. Simultaneously, the cell margin at the opposite cell pole thickened and darkened to some extent, forming a pale blue 'chromatin cap' as demonstrated by the Feulgen-positive reaction. The central part of the cap often becomes a 'spike-like structure' (Fig. 10.132F). With further development, the hemispherical component elongates

TABLE 10.11 Histochemical data for secretion in anterior (AVD), medium (MVD), and posterior (PVD) regions of vas deferens and spermatophore wall in *Callinectes danae* males

	Histochemical compound			
Secretion	Acid polysaccharides	Neutral polysaccharides	Proteins	Lipids
AVD type I	+++	−	−	−
AVD type II	−	+++	+++	
MVD granules	−	+++	+++	−
MVD matrix	−	++	+++	−
PVD	−	++	+++	−
Spermatophore wall	−	+++	+++	−

+++ = strongly positive, ++ = positive, − = negative.
From Zara et al. (2012).

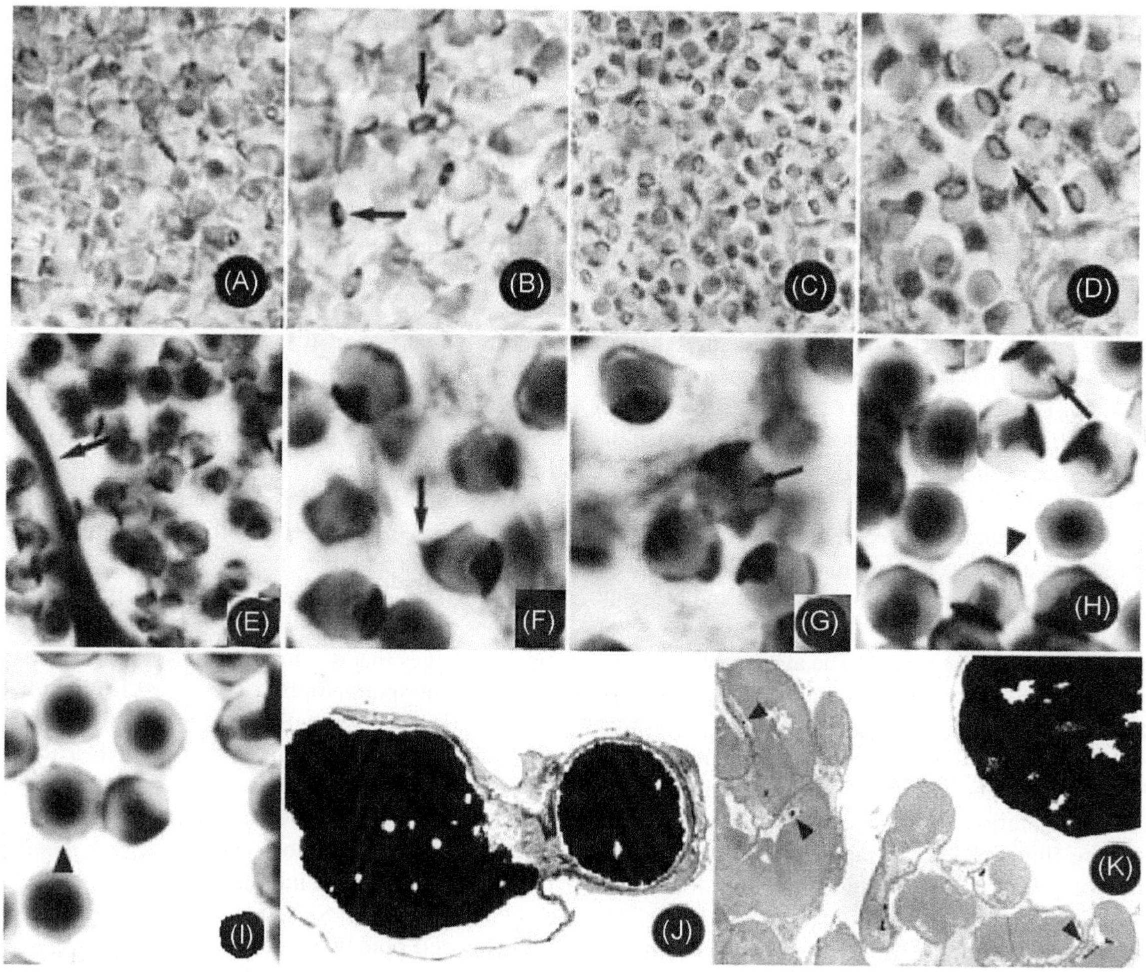

FIG. 10.132 Formation of the spermatophore of *Chionoecetes opilio (from Sainte-Marie and Sainte-Marie, 1999a,b using light microscopy)*: (A) T.S. through the vas deferens (nearest to the testis) showing 'young sperm cells', Dominici's stain. X=825; (B) enlarged part of (A) showing the *pale area* within a *dark ring (arrow)* corresponds to a slight cell invagination. X=1650; (C) this section shows that the dark rings are better developed and obvious. X=825; (D) enlarged portion of (C) showing the differentiation of the dark ring. X=1650; (E) cells enclosed in this spermatophore are more evolved spermatids; *arrow* indicates the spermatophore wall. X=1650; (F) enlarged portion showing the 'hemispherical component' with the changes of the ring and the spike *(arrow)*. X=3200; (G) enlarged portion showing the changes occurred to the hemispherical component *(arrow)*. X=3200; (H) enlarged portion of (E) showing the central 'funnel-like apparatus', at its margin with the narrow part of the apparatus, there are two small colourless rounded areas *(arrow)*, the cells are hexagonal *(arrowhead)*. X=3200; (I) enlarged portion of (E) showing the hexagonal shape of spermatids more obvious *(arrowhead)*. X=3200; (J) section filled only with dark blue IAM. X=35; (K) T.S. near that seen in A showing part of IAM-loaded structure in the upper right-hand corner, the others are sections through the vas deferens filled with pale young sperm cells, a tiny amounts of dark IAM scattered in between *(arrowheads)*. X=20.

and gives rise to a short, narrower, and relatively tubular element that joins the chromatin cap at the base of its spike (Fig. 10.132G). All these changes produced a brownish 'funnel-like apparatus' crossing the centre of the cell, with two tiny colourless rounded areas (possibly centrioles) appearing at the junction of the wide and narrow parts of the apparatus (Fig. 10.132H). The resulting cell has a hexagonal shape that was best seen in polar view (Fig. 10.132I). Such a cell is still pink and its funnel-like apparatus is less evident than its dark ring. These cells at this developmental stage are referred to as 'immature spermatids'.

In addition to the T.S. of the immature spermatids, there are one or few peculiar very large sections filled only with dark blue matter (Fig. 10.132J), referred to as 'type I amorphous matter' (IAM). This matter is PAS-positive and seemed to be homogeneous at low magnification, but higher magnification revealed that it is composed of little spheres. There is no evidence that IAM is secreted in these sections. Indeed, sperm-loaded sections of the vas deferens closest to the testis contain no detectable or only tiny amounts of IAM that are restricted to one side of the periphery of the section, often occurring in the form of a rounded mass or a single strand (Figs 10.132K and 10.133A, B). In sections somewhat far from the testis, the quantity of IAM progressively increases and IAM mixes with the young sperm cells to varying degrees. In other sections, IAM spreads broadly

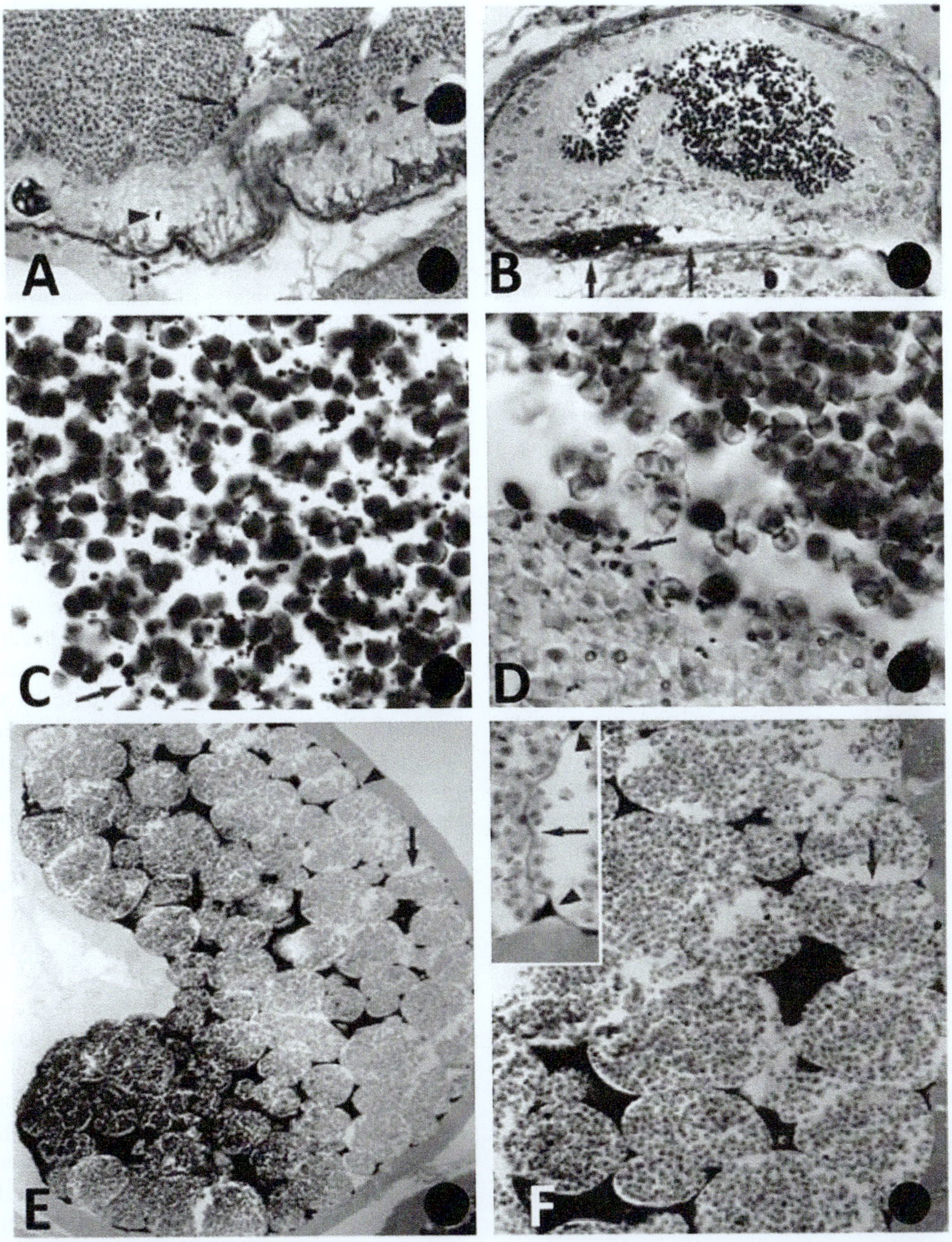

FIG. 10.133 The formation of the spermatophore of *Chionoecetes opilio:* (A) T.S. through the vas deferens filled with pale young sperm cells, *arrowheads* point to drops of IAM within the epithelium, the *arrows* indicate a small area at the periphery of the lumen which contains tiny dark droplets of IAM. X=210; (B) the lumen of this section is loaded with dark immature spermatids; beneath the epithelium, a canal *(arrows)* carries dark IAM. X=80; (C) this section shows the mixing of dark droplets of IAM *(arrow)* with darkened immature spermatids unenclosed in a spermatophore. X=825; (D) this section shows two portions: lower contains sperm cells not mixed with the material, while the upper contains dark immature spermatids mixed with drops of IAM *(arrow)*. X=825; (E) T.S through the vas deferens farther from the testis, loaded with newly formed spermatophores, the pale ones contain pale young sperm cells while the darkest ones contain dark spermatids; the *arrowheads* indicate *triangles* of dark IAM covering epithelium, Dominici's stain. X=80; (F) enlarged part indicated by an *arrow* in (E), a thin of IAM forms the basal pellicle of the newly formed spermatophore. Inset shows enlargement of the pointed site. The IAM projection extends from the tip of a peripheral triangle of IAM *(lower arrowhead)* to a tip of deep grossly stellate mass of IAM *(upper arrowhead)*.

and mixes with a large portion, or all of the cell population (Fig. 10.133C). In sections where the distribution of IAM is restricted, young sperm cells develop into immature spermatids (Fig. 10.133D).

Spermatid differentiation. The subsequent differentiation of immature spermatids into mature ones unfolded wherever a substantial quantity of IAM was mixed with the cell population, which form a thin veil at the surface of each cell. The most striking visible change during this stage is the formation of the acrosome, seen in sections stained with PAS, which typically colours the acrosome purple (the immature spermatid is PAS-negative). In immature spermatid, the funnel-like apparatus appears only as a lighter area (Fig. 10.133B). Therefore, the diagnostic feature of the mature spermatid of the snow crab is its acrosome, composed of a PAS-positive vesicle surrounding the PAS-negative funnel-like apparatus.

Both immature and mature spermatids are enclosed in spermatophores. A spermatophore wall first consists of a thin pellicle of blue IAM. There is relatively little IAM inside spermatophores, but significant amounts are present between them in the form of stellate masses (Fig. 10.133E and F). Further along, in the lumen of the vas deferens, the late stage of spermatophore formation was preceded by the introduction of a second type of amorphous matter (IIAM). The latter is PAS-negative and stained pink-red with the Dominici method. It forms a thick coating of PAS-negative surrounding the spermatophores. Its appearance partially or largely masks the PAS-positive reaction of the fine basal pellicle, rendering its observation difficult.

Sainte-Marie and Sainte-Marie (1999b) observed in the spermathecae of inseminated females a form of sperm cell yet more evolved than the mature spermatid described above. It appears that cells previously interpreted as spermatozoa were in fact spermatids. Within the contents of the spermathecae, spermatophores enclosing either immature spermatids, mature spermatids, hitherto unreported spermatozoa, or cell forms intermediate between these three types. The authors reported that the transformation of spermatids into spermatozoa was observed only in the female. The diversity of sperm cell types and the ordered placement of semen constituents within the spermatheca suggest that the sperm are partitioned for short- or long-term use.

Another study of the formation of the spermatophores of this Atlantic snow crab *Chionoecetes opilio* obtained from the same locality (islands and bays of the Atlantic, east Canada) was encountered. Beninger et al. (1988) presented a different view and completed the process of formation of the spermatophore. They described the AVD of this species as a 'complex network of ducts' (Fig. 10.134A) (the same of that of *P. pelagicus*), some of which secrete substances to form a 'pellicle'. The latter encases the 'spermatozoa' with some of the matrix in spherical spermatophores, which are stored within the matrix-filled lumen of the AVD and MVD. Phase-contrast light microscopic observations of fresh spermatophores taken from the MVD show tightly packed spermatozoa enveloped by a convoluted pellicle (Fig. 10.134B) The complex folded form of the pellicle is evident in SE micrographs (Fig. 10.134C and D). It is visible as a uniform, 2-μm thick, acellular folded membrane in the TE micrograph of Fig. 10.134E. These three observations virtually exclude the possibility that the folded nature of the spermatophoric wall is an artefact.

The female system. It is well known that the gonad of the crustacean animals, like many arthropod ones, is continuous with its coelomic gonoduct. In the female crab *Portunus*, the ovaries, contrary to the testes, constitute the whole H-shaped structure. Hence, the transverse bridge is of ovarian tissue. It marks the anterior ovary from the posterior one. Each posterior ovary bends sharply to the inside and then proceeds straight (Fig. 10.135(I) upper). The two posterior ovaries are close to each other, unlike the two anterior ones, which are widely separated. The gonoduct appears as an elongation from the corresponding posterior ovary from the ventral surface at the level of the bend. This portion of connection can never be determined in the gross anatomy even in the final stages of maturation of the system. The gonoduct extends vertically to the ventral side of the animal, but this extension can be seen in situ only after the isolation of the thoracic muscles that the gonoduct penetrates. The latter opens on the sixth thoracic sternite just on the side of the apex of the overlapping abdomen (Fig. 10.118D). This means that this gonoduct is very short. Morphologically, it is a simple narrow tube and cannot be differentiated into regions. However, the terminal (distal) part that opens to the outside is the vagina, since it is lined by ectoderm (cuticle) (Spalding, 1942; George, 1963). Therefore, the proximal part is the oviduct (coelomic duct) according to the general description given for the Brachyura by Calman (1909). Just after copulation, the gonoduct of *Portunus* dilates, forming the spermatheca.

Again, contrary to the male *Portunus*, the female crab has a definite breeding season. The current author can actually select a crab of definite maturity stage from the catch, having been greatly familiar with this species during the overall study of its spermatophore (1971–1973) and during the overall study of the female reproductive biology (1974–1979). In addition, as a citizen of Portsaid, it is possible to report the following significant observations:

1. This crab is available all year round, but its percentage in the catch decreases greatly in winter months.
2. Adult female crabs are abundant in the catch between September and November.
3. Adult soft-shelled female crabs are rare in the catch and may be absent.
4. Adult hard-shelled nonovigerous females are abundant from June to August.
5. Ovigerous females constitute most of the catch from spring until late summer with two peaks: one in May and the other in September.

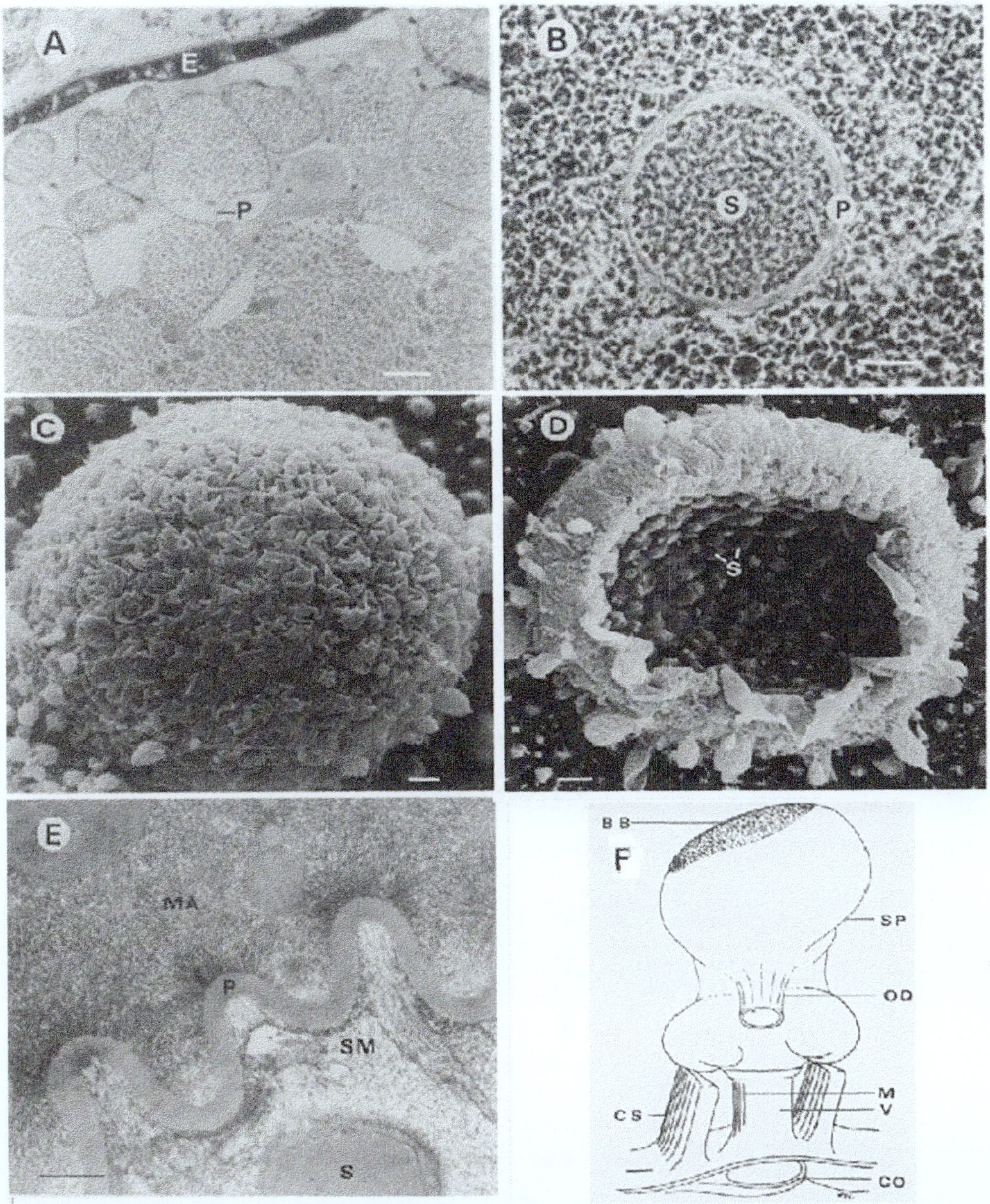

FIG. 10.134 Formation of the spermatophore of *Chionoecetes opilio (from Beninger et al., 1988)*: (A) in the region between the testis and AVD, cuboidal secretory epithelium (E) secretes the pellicle (P) which envelops the spermatozoa forming complete spermatophores, H&E. Bar = 50 μm; (B) spermatophore in fresh smear of MVD contents, notice irregular pellicle enclosing spermatozoa (S), Phase contrast optics, Bar = 30 μm; (C) SEM of a whole spermatophore taken from the MVD, notice the highly convoluted pellicle. Bar = 10 μm; (D) SEM of a ruptured spermatophore showing the packing sperm (S), peripheral globular objects are freeze-dried MVD matrix. Bar = 10 μm; (E) TEM showing the structure of spermatophore pellicle. MA = matrix, MAD, SM = spermatophore matrix. Bar = 0.5 μm; (F) external anatomy of the spermatheca of *C. opilio*, attached to inner wall of the vagina, black band (BB), insertion of coxa (CO), cuticular septum (CS), oblique muscles (M), oviduct (OD), spermatheca (SP) concave (inner) wall of cuticular vagina (V). Bar = 100 μm.

During maturation of the system, it was observed that there is a definite synchronization between the changes that occur to the ovaries and those that occur to their ducts (Fig. 10.135(II)). Through the study of these changes, more than one line of research was done as will be seen later. These changes, in addition to the physiological activities of the animal, such as moulting, copulation, dehiscence of the spermatophores, fertilization, spawning, and incubation, were taken in consideration to determine the ovarian cycle and/or the reproductive cycle. El-Shereif (1983) and Bawab and El-Sherief (1988) determined the stages of the reproductive cycle of the female crab *P. pelagicus*. These stages are as follows:

A. Soft-shelled female

Stage 1: Virgin female (precopulation), March

Stage 2: Copulated female (early stage), March

B. Hard-shelled female

Stage 3: Copulated female (late stage), April

Stage 4: Prespent female (ripe female), late April

Stage 5: Ovigerous female (May spawning), first spawning May

Stage 6: Inter spawn (ripe female, second ovarian cycle), June–August

Stage 7: Ovigerous female (September, second spawning), September

Stage 8: Nonbreeding female, winter season

Bawab and El-Sherief (1988) summarized the morphological features of the stages of the reproductive cycle in Table 10.12. These features are based on the morphological appearance of the female and the anatomical studies of the ovaries and their ducts.

Copulation. It becomes clear that **copulation** is the first step in the reproductive cycle of brachyuran female crabs. Copulation usually occurs just within 24 h of moult (soft-shelled postmoult phase), as reported in many of the previous works such as Edwards (1966), Hartnoll (1969), Fielder and Eales (1972), and Schram (1986), then recently by experimental observations of the sand crab *P. pelagicus* (Soundarapandian et al., 2013). Since copulation is not observed in nature (Banoub, 1963; Ryan, 1967a,b), the newly moulted crabs become very rare in the catch in Alexandria fisheries. In spite of Banoub's remark that the

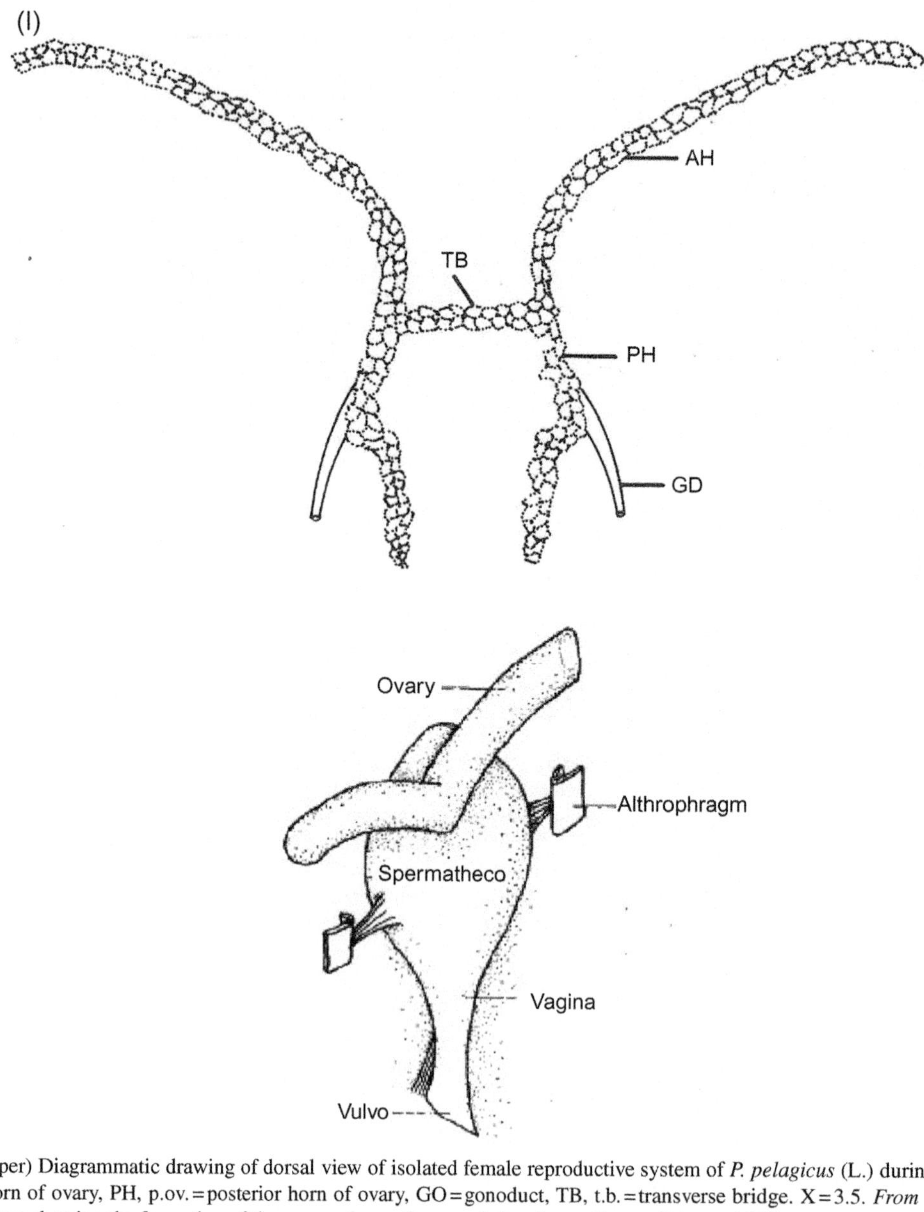

FIG. 10.135 (I) (upper) Diagrammatic drawing of dorsal view of isolated female reproductive system of *P. pelagicus* (L.) during nonbreeding season. AH, a.ov. = anterior horn of ovary, PH, p.ov. = posterior horn of ovary, GO = gonoduct, TB, t.b. = transverse bridge. X = 3.5. *From Bawab and El-Sherief (1988).* (Lower) Diagram showing the formation of the spermatheca after copulation (according to Calman, 1909).

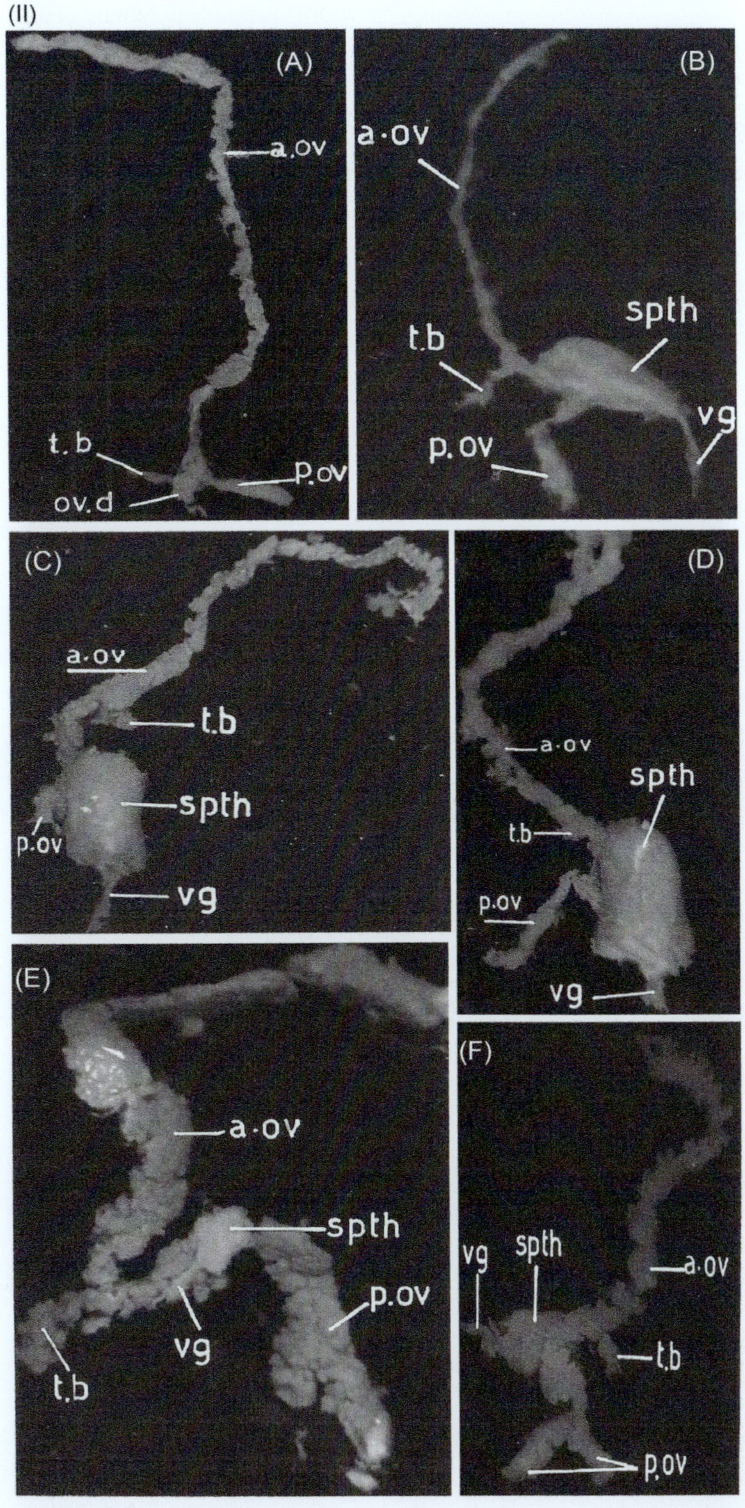

FIG. 10.135, CONT'D (II) (A–F) Dorsal view of the right side of the female reproductive system of *P. pelagicus*: (A) isolated from soft-shelled specimen obtained in March, stage 1 (virgin female). X=3; (B) isolated from soft-shelled specimen obtained in late March, stage 2 (fresh-copulated female). spth=spermatheca, vg=vagina. X=2; (C) isolated from hard-shelled specimen obtained in April, stage 3 (late copulated female). X=1.5; (D) isolated from hard-shelled specimen obtained in May, stage 5 (ovigerous female), the spermatheca forms the dome. X=2; (E) isolated from hard-shelled specimen obtained in August, stage 6 (second ovarian cycle). X=1.5; (F) isolated from hard-shelled female obtained in September, stage 7 (ovigerous female, second spawning). X=1.5. *From El-Sherief (1983).*

TABLE 10.12 Cyclic changes of the Ovaries of the Crab *Portunus pelagicus* (L.) correlated with the changes of the spermatheca during the reproductive cycle

	Representative	Ovary			
Stage and date	**Female**	**Morphology**	**Histology**	**Vital activity**	**Spermatheca**
Precopulation March	• Soft-shelled • Virgia	• Narrow, ribbon • Transparent	• Cortex 25U • Few oocytes in stroma	Multiplication and growth of cogenesis	• Small narrow • Undifferentiated • Empty
Fresh-copulation March	• Soft-shelled • Copulated	• Narrow, ribbon • Somewhat transparent	• Cortex 15U Follicles in stroma, compact	• Naturation of oocytes	• Large, pear-shaped • Male product inside.
Late-Copulation April	• Hard-chelled • Copulated	• Zigzag, ribbon • Pale yellow	• Cortes 7U • Nature and fertilized ova	• Fertilization	• Maximum size • Oval • Male product inside
Prespawn May	• Hard-shelled • Ripe	• Maximum size • Projecting lobes • Reddish orange	• Cortex 5U • Different stages of vitellogenic ova	• Vitello-genesis	• Large • Dorsal dose • Half-filled with male product
May-spawn Late May	• Hard-shelled • Ovigerous	• Large size • Yellow	• Cortes 20U • Developing oocytes • Debris	• Spawning • Second oogenesis	• Smaller size • Half-filled with male product
Interspawn August	• Hard-shelled • Ripe	• Maximum size • Reddish orange	• Cortex 7U • Nature ova and vitellogenic ova	• Second maturation • Second vitellogenesis	• Smaller • Undifferentisted. • Rather empty

TABLE 10.12 Cyclic changes of the Ovaries of the Crab *Portunus pelagicus* (L.) correlated with the changes of the spermatheca during the reproductive cycle—cont'd

	Representative	Ovary			
Stage and date	Female	Morphology	Histology	Vital activity	Spermatheca
September-spawn September	• Hard-shelled • Ovigerous	• Large size • Coloured	• Cortex 15U • Debris of ensheathed yolk spheres	• Getting rid of debris • Recover the immature condition	• Small • Undifferentiated • Empty
Nonbreeding November	• Hard-shelled • Immature reproductive system	• Narrow, Ribbon • Transparent	• Cortc 20U • Few germinal oocytes and debris	• Inactive	• Small • Empty • Differentiated into oviduct and vagina

From Bawab and El-Sherief (1987b, 1988).

fishermen observed *Callinectes sapidus* in the act of copulation in Lake Edku (east of Alexandria), the soft-shelled specimens of *Portunus* were very rare in the catch (Bawab and El-Sherief, 1988). However, Edwards (1966) cultured more than 300 of the European edible crab, *Cancer pagurus* Linnaeus, under conditions almost like nature.

It appears that this culture was a very successful one, since many of the crabs survived for 2 years. Edwards observed that there is a powerful attraction between the sexes prior to the female moult, i.e. there is courtship. He stated that the attraction between the sexes is chemical and 'some time before the moult of copulation the female is exuding a substance which is attractive to the males'. The males can detect this substance at a distance. He also reported that this attraction between the sexes has been noticed on the foreshores along the Norfolk and Yorkshire coasts, where hard males have been found attending both hard and moulted females. These crabs were usually found at the low-water mark, hiding under rocks, during July and August, which is the main moulting period on the east coast of England. Ryan (1966) evidenced experimentally the release of a sex-attractant pheromone in *Portunus sanguinolentus* (Herbst), a large edible crab of the Indo-West Pacific, obtained from Kaneohe Bay, Hawaii. He added that the obvious possible source of the attractant (pheromone) was the urine of the premoult females, which was released through the excretory pores. He described the effect of this attractant as follows:

whenever a premolt female was placed in a holding cage containing several males, the latter began a peculiar display and search behavior. Each male became active, walked about the cage on the tips of its dactyls with its body elevated at maximum height above the substratum, extended its ox-blood red chelae (female chelae are white), and attempted to pull whatever crab it came in contact with, male or female, into the precopulatory holding position. Crabs other than premolt females usually escaped.

The origin of the pheromone, its chemical nature, and the way it is detected by males at that time, remain to be detected. In contrast with crustaceans, occurrence of sex attractants in insects was well documented at that time, and so later on Barrington (1979) recorded this subject in detail.

Another observation stated by Edwards is that the females were continuously attended by a male for a period of 3–21 days before the moult, and for a further period of 1–12 days after the moult. This means that the attendance before moulting was longer. He also observed that the female was accompanied by only one male throughout the period of attendance. Edwards was considered the first to describe this period of attendance in detail, since only a brief reference to male protection of the female was made by few of the previous authors. However, Hartnoll (1969) stated that there are two basic patterns of mating in the Brachyura: the **first** in Cancridae and Portunidae in which copulation occurs very soon after the moult of the female;

courtship is a prolonged process, involving the attendance of the premoult female until ecdysis, then releases her to moult, but guards and assists her during the process. After copulation, usually the male continues to carry the female until her integument is sufficiently hardened; the **second** pattern occurs in those crabs where the female is not recently moulted at the time of copulation, a hard female and brief courtship. All species studied are of the Grapsidae and Ocypodidae. The most thoroughly studied examples are several species of the genus *Uca*.

This behaviour of attendance was referred to as 'premoult guarding', 'precopulatory guarding', and 'postcopulatory guarding' in the experimental work on the sand crab, *Portunus pelagicus* (Soundarapandian et al., 2013). These authors evidenced the courtship, attendance, and act of copulation in Brachyura. They called the copula position of the male and female crabs the **cradle-carrying position**. They stated that the males only initiate this position by approaching the pubertal females who have not shown any sort of approach towards the male. The female is protected by a male before and after her pubertal moult, by being grasped by the male's first pair of walking legs and held right side up in a cradle-carrying position under the male. They divided the mating behaviour into five phases as follows.

1. **Premoult guarding:** the hard-shelled male crab and the female crabs at the verge of premoult stage were introduced into the experimental tank. The female attracts the hard-shelled male. The male crab contacts the female by moving towards her and extending his bigger chelate in front of her, without showing any other attracting courtship. Then he catches her to form the cradle-carrying position described above. During this position, both animals face in the same direction. In general, the male is active and aggressive but the female either remains motionless or actively moves to push her body back under the male as if to aid in maintaining the copula position (Fig. 10.136A). This description was referred to as the courtship stage in the reproductive behaviour of the same species, *P. pelagicus* in view of Fielder and Eales (1972). The latter authors added that once the courtship position was adopted, much time was spent in the same location, the female either resting on the sand or almost completely buried and the male partially so. This courtship lasted about 4 days in one of their two experiments.
2. **Moulting:** at that time of the female, the pair was separated from the cradle carrying position and subsequently the female start to moult. During this time, the male crab moves around the female crab and helps her to moult by removing the shell, and protects her from others until she attains normal hardness of the exoskeleton (Fig. 10.136B). In *C. pagurus*, immediately after completing her moult, the female lay as an inert mass on the tank bottom (Edwards, 1966).
3. **Precopulatory guarding:** after moulting, the male and female crabs would form the cradle carrying position as earlier for sometime, until the mating process starts. These two phases, moulting and precopulatory, were considered one phase as precopulation moulting in the work of Fielder and Eales (1972).
4. **Copulation:** this is initiated when the female's exoskeleton is soft (information that was mentioned in all the works that dealt with this aspect). In addition, Edwards (1966) reported that for successful copulation and impregnation, the female must be in a very soft-shelled condition. During this vital event, the male crab becomes very active and rotates the female by using his walking legs, and by using the chelate, he turns over her (Fig. 10.136C). The female assists this action by positioning herself upside down beneath him and extends her abdomen allowing the male to insert his paired gonopods into her genital pores. By this time, the male and female crabs are facing in opposite direction (Fig. 10.136D). During copulation, the male often walks around with the female attached to its ventral surface, holding her with the third and fourth walking legs (Fig. 10.136E). This vital event lasts in *C. pagurus*, in some cases for 3 h, and in others for 2 h (Edwards, 1966).
5. **Postcopulatory guarding** (Fig. 10.136E)**:** the behaviour after the completion of copulation, the male crab liberates the female from mating position and they embrace in a short while to form the cradle carrying position. The latter continues for just a few hours, during which the female is inactive until she attains normal hardness of her exoskeleton.

Duration of copulation in Brachyura. There are various records of the duration of copulation. Most of these records are based on captive specimens under very artificial conditions. It is not clear whether the stated duration refers to copulation only, or includes courtship behaviour. In addition, it is not always certain that the pleopods are inserted for the entire period that the copulatory position is maintained (Hartnoll, 1969). The latter listed different durations encountered in his review. The most common is 5 min to 1 h; less than 5 min, Ryan (1966) recorded that copulation in *Portunus sanguinolentus* occurs in 2–5 min; 1–5 h; and in few brachyurans, copulation occurs over 5 h.

The impregnation of the female crab initiates two vital activities: the maturation of the ovaries and the sudden formation of the spermatheca and the plug. It is suggested that these two latter structures are formed within a few minutes and are associated with each other to a great extent.

Formation of the spermatheca. To study the formation of the spermatheca and the plug in the female crab, *P. pelagicus*, Bawab and El-Sherief (1989) first studied the original histological structure of the gonoduct, i.e. that of the nonbreeding stage (Fig. 10.135(I)), to understand the changes that occur to form these two structures. The proximal region

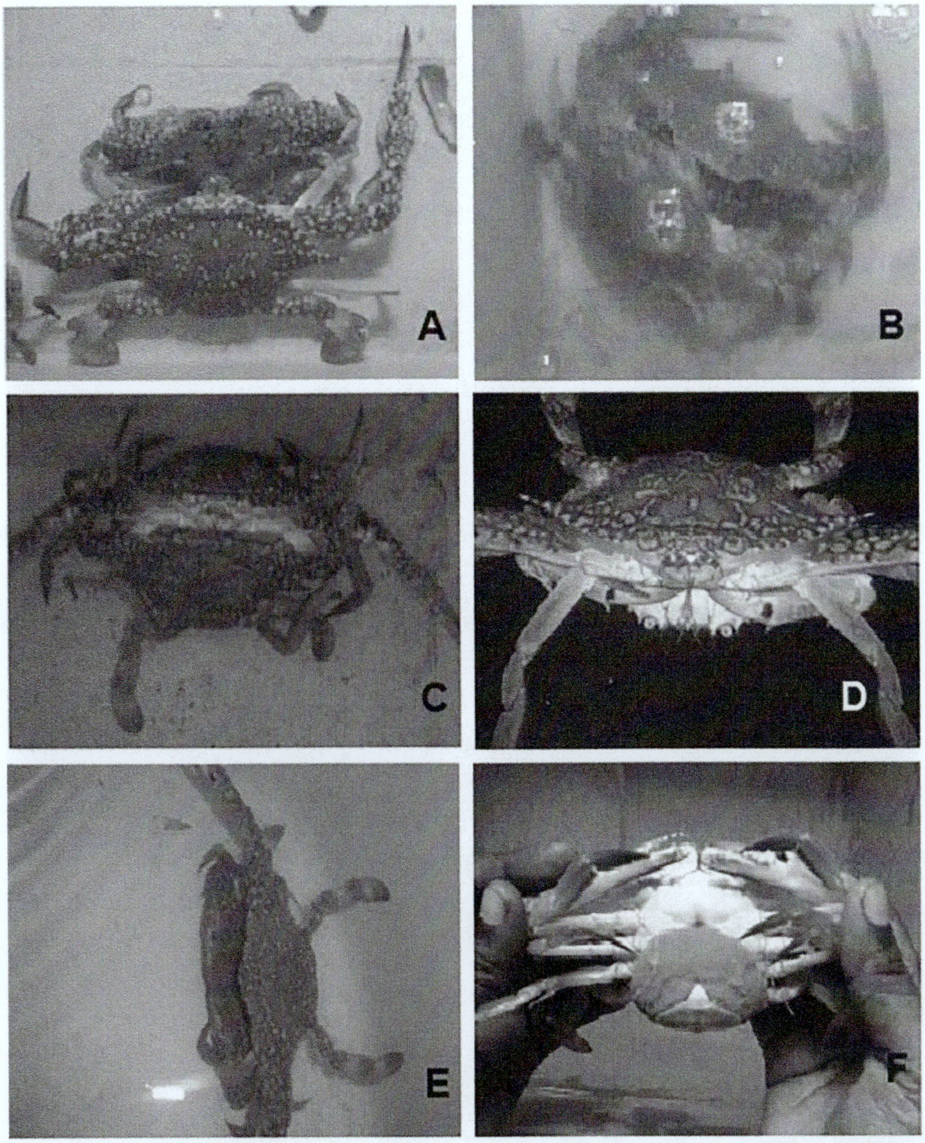

FIG. 10.136 Various positions of mating behaviour of *P. pelagicus* presented by *Soundarapandian et al. (2013)*: (A) premoult guarding; (B) moulted shell; (C) male try to rotate the female during the precopulatory guarding; (D) mating; (E) post copulatory guarding; (F) newly spawned female.

of the gonoduct is a true coelomic **oviduct**. Its lumen is a continuation of that of the ovary, and the outer covering layer is of squamous epithelium. The latter is readily observed by its red colouration when treated with Mallory's triple stain. Fig. 10.137A shows that the oviduct has the largest diameter (about 2.75 mm), the thickest wall (about 500 μm) and the narrowest lumen. The greater part of its wall consists of condensed bundles of smooth muscles, running longitudinally and obliquely, embedded in compact connective tissue with several scattered blood vessels. Mason's trichrome stain gives a well-differentiated picture for these tissues. The oviduct is lined by multicellular glandular mucosa, the cells of which are small and round. This mucosa and muscles of considerable portion of the dorsal wall strikingly invaginate deep into the lumen.

The distal region is the **vagina**, since it is lined by cuticle as mentioned above. Fig. 10.137B shows its narrower diameter of about 2 mm with a wall measuring 300–350 μm. Like the oviduct, it is highly muscular with irregularly arranged bands, but there is a pronounced complete layer of well-arranged muscular bands just beneath the epithelial lining. The latter consists of inactive columnar cells. The epithelium is lined by a thick horizontally laminated cuticle, measuring about 50 μm. It accepts the aniline blue and the light green of the two above stains, respectively, and it is negative to the iron haematoxylin of the Heidenhain's. These results indicate the presence of chitin (Yonge, 1932). The irregular folding of the lining epithelium with its cuticle characterizes the vagina. The folds of the upper side are few but deeply invaginated, and consequently the lumen

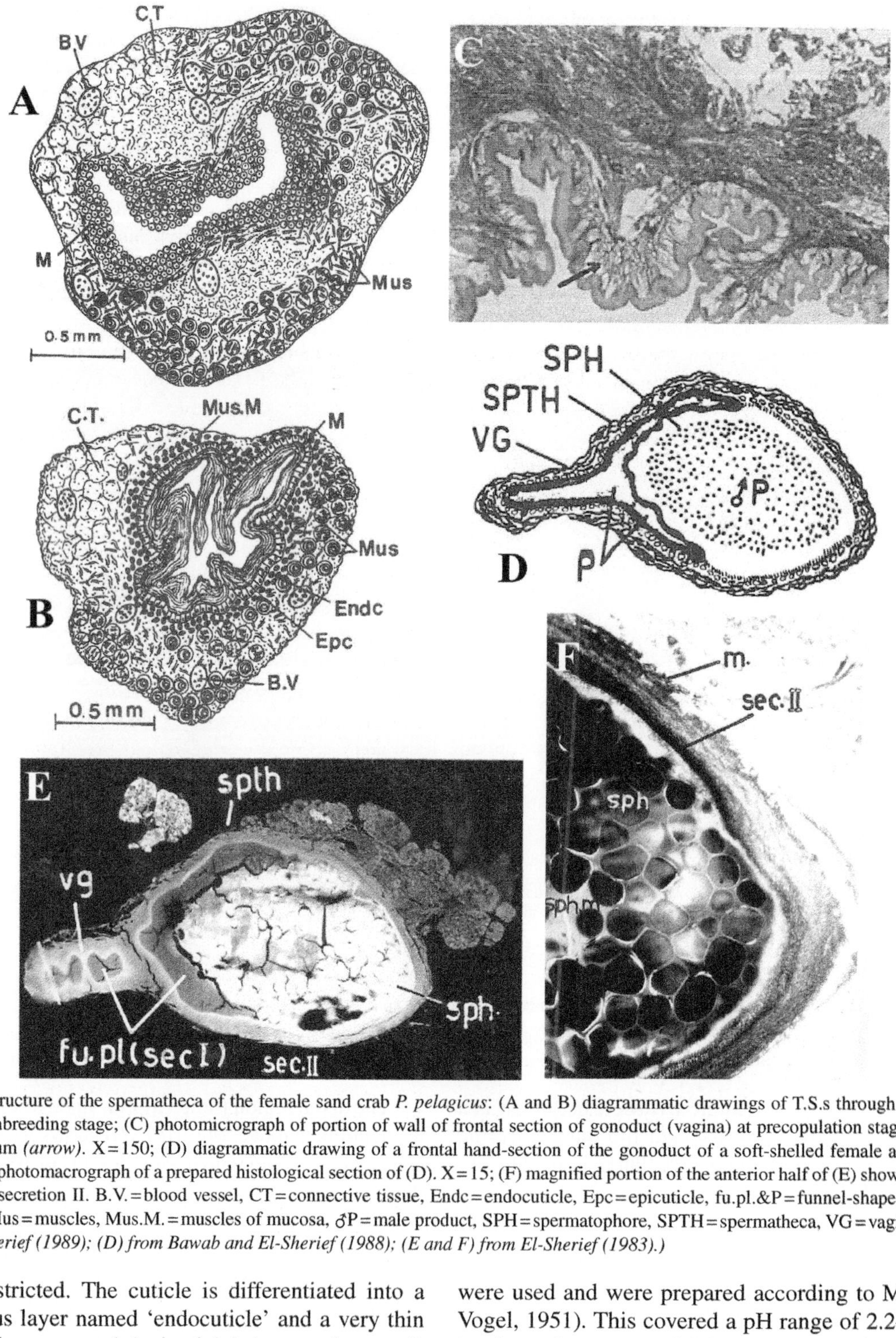

FIG. 10.137 Structure of the spermatheca of the female sand crab *P. pelagicus*: (A and B) diagrammatic drawings of T.S.s through (A) oviduct; (B) vagina during nonbreeding stage; (C) photomicrograph of portion of wall of frontal section of gonoduct (vagina) at precopulation stage showing active cuticular epithelium *(arrow)*. X = 150; (D) diagrammatic drawing of a frontal hand-section of the gonoduct of a soft-shelled female at fresh-copulated stage. X = 10; (E) photomacrograph of a prepared histological section of (D). X = 15; (F) magnified portion of the anterior half of (E) showing the epithelial wall loaded with secretion II. B.V. = blood vessel, CT = connective tissue, Endc = endocuticle, Epc = epicuticle, fu.pl.&P = funnel-shaped plug (secretion I), M = mucosa, Mus = muscles, Mus.M. = muscles of mucosa, ♂P = male product, SPH = spermatophore, SPTH = spermatheca, VG = vagina. *((A–C) From Bawab and El-Sherief (1989); (D) from Bawab and El-Sherief (1988); (E and F) from El-Sherief (1983).)*

is highly constricted. The cuticle is differentiated into a major chitinous layer named 'endocuticle' and a very thin nonchitinous layer named 'epicuticle' (nomenclature after Dennell, 1947a,b, 1960). The epicuticle is differentiated into an outer lipoid layer by Sudan black B, and an inner proteineous one by Bromphenol blue (Mazia et al., 1953). To confirm the presence of chitin, the iso-electric point (I.E.P.) was determined. The method adopted was that employed by Yonge (1932). Citric acid/phosphate buffers were used and were prepared according to McIlvaine (see Vogel, 1951). This covered a pH range of 2.2–8.0, increasing by an increment of 0.2 pH. The experiment shows that the endocuticle has an I.E.P. at pH 3.5, the same as that of the 'chitin' layer of the fore-gut of *Homarus gammarus* (Linnaeus, 1758) (Yonge, 1932, as *H. vulgaris*). In addition, it shows that the I.E.P. of the epicuticle is at pH 5.0–5.1, almost the same as that recorded before for various animals (Yonge, 1932; Dennell, 1960).

The **transitional duct** is the middle third of the gonoduct. It merges smoothly into the oviduct on one side and into the vagina on the other, without any external differentiation. The T.S. shows that its dorsal wall is a natural continuation of that of the preceding oviduct while its ventral wall has the same structure as the cuticular epithelium of the vagina. Its folds, however, are in the form of shallow crypts. It is suggested that this conspicuous structure of the ventral wall corresponds to the side pouch illustrated as the spermatheca of *C. maenas* (see Spalding, 1942), while the whole region corresponds to the elongated spermatheca of *Neptunus (= portunus) sanguinolentus* (George, 1963). The latter pulled the chitinous lining of the vagina and the spermatheca from the genital opening. George described it as a funnel-shaped structure, of which one side is longer than the other. This is strong evidence for the present histological picture.

Precopulation stage. The soft-shelled virgin female crab has a gonoduct almost similar to that described above (Fig. 10.135(II) A). The only difference shown in the histological picture that the epithelial cells become somewhat active. Those of the oviduct are heavily loaded with fuchsinophilic secretory granules. On the other hand, those of the vagina become more active. Each cell elongates, broadens (Fig. 10.137C), and is loaded with a fluid substance accepting violet colour with the two above stains.

Fresh-copulation stage. The spermatheca firstly appears in the gonoduct of this soft-shelled and freshly copulated crab. It is a pear-shaped structure (Fig. 10.135(II) B). The wide portion is the spermatheca while its stalk is the vagina (Fig. 10.137D) (Bawab and El-Sherief, 1988). The histological structure of the gonoduct of *P. pelagicus* permits the free expansion of about two-thirds (oviduct and transitional duct), forming the spermatheca at the time of copulation. This **temporary** condition resembles Dromiacea (Calman, 1909) but differs from both *C. maenas* (Spalding, 1942), *Neptunus (= portunus)* (George, 1963), and recently from *Chionoecetes opilio* (Beninger et al., 1988; Sainte-Marie and Sainte-Marie, 1998) in which the spermatheca is a definite and **permanent** pouch (Fig. 10.134F). In *P. pelagicus*, the thickness and arrangement of the muscle coat constitute the **main factors** permitting the great distension of the gonoduct forming the spermatheca. The collagen of the dense connective tissue is necessary to synthesize or repair the muscle fibres. The histological pattern of the spermatheca is completely different from its origin (oviduct) (Fig. 10.137E). It has a comparatively thin wall (Fig. 10.137F) measuring about 250 μm in thickness and its tissues are regular in arrangement. They are an outer compact connective tissue layer, a middle longitudinal muscle layer, and an inner lining mucosa. The latter is the same as that of the original oviduct, but each cell is loaded with its secretion stained deep blue with Mallory's stain. Its lumen is greatly widened and completely occupied by the male product. This product consists of the spermatophores embedded in a homogeneous matrix. The study of Bawab and El-Sherief (1989) shows that the spermatheca is built on the same basis as the uterus of higher animals (Bloom and Fawcett, 1975). The disappearance of the invaginated portion of the wall of the original oviduct after the impregnation was interpreted as **another factor** facilitating the widening of the spermatheca and regulation of its wall. In addition, the deep folds of the dorsal cuticular epithelium of the vagina were considered **a third factor**. They can expand anteriorly by their own muscle layer to compensate the missing dorsal cuticular lining of the formal transitional duct, which differentiates into the posterior half of the spermatheca.

It is worth mentioning that Hartnoll (1968) studied and discussed the structure of the female gonoduct, especially the spermatheca of some different crabs, whose genital openings open on the sixth thoracic sternite. He reported that in all members, the vulva leads via a vagina to a spermatheca, and this communicates by a short oviduct with the ovary (i.e. the spermatheca is a permanent structure). However, there are considerable differences in the form of these parts, so it became apparent that he could interpret two basic patterns: 'simple' and 'concave'. The latter may be elaborated by the presence of an operculum, which is either normally mobile or immobile. He described both patterns in detail. The simple pattern occurs in *Carcinus maenas*, which was previously describrd (Spalding, 1942). This simple pattern also occurs in *Cancer pagurus* Linnaeus (Williamson, 1904), in *Callinectes sapidus* (Cronin, 1942), and in *Portunus sanguinolentus* (Herbst) (Ryan, 1967a,b). Only *Cancer pagurus* belongs to family Canceridae, while the other three species belong to family Portunidae (FAO, WFP and IFAD, 2012). Hartnoll presented a diagrammatic drawing to explain the form and structure of the simple gonoduct (Fig. 10.135(I) lower). It is almost the same of the freshly copulated female *Portunus pelagicus* (Fig. 10.135(II) B). The latter is of a mated crab, and that of the nonmated is a short, narrow, and straight duct (Fig. 10.135(I)). If one does not consider the permanent or temporary spermatheca, then the gonoduct of *Portunus pelagicus* is of the simple pattern, like the others of family Portunidae.

Plug of the female *P. pelagicus*. Edwards (1966) reported that Williamson (1900) described how a fluid, produced from glands lining the spermatheca of the female, flowed out into the vagina when the male copulatory organ was introduced. On withdrawal of the male organ, this fluid hardened on contact with seawater to form a white structure, which plugged the opening of the oviduct. According to Williamson, two important pieces of information were obvious: this **plug**, with only few exceptions, is found in the oviducts of all impregnated female crabs to protect the male product; and this plug is made by the female crab. Later on, Spalding (1942) studied in detail, the male and female reproductive systems of *Carcinus maenas* (Linnaeus, 1758). He raised the possibility that the plug is made by the male crab and not by the female. Some subsequent authors followed Spalding's view. Bawab and El-Sherief (1989) were greatly interested in this aspect and so contributed much to the origin, nature, and formation of this plug in the female crab *P. pelagicus*. They evidenced the views of Williamson, but opposed the idea that the plug hardened on contact with seawater.

Formation of the plug. Bawab and El-Sherief (1989) reported that just after the impregnation of the male product, in the fresh-copulation stage, the posterior half of the spermatheca and the vagina are no longer lined by the cuticular epithelium. It is replaced by a multilayered mucosa, which might be a new synthesis, or a backward extension of the lining of the original oviduct. The disappearance of the cuticular epithelium may occur as a response to a definite substance secreted at the proper time. The only confirmation is the fluid substance secreted by the cuticular columnar cells observed in the precopulation stage. The activity observed at the fresh-copulation stage is that the free secretion lining the spermatheca is differentiated in its two halves. In the anterior half, this secretion forms a smooth, comparatively thin, and continuous layer in touch with the male product (Fig. 10.138A and H), which is enclosed in a cup-portion of a funnel-shaped structure (Fig. 10.137D). This secretion shows a marked tendency for acid fuchsin. Previously, it was defined as secretion II, which has no role in the formation of the plug, but is responsible for the dehiscence of the spermatophores (El-Sherief, 1983). The secretion in the posterior half constitutes a very thick, frothy layer (Fig. 10.138B), measuring about 330 μm. The distortion of the cells close to the secretory mucosa with their violet colour attained by Mallory's stain indicates that this secretion is of holocrine origin (Garven, 1965; Leonhardt, 1977). It was named secretion I (El-Sherief, 1983), which forms the cup-portion of the plug (Fig. 10.137D). The frothy layer is not in touch with the male product; the two are separated by another layer (Fig. 10.138B and H), having the same thickness, appearance, and almost the same affinity to stains as the original endocuticle. The latter still accepts the aniline blue with mere diffusion of the violet colour of Mallory's stain.

The vagina at this stage maintains its original size. However, its histological pattern changes and assumes the picture and activity of the posterior half of the spermatheca. Then, the frothy secretion together with the collapsed endocuticle cause a sort of stenosis in the lumen of the vagina forming, at the same time, the stalk of the funnel-shaped plug (Figs 10.138H and 10.137E). The levelling of the crypts of the columnar epithelium (Fig. 10.137B), owing to the enlargement of the anterior part of the vagina because of the formation of the spermatheca, gives the plug its characteristic funnel-shape. The plug firmly blocks the vagina and thus prevents the back exit of the spermatophores and/or the entrance of seawater that might cause their dehiscence.

Late-copulation stage. The gonoduct at this stage is still pear-shaped (Fig. 10.135C), but the spermatheca itself becomes oval because of its great increase in size. Its histological structure, though appearing the same as the fresh stage, is not really so, according to the critical study of Bawab and El-Sherief (1989) which proved some significant differences as follows.

1. The free secretion in the anterior half of the spermatheca is no longer definite continuous lining and its fuchsinophilic affinity decreases. It merges smoothly into the male matrix and comes in touch with the walls of the spermatophores (Fig. 10.138C).

Chemical structure of the plug.

2. The frothy layer forming the funnel-structure (plug) still attains the violet colour with each of the triple stains. It is impregnated with the secretory substance (Fig. 10.138D) which was previously loaded in the cells of the mucosa. This substance is clearly condensed in the inner parts of the frothy layer and in the neighbouring original endocuticle (Fig. 10.138E). The latter loses the characteristic colour of chitin when stained with aniline blue or light green (Yonge, 1932). Simultaneously, it accepts the deep green or greenish-red colour of Masson's stain and reacts in a markedly positive manner with Heidenhain's haematoxylin. This is a preliminary indication for the addition of protein, especially collagen (Pantin, 1948; Mahoney, 1973). However, Dennell (1960) described the exoskeleton of the arthropods as having a chitin-protein structure. The chitin is of α-type and the protein is not collagen (Rudall, 1955). On the other hand, El-Sherief (1983) studied histochemically the structure of this funnel-structure and by specific reactions proved the complete absence of tyrosine, tryptophan, arginine, cysteine, and cysteine.

Bawab and El-Sherief (1989) applied more specific histochemical reactions on sections of the gonoduct at this late-copulation stage (Table 10.13).

The funnel-structure shows distinctive results:

a. Its **protein** content is very high as is shown by the Oxidized Tannin-Azo method (OTA), which is in accordance with the results of the used stains.
b. **Lipids** are present in a considerable percentage and may exist in different complexes as shown by the lipid reactions and the intensive colouration attained by the Periodic Acid-Schiff method (PAS) (El-Sherief, 1983). This supposition is consistent with the results recorded in Table 10.13. The red colouration attained by the PAS method indicates that **phospholipids** with unsaturated bonds might be present in considerable amounts. This is supported by the colour attained by the Nile blue technique (Cain, 1947a).
c. Since pyridine dissolves phospholipids and other lipids, but not bound lipids (Cain, 1947b), the positive result

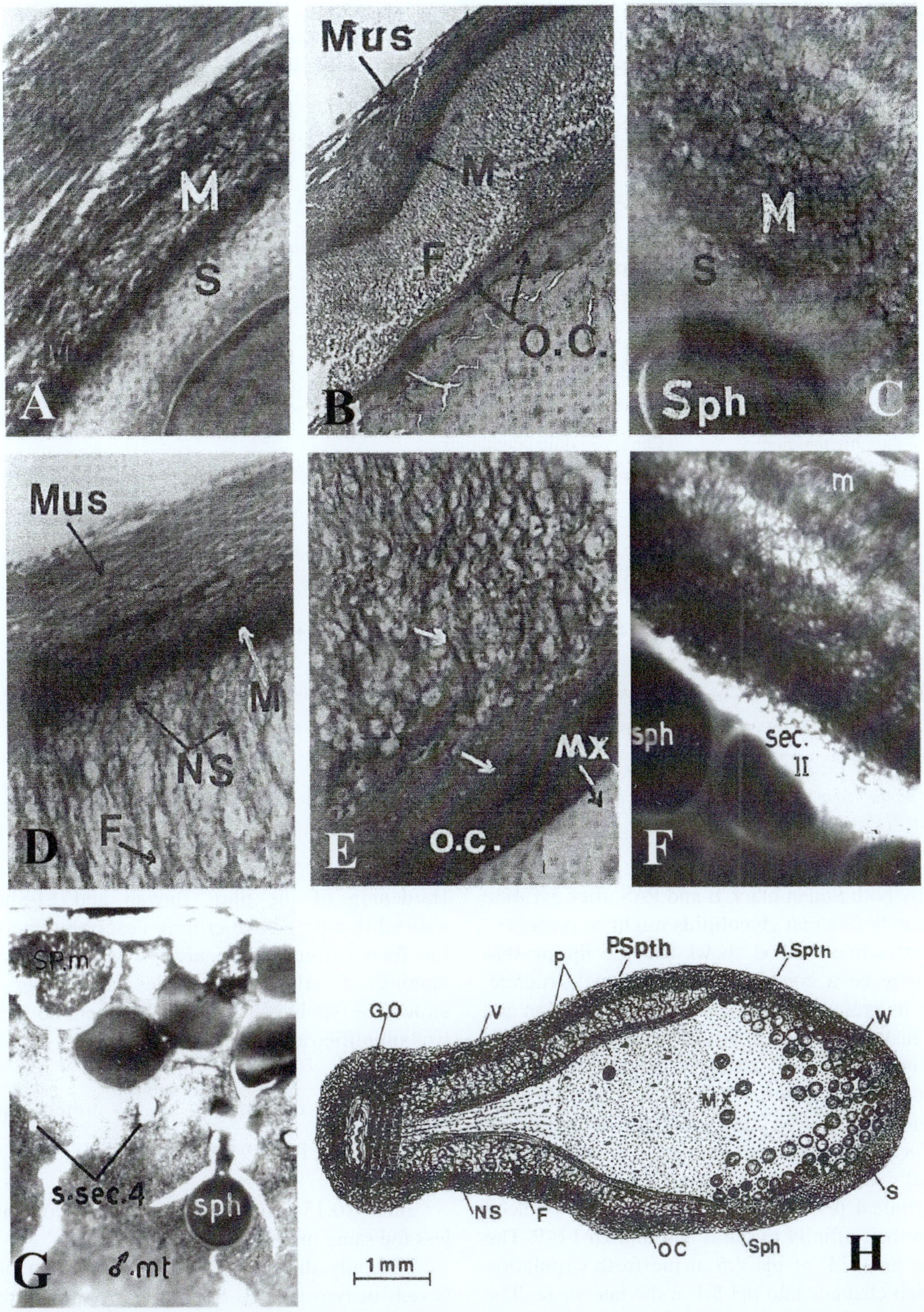

FIG. 10.138 Photomicrographs (except (H)) of T.S.s through different regions of spermatheca showing its activity: (A) enlarged portion of anterior region at fresh-copulation stage (see Fig. 10.137) showing the free secretion of the mucosa. X = 500; (B) posterior region at the same stage showing thick frothy secretion. X = 100; (C) enlarged portion of anterior region showing its free secretion in touch with spermatophores. X = 500; (D) enlarged portion of posterior region showing new secretion impregnating frothy layer. X = 200; (E) enlarged portion of posterior region showing concentration of new secretion *(arrow)* in inner parts. X = 200; (F) magnified portion of anterior region showing secretion II and the beginning of dehiscence of the spermatophores. X = 200; (G) portion beyond (F) showing the complete dehiscence of some spermatophores and the complete disappearance of secretion 4 of the male matrix. X = 100; (H) diagrammatic drawing of frontal section of the gonoduct at late copulation stage showing two components of the plug. A.Spth = anterior half of the spermatheca, F = frothy secretion, G.O. = genital opening, M = mucosa, Mus = muscles, MX = male matrix, NS = new secretion, O.C. = original cuticle, P = plug, P.Spth = posterior half of the spermatheca, S = secretion, Sec II = secretion II, sec. 4 = secretion 4, Sph = spermatophore, SP.m = sperm mass, V = vagina, W = wall. *((A–E and H) From Bawab and El-Sherief (1989); (F and G) from El-Sherief (1983).)*

TABLE 10.13 Responses to some histochemical reactions given by the two components of the plug found in the spermatheca of the female crab *Portunus pelagicus* (L.) at the late-copulation stage

		Cuticle of vagina		Plug in spermatheca	
Method	Material demonstrated	Epic.	Endoc.	Funnel-structure	Male matrix
Oxidized tannin-Azo method (OTA) (Dixon) (Pearse, 1968)	Amino-groups of proteins (θ-amino-groups)	++	–	+++	++
Sudan black B (Pearse, 1968)	Lipids	++	++	++++	++
Perforrnic acid-Schiff (Lillie) (Pearse, 1968)	Unsaturated lipids Phospholipids with unsaturated bonds	–	+	++	–
Nile Blue (Cain, 1947a)	Phospholipids	++	++	++	+++
Sudan black B after hot pyridine extraction (Baker, 1946)	Complex conjugated lipids	–	++	+	+
Periodic Acid-Schiff (Pearse, 1968) after hot pyridine ext. (Baker, 1946)	Glycolipids	–	++	++	++
Salt extraction (Holezinger and Balint) (Pearse, 1968)	Lipoproteins	–	±	+++	++
Ferric chloride (Dennell, 1947b; Krishnan, 1951)	Free orthodihydroxy phenols	+	–	++	±
Argentaffin (Lison, 1960)	Reducing aromatic substances	++	–	+++	++
Catechol (Smyth, 1954)	Polyphenol oxidase	+	–	+++	++

The same reactions were carried out on the vagina of the nonbreeding stage for comparison.
+ indicates the production of the colour of the reaction; its repetition indicates intensity of this coloration.
From Bawab and El-Sherief (1989).

obtained by both Sudan black B and PAS after pyridine extraction indicates that **glycolipids** might be present.

d. The salt extraction method shows that the **lipoprotein** complex may be a constituent of the funnel-structure. Therefore, its presence justifies its mode of secretion and its frothy histological appearance (Fig. 10.6) on the one hand, and its natural amber colour on the other.

3. The most striking feature observed during this late-copulation stage is the great similarity between the funnel-structure and the matrix engulfed in its cup. Both are naturally amber-coloured and both become hard during technical processes. Both undergo pronounced changes in their affinity to stains and in their I.E.P. The funnel has an I.E.P of pH 4.6 at the fresh-copulation stage, which changes into pH 5.1 at the late stage. The matrix has an I.E.P. of pH 3.4 at the former stage, changing into pH 5.7 at the latter. Again, both develop great resistance to acids. Gelatin-embedded frozen sections of about 20 μm thickness were treated with cold concentrated HCl acid. The frothy layer resisted for a period exceeding 7 days, while the matrix began to dissolve just after the 7th day. Both strongly reduce ammoniacal silver hydroxide (Table 10.13).

Hardening of the plug. Bawab and El-Sherief (1989) noticed that the above criteria describing the chemistry of the funnel-structure are always associated with phenolic tanning in arthropod cuticles and other invertebrate structures (Richards, 1951). The phenolic tanning system was simplified and summarized by Smyth (1954) as follows:

$$\text{Protein} + \text{diphenol} \xrightarrow{\text{Polyphenol oxidase}} \text{protein} + \text{quinone} = \text{tanned protein}\left(\text{sclerotin}\right).$$

Table 10.13 shows that the protein content is more or less the same in both components.

The only difference is that the protein of the matrix is rich in tyrosine (El-Sherief, 1983). The green colouration attained by the ferric chloride test proved that the new secretion impregnating the lipoproteins constituting the funnel-structure (Fig. 10.138D and H) is rich in the free ortho-dihydroxy phenols (Table 10.13). The matrix, however, shows a much lower reactivity with the same test. On the other hand, the catechol technique proved the presence of the polyphenol oxidase in great amounts in both components (Table 10.13). These results indicate that the two components have the requirements of phenolic tanning. This is an additional similarity providing suggestive evidence that the funnel-structure and the male matrix have similar function in the gonoduct of *P. pelagicus*. Bawab and El-Sherief (1989) mentioned above that the frothy layer is the first secretory substance, after the detached chitinous cuticle, forming the funnel-structure. In addition, the natural amber-colour of this structure provides fundamental evidence to the quinone tanning. This may explain the intensive positive reaction attained by the OTA method (Table 10.13), indicating a high protein content, especially the θ-amino-groups of lysine. Pearse (1968) reviewed and discussed the action of the tannic acid on the amino-groups of proteins. Therefore, lysine might be present as a constituent in the funnel, compensating for the absence of phenolic amino-acids. Despite the fact that Hackman (1949, 1953) and Hackman and Todd (1953) showed and reviewed the importance of tyrosine in tanning due to its phenolic properties, Hackman (1953) showed the possibility and importance of the θ-amino-groups of lysine in this reaction 'as it provides a means of cross-linking the protein chains at points other than the ends' (structural formula of lysine shown below (Murray et al., 1996)).

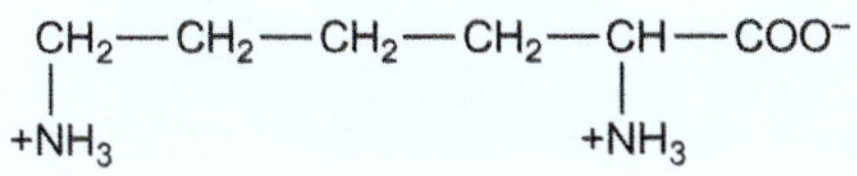

On the other hand, the lipids in the frothy layer were demonstrated to be sterols (El-Sherief, 1983). Bawab and El-Sherief (1989) added that this layer contains varieties of conjugated lipids that are of general occurrence in animal tissues and structures. The most striking is the presence of the lipoprotein complex. Malek (1952) proved its presence in some naturally tanned structures, and Blower (1951) thought that lipoids occur in the cuticle of certain myriapods. More recently, Malek and Bawab (1971) proved that lipoprotein is the substrate of quinone tanning of the spermatophore of *Penaeus trisulcatus* (= *P. 'Melicertus' kerathurus*). The consequent secretion to the lipoprotein forming the funnel is rich in dihydroxyphenols. It was suggested that these polyphenols are the precursors of the quinones responsible for the hardening of this funnel-structure. This is justified by the presence of polyphenol oxidase in the same localities of its substrate. The histological preparations show that the gonoduct of the female *portunus* is richly supplied with blood. This evidence suggests that the funnel may derive the polyphenol needed for its tanning from the blood. After being tanned and with its chitinous and lipoid nature, the funnel-structure attains a protective property as it becomes rigid, hard, resistant and impermeable (Yonge, 1936; Dennell, 1947b).

The above description of the funnel-structure of *portunus* proved that it is purely a female product. In this, it evidenced greatly the view of Williamson (1900), but of course the hardening is not caused by mere contact of water. On the other hand, this describtion provides considerable difference between *portunus* and *Carcinus*. The plug of the latter is formed by the male product impregnated in the spermatheca, the 'lower part' of which is chitinous (Spalding, 1942). In *portunus*, chitin was shown to be completely absent from its male product even when it is still in its natural sites within the male reproductive system as described before in this chapter. However, the 'upper part' of the 'sperm plug' of *Carcinus* and the matrix engulfed in the cup of the funnel-structure of *portunus* are similar. In spite of the fact that the latter is almost free from spermatophores (Fig. 10.138H), both are hard, homogeneous, hyaline in appearance, and cuticular in nature. This matrix basically is the male product impregnated in the female. El-Sherief (1983) showed that this matrix, in addition to being nonchitinous, it contains large amounts of proteins rich in tyrosine. Bawab and El-Sherief (1989) added that lysine and free dihydroxyphenol may be negated but the enzyme polyphenol oxidase is present in considerable amounts. This information provides a strong probability that this matrix undergoes, at an earlier stage, a self-tanning process as does the spermatophore of *Penaeus trisulcatus* (= *P. kerathurus*) (Malek and Bawab, 1971).

The comparable characters of the funnel-structure and its engulfed matrix, considering their relative positions, suggest that each is complementary to the other in fitting tightly to the lumen of the vagina and closing it against external factors. Consequently, the spermatheca becomes a chamber equipped for the preservation and protection of the spermatophores. Considering this view, the plug of *Purtunus pelagicus* is formed partly by secretions of the female individual and partly by the matrix impregnated by its partner. Hartnoll (1969) dealt with the plug in Brachyura and described it as 'a hard body which extends into the vagina and even protrudes from the vulva', in many crabs and in only two groups did the plug not protrude. The study of Bawab and El-Sherief (1989) indicates that *Purtunus pelagicus* is one of the latter.

It is a fact that the formation of the plug of *Purtunus pelagicus* is correlated with copulation, which is proved to be correlated with the moult (Bawab and El-Sherief, 1988). Hence, the hardening of the cuticle of the body, of the funnel-structure and of the male matrix may proceed at the same time and may have the same requirements. According to Brown (1950), the sequence of events leading to the phenolic tanning differs at least in the two components of the plug: the funnel-structure and the male matrix.

Maturation of the ovaries. It was mentioned above (p. 164) that two vital activities are initiated just after the impregnation of the male product into the temporary spermatheca of the female *Purtunus pelagicus*. Maturation of the ovaries is one of these activities. To observe carefully the changes attained by the ovaries during their maturation, those of the **nonbreeding stage** must be described. The ovaries of the female crab during winter (nonbreeding period) are immature, i.e. very small, narrow, and transparent ribbons (Fig. 10.135(I)). It is very difficult to define or isolate them, in spite of the fact that the hepatopancreas forms a suitable background underneath. A T.S. at any level (Fig. 10.139A) shows that each ovary is trilobed and covered by an 'ovarian sheath' made of squamous epithelium. The ovarian tissue is differentiated into an outer relatively thin layer of compact connective tissue measuring about 15 μm and an inner very thick one of loose connective tissue. The former is referred to as a cortex, while the latter is the 'medulla' which surrounds a central cavity. This differentiation is seen in the immature stage (Fig. 10.139B and C), and disappears as maturation proceeds. The cortex is considered the true germinal epithelium since it contains very small developing germinal cells. The cortex fluctuates in its thickness during the maturation of the ovary. On the other hand, the connective tissue of the medulla is the stroma (Threadgold, 1976); it contains scattered oocytes, usually degenerating ones, and much debris (Fig. 10.139B).

Precopulation stage. The moult of the female crab *Portunus* is a stimulus for copulation (Bawab and El-Sherief, 1987b, 1989). The latter work added that it is an activator for the maturation of the ovaries. During the precopulation stage, the ovaries are still immature in appearance, but they enlarge slightly in size. The cortex thickens,

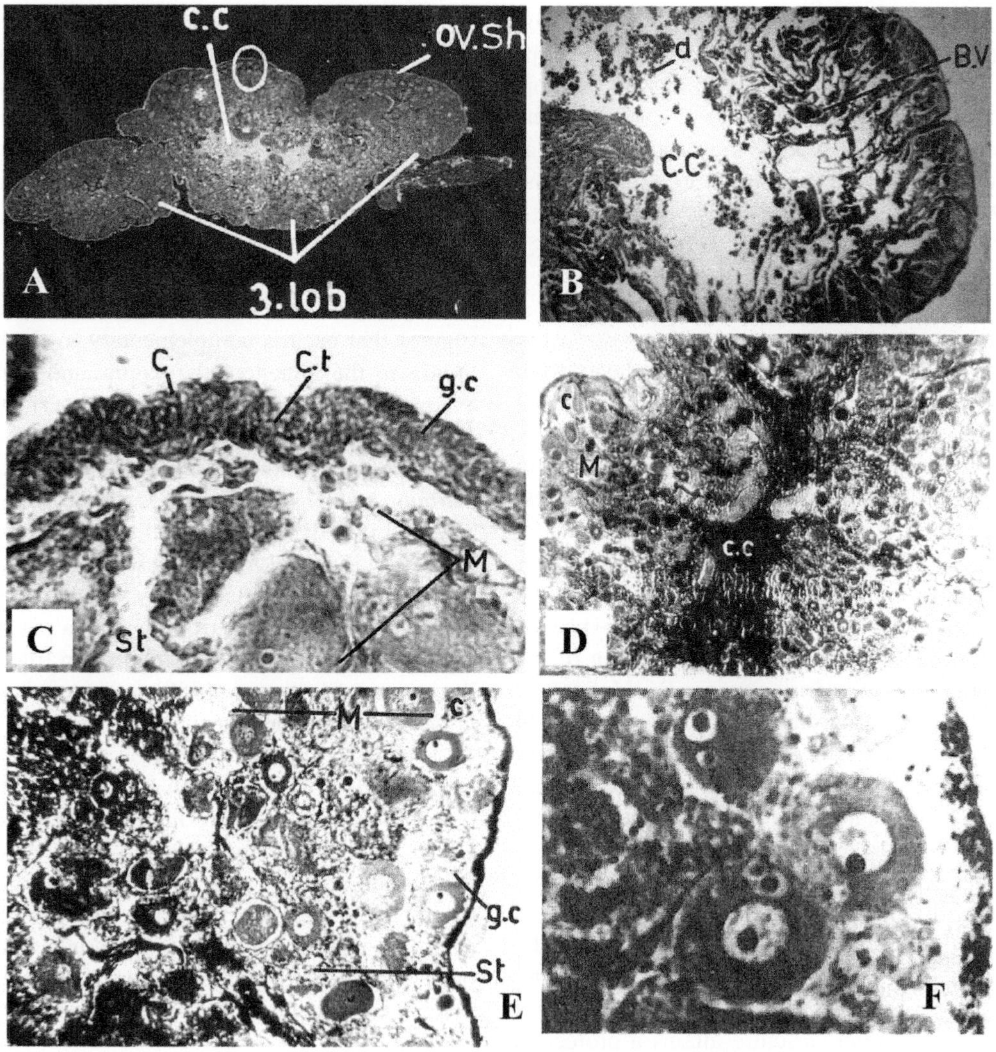

FIG. 10.139 Photomicrographs of T.S.s of the ovaries of *P. pelagicus* showing their structure: (A) through the proximal part of ovary obtained during the nonbreeding season. X=15; (B) magnification of portion marked by a *circle* in (A) showing histological structure. X=30; (C) high magnification of the same portion describing the details. X=500; (D) through the ovary of virgin female (stage 1). X=50; (E) magnified portion of (C) showing details of cortex and medulla. X=150; (F) high magnified portion of (E) showing the structure of the growing oocytes. B.V.=blood vessel, C=cortex, C.C.=central cavity, c.t.=connective tissue, d=debris, g.c.=germinal cells, 3 lob.=three lobes of the ovary, M.=medulla, ov.sh.=ovarian sheath, St=stroma. *(From El-Sherief (1983) except (F) from Bawab and El-Sherief (1987b).)*

attaining about 20–30 μm, and its cells proliferate. The stroma contains growing oocytes with diameter ranging from 65 to 75 μm, while those of their nuclei range from 25 to 20 μm (Fig. 10.139D and E). The smaller the size of the oocyte, the larger is its nucleus. The larger the size of the oocyte, the deeper is its position within the stroma (Fig. 10.139E). Each has clear and homogeneous cytoplasm and a relatively large and rounded nucleus. The latter has a central position, a granulated appearance, and one or more distinct nucleoli. Later on, these nucleoli become very close to the nuclear membrane (Fig. 10.139F). The ovary at this stage is distinctly characterized by: (1) some of the growing oocytes are in groups in the deeper parts of the stroma (Fig. 10.139D) which bulge into the central cavity. The latter becomes reduced and with irregular outline; (2) some of these oocytes, the larger in size, change their appearance and affinity to stains. Their cytoplasm becomes haematoxylinophilic, slightly evacuated and surrounded with detectable wall (Fig. 10.139F). These may be primordial germ cells, oogonia, and/or young oocytes (Giese and Pearse, 1974). This means that multiplication and growth phases of oogenesis are the characteristic features of this stage and some oocyts attain the full-grown size (75 μm) as what was recorded by Giese and Pearse (1974). However, they have not yet attained maturity. Maturation may occur during the fresh-copulation stage; (3) great number of very small cells in the cortex proliferate, each with a diameter measuring about 4 μm and with distinct nucleolus. Small groups of these cells are clearly observed in the stroma between the growing oocytes (Fig. 10.139E). At this stage of maturity, they are referred to as **cortical cells**.

Fresh-copulation stage. The ovaries increase in size and relobulates or follow a zigzag course (Fig. 10.140A). Some turbidity is noticed distributed among their tissues. The T.S.

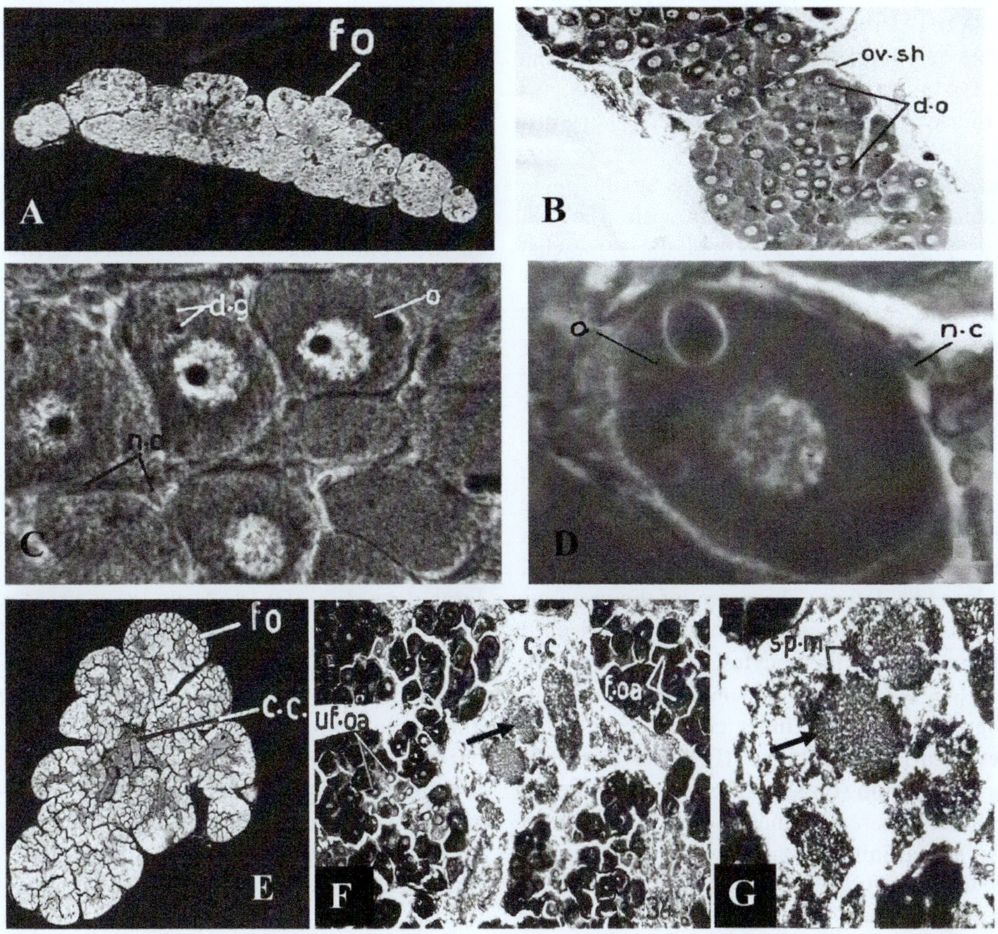

FIG. 10.140 Photomicrographs of T.S.s through different levels of the ovary of a freshly copulated female (stage 2) from (A–D): (A) showing enlargement and lobulation of the ovary. X=15; (B) magnified portion of (A) showing the compact developing oocytes. X=90; (C) high Magnification of a portion of (B) showing the structure of the oocytes at this stage. X=350; (D) more enlargement of one oocyte showing the surrounding nurse cells. X=600; (E–G) of the ovary of late copulated female (stage 3): (E) shows more enlargement and much lobulation, sperm masses in the central cavity *(arrow)*. X=15; (F) magnified portion of (E) showing the central cavity occupied by sperm masses *(arrow)*. X=50; (G) high magnification of the central cavity. X=150. c.c=central cavity, d.g=distinctive granules, d.o=developing oocyte, fo=follicles, f.oa=fertilized ova, n.c=nurse cell, o=oocyte, Ov.sh=ovarian sheath, sp.m=sperm mass, uf.oa=unfertilized ova. *(From Bawab and El-Sherief (1987b).)*

shows a distinct cortical layer measuring about 15 μm but almost without definte cells. The growing oocytes increase greatly in number, to the extent that the whole picture of the T.S. at any level of the ovary appears to be formed only of these oocytes (Fig. 10.140B). The size of each oocyte increases too, attaining a diameter of 75–95 μm. Its cytoplasm becomes granulated, still haematoxylinophilic, its wall is well defined, and its nucleolus changes neither in size nor in its affinity to stains. Most of the nucleoli are close to the nuclear envelope (Fig. 10.140C). This stage is also characterized by the appearance of the **yolk nuclei** in the cytoplasm; some oocytes may have one, others may have two or more. The yolk nuclei are in the form of a cap-like structure on one side of the nucleus, or anywhere in the cytoplasm, but usually on one side of the cell (Fig. 10.140C). Bawab and El-Sherief (1987a) referred to these structures as **directive granules**. In addition, each oocyte becomes surrounded by follicular cells, the so-called cortical cells in the previous stage, and called in the work of El-Sherief (1983) nurse cells (Fig. 10.140D). Later they form the chorion of the egg. These become the only remaining constituent in the stroma (Fig. 10.140C). Owing to the fact that the growing oocytes are very close to each other, the stroma becomes very difficult to detect.

Late-copulation stage. The ovaries increase gradually in size and their natural colour becomes yellowish. The T.S. shows numerous pronounced lobules and the somewhat narrow central cavity (Fig. 10.140E). The latter is occupied by sperm masses (Fig. 10.140F) having the same appearance and staining affinity of the dehisced spermatophores found in the spermatheca at this stage (Bawab and El-Sherief, 1989). The cortex reaches its minimum thickness, measuring 7–5 μm, while the whole area is occupied by grown-up oocytes, which are separated from each other (Fig. 10.141A). Each is somewhat rounded in outline but does not yet form a follicle. It has approximately the same size as before. The nucleus is not accurately central in position and loses its definite characters of the previous stage. The cytoplasm is differentiated by the stains used (Fig. 10.141B), but many of the oocytes still have the yolk nuclei. Each oocyte is surrounded by two well-differentiated membranes (Fig. 10.143A): the inner corresponds to the **vitelline** or **fertilization** membrane (El-Sherief, 1983); and the outer is the **chorion** (El-Sherief et al., 1986). In the middle of each lobule, there are smaller cells that differ in their appearance and affinity to stains, which penetrate in between the growing oocytes (Fig. 10.141A). Most of them are in different stages of autolysis. Hence they were given the name **interstitial cells** (Bawab and El-Sherief, 1987b); the same were also called undifferentiated ova (El-Sherief, 1983).

During the above stage (late-copulation stage), two important vital activities occur: dehiscence of the spermatophores and fertilization of the oocytes. The latter depends greatly on the former. The spermatheca during this stage is the largest and oval in appearance (Fig. 10.135(II) C). Bawab and El-Sherief (1989) illustrated and stated that the epithelial wall of the anterior portion of this spermatheca secretes a fluid substance (sec. II) which merges smoothly into the male matrix in between the spermatophores (Fig. 10.138C and F). Therefore, this portion of the spermatheca contains some of the spermatophores with **dissolved walls**, i.e. dehisced spermatophores (Fig. 10.138G). It is striking that this secretion later disappears as a definite layer and the epithelial wall retains its thickness, i.e. it is renewed. Similarly striking is that some of secretion 4 of the male matrix disappears. Therefore, the dehiscence of the spermatophores of *P. pelagicus* occurs chemically. However, the mechanism of this dehiscence and the substances used need further investigation. Interestingly, around the same period of the interpretation of the above process of dehiscence, Beninger et al. (1988) interpreted this process in the spermatheca of *Chionoecetes opillio* as possibly occurring by dilution of 'the seminal fluids/spermatophore storage matrix with seawater'. However, in *P. pelagicus*, this is impossible because of the presence of the plug.

Up until the works of Bawab and El-Sherief, no information concerning the process of **fertilization** was encountered. However, El-Sherief (1983) stated that masses of the dehisced spermatophores are clearly seen in the central cavity of the ovary. Critical observation reveals that free sperm can be detected merging within the frothy matrix inbetween the developing oocytes in the vicinity of the cavity (Fig. 10.140E–G). This matrix might be the result of the autolysis of the interstitial cells forming a suitable medium for the vitality of the sperm. It was then interpreted that fertilization occurs inside the ovarian tissue and not in its cavity or its ducts. This means that the oocyte is fertilized in its initial position in the ovary. This is confirmed by examination of other selected sections of another ovary obtained at a later date. Fig. 10.141A shows a great number of oocytes with clearly elevated membranes. The fertilized cell (Fig. 10.141B) has two distinct and well-differentiated membranes by the used stains: the inner is very thin and closely applied to the surface of the cell—that is, the vitelline membrane which becomes the fertilization membrane; and the outer is thicker—the chorion. A comparable condition was recorded previously for the crab *Menippe mercenaria* (Binford, 1913) and for the crab *Scylla* (Estampador, 1949). Cheung (1966) studied the egg membranes and their attachment in the shore crab *Carcinus maenas*, and reported that the internal fertilization of this crab occurs in the lumen of the ovary or within the oviducal region. However, this process of fertilization in Brachyura in general, and in *P. pelagicus* specifically, definitely occurs in the ovary, but needs more cytological studies.

Prespawn stage. The ovaries during this stage attain their maximum size (Fig. 10.135D) and occupy most of the body cavity of the specimen. Their natural colour undergoes a series of changes, beginning with yellow, to orange and reddish-orange. Histologically, the cortex attains the minimum thickness measuring about 5 μm. The fertilized oocytes with their well-marked membranes are much more separated from each other (Fig. 10.141C). There is no change with respect to the central cavity except that the sperm masses disappear completely. The oocytes, on the other hand, undergo marked changes in their size and cytoplasm. This, together with the natural colour changes, indicates that this is the stage of yolk formation, deutoplasmogenesis (vitellogenesis) (Fig. 10.141C and D).

Deutoplasmogenesis. This is a vital activity performed by the oocytes to change to comparatively large eggs condensed with yolk. It does not begin with this stage, and to understand the sequence of changes done during this

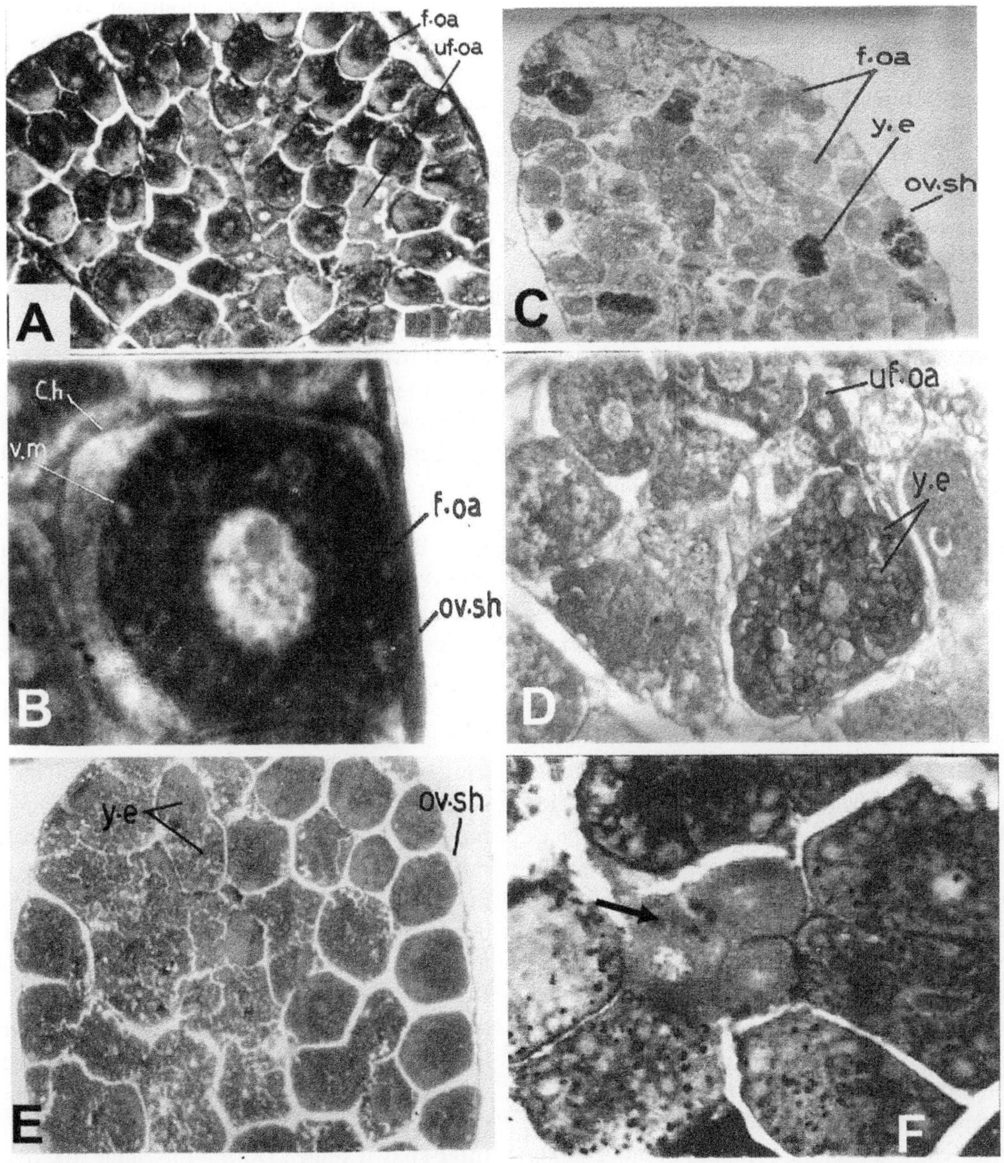

FIG. 10.141 Photomicrographs of T.S.s of the ovary of a late copulated female (stage 3): (A) showing the fertilized oocytes and the accumulation of the unfertilized ones and/or interstitial cells *(arrow)*. X = 150; (B) Highly magnified fertilized oocyte. X = 1000; (C–F) of the ovary of prespent female (stage 4); (C) shows different stages of vitellogenic oocytes. X = 150; (D) magnified portion of (C) showing late vitellogenic oocyte. X = 600; (E) portion of the ovary showing final stage of vitellogenesis, notice the periphery with oocytes having compact homogeneous yolk. X = 150; (F) magnified portion of the ovary of this stage showing the glandular interstitial cells *(arrow)*. X = 300. Ch & v.m = egg membranes, y.e = yolky egg, other lettering as in Fig. 10.140. *(From Bawab and El-Sherief (1987b).)*

process, the nature of the cytoplasm of the oocyte before this process must be recorded. Bawab and El-Sherief (1987a) studied this aspect histochemically. The striking result is the complete absence of proteins with –S–S group and lipids of steroid nature (Table 1, 1987a). The **first step** is the appearance of yolk deposition that begins nearly with the growth period of oogenesis (precopulation stage). The cytoplasm is gradually loaded with homogeneously distributed fine droplets (Fig. 10.140B), strongly positive to Sudan black B, Nile blue sulphate, and never to the protein or to carbohydrate reactions. This means that the basic component of this yolk (in oocytes of *P. pelagicus*) is of a fatty nature and explains the natural pale yellow colour attained by the ovaries in the living state. During this step and the **second step**, the oocyte increases gradually in size and is still loaded with the fat droplets in the same density and property. The yolk nuclei appear (Fig. 10.140C) (fresh-copulation stage) and are mainly of protein nature but without tryptophan, or –S–H group. Traces of natural mucopolysaccharides can be detected, but both glycogen and fat can never be traced in these nuclei.

The previous two steps of yolk deposition are during the oogonial phase, while the following third and fourth steps are postoogonial, during which each 'ovum' attains its definitive size. The **third step** is characterized by great enlargement of the oocyte and by the deposition of two different types of globules (Fig. 10.141D) that are numerous and much larger than the yolk nuclei. The globules can be differentiated by their size, appearance, and chemical nature. One type is of large translucent globules, hence their name. They accept nearly all the colours of the histochemical stains used in Table 10.14. However, the glycogen is absent from these translucent globules. The other type is of somewhat smaller and opaque globules, again hence their name. They have opposite results to almost all the specific reactions (see Table 10.14) but the glycogen is detectable in considerable amounts. The deposition of these globules is everywhere in the cell, but not at the expense or displacement of the previous fatty droplets. The latter seems to conjugate with proteins and/or carbohydrates. Therefore, the developing oocyte, at the end of this step, is heavily loaded with three types of spheres: fatty granules, translucent, and opaque globules. Accordingly, the fresh ovaries have a deep orange colour. It is worth mentioning that the yolk nuclei disappear during this step. The important feature of the **fourth step** is that the individuality of these three spheres disappears completely, i.e. all the deposited materials in the developed oocyte (egg) become a compact homogeneous substance (Fig. 10.141F) and attain the reddish-orange colour in the live state. Deutoplasmogenesis in this species *P. pelagicus* is an **autonomous intraoogonial process**. The fourth step (prespawn stage) finishes both the vitellogenesis and ovarian cycle, after which the product starts to discharge to the outside.

At the time that all the oocytes are fully occupied by the yolk constituents, they become separated from each other (Fig. 10.141E). Most of the interstitial cells of the previous stage disappear and the persisting ones differ greatly. They are larger, unnucleated, stellate, and with dendrite-like processes (Fig. 10.141F). They attain the haematoxylinophilic affinity of the follicular cells of the fresh-copulation stage. Just before spawn, at about the end of prespawn stage, some other interstitial cells appear between the yolky 'eggs' of each lobule of the ovary. They are globular, glandular in appearance, nucleated, and accept the blue and green colours of Malloy's and Masson's combined stains, respectively. In a process of holocrine secretion (Garven, 1965; Leonhardt, 1977) or a process of autolysis, they become a suitable matrix (Fig. 10.142A) clearly observed coalescing the ripe eggs forming a brood ready to discharge (Fig. 10.142B).

It is very important to describe the changes attained by the spermathecae during the final steps of this stage: the prespawn stage (Fig. 10.135D). The interior of each spermatheca changes to be equipped for the passage of the broods of the fertilized eggs. Bawab and El-Sherief (1988) stated that the spermatheca undergoes a significant change. It is no longer divided into anterior and posterior portions, but into dorsal and ventral ones. The dorsal portion elevates to form a **dome** (Fig. 10.142C), which causes the differentiation of the lumen into the dorsal chamber, which is still occupied by the undehisced spermatophores embedded in their hardened matrix, and the ventral chamber, which is almost empty. The cavity of the latter becomes continuous with that of the vagina. The histological appearance assumes that this dorsal portion is the previous posterior portion (Fig. 10.138H). The significant feature of this change is the disappearance of the plug, which would need further histochemical studies.

May-spawn stage (ovigerous female). The ovaries at this stage still have the distended and lobulated appearance of the preceding stage (same size) but lose their colour. They retain the yellowish colour and the histological pattern of the fresh-copulation stage gradually attains that of the late-copulation stage, but differs in having a more active and thicker cortex, which measures about 25 μm (Fig. 10.142E). Beside the developing oocytes found in the periphery, the deeper parts of the stroma and the central cavity become loaded with debris. The latter consists of degenerating eggs, spheres of yolk granules, and much of the frothy matrix (Fig. 10.142D).

Inter-spawn stage. During this stage, the ovaries repeat the events of both the late-copulation and the prespawn stages, i.e. they undergo the second ovarian cycle. Crabs representing these three stages are hard-shelled and nonovigerous. The ovaries cannot be differentiated except by the associated spermathecae (Fig. 10.135(II) E) (Bawab and El-Sherief, 1988).

TABLE 10.14 The chemical nature of the different granules forming the yolk in the eggs of the crab *Portunus* during the process of deutoplasmopenesis

			Deutoplasmogenesis				
			Step 2		Step 3		
Reaction	Indication	Predeuto-plasmo-genesis	Cytoplasm	Directive granules	Granules	Translucent globules	Opaque globules
Mercuric bromophenol blue (Mazia et al., 1953).	Protein	bbb	bb	bb	bbbbb	bbbb	bbb
Biuret reaction (Gabe, 1976)	Peptide linkage	v	v	vv	vv	vv	vv
Aqueos bromophenol blue (Mazia et al., 1953).	Basic protein	bb	bb	bb	bb	bbb	b
Toluidine blue (Kramer and Windrum) (Pearse, 1968).	Acidic groups (metachromatic sub.)	bb	bb	bb	bb	–	–
Xanthoproteic reaction (Pearse, 1968).	Phenyl group	oo	oo	ooo	oooo	o	oo
Millon reaction (Benaley and Gersh modification) (Pearse, 1968)	Tyrosine	PP	PP	P	PP	P	–
Nitroso naphthol method (Udenfriend, Cooper and Stoward) (Pearse, 1968)	Tyrosine	y	yy	y	yy	y	y
DMAB-nitrite method (Adams) (Pearse, 19S8)	Tryptophan	b	–	–	–	–	–
Tryptophan reaction (Adams) (Pearse, 1968)	Tryptophan	B	–	–	–	–	–
Ninhydrin Schiff reaction (Yasuma and Itchikawa) (Pearse, 1968)	NH_2 group	P	P	P	P	PP	P
Oxidized Tannin Azo method (Dixon) (Pearse, 1968)	Amino group	P	P	P	P	PP	–
Sakaguchi react ion (Pearse, 1968)	Arginine	oo	o	o	o	o	o
Ferric ferri-cyanide reaction (Pearse, 1968)	S–H	b	–	–	–	b	–
Performic acid-Alcian blue (Pearse, 1968)	S–S	–	–	–	–	–	–
PAS (Mc Manus) (Pearse, 1968)	Mucosubstance	R	R	R	R	RRRRR	RRR
Alcian blue method (Pearse, 1968)	Acid mucopolysaccharides	–	–	–	–	–	–
Alcian blue PAS (Pearse, 1968)	Acidic and neutral MPS	bb RR	bbb RR	RR	bbb	RRRRR	RRR
Best's Carmine (Pearse, 1968)	Glycogen	PP	PP	–	PPP	P	PP
Sudan black B (McManus) (Pearse, 1968)	Lipids	bb	bb	–	bbbb	bb	bbb
Sudan black B (Pearse, 1968)	Lipids	bb	–	–	bb	bbb	bbb
Nile blue sulphate (Cain, 1947a,b)	Neutral and acidic lipids	b	bb	–	bbbb	bbbb	bbbb
Liebermann Bürchardt reaction (Pearse, 1968)	Cholesterols and its esters	–	–	–	–	–	–
Schultz reaction (Weber et al.) (Pearse, 1968)	Cholesterols and its esters	–	–	–	–	–	–

The symbol indicates the production of the colour: B, black; b, blue; g, green; o, orange; P, pink; R, red; v, violet; y, yellow. The repetition of the symbol indicates the intensity of this colour.
From Bawab and EL-Sherief (1987a).

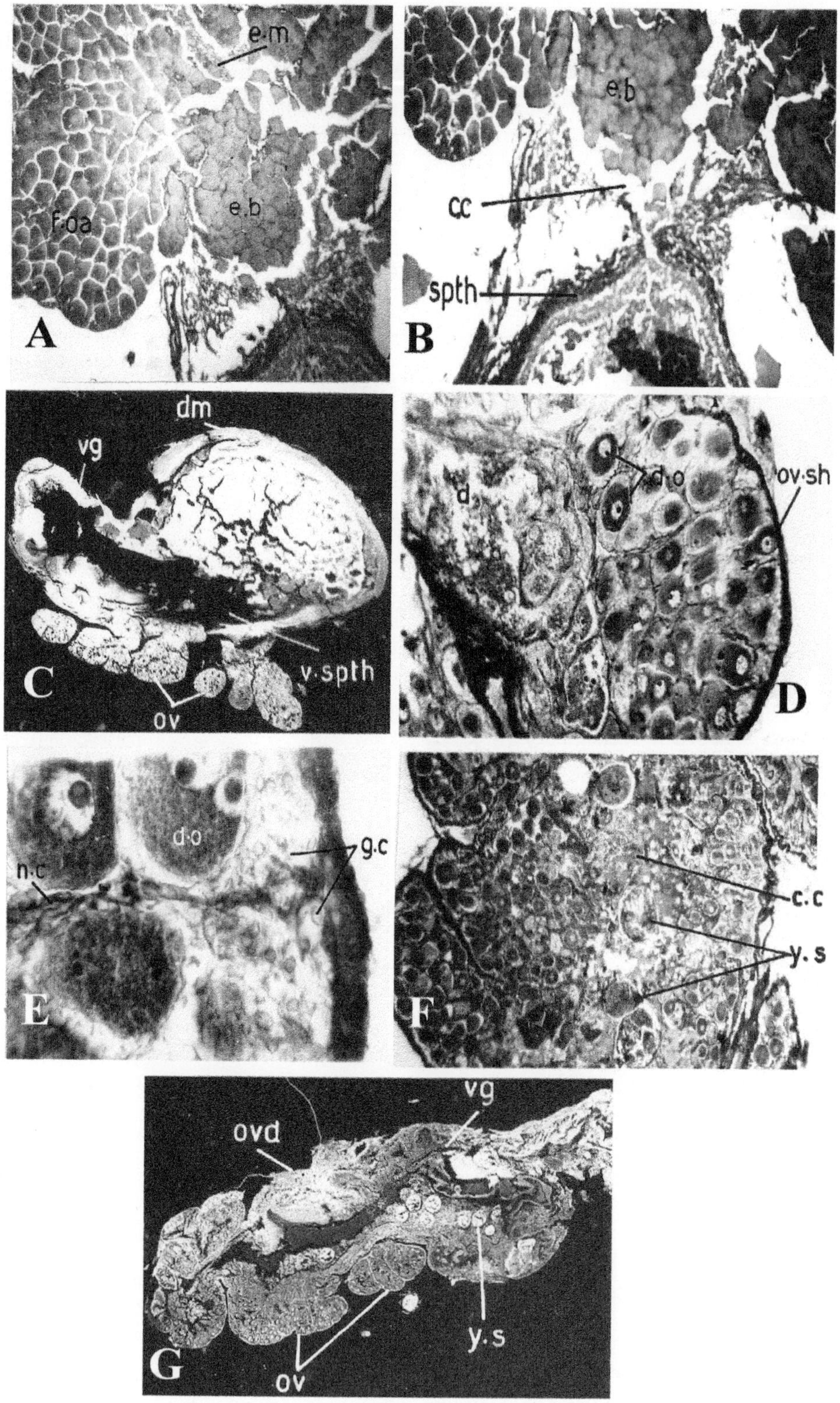

FIG. 10.142 Photomicrographs of frontal sections of the ovary connected to its spermatheca: (A) showing egg brood in the central cavity just before spawning (prespawn stage). X=75; (B) another section showing the continuation of the central cavity with the spermatheca. X=75; (C) L.S. of the gonoduct of the ovary of a prespent female showing the dome. X=150; (D) T.S. through the ovary of May-ovigerous female showing the new developing oocytes of the second cycle. X=150; (E) magnified portion of (D) showing the proliferated germinal cells from the cortex. X=400; (F) T.S. through the ovary of ovigerous female after second spawning showing its central cavity, notice its similarity with (D). X=100; (G) frontal section through the gonoduct of the above ovary showing the debris (yolk spheres). X=20. dm=dome (dorsal portion of the spermatheca), d=debris, d.o=developing oocytes, eb=egg brood, em=egg matrix, g.c=germinal cells, n.c=nurse cell, ov=ovary, ovd=gonoduct, spth=spermatheca, vg=vagina, v. spth=ventral part of the spermatheca, y.s=yolk spheres, other lettering as before. *(From Bawab and El-Sherief (1987b).)*

September-spawn stage. The ovaries decrease gradually and slowly in size to recover the appearance of the immature condition. Their natural orange colour persists for some time and then fades slowly. The cortex thickens again becoming about 15 μm. The remnants in the lobules are considered as debris, even the yolky eggs. The characteristic feature of this stage is the presence of ensheathed spheres of dense yolk granules (Fig. 10.142F), much of which are clearly seen within the cavity of the associated spermathecae (Fig. 10.142G). This may be interpreted as the female reproductive system preparing itself for the following breeding cycle.

During the reproductive cycle that was described above and each female cycle, the ovaries undergo two separate cycles: one ends during May and the other ends during September. Just at each spawning, the female carries the eggs and brood them until hatching. Table 10.12 summarizes the marked histological changes of the ovaries during these first and second cycles, differentiates between them, and correlates them with those of the spermathecae during the reproductive cycle.

Egg-membranes. The crabs of the May- and September-spawn stages are easily selected because of the eggs being carried on their pleopods (Figs 10.118E and 10.136F). They are usually found in May and in September in the catch in Alexandria. During these stages, contribution concerning the **brooding** nature of this *P. pelagicus* must be added. El-Sherief et al. (1986) studied macroscopically and microscopically the structure of eggs on the pleopods. Ripe reddish-orange large eggs can be observed macroscopically and even by the naked eye within the ovary of the prespawn stage. Smears of this ovary show that each egg is surrounded by two well-differentiated membranes. According to the view mentioned above that fertilization occurs within the ovary, the inner of these membranes, essentially vitelline, becomes the fertilization membrane and the outer is the chorion (Figs 10.141B and 10.143A). Just after the female spawns, the eggs are incubated on the pleopods. The freshly hanged egg (Fig. 10.143C) is a large spherical mass with about the same diameter attained during the prespawn stage that may reach 275 μm. Both the stereo- and phase-contrast microscopes show the surface of this egg divided into polygonal facets (Fig. 10.143C) that it is described as having a football-like appearance. The corners of each facet project outwards as folds, which are regularly displaced according to the regularity of the facets. Stained smears of this hanged egg, i.e. at this stage of incubation, supported the folding appearance of the surface (Fig. 10.143D). In addition, these smears show that each egg is enclosed within two other extra-membranes outside the chorion described in Fig. 10.143A. The whole diameter of the smeared egg with all the membranes is about the same as that of the naturally hanged one.

Histological preparation of these smeared eggs clarified the situation (Fig. 10.143E). The fertilization membrane still surrounds closely and limits the outline of the spherical egg. The chorion is differentiated into two layers; both are free of chitin as indicated by the reaction of Alcian blue after sulphation (Table 10.15). The inner layer is lipoid in nature, as indicated by the positive result of the Sudan black B stain. The outer layer is amber coloured naturally and consequently is unstable in its acceptance to the different stains. However, this outer layer of the chorion consists of protein, lipid, and neutral mucopolysaccharides (Table 10.15). On the other hand, the two extra-membranes are well differentiated from each other and from the chorion and fertilization ones. The inner extra-membrane is the thickest of all; it measures about 12.0 μm and is differentiated into two layers: the inner-most one constitutes almost the whole thickness and it is chitinous since it attains the blue colour of the aniline blue of Mallory's triple stain. This is confirmed by the reaction of Alcian blue after sulphation (Table 10.15). The outer layer of the inner extra-membrane is a thin sheet, measures about 2.7 μm in thickness. It is nonchitinous, but the protein substance is more pronounced. Owing to the chitinous nature and thickness of the inner-most layer of this extra-membrane, it is referred to as a **chitinous membrane**. The outer extra-membrane surrounds each egg and firmly attaches it to the setae of the pleopod by a separate funiculus (Fig. 10.143B). Table 10.15 shows that it consists of acid mucopolysaccharide, lipid, but no protein. It is named the attaching membrane. It is worth mentioning that this attaching membrane resembles the outer layer of the chorion in that it is naturally amber coloured, making it unstable to the stains used in Table 10.15.

The origin of these two extra-membranes needs thorough examination of the oviduct through the ovarian cycles. However, El-Sherief et al. (1986) examined by histological studies the pleopods on which the eggs are attached: first, the pleopod of the freshly incubated eggs; and second, the pleopods at the end step of incubation. Sections of the first specimen showed copious secretion inner to the cuticular wall of the pleopod (Fig. 10.143F). It is discharged through definite ductules penetrating the cuticle covering the pleopod (Fig. 10.143F). The heaviness of this secretion obscures the true picture of their origin. However, these glands can be isolated from the pleopods after fixation in 10% neutral formol. The reactions in Table 10.15 show that the nature of the secretion of these glands is fairly similar for the attaching membrane. On the other hand, the sections of the second specimen at the end of incubation showed that these dermal glands have a definite and clear structure (Fig. 10.143G), but without secretion.

It appears that the process of egg-membrane formation and the attachement of eggs to the pleopods in Decapod Crustacea is not a recent subject. Cheung (1966) reviewed the history of this subject and showed that a few works were

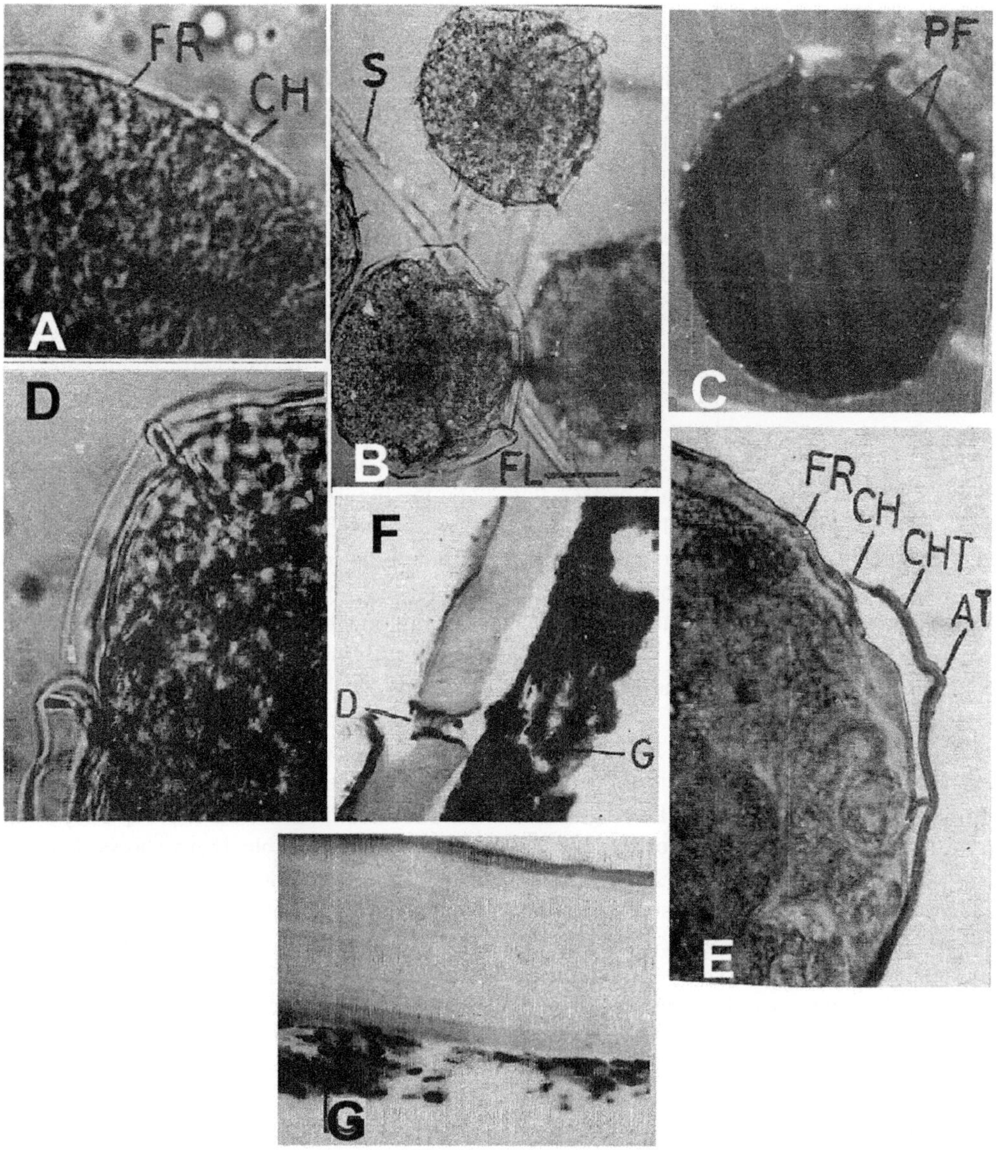

FIG. 10.143 (A) Magnified part of an egg from a smear of ovary of prespent female *(reddish-orange)*. X = 1000; (B) stained W.M. preparation of incubated eggs hanged on the setae of the abdominal appendages, showing the chitinous and attaching membranes, notice the funiculus. X = 100; (C) W.M. preparation of unstained hanged egg showing its external appearance. X = 160; (D) magnified part of stained egg of the same above stage showing the folding appearance. X = 500; (E) section through freshly incubated egg showing the four membranes. X = 400; (F) L.S. through the pleopod of freshly spawned female showing the dermal glands filled with secretion. X = 350; (G) L.S. through the pleopod of female at late stage of incubation showing the dermal glands devoid of secretion. X = 900. AT = attaching membrane, CH = chorion, CHT = chitinous layer, D = ductile, FL = funiculus, FR = fertilization membrane, G = dermal glands, PF = polygonal facets, S = setae. *(From El-Sherief et al. (1986).)*

carried out as early as 1787 and during the 19th and the 20th centuries. Cheung reported that it was believed that in lobsters and crayfish, the eggs are soon found in the 'incubation cavity', which is packed with a soft substance secreted by the 'cement glands' in the abdominal limbs. Then, the female begins to execute typical circular movements and the 'cement' gradually hardens. Serial sections of the endopodites and exopodites of *Carcinus* pleopods were examined in search of cement glands, in both mature and spent female specimens. No such rosette structures with cyclic secretory activity corresponding to the development of the ovary have been found. In addition, the abdomens of these specimens were also sectioned, but the cement glands were not found. Therefore, in Brachyura the process is different. One of the earliest workers, Williamson (1904), claimed the 'impaling' theory of egg attachment (that each egg is impaled on a hair (setae of pleopod) which has been thrust through the external egg-membrane). However, this theory was not supported by any of the later workers. The most comprehensive account of crustacean egg-membrane formation and egg

TABLE 10.15 **Staining reactions of the membranes of the freshly incubated eggs of** *Portunus pelagious* **(L.) and dermal glands of the pleopods**

Stain	Chorion		Chitinous membrane		Attaching membrane	Dermal glands
	Inner	Outer	Inner	Outer		
	0.5	0.4	0.9	2.7	12.0	
	Microns		Microns		Microns	
Heidenhain's iron haematoxylin (Pantin, 1948)	+ve	+ve	−ve	+ve	+ve	+ve
Mallory's triple stain (Pantin, 1948)	RR	−ve	bb	RR	R	RR
Masson's trichrome stain (Pantin, 1948)	RR	RR	gg	R	RR	RR
Mercuric bromphenol blue (Mazia et al., 1953)	−ve	bb	bb	bbb	−ve	b
PAS+Alcian blue (Pearse, 1968)	bb	RR	RR	RRR	bbb	RRR
Sudan Black B (Pearse, 1968)	bb	bb	b	−ve	bb	bb
Alcian blue after sulphation (Pearse, 1968)	−ve	−ve	bbb	−ve	−ve	−ve

From El-Sherif et al. (1986).

attachment has been given by Yonge (1937, 1955), as well as others. The results of these works gave Yonge the opportunity to consider that the crustacean egg basically consists of two external nonprotoplasmic membranes, in addition to a living membrane situated inside them; the inner is formed of chitin, whereas the outer is formed of epicuticular substances, in a manner analogous to the integument of the Crustacea. Afterwards, this was called **integumental theory** (Yonge's theory). Regarding the chitinous membrane (the inner membrane of Yonge), Cheung mentioned that it has been noted in *Carcinus*; it is formed long after the appearance of the fertilization membrane (during the embryonic stage). Therefore, its formation from secretion from the oviduct is impossible. El-Sherief et al. (1986) discussed this point without a clear-cut view and proposed more investigation to the oviduct of *P. pelagicus*.

Cheung (1966) studied thoroughly (experimentally, histologically, histochemically and chemically) the development of egg-membranes and egg attachment in *Carcinus maenas*, a species in which Yonge's theory was accepted without any doubts by Spalding (1942). During the investigation, evidence had accumulated that indicated an entirely different method of egg-membrane formation and egg attachment. Therefore, Cheung studied, in addition, other decapods, *Nephrops norvegicus*, *Homarus vulgaris*, and *Astacus pallipes* (freshwater) in order to help solve the problem as far as possible. In *Carcinus*, the first membrane (vitelline membrane) of the egg is acquired in the early oocyte stage, when the oocyte is first 'cut' off from the syncytial germinating zone (Fig. 10.144B (a)). It appears very thin at the beginning, and later, the small 'follicular nuclei' called accessory cells gather around this to form a 'mono-nuclear layer' (Fig. 10.144B (b)). At maturity, the vitelline membrane appears to be much thicker (Fig. 10.144B (c)), and the mono-nuclear layer finally falls apart when fully mature. While the egg is laid, this layer with its follicular cells is left behind in the ovary (Fig. 10.144B (d)). This description was not observed in *P. pelagicus* (El-Sherief et al., 1986). It is worth mentioning that Cheung, in his work, completely refused the view of the formation of the egg-membrane (chorion) from follicular cells, which was described for the egg of *P. pelagicus* (Figs 10.141B and 10.143A). However, he stated that layers 1, 2, and 3 (Fig. 10.144A (e) and (f)) are found universally in the four species studied, which were called 'trichromatic membrane'. He evidenced experimentally that layer 1 (vitelline membrane) is elastic in both eggs of *Astacus* and *Carcinus* (no mucous is produced for surrounding the eggs at spawning in *Carcinus*). Before the egg became attached to setae, no additional membrane was found. The next stage was the collection of a viscous fluid below the vitelline membrane, followed by the appearance of another layer below the fluid. Therefore, the fluid became compressed between the vitelline membrane and this new layer (later layer 3). Very soon, this layer also separated from the egg and became closely applied to the vitelline membrane with the fluid in between. The fluid space then became much reduced and the remaining fluid solidified into a uniform layer 2. The latter is so thin that it could easily be overlooked under the light microscope. Where the vitelline membrane was drawn out to form the funiculus, the solidified layer was expanded, filling the whole space.

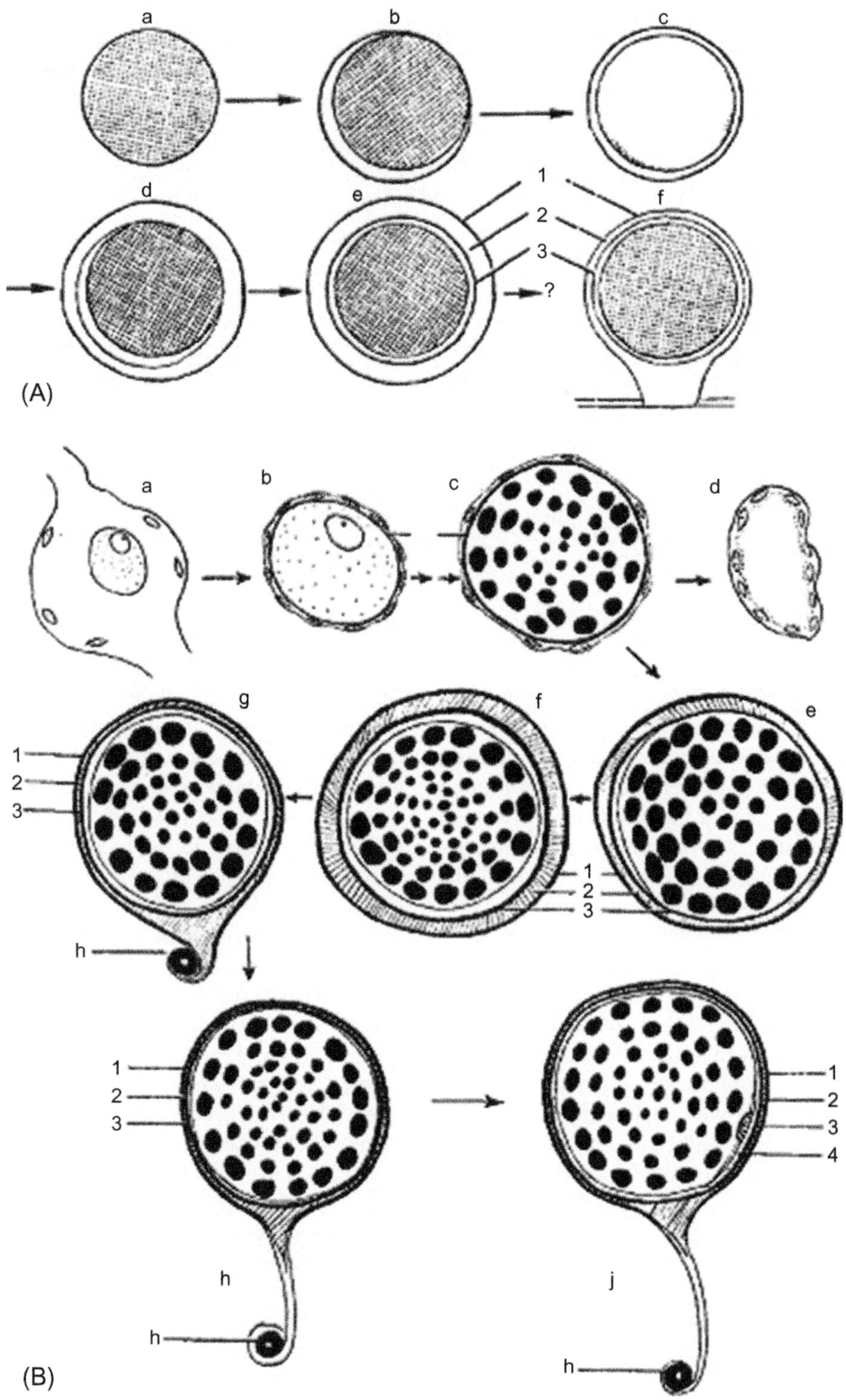

FIG. 10.144 (A) Formation of layers 1 and 3 in *Carcinus*: (a) newly laid egg; (b and c) space formed between layer 1 and yolk; (D) appearance of another layer (presumably layer 3) from yolk; (E) 1 and 3 as two eccentric layers around yolk; (F) a stage not actually observed in *Carcinus* but presumed to happen eventually; (B) Diagrammatic drawings showing hypothetical description of egg-membrane formation and egg-attachement of *Carcinus*: (a) early oocyte in immature ovary; (b) follicular nuclei surrounding oocyte; (c) mature egg in ovary; (d) empty follicle; (e and f) newly laid egg, formation of trichromatic membrane; (g and h) formation of funiculus; (j) old attached egg with eye spot showing formation of layer 4. *(From Cheung (1966).)*

However, Cheung believed at that time, in the new theory of fertilization and the formation of the fertilization membrane from vitelline membrane (see Introduction). He then stated that the trichromatic membrane might really be the fertilization membrane. From his study, he thought that all the egg-membranes are formed by the egg itself. No externally deposited membranes have ever been seen to be formed at any stage. As to exactly how the funiculus and/or the attachment of the eggs are formed, no observations have been made in his work. However, the funiculus of each egg of *Carcinus* is attached directly to a hair, coiling around it. It therefore differs from that of *Astacus pallipes.*

Embryology. It appears that studies on the embryology of brachyuran crabs are recent compared to those of other groups of decapod crustaceans, e.g. of crayfishes. However, Hartnoll and Paul (1982) proved experimentally that there were no problems in obtaining satisfactory embryonic development when berried female *Carcinus* were kept in the laboratory, and many other species of Brachyura have also successfully incubated eggs in captivity. As it will be seen in the foregoing accounts, this success becomes a fact especially with crabs living in the tropical regions of the western Indian Ocean and the Eastern Pacific. Hartnoll and Paul selected the green shore crab *Carcinus maenas* for their work, which had two basic different aims. One was to see whether the eggs could be reared in isolation from the female as well as while carried attached to her pleopods. The other was to investigate the effects of salinity and temperature on the duration and success of incubation. They collected mature nonovigerous females from local shores of Port Erin, Isle of Man, west United Kingdom. They reared them in the laboratory to obtain recently berried females. Some of the latter were used to record the duration of embryonic development and percentage hatch of eggs incubated. The eggs were removed from the other recently berried females to investigate the development of isolated eggs. Two alternative techniques were then tried. Some eggs were kept under optimum conditions such as water flow, aeration, normal salinity seawater, and temperature but without agitation. (It was observed that incubating female crab periodically agitates the eggs by movements of the abdomen and pleopods.) The other method involved agitation but the eggs were reared under a wide range of salinities from 0 to 69.6‰ and at two different temperatures, 30°C and 25°C. In this work of Hartnoll and Paul, isolated eggs would not develop in static water, most probably due to poor oxygenation and infection. Three topics of results and discussion were obtained from this work: the effects of removal of eggs from the parent crab; the effects of temperature; and the effects of salinity. For the first topic, with continuous agitation, the isolated eggs showed better survival and faster embryonic development than attached eggs under comparable conditions. Embryonic development was successful at temperature from 11°C to 25°C. The closest comparison is between the attached eggs at 24°C and the isolated eggs at 25°C, both at normal salinity. The percentage hatch was 71% for attached and 88% for isolated eggs. The entire isolated hatch moulted successfully to the first zoea. In addition, the duration of the embryonic development was 20 days for attached but only 14 days for isolated eggs. The effect of varying the salinity demonstrated that the eggs of *Carcinus* are extremely tolerant. Development to hatching occurred over the range 17.4‰ to 34.8‰, and appreciable survival beyond the prezoea occurred from 26‰ to 39‰. However, the success rate declined from around 90% at normal salinity to only about 50% at the extreme values.

Hartnoll and Paul stated that it is clear in *Carcinus* that the incubation of isolated eggs is quite practicable, and there is no reason to believe that it will not work with other species. This is in spite of the fact that they did not use developmental changes of the embryo as a factor. They used only statistical analysis of the data recorded for the environmental parameters in which the specimens of *Carcinus maenas* were reared.

The information available on the embryonic development of brachyurans focused mainly on marketable species, e.g. species of family Portunidae. The division of this period of development is based on the relative proportion of yolk compared to the animal pole and the differentiation of certain structures of the developing embryo. It was described as 10 embryonic stages. More recently, others have used a lower number of stages to distinguish embryos.

Before describing the embryological development of the different genera, Anderson (1973) chose the brachyuran crab *Macropodia longirostris* (Fabricius, 1775) to be an example to what is meant by embryonization. In *Macropodia* (Fig. 10.145), the nauplius and protozoea stages are passed during embryonic development. The stages are clearly recognizable in the embryo: the nauplius stage by its 'precocious' development of the three naupliar segments and their limb buds; the protozoea stage by its precocious development of the maxillules, maxillae, first and second maxillipeds, and abdominal segments. The embryo proceeds directly to a hatching stage equivalent to the first Mysis stage of *Penaeus*. The continued suppression of the posterior thoracic region of *Macropodia* at the hatching stage is a specialized brachyuran feature. Embryonization of development is the rule rather than the exception in the Malacostraca, and is also common in the Brachyura.

Studies on the embryology of brachyuran genera, especially the marketable species, flourish in the Indo-Pacific region due to the availability of man-made crab cultures. Arshad et al. (2006) studied fecundity, embryology, and larval development of the blue swimming crab *Portunus pelagicus* (Linnaeus, 1758). This is the major marine crab species distributed through the coastal margins of the tropical regions of the western Indian Ocean and the Eastern Pacific. Although they can be caught commercially, the specimens

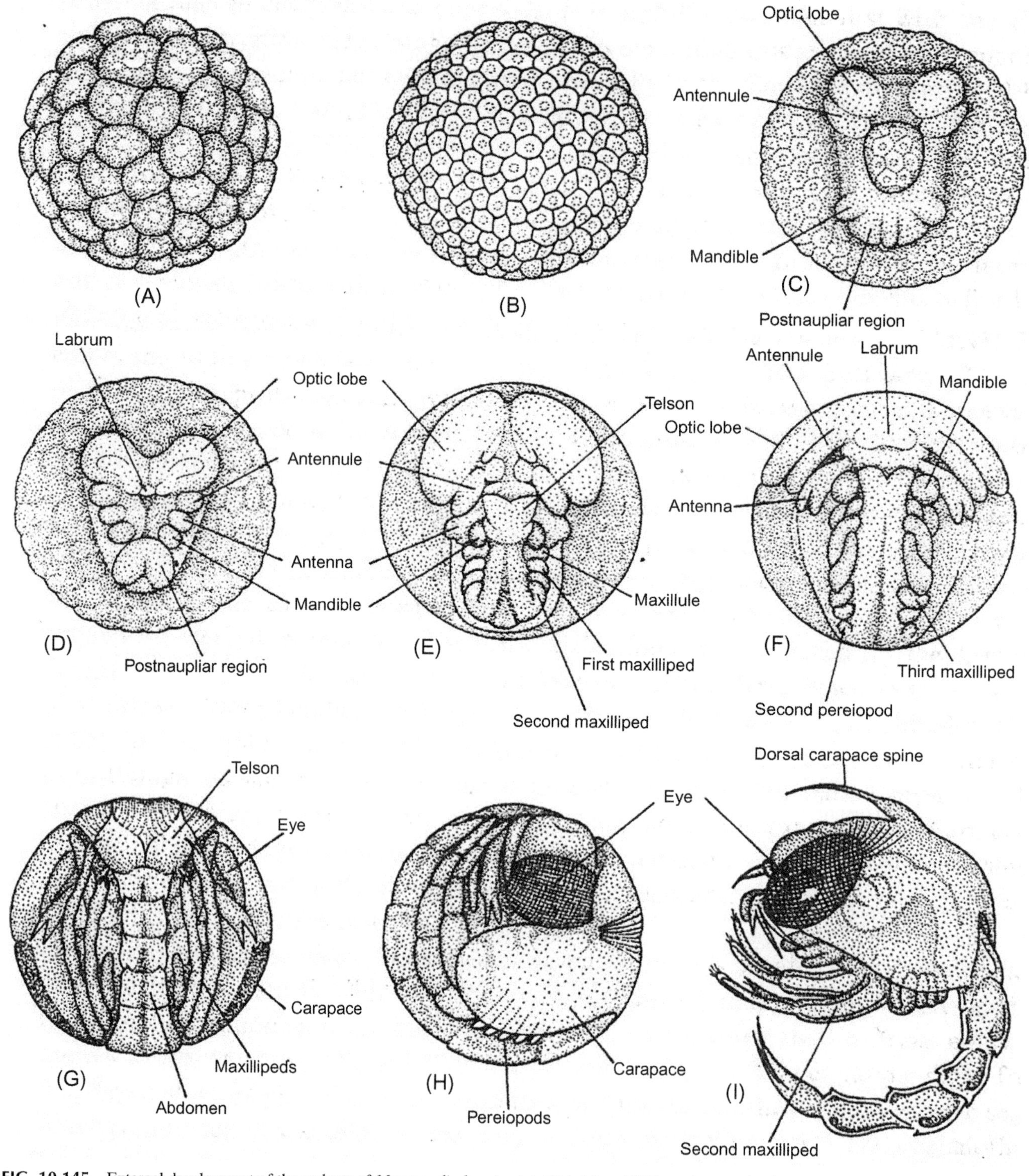

FIG. 10.145 External development of the embryo of *Macropodia longirostris* (Fabricius, 1775) marine species in Europe (family Majidae, subfamily Inachinae (Calman, 1909)) or family Majidae Samouelle, 1819 (*Schram, 1986*); (A) cleavage; (B) blastoderm; (C) nauplius rudiment; (D) formation of caudal papilla; (E and F) formation of postnaupliar segments and limbs; (G and H) completion of development; (I) newly hatched first zoea. *(After Lang and Fioroni (1971), from Anderson (1973).)*

were obtained from their natural habitat in Port Dickson, Malaysia. After being transferred to the laboratory, they were disinfected with 100 ppm formalin for 30 min before being placed in maturation and hatching tanks. To study embryology, a sample of ova from each ovigerous crab was removed every 24 h. Examination of freshly removed (living) eggs was done using optical microscope (25× magnifications). Eggs were placed in seawater using excavated slides. Viable eggs of *P. pelagicus* were spherical, yellowish-orange in colour, and somewhat translucent, ranging from 0.25 to 0.45 mm (average 0.34 ± 0.038 mm) in diameter. There was a slight increase in egg size from newly spawned ones to just prior to hatching. During development, the colour of eggs underwent cyclic change from yellowish-orange to yellow, brown, and dark grey as the yolk was used up and an outline of the embryo became visible. Alterations of the colour of eggs through embryogenesis are common in crustaceans, especially in brachyurans.

Egg development of *P. pelagicus* in the study of Arshad et al. (2006) lasted 7–8 days and was divided into nine periods (Fig. 10.146):

- Period I, day 1 morning: immediately after spawning, the egg was dark yellow-orange, the funiculus was not spiralled.
- Period II, day 1 afternoon: the egg was yellow totally divided, the free region of yolk was visible, egg funiculus spiralled.
- Period III, day 2: the egg was yellowish, free area was conspicuously larger, yolk fragmentation more evident and the embryo appeared whitish in lateral view.
- Period IV, day 3: the egg was brownish and the ocular lobes of the embryo were already visible.

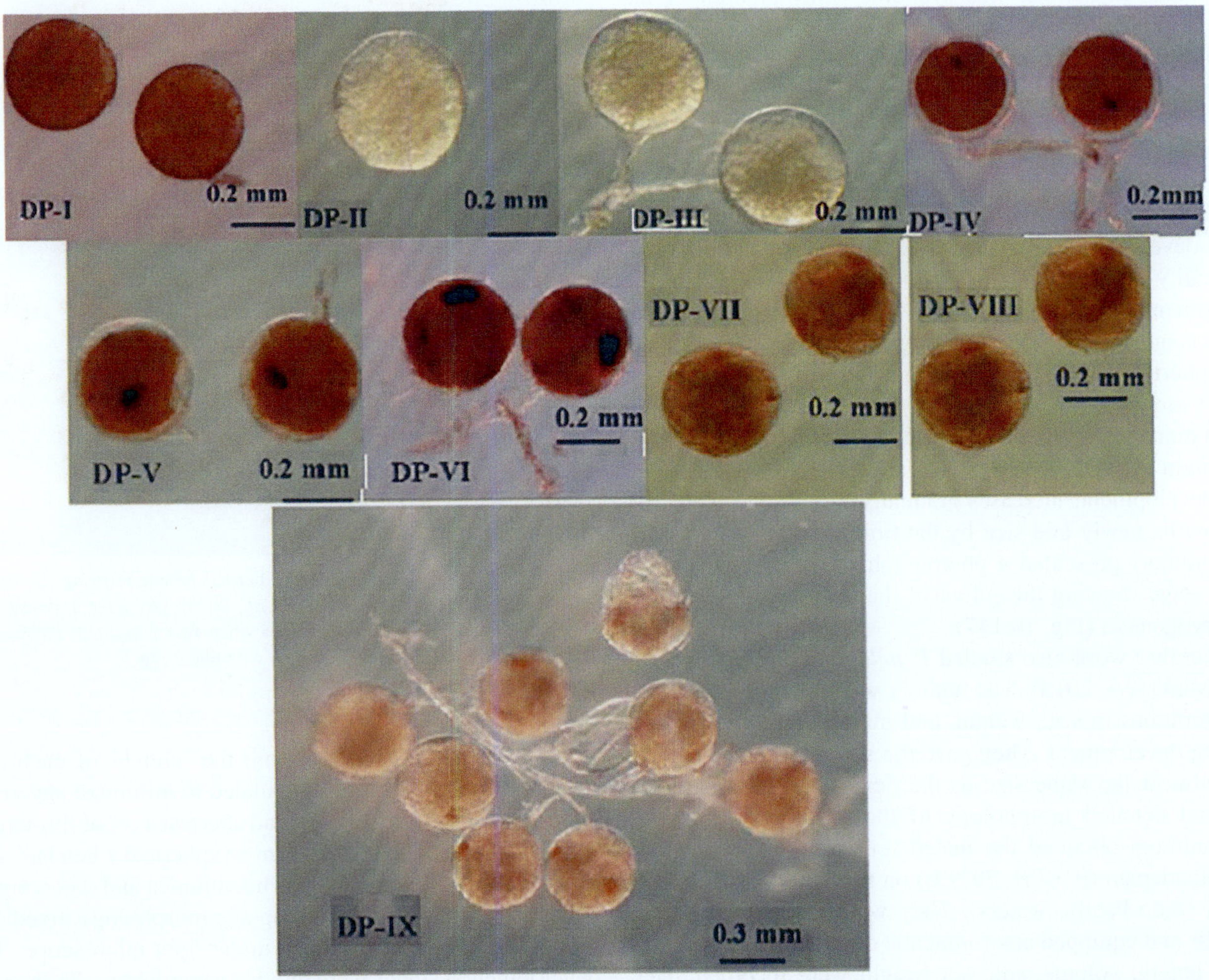

FIG. 10.146 Embryonic development of blue swimming crab *P. pelagicus (from Arshad et al., 2006)*. **DP-I**, day 1 morning, egg just after extrusion, undivided, fully filled with *dark yellow/orange yolk*, and funiculus not spiralled; **DP-II**, day 1 afternoon, free region of yolk visible, egg totally divided, funiculus spiralled; **DP-III**, day 2, free region larger, yolk fragmented, first appearance of *whitish embryo* in lateral view; **DP-IV**, day 3, *brownish* in colour, ocular lobes visible; **DP-V**, day 4, *brown*, ocular lobes larger and darkly pigmented, abdominal chromatophores visible; **DP-VI**, day 5, appendages completely formed, heart large with evident beating; **DP-VII**, day 6, *colour grey*, embryo in lateral view occupied almost all the egg volume, heart more advanced and with cardiac chromatophores, first movement of the embryo; **DP-VIII**, day 7, *grey-black*, the egg ready to hatch; **DP-IX**, day 8, hatching as first zoea.

- Period V, day 4: the egg was brown and the ocular lobes more evident and darkly pigmented, abdominal chromatophores already visible.
- Period VI, day 5: the egg was brown-black, chromatophores and appendages were completely developed, heart appeared larger with evident beating.
- Period VII, day 6: the egg was grey, laterally the embryo occupied almost all the volume of the egg, heart beating and cardiac chromatophores were more visible, and embryo began to move.
- Period VIII, day 7: the egg was grey-black and ready to hatch.
- Period IX, day 8: hatched as first zoea.

It appears from the available literature that this species *P. pelagicus* is a favourite brachyuran animal to study its embryology due to its commercial importance. Therefore, more than one work was available. Ikhwanuddin et al. (2012) studied the embryonic development of the eggs of the female crab ***Portunus pelagicus*** obtained from coastal waters of Johor, Malaysia (Indo-Pacific waters). The study was concerned only with change of the colour and size of these eggs. There are four colours of gonad, ranging from milky white for stage 1, yellowish during stage 2, orange in stage 3, and finally the ovary turned dark orange when well developed. The authors were interested more in the fecundity of the species rather than the morphological development of the embryos. They divided the process into three stages: stage 1 when the eggs are first deposited, they are spherical, bright yellow-orange, and with main diameter 0.307 mm; stage 2 in which the eggs become brown with main diameter 0.386 mm; and stage 3 with black eggs having main diameter 0.396 mm. It appears that the egg, during development, increases gradually in size and is nearly double its newly laid size by the time it is ready to hatch. The authors presented a photograph of a berried crab for each stage showing the colour of the spawned eggs during embryogenesis (Fig. 10.147).

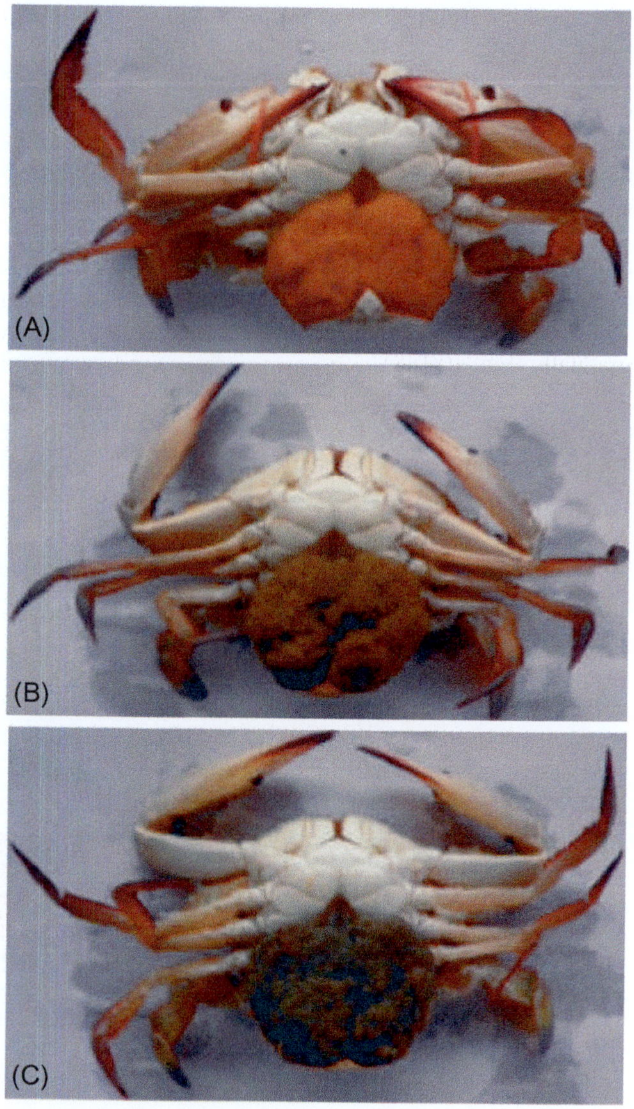

FIG. 10.147 Photographs showing berried female *P. pelagicus* during embryogenesis *(from Ikhwanuddin et al., 2012)*. (A) stage 1 shows eggs spherical and *yellow/orange* in colour when first deposited; (B) stage 2 with brown coloured eggs; (C) stage 3 with black eggs.

Another work also studied ***P. pelagicus***, in India (Ravi and Manisseri, 2013). The authors were mainly interested in alterations in size, weight, and morphology of the eggs during development. They gave the same results of colour and almost the same size as the developing eggs but additional detailed morphology of the developing embryo. The authors obtained the mated females from Palk Bay, off Mandapam (9°17′N, 79°9′E) on the south-east coast of India (Indo-Pacific waters). They were maintained under similar and equipped environmental conditions. Their carapace length, volume, and wet weight were recorded. The spawning seemed to occur during the night or early morning hours. The eggs were light yellow and laden with yolk when deposited. Based on the changes that occurred in egg morphology at 24-h intervals, embryo development was divided into eight stages. To study variations in the egg, 20 eggs were randomly taken from the 'clutch' of each animal daily and the average calculated to minimize any error. Measurements of the longest and shortest axes of this egg of *P. pelagicus* showed that it is almost spherical when laid, and throughout the development. The volumes and wet weights were also recorded. Concerning egg morphology, fixed and prepared eggs were examined under light microscope. The developmental stages were divided according to the changes occurring in the quantity of the yolk, the space occupied by the embryo, and the appearance and subsequent growth of the organs. Since embryo development is a continuous process, the term 'period' was used instead of 'stage' to describe a phase of 24 h. In addition, Arshad et al. (2006)

preferred this terminology to describe the development phases of the eggs of the same species.

In this study of Ravi and Manisseri (2013), the complete embryonic development of *P. pelagicus* took place in 8 days (incubation period at 29 ± 0.1°C (± SE)). The eggs in this study have mean diameter 0.324 ± 0.014 mm (± SD), average volume of 0.016 ± 0.002 mm^3 (± SD), and a mean wet weight of 0.048 ± 0.007 mg (± SD). The mean egg diameter increased gradually during embryogenesis from day 1 to day 6, and increased rapidly from day 6 until hatching. Egg morphology was continuous and the embryo developed gradually (Fig. 10.148).

- Period I (day 1): macrolecithal, spherical yellow eggs, the entire volume filled uniformly with yolk mass, cleavage not commenced, clutch was **light yellow** (Fig. 10.148 (1)).
- Period II (day 2): yellow eggs, small bud like invaginations could be seen mainly on one side which denoted the formation of a cluster of presumptive primordial cells, fragmented oil globule like appearance on the egg surface was more pronounced, clutch continued to be **light yellow** (Fig. 10.148 (2)).
- Period III (day 3): yolk content along the periphery at one side has been reduced by a large amount, certain

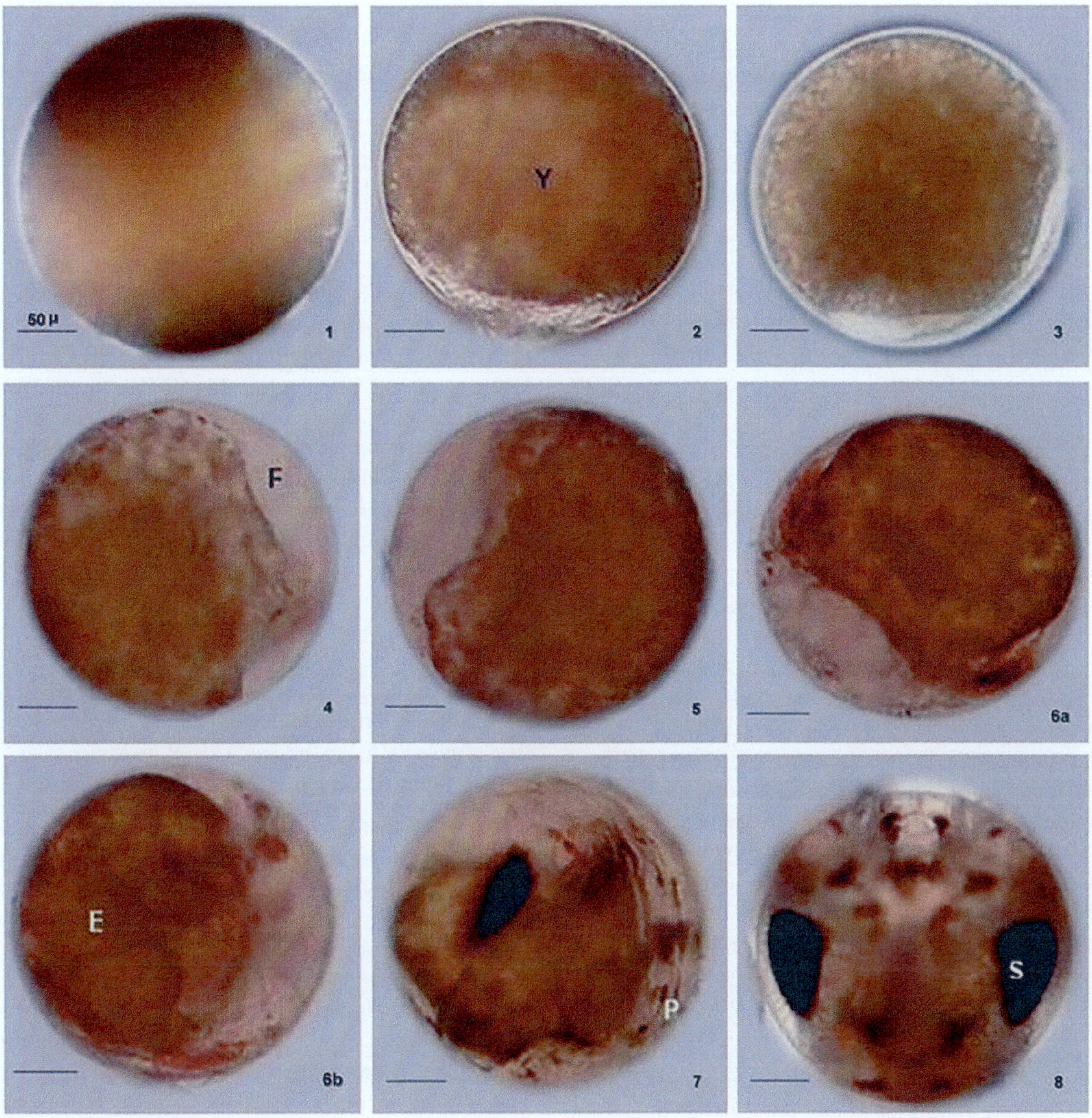

FIG. 10.148 Photomicrographs of the eggs of *Portunus pelagicus* during incubation showing morphological changes *(from Ravi and Manisseri, 2013)* days 1, 2, 3, 4, 5, 6 (6a* and 6b*), 7, and 8; E=developing embryo, F=free space, P=pigments, S=eye spot, Y=yolk, * from different angles. Scale bar=50 μm.

other peripheral regions were also transparent, and clutch became **dark yellow** (Fig. 10.148 (3)).

- Period IV (day 4): yolk content reduced by 35%, more portions of the egg were transparent, invaginations became deeper, clutch continued to be **dark yellow** (Fig. 10.148 (4)).
- Period V (day 5): not much change in the egg volume transparency, light yellowish-orange eye spots observed with difficulty, clutch changed to **yellow-orange** (Fig. 10.148 (5)).
- Period VI (day 6): egg volume transparency was more than 50%, developing embryo was fairly visible, middle of the yellow-orange eye streak gradually darkened and so could be observed, the head and the main body parts could be differentiated, pigmentation faintly visible, clutch was **light grey** (Fig. 10.148 (6a)).
- Period VII (day 7): the heart started to function, ocular lobes large with brownish-black colour and oval-shaped with developed dark core, heavy pigmentation due to the high deposition of chromatophores throughout the head region of the embryo, the latter occupied the entire space inside the egg, slight pigmentation on the abdominal segments, body parts acquired definite shapes, clutch continued to be **light grey** (Fig. 10.148 (6b)).
- Period VIII (day 8): eye spots very large and nearly black, appendages could be seen, heavy pigmentation on the head and the rest of the body, the yolk nearly used up except for a small patch in the cephalic region, heart beating vigorously, transparent embryo, clutch was **dark grey** (Fig. 10.148 (7)).
- Zoea 1 (day 9): free swimming zoea 1 larvae.

India is a country that is actively involved in crab culture and research in Indo-Pacific region. Therefore, embryological studies form a base line to develop hatchery seed production and/or technology. ***P. pelagicus*** is one of the two commercially important genera in this region, and thus many authors are interested in its embryology such as Soundarapandian. One of the works of this author with Tamizhazhagam in 2009 dealt with this species, and there was great interest in its fecundity. They obtained healthy females with yellowish-orange coloured eggs from Parangipettai, Tami Nadu and retained them in equipped tanks during the experimental period. A few eggs were removed daily for examination and description (but without presented photos or diagrams). The eggs undergo colour change as in the previous works during the gradual development, which was reported in this work to be 6–7 days. The eggs at the time of oviposition were quite distinct and large. Their incubation period was divided into six stages:

- Stage I: **Blastula** during which the egg was round, golden yellow, undeveloped and appeared as a mass of undifferentiated cells since cleavage and gastrulation were not clear. Yolk granules were denser. The diameter of the freshly laid egg was reported as 0.35–0.36 mm.
- Stage II: **Gastrula**—The eggs were still round and deep yellow or yellowish orange in colour. The space between the egg wall and the developing embryo was visible. The diameter of the developing egg increased to 0.37–0.38 mm.
- Stage III: **Eye placode**—The eggs had the same shape and colour as the previous stage, but the yolk granules were not dense. Cleavage and organogenesis were distinct and the eye spots appeared as scarlet crescent. The egg diameter was 0.39–0.40 mm.
- Stage IV: **Pigment**—The eggs became slightly elliptical with brown colour. Appendages of the developing larva were pigmented and its diameter was 0.40–0.41 mm.
- Stage V: **Heartbeat**—The eggs were dark brown or black in colour and the eyes round in shape. The colour change is caused by absorption of yellow yolk and development of dark pigments in the eyes. The heart started to beat vigorously, the diameter was 0.41–0.42 mm.
- Stage VI: Newly hatched first zoea- the freshly hatched I zoea had CL ranges from 1.05 to 1.025 mm. It swam freely in water. The fecundity rate ranged between 900,000 and 1,000,000 eggs. The hatching success of freshly hatched I zoea was 75%.

Embryology of the other commercially important portunid crab (another Indian species) ***Scylla serrata*** (Forskal) was studied (Samuel and Soundarapandian, 2010) with the same view of division of embryological development. Healthy gravid females of this mud crab *Scylla serrata* (Forskal) with early broods (yellowish orange coloured eggs) were obtained from Parangipettai, Tamilnadu coast (lat. 11°29′N and long. 79°46′E), the same locality of the above species, *P. pelagicus*. During the experimental period, optimum environmental parameters were maintained (salinity 30–35 ppt., pH 7.8–8.2, and temperature of 27–30°C). The colour and diameter of the eggs were recorded daily. These eggs of the mud crab underwent the same colour change and almost the same size change as those of *P. pelagicus*. The incubation period of this mud crab was 7–9 days. Its embryological development was divided into **five stages** (almost the same as *P. pelagicus*) and almost with the same description but with photographs:

- Stage I: **Blastula**, cleavage and gastrulation were not clear.
- Stage II: **Gastrula**, the space between the egg wall and the inner developing embryo was visible.
- Stage III: **Eye placode**, yolk granules were not dense. Segmentation and organogenesis were distinct. The eye spots appeared scarlet crescent.
- Stage IV: **Pigment**, eggs with slightly elliptical shape. Appendages of embryonic larvae were pigmented.
- Stage V: **Heartbeat**, the eggs were dark brown or black in colour. Eyes were round in shape, heartbeat was vigorous. Hatching usually occurred in the early hours of the day as all the portunid crabs. The newly hatched I zoea came out of the egg cases and swam freely in the water.

Samuel and Soundarapandian (2010) mentioned that the newly hatched larvae could be subjected to a variety of stress tests including ammonium, salinity, formalin, and starvation. The results of these tests would be definitely the factors for choosing eggs with good quality. Rearing of healthy larvae would be possible only if the embryonic development is well studied.

Another species of the above genus, ***Scylla olivacea*** (Herbst, 1796) from Malaysia region was studied by Ikhwanuddin et al. (2015). Mud crabs have been cultured in China for the past 100 years. Due to their meat quality and their large size, their culture spread throughout Asia in the last 30 years. Mud crabs of the genus *Scylla* are commercially important, as mentioned above, and are conspicuous crustaceans that provide basic source of income for coastal fishing communities throughout the Indo-Pacific region. FAO statistics shows that the mud crab aquaculture is well developed in China and Philippines, but from the available literature, it appears that the aquaculture of this genus and also genus *Portunus* is spread in the Indo-Pacific region. In an effort to develop mud crab hatchery technology, comprehensive knowledge on embryogenesis may help in obtaining healthy zoea from developing eggs.

Ikhwanuddin et al. (2015) obtained five mature specimens of the orange mud crab, *S. olivacea* from Setiu Wetlands, Terengganu, Malaysia (5.65°N, 102.77°E) which they induced to mate to produce berried female crabs. Crabs were induced to moult by automizing the appropriate walking legs by crushing the carpus or the merus; the chelae were removed by crushing the propodus. This means that this experiment begins at early stages than the above one. The authors observed the premoult, moult and postmoult attendance of the crabs. However, the culture conditions were prepared as has been designed by Samuel and Soundarapandian (2010). Samples of a minimum of 10 eggs (the fertilized egg is macrolecithal, centrolecithal, spherical and has a uniform dark olive green colour (Fig. 10.149A)) were taken from the female in the morning and night at 12-h intervals. The size and morphology of the developing embryos were almost the same as recorded for *S. serrata*. However, the authors presented photos of the spawned crabs showing the cyclic colour change of the eggs during embryogenesis. This appeared to be orange on day 1, became brown on day 4, grey on day 5, and finally dark grey on day 7 before hatching. Therefore, the duration and five stages of incubation are the same as in *S. serrata*. Only the cleavage of the egg of *S. olivacea* was recorded to begin as 2-daughter-blastomeres, increasing to more than 128 blastomeres (Fig. 10.149C). The phase between blastula and gastrula was not clearly marked until gastrulation took place by epiboly with increase of the yolk-free portion. The latter continued to increase gradually. On the seventh day prior to hatching, there were many chromatophores present, mostly dark in colour, and the yolk granules reduced in size. The heart was beating vigorously and the embryo occupied most of the available egg volume. Hatching occurred on the eighth day and before hatching, the larva moved inside the egg specifically the abdominal part. The mean length of the newly hatched I zoea was 1027.27 ± 90.01 μm. The embryonic development chronology of this species *S. olivacea* under laboratory conditions was recorded in a table.

Pinheiro and Hattori (2002) presented a different view of embryological division of a **portunid *Arenaeus cribrarius*** (Lamarck, 1818), from a different locality, from Uhatuba São Paulo, Brazil (South Atlantic Ocean). The authors maintained the ovigerous females in aquaria under controlled temperature (25 ± 1°C) salinity (35 ± 1‰) and photoperiod (12:12 h). They examined the eggs daily and divided the process into eight stages according to the egg size. They photographed the stages (publication local language), and recorded that the shape of the egg was low elliptical throughout embryogenesis and showed a size increase (24.4%).

The same authors, Pinheiro and Hattori (2003) studied the embryology of another species, the mangrove crab ***Ucides cordatus*** (Linnaeus, 1763) Family (Ocypodidae). They obtained the females bearing recently extruded egg 'batches' from different mangrove areas in Iguape, SP, Brazil. They were maintained in equipped and controlled environmental conditions: salinity (15 ± 1‰), temperature (27 ± 1°C) and photoperiod (12:12 h). Three experiments were undertaken as follows.

In **experiment 1**, egg samples were isolated daily from females to record biometry. During development, the eggs of this species undergo the series of colour change as that of the portunid *Arenaeus cribrarius*. Generally, the freshly extruded egg of *U. cordatus* is golden yellow and/or yellowish orange. It seems that this serial colour change is common in developing brachyuran eggs. As recorded above, it becomes deep orange, brown, deep brown, grey, then deep grey and/or black. In addition, the developing eggs of the two species worked on by Pinheiro and Hattori were slightly ellipsoid, with size diameter ranges from 390 to 540 μm, cleavage mesolecithal and the embryogenesis was divided into eight stages. The incubation period of *U. cordatus* passed through 19 ± 1 day at 27°C.

Experiment 2 was carried out to examine the effect of salinity on the embryonic development. The eggs were kept at 5‰, 15‰, 25‰, and 35‰, with constant aeration. Half of the eggs survived and free-swimming larvae were obtained when reared at 15‰ and the others did not survive beyond the 17th day. At the other salinities, the egg development ceased completely after the third day (5‰), while the process ceased in stage 2 (35‰), with different difficulties in each case. This experiment shows that the most suitable medium to incubate eggs is salinity 15‰. Physiologically, the adult *U. cordatus* is capable of tolerating salinities common to the mangrove environment from 2‰ to 30‰. However, eggs and larvae are very sensitive to sudden salinity alterations.

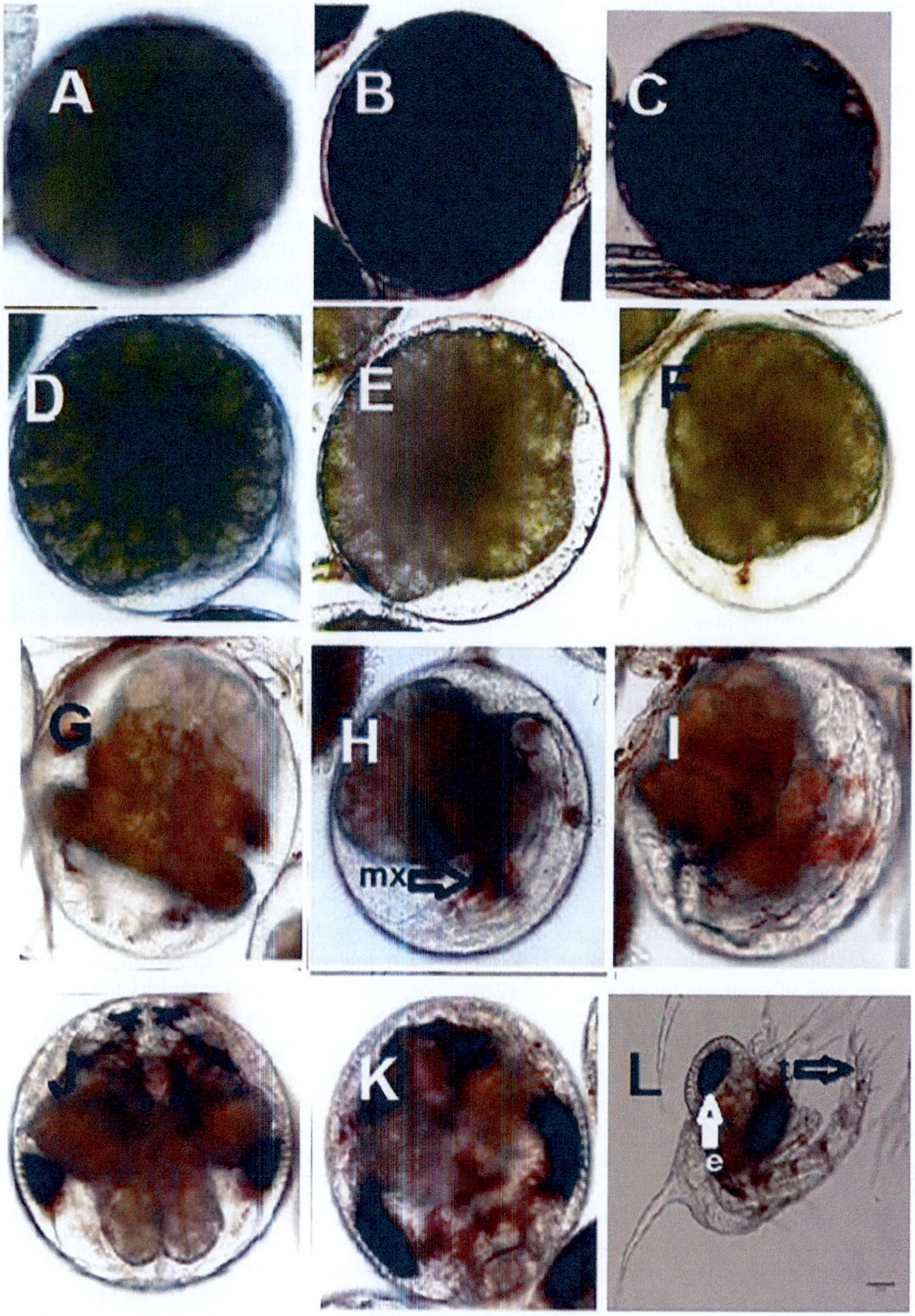

FIG. 10.149 Photomicrographs showing developmental stages of the orange mud crab *Scylla olivacea (from Ikhwanuddin et al., 2015)*, stages were identified according to Samuel and Soundarapandian (2010): (A) unfertilized egg; (B) fertilized egg before cleavage; (C) blastula; (D, E) gastrula; (F) 3 days embryo; (G and H) 4 days embryo; (I) 5 days embryo; (J) 6 days embryo; (K) 7 days embryo; (L) zoeae hatching. e = eye, mx = maxillipeds, t = telson.

In **experiment 3**, eggs were reared in one case of these salinities, e.g. in brackish (15‰) water, but no aeration was provided.

For each experiment, eggs were removed and examined under a stereomicroscope. The morphology of each embryo was recorded in a computer image analysis system, and schematic drawings were produced and identified according to the morphological changes, marked by biometric alterations (Fig. 10.150). The embryonic development of this species, *U. cordatus*, is very similar and matches the general embryonic pattern observed in many brachyuran crabs, e.g. *P. pelagicus* (Arshad et al., 2006). Therefore, the embryonic stages of *U. cordatus* can be adequately identified by verifying the morphology of embryos and chromatic patterns of eggs. Although contributing to their identification, biometric data alone are not sufficient to achieve a precise discrimination.

Diagrams of embryonic development of *Macropodia* and of *Ucides* can be used as examples to understand those of other species.

Another example of embryonic development is that of the estuarine crab *Neosarmatium indicum* (A. Milne Edwards, 1868), a freshwater inhabitant. Sarker et al. (2012) collected several pairs of *N. indicum* by hand during the spawning season, from the Nuha River mangrove swamp, southern Okinawa Island, Japan. This swamp is a unique mangrove forest densely populated by *Kandelia candel* and by crabs of the gen *Perisesarma* and *Neosarmatium indicum.* Among the collection, two pairs were captured during active courtship behaviour. They were reared separately and the environmental conditions were maintained carefully (temperature 25°C, salinity 28‰). The authors observed copulation that lasted 5–7 h. They recorded that fertilization occurred 'within the female body cavity' and the fertilized eggs were extruded within 48 h. A sample of at least 10 eggs from each female was removed every hour up to cleavage, and then every 24 h up to hatching. The events of development of these eggs were observed and described by using an optical DIC microscope equipped with a digital camera, video camera, and printer.

The fertilized egg just after spawning is macrolecithal, centrolecithal, and spherical in shape, with a diameter of 0.36 mm, that increases gradually during development reaching 0.47 mm just at hatching. It is filled with uniformly distributed yolk, dark olive in colour, and without evidence of any segmentation (Fig. 10.151A). Sarker et al. (2012) considered this description as stage 1 of the embryo development, which was completed within 16 days of incubation and divided into 16 stages. The division was mainly based on morphological differentiation of the embryo rather than on the colour change. The authors presented a table of chronology of embryological development of *N. indicum* at constant room temperature of 25°C.

- **Stage 1** (fertilized egg): 2–10 h of incubation, no polar bodies, and three distinct membranes surrounded the freshly attached eggs on pleopods. After 1 h, the first polar body-like structure was extruded and the yolk was yellowish in colour (Fig. 10.151B).
- **Stage 2** (cleavage): 12–17 h of incubation and the cleavage was holoblastic and characterized by the early appearance of a segmentation cavity (blastocoele). The first cleavage typically occurred within 11–12 h of incubation and a large cavity was formed between the two blastomeres (Fig. 10.151C); the second cleavage followed within the next hour, giving rise to four large blastomeres (Fig. 10.151D). At the end of 14 h, the third cleavage occurred, giving rise to eight blastomeres (Fig. 10.151E), and so on until the fifth cleavage, which occurred within 16–17 h. Later cleavage(s) could not be detected clearly due to the condensation of the blastomeres.
- **Stage 3** (blastula): 18–40 h of incubation, during which it was impossible to distinguish individual cells. However, the outer appearances of the developing embryos looked like the uncleaved fertilized eggs (Fig. 10.151H).
- **Stage 4** (gastrula): 41–48 h of incubation; at the end of this period, the developing embryo reached the gastrula stage (Fig. 10.151I).

Stages 5–14 involved the developing embryos.

- **Stage 5** (3–4 days): at the beginning of this period, the yolk droplets were somewhat larger and yellowish (Fig. 10.151J), a yolk-free portion of the egg with evidence of tissues was first observed at the end of this period (Fig. 10.151K).
- **Stage 6** (5 days): the additional feature was the increase of the yolk free portion in size without any observed organ (Fig. 10.151L).
- **Stage 7** (6 days): further increase in the yolk free portion was observed with evidence of distinct tissues (Fig. 10.152A).
- **Stage 8** (7 days): primordial of organs were visible in ventro-lateral position of the yolk free portion during this day, ocular, thoracic-abdominal and cephalic portion of papilla were observed (Fig. 10.152B–C).
- **Stage 9** (8–9 days): eyes like organ was visible on the eighth day of incubation (Fig. 10.152D) and at the end of the ninth day, the developing eyes increased in size (Fig. 10.152E).
- **Stage 10** (10 days): eyes completed more or less their size, and during this day the primordial of the heart was first visible (Fig. 10.152F).
- **Stage 11** (11 days): the eyes were still incomplete, the developing heart and its beating were observed. At this stage the yolk components arranged in four lobes (Fig. 10.152G).

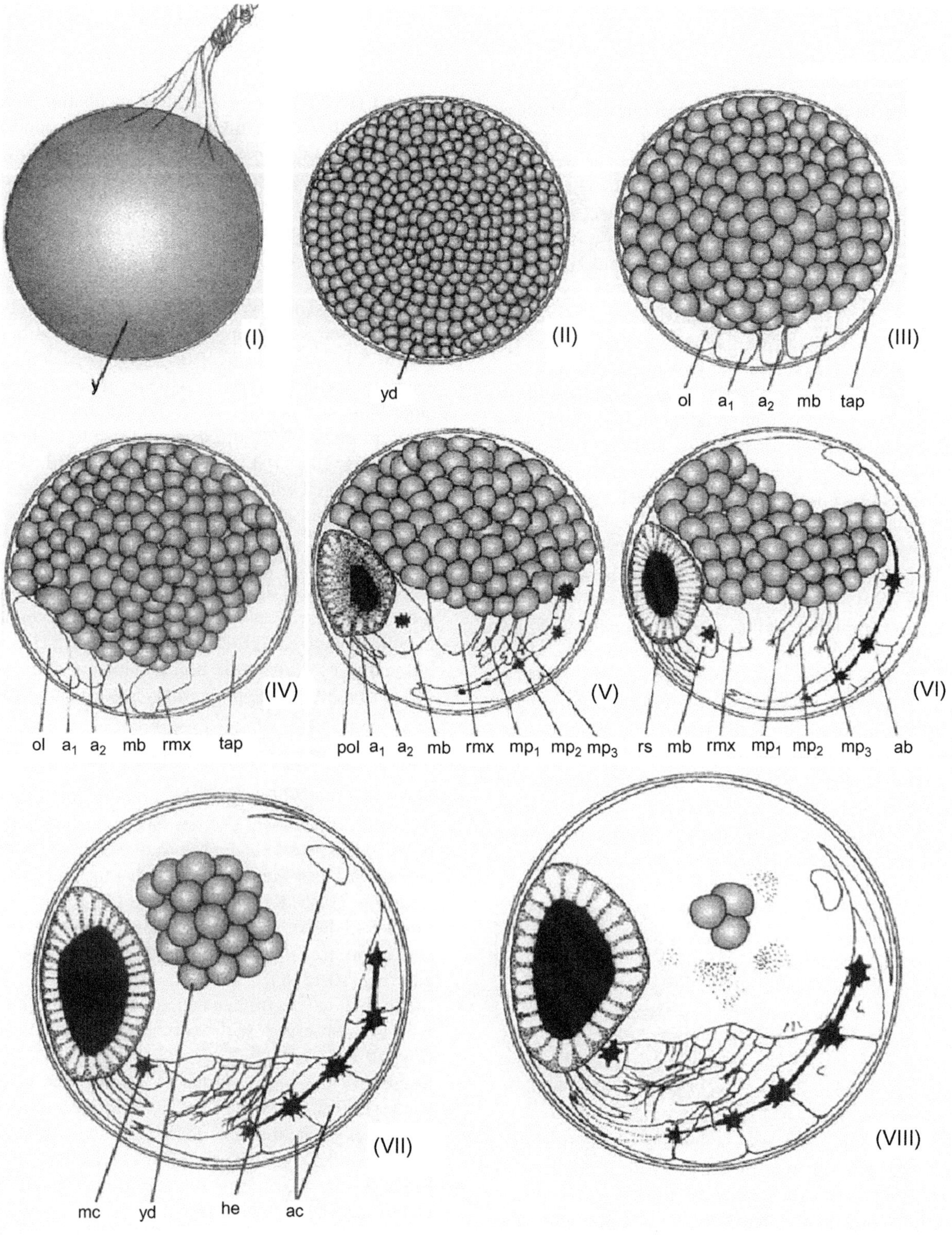

FIG. 10.150 Schematic drawings obtained by a computer image analysis system describing the embryological stages of the mangrove crab *Ucides cordatus (from Pinheiro and Hattori, 2003)*. (I) precleavage stage- egg just after extrusion, undivided, fully filled with yolk, dark 'bordeaux' and opaque, funiculus not 'spiralled'; (II) cleavage to blastula—after 2 days, egg totally divided, dark bright bordeaux, funiculus now spiralled, embryo not yet defined; (III) naupliar stage—after 5 days, fragmentation of yolk evident, embryo appeared as whitish translucent pole occupying about one-quarter egg volume, ocular lobes and three naupliar appendages appear; (IV) metanaupliar stage—8 days, embryo occupied more than ¼ dark 'ochre' egg, 6 embryonic rudiments are distinct but thoracico-abdominal process still not metamerized, heartbeats; (V) pigmented stage—10 days, embryo occupying about half of egg volume, ocular lobes more evident with black pigmentation in central area, eyes oval in lateral view, abdominal chromatophores already visible and connected, metamerization of thoracico-abdominal process; (VI) double chromatophoric bridge stage—14 days, in lateral view two bilobed yolk occupied one third of egg volume, in ventral view chromatophore pair in each abdominal somite, connected together by double chromatophoric bridge, heart now larger with evident beating; (VII) prehatching stage—16 days, in lateral view embryo occupied three-quarters of egg volume, yolk two small connected patches, ocular area ovoid and intensively pigmented, colouration of egg light ochre, beating of heart more noticeable and cardiac chromatophores visible, 6 abdominal segments and telson visible; (VIII) hatching stage—19 days, larva totally formed, in lateral view two small yolk droplets at dorsal region of carapace, having size and shape as the embryo eyes, carapace, maxillipeds, like abdominal somites evident. a_1 = antennule, a_2 = antenna, ab = abdomen, ac = abdominal chromatophores, h = heart, mb = mandible, mc = mandibular chromatophore, mp_1 to mp_3 = maxillipeds 1 to 3, ol = optic lobe, pol = pigmented optic lobe, rmx = rudiment of maxilla and maxillule, Rs = rostral spine, tap = thoracico-abdominal process, y = yolk, yd = yolk droplets.

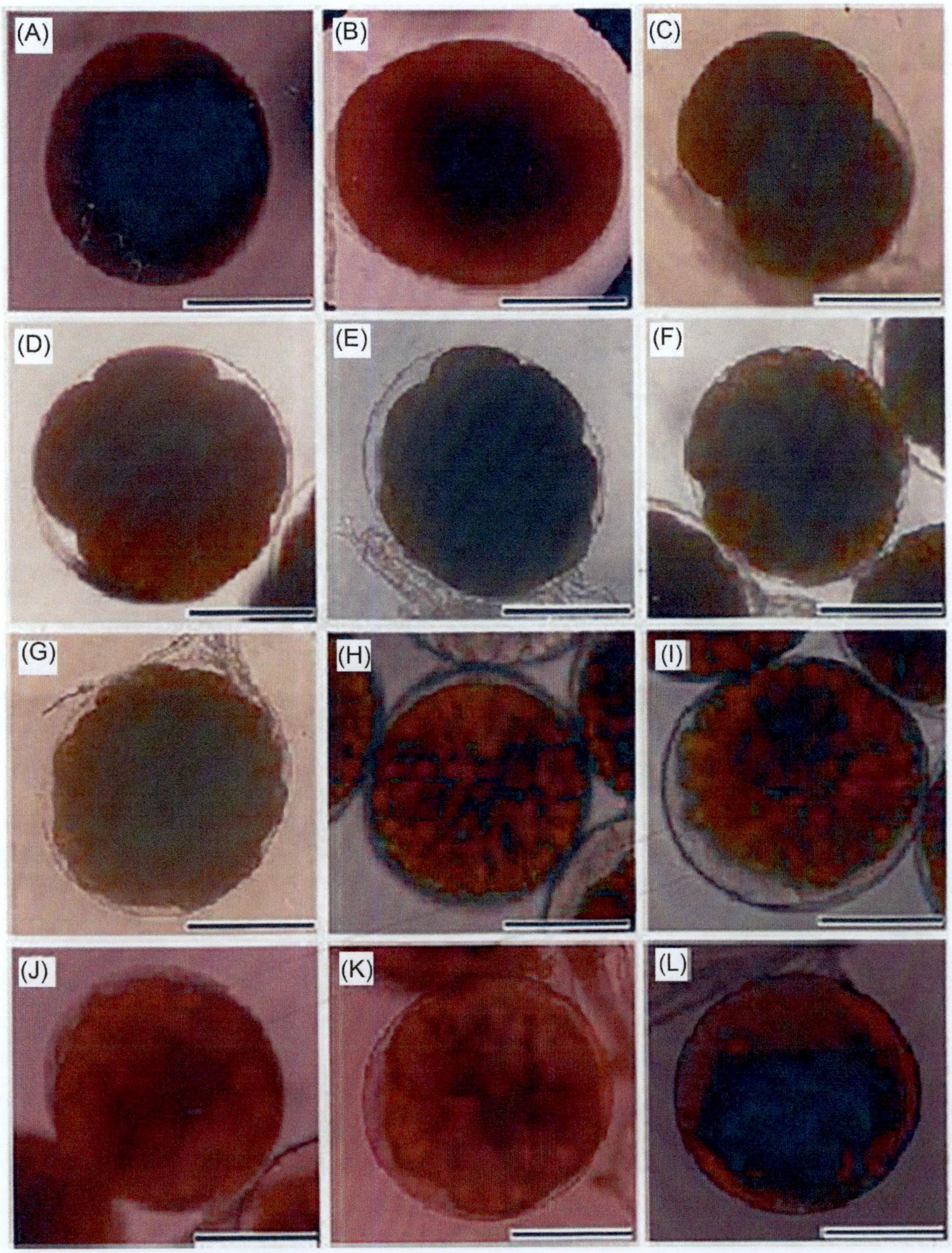

FIG. 10.151 Photomicrographs of early developmental stages (cleavage) of the egg of the estuarine crab, *Neosarmatium indicum (from Sarker et al., 2012)*: (A) unfertilized egg; (B) stage 1 (fertilized egg before cleavage); (C) stage 2 (cleavage, 2-blastomere; (D) 4-blastomere; (E) 8-blastomere; (F) 16-blastomere; (G) 32-blastomere); (H) stage 3 (blastula); (I) stage 4 (gastrula); (J and K) stage 5 (developing embryos); (L) stage 6. Scale bar = 100 μm.

- **Stage 12** (12–13 days): the eyes were more visible, the heart beating was faster and further reduction of the granules becoming two lobes (Fig. 10.152H).
- **Stage 13** (14 days): embryo has occupied half of the volume of the egg with ventral position, the eyes were larger with visible cornea and pigmentation like retina was more intense, heart completed its shape and beating faster. Primordia of antennae and mandible were more visible (Fig. 10.152I).
- **Stage 14** (15 days): about three-quarters of yolk granules was consumed, while the remaining was fragmented. The embryo grew considerably and occupied

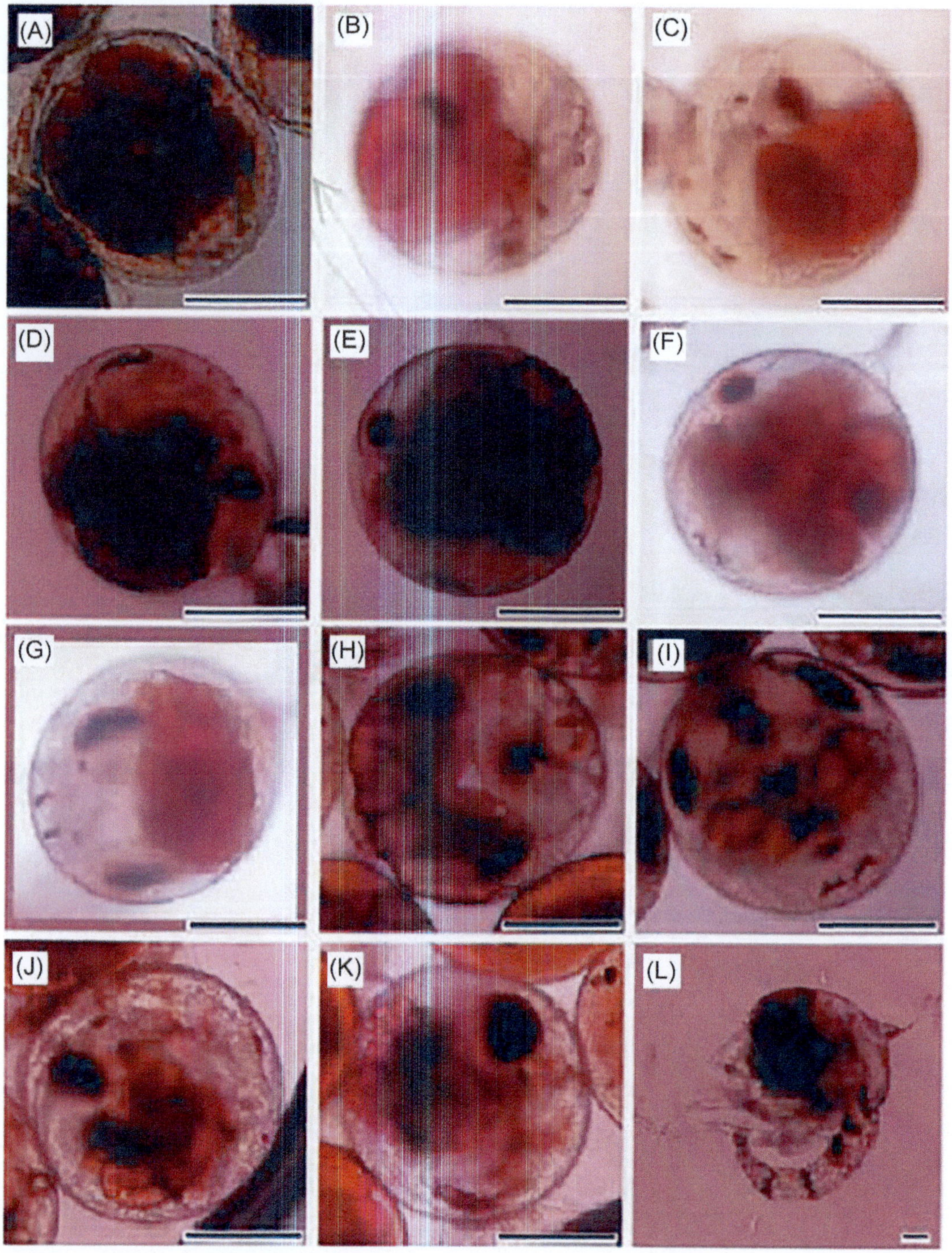

FIG. 10.152 Photomicrographs of later stages of development of the egg of *Neosarmatium indicum* from gastrula to hatching *(from Sarker et al., 2012)*: (A) stage-7; (B and C) stage-8; (D–E) stage-9; (F) stage-10; (G) stage-11; (H) stage-12; (I) stage-13; (J) stage-14; (K) stage-15 before hatching; (L) zoea stage. Scale bar = 100 μm.

the whole egg except some reserve yolk, heart dorsal in position with regular beating, ocular lobes well developed, eyes large and more complex, appendages larger but incompletely segmented with developing setae. First appearance of rostral spine was visible (Fig. 10.152J).

- **Stage 15** (before hatching) 16 days, the embryo occupied almost all the available egg volume and all yolk was almost consumed except for some traces in the cephalothorax. The heartbeat was faster and the egg membrane was transparent, so the chromatophores mostly the black ones were easily observed under the stereomicroscope. Carapace, maxillipeds, and telson were well differentiated. Frequently, the abdomen and appendages of the embryo moved within the egg membrane just before hatching (Fig. 10.152K).

Stage-first zoea (17 days of incubation period), the movement of the abdomen and appendages enabled the larva to rotate inside the membrane 180 degrees, leaving it with the telson and rostral spine facing towards the funiculus. A few hours prior to hatching, the egg reached its maximum size due to absorbing water.

Hatching. A remarkable feature of larval release is the synchorony with environmental cues such as day-night, tidal, or lunar cycles. Hatching (breakage of the outer egg membrane) usually occurs in the early hours of the day in almost all the Portunid crabs, as was recorded above. Sarker et al. (2009) suggested that in sesarmid crabs, light is a major factor inducing the release of larvae. This vital activity occurs under well-defined conditions of light, tide, and several other exogenous and endogenous factors. In all cases, all females carrying mature eggs await sunset or dusk before proceeding to the seashore. However, nocturnal release rhythms have been observed for a number of crustaceans, e.g. the fiddler crab, *Uca*. Though *P. bidens* is not nocturnal, in most cases during the experiments of Sarker et al., larval release of this species occurred in the evening and early in the morning. Precisely timed rhythms in larval release imply that hatching time is also synchronized within a batch of eggs attached to a female.

Little is known about the hatching mechanism in Crustacea. It might be mechanical, nonmechanical, or a combination of both. From field observations, Saigusa (1992) stated that this vital activity is an extremely short-lived event and suggested several possible mechanisms: (1) physical stimuli from the female during release behaviour (i.e. vigorous fanning movements of pleopods and abdomen cause breakage of the egg membrane); (2) a hatching enzyme secreted by the zoea itself or the female that digests the egg case; or (3) some other mechanism that controls the time of hatching. Hatching itself as a vital activity was the main aim of Saigusa (1992), who observed the crab *Sesarma haematocheir* in the field. This crab burrows in land surrounding small estuaries in Japan and along the coastlines of south-east Asia. In early summer, a female incubates 30,000–50,000 eggs in her folded abdomen for 1 month. She emerges at the riverside or seashore where the zoeae are released. This larval release behaviour is completed in 3–5 s and all the zoeae are released from the pleopods. Then the female returns to land and incubates the next clutch, which will appear within a few days. Each female incubates and releases larvae two or three times between July and September. This study of Saigusa (1992) was conducted at two sites on the coast of the Inland Sea, Japan: the margin of the outlet of a small river at Kasaoka, where the author observed the larval release activity, and at a similar place at Ushimado, Okayama Prefecture, 70 km east of Kasaoka. Ovigerous females approaching riverside were captured about dusk and brought to the laboratory. Small samples of eggs (embryos folded within the membranes, expected to be released that night) were removed from the female and prepared for stereomicroscopic examination. In addition, the empty egg cases remaining immediately after larval release, detached from the pleopods of females with forceps, were prepared for the SEM to evidence the presence of a hatching enzyme.

Observations by stereomicroscope indicate that the egg cases of *Sesarma haematocheir* consist of two membranes: a thick outer membrane and an inner thin membrane as in other decapods. In addition, the case of this crab has another thin membrane lining the outer membrane, which is clearly visible by SEM (Fig. 10.153A (arrow)). From SEM studies, egg hatching in *Sesarma* appears to result from mechanical rupture of the egg capsule. The pressure causing rupture comes from within the egg and not from the female. The pressure may be exerted by the swelling embryo itself, perhaps associated either with the uptake of water, or as a result from osmotic swelling of the inner membrane enclosing the embryo. Saigusa photographed some eggs enlarging in size due to their swelling (Fig. 10.153A (H1, H2, H3)) from water reaching the embryo as the female circulates water over the egg 'sponge' with her pleopods, when she moves about on land. Saigusa stated that the process of egg-membrane rupture is similar to that described for other decapods. However, he proposed another explanation for membrane breakage: that a sudden increase in excretion by the embryo leads to increased salt content and consequent influx of water surrounding the embryo. Therefore, egg hatching in *Sesarma* is not fully accounted for by either hypothesis, and the presence of a 'hatching enzyme' has not been obviously demonstrated. Hatching in this crab obviously includes temperature-dependent factors, and there are many other phenomena that still need to be studied.

Later on, osmotic hatching was suggested for a number of crustaceans. During the hatching process of *S. serrata* (Samuel and Soundarapandian, 2010), the fully developed I-zoea comes out of the egg case and swims freely in the water. It was reported that before the process, there is a period of swelling of the eggs followed by osmotic swelling of the inner egg membrane at the start of hatching. The swollen inner egg membrane then ruptures the chorion by pressure from within the embryo. The inner membrane is subsequently ruptured by mechanical action of the larval abdomen.

Another problem is the synchronization of hatching in the large number of embryos carried by the female, i.e. all the larvae are released at the same time. Saigusa (1992) stated that this problem in the female *Sesarma haematocheir* has not been solved by his work. However, in his discussion, he mentioned that improved hatching timing occurs

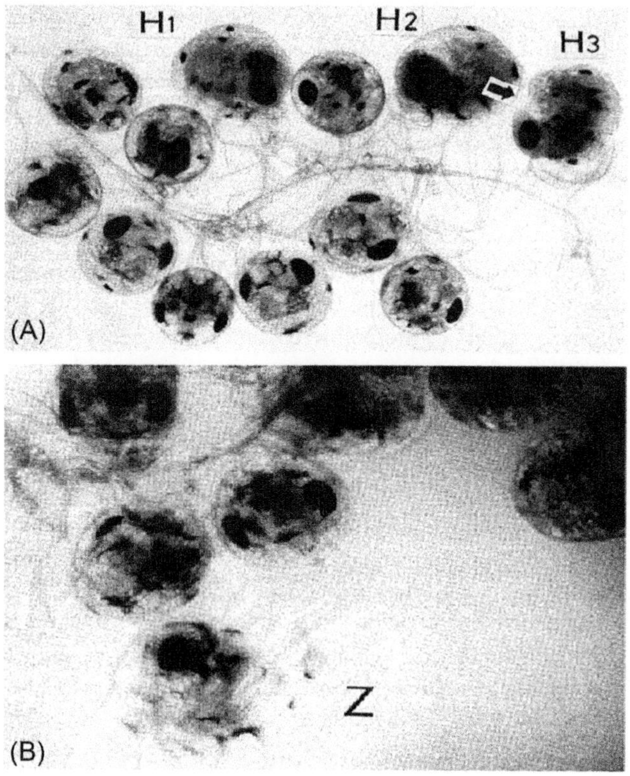

FIG. 10.153 (A) Hatching of *Sesarma haematocheir* eggs, increase in volume following rupture of the outer membrane (H1, H2, H3), *arrow* shows thin membrane enclosing the larva after bursting of the egg case, other embryos have not yet hatched; (B) zoea larva (Z) that has begun to swim after throwing off the thin membrane. *(From Saigusa (1992).)*

in turbulent water when the females rapidly and vigorously 'pump' their abdomens. Groups of larvae are released with each pump, and it was suggested that pumping behaviour itself assists in breaking the egg membrane, which enhances hatching synchrony. In his crab, *Sesarma*, this behaviour was not observed. However, he exposed detached embryos (isolated eggs) to vigorously aerated water, which caused a better synchrony of hatching but with a delay in hatch time. *Postembryonic development (larval development).* Very early in 1909, Calman stated that the Brachyuran eggs, as a rule, hatch in a zoea-stage (Fig. 10.154A). Since rudiments of the posterior thoracic limbs are frequently present, it might be more correct to call the larva a Metazoea. As a rule, a rostrum, a median dorsal, and a pair of lateral spines are present. Of these, the dorsal spine is the most constant and it is a characteristic of the zoea-stage. In many Brachyura, the larva when hatched is enclosed in a 'cuticle', which is moulted shortly after hatching, and this cuticle, in many cases, presents characters differing from those of the larva that escapes from it. In *Carcinus maenas*, for instance, this larva is succeeded by a Megalopa larva (see Fig. 10.154B and C). Very recently, Arshad et al. (2006) studied the embryology of the blue swimming crab *P. pelagicus*, as mentioned above. After hatching, actively swimming larvae were reared carefully in the laboratory under suitable conditions. Larval development was observed under a dissecting microscope before the larvae were preserved at each stage. The data of larval development were analysed descriptively. However, only the most

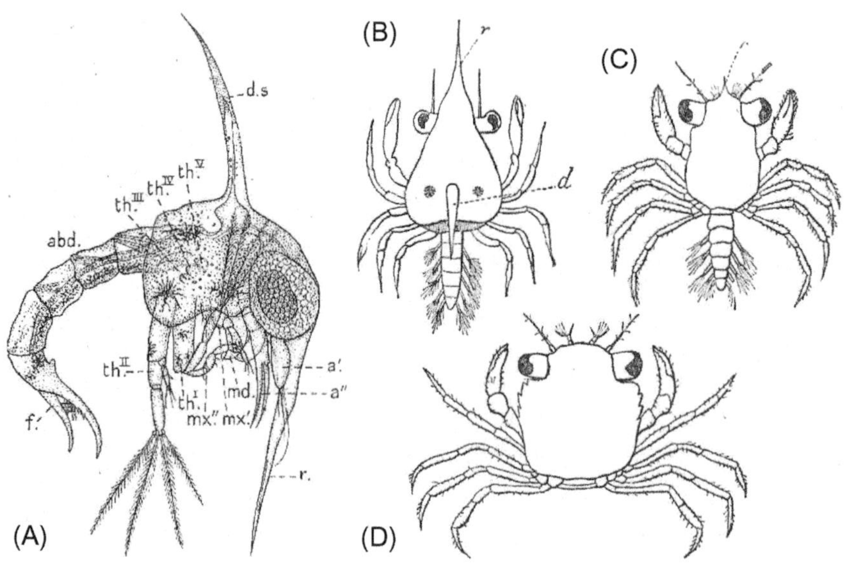

FIG. 10.154 Drawings illustrating larval development of *Carcinus maaenas (from Calman, 1909)*: (A) first zoea stage (after the first moult) (after Faxon); (B) young megalopa; (C) old megalopa; (D) post larval stage (first crab) *((B–D) after Korschelt and Heider's Embryology)*. a′ = antennule, a″ = antenna, abd = abdomen, d & d.s = dorsal spine of carapace (the so-called 'Zoea' spine), f = furcate telson, Md = mandible, mx′ = maxillula, mx″ = maxilla, r = rostrum, th^{I}, th^{II} = first and second maxillipeds (biramous and natatory), th^{III}–th^{V} = rudiments of three following thoracic appendages.

prominent features of the larval development stages were described in this work. The larval development of the blue swimming crab, *P. pelagicus*, comprised of four zoeal stages and a megalopa. The latter then metamorphosed into the crab stage (Fig. 10.155). The total duration of this larval development required a minimum of 14–19 days at 28–30 ppt. The first and second zoeal stages took 3–4 days each, the third and fourth stages took 2–3 days each, and the megalopa took 3–4 days. Moulting for the zoea and megalopa took place by a split at the dorsal boundary between the cephalothorax and the abdomen. The first crab instar spanned 15–18 days at the above salinity. All zoeal stages possessed a long rostral, a dorsal spine, and a pair of short lateral spines on the carapace (see Fig. 10.155A–D (a)). They swam with the dorsal spine usually directed anteriorly and the rostral spine ventrally.

Larval development of another group of crabs, the sesarmid crabs (Fig. 10.156), was studied by Islam and Shokita (2000). They obtained ovigerous females of *Perisesarma bidens* from the mouth of west bank of Nuha river in Manko estuary, Naha, Okinawa Island, Japan. They reared them in seawater of 27 ± 1‰ salinity with moderate aeration and temperature range from 21.3°C to 23.5°C. Hatching occurred in the morning and the larvae passed through four zoeal stages and one megalopa stage. Morphology of all larval stages and first crab stage was described and illustrated in detail, and compared with other species of sesarmid crabs. The zoeal morphological features of *P. bidens* are almost similar to other species of *Sesarma*; all differ from the above described in lacking a pair of lateral spines on carapace, but have three pairs of setae between the long telson furca.

Artificial insemination. Sarker et al., who studied the embryonic development of the estuarine crab *Neosarmatium indicum* in 2012, presented a completely different and surprising work in 2009. They artificially inseminated the unfertilized eggs of the mangrove semiterrestrial sesarmine crab *Perisesarma bidens* (De Haan) and studied their early embryonic development. As mentioned in their introduction, whether fertilization in crabs occurs internally or externally is controversial, and the possibility of artificial insemination also remains in doubt. This was the main aim of their work, to try artificial insemination and to study early development of *P. bidens*, which is of great importance to crab fisheries for artificial seed production. They selected the crabs exhibiting courtship behaviours as being reproductively active, and such active male and female crabs were captured by hand during their spawning season from May to September. This crab lives in the same mangrove forest mentioned in their other work. This species is of great importance to crab fisheries for artificial seed production. The collected crabs were transported to the laboratory and reared as pairs (one male and one female) in suitable and equipped containers filled with 80% filtered seawater (FSW). The environmental conditions were maintained: aeration, substratum, photoperiod (12 h intervals), food, and temperature (25°C). Mating behaviour was observed in the containers at noon every day in the laboratory. When copulation occurred, it continued for 0.25–4 h. Just after finishing this vital activity, the male and female were separated. The female was checked at 2–3 h intervals for laying eggs. It extruded unfertilized mature eggs 24–48 h after copulation. The eggs were immediately attached to the nonplumose setae of the endopodites of the female abdominal appendages (the same observation in *Carcinus maenas* (Cheung, 1966)). They were attached through the stalk or funiculus 5 min after laying, and were then encased in an investment coat. Mature unfertilized eggs were collected from the pleopods immediately after laying and were kept in 80% FSW. At the same time, the same female retains the spermatophores, received from the male during copulation, in its own spermatheca. Normal spermatozoa were collected from the spermatheca of this female and diluted in 80% FSW for insemination (This means that the spermatophores were dehisced inside the spermatheca as what occurs in *P. pelagicus* (Bawab and El-Sherief, 1989).)

Newly released mature eggs were spherical, centrolecithal, uniformly filled with yolk, and had an average diameter of 313.05 ± 8.41 μm. Only one layer of egg membrane was closely attached to the egg surface. After egg-laying, the egg showed no particular sign, such as polarity, or micropyle (Fig. 10.158A). The study of Sarker et al. (2009) indicated that up to 10 min after laying, newly released eggs were unfertilized. On the other hand, the spermatozoa, obtained from the spermatheca of this mated female, appeared uniformly spherical in shape, measuring 15–20 μm in diameter (Fig. 10.157A and B). Each has several slender radiating nuclear arms and a prominent acrosome overlying the nuclear mass. Semen that was examined after collection from the spermatheca contained nonmotile sperm. The above prepared unfertilized eggs and sperm were shaken well in a glass beaker for artificial insemination. Sperm motility was restored when mixed with whole eggs. After about 3–5 min, excess sperm were removed by rinsing the eggs 3–5 times with 80% FSW. These eggs were incubated in 100 mL flat cylindrical culture bottles containing 70 mL of 80% FSW at 25°C in a water bath. After insemination, eggs were taken from the culture bottle and continuously examined until cleavage began, and then at 2–6 h intervals until hatching, using a light microscope equipped with a digital camera and a video camera on a TV monitor connected to a time-lapse video set. The embryological development of these eggs was divided into many stages (Fig. 10.159).

Embryogenesis of artificially inseminated eggs. These fertilized eggs were macrolecithal, centrolecithal, spherical and with a uniform dark olive colour. No polar bodies were observed, but after 3 min of insemination, the fertilization membrane was first observed followed by the appearance

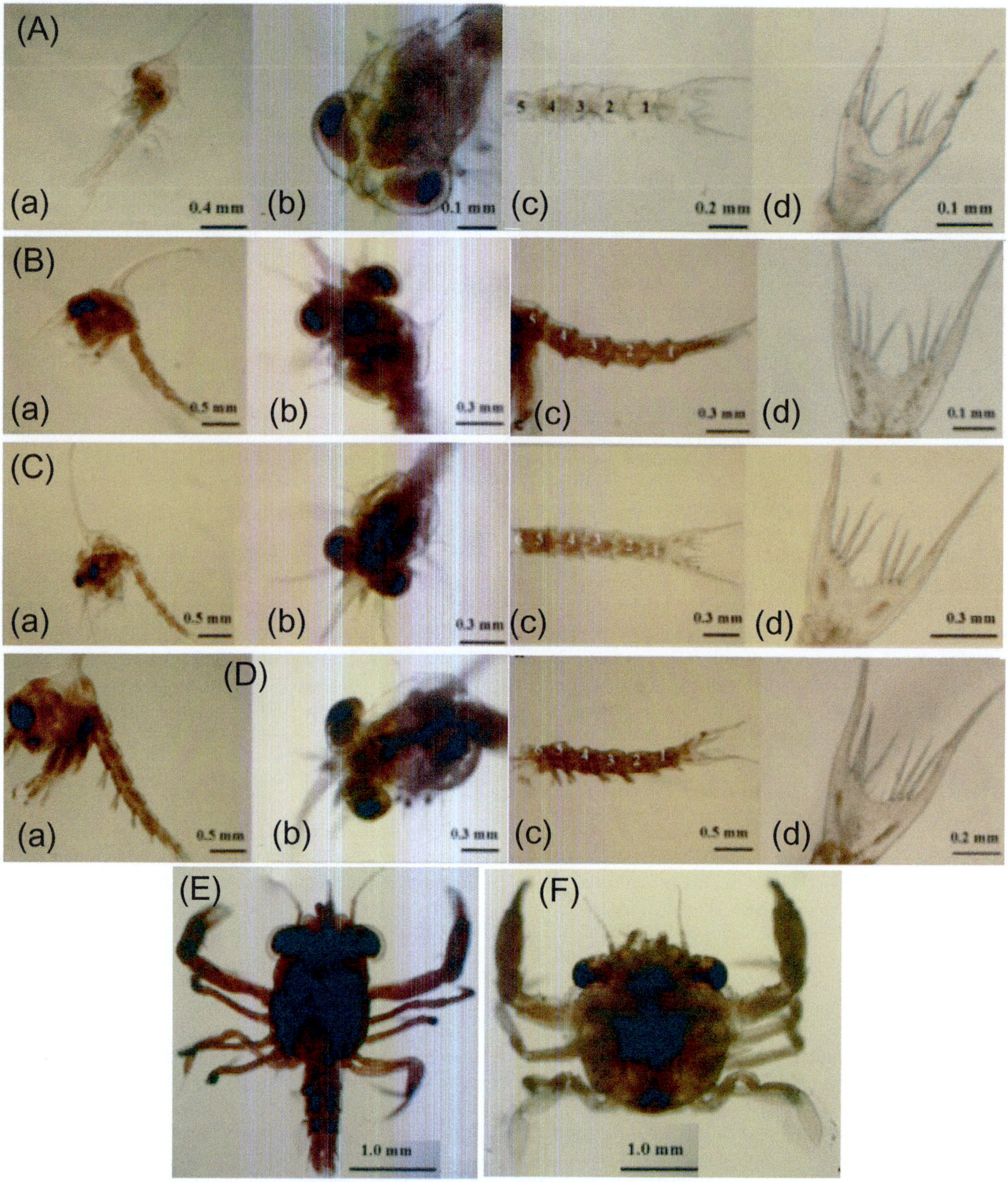

FIG. 10.155 Larval stages of *P. pelagicus*: (A–D) zoeae stages I–IV ((a) whole animal, (b) eyes sessile in (A), stalked in the others, (c) abdomen plus telson, (d) telson, three pairs of setae between the long telson furca). (A) Zoea I frontal view, abdomen five segments; (B) zoea II lateral view, a pair of short plumose setae on median margin of cleft part; (C) zoea III lateral view, abdomen six segmented with paired pleopod buds at ventral posterior end of somites 2–5; (D) zoea IV lateral view, buds well developed; (E) dorsal view of W.M. of megalopa larva; (F) dorsal view of W.M. of first crab. *(From Arshad et al. (2006).)*

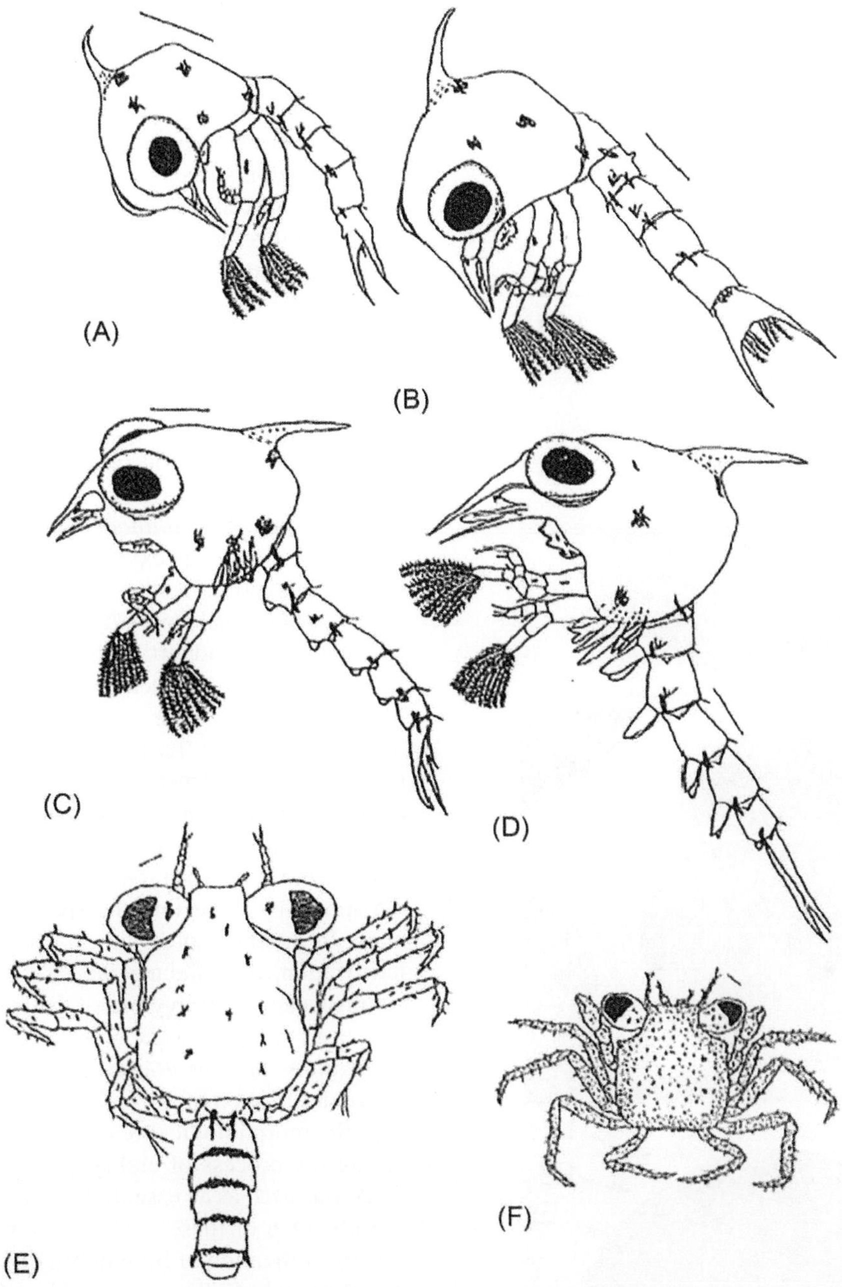

FIG. 10.156 Larval development of *Perisesarma bidens (from Islam and Shokita, 2000)*: (A) zoea I; (B) zoea II; (C) zoea III; (D) zoea IV. Scale bar=0.2 mm; (E) megalopa larva. Scale bar=0.1 mm; (F) first crab (postlarval stage). Scale bar=0.2 mm.

of the fertilization cone 5 min after insemination. The first polar body was extruded from the zygote 10–15 min and the second one 20–25 min after insemination (Fig. 10.157C and D). After 25–30 min, and at the time the yolk was yellowish, three distinct egg membranes appeared (Fig. 10.157E). Cleavage (Fig. 10.158) occurred within 23–24 h of insemination. It was equal, holoblastic, and characterized by the early occurrence of a cleavage cavity, since it appeared during the first cleavage (2-cell stage). After the fifth cleavage (32-cell stage) and due to the large density of blastomeres, the following stages of cleavage could not be clearly detected. Therefore, the phase between the blastula and gastrulation was not clearly marked, but was a gradual transformation. Gastrulation took place by epiboly. Embryonic development of the artificially inseminated eggs of *P. bidens* was completed within 17 days of incubation. Just at hatching, the embryo frequently moved its abdomen and appendages inside the egg. These internal movements of the larva caused the egg to rotate 180 degrees, allowing the telson and rostral spine to face towards the funiculus. During this time, the egg dilated due to water uptake, reaching its maximum size (0.41 mm). Hatching as a process was completed

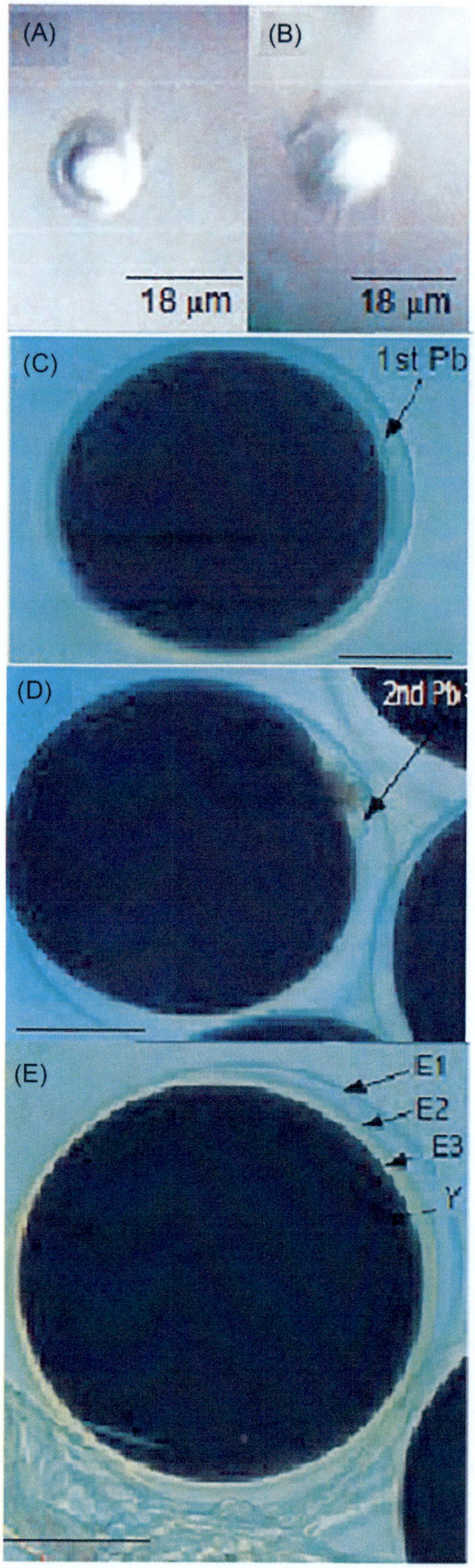

FIG. 10.157 Photomicrographs showing gametes of the mangrove crab *Perisesarma bidens (from Sarker et al., 2009)*: (A and B) Whole mount of sperm; (C) formation of the first polar body *(arrow)* of the egg; (D) formation of the second polar body *(arrow)*; (E) three distinct layers of the fertilized egg (E1, E2, and E3), Y = yolk.

within 5–40 min during the night. Some larvae were also released early in the morning as zoeae-1 (Fig. 10.159O). About 65% of the larvae hatched as zoeae-1 after 17 day of incubation. This means that the laboratory experiments of artificial insemination of *P. bidens* succeeded, since the rate of fertilization success as determined by the cleavage rate was 95%. Sarker et al. (2009) stated that two factors are essential for this success: (1) determination of the precise egg-laying time after copulation; and (2) the highest numbers of normal sperm collected from the spermatheca. These experiments evidenced the success of development of the isolated eggs of the mangrove crab *Perisesarma bidens*, as previously done on those of *Carcinus maenas* (Hartnoll and Paul, 1982).

From the discussion of this work, it is concluded that external fertilization occurs in some crabs. Males of the spiny king crab, *Paralithodes brevipes*, mate with females and fertilize the eggs externally within the brood chamber (Sato et al., 2005). In other crabs, the female releases sperm into the abdominal cavity (brood chamber), where they attach to the eggs and fertilization occurs. This means that the spermatophores were dehisced inside the spermatheca and the female released the sperm outside. Sarker et al. (2009) mentioned that the 'fertilization pattern of this crab can be regarded as incomplete internal fertilization' and considered the fertilization of their crab, *Perisesarma bidens*, of this type. For their artificial insemination, they obtained the unfertilized eggs when released by the female and collected the sperm by dissecting the spermatheca of the same female. In the opinion of the current author, this is external fertilization, since the meeting of the gametes does not occur inside any of the regions of the female reproductive system; even the sperm were collected from the spermatheca.

Utilization of lipovitellin during embryogenesis. The above description of the embryology of crabs mainly dealt with the morphological features of the egg and/or embryo during the process of embryogenesis. The latter vital activity was used as an essential basis to study other different aspects. One of these, and greatly related to the embryo, was the utilization of lipovitellin during embryogenesis of the blue crab *Callinectes sapidus* (Walker et al., 2006). It is well known that the major component of the egg yolk of most marine invertebrates is lipovitellin (LpII). It provides materials and energy for the embryos until they hatch from the egg and begin to feed. Embryos of any species are nutritionally dependent on lipids and lipoproteins stored as yolk within the eggs. In *Callinectes sapidus*, LpII is a water-soluble lipoprotein composed of approximately equal amounts of lipid (48%), protein (50%), and carbohydrate (2%). Phospholipids, primarily phosphatidyl choline, comprise the major lipid class, along with less amounts of triacylglycerols and cholesterol, and a noncovalently bound carotenoid that gives lipovitellin a characteristic yellow colour. The protein moiety is composed of two apoproteins, referred to as Protein A (107 kDa) and Protein B (78 kDa).

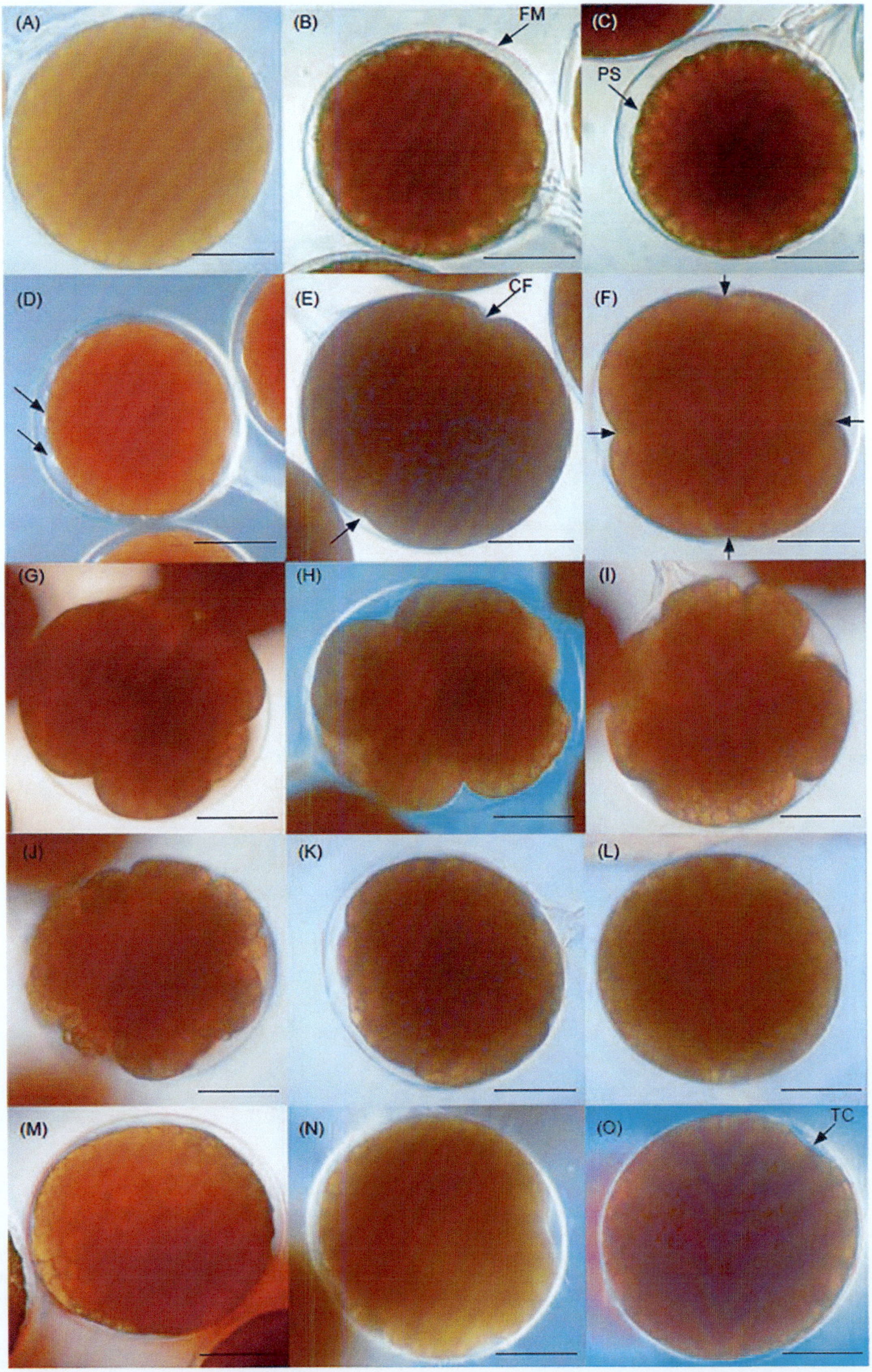

FIG. 10.158 Stages of early development (cleavage) of an artificially inseminated egg *(from Sarker et al., 2009)*. (A) Unfertilized egg of *Perisesarma bidens*; (B–D) fertilized egg before cleavage, formation of fertilization membrane (FM), perivitelline space (PS), and extrusion of polar bodies *(arrows)*; (E) 2-cell stage (*arrows* indicate the cleavage furrow passed through the animal vegetal pole (CF)); (F) second cleavage occurred 2 h next to the first producing four large, equal blastomeres (4-cell stage); (G) at 27–29 h of incubation, the third cleavage occurred and the egg reached the 8-cell stage; (H) by the fourth cleavage at 31–32 h of incubation, the egg reached the 16-cell stage; (I) within 33–35 h of incubation, the cleaved egg underwent the fifth cleavage producing the 32-cell stage; (J–L) when ~128 blastomeres were present at 44–45 h of incubation, the developing embryo reached the blastula stage, the surface of each blastomere was granular that the internal features could not be detected; (M and N) 2 days after insemination, the developing blastula reached the early gastrula stage; (O) after 3 days, there was a cluster of presumptive primordial cells beginning to form as a patch located in the ventral position (tissue cupTC). Scale bars = 100 μm.

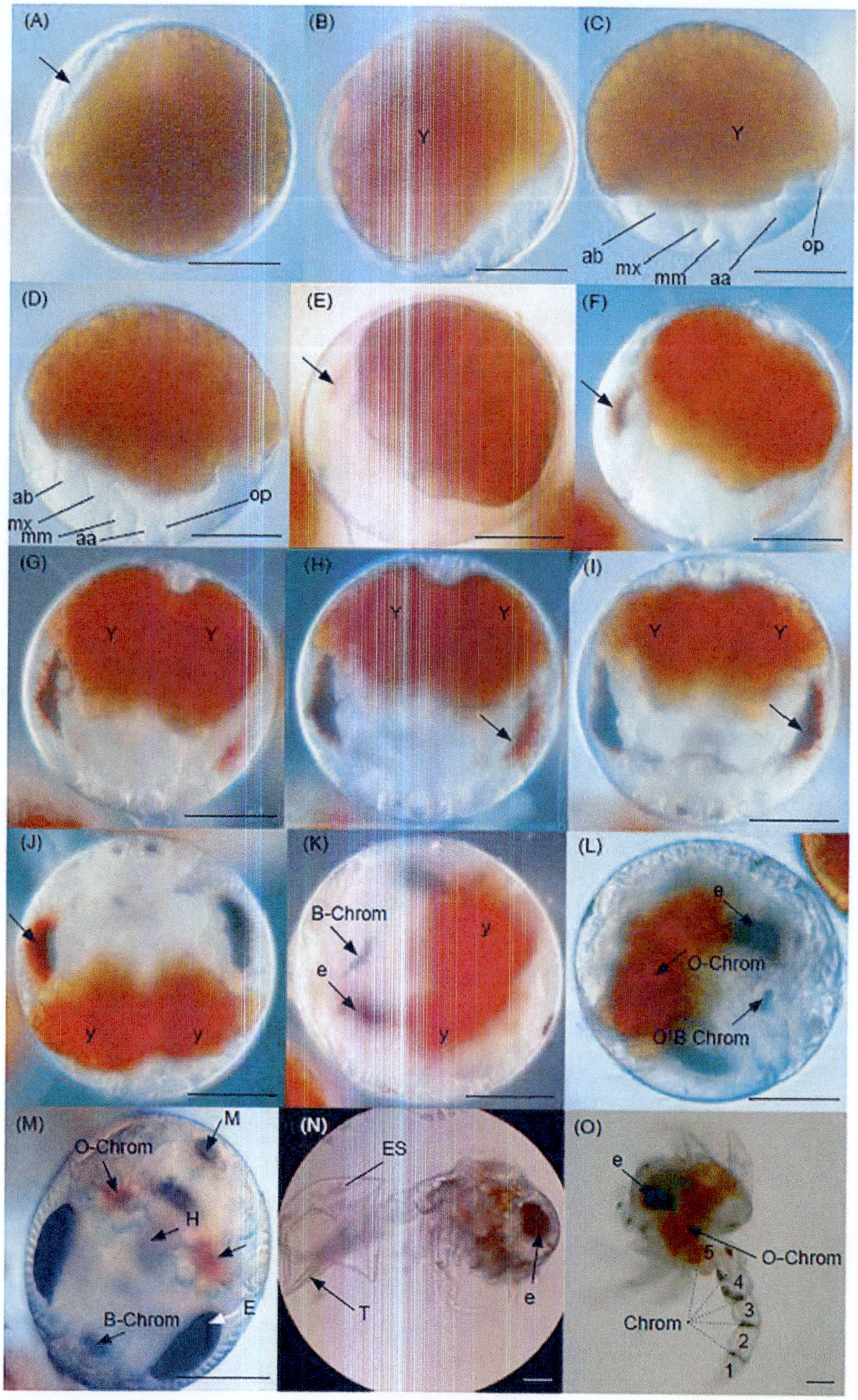

FIG. 10.159 Late stages of the egg development of embryos of *Perisesarma bidens* from gastrula to hatching *(from Sarker et al., 2009)*: (A) 4 days of incubation, droplets of yolk were larger and with lighter yellowish colour, the tissue cap grew larger, the yolk-free space *(arrow)* was observed; (B) 5 days embryo, the yolk-free portion increased in size; (C) 6 days embryo, further increase in the yolk-free portion, within which some tissue formation were distinct (buds of antennule-antenna (aa), thoracic-abdominal (ab), maxillule-maxilla (mm), maxillipeds (mx), and optical lobes (op)); (D) 7 days embryo, the previous organs were visible in the ventrolateral position of the yolk-free portion; (E) 8 days embryo, crescent-shaped eyes *(arrow)* first became visible; (F) the developing eyes were larger in size *(arrow)*; (G) 10 days of incubation, the eyes were more or less complete in size, a heart-like structure was first observed and its beat was slow at about 60–78 beats per minute (bpm); (H) 11 days, further development of the heart and its beat (100–110 bpm), yolk components were arranged in four lobes; (I) 12 days embryo, eyes were more visible *(arrow)*, heartbeat was faster, reduction of the yolk granules was observed; (J) 13 days of incubation, yolk components were arranged in two lobes, the eyes had grown and were triangular *(arrow)*, chromatophores appeared on the abdomen; (K) 14 days, embryo occupied half of the volume of the egg in the ventral position, the heart was complete and its beat was faster (120–130 bpm), the eye pigmentation was intense and its cornea was formed, the antennae and mandible were more developed; (L) 15 days, the embryo had grown and occupied the entire egg volume and the remaining yolk was divided into three patches, the heart dorsal in position, ocular lobes in their final position, eyes (e) larger and more complex, the first evidence of the rostrum; (M) 16 days, all structures of the embryo were distinct, the eyes were completely differentiated into cornea and retina, abdomen was completely segmented and each segment bore chromatophores; (N) 17 days, embryo just before hatching occupied almost all of the available egg volume, the yolk was almost completely consumed, with only traces in the cephalothorax, the carapace, maxillipeds, and telson (T) were well differentiated, there were chromatophores, mostly black (B-chrom), the egg membrane was transparent, so the pigment cells could easily be observed under a stereomicroscope; (O) first zoea hatching. Chrom = chromatophores, O-Chrom = orange chromatophores, ES = egg shell, O-B Chrom = orange-black chromatophores, Y = yolk free into the eggs, y = reserve yolk in the developing zoeae. Scale bars = 100 μm.

Walker et al. (2006) collected 'sponging' (berried) female blue crabs by trapping from estuaries in Charleston Harbour (SC) and Savannah (coastal Georgia, South-East United States). The eggs develop over a period of 10–13 days at 28°C, and pass through developmental stages with almost the same morphological features as described above. To characterize the developmental stages more closely, embryo clusters were sampled from the abdomen of these crabs daily and examined using dissecting and Differential Interference Contrast microscopes. Images were obtained photographically, scanned, and digitized; figures were then compiled using Adobe Photoshop software. Digitally scanned images of each stage were analysed using Skipper analysis software, to obtain an approximate value of yolk area relative to clear embryonic area, as well as appearance and quantity of pigment in the eyes. For the first 3 days of development at 28°C, embryo volumes are largely made up of yolk. The embryo proper is transparent, V- or comma-shaped, and lies along one edge of the yolk mass. As the embryos develop and utilize the yolk, an increasing percentage of the embryo is transparent and/or darkly pigmented (Fig. 10.160). The authors stated that one can estimate the embryonic age (stage), by observing this ratio and/or the pigmentation of the eyes.

A total of 10 monoclonal antibodies to proteins of LpII were used, to quantify LpII in different embryonic stages of *Callinectes sapidus* by ELISA (enzyme-linked immunosorbent assay). LpII could be detected at concentrations as low as 10 ng/mL. LpII decreased gradually in concentration from stage 0 to stage 5 (250–190 ng/embryo), followed by

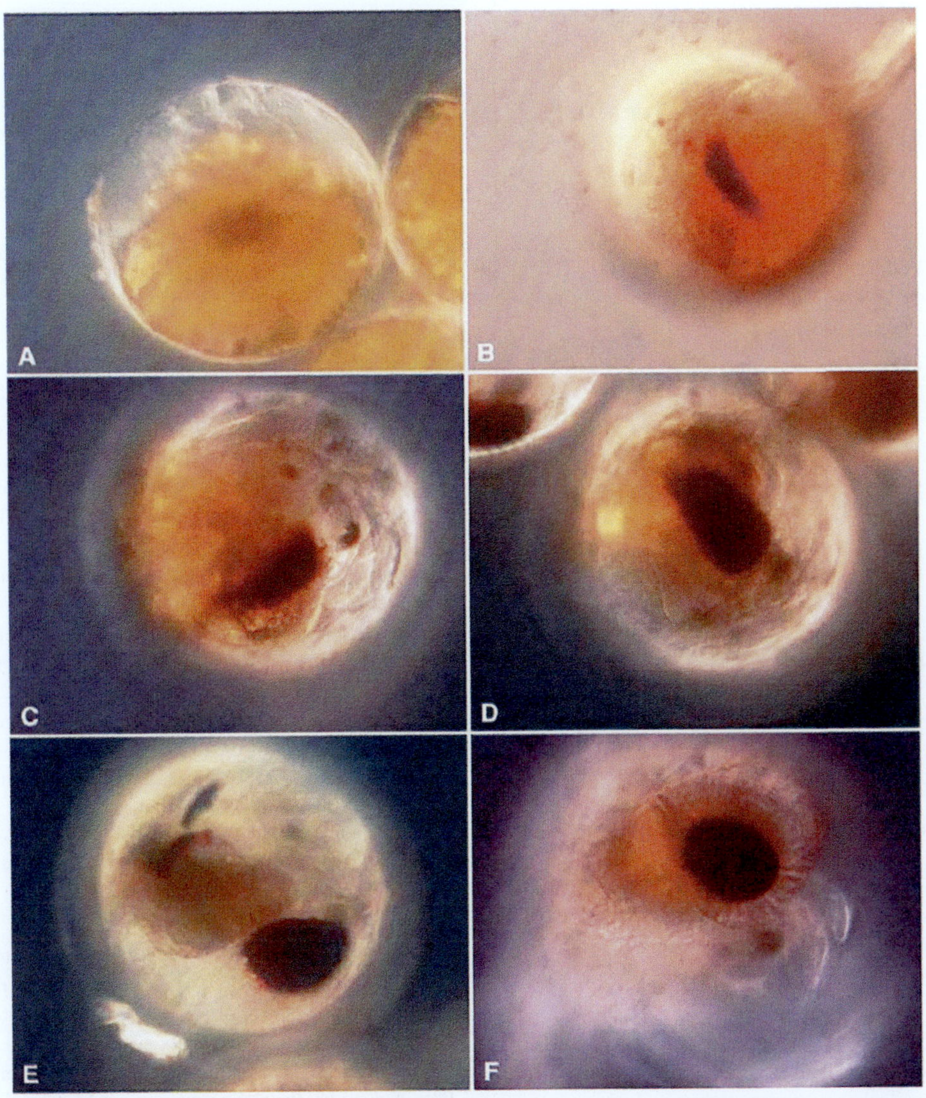

FIG. 10.160 Scanned images of stages of embryogenesis of the blue crab *Callinectes sapidus* estimating the ratio of clear embryonic area to yolk area; (A) early stage 7, the transparent comma-shaped embryo lies on one edge of the yolk mass, the pigmentation of the eye is beginning; (B) early stage 8 embryo, the pigmented eye is large and moon-shaped; (C) late stage 8 embryo, by the end of this stage the heart begins to beat visible through the chorion, the pigmented eye is larger, segmented appendages and abdomen appear; (D) stage 9 embryo; (E) stage 10 embryo, the remaining yolk lies in two paired regions along the dorsal part; (F) stage 11 embryo hatches as prezoea, a transient stage that may last minutes to hours. *(From Walker et al. (2006).)*

more rapid utilization from stage 6 to stage 11 (230–2 ng/embryo) (Fig. 10.161A). Thus, most of the LpII was utilized during embryogenesis. The prezoeae had barely detectable amounts of LpII, i.e. 2 ng/embryo.

The authors followed the degradation of proteins A and B during embryogenesis by probing immunoblots with antibodies specific for these proteins. Immunoblots of specimens from the final embryonic stages probed with the antibody against Protein B, showed intense bands at 52 kDa and 35 kDa, but no detectable band at the original 75 kDa position (Fig. 10.161B).

As embryonic development proceeds, the authors observed cells at the lateral edges of the embryonic disc, which are immunoreactive for LpII at their interface with the yolk. These cells share cytologic features with the cells of the hepatopancreas of later stages, and may also contribute to the formation of that tissue. The cells of the developing hepatopancreas appear to play an important role in the utilization of LpII.

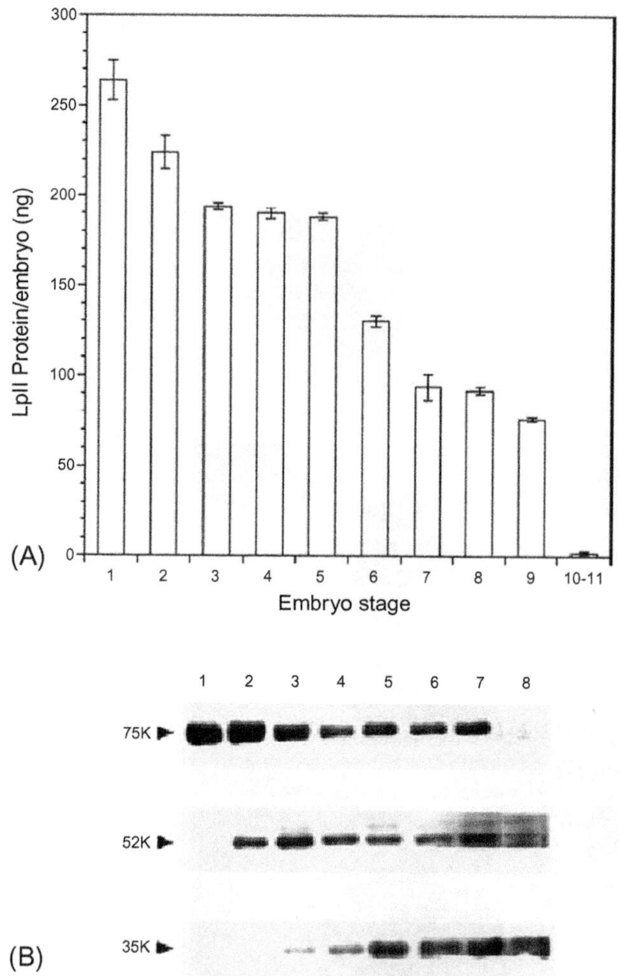

FIG. 10.161 (A) Changes in the LpII concentrations during embryogenesis of *Callinectes sapidus*. LpII was quantified by ELISA using a monoclonal antibody to protein B of LpII. Results for LpII are given as mean ± standard deviation ($n=3$) for 200 pooled embryos; (B) Immunoblot of proteins from embryo extracts (100,000 × *g* supernatant) of different embryonic stages. The lane numbers refer to extracts from: (1) mature oocytes, (2) stage 4 embryos, (3) stage 5 embryos, (4) stage 6 embryos, (5) stage 7 embryos, (6) stage 8 embryos, (7) stage 9 embryos, (8) stage 10–11 embryos. *(From Walker et al. (2006).)*

REFERENCES

Alfaro, J., Muñoz, N., Vargas, M., Komen, J., 2003. Induction of sperm activation in open and closed thelycum penaeoid shrimps. Aquaculture 216, 371–381.

Allen, B.M., 1916. Notes on the spiny lobster *Panulirus interruptus* of the California coast. Calif. Univ. Pubs. Zool. 16 (12), 139.

Alwes, F., Scholtz, G., 2006. Stages and other aspects of the embryology of the parthenogenetic Marmorkrebs (Decapoda, Reptantia, Astacida). Dev. Genes Evol. 216, 169–184.

Anderson, D.T., 1967. Larval development and segment formation in the branchiopod crustaceans *Limnadia stanleyana* and *Artemia salina*. Aust. J. Zool. 15, 47–91.

Anderson, D.T., 1969. On the embryology of the cirripede crustaceans *Tetraclita rosea* (Krauss), *T. purpurascens* (Wood), *Chthamalus antennatus* (Darwin) and *Chamaesipho columna* (Spengler) and some considerations of crustacean phylogenetic relationships. Philos. Trans. R. Soc. B 256, 183–235.

Anderson, D.T., 1973. Embryology and Phylogeny in Annelids and Arthropods. Pergamon, Oxford.

Andrews, E.A., 1906. Ontogeny of the annulus ventralis. Biol. Bull. 10 (3), 122–137.

AQUACOP, 1977. Observations sur la maturation et la réproduction en captivité des crevettes Pénei- dés en milieu tropical: 1–15 (Groupe de travail aquaculture, ICES, 10–13 May 1977, Brest, France).

Arshad, A., Efrizal, Kamarudin, M.S., Saad, C.R., 2006. Study on fecundity, embryology and larval development of blue swimming crab *Portunus pelagicus* (Linnaeus, 1758) under laboratory conditions. Res. J. Fish. Hydrobiol. 1 (1), 35–44.

Atema, J., 1995. Chemical signals in the marine enviroment – dispersal, detection, and temporal signal analysis. PNAS, 62–66.

Balasubramanian, C.P., Suseelan, C., 2000. Male reproductive system and spermatogenesis in the deep water crab *Charybdis smithii* McLeay (Brachyura: Portunidae). Indian J. Fish. 47 (4), 275–282.

Baker, J.R., 1946. Memoirs: The histoehemieal recognition of lipine. J. Cell Sci. 2 (348), 441–470.

Balss, M., 1936. The fishery grounds near Alexandria. VII. Decapoda. Notes and memoirs. Fish. Res. Dir. 15, 1.

Banoub, M.W., 1963. Survey of the blue-crab *Callinectes sapidus* (Rath.), in Lake Edku in 1960. Alex. Inst. Hydrobiol. 69, 1–18.

Barnes, R.D., 1974. Invertebrate Zoology, third ed. W.H. Saunders Company, London.

Barnes, J.M., Barnes, R.S.K., Klepal, W., 1977. Studies on the reproduction of cirripedes. 1- Introduction: copulation, release of oocytes and formation of the egg lamellae. J. Exp. Mar. Biol. Ecol. 27, 195–218.

Barrington, E.W., 1979. Invertebrate Structure and Function, second ed. The English Language Book Society and Nelson.

Bate, C.S., 1888. Report on the Crustacea Macrura collected by the H.M.S. Challenger during the years 1873–1876. Rep. Sci. Results Voyage H.M.S. Chall. Zool. 24, 1–942.

Bauer, R.T., 1987. Stomatopod grooming behaviour: functional morphology and amputation experiments in *Gonodactylus oerstedii*. J. Crustac. Biol. 7 (3), 414–432.

Bauer, R.T., Cash, C.E., 1991. Spermatophore structure and anatomy of the ejaculatory duct of *Penaeus setiferus*, *P. duorarum*, and *P. aztecus* (Crustacea: Decapoda): Homologies and functional significance. Trans. Am. Microsc. Soc. 110 (2), 144–162.

Bauer, R.T., Caskey, J.L., 2006. Flagellar setae of the second antennae in decapod shrimps: sexual dimorphism and possible role in detection of contact sex pheromones. Invertebr. Reprod. Dev. 49 (1–2), 51–60.

Bauer, R.T., Min, L.J., 1993. Spermatophores and plug substance of the marine shrimp (Crustacea: Decapoda: Penaeidae): formation in the male reproductive tract and disposition in the inseminated female. Biol. Bull. 185, 174–185.

Bawab, F.M., 1964. The Morphology and Histology of the Male Reproductive System of Penaeus (Penaeidae-Crustacea) (M.Sc. thesis). Faculty of Science, Alexandria University, Egypt.

Bawab, F.M., 1970. The Formation of the Spermatophore of *Penaeus trisulcatus* Leach (Crustacea-Decapoda) (Ph.D. thesis). Faculty of Science, Alexandria University, Egypt.

Bawab, F.M., El-Sherief, S.S., 1981. The initial formation of the sperm masses of the spermatophores of *Portunus pelagicus* L. (Crustacea - Brachyura). Bull. Fac. Sci. Alex. Univ. 21 (4), 101–118.

Bawab, F.M., El-Sherief, S.S., 1987a. The chemical constituents of the yolk formed during deutoplasmogenesis in the crab *Portunus pelagicus* (L.) (Crustacea-Brachyura). Bull. Fac. Sci. Alex. Univ. 27 (2), 198–217. Chemistry.

Bawab, F.M., El-Sherief, S.S., 1987b. Cyclic changes in the histological structure of the ovaries of *Portunus pelagicus* (L.) (Crustacea-Brachyura). Egypt. J. For. Hist. 10 (2), 297–307.

Bawab, F.M., El-Sherief, S.S., 1988. Stages of the reproductive cycle of the female crab *Portunus pelagicus* (L., 1758) based on the anatomical changes of the speratheca (Decapoda Brachyura Portunidae). Crustaceana 54 (2), 139–148.

Bawab, F.M., El-Sherief, S.S., 1989. Contributions to the study of the origin, nature and formation of the plug in the spermatheca of the female crab *Portunus pelagicus* (Linnaeus, 1766) (Decapoda, Brachyura). Crustaceana 57 (1), 9–24.

Beach, D., Talbot, P., 1987. Ultrastructural comparison of sperm from the crayfishes Cherax tenuimanus and Cherax albidus. J. Crustac. Biol. 7 (2), 205–218.

Benesch, V.R., 1969. Zur ontogenie und morphologie von *Artemia salina* L. Zool. Jb. Anat. Bd. 86, 307–458.

Beninger, P.G., Elner, R.W., Foyle, T.P., Odense, P.H., 1988. Functional anatomy of the male reproductive system and the female spermatheca in the snow crab *Chionoecetes opilio* (O. Fabricius) (Decapoda: Majidae) and hypothesis for fertilization. J. Crustac. Biol. 8 (3), 322–332.

Biffis, C., Alwes, F., Scholtz, G., 2009. Cleavage and gastrulation of the dendrobranchiate shrimp *Penaeus monodon* (Crustacea, Malacostraca, Decapoda). Arthropod Struct. Dev. 38, 527–540.

Binford, R., 1911. Notes on the life history of *Callinectes sapidus*. Johns Hopkins Univ. Cire (N.S.) 2, 14–16.

Binford, R., 1913. The germ cells and the process of fertilization in the crab, *Menippe mercenaria*. J. Morphol. 24 (2), 147–201.

Bloom, W., Fawcett, D.W., 1975. A Text Book of Histology, 10th ed., Asian edition Igaku Shoin Ltd., Tokyo.

Blower, G., 1951. A comparative study of the chilopod and diplopod cuticle. Q. J. Microsc. Sci. London 92, 141–161.

Boas, J.E.V., 1880. Studier over Decapoderns Slaegtskabsforhold. Vidensk. Selsk. Kristianaia, Skrifter 6 (5), 25–210.

Boas, J.E.V., 1883. Studien über die verwandtschaftsbeziehungen der Malakostraken. Morph. Jahrb. 8, 485–579.

Borradaille, L.A., 1901–1904. Land Crustaceans; marine Crustaceans, pts. i–vii, ix–xi, xii. In: Gardiner, J.S. (Ed.), The Fauna and Geography of the Maldive and Laccadive Archipelagoes; Being the Account of the Work Carried on and of the Collections Made by an Expedition During the Years 1899 and 1900. University Press, Cambridge.

Borradaille, L.A., 1907. On the classification of the decapod crustaceans. Ann. Mag. Nat. Hist. 19 (7), 457–486.

Borradaile, L.A., Potts, F.A., with chapters by Eastham, L.E., Saunders, J.T., 1977. The Invertebrata, fourth ed. Revised by Kerkut, G.A.Cambridge University Press.

Brisson, M.J., 1756. Regnum animale in classes IX distributum, sive Synopsis methodica sistens generalem animalium distributionem in classes IX, & duarum primarum classium, quadrupedum scilicet & cetaceorum, particularem divisionem in ordines, sectiones, genera & species. (cited in Calman, 1909).

Brown, C.H., 1950. A review of the methods available for the determination of the types of forces stabilizing structural proteins in animals. J. Cell Sci. 3 (15), 331–339.

Buhl-Mortensen, L., Høeg, J.T., 2006. Reproduction and larval development in three scalpellid barnacles, *Scalpellum scalpellum* (Linnaeus 1767), *Ornatoscalpellum stroemii* (M Sars 1859) and *Arcoscalpellum michelottianum* (Seguenza 1876) (Crustacea: Cirripedia: Thoracica): implications for reproduction and dispersal in the deep sea. Mar. Biol. 149, 829–844.

Bullough, W.S., 1981. Practical Invertebrate Anatomy, second ed. Macmillan Press Ltd, Basingstoke, London.

Burkenroad, M.D., 1981. The higher taxonomy and evolution of Decapoda (Crustacea). Trans. San Diego Soc. Nat. Hist. 19 (17), 151–268.

Burkenroad, M.D., 1984. A note on branchial formulae of Decapoda. J. Crustac. Biol. 4 (2), 277.

Cain, A.J., 1947a. The use of Nile Blue in the examination of lipoids. Q. J. Microsc. Sci., Lond. 88, 383–392.

Cain, A.J., 1947b. An examination of Baker's acid haematein for phospholipines. Q. J. Microsc. Sci., Lond. 88, 467–478.

Caldwell, R.L., 1979. Cavity occupation and defensive behaviour in the stomatopod *Gonodactylus festai*: evidence for chemically mediated individual recognition. Anim. Behav. 27, 194–201.

Caldwell, R.L., Dingle, H., 1976. Stomatopods. Sci. Am. 234, 80–89.

Calman, W.T., 1896. On the genus Anaspides and its affinities with certain fossil Crustacea. Earth Environ. Sci. Trans. Royal Soc. Edinburgh 38 (4), 787–802.

Calman, W.T., 1899. On the characters of the crustacean genus Bathynella, Vejdovsky. Zoo. J. Linn. Soc. 27 (176), 338–344.

Calman, W.T., 1904. On the classification of the Crustacea Malacostraca. Ann. Mag. Nat. Hist. 13 (7), 144–158.

Calman, W.T., 1909. Crustacea. In: Lankester, E.R. (Ed.), A Treatise on Zoology. vol. 7. Adam and Charles Black, London.

Campbell, F.L., 1929. The detection and estimation of insect chitin, and the irrelation of chitinization to hardness and pigmentation of the cuticle of the American cockroach *Periplaneta americana* L. Ann. Entomol. Soc. Am. 22, 401–426.

Carleton, H.M., 1944. Schafir's Essentials of Histology. Longmans, Green and Co, London.

Celada, J.D., De Paz, P., Gaudioso, V.R., Fernandez, R., 1987. Embryonic development of the freshwater crayfish (*Pacifastacus leniusculus* Dana): a scanning electron microscopic study. Anat. Rec. 219, 304–310.

Celada, J.D., Carral, J.M., Gonzalez, J., 1991. A study on the identification and chronology of the embryonic stages of the freshwater crayfish *Austropotamobius pallipes* (Lereboullet, 1858). Crustaceana 61 (3), 225–232.

Champion, H.F.B., 1987. The functional anatomy of the male reproductive system in *Penaeus indicus*. S. Afr. J. Zool. 22 (4), 297–307.

Chandini, T., 1989. Survival, growth and reproduction of *Daphnia carinata* (Crustacea: Cladocera) exposed to chronic cadmium stress at different food (Chlorella) levels. Environ. Pollut. 60 (1–2), 29–45.

Charniaux-Cotton, H., 1960a. La glande androgène du crustacè stomatopode *Squilla mantis*. Bull. Soc. Zool. Fr. 85, 110–114.

Charniaux-Cotton, H., 1960b. Sex determination. In: Talbot, H. (Ed.), The Physiology of Crustacea. Waterman, Academic Press, London.

Charnov, E.L., 1980. Sex allocation and local mate competition in barnacles. Mar. Biol. Lett. 2, 53–57.

Charnov, E.L., 1982. The Theory of Sex Allocation (Vol. 18). Princeton University Press, Princeton, NJ.

Charnov, E.L., 1987. Sexuality and hermaphroditism in barnacles: a natural selection approach. In: Southward, A.J. (Ed.), Barnacle Biology, vol. 5. Balkema, Rotterdam, pp. 89–103.

Cheung, T.S., 1966. The development of egg-membranes and egg attachment in the shore crab, *Carcinus maenas*, and some related decapods. J. Mar. Biol. Assoc. U.K. 46 (2), 373–400.

Choi, J.H., Hong, S.Y., 2001. Larval development of the kishi velvet shrimp, *Metapenaeopsis dalei* (Rathbun) (Decapoda: Penaeidae), reared in the laboratory. Fish. Bull. 99, 275–291.

Chow, S., Dougherty, M.M., Dougherty, W.J., Sandifer, P.A., 1991. Spermatophore formation in the white shrimps *Penaeus setiferus* and *P. vannamei*. J. Crustac. Biol. 11 (2), 201–216.

Chucholl, C., Pfeiffer, M., 2010. First evidence for an established Marmorkrebs (Decapoda, Astacida, Cambaridae) population in Southwestern Germany, in syntopic occurrence with *Orconectes limosus* (Rafinesque, 1817). Aquat. Invasions 5 (4), 405–412.

Coman, F.E., Norris, B.J., Pendrey, R.C., Preston, N.P., 2003. A simple spawning detection and alarm system for penaeid shrimp. Aquacult. Res. 34, 1359–1360.

Cronin, L.E., 1942. A Histological Study of the Development of the Ovary and Accessory Organs of the Blue Crab, *Callinectes sapidus* Rathbun (Masters thesis). University of Maryland. 37 pp..

Cuvier, G., 1829. Le regne animal, distribue d'apres son organisation, pour servir de base a l'histoire naturelle des animaux (etc.) Nouv. ed. rev. et augm. vol. 4. . Deterville.

Dahlgreen, V., Kepner, W.A., 1908. Principles of Animal Histology. Macmillan, New York.

Dall, W., Hill, B.J., Rothlisberg, P.C., Sharpless, D.J., 1990. In: Blaxter, J.H.S., Southward, A.J. (Eds.), The Biology of the Penaeidae. Advances in Marine Biology, vol. 27. Academic Press Limited, London.

Darwin, C., 1851–1854. A Monograph on the Subclass Cirripedia With Figures of All the Species. 2 vols. Svo. Ray Society, London.

Davey, K.G., 1959. Spermatophore production in *Rhodnius prolixus*. Q. J. Microsc. Sci. 100 (2), 221–230.

De Saint-Brisson, S.C., 1985. The mating behaviour of *Penaeus paulensis* Perez-Farfante (Decapoda, Penaeidea). Crustaceana 50 (1), 108–110.

Deecaraman, M., Subramoniam, T., 1980. Male reproductive tract and accessory glands of a stomatopod *Squilla holoschista*. Int. J. Invert. Reprod. 2, 175–188.

Deecaraman, M., Subramoniam, T., 1983. Mating and its effect on female reproductive physiology with special reference to the fate of male accessory sex gland secretion in the stomatopod *Squilla holoschista*. Mar. Biol. 77, 161–170.

Dennell, R., 1947a. A study of an insect cuticle: the formation of the puparium of *Sarcophaga falculata* Pand. (Diptra.). Proc. R. Soc. Lond. (B) 134, 79–110.

Dennell, R., 1947b. The occurrence and significance of phenolic hardening in the newly formed cuticle of Crustacea Decapoda. Proc. R. Soc. Lond. (B) 134, 485–503.

Dennell, R., 1960. Integument and exoskeleton. In: Waterman, T.H. (Ed.), The Physiology of Crustacea. vol. 1, pp. 449–472.

Dennell, R., Malek, S.R.A., 1955. The cuticle of the cockroach *Periplaneta Americana*. III. The hardening of the cuticle: impregnation preparatory to phenolic tanning. Proc. R. Soc. B 143, 414.ss.

Dingle, H., 1969. Ontogenetic changes in phototaxis and thigmokinesis in stomatopod larvae. Crustaceana 16, 108–110.

Dingle, H., Caldwell, R.L., 1972. Reproductive and maternal behaviour of the mantis shrimp *Gonodactylus bredini* Manning (Crustacea: Stomatopoda). Biol. Bull. 142, 417–426.

Drewes, C., 2002. Artemia franciscana. 4 pp. www.eeob.iastate.edu/faculty/DrewesC/htdocs/ARTEMIA.PDF.

Dudenhausen, E.E., Talbot, P., 1983. An ultrastructural comparison of soft and hardened spermatophores from the crayfish *Pacifastacus leniusculus* Dana. Can. J. Zool. 61, 182–194.

Edwards, E., 1966. Mating behavior in the European edible crab (*Cancer pagurus* L.). Crustaceana 10, 23–30.

El-Shereif, S.S., 1975. The Structure and Histology of the Male Reproductive System of *Portunus pelagicus* and the Formation of the Spermatophores (M.Sc. thesis). Faculty of Science, Alexandria Univ., Egypt.

El-Shereif, S.S., 1983. Cyclic Morphological, Histological and Histochemical Changes of the Female Reproductive System of the Crab *Portunus pelagicus* (L.) (Ph.D. thesis). Faculty of Science, Alexandria Univ., Egypt.

El-Shereif, S.S., 1991. Fine structure of the sperm and spermatophores of *Portunus pelagicus* (L.) (Decapoda, Brachyura). Crustaceana 61 (3), 271–279.

El-Shereif, S.S., Bawab, F.M., El-Zayat, M.A., 1986. The egg membranes of the crab *Portunus pelagicus* (L.) (Crustacea - Brachyura). Bull. Fac. Sci. Alex. Univ. 26 (4), 148–164.

El-Shereif, S.S., El-Khodary, G.M., Ghonim, A.Z., 2012. Ovarian cycle and scanning electron micrographs of the spawned egg of female mantis shrimp *Oratosquilla massavensis* (Alexandria Egypt). J. Basic Appl. Zool. 65, 116–124.

Erkan, M., Tunali, Y., Sancar-Bas, S., 2009. Male reproductive system morphology and spermatophore formation in *Astacus leptodactylus* (Eschscholtz, 1823) (Decapoda: Astacidae). J. Crustac. Biol. 29 (1), 42–50.

Estampador, E.P., 1949. Scylla. II. Comparative studies on spermatogenesis and oogenesis. Phill. J. Sci. 78 (3), 301–353.

FAO, WFP and IFAD. 2012. The State of Food Insecurity in the World 2012. Economic growth is necessary but not sufficient to accelerate reduction of hunger and malnutrition. Rome, FAO.

Fielder, D.R., Eales, A.J., 1972. Observations on courtship, mating and sexual maturity in *Portunus pelagicus* (L., 1766) (Crustacea, Portunidae). J. Nat. Hist. 6, 273–277.

Foster, B.A., 1983. Complemental males in the barnacle *Bathylasma alearum*. Aust. Mus. Mem. 18, 133–139.

Fyhn, U.E.H., Costlow, J.D., 1977. Histology and histochemistry of the ovary and oogenesis in *Balanus amphitrite* L. and *B. eburneus* Gould (Cirripedia, Crustacea). Biol. Bull. 152, 351–359.

Gabe, M., 1976. Histological techniques. Masson.

Gardiner, M.S., 1972. The Biology of Invertebrates. McGraw-Hill BookCompany, London, New York.

Garven, H.S.D., 1965. A Student's Histology, second ed. E.S. Living Stone Ltd, Edinburgh, London.

George, M.J., 1963. The anatomy of the crab *Neptunus sanguinolentus* Herbst. J. Madras Univ. B 33 (3), 289–304.

Giese, A.C., Pearse, J.S., 1974. Reproduction of Marine Invertebrates. Acoelomate and Pseudocoelomate Metazoans. vol. I. Academic Press, New York, London.

Gilbert, S.F., Raunio, A.M., 1997. Embryology, Constructing the Organism. Sinauer Associates, Inc. Publishers, Sunderland, MA.

Green, I.J., 1954. Size and reproduction in *Daphnia magna* (Crustacea: Cladocera). Proc. Zool. Soc. Lond. 124 (3), 535–545.

Grobben, C., 1876. Die Geschlechtsorgane von *Squilla mantis*, Rond. S.B. Akad. Wiss. Wien. l x xiv. Abth. i, 389–406. 1 pl.s.

Grobben, C., 1878. Beitrage zur Kenntniss der männlichen Geschlechtsorgane der Dekapoda. Arbeiten aus dem Zoologischen Institute de Universität Wien und der Zoologischen Sation in Triest, 1, 1–94.

Grobben, C., 1892. zur kenntniss des stamm baumes und des systems des Crustacea sitz. Per. Akab. Wein, ci, abth. I, 237–274.

Hackman, R.H., 1949. (Ph.D. dissertation). University of Cambridge (cited in Hackman, 1953).

Hackman, R.H., 1953. Chemistry of insect cuticle. 1. The water soluble proteins. J. Biochem. 54, 362–367.

Hackman, R.H., Todd, A.R., 1953. Some observations on the reaction of catechol derivatives with amines and amino-acids in presence of oxidizing agents. Biochem. J. 55 (4), 631–637.

Hamadi, A.A., 1995. Crayfish News 17 (2), 2.

Hamon, M., 1937. Les mecanismes produisant la dehiscence des spermatophores d' *Eupagurus prideauxi* Leach. C. R. Sci. Acad. Des Sci. Paris 204, 1504.

Hamon, M., 1941a. Histologie du canal deferent d'*Eupagurus prideauxi* Leach. C. R. Soc. Biol. Paris 135 (19–20), 1611.

Hamon, M., 1941b. Les processus secretoires du canal deferent d'*Eupagurus prideauxi* Leach. C. R. Soc. Biol. Paris 135 (19–20), 1615.

Hansen, H.J., 1895. Ordnung: Isopoda. Isopoden, Cumaceen und Stomatopoden der Plankton Expedition. In: Ergebnisse der Plankton-Expedition der Humboldt-Stiftung. vol. 2, pp. 1–48.

Harmer, S.F., Shipley, A.E., 1923. The Cambridge Natural History. Crustacea and Arachnids, vol. IV. Macmillan and Co. Limited, London.

Hartnoll, R.G., 1968. Morphology of the genital ducts in female crabs. J. Linn. Soc. (Zool.) 47 (312), 279–300.

Hartnoll, R.G., 1969. Mating in the Brachyura. Crustaceana 16, 161–181.

Hartnoll, R.G., Paul, R.G.K., 1982. The embryonic development of attached and isolated eggs of *Carcinus maenas*. Int. J. Invertebr. Reprod. 5, 247–252.

Hatziolos, M.E., Caldwell, R.L., 1983. Role reversal in courtship of stomatopod *Pseudosquilla ciliata* (Crustacea). Anim. Behav. 31, 1077–1087.

Hayashi, Y., Heckmann, L.H., Callaghan, A., Sibly, R.M., 2008. Reproduction recovery of the crustacean *Daphnia magna* after chronic exposure to ibuprofen. Ecotoxicology 17 (4), 246–251.

Heath, H., 1924. The external development of certain phyllopods. J. Morphol. 38 (4), 453–483.

Heldt, H., 1931. Observation sur la ponte, la fécundation et les premiers stades du développement de l'oeuf chez *Penaeus caramote* Risso. C. R. Sci. Acad. Sci. Paris 193, 1039–1041.

Heldt, H., 1932. Fécondation chez la crevette rose du large *Parapenaeus longirostris* Lucas. C. R. Sci. Acad. Sci. Paris 194, 2162–2163.

Heldt, H., 1938. La reproduction chez les crustacés Decapodes de la famille des Pénéides. Ann. Inst. Oceanogr. Paris 18, 31–206.

Helluy, S.M., Beltz, B.S., 1991. Embryonic development of the American lobster (*Homarus americanus*): quantitative staging and characterization of an embryonic molt cycle. Biol. Bull. 180, 355–371.

Hickman, V.V., 1937. The embryology of the Syncarid Crustacean *Anaspides tasmaniae*. Pub. and Proc. Roy. Soc. Tasm. 1936, 1–35.

Hickman, C.P., 1973. Biology of the Invertebrates, second ed. The C.V. Mosley Company, Saint Louis.

Holthuis, L.B., 1956. Notes on a collection of Crustacea Decapoda from the Great Bitter Lake, Egypt, with a list of the species of Decapoda known from the Suez Canal. Zool. Mededel. 34 (22), 301–330.

Holdich, D.M., 2002. Distribution of crayfish in Europe and some adjoining countries. Bull. Fr. Peche Piscic. 367, 611–650.

Holdich, D.M., Black, J., 2007. The spiny-cheek crayfish, *Orconectes limosus* (Rafinesque, 1817) (Crustacea: Decapoda: Cambaridae), digs into the UK. Aquat. Invasions 2 (1), 1–15.

Holthuis, L.B., 1980. Shrimps and prawns of the world. An annotated catalogue of species of interest to fisheries. FAO Fish. Synop. 1 (125).

Hubenova, T.A., Vasileva, P.L., Zaikov, A.N., 2009. Histological analysis of ovary development in narrow-clawed crayfish *Astacus leptodactylus* Esch. 1823 (Crustacea, Decapoda, Astacidae), reared in ponds in South Bulgaria. Aquacult. Aquar. Conserv. Legis. 2 (3), 261–270.

Hudinaga, M., 1942. Reproduction, development and rearing of *Penaeus japonicus* Bate. Jpn. J. Zool. 10, 305–393.

Huxley, T.H., 1880. The Crayfish. An Introduction to the Study of Zoology. C. Kegan Paul & Co., London.

Huxley, T., 1883. Inaugural address. International Fisheries Exhibition Literature 4, 1–9.

Ikhwanuddin, M., Azra, M.N., Siti-Aimuni, H., Abol-Munafi, A.B., 2012. Fecundity, embryonic and ovarian development of blue swimming crab, *Portunus pelagicus* (Linnaeus, 1758) in coastal water of Johor, Malaysia. Pak. J. Biol. Sci. 15 (15), 720–728.

Ikhwanuddin, M., Lan, S.S., Abdul Hamid, N., Fatihah Zakaria, S.N., Azra, M.N., Siti Aisah, A., Abol-Munafi, A.B., 2015. The embryonic development of orange mud crab, *Scylla olivacea* (Herbst, 1796) held in captivity. Indian J. Fish. Sci. 14 (4), 885–895.

Ingle, R.W., Thomas, W., 1974. Mating and spawning of the crayfish *Austropotamobius pallipes* (Crustacea, Astacidae). J. Zool. Lond. 173, 525–538.

Islam, M.S., Shokita, S., 2000. Larval development of a mangrove crab (*Perisesarma bidens* De Haan) (Crustacea: Brachyura: Sesarminae). Bangladesh J. Fish. Res. 4 (1), 43–56.

Jackson, H.G., 1913. Eupagurus. Liverpool Mar. Biol. Comm. Mem. 21, 1–29.

Kao, H., Chan, T., Yu, H., 1999. Ovary development of the deep-water shrimp *Aristaeomorpha foliacea* (Crustacea: Decapoda: Aristeidae) from Taiwan. Zool. Stud. 38 (4), 373–378.

Keele, C.A., Neil, E., 1973. Samson Wright's applied physiology. Am. J. Phys. Med. Rehab. 52 (1), 45.

Kelly, M.W., Sanford, E., 2010. The evolution of mating systems in barnacles. J. Exp. Mar. Biol. Ecol. 392, 37–45.

Khafage, A.R., 2000. Detailed Studies on the Structure of the Thelycum of Female Penaeids to Determine Its Role in Copulation and Fertilization (M.Sc thesis). Faculty of Science, Alexandria University.

Khafage, A.R., Bawab, F.M., 2013. Comparative study of the thelycum structure of seven penaeid species from catches of Alexandria, Egypt to correlate the structure with the function. In: Oral Presentation in Aquaculture (21–25 February) Nashville, Tennesse, USA.

Khafage, A.R., Bawab, F.M., 2014. Structure and function of coxal vesicle of the thelycum in the prawn *Penaeus kerathurus* (Decapoda: Penaeidae). In: Oral Presentation in BIT's 3rd Annual World Congress of Ocean, Division Aquaculture and Fisheries. Dalian, China.

King, J.E., 1948. A study of the reproductive organs of the commom marine shrimp, *Penaeus setiferus* (Linnaeus). Biol. Bull. 94 (3), 244–262.

Kingsley, J.S., 1892. The embryology of Limulus. J. Morphol. 7 (1), 35–68.

Krishnan, G., 1951. Phenolic tanning and pigmentation of the cuticle in Carcinus maenas. J. Cell Sci. 3 (19), 333–342.

Koca, S.B., Yigit, N.Ö., Eralp, H., 2013. Embryonic and postembryonic of freshwater cayfish (*Astacus leptodactylus* Eschscholtz, 1823). Egirdir Su Ürünleri Fakültesi Dergisi 9 (i), 21–30.

Kodama, K., Shimizu, T., Yamakawa, T., Aoki, I., 2004. Reproductive biology of the female Japanese mantis shrimp *Oratosquilla oratoria* (Stomatopoda) in relation to changes in the seasonal pattern of larval occurrence in Tokyo Bay, Japan. Fish. Sci. 70, 734–745.

Kodama, K., Shiraishi, H., Morita, M., Horiguchi, T., 2009. Reproductive biology of the Japanese mantis shrimp *Oratosquilla oratoria* (Crustacea: Stomatopoda): annual cycle of gonadal development and copulation. Mar. Biol. Res. 5, 415–426.

Komai, T., 1920. Spermatogenesis of *Squilla oratoria* de Haan. J. Morphol. 34, 307–333.

Kooda-Cisco, M.J., Talbot, P., 1982. A structural analysis of the freshly extruded spermatophore from the lobster, *Homarus americanus*. J. Morphol. 172, 193–207.

Kooda-Cisco, M.J., Talbot, P., 1983. A technique for electrically stimulating extrusion of spermatophores from the lobster, *Homarus americanus*. Aquaculture 30, 221–227.

Kücükdermenci, A., Lok, A., 2012. An investigation on ovarian development of grooved shrimp *Melicertus kerathurus* in Izmir Bay, Turkey. J. Mar. Biol. Assoc. U.K. 92 (3), 531–538.

Kumé, H., Dan, K., 1968. Invertebrate Embryology. Translated by Dan, J.C. NOLIT Publishing House, Belgrade, Yugoslavia

Kuruppu, M.M., Ekaratne, S.U.K., 1995. Life history of the brine shrimp *Artemia parthenogenetica* Sri Lanka (Crustacea: Branchiopoda: Anostraca: Artemiidae). J. South Asian Nat. Hist. 1 (2), 203–212.

Lamarck, J.B., 1802. Recherches sur l'organisation des corps vivants: précédé du discours d'ouverture du cours de zoologie donné dans le Museum d'Histoire Naturelle.

Lamsaard, S., Sriurairatana, S., Withyachumnarnkul, B., 2012. Changes in oviduct structure in the black tiger shrimp, *Penaeus monodon*, during ovarian maturation. J. Zhejiang Univ. Sci. B (Biomed. & Biotechnol.) 13 (10), 846–850.

Lankester, E.R., 1881. *Limulus,* an arachnid. Q. J. Microsc. Sci. 21, 504–548.

Latreille, P.A., 1804. Histoire naturelle, générale et particulière des crustacés et des insectes: ouvrage faisant suite aux owuvres de Leclerc de Buffon, et partie du cours complet d'histoire naturelle rédigé par CS Sonnini, membre de plusieurs Sociétés savantes (Vol. 13). Dufart.

Latreille, P.A., 1806. Genera crustaceorum et insectorum secundum ordinem naturalem in familias disposita, iconibus exemplisque plurimis explicata. (Vol. 1). A. Koenig.

Latreille, P.A., 1810. Considérations générales sur l'ordre naturel des animaux composant les classes des crustacès, des arachnides, et des insectes: avec un tableau méthodique de leurs genres, disposés en familles. F. Schoell.

Leach, W.E., 1815. Malacostraca Podophthalamata britanniae. Britanniae, London.

Lereboullet, D.A., 1860. Recherches sur le mode de fixation des oeufs aux fausses pattes abdominales dans les écrevisses. Ann. Sci. Nat. Scr. Zool. 4, 359–378.

Leonhardt, H., 1977. Human Histology, Cytology and Microanatomy. Year Book. Medical Publishers, Inc, Chicago.

Lison, L., 1953. Histochimie et cytocheimie animales. Gauthier-Villars, Paris.

López-Camps, J., Bargalló, R., Bozzo, M.G., Durfort, M., Fontarnau, R., 1981. The spermatogenesis of crustaceans. VII. Review of spermatozoon of the crayfish *Astacus astacus* (Malacostraca, Decapoda, Macrura, Reptantia). Gamete Res. 4, 65–82.

Lucić, A., Erben, R., Lacković, G., 2006. Morphological changes in Astacus astacus gonads during the reproductive cycle. Bull. Fr. Pêche Piscic. 380–381, 1183–1196.

Mahoney, R., 1973. Laboratory Techniques in Zoology. Butterworths and Co., London.

Malek, S.R.A., 1952. A lipoprotein precursor of sclerotin in the Cockroach cuticle. Nature (London) 170, 850–851.

Malek, S.R.A., Bawab, F.M., 1971. Tanning in the spermatophore of a crustacean (Penaeus trisulcatus). Experientia 27, 1098.

Malek, S.R.A., Bawab, F.M., 1974a. The formation of the spermatophore in *Penaeus kerathurus* (Forskål, 1775) (Decapoda, Penaeidae). 1-The initial formation of sperm mass. Crustaceana 26 (3), 273–285.

Malek, S.R.A., Bawab, F.M., 1974b. The formation of the spermatophore in *Penaeus kerathurus* (Forskål, 1775) (Decapoda, Penaeidae). 11-The deposition of the main layers of the body and of the wing. Crustaceana 27 (1), 73–83.

Martin, P., Dorn, N.J., Kawai, T., von der Heiden, C., Scholtz, G., 2010a. The enigmatic Marmorkrebs (marbled crayfish) is the parthenogenetic form of *Procambarus fallax* (Hagen, 1870). Contrib. Zool. 79, 107–118.

Martin, P., Shen, H., Füller, G., Scholtz, G., 2010b. The first record of the parthenogenetic Marmorkrebs (Decapoda, Astacida, Cambaridae) in the wild in Saxony (German raises) the question of its actual threat to European freshwater ecosystems. Aquat. Invasions 5 (4), 397–403.

Marshall, A.J., Williams, W.D., 1975. A Text Book Zoology: The Invertebrates. ELBS and Macmillan.

Mason, J.C., 1970. Egg-laying in the western north American crayfish, *Pasifastacus trowbridgii* (Stimpson) (Decapoda, Astacidae). Crustaceana 19, 37–44.

Matthews, D.C., 1951. The origin, development, and nature of the spermatophoric mass of the spiny lobster, *Panulirus penicillatus* (Oliver). Pac. Sci. 5 (4), 359–371.

Matthews, D.C., 1953. The development of the pedunculate spermatophore of the Hermit Crab, *Dardanus asper* (De Haan). Pac. Sci. 7 (3), 255–266.

Matthews, D.C., 1954a. The development of the spermatophoric mass of the Rock Lobster, *Parribacus antarcticus* (Lund). Pac. Sci. 8 (1), 28–34.

Matthews, D.C., 1954b. A comparative study of the spermatophores of three Scyllarid Lobsters, *Parribacus antarcticus, Scyllarides squammosus* and *Scyllarus martensii*. Q. J. Microsc. Sci. 95 (2), 205–215.

Matthews, D.C., 1954c. The origin and development of the spermatophoric mass of a Nephropsid Lobster, *Enoplometopus occidentalis* (Randall). Pac. Sci. 8 (2), 115–120.

Mazia, D.A., Brewer, A., Alfert, M., 1953. The cytochemical staining and measurements of protein with Mercury Bromphenol Blue. Biol. Bull. 104, 54–60.

Medina, A., 1994. Spermiogenesis and sperm structure in the shrimp *Parapenaeus longirostris* (Crustacea; Dendobranchiata): comparative aspects among decapods. Mar. Biol. 119 (3), 449–460.

Medina, A., Mourente, G., de la Rosa, L., Santos, I., A and Rodríguez, A., 1994. Spermatozoal ultrastructure of *Penaeus kerathurus* and *Penaeus japonicus* (Crustacea, Dendrobranchiata). Zoomorphology 114, 161–167.

Medina, A., Vila, Y., Mourente, G., Rodríguez, A., 1996. A comparative study of the ovarian development in wild and pond-reared shrimp, *Penaeus kerathurus* (Forskål, 1775). Aquaculture 148, 63–75.

Medina, A., Scelzo, M.A., Tudge, C.C., 2006. Spermatozoal ultrastructure in three Atlantic Solenocerid shrimps (Decapoda, Dendrobranchiata). J. Morphol. 267, 300–307.

Milne-Edwards, H., 1834–1840. Histoire naturelle des Crustaces (Rorert's suites a Buffon). 3 vols. and Atlas, 8 vol. Paris.

Mirheydari, S.-M., Paolucci, M., Matinfar, A., Soltani, M., 2014. Annual study on ovarian structure of *Astacus leptodactylus* (Eschscholtz, 1823) in Aras Dam Lake, Iran. Glob. Vet. 12 (6), 850–857.

Mouchet, S., 1930. Formation des spermatophores non-pedunculés des Crustacés Décapodes. C. R. Acad. Sci. Paris 191, 1090.

Mouchet, S., 1931. Spermatophores des crustacés décapods, anomures et brachyures et castration parasitaire chez quelque pagures. Ann. Sta. Océanogr. Salammbô. 6, 1–203.

Muhammad, F., Sultana, R., Shafi, M., 2014. Structure of vasa deferentia and spermatophores in *Parapenaeopsis stylifera* (H. Milne Edwards) (Decapoda: Penaeidae). Afr. J. Biotechnol. 13 (8), 983–988.

Munuswamy, N., Subramoniam, T., 1985. Studies on the oviductal secretion and ovulation in *Streptocephalus dichotomus* Baird, 1860 (Anostraca). Crustaceana 49 (2), 113–118.

Murray, R.K., Granner, D.K., Mayes, P.A., Rodwell, V.W., 1996. Harper's Biochemistry, 24th ed., Middle East edition Librairie de Liban, Appleton and Lange, Norwalk, CT/Los Altos, CA.

Newman, W.A., Zullo, V.A., Withers, T.H., 1969. Cirripedia. In: Moore, R.C. (Ed.), Treatise on Invertebrate Paleontology, Part R, *Arthropoda*. vol. 4 (1). Geol. Soc. Am. Univ. Kansas Press, Lawrence, pp. R206–R295.

Nikstrat, H., Kouba, A., Kozák, P., 2014. Post-mating morphological changes in the spermatozoon and spermatophore wall of the crayfish *Astacus leptodactylus*: Insight into a non-motile spermatozoon. Anim. Reprod. Sci. (Elsevier) 149, 325–334.

Ortmann, A.E., 1894. A study of the systematic and geographical distribution of the decapod family Atyidae Kingsley. Proc. Acad. Natl. Sci. Phila. 397–416.

Ortmann, A.E., 1896. Das System der Decapoden-Krebse. Zool. Jahrb. Abt. Syst. Geogr. Biol. Thiere 9, 409–453.

Oser, B.L., 1965. Hawk's Physiological Chemistry, 14th ed. McGraw-Hill, London.

Pantin, C.F.A., 1948. Notes on Microscopical Technique for Zoologists. Cambridge University Press, Cambridge.

Patek, S.N., Caldwell, R.L., 2005. Extreme impact and cavitation forces of a biological hammer: strike forces of the Peacock mantis shrimp *Odontodactylus schllarus*. J. Exp. Biol. 208, 3655–3664.

Pawlak, J.B., Sellars, M.J., Wood, A., Hertzler, P.L., 2010. Cleavage and gastrulation in the kuruma shrimp Penaeus (Marsupenaeus) japonicus (Bate): a revised cell lineage and identification of a presumptive germ cell marker. Dev. Growth Differ. 52, 677–692.

Pearse, A.G.E., 1968. Histochemistry. Theoretical and Applied, third ed. vol. I. Churchill, London.

Pearson, J., 1908. L.M.B.C. Memoirs on Typical British Marine Plants and Animals. XVI. Cancer. W.A. Herdman Ed., London.

Pelseneer, P., 1906. In: Lankester, E.R., Black, C. (Eds.), Mollusca: A Treatise on Zoology. London. Mem., 32. vol. V. University Press of Liverpool.

Perdichizzi, A., Pirrera, L., Micale, V., Muglia, U., Rinelli, P., 2012. A histological study of ovarian development in the giant red shrimp *Aristaeomorpha foliacea* (Crustacea: Decapoda: Aristeidae) from the southern Tyrrhenian Sea (Western Mediterranean). Sci. World J. 2012. 289608. 9 pages https://doi.org/10.1100/2012/289608.

Pflugfeder, O., 1970. Lehrbuch dez Entwicklungsgeschichte undEntwicklungsphysiologie der Teire. Veb Gustav Fischer Verlog Jena.

Pinheiro, M.A.A., Hattori, G.Y., 2002. Embriologia de *Arenaeus cribrarius* (Lamarck) (Crustacea, Brachyura, Portunidae). Rev. Bras. Zool. 19 (2), 571–583.

Pinheiro, M.A.A., Hattori, G.Y., 2003. Embryology of the mangrove crab *Ucides cordatus* (Brachyura, Ocypodidae). J. Crustac. Biol. 23 (3), 729–737.

Pongtippatee, P., Vanichviriyakit, R., Chavadei, J., Plodpai, P., Pratomchart, B., Sobhon, P., Withyachumnarnkul, B., 2007. Acrosome reaction in the sperm of the black tiger shrimp *Penaeus monodon* (Decapoda, Penaeidae). Aquac. Res. 18, 1635–1644.

Primavera, J.H., 1979. Notes on the courtship and mating brhaviour in *Penaeus monodon* Fabricius (Decapoda, Natantia). Crustaceana 37, 287–292.

Provenzano Jr., A.J., Manning, R.B., 1978. Studies on development of stomatopod crustacea II. The later larval stages of Gonodactylus oerstedii Hansen reared in the laboratory. Bull. Mar. Sci. 28 (2), 297–315.

Ramadan, M.M., 1936. Report on a collection of Stomatopoda and Decapoda from Hurghada, Red Sea. Bull. Fac. Sci. Egypt. Univ. Egypt 6, 1.

Ravi, R., Manisseri, M.K., 2013. Alterations in size, weight, and morphology of the eggs of blue swimming crab *Portunus pelagicus* Linnaeus, 1758 (Decapoda, Brachyura, Portunidae) during incubation. Turk. J. Fish. Aquat. Sci. 13, 509–515.

Reverberi, G., 1971. Experimental Embryology of Marine and Fresh-Water Invertebrates. North-Holland Publishing Company, Amsterdam, London.

Richards, R.B., 1951. Polyethylene-structure, crystallinity and properties. J. Appl. Chem. 1 (8), 370–376.

Roesel von Rosenhof, A.J., 1755. Der monatlich herausgegebenen Insecten- Belustigung 3: 305-328. Der Fluskrebs hirsigen Landes, mit seinen merkiwurdigen Eigenschaffien. Niirnberg.

Ronquillo, J.D., Saisho, T., McKinley, R.S., 2006. Early developmental stages of the green tiger prawn, *Penaeus semisulcatus* de Haan (Crustacea, Decapoda, Penaeidae). Hydrobiologia 560, 175–196.

Roosen-Runge, E.C., 1977. The Process of Spermatogenesis in Animals. Cambridge University Press, London.

Rudall, K.M., 1955. The destribution of collagen and chitin. Symp. Soc. Exp. Biol. 9, 49–71.

Russel-Hunter, W.D., 1979. A Life of Invertebrates. Macmillan Publishing Co., Inc, New York.

Ryan, E.P., 1966. Pheromone: evidence in a decapod crustacean. Science 151, 340–341.

Ryan, E.P., 1967a. Structure and function of the reproductive system of the crab *Portunus sanguinolentus* (Herbst) (Brachyura-Portunidae). I. The male system. In Proc. Symp. Crustacea. Mar. Biol. Assoc. India 2, 506–521.

Ryan, E.P., 1967b. Structure and function of the reproductive system of the crab *Portunus sanguinolentus* (Herbst) (Brachyura-Portunidae). II. The female system. In: Proc. Symp. Crustacea, part 2, 12–15 January 1965. Mar. Biol. Assoc., India, Mandapan Camp, India, pp. 522–544.

Saigusa, M., 1992. Observations on egg hatching in the estuarine crab *Sesarma haematocheir*. Pac. Sci. 46 (4), 484–494.

Sainte-Marie, G., Sainte-Marie, B., 1998. Morphology of the spermatheca, oviduct, intermediate chamber, and vagina of the adult snow crab (*Chionoecetes opilio*). Can. J. Zool. 76, 1589–1604.

Sainte-Marie, G., Sainte-Marie, B., 1999a. Reproductive products in the adult snow crab (*Chionoecetes opilio*). I. Observations on spermiogenesis and spermatophore formation in the vas deferens. Can. J. Zool. 77, 440–450.

Sainte-Marie, G., Sainte-Marie, B., 1999b. Reproductive products in the adult snow crab (*Chionoecetes opilio*). II. Multiple types of sperm cells and of spermatophores in the spermathecae of mated females. Can. J. Zool. 77, 451–462.

Sallam, W.S., 2005. Reproductive biology of the mantis shrimp *Erugosquilla massavensis* from Port Said, Egypt. Egypt J. Aquat. Biol. Fish. 9 (1), 171–183.

Samuel, N.J., Soundarapandian, P., 2010. Embryology of commercially important portunid crab *Scylla serrata* (Forskal). Curr. Res. J. Biol. Sci. 2 (1), 38–41.

Sarker, M.M., Islam, M.S., Uehara, T., 2009. Artificial insemination and early embryonic development of the mangrove crab *Perisesarma bidens* (De Haan) (Crustacea: Brachyura). Zool. Stud. 48 (5), 607–618.

Sarker, M.M., Islam, M.S., Uehara, T., 2012. Embryonic development of the estuarine crab *Neosarmatium indicum* (Crustacea: Brachyura: Sesarmidae) from the mangroves of the Okinawa Island, Japan. Univ. J. Zool. Rajshahi Univ. 31, 49–54.

Sato, T., Ashidate, M., Goshima, S., 2005. Negattive effects of delay mating on the reproductive success of female spiny king crab, *Paralithodes brovipes*. J. Crustac. Biol. 25, 105–109.

Scholtz, G., Kawai, T., 2002. Aspects of embryonic and postembryonic of the Japanese freshwater crayfish *Cambaroides japonicus* (Crustacea, Decapoda) including a hypothesis on the evolution of maternal care in the Astacida. Acta Zool. (Stockholm) 83, 203–212.

Scholtz, G., Braband, A., Tolley, L., Reimann, A., Mittmann, B., Lukhaup, C., Steuerwald, F., Vogt, G., 2003. Parthenogenesis in an outsider crayfish. Nature (London) 421, 806.

Schram, F.R., 1983a. Remipedia and crustacean phylogeny. Crustacean Issues, 1, 23–28.

Schram, F.R., 1983b. Method and madness in phylogeny. Crustacean Issues, 1, 312–350.

Schram, F.R., 1986. Crustacea. Oxford University Press, Oxford.

Shiino, S.M., 1942. Studies on the embryology of *Squilla oratoria*. Mem. Coll. Sci. Kyoto Imp. Univ. 17 (B), 11–174.

Shiino, S.M., 1968. Crustacea. In: Kumé, M., Dan, K. (Eds.), Invertebrate Embryology, The NOLIT, Publishing House, Belgrade, Yugoslavia, pp. 333–388.

Shiino, S.M., 1988. Crustacea. In: Kume, M., Dan, K. (Eds.), Invertebrate Embryology. Garland Publishing, New York, pp. 333–388.

Smith, G.M.A., Woods, M.A., Shipley, G.B.E., Worburton, M.A., Thompson, C.B., 1923. The Cambridge Natural History. (Crustaceaand Arachnida). Macmillan and Co., Limited, London.

Smyth, J.D., 1954. A technique for the histochemical demonstration of polyphenol oxidase and its application to egg-shell formation in Helminths and byssus formation in *Mytilus*. Q. J. Microsc. Sci. Lond. 95, 139–152.

Sorgeloos, P., Lavens, P., Leger, P., Tackaert, W., Versichele, D., 1986. Manual for the Culture and Use of Brine Shrimp Artemia in Aquaculture, University of Ghent, Ghent, Belgium.

Soundarapandian, P., Varadharajan, D., Anand, T., 2013. Mating behavior of sand crab, *Portunus pelagicus* (Linnaeus). Open Access Sci. Rep. 2, 685–689.

Spalding, J.F., 1942. The nature and formation of the spermatophore and sperm plug in *Carcinus maenas*. Q. J. Microsc. Sci. 83, 399–422.

Stephenson, W., Campbell, B., 1959. The Australian Portunidae. Aust. J. Mar. Fresh Water Res. 10 (1), 88.

Stewart, M.J., Stewart, P., Soonklang, N., Linthong, V., Hnn, P.J., Duan, W., Sobhon, P., 2010. Spermatogenesis in the blue swimming crab, *Portunus pelagicus*, and evidence for histones in mature sperm nuclei. Tissue Cell 42, 137–150.

Storer, T.I., Stebbins, R.C., Usinger, R.L., Nybakken, J.W., 1983. General Zoology, sixth ed. Tata McGraw-Hill Publishing Company Ltd, New Delhi.

Subramoniam, T., 1995. Light and electron microscopic studies on the seminal secretions and the vas deferens of the penaeiodean shrimp, *Sicyonia ingentis*. J. Biosci. 20 (5), 691–706.

Tatarazako, N., Oda, S., 2007. The water flea *Daphnia magna* (Crustacea, Cladocera) as a test species for screening and evaluation of chemicals with endocrine disrupting effects on crustaceans. Ecotoxicology 16, 197–203.

Tatarazako, N., Ishibashi, H., Teshima, K., Kishi, K., Arizono, K., 2004. Effects of triclosan on various aquatic organisms. Environ. Sci. 11 (2), 133–140.

Taylor, J.R.A., Patek, S.N., 2010. Ritualized fighting and biological armor: the impact mechanics of the mantis shrimp's telson. J. Exp. Biol. 213 (20), 3496–3504.

Terra, N.R., Feiden, I.R., 2003. Reproduction and survival of *Daphnia magna* Straus, 1820 (Crustacea: Cladocera) under different hardness conditions. Acta Limnol. Bras. 15 (2), 51–55.

Thompson, W., 1835. Memoirs on the metamorphosis and natural history of the Pinnotheres or pea-crabs. Entomol. Mag. 3 (6), 85–90.

Thompson, J.E., 1930. Ethnology of the Mayas of Southern and Central British Honduras. Anthropological Series, vol. 17 (2). Publications of the Field Museum of Natural History, pp. 27–213.

Threadgold, L.T., 1976. Fasciola hepatica: ultrastructure and histochemistry of the glycocalyx of the tegument. Exp. Parasitol. 39 (1), 119–134.

Türkay, M., 2001. Decapoda. In: Costello, M.J., et al. (Eds.), European Register of Marine Species: A Check-list of the Marine Species in Europe and a Bibliography of Guides to Their Identification. Collection Patrimoines Naturels, 50, pp. 284–292.

Ünis, Ç., Erkan, M.B., 2012. Morphology and Development of the female reproductive system of *Astacus leptodactylus* (Eschscholtz, 1823) (Decapoda, Astacidae). Turk. J. Zool. 36 (1), 775–784.

Verma, P.S., 1977. A Manual of Practical Zoology: Invertebrates. S. Chand and Company Ltd, Ram Nagar, New Delhi.

Vogel, A.I., 1951. A Text Book of Quantitative Inorganic Analysis—Theory and Practical, second ed. Longmans, London.

Vogt, G., 2008a. The marbled crayfish: a new modern organism for research on development, epigenetics and evolutionary biology. J. Zool. 276, 1–13.

Vogt, G., 2008b. Investigations of hatching and early post-embryonic life of freshwater crayfish by in vitro culture, behavioral analysis, and light and electron microscopy. J. Morphol. 269, 790–811.

Vogt, G., 2011. Marmorkrebs: natural crayfish clone as emerging model for various biological disciplines. J. Biosci. 36 (2), 377–382.

Vogt, G., Tolley, L., 2004. Brood care in freshwater crayfish and relationship with the offspring's sensory deficiencies. J. Morphol. 262, 566–582.

Vogt, G., Tolley, L., Scholtz, G., 2004. Life stages and reproductive components of the Marmorkrebs (Marbled Crayfish), the first parthenogenetic Decapod Crustacea. J. Morphol. 261, 286–311.

von Baldass, F., 1937. Entwicklung von *Holopedium gibberum*. Zoöl. Jahrb. Abt. Anat. Ontog. 63, 399–453 (cited in Kumé and Dan, 1968).

Walker, A., Ando, S., Smith, G.D., Lee, R.F., 2006. The utilization of lipovitellin during blue crab (*Callinectes sapidus*) embryogenesis. Comp. Biochem. Physiol. B 143, 201–208.

Wilder, M.N., Okumura, T., Tsutsui, N., 2010. Reproductive mechanisms in Crustacea focusing on selected prawn species: vitellogenin structure, processing and synthetic control. Aqua Biosci. Monogr. (ABSM) 3 (3), 1–49.

Williamson, H.C., 1900. Contributions to the life history of the edible crab (*Cancer pagurus*, Linn.). Rep. Fish. Bd. Scotland 18 (3), 77–143.

Williamson, H.C., 1904. Contributions to the life history of the edible crab (*Cancer pagurus*, Linn.) and of other decapod Crustacea. Impregnation, spawning, casting, distribution, rate of growth. Rep. Fish. Bd. Scotland 22 (3), 100–140.

Winsor, G.L., Innes, D.J., 2002. Sexual Reproduction in *Daphnia pulex* (Crustacea: Cladocera): Observations on Male Mating Behaviour and Avoidance of Inbreeding. vol. 47. Blackwell Science Ltd, 441–450.

Wolfe, A.F., 1971. A histological and histochemical study of the male reproductive system of *Artemia* (Crustacea, Branchiopoda). J. Morphol. 135, 51–70.

Wolfe, A.F., 1973. Observations on the clasping behavior of Artemia salina. Amer. Zool. 13, 472.

Wolfe, A.F., Dillman, J.F., 1990. Light and electron microscopic study of *Artemia* sperm. Am. Zool. 30 (4), 7a.

Yamaguchi, S., Ozaki, Y., Yusa, Y., Takahashi, S., 2007. Do tiny males grow up? Sperm competition and optimal resource allocation schedule of dwarf males in barnacles. J. Theor. Biol. 245, 319–328.

Yamaguchi, S., Yusa, Y., Yamato, S., Urano, S., Takahashi, S., 2008. Mating group size and evolutionarily stable pattern of sexuality in barnacles. J. Theor. Biol. 253 (1), 61–73.

Yamaguchi, S., Charnov, E.L., Sawada, K., Yusa, Y., 2012. Sexual systems and life history of barnacles: a theoretical perspective. Integr. Comp. Biol. 52 (3), 356–365.

Yamaguchi, S., Yusa, Y., Sawada, K., Takahashi, S., 2013. Sexual systems and dwarf males in barnacles: integrating life history and sex allocation theories. J. Theor. Biol. 320, 1–9.

Yano, I., 1988. Oocyte development in the kuruma prawn *Penaeus japonicus*. Mar. Biol. 99, 547–553.

Yano, I., Kanna, R.A., Oyama, R.N., Wyban, J.A., 1988. Mating behaviour in penaeid shrimp *Penaeus vannamei*. Mar. Biol. 97, 171–175.

Yonge, C.M., 1932. On the nature and permeability of chitin. I- The chitin lining the fore gut of Decapod Crustacea and the function of the tegumental glands. Proc. R. Soc. Lond. (B) 111 (772), 298–329.

Yonge, C.M., 1936. On the nature and permeability of chitin. II – The permeability of the uncalcified chitin lining the foregut of *Homarus*. Proc. R. Soc. Lond. (B) 120, 15–41.

Yonge, C.M., 1937. The nature and significance of the membranes surrounding the developing eggs of *Homarus vulgaris* and other decapods. Proc. Zool. Soc. Lond. A 107, 499–517.

Yonge, C.M., 1955. Egg attachment in *Crangon vulgaris* and ather Caridea. Proc. R. Soc. Edinb. 65, 369–400.

Yusa, Y., Takemura, M., Miyazaki, K., Watanabe, T., Yamato, S., 2010. Dwarf males of *Octolasmis warwickii* (Cirripedia: Thoracica): the first example of coexistence of males and Hermaphrodites in the Suborder Lepadomorpha. Biol. Bull. 218, 259–265.

Yusa, Y., Yoshikawa, M., Kitaura, J., Kawane, M., Ozaki, Y., Yamato, S., Høeg, J.T., 2012. Adaptive evolution of sexual systems in pedunculate barnacles. Proc. R. Soc. Lond. B Biol. Sci. 279 (1730), 959–966.

Yusli, W., Ali, M., 2010. Biological information on the mantis shrimp *Harpiosquilla raphidia* (Fabricius, 1798) (Stomatopoda, Crustacea) in Indonesia with a highlight of its reproductive aspects. J. Trop. Biol. Conserv. 7, 65–73.

Zara, F.J., Toyama, M.H., Caetano, F.H., López-Greco, L.S., 2012. Spermatogenesis, spermatophore, and seminal fluid production in the adult blue crab *Callinestes danae* (Portunidae). J. Crustac. Biol. 32 (2), 249–262.

Zehnder, H., 1934. Uber die embryonalentwicklung des flusskrebses. Acta Zoologica. 15, 261–408.

Chapter Outline

Definition **715**
General features of Mollusca **715**
Classification of Mollusca **716**
Class- Amphineura 716
Polyplacophora 716
Reproduction 716
Agametic reproduction 716
Gametic reproduction 716
Dimorphism 716
Hermaphroditism and sex ratio 716
Anatomy of the reproductive system 716
Gametogenesis 722
Oogensis 722
Spermatogenesis 723
Reproductive behaviour and spawning 727
Early development 728
Cleavage 728
Gastrulation 728
Larval development 728
Larval eye 730
Foot and pedal gland 730
The shell 730
Settlement and metamorphosis 730
Brooding and ovoviviparity 730
Later development 732
The eyes 732
The aesthetes 732
The nervous system 732
The head 732
Monoplacophora 733
Aplacophora 735
Early development 735
Larval development 738
Metamorphosis 738
Class- Gastropoda 739
Reproduction 739
Agametic reproduction 739
Gametic reproduction 739
Dimorphism 739
Hermaphroditism 739
Protandric hermaphrodites 741
Reproductive system 741
Eremina ehrenbergi 741
The hermaphrodite gonad 741
The hermaphrodite duct 744
The talon 744
Albumen gland 748
The spermoviduct 753
The male side of the system 755
The vas deferens 757
The penis 758
The genital atrium 758
The female side of the system 758
The oviduct 758
The spermatheca and spermathecal duct 758
The vagina 760
The accessory organs of the vagina 760
The mucous glands 760
The dart sac 764
The dart 764
Gametogenesis 764
Spermatogenesis 764
Structure of the fully mature spermatozoa 767
Abnormal spermatogenesis 767
Other examples of dimorphism of sperm in Mollusca 769
Functions of abnormal spermatozoa 769
Spermatophores 769
Eremina ehrenbergi 771
Oogenesis 771
The spherical oocyte 775
Gonad of ***Melampus coffeus*** 777
Back to *E. ehrenbergi* 777
1- Sperm pathway and spermatophore formation 777
Spermatophore formation 782
Copulation 782
2- Pathway of the foreign sperm 783
3- The pathway of the oocyte 783
The egg/fertilization 785
The fully formed egg 786
Back to Gastropoda in general 786
Reproductive behaviour 786
Spawning aggregation 786
Sex recognition 786
Copulation 786
Brooding behaviour 786
Spawning and rearing of youngs 786
Adalaria proxima 788
Fertilization 789
Early development 789
Cleavage 789
Cleavage of ***Trochus*** (Annelidan mode) 790
Cleavage of ***Ilyanassa*** 791
Differentiation of germ layers 793
1- Segregation of ectoderm 793
2- Differentiation of ento-mesoblast 793
3- Ecto-mesodermal cells 794
Blastulation 794
Gastrulation and gastrula stage 794
Types of gastrulation 794
Larvae (trochopore and veliger stages) 797
Shell formation 797
Foot, pedal gland, and statocysts 797

Radula sac 799
Velum 799
Head vesicle 800
Digestive system 800
Later development 800
Torsion 800
Coiling of the visceral hump 802
Detorsion 802
Echinospira larva 804
Trophon geversianus 804
References **810**

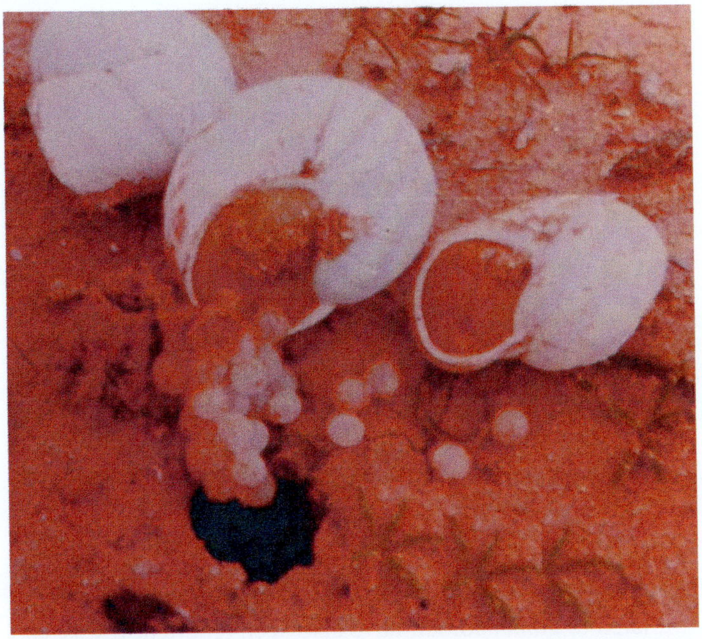

Chapter 11

Phylum Mollusca

Latin *mollis*, 'soft'
(about 128,000 sp.)

DEFINITION

Mollusca comprise the second largest animal phylum and, as such, form an extremely diverse group. Most of its animals are familiar, such as snails, slugs, clams, oysters, squids, scallops, and octopuses. These animals are characterized by more than one conspicuous phenomenon: hermaphroditism, dimorphism, formation of spermatophores, and most notably, spiral cleavage.

GENERAL FEATURES OF MOLLUSCA

- Unsegmented shizocoelomata animals (protostomata).
- The body is divided into three distinctive regions, usually a well-developed head, a large, muscular ventral foot, and a dorsal visceral hump (Fig. 11.2A).
- The mantle, which is a membranous sheet, encloses the visceral hump.
- The mantle often secretes a calcareous shell.
- The mantle encloses a mantle cavity into which opens the anus.
- The shizocoel (coelom) is primitively represented as three pairs of cavities, the renocoele, the gonocoele, and the pericardium. They do not develop in a manner suggestive of mesodermal segmentation. Initially, these cavities are intercommunicated and the reproductive cells are discharged through the renocoele. In the more advanced molluscs there is a progressive separation of the renal and the gonadal products and each discharge through their own duct (Fig. 11.3).
- Therefore, the kidneys are mesoblastic.
- It has a pair of ctenidia.
- The alimentary canal usually consists of a buccal mass, **radula**, and salivary glands, and the stomach, into which a digestive gland opens.
- The heart consists of a median ventricle and two lateral auricles.
- Arterial and venous systems often expanding into a more or less extensive haemocyanin as a respiratory pigment.
- The nervous system, consisting of a circumoesophageal ring, is often concentrated into cerebral and pleural ganglia, pedal cords and ganglia, and visceral loops, pericardium, renocoel, and gonocoel (Fig. 11.2A).
- Generally Mollusca are dioecious; however, hermaphroditism is present in the most famous gastropods: Pulmonata and Opisthobranchiata (Fig. 11.12).
- Sperm, generally long-tailed, looks like that of vertebrates (vertebrate sperm).
- Gastropods and pelecypods produce dimorphic sperm: eupyrene, apyrene (Fig. 11.26).
- Some species are characterized by discharging their sperm in spermatophores; e.g. gastropods and cephalopods (Fig. 11.22).
- Egg is moderately yolky and holoblastic, except that of cephalopods, which is telolecithal and meroblastic.
- External and internal fertilization.
- Spiral cleavage is the normal pattern for all the molluscs except the cephalopods.
- Indirect development through trochophore larva, which develops into veliger larva.

Invertebrate Embryology and Reproduction. https://doi.org/10.1016/B978-0-12-814114-4.00011-4

CLASSIFICATION OF MOLLUSCA

Class I: Amphineura (Placophora).
Subclass 1: Polyplacophora, e.g. *Chiton.*
Subclass 2: Monoplacophora, e.g. *Neoplina.*
Subclass 3: Aplacophora, e.g. *Neomenia.*
Class II: Gastropoda.
Subclass 1: Prosobranchiata, e.g. *Pterotrachea.*
Order 1: Archaeogastropoda, e.g. *Haliotis* (abalone, ormer), *Patella* (limpet), *Trochus* (top shell).
Order 2: Neritacea, e.g. *Nerita.*
Order 3: Mesogastropoda, e.g. *Littorina* (periwinkle), *Viviparous* (river snail), *Crepidula* (slipper limpet).
Order 4: Neogastropoda, e.g. *Buccinium, Busycon* (whelks).
Subclass 2: Opisthobranchiata, e.g. *Aplysia* (sea hare), *Doris* (sea lemon), *Eolis* (sea slug).
Subclass 3: Pulmonata.
Order 1: Basommatophora, e.g. *Limnaea* (pond snail).
Order 2: Stylommatophora, e.g. *Helix* (land snail), *Eremina* (desert snail).
Class III: Scaphopoda, e.g. *Dentalium, Cadulus.*
Class IV: Lamellibranchiata (pelecypoda).
Order 1: Protobranchiata, e.g. *Nucula, Yoldia.*
Order 2: Filibranchiata, e.g. *Mytilus, Anaemia, Donax.*
Order 3: Eulamellibranchiata, e.g. *Anodonta, Cardium, Unio.*
Order 4: Septibranchiata, e.g. *Poromya, Cuspidaria.*
Class V: Cephalopoda.
Order 1: Dibranchiata, e.g. *Sepia, Loligo, octopous.*
Order 2: Teterabranchiata, e.g. *Nautilus, Baculites.*

The present work will deal with each of these phenomena in its proper place.

The development of Amphineura, Gastropoda, lamellibranchiate, and Scaphopoda are more or less similar. The development of the cephalopoda is completely different: it undergoes a different pattern of cleavage and develops directly, without larvae. Hence, it is separated in its own chapter.

Class- Amphineura

Polyplacophora

Polyplacophora, or chitons (Fig. 11.1D), are common inhabitants of marine rocky shores. They can be abundant in intertidal areas, where they usually graze on encrusting algae and sessile animals. They can be recognized by their eight overlapping shell plates, known as valves (six intermediate valves and two terminal valves at the head and tail), which are firmly anchored in a tough muscular girdle (Eernisse, 2007). Chitons provide particularly favourable material for reproductive studies because they are readily available and easily collected. They are relatively large, and because the single gonad is discrete, conveniently dissected, and filled with numerous sperm or eggs when ripe. However, fossils of chitons, including some that are more than 500 million years old, suggest that, throughout their history, their lifestyle and appearance have not changed very much (Eernisse, 2007). The latter author also reported that Chitons sometimes aggregate and simultaneously spawn (e.g. the giant gumboot chiton, *Cryptochiton stelleri*).

Reproduction

Agametic reproduction. No form of agametic reproduction has been reported for chitons.

Gametic reproduction

Dimorphism. Most of chitons are dioecious. There are only a few reports of external sexual dimorphism. Pelseneer claimed that mature males of *Lepidochitona cinereus* (Linnaeus, 1767) could be recognized by noting the red colour of the testis, visible through the musculature of the foot. Other colour dimorphisms, apparently unrelated to gonad colour, have also been reported. It was said that the ventral surface of males of *Chiton tuberculatus* in Bermuda was a dull buff colour, while that of females was pink, salmon, or orange. The colour of reproductively active gonads is usually different in males and females. The gonad wall is generally dull reddish brown, as seen in small, undeveloped gonads, while the sperm are white and the oocytes are usually dark grey-green to black. Testes full of sperm are pale red, orange, or salmon pink, and ovaries full of eggs are usually dark green to black. These colour distinctions are not present in all species of chiton.

In addition to colour differences of the gonad, it was said that dissected females of *Katharina tunicata* could be distinguished easily from males by the presence of thickened glandular oviducts (shell glands).

Hermaphroditism and sex ratio. Pearse (1979) reported that there is only one confirmed report of a hermaphroditic species of chiton. It was said that, in central California, young specimens of the brooding intertidal species, *Cyanoplax dentien* (=*Tachydermon raymondi*) were females, while larger specimens had gonads containing both oocytes and sperm. Populations of many species of chitons have a significant predominance of males. It was suggested that the predominance of males in the species studied might be an adaptation to their intertidal habitat. Considerable variation in sex ratios among different samples of *Chiton pelliserpentis* was noted, with females predominating in smaller size groups and males predominating in larger size groups. It was proposed that this difference might be due to earlier sexual maturation of females than males and, perhaps, a higher mortality rate for females.

Anatomy of the reproductive system

The gonad in most species of chitons is a single lobulated structure in the middorsal region, between the gut and dorsal body wall. The gametes are transported to the outside by a special pair of gonoducts that arise from near the

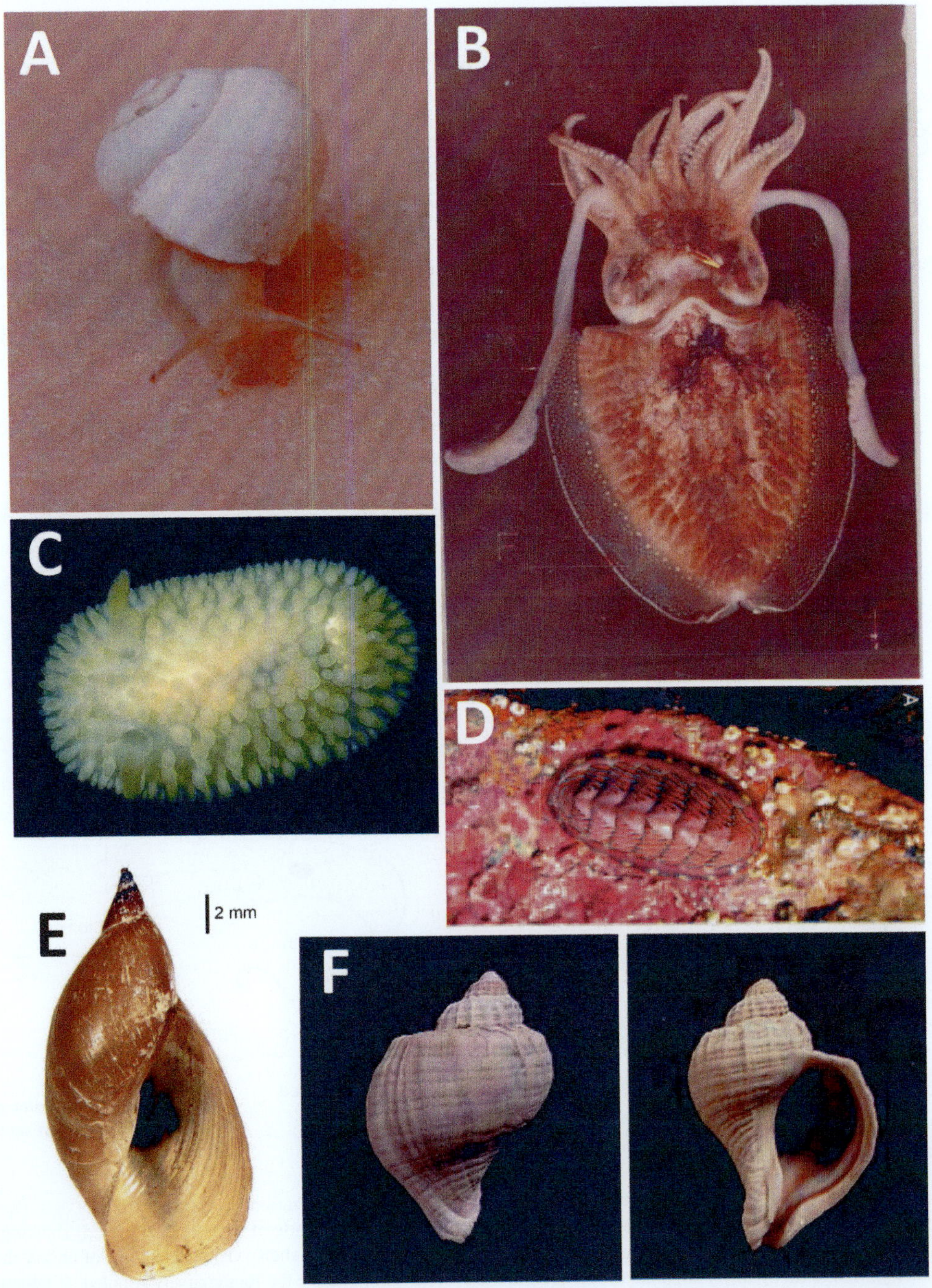

FIG. 11.1 Photographs of live specimens of some molluscan animals: (A) dorsal view of adult desert snail *Eremina ehrenbergi* Roth, 1839, Its identification according to Biggs (1959). X=2 *(from Abd El-Kerim, 1988)*; (B) dorsal view of a fresh male specimen of *Sepia officinalis*. X=⅓ *(from Ismail, 1981)*; (C) *Adalaria proxima* (Alder & Hancock, 1854), *courtesy of Bernard Picton*; (D) chiton, *Tonicella lineata* (British Columbia, Canada) feeds on crustose coralline algae. *Photograph by Eernisse (2007)*; (E) *Lymnaea acuminata* Lamarck, 1822 accepted as *Radix rufescens* (Gray, 1822) *courtesy of Vinarski, Maxim*; (F) Two views of the Shell of *Trophon geversianus* (Pallas, 1774). *(C and E) From (WoRMS); (F) from: https://wikivisually.com/wiki/Trophon#cite_note-12.*

dorsoposterior end of the gonad (Figs 11.2E and 11.4A), instead of by the nephridia. A gonopore is located in each pallial groove in front of the nephridiopore (Barnes, 1974). However, each gonoduct extends along the anterior surface of the pericardium and then curves ventrolaterally to open into the mantle groove, at level one to nine gills anterior to the excretory pore, near the boundary between the sixth and seventh shell valves (Pearse, 1979).

The female oviducts are simple ducts lined with a ciliated epithelium, but after a short distance they are lined with elongated secretory cells. The glandular portions of the oviducts probably produce much of the mucous that envelops

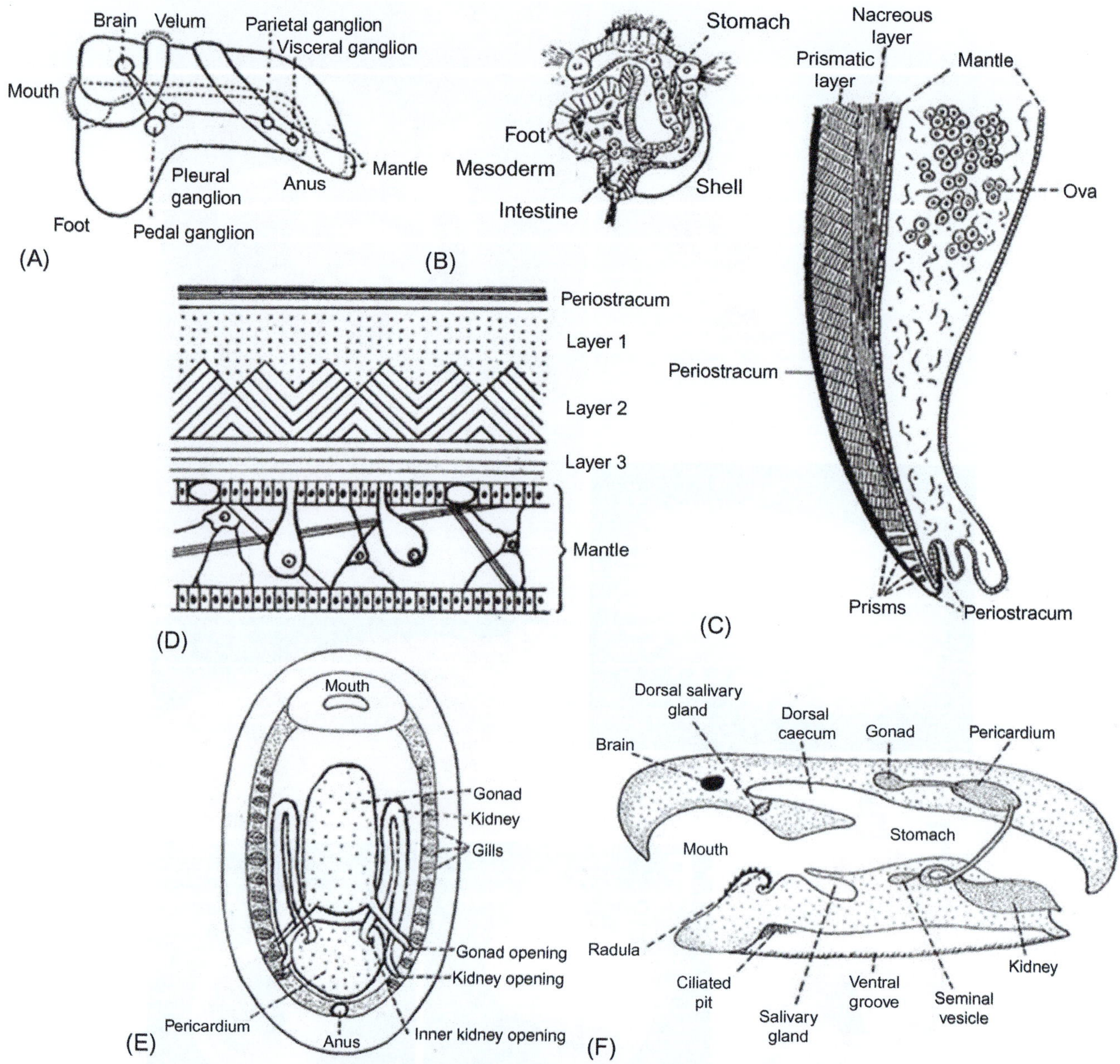

FIG. 11.2 (A) Illustration of veliger larva of *Paludina* before torsion showing molluscan organization; (B) early veliger larva of *Patella coerulea (after Patten)* showing the beginning of the formation of the shell; (C) V.S. through the edge of the mantle of *Mytilus (after Field)* showing the process of formation of the shell; (D) diagram of T.S. of generalized lamellibrach shell (example of molluscan shell); (E) ventral view of chiton showing external and internal bilateral symmetry. Mantle cavity finely stippled, the division of the coelom shown above the foot coarsely stippled; (F) diagram of the structure of an aplacophoran, *Proneomenia*. *(All from Borradaile et al. (1977).)*

the eggs when they are spawned. On the other hand, the sperm ducts in the male are simple ducts lined with a ciliated epithelium, continuous with that of the dorsal surface of the testis. Their epithelial cells are not secretory, but they often contain yellowish-brown globules (Hyman, 1967). Although it was not reported that this system stores sperm, the sperm ducts are enlarged anteriorly, just beyond the testis or at their end, forming seminal receptacles.

Avila-Poveda and Abadia-Chanona (2013) studied 'emergence', development, and maturity of the gonad of two wild chitons of family Chitonidae: *Chiton articulatus* and *C. albolineatus*. The knowledge of this gonadal ontogeny, which is one of the fundamental aspects of developmental biology, is not enough, making it difficult to be determined. However, it appears that it happens too early in the life cycle almost immediately after the metamorphic stage. The authors categorized the gonadal ontogenesis and gonad development stages into three periods ('Pw' without gonad, 'Pe' gonad emergence (origin), and 'Pf' gonadal sac formed) and four stages ('S0' gametocytogenesis, 'S1'

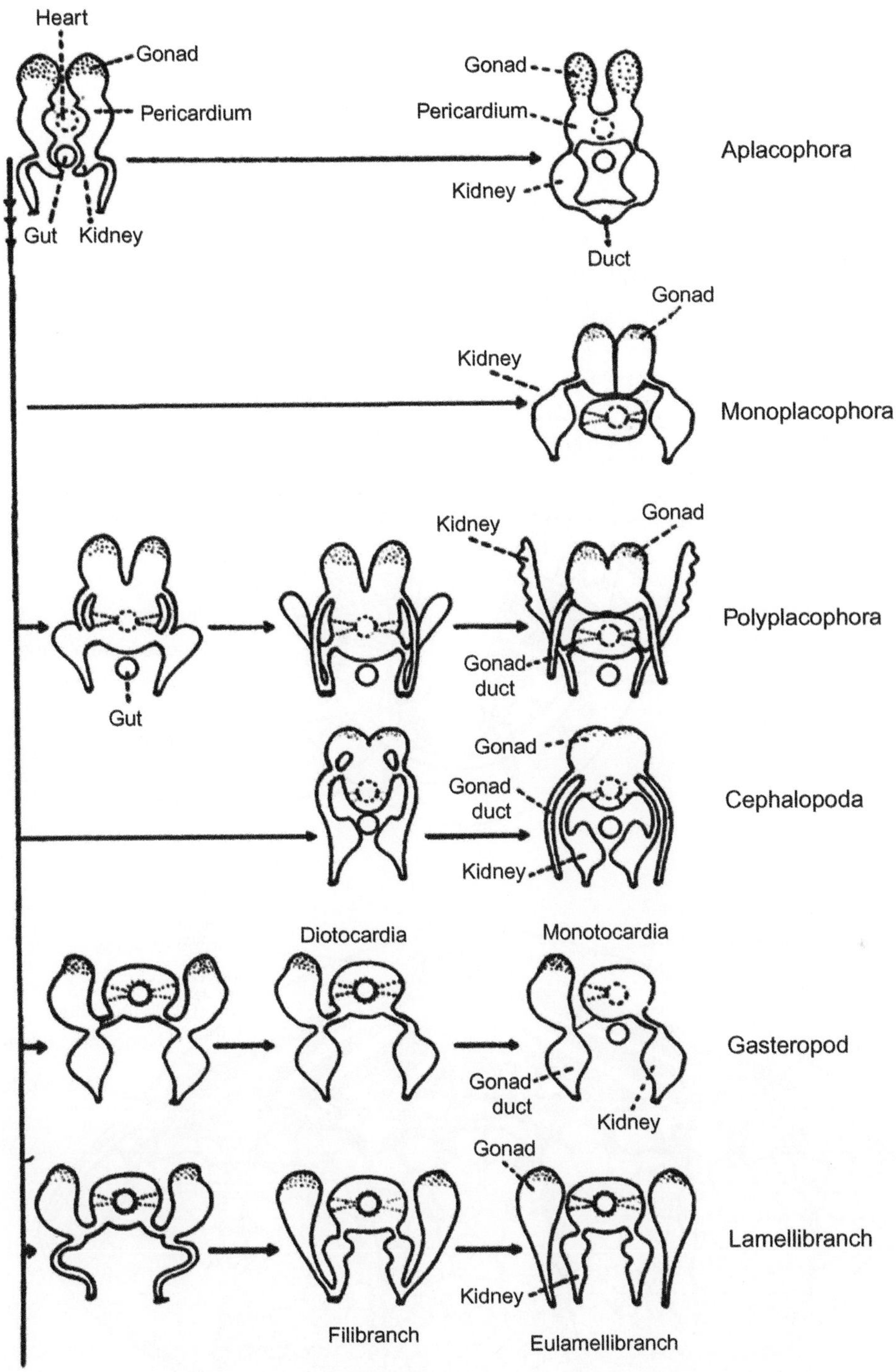

FIG. 11.3 Development and relationship of coelomic cavities in molluscs. The coelom is represented by the gonad cavity, the kidney cavity, and the pericardium. Note the tendency for the renocoele and gonocoele to separate. *(All except monoplachophora after Goodrich, from Borradaile et al. (1977).)*

gametogenesis, 'S2' mature, and 'S3' spawning), respectively. In addition, they evaluated allometric relationships between the lengths and body weight of chitons throughout early life stages. The authors suggested that there is a subtle change in chitons' body shape as they develop into adulthood, which particularly evolves in response to gonad maturation. These two chitons, particularly *Chiton articulatus* (locally known as 'dog tongue' or 'sea cockroach'), have gained regional importance and economic interest in the southern Mexican Pacific, where some restaurants have begun to offer the 'sea cockroach' to tourists as a gourmet and aphrodisiac food (Avila-Poveda and Abadia-Chanona, personal observation).

Thus, the first appearance of the gonad evolves as a projection of squamous epithelium, which extends from the

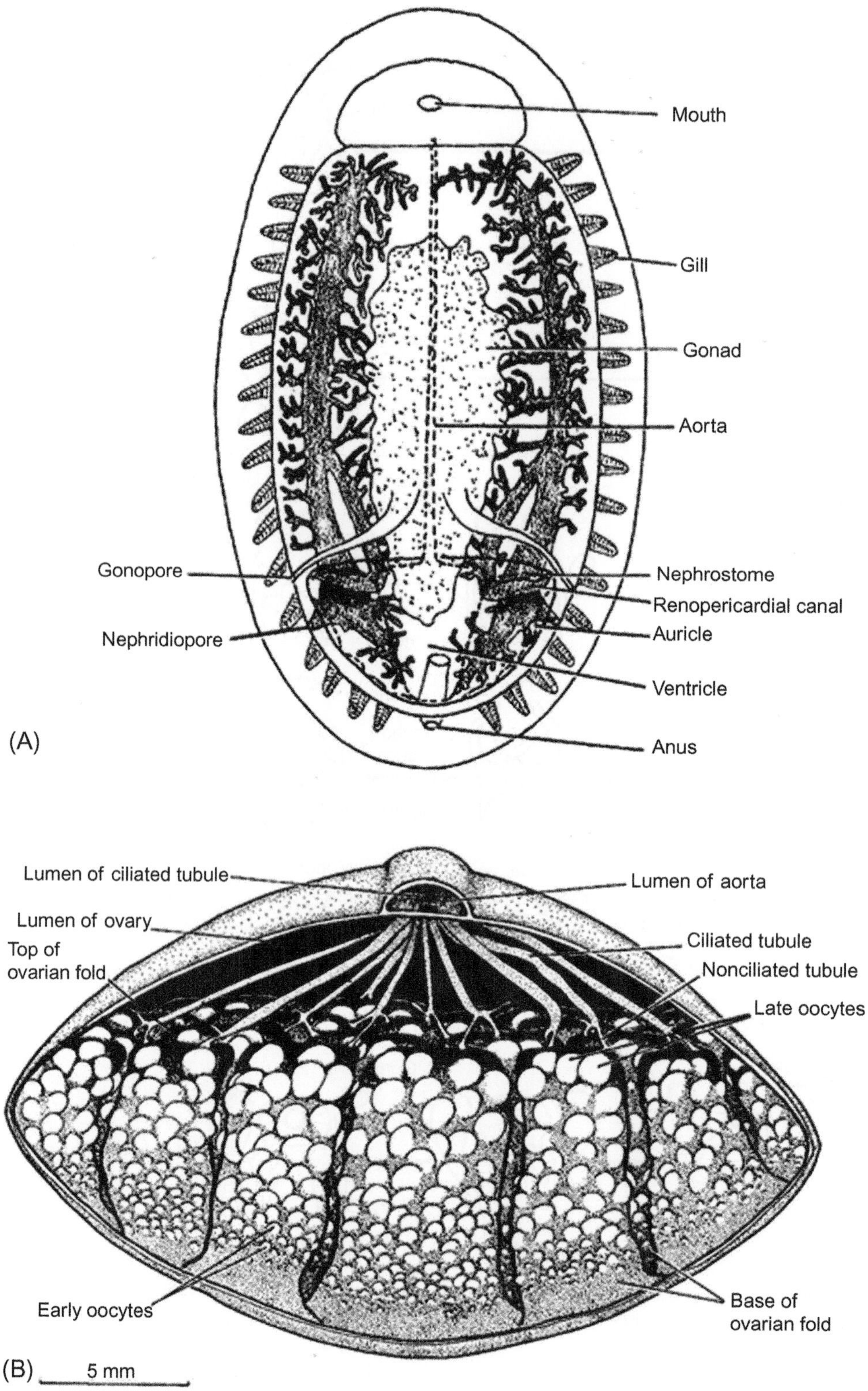

FIG. 11.4 (A) Diagram showing internal structure of a chiton *(after Lang and Haller, from Barnes, 1974)*; (B) diagrammatic cross section through the ovary and aorta of *Chiton septentriones* (Asby) showing the ovarian folds and their blood vessels ('ciliated tubules') *(After Selwood, 1968; from Pearse, 1979)*;

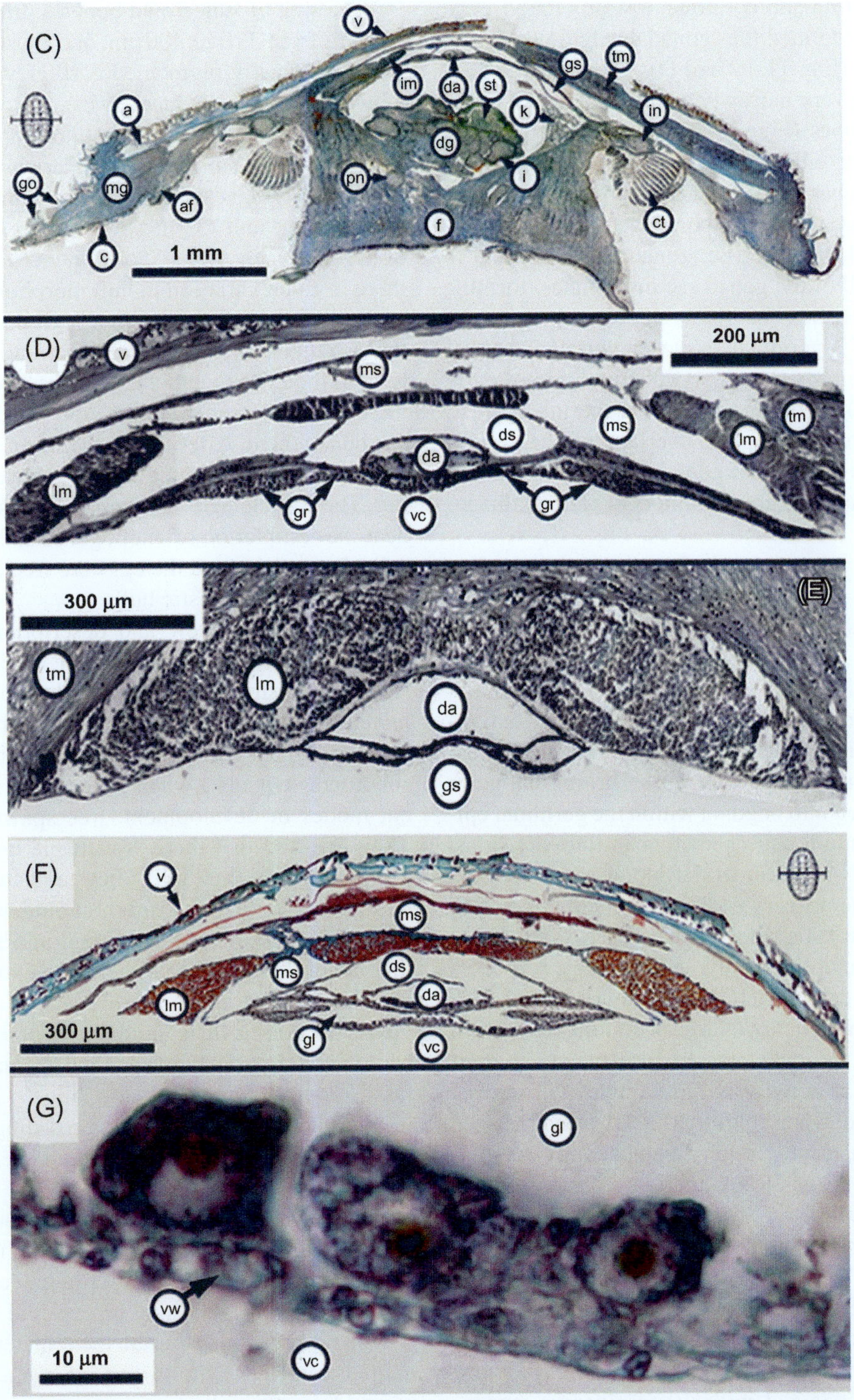

FIG. 11.4, CONT'D (C–G) T.S.s through *Chiton articulatus* to illustrate emergence, development and maturity of the gonad: (C) composite digital image in transverse section of the whole specimen through the starting period of gonad emergence; (D) magnification of the dorsal part from the whole specimen of (C); (E) magnification of the dorsal part from the whole specimen during the formation of the gonadal lobes; (F) composite digital image in transverse section of the dorsal part of the whole specimen showing the general histological anatomy during the merging of the gonadal lobes; (G) magnification of the ventral right side of the gonad sac without tissue plates but showing the rise of the early oogonia with basophilic granular cytoplasm and one nucleolus, as well as the development of the early cells that constitute the tissue plates on the ventral wall of the gonad sac. a = articulamentum, af = accessory fold, c = cuticle, ct = ctenidium, da = dorsal aorta, dg = digestive gland, ds = dorsal aorta sinus, f = foot, gl = gonadal lumen (i.e. an inner open space or cavity of the gonad), go = girdle ornamentations, gr = gonadal ridge (i.e. the precursor to the gonad), gs = gonadal sinus (i.e. cavity formed by a bending or curving from the gonadal ridge and containing chiefly blood), I = intestine, k = kidney, lm = longitudinal muscle, ln = lateral 'visceral' nerve cord, mg = mantle girdle, ms = muscle sinus, pn = pedal nerve cord, st = stomach, tm = transversal muscle, v = valve, vc = visceral cavity 'hemocoel', vw = ventral wall of gonad sac. *(C–G) From Avila-Poveda and Abadia-Chanona (2013).*

wall of the dorsal aorta and is thrown towards the visceral cavity; this could be termed the 'gonadal ridge' or the precursor to the gonad (Fig. 11.4C and D). The projection enlarges, until closing very narrowly against the dorsal aorta, forming two oval lobes (Fig. 11.4E). Then, the squamous epithelium becomes ciliated cubical epithelial cells. In turn, the two oval lobes merge and develop into the gonad sac (Fig. 11.4F and G). Finally, the gonad sac starts to expand to form a hollow space (i.e. gonadal lumen), and in turn, the ventral wall of the gonad sac invaginates, forming folds inward of the lumen (i.e. tissue plates), which, in the view of this work, is germinal tissue or germinal epithelium. Gametocytogenesis (i.e. period of formation of early oocytes (oogonia) or early spermatocytes (spermatogonia), i.e. immature gametes) takes place overlapping the periods of gonad emergence (Pe) and of gonadal sac formation (Pf) in both chiton species. The above authors explained this information by tables.

Gametogenesis

Oogensis. Chiton undergoes gametogenesis in two steps: the first to increase size, the second to undergo vitellogenesis (Fig. 11.6A). Light microscopic analysis of oogenesis in chiton has been reported by more than one scientist, including Pelseneer (1898). Pearse (1979) reviewed this aspect and reported that in *C. septentriones,* primordial germ cells are differentiated into oogonia within the germinal epithelium. Each oogonium undergoes four mitotic divisions. Those resulting cells adjacent to the blood sinuses of the ovarian folds become primary oocytes; while the overlying cells become follicle cells (Fig. 11.6B). Early oocytes are previtellogenic, develop gradually, and undergo a process of vitellogenesis: so called vitellogenic oocyte (Fig. 11.6C). The one which finishes vitellogenesis, becoming full-grown with optimum size, is called the mature oocyte (ovum) (Fig. 11.6D). Nuclear activity is most intense during the previtellogenic phase, when ribosomal RNA is synthesized and accumulated in the cytoplasm, which becomes very basophilic. Associated to this RNA, there are basic proteins, which become particularly prominent in previtellogenic oocytes. At the end of the previtellogenic stage, RNA synthesis ceases and does not begin again until after the trochophore larva is developed. However, these changes in the volume ratios of the nucleus, nucleolus, and cytoplasm are similar to those in other animals.

In another species, e.g. *Lepidocitona cinereus*, oogenesis proceeds in somewhat different steps (Fig. 11.6E–J). The vitellogenic oocyte is often characterized by lobulations formed by infoldings of the cell membrane. The dense masses of yolk nuclei disperse, and piles of annulate lamellas contact the cytoplasmic vacuoles, which become filled with mucopolysaccharides and acid lipid material. The source of yolk appears to be mainly de novo synthesis, possibly by mitochondrion transformations (Pearse, 1979).

The size of full-grown oocytes differs with the species, ranging from 110 to 400 μm, and most are filled with yolk, which is usually pigmented dark green and surrounded by a hull ('chorion') (Pearse, 1979). According to the arrangement of the hull, the eggs of chitons are grouped into three: (1) the hull is smooth and spherical (e.g. *Lepidopleurus asellus*); (2) it is made up of a number of cupules (e.g. *C. porosus*); and (3) the hull possesses numerous projections or spines (e.g. *Chaetopleura apiculata*). However, there is much variation in hull morphology among different species of chitons, and a thorough analysis of the various types remains wanting. The function of the hull might be to hold the eggs together in clumps after spawning, or hold the eggs to the substrate, or hold them in the mucous that surrounds them. After that, in addition of a chorion, a vitelline envelope is secreted by the oocyte just before spawning. The follicle cells serve mainly to shape the hull. Then, near the completion of vitellogenesis, the remaining follicle cells degenerate, and the eggs are free in the ovarian lumen with their characteristic hulls.

This was the view and description, especially of the full-grown oocyte, until 1980. In 2013, Liuzzi and Zelaya studied ultrastructurally the egg-hull of *Ischnochiton stramineus* (Sowerby, 1832). First of all, they reported that the morphology of the hull varies, not only among different suborders, but also when comparing species with different modes of development, free spawning, brooding, etc. They reported that there is valuable information about the morphology of this hull. They tabulated the form of egg-hull processes in many different chitons, each with its suborder, order, brooding condition, and the reference. From that table, it is noted that the hull of southwestern Atlantic chitons remains poorly known. They also presented a review concerning their animal and its hull, as summarized here. Sowerby (1832) described *Ischnochiton stramineus* (as *Chiton stramineus*) from Chiloe' Island (Chile), based on morphological characters, providing novel information on its brooding behaviour and the general egg morphology (studied with light microscopy). The name *Ischnochiton stramineus* becomes definite and its parental care becomes an interesting field. Liuzzi and Zelaya (2013) provided additional information on the reproductive characteristics of *I. stramineus*, while focusing on egg-hull ultrastructure.

Liuzzi and Zelaya used their material from preserved animals, and gave the history of each. The eggs were removed from the pallial grooves, and prepared for the study. One of the animals was bearing juveniles with fully developed eight-valved shells.

Eggs (Fig. 11.6N) are condensed in compact masses in the mantle groove on either side of the foot. One specimen brooded up to 1600 eggs (Fig. 11.6O). Egg hulls (hull is the chorion, as mentioned above) are formed of polygonal, perforated plates, about 30 μm in width (Fig. 11.6P). Micropores (0.2–0.5 μm; Fig. 11.6P) are randomly distributed over the entire surface of each plate, except for the area near the base of the projections (Fig. 11.6Q). Each plate has raised periphery, forming rounded, unperforated ridges (Fig. 11.6Q and P). It shows a small subcentral elevation, which is the base of a hollow and elongated projection with a spiral tip. This tip enlarges distally, forming a small ampoule (Fig. 11.6N, Q, R, and S). Each hull contains ~204 of these curled projections, which interlock with those of other eggs and remain even in developing embryos (Fig. 11.6T). Seven fully metamorphosed young, of about 6 mm length, were found in one of the studied specimens. So, the authors suggested that *I. stramineus* eggs (or at least some of them) are retained in the pallial grooves until completely developed.

Spermatogenesis. The process of spermatogenesis in different species of chitons was described using light microscopy. However, Pearse (1979) reported the detailed electron microscopic analysis of the events of this vital activity in *C. pelliserpentis*. In the testis of this species, spermatogonia are identified as large cells arranged in loose groups along the testicular folds, as stem cells or primordial germ cells. They seem to give rise to pear-shaped (7×20 μm) spermatogonia. Each contains a large nucleus (6.5–8 μm in diameter). The spermatogonia multiply, thickening the surface of the fold, forming pockets at first, then continuous bands of primary spermatocytes. The nuclei of the latter are lightly stained. Secondary spermatocytes containing densely stained nuclei (1.5–2 μm) are produced from the division of the primary ones. The secondary spermatocytes are differentiated, giving rise to small spermatids (1 μm in diameter) with little cytoplasm. Then spermiogenesis is completed as the spermatids elongate into 1×3 μm triangular heads, and develop a tail well over 10 μm long. The sperm migrate between the epithelial cells and accumulate within the testicular lumen.

The sperm of chitons (Fig. 11.5A and E) are pointed and, in many species, they have a long anterior filament, similar to that found in some aplacophoran species. It is referred to as an acrosomal filament or an acrosome. It appears that this filament contains only an extension of the nucleus. However, the sperm did not display an acrosome reaction during egg water tests. No acrosome has been observed in electron, eccentrically near the posterior end of the head, in most species studied by electron microscopy.

A more recent study on gametogenesis in two chitons, *Chiton articulatus* and *C. albolineatus,* is encountered in the work of Avila-Poveda and Abadia-Chanona (2013) (see above). They added much information to the study of gametogenosis, and divided this vital activity into:

Stage 0 (gametocytogenesis); during which the first sexual cells (oogonia and early oocytes) arise in females, exceptionally early during the end period of gonad origin, when the cuboidal epithelial cells of the gonadal sac are created, but still without germinal tissue (Fig. 11.4G). The primary oocyte is a small cell that measures from 15 to 62.5 μm in diameter. Its cytoplasm is hyaline, vacuolated, heavily basophilic, and with a translucent nucleus, having one basophilic spherical nucleolus (Fig. 11.6A). In males, on the other hand, these first sexual cells, spermatogonia, and early spermatocytes, arise later than in females, but at the same stage of gonad sac formation, i.e. when the sac is wholly expanded, forming a gonadal lumen.

Stage 1 (gametogenesis); during which secondary oocytes and spermatocytes and spermatids are formed. In females of *C. articulates,* the gonad shows asynchronous groups of oocytes, i.e. primary oocytes, secondary oocytes, and mature ova are present without a dominant group (Fig. 11.6A). The gonad of the other species, *C. albolineatus,* shows only a synchronous group of secondary oocytes without replenishment by earlier oocytes. The secondary oocyte is larger, measuring from 46 to 162 μm in diameter. Its cytoplasm is somewhat diminished of basophilia, nonvacuolated but wholly granular, with a relatively smaller spherical nucleus than early oocytes. No follicle cells surrounding them (Fig. 11.6A). The testis in males shows spermatocytes and spermatids occurring around the germinal epithelium and forming groups, and some spermatozoa beginning to fill the gonadal lumen (Fig. 11.5A).

Stage 2 (maturation); formation of mature ova and mature sperm, in both species the female shows only a synchronous group of mature ova without replenishment by earlier oocytes (Fig. 11.6A, mo). The mature ova, excluding the chorion, measure from 134 to 216 μm in diameter. The cytoplasm is acidophilic, filled with uniform protein yolk granules (vitellum), and a fully developed chorionic process with projections (i.e. chorion or hull) covers the mature ova and/or shed into the gonadal lumen (Fig. 11.6T). Males reach the mature stage at a smaller size than females and the lumen of the gonad becomes filled with sperm (Fig. 11.5B and C).

Critical information is reported in this work of Avila-Poveda and Abadia-Chanona (2013), and is worth mentioning here. Recently, it has been indicated that *Chiton articulatus* reveals an unusually high incidence of hermaphroditism, commonly being the female tissue, which occupies the greatest part of the gonad sections, in respect to male tissue. However, this study did not indicate the total length of the studied specimens. It is therefore difficult to integrate this information to recognize whether *C. articulatus* is: (1) a hermaphrodite species in the adult stage; and/or (2) whether these are just juvenile specimens that are in a process of sexual differentiation. The authors furthermore reported that the term juvenile introduces young chitons that meet the criteria from Stage 0 (gametocytogenesis) and Stage 1 (gametogenesis) and is defined as the stage between the subjuvenile and subadult stages.

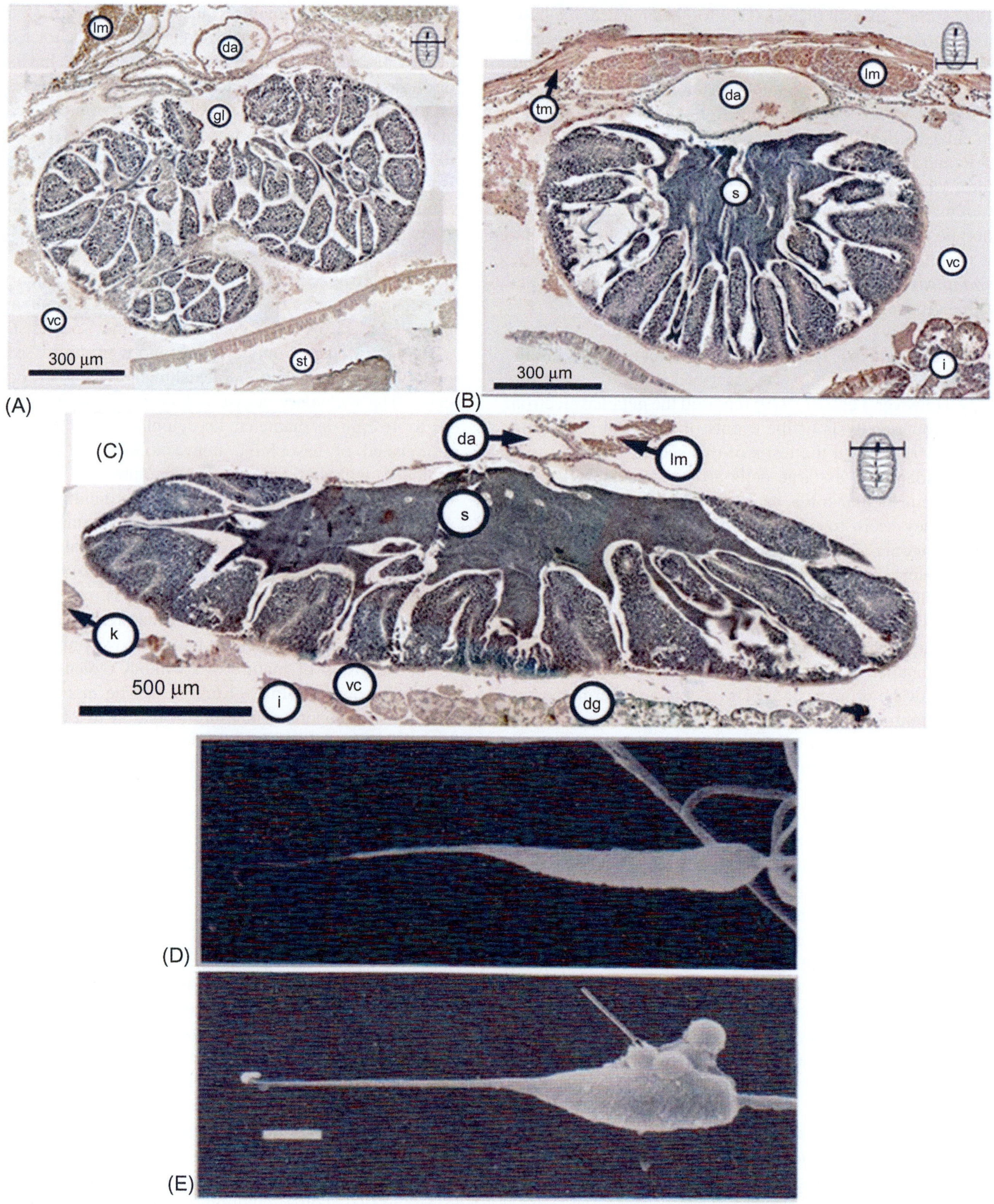

FIG. 11.5 Spermatogenesis: (A) Composite digital image in transverse section of a *Chiton albolineatus* male, with testis in the gametogenesis stage (S1) showing spermatocytes and spermatids occurring around the tissue plates and forming groups, and some spermatozoa beginning to replenish the gonadal lumen; (B and C) Composite digital images in transverse section (B) of the posterior dorsal part and (C) of the anterior dorsal part from mature *Chiton albolineatus* males, showing the gonadal lumen fully replenished with sperm. s = spermatozoa, other letters as in Fig. 11.4; (D and E) SEM of sperm taken from the testes of chitons to show the long anterior filament: (D) *Stenoplax heathiana* Berry; (E) *Mopalia muscosa* (Could), the plasmalemma is torn, revealing the spherical mitochondria *(arrow)*. *((A–C) From Avila-Poveda and Abadia-Chanona (2013) and (D and E) from Pearse (1979).)*

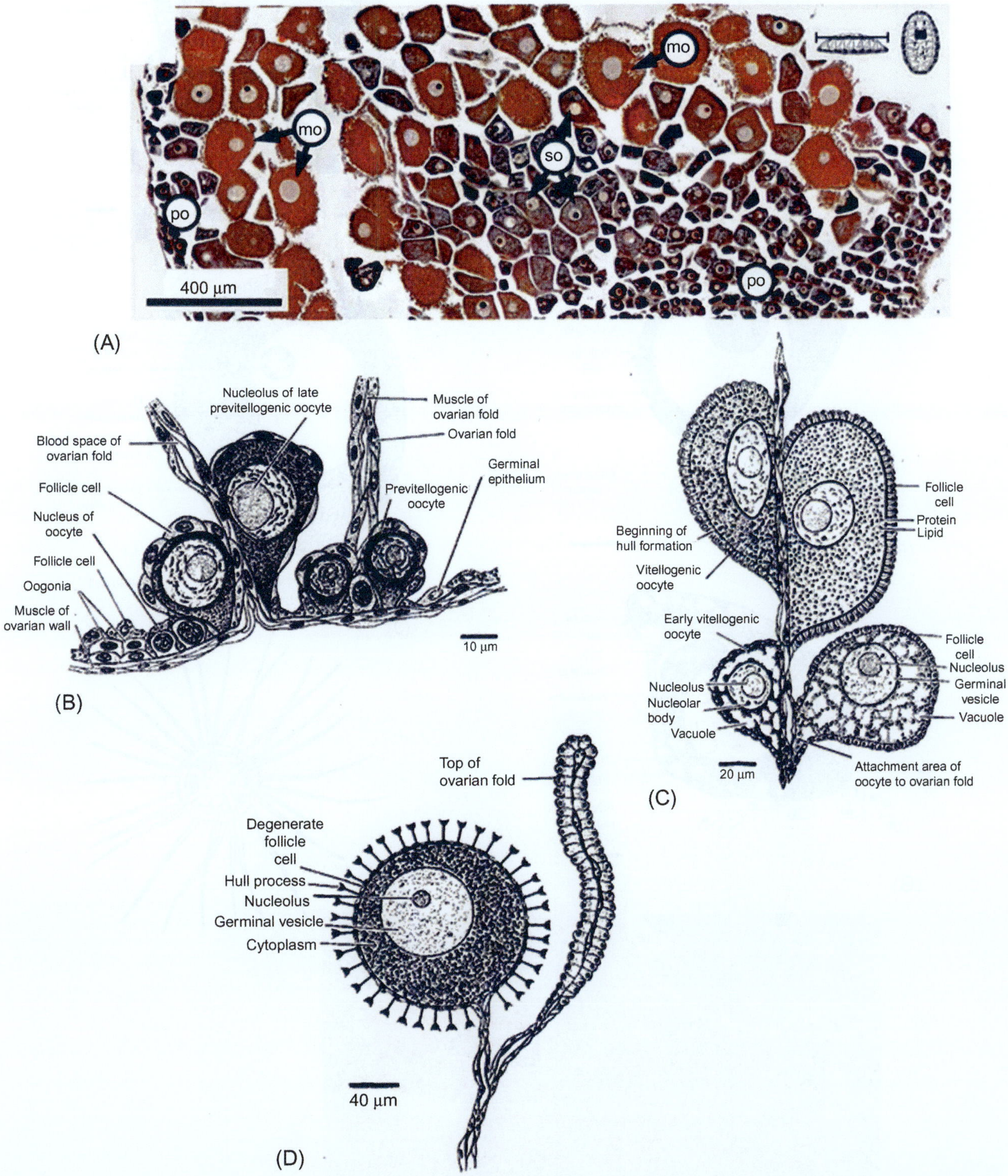

FIG. 11.6 (A) Composite digital image in frontal section of the anterior dorsal part in a female *Chiton articulates*, with the ovary in the **oogenesis** stage (S1) showing asynchronous groups of oocytes, i.e. primary oocytes, secondary oocytes, and mature ova without a dominant group. po = primary oocytes, so = secondary oocytes, mo = mature ova *(from Avila-Poveda and Abadia-Chanona, 2013)*; (B–D) schematic drawings of oocytes in different stages of development along the ovarian folds of *Chiton septentriones* (Ashby) (see Fig. 11.4B): (B) previtellogenic oocytes near the base of the ovarian folds; (C) vitellogenic oocytes along the middle portion of an ovarian fold; (D) full-grown oocyte near the top of an ovarian fold, note different scale bars *(after Selwood, 1968, from Pearse, 1979)*;

(Continued)

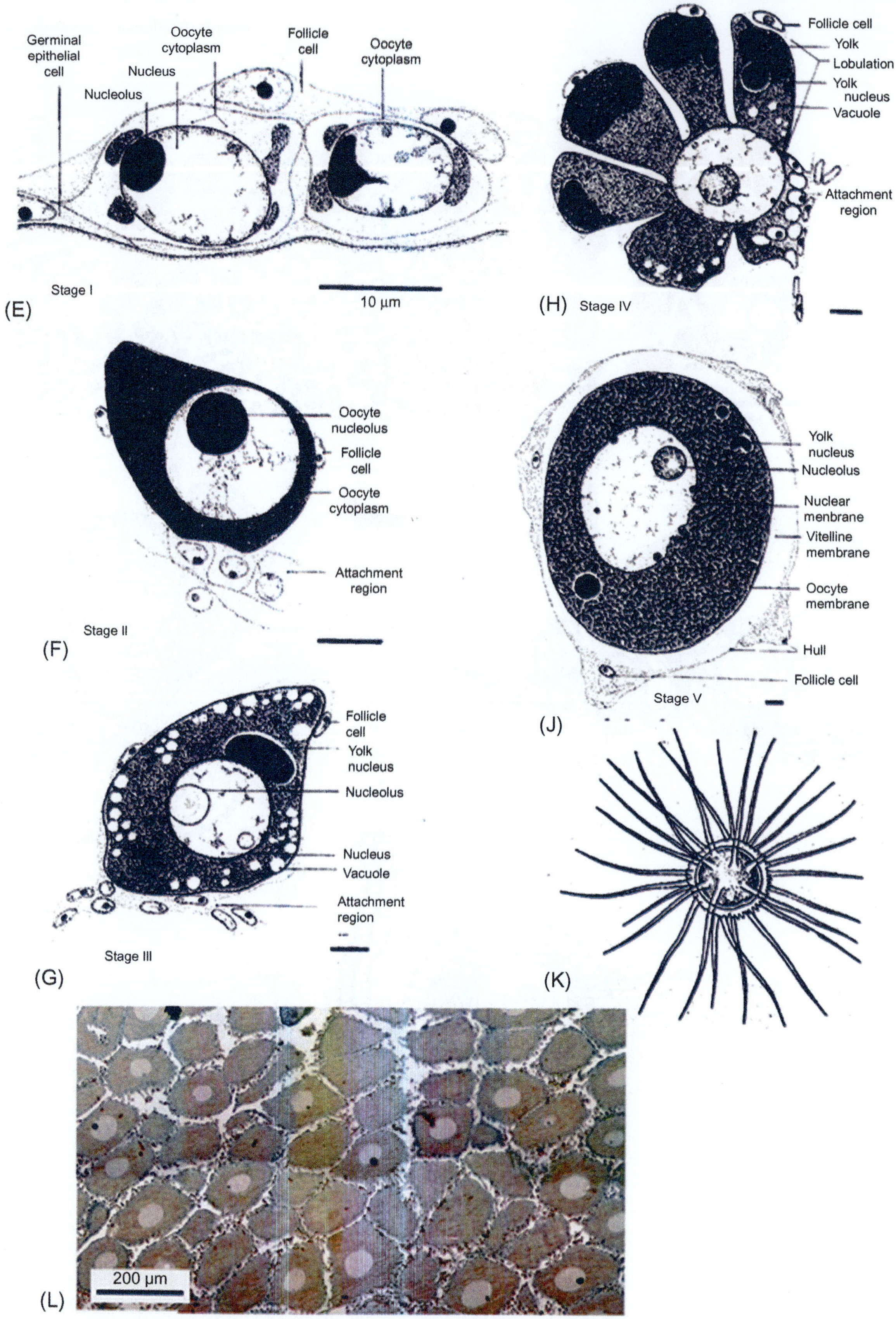

FIG. 11.6, CONT'D (E–J) schematic drawings of oocytes of *Lepidochitona cinereus* (Linné) in different stages of development: (E) previtellogenic oocytes embedded in the germinal epitelium; (F) previtellogenic oocyte protruding from the germinal epithelium; (G) previtellogenic oocyte with yolk nucleus and vacuoles; (H) vitellogenic oocyte showing characteristic lobulations; (J) full-grown oocyte free in the ovarian lumen and surrounded by the hull. Bars = 10 μm, note the different sizes *(after Richter and Götting, 1974; from Pearse, 1979)*; (K) egg of the amphineuran, *Liolophura japonica* Lischke (Taki), the chorion assumes a highly characteristic structure *(from Wada, 1968)*; (L) Mature Chiton female. Composite digital image in transverse section of an ovary of *Chiton articulatus* in mature stage (S2) showing only a synchronous group of mature ova without replenishment by earlier oocytes *(from Avila-Poveda and Abadia-Chanona, 2013)*;

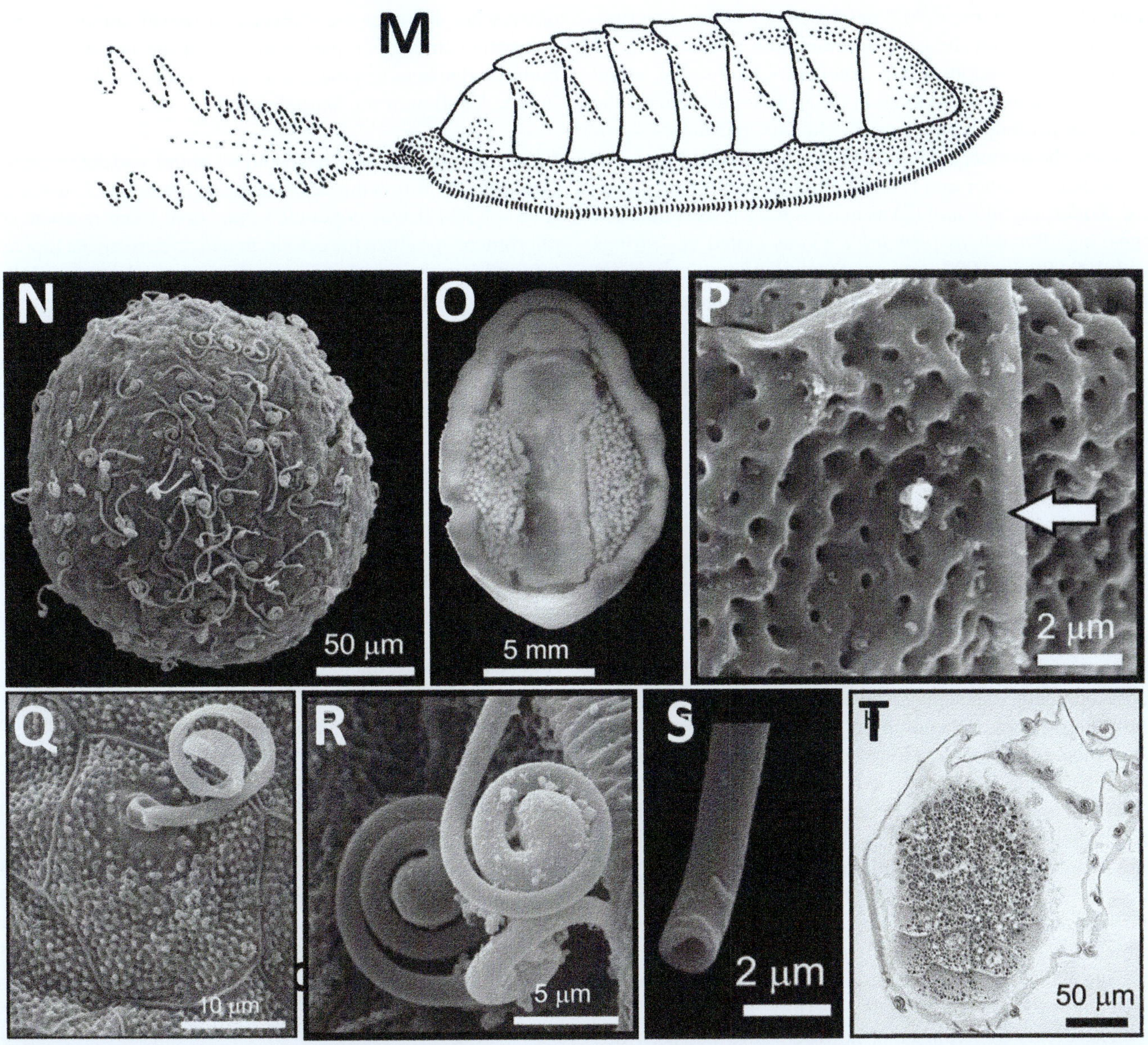

FIG. 11.6, CONT'D (M) Drawing of a spawning male specimen of *Lepidochitona cinereus* (Linné). *(After Mathews, 1956, from Pearse, 1979)*; (N–T) ultrastructural study of the egg-hull of *Ischnochiton stramineus* from Malvinas/Falkland Islands: (N) SEM showing general view of an egg; (O) reflex image of ventral view of adult specimen showing compact mass of eggs stored in the pallial groove; (P) SEM showing detail of micropores and junction between plates, *arrow* points to the rounded ridge; (Q) SEM showing details of a polygonal egg plate; (R) SEM showing details of the spiral egg hull projections; (S) SEM showing details of a hollow broken projection; (T) Optical microscope image showing histological section of a developing embryo showing curled egg hulls *(from Liuzzi and Zelaya, 2013)*.

Abadia-Chanona added that a study of reproductive biology of *C. articulatus* from Oaxaca, Mexico is ongoing, and histological observations reveal no hermaphroditism for specimens of total length 'TL' greater than 40 mm, i.e. adult stage (Abadia-Chanona, personal observation).

Reproductive behaviour and spawning. Pearse (1979) reported that there is no evidence of sexual pairing during spawning in chitons. However, it was reported that a small number of the hermaphroditic, brooding chiton *Cynoplax dentiens* were grouped together during the breeding season. However, Chitons sometimes aggregate and simultaneously spawn (e.g. the giant gumboot chiton, *Cryptochiton stelleri*) (Eernisse, 2007). Individuals of several species of chitons have been reported to become active just before spawning, in the laboratory. Many different workers interested in this field noted that these animals tended to position themselves vertically in aquariums just before spawning. Immediately before spawning, the girdle arches upwards to form a spout at the posterior end, or on both sides near the posterior end, and then the gametes are expelled in one or two streams

(Fig. 11.6M). Thorpe (1962) described egg laying as a two-stage process with the eggs accumulating in the pallial or mantle groove before being expelled. Chitons that brood their eggs merely retain them in the mantle groove until the larvae or juveniles are released. Mucous, presumably produced in the oviducts, often surrounds the expelled eggs and holds them together as: (1) small clumps of a few eggs each e.g. *Katharina tunicate;* (2) as masses with several hundred eggs, e.g. *Plaxiphora primordia*; (3) as coiled egg strings with 100,000–200,000 eggs, e.g. *Stenoplax heathiana;* or (4) as long egg strings reaching nearly 1.5 m in length, which probably contain millions of eggs, e.g. *Cryptochiton stelleri* (Eernisse, 2007). In some species, apparently, little or no mucous coats the eggs. The mucous surrounding the egg masses usually does not last long. The egg strings of *Cryptochiton stelleri* were soon broken into fragments by waves.

Early development

There are many records on the development of different species of polyplacophorans. Pearse (1979) reviewed such previous studies and reported that all these have free-swimming larvae. And although individuals of *Middendorffia polii* and *Lepidochitona cinereus* (rarely) brood their eggs, the embryos are released when they hatch as trochophores. Developmental rates may vary considerably even within the same species, depending on environmental conditions, especially temperature. However, the development of all species of chitons studied is remarkably similar and varies mainly in temporal details. Pearse (1979) established a table concerning the timing from the first cleavage to settlement, including the references. Fertilization occurs after the eggs are spawned, probably often while they are within the pallial groove. During the spawning of *Chiton tuberculatus,* for example, a thick gelatinous stream of sperm was observed passing under the girdle of a female, and then numerous eggs emerged surrounded by the sperm.

The events of fertilization, however, have not been described in detail for any species of chiton. It is unclear how the sperm penetrate the egg hull, especially since they apparently lack acrosome. Miller (1977) reported that chitons have chemotactic sperm, as also found in hydroids and ascidians. Sperm of a variety of chitons including *Katharina tunicata* and some other species were attracted to both seawater extracts (egg water) and alcoholic extracts of egg without agglutinating. Presumably, this behaviour aids sperm in locating eggs. However, unlike those of hydroids and ascidians, chiton sperm showed a complete absence of species specificity. Moreover, it was found that hybrid larvae could be formed from gametes of *Katharina tunicata, Mopalia mucosa*, and *Tonicella lineata,* but these hybrids have not been raised to maturity.

During fertilization, a space is formed between the egg hull and the cell membrane. This space is attributed to shrinkage of egg cytoplasm. Spawned eggs may remain viable for 24–40 h after release. The germinal vesicle breaks down when the eggs are spawned. However, meiosis is not completed until after fertilization, and the resulting polar bodies lodge between the egg membrane and the hull, marking the location of the animal pole.

Cleavage. Holoblastic, nearly equal, spiral, and determinant cleavage. By following cell lineage of *Stenoplax heathiana* (Fig. 11.7A) it was concluded that: first, three quartets of micromeres produce the ectoderm and its derivatives (apical tuft and larval eyes, nervous system, stomodaeum, radula, foot glands, shell valves). A 4d cell forms from one macromere, the endomesodermal cell, which produces most of the mesodermal derivatives (muscles, heart, kidneys, and gonad). Remaining of the fourth quartet (4a–4c) and macromeres form the digestive tube (pharynx, oesophagus, stomach, intestine, and their glandular diverticulum).

Gastrulation. Gastrulation (Fig. 11.7B) mainly occurs by invagination. Firstly, the macromeres invaginate, followed by the endomesodermal cell, which lodges near the blastopore and divides into 2 to form a pair of posteriolateral cells. Gastrulation by delamination was also evidenced. Endomesenchyme cells proliferate, partly filling the blastocoel. The early formation of paired coelomic pouches (mesoderm) in *Middendorffia polii* was described by Kwalevsky (1883), but in the other species this formation could not be detected until after metamorphosis. The position of the blastopore shifts from the posterior pole anteriorly along the midventral surface, as a consequence of rapid proliferation of the posteriodorsal ectodermal cells, until it is located on the midventral surface (Fig. 11.7B (a) and (b)). It sinks inwards in front of the stomodeual invagination and nearly closes. At the anterior pole of the embryo, micromeres form the tetraradiate molluscan cross, alternating with four clusters of trochoblasts and capped with the apical and rosette, which form the annelidan cross. The apical cells become the sensory cells bearing the apical tuft (Fig. 11.7C), while the trochoblasts form the prototroch, consisting of two rows of ciliated cells. The prototroch divides the embryo into pretrochal and posttrochal regions. The former, with apical tuft, is mainly larval, while the latter will form the definitive adult (Fig. 11.7D). With the formation of the prototroch, the embryo moves around within the egg hull and is nearly ready to hatch.

Larval development

The embryo hatches when the egg hull splits open by unknown means. The embryo pushes through the hull opening as a free-swimming trochophore larva, using rapid body flexions. Most chiton embryos hatch as trochophores (or as pseudotrochophores) before the appearance of the foot, shell valves, or larval eyes. But in some species (e.g. *Cryptochiton stelleri*), these structures are well developed by the time of hatching. The trochophore larvae of chitons do not have protonephridia, but they do form the radula sac,

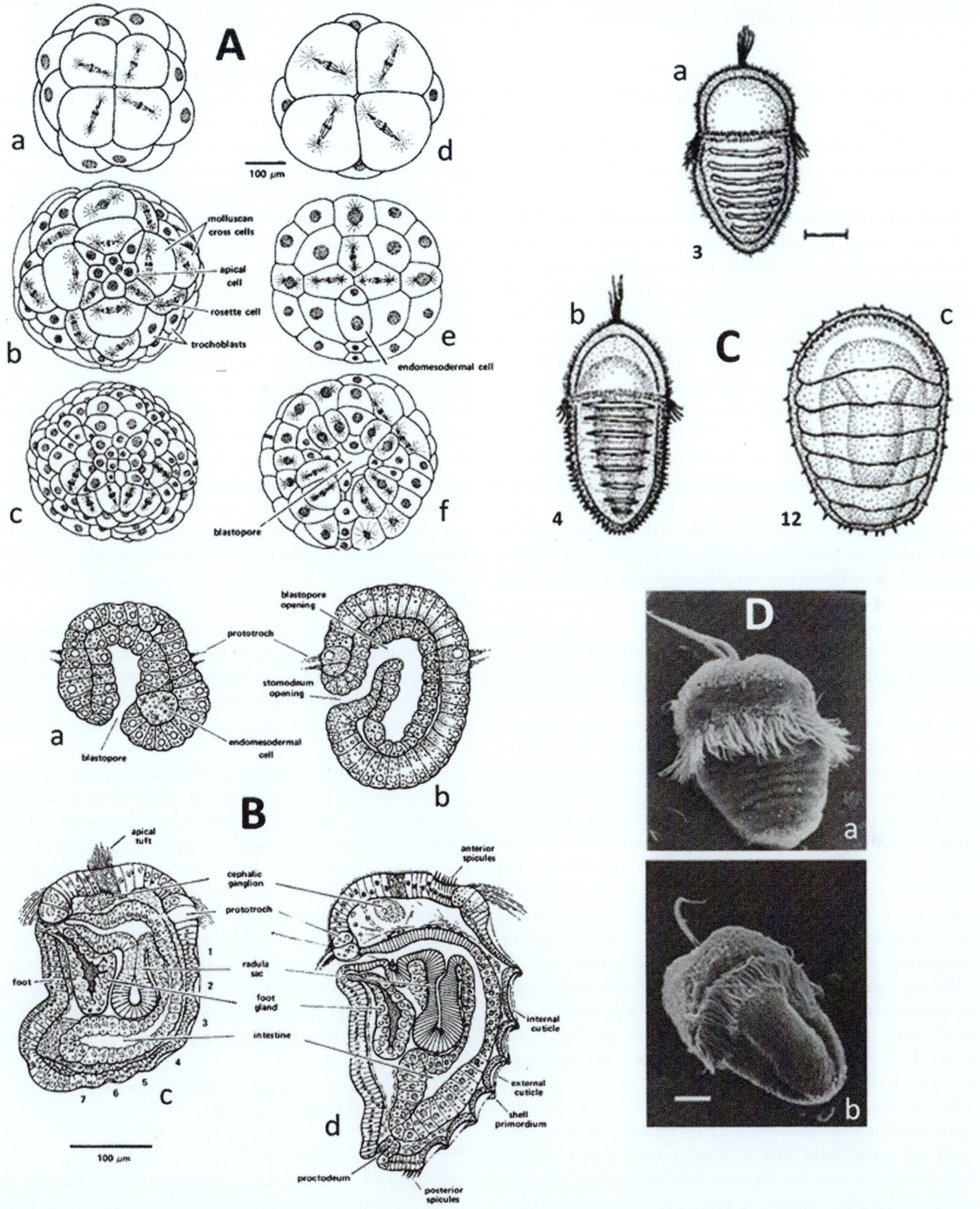

FIG. 11.7 (A) Drawings illustrating early cleavage in *Stenoplax heathiana* Berry showing development of the molluscan cross, the appearance of the endomesodermal cell (the 4d cell,), and the beginning of gastrulation. (a–c) Anterior (animal) view, (d–f) posterior (vegetal) view *(after Heath, 1899)*; (B) drawings of L.S.s of embryos and a larva of *Middendorffia polii* (Philippi) showing development of the main organ systems: (a) late gastrula with prototroch and beginning of the ventral displacement of the blastopore; (b) ventral displacement of the blastopore and stomodeal invagination complete; (c) embryo near the time of hatching with apical tuft, cephalic ganglion, radula sac, foot gland, ventral foot, and dorsal spaces marking the position of the shell valves (numbered 1–7); (d) larva near the time of settlement after loss of the apical tuft *(after Kwalevsky, 1883)*; (C) drawings of the dorsal appearance of the larva at hatching (a), and larva at settling (b), and juvenile shortly after settling of *Cryptochiton stelleri* (Middendorff), numbers give the age in days after fertilization. Bars = 100 μm; (D) SEM of *Katharina tunicata* (Wood): (a) dorsal view of 3-day-old larva showing indentations separating the future shell valves; (b) ventral view of 8-day-old larva showing a well developed foot. Bar = 50 μm (originals by F.S. Chia and R.M. O'Clair, University of Alberta). *All adapted and/or from Pearse (1979).*

pedal gland (Fig. 11.7C (a)), and other structures characteristic of molluscs.

The free-swimming trochophore stage, in general, is very short, lasting several hours or, at most, 2–3 days.

Larval eye. Pigmented cells form the two temporary larval eyes located laterally just behind the prototroch. Their photosensitivity or function has not been demonstrated, but larvae are sensitive to light, being photopositive before settling and photonegative during and after settling. The eyes gradually become indistinct after settlement, when they are covered by the expanding mantle and first or second shell valves, but they may be retained in the lining of the pallial groove well into adult life.

Foot and pedal gland. As the larva develops, the dorsal and ventral surfaces become distinct (Fig. 11.7B). The ventral surface forms the foot with tall cells bearing short cilia. At its end, the larva begins to creep on the bottom with the developed foot. Near the posterior lip of the stomodaeum, the larval gland forms as a simple invagination, and in some species, this apparently multiplies to form numerous mucous secretory glands in the adult. Together with the degeneration of the apical tuft and prototroch, the posterior part of the body grows and assumes the adult form.

The shell. One of the major distinguishing characteristics of adult chitons is the possession of a shell consisting of eight overlapping plates. The rudiments of these are also seen in the larvae. About the time when the larva changes to a bottom-crawling life, a thin cuticle forms dorsally, and transverse rows of clear cells mark the division of the shell valves. The clear cells probably secrete the cuticle, and small cells between the rows form the shell valves. In most species, the first seven shell valves form nearly simultaneously as irregularly shaped calcareous deposits, and the tegmentum forms before the articulamentum (for more details, see Pearse, 1979). The first valve forms at the level of the prototroch. The eighth shell valve may form nearly at the same time as the first seven valves (e.g. *C. olivaceus*), 1–2 days later when the larva settles, or much later, e.g. 3 days after settlement in *C. stelleri* (Fig. 11.7C (a)), or 5 days after settlement in *Tonicella lineata*, or 5 weeks after settlement in *Mopalia lignosa*. This means that the eighth shell valve may form after settlement, but the exact timing differs with the species. Hyman (1967) speculated that the initial seven plates maybe a 'phylogenetic reminiscence,' since the oldest known fossil chiton had only seven shell valves.

Considering the process of formation of the shell (secretion) itself, Borradaile et al. (1977) reported that, characteristically, the ectodermal epithelium of the mantle secretes the shell in the Mollusca, and in most of them the method is the same. The original shell is laid down by the mantle of the veliger larva (Fig. 11.2B). All extension takes place by secretion at its edge (Fig. 11.2C). The outer shell layer, the periostracum, formed of horny conchiolin, is first produced in a groove, and then the prismatic layer, largely consisting of calcite or arragonite, is secreted underneath it by the cells of the thickened edge. The innermost **nacreous layer** (also mostly calcium carbonate) is, however, formed by the cells of the whole of the mantle. Under such conditions, as occurs in the formation of pearls, this general epithelium is capable of secreting any of the three shell layers (Fig. 11.2D).

With their dorsal shell valve primordium, ventral foot, prototroch, and apical tuft, chiton larvae have a very characteristic appearance (Fig. 11.7D). They are mainly if not wholly, nonfeeding larvae, serving in short-range dispersal. However, if the larva is supplied with enough, and good, food, it develops more rapidly. The free-swimming life of the larva is relatively short, ranging from a few hours to a few days. Pearse (1979) mentioned the life span of the larvae of different species in plankton in his tables.

Settlement and metamorphosis

Settlement is gradual and, in the laboratory, the larvae may spend time dropping to the bottom of the dish and creeping, and then swimming up in the water again. There is some evidence of substrate selection. Pearse (1979) presented a review of a number of species of chitons and their substrate with their full references.

Settlement and metamorphosis occurred even when the anterior third or posterior portions of the larva were surgically removed. Without suitable substrate, larvae delayed settling for nearly a month. At settlement, the protoroch is lost (Fig. 11.7C (c)), and its disappearance marks the beginning of juvenile life. Metamorphosis mainly involves growth; some authors have suggested that there is no metamorphosis in chitons. In chitons, the larval pretrochal region gradually diminishes, as the region anterior to the first shell valve forms the anterior portion of the mantle and overgrows the mouth. Within a short time after settlement, the animal has the appearance of a tiny chiton (Fig. 11.7C (c)).

There is some evidence of substrate selection, i.e. each larva selects its suitable substrate.

Brooding and ovoviviparity

Although free spawning is most common, the females of about 5% of all chiton species brood their eggs instead (Fig. 11.8A (a)–(d)), with embryonic and larval development completed within the pallial groove of the brooding mother (Eernisse, 2007), sometimes with embryos sticking together in rod-shaped broods (Fig. 11.8A (c)). This justified the previous view, which stated that a relatively large number of chitons brood their embryos in the pallial (mantle) groove. Approximately equal numbers of embryos are held on each side of the foot, and often

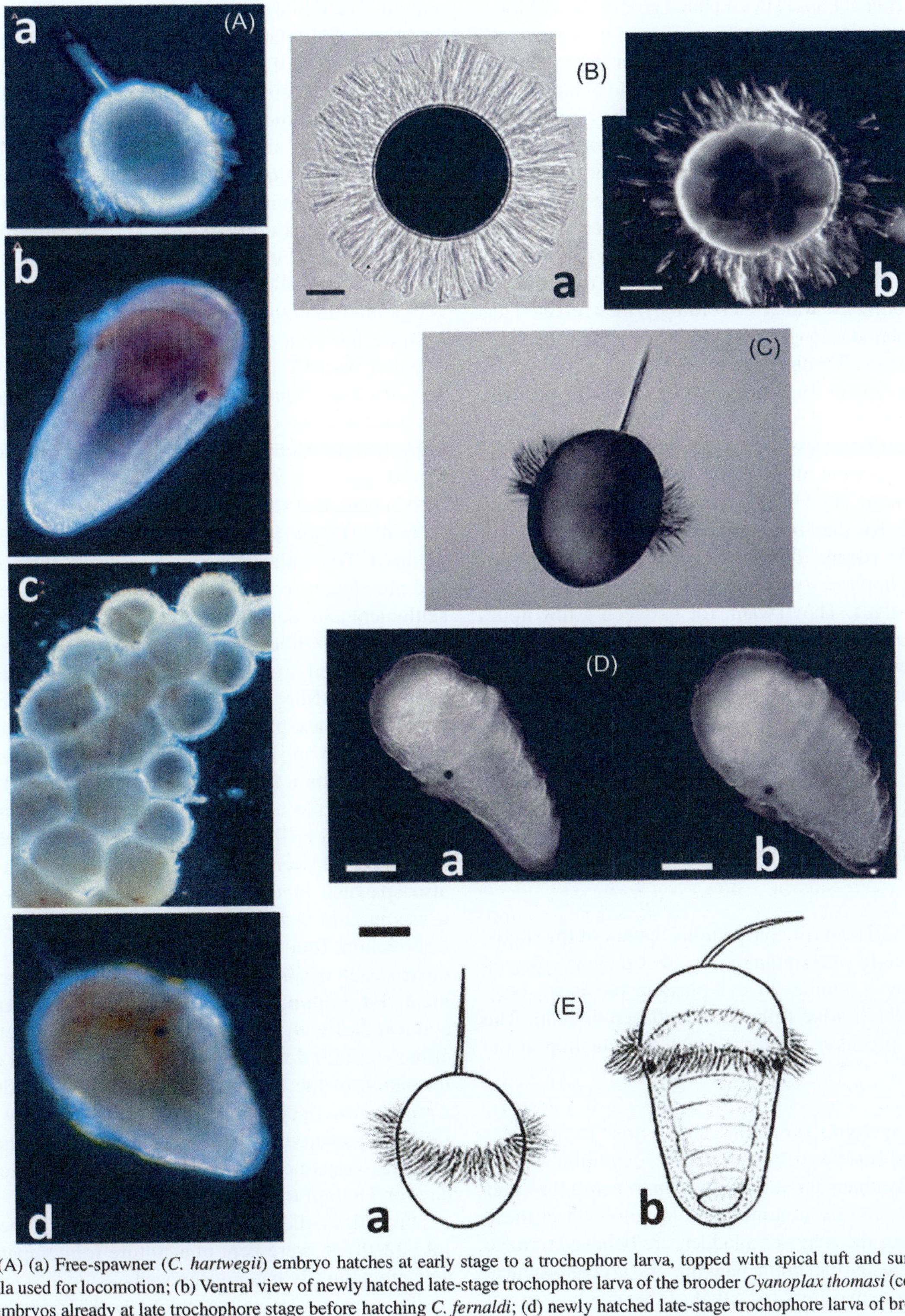

FIG. 11.8 (A) (a) Free-spawner (*C. hartwegii*) embryo hatches at early stage to a trochophore larva, topped with apical tuft and surrounded by prototroch flagella used for locomotion; (b) Ventral view of newly hatched late-stage trochophore larva of the brooder *Cyanoplax thomasi* (compare with (a)); (c) brooded embryos already at late trochophore stage before hatching *C. fernaldi*; (d) newly hatched late-stage trochophore larva of brooder *C. fernaldi (Eernisse, 2007)*; (B) Early development of *Cryptochiton stelleri*: (a) Fertilized egg with surrounding hull; (b) 16-cell stage with spiral cleavage evident. Bar = 100 μm; (C) Two-day-old trochophore larva of *Cryptochiton stelleri* just after hatching. Note the lack of dorsal plates or bumps that are the precursors to plate formation. There is no elongation of the posttrochal region, and the eyes and foot are yet to develop. Bar = 100 μm; (D) Late trochophores of *Cryptochiton stelleri*. Dorsal side is facing the upper right corner of photographs: (a) elongation of posttrochal region is visible as well as some dorsoventral flattening, foot and seven plate precursors are evident underneath body surface, yolk is still visible; (b) The seven bumps that are plate precursors are visible on outer surface of larval body, as in (a) eyes and prototroch are present. Bars = 100 μm; (E) Development of *Cryptochiton stelleri* observed in Oregon in the study of Lord (2011): (a) just after hatching (no plate formation or posttrochal elongation); (b) after settlement (shell plate rudiments are plate precursors as no calcification has occurred. Bars = 100 μm. *((B–E) From Lord (2011).)*

nearly fill all available space in the pallial groove. In addition to these brooding species, *Callistochiton viviparous* is apparently ovoviviparous, according to Plate's findings in 1899. About 15 embryos, some with 7 shell valves, were found in the ovary of a single specimen. In a more recent work, Liuzzi and Zelaya (2013) studied the hull of the egg of *Ischnochiton stramineus* (see above) and presented some information on brooding in chitons. They recorded and tabulated that, at the time of their work, the number of known species that brood eggs/juveniles in the pallial groove is 36, *I. stramineus* being one of them. They also recorded that Plate (1899) concluded that brooding only occurs during the winter. Accordingly, they revealed that the brooding period is longer than previously assumed, at least for their specimen. This is because of the finding of a specimen brooding eggs in August, and another carrying juveniles in December.

Pearse (1979) reviewed this brooding in chitons and reported that, in view of previous authors, it is an obligatory habit. Kwalevsky (1883) found that in *Middendorffia polii*, the embryos died or developed abnormally when not brooded by the parent. The young of this species were released as trochophores with a prototroch, but with developing shell valves. They swam, for between a few hours and a few days, in culture dishes before settling. Most brooding species, and the single known ovoviviparous species, apparently retain the embryos until tiny juvenile chitons, with seven or eight fully formed shell valves, are released.

Later development (Wada, 1968)

The eyes. The larval eyes disappear with the completion of the mantle and shell.

The aesthetes. The most conspicuous feature of the shells, which are of eight overlapping plates that they are pierced at the sides by a number of cytoplasmic processes (aesthetes), and which arise from the mantle epithelium. The ends of these processes perform a sensory function at the shell surface.

The nervous system. The brain ganglia arise from the radially arranged rosette cells. The paired longitudinal pleural nerve cords characteristic of the Amphineura, the pair of nerve cords, and the commissures, which connect these transversely like the rungs of a ladder, are believed to have the same origin as the brain ganglia. As a consequence, this nervous system differs from that of the annelids, in not showing metameric structure. Nervous system in adult differs from other molluscs.

The head. Furthermore, the head region is very poorly developed, without tentacles, except for some species which have tentacle like processes on the mantle in front of the mouth. The labial palp, which is formed around the mouth, and the subradular organ, which is formed in the stomodaeum in front of the radula sac as an invagination of the stomodaeal epidermis, are also thought to be sensory organs.

There is a somewhat recent study of larval development, metamorphosis, and early growth of the gumboot chiton, *Cryptochiton stelleri* (Middendorff, 1847). It is a common intertidal inhabitant on the western coast of North America, from central California to Alaska, and westward to northern Japan. It is up to 36 cm long, the largest chiton species in the world (Lord, 2011). This author collected adults from the rocky intertidal zone at South Cove, Cape Arago, Oregon. Despite the abundance of this species, small individuals (smaller than 15 cm long) are rarely found. The author observed adult *Cryptochiton stelleri* spawning naturally (without artificial cues) in the spring of both 2009 and 2010, and described the method. Four chitons were observed with released eggs near the paired gonopores in the pallial groove. This means that the author did not observe their spawning naturally. In the laboratory, these eggs were isolated and cultured. The author described, in detail, the establishment and maintenance of the culture, and the attempts to induce settlement and metamorphosis.

In another line of work, male and female *C. stelleri* were collected from the same above-mentioned site, and cultured. On May 6, 2010, the animals spawned and continued to release gametes intermittently for 2 days. The author observed and described the process of spawning in the laboratory. Some females raised their posterior ends while spawning, while others lay entirely upside down to release or showed no visible difference in behaviour while spawning. The eggs were fertilized with sperm from males that also spawned in the laboratory. It was not clear whether it was natural or artificial fertilization. About 2 days post-fertilization, four treatments were set up, with four culture dishes, each of 50 larvae in filtered seawater, for each treatment: **1st**, nothing was added to the filtered seawater, as a control; **2nd** with a small rock covered with encrusting coralline algae; **3rd** with encrusting coralline algal extract (the formation of the latter was described); **4th** with extract of *Cryptopleura* (a delicate leafy red algae eaten by juvenile *C. stelleri*), prepared as an extract of the 3rd treatment. Larvae from each treatment were examined microscopically once a day for 1 month to observe metamorphosis.

Juveniles collected at Cape Blanco (not Cape Arago) on July 22, 2009, were kept in a culture for comparison, since they are naturally metamorphosed.

The green eggs were stuck in the mucous covering the gills, towards the posterior end of the pallial groove. Both the eggs and sperm sank in seawater. The eggs were yolky, indicated by their opaque appearance, but the hull obstructed a clear view of the embryos. Fertilization was observed shortly after gametes were mixed together in culture (Fig. 11.8B (a)). Cleavage is total, spiral (Fig. 11.8(b)), began 4h 40min postfertilization. Subsequent divisions occurred ~ every 40min. The author tabulated times from the stage of mixing gametes to metamorphoses of the larva in *C. stelleri*. By 27h postfertilization, four distinct patches of cells had developed cilia. The eggs remained on the bottom of the cultural dishes until hatching. At 46h the early trochophore hatched and was characterized by an apical tuft and prototroch, which propelled the bright green larvae rapidly around the culture dish (Fig. 11.8C). After the next 2days, the posttrochal region elongated and a pair of red larval eyes appeared behind the prototroch. The foot developed as some dorso-ventral flattening (Fig. 11.8D (a)). At about 5days after fertilization (3days posthatching), the larvae developed seven shell plate rudiments and settled to the bottom of the culture dishes where they began creeping (Fig. 11.8D (b)).

The larvae of *Cryptochiton stelleri* described in this study by Lord (2011) are similar to those of most of the other described species of nonbrooding chitons. They have a short life span and are nonfeeding. The larvae look like those of other chitons, but slightly larger: 300μm at hatching (Fig. 11.8C). The appearance of only seven plates is consistent with other chiton species, since many chitons do not develop the eighth plate until well after metamorphosis.

In controls with only filtered seawater, no further development occurred, with loss of the prototroch and no plate formation (shell). This means that the larvae have no cues for metamorphosis. Some different trials occurred by the addition of different kinds of, for example, seaweeds to the culture. Metamorphosis was never induced and the larvae died in these and in the controls. On the other hand, the embryos cultured in May 2010, developed on a similar timeline, hatched 2days after fertilization. The trochophores exposed to the 3rd treatment (algal extract) and coralline algae-covered rock began metamorphosing 3days after hatching (5days postfertilization). Metamorphosis occurred after the elongation of the posttrochal region and the formation of the rudiments of the seven shell plates. The larvae were found, during this period of time, on the bottom of the dishes (Fig. 11.8E). In the control and in 4th treatment, the larvae did not metamorphose for over 2 months, and metamorphosis could still occur after this duration.

Metamorphosis was triggered solely by the addition of coralline algal extract, which has also been shown to induce other species metamorphosis. It is also possible that a coralline algal cue ensures that metamorphosis occurs at a low tidal level, where encrusting coralline algae are found. The process itself involves: (1) the loss of the prototroch and apical tuft; (2) beginning of the formation of seven shell-plates. The plates began as thin lines and slowly expanded until they covered the body of the juvenile. Shell plates continued to develop and widen as juveniles grew.

Monoplacophora

The members of this molluscan subclass were known only as fossils from the Cambrian to the Devonian periods until 1957, when the Danish 'Galathea Expedition', dredging off the Mexican west coast, made a rich haul at station 716 (Lemche, 1975). The many interesting animals found there included 10 live specimens of *Neopilina galatheae* and three extra shells from a collection at abyssal depths (3600m) of the Pacific Ocean, off Costa Rica. Lemche described this animal as bilaterally symmetrical with 'internal metamerism' (Fig. 11.9A and B) and with an external piece of shell much like that of a limpet (Fig. 11.9D). This 'shell covers the pallium and extends all over the dorsum', and the anus is postero-median. One of these specimens had kept a dextrally coiled larval shell of about 0.15mm in diameter, and forming one and a half whorls, of the type characteristic of pelagic larvae (Fig. 11.9E). Internally, there is a well-developed coelomic cavity. The nephridia open at the base of the gills and the gonad products pass to the exterior via some of the nephridia (Fig. 11.9C). The sexes are separate. The animals show a strongly marked metamerism and it is problematic whether this resembles the metamerism seen in the arm of the starfish, or the coelomic segmentation seen in the annelids. Subsequently, some other specimens were discovered, but all were very similar and were placed in the genus *Neopilina*. Because of the abyssal habitat of these living monoplacophorans, it has been difficult to secure material for study, and what is known about reproduction is primarily inferred from the descriptions of the anatomy and histology of *Neopilina galatheae* by Lemche and Wingstrand (1959). This latter work includes a table reviewing the natural condition and external features of the 10 specimens of Lemche (1957).

The discovery of more species of monoplacophorans was continued and, in 2013, Haszprunar and Ruthensteiner reported that more than 30 extant species were known, reflecting a radiation starting in the Mesozoic, whereas the origin of the clade certainly lies deep in the Palaeozoic and probably occurred already in the Lower Cambrian. These authors gave a review of the current state of knowledge on the Monoplacophora and discussed and resolved a number of questions related to this taxon, especially the seriality of some organ systems. One of these newly discovered species is *Laevipilina theresae* in 2006 by Schrödl. This author found this new monoplacophoran in a collection collected off Kapp Norwegia, Eastern Weddell Sea, at approx. 800m depth. This *Laevipilina theresae* is named after the author's daughter and it is the third species known from Antarctica. It is unique in having its apex clearly surpassing the anterior shell margin and showing two concentric bulges around the base of the apex. Schwabe (2008) presented a summary of literature records of Monoplacophora from below 2000m. Reports have been published of 11 described species of monoplacophorans from abyssal and hadal depths,

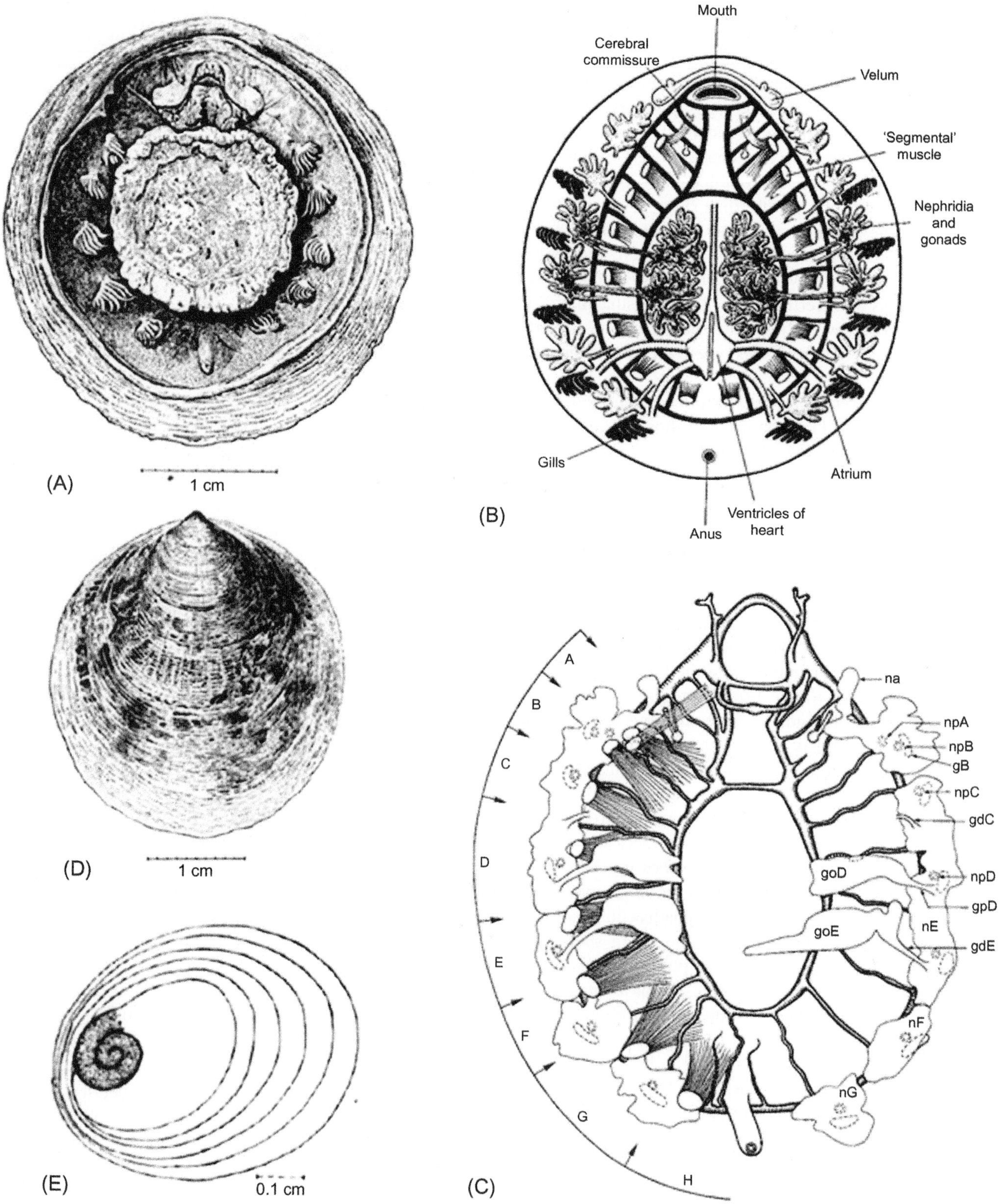

FIG. 11.9 *Neopilina galatheae*: (A) ventral view of holotype (specimen 1); (B) diagram of the segmented organs; (C) *Vema ewingi*, specimen 2, dorsal view, Immature specimen. Graphic reconstruction of nephridial complex, nephridiopores, outline of immature gonads and gonoducts. Gill bases, nervous system and muscles added for orientation. The sector sequence is indicated to the left. Dorsal view. A–H = pedal retractors A to H, gB = base of gill B, gdC-gdE = gonoducts of sectors C to E, goD and goE = gonads of sectors D and E, na = anterior nephridial diverticulum (probably referable to sector A), nE-nG = separate nephridial sacs of sectors E to G, npA-npD = nephridiopores A to D, those of E to G have no lettering; (D) dorsal view of the shell of the holotype; (E) apical part with larval shell of specimen 3. *((A, D, and E) From Lemche (1957); (B) from Borradaile et al. (1977), after Lemche and Wingstrand (1969); (C) from Wingstrand (1985).)*

as well as several records of deep water species of uncertain taxonomic position in both classes, Monoplacophora and Polyplacophora. In addition to their description, their occurrence and geographic distribution are briefly discussed (Schwabe, 2008). Although, more detailed information on monoplacophorans' morphology, anatomy, ecology, and their phylogenetic position (Lemche and Wingstrand, 1959 and others) are available, good material for embryological and molecular studies remains unavailable.

However, Wingstrand (1985) in his later work concentrated on points where the original description was insufficient or directly wrong. He also paid particular attention to the species *Vema ewingi* (Clarke and Menzies, 1959), which differs from *Neopilina galatheae* in the metameric repetition of organs. *Vema* has a more complete set of metameric organs, including six pairs of gills, seven pairs of nephridia and three pairs of gonoducts (although one is vestigial), but it still has eight pairs of retractors. He registered the names of the 10 new recent species which have been included in class Monoplacophora, family Neopilinidae, since Lemche's original description of *Neopilina galatheae* appeared in 1957. With respect to the reproductive system, he reported that the gonads of all the new specimens are immature. But in *Neopilina galatheae* and *Vema ewingi,* gonadal rudiments are visible as dark masses of small germinal nuclei, situated along the bottom of the perivisceral cavity, below the intestine. The walls of the gonads are thin even in the large mature specimens. The reconstructions of the gonads in the new specimens, therefore, only show the outline of the clusters of dark germinal cells (Fig. 11.9). As in the mature specimens of *N. galatheae*, the immature specimen has two pairs of gonadal rudiments, and two pairs of gonoducts that pass out to the pallial fold closely behind the corresponding pedal retractors, respectively. In the pallial fold each gonoduct opens into the nephridium of the same sector: the duct widens into a funnel-like terminal part that fuses with the dorsal or dorsomedial wall of the nephridial sac.

Aplacophora

The aplacophorans are simplified forms of molluscs that have many worm-like characteristics; however, they maybe differentiated from the annelids in that they show no trace of segmentation. In addition, they have a molluscan type of coelom (pericardium, gonocoele, and renocoele, Fig. 11.3). In addition, they sometimes possess a radula (Fig. 11.2F). The foot is absent or represented by a median ridge in a ventral groove and the mantle is correspondingly enlarged. There are no shell plates but only spicules. The mantle cavity may be represented by a small cloacal chamber at the posterior end, gills present (*Chaetoderma*) or absent (*Neomenia*) (Fig. 11.2F) (Borradaile et al., 1977).

Most aplacophorans are small (1–5 mm) and inhabit the deep sea, where they are not easily obtainable for developmental studies. *Epimenia babai* can reach 30 cm in length and lives in relatively shallow water, from 20 to 70 m in depth (Salvini-Plawen, 1997). Okusu (2002) reported that the neomenioid aplacophoran *Epimenia babai* from Japan is hermaphroditic and is considered to have a unique larval form among neomenioid aplacophorans in that it lacks the true test structure characteristic of neomenioid pericalymma larvae. Because of its unique development, the larval form of *E. babai* has been suggested to be an intermediate form between pericalymma and trochophore larvae. Okusu mentioned that this species was primarily described by Baba (1938, 1999), but details of morphology and cell fate during embryogenesis were not studied. So, Okusu (2002) reexamined the early and late development of this species, *E. babai*, to elucidate some of the unknown developmental features of a neomenioid aplacophoran, such as the morphology and fate of the pericalymma test.

Okusu collected adult specimens of *E. babai,* offshore near Amakusa Marine Biological Laboratory (Kumamoto, Japan) in 1999 and 2000, during the most active breeding period, between June and September. These specimens were kept and maintained carefully and perfectly in the laboratory with running seawater and at 26°C. The animals spawned spontaneously several hours after sunset or were induced to spawn by turning off the light after 12–14 h of constant light. Three individuals of *Epimenia babai* laid ~20,000 eggs over 90 consecutive nights in the laboratory. Although the time of sperm transfer is not known, fertilization is assumed to have occurred internally while the posterior ends of the animals were entwined. However, an observation that one individual, isolated for more than 90 days, laid viable fertilized eggs suggests that either *E. babai* can self-fertilize, or that sperm were held for that period in the seminal receptacles. Anyhow, the eggs were laid as pairs of egg-sheets, one pair at a time, at intervals of about an hour. The author did not observe the spawning of the male product. The author cultured the eggs of the spawned egg sheets and prepared them technically for his study. More than 1000 embryos from three adults were analysed.

Early development

Okusu studied embryogenesis of *E. babai* by both scanning electron and light microscopy. At the time of deposition, the fertilized eggs were always at the single-cell stage. No brooding was observed. They are oval, but become spherical within 30 min at a diameter of 250 μm. Each egg is enclosed in a tough transparent egg capsule 300–350 μm in diameter (Fig. 11.10A). It is opaque, yellow because of its high yolk content, and surrounded by numerous 'accessory granules'. No polar bodies were observed (Fig. 11.10B), as were described in the works of Baba.

Cleavage is spiral, unequal, and asynchronous (Figs 11.2–11.9). 1st cleavage is meridional, i.e. parallel to the animal vegetal axis, and unequal. Just before the

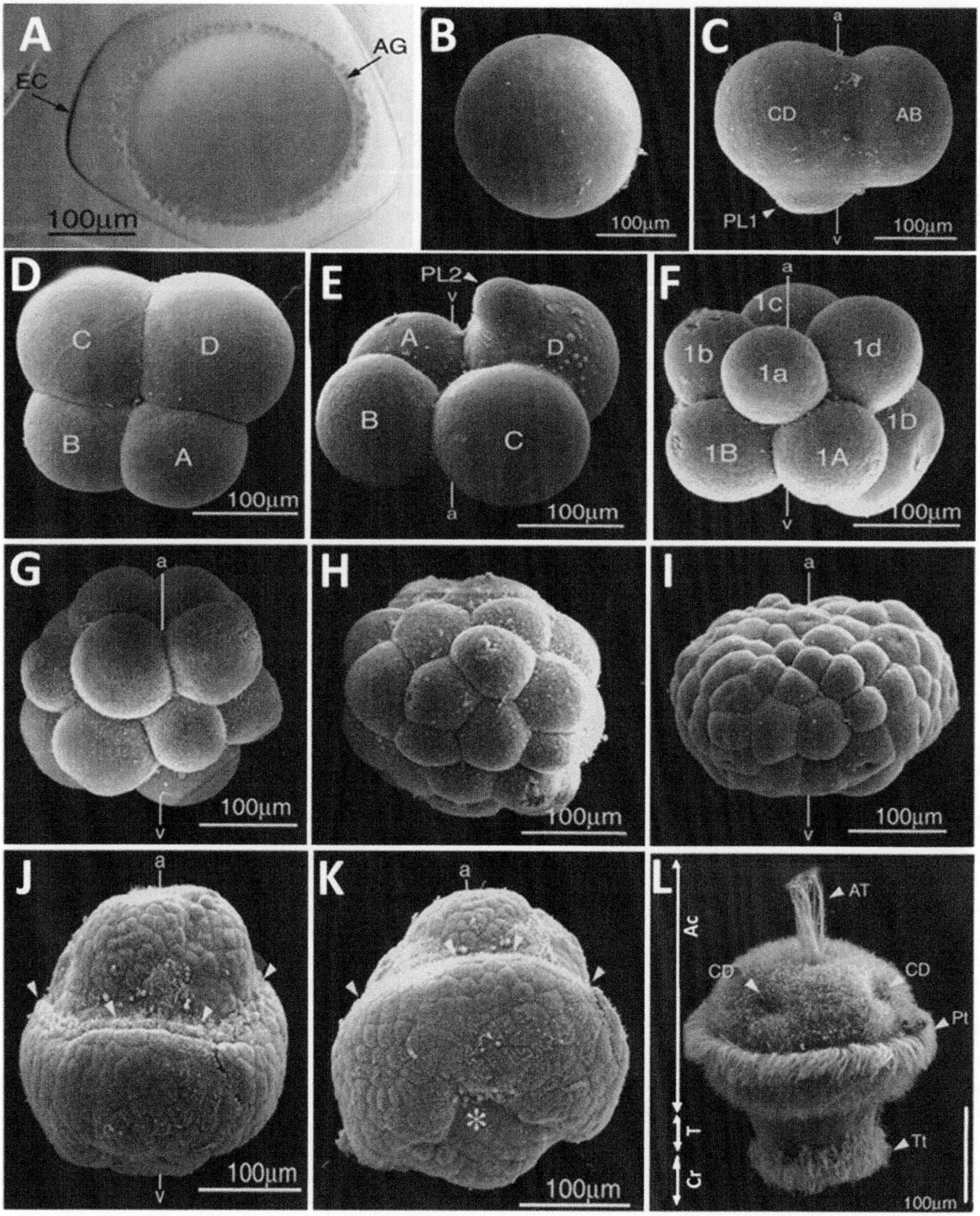

FIG. 11.10 Embryogenesis and Development of *Epimenia babai*: (A) fertilized egg (~250 μm in diameter) in a single egg capsule (EC) membrane. Numerous accessory granules (AG) were observed within the capsule; (B–K) SEM of the early embryological development of *Epimenia babai* before hatching. Animal vegetal axis (a–v) and dorsal ventral axis (D–V) are indicated; (B) a fertilized egg just after spawning, polar bodies not detected; (C) The beginning of first cleavage (lateral view). The first polar lobe (PL 1) has protruded from the vegetal side as the two cells (AB/CD) divide meridionally. The polar lobe was then absorbed by the CD cell, making it larger than AB; (D) animal view of the 4-cell embryo (A–D); (E) vegetal view of the 4-cell (A–D) embryo. The second polar lobe (PL2) protruded from the vegetal side of the D cell, and was soon absorbed by the D cell; (F) lateral view of the 8-cell stage (la-Id, 1A, IB, ID); (G) lateral view of 8- to 16-cell stage; (H) animal view of 16- to 32-cell stage; (I) lateral view of 64-cell to blastula stage. The embryo has flattened in animal-vegetal direction; (J) lateral view of gastrula. The *arrowheads* indicate ciliated presumptive prototrochal cells. The entire embryo is covered with short microvilli; (K) lateroventral view of embryo showing proliferation of the unciliated ectodermal cells dorsal to the blastopore. This outgrowth of the cells has shifted the ventral opening of the blastopore (*) towards the ventral side of the larva; (L) larva after hatching from the egg capsule (3-days after oviposition. 1-day-old larva). Ventral view of the larva showing its three distinct regions: ciliated preoral apical cap (Ac), unciliated trunk (T), and ciliated caudal region (Cr). AT = apical tuft, CD = cerebral depression, Pt = prototroch, Tt = telotroch;

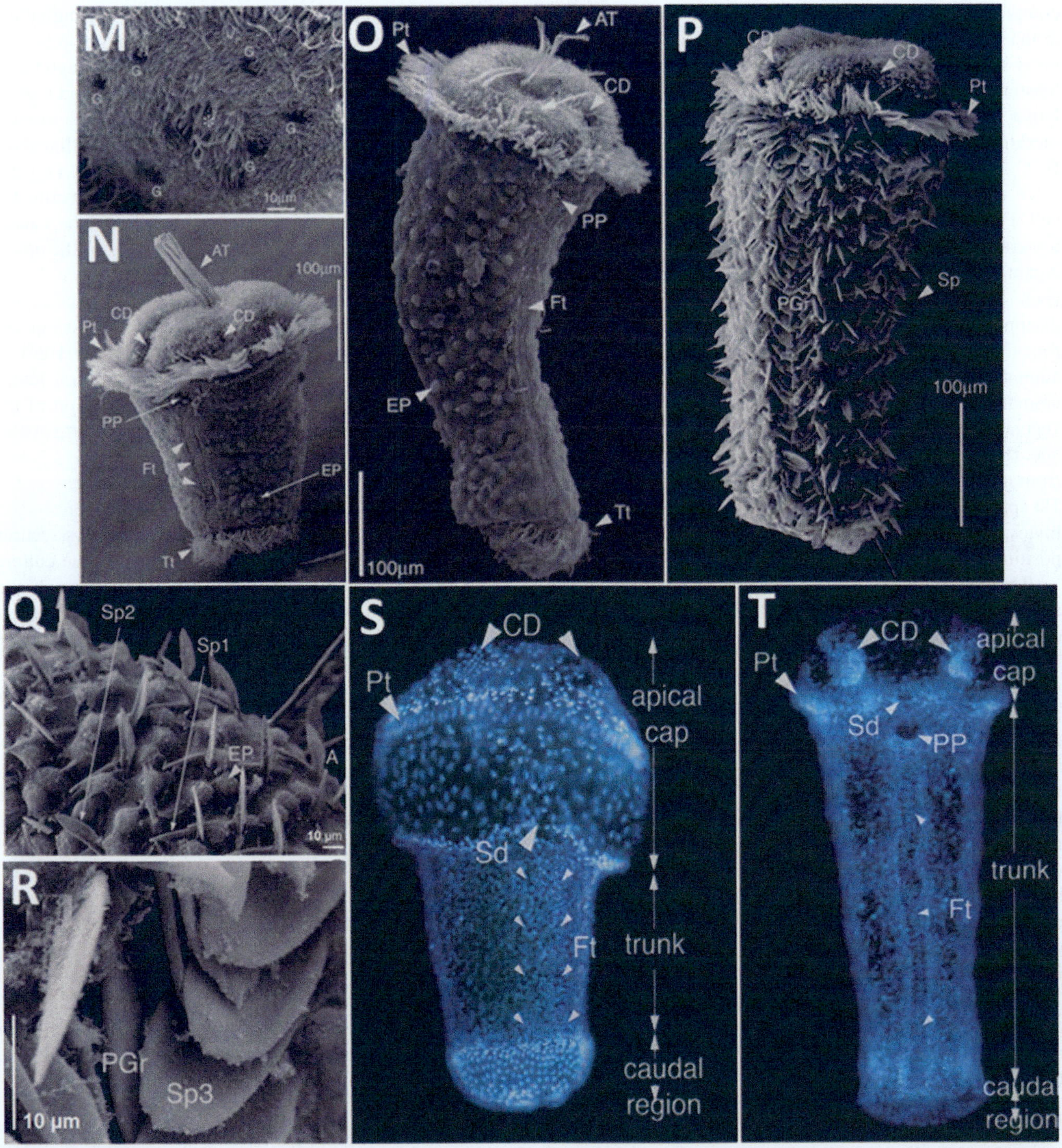

FIG. 11.10, CONT'D (M) Bottom surface of the caudal region of 1-day-old larva. The entire surface is ciliated except for scattered unciliated globules. The cilia at the periphery and in the centre of the caudal region (*) are longer than the rest of the cilia on the caudal surface. G = globules; (N) ventral view of larvae aged 4–6 days. The ciliated foot has formed midventrally on the trunk, and the pedal pit has formed at its apical end. Epidermal papillae are beginning to form on the trunk. The posttrochal region of the apical cap is diminishing as the trunk region elongates. EP = epidermal papilla, Ft = foot, PP = pedal pit; (O) (7- to 8-day-old larva), beginning of the gradual process of metamorphosis. The posttrochal region of the apical cap has completely disappeared. The trunk has become covered with numerous large epidermal papillae; (P) (9- to 11-day-old larvae), larva during the completion of metamorphosis. The larva covered by spicules on the trunk. The apical cap has diminished, and the apical tuft is being resorbed at the apical end. Sp = spicule; (Q) dorsal view of two types of the three of larval spicules right after metamorphosis. These spicules are equally abundant over the entire body of the juvenile: hollow upright spicules (Spl) lie against the body, while bladelike solid spicules (Sp2) project outwards from the body; (R) the third type of larval spicule (Sp3) is broad and solid, forming a single row on each side of the pedal groove; (S and T) Hoechst nuclear staining showing the development of larvae and the degeneration of large apical cap cells; (S) ventral view of a 3- to 4-day-old larva (cf. Fig. 11.10L and N). Cells of the apical cap are larger than cells on the trunk. Arrowheads indicate the broad ciliated foot starting to form; (T) ventral view of a 7-day-old larva (cf. (O)). The number of cells in the apical cap has decreased; posttrochal region of the apical cap has diminished. The cells in the apical cap still remain large. The foot narrows as the trunk elongates. Sd = stomodeum. *(All from Okusu (2002).)*

completion of the first cleavage, the first polar lobe appears at the vegetal side of the embryo (Fig. 11.10C). This first polar lobe is less yolky and is soon absorbed by one of the blastomeres, resulting in the CD blastomere being slightly larger than the other AB blastomere. 2nd cleavage also is meridional and unequal, resulting in the blastomeres A, B, C, D, but the cleavage itself is slightly asynchronous, since the larger CD blastomere divides later than the AB cell (Fig. 11.10D and E). The cross-furrow of the four blastomeres is visible, at this stage, but indistinct (Fig. 11.10D). Again, a second polar lobe appears at the vegetal region (Fig. 11.10E) during the CD divisions. Within 30 min, it disappears and the D blastomere becomes the largest of the four. 3rd cleavage is equatorial, i.e. perpendicular to the animal-vegetal axis, giving rise to four micromeres at the animal pole (1a–1d) and the four macromeres remain at the vegetal pole (1A–1D) (Fig. 11.10F). Since it is spiral cleavage, the four micromeres are shifted dextrally (clockwise from the animal pole) relative to the four macromeres. The 1D appears larger than the others (1A–1C). Cleavage continues, but slightly asynchronous and by alternation of the position of the micromere quartets.

The 4th cleavage is sinistral (anticlockwise from the animal pole), producing 16 cells (Fig. 11.10G). The 5th cleavage is dextral, producing 32 cells (Fig. 11.10H). The cleaved embryo reaches 64 cells about 15 h after laying, becoming a blastula. The latter is flattened along the animal-vegetal axis. This asynchronous cleavage is difficult to be followed and reaches the gastrula stage within 24 h after oviposition. Gastrulation occurs, within the capsule, by both epiboly and invagination (Fig. 11.10J). Gastrula stage shows: (1) cells between the blastopore and presumptive prototrochal cells thickened to form a broad lip around the vegetal side; (2) the surface of the gastrula is covered by patches of short cilia (Fig. 11.10J). Within the blastopore, ectodermal cells proliferate causing shifting of the blastopore opening to the future ventral side of the larva (Fig. 11.10K). Two days after oviposition, the presumptive apical tuft and prototroch appear in the apical region. The telotroch at the caudal end of the ectodermal trunk, also appears (Fig. 11.10L).

Larval development

Three days after oviposition, the larva hatches and swims forwards and spirally, using the prototroch. The 3 days old free-swimming larva has three distinct regions: apical cap, trunk and caudal region (Fig. 11.10L and S). The apical cap is the entire preoral sphere of the larva, it is completely ciliated and divided into a pretrochal and a posttrochal region. A prominent row of compound prototroch cilia at the equator of the larva divides the sphere into pre- and posttrochal regions (Fig. 11.10L). In the pretrochal hemisphere of the apical cap, most of the larval structures develop, such as the apical tuft, secretory globules, and cerebral ganglial depressions. The latter structures are formed on the pretrochal hemisphere of the apical cap, and they persist through metamorphosis (Fig. 11.10L). The trunk is an unciliated region, and gives rise to cuticle and epidermis, also to a ciliated stomodaeum formed on the ventral side of the unciliated ectodermal trunk, directly below the apical cap (Fig. 11.10S). The caudal region is covered by short cilia, but the cilia differentiate in the different parts of the region. The caudal region also contains globules, similar to those of the apical cap (Fig. 11.10M).

The 4–6-day-old larva elongates, the postrochal region of the apical cap narrows, and the cerebral depressions become deeper as the trunk grows longer (Fig. 11.10N). A foot is formed by differentiation of the midventral longitudinal band (15–20 μm across) of the trunk region of the larva, which becomes ciliated (Fig. 11.10N).

Metamorphosis

The larva begins metamorphosis at age 7–8 days, ceases its active swimming and sinks to the bottom of the culture jar, in spite of the ciliary movement of the prototroch and apical tuft. The unciliated trunk grows longer, the epidermal papillae increase in number and size, and presumptive spicules are formed (Fig. 11.10O). Although the posttrochal region of the test narrows as the trunk grows, lengthening of the trunk appears to be due to proliferation of definitive ectodermal trunk cells. Metamorphosis is completed during days 9–11. The trunk of the larva is then completely covered with epidermal papillae and spicules that extend well beyond the cuticle (Fig. 11.10P). The whole caudal region and the cap become enclosed within the trunk and covered by the posterior and anterior extensions of definitive ectoderm.

The postmetamorphic juvenile crawls on the ciliated foot. The latter is located in the midventral pedal groove. It appears that this juvenile is sensitive to vibration, but not to light. It becomes more transparent, as the original yolky yellow colour becomes internally restricted. The animals of this culture lived for up to 1 month without any particulate food source, and the gut was not differentiated in histological sections during that period.

Strikingly, the difference of the spicules in the juvenile and the adult is not only in shape, but also in distribution. First of all, the aragonitic spicules develop from the definitive ectoderm. A spicule begins as a solid tip, continues to an open-ended hollow spicule, and finally becomes a closed-ended hollow spicule. The spicules of adults are solid and bladelike along the foot groove and near the mouth, but the rest of the body is covered by hollow, needlelike spicules forming a crisscrossed meshwork within the cuticle. But in juveniles, flat bladelike spicules are distributed evenly over the entire body, along with upright and needlelike spicules

(Fig. 11.10Q and R). The juvenile pedal spicules are flat and bladelike as in the adults (Fig. 11.10Q and R), and the head region is also covered by flat, bladelike spicules.

As mentioned above, brooding was not observed in *E. babai* during this study of Okusu (2002). However, the capacity to keep embryos within the posterior mantle cavity in fast water currents cannot be refuted. Brooding of larvae in the posterior mantle cavity was previously reported in *E. babai* (Baba, 1938). During this study, *E. babai* individuals tightly curled their posterior part around the alcyonarian corals, thus pressing closed the mantle cavity opening and keeping their eggs within the mantle cavity. This means that this species, *E. babai*, and other species of *Epimenia*, in their natural habitat, live and feed on the alcyonarian corals that prefer strong swift water currents (Baba, 1938). Again, this means that brooding may take place in the natural habitat.

It remains to report that the results of this study, using scanning electron and light microscopy, semithin sections, Hoechst nuclear staining, and programmed cell death staining, enabled the author to discuss the homology and evolution of molluscan larval tests. It was suggested that the test of the neomenioid larvae is an enlarged preoral sphere of a trochophore. The test morphologies of neomenioid larvae are compared to those of pericalymma larvae of protobranch bivalves.

Class- Gastropoda

Gastropods (snails and slugs) constitute by far the largest and most diverse class of Mollusca, with an estimated 40,000 marine, 3000 freshwater, and 24,000 terrestrial species (Baur, 1994). The archaeogastropods are considered mostly like the ancestral gastropods. Thus, reproductive biology is relatively simple. Most are dioecious, broadcast spawners that release mature gametes through the right renal duct into the mantle cavity and then into the external environment. Fertilization is external, and a trochophore larva is generally formed before the veliger is reached (Webber, 1977). In prosobranchs mesogastropods, fertilization is internal through copulation, and development is partially or completely in an egg capsule that is generally externally deposited. However, many forms are:

a. Protandric hermaphrodites (e.g. *Crepidula*)
b. Parthenogenetic (e.g. *Campeloma*)

Many brood their eggs (e.g. *Cypraea*) or are viviparous (e.g. *Viviparous*), internal fertilization occurs in some species in which the male lacks a penis (e.g. *Bittium, Turritella*) and in many species in which the male produces both normal and abnormal spermatozoa. The neogastropods are considered the most advanced gastropods. They are primarily marine forms, and their reproductive biology generally follows the pattern of the mesogastropods.

Opisthobranchia have some modifications, including detorsion, and reduction or loss of the ctenidia and shell, so it becomes soft-bodied (slug-like morphology). Almost all are simultaneous hermaphrodites with extraordinary complex reproductive structures. Pulmonates indicate a widespread tendency towards protandric hermaphroditism. They provide more favourable experimental material and also claim more medical and economical importance.

Reproduction

Agametic reproduction

The only type of agametic reproduction that occurs in prosobranchs is parthenogenesis. The first experimental evidence confirming parthenogenesis was for *Potamopyrgus jenkinsi* (Boycott, 1919). Only once in this common species has a male been reported. This single male apparently underwent normal spermatogenesis.

Cytologically, the parthenogenesis of *Potamopyrgus jenkinsi* is ameiotic; that is, a single maturation division occurs during oogenesis, and meiosis does not occur. Parthenogenesis and polyembryony, are known among opisthobranchs and pulmonates.

Gametic reproduction

Dimorphism. Webber (1977) reported that the differences between male and female prosobranchs of the same species (except in family Entoconchidae) are generally not pronounced. In those species with internal fertilization (most mesogastropods and neogastropods), a penis located behind the right tentacle is often visible.

The one group of prosobranchs in which sexual dimorphism is pronounced is the parasitic family Entoconchidae. Species in this mesogastropods are endoparasites of holothuroids. Females are generally vermiform and have no shell except in larval stages (Fig. 11.11A). Males are much reduced and occur as dwarfs in the female.

One genus of this group (*Enteroxenos*) was believed to be hermaphroditic and the group has been classified by some authors as an opisthobranch group (Hyman, 1967). It seems that this phenomenon is not pronounced in both Opithobranchiata and Pulmonata.

Hermaphroditism. Most prosobranchs are dioecious; there are, however, a number of hermaphroditic species. Webber (1977) tabulated the reported hermaphrodite genera of Prosobranchiata. Hermaphroditic patterns may be simultaneous (with eggs and sperm maturing simultaneously in the same gonad), or consecutive (with either eggs or sperm in the same gonad but at different times). Species showing simultaneous hermaphroditism generally have cross-fertilization, although self-fertilization may rarely occur. Consecutive hermaphrodites may be protandric (change from male to female) or may alternate sex every other season. The best known protandric hermaphrodites are found

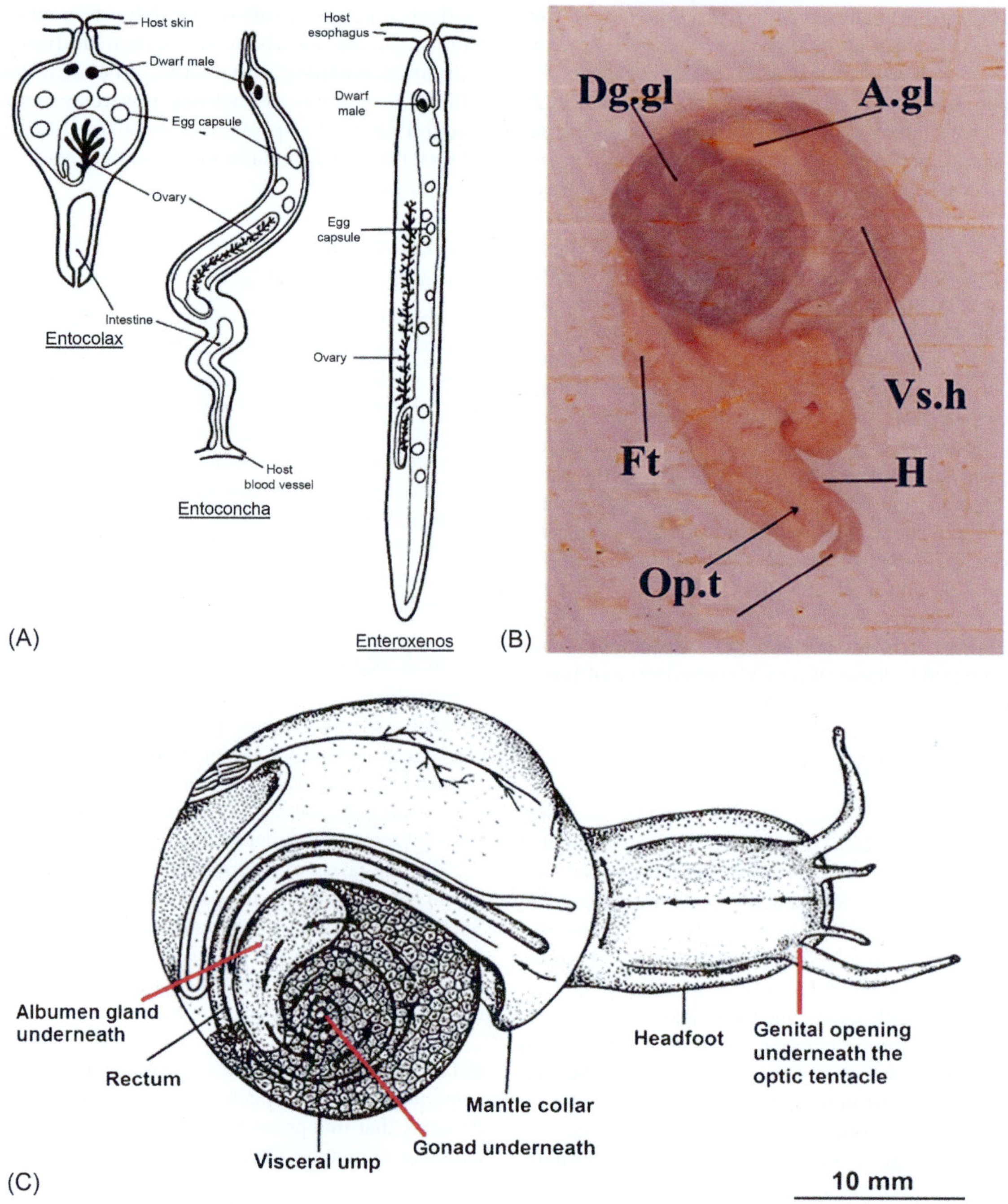

FIG. 11.11 (A) Sexual dimorphism in the family Entoconchidae *(after Lützen, 1968, from Webber, 1977)*; (B) right side view of naked snail *Eremina ehrenbergi*. The *arrow* indicates the position of the genital opening. X=about 2.5; (C) diagrammatic drawing showing the dorsal view of naked snail *Eremina* and representing the direction of the incisions for dissection *((B and C) from Abd El-Kerim, 1988)*.

in the mesogastropod family Calyptraeidae (Calyptraea, and Crepidula).

The opisthobranchs and pulmonates are mainly hermaphrodites. It has been proposed that the hermaphroditic pattern in opisthobranchs developed by the superimposing of a male reproductive system on a female one. All marine pulmonate hermaphrodites produce male and female gametes in a common gonad, like freshwater and terrestrial groups. While pulmonates, especially the nonmarine ones, are ambisexual, marine pulmonates often exhibit various degrees of sex change that are mostly (and possibly all) protandric.

Unlike those of some prosobranchs, the protandry of marine pulmonates does not involve conspicuous changes in the reproductive ducts and genitalia, which instead remain clearly hermaphroditic throughout sexual changes within the gonad.

Protandric hermaphrodites. The most widely studied prosobranch protandric hermaphrodites are in the genus *Crepidula.* The young *Crepidula* has both spermatogonia and oogonia in the gonad. *Crepidula* is unique in that individuals of some of its species are found in chains of about 12–19 animals attached one to each other's shell. The oldest and largest individual is usually attached to the substratum, the others clinging to the shell of the animal beneath. The chain twists to the right, so that the genital apertures of all individuals approximate. The smallest individual, at the summit, is a young male; the oldest and larger ones near the base are females, with transitional hermaphrodites in between. The relative proportions of males and females in a chain can influence the sexual development of newly added young males, and it is claimed that a male will remain a male as long as it is attached to a female. Chain formation has been regarded as important for breeding in a relatively sedentary species, but its main importance may be in the production of a stronger current to which all members contribute and from which all benefit.

Another notable case is that in some groups, protandric hermaphrodites may occur in only some species of a genus. For example, in the limpet *Patella, Patella vulgata, P. caerulea,* and *P. intermedia* are protandric hermaphrodites, while *P. aspera* and *P. depressa* are generally dioecious. Sex change does not occur in all the population of one species. It was recorded that 90% of the individuals of *Patella vulgata* mature as males and after a year or more change to females. In *Fissurella,* Bacci (1947) found that only 12% of the individuals underwent sex reversal, and the remainders were dioecious.

Another notable case, recorded in Marshall and Williams (1975), is that the slipper-limpet *Crepidula fornicata* is always a protandrous consecutive hermaphrodite, i.e. begins life as a male, passes through a hermaphrodite stage and finally reaches a stage of pure female. Initially the gonad has both spermatogonia and oogonia, but in the male phase, the oogonial development is suppressed. During the transition stage, the male gonad and its trait degenerate, and normally, but not always, the female gonad and its associated structures develop. *Crepidula* is unique.

Reproductive system. See Fig. 11.12 showing drawings of the isolated hermaphrodite reproductive system of some gastropods. Fig. 11.12D illustrates that of *Eremina ehrenbergi.*

Eremina ehrenbergi

Almost all the work on *Eremina ehrenbergi* was published, but all the figures presented here are the original from El-Bawab laboratory, done by Abd El-Kerim and Mehanney.

Eremina is a stylommatophoran genus that commonly inhabits the Egyptian desert. Different species of *Eremina* live in large scattered groups on the shrubs, feeding on their stems and leaves in the Sahara (the world's largest desert (Philip's, 2003)), west of Alexandria. *Eremina ehrenbergi* (Fig. 11.1A) was chosen for an overall work to study its hermaphrodite reproductive system and its reproduction. It was identified according to Biggs (1959). Some large specimens with shell diameter 2.4–3 cm and with length 2.2–2.5 cm were fixed in 10% formalin and sent to the British Museum (Natural History, Department of Zoology) for confirmation. Measurements were done according to Salib (1975) and Emberton (1985). Live adult snails were collected regularly, twice every month from July 1984 to August 1985, from different areas of El-Omayed (from El-Hammam to Gabel Kashm El-Eish). The snails were successfully kept as a culture in the laboratory. From August 1985 up till March 1989, other collections were brought at irregular intervals to renew the culture, or to use the specimens to confirm some observations during the course of the work. The culture was maintained in proper conditions due to the care provided by both Abd El-Kerim and Mehanney.

For study, the animals must be prepared by a special treatment to prevent the contraction of the head-foot and to facilitate the isolation of the reproductive system. It was found that the most convenient and suitable method for the study of the anatomy of the specimen is that preferred by Mahoney (1973). The snails were immersed in tap water previously boiled, and then cooled to 60°C. The full extended specimens were withdrawn from the shell gently by hand, or the shell itself was broken gently by a small hammer to keep the specimen in a very good condition. The snail was then transferred directly into 10% formalin or 70% alcohol for at least 2 days. The animal then becomes ready for dissection and study of the reproductive system.

The hermaphrodite gonad. Naturally, this system occupies the most dorsal part of the visceral hump. It is very long, complex and well differentiated into a number of constituents and accessories. It is an unpaired system and opens by a single very small genital opening on the right side of the head, just underneath the optic tentacle (Fig. 11.11C). The system begins with the hermaphrodite gonad, which is found just underneath the apex of the visceral hump (Fig. 11.11C) where the digestive gland is positioned. It occupies the surface of the second whorl and is covered by the first one. It can be seen externally through the transparent mantle wall. The gonad is fan-like, but small compared to the other constituents of the system (Figs 11.12D and 11.13A). The immature gonad is white in colour, and gradually acquires a creamy tinge during maturation. When very mature, it measures about 2.5–3 mm in length and 7–9 mm in width. The digestive gland furnishes suitable background for the gonad to be distinguished. Histologically, it appears as a branched, but not anastomosed, tubular mass (Fig. 11.13B and C). Each tube or its branch ends freely or separately, while connective tissue is observed in between the distal portions of these tubules, where their number is

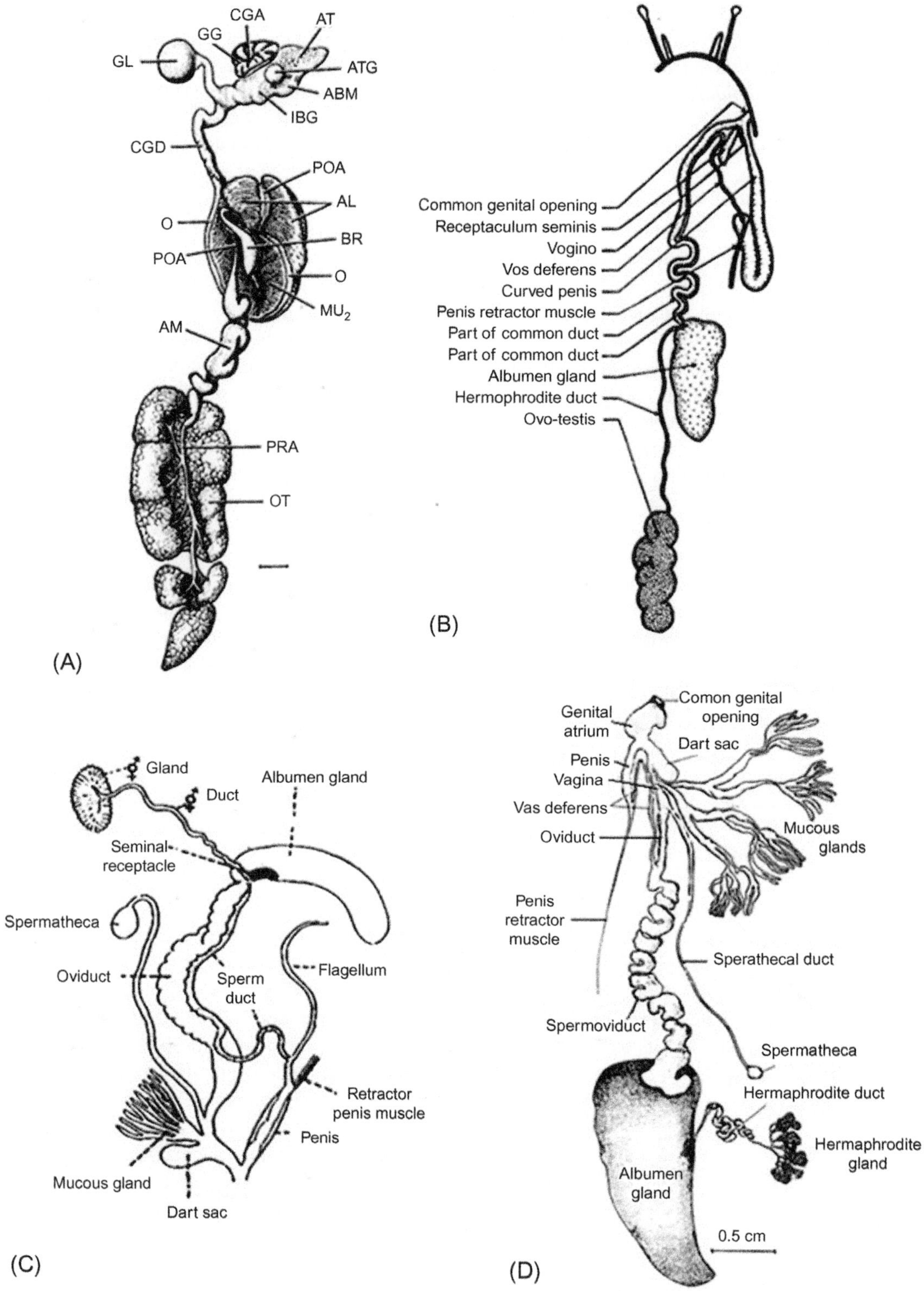

FIG. 11.12 Diagrams of hermaphrodite reproductive systems of some different molluscs: (A) left aspect of that of the anaspidean *Phyllaplysia taylori* (Opisthobranch), from life. X = roughly indicates 2 mm. ABM = atrial basal muscle, AL = albumen gland, AM = ampulla, AT = atrium, ATG = atrial gland, CGA = common genital aperture, CGD = common genital duct, GG = genital groove, GL = gametolytic gland (copulatory bursa), ISG = internal seminal groove, MU 2 = mucous gland (secondary lobe), OT = ovotestis, POA = postampullar duct, PRA = preampullar duct, SR = seminal receptacle; (B) of the slug *Limax*; (C) of *Helix pomatia*; (D) of *Eremina ehrenbergi*. *((A) From Beeman (1977), (B) from Bullough (1981), (C) from Borradaile et al. (1977), and (D) from Mehanney (1991).)*

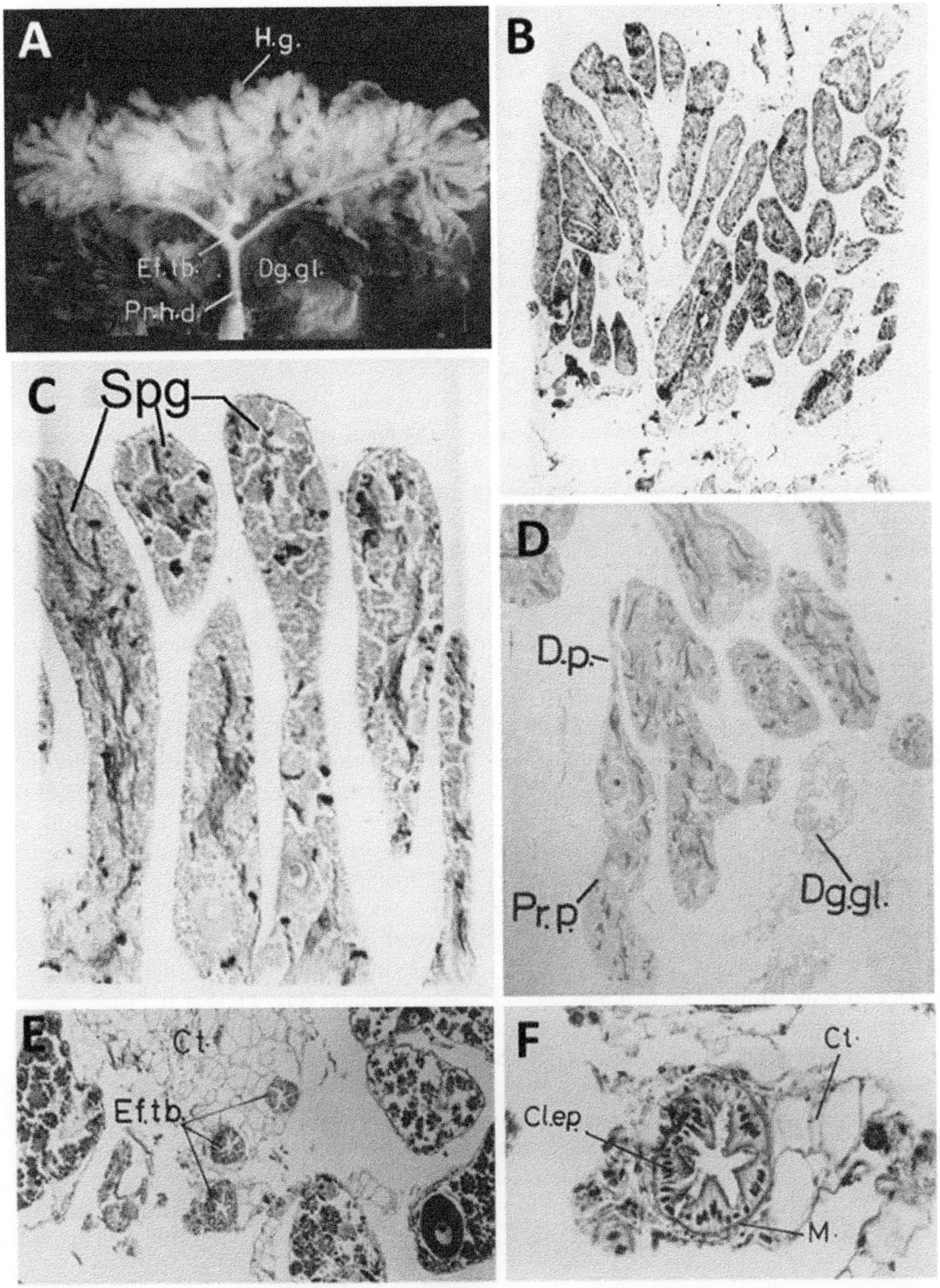

FIG. 11.13 (A) Photomacrographs of the dorsal view of the hermaphrodite gonad of *Eremina*. X=10; (B–F) photomicrographs: (B) of L.S. of the hermaphrodite gonad of *Eremina*. X=50; (C) of a magnified portion of a L.S. of the gonad prepared in October showing the dominance of spermatogenesis. X=120; (D) of a portion of L.S. of the gonad showing the differentiation of the gonadal tube. X=50; (E) of a portion of a T.S. of the distal region of the gonad showing the three efferent tubes and conn. tiss. in between. X=100; (F) enlarged of one of the efferent tubes showing its structure. X=500. Ct=conn. tiss., Cl.ep.=ciliated epithelium, D.p.=distal portion, Dg.gl=digestive gland, Ef.tb=efferent tubes, H.g=hermaphrodite gonad, M.=muscle, Pr.P.=proximal portion of gonadal tube, Pr.h.d.=proximal hermaphrodite duct, Spg=different stages of spermatogenesis.

limited (Fig. 11.13E and F). Each gonadal tubule is differentiated into the proximal glandular portion, which proliferates the gonial cells, and the distal narrower portion with wider lumen (Fig. 11.13D). (The terms 'proximal' and 'distal' are with respect to the gonad). The distal portion constitutes the passage for the product of the glandular one (Figs 11.25M and 11.29D) and is hence referred to as conducting portion.

The conducting portions of all gonadal tubes are collected distally in three main efferent tubes. The transverse section through the distal part of the gonad shows three narrow tubes (Fig. 11.13E). Each is circular in outline with a comparatively thick wall and diameter measuring about 100 μm (Fig. 11.13F). It is surrounded by a thin layer of connective tissue, which appears to be continuous with the binding tissue. Very fine muscle fibres invade the connective tissue, forming a continuous layer, which can be identified by accepting the acid fuchsin when treated with Masson's trichrome stain. Innermost is a layer of the columnar epithelial cells, with different lengths, giving the appearance of thickenings. The latter extend in the lumen as several villi. Each cell has clear cytoplasm and a basal oval, small nucleus (Fig. 11.13F). Its free edge is not smooth, and whether it is ciliated or secretory, needs ultrastructural examination.

The hermaphrodite duct. The collection of the efferent tubes is the beginning of the hermaphrodite duct (Fig. 11.13A). The latter passes downwards through the visceral hump, along the inner side of the third whorl where the digestive gland lodges, and extends untill it reaches the albumen gland (Fig. 11.12D). It is comparatively very long and its colour is chalky-white or brownish cream. The whole duct can be detected externally through the thin, transparent mantle wall after shifting the first and second whorls of the visceral hump. Along the whole length of this duct, it is surrounded by connective tissue continuous with the binding one and followed by a circular muscle layer. Morphologically and histologically, it is differentiated into three regions (Fig. 11.14A). The proximal region is a short, straight and slender tube, having the narrowest lumen. Its lining epithelium is formed of ciliated and secretory columnar cells (Fig. 11.14B and C). The middle is the longest, thickest and folded region (Fig. 11.14D), and contains a milky white viscous substance that protrudes to the outside just at puncturing by a pin any portion of this region of a dissected live specimen. Even in tap water, this substance swells greatly and the examination of a smear shows that it embeds numerous live sperm, which remain alive for a long time (Fig. 11.14H). This sperm mass in this middle region of the hermaphrodite duct indicates that this region is not only conducting like its continuations, the proximal and distal regions, but also is storage for the freshly produced sperm. So it is named vesicula seminalis (Pelseneer, 1906; Duncan, 1958, 1960a,b; Demian, 1960; Rigby, 1963; Kugler, 1965; Visser, 1981). In addition, its epithelial lining varies from very short cubical cells (Fig. 11.14F) to tall columnar cells (Fig. 11.14E), not ciliated, but secretory, since they contain strands of substance with the same affinity to stains as that heavily condensed in their free edges (Fig. 11.14E). It appears that this substance discharges merocrine (Bloom and Fawcett, 1976).

The distal hermaphrodite duct (Fig. 11.14G), looks like the proximal, in size, structure, and appearance, but its lumen is wider and empty. It extends towards the albumen gland to connect a small cylindrical, elongated, blind, claw-shaped pouch, found in the distal part of this gland in its concave surface (Figs 11.14A and 11.15A, B). This pouch is referred to as 'talon' (Gugler, 1964), however it becomes a common name referred to a lateral diverticulum greatly related to the hermaphrodite duct of the stylommatophoran snails (Berry, 1977). It corresponds to the fertilization chamber described in some different basommatophorans (Duncan, 1960a,b). In *Eremina ehrenbergi*, this talon differs in size in the different specimens. The portion of connection of hermaphrodite duct and talon is in the form of loop giving a U-shaped structure or, more precisely, a hairpin loop. The two limbs are dissimilar, since the duct is a narrow tube, while the talon is a wide pouch (Fig. 11.15B). The talon proceeds forwards, inside the albumen gland tissue, reaching its distal portion, where the spermoviduct, the voluminous and largest constituent of this reproductive system of *Eremina* (Fig. 11.12D), lodges. At this level, the distal portion of the talon connects these two elements: the spermoviduct and the albumen gland itself. The latter gland has a central narrow duct, which curves to connect the distal talon (Fig. 11.15E) at the base of the spermoviduct. At the same time, it merges with the ventral straight side of the spermoviduct (Fig. 11.15D and E).

The talon. The histological study (transverse, longitudinal, and frontal sectioning) of the albumen gland containing the talon clarified the picture and added much to the understanding of this peculiar constituent of the reproductive system of *E. ehrenbergi*. First of all, this talon, which definitely originates from the hermaphrodite duct in stylommatophora, corresponds to the so-called Carrefour of Basommatophora (Visser, 1981). This is the fertilization pouch studied by Duncan (1960a,b). Secondly, it can be compared to the description of that of two different stylommatophoran given by Rigby (1963) and Lind (1973). The former studied *Oxychilus cellarius* from family Zonitidae and the latter studied *Helix pomatia* which belongs, as *Eremina*, to family Helicidae (Solem, 1974). The three animals show common principles concerning the structure of the talon: (1) its main component is the fertilization chamber; (2) this chamber gives a blind outgrowth, the receptaculum seminis; (3) the narrow end of the talon communicates the lumens of the albumen gland and the spermoviduct.

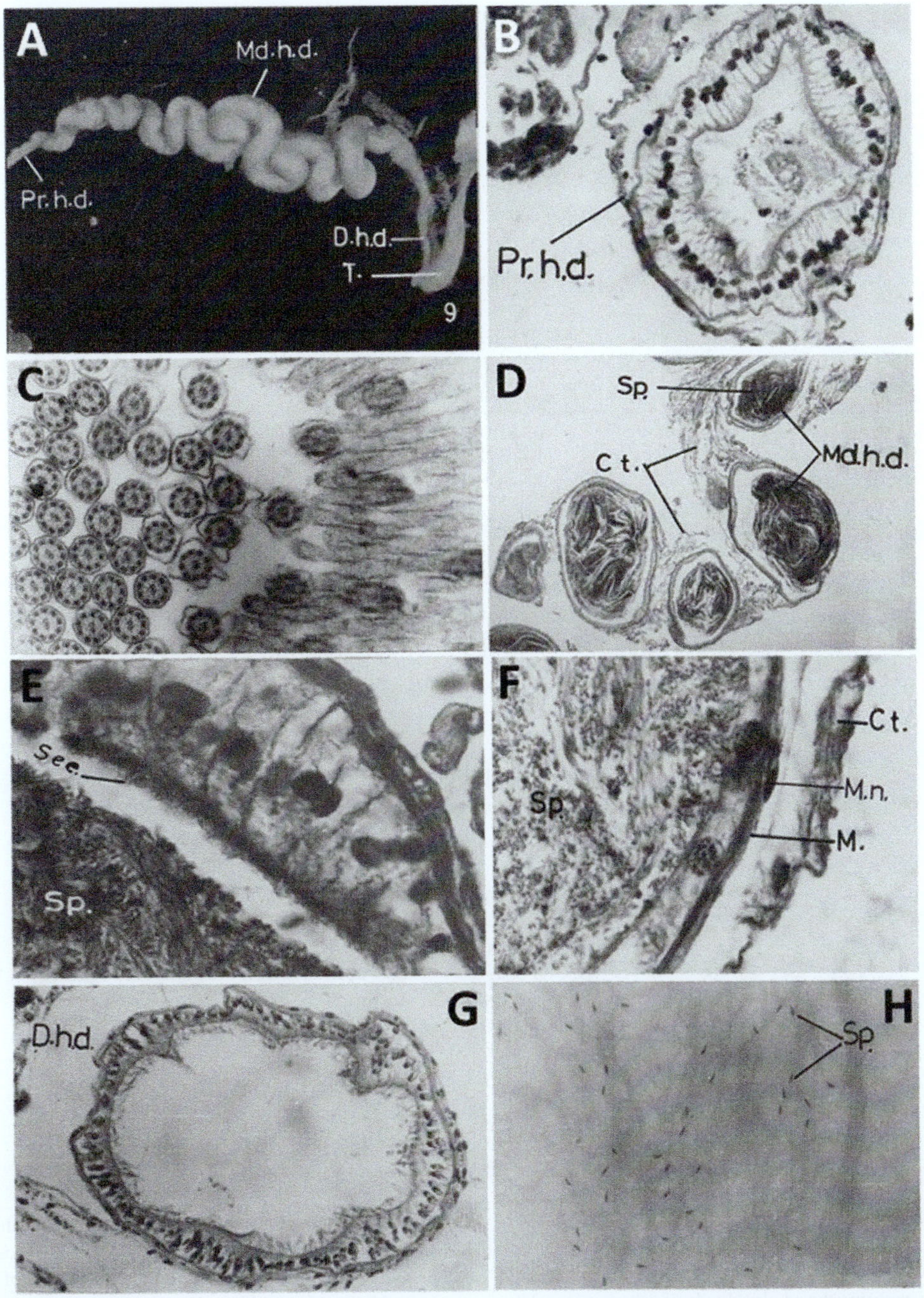

FIG. 11.14 Structure of the hermaphrodite duct (HD) of the reproductive system of *Eremina*: (A) showing its differentiation. X=10; (B) Photomicrograph of a T.S. of the proximal region of the HD. X=75; (C) TEM of a portion of the lining layer of section B, notice transversely cut cilia. Gultraldehyde osmium fixative, uranyle acetate and lead citrate stain. X=30,000; (D) L.S. through the folded middle region of the HD, notice the connective tissue binding the folds. X=50; (E–H) Photomicrographs: (E) of a highly magnified portion of the wall of the middle region of the HD shown in (D). Notice the very tall columnar epithelium and the heavily stained edges. X=1200; (F) of another magnified portion of the wall of the same region showing that the cells are cubical. X=1200; (G) of a T.S. of the distal region of the HD, notice that its cells are secretory. X=100; (H) of a smear preparation of the contents of the middle region of the HD showing fully formed sperm embedded in a medium proving that this region is the vesicula seminalis of this system of *Eremina*. X=250. Ct=connective tissue, D.h.d.=distal hermaphrodite duct, M=muscles, Md. H.d.=middle region of HD, M.n=muscle nucleus, Sec=secretion, Sp=spermatozoa.

In *Eremina*, at its beginning as a continuation of the distal hermaphrodite duct, it is a circular structure covered by a very thin layer of connective tissue. It encloses a number of different ductules, and an oblique connective tissue septum separates the upper group from the lower one (Fig. 11.16A). The latter consists of two irregular ductules, bounded by connective tissue, to maintain their position relative to each other. The largest is in the form of the letter C, enclosing the other smaller one in its concavity. The smaller appears as two incompletely separate ones. The outer is the distal

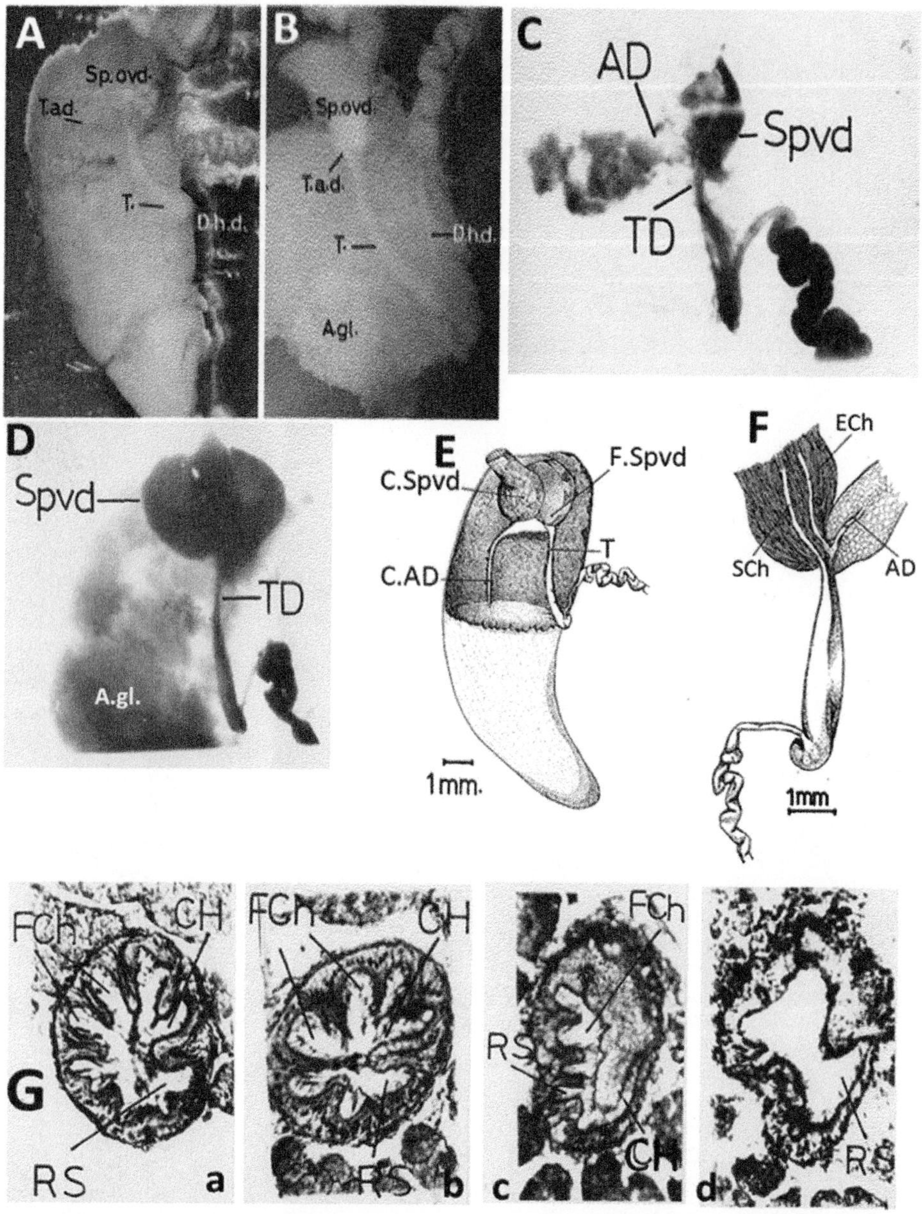

FIG. 11.15 Distal HD and talon of the reproductive system of *Eremina*: (A) photomacrograph of a W.M. of the albumen gland showing its relationship with both HD and spermoviduct. Also shows the continuation of the distal HD with the talon lodging the ventral surface of the most distal portion of the gland. Notice the connection of the distal talon with the albumen duct. X = 10; (B) dissection of the albumen gland, i.e. removal of much tissue from the ventral concave surface exposing the whole talon inside the tissue and its connection with the spermoviduct. X = 10; (C) isolated talon showing its connection; (D) dorsal view of the talon with its associated proximal portion of the spermoviduct which lodges distal albumen gland. Most of the albumen gland was removed to expose this association. X = 10; (E) diagrammatic Camera Lucida drawing of a dissected albumen gland showing the connection of the three elements in situ. Notice the central albumen duct and its continuation to connect both the talon and spermoviduct. Albumen tissue of dorsal distal half was removed; (F) diagrammatic camera Lucida drawing of ventral view of isolated talon with its associated portions, notice that the sperm- and egg-channels, inside the spermoviduct, are parallel; (G) successive T.S.s of the most end of the narrow talon showing its interior appearance: (a) beyond the origin of the receptaculum seminis from the fertilization chamber, showing the orientation of each component in the narrow talon. X = 175; (b) at the base of the spermoviduct, and the direction of the ciliated hood to this organ. X = 250; (c) continuation of the ciliated hood to be sperm-channel. X = 200; (d) constriction of the passage of the receptaculum seminis to connect the egg-channel. X = 200. AD = albumen duct, A.gl. = albumen gland, C.Spvd = coiled portion of Sp.ovd, D.h.d = distal hermaphrodite duct, ECh = egg channel, F.Sp.ovd = funnel of Sp.ovd, SCh = sperm channel, Sp.ovd = spermoviduct, T = talon, T.a.d. = Talon-albumen duct, TD = narrow talon.

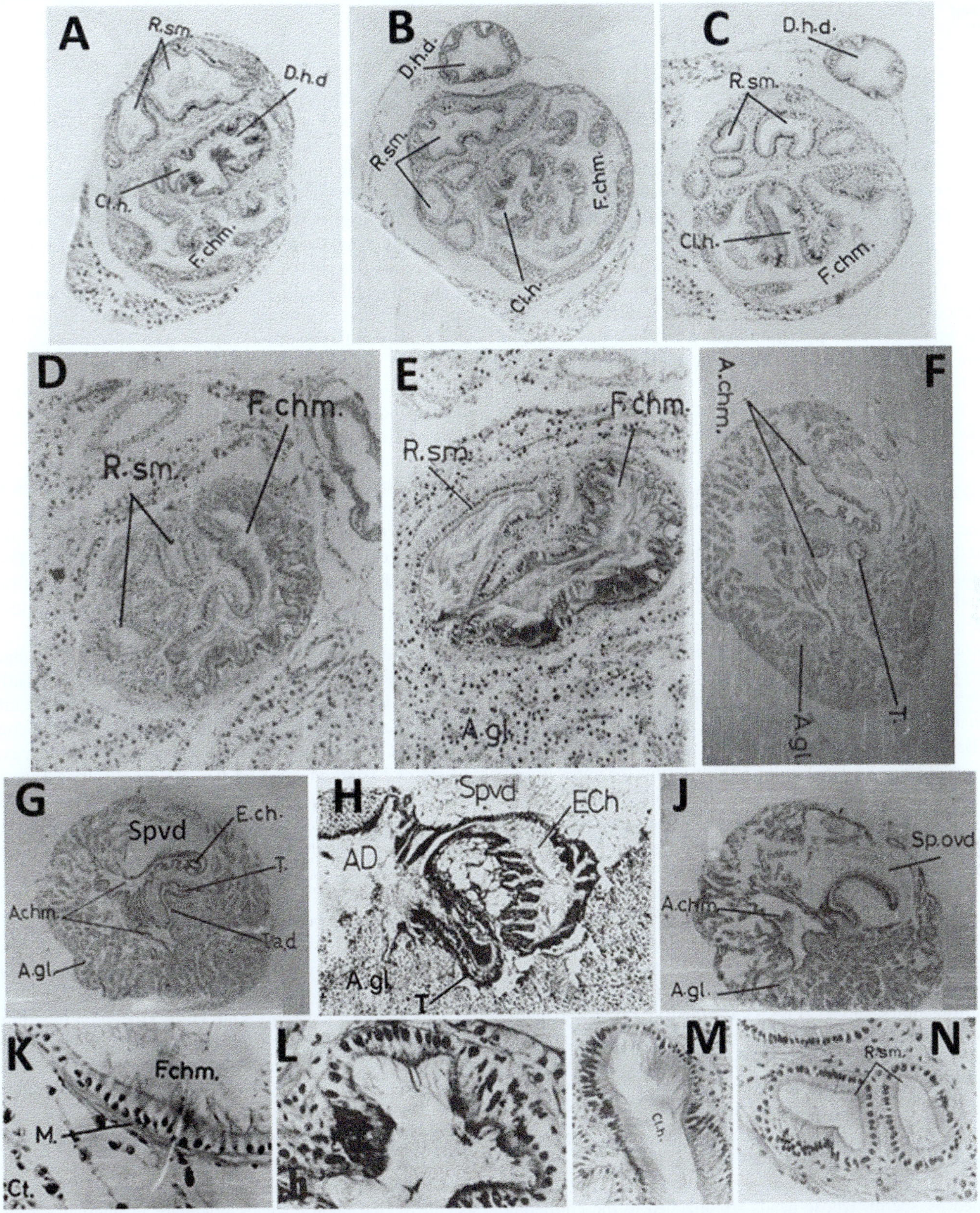

FIG. 11.16 Photomicrographs of selected sections of serial Transverse preparation of the talon of *Eremina* starting from its proximal part to its distal one, showing its structure at the levels indicated in Fig. 11.18: (A) just at the flexion of the distal hermaphrodite duct; (B) the latter appears outside the talon, and its continuation in talon is the ciliated hood; (C) at the middle of the talon, the ciliated hood appears as a deep fold of the fertilization chamber; (D) at about two thirds the length of the talon, showing the six ducts of the receptaculum seminis; (E) of a level showing the origin of receptaculum seminis from the fertilization chamber; (F–J) another serial preparation to show the relationship of the talon with the albumen gland and spermoviduct: (F) the most distal part of the talon and the additional albumen chamber; (G) showing the talon-albumen duct connecting the talon with the central albumen chamber; (H) magnified portion of a frontal S. showing the continuation of the talon with the albumen chamber and the spermoviduct; (J) showing the separation of the lumen of the spermoviduct from the albumen chamber; (K) magnified portion showing the wall of the fertilization chamber of the talon; (L) Two villi of the epithelial lining of the ciliated hood; (M) magnification of the ciliated hood; (N) magnification of the walls of the receptaculum seminis. A. chm = albumen chamber, A. gl = albumen gland, Cl. H. = ciliated hood, D.h.d = distal hermaphrodite duct, E. ch. = egg channel of spermoviduct, F. chm = fertilization chamber, R. sm. = receptaculum seminis, Sp.ovd. = spermoviduct, T. = talon, T.a d. = talon albumen duct. (A–J except H) X = 75, X of H = 120, of K, N = 700. *((K–N) From Bawab (1988).)*

hermaphrodite duct after its flexion to connect the talon. The inner is the continuation of the hermaphrodite duct inside the talon. It, with the C-shaped one, constitutes the talon. The C-shaped ductule is considered the main component of the talon. Rigby (1963) referred to the former as 'the ciliated hood' and to the latter as 'the fertilization chamber'. The ciliated hood proceeds independently in the concavity of the fertilization chamber (Fig. 11.16B). It is lined by one layer of columnar cells with various lengths (Bawab, 1988), hence the appearance of some shallow villi (Fig. 11.16L). Its lining cells are heavily loaded with dense striations, to the degree that their boundaries disappear (Fig. 11.16M); this maybe a secretory substance, or cilia, or both. They probably serve to offer the proper direction for own sperm (Bawab, 1988). The ciliated hood merges in the fertilization chamber to become a fold of its wall, at about the middle of the talon (Fig. 11.16C).

The fertilization chamber is the largest component of the talon and is lined with a folded layer of epithelium, the cells of which fluctuate between columnar and cubical cells (Fig. 11.16K). The nuclei are generally basal, elongated, and not diffused nor wholly condensed. Many of them are densely stained. Fibre-like, deeply stained extrusions extend from the free edges of the cells, indicating that they maybe secretory. A specimen prepared on 25 August showed that the lumen of this chamber is occupied by numerous eggs that appeared as being embedded in a medium of sperm (Bawab, 1988) (Fig. 11.32C).

Referring back to Fig. 11.16A, to follow the upper group of ductules of the talon, these are transverse sections of end portions of the third component of the talon, the receptaculum seminis. This component is the same as in the stylommatophoran reproductive system of *Oxychilus cellarius* (Rigby, 1963) and in the opisthobranchs (Rigby, 1965). Lind (1973) illustrated the talon in the genital tract of *Helix pomatia* and described its 'spermathecal sac' as being composed of three to five blind sacs. In *Eremina*, this component (receptaculum seminis) originates as an outgrowth of the fertilization chamber (Figs 11.16E and 11.17D). It extends blindly backwards, along about three quarters of the length of the talon, and branches and rebranches giving rise to the different numbers in the different levels (Fig. 11.16B–D). It is lined by a single unfolded layer of typical columnar epithelium (Fig. 11.16N). Each cell has a basal, somewhat rounded and condensed nucleus (Bawab, 1988). During October, this receptaculum seminis was found to be occupied by sperm, whose heads are attached to the epithelial lining (Fig. 11.32A and B). This picture was described as oriented sperm (Breckenridge and Fallil, 1973) indicating that they are foreign sperm; this epithelium maybe secretory, to nourish or activate these foreign sperm.

In general, each part of the different components of this talon is surrounded by connective tissue, in which are embedded muscle fibres. This connective tissue appears continuous with that surrounding the whole structure on one hand, and the binding tissue on the other. It differs in the two sites: that which surrounds the components is compact tissue; while the binding is loose, connective tissue (Bloom and Fawcett, 1976). Another noticeable observation is that the muscle layer is more pronounced around the fertilization chamber and the muscle fibres themselves differ in size around the different components.

Just beyond the origin of the receptaculum seminis, the talon proceeds obliquely, as a narrow duct within the albumen gland (Fig. 11.17D), to pass from the ventral surface of the albumen gland to its dorsal surface, reaching almost the end of this gland where the spermoviduct lodges upon the gland (Fig. 11.15A and B). Along this narrow talon its interior appearance changes, but its components can be discernible (Fig. 11.15G). This is well observed by careful examination of the seriality in more than one preparation. Its epithelial lining forms definite numbers of deep villi, formed by the elongation of the columnar cells and not by the folding of the layer. This epithelial layer is covered, along the whole length of the talon, by a thin layer of connective tissue that is provided with fine muscle fibres.

Albumen gland. It is appropriate to deal with the albumen gland at this point in the description, since it is greatly related to the talon, i.e. to the hermaphrodite duct, examined as was described above, and with the spermoviduct (see later). It is a voluminous organ in the reproductive system of *Eremina* (Fig. 11.12D). It occupies the curve of the 4th whorl of the visceral hump, lateral to the digestive gland and parallel to the rectum (Fig. 11.11B and C). It is a curved organ where the convex surface is the dorsal side, and the concave surface is the ventral side (Fig. 11.12D). This albumen gland always attains the yellowish or cream colouration. However, this colour, together with its size, varies greatly from one specimen to the other. Its maximum size may reach 3–5 cm in length, 2 cm in width, and 1 cm in height.

This albumen gland does not embed the talon (continuation of the distal hermaphrodite duct) only, but it also connects its most distal portion (Fig. 11.15A). This connection cannot be seen externally, but the careful removal of albumen tissue surrounding this portion exposes an elongated then curved duct that connects the talon (Fig. 11.15C and E) just neighbouring (or beneath) the spermoviduct.

Along about 250 μm, as measured in one of the prepared specimens, the talon changes gradually in internal appearance (Fig. 11.15G), till it reaches the base of the spermoviduct, where the albumen duct forms the inner edge of the SPERMOVIDUCT (Fig. 11.17A and B). This picture is well confirmed by the frontal section shown in Fig. 11.16H. At this level, the talon appears just underneath the albumen duct, but does not form the initial portion of the spermoviduct, as it appears macroscopically (Fig. 11.15E). The most important factor that follows from this critical connection is the difference between the two components, the albumen gland and spermoviduct, both histologically and in acceptance to stains. However, the nearer the albumen duct becomes to the spermoviduct, the more it becomes its inner edge properly. This portion, which is nearest, becomes curved, and that is its semilunar lumen (Fig. 11.16G). The other continuous portion is wider and still related to the albumen gland as albumen duct. At this level, the talon connects with the central albumen duct (Fig. 11.16G). This connection is well observed in the macroscopical examination and is referred to as the talon-albumen duct (Fig. 11.15A). Just at this level, or a few microns beyond, the talon changes the orientation of its villi (Fig. 11.15G (b) and (c)) and deviates that of the ciliated hood to join the neighbouring end of the curved semilunar lumen of the spermoviduct (Fig. 11.16G and H) (to be sperm-channel, see later). The lumen of spermoviduct separates completely from the albumen chambers and, at this level, the spermoviduct becomes an independent component, but still upon the albumen gland. The portion of its semilunar lumen connecting the ciliated hood of the talon proceeds as the sperm channel; while the remainder of the lumen becomes the egg channel.

The whole bulk of the narrow talon proceeds forwards for a very few microns, still underneath the base of the

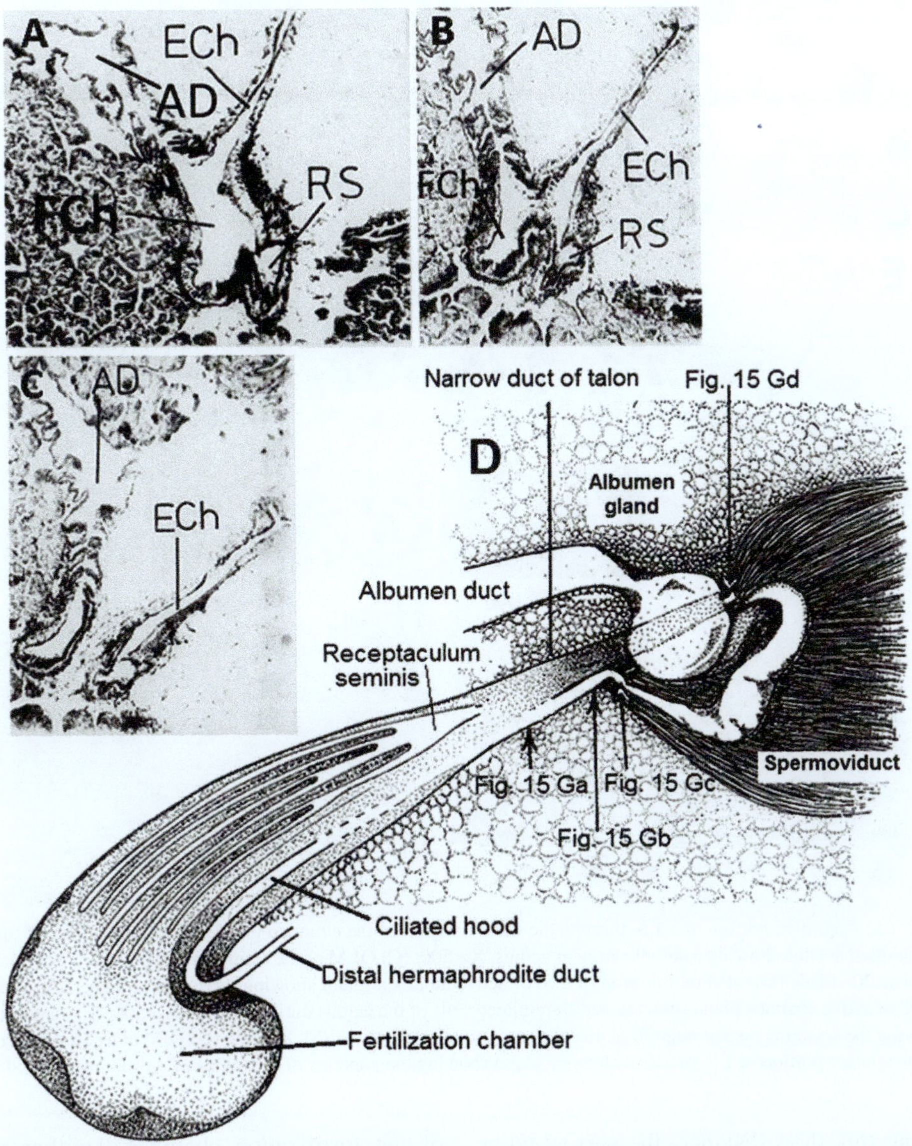

FIG. 11.17 Another preparation of serial T.S.s of the albumen gland of *Eremina* showing the talon: (A) showing the constriction of the receptaculum seminis (RS); continuation of the lumens of fertilization chamber, albumen duct (AD), and egg-channel (ECH); (B) showing the continuation of the albumen duct with the fertilization chamber; the separation of the egg-channel connected with the receptaculum seminis; (C) showing the complete separation of the egg-channel connecting the RS from the albumen duct connecting F Ch of the talon. X = 100; (D) diagrammatic illustration of the structure of the talon and its relationship with both the albumen duct and the spermoviduct;

(Continued)

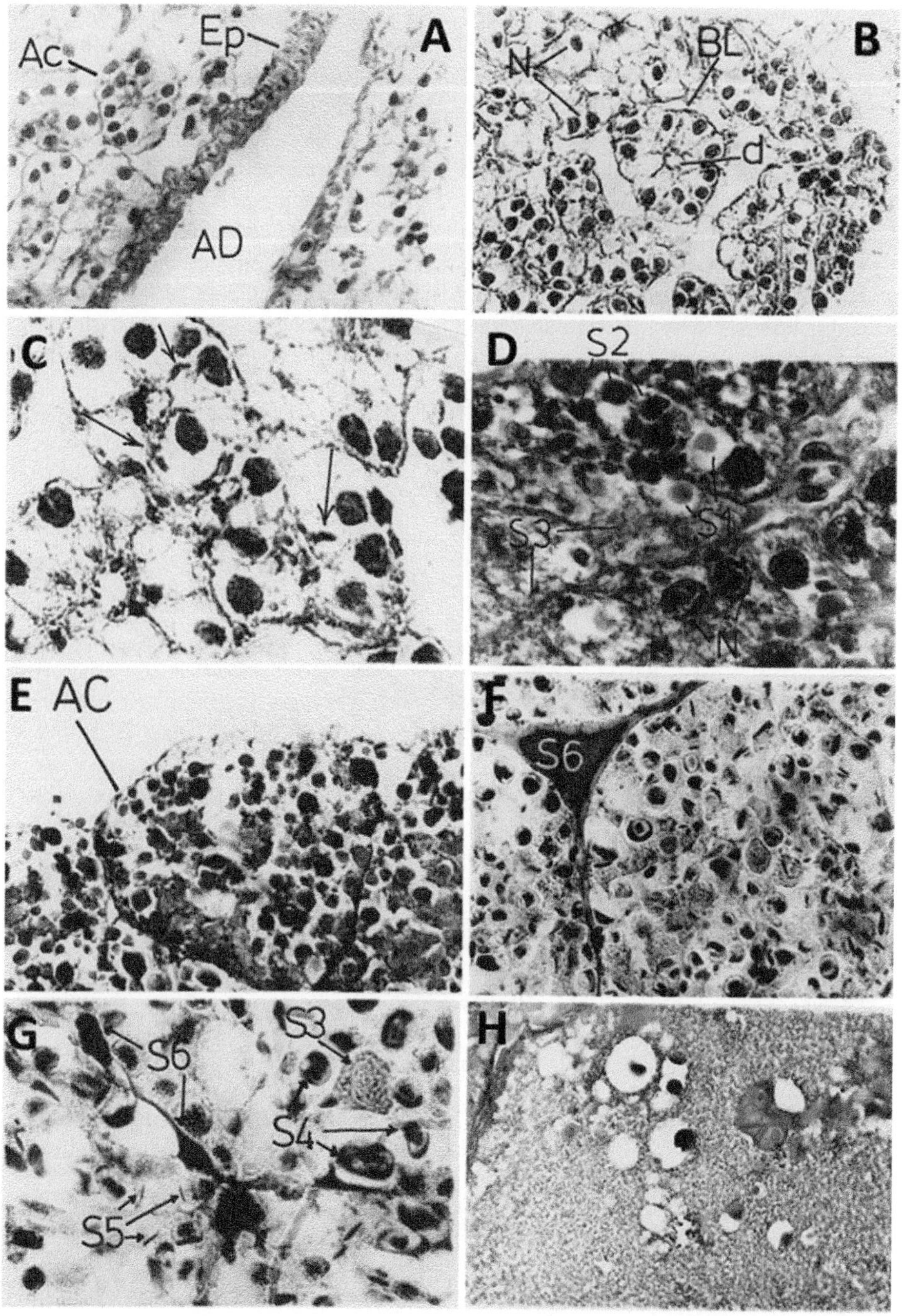

FIG. 11.17, CONT'D (A) magnified portion of a T.S. through the middle region of the albumen gland showing the lining of the central albumen duct. X=400; (B) another magnified portion showing a definite circular acinus. X=500; (C) OLM of the acini of an inactive albumen gland showing the ciliated cells (indicated by arrows). X=1500; (D) OIM of a portion of T.S. of active albumen gland showing three different secretory substances. X=1050; (E) magnified portion of T.S. of active albumen gland showing undifferentiated cells of the acinus during stage II. X=500; (F) magnified portion of T.S. of active albumen gland showing the secretion during stage III, notice the extra-acinus secretion (S6). X=700; (G) OIM of a portion of (F) showing different secretions. X=1150; (H) magnified portion of T.S. of active albumen gland showing the contents of the lumen of the albumen duct during stage III. X=500.

albumen duct. Along this short distance, the narrow talon reorients its villi again, constituting the fertilization chamber and the receptaculum seminis (Fig. 11.15G (d)). Those of the latter gradually become a constricted portion to be aside from facing the base of the albumen duct as those of the fertilization chamber. Another preparation of the albumen gland shows a different orientation of this relation. Fig. 11.17 A shows this constricted portion and the continuation of the fertilization chamber with the albumen duct, which is still continuous with the egg-cannel of the spermoviduct, at this level, the lumens of the egg-channel, albumen duct, and the fertilization chamber of the talon are continuous. Just beyond this level, the egg-channel is on its way to separate from the albumen duct (Fig. 11.17B) and the constricted receptaculum seminis becomes continuous with the egg-channel. Fig. 11.17C shows this complete separation and the complete independence of the spermoviduct from both the talon and the albumen gland.

The histological structure of the albumen gland leads to understanding its secretory function. Mehanney (1991) and Bawab et al. (1993) studied the structure and the secretory function of this albumen gland. Any section of the albumen gland shows many acini of different sizes and outlines. Comparing this picture with those in the textbooks of histology, it is a compound tubuloacinar gland. The acinus of a circular outline (Fig. 11.17 con. B) gives a simple picture to study. It consists of about 12 glandular cells enclosing a narrow ductule. This picture is the same as that of *Eobania vermiculata* and *Helix aspersa* (Bani, 1979, and Courtot and Gomot, 1982). These authors observed ultrastructurally very narrow ciliated cells in between the glandular ones. Courtot and Gomot added that these cells are very narrow, elongated, and located from the basal lamina to the free edge. They have broad and ciliated free edges at the same level as the glandular cells. However, Berry (1977) reviewed and gave an agreement that the albumen gland of pulmonates comprises columnar epithelial cells, interspersed with narrow, wedge-shaped ciliary cells. It was concluded that the secretory cells of the albumen gland are of one type (de Jong-Brink et al., 1979) and each cell can secrete many different substances. It was very difficult to identify this type of cells in the albumen gland of *Eremina* by light microscope. However, by the aid of an oil immersion lens, and with some patience, it could be observed and Fig. 11.17 continued C clarified these cells. Since one understands these cells, it could be observed even by normal magnification.

Following the difficulty encountered in preparing histological sections of the albumen gland to study the secretory function, because of the large nature of this organ, it was thought that biochemical reactions would lead to understanding its secretory substances. For this, 20 large active albumen glands were squeezed and dissolved in a medically prepared saline solution (about six times their volume) (Cole, 1944). The resulting solution was used for the biochemical tests. Table 11.1 shows the results of these tests. The results enable the use of proper stains, which in turn, gives the opportunity to understand the histochemical and histological picture. It appears that the activity of the albumen gland of *Eremina* can be identified as three successive stages, each with definite secretions identified by successive numbers to facilitate their following. The staining affinity of these secretions was described in Table 11.2.

Stage I: during this stage, the glandular cells of each acinus are well differentiated by their large nuclei, and their cytoplasm is occupied by three different secretions (Fig. 11.17 cont. D): secretion 1: large bright globules, few in number and well distributed in the cell; secretion 2: large in amount, in the form of small opaque globules that occupy almost the whole cell and; secretion 3: masses of fine granules. A remarkable observation is that few, very fine cells are scattered in the thin connective tissue coat surrounding each acinus, the function of which cannot be investigated, they are referred to as extra-acinar cells.

TABLE 11.1 Biochemical tests on the emulsion of the albumen gland of *Eremina*.

Test	Reacting group	Positive result of reaction	Observed result
Biuret	Peptide linkages	Rose	+
Ninhydrin	Any amino acid contains free carboxyl group and amino group	Blue	++
Xanthoproteic	Phenolic group	Yellow→to orange	++
Mercuric nitrite (modified millon's)	Phenol (tyrosine)	Red (pink)	++
Aldehyde	Tryptophane	Deep violet purple	+
Sulphur	Sulphide group	Brown or black	++
Arginine (Sakaguchi)	Arginine	Red	–
Molisch's reaction	Carbohydrate	Violet ring	++
Furfural test	Fructose, sucrose and pentose	Deep purple	–

From Bawab et al. (1993).

TABLE 11.2 Chemical nature of the secretory substances of the albumen gland of *Eremina*

	Stages of active albumen gland	I			II	III	
	Number of secretion	1	2	3	4	5	6
	Form of secretion						
Staining reaction	Indication	Globular	Small globules	granulated	Vaculated large globular	Rod-like crystals	Fluidy (extra-acinar)
Mallory's triple Stain		Bright blue	Blue	Violet	Pale orange	Deep violet	Bright red
Masson's Trichrome Stain		Bright green	Deep green	Deep violet	Pale red	Deep violet	Orange
Heidenhain's Iron haematoxylin		++	+++	+++	++++	++	++
Mercury—bromphenol blue (after Bonhag)	Protein	++++	++	++	–	+++	++++
Toluidine blue	Basic protein	–	++	++	++	++	+++
Performic acid Schiff	Ss	++++	++	++	–	++	–
PAS (after McManus)	Mucosubstance	++	+++	+++	++++	++++	+
Alcian blue pH2.5	Weakly acidic sulphated muco- substance	–	+++	+	++++	+++	–
Alcian blue pH1	Sulphated muco-substance	–	+++	+++	++++	–	–
Alcian blue PAS (after Mowry, 1963)	Acidic and neutral MPS	- ++	++ +++	- +++	++++ -	- ++	- +
Best's carmine	Glycogen	–	++	++++	++	+++	++
Sudan Black B (after Mcmanus, 1946)	Lipids	–	–	–	–	–	–

From Bawab et al. (1993).

Stage II: The cells of the acinus cannot be differentiated, but the acinus itself can be limited. The cytoplasm is largely masked by the secretory substances, the above three ones can be differentiated. In addition, another, secretion 4, appears, in the form of large globules (Fig. 11.17 con. G), it differs in its affinity to stains from secretion 1 (see Table 11.2). At the same time, secretion 1 becomes very difficult to be observed.

Stage III: Not only the cells of the acini which cannot be differentiated, but also the acini become so difficult to define their boundaries (Fig. 11.17 con. F). The secretion of the albumen gland appears as consisting of a heterogeneous assortment of inclusions. Critically, all the above secretions can be detected except secretion 1. A new one, Secretion 5, appears as rod-like crystals within the globules of secretion 4 (Fig. 11.17 con. G).

It is worth mentioning that all the above five different secretions produced during the three different stages of activity of the albumen gland are intra-cellular secretions. During Stage III, an additional but extra-acinar shiny, fluidy substance, secretion 6, occupying the spaces in-between the acini, appears to define the limits of the acini of the section of the albumen gland. It was difficult to define the source of this secretion. The histochemical results shown in Table 11.2, were compared with the results of the biochemical tests (Table 11.1). Each of the two confirms the other. In the lumen of the duct of the albumen gland, there is a somewhat homogeneous substance, since many of its components are not well differentiated (Fig. 11.17 con. H).

Spermoviduct. This organ passes through the mantle cavity. It is a characteristic feature for the Stylommatophora that the pallial region of the reproductive system is a single functional duct (Pelseneer, 1906). This organ of the reproductive system of *Eremina ehrenbergi* is greatly voluminous, glandular, and is the longest of this system (Fig. 11.12D). Concerning its size, it is noticed from the dissected snails of one collection, obtained on the 15th of November 1987, that it varies greatly, from voluminous (Fig. 11.18A (a)) to a thread-like organ (Fig. 11.18A (b)). Fifty specimens were dissected to study the reproductive system in general; about 25 specimens had different sizes of this spermoviduct (Mehanney, 1991).

At first, it was thought that this difference is according to the size of the system as a whole, or according to the physiological state of the system. This supposition was based on the fact that some of these snails copulate during October (Abd-El-Kerim, personal observation). However, Hammad et al. (1977) recorded that *Eremina desertorum* (=*Eremina ehrenbergi*) lays its eggs during January, February, and March; i.e. this is the breeding season. It was striking that the observation of November was repeated in another collection obtained on 24th of February 1989. This observation was greatly considered, since it might be significant. A third collection, obtained on the 9th of September 1989, confirmed this observation. The snails were aestivating, since they were completely adhering to the shrubs with the epiphragm covering their apertures. About 50 snails were dissected and the same observation was encountered. This could only be interpreted as that the spermoviduct of this *Eremina* has two different morphological sizes: a voluminous one (Fig. 11.18A (a)) that has a greatly folded pathway, and creamy or white in colour. One side is peculiar in appearance; it has a straight extension and smaller diameter. This is the male side. On the other hand, the thread-like spermoviduct (Fig. 11.18A (b)) has the same colour, but its whole extension seems to be straight. Its two components are clearly parallel to each other, but one of them, the female one, is slightly larger and wavy.

Generally, to the naked eye, the spermoviduct appears as if it emerges from the albumen gland (Fig. 11.15A, B, D, and E). It has a conspicuous appearance, since its dorsal wall is enlarged and folded, while its ventral one is a narrow, and straight, simple tube (Fig. 11.18B). Beyond the albumen gland, it extends along the midline of the body cavity. A hand section, taken through the mid-portion of an adult specimen (Fig. 11.18C), and the examination of that of an inactive small specimen show that its lumen is differentiated into a wide channel enclosed within the folded thick wall and is lined by irregularly-shaped and sized cells (Fig. 11.18F, G, and H), giving the appearance of that of the photomicrographs of the transitional stratified epithelium illustrated in the text books of histology, e.g. Bloom and Fawcett (1975); and a very narrow one, enclosed within the straight wall (Fig. 11.18C and D). The folded thick wall (Table 11.3) is considered as a gland, comparable and opposite in position, to the prostate gland of the sperm channel. It is preferred to refer to this gland as oviducal gland. Histologically, the narrow channel is the sperm channel; it contains, in some preparations, some fully formed sperm. This channel can be distinguished externally by a characteristic tissue defined as 'prostate gland' (Fig. 11.18C, D, J, and K) (Kugler, 1965; Rigby, 1963; Visser, 1981).

Bawab et al. (1995) did many different trials to describe the critical relationship between the talon, albumen gland, and spermoviduct. The talon was exposed by removing portions of the albumen gland from its ventral surface, although this was not entirely sufficient. The parts of the albumen gland including the talon were dehydrated, cleared, and examined. This preparation confirmed that the talon continues as the sperm-channel in the funnel-shaped portion of the spermoviduct. On the other hand, the egg-channel is the continuation of the essential talon after receiving the albumen duct (Fig. 11.15F).

From the above description, it becomes evident that the distal narrow talon gives rise to a definite passage, which diverges aside to continue as the sperm-channel of the spermoviduct (Fig. 11.15F). The essential portion of the distal narrow talon becomes, after receiving the albumen duct, the

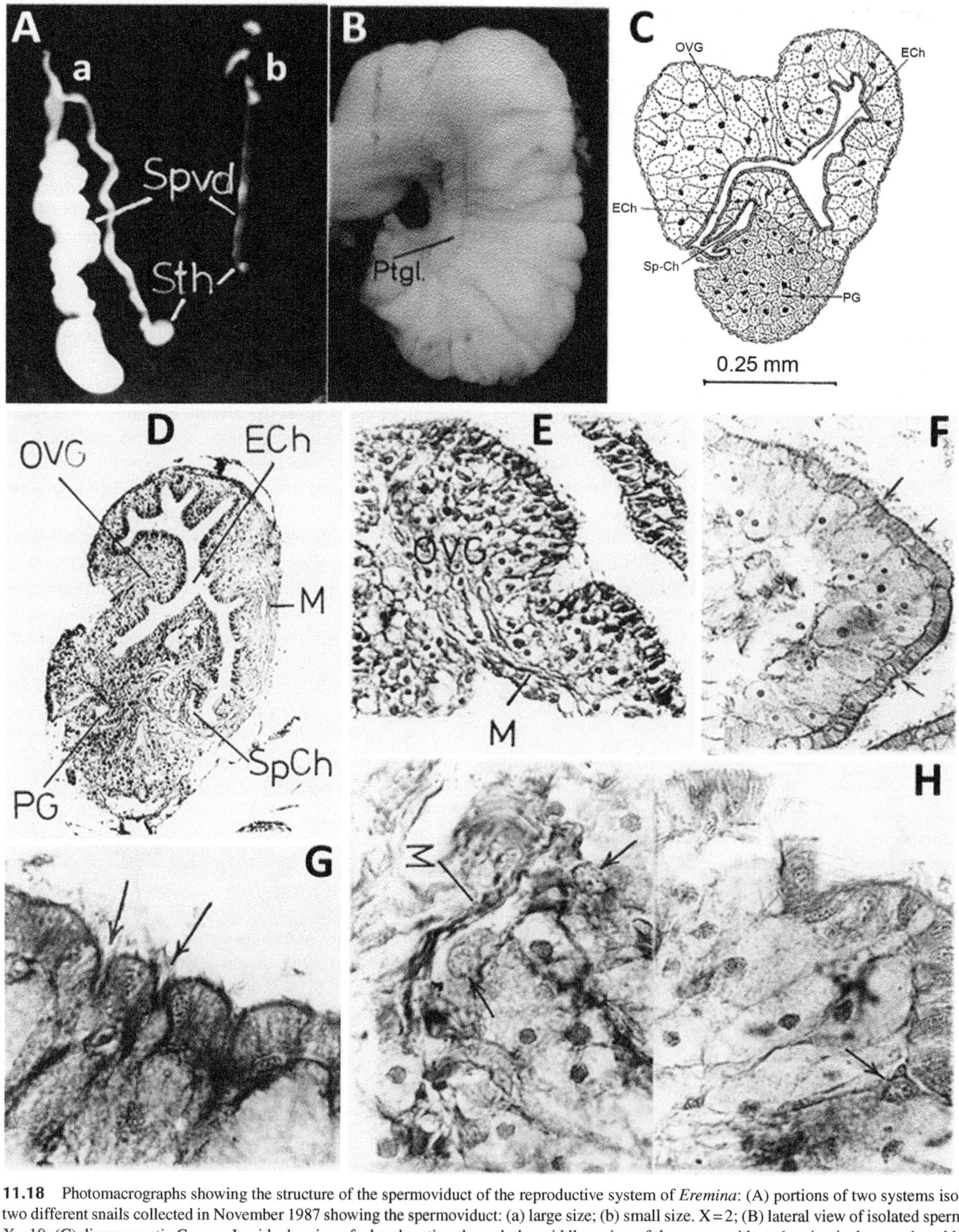

FIG. 11.18 Photomacrographs showing the structure of the spermoviduct of the reproductive system of *Eremina*: (A) portions of two systems isolated from two different snails collected in November 1987 showing the spermoviduct: (a) large size; (b) small size. X=2; (B) lateral view of isolated spermoviduct. X=10; (C) diagrammatic Camera Lucida drawing of a hand section through the middle region of the spermoviduct showing its lumen: dorsal larger egg-channel and ventral narrow sperm-channel; (D–H) histological structure of the spermoviduct: (D) T.S. through an inactive spermoviduct. X=100; (E) magnified portion of section D to show the histology of the oviducal gland. X=400; (F) magnified portion of the wall of another active spermoviduct showing the wall of the egg-channel. X=400; (G and H) Oil immersion magnification of portion of (F) showing in (G) the detailed structure of the epithelial lining. X=1500; in (H) the detailed structure of the oviducal gland, *arrows* indicate remarkable cells. X=1500;

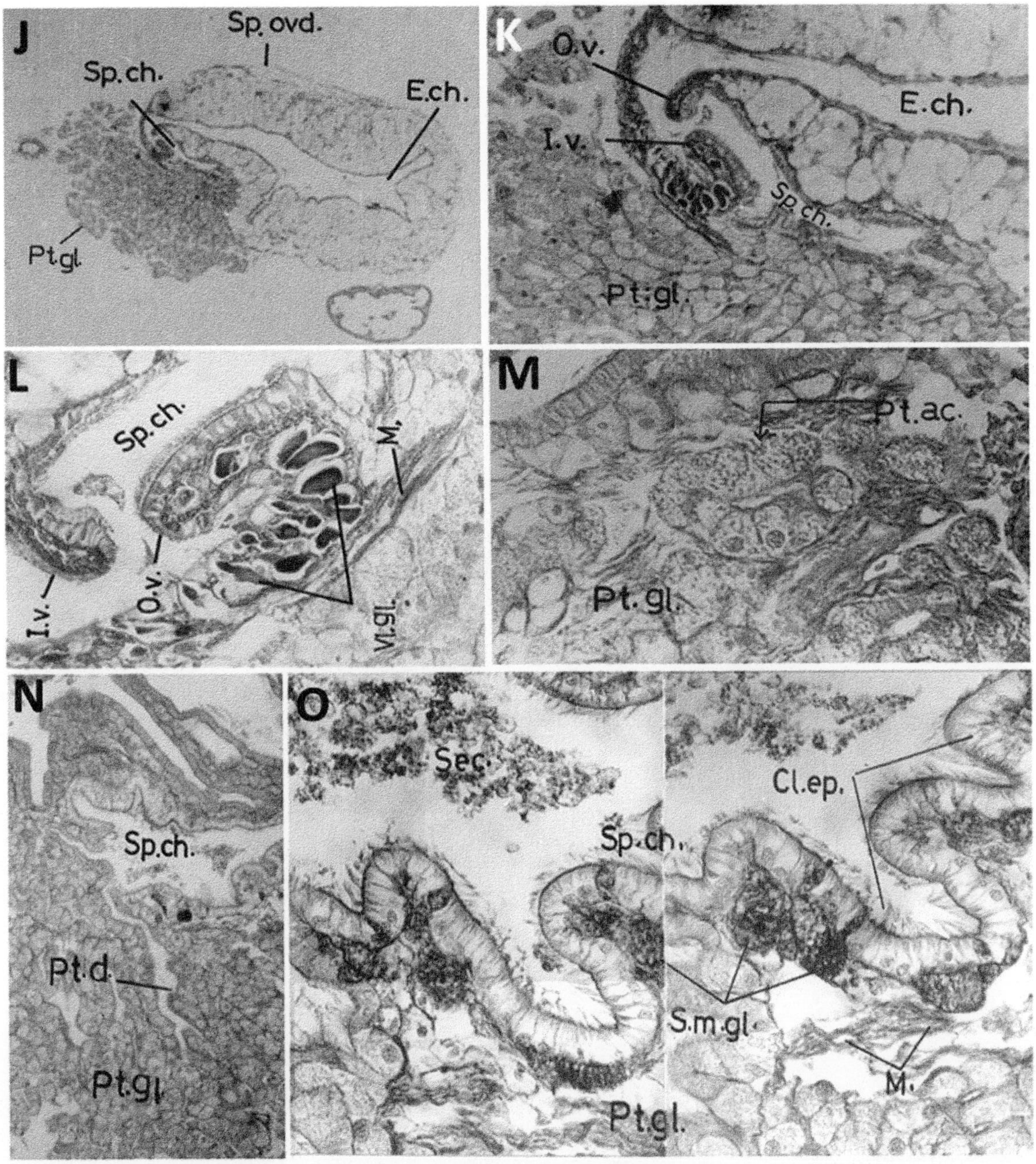

FIG. 11.18, CONT'D (J) photomicrograph of a T.S. of an active spermoviduct in its middle region. X=50; (K) magnified portion of (J) showing the two valves. X=150; (L) magnification of the portion of connection of sperm-channel with egg-channel showing the valves and the valve gland. X=500; (M) magnified portion of the prostate gland showing its acini. X=700; (N) photomicrograph of a T.S. of prostate gland showing a ductile opened in the sperm-channel. X=150; (O) a magnified portion of a T.S. showing the epithelial lining of the sperm-channel and its associated glands, X=700. Cl.ep.=columnar epithelium, ECH=egg channel, Ep.=epithelium, I.V.=inner valve, M=muscle, N=nuclei, O.V.=outer valve, Ov=oviduct, OVG=oviducal gland, PG or Pt. gl.=prostate gland, Pt. ac.=acinus of prostate gland, Pt. d.=duct of prostate gland, S.m.gl.=submucosal gland, Sec=secretion, SpcH=sperm channel, Sp.ovd.=sperm oviduct, Vd=vas deferens.

egg-channel. The two channels continue parallel to each other within the funnel-shaped portion of spermoviduct. Their lumens are slit-like, while the whole bulk is occupied by the glans constituting the walls of both sides, as it appears in all the preparations. This is the beginning of the independence of the spermoviduct as a component of this hermaphrodite system. Its end portion is the beginning of the bisexual portion of this complex system.

The male side of the system. The critical end portion of the spermoviduct (Fig. 11.19A) undergoes some gradual and synchronized changes to continue as a separate vas

TABLE 11.3 Staining affinity of the layers of the wall of the egg-channel of the spermoviduct of *Eremina*

	Inactive wall				Active wall				
	Epithelial lining		Oviducal gland cells		Epithelial lining			Oviducal gland cells	
Stain	Nuclei	Cytoplasm	Nuclei	Cytoplasm	Nuclei	Striation	Free surface	nuclei	Granules
Heidenhain's iron haematoxylin	Dark blue	Faint blue	dark blue	Faint blue	Dark blue	Blue	Blue	Black blue	Faint blue
Masson's trichrome stain	Brownish red	Green	Brownish red	Green	Dark blue	Green	Green	Dark blue	Green
Mallory's triple stain	Red	Blue	Red	Blue	Red	Orange	Blue	red	Blue

From Mehanney (1991).

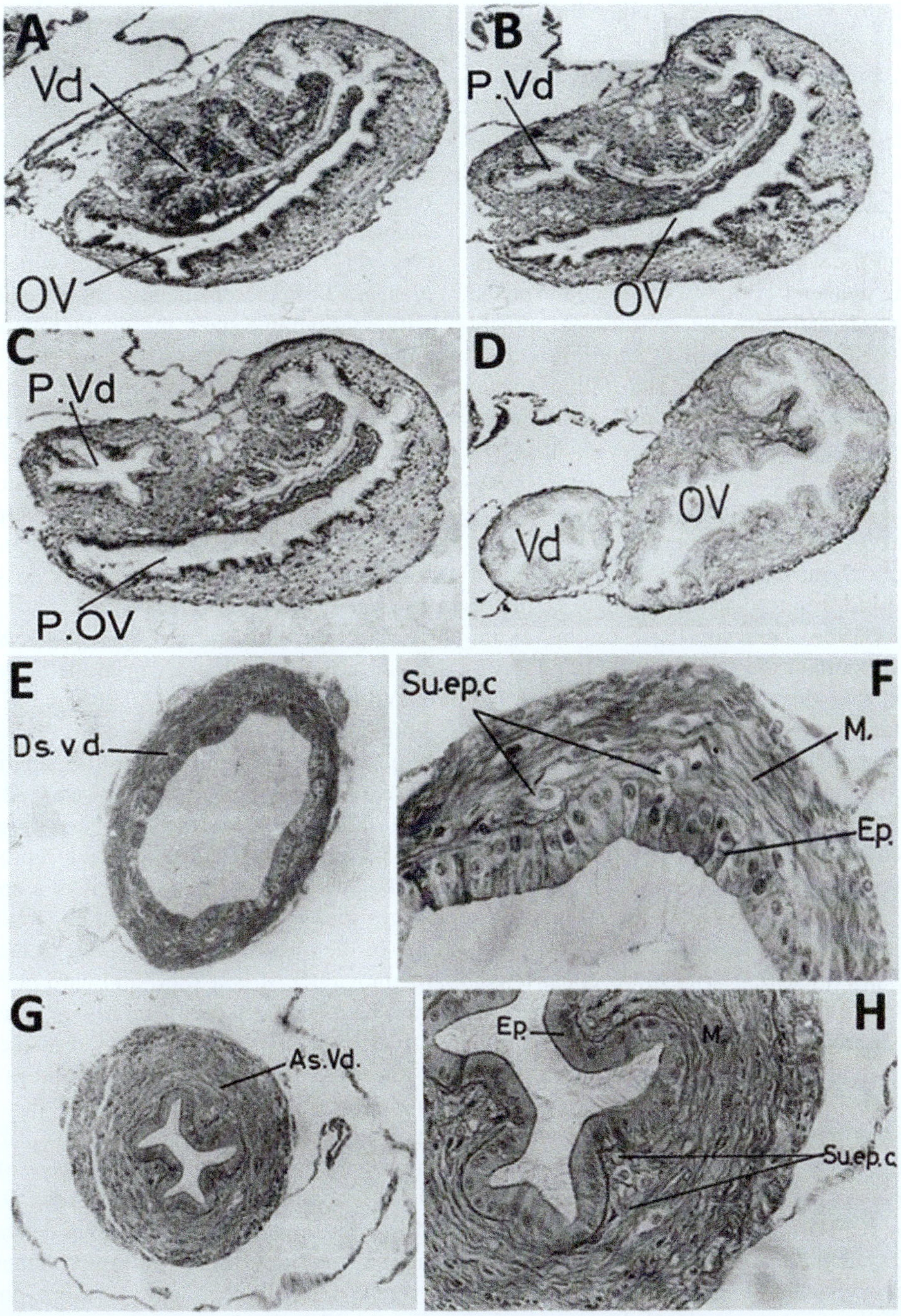

FIG. 11.19 (A–D) Serially selected T.S.s taken through the end part of the spermoviduct of *Eremina* showing the continuation of the egg-channel and sperm-channel into an oviduct and vas deferens; (E–H) photomicrographs of T.S. showing the structure of the vas deferens (vd): (E) section through the middle region of the descending vd. X = 100; (F) magnified portion of (E) showing the structure of the wall of the vd. X = 600; (G) section through the thick portion of the ascending vd. X = 100; (H) magnified portion of (G) showing the structure of the wall, striated epithelium and the subepithelial cells. X = 500. As.vd = ascending vas deferens, ds.vd = descending vas deferens, Ech = egg-channel, Ep = epithelium, M = muscle, OV = oviduct, Su.ep.c = subepithelial cells, Spch = sperm-channel, Vd = vas deferens.

deferens and oviduct (Fig. 11.19D). This portion is referred to as a transition region in the snail *Achatina fulica* (Breckenridge and Fallil, 1973). The prostate gland decreases gradually in size till it disappears completely at the point of separation.

The vas deferens. The vas deferens is the first portion (organ) beyond the spermoviduct that continues as a narrow cylindrical U-shaped structure with one limb longer than the other (Figs 11.12D and 11.23A). The longer limb is the part from its beginning till its flexion, referred to as the descending vas deferens. The other limb (continuation of the descending vas deferens after the flexion) is the shorter, ascending vas deferens. Histologically, along almost its whole length, it has the same structure. Fig. 11.19E and F presents the typical appearance of this vas deferens. The diameter of the latter increases gradually, within a very short distance forming the end portion of the ascending limb (Fig. 11.19G). This end portion differs somewhat from the whole length of the vas deferens. The increase in the diameter is due to the increase of the thickness of the muscular layer. The muscles are mostly circular in arrangement, but longitudinal, and different arrangements are present (Fig. 11.19H). The folding of the epithelial lining forms few crypts or villi and so the lumen is highly constricted (Fig. 11.19G). It appears that this epithelium is secretory.

The penis. The end portion of the ascending vas deferens flexes again to continue as a penis (Fig. 11.12D). Its end connecting with the vas deferens elongates, much as a fibre-like muscle, referred to as a 'penis retractor muscle' (Pelseneer, 1906; Duncan, 1958; Rigby, 1965; Visser, 1981). The examination of the reproductive system in situ proves that this muscle extends backwards to a point found far away, above the kidney, where it is inserted on the floor of the mantle cavity. This penis is a white tubular organ and lies in the head region. It is enclosed in a comparatively thick sheath, which can be observed by the naked eye. The longitudinally incised penis (Fig. 11.20A) shows that its lumen is continuous with that of the ascending vas deferens. Its wall is highly muscular and it is differentiated into two somewhat equal regions. The lumen of the proximal region is more constricted, and this differentiation is marked internally because the thick folds of the proximal region fade abruptly forming the distal region. This description coincides with that of *Succinea putris* (Rigby, 1965). This author referred to the proximal region as the 'penis proper' and to the distal one as the 'praeputium'.

The genital atrium. The genital atrium is the end part of both the male and female sides (Figs 11.12D and 11.23A) of the reproductive system of the snail *Eremina ehrenbergi*. It is a small common chamber with a rigid muscular wall in the normal condition, i.e. not copulated. It opens to the outside by the genital opening by which the genital product discharges.

The female side of the system

The oviduct. Fig. 11.19D shows the beginning of the oviduct. It is very narrow compared to its preceding continuation in the spermoviduct, but it is wider than the vas deferens. It proceeds as a simple tube (Fig. 11.21A) parallel to the descending vas deferens (Figs 11.12D and 11.23A), and before its continuation to form the vagina, it receives the spermathecal duct. This connection marks the beginning of the vagina. Histologically (Fig. 11.21B), it differs from all the other organs of this hermaphrodite system in that, in more than one preparation, its epithelial lining shows secretory function (Fig. 11.21C). No definite free surface was observed for the cells, since the cytoplasm appears while discharging (apocrine) its product to the lumen. This secretion itself is frothy in appearance and has the same affinity to stains as the cytoplasm of the lining epithelium (Fig. 11.21C). Also this oviduct differs by the striking presence of numerous deeply stained nuclei-like structures with considerable sizes, giving the impression that they are amoebocytes.

In another preparation, the epithelial lining of the oviduct is not as folded as in Fig. 11.21B; i.e. it has shallow folds, and the cytoplasm of the cells is clear, faintly stained, and with limited free surfaces all around the layer (Fig. 11.21D and E), with no secretion. In the opinion of the current author, this specimen is just at the finish of the aestivation and there is no source of food in the desert.

The spermatheca and spermathecal duct. The spermatheca is a common organ in the hermaphrodite system of pulmonates (Berry, 1977; Solem, 1978) that joins the female side by a duct, in between the oviduct and vagina (Fig. 11.22). Generally, it is the smallest organ of this hermaphrodite system, but its size and the length of its duct varies greatly with the species (Fig. 11.12C and D). Its size may also change with the reproductive condition of the specimen. In *Eremina*, it is a spherical mass that lodges in the mantle cavity just underneath the distal portion of the albumen gland, nearest to the beginning of the spermoviduct. It is a solid and hard element when seized by forceps or even by fingers. It is opaque and brown in colour, but this colour differs from the brownish cream to the very dark brown in the different specimens of the one collection. It is supposed that the dark brown spermatheca preserves spermatophore(s). This difference in the natural colour indicates the degree of hardening of the wall of the spermatophore (Richards, 1951; Bawab, 1970; Malek and Bawab, 1974a,b). By a delicate incision under the binuclear microscope, the spermathecal wall was gently removed and the contents were photomacrographed (Fig. 11.27E). Each spermatheca contains a capsule that is typically pear-shaped in outline. By smearing this capsule on a slide with a drop of water and examining microscopically, vibrating live sperm were observed and remained alive during the whole period of examination (several hours). This experiment was repeated for the spermathecae of all different colours and the result was the same. This indicates that the spermatheca of *Eremina*, always contains spermatophore(s).

The texture of the occupied spermatheca is hard, and very difficult to be prepared histologically as it is rendered harder during the technical preparation. However, its orientation

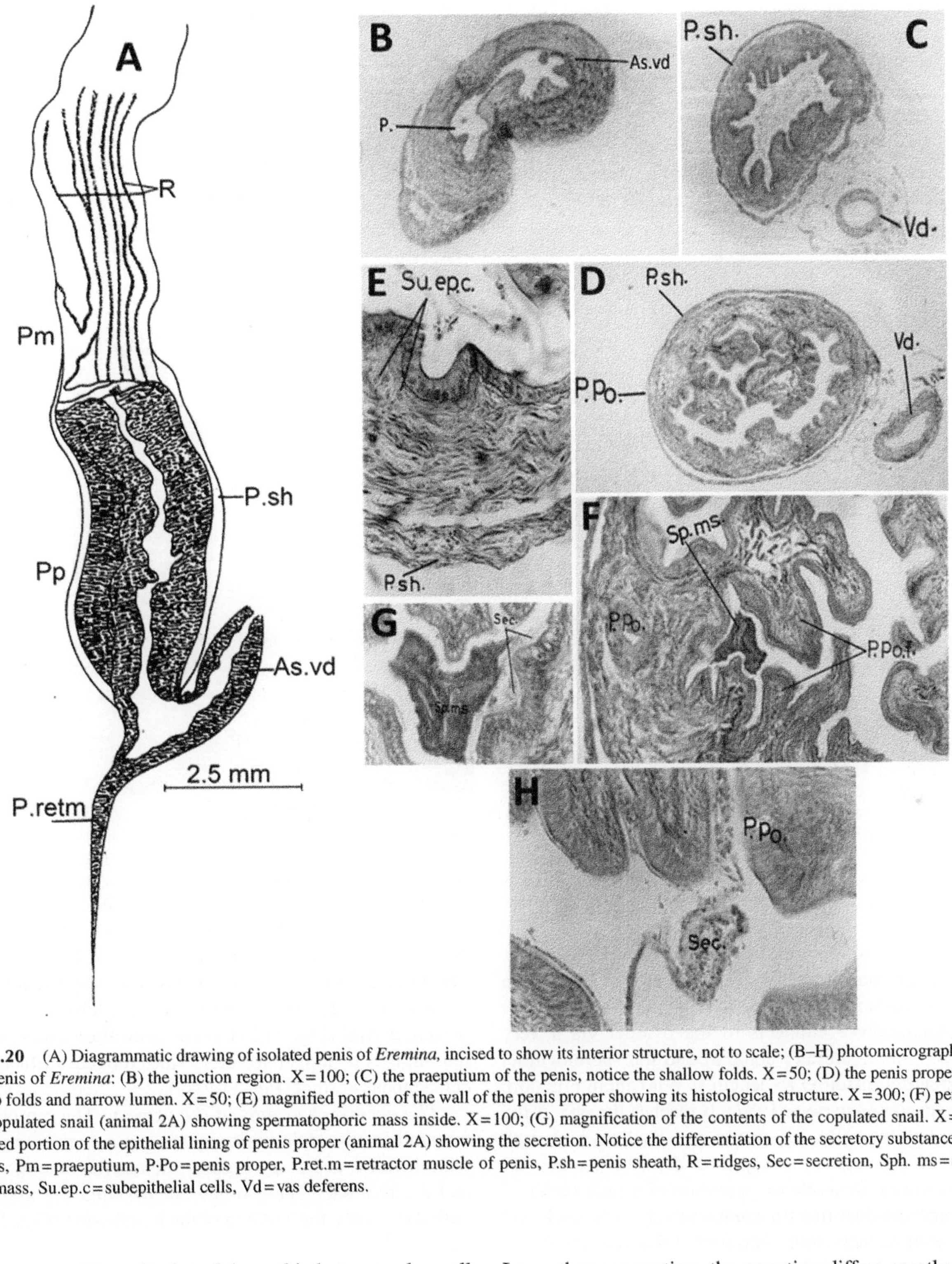

FIG. 11.20 (A) Diagrammatic drawing of isolated penis of *Eremina*, incised to show its interior structure, not to scale; (B–H) photomicrographs of T.S.s of the penis of *Eremina*: (B) the junction region. X = 100; (C) the praeputium of the penis, notice the shallow folds. X = 50; (D) the penis proper showing the deep folds and narrow lumen. X = 50; (E) magnified portion of the wall of the penis proper showing its histological structure. X = 300; (F) penis proper of precopulated snail (animal 2A) showing spermatophoric mass inside. X = 100; (G) magnification of the contents of the copulated snail. X = 500; (H) magnified portion of the epithelial lining of penis proper (animal 2A) showing the secretion. Notice the differentiation of the secretory substance. X = 350. P = penis, Pm = praeputium, P·Po = penis proper, P.ret.m = retractor muscle of penis, P.sh = penis sheath, R = ridges, Sec = secretion, Sph. ms = spermatophoric mass, Su.ep.c = subepithelial cells, Vd = vas deferens.

was done according to its duct. It has a thin but muscular wall (Fig. 11.22) and a secretory elongated columnar epithelium lining (Fig. 11.22C), enclosing a comparatively wide lumen. Some cells discharge their secretion holocrine (Leonhardt, 1977). Separate large masses of this secretion appear as accumulations in the free surface of these cells (Fig. 11.22D). In another preparation, the secretion differs greatly, since it is condensed, forming a thick granulated layer, observed as a diffused substance (Fig. 11.22E) within the lamellated wall of the spermatophore found inside the lumen (see above). This lamellated wall (Fig. 11.22B) encloses a small homogeneous mass, which is the spermatophoric mass containing

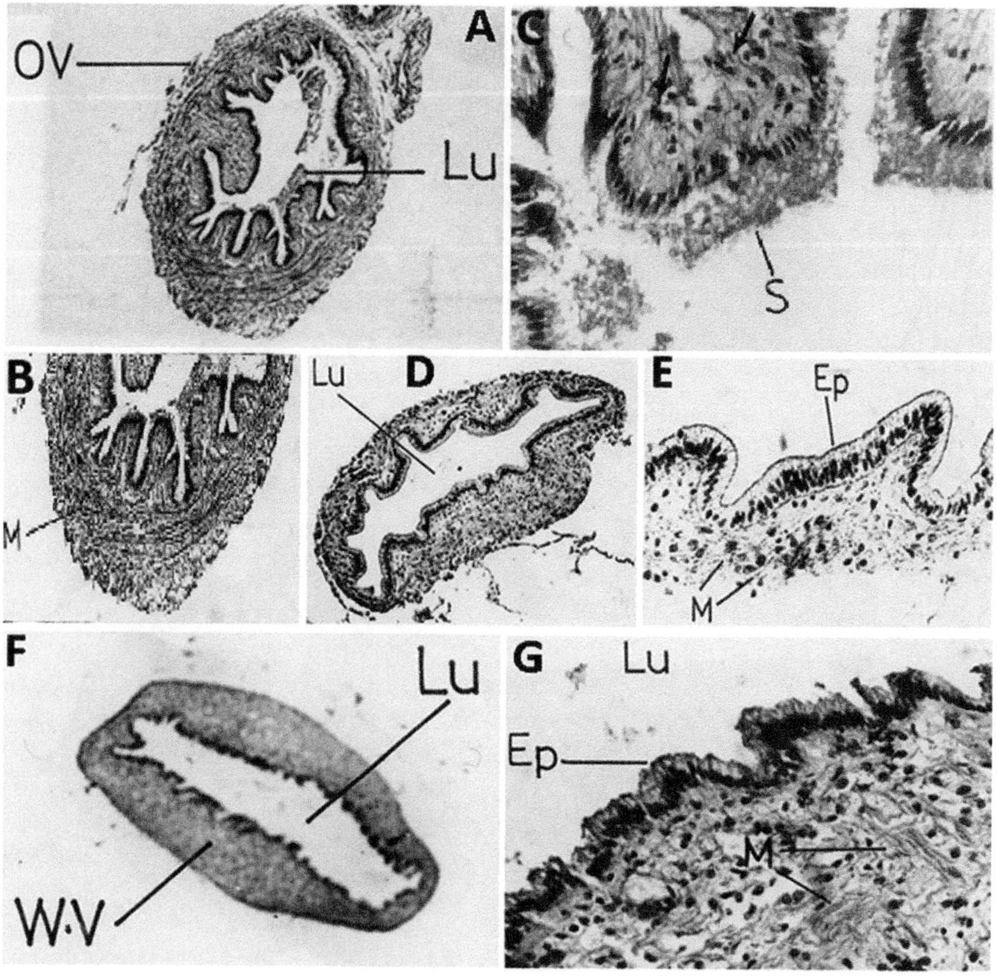

FIG. 11.21 Photomicrographs of T.S. of the oviduct of *Eremina* through its middle region: (A) whole mount of the section. X=60; (B) magnified portion of the section showing the histological structure. X=100; (C) magnification of portion of the epithelial lining showing its secretion. Notice the *arrows* that indicate the amoebocytes. X=550; (D) section through the middle region of an inactive oviduct. X=100; (E) magnified portion of the epithelial lining of (D). X=600; (F) section through the middle region of the vagina. X=50; (G) magnified portion showing the histological structure of the wall of the vagina. Notice the thick muscle layer, arranged irregularly. X=600. Ep=epithelium, Lu=lumen, M=muscle, OV=oviduct, S=secretion, W.V=wall of vagina.

sperm according to the artificially dehisced one. This spermatophore appears the same as that found in the atrium (Fig. 11.27F) although the latter is longitudinal section.

The spermatheca connects to the oviduct via a very long duct, which naturally coils around the spermoviduct while extending distally. Its length is a taxonomic feature with the species (Aparicio, 1982). It reaches the end of the oviduct. Histologically, it is the same all along its length (Fig. 11.22F), except for a very short distance just beyond the spermatheca. In another preparation of a snail obtained in February, the duct itself is comparatively very thick and the epithelial cubical lining is refolded that the lumen becomes narrow (Fig. 11.22G). In the two preparations, the lumen is empty. The duct opens laterally at the right side of the end port of the oviduct (Fig. 11.23A).

The vagina. Externally, this organ appears as a continuation of the oviduct. It is a very short organ, but important since it receives two characteristic accessory organs: the mucous glands and the dart sac (Figs 11.12D and 11.23A). The end part of the oviduct changes gradually to become the vagina. The latter differs in having wide lumen and thicker muscular wall (Fig. 11.21F). The typical columnar epithelial lining is no longer secretory since the free surfaces of its cells appear normal (Fig. 11.21G).

The accessory organs of the vagina. The reproductive system of some stylommatophoran pulmonates is provided with different accessory organs, such as the mucous glands and the dart sac. These organs in *Eremina* are associated with the vagina, the end part of the female side (Figs 11.12D and 11.23A).

The mucous glands. It appears that these separate glands are specific for individuals that belong to family Helicidae. The branching and the number of the terminal branches are used as taxonomic feature with the genus and species (Aparicio, 1982). In *Eremina,* these are two simple branched tubular glands, since they are similar to the illustrations found in

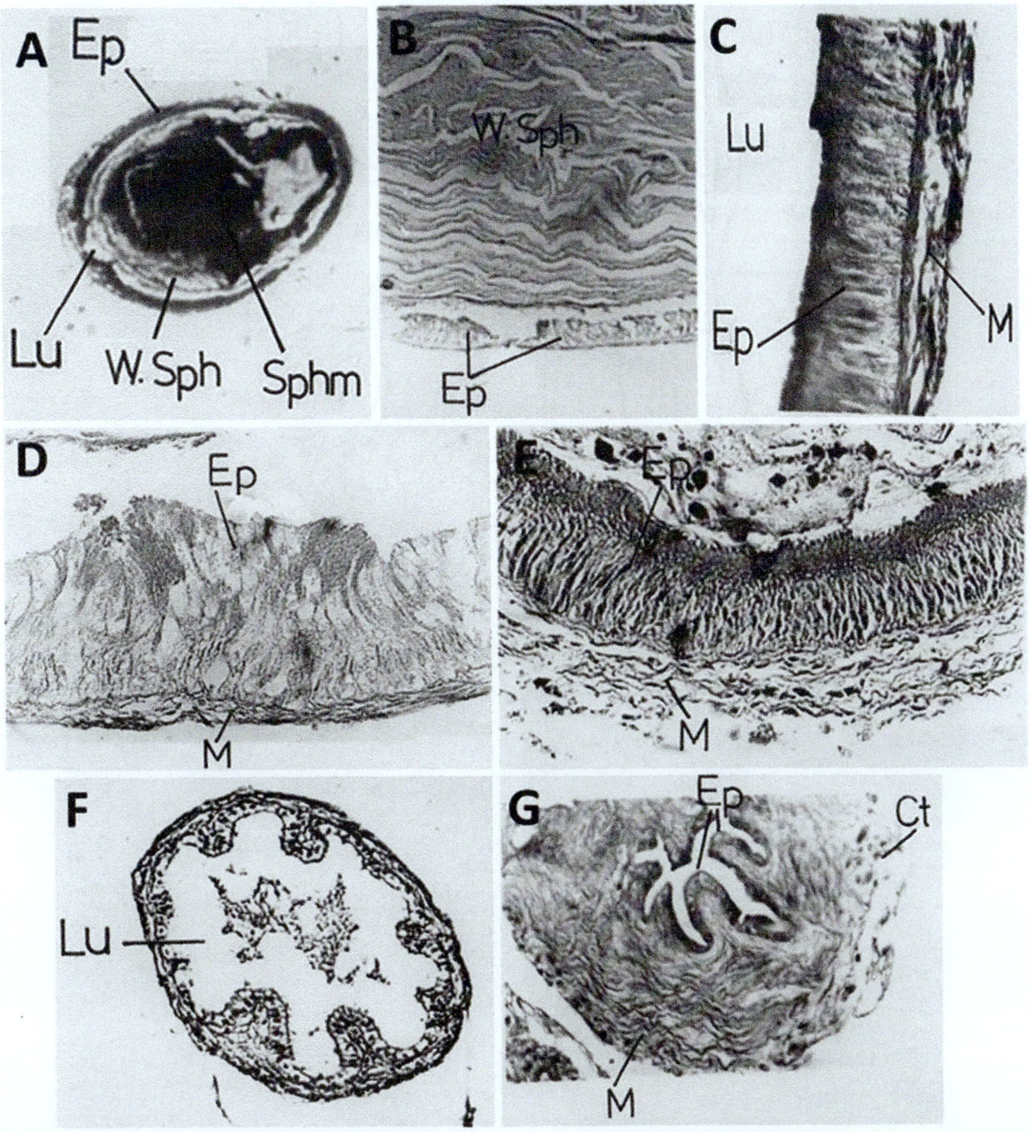

FIG. 11.22 Photomicrograph of T.S. of the spermatheca and its duct of *Eremina*: (A) section through the middle region of the spermatheca showing its contents. X=50; (B) magnified portion showing the lamellated appearance of the wall of the spermatophore inside the spermatheca. X=200; (C) magnified portion of the wall of section A showing the histological structure of the epithelial lining. Notice the dense free edge of the free surface, which indicates the merocrine secretion. X=500; (D) another portion showing holocrine secretion. X=300; (E) a third portion showing the diffused secretion. X=200; (F) section through the middle region of the spermathecal duct of small size hermaphrodite system of *Eremina*, obtained during December showing its histological structure. X=250; (G) magnified portion of section through the middle region of the spermathecal duct of a large system obtained during February. Ct=connective tissue, Ep=epithelium, Lu=lumen, Sph.m=spermatophoric mass, W.Sph=wall of spermatophore.

Bloom and Fawcett (1975). On either side of the opening of the dart sac opens the common tube of one of the two glands, i.e. the three openings are at the same level with respect to the vagina (Fig. 11.23B). The branching of each gland is regular, especially at the basal parts, each gland divides into two branches, then each branch divides into two, becoming four in number. Then, the branching seems to be irregular but peculiar; Fig. 11.23A explains this branching. The size of these glands varies according to the vital condition of the animal. It was noticed that the copulated snails have the largest size of mucous glands (Abd El-Kerim, personal communication). The histological studies confirmed that these glands are inactive during July, i.e. their epithelial lining is not secretory (Fig. 11.24A and B), while during August and October, this epithelium shows the active phase of these glands (Fig. 11.24C–F). The cells are larger, with a large nucleus, which has an active appearance. The cytoplasm, in general, is heavily loaded with secretory

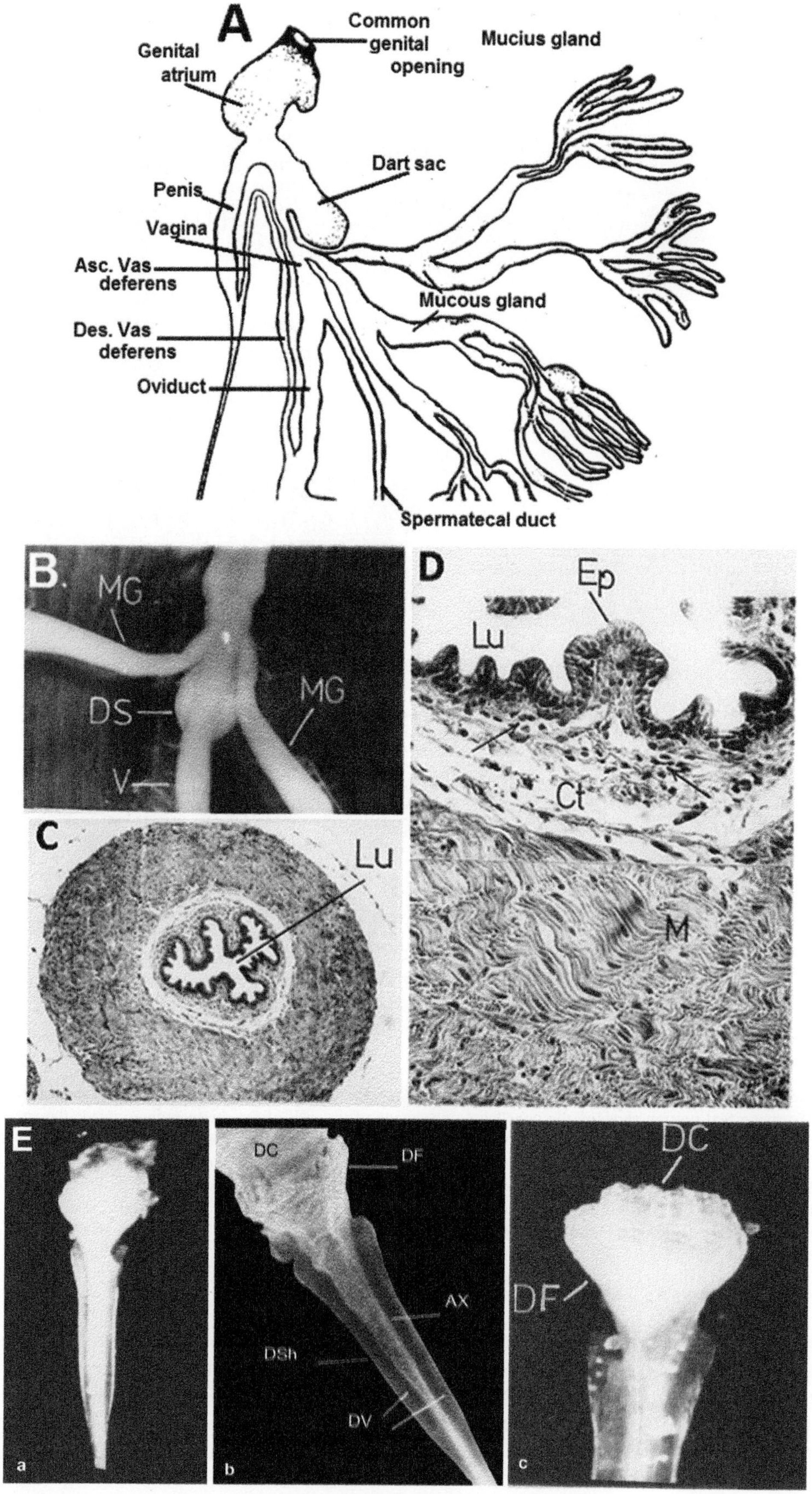

FIG. 11.23 (A) Magnification of the most distal portion of the hermaphrodite reproductive system of *Eremina* showing the connections of the organs of this region; (B) photomacrograph of the ventral view of the vagina showing the connection of the two lateral mucous glands and the dart sac. X=10; (C and D) photomicrographs showing the structure of the dart sac: (C) whole mount of a T.S. through the middle region of the sac. X=100; (D) magnified portion of (C) showing the histological structure of the wall of the sac, *arrows* indicate the wandering cells. Ct=connective tissue, Ep=epithelium, Lu=lumen, M=muscles; (E) photographs showing dart of *Eremina ehrenbergi*: (a) an isolated dart. X=20; (b) SEM of an isolated dart showing its structure. X=50; (c) magnified basal part of the dart showing the corona. X=50.

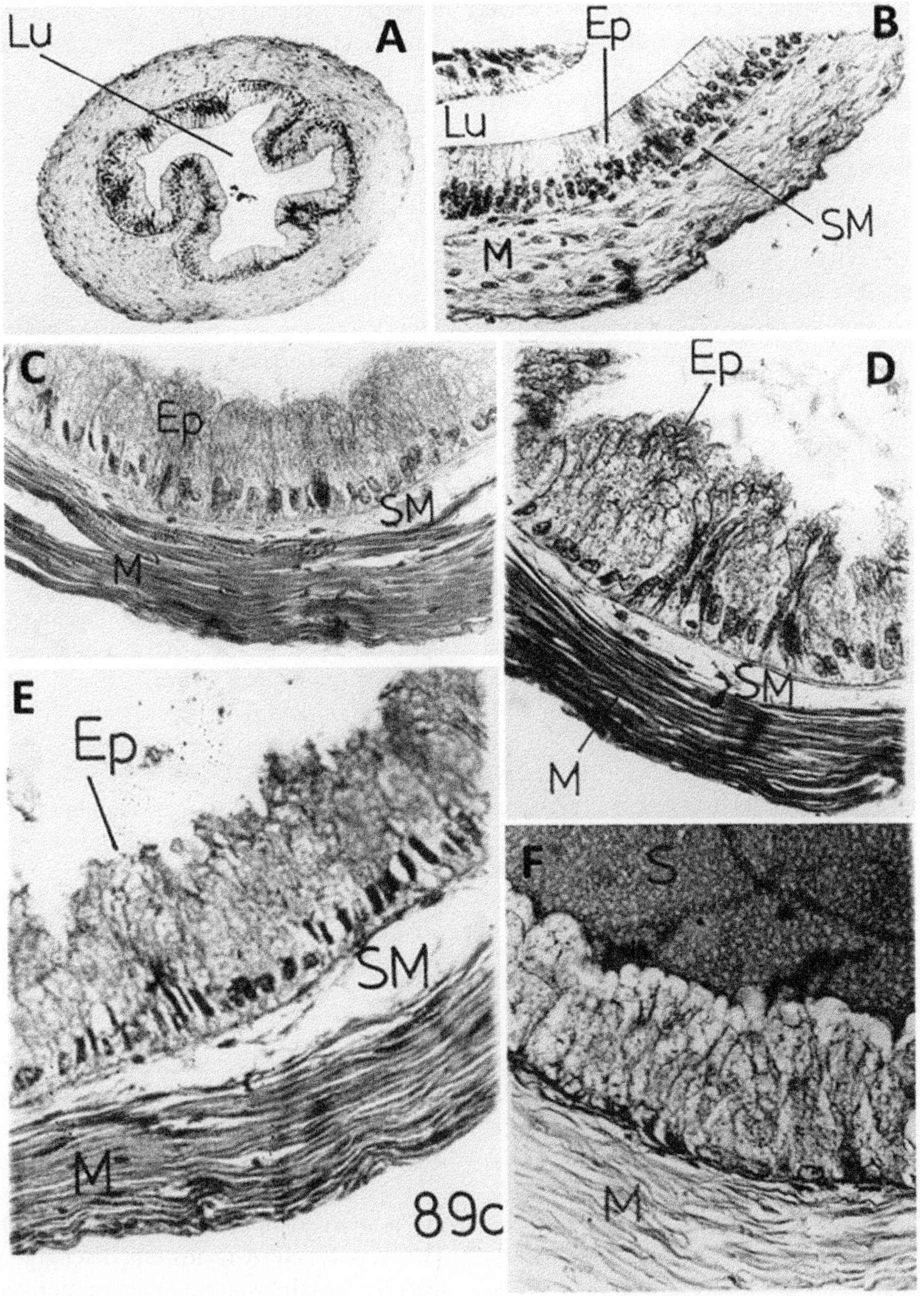

FIG. 11.24 Photomicrographs of T.S.s showing the structure of the mucous glands: (A) WM of the section of an inactive gland of a snail obtained in July. X=150; (B) magnified portion of the wall of the section, notice the smooth free edge of te epithelial lining. X=350; (C–F) showing portions of the wall of the gland of active snail showing the secretory activity of the gland: (C) shows the frothy appearance of the secretion; (D) shows the rosette-like extensions of the secretory cells; (E) showing the holocrine secretion; (F) showing the apocrine secretion. (C–F) X=500. Ep=epithelium, M=muscles, SM=submucosa.

substance, that is frothy in appearance or finely granulated (Fig. 11.24C). The latter becomes copious in many portions of the epithelial lining, to the degree that the free boundaries of the cells disappear. The cytoplasm, or more precisely the secretory substance, extends freely so that the free surface becomes irregular (Fig. 11.24D). At this step of activity, the secretory cells themselves appear in a characteristic form by grouping together in rosette-like extensions (Fig. 11.24D). The extended portions of the cells give the impression that the secretion (Fig. 11.24D) is holocrine (Leonhardt, 1977).

The preparation from October, in specific, shows another picture. The lumen is wholly filled with dense substance (Fig. 11.24F). The epithelial lining cells differ too. They are not frothy in appearance, their free surfaces are globular, and are in touch with the contents of the lumen. This picture indicates that this secretion is discharged apocrine (Leonhardt, 1977).

A striking observation is that the epithelial lining of the end branches of a specimen obtained in August has a completely different appearance. Its cells are represented only by the nuclei close to the basal lamina or by very short cells. The latter appear as if they are newly built.

The dart sac and dart. The other organ associated with the vagina is the dart sac. It is in the form of an appendix joining the vagina at its end, in between the mucous glands (Figs 11.12D and 11.23A). It is a short but swollen sac, and always hard, since it contains the calcareous dart which it secretes. Fig. 11.23C shows a section through the middle region of the sac. It is circular in outline with a very thick muscular wall and a definite, narrow lumen. The latter may be in the form of two narrow channels crossing each other, giving a cross-shaped lumen. The muscles in the wall are in different directions and occupy about ⅘ of its thickness. The epithelial lining is in columnar stratified form (Bloom and Fawcett, 1975). Just underneath the epithelial lining, small cells are condensed in the connective tissue of the submucosa. They are referred to as wandering cells (Fig. 11.23D).

It is striking that the lumen of the dart sac was always empty in the sections, contrary to the fact that it always contains a dart, since the latter can be isolated from the sac of any snail at any time of the year. In the opinion of the current author, it might be that it dissolves during the technical preparations.

The dart. The dart is a delicate, pointed, calcareous weapon, generally possessed by land snails (Tompa, 1980). It is commonly known as 'telum amoris', or 'love dart' (Morton, 1979; Mann, 1984) which is discharged by the partners with some velocity before courtship. To study the dart of *Eremina,* it was isolated as follows. The dart sac was macerated in 10% potassium hydroxide solution (Mahoney, 1973) or in Clorax (5.25% sodium hypochlorite) (Tompa, 1980; Dillaman, 1981). The isolated dart was thoroughly washed with distilled water and then dried. To be kept permanently, it was mounted on a piece of black card firmly stuck to the slide (Mehanney, 1991). The dart of *Eremina* (Fig. 11.23E (a)) measures about 3 mm. Its morphology is nearest to that of *Eobania vermiculata*, illustrated by Tompa (1980). Since *Eobania* and *Eremina* belong to the same family, Helicidae, the terminology adopted by Dillaman was used to describe the dart of *Eremina*. However, the two darts are somewhat different. Fig. 11.23(a) and (b) shows the structure of the dart of *Eremina*, it consists of the corona (basal part), flare, and the shaft. In situ, the corona is attached to the blind end of the dart sac by a voluminous muscle band. The corona is cup-shaped (Fig. 11.23E (c)) with diameter measuring about 1 mm and its edge is irregular and in the form of festoons, which gives the dart its specific form. The flare is the stem of the corona (Fig. 11.23E (b) and (c)), which joins the shaft, and measures about 0.5 mm; it is as broad as the base of the corona and gradually narrows towards the shaft. The latter almost constitutes the whole length of the dart which measures about 1.9 mm. This dart has a central axis (extension of the flare) and a vane on either side (Fig. 11.23E (b)). The shaft seems opaque and thus clearly differentiated from the vanes. It gradually narrows towards its end, which is described as a pointed end. The vane is comparatively very thin and thus transparent. It begins as a broad wing that extends along the whole length of the central axis. It gradually narrows and ends at the pointed end of the central axis.

It is appropriate to record that the SEM studies did not add much to that of the above description. Only the edge of the corona appears more obvious (Fig. 11.23E (b)).

Gametogenesis

Spermatogenesis. Spermatogenesis in prosobranchs has received considerable attention, probably because of the presence of complicated polymorphic spermatozoa in many species (Zabala et al., 2012; Giménez, 2013). Generally, its male produces a 'normal' sperm (eupyrene) capable of fertilization and (parapyrene) not fertile. Many species, as regular feature, produce 'abnormal' sperm that are irregular in chromatin content and do not play a role in fertilization. Few species produce spermatophores, and the spermatophore capsule is digested in the female after the sperm have escaped.

Beeman (1977) discussed and reviewed the literature on molluscan spermatogenesis, especially that of opisthobranchiata He reported that it is poorly known, in spite of the good description of early stages of spermatogenesis, the phase-contrast study of opisthobranch spermiogenesis, and the few excellent drawings of the stages for several opisthobranchs. Beeman reported the conclusion that the morphology of molluscan sperm was related to the biology of fertilization. In addition, he reported the proposal that the molluscan sperm is of two types: (1) sperm bearing a short compact head, which are the primitive type. This is retained by forms that discharge their spermatozoa directly into the water; (2) Sperm delivered by copulation, which displayed more elongated heads and the middle pieces were considered advanced. In the laboratory of Beeman, the first ultrastructural and phase-contrast studies on spermatogenesis of *Phyllaplysia taylori* and *Aplysia californica* showed that this reproductive activity occurs simultaneously with oogenesis in the ovotestis. This result disproved the notion that gamete isolation in such animals is a matter of protandry or 'staggered' gametogenesis. The total period of spermatogenesis in *Phyllaplysia* is about 20 days.

The morphologically mature sperm of *Phyllaplysia* has a long thread-like (about 0.3 μm diameter) nucleus wrapped in long helical turns around the anterior part of a standard

9 + 2 sperm filament or axoneme (with similarities to that of the sperm of the pulmonate *Testacella*).

Berry (1977) reviewed the works on gametogenesis in marine pulmonate Basommatophora, and in his view, it received scant attention.

In the desert snail *Eremina ehrenbergi*, Abd el-Kerim (1988) and Bawab et al. (1992a,b) studied, in detail, its male tract and the process of spermatogenesis. As mentioned above, the gonadal tubule is differentiated into a proximal wide reproductive half or generative half, and a distal narrow conducting half (Fig. 11.13C and D). The examination of the serial histological preparations of the hermaphrodite gonad (ovotestis) of different adult specimens representing different seasons reveals that this gonad undergoes spermatogenesis all year round. Moreover, it is interpreted that its maximum production of the male gametes occurs during autumn (September to November). The spermatogonia proliferate from the primordial germ cells found in the wall of the proximal half (Fig. 11.25A), so that the lumen becomes occupied by many different stages of spermatogenesis (Fig. 11.28(I) A). At any level of this portion of the tubule, one can observe clusters of rounded cells, at different steps of development, attached to the germinal wall. Each cell is large, with a diameter measuring about 14 μm, and a voluminous rounded central nucleus with a diameter measuring about 11 μm. The cytoplasm of the cell appears as a clear peripheral area

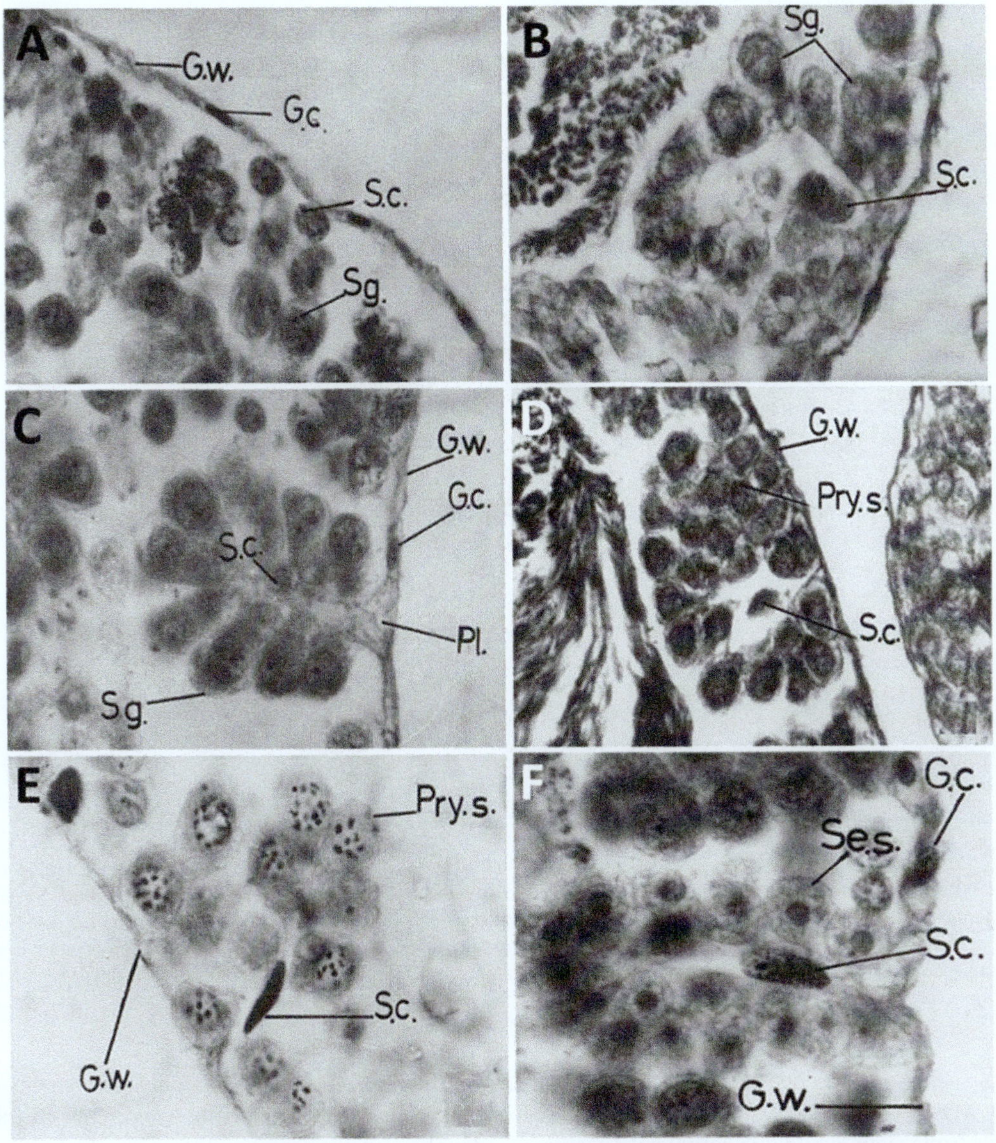

FIG. 11.25 Spermatogenesis in the hermaphrodite gonad of *Eremina*: (A) showing the germinal wall of the gonadal tube and beginning of arrangement of spermatogonia. X = 750; (B) cluster of spermatogonia. X = 1000; (C) fully arranged and formed cluster of spermatogonia. X = 1000; (D) cluster of 1^{ry} spermatocytes. X = 200; (E) cluster of 1^{ry} spermatocytes during different stages of prophase I of meiosis. X = 800; (F) fully formed cluster of 2^{ry} spermatocytes. X = 1200;

(Continued)

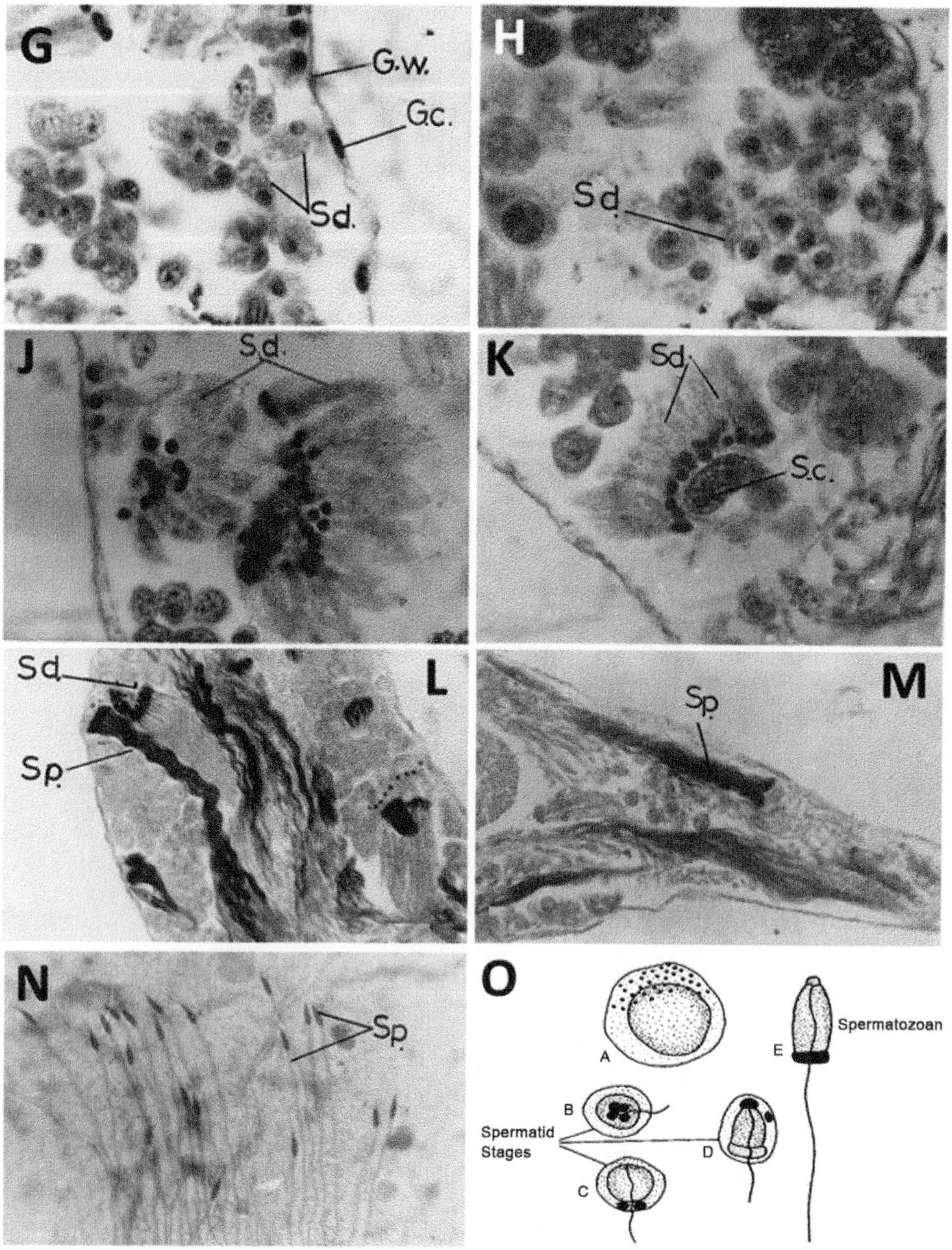

FIG. 11.25, CONT'D (G) cluster of early oval spermatids. X=900; (H) cluster of early rounded spermatids. X=1000; (J) three stages of cluster of late spermatids. X=600; (K) cluster of late spermatids during spermatiolosis with the Sertoli-cell in a typical representation. X=800; (L) clusters of metamorphosing spermatids and of spermatozoa. X=500; (M) L.S. of gonadal tube of the hermaphrodite gonad showing the distal conducting portion containing clusters of spermatozoa. X=600; (N) smear preparation from the gonad showing that the spermatozoa are fully mature in structure. X=900; (O) showing stages of spermatiolosis of the prosobranch Trochus turbinatus: (A) spermatocyte, (B–D) successive stages of spermatid, (E) spermatozoan. Scale line = 1 μm. *(After Tuzet (1930), from Webber (1977).)*

surrounding the nucleus. It accepts the colour of the acid fuchsin (AF) by Masson's trichrome stain, and the blue colour by Mallory's triple stain, while it is completely negative to the Heidenhain's iron haematoxylin (HIh). Oppositely, the nucleus has a granulated appearance and stains heavily with the acid fuchsin and attains the black blue colour of the haematoxylin. In addition, this nucleus contains small, densely stained one or two nucleoli. These cells of the clusters with the above description in different pulmonate gonads are defined as spermatogonia (Berry, 1977; Sabelli et al., 1978).

In spite of the fact that each of the three clusters is attached to the germinal wall of gonad of *Eremina*

(Fig. 11.25A), they differ in their number and arrangement of cells. However, the clusters of Fig 11.25B and C show a cytoplasmic strand extending as a pillar, around which the cells are arranged in a peculiar manner. This difference might be due to the level of cutting the cluster. The cluster of Fig. 11.25C is the typical form of the spermatogonia cluster.

Concerning the germinal wall itself, it is a peculiarity that the portion attaching the cluster has a pronounced thickness compared to the other portions at the same level. Generally, it consists of a thin layer of connective tissue in which some fine muscle fibres are scattered, distinguished by their acceptance of the AF of Masson's trichrome stain. Just beneath this layer, the gonial cells form different groups in different sizes and arrangements (Fig. 11.28A).

In each of the three clusters shown above, a different but distinct cell, which is the same size as its neighbouring spermatogonia, can be observed at one side of the cluster. Its outline cannot be definitely identified, but it was investigated as an oval cell. Its nucleus is comparatively very large, oval and with definite nuclear membrane. It accepts the heavy red colour of AF and the black blue colour of the HIh stain. This nucleus has a granulated appearance and two very small nucleoli. Bawab et al. (1992a,b) defined this cell as a 'Sertoli-like', cell following the terminology which is applied to the same cell in the gonad of the echinoderm, *Luidia clatbrata* (Hinsch and Dehn, 1979). Fig. 11.25E shows the cells of the cluster decrease slightly in size, each measures about 13 μm and its nucleus is 9 μm in diameter. The nuclei of these cells show different steps of the prophase I of meiosis (Prof. Dr. Anga Ramadan, personal communication). These are 1^{ry} spermatocytes and will give 2^{ry} spermatocytes. Clusters of the latter are shown in Fig. 11.25F. Each cluster is of 10–12 rounded smaller cells with a diameter measuring about 10 μm and a condensed acentric nucleus measuring about 5 μm. The Sertoli-like cell, the same as described above, is still in the centre of the cluster of the 2^{ry} spermatocytes (Fig. 11.25F) and with the same appearance. In addition, all the cells from the spermatogonia to the 2^{ry} spermatocytes attain the same reaction with the mentioned stains.

The germinal wall, corresponding to the clusters of the 1^{ry} spermatocytes and 2^{ry} spermatocytes, is not so pronounced and the germinal cells are not observed. Fig. 11.25G shows another cluster of rounded cells having about the same size and appearance of the 2^{ry} spermatocytes, but the nucleus of each cell shifts towards one pole, becoming typically polar in position. Gradually, the cell becomes oval, with about the same measurements, but its nucleus decreases in size, becoming 2.5 μm in diameter. The Sertoli-like cell is still found inside the cluster. Comparing this picture with that described and illustrated for mammals (Bloom and Fawcett, 1976), this is the early stage of the differentiation of the spermatids. To facilitate understanding, Bawab et al. (1992a,b) referred to the cells of these clusters as the early spermatids. The latter are considered the end of the process of spermatogenesis. Gradually they differentiate to become the late spermatids (Fig. 11.25J) beginning the process of spermiogenesis. Figs 11.25K and 11.25L show successive stages of the differentiation of the late spermatids to early spermatozoa. The heads of the latter are attached to the Sertoli-like cell. An advanced stage shows that the maturing spermatozoa become independent of this cell, which disappears completely (Fig. 11.25L). The disappearance of this Sertoli-like cell at this stage evidenced that it is for nursery during spermatogenesis and spermiogenesis, and it also has a supportive function, since its unlimited outline may extend strands intervening the gonial cells.

It is worth mentioning that the portions of the germinal wall attaching the 'bunches' of spermatids or metamorphosing sperm lack their pronunciation and germinal cells. They retain their normal form after the separation of the 'bunches'. The product, as well as the ova, are discharged to the hermaphrodite duct via the conducting portions of the gonadal tubes (Fig. 11.25M).

Structure of the fully mature spermatozoa. Smears prepared from the gonad (Fig. 11.25N) shows the structure of the fully mature sperm without any difference from that which appears in the smear prepared from the contents of the middle hermaphrodite duct (Fig. 11.14H). The mature sperm of *E. ehrenbergi* is very long and looks like that of vertebrates (Huettner, 1949). It consists of two regions: the head and the tail. Its overall length measures ~400 μm, of which the head takes up 8 μm. The head contains a condensed nucleus, which accepts the deep blue/black colour of HIh. It tapers anteriorly, forming the acrosome (Walker and MacGregor, 1968). The nucleus is ensheathed by a comparatively thin cytoplasmic layer. At this level of examination, the middle piece cannot be differentiated from the tail, except by critical examination with the oil-immersion lens. A small portion just beyond the head stains faintly, and shows the position of the middle piece. The tail, oppositely, stains deeply and constitutes most of the length of the spermatozoon. The whole structure is ensheathed by a,comparatively, very thin layer of cytoplasm (Abd El-Kerim, 1988). Fig. 11.26 shows different types of spermatozoa in gastropods.

Abnormal spermatogenesis. In prosobranchs (Webber, 1977), dimorphism of sperm occurs very frequently and has attracted much attention. This feature has demonstrated the approach to the spermatogenesis in most of the group. As a consequence, it is very difficult to compose from the literature a simple and consistent account of sperm development in prosobranchs. However, the process of formation of the abnormal sperm was investigated in different species (cytological point of view).

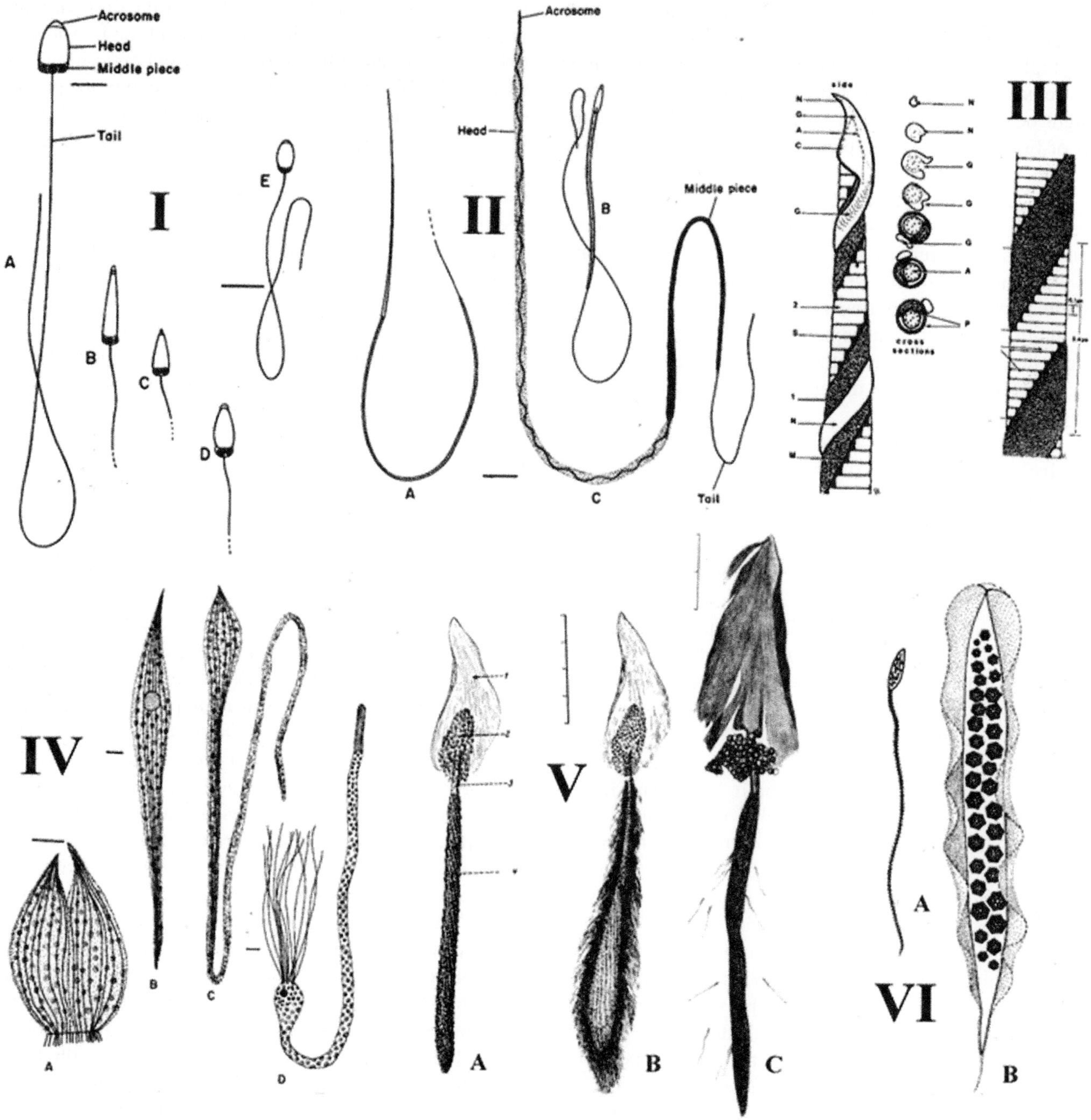

FIG. 11.26 Types of spermatozoa in gastropoda: (I) examples of archaeogastropod: (A) *Emarginula crassa*, (B) *Patella vulgata*, (C) *Acmaea testudinalis*, (D) *Gibbula cineraria*, (E) *Calliostoma militare*. Scale line for A = 2 μm, for E–B = 5 μm *(After Franzén, 1955)*; (II) examples of mesogastropod and neogastropod eupyrene (normal) spermatozoa: (A) *Rissoa incnspicua*, (B) *Hydrobia ulvae*, (C) *Nassarius nitida*. Scale line = 5 μm *((A and B) after Franzén, 1955, (C) after Tuzet, 1930)*; (III) (A) drawing of the nuclear region of the spermatozoan of *Aplysia californica*; (B) drawing of the postnuclear region of the spermatozoon of the anaspidean *Aplysia californica*. A = axoneme, C = Chapuisseptum, 1 = primary helix, 2 = secondary helix. *(After Chapuis, 1973, from Beeman, 1977)*; (IV) examples of mesogastropod and neogastropod (abnormal) spermatozoa: (A) *Columbella rustica*; (B) *Capulus ungaricus*; (C) *Aporrhais pespelicani*; (D) *Viviparus viviparous*. Scale lines for A and B = 5 μm, for C and D = 10 μm *(after Tuset, 1930). (I, II, IV) From Webber (1977)*. (V) Atypical, apyrenic spermatozoon of *janthina*, fully formed. 160: 1st Scale: 200; (A) Before fixation of typical spermatozoa, 1 = drive plate. 2 = yolk mass. 3 = connection piece, 4 = attachment piece; (B) After fixation of typical spermatozoa; (C) Fixed atypical spermatozoon of *Janthina*. The eupyrenic spermatozoa begin to attach to the attachment piece. Splitting plate slightly split. 225: 1st scale: 100 (Ankel, 1930). (VI) (A) *Oligopyrene* and (B) *eupyrene* sperm of the snail *Strombus bituberculatus*. *After Galstoff, from Gardiner (1972)*.

The deeper insight into the process of spermatogenesis shows that there are different morphological structures for the atypical spermatozoa (abnormal). Diminution or lack of chromatin was established from the normal sperm (eupyrene) with normal chromatin content. Those with subnormal chromatin are oligopyrenes, and those spermatozoa without chromatin are apyrenes.

The classical case is that of the pond-snail, *Paludina,* in which two types of spermatozoa were discovered by Siebold (1836). Its atypical sperm is oligopyrene. Generally, the atypical sperm is much larger than the normal one and occurs in low frequency.

Other examples of dimorphism of sperm in Mollusca (Webber, 1977)

1. In *Patella caerulea*, the normal sperm has a head 3 μm in diameter with a flagellum 8 μm long. The 'abnormal' sperm has a head of about 7 μm diameter with a flagellum about 7 μm long.
2. In *Buccinum undatum*, it was found that the eupyrene and abnormal sperm were morphologically identical except for the reduced quantity of chromatin in the abnormal.
3. In *Viviparus*, two types of abnormal spermatozoa were found and are vermiform and motile. The abnormal sperm is oligopyrene and multiflagellar.
4. In *Nassarius*, sperm are fusiform and capable of little movement (Pelseneer, 1935).
5. In *Clathrus*, and also in *Janthina*, *Cerithiopsis*, and *Turritella*, a giant abnormal spermatozeugmata was found; these spermatozeugmata may be as large as 900 μm and are motile.
6. In *Chenopus,* they are apyrene and have no free tails although they are motile.
7. In *Vermetus,* they are spindle-shaped, and contain large yolk platelets, have free tails, but appear completely immotile.
8. In *Strombus,* spermatozoa are equipped with undulating membrane and yolk, they are apyrene and nonmotile.
9. In *Conus mediterraneus,* they are spherical with two flagella, without yolk, apyrene, and nonmotile.
10. In *Bithynia* and closely related species, the atypical spermatozoa are morphologically very similar to the typical one, but they are oligo- or hyperpyrene and all of them arise in meiosis.
11. In *Valvata*, three different kinds of atypical spermatozoa occur that departed from normal development at various stages, all postspermatogonial.

Functions of abnormal spermatozoa. By far the most elaborate and, in fact, unique atypical sperm cell is the giant spermatozoan, also called a spermatozeugmata, of the marine snail *Janthina*. When fully mature, it is approximately 750 μm long, equipped with a large undulating membrane called a 'driving plate', a mass of granules, and a tail composed of thousands of fused flagella, which have arisen through a process of extreme multiplication of centrioles. The surface of the giant tail is modified by longitudinal rows of small depressions, into which typical spermatozoa may insert. Thus the atypical spermatozoan serves as a carrier for hundreds of typical ones.

In *Fusitriton oregonensis,* there are two different, highly differentiated atypical spermatozoa, which may have a specific function. Type A (carrier) is a tailless, apyrene cell, and possibly a nutrient source for eupyrene sperm. It is about three times the size of typical sperm, of which it carries up to a hundred on its surface. Approximately one in a hundred spermatogonia gives rise to a Type A cell. Type B (lancet) is still large, also apyrene, but does not serve as a carrier.

In *Viviparus,* the abnormal sperm has a polysaccharide sheath around its middle portion. It was suggested that this sheath provided nourishment for the eupyrene sperm in the female genital duct, by disintegrating and liberating the polysaccharide.

In *Aporrhais*, it was noticed that apyrene sperm was found in the eggs up to 20 min after fertilization, so it was suggested that 'abnormal' sperm played a part in the formation of nurse eggs that are found in egg of some species.

In other cases, the atypical forms, particularly those containing appreciable amounts of yolk, may contribute to the nourishment of the typical spermatozoa. In the majority, there is no apparent function and atypical cells die after a relatively short life. Certainly the prosobranchs offer unique approaches to the cellular and molecular systems, which determine the differentiation of the male gamete in all its stages of development. The morphologically mature sperm of *Phyllaplysia taylori* and *Aplysia californica* has turns around the anterior part of a standard 9 + 2 sperm filament of axoneme. The structure of the spermatozoan of *Phyllaplysia taylori* seems to have many similarities to that of sperm of the pulmonate *Testacella.*

Spermatophores. All three divisions of gastropods include species which discharge semen in free form, that is as sperm suspension in seminal plasma, and others in which prior to ejaculation, the spermatozoa are either aggregated into loosely assembled naked conglomerates (spermatozeugmata) or encapsulated into spermatophores.

In prosobranchs, there are some species that produce spermatozeugmata (Fig. 11.27A), and others in which the spermatozoa are encapsulated into spermatophores. *Neritina* and *Fissurella* are typical examples of animals that produce spermatophores. Those of *Neritina reclivata,* which Andrews (1936) studied in Florida, are 2.4–3 cm long and 1 mm wide (Fig. 11.27B), whereas the length of the female's body is only 2 cm; the spermatophore of this snail

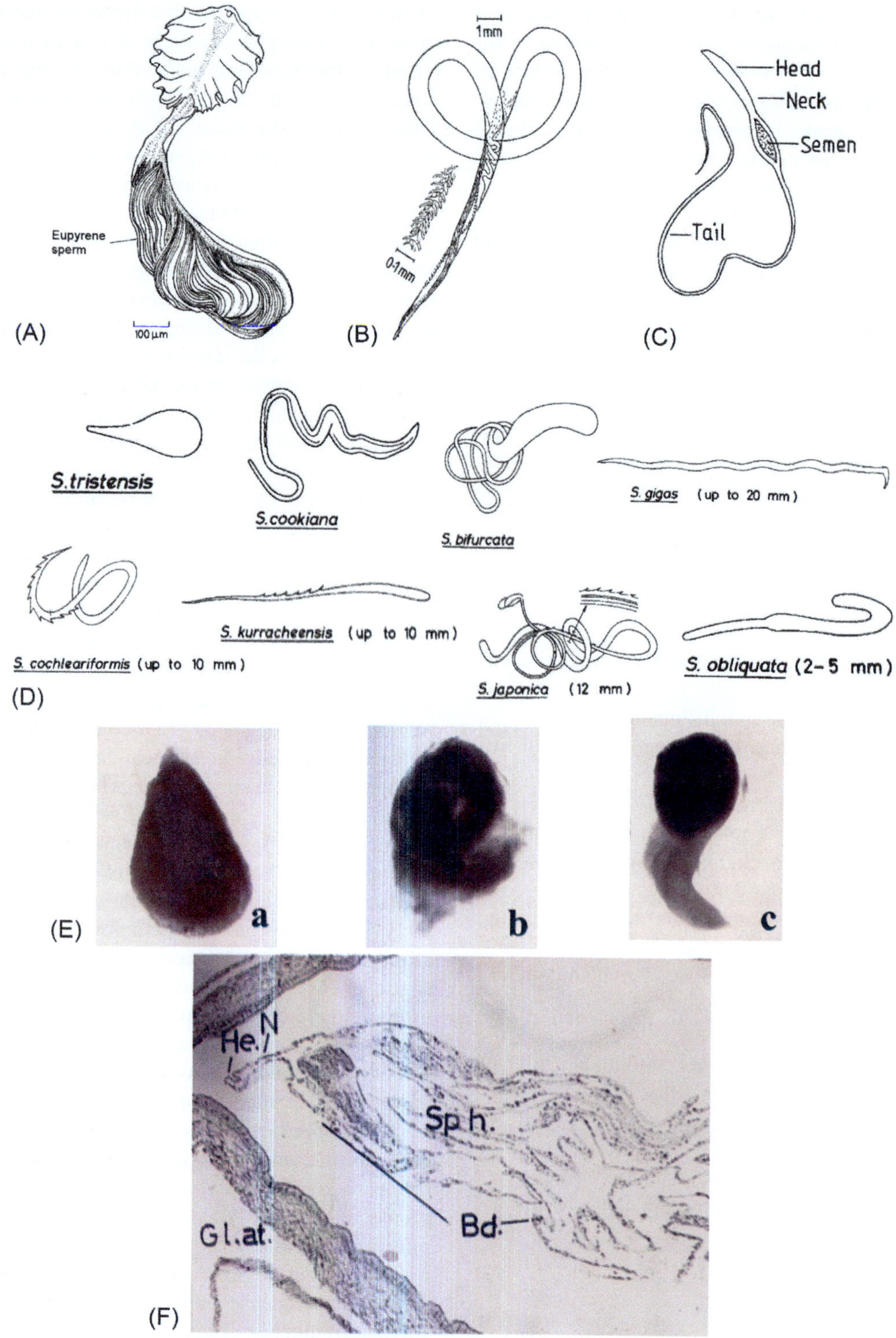

FIG. 11.27 Types of spermatophores in Gastropoda: (A) spermatozeugma of *Cerithiopsis tubercularis* carrying eupyrene spermatozoa *(after Fretter and Graham, 1962)*; (B) *Neritina reclivata*, a prosobranchia, spermatophores freshly removed from the female, and a portion with a row of triods *(redrawn from Andrews, 1936)*; (C) spermatophore of *Helix pomatia*, of natural size *(based on a drawing by Meisenheimer, 1907)*; (D) some spermatophores of *Siphonaria* sp. *(based on Hubendick, 1946–1947 and Abe, 1940)*; (E) isolated spermatophore of *Eremina ehrenbergi*, cleared: (a) just at its removal from the spermatheca (unravelled); (b) after some time; (c) ravelled spermatophore; (F) photomicrograph of L.S. of the genital atrium of copulated snail *Eremina* (Animal 1A) with spermatophore inside. X = 120. Bd = body of Sph, Gl.at = genital atrium, He = head, N = neck, Sph = spermatophore. *(A and D) From Webber, 1977 and (B and C) from Mann, 1984, (E and F) of Eremina.*

looks like a 'white turgid cylinder bent in a loop and with both attenuated ends stuck together' (Mann, 1984). The two end portions which enclose the spermatozoa are of unequal width, the wider one displaying in addition, some 'spicules, unique in being chitinoid secretions'. The latter are supposed to provide an anchoring device that keeps the tips of the spermatophore in position while the spermatozoa are being discharged and moved to the receptaculum seminis.

The slugs attach their spermatophores at various places on the animal to be impregnated. A frequent site of attachment in the visceral sac, but deposition on the head-foot complex, most commonly in its middle, was also observed. One or more spermatophores are attached to each animal. Discharge of spermatozoa is within a few days and gradually the whole of the spermatophore is emptied, making it possible for the spermatozoa to penetrate through the skin of the receiving animal. In histological sections, the changes that can be seen in the contact area, between the base of the spermatophore and the epidermis, include disappearance of nuclei in the epidermal cells, most probably resulting from cell autolysis which had been caused by the penetrating spermatozoa.

Microhedyle cryptophthalma, provides a typical example of an acochidiacean opisthobranch with two separate sexes, in which the accumulation of spermatozoa under the epidermis has been reported. The body of this slug, which inhabits the sandy surf beaches of the Mediterranean Sea, is 0.7–1.6mm long. The spermatozoa are transferred in spindle-shaped, 80–187μm long and 17–30μm wide spermatophores; each contains 200–300 spermatozoa. Attachment occurs mostly in the lateral parts of the visceral sac. Some recipients were seen with their empty capsules still attached sometime after the spermatozoa had been discharged (Mann, 1984).

It was indicated very early that, in pulmonate gastropods, the 'penis' can provide a site for sperm encapsulation into a spermatophore. The latter consists of four component parts: head portion, neck portion, semen container, and the long, filamentous tail (Fig. 11.27E (c)). The 'penis' of pulmonates is an organ involved in both copulation and spermatophore production. In the basommatophoran, *Siphonaria obliquata,* the spermatophores are hooked, and in some other species as *Siphonaria cochleariformis* and *Siphonaria kurracheesis,* they are equipped with spines. Most are less than 1 cm long, but that of *Siphonaria gigas* is nearly 2cm long and bears an angular crook at one end (Fig. 11.27D).

Eremina ehrenbergi. In the stylommatophoran *Eremina ehrenbergi*, the spermatophore consists mainly of a body and tail (Fig. 11.27E). Critically, and by unravelling the very small anterior region, it appears that this spermatophore is differentiated into a comparatively large spherical head and very short and narrow neck (Fig. 11.27E (b) and (c)). The latter extends as a thick and curved tail (Bawab et al., 1993). This described spermatophore was isolated from the very small organ, spermatheca (Fig. 11.12D). This organ in a dissected live specimen is solid and hard when seized by forceps or even by fingers. Its size changes with the reproductive state of the animal. It is always opaque and brown in colour, but this colour differs from the brownish cream to the very dark brown in the different specimens of the same collection. The current author (Bawab) supposed that the dark brown spermatheca preserves spermatophore(s). By a delicate incision under the binuclear microscope, the spermathecal wall was gently removed and the content (capsule of different outline) was photographed. It seems that the pear-shaped form is the typical form. To prove that this capsule is the spermatophore, its wall was smeared on a slide with a drop of water (tap water). Microscopically, the live sperm were shown vibrating; this means that the spermatophore was artificially dehisced. The sperm remain alive during several hours of examination. This is fully formed and mature sperm. Its description is the same as that prepared in the smear of the gonad and the smear of the vesicula seminalis (middle region of the hermaphrodite duct) mentioned above (p. 60).

Oogenesis. Although the reported studies on oogenesis in prosobranchs are fewer than those on spermatogenesis, a general pattern can be described. Stages of oogenesis are outlined in Fig. 11.28(I) upper. Oogonia arise from the germinal epithelial, divide mitotically, and have a large, spherical, clear nucleus that nearly fills the cell. Mitotic division terminates with 2^{ry} oogonia. The important phenomena that occur during oocyte development include premeiotic nuclear changes, cytoplasmic growth, and the maturation division.

In Opisthobranchs, the process of oogenesis, unlike spermatogenesis, does not seem especially unique (Beeman, 1977). However, laboratory studies on *Aplysia* have shown that these animals, like many other marine species, can be brought to reproductive condition out of season by increasing the temperature. The animals, e.g. *Phyllaplysa taylori*, appeared to have two overlapping waves of reproductive activity per year. First, a 'winter crop' with a slow early growth, a peak in mean size about July, a probable generation time of about 7–9 months. The second is a 'summer crop' with rapid growth, a peak in mean size in October, and a probable generation time of only about 3–5 months. The effect of environmental factors was not determined except that starved animals showed the expected gonadal atrophy.

Berry (1977) reported that, in Malayan ellobiids, there are no preferred regions of the ovotestis for oogenesis (Fig. 11.28(I) lower), although Pelseneer (1894) stated that in *Amphibola,* oocytes develop in one side of the gonad and spermatozoa develop on the other. He also presented another view for *Siphonaria hispida,* in which the oocytes occur more to the periphery of the gonad and spermatozoa occur nearer to the centre. Berry, himself, confirmed this picture in Malayan *S. exigua*. He described the process, in *Siphonaria hispida*, where Oogonia and their nuclei enlarge and become clear, with reticulate or scattered chromatin and prominent nucleolus. The oogonia lie among the other gonadal tissue in amoeboid fashion and eventually become stationary, surrounded by follicle cells over their free surfaces. They then grow by the deposition of yolk (vitellogenesis). Oocytes of *Siphonaria hispida* reach 150 μm in greatest diameter and their nuclei reach 60 μm.

This latter description is almost the same encountered in the work of Mehanney (1991) and Bawab et al. (1992a,b)

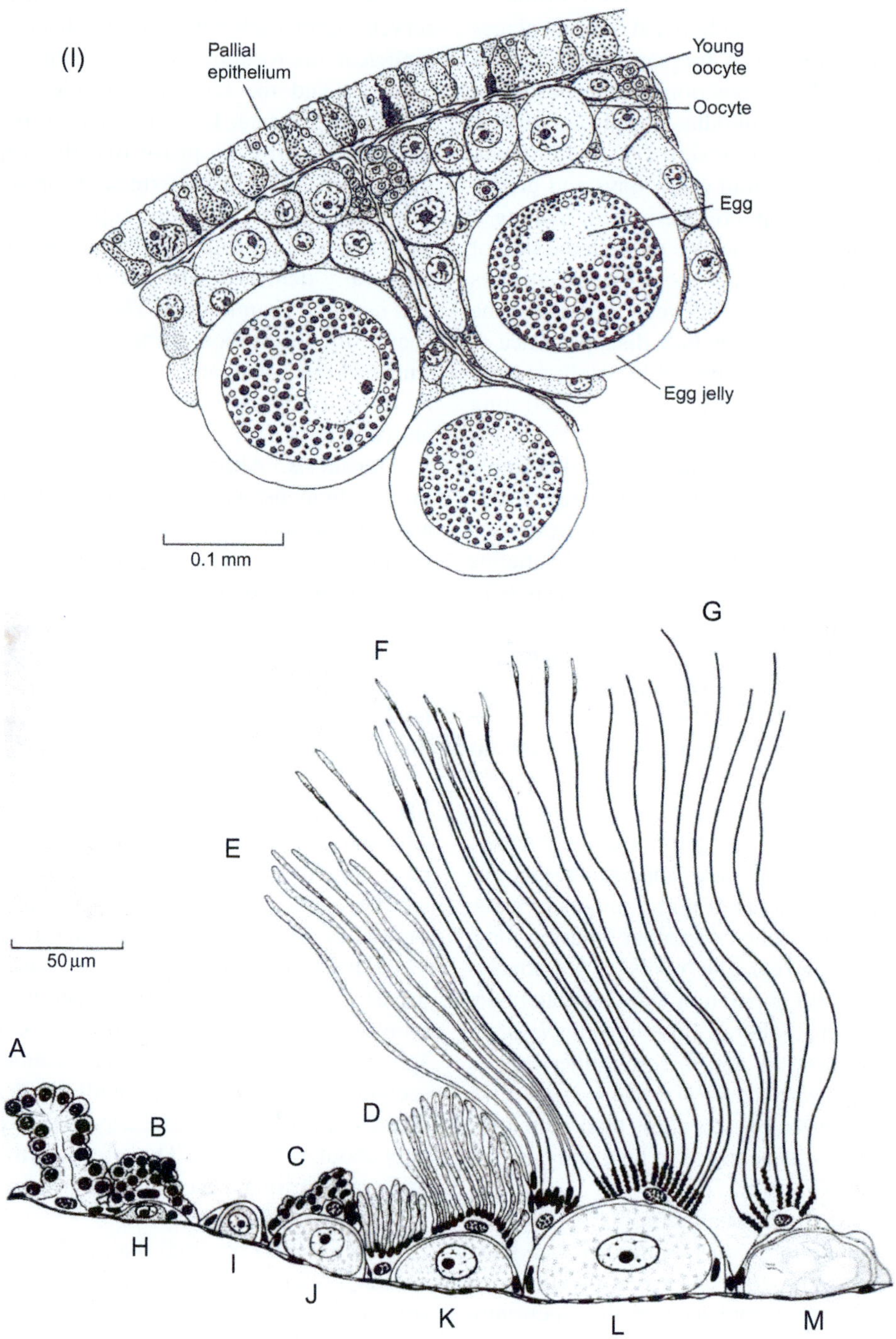

FIG. 11.28 (I) Upper photo-stages of oogenesis in *Monodonta lineata (after Fretter and Graham, 1962, from Webber, 1977)*; Lower photo-composite drawing of part of ovotestis of *Cassidula aurisfelis*: (A) spermatogonia; (B) early spermatocyte; (C) late spermatocyte; (D) early spermatid; (E) late spermatid; (F and G) spermatozoa. Stages A–G are all borne on Sertoli cells; (H–M) growing oocytes; (M) degenerating oocyte, note follicle cells covering oocytes. *From Berry (1977)*.

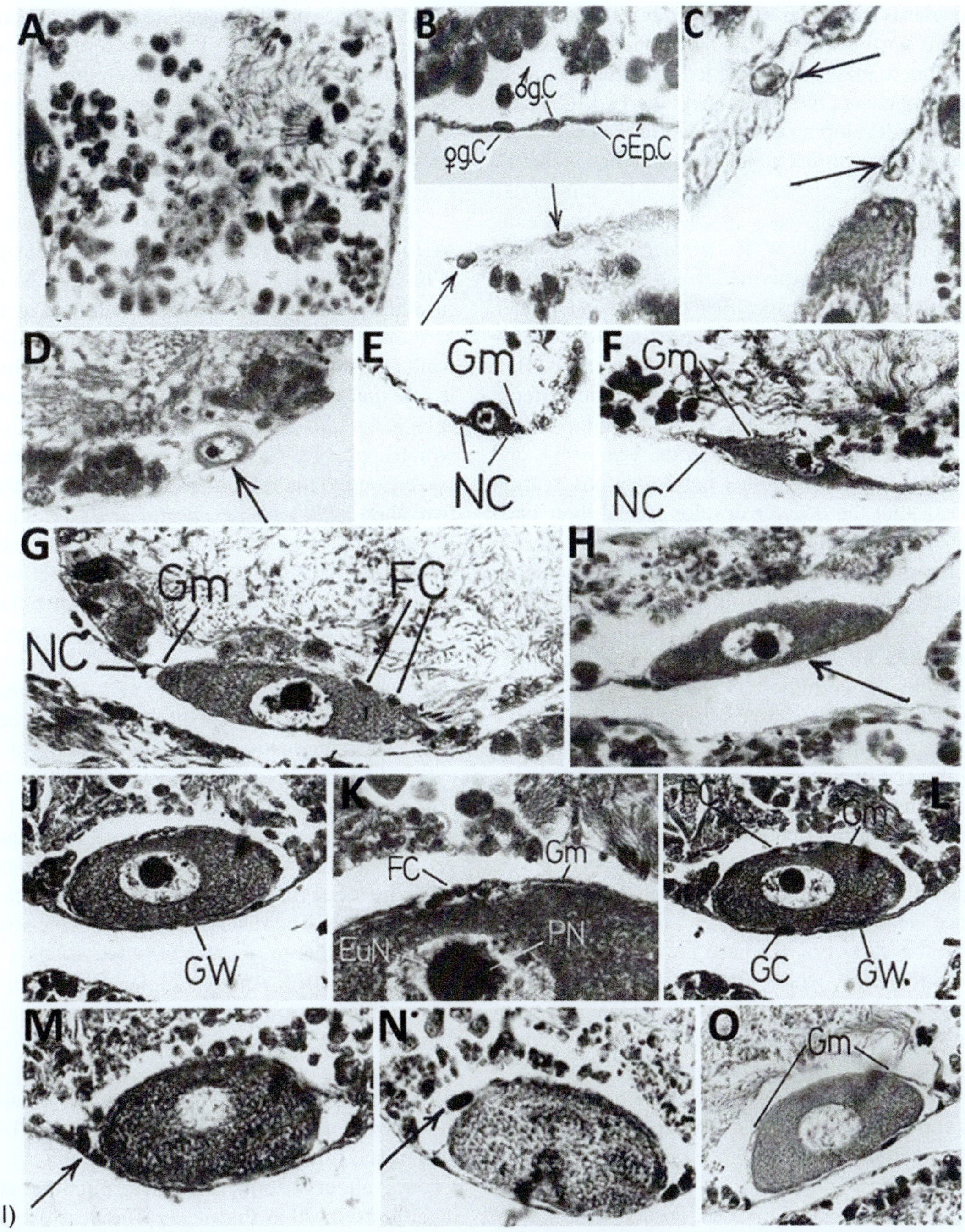

FIG. 11.28, CONT'D (II) Portions of T.S.s of hermaphrodite gonad of *E. ehrenbergi* showing stages of oogenesis: (A) germinal wall and germinal cells, note that both spermatogenesis and oogenesis are observed side by side in this portion of the gonad. X = 100; (B) oil immersion magnification (OIM) of portions of different acini showing different gonial cells in germinal wall, arrows indicate primordial of female gonial cell. X of upper portion = 1050, and of lower = 900; (C) OIM of germinal wall of different acini showing the female gonial cells, indicated by *arrows*. X = 900; (D) magnified portion showing very early oocyte, indicated by an *arrow*. X = 400; (E) another magnified portion showing early stage of oogenesis with a nurse cell on either side, prepared in July. X = 400; (F) magnified portion showing spindle-shaped oocyte. X = 300; (G) showing oocyte with optimum length of the spindle. X = 200; (H) showing the ends of the spindle become round, i.e. beginning of vertical growth of the oocyte; (J) showing the oval-shaped oocyte, still the oocyte applied to the germinal wall and attain follicle cells. X = 300; (K) OIM showing germinal wall and follicle cells of those seen in (J); (L) oval oocyte with germinal cells in the germinal wall. X = 350; (M–O) showing one oval-shaped oocyte in three successive sections within its germinal membrane. X = 400, all the oocytes from M to O prepared in July.

studying oogenesis in the desert snail *Eremina ehrenbergi*. The generative portions of the tubular gonad of this species, as mentioned above, are occupied by different stages of both spermatogenesis and oogenesis (Figs 11.13C, D and 11.28(II) A). The developing oogonia and/or oocytes are always applied to the germinal wall. The difference is that no movement of the young oocytes was detected or interpreted in any of the 36 prepared gonads of specimens representing almost every month of the year. The noticeable observation is that the metamorphosing spermatids and sperm predominate in the prepared gonad of specimen obtained in October. In other months, the number of the growing oocytes appeared to exceed that of the stages of spermatogenesis. But, the number of the developing oocytes per section differed with the different months. The peak was observed in two of the six gonads prepared in July, one in the year 1984, and the other in the year 1985. Another noticeable observation must be added, that the oocytes develop up till their optimum size and form, while attaching to the germinal wall, opposite to the developing spermatids. The latter undergo spermeogenesis while attached to the germinal wall or inside the cavity of the acinus (Fig. 11.25J–L). This means that the oocytes develop peripherally, while the sperm develop peripherally or centrally.

According to the above observations, the study of the process of oogenesis in *Eremina ehrenbergi* is more feasible during July. The two specimens from July provided successful serial transverse preparations. Practically, the gonial cells, as well as the primordia of both male and female, are difficult to distinguish from each other (Giese and Pearse, 1974; Siracusa et al., 1985). And considering that the gonad of this species is hermaphrodite, the early stages of oogenesis and even the early oocytes have never been differentiated. The gonial cells, which constitute a discontinuous layer inside the germinal wall (Fig. 11.28(II) B upper part) are more or less of the same size, which is greatly small, with a comparatively very large nucleus. It is appropriate to describe this cell as consisting only of a nucleus. It appears that this portion of germinal wall differs from that forming spermatogenesis. It contains more than one type of cell:

1. Cell with dense elongated nucleus, normal germinal or gonial cell;
2. Granulated cell, which always attains the oval appearance (Fig. 11.28(II) B lower part), irrespective of the size, it is similar to that beginning spermatogenesis. So Bawab interpreted and referred to it as female gonial cell;
3. Very few cells, each is oval with very large reticulate nucleus, with one or at most two granules. The latter accept the dark blue black colour of the HIh with the same intensity of that of the reticulate nucleus. The cytoplasm constitutes only the outline of the cell. The authors of this work stressed on this cell, since its nucleus resembles greatly that of the very young oocyte appearing in the same preparation (Fig. 11.28(II) D). Bawab referred to it as female gonial cell. It maybe oogonium or primordium of oocyte;
4. Cell that resembles the upper one, but projects more in the cavity, it is referred to as very early oocyte (Fig. 11.28(II) C).

The examination of the serial preparations of the gonads of July, show an advanced picture for the young oocyte (Fig. 11.28(II) E). This oocyte seems to be more circular in outline, but more or less the same as in the above description. It differs in being closely attached to the germinal wall, and having more cytoplasm. A noticeable observation is that this wall is quite thin and lacks cells. A third difference is that this oocyte is enclosed within a thin membrane, which appears to be separate from the germinal wall. It may correspond to the follicular membrane of the oocytes of the two slugs, *Philomycus caroliniarnis* (Kugler, 1965) and *Arion ater* (Lusis, 1961), which consists of several cells in the form of a thin layer. In *Eremina*, inner to the germinal membrane, and on either side of the oocyte, there is a somewhat small cell represented only by its dense oval nucleus (Fig. 11.28(II) E). Following the description of the two slugs, *Arion ater* (Lusis, 1961) and *Agriolimax reticulatus* (Hill and Bowen, 1976; Hill, 1977), these cells are nurse cells. This stage is the beginning of the growth of the oocyte, which again follows that which occurs in *Agriolimax reticulatus*, as it occurs along two different but successive axes: one is parallel to the germinal wall; and the other is vertical on this wall. The oocyte grows along the parallel axis with a rate that greatly exceeds that which occurs along the other axis. The result is that the oocyte attains its previous oval outline or, more precisely, attains a spindle form (Fig. 11.28(II) F). Another remarkable observation is that the cytoplasm grows at such a great rate that it retains its normal proportionality with its nucleus. It becomes granulated and deeply stained (Fig. 11.28(II) G).

This spindle oocyte grows gradually to attain the optimum size with a definite length of the parallel axis, which may reach 100 μm, while the length of the vertical axis reaches about 30 μm. The corresponding germinal wall is still so very thin that it is difficult to be differentiated. However, some very small cells with dense or diffused nuclei appear in this wall. The spindle-shaped oocyte is still enclosed within the germinal membrane, and with the nurse cell on either side (Fig. 11.28(II) H). Along this membrane few cells appear, not exceeding 3 or 4 in number. Each is oval or spherical, attaining about 6 μm in diameter and has a diffused nucleus. Oil immersion magnification shows these cells are closely attached to the oocyte. They are referred to as follicular cells, since they appear later than the nurse cells, which appear at the onset of growth phase. Such oocyte (Fig. 11.28(II) H) represents the end step of the spindle-shaped growth, i.e. the growth along the parallel axis.

The spindle-shaped oocyte continues growing gradually at this stage, along the vertical axis, with a rate that exceeds greatly that of the parallel one. Firstly, the pointed edges of the oocyte, shown in Fig. 11.28(II) H, become rounded in Fig. 11.28(II) J, with no change in the dimensions, or in the affinity to stains of both the cytoplasm and nucleus. With the increase in growth, there is an increase in the oval shape of the oocyte, i.e. its sides are more rounded (Fig. 11.28(II) L). This gradual change is accompanied by the gradual independence of the growing oocyte from the germinal wall (Fig. 11.28(II) O). The latter gradually regains its normal thickness with its invaded cells, which might be normal germinal cells or additional nurse cells. The critical examination of one oocyte (Fig. 11.28(II) L) by following it in the successive levels in the serial transverse sections of one preparation of the gonad proves that:

1. The oocyte measures about 140 μm in its parallel axis, while its vertical axis measures about 60 μm. Its germinal membrane, which is still in firm connection with its free surface, thickens and becomes well differentiated on the sides owing to their rounded form (Fig. 11.28(II) K and L);
2. The follicle cells on this membrane increase in number and appear in different positions in the different levels, confirming that they do not form a complete layer around the oocyte. The high magnification (Fig. 11.28(II) K) shows that these cells are different in outlines but with dense nuclei. Sometimes, their walls appear continuous with the germinal membrane, or appear continuous with the growing oocyte itself;
3. One or two other cells appear on either side of the oocyte, larger than the follicular cells, and differing from the previously described nurse cells (Fig. 11.28(II) M, N);
4. The oocyte itself, at this stage, has the same granulated cytoplasm and affinity to stains. Its nucleus increases the chromatin material and keeps the reticulate form or it changes to the granular form. A noticeable observation is that the enlarged nucleolus is differentiated into a huge sphere and a very small capitate portion (Fig. 11.28(II) K). According to Hill and Bowen (1976) the former is the eunucleolus and the latter is the paranucleolus.

The whole appearance of this growing oocyte and its nucleus remains the same, only its size and outline change. Gradually, the wall of the oocyte itself becomes well differentiated all around its surface. The noticeable observation is that this oocyte accepts the specific colours of both the periodic acid-Schiff and bromphenol blue reactions and its two membranes were not observed, but detected histochemically. Since the wall of the oocyte thickens gradually during its growing, i.e. during its intimate association with the follicle cells, then the latter may have a role in the formation of the oocyte membranes. This wall stains deeper than the oocyte contents when subjected to both protein and carbohydrate reactions, and portions of the periphery stain deeply with protein reactions only, indicating the deposition of vitelline membrane and then chorion (Anderson, 1969). As for the stage of vitellogenesis, the current author supervised this work and personally followed each step of study, examination, and discussion, and did not define this process.

The spherical oocyte. The growth of the oocyte continues along the vertical plane, and hence gradually attains the spherical form (Fig. 11.28(II) O). Accordingly, the oocyte undergoes some changes: (1) the more the spherical form is attained, the more the corresponding germinal wall thickens, and the more germinal cells appear in it (Fig. 11.29). The growing oocyte is still applied to this wall, but not so firmly attached as in case of the oval or spindle oocytes; (2) the germinal membrane cannot be differentiated from the free surface of the oocyte (Fig. 11.29A (a)), but it is separated on either sides; (3) the nurse cells, previously observed on the sides of the oocyte, disappear completely.

The oocyte just attains the optimum spherical form; it measures 100 μm along the parallel axis, and 80–100 μm along the vertical axis. Its cytoplasm has the same fine granulated appearance described in the lower stages. It just differs in that its outline (cell membrane) becomes densely stained and so clearly defined. Its nucleus with the nucleolus are the same as in previous stages, but the position of the nucleus is somewhat shifted to the periphery. The staining affinity of the oocyte cytoplasm is the same during its growth (see Table 11.4). Some other oocytes with the same description and dimensions are found in the same preparation, just beyond the germinal wall, i.e. inside the lumen of the acinus (Fig. 11.29B). This oocyte differs from that in Fig. 11.29A, in that it has a definitive thick wall all around. On comparing the two oocytes, the one that appears in Fig. 11.29B can be considered as the ovulated one.

In the other gonad preparation, also in July, only one oocyte was observed while passing through the conducting portion of the gonadal tubule (Fig. 11.29D) to the efferent duct. It is not spherical in outline, but it is amoeboid in form (Fig. 11.29D and E). It occupies almost the whole lumen of the conducting tubule. In this preparation, more than one amoeboid oocyte was observed attached to the germinal wall (Fig. 11.29C). The latter is thick and with the same appearance of that attaching the spherical oocyte (Fig. 11.29A (a)). However, it seems that it contains more germinal cells, one or more of these cells surrounds the amoeboid oocyte and accompanies it. The thorough histological examination proves that the growth of the oocyte is a reflection of the process of deutoplasmogenesis in general, and not for specific vitellogenesis, i.e. this oocyte found in the hermaphrodite gonad lacks yolk. In the view of the current author, no vitellogenesis occurs during this growth. However, the affinity of the spherical oocyte

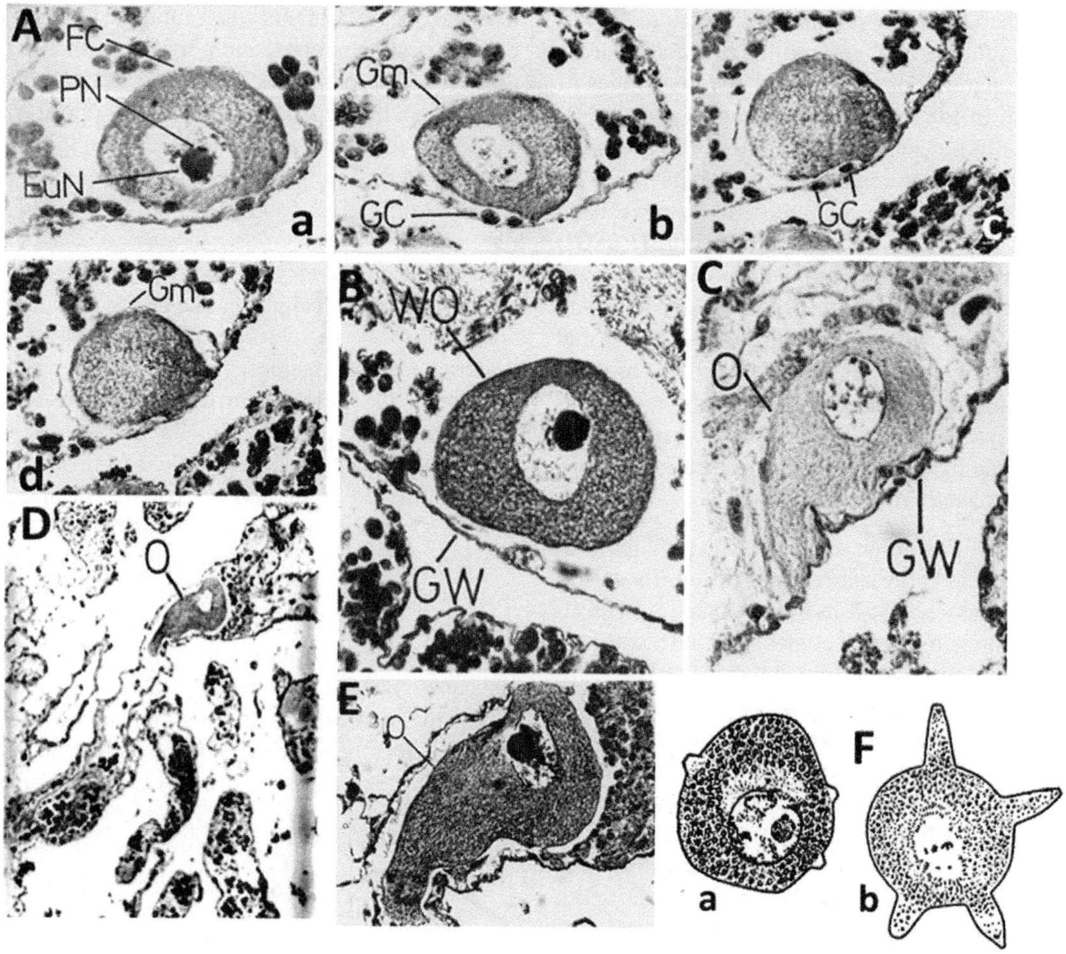

FIG. 11.29 Development of the spherical oocyte of *E. ehrenbergi* and its ovulation: (A) showing one oocyte in successive levels, (a–d). X for a=600, for c–d=350; (B) magnified portion showing oocyte to be ovulated since it is not attached to the germinal wall; (C) portion of L.S. showing amoeboid-shaped oocyte on its way to be ovulated. X=700; (D) amoeboid oocyte passing through the conducting portion of the gonadal tubule. X=100; (E) magnification of oocyte in D. X=500; (F) two stages in the maturation of the land snail *Eulota similaris stimpsoni* egg (Ikeda), protrusions through which the sperm enter the egg are formed as it matures *(from Wada, 1968)*. EuN=Eunucleolus, FC=follicle cell, GC=germinal cell, Gm=germinal membrane, GW=germinal wall, O=oocyte, PN=paranucleolus, WO=wall of oocyte.

TABLE 11.4 Staining affinity of the spherical oocyte

		Cytoplasm		Nucleus		
Stain	**Wall**	**Back ground**	**Granules**	**Nuclear membrane**	**Reticulum of nucleus**	**nucleolus**
Heidenhain's iron haematoxylin	Deep blue	–	Black/blue	Deep blue	Blue	Deep blue black
Ehrlich's haematoxylin stain	Very deep purple	Faint red	Deep purple	Deep purple	Deep purple	Deep red
Masson's trichrome stain	Dark blue	Very faint red	Blue	Dak biue	Blue	Shiny deep red
Mallory's triple stain	Deep red	Faint blue	Deep red	Blue	Blue	Shiny red
Weigert's haematoxylin and Van Gieson's stain	Reddish brown	Faint yellow	Reddish brown	Deep brown	Deep brown	Reddish brown

From Mehanney (1991).

to stains was critically examined by the oil immersion lens, see Table 11.4. In addition, the nature of the granules found in the cytoplasm of the growing oocyte was examined histochemically. They reacted only with the mercuric bromphenol blue and the periodic acid-Schiff reagents, indicating that they may consist of proteins and carbohydrates, separately or combined. But the wall of the oocyte reacted positively with the periodic acid-Schiff (PAS), Alcian blue, and with Best's carmine. At the same time, the performic acid-Schiff reagent stains many portions of this wall. This result indicates that this wall consists of glycogen and proteins.

Fig. 11.30 summarizes the process of oogenesis as described in the hermaphrodite gonad of *Eremina ehrenbergi*. The study of the hermaphrodite gonad of *Eremina* during the whole year not only aids the understanding of the processes of gametogenesis, but also provides some other striking observations:

1. Spheres of structureless contents (Fig. 11.31A) appear in gonads prepared in July (peak of oocyte production). Some masses of metamorphosing spermatids and sperm were observed, in a disordered appearance, beside these spheres. Each of the latter is invaded by definite cells without definite outline, which may originate from the germinal wall. They are spherical, oval, kidney-shaped, or angular and are deeply stained, as are the granules occupying them. For convenience, these cells are referred to as phagocytes (Mehanney, 1991).
2. The same above observation was encountered in one gonad of September (Fig. 11.31C), while the acinus contains degenerating stages of the two types of gametogenesis. Cells in great numbers were observed scattered in groups in between the contents of the acinus. Each cell is more or less rounded, densely and brightly stained, accepts both the haematoxylin and acid fuchsin (Fig. 11.31C and G). The striking observation is that these cells are well distributed in the gonads prepared in June and September, as well as in the empty gonads, e.g. that of February. At this level of work, it is not possible to investigate their origin or function and so they are named postreproductive cells.
3. The third striking observation was found in gonads prepared in months from December to June. The gonad itself, of each month, is small comparable to that of July, in spite of the fact that the system as a whole is of reasonable size. The histological appearance of any of these gonads differs completely from that of July (peak of oocyte production) (Mehanney, 1991) and from that of October (peak of sperm production) (Bawab et al., 1992a,b). Many of the acini of these gonads are very small (Fig. 11.31E and F) but are of reasonable size compared to the small size of the gonad. The acinus itself is almost empty inspite of its germinal wall being of normal thickness. The contents are degenerating oocytes, metamorphosing sperm, or masses of cells.

Concerning the first observation, it is reasonable that the acini must clean and clear the field for the production of oocytes which begins during early summer, thus the appearance of degenerating or destroyed stages of spermatogenesis. In addition, during early autumn, spermatogenesis predominates and the destroyed developing oocytes were well observed within the acini. This means that the gonad of *Eremina* produces the two types of gametes, side by side, but discharges only one type at a time. On the other hand, the appearance of the group of cells during June and during September, referred to as postreproductive cells (Figs 11.30 and 11.31), which becomes a characteristic feature in the empty acini during February–June, can be considered to be for phagocytosis.

Gonad of Melampus coffeus. It appears that the pulmonate snails are alike in their gametogenesis, since the gonad developmental stages shown in Fig. 11.31 con. of *Melampus* (Maia et al., 2012) seem to undergo the same steps of that of *Eremina*. This animal, *Melampus coffeus* (Linnaeus, 1758) (Ellobiidae), belongs to a primitive group of pulmonate molluscs, they are probably the most primitive living group of Pulmonata and they represent the transition between marine and land molluscs. Their study is a key to understand the evolution of the Gastropoda. This *M. coffeus* is a common species in mangroves and marshes of the Atlantic Ocean and can be found in North and South America from Florida to Uruguay. The spawned eggs were only found in the field during the rain, between January and June.

The spawn had spherical eggs arranged in cords, without capsules, but an envelope with a single yolk-rich embryo embedded in an albuminous material. The embryonic development occurs within that envelope and lasts 7–12 days. The authors identified the following stages of embryonic and larval development: 'Egg', 'Segmentation', 'Preveliger', and 'Veliger' (Fig. 11.39C).

Back to E. ehrenbergi. After dealing with the structure of the different organs of the hermaphrodite reproductive system of *E. ehrenbergi*, it becomes appropriate to deal with the passage and fate of its products:

Sperm pathway and spermatophore formation. The mature sperm discharge from the gonad to the efferent tubes, passing through the conducting portions of the gonad (Fig. 11.25M). Facilitating this discharge are: the lashing of the flagella of the sperm themselves; the rhythmic contractions of the muscle fibres of the wall of the gonadal and efferent tubes (Fig. 11.13E and F); the secretory substance formed by the epithelial lining of the efferent tubes; and the probable ciliation of this region. One or more factors of

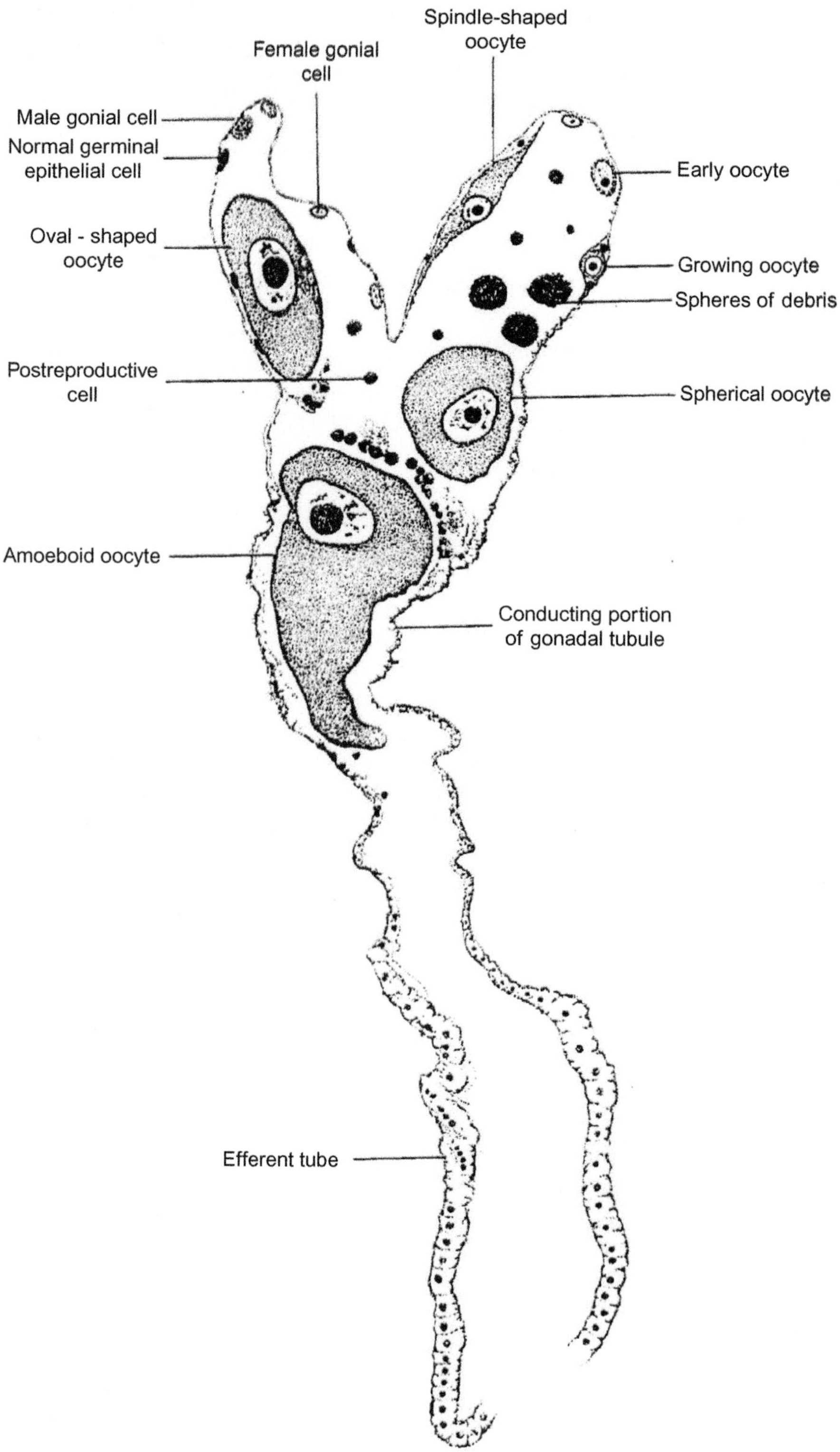

FIG. 11.30 Diagrammatic illustration summarizing the process of oogenesis in the acini of the gonad of *Eremina ehrenbergi*.

the above may share in this function. Then, the sperm pass through the proximal hermaphrodite duct. The cilia and secretion of this region facilitates this passage (Fig. 11.14B and C), till the sperm reach the middle region. Within the latter, the sperm are condensed and preserved and hence its name, the vesicular seminalis (Fig. 11.14D), proposed by many different authors. Within the vesicula seminalis (middle hermaphrodite duct), the sperm coming through its passage are transformed into sperm mass by mixing with the secretion of this region (Fig. 11.14E and F). These

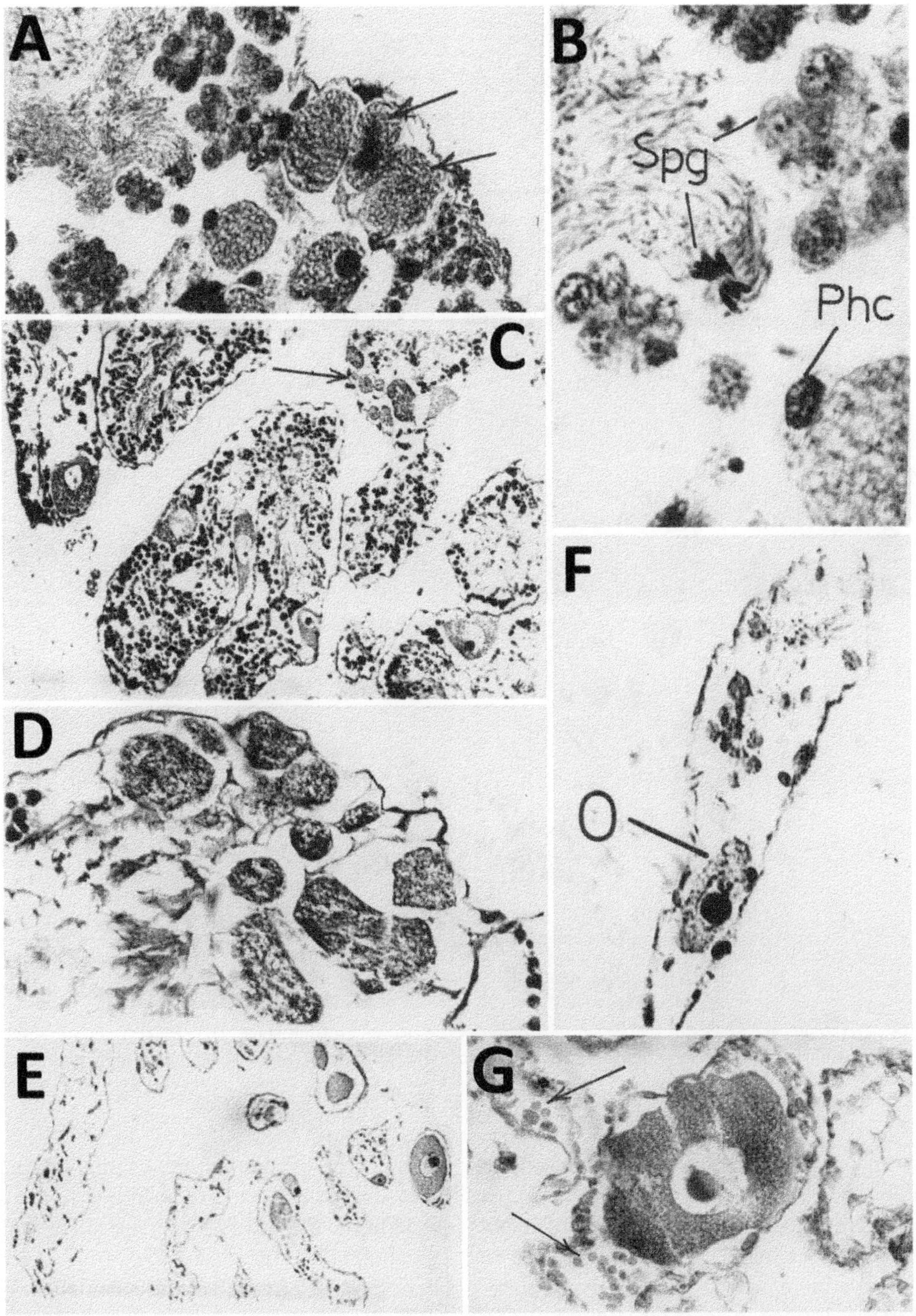

FIG. 11.31 Portions of T.S.s of gonad of *Eremina ehrenbergi* showing degeneration of the contents: (A) magnified portion of gonad of July showing spheres of debris indicated by arrows. X=500; (B) OIM of a part of (A) showing degeneration of stages of spermatogenesis and phagocyte; (C) gonad of September showing degeneration of stages of spermatogenesis and oogenesis. X=100; (D) another magnified portion showing spheres of debris. X=400; (E) small gonad of December and empty of acini. X=150; (F) magnified portion of (E) showing degenerating oocyte. X=500; (G) gonad of June showing the appearance of postreproductive cells surrounding the degenerating oocyte. X=400.

(Continued)

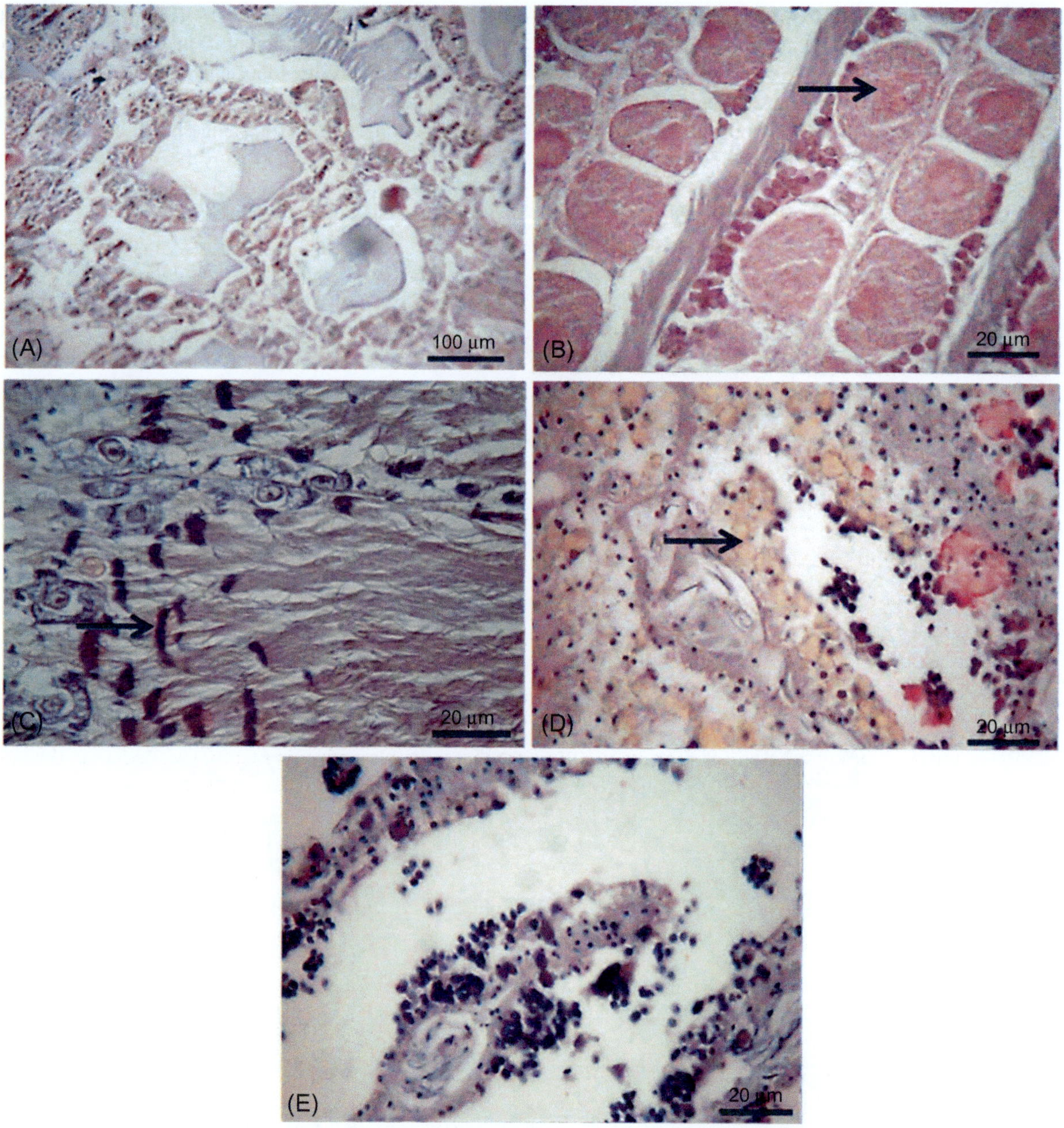

FIG. 11.31, CONT'D Developmental stages of the gonad of *Melampus coffeus*: (A) resting stage; (B) 'Final Maturation/Release' stage, showing mature oocytes; (C) 'final maturation/release' stage, showing spermatozoids; (D) 'final maturation/release' stage, showing macrophage cells; and (E) 'Exhaustion' stage. *(From Maia et al. (2012).)*

secretions, as they appear in the photos, coalesce the sperm (Fig. 11.14H), inhibit their activity, and continue their preservation. The sperm are alive but not motile or active. In the opinion of the current author, this is the initial step in the formation of the spermatophoric mass of the spermatophore, compared to that formed in *Penaeus* (Bawab, 1970; Malek and Bawab, 1974a).

A different view was recorded by Ghose (1963) and Breckenridge and Fallil (1973), who studied stylommatophoran reproductive systems. They showed that the epithelial lining of the vesicula seminalis might have phagocytic function, since the sperm are absorbed in the appropriate time. In addition, Rigby (1965) recorded the same function of the opisthobranch, *Succinia putris*. In *Eremina*, no phagocytes were observed in the vesicula seminalis. However, Morton (1967) reported that phagocytosis in Mollusca is performed by stellate amoebocytes found in the blood, which assume this power and can escape to any part of the body.

However, the main and most important function of the hermaphrodite duct, in *Eremina*, is the conduction of the products of the gonad to the following organ, which is the spermoviduct. But the albumen gland, an accessory organ, intervenes with the continuation. The end portion of the distal hermaphrodite duct reaches the surface of the albumen gland and then embeds in its tissue (Fig. 11.15A and B) as what is known as 'talon', the conspicuous and characteristic organ of the stylommatophoran reproductive system. In spite of the complicated structure of the talon, the sperm mass can pass directly to the sperm-channel of spermoviduct via the ciliated hood of the talon (Figs 11.15F and 11.17D). Reaching the sperm-channel of the spermoviduct, the sperm mass does not enter the egg-channel, since there is a histological separation of the two channels. The upper epithelial lining of this sperm-channel, as seen in the photo, at the portion of connection, protrudes inside the lumen forming a characteristic fold. The latter extends as a flap guarding this connection. The corresponding lower epithelial lining of this sperm-channel, together with the continuing part of that of the egg-channel, form an opposite fold facing the previous one. The former referred to as outer valve and the latter is the inner valve (Fig. 11.18K and L). A muscular layer is critically observed surrounding the epithelial wall of this channel and more pronounced in the vicinity of the outer valve (Fig. 11.18L).

The sperm-channel is lined by highly folded layer of epithelium, which is secretory (Fig. 11.18J and K). In addition to this secretion, there are the associated glands, the largest and most famous of which is the prostate gland. Berry (1977) reported a fairly sufficient review for the position of this gland with respect to the male duct in the three groups of Gastropoda. In *Eremina*, this gland is relatively large and without definite outline. It is a compound acinar gland with the same histological picture photographed in Bloom and Fawcett (1976) (Fig. 11.18J, K, and M). It opens in the lumen of the sperm-channel with more than one ductule and therefore it is an exocrine gland. It secretes a granular substance clearly observed within the acinar cells (Fig. 11.18N), tubules, and also in the lumen of the sperm-channel, with the same appearance and affinity to stains. Concerning this histological picture, it appears that it is the same in Stylommatophora, since there is no significant difference in the available studied snails (Ghose, 1963; Rigby, 1963; Breckenridge and Fallil, 1973; Visser, 1981). The fundamental function of this prostate gland in Gastropoda is to secrete substance(s) to 'mix' with the sperm before reaching the penis (Gardiner, 1972). It is well-known information that *Eremina* discharges its sperm encapsulated in spermatophore(s) and therefore the role of the prostate gland is not accurately determined. However, in the view of the current author, these secretions in the sperm-channel ensheathes the sperm mass, forming a spermatophoric mass ready to be encapsulated in a following organ. Beeman (1977) reported that in some cephalaspideans opisthobranchs, 'the prostate has the function of producing spermatophores, the prostate duct serving as an ejaculatory duct'. This was one of the aims of the current author to define the accurate role of the sperm-channel in the formation of the spermatophore.

In addition to the prostate gland, there are two other glands, well differentiated from the prostate one, also associated with the sperm-channel of *Eremina*. One is defined as 'submucosal gland', composed of many separate large cells heavily loaded with secretion. They scatter in the connective tissue surrounding the wall underneath the epithelial lining of the sperm-channel (Fig. 11.18O). Many of these were observed within the lining epithelial cells themselves. The position and form of these cells give the impression that they are amoeboid and can move in between the epithelial cells to discharge wholly (holocrine) in the lumen of the sperm-channel (Fig. 11.18O). The other gland is the 'valve gland', found in the submucosa of the outer valve of the sperm-channel (Fig. 11.18L). Each cell is occupied by a patch of secretion. Both the submucosal and valve glands are similar in position, affinity to stains, and in discharging their product holocrine.

Kugler (1965) studied the reproductive system of the slug *Philomycus carolinianus* histochemically, and differentiated between three glands associated with the 'spermatic groove' (sperm-channel in *Eremina*). This author interpreted these as prostate, mucous cells, and subepithelial mucous cells. The author added that the mucous cells are exocrine glands, since each discharges its secretion 'by way of a thin neck extending between the cells lining the lumen'. The position and appearance of these cells indicate that they may correspond to the submucosal gland observed in *Eremina*. But the latter is interpreted as an endocrine gland, since each cell may move by amoeboid movement to be discharged holocrine. As it is understood from the work of Kugler (1965), the accurate function of the secretion of the other gland—the 'subepithelial mucous gland'—which has no correspondence in *Eremina*, is used to bind the sperm together into a 'sperm thread'. A similar gland is present in another slug, *Athoracophorus bitentaculatus* (Burton, 1978). If this view is correctly interpreted, then the formation of the spermatophores in stylommatophora may proceed according to the method of formation of those of Crustacea (Malek and Bawab, 1974b). In this, the sperm-channel corresponds to the main vas deferens of *Penaeus*.

The third gland of *Eremina*, which is the valve gland, has no correspondence in any of the snails studied, as the available literature shows. It is supposed that this gland may secrete enzyme(s) and/or hormone for working the valve(s).

The spermatophoric mass formed in the sperm-channel passes to the vas deferens. The latter, in *Eremina*, confirms the rule that this organ forms a U-shaped pathway in stylommatophorans (Pelseneer, 1906; Visser, 1981). In *Eremina*, the term 'descending vas deferens' (Fig. 11.23A) is the limb connecting the sperm-channel, and the term 'ascending vas deferens' (Fig. 11.23A) is the limb connecting the penis (Figs 11.20A, B and 11.23A). These terms are used according to the direction of each limb. The wall of both limbs is provided with well-developed muscles and again it seems that this is the picture in the stylommatophora. This indicates that this organ is mainly for conduction. However, its epithelial lining is ciliated in some members (Rigby, 1963; Visser, 1981), and not ciliated in others (Kugler, 1965; Breckenridge and Fallil, 1973) In *Eremina*, at the level of the histological study, its epithelial lining is not ciliated and *Eremina* belongs to the second group. On the other hand, the vas deferens in *Eremina* is characterized by the striated cytoplasm of the lining epithelium; and the submucosal cells scattered in the wall. These two features may provide evidence that the vas deferens of *Eremina* is secretory. Holm (1964) described similar cells in the outer tissue surrounding the vas deferens of *Lymnaea stagnalis,* which confirms the secretory function of the vas deferens. However, no secretion was observed in the lumen, or in the epithelial lining in the histological preparations of *Eremina,* which are numerous. And the very thick muscular wall gives the impression that the contents within this vas deferens pass to the following organ by muscular movement.

It is a general picture of the Stylommatophora that the ascending vas deferens thickens before flexing to proceed as a penis. Generally, the region of flexion extends backwards as the penis retractor muscle. This portion of penis connecting both the ascending limb of vas deferens and its retractor muscle (Figs 11.12D and 11.20A) is defined as 'the junction region'. From this region (also in *Helix pomatia*, for example) originates the so-called flagellum. It is elongated, filamentous and extends backwards along the spermoviduct (Pelseneer, 1906; Meisenheimer, 1907; Lind, 1976). The presence of the flagellum and its form is an indicator for the structure of the spermatophore. It is believed that it is responsible for the formation of the so-called 'tail' of the spermatophore (Lind, 1973; Breure and Eskens, 1977; Mann, 1984). The penis of *Eremina* lacks this flagellum and hence it differs completely from *Helix pomatia*, which is generally mentioned in textbooks exemplifying gastropod, and pulmonates in specific (Berry, 1977; Mann, 1984).

The penis is a pedal organ (Duncan, 1960a,b), and the pedal parts are ectodermal in origin (Wada, 1968). Considering its structure, Rigby (1963, 1965) gave a somewhat satisfactory review in this respect. Basically it is divided into two regions: penis proper and praeputium (Fig. 11.20A), the latter is the end part of the male side of this hermaphrodite system. Pulmonate stylommatophorans differ from prosobranchs and opisthobranchs in that the praeputium only is ensheathed by a muscular sheath. However, Abd El-Kerim (1988) and Bawab et al. (1993) studied and described that the two portions of the penis are surrounded by the muscular sheath. So, the essential function of the penis is copulation. Mann (1984) added that the penis of pulmonates is involved in both copulation and spermatophore formation. It seems that Breure and Eskens (1977), who observed the formation of the spermatophore of a land snail, have an opposite view, since they reported that the penis 'does not play a role in the process of spermatophore formation'. Their observations cannot be applied on all snails, since their animal, *Drymaeus canaliculatus*' penis consists only of the praeputium, which has no role in the formation of the spermatophore.

Spermatophore formation. It is a peculiar picture that the spermatophoric mass of a spermatophore differs in its sheath, not only in the different groups, but also in the different species of the same group: In Crustacea (Bawab, 1970; El-Sherief, 1975); and in Mollusca (Ismail, 1981; Abd El-Kerim, 1988). Mann (1984) presented different spermatophores of different animal groups. In *Eremina*, the spermatophoric mass begins as a sperm mass in the vesicula seminalis (middle hermaphrodite duct), and becomes spermatophoric mass within the sperm-channel of spermoviduct, then passes through the vas deferens to the penis. Within the penis, and precisely, the penis proper, the spermatophoric mass (Fig. 11.20F and G) attains its outer 'tunics' to be a spermatophore (Fig. 11.27F). The lining epithelium of this region is folded in a specific manner so that the lumen becomes well equipped to be a mould for the spermatophore. Two of the epithelial folds are large and heavily refolded, occupying almost the whole lumen (Fig. 11.20D). The epithelial lining consists of typical columnar cells that are secretory; since they are loaded with granules, found as granulated secretion in the lumen (Fig. 11.20H). A great abundance of the so-called 'subepithelial cells' are observed scattered within the muscle layer and accurately in small masses underneath the epithelium (Fig. 11.20E). The mechanism of the completion of the formation of spermatophore cannot be described from these findings. However, Mann (1984) reported that the spermatophore of pulmonates is formed during copulation.

Copulation. As far as the findings and observations that resulted from the work on *Eremina* show, the timing of the vital activities that occur in the reproduction of this snail can be concluded. First of all, *Eremina* aestivates during August and September every year. This phenomenon was well observed during the whole period of work, which extended for about 7 years. After aestivation, the animal begins its reproductive activity by the process of spermatogenesis, which flourishes during autumn with a peak in October. On observation in the field, this snail copulates during October. Fortunately, during maintaining the culture of the snails, Abd El-Kerim observed some of the snails trying to prepare for copulation during October. Each two animals (couple) were in the position described by Lipton and Murray (1979): face to face. Three couples were marked by the letters A, B, and C and the animals of each couple were marked, e.g. 1A and 2A, etc. The moment of true copulation was not observed, maybe because of captivity, and therefore the donor and the receiver were not determined. However, the two animals of couple A were dissected, and the genital atrium of one of them (1A) was occupied by a fully formed spermatophore. On the other hand, the penis, or more precisely the penis proper, of snail (2A) was occupied by a spermatophore in its final stage of formation. At the same time, its genital atrium and spermatheca were empty. The active penis was found in specimens of Couple B. They were in a different picture, maybe because they remained in the position (copulation) for longer duration (5 h for 1B and 24 h for 2B). The spermatheca of the former (1B) was occupied by free sperm (as shown in the histological appearance of the smear done from the middle region of the hermaphrodite duct (vesicula seminalis) (Fig. 11.14H); and the receptaculum seminis of the latter (2B) was occupied by free sperm. A significant observation concerning these free sperm must be recorded that they are in a definite orientation (Fig. 11.32A and B). Concerning couple C, each of its snails was in the preparatory position and Abd El-Kerim (1988) reported a peculiar observation that the dart sac and penis of a precopulated snail, *Eremina*, snail 1C, were everted from the genital opening (Fig. 11.32G). Snail 2C did not show this picture. The dissection of both snails of couple C revealed that the male side of each was empty, as seen anatomically. However, histological study of the gonad of snail 2C evidenced that it is in its peak of activity, compared with that of the others of the three couples.

Generally, during copulation of any animal, one that is a male (donor) transfers its product to the other, which is a female (receiver). From the three couples of *Eremina* described above, the following could be concluded:

1. Each of the individuals of couple A behaved as donor and receiver at the same time, i.e. the two exchanged the spermatophores. But the vital activity did not complete, since the spermatophore of snail 1A was found in the atrium and was not transferred to the partner; while snail 2A had its spermatophore still in the penis;
2. In couple B, each of the two animals behaved as donor and receiver. Copulation was completed and the spermatophore in each was dehisced in the spermatheca. In the available literature at that time of work, there is no snail where the site of dehiscence of the spermatophore is defined. In *Eremina*, this is the first to be defined. The foreign sperm in 2B passed through their pathway (see later) to the receptaculum seminis (Fig. 11.32A);
3. In couple C, the animals were fixed during the action of copulation. In snail 1C, the penis and dart sac protruded from the genital opening (Fig. 11.32G). It appears that this couple was not in proper condition;
4. The protruded penis in snail 1C proves that the penis of *Eremina*, functions as a copulatory organ. In addition, in the snail 2A, dissected 2 h after copulation, the penis contained spermatophore in formation. This indicates that the penis of *Eremina*, shares in the formation of the spermatophore.

Pathway of the foreign sperm. Just after copulation and the snail receiving the spermatophore from the partner in the atrium, it takes its way directly to the vagina, to the oviduct, to the spermathecal duct, to the spermatheca (see Fig. 11.23A). This journey occurs very fast, and no step can be observed except that in the spermatheca. As for how this journey or passage occurs, only the muscular movements of these organs may be responsible; and/or a hormonal system may function. However, the spermatophore is retained inside the spematheca till the time of fertilization. The structure and secretory appearance of the spermatheca make it equipped to be a reservoir for a long time, and the previous views believed that secretions 'digest' the excess autosperm, allosperm, and spermatophore. The spermatheca of *Eremina* never functions with autosperm, and Meisenheimer (1907) believed that the spermathecal secretions in *Helix pomatia* 'loosen' sperm from spermatophore. Mehanney studies on *Eremina* give supportive evidence to Meisenheimer's view as for the essential function of the spermatheca. The spermatheca of *Eremina*, or spermatheca in general, is the site of preservation of the foreign spermatophore. Its secretions (Fig. 11.22C–E), which may be successive, are responsible for the dehiscence, and reactivation of its sperm to reach the receptaculum seminis in the talon. The foreign free sperm pass backwards within the spermathecal duct to its connection with the oviduct, to oviduct, and to egg-channel of the spermoviduct (Fig. 11.12D), but through a definite passage to the receptaculum seminis (RS) of the talon (Fig. 11.17B and C). Within the latter (RS) (Fig. 11.32A), these foreign sperm are still active and attach themselves to the epithelial lining, maybe for nourishment to keep their vitality (Fig. 11.32B). They remain in this position (oriented) till they are used for fertilization in the fertilization chamber (Fig. 11.32C).

The pathway of the oocyte. It was understood from the available literature that the gastropod oocyte grows, matures, and ovulates at the primary oocyte stage (Pennypacker, 1930; Kugler, 1965; Wada, 1968). When mature, it is spherical in outline, and it has both the vitelline membrane and the chorion, as mentioned above. This mature spherical oocyte passes, with amoeboid movement, through the conducting portion of the gonadal tubule (Figs 11.29D and 11.30) to the efferent tubes (Fig. 11.13E). The photomicrograph Fig. 11.29D probably provides acceptable evidence that amoeboid movement is the mode of passage of the oocyte of *Eremina*, and not the follicular cells or other modes mentioned in the literature. This oocyte passes to the hermaphrodite duct, to the talon (Fig. 11.15B). In the talon, it must pass through the

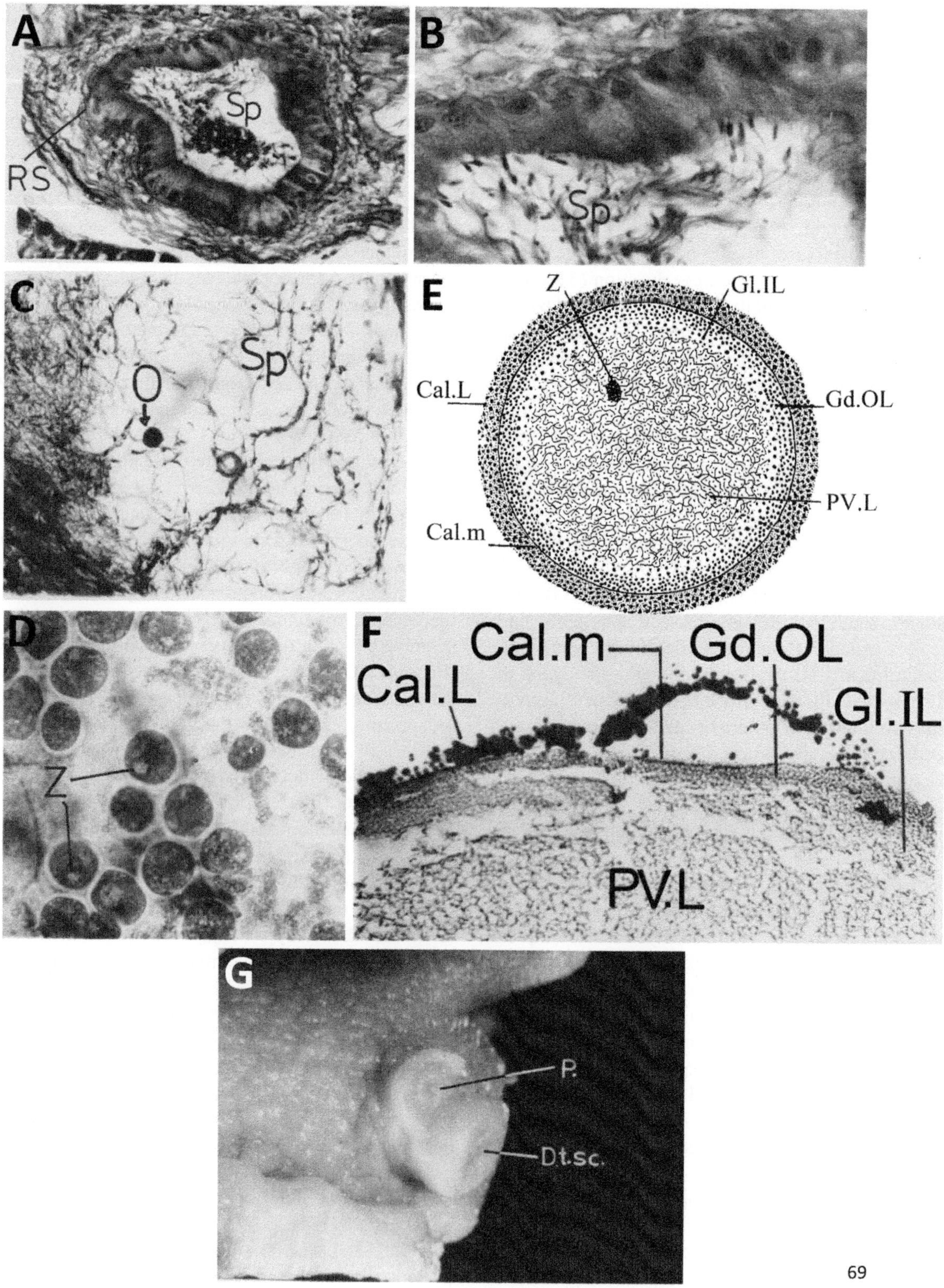

FIG. 11.32 (A) Magnified portion of a T.S. of the albumen gland of *Eremina* showing the receptaculum seminis of the talon occupied by the foreign sperm. X = 400; (B) OIM of a portion of (A) showing the oriented sperm. X = 1250; (C) magnified portion of a frontal S. of the albumen gland prepared in August showing oocyte and sperm in the fertilization chamber of the talon. X = 700; (D) OIM of the eggs found in the end part of the egg-channel of the spermoviduct. X = 850; (E) diagrammatic drawing of the laid egg showing its structure; (F) magnified portion of section of the laid egg showing the histological structure. Cal.L = calcified layer, Cal.m = calcified membrane, Gd. OL = dense granulated layer, Gl. OL = loose granulated layer, PV.L = perivitelline layer. X = 100; (G) lateral view of precopulated snail, *Eremina* (animal 1C), showing the ejaculated penis and the opened dart sac. Dt. Sc = dart sac, P = penis. X = 3.

ciliated hood. The latter does not continue as a limited duct inside the talon (Fig. 11.17D), but for the mature sperm, it becomes a definite passage (Fig. 11.16C). At the end of the limited ciliated hood (Fig. 11.17D), the spherical oocytes, by their nature, fall to the fertilization chamber. At the appropriate time, the activated sperm (foreign) in the receptaculum seminis (Fig. 11.32A and B) discharge smoothly to the fertilization chamber through the openings of its branches at different levels facilitating the oocytes to be bathed with these sperm, and fertilization occurs in this chamber (Fig. 11.32C). The foreign sperm in the corresponding component in the talon of the other animals mentioned above must be forced to pass backwards to the common entrance of the receptaculum seminis to be discharged to the fertilization chamber.

At a certain level, the talon receives the albumen duct (Figs 11.16G and 11.17A–C), or more precisely, the fertilization chamber connects both the albumen chamber and the egg-channel of the spermoviduct. In spite of the many preparations examined, it is not definite where the fertilized oocyte accepts the voluminous and different albumen secretions (Table 11.2). However, it was investigated that these secretions accompany the fertilized oocyte to the egg-channel of the spermoviduct (Fig. 11.17A–C). Within the latter, the fertilized oocytes are contaminated with the secretions of both the oviducal gland and the epithelial lining of the egg-channel, which may be glycoproteins, mucoproteins, and/or mucopolysaccharides. These may form the perivitelline membrane, may react and/or mix with those of the albumen secretions forming this perivitelline membrane (Fig. 11.32E and F), the tertiary membrane of the egg of *Eremina*. Lusis (1961) recorded that the mucoprotein of the 'oviducal part' of the spermoviduct is responsible for the formation of a corresponding membrane, but this author defined it as secondary membrane (the nomenclature and classification of egg membranes were reviewed in the work of Anderson, 1967). In addition, the secretions of the corresponding part of the spermoviduct of the *Oxychilus cellarius* (Rigby, 1963) secretes mucous, which lubricates the passage of the forming eggs, and the materials of the egg-channel are well moulded as spheres and embedded in a substance (Fig. 11.32D). Moreover, the spherical eggs are surrounded by an unstainable layer of considerable thickness, which may be the shell membrane. The work of Runham (1978) on young *Deroceras reticulatum* gives a confirmation that this duct has a definite gland, which may participate in the formation of the egg shell. On the other hand, some authors believed that the corresponding portion of the egg-channel in other stylommatophorans acts as uterus (Pelseneer, 1906; Rigby, 1963; Breckenridge and Fallil, 1973).

It appears that the egg-channel or its correspondence can share in the formation of egg membranes and the calcite one; reserve the eggs and secrete the mucous substance embedding the eggs. The different functions interpreted for the correspondence of the egg channel in the different stylomatophoran genera lead to different names referring to it. None of these names is adequate and hence the general name, egg channel, is preferable for *Eremina*.

The contents of the egg-channel (Fig. 11.32D) pass smoothly to the oviduct, which, contrary to the egg-channel, has a very thick muscular wall (Fig. 11.21B). Rigby (1963) believed that conduction is the only function of the oviduct, and its muscles keep the received contents coming from the egg-channel until the proper time of egg-laying. Another view adds a secretory function to this oviduct to finish the formation of the shell deposition around the developing eggs coming from the egg-channel (Smith, 1965, 1966; Focardi and Quattrini, 1972). Smith defined amoebocytes in the wall of the oviduct that are responsible for calcite secretion. In 1978, Runham observed 'large flask-shaped glandular cells' that open through the lining epithelium of the oviduct in young *Deroceras reticulatum*, which add materials to the periphery of the developing egg. This means that the oviduct may consist of definite secretory cells. This was not observed in the oviduct of *Eremina*, but the deeply stained cells scattered in the submucosa of the oviduct, which were assumed as amoebocytes (Fig. 11.21C), might be secretory. It was evidenced that the epithelial lining itself is secretory (Fig. 11.21C). The secretory cells are holocrine, or may be apocrine (Leonhardt, 1977) since the nuclei keep their feature during the activity of the cells and also during the discharge of the secretion. This does not mean that the amoebocytes are absent or have no role in the secretion in the oviduct of *Eremina*. However, this oviduct needs more studies to define its definite role in the formation of the developing eggs.

The contents of the oviduct pass to the vagina, which in all studied stylommatophorans is a continued passage to the oviduct. Like the oviduct, the vagina has a thick, muscular wall, but in *Eremina*, it lacks glands and secretory function. At the same time, it joins the duct of the mucous glands (Fig. 11.23A), characteristic to family Helicidae (Pelseneer, 1906; Duncan, 1960a,b; Solem, 1978; Morton, 1979). The old view (Meisenheimer, 1907), believed that the function of the secretion of these glands is associated with the individual when it behaves as a male, i.e. discharge of the spermatophore(s) during copulation. This may be confirmed by two critical and personal observations: copulation in *Eremina* occurs during October (Abd El-Kerim, 1988); and these glands attain their largest size during August to October. However, this secretion is used to cement the laid eggs in masses.

The egg/fertilization. It is appropriate to mention that no eggs were observed within the different portions of the reproductive system of *Eremina* during the period of its study, which lasted about 7 years. The oocytes were observed in the fertilization chamber of the talon (Fig. 11.32C), and the eggs in the end part of egg-channel (Fig. 11.32D). If it is conceived that fertilization occurs in the fertilization chamber, then those found in the egg-channel are zygotes surrounded with, at least, the secretion of the albumen gland. Each, of course, is not fully formed, but has a spherical shape with a diameter of about 15 µm. Histologically, the sphere has a granulated appearance deeply stained with Orange G of Mallory's triple stain. It is surrounded by a well defined thin layer, deeply stained with aniline blue of the same stain. The zygote can be observed in some of these spheres. It is conspicuously small rounded and peripheral in position.

The fully formed egg. Only the laid egg was available to be examined. The dimensions of the different layers constituting the fully formed egg could not be accurately measured, since no complete section was obtained due to the calcified membranes. However, by the stains, it was possible to describe this egg. It consists of the following:

1. The zygote is a, comparatively very small, oval cleaved embryo, embedded in a reticulate substance;
2. This substance is a very thick layer and occupies almost the whole egg. It embeds and surrounds the growing embryo. According to Bayne (1966, 1968) this is the perivitelline layer (Fig. 11.32E and F);
3. Outer to the latter is a granulated layer with a considerable thickness measuring about ¼ of the perivitelline layer. It is well differentiated into inner loosely granulated layer and outer densely granulated one (Fig. 11.32E and F);
4. The granulated layer is limited by a membranous layer which attains the typical colour of alizarine stain and therefore it maybe a calcified layer, and so it is referred to as the calcified membrane (Fig. 11.32E and F);
5. The outer most component of the egg is crystalline in appearance and has a considerable thickness. As the calcified membrane, it attains the typical colour of alizarine. It is the calcareous layer and it has the same thickness as that of the granulated layer (Fig. 11.33K).

Back to Gastropoda, In general

Reproductive behaviour

Spawning aggregation. Webber (1977) reported that in prosobranchs, *Crucibulum* and *Calyptraea,* the male of the pair was considerably smaller than the female and was found during the copulatory season attached to the shell of the female on the right side near the mantle cavity. Apparently, the female shell is the copulatory position in many prosobranchs. Pairing has also been described in broadcast spawners; e.g. some species of genus *Patella,* males and females are not found attached to each other but are close together.

Sex recognition. There is evidence of active sex recognition. The females attach to males in different ways:

1. The female releases a mucous substance and it was observed that male *Olivella* were attracted to, and followed trails of mucous of sexually mature females.
2. It was suggested that males of *Murex pomum* were attracted to females by chemotactic response. It is possible that females liberate a pheromone when sexually mature.
3. It was found that the osphradium of *Viviparus* is the organ responsible for sensing the opposite sex.
4. Another possible mechanism of sex recognition might be based on size. Many species have a size difference based on sex, and in copulating pairs, the male is often smaller.
5. Astonishing case: male–male copulation, although infrequent, has been reported. In *Littorina picta,* 7 out of 1000 and in *Littorin ascabra,* 12 out of 50 copulating pairs were male–male.

Copulation. Beeman (1977) reported that in Pulmonata, cutaneous fertilization without copulation or aphallic transfer of sperm occurs. The spermatophores are attached to the skin at numerous points on the animal to be impregnated. The spermatophores gradually empty and the sperm penetrate the skin, evidently by causing lysis of the skin cells and directed to the fertilization centre.

Brooding behaviour. Generally, there is little care for the eggs or egg cases after spawning or ovoposition. Occasionally, however, brooding of the released eggs may occur. In some cases, brooding is fairly passive and associated with the passive existence of the adult. However, *Acmaea rubella* has been observed to brood eggs in the mantle cavity. In *Calyptraea, Hipponix,* and *Stephopoma,* the spawn is attached to the substrate and covered by the mantle cavity of the foot, under the anterior part of the shell. In *Capulus,* the spawn is brooded under the female shell but is actually held in a fold of the foot. If the spawn mass is removed from protection, the embryos soon die.

Spawning and rearing of youngs. Parental care, in its broadest sense, includes the preparation of oviposition sites, the production of large, 'heavily-yolked' eggs and unfertilized nurse eggs, the care of eggs or young inside or outside the parent's body, and the provision for young before and after birth. In its narrowest sense, it refers only to the care of eggs or young when they are detached from the parent's body (Baur, 1994).

Among the Prosobranchia, the primitive Archeogastropoda lay their eggs separately or in egg masses surrounded by jelly. Except in a few ovoviviparous species, the egg of the other orders are laid separately, or in various numbers, generally less than a hundred, fixed in egg cases of characteristic shapes peculiar to each species.

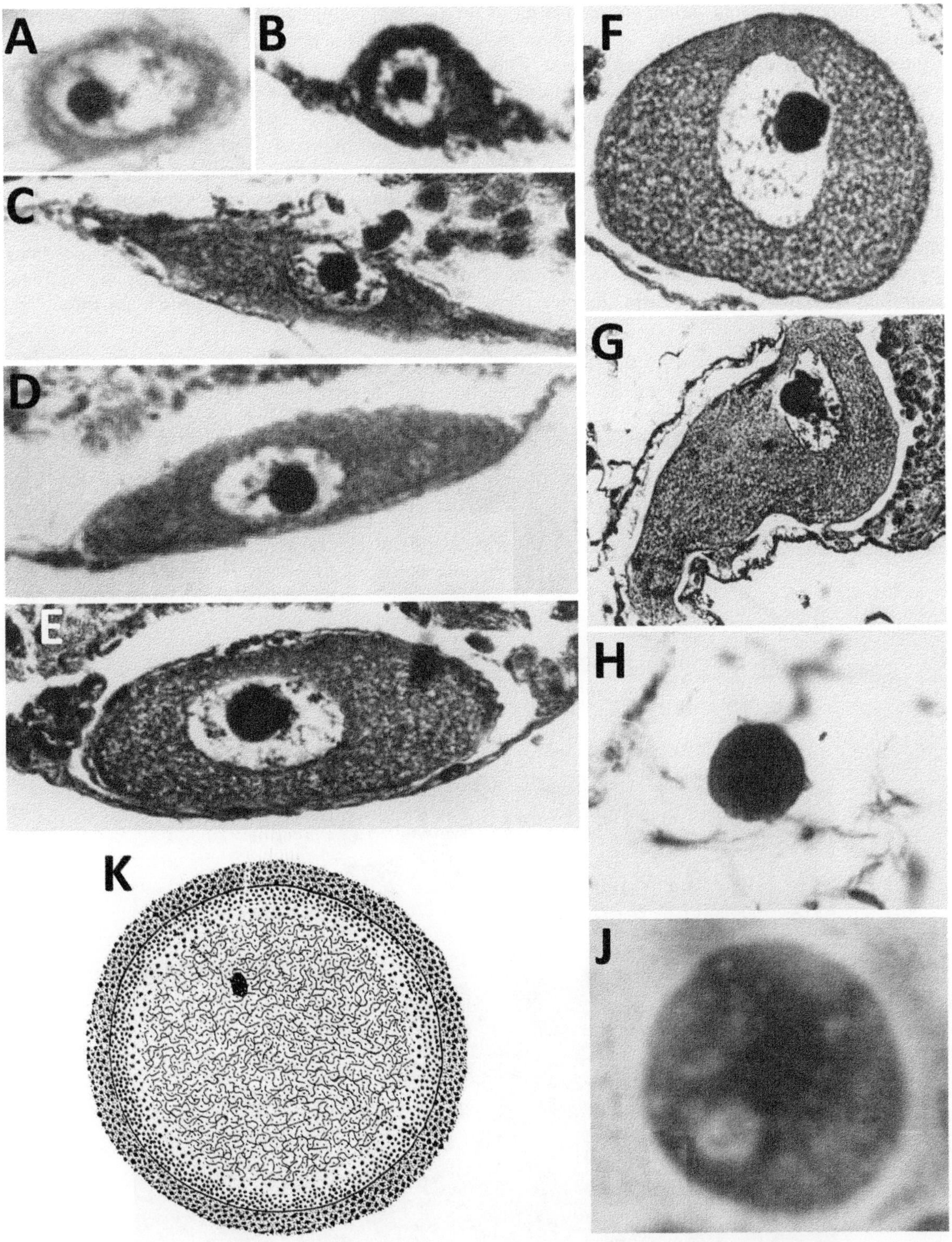

FIG. 11.33 Summary of stages of development of the oocyte of *Eremina ehrenbergi*: (A) early oocyte; (B) beginning of growth, oocyte with nurse cells; (C) spindle-shaped oocyte; (D) spindle-oocyte begins vertical growth to be spherical; (E) oval-shaped oocyte; (F) spherical oocyte; (G) amoeboid oocyte, within conducting region of gonadal tube; (H) zygote in fertilization chamber; (J) egg with perivitelline layer from end of egg-channel; (K) laid egg.

Wada (1968) reported that in *Babylonia japonica*, 40 eggs surrounded by gelatinous substance to form a single mass, are intermittently extruded from the genital pore. This mass passes along a special groove, temporarily present on the surface of the foot, to the pedal pore (also called the ventral pedal gland) on the sole of the hypopod, where it is moulded into shape and laid.

Adalaria proxima

In Opisthobranchia, Thompson (1958) selected a species of dorid nudibranch, *Adalaria proxima*, to rear small population through the complete life cycle in the laboratory. The author collected the animals from a zone of a rocky sheltered shore in the vicinity of North Wales. On this shore, *A. proxima* lives predominantly in the *Fucus serratus* (seaweed) zone. This animal, *A. proxima*, is an annual organism, breeding in early spring, beginning at the end of March and lasting approximately 8 weeks. The author observed these adult dorids copulating throughout the breeding season. The author studied and presented voluminous work dealing with the reproductive biology of this animal, *A. proxima*. First of all, he described the act of spawning (oviposition), spawn, and egg mass (Fig. 11.34A). The spawn extruded slowly from the adult, as a ribbon-shaped mass, through the genital

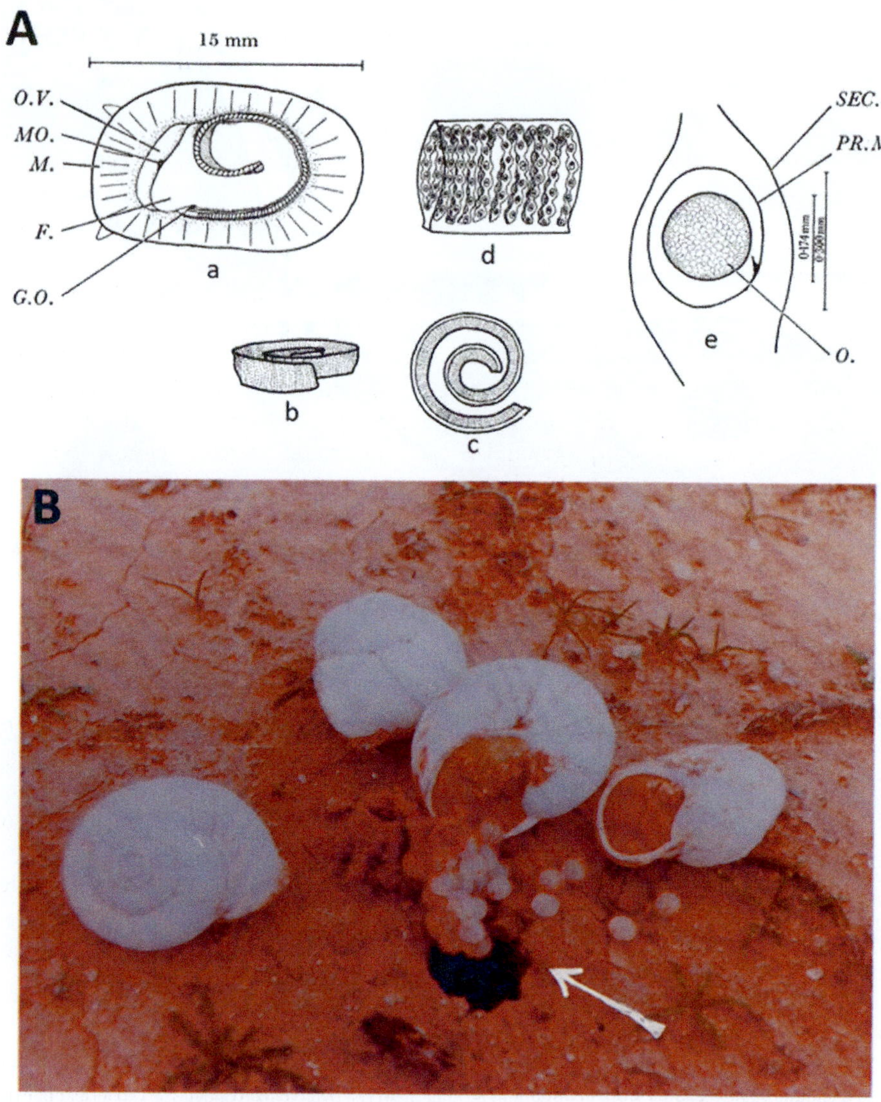

FIG. 11.34 (A) Oviposition in the opisthobranch *Adalaria proxima*: (a) mature adult in the act of spawning on a glass plate, viewed from the underside (the hatched region of the spawn ribbon is attached to the substratum); (b) completed spawn ribbon, lateral aspect; (c) completed spawn ribbon, dorsal aspect; (d) enlarged portion of spawn ribbon (diagrammatic); (e) single ovum with its envelopes. F=foot, G.O.=genital opening, M=mantle, M.O.= mouth, O=ovum, O.V.=oral veil, PR.M.=primary membrane, SEC.M.=secondary membrane *(from Thompson, 1958)*; (B) Photograph showing some snails of the pulmonate *Eremina ehrenbergi*, in the field, during the period of egg-laying (February). The *arrow* indicates a tunnel with a mass of eggs, exceptionally, out of it. It appears that the camera disturbed the snail.

aperture, which is situated antero-laterally on the right side of the body. The spawn ribbon is shaped by the slit-like genital aperture as it emerges and passes back between the mantle and the side of the foot. At this time, the jelly of the egg mass is of a sticky consistency, and pressure of the mantle and foot upon it causes it to adhere to the substratum. As the full length of the spawn ribbon is slowly extruded and pressed down, the adult moves in an anticlockwise spiral (when seen from above). The completed ribbon has the form of Fig. 11.34A (c). Oviposition of even a short ribbon takes several hours.

In nature and in culture, these spirals of spawn frequently harbour various small animals (gammarids, polychaetes, etc.). In laboratory culture, however, the majority of egg masses were laid on the walls and floor of the vessel, even when fucoids were present, and the author could not explain this. He reported that, in nature, the adults die after spawning.

This recalls what was described by Mehanney in her thesis (1991) that no egg-laying of *Eremina ehrenbergi* was observed in the laboratory. The period of egg-laying of this genus is limited (January and February) (Hammad et al., 1977). During this period in 1989, the snails were seen scattered, in great numbers, on the smooth sandy ground and not adhering to the shrubs of the field. In the morning (6–9 a.m.) of the 24th of February 1989, the snails were observed with the apertures of their shells against the surface of the ground. The accumulations of sand beside these snails differ greatly from the surrounding sand and are greatly comparable with the faeces of the snails of the maintained culture. By gentle withdrawing of one snail, it was found that the whole head-foot was extending within a tunnel (Fig. 11.34B). This tunnel and another one contained eggs, which were considered to be freshly laid. A remarkable observation was that during dissection of one of these snails, six eggs were discharged from its genital opening. Another noticeable observation is that its alimentary canal was larger than the ordinary size, and the different parts of this canal were totally full with sand grains.

The weather on that day was fine and, by critical observation in the vicinity of the upper described snails, many scattered accumulations of the faecal sand grains were found. Their removal showed the openings (mouth) of the tunnels. The tunnel itself is not vertical, but oblique, extending about 10 cm, and these eggs were found in mass at its bottom. Different masses were collected, and in the laboratory, they were found in different numbers. The largest masses were of 60 and 90 eggs. The eggs of each mass were attached to each other by a mucous substance. However, the attachment was not that firm, since some of them were easily separated from the mass.

In many species, the embryos develop inside the egg case to the veliger larval stage, before they swim out from a fixed region of the case, although forms are not rare in which the larvae develop within the case until they complete metamorphosis and attain the same morphology as the adult before they come creeping out.

There is a phenomenon that is often seen among the Murioidae, Buccinidae, and others, in which only a few eggs (15–25 of 100 in this case) complete development. The unfertilized and polyspermic eggs, and embryos, which stop developing during the early stages, are used as food by the developing veliger larvae. Furthermore, in these species the larvae hatch after passing through the veliger stage, and their sizes vary with the number of nurse eggs.

Fertilization. Molluscan eggs are all fertilized before extrusion of the first polar body, that is, during the primary oocyte stage. Among the gastropods, copulation and internal fertilization are found in all the Opisthobranchia and Pulmonata (Mesopoda), with a few exceptions that are characterized by external fertilization. Among the Prosobranchia, in some species in which the egg has chorion, it is known that the spermatozoa undergo an acrosome reaction that is believed to enable the spermatozoa to dissolve the chorion and enter the perivitelline space. At a given time after fertilization, the polar body and then the second polar body are successively extruded at the animal pole. The first polar body, unusually, does not cleave again, although there are reports of its division. At the time of the polar body formation, not only the animal pole of the egg, but also the ventral pole undergoes a form change in many cases. The sperm lies in the egg cytoplasm near the surface without much change in position until the expulsion of the second polar body. When the maturation divisions of the egg are completed, the egg and sperm pronuclei come together near the centre of the egg and unite in syngamy.

Early development

Cleavage. The molluscan egg belongs to the category of so-called 'mosaic' eggs. Their cleavage follows the spiral determinate pattern for the protostomes in general. It is total, and it is influenced by the amount of yolk contained in the egg. When the amount of yolk is small, cleavage tends to be proportionally equal, but never loses its essentially spiral character.

In Mollusca, two modes of spiral cleavage were studied:

1. Annelidan mode, exemplified by *Trochus* is the general mode;
2. In the gastropods *Nassa, Nassarius, Ilyanassa, Modiolaria* and *Aplysia,* the scaphopoda, *Detalium* and the pelecypods *Crassostrea, Pinctada, Pteria, Mytilus, Spondylus* and others, a polar lobe is formed during the early cleavages and this mode is exemplified by *Nassrias* and *Pinctada, Pteria, Mytilus, Spondylus*, and others.

Cleavage of ***Trochus*** *(Annelidan mode).* Wada (1968) illustrated and described the stages of cleavage of Trochus (Fig. 11.35). The 1st cleavage furrow appears in the animal pole and advances in a diagonally vertical direction, dividing the egg into AB and CD blastomeres. The 2nd cleavage furrow appears near the pole sides of the blastomeres, this time intersecting the first cleavage furrow at right angles. This cleavage divides AB into A and B, while CD divides into a smaller C and a larger D, to give rise to the 4-cell stage. C is about the same size as A and B. The several blastomeres of the 4-cell stage represent the four cleavage quadrants: A, lying on what will be the left side of the future embryo; B, lying anteriorly; C, at the right side.

The 3rd cleavage furrows appear in each blastomere, this time advancing in a diagonally horizontal direction and separating off the equal-sized micromeres 1a, 1b, 1c, and

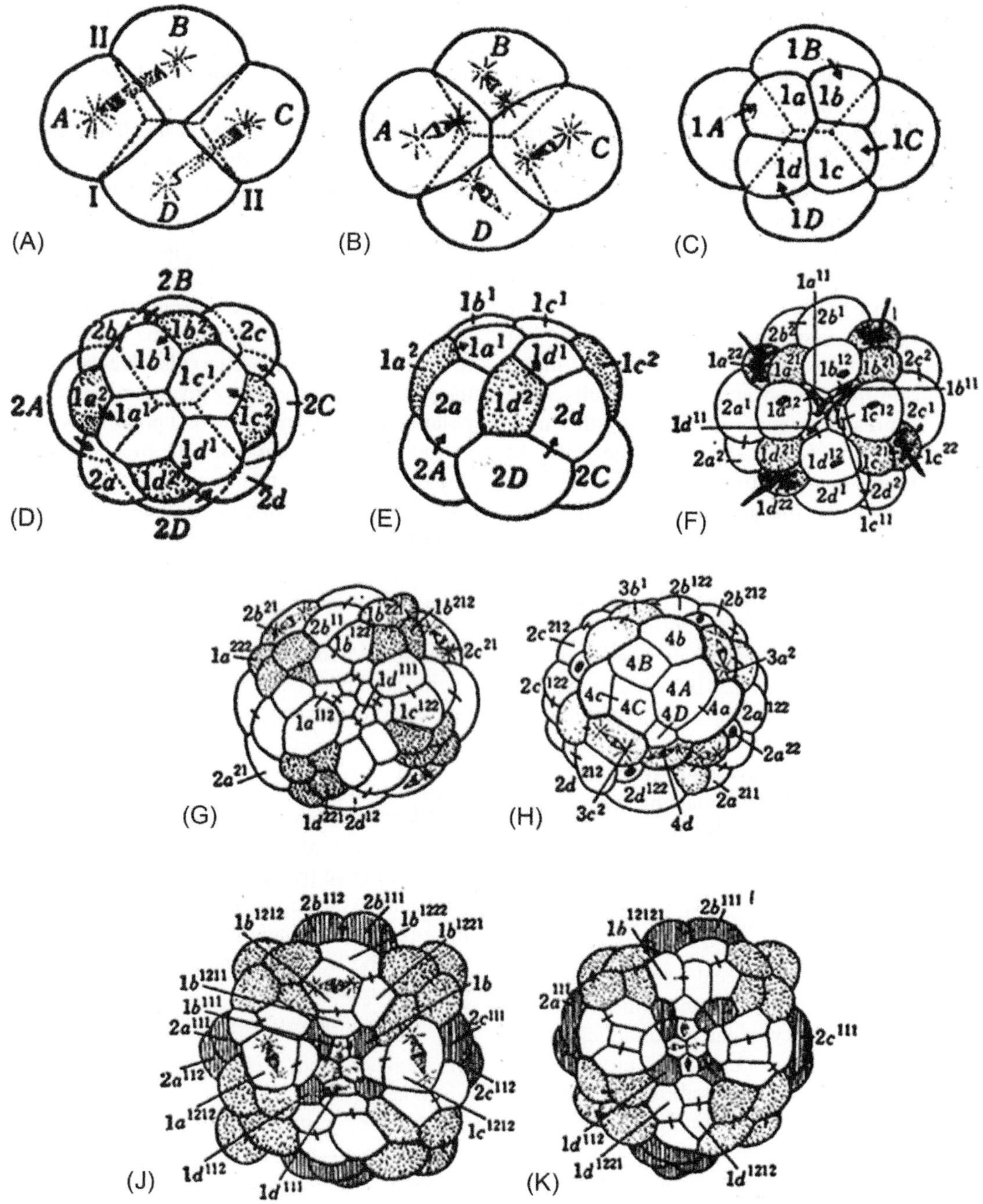

FIG. 11.35 Cleavage in *Trochus*: (A and B) 4-cell stage (from animal pole); (C) 8-cell stage (from animal pole); (D) 16-cell stage (from animal pole); (E) 16-cell stage (from side); (F) 32-cell stage (from animal pole); (G) 63-cell stage (from animal pole); (H) 81-cell stage (from vegetal pole); (J and K) more advanced stages, formation of molluscan cross. Trochoblasts (stippled), Tip cells (hatched cells at equator), peripheral rosette cells (hatched cells near animal pole), Cross cells plus rosette cells *(white). (After Robert (1902), from Wada (1968).)*

1d, diagonally upwards to form the 8-cell stage. These micromeres 1A, 1B, 1C, and 1D, forming the so-called first quartette of micromeres, which surround the animal pole. The macromeres 1A, 1B, and 1C are all of about the same size, while 1D is especially large. At the 4th cleavage, the macromeres 1A, 1B, 1C, and 1D divide obliquely towards the left and unequally, giving rise to the second micromere quartette, 2a, 2b, 2c, and 2d. At almost the same time, the micromeres of the first quartette divide equally and levotropically, becoming 8 micromeres ($1a^1$–$1d^1$) and ($1a^2$–$1d^2$, primary trochoblast) as the embryo reaches the 16-cell stage. At the 5th cleavage, the third micromere quartette is formed by dexiotropic divisions. Theoretically, the other micromere groups should simultaneously undergo dexiotropic divisions, conjointly with the special unequal cleavage of the first somatoblast, to give rise to the 32-cell stage.

After the embryo has reached the 32-cell stage, the next event is the cleavage of the macromeres (3A–3D) to form the fourth quartette (4a–4d) leaving the macromeres as (4A–4D). The third quartette divides, giving rise to ($3a^1$–$3d^1$) and ($3a^2$–$3d^2$). The second quartette divides further, to give rise to ($2a^{11}$–$2d^{11}$), ($2a^{12}$–$2d^{12}$, molluscan cross), ($2a^{21}$–$2d^{21}$), and ($2a^{22}$–$2d^{22}$). The first quartette gives rise to ($1a^{111}$–$1d^{111}$, molluscan rosette), ($1a^{112}$–$1d^{112}$), ($1a^{121}$–$1d^{121}$), ($1a^{221}$–$1d^{221}$), ($1a^{212}$–$1d^{212}$), ($1a^{122}$–$1d^{122}$), ($1a^{211}$–$1d^{211}$), and ($1a^{222}$–$1d^{222}$). Finally, the 6th cleavage gives rise to 64-blastomeres. Fig. 11.36 is the table of the cell lineage of the prosobranch *Crepidula* by Conklin (1897).

*Cleavage of **Ilyanassa obsoleta**.* Although the freshly laid egg may sometimes show a germinal vesicle, usually the first polar spindle is forming deep in the cytoplasm of the animal third of the egg, and a sperm has entered. In other cases the spindle is already at the surface.

As the first meiotic spindle of the fertilized egg moves to the surface at the animal pole, a brief and rather inconspicuous bulge at the vegetal pole comprises the first polar lobe. The first polar lobe appears, and within a few minutes the egg is again spherical and remains so for about 15 min (Fig. 11.37A–C).

The second polar lobe then appears just prior to the metaphase stage of the second meiotic division. It becomes large and conspicuous and persists, altogether, for about 50 min. The second polar body is given off approximately 20 min after the appearance of the second polar lobe, and about this

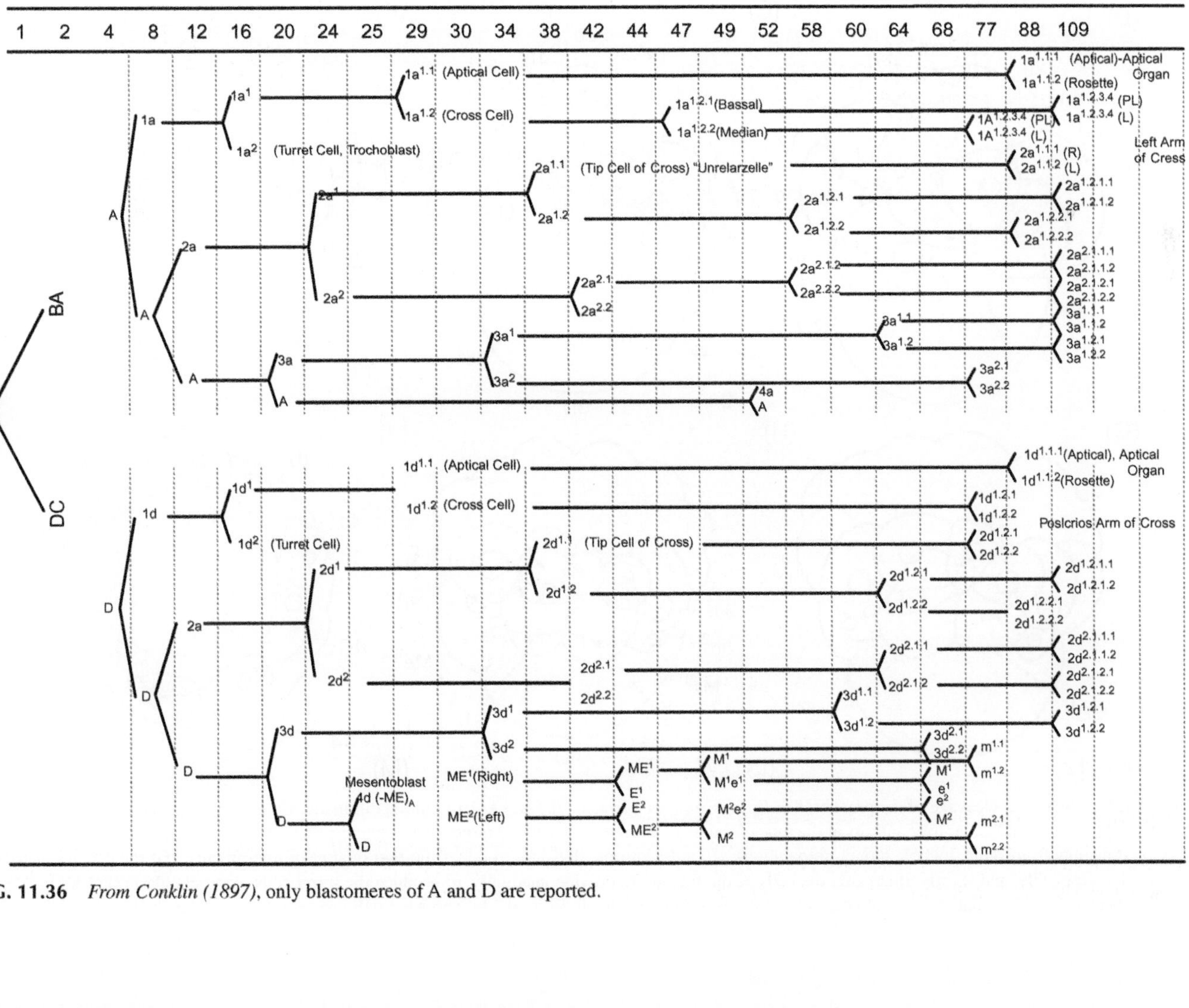

FIG. 11.36 *From Conklin (1897)*, only blastomeres of A and D are reported.

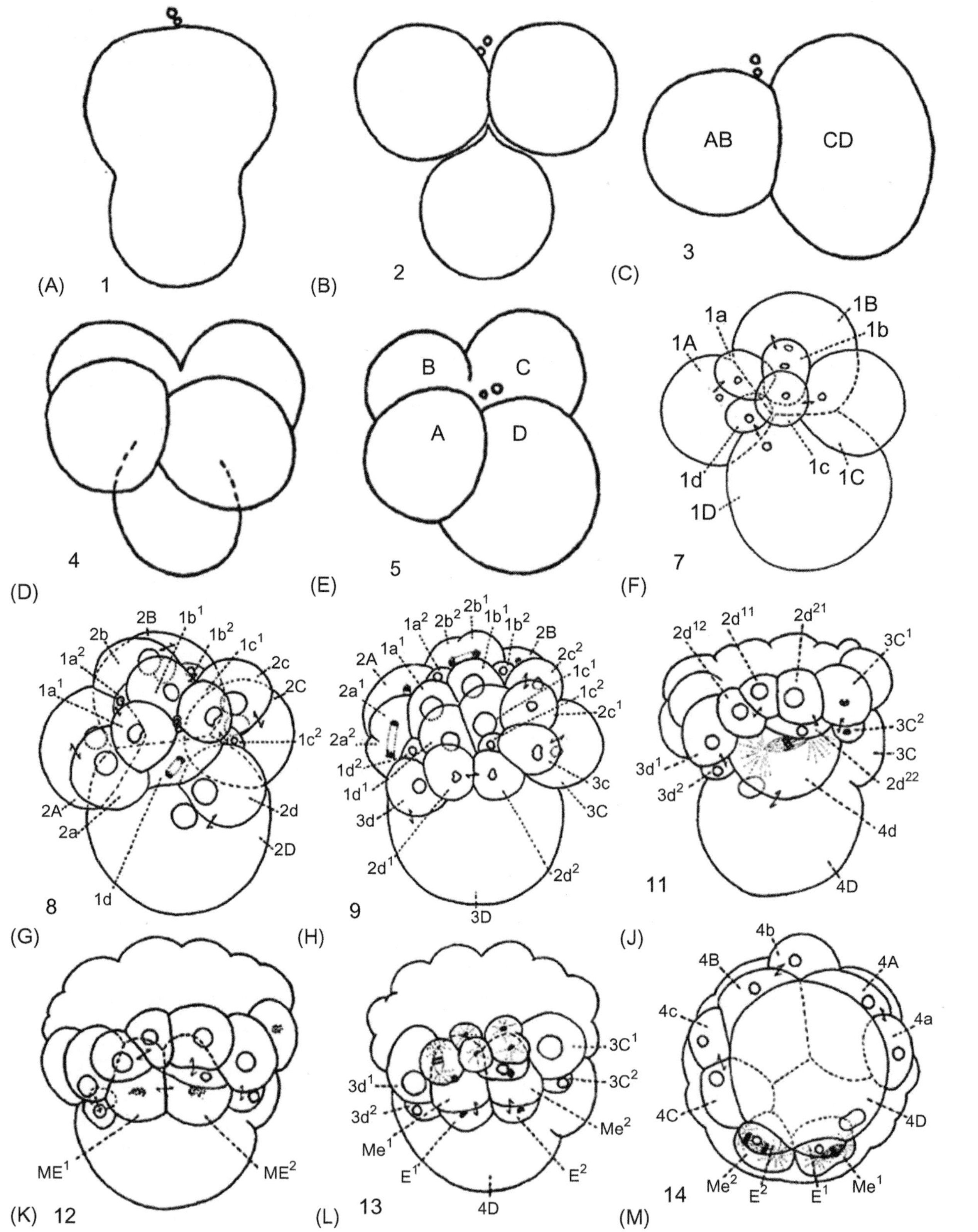

FIG. 11.37 Stages of cleavage of the prosobranch *Ilyanassa (from Clement, 1952)*: (A–F) semidiagrammatic, Camera Lucida out-lines of living eggs. X=171, showing a small polar lobe appears at the time of the first maturation division, a larger one during the second maturation division; (G–M) Camera Lucida drawings from stained whole mount preparations: (F) a dexiotropic division produces the first quartet of micromeres; (G) the second quartet is formed laeotropically, and shortly afterwards the cells of the first quartet divide unequally, to produce the turret cells, or trochoblasts (la2-ld2); (M) the egg has been oriented so that the vegetal pole is towards the observer. The division of Me' and Me2 will produce the mesoblast cells m1 and ma; (G H J) X=326.

time the lobe shows irregularities of outline not seen at the other stages. The irregularities disappear and the resorption of the lobe is followed by another spherical period.

Following the second meiotic division, the sperm nucleus, previously small and dense, enlarges and assumes a characteristic interphase condition. It then migrates to a position beneath the egg pronucleus, having meanwhile entered the prophase condition, as the egg pronucleus. During anaphase, the animal pole flattens and the egg elongates in the plane of the cleavage spindle. As the cleavage furrow forms, the lobe is further constricted from the egg, until the trefoil stage is reached. At trefoil (Fig. 11.37B), the egg consists of three spheres of roughly the same size, namely the two blastomeres and the polar lobe. A fine cytoplasmic filament may be seen extending from the upper end of the polar lobe, the notch at the bottom of the two blastomeres, but it is not evident at this time with which of the blastomeres it is absorbed. The lobe varies in different lots of eggs from about one-quarter to one-third the volume of the whole egg. It is easy to distinguish the polar lobe from a blastomere, even after they have been separated from one another. The polar lobe is very yolky in appearance, whereas the blastomeres are recognizable by a yolk-free cytoplasmic cap. The trefoil stage occurs about 3h after the time of laying and the beginning of meiosis. It soon becomes apparent that the lobe is attached to one of the blastomeres, the connection broadens and the lobe is gradually incorporated into the blastomere. In this, maybe an unequal first cleavage is achieved, the larger blastomere which has received the lobe is known as CD, and the smaller one as AB (Fig. 11.37C).

As mentioned, the polar lobe passes into the CD cell at first cleavage, and accounts for the large size of this blastomere. The lobe reappears for the fourth time prior to the second cleavage (Fig. 11.37D) and passes into the D blastomere of the 4-cell stage (Fig. 11.37E); the fourth lobe is of about the same size as the third.

The first quartette of micromeres (1a–1d) results from a dexiotropic cleavage. The 1d micromere is noticeably smaller than the other three, even though 1D is the largest of the four macromeres (Fig. 11.37F). The second quartette is formed by a levotropic division, and all cells of this group (2a–2d) are of similar size and appearance (Fig. 11.37G). Soon after the appearance of the second quartette cells, those of the first quartette subdivide to produce the small turret cells ($1a^2$–$1d^2$).

The third polar lobe makes its appearance during late prophase, while the asters are enlarging, and before the pronucleus membrane has disappeared. The interval between the disappearance of the second polar lobe and the appearance of the third is about 55min. The third polar lobe enlarges to a prominent condition during metaphase.

The third quartette cells (3a–3d) arise from a dexiotropic division of the macromeres, and are uniform in appearance, and at about the same time, the second quartette cells undergo a division (Fig. 11.37H). There are then 24 cells present. The next cells to divide are the apical cells ($1a^1$–$1c^1$). Soon thereafter the mesentoblast cell, 4d, makes its appearance. The interval between the first cleavage and the appearance of 4d is about 6–6.5h at 25–26°C. The entoblastic membrane of the fourth quartette (4a–4d) appears about 3h after the formation of 4d, meanwhile 4d will have divided equally in the plane of bilateral symmetry of the embryo, and the resulting cells will have begun the subdivisions, which eventually segregate some entoblastic cells and form the primary mesoblast bands (Fig. 11.37J and K).

Differentiation of germ layers

Segregation of ectoderm. All the material to form the cells of ectoderm is separated off from the four macromeres in the third cleavage, which are produced in a dextio-rotational-direction forming 1q (quartet) (1a–1d), as is characteristic of spiral cleavage. The fourth cleavage is sinistral, giving the 2q ($2a^1$–$2d^1$) where it is smaller in size than macromeres. The 1q is divided forming the 16-cell stage, which is also sinistral, producing ($1a^2$–$1d^2$), which occupy the animal pole region, and ($1a^2$–$1d^2$), which lie directly above the macromeres, held between the blastomeres of the 2q. ($1a^2$–$1d^2$) are called primary trochoblast and later produce the velum.

To reach the 32-cell stage, each blastomere cleaves in a dextral direction. 1q divides forming ($1a^{11}$–$1d^{11}$) which occupy the animal pole while ($2a^{12}$–$2d^{12}$) lie alternately with these, forming the (molluscan cross). ($1a^{11}$–$1d^{11}$) divide again forming ($1a^{111}$–$1d^{111}$), the apical cells or rosette cells, and ($1a^{112}$–$1d^{122}$) the peripheral rosette cross. The tip cells at the ends of the crossing arms are formed from ($2a^{11}$–$2d^{11}$).

Gastropod embryos maintain a radial symmetry as far as this stage. They are rather flattened, and a small blastocoel can be recognized inside the cell mass.

Differentiation of ento-mesoblast. From the macromeres (3A–3D), arise all the organs of endodermal nature and particularly all the mesodermal organs. Furthermore, from the 2q (*Crepidula*) or 3q cells (*Trochus, Fiona, Physa*), beside the ectodermal organs, the second mesoderm or ectomesoderm, is formed.

The division of the 3D macromeres has marked distinguishing characteristics, taking place in a sinistral direction, and producing the 4d daughter cells, which are larger than the 4D macromeres, and eventually sink into the blastocoel. The 4d blastomere is called the primordial mesodermal cell or mesoblast; since part of the ectoderm, as well, sometimes differentiates from it, it may also be called the mesentoblast, or it maybe described as the second somatoblast. In *Trochus*, the formation of 4d occurs much later, in the 64-cell stage, lagging behind the cleavage of 3A, 3B, and 3C. These three blastomeres also cleave very unequally, giving rise to the 'micromeres' 4a, 4b, and 4c, which are larger than 'macromeres' 4A, 4B, and 4C, which are called entomeres or secondary macromeres, and together with the micromeres of 5q they form the entodermal organs (digestive system). This is the symmetry division:

a. Firstly, it gives ME^2, to the left side and ME^1 to the right side, which are called teloblasts of mesodermal bands.
b. ME^2 gives rise to Me^2 & E^2; ME^1 gives rise to $Me^1$and E^1 (E^1 and E^2 are the primary enteroblasts).
c. Me^2 gives rise to m^2z^2 & M^2 e^2 and Me^1 gives rise to m^1 z^1 and M^1e^1.
d. M^1e^1 gives rise to M^1 & e^1 and M^2e^2 gives rise to M^2 & e^2 (M^1 and M^2 are the mesoblastic teloblasts and e^1 and e^2 are the secondary enteroblasts).
e. m^2 z^2 gives rise to m^2 & z^2 and m^1 z^1 gives rise to m^1 & z^1.
f. e^1 & e^2 unite with earlier formed pair E^1& E^2 to form a group of four cells.
g. M^1 & M^2 eventually divide and give rise to two irregular rows of mesoderm cells.
h. m^1 & m^2 follow the same course as M^1 & M^2, joining the latter in forming the rudiment of the mesoderm.
i. z^1& z^2 cells take care of forming its anteriormost parts.

Ecto-mesodermal cells. In prosobranch *Crepidula,* the ecto-mesoderm has its origin in three of 2q blastomeres (2a–2c). In opisthobranch *Finoa,* the rudiments can be seen after gastrulation has already begun. The $3a^{211}$, $3a^{221}$, $3b^{211}$, and $3b^{221}$, divide as they sink into the blastocoel; $3a^{2112}$, $3a^{2212}$, $3b^{2112}$, $3b^{2212}$ are pushed out of the blastopore, while the large daughter cells are gradually covered over by the growing ectoderm. These cells $3a^{2111}$, $3a^{2211}$, $3b^{2111}$, and $3b^{2211}$ are the rudiment of the ecto-mesoderm, later dividing and each forming a band composed of several cells, which lie at first in the head region of the larva.

In pulmonate *Physa* also, completely identical cells form the ecto-mesodermal rudiment. In *Planorbis,* eight cells $3a^{2212}$, $3a^{2112}$, $3b^{2112}$, $3b^{2212}$, $3c^{2112}$, $3c^{2212}$, $3d^{2212}$, and $3d^{2112}$ are said to be the ecto-mesomeres.

In Mollusca, the ecto-mesoderm is seen only during the larval period, disappearing at the time of metamorphosis.

Blastulation. In eggs with a small amount of yolk, the entomeres that lie at the vegetal side make the blastocoel, which is very small, while in eggs with a large amount of yolk, the blastula is typical stereoblastula. As it begins the blastula stage, the embryo develops cilia at regions of the body surface. The apical cilia are formed on the apical cells at the animal pole, and clumps of cilia resembling those of ctenophore comb-plate are formed on the primary trochoblasts at 4 points on the sides of the blastula. Near the statocysts, the foot ganglia differentiate from ectoderm cells. As the development proceeds, certain cells in the vicinity of the primary trochoblasts (secondary trochoblasts) also produce cilia forming a complete ring, which is called protoroch. These generally persist through veliger stage. The embryo can be seen to escape through the so-called micropyle in species that has chorion. The larvae of oysters and common mussels begin their swimming life with the vitelline membrane intact, and later gradually shed it, bit by bit.

Gastrulation and gastrula stage. In typical stereoblastula like *Crepidula,* typical epiboly can be seen, while *Paludina* shows typical emboly and there are many species which follow a process that might be called intermediate between these two.

Types of gastrulation

Invagination (emboly). This is one of the usual methods in the animal kingdom, and occurs sometimes in Mollusca. In this process, the cell layer of the posterior side of the embryo, that is, the vegetal region, bends inward the blastocoel (Fig. 11.38A and D). The portion, which thus turns inwards, is the endoderm. The part formed by the endoderm is called the archentron or primitive intestine, and its opening is the blastopore, protostomia, or primitive mouth. Even among the forms in which gastrulation takes place by invagination, there are some in which the wall of the archentron is in contact with the outer body wall, leaving no blastocoel between the two, e.g. *Phornis,* while in others, like the echinoderms, a wide blastocoel remains around the archenteron.

Epiboly. In the case of such forms as the annelids and molluscs, in which there are large blastomeres containing a great deal of yolk at the ventral side and small blastomeres distributed over the animal pole region, the cleavage of these small cells proceeds at an accelerated rate, so as to spread over and invest the large blastomeres, resulting in the formation of gastrula. Judging from later developmental results, gastrulae formed in this way preserve the same relationships as those which arise by invagination, the large cells which are enclosed being the endoderm, while the outer cell layer which surrounds them is the ectoderm. This type of gastrulation is called epiboly (Fig. 11.38B and E). However, no archenteron cavity yet appears in the internal mass of endoderm cells of these gastrulae formed by epiboly, while the part that is called the blastopore is a region at the vegetal pole where the ectoderm cells fail to cover the endodermal cell mass completely, leaving part of it exposed. But as development proceeds, the ectodermal cell layer, which encircles this blastopore, sinks inwards, forming a structure called the stomodaeum.

Multipolar proliferation (ingrowth). There is another method of endoderm formation, called multipolar proliferation (ingrowth), which closely resembles this type of delamination. In this, some of the cells constituting the blastular wall slip from various parts of this cell layer into the blastocoel, and eventually unite to form the endoderm sac (fresh-water hydra). When this releasing of cells takes place at only one side (vegetal pole) of the embryo, e.g. *Clytia,* it is called unipolar proliferation. These methods are grouped

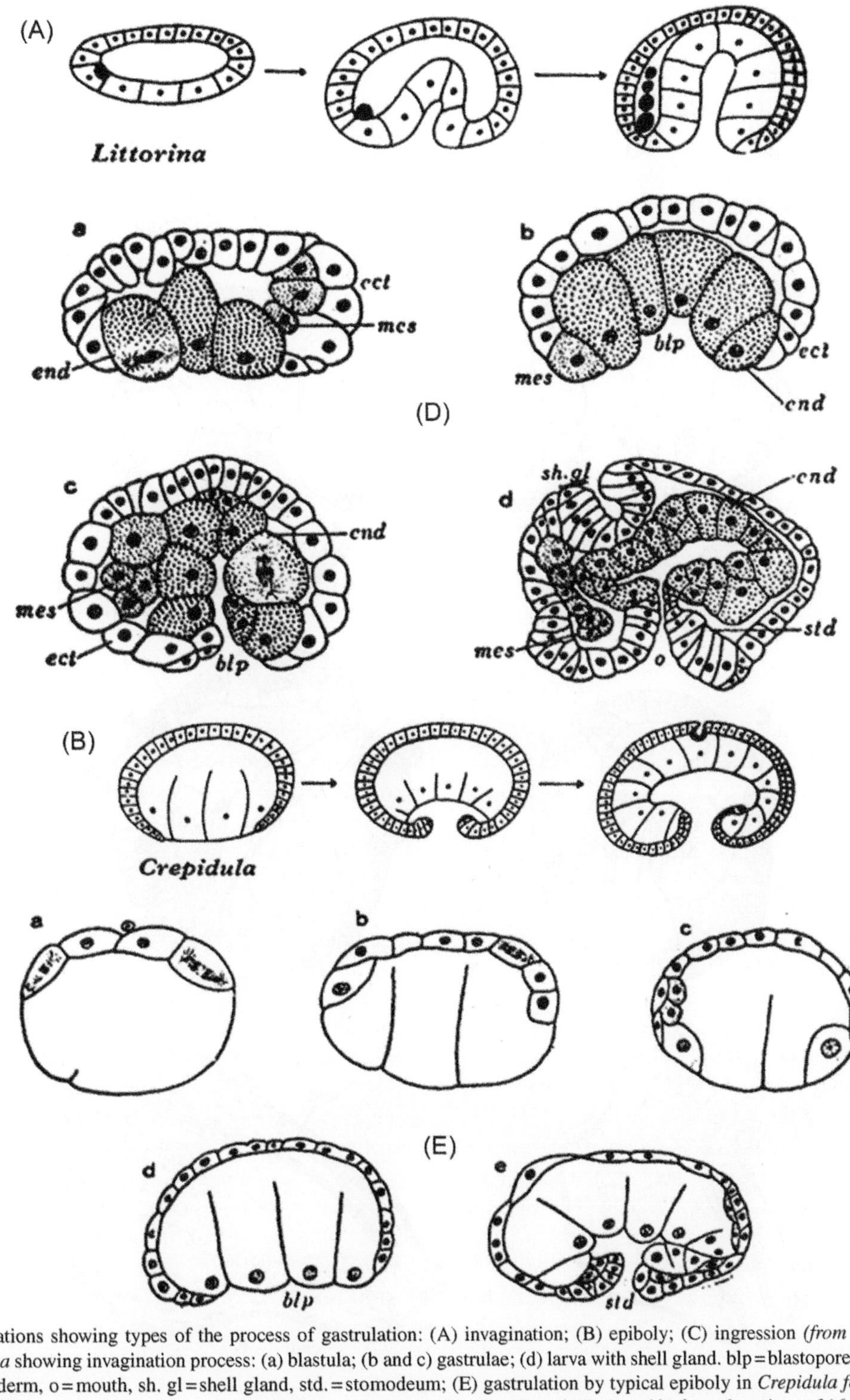

FIG. 11.38 Illustrations showing types of the process of gastrulation: (A) invagination; (B) epiboly; (C) ingression *(from Borradaile et al., 1977)*; (D) larvae of *Littorina* showing invagination process: (a) blastula; (b and c) gastrulae; (d) larva with shell gland. blp = blastopore, ect = ectoderm, end = endoderm, mes = mesoderm, o = mouth, sh. gl = shell gland, std. = stomodeum; (E) gastrulation by typical epiboly in *Crepidula formicata (after Conklin)*; (F) gastrulation in *Patella vulgata*, by the inward growth of the macromeres 4A, 4B, 4C, and 4D: (a and b) frontal sections of 16-h larva, longitudinal division of endoderm cells; (c) Parasagittal section of 16-h larva seen from right side; (d) parasagittal section of 21-h larva seen from right side. a.p. = apical plate, b.p. = position of blastopore, end. = endoderm, pr.mes. = primitive mesoderm cell, mes.bd. = mesodermal band, m.g. = incipient cavity of the mid-gut, st. = stomodaeal groove; (G) two sections of *Paludina vivipara* embryo, notice that the blastopore changes to become the anus, and the mouth is nearly formed. an = anus, end = endoderm, fg = fore-gut, mg = mid-gut, o = mouth, sh = shell, sh.gl = shell gland, ve = velum. *(D, E, and G) From Wada (1968)* and *(F) adapted from Smith (1935)*.

(Continued)

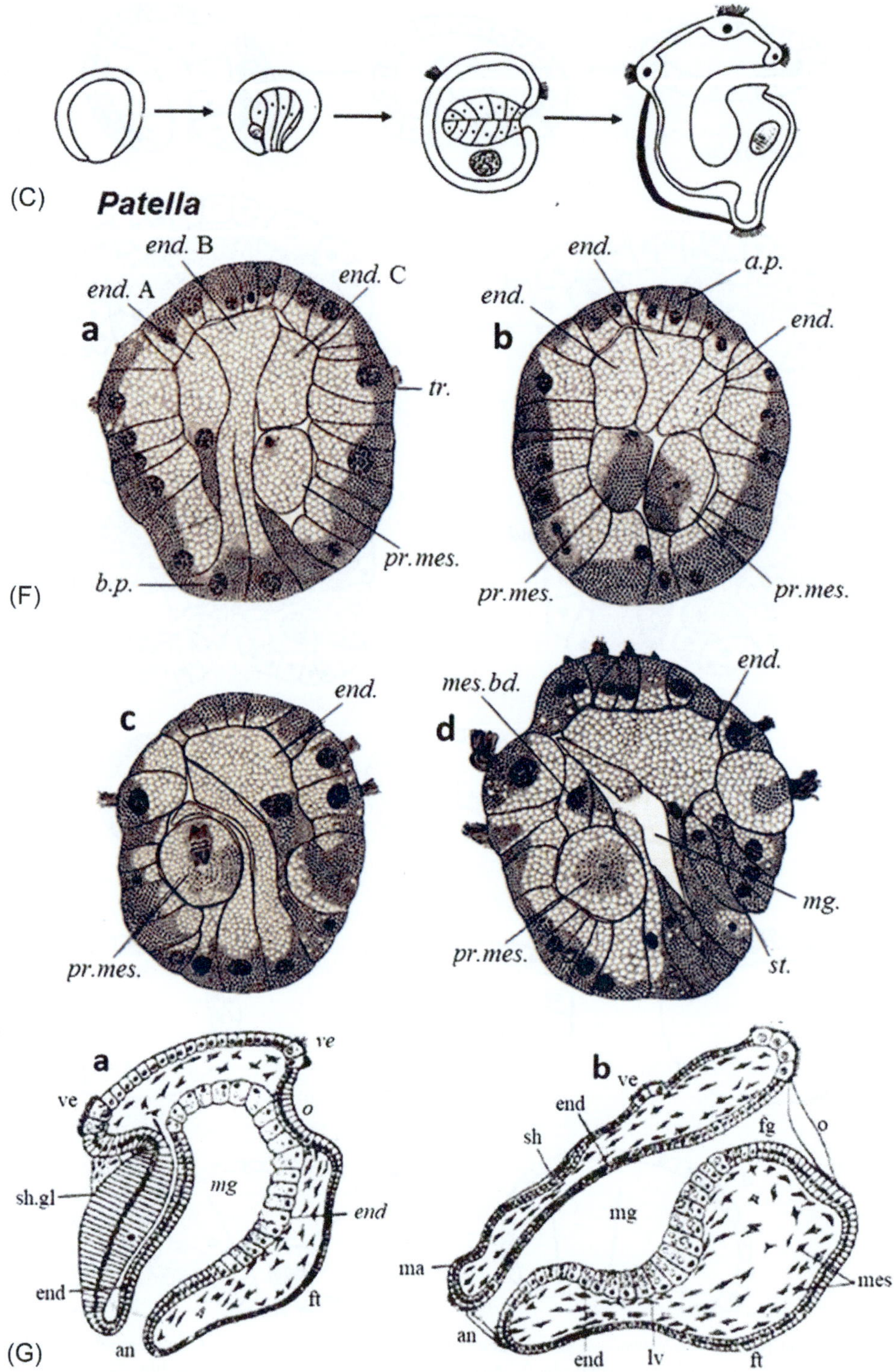

FIG. 11.38, CONT'D

together under the name of polarization, and whichever occurs, the resulting stereogastrula is solidly filled with cells, and has no archenteron or blastopore. As its development proceeds, however, an archenteron appears and a blastopore opens at the vegetal pole in the forms in which growth has taken place (Fig. 11.38C and F).

At first, a wide blastopore is seen as the result of entomeres invagination, but later, this gradually reduces to a long, narrow longitudinal slit, by the proliferation of the 2q and 3q cells, and in many species it is finally closed altogether. At this period, the posterior gastrular surface is occupied by cells of the 3q line, while the central part is composed of descendants of the 2q blastomeres. In Gastropoda, a large cell, which will form the excretory organ of the future larva, can be seen on the right side in the posterior part of the embryo. At the time when the archenteron is formed (clearly defined), the cells composing it are the macromeres 5A, 5B, 5C, and 4D, lying farthest inside, next to them are the 5a, 5b, 5c cells, and above these, $4c^2$, $4b^2$, $4a^2$, $4b^1$, and $4a^1$ lie directly below the ectoderm. In the posterior part of the body can be seen the enteroblasts E^1, E^2, e^1, and e^2, which have arisen from 4d. The ectoderm around the archenteron invaginates, forming the stomodaeum in the position of the blastopore. In some species, the blastopore remains open and changes over to a mouth. In *Paludina,* the blastopore changes into anus, and a new mouth is formed in a different position (Fig. 11.38G).

Larvae (trochopore and veliger stages)

Shell formation. Wada (1968) continued the description, and reported that shell formation in molluscs begins early in embryogenesis, during some stage of archenteron formation (Fig. 11.39). Sometimes after the invagination of the entoderm, the cells of the opposite (dorsal) side of the body proliferate vigorously, and invaginate towards the blastocoel forming the 'shell gland'. In this stage, the embryo thus shows two invaginated parts, the shell, dorsally, and the archenteron on the ventral side. Hess (1971) described this formation in *Bithynia tentaculata,* a prosobranch snail. In the young trochophore larvae at the dorsal side of the posttrochal ectoderm, the primordium of the shell gland appears as a thickening of ectodermal cells (Fig. 11.40A (b)). Later, in the centre of the thickened region, an invagination of a part of the enlarged cells takes place (Fig. 11.40A (c)). With further development, the shell gland is displaced to the left side of the embryo and enlarges more and more (Fig. 11.40A (d)). The central parts of the invagination rise again to the surface of the embryo (Fig. 11.40A (e)). The shell gland then changes from the form of a shallow trough to a circular groove, which in later stages becomes the mantle fold. The groove surrounds an area which is called the shell field. This field secretes the shell primordium in the form of a thin larval shell composed of conchioline. During development, the shell groove extends more and more (Fig. 11.40A (f)) and the shell (mantle) field finally covers large parts of the visceral hump. In *Amphibola* (pulmonata), it has been observed that the shell gland begins to secrete without invaginating, but in other gastropods invagination invariably precedes shell secretion (Wada, 1968). The shell gland is formed of cells derived from the 2d blastomere. In *Planorbis*, they are derivatives of $2d^{12}$ and $2d^{21}$.

In 1994, Eyster and Morse reported that ultrastructural information on early formation of external shells is available for only a few bivalves and gastropods. They are in agreement that most of the 'shell-secreting cells' are descendants of the 2d blastomere, but added that some 2c descendants also play a role in normal embryonic shell formation. These cells develop into an ectodermal region of the embryo called the 'shell field.' It is within the centre of this shell field that the invagination termed the 'shell gland' forms and then disappears. These authors reviewed the terminology and structure of this 'shell gland' for the past 100 years. But, they preferred to refer to the entire shell-secreting epithelium as the shell field, and the central invagination of this epithelium (i.e. the 'shell gland') will be referred to as the shell field invagination (Eyster, 1983) or SFI, a name which avoids imparting a secretory function to the invagination. Eyster and Morse added that the first ultrastructural data on molluscan shell formation in relation to the SFI were obtained from the study of gastropods. The pore of the SFI is circular, and scanning electron microscopy (SEM) studies suggest that the first shell material forms a ring surrounding the pore. Transmission electron microscopy (TEM) studies indicate that the first organic shell material lies over the pore of the SFI (Fig. 11.40B (a)), not along the SFI lumen. In the opisthobranch *Aeolidia papillosa,* the SFI lumen is prominent when the first shell material is observed. In other gastropods, the first shell material is observed only after closure of the SFI, i.e. after the cells lining the lumen become tightly apposed. TEM also shows that the first organic shell material of *A. papillosa* is visualized as two homogeneous electron dense layers having a total thickness of 20 nm (Fig. 11.40B (b), inset).

Typically, adult gastropods (and bivalves as well) secrete an organic periostracum, which lies outside the crystalline shell layers (Fig. 11.2D) and protects them from dissolution. The first organic shell layer secreted during embryogenesis is said to be a thin pellicle that becomes the periostracum. The uptake and deposition of the inorganic portion of the shell are yet to be traced in premetamorphic molluscs. In *A. papillosa*, numerous intracellular electron dense granules were seen only in cells beneath the growing edge of the shell, leading Eyster (1983) to implicate them in some role in shell formation; the granules might transport organic shell material from the Golgi complexes to the extrapallial fluid. In section, these granules are up to 200 nm long and lack any distinct internal substructure (Fig. 11.40B (b)).

Foot, pedal gland, statocysts. Furthermore, Wada described that the foot is formed between the posterior side of the blastopore and the anus. In *Planorbis* and *Trochus* it developed from the 3q cells, in *Fiona* formed from 2q, 3q. The development of the foot lags somewhat behind that of the shell gland, but rather early in the trochophore stage, the operculum is secreted on the lower surface of the foot. The formation of an operculum is peculiar to the gastropods, and even when the adult form lacks an operculum, one develops in the larva. In *Paludina,* the foot forms in front of the blastopore posterior to the point where future mouth will appear. The statocysts later receive innervation from the brain ganglia: they develop earlier than those ganglia differentiated from ectoderm cells.

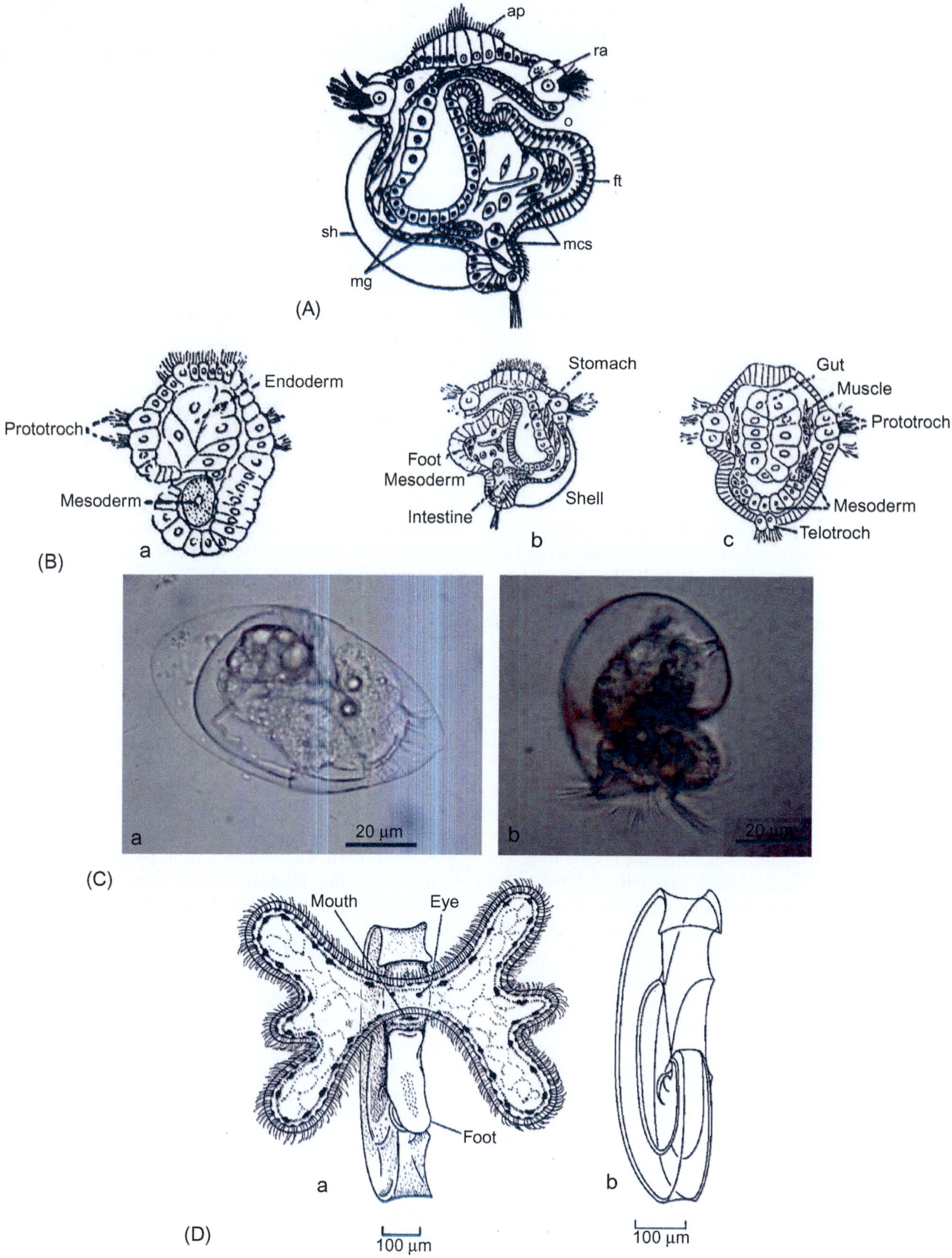

FIG. 11.39 Larvae of gastropods: (A) section of trochophore larva of *Patella (from Wada, 1968)*; (B) stages in the development of *Patella coerulea*: (a) trochophore; (b) early veliger; (c) veliger (frontal view) *(from Borradaile et al., 1977)*; (C) two stages in the development of *Melampus coffeus*: (a) preveliger; (b) veliger stage *(from Maia et al., 2012)*; (D) Echinospira larva: (a) living specimen; (b) shell. *(After Fretter and Graham, 1962, From Webber, 1977).*

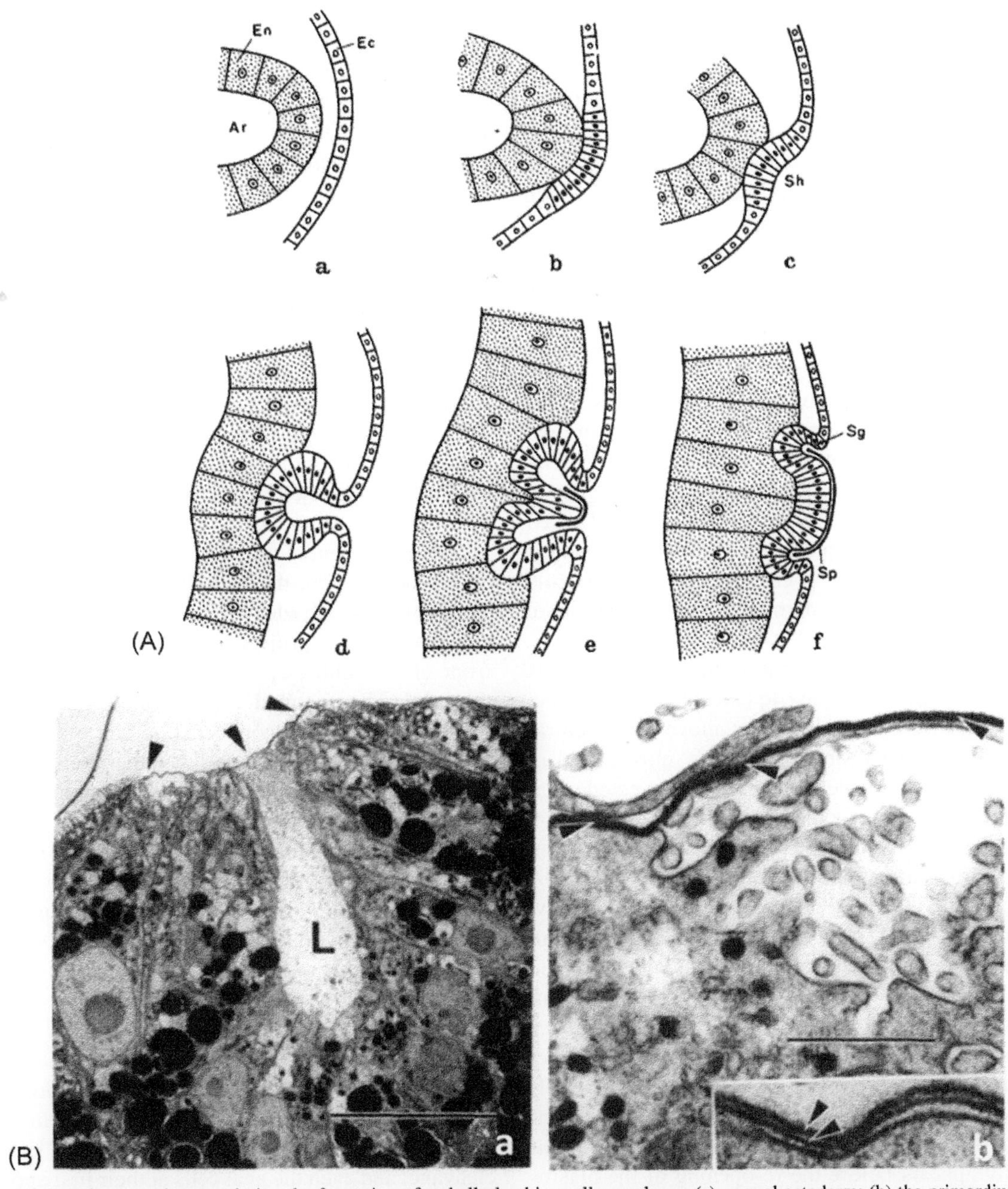

FIG. 11.40 (A) Developmental stages during the formation of a shell gland in molluscan larva: (a) normal ectoderm; (b) the primordium of the shell gland thickens; (c) the central portion of the thickened region invaginates; (d) the invagination deepens and enlarges more and more; (e) the central portion of the invagination evaginates to reach the surface of the embryo; (f) the evaginated portion changes to be a circular groove. Ar=archenteron, Ec=ectoderm, En=endoderm, Sg=shell groove, Sh=shell gland, Sp=shell primordium *(from Hess, 1971)*; (B) shell formation during molluscan embryogenesis: (a) Shell field invagination (SFI) of *Aeolidia papillosa,* showing nearly formed organic shell material (arrowheads) lying over the entrance to the SFI lumen (L), a portion of the embryonic capsule wall is shown in upper left. Bar=10 μm; (b) portion of apex of cell at the growing edge of the shell (arrowheads) in *A. papillosa,* showing characteristic electron-dense granules. The shell consists of two homogeneous electron-dense layers (inset, arrowheads), the growing edge of which is covered by long cytoplasmic extensions. Bar=0.5 μm *((B) modified after Eyster, 1983, from Eyster and Morse, 1994).*

Radula sac. The rudiment of the radula sac appears immediately after the invagination of the stomodaeum. In general, the rudiment of the radula sac develops from the cells of the stomodaeum, which grow and form a pocket extending towards the stomodaeum posteriorly. In *Physa,* it appears as a pair of invaginations behind the mouth but unrelated to the stomodaeum.

Velum. In veliger stage, the prototroch undergoes a striking growth; this expanded prototroch is then called velum. The prototroch is formed from 1q and 2q. The free-swimming period of these larvae is extremely short (several hours to several days), and as a consequence the velum is not highly developed. A typical velum is composed of two ciliated rings, one anterior to the mouth, called the preoral velum, and the other develops posterior to the mouth, which is the postoral velum. In *Crepidula* and *Planorbis,* it is said that only $1a^{22}$, $1a^{21}$, $1b^{22}$, $1b^{21}$ among the prototroch cells take part in forming the preoral velum. In addition, the tip cells of the anterior arm and the lateral arm of the molluscan cross enter into the formation of the preoral velum, while the postoral velum is mainly composed of 2a cells, to which some 1q and 3q cells are added.

Head vesicle. The part of the larva anterior to the ciliary ring of the velum is called the head vesicle. In this region, the apical organ develops, as well as the ganglia and other nerves, the eyes, and other sense organs. It is chiefly formed by 1q but in some cases 2q cell may be added to these. An apical plate, which is usually formed by the four apical cells 1a–1d occupies the anterior most part, and a large number of cilia are produced on the apical plate. In *Fiona* and *Crassostrea,* no apical plate is formed. Brain ganglia originate from the lateral ectoderm of the apical region. Cerebral ganglia are formed to the left and right of the apical plate, and are surrounded by apical cells in pulmonates. It cannot be invoked to settle the debate concerning the ectodermal or mesodermal nature of the ganglia.

Digestive system. The rudiment of the digestive tract is formed by the invagination of the 4q (except 4d), 5q and the macromeres (5A–5D). The blastopore moves in a ventral direction, becoming very small or entirely closed, and in its place a stomodaeum is nearly formed by an ectodermal invagination. This makes a connection with the archenteron (midgut) and differentiates into the foregut (oesophagus). The midgut differentiates into stomach, intestine, and digestive diverticula. The intestine is first recognized as a row of cells, which do not form a tube (as in *Fiona*), or as an evagination of the archenteron, which is tubular from the start (as in *Crepidula).* The endodermal cells derived from 4d blastomeres take part in its formation. The anus later opens at the place where the intestine comes into contact with the ectoderm. The digestive glands arise from the left and right sides of the anterior part of the midgut as a pair of foliated organs, having a lumen. The rudiment of the midgut gland usually appears before the re-opening of the blastopore, but its development is markedly retarded in embryos containing a large amount of yolk, e.g. *Buccinum*. The cells of midgut of this species contain the yolk of the nurse eggs, so the differentiation of the midgut is postponed until the veliger stage.

Later development. The important points in which the Gastropoda differ from the other molluscs in their development can be listed as:

1. The growth of the head;
2. The torsion of the posterior dorsal part of the body;
3. The spiral growth of the visceral mass and shell;
4. The secretion of the operculum on the dorsal posterior part of the foot.

In particular, the torsion of the body is a highly characteristic phenomenon that occurs in development, the following account will deal first with this subject.

Torsion. Borradaile et al. (1977) reported that in all gastropods with coiled shells, the mantle cavity is anterior, the opening is directed forwards, and the coiling of the visceral hump is directed posteriorly. But, in the development of these forms from the larva (Fig. 11.41A–D), the mantle cavity first makes its appearance behind the visceral hump, and at a particular stage the visceral hump rotates in a counterclockwise direction through an angle of 180 degrees on the rest of the body (Fig. 11.41D). This is what is known as torsion, which appears rather early in development, in the latter part of the trochophore stage. It occurs 45–46 h after fertilization in the black abalone, *Haliotis discus.* Furthermore, the process advances very rapidly, so that it is said to be completed in a few hours in *Trochus* and within a few minutes in *Acmaea,* so that it can hardly be brought about by differential growth.

Before torsion, the ctenidia and the anus point backwards, and the auricles are behind the ventricle. After torsion, the ctinidia project forwards, and the auricles are in front of the ventricle; the mantle cavity opens just behind the head. The uncoiled visceral hump has been caught in the twisting and one connective is laid over the other, one passing over the intestine and the other underneath, but both coming together near the anus and completing a figure of eight (Fig. 11.41G). This is known as chiastoneury, and constitutes an important characteristic of Prosobranchia. Muscular contractions must play their part.

Thompson (1958) through his voluminous study on biological reproduction of the Opisthobranch *Adalaria proxima*, described, discussed, and reviewed the mechanism and nature of developmental torsion and detorsion in gastropods. Firstly, he stated that torsion in *Adalaria* is no longer detectable as a mechanical process; but all the organs which are involved in ontogenetical torsion are, from their first appearance, placed in their posttorsional positions. Torsion in this species is accomplished during the later stages of cleavage, and even the anal cells at their first appearance are in the posttorsional position. Thompson distinguished five ways in which torsion is brought about, drawing examples from the various gastropods whose development has been studied:

1. 180 degrees torsion achieved by muscle contraction alone. (Found in no living forms, except possibly in *Acmaea* (Boutan, 1899) but it is possible that this is the original mechanism).
2. 180 degrees torsion achieved in two stages (Fig. 11.41E–G), an initial movement through 90 degrees, brought about by muscle contraction, followed by a slower movement through the remaining 90 degrees by differential growth processes, e.g. *Haliotis, Patella* (Smith, 1935), *Calliostoma*, etc.

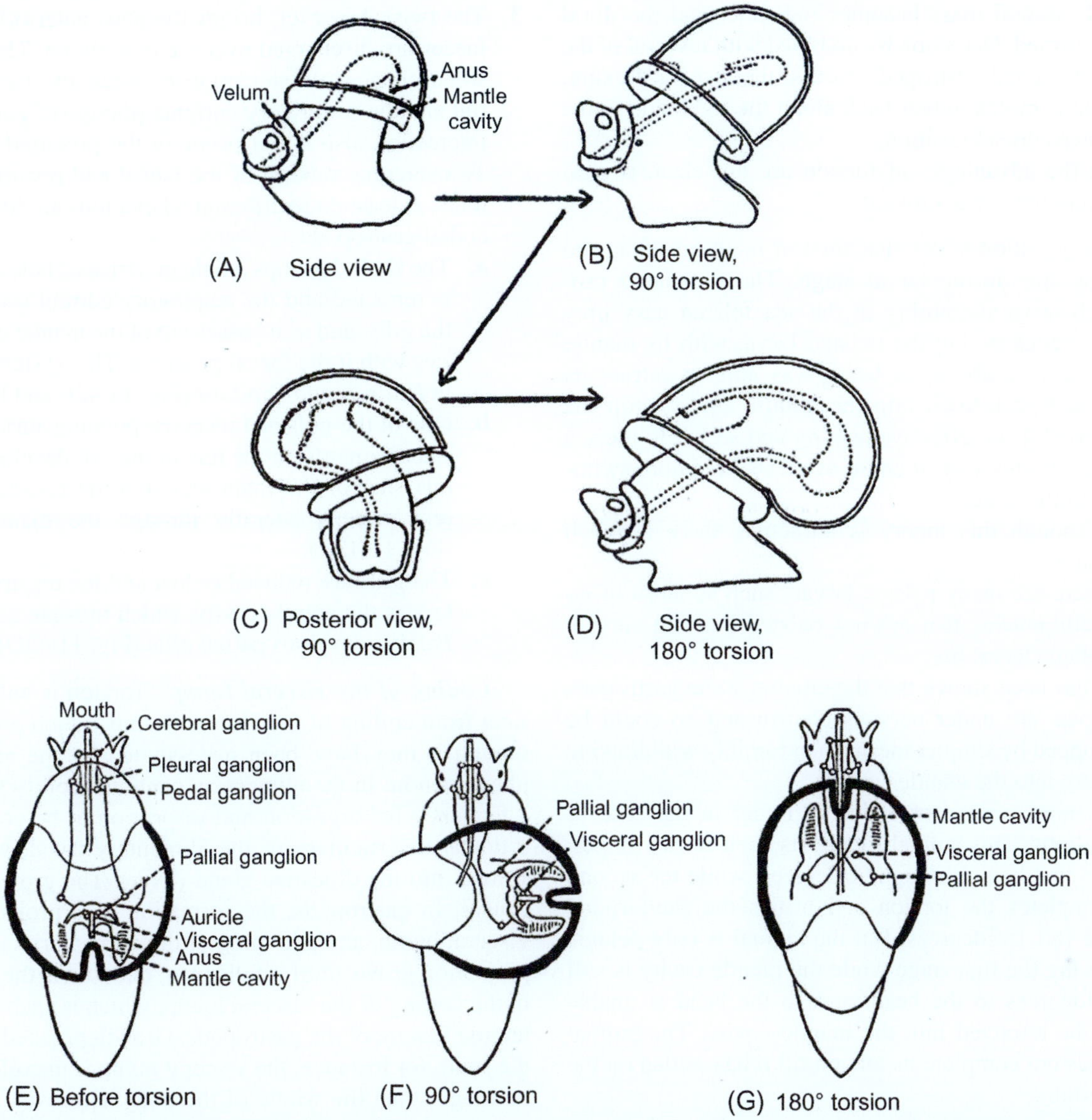

FIG. 11.41 (A–D) Illustrations to show torsion in *Paludina vivipara*: (A) almost symmetrical stage, with mantle cavity behind, but anus twisted to the right; (B) this stage shows 90 degrees of torsion anticlockwise, mantle cavity and anus to the right; (C) posterior view of B, animal with 90 degrees of torsion; (D) adult condition with 180 degrees of torsion; (E–G) diagrams to illustrate torsion, in general, seen from above. *(All after Naef, from Borradaile et al. (1977).)*

3. 180 degrees torsion achieved by differential growth processes alone, e.g. *Paludina* and *Pomatias*.
4. Torsion achieved by differential growth processes, the change in position of the anus being halted at a site approximately appropriate to the adult state, e.g. *Aplysia*, most Monotocardia.
5. Torsion no longer recognizable as a movement of the pallial complex, the organs being, from their first appearance in the embryo, in the posttorsional positions, e.g. *Adalaria*.

Torsion, then, in the veliger of *Adalaria*, although not visible as a mechanical process, would involve a movement of the pallial complex from a ventral to a dorsal position. Thompson emphasized this point because many authors have stated that developmental torsion in gastropods involves a movement of the pallial complex from a posterior to an anterior position; this is the result when translated into the adult gastropod, because of the postmetamorphal posterior extension of the foot and the consequent shifting upwards of the visceral mass, but it is surely confusing to so name the axes of the veliger that it appears to swim along the production of its own dorso-ventral axis. Metamorphosis in *Adalaria* involves no violent change in orientation, merely a general flattening of the body consequent on the loss of the shell. The distinction between head, foot, and visceral mass becomes indistinct and the dorid form is assumed. Detorsion is combined with reversal of the more fundamental gastropod process, the visceral flexure, as the anal complex moves back along the right side to the adult postero-dorsal position.

Since the advantages of torsion are not clear, several theories have been put forward:

1. One suggestion states that torsion is an advantage to pelagic life during larval stage. The untwisted prototroch larva swimming in the sea fell an easy prey to its predators, but the twisted larva, with its mantle cavity curled above its head, was able to retract its head and prototroch into the mantle cavity, stop the prototroch from effective beating and so fall to the sea bottom. In this way, it could avoid its predators swimming in the water.

   Though this theory is attractive, there is much against it:
   - **a.** There are many pelagic larvae, such as those of lamellibranchs, that are not twisted but still survive pelagic larval life;
   - **b.** It has been shown that the cilia on some gastropods larvae are under nervous control and so could be stopped by simpler means than forcibly withdrawing them into the mantle cavity;
   - **c.** A more serious objection is found in the case of Haliotis. This animal rotates its shell in two stages, the first being through 90 degrees, while the second completes the torsion and brings the shell round the full 180 degrees. But the animal is only pelagic during the first stage while the mantle cavity is still 90 degrees to the head, and so the head is unable to be retracted into the mantle cavity. The animal does not complete its torsion till it has settled on the bottom.
2. An alternative explanation is that torsion is an advantage to the adult animal. The primitive gastropod was not twisted and had its gills posteriorly on the body, tucked inside the mantle cavity. The cilia on the gills had to draw the respiratory current in from behind the animal. If the animal moved, it set up a current of water in a direction opposite to that in which it was moving and this stream would oppose the direction of the respiratory stream. It is possible that there were also weak currents flowing on the bed of the sea and that the primitive mollusc, like most animals orientated up-stream. This also would provide a stream in opposition to the respiratory current. Once the animal twists, however, all three streams would flow in the same direction and thus aid the flushing of the mantle cavity with fresh clean water (Fig. 11.42A (a)–(c)). In this way, torsion would be an adaptation present in the adult to increase the ventilation of the mantle cavity.
3. The twist, however, brings the anus anteriorly and the faeces are discharged over the oral region. There is still some chance of interaction between the faecal material and the respiratory currents (though of course, such interaction also could occur in the pretorted animals). To overcome mixing of the faecal and respiratory currents, at least three different adaptations are found in the coiled gastropods:
   - **a.** The shell develops a hole or series of holes, the anus is retracted and the respiratory current sweeps over the gills, and as it passes out of the mantle cavity carries with it the faecal products. This system is found in *Haliotis* and *Diodora* (Fig. 11.42E and F).
   - **b.** One of the gills and its corresponding auricle is lost. The filaments on the remaining gill develop only on one side of the main axis and the respiratory current sweeps laterally through the mantle cavity (Fig. 11.42C).
   - **c.** The gills are reduced or lost and the respiratory surface is the mantle cavity, which in some cases, as in *Patella*, develops pallial gills (Fig. 11.42D).

Coiling of the visceral hump. Torsion is entirely distinct from coiling of the visceral hump, which precedes it, though it may have been necessitated by the antecedent phenomenon. In the alimentary canal of molluscs, there is a tendency for digestion and resorption to be confined to a dorsal diverticulum of the alimentary canal, which develops into the digestive gland (liver). The growth of this causes, in gastropods, the formation of a projection, the visceral hump, and a looping of the alimentary canal. This projection grows until it falls over, and this is the first step in the coiling of the visceral hump, which is such a characteristic feature of the gastropods. Growth proceeds until, in the snail, for instance, the visceral hump if uncoiled would be longer than the whole of the body. However, owing to the fact that one side of the hump grows faster than the other during development, the whole organ is twisted into a compact spiral, which can be arranged so as not to interfere with the balance of the animal while crawling (Borradaile et al., 1977).

Detorsion. In most of the Opisthobranchia, this torsion secondarily undergoes a complete reversal at the time of metamorphosis; the mantle cavity becomes located in the posterior part of the body and the crossed nerves revert to their original condition. This phenomenon is called detorsion. The latter is always associated with the loss of the shell and thus the liberation of the gills from their enclosing case. Once the shell is lost, the gills become exposed to the external currents and the anterior position of the gills is no longer so great an advantage. The degree of detorsion can be correlated with the extent to which the shell is reduced (tectibranchs and pteropoda) (Borradaile et al., 1977).

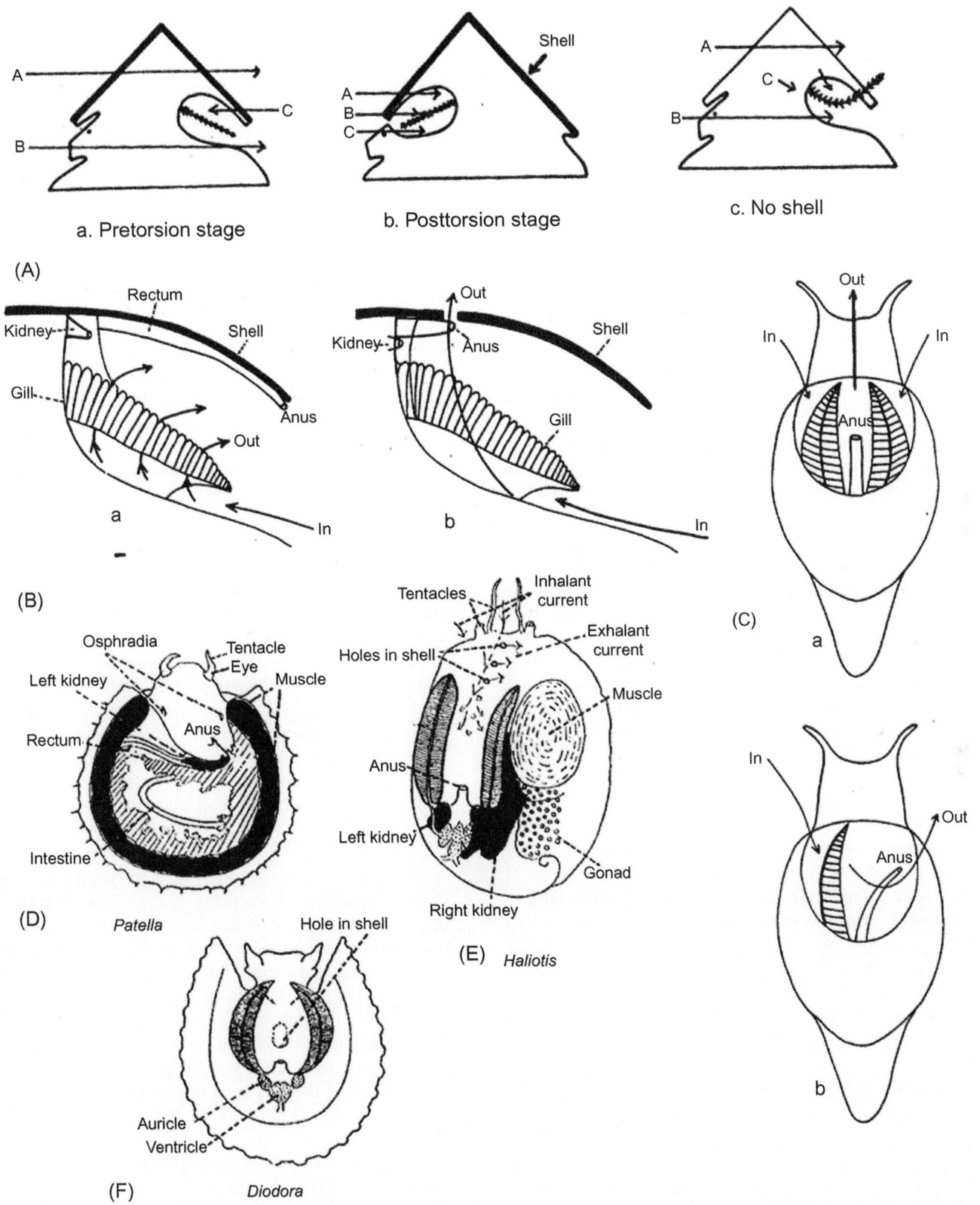

FIG. 11.42 Advantages and adaptations after torsion: (A) possible advantage of torsion: A = current in the sea, B = current due to animal's movement, C = respiratory current. In the twisted animal A and B oppose C, and do not oppose C in the posttorsion stage. In the untwisted animals, the shell is lost and the gills are once more exposed to A and B as well as C; (B) diagrams of mantle cavity: (a) primitive condition in which faeces may get on to the gills while; (b) shows condition in *Haliotis* (E) and *Diodora* (F), there is a single current through the mantle cavity carrying the faeces away from the gills; (C) currents in mantle cavity: (a) aspidobranch while; (b) pectinibranch. Note that the through-current in b carries the faeces away from the gills; (D) in *Patella*, the gills are lost and the mantle cavity develops pallial gills. *(All from Borradaile et al. (1977).)*

Echinospira larvae. Webber (1977) reported that in Prosobranchia, the third morphologically distinct larval type found is the echinospira larva. It is found in a number of mesogastropods: *Velutina velutina, Lamellaria perspicua, Lamellaria latens, Trivia monacha, Trivia arctica, Erato voluta,* and *Capulus ungaricus.* The echinospira appears to have two shells (Fig. 11.39D), the outer one transparent and separated from the inner shell by a space filled with seawater. It is believed that the seawater space aids the larva in flotation. It was suggested by Fretter and Graham (1962) that the double shell is actually the two basic layers of the shell separated; the outer, the periostratum, and the calcareous larver. The outer shell is lost at metamorphosis.

Trophon geversianus (Pallas, 1774). There are some recent accounts of *Trophon geversianus* available. This animal is a Prosobranch, Neogastropoda, Muricidae (Marshall and Williams, 1975); *Trophon geversianus* (Fig. 11.1F) shows the widest geographic distributional range (Giménez, 2013). It is a benthonic species and inhabits both rocky intertidal and shallow subtidal zones. In the south-western Atlantic, this species is reported from regions including the Malvinas (Falkland) Islands.

Giménez (2013) studied ultrastructurally the spermatogenesis and sperm (euspermatozoa and paraspemaozoa) morphology in *T. geversianus* for the first time. The eusperm, is the typical and fertile sperm, and the parasperm, is atypical and nonfertile sperm. The role of the latter has taken different explanations. One considered that the products of the paraspermatozoan breakdown function as nutrition and/or stimulation for the euspermatozoa. Fretter and Graham (1962) suggested the 'transportation' of euspermatozoa during and after sperm transfer from male to female, via mobile associations of euspermatozoa.

Giménez collected, manually, 20 specimens of *T. geversianus* at depths of 5 m at low tide, in June and December of 2010, from Golfo Nuevo, Puerto Madryn, Argentina. The gonad (testis) of this species occupies the same position as that in *Eremina ehrenbergi* and is also in contact with the digestive gland. This author described the testis as consisting of numerous spermatogenic tubules and spermatic ducts, separated from each other by connective tissue. (Fig. 11.43A). During spermatogenesis, clusters of euspermatogenic cells were observed distributed throughout the tubule. Generally, they are in small groups, in the same maturity stage, and are joined by a cytoplasmic continuity (Fig. 11.43A and B). Late euspermatocytes have the chromatin dispersed in patches throughout the nucleoplasm, the mitochondrias are located in one pole of the cell (Fig. 11.44B). The early euspermatids are roughly spherical and characterized by the condensation of the nuclear granular chromatin. The mitochondria are aggregated at the posterior pole of the cell. Then, spermeogenesis takes place and Fig. 11.44B shows the condensation of the nucleus, which is clearly correlated with nuclear elongation and a middle piece composed of five mitochondrial elements arranged around the axoneme (Fig. 11.44B). At the end of the process, the mature euspermatozoa occupy the lumen of the tubule.

The parasperm cells were observed in clusters distributed throughout the tubule (Fig. 11.43B). Paraspermatid development is characterized by, as seen by light microscopy, elongation of the cell, with the appearance of a cytoplasmic elongation at the apex of the cell and the breakdown of the nucleus into small round fragments (caryomerites) (Fig. 11.43E and F). Paraspermatocytes are connected by cytoplasmic bridges, with visible nucleoli. The paraspermatids in all stages are shown ultrastructurally, with fragmented nucleus into caryomerites, the presence of mitochondria, and internal and central axonemes and vesicles (Fig. 11.44C).

The testis of *T. geversianus* produces two types of sperm:

1. The euspermatozoa, fertile sperm; when fully formed, the sperm is composed of an acrosomal complex, nucleus, midpiece, glycogen piece and end piece, with a single incorporated axoneme. The acrosomal complex consists of a tall, conical, membrane-bound acrosomal vesicle (Fig. 11.44D). The acrosomal vesicle is approximately $N=7$ $1.14 \pm 0.08\,\mu m$ long. Apically, the vesicle membrane separates from the vesicle contents and lies close to the plasma membrane to form an electron-lucent balloon-like space, called the apical bleb (Fig. 11.44D). The acrosomal vesicle has a very deep invagination within which is subacrosomal material (Fig. 11.44E). An accessory membrane is present in the basal part of the acrosome (Fig. 11.44E).

   The nucleus is filiform with a complete and long invagination. It contains a centriolar derivative (Fig. 11.44E). The axoneme has 9+2 arrangement of microtubules and posterior to the nucleus, it is enclosed in a mitochondrial sheath, forming the midpiece region (Fig. 11.44F). The middle piece is composed of five mitochondrial elements arranged around the axoneme (Fig. 11.44B and G). Beyond the midpiece, the axoneme is associated with the glycogen piece (The glycogen composition of this has been demonstrated cytochemically, with a PAS (Periodic acid–Schiff) positive result) (Fig. 11.44H and J). The T.S. shows nine longitudinal and radiating tracts of dense granules (one tract per axonemal doublet) (Fig. 11.44J).
2. The paraspermatozoa are vermiform with multiple incorporated axonemes. They have tapered anterior and

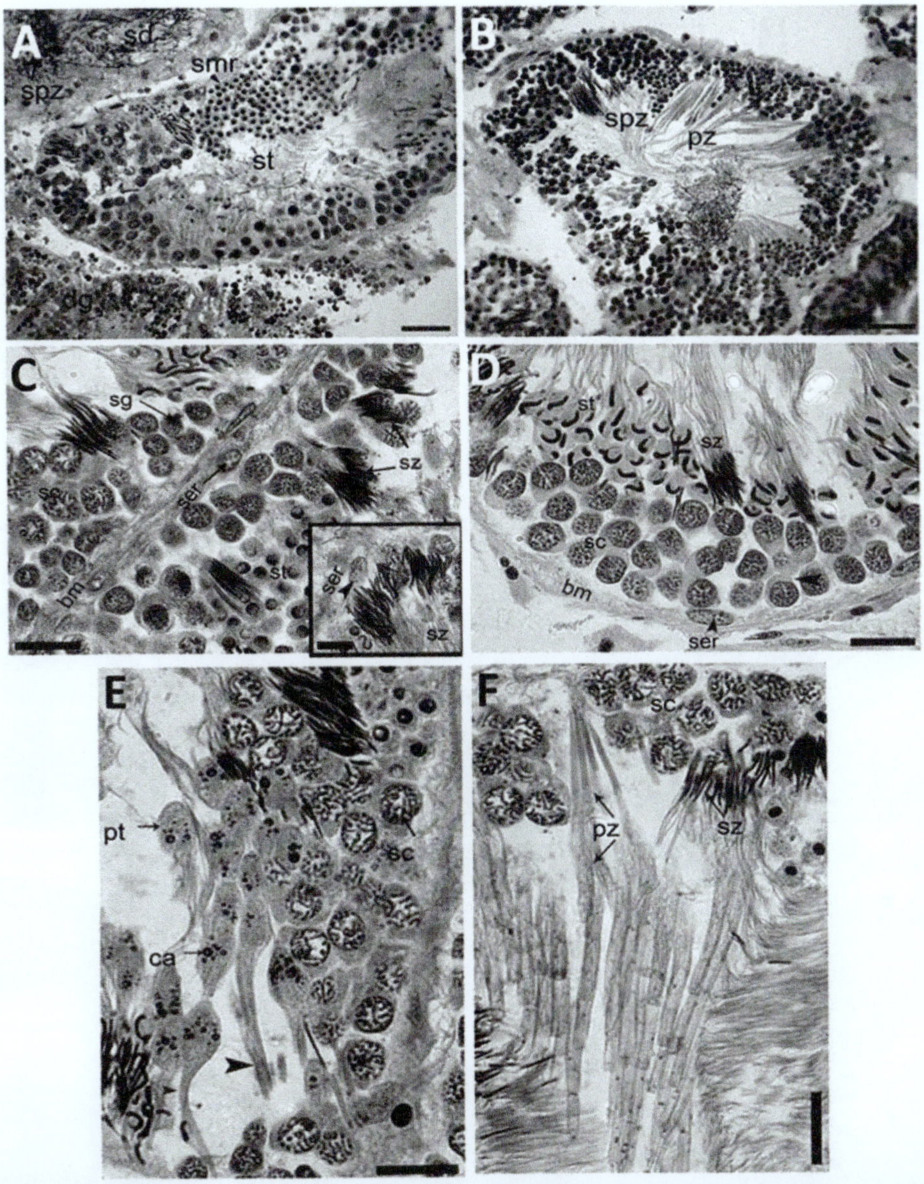

FIG. 11.43 Light microscopic sections of the testis of *Trophon geversianus* stained with haematoxylin/eosin: (A) T.S. of a spermatogenic tubule (st) surrounded by connective tissue, digestive gland (dg), and spermatogenic duct (sd) filled with mature spermatozoa (spz); (B) detail of different spermatogenic stages. Note both spermatozoa types in same tubule, spermatozoa and paraspermatozoa (pz); (C–F) sections of the testis of *Adelomelon ancilla*: (C) View of a cluster of euspermatogonia lying adjacent to the basal membrane. Note different stages of euspermatocytes and developmental stages of euspermatids. A bundle of mature euspermatozoa is visible attached to the Sertoli cell. **Inset:** Sertoli cells with spermatozoa attached; (D) Clusters of euspermatocytes connected by cytoplasmic bridges *(arrowhead)*. Note advanced stages of euspermatids and bundle of mature sperm. Note also ovoid shape of Sertoli cell nucleus; (E) Advanced paraspermatid with an elongated flagellum *(arrowhead)*; (F) Cluster of mature paraspermatozoa. Note their vermiform shape. Bm = basal membrane, ca = caryomerites, pg = paraspermatogonia, pt. = paraspermatid, pz = paraspermatozoa, sc = euspermatocytes, sg = euspermatogonia, st = euspermatid, sz = mature euspermatozoa, ser = Sertoli cell. Scale bars for A, B 50 μm, for C–F 20 μm and for C inset 10 μm.

posterior extremities and contain no discernible nucleus or nuclear derivative (Fig. 11.44K–O). In the apex of the parasperm only the axonemes and some cytoplasm persist, where each axoneme attaches to a granular deposit (Fig. 11.44K). Present near the extremities only are the peripheral axonemes (Fig. 11.44M).

In the main body region of the paraspermatozoon (Fig. 11.44L), there are:

1. 12–16 peripherally distributed axonemes lying close to, or in contact with, the inner surface of the plasma membrane. The axonemes are approximately equally separated from each other;

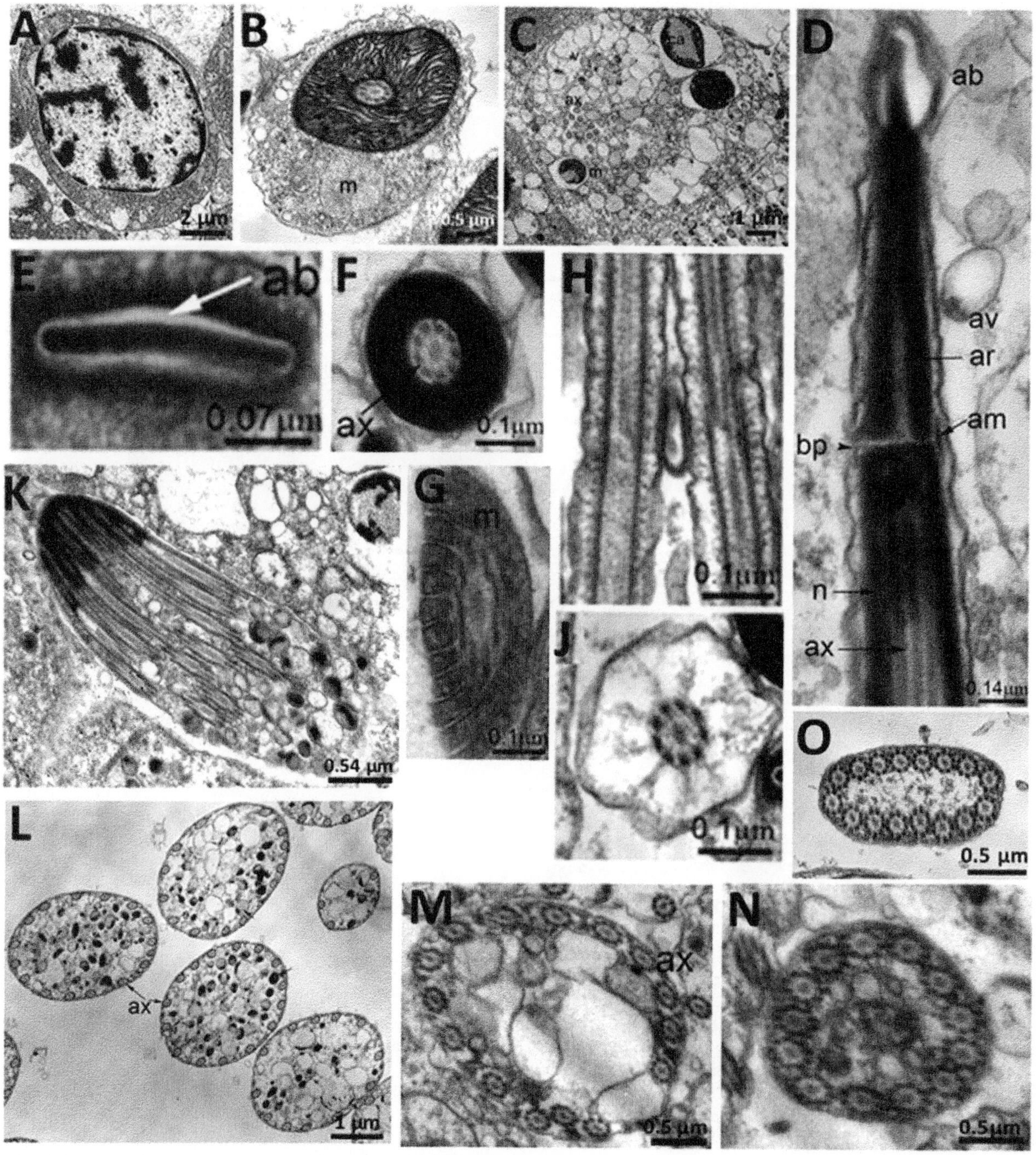

FIG. 11.44 TEM sections showing Stages of spermatogenesis of *Trophon geversianus*: (A) late spermatocyte; (B) T.S. through lamellar euspermatid stage of nucleus showing distribution of chromatin in lamellae. A well-developed Golgi complex and five rounded mitichondrias; (C) late paraspermatid. Note the vesicle excreted from a cell *(arrowhead)*. Note nuclear vesicles, which are pale and similar in density to the many clear vesicles in cytoplasm; (D–H) Euspermatozoa: (D) L.S. through acrosomal complex showing the apical bleb, subacrosomal material, acrosomal vesicle separated by the basal plate with the anterior portion of nucleus. The axoneme is continuously surrounded peripherally by nuclear material. Note the dense accessory membrane associated with the plasma membrane in the basal portion of the acrosome; (E) T.S. of the acrosomal complex from apical bleb region of the acrosomal vesicle; (F) T.S. of the nucleus. Note the disposition 9+2 of the axoneme; (G) oblique section of the midpiece showing spiralling of mitochondria; (H) L.S. of the glycogen piece showing radiating, longitudinal rows of putative glycogen granules; (J) T.S. of the glycogen piece showing the internal axoneme; (K–O) Paraspermatozoa: (K) L.S. through apex showing sheath of granular material enclosing axonemes and basal body/centriolar complexes with some dense vesicles; (L) T.S. of the main body region of a paraspermatozoon, showing 14–15 peripheral axonemes closely adherent to the plasma membrane, dense vesicles and less dense vesicles; (M–O) TS of the apex region showing a part of the peripheral axonemes. ab = apical bleb, am = accessory membrane, ar = anterior region, av. = acrosomal vesicle, ax = axonemes, bp = basal part, m = mitochondria, n = nucleus, sam = subacrosomal material, ca = caryomerites, g = Golgi complex, m = mitichondrias, n = nuclei, nu = nucleoli. *From Giménez, 2013.*

2. numerous very electron-dense vesicles;
3. occasionally, there are large round vesicles of low to moderate electron-density 'putative lipid vesicles';
4. numerous small vesicles of low electron-density, that may be mucoid deposits;
5. occasional mitochondria.

Towards the apex, the diameter of the parasperm diminishes and its shape flattens, and no vesicles or mitochondria are present (Fig. 11.44N and O). However, Giménez (2013) reported that the presence of the axoneme continuously located inside of the nucleus has not been reported before, and may prove to be a diagnostic feature of the Muricidae.

In the view of de la Barra et al. (2014), *T. geversianus* inhabits hard substrates, usually associated with bivalve populations. These authors studied oogenesis and vitellgenesis of the oocytes of this species. They collected the specimens of *T. geversianus* manually from the intertidal rocky shore at Golfo San Jose´ in Chubut, Argentina, between July 2006 and August 2007. Samples of the ovary were used monthly during this period, stained with haematoxylin and eosin, Sudan black B (to identify lipids), or Schiff's ninhydrin (to identify proteins). The sex is separate in this species, but the histological structure and position of the gonad (ovary) is the same as in *Eremina*. Each gonadal tubule is composed of germinal and accessory cells (Fig. 11.45A), and surrounded by connective tissue. The authors divided the process of oogenesis into four developmental stages:

- Stage 1; the oogonia (Fig. 11.45B). They are small spherical cells, each with a nucleus occupying almost the whole cell. They are found in groups of three or four cells (Fig. 11.45B).
- Stage 2; the previtellogenic oocytes, in which the whole cell (cytoplasm and nucleus) increases in size. No yolk is present in the cytoplasm (Fig. 11.45B).
- Stage 3: the early vitellogenic oocytes (Fig. 11.45C). These are irregularly-shaped oocytes. Some small inclusions appear in the cytoplasm, which indicate the onset of the vitellogenesis (Fig. 11.45C).
- Stage 4; the late vitellogenic oocytes, which are large irregularly shaped cells characterized by the presence of large and rounded inclusions (Fig. 11.45D).

de la Barra et al. (2014) reported that all these stages were found in every month studied (Fig. 11.45F), indicating that there is no resting phase in gametogenic cycle of Golfo San José's population. Two spawning events were registered; one between September and October 2006, and a second one between February and March 2007. During these two periods, the frequency of late vitellogenic oocytes decreases markedly, suggesting spawning events in the population (Fig. 11.45F). These events, in turn, coincided with rises in the mean surface water temperature and a 13 light-hour photoperiod.

Through digital image analysis, a lipid and a protein index (LI and PI) were calculated. Significant differences were observed in LI over the month sampled (Fig. 11.45E), contrary to what PI showed. A significant and positive correlation was found between lipid and protein content and oocytes areas, indicating that these nutrients accumulate during the entire vitellogenesis.

In each spawning event, the female lays 12 egg capsules with an average of 197 eggs each. In laboratory observations, only four eggs per capsule reached the juvenile stage, while the rest of them remained nurse eggs.

Cumplido et al. (2011) described, under laboratory conditions, the first detailed study of intracapsular development, nutrition and oxygen availability of *Trophon geversianus* from Atlantic waters. The authors collected 30 individuals (19 females and 11 males) from the rocky intertidal of Punta Cuevas, Golfo Nuevo (Puerto Madryn, Chubut, Argentina), in December 2007. They succeeded in maintaining aquaria with suitable and perfect conditions. They recorded oviposition events every day. Only recently laid egg capsules were used to study embryo development. The egg capsules were laid on the glass walls of the aquaria. Embryos within the egg capsules were classified into eight developmental stages, from egg to miniature adult.

The egg capsules of *Trophon geversianus* had on average 197.6 + 14.88 (mean + SE) uncleaved eggs (Table 11.5; Fig. 11.46A), with a range of 152–240 eggs, embedded in a transparent gelatinous matrix. Within the first 24h after egg laying, a typical holoblastic spiralian cleavage, begins with the 1st cleavage. The authors reported that this process was observed in a few eggs within the egg capsule, but stages from morula to 'trochophore' were not observed. Later, after another 24h, the nurse eggs began to disintegrate within the egg capsule (Fig. 11.46B). Size of eggs and embryos was significantly different between developmental stages, steadily and significantly increasing from early 'veliger' to hatching stage. The early veliger is characterized by the presence of a mouth and a velum and with a very few nurse eggs in the stomach (Fig. 11.46C). The intracapsular embryos began to ingest the disintegrated nurse-egg particles by the aid of the ciliated velar lobes. The embryos are transparent, so that the authors observed the ingestion of a large amount of disintegrated nurse eggs. The following stage is the late veliger, which becomes an elongated and coiled embryo. It increases in size and attains large 'prismatic' cells covering the stomach and the posterior end of the embryo (Table 11.5; Fig. 11.46D and E). A translucent shell develops during the digestion of the nurse eggs. The size of the embryo also significantly increases from late veliger to pediveliger stage (Table 11.5). The latter has reduced velum, large cephalic tentacles, basal eyes, and the foot is formed. As the foot develops, the shell grows and the velum disappears (Table 11.5; Fig. 11.46F and G).

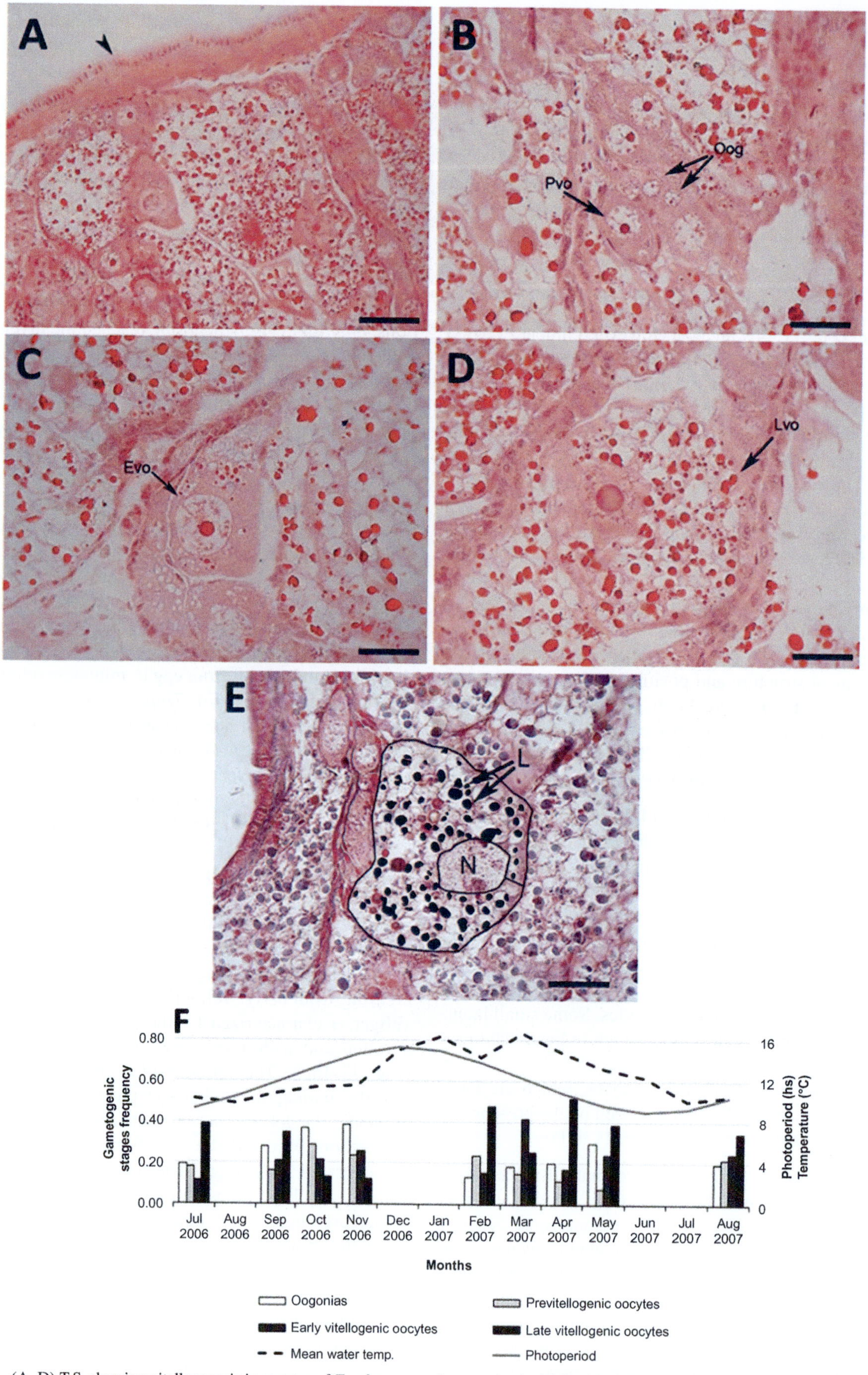

FIG. 11.45 (A–D) T.S. showing vitellogenesis in oocytes of *Trophon geversianus* stained with haematoxylin and eosin: (A) general view of the ovary, arrowhead shows simple columnar epithelium; (B) oogonia (Oog) and previtellogenic oocytes (Pvo); (C) early vitellogenic oocyte (Evo); (D) late vitellogenic oocyte (Lvo). Bars = A 100 μm, B–D 50 μm; (E) T.S. of *Trophon geversianus* ovary in paraplast sections, stained with Sudan black B, the software identifies every blue pixel within the cytoplasm. The area occupied by lipids (L) is the sum of the blue pixels. Bar = 50 μm; (F) Monthly variations in mean surface water temperature *(dashed line)*, photoperiod *(full line)*, and relative frequency of oocytes on different gametogenic stages in *Trophon geversianus* during the sampling period. *(From de la Barra et al. (2014).)*

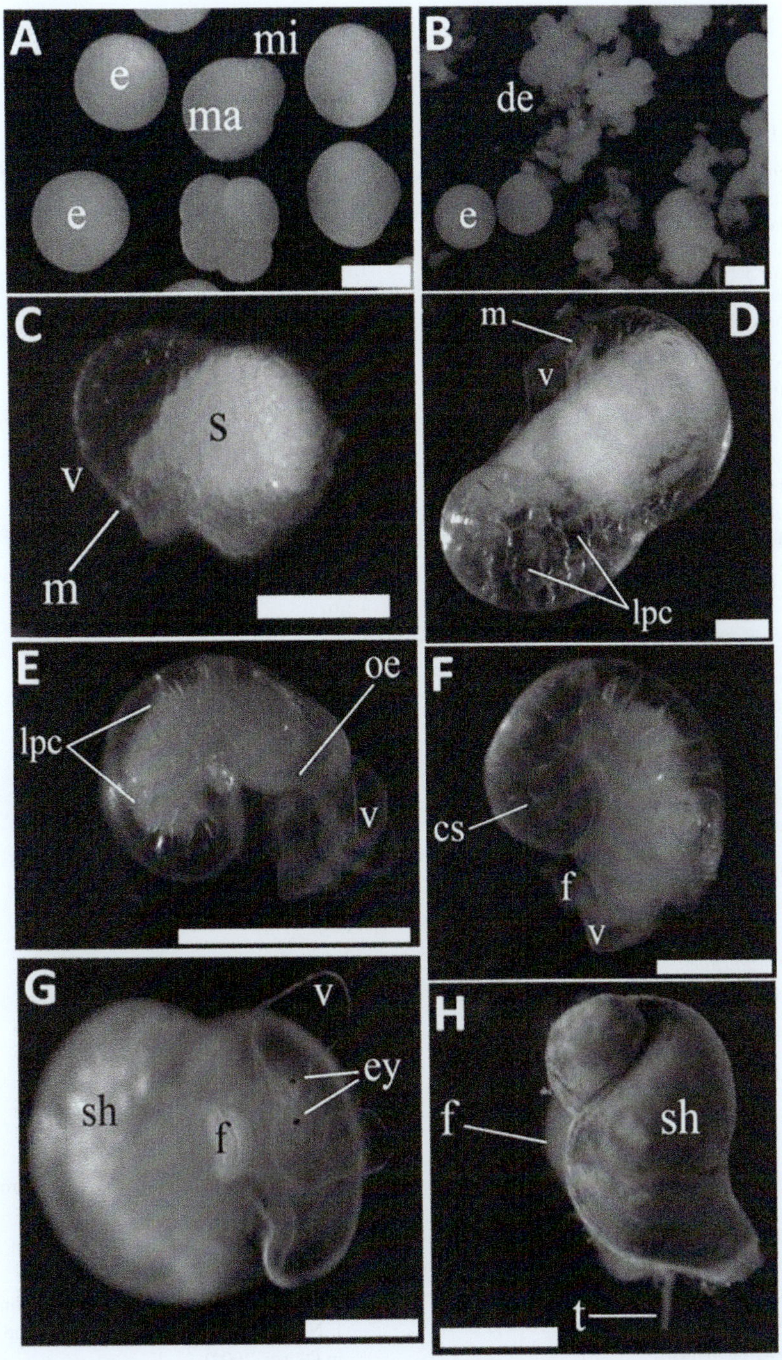

FIG. 11.46 Embryological development of *Trophon geversianus*: (A) Eggs before first cleavage and cellular cleavage stage; (B) Eggs and disaggregated nurse eggs; (C) Early 'veliger' stage with mouth, incipient velum and very few nurse eggs inside stomach; (D) Late 'veliger' stage embryo with developing velum, large 'prismatic' cells and digested nurse eggs; (E) Late 'veliger' stage with developed velum and large 'prismatic' cells; (F–G) Pediveliger stage embryo with coiled-shell, eyes and developing foot; (H) hatchling. Scale bars A–D=200 μm, E-H=1 mm. cs=coiling shell, ey=eyes, f=foot, lpc=large 'prismatic' cells, oe=oesophagus, sh=shell, t=tentacle, v=velum, e=eggs, de=disaggregated eggs, lpc=large 'prismatic' cells, ma=macromere, mi=micromere, m=mouth, s=stomach, v=velum. *(From Cumplido et al. (2011).)*

The prehatching embryo has no velum, a completely formed foot, with an operculum, with differentiation of the pedal sole tissue in the form of a split that could correspond to the accessory boring organ (ABO), and coloured embryonic shell with yolk reserves in the apical section. After 100–120 days of intracapsular development, the crawling juveniles emerged through a plug situated at the top of the egg capsule (Fig. 11.46H). The hatching juveniles exhibited a purple, calcified shell. On average, only 2.1% of the initial number of eggs developed successfully, and between 2 and 6 embryos hatched per egg capsule, which had a shell length of 2.79 + 0.03 mm.

TABLE 11.5 Embryonic intracapsular stages of *Trophon geversianus* at Punta Cuevas, Golfo Nuevo

Stage	Description	Time[a] (days)	Mean embryos size (mm)	Mean embryos per capsule	Number of egg capsule (*n*)
0 (egg)	Round and whitish, before cellular cleavage	0–1	0.27 ± 0.01 (0.24–0.32)	197.60 ± 14.88[b]	5
1 (cell cleavages)	Embryo with 2–8 macromeres and micromeres	1–12	028 ± 0.01 (0.25–0.37)	11.83 ± 0.01[b]	6
2 (early 'veliger')	Embryo with incipient velum, mouth and empty stomach	12–20	042 ± 0.03 (0.40–0.46)	9.00 ± 2.00	2
3 ('veliger')	Embryo with ciliated velum and conspicuous stomach full of eggs	20–30	1.16 ± 0.08 (0.93–1.42)	8.17 ± 1.49	6
4 (late 'veliger')	Large embryo with developed velum and large 'prismatic' cells	30–50	1.64 ± 0.04(1.39–1.99)	6.14 ± 1.19	14
5 (pediveliger)	Embryo with developing foot; velum reduced	50–70	1.92 ± 0.08 (1.52–2.43)	5.23 ± 0.79	13
6 (prehatching)	Juvenile with calcified shell and developed foot: velum resorbed	70–100	2.23 ± 0.08 (1.90–2.39)	4.30 ± 0.30	10
7 (hatching)	Crawl-away juveniles	100–120	2.79 ± 0.04 (2.61–3.05)	4.17 ± 133	12

Values represent mean size ± SE and range in parentheses.
[a]*Developmental time recorded in aquarium.*
[b]*Embryos were unable to be recognized before cellular cleavage, therefore we count and measured eggs (in general) for the first two developmental stages.*
From Cumplido et al. (2011).

REFERENCES

Abd El-Kerim, H.M., 1988. Detailed Studies on the Hermaphroditic Reproductive System of the Desert Snail *Eremina ehrenbergi* Roth, 1839 to Determine the Male Tract and Its Role in the Formation of the Spermatophores (Gastropoda - Stylommatophora) (M.Sc. thesis). Alexandria University (mostly published).

Anderson, E., 1967. The formation of primary envelope during oocyte differentiation in teleosts. J. Cell Biol. 35, 193–212.

Anderson, E., 1969. Oocyte-follicle cell differentiation into species of amphineurans (Mollusca), *Mopalia mucosa* and *Chaetopleura apiculata*. J. Morph. 129, 89–126.

Andrews, E., 1936. Spermatophores of the snail *Neritina reclivata*. J. Morphol. 60, 191–209.

Ankel, W.E., 1930. Die atypische Spermatogenese von Janthina (Prosobranchia, Ptenoglossa). Cell Tissue Res. 11 (3), 491–608.

Aparicio, M.T., 1982. Observations on the anatomy of some Helicidae from Central Spain. Malacologia 22 (1–2), 621–626.

Avila-Poveda, O.H., Abadia-Chanona, Q.Y., 2013. Emergence, development, and maturity of the gonad of two species of Chitons 'sea cockroach' (Mollusca: Polyplacophora) through the early life stages. PLoS One 8 (8), 1–12.

Baba, K., 1938. The later development of a solenogastre, *Epimenia verrucosa*. J. Dep. Agric. 6, 21–41.

Baba, K., 1999. Solenogastres (= Neomeniina, Ventroplicida). In: Mollusca I. Nakayama Shoten, Tokyo, Japan, pp. 55–107. (in Japanese). (cited in Okusu, 2002).

Bacci, G., 1947. Sex reversal in *Patella* and *Diodora*. Nature (London) 160, 94.

Bani, G., 1979. Ghiandole annesse al primo tratto delle vie genital femminili in due Elicidi (Gasteropodi Polmonoti Stylommatofori). Arch. Ital. Anat. Embriol. LXXXIV, 1–28.

Barnes, R.D., 1974. Invertebrate Zoology, third ed. W.H. Saunders Company, London.

Baur, B., 1994. Parental care in terrestrial gastropods. Experientia 50, 5–15. Birkhfiuser Verlag, CH-4010 Basel/Switzerland.

Bawab, F.M., 1970. The Formation of the Spermatophore in Penaeus (Crustacea Decapoda) (Ph.D. thesis). Faculty of Science, Alexandria University.

Bawab, F.M., 1988. Talon of the desert snail *Eremina ehrenbergi* Roth, 1839, (Stylommatophora- Gastropoda). Egypt. J. Histol. 11 (2), 363–364.

Bawab, F.M., El-Sherief, S.S., Abd El-kerim, H.A., 1992a. The gonad of the desert snail *Eremina ehrenbergi* Roth, 1839, (Stylommatophora-Gastropoda) and its role in the production of the male gametes. Funct. Dev. Morphol. 2 (2), 103–110.

Bawab, F.M., El-Sherief, S.S., Mehanney, S., 1992b. Oogenesis in the desert snail *Eremina ehrenbergi* Roth, 1839. Proc. Zool. Soc. A. R. Egypt 23 (part 1), 107–112.

Bawab, F.M., El-Sherief, S.S., Abd El-kerim, H.A., 1993. Sperm pathway and spermatophore formation in the hermaphrodite snail *Eremina ehrenbergi* Roth, 1839. Pak. J. Sci. Res. 36 (9), 426–430.

Bawab, F.M., El-Sherief, S.S., Mehanney, S., 1995. Structure and function of the female genital tract of the hermaphrodite system of *Eremina ehrenbergi* Roth, 1839. J. Egypt. Ger. Soc. Zool. 16 (D), 229–251. Invertebrate Zoology and Parasitology.

Bayne, C.J., 1966. Observations on the composition of the layers of the egg of *Agriolimax reticulatus*, the grey field slug (Pulmonata, Stylommatophora). Comp. Biochem. Physiol. 19, 317–338.

Bayne, C.J., 1968. Histochemical studies on the egg capsules of eight gastropod molluscs. Proc. Malacol. Soc. Lond. 38, 199–212.

Beeman, R.D., 1977. Gastropoda: Opisthobranch. In: Giese, Pearse (Eds.), Reproduction of Marine Invertebrates. Academic Press.

Berry, A.J., 1977. Gastropoda: Pulmonata. In: Giese, Pearse (Eds.), Reproduction of Marine Invertebrates. Academic Press.

Biggs, E.E.J., 1959. A contribution to the study of the genus *Eremina* Pfeiffer, 1885. J. Conchol. 24 (10), 332–342.

Bloom, W., Fawcett, D.W., 1975. A Text Book of Histology, 10th ed., Asian edition Igaku Shoin Ltd., Tokyo.

Borradaile, L.A., Potts, F.A., with chapters by Eastham, L.E., Saunders, J.T., 1977. The Invertebrata, fourth ed. Revised by Kerkut, G.A. Cambridge University Press.

Boutan, L., 1899. La cause principale de l'asymétrie des mollasques gastéropodes. Arch. Zool. Exp. Gén (3) 7, 203–242.

Boycott, A.E., 1919. Parthenogenesis in *Paludestrina jenkinsi*. J. Conchol. 16, 54–61 (cited in Webber, 1977).

Breckenridge, W.R., Fallil, S.F.M., 1973. Histological observations on the reproductive system of *Achatina fulica* (Gastropoda, Pulmonata, Stylommatophora). Ceylon J. Sci. 10, 85–120.

Breure, A.S.H., Eskens, A.A.C., 1977. Observations on the formation of spermatophores in a bulimulid land snail, *Drymaeus canaliculatus* (Pfeiffer, 1845) (Mollusca, Gastropoda, Pulmonata). Neth. J. Zool. 27, 271–276.

Bullough, W.S., 1981. Practical Invertebrate Anatomy, second ed. Macmillan Press Ltd, London, Basingstoke.

Burton, D.W., 1978. Anatomy, Histology and Function of the Reproductive System of the Tracheopulmonate Slug *Athoracophorus bitentaculatus*. Zool. Publ. Victoria University of Wellington (68). 17 pp.

Clarke, A.H., Menzies, R.J., 1959. *Neopilina (Vema) ewingi*, a second living species of the paleozoic class Monoplacophora. Science 129, 1026–1027 (cited in Wingstrand, 1985).

Clement, A.C., 1952. Experimental studies on germinal localization in *Ilyanassa*. 1- The role of the polar lobe in determination of the cleavage pattern and its influence in later development. J. Exp. Zool. 121, 593–625.

Cole, F.J., 1944. A history of comparative anatomy from Aristotle to the eighteenth century. Macmillan & Co. Ltd., London, p. 524.

Conklin, E.G., 1897. The embryology of *Crepidula*. A contribution to the cell lineage and early development of some marine gastropods. J. Morphol. 13 (1), 1–227.

Courtot, A.M., Gomot, L., 1982. Les stades de l'organogenése et la cytodifférenciation de la glande a'Albumen de l'escargot *Helix aspersa* Müller. Reprod. Nutr. Dev. 22 (4), 661–680.

Cumplido, M., Pappalardo, P., Fernandez, M., et al., 2011. Embryonic development, feeding and intracapsular oxygen availability in *Trophon geversianus* (Gastropoda, Muricidae). J. Molluscan Stud. 77 (4), 429–436.

de Jong-Brink, M., ter Borg, J.P., Bergamin-Sassen, M.J.M., Boer, H.H., 1979. Histology and histochemistry of the reproductive tract of the pulmonate snail, *Bulinus truncatus*, with observations on the effects of castration on its growth and histology. Int. J. Invert. Reprod. 1, 41–56.

de la Barra, P., Giménez, J., de Almeida, H.K., Arrighetti, F., 2014. Vitellogenesis and changes in lipid and protein contents in oocytes of *Trophon Geversianus* (Neogastropoda: Muricidae) in Golfo San José (Chubut, Argentina). Helgol. Mar. Res. 68 (1), 179–185.

Demian, E.S., 1960. Morphological studies on the planorbidae of Egypt. 1, On the macroscopic anatomy of *Bulinus (Bulinus) truncates* (Audoui). Ain-Shams Sci. Bull. (5), 1–84.

Dillaman, R.N., 1981. Dart formation in *Helix aspersa* (Mollusca, Gastropoda). Zoomorphology 97, 247–261.

Duncan, C.J., 1958. The anatomy and physiology of the reproductive system of the freshwater snail *Physa fontinalis* (L). Proc. Zool. Soc. Lond. 131, 55–84.

Duncan, C.J., 1960a. The evolution of the pulmonate genital system. Proc. Zool. Soc. Lond. 134, 601–609.

Duncan, C.J., 1960b. The genital system of the freshwater Basommatophora. Proc. Zool. Soc. Lond. 135, 339–356.

Eernisse, D.J., 2007. Chitons. In: Denny, M.W., Gaines, S.D. (Eds.), Encyclopedia of Tidepools and Rocky Shores. University of California Press, Berkeley, CA, pp. 127–133.

El-Sherief, S.S., 1975. The Structure and Histology of the Male Reproductive System of *Portunus pelagicus* and the Formation of the Spermatophore (M.Sc. thesis). Faculty of Science, Alexandria University.

Emberton, K.C., 1985. Seasonal changes in the reproductive gross anatomy of land snail *Triodopsis Tridentata tridentata* (Pulmonata: Polygyridae). Malacologia 26 (2), 225–239.

Eyster, L.S., 1983. Ultrastructure of early embryonic shell formation in the opisthobranch gastropod *Aeolidia papillosa*. Biol. Bull. 165, 394–408 (cited in Eyster and Morse, 1994).

Eyster, L.S., Morse, M.P., 1994. Early shell formation during molluscan embryogenesis, with new studies on the surf clam, *Spisula solidissima*. Am. Zool. 24, 871–882.

Fretter, V., Graham, A., 1962. British Prosobranch Molluscs. Their Functional Anatomy and Biology. The British Ray Society, London.

Focardi, S., Quattrini, D., 1972. Structure of the reproductive apparatus and life cycle of *Milax gagates* (Draparnaud) Mollusca, Gastropoda, Pulmonata. Boll. Zool. 39, 9–27.

Gardiner, M.S., 1972. The Biology of Invertebrates. McGraw-Hill Book Company, London, New York.

Ghose, K.C., 1963. Reproductive system of the snail *Achatina fulica*. Proc. Zool. Soc. Lond. 140, 681–695.

Giese, A.C., Pearse, J.S., 1974. Reproduction of Marine Invertebrates. In: Acoelomate and Pseudocoelomate Metazoans. vol. I. Academic Press, New York, London.

Giménez, J., 2013. Spermatogenesis and sperm morphology in *Trophon geversianus* (Gastropoda: Muricidae). J. Mar. Biol. Assoc. U.K., 1–6.

Gugler, C.W., 1964. The morphology of the talon in the land snail *Anguispira alternata*. Am. Zool. 4, 293.

Hammad, S.M., Hegazi, E.M., Ayyad, A.A., 1977. Population density and biomas of *Eremina desertorum* snail. In: Samedene Progress Report No. 3, Part I, pp. 24–42 (chapter 9).

Haszprunar, G., Ruthensteiner, B., 2013. Monoplacophora (Tryblidia)–some unanswered questions. Am. Malac. Bull. 31 (1), 189–194.

Hess, O., 1971. Fresh water gastropoda. In: Reverberi, G. (Ed.), Experimental Embryology of Marine and Fresh-Water Invertebrates. North-Holland Publishing Company, Amsterdam.

Hill, R.S., 1977. Studies on the ovotestis of the slug *Agriolimax reticulatus* (Müller). 2. The epithelia. Cell Tissue Res. 183 (1), 131–141.

Hill, R.S., Bowen, I.D., 1976. Studies on the ovotestis of the slug *Agriolimax reticulatus* (Müller) 1. The oocyte. Cell Tissue Res. 173, 465–482.

Hinsch, G.W., Dehn, P.F., 1979. Ultrastructural evidence for Sertoli-like cells in the testis of the Asteroid, *Luidia clatbrata* (Say) (Echinodermata: Platyasterida). Int. J. Invert. Rep. 1 (3), 179–185.

Holm, L.W., 1946. Histological and functional studies on the genital tract of Lymnaea stagnalis appressa Say. Trans. Am. Microsc. Soc. 65 (1), 45–68.

Huettner, A.F., 1949. Fundamentals of Comparative Embryology of the Vertebrates. The Macmillan Comp., New York.

Hyman, L.H., 1967. The Invertebrates. Mollusca I, vol. VI. McGraw-Hill, New York.

Ismail, S.Z.M., 1981. The Structure of the Spermatophore and the Male Reproductive System of *Sepia officinalis* (M.Sc. thesis). Alexandria University.

Kugler, O.E., 1965. A morphological and histochemical study of the reproductive system of the slug *Philomycus caroliniarnis* (Bosc). J. Morphol. 116, 117–132.

Kwalevsky, A., 1883. Embryogénie du *Chiton polii* (Phlippi). Ann. Mus. Hist. Nat. Marseille, Zool. 1 (5), 1–46 (cited in Pearse, 1979).

Lemche, H., 1957. A new living deep-sea Mollusc of the Cambro-Devonian class Monoplacophora. Nature 179 (4556), 413–416.

Lemche, H., 1975. The basic structural pattern of Pogonophora. Zeitschrift für zoologische Systematik und Evolutionsforschung Sonderheft 1975, 123–126.

Lemche, H., Wingstrand, K.G., 1959. The Anatomy of *Neopilina galatheae* Lemche, 1957 (Molluscs Tryblidiacea). Zoological Museum, University of Copenhagen.

Leonhardt, H., 1977. Human Histology, Cytology, and Microanatomy. Translated from German by Winstanley, D. F. Year Book Medical Publications, Inc., Chicago Thieme Publishers.

Lind, H., 1973. The functional significance of the spermatophore and the fate of spermatozoa in the genital tract of *Helix pomatia* (Gastropoda: Stylommatophora). J. Zool. (Lond.) 169, 39–64.

Lind, H., 1976. Causal and functional organization of the mating behaviour sequence in Helix pomatia (Pulmonata, Gastropoda). Behaviour 162–202.

Lipton, S., Murray, J., 1979. Courtship of land snail of genu *Partula*. Malacologia 19, 129–146.

Liuzzi, M.G., Zelaya, D.G., 2013. Egg-hull ultrastructure of *Ischnochiton stramineus* (Sowerby, 1832), a south American brooding chiton (Chitonina: Ischnochitonidae). J. Molluscan Stud. 79, 372–377.

Lord, J.P., 2011. Larval development, metamorphosis, and early growth of the gumboot chiton *Cryptochiton stelleri* (Middendorff, 1847) (Polyplacophora: Mopaliidae) on the Oregon coast. J. Molluscan Stud. 77, 182–188.

Lusis, O., 1961. Postembryonic changes in the reproductive system of the slug *Arion ater* rufus L. Proc. Zool. Soc. Lond. 137, 433–468.

Maia, R.C., Rocha-Barreira, C.A., Coutinho, R., 2012. Reproductive cycle and embryonic development of the gastropod *Melampus coffeus* (Linnaeus, 1758) (Ellobiidae) in the Brazilian northeast. Braz. J. Biol. 72 (4), 935–943.

Mahoney, R., 1973. Laboratory Techniques in Zoology. Butterworths and Co., London.

Malek, S.R.A., Bawab, F.M., 1974a. The formation of the spermatophore in *Penaeus kerathurus* (Forskal, 1775) (Decapoda, Penaeidae), I the initial formation of a sperm mass. Crustaceana 26 (3), 273–285.

Malek, S.R.A., Bawab, F.M., 1974b. The formation of the spermatophore in *Penaeus kerathurus* (Forskal, 1775) (Decapoda, Penaeidae), II the deposition of the main layers of the body and of the wing. Crustaceana 27 (1), 73–83.

Mann, T., 1984. Spermatophores. Springer-Verlag, Berlin, New York.

Marshall, A.J., Williams, W.D., 1975. A Text Book Zoology: The Invertebrates. ELBS and Macmillan.

Mehanney, S.M., 1991. Detailed Studies on the Hermaphroditic Reproductive System of the Desert Snail *Eremina ehrenbergi* Roth, 1839 (M.Sc. thesis). Alexandria University.

Meisenheimer, J., 1907. Biologie, morphologie und physiologie des Begattungs vorgangs und der eiablage von *Helix pomatia*. Zool. Jb. (Syst.) 25, 461–502.

Miller, R.L., 1977. Chemotactic behavior of the sperm of chitons (Mollusca: Polyplacophora). J. Exp. Zool. 202, 203–212.

Morton, J.E., 1967. Mollusca, fourth ed. Hutchinson University Library, London.

Morton, J.E., 1979. Mollusca, fifth ed. Hutchinson University Library, London.

Mowry, R.W., 1963. The special value of methods that color both acidic and vicinal hydroxyl groups in the histochemical study of mucins. With revised directions for the colloidal iron stain, the use of Alcian blue G8X and their combinations with the periodic acid-Schiff reaction. Ann. N. Y. Acad. Sci. 106 (2), 402–423.

Okusu, A., 2002. Embryogenesis and development of *Epimenia babai* (Mollusca Neomeniomorpha). Biol. Bull. 203, 87–103.

Pearse, J.S., 1979. Polyplacophora. In: Giese, Pearse (Eds.), Reproduction of Marine Invertebrates. vol. V. Academic Press, Inc.

Pelseneer, P., 1898. Sur la morphologie des branchies et des orifices rénaux et génitaux des chitons (1). Bulletin Scientifique de la France et de la Belgique 31, 23–30.

Pelseneer, P., 1935. Essay on zoological ethology based on the study on Mollusca [Essai d'éthologie zoologique d'après l'étude des mollusques]. Académie royale de Belgique. Classe des Sciences, Brussel, Belgium, p. 662.

Pelseneer, P., 1894. P.-J. Van Beneden malacologist: biographical note [P.-J. Van Beneden malacologiste: notice biographique]. Ann. Soc. Roy. Malac. Belgique XXIX, V–IX.

Pelseneer, P., 1906. In: Lankester, E.R., Black, C. (Eds.), Mollusca: A Treatise on Zoology. London. Mem., 32, vol. V. Universtiy Press of Liverpool.

Pennypacker, M.I., 1930. The germ cells in the hermaphroditic gland of *Polygyra appressa*. J. Morph. Physiol. 49 (2), 415–452.

Philip's, 2003. World Atlas & Gazetteer. In: Association with the Royal Geographical Society, with the Institute of British Geographers. Published in Great Britain.

Plate, L., 1899. Die Anatomie und Phylogenie der Chitonen. Zool. Jahrb. (Suppl. 5), 281–600. (cited in Pearse, 1979).

Richards, A.G., 1951. The Integument of Arthropods. Minneapolis University of Minnesota Press.

Rigby, J.E., 1963. Alimentary and reproductive systems of *Oxychilus cellarius* (Müller) (Stylommatophora). Proc. Zool. Soc. Lond. 141, 311–359.

Rigby, J.E., 1965. *Succinea putris*, a teresstial opisthobranch mollusk. Proc. Zool. Soc. Lond. 144, 445–486.

Runham, N.W., 1978. Reproduction and its control in *Deroceras reticulatum*. Malacologia 17 (2), 341–350.

Sabelli, B., Scanabissi, F.S., Merloni, M., 1978. Distribution of germ cells in the gonadic acini of *Deroceras reticulatus* (Müller) (Gastropoda, Pulmonate, Stylommatophora). Monitore Zool. Ital. (N. S.) 12, 95–106.

Salib, E.A., 1975. Morphological and Experimental Studies on the Cercaria of Human Schistosomes in Egypt (M.Sc. thesis). Ain Shams University, Cairo, Egypt.

Salvini-Plawen, L.V., 1997. Systematic revision of the Epimeniidae (Mollusca: Solenogastres). J. Molluscan Stud. 63, 131–155.

Schrödl, M., 2006. *Laevipilina theresae*, a new monoplacophoran species from Antarctica (Mollusca). Spixiana 29 (3), 225–227.

Schwabe, E., 2008. A summary of reports of abyssal and hadal Monoplacophora and Polyplacophora (Mollusca). Zootaxa, 205–222. Magnolia Press.

Siebold, T.V., 1836. Fernere Beobachtungen uber die Spermatozoen der wirbellosen Tiere. 2. Die spermatozoen der Paludina vivipara. Archiv für Anatomie und Physiologie, 232–255.

Siracusa, G., De Felici, M., Salustri, A., 1985. The proliferative and meiotic history of mammalian female germ cells. In: Metz, Monory (Eds.), Biology of Fertilization. vol. 1. Academic Press, Inc, New York.

Smith, F.G.W., 1935. The development of *Patella vulgata*. R. Soc. Lond. Ser. B 225 (520), 95–125.

Smith, B.J., 1965. The secretions of the reproductive tract of the garden slug *Arion ater*. Ann. N. Y. Acad. Sci. 118 (24), 997–1014.

Smith, B.J., 1966. Maturation of the reproductive tract of *Arion ater* (Pulmonata, Arionidae). Malacologia 4 (2), 325–349.

Solem, A., 1974. The Shell Markers Introducing Molluscs. John Wiley and Sons, London.

Solem, A., 1978. Classification of the land mollusca. In: Fretter, V., Peake, J. (Eds.), Pulmonates, vol. 2A, Academic Press, London, pp. 49–97.

Thompson, T.E., 1958. The natural history, embryology, larval biology, post-larval development of *Adalaria proxima* (Alder and Hancock) (Gastropoda, Opisthobranchia). Philos. Trans. R. Soc. Lond. B 242 (686).

Thorpe, S.R., 1962. A preliminary report on spawning and related phenomena in California chitons. Veliger 4, 202–210 (cited in Pearse, 1979).

Tompa, A.S., 1980. The ultrastructure and mineralogy of the dart from *Philomycus carolinianus* (Pulmonata: Gastropoda) with a brief survey of the occurrence of darts in land snails. Veliger 23 (1), 35–42.

Visser, M.H.C., 1981. The reproductive system of *Oopelta polypunctata* Collinge and *O. nigropunctatahisto* Morch (Pulmonata, Arionidae), with reflections upon the systematic position of the subfamily Oopeltinae. Zool. Ser. 10, 293–314.

Wada, S.K., 1968. Mollusca. In: Kumé, Dan (Eds.), Invertebrate Embryology. NOLIT Publishing House, Belgrade.

Walker, M., MacGregor, H.C., 1968. Spermatogenesis and the structure of the mature sperm in *Nucella lapillus* (L). J. Cell Sci. 3 (1), 95–104.

Webber, H.H., 1977. Gastropoda: Prosobranchia. In: Giese, Pearse (Eds.), Reproduction of Marine Invertebrates. Academic Press.

Wingstrand, K.G., 1985. On the anatomy and relationships of recent Monoplacophora. Galathea Rep. 16 (1), 94.

Zabala, S., Hermida, G.N., Giménez, J., 2012. Spermatogenesis in the marine snail *Adelomelon ancilla*. J. Molluscan Stud. 78, 52–65.

Chapter Outline

General characters **815**
Classification **815**
According to Gardiner (1972) 815
According to Borradaile et al. (1977) 816
Reproduction **816**
Agametic reproduction 816
Gametic reproduction 816
Sexual dimorphism 816
The reproductive systems 817
The male system 817
The testis 817
The gonoduct 819
I-The vas deferens 819
Proximal region 819
Main region 819
Distal region 820
II- Needham's organ 821
Spermatophores 822
Spermatophore of *Sepia officinalis* 823
Tunics 825
Connectives 827
Spermatophoric cord 827
Ejaculatory apparatus 829
Dehiscence of the spermatophore 831
Spermatozoa of *Sepia officinalis* 833
Formation of the spermatophore 833
Role of the testis (spermatogenesis) 833
Testicular duct 835
Role of the PVD 835
Role of the MVD 836
Role of the DVD 837
Role of Needham's sac 837
Hardening of the spermatophore 839
The female system 839
Buccal seminal receptacle 844
The ovary 844
Ovulation 847
Fertilization 847
The egg 848
Reproductive behaviour, mating, copulation, and spawning **850**
Development **851**
Cleavage 853
Formation of germ layers from the blastoderm 854
Organogenesis **855**
References **857**

Chapter 12

Cephalopoda (Siphonopoda)

Cuvier, G. (1817); Lankester, E.R. (1875–1893)

(Greek *kephale*, 'head', and *pous*, 'foot')

GENERAL CHARACTERS

Mollusca:

1. The body is bilaterally symmetrical. The anterior part of the foot is fused with the head; the posterior part forms an exit funnel from the mantle cavity.
2. The mantle is highly muscular and is posterior (ventral) in position.
3. The shell, if present, is internal or external and may or may not be chambered. In most recent members, it is internal and reduced or absent.
4. The head is well developed and surrounded by circumoral tentacles probably derived from the edges of the anterior part of the foot. The eyes are prominent and maybe of complex structure.
5. A radula is present, and also a pair of horny mandibles resembling a parrot's beak.
6. There are one or two pairs of ctenidia in the mantle cavity.
7. The nervous system is well developed and the chief ganglia are localized around the oesophagus to form a highly organized brain.
8. Sex is separate and there is obvious dimorphism.
9. Males produce highly developed sperm, which are described as vertebrate sperm. They are encapsulated in **spermatophores**.
10. The females produce large yolky highly telolecithal eggs.
11. All the animals are marine.
12. The genera of this group which have served as chief objects of embryological study may be classified as follows.

CLASSIFICATION

According to Gardiner (1972)

Subclass I: Nautiloidea.
Shell external, many-chambered, coiled or straight; head with numerous retractile tentacles without suckers.
Ex. *Nautilus* sp. (pearly Nautilus).

Subclass II: Coleoida (Greek *koleos*, 'sheath').
Shell internal and more or less rudimentary; head with eight arms.
Bearing suckers, some with two additional retractile tentaculate arms; single gill and excretory organ.

Order 1: Octopoda (Greek *okto*, 'eight').
Eight uniform arms, suckers are sessile without horny rims; no internal shell.
Ex. *Octopus*, *Eledone* (lesser octopus), *Argonauta* (paper nautilus).

Order 2: Sepiida (Cuttlefish).
Calcareous or horny shell, if present, limited to dorsal portion of visceral sac (except in genus *Spirula*, in which it is coiled and located at the posterior end of the cylindrical body, i.e. opposite to the head).
Ex. *Spirula*, *Sepia*, *Heteroteuthis*.

Invertebrate Embryology and Reproduction. https://doi.org/10.1016/B978-0-12-814114-4.00012-6

Order 3: Teuthidida (Greek *teuthis*, 'squid').
Squids; horny internal shell; body torpedo-shaped; pelagic swimmers.
Suborder 1: Myopsida (Greek *myein*, 'to shut', and *opis*, 'vision').
Cornea of eye almost closed.
Ex. *Loligo* (shallow water), *Sepiella*, *Sepioteuthes*.
Suborder 2: Oegopsida (Greek *oigen*, 'to open').
Cornea not closed.
Ex. *Architeuthis* (deep sea pelagic), *Todarodes* (=*Ommastrephes*) *sloani pacificus*.
Order 4: Vampyromorpha (Vampire squids).
Eight long arms united by swimming web; two small retractile arms.
Ex. *Vampyroteuthis*.

According to Borradaile et al. (1977)

Class Cephalopoda
Order I: Dibranchiata
Suborder 1: Decapoda.
Tribe a: Belemnoidea—fossils from Mesozoic rocks.
Tribe b: Sepioidea—*Spirula*, *Sepia*, *Sepiola*.
Tribe c: Oegopsida (include some gigantic forms, ex. *Architeuthis*, 60 ft long).
Tribe d: Myopsida—*Loligo*.
Suborder 2: Octopoda—*Octopus*, *Argonauta*, *Opisthoteuthis*.
Order II: Tetrabranchiata.
Suborder 1: Nautiloidea—*Nautilus*, *Orthoceras*.
Suborder 2: Ammonoidea—*Phylloceras*, *Baculites*.

REPRODUCTION

Agametic reproduction

Agametic reproduction is unknown among cephalopods (Giese and Pearse, 1977, vol. IV).

Gametic reproduction

Sexual dimorphism

Arnold and Williams-Arnold (1977) reported that in most of the decapod molluscs it is possible to distinguish male from female in sexually mature living specimens by external characteristics. The male *Loligo pealei* can be distinguished by its white testis prominently visible through the mantle in the region between the fins. In addition, the animals of similar body weight, the males are generally longer and thinner than the females, which tend to be shorter in mantle length and broader between the fins. The accessory nidamental glands of the female are often prominent as a red to orange spot when viewed through the midregion of the ventral mantle. The hectocotylus, characteristic of the male in most cephalopods, is hard to distinguish in *Loligo*, but in other species, it may have different modifications for the purpose of transferring the sperm to the female. For example, the male *Sepia officinalis* (Linnaeus, 1758) is a representative of family Sepiidae in Alexandria waters. It was accurately identified on the basis of the characteristics specified by Tompsett (1939) and 'FAO Species Identification Sheets' (Fischer, 1973). Ismail (1981) selected three large specimens, fixed them in 10% formalin and sent them to the British Museum, which confirmed the identification. The male *Sepia officinalis* has the 'hectocotylus' (Fig. 12.1B), which is the modified left fifth arm (Ismail, 1981). The modification of this arm shows itself only by the suppression of some rows of suckers at the base of the arm (Borradaile et al., 1977). In other forms, it is profoundly modified. In *Octopus*, the end of the arm is spoon-shaped and the arm is extended so as to enter the mantle cavity of the female. In other octopods, a cyst, in which the spermatophores are stored, is formed at the end of the arm. From that, a long filament protrudes. In *Philonexis* and *Argonauta* the modified arm is charged with spermatozoa, inserted into the mantle cavity of the female and then detached. This arm was described by early observers as a parasitic worm and named *Hectocotylus*. However, it was regarded as an external male genital organ by Aristotle, c.330 BC (Aristotle, 1910; Ismail, 1981). Modifications of the hectocotylus are considered as species-specific and are often used as a taxonomic character. On the other hand, the female of most species has a **seminal receptacle**, a small pouch on the ventral buccal membrane (Fig. 12.2), in which

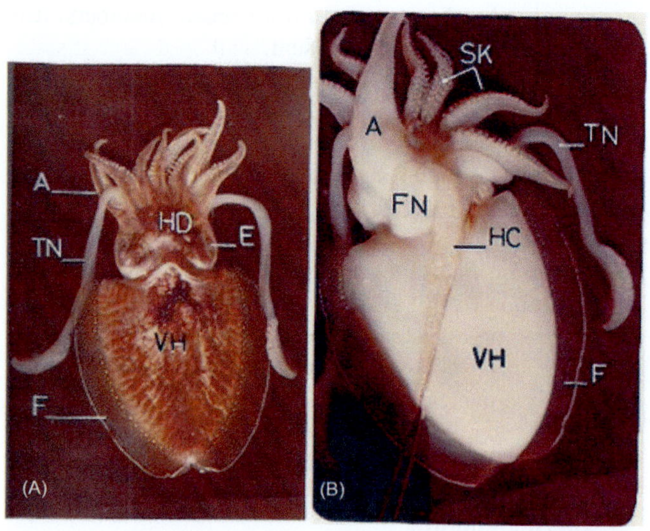

FIG. 12.1 Photographs of fresh male specimen of adult *Sepia officinalis*, showing the diagnostic features. (A) dorsal view and (B) ventral view. × one third of the natural size. A = arm, E = eye, F = fin, FN = funnel, HC = hectocotylus, HD = head, SK = sucker, TN = tentacle, VH = visceral hump. *(From Ismail (1981).)*

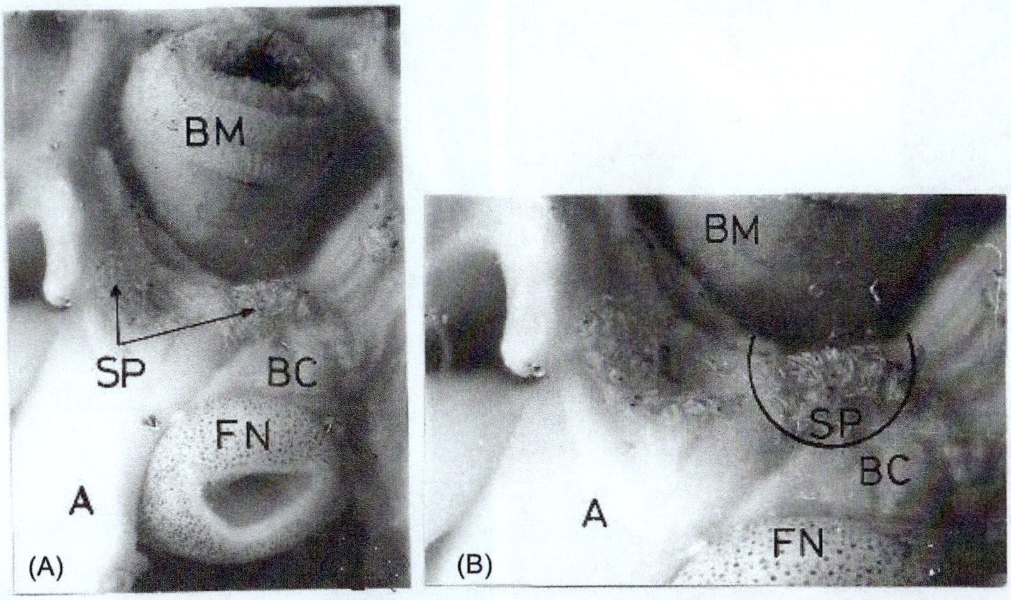

FIG. 12.2 Photographs showing bundles of spermatophores attached to the bursa copulatrix of a fresh female specimen of *Sepia officinalis*. (A) wide view, 1× and (B) magnified portion, 2×. A = arm, BC = bursa copulatrix, BM = buccal mass, FN = funnel, SPH = spermatophores. *(From Ismail (1981).)*

the male deposits its spermatophores (Bawab and Ismail, 1984a). During the mating season, this is invariably filled with spermatophores and can be seen as a small, white bulge below the mouth when the arms are spread.

The reproductive systems

The basic pattern of the reproductive system of cephalopods was first set in 1906 by Pelsenear. This system in all cephalopods is characterized by the following:

1. It is an unpaired system (Fig. 12.3), even in *Nautilus* (Haven, 1977), a primitive cephalopod, which is considered as having a paired system, has a vestigial right gonoduct, and the functional one is that of the left side.
2. There is no direct connection between the gonad and its corresponding duct.
3. The whole system is enclosed within separate coelom, the gonocoele, which is attached to the wall of the pericardium anteriorly.
4. The gonoduct opens by its own opening in the mantle cavity.
5. The entire system is dorsally placed occupying the posterior region of the mantle cavity.

The reproductive system of *Sepia officinalis* fulfils all these characters. Bawab and Ismail (1981a,b) and Ismail (1981) planned to study the structure and formation of the spermatophore of this animal, and therefore studied carefully and in detail its male reproductive system.

The male system

In a dissected fresh adult specimen of *Sepia*, this system is easily exposed since no organ of the viscera is found over or beneath it, except the ink sac (Fig. 12.3A). It is enclosed within the genital sheath. It is composed of two large components: the testis and the gonoduct.

The testis. The testis is the largest organ of this system (Figs 12.3A and 12.4A). In a mature fresh specimen, it is milky white in colour. It is ensheathed by three transparent sheaths (Fig. 12.4A). The outer one, the viscerogenital sheath, is a somewhat thick membrane and encloses the stomach as well. It therefore has to be removed to isolate the system. The middle is moderately thick and is purely genital since it encloses the whole system except the terminal part of Needham's organ. It seems that this sheath is the one described before as 'visceropericardial' (Vigelius, 1880; Tompsett, 1939) on the wrong assumption that the hearts, some of the viscera, and the gonad are enclosed within one continued cavity. The third sheath, the innermost one, is the thinnest of all, closely applied to the surface of the testis (Fig. 12.4A). It may be named as testicular sheath. Contrary to the outer and middle ones, the inner sheath cannot be removed except after fixation.

The testis has convex surfaces. It is notched in the middle of its anterior edge and this notch continues for some distance posteriorly as a superficial septum on both ventral and dorsal surfaces. At this septum, the rhythmic contractions of the testicular wall meet, well observed in live and fresh specimens. In addition, there is a very small shallow depression, which is readily visible to the naked eye only when the genital sheath is removed. It is situated posterior on the left side of the ventral surface. The **testicular pore**, previously observed in the testis of the present animal (Tompsett, 1939; Leake, 1975), is found in this depression. The opening of the PVD is directed to this depression.

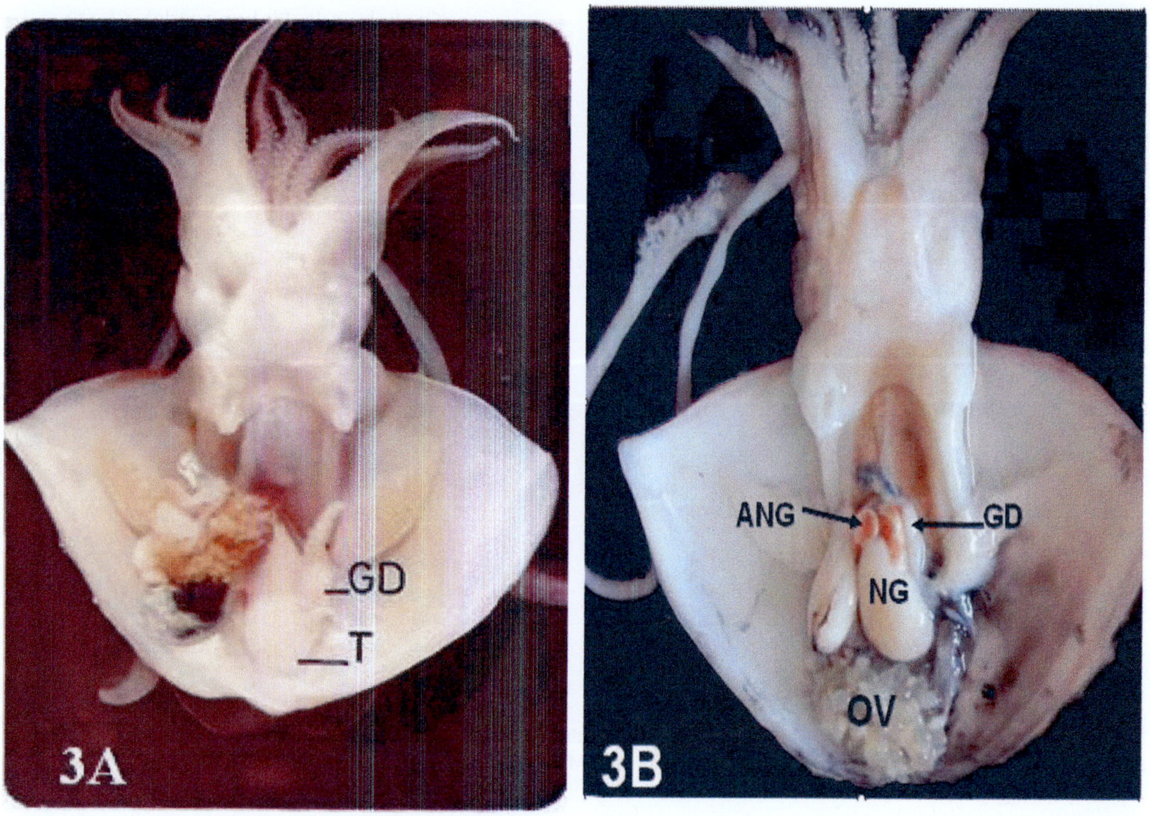

FIG. 12.3 Direct camera photos of dissected specimens of *Sepia officinalis* exposing the reproductive systems. (A) Live male, the ink sac and some of the viscera are displaced *(after Ismail, 1981)*. (B) Fresh female, the ink sac and viscera are removed. About natural size. ANG = accessory Nidamental Gland, GD = gonoduct, NG = Nidamental gland, OV = ovary, T = testis.

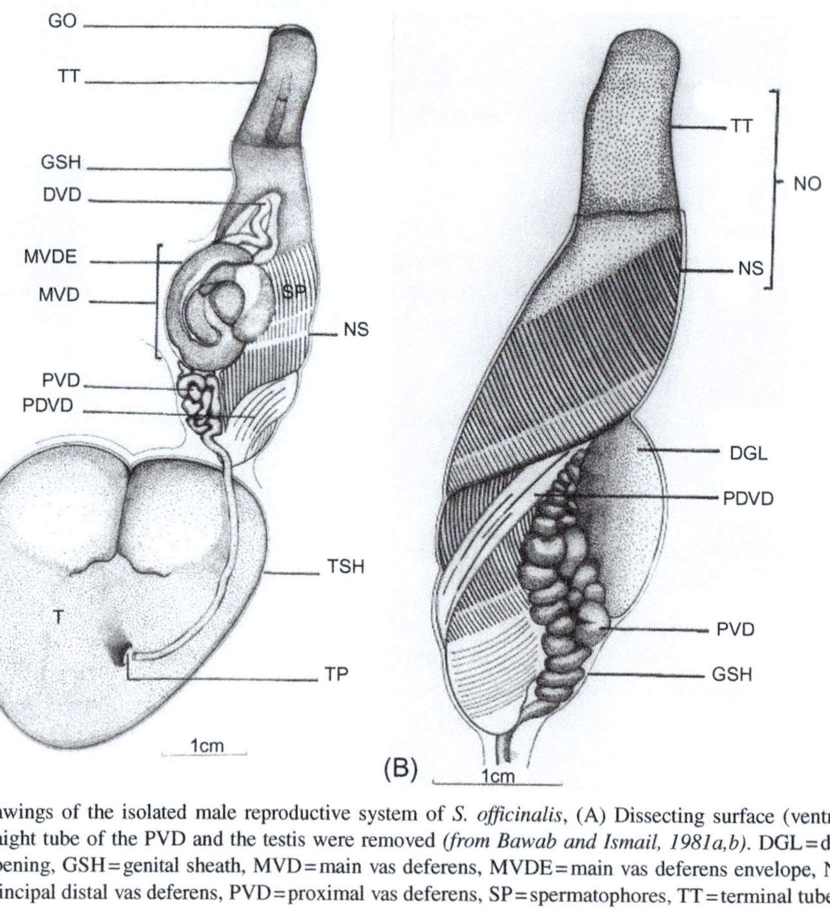

FIG. 12.4 Camera Lucida drawings of the isolated male reproductive system of *S. officinalis*, (A) Dissecting surface (ventral surface); (B) dorsal surface of the gonoduct, the straight tube of the PVD and the testis were removed *(from Bawab and Ismail, 1981a,b)*. DGL = distal gland, DVD = distal vas deferens, GO = genital opening, GSH = genital sheath, MVD = main vas deferens, MVDE = main vas deferens envelope, NO = Needham's organ, NS = Needham's sac, PDVD = principal distal vas deferens, PVD = proximal vas deferens, SP = spermatophores, TT = terminal tube of N. organ.

The gonoduct. The gonoduct of cephalopods is a very complex structure, which coils in parts, and curves and overlaps in others. When unravelled, it is very long compared to the limited area in which it is situated. This must have certain significance in connection with the manner of formation of the spermatophores. However, the gonoduct is divided morphologically and functionally into two main organs: the vas deferens and Needham's organ.

The vas deferens

The author studied and published the work on the structure and formation of the spermatophores of some crustaceans (e.g. Bawab, 1970; Malek and Bawab, 1971; Bawab and El-Sherief, 1981) and molluscs (Bawab and Ismail, 1984a). This interest allowed for the dividing of the vas deferens of these animals according to the function of its parts and applying appropriate general names. This nomenclature can be applied on the parts of any vas deferens forming these elements. Hence that of *Sepia* is divided into the proximal (PVD), main (MVD), and distal (DVD).

Proximal region. This region is the only one in agreement with the description and nomenclature of the different investigators working on different genera of cephalopods. It appears morphologically as a conducting region of two parts: a simple somewhat long tube and a massive compact coiled tube (Fig. 12.4A). The former extends in a curved pathway, mostly along the ventral surface of the testis. It begins with its simple opening, which faces the testicular opening in its depression. This observation contradicts that described before as the ciliated funnel-shaped opening (Leake, 1975). The presence of cilia needs more investigation. On the other hand, most of the coiled tube does not appear from the ventral surface of the system, since it is hidden beneath the second region (MVD). It extends dorsally along Needham's organ (Fig. 12.4B). The portion of continuation of the proximal to the main region is not visible from either surface of the system.

Main region. Seen from the ventral surface (Fig. 12.4A), the MVD appears glandular, and consists of four parts (I–IV) with different forms and diameters. This picture is justified by the histological pattern (Fig. 12.5). The diameter of each of these parts is large compared to that of the PVD and DVD. All four parts are ravelled in such a manner to form a complex disc-like mass that lodges the curvature of Needham's organ (Fig. 12.4A). This MVD is ensheathed with its own thin and transparent sheath, the **MVD sheath**. However, it is firm enough to keep this region's spherical shape, and to keep the form and lodge of each of its four parts. As far as we know, none of the previous authors observed this sheath. However, in the giant octopus a second membrane was observed after the removal of the genital one (Mann et al., 1970).

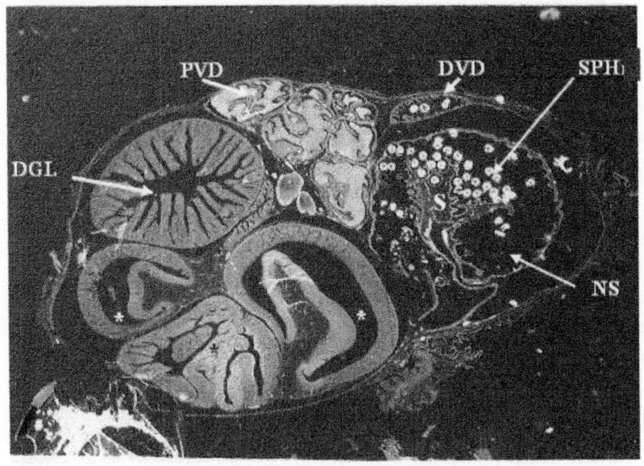

FIG. 12.5 T.S. at about the middle level of Needham's organ embedding the vas deferens. 20×. DGL = distal gland, DVD = distal vas deferens, three parts of main v.d, each indicated by star. NS = Needham's sac, PVD = proximal v.d., S = septum in Needham's sac, SPH = spermatophore.

Part I is the innermost of the spherical mass, crescent-shaped, and lodges the curvature of Needham's organ (Figs 12.4A and 12.6). Part II is a ball-like mass (Fig. 12.6), sharply distinct ventrally (Fig. 12.4A). Part III is another crescent, similar but opposite to part I. The two together form a complete circle with part II embedded in its centre. If the vas deferens is unravelled, this part II does not appear dorsally. The posterior end of the crescent part III is blind and lodges a curve of part IV (Figs 12.4A and 12.6). Tompsett (1939) named part III as 'almond' and its blind end a 'çedilla-like appendage'.

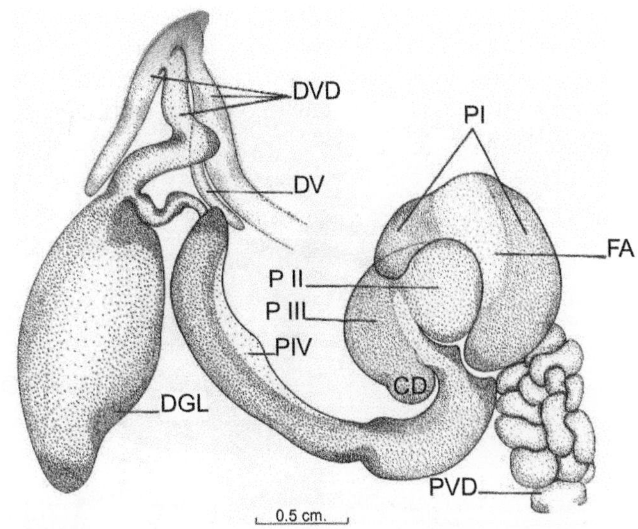

FIG. 12.6 Camera Lucida drawing of the unravelled vas deferens as seen from the dissected surface (ventral surface) showing its three definite regions. CD = cedilla of part III, DGL = distal gland, DV = diverticulum, DVD = distal vas deferens, FA = faint area of part I, PI-PIV = the four parts of main v.d., PVD = proximal v.d. *(From Bawab and Ismail (1981a,b).)*

Part IV begins as a narrow membranous side tube that arises from the inner concavity of the crescent part III (Fig. 12.6). It remains attached to the latter, but at the level of the cedilla, it becomes completely free from the wall of this part. This tube is the beginning of part IV, which then extends around the cedilla to be U-shaped tube embracing all the other parts of the MVD and making its outer tube (Figs 12.4A and 12.6). The curved portion of part IV surrounding the cedilla is stouter and naturally shorter than its outer tube. At this level of the connection with the outer tube, it constricts to decrease the diameter of the outer tube (Fig. 12.6). This figure also shows that the surface of the inner edge of this tube is fainter and/or transparent forming a hyaline, somewhat broad line indicating the presence of an internal fold. From this portion of part IV, Bawab and Ismail (1981a,b) isolated spermatophores at an advanced stage of formation. This is conceivable evidence that the MVD is the device manufacturing the spermatophores, at least in *Sepia*. In support of this, a tedious isolation of an incomplete spermatophore from the passage between part II and III of the MVD took place. Its examination proved that it is only a spermatophoric cord (see later).

It appeared that this MVD has its scientific importance in forming spermatophores. Bawab and Ismail (1981a,b) reviewed in detail the different nomenclature applied to its parts, and the valuable discussions defending some of these names. As mentioned above, general names were used to be applied to those of other animals.

The continuation of the MVD at both ends, with PVD proximally and with DVD distally, is difficult to observe, or more precisely not obvious from both surfaces of the system shown in the gross anatomy as seen in situ (Fig. 12.4A and B). Concerning the former, in the unravelled system, part I of MVD appears more voluminous and its crescent extends dorsally as a cone-like portion, which tapers gradually to be very narrow (Fig. 12.7A and B) to connect the PVD. This excitable portion is embedded in the coils of the PVD (Fig. 12.7A), the unravelling of which is almost impossible. However, it was proved that the PVD ends in the apex of this cone (Fig. 12.7B). The other distal end of MVD is part IV, which seems blind, but just before the end, the DVD originates anteriorly and an accessory duct posteriorly, from the same portion (Fig. 12.8). This duct, previously named 'small tube' (Pelseneer, 1906), 'ciliated duct' (Tompsett, 1939) or 'waste duct' (Morton, 1967), extends backwards parallel to the hyaline line of part IV as a narrow uniform tube to open in the gonocoele just at the level of the cedilla of part III, i.e. the beginning of part IV. In the ravelled system, even in the isolated one, one can never observe this duct, neither ventrally nor dorsally. To detect this duct, the system must be unravelled patiently, especially part IV. Since this duct is completely related to the MVD, Bawab and Bawab Jr. (unpublished observation) applied the name **waste duct**, preferring the view of Morton (1967).

Distal region. The previous investigators agree with its name, but restrict this name to the looped portion observed over the proximal part of Needham's organ (Fig. 12.4A and B). Ismail (1981) and Bawab and Ismail (1981a,b) patiently studied the other portions that are naturally enclosed within the triangular gelatinous mass lodging Needham's organ anterior to the MVD (Fig. 12.4A). This triangular mass was generally neglected completely or described as a 'complex appendage of the accessory gland' (Tompsett, 1939). The continuation of the DVD is critical, as its origin. Just at its origin from MVD, it proceeds as a short, very narrow wavy tube to connect the previously called the 'accessory gland' (Peterson, 1959), or the 'prostate gland' (Cuvier, 1817; Pelseneer, 1906; Mann, 1963) (Figs 12.6 and 12.8). The DVD then continues forwards within the triangular gelatinous mass as an axial wavy tube (Fig. 12.8). It connects an inverted V-shaped one, the outer limb of which is blind, while the other one

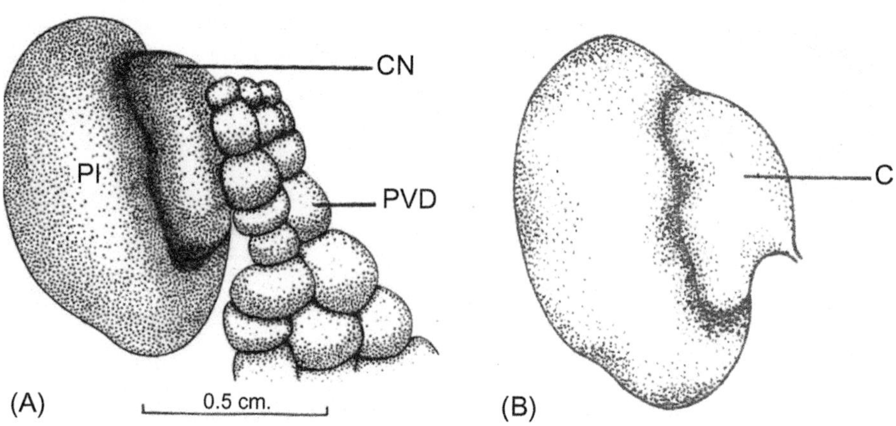

FIG. 12.7 Camera Lucida drawings showing the cone of part I of the MVD as seen from the lateral view; (A) embedded in the coils of PVD; (B) exposed cone. *(From Ismail (1981) and Bawab and Ismail (1981a,b).)*

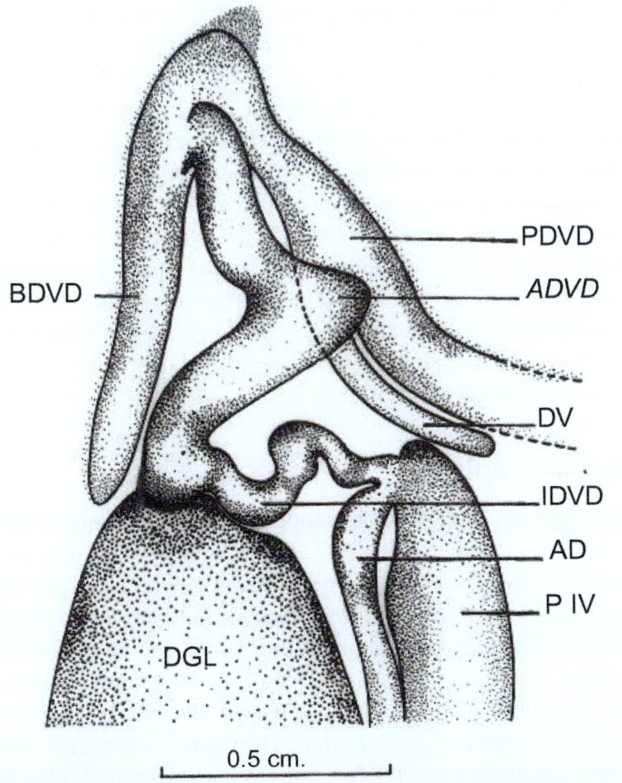

FIG. 12.8 Camera Lucida drawing showing the origin of both, the accessory duct, and the extension of the distal vas deferens (DVD) in the gelatinous triangle. Also shows the relation of the gland to the DVD. *From Ismail (1981).* AD=accessory duct, ADVD=axial DVD, BDVD=blind DVD, DGL=distal gland, DV=diverticulum, IDVD=initial DVD, P IV=part IV of MVD, PDVD=principal DVD.

connects the parallel narrow blind short tube, ending in the gelatinous mass and referred to as the **diverticulum**. The essential one extends beyond the gelatinous mass. It loops obliquely backwards over Needham's organ as to be seen dorsally (Fig. 12.4B). This is the only portion known as DVD in the previous descriptions. It loops again over Needham's organ to extend on its ventral surface to merge into its cavity (Fig. 12.4A).

The previously called 'accessory' or 'prostate' (Greek *prostate*, 'one who stands before' and is often referred to as such (Gardiner, 1972)) gland is the only gland associated to the male reproductive system of *S. officinalis*, maybe to all those of cephalopods. This gland is massive and greatly pronounced in this system (Figs 12.4B and 12.6). Critically, it is related to the DVD, so, with no doubt, it must be known as the **distal gland**. The two together, especially the gland, share in the formation of the outer cover of the spermatophores, not in the formation of the sperm mass (Bawab and Bawab Jr., unpublished). Therefore, this gland of *Sepia* is far analogous to the prostate gland of higher animals, since the latter shares in the formation of the fluid and nutrients of the semen (Keele and Neil, 1973). This is an acceptable view to contradict the previously used term 'prostate'.

Emam and Aly (2000) presented photomicrographs to describe the histological structure of the different parts of the vas deferens of the squid *Sepioteuthis lessoniana*.

Needham's organ

Needham's organ was named after its first discoverer in the 18th century (Needham, 1745). It is the end part of the gonoduct of *Sepia*. It is a spacious tube-like crescent structure (Fig. 12.9). Morphologically, it is well differentiated, even by the naked eye, into two components: an enlarged preservative sac and a comparatively short terminal conducting tube (Fig. 12.4A and B). The preservative sac is thin walled, meaning that the arranged spermatophores can clearly be seen within. It is enclosed, with the whole gonoduct, in the gonocoele. The dorsal view of this organ is characterized by two distinctive features. **First**, its anterior portion folds in an oblique manner, over its posterior portion forming a **half twist**. The latter is clearly seen passing from one side to the other at about the middle region of that surface. **Second** is that the spermatophores easily appear through its transparent wall in two oblique bundles (Fig. 12.4B). They are arranged side by side with their anterior ends (ejaculatory apparatus) directed obliquely downwards towards the blind end of the sac. This arrangement recalls that of the matchbox style (Mann, 1975). Male sexual maturity is usually considered by the number of spermatophores in Needham's sac of *Illex illecebrosus* (Coelho and O'Dor, 1984).

The blind end of the sac was studied in an allied form, *Octopus vulgaris* (Mann, 1963). It was ascertained that this portion is glandular and that is why it is free from spermatophores. On the other hand, the ventral view of Needham's organ looks different. Beside the appeared spermatophores, both its surfaces constitute suitable lodge for (almost wholly) the long but ravelled vas deferens (Fig. 12.4A and

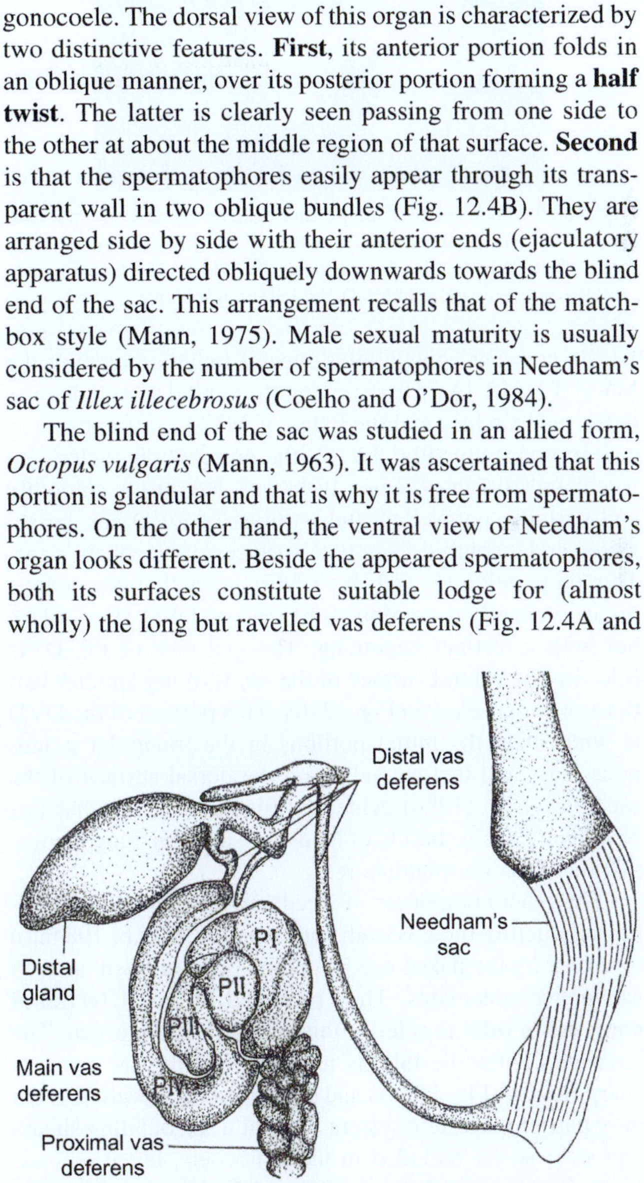

FIG. 12.9 Camera Lucida drawing of the unravelled gonoduct, especially the DVD as seen from the dissected surface (ventral surface). *(From Bawab Jr. (1996), modified after Ismail (1981).)*

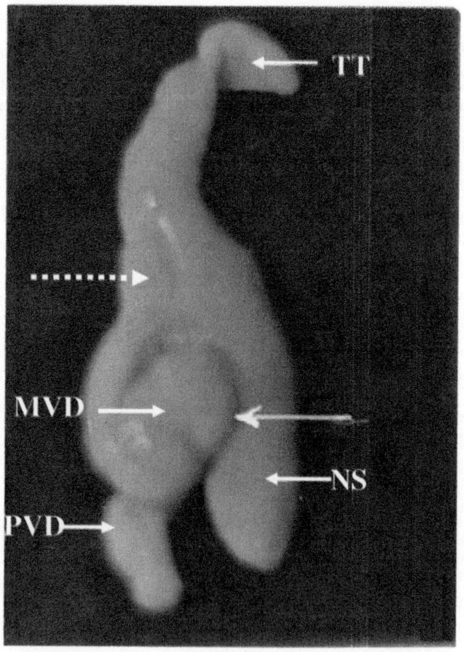

FIG. 12.10 Photograph of the isolated male gonoduct of *S. officinalis* exposed from the genital sheath *(from Bawab Jr., 1996). Dotted arrow* indicates the gelatinous triangle embedding the DVD; *straight arrow* indicates the crescentic lodge of MVD. 2×. Lettering as above.

B). Its crescent constitutes a special lodge embracing the MVD (Fig. 12.4A). On removing the genital sheath, the vas deferens dislodges and the lodge of MVD itself appears as a deep concavity (Fig. 12.10), the crescent characterizing Needham's organ of *Sepia*. If the vas deferens is smoothly and partially unravelled, the portions of DVD can be selected and displaced from both surfaces of Needham's sac. This preparation (Bawab Jr., 1996) shows that Needham's organ is merely a continued dilation of DVD (Fig. 12.9), but with a distinct beginning. The end part of the DVD folds on the ventral surface of the sac forming another half twist, which is clear in Fig. 12.4A. This portion of the DVD is wider than the initial portions in the triangular gelatinous mass and that extending on the dorsal surface of the sac. Bawab Jr. (1996) evidenced histochemically that this Needham's sac is the site of hardening and finishing the formation of the spermatophores.

The other component of Needham's organ is the terminal conducting tube. Bawab and Ismail (1981a,b, 1984a,b) observed by the naked eye its contractions in fresh, as well as, in live specimens. They preferred the term **terminal contractile tube** to refer to this end part of the organ. The terminal contractile tube is marked from the sac by very sharp flexion (Fig. 12.4A) and also by its thick wall. It is the only part of the whole system without a surrounding sheath and so it is not included in the gonocoele, but it extends freely into the mantle cavity (Fig. 12.3). It ends with a slit-like genital opening, which has two lips; the dorsal one is large, so that the slit itself is slightly shifted ventrally.

The terminal contractile tube was previously referred to as 'penis' (Tompsett, 1939; Peterson, 1959; Mann, 1963; Fields, 1965). On the other hand, many other authors did not mention this word (Bullough, 1981; Purchon, 1968; Mann et al., 1970). In addition, Morton (1967) ascertained that no cephalopod has a penis, properly so called. It is well known that penis is copulatory or an intromittent in different animals. The terminal tube of Needham's organ in *Sepia* is homologous and not analogous to this penis. This term should be rejected completely from the male system of *Sepia*. However, *Sepia* as well as all cephalopods have the hectocotylus which serves not precisely as analogous to the penis, but as an alternative.

Spermatophores

The first discovered spermatophore was that of *Sepia officinalis* (Swammerdam, 1637–1685). This species becomes mostly distinct within its group Cephalopoda in specific, and maybe within the animal kingdom in general, in being the first and the only recorded animal producing spermatophores. It remained as such for about a century, until the time of Needham (1745) who discovered and described those of the 'Calamary', i.e. *Loligo*. He named them **milt vessels**, and these were subsequently renamed **animalcules de Needham** by Cuvier (1817). Afterwards, Milne-Edwards, in 1842, studied the same elements of cephalopods and he was the first to introduce the term spermatophore.

The cephalopod spermatophore represents one of the most highly evolved and beautifully organized elements involved in invertebrate reproduction. No study and/or discussion of cephalopod development would be complete without the mention of an account of the spermatophore. The latter in cephalopods has a general plan of structure; it was believed that it is a chitinous structure (Williams, 1910). Many years later, Morton (1967) and Gardiner (1972) contributed to this view and both of them described the spermatophore of *Loligo* as a 'narrow tube of **chitin**', about two-thirds of which is filled with a dense mass of sperm (Gardiner, 1972). There is no adequate chemical evidence supporting this view in the available literature. However, Bawab Jr. (1996) and Bawab and Bawab Jr. (unpublished) proved histochemically that chitin is one of the constituents of the spermatophore of *Sepia*. It is not a simple structure. It is elongated, rod-like in form, and constitutes two scientifically significant components in a linear arrangement. According to Bawab and Ismail (1984a), and then Bawab Jr. (1996), these are a **spermatophoric cord** (SPC), and an **ejaculatory apparatus** (EA) attached together by **connectives** (beginning in a **connecting region**, CR) and are all enclosed in a rigid but transparent outer cover. The ratio of the length of each component compared to the length of the whole element differs with the species. However, it frequently varies in fine details, but it attains taxonomic significance (Marchand, 1913; Sasaki, 1929). It can be used as a taxonomic feature at the level of the species. Adam (1960) recorded that he was unable to distinguish between the two species, *Octopus macropus* and *O. aegina* by characters other than the spermatophore. The former has armed spermatophores, while the latter has unarmed ones. Likewise, Robertson (1978) established a new system of classification for gastropods based on the structure of spermatophores of different genera. In 1984, Coelho and O'Dor reported that the spermatophores have a more reliable taxonomic characteristic than other structures such as suckers, beaks, etc. More recently, Rodrigues et al. (2010) stated that the spermatophore of this group of animals have systematic value. The cephalopod spermatophore also shows another significant character. Its length differs with the different species and also with the different individuals of the same species. It is well known that this group of animals includes individuals ranging from very small to huge. Bawab Jr. (1996) surveyed spermatophore length of some species from the available literature with the aid of the recorded animal length of such species in the FAO sheets (Fischer, 1973; Fischer et al., 1987). Thereafter, Bawab and Bawab revisited the relationship between the spermatophore length and the length and/or size of its animal (Table 12.1). Indeed, this relationship was previously verified by Fields (1965), between the spermatophore length of *Loligo opalescens* and its mantle length. It was evidenced statistically (Boyle et al., 1995) that there is a strong positive relationship between spermatophore length and the body length of the males of *L. forbesi*. This fact was verified by Table 12.1, which shows that the spermatophore length is directly proportional to the length of the animal, or the mantle within the genus and/or the group.

Again, the statistical studies done by (Boyle et al., 1995) prove that there is a weak relationship between the total number of spermatophores and the body size. The terminal 'spermatophoric duct' of the giant octopus of the North Pacific, *Octopus dofleini martini*, always lodges either one or sometimes two mature 1 m long spermatophores. One can pull the spermatophore out by hand, from the live male (Mann, 1984). Usually it is possible to repeat the removal of an additional spermatophore the next day, or occasionally even in the same day. This means that this species produce very few mature spermatophores during the breeding season. On the other hand, the middle-sized animals such as mature specimens of *Sepia officinalis* from Alexandria waters, which may attain 40 cm in length, produce innumerable mature spermatophores during the breeding season. One can pull out numerous bundles of mature spermatophores from 'Needham's sac' of specimens measuring 20–25 cm in length (Bawab and Ismail, 1981a, 1984a; Bawab Jr., 1996). Other small animals, such as *Rossia macrosoma* from the eastern Mediterranean, whose mantle length ranges from 26 to 47 mm, produce a reasonably suitable number of mature spermatophores. One can find 15–103 (mean 46) spermatophores in their Needham's sac (Salman and Önsoy, 2010). Another species, *Rossia moelleri* (ML 32–41 mm) from Greenland waters (Germany) has 19–56 spermatophores in its Needham's sac (Hoving et al., 2009). Needham's sac of the mature squid, *sepiola atlantica* from the Spain Atlantic waters (ML 11.3–12.9 mm), stored 843 as a maximum number of spermatophores (Rodrigues et al., 2010). The valuable work of Nigmatullin et al. (2003) tabulated some different features of about 20 different ommastrephid squids from different regions of different oceans. This table evidenced that there is no definite relation between ML and/or size of the male mature squid and the number of spermatophores in its Needham's sac (fecundity). An opposite view was estimated for the ommastrephid short-finned squid *Illex coindetii* from the North Eastern Atlantic. It was observed that the number and length of the spermatophores tended to increase with the size of the males (Gonzalez and Guerra, 1996). These authors recorded that the number of spermatophores counted in the 'spermatophoric complex' varied between 81 in one male (ML 107 mm), and 1555 in another male (ML 345 mm).

Spermatophore of *Sepia officinalis*

The isolated spermatophores from Needham's sac of a live or fresh specimen are always in bundles and closely attached to each other. Those isolated from a specimen obtained during January are completely formed. Ismail (1981) and Bawab and Ismail (1984a) studied carefully the structure of these elements. Bawab Jr. (1996) restudied these elements, but based on their histochemical nature. That of *S. officinalis* (Figs 12.11 and 12.12) is considered one of the smallest spermatophores within cephalopods (Table 12.1). It is very narrow, measuring about 0.05–0.13 mm in diameter, cylindrical with two rounded ends, and elongated, measuring about 6–8 mm in length. Önsoy and Salman (2005) recorded that the spermatophore lengths of this species obtained from a lagoon of Aegean Sea were between 7 and 14 mm. The difference of the climate may influence this change. Alternatively, these might not be fully formed spermatophores; since Önsoy and Salman studied the reproductive biology and evidenced that the gonadosomatic indices (GSI) of males have homogeneous values throughout the cycle. However, the team of Bawab studied the hardened element of a naturally curved structure. It keeps its curvature and cannot be straightened owing to its rigid outer cover. Its length has a definite range within the species (Table 12.1). Its structure conforms to the general plan of that of cephalopods. It is quite difficult to prepare histological sections of one isolated spermatophore or even a bundle, because of its fine structure and mostly of the hard nature. The alternative was the use of transverse sections of Needham's sac. The same difficulty was encountered, yet any section shows

TABLE 12.1 Relationship of spermatophore length and animal length of some species of subclass Coleoidea, class Siphonopoda, phylum Mollusca, classification according to Gardiner (1972)

			Animal length		Spermatophore length	
Family	**Species name**	**Region**	**Length (mm)**	**Reference**	**Length (mm)**	**Reference**
Sepiolidae	*Sepiola rondeletii*	Germany	10–20	Fischer (1973)	7	Marchand (1913)
	Sepiola atlantica	Atlantic Spain	11.3–12.9	Rodrigues et al. (2010)	3.3–9.2	Rodrigues et al. (2010)
	Rossia moelleri	Greenland, Germany	32–41	Hoving et al. (2009)	9–12	Hoving et al. (2009)
	Rossia macrosoma	East Mediterranean	26–47	Salman and Önsoy (2010)	12.3–18.8	Salman and Önsoy (2010)
Sepiidae	*Sepia elegans*	Germany	40–70	Fischer et al. (1987)	3.25–3.5	Marchand (1913)
	Sepia officinalis	Liverpool	150–250	Fischer et al. (1987)	6–10	Tompsett (1939)
	Sepia officinalis	Mediterranean Alex	220–250	Bawab and Ismail (1984a,b)	8–10	Bawab and Ismail (1984a,b)
	Sepia officinalis	Aegean sea	70–140	Önsoy and Salman (2005)	7–16	Önsoy and Salman (2005)
Teuthidida	Architeuthis sp.	South Australia	2–4 m	Norman and Lu (1997)	110–200	Norman and Lu (1997)
	Septiotethis lessoniana	Gulf of Suez	20–30 cm	Emam and Aly (2000)	16 mm	Emam and Aly (2000)
Ommastrephidae	*Hyalotheuthis pelagica*	Northeast Atlantic	52–77	Nigmatullin et al. (2003)	5.0–11.5	Nigmatullin et al. (2003)
	Dosidicus gigas	East-central Pacific	130–377		17.0–31.0	
	Ommastrephes bartrainii	Northeast and west Atlantic	294–396		32.0–53.0	
	Sthenoteuthis pieropus	East-central Atlantic	107–240		15.0–42.0	
	Illex coindetii	Eastcentral Atlantic	113–205		14.0–38.0	
	Illex coindetii	Northeast Atlantic	107–345	Gonzalez and Guerra (1996)	11.1–37.7	Gonzalez and Guerra (1996)
Loliginide	*Loligo marmorae*	*Germany*			2.3–3.0	Marchand (1913)
	Loligo opalescens	Pacific Ocean	130–190	Fields (1965)	8–13	Fields (1965)
	Loligo pealii		200–300	Austin et al. (1964)	10	Gardiner (1972) and Austin et al. (1964)
	Loligo vulgaris	Germany	200–300	Fischer (1973)	12	Marchand (1913)
	Loligo forbesi	North Sea	120–450	Boyle et al. (1995)	9–25	Boyle et al. (1995)
Octopodidae	*Eledone moschata*	France	250–350	Fischer (1973)	14–19	Fort (1941)
	Eledone aldrovandi	Germany			25–26	Marchand (1913)
	Eledone cirrhosa	France	300–350	Fischer (1973)	60–70	Fort (1937)
	Octopus macropus	Germany	500–700	Fischer (1973)	46	Marchand (1913)
	Octopus vulgaris	Germany	500–700	Fischer (1973)	60–80	Marchand (1913)
	Octopus dofleini martini	North Pacific	8500 g (10–47 kg)	Storer et al. (1983) and Mann (1984	1000	Mann et al. (1966, 1970) and Mann (1984)

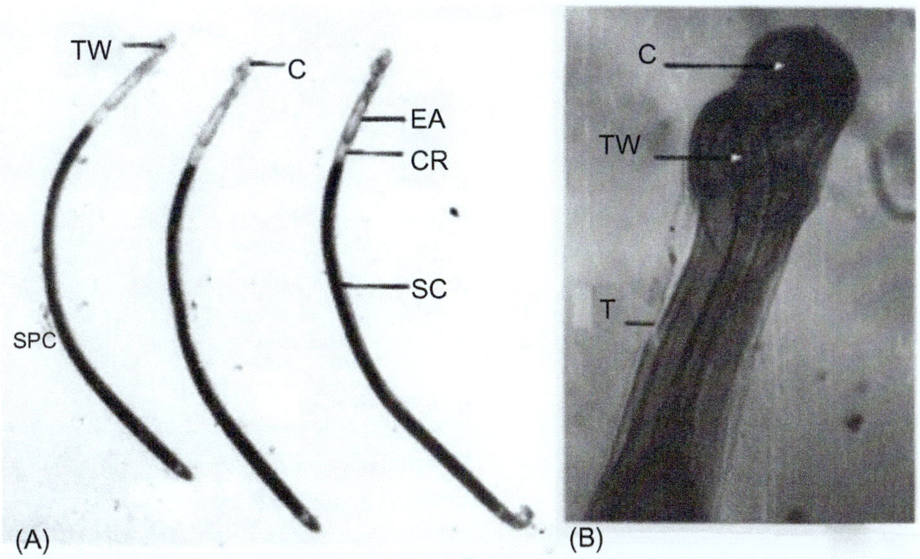

FIG. 12.11 (A) Photograph of freshly isolated spermatophores of *S. officinalis*. 15× *(from Bawab Jr., 1996)*; (B) photomicrograph of the anterior part of the spermatophore. 200× *(from Ismail, 1981)*. C=cap, CR=connecting region, EA=ejaculatory apparatus, SPC=spermatophoric cord, T=tunic, TW=twist.

fairly satisfactory sections of spermatophores at different levels (Fig. 12.13). According to the conclusions of the thorough histological and histochemical studies of Bawab Jr. (1996), some of the components of the spermatophore of *S. officinalis* were intensively resolved and so renamed. However, the components of the spermatophore are: SPC, CR, and EA, and the common constituents: the outer cover (tunics) and the connectives. These common ones are related to the whole spermatophore and will be described first.

Tunics

This term was applied to what was previously referred to as 'outer cover' (Fields, 1965; Bawab and Ismail, 1984a). This outer cover accepted much controversy; two views were presented concerning the number of tunics and/or layers forming this component. One view is that it is only one tunic (Marchand, 1913; Tompsett, 1939; and all the works of Mann et al., 1966, 1970; Mann, 1975). However, Fields and Marchand differentiated this component into three layers. Later on, Bawab and Ismail (1984a) evidenced this view by the identification of the major chemical substances constituting these layers. The thorough and detailed histochemical investigations by Bawab Jr. (1996), clarified the picture and agreed with this view, but resolved this component as double tunics. This is the second view that the spermatophore of *Sepia* is wholly enveloped by two transparent but well differentiated tunics (Fig. 12.12B). They are similar in thickness, but each has a different chemical nature. Their thickness is almost regular all around the whole length of the spermatophore. It measures about 10 μm having the same appearance all over the surface of the spermatophore. The **external tunic** is quite smooth in appearance and is differentiated into three layers. The outer most is very thin, membranous and is difficult to observe except by staining with compound and/or specific stains (Table 12.2). It is enough for the present account to record the affinities to the histological stains. The affinities to the histochemical stains and chemical reactions can be encountered in the work of Bawab Jr. (1996) and Bawab and Bawab Jr. (unpublished). Their studies were histochemical, i.e. in situ and not by extraction or by centrifugation. Mann et al. (1966, 1970) used the chemical analysis and the biochemical tests to identify the substances constituting the spermatophore of the giant octopus (*Octopus dolfeini martini*). The middle is the broadest and the inner layer has more or less the same thickness. On the other hand, the **internal tunic** is well differentiated from the outer tunic not only by its frothy appearance but also by its faint colouration with the same stain (Figs 12.12B and 12.14). The deposition of the substance forming this tunic is well observed within the lumen of the DVD attached to the dorsal surface of Needham's sac (Fig. 12.28).

It was confirmed that the spermatophore of *S. officinalis* is a chitinous structure through the application of the most general test (Alcian blue) to the most specific chitosan test (Campbell, 1929). The tunics are one of the chitinous components of this spermatophore (Bawab Jr., 1996). Chitin is a skeletal substance. It never occurs freely in nature, but is always linked with e.g. proteins, $CaCO_3$ or both (Dennell, 1949; Richards, 1951, 1970; Karlson, 1965; Pearse, 1968). It seems always to be accompanied by protein or mucoprotein in different amounts and probably of different chemical composition (Peters, 1972). Richards (1970) described this

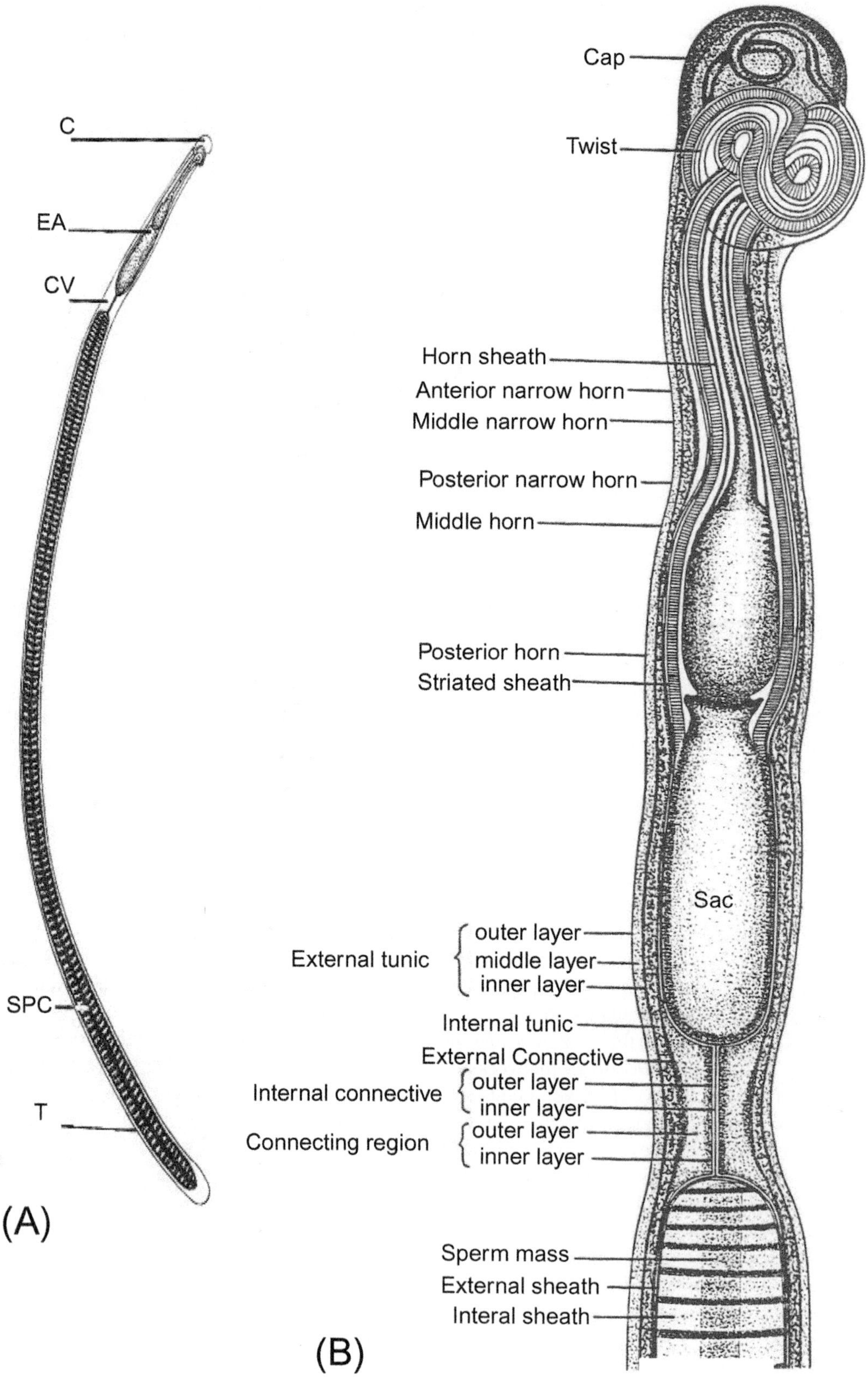

FIG. 12.12 Camera Lucida drawings of a fully formed freshly isolated spermatophore of *S. officinalis*; (A) whole mount, 40× *(from Ismail, 1981)*; (B) magnification of the ejaculatory apparatus attached with anterior portion of spermatophoric cord, 250× *(modified after Ismail, 1981, from Bawab Jr., 1996)*.

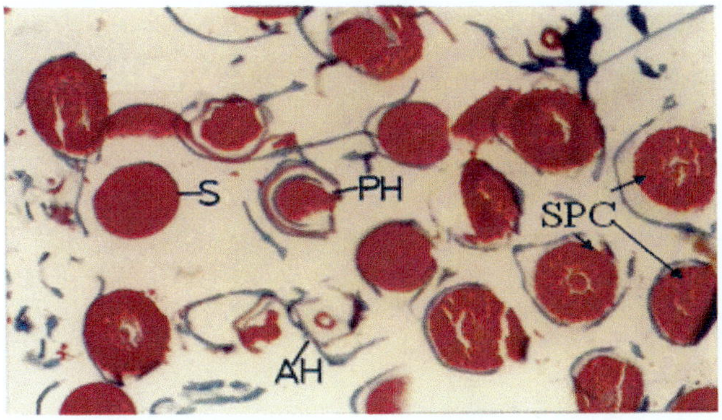

FIG. 12.13 Photomicrograph of a wide field of a portion of T.S. of Needham's sac showing sections of spermatophores at different levels, Heidenhain's Azan. 100× *(from Bawab Jr., 1996).* AH=T.S. of anterior narrow horn, PH=T.S. of posterior horn, S=T.S. of sac, SPC=T.S. of spermatophoric cord. Notice that there is no section of connecting region.

chitin as a solid, rigid, and stable structural substance, and its presence strengthens the component.

These tunics have significant importance. They extend at both ends of the spermatophore, forming two striking structures. Posteriorly, the tunics are not completely applied around the SPC, and hence there is a **free chamber** (Figs 12.12A and 12.15). Austin et al. (1964) observed thin 'streamers' of spermatozoa projecting in this chamber. This is an unreasonable interpretation because it is not in accordance with the scientific significance of the spermatophore. It appears that the substance constituting the internal tunic occupies this chamber. At the other end, just above the EA, the terminal part of the tunics elongate and coil forming the **cap**, which has its significance (see later).

Connectives

This component does not have a measurable region, but like the tunics, it encloses the whole spermatophore. In addition, like the tunics, it is more distinctive in what is known as a **connecting region** (CR) (Figs 12.11 and 12.12). The latter occupies a very short but considerable distance of the length of the spermatophore, and intervenes between the two essential components: the SPC and EA. Within this region, the true connecting material is well differentiated into axial and peripheral zones (Bawab Jr., 1996). Therefore, it was interpreted that the connectives may originate, or are precisely more pronounced in this CR, and hence the name of the latter. Macroscopically, this region appears homogeneous and the author interpreted that the basic substance of this region is that of the internal tunic, as the affinity to stains shows (Fig. 12.16). In this, it resembles the free chamber as mentioned above. To be connective, hardened material(s) impregnate definite areas of this region. The axial is narrow, filamentous, penetrating the axis of the region, densely stained and so more pronounced (Fig. 12.12B). It is not a simple filament, but it is a double-layered layer inside the other namely inner and outer layers of the axial or generally **internal connective**. At each end of this connective, its two layers together enclose the corresponding component of the spermatophore: anteriorly the EA and posteriorly the SPC. They appear surrounding each component for a considerable distance and likewise, the **external connective** is well differentiated peripherally in the CR (Fig. 12.16). It is somewhat broad, ribbon-like and densely stained. It is so pronounced that it limits the region, but is well differentiated from the internal tunic (Fig. 12.12B). It can be observed as an extended layer along the length of both the SPC and EA just underneath the internal tunic, but merging with the internal connective (Fig. 12.12B).

Spermatophoric cord

This component constitutes almost the whole length of the spermatophore; about five times that of its EA. It is an essential component. It is spiral in form and hence the striated appearance (Figs 12.11 and 12.12A). The outer tunics are closely applied to its surface along its length, that its diameter is almost the same of the spermatophore. Its posterior end is displaced from the tunics, forming a little free chamber (Fig. 12.15). The transverse section of this SPC shows well-differentiated areas (Fig. 12.17). The smooth central area constituting the core of this component is the sperm-embedded substance. It is the only area of the SPC containing sperm, since the others are sheaths and/or protectives, as confirmed by the artificial dehiscence (see later). Therefore, only the core is appropriately named the **sperm mass**; it is a coherent mass (Bawab and Ismail, 1984a). Surrounding this mass is a very thick granulated layer, the base substance of which forms a membrane-like limiting layer for the whole SPC. The granulated layer is its **internal sheath** and the membranous one is its **external sheath**. The latter is difficult to be differentiated from the connectives upper to it (Figs 12.12B and 12.17), since the central sperm mass is ensheathed by the two sheaths described just above. Yet this component is appropriately named the **spermatophoric cord**, and not merely a 'sperm mass' (Austin et al., 1964), or a 'sperm rope' (Wells and Wells, 1977; Mann et al., 1970). Confirming this view, the detection of chitin occurred only in the membranous external sheath enclosing the whole component, and glycogen is strikingly detected in the sperm mass and heavily detected in the granulated thick internal sheath (Bawab Jr., 1996). In the sperm themselves, it is used for the need of energy (Karlson, 1965; Bell et al., 1976) at the proper time. Glycogen in the above two sites is appropriate since it is useful to be consumed as a nutrient to the future activity of the sperm.

TABLE 12.2 Responses to different stains given by the layers of the different components of the spermatophore—the material does not accept the colour of the stain

Different components of the spermatophore / Stains	Spermatophorie cord					Connecting region		Ejaculatory apparatus					Common components						
	Sperm mass	Sheaths				Central area	Peripheral area	Sac		Horn			Connectives			Tunics			
		Internal (granulated)			External			Central area	Peripheral area	Body	Horn sheath	Striated sheath	Internal (axis)		External	Internal tunic (frothy)	External tunic		
		Oval granules	Granules	Subperiphery									Inner layer	Outer layer			Inner layer	Middle layer	Outer layer
1. Ehrlich's acid haematoxylin and eosin (Pantin, 1948)	Faint red	Violet	Faint red	Red	Red	Faint red	Faint red	Violet	Faint red	Violet	Faint red	Faint red	Violet	Violet	Violet	Violet	Red	Red	Red
2. Heidenhain's iron Haematoxylin (Pantin, 1948)	–	Blue-black	–	Blue-black	Blue-black	–	–	Blue-black	–	Blue-black	Blue-black	–	Blue-black	Blue-black	Blue-black	–	–	–	–
3. Safranin and light green (Lee, 1950)	Green	Pink	Faint red	Pink	Pink	Green / Reddish green	Green / Reddish green	Pink	Orange	Pink	Green	Orange	Red	Faint red	Red	Faint green	Red	Red	Red
4. Mallory's triple (Pantin, 1948)	Deep blue	Red	Orange	Reddish orange	Reddish orange	Red	Red	Red	Faint blue	Red	Red	Faint blue	Red	Red	Red	Faint blue	Blue	Blue	Blue
5. Heidenhain's Azan (Pantin, 1948)	Faint red	Red	Faint red	Red	Red	Bluish purple	Bluish purple	Red	Faint red	Red	Faint blue	Faint red	Bluish purple	Bluish purple	Bluish purple	Faint blue	Deep blue	Deep blue	Deep Blue
6. Masson's trichrome (Pantin, 1948)	Green	Blackish red	Orange	Red	Red	Green	Red	Blackish red	Faint green	Blackish red	Faint green	Faint green	Reddish green	Reddish green	Reddish green	Faint green	Green	Green	Green
7. Weigert's haematoxylin (Mahoney, 1973)	Red	Blue-black	Faint red	Blue-black	Blue-black	Yellow	Yellow	Yellow	Blue-black	Yellow	Yellow	Blue-black	Red	Red	Red		Yellow	Yellow	Blue-black

–=the material does not accept the colour of the stain.

All the stains are according to Pantin (1948), except safranin and light green according to Lee (1950) and Weigert's haematoxylin according to Mahoney (1973).

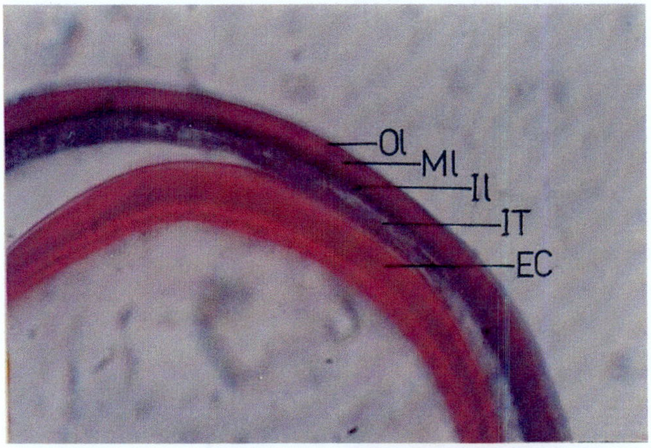

FIG. 12.14 Photomicrograph of an oil immersion magnification of a T.S. through the connecting region of the spermatophore showing the structure of the tunics, Heidenhain's Azan. 1500× *(from Bawab Jr., 1996)*. EC=external connective, IT=internal tunic, Il, Ml and Ol=inner, middle and outer layers of external tunic.

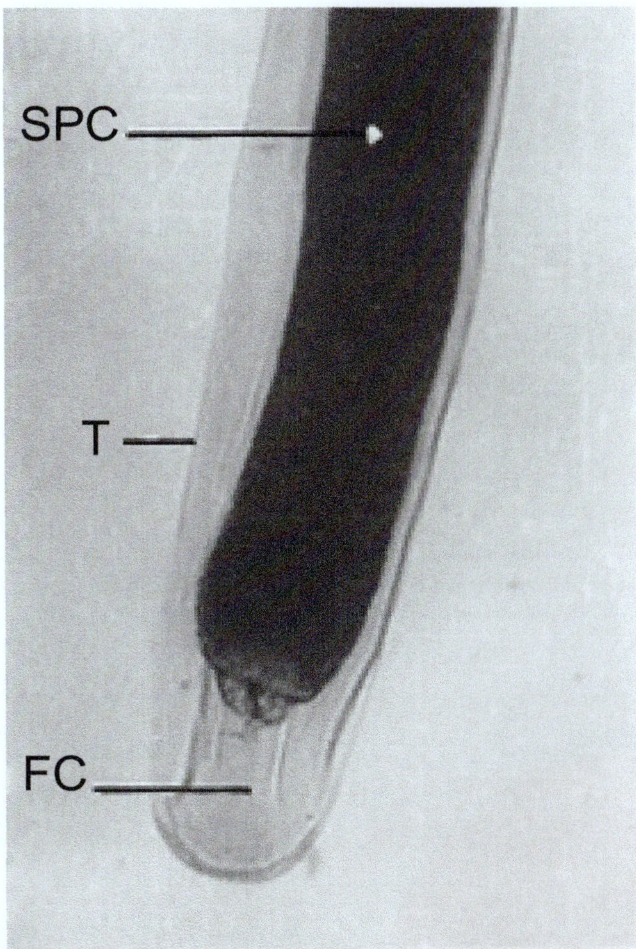

FIG. 12.15 Photomicrograph of the posterior end of the spermatophore of *S. officinalis*. 200× *(from Ismail, 1981)*. FC=free chamber; SPC=spermatophoric cord; T=tunic.

Ejaculatory apparatus

This is the second essential component of the spermatophore. It is the terminal sixth portion of the length of the spermatophore (Fig. 12.18). A distinct transverse groove divides it into two fairly equal halves. The posterior is the **sac** that just follows the CR, while the anterior is the **horn** (Fig. 12.12B). Almost all the previous investigators agreed with the term horn. The term sac was encountered in the work of Blancquaert (1925) and Fort (1937) following the nomenclature proposed by Milne-Edwards (1842). The term 'cement body' is synonymous to the term 'sac', which was proposed by Drew (1919). The body of the sac of *S. officinalis* occupies most of the area, surrounded by the connectives and the tunics outermost, i.e. having the same diameter as the SPC (Fig. 12.19A). No specific sheaths surround the sac, but a very narrow peripheral area is sharply distinct and so the distinction of the sac itself from the outer connectives (Fig. 12.19A and B). At the portion of the groove, the sac narrows slightly for the horn to begin. The horn, on the other hand, begins with the groove, and elongates with gradual decrease in its diameter for about one-third of its extension, the **basal horn**. It then decreases abruptly, becoming a tube-like horn, the **narrow horn** (Fig. 12.20). The latter ends anteriorly by coiling on itself, more than one coil forming the **twist** just underneath the cap (Fig. 12.12B). It appears that the spermatophore of *S. officinalis* is unique within encountered cephalopods, in having this twist. Tompsett (1939) described the twist but not the cap, considering that the 'outer cover' forms the thread only (see later).

To keep the uniformity of the diameter of the whole spermatophore, the horn attains its own successive sheaths, each at the proper level. Accordingly, it has a more complex picture; no one level describes the body and/or its sheaths. Again, to facilitate understanding, only two horn sheaths are concerned, each originates at a definite level. The outer one (first one), just underneath the connectives, begins to ensheath the horn from the end part of the sac (Fig. 12.21A). Critically, it is interpreted as if it is a continuation of the distinctive peripheral area of the sac and so strengthens the component as a whole. This sheath is broad, and well distinct by its natural striation, hence its name, **striated sheath** (Fig. 12.20). Similarly, but in more distinguishable manner, the peripheral but limited area of the basal horn is distinct. Just beyond the end of this basal horn, its distinct area continues as a sheath enclosing the narrow horn inner (Figs 12.12B and 12.21B) to the striated one. It is referred to as a **horn sheath**. These two sheaths, the striated and horn sheaths, are the most peculiar and are always prominent. Some other sheaths are detectable in between. Their affinity to stains is variable with the level and with the degree of change(s) that occur within Needham's sac. Bawab and Ismail (1984a) described and differentiated histochemically five sheaths surrounding this narrow horn; two of them are

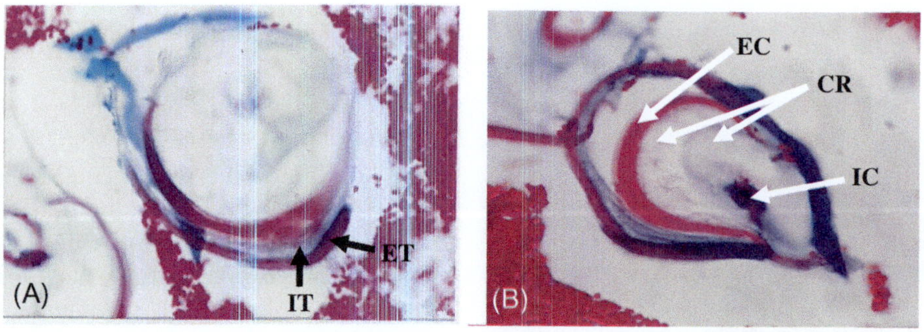

FIG. 12.16 Photomicrographs of T.S. of the spermatophore through the connecting region. 600×; (A) proper T.S.; (B) oblique one *(from Bawab Jr., 1996)*. CR = connecting region, EC = external connective, ET = external tunic, IC = internal connective, IT = internal tunic.

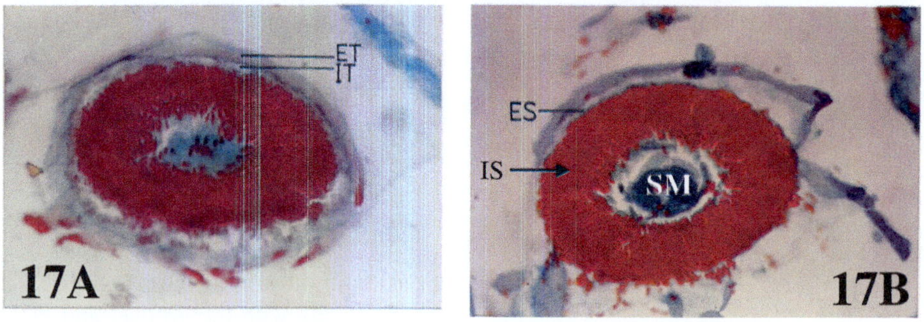

FIG. 12.17 Photomicrographs of T.S. of the spermatophore through the spermatophoric cord showing the sheaths of the latter and the differentiation of the tunics. 600× *(from Bawab Jr., 1996)*; (A) safranin and light green; (B) Mallory's triple stain. ES = external sheath, ET = external tunic, IS = internal sheath, IT = internal tunic, SM = sperm mass.

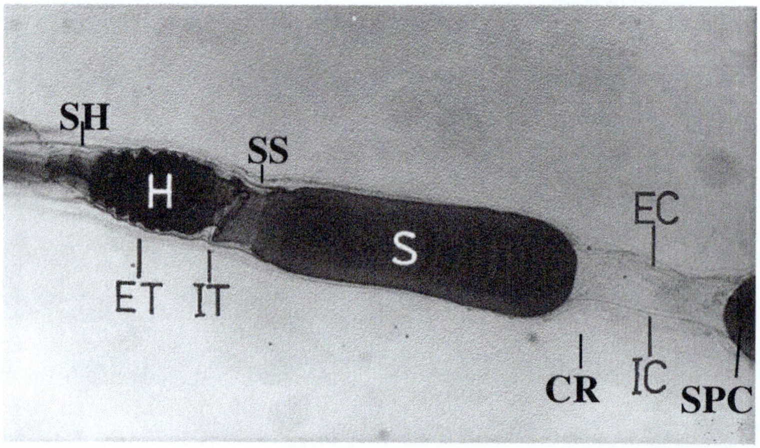

FIG. 12.18 Photomicrograph of a whole mount spermatophore stained by light green showing regions of the ejaculatory apparatus. 200× *(from Bawab Jr., 1996)*. CR = connecting region, H = horn of EA, S = sac of EA, SH = horn sheath, SS = striated sheath. Other lettering as above.

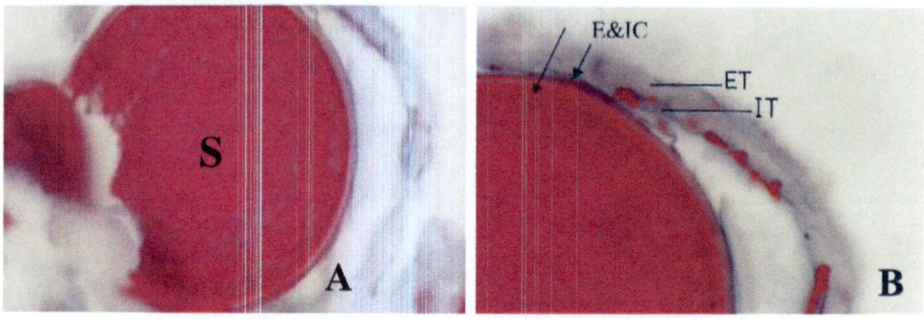

FIG. 12.19 Photomicrographs of T.C.s of the spermatophore through the sac of the EA showing histological pattern. (A) Whole section. 600×; (B) oil immersion magnification of a portion of the section. 1500×. Stain, safranin and light green *(from Bawab Jr., 1996)*. The *arrow* indicates the peripheral area of the sac. Lettering as before.

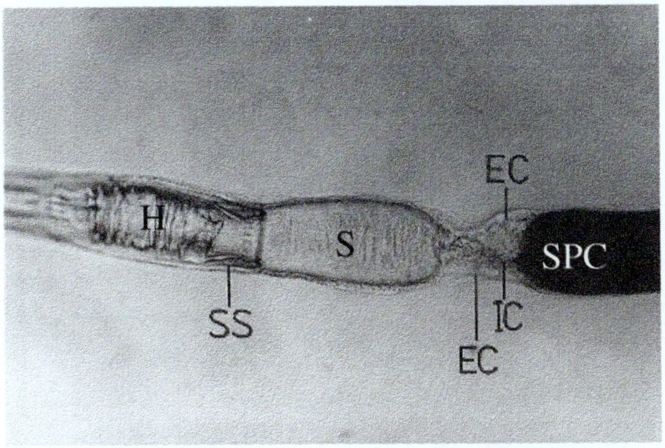

FIG. 12.20 Photomicrograph of an unstained cleared whole mount of the spermatophore. 200×. Focus adjusted to clarify external, internal connectives and striated sheath of the EA *(from Bawab Jr., 1996)*. EC = external connective, IC = internal connective, SS = striated sheath.

the upper ones. Tompsett (1939) mentioned, without any further information, that the horn has three sheaths. An opposite, but very early illustration (Marchand, 1913) shows an inner and an outer membrane surrounding the 'trompe' (= horn of the EA) of the spermatophore of the same species. That the horn is ensheathed with about five sheaths may be the most probable view (Bawab Jr., 1996).

Ejaculatory apparatus is a strong component because of its hardening. Chemically, its body is almost proteineous, and so it undergoes the process of self-tanning (Dennell and Malek, 1956). The deep yellowish-brown colour attained by this apparatus is conceivable evidence that it is a hardened component with strength. This strength is aided by the chitinous connectives and horn sheaths, which are also hardened (Bawab Jr., 1996).

Illustrations of Fig. 12.22 summarize the structure of the spermatophore at its different levels according to the histochemical results.

Dehiscence of the spermatophore

The spermatophores described above were isolated from Needham's sac of a live or fresh specimen. They can be kept fresh if immersed in seawater. However, after about 3 h, Bawab and Ismail (1984a) observed some changes that occur in the isolated spermatophore, beginning with the cap and the cap-thread, which is invisible in freshly isolated ones. These changes are presumed to be the same steps undergone by each spermatophore to dehisce inside the bursa copulatris of the female (Fig. 12.2). The steps of the artificial dehiscence are as follows (Fig. 12.23A–C):

1. The cap thread unravels and elongates, then becomes visible.
2. The connecting region extends gradually.
3. Meanwhile, the coils of the twist partially unravel.
4. The cap suddenly and vigorously vibrates, then breaks at its base.
5. Instantaneously, the EA ejects outside the tunics through the break of the cap while dragging the connectives. Still the EA connects the spermatophore at its twist.
   The last two steps—the break of the cap and the emergence of the EA—cannot be photographed because they occur instantaneously. (Video recording was not available.)
6. The connectives, in turn trail gradually and smoothly the SPC, leaving the tunics unaltered and looks like the exuvia of a moulted animal. Contrary to the ejection of the EA, that of the SPC takes a considerable time. Then, the spermatophore of *Sepia* is dehisced.

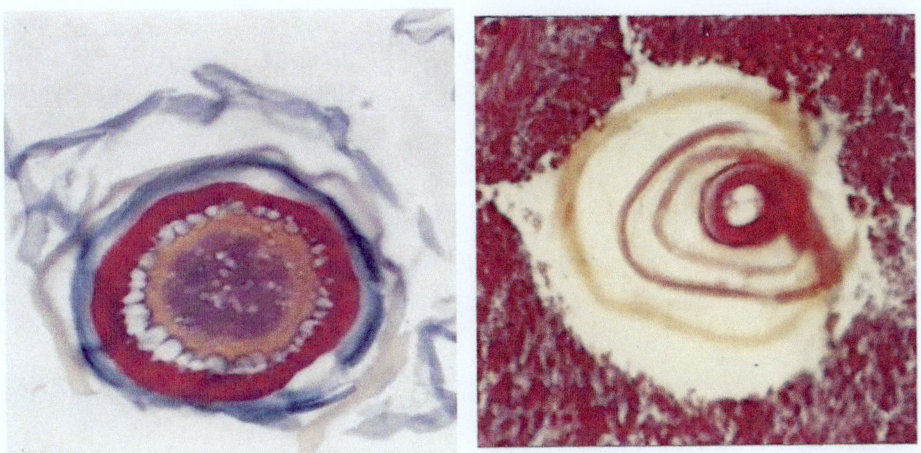

(A) (B)

FIG. 12.21 Photomicrographs of T.C.s of the spermatophore through the horn of the EA. 600×; (A) through the posterior horn, Mallory;s triple stain; (B) through the anterior horn, Mercury-Bromophenol blue. Notice the size of the body of the horn. *(From Bawab Jr. (1996).)*

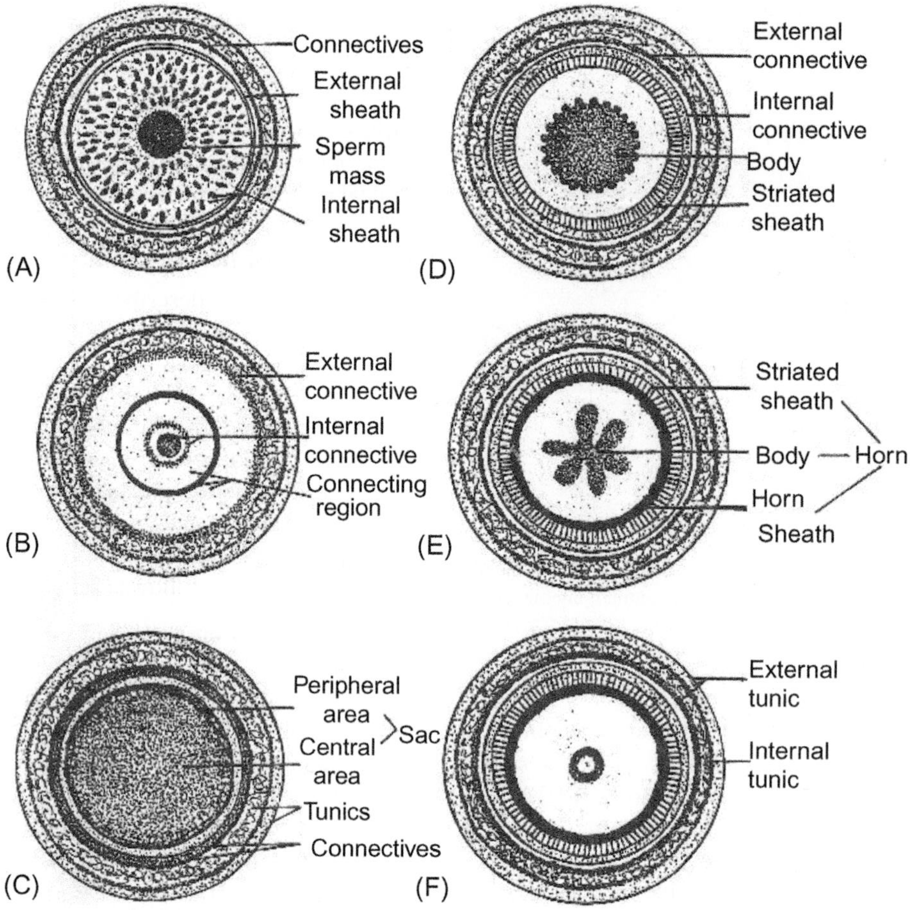

FIG. 12.22 Diagrammatic drawings of T.S.s at different levels of the spermatophore of *S. officinalis*, not to scale *(from Bawab Jr., 1996, modified from Bawab and Ismail, 1984a)*. (A) spermatophoric cord; (B) connecting region; (C) the sac of EA; (D–F) posterior, middle and anterior horn of EA.

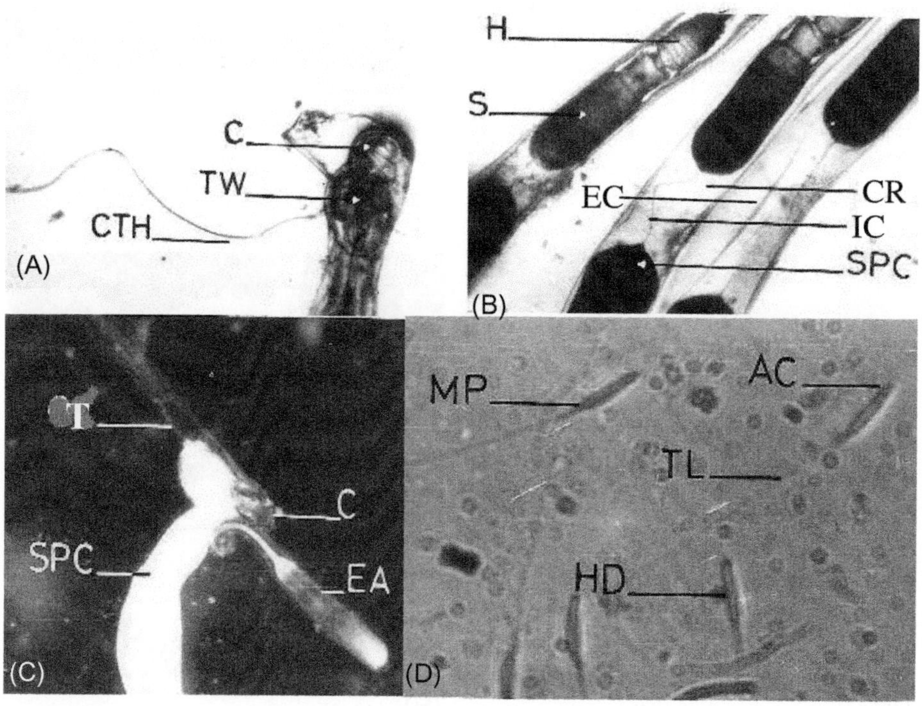

FIG. 12.23 Photomicrographs showing the successive stages of the artificial dehiscence of the fully formed spermatophore. (A) the cap thread becomes visible. 150×; (B) shows gradual extension of both the connecting region and connectives. 150×; (C) shows the gradual emergence of the spermatophoric cord which begins swelling. 50×; (D) oil immersion magnification of a smear of the dehisced SPC showing the structure of the spermatozoon. 1500×. *(From Bawab, F.M., Ismail, S.Z., 1984a.)*

7. Within a few hours, the SPC swells up due to imbibition of water. Its shape changes to a mass of hyaline substance, a smear of which shows that it embeds innumerable very active spermatozoa. It is interesting to note that these sperm obtained by the artificial dehiscence survive for several days, under suitable conditions, i.e. the sperm must be embedded in their hyaline substance, which is immersed in seawater.

A paper presented in the Scientific Council Meeting (September 1984) in a special session on squids (Coelho and O'Dor, 1984) was available after observing the results of the above experiment. These authors recorded that evidence presented that the male *Illex illecebrosus* can expel sperm independent of the copulation. Along with that, they presented scanned micrographs of biflagellate and tailless spherical spermatozoa.

1. The samples of sperm occasionally obtained from the surface of the water of the tank containing squid, to be transferred to the pool tank, as well as from the seawater of the latter itself, were in the opinion of the author, the substance of the spermatophoric cord of dehisced spermatophores. The male squid expelled spermatophores under whatever conditions. They artificially dehisced (i.e. the dehiscence did not occur in the natural site) according to the experiment mentioned above, page 19 (Bawab and Ismail, 1984a).

2. The cephalopod sperm are long tailed like those of vertebrates, which maybe biflagellate but this needs more thorough investigation. It must be considered that the samples of sperm of the above information were obtained from contaminated seawater.

Spermatozoa of *Sepia officinalis*

The cephalopod spermatozoon is comparatively large and almost closely similar to those of vertebrates. A great variety of cephalopod spermatozoa were extensively described by Fretter and Graham (1964). Franzén (1955, 1956, 1967) concluded that the morphology of the sperm of Mollusca has a certain relation to the biology of fertilization. Accordingly, he considered it as a classificatory point among this group of animals. In spite of this information, none concerned that of *Sepia*, even the work of Tompsett (1939), which was essentially interested in the study of the biology of this animal. Since the sperm constitute the principal component of the spermatophore, its description is in the appropriate position to complete the picture of that of the spermatophore. The smear of the gelatinous substance of the dehisced spermatophore was vitally stained by brilliant cresyl blue. The sperm embedded in this substance are elongated fully grown ones (Fig. 12.23D). Their essential parts are well differentiated. The head is rod-shaped, measures about 6.7 µm, and ends anteriorly with a cap-like structure, the acrosome. The head is densely stained, well differentiated from the following middle piece, which measures 2 µm. The tail, the longest component of the spermatozoon, is about six times the length of the head.

Formation of the spermatophore

From the above accounts, it appears that sperm are the essential product to be kept and protected live but inactive for a definite time, hence their encapsulation in spermatophores. The production of sperm becomes the first step in the formation of such elements.

Role of the testis (spermatogenesis)

Externally, the milky white colour indicates the maturity of the testis, which deepens gradually with the gradual increase in maturity. Histologically, Roosen-Runge (1977) recorded that he was not able to find any comprehensive accounts of the testicular histology and spermatogenesis of cephalopods. He mentioned that Thesing (1904) did not specify the reasons why he found *Loligo* and *Sepia* as unfavourable animals for the research. However, Bawab and Ismail (1984b) studied the histology and not the cytology or the molecular of the testis of *S. officinalis* (Fig. 12.24). This testis is ensheathed within three sheaths (Fig. 12.4A); the connective tissue of the testicular one is invaded with muscle fibres and blood vessels (Fig. 12.24A). The testis is too large to be prepared, so it was cut transversely into portions, well marked to keep its seriality. A histological pattern of this testis appears as any testicular tissue. The **immature follicle** (T.S. of seminiferous tubule) is blocked with the germinal epithelium and hence lacks a lumen (Fig. 12.24B). The follicles are closely applied to each other that one cannot define the contour of each. This means that the wall of each, the **membrana propria** (Fig. 12.24C and G), is greatly compressed and is detected only with the beginning of maturity. This beginning is determined by the differentiation of the germinal epithelium (Fig. 12.24C and D). One or two layers (at most) of more compact large cubical and densely stained cells constitute the lining of the membrana propria. These are the **peripheral germinal epithelium** (PGE) (Fig. 12.24D), which are well distinct from those forming the following several layers, the **essential germinal epithelium** (EGE) (Fig. 12.24B and D). The latter are also large, but rounded and faintly stained with diffused nuclei (Fig. 12.24D). In the current's author opinion, these are true **spermatogonia** in the multiplication phase; it is therefore the beginning of the process of spermatogenesis. At this level of examination, the follicle is still without lumen. Gradually in the centre of the follicle, small group of cells appear with very active nuclei, which are in different stages of division. Their number increases on the expense of the EGE. Bawab and Ismail (1984b) referred to these cells

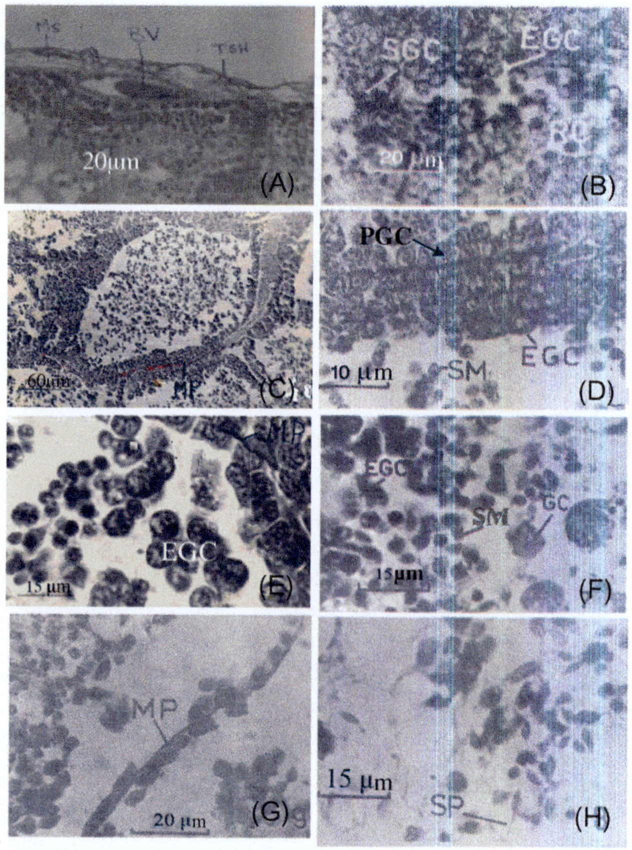

FIG. 12.24 T.S.s of the testis of *S. officinalis* showing its histological structure and the process of spermatogenesis *(from Bawab and Ismail, 1984b)*. (A) showing the nondefined testicular follicles. TSH = testicular sheath, MS = muscles, BV = blood vessel; (B) central portion of a follicle showing different types of cells. EGC = essential germ cell, RC = round cell, SGC = specialized germ cell; (C) mature follicle with central cavity and the membrana propria MP is obvious; (D) portion of the wall of the follicle in (C) showing the process of spermiation. PGC = peripheral germ cell, SM = spermatid; (E) showing a late stage of a mature follicle with complete proliferation of the EGCs; (F) showing the contents of the lumen of a mature follicle. GC = giant cell; (G) portions of neighbouring walls of fully mature follicles with degenerating germinal epithelium; (H) the contents of a fully mature follicle showing the process of spermeogenesis, rounded cells are spermatids, tailed and tailless, oval cells are metamorphosing sperm.

as **specialized germ cells** (SGC). These are **1^{ry} and/or 2^{ry} spermatocytes** (Fig. 12.24B). It is difficult to distinguish spermatogonia from 1^{ry} germ cells or from 1^{ry} spermatocytes (Giese and Pearse, 1974). Meanwhile, relatively large rounded cells with nuclei that occupy the whole cell and are so densely stained appear individually or in clusters scattered in between these SGCs. Bawab and Ismail (1984b) referred to these cells as **rounded cells** (Fig. 12.24B). At this stage of maturation, the follicle is still blocked with cells, but of different types. Gradually some of the SGCs (2^{ry} spermatocytes) give rise to very small ones with rounded or oval, densely stained nuclei. They no longer constitute tissue in the follicle but are released, causing the appearance of the central lumen. These are **spermatids** (Fig. 12.24D and F) and hence the process of **spermateolosis** begins. From this stage onwards, the follicle has different picture. The connective tissue constituting the membrana propria becomes well detectable and gradually thickens. In addition, most of the EGE cells proliferate and scatter in between the contents of the lumen (Fig. 12.24E). The latter contains stages of spermatids (those with round and oval nuclei) and metamorphosing spermatozoa (oval tailless and oval short-tailed cells) (Fig. 12.24H). The SGC (1^{ry} and 2^{ry} spermatocytes) and the rounded cells disappear completely while other cells, **giant cells** having a phagocytic appearance, appear clearly (Fig. 12.24F). Compared to the proliferated cells, they are few in number, large in size and with a different affinity to stains. Each cell contains several vacuoles in addition to some granules. They are always observed accompanying the metamorphosing spermatozoa.

In 2007, molecular studies were carried out by Martínez-Soler et al. They studied transition of nuclear proteins and chromatin structure in 'spermiogenesis' of *S. officinalis*. They showed that the pattern of spermiogenic chromatin condensation consists of a phase of 'pattering' and a phase of 'condensation', two phases that they described in detail. Some years earlier, Wouters-Tyrou et al. (1988) isolated two highly basic proteins from testis chromatin and from sperm nuclei during cuttlefish (*S. officinalis*) spermiogenesis. The first one (Protein T) is rich in arginine and is phosphorated in different levels. The second (Protein Sp) is a typical protamine that appears in a late stage of spermiogenesis and constitutes the major basic nuclear component in mature spermatozoa. In the latter study, the authors showed that protein T could be a precursor for protein Sp.

During the end stages of maturity, the membrana propria of the follicle of *S. officinalis* reaches its maximum thickness, while the PGE is on its way to disappear completely (Fig. 12.24G) and the wide lumen contains developing sperm with some debris in between. However, the highly immature and highly mature follicles can never be found together in one section. It is appropriate to mention that Gabr et al. (1998) studied maturation and fecundity of commercial cuttlefish, *S. pharaonis* and *S. dollfusi*, in the Suez Canal. They found that the maturation process for both species is similar morphologically and histologically. The few figures they presented in their work are similar to some presented in the current work on *S. officinalis*, but it must be noted that the whole processes of spermatogenesis and spermiogenesis of this latter species were described above. However, this may provide an investigation that these processes are the same within the genus or maybe within the cephalopods, in general.

On the other hand, this process of spermatogenesis differs from that of higher animals and even from some other molluscs such as that of the desert snail *Eremina ehrenbergi* (Bawab et al., 1992a) in two aspects: the nourishment of the developing spermatozoa, and their discharge to the following region. Concerning the **1st** difference, the general mode of nourishing the developing sperm is the **Sertoli cell** found in higher animals (Bloom and Fawcett, 1976) and in *Eremina ehrenbergi* (Bawab et al., 1992a). Bloom and Fawcett described this cell not only as nourishing, but also as a secretory and phagocytic cell. Bawab and Ismail (1984b) interpreted that the proliferated EGE nourishes the developing sperm in *S. officinalis*; the rounded cells may be hormonal, i.e. secretory cells; and the giant cells are phagocytic cells. The **2nd** difference is the discharge of the testicular product. It was mentioned above that the testis of cephalopods is not directly connected to its gonoduct and it has a pore on its surface. The thorough histological studies proved that all the seminiferous tubules of the testis of *S. officinalis* are oriented towards its centre (Fig. 12.25A). This picture differs from that recorded for *Octopus bimaculoides* that they are oriented towards the pore in a radiation pattern (Peterson, 1959). Thus in *Sepia*, the central follicles are always at a stage of more maturity. Spermiation takes place centrally in this position. The epithelial walls of the portions of the seminiferous tubules at this position decrease to the minimum thickness or may diminish altogether. They easily disintegrate, forming a continuous central cavity (Fig. 12.25B and C) in which the testicular product can be collected. The testis discharges this product through a definite duct connecting this central cavity with the testicular pore.

Testicular duct. In *S. officinalis*, this duct can be easily differentiated from the neighbouring follicles by its flattened outline and wide lumen. It is bounded by a thick layer of connective tissue comparable to the membrana propria. It is lined by an irregular layer of simple columnar cells, and thickens in parts and ingrows in others forming folds of different sizes and directions (Fig. 12.25D and E). It is interesting to observe numerous goblet cells (Fig. 12.25F–H) as if the epithelial lining consists only of these cells. They are in different stages of activity. They secrete mucin in the form of globules: secretion1 (Fig. 12.25F). However, at the periphery of the testis, the goblet cells decrease greatly in number and so the epithelium changes its activity. It secretes a fluid substance: secretion2 (Fig. 12.25G), which differs completely from the globules. It is appropriate to mention that the nature and affinity to stains of these substances secreted by this testicular duct (Bawab and Ismail, 1984b) are almost the same reported for the sperm-embedded substance of the SPC of the spermatophore of the same species (Bawab Jr., 1996).

Bawab and Ismail performed two experiments to prove this pathway of discharge of the testicular product.

The **first experiment** considered two factors: one is that the testicular pore opens in a depression found just facing the opening of the PVD (Fig. 12.4A); the second is the force created due to the rhythmic contractions of the testis, which can be clearly observed in the living specimens beginning at the posterior apex of this testis. They spread all over the surface of the latter and are directed towards its anterior edge to meet in the anterior septa. This action squeezes the testis, and hence presses, forcing its contents to be discharged through the pore. According to these two factors, a live male specimen was dissected in seawater. Only the ink sac was isolated to expose the reproductive system. The testis was injected with the vital stain, neutral red, at a point far away from the pore. Immediately, a strong red current of stain was observed discharging through the pore taking its way to the straight portion of the PVD, which also became red. Therefore, the testis minimizes the loss of the product in the gonocoele.

The **second experiment** was to expose the testicular duct. A fresh testis was cut transversely and a longitudinal incision was made in the lower half exposing the testicular duct inside, connecting the surface of the half to the pore. In addition, this duct was further illustrated by constructing serial sections (Fig. 12.26). It is obliquely situated since the pore is posterior and superficial in position. Therefore, this duct is naturally efficient to drain mechanically the product of the testis to the gonocoele.

In this, **the role of the testis in the formation of the spermatophore is limited to the production of sperm in a relatively fully grown state embedded in a suitable substance**. The sperm are live and motile.

Role of the PVD

The sperm-embedded substance is immediately conveyed from the gonocoele to the straight PVD. The latter is lined mainly by simple columnar epithelium enclosing a very wide lumen (Fig. 12.27A). Muscle fibres are widely distributed in its connective tissue coat. The PVD is not merely a conducting tubule, but it must be secretory in nature (Fig. 12.27B) due to the presence of a massive mucous substance. Therefore, it further assists in:

1. Nourishing the growing spermatozoa to complete their maturation
2. Coalescing the individually separated sperm; and
3. Furnishing suitable substances to convert them into a somewhat coherent sperm mass (Bawab and Ismail, 1981b)

All these functions are performed through three successive phases of activity of the PVD. The product of this activity first forms a heterogeneous mass (Fig. 12.27C) inside the coiled part of vas deferens. It is then converted into a homogeneous gelatinous mass (Figs 12.5 and 12.27D). A smear of the latter shows that it embeds numerous fully mature spermatozoa as those smeared from the completed spermatophores. Therefore, **the essential role of the PVD in the**

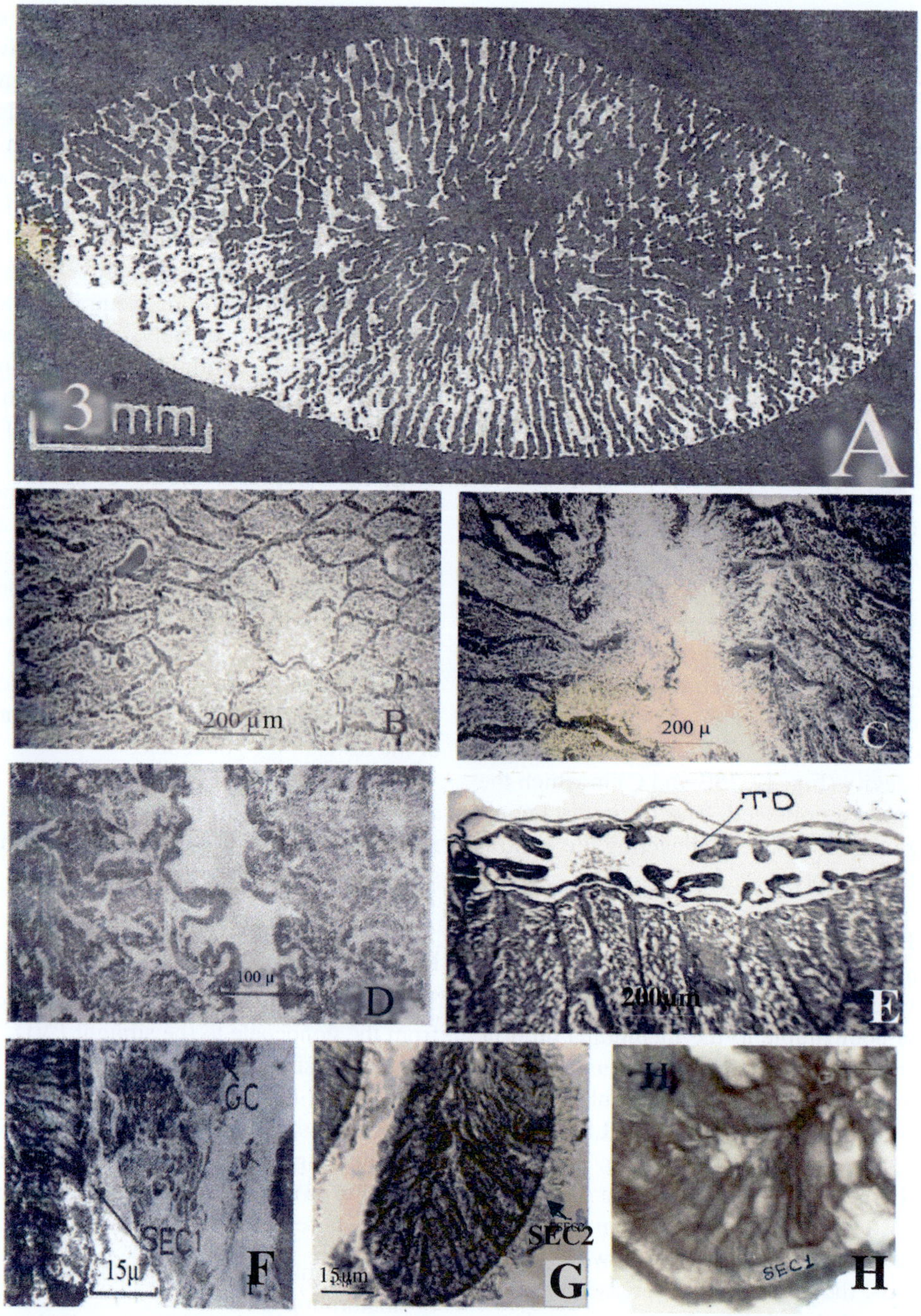

FIG. 12.25 T.S.s of the same testis showing the formation and structure of the testicular duct *(from Bawab and Ismail, 1984b)*. (A) W.M. of a transverse section of the testis showing the orientation of the seminiferous tubules with respect to the central cavity; (B) initial formation of the central cavity; (C) the central cavity; (D) T.S. through the posterior half of the testis showing the testicular duct inside; and (E) the testicular duct at the periphery; (F) eccrine extrusion of globular substance SEC1, notice the giant cell GC; (G) eccrine extrusion of the fluidy substance SEC2; (H) oil immersion magnification of the secretory Goblet cells. 1000×.

formation of the spermatophore is to complete maturation of the sperm and to convert the flow of sperm coming from the testis into a coherent mass, analogous to the semen of other animals. It not only preserves the sperm, but also keeps them live though inactive for some time.

Formerly, the epithelial cells of the PVD were described as ciliated ones. Owing to their intense secretory activity and the mode of extrusion of their product, it is appropriate to describe them as having a **brush border** or **microvilli**.

Role of the MVD

On the other hand, the glandular appearance of the MVD supports the view that it is the device manufacturing the

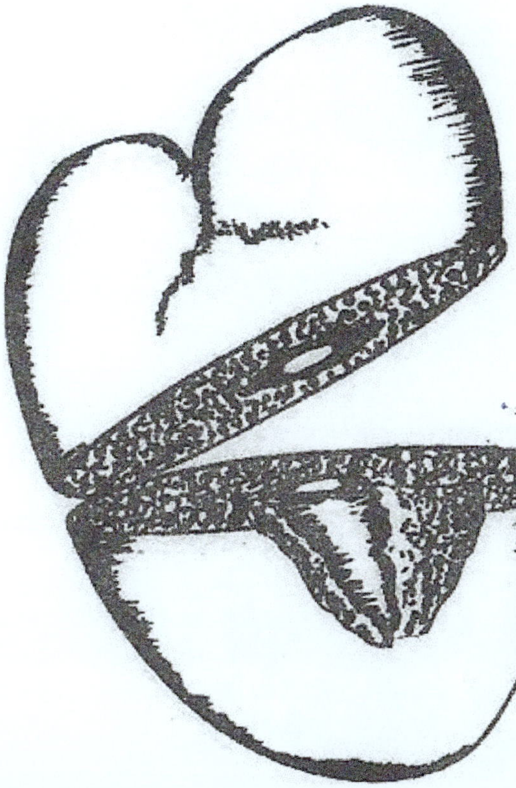

FIG. 12.26 Diagrammatic drawing of the testis cut at its middle showing the central cavity. The posterior half cut longitudinally showing the passage of the **Testicular Duct**. It was corrected by illustrated constructing serial sections. *(From Bawab and Ismail (1984b).)*

spermatophores. Each of its four parts is designed to perform a definite function of this manufacture, i.e. the form and initial portion of part I are equipped to receive the coherent sperm mass, which merges smoothly from the PVD to divide it into the characteristic size of the sperm masses of the species. Part II ensheathes each to be a spermatophoric cord. A tedious isolation of one of the latter from the passage between parts II and III (Ismail, 1981) confirms this visualization. Part III is specialized in the formation of the EA (Bawab, unpublished observation). The formation of the spermatophore of *S. officinalis* is still an ongoing line of work for Bawab and her colleagues and/or juniors. The initial portion of part IV may form the connectives and permit each SPC to connect and trail an EA. Spermatophores isolated from the tubular portion of part IV of the MVD lack only the tunics (Ismail, 1981).

Role of the DVD

The above information was later confirmed by Bawab Jr. (1996), who added that the DVD, in general, together with its distal gland, in particular, is definitely involved in the formation of the tunics. The deposition of the substance forming these tunics is well observed within the lumen of the DVD attached to Needham's sac (Fig. 12.28). Therefore, the distal gland becomes far analogous to the prostate gland of higher animals. The prostate gland in these animals shares in the formation of the fluid and nutrient material in the semen (Keele and Neil, 1973). This is an acceptable view to contradict the term 'prostate gland' (Cuvier, 1817; Mann, 1963), which was previously applied to refer to the present distal gland.

Role of Needham's sac

Ismail (1981) and Bawab and Ismail (1981a,b) studied spermatophores isolated from the DVD to be compared with those isolated from Needham's sac. The former are always soft and much longer than the latter. This means that they are completely formed in the DVD, but not fully formed, i.e. they do not have the finishing characteristics of the spermatophore. The smooth continuation of the DVD to be a spacious sac of Needham's organ facilitates the smooth entrance of those soft elements. The interior of this sac is strikingly constructed to offer these elements regular direction. Just at the entrance, a broadened septum (typhlosole) arises properly from the wall of the sac to face the coming spermatophores straight away (Bawab Jr., 1996). This septum intersects the lumen of the sac and proceeds forwards through a spiral path along the whole length of the wall. It completes one and half spirals in a forward direction towards the terminal tube. This spiral septum looks like the 'spiral valve' described for the intestine of some vertebrates (Parker and Haswell, 1974). This spiral septum with its festooned free edges and its muscles distributed in its tissues, as well as the muscles of the wall of the sac itself, is greatly involved in the completion of the formation of the spermatophores to:

1. control the entrance of the spermatophores;
2. regulate their direction. Bawab junior observed through the transparent wall of the DVD, the spermatophore with its spermatophoric cord (SPC) directed forwards to the sac, trailing its EA;
3. create peristaltic movements permitting their forward propulsion. Mann in 1963 quantitatively demonstrated a substance including 5-hydroxytryptamine (5-HT) in Needham's sac of *Octopus vulgaris*, with a remarkably high concentration in its terminal tube. It is well known that 5-HT (= serotonin or enteramine) generally increases the tone of smooth muscles wherever it occurs, and stimulates peristalsis in the intestine (Keele and Neil, 1973). The amino-acid tryptophan is obviously detected in considerable amounts as patches in the connective tissue of the spiral septum as well as of the wall of the sac. In addition, the blood supply of this organ contains this amino acid (Bawab Jr., 1996). Tryptophan is the parent substance for the production of 5-HT (Karlson, 1965; Bell et al., 1976;

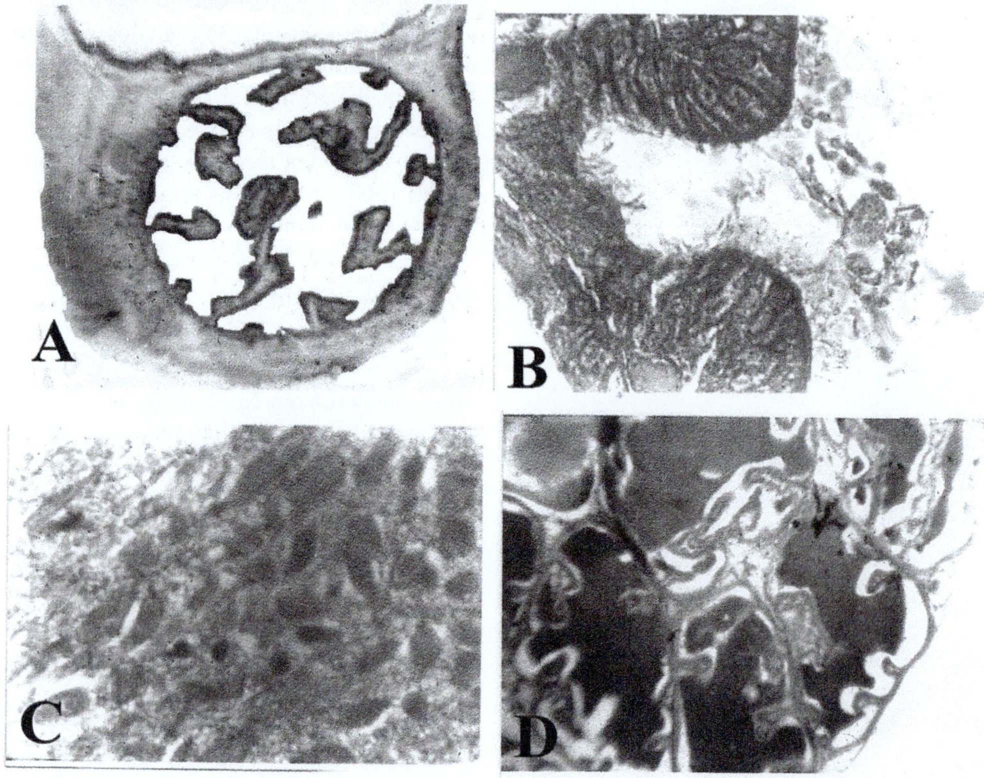

FIG. 12.27 T.S.s of PVD showing its histological structure and function. (A) Through the straight portion. 100×; (B) secretory function of PVD. 600×; (C) heterogeneous secretion in the lumen of the coiled PVD. 1000×; (D) enlarged portion of the coiled PVD showing homogeneous secretion. 100× (see Fig. 12.5). *(From Bawab and Ismail (1981b).)*

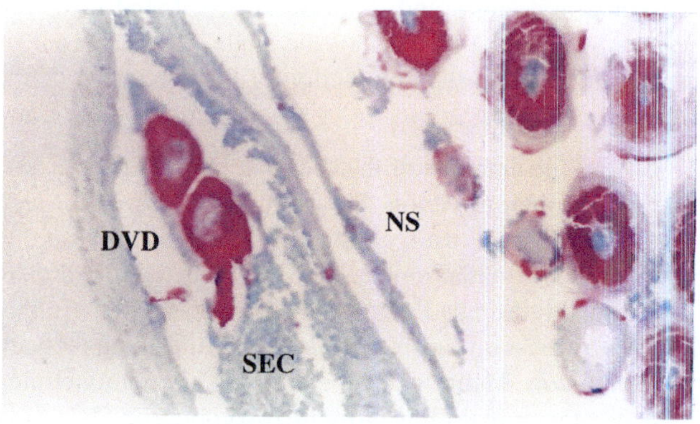

FIG. 12.28 Magnified portion of section of Needham's sac (NS) with the DVD attached to its dorsal surface showing secretory substance (SEC) in DVD, safranin and light green. 100×. *(From Bawab Jr. (1996).)*

Murray et al., 1996). If this tryptophan actually occurs in Needham's sac, especially in the blood, it becomes a strong evidence supporting the probable occurrence of 5-HT which offers a forward smooth regular passage of the spermatophores;

4. the festoons of the septum provide tubular moulds for these elements to be within, until attaining their stability and hardness;
5. then the wonderful arrangement of the spermatophores, side by side with the EA directed downwards, they form a continuous spiral thick band, which appears through the transparent wall of the sac as two or three bundles due to the folding of the sac;
6. stick the fully formed spermatophores in coherent bundles. It was well observed that the spermatophores enter Needham's sac one by one, i.e. separately, but discharge at the appropriate time as coherent bundles. Ismail (1981) presented a photograph showing these bundles in the bursa copulatrix of female *Sepia* (Fig. 12.2). Moreover, at the isolation of these elements from the sac to work on, they are always detached as groups of tightly fastened ones. They also keep this form during their fixation. Mucins were well observed and detected histochemically while discharging merocrine from the epithelial lining and merging in between the spermatophores (Bawab Jr., 1996). These mucins may be the fastening substances keeping the spermatophores as bundles in the same way of the masses or ribbons of eggs of amphibians, e.g. this probable role of Needham's sac opposes an old view (Oordt and Damste, 1939) that the secretion of glands of the hectocotylus possibly helps to attach the spermatophores to each other and to the hectocotylus during their transport into the mantle cavity of the female.

Hardening of the spermatophore

The essential and most important role of Needham's organ in the formation of the spermatophore is as follows.

1. Its equipped histological internal structure gives rise to a proper and suitable site for them to be hardened.
2. Its spiral septum forms tubular moulds for soft spermatophores to prevent curling and/or deformation. In addition, it strikingly arranges these soft elements and keeps them always in a definite pattern until accepting hardening.
3. The secretory function of its epithelial lining and that of its septum. By means of the blood, they may be provided by (at least) the free polyphenol, the enzyme(s) and/or the precursors of the latter (see Table 12.6). These substances are responsible for tanning proteins (hardening process). The phenolic tanning system was simplified and summarized by Smyth (1954) as follows:

$$\text{Protein} + \text{diphenol} \xrightarrow{\text{Polyphenol oxidase}} \text{protein} + \text{quinone}$$
$$= \text{tanned protein}\left(\text{sclerotin}\right)$$

This process (tanning) that renders the structure hard is called sclerotization (Hackman, 1953, 1971).

All the investigators previous to Bawab Jr. (1996), from Hamon (1939a,b) to Austin et al. (1964) to Mann (1963, 1975) and Mann et al. (1966, 1970, 1981), studied the chemistry of the spermatophore of some different cephalopods (*S. officinalis* was one of them) using different techniques. All agreed that the conjugated proteins and carbohydrates, in the form of glycoproteins, mucoproteins and scleroproteins were incorporated in the structure of such studied spermatophores (Table 12.3). Bawab and Ismail (1984a) evidenced experimentally and histochemically the presence of two substances in the spermatophore of *S. officinalis*: chitin (Table 12.3) and lipid. Chitin is an acid mucopolysaccharide substance belonging to the major group, carbohydrates. The application of the PAS-reaction gives a picture of this spermatophore as if it wholly constitutes this major group (Bawab Jr., 1996). This may be the cause of the old view that 'the spermatophore of Cephalopoda is chitinous structure' (Williams, 1910) and it has remained so since. On the other hand, probably, this is the first to report the presence of lipid in the spermatophore of a cephalopod (Table 12.4). The thorough and detailed histochemical studies (Bawab Jr., 1996) confirmed this view in the case of *S. officinalis*, and specifically confirmed the presence of lipoprotein. She added the presence of the basic amino acid, arginine, and the phenolic and sulphated proteins (sulphydryl, –SH, and bisulphide, –S–S) (Table 12.5). These proteins are structural proteins that accept any component stability and hardness.

Considering this chemical structure, Bawab junior concluded that the spermatophore of *S. officinalis* might be subjected to the phenolic tanning, which causes hardness, and the lipoprotein is the substrate involved. The lipid, which is of steroid type, subsequently undergoes polymerization. This phenolic process can operate enzymically by one of two methods:

1. self-tanning, the tyrosyl residues, in the presence of a phenolase enzyme, are believed to be oxidized in situ to quinones which tan the protein. The natural ambercolour of these elements provides fundamental evidence to the quinone tanning. This cross linkage of the tyrosyl residues of the protein moiety is as that which occurs in the spermatophore of *Penaeus* (Bawab, 1970; Malek and Bawab, 1971). This mode of tanning is essentially different from the;
2. tanning of the protein by the free dihydroxyphenols (according to the system of Smyth mentioned above) such as that which operate in the cuticle of arthropods and are described for the plug of the spermatheca of the crab *Portunus pelagicus* (Bawab and El-Sherief, 1989). In spite of the fact that the free dihydroxyphenol was negatively detected in such a spermatophore (Table 12.6), it can be provided by the blood. On the other hand, the enzyme phenolase was excitably detected with intense presence in the body of the EA (the most hardened component of the spermatophore of *S. officinalis*). However, the intensity of the hardening of this element is greatly beyond that of the phenolic groups of the protein moiety of such spermatophore. Sulphydryl and bisulphide occur in much more intensity (Table 12.5) than the phenolic groups. Hackman (1971) revealed that proteins are stabilized by disulphide bonds formed by a reaction of quinone with sulphydryl groups, i.e. quinone tanning (a part of sclerotization process). This tanning process is the same as that investigated in the exocuticle of the scorpion *Androctonus bicolor* through the formation of disulphide linkages (Hashem, 1973).

Therefore, the hardening and finishing of the spermatophore of *S. officinalis* occur in the end part of the gonoduct; strictly speaking, Needham's sac is the site of hardening and chemical stability of these elements. Cephalopoda is thus similar to Crustacean in which the terminal ampoule is the site of moulding and hardening of the spermatophores (Malek and Bawab, 1971). At the same time, it contradicts the view encountered in textbooks, e.g. *Invertebrate Zoology* (Barnes, 1974), that believes that 'the spermatophore is completed in a hardening gland and then in a finishing gland' and afterwards passes to be stored in Needham's sac.

The female system

The female reproductive system is comparatively simple. However, it consists of five major components: the single

TABLE 12.3 The chemical nature of the layers of the different regions of the spermatophore (carbohydrates tests)

Different regions of the spermatophore / Stains/ references	Material demonstrated	Spermiatophoric cord					Connecting region		Ejaculatory apparatus					Common components						
		Sperm mass	Sheaths				Internal layer	External layer	Sac		Horn			Connectives			Tunics			
			Internal			External			Central area	Peripheral area	Body	Horn sheath	Striated sheath	Internal (axis)		External	Internal tunic (frothy)	External tunic		
			Oval granules	Granules	Subperiphery									Inner layer	Outer layer			Inner layer	Middle layer	Outer layer
1. Periodic acid Schiff reagent (PAS) (Mc Manus; Pearse, 1968).	Mucosubstances	p	ppp	p	p	pp	p	p	p	pp	p	pppp	pp	pp	ppp	ppp	p	ppppp	ppp	ppppp
2. Toluidine blue (Kramer and Windrum; Pearse, 1968).	Mucoproteins and glycoproteins	b	b	–	b	b	–	–	–	V	–	V	V	V	V	b	V	V	V	V
3. Mucicarmine (Mayer's mucicarmine; Lee, 1950).	Mucins	R	RR	R	RRR	RRR	R	R	R	RRRR	R	RRRR	RRRR	RRR	RRR	RRR	RRRR	RRR	R	RRR
4. Alcian blue (Steedman, 1950; Pearse, 1968).	Acid mucopolysaccharides (MPS)	b	bb	b	bbb	bbb	b	b	b	bbbb	b	bbbb	bbbb	bbb	bbb	bbb	bbbb	bbb	b	bbb
5. Alcian blue-PAS (Mowry, 1963; Pearse, 1968)	Acidic and neutral MPS	p	p	p	p	p			P	b	P	b	b	–	–	b	b	P	P	bluish purple
6. Alcian blue after sulphation (Kramer and Windrum, 1954; Pearse, 1968).	Chitin	–	–	–	–	±			b	bbb	b	bbb	bbb	bbb	bbb	bbbb	bbb	–	–	bbb
7. Schultze chlor-zinc iodine reagent (Pearse, 1968).	Chitin								Rv	Rv	Rv		Rv		Rv					
8. Best's carmine (after Best, 1906; Pearse, 1968).	Glycogen	RR	RRRRR	R	RRR	RRR	R	R	RRRR	RRR	RRRR	RRR	RRR	RRR	RRR	RRR	R	RRRRR	RR	RRRRR

b=Blue, P=purple, R=red, Rv=reddish violet, V=violet, and –=the material does not accept the colour of the stain.
The repetition of each symbol indicates intensity of this colouration.

TABLE 12.4 The chemical nature of the layers of the different regions of the spermatophore (lipid tests)

Different regions of the spermatophore / Stains/ references	Material demonstrated	Spermatophoric cord					Connecting region		Ejaculatory apparatus					Common components						
		Sperm mass	Sheaths				Internal layer	External layer	Sac		Horn			Connectives			Tunics			
			Internal			External			Central area	Peripheral area	Body	Horn sheath	Striated sheath	Internal (axis)		External	Internal tunic (frothy)	External tunic		
			Oval granules	Granules	Subperiphery									Inner layer	Outer layer			Inner layer	Middle layer	Outer layer
1. Sudan black B (Mc Manus, 1946; Pearse, 1968).	Gelatin sections	bbb	bb						b	bb	b	bb	bb	bbbb	bb	bbbb	–	bbb	b	bbbbb
	Lipids																			
	Paraffin sections	bbb	bb	–	b	b	–	–	b	bbbb	b	bbbb	bbbb	bb	bb	bbbb	–	bb	b	bbbb
2. Nile blue sulphate 1% (Cain, 1947) and (Pearse, 1968).	Gelation sections		bb						bb	bbbb	bb		bbbb	bbbb	bbbb	bbbb			bbb	
	Neutral and phospholipids																			
	Paraffin sections	b	bb						bbbb	bb	bbbb	r	bb	bb	bb	bb		b	b	b
3. Nile blue sulphate 1% and 0.02% (Cain, 1947) and (Pearse, 1968).	Gelatin sections		bb						bb	bbbb	bb		bbbb	bbbb	bbbb	bbbb			bbb	
	Neutral and phospholipids																			
	Paraffin sections	bb	bb		b	b			bbbb	bbb	bbbb	r	bbb	bbb	bbb	bbb		bb	bb	bb
4. Liebermann-Bürchardt reaction (Pearse, 1968).	Gelatin sections		B		B	B						B		B	B	B				BB
	Sterols cholesterol and cholesterol esters																			
	Paraffin sections		B									B		B	B	B				BBBB
5. Schultz method (after Weber et al.; Pearse, 1968).	Gelatin sections		g		g	g						g		g	g	g				gg
	Cholesterol and cholesterol esters																			
	Paraffin sections		g									g		g	g	g				gggg

b=Blue, B=brown, g=green, r=red, and –=the material does not accept the colour of the stain.
The repetition of each symbol indicates intensity of this colouration.

TABLE 12.5 The chemical nature of the layers of the different regions of the spermatophore (protein tests)

Stains/references \ Different regions of the spermatophore	Material demonstrated	Spermatophoric cord					Connecting region		Ejaculatory apparatus					Common components						
		Sperm mass	Sheaths			External	Internal layer	External layer	Sac		Horn			Connectives			Tunics			
			Internal						Central area	Peripheral area	Body	Horn sheath	Striated sheath	Internal (axis)		External	Internal tunic (frothy)	External tunic		
			Oval granules	Granules	Subperiphery									Inner layer	Outer layer			Inner layer	Middle layer	Filter layer
1. Mercury Bromphenol blue (after Bonhag; Pearse, 1968).	Proteins	b	bb	b	b	bb	b	b	bb	bb	bb	b	b	b	b	–	b	bb	b	bb
2. Aqueous Bromphenol (Mazia et al., 1953).	Basic proteins	b	bb	–	bb	bb	–	–	–	bb	–	bb	bb	bb	bb	–	±	bb	bb	bb
3. Ninhydrin Schiff (Gabe, 1976).	Reactive NH_2 group	–	pp	–	p	p	–	–	pp	pp	pp	p	pp	pp	pp	pppp	–	pppp	pppp	pp
4. Biuret reaction (Sols' technique, 1947; Gabe, 1976).	Peptide linkage (CONH)	–	–	–	–	–	–	–	–	–	–	–	–	–	–	–	–	–	–	–
5. Sakaguchi reaction (Baker's 1947 Modification; Pearse, 1968).	Arginine (basic protein)		0	0	0	0			0	00	0		00	00	00	00	–	00	00	00
6. Xanthoproteic (Pearse, 1968).	Phenyl group (tyrosine, tryptophan and phenyl alanine)	–	–	–	–	0	–	–		0		0	0	0	0		–	–	–	–
7. Millon's reaction (Baker Modification; Pearse, 1968).	Tyrosine		P		P				p	p	p	p	p	p	p	p	–	–	–	p
8. Nitroso-Naphthol method (Udenfriend and Copper, (1952; Stosward, 1963; Pearse, 1968).	Gelatin sections					Gy				Gy		GyGy	Gy	GyGy	GyGy	GyGy	–	–	–	Gy
	Tyrosine																			
	Paraffin sections		±	±	±	Gy				Gy Gy		Gy Gy	Gy Gy	Gy Gy	Gy Gy	Gy Gy	–	–	–	Gy
9. DMAB-Nitrite method (Adams, 1957; Pearse, 1968).	Tryptophan	–	–	–	–	–			b	b	b	–	b	–	–	–	–	b	b	b
10. Tryptophan reaction (Adams, 1960; Pearse, 1968).	Tryptophan	–	–	–	–	–	–	–	–	–	–	–	–	–	–	–	–	–	–	–
11. Ferric ferricyanide method (Chévremont and Frédric, 1943; Pearse, 1968).	–SH Cysteine	–	–	–	–	–	b	b	bb	bbb	bb	bb	bbb	bb	bb	bb	bb	bbb	bbb	bb
12. Performic acid alcian blue method (Adams and Sloper, 1955, 1956; Pearse. 1968).	–S–S Cystine	–	–	–	b	bb	b	b	bb	b	bb	b	b	bbb	bbb	bbb	bb	bbb	bb	bbb

b=Blue, O=orange, p=pink, Gy=greenish yellow, and –=the material does not accept the colour of the stain.
The repetition of each symbol indicates intensity of this colouration. All these reactions were applied on paraffin sections except the nitrosonaphthol method.

TABLE 12.6 The chemical nature of the layers of the different regions of the spermatophore

Different regions of the spermatophore / Stains/ references	Material demonstrated	Spermatophoric cord					Connecting region		Ejaculatory apparatus					Common components						
		Sperm mass	Sheaths				Internal layer	External layer	Sac		Horn			Connectives			Tunics			
			Internal			External			Central area	Peripheral area	Body	Horn sheath	Striated sheath	Internal (axis)		External	Internal tunic (frothy)	External tunic		
			Oval granules	Granules	Subperiphery									Inner layer	Outer layer			Inner layer	Middle layer	Outer layer
1. Argentaffin (Lison, 1960)	Phenols, quinone, polyamines and type of sugar (reducing aromatic substances)	B	BBB	B	BBB	BB	–	–	BBBB	BB	BBBB	BBB	B	BBB	BBB	BBB	–	BBB		BBB
2. Millon's reaction (Baker Modification; Pearse, 1968).	Tyrosine		P		P				P	P	P	P	P	P	P	P	–	–	–	P
3. PAS-reaction (Mc Manus; Pearse, 1968).	Mucosubstances	*P*	*PPP*	*P*	*P*	*PP*	*P*	*P*	*P*	*PP*	*P*	*PPPP*	*PP*	*PP*	*PPP*	*PPP*	*P*	*PPPPP*	*PPP*	*PPPPP*
4. Sudan black B (Mc Manus, 1946; Pearse, 1968).	Gelatin sections	bbb	bb						b	bb	b	bb	bb	bbbb	bb	bbbb	–	bbb	b	bbbbb
	Lipids																			
	Paraffin sections	bbb	bb	–	b	b	–	–	b	bbbb	b	bbbb	bbbb	bb	bb	bbbb	–	bb	b	bbbb
5. Dilute ferric chloride (Lison, 1936; Smyth, 1954).	free O-diphenol (ortho-dihydroxyphenol) gelatin sections	–	–	–	–	–	–	–	–	–	–	–	–	–	–	–	–	–	–	–
6. Catechol (Smyth, 1954)	Polyphenol oxidase Gelatin sections	B							BBBB		BBBB							BB		BB

b=Blue, B=brown, P=pink, *p*=purple, –=the material does not accept the colour of the stain.
The repetition of each symbol indicates intensity of this colouration.

ovary, the single oviduct with thin-walled and glandular portions, the paired nidamental glands, the paired accessory nidamental glands (Fig. 12.3B) and the buccal seminal receptacle (Fig. 12.2). In *Sepia*, which is quite a popular species, the large ovary lies posteriorly in the apex of the mantle cavity, the same site of the testis. It is dorsal to the stomach, and caecum therefore does not need to be removed (Bullough, 1981). The thin-walled, glandular oviduct begins with a ciliated funnel-shaped opening. It proceeds forwards within the coelom on the left of the rectum to the mantle cavity. In the latter, it opens through a slit with large folded lips. On the side of the oviduct in *Loligo*, there is a large oviducal gland that secretes the outer eggshell (Bullough, 1981). There is another view: that in coleoids the oviducal gland produces albumen to coat the eggs (Marshal and Williams, 1975). The paired nidamental glands are prominent as white globular structures on the midline between the bases of the ctenidia. The paired accessory nidamental glands are attached to the anterior portions of the nidamental glands, becoming prominent during maturation with an orange to red colour.

Buccal seminal receptacle. Kumé and Dan (1968) after Soeda (1952) reported that the seminal receptacle could be recognized as white dots on the inner surface of the buccal membrane of *Ommastrephes sloani pacificus*. There are about 30 of these seminal receptacles on the buccal membrane, but in the fresh material, it is difficult to detect them with the naked eye. They have openings on the inner (mouth) side of the membrane, and branch internally; they are sometimes given the name of **fertilization sac**. As these become filled with spermatophores (Fig. 12.2), they take on a white opaque appearance, and form conspicuous protuberances on the outer side of the membrane. Microscopically, the receptacle of *Loligo vulgaris* is a pouch lined with columnar epithelium and goblet cells, which secrete a substance that suppresses the activity of spermatozoa (van Oordt, 1938). The spermatophores are ejaculated into the seminal receptacles during copulation and apparently may be retained there for long periods.

The ovary. During the development of the ovary, e.g. of *Sepia officinalis*, the entire organ increases in size as the stroma and oocytes grow, the latter increase in number as well as in size. During the breeding season, the ovary reaches its maximum size and is composed of groups of oocytes in all stages of development—i.e. the process of oogenesis. These oocytes cluster about a ramifying blood vessel (Fig. 12.3B). It looks like bunches of grapes packed together on a twisted and branched vine (Arnold and Williams-Arnold, 1977). The study of reproductive biology of the same species, *Sepia officinalis* obtained from a lagoon of the Aegean Sea (Önsoy and Salman, 2005) divided maturation of the ovary into five stages. The gonadosomatic indices show the ovary with two peaks, the first in March and the second in June. In another study of maturation, fecundity and seasonality of reproduction of *S. pharaonis* and *S. dollfusi* (Gabr et al., 1998), the authors described the maturation of the female system in general and the process of oogenesis, in specific. They divided the latter process into four stages. Stage I is immature, consisting of growing small follicular oocytes (Fig. 12.29A). This figure, in the view of the current author, is not only so, but it also shows the multiplication of oogonia and small primary oocytes. This view is based on two facts: in invertebrates, primary gonial cells usually go through a limited number of mitotic divisions, usually 2–6; secondly, in practice, primordial germ cells, primary and secondary

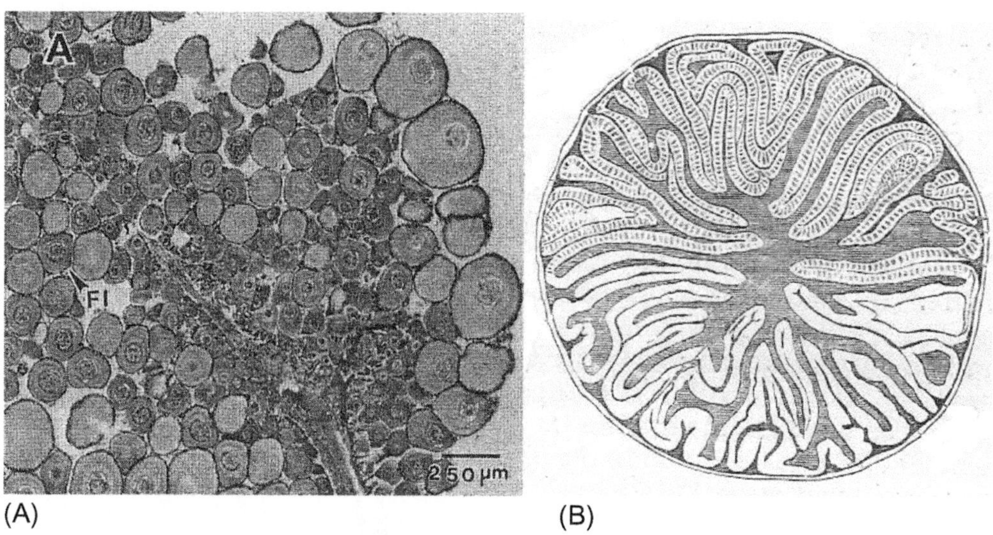

FIG. 12.29 (A) Photomicrograph of portion of T.S. of ovary of *S. pharaonis* showing immature stage, *from Gabr et al. (1998)*; (B) T.S. through mature ovarian oocyte of *S. officinalis* showing epithelial folds inside, *after Lankester from Balfour (1880)*.

gonial cells, and a young primary spermatocyte or oocyte are difficult to distinguish from each other (Giese and Pearse, 1974). In addition, stage III of Gabr et al. consists of advanced follicular oocytes especially that of the maximum size with the maximum penetration of the epithelial folds into the inside of the oocyte. The figure describing the latter oocyte is exactly the same as that of Lankester (1875) (Fig. 12.29B, see later). In spite of the fact that the principal aim of the study of these authors is fishery, but also their results on maturation highlights the significance of the earlier work done by Lankester towards the end of the 19th century.

Each oocyte of any species undergoes two growth phases during oogenesis. The second phase is a process of vitellogenesis. Very early in 1880, Balfour after Lankester (1875) described "the nutrition of the 'eggs' of *Sepia* during their growth." The eggs develop in connective-tissue pouches which give rise to the early double pedunculated capsule, cells of its inner layer soon change into epithelial layer. The region in between the two layers becomes highly vascularized and the epithelial follicle penetrates the cytoplasm, forming folds in a most remarkable manner. The folds increase in number and size with the increase of the size of the egg. When the egg reaches its optimum size, the folds project into, and nearly completely fill up, the body of the 'ovum' (Fig. 12.29B). Each fold is thoroughly supplied with blood vessels. This means that the ovum is surrounded and invaded by a sufficient amount of blood supply. During this stage, when the follicle epithelium fill up the ovum, its cells have the character similar to that of the goblet cells of a mucous membrane, and 'pour' out their 'metamorphosed protoplasm' (deutoplasm, i.e. product of vitellogenesis) into the body of the ovum. Therefore, the process of vitellogenesis in *Sepia* cannot be interpreted as an **autonomous intraoogonial process** (Threadgold, 1976). This type of follicular oocyte and its conspicuous mode of nutrition are unique in invertebrate eggs. It recalls that of a tunicate *Styela plicata*, in which a few cells invade the cytoplasm, forming a discontinuous inner layer. They are called test cells (Fig. 12.10D, Chapter 13). This tunicate follicular oocyte can be considered an intermediate type to that of Cephalopoda.

A lot more recent to that time of Lankester, Arnold and Williams-Arnold (1977) reviewed the studies on this aspect in some decapods, and *Sepia* is one of them. No striking difference was recorded, in spite of the difference in time and hence in the tools used. Rather than the observations obtained from studies on the level of TEM, only the light microscopical studies of Cowden (1968) on the development (oogenesis) of the oocyte of *Loligo brevis* deserve attention. This author divided this process into 10 developmental stages. Indeed, the description of Lankester (considering the time at which it was made) can be applied on these 10 stages and the only figure presented in Balfour (1880) (Fig. 12.29B) is exactly the same as stages VI and VII of Cowden. Both works confirm each other in describing this follicular oocyte, and the follicular epithelium is exceptional in that it invades the interior cytoplasm. The only addition of Cowden is its detailed description of the histology and cytology of the cells. However, Arnold (1971) (the same Arnold mentioned above) preferred to report the view which divides the process of oogenesis in cephalopods into its essential three phases: (1) multiplication phase, in which the folds of the follicular epithelium appear and penetrate into the developing oocyte; (2) growth phase which occurs by active secretion of the follicular folds causing the increase in size of the oocyte and the formation of the chorion. It is well-known that the cephalopod oocyte is arrested at the prophase of the 1^{ry} oocyte. This means that oogenesis stops in the ovary after the oocyte attains its growth and is assumed for fertilization, i.e. maturation is attained during fertilization. (3) The changes occur to the oocyte in the ovary after growth. Regression and disappearance of the folded epithelium complete the structure of the ovum. This was considered as the third phase of oogenesis in the above description.

The fully grown oocyte and/or egg is usually surrounded by various accessory coats, which differ with the species. Basically, all cephalopod eggs are contained in a clear to cloudy opaque chorion. Very early in 1880, Balfour, after Lankester, described 'the ova of Cephalopoda, like those of Gastropoda, are naked, being without a vitelline membrane or chorion'. The egg capsule, which is formed for them in their passage down the oviduct, is perforated in *Sepia* by a micropylar aperture. First of all, one can understand from this description that the egg of *Sepia* has chorion. On Gastropoda, Bawab et al. (1992a,b) discussed the presence of a vitelline membrane and chorion in some species, and presented indications of a probable presence of these two membranes in the oocyte of the stylommatophoran *Eremina ehrenbergi*. Arnold and Williams-Arnold (1977) described ultrastructurally the formation of the chorion of the cephalopod oocyte by the follicular cells. Firstly, it develops as a series of droplets at the interface between the follicular epithelium and the oocyte proper during stage VIII of Cowden (1968). These droplets fuse and eventually make a continuous layer surrounding the oocyte and following the exact contours of the folded epithelium (Fig. 12.30). A small gap occurs in this layer in the region of the future micropyle. A pad of modified cells from the surrounding stroma is present in this area.

The ultrastructural and cytochemical analysis of oogenesis in the squid, *Loligo pealii* (Selman and Arnold, 1977), provided unquestionable evidence detailed with a clear picture of the structure and function of follicular cells, and perhaps of Cephalopoda in general. Discussed summaries of the contributions of this work are as follows:

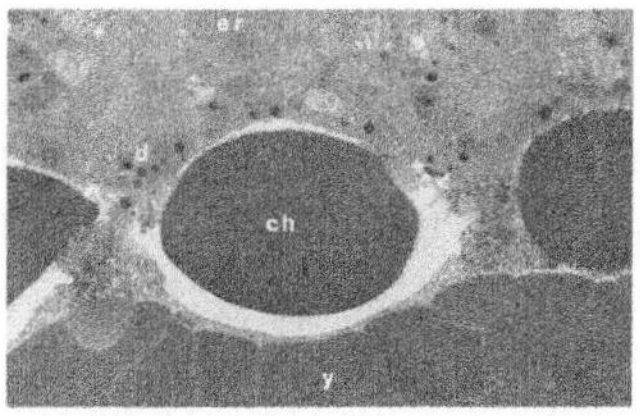

FIG. 12.30 Photomicrograph describing the formation of chorion in *Loligo* oocyte. D = dense droplets in the cytoplasm of the follicular epithelium, ch = Chorion, y = yolk, er = endoplasmic reticulum. *(After Arnold and Williams-Arnold in Giese and Pearse (1977).)*

1. The division of the process of oogenesis into only five stages according to the structure of the follicle cell.
2. The differentiation of the follicular cells, being squamous during stage I, proliferate and change to columnar during stage II, confirming the work of Cowden (1968); they continue proliferation and the cells differentiate into an active secretory syncytium during stage III.
3. Formation of the syncytium and its degeneration, the follicular cells are separated laterally from each other by only a small (~300 Å) intercellular space. During stage III, intercellular bridges connect the cytoplasm of adjacent follicle cells, which are then occasionally occluded by folds and dense material (Arnold, 1974). During the same stage, the lateral membranes separating adjacent follicle cells retract, forming follicular syncytium. Bundles of microtubules lie parallel with the retracting membrane and presumably contribute to the retraction process. After the completion of the process of vitellogenesis and formation of the chorion, the folds of the follicular syncytium are displaced by being pushed to the periphery of the follicle due to the central accumulation of yolk bodies in the oocyte, then degenerate at the end of stage V to permit the shedding of the fully grown oocyte.
4. Detailed ultrastructural description of the cytodifferentiation of the follicle cells during all stages, especially the syncytium and definition of the role of each inclusion, especially that of Golgi apparatus.
5. Vitellogenesis, yolk synthesis in cephalopods, was investigated very early, establishing that the follicular cells themselves synthesize yolk proteins, as described above (Lankester, 1875 and others, e.g. Cowden, 1968). Some other investigators believed that 'exogenous lipoprotein' can be transported from the blood to the oocyte via follicle cells. Regarding a third condition, in *L. pealii*, Selman and Arnold (1977) described the follicular syncytium as the primary site of yolk synthesis. At the beginning of the process during stage IV, the cytoplasm of the oocyte is scant and shows no ultrastructural indication of synthesis. However, the possibility of endogenous yolk synthesis occurs, i.e. the oocyte may also be involved in the formation of yolk. This is indicated by the simultaneous appearance of PAS-positive material in the Golgi complexes of the syncytium and in granules in the oocyte (carbohydrate substance) (Fig. 12.31). In a previous experimental work, Selman and Wallace (1972) evidenced by autoradiograms that the follicle cells were the major site of protein synthesis. They isolated growing oocyte-follicle cells (0.9 mm diameter) of the squid *Loligo pealei* and incubated with ^{2}H-leucine for various lengths of time. The experiment also showed that the material initially synthesized in the follicle cells appeared to be transferred subsequently into peripheral yolk granules within the oocyte. These studies show that radioactive protein precursors are first incorporated into the follicular syncytium, and then found in the yolk at the periphery. Therefore, the protein may make up the major component of yolk material in cephalopods. Another view of protein uptake is that extracellular proteins are incorporated via large, noncoated vesicles. In *L. pealii*, numerous smooth-surfaced vesicles of variable morphology reside in the ooplasm; some appear empty, whereas others contain some flocculent material. The material within these vesicles may subsequently condense and become small yolk bodies (Fig. 12.32). In this, yolk production in this species is a **heterosynthetic process**, while the consistency of yolk itself is a complex of carbohydrates and protein. Gardiner (1972) reported that in view of the chemistry, yolk is broadly classified as a

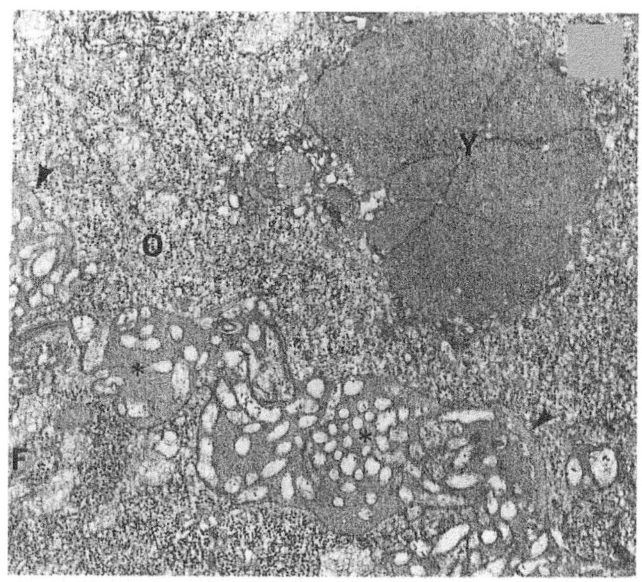

FIG. 12.31 Electron photomicrograph of follicular cells of oocyte of *Loligo pealei* showing yolk bodies in early stage IV (y) accumulated in the ooplasm; * = the extracellular space at the interface of the oocyte (O), F = follicular syncytium, *arrowheads* = extracellular fibres. 11.287×. *(From Selman and Arnold (1977).)*

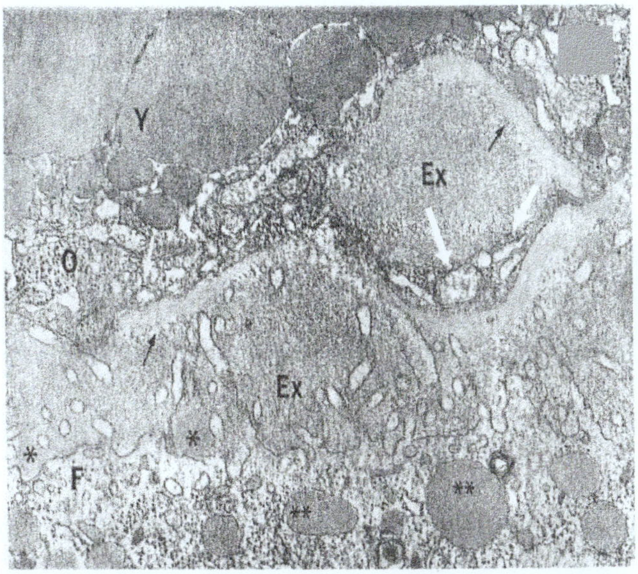

FIG. 12.32 Electron photomicrograph showing stage IV follicle at the initiation of chorion formation. O=electron dense ooplasm containing Y=yolk bodies, and *small white arrows*=small surfaced vesicles, Ex=extracellular space contains *long black arrows*=fibrous material, *=osmiophilic droplets of chorion material, F=apical syncytial contains **=secretory granules. 22,080×. *(From Selman and Arnold (1977).)*

carbohydrate, fatty or 'protein'. Eggs of different types contain more or less one type than another. In most cases, a carbohydrate yolk is associated with a protein in a mucopolysaccaride complex. The above description of the process of vitellogenesis shows that the yolk of eggs of *L. pealii* may be of that last type.

6. Secondary membrane (chorion): it is well known that the follicle cells play a major role in the formation of this chorion, in the eggs of any animal having this membrane. In *L. pealii*, slightly subsequent to the initial appearance of yolk bodies in the oocyte during stage IV, PAS-positive material begins to accumulate in the interface region as discrete droplets. Ultrastructurally, these droplets are more electron-dense than the surrounding extracellular material and subsequently fuse to form the thick secondary membrane. However, stains indicate that both protein and polysaccharide constitute the material of this chorion. The formation of the chorion continues during stage V (Fig. 12.32) and it becomes a thick (0.75 mm), continuous homogeneous layer which is distended to envelop the oocyte smoothly.
7. Ovulation of the fully grown oocyte, at the end of stage IV, yolk bodies of variable sizes which fuse to polygonal bodies accumulate in the oocyte pushing the ooplasm to the periphery where it remains as a layer averaging 30 μm thick at the oocyte surface. At the same time, the folded follicular syncytium is displaced peripherally to surround the smooth-surfaced oocyte. During stage V, the chorion completely encloses the oocyte and the latter is shed by an active contraction of the follicle (Arnold and Williams-Arnold, 1977). The approximate size of this ovulated egg is 1.6 mm × 1.0 mm (Selman and Arnold, 1977).

Ovulation. When the disappearance and complete absorption of the folds is completed in the ovum of *Sepia* (Balfour, 1880), its surface assumes a perfectly regular outline. Then the 'capsule' of the ovum bursts at the opposite pole to the peduncle, and the ovum falls into the oviduct. In *Loligo pealii*, when the egg has completed development, it is shed into the lumen surrounding it by the contraction of the follicle (Arnold and Williams-Arnold, 1977). These authors described and presented serial figures showing the process of ovulation (Fig. 12.33), i.e. 'an oocyte being extruded and the subsequent collapse of the follicle'. This vital activity occurred and completed in 3:30 min. There is a series of very slow rhythmic contractions, which pass over the surface of the follicle from the distal to proximal end. The ovulated oocyte has usually been distorted by this process and slowly begins to resume a more ovate shape as it is passed through the oviduct.

Fertilization. The events of this vital activity were described early in 1930 by Hoadley in *L. peelii*, and described the penetration of the sperm to the micropyle. However, the fertilization in cephalopods is always internal. Therefore, copulation must precede this activity.

The process of fertilization itself apparently occurs in the oviduct. Jelly-covered eggs isolated from the lower part of the oviduct have been artificially fertilized (Fig. 12.34). It was observed that one to several spermatozoa may penetrate the micropyle, but only one actually penetrates the egg surface. The head and midpiece of the sperm can be found lying directly below the micropyle. The immotile tail can be seen projecting into the intrachorionic space at least until first cleavage. At once, two events occur: the 1^{ry} oocyte resumes meiosis and the first polar body is given off; and the cytoplasm streams to establish an enlarged blastodisc directly below the micropyle. After the discharge of the first polar body, the second one is formed and the remainder of the oocyte nucleus forms the female pronucleus. The timing of these events in the artificial fertilization of *Ommastrephes* (Kumé and Dan, 1968) can be considered the same of that normal fertilization of *L. pealii* (Arnold and Williams-Arnold, 1977). During this time, the sperm head and midpiece rotate around each other and continue to move about in the vicinity of the micropyle. After the completion of the formation of the two pronuclei, the male one becomes practically identical with that of the egg. They approach each other and unite in syngamy about 140 min after sperm entry. Two phenomena accompany this cephalopod fertilization:

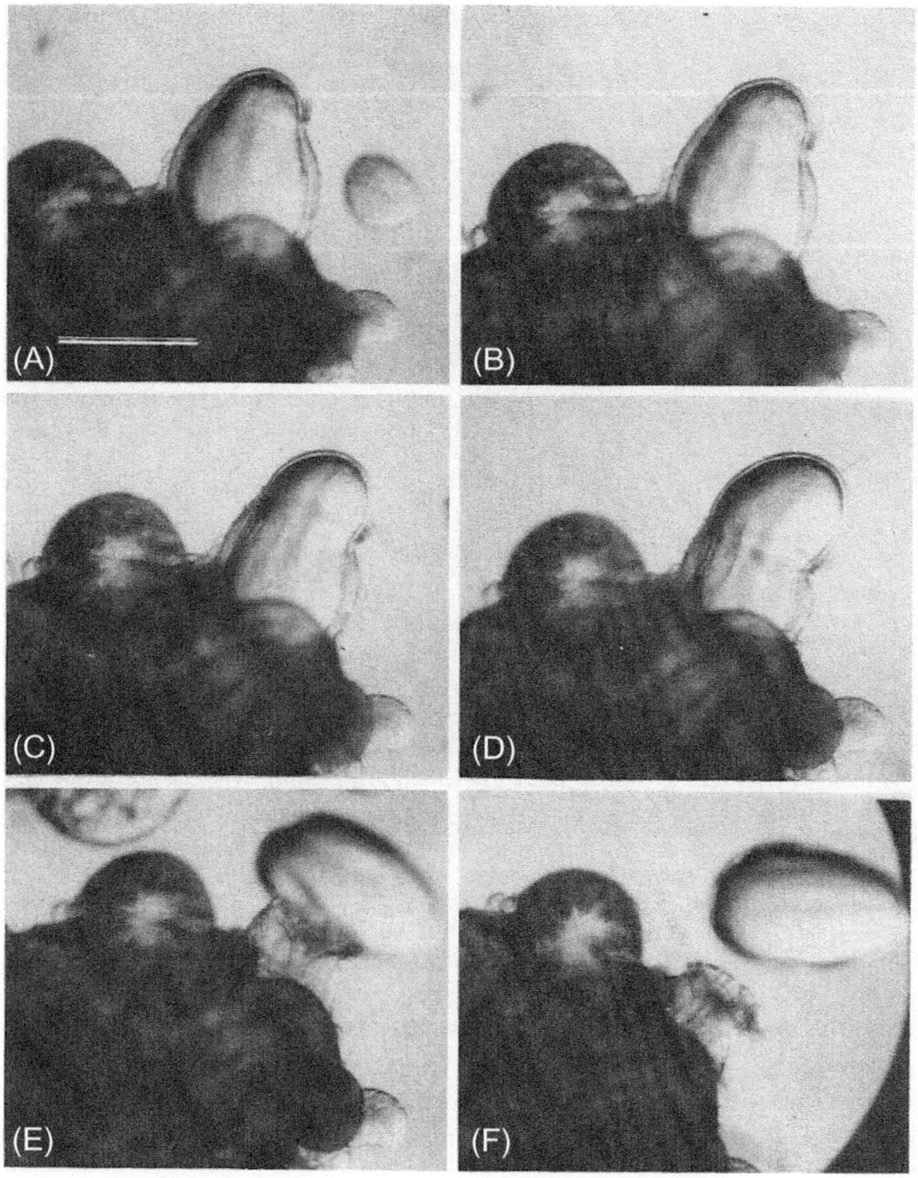

FIG. 12.33 Steps of shedding of a full-grown oocyte from an isolated piece of ovary of *Loligo pealei*. Scale line 1 mm. (A) The tip of the follicle has broken down following a period of slow rhythmic contractions. A wave of contractions is passing down the oocyte 0:00 min; (B) the constriction has passed farther down the follicle and the opening has widened 0:50; (C) further constriction and extension of the oocyte. The wall of the follicle has thickened 1:10; (D) the sides of the oocyte are deeply indented by increasing constriction 2:50; (E) the follicle and ovum a few seconds after shedding 2:55; (F) the follicle wall continues to constrict after shedding and surface of the ovum begins to smooth out 3:30. *(From Arnold and Williams-Arnold in Giese and Pearse (1977).)*

1. block to polyspermy since there is no evidence of supernumerary sperm in the zygote; and no reference to diaspermic or polyspermic development are encountered. However, the nature of this block is unknown; and
2. the polar bodies persist for some time and can divide in the egg of *L. pealii* more than once. Five polar bodies were observed at the animal pole, even during the first cleavage furrow.

The egg. The eggs, before spawning and just after fertilization, accept, outer to the chorion, jelly membranes in various amounts. This jelly substance derived from the oviducal, nedamental and accessory nedamental glands. In *S. officinalis*, the eggs are large and the chorion is coated with a thin layer of jelly attached by a strand of jelly, which the female forms around rod-like structure. The egg jelly of *S. officinalis* is soft and the female ejects ink over the newly laid eggs so they are black and opaque (for different jelly coats of cephalopod eggs, see Fig. 12.10 in Arnold and Williams-Arnold, 1977).

The fully mature cephalopod egg is a large, heavily yolked telolecithal and meroblastic egg. It is yellowish, i.e. amber-coloured, generally ellipsoidal or spherical and covered with a jelly coat(s). However, the shape and

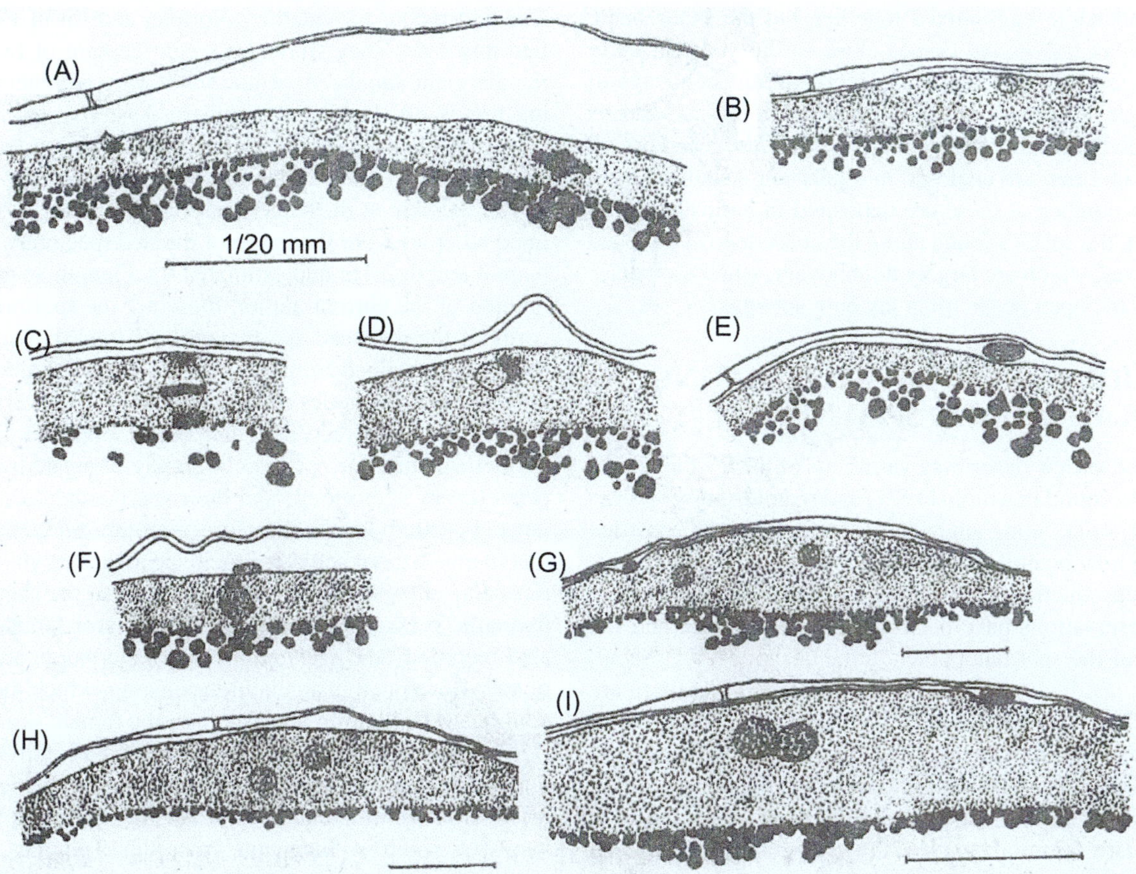

FIG. 12.34 Fertilization and fusion of pronuclei in fertilized egg of *Ommastrephes Sloani pacificus (after Soeda, 1952 from Kumé and Dan, 1968).* (A) Sperm has entered through micropyle, first meiotic spindle at metaphase; (B) sperm pronucleus, moving away from micropyle; (C) rotation of first meiotic spindle; (D) migration of sperm pronucleus; (E) formation of first polar body; (F) formation of second polar body; (G and H) egg and sperm pronuclei approach each other; (I) fusion of pronuclei near micropyle.

size of this egg differ with the species, but do not correlate with the taxonomic position or the size of the adult animal. For example, the egg of *Sepia* may reach exceptional size, about 2 cm in diameter (Gardiner, 1972). However, the egg diameter of *Sepia officinalis* obtained from Homa Lagoon (Turkey) measures 7–9 mm (Önsoy and Salman, 2005). The diameter of the eggs of *S. pharaonis* and *S. dollfusi* obtained from the Suez Canal, Bitter Lake reaches 6–15 mm (Gabr et al., 1998). Ripe eggs from the oviduct of the bobtail squid *Rossia macrosoma* vary between 5 and 10 mm in diameter (Salman and Önsoy, 2010). The mantle length of the animal itself is varied between 37 and 62 mm. Usually, larger mature females have larger eggs in their oviducts. The eggs in oviducts and the vitellogenic oocytes in the ovary of ommastrephid short-finned squid *Illex coindetii* from the Northeast Atlantic have similar diameters, ranging from 0.8 to 1.3 mm (Gonzalez and Guerra, 1996). Many other types of cephalopod eggs were reviewed by Arnold (1971).

Arnold (1971) reported that there is some brooding of the cephalopod eggs by the female. In *Sepioteuthis*, three or four eggs are contained in a common complicated jelly layer forming a capsule. In *Loligo*, there may be up to 200 eggs in an egg string, the structure of which is fairly complex. These egg strings are attached to the substrate or more commonly to other egg strings until a large common egg mass is built-up by several females over a period of a few days. The egg mass of *L. opalescens* was observed to be 40 ft in diameter. The synchrony and rate of laying eggs make *O. vulgaris* an attractive animal from an experimental standpoint. The brooding female, after depositing as many as 100,000 eggs over a period of a few weeks, undergoes behavioural changes, refuses to feed and eventually dies. It appears that this is a natural event and that death is the typical consequence of egg laying or mating in cephalopods.

As for the Japanese species, Kumé and Dan (1968) reported that *Sepiella maindroni* and *Sepia esculenta* would come in spring, into the bays along the coast from Tokyo towards the southwest, and lay their eggs on seaweeds and other objects in shallow water. The egg of *Sepiella* is surrounded by a jet-black, strong, rubber-like substance, which sticks it to the substrate; the *Sepia* eggs are also attached separately by their white gelatinous coats often covered with sand grains. These two species spawn in the same localities and at about the same time, so it is quite usual to find black and white eggs mixed together. On the other hand, *Sepioteuthis lessoniana* comes close to the coast in early summer to spawn, the eggs of which are laid in strings of about eight, surrounded by jelly. The eggs of *Loligo bleekeri* are somewhat smaller than those of *S. lessoniana* and 50–60 of them in a row are enclosed in a jelly mass shaped like a finger; a number of these are deposited in a cluster. It appears that the species found along the coast deposit attached egg masses, which are largely nonbuoyant, whereas species living in the open ocean often lay buoyant eggs.

REPRODUCTIVE BEHAVIOUR, MATING, COPULATION, AND SPAWNING

A detailed review describing variations of these vital activities can be found in Arnold (1971) and Arnold and Williams-Arnold (1977). Some amazing observations were recorded showing how beautifully these animals undergo courtships during reproduction. Concerning the process of copulation, many observations have been made on copulation and the transfer of the spermatophores in the decapods, e.g. *Sepia*, *Sepioteuthis*. In *L. pealii*, van Oordt recorded that there are two methods of copulation in this American squid. **First**, the spermatophores are inserted into the mantle cavity of only those females that are about to deposit their eggs. In this case, **the animals are paralleling each other**, which is the far more common position; the hectocotylus of the male seizes the spermatophores which appear in the opening of its funnel, and with a rapid sweep, these are immediately inserted into the mantle cavity of the female, frequently on the distal end of the oviduct. After about 5–6 s, this arm is withdrawn, the animals separate and, in further 5–6 s, the empty tunics pass out of the funnel. **Second**, the female of *L. pealii* may not yet be ready to deposit the eggs, but is capable of copulation. According to recordings in the work of van Oordt, the sperm masses of the spermatophores are 'ejaculated', i.e. the spermatophores are dehisced (this is not in view of the current author) and then attach to a receiving depression of the buccal membrane of the female (seminal receptacle below the mouth of the female) (see Fig. 12.2). Copulation in this method occurs when the animals are in a position **head to head**. When the animals separate, the empty tunics of the 'ejaculated' spermatophores are held for several minutes between the arms of the female and are then finally dropped.

Very recently, the extensive review on reproduction in oceanic cephalopods, and mating, spawning and death in these animals provide an unexpected mode of transfer of such spermatophores. Nesis (1995) stated that the cephalopods, which lack hectocotyli, transfer their spermatophores by a long penis. He also stated that "one of the most intriguing puzzles of spermatophore transfer is the rather common case of spermatophore embedding in the outer surface of the mantle -- in special sites or simply in almost every place." The first record provided by Norman and Lu in 1997 verified this case. They studied a recent capture of two mated female giant squids (*Architeuthis*) during commercial fishing trawls off Tasmania, Southern Australia, at depths of 500–1000 m. They observed in this almost 'mythical' deep-sea creature's 'sperm packages' (spermatophores) embedded in the skin at the same level on both ventral arms (and in no other arms) of the larger of the two specimens (220 kg, mantle length 2.4 m and estimated total length 15 m). In the opinion of the current author, these are the spermatophoric cords of the dehisced spermatophores (see above). All the specimens obtained from these waters were females; little is known about the males in these southern oceans. However, Norman and Lu recorded that males obtained from the northern hemisphere possess a greatly elongated muscular penis (up to 92 cm long) and their spermatophores are also large (11–20 cm long). Therefore, Norman and Lu suggested that the male giant squids of their specimens might use their muscular elongated penis to 'inject' these packages under hydraulic pressure directly into the arms of females. They also recorded that *Architeuthis* spermatophores are exceptionally enveloped in a gelatinous coating, which may act as a lubricant facilitating injection into the female skin.

Jackson and Jackson (2004) examined 14 individuals of mated female onychoteuthid squid *Moroteuthis ingens* collected from commercial deep water fishing vessel trawling in waters south of Tasmania, Australia. Trawling depth is about the same recorded by Norman and Lu (1997). Many of the specimens were mature and spent adults, collected between January and March in 2002 and 2003. These authors tabulated all the information of their specimens. The most important points were both the size of the individual itself and the sites of spermatophore placement. The mantle length ranged from 467 mm to 578 mm and the weight ranged from 1699 to 4896 g. Spermatophores were found embedded in the external mantle, but more commonly in the internal mantle just under the thin mesentery, or on various sites within the visceral mass and sometimes very near to the oviducal glands. Moreover, these authors collected, during January 2003, a pair of copulated specimens, a male (380 mm ML) and a female (485 mm ML), which were strongly physically attached and difficult to separate. On dissection of the female, they found that it was immature and found the spermatophores in the mentioned sites. In addition, there were large numbers of empty tunics lying freely in the mantle cavity. On the other hand, the dissected male was spent with a ribbon-like testis and only a few spermatophores left in the penis. The latter was extended freely beyond the mantle margin. Nesis (1995) interpreted that a head-to-head position is employed by species that use the penis for direct spermatophore transfer into the female tissues.

It is not known what structure or process is responsible for this implantation. Several hypotheses have been proposed to explain this peculiar process by Hoving and Laptikhovsky (2007). These authors recorded three non-convincing hypotheses, which are based solely on morphological observations of preserved male and female squids. Thus they presented an experimental study using the deep-sea squid *Moroteuthis ingens*, the same species previously examined by Jackson and Jackson (2004). The experiment showed for the first time that spermatophores have an autonomous mechanism that enables them to migrate into body tissue. On board the research boat *Dorada* of the Falkland Islands Government Fisheries Department, the authors isolated spermatophores from the penis of a 'moribund' squid and placed them on the outer surface of another dead large specimen with a mantle length of about 300mm. With forceps, they gently pressed on the cap of the spermatophores to induce artificial dehiscence. They observed the ejaculation of the ejaculatory apparatus out of the tunics of the spermatophores while the specimen was submerged in seawater. It appears that these spermatophores had undergone the same changes that occurred in those of *Sepia officinalis* during the artificial dehiscence (Bawab and Ismail, 1984a,b). The ejaculatory apparatus of the spermatophores of *M. ingens* then dragged the other parts that the spermatophoric cords became out of the tunics. The authors referred to those uncovered spermatophoric cords as 'spermatangia'. After a few minutes, these spermatangia had autonomously embedded themselves into the tissue of the specimen wherever the spermatophores were deposited. The authors tried to explain how this penetration occurred, but this needs further discussion of the chemistry of the spermatophore itself. This process may be an adaptation for species that have no seminal receptacles. The observations of Hoving and Laptikhvosky also clarified the role of the penis. In addition to transferring spermatophores, it initiates the spermatophoric reaction by pressing the cap or by pulling its thread. Another role is probably the precise deposition of the spermatophores on the female before releasing it.

Hoving with another team (2009) studied experimentally spermatophore implantation in another species, *Rossia moelleri* collected from Greenland waters (Germany). They described this species as follows:

1. the male has a hectocotylsed arm;
2. the female lacks bursa copulatrix; and
3. the male implanted spermatophores mainly on the left side of the female body, the same site of the opening of the oviduct, in addition, on other sites, especially different regions of the mantle

Hoving et al. performed their first experiment to confirm that the spermatophores of this species are able to implant into tissue autonomously, even after being kept deep frozen for as long as 4years. In artificial seawater (35%; 8°C), the spermatangia were successfully implanted into female mantle tissue. This experiment refreshes the memory and recalls the information that spermatophores of a crustacean, or maybe of *Sepia*, were kept frozen for as long as 15years and, when dehisced, the sperm restored their motility and activity in seawater. Hoving et al. performed their second experiment to measure the force necessary to penetrate the mantle tissue. They explained, in detail, the method and instruments used. They analysed the spermatophoric reaction and calculated the force used for this implantation.

The result of the first experiment verified that the autonomous mechanism of spermatophore implantation firstly observed in *Moroteuthis ingens* (Hoving and Laptikhovsky, 2007) occurs in other cephalopods. In the cephalopod, which lacks bursa copulatrix, the male transfers its spermatophores by implantation. This process, as described for *R. moelleri*, is found in all members of subfamily Rossiinae, two sepiolid subfamilies Heteroteuthinae and Sepiolinae. From the second experiment, the authors estimated that the autonomous implantation of spermatangia of *Rossia moelleri* necessitates not only the mechanical mechanism, but probably the chemical mechanism as well.

DEVELOPMENT

The development of the most common European squids, *S. officinalis* and *L. vulgaris*, has been an object of study for many years. The ease with which the eggs of *Sepia*, *Sepiella* and *Loligo* can be collected in Japan as well has made them objects of interest to many investigators. Since the sexually mature *L. bleekeri* can readily spawn in captivity, their eggs make a convenient material for embryological research. In addition, *Octopus vulgaris* are easy to keep in an aquarium, where they often spawn. In spite of this it has been noted that 'there has been a surprising lack of experimental work done on cephalopod embryos compared to the rest of the Mollusca' (Giese and Pearse, 1977). However, the histological studies started very early, e.g. Balfour (1880) recorded the description of the cleavage of *Sepia* and *Loligo* done by N. Bobretzky and presented a diagram showing a late stage of that of *Loligo*; also, the cleavage of *Sepia* and the development of *Loligo* were described by Vialleton in 1888 (Fig. 12.35A and B). The latter work was recorded in Parker and Haswell (1943a, b) and then in the work of Pflugfeder (1970). Two other works dealing with the same items were encountered. No significant difference was observed between them, or between the above old one. Only, the picture becomes clearer and/or new observation(s) are added due to the new tools used. The **first one** was carried out by Soeda (1952, 1954) on the Japanese common squid *Ommastrephes* (=*Todarodes*) *sloani pacificus*. Soeda studied artificial fertilization and cleavage, obtained eggs by gently pressing the ripe oviducts by a glass rod, and obtained spermatozoa in proper condition, possibly by squeezing the seminal receptacles. Further developmental stages were studied on *L. vulgaris* (Kumé and Dan, 1968). The **second** was the

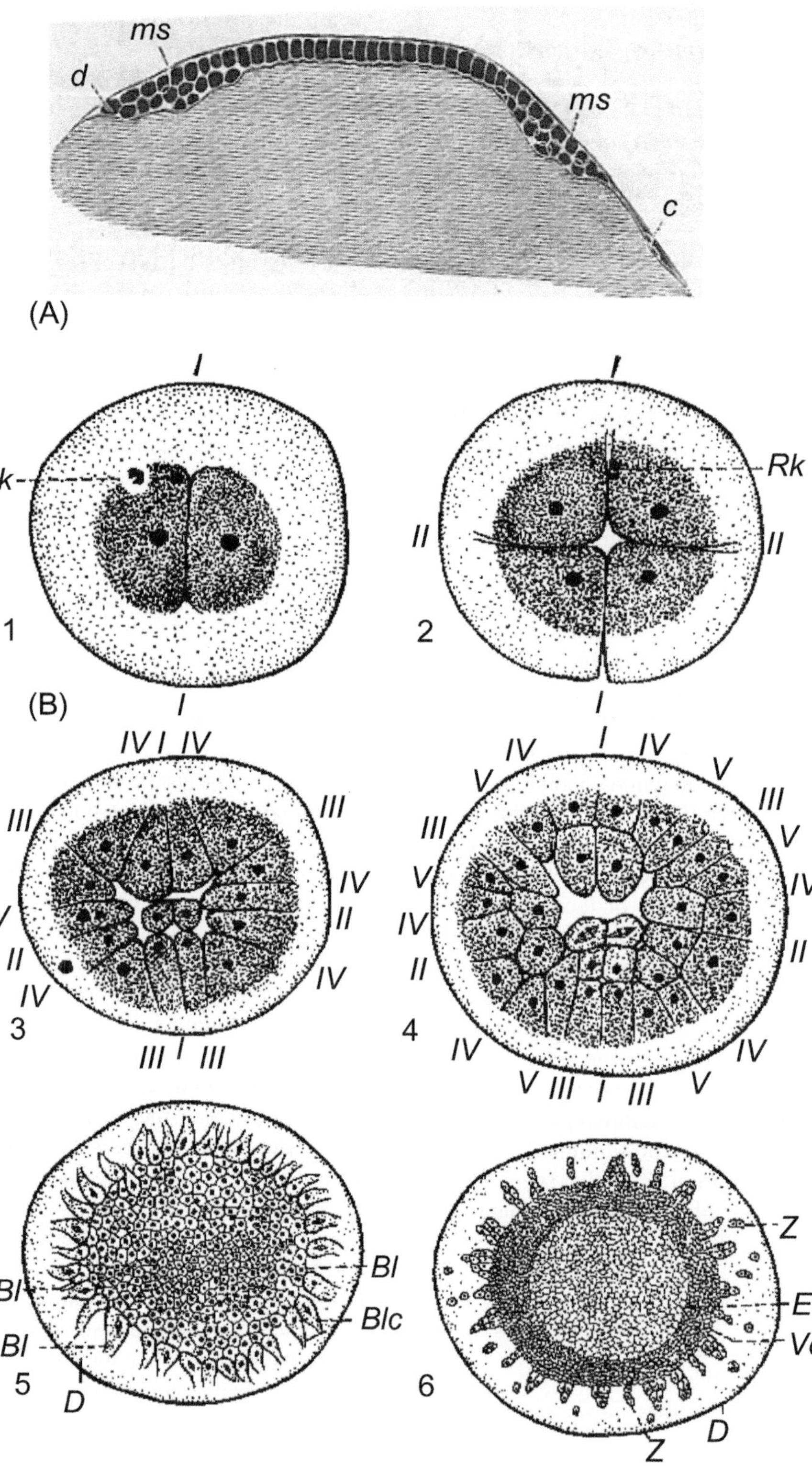

FIG. 12.35 Cleavage of cephalopod ovum: (A) T.S. through the blastoderm of a *Loligo* ovum at the beginning of the fourth day *(after Bobretsky from Balfour, 1880)*. ms = mesoblast, d = cell at the edge of the blastoderm, c = one of the segmentation cells; (B) 1–6 cleavage of the egg of *S. officinalis*: 1 = two blastomere stage; 2 = four blastomere stage; 3 and 4 = third to fifth cleavage; 5 = surface view of blastoderm of *Sepia* ovum at a late stage of segmentation; 6 = surface view of blastoderm with thick zone in the middle region. E = single-layered, Vd = thickened zone, Z = dissolving cell groups *(from Pflugfelder, 1970)*; (C–E) T.S. through the edge of the blastoderm of *Sepia* at three successive stages. Bl = blastoderm, yk = yolk, yk.ep = yolk epithelium. *(B–E) After Vialleton, from Parker and Haswell (1943a, b) and Pflugfelder (1970).*

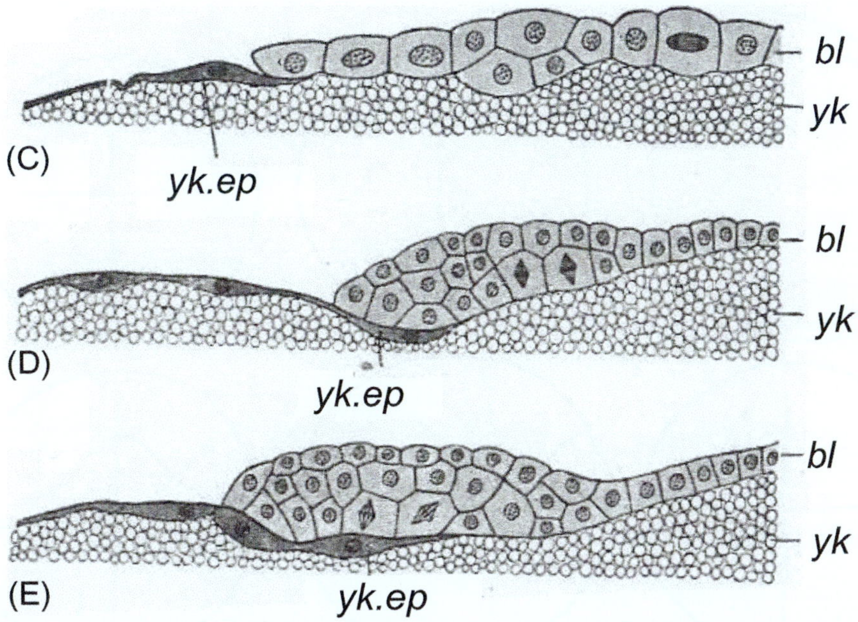

FIG. 12.35, CONT'D

work of Arnold (1971) in which the author described the whole story of the development of the squid *Loligo pealii* from fertilization until hatching the young. Arnold had another work with Williams-Arnold on Cephalopoda in Giese and Pearse (1977). He published about 13 articles on different aspects of the development of the cephalopod *Loligo pealii*. He is therefore most familiar with this animal, and became proficient in its development and in development of cephalopods in general. He reported that the pattern of development is uniform throughout the cephalopods.

Cleavage

The heavily yolked telolecithal cephalopod egg (Fig. 12.21E, introduction) greatly influences its development and sets this group apart from the rest of the Mollusca. It greatly modifies the cleavage, which becomes meroblastic (incomplete discoidal cleavage) and the typical molluscan spiral cleavage is not present. Cleavage begins about 3 h after sperm entry and this period lasts, in *L. pealii* approximately 12 h at 18°C (Arnold and Williams-Arnold, 1977). Fusion of the pronuclei results in formation of the first mitotic figure and cleavage occurs. During this time, an accumulation of clear cytoplasm has occurred at the animal pole to form the blastodisc, in which all subsequent development will take place. (It must be remembered that the animal pole corresponds to the future functional posterior end of the animal, while the vegetal pole corresponds to the location of the foot or functional anterior of the swimming decapod.) The zygote also undergoes a sharp change at this time, by becoming somewhat less ovate, and the intrachorionic space seems to expand. The **first** cleavage furrow appears as a slightly flattened area of egg surface on a line between the micropyle and the polar bodies at the polar pole. It suddenly appears a few minutes later, as a surface line several micrometres in length. It lengthens as it deepens until the edges of the blastodisc are reached laterally and the yolk-cytoplasm interface is reached basally; i.e. it includes more than two-thirds of the egg diameter. About 1 h later, the **second** cleavage furrow forms slightly subapically, at a right angle to the first. It begins as the first one, then it elongates until it becomes as long as the first. The base of the second furrow is attached to the base of the first and causes the latter to be temporarily slightly deviated as it forms. The **third** furrow sets in perpendicular to the second, i.e. parallel to the first. It displays the bilaterally symmetrical nature of the cleavage pattern. The **fourth** cleavage furrow essentially parallels the second furrow and cuts free two groups of micromeres, a central small group and a surrounding syncytium attached by the distal margins. Cleavage continues to separate these two groups of cells until a central layer is formed, surrounded by a peripheral syncytium (Fig. 12.35). The nuclei of the latter are partially separated by membranes, which extend centrifugally. These are the blastomeres of Vialleton (1888) (Arnold and Williams-Arnold, 1977).

Several aspects of cleavage should be noted. Soeda studied the artificial fertilization of the squid *Ommastrephes* and described its cleavage (Fig. 12.36). No significant differences can be observed and the upper description can be considered the same as that of Soeda. The latter author mentioned that the fourth cleavage results in the formation of two centrally located micromeres, and after that, the rate

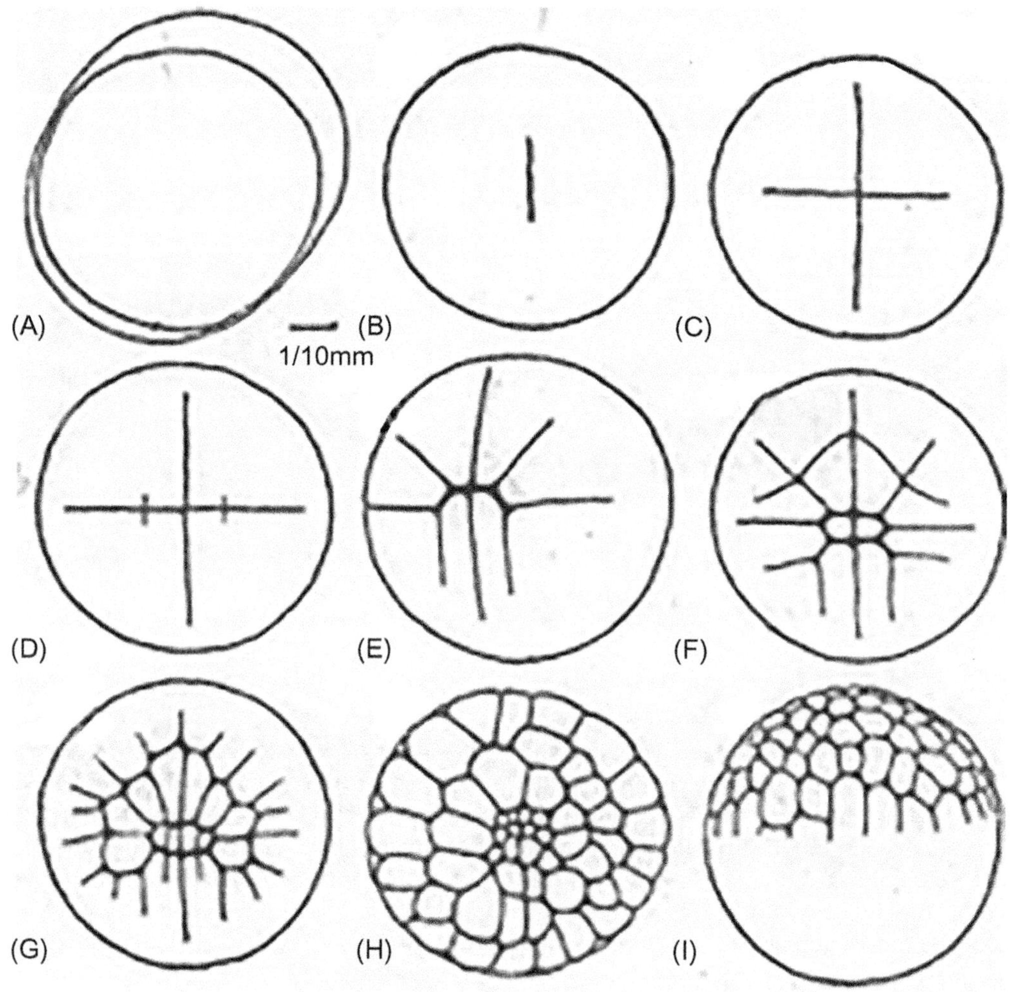

FIG. 12.36 Cleavage in the ovum of *O. sloani pacificus (after Soeda, 1952 from Kumé and Dan, 1968).* (A) Fertilized egg; (B) 45 min after fertilization; (C) completion of second cleavage; (D) 3 h after fertilization; (E) completion of third cleavage; (F) 4 h after fertilization; (G) fifth cleavage (not 4 micromeres); (H) about 6.5 h; (I) about 8.5 h after fertilization.

of cleavage begins to show local differences. The increase in the number of furrows makes it difficult to follow the process by simply observing the surface of the embryo. However, 32–64 cells can be seen, and the number of central micromeres is clearly eight. Continued cleavage produces a single-layered blastoderm (blastodisc). In a cross section of this stage, it is difficult to distinguish between the micromeres and macromeres; simply there is a single layer of blastomeres overlying the animal half of the egg. There is no formation of a yolk epithelium. Afterwards, a longitudinal section of the embryo shows a blastula stage (Fig. 12.35A and B). The latter has a single layer of blastomeres at the centre, encircled by a double-layered area. It becomes clear that the observed difference is the timing of the steps of the process. However, one can consider the species, climate, etc. The advantage of the work of Arnold (1971) is that he presented electron micrographs of the cleavage furrows and other cleavage stages.

Formation of germ layers from the blastoderm

Kumé and Dan (1968) reported that at one point of the edge of the blastodisc, very active cell proliferation takes place. This is the site of the future anus. As the result of this proliferation, and immediately beneath the blastoderm, the endoderm is being formed, at first in a crescent-shaped area. It then gradually enlarges and becomes circular. The embryo thus acquires two layers, the **ectoderm** of the blastodisc overlying the newly formed **endoderm**. The ectodermal cells of the anal side proliferate laterally, and extend towards the anterior in the space between the ecto- and endodermal layers, to give rise to the mesoderm. Another description was found in the work of Arnold (1971). He reported that the blastoderm is a single layer of cells made up of blastomeres in the centre and surrounded by a ring of peripheral ring of cells (termed blastocones by Vialleton).

The blastomeres are separated from the underlying yolk by a plasma membrane and an irregular space with many areas of close junction existing between the cell and the yolk plasma membrane. The blastocones are continuous with the yolk below, and with each other at their vegetal pole end. The *L. pealii* embryo at this stage is composed of about 60–100 cells.

In the cephalopod embryo the yolk becomes covered by a special syncytial layer of cells called the yolk epithelium (Lankester, 1875), which is digestive in function and morphogenetic in significance. There has been much speculation as to the origin of the nuclei of this layer. In the view of Arnold (1971), and according to his personal experience with this group of animals, the blastocones continue to divide and apparently some of the nuclei from the blastocones become incorporated into the layer beneath the blastoderm itself. How this occurs is not clear, but a nucleated ring of syncytial cytoplasm is established just inside the margin of the spreading blastoderm. Moreover, some blastocones continue to divide off blastomeres at their apical ends, the nuclei of which are diffused, lightly staining and have a very characteristic appearance in the EM. These characteristics are retained by the yolk epithelium throughout all of the embryonic development. As mentioned above, the yolk epithelium has a digestive function, which is important in relation to the breakdown of the platelet. The yolk epithelium continues to expand beneath the margin of the blastoderm. Then, a third layer of cells appears which is derived by marginal mitosis of the blastoderm cells themselves. The daughter cells would tend to expand the blastoderm. In this way, an intermediate layer of cells is established without the necessity of invoking cellular movements or 'gastrulation'. There have been many varying opinions as to the origin of these cells. One of them is that of Lankester (1875), who used the term 'primitive streak' to designate an area of supposed invagination, but no evidence available that supports this view or the others. Other authors, e.g. Watase (1888), had another view. This layer is derived by delamination of the upper layer. The micrographs presented by Arnold (1971) confirm this view.

The multilayered blastoderm continues to expand primarily by marginal division until the equator of the egg is reached. This shows the future organogenic region of the embryo. The remaining yolk mass becomes cellulated with a double layer of cells: the outer single layer and the inner syncytial yolk epithelium. This vegetal pole will become the external yolk sac in later development. The relative fate of the three cell layers was strongly evidenced. The inner syncytial yolk epithelium is homologous to the **endoderm**. The outer layer of cells derived by spreading of the surface of the blastoderm is then **ectodermal**. The cells have long cytoplasmic strands that connect to the inner layer. The middle cellular layer derived by delamination is the **mesodermal** or mesondodermal layer, in which many of the cells are interconnected by large cytoplasmic bridges. The entire complex structure spreads towards the vegetal pole until all the original yolk is enclosed in cells, as mentioned above, at the beginning of organogenesis.

ORGANOGENESIS

All the organs arise from the two layers of cells on the yolk epithelium. It is mentioned above that the inner syncytial yolk epithelium is digestive in function, but it discards during later development and hatching. In *Loligo pealei* (Fig. 12.37), the first embryonic organs appear as a thickening of the outer cell layers (from cuboidal to columnar) and there is a slight equatorial constriction (Fig. 12.37A). The latter separates the future **external yolk sac** from the **embryo** proper. The external yolk sac develops into a large haemal space transversed by muscle strands, which by rhythmic contractions circulate the blood through the developing vessels until the heart eventually takes its function (Arnold and Williams-Arnold, 1977). In the meantime, at the surface, at about the middle of the blastoderm, a thickening of a cap-like shape appears, the edges of which become raised above the general level of the blastoderm; this is the rudiment of the **mantle**. On the surface of this, is a developed depression, which subsequently forms a closed sac—the **shell-sac**, within which the **shell** will be formed. This does not correspond to the shell gland seen in the embryos of other molluscs, as was early pointed out by Lankester (1875) (Kumé and Dan, 1968). Below the mantle (i.e. nearer to the vegetal pole), two elevations appear, each with a pit-like depression, which are the rudiments of the **eyes** (Fig. 12.37A). The elevations become more and more prominent and the eyes themselves are formed from just a part of these elevations; each is a pit, which subsequently becomes closed to form a vesicle: the **optic vesicle**. Later, an ingrowth of a specialized group of cells in the outer surface of the eye primordium gives rise to the lens. The iris and cornea are formed by further surface invaginations. The inner wall of the optic vesicle develops directly into the retina, which has the optic ganglion directly internal to it. Further details of eye development can be found in Arnold (1971). Still nearer to the vegetal pole, slightly above the equatorial constriction, a series of paired elevations appears: the rudiments of the **arms** (Fig. 12.37B–J). They condense into the arm primordia, which grow outwards and differentiate small sucker buds on their inner surface.

After the complete enclosure of the yolk by the blastoderm, the **mouth** (Fig. 12.37B) is developed as an oval depression between the rudiments of the eyes. A row of cilia, which is developed in the neighbourhood of the mouth in some forms, perhaps represents the velum or the oral circlet of other molluscan embryos. Immediately in front of the edge of the mantle, two short ridges appear; these are the beginnings of the **gills** (Fig. 12.37C), which increase

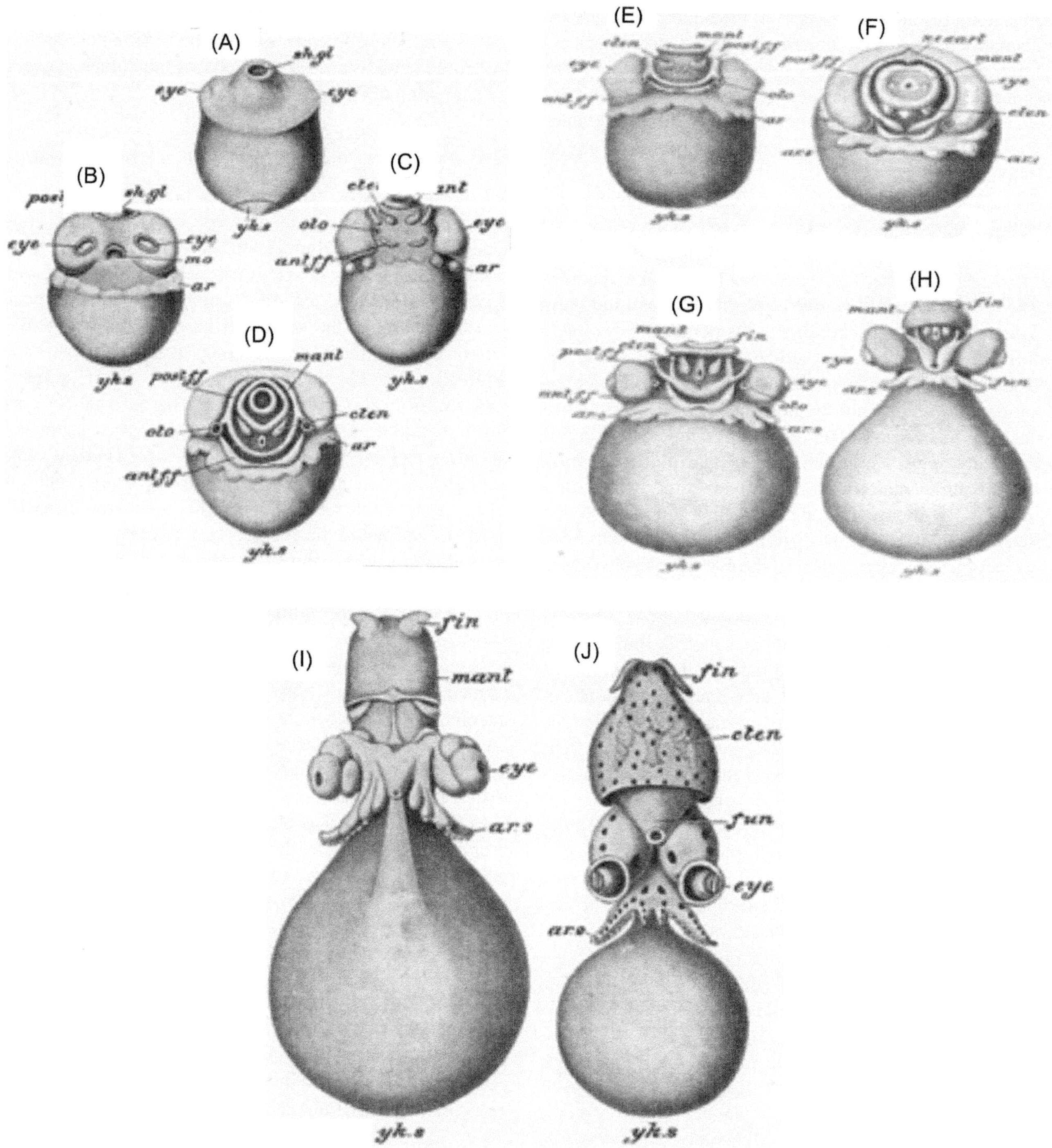

FIG. 12.37 Development of cephalopod embryo (*Loligo*) *(after Korschelt and Heider from Parker and Haswell, 1943a, b)*. (A–D) Early stages: (A) development of eyes and shell-gland rudiments; (B) later embryo from the oral side; (C, D) from the anal side; (E) later stage from the funnel side; (F) the same obliquely from above; (G, H) later stages than (E, F) from the same sides, respectively; (I, J) two successive later stages seen from the funnel side. Ant.f.f. = anterior funnel fold, ar = rudiments of arms, cten = ctenidium, eye = eye, mo = mouth, mant = rudiment of mantle, oto = statocyst, post.f.f. = posterior funnel fold, sh.gl. = shell-gland, yk.s. = yolk sac.

in size and become two flaps of tissue, and each separates into gill filaments. A branchial heart develops at the base of each gill. At the same time a pair of folds appear—the posterior funnel-folds—which are formed between these and the eyes. These are the first rudiments of the **funnel** (Fig. 12.37B–H); the greater part of which is formed from a second pair of folds—the anterior funnel-folds—developed further forwards. The mantle eventually grows downwards to cover the funnel, so only its distal end projects beyond the mantle cavity. Lateral outgrowths of the mantle have already given rise to the rudiments of the fins. Behind the anterior funnel-folds, two pit-like depressions appear, which subsequently develop into the **statocysts** (Fig. 12.37C–G). They become incorporated into the head behind the brain during development. They undergo an invagination, and the wall of the vesicle secretes small granules, which fuse together to form the statolith attached to the sensory hairs.

In respect to the **alimentary canal**, the fore-gut and hind-gut develop as invaginations from the surface and grow towards the formal animal pole. The midgut arises independently as a bipartite tube, which surrounds the internal yolk sac but fuses into a simple structure in later development. The fore-gut buds off the anterior and posterior **salivary glands**, and the hind-gut gives rise to the ink sac in a similar fashion. The anus appears as an aperture situated on a little papilla: the **anal papilla**.

As the embryo increases in size, the yolk is gradually absorbed, and the yolk sac decreases in bulk, until, when the embryo leaves the egg, it has almost completely disappeared. The embryo hatches by applying its **Hoyle organ** to the inside of the chorion and digesting a hole inside it. Hoyle's organ is composed of specialized gland cells on the dorsal posterior mantle of later stages of development. The development of these cells was described as to become rather typical goblet cells. It has shown that once the larvae have hatched, it disappears (Arnold, 1971). Properly speaking, there are no larval forms in the Decapoda–Cephalopoda. Development is direct to a miniature adult, which may or may not enter a planktonic type of existence before attaining the typical lifestyle. This direct mode of development is very unusual among molluscs, but not unexpected for animals with such large eggs.

REFERENCES

Adam, W., 1960. Cephalopoda from the Gulf of Aquaba, Ministry of Agriculture, Division of Fisheries. Halfa Bull. 26 (16), 1–26.

Aristotle, 1910. ca. 384–330 BC: 'Historia animalium' (translated by D'Arcy Thompson). Oxford University Press (Clarendon), London/New York.

Arnold, J.M., 1971. Cephalopods. In: Reverberi, G. (Ed.), Experimental Embryology of Marine and Freshwater Invertebrates, North-Holland Pub. Co., Amsterdam (Chapter 10).

Arnold, J.M., 1974. Intercellular bridges in somatic cells: cytoplasmic continuity of blastoderm cells of *Loligo pealei*. Differentiation 2, 335–341.

Arnold, J.M., Williams-Arnold, L.D., 1977. Cephalopoda: Decapoda. In: Giese, Pearse (Eds.), Reproduction of Marine Invertebrates. vol. IV. Academic Press, New York, pp. 243–290.

Austin, C.R., Lutwak-Mann, C., Mann, T., 1964. Spermatophores and spermatozoa of the squid *Loligo pealii*. Proc. R. Soc. Lond. B 161, 143–152.

Balfour, F.M., 1880. A Treatise on Comparative Embryology. vol. I. Macmillan and Co, London.

Barnes, R.D., 1974. Invertebrate Zoology, third ed. W.H. Saunders Company, London.

Bawab, F.M., 1970. The Formation of the Spermatophore in *Penaeus trisulcatus* Leach (Crustacea, Decapoda) (Ph.D. thesis). Faculty of Science, Alexandria University, Egypt.

Bawab, F.M., Bawab Jr., G.M., unpublished. Roles of Needham's organ of *Sepia officinalis* (Linnaeus, 1758) in the formation of the spermatophore (Mollusca—Cephalopoda).

Bawab, F.M., El-Sherief, S.S., 1981. The initial formation of the sperm masses of the spermatophores of *Portunus pelagicus* L. (Crustacea-Brachyura). Bull. Fac. Sci. Alex. Univ. 21 (4), 101–118.

Bawab, F.M., El-Sherief, S.S., 1989. Contributions to the study of the origin, nature and formation of the plug in the Spermatheca of the female crab *Portunus pelagicus* (Linnaeus, 1799) (Decapoda-Brachyura). Crustaceana 57 (1), 9–24.

Bawab, F.M., Ismail, S.Z., 1981a. The structure of the reproductive system of *Sepia officinalis*. Bull. Fac. Sci. Alex. Univ. 21 (4), 119–134.

Bawab, F.M., Ismail, S.Z., 1981b. The role of the testis and the proximal vas deferens in the formation of the spermatophore of *Sepia officinalis* (Mollusca – Cephalopoda). Presented in the First Indo-Pacific Symposium on Invertebrate Reproduction (27–29 November 1981). Marathwada Univ. Aurangabad, India.

Bawab, F.M., Ismail, S.Z., 1984a. Structure of the Spermatophore of *Sepia officinalis*. Bull. Fac. Sci. Alex. Univ. 24 (4), 274–310. G14.

Bawab, F.M., Ismail, S.Z., 1984b. The formation of the spermatophore of *Sepia officinalis*. 1- Role of the testis. Bull. Fac. Sci. Alex. Univ. 24 (4), 311–325.

Bawab Jr., G.M., 1996. Histological and Histochemical Studies on Needham's Organ to Determine its Role in the Formation of the Spermatophores of *Sepia officinalis* (Linnaeus, 1758) (Mollusca-Cephalopoda) (M.Sc. thesis). Faculty of Science, Alexandria University, Egypt.

Bawab, F.M., El-Sherief, S.S., Abd-El-Kerim, H.M., 1992a. The gonad of the desert snail *Eremina ehrenbergi* Roth, 1839 (Stylommatophora-Gastropoda) and its role in the production of the male gametes. Funct. Dev. Morphol. 2 (2), 103–110.

Bawab, F.M., El-Sherief, S.S., Mehanney, S., 1992b. Oogenesis in the desert snail *Eremina ehrenbergi* Roth, 1839. Proc. Zool. Soc. A.R. Egypt. 23 (pt 1), 107–113.

Bell, G.H., Emslie-Smith, D., Paterson, C.R., 1976. Text Book of Physiology and Biochemistry, ninth ed. The ELBS and Churchill Livingstone.

Blancquaert, T., 1925. L'origine et la formation des spermatophores chez les céphalopods décapods. Cellule 36, 315–356.

Bloom, W., Fawcett, D.W., 1976. A Textbook of Histology. 10th ed., Asian ed. Igaku Shoin Ltd. Tokyo.

Borradaile, L.A., Potts, F.A., with chapters by Eastham, L.E., Saunders, J.T., 1977. The Invertebrata, fourth ed. Revised by Kerkut, G.A. Cambridge University Press.

Boyle, P.R., Pierce, G.J., Hastie, L.C., 1995. Flexible reproductive strategies in the squid *Loligo forbesi*. Mar. Biol. 121, 501–508.

Bullough, W.S., 1981. Practical Invertebrate Anatomy, second ed. Macmillan Press Ltd, London, Basingstoke.

Campbell, F.L., 1929. The detection and estimation of insect chitin; and the irrelation of 'chitinization' to hardness and pigmentation of the cuticule of the American cockroach *Periplaneta Americana* L. Ann. Entomol. Soc. Am. 22 (3), 405–426.

Coelho, M.L., O'Dor, R.K., 1984. Evaluation of male reproductive features in *Illex illecebrosus* for maturity staging. In: Scientific Council Meeting – September 1984. Special Session on Squids. Serial No. N898 NAFO Canada.

Cowden, R.R., 1968. Cytological and cytochemical studies of oocyte development and development of the follicular epithelium of the squid *Loligo brevis*. Acta Embryol. Morphol. Exp. 10, 160–173.

Cuvier, G., 1817. Memoires pour seriver a l' histoire et l' anatomie des mollusques. Deterville, Paris.

Dennell, R., 1949. Earthworm chaetae. Nature 164, 370.

Dennell, R., Malek, S.R.A., 1956. The cuticle of the cockroach *Periplaneta Americana*. V. The chemical resistance of the impregnating material of the cuticle, and the 'self-tanning' of its protein component. Proc. R. Soc. B 144, 545.

Drew, G.A., 1919. Sexual activities of the squid, *Loligo pealii* (Les.). II. The spermatophore, its structure, ejaculation and formation. J. Morphol. 32 (2), 379–435.

Emam, W.M., Aly, R.H., 2000. The male reproductive system of the reef squid *Sepioteuthis lessoniana* Lesson, 1830 (Mollusca: Cephalopoda) from the Gulf of Suez, Egypt. J. Aquat. Biol. Fish. 4 (1), 265–295.

Fields, W.G., 1965. The structure, development, food relations, reproduction and life history of the squid *Loligo opalescens* Berry. In: Fish Bulletin 131. State of California, Dept. of Fish and Game, pp. 1–108.

Fischer, W., 1973. FAO Species Identification Sheets for Fishery Purpodes'. Mediterranean and Black Sea (Fishing Area 37). vol. II.

Fischer, W., Schneider, M., Bauchot, M. L. (1987). [FAO species identification sheets for fishery purposes: Mediterranean and Black Sea, Fishing areas 37. Revision 1. v. 1: Vegetation and invertebrates.-v. 2: Vertebrates]. [French].

Fort, G., 1937. Le spermatophore des céphalopods. Etude du spermatophore d' *Eledone cirrhosa* (Lamarch, 1799). Bull. Boil. Fr. 75, 357–373.

Fort, G., 1941. Le spermatophore des Cephalopodes. Etude du spermatophore d'*Eledone moschata* (Lamarck 1799). Bull. Biol. Fr. Belg. 75, 249–256.

Franzén, Å., 1955. Comparative morphological investigations into the spermiogenesis among Mollusca. Zool. Bidr. Uppsala 30, 399–456.

Franzén, Å., 1956. On spermiogenesis, morphology of the spermatozoon, and biology of fertilization among invertebrates. Zool. Bidr. Uppsala 31, 356–482.

Franzén, Å., 1967. Spermatogenesis and spermatozoa of the Cephalopoda. Ark. Zool. 19, 323–334.

Fretter, V., Graham, A., 1964. Reproduction. In: Wilbur, K.M., Yonge, C.M. (Eds.), Physiology of Mollusca. vol. 1. Academic Press, New York, pp. 127–164.

Gabr, H.R., Hanlon, R.T., Hanafy, M.H., El-Etreby, S.G., 1998. Maturation, fecundity and seasonality of reproduction of two commercially valuable cuttlefish, *Sepia pharaonis* and *S. dollfusi*, in the Suez Canal. Fish. Res. 36, 99–115.

Gardiner, M.S., 1972. The Biology of Invertebrates. McGraw-Hill Book Company, London, New York.

Giese, A.C., Pearse, J.S., 1974. Reproduction of Marine Invertebrates. Acoelomate and Pseudocoelomate Metazoans. vol. I, Academic Press, London.

Giese, A.C., Pearse, J.S., 1977. Reproduction of Marine Invertebrates. Molluscs: Gastropods and Cephalopods. vol. IV, Academic Press, New York, London.

Gonzalez, A.F., Guerra, A., 1996. Reproductive biology of the short-finned squid *Illex coindetii* (Cephalopoda, Ommastrephidae) of the NorthEast Atlantic. Sarsia 81, 107–118 (Bergen).

Hackman, R.H., 1953. Chemistry of insect cuticle. 3. Hardening and darkening of cuticle. Biochem. J. 54, 371.

Hackman, R.H., 1971. The integument of Arthropoda. In: Chemical Zoology'. vol. VI. Academic Press, New York/London (Chapter 1).

Hamon, M., 1939a. Les constituents chimques des envelopes des spermatophores de Céphalopodes. C. R. Hebd. Séanc. Acad. Sci. Paris 208, 387–389.

Hamon, M., 1939b. Caractérization de quelques acides a-aminés entrant dans la constitution des spermatophores des Céphalopodes. C. R. Hebd. Séanc. Acad. Sci. Paris 208, 835–837.

Hashem, H.O., 1973. A Study of Sulphur Containing Cuticle of a Scorpion, *Androctonus bicolor* (H&E) (Ph.D. thesis). University of Alexandria, Egypt.

Haven, N., 1977. Cephalopoda: Nautiloidea. In: Giese, Pearse (Eds.), Reproduction of Marine Invertebrates. vol. IV. Academic Press, New York, pp. 227–241.

Hoadley, L., 1930. Polocyte formation and the cleavage of the polar body in *Loligo* and *Chaetopterus*. Biol. Bull. 58, 256–264.

Hoving, H.J.T., Laptikhovsky, V., 2007. Getting under the skin: autonomous implantation of squid spermatophores. Biol. Bull. 212, 177–179.

Hoving, H.J.T., Nauwelaerts, S., Van Genne, B., Stamhuis, E.J., Zumholz, K., 2009. Spermatophore implantation in *Rossia moelleri* Steenstrup, 1856 (Sepiolidae—Cephalopoda). J. Exp. Mar. Biol. Ecol. 372, 75–81.

Ismail, S.Z., 1981. The Structure of the Spermatophore and the Male Reproductive System of *Sepia officinalis* (M.Sc. thesis). Faculty of Science, Alex. Univ., Egypt.

Jackson, G.D., Jackson, C.H., 2004. Mating and spermatophore placement in the onychoteuthid squid *Moroteuthis ingens*. J. Mar. Biol. Assoc. UK 84 (4), 783–784.

Karlson, P., 1965. Introduction to Modern Biochemistry, second ed. Translated by Doering, C.H., Academic Press, New York, London.

Keele, C.A., Neil, E., 1973. Sanson Wright's Applied Physiology, 12th ed. Oxford University Press.

Kume, H., Dan, K., 1968. Invertebrate Embryology. Translated by Dan, J.C. NOLIT Publishing House, Belgrade, Yugoslavia.

Lankester, E.R., 1875. Observation on the development of Cephalopoda. J. Microsc. Sci. 15 (cited in Ishikawa 1968).

Lankester, E.R., 1893. Note on the coelom and vascular system of Mollusca and Arthropoda. Q. J. Microsc. Sci. 34, 427–432.

Leake, L.D., 1975. Comparative Histology, An Introduction to the Microscopic Structure of Animals. Academic Press, London.

Lee, B., 1950. The Microtomist's Vade-Mecum, 11th ed. Churchill, London.

Malek, S.R.A., Bawab, F. M., 1971. Tanning in the spermatophore of a crustacean *Penaeus trisulcatus* (= *P. kerathurus*). Exp. Dermatol. 27, 1098.

Mahoney, R., 1973. Laboratory Techniques in Zoology. Butterworths and Co, London.

Mann, T., 1963. 5-Hydroxytryptamine in the spermatophoric sac of the octopus. Nature 199, 1066–1067.

Mann, T., 1975. Spermatophores. In: Duckett, Racey (Eds.), Biology of the Male Gamete. vol. 7. Linnean Society of London, Academic Press, pp. 417–421.

Mann, T., Martin Jr., A.W., Thiersch, J.B., 1966. Spermatophores and spermatophoric reaction in the giant octopus of the North Pacific, *Octopus dofleini martini*. Nature 211, 1279–1282.

Mann, T., Martin Jr., A.W., Thiersch, J.B., 1970. Male reproductive tract, spermatophores and spermatophoric reaction in the giant octopus of the North Pacific, *Octopus dofleini martini*. Proc. R. Soc. Lond. B 175, 31–61.

Mann, T., Martin Jr., A.W., Thiersch, J.B., 1981. Changes in the spermatophoric plasma during spermatophore development and during the spermatophoric reaction in the giant octopus of the North Pacific, *Octopus dofleini martini*. Mar Biol (Berl) 63, 121–127.

Mann, T., 1984. Spermatophores. Springer-Verlag, Berlin, New York.

Marchand, W., 1913. Studies Über Cephalopoden. II. Über die Spermatophoren. Zoologica (NY) 26, 169–200.

Marshall, A.J., Williams, W.D., 1975. A Text Book Zoology: The Invertebrates. ELBS and Macmillan.

MartÍnez-Soler, F., Kurtz, K., Ausió, J., Chiva, M., 2007. Transition of nuclear proteins and chromatin structure in spermiogenesis of *S. officinalis*. Mol. Reprod. Dev. 74 (3), 360–370.

Milne-Edwards, H., 1842. Observations sur la structure et les functions de quelques Zoophytes, Mollusques, et Crustacées des côtes de France. IV. Sur les spermatophores de Céphalopodes. Ann. Sci. Nat. 18 (2), 331–350. *Zoologie*.

Morton, J.E., 1967. Molluscs, fourth ed. Hutchinson University Library, London.

Murray, R.K., Granner, D.K., Mayes, P.A., Rodwell, V.W., 1996. Harper's Biochemistry. 24th ed., Middle East edition Librairie de Liban, Appleton and Lange, Norwalk, CT/Los Altos, CA.

Needham, J.T., 1745. New Microscopical Discoveries. Printed for F. Needham, over-against Gray's Inn in Holborn, London, 47–49 and 56–57.

Nesis, K.N., 1995. Mating, spawning and death in oceanic cephalopods: a review. Ruthenica 6 (1), 23–64.

Nigmatullin, C.M., Sabirov, R.M., Zalygalin, V.P., 2003. Ontogenetic aspects of morphology, size, structure and production of spermatophores in ommastrephid squids: an overview. Berliner Paläobiol. Abh. Berlin. 03, 225–240.

Norman, M.D., Lu, C.C., 1997. Sex in giant squid. Nature 389, 683–684.

Önsoy, B., Salman, A., 2005. Reproductive biology of the common cuttlefish *Sepia officinalis* L. (Sepiida: Cephalopoda) in the Aegean Sea. Turk. J. Vet. Anim. Sci. 29, 613–619.

Oordt, G.V., Damste, P.H., 1939. Experimental modification of the sexual cycle and moult in the Greenfinch. Acta brev. neerl. Physiol. 9, 140–143.

Pantin, C.F.A., 1948. Notes on Microscopical Technique for Zoologists. University Press, Cambridge.

Parker, T.J., Haswell, W.A., 1943a. A Text Book of Zoology, sixth ed. vol. I. Macmillan and Co., Limited, London.

Parker, T.J., Haswell, W.A., 1943b. A Text Book of Zoology, fourth ed. vol. I. Macmillan and Co., Limited, London.

Parker, T.J., Haswell, W.A., 1974. A Text book of Zoology. Vertebates, vol. 2, seventh ed. English Language Book Society and Macmillan.

Pearse, A.G.E., 1968. Histochemistry. Theoretical and Applied, third ed. vol. I. Churchill, London.

Pelseneer, P., 1906. In: Lankester, E.R., Black, C. (Eds.), Mollusca: A Treatise on Zoology. London. Mem., 32, vol. V. Universtiy Press of Liverpool.

Peters, W., 1972. Occurrence of chitin in Mollusca. Comp. Biochem. Physiol. 41B, 541–550.

Peterson, R.P., 1959. The anatomy and histology of the reproductive system of *Octopus bimaculoides*. J. Morphol. 104 (1), 61–88.

Pflugfeder, O., 1970. Lehrbuch dez Entwicklungsgeschichte und Entwicklungsphysiologie der Teire. Veb Gustav Fischer Verlog Jena.

Purchon, R.D., 1968. The Biology of Mollusca. Pergamon Press, Oxford.

Richards, A.G., 1951. The Integument of Arthropods. Minneapolis University of Minnesota Press.

Richards, A.G., 1970. Chitin. In: Gray, P. (Ed.), The Encyclopedia of the Biological Sciences, second ed. Van Nostrand Reinhold Company.

Robertson, R., 1978. Spermatophores of six eastern north American pyramidellid gastropods and their systematic significance (with the new genus Boonea). Biol. Bull. 155, 360–382.

Rodrigues, M., Troncoso, J.S., Guerra, Á., 2010. Spermatophore components in the Atlantic bobtail squid *Sepiola atlantica* (Cephalopoda: Sepiolidae). In: XVI Simposio Iberico de Estudios de Biologia Marina. Atlantic (Espanha).

Roosen-Runge, E.C., 1977. The Process of Spermatogenesis in Animals. Cambridge University Press, London.

Salman, A., Önsoy, B., 2010. Reproductive biology of the bobtail squid *Rossia macrosoma* (Cephalopoda: Sepiolidea) from the eastern Mediterranean. Turk. J. Fish. Aquat. Sci. 10, 81–86.

Sasaki, M., 1929. A monograph of the Dibranchiata cephalopods of the Japanese and adjacent waters. J. Fac. Sci. Univ. Tokyo Zool. Suppl. 20, 1–357 (cited in Kumé and Dan, 1968).

Selman, K., Arnold, J.M., 1977. An ultrastructural and cytochemical analysis of oogenesis in the squid, *Loligo pealei*. J. Morphol. 152, 381–400.

Selman, K., Wallace, R.A., 1972. A role of follicle cells during vitellogenesis in the squid *Loligo pealei*. Biol. Bull. (Woods Hole, MA) 143, 477.

Smyth, J.D., 1954. A technique for histochemical demonstration of polyphenol oxidase and its application to egg-shell formation in helminth and byssus formation in *Mytlus*. Q. J. Microsc. Sci. 95 (2), 139–152.

Soeda, J., 1952. On the artificial fertilization and early cleavage of the *Ommastrephes* egg. Hokkaido Reg. Fish. Res. Rep. 5 (in Japanese) (cited in Ishikawa, 1968).

Soeda, J., 1954. On the fertilization and shape of the *Ommastrephes* egg. Hokkaido Reg. Fish. Res. Rep. 11 (in Japanese) (cited in Ishikawa, 1968).

Storer, T.I., Stebbins, R.C., Usinger, R.L., Nybakken, J.W., 1983. General Zoology, sixth ed. Tata McGraw-Hill Publishing Company Ltd, New Delhi.

Swammerdam, J., 1637–1685. Biblia Naturae (Bybel der Nature) was Published Posthumously, in 1737. Isaak Severinus, Boudowyn Vander and Pieter Vander, Leyden. The English translation by Thomas Flloyd, The Book of Nature, was published in 1758, London.

Threadgold, L.T., 1976. The Ultrastructure of the Animal Cell, second ed. Pergamon Press, New York.

Thesing, C., 1904. Beiträge zur Spermatogenese der Cephalopoden. Z. Wiss. Zool. 76, 94–136.

Tompsett, D.H., 1939. Sepia. In: Liverpool Mar. Biol. Com. Mem. vol. 32. University Press of Liverpool.

van Oordt, G.J., 1938. The spermatheca of *Loligo vulgaris*. 1- Structure of the spermatheca and function of its unicellular glands. Q. J. Microsc. Sci. 80 (320: part IV), 593–599.

Vigelius, W.J., 1880. Über das Excretions system der Cephalopoden Niederland. Arch. Zool. Bd. 2, 115–184.

Watase, S., 1888. Observations on the development of Cephalopods: homology of the germ layer. Stud. Johns Hoppkins Biol. Lab. 4, 165–183.

Wells, M.J., Wells, J., 1977. Cephalopoda: Octopoda. In: Giese, Pearse (Eds.), Reproduction of Marine Invertebrates. Academic Press, New York, pp. 291–336.

Williams, L.W., 1910. The Anatomy of the Common Squid, *Loligo pealei*. Leiden, Brill, E.J., Lesuer.

Wouters-Tyrou, D., Martin-Ponthieu, A., Richard, A., Sautière, P., 1988. Characterization of the nuclear basic proteins specific of spermiogenesis in cuttlefish, *Sepia officinalis*. Biochim. Biophys. Acta Protein Struct. Mol. Enzymol. 953, 86–94.

Phylum Chordata

Chapter Outline

Definition **863**
General characters **863**
Class ascidiacea **864**
Characteristics 865
Classification 865
Order 1- Enterogona 865
Order 2- Pleurogona 865
Reproduction 865
Agametic reproduction 865
Gametic reproduction 867
Gametogenesis 868
Spermatogenesis 868
The sperm 868
Oogenesis 868
The full grown oocyte 870
Spawning 874
Fertilization 874
Early development (embryogenesis) 877
I. Maturation and ooplasmic segregation 878
a. The completion of meiosis of the egg 878
b. The development of the sperm pronucleus 878
c. Ooplasmic segregation 878
d. Syngamy, or pronuclear fusion 878
II. Cleavage (second phase of embryogenesis) 880
III. Gastrulation and neurulation 884
Gastrulation 884
Neurulation 886
IV. Morphogenesis and cell differentiation 886
Hatching 886
Structure of fully formed larva 888
Metamorphosis 888
Differentiation of adult organs 889
Endostyle 889
Atrium 889
Alimentary canal 889
Nervous system 890
Time required for development 890
Experimental studies 890
References **898**

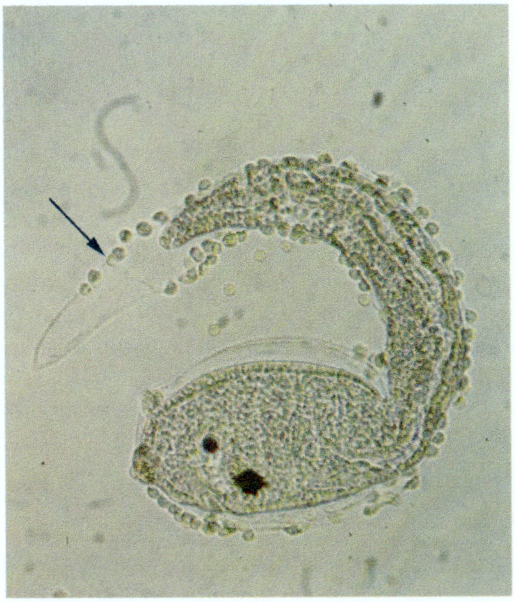

Chapter 13

Subphylum Urochordata (Tunicata)

DEFINITION

It is well known that the chordates constitute the largest phylum of deuterostomes, but most chordates are vertebrates and are outside the scope of this book. Two subphyla, the Urochordata and Cephalochordata, lack a backbone but possess the three distinguishing chordate characteristics—at some time in the life cycle, there can be found a notochord, a dorsal hollow nerve cord and pharyngeal clefts. The urochordates and cephalochordates are known as protochordates because of their simple body plans relative to those of the vertebrates.

Which of the protochordates is most closely related to the vertebrates? Studies have been made in reconstructing the evolutionary history of organisms using DNA sequences. Fig. 13.1 shows a chordate evolutionary tree based on DNA sequences. According to this tree, the tunicates diverged from the chordate line before the appearance of the cephalochordates and vertebrates. Therefore, the cephalochordates appear to be most closely related to the vertebrates, while the tunicates are the next closest vertebrate relative.

Since cephalochordates (e.g. *Amphioxus*) initiate most of the vertebrate embryology courses, this group is not treated here. In spite of the fact that the members of the urochordates are less familiar, they deserve some attention, as they are quite common marine animals.

Adult urochordates, commonly known as tunicates, little resemble other chordates. About 1300 species are known and are reasonably studied. Tunicates have been popular subjects in developmental biology for more than a century and they were used in experimental embryology. They are soft-bodied animals encased in a gelatinous or leathery tunic and hence the name of the group, **Tunicata**. They were known to Aristotle in ancient Greece, more than 2000 years ago, he gave a short account of a Simple Ascidian under the name of **Teihywm** (**Thalia**). This name is still used for a class of the group. When naturalists began to classify animals, the tunicates were considered to be molluscs because of their soft bodies and filter-feeding life styles. However, the tadpole larva of these animals was discovered in 1841 by H. Milne-Edwards, and in 1847 by Van Beneden (Kumé and Dan, 1968), and was thoroughly studied by the Russian biologist A. Kowalevsky in 1866. The latter showed that tunicates have a tailed larva with a dorsal nervous system and notochord, the unmistakable features of a chordate. In most tunicates, these chordate features disappear during metamorphosis, and the adult bears no resemblance to other chordates. Kowalevsky's important discovery sparked interest in their development. On the other hand, H. Milne-Edwards was one of the pioneers (perhaps the most famous), who studied and was greatly interested in the field of Tunicata. In 1826, he and Audouin made a series of observations on living Compound Ascidians. Among other discoveries, they found the free-swimming tailed larva, and traced its development into the young Ascidian. Milne-Edwards also founded the group of 'social' Ascidians, now known as the Clavelinidm, and gave a classification of the Compound Ascidians, which was universally accepted for many years. For his voluminous and valuable contributions, Henri Milne-Edwards' name was honoured in about 50 names of genera and species.

GENERAL CHARACTERS

- Most of the tunicates are somewhat barrel-shaped attached by one end to the substratum.

Invertebrate Embryology and Reproduction. https://doi.org/10.1016/B978-0-12-814114-4.00013-8

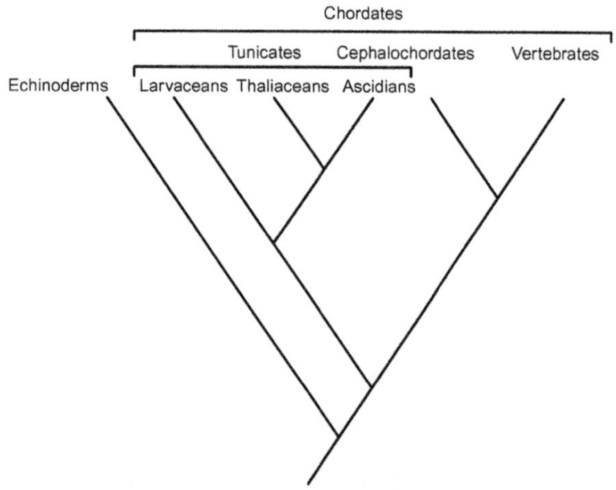

FIG. 13.1 A chordate evolutionary tree inferred from small ribosomal DNA sequences. *(From Gilbert and Raunio (1997).)*

- The body is not regionally divided and the coelom is absent.
- There is a notochord only in the tail of the larva, which is lost in the adult (except larvacea).
- There is a postanal tail in the larva. The larva looks like a microscopic tadpole, and possesses distinct chordate characteristics.
- There is an atrium into which the pharyngeal clefts open in the adult.
- The central nervous system is separated from the epidermis. In the adult, it is reduced to a single, solid, elongated ganglion, whereas in the larva it is more prominent and hollow.
- There is typically a well-developed blood system with a ventral heart through which the direction of flow of the blood is reversed from time to time.
- The body of the adult is enclosed in a test (tunic) usually composed of tunicin, a substance related to cellulose.
- Most of the members are hermaphrodites.
- The tunicates consist of three classes: larvacea, ascidiacea, and thaliacea.

The ascidians contain the majority of species and the most common and typical tunicates. The development of the members of this class has been most thoroughly studied, in contrast to that of the thaliaceans and larvaceans. The animals of the two latter groups inhabit the open ocean and so they are difficult to collect and transport to laboratories. Thus, the study of their development is poor. On the other hand, ascidians are popular as research animals in developmental biology because of the following:

1. They are easy to collect since they are sessile and usually live near the seashore.
2. Ascidian embryos and larvae have a small number of cells, e.g. the initial gastrula contains only about 110 cells, while the tadpole larva consists of a few thousand cells and only six different tissues.
3. Some ascidians have transparent embryos or large embryos; others have coloured cytoplasm.
4. Ascidians develop rapidly; rapid development is an advantage for experimental analysis.
5. Some ascidians have eggs with coloured cytoplasmic regions, which facilitates following cell fates during cleavage.
6. As simple chordates, ascidians may provide basic information about the development of more complex chordates.
7. Ascidians have small genomes that facilitate cloning genes involved in developmental processes.
8. Some closely related ascidian species have evolved different modes of development. This permits the study of evolution of development.

There are also some limitations to using ascidians as an experimental system:

1. Most ascidians have restricted breeding seasons, and so living embryos can be obtained only at certain times of the year.
2. Their genetic analysis is usually not available, as it is in *Drosophila, Caenorhabditis elegans* and the mouse.

To find an ascidian species with attractive principal features for research work, Jeffery and Swalla (1997) tabulated the species used for this purpose (Table 13.1).

CLASS ASCIDIACEA

Ascidians, often called sea squirts, are common marine invertebrates throughout the world. The majority is found in littoral waters; they attach to rocks, shells, pilings and ship bottoms, or are sometimes fixed in mud and sand by filaments or a stalk.

TABLE 13.1 Ascidians used in developmental biology

Genus/species	Family	Major attribute
Ciona sp.	Cionidae	Rapid development; long breeding season
Phallusia mammallata	Ascidiidae	Transparent embryos
Ascidia sp.	Ascidiidae	Transparent embryos
Styela sp.	Styelidae	Coloured cytoplasm
Botryllus sp.	Botryllidae	Colony formation agametic reproduction
Boltenia villosa	Pyuridae	Coloured cytoplasm
Halocynthia roretzi	Pyuridae	Large embryos
Molgula citrina, M. oclata, M. occulta	Molgulidae	Evolutionary studies

Characteristics

- The adult has lost the larval tail and is variable in form.
- The test is permanent and thick.
- There are several pharyngeal clefts, usually subdivided by longitudinal and transverse bars.
- There is an atrium opening dorsally.
- The nervous system is reduced to a single ganglion.
- The adult is always sedentary and often colonial.
- Budding usually occurs in colonial forms.

Classification

The system of classification of ascidians currently in use is that of Huus (1937). This system roughly divides the class into two orders:

Order 1- Enterogona

- The gonad is unpaired and lies close to the intestine.
- The body may or may not be divided into 'thorax' and 'abdomen'.
- The atrial cavity develops from paired lateral rudiments that later fuse mid-dorsally.
- Examples: *Ciona, Phallusia, Ascidia, Pyura, Halocynthia, Ascidiella.*

Order 2- Pleurogona

- The gonads are normally paired and always lie in the lateral mantle wall.
- The body is never divided.
- The atrial cavity is developed from an originally single median dorsal invagination of the ectoderm.
- Examples: *Styela, Botryllus, Boltenia.*

This division does not draw a major line of distinction between the single and compound species. Therefore, it is convenient, from the embryological point of view, to differentiate, following Savigny (1816), between simple ascidians (**Monoascidia**) which reproduce only gametically, and the compound ascidians (**Synascidia**) which show both gametic and agametic reproduction. The mode of gametic reproduction of both types of ascidians is essentially the same. However, the conspicuously yolky eggs of some of the compound species follow a slightly different pattern, and the formation of colonies is by budding.

The simple ascidians (Monoascdia) include both oviparous and viviparous forms. Among the oviparous are *Ciona*, *Ascidia*, *Phallusia Styela*, *Pyura* and *Molgula*, and the Japanese species *Halocynthia (Styela) roretzi* (Drasche) and *Chelyosoma siboja*. Among the viviparous forms are *Polycarpa*, *Boltenia*, *Molgula* and *Corella*. The fertilized eggs of the viviparous forms develop in the cloacal cavity.

Reproduction

Agametic reproduction

Regeneration and agametic reproduction are highly developed in colonial ascidians but absent in the solitary ones. Agametic reproduction takes place by means of budding, but is complex and exceedingly variable (Fig. 13.2). In fact, ascidians display the greatest variation in agametic budding among all groups of metazoans.

A tunicate bud is called a blastozooid and originates in different parts of the body in different groups of ascidians. With the variation of the site of bud formation, there is a corresponding variation in the germinal tissues included in the bud. The most primitive type of budding appears in species of *Clavelina* and other forms (Barnes, 1974), in which the bud arises from the stolon (Fig. 13.2A). The bud first appears as a swelling of the stolon filled with reserve food material, and may not develop until the stolon breaks away from the parent, or the parent degenerates. In *Clavelina* and other members of this family, epidermis and mesenchyme comprise the germinal tissues of such stoloniferous buds. In the colonial Polyclinidae, which have a long postabdomen bearing an extension of the epicardia and a cord of tissue from the gonads, the postabdomen breaks up within the tunic like a string of sausages (Fig. 13.2F). Each section becomes a bud, with the epidermis, genital cord and epicardial tissue acting as germinal tissues. The blastozooids reorganize in the base of the old tunic to form a new colony.

In the opinion of Berrill (1975), ascidians are with little doubt the most primitive of the three classes of tunicates. Even though the agametic reproductive processes in this group lead only to colony formation or growth, they do represent true agametic reproduction so far as the individual constituent zooids are concerned. However, as it relates to colony formation, Berrill reported that the process of budding in ascidians varies according to the general morphology of the organism:

1. Elongated thorax and a stalk-like abdomen. The latter contains the looped intestine, heart and gonads. In some, the body wall extends beyond the intestine as a postabdomen containing gonads and heart; and
2. **Holosomatous**, in which the intestinal loop, heart and gonads within the intestinal loop lie at one side of the thorax

In most merosomatous forms, a series of epidermal constrictions (strobilation) divide the abdomen or postabdomen into a series of short pieces capable of reconstituting a whole organism by a combination of reorganization and regeneration. The particular course taken depends on the character of the tissues contained within a given epidermal envelope. In species where there is no postabdomen, strobilation is confined to the abdomen between the base of the thorax and the stomach (Fig. 13.2). In most other species, strobilation

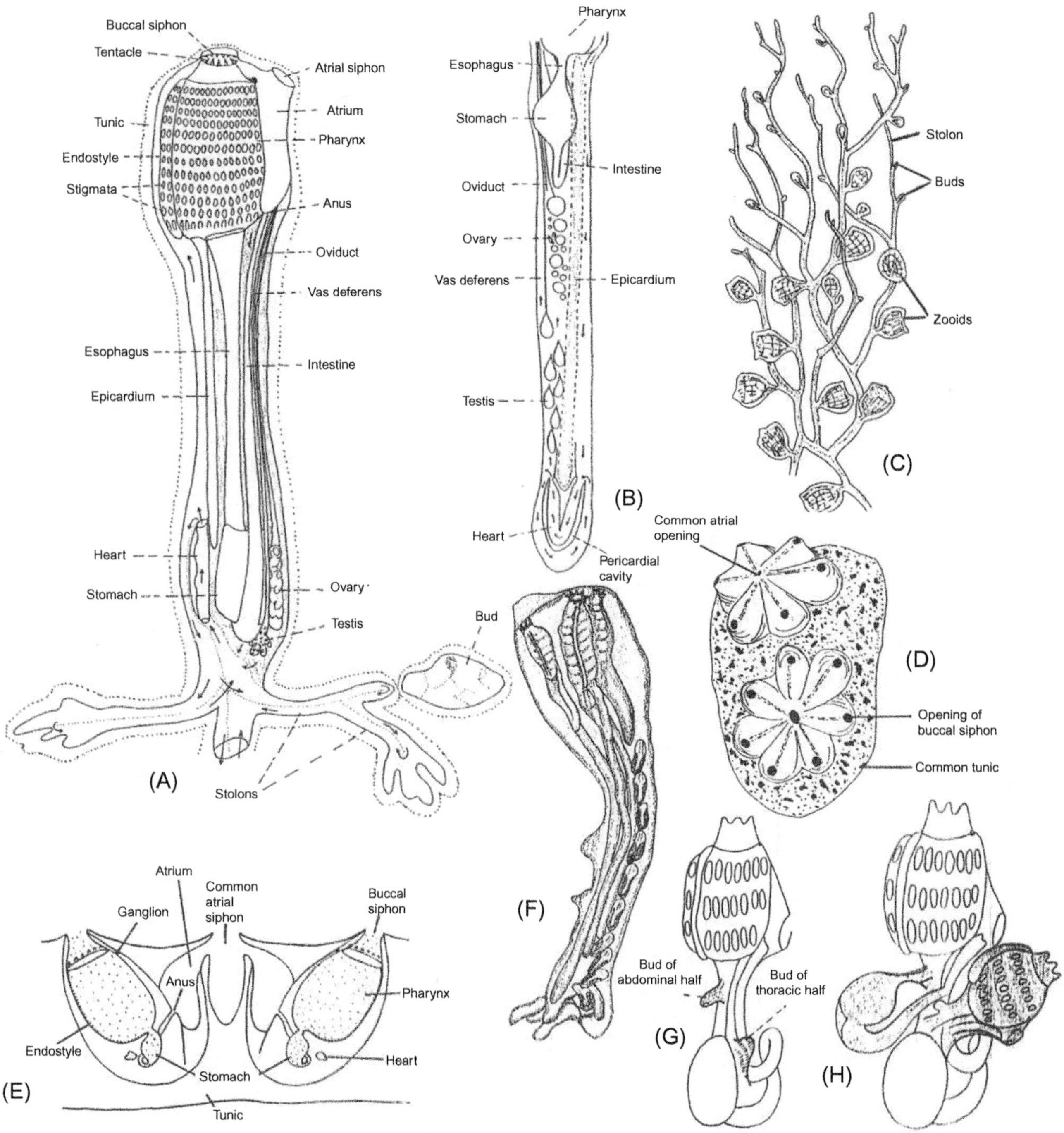

FIG. 13.2 Diagrammatic drawings illustrating agametic reproduction in some tunicates. (A) *Clavelina*, an ascidian with a long abdominal region *(after Brien)*; (B) abdominal and postabdominal region of *Sydnium (after Berrill from Brien), arrows* indicate paths of circulating blood; (C) Colony of *Perophora viridis (after Miner)*; (D) *Botryllus schlosseri*, a compound ascidian *(after Milne-Edwards from Yonge)*; (E) vertical section through *Botryllus (after Delage and Hérouud)*; (F) budding in the colonial *Circinalium concrescens*, a polyclinid. The long postabdomen becomes sectioned into buds within the common tunic of the colony *(after Brien)*; (G and H) budding in the Didemnldae; (G) two buds develop in abdominal region of parent body; (H) buds fuse, one forming the abdominal half of new individual and one forming the thoracic half *(after Salfi from Brien). All the diagrams are from Barnes (1974).*

affects only the postabdomen (Fig. 13.2). In the first group, a segment of the intestine is included, in the second it is not. (For details and more information, refer to Berrill, 1975.)

In a number of families, budding is precocious and begins at the larval stage. For example, in the Didemenidae, a few hours after settling of the larva, two buds form in the abdominal region (Fig. 13.2G). The two buds fuse, one forming the thoracic half and one the abdominal half of the new individual. In the Botryllidae, which forms the star-shaped colonies (Fig. 13.2D and E), a single bud develops on the larva. After fixation, the larva dies, and the bud develops as the initial member of the colony. The star-shaped pattern of the adult colony results from geometrical secondary budding.

Gametic reproduction

Ascidians are without exception hermaphroditic (Berrill, 1950). The gonads of some species were carefully studied by one of my students, now colleague, Saad (1995, 2002). From order Enterogona, he studied *Ciona intestinalis* (Linnaeus, 1767), *Ascidiella aspersa* (Müller, 1776) and *Phallusia mammillata* (Cuvier, 1815), in which the gonads are unpaired and it is always only one gonad/individual. From order Pleurogona, on the other hand, he studied *Styela plicata* (Lesieur, 1823) and *S. partita* (Stimpson, 1852), in which the gonads are normally paired, though it is not always the case. The number of gonads differs with the species and may differ in the same species. In *S. partita*, the gonads are two pairs, while in *S. plicata*, there is no definite number, as recorded by different authors. There are four gonads in the right side of the specimen and one or maximally two in the left side (Fig. 13.3). The most important organs that are greatly related to the gonads are the atrium and the branchial sac. The gonads are arranged asymmetrically, somewhat transversely underneath and enclosed in either sides of the atrial cavity. They are embedded in between the muscular wall of the mantle and the outer epithelial wall of the atrium bulging or budding into the atrial cavity (Fig. 13.4). The size and natural colour of the gonads differ with the degree of their maturity. Each gonad contains distinct regions devoted either to sperm or egg production, i.e. contains an ovary and a testis (Fig. 13.3), but the ovary constitutes the largest portion of the gonad.

Accordingly, an opinion was raised that some species undergo the phenomenon of **protandry**. Millar (1952) studied the annual growth and reproductive cycle of two compound and two simple ascidians. He concluded that the simple ascidians *Ascidiella aspersa* and *Ciona intestinalis* are protandrous. Sabbadin (1957) studied *Ciona*, *Molgula manhattensis* and *S. plicata*, and concluded that each of them is slightly protandrous. This means that protandry may be a common phenomenon within simple ascidians. However, Saad (1995, 2002) proved histologically that the germinal epithelium producing one type of gametes is completely separate from that producing the other type in *S. plicata*. This case is opposite to the case of the well-studied gastropod *Eremina ehrenbergi* (see later the chapter of Mollusca since protandry is common within many gastropods).

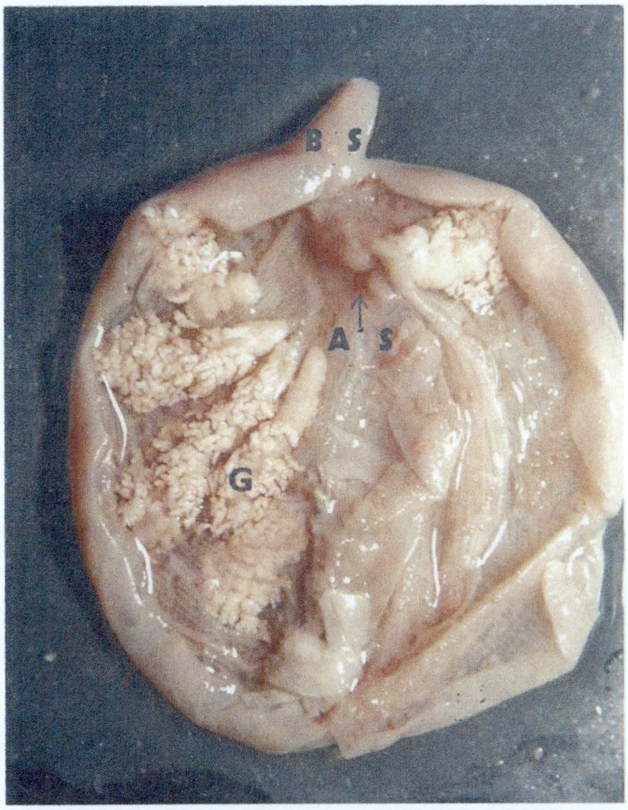

FIG. 13.3 Photograph of dissected specimen of *Styela plicata* obtained during September, about the natural size. The gonads in both sides are exposed after the complete removal of the branchial sac. AS = atrial siphon, BS = branchial siphon, G = gonad. *(From Bawab et al. (2012).)*

For convenience, enterogoniate will be indicated by EG and pleurogoneate by PG.

In respect to the breeding season of simple ascidians, it seems that there is no definite season for some species, e.g. the EG *Ascidia*, *Molgula*, *Ciona*, and *Eugyra* (Hirai, 1968). These species become sexually mature and their breeding activity falls off somewhat during the cold months, practically, they breed all the year round. In 2002, Saad and in 2008, Michael et al. confirmed that *Ciona intestinalis* breeds all year round. Their results were based on their field observations and anatomical and histological studies of the structure of the gonads.

On the other hand, Hirai (1968) reported that the breeding of some other species is limited to the warm months. The PG *Styela partita* reproduces in June to September, while the EG *Pyura vittata* reproduces in summer. In Japan, the EG *Halocynthia (Styela) roretzi* (Michael et al., 2008) spawns in Miyagi Prefecture from December to February. In the same location but in Onagawa Bay, *Styela clava*

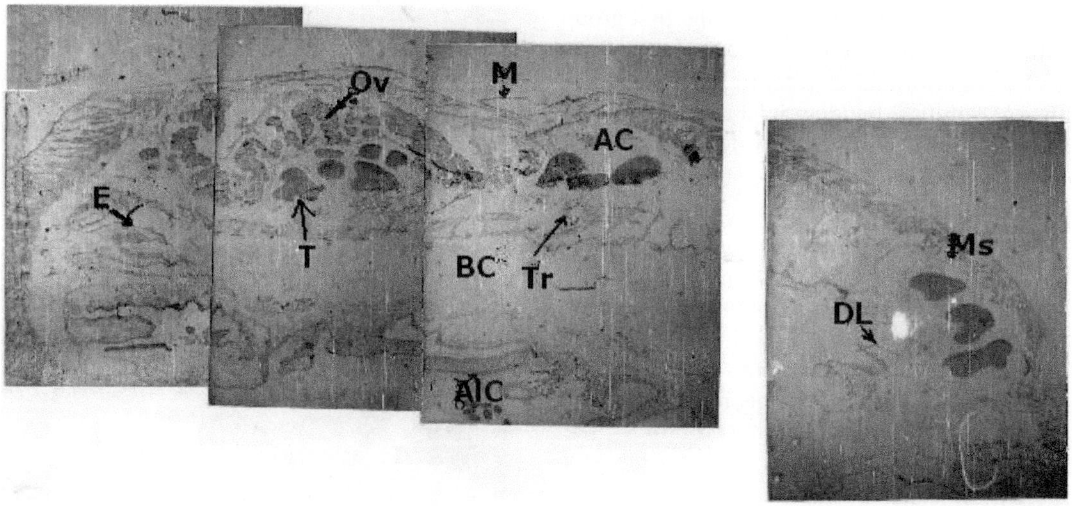

FIG. 13.4 A montage of photomicrographs of portions of one transverse section of *Styela plicata* showing the relationship between its internal organs. The specimen obtained during September. 60×. Ac=atrial cavity, Alc=alimentary canal, BC=branchial cavity (pharynx), DL=dorsal lamina, E=endostyle, M=mantle, Msc=muscles, Ov=ovary, T=testis, Tr=trabecular. Note, the section is so enlarged, even by one-k objective lens. Orientation of this section is according to the animal. *(From Bawab et al. (2012).)*

spawns during July. In addition, on Sweden's Skagerrak coast, it starts by July at the latest, and continues throughout the summer with a peak in August and September (Michael et al., 2008). At Misaki in Kanagawa Prefecture, *Styela plicata* spawns in June. In the waters of Alexandria, Egypt (south-east of the Mediterranean Sea) (Saad, 1995, 2002; Michael et al., 2008) observed and collected specimens of *S. plicata* and *S. partita* with full ripe gonads from September to February (autumn and winter) with a peak in September–October. Great numbers of highly mature specimens of *S. plicata* were collected during the latter period in two successive years: 1989, 1990 (Saad, 1995). The above review shows, that the breeding season of ascidians differs, not only with the species, but also the locality and/or the climate constitute two essential factors greatly affecting the breeding, controlling its season and the duration of this season.

Gametogenesis

Spermatogenesis. Histological appearance of the process of gametogenesis was described in some ascidians, *S. plicata*, *S. partita*, and *C. intestinalis* (Saad, 1995, 2002). The picture of the testicular region of the gonads of the two species of *Styela* is more or less the same. Stages of spermatogenesis are well pronounced during March–June (Fig. 13.5A). Almost all the germinal wall of the follicle (T.S. of seminiferous tubule) is occupied by many rows of cells, the size of which decreases from the periphery to the centre of the follicle. In July, the follicles show advanced stage of development since the zone of spermeogenesis dominates (Fig. 13.5B) and the smaller spermatids and/or developing sperm occupy the centre of the follicle where they may be preserved (Fig. 13.5C). In December, the follicles appear without a wall and with undistinguishable central mass of contents mingled with some large cells (Fig. 13.5D). It was interpreted that the central mass is formed of the remnants following the discharge of the sperm. The large cells then phagocytose these remnants to prepare the follicle for the proceeding cycle.

The sperm. As in most animals, the ascidian **sperm** are adapted for motility, transmission of the haploid genome to the egg and a short life span. However, these sperms are unique in the absence of the midpiece (note Fig. 13.14, Introduction). Each ascidian sperm consists of a head and a tail. The head contains the nucleus, an acrosome, centrosomal material and a single large mitochondrion, which is displaced to one side of the nucleus (Fig. 13.6). This single mitochondrion is formed by the fusion of numerous individual mitochondria in the spermatid cytoplasm. As the volume of the cytoplasm is reduced, the nucleus flattens and elongates. The change in the nuclear shape is mediated by bundles of microtubules, which are located in the cytoplasm adjacent to the nuclear envelope. Within the nucleus, the dispersed chromatin becomes condensed and compact into lamellae. This process is controlled by a change of chromosomal proteins from histones to protamines. The latter proteins are small basic nuclear proteins that cross-link DNA into tight packages. The tail of the sperm contains an axoneme, exhibiting a 9+2 pattern of microtubules. The mature sperm are stored in the lumen of the seminiferous tubules and pass out of the testes through the vas deferens during spawning (Gilbert and Raunio, 1997).

Oogenesis. The **eggs**, the other gamete, are produced in the ovarian region. Oogenesis proceeds in its definite phases. In *Styela plicata* and *S. partita* (Saad, 1995, 2002), for example,

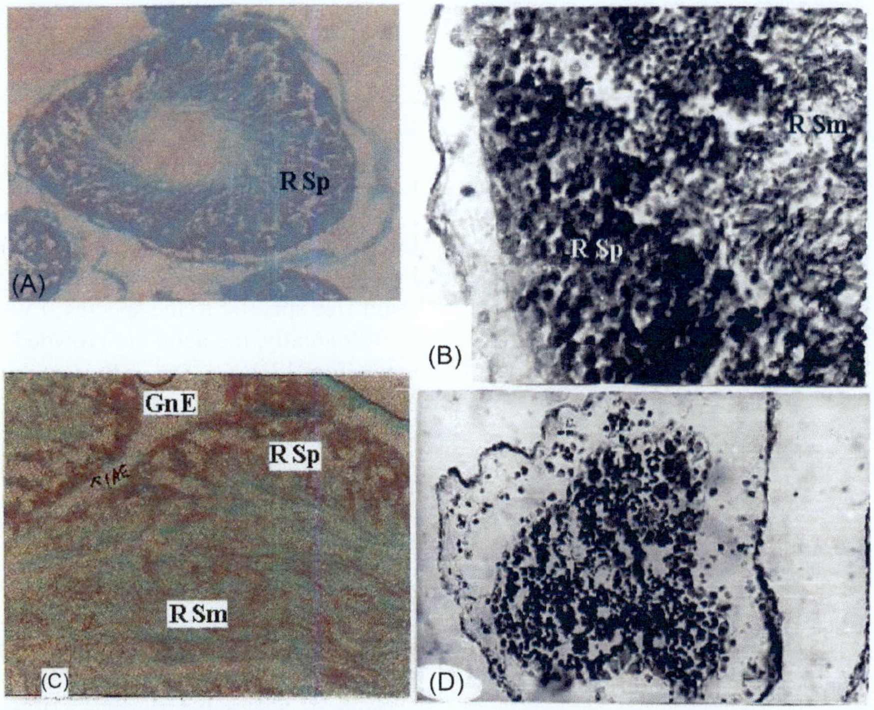

FIG. 13.5 Photomicrographs of transverse sections showing testicular follicles of *Styela* during spermatogenesis: (A) follicle at the beginning of the process. 100× *(after Saad, 2002)*; (B) oil immersion magnification of a portion of a follicle showing spermatogenesis and spermeogenesis. 1000× *(after Bawab et al.)*; (C) oil immersion magnification of a portion of a follicle showing developing sperm. 1000× *(after Saad, 2002)*; (D) a magnified testicular follicle during December after discharging its product. *(After Bawab et al.)*. (A) and (C) of *S. partita*; (B) and (D) of *S. plicata*.

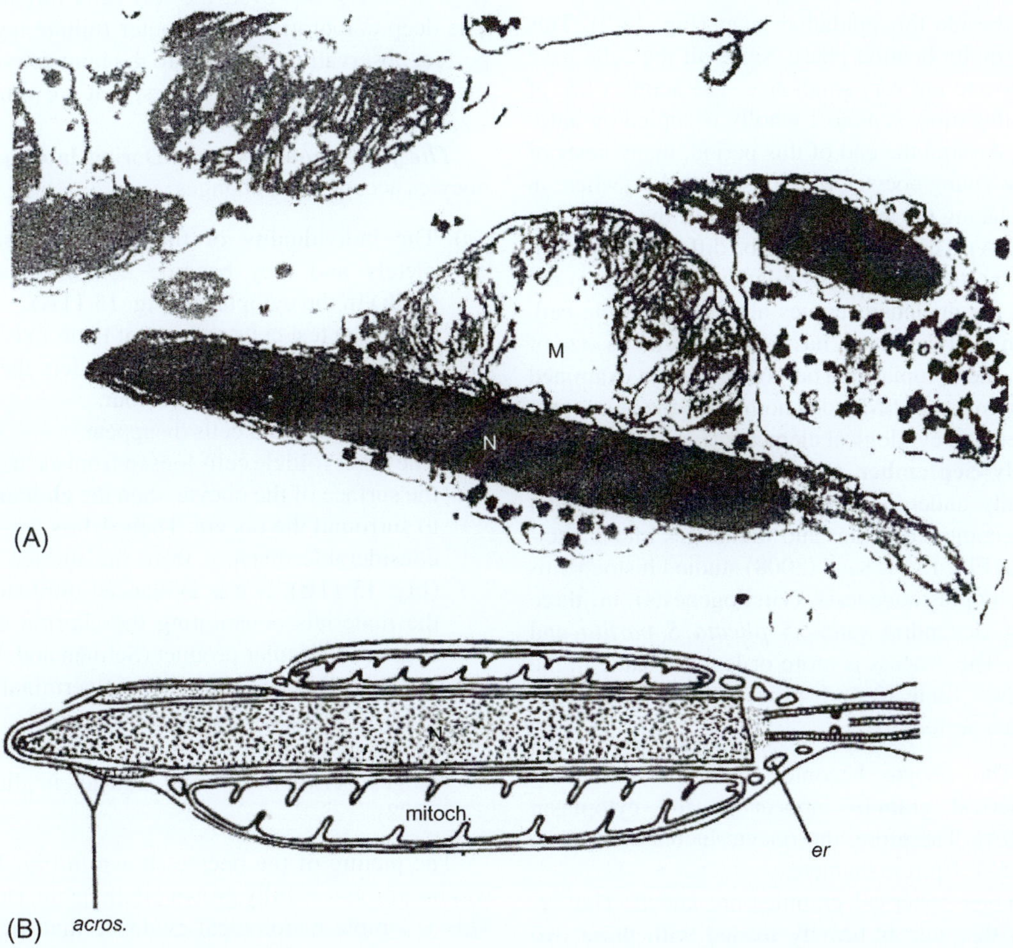

FIG. 13.6 (A) An electron micrograph showing the head of an ascidian sperm. N = nucleus, M = mitochondria *(after Fukomoto, 1990, from Gilbert and Raunio, 1997)*; (B) Diagram of a longitudinal section of a sperm of *Ascidia nigra*. A single gigantic mitochondrion (mitoch) is situated in the head of the sperm surrounding the nucleus (N). acros = acrosome, er = endoplasmic reticulum. *(Redrawn from Ursprung and Schabtach, 1964; from Reverberi, 1971.)*

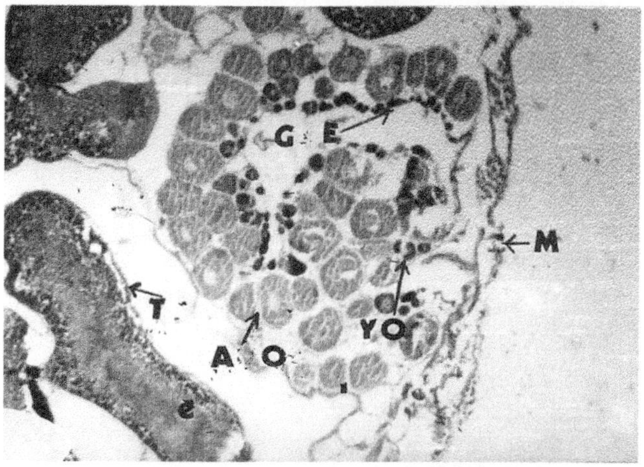

FIG. 13.7 Photomicrograph of a portion of a T.S. of *Styela plicata* showing the structure of the gonad during March. The acinus of the ovary is wholly occupied by autolized oocytes. 150× *(after Bawab et al., 2012).* AO = autolized oocytes, GE = germinal epithelium, M = mantle, T = testicular follicle, YO = oogonia and/or young oocyte.

the germinal epithelium was observed in its highest activity during **March–June**. Comparatively, a large number of oogonia was observed proliferating from, and linearly arranged on, or beside this epithelial layer (Fig. 13.7). This represents the multiplication phase, since all the cells have the same shape and are very small size. The acinus (T.S. of the ovary), at this time, is almost wholly occupied by autolysed oocytes. Around the end of this period, many nests of oogonia and/or young oocytes were observed everywhere in the acinus, replacing the autolysed oocytes of the preceding cycle (Fig. 13.8A). The oogonium can be differentiated from young oocyte by its huge nucleus relative to the size of the cell itself, i.e. the nucleus occupies almost the whole cell. The oocyte, on the other hand, has a normal relative size of the nucleus to the cytoplasm. The young oocyte examined by electron microscope (EM) adds nothing to the above picture, except detailed cytological elements (Fig. 13.8B).

During **July–September**, the growth phase occurs. The oocyte gradually undergoes two synchronized major vital activities: it becomes follicular and it changes into a vitellogenic oocyte (Fig. 13.9). Saad (2008) studied histochemichally the deutoplasmogenesis (vitellogenesis) in three ascidians from Alexandria waters, *S. plicata*, *S. partita*, and *C. intestinalis*. The process is more or less the same in the three species (see Table 13.2). It proceeds in three definite steps during this period.

- **Step 1**: The oocyte becomes follicular and very small spherical granules appear in the cytoplasm (Fig. 13.10A). Therefore, the oocyte increases in size, reaching 75–130 μm in diameter.
- **Step 2**: Larger spherical granules are added. The cytoplasm at this step is heavily loaded with these two components that their identification becomes difficult (Fig. 13.10B).
- **Step 3**: A third component in the form of patches appears during the end of August and early September.

During this latter month and after the third step, the ovaries attain their maximum size as it appears in Fig. 13.3, since each follicular oocyte attains its optimum size specific to the species, 150–200 μm in diameter. Histologically, the acini are crowded with fully grown oocytes (Fig. 13.10C), and each is enclosed within three well differentiated follicles (Fig. 13.10D) as indicated by the stains used (Saad, 2008):

1. The outer most is a thin membranous layer formed of simple squamous epithelium, referred to as outer follicle layer.
2. Just beneath the latter and above the plasma membrane of the cell, the inner follicle layer consists of relatively large cubical cells with clear cytoplasm.
3. Within the peripheral area of the cytoplasm of the oocyte, the **test cells** appear in a discontinuous layer.

Each cell is large cubical or rounded, more or less similar to that of the inner follicle. The synchronization of the formation of the three follicle layers was not determined (Fig. 13.10B); however, the test cells might be the latest. The deep colouration of the outer follicle layer permits its perfect observation, contrary to the faint colouration of both the inner follicle and test layers, that are similar, except in size and site.

The fully grown oocyte. During **late September**, the oocytes accept more changes:

(a) The individuality of the granules disappears completely and they become homogeneous component (**yolk**) in the cytoplasm (Fig. 13.11A).
(b) All of the test cells or most of them fade and intermingle with the yolk. The latter renders the fresh ovaries the deep reddish-orange colour.
(c) The outer follicle cells disappear.
(d) The inner follicle cells loosen from each other and from the surface of the oocyte, then the **chorion** is deposited to surround the oocyte. Their debris was observed at a considerable distance from the surface of the oocyte (Fig. 13.11B), It was evidenced ultra-structurally that the materials constituting the chorion of cephalopod eggs are follicular product (Selman and Arnold, 1977).
(e) A few cells from the adjacent germinal wall become gradually intact with the above ones.
(f) The oocyte attains its optimum size; in *S. plicata* it ranges between 180 and 200 μm in diameter (Saad, 2008).

The picture of the oocyte shown in Fig. 13.11A and B was interpreted as fully grown vitellogenic, isolecithal eggs. This is simple histological evidence that the ascidian egg is isolecithal, while Bates (2004) provided experimental evidence by ultraviolet irradiation of eggs and blastomeres.

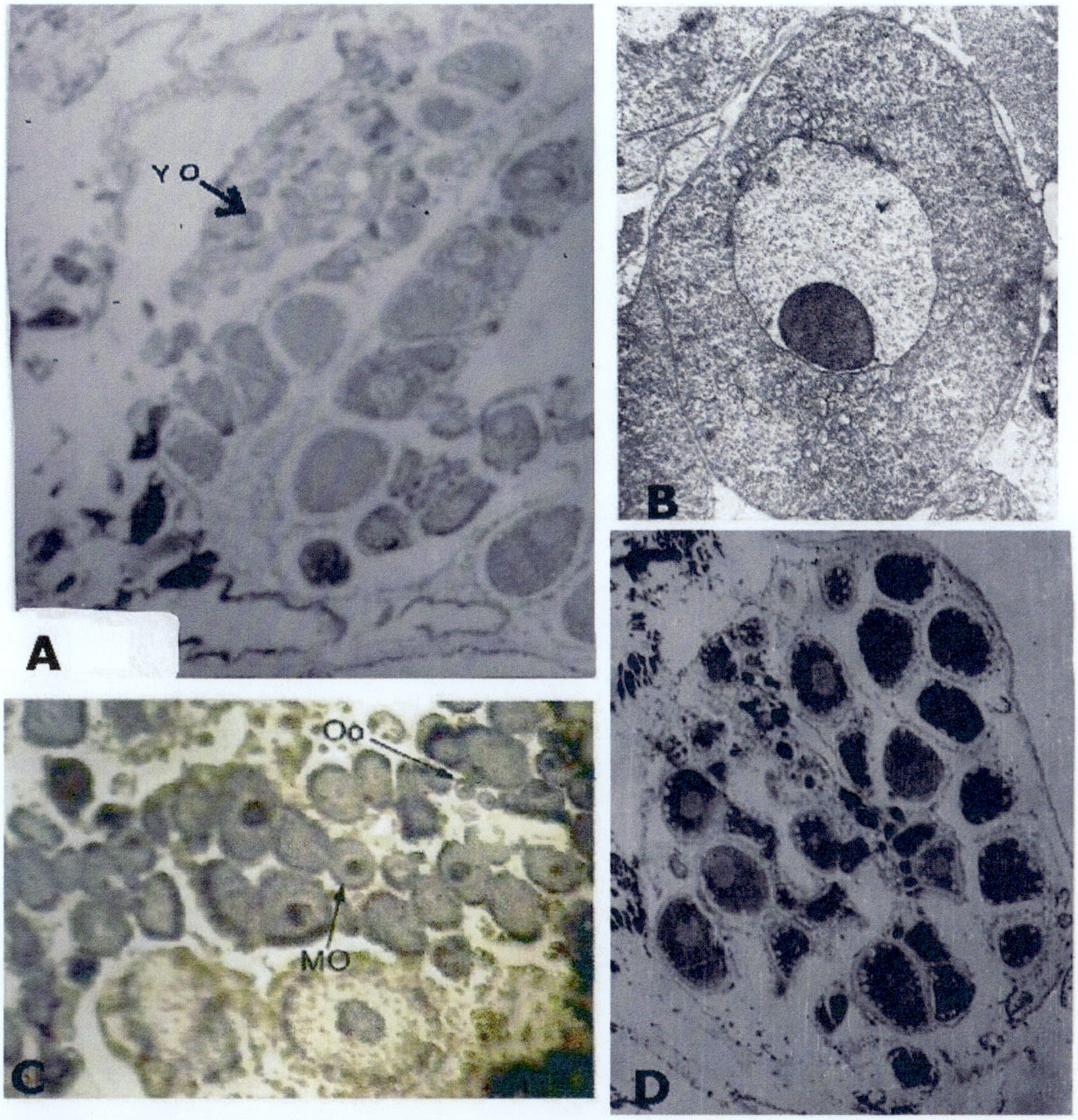

FIG. 13.8 Shows development of a tunicate oocyte. (A and D) Photomicrographs of portions of T.S. of *S. plicata (from Bawab et al., 2012)*; (C) photomicrograph of a portion of T.S. of ovary of *Ciona*. 400× *(from Saad, 2008)*; (B) original electron micrograph showing young oocyte of *Ciona*; note the large germinal vesicle, the eccentrically placed nucleolus, the cytoplasm crowded with vesicles of endoplasmic reticulum and mitochondria *(from Reverberi, 1971)*. The autolized oocytes seen in Fig. 13.7 are replaced by developing oocytes. Oo = oogonia, YO = young oocyte and MO = moderate-sized oocyte in (A) and (C). (D) shows follicular oocytes.

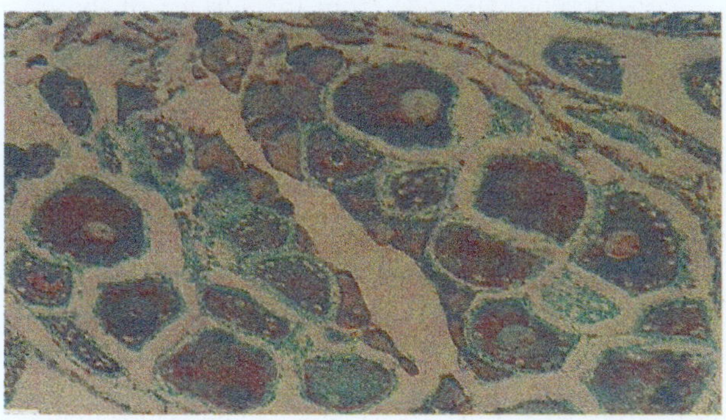

FIG. 13.9 Photomicrograph of a portion of a T.S. of *S. partita* showing follicular oocytes in the ovary of specimen obtained during July–August. The picture is almost the same of that of *S. plicata* during the same period as Fig. 13.8D. *(From Saad (2002).)*

Those oocytes (Fig. 13.11A and B) are on their way to be ovulated and to be discharged from the ovary. These two stages are more advanced, contrary to what is explained as a mature one in most of the literature. Morgan (1927), Tucker (1942), Hirai (1968) (Fig. 13.12) and Reverberi (1971), in their reviews, described and/or illustrated the ripe oocyte, in some species, as it retains its follicles especially the test cells. However, Morgan added that the test cells are later expelled from the egg and take no part in its development. It is worth mentioning that Saad (1995) studied the material by preparing the whole animal with its gonads intact, twice a month for a whole year, that involved a complete breeding cycle.

TABLE 13.2 Morphological nature of the cytoplasm and the nucleus of the oocyte of *Styela partita* at its different stages of development.

	Stages of oocyte development									
	Oogonia		Previtellogenic (young) oocyte		Vitellogenic oocyte				Postvitellogenic oocyte	
Stain references	Cytoplasm	Nucleus	Cytoplasm	Nucleus	Step 1 small granules	Step 2 large granules	Step 3 patches	Nucleus	Yolk material (homogenous)	Nucleus
Erlich hamatoxylin and eosin (Pearse, 1968)	Red	Blue	Blue	Deep blue	Deep dark blue	Opaque blue	Light blue	Deep blue	Blue	Deep blue
Heidenhain's iron hamatoxylin (Pearse, 1968)	Red	Blue	Red	Deep blue	Deep dark blue	Opaque blue	Light blue	Deep blue	Blue	Deep blue
Haemalum and eosin (Pearse, 1968)	Red	Deep blue	Dark red	Deep blue	Deep dark blue	Opaque blue	Light blue	Deep blue	Red	Deep blue
Mallory triple stain (Pearse, 1968)	Red	Deep blue	Dark red	Deep blue	Deep dark blue	Opaque blue	Light blue	Deep blue	Red	Deep blue
Masson trichrome stain (Pearse, 1968)	Red	Deep blue	Dark red	Deep blue	Deep dark blue	Opaque blue	Light blue	Deep blue	Red	Deep blue
Weigerert's haematoxylin and van gieson stain (Pearse, 1968)	Red	Deep blue	Dark red	Deep blue	Deep dark blue	Opaque blue	Light blue	Deep blue	Red	Deep blue
Alcian blue and eosin (Pearse, 1968)	Red	Blue	Red	Dark red	Deep dark blue	Opaque blue	Light blue	Deep red	Red	Deep blue
Safranin and light green (Pearse, 1968)	Faint green	Green	Light green	Dark green	Deep dark green	Opaque blue	Light green	Deep green	Green	Deep green
Modified azan stain (Pearse, 1968)	Blue	Blue	Blue	Deep blue	Deep dark blue	Opaque blue	Light blue	Deep blue	Blue	Deep blue

From Saad (2008).

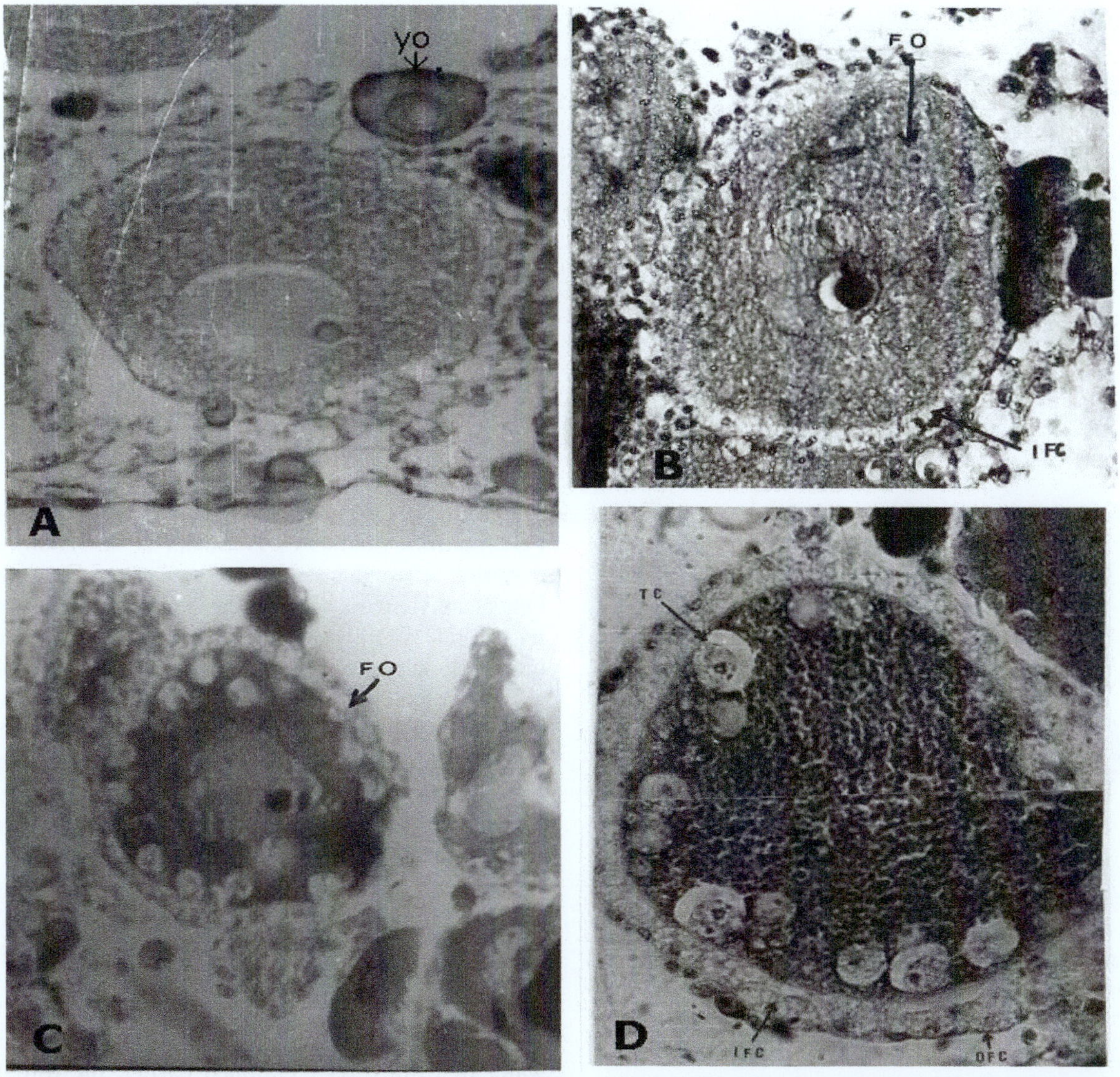

FIG. 13.10 Photomicrographs of portions of T.S. of *S. plicata* showing development of the oocyte to become follicular and vitellogenic, (A), (B) and (C) of specimens obtained during July–August. X for (A) and (B) 400, for C 250. (A) shows fully grown oocyte almost at the beginning of vitellogenesis; note its great resemblance to that of *Ciona* (Fig. 13.8B); (B) shows the formation of layer of inner follicle cells; (C) shows the completion of formation of follicular system; (D) complete follicular oocyte of specimen obtained during September, but before the end of vitellogenesis. 1000×. FO = follicular oocyte, IFC = inner follicular cell, OFC = outer follicle cell, TC = test cell, YO = young oocyte. *(From Bawab et al. (2012).)*

Since the appearance of test cells is a unique character for the oocyte of ascidians, it becomes an attractive subject to be discussed. However, neither definite nor acceptable function was recorded in the references for these test cells of any of the tunicates so far studied. Hirai (1968) recorded that the test cells are thought by some to have a secretory function, and by others to be nutritive. It was also suggested that they might produce enzymes. Seshachar and Rao (1963) studied these cells of the oocyte of the tunicate *Pyura squamulosa* in view of cytochemistry. It is rather interesting and astonishing that most of the reactions used in this work gave the same results recorded in the study of the chemical constituents of the yolk formed during the deutoplasmogenesis in the crab *Portunus pelagicus* (Bawab and El-Sherief, 1987). This may propose that these test cells have a role in the process of vitellogenesis in the oocyte of *S. plicata*. Their appearance later than the follicle layers, during the formation of the yolk and their fading during the late stages of the homogeneity of this yolk greatly evidenced this proposal (Fig. 13.11A and B). In addition, their almost complete disappearance after the process is reasonably interpreted as the test cells are specialized for vitellogenesis. Saad (2008) studied this process histochemically (Table 13.3), maybe for the first time in tunicates. He evidenced that the test cells, at least in this animal, *S. plicata*, are responsible for or sharing in the formation of the

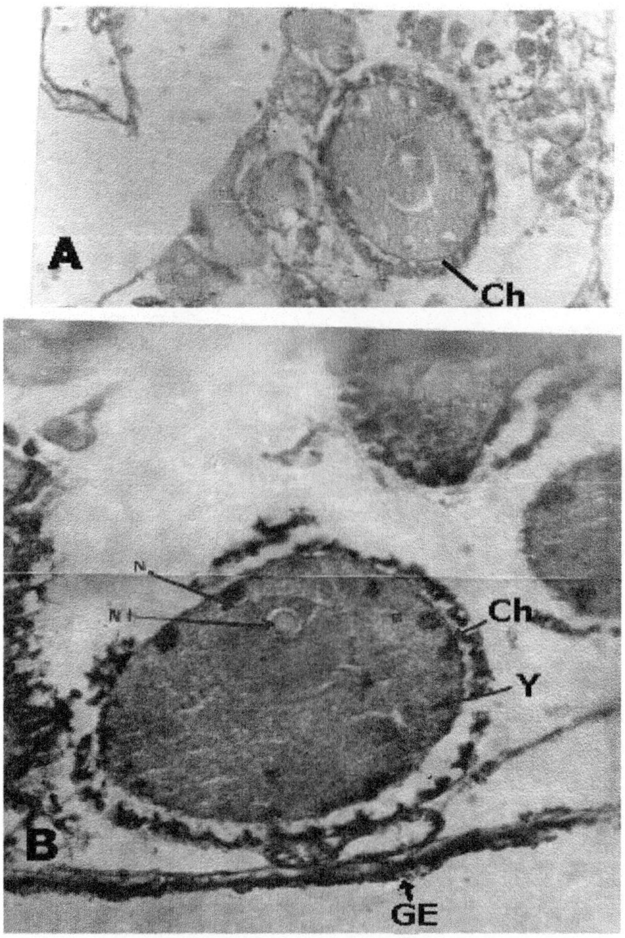

FIG. 13.11 Photomicrographs of magnified portions of T. S. of *S. plicata* obtained during the beginning of September (A) and end of September (B) showing full mature isolecithal oocytes enclosed in secondary membrane. Ch=chorion, GE=germinal epithelium, N=nucleus, Nl=nucleolus, Y=yolk. (B) shows oocyte on its way to be ovulated. *(After Bawab et al. (2012).)*

process. Previously, the electron microscopic studies on the other hand, made some important contributions concerning the function of the test cells. It was concluded that the test cells could be considered to correspond to the cortical granules observed in many types of eggs (for review see Reverberi, 1971). This means that they are secretory and so the above and earlier proposed functions can be accepted.

Spawning. Spawning is triggered by changes in the photoperiod. In some ascidians, a short pulse of light following an extending dark period triggers spawning. These species spawn in the morning. In other species, spawning is initiated by a longer light period following darkness. These species spawn at dusk. In the laboratory, spawning can be attained by adjusting the light–dark cycle. In this manner, sperm and eggs can be obtained for experimental purposes at any time during the breeding season (Gilbert and Raunio, 1997). An earlier method was reported by Hirai, 1968. It is possible to induce shedding in an individual that has spawned 2 or 3 days previously by injecting about 3 mL of 1/2 M KCl into the reproductive tract. In forms such as *C. intestinalis*, which have long genital ducts, the ripe gametes can be seen within the separate male and female ducts, and with care can be removed separately. Later, they were mixed at will to give artificial fertilization. In a recent publication (Hofmann et al., 2008), it was recorded that they obtained live gametes from *Ascidiella aspersa* and *Phallusia mammillata.* The live animals were dissected in seawater under a dissecting microscope. Eggs and sperms were obtained directly from the gonoducts using strictly separate micropipettes. Oocytes were placed in large Petri dishes (30 cm diameter). Subsequent cross-fertilization was carried out using small volumes of diluted sperm to prevent polyspermy. To ensure the success of this method, antibiotic must be added to the used seawater.

Fertilization. Studies of Hofmann et al. (2008) and the majority of ascidians studies show external fertilization. Others are viviparous, brood their embryos in the atrial cavity and release fully developed larvae. It is recorded in the introduction, that fertilization is divided into the following successive events: sperm attraction to the egg; sperm activation; sperm binding and penetration of the egg surface; and sperm-egg fusion. It appears that the steps of the process of fertilization in ascidians are the same as in sea urchins (Jeffery and Swalla, 1997). In addition, ascidian eggs have mechanisms to prevent polyspermy.

In ascidians, the vitelline envelope plays a major role in fertilization. However, *Phallusia* eggs lacking this vitelline envelope can be fertilized if sperm are first exposed to other eggs with follicle cells. Sperm activation may be achieved by interaction of the sperm with follicle cells or with a substance they secrete into seawater (like egg water of Lillie). As mentioned earlier, the follicle cells secrete materials that attract sperm to the egg (Miller, 1982). The molecules responsible for sperm chemotaxis are unknown. It was evidenced experimentally that the follicle cells also function in sperm activation. This means that the follicle cells are responsible for the two first events of fertilization in ascidian eggs (Jeffery and Swalla), or at least some of them. The third event is that the sperm must bind and penetrate the vitelline envelope to reach the egg surface. Sperm binding occurs at clefts between the bases of the follicle cells (Fig. 13.13), where parts of the underlying chorion are exposed to the environment. The general mechanism of sperm binding is the formation of a macromolecular complex between a sperm surface enzyme and a sugar residue in the chorion. The sugar-binding sites are probably parts of larger glycoprotein molecules. In *Ciona*, sperm bind to L-fucose residues in the chorion. This binding is mediated by the formation of a complex with α-L-fucosidase, which is present on the sperm surface. In *Phallusia* and *Ascidia*, the sperm binding site is *N*-acetylglucosamine.

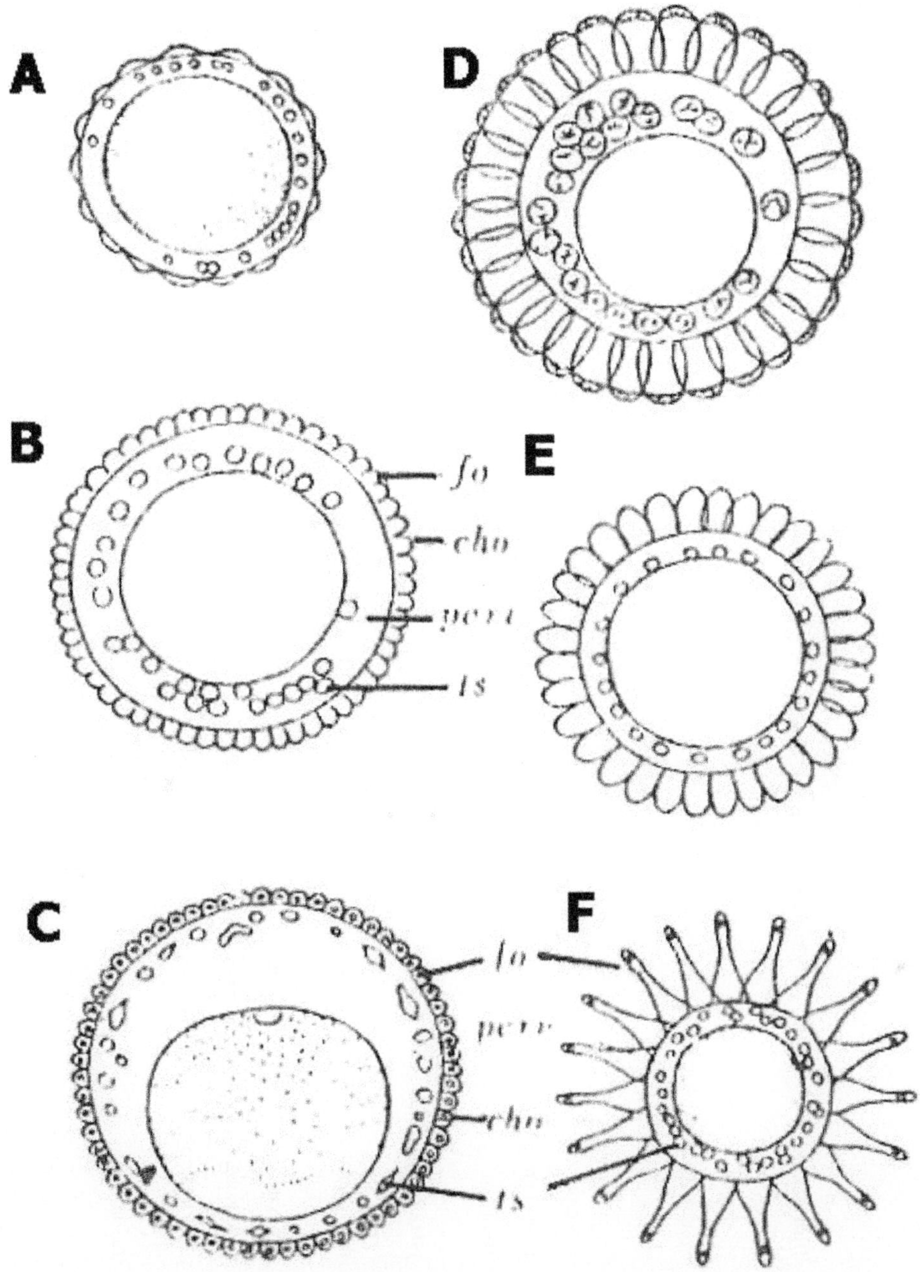

FIG. 13.12 Mature eggs of several simple ascidians. (A) *Styelopsis grossularia*, viviparous form with flattened follicle cells (Berrill, 1950); (B) *Chelyosoma siboja* (Hirai); (C) *Halocynthia roretzi* (Drasche) *(Hirai, 1941a,b)*; (D) *Corella parallelogramma* (Knaben, 1936); (E) *C. japonica*, var. *asamusi (Hirai, 1941a,b)*; (F) *Ciona intestinalis* (Berrill, 1947). cho=chorion, fo=follicle cell, perv=perivitelline space, ts=test cell. *(From Kumé and Dan (1968).)*

Many sperm are released in close proximity to the eggs during spawning. These eggs are usually fertilized by only one single sperm, suggesting that there are effective means to prevent polyspermy. In Ascidia, *N*-acetylglycosaminidase located on the egg surface is released into seawater as a response to the first sperm that binds to the chorion. This enzyme destroys the unoccupied *N*-acetylglucosamine binding sites in the chorion, prohibiting the binding of additional sperm. An additional block to polyspermy may be mediated by electrical depolarization of the egg plasma membrane. The relationship between the two modes of block to polyspermy remains to be investigated.

After the sperm binds to the vitelline envelope, it must penetrate the chorion to reach the egg surface. This process is mediated by proteolytic enzymes, which digest a hole in the chorion. This is confirmed experimentally. As mentioned previously, the head of the ascidian sperm contains a small **acrosome**. It undergoes its reaction similar to what

TABLE 13.3 The chemical nature of the different components forming the yolk in the eggs of *Ciona intestinalis* during the process of deutoplasmogenesis

		Stages of oocyte development				
		Previtellogenic (young) oocyte cell size 100 μm nucleus size 35 μm	Vitellogenic oocyte cell size 120–180 μm Nucleus size 75 μm			Postvitellogenic oocyte cell size 200 μm Nucleus size 80 μm
			Cytoplasm			
Histochemical reaction and references	**Indication**	**Cytoplasm**	**Step 1 Small granules**	**Step 2 Large granules**	**Step 3 patches**	**Yolk material (homogeneous)**
Mercury brompheno blue (Mazia et al., 1963)	Protein	bb	bbb	bbbbb	bbbb	bbbbb
Biuret reactiron (Gabe, 1976)	Peptide Linkage	vv	vvv	vvvv	vvvv	vvvv
Aqueous bromphenol blue (Mazia et al., 1963)	Acidic protein	bb	bbb	bbbbb	bbbbb	bbbbb
Toluidine blue Pearse, 1968)	Basic protein	bb	bbbb	bbbb	bbbbb	bbbbbb
Xanthoproteic reaction (Pearse, 1968)	Phenyl group	oo	oo	oo	oooo	oooo
Millon's reaction (Baker modification) (Pearse, 1968)	Tyrosine	pp	ppp	pppp	pppp	pppp
Ninhydrin Schiff reaction (Yasuma and Itchikawa) (Gabe, 1976)	NH2 group	pp	pp	ppp	pppp	pppp
Oxidized Tannin Azo method (Dixon) (Pearse, 1968)	Amino group	pp	pp	ppp	pppp	pppp
Sakaguchi reaction (Pearse, 1968)	Arginine	oo	oo	ooo	oooo	oooo
PAS (Pearse, 1968)	Mucosubs-tance	RR	RRR	RRRR	RRRRR	RRRRR
Alcian blue (Pearse, 1968)	Acid mucop-olysacch-arides	RR	RRR	RRR	RRRR	RRRRR
Alcian blue and PAS (Pearse, 1968)	Acidic and neutral MPS	B RR	BB RR	RRR	RRRR	RRRRR
Best's Carmine (Pearse, 1968)	Glycogen	pp	ppp	pppp	pppppp	pppppp
Sudan black B (Pearse, 1968)	Lipids	–	BbBbBb	BbBbBb	BbBbBbBb	BbBbBbBbBb
Nile blue sulphate (Cain, 1947)	Neutral and acidic lipid	–	bb	bbbb	bbbbbb	bbbbbb
Liebermann Bürchardt reaction (Pearse, 1968)	Cholesterols and its esters	–	–	–	–	–

From Saad (2008).

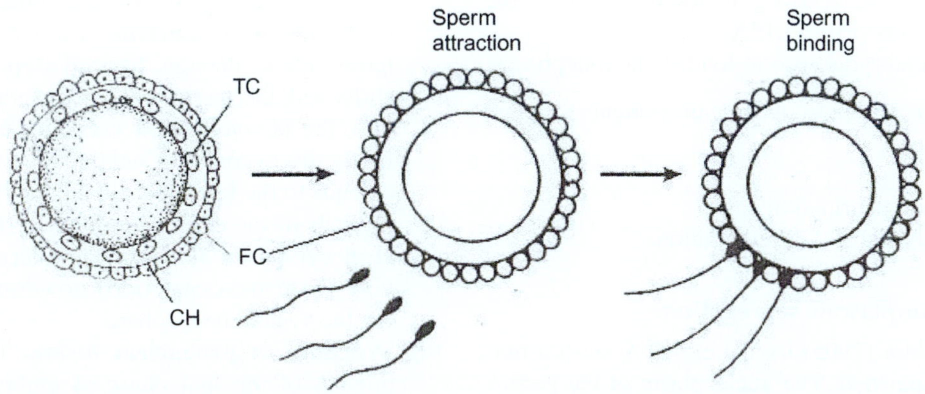

FIG. 13.13 Steps of ascidian fertilization *(adapted from Gilbert and Raunio, 1997)*. CH = chorion, FC = follicle cell, TC = test cell.

is described in the introduction. So it penetrates the chorion, the mitochondrion slides posteriorly out of its head and along the axoneme to the tip of its tail, where it is discarded (Fig. 13.14). The sliding mitochondrion may propel the sperm through the chorion. Ascidians are one of the few animals in which sperm mitochondria do not enter the egg, so after fertilization, this egg has exclusively maternal mitochondria.

It is well known that most ascidians are hermaphrodites. However, self-fertilization is undesirable. It was suggested that self-fertilization is controlled by the vitelline envelope. The results of the studies could not distinguish whether the follicle cells or the chorion is the agent that prevented self-fertilization. Further experimental studies provide suggestions that the chorion is responsible for self-sterility. However, self-sterility is established late during oogenesis, accompanied by the appearance of a thin layer of dense material on the outer surface of the chorion, but little is known about the molecules that are responsible for this event (i.e. preventing self-fertilization). Morgan (1927) reported that in several species, e.g. *Cynthia* (= *Styela*; Morgan, 1927), self-fertilization is known to take place, but in another species, *Ciona intestinalis*, self-sterility occurs. On the other hand, *S. partita* is sometimes self-fertile, and *Molgula manhattensis* shows both types of fertilization (Morgan, 1942). *Halocynthia (Styela) roretzi* and *Chelyosoma siboja* are sometimes self-fertile in very low percentages (Hirai, 1968).

After penetrating the chorion, sperm enter the perivitelline space and eventually bind and fuse with the egg. As in the case of sea urchins, ascidian sperms protrude an **apical process** that binds to the egg surface. This event probably forms the final step in the acrosome reaction. During the sperm-egg fusion, the centrosomal material enters the egg cytoplasm along with the nucleus. The centrosome later generates the sperm aster, which is required for cleavage. There is no elevation of a fertilization membrane, such as in fertilized sea urchin eggs. However, in some species, the perivitelline space expands or the chorion is hardened after fertilization. These events may also be related to the prevention of polyspermy.

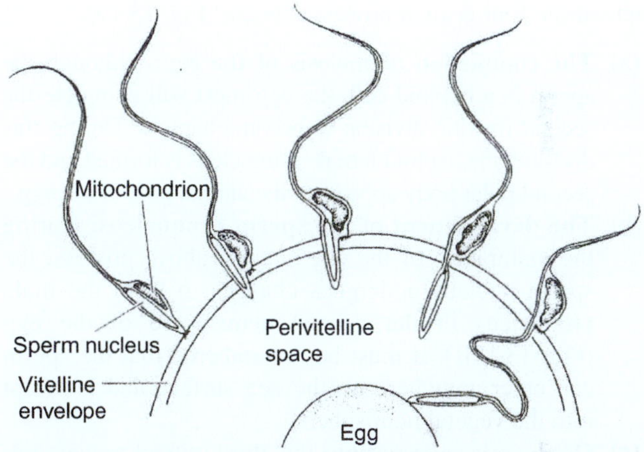

FIG. 13.14 Movement of the mitochondrion during sperm penetration of the chorion. *(After Lambert and Koch (1988), from Gilbert and Raunio (1997).)*

Early development (embryogenesis)

The fertilization and embryogenesis of the ascidian egg was in the scope of some ancient authors, from Van Beneden (1846) to Kowalevsky (1866, 1871). A more recent and detailed account was given by Conklin (1905a,b), of *Styela* (= *Cynthia*) *partita*. His drawings are depicted in most of the literature nowadays. Morgan (1927) reported that the ascidian egg, when laid, has already formed its first polar spindle. It is waiting in the resting condition for the entrance of the spermatozoon to begin a new set of changes, i.e. the egg is arrested at first metaphase. However, in a more recent publication (Jeffery and Swalla, 1997), it was reported that the ascidian egg arrested at metaphase II, so the unfertilized egg is actually a secondary oocyte (Fig. 13.9, chapter introduction). Morgan also described that the egg of *Styela* contains many scattered, yellow granules, especially in the more superficial layers (Fig. 13.15A).

Thus, ascidian embryogenesis is divided into four phases:

I. egg maturation and protoplasmic movements (ooplasmic segregation);
II. cleavage;
III. gastrulation and neurulation; and.
IV. morphogenesis and cell differentiation.

Maturation and ooplasmic segregation

According to Conklin (1905a,b), the egg of *S. partita* matures after being spawned. The nucleoplasm of the germinal vesicle moves towards the animal pole where it forms a clear cap over this region (Fig. 13.15A). Fertilization, then initiates and activates the egg for development. During this activation, four critical processes occur (Fig. 13.15):

(a) **The completion of meiosis of the egg:** Although the sperm is a haploid cell, the egg must still complete the second meiotic division to become haploid. During this division, the haploid female pronucleus is formed and the second polar body appears at the animal pole of the egg.

(b) **The development of the sperm pronucleus:** During the maturation of the egg, i.e. the above process, the sperm nucleus undergoes changes to form the male pronucleus in the vegetal hemisphere of the egg (Fig. 13.15B). It must be remembered that the sperm can enter anywhere on the egg surface and is swept into the vegetal hemisphere.

(c) **Ooplasmic segregation:** this third critical process occurs at the same time of the two previous processes, i.e. the three processes are not successive, they occur synchronously. Ooplasmic segregation, an exciting shift in egg cytoplasm regions, is initiated by sperm entry. Ooplasmic segregation in *Styela* and *Boltenia* eggs is a very fine process since they have coloured myoplasms. Ooplasmic segregation is undergone in two steps. **First step**: the clear cytoplasm, which has been capping the animal pole region, and the peripheral layer of yellow granules descend together to the vegetal hemisphere (Fig. 13.15B), where they form a cap that includes the sperm head. This downward streaming is so violent that it is said to carry the test cells along with it, so that they accumulate at the vegetal hemisphere in some cases. A small mass of hyaline cytoplasm is left surrounding the maturation spindle at the animal hemisphere and the grey yolk granules gather around it (Fig. 13.15B). The two kinds of protoplasm that descended to the vegetal hemisphere, become arranged so that the clear cytoplasm spreads out over the plasm containing the yellow pigment granules. The sperm head proceeds from its point of entrance through the clear area until it reaches the yolk region (Fig. 13.15C). There, it changes its course, and then it is in the form of the **sperm pronucleus**; it moves towards the posterior side of the egg. **Second step**: most of the yolk shifts with the sperm pronucleus forming, in some species, the famous **yellow crescent**, e.g. in *Styela* eggs firstly discovered by Conklin (1905a,b). This crescent extends to the left and right below the equator, including half of the egg circumference (Fig. 13.15D). The clear cytoplasm also moves posteriorly and the grey yolk plasm moves anteriorly downwards, finally reaching the vegetal hemisphere.

(d) **Syngamy, or pronuclear fusion:** The fourth critical process of the first phase of embryogenesis, occurs during the second step of ooplasmic segregation. The sperm aster, a radial system of microtubules is formed in the vegetal hemisphere. The sperm aster is probably responsible of localization the yellow crescent during the second step of ooplasmic segregation. As this latter process and the formation of the aster are completed, the female and male pronuclei fuse to form the diploid nucleus of the zygote.

The final appearance of the egg, e.g. egg of *S. partita*, is a yellow crescent and a clear protoplasm that appears in the posterior part of the egg at fertilization; in addition other parts of the cytoplasm also form definite regions, so that five different regions, each with its specific appearance, can be distinguished before the first cleavage. These are: the posterior yellow crescent and clear protoplasm; a grey crescent in the anterior part; the grey yolk at the vegetal side; and a region of clear yolk at the animal side (Conklin, 1905a,b). Similar regions resulting from ooplasmic segregation can also be observed in other species but with different characters, e.g. the Japanese species, *Halocynthia (Styela) roretzi* (Hirai, 1941a,b). As the result of such region formation, the uncleaved egg of ascidians exhibits bilateral symmetry. Sections of the unfertilized egg of *Ascidiella scabra* show that the distribution of the cytoplasm is already bilaterally symmetrical before fertilization (Kumé and Dan, 1968).

The axes of the tunicate larva are among its earliest commitments. The unfertilized egg shows **radial symmetry**. Indeed, all of its embryonic axes are determined by the cytoplasm of the zygote prior to the first cleavage. Their appearance is an important consequence of ooplasmic segregation. The first axis to be determined is the **dorso-ventral axis**. It is defined by the cap of cytoplasm at the ventral pole. It approximates the animal-vegetal axis. The cap, the vegetal pole region, defines the future dorsal side of the larva and the site where gastrulation is initiated. The second axis to appear is the **anterior–posterior axis**, which is also determined during the migration of the oocyte cytoplasm. This axis lies perpendicular to the dorso-ventral axis. The yellow crescent is the region of the egg that will become the

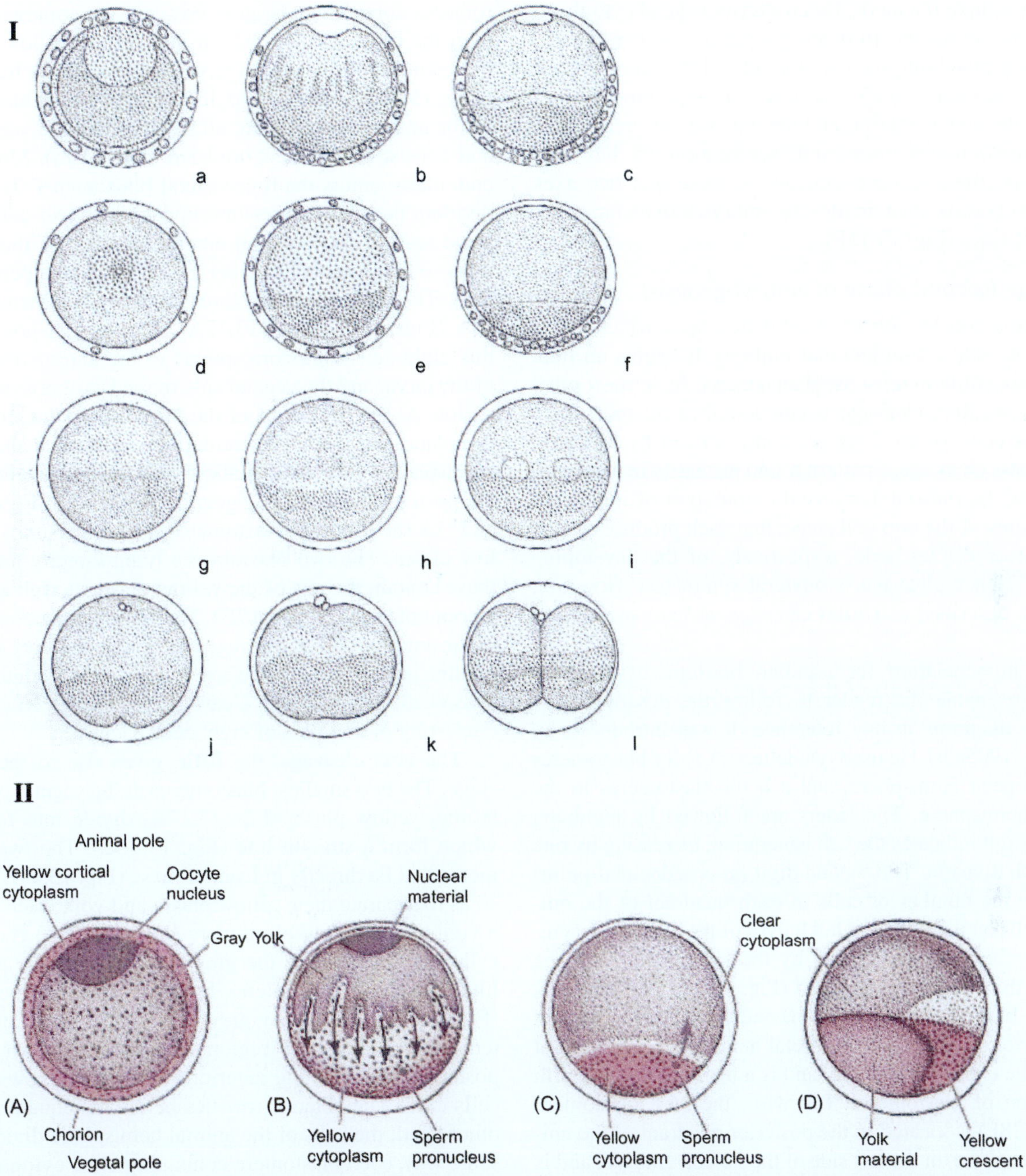

FIG. 13.15 Movement of the materials of the oocyte of *Styela (Cynthia) partita* during fertilization (ooplasmic segregation). *(I) After Conklin (1905a,b), from Morgan (1927)*: (a) unfertilized egg contains many scattered yellow granules, especially in the more superficial layers. The sperm enters at or near the vegetal pole; the egg completes its maturation, extruding the 1st, then the 2nd polar bodies. During that time, the yellow granules have begun to shift towards the vegetal pole (b–f). The ring is thickened on one side, and finally assumes the form of a crescent, whose thickest part indicates the plane in which the first division takes place (g–i). The movement of the yellow granules serves as an index of the extensive cytoplasmic changes through which the granules are carried along. These rearrangements are intimately bound up with the formation of the future median axis of the embryo. *(II) After Conklin (1905a,b), from Gilbert (2000)*: (A) before fertilization, yellow cortical cytoplasm surrounds the grey yolky inner cytoplasm; (B) after sperm entry (in the vegetal hemisphere of the oocyte), the yellow cortical cytoplasm and the clear cytoplasm derived from the breakdown of the oocyte nucleus stream vegetally towards the sperm; (C) as the sperm pronucleus migrates animally towards the newly formed egg pronucleus, the yellow and clear cytoplasms move with it; (D) the final positions of the clear and yellow cytoplasms mark the locations where cells give rise to mesenchyme and muscles, respectively.

posterior side of the larva. This posterior vegetal cytoplasm, PVC, was 'dominant' to other cytoplasms, so that when it was transplanted into the anterior vegetal region of zygotes that had had their own PVC removed, the anterior of the cell became the new posterior, and the axis was reversed. After the completion of the ooplasmic segregation, the left–right axis is specified as a consequence of these first two axes. The first cleavage then divides the embryo into its future left and right sides (Fig. 13.15F).

Cleavage (second phase of embryogenesis)

Cleavage occurs as a series of rapid cleavages, which divide the zygote into a multicellular embryo. It begins an hour after fertilization in most ascidian species. As in most other animals, ascidian cleavage is characterized by short, synchronous cell cycles. They are characterized by **bilateral holoblastic cleavage**, a pattern found primarily in tunicates. Tung (1932) confirmed, by vital colouration of individual blastomeres of the two-cell stage, that each produces either the right or the left side, respectively, of the developing embryo. Hence cleavage is bilateral symmetric. However, it can be described as **radial** cleavage, at least in its early steps.

The nomenclature for ascidian blastomeres must be known to enable the reader to follow the description of cleavage diagrams in any reference. It was introduced by Conklin (1905a,b). He used two letters: A B for blastomeres in the vegetal hemisphere, and a b for blastomeres in the animal hemisphere. The letters are followed by numbers; the first digit indicates the cell generation, increasing by one after each division. The second digit (after a decimal point) indicates the number of cells in each quadrant of the embryo. Cells on the right and left halves of the bilaterally symmetric embryo are designated by the same letters, but the cells on the right are underlined (Fig. 13.18). For example, the *A4.1* blastomere (Figs 13.16B and 13.18C) is located in the anterior quadrant of the vegetal hemisphere on the right side of the eight-cell embryo, and is a member of the fourth generation of blastomeres. Likewise, the b6.8 blastomere (Fig. 13.18F) is located in the posterior quadrant of the animal hemisphere on the left side of the 32-cell embryo, and is a number of the sixth generation of blastomeres.

The most striking phenomenon in this type of cleavage is the **first** cleavage plane. It is **perpendicular**, establishes the earliest axis of symmetry in the embryo, separating it equally into its future right and left sides (Fig. 13.15F). The half-embryo formed on one side of the first cleavage plane is the mirror image of the half-embryo on the other side. This cleavage bisects the yellow crescent and other regions (Fig. 13.15F), and at its end, the nuclei surrounded by hyaline cytoplasm move towards the animal pole, pushing aside the yolk which fills this region. Each successive division orients itself to this plane of symmetry. The **second** cleavage is also **perpendicular** and at right angle to the first, giving rise to the four-cell embryo. In this stage, the anterior blastomeres containing the grey crescent are somewhat larger than the posterior two, which include the yellow crescent material. The **third** cleavage, eight-cell stage, is **horizontal** (Figs 13.16B, 13.17A, and 13.18C). The four blastomeres of the animal pole side are all smaller than the vegetal four and consist of hyaline cytoplasm (ectoplasm). Most of the endoplasm enters the four vegetal blastomeres. The planes dividing the blastomeres into upper and lower cells in this stage tend to slant (slope) downward towards the anterior side, so that they fail to meet the second plane perpendicularly. The result is an H-shaped pattern when seen from the side (Figs 13.16B and 13.17A). The animal pole side of this eight-cell stage corresponds to the antero-ventral part of the larva, and the vegetal side to the larval postero-dorsal region. At the beginning of the **fourth** cleavage, the cleavage planes are oriented differently in the animal and vegetal hemispheres. The two smallest blastomeres are formed in the posterior part of the vegetal region (B5.2) (Figs 13.17B and 13.18E); these blastomeres consist mainly of yellow plasm. The two blastomeres lying directly anterior to these contain the rest of the yellow plasm as well as a large amount of yolk (Fig. 13.17B). The four blastomeres located in the anterior part of the vegetal side contain grey yolk, but hyaline plasm can also be seen on the anterior edge of each. This cleavage, the fourth cleavage, produces a 16-cell stage consisting of two tiers of eight cells each.

The next cleavage, the **fifth**, gives rise to the 32-cell stage. The two smallest blastomeres at the vegetal pole containing yellow plasm (Fig. 13.17B) divide into four cells which form a straight line (Fig. 13.16C). The two blastomeres that lie directly in front of these (Fig. 13.17B) cleave, so as to separate their yellow plasm and yolk, each forming a yellow and a grey blastomere (Fig. 13.16C). The yellow cells lie peripheral to the grey ones, joining the posterior line of yellow blastomeres bringing their number to six (Fig. 13.16C). The four grey yolk blastomeres at the anterior part of vegetal region divide across their antero-posterior axis, forming anterior and posterior rows of four cells each. The anterior row lies below the equator. On the other hand, the cells of the animal hemisphere divide in the same way, each blastomere containing clear cytoplasm and a small amount of yolk. These cells of the animal half are columnar, while those of the vegetal half are wide and flattened. However, these polar differences in the shape of the blastomeres are later reversed completely.

In this fifth stage, the blastomeres gradually become separated from each other at the centre of the embryo, giving rise to the blastocoel. After this cleavage, the synchrony begins to fade. However, by the **sixth** cleavage, 64-cell stage (Figs 13.17C, D and 13.18H, I), the cytoplasmic regions of the egg have been distributed to different blastomeres. The ectoplasm (epidermis) is located in the animal hemisphere blastomeres, and the endoderm in the vegetal hemisphere blastomeres. The myoplasm is located in the posterior vegetal hemisphere blastomeres (Fig. 13.19A and D–G).

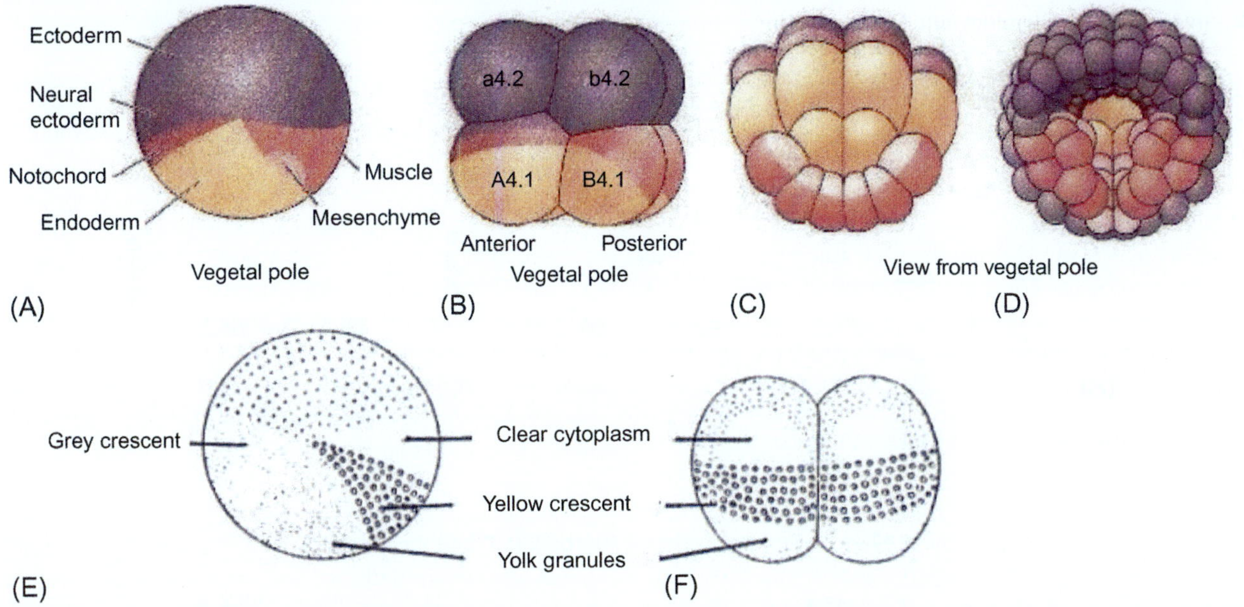

FIG. 13.16 Bilateral symmetry in the egg of the tunicate *Styela partita*: (A) uncleaved egg, the regions of cytoplasm destined to form particular organs are labelled here and coded by colour throughout the diagrams *(after Balinsky, 1981)*; (B) eight-cell embryo, showing the blastomeres and the fates of various cells. The embryo can be viewed as two four-cell halves; from here on, each division on the right side of the embryo has a mirror-image division on the left; (C), (D) views of later embryos from the vegetal pole. *(A)–(D) are from Gilbert (2000)*; (E) the same of (A), showing the natural left side view of egg immediately before first cleavage; (F) natural posterior view of two-cell stage. *((E) and (F) are after Conklin, 1927, from Kumé and Dan, 1968.)*

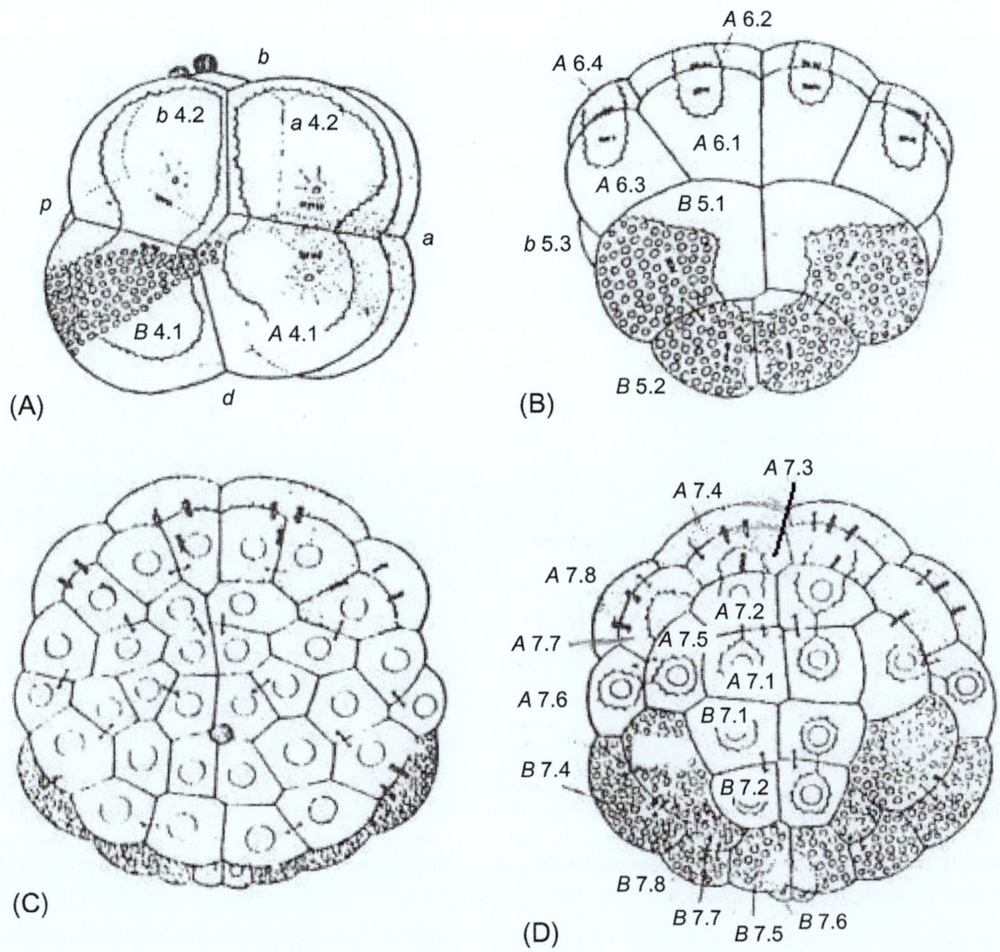

FIG. 13.17 Cleavage in *Styela partita:* (A) right side view of eight-cell stage; note that this view is the opposite to that of Fig. 13.16B; (B) dorsal view of a transition stage from 20- to 24-cell stage. Cytoplasm derived from grey crescent is contained in *A* 6.2- and *A* 6.4- blastomeres; (C), (D) are ventral and dorsal views of 64-cell stage, *A* 7.4 and *A* 7.8 are presumptive neural plate areas. *A* 7.3 and *A* 7.7 are presumptive notochore areas. *B* 7.4 and *B* 7.8 are presumptive muscle areas. Other cells containing yellow cytoplasm, together with *A* 7.6, develop into mesenchyme. Remaining ventral region develops into endoderm. Extra-crescent blastomeres of ventral side will form ectoderm. *(After Conklin (1905a,b), from Kumé and Dan (1968).)*

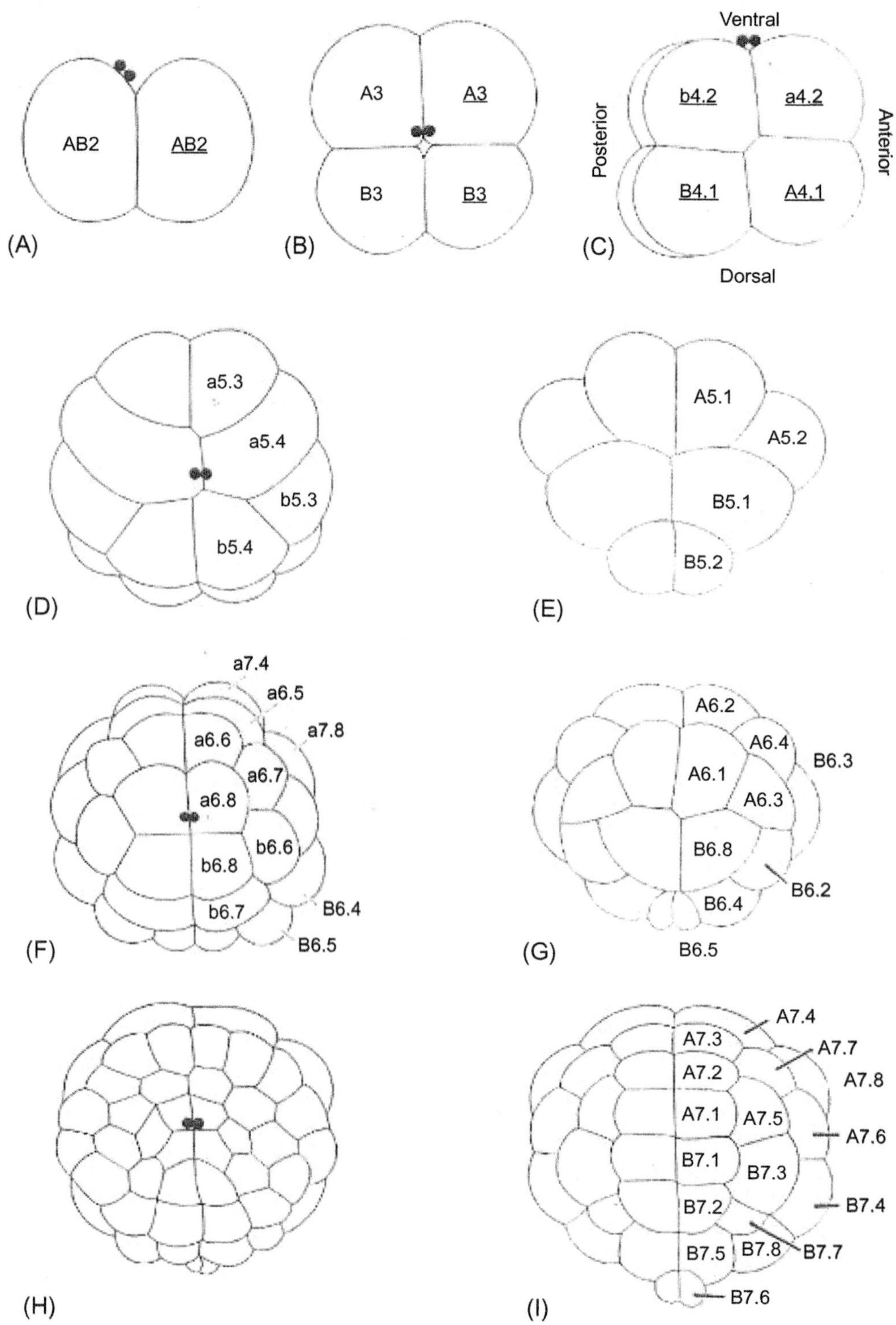

FIG. 13.18 Surface views of cleavage of the ascidian embryo showing blastomere nomenclature. (A–C) Two-cell to eight-cell embryos; note that (C) is the same of Fig. 13.17A. (D–E) 16-cell embryos viewed from the animal (D) and vegetal (E) poles. (F, G) 32-cell embryos viewed from the animal (F) and vegetal (G) poles. (H, I) 64-cell embryos viewed from the animal (H) and vegetal (I) poles. *(After Venuti and Jeffery (1989), from Gilbert and Raunio (1997).)*

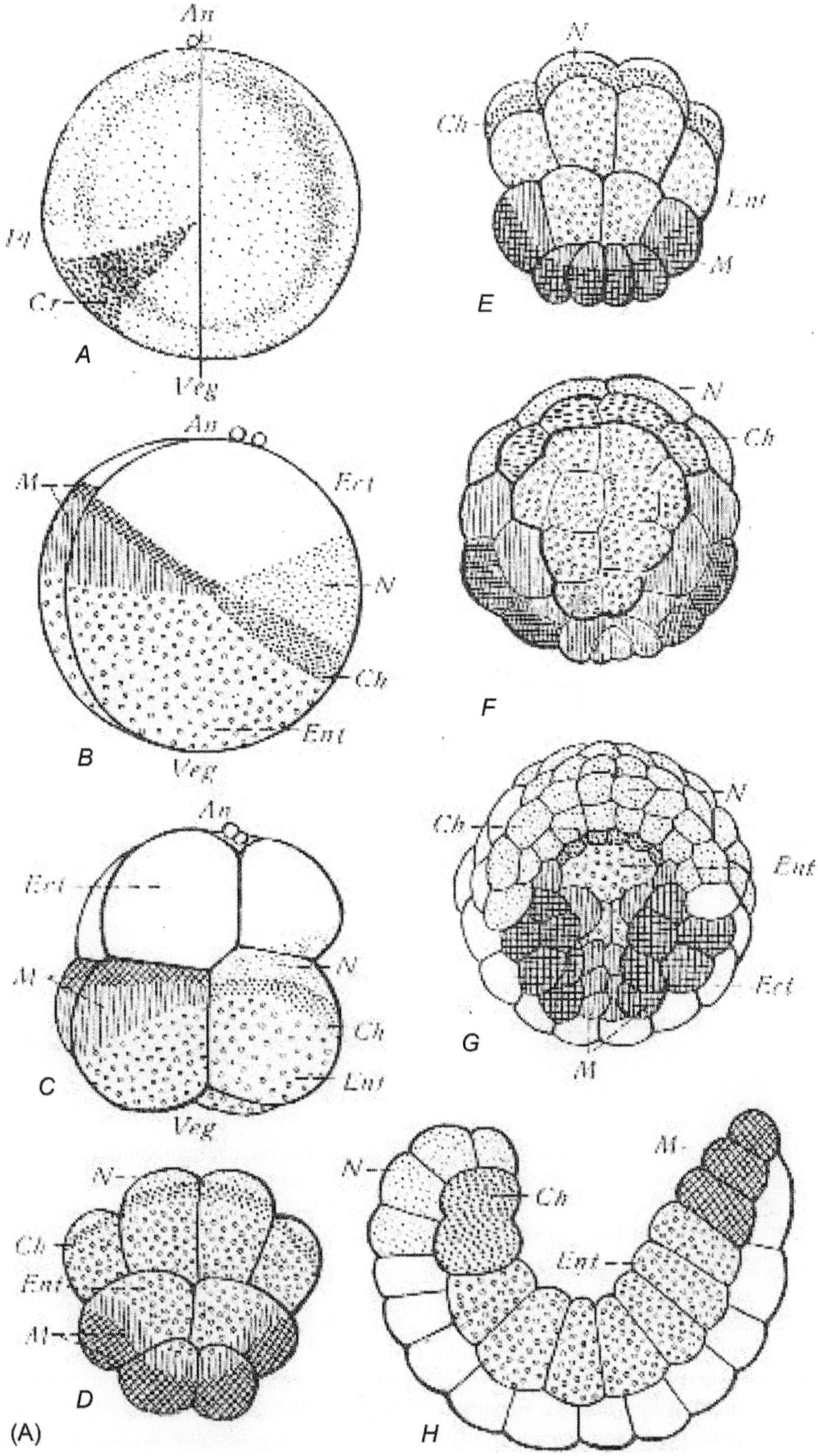

FIG. 13.19 (A) Cleavage and gastrulation in *Styela partita.* (A) showing the animal-vegetal axis of the egg; (B) two-cell stage; (C) eight-cell stage; (D) 16-cell stage; (E) 32-cell stage; (F) 64-cell stage, seen from vegetative pole. Early gastrula, as seen from vegetal pole, with T-shaped blastopore; (H) Sagittal section showing complete gastrulation of the internal cells. *After Conklin, from Pflugfeder, 1970.* An = animal pole, Ch = notochord, Cr = grey crescent, Ect = ectoderm, Ent = endoderm, M = mesoderm, N = neural plate, Pl = clear cytoplasm, Veg = vegetative pole. (B) Cell divisions. A linear version of the fate map, showing the fates of each cell of the embryo (the cell-lineage in a tunicate embryo up to 64-stage). *(After Conklin (1905a,b), and Nishida (1987), from Gilbert (2000).)*

(Continued)

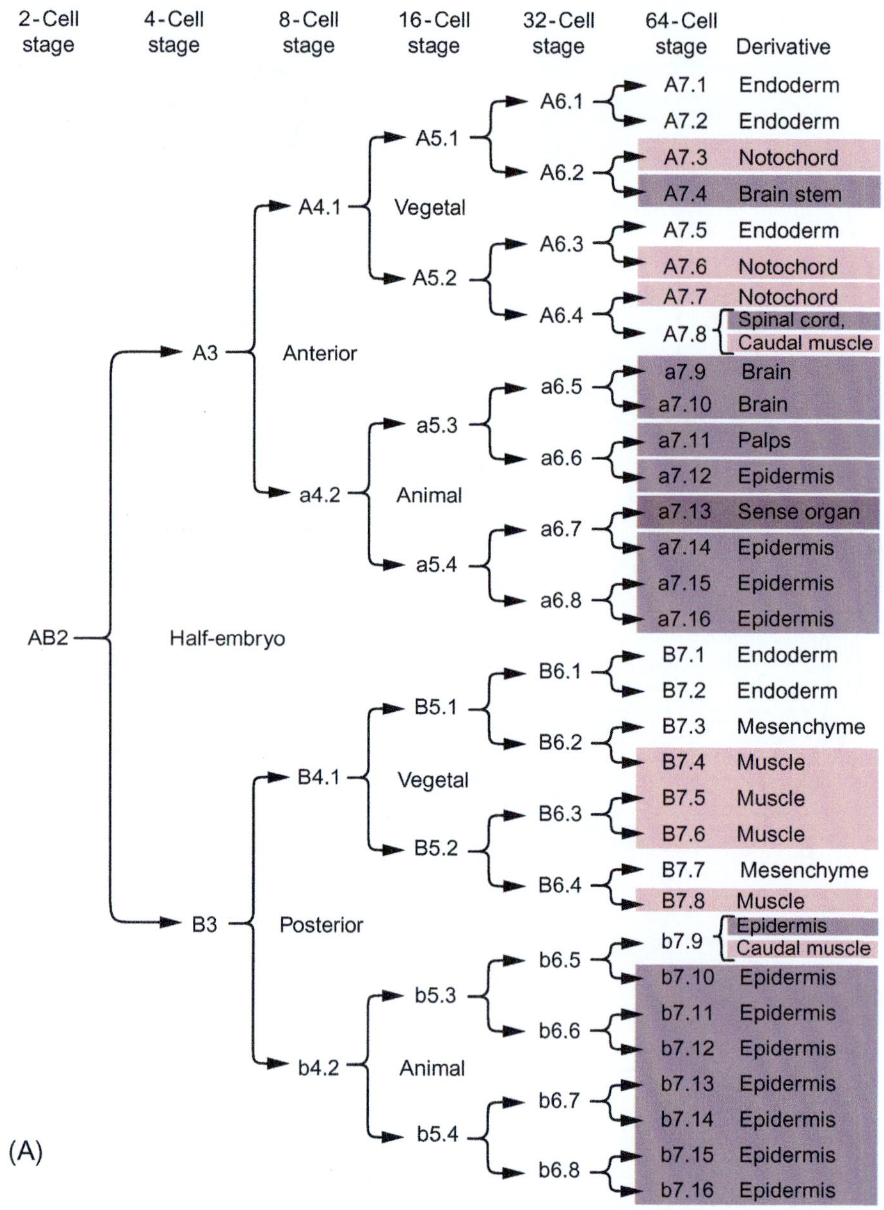

FIG. 13.19, CONT'D

On the other hand, Fig. 13.19B shows the cell lineage up to the 64-cell stage. Blastulation occurs during the late cleavage which continues through the 64-cell stage (sixth) to 110-cell stage (seventh) cleavages, when gastrulation begins (Figs 13.16D and 13.19G).

Gastrulation and neurulation

Gastrulation. Gastrulation begins 4–5 h after fertilization, as a series of cell movements in the three germ layers which convert the blastula to a multilayered embryo (gastrula). These movements are in three processes: invagination of the endoderm; involution of the mesoderm; and epiboly of the ectoderm. Firstly, **invagination** occurs by the internal movement of a sheet of cells (the endoderm) based on changes in their shape. The ascidian blastula is slightly flattened in the vegetal hemisphere. The most extreme changes occur in the four largest endoderm cells (B7.1, *B7.1*, A7.1 and *A7.1*) (Fig. 13.18I), which appear to lead the process of invagination. Inside the embryo, the invaginated endoderm cells migrate anteriorly beneath the ectoderm (Fig. 13.19A and H). After the process of invagination, the blastopore is bordered by mesoderm cells on all sides (Figs 13.19A, G–H and 13.20A, B). The fate map of the eight-cell stage shows that the presumptive notochord cells are located just upper to the endoderm (Fig. 13.19A and C); during gastrulation, they become located at the anterior lip of the blastopore (Fig. 13.19A, G, and H). On the other hand, the presumptive mesenchyme cells are located at the lateral lips of the blastopore, and the presumptive muscle cells at the posterior lip (Figs 13.19A, G and 13.20A, B).

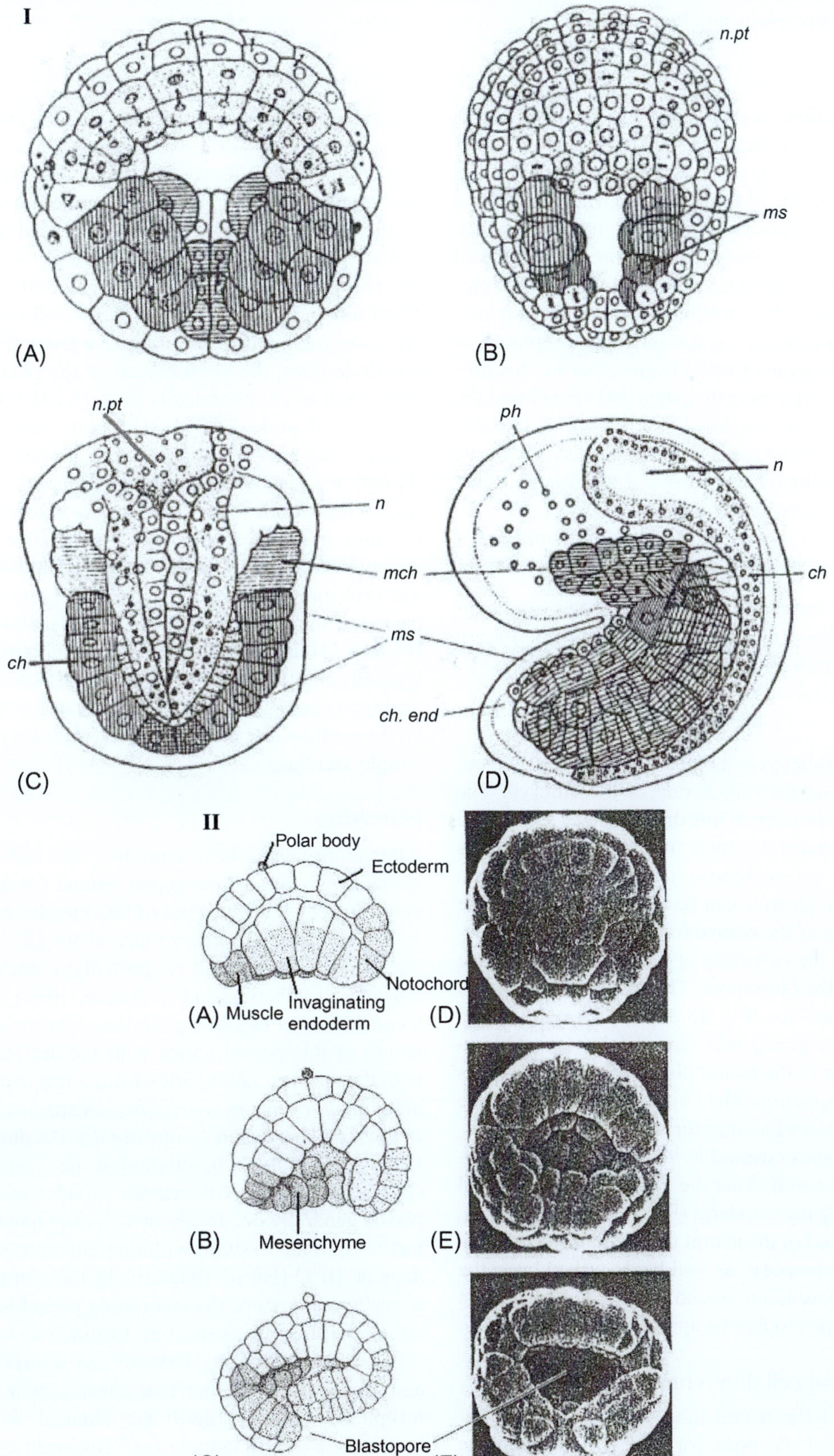

FIG. 13.20 (I) Gastrulation and early gastrula of *Styela partita:* (A) very early gastrula (180-cell stage), showing T-shaped blastopore; (B) later stage; dorsal lip grows backward, laterally-lying muscle cells draw closer to median line and blastopore narrows ventrally in posterior half of embryo; (C) dorsal view of young embryo at formation of neural tube; (D) left side view of young embryo, showing orientation of three rows of muscle cells and other organ rudiments. Ch = notochord, mch = mesenchyme, ms = muscle cell (hatched), n = neural tube, n. pt = neural plate (stippled), ph = pharyngeal region. *After Conklin (1905a,b), from Kumé and Dan (1968).* (II) Processes of gastrulation in the tunicate. Cross sections (A–C) and scanning electron micrographs viewed from the vegetal pole (D–F). (A) and (D) illustrate the invagination of the endoderm; the involution of the mesoderm (B, E); and the epiboly of the ectoderm (C, F). Cell fate is coloured in Fig. 13.16. *((A–C) After Conklin (1905), from Morgan, 1927; (D–F) after Jeffery and Swalla, 1998, from Gilbert, 2000.)*

Secondly, **involution** occurs by the mesoderm, in which a sheet of migratory mesodermal cells moves over a sheet of stationary ectodermal cells (Fig. 13.19A and H). The mesoderm cells involute over the lips of the blastopore (Fig. 13.20(II) B and E) in an anterior to posterior sequence; the presumptive notochord cells disappears into the blastopore first, followed by the presumptive mesenchyme cells, and then by the presumptive muscle cells. After the completion of invagination of endoderm, and involution of mesoderm, the epiboly of ectoderm (third movement of cells) begins. During this process, the columnar ectoderm cells flatten and spread into the vegetal hemisphere. Similar to involution, epiboly is initiated in the anterior region of the gastrula and proceeds posteriorly. Fig. 13.20(II) shows the whole process of gastrulation illustrated by scanning electron micrographs.

At the end of gastrulation (Fig. 13.20C), the embryo begins to elongate along its anteroposterior axis. This leads to:

- the extension of the neural plate (several rows of columnar ectoderm cells on the mid-dorsal side of the embryo);
- the anterior migration of the endoderm cells; and
- the posterior migration of the mesoderm cells.

Neurulation. Neurulation is the process in which the neural tube is formed on the dorsal side of the embryo. The ectoderm along the mid-dorsal line differentiates as a neural plate (Fig. 13.21A and C), which sinks inwards and folds (Fig. 13.21B) along the anteroposterior axis. The time when the blastopore of the gastrula can be seen on the dorsal side marks the beginning of the **neural fold** formation. The neural folds arise from the two sides of the neural plate, in the region posterior to the blastopore. These neural folds eventually meet at the mid-line (Fig. 13.21B) to form the neural tube. However, these neural folds are widely separated anteriorly. The folding of the neural plate occurs by a process that resembles invagination. The closure of the neural tube occurs in a posterior to the anterior wave. In this process, the blastopore becomes covered by the neural folds, forming the **neurenteric canal**. After the formation of the neural tube, it is covered by the ectoderm (Fig. 13.20(II) C and F). The anterior-most end of the neural tube remains open for a short time as the **neuropore**. As will be described later, the neuropore in simple ascidians is said to close temporarily in the tadpole larva stage, to reopen again at metamorphosis.

Morphogenesis and cell differentiation

The anterior part of the neural tube will expand and becomes the brain, and the posterior part will become the **spinal cord**. The chorda cells, which were first, in the form of a plate (Fig. 13.19A, B–H), become arranged to form a groove as the tail elongates (Fig. 13.21D), and as the tail grows longer they elongate. This stained part in *Halocynthia (Styela) roretzi* forms the ventral side of the cylindrical notochord after the completion of tail formation. The posterior part of the endoderm lies along the underside of the notochord, where it forms a ventral cord (sub-chordal endodermal cord). After a short time, these cells become dispersed. In the dorsal part of the developing larva, the rudiments of the **atrial** or **peribranchial cavities** appear in the PG animals as a single ectodermal invagination bifurcating ventrally. The walls of these cavities unite with the pharyngeal wall and differentiate into the **gill slits**. A blind tube arises from the posterior wall of the pharynx to form the rudiment of the **digestive tract** (Fig. 13.21E and F). Underlying the ventral wall of the pharynx, there is a wide band of columnar cells (Fig. 13.21E), which develops into the **endostyle** (Fig. 13.21F). In the brain-vesicle, an **eye** (ocellus) appears, having pigment granules and a lens, although this may be vestigial or lacking in some species. A **balancing organ** (static or otocyst) is formed, which also contains a spherical pigmented concretion. A thickening in the posterior wall of the brain vesicle develops into the **visceral ganglion** (Fig. 13.22A). The anterior part of the trunk becomes considerably distended and is called the chin (Fig. 13.21E and F). The front end of this structure develops typically into three processes: the **adhesive papillae**, consisting of ectodermal secretory cells; one of these is situated on the midline and the others to the left and right of it in the simple ascidians.

Hatching

After embryogenesis is complete, the tail of the embryo elongates so that it is wrapped around the trunk within the chorion. It then hatches out of this membrane.

Hatching usually takes place about 12–36 h after fertilization and is mediated by proteolytic enzymes, hatching enzymes (Lübbering and Hofmann, 1995) secreted by the larva. The latter begins a period of vigorous swimming. The length of this period varies with the individual as well as with the species, i.e. it varies from a few minutes to several hours and/or days in ovoviviparous species (Grave, 1944). In the simple ascidian *C. intestinalis*, the duration of swimming is generally 12 h, although it may last as long as 36 h (Berrill, 1950). In *Halocynthia (Styela) roretzi*, metamorphosis generally occurs about 12 h after hatching, but many larvae continue to swim without metamorphosis for several days at 10°C (Hirai, 1941a,b). In the compound ascidian *Botryllus schlosseri*, the swimming period is usually 1–3 h, although it may be as short as 30 min or as long as 27 h.

The free-swimming larva of sea squirts reacts to light and gravity in a manner that ensures their dispersal. The receptors to these stimuli are situated within the cerebral vesicle of the tadpole (see Hofmann et al., 2008). In 1986, Holmberg stated that these receptor cells are innervated by only one neuron and not by two as was previously believed. Olsson et al. (1990) recorded that receptors in the larvae of *C. intestinalis*, *Phallusia mammillata* and *Ascidia nigra* consist of a unicellular otolith and an ocellus made up from about 10 cells. Torrence and Cloney (1982) studied the receptors of the larva of *Oikopleura* (Tunicata,

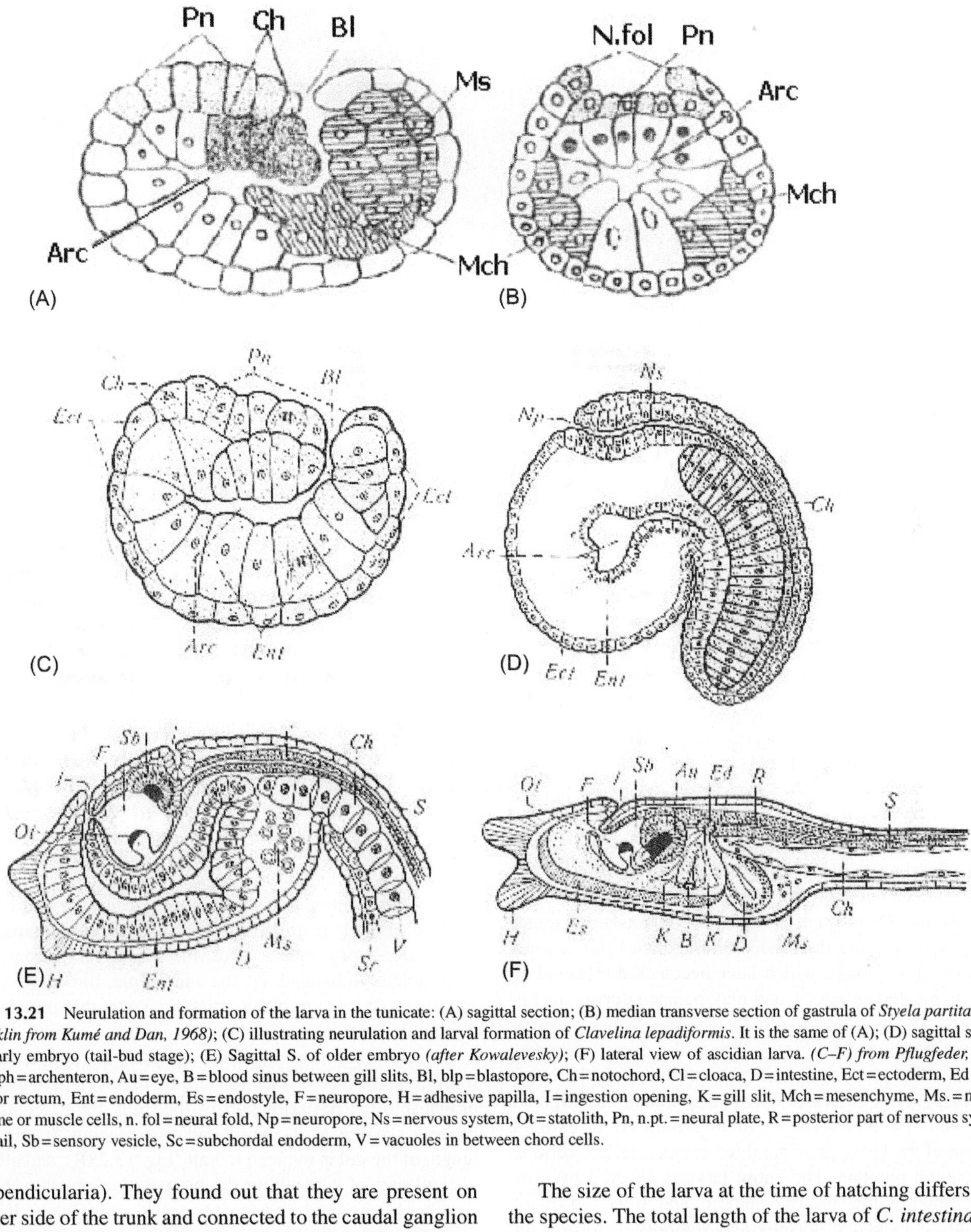

FIG. 13.21 Neurulation and formation of the larva in the tunicate: (A) sagittal section; (B) median transverse section of gastrula of *Styela partita (after Conklin from Kumé and Dan, 1968)*; (C) illustrating neurulation and larval formation of *Clavelina lepadiformis*. It is the same of (A); (D) sagittal section of early embryo (tail-bud stage); (E) Sagittal S. of older embryo *(after Kowalevesky)*; (F) lateral view of ascidian larva. *(C–F) from Pflugfeder, 1970.* Arc,ph = archenteron, Au = eye, B = blood sinus between gill slits, Bl, blp = blastopore, Ch = notochord, Cl = cloaca, D = intestine, Ect = ectoderm, Ed = hind gut or rectum, Ent = endoderm, Es = endostyle, F = neuropore, H = adhesive papilla, I = ingestion opening, K = gill slit, Mch = mesenchyme, Ms. = mesenchyme or muscle cells, n. fol = neural fold, Np = neuropore, Ns = nervous system, Ot = statolith, Pn, n.pt. = neural plate, R = posterior part of nervous system, S = tail, Sb = sensory vesicle, Sc = subchordal endoderm, V = vacuoles in between chord cells.

Appendicularia). They found out that they are present on either side of the trunk and connected to the caudal ganglion by the axons of cerebral cells, which form electrical synapses (gap junction). These authors added that this sensory system of this animal is comparable with a similar caudal sensory system in the tadpole larva of *Diplosoma macdonaldi*. Jang et al. (2005), using wild type and mutant larvae of *C. savignyi*, demonstrated that pigmentation in the tadpole's sensory organs is essential for normal behaviour.

The size of the larva at the time of hatching differs with the species. The total length of the larva of *C. intestinalis* is about 0.7 mm, in *Molgula manhattensis* 0.6 mm, in *Ascidia zara* 0.5 mm, in *Halocynthia (Styela) roretzi* 1.5 mm, in *S. clava* 1.0 mm, in *Corella japonica* 0.7 mm, in *Chelyosoma siboja* 1.0 mm, in *Eudistoma digitatum* 11 mm and in the giant tadpole of *Ecteniascidia turbinate*, it is 12.5 mm. The two latter colonial species are from Hofmann et al. (2008).

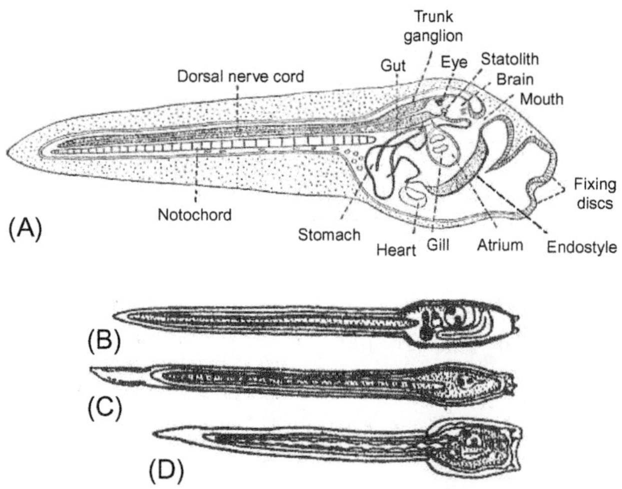

FIG. 13.22 Diagrammatic drawings of whole mounts of types of ascidian tadpole larvae: (A) of the typical larva *(from Borradaile et al., 1977)*; (B) tadpole of *Ciona intestinalis (from Berrill, 1947)*; (C) tadpole of *Halocynthia roretzi*; (D) tadpole of *Corella japonica ((C and D) from Hirai, 1951)*.

The ascidian larva is a nonfeeding larva that is important for dispersal of the sessile adult.

Structure of fully formed larva

Complete development leads to an elongated microscopic larva, called an **appendicularia** or more commonly a **tadpole** larva (Fig. 13.22). The ascidian larva has been termed a tadpole as it morphologically resembles the tadpole larva of anuran amphibians. It clearly shows characteristics of a chordate at an early developmental stage, in particular a linear row of 40 cells in the tail forming the notochord as revealed by Kowalevsky (1866). The presence of notochord in the tail led to the taxon named Urochordata as a synonym of the Chordata. A distinct tail represents the posterior half of the larva and contains the notochord and the neural tube. Dorsally, the anterior half of the larva contains a pigmented cup, a statocyst and the dilated end becomes the cerebral ganglion. The mouth, which later becomes the buccal siphon, is located anteriorly and may be open during the larval stage. The mouth leads into the pharynx, which in turn is followed by a twisted digestive loop with a dorsally directed intestine. The pharynx contains a ventral endostyle, but at first the pharynx possesses only two stigmata. The stigmata empty into a small pocket, the future atrium. At the extreme anterior of the larva, there are three ectodermal projections: the fixation papillae. The entire larva is covered by a tunic secreted by the surface ectoderm, and in the tail region the tunic is extended dorsally and ventrally to form a fin. The tail is the larval locomotor organ, and movement is brought about by special muscle bands.

Metamorphosis

Metamorphosis involves numerous rapid and coordinated morphogenetic movements, physiological changes and cellular rearrangements, and also significant degradation processes, which appear to be initiated by the time of settlement. Hoffmann et al. (2008) reviewed and reported the views of previous authors concerning ascidian metamorphosis. They listed 10 principal progressive and regressive metamorphic events. The following ones are the most salient: secretion of adhesives by anterior trunk papillae; resorption of the tail; loss of the outer cuticular layer of the tunic; rotation of the visceral organs through an arc of about 90 degrees; expansion of the branchial basket with concomitant elongation of the juvenile; retraction of the sensory vesicles; phagocytosis of visceral ganglion, sensory organs, and cells of the axial complex; and release of organ rudiments from arrested developmental state. In addition, they pointed out that the time course of these events may vary considerably, adhesion and tail resorption may occur in seconds to minutes, while trunk reorganization and phagocytosis may take hours or days. However, the onset of metamorphosis is indicated by the resorption of the tail. It is a complex process in which the larval tissues are destroyed or remoulded and replaced by adult tissues and organs. Most importantly, the adult develops gonads for gametic reproduction to produce the next generations. Metamorphosis from a tadpole larva to a feeding adult usually takes at least 5 days. For a detailed description, this process can be divided into four separate events. The first is larval attachment to a suitable substratum. Second, the tail is resorbed by contraction of the tail epidermis. Third, the larva rotates 90 degrees and begins tissue remoulding. Fourth, after several days of development, the newly formed siphons open and the juvenile begins feeding.

First: when its free-swimming period is over, the tadpole larva secretes an adhesive substance from its adhesive papillae and attaches to some object (Fig. 13.23A); its movements become sluggish and it begins the process of metamorphosis. The resulting changes consist of the disappearance of the larval tail and adhesive papillae, accompanied by the development of **attaching organs** or **ampullae**; the brain-vesicle is transformed into the **cerebral ganglion** and its accessory organs. Thus, the special larval character is completely changed. At the same time, the adult organ rudiments proceed to differentiate in a way that transforms the motile larva into a fixed sessile adult.

Second: The most conspicuous phenomenon as metamorphosis begins is the shrivelling of the tail, which is drawn into the posterior part of the trunk by a folding-up process starting at its basal end. After about 30 min the length of the tail is reduced to half (Fig. 13.23B), and all but a small part of it is within the body within about an hour. Another hour is required for it to vanish completely. Inside the body, it bends three or four times and lies folded up in the posterior region. While the tail is shrinking, the epidermis in the neighbourhood of the adhesive palps thickens and expands, giving rise to three ampullae. No change in the arrangement of the various organs can be observed until resorption of the tail is complete. About 5 h after the beginning of metamorphosis, the third event occurs.

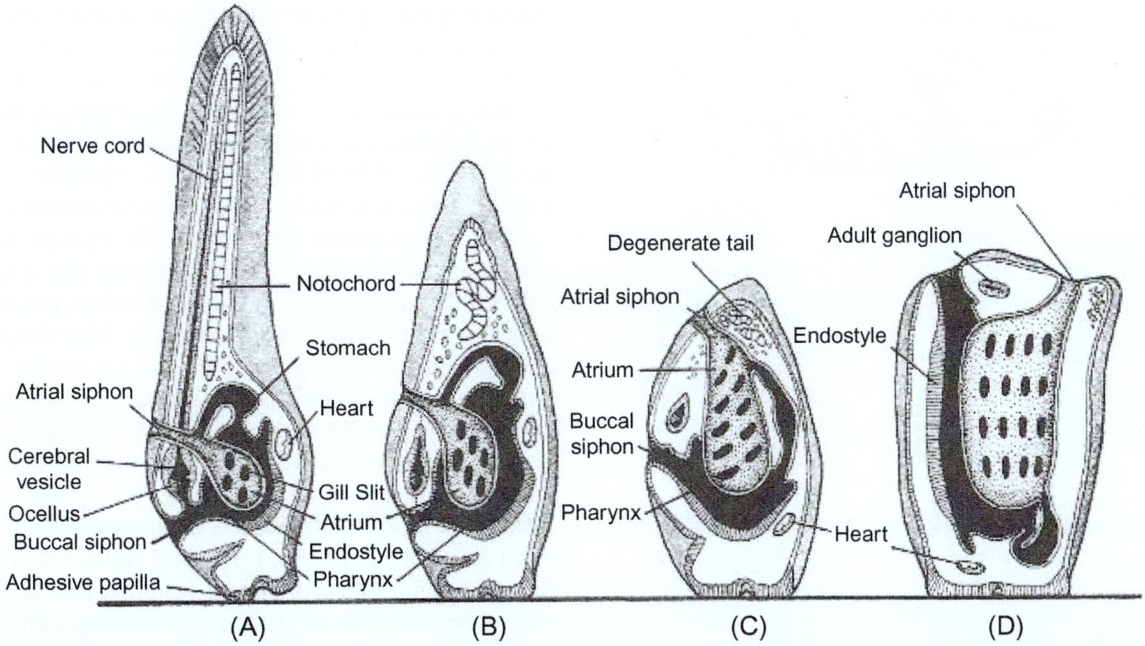

FIG. 13.23 Diagrammatic drawings showing metamorphosis of urochorodate tadpole larva *(from Barnes, 1974)*. (A) lateral view, which has just attached to the substratum by the anterior end; (B) the tail is retracted and resorbed; (C) the larval organs rotate 90 degrees and tissue remodelling continues; (D) a young individual just after metamorphosis, after several days of development, the functional siphons open and the juvenile begins feeding. *Modified after Seeliger from Brien (1948).*

Third: as a result of the rapid growth of the chin region (the area between the adhesive papillae and the mouth), the entire body is turned 90 degrees (according to Kumé and Dan (1968) and Gilbert and Raunio (1997)). The mouth or the buccal siphon is carried backwards to open at the end opposite to that of the attachment, and all other organs are rotated in the same manner. The sensory organ, which is located dorsally in the larva, turns towards the posterior part of the larval axis, and finally makes a revolution of about 90 degrees and comes to lie in a position posterior in relation to the adhesive papillae, but at the anterior end of the postmetamorphosis body axis. Little is known about the cues or mechanism(s) underlying this process.

Fourth: during the fourth event, the tail epidermis contracts causing the notochord, nerve cord and muscles to be bent together and brought inside the trunk. There, they fragment and are consumed by wandering phagocytic cells; the tail epidermis invaginates to form pockets. While the tail, thus being resorbed, the front end of the trunk elongates, and forms the centre around which the anterior part of the body grows, so that it executes a turn of 90 degrees posteriorly, or towards the former tail position.

Differentiation of adult organs

Endostyle. The rudiment of the endostyle lies in the anterior part of the endodermal organ rudiments in the larva (Fig. 13.22A). As the result of the body rotation, the endostyle is gradually shifted posteriorly from its larval position, finally turning through 90 degrees. The mouth, which was first situated at the anterior most end of the endostyle, is thus moved posteriorly, as the endostyle turns to its definitive adult position. Since the position of the mouth marks the anterior end of the body, this means that the adult body axis is exactly opposite to that of the larva. The endostyle thus lies posterior to the mouth and elongates in a direction parallel to the long axis of the body. These changes appear externally, and in the meantime, adult internal organs differentiate inside the body.

Atrium. The ectoderm of the dorsal side of the larva forms either a single or paired atrial invaginations that fuse with the pharynx to produce gill slits. In *C. intestinalis* a pair of lateral invaginations is formed. Each gives rise, simultaneously, to two gill slits (Fig. 13.24) without the formation of two corresponding outgrowths of the pharyngeal wall. First an atrial invagination is formed on each side, and then the pharyngeal wall adjacent to its inner end also invaginates. Both invaginations deepen until they become fused, then break through to form gill slits. By the growth of the pharynx and the differentiation of the internal organs, the formation of more gill slits arises successively. When the metamorphosis is about to complete in *C. intestinalis*, the heart and gill slit become active; this is the stage in which the paired gill slit primordia first divides into a left and right parts, and the atrial cavity still has a pair of openings, one on each side of the body.

Alimentary canal. From the posterior part of the pharynx, an intestine takes shape. In the basal part of this, a spherical sac develops into a stomach. The pyloric gland in *Ciona* differentiates as a coecum in the region where the stomach joins the intestine. This rudiment grows and elongates; its tip

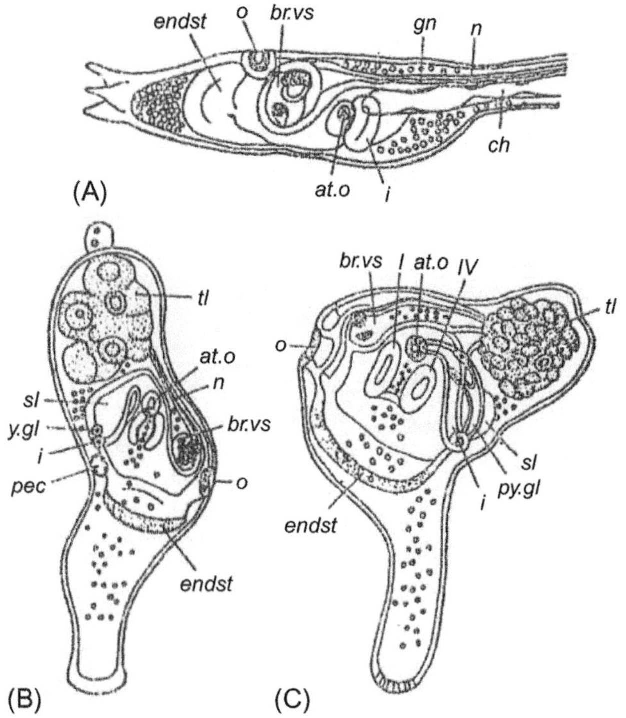

FIG. 13.24 Metamorphosis in *Ciona intestinalis* I *(after Willey, 1893a,b, from Kumé and Dan, 1968)*. (A) tadpole larva; (B), (C) successive stages in metamorphosis. I, IV = first and fourth gill slits, at.o = atrial canal, br. *Vs.* = brain vesicle, ch = notochord, endost = endostyle, gn = somatic ganglion, i = intestine, n = neural tube, o = mouth, pec = pericardium, py.gl = pyrolic gland, st = stomach, tl = tail.

first bifurcates and then re-divides many times to produce the complicated structure of the pyloric gland. The latter surrounds the digestive tract from the stomach, nearly to the anal region with its finger-shaped branches (Figs 13.24P and 13.25).

The nervous system. Hirai (1968) reported that the metamorphosis of the nervous system, which accompanies the formation of the subneural gland in *C. intestinalis*, was described very early by Willey (1894). It proceeds in five stages: **Stage 1** occurs while the larva still enclosed by the chorion. During this time, the rudiment of the neurohypophysial canal is formed as a cavity in the left side wall of the anterior part of the brain vesicle (imagine it in Fig. 13.21E); its posterior part connects the cavity of the brain vesicle. The anterior end of its cavity is a blind tube, the rudiment of the **neuro-hypophysial canal** (Fig. 13.26A and B). To follow this description, one must remember the statement: 'the neural gland, which lies on the ventral side of the nerve ganglion in the adult ascidian, has sometimes been correlated with the hypophysis of the Craniata' (Parker and Hazwell, 1974). **Stage 2** proceeds after the hatching of the larva, which becomes fixed on a substratum. The rudiment of the neuro-hypophysial canal becomes prominent and begins to separate from the brain vesicle. At this time, the blind tube approaches the midline of the brain vesicle. In **stage 3**, the blind tube develops an opening into the base of stomodaeal rudiment. At about this time, the brain vesicle is in the process of disappearing. The opening of the connection between the hypophysial canal and the neuropore means that the temporarily closed neuropore reopens, after the formation of the stomodaeum. The sagittal section of an attached larva (Fig. 13.26C) shows the relation between the nervous system and the other organs. The hypophysial canal can be seen to open into the stomodaeum. Fig. 13.26D, an enlargement of the hypophysial canal region, shows the canal to be surrounded by columnar cells. Near the posterior end its cavity becomes irregular, and the cellular wall of its upper part is composed of numerous cell masses. This marks the beginning of adult ganglion formation. During **stage 4**, while the first four gill slits are being formed, the ganglion grows and increases in size (Fig. 13.26D). The lumen of the larval neural tube is being obliterated; the ganglion acquires a large, solid structure. Anterior to the latter, there is a funnel-shaped passage leading to the stomodaeal part of the branchial sac. Below this opening, a ciliated process appears on the dorsal side of the branchial sac; this is the rudiment of the adult ciliated groove (Fig. 13.26E).

In **stage 5**, 'in the younger immature adults, anteriorly the duct of the hypophosis expands into a funnel-shaped dilation, which in its turn opens into the branchial sac at the end of a papilliform prominence. The latter projects boldly into the branchial chamber' (Willey, 1894). In this stage, the hypophysis and the ganglion, have been gradually differentiating themselves from the common neuro-hypophysial tube, which lies closely applied to, but at the same time distinct from, the ganglion; glandular tissue is seen to be developing from its ventral wall. The hypophysis rudiment includes a lumen, which is reduced in the anterior part. The cerebral ganglion arises by proliferation and constriction from the dorsal wall of the neuro-hypophysial tube. 'The anterior portion of the hypophysis, including the funnel-shaped dilation and the duct, is derived from a secondary evagination from the stomodaeal region of the branchial sac' (Willey, 1894).

Time required for development

In *Ciona intestinalis*, both eggs and sperm mature within their respective ducts, and fertilization takes place externally. The fertilized egg begins to cleave within about 1 h, and reaches the tadpole larva stage 25 h after fertilization at 16°C. The swimming stage lasts for about 12 h, although it may be as short as 6 h, or as long as 36 h. Postmetamorphosis larvae can be raised to maturity in the laboratory with diatoms as food. The egg is green or yellowish-green, with a diameter of 0.16 mm; artificial fertilization is possible (Morgan, 1945).

Experimental studies

Meticulous observation of ascidian embryogenesis dates back to Krohn (1852), Kowalevsky (1866) and particularly to

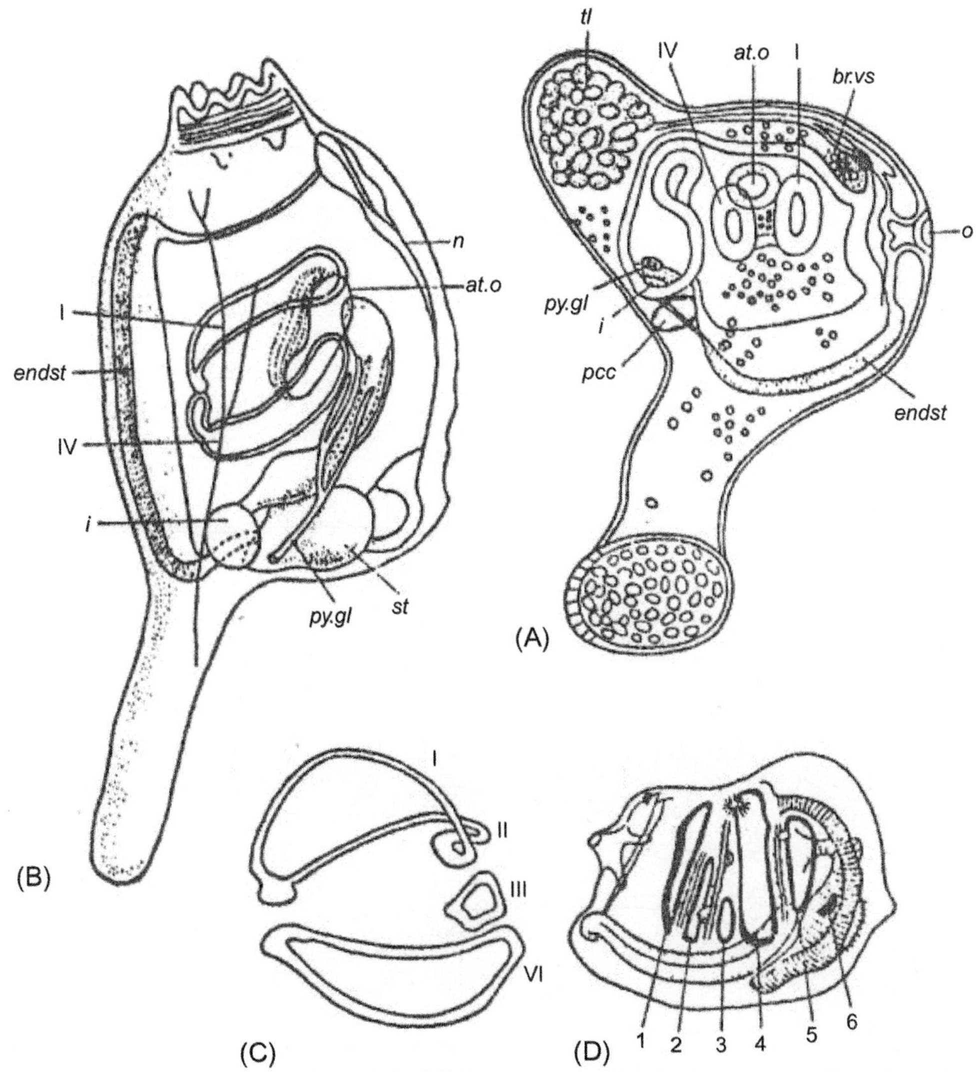

FIG. 13.25 Metamorphosis in *C. intestinalis* II (Willey), and *Phallusia scabroides* (Korschelt Heider). (A), (B) later stages in metamorphosis of *C. intestinalis* larva; (C) differentiation of gill slits in larva of *C. intestinalis*; (D) differentiation of gill slits in larva of *Phallusia scabroides*. Abbreviations as above. *(From Kumé and Dan (1968).)*

Conklin (1905a,b). The latter established the cell lineage and fate map of *Styela plicata*. More recent studies of early development under constant temperature conditions and along a precise time scale are available for the two EG ascidians: *Ascidiella aspersa* (Müller, 1776) (Niermann-Kerkenberg and Hofmann, 1989) and for *Phallusia mammillata* (Cuvier, 1815) (Wahl und Lafargue, 1992). The observations of Hofmann et al. (2008), carried out on the same two species, match very well the data of the above published two works. They confirm that embryonic development is very similar, both morphologically and with regard to the time course, in the two species and they serve well as a prerequisite for further experimental studies. Hofmann et al. aimed to study metamorphosis and the possible impact of defined chemicals in eliciting larval-juvenile transformation. The ascidians: *A. aspersa* were collected from the North Sea, Island of Helgoland (Germany) and *P. mammillata* were collected from Banyuls-sur-Mer (France). The former was treated in the laboratory at the Ruhr-University Bochum. As specimens of *A. aspersa* were more readily available, this species has been chosen for all subsequent experiments related to larval metamorphosis. Both species show rapid embryonic development, which leads to formation of motile tadpole-like larvae. Suitable live specimens were chosen and dissected in seawater under a dissecting microscope to obtain gametes. Eggs and sperm were obtained directly from the gonoducts using strictly separate micropipettes. This means that the gametes were obtained surgically from the gonoducts. Both species are known to be self-fertile, but eggs were routinely inseminated with sperm from a different animal.

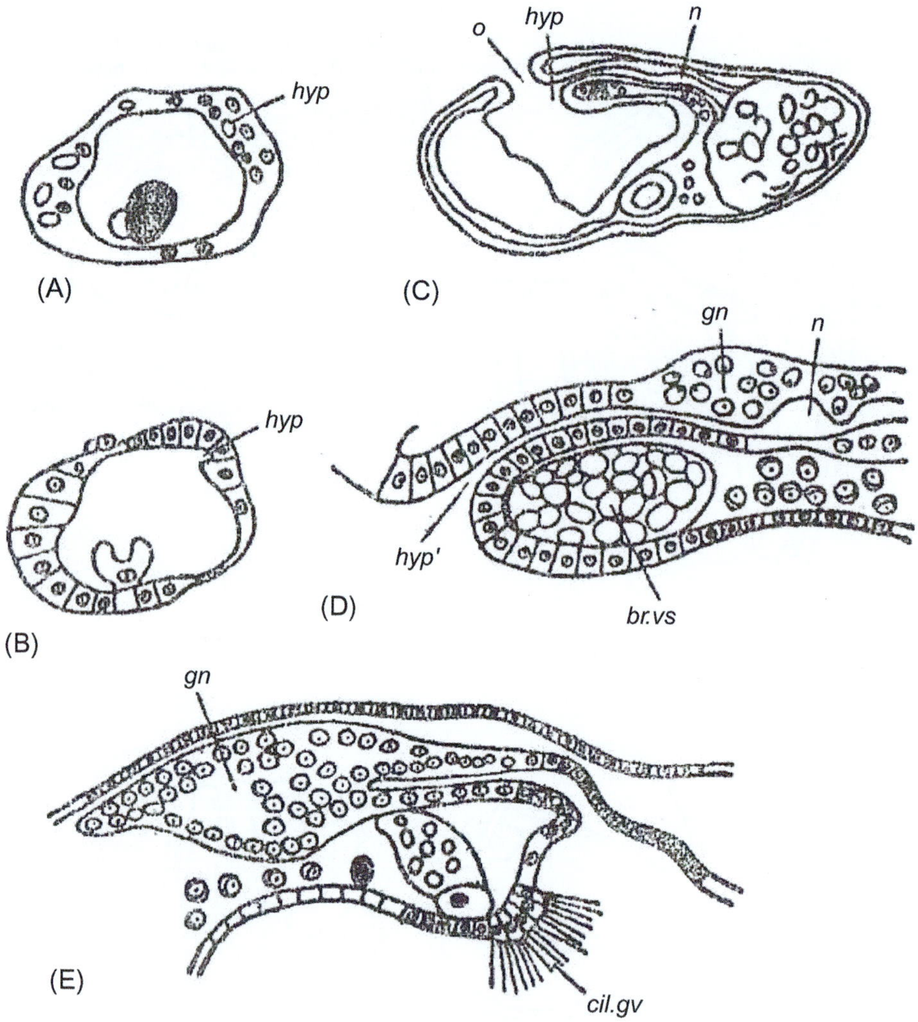

FIG. 13.26 Formation of nervous system in *Ciona intestinalis (after Willey, 1893a,b, from Kumé and Dan, 1968)*: (A) formation of neuro-hypophysial canal in side wall of brain vesicle; (B) opening of neuro-hypophysial canal into brain vesicle; (C) sagittal section showing the relation between the N.S. and the other organs; (D) differentiation of ganglion in dorsal wall of neuro-hypophysial canal (enlargement of the hypophysial canal region); (E) ganglion at more advanced stage. br.vs=vestige of degenerating larval brain vesicle, cil.gv=rudiment of ciliated groove, gn=ganglion rudiment, hyp=neuro-hypophysial canal, hyp'=opening of neuro-hypophysial canal into gill sac, n=neural tube, o=mouth.

Mature oocytes of *A. aspersa* and *P. mammillata* were transferred from the oviduct to large Petri dishes and then artificially fertilized with **heterologous** sperm. This cross-fertilization was carried out using small volumes of diluted sperm to prevent polyspermy. The examination of mature oocyte reveals that it is enclosed in a three-layered envelope. One layer of cells is directly opposed to the oolemma, the test cells. It is separated from the acellular vitelline envelope or chorion by a large perivitelline space. A layer of vacuolated larger cells inserts at the outer surface of the chorion, the follicle cells. In *A. aspersa*, the test cells are evenly spaced and the large follicle cells form a tight outer layer. In *P. mammillata*, test cells are arranged in a less orderly fashion and the follicle cells, which detach more easily upon handling, are significantly smaller (Fig. 13.27). This relatively large envelope encloses an oocyte measuring only

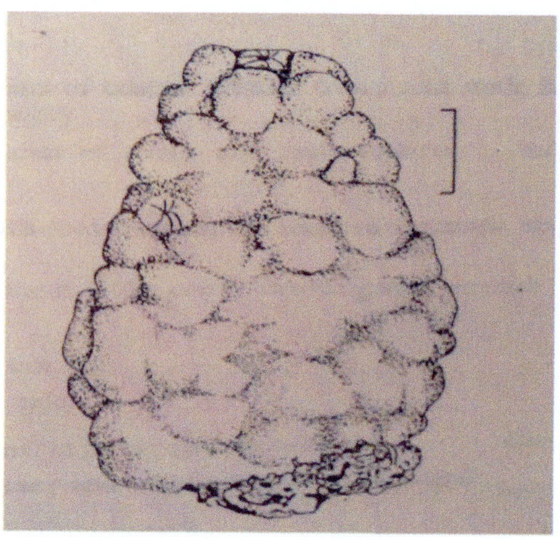

Diagrammatic drawing of an adult animal *(from Millar, 1970)*

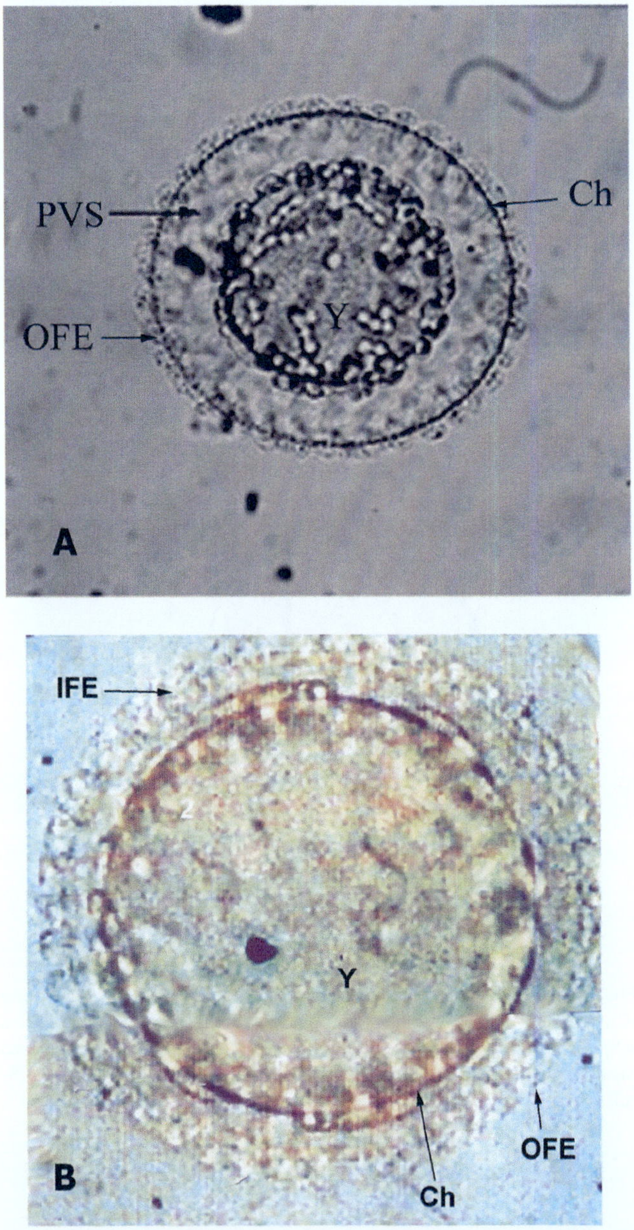

FIG. 13.27 Experimental development of *Phallusia mammillata* (Cuvier, 1815). Phase contrast photomicrographs of whole mounts of live stages of development of the animal. *From Hofmann et al. (2008).* Fertilized egg: (A) fixed material at lower magnification to provide better overview of ovular envelopes. 250×; (B) live fertilized egg. 400×. Ch = chorion, IFE = inner follicular epithelium, OFE = outer follicular epithelium, PVS = perivitelline space, Y = yolk material.

about 75 μm in diameter. It is moderately yolk rich, seemingly isolecithal in the unfertilized condition, and devoid of coloured pigments in both species. Sperm entry launches contraction waves of the egg cortex leading to specific alterations of the cell shape, which can easily be observed with a microscope (Niermann-Kerkenberg and Hofmann, 1989). Concomitant with these cortical contractions, components of the cytoplasm become shifted and rearranged in a process termed **ooplasmic segregation** (see above). At the same time, meiotic divisions are completed with two or sometimes three polar bodies appearing at the animal pole. Male and female pronuclei merge to yield the zygote nucleus, which then enters the first division cycle. Fertilized eggs were maintained in an incubator or temperature conditioned room at a constant temperature of 20°C. The egg then undergoes holoblastic cleavage as follows:

1. This cleavage starts with a first meridional cleavage furrow proceeding from the animal to the vegetal pole and resulting in **two** equal and identical blastomeres (after 2 h of fertilization). Cytoplasmic segregation continues with yolk granules concentrating in the vegetal hemisphere of the egg (Fig. 13.28).

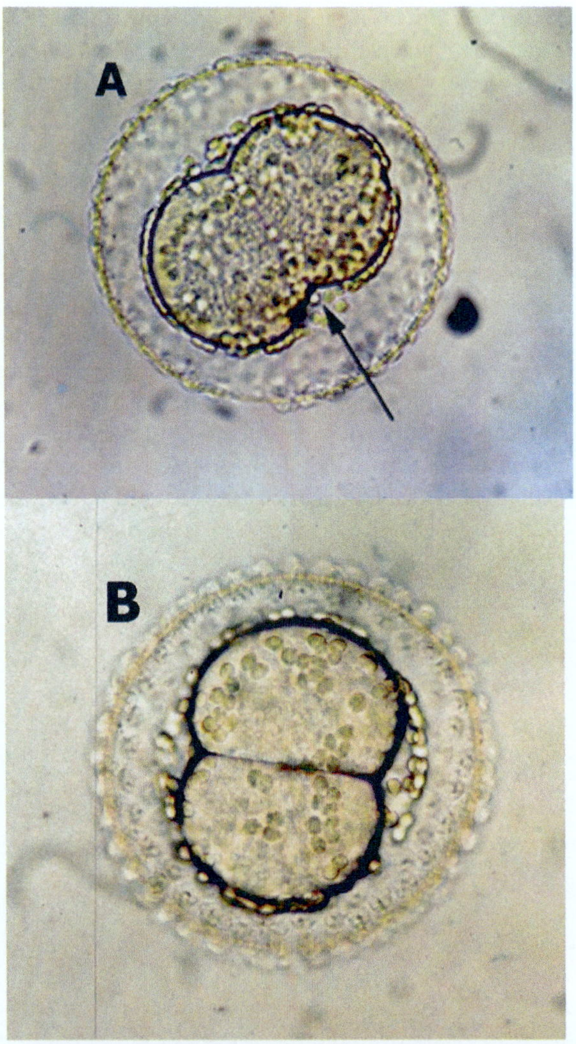

FIG. 13.28 Experimental development of *Phallusia mammillata* (Cuvier, 1815). Phase contrast photomicrographs of whole mounts of live stages of development of the animal. *From Hofmann et al. (2008).* The first cleavage: (A) the onset of division. The first cleavage furrow is indicated by an arrow. 250×; (B) two-cell stage, the first division completed. 100×.

2. The second cleavage furrow is also meridional but perpendicular to the first one, resulting, in the present species, in **four** approximately equal blastomeres after 2.15 h of fertilization (Fig. 13.29A).
3. The third cleavage furrow is transverse or latitudinal. It transects the first and second furrows medially and is shifted to the animal hemisphere resulting in **four pairs** of unequal blastomeres, after 2.30 h of fertilization (Fig. 13.29B).
4. The fourth cleavage leads to divisions at the same time, which are differently oriented in the animal and vegetal tiers. A **16-cell** stage results, showing in animal view a prominent longitudinal furrow that indicates the axis of bilateral symmetry, after 3.40 h of fertilization (Fig. 13.30A).
5. The fifth cleavage generates the **32-cell** stage, which shows different sizes and patterns of the animal and vegetal hemispheres (Fig. 13.30B and C).
6. After the fifth cleavage, the division pattern gets complicated and gradually becomes asynchronous. However, the **64-cell** stage could be estimated (Fig. 13.31A). The blastomeres gradually reduce in size, forming a compact sphere. After cleavage, the resulting blastomeres move to the periphery of the sphere, where they arrange themselves forming a blastoderm. The latter encloses a rudimentary cavity, the **blastocoel**. This stage of development is referred to as the **blastula** stage, after 4 h of fertilization.
7. The blastomeres at one pole of this sphere, the vegetal pole, begin to invaginate. Invagination proceeds, transforming the blastula into a double cell-layered **gastrula**, after 4.30 h of fertilization (Fig. 13.31B). Thus a new cavity appears, **archenteron** or gastrocoal, which increases gradually in size on the expense of the size of the blastocoel. The archenteron has an aperture towards the perivitelline space, which is referred to as the **blastopore**.

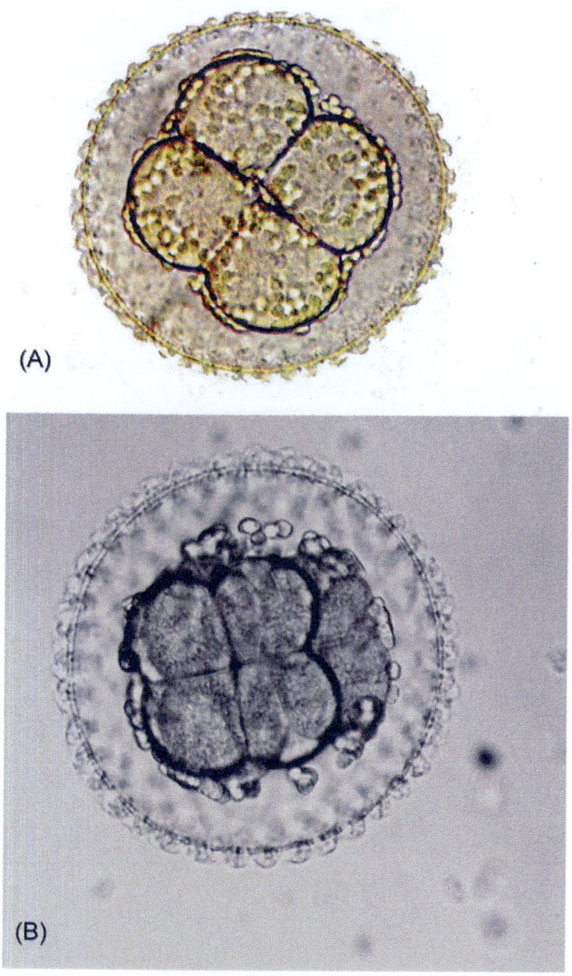

FIG. 13.29 Experimental development of *Phallusia mammillata* (Cuvier, 1815). Phase contrast photomicrographs of whole mounts of live stages of development of the animal. *From Hofmann et al. (2008).* Fixed material: (A) second cleavage (4-cell stage). 100×; (B) third cleavage (8-cell stage). 250×.

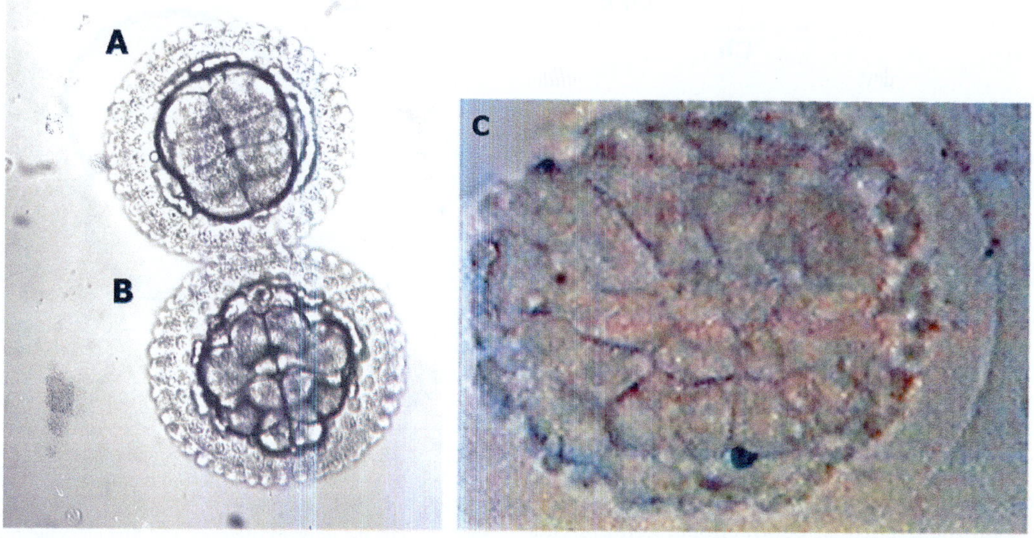

FIG. 13.30 Experimental development of *Phallusia mammillata* (Cuvier, 1815). Phase contrast photomicrographs of whole mounts of live stages of development of the animal. *From Hofmann et al. (2008).* (A and B) Fixed material: (A) fourth cleavage (16-cell stage), note the longitudinal furrow indicating left–right symmetry at the 16-cell stage; (B) fifth cleavage (32-cell stage). 100×; (C) live 32-cell stage (from a black/white film) 400×.

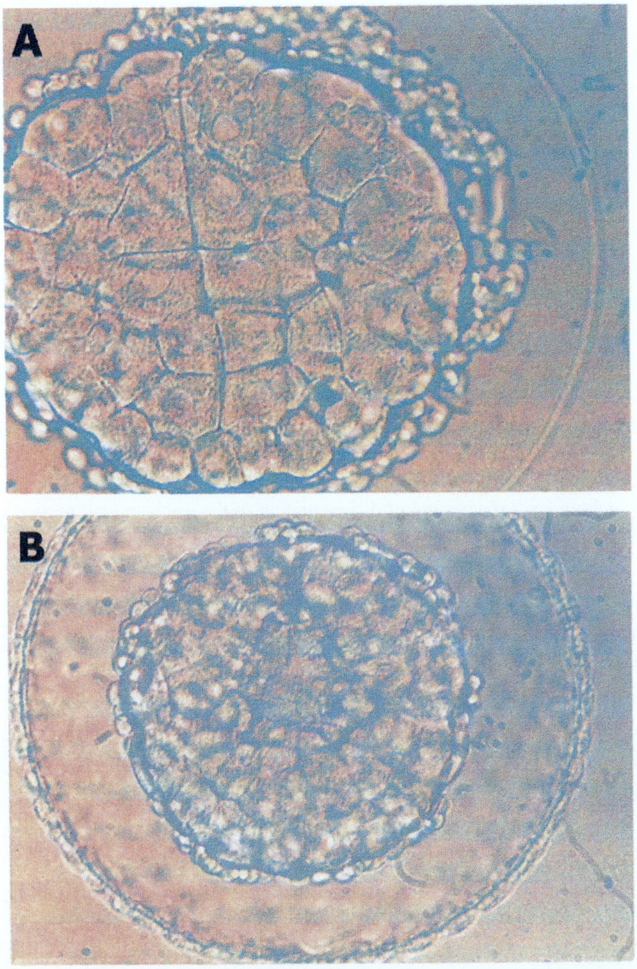

FIG. 13.31 Experimental development of *Phallusia mammillata* (Cuvier, 1815). Phase contrast photomicrographs of whole mounts of live stages of development of the animal. *From Hofmann et al. (2008).* Live material: (A) 64-cell stage; (B) gastrula stage. 400×.

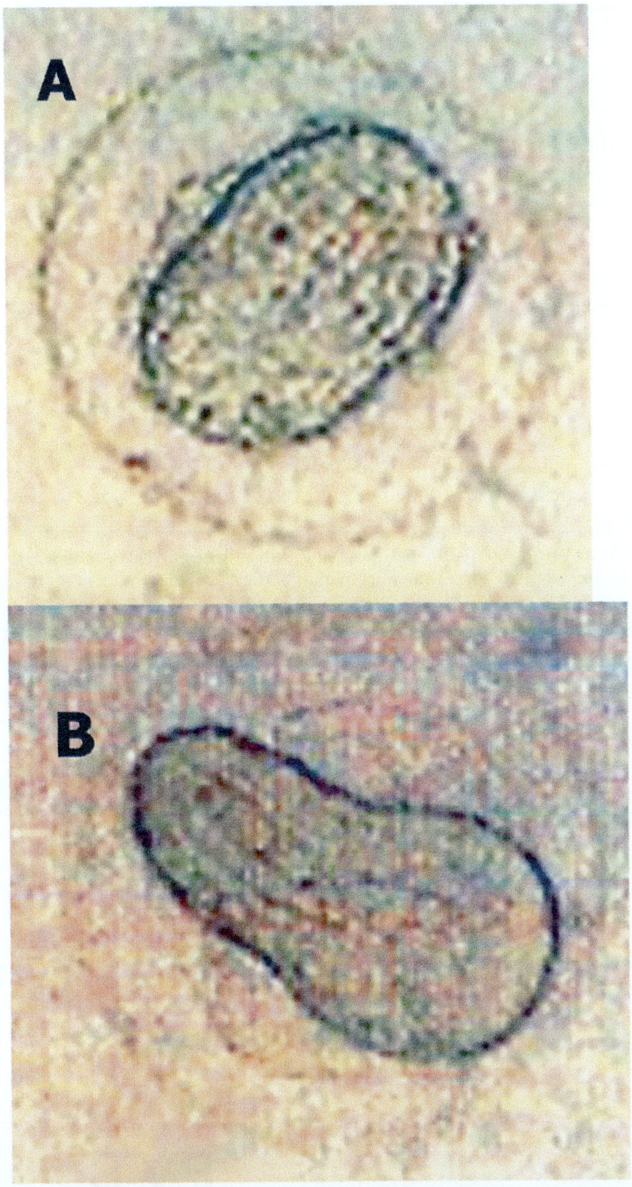

FIG. 13.32 Experimental development of *Phallusia mammillata* (Cuvier, 1815). Phase contrast photomicrographs of whole mounts of live stages of development of the animal. *From Hofmann et al. (2008).* (A) Neurula stage; (B) tailbud stage. 400×.

8. At the dorsal side of the gastrula, precisely in its median portion, rows of neuroectodermal cells form the so-called neural plate, which then invaginates to form the tube-like rudiment of the **neural system**. This stage is termed **neurula** after 7 h of fertilization (Fig. 13.32A).
9. The slightly elongated neurula develops into an **embryo** in which a more compact anterior portion becomes segregated from a slender tail portion. This is called the **tailbud** stage, and occurs after 9 h of fertilization (Fig. 13.32B).
10. Still inside the chorion, the embryonic cells of the tailbud stage become arranged according to their prospective fate and progressively differentiate within the emerging larval organism. The surface of the late embryo remains covered by the test cells, but these become lifted off the epidermis when the larval tunic is secreted. The larva appears fully developed with the elongated tail bent back over the anterior head-trunk portion after 11 h of fertilization (Fig. 13.33). Prominent features are the notochord accompanied by longitudinal muscle strands in the tail, also by the dorsal neural tube and a ventral strand of endoderm. Furthermore, the brain vesicle shows the pigmented dorsal ocellus and the ventral statolith in anterior portion. At the anterior end of the developing larva, the rudiments of the adhesive papillae are visible (Fig. 13.34).
11. Hatching of the larva takes place assisted by a hatching enzyme and vigorous beats of the tail starting about 15 h after fertilization and completing by 20 h after

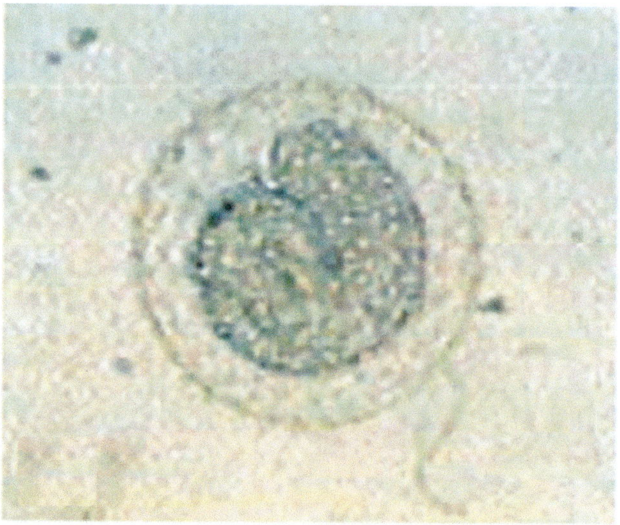

FIG. 13.33 Experimental development of *Phallusia mammillata* (Cuvier, 1815). Phase contrast photomicrographs of whole mounts of live stages of development of the animal. *From Hofmann et al. (2008).* Developing larval stage, early stage. 250×.

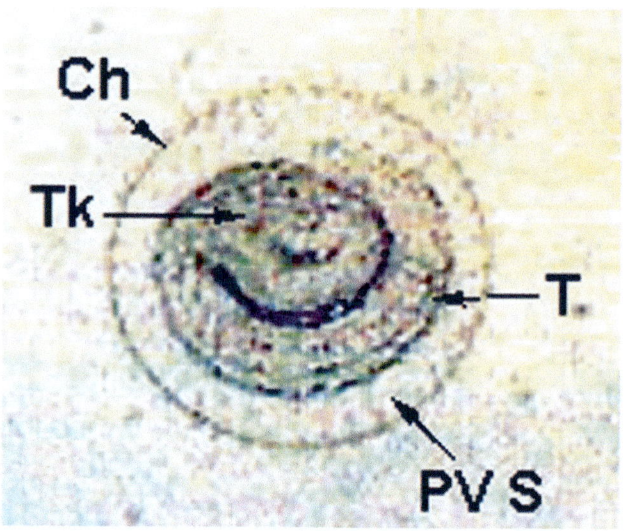

FIG. 13.34 Experimental development of *Phallusia mammillata* (Cuvier, 1815). Phase contrast photomicrographs of whole mounts of live stages of development of the animal. *From Hofmann et al. (2008).* Development and metamorphosis of the larva: advanced larval stage, but still inside the chorion. 250×.

fertilization at the latest. The larva at the moment of hatching is very small in size; it is smaller than that of *C. intestinalis*. It measures about 0.4 mm in length. It may remain in the positively phototactic and negatively geotactic swimming state in the water column for a number of hours or days. The hatched larva of *P. mammillata* still carries test cells adhering to the tunic (Fig. 13.35).

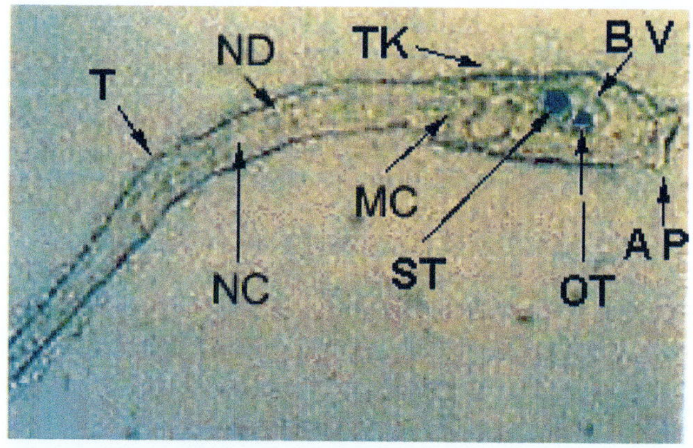

FIG. 13.35 Experimental development of *Phallusia mammillata* (Cuvier, 1815). Phase contrast photomicrographs of whole mounts of live stages of development of the animal. *From Hofmann et al. (2008).* Hatching larva (beginning of differentiation of the trunk and tail regions), the tail is only partially shown. 400×.

Metamorphosis

Transformation of the swimming larva into the sessile juvenile ascidian was studied experimentally. The experiments were carried out on newly hatched *A. aspersa* larvae. Taking the advantage of the very small size of these tadpole larvae, the experiments were carried out in wells. Twenty-four well tissue culture plates with each well accommodating a volume of up to 2.2 mL culture medium. The plates can be easily viewed under a dissection microscope as well as under a compound microscope using a 10× objective. Fifteen larvae were carefully pipetted into each well provided with solutions of one of the following chemicals: NH_4Cl, NH_4, C_8, serotonin, acetylcholine, RF-amide and LW-amide. Controls were run in filtered seawater and in seawater containing antibiotics. All samples were maintained at a constant temperature of 20°C, examined carefully and labelled for progress of metamorphosis at intervals of 24 h for a total of 120 h.

Observations in vitro have shown that the larva begins metamorphosis by a switch towards photophobic and positively geotactic behaviour, and attachment is not a prerequisite. However, retraction and coiling of the axial structures in the tail within the integument, followed by accumulation of the degrading cellular material at the posterior end of the trunk is the subsequent event. Test cells detach and the tunic of the tail portion separates from the remainder of the larval body. The anlagen of the internal organs of the emerging juvenile ascidian undergo a rotation of about 90 degrees as compared to the antero-posterior axis of the larva. The hitherto nonfunctional alimentary tract with the anterior gill chamber and ventral endostyle, then open via an inhalant siphon distally and evacuate transiently via two exhalant siphons. The latter will merge and form a single one over time (Berrill, 1947). The larval brain vesicle becomes transformed into the dorsal ganglion connected with the tubular neural gland, while the larval sensory organs are degraded. The melanin of their pigment cells may persist for a while.

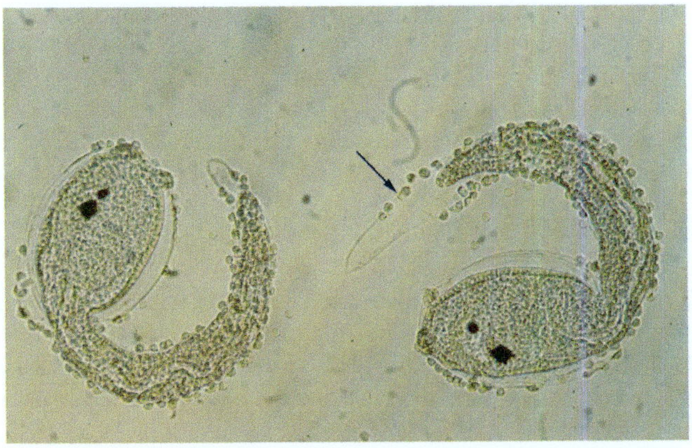

FIG. 13.36 Experimental development of *Phallusia mammillata* (Cuvier, 1815). Phase contrast photomicrographs of whole mounts of live stages of development of the animal. *From Hofmann et al. (2008).* Stage 1 metamorphosis (early tail resorption is indicated by an *arrow*). 250×.

Three different metamorphic stages were considered:

- **Stage 1**—hatched, larva showing all untransformed larval characteristics, exhibiting a long tail, and strong movements (Fig. 13.36).
- **Stage 2**—larva has a short tail. The larva shows significant retraction and coiling of the axial structures of the tail; degrading cells assemble at the posterior part of the trunk. The pigmented sensory organs within the brain vesicle are still prominent (Fig. 13.37).
- **Stage 3**—metamorphosed larva in the form of larval head-trunk region with tail structures fully resorbed. Degrading sensory pigments are visible, and tunic layer present. Rotation of the anlagen of alimentary tract and branchial basket completed, growth and further differentiation of the juvenile ascidian underway (Fig. 13.38).

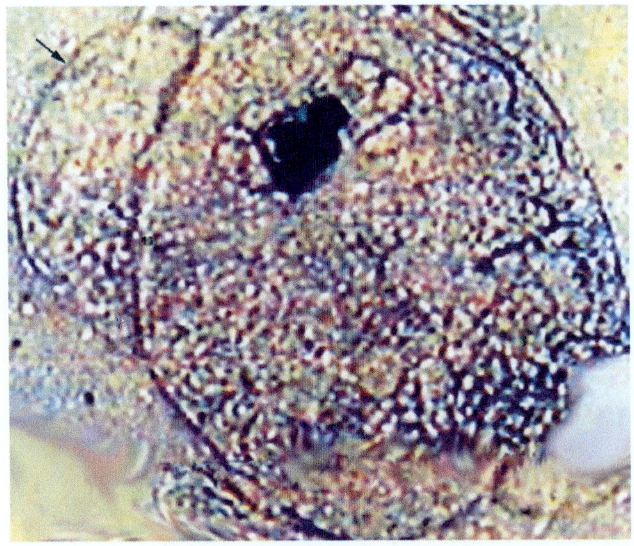

FIG. 13.38 Experimental development of *Phallusia mammillata* (Cuvier, 1815). Phase contrast photomicrographs of whole mounts of live stages of development of the animal. *From Hofmann et al. (2008).* Stage 3 metamorphosis, newly metamorphosed larva. 400×. NC = notochord, ND = nerve cord, OT = otolith, ST = statolith, T = tail, TK = trunk, other lettering as before.

In all the performed trials and controls, metamorphosis to stage 3 occurred. Differences in efficiency on both altering time course and proportion of larva-to-juvenile transition by chemical treatment were gradual. Indeed, both values were somewhat lower in the controls. As RF-amide and LW-amide are known as neuropeptides with morphogenetic functions, e.g. in the hydroid *Hydractinia echinata*, fixed whole-mount preparations of larval and metamorphic stages of *A. aspersa* were subjected to tests for both anti-RF-amide and anti-LW-amide immunoreactivity in the larval and transforming nervous system. For more details, refer to Saad (2002) and Hofmann et al. (2008).

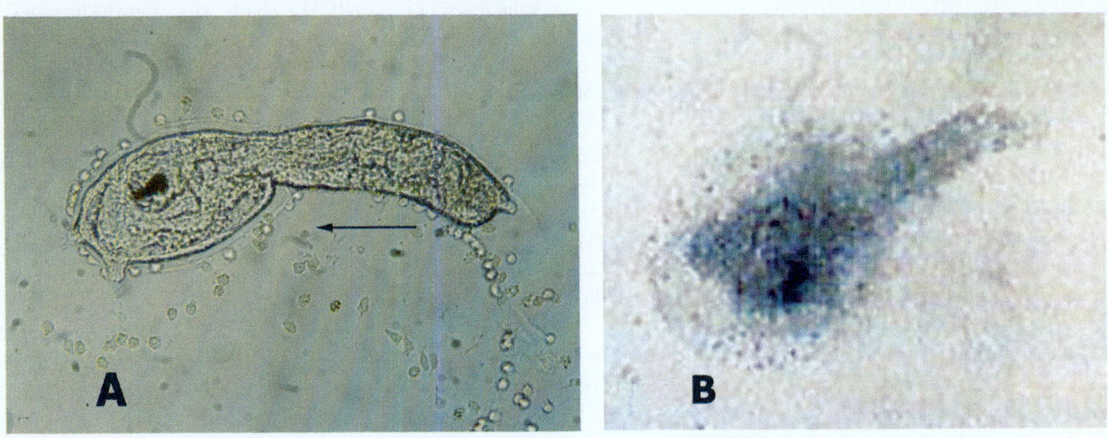

FIG. 13.37 Experimental development of *Phallusia mammillata* (Cuvier, 1815). Phase contrast photomicrographs of whole mounts of live stages of development of the animal. *From Hofmann et al. (2008).* Stage 2 metamorphosis. 250×: (A) advanced tail resorption, the direction of tail phagocytosis is indicated by an arrow; (B) partially metamorphosed larval stage (note the test cells surrounding the larva and the chorion and the inner follicular cells on the periphery).

REFERENCES

Barnes, R.D., 1974. Invertebrate Zoology, third ed. W.H. Saunders Company, London.

Bates, R.W., 2004. Ultraviolet irradiation of egg and blastomere isolation experiments suggests that gastrulation in the direct developing ascidian, *Molgula pasifica*, requires localized cytoplasmic determinants in the egg and cell signaling beginning at the two cell stage. Evol. Dev. 6 (3), 180–185.

Bawab, F.M., El-Sherief, S.S., 1987. The chemical constituents of the yolk formed during deutoplasmogenesis in the crab *Portunus pelagicus* (L.) (Crustacea-Brachyura). Bull. Fac. Sci. Alex. Univ. 27 (2), 198–217. (chemistry).

Bawab, F.M., Michael, M.I., Khalil, S.H., Saad, G.A., 2012. Studies on the reproductive and nervous systems of the Protochordate *Styela plicata* (Lesueur, 1823) to correlate the nervous activity on reproduction. Egypt. J. Aquat. Res. 38 (234), 125–145.

Berrill, N.J., 1947. The development and growth of *Ciona*. J. Mar. Biol. Assoc. UK 26, 616–625.

Berrill, N.J., 1950. The Tunicata with an Account of the British Species. Ray Society Publications (133), pp. 1–354.

Berrill, N.J., 1975. Chordata: Tunicata. In: Gease, Pearse (Eds.), Reproduction of Marine Invertebrates, vol. II. Academic Press.

Borradaile, L.A., Potts, F.A.. with chapters by Eastham, L.E., Saunders, J.T., 1977. The Invertebrata. fourth ed. Revised by Kerkut.G.A.Cambridge University Press.

Brien, P., 1948. Embranchement des Tuniciers. In: Grasse, P. (Ed.), Traite de zoologie, Vol. II, Echinodermes, Stomocordes, Procordes. Masson et Cie, Paris, pp. 553–930.

Conklin, E.G., 1905a. The organization and cell lineage of the ascidian egg. J. Acad. Sci. (Philadelphia), 1–119.

Conklin, E.G., 1905b. Organ forming substances in the egg of ascidians. Biol. Bull. 8, 205–230.

Gilbert, S.F., 2000. Developmental Biology, sixth ed. Sinauer Association, Inc. Publishers, Sunderland, MA.

Gilbert, S.F., Raunio, A.M., 1997. Embryology. In: Constructing the Organism. Sinauer Associates, Inc. Publishers, Sunderland, MA.

Grave, C., 1944. The larva of *Styela* (*Cynthia*) *partita*: structure, activities and duration of life. J. Morphol. 75, 173–188.

Hirai, E., 1941a. An outline of the development of *Cynthia roretzi* Drasche. Sci. Rep. Tohoku Imp. Univ. Biol 16, 257–261.

Hirai, E., 1941b. The early development of *Cynthia roretzi*. Sci. Rep. Tohoku Imp. Univ. Biol 16, 217–232.

Hirai, E., 1951. A comparative study on the structures of the tadpoles of ascidians. Sci. Rep. Tohoku Univ. Biol. 19, 79–87.

Hirai, E., Tunicata, 1968. In: Kumé, Dan (Eds.), Invertebrate Embryology. Translated by Dan, J. C. NOLIT Publishing House, Belgrade, Yugoslavia.

Hofmann, D.K., Michael, M.I., Khalil, S.H., El-Bawab, F.M., Saad, G.A., 2008. Larval metamorphosis in *Ascidiella aspersa* (Müller, 1776) and *Phallusia mammillata* (Cuvier, 1815) Urochordata, Ascidiacea – an experimental study including an immunocytochemical approach. In: Proc. 5th Int. Con. Biol. Sci. (Zool.), vol. 5. pp. 235–248.

Holmberg, K., 1986. The neural connection between the Langerhans receptor cells and the central nervous system in *Oikopleura dioica* (Appendicularia). Zoomorphology 106 (1), 31–34.

Huus, J., 1937. Ascidiacea. Handbuch der Zoologie. 5, 545–672.

Jang, D., Tresser, J.W., Horie, T., Tsuda, M., Smith, W.C., 2005. Pigmentation in the sensory organs of the ascidian larva is essential for normal behavior. J. Exp. Biol. 208, 433–438.

Jeffery, W.R., Swalla, B.J., 1997. Tunicates. In: Gilbert, Raunio (Eds.), Embryology, Constructing the Organism. Sinauer Associates, Inc. Publishers.

Kowalevsky, A., 1866. Entwieklungsgesehichte der einfachen Aschidien. Mém. Acad. Imp. Sc. St. Pétersbourg 7, 1–19.

Kowalevsky, A., 1871. Entwicklungsgeschichte der Amphioxus lanceolatus. Idem xi.

Krohn, A., 1852. Über die Entwicklung der Ascidien. Virchows Arch. 6, 312–333. Plate 8.

Kume, H., Dan, K., 1968. Invertebrate Embryology. Translated by Dan, J.C. NOLIT Publishing House, Belgrade, Yugoslavia.

Lübbering, B., Hofmann, D.K., 1995. Do test cells secrete a hatching enzyme in *Ascidiella aspersa* (Tunicata, Ascidiacea)? Helgoländer Meeresun. 49, 137–143.

Michael, M.I., Hofmann, D.K., Khalil, S.H., El-Bawab, F.M., Saad, G.A., 2008. Comparative studies of the nervous and reproductive systems of *Styela plicata*, *S. partita*, and *Ciona intestinalis* (Urochordata, Ascidiacea) collected from the Mediterranean Sea near Alexandria, Egypt. In: Proc. 5th Int. Con. Biol. Sci. (Zool.), vol. 5. pp. 330–347.

Millar, R.H., 1952. The annual growth and reproductive cycle in four ascidians. J. Mar. Biol. Ass. 31, 41–61.

Millar, R.H., 1970. British Ascidians (Tunicata-Ascdiacea). Keys and Notes for the Identification of the Species. Published for the Linnean Society of London by Academic Press.

Miller, R.L., 1982. Sperm chemotaxis in ascidians. Am. Zool. 22, 827–840.

Morgan, T.H., 1942. Cross and self fertilization in the ascidian *Molgula manhattensis*. Biol. Bull. 82 (1), 172–177.

Morgan, T.H., 1945. The conditions that lead to normal or abnormal development of *Ciona*. Biol. Bull. 88, 50–62.

Niermann-Kerkenberg, E., Hofmann, D.K., 1989. Fertilization and normal development in *Ascidiella aspersa* (Tunicata) studied with Nomarski-optics. Helgoländer Meeresun. 43, 245–258.

Olsson, R., Holmberg, K., Lilliemarck, Y., 1990. Fine structure of the brain nerves of *Oikopleura dioica* (Urochordata-Appendicularia). Zoomorphology 110, 1–7.

Parker, T.J., Haswell, W.A., 1974. Textbook of Zoology. Vol. I. Invertebrates. (Williams, W. D. (William David) & Marshall, A. J. "Jock" — Editors). Published by Macmillan, London, pp. 779.

Pearse, A.G.E., 1968. Histochemistry. Theoretical and Applied, third ed., vol. I. Churchill, London.

Pflugfeder, O., 1970. Lehrbuch dez Entwicklungsgeschichte und Entwicklungsphysiologie der Teire. Veb Gustav Fischer Verlog Jena.

Reverberi, G., 1971. Experimental Embryology of Marine and Fresh-Water Invertebrates. North-Holland Publishing Company, Amsterdam, London.

Sabbadin, A., 1957. Cyclobiologico di *Ciona intestinalis* (L.), *Molgula manhattensis* (Dekay) e *Styela plicata* (Lesueur). Nella laguna veneta. Archivio di Oceanografia a Limnologia. 11, 1–30.

Saad, G.A., 1995. Experimental and histological studies on the nervous and reproductive systems of the protochordate *Styela plicata* (Lesueur). Urochordata-Ascidiacea. (M.Sc. thesis) Faculty of Science, Alexandria University.

Saad, G.A., 2002. Comparative Studies of the Nervous and Reproductive Systems of some Species of Urochordates with Emphasis of the Nervous System on Reproduction and Larval Metamorphosis. (Ph. D. Thesis).

Saad, G.A., 2008. Histological and histochemical studies of the process of deutoplasmogenesis (vitellogenesis) in the oocyte of *Styela plicata* (Lesuaer, 1823) *S. partita* (Stimpson, 1852) and *Ciona intestinalis* (Linnaeus, 1767). Urochordata-Ascidiacea. Egypt. J. Aquat. Res. 34 (2), 387–411.

Savigny, J.C., 1816. Recherches anatomiques sur les ascidies composes et sur les ascidies simples—Système de la classe des Ascidies. In: Mémoires sur les Animaux sans Vertèbres, Pt 2. G. Dufour, Paris, pp. 1–239.

Selman, K., Arnold, J.M., 1977. An ultrastructural and cytochemical analysis of oogenesis in the squid, *Loligo pealei*. J. Morphol. 152 (3), 381–400.

Seshachar, B.R., Rao, V.S.R., 1963. Cytochemistry of the test cells in the eggs of *Pyura squamulosa* (Tunicata, Pleurogonia). Q. J. Microsc. Sci. 104 (4), 459–464.

Torrence, S.A., Cloney, R.A., 1982. Nervous system of ascidian larvae: caudal primary sensory neurons. Zoomorphology 99, 103–115.

Tung, T.C., 1932. Expériences de coloration vitale sur l'oeuf d' *Ascidiella aspersa*. Arch. De Biol. 43, 451–469.

Van Beneden, P.J., 1846. Reecherclies Sur Pembryogenie, I'Anat. et la Physiol, des Ascidies simples. Mem. Acad. Boy., Belgique.

Wahl, M., Lafargue, F., 1992. Ascidien: Entwicklung. In: Emschermann, P., Hoffrichter, O., Körner, H., Zissler, D. (Eds.), Meeresbiologische Exkursion: Beobachtung und Experiment, Abschnitt 1.3, vol. 28. Gustav Fischer, Stuttgart, pp. 163–168.

Willey, A., 1893a. Studies on the Protochordata I. On the origin of the branchial stigmata, preoral lobe, endostyle, atrial cavities, etc., in *Ciona intestinalis* L. with remarks on *Clavelina lipadiformis*. Q. J. Microsc. Sci. 34, 317–360. (cited in Hirai, 1968).

Willey, A., 1893b. Studies on the Protochordata II. The development of the neuro-hypophysial system in *Ciona intestinalis* L. and *Clavelina lipadiformis*. Q. J. Microsc. Sci. 35, 295–316. (cited in Hirai, 1968).

Further Reading

Alberts, B., Johnson, A., Lewis, J., Walter, P., Raff, M., Roberts, K., 2002. Molecular Biology of the Cell, fourth ed. Garland Science, Taylor and Francis Group.

Al-Hussaini, A., Demian, E.S., 1967. Practical Animal Morphology. Macmillan & Co Ltd, London.

Anderson, D.T., 1973. Embryology and Phylogeny in Annelids and Arthropods. Perganon Press, Oxford, New York, Sydney. 495 p.

Aparicio, M.T., 1937. Certain reproductive organs in the Neritidae. Malacologia 61, 525–549.

Aristotle, 1910. The Works of Aristotle. Translated into English under the editorship of J.A. Smith and W.D. Ross. vol. IV. Historia Animalium, by D'Arcy Wentworth Thompson. Clarendon Press, Oxford.

Armstrong, W.P., 1999. Watermelon snow, a strange phenomenon caused by algal cells of the Chlorophyta. In: Wayne's Word Noteworthy Plants.

Balfour, F.M., 1880. A Treatise on Comparative Embryology, vol. I. Macmillan and Co, London.

Barnes, R.D., 1974. Invertebrate Zoology, third ed. W.H. Saunders Company, London.

Barrington, E.W., 1979. Invertebrate Structure and Function, second ed. The English Language Book Society and Nelson.

Bell, G.H., Emisle-Smith, D., Paterson, C.R., 1976. Text Book of Physiology and Biochemistry, ninth ed. The English Language Book Society and Churchill Livingstone.

Bloom, W., Fawcett, D.W., 1975. A Text Book of Histology, 10th ed., Asian edition Igaku Shoin Ltd., Tokyo.

Boas, J.E.V., 1886. Kleinere carcinologische Mittheilungen: 1, Eine neue Art der Gattung Apseudes. Zool. Jahrb. II, 109–116.

Borradaille, L.A., 1917. On the structure and function of the mouth-parts of the palaemonid prawns. Proc. Zool. Soc. Lond. 87 (1), 37–72.

Borradaille, L.A., 1926. Notes upon crustacean limbs. Ann. Mag. Nat. Hist. 17 (9), 193–213.

Borradaile, L.A., Potts, F.A., with chapters by Eastham, L.E., Saunders, J.T., 1977. The Invertebrata, fourth ed. Revised by Kerkut, G.A. Cambridge University Press.

Bullough, W.S., 1981. Practical Invertebrate Anatomy, second ed. Macmillan Press Ltd, London, Basingstoke.

Burighel, P., Cloney, R.A., 1997. The invertebrate chordates. In: Microscopic Anatomy of Invertebrates. Urochordata: Ascidiacea. Wiley-Liss (Chapter 4).

Calman, W.T., 1907. On new or rare Crustacea of the order Cumacea from the collection of the Copenhagen Museum: Part I. Trans. Zool. Soc. Lond. xviii, 1–58.

Calman, W.T., 1909. In: Sir Lankester (Eds.), Treatise on Zoology. Part VII. Adam and Charles Black, London.

Carleton, H.M., 1944. Schafir's Essentials of Histology. Longmans, Green and Co, London.

Carré, D., Sardet, C., 1984. Fertilization and early development in *Beroe ovata*. Dev. Biol. 105, 188–195.

Charniaux-Cotton, H., 1985. Vitellogenesis and its control in malacostracan Crustacea. Am. Zool. 25, 197–206.

Cheng, T.C., 1964. The Biology of Animal Parasites. W.B. Saunders Company, Philadelphia, London.

Claus, C., 1872. Uööber den Bau und die systemische Stellung von *Nebalia*. Z. Wiss. Zool. 22, 323–330.

Claus, C., 1888. Uber den organismus der Nebaliden und die systematische Stellung der Leptostraken. Arb. Zool. Inst. Wien 8, 1–148.

Cole, S.W., 1944. Practical Physiological Chemistry, eigth ed. W. Heffer and Sons, Cambridge, England.

Duncan, C.J., 1959. The life cycle and ecology of the freshwater snail *Physa fontinalis* (L). J. Anim. Ecol. 28, 97–118.

El-Banhawy, H.A., Demian, E.S., Shalaby, A.A., Roshdy, H.A., Saoud, M.F., Said, E.E.D., 1975. Text Book of Zoology. Dar Al-Maaref, Cairo.

El-Sherief, S.S., Bawab, F.M., El-Zayat, M.A., 1986. The egg membranes of the crab *Portunus pelagicus* (L.) (Crustacea-Brachyura). Bull. Fac. Sci. Alex. Univ. 26 (4), 148–164.

Encyclopedia Americana. International edition Birmingham to Burlington. 1982. Grolier Incorporated. Complete in thirty volumes. First published in 1829.

Ereskovsky, A.V., 2006. A new species of *Oscarella* (Demospongiae, Plakinidae) from the western sea of Japan. Zootaxa 1376, 37–51.

Foerder, C.A., Klebanoff, S.J., Shaprio, B.M., 1978. Hydrogen peroxide production, chemiluminescence, and the respiratory burst of fertilization: interrelated events in early sea urchin development. Proc. Natl. Acad. Sci. U. S. A. 75 (7), 3183–3187.

Fretter, V., Graham, A., 1962. British Prosobranch Molluscs. Their Functional Anatomy and Biology. The British Ray Society, London.

Fretter, V., Graham, A., 1964. Reproduction. In: Wlbur, K.H., Yonge, C.M. (Eds.), Physiology of Mollusca, vol. I. Academic Press, New York, pp. 127–164.

Gardiner, M.S., 1972. The Biology of Invertebrates. McGraw-Hill Book Company, London, New York.

Garven, H.S.D., 1965. A Student's Histology, second ed. Livingstone Ltd., Edinburgh, London.

Giese, A.C., Pearse, J.S., 1974. Reproduction of Marine Invertebrates. Acoelomate and Pseudocoelomate Metazoans, vol. I. Academic Press, New York, London.

Giese, A.C., Pearse, J.S., 1975a. Reproduction of Marine Invertebrates. Entoprocts and Lesser Coelomates, vol. II. Academic Press, New York, London.

Giese, A.C., Pearse, J.S., 1975b. Reproduction of Marine Invertebrates. Annelids and Echiurans, vol. III. Academic Press, New York, London.

Giese, A.C., Pearse, J.S., 1977. Reproduction of Marine Invertebrates. Molluscs: Gastropods and Cephalopods, vol. IV. Academic Press, New York, London.

Giese, A.C., Pearse, J.S., 1979. Reproduction of Marine Invertebrates. Molluscs: Pelecypods and Lesser Classes, vol. V. Academic Press, New York, London.

Gilbert, S.F., 2000. Developmental Biology, sixth ed. Sinauer Association, Inc., Publishers, Sunderland, MA.

Gilbert, S.F., Raunio, A.M., 1997. Embryology, Constructing the Organism. Sinauer Associates, Inc. Publishers, Sunderland, MA.

Goldstein, D.J., 1962. Correlation of size of dye particle and density of substrate, with special reference to mucin staining. Stain. Technol. 37, 79–93.

Gray, P. (Ed.), 1971. Encyclopedia of the Biological Sciences. second ed. Van Nostrand Reinhold Company, pp. 1–20.

Grove, A.J., Newell, G.E., 1955. Animal Biology, fourth ed., reprint. University Tutorial Press Ltd, London.

Hackman, R.H., 1971. The integument of arthropoda. In: Florkin, M., Scheer, B.T. (Eds.), Chemical Zoology, vol. VI. Academic Press, New York, London.

Harmer, S.F., Shipley, A.E., 1923. The Cambridge Natural History. Crustacea and Arachnids, vol. IV. Macmillan and Co. Limited, London.

Hickman, C.P., 1973. Biology of the Invertebrates, second ed. The C.V. Mosley Company, Saint Louis.

Hirai, E., Tsubata, B., 1956. On the spawning of an ascidian, *Halocynthia roretzi* (Drasche). Bull. Mar. Biol. Sta. Asamusi 8, 1–4.

Ho, K.M., Kasha, K.J., 1975. Genetic control of chromosome *elimination* during haploid formation in barley. Genetics 81, 263–275.

Hyman, L.H., 1937. Reproductive system and copulation in *Amphiscolops langerhansi* (Turbellaria Acoela). Biol. Bull. 72, 319–326.

Hyman, L.H., 1940. The Invertebrates, Protozoa Through Ctenophora, first ed. McGraw-Hill Book Company, Inc., New York, London. Third expression.

Hyman, L.H., 1944a. A new Hawaiian polyclad flatworm associated with Teredo. Occas. Pap. Bernice P. Bishop Mus. Honolulu 18, 73–75.

Hyman, L.H., 1944b. Marine Turbellaria from the Atlantic coast of North America. Am. Mus. Novit. 1266, 1–15.

Hyman, L.H., 1951. The Invertebrates: Acanthocephala, Ashelminthes and Entoprocta. The Pseudocoelomate Bilateria, vol. III. McGraw-Hill Book Company, Inc., New York, Toronto, London.

Hyman, L.H., 1967. The Invertebrates. Mollusca I, vol. VI. McGraw-Hill, New York.

Invertébrés: Beauté du Monde Animal, Tome XII. 1974. Librairie Larousse, Paris.

Jeffery, W.R., Swalla, B.J., 1990. The myoplasm of ascidian eggs: a localized cytoskeletal domain with multiple roles in embryonic development. Semin. Dev. Biol. 1, 253–261.

Johnston, P.G., Watson, C.M., Adams, M., Paull, D.J., 2002. Sex chromosome elimination, X chromosome inactivation and reactivation in the southern brown bandicoot *Isoodon obesulus* (Marsupialia: Peramelidae). Cytogenet. Genome Res. 99 (1-4), 119–124.

Karlson, P., 1965. Introduction to Modern Biochemistry, second ed. Translated by Doering, C.H. Academic Press, New York, London.

Keele, C.A., Neil, E., 1973. Sanson Wright's Applied Physiology, 12th ed. Oxford University Press.

Kirk, D.L., Birchem, R., King, N., 1986. The extracellular matrix of *Volvox*: a comparative study and proposed system of nomenclature. J. Cell Sci. 80, 207–231.

Kochert, G., 1968. Differentiation of reproductive cells in *Volvox carteri*. J. Protozool. 15 (3), 438–452.

Kume, H., Dan, K., 1968. Invertebrate Embryology. Translated by Dan, J.C. NOLIT Publishing House, Belgrade, Yugoslavia.

Leake, L.D., 1975. Comparative Histology, An Introduction to the Microscopic Structures of Animals. Academic Press, London.

Lee, B., 1950. The Microtomist's Vade-Mecum, 11th ed. Churchill, London.

Lexicon Universal Encyclopedia. 1983. Lexicon Publications, Inc., New York. Parts 1–21.

Lillie, F.R., 1913a. Studies of fertilization. V. The behavior of the spermatozoa of *Nereis* and *Arbacia* with special reference to egg extractives. J. Exp. Zool. 14, 515–574.

Lillie, F.R., 1913b. The mechanism of fertilization. Science 38 (980), 524–528.

Magruder, W.H., Hunt, J.W., 1979. Seaweeds of Hawaii: A Photographic Identification Guide. The Oriental Publishing Company, Honolulu, Hawaii, 116 pp.

Mahoney, R., 1973. Laboratory Techniques in Zoology. Butterworths and Co., London.

Mann, T., 1975. Spermatophores. In: Duckett, J.G., Racey, P.A. (Eds.), Biology of Male Gamete, vol. 7. Linnean Society of London, Academic Press, pp. 417–421.

Mann, T., 1984. Spermatophores. Springer-Verlag, Berlin, New York.

Marshall, A.J., Williams, W.D., 1975. A Text Book Zoology: The Invertebrates. ELBS and Macmillan.

Martindale, M.Q., Baxevanis, A.D., 2013. The genome of the ctenophore *Mnemiopsis leidyi* and its implications for cell type evolution. Science 342, 124–259. https://doi.org/10.1126/science.1242592.

Metz, C.B., Monroy, A., 1985a. Biology of Fertilization. Model Systems and Oogenesis, vol. I. Academic Press, Inc, New York, London.

Metz, C.B., Monroy, A., 1985b. Biology of Fertilization. Biology of the Sperm, vol. II. Academic Press, Inc., New York, London.

Metz, C.B., Monroy, A., 1985c. Biology of Fertilization. The Fertilization Response of the Egg, vol. III. Academic Press, Inc., New York, London.

Milne-Edwards, H., 1840. Histoire naturelle des crustacés: comprenant l'anatomie, la physiologie et la classification de ces animaux. vol. 3. Librairie encyclopédique de Roret.

Milne-Edwards, H., 1842. Observations sur la structure at les fonctions de quelques Zoologies, Mollusques at Crustacées des câtes de France. Sur les Spermatophores de Céphalopodes. Ann. Sci. Nat. 18 (2) (Zoologie).

Milne-Edwards, H., 1843. Note sur deux Crustacés fossiles de l'ordre des isopodes. Ann. Sci. Nat. 2e ser. (Zool.) 20, 326–329.

Minchin, M.A., Fowler, B.A., Bourne, M.A., 1900. In: Lankester, E.R. (Ed.), A Treatise on Zoology. Part II: The Porifera and Coelen. Adam and Charles Black, London.

Morgan, T.H., 1927. Experimental Embryology. Columbia University Press.

Morton, J.E., 1967. Mollusca, fourth ed. Hutchinson University Library, London.

Murray, R.K., Granner, D.K., Mayes, P.A., Rodwell, V.W., 1996. Harper's Biochemistry, 24th ed., Middle East edition Librairie de Liban, Appleton and Lange, Norwalk, CT/Los Altos, CA.

Needham, J., 1934. A History of Embryology. Cambridge University Press, Cambridge, England.

Needham, J., 1945. New Microscopical Discoveries. F. Needham, London.

Nozaki, H., 2003. Origin and evolution of the genera *Pleodorina* and *Volvox*. Biologia, Bratislava 58, 425–431.

Pantin, C.F.A., 1948. Notes on Microscopical Technique for Zoologists. University Press, Cambridge.

Parker, T.J., Haswell, W.A., 1943. A Text Book of Zoology, fourth ed., vol. I. Macmillan and Co., Limited, London.

Parker, T.J., Haswell, W.A., 1974. A Text book of Zoology. Vertebates, vol. 2, seventh ed. English Language Book Society and Macmillan.

Pearse, A.G.E., 1968. Histochemistry. Theoretical and Applied, third ed., vol. I. Churchill, London.

Pelseneer, P., 1906. In: Lankester, E.R., Black, C. (Eds.), Mollusca: A Treatise on Zoology. London. Mem., 32, vol. V. Universtiy Press of Liverpool.

Pflugfeder, O., 1970. Lehrbuch dez Entwicklungsgeschichte und Entwicklungsphysiologie der Teire. Veb Gustav Fischer Verlog Jena.

Philip's, 2003. World Atlas & Gazetteer. In Association with the Royal Geographical Society, with the Institute of British Geographers. Published in Great Britain.

Pickett-Heaps, J.D., 1975. Green Algae, Structure, Reproduction & Evolution in Selected Genera, first ed. Sinauer Associates Inc., Publishers, Sunderland, MA.

Prosser, C.L., 1973. Comparative Animal Physiology, third ed. W. B. Saunders Company, Philadelphia, London.

Purchon, R.D., 1968. The Biology of Mollusca. Pergamon Press, Oxford.

Rajan, S.S., 2001. Introduction to Algae, first ed. Anmol Publication Pvt Ltd, New Delhi.

Reverberi, G., 1971. Experimental Embryology of Marine and Fresh-Water Invertebrates. North-Holland Publishing Company, Amsterdam, London.

Richards, A.G., 1951. The Integument of Arthropods. Minneapolis University of Minnesota Press.

Roosen-Runge, E.C., 1977. The Process of Spermatogenesis in Animals. Cambridge University Press, London.

Russel-Hunter, W.D., 1979. A Life of Invertebrates. Macmillan Publishing Co., Inc, New York.

Smith, G.M.A., Woods, M.A., Shipley, G.B.E., Worburton, M.A., Thompson, C.B., 1923. The Cambridge Natural History. (Crustacea and Arachnida). Macmillan and Co., Limited, London.

Soliman, G.N., 1996. Invertebrate Zoology, The Mid-eastern Invertebrate Fauna. Part I. The Palm Press, Cairo.

Soundarapandian, P., Tamizhazhagan, T., 2009. Embryonic development of commercially important swimming crab *Portunus pelagicus* (Linnaeus). Curr. Res. J. Biol. Sci. 1 (3), 106–108.

Storer, T.I., Stebbins, R.C., Usinger, R.L., Nybakken, J.W., 1983. General Zoology, sixth ed. Tata McGraw-Hill Publishing Company Ltd, New Delhi.

Taki, 1933. Gastropoda, Pelecypoda. Iwanami Biol. Course. Iwanami Shoten, Tokyo (cited in Wada, 1968).

Threadgold, L.T., 1976. The Ultrastructure of the Animal Cell, second ed. Pergamon Press, New York.

Tompsett, D.H., 1939. Sepia. Liverpool Marine Biology Committee.

Verma, P.S., 1977. A Manual of Practical Zoology: Invertebrates. S. Chand and Company Ltd, Ram Nagar, New Delhi.

Villee, C.A., Walker, J.R.W.F., Smith, F.E., 1964. General Zoology. W.B. Saunders Company, London.

Vogel, A.I., 1951. A Text Book of Quantitative Inorganic Analysis-Theory and Practical, second ed. Longmans, London.

Waterman, T.H., 1960. The Physiology of Crustacea. Academic Press, New York, London.

West, A.B., Lambert, C.C., 1975. Control of spawning in the tunicate *Styela plicata* by variations in a natural light regime. J. Exp. Zool. 195, 203–270.

Index

Note: Page numbers followed by *f* indicate figures and *t* indicate tables.

A

Abranchiata, 481
Achatina fulica, 755–757
Acmaea rubella, 786
Acoela, 340–345
  Convolutriloba longifissura, 343–345, 346–347*f*
  Neochildia fusca, 343, 345*f*
  Polychoerus carmelensis, 341–345, 344*f*
Acotylea, 340
Acropora, 297–301
  cleavage, 298
  embryological development, 299*f*
  gastrulation, 298
  spawning, 298
  stomodaeum and gastrodermis, formation of, 298–300
Acrosomal complex, 32–34
Acrosomal reaction, 49–51
Acrosome, 32–34, 35*f*, 875–877
Actinophrys, 89–91
Actinophyridae, 89
  morphology, 89–90, 90*f*
  reproduction, 90–91, 91–92*f*
Actinopoda, 89*t*
Actinula, 194
Action-at-a-distance, 49
Adalaria proxima, 717*f*, 788–809, 788*f*
Aeolidia papillosa, 797, 799*f*
Aequorin, 57–58
Aesthetes, 732
Aflagellated spermatozoons, 38
Agamermis decaudata, 435*f*
Agametic method, 138
Agametic reproduction, 72, 716, 816, 865–867
  Alatina cf. *moseri*, 250
  Antipatharia, 255
  Cassiopea andromeda, 228–229, 230–231*f*
  Cnidaria, 177–180, 178*f*
    Anthozoa, 180
    Hydrozoa, 177–179, 185*f*
    Scyphozoa, 179–180
    types of, 179
  Heteroxenia fuscescens
    budding, 266
    pinnitomy, 267
    transverse fission, 266–267, 270*f*
  Platyhelminthes, 345–348, 348*f*
  Porifera
    budding, 111
    external bud formation, 111–114, 111–114*f*
    fragmentation, 116
    internal buds, 114–116
    reduction bodies, 116
Agglutination, 45–46
Agriolimax reticulatus, 774
Alatina cf. *moseri*, 248–251, 249*f*
  agametic reproduction, 250
  cyst formation, 250–251
  metamorphosis, 251
  young medusa, 251
Alcyonacea, 264–276, 265*f*
  Alcyonium spp.
    A. coralloides, 264–266
    A. digitatum, 264
  Capnella gaboensis, 273–276, 275*f*
  Heteroxenia fuscescens, 266–273
  Xenia umbellata, 273, 274*f*
Alcyonium spp.
  A. coralloides, 264–266
  A. digitatum, 264
Alecithal oocyte, 40
Alfa-actinin, 35
Alimentary canal, 857, 889–890
Allolobophora caliginosa, 94
Alteromonas espejiana, 186–187
14-amino acid peptide, 49
Amnion, 403
Amoeboid cells, 403
Amphiblastula larva, 135, 166–167
Amphimixis, 89, 91, 92*f*
Amphineura
  Aplacophora, 735–739
  Monoplacophora, 733–735
  Polyplacophora, 716–733
Anabiosis, 455
Anal papilla, 857
Anaplectus granulosus, 435*f*
Ancylostoma caninum, 441
Androdioecy, 506–507
Androgamone 1, 31
Animal oocytes, meiotic arrest in, 29–30
Anisogamous reproduction, 77, 80*f*
Annelids, 53–55
Anomura, 479–480
Anopla, 390
Anostraca
  Artemia, 485–493
  early development
    blastula, 489–490
    cleavage, 489
    gastrulation, 490–491
    hatched nauplius, 491–493
    parthenogenesis, 493
  morphology, 484
  reproductive system, 484
Anostracans, 484, 484*f*, 488*f*, 489
Anterior–posterior axis, 878–880
Antheridia, 82
Antheridium, 77
Anthoplexaura dimorpha, 279
Anthozoa, 176–177, 180, 253–301
  Ceriantharia, 260–262
    Cerianthus iloydii, 262
    Pachycerianthus multiplicatus, 260–262, 261*f*
  Ceriantipatharia
    Antipatharia, 253–260
    Ceriantharia, 260–262
  Octocorallia
    Alcyonacea, 264–276, 265*f*
    Gorgonacea, 276–279, 278*f*
    Stolonifera, 262–264
  (*Octocorallia see* Octocorallia)
  types of, 254*f*
Anticoma typica, 432, 435*f*
Antipatharia
  agametic reproduction, 255
  Antipathes, 258–260, 259*f*
  Cirrhipathes anguina, 255–258
  gametic reproduction, 255
Antipathes spp., 258–260, 259*f*
  A. griggi, 258–260, 259*f*
Anus, 331–333
Apertural males, 506–507
Aphelenchoides, 431
Aplacophora, 735–739
  early development, 735–738
  larval development, 738
  metamorphosis, 738–739
Aplanospore formation, 74
Aplysia californica, 769
Aplysilla rosea, 117–118
Apodemes, 590
Aporrhais, 769
Appendicularia, 888
Aptera, 477
Apyrene, 39
Arachnid, 477
Arbacia spp.
  A. punctulata, 37*f*
  A. zygote, 59*f*
Archaeocytes, 110
Archenteron, 62, 894

Archeophora, 340
Architomy, 346
Arcoscalpellum michelottianum, 506*f*, 507–508
Arenaeus cribrarius, 689
Aristaeomorpha foliacea, 560*f*
atresic stage, 574
early vitellogenic stage, 574
late primary stage, 574
late vitellogenic stage, 574
oogonia and early primary stage, 573–574
postvitellogenic stage, 574
Aristeus coruscans, 532
Aristotle, 4
Armadillidium vulgare, 445
Artemia spp., 485–493
A. parthenogenetica, 493, 494*f*
A. salina, 484–485, 485*f*
hatching, 492*f*
postembryonic development, 493*f*
embryology, 489–493
hatched nauplius, 491–493
mating, 488*f*
parthenogenesis, 493
reproductive system, 485–486
female, 486–489
male, 485–486, 487–488f
winter eggs, 489
Artificial insemination, 697
Asbestopluma occidentalis, 157–162, 159*f*, 161*f*
Ascandra minchini, 137, 139*f*
Ascaris spp.
A. lumbricoides, 437–438, 440–441, 445, 446*f*
A. megalocephala, 19–20, 433*f*, 434–435, 444*f*, 445, 449*f*
A. suum, 437, 438*f*, 443
type of fertilization, 58–59
Ascidiacea, 864–865
Ascidiella aspersa, 867, 874, 890–891
Asexual reproduction, 90–91
Chlamydomonas, 72–75, 74–76*f*
Paramecium, 96, 98*f*
types of, 72
Volvox, 81–82, 85*f*
Astacidae, 590–629
sperm, 597–598
spermatophore formation, 595–597
up-to-date embryological studies on, 609–611
Astacopsis franklinii, 590
Astacus spp., 590–629
A. astacus, 597–598
spawning, 602–603
spermatogenesis in, 604–605
A. fluviatilis, 497*f*, 536, 590–592, 593*f*, 606–620
brooding, 603–604
A. leptodactylus, 590–592, 593*f*
female reproductive system, 598, 599*f*
reproductive system, 594*f*
spermatophore formation, 593–597
A. pallipes, 604
egg membranes in, 601*f*
brooding, 603–604
copulation, 598–602
development, 604–605
early embryological studies, 606
gametes formation, 604–605
gastrulation, 606
hatching, 620–629
morphology and sexual dimorphism, 590
parthenogenesis, 628–629
postembryonic development, 620–624
reproductive systems, 590–598
spawning, 602–603
Asterina pectinifera, deutoplasmogenesis in, 29, 30*f*
Astrangia danae, 295–297
fertilization and development, 297
gametes, 295–297
histological preparations of tissues, 296*f*
oogenesis, 295
spawning, 295
spermatogenesis, 295
Atrium, 758, 889
Aurelia aurita, 214*f*, 220–223
Austropotamobius pallipes, 602–603
copulation, 598–602
embryonic stages of, 610*f*
mating process, 600*f*
Autogamy, 90–91, 100
Autonomous intraoogonial process, 676, 845
Autumn, 115
Auxocytes, 271
Axial filament, 35
Axoneme, 35, 804–805, 806*f*
Axpodia, 89

B

Babylonia japonica, 788
Balanus spp., 509–510*f*, 518
B. amphitrite, 505, 509–510*f*, 511
B. balanoides, 511
B. balanus, 511
B. eburneus, 505, 511
B. nubilis, 517*f*
B. tintinnbulum, 505*f*
Bargmannia elongata, 208–210, 209*f*
deep-sea siphonophore, 208, 209*f*
growth zone and budding sequence, 208–210
siphonosome cormidial organization, 208
Basal Trachymedusae hypothesis, 212
Bathylasma alearum, 505–506
Bdellomorph, 390
Beroidea, 311
Biflagellated sperm, 38–39, 39*f*
Bilateral holoblastic cleavage, 880
Bindin, 49–51
Bithynia, 769
Blastocoele, 405
Blastoderm, 606
Blastopore, 415, 894
Blastostyles, 187–188
Blastulation, 61, 63*f*, 401, 450, 794
Anostraca, 489–490
Cnidarian embryogenesis, 183
development, 61, 63*f*
Diplostraca, 501
Hydra vulgaris, 205
Blind pouch, 536
Boas, 481
Bolinopsis microptera, 315, 319*f*
Borradaile, 481
Bosmina longispina, 495*f*
Botryllus schlosseri, 866*f*, 886
Bougainvillea spp.
B. fructuosa, 191
B. multitentaculata, 181, 182*f*
Boveri, Theodor Heinrich, 10
Box jellyfish, 239
Brachyura, 479–480
copulation in, 664
definition, 631
habits and habitat, 631–632
morphology and sexual dimorphism, 632–633
reproductive systems, 633–704
male system, 633–658
mature follicle, 641–643
spermatozoa, 643–647
sperm masses, 641
spermatophore, 647–650
Brain, 376–377
Branchiopoda
Anostraca, 484
Diplostraca, 493–495
Brooding
Astacus spp., 603–604
Astacus fluviatilis, 603–604
Cirripedia, 509–511
Cnidaria, 183
Gastropoda, 786
Hormiphora plumosa, 313, 314*f*
Mnemiopsis leidyi, 335–336
Polyplacophora, 730–732
Buccal seminal receptacle, 844
Buccinum undatum, 769
Bud/budding, 111, 179–180
Bythotrephes longimanus, 495*f*

C

Caenorhabditis elegans, 20, 437, 438*f*
dauerlarva, 465
development, 462–465
embryogenesis, 462–463
gametes, 462
gametogenesis, 458–462
gastrulation, 464
gonadogenesis, 458
hermaphrodite, 458
historical review, 455–456
nematode embryology, 465–469
nomenclature, 455
polarity, 464–465
postembryonic development, 465
reproductive system, 457–458
sex differentiation, 457
Calcarea
Calcaronea, 131–137, 132*f*
amphiblastula larva, 135

inversion, 135
metamorphosis and settlement, 135
Sycon raphanus, development of, 135–137
Calcinea, 137–138
calcinea larva, 137–138, 138–140*f*, 143–145*f*, 147*f*, 150–151*f*, 153–155*f*, 157*f*, 160–161*f*, 163–164*f*
Hexactinellida, 138–148
demospongiae (*see* Demospongiae)
(*Oopsacas minuta see Oopsacas minuta*)
reproduction, 138–141
Calcaronea, 131–137, 132*f*
amphiblastula larva, 135
inversion, 135
metamorphosis and settlement, 135
Sycon raphanus, development of, 135–137
Calciblastula, 137–138
Callinectes spp.
C. danae, 653–655
C. sapidus, 631–632, 647, 700, 703–704, 703–704*f*
Callistochiton viviparous, 730–732
Callyspongia diffusa, 115–116, 162, 165
Calyconecta, 206
Cambaroides japonicus, 611–613, 614*f*
Camera lucida, 21*f*
Capnella gaboensis, 273–276, 275*f*
Carapace, 482
Carcenonemertes epialti, 390–396
Carcinus maenas, 554, 604, 633–634, 647–650, 653, 696–697, 696*f*
Caridea, 481
Caridomorpha, 481
Carnivorous sponge, embryogenesis of, 160*f*
Carybdea spp.
C. marsupialis, 241–243
C. sivickisi, 243–247
courtship, spermatophore formation and sperm-transfer, 246
female gonad and putative spermathecae, histology of, 246
settlement and development of embryos, 247
sexual maturation, 243–246
spermatophore and fertilization, dehiscence of, 246
Cassidula aurisfelis, 772–773*f*
Cassiopea andromeda, 232
agametic reproduction, 228–229, 230–231*f*
development, 229–232
strobilation, 232–233
Cell adhesion, 49–52, 50–52*f*, 99
Cell–cell interaction, 69–70
Cell theory, 8, 10
Centrioles, 34–35
Centrolecithal oocyte, 40
Cephalopoda
blastoderm, 854–855
buccal seminal receptacle, 844
characters, 815
classification, 815–816
cleavage, 853–854
connectives, 827
copulation, 850–851
development, 851–855
distal region, 820–821
DVD, 837
egg, 848–850
fertilization, 847–848
gonoduct, 819
main region, 819–820
mating, 850–851
MVD, 836–837
organogenesis, 855–857
ovary, 844–850
ovulation, 847
proximal region, 819
PVD, 835–836
reproduction, 816–850
agametic, 816
gametic, 816–822
reproductive system, 817–822
sexual dimorphism, 816–817
reproductive behaviour, 850–851
spawning, 850–851
spermatophores, 822–850
dehiscence, 831–833
formation, 833–838
hardening, 839–850
spermatozoa, 833
testicular duct, 835
tunics, 825–827
vas deferens, 819–821
waste duct, 820
Cephalothorax, 482
Cephalothrix rufifrons, 415
Ceractinomorpha, 110
Cerebral ganglion, 888
Cerebratulus spp., 394–395
C. lacteus, 390–394, 392–393f, 396
Ceriantharia, 260–262
Cerianthus lloydii, 262
Pachycerianthus multiplicatus, 260–262, 261*f*
Cerianthus iloydii, 262
Ceriantipatharia, 176
Cerithiopsis tubercularis, 770*f*
Cestidea, 310
Cestoda, 340–341
Charybdea spp.
C. marsupialis, 241
C. rastonii, 242*f*
Charybdis spp.
C. japonica, 505*f*, 507
C. smithii, 648–650
Chemotaxis, 49
Chenopus, 769
Cherax spp.
C. albidus, 590, 597–598, 597*f*
C. tenuimanus, 590, 597–598, 597*f*
Chick developmental anatomy, 5*f*
Childia spinosa, 349*f*
Chionoecetes opilio, 641–643, 648–650
Chirocephalopsis bundyi, 484
Chirocephalus diaphanous, 484
Chitin, 653
Chitinous membrane, 679
Chitons, 716
Chiton spp.
C. albolineatus, 718–719, 724*f*
C. articulatus, 718–719, 725–727*f*
C. pelliserpentis, 716
C. septentriones, 720–721*f*, 722, 725–727*f*
C. tuberculatus, 716
Chlamydomonas, 69–79
asexual reproduction, 72–75, 75–76*f*
aplanospore formation, 74, 76*f*
palmella stage, 74–75
zoospore formation/binary fission, 72–74, 74–76*f*
C. eugametos, 73*f*
C. moewusii, 72*f*
C. pulvisculus, 71
C. reinhardtii, 71, 71*f*, 79*f*
C. smithii, 79*f*
habitat, 71
morphology, 71–72, 72–73*f*
sexual reproduction, 75–77, 77–79*f*
anisogamous reproduction, 77, 80*f*
isogamous reproduction, 76–77, 78–79*f*
oogamous reproduction, 77
zygote and zygospore, 77–79
Chondrophora, 176
Chondrosia reniformis, 108*f*
Chorion, 674
Chromadorina epidemos, 430
Chromatin diminution, 19–20
Chromosome elimination, 20–21, 20–22*f*
Chromosome theory, 8
Chrysaora quinquecirrha, 224
Ciliophora, 95–96, 95*t*
habitat, 96
Paramecium, 96–100
agglutination, 99, 99*f*
asexual reproduction, 96, 98*f*
holdfast union, 99
morphology, 96
new cell line development, 100
paroral union, 99
reduced haploid genome, meiotically fusion, 99–100
sexual reproduction, 97–98
Cinctoblastula, 164–165
Ciona intestinalis, 867, 876*t*, 888*f*, 890*f*
Circinalium concrescens, 866*f*
Cirrhipathes anguina, 255–258
mature female polyp, 258
mature male polyp, 255–258, 257*f*
spawning, 258
Cirripedia
brooding, 509–511
cleavage, 511–514
copulation, 508–509
development, 511–518
differentiation of embryo, 516
fertilization, 509–511
gastrulation, 516
metamorphosis of larvae, 516–518
presumptive areas, 514
0presumptive ectoderm, 516
presumptive mesoderm, 516

Cirripedia *(Continued)*
presumptive midgut, 514–515
presumptive stomodaeum, 515
proctodaeum, 515
reproductive system, 503–508
thoracica, 503
Cladocera, 493, 495*f*
Clathrina contorta, 138*f*
Clathrus, 769
Clavelina, 866*f*
Clavularia hamra, 262, 263*f*
Cleavage, 354, 789–793, 853–854, 880–884
Acropora, 298
Cnidarian embryogenesis, 183
development, 60–61, 61*f*
fifth, 354–356
first, 142–143, 354
fourth, 354
Hormiphora plumosa, 323–326, 324–325*f*
Hydra vulgaris, 205
Nematoda
chromatin diminution, 447–448
one-to two-cell stage, 447
two-to four-cell stage, 447
four-to eight-cell stage, 449–450
fifth cleavage, 450
eight-cell to 16-cell stage, 450
Nematostella vectensis, 284
in *Oopsacas minuta*, 142–143
polyaxial, 137
radial, 60, 880
second, 354
spiral, 60
third, 142–143, 354
Cliona spp.
C. stationis, 163, 164*f*
C. vastifica, 111–113
Cnidaria
agametic reproduction, 177–180, 178*f*
Anthozoa, 180
Hydrozoa, 177–179, 185*f*
Scyphozoa, 179–180
types of, 179
Anthozoa (*see* Anthozoa)
body plan of, 177, 177*f*
characteristics, 173, 174–175*f*
classification, 175–177
Anthozoa, 176–177
Cubozoa, 176
Hydrozoa, 175–176
Scyphozoa, 176
Cubozoa (*see* Cubozoa)
definition, 173
gametic reproduction, 180–187
brooding, 183
embryogenesis (*see* Cnidarian embryogenesis)
gametes, 181
gametogenesis, 181
germ cells, origin of, 180
gonadogenesis, 180–181
spawning, 181–183
Hydrozoa, 187–212
Bargmannia elongata, 208–210, 209*f*
Bougainvillea fructuosa, 191
Eucheilota paradoxica, 189–191, 190*f*
Eudendrium racemosum, 195–199, 197*f*
Hydra vulgaris, 201–206
Nanomia bijuga, 210–212, 211*f*
Obelia geniculata, 187–189
Tubularia mesembryanthemum, 191–195, 193*f*
regeneration, 180
Scyphozoa, 176, 179–180
Discomedusae (*see* Discomedusae)
Medusozoa Peterso, 212–238, 213–214*f*
Stauromedusa, 215–218
Coccidia, 91*t*
Coeloblastula, 61, 163*f*, 167
Coelom, 401–403
Coenobium, 81
Coleoida, 815–816
Collecting duct, 653–655
Colonial forms, 69
Comb plates, 327–331
Comparative phase, 7–9
Complemental males, 505–506
Composite egg, 380–385
Conchostraca, 493
Conjugation, 96, 99–100*f*
Conklin, Edwin Grant, 12
Connecting region (CR), 822–823, 827
Connectives, 827
Connective tissue (CT), 536
Conus mediterraneus, 769
Convolutriloba longifissura, 343–345, 346–347*f*
Copula sivickisi, 247–248
female gonad with spermatozeugmata, 248
fertilization, 248
gonads structure, 247
spermatozeugmata, 247
spermatozoon, 247
Copulation, 182–183, 782–783, 786
Corella japonica, 887–888, 888*f*
Coronatae, 219–220
Cortex, 36
Cortical cells, 672–673
Cortical cristae, 574
Cortical granules (CG), 36–37, 52*f*, 55–60
calcium in, 57–58
third substance of, 56–57
Cortical rod protein (CRP), 574
Cortical rods, 569
Corticium candelabrum, 121, 122*f*, 124*f*, 130*f*, 163*f*
Corticogenesis, 319
Cotylea, 340
Cotylorhiza tuberculata, 233–236, 234*f*
embryology, 235
embryos release, 235
fertilization, 235
oogenesis, 235
planula, 235
polyp, 235
reproductive system, 233
scyphistoma, 235–236
settlement, 235
spermatogenesis, 233–235
Courtship, 246, 523–524
Covalent cross-linking, 56–57
Cradle-carrying position, 664
Crambe crambe, 121–123
Crepidula spp., 741
C. fornicata, 741
Cribrinopsis japonica, 290–291
Cruspedacusta sowerbyi, 199
Crustaceans
Arthropoda, 477–478, 482
Branchiopoda, 484–503
Cirripedia, 503–511
classification of, 483–704
general characters, 482–483
historical review, 477–482
Malacostraca, 518–704
Cryptochiton stelleri, 727–728, 729*f*, 731*f*, 732–733
Ctenophores
classification
Hormiphora plumosa, 311, 312*f*
Mnemiopsis leidyi, 326–327
Nuda, 311
Tentaculata, 310
definition, 309, 310*f*
distinctive features, 309–310
Ctenophore theory, 341
Ctenoplana korotneffi, 314–315
Cubomedusa, 239
Cubozoa, 176, 239–253, 240*f*
Alatina cf. *moseri*, 248–251, 249*f*
agametic reproduction, 250
cyst formation, 250–251
metamorphosis, 251
young medusa, 251
Carybdea marsupialis, 241–243
Carybdea sivickisi, 243–247
courtship, spermatophore formation and sperm-transfer, 246
female gonad and putative spermathecae, histology of, 246
settlement and development of embryos, 247
sexual maturation, 243–246
spermatophore and fertilization, dehiscence of, 246
Charybdaea marsupialis, 241
Copula sivickisi, 247–248
female gonad with spermatozeugmata, 248
fertilization, 248
gonads structure, 247
spermatozeugmata, 247
spermatozoon, 247
Morbakka virulenta, 251–253, 252*f*
Tripedalia cystophora, 239–241, 240*f*
Cyanea spp.
C. arctica, 212
C. capillata, 212–215
Cyanoplax spp.
C. dentien, 716, 727–728
C. thomasi, 731*f*
Cyclestheria hislopi, 503
Cyclomorphosis, 495
in *Daphnia*, 497*f*
Cyclops vernalis, 558
Cydippidea, 310, 326
Cypris larva, 516–518
Cystonectea, 207

D

Dalyellid flatworm, 366–369, 366*f*
Daphnella brachyuran, 497*f*
Daphnia spp., 493–494
 cyclomorphosis in, 497*f*
 D. magna, 499
 D. pulex, 498*f*
 embryology, 499–501
 morphology, 494–495
 reproductive system, 495–503
Dart sac, 764
Dauer larva, 465, 466*f*
Daughter colonies, 81
4d cell, 356
De Beer, Sir Gavin, 9
Decapoda, 479–480, 530–704, 816
Delamination, 64, 403
Demospongiae
 Donatia deformis, 149
 Tetractinomorpha, 148–149
 viviparous sponges, 149–162
  carnivorous *Asbestopluma occidentalis*, 157–162
  Plakina monolopha, 150–152
  Spongia officinalis, 152–154
  Stylocordyla borealis, 155–157
Dendraster excentricus, 47
Dendrobranchiata, 481, 583
Dendrocoelum lacteum, 349*f*, 380–385, 381–382*f*
Deontostoma californicum, 428–429f, 432, 435–436, 445
Dermatoxys veligera, 445, 446*f*
Deroceras reticulatum, 785
Descriptive phase, 4–7
Desmolaimus zeelandicus, 435*f*
Desor larva, 410
Determinate, 60
Detorsion, 802–809
Deutoplasmic oocytes, 25–29
Deutoplasmogenesis, 24–29
 Asterina pectinifera, 29, 30*f*
 Portunus pelagicus, 24–29
  deutoplasmic oocytes, 25–29
  predeutoplasmic oocyte, 24–25, 25–26*f*, 27–28t
Development, 320–326, 342–345, 354–359
 Astrangia danae, 297
 blastulation, 61, 63*f*
 cleavage, 60–61, 61*f*
 early development, 60–64, 789–797
 gastrulation, 61–64
 Nematoda, 445–451
 Penaeus kerathurus, 563–590
 Porifera, 131
 Portunus pelagicus, 696–697
 sea urchin, 14*f*
Developmental force, 7
Dibranchiata, 816
Differential interference contrast (DIC), 322–323
Difflugia spp., 88–89, 88*f*
Difflugiidae, 88
Dil, 12
Diminution, 19–20
Dioecy, 506–507
Diplosoma macdonaldi, 886–887
Diplostraca, 493–495
 blastula, 501
 cleavage, 499–501
 differentiation of embryo, 502–503
 embryology, 499–501
 gastrulation, 501–502
 morphology, 494–495
 presumptive areas of blastula, 501
 presumptive ectoderm, 501
 presumptive mesoderm, 501
 presumptive stomodaeum, 501–503
 proctodaeum, 501–503
 reproductive system, 495–503
Dipsastraea (Favia) speciosa, 300–301, 302–303*f*
Directive granules, 673–674
Discogastrula, 64
Discomedusae, 176
 Coronatae, 219–220
 Rhizostomae
  Cassiopea andromeda, 228–233, 230–231*f*
  Cotylorhiza tuberculata, 233–236, 234*f*
  Lychnorhiza lucerna, 236–238, 237*f*
  Mastigias, 224–228, 227*f*
 Semaeostomeae
  Aurelia aurita, 220–223
  larval and onset of metamorphosis, settlement of, 224
  planula larva, 223–224
  Semaeostomeae Scyphozoa, 224
  tentacles and gastric pouches, formation of, 224
Dissolved walls, 674
Distal gland, 821
Distal vas deferens (DVD), 536–539, 593–595, 641, 650, 837
Ditylenchus triformis, 453, 453–454*f*, 455
Dityrosine cross-links, 56–57
Diverticulum, 820–821
Division of labour, 84–87
Donatia deformis, 111–113, 149
Dorsal rudiment, 406–410
Dorso-ventral axis, 878–880
Double eggs, 397–398
Driesch, Hans Adolf Eduard, 10
Driving plate, 40
Dryodora glandiformis, 317*f*, 320
Dugesia benazzii, 352
Dwarf males, 505–508, 506*f*

E

Early responses, egg metabolism activation, 59–60
Echiuroids, 53–55
Ectoderm, 62
Ectomesoderm, 64
Egg, 181
 maturation, stages of, 24*f*
 membranes, 40–42, 42*f*, 679–683
Egg water, 45–46
Eimeria spp., 92–94, 93*f*
 E. schubbergi, 92–94
Ejaculatory apparatus (EA), 822–823, 829–831
Electrical polyspermy block, 53–55
Electron-dense material (EDM), 597*f*
Emboly, 62
Embryogenesis, 153–154, 158–159, 877–886
 carnivorous sponge, 160*f*
 Cnidarian
  blastulation, 183
  cleavage, 183
  fertilization, 183
  gastrulation, 183–186
  larval forms, 186
  larval settling, 186
  postembryonic development, 186–187
 Hydra vulgaris, 203*f*
 Oopsacas minuta, 142–146
 Tubularia mesembryanthemum, 193*f*, 194–195
Embryology, 3
 cell theory, 8
 chromosome theory, 8
 genetics and gene theory, 8–9
 history of, 4
 phylogeny and recapitulation theory, 9
 principles of, 3–4
 retention theory, 9
Embryonic pharynx, 380
Embryotheca, 205
Endoderm, 62
Endomixis, 90–91, 100, 101*f*
Endostyle, 889
Enopla, 390
Enoplus communis, 432–433, 435*f*
Enterobius vermicularis, 428–429*f*, 445, 446*f*
Enterogona, 865
Ento-mesoblast, 793–794
Envelope, 321*f*
Eobania vermiculata, 751
Ephippium, 498
Ephydatia fluviatilis, 117, 119, 124*f*, 165
Epiactis japonica, 287–289, 289*f*
Epiboly, 62–64, 794
Epidermis, 376–377
Epifluorescence optics, 322–323
Epigenesis, 4
Epimenia babai, 735–738, 736–737f
Epimorphosis, 377
Epithelium, 403
Equatorial band, 146
Eremina spp.
 albumen gland of, 751–752*t*
 E. ehrenbergi, 740*f*, 741–788, 867
  copulation, 782–783
  foreign sperm pathway, 783
  oocyte pathway, 783–788
  spermatophore formation, 782
  sperm pathway, 777–783
 female side of system, 758–764
 male side of system, 755–758
 spermoviduct, 753–755, 756*t*
Erugosquilla massavensis, 523, 527*f*
Essential germinal epithelium (EGE), 833–834
Ethusa mascarone, 497*f*
Euastacus serratus, 590
Eucarida, 530
Eucheilota paradoxica, 189–191, 190*f*
Eucopidae, 481

Eudendrium spp., 17, 18*f*
E. racemosum, 184–186, 195–199, 197*f*
Gonionemus depressum, development of, 199–201
life cycle, 197*f*
metamorphosis, 198–199, 200*f*
Eugregarinaria, 91*t*
Eumalacostraca, 478–479
Eumetazoa. *See* Cnidaria
Eupagurus spp.
E. bernhardus, 186–187
E. prideauxi, 643, 648–650
Eupyrene, 39
Eurycercus glacialis, 495*f*
Evacuated follicle, 647
Evadne nordmanni, 495*f*
Excretory pores, 331–333
Exocytosis, 37–38
Exoskeleton, 482
Experimental phase, 9–15
External yolk sac, 855

F

Farrea sollasii, 138, 140*f*, 141
Female colonies, 266
Female pronucleus, 58, 321*f*
Fertilization, 43–60, 847–848, 874–877
Astrangia danae, 297
Carybdea sivickisi, 246
Copula sivickisi, 248
Cotylorhiza tuberculata, 235
embryogenesis Cnidarian, 183
Gastropoda, 785–789
Heteroxenia fuscescens, 271
Hormiphora plumosa, 320–323
Hydra vulgaris, 204–206
membrane, 41, 45
membrane formation, 47–49, 48*f*
Nematoda, 441–445
Nematostella vectensis, 284
Nemertinea, 396
Penaeus kerathurus, 575
Platyhelminthes, 353–354
Porifera, 129–131
in sea urchin eggs, 44–47, 46*f*
sea urchin type of, 58
up-to-date concept, process, 49–56
cell adhesion, 49–52, 50–52*f*
cortical granule reaction, calcium in, 57–58
echiuroids, annelids, and molluscs, 53–55
egg metabolism, activation of, 59–60
fertilization envelope, elevation of, 55
haploid nuclei fusion, egg and sperm, 58–59
hardening of fertilization envelope, 56–57, 57*f*
inactivate sperm, peroxide release to, 56
membrane potential, ion-dependent changes in, 53, 54*f*
secretory products, cortical granules, 55–56
Fertilization cone, 44–45
Fertilization membrane, 674
Fertilization sac, 844
Fertilizin, 45–46
Fertilizing spermatozoon, 44–45
Fifth larva formation, 146
Filtering apparatus, 482
First larva formation, 146
First mitosis, 321*f*
First phase, *Eucheilota paradoxica*, 189
First successful sperm-egg reaction, 53
Fission, 180
Flagellated spermatozoon, 38
Four haploid zoospores, 79
Fourth larva formation, 146
Fragmentation, 111–112*f*, 116
Fresh-copulation stage, 662–663*t*, 667–668, 669*f*, 673–674
Fringe, 76–77
Fucus serratus, 788–789
Fura-2, 57–58
Fusitriton oregonensis, 769
Fusogenic proteins, 52

G

Galathea squamifera, 497*f*
Galaxea fascicularis, 300, 301*f*
Gametes, 17, 94–95, 511–518
Astrangia danae, 295–297
Cnidaria, 181
formation, gametogenesis, 22–31, 23*f*
animal oocytes, meiotic arrest in, 29–30
egg, senescence of, 31
oocyte growth, 24–29
spermatozoa, senescence of, 31
isogametes, 94–95
macrogametes, 77
maturation of
oocyte maturation, 42
spermatozoon maturation, 43
microgametes, 77
Polychoerus carmelensis, 341–342
sperm cell evolution (*see* Sperm cell)
structure of, 31–38
types of
egg, 40–42
sperm, 38–40
Gametic reproduction, 72, 313–326, 716, 816–822, 867–877
Antipatharia, 255
Heteroxenia fuscescens, 267
embryology, 271–273
fertilization, 271
ovum, 271
spermaries, 267–271
Hexactinellida, 138
Platyhelminthes, 348–354
Porifera
Calcarea (*see* Calcarea)
development, 131, 131*f*
fertilization, 129–131
gametogenesis, 118–131, 123–124*f*
germ cells, origin of, 117–118
gonads, 117
oogenesis, 123–128, 126–128*f*
reproductive behaviour and spawning, 128–129, 130*f*
sexual differentiation, 117
spermatogenesis, 119–123, 120*f*, 122*f*
Gametocytes, 94–95
Gametogenesis, 17, 314–320, 350–353, 722–728, 764–777, 868–877
Anthoplexaura dimorpha, 279
Cnidaria, 181
gamete formation, 22–31, 23*f*
animal oocytes, meiotic arrest in, 29–30
life span of, 30–31
oocyte growth, 24–29
gametes, structure of (*see* Gametes)
germ cell, origin of
Eudendrium sp., 17
In *Hydra* sp., 17–18, 18*f*
Obelia sp., 17
primordial gonocytes, natural phenomena identification (*see* Primordial gonocytes)
maturation of gametes
oocyte maturation, 42
spermatozoon maturation, 43
Porifera, 118–131, 123–124*f*
types of gametes
egg, 40–42
sperm, 38–40
Urochordata
fertilization, 874–877
fully grown oocyte, 870–874
oogenesis, 868–874
spawning, 874
sperm, 868
spermatogenesis, 868
Gamocyst, 94–95
Gamontocyst, 94–95
Gamonts, 90–91, 94–95
Gastric mill, 482
Gastrodes spp.
G. parasiticum, 315*f*, 326
oogenesis in, 315*f*, 320
Gastropoda
albumen gland, 748–753
blastulation, 794
dart, 764
dart sac, 764
early development
cleavage, 789–793
ectoderm, 793
ecto-mesodermal cells, 794
ento-mesoblast, 793–794
egg
brooding behaviour, 786
copulation, 786
fully formed egg, 786
sex recognition, 786
spawning aggregation, 786
spawning and rearing of young, 786–788
fertilization, 785–789
gametogenesis, 764–777
abnormal spermatogenesis, 767–769
abnormal spermatozoa, 769
dimorphism of sperm, 769
oogenesis, 771–775
spermatogenesis, 764–769
spermatophores, 769–771

spermatozoa, 767
gastrulation
epiboly, 794
invagination, 794
multipolar proliferation, 794–797
genital atrium, 758
hermaphrodite duct, 744–753
larvae
digestive system, 800
foot, 797
head vesicle, 800
pedal gland, 797
radula sac, 799
shell formation, 797
statocysts, 797
velum, 799–800
later development
detorsion, 802–809
echinospira larvae, 804
torsion, 800–802
visceral hump, 802
mucous glands, 760–764
oviduct, 758
penis, 758
reproduction
agametic, 739
dimorphism, 739
gametic, 739–741
hermaphroditism, 739–740
protandric hermaphrodites, 741
reproductive system, 741
spermatheca, 758–760
spermathecal duct, 758–760
spermoviduct, 753–755
spherical oocyte, 775–777, 776*t*
talon, 744–748
vagina, 760–764
vas deferens, 757–758
Gastrulation, 61–64, 183–186, 794–797, 884–886
Acropora, 298
Anostraca, 490–491
Cnidarian, 183–186
development, 61–64
differentiation of embryo, 491
Hormiphora plumosa, 326
Hydra vulgaris, 205
Polycladida, 356, 356*f*, 362*f*
Gazellostrongylus lerouxi, 428–429*f*
Geigy's method, 22*f*
Gemmule formation, 114–116
Gene theory, 8–9
Genetics, 8–9
Genitalia, 376–377
Germ-cell determination, localized cytoplasmic factors in, 21–22
Germinal cells, 641
Germination, 115
Germplasm, 9–10
Giant cells, 833–834
Gieysztoria superba, 366–369, 366*f*
juvenile of, 368–369
Gills, 482
Girardia tigrina, 380–385, 382–384*f*
Glycolipids, 668–670
Gonadogenesis, Cnidaria, 180–181
Gonads, 314, 777
Anthoplexaura dimorpha, 279
Porifera, 117
Gonial cells, 22
Gonidia, 81
Gonionemus depressum, 199–201
Gonodactylus spp.
G. bredini, 523*f*, 524, 528
G. festai, 520
G. oerstedii, 520, 528, 531*f*
Gonodendron, 210
Gonoduct, 819
Gonophore, 191–194
Gorgonacea
Anthoplexaura dimorpha, 279
Junceella fragilis, 276–279
sea fan and corallium, 276, 278*f*
Gotte's larva, 359
Grantia compressa, 117, 118*f*, 162
Green glands, 482
Gut, 376–377

H

Habits/habitat, 71, 81, 590
Haeckel, Ernst Heinrich, 9
Halcampa duodecimcirrata, 286–287, 287*f*
Halichoanolainus robustus, 435*f*
Haliclona spp.
H. ecbasis, 117
H. limbata, 165
H. loosanoffi, 115–116
Haliclystus spp.
H. octoradiatus, 215–218
H. sanjuanensis, 215
Haliotis discus, 800
Halisarca spp., 117, 165
H. dujardini, 119
Halocynthia roretzi, 888*f*
Haploid cell, 71
Harpiosquilla raphidia, 523
Hartsoeker, Nicolas, 6, 6*f*
Harvey, William, 4
Hatching, 886–888
Artemia spp., 492*f*
Astacus spp., 620–629
Hydra spp., 206
Mnemiopsis leidyi, 335
Nematoda, 451–452
Portunus spp., 695–696
Urochordata, 886–888
Head-to-head agglutination, 46
Head vesicle, 800
Heligmosomoides polygyrus, 438*f*
Helix spp.
H. aspersa, 751
H. pomatia, 744, 748
Hermaphrodite duct, 744–753
Hermaphroditism, 431–432, 739–740
Heteonemertea, 397–398
Heterakis spp.
H. gallinae, 436*f*
H. gallinarum, 430
Heterodera spp.
H. marioni, 435*f*
H. schachtii, 445
Heteromyenia, 113–114*f*
Heteronemertea, 415
Heterosynthetic process, 846–847
Heterothallic, 75–76
Heteroxenia fuscescens, 266–273, 270*f*
agametic reproduction
budding, 266
pinnitomy, 267
transverse fission, 266–267, 270*f*
gametic reproduction, 267
embryology, 271–273
fertilization, 271
ovum, 271
spermaries, 267–271
Hexacorallia, 176–177
Actiniaria (*see* Sea anemones)
Scleractinia, 291
Acropora, 297–301
Astrangia danae, 295–297
Galaxea fascicularis, 300, 301*f*
true/stony corals, 291–301, 292*f*, 294*f*
sea anemones and stony corals, 280
Hexactinellida, 138–148
Demospongiae, 148–149
Asbestopluma occidentalis, 157–162
Donatia deformis, 149
Plakina monolopha, 150–152
Spongia officinalis, 152–154
Stylocordyla borealis, 155–157
Tetractinomorpha, 148–149
Oopsacas minuta, 141–148, 142–145*f*
embryogenesis, 142–146
fully formed larva, 146–148
larva formation, 146
oogenesis, 141–142
reproduction, 141
spermatogenesis, 141
reproduction, 138–141
Hippospongia communis, 118–119
His, Wilhelm, 9
Holdfast union, 99
Holopedium, 499
reproductive system, 495–503
Holosomatous, 865
Homardiens, 479–480
Homarus spp.
H. americanus, 595, 596*f*, 610–611, 612*f*
H. vulgaris, 604
Homoscleromorpha, 110, 151*f*
Homothallic, 75–76
Homothetogenic, 96
Hoplocarida, 518–530
courtship and copulation, 523–524
definition, 518
embryology
early, 528
gametes, 524–527
postembryology, 528–530
habitat and behaviour, 520
history, 518–520
mature stage, 527
morphology, 520–521

Hoplocarida *(Continued)*
oocyte stage, 526
remature oocyte, 526–527
reproductive systems, 521–522
sexual dimorphism, 521
sperm cord formation, 522–523
vitellogenic oocyte, 526
Hoplonemertea, 415
Hoploplana inquilina, 358–359, 360–361*f*
Hormiphora plumosa, 311, 312*f*, 334*f*
external morphology, striking and characteristic feature in, 311
reproduction
brooding, 313, 314*f*
cleavage, 323–326, 324–325*f*
embryonic development, 320, 324–325*f*
fertilization, 320–323, 321*f*
Gastrodes, oogenesis in, 315*f*, 320
gastrulation, 326
germ cells and gonads, origin of, 313
gonads, 314
larval types, 326
mature sperm, 315
oogenesis, 315–320, 318–319*f*
regeneration, 311–313, 316*f*
spawning, 320
spermatogenesis, 314–315, 317*f*
Horn sheath, 829–831
Hoyle organ, 857
Hyaline layer, 55
Hyaline matrix, 19
Hydractinia echinata, 186
Hydra spp., 17–18, 18*f*, 174–175*f*
H. carnea, 204–206
H. fusca, 205
H. vulgaris, 201–206, 203*f*
cleavage and blastulation, 205
cyst formation, 205–206
embryogenesis and development of, 203*f*
embryo, hatching of, 206
fertilization, 204–206
gametogensis, 204
gastrulation, 205
Siphonophora, 206–212, 207*f*
nematocysts of, 175*f*
Hydroida, 175–176
Hydrozoa, 175–176, 187–212
Bargmannia elongata, 208–210, 209*f*
Bougainvillea fructuosa, 191
Eucheilota paradoxica, 189–191, 190*f*
Eudendrium racemosum, 195–199, 197*f*
Hydra vulgaris, 201–206, 203*f*
Nanomia bijuga, 210–212, 211*f*
Obelia geniculata, 187–189
Tubularia mesembryanthemum, 191–195, 193*f*
Hyostrongylus rubidus, 430
Hypodermic injection, 353
Hysterothylacium reliquens, 427*f*, 430–431, 433*f*

I

Ibla, 505–506
Illex spp.
I. coindetii, 823
I. illecebrosus, 821, 833
Ilyanassa obsoleta, 791–793
Imaginal disc, 403
Immature follicle, 641, 833–834
Immaturity period, 98
Indeterminate, 60
Ingression, 64
Integumental theory, 679–681
Intermediate development, 358
Interstitial cells, 18, 674
Inter-spawn stage, 676
Intestina, 423
Intromittent organ, 641
Invagination, 62, 794
Inversion, 81–82
Invertebrate phyla, 3
Involution, 64
Ischnochiton stramineus, 722, 725–727*f*, 730–732
Isogametes, 94–95
Isogamous mating, 79*f*
Isogamous reproduction, 76–77, 78–79*f*
Isogamy, 76–77
Isolecithal oocyte, 40
Iwata's larva, 410–411

J

Junceella fragilis, 276–279, 278*f*

K

Katharina tunicata, 716, 727–728, 729*f*
Kato's larva, 359–361
32-kDa glycoprotein, 82
Kronborgia caridicola, 353–354

L

Laevipilina theresae, 733–735
Larval forms, 186
Larval settling, 186
Late-copulation stage, 662–663*t*, 668–671, 670*t*, 674
Late responses, egg metabolism activation, 59–60
Latona setifera, 497*f*
Leander aspersus, 497*f*
Lepas spp., 518, 518*f*
L. anatifera, 505*f*, 514
Lepidochitona cinereus, 716, 722, 725–727*f*, 728
Leptodora, 494–495
Leptostraca, 478
Leucandra blanca, 165–166
Leucosolenia spp.
L. complicata, 132*f*
L. variabilis, 162, 166
Life history theory, 508
Limited chromosomes, 20
Limnadia, larval development and metamorphosis of, 502, 504*f*
Lineus spp.
L. alborostratus, 396
L. ruber, 394, 411–415
L. toquatus, 399
L. torquatus, 392–393*f*
Lipids, 653, 668
Lipoprotein, 670
Lipovitellin, 700–704
Litopenaeus occidentalis, 567–568*f*
Little green balls, 74–75
Lobata, 310
Loligo spp.
L. bleekeri, 849–850
L. opalescens, 822–823
L. pealei, 816–817, 846–847, 846*f*
Lucernaria janetae, 215
Luther's larva, 359
Lychnorhiza lucerna, 236–238, 237*f*
Lymnaea spp.
L. acuminata, 717*f*
L. stagnalis, 781–782
Lyrocteis imperatoris, 313

M

Macrobrachium rosenbergii, 558
Macrogametes, 77
Macromeres, 356
Macronucleus, 95–96
Macropodia longirostris, 683, 684*f*
Macrostomum spp., 372–377, 373–374f
Macrura, 479–480
Malacostraca
Eucarida, 530
habits and habitat, 590
Hoplocarida, 518–530
Male colonies, 264–266
Malpighi, Marcello, 5*f*, 6
Marginal cells, 641
Marine spermatozoon, 32–35
Maritigrella crozieri, 361–363
Marsupenaeus japonicus, 575
Mastigias, 224–228, 227*f*
Mastigophora, 69–87, 70*f*
Chlamydomonas, 69–79
asexual reproduction, 72–75, 75–76*f*
habitat, 71
morphology, 71–72, 72–73*f*
sexual reproduction, 75–77, 77–79*f*
zygote and zygospore, 77–79
colonial forms, 69
solitary forms, 69
Volvox, 79–87
asexual reproduction, 81–82, 83*f*, 85*f*
features of, 84–87
habitat, 81
morphology, 81, 84*f*
sexual reproduction, 82–84, 85–86*f*
Mating reaction, 99
Mating type-specific components, 77
Mature oocyte, 128
May-spawn stage, 676
Medial vas deferens (MVD), 836–837
Megalospheric proloculus, 89
Meiotically, zygotes, 79
Meiotic arrest, in animal oocytes, 29–30
Melampus coffeus, 777, 798*f*
Membrana propria, 833–834

Membrane lamellar complex (MLC), 597*f*
Mermis subnigrescens, 427
Mesostoma spp., 366
  M. lingua, 369–375, 370–371*f*
Metagamic phase, 96
Metamorphosis, 135, 162, 186–187, 888–889
Metanauplii, 516–518
Metapenaeus monoceros, 532
Metazoa, 72
Metoncholaimu spritiuris, 435*f*
Microciona spp., 162
  M. coccinea, 167
  M. microjoanna, 124*f*
  M. prolifera, 117
Microgametes, 77
Microhedyle cryptophthalma, 771
Micronucleus, 95–96
Microscope, discovery of, 5–7
Microspheric proloculus, 89
Microstomum lineare, 346
Microvilli, 51–52
Micrura alaskensis, 390–395, 405–410
  blastula, 406
  cleavage, 406, 407–408*f*
  embryonic development, 406
  juvenile development, 406
  juvenile rudiment development, 406–410
  nervous system, 410
  pilidium larva, 406
Micrura caeca, 403
Middendorffia polii, 728, 729*f*, 732
Middle vas deferens (MVD), 593–595, 637–641, 650
Mitella mitella, 511–513, 512*f*, 514*t*
Mitochondria, spermatozoa, 35
Mitotic process, 96
Mnemiopsis leidyi, 326–327, 328*f*
  brooding, 335–336
  hatching, 335
  organogenesis
    comb plates, 327–331, 329–331*f*
    gastrovascular system, 335
    nervous system, 335
    sense organ, 334–335, 334*f*
    stomodeal pharynx, 331–334, 332*f*
Moina spp.
  M. brachycephala, 495*f*
  M. paradoxa, 497*f*
  reproductive system, 495–503
Mollusca, 53–55
  Amphineura, 716–739
  classification, 716–809
  features, 715
  Gastropoda, 739–809
Monaxonida, 114–115
Monoascidia, 865
Monocystis, 94–95, 95*f*
Monodonta lineata, 772–773*f*
Monoplacophora, 733–735
Monopodial growth, 189
Mopalia spp.
  M. lignosa, 730
  M. muscosa, 724*f*
Morbakka virulenta, 251–253, 252*f*
Morgan, Thomas Hunt, 8–10
Moroteuthis ingens, 850
Morphallaxis, 377–380
Morphogenesis, 886
mt(+), 76–77
mt(-), 76–77
Mucoid substance (MS), 597*f*
Mucopolysaccharides, 55
Mucous glands, 760–764
Müller's larva
  formation of, 356–359
  metamorphosis of, 358
Multiple fission, 94–95
Multipolar ingression, 64
Myopsida, 816

N

Nanomia bijuga, 210–212, 211*f*
Nassarius, 769
Natantia, 481
Nauplius stage, 477–478, 483
Nausithoë aurea, 217*f*, 219–220
Nautiloidea, 815
Nebalia, 478
Nectonemertes spp., 390–394
  N. mirabilis, 394
Needham's organ, 821–822
Needham's sac, 837–838
Nematoda
  anabiosis, 455
  classification, 424–426
  cleavage
    chromatin diminution, 447–448
    one-to two-cell stage, 447
    two-to four-cell stage, 447
    four-to eight-cell stage, 449–450
    fifth cleavage, 450
    eight-cell to 16-cell stage, 450
  copulation, 441
  definition, 423
  development, 445–451
  early embryology, 447
  embryological peculiarities, 452
  fertilization, 441–445
  gastrulation, 450–451
  general characters, 424
  hatching, 451–452
  historical review, 423–424
  life cycles, 453–455
  phylum, 425
  postembryonic development, 452–455
  reproduction
    agametic, 426–427
    gametic, 430–445
    gametogenesis, 434–441
    germ cells and gonads, 433–434
    hermaphroditism, 431–432
    intersexes, 431
    oogenesis, 437–441
    parthenogenesis, 427–430
    sex attractants, 441
    sexual dimorphism, 430
    spermatogenesis, 434–437
    spicules, 430–431
Nematostella vectensis, 280–285
  blastula, 284–285
  cleavage, 284
  fertilization, 284
  spawning, 280–284
Nemertinea
  blastula, 401
  body wall formation, 403–405
  classification, 390
  cleavage, 399–401
  direct development, 415–418
  distinctive features, 389–390
  double eggs in, 397–398
  early development, 398–401
  early juvenile, 415
  fertilization, 396
  fusion of discs, 404–405
  gametogenesis, 394–396
  gastrulation, 401
  Iwata's larva, 410–411
  Lineus ruber, 411–415
  Micrura alaskensis, 405–410
  obtaining material, 397
  oogenesis, 395–396
  pilidium larva, 401–405
  polyspermy, 397
  proboscis, 414–415
  Procephalothrix simulus, 416–418
  reproductive system, 390–394
  Schmidt's larva
    metamorphosis, 411–414
    structure, 411
  spawning, 396
  spermatogenesis, 394–395
Neoascaris vitulorum, 426*f*
Neochildia fusca, 343, 345*f*
Neofibularia nolitangere, 128–129
Neoophora, 340
Neopilina galatheae, 733, 734*f*, 735
Neosarmatium indicum, 691, 693–694*f*, 697
Nephridia, 405
Neptunus sanguinolentus, 647–650
Neritina reclivata, 769–771
Nervous system, 173, 310, 335, 405, 410, 732, 890
Nests, 137
Neuro-hypophysial canal, 890
Neurulation, 884–886, 887*f*
Nitzschia curvilineata, 372
Nonmotile zygote, 77
Nonsoluble 30,500-Da protein, 51
Notoplana humilis, 354, 363–366, 365*f*
Nuda, 311
Nutrition, of embryos, 162

O

Obelia spp., 17, 174*f*, 187*f*, 188
  O. geniculata, 187–189
    development, 188–189
    gametic reproduction, 188
    germ cell origin, 188
    medusa, development of, 188–189
    medusa, structure of, 188

Octocorallia, 176
Alcyonacea, 264–276, 265*f*
Alcyonium coralloides, 264–266
Alcyonium digitatum, 264
Capnella gaboensis, 273–276, 275*f*
Heteroxenia fuscescens, 266–273
Xenia umbellata, 273, 274*f*
Gorgonacea
Anthoplexaura dimorpha, 279
Junceella fragilis, 276–279
sea fan and corallium, 276, 278*f*
Stolonifera, 262–264
Octolasmis warwickii, 505*f*, 507
Octopoda, 815
Octopus vulgaris, 821–822, 837–838, 851–853
Odontodactylus scyllarus, 519*f*, 520, 523*f*
Oegopsida, 816
Oligopyrene, 39
Oocyst, 94–95
Oocyte, 775–777, 783–788
maturation, 42
secondary, 23–24
Oocyte growth, 24–29, 870–874
Asterina pectinifera, deutoplasmogenesis in, 29, 30*f*
Portunus pelagicus, deutoplasmogenesis in, 24–29
deutoplasmic oocytes, 25–29
predeutoplasmic oocyte, 24–25, 25–26*f*, 27–28t
Oogamous reproduction, 77, 82
Oogenesis, 158–159, 279, 315–320, 352–353, 569, 605, 771–775, 868–874
Astrangia danae, 295
Cnidaria, 181
Cotylorhiza tuberculata, 235
in Gastrodes spp., 315*f*, 320
Oopsacas minuta, 141–142
Porifera, 123–128, 126–128*f*
Ooplasmic segregation, 878–880, 879*f*, 892–893
Oopsacas minuta, 141–148, 142–145*f*
embryogenesis, 142–146
fully formed larva, 146–148
larva formation, 146
oogenesis, 141–142
reproduction, 141
spermatogenesis, 141
Oospore, 83–84
Optic vesicle, 855
Oratosquilla spp.
O. massavensis, 523, 525–527, 527*f*
O. oratoria, 497*f*, 521–526, 522*f*, 526–527*f*, 528
Orconectes limosus, 591–592*f*
Ornatoscalpellum stroemii, 506*f*
Ortmann, 481
Oscarella spp., 114, 117
O. lobularis, 119, 124*f*, 166
O. nathaliae, 151–152
Ovary development, in sexually mature crayfish, 605
Oviparous sponges, 162–163
Ovoperoxidase, 56–57
Oxychilus cellarius, 744, 748, 785

P

Pachycerianthus multiplicatus, 260–262, 261*f*
Pacifastacus spp.
P. leniusculus, 595–597, 596*f*
P. trowbridgii, 602–603
Palaemonids, 480–481
Palaeonemertini, 390
Palmella stage, 74–75
Panulirus penicillatus, 643
Paralithodes brevipes, 700
Paramecium spp., 96–100
agglutination, 99, 99*f*
asexual reproduction, 96, 98*f*
holdfast union, 99
morphology, 96
new cell line development, 100
paroral union, 99
P. aurelia complex, 98
P. aurelia, mating types of, 97–98
P. caudatum, 96
reduced haploid genome, meiotically fusion, 99–100
sexual reproduction, 97–98
Parapenaeopsis stylifera, 539–541
Parapenaeus spp.
P. japonicus, 575–588
P. longirostris, 566*f*, 575, 588–590
P. semisulcatus, 575, 577–578*f*
Parascaris, 19
P. equorum, 437–438, 443, 445, 446*f*, 447–448
Paratomy, 346
Paratylenchus scribneri, 428–430
Parazoa
biological associations, 167
larvae types, 164–167
amphiblastula larva, 166–167
parenchymula larva, 165–166
larval behaviour and settling, 162
larval release, 162
metamorphosis, 162
nutrition of embryos, 162
oviparous sponges, 162–163
Porifera
agametic reproduction, 111–116
characteristics, 107–109
classification, 109–110
definition, 107
embryological characteristics of porifera, 110
evolution theory, 107
factors, 116
gametic reproduction, 117–131
sponge and body planes, structure of, 109*f*
sponges, morphology and physiology of, 107
Parenchymula, 165–166
Parthenogenetically, 91
Patella spp.
P. caerulea, 769
P. vulgata, 741
Pedal gland, 730, 797
Pedal laceration, 180
Pellicle, 96
Penaeidae, 481, 532
Penaeoidea
definition, 531
habitat, 531–532
Penaeus spp.
P. japonicus
chromatin nucleolus stage, 569
early perinucleolus stage, 569
late perinucleolus stage, 569
maturation stage, 571–573
oil globule stage, 569
oocyte development, 571–573, 571–573*f*
prematuration stage, 571
synapsis stage, 569
yolk granule stage, 571
yolkless stage, 571
P. kerathurus, 532, 653
argentaffin material, in spermatophore, 556–557
ascending limb, 543
carbohydrates distribution, 556
chitin, absence, 554–555
copulation, 560–563
descending limb, 547–548
development, 563–590
distal vas deferens, 548–549
early embryology, 575–588
female reproductive systems, 557–590
fertilization, 575
gametes, 563–574
larval development, 588–590
lipids distribution, 555
main layers deposition of body and wing, 543–546
male reproductive systems, 532–557, 537–538*f*
morphology and sexual dimorphism, 532
proteins distribution, 555–556
spawning, 574–575
spermatophore (*see* Spermatophore)
sperm mass formation, 542–543
terminal ampoule, 549–553
testes, 532–536
wing neck, 550–551
P. monodon, 560–563, 575–588
P. setiferus, 554, 653
P. trisulcatus, 595–597
P. vannamei, 560–563
Pennant, 477
Pennaria tiarella, 184, 184*f*
Perforatorium, 32–34
Period of sexual maturity, 98
Peripheral germinal epithelium, 833–834
Perisesarma bidens, 697–700, 699–702*f*
Perophora viridis, 866*f*
Peroxidase, 56–57
Petrosia ficiformis, 117–118, 118*f*, 121, 130*f*, 162, 164*f*
Phallusia mammillata, 874, 890–897
Pharyngeal axis (PA), 332*f*
Pharynx, 376–377
Pheromone, 82
Phialopore, 81–82
Philometra amazonica, 424

Philomycus carolinianus, 774, 781
Phospholipids, 668
Phyllaplysia taylori, 742*f*, 769, 771
Phyllobranchiata, 481
Phylogeny and recapitulation theory, 9
Phylum, 3, 477, 482
Physalia physalis, 207*f*
Physonectae, 206–207
Phytomonadina, 70*f*
Pilidium gyrans, 397–398
Pilidium larva, 401–405
  alimentary canal, 405
  body cavity, 405
  body wall, 405
  metamorphosis, 403–405
  muscles, 405
  nervous system, 405
  structure of, 403
Plakea, 81–82
Plakina spp.
  P. jamaicensis, 151–152
  P. kanaky, 150*f*, 152, 153*f*
  P. monolopha, 150–152, 151*f*
  P. trilopha, 152, 154*f*
Planetary movements, 81
Planula theory, 326, 341
Planuloid buds, 179
Plasma membrane, 36
Platyctenea, 310
Platyhelminthes, 339
  Acoela, 341–345
    Convolutriloba longifissura, 343–345, 346–347*f*
    Neochildia fusca, 343, 345*f*
    Polychoerus carmelensis, 341–345, 344*f*
  agametic reproduction, 345–348, 348*f*
  classification
    Cestoda, 340–341
    Trematoda, 340
    Turbellaria, 340
  Ctenophore theory, 341
  definition, 339
  Dendrocoelum lacteum, 380–385, 381–382*f*
  distinctive features, 339
  gametic reproduction, 348–354
  gametogenesis
    polymorphism of sperm, 352
    spermatogenesis, 350–352, 351*f*
    spermatozoa, 350–352
    Turbellaria, sperm transfer mechanisms, 352
  Girardia tigrina, 380–385, 382–384*f*
  oogenesis, 352–353
    copulation and/or fertilization, 353–354
    oocytes, 353
    oviposition, 353–354
  Planula theory, 341
  Plug, 667–671
  Polarity, 464–465
  Polycladida
    flatworms larvae, 359–366, 363*f*
    formation, Müller's larva, 356–359
    gastrulation, 356, 356*f*, 362*f*
    Götte's larva, 359
    Hoploplana inquilina, 358–359, 360–361*f*
    metamorphosis, Müller's larva, 358
    Notoplana humilis, 363–366, 365*f*
    simple egg, cleavage of, 354–356, 355*f*
  reproductive system, 348–350, 349*f*
  Rhabdocoela
    Gieysztoria superba, 366–369, 366*f*, 372
    Macrostomum sp., 372–377, 373–374f
    Mesostoma lingua, 369–372, 370–371*f*
  Tricladida, 377–385, 377*t*
    epimorphosis, 377
    morphallaxis, 377–380, 378*f*
Plaxiphora primordia, 727–728
Pleocyematan, 569
Pleurobrachia bachei, 333–334
Pleurogona, 865
Plugged junctions, 143
Podocysts, 179
Podon intermedius, 495*f*
Polyaxial cleavage, 137
Polycelis tenuis, 377–378
Polychaerus carmelensis, 341–345, 344*f*, 352
  gametes morphology, 341–342
  mating, 342
Polycladida, 340
  flatworms larvae, 359–366, 363*f*
  formation, Müller's larva, 356–359
  gastrulation, 356, 356*f*, 362*f*
  Götte's larva, 359
  Hoploplana inquilina, 358–359, 360–361*f*
  metamorphosis, Müller's larva, 358
  Notoplana humilis, 363–366, 365*f*
  simple egg, cleavage of, 354–356, 355*f*
Polydisc strobilation, 223
Polymastia spp.
  P. mammillaris, 117, 119
  P. robusta, 163, 167
Polymorphic spermatozoa, 39
Polyp
  Cotylorhiza tuberculata, 235
  lower part of, 266–267
  mature female, 258
  mature male, 255–258, 257*f*
  upper part of, 266–267
Polyphemus
  gastrulation, 501–502
  P. pediculus, 495*f*, 497–498*f*, 499–500, 500*t*
  presumptive ectoderm, 501
  presumptive mesoderm, 501
  proctodaeum, 501
  reproductive system, 495–503
Polyplacophora, 716–733
  brooding, 730–732
  cleavage, 728
  dimorphism, 716
  early development, 728
  gametogenesis, 722–728
  gastrulation, 728
  hermaphroditism, 716
  larval development, 728–730
    foot and pedal gland, 730
    larval eye, 730
    shell, 730
  later development
    aesthetes, 732
    eyes, 732
    head, 732–733
    nervous system, 732
  metamorphosis, 730
  oogensis, 722–723
  ovoviviparity, 730–732
  reproduction
    agametic, 716
    gametic, 716
  reproductive behaviour, 727–728
  reproductive system, 716–722
  settlement, 730
  sex ratio, 716
  spawning, 727–728
  spermatogenesis, 723–728
Polyspermy, 397
  rapid prevention of, 53
  slower mechanical block to, 55
Portuguese Man o' War, 207*f*
Portunus spp.
  P. pelagicus, 24–29, 497*f*, 526, 631–632, 631*f*
    artificial insemination, 697
    chemical composition, spermatophore, 653
    copulation, 660–664
    cyclic changes of ovaries, 662–663*t*
    deutoplasmic oocytes, 25–29
    deutoplasmogenesis, 675–676
    egg-membranes, 679–683
    embryogenesis of artificially inseminated eggs, 697–700
    embryology, 683–695
    female system, 658–704
    formation, spermatophore, 641–650
    fresh-copulation stage, 667–668, 673–674
    hatching, 695–696
    interspawn stage, 676
    larval stages, 698*f*
    late-copulation stage, 668–671, 674
    lipovitellin, 700–704
    male reproductive system, 633–658
    mature follicle, 641–643
    may-spawn stage, 676
    moulding, spermatophore, 650–651
    moulting, 664
    nonbreeding stage, 671–672
    ovaries maturation, 671–683
    oviduct, 664–665
    plug, 667–668, 670–671
    postcopulatory guarding, 664
    postembryonic development, 696–697
    precopulation stage, 667, 672–673
    precopulatory guarding, 664
    predeutoplasmic oocyte, 24–25, 25–26*f*, 27–28t
    premoult guarding, 664
    prespawn stage, 675–676
    september-spawn stage, 679–683
    spermatheca formation, 664–671
    spermatid differentiation, 658
    spermatozoa, 643–647
    sperm masses, 641
    structure, spermatophore, 651–653
    vagina, 665–666
  P. sanguinolentus, 524, 663–664

Postzygotic, 83–84
Potamopyrgus jenkinsi, 739
Pre-copulation stage, 662–663*t*, 666*f*, 667, 672–673
Predeutoplasmic oocyte, 24–25, 25–26*f*, 27–28t
Preformation, 6
Pre-spawning stage, 602–603, 662–663*t*, 675–676, 678*f*
Presumptive ectoderm, 490
Presumptive mesoderm, 490
Presumptive midgut, 489–490
Presumptive proctodaeum, 490
Presumptive stomodaeum, 490
Previtellogenesis, 315
Primary delamination, 64
Primary yolk pyramids, 606
Primordial gonocytes, 18–19
  cytoplasm, germ-cell determination, 21–22
  nucleus, features
    chromatin diminution, 19–20
    chromosome elimination, 20–21, 20–22*f*
Pristionchus aerivora, 442*f*
Proboscis, 405–410, 414–415
Procambarus spp.
  P. clarkii, 590
  P. fallax, 613–620, 615*f*
Procephalothrix simulus, 416–418, 417*f*
Progamic phase, 96
Prosorhochmus, 416
Protandric simultaneous hermaphroditism, 506–507
Proteases, 55
Protective outer wall, 83–84
Proteins, 653, 668
Protostomia, 62
Protozoa, 72
  Ciliophora, 95–96
    habitat, 96
    Paramecium, 96–100
  Mastigophora, 69–87, 70*f*
    (*Chlamydomonas see Chlamydomonas*)
    (*Volvox see Volvox*)
  Sarcodina, 87–88
    Actinophyridae, 89
    Difflugia sp., 88–89, 88*f*
  Sporozoa, 91–95
    Eimeria spp., 92–94, 93*f*
    monocystis, 94–95, 95*f*
Proximal vas deferens (PVD), 536, 539–541, 593–595, 634–637, 647–650, 835–836
Pseudalius inflexus, 433*f*
Pseudocnids, 389
Pseudomorula, 194–195
Pseudonavicella, 94–95
Pseudosquilla ciliata, 523
Pseudotransverse, 72–73
Ptera, 477
Pycnogonida, 478
Pylome, 88–89
Pylorus, 415

Q

Quadrant, 60
Quadriflagellate zygote, 77

R

Radial cleavage, 60, 880
Radial symmetry, 878–880
Radula sac, 799
Rathke, Martin Heinrich, 8
Recapitulation theory, 9
Reception cone, 43–44
Regeneration, 111–114, 117, 311–313
Regulative development, 11*f*
Reniera simulans, 119
Reptantia, 590
Resact, 49
Resting eggs, 498
Resting volvocids, 83–84
Retention theory, 9
Rhabdias bufonis, 447
Rhabdocoela, 340
  Gieysztoria superba, 366–369, 366*f*, 372
  Macrostomum spp., 372–377, 373–374f
  Mesostoma lingua, 369–372, 370–371*f*
Rhabdoids, 389
Rhaptothyreus typicus, 431
Rhizopoda, 87*t*
Rhizostomae
  Cassiopea andromeda, 228–233, 230–231*f*
  Cotylorhiza tuberculata, 233–236, 234*f*
  Lychnorhiza lucerna, 236–238, 237*f*
  Mastigias, 224–228, 227*f*
Rhodopsin, 72
Rimapenaeus similis, 562–563, 563*f*
Rossia macrosoma, 823
Rotylenchulus reniformis, 449*f*
Roux, Wilhelm, 10

S

Sabatieria hilarula, 435*f*
Sagartia troglodytes, 285–286, 286*f*
Salicoques, 480–481
Salivary glands, 857
Sarcodina, 87–88
  Actinophyridae, 89
    morphology, 89–90, 90*f*
    reproduction, 90–91, 91–92*f*
  Difflugiidae, 88–89, 88*f*
Scalpellum spp., 505–506
  S. gigas, 506*f*
  S. peronii, 506*f*, 507
  S. scalpellum, 505*f*, 507–508
  S. vulgare, 505–506*f*
Schmidt's larva, 410–415
  metamorphosis, 411–414
  structure of, 411
Sciara spp., 20–21
  S. coprophila, 20–21, 20*f*
  S. ocellaris, 21
  S. pauciseta, 20–21
Scleractinia, 291
  Acropora, 297–301
  Astrangia danae, 295–297
  Galaxea fascicularis, 300, 301*f*
  true/stony corals, 291–301, 292*f*, 294*f*
Sclerospongia, 109–110
Scylla spp.
  S. olivacea, 689
  S. serrata, 688
Scyphozoa, 176, 179–180
  Discomedusae (*see* Discomedusae)
  Medusozoa Peterso, 212–238, 213–214*f*
  Stauromedusa, 215–218
    Stylocoronella riedli, 218
  types of, 213*f*
Sea anemones, 280, 283–284*f*
  Cribrinopsis japonica, 290–291
  Epiactis japonica, 287–289, 289*f*
  Halcampa duodecimcirrata, 286–287, 287*f*
  Nematostella vectensis, 280–285
    blastula, 284–285
    cleavage, 284
    fertilization, 284
    spawning, 280–284
  Sagartia troglodytes, 285–286, 286*f*
  Tealia crassicornis, 289–290
Sea urchin
  development, 14*f*
  type of fertilization, 58
Sea wasp, 241
Secondary delamination, 64
Secondary oocytes, 23–24
Secondary yolk pyramids, 606–609
Second cleavage, 354
  in *Oopsacas minuta*, 142–143
Second larva formation, 146
Second phase, *Eucheilota paradoxica*, 189
Semaeostomeae
  Aurelia aurita, 220–223
  larval and onset of metamorphosis, settlement of, 224
  planula larva, 223–224
  Scyphozoa, 224
  Semaeostomeae Scyphozoa, 224
  tentacles and gastric pouches, formation of, 224
Seminal receptacle, 816–817
Seminiferous tubule, 32*f*
Senescence, 31
Sense organ, 334–335
Sepia spp.
  S. esculenta, 849–850
  S. officinalis
    agametic reproduction, 816
    copulation, 850–851
    dehiscence, spermatophores, 831–833
    formation, spermatophores, 833–838
    gametic reproduction, 816–822
    hardening, spermatophores, 839–850
    mating, 850–851
    organogenesis, 855–857
    reproductive behaviour, 850–851
    reproductive system, 817–822
    sexual dimorphism, 816–817
    spawning, 850–851
    spermatozoa, spermatophores, 833

Sepiella maindroni, 849–850
Sepiida, 815
Sepiola atlantica, 823
Sepioteuthis lessoniana, 821, 849–850
Sergestidae, 481
Seriata, 340
Sertoli cell, 834–835
Sesarma haematocheir, 695–696, 696*f*
Sex(es), 97–98
Sex allocation theory, 508
Sex attractant, 441, 562, 663
Sex-inducing pheromone, 82
Sexual differentiation, Porifera, 117
Sexual dimorphism, 483
Sexual reproduction, 75–77
  Paramecium spp., 97–98
  types of, 72
  Volvox, 82–84, 85–86*f*
Shell glands, 482
Shells/shell, 88–89, 734*f*
Short-flagellated, 38
Shoulder-to-shoulder, 77
Shrimp ovarian peritrophin (SOP), 574
Sicyonia ingentis, 539–541, 567–568*f*
Sida crystalline, 495*f*
Simocephalus, reproductive system, 495–503
Simultaneous hermaphroditism, 506–507
Siphonaria spp.
  S. cochleariformis, 771
  S. hispida, 771–772
  S. kurracheesis, 771
  S. obliquata, 771
Siphonophora, 176, 206–212, 207*f*
Siphosome, 208–210
Snow algae, 71
Solenocera spp.
  S. Africana, 567–568*f*
  S. membranacea, 567–568*f*
Solitary forms, 69
Somatoplasm, 9–10
Soybean tyrosine inhibitor (SBTI), 37
Spallanzani, Lazzaro, 43
Spawning, 22–23, 42, 128–129, 130*f*, 181–183, 248–250, 258, 262–264, 263*f*, 279–284, 283–284*f*, 294*f*, 295, 298, 320, 396, 574–575, 602–603, 727–728, 786–788, 850–851, 874
Specialized germ cells (SGC), 833–834
Species-specific attraction, 49
Species-specific recognition, 51
Spemann, Hans, 12
Spermateolosis, 833–834
Spermatheca, 243, 454–455, 459*f*, 460, 613, 658, 662–663t, 664–671, 674, 678*f*, 758–760, 783
Spermatids, 833–834
Spermatogenesis, 279, 314–315, 317*f*, 350–352, 563–564, 723–728, 764–769, 868
  in *Astacus astacus*, 604–605
  Cnidaria, 181
  Porifera, 119–123, 120*f*, 122*f*, 125*f*
Spermatogonia, 833–834
Spermatophore, 353, 483, 782
  Penaeus kerathurus
    argentaffin material in, 556–557
    chemical composition, 554–557
    external appearance, 553–554
    formation, 541–553
    hardening, 557
    preliminary connection, 546
  Portunus pelagicus
    chemical composition, 653
    formation, 641–650
    moulding, 650–651
    structure, 651–653
  Sepia officinalis, 822–850
    dehiscence, 831–833
    formation, 833–838
    hardening, 839–850
    spermatozoa, 833
Spermatophoric cord (SPC), 822–823, 827
Spermatophoric duct, 543
Spermatozoa, 31–32, 34*f*, 350–352, 486, 497*f*, 767, 769
Spermatozoon maturation, 43
Sperm cell, 31–35, 32–33*f*, 181
  agglutination, 45–46
  entry, 321*f*
  mature egg, 36–38
  packets, 82
  typical marine spermatozoon, 32–35
Sperm lysins, 49–51
Sperm mass, 827
Spermoviduct, 42, 42*f*, 744, 746–747*f*, 748–749, 753–757, 756*t*, 757*f*, 780–781, 783, 785
Sperm pronucleus, 58, 878
Spherical mass, 634–637
Spicules, 109–113, 137, 139–141, 140*f*, 149, 158*f*, 166, 276, 428–429*f*, 430–431, 441, 736–737*f*, 738–739
Spindle-shaped schizozoites, 92–94
Spiral arrangement, 60
Spiral cleavage, 60
Spongia officinalis, 152–154, 155–156*f*, 167
Spongilla lacustris, 115–117, 119, 122*f*, 124*f*
Spore, 72–73
Sporoblasts, 94–95
Sporocyst, 94–95
Sporozoa, 91–95
  Eimeria spp., 92–94, 93*f*
  monocystis, 94–95, 95*f*
Sporozoites, 94–95
Squilla spp., 520
  S. armata, 528
  S. empusa, 519*f*, 528
  S. holoschista, 521–522, 521–522*f*, 524, 525*f*
  S. mantis, 518–522, 519*f*
  S. oratoria, 497*f*, 524–525, 528, 529–530*f*
Stalked jellyfish, 215
Statocysts, 483, 797
Stauromedusae, 176, 215–218
Stenoplax heathiana, 724*f*, 727–728, 729*f*
Stercoraliss tercoralis, 428–429*f*
Stereoblastula, 61
Stichostemma grandis, 397–398
Stolonifera, 262–264
Stomatopoda, 518–522, 519*f*, 528
Stomodeal pharynx, 331–334, 332*f*
Stomodeum, 62–64
Streptocephalus dichotomus, 484, 486–488*f*, 488–489
Striated sheath, 829–831
Strombus, 769
Strongylocentrotus purpuratus, 14*f*
Structuralization, 56–57
Styela plicata, 867–868*f*
Stylochoplana parasitica, 339
Stylochus spp., 359
  S. uniporus, 359
Stylocordyla borealis, 155–157, 157–158*f*
Stylocoronella riedli, 218
Succinia putris, 780
Summer eggs, 498
Sun animalcule, 89–90
Superficial cleavage, 483
Sycon spp., 136*f*
  S. raphanus, 134*f*, 136*f*, 162, 166
Symmetrogenic fission, 72–73, 96
Sympodial growth, 189
Synascidia, 865
Syngamy process, 46*f*, 94–95, 878
Syngen, 98

T

Tadpole larva, 888
Talon, 744–751, 749–750f, 753, 780–781, 783–785
Tealia crassicornis, 289–290
Telolecithal oocyte, 40
Telosporidea, 91*t*
Telson, 482
Temnocephala, 340
Tentaculata, 310
Testicular ductules, 641–643
Testicular pore, 817
Tethya aurantium, 163, 165
Tetilla=(Tethya) serica, 163
Tetrabranchiata, 816
Tetraclita rosea, 511–513, 513*f*
Tetractinomorpha, 110, 148–149
Teuthidida, 816
Thaumaioscyphus distinctus, 221–222*f*
Theoria Generationis, 6–7
Third larva formation, 146
Third megalospheric forms, 90*f*
Third phase, *Eucheilota paradoxica*, 189
Thoracica, 503
Thysanozoon brocchi, 353
Tjalfiella tristoma, 313
Tobrilus gracilis, 435*f*
Toenioplana teredini, 339
Tonicella lineata, 730
Torsion, 800–802, 801*f*, 803*f*
Trabecular reticulum, 146
Trachylina, 176
Trachypenaeus similis, 539–541, 567–568*f*
Transitional duct, 667

Transition membrane, 55
Transverse fission, 179
Trematoda, 340
Trichimella, 148
Trichinella spiralis, 441
Trichobranchiata, 481
Trichuris suis, 428–429*f*, 430, 446*f*
Tricladida, 340, 377–385, 377*t*
  epimorphosis, 377
  morphallaxis, 377–380, 378*f*
Trilobita, 478
Tripedalia cystophora, 239–241, 240*f*
Triploblastica. *See* Platyhelminthes
Trochopore and veliger stages, 797–800, 798*f*
Trochus, 790–791
Trophon geversianus, 804–809
Trophozoite, 91–92, 94
True/definitive mesoderm, 64
Tubularia spp.
  T. indivisa, 505*f*
  T. larynx, 194
  T. mesembryanthemum, 191–195, 193*f*
    actinula larva, 195
    embryogenesis, 193*f*, 194–195
Tubutanus polymorphus, 394
Tunicata, 863
Tunics, 782, 825–827, 829*f*, 831, 837
Turbatrix aceti, 447, 449–450
Turbellaria, 340–341, 352

U

Ucides cordatus, 689, 692*f*
Unipolar ingression, 64
Urochordata
  alimentary canal, 889–890
  Ascidiacea, 864–865
  atrium, 889
  cell differentiation, 886
  characteristics, 863–897
  classification, 865
  cleavage, 880–884
  early development, 877–886
  endostyle, 889
  Enterogona, 865
  experimental studies, 890–897
  fully formed larva, 888–890
  gametogenesis
    fertilization, 874–877
    fully grown oocyte, 870–874
    oogenesis, 868–874
    spawning, 874
    sperm, 868
    spermatogenesis, 868
  gastrulation, 884–886
  hatching, 886–888
  maturation, 878–880
  metamorphosis, 888–889, 896–897
  morphogenesis, 886
  nervous system, 890
  neurulation, 886
  ooplasmic segregation, 878–880
  Pleurogona, 865
  reproduction, 865–897
    agametic, 865–867
    gametic, 867–877
  spinal cord, 886
  time requirement for development, 890
  visceral ganglion, 886

V

Valvata, 769
Vampyromorpha, 816
van Leeuwenhoek, Anton, 5, 31–32, 43
Vas deferens, 536, 633–634
Vas efferens, 653–655
Velum, 799–800, 809*f*
Vema ewingi, 734*f*, 735
Vermetus, 769
Verongia archeri, 128–129
Vertebrate sperm, 38
Vis-à-vis, 77
Viscosia carnleyensis, 445
Visual purple, 72
Vitelline membrane, 36, 41, 674
Vitellogenesis, 24–29, 123–125
Viviparous sponges, 117, 119, 125–126, 149–162
Viviparus, 769
Volvox spp., 79–87
  asexual reproduction, 81–82, 83*f*, 85*f*
  features of, 84–87
  habitat, 81
  morphology, 81, 84*f*
  sexual reproduction, 82–84, 85–86*f*
  V. carteri, 82, 85*f*
von Baer, Karl Ernst, 7
von Pander, Heinrich Christian, 7

W

Watermelon snow, 71
Weismann, August, 9
Wing duct, 543
Winter eggs, 498
Wolff, Friedrich, 6

X

X-chromosomes, 21
Xenia umbellata, 273, 274*f*
Xiphopenaeus riveti, 567–568*f*
Xiphosura, 478

Y

Yolk nuclei, 673–674

Z

Zooids, 206
Zoospore formation/binary fission, *Chlamydomonas*, 72–74, 74–76*f*
Zostera marina, 435–436
Zygocyst, 94–95
Zygospore, 77
  Chlamydomonas, 77–79
Zygote, 20, 45, 94–95
  Chlamydomonas, 77–79
  nonmotile, 77
  quadriflagellate, 77

CPI Antony Rowe
Eastbourne, UK
December 14, 2022